BEST AND TAYLOR'S
Physiological Basis of Medical Practice

TWELFTH EDITION

BEST AND TAYLOR'S
Physiological Basis of Medical Practice

TWELFTH EDITION

EDITED BY

John B. West, M.D., Ph.D., D.Sc.

Professor of Physiology and Medicine
University of California, San Diego
School of Medicine
La Jolla, California

Williams & Wilkins

BALTIMORE • PHILADELPHIA • HONG KONG
LONDON • MUNICH • SYDNEY • TOKYO
A WAVERLY COMPANY

Editor: Timothy S. Satterfield
Associate Editor: Linda Napora
Copy Editor: Shelley Potler
Designer: Norman Och
Illustration Planner: Lorraine Wrzosek
Production Coordinator: Adèle Boyd-Lanham

Copyright © 1990
Williams & Wilkins
428 East Preston Street
Baltimore, MD 21202, U.S.A.

Accurate indications, adverse reactions, and dosage schedules for drugs are provided in this book, but it is possible that they may change. The reader is urged to review the package information data of the manufacturers of the medications mentioned.

Made in the United States of America

First Edition, 1937
Second Edition, 1939
Third Edition, 1943
Fourth Edition, 1945
Fifth Edition, 1950
Sixth Edition, 1955
Seventh Edition, 1961
Eighth Edition, 1966
Ninth Edition, 1973
Tenth Edition, 1979
Eleventh Edition, 1985

Spanish Editions, 1939, 1941, 1943, 1947, 1954, 1982, 1986
Portuguese Editions, 1940, 1945, 1976, 1987
Rumanian Editions, 1958, 1966
Italian Editions, 1959, 1979
Polish Editions, 1960, 1971
Taiwan Editions, 1980, 1982
Indian Edition, 1967
Asian Edition, 1967
Philippine Edition (English), 1981

Library of Congress Cataloging in Publication Data

Best, Charles Herbert, 1899–1978.
[Physiological basis of medical practice]
Best and Taylor's physiological basis of medical practice.—12th
ed. / edited by John B. West.
p.   cm.
Includes bibliographies and index.
ISBN 0-683-08947-1
1. Human physiology.   I. Taylor, Norman Burke, 1885–
II. West, John B. (John Burnard)   III. Title.   IV. Title:
Physiological basis of medical practice.
[DNLM: 1. Physiology.   QT 104 B561p]
QP34.5.B47   1990
612—dc20
DNLM/DLC
for Library of Congress   89-16562
                                                                      CIP

10 9 8 7

Preface to the Twelfth Edition

The preface to the first edition was written more than 50 years ago, and the same principles still ring true. Extensive changes have been made in this 12th edition. The section on the gastrointestinal system has been completely rewritten and we welcome the new section editor, Stephen J. Pandol, M.D. All parts of the book have been thoroughly revised and brought up to date. The chapters on respiration have been completely recast. Both the sections on blood and on neurophysiology have been considerably shortened so that the whole volume is now about 15% leaner than the 11th edition. We sympathize with the modern-day student who is continually presented with exciting new advances, but less commonly informed about what he can omit. Thus, this new edition represents a major change.

As in the 11th edition, the guiding principle has been to produce a book that emphasizes the clinical relevance of physiology, yet is authoritative with appropriate scientific rigor. In general, the section editors are people who teach first-year medical students, but who also have first-hand knowledge of the issues that are important in clinical medicine. Our aim was to produce a text that is clearly written, didactic, not necessarily encyclopedic, and that recognizes the limited time available to medical students.

During the 5 years that have elapsed since the 11th edition, spectacular advances have been made in molecular and cell biology. In some medical schools, this has resulted in a reduced emphasis on the study and teaching of organ physiology. However, this topic remains one of the cornerstones of the intelligent practice of clinical medicine, and is no less important than it was 50 years ago when the first edition of this textbook was produced. While acknowledging the importance of modern molecular and cell biology, we reaffirm the importance of physiology in the day-to-day practice of medicine.

We wish to thank Michelle Lambert for nearly all the new illustrations, and the editors and the staff of Williams & Wilkins for their cooperation.

For the editors

November 1989 J.B.W.

Preface to the First Edition

Physiology is a science in its own right and the laboratory worker who pursues his research quite detached from medical problems need offer no apology for his academic outlook. Indeed some of the most valuable contributions to medical science have been the outcome of laboratory studies whose applications could not have been foreseen. Nevertheless, we feel that the teacher of physiology in a medical school owes it to his students, whose ultimate interest it must be conceded is in the diagnosis and treatment of disease, to emphasize those aspects of the subject which will throw light upon disorders of function. The physiologist can in this way play a part in giving the student and practitioner a vantage point from which he may gain a rational view of pathological processes.

We have endeavored to write a book which will serve to link the laboratory and the clinic, and which will therefore promote continuity of physiological teaching throughout the pre-clinical and clinical years of the undergraduate course. It is hoped that when the principles underlying diseased states are pointed out to the medical student, and he is shown how a knowledge of such principles aids in the interpretation of symptoms or in directing treatment, he will take a keener interest in physiological studies. When such studies are restricted to the classical aspects of the subject, apparently remote from clinical application, the student is likely to regard them only as a task which his teachers in their inscrutable wisdom have condemned him to perform. Too often he gains the idea, from such a course, that physiology is of very limited utility and comes to believe that, having once passed into the clinical years, most of what he has "crammed" for examina-

tion purposes may be forgotten without detriment to his more purely medical studies. Unfortunately, he does not always realize at this stage in his education how great has been the part which physiological discoveries have played in the progress of medicine, and that the practice of today has evolved from the "theories" of yesterday.

Many physiological problems can be approached only through animal experimentation. Advances in many fields, most notably in those of carbohydrate metabolism, nutrition, and endocrinology, bear witness to the fertility of this method of research. On the other hand, many problems can be elucidated only by observations upon man, and physiology has gained much from clinical research. The normal human subject as an experimental animal possesses unique advantages for many types of investigation; and in disease, nature produces abnormalities of structure and function which the physiological laboratory can imitate only in the crudest way. Within recent years the clinical physiologist, fully realizing these advantages and the opportunities afforded by the hospital wards, has contributed very largely to physiological knowledge. In many instances, clinical research has not only revealed the true nature of the underlying process in disease, but has cast a light into some dark corner of physiology as well; several examples of clinical investigation which have pointed the way to the physiologist could be cited. In the last century, knowledge of the processes of disease was sought mainly in studies of morbid *anatomy;* biochemistry was in its infancy and many of the procedures now commonly employed for the investigation of the human subject had not been devised. Today, the student of scientific medicine is directing his attention more and more to the study of morbid *physiology* in his efforts to solve clinical problems. This newer outlook has borne fruit in many fields. It has had the beneficent result of drawing the clinic and the physiological and biochemical laboratories onto common ground from which it has often been possible to launch a joint attack upon disease. We feel that this modern trend in the field of research should be reflected in the teaching of medical students, and have therefore given greater prominence to clinical aspects of the subject than is usual in physiological texts.

In order to understand the function of an organ it is usually essential to have a knowledge of its structure. For this reason we have followed the plan of preceding the account of the physiology of a part by a short description of its morphology and, in many instances, of its nerve and blood supply. The architecture and functions of the central nervous system are so intimately related that some space has been devoted to a description of the more important fiber tracts and grey masses of the cerebrum, cerebellum and spinal cord.

We wish to thank our colleagues in physiology, biochemistry and anatomy whom we have drawn upon on so many occasions for information and advice; without their generous help the undertaking would have been an almost impossible one. We are also deeply grateful for the unstinted assistance which we have received from our friends on the clinical staff, several of whom have read parts of the text in manuscript or in proof.

C.H.B.
October 15, 1936   N.B.T.

Section Editors

General Physiological Processes
Madeleine A. Kirchberger, Ph.D., and Irving L. Schwartz, M.D.

Cardiovascular System
John Ross, Jr., M.D.

Blood
Samuel I. Rapaport, M.D.

Body Fluids and Renal Function
Nora Laiken, Ph.D., and Darrell D. Fanestil, M.D.

Respiration
John B. West, M.D., Ph.D., D.Sc.

Gastrointestinal System
Stephen J. Pandol, M.D.

Metabolism
Daniel Steinberg, M.D., Ph.D.

Endocrine System
Gordon N. Gill, M.D.

Neurophysiology
Robert B. Livingston, M.D.

Contributors

SECTION 1

General Physiological Processes

Madeleine A. Kirchberger, Ph.D., *Editor*
Associate Professor of Physiology and
    Biophysics
Mount Sinai School of Medicine
New York, New York

Irving L. Schwartz, M.D., *Editor*
Harold and Golden Lamport
    Distinguished Service Professor
Mount Sinai School of Medicine
New York, New York

Sandra K. Masur, Ph.D.
Associate Professor of Physiology and
    Biophysics
Mount Sinai School of Medicine
New York, New York

Victor L. Schuster, M.D.
Associate Professor of Medicine
Albert Einstein College of Medicine
Bronx, New York

George J. Siegel, M.D.
Professor of Neurology
The University of Michigan
School of Medicine
Ann Arbor, Michigan

SECTION 2

Cardiovascular System

John Ross, Jr., M.D., *Editor*
Professor of Medicine
University of California, San Diego
School of Medicine
La Jolla, California

James W. Covell, M.D.
Professor of Medicine
University of California, San Diego
School of Medicine
La Jolla, California

Gregory K. Feld, M.D.
Assistant Professor of Medicine
University of California, San Diego
School of Medicine
La Jolla, California

Geert Schmid-Schoenbein, Ph.D.
Associate Professor of Biomedical
    Engineering
University of California, San Diego
La Jolla, California

SECTION 3

Blood

Samuel I. Rapaport, M.D., *Editor*
Professor of Medicine and Pathology
University of California, San Diego
School of Medicine
La Jolla, California

Helen M. Ranney, M.D.
Professor of Medicine
University of California, San Diego
School of Medicine
La Jolla, California

Raymond Taetle, M.D.
Associate Professor of Pathology and
    Medicine
University of California, San Diego
School of Medicine
La Jolla, California

SECTION 4

Body Fluids and Renal Function

Nora D. Laiken, Ph.D., *Editor*
Lecturer in Medicine and Assistant Dean
University of California, San Diego
School of Medicine
La Jolla, California

Darrell D. Fanestil, M.D., *Editor*
Professor of Medicine
University of California, San Diego
School of Medicine
La Jolla, California

SECTION 5

Respiration

John B. West, M.D., Ph.D., D.Sc., *Editor*
Professor of Physiology and Medicine
University of California, San Diego
School of Medicine
La Jolla, California

SECTION 6

Gastrointestinal System

Stephen J. Pandol, M.D., *Editor*
Associate Professor of Medicine
University of California, San Diego
School of Medicine
La Jolla, California

James Christensen, M.D.
Professor of Medicine
University of Iowa
College of Medicine
Iowa City, Iowa

Kiertisin Dharmsathaphorn, M.D.
Associate Professor of Medicine
University of California, San Diego
School of Medicine
La Jolla, California

Alan F. Hofmann, M.D.
Professor of Medicine
University of California, San Diego
School of Medicine
La Jolla, California

Jon I. Isenberg, M.D.
Head, Gastroenterology Division
Professor of Medicine
University of California, San Diego
School of Medicine
La Jolla, California

SECTION 7

Metabolism

Daniel Steinberg, M.D., Ph.D., *Editor*
Professor of Medicine
University of California, San Diego
School of Medicine
La Jolla, California

SECTION 8

Endocrine System

Gordon N. Gill, M.D., *Editor*
Professor of Medicine
University of California, San Diego
School of Medicine
La Jolla, California

Robert A. Brace, Ph.D.
Professor of Reproductive Medicine
University of California, San Diego

School of Medicine
La Jolla, California

Bart C. J. N. Fauser, M.D., Ph.D.
Assistant Professor of Obstetrics and
    Gynecology
Dijkzigt University Hospital
Rotterdam, The Netherlands

Aaron J. W. Hsueh, Ph.D.
Professor of Reproductive Medicine
University of California, San Diego
School of Medicine
La Jolla, California

James H. Liu, M.D.
Associate Professor of Obstetrics and
    Gynecology
University of Cincinnati
College of Medicine
Cincinnati, Ohio

Leslie Myatt, Ph.D.
Associate Professor of Obstetrics and
    Gynecology
University of Cincinnati
College of Medicine
Cincinnati, Ohio

Robert W. Rebar, M.D.
Professor of Obstetrics and Gynecology
University of Cincinnati
College of Medicine
Cincinnati, Ohio

Robert Resnick, M.D.
Professor of Reproductive Medicine
University of California, San Diego
School of Medicine
La Jolla, California

Michael A. Thomas, M.D.
Assistant Professor of Obstetrics and
    Gynecology
University of Cincinnati
College of Medicine
Cincinnati, Ohio

SECTION 9

Neurophysiology

Robert B. Livingston, M.D., *Editor*
Professor of Neurosciences
University of California, San Diego
School of Medicine
La Jolla, California

Contents

SECTION 1: GENERAL PHYSIOLOGICAL PROCESSES

edited by Madeleine A. Kirchberger and Irving L. Schwartz

4. Excitation and Contraction of Skeletal Muscle   62
Madeleine A. Kirchberger

SECTION 2: CARDIOVASCULAR SYSTEM

edited by John Ross, Jr.

SECTION 3: BLOOD

edited by Samuel I. Rapaport

SECTION 4: BODY FLUIDS AND RENAL FUNCTION

edited by Nora D. Laiken and Darrell D. Fanestil

SECTION 5: **RESPIRATION**

edited by John B. West

SECTION 6: GASTROINTESTINAL SYSTEM

edited by Stephen J. Pandol

SECTION 7: METABOLISM

edited by Daniel Steinberg

SECTION 8: ENDOCRINE SYSTEM

edited by Gordon N. Gill

SECTION 9: NEUROPHYSIOLOGY

edited by Robert B. Livingston

SECTION 1

General Physiological Processes

CHAPTER 1

By Way of Introduction: The Cell

The science of mammalian physiology involves the study of dynamic interrelationships that exist among cells, tissues, and organs, and reaches ultimately to the level of the organism as a whole. The cell is the smallest functional unit and is itself composed of organelles. In this chapter, we shall focus on the structure and function of these organelles.

One group of organelles is bounded by a limiting membrane; a second group is not so delimited. The former group encompasses the nucleus, endoplasmic reticulum, Golgi apparatus, lysosomes, peroxisomes, and mitochondria—and in this category we also include the plasma (outer) membrane of the cell as a whole. The group of organelles that are not bounded by a membrane includes the chromosomes, nucleoli, microtubules, ribosomes, microfilaments, and centrioles.

Although there was some knowledge of the cell boundary and subcellular organelles prior to the advent of the electron microscope, the complicated ultrastructure of intracellular membrane systems was unforeseen (Fig. 1.1). These membrane systems provide closed compartments within the cell with distinctive environments because they maintain higher concentrations of ions (e.g., low pH in lysosomes) or enzymes (the tricarboxylic acid cycle enzymes in mitochondria, phosphatases in the Golgi apparatus). In addition, the membrane itself consists of microdomains in which proteins or lipids are arranged in an ordered sequence resulting in a functionally meaningful pattern of enzymes, co-factors, or carriers.

CELL MEMBRANE

The cell membrane is a permeability barrier. If a cell is placed in hypotonic solution and if it contains molecules which cannot penetrate its outer membrane, it will swell; conversely, it will shrink if placed in a hypertonic medium. In both instances water moves down its concentration gradient (Chapter 2). Thus, the cell behaves as an osmometer. Nonpolar molecules (gases, lipids) move freely across the membrane; polar molecules penetrate the membrane much less readily and indeed it is the selective permeability of the plasma membrane to certain ions which determines the excitability characteristics of nerve and muscle cells.

Chemical Composition

Although the general chemical nature of the membrane found at the cell boundary was predicted on the basis of physiological data, the detailed molecular structure of the membrane is not yet known. Models of cell membranes prepared by combining their lipid and protein constituents (partly known from chemical analysis of purified cell membrane preparations) exhibit some physiological characteristics similar to those of natural membranes. The role of lipid-lipid interactions in membrane structure has been at the center of attention because such interactions can explain much of the presently known phenomena of membrane transport. Quantitative studies of isolated cell membranes revealed that enough lipid is present to be arranged as a bilayer coating the cell. Artificial mixtures of extracted cellular polar lipids (lecithin, phospholipid, and steroids) under appropriate conditions will form a bimolecular layer spontaneously. Presumably the polar (hydrophilic) ends of the lipids form the two outer borders, making them available for interaction with other polar molecules such as proteins (Fig. 1.2). On a weight basis, membranes contain a significantly larger amount of protein than lipid (ratio up to 4:1); however, due to the high molecular weight of proteins, this relationship is reversed on a molar basis (protein to lipid ratio ranging from 1:100 to 10:100). Some proteins are associated peripherally with one of the polar surfaces of the lipid bilayer. Other proteins, the integral membrane proteins, are not restricted to the surfaces of the plasma membranes but extend into the bimolecular lipid layer (Fig. 1.2).

In addition to lipid and protein, carbohydrate is associated with the cell membrane as lipopolysaccharide and as protein-polysaccharide. The carbo-

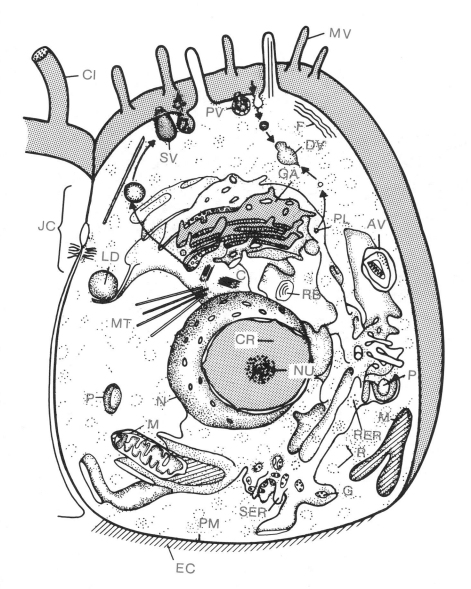

Figure 1.1. Schematic diagram of a cell and its organelles drawn to reveal their three-dimensional structure. *AV*, autophagic vacuole; *C*, centriole; *CI*, cilium; *CR*, chromatin; *DV*, digestion vacuole; *EC*, extracellular coat (e.g., basement membrane); *F*, microfilaments; *G*, glycogen; *GA*, Golgi apparatus; *JC*, junctional complex; *LD*, lipid droplet; *M*, mitochondrion; *MT*, microtubules; *MV*, microvillus; *N*, nucleus; *NU*, nucleolus; *P*, peroxisome; *PL*, primary lysosome; *PM*, plasma membrane; *PV*, pinocytic vesicle; *R*, ribosomes and polysomes; *RB*, residual body; *RER*, rough endoplasmic reticulum; *SER*, smooth endoplasmic reticulum; *SV*, secretion vacuole. The organelles have been drawn only roughly to scale. The sizes and relative amounts of different organelles can vary considerably from one cell type to another. (Adapted from Novikoff and Holtzman, 1976.)

hydrate moieties of the membrane serve to modify the electric charge at its surface and provide specific surface-binding sites. Cytochemically demonstrable polysaccharide-protein complex is associated with many cell surfaces as an extracellular layer. In some places, particularly at luminal surfaces, this layer forms a fuzzy coat—often referred to as a glycocalyx (Fig. 1.3)—which may act as a crude filter and/or facilitate the attachment of molecules for endocytic (see below) transport across the cell membrane.

Structure of the Unit Membrane

A pattern generally found in almost all cellular membranes prepared for microscopy by conventional techniques consists of three layers, i.e., two electron-dense layers on either side of a single electron-lucent layer (Fig. 1.3). This has been termed a *unit membrane,* or a *three-layered membrane.* Electron microscopic examination of osmium tetroxide-fixed sectioned tissue shows the cell membrane to be 7–10 nm wide. The electron-lucent line in the unit membrane is thought to represent the lipid layer. Hydrophobic bonding in the lipid bilayer region may make it inaccessible to osmium deposition. Thus, the two electron-dense lines would result from deposition of osmium at the surfaces of this bilayer. There is much physiological evidence to suggest that the lipid bilayer is interrupted by proteins which, as hydrophilic molecules (Fig. 1.2), connect the two outer surfaces of the membrane and provide transmembrane channels (a few Å in width) for transfer of water molecules and ions (Chapter 2). The protein components of the plasma

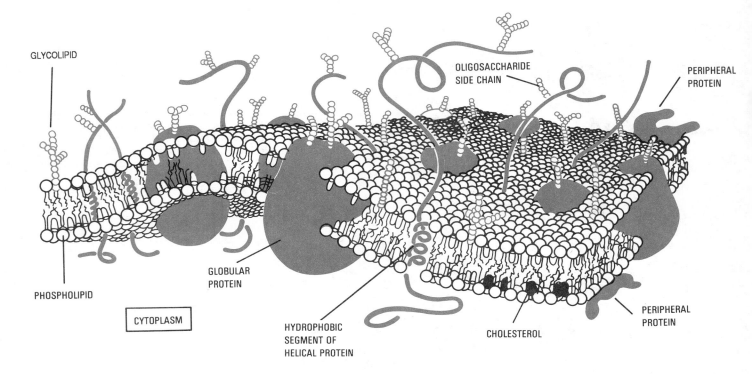

GLYCOLIPID

OLIGOSACCHARIDE SIDE CHAIN

PERIPHERAL PROTEIN

PHOSPHOLIPID

GLOBULAR PROTEIN

CYTOPLASM

HYDROPHOBIC SEGMENT OF HELICAL PROTEIN

CHOLESTEROL

PERIPHERAL PROTEIN

Figure 1.2. Diagram of a three-dimensional model of a cell membrane segment showing the phospholipid bilayer in which cholesterol and proteins are embedded. Presumably the hydrophobic portions of certain globular and helical proteins allow their integration within the hydrophobic lipid regions. Both lipid and peripheral protein components and associated groups of either surface, cytoplasmic or extracellular, may differ from one another. (Modified from Bretscher, 1985.)

membrane are either tightly associated with it *(integral)* or more readily dissociated from it *(peripheral)*.

Additional electron microscopic structural information comes from unfixed membranes studied by the freeze-etch technique. When rapidly frozen tissue is fractured by a sharp knife, the membranes tend to fracture along the middle layer of their bimolecular lipid leaflets. The evaporation of a thin layer of carbon or platinum onto the exposed surface produces a replica which is viewed in the electron microscope. Intramembrane particles are seen mainly in replicas of membranes having integral proteins (and also in various lipid-cholesterol mixtures).

In most metabolically active membranes, repeating structures are visible at various intervals within the layer. These structures may represent proteins that extend through much of the thickness of the membrane (Fig. 1.3*B*). This interpretation is supported by evidence from other techniques which label the outer portions of integral proteins at either the outside or inside surface. The positions of the intramembranous particles may be more or less stable. With varying physiological conditions the distribution pattern of these particles may shift within a membrane. This lateral mobility of components, e.g., proteins, within the membrane is seen if cells of different origins are caused to fuse. Rather than a patchwork of the two original membrane particle patterns the result is an intermixture of both characteristics. Hormone effects resulting in transmembrane signaling may depend on this mobility of membrane proteins: for example, formation of a receptor dimer may be required for activation—and receptor and integral membrane proteins may need to associate in order to activate adenylate cyclase and thereby increase the level of the intracellular second messenger, cAMP.

Although the physiology of the movement of molecules across membranes is dealt with in Chapter 2, we will here indicate a few selected examples of transport phenomena correlated with structural features of the cell membrane. The lipid bilayer serves as a transport medium for lipid-soluble molecules to gain entry into the cell, whereas protein-lined hydrophilic "pores" probably provide channels for diffusion of polar entities such as water and ions. Surprisingly, certain lipid bilayers show rates of movement sufficient to account for osmosis. The

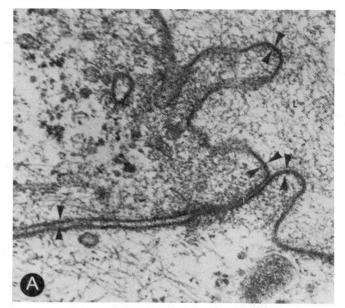

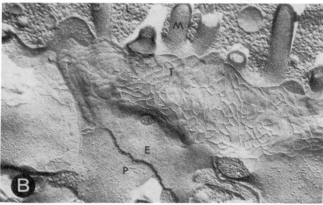

Figure 1.3. *A,* Electron micrograph of portions of epithelial cells bordering the lumen (*L*) of the toad urinary bladder, as seen in a thin section. The three-layered membrane structure of two-cell membranes is seen between the *arrowheads.* In the tight junction (*T*) between the cells the external space is obliterated between the two cell membranes so that a five-layered image is seen (compare with *B*). The "fuzz" in the lumen is associated with the cell surface (see text) (×38,000) (S. Masur). *B,* Toad urinary bladder epithelial cells prepared by the freeze-fracture technique. The fracture has resulted in exposure of the inner or protoplasmic (*P*) leaflet of the membrane in which numerous particles are visible. Fewer particles are seen in the extracellular (*E*) fracture face. The fused region (*central dark line* in *A*) of the tight junction (*T*) appears as weblike protuberances and complementary grooves. The extensive tight junction presumably prevents the intercellular movement of most materials out of the bladder lumen (*L*). A microvillus is seen at *M* (×33,000). (Courtesy of J. B. Wade.)

permeability to water, however, seems to be inversely related to the proportion of cholesterol in the phospholipid membranes.

Endocytosis and Exocytosis

Membrane components are subject to continual turnover. In certain cell types, portions of the membrane invaginate into the cell and pinch off to form the boundary of an intracellular vesicle, vacuole, or tubule. External material is carried into the cell by this process, referred to as *endocytosis* (Fig. 1.1). This material and its enclosing membrane may fuse with lysosomes, or after delivering endocytosed material to an intracellular endosome (see Fig.1.6), the specialized endocytic vesicle membrane may return to the plasma membrane. The fusion to the plasma membrane with membranes of intracellular origin is termed *exocytosis*. When secretory granules exocytose at the cell membranes they release their internal material to the outside of the cell but their membranes are retrieved as endocytic vesicles. These vesicles may, in fact, return to their intracellular origin, the Golgi area. As a result there is a flow of membrane and of material enclosed in membrane-delimited spaces between the surface and intracellular compartments. Similar exchanges seem to occur between certain intracellular organelles, notably the endoplasmic reticulum and the Golgi apparatus. In the various processes involving formation of vesicles (e.g., pinocytic endocytosis) there is often a "bristle" coating, presumably clathrin or a related protein, on the cytoplasmic side of the vesicle membrane (see Figs. 1.1 and 1.6). Exocytosis and incorporation into the surface of intracellular membrane-containing transport units (e.g., channels or transporters) may be the structural basis for rapid changes of cell permeability by increasing the numbers of cell membrane transporters in response to hormonal stimulation.

Structural Aspects of Increasing the Cell Surface

Stable evaginations and invaginations of the cell membrane are important elements in providing a dramatic increase in surface area contact between cell and environment. Microvilli, finger-like evaginations, are generally associated with cell surfaces involved in absorption processes, such as in the intestine and kidney (Fig. 1.3). Conversely, in striated muscle, one finds an invagination of the cell membrane, the transverse tubule, associated with each sarcomere. Since the transverse tubules are continuous with the surface membrane they provide a direct route for the communication of alterations at the cell surface to the contractile system deep within the muscle fiber (Chapter 4).

THE NUCLEUS

Function of Nucleus

The nucleus has two principal functions: replication of deoxyribonucleic acid (DNA) and synthesis of ribosomal, messenger, and transfer ribonucleic acids (RNAs). Because it is best understood, we shall discuss in some detail ribosomal RNA synthesis which occurs in the nucleolus.

Each nucleus possesses one or more *nucleoli*, not delimited by membranes. Each nucleolus consists of a roughly spherical, dense array of fibrils and granules rich in RNA. Often the nucleolus is found in intimate association with special regions of DNA (known as nucleolar organizer regions) which are presumed to carry the information for *ribosomal RNA*. Nucleolar RNA (45S predominantly) is almost certainly a precursor form of ribosomal RNA found in the cytoplasm; if one "labels" RNA synthesized in the nucleolus with radioactive nucleotides, labeled RNA molecules are subsequently detected in the cytoplasm. These RNA molecules complex with protein and form, respectively, a 30S and a 60S subunit. One small and one large subunit combine in the cytoplasm to form a *ribosome* (15–25 nm in diameter). There may be several million ribosomes in a given cell.

The specific function of ribosomal RNA is not well understood; generally speaking, however, all three types of RNA are involved in the translation of genetic information contained in the DNA molecule into specific proteins that are synthesized in the cytoplasm. The ribosomes interact in the process of protein synthesis with two other types of RNA: large *messenger RNA* molecules (mRNAs) determine the sequence of amino acids in proteins by specifying the order of attachment of the small *transfer RNAs* carrying the appropriate amino acids. Our belief that DNA is the template for these RNAs is derived largely from experiments with procaryotic (bacterial) cells.

As the information in a messenger RNA molecule is being "read," several ribosomes attach via their smaller subunits to the mRNA. The combination of an mRNA and its attached ribosomes is referred to as a *polysome*. Each ribosome of a polysome synthesizes a polypeptide chain, so that several chains will be produced simultaneously by a polysome. The nascent peptide seems to be attached to the larger ribosomal subunit; completed protein is released to the cytoplasm. An average polypeptide may be synthesized in 10–20 s.

Morphology of the Nucleus

The interphase nucleus is readily seen in the light microscope as a spheroidal body with a "suggestion" of internal organization (Fig. 1.1). The DNA-containing material can be specifically stained. The nuclear chromatin can be resolved into two types: euchromatin (loosely coiled) and heterochromatin (compact). It seems likely that the euchromatic regions are more active in the transcription process than the heterochromatic regions, i.e., there is little demonstrable RNA synthesis in chromosomes that are largely or entirely composed of heterochromatin, as in the case of sperm cells, polymorphonuclear leukocytes, and the Barr body (one of the X chromosomes of female cells). The association of euchromatin with active transcription may account for some of the selective genetic expression associated with characteristic chromosomal uncoiling patterns found in different tissues within the same organism or at different developmental stages in the same tissue.

Chemical analysis has shown that the chromosome consists of DNA associated with basic proteins (histones) and with other (nonhistone) protein. It has been speculated that the complexing of histone with DNA may have a protective or structural function (preventing alteration or denaturation of the DNA, controlling coiling, etc.), or the histone may have a repressor function (interfering with the template activity of DNA). The amounts of RNA and nonbasic nuclear protein seem to vary in parallel with the metabolic activity of the cell—e.g., sperm cell nuclei have essentially neither RNA nor nonbasic protein.

Isolated chromosomes studied by electron microscopy appear as masses of fibers around 25 nm in diameter, or may have a beaded look with periodic DNA coiling around histone groups (nucleosomes). An individual DNA double helix coated with protein measures less than 5 nm, and while it is known that the fibers of chromosomes are coiled, the nature of the packaging of nucleic acid and proteins is yet to be described. Nevertheless, there are theories, consistent with current evidence, suggesting that a single chromosome contains one, or at most a very few, extremely elongated DNA molecules complexed with protein and coiled into a fiber structure which is seen in the electron microscope.

INTRACELLULAR MEMBRANE SYSTEMS

Nuclear Envelope

The boundary of the nucleus, the nuclear envelope, is a double membrane complex (Fig. 1.4). Each membrane is approximately 7–8 nm thick. The envelope, a flattened sac with an enclosed perinuclear space, resembles the rough endoplasmic reticulum (ER) (see below): (1) the cytoplasmic surface of the outer (cytoplasmic) nuclear membrane has granules which appear to be ribosomes; (2) direct continuities are seen between the cytoplasmic portion of the nuclear membrane and the ER (Fig. 1.4); and (3) the presence of certain enzymes can be demonstrated cytochemically in both the perinuclear space and the cisternae of the ER.

The inner surface of the nuclear membrane is often associated with chromatin and an "internal dense lamella"; the latter may provide some rigidity to the structure. The inner and outer membranes of the envelope join at intervals to form "pores" tens of nanometers in diameter (Fig. 1.4).

How does a "directive" of the nucleus reach the cytoplasm or, conversely, how do cytoplasmic and other external feedback messages reach the nucleus? Non-nuclear substances can act as inducers or repressors of the synthesis of specific proteins in the cytoplasm. This almost certainly requires interaction with genes. Furthermore, most gene products (e.g., mRNA) must leave the nucleus and enter the cytoplasm to express their effects. Permeability properties of the nucleus are too complex to be explained by simple holes. The morphology of the nuclear boundary provides for two alternative routes for the transfer of information—either across membranes of the perinuclear sac or through "pores." The pores are often referred to as "annuli" to emphasize that they are not simple holes but rather organized regions: often pores are seen which contain a diaphragm or plug. In addi-

tion, the membrane adjacent to the pore may show morphological traces of special organization.

To date, the morphological evidence in support of the physiological and biochemical data on transnuclear transport through pores rests mainly on a few observations, such as the movement of electron-dense material (thought to be RNA-containing granules) through the nuclear pores in the insect salivary gland and some other tissues, and the movement of a marker (colloidal gold) into the nucleus when it is injected into the cytoplasm of ameba. Clear morphological evidence on passage of material across the nuclear membranes, as distinct from the pores, is not available.

Endoplasmic Reticulum

Often an integrated biochemical and ultrastructural investigation (involving cell disruption, isolation, and analysis of a homogeneous organelle population) leads to the clearest understanding of organelle function in situ. From such studies, in a variety of cell types, a fraction of membrane-delimited vesicles, referred to as *microsomes*, is recovered. The microsomes perform several functions, including the provision of a base for the attachment of ribosomes, the biosynthesis of lipids, and, in the case of striated muscle, the accumulation and release of calcium. In the intact cell, microsomal

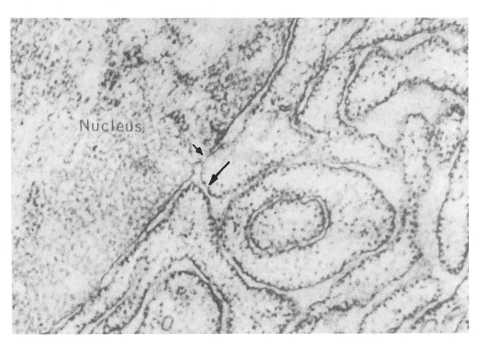

Figure 1.4. Electron micrograph of a portion of a serous cell of mouse salivary gland. The rough endoplasmic reticulum *(large arrow)* of the cell is continuous with the outer membrane of the nuclear envelope. A nuclear pore is indicated by the *small arrow* (osmium fixation; ×84,000.) (From Fawcett, 1966.)

vesicles are not found as such; rather, one observes a tubular network known as the ER (Figs. 1.1 and 1.4). It is assumed that the majority of microsomal vesicles represent a reproducible, preparative artifact arising during cell disruption by the shearing into fragments and closing up of the tubules and sacs of the ER.

ROUGH ER

As noted above, the ER (and/or the microsomes derived from the ER) can provide a base for the attachment of ribosomes (Fig. 1.5A). Such ribosome-carrying ER is referred to as *rough ER*. Microsomal vesicles derived from this rough ER were found to be capable of protein synthesis, the newly synthesized protein appearing in the vesicle lumen. In the rough ER in situ the nascent protein likewise can be demonstrated within the reticulum lumen. This unidirectional passage into the lumen is thought to result from the folding of the original 0.5- to 1.0-nm wide protein into a three-dimensional structure which is large enough to be retained.

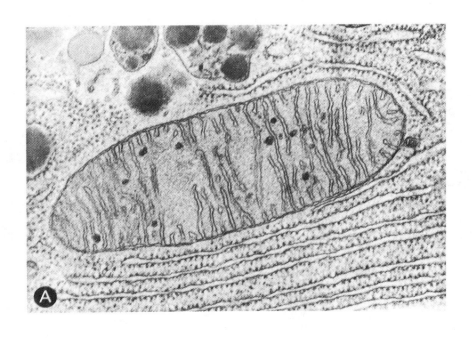

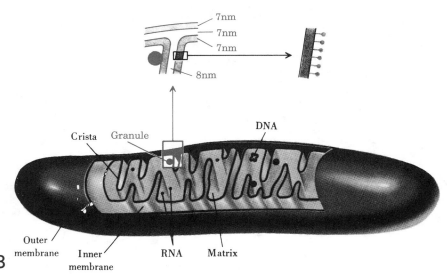

Figure 1.5. *A,* Electron micrograph of a portion of a pancreatic cell showing the mitochondrial structures labeled in the diagrammatic representation in *B.* A parallel array of cisternae of rough endoplasmic reticulum is also evident ($\times$60,500). (From Fawcett, 1966.) *B,* Diagrammatic representation of a mitochondrion. (Data from Ornstein, Palade, Sjostrand, Fernandez-Moran, and others. From Novikoff and Holtzman, 1976.)

The rough ER seems to grow by synthesizing more of itself. The newly made rough ER may lose its ribosomes and thus become converted to *smooth ER*. The relative proportions of rough and smooth ER vary within different cells; for example, the rough ER is extensive in cells which specialize in synthesizing protein for export, while the smooth ER is extensive in steroid-secreting cells.

SMOOTH ER

As noted above, the ER which lacks ribosomes is referred to as the smooth ER. The membrane of the ER carries enzymes which are important in several biosynthetic pathways. For example, the enzymes required for the synthesis of steroid are found in microsomal fractions of steroid-secreting cells. Enzymes involved in triglyceride synthesis as well as phospholipid synthesis are also found in this fraction, the phospholipid sometimes appearing in the ER as small fat droplets. In liver cells, important drug-degrading enzymes are associated with the smooth ER. Also, in the hepatocytes, the close spatial relationship of the smooth ER with glycogen, the major storage form of glucose, suggests that the smooth ER may function in glycogen metabolism. In muscle, the smooth ER (sarcoplasmic reticulum) controls the local concentration of calcium ions near the contractile machinery and thereby influences the contraction and relaxation process (Chapter 4).

Golgi Apparatus

The Golgi apparatus is believed to be a site for the concentration of protein and polysaccharide. It is also a site for completion of the synthesis of the carbohydrate moiety of glycoprotein, e.g., the synthesis of the carbohydrate moieties of thyroglobulin and immunoglobulin begins in the ER, but the terminal sugars are added in the Golgi apparatus. In the case of synthesis of polysaccharides destined for secretion, the precursors are first seen in the Golgi apparatus. Therefore, the apparatus is believed to be the site of synthesis and packaging of polysaccharides for secretion. These products are usually packaged as "granules" within Golgi-derived vacuoles or vesicles which then migrate away from the Golgi apparatus. The enzymes involved in the polymerization of polysaccharide or addition of carbohydrate to protein, glycosyl transferases, have recently been used as marker enzymes for the biochemical isolation of the Golgi apparatus.

The Golgi apparatus consists of a stack of several membranous saccules with associated vacuoles and vesicles (Fig. 1.6). The ER in some cell types is assumed to contribute to the "forming face" or

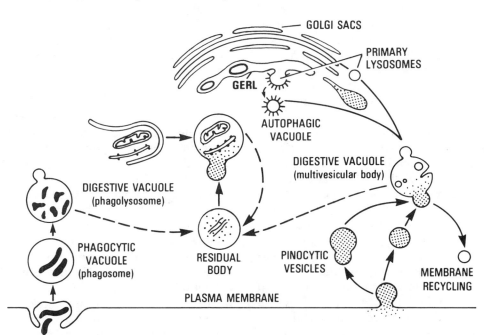

Figure 1.6 Diagram suggesting probable interrelations of some organelles in the membrane-bound transport of lysosomal hydrolases and the formation of lysosomes. Illustrated are: (1) Fusion of primary lysosomes (Golgi vesicles) with phagocytic and pinocytic vesicles resulting in the formation of digestive vacuoles (multivesicular bodies). Some of the vesicles are "coated" (see text). (2) Vesicles and autophagic vacuoles are among the structures that may form from closely interrelated components of the ER and Golgi apparatus (GERL). These membranes are also called the trans-Golgi network. (3) Formation of residual bodies by the accumulation of indigestible residues within lysosomes. (From Holzman and Novikoff, 1983.)

"outer" surface of the Golgi apparatus. Within the stacked membranes of the Golgi apparatus materials are concentrated as they pass from the saccules on the "outer" surface to those forming the "inner" surface of the apparatus (Fig. 1.1). In addition, while passing through the Golgi cisternae, these luminal proteins are modified covalently by removal and addition of specific sugars. In exocrine and endocrine cells the mature secretory granules are generally found in association with the "inner" saccules.

Lysosomes

Lysosomes have been found in virtually all animal cells which have been studied. As organelles they are best defined by biochemical and cytochemical criteria: a lysosome is a membrane-delimited body containing demonstrable acid-hydrolase activity and an intravesicular pH of 5–6. Over 30 acid hydrolases are known to occur in the lysosomes; these enzymes can digest essentially all macromolecules. Material to be digested becomes enclosed within lysosomal membranes permitting isolated, controlled degradation. A proton-translocating ATPase maintains the low intravesicular pH. There are numerous findings suggesting that release of hydrolases from the lysosome into the cell may be important in various pathological states. In silicosis (miner's disease) it is believed that macrophages of the lung take up silica into phagocytic vacuoles which, upon fusion with lysosomes (Fig. 1.6), make the lysosomal membranes leaky. In some inflammations, hydrolases may be released at the surface of the phagocytic cells and affect the adjacent tissues. In many cases these findings are yet to be fully evaluated.

Microscopic identification of lysosomes often consists of demonstration of acid phosphatase activity within a membrane-delimited body; the assumption is that other lysosomal hydrolases are also present. Morphologically, the lysosomes are a motley group of subcellular bodies (Fig. 1.6). Their appearance depends largely on the origin of the enclosed material which is destined for intracellular digestion by the lysosomal hydrolases. In polymorphonuclear white blood cells, Golgi-derived lysosomal granules fuse with phagocytic vacuoles formed as a result of endocytosis of foreign material. *Autophagic* vacuoles are lysosomes which contain bits of the cell's own substance which have been separated along with the hydrolases from the rest of the cytoplasm within a membrane-delimited space. Autophagia, which may be enhanced by stress, is hypothesized to be important for the turnover of some cell constituents.

The degraded soluble products of lysosomal hydrolysis can either enter the anabolic pool to be reused in biosynthesis or to be secreted. An example of the latter is the secretion of thyroid hormone: thyroid colloid travels from the follicle lumen within endocytic vacuoles which then fuse with the lysosomes; the colloid is hydrolyzed and thyroxine is released. It is also noteworthy that indigestible residues accumulate within lysosomes, a phenomenon which accompanies the aging process in neural and other cells. The accumulation of lipid deposits in blood vessels may be one factor contributing to the development of atherosclerosis.

Peroxisomes

The peroxisomes constitute another group of membrane-delimited bodies. They are often associated with the ER. They are concerned with the metabolism of peroxide: peroxisomes contain enzymes (such as catalase) which destroy hydrogen peroxide and other enzymes which produce hydrogen peroxide (such as D-amino acid oxidase and, in some species, urate oxidase). Peroxisomes have, thus far, been found in essentially all cell types, of which liver and kidney are among the best established examples. The function of peroxisomes is currently under active investigation, and there is evidence that, in some species, they are involved in carbohydrate synthesis from fat and in the degradation of purines and fats, as well as in the detoxification of hydrogen peroxide. New peroxisomes are apparently formed by division of preexisting peroxisomes.

Mitochondria

The early cytologists noted that mitochondria were closely associated with motile processes and were situated in regions of intense metabolic activity such as sites of active transport. Biochemists have since shown that two major metabolic pathways, the tricarboxylic acid cycle and the electron transport chain, are situated within the mitochondrion; thus, this organelle is involved in the metabolism of lipids, amino acids, and carbohydrates.

The "typical" mitochondrion (Fig. 1.5) has a length of 5–10 μm and a diameter of 0.5–1.0 μm. It is bounded externally by two lipoprotein membranes (each about 7 nm thick), the inner one of which is thrown into folds termed cristae or tubules. Within the inner membrane is a matrix containing granules, RNA and DNA. The DNA is a small circular molecule and is thought to provide some, but not all, of the necessary information for replication of the mitochondria. The RNA is responsible for synthesis of a few of the proteins of the mitochondrion; the majority of the mitochondrial enzymes are synthesized on cytoplasmic ribosomes under the direction of the nuclear DNA. From physiological and biochemical evidence, it is generally believed that the respiratory enzymes and the components of oxidative phosphorylation are associated in an ordered array on the inner membrane. The order is thought to promote the sequential interaction of substrates and enzymes in these multienzyme systems with concomitant conservation of energy. Therefore, much effort has been directed toward isolation of modular physiological units from inner membrane fractions. The fact that multienzyme assemblies can be obtained from mitochondria has been encouraging. Moreover, morphological studies support the multienzyme assembly hypothesis. First, the respiratory activity of mitochondria is roughly proportional to the amount of inner membrane—a finding which could be explained by presuming an increased number of repeating units or respiratory assemblies associated with the increased membrane area. Second, a repetitive array of particles found in certain preparations suggests a similarly repetitive assemblage of inner membrane enzyme systems. These regularly spaced particles are attached by small stalks to the inner membrane of unfixed and negatively stained isolated mitochondria. (In negative staining, the surface details of the material under study appear as light objects against a dark background.) Surprisingly, after isolation, the stalked particles contained an ATPase predominantly; however, it is suspected that the equilibrium of the reaction catalyzed by the ATPase in the stalked particles is reversed in the intact mitochondrial oxidative phosphorylation system, and that it couples phosphorylation of adenosine diphosphate (ADP) with electron transport. This suggestion is supported by the fact that dissociation of the particle from submitochondrial preparations abolishes the ability of the preparation to carry out oxidative phosphorylation; when the particles are added back, oxidative phosphorylation returns too. These particles are closely associated with the assemblage of enzymes and cytochromes that are distributed in the inner membrane and that form the electron transport system of the cell. Normally these two essential processes, namely, ATP formation and electron transport, are tightly coupled. The morphology of the inner membrane enzyme system remains a matter of current dispute.

The outer membrane of the mitochondrion has a different enzyme content, a larger percentage of lipids, and is more permeable to simple sugars than the inner membrane. In addition to the demonstrably membrane-anchored enzymes of both inner and outer membranes, certain enzymes (such as those of the tricarboxylic acid cycle) are solubilized after disruption of the mitochondria. These enzymes are presumed to be situated in the matrix or possibly loosely attached to a mitochondrial membrane.

Physiologists can clearly define the metabolic state of the mitochondria in terms of electron transport and oxidative phosphorylation. Careful electron microscopic study of isolated mitochondria, and in a few cases of mitochondria in situ, has resulted in the identification of two functionally related states: (1) the condensed state in which the matrix appears dense and the space between the inner and outer membrane is enlarged; this state is seen when oxidative phosphorylation is proceeding at a rapid rate under conditions of excess ADP and inorganic phosphate, and (2) the orthodox state (see Fig. 1.5A) when ADP and P_i are rate-limiting in oxidative phosphorylation. It is hoped that this kind of structural alteration can be explained eventually in terms of the interaction of mitochondrial macromolecules.

MICROTUBULES AND MICROFILAMENTS

The asymmetry observed in certain cell types is sharply at variance with the picture of an idealized cell in which all the organelles surround a central nucleus symmetrically. A nerve cell in which the axon runs several feet as an extension of the perikaryon, an elongated muscle cell, and a squamous (or columnar) epithelial cell all exemplify such asymmetry. Likewise, the nonrandom (asymmetric) movement of subcellular elements is exemplified by transport of neurosecretory products (axonal flow), sliding myofilaments in muscle contraction, and chromosomal movement in cell division. Electron microscopy of aldehyde-fixed tissue has revealed a morphological basis for asymmetric structure and movement in the form of entities referred to as *microtubules* and *microfilaments;* these structures are not membrane-delimited.

Microtubules and Centrioles

In cross-section microtubules are 20–30 nm in diameter and may be followed for several microns in longitudinal sections. They are found in many regions in which phase-contrast and polarizing microscopy had previously demonstrated the presence of formed oriented, elongated elements. Microtubules are often associated with oriented movement, e.g., axonal transport of neurotransmitter from the nerve cell body (where synthesis occurs) to the synaptic terminal where transmitter is released (see Chapter 3). In nerve cells microtubule subunits are also synthesized in the cell body but assembled in the axon.

One of the best established examples of microtubule association with movement in all cells is chromosomal movement in the mitotic spindle. The mitotic poles—toward which the microtubules of the spindle orient and toward which the chromosomes move—have centrioles, usually two per cell. In the interphase cell the pair of centrioles is generally found with long axes at right angles to each other (Fig 1.1). Each centriole is a cylinder 0.15 μm in diameter and 0.5 μm in length, composed of nine sets of microtubule-like elements (Fig. 1.7).

The organization of microtubules can be disrupted by physical means (freezing or high pressure) or by chemical treatment, especially with colchicine. When this is done, motion is inhibited and

some of the structure collapses. Therefore, the affinity of microtubular protein for colchicine serves as a means of identifying microtubular protein in a cell fraction. Isolated and disrupted microtubules yield protein subunits of approximately 6×10^4 daltons. These appear to be globular subunits which may be arranged in a helical fashion to form the microtubules. In normal cells the microtubular protein appears to be present in a form which is assembled into tubules (e.g., for the formation of the mitotic spindle) under appropriate, but as yet not understood, stimulation.

Cilia and Flagella

Microtubule-like structures may also be organized into organelles as diagrammed in Figure 1.7. Cilia and flagella are rapidly beating cell processes which extend 10–200 μm from the cell and are surrounded by a membrane which is continuous with the plasma membrane. The intracellular basal bodies of cilia and flagella are also composed of microtubular structures arranged in the pattern of nine basic units (often referred to as "9+0"); they are widely assumed to be an alternate form of the centriole.

Cilia and flagella generally have, in addition to the basic nine outer sets, a central pair of microtubules ("9+2"). Good evidence suggests that the sliding of tubules within a doublet is the motile force. The process of beating requires cellular energy as indicated by the findings that exogenous ATP can cause beating in isolated cilia and flagella and that the "arms" of the nine sets of microtubules contain an ATPase. The tubules of the cilia are composed of molecules similar to those of the other cellular microtubules.

Microfilaments

Microfilaments are a heterogeneous class of long, thin nontubular structures. Thin microfilaments, 5–7 nm in diameter, are made of actin. Among the most commonly seen microfilaments are those which appear to serve as the structural core of microvilli. Also frequently encountered are tonofilaments on the intracellular side of desmosomes (see below). The best example of association of microfilaments with motion is the extensively developed myofilament system which forms the basis of muscle contraction (Chapter 4). The myofilament proteins, actin and myosin, have now been localized in many other cell types as a result of improved techniques for the cellular localization of specific proteins. Results of the application of antibody to actin have implicated some form of this protein as a constant component of thin (5-nm diame-

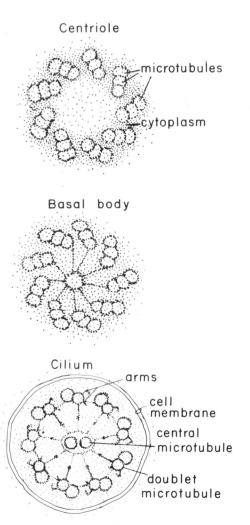

Figure 1.7 Diagrammatic representations of organelles composed of microtubules. (Adapted from Fawcett, 1966; Gibbons, 1967; Satir, 1974.)

ter) microfilaments. Likewise a myosin-like protein, associated with transiently formed thick microfilaments (10 nm), is found in virtually all cells.

JUNCTIONAL COMPLEX

We have thus far limited our view of the cell to those entities circumscribed by and including its boundary, the plasma membrane. Cells rarely are continuous with one another, usually a space of 10–20 nm separates them. Cells are associated in tissues by various means; the best described is the *junctional complex* in epithelial cells (Fig. 1.1). In this complex the plasma membranes of two adjacent cells contribute to specialized attachment sites: a tight junction *(zonula occludens)*, a desmosome *(macula adherens)*, and between these two usually a less well-defined *zonula adherens* where the two membranes are separated by a constant 20-nm space. The desmosome may contain organized extracellular material between the two cell membranes, which may be seen as an additional dense line in parallel with the membranes.

In the region of the tight junction, the outermost layers of the two cellular unit membranes appear to be very closely associated or fused with one another (Fig. 1.1); externally applied tracer molecules (such as ferritin, an iron-containing electron-dense protein) cannot penetrate between the cells at the tight junction. Movement across epithelial cell layers with tight junctions requires a pathway through cells rather than around them. Certain cells are considered to be electronically coupled in that the usual insulation effect on passage of an applied electric current between cells is greatly reduced at a specialized junctional area. This area has been called a *gap junction* because in a thin section it appears to have many small regions of contact between the plasma membranes with obliteration of (gaps in) the adjacent dense lines. Communication between and coordination of the individual cells of cardiac muscle may be effected via such gap junctions.

CHAPTER 2

Properties and Functions of Cell Membranes

OVERVIEW OF FUNCTION

Structure of Cell Membranes

The structure of biological membranes was discussed in some detail in Chapter 1. In this chapter we present only a brief review of structure; the main emphasis here is on function. The current best model of membrane structure, termed the fluid mosaic model, was put forth by Singer and Nicholson in 1972. Depicted in Figure 1.8, it has two essential components, a lipid bilayer and protein. The lipid bilayer is composed of phospholipids that are oriented with polar head groups toward the aqueous environment and with acyl chains toward the hydrophobic interior. The lipid molecules exhibit translation, vibration, and rotation, but are confined to the bilayer. The lipid bilayer serves as a major barrier to movement of solutes across the membrane. Membrane proteins are immersed to greater or lesser extents in the lipid bilayer and exhibit essentially free lateral motion. Integral membrane proteins have one or more hydrophobic domains embedded in the interior of the lipid bilayer. Proteins that have polar regions on each side of the bilayer are termed "transmembrane proteins." For the purposes of this chapter we will concentrate on this type of integral membrane proteins, as such proteins serve as a major route for transmembrane movement of various solutes. Indeed, for each of the specific transport proteins that will be discussed in this chapter, the primary amino acid sequence is known from cDNA cloning, and each is a transmembrane (thus, integral membrane) protein.

Other membrane proteins, called "extrinsic proteins," do not contain hydrophobic segments embedded in the lipid bilayer interior. Generally not involved in membrane transport, they are loosely associated with the lipid bilayer or are attached to the integral membrane proteins.

In this chapter, the term "cell membranes" refers both to the plasma membrane, which surrounds the cell, and to organellar membranes, which form organellar compartments within the cell interior.

General Functions of Cell Membranes

All cell membranes have three general functions. The first function, which is the focus of this chapter, is to *establish, maintain, or vary in a controlled fashion the concentration of electrolytes, nonelectrolytes, or water between two separate aqueous compartments*. This separation is accomplished by a series of solute *pumps* and solute/solvent *leaks*.

The concept of a pump is fairly easy to understand. A pump turns out to be an integral membrane protein that moves a solute against a concentration gradient by using energy.

The concept of leak, on the other hand, may not be so immediately easy to grasp. As used here, the term leak means any mechanism or route (other than a pump) by which a solute or solvent (water) molecule crosses a membrane down a concentration gradient. The student should get past any notion of a leak as something bad, or as a nonspecific hole in the membrane. As we use the term here, *leak pathways are usually specific proteins that facilitate useful movements of solute and solvent molecules*.

Two other functions of cell membranes are mentioned here for completeness. Membranes are a site of *signal transduction* as, for example, when a peptide hormone outside the cell binds to a cell surface receptor, activating an intracellular phosphorylation cascade. Membranes also serve as *surfaces favoring certain molecular interactions*, e.g., covalent bonding of a polar molecule (glycerol phosphate) to an amphipathic molecule (stearoyl coenzyme A) using the endoplasmic reticulum membrane as a catalytic surface. These two functions will not be further discussed.

General Overview: Establishment, Dissipation, and Maintenance of Solute/Solvent Gradients Across Cell Membranes

As mentioned above, the major function of cell membranes is to separate two aqueous compartments of dissimilar composition, and generally to keep the compositions of the two compartments dis-

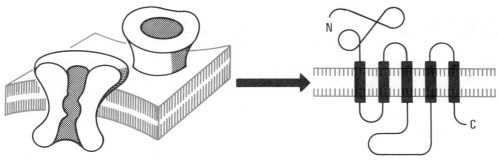

Figure 1.8. Fluid-mosaic model of membrane structure shown in three dimensions on the *left* and in cross-section on the *right*. On the *left,* two globular integral membrane proteins that might be pores, channels, or carriers are depicted by the solid bodies in the lipid bilayer. A cross-sectional representation of one of these proteins is shown on the right for comparison with Figures 1.9, 1.15, and 1.19.

similar. The establishment of solute concentration gradients across cell membranes by pumps requires energy. The energy for this separation derives from metabolism and is packaged for delivery to cell membrane ion pumps in the form of ATP. Several integral membrane proteins have now been described that concentrate ions across a membrane by cleaving ATP to ADP and inorganic phosphate (P_i) (discussed more fully below). At the same time that ATP-utilizing pumps concentrate solutes from one compartment into another across a membrane, the solutes tend to leak back across the membrane down the concentration gradient. As mentioned above, this back-leak can occur via one of several routes, including movement directly through the lipid bilayer ("simple diffusion"), or via interaction with one or more integral membrane proteins ("facilitated diffusion"). The steady-state concentration gradient that can be achieved across the membrane is a function of the relative rates of the various pump and leak components. High pump rates relative to leak rates generate large gradients; if leaks are large relative to pump rates, gradients are small. Table 1.1 shows the steady-state concentrations of several different ions within the cell and blood. It is clear that pump/leak systems can generate variable (and substantial) gradients across membranes.

Table 1.2 lists eight examples of pump/leak mechanisms involved in membrane transport. This table serves as the outline for most of this chapter. The reader should refer back to this table frequently as each mechanism is discussed. Each specific example will be used to illustrate a general concept of membrane transport; these concepts are listed in the right-hand column of Table 1.2.

SPECIFIC PUMP/LEAK MECHANISMS IN MEMBRANE TRANSPORT

ATPase Pumps: Primary Active Transport

The best-characterized ion-translocating ATPase is the Na/K-ATPase. The enzyme is a transmembrane protein. Several features of its structure, derived from cDNA cloning and more classical biochemical approaches, are depicted in Figure 1.9. The pump has two subunits, α and β, that form a dimer. The dimer translocates 3 Na^+ ions from the cytoplasm to the cell exterior and 2 K^+ ions from the cell exterior to the cytoplasm, cleaving one ATP to ADP and P_i in the process. The enzyme can be inhibited by ouabain, a compound related to the cardiac glycoside digitalis. Figure 1.9 shows that, typical of integral membrane proteins, both the β and α subunits have hydrophobic sequences (1 and 6–8, respectively) embedded in the lipid bilayer, and each has hydrophilic sequences on both sides of the membrane. Because the Na^+/K^+ stoichiometry is 3:2, each "turnover" of the pump transfers one positive charge out of the cell. The consequences of such charge transfer will be dealt with later in this chapter. The important point here, however, is that *K^+ is concentrated within the cell and Na^+ is concentrated outside of the cell.*

There exist two other general categories of ion-translocating ATPases. Proton-ATPases are present in endosomes, lysosomes, mitochondria, and the

Table 1.1.
Ionic Gradients Generated by Pump/Leak Systems[a]

Ion	Concentrations		Blood:Cytoplasm Gradient
	Blood	Cytoplasm	
Na^+	145 mM	12 mM	12:1
K^+	4 mM	140 mM	1:35
H^+	40 nM	100 nM	1:2.5
Cl^-	115 mM	4 mM	29:1
HCO_3^-	25 mM	10 mM	2.5:1
Mg^{2+}	1.5 mM	0.8 mM	1.9:1
Ca^{2+}	1.8 mM	100 nM	18,000:1

[a]Ca^{2+} and Mg^{2+} concentrations are free (unbound) concentrations. Intracellular values are representative, e.g., erythrocyte [Mg^{2+}]. Intracellular anionic proteins (not shown) balance intracellular cations.

Table 1.2.
Representative Pump/Leak Mechanisms in Membrane Transport

General Category	Specific Mechanism	Species Pumped or Leaked	Representative Membrane Protein	Energy Source	Charge Transfer	General Membrane Transport Concept
Pumps	Translocating ATPases	Na^+ and K^+ pumped	Na/K-ATPases[a]	ATP	Yes (+)	Primary active transport
Leaks: Simple Diffusion	Solute diffusion	Glycerol leaked	None	Δ [Glycerol]	No	Simple diffusion
	Water diffusion	H_2O leaked	None	Δ [THO]	No	Use of tracers
Leaks: Osmosis	Osmotic water flow	H_2O leaked	"Pore"	Δ [H_2O] and Δ impermeant solute	No	Osmosis
Leaks: Facilitated Diffusion	Simple carrier	Glucose leaked	Erythrocyte glucose transporter[a]	Δ [Glucose]	No	Carrier-mediated transport
	Countertransport	HCO_3^-	Erythrocyte Band 3[a]	Δ [HCO_3]	No	Carrier-mediated exchange
	Cotransport	Na^+ leaked (glucose pumped)	Na^+-dependent[a] glucose transporter	Δ [Na^+]	Yes (+)	Secondary active transport
	Ion channels	Na^+ leaked	Brain Na^+ channel[a]	Δ [Na^+]	Yes (+)	Electrical potentials and current

[a]Primary amino acid sequences have been deduced from cDNA cloning; models shown in text. "Δ" = difference in concentration of substance across the membrane.

plasma membranes of certain epithelial cells. There are probably several forms of H^+-ATPases. The resulting H^+ gradients play a role in the degradation of proteins, digestion of ingested food, and elimination from cells of the protons generated by metabolism. Calcium-ATPases are found in the sarcoplasmic reticulum, certain plasma membranes, and other organelles. These pumps create 10,000:1 calcium gradients across these membranes. Control of the calcium leak step is important in excitation-contraction coupling in muscles and in signal transduction across many plasma membranes.

All three types of ion-translocating pumps are *reversible*. If one creates a high enough concentration gradient of the transported species oriented opposite in direction to that of normal pumping, one can generate ATP from ADP and P_i. Indeed, H^+ gradients across mitochondrial membranes are derived

from metabolism and generate ATP from ADP and P_i using an "H^+-ATPase."

These ion-translocating ATPase pumps perform *active transport*. Active transport can be defined as the movement of solute molecules against a concentration gradient by an energy-dependent mechanism. (In the case of charged electrolytes, the movement is against an "electrochemical gradient." The distinction between an electrochemical gradient for electrolytes and a chemical gradient for nonelectrolytes will be discussed in the second section of this chapter.) Active transport comes in two basic forms. The ion pumps discussed here perform *primary active transport*: the movement of a solute across a membrane against a concentration (or electrochemical) gradient is *directly* powered by the hydrolysis of ATP. The other form of active transport, secondary active, utilizes the potential energy

Figure 1.9. Model of Na/K-ATPase orientation in the membrane based on cDNA cloning. The α subunit has five binding sites: (1) Na^+ binding site, (2) K^+ binding site, (3) ouabain binding site, (4) phosphorylation site, (5) ATP binding site. Amino acid sequences predict 6–8 membrane-spanning segments. It is not clear whether the carboxyl terminus is intra- or extracellular. The β subunit has one membrane-spanning segment, a short cytoplasmic amino terminus, and three potential external glycosylation sites (CHO). (Adapted from Rossier et al., 1987.)

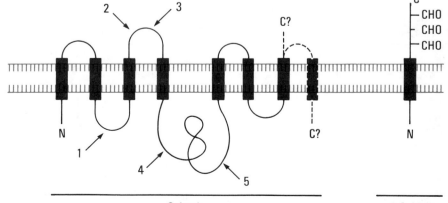

present in the concentration gradients produced by ion-translocating ATPases. Here, energy dissipated by the movement of one solute *down* a gradient is used to move another solute *up* a gradient. No ATP is directly hydrolyzed in the process. Secondary active transport will also be discussed in more detail below.

Thus, ion-translocating ATPases generate concentration gradients using ATP. *Concentration gradients form the basis for all other membrane transport processes.* Solute and solvent molecules leak across cell membranes down concentration gradients by several different mechanisms. These various leak mechanisms will now be discussed in turn.

Solute Leaks via Simple Diffusion

Table 1.2 gives an example of a solute, glycerol, that, within experimental error, appears to cross cell membranes via the process of "simple diffusion." If glycerol is placed in an aqueous solution on one side of a cell membrane at a concentration higher than that on the other side, within a short time the two concentrations will equalize. How does this occur?

Diffusion of molecules in solution from an area of high concentration to an area of lower (or zero) concentration results directly from the Brownian motion of these molecules. This process is easiest to conceptualize if one considers the following experiment (Fig. 1.10). Two fluid compartments are separated by a completely impermeable barrier. The left-hand compartment consists of a well-mixed solution of solute X in water at some finite concentration. The right-hand compartment contains only water without solute. At time zero, the barrier is quickly removed without causing turbulence and the two solutions meet each other at a planar interface. What happens next is depicted in Figure 1.10. Each individual solute particle undergoes random jumping movements via Brownian motion. Consider especially the two solute molecules closest to the liquid-liquid interface at time zero (top) in Figure 1.10. There is a finite chance that, within a given time, one of these molecules will randomly cross the interface from left to right. On the contrary, at time zero there is no chance for a solute molecule to move from right to left across the interface, simply because no solute molecules exist in the right compartment. Einstein showed mathematically that such random motion of individual solute molecules results in a process that, at the macroscopic level, we call "diffusion down a concentration gradient." Eventually, the solute concentrations in the two compartments will equalize. Such a state is more disordered than that at time zero, i.e., the entropy

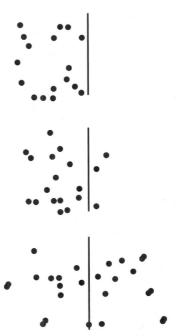

Figure 1.10. Computer simulation of the diffusion process. The *vertical line* is a planar interface that separates a region containing solute molecules *(left)* from one devoid of solute *(right)*. Each molecule is allowed only random motion with time by the computer program. Shown are solute distribution patterns at time zero *(top)*, an intermediate position *(middle)*, and at equilibrium *(bottom)*. (Adapted from W. D. Stein, 1986.)

of the system has been increased, and so diffusion is energetically favored. The "diffusion" in Figure 1.10 was modeled on a computer simply by imparting Brownian motion to each dot and allowing time to pass.

The diffusion rate or flux (J) of solute molecules across the interface is a function of: (1) the area of the interface (A); (2) the difference in the concentration (c) of the solute across the width (x) of the interface, dc/dx; and (3) a proportionality constant called the diffusion coefficient (D) that varies directly with the speed of random motion of the solute in the solvent (water). *Fick's first law of diffusion* states that

$$J = -DA \frac{dc}{dx} \qquad (1)$$

The minus sign exists because solute molecules diffuse from a region of high concentration (large "c" term) to a region of low concentration (small "c" term) as a function of distance, x. Therefore, referenced to the region of high solute concentration, dc/dx is always negative.

The diffusion of glycerol across a cell membrane can be explained in a similar fashion, with the lipid bilayer substituting for the liquid-liquid interface of Figure 1.10. Now, however, in order to calculate the *diffusion rate (flux) through the membrane, J^m*, one must know the diffusion coefficient for the solute within the membrane, D^m, as well as the concentration gradient within the membrane, dc^m/dx^m. Then:

$$J^m = -D^m A^m \frac{dc^m}{dx^m} \qquad (2)$$

The term dc^m/dx^m is a function of the tendency of a given solute to dissolve into the lipid bilayer from the aqueous solution. Glycerol's three OH^- groups constitute three hydrophilic centers that can form hydrogen bonds with water, i.e., glycerol is in a more favorable energetic state in water than in an apolar, hydrophobic lipid bilayer. The tendency of a given solute to distribute in oil versus water is given by the *partition coefficient, β*. Glycerol has a low value of β, i.e., it is much more soluble in water than in oil. Higher values of β indicate a greater lipid:water solubility ratio. Figure 1.11 relates β to the concentration profile within the membrane. It is clear that a solute with a low β has a low value of dc^m/dx^m, and thus a low rate of diffusion through the lipid bilayer.

A useful term that lumps together β, D^m, and the

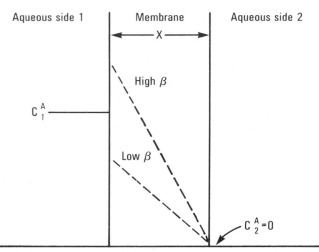

Figure 1.11. Concentration profile of two theoretical solutes within the lipid bilayer. Solute is present in aqueous compartment 1 at concentration C_1^A. The concentration in aqueous compartment 2, C_2^A is zero. A solute that partitions preferentially into oil as opposed to water is shown by the "high β" line. A solute (such as glycerol) that is more water-soluble than lipid-soluble is shown by the "low β" line. The slope dc^m/dx^m has a larger negative value for the "high β" solute than for the "low β" solute, thus the diffusional flux across the membrane of the solute with a high β value is higher than the flux of the solute with a low β.

thickness, x, of the membrane is the *permeability coefficient, P*:

$$P = \frac{\beta D^m}{x} \qquad (3)$$

For a standard membrane thickness, P is thus proportional to a solute's lipid solubility (β) and its diffusion coefficient in a lipid solvent, D^m.

It is easy to show that solute diffusion (flux) across a lipid bilayer follows a form of Fick's law in which Δc is the concentration difference between the two aqueous compartments:

$$J = -PA \, \Delta c \qquad (4)$$

Equation 4 predicts that, at a constant P and A, J will vary linearly as a function of Δc. This linear relationship is a hallmark of diffusion occurring either in free solution or across a lipid bilayer, as opposed to the nonlinear relationship that is seen with facilitated diffusion mechanisms (below).

Returning to Table 1.2, then, we see that the "leak" flux of glycerol down a concentration gradient across a cell membrane occurs by simple diffusion. No membrane protein is involved in the process. The source of energy is simply the glycerol concentration gradient imposed by the investigator. The process is electroneutral; no charge is transferred.

Do physiologically important solutes cross cell membranes by simple diffusion? The answer is "yes," although it can be quite difficult to determine exactly how much this mechanism contributes to total solute flux across a membrane. The more one studies solute flux, the more one tends to discover nondiffusional leak pathways. This does not negate the fact, however, that such simple diffusional pathways exist.

Leak of Water via Simple Diffusion: Use of Tracers

In the previous section we discussed the physical basis for diffusion of solute molecules either free in solution or across a lipid bilayer, that is, Brownian motion from a region of high concentration to one of low concentration. One can use the same concepts to study the diffusion of water molecules within the solvent, i.e., within water itself. How? One simply makes a small percentage of water molecules distinguishable from all the other water molecules using tritiated (^3H) water (^3H-O-H, abbreviated THO).

Reconsider the experiment described in Figure 1.10, but instead of solute molecules in one solution, imagine that a few of the myriad water molecules in the lefthand solution are radioactive THO. Their concentration can be determined using scintillation counting. At time zero after removing the hypothetical barrier, these THO molecules begin to diffuse from left to right as if they were solute. This occurs because *there is a concentration gradient for the tracer*. The equation describing diffusional flux of radioactive tracer is the same as Fick's Law for a solute:

$$J^*(l \rightarrow r) = -DA \frac{dc^*}{dx} \qquad (5)$$

where $J^*(l \rightarrow r)$ is the flux of *tracer water molecules* from left to right and c^* is the concentration of *tracer* water molecules in the nonradioactive water solvent.

In using tracers to study transport, one assumes that the tracer behaves in all respects as does the nonradioactive moiety, i.e., that ^{3}H-O-H and H-O-H behave identically. The evidence available supports this assumption.

If one can envision the experiment in Figure 1.10 using THO, one can then study true water diffusion across a lipid bilayer as in Figure 1.11 using THO. In the same way that the oil:water partition coefficient, β, for a solute describes its tendency to reside in either water or lipid phases, so does β for THO measure the tendency of water molecules to enter the lipid bilayer. β for THO is about 2.5 fold less than β for glycerol, i.e., water molecules prefer the company of other water molecules over oil to an even greater extent than do glycerol molecules.

The permeability coefficient of a synthetic lipid bilayer to THO is fairly low. Interestingly, it is roughly the same as that of several plasma membranes whose specialized function is to minimize water flow (e.g., some water-impermeable kidney tubule membranes). Such low permeability values appear to represent the best achievable ability of biomembranes to compartmentalize water. For most other (perhaps all other) cell membranes, additional mechanisms are available for transporting water at higher rates across the membrane. One very important water transport mechanism, osmosis, is discussed next.

Leaks: Osmosis

Osmosis is a mechanism producing large water fluxes across a membrane. In order for osmosis to occur, two conditions must be met: (1) a solute must be present at a higher concentration on one side of the membrane than the other; and (2) that solute must have a lower diffusional permeability, P,

across the membrane than P for H_2O. Ions represent the typical solutes causing osmotic water flow. Ions meet both of these criteria: (1) they are concentrated across cell membranes (Table 1.1) by ion-translocating ATPase pumps as discussed above, and (2) they exhibit very low diffusional permeability coefficients relative to water. Table 1.3 shows that P for Na^+ and K^+ across erythrocyte membranes is many orders of magnitude lower than P for water (THO). If ions, e.g., NaCl, are concentrated on one side of a cell membrane by pumps, water will flow toward the side of the higher NaCl concentration by osmosis.

One of the first demonstrations of osmosis used an arrangement of a thistle tube, a parchment membrane, and sucrose as shown in Figure 1.12. Parchment membrane is essentially impermeable to sucrose, but is freely permeable to water. A solution of sucrose in water causes osmotic water flow across the parchment membrane toward the sucrose side, because the conditions for osmosis have been met (concentration difference of a solute with a lower membrane permeability than that of water). In fact, enough water flows in this arrangement to raise a column of the sucrose solution up the narrow portion of the thistle tube, graphic evidence that *osmosis produces hydrostatic pressure*.

What causes osmosis? Theories of osmosis at the microscopic level work best if one postulates a membrane "pore" that allows water molecules, but not solute molecules, to pass through. This is shown schematically in Figure 1.13. There must exist a critical, microscopic planar interface at the junction of the mouth of the water-filled pore and the solution containing the impermeant solute. At this interface, water molecules exist in a higher concentration at the mouth of the pore than in the adjacent sucrose solution (where a portion of the solution volume is occupied by sucrose molecules). By the same process of Brownian motion that accounts for simple diffusion, as discussed above, a water molecule jumps at random from the mouth of the pore

Table 1.3.
Orders of Magnitude of the Permeability Coefficients of Erythrocyte Membranes[a]

Substance	Permeability Coefficient, cm/s	Source
Water	10^{-2}	Bovine
Urea	10^{-4}	Bovine
Cl^-	10^{-4}	Human
K^+	10^{-8}	Human
Na^+	10^{-10}	Human

[a]For references see Stein, 1986.

Figure 1.12. Demonstration of osmosis. A parchment membrane that is impermeable to sucrose but permeable to water is stretched over the wide end of a thistle tube. A sucrose/water solution is introduced inside the tube. The tube is lowered into water so that the top of the sucrose solution is even with the top of the water *(left)*. After some time *(right)*, water has flowed into the thistle tube, raising the sucrose solution up by a given hydrostatic pressure, ΔP (cm H_2O).

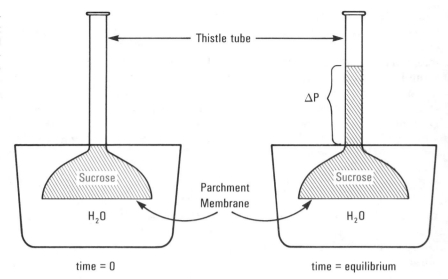

into the sucrose solution. The movement of one such water molecule out of the pore creates a momentary vacancy in the pore itself, i.e., the number of water molecules, n, within the pore is temporarily lowered. From the well-known gas law as applied to water in the pore:

$$PV = nRT \qquad (6)$$

a drop in n produces a drop in pressure, P, within the pore (R, the gas constant; T, the temperature; and V, the volume of the pore are assumed to be constant). A gradient of pressure within the pore develops that drives *bulk water flow,* not diffusional water flow, through it from left to right.

It is important to understand that the rate of

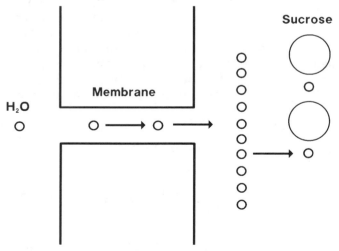

Figure 1.13. Model for osmotic water flow through a semipermeable membrane. Sucrose molecules, represented by *large circles,* are present in the righthand solution and cannot penetrate the pore. Water molecules *(small circles)* move from left to right through the pore.

water flow produced by osmosis is much larger than can be accounted for by simple diffusion of water alone. In an experimental arrangement such as that in Figure 1.12, one can substitute THO for sucrose and measure tracer appearance in the lower chamber. One finds that the rate of THO movement across the membrane by diffusion is several hundred times lower than the rate of osmotic water flow measured using sucrose. This suggests that the artificial membrane must contain the microscopic equivalent of the sucrose-impermeable "pore" shown in Figure 1.13. In contrast, artificial lipid bilayers show essentially equivalent rates of diffusional and osmotic flow, consistent with the absence of microscopic "pores" in a pure lipid bilayer.

The exact biochemical and structural nature of these pores in cell membranes remains to be determined. No "water pore" has yet been purified, nor has its cDNA been cloned. Yet it seems clear that such pores must exist. There is some evidence that water molecules move through integral membrane proteins intended for other purposes (e.g., ion channels or carriers) as if the interior of these proteins represented water "pores."

Let us now turn to a useful mathematical description of osmosis that takes into account the inability of solute molecules to enter the water pore. By analogy with Fick's law of diffusion in which flux = permeability $\times$ concentration gradient, osmotic water flow can be described

$$J_v = L_p \, \Delta \pi \qquad (7)$$

where J_v is net volume flux, L_p is a permeability-like term called the hydraulic conductivity, and $\Delta \pi$ is the difference in osmotic pressure across the membrane. Osmotic pressure here is measured in the standard way using colligative properties, e.g., by freezing-point depression. As long as the solute(s) making up $\Delta\pi$ has a very low permeability across the membrane, this equation is adequate to predict the rate of osmotic water flow.

In considering Figure 1.13, however, it is clear that, if solute were to move through the membrane from right to left, this would reduce the solute concentration at the righthand mouth of the pore, raise it at the lefthand mouth, and diminish the forces favoring left-to-right osmotic water flow. In the case of the thistle tube in Figure 1.12, the hydrostatic pressure would be less than that achieved with a completely impermeable solute such as sucrose. In the extreme, if a solute were as permeable through the membrane as water, it would readily equilibrate across the parchment membrane, equalizing its own concentration, producing no osmotic water flow whatsoever.

The interplay of solute permeability and osmotic water flow can be dealt with mathematically by introducing a term called the *reflection coefficient:*

$$J_v = \sigma L_p \Delta\pi \qquad (8)$$

If a solute is completely impermeant across the membrane, i.e., it is "reflected" away from the membrane, $\sigma = 1$ and the maximum J_v for any $\Delta\pi$ is achieved. If, on the other hand, a solute permeates the membrane as well as THO, then it is not "reflected" at all, $\sigma = 0$ and J_v is zero.

The concept of the reflection coefficient is biologically important in determining whether a given solute is likely to cause osmotic water movement across cells, and thus produce changes in cell volume. Consider two separate red blood cells bathed in plasma-like artificial solutions (Fig. 1.14). Suppose that we abruptly add a hypothetical solute of $\sigma = 1$ to the first (top), and an equimolar amount of a different solute with $\sigma = 0$ to the second (bottom). The first will cause osmotic water flow out of the cell, and thus decrease cell volume. The second will cause no change in cell volume.

Such an experiment illustrates the difference between the concepts of *osmolarity* and *tonicity.* Osmolarity is a function of the number of solute particles in a solution; it is measured using colligative properties such as freezing-point depression or vapor-pressure elevation. In contrast, tonicity is a function of how well a given solute causes osmosis across cell membranes. In the example just given, addition of impermeant solute ($\sigma = 1$) to the solution bathing red cells makes that solution *hyperosmolar* and *hypertonic* relative to the cell interior. Therefore, the cells shrink. In the second case, addi-

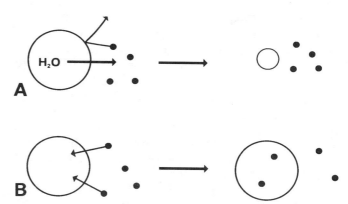

Figure 1.14. Osmolarity versus tonicity. Two different cells, *A* and *B*, are depicted by the *circles*. At time zero, four solute molecules are added outside each of the cells, raising the osmolarity accordingly. *A,* The reflection coefficient, σ, for the solute is 1. Osmotic water flow out of the cell occurs, causing cell shrinkage. The hyperosmolar solution created by the solute addition is thus a *hypertonic* solution. *B,* The same number of solute molecules are added, but now $\sigma = 0$. At time zero, the extracellular osmolarity is just as high as in *A* above. However, no osmosis and no cell shrinkage occurs. The hyperosmolar solution is thus *isotonic.*

tion of a completely permeant solute ($\sigma = 0$) creates a *hyperosmolar, isotonic* solution. In this case, although the concentration of solute has increased, no change in cell volume occurs.

In this section we have seen that osmosis represents a special kind of water leak capable of moving large volumes of water quickly across cell membranes. Because no ATP-utilizing "water pump" has ever been found (it turns out that such a pump would be energetically too expensive), osmotic water flow represents passive transport driven by solute concentration gradients. Indeed, when a large rate of water transport across cell membranes is required (e.g., in kidney tubules), this is accomplished by pumping ions, creating ionic gradients, and driving osmosis.

Leaks: Facilitated Diffusion: Carrier-Mediated Transport

We previously established that solute molecules can cross cell membranes by simple diffusion, but that the rate of this process depends heavily upon the solubility and mobility of the solute molecule in the lipid bilayer. Many biologically important hydrophilic solutes need to cross cell membranes rapidly; simple diffusion is inadequate to meet this need. For such solutes the rate of passive (non-pump-mediated) flux across membranes is speeded through the process of *facilitated diffusion.* This process uses an integral membrane protein to provide a relatively specific, "aqueous" route for such solutes across the lipid bilayer. The transport of glucose across red blood cell membranes typifies facilitated diffusion. The protein responsible, the glucose transporter, has been sequenced by cDNA cloning methods; the proposed general structure vis-à-vis the lipid bilayer is shown in Figure 1.15. As with the Na/K-ATPase, we see that the glucose transporter is an integral membrane protein with hydrophobic sequences embedded in the lipid bilayer and both internal and external hydrophilic domains in the aqueous environment. The carrier binds glucose on one side of the membrane and translocates it to the other.

The glucose transporter exhibits the classic characteristics of a membrane carrier system: (1) the flux, J, down a concentration gradient is too large to be accounted for by simple diffusion based on an estimated solute mobility in the membrane (D^m) or lipid solubility (β); (2) Fick's law is not obeyed, i.e., flux does *not* vary linearly with the concentration gradient, Δc. Rather, J versus Δc shows saturation (enzyme-like) kinetics; (3) transport can be inhibited by structural analogues, as seen in competitive inhibition of enzyme activity; and (4) addition of substrate to the opposite side of the membrane (so-called "trans" side) accelerates flux from the "cis" to the "trans" side of the membrane.

The concentration dependence of solute flux in one direction across a membrane for a typical carrier-mediated system is illustrated in Figure 1.16. Because the carrier protein is embedded in a lipid bilayer which itself allows a finite (albeit low) rate of simple diffusion, one must subtract out this latter (linear) component from the total J. The remaining flux shows Michaelis-Menten type kinetics:

$$J(cis \rightarrow trans) = \frac{J_{max} \cdot \Delta c}{K_m + \Delta c} \qquad (9)$$

where J_{max} is the maximum achievable $J(cis \rightarrow trans)$ and K_m is the affinity constant.

The analogy with enzymes is a very appropriate one. Both transmembrane solute carriers and enzymes: (1) show K_m values ranging from the micromolar to the molar range; and (2) have "turnover" numbers ranging from 10^2–10^6/s. It also seems clear that, in the same way that enzymes undergo small conformational changes after binding substrate, so do carriers undergo structural change in the course of translocating solute across membranes.

For years there has been debate about the exact molecular mechanisms of such conformational changes and solute translocation. The possibilities have included: (1) that the transport protein binds solute on one side, turns a flip-flop in the membrane (like a revolving door), and unloads on the other side; (2) after loading on one side, the protein diffuses across the bilayer like a ferryboat, unloading on the other side; (3) the protein represents a fairly rigid tube spanning the lipid bilayer. It binds solute on one side and then undergoes a small conformational change during which the solute molecule "squirts" through the tube and is unloaded on the other side. This third mechanism now appears to be the correct one. Figure 1.17A shows a schematic of glucose translocation. After binding glucose on the "cis" side, the carrier unloads on the "trans" side. If trans glucose concentrations are low, the carrier then returns unloaded to the cis side again.

This kinetic model suggests that the glucose transporter might exist three dimensionally almost like a pore in the membrane. In fact, the protein structure in Figure 1.15A is misleading because it is laid out flat; probably three dimensionally, the membrane-spanning domains wrap around to form the walls of a cylinder, like wooden staves of a barrel.

Countertransport: Carrier-Mediated Exchange

An alternative mode of operation of a solute carrier is that the carrier cannot return from the trans to the cis side unloaded; only a loaded carrier can return. Such a carrier, illustrated in Figure 1.17B, mediates only *solute exchange*. The best-studied exchange protein is the Cl-HCO_3 exchanger of red cells, also known as "band 3" protein. (The membrane proteins of the erythrocyte were named by their relative positions, detectable as bands, after electrophoresis on polyacrylamide gels.) Band 3 is crucial for movement of CO_2 produced by tissues into the lungs for exhalation. In capillaries and venules, CO_2 diffuses into erythrocytes where it is converted to HCO_3^- by carbonic anhydrase. HCO_3^- then leaves the red cell via band 3 in exchange for Cl^-, which enters the cell. The whole process is reversed in the lung.

Like the other membrane proteins discussed, band 3 has been sequenced using recombinant DNA techniques. A model of its orientation is shown in Figure 1.15B; the familiar structural themes of a transmembrane protein are obvious. Band 3 has been extensively characterized kinetically. These studies have revealed three general results about ion exchange by band 3: (1) It is "electroneutral" (i.e., the carrier never slips trans-to-cis unloaded). Thus, exchange of anions is obligatorily 1:1. (2) The "squirt-through" translocation mechanism as described above is the correct one. The "revolving door" and "ferryboat" models are wrong. (3) Several different anions can bind to either the cis or trans binding sites, all with different K_ms. It is the

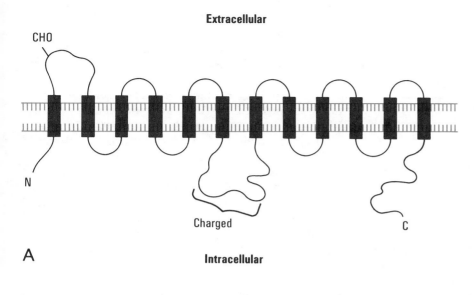

Extracellular

Charged

A   **Intracellular**

Figure 1.15. Proposed model for the orientation of three transmembrane solute carrier proteins in the lipid bilayer. *A,* Erythrocyte glucose transporter. *N* = amino terminus, *C* = carboxyl terminus. Twelve putative membrane-spanning domains are shown as *rectangles.* The intracellular hydrophilic segment marked "charged" contains numerous positively and negatively charged amino acids. *CHO* = glycosylation site. (Adapted from Mueckler, 1985.) *B,* Band 3 orientation in the erythrocyte plasma membranes. Twelve membrane-spanning segments are shown with a possible 13th. *N* = amino terminus, *C* = carboxyl terminus (position not yet determined). *CHO* = oligosaccharide attachment site. *S* = binding site for stilbene disulforates, inhibitors of anion exchange. (Adapted from Kopito and Lodish, 1985.) *C,* Na⁺-glucose cotransporter. There are 11 membrane-spanning regions, two potential glycosylation sites (*CHO*), and a highly charged hydrophilic section near the C-terminus that may play a role in glucose binding. (Adapted from Hediger et al., 1987.)

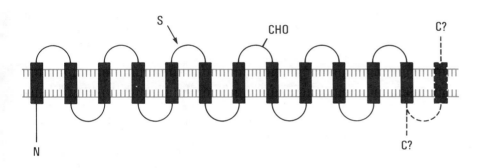

Extracellular

B   **Intracellular**

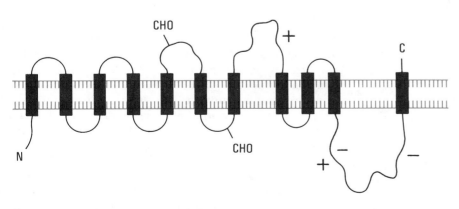

Extracellular

C   **Intracellular**

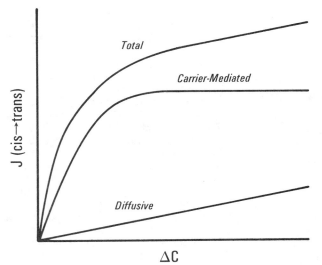

Figure 1.16. Typical concentration-dependence (Δc) of flux (J) for a carrier-mediated (facilitated-diffusion) process. Radioactive solute is added to one side ("cis" side) and its disappearance from that side (and/or its appearance on the opposite, or "trans" side) is measured. The flux shown is thus a unidirectional one, J(cis → trans). Total flux shows non-Michaelis-Menten kinetics, because it is actually the sum of two processes. A small (simple) diffusive flux, shown at the bottom, varies linearly with Δc. When this linear component is subtracted from the total flux, the remaining flux, the carrier-mediated flux, shows typical saturation kinetics.

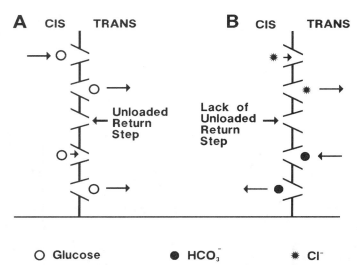

Figure 1.17. Models of two types of carrier-mediated transport. *A*, Simple carrier similar to erythrocyte glucose transporter. Solute (glucose) binds to protein site on one side (*cis*), is translocated across the lipid bilayer, and unloaded (*trans*). The carrier returns unloaded (although it can also return loaded if the trans glucose concentration is high). *B*, Exchange carrier such as the band 3 Cl-HCO$_3^-$ exchanger of erythrocytes. Cl$^-$ binds to a cis site and is translocated across. This carrier will not return unloaded; it remains "facing" the trans side until it binds a solute. In this example it binds HCO$_3^-$ trans (although it could also bind Cl$^-$ trans).

relative cis and trans concentrations of Cl$^-$ versus HCO$_3^-$, plus the various K_ms at each site for each ion, that determine overall transport rates by band 3. Despite this complexity, it should be kept in mind that, in both peripheral tissues and in lung capillaries, it is a HCO$_3^-$ gradient across the cell membrane that is dissipated by this particular leak pathway. HCO$_3^-$ movement is passive, i.e., downhill, and occurs many orders of magnitude faster than would be the case via simple diffusion without band 3.

Another ion exchanger of major importance is the Na-Ca exchanger. Although the exact stoichiometry of exchange is not clear, it appears to be about 3–5 Na$^+$ ions exchanged for each Ca^{2+} ion. Thus, the exchanger is probably electrogenic. Na-Ca exchange has been found in most excitable membranes, including those of heart and brain. In the heart, Na$^+$-Ca^{2+} exchange might serve to maintain a low concentration of free intracellular calcium or to bring Ca^{2+} into the cell during the systolic phase of contraction, or both.

Finally, a 1:1 electroneutral Na$^+$-H$^+$ exchanger appears to be nearly ubiquitous in all cells. This exchanger plays a crucial role in many cellular processes, including proliferation, differentiation, and hormone-responsiveness.

Electrogenic Cotransport: Secondary Active Transport

We have seen that glucose molecules, as well as HCO$_3^-$ and Cl$^-$ ions, can move across membranes via proteins. Certain transporters combine movement of more than one molecule in a given direction. This is called *cotransport,* or symport. Both solutes must bind to the transporter on one side of the membrane; only then can *both* be translocated across.

Except for certain systems in yeast, virtually every cotransporter described to date in eukaryotic cells has the Na$^+$ ion as one of the transported solutes. There is good reason for this. Recall from Table 1.1 and the discussion of ion-translocating ATPases that most eukaryotic cells have a low intracellular Na$^+$ concentration, due to the Na/K-ATPase pumping activity. Thus, there is constantly an inwardly directed Na$^+$ gradient across the plasma membrane (Table 1.1). Na$^+$ ions tend to diffuse down this gradient into the cell. Essentially all cotransporters couple Na$^+$ transport to that of another solute. In the process, Na$^+$ moves down its concentration gradient *but the cotransported solute is actually pumped up a gradient.* This process is termed *secondary active transport* of the cotransported solute. It does require energy, but the transport step is once-removed from ATP hydrolysis. The energy comes directly from the Na$^+$ gradient. Table

1.2 gives an example of Na^+-coupled solute transport: electrogenic Na^+-glucose cotransport. One Na^+ is "leaked," i.e., moves passively down its gradient, and one glucose is "pumped," i.e., moved up a gradient. One positive charge is transferred across the membrane in the process. The cDNA for the Na^+-dependent glucose transporter has also been cloned; the familiar elements are shown in Figure 1.15C. (Of interest, this transport protein has no sequence homology with the erythrocyte (nonelectrogenic) glucose carrier.)

The important point is that Na^+-linked cotransport represents an efficient way to move numerous and varied solutes *up* concentration gradients into cells. Table 1.4 lists several such systems. Some of these cotransporters, such as Na^+-glucose cotransport, are electrogenic. Others, such as $Na^+K^+2Cl^-$ cotransport, are electroneutral.

Leaks: Ion Channels

We have seen that hydrophilic solutes permeate lipid bilayers poorly, but that their flux across the membrane can be dramatically increased by the use of carrier proteins, i.e., "facilitated diffusion." One special form of facilitated diffusion takes place across integral membrane proteins called *channels*. Channels transport ions at very fast rates. Table 1.5 shows the maximum number of solute molecules that can cross a membrane (natural or artificial) via various membrane proteins. It is clear that pumps move solute the most slowly at several hundred ions per second. Carriers are intermediate at about 10,000 molecules per second. Channels provide the highest flux rates. One naturally occurring channel, the brain Na channel, transports about 10 million ions per second. An artificial channel (the gramicidin channel) under optimal conditions can transport 200 million cesium ions per second! Why do cell membranes have channels with such high transport rates?

First, one might envision that channels are needed in order to transfer charge across membranes, and thus set up electrical potentials. However, we have seen that Na^+-cotransport systems can also transfer charge and thus set up potentials.

Table 1.4.
Partial List of Na^+ Cotransport Systems Present in Mammalian Cells[a]

Ions	Sugars	Amino Acids	Other
Na^+/Cl^-	$Na^+/glucose$	$Na^+/alanine$	$Na^+/lactate$
$Na^+/K^+/2\,Cl^-$	$Na^+/galactose$	$Na^+/glycine$	$Na^+/cholate$
Na^+/PO_4^-		$Na^+/serine$	$Na^+/citrate$
$2Na^+/PO_4^=$		$Na^+/cysteine$	$Na^+/folate$

[a]For references see Stein, 1986.

Table 1.5.
Maximum Number of Solute Molecules Transported per Second per Transport Protein

Protein	Class	Solute	Molecules/second
Na/K-ATPase	Pump	Na^+	5×10^2
Ca-ATPase	Pump	Ca^{2+}	2×10^2
Band 3	Carrier	Cl^-	5×10^4
Glucose carrier	Carrier	Glucose	1×10^4
Na channel	Channel	Na^+	1×10^7
Gramicidin	Channel	Cs^+	2×10^{8a}

[a]20 pS conductance, 100 mV driving force. (Adapted from Hille, 1984.)

So a channel is not necessary just to create a membrane potential. Second, channels mediate very high fluxes. Are channels needed to achieve such high fluxes? No. One can also achieve high flux rates, even with pumps, simply by increasing the density of pumps or carriers in the membrane (for example, band 3 is present at 10^6 copies per red cell, producing an extremely large Cl^- flux per unit membrane area). Neither of these needs would alone appear to require a channel.

The special property of channels, particularly for excitable cells, *is that they open and close very quickly*, i.e., their transport rate can be controlled over a range of milliseconds. This rapid control, combined with the very high flux rates inherent in channels, has two major functions: (1) For calcium channels, calcium can be essentially instantaneously released from intracellular stores so as to produce rapidly controlled events such as muscle contraction, fusion of vesicles with the plasma membrane, etc. (2) For other channels, such as brain Na^+ channels, rapid channel opening suddenly changes the electrical potential across the cell membrane. This produces crucial effects such as initiating or propagating a nerve action potential.

The rapid control of ion flux through channels is best illustrated using the technique of patch-clamping. By this method a single ion channel can be studied in isolation. As shown in Figure 1.18, ion flux is measured as current through the single channel. This particular example shows opening of a K^+ channel by application of adrenaline to trachea epithelial cells. Patch clamping has given the most graphic evidence that channel opening or closing is regulated according to the needs of the cell.

The cDNA for one of several Na^+ channels in the brain has now been cloned. The full-length cDNA sequence shows four units of internally repeated sequence (I-IV). The model for this channel is shown in Figure 1.19. It is apparent that this particular model represents the most three dimensional view of an integral membrane protein of those we have considered. The general similarity in structure

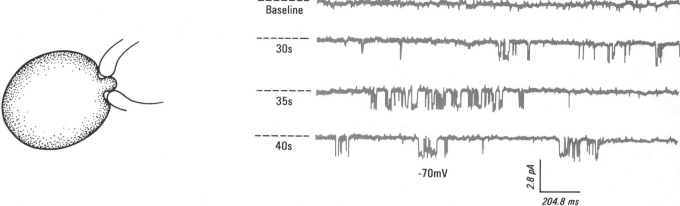

Figure 1.18. Electrical activity from a single potassium channel on a trachea epithelial cell studied by the patch clamp technique. The *left side* shows a single cell with a patch-clamp electrode attached, a so-called "cell-attached patch." The lumen of the electrode is attached to an amplifier and recording device so as to generate the tracing on the right. On the right, the *dashed line* represents current flow through the patch when all channels are closed. The top sweep, left to right, was taken under nonstimulated, baseline conditions. Although some elec-

trical noise is apparent, there are no channel openings. The bottom three sweeps represent channel activity in response to epinephrine. Downward deflections represent channel openings that pass about 2.8 picoamperes of current per opening. The membrane voltage is −70 mV. Although the bottom three sweeps show variable degrees of channel opening, the important point is that channels now open in response to epinephrine. (Courtesy of M. J. Welsh.)

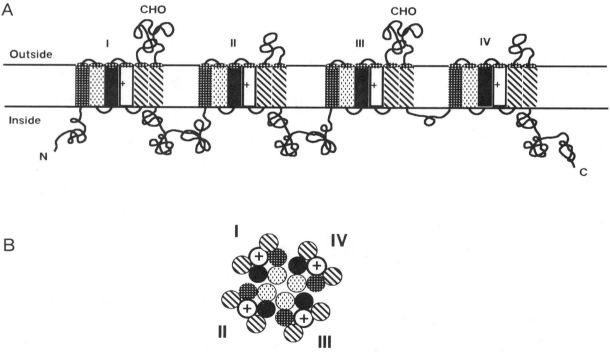

Figure 1.19. Proposed transmembrane topology of the Na⁺ channel viewed parallel (*A*) and perpendicular (*B*) to the plane of the membrane. In *A*, four units of homology spanning the membrane are displayed linearly. Segments S1–S6 in each repeat (numbered I–IV) are indicated by cylinders as follows: S1, *cross-hatched;* S2, *stippled;* S3, *solid;* S4, indicated by a *plus sign;* S5 and S6, *hatched.* Putative glycosylation

sites (*CHO*) are noted. In *B*, the ionic channel is represented as a central pore surrounded by four units of homology. Segments S1–S6 in each repeat are represented by *circles* as in *A*. (Redrawn with permission from Noda et al., Nature 320:188–192, 1986. Copyright © 1986 Macmillan Magazines Ltd.)

between pumps (Fig. 1.9), carriers (Fig. 1.15), and channels (Fig. 1.18) suggests that the difference in function is subtle; all provide a "pore" through the membrane, and all undergo conformational changes that control the rate of solute (ion) flow through the "pore."

Besides patch clamping, ion movement through channels can be studied in another way. Radioactive ions move through channels. The tracer fluxes through channels show saturation kinetics and competition for transport by nontracer ions. That is, tracer flux shows Michaelis-Menten properties (shown previously, Fig. 1.16). Moreover, patch clamp experiments also show that the electrical conductance through a given channel saturates as the concentration of the transported ion increases. Both of these pieces of evidence suggest that ions do not diffuse freely within the channel as in aqueous solution. Rather, as with substrate binding to enzyme catalytic sites, ions must bind to specific sites on or in the channel. Such binding sites account for the selectivity of a channel, i.e., make it Na^+-selective, Cl^--selective, etc.

In the preceding sections we have discussed many aspects of basic membrane transport. However, most of the examples have involved electroneutral transport. Because electrical potentials across membranes are so important, in the remainder of this chapter we focus on the electrical properties of cell membranes.

ELECTRICAL PROPERTIES AND MEMBRANE POTENTIALS

Capacitance, Charge Transfer, and Ohm's Law

A parallel-plate capacitor is made up of two conductive surfaces, or plates, separated by a relatively nonconductive medium. Passage of current across the capacitor removes charge from one plate and stores it on the other. The capacitance, C, is a measure of how much charge, Q, is required to set up a given potential difference, E, between the two plates:

$$C = \frac{Q}{E} \qquad (10)$$

Pure lipid bilayers show capacitances of around 0.8 microFarads per square cm ($\mu F/cm^2$), whereas cell membrane capacitances are around 1.0 $\mu F/cm^2$. These values are large as capacitors go, i.e., cell membranes can generate a large membrane voltage with the movement of relatively few ions (charges) across the membrane. In fact, the movement of 6000 ions over a membrane area of 1 μm^2 will generate a potential difference across the membrane of 100 mV.

How do potentials develop across membranes? Figure 1.20*A* shows a theoretical model in which a membrane containing K^+ channels (perhaps similar to those in Fig. 1.18) separates two aqueous compartments. The membrane is permeable to K^+ but not to anions, and the K^+ concentration of the "In" (intracellular) compartment is 10 fold that of the "Out" (extracellular) compartment. K^+ ions tend to diffuse by Brownian motion in both directions across the membrane. However, because the K^+ concentration begins higher in the left ("In") compartment, there is greater net movement of K^+ from left to right.

So far, the situation depicted in Figure 1.20 seems very similar to facilitated diffusion as discussed for many solutes above. If K^+ were not a charged solute, it would equilibrate by diffusion until the K^+ concentrations were equal in the two compartments. Because K^+ is charged, however, the diffusion of K^+ down its concentration gradient produces separation of charge. Thus, K^+ ions will undergo net diffusion from left to right until the membrane is charged to a certain degree, with the "In" side negative relative to the "Out" side. At this point, the electrical potential across the membrane will prevent further net movement of K^+ ions. That is, negative charge along the "In" side of the membrane will retard further movement of K^+ through channels from "In" to "Out." Very few K^+ ions need to move from left to right to create a substantial electrical potential across the membrane. In fact, there is so little net movement of K^+ from left to right in this particular example that the K^+ concentrations on the left and right sides are imperceptibly changed. It should be emphasized that, once the membrane has become fully charged, only the *net* movement of K^+ from left to right stops; individual K^+ ions are still crossing the membrane at equal rates in both directions, driven by Brownian motion.

How much voltage is generated before net K^+ movement stops? The equation describing the voltage that develops is a famous one in membrane biology, the *Nernst* equation:

$$E = -\frac{RT}{zF} \ln \frac{[K]_{in}}{[K]_{out}} \qquad (11)$$

where E is the *equilibrium potential* for K^+ in millivolts, R is the gas constant (8.31 volts × coulombs per degree Kelvin per mole), z is the valence (here,

Figure 1.20. Two examples of the development of equilibrium potentials across ion-selective membranes. *A*, A membrane separates two compartments. The left, or "In" compartment contains 100 mM K⁺. The right, or "Out" compartment contains 10 mM K⁺. The membrane is freely permeable to K⁺ but not to the accompanying anion. Net K⁺ movement from left to right creates a membrane voltage, left side negative. *B*, The same example as in *A*, except that Na⁺ has been substituted for K⁺ and the concentrations reversed. This arrangement causes right-to-left net Na⁺ movement and the development of a membrane voltage that is now positive on the left side.

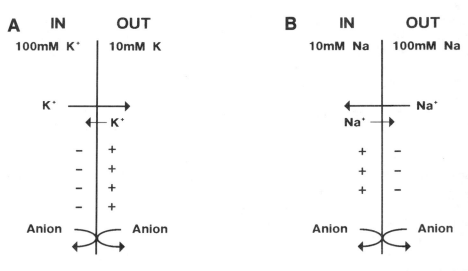

K^+-Selective Membrane $\qquad$ Na^+-Selective Membrane

plus one), F is the Faraday constant (9.65×10^{-4} coulombs per mole), and $[K]_{in}$ and $[K]_{out}$ are K⁺ concentrations at equilibrium in the two compartments. It is handy to know that the natural logarithm (1n) is equal to 2.3 times the log 10, and that for z = 1, the value of RT/zF at body temperature (37°C) is 26.7 mV. Thus, the entire term 2.3 RT/zF equals 61.5 mV at 37°C. For the example given above:

$$E_K = -(61.5)\log 10 = -61.5 \text{ mV}$$

In the model shown in Figure 1.20A, because K⁺ is the only permeant ion, a minute transfer of K⁺ ions across the membrane has created a membrane voltage equal to the equilibrium voltage of K⁺, i.e., 61.5 mV. This situation is quite similar to that of resting nerve cells in which the major membrane ionic permeability is to K⁺ and the intracellular K⁺ concentration is kept high relative to that outside the cell by the Na/K-ATPase.

Figure 1.20B shows the same membrane but with two important differences. First, the membrane now contains only Na⁺-conductive channels. Second, the Na⁺ concentration gradient is opposite that of K⁺ in Figure 1.20A. This Na⁺ concentration gradient is also maintained by the Na-K-pump. Now if we calculate E_{Na}, we find +61.5 mV (left side positive). This situation is analogous to that across a nerve cell membrane at the peak of the action potential, when the membrane is essentially only permeable to Na⁺.

Most cells maintain a membrane voltage of −70 to −100 mV, inside negative. This voltage essentially represents the sum of the equilibrium contributions by all permeant ions, weighted according to their relative conductive permeabilities. For nerve cells containing primarily K⁺, Na⁺, and Cl⁻ conductances, this concept is expressed by the *Goldman equation:*

$$V_m = \frac{RT}{F} \ln \frac{P_K C_K^0 + P_{Na} C_{Na}^0 + P_{Cl} C_{Cl}^i}{P_K C_K^i + P_{Na} C_{Na}^i + P_{Cl} C_{Cl}^0} \qquad (12)$$

where V_m = membrane voltage, P = conductive permeability to a given ion, C_x^0 = concentration of ion x outside, C_x^i = concentration of x inside.

One can get an idea of whether a given ion is actively transported by determining whether there is a steady-state *electrochemical gradient* for the ion across the membrane. The electrochemical gradient, $\Delta\tilde{\mu}$, for ion x is given by:

$$\Delta\tilde{\mu} = E_x + V_m \qquad (13)$$

Clearly, when the equilibrium potential, E_x, and the membrane voltage, V_m, are equal and opposite, there is no electrochemical gradient ($\Delta\tilde{\mu} = 0$). If $\Delta\tilde{\mu} \neq 0$ and the membrane definitely contains a conductance for ion x, then x is actively transported. If $\Delta\tilde{\mu} = 0$, then x is probably passively distributed across the membrane (although not definitely so).

Finally, with a given membrane potential one can determine how much current will flow through a membrane channel when it opens. This is simply given by *Ohm's Law* ($V = I/G$) where V = membrane voltage (usually in mV), G = channel conductance (units of picoSiemens), and I = current flowing through the channel (in picoamperes).

Donnan Equilibrium

The last topic dealing with electrical potentials is an interesting and biologically important one, the phenomenon known as the Donnan (or Gibbs-Donnan) equilibrium. The Donnan equilibrium represents the concentrating of ions and the generating of a membrane potential that occurs solely because *an impermeant anion is present on one side of a membrane but not the other*. Notice that this situation differs from the model just developed in Figure 1.20 for the Nernst equation in which there were no permeable anions on either side of the membrane. Because all cells contain impermeable, anionic proteins and nucleotides, the Donnan effect is ubiquitous.

To understand the Donnan equilibrium, we set up two compartments separated by a membrane, as shown in Figure 1.21. The membrane is permeable to Na^+ and Cl^-, but not to the anionic protein, P^-. For the moment, consider that the membrane is also water-impermeable. At time zero (Fig. 1.21A), the Na^+P^- concentration in the left compartment is equal to the Na^+Cl^- concentration in the right compartment. Also, if one measures the Na^+ and Cl^- concentrations individually in the right compartment, within the limits of detection their concentrations are equal, i.e., we say that, "macroscopically," electroneutrality exists. At time zero, before any ion has moved across the membrane, there is no potential across it.

Now (Fig. 1.21B) we allow Na^+ and Cl^- ions to diffuse (P^- is impermeant and cannot diffuse). Because Na^+ is present at equal concentrations on the two sides, no net Na^+ diffusion occurs. Only Cl^- diffuses down its concentration gradient from right to left.

With the movement of only a few thousand Cl^- ions per square micron of membrane (a change too small to be measurable by available techniques), a substantial membrane voltage develops, left side

negative. This voltage now has two effects: (1) it begins to drive Na^+ ions across the membrane from right to left (Fig. 1.21B), and (2) it opposes further Cl^- movement from right to left. Ultimately, an *equilibrium* is reached between oppositely directed chemical gradients for Na^+ and Cl^- and the membrane voltage (Fig. 1.21C). Again, within the limits of detection, electroneutrality will be maintained within each compartment (concentration of anions = concentration of cations).

We stated above that, in a passive system with conductance present in the membrane for the ion of interest, passive distribution of ion x results in $E_x = V_m$. In the Donnan equilibrium, Na^+ and Cl^- conductances are present, and no energy is put in. Thus, at equilibrium $E_{Na} = V_m$ and $E_{Cl} = V_m$, or $E_{Na} = E_{Cl}$. Therefore,

$$E_{Na} = -\frac{RT}{zF} \ln \frac{[Na]_l}{[Na]_r} = -\frac{RT}{zF} \ln \frac{[Cl]_r}{[Cl]_l} \qquad (14)$$

Dividing by $-RT/zF$ yields

$$\ln \frac{[Na]_l}{[Na]_r} = \ln \frac{[Cl]_r}{[Cl]_l} \qquad (15)$$

Taking antilogs

$$\frac{[Na]_l}{[Na]_r} = \frac{[Cl]_r}{[Cl]_l} \qquad (16)$$

or

$$[Na]_l[Cl]_l = [Na]_r[Cl]_r \qquad (17)$$

This equation is known as the *Donnan relationship*.

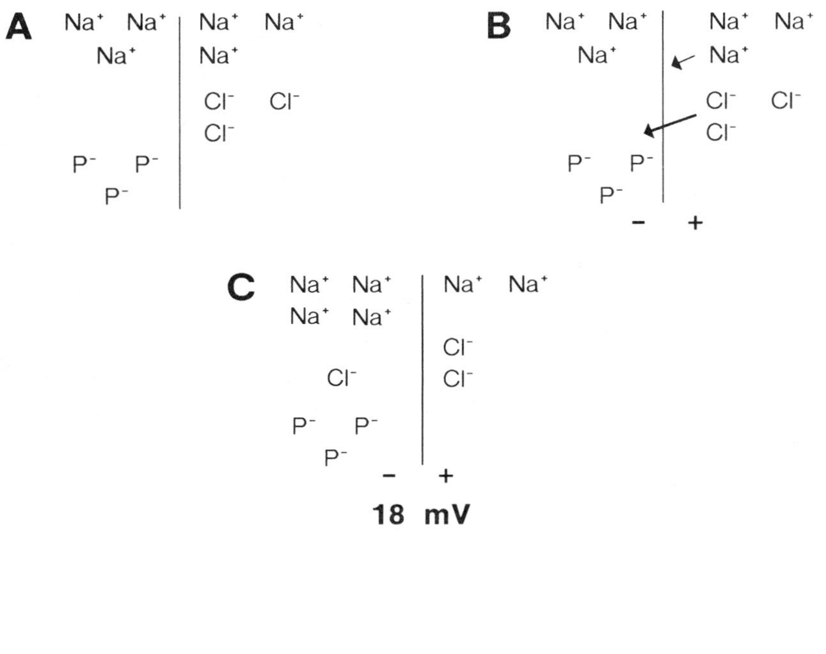

Figure 1.21. Generation of the Gibbs-Donnan equilibrium. The membrane is permeable to Na^+ and Cl^-, but not to anionic protein, P^-. *A,* Time zero. Electroneutrality on each side. *B,* A fraction of a second later, Cl^- is diffusing down its concentration gradient from right to left. This separates charge, making the membrane negative on the left side. The voltage across the membrane secondarily drives Na^+ diffusion right-to-left. *C,* At equilibrium, NaCl has accumulated on the left; the equilibrium voltage is 18 mV, left side negative.

One can also calculate the equilibrium concentrations of Na^+ and Cl^-. To preserve electroneutrality

$$[Na]_r = [Cl]_r = [Cl]_r^0 - [Cl]_l \qquad (18)$$

where $[Cl]_r^0$ is the starting Cl concentration in the right compartment. Also,

$$[Na]_l = [Na]_l^0 + [Cl]_l \qquad (19)$$

where $[Na]_l^0$ is the starting Na concentration in the left compartment. From *equation 17*

$$[Na]_l[Cl]_l = [Na]_r[Cl]_r$$

we can substitute *18* and *19* into *17* to get

$$([Na]_l^0 + [Cl]_l)([Cl]_l) = ([Cl]_r^0 - [Cl]_l)([Cl]_r^0 - [Cl]_l) \qquad (20)$$

Knowing $[Na]_l^0$ and $[Cl]_r^0$, one can solve *equation 20* for $[Cl]_l$. In Figure 1.21A we showed 3 Na^+ ions in the left compartment at time zero and 3 Cl^- ions in the right compartment at time zero. For simplicity, let $[Na]_l^0 = 3M$ and $[Cl]_r^0 = 3M$. From *equation 20*, $[Cl]_l = 1M$.

One can now determine the membrane voltage that has developed. Recall that, at equilibrium $E_{Na} = E_{Cl} = V_m$, so that V_m can be determined from *either* E_{Na} or E_{Cl}. Let us use E_{Cl}. We know that $[Cl]_l = 1M$. From *equation 18* $[Cl]_r = [Cl]_r^0 - [Cl]_l$, so $[Cl]_r = 3M - 1M = 2M$. Thus, from the Nernst equation:

$$E_{Cl} = V_m = -\frac{RT}{zF} \ln \frac{[Cl]_r}{[Cl]_l} \qquad (21)$$

At 37°C, $V_m = -18$ mV.

When we began developing the Donnan equilibrium, we treated the membrane as water-impermeable. Actually, cell membrane and capillaries are permeable to water. Therefore, it is important to appreciate that, at equilibrium, the total solute concentration on the left in Figure 1.21C is $[Na]_l + [Cl]_l + [P^-] = 4M + 1M + 3M = 8M$, whereas that on the right is $[Na]_r + [Cl]_r = 2M + 2M = 4M$. This 2:1 left-to-right solute concentration gradient would drive substantial osmotic water flow if mechanisms were not present to counteract such flow. Indeed, because cells contain impermeable anions (proteins and nucleotides), the Donnan effect *does* tend to produce osmotic water flow into cells with resultant cell swelling. Under normal circumstances, this tendency appears to be offset by active extrusion of ions, e.g., by the Na/K-ATPase. However, if depletion of cellular ATP pools occurs, cells will swell, probably by stopping the Na pump and allowing the Donnan effect to express itself.

In the case of capillary beds of the body, a Donnan equilibrium occurs owing to the presence of blood proteins, to which most capillaries are essentially impermeable. By analogy to Figure 1.21, the Na^+ concentration in the capillary plasma is greater than that of the Na^+ concentration in the interstitial fluid surrounding the capillary. Likewise, the Cl^- concentration in the capillary is less than the Cl^- concentration in the interstitial fluid. The resulting osmotic gradient tending to move water into the capillary lumen from the surrounding interstitial fluid is counteracted by hydrostatic pressure within the capillary (derived from the energy of cardiac contraction). Thus, the Donnan phenomenon is a comprehensive equilibrium involving a balancing of osmotic, electrochemical, and hydrostatic driving forces.

CHAPTER 3

Excitation, Conduction, and Transmission of the Nerve Impulse

The nervous system is a complex array of specialized structures which serve to receive, store, and transmit information—thereby integrating the activities of spatially separated cells, tissues, and organs and making it possible for a multicellular organism to function as a coordinated unit in terms of growth, development, and the ability to do work and to adapt to changes in the environment. Our efforts to understand the function of the nervous system involve the concepts and languages of many disciplines, ranging from mathematics and physics to the physiology and behavior of human beings in all of their complexity. In this chapter, however, we will be concerned primarily with the generation and propagation of the impulses which constitute the main currency of neural transactions.

STRUCTURE OF NERVOUS TISSUE

Neuron

The neuron consists of a body (soma, perikaryon) and two types of processes—the dendrite and the axon (Fig. 1.22a). In vertebrates, the bodies of the nerve cells lie within the grey matter of the central nervous system or in outlying ganglia, e.g., posterior spinal root, cranial, or autonomic ganglia. The white matter of the brain and spinal cord and of the peripheral nerves is composed of bundles of nerve fibers. The core of each nerve fiber is formed by a process of a nerve cell, and many of them are surrounded by a sheath of myelin which gives them a white appearance. The grey matter receives a rich blood supply from the vessels of the pia mater; the blood supply to the white substance is much less profuse.

There are a number of different types of nerve cell; those in which axon and dendrite arise by a common stem are called unipolar, and those in which the axon and the dendrite or dendrites spring from opposite or at least different parts of the soma are called bipolar or multipolar. The cell bodies or somata are of various sizes and forms—stellate, round, pyramidal, fusiform, etc.

After fixation and staining by special techniques, various structures are seen in the cytoplasm or perikaryon of the nerve cell body: (1) neurofibrils; (2) Nissl bodies or tigroid substance; (3) Golgi apparatus; (4) mitochondria; (5) ribosomes; and (6) the endoplasmic reticulum (see also Chapter 1). Electron microscopy has revealed much of the detailed structure of these intracellular entities. Their structure and function appear to be the same in all cells studied so far. Mitochondria, found along the entire length of the axon, contain all the enzymes required for the respiratory activity of the cell, and are, therefore, responsible for those functions dependent upon aerobic metabolism. The neurofibrils appear as fine filaments which stream through the cytoplasm from dendrites to axon (Fig. 1.22b); they enter the latter process and extend to its terminations. The Nissl bodies composed of ribonucleic acid and polysomes are granular masses stainable with basic dyes and occur in the perikaryon and dendrites but not in the axon. They give a striped or tigroid appearance to the cell. They are absent from the region of origin of the axon (axon hillock) and vary in size and number with the state of the neuron; they undergo disintegration (chromatolysis) in a fatigued or injured cell or in one whose axon has been sectioned. This means that the synthetic machinery of the perikaryon is at least partially regulated by events in the peripheral processes of the cell. The nature of this feedback control is not known. The internal reticular apparatus of Golgi is a coarse network seen within the cells when special methods—e.g., impregnation with silver chromate—are employed which leave the Nissl bodies and the neurofibrils invisible.

The nucleus of the nerve cell contains one and sometimes two nucleoli but, as a rule, no centrosome. The absence of a centrosome indicates that the highly specialized nerve cell has lost its power

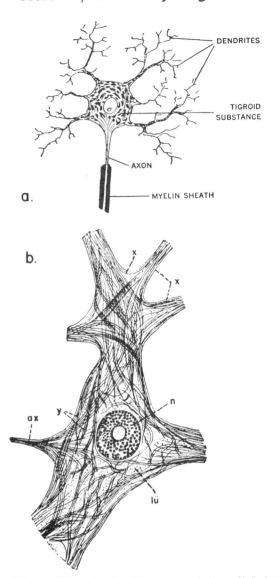

DENDRITES

TIGROID
SUBSTANCE

AXON

MYELIN SHEATH

a.

b.

x

x

ax

y

n

lü

Figure 1.22. *a,* Different parts of the neuron. *b,* Neurofibrils in a cell from the anterior grey column of the human spinal cord. *ax,* axon; *lü,* interfibrillar spaces; *n,* nucleus; *x,* neurofibrils passing from one dendrite to another; *y,* neurofibrils passing through the body of the cell.

of division. Nerve cells once destroyed are replaced merely by neuroglia.

The linking together of neurons to form conducting pathways is effected by the contact (but not union) of the axon terminal of one nerve cell with the body, dendritic process, or in some instances, the axon of another. Such a junction, without anatomical continuity, is called a *synapse.*

Though the nerve cell frequently possesses more than one dendrite, the axon is single. The axon may be long and contribute to one of the tracts of the central nervous system forming the white matter, or it may terminate as a peripheral nerve fiber. Such cells are referred to as Golgi I type. In the Golgi II

type cell, the axon is short and ends within the grey matter by making contact with another neuron. The axon arises from a small elevation on the surface of the cell body, the axon hillock. It may give off short collateral branches or run as an unbranched fiber, not dividing until it has reached its destination. The dendrite is the receptive process of the neuron; the axon is the discharging process, i.e., the former transmits the impulse toward the cell body and the latter away from the cell body. Within the central nervous system the dendrite is usually short and possesses many branches, but in the peripheral sensory nerves it is comparable in length with an axon. Nerve fibers which carry impulses to the central nervous system are termed *afferent;* those conveying impulses from the central nervous system to the periphery are called *efferent.* A mixed nerve contains fibers of both types. The fact that conduction is unidirectional is a consequence of the properties of synapses, for nerve fibers per se can be made to conduct in either direction.

As with other cells, the cytoplasm of the soma is enveloped by a plasma or unit membrane composed of a bimolecular leaflet of lipid material covered and penetrated by protein. This extends over the processes of the nerve cell. The electrical properties of nerve depend upon the plasma membrane.

Nerve Fiber

The white matter of the central nervous system and the peripheral nerves are composed of thousands of individual nerve fibers. Within the grey matter the axons are enclosed only by the plasma membrane; but upon leaving the grey substance they acquire a sheath of lipid material called myelin. The myelin sheath is associated with somatic nerves of large diameter and generally not with nerve fibers of small diameter. Hence, we speak of myelinated (medullated) or unmyelinated (unmedullated) nerves. Myelinated fibers, usually larger than 1 μm in diameter, conduct faster than the smaller unmyelinated fibers.

The myelin sheath of the somatic nerves consists of compressed layers of Schwann cells spiraling concentrically to form a wrapping around the axons. The outer layer of the myelin contains the flattened nuclei of the Schwann cells and is termed the neurilemma or sheath of Schwann. Myelinated nerves appear as if constricted at regular intervals along their course. This appearance is due to the absence of myelin at these points and the dipping inwards of the neurilemma; these points are known as the nodes of Ranvier. The segments between the nodes vary in length in different nerves, but the usual internodal distance is about 1 mm. Each internodal segment of the neurilemma consists of a single Schwann cell. The fibers forming the white matter of the central nervous system and optic nerves have no neurilemma; this membrane is replaced by glial cells which produce the myelin sheaths as the Schwann cells do peripherally.

Compared with the myelin sheath the neurilemma in the region of the node of Ranvier under appropriate conditions becomes highly permeable to sodium and potassium ions. There is good evidence that the conduction of the impulse along a myelinated nerve is a "leaping" from node to node over the intersegmental regions rather than a continuous process. This type of conduction is called saltatory (from the Latin word *saltus*, a leaping).

MYELINATION OF FIBER TRACTS IN THE CENTRAL NERVOUS SYSTEM

The nerve fibers in the various conducting pathways receive their myelin sheaths at different ages, and it is generally believed that the myelination of a given tract and the time at which it commences to function coincide. The sensory tracts become myelinated first, those of the posterior columns of the spinal cord between the 4th and 5th months of fetal life (human). The spinocerebellar tracts are myelinated later, and the motor paths, e.g., corticospinal (pyramidal) tracts, do not begin to be invested with myelin sheaths until the 2nd month of life and are not completely myelinated until about the 2nd year, about the time when the child has learned to walk. The fibers of association paths, for the most part, myelinate at still later dates. The high insulating property of myelin, in addition to confining the nerve impulse to individual fibers and thus preventing cross-stimulation of adjacent axons, serves to increase the rate of conduction.

The neuroglia or "glia" (Greek, "glue") is a special type of interstitial tissue. Its cells are of three kinds, astrocytes, oligodendrocytes, and microglia. The microglia appear to have a phagocytic role, since they wander into the central nervous system from the blood vessels and increase in number during inflammatory processes. Astrocytic processes are found abutting blood vessels and investing synaptic structures, neuronal bodies, and neuronal processes. Although the functions of astrocytes are not precisely known, putative roles include involvement in support, transport mechanisms, inflammatory and reparative reactions, and isolation of neuronal elements. The oligodendroglia are responsible for the formation of the myelin layers around axons within the central nervous system.

Degeneration and Regeneration of Nerve

When a peripheral nerve is cut, the part of the nerve separated from the cell body shows a series of chemical and physical degenerative changes. At the same time the fibers of the proximal stump of the nerve, those still attached to their cell bodies, grow distally toward the separated part of the nerve; these changes constitute the process of regeneration (Fig. 1.23).

DEGENERATION

The degenerative period may be divided into an early and late phase. Shortly after the nerve has been sectioned, the axon swells, the myelin sheath begins to form bead-like structures, and a series of enlargements (round, fatty fragments) appear. The axon breaks up, and its parts are devoured by macrophages. Up to 3 days after section, the distal nerve will continue to conduct an impulse. Changes in the action potential can be observed as early as 2 days after section. After the 3rd day, the ability of the nerve to conduct has seriously deteriorated and after the 5th day, an impulse can no longer be evoked. The first period (3–5 days) is one in which the obvious changes in nerve are largely structural. Changes in the ultrastructure of the myelin sheath can be shown to occur during this period, as well as changes in the endoplasmic reticulum, mitochondria, neurofibrils, and the plasma membrane.

Up to the 8th day, little or no changes in lipid histochemistry may be detected, although changes in

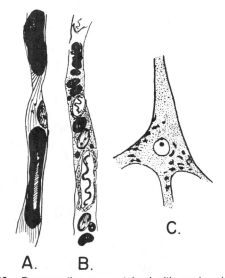

Figure 1.23. Degenerating nerve stained with osmic acid. *A* shows appearance of distal segment of nerve fiber 2 days after section; note large masses of myelin derived from medullary sheath. *B*, 5 days after section, smaller myelin particles together with droplets of fatty acids and fragmented neurofibrils. *C*, Retrograde degeneration in cell body, disintegration of Nissl bodies.

nerve cholinesterase and failure in the ability to synthesize acetylcholine have been demonstrated about the 3rd day. The lipids appear to maintain their original chemical structure. From the 8th to the 32nd day after section, the myelin gradually disappears, while Schwann cells and macrophages have both increased greatly in number. Why the myelin breaks down is not clear. It has been suggested that the macrophages and Schwann cells secrete enzymes which aid in the destruction of myelin. The principal myelin lipids are free cholesterol, and two lipids containing sphingosine: cerebroside and sphingomyelin. At the time of the disappearance of the myelin, cholesterol esters appear in large quantities, and free cholesterol disappears.

The changes just described are generally known as Wallerian degeneration. In addition, there are other changes that occur in the neuron on the proximal side of the section (retrograde degeneration). The nerve fiber as far centrally as the first node of Ranvier shows changes similar in nature to those just described. In the cell body itself, swelling of the cytoplasm and nucleus occurs and the Nissl granules undergo disintegration (chromatolysis). Atrophy of the cell body may ultimately result.

REGENERATION

Following section of a nerve, the fibers in the central stump begin to send out branches consisting of outgrowths of the axon near the cut tip. Up to 50 branches may sprout from a cut axon. At the same time, there is a rapid proliferation of the Schwann cells. If a gap larger than 3 mm exists between the central and peripheral stumps, the fibers tend to intermesh and form a tumor-like swelling called a neuroma. In such an event regeneration will probably never occur. For this reason it is necessary to join accurately by suture the proximal and peripheral stumps. If the neuroma is composed of sensory fibers, it may be very painful to pressure, often a troublesome complication following amputation. When peripheral degeneration has proceeded for enough time so that the peripheral stump contains only empty neurilemmal tubes, one of the outgrowing sprouts enters a tube and grows to form a new axon. Only one of the many sprouts enters an empty tube. The ability of axons to grow into a peripheral stump has led some observers to invoke a special chemical attraction between the nerve fiber and the terminal organ. This is the doctrine of neurotropism.

The rate of growth of regenerating nerve is from 1–4 mm/day. It has been established, in a variety of experiments, that there is a flow of axoplasm down the normal nerve at about the same rate.

PRINCIPLES OF BIOELECTRICITY

It will be recalled that the practical unit for quantitation of charge is the coulomb which equals 6.2×10^{17} electrons. A positive charge results from the removal of electrons from a neutral body, as when one electron is removed from the neutral sodium atom to form a sodium ion.

An electric field exists in the space around a charge and extends to infinity. The strength of this field is measured by the force which is exerted on a unit charge placed at any point in the field. The intensity of the field decreases inversely as the square of the distance from the charge (Coulomb's law). In order to move a charge against the field, work must be performed. Thus, it takes work to move a positive charge up to a region where another positive charge is situated, and the potential at any point in the field is defined as the work that must be done on a unit positive charge to move it up to that point from an infinite distance. This is the absolute definition of potential which is seldom necessary in practice. The significant quantity in experimental work is the difference in potential between two points, which is defined as the work necessary to move a unit charge from one point to the other. The work is measured in joules. The practical unit of potential difference is the volt, i.e., the potential difference against which 1 joule of work is done in the transfer of 1 coulomb.

Origin of Potentials

A difference in potential is associated with a flow of current in a medium (e.g., electrons in metal conductors and charge-carrying ions in solutions of electrolytes). The current flow requires energy since it encounters resistance in the medium. In order to produce electrical currents, special sources of electrical energy are available, such as the battery and generator. Such sources of energy are said to produce an electromotive force or emf. In solutions, electrical energy is produced by chemical reactions which result in the separation of charge at electrodes. Other sources of energy produce separation of charge by the flow of ions from solutions of high concentration, such separation of charge itself constituting a form of stored energy which is manifest as a difference in electrical potential.

Two sources of energy are important in understanding the origin of biological potentials; these sources give rise to emfs called *electrode* or *concentration* potentials and *diffusion* or *membrane* potentials. Concentration potentials arise in every measurement of potential difference in solution and are generally to be avoided in biological measurements. They arise whenever an electrode is dipped into a solution containing an ion in common with the electrode. Thus, an electrode made of silver and coated with silver chloride inserted into a solution of sodium chloride will produce a potential difference between the electrode and the solution. The silver dissolves in the solution as silver ion, Ag^+, leaving the electrode negative. The positive Ag^+ ion remains at the electrode, forming a double layer of charge with the negative electrode. Charge is therefore separated at the electrodes. If two such electrodes are set up, each dipping into a different concentration of sodium chloride (a concentration cell), double layers will be set up at each electrode, but the double layer at one electrode will be more charged. When two such electrodes are connected by a wire, a current will flow. The emf produced between the two electrodes can be calculated from the Nernst equation for the concentration cell,

$$E = 2.303 \ (RT/F)\log(C_1/C_2)$$

where the electrolyte is univalent. R is the gas constant (equal to 1.99 cal/mole/°C, or 8.3 joules/mole/°C), T absolute temperature (°K), F the Faraday (23,050 cal/v/mole, 96,494 coul/mole), and C_1 and C_2 are the concentrations of electrolyte in each solution. Note, that according to the formula if the concentrations are equal in each solution, the emf is zero. The method of producing a concentration cell is shown in Figure 1.24.

DIFFUSION POTENTIALS

Let us place two solutions containing different amounts of NaCl in contact by means of membrane freely permeable to the electrolyte. An emf whose origin is quite different from that of the electrode potentials we have discussed will arise at the membrane. The potential difference arises from the diffusion of ions across the membrane or at any junction between such solutions. They are therefore called *diffusion, membrane,* or *junction* potentials. The sodium chloride in the solution of higher con-

centration tends to diffuse across the membrane into the other solution, the sodium ion diffusing faster than the chloride ion. The emf at the membrane depends upon the relative rates (mobilities) at which the sodium and chloride ions move and the relative concentrations of sodium chloride. Since Na^+ diffuses faster than Cl^-, a separation of charge is produced; the solution into which a new flow of Na^+ occurs becoming positive. At biological membranes, somewhat special hindrances to the movement of ions are present. Potassium and chloride usually diffuse at the same rate in solution; however, the ability of chloride to move across the cell membrane is more limited than that of potassium. Under these circumstances, potassium diffuses ahead of chloride, and an excess of potassium ions appears on one side of the membrane. Although this excess is very small, it is sufficient to set up an emf or potential difference, so that the solution containing the lesser concentration of potassium chloride becomes positive to that containing the higher concentration. There is, in effect, a greater loss of potassium from the more concentrated to the less concentrated solution. The more concentrated solution becomes negative, since it has lost positive charge. The concentration of K^+ increases until the emf has become sufficiently negative to keep any further K^+ from diffusing across the membrane; this is the equilibrium potential for K^+. The diffusion potential at a membrane is given by a formula similar to the Nernst equation in which the latter has been modified to take into account the mobilities of the ions. The equation is:

$$E = [(u - v)/(u + v)]2.303 \ (RT/F)\log(C_1/C_2)$$

where the electrolyte is univalent, u and v are the mobilities of the cation and anion, respectively, i.e., the rates at which the ions move under unit field strength in solution. The remaining symbols are the same as those in the Nernst equation. Note that when the anion and cation mobilities are equal, $u = v$, and the emf is zero.

TRANSMEMBRANE POTENTIAL

Methods are available for measuring the potential difference across the cell membrane at rest and during the activity associated with impulse generation and propagation. In some cells, such as the giant axon of the squid, it is possible to insert an electrode into the body of the axon down its length and, by placing another electrode outside the cell, to measure the transmembrane potential directly. In other cells, where this technique is not feasible, another method is used. A glass capillary is drawn

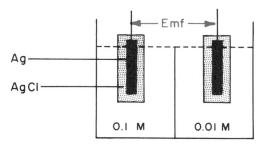

Figure 1.24. A concentration cell. Both solutions contain sodium chloride. The partition between the compartments is freely permeable to water and salt.

out to a fine tip less than 1 μm in diameter. The capillary is filled with saturated KCl or NaCl and serves as a microelectrode. It is inserted directly into the soma, muscle fiber, or even peripheral nerve fiber. The potential difference is measured between the microelectrode and a large, nonpolarizable electrode (connected via a KCl bridge) located outside the cell; this is the transmembrane potential. Such a method always involves the possibility that the injury caused by the puncture will result in a gradual fall in the potential and death of the cell. The membrane potentials of a variety of cells are shown in Table 1.6; they range from 61–94 mv, with the cell interior negative in relation to the external medium.

In recent years, two approaches have been developed for the assessment of the contributions to transmembrane current flow of individual ionic channels. The first of these methods involves analysis of the minute variations (noise analysis) in the conductance of a large segment of cell membrane (Begenisich and Stevens, 1975; Conti et al., 1976); the second of these methods involves the measurement of the ionic current flow through very small patches of cell membrane (Neher et al., 1978; Hamill et al., 1981).

Theory of the Membrane Potential

The original theory of the membrane potential, largely valid today in broad outline, was first enunciated by Bernstein (1912) at the turn of this century. It is well known that the concentration of potassium is much higher inside cells than outside. Table 1.7 gives the value of the ratio of inside to

Table 1.6.
Resting and Action Potentials

Tissue	Resting Potential, mv	Action Potential, mv
Loligo axon	61	96
Sepia axon	62	122
Carcinus axon	82	134
Frog myelinated nerve fiber	71	116
Frog striated muscle fiber	88	119
Frog cardiac muscle fiber	70	90
Dog cardiac muscle fiber	90	121
Kid cardiac muscle fiber	94	135

outside concentrations for potassium and other important ions in various tissues. The ratios for K^+ vary from 23–68. Bernstein maintained that the membrane potential was the result of the outward diffusion of potassium ions from the cells. Since the resting potential is a diffusion potential in which the mobility of the chloride anion was taken as zero and sodium was also presumed to be unable to penetrate the cell, the magnitude of the membrane potential can be calculated from the formula for the diffusion potential in which $v = 0$. The formula reduces to that of Nernst (Chapter 2, *equation 11*). Partial verification for this theory has been obtained by Hodgkin and Huxley (1952b) in several experiments in which the external potassium concentration was varied and resting potential of the squid axon measured. It was shown that the resting potential varied directly with the logarithm of the external concentration of potassium over a wide range. Table 1.7 shows that the sodium is largely present outside cells. The resting potential, V_r, was

Table 1.7.
Ionic Content of Nerve and Muscle Cells

Tissue	Sodium In, mM	Sodium Out, mM	Sodium Ratio	Potassium In, mM	Potassium Out, mM	Potassium Ratio	Chloride In, mM	Chloride Out, mM	Chloride Ratio
Carcinus nerve		460		380	10	38		540	
Carcinus nerve		460		230	10	23		540	
Frog nerve (Nov.)	37	120	0.31	110	2.5	44		120	
Frog nerve (Mar.)				170		68			
Frog sartorius muscle	15	120	0.12	125	2.5	50	1.2	120	0.01
Frog sartorius muscle	26	120	0.22	115	2.5	46	11	120	0.092
Rat cardiac muscle	13	150	0.087	140	2.7	52		140	
Dog skeletal muscle	12	150	0.08	140	2.7	48		140	

Ionic fluxes of resting membrane

Tissue				pmol cm^{-2} sec^{-1}					
Sepia axon	61	31		17	58				
				(11)	(33)				
Carcinus axon				19	22				
Frog sartorius muscle	13	16		7	5				
Frog sartorius muscle		5–10			20				
Frog abdominal muscle		5		10	10				
Frog ext. long. dig. IV muscle[a]				4	5				

[a]Extensor longus digitorum.

shown to be independent of the external sodium concentration [$Na_{outside}$] in a series of experiments on the squid axon in which [$Na_{outside}$] was varied over a wide range with no effect on V_r. Similar findings have been encountered in experiments on frog cardiac and skeletal muscle.

Ionic Distributions and the Membrane Potential

If no other ion but potassium could penetrate the membrane, the membrane potential would, as noted above, be given by the equation for a diffusion potential in which the anion mobility is zero. The resulting equation is then equal to the Nernst equation for a concentration cell. Actually, however, potassium does not pass through the membrane alone; there is always some transmembrane flux of the sodium and chloride ions. Two schemes for the membrane potential have been proposed to take into account the flow through the membrane of ions in addition to potassium. These are the Goldman equation and the Hodgkin-Huxley equivalent circuit for the resting membrane. Tests of these equations are usually made by plotting the membrane potential against the external potassium concentration and observing how closely the equations agree with the experimentally derived curve. All the equations, the Nernst relation included, are satisfactory over some range of potassium concentrations.

In order to understand the derivation and significance of the Nernst equation, let us consider the problem of moving an ion in a solution in which a potential difference is present. Work must be done with or against two forces, the electrical field, and any concentration difference in the ions. Thus, both "electrical" and "concentration" work must be performed. The electrical work (in joules) necessary to move 1 mole of a univalent ion (i.e., one equivalent) against a potential difference, E (in volts), is given by the expression: work = FE, where F is the Faraday (96,494 coul/mole). The concentration work required to move a mole of ion from a concentration C_1 to a higher concentration C_2 is given by the expression: work = $2.303\ RT \log C_2/C_1$, where R is the gas constant and T the absolute temperature. The total work is given by the sum of the electrical and the concentration work, or work = $FE + 2.303\ RT \log C_2/C_1$. When the system is at equilibrium, the work required is zero (this is the definition of thermodynamic equilibrium), and the potential difference is given by the Nernst equation,

$$E = 2.303(RT/F)\log(C_1/C_2)$$

The physical interpretation of this equation is that the tendency of an ion to diffuse down its concentration gradient is countered by the buildup of an electric field at the junction of the two solutions. The direction of the field is such as to hold back the ion from further movement. As noted in Chapter 2, it is necessary to move only a very minute number of ions in order to produce the restraint required to obtain equilibrium of electric and concentration gradients. As an example of an application of the Nernst equation, we calculate the transmembrane potential difference which might exist in frog muscle on the assumption that the muscle cell membrane is permeable only to K^+. The internal concentration of potassium is 155 meq/liter; the external concentration is 4 meq/liter. Substituting these values into the Nernst equation at a temperature of 27°C, we obtain for the potential difference,

$$E = 2.303(8.3 \times 300/96,500)\log(4/155)$$
$$= -95\ mv$$

The Goldman equation will not be derived here. However, in the derivation, the assumptions are made that the total flow of current through the membrane is zero, that is, that the flow of negative charges is equal to the flow of positive charges, and also that the drop in potential across the membrane is linear. Using these assumptions it can be shown that the resting potential will be given by the relation

$$E = 2.303 \frac{RT}{F} \log \frac{P_K C_K^O + P_{Na} C_{Na}^O + P_{Cl} C_{Cl}}{P_K C_K + P_{Na} C_{Na} + P_{Cl} C_{Cl}^O}.$$

In this equation R, T, and F have their usual significance. P represents the permeability of the membrane to the ion and the superscript "O" above the concentration, i.e., C^O, represents the concentration outside the cell, the unlabeled C, the concentration within the cell. When the permeabilities to Na^+ and Cl^- are taken to be zero, the equation reduces to the Nernst equation. If permeabilities are assumed to have the ratio,

$$P_K:P_{Na}:P_{Cl}::1:0.04:0.45,$$

then the emf of the membrane as given by the equation agrees rather well with the value measured in the squid axon, and the resting membrane potential agrees well with the calculated value over a more than 50-fold variation of concentration of external potassium.

The Hodgkin-Huxley equivalent circuit constitutes a third approach to the analysis of the nerve cell membrane potential. The membrane is assumed to contain separate channels through which each of the ions passes without interference from the others. The total electrical current flow through the membrane is again assumed to be zero. Each ion, in passing through its channel, encounters resistance to its movement through the membrane. Three such channels are shown, one for Na^+, K^+, and Cl^- (Fig. 1.25); each is represented by a battery (equilibrium potential) whose emf is calculated from the Nernst relation and whose opposition to current flow is shown as an electrical resistance. Often, the reciprocal of resistance, the conductance (g), is used to represent the channel permeability. If the total current from the membrane is zero, it can be shown that the membrane potential, E, resulting from the three emfs in parallel is given by the equation

$$E = \frac{E_K g_K + E_{Na} g_{Na} + E_{Cl} g_{Cl}}{g_{Na} + g_K + g_{Cl}} .$$

The gs represent the conductance of the membrane; they correspond approximately to the permeabilities to each ion. The emfs are the equilibrium potentials for each ion. Thus, in frog muscle the equilibrium potential for potassium is -95 mv, for sodium $+65$ mv, and for chloride -90 mv, values which have been calculated for each ion from the Nernst equation and the concentration of ions inside and outside the cell. Assuming further that the potassium conductance is 100 times greater than the Na and Cl conductances (i.e., that $g_K = 100 g_{Na} = 100 g_{Cl}$), which is approximately true for frog muscle, we obtain for the emf across the membrane,

$$\frac{-95(g) + 65(g/100) + (-90)(g/100)}{g + g/100 + g/100}$$

or -95 mv, a value very close to the actual membrane potential. The Hodgkin-Huxley equivalent circuit will prove very important in discussing the mechanism of the action potential in later sections.

Electrogenic Na-K Pump Produces a Portion of the Membrane Potential

A small proportion of the resting membrane potential, about 4 mV, is produced by the unequal exchange of 3 Na^+ for 2 K^+ per mol ATP hydrolyzed

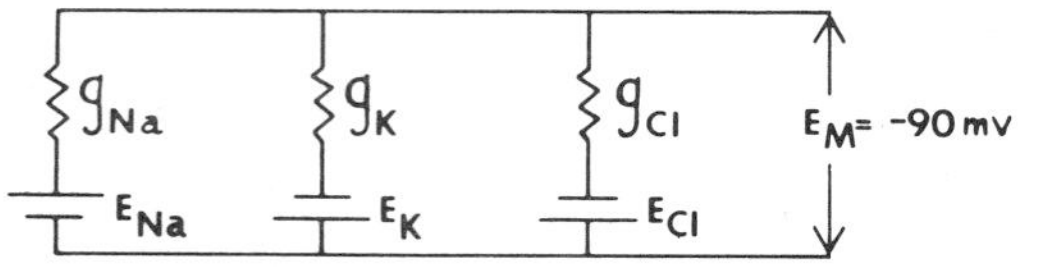

to ADP $+ P_i$ in the active transport mediated by Na/K-ATPase (Na-K pump). In turn, since there is net extrusion of one positive charge per mol ATP hydrolyzed, the pump rate is partly sensitive to the membrane potential, probably through a voltage-sensitive conformational reaction of the pump protein (Goldshlegger et al., 1987). This portion of the Na-K pump activity, termed *electrogenic,* results in hyperpolarization (or increased inside negativity) of the membrane. When the membrane is depolarized, the pump rate increases and when the membrane is repolarized, the pump rate decreases. Conversely, conditions that produce stimulation of the pump, such as increased intracellular Na^+, tend to hyperpolarize the membrane. Pump poisons such as ouabain, or decreasing temperature which reduces the pump rate, produce an immediate depolarization to an extent related to the electrogenic component.

EXCITABILITY AND CONDUCTIVITY OF THE NERVE FIBER

The generation and transmission of a nerve impulse involves two conceptually independent but operationally related processes, excitation and conduction.

When a nerve is stimulated, electrical events which are not propagated occur in the membrane in the vicinity of the electrodes. If the local events have particular characteristics (described below), the membrane potential undergoes an abrupt change, termed the *action potential,* which is self-propagated along the axon. Excitation refers to the events leading to the generation of an action potential; conduction refers to the propagation of the action potential which proceeds away from the site of excitation, much as a wave travels in a taut string. It is fundamental that nerve may be excited anywhere along its length and propagation is away from the point of stimulation in both directions. However, when a junction (synapse) intervenes between nerve fibers in the mammalian nervous system, then propagation can only continue in one direction. Experimentally many different kinds of stimuli (electrical, thermal, mechanical, chemical) may be utilized to excite nerve or muscle; however, all of those stimuli operate by depolarizing the nerve fiber. Since electrical stimuli of any intensity, shape (wave form), and duration may be easily produced both accurately and repetitively, this form of stimulation is used universally to study the phenomena of excitation and propagation. Moreover, the nerve impulse is electrical in nature, and many of its effects as an excitatory agent can be simulated by the electrical stimulus.

Figure 1.25. The Hodgkin-Huxley equivalent circuit for the membrane. The symbol E represents the "equilibrium" potential as determined for each ion from the Nernst equation. The symbol g represents the conductance of each ion.

Characteristics of the Stimulus

Nerve responds to electrical stimulation provided that the electrical stimulus fulfills certain specific criteria. It must be of sufficient *intensity* and *duration* to reduce the transmembrane potential from its resting value to a critical voltage which, when achieved, results in the development of a propagated impulse. This reduction in membrane potential is termed *depolarization,* and the critical voltage required for impulse propagation is termed the *threshold voltage* or *critical firing potential.* The relationship between the intensity and duration of the initial stimulus is further discussed below. A current just adequate to cause an impulse is called a *threshold* stimulus. Intensities below threshold are referred to as *subliminal.* Thresholds vary only slightly if the temperature and external ionic composition are maintained constant.

Another characteristic of the stimulus is its rate of rise. If the current is increased too slowly, the nerve will not respond. Figure 1.26 shows two linearly rising currents, one of which is able to reach threshold. The other current rises too slowly, and the nerve is able to accommodate to the passage of the current. *Accommodation,* therefore, consists of a rise in threshold of the tissue during stimulation. To minimize accommodations, it is convenient to employ stimulus currents which rise extremely rapidly. Two such stimuli are shown in Figure 1.27, the square wave and the exponential pulse.

Even the rapidly rising pulses of Figure 1.27, if too short in duration, would not result in an impulse. The two properties, intensity and duration, obviously interact, and it is therefore important to be aware of the relationship between the threshold stimulus intensity and the duration of the stimulus—the *strength-duration* relationship.

For this purpose, the following kind of experi-

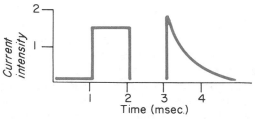

Figure 1.27. Two commonly used stimuli: a rectangular pulse *(left)* and an exponential pulse *(right).*

ment is usually performed. A stimulus of a fixed duration, e.g., 1 ms, is applied to a nerve through two electrodes, one of which is the cathode $(-)$, the other the anode $(+)$. The threshold stimulus is determined by increasing the current until a response is obtained—at the particular duration selected. A series of other durations are selected and the thresholds determined. The thresholds obtained at each duration are plotted as a function of the duration. The curve so obtained is called the strength-duration relationship for nerve (Fig. 1.28).

The curve is accurately described over most of its course by the empirical relationhip $I = I_0(1 - e^{-kt})^{-1}$, where I_0 and k are constants, t is the stimulus duration, and I the threshold current. For short durations, in which accommodation is presumably slight, the relation is approximated by the equation, $It = $ constant. This relationship may be interpreted as follows: The current I is the charge per unit time which is placed on the membrane in a time t. The product of the current and time is, therefore, equivalent to a constant charge. The equation implies that a critical amount of charge must be placed on the membrane, whatever the current or duration.

What is the meaning of this critical amount of charge? Placing a charge on the membrane by an appropriately oriented stimulus current reduces the net charge on the membrane, i.e., part of the charge on the membrane is neutralized. Such a partial neutralization of charge is equivalent to a *depolarizing*

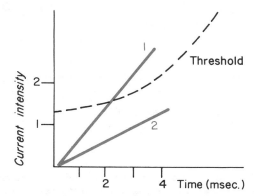

Figure 1.26. A schematic representation of the effect of rate of rise of current. Curves *1* and *2* represent two stimuli with different rates of rise. Stimulus 2 never reaches threshold *(dashed line).* Stimulus 1 attains threshold.

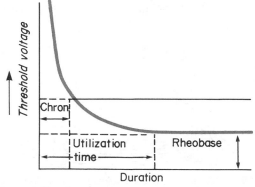

Figure 1.28. Strength-duration curve; *chron,* chronaxie.

of the membrane, i.e., to a reduction in the membrane potential from its resting negative value toward 0. Increased polarization or *hyperpolarizing* results in an increase in the membrane potential, i.e., increasing negativity.

A feature of excitation discovered quite early was the fact that the nerve impulse originated at the cathodal stimulating electrode, i.e., at the negative stimulating electrode. Figure 1.29 illustrates the lines of current that flow into and out of the electrodes. At the cathode, the current flows outward through the membrane; at the anode, where current is flowing inward, excitation is hindered. Continuous current flowing through the nerve depolarizes the membrane in the vicinity of the cathode and hyperpolarizes it in the vicinity of the anode. Thus, it is easier to stimulate a nerve in the vicinity of the cathode because the membrane potential has been lowered; conversely, at the anode, the membrane potential is increased and higher currents are required to excite the underlying tissue. These effects of direct current on the excitability of nerve and muscle have been referred to as electrotonic phenomena; the depolarization at the cathode as *catelectrotonus*, and the hyperpolarization at the anode *anelectrotonus*.

Summation of the Local Excitatory State

The application of a brief subthreshold stimulus has a residual effect on nerve, even though an impulse is not elicited. This residual effect is revealed by the fact that the application of a second subthreshold stimulus within a millisecond can elicit a response. The longer the interval between the two stimuli, the more intense the second stimulus must be to yield a response. The important point, however, is the fact that two subthreshold stimuli can sum their effects on the nerve membrane so that a response is evoked. The interpretation of this experiment is that the first subthreshold stimulus produces a change in the membrane potential which lasts for a millisecond or more and that this change facilitates the effect of a second stimulus.

The change in the membrane caused by the first stimulus is referred to as the "local excitatory state" in the vicinity of the electrode. It is a nonpropagated response of the membrane which is associated with the depolarization of the membrane in the region of the cathode (see below).

LOCAL EXCITATORY STATE

The experiments on subthreshold summation and the electrotonic properties of nerve led to the expectation that the membrane potential should be decreased in the vicinity of the cathode, and indeed, this expectation has been affirmed by direct experiments in which the changes in membrane potential were measured in the vicinity of the cathode and anode.

It is necessary first to consider the purely passive changes in the membrane potential which result from the fact that nerve has the properties of an electric cable, namely, electrical resistance and capacitance (the ability to store charge). The capacitance, C, is defined as the charge, Q, which must be placed on two surfaces in order to produce a unit potential difference, V, between them ($C = Q/V$). The units of C are farads when Q is expressed in coulombs and V in volts. The capacitance of the nerve membrane is 1 $\mu f/cm^2$.

For each unit length of nerve, there is an external electrical resistance to current flow through the extracellular medium and an intracellular resistance to flow through the cytoplasm. The transmembrane resistance constitutes a third electrical resistance. Each unit length of nerve may be considered as an electrical cable (Fig. 1.30) consisting of these resistances and a condenser (the capacitance). If a stimulating current is passed through a section of nerve, its condensers will charge, but the most distant ones will be least charged since more external and internal resistance is included between them and the source of current. If one were to measure the charging process at any moment, the voltage

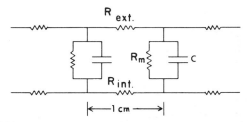

Figure 1.30. Equivalent circuit of a nerve (radius 1 μm) considered as an electric cable. $R_{ext.}$ is the resistance of 1 cm of the external medium; generally low. $R_{int.}$ is the resistance of 1 cm of the internal axoplasm (10^7 Ω) and R_m is the resistance of 1 cm of the membrane to radial currents—of the order of 10^{10} Ω. The capacitance, C, is 1 $\mu f/cm^2$ of nerve surface.

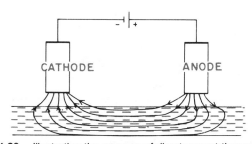

Figure 1.29. Illustrating the passage of direct current through tissue. Anode is defined as the electrode which sends current *into* tissue.

would be highest at the electrodes and would decrease exponentially as one proceeded away from the stimulating electrodes (Fig. 1.31). This charging process is almost instantaneous and represents a nonpropagated buildup of charge along the nerve. It extends for some distance along the nerve, but it is never propagated as a wave.

When nerve is stimulated, two electrical events take place. The membrane charges passively as a cable and, at the same time, the nerve begins to react physiologically at the cathode. Events at the anode are passive, and since excitation does not occur there, the time course of charging at the anode may be taken as that for the passive or physical charging of nerve. At the cathode, passive charging occurs, but in addition, an active process occurs which we have called the local excitatory process. If the passive process at the cathode is subtracted graphically from the overall recorded response, the local excitatory process should be obtained. The passive physical process at the anode and cathode are proportional to the stimulating current. In order to measure the local responses, an experiment is performed in which the potential difference between electrodes placed at the anode and cathode is measured with respect to a distant electrode. Figure 1.32 shows the results of the potential measurements and also the result of subtracting the passive response from the overall response to obtain the local potential which corresponds to an active, nonlinear process at the electrode.

The significance of the local potential is that it demonstrates the time course of the depolarization at the cathode. When the depolarization at this electrode reaches a critical value an action potential will be initiated which will propagate away from the electrode. Such local responses can be found in many tissues. For example, receptors must be depolarized before they give rise to an action potential; the potential representing the local excitatory state of the receptor is called the generator potential. An active process also occurs at the neuromuscular end-plate which is called the end-plate potential; this depolarization initiates the propagated action

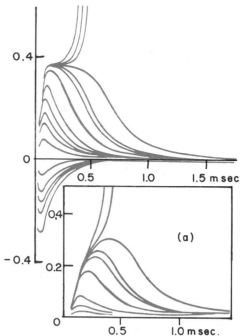

Figure 1.32. Electrical changes at stimulating electrode produced by shocks with relative strengths, successively from above, 1.00 (upper 6 curves), 0.96, 0.85, 0.71, 0.57, 0.43, 0.21, −0.21, −0.43, −0.57, −0.71, −1.00. The ordinate scale gives the potential as a fraction of the propagated spike, which was about 40 mv in amplitude. The 0.96 curve is thicker than the others, because the local response has begun to fluctuate very slightly at this strength. The width of the line indicates the extent of fluctuation. *Inset (a)*, Responses produced by shocks with strengths, successively from above, 1.00 (upper 5 curves), 0.96, 0.85, 0.71, 0.57; obtained from curves in upper figure by subtracting anodic changes from corresponding cathodic curves. Ordinate as above. (From Hodgkin, 1939.)

potential of muscle. At the synapses of neurons, a local, nonpropagated potential (EPSP) may be recorded which gives rise to the nerve impulse of the neuron. The critical event in the excitation of all these cells is the local, nonpropagated depolarization. When this depolarization reaches threshold magnitude, the explosive, propagated change in membrane potential, called the nerve impulse or action potential, is set off.

NERVE IMPULSE

Recording the Action Potential

The externally recorded action potential may be obtained by stimulating a nerve at one end and picking up the responses of the nerve some distance away with two recording electrodes and a recording device. Figure 1.33 shows the arrangement by which the action potential of nerve is usually recorded with a cathode ray oscilloscope. Unlike any instrument previously employed for this purpose (e.g., the string galvanometer or the capillary

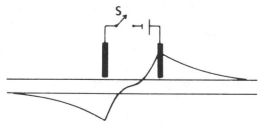

Figure 1.31. Illustration of the passive, electrotonic potential along the nerve resulting from the passage of direct current into the nerve.

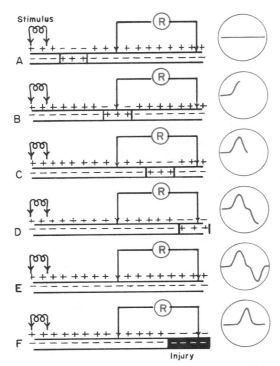

Figure 1.33. Passage of action potential down a nerve. Stimulus is applied at left. Cathode ray oscilloscope is indicated at *R*. The recorded potential is shown at the face of the cathode ray oscilloscope at the extreme right of each figure. A monophasic potential is shown in figure *F*.

electrometer), the moving part of the oscilloscope is a stream of electrons and consequently has virtually no mass, and thus, no inertia. It is, therefore, capable of recording very rapid changes in electrical potential. The instrument consists of an evacuated tube; an electron stream from a hot cathode strikes a fluorescent screen upon which it produces a spot of light. On either side of the electron stream is placed a vertical plate. A potential difference is created between the pair of plates; the electric field set up across the path of the stream deflects it horizontally and sweeps it across the screen. The spot of light is converted into a horizontal streak. By means of a sweep oscillator the horizontal deflections are repeated many times per second. It is this latter set of plates into which the nerve action potential is fed so that a vertical deflection of the electron stream results with production of a standing wave which is photographed and a permanent record thus obtained. The speed of the horizontal movement of the spot of light enables the time scale to be calculated; this can be varied by altering the potential applied to the vertical pair of plates. The magnitude of the action potential is determined from the height of the wave. Before reaching the recording system, the action current is amplified

several thousand times by passing it through an amplifier. This is necessary because the electron stream requires about 50 v to cause a deflection of 1 cm on the face of the tube.

The response on the cathode ray oscilloscope shows up as a diphasic variation of potential. The cathode ray oscilloscope records the potential under one recording electrode as a function of time with respect to the other recording electrode. When the nerve is stimulated, an electrical impulse is generated which travels along the fiber. The arrival of this impulse at the first electrode causes the oscilloscope beam to deflect in one direction. When the impulse passes between the electrodes, no recorded potential is observed. But when the impulse reaches the second electrode, a potential is recorded which is opposite in polarity to the first and accordingly, causes the oscilloscope beam to deflect in the opposite direction. We, therefore, say that there are two phases present in the recorded action potential. It is quite clear that one can eliminate the second phase from the recording by preventing the negativity of the impulse from reaching the second electrode. This can be done by crushing the nerve either in the region between the two electrodes or directly under the second electrode. Of course, if the crush is at the first electrode, no potential will be recorded at all, since the action potential cannot pass a dead region of nerve. The potential recorded when the nerve is crushed at the second electrode is called a monophasic potential and is shown also in the accompanying figure (Fig. 1.33*F*).

COMPOUND NATURE OF ACTION CURRENT RECORDED FROM A NERVE TRUNK

Erlanger and Gasser (1937) studied the action potential of mixed nerve trunks by means of the cathode ray oscillograph. They showed that the recorded "spike" is actually compounded of the individual spikes of many axons which were classified into three main types of nerve fiber—referred to as the A, B, and C groups (Fig. 1.34). Several properties of nerve are correlated with the diameters of the fibers: the larger the fiber diameter, the greater the conduction velocity; the greater the magnitude of electrical response, the lower the threshold of excitation and the shorter the duration of response and the refractory period. The relationship of conduction velocity to diameter of the nerve fiber is a linear one (Fig. 1.35). The amplitude of the externally recorded potential is also linearly related to the fiber diameter.

The A group is composed of the largest fibers, 1–20 μm in diameter, with conduction rates from 5 m/s or less for the smallest fiber to 100 m/s for the

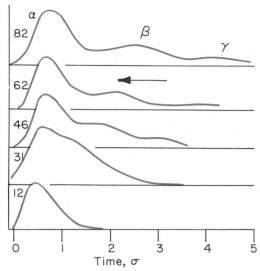

Figure 1.34. Cathode ray oscillograph records of the action currents in the sciatic nerve of the bullfrog after conduction from the point of stimulation through the distances (in millimeters) shown at the left. The action potentials might be compared to runners in a race who become separated along the course as the faster contestants outstrip the slower; thus, in a record at 82 mm from the point of stimulation, three waves are shown, whereas at 12 mm the potentials are fused, and only one large wave appears. (Adapted from Erlanger and Gasser, 1937.)

largest. The fibers of the A group are all myelinated, and both sensory and motor in function, and are found in such somatic nerves as the sciatic and saphenous nerves.

The B fibers are myelinated and have diameters from 1–3 μm and conduction velocities from about 3–14 m/s. The B fibers are found solely in preganglionic autonomic nerves. The C group, composed of the smallest fibers (less than 1 μm in diameter), are unmyelinated and have a conduction rate of around 2 m/s or less; many are found in cutaneous and visceral nerves. They have a high threshold, 30 fold

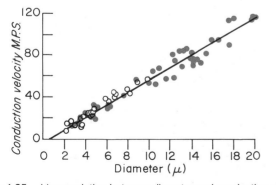

Figure 1.35. Linear relation between diameter and conduction velocity of mammalian nerve fibers. Each point represents a determination of the maximum conduction velocity in meters per second and of the diameter in microns of the largest fiber of an individual nerve. *Dots,* adult nerves; *circles,* immature nerves. (Adapted from Hursh, 1939.)

that of the A group. The A group of fibers makes by far the greatest quantitative contribution to the compound action potential, and the C group the least. The electrical potentials recorded from both A and C fibers exhibit slow variations in potential following the action potential, negative and positive after-potentials, but the B group does not exhibit a negative after-potential with a single response (though a negative after-potential does appear upon repetitive stimulation). The B fibers are the most susceptible to asphyxia, the C fibers, the least so.

The linear relationship between fiber diameter and conduction velocity holds also for growing nerves of young animals. During growth the diameters of the nerve fibers enlarge. Conduction velocity increases proportionately so that the time taken for an impulse to travel from the toes of a kitten a few days old to the spinal cord is the same as for a full-grown cat. Thus, the kitten and the cat react to stimulation with about equal promptness.

The diameters of the regenerating fibers in a sectioned or crushed nerve also enlarge gradually, and conduction velocities increase accordingly, the relationship again being a linear one. The maximum conduction velocity is not reached until maximum diameter of the fiber is attained. If the axons of the nerve fibers alone are interrupted, the sheaths of the nerve fibers remaining intact, the diameters and conduction velocities may reach those of the normal nerve; this rarely occurs if the nerve has been completely severed.

Conduction Rates

The velocity of the nerve impulse varies in different nerve fibers in accordance with their diameters, the thicker fibers conducting more rapidly than the fibers of smaller diameter. In the large afferents from muscle spindles of the mammals, the rate is from 80–120 m/s. Sensory nerves of the skin being of smaller diameter have slower conduction rates. Nonmedullated fibers conduct more slowly than medullated fibers. Some of the fibers subserving pain sensation and those of the sympathetic nervous system have a very slow conduction rate.

The following table gives approximate conduction rates in nerves of several different animals:

Medullated nerve, mammal, 37°C: 120 m/s
Medullated nerve, dogfish, 20°C: 35 m/s
Medullated nerve, frog, 20°C: 30 m/s
Nonmedullated nerve, crab, 22°C: 1.5 m/s
Nonmedullated nerve, mammal, 37°C: 1 m/s
Nonmedullated nerve, olfactory of pike: 20°C: 0.2 m/s
Nonmedullated nerve, in fishing filament of physalia, 26°C: 0.12 m/s
Nonmedullated nerve, in *Anadon*: 0.05 m/s
Compare the velocity of sound in air at 0°C: 331 m/s

By an indirect method of measurement the rates of conduction in various human postganglionic sympathetic nerves have been found to be from 0.85–2.30 m/s. The lower figures were obtained for the nerves of the leg, the higher ones for the nerves of the chest.

ACTION CURRENTS AND EXCITATION

The externally recorded action potential represents a variation in electrical potential along the nerve. An alternative way of looking at the nerve impulse is as a set of currents flowing out of the membrane ahead of the area of greatest depolarization, or more particularly, as the movements of these currents along the nerve (Fig. 1.36). The term action current is as appropriate as action potential. The advantage of understanding the currents of nerve as well as the potential becomes apparent when one considers how the self-exciting properties of the nerve impulse arise; for it is the currents themselves which act to depolarize the nerve. The analogy is often made that the nerve acts like a fuse along which the ignition progresses. Action currents leave the nerve ahead of the region of depolarization, acting as a virtual cathode since they have the direction of currents flowing into a cathodal electrode. The hypothesis that action currents act to depolarize the region ahead of them was substantiated by experiments in which it was demonstrated that if action currents were allowed to enter but not to excite a region beyond a narcotized stretch of the nerve, then the action currents gave rise to two phenomena beyond the blocked region: an increase in excitability and a depolarization of the fiber. Neither the increase in excitability nor the depolarization were large enough to set up an impulse; the block diminished the intensity of the currents, but it was clear from this experiment that the local currents generated by the action potential (as distinct from the action potential itself) could cause depolarization and, therefore, a change in excitability.

Another hypothesis which is a part of the general theory of self-excitation holds that in order for a propagated nerve impulse to be generated, not only is a depolarization necessary, but there must also be an increase in permeability to ions at the depolarized region. Thus, depolarization brings about an increased ion flow through the "not yet excited" membrane bordering on the depolarized region. In other words, during the action potential a path is opened in the membrane through which the charge from resting membrane can flow. Thus, by discharging resting membrane ahead of itself, the action potential becomes self-propagating. Evidence for this characteristic of self-propagating activity was obtained by Curtis and Cole (1938) who showed that there was an increased permeability to ions

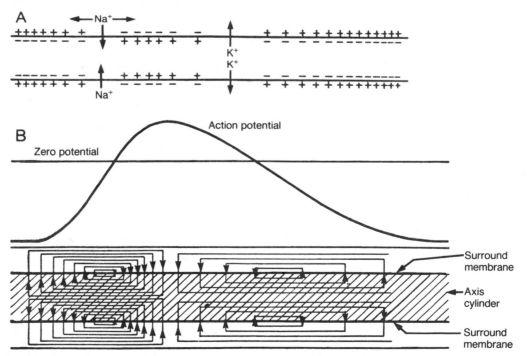

Figure 1.36. *A*, The movement of sodium and potassium during the action potential, traveling in the direction of the *arrow*. The charge in each region is also shown. *B*, Upper curve shows the potential distribution of the impulse along the nerve. The lower part shows the flow of currents in the external medium and within the fiber. There is a reversal of potential during the spike. (Adapted from Eccles, 1953.)

during activity. This was demonstrated as a decrease in transverse resistance of the membrane from a resting value of 1000 Ω/cm^2 to 25 Ω/cm^2 during the rise of the action potential in the squid axon. Decreases in membrane resistance have also been demonstrated in other tissues, such as muscle, during propagated activity.

"All-or-None" Principle

A stimulus which is just capable of exciting a nerve fiber (threshold stimulus) sets up an impulse which is no different from one set up by a much stronger stimulus. The impulse generated by the weak stimulus is conducted just as rapidly and is equal in magnitude to that generated by the strong stimulus when judged by the action current developed or the mechanical response of the muscle it innervates. Thus, the propagated disturbance established in a single nerve fiber cannot be varied by grading the intensity or duration of the stimulus, i.e., the nerve fiber under a given set of conditions gives a maximal response or no response at all. To make use again of the train of gunpowder analogy—the flame of a match applied to the powder fuse will start a traveling spark no less intense than one started by the flame of a torch. The restoration of the strength of the impulse to its original value after passing from a narcotized region into normal nerve also shows the "all-or-none" nature of nervous conduction. The well-known fact that a strong stimulus applied to a nerve *trunk* causes a compound action current of greater amplitude, and a greater muscular response than a weaker stimulus is due to the fact that the nerve trunk is composed of many fibers, each of which supplies a group of muscle fibers. The weak stimulus excites only a proportion of the units of the nerve trunk, whereas a maximal stimulus excites them all. For example, the cutaneous dorsi muscle of the frog is supplied by a nerve which contains only 8 or 9 fibers; each of these innervates about 20 muscle fibers. When the nerve was stimulated by shocks of gradually increasing intensity, the muscular responses did not show a similar continuous rise in amplitude; on the contrary, the responses of the muscle increased in a series of well-defined steps. In other words, increasing the stimulus intensity produced no effect for a time upon the amplitude of the muscular response, but then a slight increase in strength of stimulus produced a sudden rise in amplitude. The steps which were never greater in number than the number of fibers in the nerve were due to additional fibers becoming excited as the strength of stimulus reached a certain value.

It must also be remembered that the "all-or-none"

principle applies only for the condition of the nerve at the point where, and the moment when, the impulse arises. A stimulus which will give rise to a response of a certain magnitude under one condition of the nerve may give a much smaller response under other conditions, e.g., during the relative refractory period (see below), narcosis, oxygen lack, etc.

Absolute and Relative Refractory Periods of Nerve

For a brief interval following the passage of an impulse along the nerve fiber, a second stimulus, however strong, is unable to evoke a response. This interval is called the absolute refractory period. In a frog's sciatic nerve at a temperature of about 15°C the absolute refractory period has a duration of between 2 and 3 ms. Its duration is roughly the same as the action potential "spike." It is much shorter in mammalian nerve (0.4–1.0 ms in large medullated nerve fibers).

The period during which the nerve is absolutely refractory is succeeded by one in which the nerve, though it will not respond to a stimulus of the same strength as it did before the passage of the impulse, will respond to a somewhat stronger one. The excitability of the nerve gradually increases and the strength of stimulus necessary for excitation becomes progressively less (Fig. 1.37). In the end, the restoration of excitability is complete, and the nerve responds to a stimulus of no greater strength than that which is capable of exciting a resting nerve. The period following the absolute refractory phase and during which the excitability gradually rises to normal is called the relative refractory period. The time required for the excitability of the nerve to return to about 95% of its resting level ranged from 10–30 ms (full recovery may not be attained until the lapse of 100 ms). It should be pointed out that the failure of the nerve to conduct a second impulse is not due simply to lowered excitability at the point in the nerve where the original

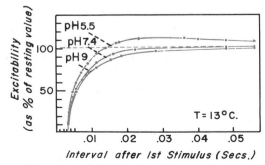

Figure 1.37. Recovery of excitability in nerve perfused with fluids of different pH. (Adapted from Adrian, 1935.)

stimulus was applied, for during the absolute refractory period a stimulus applied to any other point upon the nerve likewise fails to set up an impulse. The passage of the impulse along the nerve leaves in its wake a change of state in the membrane organization. A certain time is required for the changes associated with the passage of the impulse to become reversed and the nerve restored to its resting condition.

The refractory period renders a continuous excitatory state of the nerve impossible. Fusion or summation of impulses does not occur. The refractory period obviously must also set an upper limit on the frequency of the impulses. In the mammal the absolute refractory period is about 0.5 ms. The intervals between impulses cannot be shorter than the absolute refractory period; the maximum impulse frequency is around 1000/s. At this rate, the impulses are traveling in the *relative* refractory period of their predecessors and are weaker and more slowly conducted. In frog nerve with its refractory period of from 2–3 ms, the maximal impulse frequency is between 250 and 300/s.

Classical Mechanism of the Action Potential

The mechanism by which the action potential is produced has not been examined, but the role of the currents in exciting a region ahead of the approaching nerve impulse was noted. It is now appropriate to consider the origin of the action potential in terms of the prevalent ions, namely, sodium and potassium. The resting potential was shown to arise from a diffusion of potassium ions across the cell membrane. The action potential has a more complex origin. Recent investigations have demonstrated that the action potential results chiefly from the movement of sodium ions into the nerve impulse, and that there is a movement outward of potassium ions later during the fall of the action potential.

At the turn of the century, the action potential was attributed to a complete depolarization of the membrane, creating an area more negative than other parts of the nerve into which the ionic currents flowed. This concept, first enunciated by Bernstein (1912), could not be tested with the methods and tissues available until 1940. At that time two groups, one in England and the other in America, performed the critical experiments which led to a reexamination of the Bernstein concept. These experiments utilized the giant axon of the squid to make a direct measurement of the transmembrane potential. This axon is 0.1 mm or more in diameter, so that it is possible to insert one electrode directly

into it along its length and to place another electrode outside. The resting potential was measured and the nerve stimulated to obtain an action potential. The results of such an experiment are illustrated in Figure 1.38. According to the classical concept, the action potential arises from the fall of the membrane potential toward zero, the nerve being completely depolarized when the membrane potential reaches zero. In the case of the squid axon instead of a simple depolarization of the membrane, an "overshoot" of the potential past the zero baseline was noted. Such an overshoot represents a reversal of the potential from the resting state so that during the peak of the action potential, the inside of the cell becomes about 50–60 mv positive to the outside (Fig. 1.38). This is an astonishing result for, on first consideration, there does not appear to be any mechanism by which the potential difference could invert. Bernstein had considered the resting potential a diffusion potential resulting from the permeability of nerve to potassium and the impermeability to sodium and chloride, and the electrochemical events during the action potential were attributed to an equal flow of anion and cation other than potassium into the nerve fibers, so that the diffusion potential across the membrane disappeared when the sodium and chloride flows were permitted to take place. (In a diffusion cell the emf is proportional to the difference of the mobilities of cations and anions, and if these are equal, the emf is zero.) This hypothesis, while showing how the membrane potential could become zero, did not appear to contain an explanation of how the reversal of potential (which has been referred to as overshoot) might arise.

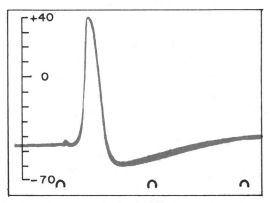

Figure 1.38.   Action potential recorded with an internal electrode from a squid giant axon. The scale shows the internal potential in millivolts relative to the outside bath. Time marks are in 2-ms intervals. (Adapted from Hodgkin and Huxley, 1945.)

MODERN CONCEPT OF THE ACTION POTENTIAL

Closer consideration of the details of ionic flow pointed to sodium as the ion which might give rise to the reversal of potential. Since the concentration of sodium is higher outside than inside the nerve, a flow of sodium inward would tend to make the inside positive with respect to the outside. The experimental result establishing that sodium is the most important cation involved in both the production of the action potential and of the overshoot is simple. The squid axon is normally surrounded by sea water osmotically equivalent to a 0.3 M sodium chloride solution. If the concentration of the sodium in the sea water is changed so that less sodium is present, the amplitude of the action potential will diminish. In such experiments, the resting potential is unaffected by altering the sodium concentration. Accordingly, it was concluded that sodium is essential to the production of the action potential including the overshoot. Figure 1.39 illustrates the type of result obtained from such substitution experiments.

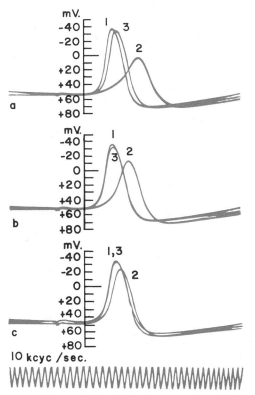

Figure 1.39. Single action potentials recorded from a squid giant axon. *1*, Normal action potential in sea water. *2*, Action potential after equilibration in medium with altered sodium. *a* is 0.33 times normal, *b* is 0.5 times normal, and *c* is 0.71 times normal in concentration. *3*, Action potential after return to normal sodium solution.

Note that in applying half sea water, the osmotic lack of sodium is balanced by substituting an osmotically equivalent amount of a nonelectrolyte such as sucrose or an electrolyte such as choline chloride.

Voltage "Clamping"

Various experiments were carried out to determine the dependence of the sodium and potassium ionic currents on the membrane potential. Unfortunately, it is not possible to measure one ionic current flow during a depolarization of nerve directly and independently of another. The changes in current flow affect the membrane potential, which in turn affects the current flow. This concept of a mutual influence of current and potential is important for the understanding of the mechanism of the action potential, but it is necessary to eliminate the interaction operationally if one wishes to study the effect of a change in membrane potential on the ionic fluxes. Such independence of current and potential is achieved by depolarizing the membrane to a given value and then subsequently maintaining the potential at that value by provision of current via an external circuit. The method is referred to as the *voltage clamp method* because the membrane potential is "clamped" (fixed) at an invariant value. Thus, it became possible to investigate the ionic flows resulting from a given step of depolarization. Note that once the potential has been clamped, there will be no flow of current into the capacitive element of the membrane, since to put charge on a condenser requires a changing potential. The current, therefore, flows only through the external, internal, and transmembrane resistances noted above.

In such voltage clamp experiments it has been observed that the current may change while the depolarization is being maintained. This appears to violate Ohm's law, which states that current flow in a system depends upon the voltage ($I = E/R$), and the voltage across the membrane is held constant in these experiments. The explanation of these findings rests in the fact that the resistance or permeability of the membrane changes during a voltage clamp experiment, and thus, although the potential is constant, extensive changes in current occur.

The effect of a small clamp on the current flow through the membrane is shown in Figure 1.40. The depolarizing clamp causes a diphasic current which is first directed inward (into the nerve fiber) and then outward. If a solution which lacks sodium is substituted for the outside bath, the initial inward current disappears, but the outward current remains unaffected. The early, inward current is,

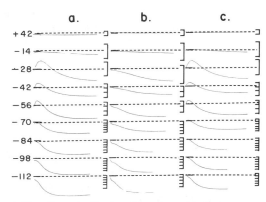

Figure 1.40. Currents flowing through squid giant axon membrane during a steady voltage clamp. Figures on left are clamped values of the membrane potential in millivolts relative to resting potential (outside minus inside). Minus values represent depolarizations. Columns *a* and *c*, axon in sea water. Column *b*, axon in choline sea water without sodium. Vertical scale, 1 div = 0.5 ma/cm³. Time dots are 1 ms apart. (Adapted from Hodgkin and Huxley, 1952d.)

therefore, attributed to sodium inflow. The dependence of the sodium current on the amount of depolarization is revealed by the experiments in which progressively greater amounts of depolarization are used (Fig. 1.40). As the membrane potential is decreased (greater clamp voltage), the inward current increases. With further increases in the depolarization, however, the sodium current decreases in magnitude and ultimately disappears.

The results of a brief depolarization or clamp may be outlined as follows: (1) There occurs, initially, a brief inward current resulting from an inward movement of sodium ions, followed by an outward flow of potassium ions. (2) The amount of sodium flowing depends upon the extent of depolarization; over a rather broad range it is seen that the greater the depolarization, the larger the sodium current. (3) The potassium current is relatively unaffected by the depolarization but is greatly affected by a hyperpolarization of the membrane. Increasing the membrane potential from, for example, −60 to −80 mv increases the potassium current.

Hodgkin-Huxley Theory

To understand the mechanism of the action potential, it is necessary to examine the results of the voltage clamp experiments in terms of a detailed theory—evolved by A. Hodgkin (1964) and A. F. Huxley (1964)—which explains many of the phenomena of excitation and conduction in nerve and muscle. According to this theory, the nerve membrane is represented as containing three channels through which sodium, potassium, and chloride ions may move independently. The individual ions are forced through the membrane by their electrochemical potential gradients. A schematic circuit

representing the three-channel hypothesis is shown in Figure 1.41. The value of each emf is given by the Nernst equation, the three emfs being equivalent to the three "equilibrium" potentials (as discussed above in the section on the membrane potential). It is important to understand that the network is only an electrical "equivalent" and that the channels are not physically located one next to the other. Each of the channels represents a group of many identical channels in a unit area of the membrane lumped together schematically as one channel. The resistance signifies the opposition encountered by each ion in the unit area of membrane. The reciprocal of the resistance, the conductance (g), is proportional to the permeability of the membrane to the ion.

The sequence of ionic events in an action potential may now be described as follows: A depolarizing voltage (stimulus) is applied to the membrane and the sodium influx increases in the direction of its concentration gradient. If this Na^+ influx is greater than the flux of potassium and chloride (both of which constitute current flowing in the opposite direction), then the net sodium entry causes a change in the membrane potential. But, as noted above, the change in potential across the membrane brings about an increased sodium influx, which in turn leads to a further decrease in the membrane potential. Therefore, the influx of sodium builds up quite rapidly. This mutual effect of sodium influx and membrane potential constitutes the *regenerative* factor in impulse transmission and is an example of a positive feedback process. In terms of the equivalent circuit of Figure 1.41, the action potential develops in the following way. The increase in permeability to sodium is essentially a decrease in resistance. The potential

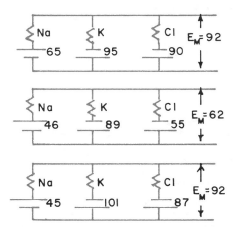

Figure 1.41. Equivalent circuits for three membranes. *Upper* is for mammalian skeletal muscle fiber; *middle* is for squid giant axon; and *lower* is for frog sartorius muscle fiber. E_M is resting potential; all values are in millivolts.

difference across the membrane, which is the resultant of the three emfs in parallel, will approach that emf which has the lowest resistance in series with it. The resistance to potassium and chloride does not change. The membrane potential therefore tends to approach the equilibrium potential of the sodium ion which, in volts, is

$$E_{\text{Na}} = 0.058 \log(\text{Na}_{\text{outside}}/\text{Na}_{\text{inside}}).$$

The equilibrium potential of the sodium ion is approximated but never quite reached by the peak of the overshoot. It is possible that an increased efflux of potassium and an active process of sodium extrusion accounts for the difference between the experimentally observed peak of the overshoot and the Na^+ equilibrium potential, i.e., the theoretically expected peak.

The rapid rise in sodium current does not continue for very long and, as shown by the clamp experiments, the sodium currents are rapidly followed by a potassium efflux from the axon. The sodium influx is terminated by a process called *sodium inactivation* which develops during the increase in sodium flux. Evidence for the existence of such an inactivation process has been adduced in clamp experiments of another type. For example, if one applies a clamp of the order of 10 mv, this voltage change is insufficient to cause a sodium influx, yet it can nevertheless set into action the process of inactivation—as demonstrated by the fact that, following the small depolarization, a larger depolarization of about 40 mv does not evoke the magnitude of sodium current that would have developed if the smaller voltage had not been applied initially. In other words the small initial voltage clamp rendered the larger subsequent voltage clamp less effective in changing the permeability of the membrane to Na^+. Thus, by following small clamps by larger clamps at various intervals, the time course of development of the inactivation can be ascertained (Fig. 1.42). Thus, the inactivation process turns off the sodium current and thereby contributes toward restoration of the resting state.

A second process also aids in the restoration of the resting membrane potential. It was noted that a potassium efflux follows the sodium influx. The time course of efflux of potassium ions corresponds to the fall of the action potential. The increase in conductance to potassium means that the change in potential across the membrane is now being driven by the electrochemical potential due to the consequence of the increased permeability to potassium ions (sometimes referred to as rectification) tends to restore the membrane potential to its original state of inter-

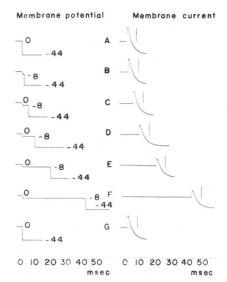

Figure 1.42. Development of "inactivation" during constant depolarization of 8 mv. *Left column,* displacement of membrane potential from the resting value in mv. *Right column,* membrane currents as a function of time for the displacement in the left column. The *vertical lines* show the sodium current expected in the absence of the conditioning step. (Adapted from Hodgkin and Huxley, 1952d.)

nal negativity. Figure 1.43 shows the time course of the increase in conductance of potassium in relation to the sodium inactivation process.

Figure 1.39 shows that after the inactivation, the membrane potential becomes more negative than the resting potential (hyperpolarization), taking many ms or up to s under some conditions to return to the resting potential. This hyperpolarization (originally called positive after-potential) is related to two factors: excess K^+ conductance that persists before the K^+ channels are inactivated at the end of the action potential and the electrogenic portion of the Na-K pump before its rate is slowed by the repolarization of the membrane.

Under some conditions, particularly after repeated frequent action potentials, the repolariza-

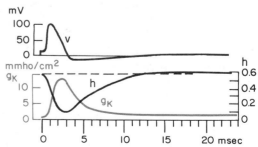

Figure 1.43. Time course of inactivation (*h*) and potassium conductance (g_K) during a nonpropagated action potential. *Upper curve* is action potential in response to a 15 mv initial depolarization. (Adapted from Hodgkin and Huxley, 1952d.)

tion is slowed before the membrane potential reaches the resting level (originally called negative after-potential) which is probably related to accumulation of K^+ at the outer surface of the membrane.

Both the sodium and potassium flows during the phase of increased permeability are in the direction of their concentration gradients. The flow of ions during the action potential is passive and depends upon the concentration gradient and on interactions among ions and with the channel (Hille, 1978). The production of a nerve impulse would be expected, therefore, to be independent of metabolism. This deduction is substantiated by experiments in which both aerobic and anaerobic processes are blocked by inhibitors. Nerves may continue to conduct for hours, even when stimulated at high rates under the influence of metabolic inhibitors. The increased permeability to sodium and potassium persists, although the means for pumping sodium and potassium have been cut off. This is not to say that transport processes are not important for maintaining the ionic inequalities. If nerve is continually stimulated while being treated with an inhibitor, it fails to pump out the sodium which accumulates inside the nerve both during the resting state and during excitation—and it also fails to reaccumulate potassium to restore its resting intracellular K^+ concentration. Ultimately, therefore, the nerve will depolarize as the result of potassium loss and/or sodium influx. Thus, the ion pumps and associated metabolic activity which maintain the steady-state ionic gradients across the nerve cell membrane are necessary in the long run, but neither the resting potential nor the action potential depends primarily upon metabolism for their production.

How does this theory of the action potential explain other important phenomena of nerve, such as refractoriness and accommodation? The absolute refractory period is the result of the inability of nerve to respond to a stimulus no matter how intense. This inability corresponds to the period during the action potential when sodium influx is completely inactivated. It is impossible to turn on the sodium influx machinery in any degree whatever at the height of the inactivation process. During the relative refractory period, the nerve is under the influence of partial sodium inactivation and increased potassium efflux. The potassium efflux tends to maintain the membrane potential at the original unexcited level. The restorative processes of sodium inactivation and potassium efflux (rectification) must be opposed by very large depolarizations in order to get sufficient sodium influx to start regeneration. Similarly, accommodation,

which is a rise in threshold during the persistent application of a linearly rising current, may be explained as the result of growth of inactivation at the cathode of an exciting current, much as inactivation developed in the clamp experiments which demonstrated the existence of the process.

Little is known concerning the molecular events within the nerve membrane that are triggered by depolarization and which lead to increased sodium conductance. It is clear, however, that these events are reversible.

As noted above, the excitatory process of axon and muscle membranes involves voltage-dependent transient increases in permeability to Na^+ and K^+ ions in specific channels for each. In mammalian myelinated nerve, the relative importance of potassium channels diminishes; for example, rabbit sciatic nerve nodal membrane lacks K^+ channels and depends on inactivation of Na^+ channels and leakage conductance for repolarization.

Ca^{2+} channels also exist to account for Ca^{2+} action potentials in certain locations (see below). While the steady-state membrane conductance to specific cations is a function of the membrane potential, the average conductance of single channels does not depend on the membrane potential. The simplest interpretation of this apparent paradox is that the single channels have two conductance states, open or closed, and the number of open channels depends on the membrane potential.

The opening of a channel is itself associated with a small current called the "gating current" (Armstrong, 1975; Armstrong and Bezanilla, 1977). This gating current has been measured for Na^+ channels in squid axon and frog nerve node of Ranvier by blocking the Na^+ channel with tetrodotoxin and the K^+ channel with tetraethylammonium, and then determining the current change associated with one or more step increments in the clamped membrane voltage. Tetrodotoxin, while blocking the entry into the Na^+ channel, does not prevent the opening and closing of the gate and, under this condition, the gating current is a large fraction of the total charge displacement. The gating mechanism behaves as if three or four charged particles cooperate in the opening of each channel (Keynes, 1979). The energy source for the gate opening seems to be the electric field; no chemical sources have been identified. The closing of the channels may involve a further rearrangement of the gating particles or a separate and independent set of blocking components. Hille (1978) has estimated that the cross-sectional dimensions of the Na^+ channel are 3×5 Å. In frog node of Ranvier there are about 3000 sodium channels per μm^2. The single channel conductance is 3–8 pmho. A value of 5 pmho corresponds to 3×10^6 ions per second for a 100-mv driving force (because of this high rate the channel is considered to be an aqueous pathway); this ion flux is far in excess of turnover rates for carriers. However, this is less than the theoretical maximum estimated for free diffusion, and there is saturation at high concentrations of sodium ion. Therefore, there appear to be interactions between ions and the channels.

Channel Gating

The gating currents result from charge movement across the plane of the membrane. Since Na^+ conductance channel polypeptides have been isolated and sequenced, it is possible to construct channel models in which gating currents and conductance pores may be assigned to specific amino acid sequences or regions of the polypeptides (Noda et al., 1986). The gating charges are related to charged or dipolar groups on amino acids of the polypeptide channel. Their movement is sensitive to the electric field and is associated with conformational changes in the channel that permit or block the passage of the respective cations.

Hodgkin and Huxley proposed a kinetic model for the opening and closing of the Na^+ and K^+ channels. This model may be described as follows:

In this model, depolarization of the membrane causes voltage-sensitive conformational reactions to proceed to the right while repolarization or hyperpolarization causes these reactions to proceed to the left. Na^+ channels may exist in at least three conformational states. Na^+ channels that are closed at resting membrane potentials open rapidly upon depolarization allowing Na^+ entry that further depolarizes the membrane and potentiates additional Na^+ channel opening *(regeneration)*. The open channels change slowly to an inactivated conformation which slowly reverts to the resting closed state only after the membrane is repolarized. The recovery from inactivation probably corresponds to the absolute refractory period. As the membrane depolarizes, the voltage-sensitive K^+ channels open slowly, permitting K^+ to flow out. The K^+ efflux participates together with inactiva-

tion of the Na^+ channels in repolarizing the membrane. *Accommodation* is related to the slow inactivation of Na^+ channels when the membrane is maintained at an intermediary level of depolarization. It is presently believed that the amino acid sequences for Na^+ channel opening are at the extracellular surface and that the inactivation is related to regions of the ion channels exposed at the cytoplasmic surface. The inactivation occurs slightly slower than the opening of channels thus permitting Na^+ influx before the channel is closed (Hille and Catterall, 1989).

Increasing calcium ions at the extracellular surface of the membrane raises the threshold for excitation (i.e., it shifts the critical firing potential to a more depolarized level) while decreasing the calcium ion concentration lowers the threshold for excitation by shifting the critical firing potential closer to the resting potential. Ca^{2+} may exert these effects by combining with membrane surface-negative charges near the voltage-sensing component of the channel gate.

Recently, attention has been drawn to voltage-dependent calcium currents, i.e., calcium action potentials (Llinas and Walton, 1980) which have been identified in neuronal cell bodies, dendrites, and presynaptic terminals. Calcium action potentials are much slower than sodium action potentials; they are not blocked by tetrodotoxin or removal of extracellular sodium. They are blocked by removal of extracellular calcium and by calcium channel blockers. Also, they are associated in many cases with a calcium-activated increase in potassium conductance. The calcium action potential and the calcium-activated potassium conductance are components of a negative feedback mechanism which influences both membrane excitability and the level of intracellular calcium. The calcium-activated increase in potassium permeability operates to terminate the calcium action potential and thereby to reduce the influx of calcium ions (Llinas and Walton, 1980; Thompson and Aldrich, 1980). The role of intracellular Ca^{2+} in transmitter release will be considered below.

There are at least three types of voltage-sensitive calcium ion channels (VSCC) that can be distinguished by pharmacological and physiological means.

L channels are very sensitive to the dihydropyridine (DHP) class of drugs (such as nitrendipine) and to blockage by Cd^{2+} and the toxin from the marine snail *Conus geographicus* (ω-CgTx). N channels are very sensitive to Cd^{2+} and ω-CgTx but insensitive to DHP while T channels are weakly sensitive to Cd^{2+} and ω-CgTx and insensitive to DHP. In cultured chick dorsal root ganglion cells, the three types of channels contribute to the overall Ca^{2+} current. Small depolarizing steps give rise to small Ca^{2+} transient currents that are relatively low in sensitivity to Cd^{2+} and that are not blocked by DHP or ω-CgTx. Strong depolarizations produce an intermediate level of inactivating current that is blocked by Cd^{2+} and not modulated by DHP in addition to a nonactivating (over hundreds of ms) current that is modulated by DHP and blocked by Cd^{2+} as well as by ω-CgTx. The L channels (noninactivating Ca^{2+} current) are believed mainly localized in neuron cell bodies and postsynaptic membranes. N channels (intermediate inactivating Ca^{2+} current) appear localized to nerve terminals where their Ca^{2+} currents may mediate neurotransmitter release (Miller, 1987).

A number of membrane-bound signaling molecules permit chemical or electrical stimuli to produce a wide variety of physiological responses in different classes or subcellular regions of neurons.

Guanine nucleotide-binding proteins may function in regulating or coupling receptor effects onto voltage-dependent ion channels in neuronal membranes or onto receptor-linked ion channels in postsynpatic membranes (Logothetis et al., 1987; Hescheler et al., 1987).

Transformations in neuronal firing patterns and synaptic efficacy may be regulated or induced by activation of second messenger systems whose final pathway is phosphorylation of various membrane proteins or enzymes of neurotransmitter metabolism. These systems include (1) the cyclic nucleotide-dependent protein kinases, (2) Ca^{2+}-calmodulin-dependent protein kinases, (3) Ca^{2+}-diacylglycerol-dependent protein kinase C. Presently there is evidence for involvement of protein kinase/phosphatase systems with Na^+, K^+, and Ca^{2+} voltage-dependent channels, neurotransmitter-dependent receptor and channel proteins, and synaptic vesicle-associated proteins (synapsin I) (Nestler and Greengard, 1989).

NERVE METABOLISM

Intermediary Metabolism

Nerve, like other tissues, contains the enzymatic apparatus for glycolysis, the citric acid (Krebs) cycle, and the electron transport system and, thus, can generate and store energy in the form of adenosine triphosphate (ATP). The pathway of glycolysis is concerned with the breakdown of glucose to pyruvate and/or lactate. Only a small amount of ATP is produced in this pathway which is anaerobic. The major part of the ATP is produced aerobically in the Krebs cycle and the electron transport

system (for details, see discussion of the energy-generating systems of cells in Chapter 4).

Under anaerobic conditions, e.g., in an atmosphere of nitrogen, resting nerve converts glucose to lactate quantitatively. However, the nerve at rest produces very little lactic acid aerobically, although it has been shown that there is a clear-cut increase in lactic acid production in an oxygen atmosphere (aerobic glycolysis) when the nerve is stimulated at 100/s.

Nerve conduction continues under anaerobic conditions (nitrogen atmosphere); for, as noted previously, the action potential does not depend immediately upon metabolism. Ultimately, of course, conduction will fail under anaerobic conditions because of the gradual accumulation of sodium and consequent depolarization of the nerve fiber.

Further evidence for the existence of glycolytic systems in nerve comes from the effect of metabolic inhibitors on nerve metabolism. Iodoacetic acid, which interferes with glycolysis, causes a decrease in the oxygen consumption of nerve. At the same time the effect of the inhibitor may be slowed by the administration of sodium lactate, the latter probably forming pyruvate, which is then oxidized. The cofactor necessary for the formation of pyruvic acid from lactic acid, nicotinamide adenine dinucleotide, is found in frog nerve. It has also been shown that the administration of glucose partially prevents the loss of potassium and gain of sodium resulting from anoxia of nerve.

Evidence for the presence of the Krebs cycle in nerve comes from the effect of inhibitors on the oxygen consumption of frog nerve. Methylfluoracetate, an inhibitor which acts in the Krebs cycle, inhibits the rate of oxygen uptake and ultimately causes a failure of conduction. If the nerve is preequilibrated with sodium fumarate or sodium succinate, the inhibition is overcome, and the nerve regains its ability to conduct and to consume oxygen. Fumarate and succinate are intermediates in the Krebs cycle and serve to keep going the portion of the cycle devoted to converting succinate or fumarate to carbon dioxide and water (see energy-generating systems, Chapter 4).

The importance of the electron transport scheme has been demonstrated by experiments in which the cytochrome oxidase of nerve has been inactivated by carbon monoxide. When nerve is poisoned with carbon monoxide, the cytochrome oxidase fails to function, and molecular oxygen cannot be reduced for its ultimate reaction with metabolically derived hydrogen to form water. The nerve behaves as though it were in nitrogen. However, it is possible to dissociate the carbon monoxide from the cytochrome oxidase by shining visible light on the nerve. The nerve then regains its ability to consume oxygen and will conduct an action potential.

Resting and Active Metabolism

In the resting state, the oxygen consumption of frog sciatic nerve is quite low, 30–40 mm^3/g wet weight/h in contrast to that of mammalian nerve which ranges from 200–300 mm^3/g/h. These values may be compared with those of brain which is of the order of 2,000 mm^3/g/h. Corresponding to the consumption of oxygen, there is a resting release of heat by nerve which amounts to 0.15 cal/g/h for frog sciatic nerve.

The oxygen consumption of nerve in vitro is accompanied by a release of carbon dioxide such that the respiratory quotient (moles CO_2/moles O_2) is 0.8. This value indicates that the fuel of nerve is not exclusively glucose which would yield a respiratory quotient (RQ) of 1, as indeed proves to be the case when the RQ of mammalian brain is measured in vivo. During excitation the respiratory quotient of the extra oxygen consumption of nerve changes to 0.9, suggesting that the recovery processes following excitation use a different substrate than is used during the resting state.

How much of the resting heat production (metabolism) of nerve can be attributed to the heat production of the process pumping Na$^+$ out of nerve and K$^+$ into the cell during the resting state? One can calculate the energy required to extrude sodium against both the electrical force in the membrane and the concentration gradient. A similar calculation may be made of potassium uptake. If the process is assumed to be inefficient and to produce heat equal to the work done (50% mechanical efficiency) an estimate of the heat production can be obtained. In the case of sodium, it is necessary to pump the ion against both the electric field and its concentration gradient, since the ion is positive and it must be forced out against the negative attraction of the interior. The electric work required is given by *EF* (ionic flux). The concentration work is obtained from the relationship 2.303 *RT* log C_2/C_1, where the symbols have their usual significance. Substituting the values for frog nerve of $E = 0.070$ v, $F = 23,050$ cal/v/mole, $C_2/C_1 = 3$ (sodium), and a sodium flux of 10^{-5} mole/g/h, one obtains an electrical work of 0.015 cal/g/h and a concentration work of 0.007 cal/g/h. A similar calculation for potassium shows that the energy necessary to pump this ion into nerve is 0.019 cal/g/h. The work necessary to pump both sodium and potassium is 0.041 cal/g/h. If the pump has a mechanical efficiency of 50%, then its operation—complete with inefficiency—accounts for 0.082 cal or about 50% of the total energy con-

sumption of 0.15 cal/g/h noted above. The value of 0.041 cal/g/h corresponds to 4.1 kcal/mole of Na$^+$ and K$^+$ exchanged. Thus, if 8–10 kcal/mole is taken as the free energy of ATP hydrolysis, it is seen that approximately 2–3 moles of Na$^+$ can be exchanged for 2–3 moles of K$^+$ for each mole of ATP converted to adenosine diphosphate plus inorganic phosphate.

When nerve is stimulated, there is an increase in the oxygen consumption and an increased loss of heat. The extra oxygen consumption may double. The increase in heat production and oxygen consumption parallel one another and depend upon the frequency of stimulation. The increase in oxygen consumption rises gradually with increase in frequency of stimulation, leveling off at about 100 impulses/s. The extra heat which accompanies short tetanic stimulation has been shown by Hill (1938, 1959) and co-workers to consist of two phases: an early one lasting only 2–3 s is called the initial heat, which amounts to a few percent of the total extra heat; this is followed by the remainder of the heat production which is called the delayed heat. The latter is presumably concerned with the recovery process in nerve involving the pumping of ions. The extra oxygen consumption of nerve may be eliminated by inhibitors (azide and methylfluoracetate) which have no effect on the action potential.

Abbott et al. (1958), using improved thermal methods, have been able to measure the heat output resulting from a single shock to a crab nerve. They found that the initial heat accompanying an action potential occurred in two phases, an early, rapid portion, 1 ms in duration, which was positive (exothermic), and a delayed, negative (endothermic) heat of absorption. The actual heats during an impulse were $+14 \times 10^{-16}$ and -12×10^{-16} cal/g/impulse. The difference gives the net initial heat production. Calculations by Hodgkin indicate that the initial heat can be accounted for by the heat production of the action currents flowing within and without the nerve. But other processes such as heat produced by the mixing of sodium entering nerve may also account for the initial heat.

Thus, active cation transport is the major energy-consuming process in specialized tissues such as nerve. McIlwain (1966) has shown that electrical pulsation of brain slices in vitro results in cation-dependent respiratory stimulation accompanied by acceleration of active cation fluxes. About 30% of the increased respiration may be accounted for by the increased cation flux. The maintenance or restoration of cation gradients in nerve, as well as other tissues, has been correlated with a sodium-and-potassium-stimulated adenosine triphosphatase present in membranes. This enzyme utilizes the chemical bond energy in ATP to translocate Na^+ and K^+ across the membrane against their respective concentration gradients.

SYNAPTIC TRANSMISSION

In the nervous system of mammals, the formation of conducting pathways does not involve direct anatomical continuity between one neuron and another or between a neuron and a cell of an effector organ. An axon gives rise to many expanded terminal branches (presynaptic terminal boutons). The postsynaptic component of the synapse may be formed by any part of the surface of the second neuron, with the exception generally of the axon hillocks; it usually is a dendrite (axodendritic synapse), but it may be part of the cell body (axosomatic synapse) or part of the membrane of another axon (axoaxonic synapse). A single neuron may be involved in many thousands of synaptic connections, but in every case the impulse transmission can occur only in one direction. In the mammalian central nervous system there is a cleft of about 20 nm (the *synaptic cleft*) separating the presynaptic axon terminal (the *presynaptic membrane*) and the surface of the postsynaptic component (the *postsynaptic membrane*).

At the sites of apposition, both the presynaptic and postsynaptic membranes exhibit a plaque of electron-dense material on their cytoplasmic surfaces. Beneath this plaque on the presynaptic membrane and within the presynaptic bouton, there are vesicles ranging in diameter from 10–50 nm and often numerous mitochondria.

The action potential does not cross the synaptic cleft but instead causes the release of a transmitter substance which is stored in the vesicles of the presynaptic terminals. The release of transmitter requires the entry of calcium ions into the presynaptic terminal from the extracellular milieu upon arrival of the action potential. The transmitter molecule then crosses the gap, attaches to "*receptors*" on the postsynaptic membrane, which is then either *excited* or *inhibited* depending on which ionic channels are opened by the binding of the transmitter to the postsynaptic receptor (see below).

Figure 1.44 shows by autoradiography the distribution of three different receptors in human cerebrum.

Chemical transmission was first demonstrated by Otto Loewi in a classical experiment in which the heart of a frog was slowed by perfusing it with blood taken from the heart of a second frog that had been slowed as a result of vagal stimulation. Loewi (1921) concluded that the vagal stimulation had caused the release of a substance identified as acetylcholine—which altered the excitability of cardiac muscle.

Excitatory Postsynaptic Potential (EPSP)

Eccles (1953) and co-workers have inserted finely drawn-out microelectrodes made of glass and filled with KCl or NaCl into the cell body of motoneurons of the spinal cord. When the microelectrode is inserted, a resting potential of about 70 mv is observed, with the interior of the cell negative to the outside. If the motoneuron is antidromically excited, i.e., by stimulating the axon of the motoneuron, the motoneuron is excited without the intervention of the synapse. An action potential with an overshoot of about +20 mv appears. Thus, antidromic excitation gives rise to an action potential in the postsynaptic cell, the motoneuron, which is no different in its general characteristics than that obtained from squid axon or the muscle fiber.

The events of interest in synaptic excitation, however, occur when excitation is delivered orthodromically, i.e., to the afferent neuron of the dorsal root, so that the presynaptic endings of this neuron are excited. In such experiments in which a synapse is present, an additional potential appears in recording from the postsynaptic cell which is a sign of postsynaptic depolarization. Figure 1.45 shows the response of the motoneuron to a presynaptic volley. The slow potential change which appears just before the action potential is called the excitatory postsynaptic potential or EPSP. It consists of a depolarization of the membrane which may last for 20 ms and can best be seen when the extent of depolarization is insufficient to give rise to an action potential. When the membrane potential falls to a critical value of about −60 mv, the motoneuron fires, and an action potential appears superimposed on the EPSP. Temporal summation of such EPSPs is also possible. If two subliminal volleys are sent in over the same nerve, each volley produces an effect which is manifested by an EPSP. The EPSPs will then sum, and if the critical level of depolarization is reached, an impulse will be set off.

The EPSP is monophasic and nonpropagating; it represents a depolarization which is localized to the soma of the motoneuron. Unlike the action potential, the EPSP is a potential which is not all or none in character, since it can be augmented simply by increasing the intensity of the input volley. Moreover, the EPSPs of different inputs can sum on a postsynaptic cell to produce a greater depolarization. These characteristics show that the EPSP is produced in a process which is fundamentally different from that of the action potential.

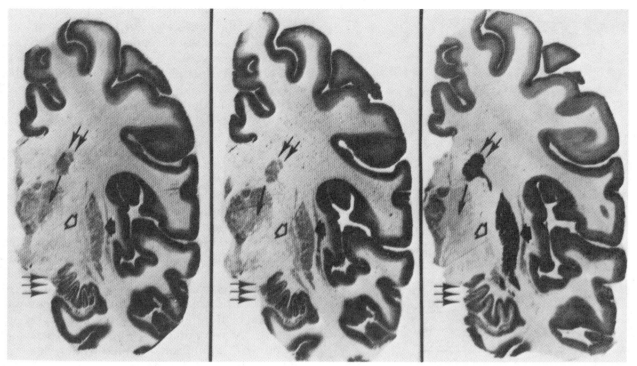

Figure 1.44. Autoradiograms of the regional localization of GABA (γ-aminobutyric acid), benzodiazepine, and muscarinic cholinergic receptors in 30-μm coronal frozen sections of normal human brain. Images were made onto tritium-sensitive film as described by Penney and Young, 1982. Images reflect the binding of tritiated ligands to membrane receptor molecules in the tissue sections. The darker regions of the image reflect higher amounts of receptor binding; the lighter regions, lower amounts. *Left,* [³H]muscimol binding to GABA receptors. Muscimol is a potent GABA agonist which binds to GABA receptors with high affinity. Binding is highest in cerebral cortex and hippocampus, intermediate in putamen and thalamus, much less in globus pallidus, and negligible in white matter. *Middle,* [³H]-flunitrazepam binding to benzodiazepine receptors. Flunitrazepam is a potent benzodiazepine agonist. Benzodiazepines are drugs used as antianxiety and muscle relaxant agents and have been shown to interact with specific brain receptors closely linked to GABA receptors in the membranes. Binding is very similar in distribution to that seen with [³H]muscimol. *Right,* [³H]quinuclidinylbenzilate (QNB) binding to muscarinic cholinergic receptors. QNB is a potent muscarinic antagonist, similar to but more potent than atropine. Binding is highest in caudate and putamen, intermediate in cerebral cortex and hippocampus, less in thalamus, little in globus pallidus, and negligible in white matter. *Narrow dark arrow,* thalmus; *open arrow,* globus pallidus; *closed arrow,* putamen; *double arrow,* caudate; *triple arrow,* hippocampus. The claustrum is the narrow band just lateral to the putamen in all images. (Courtesy of J. B. Penney and A. B. Young.)

Several lines of evidence indicate that the EPSP arises from an influx of all ions through the postsynaptic membrane into the cell. One group of experiments concerns setting the membrane at a given membrane potential by passing current across the membrane. A double-barreled microelectrode was inserted into a motoneuron. One electrode was used to depolarize or hyperpolarize the neuron with a direct current, the other to record the EPSP. When the membrane potential was increased, the amplitude and rate of rise of the EPSP increased. Conversely, when the membrane potential was decreased, the EPSP decreased. Reversal of the direction of the EPSP occurred when the membrane potential was set up at zero volts, which is therefore the equilibrium value for the EPSP. The explanation of this effect goes back to the experiments on the squid axon. It will be recalled that when the axon was clamped at the equilibrium potential of an ion, a reversal of current flow occurred through the membrane above and below this value of clamp voltage. We may imagine that the motoneuron postsynaptic membrane contains channels through which ions can flow to cause the development of an EPSP. These channels are additional to those serving as producers of the action potential. A diagram of the equivalent circuit for such channels is shown in Figure 1.46. When the membrane potential is set at zero, the equilibrium potential of the EPSP, a change in resistance of this channel can have no effect on the recorded potential. No EPSP will develop. Under normal circumstances the membrane potential is determined by the equilibrium potential for potassium, but when the EPSP channels decrease in resistance, the potential recorded tends to approach the equilibrium potential for the EPSP. Now the only process which can have an equilibrium potential of zero is one in which the

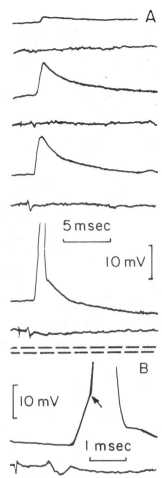

A

5 msec

10 mV

10 mV

1 msec

B

Figure 1.45. Intracellular potentials set up in a biceps-semitendinosus neuron by various sizes of volleys in the afferent nerve (lower records). *A* shows synaptic potentials (upper records) of graded size, the largest setting up an action potential. In *B* a faster record of this response is shown. Note spike arising at *arrow* from more slowly rising synaptic potential. (From Brock et al., 1952.)

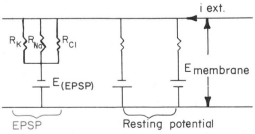

Figure 1.46. Equivalent circuit of motoneuron membrane for excitatory postsynaptic potential *(EPSP)*. Left circuit represents channels giving rise to EPSP. Right circuits represent circuit of polarized membrane.

membrane is freely permeable to all ions. This process will have an equilibrium potential of zero because it is a diffusion potential, and the value of the emf developed in a diffusion potential depends directly upon the difference of mobilities of the cations and anions. During the production of the EPSP it is presumed that the mobilities are equal so that the process has an equilibrium potential of zero.

Inhibitory Potentials

If a monosynaptic reflex is evoked by orthodromic excitation, an inhibition of the reflex can be produced by stimulating the afferent nerve from an ipsilateral antagonistic muscle as described more fully in the section on inhibition. The effect of the inhibitory stimulus is to cause a hyperpolarization of the motoneuronal postsynaptic membrane in a direction opposite to that of the EPSP (Fig. 1.47). This hyperpolarization consequent upon an inhibitory stimulus is called the inhibitory postsynaptic potential or IPSP. Its time course is the same as that of EPSP, of the order of 20 ms, and corresponds to the curve of inhibition obtained from reflex studies. Indeed, the IPSP is a mirror image of the EPSP in the same motoneuron. The IPSP, being a hyperpolarization of the membrane, renders it less excitable. Any EPSP which occurs during an IPSP will generate currents which will be less effective in causing excitation of the cell.

The problem of determining the nature of the current responsible for the IPSP has been attacked by the same procedures described for the EPSP. Using double-barreled microelectrodes, it is found that setting the membrane potential at −80 mv will

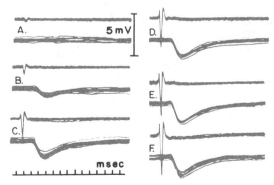

Figure 1.47. Lower records give intracellular responses of biceps-semitendinosus motoneuron to a quadriceps volley of progressively increasing size, as is shown by the upper records which are recorded from the L6 dorsal root by a surface electrode (downward deflection signaling negativity). Note three gradations in the size of the IPSP; from *A* to *B*, from *B* to *C*, and from *D* to *E*. Voltage scale gives 5 mv for intracellular records, downward deflection indicating membrane hyperpolarization. (From Coombs et al., 1955.)

cause the disappearance of the IPSP. The potential −80 mv is the average of that determined by potassium (−70) and chloride (−90), suggesting that the IPSP results from an increased flow of K⁺ and Cl⁻, making the inside more negative. Further evidence, however, points to chloride as the principal ion involved in producing the IPSP. The method of electrophoretic injection has been used to demonstrate this. One barrel of a double microelectrode is filled with KCl and made negative so that this electrode effectively drives chloride ions into the cell; it is a hyperpolarizing current since it is internally negative. The other barrel electrode records the IPSP. It is found that the injection of chloride into the cell converts the IPSP, elicited by a group I_a afferent volley, from a hyperpolarization to a depolarization (EPSP). The injection of chloride caused an increased internal concentration of chloride. The transmitter causing inhibition results in an increased permeability to chloride, but the chloride ionic gradient is now opposite to its normal direction so that a depolarization will occur instead of a hyperpolarization when the membrane becomes especially permeable to chloride. Chloride now flows out and makes the inside of the cell less negative. A number of anions were injected internally, and all those with less than a certain hydrated ion diameter caused a reversal of the IPSP to an EPSP. Apparently, the inhibitory substance opens pores which allow ions below a certain size to pass. Since chloride is the only anion which under in vivo conditions exists in a high enough concentration to flow in during such an inhibition, it is presumed that chloride is the main contributor to the inhibitory currents.

TYPES OF INHIBITION

Some restraint must be placed upon the ability of muscle and neural circuits to respond. In reaching for an object, for example, the muscle must be controlled if the movement is to be accurate and the muscle is not to overshoot the mark. Certainly reciprocal innervation of some type must often be employed so that when a muscle is activated its antagonist will be inhibited; otherwise, persistent opposition of an undesired sort will be encountered. These two reasons give some indication of the usefulness of inhibition in motor movement. A third less obvious necessity for inhibition arises from the very complexity of neural activity in which some form of "negative feedback" or inhibition is necessary in order to keep the complex neural networks from overactivity. An abnormal form of such activity is observed, for example, in strychnine poisoning in which inhibitory neurons have been shown to be inactivated. In a strychninized animal any stimulus leads to persistent neural activity and convulsion. Again, in the disorder known as parkinsonism, periodic activity in the form of a tremor of a limb manifests itself, presumably as a result of injury to inhibitory systems which restrain normal motor activity.

Three kinds of inhibition have been extensively studied: (1) direct; (2) presynaptic; and (3) Renshaw cell or recurrent inhibition. A fourth type of inhibition, called indirect inhibition and extensively discussed by Sherrington, is not now considered to be inhibition per se but a form of occlusion.

Direct Inhibition

The direct form of inhibition can be observed by evoking a monosynaptic reflex and then depressing the amplitude of the reflex by stimulating an ipsilateral skin nerve (type II fibers). It is necessary to stimulate the inhibitory nerve shortly before eliciting the reflex. By varying the interval between the inhibitory and excitatory stimulus, the time course of direct inhibition can be obtained. The inhibitory nerve is stimulated first. A curve of the form of Figure 1.48 is obtained in which maximum inhibition is obtained at an interval of 0.5 ms between the two stimuli and with a delay lasting for about 10 ms. The interval 0.5 ms is equivalent to one synaptic delay so that at least one inhibitory interneuron is interposed between the afferent neuron and the motoneuron. It is believed at present that this interneuron in the inhibitory pathway possesses the ability to secrete a substance at its terminals which hyperpolarizes the membrane and leads to an IPSP. Thus, a neuron which is normally excitatory to other neurons may exert an inhibitory action by the interposition of an inhibitory neuron.

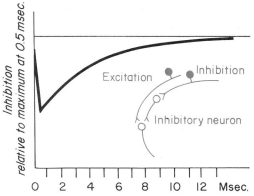

Figure 1.48. Time course of direct inhibition. *Insert* shows circuit containing inhibitory neuron. (Adapted from Lloyd, 1946.)

Direct inhibition is very susceptible to strychnine injection which can completely abolish it. This effect of strychnine is the basis for the explanation of its convulsant activity—all inhibition of the direct type having been removed, the slightest stimulus causes a tremendous extensor response.

Presynaptic Inhibition

Early in the studies of the electrical responses evoked by reflex action, electrodes were placed on the surface of the spinal cord or on dorsal roots, and large potentials which were positive to distant regions, called P waves, were recorded. It was suspected that this positivity might be the sign of some inhibitory presynaptic activity going on in the cord. To demonstrate presynaptic inhibition, a reflex response is elicited in extensor motoneurons. This reflex will be inhibited if it is preceded by stimuli from any nerve entering the cord. Group I_a and I_b fibers are most effective in causing the inhibition, but groups II and III will also serve. The time course of such an inhibition is shown in Figure 1.48. Unlike direct inhibition, maximum inhibition is observed when the inhibitory stimulus precedes the excitatory by about 20 ms. The inhibition may endure for as long as 200–300 ms and is extremely resistant to strychnine. Several lines of evidence indicate that this inhibition is presynaptic. Recording within the motoneuron at the time that inhibition is produced shows no changes in the membrane potential. Neither a hyperpolarization nor a depolariza-

tion is produced by an inhibitory volley. Instead, it is noted that the magnitude of the EPSP which is set up by the reflex volley is diminished by the inhibitory stimulus and that the magnitude of the EPSP parallels the curve of inhibition (Fig. 1.49). It has also been shown that there is a depolarization of the presynaptic fine terminals entering the dorsal region of the cord. Microelectrodes are inserted into and just outside a nerve fiber. The undesirable potential of neighboring neurons which is recorded inside a nerve fiber together with the membrane potential, may be obviated by subtracting the externally recorded potential from the internally recorded to obtain the true membrane potential. It is found that the time course of the depolarization produced by inhibitory volleys parallels the time course of inhibition.

Although the mechanism is not clear, depolarization in axoaxonal synapses on presynaptic terminals may reduce the amount of excitatory transmitter released from the presynaptic terminals, thus decreasing the EPSP amplitude. Since long delays are involved in presynaptic inhibition, many interneurons are interposed between the first and last neurons of the chain.

Renshaw Cell Inhibition

In 1946, Renshaw discovered that a volley of impulses delivered to motor axons causes an inhibition of all types of motoneurons at the segmental level. This type of inhibition has therefore been

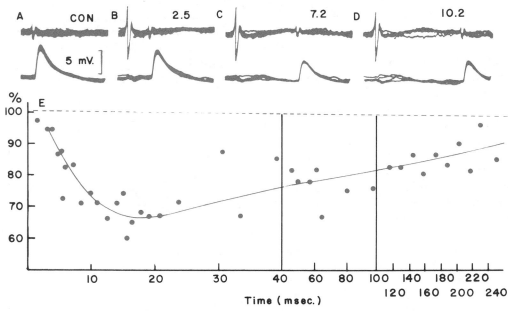

Figure 1.49. Time course of presynaptic inhibition. (*E*). Inhibitory stimulus is maximal group I volley in biceps-semitendinosus nerve. Monosynaptic EPSPs were evoked at various intervals after the inhib-

itory stimulus by stimulating the gastrocnemius-soleus nerve maximally. *A* shows control EPSP; *B*, *C*, and *D* are the EPSP at intervals marked above the records.

called antidromic because the inhibition may be evoked by firing backwards over the motor roots into the spinal cord. When such antidromic excitation was used, an afterdischarge of quite high frequency was observed in microelectrode recordings from the ventral horn. It was also shown that the discharge did not occur in the motoneuron but in neighboring cells near the motoneuron—in cells which discharge with a high frequency when the antidromic excitation occurs. At the same time the motoneuron is inhibited. These neighboring cells are therefore believed to cause the inhibition. The motoneuron displays a hyperpolarization which has all the characteristics of an inhibitory postsynaptic potential, and the IPSP lasts for the period of time corresponding to the discharge in the neighboring cells which are called Renshaw cells. An anatomical pathway has been suggested to explain these results (Fig. 1.50). The motoneuron gives off a collateral to the Renshaw cells; the Renshaw cell axon returns to the motoneuron and inhibits it.

The repetitive discharge of the Renshaw cell is presumed to be caused by an accumulation of acetylcholine at the junction between the motoneuron collateral fiber and the Renshaw cell. It has been shown, for example, that the discharge of the Renshaw cell may be prolonged by anticholinesterases, which allow acetylcholine to accumulate. It is also possible to inhibit the discharge with β-dihydroerythroidine, a substance which blocks the action of acetylcholine. The functional significance of this pathway is not clear. It appears to serve as a general synaptic inhibitor and may act to limit the frequencies of impulses going to the motor end-plate.

ELECTRICAL AND CHEMICAL TRANSMISSION

The events at a synapse may be interpreted according to either a chemical or electrical theory of synaptic transmission. The chemical theory of transmission attributes the response of the postsynaptic cell to a chemical substance (transmitter)

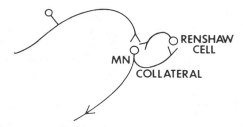

Figure 1.50.   Schematic drawing of recurrent inhibition.

released from the presynaptic nerve terminals. The electrical mechanism of transmission presupposes the activation of postsynaptic neurons by electrical currents flowing out from the presynaptic terminals into the postsynaptic cell. Examples of each of these types of transmission may be found in various kinds of synapses among vertebrates and invertebrates. It may be stated, however, that no firm evidence of electrical transmission at mammalian synapses has as yet been obtained.

The present-day understanding of chemical transmission is based primarily on studies of the neuromuscular junction because of its accessibility to investigation. The chemical mechanism of transmitter release and action may be schematically outlined as in Figure 1.51.

Cholinergic Transmission

The action potential, upon invasion of the presynaptic terminal, causes release of transmitter material. In the case of neuromuscular junction in which the transmitter is acetylcholine, this substance is stored in clear vesicles in the presynaptic nerve terminal. In their classical experiments with frog muscle fibers, Fatt and Katz (1952) observed that under certain conditions the muscle end-plate (postsynaptic membrane) exhibited intermittent, spontaneous miniature potentials in a random manner. These miniature end-plate potentials (MEPPs), of insufficient amplitude to depolarize the membrane to threshold, were related to random release of acetyl-

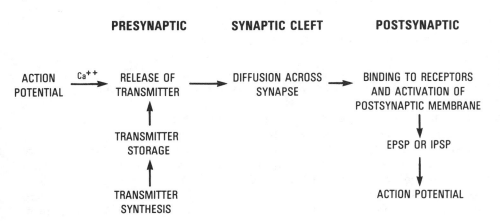

Figure 1.51. Outline of processes involved in storage and release of chemical transmitter. It should be noted that some transmitters have receptors in the presynaptic membrane as well as in the postsynaptic membrane; occupation and activation of these presynaptic receptors serve to self-inhibit and thereby modulate release.

choline packets from the presynaptic vesicles. When the external Ca^{2+} concentration was reduced, the amplitude of the end-plate potentials decreased in a stepwise manner; under various conditions, the amplitudes were integral multiples of the miniature components. However, graded ionophoretic application of acetylcholine directly onto the end-plate produced a continuously graded potential change, thus indicating that the stepwise electrical changes were not due to the nature of the receptor response, but, rather, to the nature of the transmitter release. Their conclusion was that the transmitter is released in quanta or packets of uniform size. Depolarization of the presynaptic terminal increased the frequency of quanta release, thus allowing the MEPPs to summate and reach threshold for firing of an action potential in the postsynaptic cell. Other studies have shown that entry of Ca^{2+} is essential for the increased quantal release during the action potential and that external Mg^{2+} antagonizes the Ca^{2+} action; however, the detailed mechanism for the coupling of excitation to secretion of transmitter is not known.

The released acetylcholine diffuses across the synaptic cleft in a very short time (<0.6 ms) to attach to receptor sites on the postsynaptic membrane. Specific acetylcholine receptor molecules can be extracted and purified from muscle end-plates (Changeux, 1981). The mechanism by which attachment of acetylcholine to its receptor alters the channel permeability in producing the end-plate potential probably involves rearrangement of the receptor structure. The acetylcholine receptor complex, consisting of five polypeptide subunits, contains the acetylcholine binding sites and the ionic channels. The action of acetylcholine is terminated by its hydrolysis catalyzed by acetylcholinesterase present in the postsynaptic membrane. Numerous observations indicate that the quantal nature of transmission is applicable to interneuronal synapses in many portions of the vertebrate nervous system and also to other transmitters, for example, norepinephrine. In the nervous system, combination of the transmitter with receptor gives rise to the EPSP or IPSP. However, direct demonstrations of specific transmitters and their mode of release in the mammalian central nervous system have not yet been achieved.

The basis for establishing that chemical transmission occurs involves the following purely chemical criteria. A transmitter substance must be found (1) which is produced presynaptically, that is, the pertinent enzyme systems for its production are present in the presynaptic terminals; (2) which is released and detected when the presynaptic terminals are stimulated; (3) the action of which is dissipated by a specific enzyme (as in the case of acetylcholinesterase which destroys acetylcholine) or by reuptake (as in the case of norepinephrine) or by simple diffusion away from the postsynaptic membrane; and (4) the action of which, when applied directly onto the postsynaptic membrane, mimics the effects of presynaptic nerve stimulation. In the case of acetylcholine, the transmitter at the neuromuscular junction, such criteria have been met.

Additionally, acetylcholine has been shown to be a transmitter in autonomic ganglia, postganglionic parasympathetic terminals, certain postganglionic sympathetic terminals (sweat glands and vasodilator fibers), and in motoneuron collateral-Renshaw cell synapses in the spinal cord. Examples of some of the evidence bearing on the role of acetylcholine as a transmitter in the central nervous system are: (1) choline acetylase and vesicles containing acetylcholine are present in nerve endings from many nuclei of the nervous system; (2) acetylcholine release has been detected in the cerebral cortex after excitation of a sensory nerve, but only in eserinized preparations in which acetylcholine cannot be destroyed; (3) acetylcholinesterase is abundantly present in neural tissue; and (4) injection of acetylcholine either directly or electrophoretically will activate many cells in the nervous system.

Other Transmitters

There is substantial evidence that norepinephrine is the transmitter released at postganglionic sympathetic nerve terminals. The inactivation of norepinephrine action involves reuptake into the presynaptic terminal. Within the central nervous system norepinephrine is found abundantly in the hypothalamus and brainstem reticular formation where it may have a transmitter role. On the other hand, dopamine, a precursor of norepinephrine, is abundant in the corpus striatum, where it is thought to act as a transmitter. Histochemical fluorescent studies have shown that dopamine is present in nerve terminals within the corpus striatum that derive from cell bodies located in the substantia nigra. In Parkinson's disease, which is associated with symptoms of basal ganglia dysfunction, there is degeneration of this biochemically defined nigra-striatal path and depletion of dopamine in the corpus striatum. Administration of L-dopa, which is decarboxylated to form dopamine in vivo, ameliorates the symptoms of parkinsonism. Both catecholamines, dopamine and norepinephrine, are stored in granular vesicles in presynaptic nerve terminals.

A number of other substances have been suggested as possible synaptic transmitters and inhibitors. Serotonergic axons extend from median raphe nuclei in the brainstem to many areas of the brain, but particularly to the hypothalamus. Evidence is accumulating that γ-aminobutyric acid is an inhibitory transmitter in many portions of the central nervous system and that glycine acts as an inhibitory transmitter in the spinal cord. In Huntington's disease, another type of disorder of the basal ganglia, neurons containing γ-aminobutyrate are affected (Perry et al., 1973). Glutamic acid and aspartic acid have been suggested as important excitatory transmitters. These are as effective as acetylcholine in activating neurons when injected into the cells electrophoretically. These amino acids, however, have not been found associated with vesicles, and mechanisms of their release are not known. Postulated inactivation mechanisms involve reuptake into synaptic and possibly glial membranes. A number of peptides (e.g., substance P, neurotensin, endorphins, enkephalins, thyrotropin-releasing hormone, luteinizing hormone-releasing hormone, angiotensin, bradykinin, bombesin, oxytocin, vasopressin, melanocyte-stimulating hormone (MSH), MSH release-inhibiting factor, adrenocorticotrophic hormone, somatostatin) and other substances (e.g., histamine, prostaglandins, and purines) found in nerve tissues may also play roles in transmission or modulation of synaptic function, but their precise roles remain to be elaborated (Hedquist, 1974; Hughes et al., 1975; Takahashi and Otsuka, 1975). The interested student is referred to section II of a recent text on neurochemistry (Siegel et al., 1989) for additional reading on synaptic function.

Greengard (1976) has suggested that cyclic nucleotides (adenosine 3',5'-phosphate, guanosine 3',5'-phosphate), which serve in many tissues to translate hormonal messages into intracellular responses, may also mediate certain postsynaptic cell responses to neurotransmitters.

There are other indications that chemical transmission takes place at mammalian synapses. Some synaptic delay would be anticipated if the release, diffusion, and attachment of a substance were factors important in transmission. Thus, the finding of a synaptic delay at vertebrate synapses of 0.5 ms is consistent with a chemical theory. A second phenomenon which is consistent with the chemical theory is the fatigability of synapses when rates of stimulation are increased. At 40 or 50 impulses/s, many synapses within the nervous system fail to transmit an impulse. According to the chemical theory, such a failure would be in accord with the expectation that there will be a frequency of stimulation above which the synthesis of transmitter is unable to keep pace with its release from the terminals. At such high frequencies, the amount of transmitter released per impulse would decrease and fail to activate the postsynaptic membrane. A third, well-known observation is the susceptibility of synapses to anoxia and metabolic inhibitors. Thus, cutting off the supply of oxygen interferes with the synthesis of ATP and other chemical materials necessary for the manufacture of the transmitter substance. The supply of transmitter diminishes and the amount released per impulse again becomes inadequate to activate the postsynaptic membrane. A fourth phenomenon is the repetitive discharge of neurons as noted in the case of the Renshaw cell where a repetitive discharge ensues from a single shock. This repetitive discharge is the result of a prolonged depolarization lasting more than 100 ms. The prolonged depolarization is attributable to the accumulation of the transmitter substance at the synapse.

Electrical Transmission

Criteria for the presence of electrical transmission are essentially the opposite of those for chemical transmission. Thus a system which transmits impulses across junctions at high frequencies with no delay and in which anoxia and metabolic inhibitors have little or no effect is presumed to be an electrical synapse. Such systems behave, indeed, as though no junction were present, although there may be an anatomical junction with little or no space between the presynaptic and the postsynaptic membranes. Such electrical junctions have not yet been discovered in the mammalian nervous system, but artificial synapses called "ephapses" have been created at junctions between nerve fibers so that an incoming electrical impulse will either modify the excitability in a contiguous nerve or actually set off an action potential. Among the invertebrates, several examples of purely electrical junctions have been discovered. The giant motor synapses of the crayfish are designed to transmit electrically in one direction only. Simultaneous recording from both presynaptic and postsynaptic cells has demonstrated that practically no delay (about 0.1 ms) occurs. The postsynaptic response has the shape of an EPSP, and spikes may be generated from sufficiently strong EPSPs. Such synapses appear to act as rectifiers, i.e., they pass current more easily in one direction than the other.

Excitation and Contraction of Skeletal Muscle

GENERAL CHARACTERISTICS OF MUSCLE

Of the different types of muscle tissue, only skeletal muscle will be discussed in this chapter. A thorough understanding of skeletal muscle physiology will provide a basis for subsequent study of cardiac and smooth muscle. Both skeletal and cardiac muscle are called striated muscle because of a repetitive pattern of light and dark bands seen along the length of the muscle cell under the light microscope. Much of our knowledge of mammalian skeletal muscle stems from studies of frog skeletal muscle to which frequent reference will be made.

The main functions of skeletal muscle tissue are development of tension and shortening. The nervous system coordinates the activity of various muscles and of different parts of one or more muscles to produce useful movements and postures. The effect of muscle activity is transferred to the skeleton by means of tendons; the translational displacements between various parts of the muscle mass, displacements that are associated with movement, are facilitated by the interposition of connective tissue septa. The latter structures are present at places where the translational displacements are most pronounced.

The tension developed by a muscle in the body is graded and adjusted to the load. If contraction is associated with shortening, the tension is adjusted both to the load and to the velocity of the shortening. The graded response of muscle is due to variation in the degree of activation of the tissue by the motor nerves.

STRUCTURE

Myofibers, Myofibrils, and Myofilaments

Skeletal muscle is composed of numerous parallel elongated cells referred to as *muscle fibers* or *myofibers*. These are about 10–100 μm in diameter and vary with the length of the muscle, often extending its entire length. Under the electron microscope, the subcellular structure of the skeletal muscle fibers is seen to be composed of smaller fibrous structures 1 μm in diameter, *myofibrils*, which are separated by cytoplasm and arranged in parallel along the long axis of the cell (Fig. 1.52). Each myofibril is further subdivided into thick and thin *myofilaments* which are referred to simply as *thick* and *thin filaments*. Thin filaments are about 7 nm wide and 1.0 μm long, and thick filaments are about 10–14 nm wide (in mammals) and 1.6 μm long. The arrangement of the thick and thin filaments produces the cross-striated appearance of the muscle, which results from a regular repetition of dense cross-bands (1.6 μm in length) separated by less dense bands. The dense cross-bands, referred to as A bands because they are strongly anisotropic,* contain the thick filaments arranged neatly in parallel. The less dense segments, the *I bands*, contain the thin filaments, which extend symmetrically in opposite directions from a dense thin line, the *Z line*. The Z line appears in embryonic muscle before the thin filaments develop and is comprised of a lattice-like protein, intertwined with the thin filaments which extend perpendicularly from the Z line for 1 μm on either side. The term I band is based on the fact that this zone was originally considered to be isotropic. Although it is now recognized to be weakly birefringent, it is very much less birefringent than the A band.

The width of the I band varies with the degree of stretch or shortening of the muscle fiber (Fig. 1.53). Since the length of the individual thin filaments is 1 μm and there are two sets of such filaments extending from the Z line, which is 0.05 μm wide, the total width of the Z line-thin filament complex is 2.05 μm. The Z-line structure contributes toward keeping the thin filaments arranged in register and with a regular spacing. The gap between the terminations of

---

*Optically anisotropic substances have different refractive indices for different polarization planes of the incident light and are birefringent. Optically isotropic substances have a single index of refraction.

a.

b.

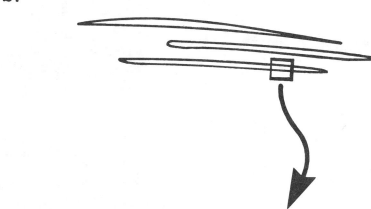

c.

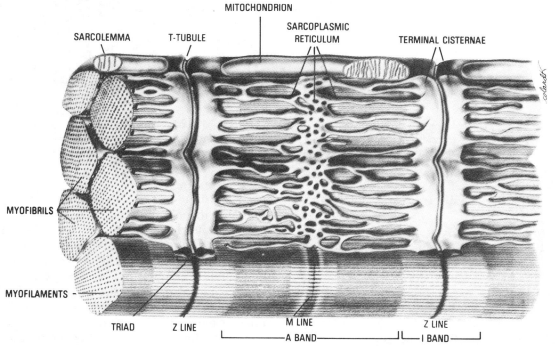

Figure 1.52. Structure of skeletal muscle. *a,* Whole muscle. *b,* Muscle fibers. *c,* Schematic representation of a segment of a skeletal muscle fiber. (Adapted from Fawcett and McNutt, 1969.)

the thin filaments is called the *H zone*, and the darker area in the center of the H zone, the *M line*.

Sarcomere

The molecular basis for the difference in isotropicity between A and I bands described above becomes evident from an understanding of the structure of the *sarcomere*, the fundamental contractile unit of muscle. A sarcomere consists of the region between two consecutive Z lines; thus, this unit consists of one A band and one-half I band at each end of the A band. Its length at rest varies

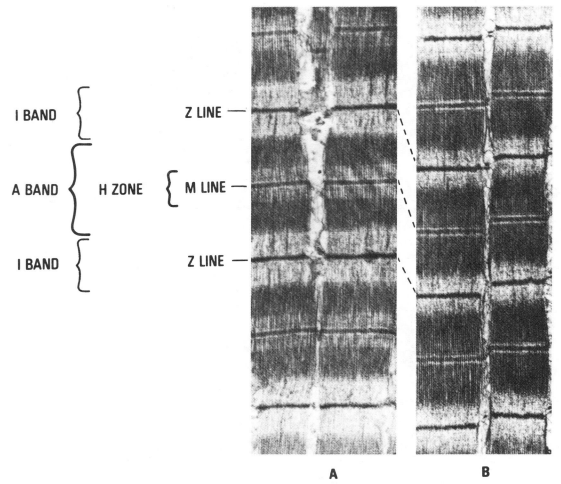

I BAND

A BAND

H ZONE

I BAND

Z LINE —

M LINE —

Z LINE —

A

B

Figure 1.53. Electron micrographs of muscle (frog sartorius) fixed at different stages of shortening. *A,* Muscle had shortened only very slightly and a wide H zone is still visible (×15,500). *B,* As in *A* but the muscle was allowed to shorten further so that a very short H zone (and a shorter I band) is visible (× approximately 15,500). (From Huxley, 1964.)

between 2.0 and 2.6 μm in frog and mammalian muscle.

At the normal length of the muscle in situ (rest length) the two sets of filaments in most muscles interdigitate with a rather extensive zone of overlap at each end of the A band (Fig. 1.54*A*). If a muscle is allowed to shorten, the relationship between the half-width of the I bands, the width of the overlap zone, and the width of the H band is the same as described above (Fig. 1.53). At sarcomere lengths less than 2 μm, that is, less than the sum of the lengths of the two sets of thin filaments in the sarcomere, a somewhat more dense zone appears in the middle of the A band. This phenomenon is due to a double overlap of the thin filaments (Fig. 1.54*D*). An increased density seen in the region of the M line is called a contraction band, specifically a CM band. A second contraction band (CZ band) is seen at sarcomere lengths of less than 1.5 μm. This band may be due to the penetration of the Z line by the ends of

the thick filaments or to ends of the thin filaments reversing direction at the Z line.

If the muscle is stretched, the zone of overlap decreases in width in proportion to the increase in half-width of the I band. The zone in the middle of the A band bounded by the two zones of overlap, the H zone, also increases in width, this increase corresponding to the decrease in width of both zones of overlap. This means that when a muscle is stretched, the thin filaments are being pulled out of the A band, while the thick filaments remain constant (Fig. 1.54*B* and *C*). At a sarcomere length of about 3.6 μm, the thin filaments have been pulled out completely, and the two sets of filaments appear now in an end-to-end arrangement.

In accord with the above considerations, the microscopic patterns seen in cross-sections of muscle fibers depend on the level of the section (Fig. 1.55). Near the Z line, only thin filaments are observed, whereas at the site of overlap of thick

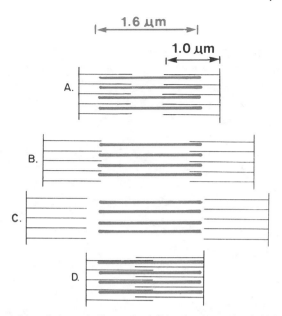

Figure 1.54. Schematic illustration of the arrangement of thick filaments and thin filaments in the sarcomere at various degrees of stretch and shortening. *A* and *B* show different degrees of overlap of the two types of filament, *C* illustrates the case when the thin filaments have been completely pulled out from the A band with a gap between A band and the thin filaments. In the extensively shortened sarcomere in *D* a zone of double overlap of thin filaments has developed in the center of the A band.

and thin filaments, each thick filament is surrounded by six thin filaments and each thin filament by three thick filaments. A cross-section through the M line shows the thin connections between adjoining thick filaments. A representative electron micrograph of a cross-section through several myofibrils is seen in Figure 1.56.

Sarcolemma

The sarcolemma is the outer membrane surrounding each muscle fiber. It consists of the *plasma membrane* proper (plasmalemma) and a *basement membrane*. The plasma membrane has the general double membrane structure of most biological membranes while the basement membrane on the outer surface of the plasma membrane consists of a thin mucopolysaccharide-rich coating containing fine collagen fibrils, which fuse with the tendons at the ends of the muscle. The main function of the plasma membrane in muscle contraction is to conduct the wave of depolarization originating at the motor endplate over the entire cell surface to initiate contraction. The plasma membrane of skeletal muscle fibers, in contrast to cardiac muscle fibers and many other cell types, contains no gap junctions or tight junctions. This property promotes electrical sepa-

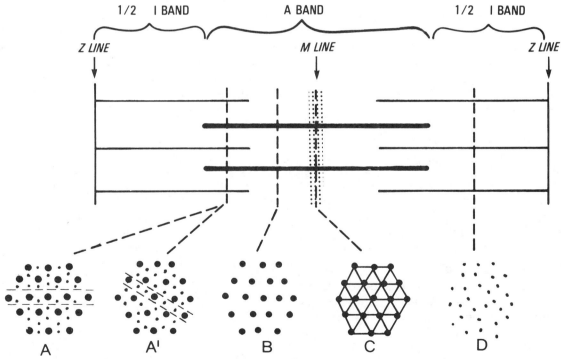

Figure 1.55. Diagram illustrating the arrangement of the thick and thin filaments in a sarcomere as they would appear in a longitudinal section of a muscle fiber. Immediately below are diagrams of transverse sections taken through the A band where the thick and thin filaments overlap (*A* and *A'*); the A band where there is no overlap, i.e., H zone, (*B*); the M line (*C*); and the I band (*D*). If a longitudinal section were cut in the plane which is indicated by the *broken lines* in (*A*), it would show one (apparently) thin filament between each two thick ones. If it were cut in the other plane shown by *broken lines* in (*A'*), it would show two thin filaments between each two thick ones. (Adapted from Huxley and Hanson, 1960.)

Figure 1.56. Cross-section through several myofibrils, at the level of the A band, illustrating the relative disposition of thin and thick filaments in the sarcomere. At right, thin filaments occupy a trigonal position in the hexagonal lattice of thick filaments (*A*). At left, in the H zone (*H*), only thick filaments are present. Cross-links join thick filaments to each other at the M line (*M*), located in the center of the sarcomere. (×33,000). (From Franzini-Armstrong and Peachey, 1981.)

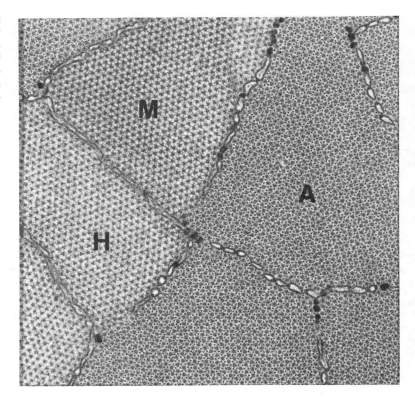

ration of the skeletal muscle cells from one another. Besides shielding the muscle fiber, the sarcolemma imparts a characteristic resistance to stretch to the muscle as a whole. Because this resistance is anatomically disposed in parallel with the contractile elements, it is usually lumped together with the resistance contributed by the connective tissue surrounding the muscle fibers, and both are referred to as *parallel elastic elements.*

Tubular extensions of the sarcolemma, called *T (transverse)-tubules* or *sarcotubules*, extend deep into the fiber at the level of either the Z line or at the junction of the A-I bands, depending on the type of muscle fiber or the animal species. The T-tubules, about 0.03 μm in diameter, allow a wave of depolarization traveling along the sarcolemma during muscle excitation to pass rapidly into the fiber so that deep lying myofibrils may be rapidly activated.

Sarcoplasm

The sarcoplasm of a muscle fiber consists of the contents of the sarcolemma, excluding the proteins of the contractile elements and nuclei. It contains the usual cytoplasmic organelles including mitochondria (sarcosomes), sarcoplasmic reticulum, and Golgi apparatus.

Sarcoplasmic Reticulum

The sarcoplasmic reticulum is an elaborately anastomosing tubular network which surrounds the myofibrils and runs parallel to the myofilaments. The slender tubules, approximately 0.04 μm in diameter, may extend the full length of the sarcomere and end in dilated structures called *terminal cisternae*, which lie on opposite sides of the T-tubules. The sarcoplasmic reticulum can thus be divided into the longitudinal sarcoplasmic reticulum and the terminal cisternae. A group of one T-tubule and two terminal cisternae is called a *triad* because of the triple spaces seen in cross-sections of these structures (Fig. 1.57). Densities called "feet" are located in the narrow space between the terminal cisternae and the T-tubules (about 12–14 nm). These structures appear to play a role in excitation-contraction coupling and are probably calcium ion channels. The sarcoplasmic reticulum immediately adjacent to the T-tubules is also called *junctional sarcoplasmic reticulum.* In this terminology, the remaining sarcoplasmic reticulum is called free sarcoplasmic reticulum because it is not closely associated with another membrane.

The functions of the sarcoplasmic reticulum are the release of calcium (perhaps through the foot structures) during muscle contraction and the sequestration and storage of calcium during muscle relaxation. Some of the earliest evidence for the function of the sarcoplasmic reticulum came from the studies in frog skeletal muscle by Huxley and Taylor, who found that following exploratory stim-

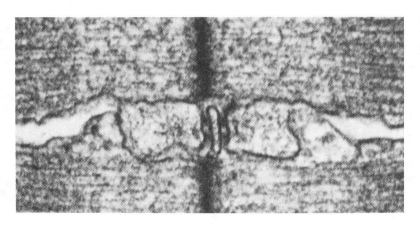

Figure 1.57. Electron micrograph of a longitudinal section of a muscle fiber showing a triad at the level of the Z line. Two "feet" are visible on each side of the T-tubule ($\times$ 120,000). (From Franzini-Armstrong, 1970.)

ulation with minute electrical currents applied with a microelectrode at intervals of 1 μm along the length of the sarcolemmal membrane, a local contraction was elicited *only* when the stimulatory microelectrode was placed at the level of the Z line at which are found (in frog skeletal muscle) the openings of the T-tubules. These findings revealed an important relationship between the T-tubules and the adjoining cisternae of the triad and also between the triad as a whole and the contractile elements (myofilaments) of the muscle fiber. Somewhat later, in an autoradiographic study by Winegrad which allowed localization of radiolabeled calcium during muscle contraction and relaxation, evidence was obtained to indicate that during muscle excitation, calcium stored in the terminal cisternae is released to allow binding to troponin (see below) and interaction of the thick and thin filaments. Muscle relaxation was correlated with the sequestration of the calcium at the level of the A band by the longitudinal sarcoplasmic reticulum.

Nuclei

Skeletal muscle fibers are multinucleated. In embryonic tissue the nuclei are located in the center of the fiber; however, in the course of differentiation, they come to occupy a peripheral position.

PROTEINS OF THE CONTRACTILE ELEMENTS

Thin Filaments

Thin filaments are composed primarily of three types of protein: actin, tropomyosin, and troponin in a ratio of 7:1:1. The so-called "functional unit" necessary for relaxation consists of 7 actin monomers, 1 tropomyosin, and 1 troponin complex.

ACTIN

In vitro, actin molecules exist in two states: G-actin and F-actin. G-actin is a monomeric globular protein with a molecular weight of 42,000 and a

diameter of 4–5 nm. Each monomer contains binding sites for other actin monomers, myosin, tropomyosin, troponin I, ATP, and cations. F-actin, a fibrous polymer consisting of approximately 300 or more monomers of G-actin, is formed in vitro when salts at physiological ionic strength are added to G-actin monomers. The link between actin monomers in G-actin is noncovalent, probably due to hydrophobic interaction. The basic structure of the thin filament consists of two strands of F-actin polymers intertwined in the conformation of a double stranded helix (Fig. 1.58).

TROPOMYOSIN

Tropomyosin is an elongated protein which consists of two α-helical chains, each with a molecular weight of approximately 35,000, wound around each other to form a coiled-coil. This coiled-coil lies in each of the two grooves, 180° apart, formed by the double stranded helix consisting of F-actin (Fig. 1.59). Each molecule of tropomyosin extends over seven actin monomers.

TROPONIN

Troponin consists of a complex of three separate proteins: troponin-T (abbreviated TN-T), troponin-

a.

b.

Figure 1.58. Structure of F-actin. *a,* Model showing double stranded helix of F-actin. F-actin is a polymer of an indeterminate number of G-actin monomers. *b,* Electron micrograph of a segment of an F-actin filament. (Approximately $\times$230,000.) (Courtesy of James Spudich.)

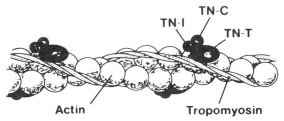

Figure 1.59. Model of thin filament of muscle. Thin filament consists of a double stranded helix of F-actin, two double stranded helices of tropomyosin, and a troponin complex, consisting of TN-T, TN-C, and TN-I, located at each group of seven actin monomers. (Reprinted with permission from McCubbin and Kay, 1980. Copyright 1980 American Chemical Society; based on model by Ebashi et al., 1969.)

C (TN-C), and troponin-I (TN-I) (Fig. 1.59). Each troponin complex is bound to a tropomyosin molecule. While only actin and myosin are directly involved in tension generation, the tropomyosin and troponin complex are known to regulate the actin-myosin interaction, hence are called regulatory proteins.

TN-T (molecular weight, 30,000) binds the other two troponin subunits to tropomyosin. TN-C (molecular weight, 18,000) is the calcium acceptor protein of the troponin complex. It binds to TN-T and to TN-I (molecular weight, 22,000), which is believed to induce the inhibitory conformation of the actin-tropomyosin filament. If indeed the actin is masked during relaxation (this is the traditional viewpoint), then tropomyosin masks the actin when tropomyosin is in its inhibitory conformation. TN-I probably participates in shifting tropomyosin to this position. Strictly speaking, TN-I is not the prime mover, since there is 1 TN-I molecule per 7 actin monomers, and TN-I is too small to "mask" 7 actin monomers.

OTHER PROTEINS ASSOCIATED WITH THIN FILAMENTS

α-Actinin is a protein localized at the level of the Z line. It may play a role in the attachment of F-actin to the Z line. Two other proteins at the Z line, *desmin* and *vimentin*, may serve a similar function.

Thick Filaments

Thick filaments consist primarily of myosin and to a lesser extent of various other proteins.

MYOSIN

Myosin is a dimer with a molecular weight of 480,000 measured under nondenaturing conditions. It consists of two globular heads, which hydrolyze ATP and interact with actin, and a rod-like region, which confers stability to the molecule. In in vitro studies, purified myosin may be treated with proteolytic enzymes in order to study the properties of

different segments of the molecule (Fig. 1.60). Treatment of myosin with trypsin splits the protein into two components: light meromyosin (LMM) (molecular weight, 140,000–160,000), which consists of the tail (rod-like) part of the native molecule, and heavy meromyosin (HMM) (molecular weight, 340,000–380,000), the primarily globular end of the native molecule, to which is attached a segment of the rod-like portion. Papain cleaves HMM or myosin into a globular protein, S_1, and a rod-like protein, S_2, or myosin rod consisting of LMM and S_2. Both HMM-S_2 (molecular weight 60,000) and LMM are insoluble in solutions of low ionic strength, whereas HMM-S_1 is soluble under these conditions. Myosin adenosine triphosphatase (ATPase) activity, i.e., the ability to hydrolyze ATP, is retained in HMM and in HMM-S_1 obtained with papain; LMM and HMM-S_2 are totally inert in

a.

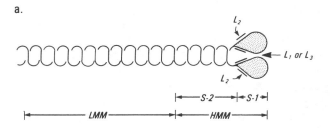

b.

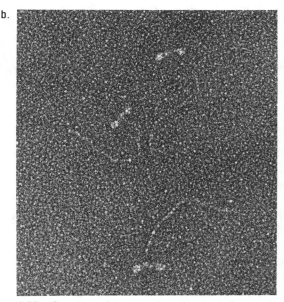

Figure 1.60. Structure of the myosin molecule. *a,* Schematic representation: The molecule is a dimer consisting of two globular heads joined to a rod-like tail region. Hinge-regions occur at the head-tail junction and in the tail near the site of proteolytic cleavage of the molecule into LMM and HMM. The light chains (L_1, L_2, and L_3) are associated with the myosin heads. *b,* Electron micrograph of shadow cast myosin molecules magnified approximately 271,000 times. (Courtesy of S. S. Margossian and H. S. Slater.)

this respect. The ATPase activity of myosin obtained from different types of skeletal muscle correlates with the shortening velocity of the particular muscle type.

When column-purified myosin obtained from fast skeletal muscle is examined by sodium dodecyl sulfate-polyacrylamide gel electrophoresis (denaturing conditions), it is found to consist of heavy chains (molecular weight, 200,000) and three light chains called L_1 (molecular weight, 25,000), L_2 (molecular weight, 18,000), and L_3 (molecular weight, 16,000). These light chains are bound noncovalently to the globular heads (S_1) in the neck region near the junction with S_2.

The L_2 light chains, two per myosin molecule, one on each head, are also called DTNB light chains because they can be dissociated with 5,5'-dithiobis-(2-nitrobenzoic acid) (DTNB). The L_2 light chain is not essential for either the myosin or actin-activated ATPase activity, but modulates the latter (Pemrick, 1980). The other two light chains, L_1 and L_3, are called the alkali light chains because they may be removed with alkali treatment. It is thought that the L_3 light chain is derived from the L_1 light chain since the entire amino acid sequence of L_3 is contained in a 16,000 molecular weight C-terminal stretch of L_1. Traditionally, removal of these light chains has resulted in loss of ATPase activity; hence they have been called essential light chains. Recent technology, however, has made it possible to demonstrate ATPase activity in the absence of alkali light chains (Sivaramakrishnam and Burke, 1982).

Myosin molecules will aggregate in a particular pattern to form filaments, similar to the thick filaments seen in intact muscle, when the ionic strength of the medium is reduced, e.g., from 0.5 M to 0.1 M KCl (Fig. 1.61). The morphology of the thick filaments agrees with the assumption that the myosin molecules aggregate with their globular ends directed toward the ends of the filaments. The middle smooth shaft corresponds to a close packing of the rod-shaped part of the molecules, and the series of projections correspond to their globular ends.

Skeletal muscle myosin is heterogenic or polymorphic with respect to both the heavy and light chains. Various isoforms of myosin can be identified in different muscle types by certain types of gel electrophoresis or by immunocytochemical methods. These methods have been used to distinguish three isomyosins in fast skeletal muscle fibers and two isomyosins in slow fibers, all with the same molecular weight. The presence of these isomyosins correlates with the ATPase activity. In fast fibers,

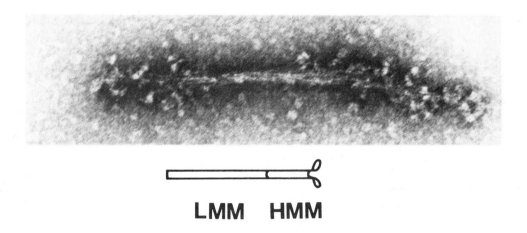

LMM   HMM

Figure 1.61. Electron micrograph of a synthetic thick filament formed by aggregation of purified myosin in 0.1 M KCl. Projecting cross-bridges can be seen. Filaments containing light meromyosin alone (not shown) have no projections. (Approximately ×162,000). (From Huxley, 1969.) Shown immediately below in order to illustrate the pattern of aggregation are diagrams of a single myosin molecule and a synthetic thick filament. *LMM,* light meromyosin; *HMM,* heavy meromyosin.

there are three light chains, LC_1^f, LC_2^f, and LC_3^f whereas in slow fibers, there are three different forms of light chains designated LC_{1a}^s, LC_{1b}^s, and LC_2^s. The possibility of fast and slow myosins existing in the same fiber cannot be excluded particularly during states of transition such as occur in cross-innervation experiments, during maturation, in response to exercise or changes in hormonal status. For example, changes in circulating levels of thyroid hormone are known to affect muscle phenotype.

OTHER PROTEINS ASSOCIATED WITH THICK FILAMENTS

A number of proteins in addition to myosin are associated with the thick filaments. C-protein, the best studied of these proteins, is associated with the A band on either side of the M line and binds to myosin or light meromyosin at physiologic ionic strength, indicating interaction with the thick-filament backbone. There are three identifiable components of the M-band region. They are creatine kinase, which allows rapid regeneration of ATP during contraction, myomesin, and M protein. Some of the nonmyosin thick filament proteins may help to keep the filaments in register within the sarcomere.

In molluscan catch muscle, the thick filaments consist of a core of rod-like proteins called *paramyosin* and a surface layer of myosin molecules. The paramyosin participates in the maintenance of a very high tension by the muscle with little expenditure of ATP.

In Vitro Studies of the Contractile Process

The properties of the contractile proteins can be studied in a number of in vitro systems. Purified myosin can be used to study its ATPase activity. Myosin ATPase can be activated by Mg^{2+}, K^+ (plus EDTA), NH_4^+, or Ca^{2+}. All of these ions can be utilized as all bind to ATP; however, the physiological substrate of myosin is MgATP. Greatest activation is obtained with the physiological activator actin in the presence of MgATP. ATPase activity may also be studied in preparations of HMM or the S_1 subfragment of HMM.

Actin and myosin combine to form actomyosin, a highly viscous complex or thick gel. If the actin filament contains troponin and tropomyosin, it is called *"regulated actin,"* which means that the interaction of actin and myosin in the presence of ATP is Ca^{2+} sensitive. In the absence of Ca^{2+}, troponin is calcium free and tropomyosin is in its inhibitory postion on actin (see above). As a result, little or no interaction occurs between actin and myosin. This in vitro situation is analogous to muscle in the relaxed state. In the presence of Ca^{2+}, regulated actin is no longer exerting an inhibitory influence on the system, actin-myosin interaction is optimal, and the in vitro situation is similar to muscle contraction.

Actin-myosin interaction can be followed in a spectrophotometer due to changes in light scattering associated with ATP-mediated changes in the actomyosin gel. The addition of ATP to an actomyosin gel first dissociates the actin and myosin resulting in a clearing phase (minimal light scattering), which is very long (time scale of hours) in the absence of Ca^{2+} and fairly short in the presence of Ca^{2+} (3–5 minutes) since ATP is rapidly hydrolyzed only in the latter case. When the ATP level falls below a critical value, a phenomenon known as *superprecipitation* (increase in light scattering) occurs due to shrinking and loss of water content in the actomyosin gel. Superprecipitation is believed to be analogous to an elementary contractile process.

The contractile process can also be studied in various preparations of muscle fibers. In early studies, glycerinated fibers were frequently used. Prolonged treatment of muscle in 40–50% glycerol results in loss of cellular membranes, including the sarcoplasmic reticulum and the regulatory proteins of the thin filament (unless precautions are taken to retain the latter) while leaving the contractile proteins (actin and myosin) intact. Addition of ATP causes contraction of the fibers. Later it became possible to isolate intact fibers by careful dissection from the muscle. The latest advance in the study of single fibers has been the development of the "skinned fiber" preparation. These are fibers from which the sarcolemma has been mechanically removed. It is, therefore, possible to vary directly the composition of the medium surrounding the myofibers without the permeability barrier imposed by the surface membrane.

THEORIES OF CONTRACTION

Historical Development

From the 1840s until the 1920s, the *viscoelastic theory*, also called the *new elastic body theory*, was postulated to explain muscle contraction. According to this theory, muscle acts like a stretched spring (or new elastic body) contained in a viscous medium. The amount of energy to be released upon contraction would depend on how much the muscle or spring was stretched; when released, the muscle or spring would liberate all of its energy in an all-or-none fashion irrespective of the work done. Thus it had been believed that a preset amount of additional energy is fed into the contractile machinery when a muscle contracts (in association with the assumption by the muscle of new elastic characteristics), and that this present increment of energy could be proportioned between work and heat, the total amount of energy (work + heat) remaining constant. That this was not the case was demonstrated in the 1920s by Fenn, who observed that the total energy released by a muscle (work + heat) increases as muscle work increases. This finding became known as the *Fenn Effect.*

Fenn's finding indicated that energy release is determined not only by the activation process but also by the work load imposed on the muscle. Thus the mechanical function was shown to control chemical reactions in muscle cells, the molecular mechanism being finely attuned to the work demand, i.e., the performance of a given increment of muscular work calls for the generation and delivery of a commensurate increment of energy. While these findings add little insight into the molecular mechanism of muscle contraction, they are significant because they showed that the viscoelastic theory of muscle contraction, a view held for 80 years, was incorrect.

Since the identification of actin and myosin as the contractile proteins, various other theories of contraction and relaxation have been postulated. According to the *continuous filament theory,* during contraction actin and myosin combined to form one continuous filament, which underwent folding and thereby shortening. The folding was postulated to be due either to thermal agitation or to loss of water molecules from the intrinsic chemical structure of fibers.

Observations by interference microscopy did not support the continuous filament theory, because actin and myosin appear in regularly spaced arrays. After contraction, the lengths of thick and thin filaments were observed to be unaltered and only their relative position changed. Another theory, known as the *sliding filament theory,* had been proposed independently in 1954 by A. F. Huxley and by H. E. Huxley and their collaborators. As more information became available based on electron microscopic observations, X-ray diffraction patterns of muscle fibers both at rest and in the contracted state, and biochemical data on the contractile proteins, the sliding filament theory was transformed into the *cross-bridge theory* of muscle contraction. The original Huxley theories have been extended several times in subsequent versions such as the Huxley-Simmons model of muscle contraction. These theories are described below.

Sliding Filament and Cross-Bridge Theories

The sliding filament theory states that muscle contraction is the result of two overlapping sets of filaments sliding past each other. Specifically, the thin filaments at each end of the sarcomere move in opposite directions toward the center and between the thick filaments to which they are linked by cross-bridges. Polarity of the thick filaments, as described below, is thus a requisite for contraction.

The molecular basis for the sliding motion of the filaments became evident with the elucidation of the structure of actin and myosin. The globular heads of myosin form cross-bridges with the actin monomers, hence the newer designation cross-bridge theory.

Huxley (1969) suggested that the cross-bridges move to and fro, first attaching to the thin filaments and pulling them toward the center of the A band during the cycle and then detaching again prior to their return stroke—similar to the action of a ratchet. The cross-bridge theory is thus also known as the *ratchet theory* of muscle contraction. The cross-bridges are attached to thin filaments while force is being developed and serve as the agents through which the mechanical force is transmitted.

Each cross-bridge consists of the globular heads of the myosin (S_1-subfragment) and an α-helical tail (S_2-subfragment) by which the cross-bridges attach to the backbone of the thick filament (compare Figs. 1.60A, 1.64, and 1.66; see also Huxley and Kress, 1985). Myosin molecules are arranged in the thick filaments with a definite structural polarity so that the heads of the molecules are always directed away from the midpoint of the filament. Thus, all cross-bridges in one-half of an A band have the same orientation (polarity). This polarity is reversed in the opposite half of the A band. The actin monomers are also oriented oppositely on either side of their attachments to the Z line. During contraction the sets of thin filaments in each half sarcomere are drawn toward the center of the A band and subjected to sliding forces oriented in opposite directions.

Using X-ray diffraction H. E. Huxley had initially predicted that the contractile components (later identified as actin and myosin) must have different structures. Further study of X-ray diffraction patterns indicated that in striated muscle of vertebrates the cross-bridge projections on the thick filaments are arranged so that there are three pairs of cross-bridges per 360° turn (Fig. 1.62). Each pair of bridges projects out from the myosin backbone in opposite directions, so that the two bridges in any pair form an angle of 180° with each other, i.e., any pair of cross-bridges form a straight line at any given level. Taking any pair of cross-bridges as a reference, the nearest neighbor pair of bridges on either side occurs at a distance of 14.3 nm and is rotated relative to the reference pair by 120°. This arrangement continues so that the full pattern

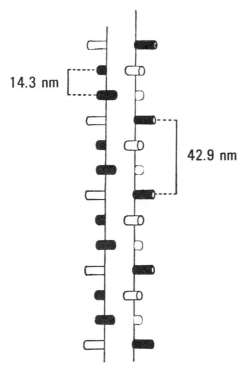

Figure 1.62. Diagram of cross-bridge arrangement on thick myosin-containing filaments of frog sartorius muscle which would account for the observed X-ray pattern. (Adapted from Huxley, 1969.)

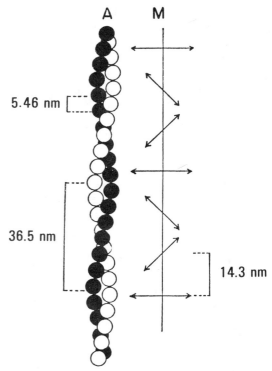

Figure 1.63. Diagram of the arrangement of G-actin monomers in thin (actin (*A*)) filaments derived from X-ray diffraction and electron-microscope observations. Both the pitch of the helix and the subunit repeat differ from those of the thick (myosin (*M*)) filaments, indicated schematically alongside. Thus, cross-bridges between filaments would act asynchronously, and a sequence of them would develop a fairly steady force as the filaments moved. (Adapted from Huxley, 1969.)

repeats itself at intervals of 3×14.3 nm or 42.9 nm.

The X-ray diffraction patterns of the thin filaments show a double helical structure, with subunits (the G-actin monomers) repeating at 5.46 nm intervals along each strand. The position of any subunit on one strand is staggered relative to the position of its nearest neighbor subunit on the other strand by half a subunit period, i.e., 2.73 nm. The chains twist around each other with cross-over points at 36- to 37-nm intervals, so that the pitch of the helix formed by either of the two chains is 72–74 nm (Fig. 1.63).

On the basis of these observations, a model (Fig. 1.64) was suggested in which the cross-bridges are attached to the backbone of the thick filaments and extend outward beyond the myosin backbone by a short subunit tail. This tail is attached to the thick filament only at one end. This type of attachment makes it possible for the cross-bridges to move away from the backbone toward the thin filament. As noted previously, the junction between the heavy meromyosin tail to the light meromyosin of the thick filament backbone is susceptible to trypsin digestion and, therefore, probably represents a non-helical flexible region of the molecule. Similarly, the junction between the linear part of the HMM tail with the globular head is also susceptible to enzy-

matic digestion, and this site of attachment probably also constitutes a flexible site. Therefore, the presence of two flexible points on the tail of the cross-bridge would give the HMM head the mobility

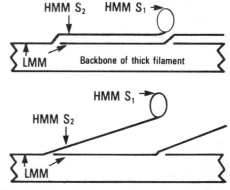

Figure 1.64. Suggested behavior of myosin molecules in the thick filaments. The light meromyosin (*LMM*) part of the molecule is bonded into the backbone of the filament, while the linear portion of the heavy meromyosin (*HMM*) component can tilt further out from the filament (by bending at the HMM-LMM junction), allowing the globular part of HMM (that is, the S₁ fragment) to attach to actin over a range of different side spacings, while maintaining the same orientation. (From Huxley, 1969.)

needed for it to become properly oriented toward the thin filament during contraction. Because of these flexible points or hinge regions, the cross-bridge projections may change either in orientation or length at different sarcomere lengths when the lateral distance between thick and thin filaments changes (Fig. 1.65). Because the lateral distance between thick and thin filaments increases as the fiber shortens, the volume of the muscle fiber remains constant during contraction.

On the basis of both mechanical and biochemical considerations, the flexible tail of the cross-bridge cannot itself account for the development of tension during contraction, inasmuch as the ATPase activity of HMM remains with the globular head after the linear tail is removed by enzymatic treatment. However, the orientation of the linear tail to the thick filament backbone suggests that the tail serves to sustain the tension which is developed by the globular end of the HMM cross-bridge. Observations of interactions between cross-bridges and thin filaments indicate that globular HMM or S_1 binds tightly to actin at least during part of the contraction cycle. According to the Huxley-Simmons model of muscle contraction (see Harrington, 1979), tilting of the globular HMM head of the cross-bridge is associated with a simultaneous stretching of the spring-like elastic component in S_2; retraction of the elastic component produces the power stroke or movement. This model of muscle contraction is depicted in Figure 1.66.

During the cyclic attachment and detachment of the cross-bridges, ATP is hydrolyzed by a series of biochemical reactions. Figure 1.67 shows a scheme of reactions which can be formulated based on in vitro studies. Actin and myosin in the absence of ATP are bound (AM) due to the very high affinity of myosin for actin (association constant 10^7 M^{-1}) (step 11). AM is called a rigor complex. If ATP [which may bind to both M (step 7) and AM (step 1)] is then introduced into the system, ATP dissociates the AM complex (step 2) because the affinity of M-ATP for actin is much weaker (10^4 M^{-1}) than the affinity of M (without ATP) for actin (10^7 M^{-1}). ATP hydrolysis can occur both with myosin bound to actin (step 10) and with myosin dissociated from actin (step 3). At physiological ionic strength, Mg^{2+}, and MgATP concentrations, the reaction sequence proceeds mainly from steps 1–6. This is because actin affinity is low with ATP bound to myosin. The cross-bridge states prior to P_i liberation are termed weakly bound cross-bridge states. Following P_i liberation, the affinity of myosin for actin increases (to 10^6 M^{-1} at step 5) to generate the strongly bound cross-bridge states. It is generally assumed that the energy derived from the hydrolysis of ATP at step

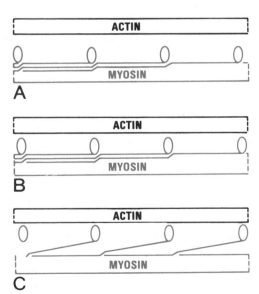

Figure 1.65. Diagram showing relative positions of filaments and cross-bridges at two different interfilament spacings [(*A*) 25 nm and (*B*) 20 nm] corresponding, in frog sartorius muscle, to sarcomere lengths of ~2.0 and ~3.1 μm. The X-ray diagram (not shown) suggests that in a relaxed muscle the cross-bridges do not project very far toward the thin (actin) filaments. During contraction or in rigor, the cross-bridges could attach to the actin filaments by bending at two flexible junctions, as shown in (*C*). (From Huxley, 1969.)

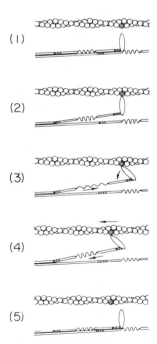

Figure 1.66. Huxley-Simmons model of muscle contraction. (*1*) resting state, (*2*) attachment of S_1 to actin (thin) filament, (*3*) rotation of S-1 while it is attached to the actin filament and simultaneous stretching of the spring-like elastic component in S_2, (*4*) power stroke resulting from retraction of elastic component, (*5*) return of cross-bridge to resting state. (From Harrington, 1979.)

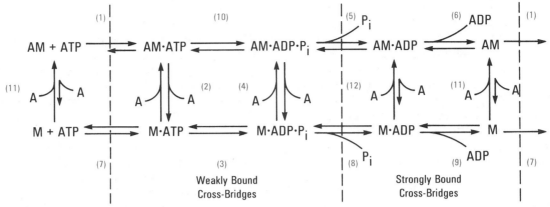

Figure 1.67. Simplified scheme of reactions of hydrolysis of ATP by myosin based on in vitro studies with HMM or S_1-subfragment. (Modified Lymn-Taylor, 1971; model as described by Eisenberg and Hill, 1985; and Hibberd and Trentham, 1986.) Favored directions of some of the reactions are indicated by *longer arrows*. The numbered reactions are discussed in the text. (Courtesy of S. Pemrick.)

10 is utilized for force generation or movement of the cross-bridges (i.e., the power stroke) at steps 5 or 6 and is associated with a conformational change in myosin (see step 4 in Fig. 1.66). Step 5 is the rate-limiting step in the Lymn-Taylor (1971) model.

In the absence of actin, ATP is hydrolyzed at step 3 in the sequence $7 \rightarrow 3 \rightarrow 8 \rightarrow 9$, and the rate-limiting step is 8. Actin is termed a cofactor of myosin ATPase, not only because it changes the rate-limiting step, but also because actin increases, by a factor of 100 to 200 fold, the rate of release of ADP and P_i from actomyosin (at steps 5 and 6). Under normal conditions ATP is present in the cell in millimolar concentrations and therefore does not limit this cyclic process, which is instead controlled by Ca^{2+} at steps 4 and 5.

In the absence of Ca^{2+}, an inhibitory conformation of regulated actin may prevent formation at step 4 of the $AM \cdot ADP \cdot P_i$ complex. The physical (steric) blocking of myosin binding to actin by troponin-tropomyosin in the absence of Ca^{2+} has been the traditional view until recently when it became possible to demonstrate that 10–20% of the "heads" of myosin hydrolyze ATP at step 10. This introduces the concept that relaxation need not be associated with dissociation of actin and myosin but rather with a particular conformation of myosin bound to actin and with ability of actin to respond to this conformation of myosin (Stein et al., 1979; Murray et al., 1981). Thus myosin can bind weakly to actin in the absence of Ca^{2+}, and the transition from weakly bound to strongly bound cross-bridges (step 5) is inferred also to be Ca^{2+} sensitive, i.e., it may be regulated by the Ca^{2+} concentration and the Ca^{2+} affinity of troponin. Although troponin-tropomyosin is in its inhibitory conformation, some weakly bound cross-bridges may persist, hence the physical blocking of myosin binding to actin in the absence of Ca^{2+} may not be complete as was traditionally assumed.

When all the ATP in the muscle is lost and the muscle goes into rigor, as is seen in *rigor mortis*, it is believed that a large portion of the cross-bridges become attached to thin filaments. Thus, the effect of ATP on actin-myosin interaction is two fold: (1) to provide energy for movement, and (2) to reduce the affinity betwen actin and myosin to allow cyclic interaction of the two proteins. The latter function of ATP is often called its *plasticizing effect.*

MECHANICAL CHARACTERISTICS OF MUSCLE

Single Muscle Twitch

The contraction-relaxation of a skeletal muscle in response to a single stimulus is called a *twitch* (Fig. 1.68). When a muscle is activated by a single stimulus while extended by a moderate load, there is a brief lag between the arrival of the stimulus and the initiation of tension development. This lag is referred to as the *latent period,* which lasts approximately 2–4 ms. The initial part of the latent period is due to the spread of the action potential along the plasma membrane and T-tubules and transmission of the signal to the sarcoplasmic reticulum to cause calcium release. When the tension has reached a value exceeding the stretching force exerted by the load, the muscle shortens rapidly. This development of contractile force is transient, and as the contractile force decreases, the muscle lengthens and returns to its relaxed condition.

During the latent period, the muscle may lengthen slightly, a phenomenon called *latency relaxation.* The subcellular events during this period are unknown, although it has been suggested that

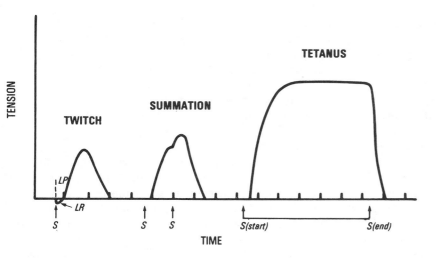

Figure 1.68. Single muscle twitch, summation, and tetanus. Each division on the time scale represents 5 ms. *S,* stimulus; *LP,* latent period; *LR,* latency relaxation. Latency relaxation may or may not be observed, i.e., it is not routinely recorded under normal experimental conditions. High frequency stimulation is necessary to produce tetanic tension, which, although not shown above, may begin to fall despite continued stimulation due to muscle fatigue.

latency relaxation reflects the disruption of long-lasting cross-bridge bonding between thick and thin filaments just prior to the initiation of the more rapid "make and break" cross-bridge bonding associated with the sliding process. Latency relaxation is most marked at longer lengths of muscle and is not at all evident when the muscle length is substantially (10% or more) below the normal body length.

Summation and Tetanus

If a skeletal muscle is stimulated and a second stimulus is applied before relaxation is complete, a second contraction, which develops a greater tension, is fused to the first contraction. This is called *summation.* A possible explanation for the increase in tension may be that Ca^{2+}, which would normally have been sequestered by the sarcoplasmic reticulum, remains in the sarcoplasm from the previous contraction and together with additional Ca^{2+} released with the second stimulus constitutes more activator Ca^{2+} than would be available if relaxation had been complete.

If the stimulus is repeated at a sufficiently high rate, the muscle will not relax between each stimulus but rather will remain in a contracted state which is referred to as *tetanus.* The plateau of such a tetanic contraction exceeds the peak of a single twitch (Fig. 1.68). If the stimulus is of an intensity high enough to activate all of the fibers within the muscle, the maximum contraction of which the muscle is capable is produced. This occurs only rarely physiologically as most routine daily tasks do not require a maximum voluntary contraction of skeletal muscle. Tetanus, whether of a single fiber or an intact muscle, cannot be maintained for prolonged periods due to fatigue, the exact source of which remains unknown.

Changes in the Excitability of Muscle During the Contraction-Relaxation Cycle

The action potential of a skeletal muscle fiber is similar to that of a nerve except that it is about 5 ms in duration instead of 2 ms. Like the nerve action potential, it has an *absolute refractory period* (also called effective refractory period) and a *relative refractory period* (see Chapter 3). The action potential of a skeletal muscle fiber typically has a long negative after-potential during which time the fiber is hyperexcitable. This is called the *supernormal period.*

Refractoriness of the plasma membrane plays little role in the physiological properties of skeletal muscle. This is because the action potential is very brief and the refractory period is over before the mechanical response even begins (Fig. 1.69). Therefore it is possible to tetanize a skeletal muscle fiber. If a second stimulus is applied soon after an initial

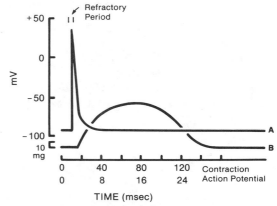

Figure 1.69. Electrical (*A*) and mechanical (*B*) responses of a fast-twitch skeletal muscle fiber. Because the refractory period of the action potential is over before the mechanical response of the fiber even begins, skeletal muscle can be tetanized by rapid stimulation above a certain critical frequency.

stimulus, its action potential will produce a second twitch before the mechanical response to the first stimulus is complete, resulting in fusion of twitches.

Motor Unit

In normal skeletal muscles, fibers probably never contract as isolated individuals. Instead several of them contract at almost the same moment, all being supplied by branches of the axon of one spinal motor neuron. In other words, when the motor neuron is excited by a stimulus at or about threshold, all muscle fibers innervated by the neuron will be activated. The single motor neuron and the group of muscle fibers which it innervates are called a *motor unit* (Fig. 1.70); this is the smallest part of the muscle that can be made to contract independently.

The muscle fibers of a motor unit normally contract sharply upon the arrival of impulses at frequencies ranging up to 50/s. This frequency seems to be the upper physiological limit for axonal propagation of motor neurons in most mammals including man. However, motor units may be deliberately fired at much slower frequencies, even to as low as single isolated contractions at will. Normal human beings can learn easily to provide various frequencies of impulses—usually below 16/s—from their spinal motor neurons.

The number of striated muscle fibers that are served by one axon, i.e., the number in a motor unit, varies widely. The size of the motor units in a muscle is correlated with the precision with which the tension developed by the muscle is graded. Generally, muscles controlling fine movements and adjustments (such as those attached to the ossicles of the ear and of the eyeball, larynx, and pharynx) have the smallest number of muscle fibers per motor unit. The muscles that move the eye have less than 10 fibers per unit; the muscles of the middle ear, 10–125; the laryngeal muscles 2–3; and the pharyngeal muscles have 2–6. These are all rather small and delicate muscles and they control fine or delicate movements. On the other hand, large, coarse-acting muscles have motor units with many muscle fibers, e.g., the human gastrocnemius has 2,000 or more.

Even the largest bundles of muscle fibers are quite small, and so a strong contraction of a skeletal muscle requires the participation of many motor units. Further, there is a complete asynchrony of the motor unit contractions imposed by asynchronous volleys of impulses coming down the many axons. Thus, with motor units contracting and relaxing at differing rates, a smooth pull results. In certain disturbances the contractions become synchronized, resulting in a visible tremor.

The fibers in a motor unit may be scattered and intermingled with fibers of other units. Thus, the individual muscle bundles seen in routine histology do not correspond to individual motor units as such. In rat muscle, the fibers of a motor unit are widely scattered. In man, a similar condition is probable; the spike potentials of each motor unit in the biceps brachii are localized to an approximately circular region of 5 mm diameter in which the fibers of the unit are confined. However, the potentials can be traced in their spread to over 20 mm distance; thus, the area of 5 mm diameter includes many overlapping motor units.

Gradation of Muscle Tension: All-or-None Response

The gradation of the tension developed by skeletal muscle depends on variation in the number of motor units activated and on the frequency of stimulation. Stimulation at *threshold intensity* (minimum effective stimulation) evokes a minimal twitch tension. There is one response to one stimulus. If the intensity of the stimulus is increased stepwise, the amplitude of the twitch tension will likewise increase. However, with continued increase in the intensity of stimulation, a maximal contractile response is attained, after which a further increase in the intensity of stimulation no longer results in a change in the amplitude of the twitch tension (Fig. 1.71).

The explanation for these responses lies in the fact that with low intensity stimuli, only the most excitable nerve fibers respond and activate their

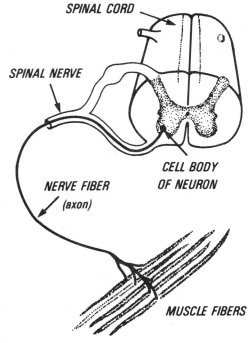

Figure 1.70.   Diagram of a motor unit. (From Basmajian, 1978.)

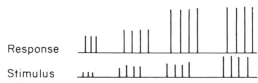

Figure 1.71. Recruitment of motor units with increasing stimulus intensity. The response of the individual unit is "all-or-none," but the threshold of excitability differs among the units.

respective motor units, but as the intensity of the stimuli is increased, more motor units are recruited, until finally all motor units are responding, at which time the twitch tension is maximal at a particular frequency of stimulation, and therefore an increase in intensity of the stimuli can no longer elicit an increased response. The individual muscle fibers have long been believed to respond to activation in an *"all-or-none"* manner, that is, to exhibit contractions of uniform intensity once their particular threshold of excitability had been reached. In seeming conflict with the classical "all-or-none" concept of muscle fiber contraction are the findings of Costantin and Taylor (1973) which indicate that the contraction of single muscle fibers can be graded by varying the magnitude of membrane depolarization, resulting in activation of variable numbers of myofibrils within the fibers. However, in the muscle in situ it is not possible to vary the magnitude of action potential, which remains all or none thus accounting for the all-or-none property of the contractile response of a skeletal muscle fiber to neural stimulation.

Elastic Elements and Force Generation

Skeletal muscle may be thought of as a two-component assembly consisting of an elastic element in series with a viscous contractile element, and a second elastic element in parallel with the two-component assembly (Fig. 1.72). The *parallel elastic element* has already been identified as consisting in part largely of the sarcolemma. It appears, moreover, to be intracellular as well in view of its persistence in isolated skinned fibers. This observation is consistent with the recent finding of a highly elastic high molecular weight protein, called *titin* or *tubulin*, which forms a net-like structure around the

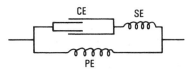

Figure 1.72. Model of skeletal muscle fiber to show contractile and elastic elements. The contractile element (*CE*) is in series with a series elastic element (*SE*). The CE and SE are in parallel with the parallel elastic element (*PE*).

thick and thin filaments. The *series elastic element* is now thought to reside in the hinge regions of myosin as described below.

If a muscle is rapidly stretched during the early part of a contraction, it is capable of developing a considerably greater tension than that recorded under normal conditions (i.e., without stretching). This property has been interpreted to reveal the presence in the muscle of an elastic element coupled in series with the contractile elements. The *quick stretch experiments* also show that the contractile machinery is activated already during the latent period and that this activation, which A. V. Hill (1927) referred to as the "active state" (see below), very rapidly reaches a maximum which is maintained for only a short period of time.

The existence of a series elastic element can also be demonstrated by *quickly releasing* one point of fixation of an isometrically contracting muscle and allowing it to shorten before again fixing it at a new (shorter) length. Immediately after release (even if the muscle is shortened by as little as 1 mm) the tension falls abruptly to zero and then redevelops gradually. This observation suggests that the initial loss of tension is due to elastic recoil of a series elastic element in the sarcomere and that the subsequent slower redevelopment of tension is the result of shortening of the contractile apparatus per se.

A. V. Hill (1938) believed that much of the elasticity resided in the tendons and in the connective tissues in series with the contractile elements. Later, however, A. F. Huxley and Simmons (1973) demonstrated that even in an isolated single skeletal muscle fiber, devoid of any tendinous attachment, quick release results in an initial abrupt loss of tension followed by a slow redevelopment of tension. Since the degree of the slow tension development depends on the precise instant of the quick release in the contraction-relaxation cycle, A. F. Huxley (1974) has suggested that the site of series elasticity as well as the site of tension generation are both in the sarcomere. These observations support the idea that the cross-bridges themselves are the sites of tension generation in muscle—indeed, that they are independent force generators—and, therefore, that the degree of overlap between the thick and thin filaments is directly proportional to the level of tension generated in a twitch (see below).

Resting Tension

It has been observed that muscle which is neither stimulated nor stretched beyond the normal body length still maintains some degree of tension, a phenomenon which is revealed by the simple fact that if a tendon is cut in vivo its associated muscle shortens. D. K. Hill (1970) had obtained evidence to suggest that resting tension is due to residual long-lasting interactions between thick and thin filaments in the absence of Ca^{2+}. A newer view is that some of the resting tension may be due to cross-bridges which somehow have been able to complete the power stroke *at low Ca²⁺* (step 5 in Fig. 1.67). This, however, cannot explain the sharply rising resting tension at long sarcomere lengths where cross-bridge formation is not possible (see text below and Fig. 1.74). Here the resting tension appears to be due to the elastic forces of noncontractile elements such as the sarcolemma and elastic fibers within the myofibers. Thus the highly elastic protein *titin* could contribute to the high resting tension seen at long sarcomere lengths.

Isometric Contraction

When the two ends of a muscle are held at fixed points, stimulation causes the development of force without change in muscle length called isometric tension. Isometric tension can be recorded in either a single muscle fiber or a whole muscle, both of which can also be tetanized isometrically. In the whole muscle, if a stimulus of an intensity sufficient to activate all of the fibers within the muscle is given at a high enough (critical) frequency, a sustained maximal isometric contraction called maximal isometric tetanus is developed.

ISOMETRIC LENGTH-TENSION CURVE

Isometric tension depends not only on the frequency of fiber or muscle stimulation but also on the initial length of the fiber or muscle. The length-tension relationship can be explained in terms of the sliding filament theory of muscle contraction. The length of the muscle prior to activation is called the *initial length* (L_i) or *rest* length. With L_i at considerably less than normal body length, the thin filaments overlap each other in the center of the sarcomere (Fig. 1.73) to such an extent as to reduce the number of actin sites available for interaction with myosin. As the sarcomere is stretched, the "abnormal" overlap of the thin filaments is progressively reduced, and more active sites on the actin become available for linkage to the myosin cross-bridges. The length at which tension becomes maximal is defined as *optimal length* (L_o). The term maximum length (L_{max}) is sometimes used in the same context as L_o, particularly in cardiac muscle physiology. L_o refers to the plateau region of the length-tension curve for a particular muscle. In frog muscle it is between 2.0 and 2.2 μm. It is usually the normal

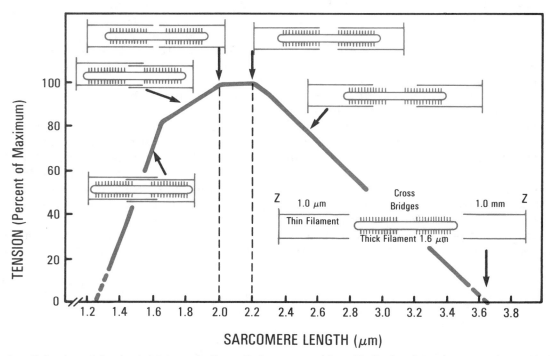

Figure 1.73. Length-tension relation in skeletal muscle fibers. *Ordinate,* Isometric tetanic tension developed (total tension during tetanus minus resting tension). *Abscissa,* Striation spacing in micrometers. Diagrams alongside the length-tension curve show critical stages in the degree of overlap of the thick and thin filaments. Optimum overlap is between 2.0 and 2.2 μm. (Adapted from Gorden et al., 1966.)

length of muscle in the body (normal body length). In this condition, the position of the thick and thin filaments is such as to provide a maximum number of active sites for interaction (Fig. 1.73). As the muscle is further stretched beyond the optimal length the overlap of thick and thin filaments is again progressively reduced, and contractile tension diminishes until a point is reached at which there is no overlap (Fig. 1.73), and, consequently, no active tension can be developed. This occurs at sarcomere lengths greater than 3.65 μm.

It should be appreciated that Figure 1.73 shows the level of *active isometric tetanic tension* development at sarcomere lengths ranging from 1.2–3.6 μm. The change in *resting tension* at various sarcomere lengths is represented in a general sense by the curve shown in Figure 1.74. The *total tension* developed by the muscle in the sum of the resting tension and the active tension as the sarcomere length is increased. The plot of total tension against sarcomere length generates an N-shaped curve rising steeply at long sarcomere length, reflecting the dominance of resting tension.

For many years the interaction of the thick and thin filaments has been the basis for an explanation of the length-tension curve. More recent studies indicate that the calcium sensitivity of the thin filaments may also be length-dependent and may be an important factor in the ascending limb of the length-tension curve.

Isotonic Contraction

When one end of the muscle is free, stimulation produces shortening while exerting a constant force. This is called an isotonic contraction. The degree of shortening of a muscle bears a direct rela-

tionship to the load upon which it exerts force. A progressive increase in the frequency of stimulation at above-threshold levels will evoke a partial and then a complete state of tetanus which, under conditions of isotonic contraction, is characterized by sustained maximal shortening. The additional shortening observed during tetanus as compared with the shortening observed during a single twitch indicates that in the latter circumstance the muscle is not fully activated.

Force-Velocity Relationship

If a muscle is contracting isotonically (i.e., shortening against a load), the velocity of contraction is inversely related to the load which the muscle is lifting. If the muscle is unloaded, it shortens with maximum velocity, and as the load increases, the velocity of shortening declines. When the load reaches a maximum (which is equal to the maximum isometric tension the muscle can develop), the muscle no longer shortens. The relationship between load and the velocity of shortening is hyperbolic as shown in Figure 1.75. Hill (1938) adduced the following empirical equation on the basis of data obtained from frog sartorius muscle tested at 0°C:

$$(P + a)(V + b) = \text{constant} \qquad (1)$$

where P is the load applied to the muscle, V is the velocity of shortening, and a and b are constants with dimensions of force and velocity, respectively. This equation states that force is inversely related to velocity. It can be seen from Figure 1.75 that when $P = P_0$, $V = 0$, P_0 being defined as the maximum isometric tension that the muscle can develop; therefore

$$(P_0 + a)b = \text{constant} \qquad (2)$$

Therefore

$$(P + a)(V + b) = (P_0 + a)b \qquad (3)$$

Eq. 3 can be rearranged to take the following form, which is known as the Hill equation:

$$(P + a)V = (P_0 - P)b. \qquad (4)$$

The constant a is independent of temperature while b is temperature-dependent, exhibiting a Q_{10} ranging from 2.0–2.5. Thus, b appears to be related to chemical processes underlying the mechanism of shortening. The value of b varies from muscle to muscle but is similar for the same type of muscle,

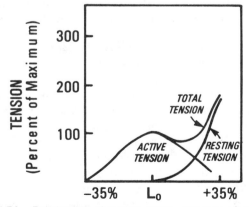

Figure 1.74. Relationship of resting, active, and total tensions in a skeletal muscle. Total tension is the sum of the active tension developed during contraction plus the resting tension in the muscle prior to stimulation. L_o, optimal muscle length (Adapted from Spiro and Sonnenblick, 1964.)

Figure 1.75. Force-velocity data from a typical soleus muscle. *Open circles* are experimentally determined points. *Solid line* was obtained as regression fitted to data plotted *(inset)* using linearized form of Hill equation. *Dashed line* is extrapolation to value of V_{max} computed from equation for $P = 0$. Velocity is given in fiber lengths (l_0)/s. *P*, force; *V*, velocity; *a* and *b*, constants. Note that P is force normalized with respect to cross-sectional area (stress). (From Murphy and Beardsly, 1974.)

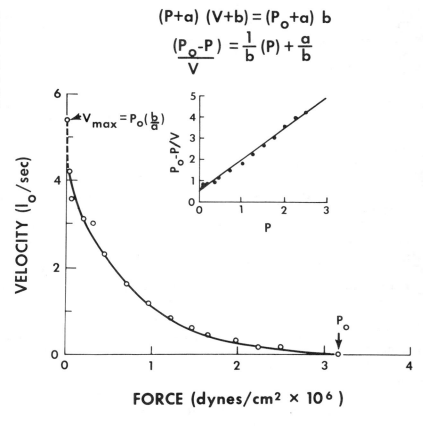

$$(P+a)(V+b) = (P_0+a)\,b$$

$$\frac{(P_0-P)}{V} = \frac{1}{b}(P) + \frac{a}{b}$$

$$V_{max} = P_0\left(\frac{b}{a}\right)$$

FORCE (dynes/cm² × 10⁶)

i.e., fast or slow twitch skeletal or cardiac or smooth muscle. Force-velocity curves are useful for comparing different muscles with respect to V_{max}, a, and b. The constants a and b may be derived from the linearized form of the Hill equation (see inset to Fig. 1.75) in which the slope is equal to $1/b$ and the y intercept is a/b. V_{max}, which correlates with the rate of myosin ATPase activity, may be derived from *Eq. 3*. From *Eq. 3* it follows that when

$$P = 0 \quad \text{and} \quad V = V_{max},$$

then

$$a \cdot V_{max} = b \cdot P_0$$

and, rearranging,

$$V_{max} = (b/a) \cdot P_0. \tag{5}$$

Concept of Active State

A. V. Hill postulated that in the simple twitch response all the contractile elements undergo a rapid transition from a resting state to an activated state which persists for a variable period of time, and then declines before peak contractile tension is reached. Observations obtained from "quick stretch" experiments provided the basis for the formulation of the concept of the "active state." The contour of the intensity of the active state (I_{as}) can be obtained by determining the maximum load that a muscle can sustain after a small quick stretch at various times after stimulation. Shortly after a muscle starts to contract (about 20 ms after stimulation for frog muscle), a quick stretch demonstrates that, in fact, the load-bearing capacity (i.e., the capacity to develop tension) of the contractile element has reached its peak. Thereafter I_{as} declines rapidly. The maximal intensity of the active state was originally defined as the maximal isometric tension which the contractile elements of muscle can develop independently of the effects of the elastic elements. Only a part of this maximal tension is actually manifest in a twitch because of the damping effect of series elastic elements; i.e., the "spring-like" quality of the series elastic elements initially absorbs the energy developed during the onset of the active state in the sarcomere.

The quick stretch method for mapping the time course of the active state provided only discontinuous glimpses of the total time course. A new method was developed which does not require any manipulation of the muscle and which provides a continuous map of the active state contour. In this new technique, tension is transduced into an electrical signal which, in turn, is differentiated so that the change in muscle tension as a function of time, dP/dt, can be recorded for a muscle twitching isometrically (Fig. 1.76). Inasmuch as the stretching of the elastic elements (which develops the isometric tension) is a reflection of the state of activity of the contractile elements, the dP/dt curve in effect indicates the instantaneous changes in the contour of the active state.

Findings based on measurements of the active state of muscle can be readily understood in terms of the availability of Ca^{2+} to troponin. Thus we know now that (1) intracellular Ca^{2+} concentrations are highest when maximum I_{as} is attained; (2) Ca^{2+} is rapidly sequestered by the Ca^{2+}-activated calcium pump of the sarcoplasmic reticulum, resulting in a decline in the active state; (3) factors which affect the amount and duration of Ca^{2+} availability affect the intensity and duration of the active state; and (4) various conditions which increase contractile tension affect either the peak or the duration of the active state.

Muscle Relaxation

For many years, a physiological factor which reverses the contractile process was sought. In the early 1950s, Marsh observed that a shrunken actomyosin gel could be made to take up water by the addition of a supernatant fluid which had been recovered from the centrifugation of ATP-treated muscle homogenates. This muscle extract could relax contracted glycerinated muscle fibers. The

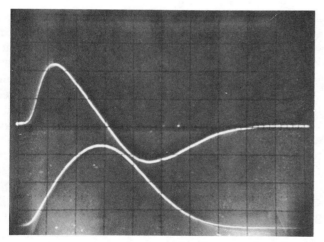

Figure 1.76. A technique for mapping the contour of the active state. The monophasic *(bottom)* curve is a recording of isometric tension in rat tibialis anterior muscle measured at 22°C. The biphasic *(top)* curve is a recording of dP/dt, the rate of change in the development and release of tension; 20 ms per division. Peak + dP/dt corresponds to the peak in cytoplasmic Ca^{2+}. (Courtesy of Z. Penefsky.)

unknown component responsible for these effects was called "relaxing factor." It was later found to sediment in a moderate gravitational field, and it therefore appeared to be particulate in nature.

Many investigators attempted to find a small molecule which, on release from the particulate, "relaxed" the ATP-contracted actomyosin system. However, several years elapsed before it became apparent that the relaxing effect could be due to the removal of an element from the contracted actomyosin system, and therefore that the relaxing factor could be the system involved in the removal and sequestration of this element.

In the mid-1950s it was observed that the addition of ethylenediaminetetraacetic acid (a Ca^{2+} chelator) in the presence of ATP mimicked the physiological action of the relaxing factor; but it was not until 1959 that Weber demonstrated that the ATP-actomyosin system depended on minute amounts of Ca^{2+} for ATPase activity.

The effect of the relaxing factor was then conceived to be a chelating system which was responsible for the removal of an agent, Ca^{2+}, which regulated the actin-myosin interaction. The relaxing factor was identified on further study as the fragmented sarcoplasmic reticulum recovered from a muscle homogenate. It was shown that these vesicular fragments could indeed concentrate Ca^{2+} against very high gradients.

Much of the protein of sarcoplasmic reticular membranes consists of the *calcium pump protein,* a 100,000-dalton protein that has Ca^{2+}-activated ATPase activity. The calcium pump protein transports Ca^{2+} present in the sarcoplasm during contraction into the sarcoplasmic reticular tubules causing muscle relaxation. For every two molecules of Ca^{2+} transported, one molecule of ATP is hydrolyzed. The rapid uptake of calcium by the sarcoplasmic reticulum lowers the sarcoplasmic Ca^{2+} from approximately 10 μM during the height of contraction to 0.1 μM or less, allowing the muscle to relax. The calcium is sequestered within the tubules where the total calcium concentration is in the millimolar range. The sequestered calcium is presumed to move from the longitudinal component of the sarcoplasmic reticulum to the terminal cisternae, which are known to contain a calcium-binding protein called *calsequestrin,* which can bind large quantities of calcium. This calcium becomes available for release from the sarcoplasmic reticulum for the next contraction.

Events leading to muscle relaxation in the intact muscle may be summarized as follows. The sarcoplasmic reticular calcium pump is activated by the Ca^{2+} present in the sarcoplasm during contraction.

Since the free Ca^{2+} is in equilibrium with the calcium bound to TN-C, the latter is dissociated from the TN-C-Ca complex as Ca^{2+} ions and removed from the sarcoplasm. Actin-myosin interaction is now inhibited, and the contractile elements return to their resting state. At this point all of the Ca^{2+} that had been released from the sarcoplasmic reticulum is presumed to be resequestered. Some studies suggest that a protein called *parvalbumin* may serve as an intermediate calcium-binding protein in the movement of Ca^{2+} between the myofibrils and the sarcoplasmic reticulum.

Staircase

In 1871 Bowditch, working with the frog heart, observed that if the heart had been inactive for a period of time a sudden series of stimulations induced a corresponding series of contractions that increased in amplitude until a steady-state was reached. This phenomenon was called "staircase" or "Treppe." If the rate of stimulation of the heart was suddenly decreased, a series of consecutively declining contractions was obtained, until a second steady-state was reached (negative staircase or negative Treppe). The same phenomena are observed in skeletal muscle, but because of the shorter refractory period in this tissue the rate of stimulation must be much higher to elicit staircase effects.

Staircase phenomena are believed to be the result of changes in the intracellular distribution of Ca^{2+} with changes in frequency. A positive staircase may be the result of more Ca^{2+} being available to the contractile proteins. A negative staircase could be the result of a lag in calcium movement within the sarcoplasmic reticulum, and therefore less calcium would be available for subsequent contractions. The physiological significance of staircase phenomena, demonstrated in isolated skeletal muscle, is not known.

Posttetanic Potentiation

For a brief period of time after cessation of tetanic stimulation of a fast muscle (such as the extensor digitorum longus) a single stimulus induces a twitch tension of higher amplitude than that elicited in a muscle not subjected to previous tetanization. This "posttetanic potentiation" (PTP) declines exponentially and is no longer manifest after the fifth and sixth posttetanic twitch. The phenomenon of PTP may reflect a transient increase in the level of calcium in the cytoplasm during the tetanization.

Contracture

Agents which lead to a sustained elevation of cytosolic calcium (by promoting release of calcium or by inhibiting reaccumulation of calcium by the sarcoplasmic reticulum) can induce a state of prolonged contraction *in the absence of action potentials*. This phenomenon is referred to as contracture. If muscle cell membranes are depolarized by high concentrations of K^+, a state of contracture will ensue because the depolarization causes release of calcium.

Clinically, a contracture analogous to pharmacological contracture is produced in McArdle's phosphorylase deficiency, a condition characterized by an inability of muscle to relax after strenuous exercise. The reason for this appears to be that the muscle cannot utilize glycogen and produce ATP, which is necessary for calcium uptake by the sarcoplasmic reticulum.

ACTIVATION OF MUSCLE

Neuromuscular Transmission

The activity of muscle is controlled by the central nervous system through the motor innervation of the muscle fibers. Each motor nerve fiber splits up into a number of branches that make contact with the surface of the individual muscle fibers via several bulb-shaped endings. These endings are arranged in a group, and with a specialized structure of the surface of the muscle fiber, they form an entity which is referred to as a *neuromuscular junction, myoneural junction,* or *motor end-plate* (Fig. 1.77).

Development of End-Plate Potentials and Activation of Skeletal Muscle Fibers

The action potential conducted along the nerve fiber enhances the endocytic release into the neuromuscular "cleft" of acetylcholine (ACh) from packets (vesicles) located in the nerve endings. A small number of vesicles release their contents intermittently from unstimulated nerve endings, thereby accounting for the quantified subthreshold *miniature end-plate potentials* (MEPP) described below and in Chapter 3. When an action potential reaches the nerve terminals in the region of a motor end-plate, there is an enhanced permeability to Ca^{2+} ions which increases the exocytic release of ACh from several hundred vesicles at the presynaptic (prejunctional) membrane—so that the number of ACh molecules that diffuses across the junctional gap to react with specific ACh receptor protein in the postjunctional membrane equals or exceeds the threshold amount needed for induction of an action potential in the muscle fiber. Excess ACh is rapidly inactivated via hydrolysis by acetylcholinesterase, which is present on the surface of the postjunctional membrane.

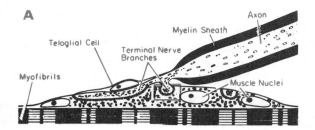

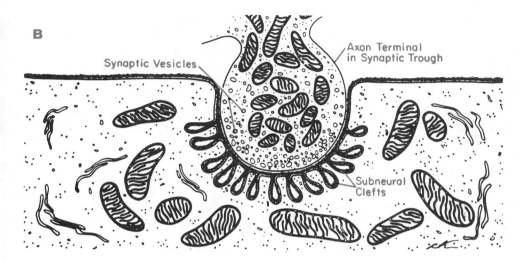

Figure 1.77. Motor end-plate for neuromuscular transmission in striated muscle. *A*, Longitudinal section. *B*, Electron microscopic appearance of the blue rectangular area indicated in *A* showing the ultrastructure of the junction between an axon terminal and the muscle fiber membrane. The invagination of the muscle fiber plasma membrane (sarcolemma) forms the synaptic trough into which the axon terminal protrudes. The space between the plasma membrane of the axon terminal and the invaginated sarcolemma is referred to as the neuromuscular cleft or synaptic gap. The invaginated sarcolemma (postjunctional or postsynaptic membrane) has many folds (subneural clefts) which greatly increase its surface area. Acetylcholine is stored in the synaptic vesicles located in the axon terminal. Acetylcholine receptor protein and acetylcholinesterase are both associated with the postjunctional membrane. (From Bloom and Fawcett, 1975.)

The receptor density is many times greater at the neuromuscular junction than at any other site on the sarcolemmal membrane. Thus the structural and chemical specificity of the ACh receptor explains why ACh is a very much more effective agonist when applied to the myoneural junction than when applied elsewhere on the sarcolemma. The innervation of skeletal muscle plays an important role in the differential sensitivity of various regions of the sarcolemma to ACh. Following a section of a motor nerve the total number of receptors is increased more than 10 fold, and the ACh receptor material, normally concentrated at the end-plate, becomes dispersed over the entire surface of the sarcolemma which now becomes equally sensitive to ACh. After the motor nerve has regenerated, however, and functional innervation of the muscle is reestablished, the accumulation of receptor decreases throughout the sarcolemma and the sensitivity of the nonjunctional portion of the sarcolemma diminishes.

The ACh receptor molecule is composed of five subunits with a molecular weight in the range of 40,000–70,000. At rest the electrical potential difference across the muscle cell membrane (i.e., the resting potential) is approximately −90 mV (inside negative). Because of the difference in concentration of Na^+ ions (high outside the muscle cell) and K^+ ions (high inside the muscle cell) and because the formation of an ACh-receptor complex results in an increase in the ionic permeability of the postjunctional membrane to small cations, the release of threshold amounts of ACh into the neuromuscular junctional cleft gives rise to a sudden influx of Na^+ ions and an efflux of K^+ ions through the plasma membrane. This short-circuits the adjacent parts of the plasma membrane and causes a drop in the membrane potential to the level where the membrane becomes electrically excited (i.e., the membrane potential rises to the critical firing level, approximately −50 mV), and a wave of depolarization spreads along the muscle fiber (in the same manner as the nerve impulse is propagated along an axon). The electrical response developed as a consequence of the acetylcholine effect on membrane permeability can be recorded as an action potential with the inside of the membrane briefly becoming positive. The conduction of one impulse involves only a minute amount of Na^+ ions (about 4×10^{-12} moles/cm^2) entering the muscle fiber and an equivalent leakage of K^+ ions to the extracellular space. Through a recovery mechanism, involving active transport processes, the ion concentration differences between the muscle cell interior and the extracellular space is reestablished by Na^+ extrusion and K^+ uptake.

The activity of acetylcholinesterase rapidly abolishes the effect of acetylcholine. The electrical response of the plasma membrane of the muscle fiber is therefore efficiently controlled by the nerve impulses reaching the motor end-plate. Even without any such impulses, there are, as noted above, minute spike potentials developed at the motor end-plate region of the muscle plasma membrane (MEPP), which are due to a spontaneous release of ACh from ACh-containing vesicles in the nerve terminals of the neuromuscular junction. Each MEPP is the result of the release of ACh from one vesicle, which contains a fixed number of ACh molecules. Hence, ACh is released in discrete packets or *quanta*. The concentration of ACh that results from the spontaneous release of ACh producing a MEPP is generally not high enough to produce a sufficient permeability change in the plasma membrane of the muscle fiber to raise the membrane potential to -50 mv, the threshold for the development of a propagated electric response. The minute spike potentials therefore remain localized to the end-plate region.

Actions of Drugs on the Neuromuscular Junction

DRUGS THAT MIMIC ACh

Drugs, such as carbamylcholine and nicotine, mimic the effect of ACh at the motor end-plate. However, because these drugs are either not inactivated or are very slowly inactivated by cholinesterase, their action is much more prolonged than that of ACh. In moderate dosage these drugs will cause a muscle spasm: in high dosage, however, they induce a state of partial or complete paralysis (depolarizing block) similar to the "desensitization" phenomenon observed when the neuromuscular junction is exposed to ACh for an extended period. This type of diminution or ablation of response has been observed in other receptor systems and is believed to be due to a gradual conversion of the transmitter-receptor complex from an active to an inactive form. The inactive complex then dissociates to generate an inactive or "desensitized" free receptor which only slowly reverts to the normal, active receptor.

DRUGS THAT BLOCK TRANSMISSION

Drugs such as *d*-tubocurarine and its congeners bind to the acetylcholine receptor with high affinity, but the curare-receptor complex does not alter the ionic permeability of the postjunctional membrane but rather prevents ACh from binding. Thus curare or *d*-tubocurarine diminishes or blocks neuromuscular transmission by acting as a competitive antagonist of ACh at the motor end-plate, an effect which can cause a fatal paralysis of the muscles needed for respiration.

DRUGS THAT INACTIVATE CHOLINESTERASE AND THEREFORE STIMULATE

Yet certain other drugs—e.g., physostigmine, neostigmine, and diisopropyl fluorophosphate (DFP)—inactivate acetylcholinesterase and thus permit ACh to accumulate within the neuromuscular cleft and in turn to stimulate neuromuscular transmission excessively. These drugs initially induce a persistent contraction (spasm) but also can cause depolarizing block at higher doses. DFP is particularly dangerous in this respect because it inactivates cholinesterase virtually irreversibly (i.e., for several weeks), whereas physostigmine and neostigmine remain bound to cholinesterase only for a few hours—after which time they dissociate from the enzyme which then becomes fully capable of hydrolyzing ACh. These latter drugs are used for the treatment of myasthenia gravis, a disease in which neuromuscular transmission is defective because of impairment of receptor function by antireceptor antibodies which react with the postjunctional membrane. The rationale for this treatment, therefore, is simply to inhibit the enzymatic action of ACh esterase so that sufficient amounts of ACh can accumulate to restore a normal level of neuromuscular transmission despite the inadequate numbers of functional receptors at myasthenic neuromuscular junctions.

Electrical Stimulation via the Sarcolemma

Although the normal path of the electrical activity is via the neuromuscular junction, it is possible to stimulate the sarcolemma directly. If the intensity of the stimulus is subthreshold, a brief local depolarization can be induced, which will reverse upon the removal of the stimulus. However, if the stimulus depolarizes the membrane to the critical firing potential (threshold), an action potential will be observed. Whereas the subthreshold response is decremental and does not propagate far from the site of stimulation, the action potential is self-regenerative and will be conducted in both directions away from the site of stimulation.

Action potentials recorded from fast skeletal muscles depolarize the membrane rapidly, overshoot, and then repolarize rapidly to a level of approximately -90 mV. At this level of membrane potential the repolarization process develops a relatively long time constant. The membrane voltage during this latter period of slower repolarization is referred to as a *negative after-potential*, which gradually returns toward the normal resting potential, also similar to the response of nerve fibers. The contribution of various sites on the sarcolemma (surface membrane, transverse tubules) to the ionic fluxes which determine the shape of the action potential has been investigated extensively, and it has been shown that depolarization of the muscle cell membrane is largely due to Na^+ ion influx in most species. Although the transverse tubules appear to be an extension of the sarcolemma, the membranes may have different conductance properties, as suggested by the analysis of the action potential recorded after the T-tubules have been destroyed with hypertonic Ringer's glycerol solution and the analysis of membrane currents after voltage clamping.

The usual rapid repolarization that follows a spike generated by the muscle membrane is due to a delayed rectification of the membrane. Inactivation of the conductance of Na^+ ions and increase in the conductance of K^+ ions are responsible for the quick phase of repolarization, following which there is an early and late negative after-potential, with the membrane remaining slightly depolarized for about 0.25 s.

The origin of the fast phase of repolarization (delayed rectification) seems to be on the surface membrane, whereas the after-potential seems to originate mostly in the transverse tubules. It has been shown that if the transverse tubules are destroyed by exposure to hypertonic solutions, muscle will no longer contract in response to stimulation (excitation-contraction uncoupling); in addition, the total membrane capacitance is reduced from 6.5 to 2.6 $\mu F/cm$ and the after-potentials will disappear, although the earlier features of the action potential (spike and the rapid phase of repolarization) persist.

The after-potentials may be due to K^+ ion accumulation in the tubules. Using rubidium in place of potassium, it has been observed that the after-potentials can be suppressed.

The contribution of Cl^- ions to the ionic currents of the muscle cell membrane has long been controversial. It is now known that chloride conductance (G_{Cl}) is pH-sensitive. At values of pH of about 5.6, G_{Cl} is inhibited. Under such conditions it has been shown that G_K in the tubules changes in direction during the after-potential; this anomalous rectification seems to be a property of the surface membrane; whereas the slow rectification (which accounts for the return from maximal hyperpolarization to the resting potential at the end of the

action potential cycle) seems to take place within the tubules (which occupy a total of 0.002–0.003% of the fiber volume).

Recapitulation of the Role of Mono- and Divalent Ions

1. K^+ ion, the major component of the intracellular fluid, functions chiefly to maintain the resting potential, which approximates the K^+ equilibrium potential. If the concentration of K^+ in the extracellular fluid is increased, resting membrane potential will decline according to the Nernst equation $E = -RT/F \ln [K_I]/[K_E]$, where K_I is the intracellular K^+ concentration and K_E is the extracellular K^+ concentration.

2. Na^+ ion, the major component of the extracellular fluid, functions to maintain the osmotic pressure (osmolality) of the extracellular compartment. The permeability of the muscle cell membrane to Na^+ is very low at rest but increases markedly following stimulation. Thus, Na^+ ions "rush" into the cell, and these positive charges depolarize the membrane. Because Na^+ permeability is voltage-dependent, more Na^+ ions "rush" into the cell as depolarization progresses in what amounts to an autocatalytic, positive feedback process which is propagated along the muscle membrane. This self-regenerative process is qualitatively identical to the action potential of nerve. For a brief instant the inside is positive with respect to the outside. At this point (the peak of the overshoot) the membrane potential approaches the Na^+ equilibrium potential. However, the increased Na^+ permeability quickly reverses, permeability to K^+ increases, and the repolarization process is initiated.

3. The role of Ca^{2+} ions in muscle physiology will be discussed below in connection with the relaxation of muscle and under the heading of "Excitation-Contraction Coupling."

4. Mg^{2+} ions are important in many enzyme reactions, including those of glucose metabolism and myosin ATPase activity. The cytoplasmic concentration of Mg^{2+} has been estimated to be in the millimolar range.

5. Cl^-, the major anion in the extracellular fluid, passively follows the cations moving across the muscle cell membrane distributing according to the Gibbs-Donnan equilibrium.

Regulatory Role of Calcium

Ca^{2+} is needed to release the inhibition of ATP hydrolysis by myosin in the presence of regulated actin. In the absence of Ca^{2+} (intracellular concentration less than 0.1 μM), tropomyosin sterically blocks the attachment of the myosin head to the actin according to one hypothesis (Fig. 1.78). This steric hindrance by the elongate tropomyosin molecule occurs along the entire length of the F-actin. In addition, at each seven actin monomers of F-actin, the TN-I is bound to both actin and tropomyosin in the absence of Ca^{2+}. In the presence of Ca^{2+}, when Ca^{2+} binds to the TN-C, there is a conformational change in the troponin complex whereby the TN-I dissociates from the actin and the tropomyosin. This allows the tropomyosin to move closer to the groove of the two-stranded actin helix. The tropomyosin now no longer blocks the interaction of myosin with actin, and muscle contraction occurs. Recently acquired evidence suggests that troponin-tropomyosin in the absence of Ca^{2+} prevents only the strongly attached state of myosin but not the weakly attached state (compare Fig. 1.67).

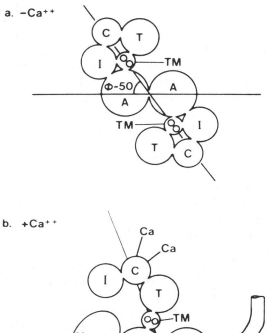

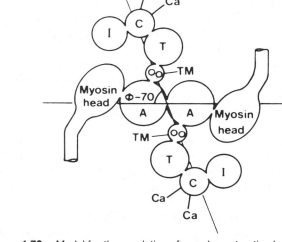

Figure 1.78. Model for the regulation of muscle contraction by Ca^{2+}. Abbreviations are as follows: *A*, actin; *TM*, tropomyosin; *I*, troponin I; *C*, troponin C; *T*, troponin T. (Reprinted from McCubbin and Kay, 1980. Copyright 1980 American Chemical Society; adapted from Potter and Gergely, 1974.)

Excitation-Contraction Coupling

From the foregoing it is clear that the stimulation of the muscle sets up a process which starts on the sarcolemma and ends in the contractile machinery. Electron micrographs have shown that the closed end of the T-tubules are in close proximity (5 nm) to the terminal cisternae. Evidence by Chandler and Schneider (1976) suggests that when an action potential reaches the T-tubules, a charge transfer occurs which in some manner leads to a conformational change in the membrane of the cisternae and thereby to an opening of ion channels. Ionic currents, thus generated, are presumed to induce the release of Ca^{2+} ions from the cisternae into the sarcoplasm.

Release of the Ca^{2+} into the sarcoplasm has been demonstrated by injecting aequorin into muscle. This substance (which can be extracted from jelly fish) luminesces in the presence of Ca^{2+} ions with an intensity that is proportional to the Ca^{2+} ion concentration. After stimulation, luminescence increases, reaches a maximum during the course of tension development, and decreases to a minimum by the time peak contractile tension is developed.

The entire process of excitation-contraction can be abbreviated as follows: Stimulus → Depolarization of the sarcolemma → Action potential initiated and propagated along the T-tubules → Calcium released from the SR system → Calcium ions diffuse and attach to the binding sites on troponin C → Inhibitory effect of troponin I on the interaction of actin and myosin is removed → thin filaments slide along thick filaments shortening the sarcomere. This series of events is fully reversible under normal conditions. Following repolarization there is an increase in the uptake of calcium ions by the sarcoplasmic reticulum. As this sequestration proceeds, more Ca^{2+} ions dissociate from their binding sites on troponin C resulting in a restoration of acto-myosin-inhibiting property of troponin I and, accordingly, in a return of the thick and thin filaments to the resting state (relaxation). Indeed, there is evidence that the rate of muscle relaxation is proportional to the removal of calcium from the contractile protein.

HEAT PRODUCTION DURING THE CONTRACTION-RELAXATION CYCLE

Muscle contraction is associated with heat production, a sign of the chemical and physical events associated with contraction. The generation of heat by muscle can be measured by an elaborate system of thermocouples coupled in series to form a thermopile.

In 1920, Hill and Hartree described four different heats in an isometrically contracting muscle: initial heat, maintenance heat, relaxation heat, and recovery heat. Since these early studies our understanding of the basic metabolic processes underlying the production of heat during the contraction cycle has vastly expanded and the original terminology has been modified to reflect modern concepts. Therefore the precise definitions of the original terms for the different heats will not be given; the general sense of these terms will become apparent in the following discussion of muscle energetics.

The initial, rapid liberation of heat during an isometric twitch or tetanus has been called *activation heat* by some muscle physiologists. It was associated with the development of the active state and is derived from initial processes other than cross-bridge cycling. These processes include the release of calcium from the terminal cisternae and binding of calcium to sarcoplasmic proteins, primarily troponin and parvalbumin, which is found in fish, frogs and other amphibians, and to a lesser extent in mammalians.

Hill and Hartree had originally observed a steady rate of heat production in an isometric tetanus following the initial heat production and called it "maintenance heat." Later investigators divided heat production during an isometric tetanus into a *labile maintenance heat* and a *stable maintenance heat*. The labile heat is ascribed to processes independent of filament overlap (such as the binding of calcium to sarcoplasmic proteins), and the stable maintenance heat primarily to cross-bridge turnover and reuptake of calcium by the sarcoplasmic reticulum.

When muscle shortens, there is an increase in the cycling rate of the cross-bridges as well as an increase in heat production. A. V. Hill stated that the *"shortening heat"* is the additional heat produced during shortening above the "maintenance heat" produced by an isometrically contracting muscle. Based on thermodynamic principles, the rate of energy liberation during muscle shortening must equal the sum of the heat produced per unit of distance shortened and the work done per unit of time.

Resting muscle has rubber-like elasticity, that is, it shortens when warmed and lengthens when cooled. In activated muscle, the elasticity is no longer rubber-like but spring-like. The change in elastic properties as the muscle contracts is associated with the absorption of heat. An equivalent amount of heat, called *thermoelastic heat*, is released during muscle relaxation. Thus no net loss or gain of heat occurs as a result of changes in muscle elasticity during a full contraction-relaxation cycle. A correction for thermoelastic heat, however, must be made in measurements involving less than a complete cycle. Thermoelastic heat released during muscle relaxation can account for the *"relaxation heat"* originally described by Hill and Hartree.

The shortening heat that is obtained in a rapidly shortening muscle after making the appropriate corrections is found to contain a significant amount of "unexplained heat." Unexplained heat is heat that cannot be accounted for by the hydrolysis of high-energy phosphates (see below). An interesting problem in muscle physiology concerns the relationship of this unexplained heat production during muscle shortening to cross-bridge intermediates such as are shown in Figure 1.67. The unexplained heat produced during rapid shortening might be attributed to rapid detachment and reattachment of cross-bridges to the thin filaments in the absence of ATP hydrolysis with the liberation of heat (e.g., step 12 and the reverse reaction in Fig. 1.67).

The energy for muscle contraction is derived from chemical reactions in the muscle fiber. In "glycolytic" muscles, the main energy source is glycogen. This energy is made available without any oxygen being consumed, even when oxygen is present. The oxidative (aerobic) chemical reactions are therefore not directly associated with muscle contraction but with the recovery processes which operate to provide energy in a form that is readily available to the contractile machinery. Heat liberated as a result of these recovery processes was called *"recovery heat"* by Hill and Hartree (1920).

ENERGY-GENERATING SYSTEMS

Adenosine Phosphates

The compound that appears to be ultimately involved in providing energy to the contractile machinery is adenosine triphosphate. In this process, energy is utilized in a reaction coupled with the hydrolytic splitting of the terminal phosphate group from ATP:

$$ATP \rightarrow ADP + P_i$$

This reaction involves a liberation of about 7,300 cal/mole of ATP. The bond between the terminal phosphate group and the neighboring phosphate group is often referred to as a "high energy" or "energy-rich" bond designated by the symbol $\frown$. It must be realized when using these terms that the energy is contained in the molecule as a whole in the case of so-called "high energy" compounds, such as

ATP. Much of the free energy is released when these compounds are hydrolyzed. The structure of adenosine triphosphate can be written as follows:

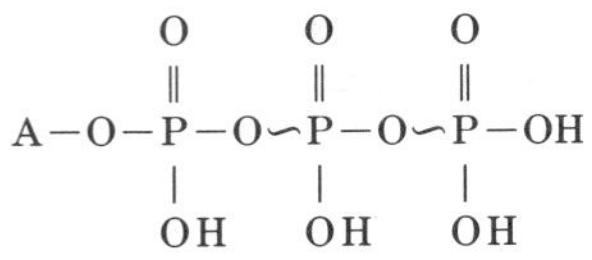

$$
\begin{array}{ccccc}
 & O & & O & & O \\
 & \parallel & & \parallel & & \parallel \\
A-O-P & -O- & P & -O \sim & P & -OH \\
 & | & & | & & | \\
 & OH & & OH & & OH
\end{array}
$$

The ATPase activity of myosin permits the hydrolysis of ATP to take place at the contractile machinery itself where energy is transferred from ATP to the myosin-actin system to allow muscle contraction.

The restoration of ATP involves phosphorylation of ADP. Prompt resynthesis of ATP is achieved by phosphate group transfer from creatine phosphate to ADP, catalyzed by creatine kinase (Fig. 1.79). Creatine phosphate, depleted in this process, is ultimately regenerated by the transfer of phosphate from ATP to creatine. In this way a second high-energy compound, creatine phosphate, is built up in the muscle during recovery. The ATP and the creatine phosphate represent energy stored in a form that can be utilized much more rapidly than would be possible if ATP were to be replenished by the slower process of oxidative phosphorylation.

Since the reactions associated with the contraction, the hydrolysis of ATP, and the transfer of phosphate from creatine phosphate to ADP are nonoxidative processes, it becomes obvious that no oxygen is consumed during contraction provided that no recovery metabolism is maintained in parallel. Oxygen is consumed during the recovery phase in connection with oxidative phosphorylation of ADP.

The energy required for the formation of the high energy ATP and creatine phosphate is supplied by glycolysis and by the oxidation of acetyl CoA (derived from pyruvic acid) in the citric acid cycle with the associated oxidative phosphorylation by the respiratory chain. Glucose metabolism via the pentose shunt pathway is negligible in skeletal muscle. In addition to glycogen, fatty acids and amino acids can furnish fuel for the muscle, and in resting muscle, glycogen breakdown is not required to maintain metabolism. Muscle metabolism can be maintained by the uptake of carbohydrates and other compounds from the blood. In resting muscle, carbohydrates are responsible for about 60% of the energy requirements of the tissue.

Glycogenesis

The synthesis of glycogen from glucose is called glycogenesis. Muscle and liver cells are able to store large amounts of glucose in the form of glycogen. The glycogen is stored in complexes called *glycogen particles*, which are found in the cytoplasm surrounding the myofibers. In addition to the stored glycogen, the glycogen particles contain the enzymes necessary for glycogen synthesis and breakdown to hexose. The first step in glycogenesis involves the phosphorylation of glucose to glucose 6-phosphate, a reaction catalyzed by *hexokinase* and associated with utilization of energy-rich phosphate in the form of ATP.

$$
\text{Glucose} + \text{ATP} \xrightarrow{\text{Mg}^{2+}} \text{glucose 6-phosphate} + \text{ADP}
$$

The product, glucose 6-phosphate, is trapped within the cell as cell membranes are poorly permeable to the compound. It is converted to glucose 1-phosphate by the enzyme *phosphoglucomutase*, which requires Mg^{2+} and glucose 1,6-diphosphate as cofactors (Fig. 1.80). Glucose 1-phosphate is then activated by reacting with uridine triphosphate (UTP) to form UDP-glucose in a reaction catalyzed by *UDP-glucose pyrophosphorylase*. The glucose moiety of UDP-glucose is transferred by the enzyme *glycogen synthase* (UDP-glucose glycogen-transglucosylase) to a preexisting glycogen molecule (primer) to which it is attached in a 1,4-linkage involving the 1 carbon and 4 carbon of adjoining glucose molecules. The formation of such bonds

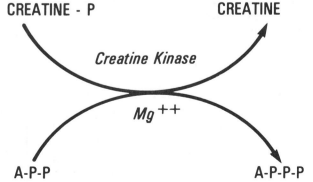

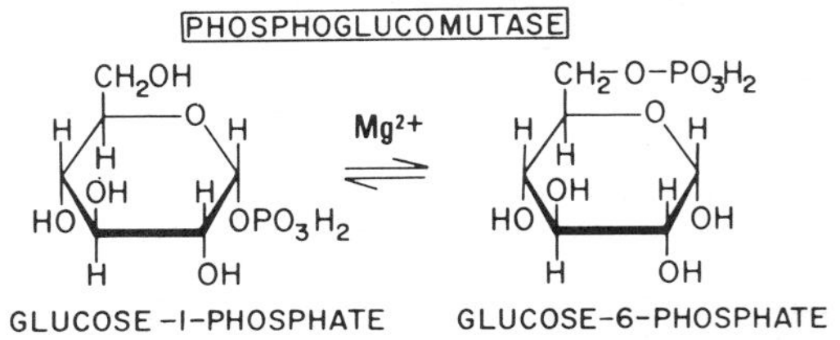

CREATINE - P　　　　　　　　**CREATINE**

Creatine Kinase

Mg^{++}

A-P-P　　　　　　　　**A-P-P-P**

Figure 1.79. Resynthesis of ATP with phosphate from creatine phosphate.

PHOSPHOGLUCOMUTASE

GLUCOSE –1–PHOSPHATE　　　GLUCOSE–6–PHOSPHATE

Figure 1.80. Interconversion of glucose phosphates by the enzyme phosphoglucomutase.

gives rise to linear chains of glucose residues. At intervals, chains of 8–12 glucose residues are picked up and transferred to make a branch which is attached to the main chain of glucose residues by means of a 1,6 linkage. This reaction is catalyzed by the *"branching enzyme"* (amylo-1,4-1,6 transglucosylase). In this way, a branch point is established, and the glycogen molecule grows to a large, highly branched structure (Fig. 1.81), with a molecular weight that ranges from 240,000 to 10^7.

The rate of glycogen synthesis is controlled by a regulatory control mechanism. The rate-limiting enzyme in this process is glycogen synthase, which exists in two forms called glycogen synthase D (D, dependent) and glycogen synthase I (I, independent). Glycogen synthase D, the less active form, is dependent on glucose 6-phosphate for activity, whereas glycogen synthase I, the more active form, is independent. In order for glycogen synthase D to be converted to glycogen synthase I, it may be dephosphorylated by a *glycogen synthase phosphatase*. Conversely, for conversion of glycogen synthase I to glycogen synthase D, the latter must be phosphorylated by cyclic AMP-dependent protein kinase. The enzyme is activated when cellular cyclic AMP concentrations are increased as a result of the stimulation of adenylate cyclase by epinephrine. Epinephrine is released from the adrenals in response to stress when energy demands by the muscle may be increased, as during the "fight or flight" response. At such a time it would be inappropriate for the muscle to store available glucose in the form of glycogen when it is needed for energy.

Glycogenolysis

Glycogenolysis is the breakdown of glycogen to glucose. During glycogenolysis (Fig. 1.82), glycogen is first partially degraded to glucose 1-phosphate; this degradation is catalyzed by *glycogen phosphorylase*, an enzyme whose activity is confined to the 1,4 linkages. All of the residues on the branches except for the proximal residue are transferred to

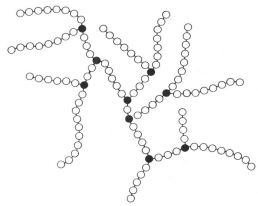

Figure 1.81. Part of a branched glycogen molecule with each circle representing one glucose residue. *Solid circles* represent branch points of glucose residues whose 6-carbon is joined in 1,6-linkages of glucose residues, whereas *open circles* represent linear chains of glucose residues consisting of 1,4-linkages.

the main chain by the debranching enzyme. The 1,6 linkage involving the proximal residue is then split by *amylo-(1,6)-glucosidase*, releasing this branch point residue as free glucose. Thereafter the remaining molecule can be degraded completely by glycogen phosphorylase.

Glucose 1-phosphate is converted to glucose 6-phosphate in a reaction catalyzed by *phosphoglucomutase*, the same enzyme utilized in glycogenesis (see above). The glucose 6-phosphate enters the main pathway of glycolysis to form pyruvic acid followed by either complete oxidation via the citric acid cycle and its associated oxidative phosphorylation system or conversion to lactic acid under anaerobic conditions. The lactic acid is then reconverted to pyruvic acid and metabolized further when O_2 is again available to the muscle or is metabolized in the liver. Glucose 6-phosphate cannot be converted to free glucose in muscle tissue because it, unlike the liver, kidney, and small intestine, lacks the enzyme glucose 6-phosphatase necessary for this conversion.

The rate-limiting step in glycogenolysis is at the level of glycogen phosphorylase, and an elaborate

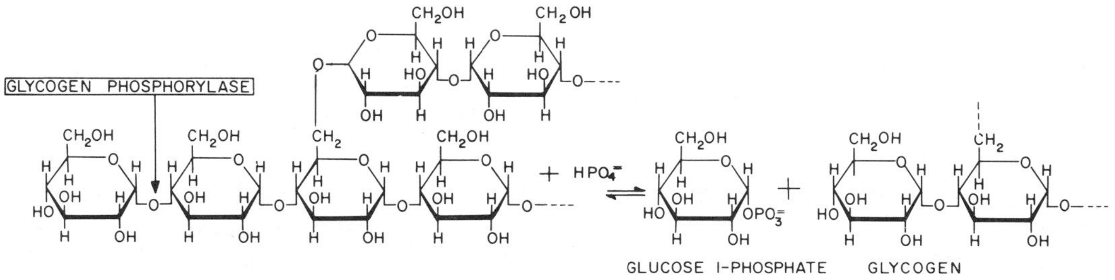

Figure 1.82. Glycogenolysis. Glycogen is first degraded by glycogen phosphorylase, which acts at 1,4-linkages.

control mechanism exists. Glycogen phosphorylase exists in an active form called *phosphorylase a* and a less active form called *phosphorylase b*. These two forms of phosphorylase differ in several respects. Phosphorylase *b* has a requirement for AMP, whereas phosphorylase *a* does not. Phosphorylase *a* exists as a tetramer which has been phosphorylated on serine residues by another enzyme called *phosphorylase b kinase* (see below). When these phosphate groups are removed by the action of an enzyme called *phosphorylase phosphatase*, there is a rearrangement of the enzyme to a dimeric form. Phosphorylase *b* is the enzyme form present in resting muscle.

Phosphorylase *b* kinase requires Ca^{2+} for activation. The Ca^{2+} requirement reflects the fact that calmodulin, a calcium-binding protein, is a subunit of phosphorylase *b* kinase. During muscle activation when cytosolic Ca^{2+} rises, phosphorylase *b* kinase becomes activated and thereby increases glycogenolysis, which supplies energy for contraction. Additional activation of phosphorylase *b* kinase may occur when it too, like phosphorylase *b*, is phosphorylated, in this case by cyclic AMP-dependent protein kinase, which can also be called phosphorylase *b* kinase kinase when phosphorylase *b* kinase is its substrate. When the phosphorylase system is activated by phosphorylation, the synthase is inactivated by the very same process. This biochemical switch permits glycogen breakdown to occur without the concurrent competition of glycogen synthesis.

The scheme in Figure 1.83 summarizes the pathways of glycogenesis and glycogenolysis.

Glycolysis

Glycolysis is the degradation of glucose to pyruvic acid. It occurs in the sarcoplasm, and the enzymes involved are part of the easily extracted soluble proteins of muscle. In glycolysis, also known as the *Embden-Meyerhof pathway* of glucose metabolism, two molecules of ATP are formed per molecule of glucose degraded.

$$\text{Glucose} + 2\ \text{ADP} + 2\ \text{P}_\text{i} \rightarrow 2\ \text{pyruvic acid} + 2\ \text{ATP}$$

The conversion of free glucose to glucose 6-phosphate is bypassed when glycolysis follows glycogenolysis because glucose 6-phosphate is then formed directly from glucose 1-phosphate in a reaction not involving utilization of phosphate-bond energy. The net gain of glycolysis will, therefore, in this case, be 3 moles of ATP per mole of glucose residues in glycogen. When free glucose is derived from the bloodstream rather than by glycogenolysis

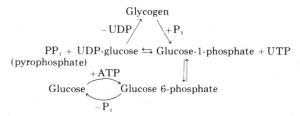

Figure 1.83. Summary of pathways of glycogenesis and glycogenolysis.

within the muscle, it is converted to glucose 6-phosphate by hexokinase in the reaction described under Glycogenesis.

Hexokinase and numerous other enzymes in glucose metabolism require Mg^{2+} or other metals as cofactors for activity. These cofactors are indicated in Figure 1.84, which summarizes the reaction steps in glycolysis.

Glucose 6-phosphate is converted to fructose 6-phosphate in a reaction catalyzed by *phosphoglucose isomerase*. A second phosphate group is introduced at the 1-carbon of fructose 6-phosphate to yield fructose 1,6-diphosphate. The phosphate group is transferred from ATP, and the transfer is catalyzed by the enzyme *phosphofructokinase*. Since 1 mole of ATP is hydrolyzed per mole of fructose 1,6-diphosphate formed, this reaction involves expenditure of energy.

In the next step, the 6-carbon compound fructose 1,6-diphosphate is split between carbons 3 and 4 to yield two 3-carbon compounds, dihydroxyacetone phosphate and D-glyceraldehyde 3-phosphate. The enzyme *aldolase* catalyzes this cleavage.

The two triosephosphates formed are interconvertible by isomerization, and dihydroxyacetone phosphate is converted to D-glyceraldehyde 3-phosphate in a reaction catalyzed by *triosephosphate isomerase*.

D-Glyceraldehyde 3-phosphate is then converted to 3-phosphoglyceric acid. The formation of 3-phosphoglyceric acid consists of an oxidation of an aldehyde to an acid and takes place in the following steps:

(1) 3-Phosphoglyceraldehyde + NAD^+ + P_i → 1,3-diphosphoglyceric acid + NADH, catalyzed by *phosphoglyceraldehyde dehydrogenase*, and
(2) 1,3-Diphosphoglyceric acid + ADP → 3-phosphoglyceric acid + ATP, catalyzed by *phosphoglyceric acid kinase*.

It is seen that the energy of the oxidation of the aldehyde of 3-phosphoglyceraldehyde to the carboxylic acid of 3-phosphoglyceric acid has been preserved in the formation of an anhydride bond in the 1-position of 1,3-diphosphoglyceric acid and uti-

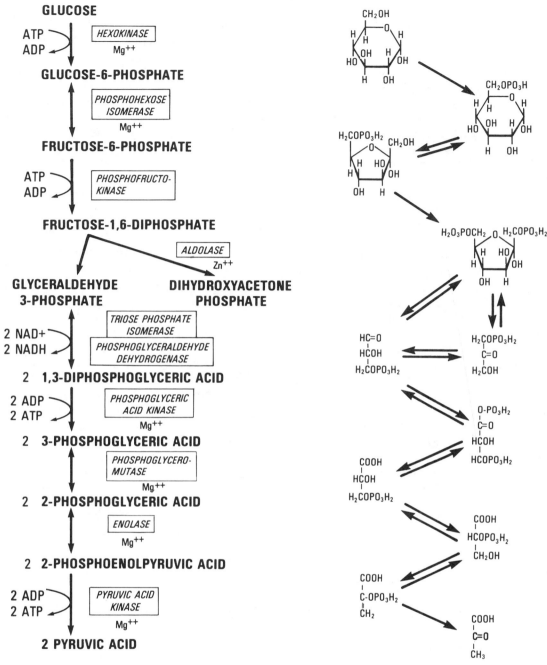

Figure 1.84. Glycolysis. All steps are reversible except the conversions of glucose to glucose 6-phosphate, of 2-phosphoenol pyruvic acid to pyruvic acid, and of fructose-6-phosphate to fructose-1,6-diphosphate.

lized to form ATP. Intact sulfydryl groups in phosphoglyceraldehyde dehydrogenase are critical for enzyme activity. The inhibition of glycolysis by iodoacetate is due to its reaction with these sulfhydryl groups. This is an example of the coupling of an energy-releasing oxidation with phosphorylation of ADP in order to retain energy in a chemically available form, thus preventing it from being dissipated as heat. Oxidation of NADH is also coupled to

ADP phosphorylation and yields 3 moles of ATP per mole of NADH (see below).

The 3-phosphoglyceric acid is transformed into 2-phosphoglyceric acid in an intermediate step in which 2,3-diphosphoglyceric acid is utilized as a cofactor. The overall reaction is catalyzed by *phosphoglyceromutase* and is similar in mechanism to that involving glucose 1-phosphate and glucose 6-phosphate.

2-Phosphoglyceric acid is converted by dehydration to phosphoenolpyruvic acid by *enolase*. Fluoride, a potent inhibitor of glycolysis, acts on this enzyme. Phosphoenolpyruvic acid is a high-energy compound. A transfer of its phosphate group to ADP yields one molecule of ATP plus pyruvic acid in a reaction catalyzed by pyruvic acid or *pyruvate kinase*. This is the last reaction during glycolysis that can furnish energy for the synthesis of ATP.

NADH formed as a result of conversion of 3-phosphoglyceraldehyde to 3-phosphoglyceric acid can be oxidized to NAD^+ in one of two ways. In the presence of O_2, the NADH is oxidized by a transfer of electrons to the cytochrome chain in the mitochondria. This process is described further under Oxidative Phosphorylation. In the absence of O_2, NADH oxidation is coupled to lactic acid formation. The coupling occurs in contracting muscle with normal blood supply when muscle activity exceeds a certain intensity and O_2 becomes limiting relative to demand. Lactic acid can thus be formed under both aerobic and anaerobic conditions. The amount of lactic acid formed represents an *oxygen debt* which can be repaid during the recovery period of the muscle. Alternatively, if the muscle contraction continues, lactic acid is released into the bloodstream and carried to the liver where it is converted to glucose 6-phosphate via gluconeogenesis. The glucose is then released by the action of the liver's glucose 6-phosphatase to be used as energy by the muscle. The conversion of lactic acid to glucose by liver which is returned to the muscle is called the *Cori cycle*.

The coupling of the reactions leading to restoration of NAD^+ can be summarized as shown in Figure 1.85. No oxygen, however, is consumed during glycolysis, and the presence of oxygen is not required.

The main control point in glycolysis occurs at the level of phosphofructokinase, which is a rate-limiting enzyme. The activity of this enzyme is increased by fructose 6-phosphate, ADP, AMP, and inorganic phosphate. Recently, fructose 2,6-diphosphate has been found to be a powerful physiological regulator, producing stimulation. Epinephrine also activates phosphofructokinase because it stimulates the production of cyclic AMP which, in turn, helps activate an enzyme catalyzing the production of the potent activator fructose 2,6-diphosphate. On the other hand, ATP at high concentrations and phosphocreatine inhibit phosphofructokinase activity. A particularly important inhibitor is citrate, which is produced by aerobic metabolism via the citric acid cycle. Therefore, when oxygen is present, glycolysis is slowed reflecting the more efficient production of ATP by oxidative metabolism.

Citric Acid Cycle

The citric acid cycle (tricarboxylic cycle or Krebs cycle), occurring in the mitochondria, represents a cyclic sequence of chemical reactions by which acetic acid bound to coenzyme A (CoA) to form acetyl coenzyme A is oxidized to CO_2 and H_2O and CoA is re-formed (Figs. 1.86 and 1.87). Pyruvic acid is oxidized in a series of reactions involving several enzymes and four cofactors which form a complex constituting *pyruvic acid dehydrogenase*. The net effect of these reactions is summarized in Figure 1.88.

The acetyl CoA is an energy-rich compound and, in addition to glycolysis, can also be formed from fatty acids and certain amino acids. The citric acid cycle, therefore, represents a common pathway in the oxidative degradation of carbohydrates, fats, and proteins. Glucose and fatty acid metabolism, however, are the main sources of acetyl CoA.

In the citric acid cycle, acetyl CoA reacts with the 4-carbon oxaloacetic acid to form the 6-carbon citric acid, a reaction catalyzed by citrate synthase, formerly referred to as the *"condensing enzyme."* Cit-

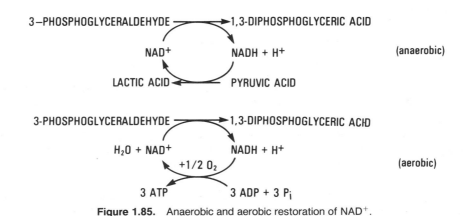

Figure 1.85. Anaerobic and aerobic restoration of NAD^+.

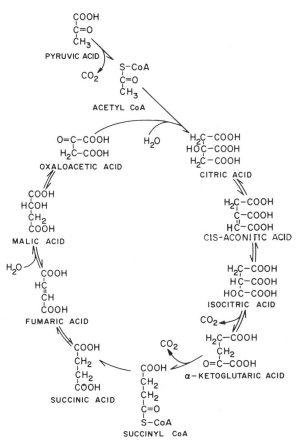

Figure 1.86. The citric acid cycle.

ric acid is converted to its isomer, isocitric acid. Oxidative decarboxylation of isocitric acid results in the formation of α-ketoglutarate (plus CO_2) in a series of reactions analogous to the pyruvate dehydrogenase reactions (see above). Succinyl CoA (plus CO_2) is then formed from α-ketoglutarate and CoA. The energy of the thioester bond in succinyl CoA is used to form guanosine 5'-triphosphate (GTP) from guanosine 5'-diphosphate (GDP) in a reaction producing free succinate and free CoA. Succinate is oxidized to form fumarate and reduced flavin adenine dinucleotide ($FADH_2$). Hydration of fumarate to malate and finally oxidation of malate to form oxaloacetate complete the cycle. One molecule of the 2-carbon acetyl fragment is oxidized during the cycle, with two molecules of CO_2 being produced and four hydrogen atoms removed. The next exchange during one cycle is shown in the following scheme:

$$CH_3COOH + 2\,O_2 \rightarrow 2\,CO_2 + 2\,H_2O + energy$$

The five energy-yielding reactions involve three steps in which NADH is produced, one step in which $FADH_2$ is produced and one substrate-level phosphorylation, where GTP is produced and subsequently converted to ATP by a kinase.

(1) Isocitrate $+$ NAD^+ $\rightarrow$ α-ketoglutarate $+$ CO_2 $+$ NADH $+$ H^+
(2) α-Ketoglutarate $+$ NAD^+ $+$ CoA $\rightarrow$ succinyl CoA $+$ CO_2 $+$ NADH $+$ H^+
(3) Succinyl CoA $+$ GDP $+$ P_i $\rightarrow$ succinate $+$ CoA $+$ GTP
(4) Succinate $+$ FAD $\rightarrow$ fumarate $+$ $FADH_2$
(5) Malate $+$ NAD^+ $\rightarrow$ oxaloacetate $+$ NADH $+$ H^+

Oxidative Phosphorylation

The reactions in the citric acid cycle are coupled to the respiratory chain, consisting of a sequence of hydrogen and electron carriers also located within the mitochondria. These reactions are shown in Figure 1.87. The respiratory chain is the site of the oxidative phosphorylation of ADP to ATP.

The electron carriers are iron-containing hemoproteins or cytochromes, which transfer electrons by alternating between reduced and oxidized states, the iron switching between ferrous and ferric forms (Fig. 1.89). An intermediate step involving the removal of hydrogen from NADH by NADH dehydrogenase (which is a flavoprotein) precedes the passage of electrons along the series of cytochromes in the carrier chain. The flavoprotein is oxidized by ubiquinone (coenzyme Q, CoQ), a lipid-soluble quinone acting as a coenzyme. Ubiquinone, which can drift laterally in the membrane, shuttles redox equivalents between $FMNH_2$ and cytochrome b (Cyt b) and between $FADH_2$ and Cyt b, all of which are membrane bound. In the step involving oxidation of ubiquinone, the electron from hydrogen is transferred from ubiquinone to the first cytochrome in the cytochrome series, Cyt b, and the proton is donated from the membrane to the medium. Electrons are now passed along the chain of cytochromes, involving in addition to cytochrome b, cytochrome c_1 (Cyt c_1), cytochrome c (Cyt c), cytochrome a (Cyt a), and cytochrome a_3 (Cyt a_3). The complex consisting of Cyt a plus Cyt a_3 is called cytochrome oxidase, which can react directly with oxygen. When oxygen is reduced by cytochrome oxidase, hydroxyl ions are formed that combine with the protons generated at the oxidation of ubiquinone to form water. The net effects of the electron transport along the respiratory chain are illustrated in Figure 1.87.

During the electron transfer along the respiratory chain, starting with NADH, three ADP molecules are phosphorylated to yield three ATP molecules per atom of oxygen. This is expressed as a P/O ratio of oxidative phosphorylation of 3. The energy for the formation of ATP is derived at three points

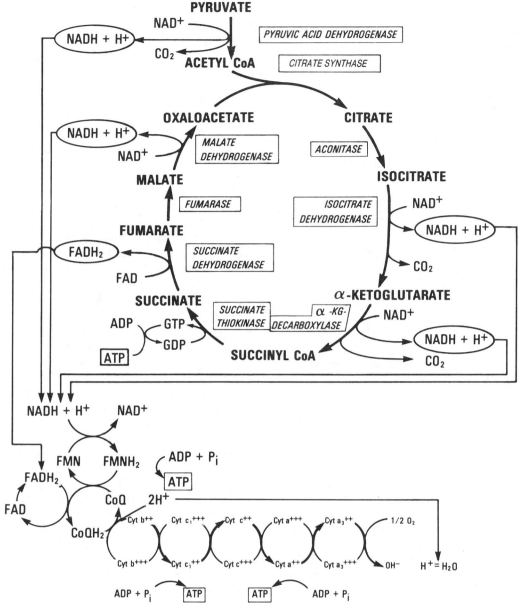

Figure 1.87.   The citric acid cycle and the electron transport chain.

along the respiratory chain; these three points are considered to be in the segment between NADH and cytochrome b, between cytochromes b and c, and at the oxidation of cytochrome c by *cytochrome oxidase*. The minimum difference in oxidation-reduc-

tion potential which is required to provide about 7,300 cal/mole for phosphorylation of ADP is 0.2 V. At the fourth energy-yielding reaction in the citric-acid cycle described above, succinate is oxidized to fumarate by succinate dehydrogenase. Since the

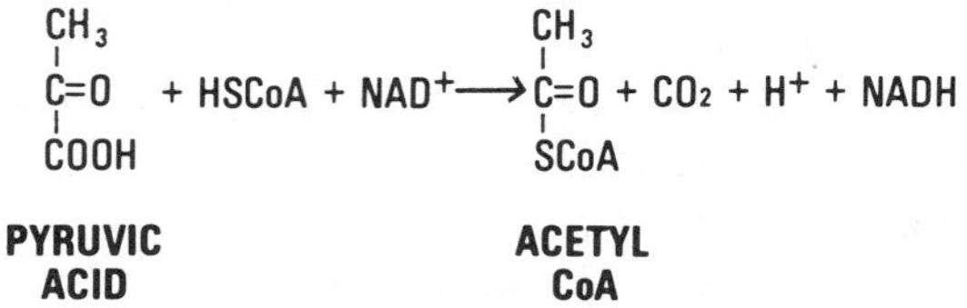

$$\underset{\substack{\text{PYRUVIC}\\\text{ACID}}}{\underset{\substack{|\\\text{COOH}}}{\overset{\substack{\text{CH}_3\\|}}{\text{C=O}}}} + \text{HSCoA} + \text{NAD}^+ \longrightarrow \underset{\substack{\text{ACETYL}\\\text{CoA}}}{\underset{\substack{|\\\text{SCoA}}}{\overset{\substack{\text{CH}_3\\|}}{\text{C=O}}}} + \text{CO}_2 + \text{H}^+ + \text{NADH}$$

Figure 1.88.   Reaction by which acetyl CoA is produced as pyruvic acid is oxidized.

A. $\text{CoQH}_2 + 2\,\text{Fe}^{+++} \longrightarrow \text{CoQ} + 2\,\text{H}^+ + \text{Fe}^{++}$

B. $\frac{1}{2}\,\text{O}_2 + \text{H}_2\text{O} + 2\,\text{Fe}^{++} \longrightarrow 2\,\text{OH}^- + 2\,\text{Fe}^{+++}$

Net. $\text{CoQH}_2 + \frac{1}{2}\,\text{O}_2 \longrightarrow \text{CoQ} + \text{H}_2\text{O}$

Figure 1.89.   Summary of reactions in which ubiquinone (CoQH$_2$), oxygen, and cytochrome oxidase interact to form water.

oxidation of NADH is bypassed, the P/O ratio is 2 instead of 3.

The exact mechanism by which the phosphorylation of ADP is coupled to the electron transport along the respiratory chain is unknown. However, an abundance of evidence has been obtained in favor of the *chemiosmotic hypothesis* first postulated by Peter Mitchell in 1961. According to this hypothesis, energy is transferred to ATP from a proton-motive gradient consisting of a charge gradient and a pH gradient across the mitochondrial membrane by an ATPase complex—i.e., by a H^+ pump that is being driven "backwards" by the gradient. If this coupling mechanism is inhibited by drugs, respiration can, under experimental conditions, proceed without the generation of ATP, indicating that the electron transport system can remain intact when the phosphorylation is uncoupled. Under physiological conditions, however, respiration and phosphorylation are tightly coupled, and the rate-limiting factors in these processes are the concentrations of ADP and inorganic phosphate. The ADP concentration, on the other hand, is determined by the rate at which ATP is utilized. This represents a self-regulatory mechanism by which the rate of energy generation is adjusted to the requirements of the cell.

The mitochondrial matrix contains the enzymes for the oxidative decarboxylation of pyruvate and the reactions of the citric acid cycle, whereas the components of the respiratory chain are arranged in proper sequence in the inner membranes of the mitochondria. Another type of constraint which may be imposed upon the mitochondrial membrane is the orientation of the binding sites of the electron carriers. For example, the binding sites for NADH dehydrogenase and succinic acid dehydrogenase are assumed to face the matrix at the surface of the membrane elements. It has been speculated that the electron transfer might be facilitated by the close arrangement of the components of the respiratory chain, which would allow transfer with small changes in orientation of the individual electron carriers.

Finally it should be noted that the mitochondria are impermeable to NADH. Therefore special energy-requiring mechanisms, called substrate shuttles, exist to transfer the reducing equivalents of NADH produced during glycolysis from the cytoplasm into the mitochondria.

Energy Yield in Glucose Oxidation

Table 1.8 summarizes the energy-yielding and energy-requiring reactions during glucose degradation. In reactions 4, 5, and 9, ATP is formed from ADP participating in the reaction. In the other reactions except reaction 10, ATP generation is a consequence of NADH oxidation in the respiratory system of mitochondria.

$$NADH + H^+ + \tfrac{1}{2}O_2 + 3\,ADP + 3\,P_i \rightarrow NAD^+ + H_2O + 3\,ATP$$

NAD is bypassed in reaction 10 and hydrogen is transferred to FAD, and ATP is formed in connection with a coupling of $FADH_2$ oxidation and ADP phosphorylation according to the following net reaction:

$$FADH_2 + \tfrac{1}{2}O_2 + 2\,ADP + 2\,P_i \rightarrow FAD + H_2O + 2\,ATP$$

In reactions 6, 7, and 8, 1 mole of CO_2 is formed for each mole of NAD formed. In glycolytic degradation of glucose equivalents derived from glycogenolysis reaction 1 is not involved, and the total number of

Table 1.8.
Energy Yield in Glucose Degradation

Reaction	ATP Yield
(1) Glucose + ATP → glucose 6-phosphate + ADP	−1
(2) Fructose 6-phosphate + ATP → fructose 1,6-diphosphate + ADP	−1
(3) 2 D-Glyceraldehyde 3-phosphate + 2 NAD^+ + 2 P_i ⇌ 2 1,3-diphosphoglyceric acid + 2 NADH + 2 H^+	+6[a]
(4) 2 1,3-Diphosphoglyceric acid + 2 ADP ⇌ 2,3-phosphoglyceric acid + 2 ATP	+2
(5) 2 Phosphoenolpyruvic acid + 2 ADP → 2 pyruvic acid + 2 ATP	+2
(6) 2 Pyruvic acid + 2 CoA + 2 NAD^+ → 2 acetyl CoA + 2 CO_2 + 2 NADH + 2 H^+	+6
(7) 2 Isocitric acid + 2 NAD^+ ⇌ 2 α-ketoglutaric acid + 2 CO_2 + 2 NADH + 2 H^+	+6
(8) 2 α-Ketoglutaric acid + 2 CoA + NAD^+ → 2 succinyl CoA + 2 NADH + 2 H^+ + 2 CO_2	+6
(9) 2 Succinyl CoA + 2 GDP + 2 P_i → 2 succinic acid + 2 CoA + 2 GTP	+2
[2 GTP + 2 ADP ⇌ 2 GDP + 2 ATP]	+4
(10) 2 Succinic acid + 2 FAD ⇌ 2 fumaric acid + 2 $FADH_2$	
(11) 2 Malic acid + 2 NAD^+ ⇌ 2 oxaloacetate acid + 2 NADH + 2 H^+	+6
Total	+38

[a]Substrate shuttles transferring the reducing equivalents of NADH from the cytoplasm into the mitochondria require energy. Therefore the actual ATP yield at step 3 is only +4 or +5.

moles of ATP formed per mole of glucose equivalents is, therefore, 39.

Total oxidation of glucose to CO_2 and H_2O yields 686,000 cal/mole, but only 56,000 cal/mole are produced when the glucose degradation does not proceed further than to lactic acid. The efficiency of glycolysis, that is, of the energy-generating reactions under anaerobic conditions, can be calculated by assuming that the phosphorylation of three ADP molecules represents about 24,000 cal/mole, as greater than 8,000 cal/mole are released when phosphate is transferred from high-energy compounds, e.g., phosphoenolypyruvate, to ADP. This is $24,000/56,000 \times 100\%$, or about 43% of the potentially available energy during glycolysis following glycogenolysis. In comparison with the energy released during complete oxidation of glucose—686,000 cal/mole—the energy stored in three molecules of ATP will represent $24,000/686,000 \times 100\%$ or about 3% of the total energy available as chemical energy in glucose. This shows the low energy yields of anaerobic energy metabolism. In comparison, 39 molecules of ADP are phosphorylated in the aerobic degradation of a glucose equivalent derived from glycogenolysis, representing a free energy storage of about 312,000 cal. The efficiency of the oxidative degradation of glucose is therefore $312,000/686,000 \times 100\%$ or about 45%, which is in good agreement with the efficiency estimated from measurements of the recovery heat.

ELECTROMYOGRAPHY

Motor Unit Potential

During the normal twitch of a muscle fiber, a minute electrical potential is generated, which is dissipated into the surrounding tissue and hence can be detected on the skin above the muscle. The duration of the action potential associated with this twitch is about 1–4 ms. Since all the muscle fibers of a motor unit do not contract at exactly the same time—some being delayed for several milliseconds—the electrical potential developed by the single twitch of all the fibers in the motor unit is prolonged. The electrical result of the motor-unit twitch, then, is an electrical discharge or motor-unit potential lasting about 5–8 ms (and often as long as 12 ms). The majority of these motor-unit potentials have an amplitude of around 0.5 mV.

Motor-unit potentials may be detected by an electromyograph. Basically, an electromyograph is a high-gain amplifier to which electrodes are connected. The electrodes are placed either on the skin above the muscle ("skin" electrodes) or into the muscle (needle or wire electrodes). When the motor-

unit potentials obtained by electromyography are displayed on a cathode ray oscilloscope, the result is a sharp spike that is most often biphasic (Fig. 1.90), but it may also have a more complex form, depending on physical and other factors.

Generally, the larger the motor-unit potential registered, the larger is the motor unit producing it. However, complicating factors, such as distance of the unit from the electrodes and the types of electrodes and equipment used, determine the final size and pattern of the individual motor-unit potential that is recorded. There are, furthermore, characteristic variations with different normal muscles, e.g., the smallness of potentials in facial muscles as compared with those in muscles of the extremity.

Clinical Application

Electromyography is most useful for distinguishing nerve from muscle disease. Typically, electromyograms are obtained (a) at rest, when normally there is no spontaneous muscle activity, (b) during slight muscle contraction in order to assess the size and duration of activity of motor units, thus enabling differentiation between myopathic and neuropathic diseases, and (c) during maximal muscle activity in order to determine if there is abnormal recruitment of motor units, which, again, can differentiate whether the problem is myopathic or neuropathic.

According to the *Henneman principle* (Henneman et al., 1974), under normal conditions, the smaller potentials appear first with a slight contraction. As the force is increased, larger and larger potentials are recruited, this being the normal pattern of recruitment. This normal pattern is absent in cases of partial paralysis due to injuries or lesions of the lower motor neuron, i.e., the small

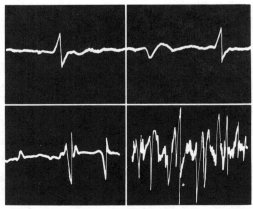

Figure 1.90. Normal electromyograms. The single potential in the upper corner has a measured amplitude of 0.8 mV+ and duration of 7 ms.

potentials never appear, apparently because only the larger motor units have survived.

Two common abnormalities that may be detected at rest when normal muscle is quiescent are fibrillation and fasciculation. *Fibrillation* is the contraction of single muscle cells which have become completely dissociated from nervous control due to destruction of a motor nerve and subsequent degeneration of the distal part of the axon by disease or wilting of the nerve. Fibrillation occurs for several weeks and then ceases as the muscle cells atrophy. Because the motor unit has been disrupted, there can be no summation of muscle fibers. Therefore, the potentials must be measured from single muscle cells with the aid of bipolar needle electrodes inserted into the muscle.

Fasciculation is the contraction of groups of muscle cells integrated by a single axon into a motor unit. Fasciculation occurs when an anterior motor neuron is destroyed, as in motor neuron disease or the axon is severed. Spontaneous impulses will arise as the distal axon atrophies. Until atrophy is complete, the muscle cells composing the motor unit will contract in response to these spontaneous impulses. These contractions are therefore strong enough to allow measurement of potentials with skin electrodes.

FUNCTION OF WHOLE SKELETAL MUSCLE

Gross Organization

Approximately 600 skeletal muscles, composed of millions of muscle fibers, account for 40–45% of the human body weight. Surrounding individual muscle fibers is connective tissue called the *endomysium,* which contains small blood vessels and nerves. Groups of muscle fibers, collected into bundles of fascicles, are bound together by a more dense layer of collagenous and elastic fibers called the *perimysium.* Finally the connective tissue which binds the fascicles into definitive muscle is called the *epimysium.* The connective tissues of all these three layers are actually in continuity with each other. At the ends of the elongated muscle, the connective tissue forms a common bundle of fibers called the *tendon.*

The arrangement of the muscle fibers in a muscle is variable (see Fig. 1.91) and has been classifed as follows:

1) Muscle fibers parallel to the long axis of the muscle (e.g., the strap-like sartorius, the pectineus, and the fusiform biceps brachii and flexor carpi radialis).
2) Muscle fibers oblique to the long axis of the mus-

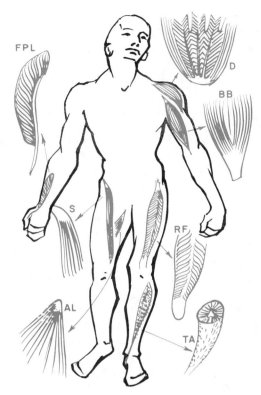

Figure 1.91.   The form and distribution of several varieties of striated skeletal muscle: the pennate, flexor pollicis longus (*FPL*); multipennate, deltoid (*D*); fusiform, biceps brachii (*BB*), bipennate, rectus femoris (*RF*); circumpennate, tibialis anterior (*TA*), radial, adductor longus (*AL*), and strap-like sartorius (*S*). (The male figure redrawn from the First Muscle Tabule in Fabrica by Vesalius.)

cle. Muscles whose fibers are arranged so that they insert obliquely into a tendon include several varieties: (1) unipennate (e.g., flexor pollicis longus); (2) bipennate (e.g., rectus femoris); (3) circumpennate (e.g., tibialis anterior); and (4) multipennate (e.g., deltoid).
3) Muscle fibers arranged in a radial or triangular shape (e.g., adductor longus, pectoralis minor).

Muscles whose component fibers are in parallel array are able to shorten over greater distances than muscles whose fibers are organized in the pennate manner. The penniform arrangement of muscle fibers is associated with those muscles which exert great force rapidly over a short distance.

In order for muscles to perform their function, they are attached, for the most part, from one bone across a joint or joints to another bone by *tendons* or flattened sheets of connective tissue called *aponeuroses.* The tendons and aponeuroses are formed of collagen, which is flexible but practically inextensible.

The points of muscular attachment are called the origin and the insertion. The *origin* of a muscle is usually the fixed or proximal attachment, while the *insertion* is the distal attachment to the bone which is moved. There are exceptions and in many cases the designation of the origin and insertion of a muscle has become a matter of anatomical convention. The force developed by an uncontrolled maximal contraction of a whole muscle is of little practical use in the movements of man and animal. Therefore, the force and movement generated in a given circumstance by a given muscle depends upon the time-integrated individual activities of its component motor units. For fine control of movement one or more motor units are employed. As more force is required the number of stimuli carried by each motor unit nerve is increased, along with the recruitment of additional motor units. The grading of muscular activity is the result, therefore, of asynchronous firing of the motor units of a whole muscle.

In considering the function of muscles in man, it should be noted that no muscle acts alone even in the simplest movements, i.e., a variety of muscles which are described as agonists, synergists, and antagonists are involved in each action. The muscle which produces the movement is termed the *agonist*, while the muscle which opposes the movement is termed the *antagonist*. The *synergists* are muscles which act together to produce a movement which no muscle could produce alone. An example of synergistic function can be seen in the adduction of the hand at the wrist. The flexor carpi ulnaris will both flex and adduct the hand. To produce adduction alone, the extensor carpi ulnaris must be brought into action to offset flexion and yet permit adduction of the hand by the flexor carpi ulnaris. Examples of antagonistic muscles are the biceps brachii and triceps brachii, which are mutually antagonistic with respect to movement at the elbow joint.

The antagonist is usually reflexly inhibited when the agonist is brought into action, e.g., if one holds a weight on the palm of the hand with the elbow flexed perpendicular to the body, the biceps brachii (in this case the agonist) develops tension while the triceps brachii (the antagonist) is flaccid. This action is reversed—the triceps being contracted and the biceps relaxed—when the hand presses down onto a table with the arm flexed. When however, the arm is extended and the elbow joint is fixed, both the biceps and the triceps are contracted. Large, translational movements may occur between muscles that exhibit antagonistic effects, such as flexors and extensors, and such muscles are also frequently separated by extensive connective tissue septa.

Not all striated muscles are associated with the movement of skeletal parts at joints or with the fixation or maintenance of joint stability as related to posture. Muscles serve as *sphincters* when they encircle an orifice (e.g., eyelid, lips, and anus). Striated muscle is found in the upper two-thirds of the esophagus in a tubular arrangement which aids in the swallowing mechanism. The diaphragm is a thin muscular sheet which upon contraction serves to enlarge the thorax and compress the abdomen, effects which are associated with inspiration. This action is reversed during expiration when the muscle relaxes. In addition, there are certain cutaneous muscles which have at least one of their attachments to the skin, e.g., the mimetic facial muscles and the panniculus carnosus of the trunk.

Force Transmission to the Skeleton

The force generated by a muscle is transmitted to the skeleton by means of its *tendons*. The submicroscopic collagen filaments of tendons extend along the surface of the individual muscle fibers at the tapering ends of the muscle fibers. The surface area of the muscle fibers is fairly large here due to longitudinal folds of the plasma membrane. There is close contact between collagen filaments and the muscle fiber surface.

The shape of a tendon varies with the shape of the muscle to which it is attached, but the muscle fibers are always oriented at an angle to the main direction of the tendon. Muscles in which the angle is relatively large are called pennate muscles. This angle, which never exceeds 10–20°, serves to prevent fraying of the tendon when the diameter of the fibers increases in connection with contraction and shortening.

A tendon may form one or several flat sheets at the region of contact with the muscle, and the muscle fibers can impinge on one or both sides of such a sheet. The length of the fibers extending between tendons is usually considerably shorter than the overall length of the muscle. The distance bridged by the muscle fibers is related to the range of shortening under physiological conditions and to the fact that under these conditions the maximal shortening of individual muscle fibers is kept within a certain limit, viz., about 30% of the normal length of the muscle in the body. This is presumably an adjustment to the relationship between the maximal tension developed by a muscle as a function of its length.

In places where a large force is required but where the range of shortening is limited, muscles are characterized by multipennate structure, with the tendons forming several septa arranged in two planes. In this case the effective cross-section of the muscle is considerably larger than the anatomical cross-section. The septa of the tendons in the human deltoid muscle were found to develop gradually during embryological growth and to be reduced in "old age." It is therefore likely that the gross structure of a muscle is dynamically adjusted to its functional requirements as is its cross-sectional area.

Types of Skeletal Muscle Fibers

It has been recognized for many years that muscles vary in color, not only from species to species but also within the same animal. Striking examples of this are the crimson red pectoralis of the pigeon as contrasted with the stark white chicken pectoralis and the red soleus lying near the white gatrocnemius in the lower limb of animals and man. Using morphological, histochemical, biochemical, and physiological properties as criteria, one is able to distinguish several distinct muscle types in the animal kingdom. The voluntary skeletal muscles of mammals consist wholly of twitch-type fibers, which can be characterized by their ability to produce (on nerve stimulation) a propagated action potential and to respond with fast shortening or tension development. Other types of muscle fibers have evolved for specific purposes in other species and include insect flight muscle (which will not be described further) and slow fibers, which produce tonic (as opposed to phasic) contractions. Slow fibers are widely distributed in the animal kingdom, but in mammals they are found only in the extrinsic muscles of the ear and in the extraocular muscles, which have both slow and twitch types.

SLOW-TYPE FIBERS

Although students of human physiology will be concerned primarily with the twitch-type muscle of the skeletal musculature (see below), various properties of slow-type fibers will be given in order to compare them with twitch-type fibers. The slow fibers differ fundamentally from the twitch fibers because they have lower levels of resting potential (-50 to -70 mV as compared to -80 to -90 mV for fast fibers), they do not produce propagated action potentials, they exhibit a longer latency of response (6–8 ms as compared to 2–4 ms for fast fibers), and they do not respond to stimulation by rapid shortening or rapid tension development. Only repeated stimulation results in a significant rise in tension by these slow muscle fibers. In addition, the slow muscle fibers are innervated by numerous nerve endings (the *en grappe* type), distributed widely over the whole fiber length, whereas the twitch fibers are innervated by a single discrete, localized nerve ending (the *en plaque* type) with a distinct neuromuscular junction. The numerous areas of nerve contact, which locally activate the contractile apparatus of the fiber, compensate for the inability of a slow fiber to produce propagated muscle impulses.

The slow-type muscle system has been studied in greatest detail in the frog. It produces the prolonged contraction during the male's amplexus, which may last for days during the mating season. In this animal, tension development by slow fibers following reflex activation is similar to tension development after direct nerve stimulation (provided that the direct stimulation is of sufficient intensity to activate most of the small nerve fibers serving the muscle). The motor units of the slow fiber system may be normally activated, therefore, by a synchronous high frequency discharge from the spinal cord. In the frog, the nerve fibers to the slow muscle fibers are small and slow conducting (2–8 m/s), while the nerve fibers to the twitch fibers are large and fast conducting (8–40 m/s). No overlap of innervation between the two muscle groups has been found. The slow muscle fibers are reflexly activated even during periods when the twitch fibers are at rest. The slow fibers can also be activated during twitch fiber activity and can maintain large tensions originally produced by the twitch fibers over very long periods by virtue of low frequency stimulation. A functionally useful synergism has been demonstrated by the twitch and slow fibers in the frog iliofibularis muscle. First, small nerve stimulation at 30/s results in a smooth rise in tension. If a burst of twitch activity is added, via the large nerve system, the final tension developed exceeds that which would have been developed without the twitch activity. Second, when stimulation ceases there is a very gradual relaxation of the generated tension. If a burst of twitch activity is initiated during the slow relaxation phase, the residual tension in the slow fibers collapses rapidly.

Slow fibers and twitch fibers have several similarities and some major differences in their fine structure. In both muscle types the myofilaments of the myofibrils are of the same length, even though sarcomere lengths may differ. However, the slow fibers have no discrete M lines, the H zones are poorly defined, and the Z lines are thick. Although the myofibrils tend to be large and vaguely defined in cross-section, this cannot be considered as the sole criterion for slow fibers.

A striking difference between twitch and slow fibers is the absence or poor development of the transverse tubular or T system in the slow fiber: In the twitch fiber the T-tubule is in continuity with the sarcolemma; but, in those slow fibers in which a T system can be recognized, the small delicate tubules have no specific localization with respect to the sacrolemma or the contractile apparatus.

The motor nerve endings on slow muscle fibers, as revealed by electron microscopic studies, are readily distinguishable from those of twitch muscle fibers. The specific region of nerve-muscle contact is small compared with that on twitch fibers. In addition, there are no junctional folds in the muscle membrane or sarcolemma in the region of the nerve ending. Such junctional folds are a constant feature of the motor end-plate in twitch fibers.

TWITCH-TYPE FIBERS

Classification

Twitch-type fibers can be further classified into three distinct subtypes which have been variously designated through the years and now frequently are referred to as fast fatigable, slow fatigable, and fast fatigue-resistant (Table 1.9). Fast fatigable fibers, also called white fibers, have high myofibrillar ATPase activity, high glycolytic enzyme activities, an intermediate glycogen content, a low myoglobin content, and a small mitochondrial content. They produce a fast twitch and fatigue rapidly. The latter property is related to their limited glycogen content and low capacity for oxidative metabolism.

Fast fatigue-resistant fibers also have high myofibrillar ATPase activity, a high glycogen content, and an intermediate level of glycolytic enzymes but have a large mitochondrial content and a high myoglobin content, related to their resistance to fatigue. Slow fatigue-resistant fibers have low myofibrillar ATPase activity, low glycogen content but a high myoglobin content, high mitochondrial oxidative enzyme activites, and an intermediate mitochondrial content. The latter two fiber types are red in color. Thus red fibers can be either fast or slow contracting.

The red color of muscle is due to *myoglobin* which is related to the hemoglobin of red blood cells. Myoglobin from skeletal muscles has a molecular weight of 16,700 and contains 1 heme (iron-porphyrin complex) per molecule. The hemoglobin of the blood is composed of 4 hemes/molecule and differs also in the protein globin. Myoglobin has a greater affinity for oxygen than does hemoglobin and may serve as an oxygen reservoir and facilitate the transport of oxygen within the muscle fiber. In addition to its high affinity for oxygen, myoglobin loads and unloads its oxygen with great speed. Myoglobin is probably synthesized within the muscle fiber much as hemoglobin is synthesized within the red blood cell. A direct relationship has been found between the myoglobin content of muscle and the rate of blood flowing through a given muscle mass, usually

Table 1.9.
Morphological and Histochemical Types of Twitch Fibers in Mammalian Skeletal Muscle[a]

	Fast fatigable	Slow fatigue-resistant	Fast fatigue-resistant
Classifications			
Burke et al. (1973)	Fast fatigable	Slow fatigue-resistant	Fast fatigue-resistant
Other designations in the literature:			
(a)	Fast-twitch white	Slow-twitch intermediate	Fast-twitch red
(b)	White	Medium	Red
(c)	A	B	C
(d)	II	I	II
(e)	I	III	II
(f)	White	Intermediate	Red
(g)	A	C	B
Morphological properties			
Mitochondrial content	Small	Intermediate	Large
Z line	Narrow	Intermediate	Broad
Fiber diameter	Large	Intermediate	Small
Neuromuscular junction	Large and complex	Intermediate	Small and simple
Histochemical properties			
Distribution of succinic dehydrogenase	Even network	Even network	Predominantly subsarcolemmal
Oxidative enzyme activities	Low	High or intermediate	Intermediate or high
Mitochondrial ATPase	Low	Intermediate	High
Glycolytic activities	High	Low or variable	Intermediate or low
Myoglobin content	Low	High	High
Glycogen content	Intermediate	Low	High
Myofibrillar ATPase at pH 9.4	High	Low or variable	High
pH sensitivity of myofibrillar ATPase	Acid labile, alkali stable	Acid stable, alkali labile	Acid labile, alkali stable
Formaldehyde sensitivity of myofibrillar ATPase	Sensitive		Stable

[a]Table, in modified form, and pertinent references (not shown) are from Close, 1972.

referred to as the fractional blood flow and expressed as the percentage of cardiac output perfusing 100 g of muscle.

The *density of the capillary network* is highest in the slow, fatigue-resistant fibers, lowest in fast fatigable fibers, and intermediate in fast fatigue-resistant fibers. Myocardial fibers and the external oculomotor fibers have a denser capillary network than other muscles. Twitch time is proportional to the fractional blood flow (thus, red muscles twitch more slowly than white muscles).

Metabolism of the red fibers is mostly aerobic, while white muscles can also function anaerobically as illustrated by the increase in the lactic acid content in the veins draining an exercising muscle. Such anaerobic metabolism leads to "oxygen debt," from which the muscle must ultimately recover by oxidative replenishment of its energy stores. White fast muscles undergo only short bursts of activity and possess little myoglobin. It should be recalled, however, that most muscles are a mixture of both red (slow and fast fatigue-resistant and white fast fatigable) fibers.

Both white and red fast-twitch fibers not only are able to contract faster than red slow-twitch fibers, but also they relax faster. The shorter relaxation time of either white or red fast-twitch fibers correlates with a faster rate of calcium transport by the sarcoplasmic reticulum.

Heterogeneity of Muscle

The above fiber types have been well characterized in the cat, where some muscles consist almost exclusively of one fiber type. In the human, most skeletal muscles are mixed, that is, they contain different fiber types in various percentages, although a single fiber type may predominate in some muscles. For example, the human soleus consists largely of slow fatigue-resistant fibers. The anatomically equivalent muscle may consist of different percentages of fibers in various subspecies of the same animal. Thus the leg muscles are primarily red in the rabbit but primarily white in the hare.

Whole muscles provide a spectrum of speeds of contraction and relaxation as shown in Figure 1.92. It can be seen that the contraction time of an extraocular muscle (7.5 ms at 37°C) is five times faster than that of the gastrocnemius muscle (40 ms) and 12 times faster than that of the soleus muscle (90 ms). The extraocular muscles and the soleus represent the fast and slow extremes in most mammals. It is important to remember that whole muscles are not normally activated synchronously to yield the maximal twitch response of the kind recorded in Figure 1.92. Furthermore, the twitch response of

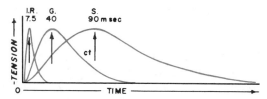

Figure 1.92. Illustration of the isometric contraction-relaxation (twitch) curves for three mammalian skeletal muscles. The lateral rectus (*l.r.*), gastrocnemius (*G*), and soleus (*S*) represent the extremes (fast and slow) and midrange in a spectrum of contraction-relaxation times for mammalian muscles. The *arrows* indicate peak tension development which corresponds to the contraction time for each muscle (lateral rectus, 7.5 ms; gastrocnemius, 40 ms; and soleus, 90 ms.)

isolated whole muscles does not reveal the heterogeneous nature of the tissue either in functional or structural terms. The gastrocnemius illustrates the heterogeneity of muscle. The white gastrocnemius muscle of the cat is in the midrange of muscle speed (contraction time is about 40 ms and contraction-relaxation time about 160 ms). The contraction time of the motor units which comprise the gastrocnemius vary from 17.8–129 ms, with a distribution peak between 30 and 40 ms. The peak distribution corresponds to the contraction time for the twitch of the isolated whole muscle.

Effect of Exercise

The cross-sectional area of a muscle increases when the muscle is forced repeatedly to develop maximal or close-to-maximal tension, as in regular exercise or work that requires forceful muscle action. This increase is due to an increase of the cross-sectional area of the individual muscle fibers (*hypertrophy*) but not to a formation of new fibers (*hyperplasia*). Hypertrophy will persist and increase only if the work load is continuously increased during a training period. If the load is kept constant, the growth of the muscle stops when its strength has been adjusted to the load.

Muscles exposed to sustained rhythmic activity over fairly long periods of time at a load below maximum show an increase in the density of the network of blood capillaries extending between the muscle fibers. This appears as an adjustment to the requirements for increased oxygenation associated with this form of activity. In animals, experiments have shown the effect to be more pronounced in the young than in adults.

Effects of Hormones on Muscle

Key effects of hormones on skeletal muscle are outlined only briefly below, as these are discussed in greater detail in chapters on the endocrine system.

Insulin is necessary for the uptake of glucose by the muscle fiber and other, but not all, cells. It acts to enhance the rate of transport of glucose across the cell membrane, which is the rate-limiting step in the entry of glucose into the muscle cell.

Glucocorticoids, as exemplified by cortisol, mobilize amino acids from extrahepatic tissues, primarily from muscle, by producing a decrease in protein synthesis. The amino acids are then carried via the bloodstream to the liver where they participate in gluconeogenesis, which is also stimulated by cortisol. In the normal state, glucocorticoids maintain blood sugar and glycogen content of liver and muscle. In Cushing's syndrome and other conditions such as clinical treatment with steroids where there is prolonged elevation of circulating levels of glucocorticoids, glucocorticoids produce a diabetes-like state because of their ability to raise blood sugar and to decrease the sensitivity of muscle and other insulin-sensitive tissues to insulin. Glucocorticoids thus are said to have a diabetogenic effect. Mobilization of amino acids from muscle may result in a generalized decrease in muscle mass and may contribute to the muscle wasting which is commonly seen in Cushing's syndrome (cortisol excess).

Testosterone has a marked anabolic effect on muscle due to its ability to stimulate protein synthesis.

Epinephrine has already been described as an important regulator of glycogen synthesis and glycogenolysis because of its ability to stimulate cyclic AMP formation in the cell. In white fast-contracting skeletal muscle, it produces a slight inotropic effect, i.e., it enhances twitch tension. This effect requires pharmacological doses of epinephrine and therefore probably does not occur under normal physiological conditions. In red slow-contracting skeletal muscle, epinephrine in physiological doses causes a decrease in twitch tension and an increase in the rate of muscle relaxation. The latter effect appears to be on the sarcoplasmic reticulum of slow-contracting skeletal muscle, which like cardiac sarcoplasmic reticulum, contains a low molecular weight protein called phospholamban. When phospholamban is phosphorylated by cyclic AMP-dependent protein kinase, it increases the activity of the calcium pump (Kirchberger and Tada, 1976).

Thyroid hormone, as mentioned above, is known to affect the myosin phenotype or isozyme pattern in a particular muscle. Clinically, hyperthyroidism produces a myopathy which is characterized by muscle weakness and fatigability. The symptoms may be related to a change in skeletal muscle phenotype either as a result of direct effect of thyroid hormone or its lack on the myosin isozyme composition of the muscle or an indirect effect due to a change in the pattern of neural activation of the muscle with a resultant loss of fast-twitch fibers (Johnson et al., 1980). In the latter case, a change in myosin light chains would be expected to occur based on our knowledge gained from cross-innervation experiments. In cardiac muscle, thyroid hormone produces isozymes of myosin which differ with respect to the heavy chain.

BIBLIOGRAPHY

ABBOTT, B. C., A. V. HILL, AND J. V. HOWARTH. The positive and negative heat production associated with a nerve impulse. *Proc. Roy. Soc. (London) Ser. B.* 148: 149–187, 1958.

ADRIAN, E. D. *The Mechanism of Nervous Action.* Philadelphia: Univ. Pennsylvania Press, 1935 (Reprinted, 1959).

ADRIAN, R. H. The effect of internal and external potassium concentration on the membrane potential of frog muscle. *J. Physiol. (London)* 133: 631–658, 1956.

ADRIAN, R. H., L. L. COSTANTIN, AND L. D. PEACHEY. Radial spread of contraction in frog muscle fibres. *J Physiol. (London)* 204: 231–257, 1969.

ALBERTS, B., D. BRAY, J. LEWIS, M. RAFF, K. ROBERTS, AND J. D. WATSON. *Molecular Biology of the Cell.* New York & London: Garland Publishing Inc., 1989.

ARMSTRONG, C. M. Ionic pores, gates, and gating currents. *Qtr. Rev. Biophys.* 7: 179–210, 1975.

ARMSTRONG, C. M., AND F. BEZANILLA. Inactivation of the sodium channel. II. Gating current experiments. *J. Gen. Physiol.* 70: 567–570, 1977.

BÁRÁNY, M. ATPase activity of myosin correlated with speed of muscle shortening. *J. Gen. Physiol.* 208: 197–218, 1967.

BARCENAS-RUIZ, L, D. J. BEUCKELMANN, AND W. G. WIER. Sodium-calcium exchange in heart: membrane currents and changes in [Ca]$_i$. *Science* 238: 1720–1722, 1987.

BASMAJIAN, J. V. *Muscles Alive: Their Functions Revealed by Electromyography,* Ed. 4. Baltimore: Williams & Wilkins, 1978.

BEGENISICH, T., AND C. F. STEVENS. How many conductance states do potassium channels have? *Biophys. J.* 15: 843–846, 1975.

BERNSTEIN, J. *Elektrobiologie.* Braunschweig, Germany: Friedr. Vieweg Sohn, 1912.

BLOOM, W., AND D. W. FAWCETT. *A Textbook of Histology,* Ed. 2. Philadelphia: W. B. Saunders, 1975.

BOWDITCH, H. P. Über die Eigenthumlichkeiten der Reizbarkeit, welche die Muskelfasern des Herzens zeigen. *Arb. Physiol. Anstalt, Leipzig* 6: 139–176, 1871.

BROCK, L. G., J. S. COOMBS, AND J. C. ECCLES. The recording of potentials from motoneurons with an intracellular electrode. *J. Physiol. (London)* 117: 431–460, 1952.

BRODSKY, W. A., A. E. SHAMOO, AND I. L. SCHWARTZ. Dissipative transport processes. In: *Handbook of Neurochemistry,* edited by A. Lajtha. New York: Plenum Press, 1971, vol. V, pt. B, chap. 20, p. 645–681.

BRONNER, F., AND A. KLEINZELLER (eds.). *Current Topics in Membranes and Transport.* New York: Academic Press, vols. 1–16, 1970–1982.

BROOKHAVEN NATIONAL LABORATORY. *Structure, Function and Evolution in Proteins,* report of a symposium. Brookhaven Symposia in Biology, 21, vol. I and II, 1969.

BURKE, R. E., D. N. LEVINE, P. TSAIRIS, AND F. E. ZAJAC, III. Physiological types and histochemical profiles in motor units of the cat gastrocnemius. *J. Physiol. (London)* 234: 723–748, 1973.

BURKE, R. E., AND P. TSAIRIS. The correlation of physiological properties with histochemical characteristics in single muscle units. *Ann. N.Y. Acad. Sci.* 228: 145–159, 1974.

CALDWELL, P. C., A. L. HODGKIN, R. D. KEYNES, AND T. L. SHAW. The effects of injecting "energy-rich" phosphate compounds on the active transport of ions in the giant axons of *Loligo. J. Physiol. (London)* 152: 561–590, 1960.

CHANDLER, W. K., AND M. F. SCHNEIDER. Time-course of potential spread along a skeletal muscle fiber under voltage clamp. *J. Gen. Physiol.* 67: 165–184, 1976.

CHANGEUX, J.-P. The acetylcholine receptor: An "allosteric" membrane protein. *Harvey Lectures* 75: 85–255, 1981.

CHAPMAN, D., AND D. F. H. WALLACH (eds.). *Biological Membranes.* London: Academic Press, 1968–1976, vols. 1–3.

CLOSE, R. I. Dynamic properties of skeletal muscle. *Physiol. Rev.* 52: 129–197, 1972.

COLE, K. S., AND J. W. MOORE. Ionic current measurements in the squid giant axon membrane. *J. Gen. Physiol.* 44: 123–167, 1960.

COLLANDER, R. Cell membranes: their resistance to penetration and their capacity for transport. In: *Plant Physiology,* edited by F. C. Steward. New York: Academic Press, 1959, vol. II.

CONTI, F., B. HILLE, B. NEUMCKE, W. NONNER, AND R. STÄMPFLI. Measurement of the conductance of the sodium channel from current fluctuations at the node of Ranvier. *J. Physiol. (London)* 262: 699–727, 1976.

COOMBS, J. S., J. C. ECCLES, AND P. FATT. The inhibitory suppression of reflex discharges from motoneurones. *J. Physiol. (London)* 130: 396–413, 1955.

COSTANTIN, L. L., AND S. R. TAYLOR. Graded activation in frog muscle fibres. *J. Gen. Physiol.* 61: 424–443, 1973.

CURTIS, H. J., AND K. S. COLE. Transverse electric impedance of the squid giant axon. *J. Gen. Physiol.* 21: 757–765, 1938.

DAHL, J. L., AND L. E. HOKIN. The sodium-potassium adenosinetriphosphatase. *Ann. Rev. Biochem.* 43: 327–356, 1974.

DARNELL, J., H. LODISH, AND D. BALTIMORE. *Molecular Cell Biology.* New York: Scientific American Books, Inc. 1986.

DINNO, M. A., AND A. B. CALLAHAN (eds.). *Membrane Biophysics: Structure and Function in Epithelia.* New York: Alan R. Liss, 1981.

DOWBEN, R. M. *General Physiology.* New York: Harper & Row, 1969.

EBASHI, S., M. ENDO, AND I. OHTSUKI. Control of muscle contraction. *Qtr. Rev. Biophys.* 2: 351–385, 1969.

ECCLES, J. C. *The Neurophysiological Basis of Mind.* London: Oxford Univ. Press, 1953.

EINSTEIN, A. *Investigations on the Theory of the Brownian Movement.* New York: Dover Publications, 1926.

EISENBERG, E., AND T. L. HILL. Muscle contraction and free energy transduction in biological systems. *Science* 227: 999–1006, 1985.

EISNER, D. A., AND W. J. LEDERER. Na-Ca exchanger: stoichiometry and electrogenicity. *Am. J. Physiol.* 248: C189–C202, 1985.

ELLIOT, A., AND G. OFFER. Shape and flexibility of the myosin molecule. *J. Mol. Biol.* 123: 505–519, 1978.

ERLANGER, J., AND H. S. GASSER. *Electrical Signs of Nervous Activity.* Philadelphia: Univ. Pennsylvania Press, 1937.

FATT, P., AND B. KATZ. Spontaneous subthreshold activity of motor nerve endings. *J. Physiol. (London)* 117: 109–128, 1952.

FAWCETT, D. *The Cell, an Atlas of Fine Structure.* Philadelphia: W. B. Saunders, 1966.

FAWCETT, D. W., AND N. S. McNUTT. The ultrastructure of the cat myocardium. I. Ventricular papillary muscle. *J. Cell Biol.* 42: 1–45, 1969.

FENN, W. O. A quantitative comparison between the energy liberated and the work performed by the isolated sartorius muscle of the frog. *J. Physiol. (London)* 58: 175–203, 1923.

FENN, W. O., AND B. S. MARSH. Muscular force at different speeds of shortening. *J. Physiol. (London)* 85: 277–297, 1935.

FINEAN, J. B., R. COLEMAN, AND R. H. MITCHELL. *Membranes and Their Cellular Functions,* Ed. 2. New York: Halsted Press, John Wiley & Sons, 1978.

FORRESTER, T., AND H. SCHMIDT. An electrophysiological investigation of the slow fibre system in the frog rectus abdominis muscle. *J. Physiol. (London)* 207: 477–491, 1970.

FRANKENHAEUSER, B., AND A. L. HODGKIN. The after-effects of impulses of the giant nerve fibers of *Loligo. J. Physiol. (London)* 131: 341–376, 1956.

FRANZINI-ARMSTRONG, C. Membranous system in muscle fibers. In: *The Structure and Function of Muscle,* edited by G. H. Bourne. New York: Academic Press, 1973, vol. 2, pt. 2, p. 531–619.

FRANZINI-ARMSTRONG, C. Studies of the triad. I. Structure of the junction in frog twitch fibers. *J. Cell Biol.* 47: 488–499, 1970.

FRANZINI-ARMSTRONG, C., AND L. D.

PEACHEY. Striated muscle—contraction and control mechanisms. *J. Cell Biol.* 91: 166s–186s, 1981.

FRYE L. D., AND M. EDIDIN. The rapid intermixing of cell surface antigens after formation of mouse-human heterokaryons. *J. Cell Sci.* 7: 319–335, 1970

GALL, J. G., K. R. PORTER, AND P. SIEKEVITZ (eds.). Discovery in cell biology. *J. Cell Biol.* 91: 1s–306s, 1981.

GEREN, B. B. The formation from the Schwann cell surface of myelin in the peripheral nerves of chick embryos. *Exp. Cell. Res.* 7: 558–562, 1954.

GIBBONS, I. R. The organization of cilia and flagella. In: *Molecular Organization and Biological Function*, edited by J. M. Allen. New York: Harper and Row, 1967, Chap. 8, p. 211–237.

GOLDMAN, Y. E. Special topic: Molecular mechanism of muscle contraction. *Ann. Rev. Physiol.* 49: 629–636, 1987.

GOLDSHLEGGER, R., S. J. D. KARLISH, A. REPHAELI, AND W. D. STEIN. The effect of membrane potential on the mammalian Na^+, K^+-pump reconstituted into phospholipid vesicles, *J. Physiol. (London)* 387: 331–355, 1987.

GORDON, A. M., AND E. B. RIDGWAY. Length-dependent electromechanical coupling in single muscle fibers. *J. Gen. Physiol.* 68: 653–669, 1976.

GORDON, A. M., A. F. HUXLEY, AND F. J. JULIAN. The variation in isometric tension with sarcomere length in vertebrate muscle fibres. *J. Physiol. (London)* 184: 170–192, 1966.

GRANGER, B. L., AND E. LAZARIDES. Desmin and vimentin coexist at the periphery of the myofibril Z disc. *Cell* 18: 1053–1063, 1979.

GREASER, M. L., AND J. GERGELY. Reconstituion of troponin activity from three protein components. *J. Biol. Chem.* 246: 4226–4233, 1971.

GREENGARD, P. Possible role for cyclic nucleotides and phosphorylated membrane proteins in postsynaptic actions of neurotransmitters. *Nature (London)* 260: 101–108, 1976.

GUIDOTTI, G. Membrane proteins. *Ann. Rev. Biochem.* 41: 731–752, 1972.

HALL, J. L., AND D. A. BAKER. *Cell Membranes and Ion Transport.* New York: Longman, 1977.

HAMILL, O. P., A. MARTY, E. HEHER, B. SAKMANN, AND F. J. SIGWORTH. Improved patch-clamp techniques for high-resolution current recordings from cells and cell-free membrane patches. *Pflügers Arch.* 391: 85–100, 1981.

HARRINGTON, W. F. Contractile proteins of muscle. In: *The Proteins*, Ed. 3, edited by H. Neurath and R. L. Hill. New York: Academic Press, 1979, vol. IV, chap. 3, p. 245–409.

HEDIGER, M. A., M. J. CODY, T. S. IKADA, AND E. M. WRIGHT. Expression cloning and cDNA sequencing of the Na^+/glucose co-transporter. *Nature* 330: 379–381, 1987.

HEDQVISTK, P. Role of the α-receptor in the control of noradrenaline release from sympathetic nerves. *Acta Physiol. Scand.* 90: 158–165, 1974.

HEINZ, E. *Electrical Potentials in Biological Membrane Transport.* Heidelberg: Springer-Verlag, 1981.

HEINZ, E. *Mechanics and Energetics of Biological Transport.* Berlin: Springer-Verlag, 1978.

HENNEMAN, E., H. P. CLAMANN, J. D. GILLIES, AND R. D. SKINNER. Rank order of motoneurons within a pool: law of combination. *J. Neurophysiol.* 37: 1338–1349, 1974.

HESCHELER, J., W. ROSENTHAL, W. TRAUTWEIN, AND G. SCHULTZ. The GTP-binding protein, G_O, regulates neuronal calcium channels. *Nature*, 325: 445–446, 1987.

HIBBERD, M. G., AND D. R. TRENTHAM. Relationships between chemical and mechanical events during muscular contraction. *Annie. Rev. Biophys. Biophys. Chem.* 15: 119–161, 1986.

HILL, A. V., AND W. HARTREE. The four phases of heat production of muscle. *J. Physiol. (London)* 54: 84–128, 1920.

HILL, A. V. *Muscular Movement in Man.* New York: McGraw-Hill, 1927.

HILL, A. V. The heat of shortening and the dynamic constants of muscle. *Proc. R. Soc. (London) Ser. B* 126: 136–195, 1938

HILL, A. V. The abrupt transition from rest to activity in muscle. *Proc. Roy. Soc. B*, 136: 399–419, 1949.

HILL, A. V. *First and Last Experiments in Muscle Mechanics.* London: Cambridge Univ. Press, 1970.

HILL, D. K. The effect of temperature on the resting tension of frog's muscle in hypertonic solutions. *J. Physiol. (London)* 208: 741–756, 1970.

HILLE, B. Pharmacological modifications of the sodium channels of frog nerve. *J. Gen. Physiol.* 51: 199–219, 1968.

HILLE, B. Ionic channels in excitable membranes: current problems and biophysical approaches. *Biophys. J.* 22: 283–294, 1978.

HILLE, B. *Ionic Channels of Excitable Membranes.* Sunderland, MA: Sinauer Assoc., 1984.

HILLE, B., AND W. A. CATTERALL. Electrical excitability and ionic channels. In: *Basic Neurochemistry*, Ed. 4. edited by Siegel, G. J., B. W. Agranoff, R. W. Albers, and P. Molinoff. New York: Raven Press, p. 71–90, 1989.

HODGKIN. A. L. The subthreshold potentials in a crustacean nerve fiber. *Proc. Roy. Soc. (London) Ser. B* 126: 87–121, 1939.

HODGKIN, A. L. Ionic movements and electrical activity in giant nerve fibres. *Proc. Roy. Soc. (London) Ser. B* 148: 1–37, 1957.

HODGKIN, A. L. The ionic basis of nervous conduction. *Science* 145: 1148–1154, 1964.

HODGKIN, A. L. *The Conduction of the Nerve Impulse.* Springfield, IL: Charles C Thomas, 1964.

HODGKIN, A. L., AND A. F. HUXLEY. Resting and action potentials in single nerve fibres. *J. Physiol. (London)* 104: 176–195, 1945.

HODGKIN, A. L., AND A. F. HUXLEY. Currents carried by sodium and potassium ions through the membrane of the giant axon of *Loligo. J. Physiol. (London)* 116: 449–472, 1952a.

HODGKIN, A. L., AND A. F. HUXLEY. The components of membrane conductance in the giant axon of *Loligo. J. Physiol. (London)* 116: 473–496, 1952b.

HODGKIN, A. L., AND A. F. HUXLEY. The dual effect of membrane potential on sodium conductance in the giant axon of *Loligo. J. Physiol. (London)* 116: 497–506, 1952c.

HODGKIN, A. L., AND A. F. HUXLEY. A quantitative description of membrane current and its application to conduction and excitation in nerve. *J. Physiol. (London)* 117: 500–544, 1952d.

HODGKIN, A. L., AND R. D. KEYNES. The mobility and diffusion coefficient of potassium in giant axons from *Sepia. J. Physiol. (London)* 119: 513–528, 1953.

HODGKIN, A. L., AND R. D. KEYNES. Active transport of cations in giant axons from *Sepia* and *Loligo. J. Physiol. (London)* 128: 28–60, 1955.

HODGKIN, A. L., AND R. D. KEYNES. Experiments on the injection of substances into squid giant axons by means of a microsyringe. *J. Physiol. (London)* 131: 592–616, 1956.

HODGKIN, A. L., A. F. HUXLEY, AND B. KATZ. Measurement of current-voltage relations in the membrane of the giant axon of *Loligo. J. Physiol. (London)* 116: 424–448, 1952.

HOFFMAN, J. F. (ed.) *Membrane Transport Processes.* New York: Raven Press, 1977, vol. 1.

HOLTER, H. Pinocytosis. *Int. Rev. Cytol.* 8: 481–504, 1959.

HOLTZMAN, E., AND A. B. NOVIKOFF. *Cells and Organelles*, Ed. 3. New York: Holt, Rinehart and Winston, 1983.

HOMSHER, E. Muscle enthalpy production and its relationship to actomyosin ATPase. *Ann. Rev. Physiol.* 49: 673–690, 1987.

HUBBELL, W. L., AND H. M. McCONNELL. Spin-label studies of the excitable membranes of nerve and muscle. *Proc. Natl. Acad. Sci. U.S.A.* 61: 12–16, 1968.

HUBBELL, W. L., AND H. M. McCONNELL. Molecular motion in spin-labeled phospholipids and membranes. *J. Am. Chem. Soc.* 93: 314–326, 1971.

HUGHES, J., T. W. SMITH, H. W. KOSTERLITZ, L. A. FOTHERGILL, B. A. MORGAN, AND H. R. MORRIS. Identification of two related pentapeptides from the brain with potent opiate agonist activity. *Nature (London)* 258: 577–579, 1975.

HURSH, J. B. Conduction velocity and

diameter of nerve fibers. *Am. J. Physiol.* 127: 131–139, 1939.

HUXLEY, A. F. Electrical processes in nerve conduction. In: *Ion Transport Across Membranes*, edited by H. T. Clarke and D. Nachmansohn. New York: Academic Press, 1954, p. 23–24.

HUXLEY, A. F., AND R. NIEDERGERKE. Structural changes in muscle during contraction. *Nature* 173: 971–973, 1954.

HUXLEY, A. F. Excitation and conduction in nerve: quantitative analysis. *Science* 145: 1154–1159, 1964.

HUXLEY, A. F. Muscular contraction. *J. Physiol. (London)* 243: 1–43, 1974.

HUXLEY, A. F., AND R. M. SIMMONS. Mechanical transients and the origin of muscular force. *Cold Spring Harbor Symp. Quant. Biol.* 37: 669–686, 1973.

HUXLEY, A. F., AND R. STÄMPFLI. Evidence for saltatory conduction in peripheral myelinated nerve fibres. *J. Physiol. (London)* 108: 315–339, 1949.

HUXLEY, A. F., AND R. STÄMPFLI. Direct determination of membrane resting potential and action potential in single myelinated nerve fibres. *J. Physiol. (London)* 112: 476–495, 1951.

HUXLEY, A. F., AND R. E. TAYLOR. Local activation of striated muscle fibres. *J. Physiol. (London)* 144: 426–441, 1958.

HUXLEY, H. E. Electron microscope studies on the structure of natural and synthetic protein filaments from striated muscle. *J. Molec. Biol.* 7: 281–308, 1963.

HUXLEY, H. E. Structural evidence concerning the mechanism of contraction in striated muscle. In: *Muscle: Proceedings of a Symposium Held at the Faculty of Medicine, University of Alberta*, edited by W. M. Paul, E. E. Daniel, C. M. Kay, and G. Monckton. Oxford: Pergamon, p. 3–28, 1964.

HUXLEY, H. E. The mechanism of muscular contraction. *Science* 164: 1356–1366, 1969.

HUXLEY, H., AND J. HANSON. Changes in the cross-striations of muscle during contraction and stretch and their structural interpretation. *Nature* 173: 973–976, 1954.

HUXLEY, H. E., AND J. HANSON. The molecular basis of contraction in cross-striated muscles. In: *The Structure and Function of Muscle*, edited by G. H. Bourne. New York: Academic Press, 1960, vol. 1, chap. 7, p. 183–227.

HUXLEY, H. E., AND M. KRESS. Cross-bridge behaviour during muscle contraction. *J. Muscle Res. Cell Motil.* 6: 153–161, 1985.

JOHNSON, M. A., F. L. MASTAGALIA, AND A. G. MONTGOMERY. Changes in myosin light chains in the rat soleus after thyroidectomy. *FEBS Lett.* 110: 230–235, 1980.

KAPLAN, D. M., AND R. S. CRIDDLE. Membrane structural proteins. *Physiol. Rev.* 51: 249–273, 1971.

KATZ, B. The relation between force and speed in muscular contraction. *J. Physiol. (London)* 96: 45–64, 1939.

KATZ, B. *Nerve, Muscle and Synapse.* New York: McGraw-Hill, 1966.

KATZ, B. *The Release of Neural Transmitter Substances.* Springfield, IL: Charles C Thomas, 1969.

KATZ, A. M., AND A. J. BRADY. Mechanical and biochemical correlates of cardiac contraction (I). *Mod. Concepts Cardiovasc. Dis.* 40: 39–43, 1971.

KENDREW, J. C. Side-chain interactions in myoglobin. *Brookhaven Symp. Biol.* 15: 216–228, 1962.

KEYNES, R. D. The ionic movements during nervous activity. *J. Physiol. (London)* 114: 119–150, 1951.

KEYNES, R. D. Ion channels in the nerve-cell membrane. *Sci. Am.* 240: 126–135, 1979.

KINNE, R., AND I. L. SCHWARTZ. Resolution of the epithelial cell envelope into luminal and contraluminal plasma membranes as a tool for the analysis of transport processes and hormone action. In: *Membranes and Disease*, edited by L. Bolis, J. F. Hoffman, and A. Leaf. New York: Raven Press, 1976.

KINNE, R., AND I. L. SCHWARTZ. Isolated membrane vesicles in the evaluation of the nature, localization, and regulation of renal transport processes. *Kidney Internat.* 14: 547–556, 1978.

KIRCHBERGER, M. A., AND M. TADA. Effects of adenosine 3′:5′ -monophosphate-dependent protein kinase on sarcoplasmic reticulum isolated from cardiac and slow and fast contracting skeletal muscles. *J. Biol. Chem.* 251: 725–729, 1976.

KNAPPEIS, G. G., AND F. CARLSEN. The ultrastructure of the M line in skeletal muscle. *J. Cell. Biol.* 38: 202–211, 1968.

KOPITO, R. R., AND H. F. LODISH. Primary structure and transmembrane orientation of the murine anion exchange protein. *Nature* 316: 234–238, 1985.

KUSHNER, D. Self-assembly of biological structure. *Bact. Rev.* 33: 302–345, 1969.

KYTE, J. Purification of the sodium- and potassium-dependent adenosine triphosphatase from canine renal medulla. *J. Biol. Chem.* 246: 4157–4165, 1971.

LLINÁS, R., AND K. WALTON. In: *The Cell Surface and Neuronal Function*, edited by C. W. Cotman, G. Poste, and G. L. Nicolson. New York: Elsevier/North-Holland Biomedical Press, 1980, chap. 3, p. 87–118.

LLOYD, D. P. C. Facilitation and inhibition of spinal motoneurones. *J. Neurophysiol.* 9: 421–438, 1946.

LOEWI, O. Über humoral Übertagbarkeit der Herznervenwirkung. I. Mitteilung. *Pflügers Arch.* 189: 239–242, 1921.

LOGOTHETIS, D. E., Y. KURACHI, J. GALPER, E. J. NEER, AND D. E. CLAPHAM. The β_γ subunits of GTP-binding proteins activate the muscarinic K$^+$

channel in the heart. *Nature* 325: 321–326, 1987.

LYMN, R. W., AND W. W. TAYLOR. Mechanisms of adenosinetriphosphate hydrolysis by actomyosin. *Biochemistry* 10: 4617–4624, 1971.

MACPHERSON, L., AND D. R. WILKIE. The duration of the active state in a muscle twitch. *J. Physiol. (London)* 124: 292–299, 1954.

McCUBBIN, W. D., AND C. M. KAY. Calcium-induced conformational changes in the troponin-tropomyosin complexes of skeletal and cardiac muscles and their roles in the regulation of contraction-relaxation. *Acct. Chem. Res.* 13: 185–192, 1980.

McPHEDRAN, A. M., R. B. WUERKER, AND E. HENNEMAN. Properties of motor units in a homogenous red muscle (soleus) of the cat. *J. Neurophysiol.* 28: 71–84, 1965.

MARGOULIES, M. (ed.). *Protein and Polypeptide Hormones.* Amsterdam: Excerpta Medica Foundation, 1968.

MILLER, R. J. Multiple calcium channels and neuronal functions. *Science* 235: 46–52, 1987.

MITCHELL, P. Vectorial chemistry and the molecular mechanisms of chemiosmotic coupling: power transmission by proticity. *Trans. Biochem. Soc.* 4: 662–663, 1976.

MITCHELL, P. Keilin's respiratory chain concept and its chemiosmotic consequences. *Science* 206: 1148–1159, 1979.

MORGAN, D. L., AND U. PROSKE. Vertebrate slow muscle: its structure, pattern of innervation, and mechanical properties. *Physiol. Rev.* 64: 103–169, 1984.

MUECKLER, M., C. CARUSO, S. A. BALDWIN, M. PANICO, I. BLENCH, H. R. MORRIS, W. J. ALLARD, G. E. LIENHARD, AND H. F. LODISH. Sequence and structure of a human glucose transporter. *Science* 229: 941–945, 1985.

MULIERI, L. A., AND N. R. ALPERT. Activation heat and latency relaxation in relation to calcium movement in skeletal and cardiac muscle. *Can. J. Physiol. Pharmacol.* 60: 415–588, 1982.

MURER, H., AND R. KINNE. The use of membrane vesicles to study epithelial transport processes. *J. Membrane Biol.* 55: 81–95, 1980.

MURPHY, R. A., AND A. C. BEARDSLEY. Mechanical properties of the cat soleus muscle in situ. *Am. J. Physiol.* 227: 1008–1013, 1974.

MURRAY, J. M., AND A. WEBER. The cooperative action of muscle proteins. *Sci. Am.* 230: 58–71, 1974.

MURRAY, J. M., A. WEBER, AND M. K. KNOX. Myosin sulfragment 1 binding to relaxed actin filaments and steric model of relaxation. *Biochemistry* 20: 641–649, 1981.

NEHER, E., B. SAKMANN, AND J. H. STEINBACH. The extracellular patch clamp: a method for resolving currents through individual open channels in biological membranes. *Pflügers Arch.* 375: 219–228, 1978.

NERNST, W. Zür Theorie des elektrischen Reizes. *Pflügers Arch.* 122: 275–314, 1908.

NESTLER, E. J., AND P. GREENGARD. Protein phosphorylation and the regulation of neuronal function. In: *Basic Neurochemistry*, Ed. 4, edited by Siegel, G. J., B. W. Agranoff, R. W. Albers, and P. Molinoff. New York: Raven Press, p. 373–398, 1989.

NODA, M., T. IKEDA, T. KAYANO, H. SUZUKI, H. TAKESHIMA, M. KURASAKI, H. TAKAHASHI, AND S. NUMA. Existence of distinct sodium channel messenger RNAs in rat brain. *Nature* 320: 188–192, 1986.

NORRIS, F. H., JR., AND R. L. IRWIN. Motor unit area in a rat muscle. *Am. J. Physiol.* 200: 944–946, 1961.

NOVIKOFF, A. B., AND E. HOLTZMAN. *Cells and Organelles*, 2nd ed. New York: Holt, Rinehart and Winston, 1976.

PAGE, S. G. Comparison of the fine structures of frog slow and twitch muscle fibres. *J. Cell. Biol.* 26: 477–497, 1965.

PEACHEY, L. D. The sarcoplasmic reticulum and transverse tubules of the frogs's sartorius. *J. Cell Biol.* 25: 209–231, 1965.

PEMRICK, S. M. (1980). The phosphorylated L_2 light chain of skeletal myosin is a modifier of the actomyosin ATPase. *J. Biol. Chem.* 255: 8836–8841, 1980.

PENEFSKY, Z. P. A model active state in cardiac muscle: A study of the first derivative of isometric tension. In: *Research in Physiology*, edited by F. F. Kao, K. Koizumi, and M. Vassalle. Bologna: Aulo Gaggi Publisher, 1971, p. 239–251.

PENNEY, J. B., AND A. B. YOUNG. Quantitative autoradiography of neurotransmitter receptors in Huntington disease. *Neurology* 32: 1391–1395, 1982.

PERRY, T. L., S. HANSEN, AND M. KLOSTER. Huntington's chorea: deficiency of γ-amino-butyric acid in brain. *N. Engl. J. Med.* 228: 337–342, 1973.

PETERS, A., S. L. PALAY, AND H. DEF. WEBSTER. *The Fine Structure of the Nervous System: The Neurons and Supporting Cells.* Philadelphia: W. B. Saunders, 1976.

PINTO DA SILVA, P., AND D. BRANTON. Membrane splitting in freeze-etching. *J. Cell Biol.* 45: 598–605, 1970.

PORTER, K. R., AND A. B. NOVIKOFF. The 1974 Nobel Prize for Physiology or Medicine. *Science* 186: 516–520, 1974.

PORTER, K. R., AND G. E. PALADE. Studies on the endoplasmic reticulum. III. Its form and distribution in striated muscle cells. *J. Biophys. Biochem. Cytol.* 3: 269–300, 1957.

POST, R. L., C. R. MERRITT, C. R. KINSOLVING, AND C. D. ALBRIGHT. Membrane adenosine triphosphatase as a participant in the active transport of sodium and potassium in the human erythrocyte. *J. Biol. Chem.* 235: 1796–1802, 1960.

POTTER, J. D., AND J. GERGELY. Troponin, tropomyosin, and actin interactions in the Ca^{2+} regulation of muscle contraction. *Biochemistry* 13: 2697–2703, 1974.

PREWITT, M. A., AND B. SAFALSKY. Enzymic and histochemical changes in fast and slow muscles after cross innervation. *Am. J. Physiol.* 218: 69–74, 1970.

RALL, J. A. Sense and nonsense about the Fenn effect. *Am. J. Physiol.* 242: H1–H6, 1982.

REIS, D. J., AND G. F. WOOTEN. The relationship of blood flow to myoglobin, capillary density, and twitch characteristics in red and white skeletal muscle in cat. *J. Physiol. (London)* 210: 121–135, 1970.

RENSHAW, B. Central effects of centripetal impulses in axons of spinal ventral roots. *J. Neurophysiol.* 9: 191–204, 1946.

ROBERTSON, J. D. New observations on the ultrastructure of the membranes of frog peripheral nerve fibers. *J. Biophys. Biochem. Cytol.* 3: 1043–1047, 1957.

ROBERTSON, J. D. Unit membranes: a review with recent new studies of experimental alterations and a new subunit structure in synaptic membranes. In: *Cellular Membranes in Development*, edited by M. Locke. New York: Academic Press, 1964, p. 1–81.

ROSSIER, B. C., K. GEERING, AND J. P. KRAEHENBUHL. Regulation of the sodium pump: How and why? *TIBS* 12: 483–487, 1987.

SALKOFF, L., A. BUTLER, A. WEI, N. SCAVARDA, K. BAKER, D. PAVRON, AND C. SMITH. Molecular biology of the voltage-gated sodium channel. *Trends in Neuroscience* 10: 522–527, 1987.

SATIR, P. How cilia move. *Sci. Am.* 231: 44–52, 1974.

SCHAFER, J. A., AND T. E. ANDREOLI. Principles of water and nonelectrolyte transport across membranes. In: *Physiology of Membrane Disorders*. edited by Andreoli, T. E., J. F. Hoffman, D. D. Fanestil, and S. G. Schultz, New York: Plenum, 1986.

SCHULTZ, S. G. *Basic Principles of Membrane Transport.* London: Cambridge University Press, 1980.

SIEGEL, G. J., AND R. W. ALBERS. Nucleosidotriphosphate phosphohydrolases. In: *Handbook of Neurochemistry*, edited by A. Lajtha. New York: Plenum Press, 1970, vol. IV, chap. 2, p. 13–44.

SIEGEL, G., B. W. AGRANOFF, R. W. ALBERS, AND P. MOLINOFF (eds.). *Basic Neurochemistry: Molecular, Cellular, and Medical Aspects*, Ed. 4. New York: Raven Press, 1989.

SINGER, S. J. The molecular organization of membranes. *Ann. Rev. Biochem.* 43: 805–833, 1974.

SINGER, S. J., AND G. L. NICOLSON. The fluid mosaic model of the structure of cell membranes. *Science* 175: 720–731, 1972.

SIVARAMAKRISHNAN, M., AND M. BURKE. The free heavy chain of vertebrate skeletal myosin subfragment 1 shows full enzymatic activity. *J. Biol. Chem.* 257: 1102–1105, 1982.

SKOU, J. C. The influence of some cations on an adenosine triphosphatase from peripheral nerve. *Biochim. Biophys. Acta* 23: 394–401, 1957.

SPIRO, D., AND E. H. SONNENBLICK. Comparison of the ultrastructural basis of the contractile process in heart and skeletal muscle. *Circ. Res.* 15 (Suppl. II): 14–36, 1964.

STARR, R., AND G. OFFER. Polypeptide chains of intermediate molecular weight in myosin preparations. *FEBS Lett.* 15: 40–44, 1971.

STEIN, L. A., R. P. SCHWARZ, JR., P. B. CHOCK, AND E. EISENBERG. Mechanism of actomyosin adenosine triphosphatase. Evidence that adenosine 5'-triphosphate hydrolysis can occur without disruption of the actomyosin complex. *Biochemistry* 18: 3895–3909, 1979.

STEIN, L. A., P. B. CHOCK, AND E. EISENBERG. The rate-limiting step in the actomyosin adenosinetriphosphatase cycle. *Biochemistry* 23: 1555–1563, 1984.

STEIN, W. D. *Transport and Diffusion Across Cell Membranes.* New York: Academic Press, 1986.

SWEADNER, K. J., AND S. M. GOLDIN. Active transport of sodium and potassium ions: Mechanism, function and regulation. *N. Engl. J. Med.* 302: 777–783, 1980.

TAKAHASHI, T., AND M. OTSUKA. Regional distribution of Substance P in the spinal cord and nerve roots of the cat, and the effects of dorsal root section. *Brain Research* 87: 1–11, 1975.

TAYLOR, J. H. (ed.) *Molecular Genetics.* New York: Academic Press, 1963 and 1967, pts. I and II.

THOMPSON, S. H., AND R. W. ALDRICH. In: *The Cell Surface and Neuronal Function*, edited by Cotman, C. W., G. Poste, and G. L. Nicolson. New York: Elsevier/North-Holland Biomedical Press, 1980, chap. 2, p. 49–85.

TOSTESON, D. C. (ed.). *Membrane Transport in Biology Series.* Berlin: Springer-Verlag, 1978 and 1979, vols. I and II.

TOSTESON, D. C., Y. A. OVCHINNIKOV, AND R. LATORRE (eds.). *Membrane Transport Processes.* New York: Raven Press, 1977, vol. 2.

VERATTI, E. Ricerche sulla fine struttura della fibra muscolare striata. *Arch. Ital. Biol.* 37: 449–454, 1902.

WEBER, A. On the role of calcium in the activity of adenosine 5'-triphosphate hydrolysis of actomyosin. *J. Biol. Chem.* 234: 2764–2769, 1959.

WEISSMANN, G., AND R. CLAIBORNE (eds). *Cell Membranes: Biochemistry,*

Cell Biology and Pathology. New York: Hospital Practice Publishing Co., 1975.

WELSH, M. J., AND C. M. LIEDTKE. Chloride and potassium channels in cystic fibrosis airway epithelia. *Nature* 322: 467–470, 1986.

WINEGRAD, S. Autoradiographic studies of intracellular calcium in frog skeletal muscle. *J. Gen. Physiol.* 48: 455–479, 1965.

WINKELMANN, D. A., AND S. LOWEY. Probing myosin head structure with monoclonal antibodies. *J. Molec. Biol.* 188: 595–610, 1986.

WOLFE, S. L. *Biology of the Cell*, Ed. 2. Belmont, CA: Wadsworth Publishing Co., 1981.

WUERKER, R. B., A. M. McPHEDRAN, AND E. HENNEMAN. Properties of motor units in a heterogeneous pale muscle (M. gastrocnemius) of cat. *J. Neurophysiol.* 28: 85–99, 1965.

WYSSBROD, H. R., W. N. SCOTT, W. A. BRODSKY, AND I. L. SCHWARTZ. Carrier-mediated transport processes. In: *Handbook of Neurochemistry*, edited by A. Lajtha. New York: Plenum Press, 1971, vol. V, pt. B, chap. 21, p. 683–819.

YOUNG, J. Z. The functional repair of nervous tissue. *Physiol. Rev.* 22: 318–374, 1942.

SECTION 2

Cardiovascular System

CHAPTER 5

Introduction to the Cardiovascular System

The heart and circulation constitute primarily a transport and exchange system. Large blood vessels in the circulatory loop provide the pathways for distributing and collecting blood-borne materials. The heart generates the energy for moving the blood through the circuit, and in the various organs and tissues, exchange of oxygen, carbon dioxide, and other metabolites occurs across the extremely thin walls of the capillaries. (For the historical development of concepts about the heart and circulation, see Fishman and Richards, 1964.)

An understanding of certain principles concerned with heart and circulatory function is fundamental to an appreciation of disease states, as well as of the action of various pharmacological agents. Therefore, as we examine the physiology of the normal heart and circulation it will be useful to reinforce these physiological principles with some examples of the derangements that can occur in human disease. Physicians encounter a great deal of heart and circulatory disease; for example, nearly 60 million people in the United States have cardiovascular disease, and about ½ of a million people die each year from heart attack alone. Cardiovascular disease is responsible for approximately ½ of all male deaths between the ages of 35 and 64 yr, far more than accounted for by cancer. Such figures emphasize not only the importance to the physician of obtaining an appreciation of basic cardiocirculatory phenomena, but also the need for additional research in this area.

In this chapter the general components of the heart and circulation will be briefly and simply described. The details of blood vessel and cardiac structure and function will be considered in the ensuing chapters, and the regulation of certain special circulations of the body (such as those supplying the brain, the kidney, the lung, the gastrointestinal tract, and the fetal circulation) are presented in other sections of this book.

GENERAL COMPONENTS OF THE CIRCULATORY SYSTEM

Each type of blood vessel has a different function, and whereas some organs have a highly specialized circulatory supply (such as the kidney and liver), the basic components supplying most tissues and organs consist of: artery→arteriole→capillary→venule→vein.

Arteries

The arteries serve to deliver oxygenated blood to the various organs. They have a relatively thick muscular wall and are sufficiently distensible to smooth the pulsations generated by the heart (Chapter 7). The channel of the arteries ranges from several cm in diameter (the aorta, the largest artery, which directly leaves the heart) through progressive branchings down to tiny arteries 1 mm or less in diameter.

The arteries are particularly susceptible to atherosclerosis, a buildup of cholesterol and other lipids in the artery walls, which can produce narrowed regions in the channel. When such a diseased artery supplies the brain, for example, a stroke can ensue.

Arterioles

The smallest branches of the arteries are the arterioles, small thick-walled muscular vessels which regulate the resistance to flow through the various organs and tissues of the body, thereby controlling the distribution and rate of blood flow to these regions. The ratio of the thickness of the vessel wall to the diameter of the vessel is high, and contraction or relaxation of the smooth muscle in the wall thereby allows them to serve as "stopcocks" which can readily control the vessel caliber (Fig. 2.1). The state of contraction of this "vascular smooth muscle" is regulated by several factors, including nerve impulses which reach nerve terminals lying within the smooth muscle, circulating vasoactive hor-

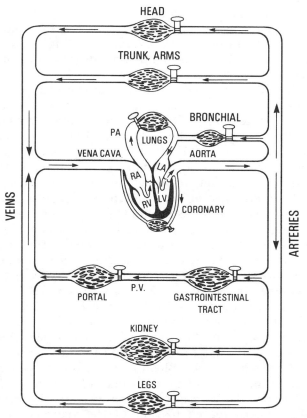

Figure 2.1. The pulmonary and systemic circulations. Note that the pulmonary circulation is connected in series to the remainder of the circulation (blood flows first through the chambers of the right heart and lungs, and then through chambers of the left heart). The organ systems operate in parallel from the systemic circulation. The small bronchial arteries supply oxygenated blood to the tissue of the lungs. *PA*, pulmonary artery; *RA*, right atrium; *LA*, left atrium; *LV*, left ventricle; *RV*, right ventricle; *PV*, portal vein.

mones, and metabolites carried in the blood; it can also be affected by a variety of pharmacological agents. The degree of contraction of these sphincter-like arterioles not only regulates the amount of blood flow to an organ, but when a sufficiently large number of arterioles are involved the pressure is affected in the large arteries behind them (Fig. 2.1). This effect can be likened to constriction or release of a screw-clamp placed on a garden hose; a change in resistance will raise or lower the pressure in the hose between the open faucet and the screw-clamp.

A disease which involves the arterioles is high blood pressure, or hypertension (see Chapter 16). The "tone" or state of contraction of the smooth muscle of the arterioles is increased in this disorder, and since the heart tends to maintain its output of blood to meet the needs of the body, the increased resistance to flow provided by the arterioles throughout the body elevates the pressure in the large arteries, with a number of unfavorable consequences.

Capillaries

The blood vessels continue to divide into smaller and smaller branches to reach the smallest unit, the capillary, a tiny thin-walled vessel through which exchange of materials occurs by diffusion and ultra-filtration. The distribution of blood flow through the capillary bed of an organ is further controlled in part by another set of tiny sphincter vessels, the precapillary sphincters, which contain some smooth muscle (Chapter 6). The thin wall and small cross-sectional area of the individual capillaries are ideally suited for exchange. The capillary channel (lumen) is approximately 5–7 μm in diameter, just sufficient to accommodate a red blood cell (see Fig. 2.17). The volume of the capillary bed is small (4% of the total blood volume), but the vast number of capillaries provides a large total cross-sectional area, which leads to a very slow rate of blood flow in that segment of the circulation, thereby further favoring transcapillary exchange.

Alterations in the capillaries may occur during disease states. For example, an increase in the permeability of the capillary wall can lead to leakage of intravascular proteins, fluid, and other substances out of the vascular bed (see Shock, Chapter 18).

Venules

These small vessels have a relatively large channel and a thinner muscular wall than the arteries, and they form the low pressure collecting system for the venous (relatively deoxygenated) blood as it leaves the capillaries. The venules, like the arterioles, are innervated by the autonomic nervous system, and the smooth muscle in their walls can contract or relax, thereby contributing to the pressure in the capillary bed as well as to the overall size (capacitance) of the venous compartment.

Veins

The venules progressively join together to form larger veins which eventually merge into the two largest veins in the body, the venae cavae (superior and inferior vena cava). These two large, thin-walled vessels return venous blood directly to the heart. The veins contain the largest proportion of blood within the circulaton (about 60%) and thereby serve as a blood reservoir; they can accommodate large changes in blood volume with rather small changes in pressure and are much more distensible than the arteries (Chapter 7).

Certain disease processes can primarily affect the veins. For example, inflammation of the wall of a vein can lead to the formation of a blood clot on the wall; eventually, it may dislodge and travel to the lungs. Such "pulmonary emboli" are unable to pass through the capillaries in the lung and can cause serious obstruction to blood flow.

Lymphatics

The lymphatic vessels are tiny, thin-walled channels that are not connected in sequence with the main blood vessels of the body, but rather serve as an additional drainage and transport system. They form a branching network throughout the tissues and organs of the body, and the fluid contained within them (lymph) also passes through the lymph nodes and is eventually channeled into a large vein. The ends of the lymphatic vessels are in direct communication with the extravascular fluid of the various organs, including the gastrointestinal tract, and they serve to return proteins, fats, and other substances to the circulation (Chapter 6).

Many diseases can involve the lymphatics; infection or scarring with obstruction of lymphatics can lead to swelling of the tissues (edema), and cancer cells are often seeded via the lymphatics into the lymph nodes.

PULMONARY AND SYSTEMIC CIRCULATIONS

There are two separate major circulations of the body which are linked in sequence, so that blood flows through the circulation to the body and then to the lungs, before returning again to the body.

The path of blood flow through these two circulations can be traced in Figure 2.1, the venous blood returning from the body being pumped by the right side of the heart into the pulmonary artery to the lungs (an unusual "artery" in that it contains venous blood) and, from there, into the pulmonary capillaries within the lung where uptake of oxygen and release of carbon dioxide occur. The blood then flows to the pulmonary veins (again, unusual "veins" in that they contain oxygenated blood) and back to the left-sided chambers of the heart, where it is pumped to the arteries and capillaries of the rest of the body.

The pulmonary circulation (sometimes called the "central" or "lesser" circulation) carries a relatively low pressure (about one-fifth that in the systemic arteries, with a mean pressure of 15–20 mmHg in the pulmonary artery) and serves a special function in providing a low resistance pathway for the entire output of blood from the body to traverse the lungs (Section 5, Chapter 35). The pulmonary circulation, in addition to gas exchange, serves a secondary function as a blood volume reservoir for the left heart (it contains about 10% of the total blood volume), and its high distensibility allows it to accommodate readily the large increases in blood flow that can occur (for example, during physical exercise) without manifesting a substantial change in pressure.

The systemic circulation (sometimes called the "peripheral" circulation) also has special properties (Chapters 6 and 7) and provides a high pressure source of oxygenated blood to supply the organs of the body (Fig. 2.1). It also drains the venous blood back to the right side of the heart; as mentioned, the veins serve an important volume reserve function and contain most of the blood in the circulation.

The mean pressures within the components of the systemic circulation drop progressively as blood flows from the arteries to the veins, with the highest pressure existing in the systemic arteries and the lowest in the veins. In the capillary bed, the pressure has fallen sufficiently to give a mean hydrostatic pressure that is in relative equilibrium with the plasma osmotic pressure (Chapter 6), thereby allowing fluid movement out of the thin-walled capillaries to be balanced by fluid gain due to the osmotic effect. Blood flow is slow in the capillaries, and an analogy to the marked slowing effect of the very large cross-sectional area of the capillary bed, compared to that of the arteries, is to allow water from a rapidly running hose to flow over a large screen; the high velocity of flow through the hose abruptly slows as filming over the screen takes place (see relation between volume flow and flow velocity, Chapter 6).

FEATURES OF OVERALL CIRCULATORY ORGANIZATION AND CONTROL

There are numerous miniature circulations to each of the organs and tissues of the body. These circulations operate *in parallel* off the systemic arteries, with the larger arteries serving as a "pressure reservoir" for oxygenated blood (Fig. 2.1). The regulation of blood flow through each organ is dependent upon its moment-to-moment function and metabolic needs. For example, the blood flow through a skeletal muscle is largely determined by local factors related to the varying metabolic requirements of the muscle as it alters its contractile work. If the leg muscle contracts repeatedly while running, the smooth muscle in the walls of the arterioles supplying the leg will relax to allow an increase in blood flow. In this manner, a series of "variable stopcocks" (the arteriolar resistance vessels) can regulate flow to each organ, operating semi-independently and in parallel off the large arteries (Fig. 2.1).

The general resistance equation: a simplified

equation (discussed in more detail in Chapter 7) which describes these pressure-flow relations can be termed the "general resistance equation":

Flow =

$$\frac{\text{driving pressure or pressure difference across a vascular bed}}{\text{vascular resistance}}$$

This equation, like Ohm's law of electrical circuits (current flow = voltage/resistance), makes intuitive sense. Again considering the garden hose analogy, the rate of flow will be directly proportional to the driving pressure from the faucet and inversely proportional to the amount of resistance provided by the screw clamp on the hose; as pressure increases with resistance constant, the flow will increase. As vascular resistance increases, in order to hold the pressure constant the flow from the faucet must be decreased; or, if flow is maintained constant as the screwclamp is tightened, the driving pressure must increase.

Regulation of the Systemic Arterial Pressure

The systemic arterial pressure (SAP) is held *relatively constant* by means of reflexes from the autonomic nervous system (Chapter 16) which control both the function of the heart and, hence, the cardiac output (Q) as well as the *total* peripheral vascular resistance (TPVR). In this context, the general resistance equation may be arranged as follows:

$$SAP = Q \times TPVR$$

The TPVR, of course, represents the net effect of the resistance in *all* of the organ beds of the systemic circulation. When sympathetic nerve impulse traffic increases (as during a change from supine to upright posture) norepinephrine stored in the nerve terminals around the arterioles is released. It then acts on smooth muscle cell receptors (α-receptors), causing them to contract (vasoconstriction), thereby increasing vascular resistance. Alternatively, when the blood pressure rises, sympathetic impulses decrease (through a feedback reflex), and vascular resistance decreases, tending to restore the blood pressure toward normal. Thus, through feedback mechanisms the blood pressure in the systemic arteries tends to be protected at a relatively constant level, so that an adequate pressure head is provided for perfusion of the various organs of the body.

Using the above resistance equation as applied to the regulation of arterial pressure, one can understand why the blood pressure does not drop markedly and remain low when we assume the erect posture from a lying position. Since it is known that a large volume of blood is transiently sequestered in the lower extremities and abdomen, such storage of blood should lead to a decrease in the return of blood to the heart and thereby to a drop in pressure. However, the pressure drops only mildly as the cardiac output (Q) initially falls because there is a prompt reflex increase in vascular resistance (TPVR) produced by the autonomic nervous system. Also, reflex neutral stimulation of the heart leads to an increased heart rate and augmented force of the heartbeat, and reflex contraction of the veins further tends to restore the cardiac output toward normal.

Local Organ Blood Flow

The blood flow to an organ (local Q), as mentioned earlier, is largely controlled by organ metabolism, the products of which affect local vascular resistance (local VR). The general resistance equation may be restated as follows:

$$\text{Local } Q = SAP/\text{Local } VR$$

As mentioned, the systemic arterial pressure is held relatively constant by overall circulatory reflexes. Therefore, in effect, each organ bed takes the flow of blood that it requires by regulating its local VR; as organ flow requirements increase at a constant SAP, local VR decreases, and local Q must increase.

GENERAL FEATURES OF THE HEART

The heart is a four-chambered organ composed of a special type of muscle called "myocardium" (see James et al. (1928) and Netter (1969) for details of cardiac anatomy). There are two separate pumps, the right ventricle and the left ventricle, each of which is supplied with a "booster pump"—the right atrium and the left atrium (the atria have earlike appendages and are sometimes called "auricles"). As we shall see in Chapter 8, the electrical excitation of the heart muscle is organized so that the atrial booster pumps are timed to contract just prior to the contraction of the ventricles; thus, the two atria contract nearly together, helping to fill the main pumping chambers just before the ventricles contract to eject blood into the pulmonary artery and the aorta.

Pericardium

A thin, membranous sac envelops the heart, and the space between this sac and the heart contains a small amount of fluid which serves to lubricate the heart surface as it moves during contraction. The pericardium is composed primarily of inextensible fibrous tissue which, under normal conditions, is relatively flaccid, but which, if it is acutely overstretched, becomes quite stiff. Thus, during acute overtransfusion of blood or other forms of rapid fluid overload, the pericardium can limit overdistension of the heart (Chapter 18). The pericardial sac also can become inflamed, a condition called "pericarditis." If this condition becomes chronic, the pericardium may become sufficiently thickened to limit normal filling of the heart.

Five Cardiac Subsystems

The heart may be conveniently considered as five subsystems, each of which will be considered in more detail in subsequent chapters. Cardioactive drugs and various disease processes tend to affect each of these subsystems somewhat differently.

Electrical Pacemaker and the Conducting System of the Heart

As shown in Figure 2.2, the self-firing pacemaker of the heart (sinoatrial node) is situated in the right atrium; the electrical impulse it generates travels into the two atria, which are stimulated to contract. There are some specialized conducting pathways within and between the atria. Then the impulses

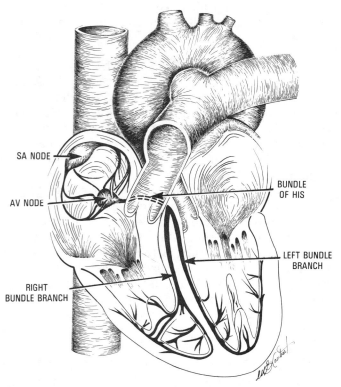

Figure 2.2.   Diagram of the electrical pacemaking and conducting system of the heart. *SA*, sinoatrial node; *AV*, atrioventricular node.

reach the main specialized conduction system, consisting of the atrioventricular node or junction which delays the impulse; it then travels down a rapid conduction system (the bundle branches) to the two ventricles, causing them to contract.

The pacemaker and specialized conduction tissues are influenced by the intrinsic innervation to the heart (the autonomic nervous system), which controls the heart rate and the velocity with which impulses are conducted through the specialized system.

Disorders of the heart's rhythm are common; disease of the pacemaker can result in rapid or irregular heart rhythms, whereas damage to the atrioventricular junction or specialized conduction system to the ventricles can result in faulty conduction of the electrical impulse to the ventricles from the atrium, so-called "heart block" (Chapter 9).

Heart Muscle

The muscle of the heart (myocardium) is composed of branched interconnected muscle fibers (cells) which are structured to rapidly transmit the electrical impulse from cell to cell. The walls of the two atria are thin (approximately 2 mm thick in the human heart) (Fig. 2.3, *left*). The wall of the right ventricle in the normal human heart is about 3–4 mm in thickness, and the left ventricle measures about 8–9 mm (Fig. 2.3, *left*), the difference in wall thickness relating to the higher pressure that the left ventricle is required to generate. The muscle bundles of the ventricles tend to wind spirally around the pumping chambers, originating and ending their insertion onto the fibrous portion (annuli fibrosi) of the heart surrounding the heart valves. In the left ventricle, the muscle is arranged across the wall in a particular manner, with the fibers winding in a counterclockwise direction at the inner wall, in a horizontal (circumferential) direction in the midwall, and in a clockwise direction at the outer wall. This is shown in Figure 2.4 as the deviation of fiber angles from the ventricular long axis. It can be seen that most of the fibers lying in the center of the left ventricular wall run at right angles to the long axis, whereas those near the inner and outer surfaces run longitudinally, more parallel with the long axis (Fig. 2.4) (see Streeter et al., 1969; Streeter, 1979).

The force and speed of contraction of the heart muscle is variable and is affected by such factors as the length of the fibers, and by stimulation of the nerve supply to the heart (adrenergic nerve terminals release norepinephrine, a powerful stimulant of contraction, into the myocardium).

Many diseases directly attack the heart muscle

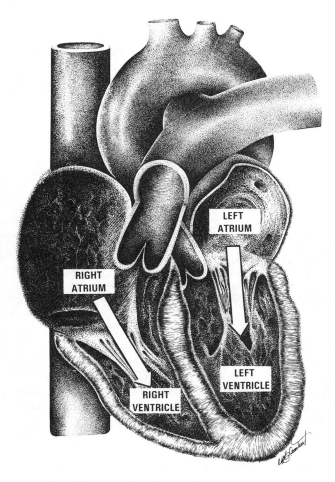

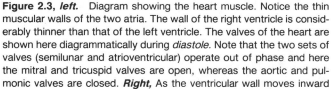

Figure 2.3, *left*. Diagram showing the heart muscle. Notice the thin muscular walls of the two atria. The wall of the right ventricle is considerably thinner than that of the left ventricle. The valves of the heart are shown here diagrammatically during *diastole*. Note that the two sets of valves (semilunar and atrioventricular) operate out of phase and here the mitral and tricuspid valves are open, whereas the aortic and pulmonic valves are closed. ***Right,*** As the ventricular wall moves inward to eject blood during *systole,* the atrioventricular valves (mitral and tricuspid) are shut and the semilunar valves (aortic and pulmonic) open, whereas during ventricular filling, opening and closure are reversed. Illustration is at a time just after the onset of ventricular ejection, before significant emptying of the ventricles has occurred into the aorta and pulmonary artery. For further discussion see text.

(for example, viral infestation is termed "myocarditis"), and other chronic forms of myocardial inflammation may accompany diseases such as rheumatic fever. Extensive damage to the heart muscle results in heart failure (Chapter 18).

Valves of the Heart

There are four one-way valves within the heart. Two of them lie between the ventricles and the great arteries and are called semilunar valves (Fig. 2.5); the aortic valve prevents blood from leaking backward from the aorta into the left ventricle, and the pulmonic valve performs a similar function between the pulmonary artery and the right ventricle (Fig. 2.3, *left*). The two other valves, called atrioventricular valves (Fig. 2.5), prevent leakage of blood backward from the ventricles into the atria

when the ventricles are contracting to eject blood into the great vessels (Fig. 2.3, *right*); the two-leaflet mitral valve serves this function between the left ventricle and left atrium, and the three-leaflet tricuspid valve prevents backward leakage of blood from the right ventricle into the right atrium (Fig. 2.5).

The two sets of valves, semilunar and atrioventricular, function out of sequence, that is, when one set is open, the other is closed. As shown in Figure 2.3, the aortic and pulmonic valves are closed when the ventricles are in the resting phase of the heart beat, called *diastole*, whereas the atrioventricular valves are open during diastole, allowing the two pumping chambers to fill from the atria (Fig. 2.3, *left*). During the contraction phase of the heartbeat, called *systole*, the ventricles develop pressure and

ENDOCARDIUM

DECILE NO.

MID-WALL

100μ

EPICARDIUM

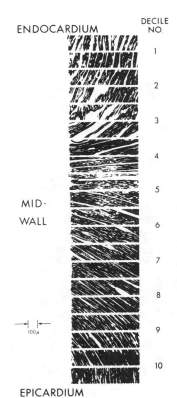

Figure 2.4. Myocardial fiber directions across the wall of the dog's left ventricle form endocardium (the inner wall, decile 1) to epicardium (the outer wall, decile 10). The fibers are viewed as if facing the front of the heart, and it can be seen that the bulk of the fibers in the midwall tend to run in a circumferential (horizontal) direction, whereas fibers in the inner (endocardial) and outer (epicardial) layers tend to run more vertically, along the long axis of the heart. (From Streeter et al., 1969.)

CUSPS OF SEMILUNAR VALVE OF PULMONARY ARTERY

CUSPS OF SEMILUNAR VALVE OF AORTA

ANNULI FIBROSI

MITRAL VALVE

TRICUSPID VALVE

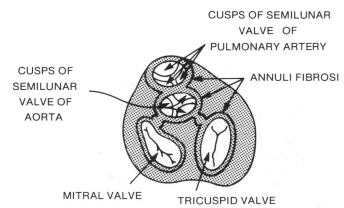

Figure 2.5. View of the heart, above, with great vessels and atria removed. There is a fibrous skeleton connecting all four valves, with an annulus fibrosis surrounding each set of valve leaflets. At the center is the aortic valve with three semilunar cusps, and above it is the similar pulmonic valve. The bicuspid mitral valve separates the left atrium from the left ventricle (atrioventricular valve), and the three-leaflet tricuspid atrioventricular valve separating the right atrium and ventricle is also shown.

eject blood, and the atrioventricular (mitral and tricuspid) valves snap shut, while the semilunar (aortic and pulmonic) valves open (Fig. 2.3, *right*).

Understanding the function of the valves becomes important in considering the significance of normal and abnormal heart sounds, as they are heard over the chest wall through the stethoscope (Chapter 13).

A number of disease states can affect one or more valves of the heart. Such disorders can be acquired,

RIGHT CORONARY ARTERY

LEFT MAIN CORONARY ARTERY

CIRCUMFLEX CORONARY ARTERY

CORONARY VEIN

LEFT ANTERIOR DESCENDING CORONARY ARTERY

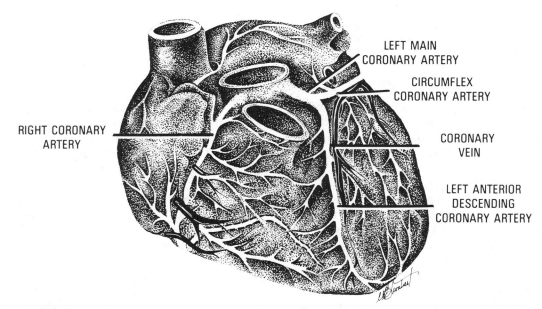

Figure 2.6. Simplified scheme of the coronary arteries. In most human subjects, the left coronary artery divides into two branches supplying the front wall of the heart (anterior descending coronary artery) and the lateral wall of the left ventricle (circumflex coronary artery),

whereas the single right coronary artery provides branches to the right ventricle and then circles backward to supply the posterior wall of the left ventricle *(dashed lines)*.

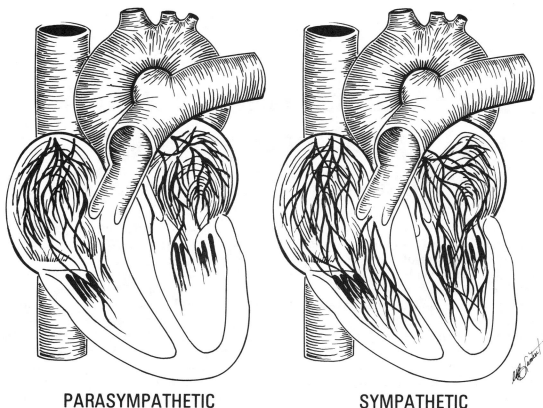

PARASYMPATHETIC

SYMPATHETIC

Figure 2.7. Diagrammatic representation of the autonomic nerve supply to the heart. There is rich innervation of the SA and AV nodes and of both atrial and ventricular myocardium by the sympathetic nervous system. The parasympathetic system predominantly supplies the SA and AV nodes and the atrial myocardium with only a few branches to the ventricles.

as with rheumatic fever (which leads to valve scarring), or due to a congenital deformity. Either can lead to valve leakage, or to valve narrowing with partial obstruction to blood flow.

Coronary Circulation

The coronary arteries are the first branches to arise from the aorta, and flow through these vessels supplies oxygen and nutrients to the heart itself (Fig. 2.6). Since the heart is an active organ, the coronary circulation provides a rich blood supply to the myocardium.

Disease of the coronary arteries, most often due to atherosclerosis, can cause impairment of the blood supply to the heart. When a coronary artery closes off completely, a heart attack (death of the heart muscle) usually occurs rapidly, often with fatal consequences (Chapter 15).

Autonomic Innervation of the Heart

Sympathetic nerve fibers (adrenergic nerve endings containing norepinephrine) ramify in the heart muscle of the atria and ventricles, as well as in the sinus node and the atrioventricular junction (Fig. 2.7). Stimulation of this system speeds up the electrical pacemaker, facilitates conduction through the specialized conduction system, and increases the force of heart muscle contraction. Parasympathetic fibers (cholinergic nerve endings containing acetylcholine) ramify in the pacemaker, the atrioventricular junction, and in the muscle of the two atria (there is little cholinergic innervation of the ventricles) (Fig. 2.7). Stimulation of the parasympathetic nervous system slows the pacemaker, delays conduction through the specialized conduction system, and reduces the force of the contraction of the atria (Chapter 16).

Under normal resting conditions, the parasympathetic and sympathetic systems interact to regulate the heartbeat at a rate of about 70/min. Activation of the sympathetic nervous system by emotional or physical stress (see Section 9) increases the heart rate and the force of the heartbeat, thereby augmenting the output of blood from the heart. Simultaneously, sympathetic stimulation of the peripheral arteries and veins also causes release of norepinephrine into the walls of those vessels and into the circulation from the adrenal glands, thereby raising the blood pressure and increasing the return of venous blood to the heart (Chapter 16).

CHAPTER 6

Structure-Function Relations in the Peripheral Circulation

In order to understand the function of the circulation and the dynamics of pressure and flow throughout the peripheral vascular bed, it is necessary to consider the structural characteristics of various vascular components in relation to their function. The general arrangement of the system of conduits that constitute the circulation can be diagrammed as shown in Figure 2.8. On the *left*, large phasic pressure changes are generated by the left ventricle as it delivers the stroke volume into the aorta. The elastic properties of the large arteries and aorta allow them to expand rapidly as the volume of blood enters the arterial bed. This expansion of the large arteries during systole allows them to store blood transiently, which is then passed onward by recoil during diastole (the arteries acting much like an accessory pump, thereby maintaining blood flow during diastole); this process has been termed the "Windkessel Effect" (Fig. 2.8). Thus, the phasic character of blood flow during systole as it crosses the aortic valve is progressively modified as it moves further down the arterial system, to yield a more and more steady flow pattern.

The small arteries and arterioles determine the systemic vascular resistance. The most distal vessels in the arterial distribution system (the tiny terminal arterioles and small arcade arterioles) open and close to regulate blood flow through the capillaries; these "sphincter vessels," terminal arterioles that can close their lumen by infolding of the endothelial cells, shown in Fig. 2.8 (Schmid-Schönbein and Murakami, 1985) therefore determine the area of the capillary bed perfused. The postcapillary resistance vessels (small venules) help to regulate capillary hydrostatic pressure and thereby affect fluid exchange in the capillaries. These vessels are numerous and therefore have small pressure drops. Finally, the larger venules and veins of the systemic circulation contain most of the blood volume, and in some veins relatively minor changes in the diameter of these vessels can modify the return of blood to the heart by changing their capacity to store blood. (These veins are therefore termed the "capacitance vessels" [Fig. 2.8].)

The terms resistance and capacitance vessels represent idealizations. Every blood vessel is distensible and therefore has capacitance and every vessel has viscous pressure drops and therefore has resistance. A highly distensible vein, as in the skin, can distend enormously and reduce its resistance, but when it is collapsed its resistance is high and may exceed that of a resistance vessel of comparable size. In some regions of the circulation, arteries are more distensible than veins (e.g., in skeletal muscle), but major arteries are stiffer than veins.

The progressive fall of intravascular pressure in the normal circulation is also shown in Figure 2.8, beginning with the pump (left ventricle) (Joyner and Davis, 1987). The precapillary resistance vessels (arterioles and the precapillary sphincter region) provide the largest pressure drop, from a mean pressure of about 85 mm Hg in the aorta to 38 mm Hg (Fig. 2.8). Across the capillary vessels, the pressure drops from about 38 mm Hg to about 25 mm Hg (in the postcapillary venules), and the remainder of the pressure fall occurs in the postcapillary resistance venules to reach the normal venous pressure of 5 mm Hg in the large capacitance veins. In terms of vascular resistance, about 70% lies in the precapillary area, 10% in the capillaries, and 20% in the postcapillary region (Mellander and Johansson, 1968).

The *mean velocities of blood flow* through the various segments can be deduced from the geometry of the systemic circulation, since at any point the mean linear velocity equals total volume flow divided by the cross-sectional area (CSA) of the vascular bed:

$$\text{Velocity (cm/s)} = \frac{\text{Volume flow (cm}^3\text{/s)}}{\text{CSA (cm}^2\text{)}}$$

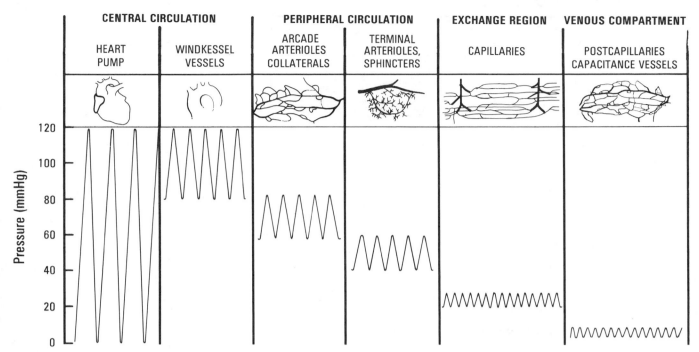

Figure 2.8. Blood pressures in different regions of the systemic circulation. The greatest pressure drop and reduction of pulse pressure occurs in arteries and arterioles. Note that arterioles and venules in most peripheral organs seem to form meshworks.

The smallest cross-sectional area which receives the entire output of blood from the heart is in the aorta (about 3.2 cm² in man) and, accordingly, the mean flow velocity is highest in that vessel. For example, if the output of blood from the normal resting heart (cardiac output) is about 6 liters/min (100 cm³/s) divided by 3.2 cm² equals a mean velocity of 31 cm/s. The peak aortic flow velocity and other vessel diameters and velocities are shown in Table 2.1. As the arteries progressively divide into smaller and smaller vessels the total cross-sectional area increases, and flow velocity correspondingly falls (Fig. 2.9). Each capillary is tiny (Table 2.1), but since the overall capillary bed contains many billions of vessels, it has a total cross-sectional area several hundred times that of the aorta (cross-sec-

tional area = 2000 cm²), and hence the mean blood flow velocity falls several hundred fold, yielding the slowest mean flow velocity in capillaries (approximately 0.05 cm/s). This slowing of blood velocity in the capillaries (Fig. 2.9) provides a highly favorable setting for exchange of gases and metabolites across the short length of the capillaries. They have large contact area with the tissue and still have a small pulsatile pressure. The two venae cavae exhibit a mean velocity about one-half that of the descending aorta (Table 2.1).

The average *distribution of the blood volume* within the various segments of the circulation is shown in Table 2.2 and Figure 2.9. The total volume of blood contained in a normal human subject is about 7% of body weight (5 liters in a 75-kg man), so that the total blood volume must circulate about once per minute. Despite a large cross-sectional area, the numerous capillaries are short and of small caliber, and contain only about 4–6% of the total blood volume; the systemic arteries contain about 12–14% of the blood volume, whereas the veins (mainly the small veins) contain nearly two-thirds (64%) of the total blood volume, emphasizing their role as a reservoir for blood. About 9–10% of the blood volume is contained in the pulmonary circulation, and 6–7% is within the heart itself during diastole (Table 2.2) (Milnor, 1968).

The *distribution of the cardiac output* to the various major organ systems of the body is shown sche-

Table 2.1.
Vessel Diameter, Blood Velocity, and Reynolds Number for the Systemic Circulation of Man

Structure	Diameter, cm	Blood Velocity, cm/s	Tube Reynolds[a] Number
Ascending aorta	2.0–3.2	63[b]	3,600–5,800
Descending aorta	1.6–2.0	27[b]	1,200–1,500
Large arteries	0.2–0.6	20–50[b]	110–850
Capillaries	0.0005–0.001	0.05–0.1[c]	0.0007–0.003
Large veins	0.5–1.0	15–20[c]	210–570
Venae cavae	2.0	11–16[c]	630–900

(From Whitmore, 1968). [a]Assuming viscosity of blood is 0.035 poise. [b]Mean peak value. [c]Mean velocity over indefinite period of time.

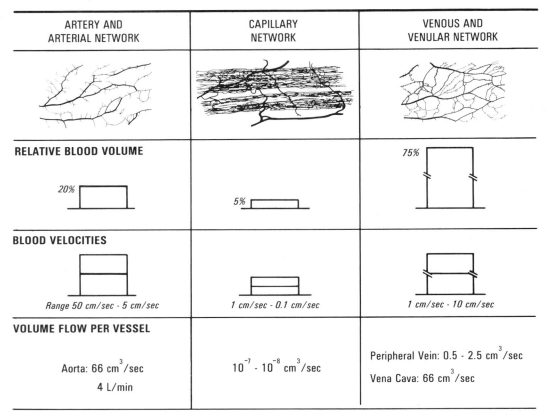

ARTERY AND ARTERIAL NETWORK	CAPILLARY NETWORK	VENOUS AND VENULAR NETWORK
RELATIVE BLOOD VOLUME 20%	5%	75%
BLOOD VELOCITIES *Range 50 cm/sec - 5 cm/sec*	*1 cm/sec - 0.1 cm/sec*	*1 cm/sec - 10 cm/sec*
VOLUME FLOW PER VESSEL Aorta: 66 cm^3/sec 4 L/min	$10^{-7} - 10^{-8}$ cm^3/sec	Peripheral Vein: 0.5 - 2.5 cm^3/sec Vena Cava: 66 cm^3/sec

Figure 2.9. The magnitude of blood volume, blood velocity, and volume flow per vessel. The volume flow changes over many orders of magnitude in the circulation due, in part, to vessel diameter variations.

matically in Figure 2.10. The arrangement of these circulations "in parallel" off the main arterial bed is evident. In the resting state, approximately 15% of the cardiac output goes to the brain, 15% to the muscles, 30% to the gastrointestinal tract, and 20% to the kidneys; the coronary circulation takes about

Table 2.2.
Estimated Distribution of Blood in Vascular System of Hypothetical Adult Man[a]

Region	Volume ml	%
Heart (diastole)	360	7.2
Pulmonary		
Arteries	130 ⎫	2.6 ⎫
Capillaries	110 ⎬ 440	2.2 ⎬ 8.8
Veins	200 ⎭	4.0 ⎭
Systemic		
Aorta and large arteries	300 ⎫	6.0 ⎫
Small arteries	400 ⎪	8.0 ⎪
Capillaries	300 ⎬ 4,200	6.0 ⎬ 84.0
Small veins	2,300 ⎪	46.0 ⎪
Large veins	900 ⎭	18.00 ⎭
	5,000	100

(From Milnor, 1968). [a]Age, 40 years; weight, 75 kg; surface area, 1.86 m^2.

5%, and the skin and bones 10%. During exercise, of course, the proportion going to the coronary circulation and the working skeletal muscles increases markedly. Notice that the arterial blood supply which nourishes the lung tissue (bronchial arteries, a circulation distinct from the blood flow through the pulmonary artery) recirculates directly back to the left atrium and left ventricle, so that the left ventricle pumps slightly more blood than the right ventricle (less than 3% of the cardiac output) (Fig. 2.10).

In the systemic circulation the distribution of blood within organs and among organs is achieved via a meshwork of arterial connections (also denoted in the microcirculation as arcade arterioles and in some organs as collateral or a plexus). This meshwork assures that the capillary flow in the most distal regions of the circulation is the same as in proximal vessels.

VASCULAR SMOOTH MUSCLE

The presence of smooth muscle cells in the walls of the arteries, arterioles, and veins give these vessels the capability of changing their caliber. The smooth muscle cells are in general arranged circum-

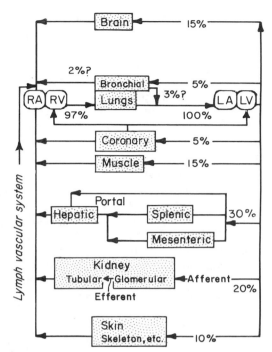

Figure 2.10. Stylized block diagrams of the circulatory system. A scheme of the circulatory system including the major vascular components (heart, arteries, microcirculation, veins) which distribute to and collect blood from the tissues. The special circulatory beds to be discussed separately are indicated and the relationship to the lymphatic collecting duct system represented. *Solid lines* depict arteries and veins, *arrowheads* show the direction of flow, and *shaded blocks* indicate the microcirculation (arterioles, capillaries, and venules). Approximate percentages of cardiac output going to the various circulatory beds are indicated near each line.

nally (which may be somewhat analogous to the Z lines of skeletal muscle, see Section 1, Chapter 4). Sarcoplasmic reticulum is present in smooth muscle, although it is less well developed than in skeletal or cardiac tissue. Smooth muscle cells appear to have a rather low resting transmembrane potential (-40 to -60 mV), and excitation spikes occur due to induced alterations in membrane potential (Somlyo and Somlyo, 1981). In some types of vascular smooth muscle, spontaneous action potentials are observed as a consequence of changing resting membrane potential.

Arteries down to the precapillary vessels and large veins are innervated by adrenergic nerve fibers, which tend to be located in the outer muscular wall or at the boundary with the outer connective tissue coat. The innervation of the veins is somewhat less dense than that of the arteries. Arterioles about 50 μm in diameter have seven or eight nerve fiber bundles giving rise to a network that surrounds the vessel. The discharge rate of the autonomic nerves to the arterioles appears to be sufficient to maintain a resting level of contraction or tone. There appears to be close contact of nerve terminals with the muscle cells in small arteries, while in larger arteries the neurotransmitter norepinephrine released from neural storage vesicles may encounter large diffusion distances. There is some evidence that "key cells" once activated by neurotransmitter may activate cells toward the inner wall and along arteries by initiating propagated action potentials, which travel via the nexus junctions. In different parts of the vascular tree removal of autonomic vasoconstrictor fibers has different effects; certain regions of the skin are under powerful sympathetic control and show little intrinsic muscle tone after denervation, whereas regions such as the brain, heart, and skeletal muscle exhibit autoregulation (see below), tend to override neural influences metabolically, and exhibit a strong degree of intrinsic tone after sympathectomy (Johnson, 1980).

The sarcoplasmic reticulum appears to be the major calcium storage site in vascular smooth muscle, with contraction and relaxation under cytoplasmic calcium control (Johansson, 1981). Thus, the levels of intracellular calcium appear to determine the contractile state of vascular smooth muscle. (This does not appear to operate through the troponin-tropomyosin mechanism, and there is some evidence that calcium control of myosin-light chain phosphorylation is involved through activation of light chain kinase; this may be the site of cyclic-AMP action.) Current evidence indicates that besides the temperature many vasoactive sub-

ferentially around the vessel, and are found in varying numbers in different components of the circulation (for general reviews on vascular smooth muscle see Bulbring et al., 1970; Bohr, 1978; Johansson, 1981). The smooth muscle occupies the middle portion of the blood vessel wall, tending to be arranged in a spiral or a helical manner in larger vessels, with a dominantly circumferential direction in the small arterioles. Smooth muscle cells vary widely in size among species (in the pig, for example, they average 100 μm long and 3 μm in diameter). They are generally narrow and tend to be arranged in muscle bundles, which can respond as an "effector unit." Nexus junctions (similar to "gap junctions" of striated muscle) have been identified by electron miscroscopy and appear to represent sites for electrical coupling between cells; other areas of the cell-to-cell junctions appear to serve for tension transmission. The cells are packed with myofilaments, which lie largely parallel to the long axis; their arrangement is less uniform than in skeletal muscle, and the myofilaments appear to attach to the plasma membrane and to "dark bodies" inter-

stances and substances elaborated by endothelial cells (endothelial relaxing factor [Fridovich et al., 1987; Furchgott and Zawadzki, 1980; Ignarro et al., 1989]) operate to alter the force of smooth muscle contraction through changes in intracellular calcium. Norepinephrine and some other agents that stimulate contraction also may produce greater depolarization and increase the frequency of spike potentials. However, pharmacological effects clearly can occur without changes in membrane depolarization, and it is considered likely that alterations in membrane permeability (such as to calcium) are important. Locally reduced metabolites also affect vasomotor tone including CO_2, lactic acid, ATP breakdown products (such as adenosine), and other potent vasoactive substances such as histamine, prostaglandin, and serotonin (Somlyo and Somlyo, 1981). However, their role in regulating basal vascular tone is uncertain.

Smooth muscle exhibits a length-active tension relation (enhanced stretch increasing the force of contraction). In this connection, transmural distending pressure within the vessel is important, either increased pressure or pulsatile pressure producing a myogenic vascular contraction (increased tone); the mechanism of this response is not clearly established. As mentioned, some smooth muscles have inherent myogenic tone as a result of electric spike discharge or nonelectrogenic mechanisms.

With long-standing increases in transmural pressure, vascular smooth muscle (like other muscles) undergoes hypertrophy (increased cell size), with thickening of the muscular layer. This is found predominantly in the heart and the central arteries but to a lesser degree in the microcirculation. In such thickened arteries, a given degree of smooth muscle shortening produces a greater reduction in luminal diameter, leading to hyperreactivity of the arteries in hypertension.

TENSION IN VASCULAR WALLS

Calculation of tension in blood vessels is based on the principle of mechanical equilibrium. Consider a blood vessel with internal radius r and wall thickness h. The internal pressure P is balanced by an average stress σ (force/unit wall area in circumferential direction) such that:

$$\sigma = \frac{rP}{h}$$

The resultant tension in the wall (force per unit wall length, expressed as dyne/cm) is:

$$T = \sigma h = rP$$

This law indicates that the smaller the radius the greater the mechanical advantage in terms of the tension that must be borne by the vessel wall. Thus, a capillary with its very small diameter (7 μm) and relatively thick wall (1 μm) can support an intravascular pressure of 25 mm Hg with a stress of only 2×10^4 N/m^2, whereas the aorta with its large radius and relatively thin wall carries a tension in the wall of 16×10^4 N/m^2 (Fig. 2.11A). This is also illustrated using a balloon of nonuniform shape (Fig. 2.11C); the tension in the wall is much lower at the small end of the balloon than in the central, dilated portion. When the wall is relatively thick, as in the arterioles, a relatively lower wall tension exists, and the smooth muscle bears a smaller load despite a relatively high intravascular pressure; this provides a mechanical advantage for changing blood vessel diameter in these small vessels. The equations above also have importance in considering aortic aneurysm (a large saccular bulging area, usually commencing in an area of disease in the aortic wall). In an aneurysm, any rise in blood pressure produces an increase in radius, with further thinning (reduction of h) of the aortic wall, thereby resulting in a further rise in wall stress σ; this sequence leads to further stretching, etc. Thus, an aneurysm can become unstable, and expansion eventually leads to rupture of the aorta in some individuals (see also influence of lateral pressure in locally dilated vessels, Chapter 7).

LARGE ARTERIES AND VEINS

The relative sizes and structures of various components of the vascular system are summarized in Figure 2.11A (Schwartz et al., 1981). The upper numbers represent the diameter of the lumen and the wall thickness (Burton, 1954). It can be seen that the channel of the aorta is very large (2.5 cm), with a relatively thin wall (2 mm). The ratio of wall thickness to lumen diameter rapidly increases as the smaller arterioles are reached. Notice also that the proportion of muscle increases as the arteries become smaller, whereas the relative proportion of elastic tissue decreases. The outer layer of connective tissue around a blood vessel is termed the *adventitia*, the middle muscular layer is called the *media*, and in most arteries there are one or more *elastic layers*, one of which lies inside the muscular wall.

All blood vessels are lined by a single layer of cells, the *endothelium*. In the arteries and the veins the cells of the endothelium are thicker than in capillaries (see below), and there tend to be few fenestrations between cells. The important role of the

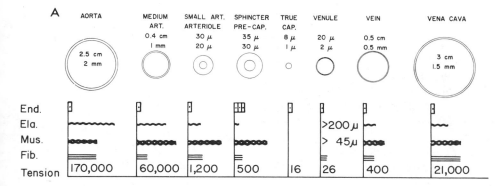

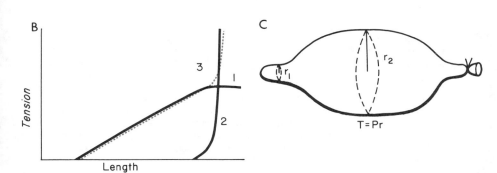

Figure 2.11. Illustrated are the lumen diameters, wall thickness, relative amounts of different tissues, and wall tensions (dynes/cm²) in the various blood vessels. *Lower left:* tension-length diagrams of elastic (*1*) and collagen fibers (*2*) presented alone and in combination (*3*). *Lower right:* balloon demonstrating pressure-tension relation. The pressure (*P*) within the balloon is equally transmitted to all parts of the contained air (Pascal's law), and the tension (*T*) in the balloon varies with the radius (*r*) in that portion, Laplace's law. (Adapted from Burton, 1954, Wolfe, 1952.)

endothelium in preventing intravascular blood clotting is discussed in Section 3, Chapter 24.

The blood vessels, like many biological tissues, do not obey Hooke's law when passively stretched. (Hooke's law states that when purely elastic materials are stretched, the tension is linearly proportional to the degree of elongation.) The more blood vessels are stretched the more they resist stretch; thus, blood vessels develop a higher tension the more they are distended, becoming progressively stiffer or less compliant (Fronek and Fung, 1980). Blood vessels are also *viscoelastic*, i.e., at any length the wall tension depends on time and vice versa. The pressure-volume curve of a large artery such as the aorta is shown in Figure 2.12. Notice the relatively steep rise in pressure as volume is changed, giving a nearly exponential relation. This shape of the curve reflects, in large part, initial stretching of the elastic fibers, which are partially folded and tend to obey Hooke's law up to their point of yield (Fig. 2.11*B*, curve 1). At that point, the collagen fibers (fibrous tissue) have been stretched from loose curved shapes into straight fibers as the vessel wall becomes stiffer; these fibers are very stiff with small changes in length causing large changes in tension (Fig. 2.11*B*, curve 2). The resultant of these two curves is a sum of forces in the two types of fibers, the lower flatter portion

being determined largely by elastic fibers (as well as muscle) and the upper steep portion being determined by the nondistensible collagen (Fig. 2.11*B*, curve 3) (Burton, 1954; 1972). The pressure-volume characteristics of larger arteries such as the aorta tend to become stiffer with age (*dotted line*, Fig. 2.12). This alteration would explain how a larger

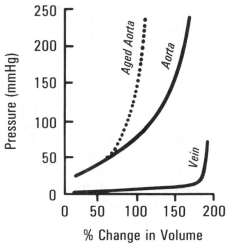

Figure 2.12. Compliance of central blood vessels. Not all arteries will harden during aging; some arteries actually may become softer with age, to the point of arterial aneurysm.

pulse pressure results from a normal stroke volume in older subjects (Chapter 7).

As shown in Figure 2.11*A*, the larger veins and venae cavae tend to have a large lumen and a relatively thin wall. The pressure-volume curve of veins tends to be quite flat over the physiological range of pressures, that is, large changes in the volume of blood contained in the reservoirs can occur with minimal change in the pressure, a property valuable to the capacitance function of the venous bed (Fig. 2.12); eventually, at high distending pressures, the venous bed will also become stiff (Fig. 2.12).

The walls of the veins contain smooth muscle, and the veins are innervated down to the level of the venules. The details of the innervation are organ dependent. Reflex changes in sympathetic tone affect the caliber of the veins and thereby the size of the storage compartment (Chapter 16), as well as to some degree the compliance of their walls.

ARTERIOLES

The arterioles have a high muscle content, and the ratio of wall thickness to channel diameter is high, thereby favoring their role in regulating vascular resistance (Fig. 2.11*A*). The arterioles are innervated by the autonomic nerves. Several phenomena, which occur predominantly in arteriolar beds, deserve discussion.

Critical Closure

When an organ, such as skeletal muscle or the coronary circulation, are perfused at low pressure, one can see that the flow stops although the arterial pressure is larger than the venous pressure. This pressure has been designated as "critical closing pressure" (Burton, 1972) or also "zero flow pressure." The critical closing pressure can have different origins, it may be determined by collateral inflow to the arteriolar meshwork; in certain situations it may be due to Rouleaux formation by red cells in the blood, to external high tissue compression compressing microvessels, or to high vascular smooth muscle tone with closure of small arterioles (as seen in terminal arterioles). It can also be caused during transient flow changes, since the pressure does not immediately follow the flow in distensible blood vessels. A series of different zero-flow intercepts can be produced as the critical closing pressure of a given vascular bed is changed by varying the vasomotor tone with different levels of sympathetic nerve stimulation (Fig. 2.13). It can also be seen from this diagram, that critical closure can occur at a transmural pressure of 30 mm Hg, for example, when there is a relatively high degree of sympathetic nerve stimulation, whereas with low

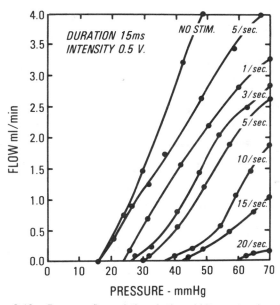

Figure 2.13. Pressure-flow relations in the rabbit ear at various levels of stimulation of the sympathetic nerve supply. Note that increased frequency of sympathetic stimulation produces vasoconstriction, with a shift of the curves to the right and a decreased slope, together with a higher pressure intercept (higher critical closing pressure). (From Girling, 1952.)

sympathetic tone substantial flow remains at this pressure level (Girling, 1952).

Autoregulation

Most vascular beds, particularly those with relatively low neurogenic control (such as the coronary, skeletal muscle, and cerebral circulations), exhibit "autoregulation" (for general discussion, see Johnson, 1964, 1980; Haddy and Scott, 1977); that is, regardless of changes in perfusion pressure over a substantial range, the local vascular resistance will change to maintain the blood flow relatively constant (see also discussion of this phenomenon in the coronary circulation, Chapter 15). Thus, as long as the metabolic requirements of the tissue supplied by that vascular bed remain constant, the blood flow will be regulated by local mechanisms. This type of autoregulatory response is diagrammed in Figure 2.14 *(solid line)*; the curve has a sigmoid shape with a central relatively flat portion where flow is regulated, a steep portion at the lower end of the pressure-flow curve where autoregulation fails, and another region at the high end of the pressure-flow curve where autoregulation is incomplete (Fig. 2.14). The shape of the curve and the position of the lower "knee" (the point at which flow starts to fall) differ in various tissues, and the speed of the autoregulatory response also may differ (for example, in the heart it takes 8–10 s for autoregulation to occur after the pressure is suddenly changed).

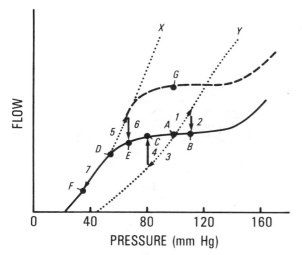

Figure 2.14. Diagram of autoregulation produced by changing the perfusion pressure in a vascular bed, while keeping the metabolic activity constant *(solid line)*. Autoregulation at a higher level of metabolic activity is shown by the *dashed line*. The *dotted lines* show pressure-flow relations prior to the autoregulatory response. For further discussion see text.

In Figure 2.14, two pressure-flow curves in the steady state at different levels of vasomotor tone in the absence of autoregulation are shown by the *dotted lines* (X and Y), and the autoregulated curve is shown by the *solid line*. For example, at a perfusion pressure of about 95 mm Hg, at a given level of resting metabolic function of the muscle, blood flow is regulated at point A (if the tissue were metabolically active so that its oxygen consumption were increased, flow at that pressure would be metabolically regulated to a higher level, so-called "active hyperemia" point G, the *dashed line* indicating that the curve is shifted upward) (Stainsby, 1962). If from *point A* the perfusion pressure is suddenly increased to 110 mm Hg, flow will initially rise along curve Y *(arrow 1)*, but active constriction of the muscle of the arterioles will then occur, increasing the vascular resistance, so that flow is regulated *(arrow 2)* back to nearly the same level but at the higher perfusion pressure *(point B)*. Conversely, if the perfusion pressure is suddenly dropped from 95 mm Hg to 80 mm Hg flow will initially fall along curve Y *(arrow 3)*, but then smooth muscle relaxation occurs, lowering the vascular resistance and returning flow to near the normal level *(arrow 4, point C)* but at the lower perfusion pressure. Curve X (Fig. 2.14) represents the unregulated pressure-flow curve when the arteriolar bed is *maximally dilated* (its lowest vascular tone). *Point D* is situated on the lower knee of the autoregulated curve; if perfusion pressure is raised to about 65 mm Hg, the flow will initially follow the curve of maximum dilatation *(arrow 5)* and then return to a slightly

increased flow at *point E (arrow 6)*. On the other hand, if at that point perfusion pressure is lowered *below point D*, e.g., to 35 mm Hg, since the autoregulatory reserve is exhausted, flow will *fall* as perfusion pressure is lowered *(arrow 7 to point F)*, i.e., the pressure-flow relation follows curve X.

The mechanism of autoregulation remains uncertain (Johnson, 1980). Four mechanisms have been proposed: *myogenic, metabolic, tissue pressure*, and *flow-dependent dilation*. Some investigators believe that a myogenic response related to the degree of stretch of the smooth muscle is the predominant factor producing alterations in vascular resistance during autoregulation. In fact, an isolated vascular muscle strip will, when stretched, contract to a length shorter than that existing prior to the stretch. Others believe that a chemical mediator is likely to be the dominant factor; thus, under conditions of constant oxygen consumption if a vasodilator metabolite were being produced at a constant rate, transiently increasing the flow would wash out the metabolite at a more rapid rate and reduce the concentration of the vasodilator, thereby increasing vascular resistance; conversely, if perfusion pressure were dropped, the vasodilator would wash out at a reduced rate, increase its concentration, and decrease vascular resistance. Such chemical mediators undoubtedly function during active hyperemia (muscle contraction), but their nature remains unknown in skeletal muscle and in most other tissues; much information suggests that in the coronary circulation adenosine is important (see Chapter 15). A third potential mechanism in autoregulation is a change in tissue pressure, at least in some tissues (such as the kidney) in which the vascular bed lies within an encapsulated organ. Thus, increased perfusion pressure might result in filtration of fluid and an increase in tissue pressure, which could then tend to compress the arterioles and reduce flow back toward its previous level (and vice versa). In arteries and arcade arterioles a flow-dependent dilation can be observed which seems to occur at constant pressure. It is possible that during autoregulation, all four mechanisms may participate, their relative importance depending on the region of the circulation and the tissue involved.

Some of the most significant experiments indicating a role for myogenic phenomena are those showing an increase in precapillary resistance in response to elevation of the *venous* pressure which decreases flow. The myogenic response of the arteriolar smooth muscle to stretch appears to act in concert with the metabolic control, but metabolic control may override the myogenic control (Meininger et al., 1987). For example, in the presence of high tissue P_{O_2} in muscle, the arterioles will constrict irrespective of the vascular pressure (Duling and Klitzman, 1980). Direct observations of arterioles in the microcirculation have also supported the presence of myogenic responses, showing decreases in diameter in response to increased arterial pressure, and arteriolar dilation in response to a lowered pulse pressure. In fact, studies on pulse pressure effects have suggested that about one-third of the basal tone of arterioles in skeletal muscle of cats may be of myogenic origin stimulated through pulsatile effects on the arterioles (Grande et al., 1979).

MICROCIRCULATION

For practical purposes we shall consider the microcirculation as comprised of the smallest arterioles and the precapillary sphincter region, the capillaries, and the venules (for general reviews see Landis, 1964; Zweifach, 1973; Fung, 1977; Wiedeman et al., 1981, Renkin and Michel, 1984; Baker and Nastuck, 1986).

Capillaries

The structure of capillary endothelium varies widely in different organs, being relatively continuous in the brain and heavily fenestrated in the liver (for review see Karnovsky, 1968). The special structure of capillaries in the kidney and lung are described in Chapters 26 and 35. The capillaries vary in diameter from 3–8 μm (depending on the species; ordinarily, they are about the size of a red blood cell or slightly smaller), and they are lined by a single layer of endothelial cells (Fig. 2.15). The cells of the capillary endothelium are thinner than those of the arteries and veins and, in addition to fenestrations, in some tissues they may exhibit clefts between cells (Fig. 2.16); they also contain vesicles (Fig. 2.16) which, in rare instances, form tubular channels entirely across the endothelial cells. The capillary endothelium in the heart and skeletal muscle is about 1 μm thick (except at the nucleus where it is slightly thicker) and shows intercellular clefts and vesicles, but no fenestrations. These clefts or slits (sometimes called aqueous pores) average about 2.5 nm in width, and show

Figure 2.15. Section of capillary wall stained with silver nitrate to demonstrate the intercellular cement. Nuclei of the cells stained with hematoxylin.

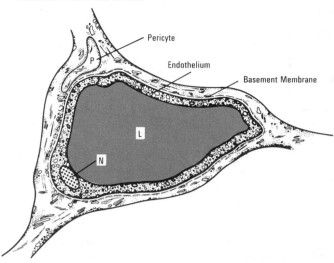

Figure 2.16. Ultrastructural image of skeletal muscle capillary. The endothelium contains numerous vesicles, microtubules, and microfilaments and is surrounded by a continuous basement membrane. The interstitium typically contains pericytes and collagen fibers. *N* = nucleus, *L* = lumen of vessel.

narrowed areas (tight junctions) which may limit large molecule movement (Fig. 2.16). The clefts have been calculated to be relatively few (less than 0.2% of the capillary surface area) and allow bulk flow of water and ready diffusion of smaller, water-soluble substances such as NaCl (except in the brain). As molecular size approaches pore size, however, diffusion is increasingly limited (Landis and Pappenheimer, 1963) (see subsequent section on transcapillary exchange). In the transport of large molecules there is controversy about the relative importance of fenestrations and vesicles (the latter process is termed *pinocytosis* or vesicular transfer).

The basement membrane which lies outside the endothelial cells may be particularly important in limiting permeability in some tissues, but in many capillary beds overall permeability is affected both by the endothelium and the basement membrane. Some rigidity to the single layer of endothelial cells of the capillaries is also provided by the basement membrane. Outside the basement membrane is the "ground substance," composed largely of mucopolysaccharides, which also may affect diffusion properties across the capillary wall. The capillaries contain no smooth muscle and are not subject to active contraction although endothelial cells may change shapes with a pressure reduction.

The *precapillary region* consists of small blood vessels that retain a single smooth muscle coat. This region functionally controls blood flow distribution and hence the area of the capillary bed perfused. It encompasses a relatively wide range of vessel sizes, including small arcade arterioles, terminal arterioles (in some organs like the mesentery they have also been designated as metaarterioles) (Fig. 2.17), and precapillary arterioles (9–16 μm) which contain only a single layer of smooth muscle cells and give rise directly to several capillary branches in many tissues. The precapillary sphincters are smaller and usually serve only one or two capillaries. Some of the metaarterioles give rise to so-called "thoroughfare channels" which in some tissues allow partial or complete bypass of the capillary bed (Zweifach, 1973). Thoroughfare or preferential channels in some beds, such as the mesentery, offer lower resistance arteriovenous connections and also exhibit lateral sphincters serving capillaries (Fig. 2.17). However, in other beds (particularly skeletal muscle) thoroughfare channels are sparse or absent, and the arcade arterioles branch into terminal arterioles (about 15 μm in diameter) which then subdivide into many capillaries (Schmid-Schönbein et al., 1986). There is a variety of branching patterns, lengths, diameters, and flow velocities between (and within) capillary beds (for example, see Klitzman and Johnson, 1982).

The *postcapillary venules* (15–20 μm) contain pericytes (cells which lie outside the endothelium, whose function may be related to endothelial cell mitosis but is not well established) and are unable to regulate their caliber; they constitute the main postcapillary resistance region, although it is probable that postcapillary resistance control lies also in larger venules up to 300 μm in diameter. The collecting venules (35–45 μm) contain no smooth muscle and they, in turn, join the large venules (up to 1000 μm), which do have smooth muscle and begin to serve the capacitance function performed by larger venules and veins.

Control of the Microcirculation

A variety of methods has been used to study the microcirculation, including electron microscopy, high power study of living tissue (vital microscopy), high speed cinemicrophotography, fluorescence or television microscopy (by which red cell velocities and vessel diameters can be measured, red cell velocities on the order of a 1 mm/s having been measured in a number of tissues in animals as well as in man), microtechniques for determining intravascular pressure, hematocrit, and local oxygen partial pressure (Goodman et al., 1974; Wiedeman, 1963; Wiederhielm and Weston, 1973; Baker and Nastuck, 1986). Such techniques have allowed measurement of capillary density as well as the dynamic behavior of the microcirculation; for example, as mentioned local contractile responses to passive stretch have been demonstrated in some tissues. Microscopic studies using fluorescent staining of nerve terminals have shown adrenergic innervation down to the level of the precapillary sphincter in some tissues, and direct stimulation of the perivascular nerves has allowed measurements of contractile responses in both the small arteries and veins. β-Adrenergically mediated dilatation of precapillary sphincters has also been demonstrated directly by injection of β-agonists, and a variety of potent, naturally occurring vasoactive substances has been shown to have direct or indirect effects on the microcirculation. These include CO_2 (a particularly potent regulator of the cerebral circulation), P_{O_2}, pH, and polypeptides such as bradykinin which are released from lysosomes or by activation of the blood components during tissue injury. Certain prostaglandins and histamine are also potent vasodilators.

There is no agreement as to which local factors regulate flow through the capillary bed in most tissues. Many of the chemical mediators mentioned above, as well as adenosine, have been considered. Some factor related to oxygen metabolism undoubtedly is involved, and increasing local tissue metabolism produces increased flow. In addition, decreasing the P_{O_2} in the arterial blood enhances local perfusion, whereas increasing P_{O_2} is associated with diminished capillary density and a decrease in perfusion (Duling and Klitzman, 1980). Thus, capillary density is sensitive to tissue P_{O_2} in most tissues, although the related direct chemical mediators remain unknown (Duling, 1980). The tissue P_{O_2}, in turn, is regulated by the capillary P_{O_2} and the rate of tissue O_2 consumption (VO_2); these factors together with the intercapillary distance determine the gradient for O_2 diffusion (see next section). Thus, regulation of flow and capillary density (surface area) are highly important control mechanisms in the microcirculation. The regulation of resistance vessel caliber in the pre- and postcapillary regions

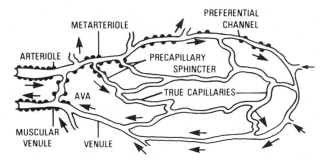

Figure 2.17. Terminal vascular bed, with preferential channels. *AVA*, arteriovenous anastamosis. (Adapted from Zweifach, 1949, as redrawn by Burton, 1972.)

also affects capillary pressure, thereby influencing transcapillary filtration, as further discussed in the next section.

TRANSCAPILLARY EXCHANGE

The capillary endothelium is relatively impermeable to proteins and large molecules. Substances such as crystalloids and gases in solution which are lipid soluble are relatively freely diffusible. Substances which are not lipid soluble have to follow aqueous channels and their transport is greatly retarded. Delivery or uptake of diffusion-limited substances are importantly related to the capillary transit time. Hydrostatic pressure and colloid osmotic pressure largely govern the movement of fluid in and out of the capillaries, whereas larger molecules move by several processes. It is convenient to consider transcapillary transport in terms of three processes: diffusion, filtration, and large molecule movement. (For reviews see Landis, 1964; Landis and Pappenheimer, 1963; Curry, 1986.)

Diffusion

The driving force for diffusional transport is a concentration gradient. For small molecular weight substances this is by far the major process by which nutrients and metabolic products cross the capillary membrane. The rate of movement of solutes, and the fluid carrying them, is far greater than that transferred by filtration. Small, lipid-soluble molecules (such as CO_2 and O_2) move freely by diffusion across the endothelial cells, as well as through the capillary pores. Water and water-soluble materials are believed to cross mainly through water channels along the microscopic clefts (pores or slits) between endothelial cells. For small, water-soluble molecules such as NaCl and glucose pore size is not an important consideration, and the rate of delivery of the substance to the capillary bed (blood flow rate) primarily determines their transport; movement of these substances is *flow-limited.* Larger molecules (such as sucrose, 358 daltons) begin to be limited by pore size and macromolecules, such as antibodies, are not effectively transported by diffusion. In general, such lipid-insoluble *diffusion-limited* substances show diffusion rates that are inversely related to their molecular size.

There are five main factors which affect diffusion between the intravascular and tissue compartments. These include: (1) the *intercapillary distance;* (2) the *blood flow;* (3) the *concentration gradient* for the solute; (4) the *capillary permeability;* (5) the *capillary surface area.*

The effects of intercapillary distance and blood

flow rate on O_2 transport serve as a useful example for other readily diffusible substances (although O_2 delivery is complicated by the equilibration of O_2 in solution with hemoglobin*; see discussion in Section 5, Chapter 37). The diffusion of O_2 out of a capillary is dependent on the radial gradient from the vessel interior to the tissue, the intravascular O_2 partial pressure, and the consumption of O_2 in the tissue.

Since O_2 is continuously extracted from the blood under normal circumstances, the local P_{O_2} drops continuously from the arterial to the venous end of the capillary. This drop in P_{O_2} depends on flow rate, capillary length, the tissue O_2 consumption ($\dot{V}O_2$), and myoglobin binding.

The *radial gradient* has traditionally been analyzed using Krogh's tissue cylinder concept (Krogh, 1919; 1959), which considers some of the factors responsible for radial diffusion from a capillary. The Krogh equation employs the radius of the capillary and the radius of the tissue cylinder around it, together with the capillary concentration or partial pressure of a substance and its diffusion coefficient, to calculate the concentration or partial pressure of the substance at a given distance from the capillary† (see Fig. 5.20, Section 5, Chapter 37).

The radial gradient can be calculated at any point along the capillary. Obviously, if two capillaries are far apart, tissue P_{O_2} can fall to a very low value between them, whereas if the intercapillary distance is small (capillary density and surface area high), the tissue P_{O_2} can be well maintained. Under conditions such as exercise in skeletal muscle, the high $\dot{V}O_2$ may produce a low tissue P_{O_2}, despite marked capillary perfusion. It has been hypothesized that at high $\dot{V}O_2$ in heart muscle, the great variability of capillary lengths and flow velocities in capillaries can result in inhomogeneity of tissue P_{O_2}; under some conditions this can lead to "islands" where tissue P_{O_2} is insufficient to maintain mitochondrial respiration, so-called "tissue hypoxia,"

---

*The level of hemoglobin (Hgb) is an important determinant of O_2 delivery, as well as the characteristics of O_2-Hgb dissociation. O_2 begins to diffuse through the arterial wall before it reaches the capillary, but most of the O_2 is unloaded from Hgb when the capillary P_{O_2} falls below 60 mm Hg, as the dissociation curve steepens, and small further decreases in P_{O_2} then result in large reductions in Hgb saturation (see Fig. 5.12, Section 5, Chapter 37).

---

†For oxygen

$$\text{mean } P_{TO2} = P_{capO2} - A\,\frac{VO_2 R^2}{4D}$$

where P_T and P_{cap} = partial pressures in tissue and capillary, A = area of the tissue cylinder outside the capillary, R = radius of the cylinder from the center of the capillary (related directly to intercapillary distance), and D = O_2 diffusion coefficient.

and decreased tissue function or cell damage can result (mitochondrial function decreases at tissue P_{O_2} below 0.1 mm Hg) (Honig and Bourdeau-Martini, 1975).

Fick's law of diffusion considers the remaining three variables (concentration gradient, capillary permeability, and surface area) as follows:

$$M = D\frac{A}{T}(C_i - C_o)$$

where M = the diffusion flux of the material, D = the free diffusion coefficient for that substance (which is related to its molecular weight). A = capillary surface area, T = the thickness of the membrane (to which diffusion rate is inversely proportional), and $C_i - C_o$ = the concentration difference for the substance from inside to outside of the vessel (see further discussion relative to the lung, Section 5, Chapter 35).

For a thin-walled membrane, the values for D, T, and A are usually replaced by PS (in which P = permeability and S = surface area of the capillary), and the equation then becomes:

$$M = PS(C_i - C_o)$$

Since the permeability rarely changes (except under pathological conditions), PS becomes proportional to the capillary surface area under many conditions. Some substances (such as O_2) begin to diffuse out of the arteries well before the true capillary vessels are reached, but all demonstrate a progressive fall of the concentration gradient as they pass from the arterial to the venous end of the capillary bed. Of course, CO_2 concentration is higher outside the capillaries than within, and this substance moves into the capillary in accordance with its concentration difference.

Most metabolites, nutrients, and other common substances (such as urea and glucose) are quite readily diffusible, and only small concentration differences are required for their transport. For most of these flow-limited (not diffusion-limited) substances, their transport can then be determined from a modified Fick equation:

$$M = Q(C_a - C_v)$$

where Q = the blood flow rate and $C_a - C_v$ = arteriovenous concentration difference. This equation is valid in situations where tissue and capillary concentrations reach equilibrium during passage of the blood.

Filtration

Fluid movement results from a hydrostatic or osmotic pressure difference across a membrane (actually causing bulk flow of fluid, a process distinct from diffusion which depends on concentration gradients). Such bulk fluid movement occurs across the endothelium and in the interstitium. The plasma proteins, primarily albumin and globulin, are in high concentration within the blood compartment (6.0 − 8.0 g/100 ml or g/dl), whereas protein concentrations in the interstitial fluid are relatively low (0.7 − 2.0 g/100 ml). The proteins (with hyaluronic acid in solution in the interstitial space) are restricted in free transport across the endothelium and therefore produce a *colloid osmotic pressure* (oncotic pressure) which is determined primarily by the number of molecules in solution. Thus, about 75% of the total oncotic pressure of the plasma results from albumin which has a relatively small molecular weight (69,000), and the remainder is due largely to globulins. It should be pointed out that some proteins can and do leak through pores in the capillary wall or are moved by vesicular transport, and these proteins and other substances that filter out of the capillary wall are returned to the circulation by the lymphatic system (see below).

Several factors determine the rate at which fluid is filtered across the capillary membrane. These include:

1. The *filtration coefficient* for a given capillary wall. This depends on the tissue (for example, this coefficient for water is much higher in the kidney than in skeletal muscle), and the coefficient also may vary under different physiological conditions. In many organs the filtration coefficient is lower on the arterial end than in venules and it depends on the properties of the endothelium. Inflammatory substances such as histamine and others greatly enhance the filtration coefficient across the endothelium.

2. The *capillary hydrostatic pressure*. There is, of course, a gradient of pressure from the arterial to the venous end of the capillary (Fig. 2.18), and the intracapillary pressure, strongly influenced by microvascular network anatomy due to interconnection, can be modified by changes in the caliber of the precapillary as well as the postcapillary resistance vessels.

3. The *interstitial fluid hydrostatic pressure*. The level of hydrostatic pressure in the interstitial fluid is controversial, both small positive and small negative values having been reported under normal conditions (the average value is probably close to zero) (Guyton, 1963). Under abnormal conditions,

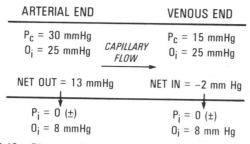

ARTERIAL END		VENOUS END
P_c = 30 mmHg	CAPILLARY	P_c = 15 mmHg
O_i = 25 mmHg	FLOW →	O_i = 25 mmHg
NET OUT = 13 mmHg		NET IN = –2 mm Hg
↓		↓
P_i = 0 (±)		P_i = 0 (±)
O_i = 8 mmHg		O_i = 8 mm Hg

Figure 2.18. Diagram of a capillary showing the effects of hypothetical plasma and tissue hydrostatic and oncotic pressures on transcapillary fluid movement. For further discussion, see text. *P*, pressure; *O*, oncotic pressure in the capillary (*c*) and interstitial (*i*) regions.

however, this pressure may become substantially positive (such as with severe swelling [edema] of the legs or other tissues).

4. The *colloid osmotic pressure of the plasma*. This is determined by the plasma protein concentration. The colloid osmotic pressure tends to draw water into the capillary in accordance with its concentration. Of course, in disease states where the serum albumin is low, this pressure can become abnormally reduced.

5. The *colloid osmotic pressure of the interstitial fluid* (discussed earlier).

An equation developed by E. H. Starling modified to include interstitial colloidal pressure, has been applied to describe ultrafiltration across a single blood capillary. It should be understood, however, that the capillary beds of various organs differ markedly in their filtration properties (for example, the kidney *filters* throughout its entire capillary length, whereas *absorption* may occur predominantly in some capillary beds in the intestinal wall, and permeability may vary in portions of the same capillary bed). In the general scheme shown in Figure 2.18, filtration predominantly occurs at the arterial end of the capillary and absorption at the venous end. *Starling's law of ultrafiltration* states that:

$$FM = K(P_c + \pi_i - P_i - \pi_c)$$

In this expression *FM* (ml/time/unit pressure) is the total fluid motion (over the entire extent of an organ). For such a definition *K* depends on capillary surface. The same equation, however, can also be formulated on a per area basis. In this case *FM* is local flux per unit area (ml/time/unit area/unit pressure). The *K* is also expressed on a per area basis and is independent of capillary surface area, recruitment, etc., where *FM* = fluid movement (+ outward, − inward), *K* = filtration coefficient of capillary wall, P_c = hydrostatic pressure in the cap-

illaries, P_i = hydrostatic pressure in the interstitial fluid, π_c = oncotic pressure in the plasma in the capillary, and π_i = oncotic pressure in the interstitial fluid.

Thus, net filtration rate will be determined by the difference between the forces tending to move fluid outward (P_c and π_i) and forces tending to move fluid inward (P_i and π_c). If the coefficient is ignored, it can be seen from Figure 2.18 that at the arterial end of this particular capillary the net driving pressure outward is about 38 − 25 mm Hg = 13 mm Hg (causing bulk flow of water toward the interstitial fluid [*arrows*]), whereas toward the venous end of the capillary the net driving pressure is inward 23 − 25 mm Hg = − 2 mm Hg (causing fluid movement into the capillary [*arrows*]). In skin, the outward pressure at the arterial end has been calculated at +7 mm Hg and the inward pressure at the venous end at −9 mm Hg (Landis and Pappenheimer, 1963).

A more important controlling mechanism than this delicate hydrostatic-osmotic balance in the individual capillaries of many tissues, however, is the presence of *vasomotion* in the precapillary beds. Thus, direct measurements in the microcirculation have shown spontaneous, relatively prolonged periods of contraction or even closure of terminal arterioles or precapillary sphincters, producing a lower capillary hydrostatic pressure (5–10 mm Hg); this type of behavior dominates in some tissues, such as skin and muscle, so that the oncotic pressure in the plasma favors fluid reabsorption. Occasionally, the precapillary vessels spontaneously open widely, and capillary pressures as high as 40–50 mm Hg (favoring net filtration) are transiently recorded (Wiederhielm and Weston, 1973; Colantuoni et al., 1984).

As shown in Figure 2.19, an increase in either arterial or venous pressure will increase capillary hydrostatic pressure, and vice versa. Capillary pressure (P_c) is controlled both by the precapillary resistance (R_1) and the postcapillary resistance (R_2). An increase in arteriolar (precapillary) resistance (R_1) will lower capillary pressure, whereas an

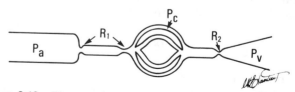

Figure 2.19. Diagram of an artery, vein, and a microcirculatory unit. P_a equals pressure in the artery. P_v equals pressure in the vein. Resistance (R_1) in the arteriole and precapillary resistance vessel is shown together with that of the postcapillary venule (R_2). *PC*, capillary pressure.

increase in venous (postcapillary) resistance (R_2) will increase capillary pressure. The normal ratio of $R_1/R_2 = 4/1$ (that is, 70–80% of the resistance is in the precapillary region). Since the venous resistance is much lower than the arterial resistance, an increase in venous pressure has a substantially larger effect on capillary hydrostatic pressure than does a change in arterial pressure. If R_1 over R_2 increases it will favor net reabsorption, whereas if it decreases net filtration will be favored. Both of these resistances can operate independently to maintain the R_1/R_2 ratio relatively constant. For example, when the venous pressure rises, as upon assuming the standing position, capillary pressure will tend to rise; however, the increased transmural pressure results in myogenic constriction of the precapillary vessels (arterioles and sphincters), which may actually close capillary beds and reduce the capillary surface available for filtration; therefore, fluid accumulation (edema) usually does not occur under these conditions (see subsequent discussion of edema).

It should be pointed out that the filtration coefficient, K, in the Starling equation, which importantly influences the movement of fluid, is affected by the capillary surface area as well as the permeability of the membrane; the surface area may be greatly changed during capillary "recruitment" or by closure of the precapillary sphincters. If the other values in the Starling equation are known, and membrane permeability is assumed not to change, the coefficient can be calculated and will provide some measure of the relative capillary surface area (the "capillary filtration coefficient," Landis and Pappenheimer, 1963).

Large Molecule Movement

Such movement can occur possibly by vesicular transport or by convection directly through capillary fenestrations. In vesicular transport, enclosure of plasma within vesicles occurs and the vesicle then moves from one side of the cell membrane to the other, thereby transporting large molecules (such as proteins) from the inner to the outer membrane surface. This process has also been called micropinocytosis; its relative importance in many tissues remains controversial. Fenestrae ("windows") which allow large molecule movements exist in many capillary membranes, although in certain tissues such as skeletal muscle and nervous tissue fenestrae are few. Fenestrated endothelial cells are found in the intestine, kidney, endocrine organs, and very large vascular openings are present in the liver and bone marrow. As mentioned, even in capillaries without apparent fenestrae, some protein

"leak" occurs, and this has been theoretically accounted for by a very few large clefts (Landis and Pappenheimer, 1963).

LYMPHATICS

Structure of the Lymphatic System

Lymphatics are present in most tissues in close association with the blood vessels of the microcirculation and the tissue parenchyma (Fig. 2.20). One of their functions, to return excess tissue fluid and proteins to the intravascular compartment, is intimately related to the movement of these substances across the capillary endothelium. The role of osmotic pressure and hydrostatic pressure in controlling the flow of fluids out of the vascular system, is discussed elsewhere previously. As shown in Figure 2.20, the knob-like ends of lymphatic capillaries in a gland are shown almost in contact with blood capillaries. The direction of flow of fluids and solutes out of the arterial capillary and into tissue spaces and back into the venous capillaries and into lymphatic capillaries is shown by the arrows.

By combining the injection of colloidal materials and electron microscopy, fundamental observations have been made about the structure and function of the lymphatic capillary (Leak and Burke, 1968, 1968; Cliff and Nicoll, 1970; Leak, 1976; Skalak et al., 1984). The endothelium of the lymphatic capillaries is similar to that of the blood capillaries. There is little or no basement membrane. The diameters of the lymphatic capillaries vary widely from dilated terminal capillary bulbs to structures that are either very narrow and slit-like, or dilated cap-

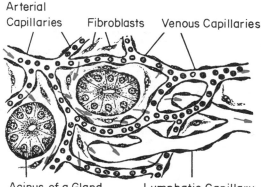

Figure 2.20. Diagram showing the relationship of lymphatic capillaries to blood capillaries and to the tissue fluids around the acini of a gland. Similar relationships exist in most of the organs of the body. *Arrows* indicate the direction of flow of fluid leaving the arterial capillaries, permeating the connective tissue spaces as tissue fluid, and re-entering the blood capillaries on the venous side. The lymphatic capillaries supplement the venous capillaries in the drainage of fluid from the tissues to the circulatory system. (From Copenhaver, 1967.)

illaries, depending upon their state of function and the organ. Pinocytosis has been clearly demonstrated; it is one method of transportation of material from the tissue spaces into the lymph capillaries. The electron micrographic studies of the lymphatic endothelium show that in some areas away from the nucleus, the endothelium is very attenuated, being in some places little more than the two membranes in width. There is considerable regional specialization of terminal lymphatic structure, but in most tissues the blind-ended channels tend to interconnect, and usually there are wide gaps or fenestrae between the cells of the lymphatic walls (Zweifach, 1973). In addition, there are collagenous anchoring filaments attached to lymphatic endothelium and extending out to unite with collagenous bundles in the adjoining tissue. The detail of this structure is shown in Figure 2.21. The lymphatic channels in skeletal muscle of the rat have been investigated in detail (Skalak et al., 1984). There, it was discovered that lymphatic channels are in close association with the arcade arterioles, in places almost completely surrounding them (Fig. 2.22). Whereas the lymphatic channels are of irregular shape and their walls consist of a single endothelial lining, the adjacent arteriole has the usual circular cross-section with a smooth muscle coat, capable of contracting against high pressure gradi-

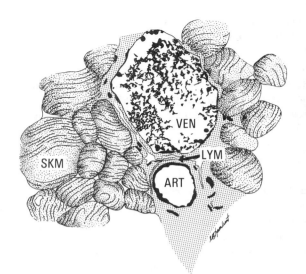

Figure 2.22. Lymphatic channels in skeletal muscle between an arteriole *(ART)* and venule *(VEN)*. The lymphatic has a continuous endothelium and is of irregular shape. If the arteriole is dilated, the lymphatic is virtually compressed, as shown, but as the arteriole contracts the lymphatic channel is stretched open. This suggests that arterial vasomotion or pressure pulsation may serve as a lymph pump mechanism in this organ. Arteriole, venule, and lymph vessels are surrounded by skeletal muscle fibers *(SKM)*. Skeletal muscle contraction may also compress and expand these lymphatics for lymph fluid formation.

ents. The lymphoarteriolar pair is deeply embedded in muscle fibers (Fig. 2.22). Contraction of the arteriole leads to expansion of the lymphatic channel, whereas dilation causes compression of the lymphatics. These results suggest that arteriolar vasomotion, arteriolar pulsation as well as muscle contraction all serve to compress and expand lymph terminals and thereby serve as a lymph pump. Retrograde flow is prevented by lymph valves. One of the secondary effects of the heart beat may be to pump lymph fluid. Such pairing of arteries and lymphatics can also be observed in the lung, the heart and many other organs.

When fluid accumulates in the interstitial spaces by transudation (or is injected), one can visualize that this fluid will occupy space between the lymphatic capillary and tissue collagenous fibers. Since the fibers attached to the lymphatic endothelium and the tissue collagen are inelastic, one would expect the overlapping endothelium to be pulled apart so as to provide large pores through which the edema fluid under greater pressure can flow into the lumen of the lymphatic capillary. The combination of a mechanism to open up the spaces between endothelial cells and the process of pinocytosis gives two mechanisms by which fluid and its solutes can be rapidly transported from blood capillaries through the interstitial spaces and then into the lymph capillaries. In the bat wing, Cliff and

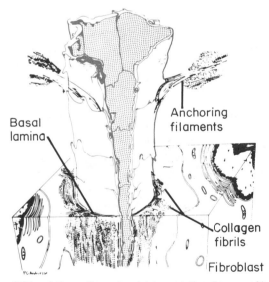

Figure 2.21. A three-dimensional, interpretative diagram of lymphatic capillary that was reconstructed from collated electron micrographs. The three-dimensional relation of the lymphatic capillary to the surrounding connective tissue area is illustrated. The lymphatic anchoring filaments appear to originate from the endothelial cells and extend among collagen bundles, elastic fibers, and cells of the adjoining tissue area, thus providing a firm connection between the lymphatic capillary wall and the surrounding connective tissue. Irregular basal lamina and collagen fibers are as marked. (From Leak and Burke, 1968.)

Nicoll (1970) have shown that the terminal lymphatic capillary bulbs enter into collecting ducts that possess a contractile wall with smooth muscle, a feature found in few other mammalian organs. These lymphatic channels are usually not paired with arterioles, they show spontaneous contractility, and have a smooth muscle and valves to prevent retrograde flow. The smooth muscle of such lymph channels is strongly myogenic. These emerge into transport lymphatic channels with very active contractility and valves. The transport lymphatics combine into larger and larger lymphatic vessels propelling the lymph to and through the lymph nodes and toward the great veins. Ultimately major lymphatic vessels are formed in the extremities and within the viscera, and then merge into the right lymphatic and thoracic ducts. From these, lymph pours into the bloodstream by way of the right and left subclavian veins.

In the course of flowing from the periphery to the entrance in the blood vessels, the lymph flows through one or more lymph nodes. As the lymphatics approach the node, they break up into finer channels called the afferent lymphatics. These penetrate into the sinuses of the cortex of the lymph node, then through the cortex and the medullary sinuses which are lined by phagocytic cells. One or more large efferent lymphatics formed by confluence of the numerous medullary lymphatics emerge from the hilum of the lymph node and progress onward to join the major lymphatic vessels.

In the walls of the abdominal cavity the lymphatics are most abundant on the undersurface of the diaphragm, where the greatest lymphatic absorption takes place from the peritoneal cavity. There is also substantial absorption into the lymphatics of the omentum. Lymphatic systems exist in all organs with the exception of the central nervous system and the cornea. Protein is not readily absorbed from the interior of the alveoli of the lung into the lymph capillaries because of the poor lymphatic supply, or possibly because it cannot readily penetrate the alveolar wall. Water passes readily into the lung capillaries, however, because of the low hydrostatic pressure within them.

Flow of Substances Between Blood and Lymph

There is a rapid and continual exchange between the intravascular and extravascular compartments by diffusion. Transcapillary exchange by filtration takes place on an exceptionally large scale without any perturbation to the circulating blood volume or the electrolyte balance; this movement of fluid into and out of the blood capillaries depends on the balance of hydrostatic and osmotic pressures acting across the capillary endothelium. The osmotic force exerted by the circulating plasma protein molecules within the blood capillaries counterbalances the hydrostatic pressure in the capillaries. The capillary wall, however, is not completely impermeable to plasma proteins, and all protein molecules leak through to a certain extent. Proteins are found in the interstitial fluid and lymph in lower concentration than in plasma (Yoffey and Courtice, 1956). The accumulation of proteins in the interstitial fluid, with an increasing osmotic pressure that would disrupt the balance of forces controlling the exchange of fluid across the capillary membrane, is precluded by the rapid flow of protein molecules directly into the lymphatics, presumably by flow through the gaps in the lymphatic capillary endothelium during local expansion of terminal lymphatics, either due to contracting arterioles or elastic recoil of surrounding tissue components.

The escape of plasma proteins into interstitial fluid varies with the tissue. It has been shown that a quantity equal to roughly 50–100% of the circulating plasma proteins escapes across the blood capillary membrane and reenters the blood through the lymphatics each day (Yoffey and Courtice, 1956).

Composition and Flow of Lymph

The *protein concentration* of lymph is always lower than that of plasma, even though there is considerable variation between lymphatics, depending upon the organ. As oral or intravenous fluid intake is increased, the protein concentration of lymph decreases. Recent experimental observations suggest that lymph/proteins are modulated during passage through the lymph nodes according to the hydrostatic and oncotic pressure in the nodal vasculature.

Lymph contains all of the *coagulation factors*, and clots, although less readily than blood plasma. There is a much higher concentration of some coagulation factors in the hepatic lymph in contrast to peripheral lymph, because these factors are made within the liver. Antibodies are found in the lymph, and will be discussed in more detail in Chapter 22. Almost all of the enzymes that are found in plasma are also found in lymph to a lesser extent.

The *electrolyte concentration* in lymph is not essentially different from that of plasma. The total cation concentration is slightly lower in the lymph than in the plasma, and chloride and bicarbonate levels tend to be higher. The direction of these differences in concentration is consistent with the Gibbs-Donnan equilibrium operating on two phases whose concentrations of nondiffusible ions (proteins) differ.

Of the plasma *lipids,* cholesterol and phospholipid being mainly associated with protein as lipoprotein, are present in the lymph in concentrations which vary with the level of protein in the lymph (Yoffey and Courtice, 1956). Neutral fat in the form of chylomicrons depends on the degree of fat absorption from the gastrointestinal tract. Immediately after meals there are large quantities of lipoproteins and fats in the lymph coming from the gastrointestinal tract. Between meals, the fat in the thoracic duct drops to low levels.

The other constituents of the plasma that are readily diffusible, nonprotein substances, and nonelectrolytes are present in the lymph in concentrations approximately the same as in the plasma.

The lymph contains *cells.* Lymphocytes of all sizes and degrees of maturity are the most numerous. A rare monocyte and macrophage is found. Platelets are not observed. Red cells, when present, indicate the degree of bleeding into the tissues as a result of injury or shock, or when found in the thoracic duct suggest the presence of intestinal parasites. Granulocytes are found in the lymph draining from areas of infection. Plasma cells are also found in small numbers. The concentration of lymphocytes in efferent lymph leaving a lymph node is always much greater than that in afferent lymph.

Any condition that increases the outpouring of fluid from the capillaries into the tissues will increase the flow of lymph if there is no obstruction. The factors influencing the flow are the pulse pressure, the contractility of the collecting lymph vessels, activity of the skeletal muscles, peristalsis, and a differential pressure between interstitial spaces and lymphatics. Presumably, the retention of fluid inside of the lymph capillary is in part due to the extraordinary low outflow resistance along the lymphatic ducts and the relative high resistance for fluid to return into the interstitial space across the lymphatic endothelium. Propulsion of lymph from one region to another is brought about by the extrinsic and intrinsic forces described earlier. The lymph pressure in the vessels will rise and the lymph will move centrally to a region where the pressure is lower. The numerous valves within the lymphatics prevent retrograde flow of the lymph.

Functions of the Lymphatic System

A most important function is obviously the return of protein, water, and electrolytes from tissue spaces to the blood. The lymphatics are exceptionally important in the absorption of nutrients, particularly fats, from the gastrointestinal tract.

The lymphatics serve also as a transport mechanism to remove red blood cells that have been lost into the tissues as the result of hemorrhage, or bacteria that may have invaded the tissues. When there is an infection in a distant part of the body, the regional lymph node becomes inflamed as a result of the localization of bacteria or toxins carried in the lymph to the gland. The lymph nodes contain a very efficient filtration system in the cortical and medullary sinuses; these are lined with phagocytic cells that engulf bacteria and red cells or other particulate material. The efficacy of the filtration system can be demonstrated by the direct perfusion of pathogenic bacteria into lymphatics that are afferent to a lymph node. Simultaneous culturing of the efferent lymphatic will show it to be sterile. The capability of the lymph node to filter out pathogenic bacteria can be overwhelmed, however.

Conditions that Increase the Lymph Flow

INCREASE IN CAPILLARY PRESSURE

Landis and Gibbon (1933) found that filtration from the capillaries shows a definite increase when the venous pressure rises above 12–15 cm H_2O. The rate of filtration from the capillaries is directly proportional to the increase in venous pressure (Fig. 2.23). In their experiments, when the venous pressure was increased to any given level the filtration

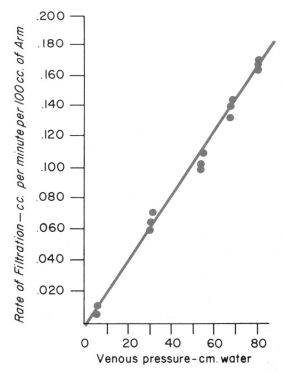

Figure 2.23. Rates of filtration produced during 30 min by venous pressures between 20 and 80 cm of H_2O. (Adapted from Landis and Gibbon, 1933.)

rate increased rapidly at first, but gradually slowed and finally ceased. This falling off in the filtration rate they ascribed to a rise in extracapillary pressure due to the fluid accumulation, which opposed the hydrostatic pressure within the blood capillary. The accumulation of extracellular fluid may be expected therefore to be greater in regions that are loose in texture and where the skin is readily stretched. In persons with firm, resistant skin, edema, for the same reasons, is later in appearing and less pronounced.

Increased pressure in the portal vein or in the hepatic veins produced, for example, by obstruction causes increased filtration into the tissues of the abdominal viscera and a great increase in the volume of lymph flowing through the thoracic duct. Increase in arterial pressure, on the other hand, does not increase filtration in animals until the pressure reaches around 300 mm Hg. Pulse pressure has a pronounced influence on lymph transport. If the rabbit ear is perfused at steady pressure no lymph fluid is formed. If, however, a pulse pressure is superimposed on the mean pressure, lymph transport is intiated.

INCREASE IN CAPILLARY SURFACE AREA

This increases the leakage of protein and fluid. It may follow any change that causes distension of the capillary vessels, e.g., (1) increase in capillary pressure; (2) increase in the local temperature; or (3) infusion of fluid.

INCREASE IN CAPILLARY PERMEABILITY

(1) *Rise in temperature* locally may increase capillary permeability. (2) *Capillary poisons* such as peptone increase the flow of lymph from the thoracic duct, probably as a result of their injurious effect upon the abdominal capillaries. Other substances that increase lymph flow in this way are extracts of strawberries, crayfish, mussels, leeches, histamine, and foreign proteins. To what extent this is due to a change in permeability and to what extent it is a consequence of capillary dilation is not known. (3) *Reduced oxygen supply* to the tissues increases lymph flow, probably because of dilation of blood vessels but possibly also through damage to the capillary endothelium.

HYPERTONIC SOLUTIONS

The intravenous injection of a concentrated solution of glucose, sodium sulfate, or sodium chloride causes an increased flow of lymph from the thoracic duct. These substances in concentrated solution may increase permeability of the capillary wall but also exert an osmotic effect that alters the normal equilibrium between the extravascular and intravascular fluids. Water at first enters the plasma from the tissue spaces, particularly of the muscles and subcutaneous tissues of the limbs, which in consequence show a fall in volume; the brain shrinks. Thus, removal of fluid may actually extend to the fluids within the cells which undergo shrinkage; general desiccation of the tissue may result. The blood volume for the time is greatly augmented, the excess fluid for the most part being accommodated in the capacious capillary and venous areas of the abdomen. As the electrolyte injected moves out of the plasma and into the interstitial fluid it carries water with it. The viscera—liver, kidneys, spleen, and intestines—then increase in volume, due not only to the distension of their vascular beds, but more importantly to the great outpouring of fluid as well as of protein that occurs through the capillary walls. This fluid then swells the volume of lymph in the thoracic duct. In this way these substances produce a redistribution of fluid.

The injection of isotonic saline will also increase the passage of protein and fluid from the capillaries and increase lymph flow. This too is probably a consequence of the changes in osmotic and hydrostatic pressures in the capillaries.

INCREASED FUNCTIONAL ACTIVITY

When a gland or muscle becomes active, an increase in lymph flow occurs which starts a little after the commencement of the secretory or contractile response, but is nearly synchronous with the increased metabolism resulting from the activity. The increased flow is ascribed to (1) formation of metabolites that increase the osmotic effect of the tissue fluids and cause more fluid to leave the vessels; (2) vasodilation, with increased capillary pressure and increased fluid and protein leakage.

During rest, the flow along the lymph vessels of the muscles and subcutaneous tissues is slight, and the protein content of the lymph is high. During activity, the protein concentrations fall, since less transuded water undergoes reabsorption into the blood and more is carried away by the lymph channels. This may occur even though the leakage of protein from the vessels is greater. The contracting muscles exert a pumping effect upon the lymph, driving it along the lymphatic vessels.

MASSAGE AND TISSUE MOVEMENT

These act to a certain extent like muscular activity. As discussed earlier, they augment the blood flow, capillary pressure, and capillary surface and so increase lymph formation, presumably by rhythmically expanding and compressing the terminal lymph channels in the tissue. The manipulations and movements of the muscles serve to propel the lymph along the lymphatic channels. Thus, walking, respiration, gentle skin massage, and exercise serve to increase lymph flow. Anesthesia reduces lymph flow rates.

EDEMA

Edema is a term applied to an excessive accumulation of fluids in the tissue spaces and is due to a disturbance in the mechanisms of fluid interchange, which has been considered in the preceding pages. Instead of there being a perfect balance between the inward and outward flow of fluid through the capillary membrane, absorption is exceeded by transudation. The particular factor or factors of the mechanism that are disordered are not always clear, and a satisfactory explanation of all forms of edema cannot be given. But from previous discussions it is evident that the following factors will tend to increase the volume of interstitial fluid: (1) reduction in the protein concentration of plasma (edema commences when the serum albumin concentration has fallen below 2.5 g/dl); (2) a general or a local rise in capillary blood pressure; (3) increased permeability of the capillary membrane; (4) increase in the filtering surface as when the capillaries dilate; and (5) obstruction of the lymph channels, cessation of vasomotion and elimination of pulse pressure.

There is a tendency for the accumulation of edema fluid to progress so far and then become stationary, provided that the conditions producing it remain constant, since eventually the interstitial pressure rises to reach a critical value and limits further transudation of fluid. Several factors operate to limit edema formation even under normal conditions. In fact, the venous pressure can rise to 10 mm Hg without obvious edema, and some fall in the plasma proteins also can occur without edema formation. Factors preventing edema include: (1) With slight edema, interstitial fluid hydrostatic pressures rises, and there is dilution of interstitial proteins (both effects would counteract further fluid filtration). (2) A rise in venous pressure causes myogenic arteriolar constriction (see "Autoregulation," Fig. 2.14), which will lower capillary pressure and decrease capillary surface area. (3) As tissue pressure rises with mild edema, the tethered lymphatics are widely open and, as discussed, there is a marked increase in lymph flow. (4) Finally, as the connective tissue and skin are stretched by edema toward their elastic limit, further edema formation will be restricted.

Since edema is only a symptom of some primary condition it may have a variety of causes, according to the particular disease with which it is associated.

1. *Cardiac Edema.* In congestive heart failure there is both an increase in extracellular fluid and in salt retention. The volume of this extracellular fluid can be reduced by restriction of the salt intake or by the administration of diuretics that remove both water and salt; or it can be discharged by improving cardiac action by digitalis.

2. *Mechanical Obstruction of Veins.* When the main veins leading from a part are obstructed by any mechanism, as in cirrhosis of the liver, thrombosis, etc., an increased transudation of fluid occurs. This is due in part to the rise in intracapillary pressure and an increase in the filtering surface, but the permeability of the capillary wall may also be increased.

3. *Edema Due to Renal Disease.* In chronic nephritis, edema is not usually pronounced unless the heart is failing; however, in the nephrotic syndrome it is an outstanding feature. In this condition a reduction in plasma protein as a consequence of the loss of protein in the urine leads to the passage of an abnormally large volume of fluid from the capillaries throughout the body. This in turn is probably responsible in some way for renal retention of salt and as a consequence of this, retention of water.

4. *Inflammatory Edema.* In this type, several factors combine to produce the fluid infiltration of the tissues. Increased capillary pressure occurs, due to dilation of the vessels and local slowing of the bloodstream as well as to thrombosis and obstruction of the returning veins. The lymphatics for a variable distance from the inflammatory area are damaged by bacterial toxin or granulocytes accumulating in the microcirculation so that a fluid with a high protein content escapes from the vessels. The edema is localized to an area of varying extent surrounding the injured site.

5. *Edema Caused by Malnutrition or Toxic Substances.* Edema may occur in the anemias or in conditions in which the general nutrition of the body suffers. When the diet is deficient in vitamins, or there is too little fat or protein in the diet, edema may occur, as in beriberi, scurvy, "war edema," or in the faulty nutrition of infants. In animals, edematous conditions have actually been induced by general underfeeding, or by a diet deficient in fat and soluble vitamins, or by one deficient in protein alone. The factors

responsible for the increased transudation in these cases are not always clear, but in others there is a marked lowering of plasma protein (more particularly in the albumin), which alone is sufficient to account for the edema.

Certain chemical substances such as arsenic, salts of heavy metals, and the toxins of certain infectious diseases, such as diphtheria, acute nephritis, etc., are known to act as capillary poisons and apparently cause edema in this way. Histamine causes local edema at the point of injection by inducing capillary dilation and increased permeability of the membrane.

6. *Chronic Lymphatic Obstruction.* Widespread obstruction of the lymph vessels may result from congenital or familial disorders of the lymph vessels or after lymphectomy. Acquired lymphedema results from obstruction of the lymph channels by cancer, scars, operative removal of lymph nodes, and fibrosis caused by X-ray therapy. It may also follow a low grade lymphangitis from parasites such as filariasis.

CHAPTER 7

Dynamics of the Peripheral Circulation

William Harvey (English physician, 1578–1657) largely dispelled the idea that blood ebbed and flowed through the arteries and veins. After studying in Padua under Fabricius, whose study of the venous valves apparently first generated the idea that blood flowed through the veins toward the heart, Harvey arrived at the conclusion that the blood must circulate around and around. In a series of experiments in animals involving vessel ligations and measurements of the volume of blood flowing from severed vessels, he concluded that "the movement of the blood is constantly in a circle, and is brought about by the beat of the heart." He also measured the volume of blood contained in the ventricles and calculated that the amount ejected by the ventricles in ½ hour was greatly in excess of the volume of blood contained in the entire body, thereby reinforcing his conclusion that the blood must circulate (despite the fact that the microscope was just coming into use and capillaries had not yet been demonstrated). These discoveries were summarized in a classic book published by Harvey in 1628 (see translation, 1928).

In this chapter, we consider primarily those features of the blood itself and of the peripheral circulation which are responsible for the patterns of pressure and blood flow through the arteries and veins consequent to the pumping action of the heart. Such properties of the pulmonary circulation are considered in Section 5, Chapter 36.

Certain physical properties of the blood are important in understanding the pressure and flow characteristics of the circulation. (The science of rheology is concerned with investigation of the deformation and flow of matter.) A number of different types of cells and formed elements are carried in the blood plasma (see Section 3, Chapter 19). With a normal red blood cell (erythrocyte) count of 5 million/mm^3 and a normal white blood cell count of approximately 7000/mm^3, the red cells constitute about 99.9% of the circulating cell population. Indeed, the red cells are so numerous that normally they occupy about 40% of the volume of the blood, a value expressed as a percentage termed the *hematocrit*. Blood plasma is a viscous fluid. The red blood cells modify this property so that blood assumes anomalous viscous properties (see below). In spite of their relatively small numbers, white blood cells play an important role in blood flow through the microcirculation and capillaries, especially in low flow states (ischemia) (Schmid-Schönbein and Engler, 1987).

The main forces that act on the blood in the circulation are listed below.

f_a = inertial force due to acceleration, including two types: (1) time acceleration due to the pulsatile ejection from the heart and (2) spatial acceleration, e.g., at entry into a constriction

f_v = viscous force due to fluid viscosity

f_p = pressure force, the driving mechanism in blood vessels; the energy is generated by the heart

f_g = gravitational force

According to Newton's law of motion

$$f_a = f_v + f_p + f_g. \qquad (1)$$

This is the fundamental law governing the motion of blood. The inertial and gravitational forces depend on the density of the blood, whereas the viscous force depends on blood viscosity. The magnitude of these forces varies widely throughout the circulation, with the exception of the gravitational force, which is constant on earth. An estimate of the relative magnitude of inertial and viscous forces is provided by Reynold's number, N_R.

$$N_R = \frac{\text{Inertial force}}{\text{Viscous force}} \qquad (2)$$

N_R is of the order of 1000 in large arteries, indicating that in these vessels the inertial forces are highly important. In the capillaries, however, $N_R = 10^{-3}$, indicating that inertial forces are negligible (for review, see Fung, 1984).

BLOOD VISCOSITY

Throughout most of the circulation, blood flow is nonturbulent (except at the valves and certain other areas, as well in abnormally narrowed regions produced by atherosclerosis where random pressure and velocity fluctuations can be observed); that is, blood flows through the vessels in an "orderly" manner, with flow velocity zero at the vessel wall and progressively increasing toward the center of the vessel. In spite of being laminar (i.e., not turbulent), the streamlines of flow are complex, with eddy formation, transient velocity fluctuations, recirculation zones, boundary layer formation in large vessels, etc. One of the most important forces determining these streamlines is the viscous force. It arises in any fluid because two layers of fluid with different velocities experience a friction or resistance to slippage. This feature varies from fluid to fluid and is due to a property of a particular fluid: its *viscosity* (for review see Cokelet et al., 1980).

A variety of instruments (viscometers) can determine the viscosity of fluids, and one method of measurement involves the passage of fluid at a known pressure through a tube of precisely known dimensions; the viscosity of the fluid can then be calculated from the flow rate using the Poiseuille equation (see below). A *Newtonian fluid*, such as water, is defined as one which maintains a *constant viscosity* at any flow velocity, whereas a non-Newtonian fluid changes viscosity at different flow velocities in such a tube. In a complex non-Newtonian suspension such as blood, the viscosity is most importantly determined by the red blood cells but also to a small degree by the plasma proteins in solution. In such a suspension, the apparent viscosity of blood *changes* considerably, depending upon the velocity of blood flow, the hematocrit, and the size of the blood vessel.

Viscosity, η, is defined as the ratio of shear stress (a tangential force per unit area, F/A, resisting or pushing flow) to shear rate, $\Delta V/\Delta X$ (the change of velocity ΔV between two neighboring fluid layers divided by their distance ΔX, measured normal to ΔV). Thus

$$\eta = \frac{\text{Shear stress}}{\text{Shear rate}} = \frac{F/A}{\Delta V/\Delta X} \qquad (3)$$

For a Newtonian fluid flowing through a cylindrical tube and exhibiting laminar flow, solution of *Eq. 3* after neglecting the inertial force, f_a, and gravitational force, f_g, yields a parabolic velocity profile across the channel, with the fastest velocity in the center and a velocity assumed to be 0 at the boundary with the wall (Fig. 2.24). With normal laminar flow of blood in arteries and veins, the profile is not strictly parabolic because the inertial force is not negligible and the geometry of the blood vessels is considerably more complex than that of straight cylindrical tubes; there are bifurcations, curvature, and, to a limited degree, taper.

The dimensions of viscosity are measured in dynes/cm^2 divided by (cm/s)/cm, a unit termed the *poise* (after Poiseuille). The practical unit 0.01 poise is termed the *centipoise*, and water at room temperature has a viscosity of about 1 centipoise. For convenience, the viscosity of various fluids is often related to that of water at the same temperature, and this relation is termed the *relative viscosity*. At a normal hematocrit, blood has a viscosity 3 or 4 times higher than that of water (relative viscosity of 3 to 4).

Plasma behaves like a Newtonian fluid but, because of the suspended red cells, changes in whole blood viscosity are markedly nonlinear with alterations in flow velocity. Thus, as flow velocity decreases, viscosity increases, and vice versa (Fig. 2.25). The shape of this viscosity-velocity relationship is determined entirely by the behavior of individual red blood cells (Chien et al., 1966; Chien, 1975). At shear rates below 1 (Fig. 2.25), "rouleaux" formation occurs, with increasing aggregates of stacked red cells causing increasing viscosity (the blood proteins fibrinogen and globulin promote this, and the effect is lost in their absence). At increasing shear rates over 1 (Fig. 2.25), less marked changes (decreases) in viscosity occur, which appear to relate to deformation of the red cells (primarily to elongation and longitudinal orientation of the membrane). Red cell deformation is

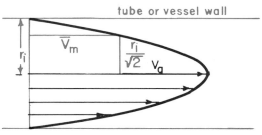

Figure 2.24. Parabolic flow velocity profile plot. The plot represents calculation of flow velocity (*V*) at various radial distances (*r*) from the center of the tube to the wall (where flow is assumed to be stationary, *V* = 0). Flow is fastest in the central axial stream (*V*$_a$) and decreases to 0 at the wall. At a distance $r/\sqrt{2}$ from the center of the tube, the flow velocity is half that along the central axis and equals the mean flow velocity (*V*$_m$) along the tube. The arrows indicate relative flow velocities at various radial distances (*r*) from the central axis.

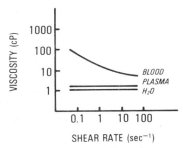

Figure 2.25. Relationship between the log of viscosity and the shear rate (related to blood flow velocity). Notice the nonlinear relation in whole blood and the essentially flat (Newtonian) relationship in plasma and water (the viscosity of plasma being slightly higher than that of water). (Adapted from Chien, 1966.)

associated with reduced viscous dissipation in the plasma, the red cell membrane, and the hemoglobin solution; this progressive decrease in viscosity is prevented by making red cells nondeformable (see Chien et al., 1966). It can be seen from Fig. 2.25 that, in the high velocity gradients occurring in large blood vessels (see Chapter 6), the curve is relatively flat, with viscosity changing little with flow velocity changes (relatively Newtonian behavior). At slower flow velocities, the shear-dependent increase in viscosity is marked (viscosity effects may become important in the veins, particularly when the cardiac output is very low). In certain disease states, such as sickle cell anemia, the red blood cells become less deformable due to polymerization and gel formation of the hemoglobin, and this can lead to an increase in blood viscosity. The viscosity is also increased when abnormally high concentrations of proteins are present in the plasma, as occurs in some disease states (Dintenfass, 1976).

The viscosity of blood is also markedly dependent upon the hematocrit, and this effect, as determined in vitro, is diagrammed in Figure 2.26. The effect of

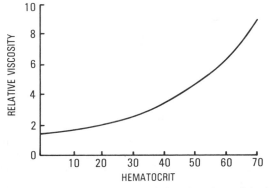

Figure 2.26. Relationship between relative viscosity and the hematocrit in whole blood at constant shear rate. Notice that at a normal hematocrit, blood is three to four times as viscous as water and, at higher hematocrits, the viscosity increases sharply.

hematocrit on blood viscosity increases sharply above a hematocrit of about 50%, as in diseases associated with overproduction of red blood cells (i.e., polycythemia vera, where hematocrits of 60–70% can occur) or in individuals living at high altitudes. In such conditions, substantial increases in the blood viscosity can occur with unfavorable effects on blood flow and the work of the heart. Relative blood viscosity in the intact circulation (as in the hind limb of a dog) is actually considerably lower over a range of hematocrits than that measured in glass tubes, although the shape of the curves is somewhat similar.

The origin of this phenomenon lies in an important effect of vessel size on viscosity, i.e., a decrease in viscosity occurring at tube radii below 0.5–1.0 mm (Fig. 2.27). This phenomenon is known as the Fahraeus-Lindqvist effect (1931), and it tends to favor flow through the capillaries, as well as through the small arterioles. This effect may, in large part, explain the lower blood viscosity measured in vivo (mentioned above). The reason for the effect is complex (see Fung, 1977 and 1981), but it has been ascribed at least in part to a phenomenon causing the layers of blood next to the vessel wall to be relatively cell-free and leading to lower hematocrits in small blood vessels. Thus, there is a tendency for red cells to move toward the higher velocity region in the center of a tube, increasing the hematocrit there while the hematocrit decreases near the walls; the cell-poor layer near the wall becomes proportionally larger as the vessel diameter becomes smaller. The rate of flow of the plasma through blood vessels is substantially slower than the rate of movement of the red blood cells. This phenomenon results from the fact, mentioned above, that the central (axial) portion of the bloodstream contains a greater proportion of red cells moving at a greater velocity. Therefore, the mean transit time of the plasma, which moves more uniformly across the wall, is slower. In microvessels, side branches can therefore have a much lower hematocrit, a phenomenon sometimes called *plasma skimming.* In general, the portioning of cells and plasma at small bifurcations is unequal, with more red cells passing into the faster daughter channel (Schmid-Schönbein et al., 1980). Indeed, in the capillaries of skeletal muscle, for example, the hematocrit is well below 20%, compared to 40–45% in the large vessels (see Klitzman and Johnson, 1982). The lowering of hematocrit results in a blood viscosity closer to that of plasma, and undoubtedly it is an important component of the fall of viscosity *(Fahraeus-Lindqvist effect)* in small vessels (Fig. 2.27). Without this effect, the overall slowing of blood

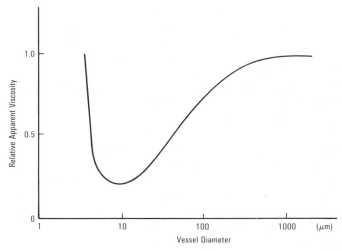

Figure 2.27. The reduction of relative apparent viscosity (apparent viscosity/plasma viscosity) with vessel diameter. This is due to radial displacement of red cells toward the lumen center and faster movement of the cells versus the plasma. In small capillaries, the cells are highly squeezed against the endothelium so that the apparent viscosity rises again.

velocity in the capillaries and adherence of red cells into aggregates at slow velocities would markedly enhance capillary blood viscosity.

The red blood cell, which is shaped like a biconcave disc, is able to deform and to squeeze through openings much smaller than its normal diameter (about 8.5 μm). As we have seen, this immense *deformability of the red blood cell* membrane is an extremely important property and allows the red cells to crowd readily through complex, narrow capillary networks with only relatively modest increases in flow resistance (Lipowsky et al., 1980; Schmid-Schönbein, 1976; Cokelet et al., 1980). Although the red cell membrane bends and shears easily, it will not tolerate *stretch* and, if the volume of the red blood cell is increased, the cell tends to become more spherical and to break easily (a property that becomes important in certain types of anemia). (For a general discussion of red cell structure and function, see Section 3, Chapter 24).

In contrast to the highly deformable red cells, the white cells or leukocytes are much stiffer and can impose a high resistance in capillaries when they have to undergo deformation for passage. The white cells are spherical and have a volume 2–3 times larger than that of the thin, disc-shaped red cells, with a diameter of 6 μm (for lymphocytes) to 8 μm (granulocytes and monocytes). On hematological blood smears, leukocytes are compressed to thin, pancake shapes, the diameter of which far exceeds their undeformed spherical diameter. Under low flow conditions (such as shock or partial obstruction of a blood vessel), the larger granulo-

cytes and monocytes may not pass through individual capillaries. This can lead to obstruction of these capillaries and adhesion between the endothelial cells lining the vessel and the granulocytes and may be accompanied by tissue damage due to oxygen radical formation and lysosomal enzyme release. Granulocyte capillary plugging has been determined to be the cause of the capillary no-reflow phenomenon with reperfusion after a period of arterial occlusion in the heart and in other organs (Schmid-Schönbein and Engler, 1987).

VASCULAR RESISTANCE

The simplest expression of resistance to flow through a hydraulic circuit is like Ohm's law for electrical circuits, in which $I = E/R$ (where I = current flow, E = voltage, and R = resistance of the circuit). The analogy in the vascular bed is the general resistance equation mentioned in Chapter 5:

$$Q = \frac{P_A - P_V}{R}$$

where Q = blood flow, P_A and P_V = pressure (P) difference across the vascular bed between artery (A) and vein (V), and R = resistance in units of pressure/flow rate, e.g., mm Hg/ml/min. The application of this equation for defining blood flow to an organ (also discussed in Chapter 5) is:

$$Q = \frac{P_A - P_V}{\text{Local VR}}$$

where local VR = organ or tissue vascular resistance. This equation is useful for describing the total vascular resistance across all systemic organ beds of the body, which maintains the blood pressure, rearranged as:

$$\text{SAP} = \text{CO} \times \text{TPVR}$$

where SAP = systemic arterial pressure, CO = cardiac output, and TPVR = total peripheral vascular resistance. In this equation the venous pressure, which is small compared to SAP, is neglected.

When resistances are arranged in series according to the laws of electrical circuits, the total resistance to flow is the sum of the resistances (Fig. 2.28, *left*). When resistances occur in parallel, preferential flow occurs through the pathways of least resistance and the total resistance is less than the resistance of any individual vessel, being the reciprocal of the sum of the reciprocals of individual resistance (Fig. 2.28, *right panel*). Conductance, the reciprocal of resistance, is used as a measure of the ability of a blood

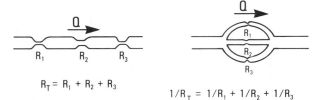

$$R_T = R_1 + R_2 + R_3$$

$$1/R_T = 1/R_1 + 1/R_2 + 1/R_3$$

Figure 2.28. *Left panel,* Diagram showing the calculation of total resistance (R_T) and various resistances aligned in *series* as blood flows through a vascular circuit. *Right panel,* Diagram showing the calculation of total resistance (R_T) as blood flows through various resistances aligned in a *parallel* circuit.

vessel to accept blood flow at a given pressure difference, expressed as ml/second/mm Hg:

$$C = \frac{1}{R}$$

where C = conductance and R = resistance. The total conductance (C_T, which is equal to 1/TPVR) is therefore equal to the sum of the individual conductances in each segment for a given pressure drop:

$$C_T = C_1 + C_2 + C_3$$

In parallel circuit, adding a resistance pathway (even if it is higher than any of the individual resistances in the circuit) will add another (small) conductance and so will raise the total conductance while lowering total resistance. (Such calculations also explain how a very large number of high resistance units, such as capillary beds, despite their smaller radii, can have a lower total resistance than the arteriolar bed.) If one resistance in a parallel circuit drops substantially below the others, the total conductance will rise and the total resistance will fall since, from the equation in Figure 2.28, total resistance in a parallel circuit must always be less than that of any individual resistance.*

In a circuit such as the peripheral circulation, where a great many organs are arranged in parallel, modest increases or decreases of resistance in a given organ will not have a major impact on total vascular or blood pressure. However, tremendous differences in flow rates exist through the various organs at rest (and, therefore, there are large differences in regional resistances since perfusion pressure is the same). The flow differences that would

exist between organs during theoretical maximum flow rates to each organ are shown in Figure 2.29, and the potential maximal flows are shown in the accompanying table. With maximal dilation of all beds, total flow would be extremely high (38 liters, exceeding the limit for cardiac output), so that some beds must undergo compensatory vasoconstriction (by reflex control) when others dilate.

While these questions are highly useful for clarifying interrelationships among pressure, flow, and resistance (and for making calculations in the clinical setting; see Chapter 13), it is important to understand in more detail the factors that are responsible for vascular resistance in the circulation, since disturbance of these factors can occur in disease states. Thus, resistance to flow may be altered by a number of factors, including the characteristics of blood flow, the diameter of the vessels

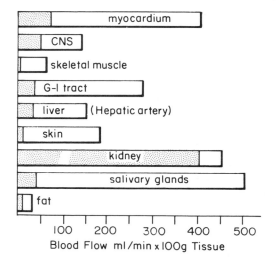

	Rest. blood flow (1/min.)	Max. blood flow (1/min.)	Organ weight (kg)
Myocardium	0.21	1.2	0.3
CNS	0.75	2.1	1.5
Skeletal muscle	0.75	18.0	30.0
G-I tract	0.7	5.5	2.0
Liver (hep. artery)	0.5	3.0	1.7
Skin	0.2	3.8	2.1
Kidney	1.2	1.4	0.3
Salivary glands	0.02	0.25	0.05
Fat	0.8	3.0	10.0
Total ≈	5.1	38.0	48.0

Figure 2.29. *Above,* Bar graph representing approximate blood flows in various organs at maximal vasodilatation *(total bar)* and at "rest" *(hatched areas)* in ml/min × 100 g of tissue at perfusion pressure of 100 mm Hg. *Below,* Approximate values for regional blood flows in a 70-kg man at "rest" and at maximal dilatation, deduced from organ weights. The organs included comprise about 70% of total body weight. (Redrawn from Mellander and Johansson, 1968.)

---

*For example, if a parallel circuit has two pathways, with each having a resistance of 4 units, the total resistance is 2 units (¼ + ¼ = ½). If the resistance in one limb falls markedly to 1 unit (a fourfold drop), the total conductance will be 0.25 + 1 = 1.25, and the total resistance (reciprocal of total conductance) becomes 0.8 unit.

in question, the distending pressure in the vessels, and the viscosity of the blood.

Poiseuille's Law

The various factors influencing resistance emerge from an analysis of the Poiseuille equation, developed from a study of steady laminar flow by a Newtonian liquid in narrow rigid glass tubes (this is the case of a balance between f_p and f_v only in *Eq. 1* above).

The law, developed by J. L. M. Poiseuille, French physicist and physician (Poiseuille, 1846), states that

$$Q = \pi \frac{(P_1 - P_2)r^4}{8\eta L}$$

where Q = flow rate, $P_1 - P_2$ = the pressure difference across the circuit, r = radius of the tube, η = viscosity of the liquid, and L = length of the tube.

The Poiseuille equation illustrates the important relationship between vascular resistance and (1) the length of the tube and (2) blood viscosity; thus, if *either one doubles, the flow will be halved* (or if flow stays constant, the pressure gradient will double). The equation also emphasizes the enormous influence of vessel radius; *if the radius doubles, the flow will increase 16-fold* if other factors are constant.

Since the experiments were originally performed using a Newtonian fluid, if the caliber and length of the tube are kept constant, the viscosity will not vary as the flow is changed and, hence, the resistance remains constant. However, if those factors which influence resistance are considered separately, the resistance is seen to be directly proportional to the viscosity and length of the tube and inversely proportional to π and the 4th power of the radius:

$$R = \frac{8L\eta}{\pi r^4}$$

This equation contains a constant, $8/\pi$; a geometric component, L/r^4; and the viscosity component, η. In selected organs where these factors were measured, it was found that, whereas Poiseuille's formula is indeed a good first approximation of normal flow in small blood vessels, other factors can enter into this relationship as well. Among the additional factors are vessel distensibility, the aggregation of blood cells, cell adhesion to the endothelium, and local protrusions of endothelial cells (e.g., in contracting arterioles or in capillaries during pressure reduction). For calculation of vascular resistances in the intact circulation and the normal range of values in humans, see Chapter 13.

Resistance in Distensible Tubes

An important difference between Poiseuille's experimental preparation and the circulation is the fact that blood vessels are distensible. In nondistensible tubes using Newtonian fluids, as pressure increases flow increases linearly, and the resistance (the slope of the pressure-flow plot) remains constant (Fig. 2.30, *dashed lines*). On the other hand, in a distensible vascular bed (with no intrinsic vasomotor responses) the pressure-flow relation is nonlinear; as the pressure difference increases the flow is augmented in a nearly exponential manner as the blood vessel is distended (Fig. 2.30, *right-hand solid line*), and as the pressure rises the resistance falls (Fig. 2.30, *left-hand solid line*). The more distensible the vessel, the more the pressure-flow curve shifts toward higher flows. In beds with higher degrees of vasomotor tone (but without autoregulation), the pressure-flow curves are progressively shifted to the right and flattened.

TURBULENT BLOOD FLOW

When the blood flow rate exceeds a certain velocity, laminar flow no longer occurs, the velocity profile becomes blunted, frictional losses within the

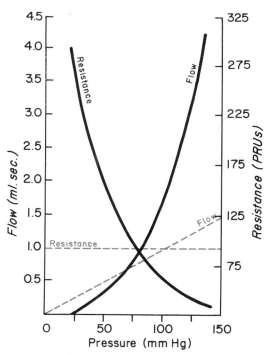

Figure 2.30. The relationships of flow, pressure, and resistance in nondistensible tubes using Newtonian fluids *(broken lines)* at constant temperature (Poiseuille's law) and blood in vessels *(solid lines)*. See text for details.

blood become high, and the resistance to flow rises. (This may occur, for example, in blood traversing a narrowed arterial area produced by atherosclerosis.) The appearance of such turbulent flow is related to the blood velocity, but also to the diameter of the vessel and the density and viscosity of the fluid. The Reynolds number (a dimensionless variable, mentioned above in relation to inertial and viscous force) serves as a useful indicator for the transition of laminar flow to turbulent flow. It can be calculated from these variables:

$$N_R = \frac{VD\rho}{\eta}$$

where N_R = Reynolds number, V = mean velocity, D = tube diameter, ρ = fluid density, and η = fluid viscosity.

Turbulent flow usually occurs when the Reynolds number exceeds a critical value at about 3000. Since the viscosity of blood is relatively high, the Reynolds number for turbulent flow is not exceeded in most parts of the circulation (see Table 2.1, Chapter 6). Laminar flow is silent and does not generate an audible signal but, with the onset of turbulent flow, random pressure fluctuations and sounds occur. In the vascular bed, these sounds may be heard with a stethoscope under certain circumstances (such as beyond a constricting vascular lesion) as vascular bruits or tapping sounds beyond the cuff of a sphygmomanometer (Korotkoff sounds) (see Fig. 2.133, Chapter 13). Also, at very high cardiac outputs, the Reynolds number may be high enough across the aortic and pulmonic valves to produce systolic heart murmurs, even in the normal circulation. It is also evident that at large vessel diameters and high flow velocities, when there is a low blood viscosity, as in anemia, the critical Reynolds number is particularly likely to be exceeded.

As flow increases, if the critical Reynolds number is exceeded, the viscous and kinetic losses of energy due to the resulting turbulent flow will produce a decrease in the flow at the same pressure difference, a factor which can affect the blood flow through diseased blood vessels. Turbulent flow with increased shear stress and strain acting at the vascular wall are important in promoting atherosclerotic lesions at certain points in the arteries (see Fry, 1968).

DETERMINANTS OF THE MEAN ARTERIAL PRESSURE

The static pressure in a closed container refers to the force exerted against a unit area of the container's wall, although typically when pressure is measured in the vascular system we refer to the height of a column of fluid supported by the force within a blood vessel. For example, when Stephen Hales (Hales, 1733) measured the arterial blood pressure in a horse, he ligated a cannula into an artery, affixed it to a glass tube 9 feet in length, and found that the column of blood from the living horse rose about 8 feet above the level of the heart. Thus, the weight of the column of blood provided a measure of the blood pressure. A mercury manometer is more convenient for measuring pressure, and it is customary to report the height to which a column of mercury (Hg) is pushed by the intravascular pressure, a column 100 mm high reflecting blood pressure of 100 mm Hg. Sometimes, pressure in the veins is measured in centimeters of water: 1 mm Hg = 1.36 cm H_2O. Commonly, for experimental purposes and during cardiac catheterization in humans, we use electronic gauges (see Chapter 13) which measure the pressure wave forms and the mean pressures by means of conductance coils or by a strain gauge attached to a rigid membrane (Chapter 13), and these gauges are calibrated against a mercury manometer.

Kinetic and Potential Energy

When blood is flowing in large arteries or veins, the total fluid energy (E) per unit volume of blood equals the sum of the pressure, plus a factor related to the influence of gravity, plus the kinetic energy:

$$E = P + \rho gh + \frac{1}{2}\rho V^2$$

where P is static pressure expressed as force (dynes/cm^2), ρ = fluid density in g/cm^3, g = the earth's gravity acceleration constant of 980 cm/s^2, h = the height of the fluid column above or below a reference level, and V = flow velocity in cm/s. In this equation, energy dissipation due to viscosity is being neglected. Therefore, the equation does not apply to flow in capillaries.

If all parts of the fluid system are at the same level, the gravitational factor can be ignored (this is the case in the supine but not the erect posture). The kinetic energy, which is proportional to the square of the velocity, is often ignored but can become important under certain circumstances; for example, although blood normally flows from a higher pressure to a lower pressure, if the kinetic energy component (velocity) is sufficiently high, blood occasionally can flow from a lower to a higher pressure. This constitutes an important mechanism for closure of valves in the heart (Fung, 1984). The driving forces for forward flow are not directly affected by the different heights of various portions of the circulation in the erect body position; thus, as pointed out by A. C. Burton (Burton, 1972), like the principle of the siphon, it is only the *difference* in the total pressure and energy at one point in the circulation from that in another that determines flow, not whether the flow is uphill or downhill. However, if the static pressure in the arterial bed is measured with a manometer in the erect posture, the pressure will differ by the height of the fluid column above or below the heart; thus, if the pressure at the heart level is 100 mm Hg one would predict a pressure in the toes of about 185 mm Hg and in the head of about 50 to 60 mm Hg. This factor is even more important in the distensible veins, as discussed subsequently.

As described in Chapter 6, the linear velocity of blood flow (V) is inversely proportional to the cross-sectional area (CSA) of the vascular stream ($V = Q/$CSA), and the kinetic energy component of the flowing blood (which is a function of V^2) can importantly affect the pressure, depending upon how pressure is measured. For example, if the orifice of the pressure-measuring tube faces upstream toward the oncoming flow (Fig. 2.31*b*), the pressure will be higher than if the orifice faces downstream because of the kinetic energy component. The pressure measured facing the flow, sometimes called the *end pressure*, therefore indicates the total fluid energy. On the other hand, when perssure is measured by a *laterally* positioned tube, it will vary depending on the velocity of flow. The *Bernoulli principle* applies to the conservation of energy in flowing fluids and, in simplest terms, indicates that the lateral pressure in a stream flowing at high velocity is less than when it flows at low velocity (Fig. 2.31). This occurs because more of the total energy is dissipated as kinetic energy ($\frac{1}{2}\rho V^2$) in the faster-moving stream. Thus, as shown in Figure 2.31, when the diameter of a vessel widens and velocity slows, the kinetic

energy term is decreased, and the pressure is higher than at *e*, whereas at *d* the pressure is lower due to higher velocity. Increased lateral pressure in regions of arteries that are abnormally enlarged may promote aneurysm formation, in addition to those factors related to wall tension (Chapter 6).

Because of velocity changes and progressive frictional energy losses, the intravascular pressure tends to be different at every point in the circulation. In the aorta, the lateral pressure is quite dependent on kinetic energy losses. However, as the circulation branches into smaller and smaller arteries and velocity slows, the kinetic energy component becomes relatively unimportant, and the pressure drop in the smaller vessels is mainly due to frictional loss (as between *c* and *e*, Fig. 2.31).

At normal cardiac output, the kinetic energy has been calculated to be less than 5% of the total fluid energy at the peak systolic pressure in the aorta; however, at high cardiac outputs, since kinetic energy is proportional to the square of velocity, it may make up nearly 15% of the total energy. Under such conditions this factor could significantly affect the total energy output and total energy requirements of the heart and should be included in calculations of total cardiac work.

PHASIC ARTERIAL PULSE WAVES AND THE ARTERIAL PRESSURE PULSE

The stroke volume is delivered into the aorta at high velocity and, after the completion of left ventricular ejection, there is a brief period of slight reverse flow which ceases as the aortic valve closes; the flow then remains at zero just above the aortic valve during diastole until the next contraction (Fig. 2.32). The distensibility of the arterial system "smooths" the flow pulse as it traverses the arterial

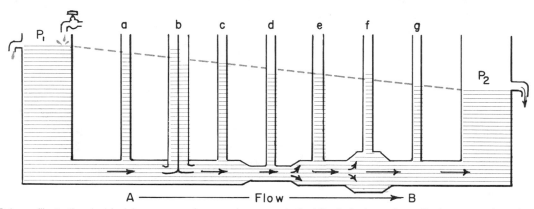

Figure 2.31. Schema illustrating, in principle, pressure changes along a tube AB through which liquid flows steadily owing to the pressure difference $P_1 - P_2$. Pressure is measured by the fluid levels in the side tubes *a–g*. Tube diameter changes at two locations, *d* and *f*. The *dashed line* indicates a theoretical pressure drop along such a tube if there were no changes in tube diameter. The Bernoulli principle is depicted by the pressure levels in *d*, as compared with *c* and *e*, and in *f*, as compared with *e* and *g*.

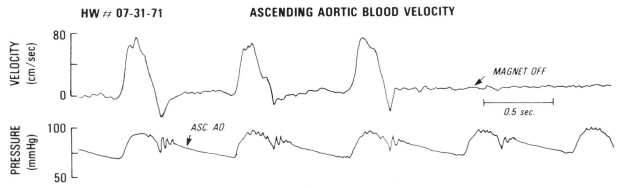

Figure 2.32. Aortic flow velocity recorded with an electromagnetic catheter tip velocity meter in the ascending aorta of a human subject. High fidelity pressure is also recorded in the ascending aorta. Zero flow velocity is indicated with the magnet off. Tracing was obtained by Gabe et al. (1969).

bed, and it also effectively lowers the systolic pressure that would be developed if the ventricle were to eject the same volume of blood into a rigid tube; this tends to decrease the workload of the heart during systole, and the energy stored in the stretched wall of the arteries during ejection is then returned as elastic recoil with further flow and pressure generation during diastole by this "passive" mechanism (see "Windkessel Effect," Fig. 2.8). Thus, beyond the initial part of the aorta, flow is maintained above zero between heart contractions.

The static pressure-volume relation of the aorta is discussed in Chapter 6 (see Fig. 2.12), and in a young human subject a 100% change in volume produces approximately a 50-mm Hg change in pressure; in older subjects the aorta is less compliant (stiffer), and the same change in volume results in a large increase in pressure. However, the pressure developed in the arterial bed in response to delivery of the stroke volume by the heart is dependent not only on the static elastic properties (compliance or stiffness) of the aorta and arterial bed, but also upon their dynamic stiffness, that is, the elastic properties are also dependent upon the rate of delivery of the stroke volume, and such frequency-dependent responses imply that the aorta also has viscoelastic properties, discussed further below.

The *pulse pressure* is defined as the difference between the peak systolic and minimum diastolic pressure, measured at any point in the systemic arterial circulation (Fig. 2.33). The *mean pressure* in the arterial circulation can be determined by electronic integration of the pressure signal obtained with a pressure transducer, or it can be determined from the recorded pressure wave form by planimetric integration of the area under the systolic and diastolic pressure waves (Fig. 2.33). If only the systolic and diastolic pressures are known (as by measurement with a sphygmomanometer), the mean pressure can be approximated as the diastolic pressure plus one-third of the pulse pressure (Fig. 2.33).

Under steady state conditions, the *mean* arterial pressure (MAP) is determined only by the cardiac output and the total peripheral vascular resistance (MAP = CO × TPVR), and the dynamic properties of the vessels have little influence. However, with sudden changes in cardiac output, the *time* required for attaining a new level of steady state MAP will depend upon the compliance (or capacitance) of the arterial bed. For example, at a fixed peripheral vascular resistance, if the cardiac output is suddenly

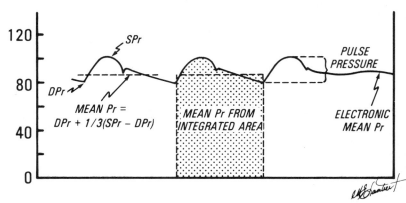

Figure 2.33. Various methods for determining the mean arterial pressure from the arterial pressure pulse. *SPr*, systolic pressure; *DPr*, diastolic pressure. For further discussion, see text.

increased in the presence of a very compliant (high capacitance) arterial bed, time will be required to fill the arterial bed and to reach a new steady state level of arterial pressure; conversely, if the system is noncompliant, the new level of pressure will be reached almost immediately.

Arterial Pressure Wave Forms

The shape of the arterial pressure wave forms changes substantially when it is measured in different portions of the arterial bed; for example, the peak systolic pressure is higher and the pulse pressure is larger in the femoral artery of the leg than in the thoracic aorta (within the chest or thorax).

In the intact circulation, from points of branching and from the tapering of vessels into small resistance vessels, pulse waves are reflected backward toward the central aorta, and these returning pressure waves interact with the oncoming waves. For example, if the aorta were cross-clamped at some distance from the heart, the reflected wave could double the amplitude of the oncoming wave (assuming no energy loss). On the other hand, if sufficient lag occurred that the reflected waves were ½ cycle (180°) out of phase, the reflected wave would subtract from the oncoming wave. The speed of a harmonic wave, C, in an elastic tube with internal radius r_i, wall thickness h, and elastic modulus E (the instantaneous stiffness of the vessel wall) is

$$C = \sqrt{\frac{Eh}{2\rho r_i}}$$

where ρ is the blood density. One of the important features of pulse wave transmission is related to the compliance of the arterial bed. The smaller blood vessels tend to be less compliant (stiffer with higher E) than the aorta, and this reduced compliance results in an increased velocity of pulse-wave transmission toward the peripheral vessels. Also, as pressure increases, arterial compliance decreases (higher E) (see Fig. 2.12, Chapter 6); therefore, the peaks of the pressure waves tend to travel faster than the lowest (diastolic) points of pressure so, as the wave form moves peripherally, there is a progressive movement of the peak toward the onset of the pressure pulse (a reduction in time to peak pressure).

Without such phenomena, the viscous forces in the blood vessel walls and the blood would tend to produce a *decrease* in the amplitude of the traveling pressure pulse wave, rather than the observed increase (Fig. 2.34). (Pulse waves are often analyzed in terms of a major first harmonic component and other higher frequency harmonics using Fou-

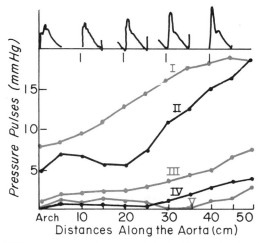

Figure 2.34. Pressure pulses recorded along the aorta of a dog from the arch to the femoral artery, with distance in centimeters. The pulse pressure increases, the dicrotic notch disappears, and a secondary wave develops on the descending limb. In the *panel below* are the first five harmonic components of these waves. This is the amplification referred to in the text. (From Taylor, 1966.)

rier analysis; see Pepine et al., 1978). In fact, high frequency oscillations, such as the incisura and positive oscillation (dicrotic wave) that follow aortic valve closure (Fig. 2.32), do tend to disappear as they are transmitted down the aorta. The contour of the pressure pulse as it moves toward the periphery is shown in Figure 2.34. From the arch of the aorta (above the aortic valve) down to the point where the aorta branches into the femoral arteries, there is a progressive increase in the amplitude of the systolic pressure and the pulse pressure, the time to the peak pressure diminishes, the systolic component narrows, and a secondary positive wave appears. Note the increasing amplitude of several of the harmonics of the pulse (particularly harmonics I and II) as the distance from the arch of the aorta toward the periphery increases (Fig. 2.34). It seems likely that these changes in the pulse wave are due to the several factors discussed above, including amplification by reflected waves, tapering of vessels, decreased compliance from central to peripheral vessels, and effects due to the more rapid travel of the wave at peak pressure. The secondary wave does not appear to be due to the dicrotic wave (which, as mentioned, is poorly transmitted), but rather to an initial central and second peripheral reflection of the primary systolic wave.

Determinants of the Arterial Pulse Pressure

Several factors importantly affect the magnitude of the *pulse pressure* measured in a given circulatory state (in addition to the location at which the pressure wave is measured, discussed above). These

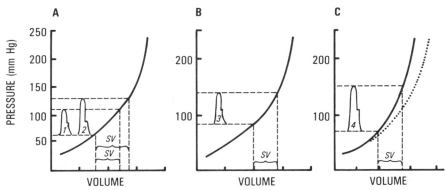

Figure 2.35. Diagrammatic pressure-volume relations of the aorta in human subjects (*curved lines* in each panel). *SV*, stroke volume: In *panel A,* increasing the stroke volume delivered into the aorta increases the arterial pulse pressure (*beat 1* and *beat 2*). In *panel B,* when the arterial pressure is elevated, the same stroke volume as shown in *beat 1* of *panel A* results in a larger pulse pressure (*beat 3*) because the aorta is operating on a steeper portion of its pressure-volume curve. In *panel C,* the normal pressure-volume relation shown in *panels A* and *B* is indicated by the *curved dotted line,* and the pressure-volume relationship seen in some elderly subjects, which exhibits decreased aortic compliance (increased stiffness), is indicated by the *solid curved line.* Under these conditions, the same stroke volume as in *beat 3* of *panel B* causes an increased pulse pressure (*beat 4*).

are the stroke volume, the aortic compliance (or capacitance, the inverse slope of the pressure-volume relation, dV/dP), and the diastolic flow or "runoff" from the aorta. The stroke volume clearly can affect the pulse pressure and, as shown in the aorta of a normal middle-aged individual, a larger stroke volume will increase the pulse pressure (Fig. 2.35, *panel A, beat 1* to *beat 2*). If the mean arterial pressure in the same individual is acutely and markedly elevated, the artery will be stretched to a stiffer portion of its pressure-volume curve (reduced compliance), and under these conditions the same stroke volume will cause a larger pulse pressure with a proportionally greater increase in systolic pressure than diastolic pressure (Fig. 2.35, *panel B, beat 3*). When aortic compliance decreases, as with age, the entire pressure-volume relationship is steeper (reduced compliance), and under these conditions the same (normal) stroke volume as in *beat 1 (panel A)* can result in an increased pulse pressure (Fig. 2.35, *panel C, beat 4*).

If the heart rate is slowed while the cardiac output remains relatively constant (as may occur with complete heart block, Chapter 9), the stroke volume rises and the pulse pressure increases (Figs. 2.35*A* and 2.36, *beat 2*); conversely, if the heart rate is increased (as by electrical pacing) and the cardiac output is unchanged (which is the usual response in the normal circulation), the stroke volume and pulse pressure diminish, although the mean arterial pressure remains constant. An isolated decrease in aortic compliance produced experimentally causes an increase in peak pressure and an increased pulse pressure (Fig. 2.36, *beat 3*). The effects of chronic hypertension are also shown in Figure 2.36 *(beat 4).* In most cases of hypertension the cardiac output

remains normal, and the rate of fall of pressure (diastolic runoff) may not change appreciably, although the increased peripheral vascular resistance at an unchanged cardiac output produces higher mean and diastolic pressures. If the hypertension is severe enough to place the aorta on a steeper portion of its pressure-volume relation, the pulse pressure will rise as well (Fig. 2.36, *beat 4*). Finally, the pulse pressure can be importantly affected by abnormally rapid diastolic runoff. This may occur with aortic regurgitation, in which the pulse pressure increases for two reasons: first, the stroke volume is greatly increased (the ventricle must eject the amount of blood that filled the ventricle from the left atrium, as well as that which regurgitated backward across the aortic valve during the previous diastole) and, second, the diastolic pressure falls rapidly in the aorta because of the combination of normal runoff into the peripheral circulation and rapid retrograde filling of the left

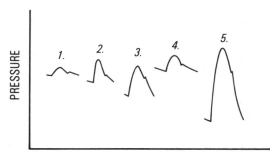

Figure 2.36. Factors affecting the arterial pulse pressure and the arterial wave form. Beat 1 is normal. Beat 2 shows the effects of an increased stroke volume. Beat 3 illustrates the effects of decreased aortic compliance. Beat 4 shows the effect of elevating the mean arterial pressure. Beat 5 indicates the consequences of aortic valve regurgitation. For further discussion, see text.

ventricle from the aorta (Fig. 2.36, *beat 5*). A similar change in the pulse pressure occurs in arteriovenous fistula, in which the fistula provides a "short circuit" connecting an artery and a vein, resulting in high cardiac output (increased stroke volume) and abnormally rapid diastolic runoff through the fistula.

PULSATILE PRESSURE-FLOW RELATIONS

With pulsatile flow in a stiff tube, the flow velocity wave lags behind the driving pressure wave and, as the frequency of the pressure pulse increases, the delay between the two pulses decreases while the amplitude of the flow velocity wave falls. In addition to frequency, a number of other factors influence pressure-flow relations in the vascular bed (including the blood viscosity and vessel characteristics) (see McDonald, 1974; Fung, 1984).

The driving pressure generated by the ventricle accelerates the column of blood in the aorta (flow leads pressure at the origin of the aorta), and once the blood starts to move through the arteries it will continue to move as long as there is a sufficient pressure difference (the term *pressure gradient* is more properly used to refer to a rate of change of pressure per unit distance). The blood will cease to move when the pressure difference is zero. The traveling pressure wave (which precedes the flow velocity wave in the peripheral arteries) can be palpated in the arteries, and its transmission time can be estimated by noting the more rapid arrival of the wave in the carotid artery than in the more distant femoral artery. Transmission of the pressure pulse produces an oscillating pressure difference (Fig. 2.37) and, if pressure is recorded at two different sites, one upstream and the other a short distance downstream, the pressure difference is positive during the upstroke of the upstream pressure pulse and negative when the delayed pressure wave arrives downstream (Fig. 2.37). The flow generated by the positive pressure difference lags behind to some degree (Fig. 2.37), as illustrated here by pressure and flow in a peripheral artery. In this case, the negative pressure wave may actually produce a small reversal of flow after the initial pulse.

Impedance

The idea of impedance in pulsatile circuits is concerned with the ratio of pressure to flow, and therefore it represents the "resistance" to *pulsatile flow*. When this ratio is calculated in a nonpulsatile circuit (analogous to a direct current flow), the ratio equals the resistance. The analogy to hydraulic impedance in electrical terms is also available for alternating current flow; instead of Ohm's law,

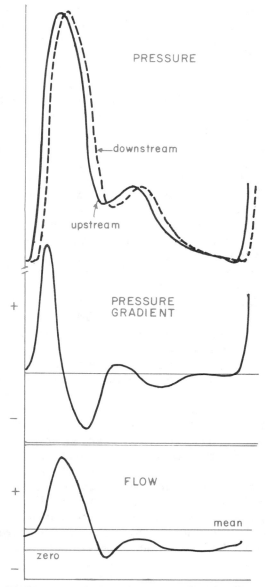

Figure 2.37. Generation of a pressure gradient by a traveling pulse wave and resultant oscillatory flow. Note small reversal of flow (below zero). Example from the femoral artery of a dog. (From Taylor, 1966.)

when dealing with a capacitive circuit the impedance (Z) is determined by the ratio of the sinusoidal voltage to the sinusoidal current flow, and the magnitude of the impedance will vary as a function of the frequency of the voltage wave, the capacitance, and other elements in the circuit. Similarly, in a hydraulic circuit when the flow is pulsatile the impedance includes not only the opposition to flow afforded by friction but also that due to the vascular compliance and the blood mass. In practice, the arterial impedance is determined at zero frequency (steady flow) and over a range of frequencies. (Analysis of pressure and flow tracings is in the time domain, whereas analysis of impedance is in the frequency domain.)† Changing the compliance of the aorta would be expected to alter the impedance (for example, clamping the aorta increases impedance).

Impedance Effects on Ventricular Function

It is of interest to compare the effects on left ventricular function of a change in peripheral vascular resistance vs. those of a change in aortic compliance. In the time domain, if pressures and flows are analyzed when peripheral vascular resistance is increased by a vasopressor drug while the left ventricular end-diastolic pressure is held constant, a marked increase in peak systolic and mean aortic and left ventricular pressure and a decrease in the stroke volume occur. If the capacitance or compliance against which the left ventricle ejects is now decreased without changing the mean aortic pressure (an artificial system with an air-buffered reservoir can be used), peak systolic pressure rises slightly, diastolic pressure falls markedly, and, again, the stroke volume is markedly reduced. The contractility (inotropic state) of heart muscle is unchanged under these conditions, but the percentage of volume ejected by the ventricle falls in both situations because of the altered loading conditions (see Chapter 12). The vasopressor drug results in an increase in resistance, but there is probably little change in aortic properties. On the other hand, with decreased aortic compliance, resistance does not change, but impedance is increased. Such studies serve to illustrate the complex interaction between the pump and the peripheral circulation (Milnor, 1975; Noordergraff, 1978) (see further discussion in Chapter 12 of impedance effects on left ventricular function).

DETERMINANTS OF THE VENOUS PRESSURE

A number of properties of the veins directly or indirectly influence the venous pressure (for reviews, see Shepherd and Vanhoutte (1975) and Fung (1984). These properties include their high compliance (see Fig. 2.12, Chapter 6) and therefore the large volume of blood that they store (64% of the total blood volume and about 80% of the systemic blood volume, Table 2.2), their tone as affected by adrenergic nerve impulses, and their potential collapse, which can occur in larger veins at several locations (depending on their distending pressure and external pressure or on the degree of

---

†Calculations of aortic impedance require Fourier analysis of high fidelity aorta pressure and flow data (Pepine et al., 1978).

tethering to the surrounding tissue). The abdominal viscera may compress the veins, for example, causing partial collapse; also, during normal inspiration, as the blood is drawn into the chest by the negative intrathoracic pressure, the neck veins tend to collapse as vena caval pressure falls (see Fig. 2.127).

In the supine position the "central" mean venous pressure in normal subjects is 0–4 mm Hg (up to about 6.0 cm H_2O) using the right atrium as the reference point. This reference point is largely independent of body position because the right ventricle, as a demand pump (i.e., expelling whatever volume flow it receives (see Chapter 12 on Starling's law)), tends to maintain the atrial pressure constant (Buyton and Jones, 1973). This level of pressure in the right atrium affects the pressure in the peripheral veins. Normally, in the small veins the pressure is 8–9 mm Hg higher than in the right atrium; therefore, if the right atrial pressure rises, the peripheral venous pressure also rises, maintaining the pressure difference favoring forward flow. The right atrial pressure, in turn, is affected by the function of the right ventricle. If the right ventricle fails, right atrial pressure will rise and tend to impede venous return, whereas good ventricular function will maintain the right atrial pressure at low values.

Other factors can affect the right atrial pressure, including the blood volume, which influences the venous return (increased blood volume increases venous return, and vice versa; see Chapter 17). The venous tone and, hence, the capacitance of the peripheral venous bed are under autonomic control (Chapter 16), and sympathetic stimulation increases peripheral venous tone, displaces blood centrally, and can thereby raise the right atrial and venous pressures; withdrawal of sympathetic tone, on the other hand, leads to increased venous capacitance and can lower the venous pressure. Other factors influencing venous return are discussed in Chapter 17.

GRAVITATIONAL EFFECTS OF POSTURE

If a normal adult is placed on a tilt table which is then abruptly moved from the supine to the upright position and if the leg muscles are not contracted, the pressure in a large leg vein near the foot will promptly rise from a few mm Hg to that of the weight of the blood volume in the veins; the column is about 4 feet or 120 cm in height (the column below the heart) and, therefore, will yield a pressure in the foot of almost 90 mm Hg (1 cm of water or blood = 1.36 mm Hg) (Fig. 2.38). Conversely, if the right atrial pressure is 1 or 2 mm Hg, the 50-cm distance above the right atrium of a vein in the head

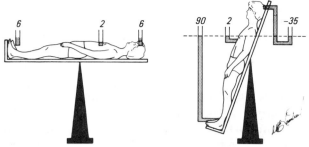

Figure 2.38. *Left panel,* Diagram of the venous pressures (mm Hg) in a normal, supine human subject. The venous pressure in the legs and head is slightly higher than in the central chest. *Right panel,* Following sudden upward rotation on a tilt table, the venous pressure rises to 90 mm Hg in the legs but remains at 2 mm Hg near the heart; it becomes 35 mm Hg negative in the head. For further discussion, see text.

in the upright tilt will result in a negative pressure in the head vein of more than −30 mm Hg (Fig. 2.38) (the veins of the cerebral circulation are normally supported externally and do not collapse under pressures). The pressures in neck and arm veins will collapse under upright conditions, although the position of the arms is important (if the arm is held downward, the hand will be below the right atrium and will exhibit a positive venous pressure). As described earlier, the high positive pressure in the feet as a subject is tilted upward produces the same increment of pressure in the arteries as in the veins, and the same difference in pressure remains between these two components of the circulation; therefore, since the total energy for forward flow is unchanged, flow will continue to the foot. However, as the venous pressure in the legs rises, the compliant veins are stretched, and a substantial volume of blood becomes sequestered in the legs and the abdomen, so that the venous return to the heart transiently decreases and the cardiac output drops markedly before compensatory reflex responses occur.

If the individual remains in the upright posture without moving the legs, the sustained high pressure in the veins will lead to a high capillary pressure, and edema (swelling) of the legs will occur after many minutes (indeed, soldiers on drill who stand for long periods at attention, without movement, can develop leg swelling). Ordinarily, however, assumption of the erect posture is associated with tensing of the muscles in the legs and is followed by walking. The peripheral veins have sets of one-way valves at frequent intervals. As the veins are compressed by neighboring muscles and emptied, the valves prevent backflow of blood into the veins from above. Thus, the venous valves interrupt the continuous column of blood in the veins. This mechanism acts as a *muscle pump,* which results in a prompt marked lowering of the venous pressure and effective filtration pressure in the lower extremities to 8–10 mm Hg upon walking (Stegall, 1966).

Sometimes, inflammatory processes in the veins or the formation of blood clots leads to damage or malfunction of the venous valves (thrombophlebitis). Under these circumstances the venous pressure may remain high, resulting in further stretching of the venous valves, enlargement of the veins (varicose veins), and constantly elevated pressures in the leg veins in the erect posture as well as during walking. This, in turn, causes chronic edema, reduced microvascular perfusion, and sometimes capillary occlusion by granulocytes in the ankles and lower legs.

Flow and Pressure Waves in the Veins

The characteristics of the pressure pulses in the right atrium and large systemic veins are described in Chapter 13; the a, c, and v waves are transmitted into visible venous pulsations in the neck from the right atrium, with some delay. When blood flow is recorded in the vena cava, there is a large positive flow wave during right ventricular systole, as the ejecting right ventricle moves away from the right atrium to produce a relative increase of atrial volume, and increased inflow of blood into the right atrium occurs (the tricuspid valve is closed) to produce the v wave in the pressure tracing. Flow again increases early in diastole as the tricuspid valve opens and blood fills the relaxing right ventricle. There is some evidence that increased inflow at this time is also related to restoring forces in the right ventricle, giving a suction effect (Hori et al., 1982).

METHODS FOR ASSESSING PERIPHERAL BLOOD VESSELS

In diagnosing peripheral vascular disease in humans (such as of the cerebral arteries or of the arteries and veins to a leg) a number of methods can be employed. Angiography, with injection of an X-ray-opaque liquid, allows X-ray visualization of regions of blockage in arteries or veins (see Chapter 13). Indirect measurements of blood flow in arteries can be made. For example, using transcutaneous Doppler flow probes (reflected ultrasound waves from the moving blood), abnormal flow wave patterns or loss of reactive hyperemia can be detected in limbs with atherosclerotic lesions. Venous occlusion plethysmography, in which the change in volume or circumference of a limb is measured while a cuff is inflated to obstruct venous outflow (not arterial inflow), allows measurement of the blood flow rate; release of the cuff can provide information on whether there is obstruction to venous outflow by thrombophlebitis (for review of noninvasive techniques, see Fronek, 1988).

CIRCULATION TO THE SKIN

The blood supply to the skin serves two main purposes: the relatively small nutrient flow, which is primarily controlled locally, and the potentially large flow concerned with heat loss regulation, which is predominantly under reflex control. Thus, most of the blood flow to the skin is neurogenically rather than metabolically controlled. (For reviews of skin functions and the skin circulation see Nicoll and Cortese, 1972; Helvig and Mostofi, 1980.)

In the skin, the arterioles form a rich plexus and, upon approaching the bases of the papillae (the layer of the corium immediately underlying the epidermis), turn horizontally and give rise to metarterioles from which originate, in turn, hairpin-shaped endothelial tubes—the capillary loops. The proximal or arterial limb of the capillary loop ascends in the papilla and then turns upon itself to form the venous limb. The latter, on reaching the base of the papilla, joins with the venous limbs of neighboring loops to form a collecting venule. The collecting venules anastomose with one another to form an extensive plexus—the subpapillary venous plexus—which runs horizontally beneath the bases of the papillae and drains into deeper veins. The capillary loops can be seen readily in the living skin under the low power of the microscope.

The color of the skin is dependent upon blood in the capillary loops and the subpapillary venous plexus. The vessels of the plexus, though more deeply placed, present a greater area parallel to the skin, whereas the capillary loops are disposed chiefly at right angles to the skin surface. When little blood is contained in the superficial vessels, the skin is unusually pale and more transparent, and the deeper venous plexuses then contribute largely to the color of the skin, often adding a leaden tint to the pallor. When the overlying vessels are open and the skin is well supplied with blood, these deeper vessels are hidden from view.

The hue of the skin, i.e., the dominance of the reddish or of the bluish hue, depends upon the extent to which the oxyhemoglobin becomes reduced during the passage of the blood through the cutaneous vessels. The degree of reduction of hemoglobin will depend, as a rule, upon the rate of blood flow. With rapid flow, the blood is more arterial in character; with slow flow, more venous, and the color of the skin becomes bluish.

Sympathetic adrenergic vasoconstrictor fibers, which are tonically active, innervate the cutaneous vessels. The vascular smooth muscle of cutaneous nutrient arterioles has both α- and β-adrenergic receptors, while the arteriovenous anastomoses have only α-receptors (Korner, 1974). The physiological importance of cutaneous β-receptors is not known. In the hands and feet a maximal cutaneous blood flow results from sympathetic blockade, but in the forearm sympathectomy increases flow only if skin and core temperatures are low.

Body warming elicits a reflex active vasodilation in the hand or foot, in addition to a release of vasoconstrictor tone (a decreased number of impulses traveling over the sympathatic nerve fibers) (Shepherd, 1963). The reflex may originate either in cutaneous receptors or by central nervous system stimulation. The vessels themselves are also sensitive to warm temperature, for the dilation following blocking of vasomotor nerves can be augmented by local heating of the hand.

Arteriovenous (AV) anastomoses, which exist in some areas of the skin, consist of communications between smaller arteries and arterioles and the corresponding venous channels, through which the blood may be shunted to the venous plexuses and the capillary areas short circuited. These AV anastomoses do not appear to be under metabolic control but, rather, are governed chiefly by reflex influences from temperature receptors and from central nervous system centers regulating heat loss by the skin (see below).

The superficial large and small veins of the skin have a nerve supply and respond by constriction to circulating epinephrine and norepinephrine. The small veins may contract separately and independently from the large veins or arterial vessels during sympathetic nerve stimulation. When the veins constrict, an increase in pressure results. Thus, a reduced blood flow may occur through the vascular bed without any obstruction or constriction of the arterial tree. In humans, forearm cutaneous vein segments isolated in situ by compression have been demonstrated to constrict following a variety of normal and noxious stimuli. This venoconstriction was blocked by chemical or surgical sympathectomy or infiltration of an anesthetic solution around the vein. Venous tone was decreased during vasovagal syncope and by stroking the skin over the vein. The fact that cutaneous veins constrict when cooled and are not affected by warming is evidence for the existence of only a vasoconstrictor nervous supply. The neurogenic control of the skin circulation is discussed further in Chapter 16.

Role of Skin Circulation in Heat Transfer

The blood vessels of the skin have a large capacity to alter flow and change heat loss through the outer layers of the skin, and with maximum vasodilation the skin can lose 7 or 8 times as much heat as during full vasoconstriction; the rate of blood flow to the skin at room temperature is many times the relatively small nutrient flow requirement. The nutrient circulation does exhibit both reactive hyperemia and autoregulation, and the skin circulation also participates in the usual baroreceptor reflexes. However, during changes in temperature the skin vascular resistance comes under other reflex control, particularly the heavily innervated AV anastomoses located primarily in the hands, feet, ears, and face (which, combined, present a large skin surface area). Total skin blood flow during marked cooling can be very low, and during extreme heating it can reach several liters per minute. In the latter circumstance, compensatory vasoconstriction must occur in other vascular beds if the blood pressure is to be maintained.

The AV anastomoses receive adrenergic sympathetic fibers which are largely under the control of the central nervous system. There is also evidence (discussed below) that, when sweating occurs at higher temperatures, chemically induced vasodilation has an important role. The center controlling heat transfer is located in the anterior hypothalamus. Direct cooling of this center or perfusion with cool blood produces cutaneous vasoconstriction, whereas heating or perfusing the center with warm blood produces sweating and cutaneous vasodilation.

Local cooling of the skin appears to cause local reflex vasoconstriction even in the absence of a change in core temperature (possibly due to increased sensitivity of the vessels to adrenergic nerve impulses, as well as a local cord reflex). That such vasoconstriction is reflex and may involve local temperature receptors is illustrated by the response to immersion of one hand in cold water when the circulation to that arm is occluded: cutaneous vasoconstriction results. When the circulation to the immersed hand is intact and the cooling is sustained, the blood temperature is lowered and leads to direct cooling of the hypothalamic center, with further reflex vasoconstriction. With prolonged severe cold exposure, local vasodilation of the skin vessels occurs despite sustained low total skin flow. (This response is responsible for the rubor or redness of face in severe cold and may serve a protective function against freezing).

Application of local heat causes local vasodilation, with opening both of the ordinary resistance vessels and of the AV anastomoses. As warmed blood reaches the hypothalamus, reflex vasodilation occurs elsewhere in the skin. The mechanism of vasodilation is not entirely clear, although when sweating is initiated by the hypothalamic center at higher temperatures there is evidence that a local vasodilator may be involved. One postulated mechanism includes sympathetic cholinergic activation of the sweat glands, which release an enzyme (kallikrein) which, in turn, induces the formation of the potent vasodilator bradykinin. Blockade of bradykinin production need not greatly impair this response, however, and other local products may be involved. This mechanism may also play a role in the central nervous system-mediated blushing response produced by anger or anxiety.

Vascular Responses of Skin to Stimulation by Mechanical and Other Agencies

WHITE REACTION

If the surface of the skin is stroked lightly with a blunt "pointed" instrument, a line of pallor appears in 15–20 s which traces the path taken by the instrument. The line attains its maximal intensity in ½–1 min and then gradually fades to disappear in 3–5 min. The white reaction proper is due to direct stimulation of vessel walls and has no nervous basis. It has been shown to be due to the tension exerted upon the walls of the minute venous vessels—collecting venules and especially those of the subpapillary venous plexus—as they respond to the stimulus by contraction.

TRIPLE RESPONSE

This comprises (1) the red reaction, (2) the flare, and (3) the wheal.

Red Reaction

If the instrument is drawn more firmly across the skin, especially of the forearm or back, a red instead of a white band appears after a somewhat shorter latent period (3–15 s), reaches its maximum in ½ to 1 min, and then gradually fades. Like the white reaction it is strictly localized to the line of stroke; it is due to dilation of the venular vessels. On either side of this, a pale area of capillary constriction (white reaction) may appear. The red reaction can be induced in its full intensity in the skin from which the circulation has been occluded, so it is due to active dilation of the venular vessels and not merely a passive result of arteriolar dilation. The red reaction is not dependent upon nervous mechanisms since it occurs after section and degeneration of the cutaneous nerves.

Spreading Flush or Flare

If the stimulus is a bit stronger, the reddening of the skin is not confined to the line of stroke but surrounds it for a variable distance (1–10 cm) according to the intensity of the injury inflicted. The temperature in the suffused area is definitely raised. This flare reaction appears a few seconds (15–20 s) after the local red line and fades sooner. It is due to dilation of the arterioles and venules, since it does not appear after the circulation of the part has been occluded by means of a tourniquet. Unlike the red reaction, however, the flare is dependent upon local nervous mechanisms (axon reflex). It occurs after the nerves are divided but not after they have degenerated.

Local Edema or Wheal

When the stimulus is still more intense, the skin along the line of the injury becomes blanched and raised above the surrounding area to a height of 1 or 2 mm or even more. Such a wheal or welt can be produced in a normal person by the lash of a whip and other types of strong, localized stimulation. In susceptible individuals, even light stimulation, such as drawing a pencil with moderate pressure over the skin of the back, will produce linear wheals surrounded by a diffuse red halo along the pencil's track. In this way letters or other designs may be embossed upon the skin. This phenomenon is spoken of as *dermographism.* The wheal makes its appearance 1–3 min after the injury and is at its maximum in 3–5 min. It is preceded by, but then replaces, the usual red reaction; it is surrounded by the flare described above. The raised patch at first is clearly demarcated, but as time passes it increases in width and decreases in height, loses its sharpness and finally, although perhaps not for some hours, disappears. The wheal is due to the transudation of fluid from the minute vessels involved previously in the red reaction; it is, therefore, a localized edema. Increased permeability of the capillary wall is judged to be the immediate cause. Wheal production does not depend upon a nervous mechanism.

The triple response has been attributed to the release of some diffusible substance by the injured cells. Injection of histamine produces a similar reaction, but whether histamine or some other vasoactive substance (e.g., ATP, bradykinin, etc.) is involved is not known.

CIRCULATION IN SKELETAL MUSCLE

Blood flow rate in resting mammalian skeletal muscle ranges from 2–5 ml/min/100 g in humans to 10–20 ml/min/100 g in cats and rabbits (Korner, 1974). One reason for this difference is the proportion of red and white fibers in the muscle groups studied, red muscle having about double the resting flow of white muscle. Skeletal muscle in general undergoes active and widely ranging fluctuations in blood flow.

Skeletal muscle is the major component of the body mass, over 40% of normal body weight. The $\dot{V}O_2$ and blood flow to skeletal muscle are low at rest, but both increase markedly during muscular exercise. The circulation to skeletal muscle is particularly suited to such a response and exhibits a high level of basal vascular tone, considerable sympathetic tone at rest, and pronounced autoregulation manifesting a high degree of local control (see Chapter 6). Also, at rest, skeletal muscle has a large capillary reserve, with most of the capillaries being functionally closed (i.e., no red cells flow through these channels because they are diverted at upstream bifurcations; anatomically, channels are present) by cyclic contractions of the precapillary vessels, and a four- to five-fold increase in the number of capillaries can occur during exercise. Finally, there is considerable reserve for the extraction of oxygen during exercise, since the venous blood draining resting muscle has a high O_2 content. Further reserves for O_2 are provided by myoglobin in red skeletal muscle, which has a high aerobic capacity. In contrast, white or mixed muscle has a high anaerobic reserve, with increased content of anaerobic enzymes, relatively fewer mitochondria, and the absence of myoglobin.

To understand blood flow in skeletal muscle it is useful to understand its vascular and microvascular anatomy. Most muscles have multiple arterial and arteriolar inflows originating at the central arteries. These inflows feed into a network of arterioles (arcades, plexus) (see Fig. 2.8) spanning the entire extent of the muscle volume. The connections to the capillaries are provided by regular side branches of the arcade arterioles. These terminal arterioles are deeply embedded between muscle fibers, can regularly close their lumens by infolding of endothelial cells, and are still innervated but also subject to strong local control mechanisms elaborated by the tissue parenchyma. Terminal arterioles feed into a bundle arrangement of capillaries with alternating sequence of terminal arterioles and collecting venules draining the local segment of the capillary bundle (Fig. 2.9). Collecting venules drain into a dense network of arcade venules which, in turn, have numerous outflow channels to the central veins (Schmid-Schönbein et al., 1985). This connecting arrangement of arcade vessels on the arterial and venous sides has many influences on blood flow in skeletal muscle. For example, it is one of the key mechanisms for ensuring homogeneous capillary perfusion of the large muscles, irrespective of their location within the organ. Because the arcade vessels are interconnected with arcade vessels in neighboring tissues, they ensure a baseline perfusion of the muscle, even if capillaries may not be perfused, since flow can occur only through the arcades. (For reviews on the circulation to skeletal muscle see Shepherd, 1963; Korner, 1974; Olsson, 1981; Schmid-Schönbein et al., 1985).

Neural Control of Muscle Blood Flow

SYMPATHETIC VASOCONSTRICTOR FIBERS

Control of muscle blood flow over a wide range is effected by variations in α-adrenergic sympathetic tone (see α- and β-receptors, Chapter 16) acting on the resistance vessels. This forms the basis of most reflex regulation of vascular resistance in skeletal muscle. The transmitter substance is norepinephrine, which stimulates α-receptors in vascular smooth muscle. Maximal stimulation of these fibers doubles vascular resistance in red muscle and increases it sevenfold in white muscle (Korner, 1974). These fibers also produce venoconstriction. Following sympathectomy, muscle blood flow increases transiently and then returns to control values owing to intrinsic regulation. During exercise, the effect of α-adrenergic stimulation on the arterioles is markedly reduced or abolished by the effects of intrinsic vasodilator mechanisms (see below). Venoconstriction, however, is well maintained (Mellander and Johansson, 1968). Sympathetic vasoconstrictor fibers are primarily concerned in postural and other reflex, homeostatic, hemodynamic adjustments, in regulation of pre- to postcapillary resistance, and in changes in the caliber of capacitance vessels (veins). The vasoconstriction they evoke is reflexly influenced by arterial baroreceptors, chemoreceptors, and cardiac baroreceptors (see Chapter 16).

SYMPATHETIC VASODILATOR FIBERS

Sympathetic nerve stimulation after α-adrenergic blockade produces in skeletal muscle a vasodilation which is blocked by atropine. The transmitter substance is thought to be acetylcholine acting on muscarinic receptors. This sympathetic outflow is not activated by the baro- and chemoreceptors mentioned above. It appears to originate in the cerebral cortex via fibers that have connections in the hypothalamus. Such sympathetic vasodilation, blocked by atropine, can be demonstrated in the dog, cat, fox, sheep, and goat, but not in primates, monkey, or rat. It has been postulated to occur in humans during stressful mental activity and vasovagal fainting (Granger et al., 1984). Such vasodilator responses are transient, and resistance returns to control levels in about 1 min, despite continued nerve stimulation. This type of active vasodilation is thought to be part of a centrally integrated neural mechanism that controls cardiovascular adjustments (increased heart rate, blood pressure, and muscle blood flow) in preparation for exercise or the defense reaction, although such a role has not been clearly demonstrated in human subjects.

The effects of circulating catecholamines on muscle blood flow are explained on the basis that both α- and β-receptors exist in the vascular smooth muscles. α-Adrenergic receptors are innervated by the sympathetic vasoconstrictor nerves and respond to circulating norepinephrine, which has little β_2 stimulating activity. β-Adrenergic receptors cause vasodilation. Epinephrine stimulates both α- and β-receptors. Injected intravenously in humans it initiates a five- or sixfold increase in muscle blood flow, which later returns to a value only twice the control level despite continued infusion. Sympathectomy does not abolish this increase; thus, it is not neurogenic. Intraarterial injections also cause vasodilation, but flow rapidly returns to or below the control level. When α-adrenergic receptors are blocked pharmacologically, epinephrine produces a sustained increase in blood flow. Thus, epinephrine stimulates α-receptors to cause vasoconstriction and β-receptors to cause vasodilation, and its overall effect is the sum of these opposing actions. Since the β-receptors in muscle appear to have the lower threshold of the two types, the usual effect of physiological amounts of epinephrine is to increase muscle blood flow, and this is one mechanism of vasodilation during exercise when circulating epinephrine is increased.

Local Control of Resistance Vessels in Skeletal Muscle

The neurally mediated changes in muscle blood flow are small in contrast to those caused by local intrinsic mechanisms. The perfused muscle bed exhibits autoregulation which minimizes changes in flow caused by alterations in perfusion pressure (Johnson, 1980). Autoregulation is present in muscle both at rest and during exercise.

The hyperemia that accompanies muscle contractions is believed to be secondary to local chemical factors rather than neurally mediated for the following reasons: (1) In a curarized muscle the direct stimulation of the muscle elicits an increased blood flow, while motor nerve excitation does not. (2) Humans with sympathectomized extremities have no muscular disabilities during exercise. (3) Local or reflex warming or cooling has little effect on this intrinsic vasomotor tone in muscles. It has been suggested, therefore, that products of muscle cell metabolism mainly act directly on the smooth muscle cells of vessels.

Thus, local metabolites override neural control during exercise, although the precise mechanism is unknown. Low P_{O_2}, increased P_{CO_2} tension, lactic acid, hydrogen ions, bradykinin, histamine, acetylcholine, adenosine, and potassium ions have all been suggested as local determinants of exercise vasodilation (Chapter 6). Most of the metabolites have been tested by infusion into the muscular vascular bed or by determining their concentration in venous blood leaving the muscle following exercise hyperemia. Each one has received only a little support as the dilator agent, but a combination of two or more of these factors may provide the answer. The vascular bed of the exercising hind limb of the dog becomes less responsive to sympathetic stimulation as the strength of contraction and oxygen uptake increase, although the blood flow of exercising skeletal muscle in the dog, except during maximal exercise, is reduced by sympathetic nerve stimulation, by intraarterial norepinephrine or epinephrine infusions, or by carotid sinus stimulation. The hypothalamic vasodilator sympathetic pathway, however, does not affect exercising muscle (see Chapters 16 and 17 for further discussion of circulatory response to exercise).

With rhythmic contraction of calf or forearm muscles, blood flow is rapidly increased both during and immediately following exercise. An increased flow (vasodilator metabolite-mediated) alternating with decreased flow (mechanical compression of vessels) parallels the relaxation and contraction of the muscle (Fig. 2.39). Mechanical obstruction, metabolite formation, and the action of muscles on the venous flow are probably all intricately involved in the control of muscular blood flow during exercise.

Reactive Hyperemia

Following a period of complete occlusion of the arterial supply to a limb or following vigorous contractions (Fig. 2.39*B*), the blood flow at first increases markedly and then later returns to the control level. This phenomenon has been named *reactive hyperemia.* The response occurs both in skin and muscle and in the coronary circulation (Chapter 15) and can be decreased in the skin by cooling, epinephrine, or tobacco smoking. The amount and duration of reactive hyperemia correlate with the previous length of arterial occlusion. This has been shown to be true for occlusion periods up to 10 min.

The blood flow or oxygen debt incurred during the period of occlusion can be calculated by multiplying the control blood flow or the oxygen usage before circulatory arrest by the duration of occlusion. This assumes that the blood flow and oxygen usage of the muscle would have remained at the previous control rates if the flow had not been occluded and that the metabolic rate of the muscle is not affected by the circulatory arrest. In the resting isolated gracilis muscle of the dog as well as in the human forearm, the oxygen debt is approximately repaid for different periods of occlusion, but the blood flow debt repayment ranges anywhere from 50–200% of that expected. The phenomenon of reactive hyperemia has also been demonstrated in dog's hind limb during exercise, and such flow debts are underpaid or barely repaid. Oxygen debts are entirely or partially repaid depending upon the level of muscle performance and the duration of the ischemic period.

In resting muscle and during light exercise both increased blood flow and oxygen extraction are involved in repayment of the oxygen debts, but during medium and strong exercise the increased blood flow is most important. During strong exercise, oxygen debts are often not repaid in the presence of a decreased arteriovenous oxygen difference; therefore, oxygen extraction is the limiting factor. The idea of oxygen debt is, although attractive, rather hypothetical. A rigorous proof for its role as control mechanism in reactive hyperemia is not available currently.

The metabolites believed to cause exercise hyperemia have also been postulated as eliciting reactive hyperemia. Since reactive hyperemia occurs in sympathectomized and denervated limbs, nervous system control is considered unimportant. The mechanical effect of the lack of pressure in the vessels during the period of occlusion has also been demonstrated to play a role in the dilation. (See Chapter 6 for further discussion of myogenic vs. metabolic theories.)

Arteriovenous Anastomoses in Skeletal Muscle

The question of whether AV shunts actually exist as functional units in skeletal muscle is still unset-

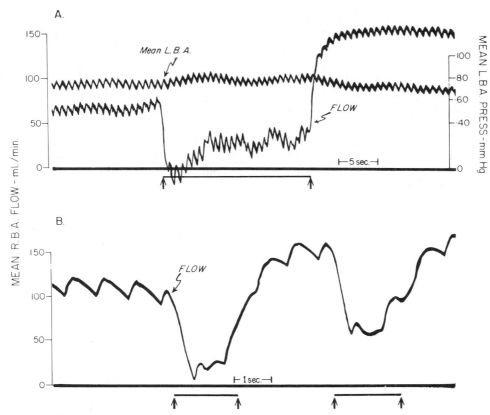

Figure 2.39. Flow in the brachial artery of a conscious human subject measured with an electromagnetic flowmeter during cardiac catheterization. *Panel A* shows mean pressure in the opposite *(left)* brachial artery *(L.B.A.),* together with mean blood flow during sustained handgrip using the right arm *(vertical arrows* and *bar below);* flow drops sharply due to mechanical squeezing of the vessels by muscle con- traction, rises gradually during continued contraction, and exhibits reactive hyperemia after cessation of handgrip. *Panel B* shows tracings at rapid paper speed illustrating the effects of two brief handgrips *(vertical arrows)* on phasic brachial artery blood flow. (From Gault et al., 1966.)

tled. Microscopic studies on the circulation of rat skeletal muscle reveal many communications between small arteries and veins which do not enter the muscle proper but instead occur in adjacent connective tissue, possibly the fascia or other neighboring organs. This is possible because the arcade arterioles in muscle often have connections to the arcade system in these neighboring organs. During inactivity, most of the flow is through these anastomoses and not to the muscle fiber capillaries. When the rats are bled in small amounts, the muscular arterioles close and flow in the capillaries ceases, while it persists in the arcade arterioles and venules and anastomotic channels. That such communications function in humans, however, and possess a means of shunt regulation has not been demonstrated. Indirect evidence for the existence of functional anastomoses is the decreased oxygen consumption and increased lactic acid production during vasodilation; such findings would occur if capillary flow were reduced. The disappearance rate of a radioisotope from a depot in skeletal mus-

cle is thought to represent only capillary blood flow and not shunt flow. During emotional stress, total forearm blood flow increases (plethysmograph), but radioisotope clearance from forearm muscle does not change. Such studies have led to the concept of a dual circulation in resting muscle; one circulatory pathway has been referred to as "nutritional" and the other as "nonnutritional."

Collateral Blood Flow

When the major artery to a vascular bed is occluded (as in atherosclerosis with thrombotic occlusion), smaller vessels arising from arteries above the occlusion supply the area with blood. This collateral flow may prove adequate for all functions. For example, some patients withstand a sudden embolic occlusion of the femoral artery with no adverse effects to the limb. Since collateral arteries function within seconds of arterial occlusion, it is thought that these vessels already exist (i.e., the artery meshwork) but then increase in size following arterial occlusion. In patients with long-standing thrombosis of the aortic bifurcation, aortography often reveals vessels as large as the femoral artery arising from the mesenteric arteries of the gut and supplying the limbs. In normal young subjects, plethysmographic studies of calf blood flow show that mechanical compression of the femoral artery at first decreases flow to one-sixth of the previous resting level, with recovery to normal flow within 6 min despite continued compression. The recovered flow is non-, or only slightly, pulsatile.

Reflex warming of the body and ganglionic blockade increase the collateral blood flow and thus demonstrate that it is mediated in part by the sympathetic nervous system. Exercise of the leg usually increases the collateral flow, but the increase may be due to a rise in systemic blood pressure.

When the external iliac or femoral artery is acutely occluded in dogs or cats, the flow drops to one-third of the previous resting level and then gradually rises to 50–100% of normal. In dogs with acute or chronic arterial occlusions, the blood pressure below the obstruction initially falls to low levels but approaches normal in a few weeks.

CHAPTER 8

Electrical Impulse Formation and Conduction in the Heart

The electrical impulses that drive the heart originate in a group of pacemaker cells lying in the sino-atrial (SA) node, and these action potentials then spread rapidly to both the right and left atrium (depolarizing them to contract). The wave of impulses then enters a conduction system made up of specialized cells, which gathers the impulses at the atrioventricular (AV) junction. There they are delayed (which allows atrial contraction to boost the filling of the ventricles prior to the onset of their contraction), and the wave then enters rapidly conducting fibers of the specialized conduction system. These carry the depolarizing wave to the left and right ventricles, causing them to contract (slightly out of sequence, the left side before the right) (see Fig. 2.4, Chapter 5).

In this chapter we will first discuss the genesis of electrical activity and electrical properties of several types of cardiac cells, since these properties are important in understanding the sequence of electrical activation, as well as how disorders of impulse formation and conduction can occur. Then, the functional anatomy and physiology of the various components of the normal cardiac conduction system and the normal sequence of cardiac excitation will be described. This sequence is due largely to differences in the ability of cells to conduct impulses and is responsible for the form of the normal electrocardiogram (Chapter 9).

BASIS OF THE RESTING MEMBRANE POTENTIAL OF HEART CELLS AND THE CARDIAC PACEMAKER

The genesis of resting cellular transmembrane potentials is complex, and its theoretical basis is detailed in Section 1, Chapter 3. It will be useful here to consider briefly the major ionic movements that produce the resting membrane potential in heart muscle cells, specialized conducting cells, and pacemaker cells. (For more detailed reviews of current concepts concerning this topic see Brady, 1974;

Fozzard, 1979; Noble, Fozzard and Arnsdorf, 1986; Wallace, 1986.)

Resting Membrane Potential of Cardiac Muscle Cells

If the cell membrane or sarcolemma of a resting muscle cell in the ventricle is punctured by means of a microelectrode, a potential difference will be recorded across the cell wall, with the inside of the cell carrying a negative potential relative to the outside. The level of this negative potential differs somewhat in various regions of the heart; in resting ventricular muscle (which is not receiving action potentials from a pacemaker) it is about -90 mV.

The basis for the resting transmembrane potential rests primarily on the selective permeability of the cell membrane to various ions. There are well known differences in concentrations of anions and cations between the intracellular and extracellular fluid compartments, Na^+ and Cl^- forming the major cation and anion in the extracellular fluid, whereas in the intracellular fluid K^+ is the major cation and the anion charge is carried by Cl^-, organic acids, proteins, phosphates, and sulfates. These ionic compositions can be mimicked in an experimental chamber in which an impermeable membrane has been placed to represent the cell boundary; under these conditions, the sum of the positive and negative charges on either side of the membrane will be equal, maintaining electrical balance, and no potential difference will be measured with a voltmeter placed across the membrane. However, if the membrane is made permeable to one ion only, that ion will attempt to cross the membrane down its concentration gradient. If the ion is K^+, for example, ions will commence to move across the membrane from the region of high K^+ concentration (the side of the chamber representing the intracellular fluid). As a few K^+ ions cross the membrane, the intracellular side will become electrically negative in the region close to the membrane (since positive charges are leaving). The buildup of positive charge on the other side of the membrane (representing the extracellular fluid) will counteract the tendency for K^+ ions to move down their concentration gradient, however, eventually causing the ion movement to cease. This sequence of events involves only a very *few* ions close to the membrane and will not result in measurable changes in the *concentration* of ions on either side. The potential difference existing across a semipermeable membrane when the movement of ions has ceased can be called the equilibrium potential for that ion (in this case for K^+).

In cardiac muscle, the resting transmembrane potential is, in fact, predominantly determined by selective permeability to K^+, with relatively slight permeability to Na^+ and less to Cl^-. The equilibrium potential for any ion can be calculated, provided its intra- and extracellular concentrations are known, using the Nernst equation.

$$V_{eq} = -\frac{61.5}{Z} \log \frac{[C]_i}{[C]_o}$$

where V_{eq} = equilibrium potential for a given ion, Z is the charge (valence) of an ion, and $[C]_o$ and $[C]_i$ represent its extracellular and intracellular concentrations, respectively (see Section 1, Chapter 3 for the more general form of the Nernst equation and the derivation of the constant from the gas constant (R), Faraday's constant (F) and T, the absolute temperature). The equilibrium potential for a given ion can be either positive or negative, depending on its concentration difference, and the logarithmic relation means that relatively large changes in the ratio of ion concentrations inside and outside the cell result in relatively small changes in the resting transmembrane potential.

When the Nernst equation is employed to calculate the resting transmembrane potential for K^+ ($Z = 1$) it is written as:

$$V_{eq} K = -61.5 \log \frac{[K^+]_i}{[K^+]_o}$$

Table 2.3 shows the relative concentrations for ions across the cell membrane in heart cells, and, by sub-stituting 150 mM and 4 mM as the normal concentrations for K^+ inside and outside the cell, a resting membrane potential of -97 mV can be calculated. If the Nernst equation is used to predict equilibrium potentials for K^+ in an ideal system in which the $[K^+]_i$ is held constant and the $[K^+]_o$ is varied, the Nernst equation predicts an inverse relation between the level of transmembrane potential and increasing extracellular $[K^+]$; that is, the cell becomes relatively more depolarized, or less negative, as the $[K^+]_o$ rises (Fig. 2.40). If the transmembrane potential is actually recorded in a resting muscle cell using an intracellular microelectrode, when the external $[K^+]$ is varied, a relation close to that predicted by the Nernst equation is observed; however, the measured curve *(solid line)* lies closer to the theoretical relation *(dashed line)* at high $[K^+]_o$ concentrations and deviates to some degree at lower concentrations, with a resting value at a normal $[K^+]_o$ of 4 mM close to -90 mV *(dotted lines, Fig. 2.40)*.

The cell membrane, as mentioned, is permeable not only to K^+ but also to Na^+, and the latter ion can also influence the resting membrane potential. Even though wide fluctuations of extracellular Na^+ alone have only a small effect on the resting membrane potential (because of the dominance of the K^+ permeability effect), the deviation from the theoretical line can nevertheless be explained in part by some degree of membrane permeability to Na^+. Also, at lower K^+ concentrations, selective permeability to K^+ appears to decrease. When the Nernst equation is applied to calculate the equilibrium potential for Na^+, considering the membrane as selectively permeable to that cation alone, a positive potential of $+42$ mV can be calculated (using intracellular and

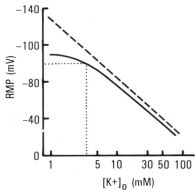

Figure 2.40. Relations between the resting membrane potential *(RMP)* and the extracellular concentration of potassium, $[K^+]_o$, in cardiac cells. The *dashed straight line* indicates the theoretical relation predicted by the Nernst equation, and the *solid line* the relation as actually measured. The *dotted lines* show that at a normal potassium concentration of 4.0 mM the RMP is approximately -90 mV.

Table 2.3.
Concentrations of Selected Ions in Mammalian Cardiac Muscle

Ion	Concentration	
	Extracellular	Intracellular
K^+	4 mM	150–160 mM
Na^+	145 mM	15 mM
Ca^{2+}	2 mM	10^{-7} M
Cl^-	120 mM	5 mM

extracellular concentrations of 30 and 145 mM, respectively). Thus, Na^+ tends to move from its high concentration outside the cell toward the inside of the cell along its concentration gradient, leaving the inner side of the membrane relatively positive compared to the extracellular fluid. As mentioned, the small leakage of Na^+ into the cell is primarily responsible for the shift downward of the actual curve (Fig. 2.40), causing the interior of the cell to be less negative than predicted by the Nernst equation. This small inward sodium movement is continuously balanced by a small outward K^+ movement, so that there is no change in the transmembrane potential with time when the cardiac muscle cell is at rest (see Fig. 2.41, phase 4). (This is not the case in pacemaker cells.)

In order to calculate the *combined* effects of various ions on the transmembrane potential when the membrane is permeable to more than one ion, the relative permeability of the membrane to each ion must be considered and the *Goldman-Hodgkin-Katz equation* can be employed (for the general form of this equation see Section 1, Chapter 3). For two or more ions:

$$V_m = -61.5 \log \frac{P_K [K^+]_i + P_{Na} [Na^+]_i + P_{Cl} [Cl^-]_o}{P_K [K^+]_o + P_{Na} [Na^+]_o + P_{Cl} [Cl^-]_i}$$

Or, for two ions:

$$V_m = -61.5 \log \frac{[K^+]_i}{[K^+]_o} + \frac{[Na^+]_i (P_{Na}/P_K)}{[Na^+]_o (P_{Na}/P_K)}$$

where V_m = transmembrane potential and P = the permeability for each ion (defined as the quantity of the ion that diffuses across each unit area of membrane per unit of concentration gradient per unit of membrane thickness).

This constant-field equation makes clear the importance of the *ratio* of permeabilities of ions in determining the transmembrane potential. When the selective permeability to K^+ is much higher than that to Na^+ (as is the case with the normal *resting* cell), the V_m approaches that calculated by the Nernst equation for K^+.

Resting Membrane Potential of Pacemaker Cells

Cells in the cardiac pacemaker region (SA node) as well as in the AV junction and the His-Purkinje network (the specialized conducting system) do not exhibit constant resting potentials, but are capable of *spontaneous diastolic depolarization*. This phenomenon has been most fully studied in the large cells which compose the rapid conducting pathway (Purkinje cells, see below), and in these cells the

mechanism for this gradual fall in the negative resting potential toward zero (it becomes progressively less negative, see Fig. 2.42C and D) is a gradual *decrease* in K^+ permeability, while the permeability to other ions remains constant. As seen from the Goldman equation for two ions, a decreasing value of P_K will increase the ratio of P_{Na}/P_K, thereby enhancing the effect of the Na^+ gradient on V_m and producing a relative increase of positive charge inside the cell due to the Na^+ leak (hence a progressively less negative transmembrane potential). Also, because of the decreasing K^+ permeability there is less K^+ movement outward, and the movement of Na^+ inward is therefore unbalanced.

This phenomenon may also be responsible, at least in part, for the gradual depolarization of resting cells in the SA node and AV junction, although the complex anatomy and small size of cells in these regions have hindered elucidation of mechanisms (Vassalle, 1977). These pacemaker cells may well have a larger permeability to Na^+ than muscle cells, as evidenced by lower (less negative) levels of resting transmembrane potential, and it has also been suggested that Ca^{2+} permeability may be significantly higher than in muscle cells. It has been further postulated that gradually increasing Ca^{2+} permeability may play a role in the diastolic depolarization of cells in the SA node and AV junction. (For example, Ca^{2+} antagonist drugs markedly reduce progressive diastolic depolarization in those regions.)

Cells in several regions (SA node, AV node, His-Purkinje network) can serve as pacemakers, that is, exhibit *automaticity*. Recent evidence suggests that in Purkinje cells there is a special voltage-dependent channel activated at negative potentials. This channel can carry either Na^+ or K^+ inward current and is mainly responsible for automaticity in Purkinje cells, with the Na^+ current being dominant at more negative resting potentials (Zipes, 1988). In the sinus node, automaticity appears to be less dependent on this current (Zipes, 1988), with pacemaker activity being mainly dependent not only on decreasing K^+ permeability (phase 4) but also on activation of the Ca^{2+} slow channel to generate the pacemaker impulse (see below). However, the origins of the sinus pacemaker currents are not entirely settled (Fozzard and Arnsdorf, 1986).

ACTION POTENTIALS IN CARDIAC TISSUE

When pacemaker cells spontaneously depolarize, they eventually reach their threshold potential and produce a spontaneous *action potential*, which completely depolarizes the cell; in such cells, action potentials typically occur at a regular rate, depending on the rate of diastolic depolarization. As in nerve fibers (see Section 1, Chapter 3), once an action potential is initiated in a given cell it will travel to adjacent cells, and if the magnitude of the action potential is sufficient to exceed the *threshold potential* of those cells a response will be triggered, which rapidly results in the complete depolarization of the adjacent cells (the triggered response is called a "propagated" action potential). Such action potentials result from rapid *changes in membrane permeability* to several ions and, of course, they provide the stimulus which carries the pacemaker potential through the specialized conduction system and initiates the contraction of heart muscle.

In pacemaker cells, the property of spontaneous diastolic depolarization followed by action potentials is referred to as *automaticity*. The threshold potential is about -70 mV in ventricular muscle cells (Fig. 2.41), although it varies somewhat depending on physiological conditions or on the presence of certain drugs. In atrial muscle, the threshold potential is 30–60 mV. An example of action potentials actually recorded from a spontaneously firing cell in the ventricular conducting system is shown in Figure 2.42*D* (for full discussion of automaticity see Vassalle, 1977).

Fast-Response Action Potentials

Fast-response action potentials are characterized by relatively high (more negative) and relatively constant resting membrane potentials and by a very rapid onset of the action potential. Cells in the atrial and ventricular muscle, specialized tracts between the two atria, and parts of the conduction system exhibit this type of action potential (Fig. 2.41 and 2.42*A* and *B*).

Such tissues exhibit rapid (muscle) or very rapid (specialized conduction tissue) conduction velocity, in contrast to the slow conduction in the atrioventricular junction. The rapid upstroke is termed phase 0, the recovery of the initial overshoot to a positive membrane potential phase 1, the plateau period phase 2, repolarization phase 3, and the resting membrane potential phase 4 (Fig. 2.41). These phases of the action potential are related to marked and rapid changes of ion permeability through separate ionic channels in the cell membrane. The passage of ions across these channels is considered to be largely controlled by "gates" which open or close in response to changes in membrane potential. Experimentally, it is possible to measure, as electrical currents, the ion fluxes across the membrane by maintaining the membrane voltage constant (so-called "voltage clamping," see Section 1, Chapter 3).

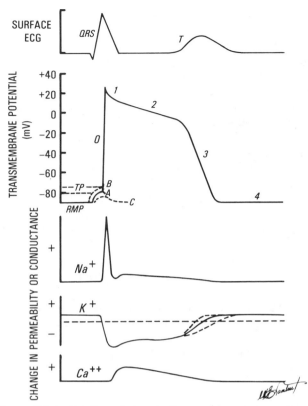

Figure 2.41. Relations between the surface electrocardiogram *(ECG)*, the transmembrane potential, and diagrams of the relative changes in ion permeability (or conductance) that are responsible for the action potential. The (0–4) phases of a typical fast action potential are indicated, the resting membrane potential *(RMP)* being phase 4. The threshold potential *(TP)* is also shown; *C* represents a subthreshold stimulus (nonpropagated); *A* a threshold stimulus, and *B* is a threshold stimulus at a lower (less negative) threshold potential *(upper dashed line)*. The large rapid increase in Na$^+$ conductance during phase 0 is shown, together with the sustained small increase in Na$^+$ conductance (slow Na$^+$ current) during phase 2 of the action potential. The large, less rapid drop in K$^+$ conductance during phase 0 of the action potential is indicated and the *dashed lines* of altered K$^+$ conductance during phase 3 indicate induced changes in P_K. The *dashed horizontal line* indicates that K$^+$ conductance is above zero at rest. The increased Ca^{2+} conductance during phase 2 of the action potential is shown. For further discussion, see text.

The *conductance* of the cell $(g)^*$ is generally measured in such experiments; it provides an electrophysiological measure that is related to membrane permeability (as permeability increases, conductance increases). (For detailed discussion or methods and current concepts of the action potential see Noble, 1984, and Arnsdorf, 1986.)

Phase 0, the rapid upstroke which follows sudden depolarization of the cell to its threshold potential, results from a voltage-dependent very rapid

---

*Conductance is the reciprocal of the electrical resistance of a membrane $(g = 1/R)$ and expresses the ease with which an ion driven by an electrical potential penetrates a membrane.

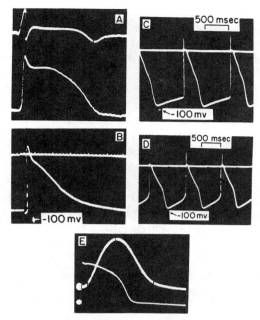

Figure 2.42. *A (upper left panel),* Transmembrane action potential of a single ventricular fiber *(lower curve),* and simultaneous electrogram of the intact heart in situ. The upstroke of the action potential coincides with the R wave of the electrogram, the plateau with the ST segment, and the phase 3 repolarization with the T wave. (Adapted from Brooks et al., 1955.) *B,* Normal transmembrane potential recorded from a single fiber of the intact atrium. (Adapted from Brooks et al., 1955.) *C,* Transmembrane action potential of a single Purkinje fiber in a nonpacemaker activity (paced from above). Note slow loss of resting potential during diastole. (Adapted from Brooks et al., 1955.) *D,* Spontaneous activity recorded from Purkinje fiber in a pacemaker area. Note slow depolarization during diastole and then abrupt onset of spontaneous depolarization. (Adapted from Brooks et al., 1955.) *E,* Temporal relation between action potential *(lower tracing)* and mechanical contraction (tension, *upper tracing*) recorded from a papillary muscle. (Adapted from Dudel and Trautwein, 1954.)

increase in sodium permeability (P_{Na}) and conductance (g_{Na}). Movement through these fast sodium channels, which is thought to represent sudden opening of pores or gates in the membrane, is favored both by the electrostatic charge across the membrane (positive outside, negative inside) and by the large concentration gradient from outside to inside for the positively charged sodium ion. A sharp drop in P_K also begins during phase 0 of the action potential (Fig. 2.41).

One model for the action of these pores (developed in nerve tissue) includes two electrically charged gates on each pore that can open and close (Hodgkin, 1964; Huxley, 1964). In such a model, the change in membrane potential from −90 to about −60 mV affects the conformation of one of the Na^+ gates and opens the fast Na^+ channel, lowering the transmembrane potential; this, in turn, opens more channels and/or further widens the opening in each gate; this sequence has been termed a *regenerative process,* a positive feedback mechanism which rap-

idly drives the action potential to a positive value at the end of phase 0 (Fig. 2.41). The inward Na^+ current is rapidly slowed as the electrical gradient across the cell membrane reverses, the inward current continuing at a positive transmembrane potential only because of the large concentration gradient for Na^+ which persists (recall that only a small number of ions actually traverse the membrane relative to the total number of ions available in the extracellular space). The "positive overshoot" of the action potential ceases at about +20 mV, and in the gating model this "turnoff" of phase 0 is considered to result from closure of the second set of gates on each channel, which respond to a change in membrane potential in an opposite manner to the first set, i.e., they progressively close as the transmembrane potential becomes less negative; this is a slower process than the opening of the first set of gates, however, so that phase 0 depolarization is rapid. As predicted by the Goldman equation, during the overshoot the transmembrane potential tends toward, but does not reach, the Na^+ equilibrium potential, because these gates close before the membrane potential can reach the Na^+ equilibrium potential (+42 mV). As might be anticipated, when the Na^+ channels are open the transmembrane potential becomes very sensitive to the external Na^+ concentration (see Goldman equation); V_m is then closely related to the log Na^+ (just as the resting transmembrane potential is related to the log $[K^+]_o$ when the K^+ channels are partially open in phase 4).

Phase 0 of the fast action potential can be specifically blocked by tetrodotoxin, an agent which blocks only the Na^+ fast channels and, under these conditions, the rapid upstroke is replaced by a slow-response action potential (see below). The rapid spike during phase 0 of a fast-response action potential usually results in a propagated depolarization of the entire cell (an all-or-nothing response), as well as cell-to-cell propagation as discussed subsequently.

Phase 1 of the action potential starts the repolarization process of the membrane (Fig. 2.41). Its mechanism is not fully elucidated, although closing of the sodium channels is an important factor. The inward movement of chloride ions may also play some role in this phase of the action potential. Phase 1 is particularly prominent in Purkinje cells.

Phase 2, the so-called "plateau" of the action potential, results from several mechanisms. It is an important feature of fibers in the rapid conduction system to the ventricles as well as of the ventricular muscle itself, and it prolongs the duration of the action potential to 200–300 ms. During that period it is not possible to initiate another action potential because the fast Na^+ channels are inactivated, and therefore the heart is not subject to tetanic contracture. (It is "refractory," see below.) The action potential in cardiac muscle contrasts with the brief action potential of skeletal muscle, which readily permits summated contractions (Section 1, Chapter 4).

Two mechanisms appear to be largely responsible for maintaining the plateau of the action potential: a slow inward Ca^{2+} current and/or a slow inward Na^+ current. Voltage-dependent activation of the slow Ca^{2+} channel with increased Ca^{2+} permeability and conductance appears to be most important for maintenance of the plateau (for review, see Reuter, 1974, Zipes, 1988). The Ca^{2+} channel opens when the transmembrane potential depolarizes to about -35 mV, and some increase in Na^+ permeability (apparently through Na^+ slow channels) also occurs at this point. The slow inward movement of Ca^{2+} and Na^+ ions is counterbalanced by movement of K^+ outward along its concentration gradient; following the initial drop in P_K it gradually increases, and this outward K^+ movement appears to remain sufficiently large to balance the inward currents (Fig. 2.41), thereby maintaining a relatively steady plateau.

The inward Ca^{2+} current is important in influencing heart muscle contraction (Chapter 10). It can be blocked by Ca^{2+} antagonists such as verapamil (which appears to close Ca^{2+} slow channels) and by Mn^{2+}. Movement of Ca^{2+} through the slow channels is enhanced by catecholamines, and it is increased by elevated $[Ca^{2+}]_o$ (increased gradient). Such channels have also been considered to be activated and inactivated by the movement of two gates, as with the fast Na^+ channel. When the calcium slow channels are blocked, some plateau of the transmembrane potential is still observed due to the inward slow Na^+ current and the outward movement of K^+, but the plateau is abbreviated and occurs at a more negative transmembrane potential.

Phase 3 of the action potential involves relatively rapid repolarization of the cell membrane. It commences with inactivation of the Ca^{2+} and Na^+ slow channels and a relatively rapid increase in P_K (or g_K), which produces a rapid outward movement of K^+ and also contributes to restoration of the relatively negative cell interior (Fig. 2.41). This change in the K^+ channel also appears to be voltage-dependent, and it progressively increases as the transmembrane potential becomes more negative, leading to a regenerative recovery process with activation of a K^+ outward current.

The P_K during phase 3 is sensitive to a number of factors which can alter the duration of the action potential (Fig. 2.41). For example, the drug lido-caine or an increased $[K^+]_o$ will increase P_K; also, acetylcholine increases P_K in atrial muscle (which has acetylcholine receptors), and all of these stimuli shorten the action potential duration. A number of other influences also can decrease action potential duration, including epinephrine and norepinephrine, digitalis, and increased frequency of contraction. Other drugs, such as quinidine, as well as decreased heart rate or lowered $[K^+]_o$, decrease P_K and conductance and increase the action potential duration (Fig. 2.41).

Phase 4, the rest phase, was discussed previously. It is clear that active membrane pumps must remove the Na^+ that enters the cell during the action potential, as well as that which enters by diffusion during diastole. The energy-dependent membrane Na^+/K^+-ATPase removes these Na^+ ions from within the cell, and the K^+ ions which left the cell during repolarization are also returned by this pump, both ions being moved against their concentration gradients (see Schwartz, 1974). Ca^{2+} entering the cell during the action potential is exchanged for Na^+ during the rest phase of the cycle and by mechanisms (Chapter 10).

A number of factors can modify the diastolic threshold potential, making a larger or smaller triggering potential necessary (Fig. 2.41). Certain drugs, such as quinidine, raise the threshold potential relative to the resting potential (make it less negative) and therefore the cells are more difficult to depolarize (less excitable), requiring a larger triggering potential (Fig. 2.41*B*). Catecholamines make the threshold potential more negative relative to the resting potential, i.e., less change in the diastolic potential is required to trigger an action potential.

Fast response cells tend to exhibit inactivation of the fast channels at less negative membrane potentials (an effect that can be produced by increased $[K^+]_o$), so that very slow depolarization of the cell toward its threshold may cause inactivation of impulse propagation.

Slow-Response Action Potentials

Cells with action potentials of this type are found in the SA and AV nodes and in the AV junctional region, where they may be of either pacemaker or nonpacemaker type. Also, fast-response cells have the capability of being converted to slow-response fibers when they are partially depolarized to a membrane potential well below the threshold for the fast response (occasionally this occurs in injured cardiac tissue, as during ischemia, see Chapter 15). The general form of the slow-response action potential is illustrated in Figure 2.43. It shows a lower resting membrane potential and a lower (less nega-

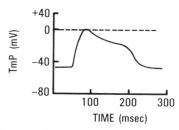

Figure 2.43. Diagram of a typical slow-response action potential of a cardiac cell. Notice the slow upstroke, lack of a phase 1, and the shorter action potential duration than the fast response shown in Figure 2.41. *TmP*, transmembrane potential.

tive) threshold potential than the fast fibers. Also, there is a slower upstroke, with loss of the spike of phase 0, and the amplitude of the action potential is less, with only a small positive phase. Although a plateau is present, repolarization is fairly rapid, leading to a shorter action potential duration than in fast fibers (Fig. 2.43).

Slow-response action potentials are thought to be mediated primarily by slow inward Ca^{2+} and Na^+ currents. Like the gates in fast-response cells, cells in the SA and AV nodes appear to have Ca^{2+} and Na^+ gates that exhibit voltage-dependent inactivation in the later phases of the action potential. The Ca^{2+} slow channels are mainly responsible for phase 0 in these cells, since they are inhibited by Ca^{2+} antagonists such as verapamil and by Mn^{2+} and are insensitive to tetrodotoxin (Zipes, 1988). When fast-response Purkinje cells are converted to the slow response, either the slow Na^+ channel or the Ca^{2+} channel may be operative (see also previous discussion under "Resting Membrane Potential of Pacemaker Cells").

MEMBRANE PROPERTIES

In addition to changes in the threshold potential (which can change the excitability of the cell), a number of other properties of the cell membrane contribute to the ability of the heart's pacemaker and conduction system effectively to conduct to the muscle and to depolarize it. These properties relate primarily to the time of recovery of excitability of the cell membrane (that is, how long it is refractory to an outside stimulus), the type of response it exhibits (fast or slow), and the conducting properties of cells and cell networks.

Refractoriness

In single cells and in groups of cells, time is required for the cell to recover partial and full excitability during the repolarization process. These periods have been divided into several segments (Hoffman, 1969).

The absolute refractory period can be determined in single cells and constitutes that period during

which the membrane cannot be reexcited by an outside stimulus, regardless of the level of external voltage applied (Fig. 2.44). In networks of cells, the absolute refractory period cannot be accurately determined because of different recovery times of various cells in the network, and the *effective* refractory period is usually determined for such cell networks.

The effective refractory period in a cell or cell network constitutes that period during which only a local response can be produced by a *larger than normal* depolarizing stimulus (Fig. 2.44). Thus, during the effective refractory period the membrane can respond, but a propagated action potential that will carry the impulse throughout the cell network cannot be generated.

The relative refractory period commences at the end of the effective refractory period and constitutes that time interval late in the action potential during which a *propagated* action potential can be generated but with a depolarizing stimulus that is *larger* than normal (Fig. 2.44).

The supernormal period is a short interval during which the cell is *more* excitable than normal; that is, a weaker than usual depolarizing stimulus

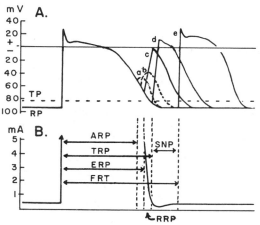

Figure 2.44. *A,* Normal transmembrane action potential and responses to series of stimuli applied during and after end of repolarization. *B,* Approximate durations of absolute refractory period *(ARP),* total refractory period *(TRP),* effective refractory period *(ERP),* full recovery time *(FRT),* supernormal period *(SNP),* and relative refractory period *(RRP).* In *A,* responses shown as *dotted lines* (a and b) are graded responses which do not propagate. Response *c* is earliest propagated response and defines end of effective refractory period. Response *d* is elicited at the time when transmembrane potential is close to level of threshold potential *(TP)* and occurs during supernormal period. Response *e* is elicited after end of repolarization and is normal in terms of rising velocity and amplitude, defining end of full recovery time. Changes in threshold shown in *B* are related to an arbitrary scale of current strength required. *Curve* shows onset of inexcitability coincident with phase 0 of transmembrane action potential, gradual decrease in threshold during phase 3 of repolarization, and restoration of full excitability after end of phase 3 of transmembrane action potential. (Redrawn from Hoffman, 1969.)

can initiate a propagated action potential (Fig. 2.44).

The full recovery time constitutes the period from the onset of the action potential to the end of the supernormal period.

In cells of this type (generally rapidly conducting) the recovery of excitability (or refractoriness) is mainly *voltage-dependent*, as discussed further below, whereas in slow-response cells (generally associated with slow conduction velocity) recovery is mainly *time-dependent*. Thus, in cells of the *slow-response* type, repolarization of the cell to its resting potential does not necessarily coincide with the recovery of excitability. In slow fibers such as in the SA node and AV junction, the refractory period can extend *well beyond* the action potential into the rest period (phase 4), causing a marked delay in the cell's ability to respond to an early stimulus.

Different groups of cells in the cardiac conduction system have different refractory periods, and a method for studying these refractory periods in the whole heart is discussed in a subsequent section. The refractory periods of various muscle cell types can be altered by physiological stimuli (such as the catecholamines) and by various pharmacological agents. The relatively long refractory periods of the cardiac conduction system and of heart muscle (which are due to the long plateau of these action potentials) are responsible for inability to produce a sustained contraction of the heart by a series of rapid impulses (which sometimes can occur during abnormal cardiac rhythms) and also prevent single extra depolarizations (so-called premature depolarizations), which occur spontaneously, from occurring so early that one contraction superimposes on another.

A number of drugs that are used to treat cardiac rhythm disturbances affect the refractory periods of heart cells. In particular, quinidine and like compounds increase the action potential duration and increase even more the duration of the effective refractory period of cells in the rapid conduction tracts and myocardium (thereby prolonging the period during the action potential when extra responses cannot be propagated). Lidocaine, on the other hand, shortens the action potential duration, but it has less shortening effect on the effective refractory period, so that, again, the ratio ERP/APD (effective refractory period/action potential duration) is increased, and a similar effect on extra impulses or rapid rhythms ensues.

Membrane Responsiveness

During the latter part of the effective refractory period in fast fibers, a large depolarizing stimulus can produce a nonpropagated response which has a slow intial rate of rise, and during the relative refractory period propagated action potentials can be generated (by larger depolarizing stimuli than normal) which, again, have a slower than normal upstroke (phase 0) (Fig. 2.44, *a–d*).

In cells exhibiting voltage-dependent refractoriness, the maximum rate of rise of the action potential (dV/dt or $\dot{V}_{max}$) is related to the membrane potential from which it is initiated, so that during the late downslope of the action potential, as the transmembrane potential becomes progressively more negative, $\dot{V}_{max}$ increases along a sigmoid curve (Fig. 2.45). This so-called *"membrane response curve"* is important in that ability of an impulse to be conducted depends on its initial upstroke velocity (the faster $\dot{V}_{max}$ the more readily it is conducted, see below) and, in addition, the response curve can be shifted by certain drugs. For example, the antiarrhythmic agent quinidine decreases membrane responsiveness by shifting the entire membrane response curve to the right and downward, so that $\dot{V}_{max}$ is reduced at any level of transmembrane potential (Fig. 2.45, *A* to *B*).

Impulse Conduction

The conduction of action potentials has been studied in detail in nerve fibers (Section 1, Chapter 3), and similar processes appear to occur in cardiac cells. Local current flow occurs through the electrically conductive solutions on either side of the cell membrane, and the local current flowing at the "front" produced by a local zone of depolarization can spread progressively over the entire cell. If the impulse is propagated, it will also spread to adjacent cells.

Several factors determine the speed and effectiveness with which an action potential is conducted. The amplitude and rate of change of the action potential ($\dot{V}_{max}$) is important, with a more rapid phase 0 increasing conduction velocity. The magni-

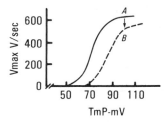

Figure 2.45. Membrane response curves for cardiac cells. *Vmax,* the maximum upstroke velocity of phase 0 of the action potential in volts per second. *TmP* is the negative transmembrane potential in millivolts. Notice the sigmoid relation in normal cardiac cells *(curve A)*, with upstroke velocity increasing with more negative transmembrane potentials. This relationship is shifted to the right by antiarrhythmic agents such as quinidine *(dashed curve B)*, so that at any level of transmembrane potential V_{max} is reduced and conduction velocity is slowed.

tude of the potential and its rate of rise will also determine whether or not it reaches the threshold of adjacent cells. The anatomy of the conducting cells is also important; an increased diameter of the cells results in an increased conduction velocity. When a cell network is organized so that cell size progressively decreases and multiple interconnections exist, conduction velocity is slowed.

Since the resting membrane potential influences $\dot{V}_{max}$, this potential has an indirect effect on conduction velocity. Increases in $[K^+]_o$, by depolarizing the membrane, can result in a marked reduction in $\dot{V}_{max}$. Drugs such as quinidine affect Na^+ conductance and shift the membrane responsiveness relationship, thereby also slowing $\dot{V}_{max}$. If the conducting cells are of the slow-response type (as in the SA or AV nodes), the upstroke velocity is slow and conduction velocity is low.

The so-called "cable properties" of the conduction system and cardiac muscle cells also determine how well the impulse is carried, and these properties in nerve fibers are discussed in detail in Section

1, Chapter 3. As with flow through a cable or wire, conduction velocity is directly related to fiber radius. The other passive cable properties can be determined by applying a current at a given point and determining its spread along the cell (Fig. 2.46). Because of the complex cell structure and cell networks, such analysis is difficult in the heart. In simpler models however, an equivalent electrical circuit can be used to represent the membrane resistance and capacitance, the extracellular resistance, and the intracellular (myoplasmic) resistance for each segment of the cell (Fig. 2.46). With application of a subthreshold square wave current, the induced local potential increases slowly as the capacitance is charged, and the magnitude of the induced (subthreshold) potential then decreases as its distance from the stimulation point increases (Fig. 2.46). These changes are determined by the input resistance of the tissue and its space and time constants (consideration of which is beyond the scope of this discussion). A subthreshold current flow precedes a propagated action potential, and the

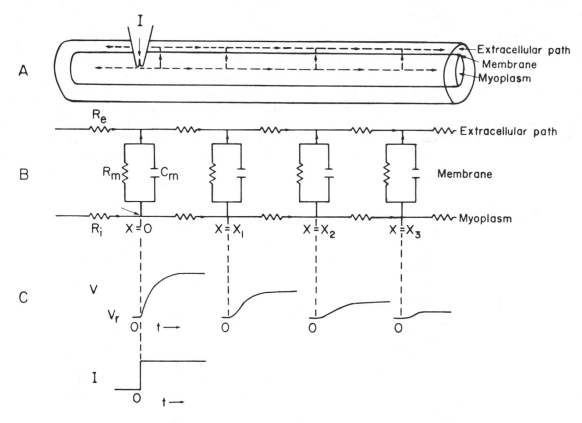

Figure 2.46. Diagram of factors influencing cable properties of cardiac cells. *Panel A,* Diagram of an elongated cardiac cell showing its cell membrane and intracellular contents (myoplasm), surrounded by a cylinder of extracellular fluid (extracellular path). A stimulating electrode supplies current flow from a current source (*I*). *Panel B,* Equivalent circuit of the cell showing repeating resistance-capacitance units along the cell membrane (R_m and C_m, respectively) at distances $X = 0$, X_1, X_2, X_3, together with resistances in the extracellular path (R_o) and resistances in the intracellular path or myoplasm (R_i). *Panel C,* Membrane voltage responses (superimposed on the resting potential, V_R), as a function of time at the various distances along the membrane in response to a subthreshold square wave current (*I*) supplied at time 0 *(lower tracing)*. (From Brady, 1974.)

membrane capacitive effects as well as the cell properties and the resistance of the extracellular fluid determine the decay of a subthreshold stimulus as it moves down a cell or fiber. (For further details of cable theory see Brady, 1974; Wallace, 1986; Fozzard and Arnsdorf, 1986.)

ANATOMY AND PHYSIOLOGY OF THE CARDIAC PACEMAKING AND CONDUCTION SYSTEMS

The characteristics of action potentials (amplitude and duration) and fiber diameters in several species are shown in Table 2.4. Cell diameters and conduction velocities in humans are shown in Table 2.5.

SA Node

The sinoatrial (SA) node is situated near the junction of the superior vena cava and the wall of the right atrium (Fig. 2.47). In humans, it is about 1.5 cm long and 0.5 cm wide. Its cells contain a few myofilaments, and the central stellate cells (so-called P cells) that probably constitute the pacemaker cells contain fewer myofilaments than the adjacent cells, which may serve to conduct the impulse out of the node (Schlant and Silverman, 1986).

Sinus node action potentials are difficult to record, but appear to exhibit a resting membrane potential of about -60 mV, and the pacemaker cells show diastolic depolarization (gradual increase in slope during phase 4). The action potential reaches threshold at about -40 mV; the upstroke velocity

is slower than in fast-response cells and does not exhibit a sustained plateau (Fig. 2.48). As mentioned earlier, SA-nodal pacemaker cells exhibit spontaneous diastolic depolarization probably mainly through a progressive reduction in K^+ permeability, and a change in Ca^{2+} permeability produces an associated inward Ca^{2+} current. Thus, Ca^{2+} antagonists such as verapamil slow the rate of diastolic depolarization and the sinus rate.

A number of factors in addition to verapamil can affect the rate of firing of the SA node. These include atropine, which speeds phase 4 depolarization by blocking acetylcholine effects (acetylcholine slows phase 4), and catecholamines, which increase the firing rate (an effect counteracted by β-adrenergic blocking drugs). The various mechanisms by which the frequency of the cardiac pacemaker can theoretically be affected are summarized in Figure 2.48 and include changing the level of the initial diastolic potential (a more negative initial potential slows the sinus rate, Fig. 2.48, a to d, and vice versa), altering the slope of diastolic depolarization (a to c), and changing the threshold potential (a to b) (Hoffman et al., 1966; Vassalle, 1977; Zipes, 1988).

Under normal conditions, the rate of firing of the SA node is under the control of the autonomic nervous system through the interplay of reflexly induced release of acetylcholine from the vagus nerve terminals (parasympathetic system) and release of norepinephrine from the sympathetic nerve terminals (both of these autonomic divisions heavily innervate the SA and AV nodes). Acetylcho-

Table 2.4.
Cardiac Resting and Action Potentials[a]

Site	Species	Fiber Diameter, μm	MRP, mV	AP, mV	D_{AP},[b]/ms
Sinus node	Frog	5	40	53	200
	Rabbit		60	66	200
AV node	Dog		53	58	350
Atrium	Frog	5	70	90	400
	Rabbit	10	78	92	150
	Dog	10	85	105	200
	Human	10	70	75	
Ventricle	Frog	10	80	95	600
	Dog	16	80	100	400
	Human	16	87	115	200
Purkinje fiber	Dog	30	90	120	300
	Sheep	75	94	130	400
Skeletal muscle[c]	Frog	130	88	120	1.5
Unmyelinated nerve fiber[c]	Squid	400	65	85	0.4

(From Hecht, 1968.) [a]MRP, membrane resting potential; AP, action potential; D_{AP}, duration of action potential. [b]The values for D_{AP} depend on temperature and heart rate—some average values given show the difference in fiber size and in duration of the action potential in comparison to cardiac tissue. [c]Values for skeletal muscle and nerve fiber are given for comparison; note particularly the differences in fiber size and in duration of the action potential in comparison to cardiac tissue.

Table 2.5.
Fiber Diameter and Conduction Velocity in Human Hearts

Site	Fiber Diameter, μm	Conduction Velocity, m/s
Sinus node	2–7	
Atrium	3–17	0.8–1.0
AV fibers (about 5 mm long)	3–11	0.05
His bundle branch	9–18	2.0
Ventricular myocardium	10–12	
Subendocardium		1
Subepicardium		0.4–1

(From Bauereisen, 1970.)

line produces both increased polarization (hyperpolarization) of the membrane and a reduction in the slope of phase 4 by increasing membrane permeability to K^+, thereby slowing the rate of the SA node. (Increased P_K results in a larger K^+ leak and a more negative cell interior, thereby partially offsetting the depolarizing inward Na^+ leak). Increased catecholamine levels can result from either norepinephrine released from nerve endings or circulating epinephrine and norepinephrine released from the adrenal glands. The mechanism of their action on phase 4 of the SA node is uncertain, but these hormones may increase Ca^{2+} permeability, thereby increasing the rate of Ca^{2+} influx and the slope of phase 4 depolarization.

If pacemaker cells are driven at a more rapid rate than their natural frequency, they become hyperpolarized, and if the driving frequency is then abruptly slowed the pacemaker cells will exhibit a long interval before the next spontaneous depolarization. This mechanism is called *overdrive suppression*, and it is important in maintaining control

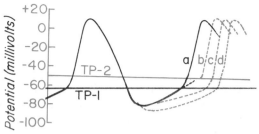

Figure 2.48. Diagram of transmembrane potential from a sinoatrial nodal pacemaker cell. On the *right* a series of succeeding pacemaker potentials depicts the mechanisms by which the discharge rate of an automatic cell may be altered. *a*, The basic rate. *b*, The delaying effect of increasing threshold potential from TP-1 to TP-2. *c*, The slope of diastolic depolarization is decreased without a change in threshold and delays firing. *d*, The delaying effect of increasing the maximum level of diastolic depolarization (hyperpolarization) with no changes in threshold potential or diastolic depolarization slope (From Hoffman et al., 1966.)

of the normal heart by the sinus node, since slower potential pacemakers elsewhere in the heart are suppressed. If the heart is artificially driven by an external electrical pacemaker, hyperpolarization of the SA node tends to occur. If the external pacemaking is then stopped, the next natural SA node cycle length will therefore be substantially prolonged. This mechanism is called "overdrive suppression of the SA node," and it also applies to other potential pacemaker sites within the heart. Sometimes such electrical pacing is used to suppress abnormal rhythms originating in the ventricles or other regions of the heart.

Interatrial or Internodal Tracts

These constitute specialized pathways of preferential conduction from the sinus node to both atria and to the AV junction (Schlant and Silverman, 1986). Three major routes can be identified (Fig. 2.47). The anterior internodal pathway sends fibers both to the AV node and to the left atrium (the latter branch is called Bachmann's bundle), the middle internodal tract goes only to the AV node, and the posterior internodal tract goes behind the superior vena cava to send a few branches to the left atrium but then continues to the AV node (Fig. 2.47). The cells in these tracts appear to constitute a mixture of ordinary muscle cells and cells which resemble the Purkinje fibers of the specialized conduction system (see below). Purkinje fibers and specialized tracts can conduct impulses at about 2.0 m/s compared with conduction velocities of 0.8–1.0 m/s in atrial muscle (Table 2.5). There is some evidence that cells in the internodal tracts can exhibit automaticity under some circumstances and that, rarely, they can conduct to the AV junction without depolarizing the atria.

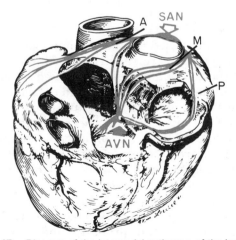

Figure 2.47. Diagram of the internodal pathways of the human heart viewed from the back. *A*, Anterior internodal pathway which sends fibers to the atrioventricular node *(AVN)* and interatrial fibers to the left atrium (Bachmann's bundle); *M*, middle internodal tract; *P*, posterior internodal tract. (From James, 1970.)

AV Junction

The cells in the atrioventricular node resemble those in the SA node. The AV node is positioned at the lower part of the septum dividing the two atria (Fig. 2.49). Three functional regions have been identified within the region called the *AV junction:* the nodal region proper, the atrionodal region which lies above the AV node, and the nodal-His region which is situated below the node toward the ventricles; from the latter region the bundle of His (see below) passes toward the interventricular septum (Fig. 2.49) (Schlant and Silverman, 1986). It is in the AV junction that impulses traversing the atria are collected and delayed for a period of time (normally up to 200 ms), so that contraction of the atria precedes activation of the ventricles.

In the AV node, the fiber diameters are small, and there are multiple sub-branches, which contribute to the slow rate of conduction and the tendency for an impulse to die out in this region; moreover, the action potentials are of the slow-response type. The slow recovery of such cells contributes to their inability to be depolarized again following a second rapid stimulus, making this region subject to complete block of impulse transmission. Also, recovery of excitability may initially occur in one direction

only, leading to so-called "retrograde block" at a time when forward conduction has recovered. The ability of the AV junction to slow and to block rapid impulses is called *decremental conduction,* which reflects the high degree of refractoriness of the AV node and its very slow conduction velocity (0.01–0.10 m/s). It is in the atrionodal region, as well as in the node itself, that the major conduction delay occurs in the passage of the impulse through the AV junction. Action potentials in the nodal-His region are transitional between those in the AV node and the bundle of His.

There are automatic pacemaker cells in all three regions of the AV junction (the atrionodal region—sometimes termed paranodal fibers, the AV node itself, and the His region) (Tse, 1984). Pacemaker cells in all three regions are accelerated by epinephrine and slowed by acetylcholine and, under abnormal circumstances (for example, slowed sinus rhythm or excess autonomic discharge), they can become the dominant cardiac pacemaker.

A number of factors affect conduction through the AV junction. Acetylcholine (ACh) released from vagal nerve endings slows conduction through the AV node by decreasing conduction velocity and increasing the refractory period (more pronounced decremental conduction). Drugs such as digitalis (which has an indirect effect through central vagal stimulation), drugs which inhibit acetylcholinesterase and thereby block ACh breakdown (such as edrophonium), and the Ca^{2+} antagonist verapamil, all slow conduction. Atropine (which blocks ACh) and quinidine (which inhibits vagal effects) speed AV conduction, as do the catecholamines. Thus, norepinephrine released from nerve terminals in the AV junction or circulating catecholamines increase the amplitude and upstroke velocity of AV nodal action potentials.

When an increased number of impulses arrives at the AV junction, its refractoriness is increased, and many of the rapid impulses are blocked. This means that, when the atria are firing at an extremely rapid rate due to an abnormal rhythm (such as atrial fibrillation), the ventricles respond at a much slower frequency. Usually conduction from the ventricles backward to the atria (so-called retrograde conduction) is blocked at a lower frequency rate than is antegrade conduction.

Increased refractoriness in response to an increased number of stimuli received by the AV junction is an important property, and it has significant effects in the presence of certain rapid rhythm disorders (Chapter 9). Decremental conduction implies that early impulses may *partially* penetrate the junction (so-called "concealed-conduction"), so

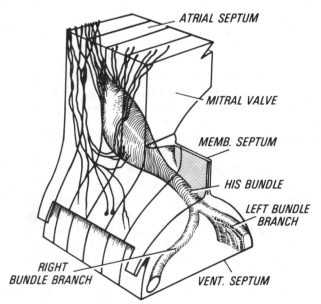

Figure 2.49. Schema of the atrioventricular node and His bundle. *ATRIAL SEPTUM,* interatrial septum; *VENT. SEPTUM,* interventricular septum; *MEMB. SEPTUM;* membranous portion of the interventricular septum. Fibers from internodal tracts are shown entering the AV node. The bundle of His runs along the lower margin of the membranous interventricular septum. It gives rise to a single right bundle branch, and then multiple left branches form a vertical sheet down the left septal subendocardium to form the origin of the left bundle branch. (From James, 1970.)

that the next impulse finds the cells of the junction only partially recovered, i.e., they remain in a more refractory state than previously. In fact, delivery of a very early electrical stimulus by an electrical pacemaker or increasingly rapid electrical pacing of the atrium up to the point of AV block provides a means of determining the refractory period of the AV junction (see subsequent section). In contrast to *electrical* pacing, however, when the SA nodal rate increases under *normal* conditions (as with exercise or excitement) the refractoriness of the AV junction does *not* increase. In fact, it decreases because the increase in heart rate under these circumstances is due to activation of the sympathetic nervous system and withdrawal of parasympathetic influence (see Chapter 16), which simultaneously affect the AV junction to improve conduction.

His-Purkinje System and Bundle Branches

As shown in Figure 2.49, a large bundle of specialized fibers originates from the lower end of the AV junction. This is called the bundle of His (described in 1893 by W. His, Jr., Leipzig). These fibers course on the right side of the upper portion of the intraventricular septum for a distance of over 1.0 cm and then divide into the *right bundle branch* and *left bundle branch* (Fig. 2.49), which supply the right and left ventricles, respectively; these bundle branches reach the ventricular walls and then ramify over the inner walls, branching into smaller bundles termed *Purkinje fibers* (see Fig. 2.4, Chapter 5). This network of small bundles is present throughout the subendocardial regions of both chambers. The cells in the His-Purkinje system (so-called *Purkinje cells*) are the largest cells in the heart and contain only a few muscle fibers; they also exhibit longer nexus junctions (low impedance cell-to-cell connections) than working myocardial cells, thereby favoring rapid impulse conduction (Legato, 1973). In the human heart, the right bundle branch is longer and considerably thinner than the left bundle branch (Figs. 2.4 and 2.49). The latter begins as a thick bundle of fibers, which then divides into two divisions, a thinner anterior division coursing to the front wall of the left ventricle and a thicker posterior division which supplies the posterolateral wall of the left ventricle. Each of these regions of the specialized conduction system can be damaged in certain disease states, leading to block in the transmission of the impulse to various regions of the heart.

The action potentials recorded in the His-Purkinje system resemble those in ventricular muscle cells but are of longer duration; they are fast-response action potentials, with a dip during phase 1 and a prominent plateau. The characteristics of these action potentials, coupled with a large cell (fiber) diameter, result in very rapid conduction velocity from the AV node through this system (2.0–4.0 m/s). Some cells in the His-Purkinje system can exhibit spontaneous diastolic depolarization (Fig. 2.42*D*), and there are also variations in the duration of the action potentials, with cells in the Purkinje fiber network near the ventricles exhibiting longer action potentials than cells higher in the His-Purkinje system (see Fig. 2.50). These so-called "gate cells" therefore can slow or block rapid sequences of action potentials before they reach the ventricles; their effective refractory period tends to prolong with increased heart rate, and early premature depolarizations originating in the ventricles may also be extinguished by the gate regions.

As might be anticipated, the conduction velocity of action potentials and the refractory periods of

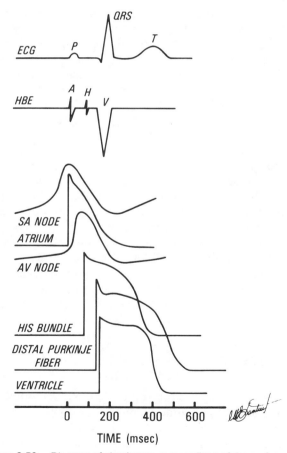

Figure 2.50. Diagram of simultaneous recordings of the surface electrocardiogram *(ECG)*, the His bundle electrogram *(HBE)* (which shows activity recorded from the atrium (*A*), the His spike (*H*), and the ventricle (*V*), and corresponding intracellular action potentials at various regions in the pacemaking and conducting system and myocardium. (For further discussion, see text.)

cells in the His-Purkinje system can be affected by a number of drugs and other stimuli.

Atrial and Ventricular Muscle

The muscle of both the atrium and the ventricle consists of relatively small diameter cells (10–15 μm) arranged in a branching network (see Fig. 2.78*A*, Chapter 10). There are special regions in the intercalated discs which separate the cells called nexus or gap junctions (mentioned above), which allow rapid electrical transmission of the impulse from cell to cell (other regions of the intercalated disc are thick and serve to transmit tension between cells) (see Fig. 2.80*A*, Chapter 10). Thus, the myocardium is a *functional syncytium*, through which the depolarizing wave front can spread relatively rapidly throughout the muscle. However, conduction in atrial and ventricular muscle is much slower than in the His-Purkinje system (Table 2.5).

Action potentials in the atrium tend to exhibit a somewhat shorter and steeper plateau (phase 2) than those in ventricular myocardium, and repolarization is slower (Figs. 2.42 and 2.50). Since the atrium is heavily innervated by the vagus nerve (in contrast to the ventricles), vagal stimulation with release of ACh can increase K^+ permeability and hence shorten the duration of the refractory period (see Fig. 2.41); it may also decrease the Ca^{2+} inward current and shorten the duration of phase 2 of the action potential. In addition, the increased P_K produced by ACh hyperpolarizes the cells, so that vagal stimulation results in a faster phase 0 and improved conduction through cells that are less refractory.

Digitalis, which indirectly stimulates the vagus, can increase the rate of an atrial rhythm abnormality or prevent an abnormality by such effects on the atrial muscle. Quinidine, on the other hand, decreases conduction velocity in atrial and ventricular muscle by shifting the membrane response curve, and it also tends to extinguish premature impulses primarily by increasing the effective refractory period in muscle as well as in Purkinje cells.

SEQUENCE OF NORMAL CARDIAC EXCITATION

A special catheter having multiple electrodes can be passed into the venous system and positioned so that one or more electrodes are located in the right atrium, another is placed on the relatively exposed portion of the bundle of His, and an electrode at the catheter tip is situated in the right ventricle. Depolarization of the atrium and ventricles can then be easily recorded and, with electronic amplification, a small electrical spike can also be recorded as the impulse travels through the His bundle (Fig. 2.50).

(The latter event is not detectable on the ordinary electrocardiogram but can sometimes be recorded on high gain computer-processed, averaged beats.) The relation of these events to the electrocardiogram (ECG) recorded on the body surface (to be discussed in the next chapter), is shown in Figure 2.50. The P wave (atrial depolarization) is followed by a pause and then by electrical depolarization of the ventricles (the QRS complex); this pause between the onset of the P wave and the onset of the QRS complex (the P–R interval) constitutes the delay in the AV junction, and it is during the latter part of this interval that the *His bundle electrogram* is normally recorded (Fig. 2.50). The T wave (Fig. 2.50) reflects ventricular repolarization.

Representative action potentials from the various regions of the heart during the entire sequence of its electrical excitation are summarized diagrammatically in Figure 2.50. Because of its very small size, action potentials from the SA node cannot be recorded on the ECG. (Like the His bundle, such small groups of cells are electrically silent on the body surface ECG, although direct recordings have now been made in experimental animals.) The impulse from the SA node spreads relatively slowly through the atrial muscle to produce the P wave, is delayed in the AV junction, and then rapidly traverses the His-Purkinje system to reach first the left ventricle, followed shortly by the right ventricle (conduction through the right bundle branch is slightly slower). The impulse subsequently travels through and depolarizes the ventricular muscle at a slower velocity, the duration of the QRS complex being much longer than that of the His bundle spike. The T wave of ventricular repolarization corresponds to repolarization of ventricular muscle cells (phase 3).

Testing the Refractory Period of the Conduction System

Using the special catheter described above for recording atrial, ventricular, and His bundle potentials, it is possible to determine the refractory periods of the various regions of the normal heart by electrically pacing the heart (see Wit et al., 1970, for details). The use of this technique to determine the refractory periods of the AV node and His-Purkinje system is illustrated in Figure 2.51. The atrium is electrically paced and a second depolarization A_2 is applied at increasingly shorter intervals; as this is done, the interval between the first and second beats, A_1–A_2, can be plotted on the abscissa, and the corresponding response of the ventricles (the V_1–V_2 interval) can be plotted on the ordinate. Since the refractoriness of the AV node increases with increasing frequency of stimulation, as A_1–A_2

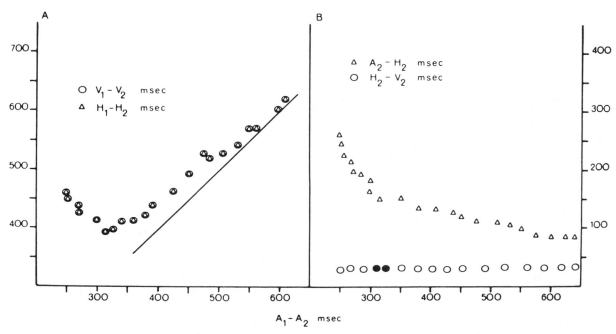

Figure 2.51. Atrioventricular conduction in the human heart studied by applying progressively earlier electrical stimuli to the atrium. *Panel A,* The intervals between an initial and a paced, premature atrial stimulus (A_1–A_2) are shown on the abscissa, and on the ordinate are shown the intervals between the initial and premature ventricular complexes (V_1–V_2, *open circles*) and the initial and premature His spikes (H_1–H_2, *open triangles*); the latter two are identical and deviate from the line of identity at progressively shorter A_1–A_2 intervals, until conduction ceases at an interval of about 250 ms (for further discussion see text). *Panel B,* The corresponding responses of the A_2–H_2 interval are shown by the *open triangles* and indicate progressive prolongation of conduction in the AV junction as A_1–A_2 shortens. The constant H_2–V_2 interval *(open circles)* indicates lack of any conduction delay in the His bundle as A_1–A_2 is shortened. (From Wit et al., 1970.)

decreases a point is reached where the ventricular response deviates from the line of identity, and V_1–V_2 fails to decrease as A_1–A_2 is further diminished (Fig. 2.51*A*). This response indicates that the P–R interval is progressively prolonging at shorter stimulus intervals. As the A_1–A_2 interval is decreased even further the interval between V_1 and V_2 starts to *increase* and, with further shortening of A_1–A_2, conduction to the ventricle fails entirely (Fig. 2.51*A*). The longest A_1–A_2 interval at which ventricular contraction still occurs constitutes the *effective refractory period* of the AV junction.

The interval between the onset of the atrial impulse and the His bundle spike (the A–H interval) and the interval from the His bundle spike to the onset of ventricular depolarization (the H–V interval) can also be measured. In normal humans, the A–H interval is 60–130 ms, and the H–V interval is 35–55 ms. As A_1–A_2 is decreased by the pacemaker virtually all of the increasing delay occurs in the A–H interval (the AV junction), whereas the H–V interval (H_2–V_2) stays quite constant (Fig. 2.51*B*). Thus, the His-Purkinje system normally does not contribute to the increasing delay between the P wave and the QRS (increasing P–R interval) produced by pacing at a rapid rate. However, in disease states the His-Purkinje system can function abnor-

mally, and progressive lengthening of the H–V interval as well as the A–H interval can be observed with the above test; sometimes, H–V conduction even fails before A–H conduction. Therefore, complete heart block (Chapter 9) can sometimes be due to failure to conduct through the His-Purkinje system, rather than the AV junction.

HIERARCHY OF CARDIAC PACEMAKERS

A number of potential pacemakers exists in the cardiac conduction system. Normally, the SA node is the fastest pacemaker (70/min at rest) and it dominates the other pacemakers by producing over-drive suppression. However, if the sinus node is markedly slowed (as by intense vagal stimulation or by disease), a pacemaker lower in the heart (often in the AV junction) will commence to fire at its natural rate, which is slower than that of the SA node (50–60/min). If there is failure of both the SA node and the AV junctional pacemaker (or if complete heart block is present), cells in the His-Purkinje system will usually drive the normal heart at a much slower rate (30–40/min). Sometimes, however, particularly in patients with heart disease, this escape frequency may be insufficient to maintain cardiac output and blood pressure at levels compatible with life.

CHAPTER 9

Electrocardiography and Disorders of Cardiac Rhythm

The electrical potentials generated by the heart can be detected by means of an electrode placed directly on the surface of the heart and, when amplified, such a recording is called an "electrogram." Since the tissues of the body contain electrolytes and are conductive, these electrical potentials spread throughout the body, and even though the changes in electrical potential become greatly attenuated they can be detected at the skin. A modern recording instrument (electrocardiograph) amplifies and records signals on moving paper, transcribing a voltage-time signal called the *electrocardiogram*. In clinical medicine, the electrocardiogram is useful for many purposes, including identification of abnormal spread of excitation through the conduction system and heart muscle, detection of changes in the size of the cardiac chambers (which alter the amplitude of the electrical signals), identification of damage to the heart (such as after a heart attack), and assessment of abnormally rapid, slow, or irregular heart rhythms.

CARDIAC DIPOLE

As electrical depolarization spreads across a muscle, there is an abrupt reversal of charge at the wave front as the outside of the cell changes from relatively positive to relatively negative. This wave front may be thought of as a moving dipole, consisting of positive and negative poles closely spaced together and situated in a volume conductor (Fig. 2.52). When such a dipole exists, current flow occurs, by convention, from the positive to the negative pole (actually cations flow toward the negative pole and anions and electrons toward the positive pole); the most dense current flow is close to the poles, with current density decreasing as distance from the poles increases (Fig. 2.52, *solid lines*). Each pole also generates a potential field in three dimensions, with isopotential lines arranged as shown in Figure 2.52 (*dashed lines* show isopotential lines in two dimensions). The *central vertical line* midway between the two poles is at zero poten-

tial, and the magnitude of the voltage at any distance from the dipole can be calculated; along the axis of the dipole (the *horizontal line* traversing both poles), the voltage is inversely related to the square of the distance from the center of the dipole and, when off the axis of the dipole, the potential varies with the cosine of the angle from the axis of the dipole. Also, the potential at *any* location varies with the strength of the dipole.

If a piece of isolated cardiac muscle is placed in a volume conductor and arranged so that it can be electrically stimulated (depolarized) at one end, while the positive pole of a recording galvanometer is placed at the other end, a characteristic tracing is obtained as the dipole of the depolarizing wave front moves across the muscle toward the recording electrode (Fig. 2.53). By convention, a positive deflection is recorded when the positive recording pole faces the positive pole of an approaching dipole (in this arrangement, the other electrode of the galvanometer is placed at a distant ground site where it is near zero potential). As shown in Figure 2.53 (*panel A*), no potential difference is recorded between the electrodes when the muscle is at rest. Following stimulation of the end of the muscle (*panels B and C*), as the potential on the outside of the muscle is reversed and the dipole is propagated across the muscle, a progressively larger positive deflection is recorded as the isopotential lines of progressively larger potential approach the recording electrode (*E*). In *panels D and E* (Fig. 2.53), the maximum potential is reached just as the muscle is entirely depolarized and then, as the zero potential line at the center of the dipole reaches the end of the muscle, the recorded potential difference returns to zero. This completes the depolarization process, which is rapid and displays a sharp positive spike (which corresponds to the QRS of the electrocardiogram and the rapid upstroke (phase 0) of the action potential).

During phase 2 (the plateau) of the action potential, the muscle remains completely depolarized,

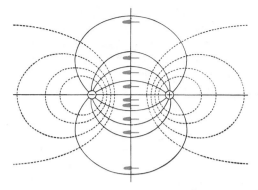

Figure 2.52. Electrical field of a dipole in a volume conductor. *Solid lines* are current flow. *Dashed lines* are isopotential lines.

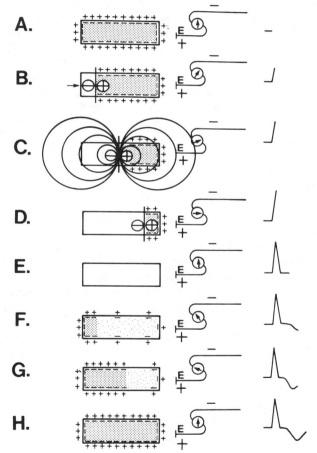

Figure 2.53. Depolarization and repolarization of a strip of myocardium immersed in a volume conductor. Electrode *E* is connected to the positive pole of a recording galvanometer which is paired with a distant electrode at near zero potential. The advancing dipole at the border between activated and inactivated muscle produces an electrical field which moves toward the electrode (*panel B*). The electrode is intersected by isopotential lines of increasing force (*panel C*) until all difference of potential is extinguished by completion of depolarization (*panel E*). Recovery occurs in all parts of the muscle nearly simultaneously but is completed at the stimulated end first (*panel G*), thereby producing relative negativity at the side facing the electrode and giving rise to a negative repolarization wave. For further discussion, see text.

and the recording line remains isoelectric (Fig. 2.53*E*). Repolarization proceeds more slowly, and recovery begins simultaneously but is completed earliest at the end where stimulation began. Therefore, the regaining of positive charge is most marked at the end away from the electrode, with movement of the repolarization wave front in the manner of a reverse dipole. Thus, the relatively negative wave front of the dipole approaches the electrode, leading to a progressively increasing negative deflection *(panel F)* which then declines as the inhomogeneous recovery process is slowly completed *(panels G and H)*.

If the positive recording electrode is placed at the midpoint of the muscle (electrode E_2, Fig. 2.54) it can be surmised that tracings of the configuration shown in the *right-hand column* of recordings will be obtained; the depolarization tracing initially will be positive, then will reach zero potential midway in the depolarization spike (as the zero isopotential point of the dipole passes under electrode E_2), and,

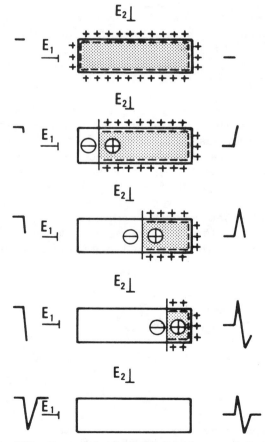

Figure 2.54. Form of complexes inscribed from an electrode placed at position E_1 or E_2 during depolarization of a strip of myocardium suspended in a volume conductor. Each electrode is paired with a distant indifferent electrode. Tracings on the *left* are related to E_1 while tracings on the *right* relate to the E_2 electrode.

finally, will become negative as the negative end of the dipole moves away to the opposite end of the muscle. The potential differences recorded when the electrode is placed at the opposite end of the muscle (compared to Fig. 2.53) are also shown in Fig. 2.54 (electrode E₁); a negative complex is evident (Fig. 2.54, *left-hand column* of recordings).

ELECTRICAL DEPOLARIZATION AND REPOLARIZATION OF THE HEART

The chambers of the entire heart contain millions of cells that are depolarized nearly simultaneously. First the atria (P wave) and then the ventricles (QRS complex) are rapidly depolarized, with both sets of chambers depolarizing slightly out of sequence, and then they repolarize more slowly with greater dispersion of the recovery of excitability (see Fig. 2.50). At any instant in time, the entire electromotive force generated by the heart can be thought of as a dipole centered in the middle of a large volume conductor (the body), and it is the sequence of these instantaneous dipoles recorded from the body surface throughout the cardiac cycle that makes up the electrocardiogram. Because of the shape of the heart and the sequence of its activation, the net voltage measured at the body surface at any moment is the sum of forces acting in *many* directions. Thus, there are large *cancellation* effects; for example, when the wave front is heading toward the posterior wall of the left ventricle it is simultaneously spreading toward its anterior wall. The *net* magnitude and direction of the voltage will then depend on the relative numbers of cells being depolarized in various regions of the heart at any moment. This process results in an electromotive force that has both magnitude and direction at any instant, that is, it constitutes a *vector*. The *magnitude* of the vector will depend largely on the *number* of cells that are being depolarized at any instant (and the distance of the heart from the recording electrode), whereas the *direction* of the vector will depend on which anatomic regions of the heart (having different locations and muscle masses) are dominant at that instant.

Atrial Depolarization

As shown in Figure 2.55, we can show the sequence of atrial depolarization (the P wave) by a series of instantaneous vectors presented by arrows that have both magnitude and direction. By convention, the *point* of the arrow represents the *positive* end of the dipole. The arrows change in amplitude and direction (right to left, up (superior) or down (inferior), and front (anterior) or back (posterior)) with all directions being given as if the observer is *facing the heart*, with right being to the observer's left. For convenience, the origin of each vector arrow is placed at a central point. As the impulse sweeps over the right atrium from the sinoatrial node, its amplitude initially is small and then increases as the wave spreads initially anteriorly and inferiorly and then from right to left (Fig. 2.55) during right atrial depolarization. The vector then moves to the left and posteriorly as the left atrium is depolarized (Fig. 2.55). The vector arrows are small because the muscle of the atrial walls is thin (correspondingly, the P wave on the electrocardiogram is normally a low amplitude wave).

These instantaneous vectors can be conveniently summarized as three vectors: the initial P vector, which points to the right and anteriorly (Fig. 2.55); the average or mean P vector, which is directed to the left and anteriorly (Fig. 2.55); and the terminal P vector, which is small and directed to the left and posteriorly (Fig. 2.55). Repolarization of the atria is not well seen on the electrocardiogram, usually

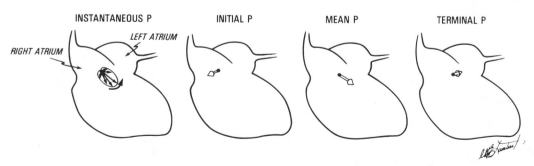

Figure 2.55. Diagram showing the heart and vectors of atrial depolarization originating from a point at the center of the heart. The left diagram shows a series of instantaneous, sequential P wave vectors, initially directed vertically, then from left to right, and finally superiorly (the tips of the arrows are connected to form a loop). The right three diagrams show that the initial average vector points toward the right, inferiorly and anteriorly; the mean P wave vector points toward the left and anteriorly, and the terminal P wave vector is directed toward the left and posteriorly.

being obscured by the onset of the much larger voltage of *ventricular depolarization.*

Ventricular Depolarization

Passage of the electrical wave front into the atrioventricular (AV) node and subsequent entry into the His bundle are electrically silent events in the standard electrocardiogram. Then, as the muscle of the ventricles depolarizes, the process can be represented by a sequence of instantaneous vectors which compose the QRS complex (Fig. 2.56). The first impulses that leave the His bundle reach the left side of the muscle of the septum (partition) between the two ventricles, leading to a small initial vector that is directed anteriorly and commonly toward the right (Fig. 2.56). Since the Purkinje fibers lie on the inner wall of the ventricles, the wave front crosses the interventricular septum from the inside surface of the left ventricle outward; this direction of depolarization from the inner (endocardial) surface to the outer (epicardial) surface subsequently occurs elsewhere in the ventricular walls. As the wave front spreads over the two ventricles, the left ventricle, because of its much larger mass, dominates the magnitude and direction of the vectors. Since the left ventricle, is situated *behind* (posterior to) the right ventricle, the instantaneous vectors which are initially to the right quickly become posterior and show increasing magnitude in the leftward direction (Fig. 2.56). Thus, the simultaneous depolarization of the thin-walled right ventricle is not seen, with its anteriorly directed forces being cancelled by the much larger forces originating from the thick-walled left ventricle. The instantaneous vectors remain leftward and posterior for a period, and then as depolarization of the heart is completed the normal vectors often point somewhat to the right (and either posteriorly or anteriorly), since the last region in the heart to be depolarized lies centrally in the region of the mitral and aortic valves (Fig. 2.56).

This series of instantaneous vectors representing ventricular depolarization can be summarized as three vectors: the initial QRS forces, which represent mainly depolarization of the interventricular septum and are directed to the right and anteriorly (Fig. 2.56); the mean QRS vector, which is large and directed to the left and posteriorly (Fig. 2.56); and the terminal QRS vector, which is often directed posteriorly and superiorly (Fig. 2.56) (there can be considerable variability, however, in the superior or interior direction of both the initial and terminal forces of the normal heart).

Ventricular Repolarization

If ventricular repolarization began at the same location and traveled in the same direction as the depolarization wave, that is, from the endocardium (where the Purkinje fibers are situated) across the wall to the epicardium, we would expect to see a slow negative wave occurring in the direction opposite to the depolarization wave. Thus, it would be directed superiorly to the right and anteriorly (Fig. 2.57, *left*), with the mean T wave vector being opposite to the mean QRS vector. However, this is *not* the case in the whole heart, because repolarization occurs in the direction opposite the QRS vector. Thus, repolarization occurs from epicardium to endocardium, leading to a T wave having the *same* direction as the depolarization wave (Fig. 2.57, *center* and *right*). However, repolarization is a slower process, fewer cells are being repolarized at any moment, and the repolarization wave (T wave) is longer (which could be displayed as more vectors per unit time) and is of lower amplitude than the depolarization wave. One theory for the opposite direction of repolarization holds that the buildup of pressure by contraction of the left ventricle compresses the subendocardial (inner wall) tissue, thereby delaying repolarization in that area.

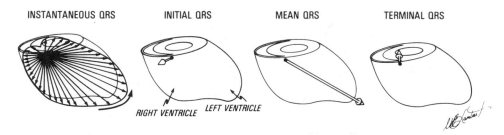

| INSTANTANEOUS QRS | INITIAL QRS | MEAN QRS | TERMINAL QRS |

RIGHT VENTRICLE   LEFT VENTRICLE

Figure 2.56. Diagram of the vectors of ventricular depolarization (the atria are removed). The instantaneous, sequential QRS vectors are represented in the left diagram as a series of arrows originating from a central point in the heart, their tips being connected by a line to form a loop. In the right three diagrams the initial average QRS vector constituting septal depolarization is directed rightward and anteriorly; the mean QRS vector is directed leftward and posteriorly; and the terminal QRS vector is directed superiorly and posteriorly.

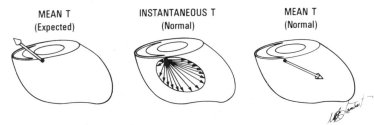

| MEAN T (Expected) | INSTANTANEOUS T (Normal) | MEAN T (Normal) |

Figure 2.57. Diagram of the vectors of ventricular repolarization showing, in the *center panel,* instantaneous, sequential T wave vectors. The *left diagram* shows that the expected mean T wave vector would be 180° opposite to the mean QRS vector (see Fig. 2.56), if repolarization followed the same pattern as depolarization (see also Fig. 2.53). However, the normal mean T wave vector *(right panel)* is directed toward the left and posteriorly, in the same direction as the mean QRS vector (for further discussion, see text).

RECORDING THE ELECTROCARDIOGRAM

For recording the electrocardiogram (ECG), metallic leads attached to wires are placed on the skin using electrolyte paste, and by arranging the electronic recording circuits appropriately it is possible to "look" at the cardiac vectors from different vantage points. A recording stylus moves up and down as the amplitude of the instantaneous vectors changes while the recording paper moves at a constant speed, allowing measurement of the duration of various events. In practice, pairs of recording electrodes or a single exploring electrode (which are termed electrocardiographic "leads") are arranged to view the heart both in the *frontal plane* (a plane passed vertically through the body as if directly facing it) (Fig. 2.58) and in the horizontal or *transverse plane* (a plane transecting the body) (Fig. 2.58).

Frontal Plane Electrocardiogram

In the early part of this century the development of an improved string galvanometer led Einthoven (W. Einthoven, Dutch physiologist, 1860–1927) to develop a three-lead system which is still in use for recording the ECG. In this system, electrodes are placed on the right arm, the left arm, and the left leg. An additional electrode is attached to the right leg and is grounded. These leads are connected in such a way that, for practical purposes, they form an equilateral triangle surrounding the heart in the frontal plane at the body surface (Fig. 2.59*A*) ("Einthoven's triangle"). (The leads are shown as if they are positioned on the two shoulders and the lower chest, although they are actually placed on the arms and legs; the effect is the same). *Lead I* measures the potential difference between the right and left sides of the chest, with the *positive pole* of the recording system *toward the left arm* (Fig. 2.59). *Lead II* records between the right arm and the left foot with its *positive pole* being toward the *left leg* (Fig. 2.59), and *lead III* is arranged to record between the left arm and the foot with its *positive pole* also being toward the *left leg* (Fig. 2.59). These leads form a simplified equivalent circuit based on the (not strictly true) assumptions that these anatomic locations are situated at a great distance from the heart,

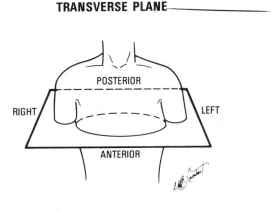

FRONTAL PLANE

TRANSVERSE PLANE

Figure 2.58. Planes transecting the body which are represented on the standard electrocardiogram. The frontal plane *(left)* passes vertically through the body, and the transverse (horizontal) plane transects the body through the chest. Note that the left and right sides of the planes are labeled as if the observer is facing the subject, that is, the subject's left is to the observer's right.

FRONTAL PLANE

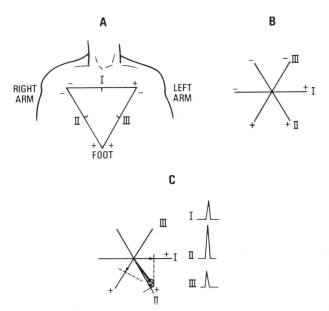

Figure 2.59. The triaxial reference system. Einthoven's triangle is represented in *panel A,* and in *panel B* the three lead axes are rearranged to cross a central point. In *panel C,* a QRS vector is drawn together with its projections on the standard limb leads I, II, and III.

equidistant from the central cardiac dipole and each other and in the same plane, and that the fluids and body tissues act as a homogeneous conductive medium.

These three leads may then be centered into a *triaxial reference system* in which the bisected lines cross a central point and are positive or negative on either side of this point (Fig. 2.59*B*). Left and right always refers to the left and right sides of the body as viewed by an observer facing the torso (Fig. 2.59*A*). If we now place a vector into this triaxial reference system (Fig. 2.59*C*), the direction and magnitude of the *projection* of the vector on any lead will determine its electrocardiographic voltage in that lead. The large vector arrow has a posterior direction, but this cannot be detected in the frontal plane, since only the magnitude of the vector projected on the frontal plane leads is being recorded. Thus, if a light were thrown from below straight upward toward the *large arrow,* the *shadow* of the arrow would be projected on lead I as an arrow (Fig. 2.59*C*). A perpendicular dropped from the tip of the arrow to the lead I axis, since the light rays are perpendicular to the lead axis, would give the magnitude of the vector in lead I. To determine the projection of the vector on lead II, a light would be thrown from the left shoulder perpendicular to that lead, and an arrow would be projected on lead II (Fig. 2.59*C*). Similarly, if a light were thrown from

below toward the right shoulder so that its rays were perpendicular to the lead III axis, the shadow would then be projected on lead III as an arrow (Fig. 2.59*C*).

At any instant (such as the peak of the depolarization wave), "Einthoven's law" states that voltages will conform to the following equation:

$$\text{lead II} = \text{lead I} + \text{lead III*}$$

This equation provides a means of checking whether the leads are properly attached to the arms and legs.

As discussed earlier, by convention any dipole that is moving toward the *positive* recording pole creates an *upright* or *positive* deflection on the electrocardiographic tracing, and the arrow tip of a vector represents its positive end. Therefore, the deflection of the pen recording the mean vectors shown in Figure 2.59*C* will yield a positive wave in each lead (notice the position of the positive poles of leads I, II, and III), and the magnitude of the waves will reflect the lengths of the arrow projected on each lead, with the largest magnitude being in lead II (Fig. 2.59*C*).

The standard "limb leads" are bipolar, measuring the potential differences between two sites. The ECG in the frontal plane has three leads added to these three standard leads. These extra leads allow other vantage points from which to view the cardiac vectors, and they are called the *augmented unipolar limb leads.* The augmented limb leads are more amplified than the standard limb leads, and to equate them the amplitude of the standard leads must be multiplied by 1.15. They are unipolar because the potential is compared to zero, not to the potential at another site. They are recorded by connecting the three standard limb terminals together through resistances and using this central point as a zero reference electrode; the other (positive) pole then becomes equivalent to an "exploring electrode." The augmented limb leads are arranged so that the positive electrode is positioned at three locations, equidistant between leads I, II, and III. One has its positive pole at the right upper chest (lead aVR), one at the left upper chest (lead aVL), and one at the foot (lead aVF), as shown in Figure

---

*According to Kirchoff's second law of circuits, the potentials measured sequentially in any three bipolar leads completing a closed circuit would be: I + II + III = 0. However, Einthoven arranged lead II so that in the normal electrocardiogram all three leads would exhibit upright QRS deflections, so that with Einthoven's system Kirchoff's law becomes I − II + III = 0, or II = I + III. Thus, at any instant, this equation holds for the potential differences recorded simultaneously by these three leads.

2.60. Thus, they lie between the standard leads giving a *hexaxial reference system*. Notice that lead aVR is positive superiorly and to the right and negative inferiorly and to the left, lead aVL is positive superiorly and to the left and negative inferiorly and to the right, and aVF is positive toward the feet and negative superiorly (Fig. 2.60). If we now place a vector in this system (Fig. 2.60) and project its shadows on leads aVF and aVR, the deflections will be equal, but the arrow on aVR is now projecting toward the negative pole, so that the ECG deflection is negative in that lead (Fig. 2.60), while it points toward the positive pole of aVF and is upright in that lead. Note that the vector is perpendicular to lead aVL, so that when a light is projected perpendicular to lead aVL, no shadow is cast and no deflection is recorded (i.e., the tracing is "isoelectric"). Keep in mind that aVR is negative in the midst of the positive poles of several leads and positive in the midst of negative poles (Fig. 2.60).

Transverse (Horizontal) Plane

There are six leads recorded in the transverse plane with the usual ECG (total of 12 leads), and they provide information on how the instantaneous cardiac vectors are directed anteriorly and posteriorly (Fig. 2.58). These are termed the *precordial leads* (because the electrodes are arranged on the front wall of the chest, over the heart or precordium). Like the augmented limb leads, these electrodes also act as unipolar exploring electrodes because they are connected to a central ground, which remains near zero potential. In Figure 2.61, the chest is represented as a cylinder with us looking down from above. The six leads, called V_1, V_2, V_3, V_4, V_5, and V_6 are shown diagrammatically, and all of their positive terminals are positioned on the front of the chest. Therefore, vectors which project

TRANSVERSE PLANE

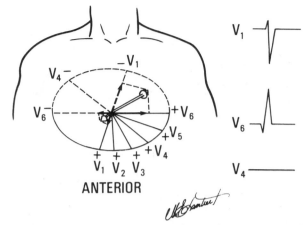

Figure 2.61. The six precordial leads, V_{1-6}, of the transverse plane lead system. Initial and mean QRS vectors are projected on selected leads of this system, giving a negative mean vector in V_1, a positive mean vector in V_6, and an isoelectric tracing in V_4 (since the mean vector is perpendicular to this lead). For further discussion, see text.

onto the positive one-half of each lead axis will yield a positive ECG deflection and vectors which project onto the negative one-half will produce a negative deflection. The *initial* and *mean* vectors of ventricular depolarization previously described (see Fig. 2.56) are also drawn in Figure 2.61, and the approach employed with the frontal plane lead system can be used to determine how these vectors are projected on the transverse plane leads. In V_1, the normal initial vector is directed anteriorly and to the right (septal depolarization), whereas the mean vector of ventricular depolarization is directed posteriorly and to the left (Fig. 2.61). Thus, when these vectors are projected on lead V_1 a small upward (positive) wave is first seen, followed by large negative deflection (Fig. 2.61). On the other hand, when the same vectors are projected onto lead V_6, the initial force is projected on the negative side of V_6, so that an initial small downward (negative) deflection is seen, followed by a large positive deflection as the arrow is projected on the positive one-half of that lead (Fig. 2.61). Lead V_4 is also shown, and the two arrows are nearly perpendicular to that lead axis, so that no projection occurs on this lead and zero deflection is recorded (Fig. 2.61).

In practice, the precordial leads are arranged in accordance with the spaces between the ribs, with the electrode being sequentially applied at the locations shown in Figure 2.62. Notice that V_6 lies in the frontal plane, so that deflections in this lead will resemble those recorded from standard lead I.

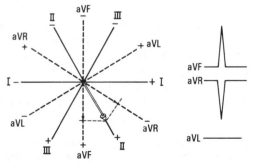

Figure 2.60. The hexaxial reference system showing the addition of the augmented limb leads (aVR, aVL, and aVF) to the standard limb leads I, II, and III. A QRS vector is shown projected onto the augmented limb leads. For further discussion, see text.

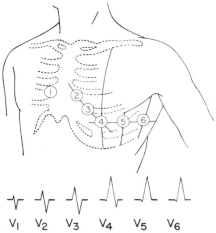

Figure 2.62. Positions of the chest electrodes for six standard precordial leads. V_1, fourth intercostal space to right of sternum. V_2, fourth intercostal space at left of sternum. V_4, fifth interspace and midclavicular line. V_3, midway between V_2, V_4. V_5, anterior axillary line, same level as V_4. V_6, midaxillary line, same level as V_4 and V_5. Below is shown general form of QRS in normal precordial leads. V_3 represents transition from predominant negativity on the *right* to predominant positivity on *left*. (From Lipman and Massie, 1959.)

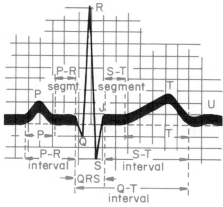

Figure 2.63. A single electrocardiographic complex giving nomenclature of the deflections and the intervals. (From Burch and Winsor, 1968.)

FORM OF THE ELECTROCARDIOGRAM

In analyzing a standard ECG tracing, it is necessary to know the voltage calibration and the significance of the "time lines" on the recording paper (which moves at 25 mm/s), as well as some terminology that has been devised to identify the various ECG waves. As shown in Figure 2.63 (a recording of lead I in a normal individual), each horizontal small box represents 0.04 s (40 ms), and the time lines of the larger boxes (made up of five small boxes) indicate 0.2 s. The atrial or ventricular rates can be measured by counting the number of small boxes between P waves or QRS complexes and dividing the number of small boxes into 1500 or the number of large boxes into 300. In the vertical direction, the ECG is calibrated so that 1 mV causes a deflection of 10 small boxes (10 mm).

The wave of atrial depolarization is called the *P wave*, the wave of ventricular depolarization the *QRS complex* (Fig. 2.63). When the initial wave of the QRS complex is negative, it is called a Q wave. The Q wave is usually followed by a positive deflection, the R wave, and then by a small negative deflection, the S wave, to form the *QRS complex*, although not all of these waves need be present in any single lead to apply this term. If a second positive deflection follows the S wave, it is labeled R′, and if a second S wave follows the R′, it is called S′. The first positive deflection is called the R wave, whether or not it is preceded by a Q wave. The wave

of repolarization is called the *T wave,* and sometimes this is followed by a small afterpotential called the *U wave* (see Fig. 2.63). The electrocardiograph is operated through an alternating current circuit. Therefore, when no voltage *change* is occurring the baseline is flat, as between the P wave and the QRS complex and during the plateau of the ventricular action potential (the *ST segment*) (Fig. 2.63) (see also Fig. 2.42*A*, Chapter 8, showing a simultaneous recording of an action potential and an electrogram).

Various other measurements made on the ECG have particular importance in abnormal states; these are indicated in Figure 2.63. The height and width of the P wave are measured because either or both may be abnormal when the atria are enlarged. The interval from the onset of the P wave to the onset of the QRS complex, the *P-R interval,* is important in that it reflects conduction from the atria to the ventricles through the AV junction and His-Purkinje system (it also includes the time of atrial depolarization; the potential from the sinoatrial (SA) node is not recorded on the surface ECG, nor is the potential from the His bundle). The P-R interval normally is about 0.14 s but may range between 0.12 and 0.2 s in normal individuals, depending largely on the heart rate and the tone of the autonomic nervous system. It can be lengthened by functional or anatomic block at any point along the specialized conduction system. The *duration of the QRS complex* normally measures about 0.08 s (normal is up to 0.09 s). The *Q-T interval* is measured from the onset of the QRS complex to the end of the T wave (Fig. 2.63). The Q-T interval is normally 0.26 to 0.45 s, and it may be prolonged or shortened by changes in heart rate, by alterations in electrolyte concentrations, or by various drugs that affect the rate of ventricular repolarization. Often, the Q-T interval is corrected for heart rate (Q-T$_c$) by dividing it by the square root of the interval between R waves.

NORMAL 12-LEAD ELECTROCARDIOGRAM

In considering the form of the normal ECG, it is useful to examine the initial, terminal, and mean vectors for the atria and ventricles (shown in Figs. 2.55 and 2.56) and to project the *loop* connecting the tips of these arrows onto the frontal plane leads (the hexaxial reference system), as well as onto the transverse plane leads (the unipolar precordial leads). (For further reading on electrocardiography see Goldberger and Goldberger, 1986, and Castellanos and Myerburg, 1986.)

P Wave

The mean vector of the P loop in the frontal plane generally points to the left parallel to lead II, giving an upright deflection in that lead (Fig. 2.55). In the transverse plane the P wave is mostly anterior, giving an initial positive deflection in lead V$_1$, followed by a small negative (posterior) deflection due to left atrial depolarization (Fig. 2.55) (the left atrium is behind the right). If the atrium is depolarized from an abnormal location (other than the SA node), such

as near the inferior vena cava, the depolarization wave can spread upward and to the right, instead of inferiorly and to the left, giving negative P waves in leads I and II. If the right atrium is enlarged, the amplitude of the P wave in the frontal plane leads is increased, but if the left atrium is enlarged there may be a prominent negative deflection in the second half of the P wave seen in lead V$_1$ (since the left atrium is situated posteriorly and depolarizes after the right), with some prolongation of the total duration of the P wave.

QRS Complex

Using the initial and terminal forces together with the mean QRS vector, the normal QRS loop is plotted in the frontal plane in Figure 2.64 *(left)*. The projections on lead I show a small negative wave (Q wave) followed by a large R wave. The projections on lead II show a small Q wave, a large R wave, and a small negative S wave (toward the negative pole of lead II). The projections on lead III show a relatively low amplitude QRS complex, with a small R wave followed by an S wave (as the loop goes first toward the positive pole of lead III and then toward the negative pole). As the series of instantaneous vectors occurs throughout the cardiac cycle, connection of their arrow points gives a time sequence

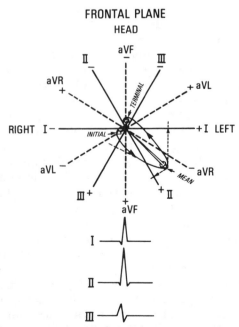

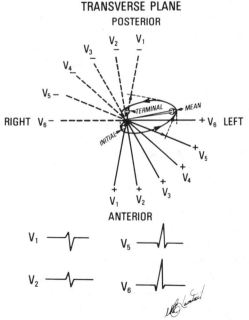

Figure 2.64. The normal QRS complex. Projection of initial, terminal, and mean normal QRS vectors, and the vector loop connecting the tips of these vectors, projected on the hexaxial lead system in the frontal plane *(left panel)* and the precordial lead system in the transverse plane

(right panel). Notice the counterclockwise direction of the heart-shaped vector loop in both the frontal and transverse planes. For further discussion, see text.

and a direction to the inscription of the loop, as shown in Figure 2.64 (see also Fig. 2.56). In this example, the loop travels counterclockwise in the frontal plane. Note that the normal QRS loop tends to be "heart shaped."

The projection of these three QRS vectors on the transverse plane, together with the QRS loop, is shown in Figure 2.64 *(right)*. The projection on V_1 shows a small initial R wave, as the initial septal forces approach the positive pole of V_1, followed by a deep S wave as the dominant left ventricular forces of the mean and terminal vectors project on the negative pole of V_1 (as mentioned, the positive and negative regions are determined by transecting each lead axis by a perpendicular line at its midpoint). In lead V_2 there is a slightly larger R wave as the loop moves toward the positive pole of that lead and an equal S wave as the loop moves counterclockwise to project on the negative half of lead V_2. In this example, lead V_2 is approximately at the "transition point" between a negative and a positive QRS complex, i.e., that lead has the most nearly *equal* positive and negative deflections. The R wave then becomes progressively larger (and the S wave smaller) in the more lateral V leads, so that in lead V_6 a small initial Q wave (septal depolarization) now projects on the negative pole; there is a large R wave as the mean vector projects on the positive pole of that lead and a small terminal negative S wave.

A normal 12-lead electrocardiogram is reproduced in Figure 2.65.

Mean Electrical Axis of the QRS Complex in the Frontal Plane

This is an important descriptor of the ECG, since it can be abnormal in a variety of cardiac disorders. It is determined by the anatomic position of the heart (vertical or horizontal, with the latter more likely in obese individuals) and by the direction of electrical depolarization of the ventricles. The *mean QRS axis* is determined from the mean QRS vector and is reported in degrees, based on the hexaxial reference system (Fig. 2.66). The normal mean QRS frontal plane axis lies between −30° and + 100°. If the mean QRS axis is −30° or more negative, there is *left axis deviation*. If the mean QRS axis lies at or to the right of 100°, there is *right axis deviation* (Fig. 2.66A).

Two simple approaches can be used to determine the mean QRS axis. First, study the six limb leads to determine in which single lead the QRS complex is most nearly isoelectric (i.e., the negative and positive areas are most equal) and in which lead the QRS deflection is largest. For example, in Figure 2.66A, lead III shows a large upright deflection because the loop is wholly on its positive side, whereas lead aVR shows equal positive and negative deflections, the positive deflection being first upright toward the positive pole of aVR and then

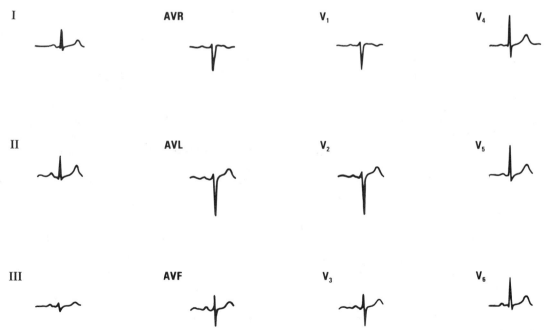

Figure 2.65. Normal sinus rhythm as seen in this normal 12-lead electrocardiograhic recording is exemplified by a regular rhythm initiated by a normal P wave, followed by a normal P-R interval, QRS complex, and T wave.

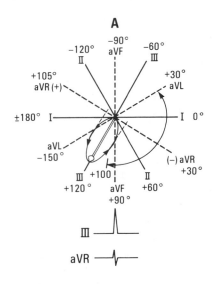

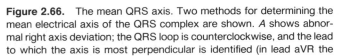

Figure 2.66. The mean QRS axis. Two methods for determining the mean electrical axis of the QRS complex are shown. *A* shows abnormal right axis deviation; the QRS loop is counterclockwise, and the lead to which the axis is most perpendicular is identified (in lead aVR the positive and negative areas of the QRS complex are most nearly equal). *B,* There is left axis deviation, and the two leads on which the QRS complex projects equally are identified (here leads III and aVL).

downward, since there is a clockwise loop passing first on the positive side and then on the negative side of aVR. Since the loop is perpendicular to aVR, the mean electrical axis is +120°, indicating right axis deviation. Therefore, with this approach the lead to which the QRS is most perpendicular (isoelectric) is determined, and the lead to which it is most parallel (largest deflection area) aids in the identification.

A second method for determining the mean QRS axis is to find two leads on which the QRS projects equally; the mean QRS axis then lies equidistant between the two leads. In Figure 2.66*B*, the mean electrical axis of the loop projects equally large negative and positive deflections on leads III and aVL, respectively. In this instance, the axis is −45°, indicating left axis deviation.

VECTORCARDIOGRAM

Given the satndard ECG, it is evident that, if the initial, mean, and terminal vector arrows are determined from the 12 scalar leads, a vector loop can be constructed in the frontal and horizontal planes. However, there is a type of electrocardiography called *vectorcardiography* which directly records vector loops using leads to record in the frontal, transverse, and sagittal planes (the latter plane transects the body vertically from front to back, at right angles to the frontal plane; see Fig. 2.58). Vectorcardiography is sometimes used in the clinical

setting to assist in the analysis of complex QRS patterns. A variety of lead systems has been developed (using numerous electrodes placed on the body torso, such as the Frank system involving eight leads) in an effort to obtain three leads that are as close as possible to being orthogonal (at right angles) and equidistant from the heart. However, because of the complex shape of the body, no perfect lead system has yet been devised.

As the vector loops are inscribed (P, QRS, and T), a series of drop-shaped marks can be inscribed on a persistence oscilloscope and the image photographed, or it can be recorded directly. Each mark occurs at a given time interval (for example, 0.025 s), and the trailing edge of the drop is narrow, allowing ready determination of the duration and direction of various portions of the vector loops simply by counting the number of marks and noting the direction of the marks (for details, see Chou et al., 1974). The vectorcardiogram is not useful for analyzing the heart's rhythm.

ABNORMALITIES OF THE QRS AMPLITUDE

Differences in amplitude of the QRS complexes from individual to individual can represent a gain or a loss of muscle mass, differences in the distance of the heart from the chest wall, or an accumulation of fluid in the pericardial sac (this produces low voltage or low amplitude QRS complexes). The amplitude of QRS complexes also can be increased by slowed conduction through the ventricles, which can lead to an abnormal direction and larger than normal amplitude of the depolarization wave due primarily to loss of simultaneous opposing forces when the affected force is late (see subsequent section). When increased QRS amplitude is due to increased muscle mass, it is called *ventricular hypertrophy* (see also Chapter 18). Hypertrophy of the left ventricle can occur, for example, with long-standing high blood pressure or with valvular heart disease (the left ventricular wall thickens and the chamber size may also enlarge). Hypertrophy of the right ventricle can occur when the pressure in the pulmonary artery is elevated due to lung disease or when there is a congenital heart defect.

Right Ventricular Hypertrophy

Since the right ventricle lies anterior to and to the right of the left ventricle, hypertrophy of that chamber would be expected to produce deviation of the mean electrical axis toward the right with an increase in the anterior forces. An example of right ventricular hypertrophy is shown in Figure 2.67 *(upper panels)*. Notice that the mean electrical axis is approximately isoelectric to lead aVR and negative in lead I, giving a mean QRS axis of +120° (right axis deviation). Notice also the tall R wave in lead V_1, which is abnormal and indicates a prominent anterior vector in the transverse plane. A valuable exercise is to plot the approximate frontal and horizontal plane QRS loops based on the initial, mean, and terminal forces of this ECG.

Left Ventricular Hypertrophy

This abnormality tends to produce a magnification of forces along the direction of the usual mean QRS vector generated primarily by the left ventri-

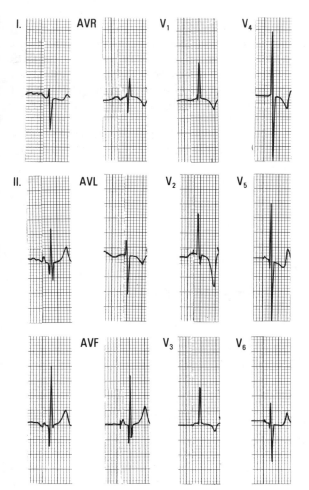

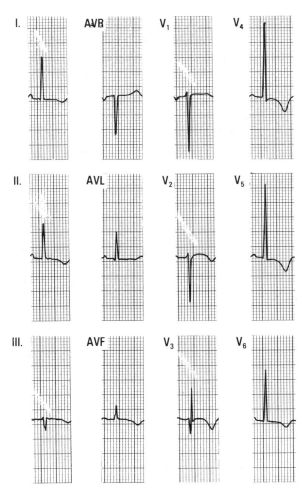

Figure 2.67. Electrocardiographic patterns of ventricular hypertrophy. Right ventricular hypertrophy *(left panels)*. The mean and terminal forces are shifted to the right and anteriorly. Left ventricular hypertrophy *(right panels)*. The initial and mean forces are shifted to the left and posteriorly.

cle. An ECG showing left ventricular hypertrophy is reproduced in Figure 2.67 *(lower panels)*. The mean QRS axis points to the left, with the largest projection negatively on lead III, while aVL and aVF show equal positive and negative amplitudes; therefore, the mean QRS axis is approximately −60° (left axis deviation). The QRS voltage is larger than normal, being largest in V_2 in the transverse plane (large S wave). Again, a useful exercise is to plot the initial, terminal, and mean vectors and to draw the QRS loops in the frontal and horizontal planes.

Notice that the T waves are abnormally directed in this tracing, being opposite to the QRS in leads I, V_4, V_5, and V_6, as well as aVL. This indicates that, in this condition, repolarization is following an abnormal direction, from endocardium to epicardium. These T wave changes are secondary to the hypertrophy, since the QRS is of relatively normal width and shows no evidence of damage to the muscle (see below).

MYOCARDIAL INFARCTION

When an area of heart muscle is damaged secondary to occlusion of a coronary artery, within a few hours the muscle cells in the involved region tend to die and depolarize completely (see Chapter 15). This leaves an electrical "window," with unopposed voltages generated by other regions of the heart producing abnormal deviation of the instantaneous vectors of the heart. Typically, death of the heart muscle ("myocardial infarction") leads to an initial vector pointing *away* from the location of the infarct, due to depolarization of normal muscle on the opposite side of the heart (away from the damaged area). There is also loss of QRS voltage due to the death of tissue. These changes often persist in patients who recover from the heart attack because of the formation of a scar in the involved region (see Goldberger and Goldberger, 1986).

Anterior Myocardial Infarction

The QRS complex is deformed by an anterior myocardial infarction, as shown in the simplified diagram in Figure 2.68. The location of the damaged area is indicated by the *black zones* on the frontal view and the transverse section. In the *frontal plane*, a small anterior infarct does not greatly affect the balance of the QRS forces since it is centered in the middle of the heart; therefore, the initial and mean QRS vectors, if plotted on the triaxial lead system, would result in relatively normal QRS complexes in leads I, II, and III. However, in the *transverse plane* (Fig. 2.68), the infarct can be seen to involve the interventricular septum and the front wall of the left ventricle and to disturb the QRS greatly. The initial vector on the transverse lead system points directly posteriorly, as the normal anterior septal force disappears and the unopposed normal forces on the posterior wall of the heart dominate. This results in loss of the normal R wave in V_1 to V_3, with abnormal Q waves in those leads, whereas lead V_6 remains relatively normal (Fig. 2.68).

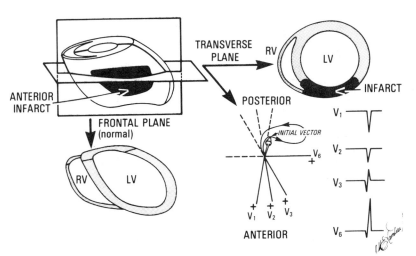

Figure 2.68. Diagrammatic representation of anterior myocardial infarction (infarction of the anterior wall of the left ventricle). As shown in the *upper left* and *lower left* diagrams, the frontal plane does not pass through the infarct, and therefore the frontal plane leads are not greatly affected (a larger infarct would affect these leads). However, the transverse plane passes through the infarct region (*upper left* and *upper right* diagrams). This results in an initial vector in the precordial leads (in the transverse plane) which is directed posteriorly, in a direction opposite to normal resulting in Q waves in leads V_1, V_2, and V_3.

Inferior Myocardial Infarction

In Figure 2.69, damage to the inferior wall of the left ventricle by coronary occlusion is shown by the *black zone* on the frontal view, so that the frontal plane vectors are importantly affected (Fig. 2.69), but a transverse plane through the middle of the heart does not reveal the infarct (Fig. 2.69). The initial vector in the frontal plane points upward and to the left (away from the infarct), producing a Q wave in leads II, III, and aVF, since it projects on the negative poles of these three leads (Fig. 2.69). Thus, unopposed normal forces on the high left lateral wall of the heart dominate the early portion of the QRS complex in inferior myocardial infarction. In contrast, if the infarct is small, the QRS in leads recorded from the transverse plane can appear relatively normal.

A variety of other infarct locations resulting from the obstruction of various coronary arteries or their branches produce different locations of Q waves in specific leads.

ABNORMAL CARDIAC REPOLARIZATION

As discussed in Chapter 8, a number of factors affect the rate of repolarization of individual cardiac muscle cells (phase 3 of the action potential), and such effects in the whole heart are reflected by changes in the T waves and ST segments of the ECG (see Fig. 2.63). For example, the T wave can be altered by an abnormal pathway of depolarization (inverted T waves in Fig. 2.67), the Q-T interval is affected by an abnormal duration of repolarization, and the T wave and ST segment are influenced by electrolyte abnormalities, drugs, or tissue damage. These changes may affect repolarization directly or through their effect on depolarization (see Castellanos and Myerburg, 1986).

Depression and Elevation of the ST Segments

Depression of the ST segments can occur transiently during angina pectoris (chest pain), due to inadequate coronary blood flow to the inner (subendocardial) wall of the left ventricle, causing partial depolarization of the deep layers of myocardium and leading to a potential difference between them and the uninjured superficial layers (see Chapter 15). During the acute phase of myocardial infarction, *elevation* of the ST segments occurs in those leads showing the area of damage (Q waves), which generally involves nearly the full thickness of the left ventricular wall. The explanation of ST segment elevation is based on the supposition that a boundary exists between a region of normal and damaged cells and that an abnormal current could flow between the two regions.

According to one theory (the theory of diastolic current of injury), the injured region is considered to be partially or completely depolarized at rest (with the damaged area therefore appearing electronegative with respect to the normal regions during diastole). A "current of injury" therefore flows

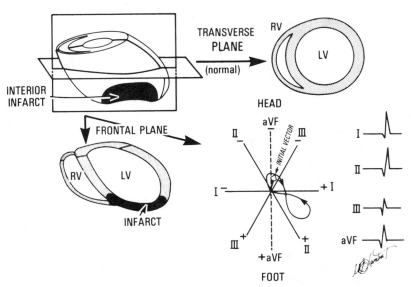

Figure 2.69. Diagrammatic representation of inferior myocardial infarction (infarction of the inferior or diaphragmatic wall of the left ventricle). The *left upper* and *right upper* diagrams show that the transverse plane does not pass through the infarct, and therefore the precordial leads remain normal. The *left upper* and *left lower* diagrams show that the frontal plane transects the infarct. Therefore, in the hexaxial lead system, the initial QRS vector is abnormally directed superiorly and to the left, giving Q waves in leads II, III, and aVF.

during electrical diastole (Fig. 2.70, *upper left panel*). This current disappears when the entire heart is fully depolarized during electrical systole (Fig. 2.70 *upper right panel*). Partial diastolic depolarization has been documented by direct intracellular recordings from cells within a zone of ischemia, at a time when ST segment elevations were evident on an epicardial electrogram. In the ECG recorded from the body surface, the AC electronic circuitry compensates for this baseline shift, which otherwise would be recorded as a depression of the segment between the end of the T wave and the onset of the P wave (Figure 2.70, *upper left*). Therefore, with this explanation, the ST segment elevation during cardiac depolarization on the surface ECG is only apparent.

Another theory (systolic current of injury hypothesis) holds that a so-called "primary" or "true" ST segment elevation occurs during electrical systole due to failure of the injured area to depolarize adequately (or as a consequence of early repolarization) (Fig. 2.70, *lower panels*). According to this theory, the injury area is considered to be nor-

mally polarized at rest, and it remains polarized during electrical systole (phase 2 of the action potential), so that there is current flow during electrical systole, with the injured region electropositive with respect to the depolarized normal zones (Fig. 2.70, *lower right panel*). This mechanism may play some role, but most of the evidence suggests that a diastolic current of injury is the dominant factor (for review, see Ross, 1976a). The effect most probably relates to loss of intracellular potassium causing local elevation of extracellular $[K^+]$, which partially depolarizes the involved membranes during diastole.

Abnormal T Waves

T waves opposite in direction to the QRS often occur with ventricular hypertrophy due to abnormal spread of the repolarization process (see Fig. 2.67), and such T waves also can be seen when there is an abnormal duration of repolarization, as with block in one of the bundle branches of the left ventricle or right ventricle (see below). In the bundle branch block, the depolarization wave spreads

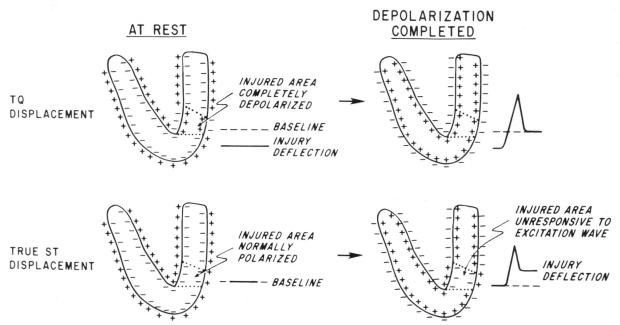

Figure 2.70. Diagram of potential mechanisms for ST segment displacement. In the upper two diagrams, at rest an injury current is produced in a lead facing the damaged area of the left ventricle *(dotted lines)* caused by partial depolarization of the cells (negative outer wall) during electrical diastole. This causes a displacement of the baseline at rest between the end of the T wave and the onset of the QRS (TQ displacement). During ventricular depolarization (the plateau phase of the electrogram), the transmembrane potential becomes zero (all cells are depolarized), and it appears as if the ST segment is elevated.

Another theory *(lower two panels)* holds that the injured area is normally polarized during the rest phase of the cycle, so that the baseline is not displaced. However, during ventricular depolarization the injured area is not depolarized, so that a relatively positive area faces the recording electrode during the action potential *(right lower panel)*, resulting in a "true" ST segment elevation. (Abnormally early repolarization of the injured area, while other cells are still depolarized, could also produce such an effect). (From Ross, 1976.)

slowly through the muscle of the chamber supplied by the involved bundle branch, rather than through the specialized conduction system, and the wave of repolarization is similarly delayed and dispersed, spreading abnormally from endocardium to epicardium. Under these circumstances, the T wave often is very large as well, as it spreads slowly over the involved ventricle; this relates to the lack of normal canceling forces, which occurred earlier as the opposite ventricle was normally depolarized and repolarized. A variety of other T wave changes occur in disease of the heart muscle and of the pericardium and in myocardial infarction, but discussion of these is beyond the scope of this text.

Drugs and Electrolytes

Digitalis can alter the ST segments and T waves (Fig. 2.71), and antiarrhythmic drugs of the quinidine type reduce P_K, thereby prolonging the action potential duration and lengthening the Q-T interval (Fig. 2.71). High serum Ca^{2+} concentration shortens phase 2 of the action potential and the Q-T interval; low Ca^{2+} decreases P_K, lengthens phase 2, and slows the rate of repolarization, thereby prolonging the Q-T interval (Fig. 2.71). Low serum K^+ reduces P_K and reduces the amplitude of the T wave, with prolongation of the Q-T interval and appearance of a prominent U wave, which may resemble the T wave (Fig. 2.71). High serum K^+ shortens action potential duration as P_K is increased, leading to a short Q-T interval with peaked, tall T waves (Fig. 2.71). With a very high concentration of serum K^+, sufficient membrane depolarization takes place to impair conduction; the P-R interval prolongs and the QRS widens to merge with the tall peaked T wave (Fig. 2.71).

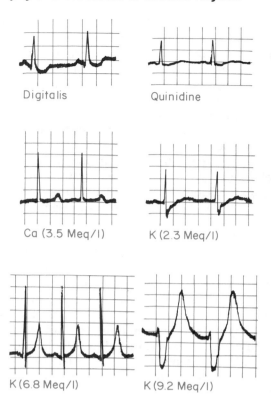

Figure 2.71. Patterns on the electrocardiogram of digitalis and quinidine effect, low serum calcium, low serum potassium, and high serum potassium.

ABNORMALITIES OF CARDIAC RHYTHM AND CONDUCTION

When the P wave is of normal configuration, each P wave is conducted to the ventricles, and the resting heart rate is within the range of 60–100 beats/min, *normal sinus rhythm* is said to exist (Fig. 2.65). Regardless of the mechanism of rhythm disturbance, when the heart rate exceeds 100 beats/min *tachycardia* is present, and when it is below 60 beats/min *bradycardia* is present. There is normally a fluctuation of heart rate that is synchronous with the respiratory cycle; the pulse accelerates slightly during an inspiration, and slows upon expiration. This normal response is called *sinus arrhythmia*. The mechanism is related to stimulation of stretch receptors in the aorta and carotid arteries as the blood pressure changes and perhaps to stimulation of receptors in the heart itself as well, which reflexly affect the heart rate, as described in Chapter 16.) (For reviews of arrhythmias, see Marriott and Myerburg, 1986, and Zipes, 1988.)

Mechanisms of Abnormal Cardiac Impulse Formation

Abnormal cardiac rhythms ("called arrhythmias or dysrhythmias") can be produced by a number of mechanisms. Such mechanisms are not fully understood in the whole heart in all instances. However, certain phenomena appear important in causing some clinical arrhythmias, particularly the rapid arrhythmias or "tachycardias." These mechanisms include: (1) automacity, (2) triggered activity, and (3) reentry (Wit and Cranefield, 1977; Cranefield, 1977; Mines, 1914).

Automaticity, which in normal pacemaker tissues is usually due to phase 4 depolarization of the membrane (see Chapter 8), may become enhanced or abnormal and lead to certain arrhythmias. One possible example of such an arrhythmia is known as an "accelerated idioventricular rhythm," in which a focus in the ventricle firing more rapidly than the sinus node takes over the heart rhythm. Such abnormal automaticity may occur in certain settings including ischemia or hypoxia or the presence of elevated cathecholamine levels or certain drugs. However, these arrhythmias are rarely very rapid and probably do not account for any serious clinical tachycardias.

Triggered activity, in contrast, is due to the development of afterpotentials (early or late) following an action potential (see Chapter 8), which reach threshold and cause a subsequent action potential or series of action potentials. An example of an arrhythmia which may be caused by this mechanism is called "torsade de pointes," a potentially serious rapid ventricular arrhythmia. However, as the term "triggered" implies, this phenomenon is not observed unless a spontaneous rhythm initiates such activity. Triggered activity is also more likely to be produced under certain conditions, including rapid heart rates, high catecholamine levels, and the presence of certain drugs (e.g., antiarrhythmic agents) or electrolyte disturbances (hypokalemia).

Reentry, however, is the most likely mechanism of the majority of clinical arrhythmias, including single premature depolarizations (called premature beats) of either the atria or the ventricles and the sustained tachycardias. Reentry may occur at a number of sites including the sinus node, atria, AV node, His-Purkinje system, and ventricles. Certain conditions or abnormalities of the myocardium are generally believed to be required for reentry to occur. These include (1) anatomic obstacles to conduction or localized slowed conduction with or without unidirectional block or (2) abnormal dispersion of refractoriness in areas of adjacent tissue resulting in localized areas of slowed conduction or block.

One model commonly used to study ventricular dysrhythmias due to reentry has been a branching Purkinje fiber attached to the ventricular myocardium (Fig. 2.72). In this diagram, the impulse normally would travel through both Purkinje fiber bundles (*A* and *B*) from the main bundle (*MB*) to depolarize the ventricular muscle (*VM*) rapidly and simultaneously. As shown in Figure 2.50 (Chapter 8), action potentials are of longer duration in the His-Purkinje system than in the ventricular muscle, so that the specialized fibers are normally refractory to reentry from the muscle. The *stippled area* in Figure 2.72 represents a zone of abnormal conduction in branch *B*, which is refractory to the ante-

grade impulse and markedly slows or blocks it. The impulse now travels only through branch *A* to depolarize the ventricle, but as the impulse travels through the muscle it finds branch *B* not refractoy, and it enters this branch retrogradely (Fig. 2.72) (branch *B* exhibits unidirectional block). The impulse is then able to travel through the *stippled area* in a retrograde manner to provide early reexcitation (reentry) of the main bundle, and it also may reenter branch *A* in a circular movement. The impulses recorded from cells directly in such branches are shown in the *upper right tracings* of Figure 2.72, and the ECG (recorded below) shows a wide, early QRS complex following each normal QRS (the latter are preceded by P waves). The broad QRS indicates a ventricular ectopic beat (premature ventricular depolarization), with the early impulse widened because it must travel slowly by an abnormal route. Such reentry mechanisms can also be rapid and repetitive (a circus movement) and can take over the ventricular rhythm. Several models of

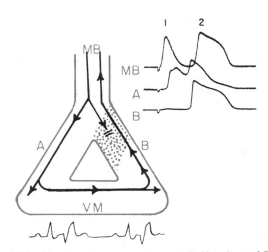

Figure 2.72. Reentry phenomenon as recorded in a loop of Purkinje fiber bundles (*A* and *B*) and ventricular muscle *(VM)*. The *shaded area* in *branch B* represents an area of unidirectional block. Conduction through this area along B toward VM is blocked, but an impulse arriving later from VM to B can be conducted retrogradely. Conduction throughout the loop is slowed. At *top right*, transmembrane action potentials from a similar arrangement of canine Purkinje fibers are shown. Conduction was delayed with elevated K^+ and catecholamines. *MB*, action potential from the main bundle leading into the loop. Action potential *1* in trace MB results from antegrade conduction in this bundle. *A*, Action potential recorded as the impulse propagates through branch A. *B*, Action potential recorded as the impulse is conduced retrogradely through branch B. Action potential *2* in MB is the retrogradely conduced reentrant impulse which is recorded further up MB and appears later than the second (reentry) impulse in branch A. The *bottom trace* shows coupled ventricular ectopic systoles following each normally conducted complex which might result from this type of reentry. (From Wit et al., 1974.)

reentrant atrial dysrhythmias have also been studied in the dog and rabbit; these models tend to support the requirements for reentry noted above (Allessie et al., 1973; Rosenbleuth and Garcia, 1947; Frame et al., 1986; Feld et al., 1986).

Characteristics and Mechanisms of Clinical Cardiac Arrhythmias

Atrial dysrhythmias can be of many types and several will be briefly discussed here. An atrial premature contraction is identified as an earlier than normal P wave, often of unusual shape, which is usually followed by a normal QRS complex (Fig. 2.73). Because the premature atrial contraction depolarizes and resets the sinus node, the subsequent normal atrial impulse then occurs at the normal sinus interval, so that there is no pause after the premature beat (compare this with the ventricular premature contraction also seen in Fig. 2.73).

Atrial tachycardias (also called supraventricular tachycardias) may occur as a result of one of several reentrant mechanisms. Often they are paroxysmal (i.e., sudden in onset and offset). The commonest form of supraventricular tachycardia (SVT) occurs when abnormal conduction is present in the AV node, resulting in two potential pathways, one with slow conduction and a short refractory period and the other with fast conduction and a long refractory period (Fig. 2.74). During sinus rhythm, impulse conduction through the AV node is predominantly through the fast pathway. However, if a premature atrial contraction occurs, the impulse blocks in the fast pathway (due to the long refractory period), conducts down the slow pathway anterogradely, and then reenters the fast pathway retrogradely. This may result in a single atrial reentrant beat or, if the anterograde pathway has fully recovered, a sustained tachycardia (i.e., slow-fast AV nodal SVT) may develop (Josephson and Kastor, 1977).

In a similar fashion, patients with a congenital accessory pathway or bypass tract connecting the atrium and ventricle (Fig. 2.75) are also prone to develop SVT. This condition is called the Wolf-Parkinson-White syndrome (Gallagher et al., 1978) The accessory pathway conducts rapidly and often has a refractory period shorter than that of the AV node. During sinus rhythm, impulses conduct anterogradely through both the AV node and the accessory pathway (unless the accessory pathway is "concealed," in which case only retrograde conduction can occur). This produces a typical delta wave on the QRS complex and a short P-R interval in the surface ECG, resulting from early activation of the

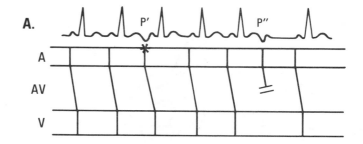

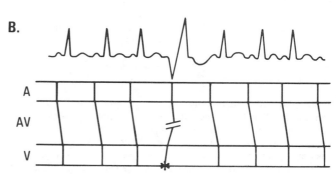

Figure 2.73. Ladder diagrams of atrial and ventricular premature contractions. During sinus rhythm the ladder diagrams show normal conduction pathways and times through the atrium (*A*), AV junction (*AV*), and ventricular myocardium (*V*). *A,* An atrial premature contraction, shown by an asterisk (*) on the ladder diagram and a (*P′*) on the electrocardiogram, is characterized by its early timing and an abnormal morphology, indicating that its origin is from a site other than the sinus node. Following this premature beat there is slightly delayed AV conduction, and the next beat is initiated by a normal sinus P wave at a normal interval. As a result of prior resetting of the sinus node by the premature atrial contraction there is no compensatory pause (see text). A second, slightly earlier premature atrial contraction (*P″*) fails to conduct through the AV junction, which is still refractory. *B,* A ventricular premature contraction, shown by an asterisk (*) on the ladder diagram and a (*V*) on the electrocardiogram, is characterized by an early QRS complex with a wide abnormal morphology. The premature beat originates in the ventricle and spreads slowly in a retrograde manner, partially penetrating the AV junction. There it collides with the normal sinus P wave (which is obscured in the abnormal ventricular complex), thereby blocking anterograde conduction of the normal P wave through the AV junction. The next P wave then occurs on schedule exactly two cycles after the last normally conducted P wave, yielding a full compensatory pause. For further discussion see text.

ventricle over the accessory pathway (Fig. 2.75). This phenomenon is called preexcitation. If a premature atrial or ventricular depolarization occurs with a critical timing, a reentrant tachycardia may be initiated, resulting in a SVT with a normal QRS (i.e., no delta wave). This is due to the fact that during the tachycardia the accessory pathway is con-

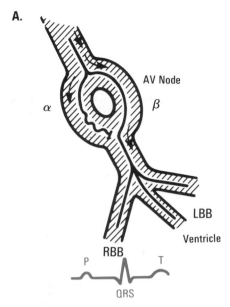

Figure 2.74. AV nodal reentry. *A,* The AV node may develop more than one functional pathway with different electrophysiological characteristics. Usually a slowly conducting pathway with a short refractory period (α) may be demonstrated in association with a rapidly conducting pathway with a long refractory period. (β). During sinus rhythm the impulse normally conducts down the fast pathway. *LBB,* left bundle branch; *RBB,* right bundle branch. *B,* However, following an appropriately timed premature atrial contraction the impulse blocks in the fast pathway (due to its long refractory period), conducts antegradely down the slow pathway, and then subsequently conducts retrogradely through the fast pathway if it has sufficiently repolarized, causing a sustained supraventricular tachycardia (note the negative retrograde P waves).

ducting only retrogradely and the AV node anterogradely (Fig. 2.75).

Supraventricular tachycardia may also occur as a result of reentry localized to a portion of the atrium only (i.e., sinus node reentry, interatrial reentry) or due to automatic or triggered activity located in the atrium (Wu, 1975; Wit and Cranefield, 1977; Cranefield, 1977).

The rates of SVT are usually between 100 to 240 beats/min. Vagal nerve stimulation (i.e., carotid massage) may slow or terminate AV nodal SVT or accessory pathway SVT, whereas it usually will not terminate interatrial reentrant, triggered, or automatic tachycardias. Atropine or catecholamines may aggravate the SVTs.

Therapy for the supraventricular tachycardias involves either slowing of conduction or prolongation of refractoriness in the normal pathway and/or the abnormal pathway or a combination of both effects. Therefore, drugs which prolong AV nodal conduction or refractoriness (e.g., calcium antagonists, beta-blockers, digitalis, amiodarone) are often effective in AV nodal SVT. In contrast, drugs or combinations of drugs which depress conduction and prolong refractoriness first in the accessory pathway (quinidine, procainamide, amiodarone) and

then in the AV node are safer and more effective in suppressing accessory pathway SVT. Triggered or automatic SVT is often controlled with a beta-blocker or calcium antagonist or a drug which markedly depresses conduction velocity and produces exit block from the focus (e.g., quinidine or flecainide).

Atrial flutter (Fig. 2.76*A*) is an even more rapid atrial dysrhythmia, which is most likely due to a reentrant wavefront around a central anatomic obstacle such as the tricuspid valve or vena cavae (Mines, 1914; Allessie et al., 1973; Boineau et al., 1980; Waldo, 1987). This dysrhythmia tends to produce a "sawtooth" baseline depicting a regular atrial rate of 240–350 beats/min (usually 300/min). The many impulses produce increased refractoriness in the AV node so that the ventricles usually beat at a slower rate, often at 150/min or in a 2:1 ratio.

Atrial fibrillation (Fig. 2.76*B*) is a common dysrhythmia and, provided the ventricular response is not too fast, it is quite compatible with a long life. Usually it occurs in the presence of heart disease, often with enlargement of one or both atria, but sometimes it is seen in "paroxysmal" form in otherwise normal subjects. Fibrillation of the atria is

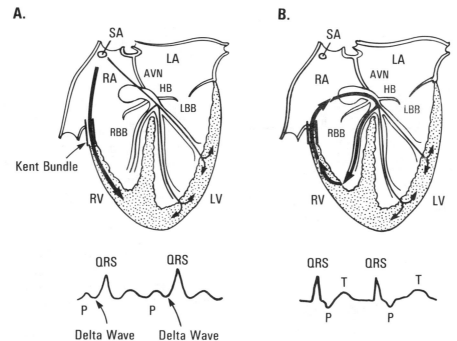

Figure 2.75. Atrioventricular reentry due to an accessory pathway in the Wolff-Parkinson-White (WPW) syndrome. *AVN,* atrioventricular node; *HB,* His bundle; *LA,* left atrium; *LBB,* left bundle branch; *LV,* left ventricle; *RA,* right atrium; *RBB,* right bundle branch; *RV,* right ventricle; *SA,* sinoatrial node. *A,* In the WPW syndrome a congenital abnormal connection (Kent bundle) between the atrium and ventricle persists, allowing electrical impulses to bypass the AV node and conduct rapidly to the ventricles. During sinus rhythm the typical delta wave of preexcitation and a short P-R interval are seen, due to simultaneous activation of the ventricle through both the accessory pathway and the AV node. *B,* A critically timed premature atrial contraction may lead to an impulse traveling only through the AV node in the normal direction and then backward over the accessory pathway, resulting in a reentrant supraventricular tachycardia. Note the normal QRS during tachycardia due to the absence of preexcitation, and the negative P waves which follow the QRS indicating retrograde depolarization of the atria. The delta wave will return upon cessation of their tachycardia.

probably related to multiple small reentry loops (sometimes called "microreentry") producing completely disordered, very rapid depolarizations of the atria; this causes them simply to quiver, rather than to contract effectively (with loss of the atrial booster pump). These depolarizations of atrial fibrillation occur at rates of 500–800/min, sometimes being difficult to discern on the ECG baseline (Fig. 2.76*B*) and, being totally irregular, they bombard the AV junction to cause varying degrees of decremental conduction, with most impulses failing to penetrate to the ventricles (see also discussion of AV node refractoriness in Chapter 8). The result is a totally irregular ventricular response which can actually be slower than that in atrial flutter because the more rapid bombardment of the AV junction in atrial fibrillation increases AV node refractoriness (Chapter 8). Once through the AV junction, the impulses travel by the normal conduction system; the QRS complexes therefore appear normal. The arterial *pulse* is irregular and of varying amplitude, not only because of the different times available between beats for ventricular filling (varying end-diastolic volume causes a variable stroke volume) but also because of the effect of changing rate on the force of contraction (force-frequency relation; see Chapter 12).

Atrial flutter and fibrillation are often treated acutely by attempting to slow the ventricular rate to the normal range with digitalis (which exerts both a direct and a central vagal effect on the AV node, increasing its refractoriness), a beta-blocker (which antagonizes sympathetic effects on the AV node), or a calcium antagonist (which directly depresses AV node conduction). They can also be terminated by applying a brief electrical shock across the chest (this approach can also be used to treat other dysrhythmias). This electrical "countershock" momentarily depolarizes all of the cells in the heart, subsequently allowing the sinus node to discharge and conduct normally, thereby abolishing the dysrhythmia. Atrial flutter may also be acutely converted to normal by transiently pacing the atrium faster than the flutter rate (i.e., overdrive pacing using an electrode catheter inserted into the atrium through a vein). Recurrence of atrial flutter

A. Atrial Flutter

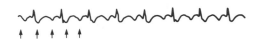

B. Atrial Fibrillation

C. Ventricular Tachycardia

D. Ventricular Fibrillation

Figure 2.76. *A,* Atrial flutter. The characteristic saw-toothed flutter waves *(arrows)* due to rapid atrial contraction are seen in this figure. The atrial rate during atrial contraction is usually 300 beats/min (ranging from 240–350). The ventricular response may be regular or irregular but is usually exactly one-half of the atrial rate (e.g., a ventricular rate of 150 beats/min may be seen if the atrial flutter rate is 300, due to 2:1 conduction through the AV node, as shown in the tracing). *B,* Atrial fibrillation. A fine or coarse irregular baseline is seen between QRS complexes during atrial fibrillation, indicating the lack of regular or organized atrial contraction. The ventricular rate is irregularly irregular due to the rapid irregular stimulation of the AV node with electrical impulses from the atrium. *C,* Ventricular tachycardia. A rapid regular rhythm with a wide and abnormal QRS complex characterizes ventricular tachycardia. The rate of the tachycardia ranges from 100–240 beats/min but is most commonly around 150–180 beats/min. This arrhythmia may be sustained and degenerate into ventricular fibrillation. *D,* Ventricular fibrillation. A chaotic low amplitude waveform characterizes ventricular fibrillation on the electrocardiogram. There is no regular rhythmic electrical or mechanical activity of the heart during this arrhythmia, which if not treated immediately is usually fatal.

or fibrillation may be prevented in many (not all) cases by the use of antiarrhythmic drugs such as quinidine, procainamide, β-blockers, or amiodarone. Digitalis alone is usually not effective in preventing recurrences of these dysrhythmias.

Ventricular dysrhythmias are also of many types, and only three important abnormalities will be discussed. *Ventricular premature beats* can be single or multiple, can originate in one or more different locations (hence, giving different configurations to the QRS), or may occur regularly after every normal beat (bigeminy). The most common type of ventricular premature beat occurs at a fixed period after the prior normal QRS (suggesting either triggered

automaticity or a reentry mechanism). Such a beat is shown in Figure 2.73*B*; the impulse arises from the ectopic focus or reentry site in the ventricles or Purkinje system and spreads slowly through the ventricular muscle, giving a wide and bizarre QRS configuration. Figure 2.73*B* illustrates normal conduction from the SA node through the atrium (*A*), corresponding to the P waves, conduction through the AV node, and ventricular (*V*) depolarization occurring rapidly to give normal QRS complexes (first three normal cycles). A ventricular premature depolarization then occurs (VPD, often called also a ventricular premature contraction, or VPC), and spreads in retrograde fashion, giving the widened QRS, which is closely "coupled" to the third normal QRS. Notice that the *next* sinus impulse occurs on schedule (*P*) but that its P wave is obscured by the premature beat (*), and the normal QRS complex does not occur because the impulse enters the AV junction at a time when it is refractory because of previous retrograde penetration of the AV junction by the premature ventricular impulse (Fig. 2.73*B*). The following P wave again occurs on schedule and is followed by a normal QRS. Notice that, in contrast to a premature atrial beat, which depolarizes the atrium and "resets" the SA node (Fig. 2.73*A*), the sinoatrial node usually is *not* reset after a ventricular premature beat. Therefore, the interval between the two normal QRS cycles bracketing the premature beat is equal to exactly two normal cycle lengths, leading to a so-called "compensatory pause" after the premature ventricular contraction. A similar phenomenon is seen with an AV junctional premature beat which fails to depolarize the atrium. In some settings, such as after a heart attack, ventricular premature beats can lead to more serious ventricular rhythm disturbances (see below), and they can be suppressed by chronic oral administration of an antiarrhythmic drug.

Ventricular tachycardia is a rapid series of regular ventricular beats showing a widened bizarre QRS morphology, that is a series of rapid ventricular premature beats (Fig. 2.76*C*). The mechanism underlying this dysrhythmia is usually reentry due to delayed conduction in ischemic or infarcted myocardium or the specialized conduction system (Josephson et al., 1978). Ventricular tachycardia often occurs at a rate of 140–170 beats/min (range, 100–240/min), and it may be sustained or paroxysmal. Because of the abnormal sequence of contraction of the ventricles and the lack of synchronized atrial impulses, the blood pressure and cardiac output may drop sharply during this dysrhythmia and require immediate treatment with an electrical countershock applied to the chest wall (i.e., electrical cardioversion).

Ventricular tachycardia is particularly dangerous because it will often degenerate to ventricular fibrillation. Episodes of ventricular tachycardia, occurring for excample after a heart attack, may be acutely treated by drugs (e.g., lidocaine, quinidine, or procainamide) which prolong the ratio of the effective refractory period to the action potential duration (i.e., they prolong the time-dependent refractoriness), depress conduction velocity, and decrease membrane responsiveness (see Fig. 2.45). However, several newer antiarrhythmic drugs (e.g., amiodarone, sotalol, or *N*-acetylprocainamide), which prolong the action potential duration and refractory period equally (i.e., they prolong the voltage-dependent refractoriness), may be more effective for both the acute treatment and chronic suppression of ventricular tachycardia.

Ventricular Fibrillation is a totally disorganized, very rapid rhythm of the ventricles similar to fibrillation of the atria (Fig. 2.76*D*). It leads to totally ineffective pumping of the ventricles (which quiver at a rate of several hundred times/min), and it is rapidly fatal, if not treated promptly, since brain death will occur within a few minutes. Ventricular fibrillation (as well as complete cessation of ventricular activity) is termed "cardiac arrest." When this occurs cardiopulmonary resuscitation almost always will be required, using mouth-to-mouth breathing and chest compression to support the circulation, followed by *electrical countershock* with an electrical defibrillator. Certain drugs may also help facilitate electrical defibrillation, including intravenous epinephrine, lidocaine, or bretyllium.

Abnormalities of Conduction through the Specialized System

Partial or complete "block" can occur at any location in the specialized conduction system, and only the most important types will be mentioned briefly (Wit et al., 1970).

Disease of the SA node may involve "sinus arrest" or occasionally "exit block" from the sinus node in which the impulse is unable to penetrate outside the nodal region. These two conditions lead to absence of P waves on the electrocardiogram and, generally, the AV junction takes over as the heart's pacemaker.

Partial and complete heart block are due to disease of the AV junction or His-Purkinje system which impairs conduction between the atria and ventricles. "First degree AV block" refers to simple prolongation of the P-R interval beyond 0.2 s. Such

block is usually in the AV junction, but sometimes it is due to delayed conduction through the His-Purkinje system. "Second degree AV block" occurs when *some* of the impulses from the atria do not pass through the AV junction of His-Purkinje system to the ventricles. The P-R interval is usually prolonged and, sometimes, every other QRS complex is "dropped" to produce 2:1 AV block; sometimes, there is progressive prolongation of the interval, followed by a dropped QRS, the so-called "Wenckebach type" of second degree AV block (Fig. 2.77). In *complete heart block* or "third degree AV block" there is *no* conduction from the atria to the ventricles (Fig. 2.77). In this condition, the ventricular rate is usually slow (about 40/min) and the QRS is widened, since it originates from a pacemaker in the Purkinje fibers, while the nonconducted P waves proceed at a normal rate. Patients who survive with this abnormality may be quite limited and may be subject to fainting spells ("Stokes-Adams attacks," see Chapter 8 for the differentiation of impaired conduction due to block in the AV junction from that due to block in the His-Purkinje system, using His bundle electrocardiography).

The usual treatment for serious disease of the SA node, for permanent severe second degree AV block, or for complete heart block is the implantation of an electronic cardiac pacemaker with its tip positioned in the right ventricle. The pacemaker is set to fire impulses at a normal rate. For patients in whom complete heart block is intermittent, pacemakers are available which fire impulses at a regular rate only when the patient's own pacemaker slows below a preset rate (so-called "demand pacemakers"). In patients with normal or abnormal SA node function but complete AV block, dual-chamber pacemakers are also available which sense and can pace the atrium and pace the ventricle, thus main-

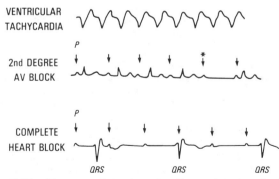

Figure 2.77. Diagrammatic tracings showing another example of ventricular tachycardia *(upper panel)*, second degree (2°) AV block of the Wenkebach type *(middle panel; asterisk* shows nonconducted P wave), and third degree (3°) AV block (complete heart block, *lower panel).*

taining the normal AV synchrony. This may be particularly helpful in patients with impaired ventricular function. Pacemakers are also available for patients with abnormal SA nodes and complete AV block which will electronically respond with a varying heart rate depending on the patient's activity and energy demands. These pacemakers sense changes in the patient's activity, temperature, blood pH, or oxygen saturation and a variety of other physiological measures and adjust the heart rate accordingly.

Right bundle branch block occurs when there is damage to the right branch of the specialized conduction system. It results in a wide QRS complex (0.12 s or more) because the right ventricular depolarization wave must pass slowly through the muscle, rather than through the conduction system. The vectors and representative scalar leads for this abnormality are shown in Figure 2.78. Since the initial septal forces in the normal ECG originate prior to depolarization of the right bundle, the initial septal depolarization is normal (vector 1, Fig. 2.78), and the QRS forces are normally directed initially toward the left ventricle (the left bundle branch is not involved) (vector 2, Fig. 2.78). However, the impulse then spreads relatively slowly through the muscle to reach the right ventricle, circumventing the damaged right bundle branch, and this causes an abnormal rightward and anterior force which is

TRANSVERSE PLANE

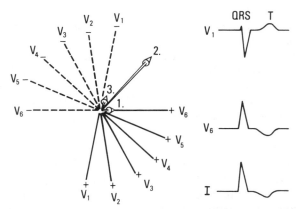

Figure 2.79. Left bundle branch block. This conduction disturbance is diagrammed in the transverse plane. The abnormal initial vector (*1*) is directed to the left, and delayed, unopposed depolarization of the left ventricle then causes a larger than normal mean and terminal QRS vector (*2* and *3*) directed to the left and posteriorly. This results in a broad QRS with an abnormally small (or absent) R wave in lead V_1 and an absent septal Q wave in lead V_6. Lead V_6 resembles the tracing in lead I on the frontal plane.

large and unopposed late in the QRS complex, giving a wide S wave (vector 3, Fig. 2.78). The loop in the transverse plane shows a normal initial vector and a large, late anterior and rightward force leading to an rSR' in lead V_1 (conventionally, as mentioned earlier, the second of two waves is marked "prime"; the smaller and larger of two waves are written with small and large letters, respectively) (Fig. 2.78). Lead V_6, like the frontal plane loop in lead I, shows a terminal vector which is slowed and directed to the right, giving a large, late S wave (Fig. 2.78).

Left bundle branch block results from damage to the left branch of the conduction system and also results in a broadened QRS complex (0.12 s or more). As shown in Figure 2.79 the initial vector is abnormal, since septal and left ventricular depolarization must now originate from the right ventricle (the right bundle branch is intact), and the impulse then spreads from right to left and posteriorly (vector 1, Fig. 2.79). The prolonged late conduction through the muscle of the left ventricle leads to unopposed forces and an enlarged mean vector (vector 2) directed more posteriorly than normal. The result is no R wave (or only a tiny R wave) in lead V_1 with a deep wide S wave, and the absence of a septal Q wave in lead V_6 with a tall wide R wave (Fig. 2.79). In the frontal plane, the tracing in lead I resembles that of lead V_6 (Fig. 2.79).

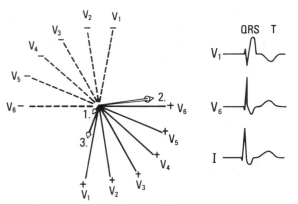

Figure 2.78. Right bundle branch block. This conduction disturbance is diagrammed in the transverse plane, showing three ECG leads. In the initial vector (*1*) septal depolarization is normally directed anteriorly to the right and the mean QRS vector (*2*) is also directed normally. However, the terminal vector (*3*) is delayed as the right ventricle is depolarized by spreading through the muscle rather than through the right bundle branch. It is unopposed and therefore a large terminal force is directed anteriorly and to the right, resulting in a large late positive wave in lead V_1 (rSR') and a late negative S wave in lead V_6. Since both lead V_6 and lead I are in the frontal plane, there is also a broad QRS in lead I due to a wide S wave representing the delayed rightward depolarization of the right ventricle.

CHAPTER 10

Cardiac Muscle: Cardiac Structure-Function Relations and Excitation-Contraction Coupling

The structure of heart muscle (myocardium) allows its function to be regulated by several intrinsic mechanisms (in addition to the autonomic nervous system). Before considering how various forms of physiological stress or drugs can affect the whole heart, it is important to understand certain of these basic structure-function relations in *isolated cardiac muscle*, as well as how various factors affect its performance (Chapter 11). The heartbeat is largely controlled by "involuntary" mechanisms, some of which are intrinsic to the muscle itself and independent of neurohumoral influences. Thus, the force of contraction can be modified by such stimuli as the resting length of the muscle, the frequency of electrical stimulation, and the concentration of various ions. These intrinsic mechanisms that govern cardiac muscle contraction are the subject of this chapter.

ANATOMY OF MYOCARDIUM

The muscular walls of the mammalian heart are composed of a branching network of muscle fibers or cells. The working muscle cells in the human heart are approximately 10–20 μm wide and 50–100 μm long (Fig. 2.80A). Each muscle cell is dominantly occupied by bundles of myofilaments, each bundle being organized into a series of repeating subunits called sarcomeres. These band-like patterns are responsible for the striated nature of both skeletal and cardiac muscle, as observed under the light microscope. By electron microscopy, the sarcomere can be seen to be composed of thick and thin myofilaments, the thick myosin filaments forming the dark A band and the thin actin filaments constituting the light I bands. The actin filaments in the I bands of each adjacent sarcomere are anchored at the Z line (Fig. 2.80B and C). On cross-section through the A band, it can be seen that there is an overlap of thick and thin filaments, the thin filaments forming a hexagonal array around each myosin filament, whereas a cross-section through the I band shows only thin filaments (Fig. 2.80D). From the ends of the two Z lines bounding each sarcomere, the thin filaments protrude toward the center of the sarcomere, each measuring about 1 μm in length, while the myosin filament is approximately 1.5 μm in length; the center of the myosin filament shows a widened area which connects adjacent thick filaments to form the M line (Fig. 2.80C). (For review of ultrastructure see Legato, 1973; McNutt and Fawcett, 1974; Sommer and Johnson, 1979; Schlant and Silverman, 1986). The thick and thin filaments slide back and forth during contraction and relaxation, much as two hair brushes interdigitate when pushed together. The molecular structure of cardiac actin and myosin resembles that of skeletal muscle (for details, see Section 1, Chapter 4), (Katz, 1977). Other cardiac cells, including the cells in the nodal regions and the Purkinje cells of the conduction system, contain relatively few myofibrils (James et al., 1982).

The myocardial cells are connected to one another by special junctions called intercalated discs (Fig. 2.81). The intercalated disc exhibits a differing structure during its course between cells, the thick transverse portions (fascia adherens) being composed of Z-substance and undoubtedly serving to transmit force from cell to cell, whereas other (longitudinally placed) regions are often much thinner and in some regions exhibit so-called "nexus junctions" (Fig. 2.81). The latter regions have a particular structure shown on freeze cleaving with scanning electron microscopy to consist of a series of interdigitating pits and projections, each of which contains a hole; it is thought that these areas allow communication of cytoplasm from cell to cell and serve as low impedance pathways for rapid transmission of the electrical impulse throughout the network of heart muscle cells (McNutt and Weinstein, 1973). The desmosomes (Fig. 2.81) may serve a similar function. The branched structure of cardiac muscle facilitates this process, in contrast to skeletal muscle, which is composed of nonbranched

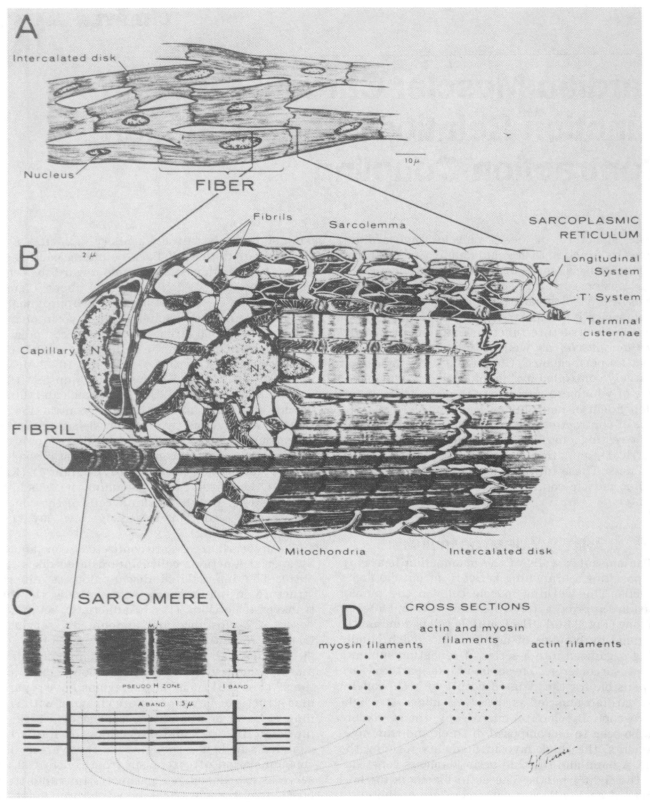

Figure 2.80. Diagram of myocardial structure. *Panel A* shows the branching network of myocardial cells together with the intercalated discs separating individual cells (fibers). The two *vertical bars* show the section of the cell with the tip of its nucleus magnified in *panel B* to show the myofibrils and transverse and longitudinal tubules (the sarcoplasmic reticulum). In *panel C* a sarcomeric unit of a myofibril is further magnified to show that the dark and light striations constituting the A and I bands, respectively, result from the arrangement of the actin and myosin filaments. In *panel D*, three cross-sections through the sarcomere are shown. One through the light I band shows only actin filaments in a hexagonal array, one through the A band *(center)* shows thick myosin filaments in the central regions of the hexagonal actin filament arrays, and a section through the light zone adjacent to the central M line *(panel C)* shows only myosin filaments. (From Braunwald et al., 1976.)

Figure 2.81. Electron micrograph of portions of two cardiac muscle cells showing in particular the intercalated disc. *S,* sarcomere; *ECS,* extracellular space (*large vertical arrowhead* shows origin of the boundary between the two adjacent cells); *Pm,* plasma membrane; *Pl,* internal plasma membrane; *M,* mitochondrion; *G,* Golgi apparatus. The features of the intercalated disc, which runs in both transverse and longitudinal directions relative to the long axis of the cell, are as follows: *FA,* fascia adherens (the thickened region of the intercalated disc into which the myofilaments insert and which serves to transmit force from cell to cell); *NSP,* a widened portion of the intercalated disk; *D,* desmosome; *N,* the nexus junction (the so-called "gap junction," which appears to serve for electrical conduction from cell to cell). (From Legato, 1973.)

muscle cell (motor) units which are activated by voluntary nerve endings (Section 1, Chapter 4).

The myocardium is rich in mitochondria, which compose 25–30% of the myocardium (many more than in skeletal muscle), and glycogen stores are plentiful (Figs. 2.80 and 2.81).

SARCOMERE LENGTH-TENSION RELATION

The sliding filament model for muscle was first developed in 1958 for frog skeletal muscle by Huxley and Niedergierke (for review see Huxley, 1974). Much experimental evidence supports this model, which states that during contraction and relaxation of muscle the myofilaments remain at *constant length* as they slide back and forth between one another.

Relation Between Active Tension and Length

It has been observed in single frog skeletal muscle fibers that there is a relation between the length to which the sarcomeres within the fiber are stretched under resting conditions and the tension developed by the fiber when it is then stimulated electrically to contract. The relation shows that as sarcomere length (the distance from Z line to Z line) is increased from about 1.7 to 2.2 μm the developed tension increases (the so-called "ascending limb") (Fig. 2.82); between 2 and 2.2 μm there is a plateau of the curve, and between 2.2 and 3.5 μm tension falls as sarcomere length increases (the so-called "descending limb") (Fig. 2.82). As in skeletal muscle, a sarcomere length-tension relationship can also be demonstrated in intact heart muscle (Fig. 2.83), but a descending limb is not evident in living muscle (Julian and Solling, 1975). This finding is related

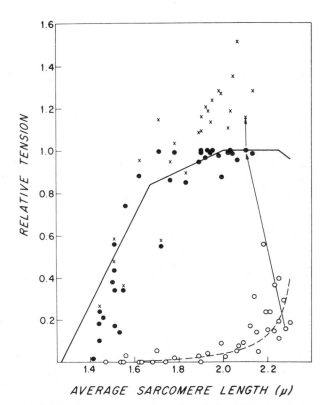

Figure 2.83. Sarcomere length-tension relation in living papillary muscles of the rat. The resting length-tension relation is shown by the *open circles*, the total active tension developed by the *X*s, of a descending limb of the length-active tension relation. (From Julian and Solling, 1975.)

primarily to the greater stiffness of heart muscle in the relaxed, unstimulated state, as discussed further below.

The shape of the sarcomere length-tension relation in skeletal muscle has been attributed to the degree of overlap of active sites or cross-bridges on the thick and thin filaments; thus, it is known that the central 0.2-μm region of the thick filament is free of active sites, thereby explaining the plateau of the length-tension relation in single fibers (Fig. 2.82). The mechanism for the descending limb observed in isolated skeletal muscle can be explained by the progressive decrease in overlap of the regions of active sites on the thick and thin filaments as the actin filaments are progressively disengaged (Fig. 2.82), and tension reaches zero in single fibers at a sarcomere length of approximately 3.5 μm where no overlap exists (Fig. 2.82).

On the other hand, the explanation for the ascending limb in both skeletal and cardiac muscle is much less clear. In the region of sarcomere lengths between 1.5 and 2.0 μm, there is "double overlap" of thin filaments with thick filaments, since below 2.0 μm these 1-μm-long thin filaments cross the M line and interdigitate with each other

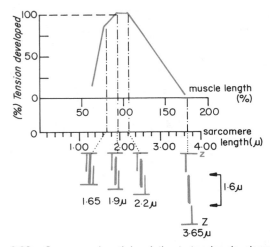

Figure 2.82. Sarcomere length in relation to tension development for skeletal muscle. Each thin (actin) filament is 1.0 μm in length, and in frog skeletal muscle the thick (myosin) filaments are 1.65 μm in length. Developed tension is constant between sarcomere lengths of 2.0 and 2.2 μm and falls at sarcomere lengths below 2.0 μm or above 2.2 μm. (Redrawn from Hanson and Lowy, 1965.)

(Fig. 2.82). This double overlap can be seen as a darker region in the A band at short sarcomere lengths on electron micrographs. The rise of tension on the ascending limb has been attributed to lessening of interference with tension development by progressively decreasing extent of double overlap of the actin filaments; also, the possibility of wider separation between thick and thin filaments due to cell shape changes with increasing distance in shorter cells has been postulated to play a role.

However, there is now evidence, particularly in isolated skeletal muscle, that, as muscle length is increased along the ascending limb of function, there is increased activation of the muscle and vice versa. Unlike skeletal muscle, intact cardiac muscle is not fully activated during contraction. When it is fully activated, as by high Ca^{2+}, the ascending limb of the sarcomere length-tension relation is relatively flat like that of skeletal muscle, but in the intact state the partially activated muscle exhibits a much steeper length-tension curve, suggesting that mechanisms in addition to simple changes in the overlap of the myofilaments are operative. Evidence is now available in cardiac muscle indicating that the increased steepness of the sarcomere length-tension relation is due both to a delayed increase in Ca^{2+} release from the sarcoplasmic reticulum following an increase in resting muscle length (and vice versa) and perhaps more importantly to increased *sensitivity* of the myofilaments to Ca^{2+}. This may be due to a change in the affinity of troponin for Ca^{2+}, although this mechanism is not yet established. Thus, the change in muscle performance resulting from changing resting muscle is related not only to the Ca^{2+} supply but also to the Ca^{2+} sensitivity of the myofibrils, and this so-called "length-dependent activation" may be a more important determinant of cardiac muscle contraction than the degree of overlap of active sites on the thick and thin filaments (for reviews, see Allen and Kentish, 1985; LaKatta, 1987) (see Fig. 2.86).

Regardless of the precise mechanisms for the ascending limb, the phenomenon of increasing sarcomere length associated with increased force of muscle contraction forms the ultrastructural basis for the Frank-Starling relation or "Starling's law of the heart," one of the intrinsic mechanisms regulating the function of the whole heart.

Relations Between Passive (Resting) Tension and Muscle Length

In skeletal muscle, tension does not rise appreciably as the resting muscle is stretched until sarcomere lengths exceed approximately 3 μm (Spiro and Sonnenblick, 1964). However, in isolated cardiac muscle (as well as in the whole heart), as the resting

muscle is passively stretched, tension in the muscle initially rises slowly and then increases rapidly above sarcomere lengths of 1.8 to 2 μm (Fig. 2.83, *open circles*). Resting tension in heart muscle is relatively low at physiological muscle lengths (sarcomeres below 2.2 μm), the unactivated resting muscle being freely extensible, but the muscle becomes extremely stiff when it is stretched further (Spiro and Sonnenblick, 1964; Spotnitz et al., 1966). Thus, there is limit in the degree to which cardiac muscle can be stretched, since at sarcomere lengths of 2.3 to 2.4 μm any further stretch is resisted by the very high resting force (Fig. 2.84).

As we shall see, this is a useful property since it prevents overdistension of the heart during changes in the return of blood to the heart (otherwise, the heart might readily be forced into a "descending limb" of function). This ability to develop high resting tension may relate to the connective tissue content of heart muscle and its branching interconnecting structure (which differ from skeletal muscle). A complex extracellular col-

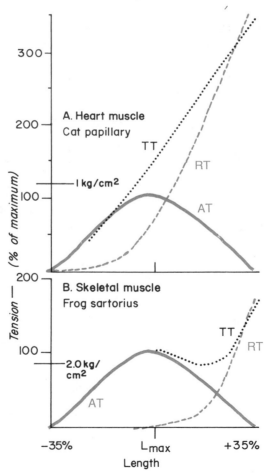

Figure 2.84. Length-tension diagrams for heart (*A*) and skeletal (*B*) muscle. *RT,* resting tension; *AT,* actively developed tension; *TT,* total tension (AT + RT). (Redrawn from Spiro and Sonnenblick, 1964.)

lagen matrix connecting and tethering myocytes to each other, and to capillaries, has been demonstrated in heart muscle by scanning electron microscopy (Caulfield and Borg, 1979). In skeletal muscle which is attached to bone, the resting tension is still low at long sarcomere lengths, and a descending limb of active tension is readily produced (Fig. 2.84). Other structural and functional differences between skeletal and cardiac muscle are shown in Table 2.6 (for review of skeletal muscle physiology see Section 1, Chapter 4).

EXCITATION-CONTRACTION COUPLING

The German physiologist Ringer discovered in the 19th century that, when calcium was removed from the medium bathing an isolated frog heart, electrical activity recorded by an electrogram from the heart was not affected, but contraction of the heart ceased. The calcium ion (Ca^{2+}) is now known to be the key link in the coupling of electrical excitation of muscle to its mechanical contraction (see also Section 1, Chapter 4). Subsequently, it was shown that local electrical stimulation of the sarcolemma of a muscle fiber at a region near the Z line produced local contraction of adjacent sarcomeres (Huxley, 1959) and also that insertion of a micropipette through the sarcolemma with local infusion of calcium within the cell caused local contraction of nearby myofibrils. (For general reviews on calcium and excitation-contraction coupling in muscle see Ebashi, 1976; Winegrad, 1979; Braunwald et al., 1988.)

Further insight into how the concentration of calcium affects contraction is provided by studying the adenosine triphosphatase (ATPase) activity of reconstituted cardiac actomyosin made with actin containing the tropomyosin/troponin complex (Fig. 2.85). When the calcium concentration is changed from about 10^{-7} to 10^{-5} M, myosin ATPase is markedly stimulated. This contrasts with the calcium sensitivity of myosin ATPase alone (Fig. 2.85) and illustrates that troponin and tropomyosin confer "calcium sensitivity" to the ATPase (Ebashi, 1976). As described in Section 1, Chapter 4, the long rod-shaped molecule of tropomyosin is anchored to actin along with the globular protein troponin and, when calcium is bound to troponin, the tropomyosin undergoes a conformational change which shifts it into the groove between the two actin strands, thereby uncovering active sites and allowing myosin filaments to form reversible bonds with actin filaments. The calcium binds to one component of troponin (TN-C) which releases another troponin component (TN-I, the inhibitory component) from actin, causing tropomyosin to shift to a nonblocking position. The precise mechanisms by which adenosine triphosphate (ATP) is hydrolyzed by the ATPase bound to myosin as active sites form and break and how this interaction is transmitted into mechanical work of the muscle remain to be fully established. (For reviews see Noble and Pollack (1977) and Gevers (1986).) However, during contraction the myosin cross-bridge changes its angle with release of adenosine diphosphate and phosphorus, and the cross-bridge bond to actin is broken when ATP again binds to myosin (thus, ATP is required for relaxation). For further details see Huxley (1974) and Katz (1970).

The importance of calcium concentration in the contraction of cardiac muscle is further demonstrated by use of muscle fiber preparations from which the sarcolemma is removed. If a strip of isolated cardiac muscle is placed in glycerol for many hours (alternatively, muscle fibers can be "skinned" by a detergent), most of the membranes and low molecular weight compounds including ATP go into

---

Table 2.6.
A Comparison Between Skeletal and Cardiac Muscle

	Skeletal Muscle	Cardiac Muscle
1. Innervation and structure	a. Motor end-plates (Ach) and motor units b. No autonomic fibers to muscle c. Small T tubules	a. Functional syncytium with propagated impulse throughout b. Sympathetic and vagal fibers to muscle c. Large T tubules and cell-to-cell electrical transmission
2. Alterations in performance	a. Length-tension relation b. Ability to summate contractions c. Force-velocity relation d. Recruitment of more fibers e. Does not alter inotropic state (except temperature) f. Contractility not responsive to external Ca^{2+} and drugs	a. Length-tension relation b. Cannot summate contractions c. Force-velocity relation d. Cannot recruit more fibers e. Inotropic state readily altered (can shift length-tension and force-velocity curves) f. Contractility responsive to external Ca^{2+}, drugs, catecholamines
3. Stress-strain curve resting muscle	a. Compliant at normal resting length	a. Relatively noncompliant at normal resting length

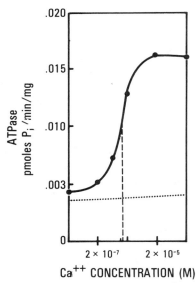

Figure 2.85. Calcium sensitivity of reconstituted cardiac actomyosin which includes the tropomyosin/troponin complex *(solid line)* expressed as ATPase activity. Note the marked increase in ATPase activity between calcium (Ca^{2+}) concentrations of 2×10^{-7} and 2×10^{-5} M. The *dotted line* shows the lack of calcium sensitivity of myosin alone, and the *vertical dashed line* indicates the calcium concentration at half-maximum activation. (Redrawn from Katz, 1970.)

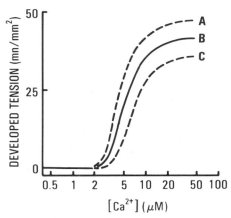

Figure 2.86. Relation between Ca^{2+} and developed tension in skinned cardiac muscle showing changes in calcium sensitivity at different resting muscle and sarcomere lengths. The *middle curve (solid line, B)* shows the increasing tension developed at a fixed resting sarcomere length of 1.9 μm as Ca^{2+} is increased progressively from below the level needed for activation up to maximum muscle activation. The *upper curve (dashed line, A)* shows the effect of the same range of calcium concentrations at an increased resting muscle length (2.0 μm), and the *lower curve (dashed line, C)* shows the effects at a decreased resting sarcomere length (1.8 μm). (Modified from Allen and Kentish, 1985.)

solution, whereas the contractile proteins remain in relatively normal structural relationships. If a skinned or glycerinated muscle is then bathed in buffer containing magnesium and ATP, it does not contract as long as the calcium remains low. However, as the calcium concentration is progressively increased the tension is elevated (Fig. 2.86), with the sigmoid-shaped curve showing the most rapid increase of tension development between 10^{-5} and 10^{-7} M calcium (resembling the sigmoid curve of ATPase activation by calcium). Such curves can be used to demonstrate altered sensitivity of the myofilaments to Ca^{2+} due to changing muscle length, mentioned above. Thus, displacement of the curve upward and to the left is observed at increased muscle (and sarcomere) lengths (increased sensitivity), and it is displaced downward at shorter muscle lengths (Fig. 2.86) (Kentish et al., 1983).

"CALCIUM CYCLE" IN CARDIAC MUSCLE

The structural basis for the transmission of the electrical impulse to the interior of the myocardial cell, where it promotes calcium release, is shown in Figure 2.80*B*. From the surface of the cell, narrow, hollow tubes which are direct extensions of the sarcolemma project inward, penetrating the intracellular space. These so-called transverse tubules (T tubules) abut on the Z bands of the sarcomeres and

carry electrical depolarization inward. In heart muscle, the T tubules are much larger than in skeletal muscle (consonant with the important role for transmitted electrical activation in cardiac muscle).

The sarcoplasmic reticulum constitutes a separate system of tubules; this network of small interconnecting tubules surrounds the sarcomeres (Fig. 2.80*B*, longitudinal system) and contains a high concentration of calcium due to the activity of an ATP-utilizing calcium pump. The sarcoplasmic reticulum does not connect directly with the sarcolemma or the T tubules, which are continuous with the sarcolemma, but its flattened saccular structures (cisternae or lateral sacs) abut the sarcolemma and T tubules (Fig. 2.87). In cardiac muscle, where one or two lateral sarcotubules wrap around a transverse tubule, the appearance in cross-section is that of a "diad" or "triad," whereas the skeletal muscle triad shows larger lateral sacs and smaller transverse tubules (Fig. 2.87).

TRIAD
CARDIAC MUSCLE

TRIAD
SKELETAL MUSCLE

Figure 2.87. Diagrammatic concepts of triadic and diadic relationships in cardiac and skeletal muscle sarcomeres. *LS,* lateral sac. (Redrawn from Legato, 1969.)

The movements of calcium during a single cardiac cycle are presented diagrammatically in Figure 2.88. In brief, a relatively small amount of calcium moves inward across the sarcolemma during the action potential, most of the calcium needed for muscle activation being rapidly released from the sarcoplasmic reticulum. This release mainly from the sarcoplasmic reticulum is triggered by the transsarcolemmal Ca^{2+} influx (Fig. 2.88) and raises the intracellular Ca^{2+} concentration sufficiently to bring about muscle contraction. Following contraction the calcium is rapidly reaccumulated by the sarcoplasmic reticulum, causing muscle relaxation, and calcium also moves out of the cell across the sarcolemma in exchange for sodium during diastole and by other mechanisms discussed below.

Inward Transsarcolemmal Calcium Movement

Depolarization and repolarization during an action potential produce a sequence of ion fluxes involving primarily sodium, potassium, and calcium (Chapter 8). First, sodium enters (fast inward current), followed by an influx of calcium (slow inward current) (Reuter, 1974). During the rest phase of the cardiac cycle, the Ca^{2+} concentration is high outside the cell (10^{-3} M or more) relative to the very low free Ca^{2+} concentration within the cell (although the mean intracellular calcium content is much higher due to that stored in the sarcoplasmic reticulum and bound elsewhere). Therefore, when the voltage-dependent Ca^{2+} channels are opened during the action potential (Fig. 2.88), Ca^{2+} flows down its concentration gradient, entering the cell through these channels in the cell membrane (including those in the T tubules). There are Ca^{2+} binding sites on the inner surface of the sarcolemma which are important in allowing rapid Ca^{2+} release (lanthanum competes for these sites) (Langer, 1978). Toward the end of the action potential, calcium influx ceases and repolarization is accompanied by potassium efflux. The Na^+/K^+-ATPase (the "sodium pump") in the sarcolemma is activated when both Na^+ and K^+ are present, uses ATP, and maintains a negative electrical potential during diastole by carrying three Na^+ ions to the outside of the cell for two K^+ ions moving back to the interior of the cell. Although Ca^{2+} enters the cell during the action potential and contributes to the intracellular pool, the amount entering is much less than that needed to activate contraction (approximately

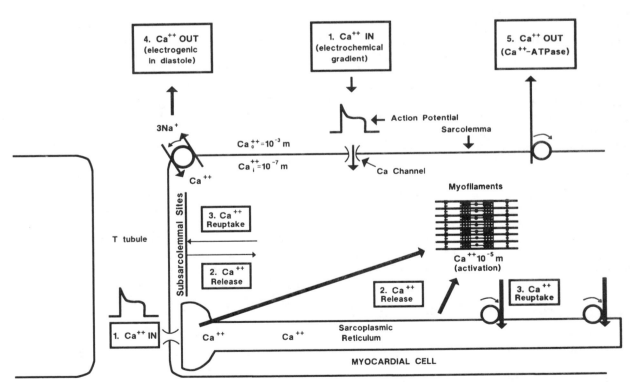

Figure 2.88. Diagram of Ca^{2+} movements during the cardiac cycle. *1,* Inward calcium movement across the sarcolemma during phase 2 of the action potential. *2,* These events trigger Ca^{2+} release from the sarcoplasmic reticulum and probably from subsarcolemmal sites. *3,* Ca^{2+} reuptake is predominantly into the sarcoplasmic reticulum by an energy-dependent pump, but some Ca^{2+} is also rebound at subsarcolemmal binding sites. *4,* Extrusion of Ca^{2+} occurs by Na^+/Ca^{2+} exchange across the sarcolemma and also by a sarcolemmal Ca^{2+} pump (*5*).

20%). However, as this calcium moves in across the sarcolemma (including the T tubules) (Fig. 2.86) it plays a significant role in contributing to the pool in the sarcoplasmic reticulum and in stimulating release of calcium from the sarcoplasmic reticulum (see below).

Release of Calcium from the Sarcoplasmic Reticulum and Cardiac Contraction

The sarcoplasmic reticulum contains much more calcium than is necessary for contraction, and only part of it is released with each cardiac cycle. Exactly how calcium release is triggered from the sarcoplasmic reticulum is not known. It has been postulated that there may be electrical depolarization of the lateral sacs. In addition, stimulation of calcium release from the sarcoplasmic reticulum by a small increase in calcium (so-called "regenerative" calcium release or "calcium-generated calcium release") has been shown to be operative (Fabiato and Fabiato, 1979) (Fig. 2.88). Negatively charged subsarcolemmal binding sites also appear to provide other intracellular rapid release sites for loosely bound calcium (Fig. 2.88).

In regulation of the heart muscle contraction, the concept of *incomplete activation* applies; that is, much more calcium is available in the sarcoplasmic reticulum (and on rapid release sites) than is actually used during a given contraction. Thus, as mentioned earlier, the myofibrils are not fully activated to achieve maximum tension development under normal conditions. Therefore, a substantial "contractile reserve" is available which may be called upon under altered conditions.

Use of bioluminescent proteins such as aequorin, which emit light when combining (reversibly) with Ca^{2+}, has elucidated mechanisms of excitation-contraction coupling. Aequorin can be loaded into individual cells and strips of cardiac muscle by microinjection, thereby allowing continuous recording of rapid intracellular changes in free Ca^{2+} concentration by light emission, together with the tension developed during electrically simulated muscle contractions. Such tracings (Fig. 2.89, *control*) show a rapid increase in myoplasmic "free Ca^{2+}" concentration shortly after the onset of the action potential, which reaches an early peak well before contractile tension is maximal and then decays more slowly (Blinks, 1986). Similar tracings have been obtained in pieces of human heart muscle obtained at the time of open heart surgery (Morgan and Morgan, 1984). Such measurements confirm directly the importance of the Ca^{2+} transient in initiating contraction, and its magnitude is important in *modulating* the force of cardiac contraction (see below).

Calcium Reuptake by the Sarcoplasmic Reticulum and Cardiac Relaxation

Reuptake of calcium by the sarcoplasmic reticulum (Fig. 2.88) is an active ATP-dependent process (Endo, 1977) and is responsible for the termination of contraction by removing Ca^{2+} from troponin-C. By this mechanism and the termination of the Ca^{2+} inward current, the so-called "cardiac relaxing system," the calcium concentration is rapidly lowered to below 10^{-7} M.

There is evidence in skeletal muscle that calcium reuptake occurs primarily in the longitudinal por-

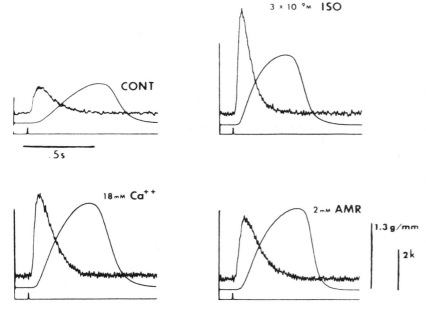

Figure 2.89. Light signals emitted by aequorin injected into cat papillary muscles, together with simultaneous recordings of isometric tension development. Light calibration (2K) indicates counts per second with a photon-counting system. Tension calibration is in grams/mm². *CONT,* control; *ISO,* isoproterenol; *Ca²⁺,* calcium; *AMR,* amrinone. The calcium transient *(upper tracing),* which follows electrical activation; *(lower tracing* shows timing of electrical stimulus), which is responsible for excitation-contraction coupling, is clearly seen in the control tracing. Note the increased light signal (indicating enhanced intracellular Ca²⁺ concentration) and increased tension development *(middle tracing)* compared to the control condition in the other panels, obtained during the steady state after administration of the various positive inotropic stimuli (isoproterenol, Ca²⁺, and amrinone). Each tracing represents a number of average responses during a steady state. (Modified from Morgan and Morgan, 1984.)

tions of the sarcoplasmic reticulum, from which it then moves to the lateral sacs for storage. However, the lateral sacs are much larger in skeletal muscle than in cardiac tissue, and it is uncertain whether this movement occurs in the heart.

The protein phospholamban, a Ca^{2+}-stimulated Mg-ATPase which spans the membrane of the sarcoplasmic reticulum, is responsible for the rapid reuptake of Ca^{2+} from the cytoplasm into the sarcoplasmic reticulum. This process occurs against a concentration gradient and requires ATP. Sarcoplasmic reticular Ca^{2+} reuptake is mainly responsible for muscle relaxation (Fig. 2.88). Relaxation is further promoted by muscle shortening during cardiac ejection, which causes decreased binding of Ca^{2+} to troponin (LaKatta, 1987).

Calcium Extrusion from the Cell (Na-Ca Exchange)

An exchange mechanism exists in the sarcolemma which *removes* calcium from the cell during the resting (diastolic) phase of the cardiac cycle (Fig. 2.88) (Langer, 1971; Reuter, 1974). Calcium concentration is relatively high outside the cell, producing a concentration gradient for calcium movement inward during the resting phase of the cycle, since intracellular calcium ion concentration is maintained at a low level by the sarcoplasmic reticulum. Na^+ competes for Ca^{2+} on the exchanger, three Na^+ ions exchanging for one Ca^{2+} ion (Caroni et al., 1980), which is therefore an electrogenic process (Reeves and Sutko, 1980). The energy for this exchange is provided mainly by the Na^+ gradient from inside to outside the cell generated by sarcolemmal Na^+/K^+-ATPase, which maintains a high intracellular Na^+ concentration (Fig. 2.88). The sarcolemmal exchange mechanism is important for two reasons: first, for maintaining an outward movement of calcium to counterbalance the inward gradient during diastole and, second, to remove calcium that enters during the action potential (Fig. 2.88). This exchange mechanism must be considered when analyzing the effects of changes in cardiac rate and other influences on transsarcolemmal calcium movements.

Any intervention which *increases the sodium gradient* will accelerate the carrier operation and *increase calcium removal* from the cell, whereas *reduction of the sodium gradient* slows the carrier and *results in calcium accumulation* inside the cell (Reuter, 1974). For example, when $[Na]_i$ increases (as occurs with inhibition of sodium/potassium-ATPase produced by digitalis), since sodium normally is higher outside the cell the transsarcolemmal gradient is reduced, the carrier slows, and $[Ca^{2+}]_i$ increases. This increased intracellular cal-

cium increases contraction, as discussed below. The same carrier mechanism operates in the giant squid axon; in this large cell, when sodium is placed directly in the interior of the cell with a micropipette (decreasing the transmembrane sodium gradient), a measurable decrease in calcium efflux follows. When $[Na]_o$ is lowered, the sodium gradient is also reduced, carrier operation slows, and, again, $[Ca^{2+}]_i$ accumulates. Increasing calcium concentration outside the cell can also increase intracellular calcium accumulation.

In addition, there is a sarcolemmal Ca^{2+} pump (Fig. 2.88), a Ca^{2+}-stimulated ATPase, which requires Mg-ATP (Caroni and Carafoli, 1981). Activation of this pump is also involved in the efflux of Ca^{2+} from the cell. It is inhibited by digitalis and may be important in other drug actions. The magnitude of its contribution to overall Ca^{2+} efflux relative to that of the Na^+-Ca^{2+} exchange mechanism is unclear at present. Slow uptake and release of Ca^{2+} by the mitochondria also occurs but may not be important in regulation of contraction.

INOTROPIC MECHANISMS IN HEART MUSCLE

Suppose a piece of isolated cardiac muscle (such as the papillary muscle of the cat) is maintained in an oxygenated bath and held at a fixed length (isometric contraction), while active tension development following electrical stimulation is measured with each twitch. Then we can very simply define a *positive inotropic influence* as one that produces an increase in the peak force developed during contraction and a *negative inotropic influence* as a decrease in the force generated under similar conditions. As we shall learn in the next chapter, a positive inotropic stimulus also increases the rate of force development and the rate of fall of force (relaxation), whereas a negative inotropic influence produces opposite effects. As mentioned earlier, the degree of activation of mammalian cardiac muscle by calcium normally is well below maximum, giving a considerable reserve for modulation of the inotropic state of heart muscle.

A number of mechanisms can affect the level of the inotropic state, thereby increasing or decreasing the force of contraction of the muscle. These include changes in the action potential itself, alterations of ion concentrations, and effects of a variety of drugs and hormones.

Action Potential Changes

When an isolated cardiac muscle is arranged so that the voltage across the membrane is held constant and the effects of various interventions on current flow across the membrane are measured (so-called "voltage clamping"), the inward calcium current during the plateau phase of the action potential can be measured. This phase of the action potential provides a "calcium gate," and it can be demonstrated that if the duration of the plateau phase is increased, or the degree of depolarization is greater, or the external calcium concentration is raised, a positive inotropic effect is produced on contraction. Conversely, shortening of the plateau phase of depolarization, a reduced degree of depolarization, or lowering of external calcium has a negative inotropic effect, leading to less active force development (Wood et al., 1969; Reuter, 1974). Certain hormones and drugs may also affect the amount of calcium entering the cell during the time of the slow inward calcium current (see below).

Effects of Changing External Ion Concentrations

As discussed, increasing external calcium concentration results in increased tension development both in isolated heart muscle and in the whole heart. Reducing external calcium diminishes the force of contraction, and lowering external sodium concentration has opposite effects (reducing the sodium gradient and promoting calcium accumulation, as discussed earlier); in fact, in the frog heart, when external calcium is markedly reduced to produce dissociation between electrical activity and mechanical contraction, if external sodium concentration is then lowered the dissociation is corrected and contraction returns. Reduction of external potassium concentration results indirectly in an increase of intracellular calcium and a positive inotropic effect. This occurs as a consequence of the ensuing decrease in intracellular potassium with a compensatory increase in intracellular sodium, leading to a decrease in the sodium gradient and decreased efflux of calcium.

It is noteworthy that in skeletal muscle, which contains a much more extensive sarcoplasmic reticulum and depends very little on transmembrane calcium influx during contraction, changing the external calcium concentration has little or no effect on the force of contraction.

Time- and Frequency-dependent Changes in Cardiac Contraction

In heart muscle, the frequency and sequence with which action potentials are generated can importantly affect the force of active contraction. These phenomena are related to a number of factors already discussed that are concerned with calcium flux and storage, and the effects are readily demon-strable in the whole heart as well as in isolated muscle. (For review, see Johnson, 1979).

Recovery of the ability to contract after an active twitch (after the electrical refractory period) has been observed to be progressive (Siebens et al., 1959). When electrically stimulated single beats are produced at progressively increasing intervals after the refractory period, the earlier (premature) beats are weaker (Fig. 2.90). Such reduced contractions are due, at least in part, to lack of full recovery of calcium stores in the sarcoplasmic reticulum and other rapid release sites. Thus, a smaller than normal release of calcium can be presumed to occur with each of these early contractions.

Changes in the frequency of action potentials per minute in cardiac muscle produce an initial slight drop in developed force which is then followed by a progressive rise in the force of contraction over the course of several beats to a new higher level of force at steady state. This response constitutes the so-called *"Bowditch staircase"* or *"force treppe"* and represents a positive inotropic effect (Fig. 2.91) (Bowditch, 1871). Conversely, abruptly slowing the frequency of contraction results in an initially slightly higher force of contraction, followed over the course of several beats by a new steady state level at a lower level of force (a *negative staircase*; i.e., a negative inotropic effect of decreased frequency). These effects have also been termed "force-frequency" relations.

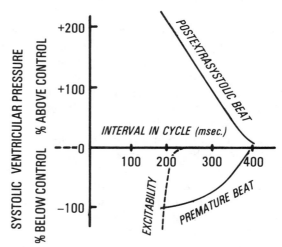

Figure 2.90 Recovery of ability to contract (systolic ventricular pressure development) in a premature beat as a function of time in the cycle after the previous beat, showing that the earlier in the cycle the weaker the premature beat. The earlier the premature beat the more forceful the postextrasystolic beat, as shown by the inverse relation between lengthening of the interval in the cycle and systolic pressure development by the postextrasystolic beat following the premature beat *(top line)*. The *dashed line* shows the recovery of electrical excitability. (Redrawn from Hoffman et al., 1956.)

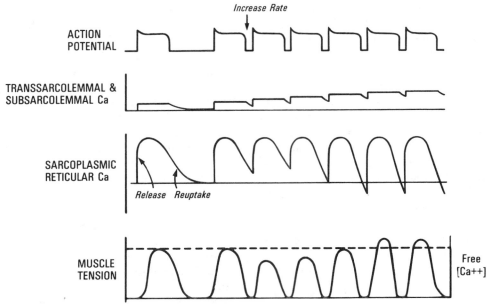

Figure 2.91.   Bowditch staircase (force-frequency relation). Hypothetical diagram of how changes in calcium uptake and release might result in a positive inotropic effect due to increased frequency of contraction of isolated cardiac muscle. The action potential of the muscle is shown at the *top* and the isometric muscle tension is shown in the bottom *tracing*. With increased rate, an increased amount of calcium per beat traverses the sarcolemma during the action potential per unit time (increased number of action potentials) and may result in more calcium storage in the intracellular binding sites; a lag in the extrusion of calcium by the sodium-calcium exchange mechanism may also occur *(second line)*. The release and reuptake of sarcoplasmic reticular calcium is shown in the *third line;* with increased rate, initially less calcium is released since reuptake is not complete and peak tension falls but, as intracellular calcium stores increase, the rate of uptake and the amount of available calcium stored in the sarcoplasmic reticulum increase *(third line)*. Therefore, the muscle tension *(fourth line)* and the amount of free calcium released per beat *(fourth line, right)* increase to a new steady-state.

Changes in calcium movement during these maneuvers are complex, and several factors are undoubtedly involved, as diagrammed in Figure 2.91. During the initial few contractions after frequency is increased, tension falls very transiently by the mechanism alluded to in the section on recovery of ability to contract. With increased frequency, there is less time available per minute during the rest phases of the cycle for electroneutral calcium extrusion by the sarcolemma, and it has also been postulated that a "lag" in the outward pumping of sodium by membrane pump (which is unable to keep up with the suddenly increased sodium load) produces a decrease in the sodium gradient and further enhances the accumulation of intracellular calcium. This is shown in simplified diagrammatic form in Figure 2.91 (Langer, 1971). These factors, coupled with the increased number of action potentials per minute, tend to produce enhanced loading of the sarcoplasmic reticulum and other rapid-release sites which reaches a steady state over the course of several contractions, resulting in a higher free Ca^{2+} concentration and more force development after each action potential (Fig. 2.91).

Postextrasystolic potentiation is a phenomenon which *follows* the occurrence of a premature electrical impulse early after a preceding contraction (in the whole heart, a spontaneous "extrasystole" may occur at this time). This produces a weak beat, followed by a marked augmentation of the active force in *the beat following the extra beat* (Hoffman et al., 1956). This positive inotropic effect then dissipates over the course of the next two or three cycles (Fig. 2.92). The degree of potentiation is related directly to the degree of prematurity of the extra beat (Fig. 2.90) (Siebens et al., 1959). Again, the changes in calcium movements responsible for this phenomenon are complex. In the whole heart, the early electrical impulse is followed by a *pause* because the next *normal* impulse reaches the heart when it is electrically refractory (Fig. 2.92). This in itself enhances the amount of calcium stored and released on the subsequent contraction, but the early electrical impulse also causes early release of an extra bolus of calcium which, along with the long pause before the next beat, leads to increased calcium uptake by the sarcoplasmic reticulum by the time of the next beat, resulting in increased contraction. In the whole heart an *additional* mechanism is involved in postextrasystolic potentiation:

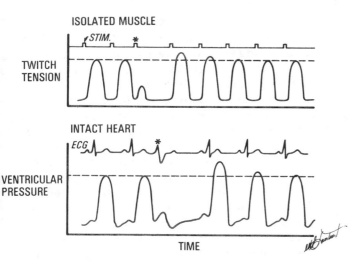

ISOLATED MUSCLE

TWITCH TENSION

INTACT HEART

VENTRICULAR PRESSURE

TIME

Figure 2.92. Postextrasystolic potentiation. *Upper tracing* shows isometric contractions of isolated cardiac muscle (electrical stimulus artifact *(STIM)* in which the third stimulus *(asterisk)* is premature. This produces a weak beat but is followed by a potentiated beat, and peak isometric force then returns to normal over three cardiac cycles. End-diastolic tension is unchanged under these conditions. In the *lower tracing*, the left ventricular pressure pulse recorded in the intact left ventricle is shown together with the electrocardiogram *(ECG)*. The third beat, a ventricular premature depolarization *(asterisk)*, produces an early beat that develops little pressure and is followed by a compensatory pause which results in an increased end-diastolic pressure on the following beat and potentiation of the systolic pressure. Two mechanisms are therefore operative in postextrasystolic potentiation under these conditions: the effect of premature electrical stimulation and the added cardiac filling of the postextrasystolic beat.

the compensatory pause following the extrasystole (see Chapter 12) results in increased filling and increased stretch on the fibers, which also results in an increased force of contraction (Fig. 2.92).

Calcium Channels, Hormones, and Drugs

Ion channels in the sarcolemma are proteins spanning the lipid bilayer which contain pores capable of being opened or closed by several mechanisms. We have mentioned the voltage-operated Ca^{2+} channel, which contains electrically charged portions that control the gating (opening and closing) of the channel in response to voltage sensors (Reuter, 1984). Other Ca^{2+} channels in the cell membrane, and the voltage-operated channels themselves, are responsive to neurotransmitters (such as norepinephrine), circulating hormones such as catecholamines, and drugs, the so-called ligand-operated or receptor-dependent Ca^{2+} channels; such channels can be gated by a number of mechanisms including direct binding of the ligand to a receptor on the channel, causing a conformational change of the channel, or secondary metabolic changes such as phosphorylation of the channel (Reuter, 1987). Such influences on the calcium channels play an important role in regulating the myocardial inotropic state and hence cardiac performance. The patch clamp technique, by which a region of cell membrane occupied by one or more receptor channels is studied with a tiny suction pipette, allows recording of local ionic currents flowing through these channels and hence indicates whether the channel is open or shut (Hamill et al., 1981).

When a β_1-adrenergic agonist such as isoproterenol (or norepinephrine) binds to its receptor on the sarcolemma, adenylate cyclase is activated (via a guanosine triphosphate-binding protein) which, in turn, causes cyclic adenosine monophosphate (AMP) formation; this binds to protein kinase with subsequent release of a catalytic subunit which phosphorylates the calcium channel (Curtis and Catterall, 1985). This phosphorylation markedly increases the time during which the Ca^{2+} channel is open, as measured by the patch clamp technique (Reuter, 1987). The net result is increased influx of Ca^{2+} into the cell, more storage in the sarcoplasmic reticulum, and a marked increase in the Ca^{2+} transient leading to augmented force of contraction (Fig. 2.89, *isoproterenol*). An increase of cyclic AMP-dependent protein kinase produced by catecholamines *also* increases the Ca^{2+} sensitivity of the Ca^{2+}-ATPase of the sarcoplasmic reticulum, thereby speeding Ca^{2+} reuptake and the rate of muscle relaxation (Katz, 1977).

Drugs such as the xanthines which inhibit phosphodiesterase lead to increased contractility by inhibiting the breakdown of cyclic AMP, and some newer positive inotropic drugs, such as amrinone, appear to mediate their stimulating effects on contraction mainly by inhibiting an isoenzyme of phosphodiesterase (Scholz and Meger, 1986), thereby increasing cyclic AMP and causing increased intracellular Ca^{2+} (Fig. 2.89, *amrinone*).

An increase of extracellular Ca^{2+} causes an increase in the Ca^{2+} transient and also an increase in the force of muscle contraction (Fig. 2.89).

Digitalis glycosides, as mentioned, inhibit the sodium/potassium-ATPase in the sarcolemma, slow sodium extrusion by this active pump and therefore diminish the sodium gradient (Schwartz, 1976); this leads to increased intracellular calcium and a positive inotropic effect (increased peak isometric force).

SUMMARY OF THE ROLE OF CALCIUM IN INOTROPIC STATE CHANGES

Based on studies in isolated cardiac muscle, Wood et al. (1969) proposed definitions of the relationships between calcium and inotropic state changes which may be modified as follows:

1. The amount of calcium bound to rapid release sites in the sarcoplasmic reticulum and subsarcolemmal regions just prior to the action potential determines the level of inotropic state or contractility of the following muscle contraction (amount and rate of tension development).
2. The amount of calcium bound at these sites at any moment depends on many factors, including:

a. The characteristics and shape of prior action potentials (which affect the influx of calcium ions during the slow inward calcium current). In the whole heart, however, such changes in action potential shape are relatively minor, and the major changes due to electrical phenomena result from mechanism b below.
b. Prior intersystolic intervals, that is, the frequency of contraction or "force-frequency relation" and postextrasystolic potentiation.
c. The effect of ions on the *outward* calcium transport system (sodium-calcium exchange). This includes such factors as the external calcium and sodium concentrations.
d. Cyclic AMP-mediated effects on calcium channels, modifying transmembrane calcium influx.

CHAPTER 11

Mechanical Performance of Isolated Cardiac Muscle

The performance of the whole heart is importantly affected by the loads imposed upon it due to changes in the level of blood pressure and the return of blood from the various organs. Such changes in loading affect the performance of the muscle composing the whole heart in the same way that they influence an isolated piece of muscle subjected to varying loading conditions. In addition to the intrinsic factors discussed in the last chapter, a number of extrinsic factors (also largely involuntary) can affect the performance of the intact heart, including sympathetic (adrenergic) and parasympathetic (cholinergic) nervous stimuli and catecholamines circulating in the blood, which influence myocardial inotropic state, blood pressure, and the heart rate. All of these responses can be better appreciated if the responses of isolated muscle to altered loading conditions and inotropic state are understood (Sonnenblick, 1962).

The term "active state" was sometimes used in the past to describe muscle activation and the level of inotropic state and may be considered in terms of the presence of a sufficient level of free calcium to trigger and maintain contraction. The time course of the "active state" does not follow exactly the level of tension developed by the muscle (Edman, 1968), since muscle activation onset precedes tension development. In addition, the level of tension builds up more slowly than the level of "active state," since tension development in the muscle is delayed by the stretching of elastic components within the muscle and its attachments.* A muscle can perform

only a few tasks: it can develop force or tension, it can shorten with no load, or it can shorten and lift a load (perform work). (For reviews on the mechanics of muscle contraction see Sonnenblick, 1962, 1966; Braunwald et al., 1976; Fung, 1981; Brady, 1984).

ISOMETRIC CONTRACTION (TWITCH) OF ISOLATED CARDIAC MUSCLE

The type of contraction of isolated muscle that is simplest to understand is the so-called *isometric contraction,* in which the muscle length is held fixed throughout contraction (Fig. 2.94). Under these conditions, when the resting muscle is electrically stimulated to contract, it develops force, but it cannot shorten (an analogy is an unsuccessful attempt to lift a heavy object with one arm, which results in force development by the biceps muscle but no movement). Thus, tension rises, peaks, and then falls as the muscle is activated and relaxes (Fig. 2.94).

In skeletal muscle a twitch can also be produced by a brief pulse of rapid electrical impulses, or (in contrast to cardiac muscle which is electrically refractory for a time following an electrical impulse) a sustained delivery of rapid pulses can produce a prolonged muscle contraction, a so-called tetanic contraction.

Effects of Changing Muscle Length

When the resting length of cardiac muscle is increased by adding more preload, more active force is developed if the muscle is then stimulated electrically to contract isometrically. Conversely, if the resting muscle length is reduced (muscle shortened by reducing preload), less active tension is developed (Fig. 2.95). (See also ascending limb of sarcomere length-tension curve, Figs. 2.82 and 2.83). The time to peak tension remains the same in such contractions and, since peak tension increases, the speed of tension development (dT/dt) also increases (Fig. 2.95).

---

*Muscle models are beyond the scope of this discussion, but for purposes of describing the differences between active state and time course of tension development, a two-component model for muscle may be mentioned. Muscle behaves as if a passive, undamped elastic element is attached in series to the contractile elements (CEs) (Fig. 2.93). This "series elastic" element (SE) transmits force to the attachments of the muscle. The time required to stretch the SE delays and modifies the characteristics of tension development (Fig. 2.93). In some models for muscle, a parallel elastic element which bears the resting force on the muscle is also added.

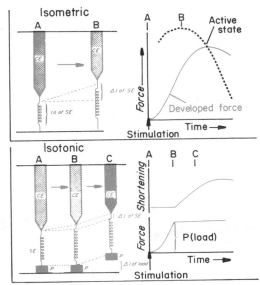

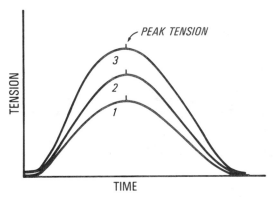

Figure 2.95. Diagram of isometric twitches from electrically stimulated isolated cardiac muscle produced by progressively increasing preloads. Twitches obtained at different times are superimposed. Notice the increase in resting muscle tension and peak tension development in beats 1–3. As peak tension rises, the time to peak tension remains constant.

Figure 2.93. Sequence of events during contraction. *Upper panel,* Isometric contraction showing events in the Hill model *(left)* and curves of force development *(solid line)* and active state *(dashed line).* A, Initial resting state. B, An instant during active contraction. The difference between the active state curve and the developed force curve is owing to the time needed for contractile element (*CE*) shortening to stretch series elastic element (*SE*). *Lower panel,* Afterloaded isotonic contraction *(left)* and curves of force and shortening *(right)* as functions of time. (Redrawn from Sonnenblick, 1966.)

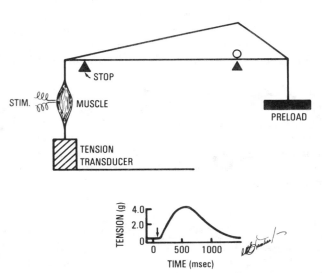

Figure 2.94. Study of isolated cardiac muscle using a lever arm arranged so that the muscle will contract isometrically. The muscle is first stretched to a given resting length by a given preload (weight), and a stop is then positioned so that when the muscle develops force it cannot shorten and move the lever arm; therefore, contraction is isometric when the muscle is then stimulated electrically to contract (*STIM*). Tension developed by the muscle is recorded by a transducer, and the isometric twitch following electrical stimulation *(arrow, lower panel)* is shown as a function of time.

This length-tension response is partly related to a change in the degree of overlap of active sites on the myofilaments although, as discussed earlier, it is possible that changes in the level of activation by calcium also play a role under these conditions (Jewell, 1977; Allen and Kentish, 1985).

Effects of Changing Inotropic State

Intensity of the active state or inotropic state of cardiac muscle (terms usually synonymous with "contractility," sometimes called "contractile state") is related in many situations to the availability of calcium and perhaps thereby to the number and frequency of active bond formations between the myofilaments. Changes in the level of inotropic state are reflected by changes in the peak tension developed in the isometric twitch (Fig. 2.96). Also, very pronounced changes occur in the *rate* of tension development and the rate of relaxation (Brutsaert et al., 1978), associated with changes in the *duration* of contraction (Fig. 2.96). Such changes due to alterations in inotropic state can occur when there is *no* change in resting muscle length. Positive inotropic stimuli (such as norepinephrine and digitalis) produce an increase in peak isometric tension, augmented velocity of both tension development and relaxation (decreased time to peak tension), and shortening of the duration of contraction (*contraction A,* Fig. 2.96). Negative inotropic stimuli (such as some anesthetic agents) produce a decreased peak tension, slowed velocity of tension development and increased time to peak tension, prolongation of the duration of isometric

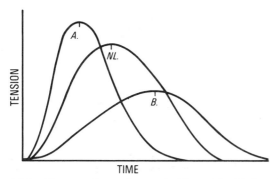

Figure 2.96. Diagram of superimposed of isometric twitches of isolated cardiac muscle initiated from the same resting muscle length and tension. *NL.*, normal twitch. *A.* shows the effect of a positive inotropic stimulus, with an increased peak tension and rate of tension development and shortening of the duration of the twitch. *B.*, Effects of a negative inotropic intervention, which depresses peak tension and rate of tension development and prolongs the duration of the twitch.

twitches, and delayed rate of relaxation (*contraction B*, Fig. 2.96).

The effects of inotropic state, or contractility, are somewhat difficult to separate conceptually from those due to the loading conditions on the heart. Thus, the degree of cardiac muscle activation by Ca^{2+} is now known to be intimately related to several intrinsic factors, including deactivation produced by active muscle shortening and length-dependent activation (discussed earlier), and nonuniformity of sarcomere lengths in isolated muscles and the whole heart also introduces complexities (Brady, 1984; LaKatta, 1987).

ISOTONIC CONTRACTION OF ISOLATED CARDIAC MUSCLE

A piece of cardiac (or skeletal) muscle can be arranged in a muscle bath using a lever arm and weights to produce a so-called isotonic contraction (Fig. 2.97). One end of the muscle is attached to a force transducer and the other end to the lever arm (which operates on a fulcrum). The opposite end of the lever arm is attached to a pan upon which various weights can be added. The following steps are used to produce an isotonic contraction.

1. A small weight (termed the "preload") is placed in the pan to produce a given degree of stretch of the resting muscle.
2. A stop is then positioned at the lever arm (Fig. 2.97) to prevent further stretching of the resting muscle when additional weights are added.
3. More weights are added to the pan (this added weight is not transmitted to the resting muscle). The total weight in the pan will then be encountered by

the muscle only when it is stimulated to contract and develop force. If this total load is not too much for the muscle to lift at a given degree of resting stretch, the pan will be lifted, and this weight when it is lifted in the air is termed the "afterload" (the load *after* the muscle starts to shorten and lift the weights in the pan, which include the preload). The ensuing contraction (Fig. 2.97) is called *isotonic* because, once the weight is lifted into the air, the muscle will shorten and then lengthen while bearing the same load (hence, isotonic).

The first portion of an isotonic contraction is isometric (since sufficient force must first be developed by the muscle to lift the afterload), and the last portion of the contraction (after the muscle has relaxed sufficiently to reposition the lever on its stop) is *also* isometric as the force falls back to the level of the preload (Fig. 2.97). Of course, if the weight is too heavy for the muscle to lift at a given preload, an isometric twitch will ensue; by definition, such a contraction is not afterloaded.

The afterloaded isotonic contraction in isolated muscle resembles to some extent the beat of the

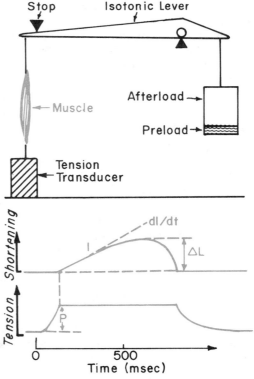

Figure 2.97. *Upper panel*, Arrangement for study of mechanics in isolated papillary muscle. *Lower panel*, Records of shortening and tension of afterloaded isotonic contraction. *P*, afterload; *dl/dt*, initial shortening velocity. (Redrawn from Sonnenblick et al., 1970.)

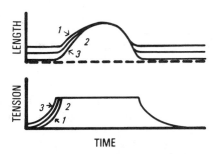

Figure 2.98. Diagram of superimposed isotonic contractions of isolated cardiac muscle in which the preload is progressively increased (beats 1–3) while the afterload is kept constant. With higher preloads, length changes (expressed as shortening from increasing resting length) become progressively larger, and the initial velocity of shortening also increases (isometric relaxation is also delayed at higher preloads, not shown).

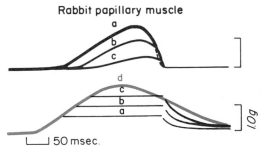

Figure 2.99. Afterloaded contractions. Relationship among papillary muscle afterload, shortening, and duration of mechanical activity. Isotonic contraction curves, *a, b, c* (*upper,* shortening scale) superimposed on the same time base with isometric myogram, *c, b, a* (*lower,* force scale). *Curve d* is the ordinary isometric contraction myogram. With progressively higher afterloads (*a, b, c, lower panel*) the extent and velocity of shortening decrease (*a, b, c, upper panel*). The more the muscle is allowed to shorten during the isotonic phase, the briefer is the isometric portion of the myogram. (Redrawn from Edman, 1968.)

intact heart, which first must develop pressure to open the valves to the pulmonary artery and aorta before it can eject blood. However, in the whole heart the load is not isotonic, since it varies as aortic pressure changes during the course of ventricular ejection. Another difference is that the ventricle ejects its contents and then relaxes at a smaller volume or muscle length, whereas the isolated muscle lifts the weight and returns it to the previous position, lengthening as it relaxes back to its initial resting muscle length. (See subsequent discussion and Fig. 2.103.)

In a single isotonic contraction, by measuring the movement of the lever arm (such as with photoelectric device), muscle length (L) can be continuously recorded, and the amount or extent of shortening (ΔL) and the rate of change of muscle length (dL/dt) also can be determined (Fig. 2.97). When only the *preload* is changed in a series of isotonic contractions, while the afterload remains unchanged, both the amount and velocity of shortening increase as the preload is increased, and they decrease as preload is reduced (Fig. 2.98).

When only the *afterload* of an isotonic contraction is changed while the preload is maintained constant, the performance of the muscle is also importantly affected, the extent and velocity of shortening both decreasing as the afterload is increased, whereas performance increases as the afterload is decreased (Fig. 2.99); thus, as might be predicted, when one attempts to lift a heavy load it cannot be lifted very far or fast, whereas if the load is lightened it is possible to lift it a greater distance at a faster speed.

FUNCTION CURVES OF CARDIAC MUSCLE

The mechanical loading conditions on an isolated cardiac muscle can be varied over a wide range,

while the muscle's contractile responses are recorded at each level of altered preload or afterload. From these responses, "function curves" can be constructed, and such curves can also be performed after an inotropic stimulus. Function curves of this type are particularly relevant to the study of function of the whole heart in the intact circulation, when loading conditions are often changing, and various pharmacological or biochemical factors can also influence heart performance. We will consider one passive curve and several "active" function curves of cardiac muscle.

Passive Length-Tension Curve

When the length of a *resting* muscle is progressively increased in a stepwise manner by progressively adding weights in preload increments, the tension rises at each step to yield the passive or resting length-tension curve (Fig. 2.100). For practical purposes, we can consider that the shape or position of this curve is *not altered* appreciably by changes in mechanical loading conditions or the ino-

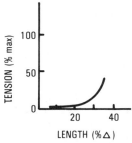

Figure 2.100. The passive of resting length-tension relation of isolated cardiac muscle.

tropic state of the muscle during a given experiment (see also Fig. 2.83).†

Isometric Length-Tension Curve and End-Systolic Length-Tension Relation

When a muscle is stimulated to contract isometrically and each contraction (or series of contractions) is produced at a progressively increased resting muscle length (higher preload), the active length-tension curve can be obtained by plotting the peak twitch tension for each twitch vs. the associated resting muscle length (Fig. 2.101). Total active tension *(dotted line)* includes the preload *(dashed line,* Fig. 2.101); "developed" tension is sometimes used in which the preload is not included (Fig. 2.101, *solid line*). The isometric length-tension curve corresponds to the ascending limb of the sarcomere length-tension curve (Figs. 2.82 and 2.83).

The isometric length-tension curve also provides a framework for isotonic contractions. Experiments in isolated muscle have shown that the length and tension at the *end* of isotonic contractions against the same afterload but from different preloads tend to be relatively independent under most circumstances of the initial length (preload) of the contraction (Fig. 2.102), but length and tension at the end of shortening are affected by the level of afterload (regardless of the initial length) (Fig. 2.102) (Downing and Sonnenblick, 1964). Thus, the tension developed by a muscle at the end of its shortening phase tends to be the *same* as the tension it would develop if it contracted isometrically from the same (shorter) resting muscle length as that at the end of shortening (Fig. 2.102). In other words, the entire isometric length-tension curve provides the *limit* for isotonic shortening, independent of preload and afterload. This *end-systolic length-tension relation* determined from isotonic contractions alone therefore is highly useful, since it represents points lying close to the *isometric length-tension curve.*

In considering such isotonic contractions in isolated muscle, it should be understood that the muscle shortens to its minimum length and then lengthens again to its original resting length while bearing

---

†This point is somewhat controversial; some investigators having reported small shifts in the passive length-tension curve under certain conditions, such as changes in muscle temperature. Resting cardiac muscle and the heart also exhibit certain time-dependent plastic or viscous properties when suddenly stretched (stress-relaxation) or suddenly loaded without a stop mechanism (creep), which are beyond the scope of this discussion (LeWinter et al., 1979). Chronic changes in the muscle (as produced by scar or hypertrophy) can markedly alter the shape or position of the passive length-tension relation.

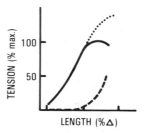

Figure 2.101. The active length-tension relation of isolated cardiac muscle. The passive length-tension relation is indicated by the lower *dashed curve*, the developed active tension by the *solid line*, and the total tension (developed tension + passive tension) is indicated by the *dotted line*.

the load (Fig. 2.103), in contrast to the mode of contraction of the heart, mentioned earlier.

Length-Shortening Curve

As mentioned, the extent of shortening of the muscle depends upon the amount of stretch of the resting muscle when the afterload is kept constant

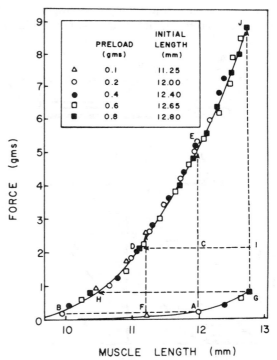

Figure 2.102. Relation between force or tension and muscle length in the papillary muscle of the cat. The resting length-tension curve is shown over a range of preloads and muscle lengths (indicated in the *inset*), and active force is plotted at increasing afterloads at a very low level (*A* to *B*), at higher levels (*G* to *H* and *C* to *D*), and for isometric contractions (*F* to *D*, *A* to *E*, and *G* to *J*). Regardless of initial muscle length, the active isometric length-tension curve tends to form the limit of shortening for isotonic contractions. (From Downing and Sonnenblick 1964.)

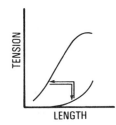

Figure 2.103. Passive and isometric length-tension curves for isolated cardiac muscle, together with the pathway of an isotonic contraction *(arrows)*. The contraction first develops tension and then shortens to the length-tension curve; it then lengthens while still bearing the afterload and relaxes isometrically.

(Fig. 2.104*A*). This can be represented on the length-tension diagram as progressively increased shortening at the same afterload, as increasing preloads cause increasing resting muscle lengths (Fig. 2.104*B*). From such data, a length shortening relation showing a positive slope (increased resting length causing increased shortening) can be plotted for any single level of afterload (Fig. 2.104*C*).

Force-Velocity Curve

This curve is determined by plotting the initial velocity of shortening, dl/dt, in a series of isotonic

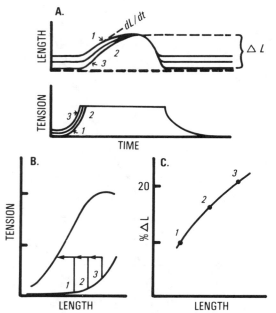

Figure 2.104. *Panel A.* Three isotonic contractions at the same afterload with increasing preloads showing progressively increasing extent of shortening (ΔL) and increased initial velocity of shortening (*dL/dt*, the slope of the initial portion of the shortening tracings). *Panel B,* Representation of *contractions 1, 2,* and *3* shown in *panel A* on the length-tension diagram, showing that, as resting muscle length increases in contractions developing the same afterload, active shortening progressively increases. *Panel C,* The relation between increasing muscle length and percentage of active shortening of the muscle (%ΔL) is plotted as a function curve.

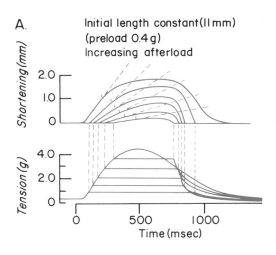

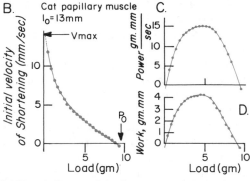

Figure 2.105. *A,* Curves of the time course of shortening and tension with increasing afterload. The initial velocity of shortening (represented by the slope of the *dashed lines* in the *top curves*) decreases with increasing afterloads. *B,* Force-velocity curve (cat papillary muscle) showing this inverse relation between afterload and initial velocity of shortening. *C,* The relation between power (load × velocity = $P \cdot \left(\frac{dl}{dt}\right)$) and load (*P*). *D,* The relation between work (load × distance = $\int_{l0}^{l1} P \cdot dl = P \cdot \Delta L$) and load. (Redrawn from Sonnenblick, 1966.)

contractions produced against progressively varied levels of afterload (Fig. 2.105*A*), the force-velocity relation being determined with each contraction beginning at the *same* resting preload and length. As the afterload is progressively increased, the velocity of muscle shortening progressively declines (Fig. 2.105*A*) until, finally, the muscle is unable to lift the weight (at that particular preload) and an isometric contraction ensues. Several features of the force-velocity curve (Fig. 2.105*B*) have been described:

1. It exhibits an *inverse* curvilinear relation. In skeletal muscle, the curve is hyperbolic and has been described by an empirical equation (Hill, 1939).
2. P_0 is the isometric tension and velocity = zero (Fig. 2.105*B*), which corresponds to one end of

the force-velocity curve. This point is common to one point on the isometric active length-tension curve at that particular resting muscle length. P_0 is altered by changing the level of inotropic state.

3. V_{max} is the velocity of muscle shortening at zero load (Fig. 2.105B). It is usually determined by extrapolation, since a muscle contracting with only its preload is not operating at zero load. V_{max} in cardiac muscle is increased by positive inotropic stimuli and decreased by negative inotropic stimuli and decreased by negative inotropic influences.‡

Force-Shortening Curve

When the resting muscle length is held constant and a series of contractions produced as the afterload is progressively increased, the extent of shortening progressively diminishes (Fig. 2.105A), yielding an inverse relation between afterload and active shortening. This may also be shown within the framework of the length-tension diagram: as afterload increases from the same preload, shortening decreases and vice versa (Fig. 2.107, *left*). Also, these data can be plotted as an inverse relation between increasing afterload and decreasing ΔL (Fig. 2.107, *right*), the force-shortening curve.

There is an unusual relationship between the afterload and the *work* performed by a muscle, where work = afterload × extent of shortening (Fig. 2.105D). Work is zero at the left end of this curve, since a large amount of shortening against zero load produces no work; the curve then rises, as the product of work and shortening increases, and

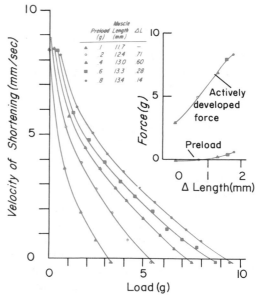

Figure 2.106. Force-velocity curves of cat papillary muscle with increasing preload which increased initial fiber length (*inset* represents length-tension curve at these preloads). Increasing preload (fiber length) does not appear to alter V_{max} but increases actively developed force at P_0 and at other loads during shortening. (Redrawn from Sonnenblick et al., 1970.)

subsequently it falls as the afterload rises further, reaching zero again at P_0, when contraction is isometric and no shortening occurs (Fig. 2.105D). Thus, the work (and power or rate of work performance, Fig. 2.105C) of the muscle is highly *load-dependent*.

Four Mechanisms That Affect the Performance of Isolated Cardiac Muscle

For practical purposes, four major factors can affect the force, speed, and extent of shortening of cardiac muscle although they are, of course, interrelated: (1) resting muscle length (preload); (2) the level of afterload; (3) the inotropic state ("contrac-

‡There is evidence that V_{max} of isolated cardiac muscle is little affected by altering resting muscle length (Fig. 2.106). It has been suggested that this might be due to an increased or decreased *number* of active sites available when muscle length is changed without alterations of the intrinsic rate or turnover of the active sites in the *unloaded* contracting muscle. (An analogy can be made to a group of runners pacing together at maximum running rate; adding more runners of like ability will not increase the speed of the group). On the other hand, as discussed in previous sections, a change in the preload *does* affect the extent and velocity of shortening of the *loaded* muscle and, when force-velocity curves are repeated at higher levels of preload, this effect results in skewing of the force-velocity curve upward and to the right with a relatively constant V_{max} obtained by extrapolation (Fig. 2.106). This displacement could also be predicted from the fact that peak isometric force is increased by increasing preload (Fig. 2.106, *inset*). Thus, at any level of afterload (other than zero) the muscle is faster and can shorten more at a higher preload, and vice versa (Fig. 2.106). (A large number of runners can jointly carry a heavy load further and faster than a few runners.) These concepts are oversimplified, however, since calcium sensitivity of the myofilaments changes with resting muscle length, as described earlier, and measurements of true $\dot{V}_{max}$ in cardiac muscle are extremely difficult because of nonuniformities and viscous and elastic properties of the muscle.

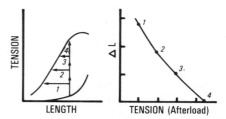

Figure 2.107. Effects of increasing afterload from a constant preload plotted on the length-tension diagram *(left panel)*. *Beats 1, 2,* and *3* show effects of increasing afterload, and *beat 4* is an isometric contraction. In the *right panel,* the extent of shortening (ΔL) in the same three beats is plotted against the tension or afterload, and an inverse relation is described.

tility"); and (4) the frequency of contraction. Contraction frequency clearly can affect the amount of work done *per minute,* as the number of contractions per minute changes. In addition, it can exert a positive or negative inotropic effect through the force-frequency relation, discussed earlier.

IDENTIFICATION OF CHANGES IN INOTROPIC STATE

A change in resting muscle length is not defined as a change of inotropic state, even though it alters muscle performance. Likewise, a change in the afterload also alters muscle performance but is not usually defined as a change in the level of inotropic state. For practical purposes, even though these two factors can alter calcium sensitivity and its availability to the myofilaments, they may usefully be considered as *mechanical* determinants of muscle performance. A change in the inotropic state of the myocardium produces a change in the performance of heart muscle (force development, velocity, extent of shortening) that can be relatively *independent* of alterations in the preload and the afterload. We can now *define* a change in inotropic state as a *shift* of an entire function curve of muscle. Thus, in cardiac muscle a shift or displacement of the force-velocity, force-shortening, length-active tension, or length-shortening curve can occur through alterations in the level of inotropic state, and the myocardium thereby can *modulate* its performance through a variety of biochemical and neurohumoral stimuli. By constructing a function curve before and after an inotropic intervention, it can be determined whether the intervention had a positive or a negative inotropic effect. The most useful single curve for identifying such changes is the isometric length-tension curve and the framework that it provides for studying the extent of shortening through the length-shortening and force-shortening relations (Fig. 2.108).

A positive inotropic stimulus shifts the length-active tension curve upward and to the left so that for any resting muscle length more peak tension is developed isometrically (Fig. 2.108). More shortening occurs in isotonic contractions from the same preloads against any given level of afterload (Fig. 2.108), leading to an upward shift of the curve relating resting length to active shortening (Fig. 2.109). Also, more shortening can occur at any level of afterload (Fig. 2.108) (leading to a shift upward of the force-shortening relation) (Fig. 2.109). Opposite shifts on these curves occur with a negative inotropic stimulus (Figs. 2.108 and 2.109). In addition,

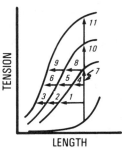

Figure 2.108. Effects of alterations of intropic state on isometric and isotonic contractions from the same resting muscle length. *Beat 10* represents an isometric contraction at the normal level of contractility, which forms a point on the isometric active length-tension relation. *Beat 7* is an isometric contraction at a depressed level of inotropic state, which reaches the depressed isometric length-tension relation, and *beat 11* is an isometric contraction in the presence of increased inotropic state, which reaches an elevated isometric length-active tension curve. At the same level of afterload *beats 1* and *3* exhibit less and more active shortening, respectively, than does *beat 2* (the control beat), and this is the case at any given level of afterload. Notice that *beat 9* shortens more than *beat 8*, while the depressed muscle is unable to shorten at that afterload.

the force-velocity relation is shifted in a characteristic parallel manner so that *both* the velocity at zero load (V_{max}) and the peak isometric tension at zero velocity (P_0) are increased with a positive inotropic stimulus (*graph 2,* Fig. 2.109). An example of force-velocity curves actually recorded from the isolated papillary muscle of the cat before and after addition to the muscle bath of the positive inotropic agent acetylstrophanthidin (a digitalis preparation) is shown in Figure 2.110.

FACTORS AFFECTING THE LEVEL OF INOTROPIC STATE

Inotropic state, the intrinsic level of myocardial contractility, is altered by many factors. These may now be summarized as the following.

Intrinsic Factors

1. The force-frequency relation provides a positive inotropic stimulus with increased frequency of contraction and vice versa. Its effect and the positive inotropic effect of postextrasystolic potentiation are mediated through variations in calcium release and/or reuptake.
2. Insufficient blood flow (ischemia) produces a negative inotropic effect due to oxygen lack, acidosis, and other factors.
3. Chronic myocardial depression or "myocardial failure" reflects a negative inotropic state due to damage to the heart muscle.

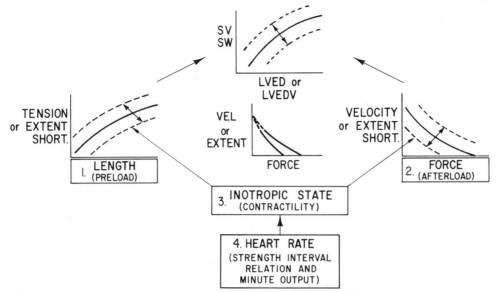

Figure 2.109. Factors influencing the performance of isolated muscle and the whole heart. The effects of the two "mechanical" determinants of performance, the resting muscle length (preload) and the tension or force (afterload), are represented by *graphs 1* and *2,* respectively. The preload affects isometric tension as well as the extent and velocity of shortening, and the afterload affects the extent and velocity of shortening. Upon these two factors plays the third important determinant, the inotropic state (*3*), which can shift the function curves of the muscle up (positive inotropic influence) or down (negative inotropic influ-ence) (*graphs 1* and *2*). The frequency of contraction or heart rate (*4*) affect the inotropic state through the force-frequency relation, but it also affects the work per minute or the output of the heart. The force-velocity or force-shortening curve also is shifted in the characteristic manner by changes in preload (shown in the *center insert;* see also Fig. 2.106). The *top panel* indicates that function curves of the whole heart relating left ventricular end-diastolic pressure or volume (*LVED* or *LVEDV*) to the stroke volume and stroke work (*SV* and *SW*) are also shifted by positive and negative inotropic influences.

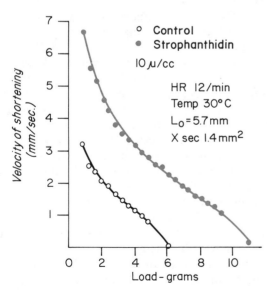

Figure 2.110. Strophanthidin effect on force-velocity relation (strophanthidin is a digitalis preparation). Both V_{max} and maximal force of contraction are increased *HR*, heart rate. (Redrawn from Sonnenblick et al., 1970.)

Extrinsic Factors

1. Local release of norepinephrine into the myocardium due to activation of the adrenergic nerve acts as a positive inotropic stimulus.
2. Release of acetylcholine due to activation of the parasympathetic nerve acts as a negative inotropic stimulus in atrial muscle and to a small extent in the ventricular muscle as well, although the innervation of the ventricles by the parasympathetic nervous system is sparse.
3. Pharmacological agents of many types affect the level of inotropic state: digitalis, amrinone, and sympathomimetic agents such as isoproterenol have a positive inotropic influence, whereas many antiarrhythmic drugs, barbiturates, calcium antagonists, and anesthetic agents have a negative inotropic effect.
4. Increases in extracellular calcium concentration produce a positive inotropic effect on cardiac (not skeletal) muscle, and vice versa.
5. Certain hormones other than epinephrine and norepinephrine, such as glucagon and thyroid hormone, have positive inotropic effects.

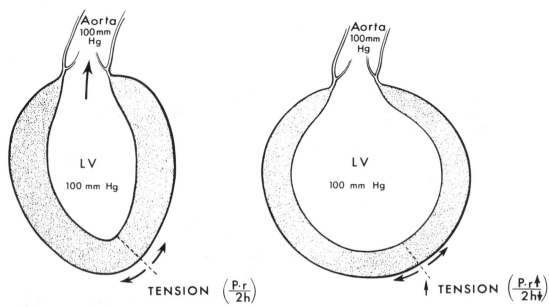

Figure 2.111. Afterload = wall tension (stress). Diagram showing how the systolic wall tension or wall stress (which represents the afterload in the myocardial fibers during left ventricular ejection) is affected by the geometry of the left ventricle (*LV*). A simplified Laplace relation relating pressure (*P*) to ventricular radius (*r*) and wall thickness (*h*) is shown. The acutely dilated ventricle on the right is developing the same systolic aortic and left ventricular pressure as the normal chamber on the left (100 mm Hg). However, since the radius of the chamber on the right is increased and its wall thickness is decreased, the wall tension is elevated compared to the chamber on the left.

ANALOGIES FROM ISOLATED MUSCLE TO THE WHOLE HEART

The mechanisms described above which influence the performance of isolated muscle also affect the performance of the whole heart. Just as the effects of these factors can be detected by examining the various function curves of muscle, by making certain analogies and assumptions the performance of the ventricles in the intact heart can be understood within a similar framework (Ross et al., 1966a). Thus, if the physiology of isolated cardiac muscle is clearly understood, insight is gained into the function of the whole heart.

In the research setting, it is possible to measure pressure in the heart and, by knowing the dimensions of the heart and its wall thickness, the force or tension developed in the myocardial wall during systole can be calculated (analogous to the afterload in an isolated muscle). Likewise, by assuming a spherical or other simplified model (such as an ellipsoid) for the shape of the left ventricle, it is possible to calculate the circumference of the chamber or its diastolic volume (analogous to resting muscle length); the end diastolic pressure in the ventricle also can be equated to the preload.

To estimate the force in the myocardial wall, one can use a simplified equation based on the law of Laplace as applied to a sphere:

$$T = \frac{P \cdot r}{2h}$$

where T = mean wall stress (analogous to afterload), P = intraventricular pressure, r = ventricular radius, and h = the wall thickness (Burton, 1957; Badeer, 1963; for more complex analyses, see Mirsky, 1974). As the heart enlarges, even if the systolic pressure in the chamber (during ejection) stays constant, the afterload (tension) will increase, since radius increases and the wall thickness must fall as the ventricle is stretched (muscle mass stays constant) (Fig. 2.111). As the left ventricle becomes smaller during the ejection of blood into the aorta, its wall thickens and, despite a small rise in the aortic pressure during ejection, the afterload or wall force will fall during ejection since it is directly proportional to chamber diameter and inversely proportional to wall thickness (that is, the normal heart "unloads itself" as it ejects).

The stroke volume (the volume of blood ejected per beat) is often measured when studying the function of the whole heart, and the stroke volume can be related to the shortening of muscle of the wall by using the equation for a sphere (volume = $4/3 \ \pi r^3$) to calculate the change in radius or to calculate circumferential shortening (circumference = $2\pi r$). Using calculations of this type, length-tension, force-velocity, force-shortening, and other relations have been determined in hearts subjected to various loading conditions and shifts in the function curves described due to changes in inotropic state (Ross et al., 1966a).

Such analogies between the function of the whole heart and that of the isolated muscle are highly useful for understanding how the performance of the heart is altered by various physiological responses, disease conditions, and pharmacological agents. They are also helpful for understanding various simplified approaches currently used for examining heart function, which often employ systolic aortic or ventricular pressure (as an index of afterload), end-systolic pressure-volume relations, and analysis of the relations between end-diastolic volume and the stroke volume.

CHAPTER 12

The Cardiac Pump

We can now consider how the heart, as a hollow muscular organ, can meet its obligations as a pump and respond to changes in venous return and loading conditions by altering ventricular volume, heart rate, and the inotropic state of the myocardium. For practical purposes, in this chapter we will consider the heart as it functions in isolated heart and heart-lung preparations, in which the influence of the brain and neurohumoral factors which affect the intact circulation are removed. In such a preparation, a reservoir or pump is available for rapid blood transfusion or removal of blood from the circulation, and a Starling resistor allows regulation of the aortic pressure and outflow resistance (Fig. 2.112) (Sarnoff et al., 1958a).

EVENTS OF THE VENTRICULAR CYCLE

The sequence of events during the entire cardiac cycle, including both the right and left atria and ventricles, will be described in Chapter 13. For the purposes of this discussion, we will consider the sequence of events only in a single chamber, the *left ventricle* (Wiggers, 1954). Electrical depolarization of the heart muscle is, of course, followed by the mechanical events of atrial and ventricular contraction. The electrical depolarization of atrial muscle, recorded by the electrocardiogram as the P wave, is followed by contraction of the right and the left atria almost together (slightly out of sequence), and after a delay in the atrioventricular node the QRS complex is followed by contraction of both ventricles (again, slightly out of sequence).

The pressure changes within the left ventricle, the flow of blood into and out of that chamber, and the ventricular volume changes during the cardiac cycle are summarized in Figure 2.113. Commencing at the QRS, the ventricle begins to contract after a brief delay (about 50 ms) and to develop pressure as it squeezes against the blood held within it. This causes deceleration of blood flowing into the ventricle through the mitral valve, leading to closure of the valve, an event which is further assisted as the

pressure continues to rise rapidly in the ventricle, while that in the left atrium stays relatively constant. Following closure of the atrioventricular (mitral) valve, as the ventricle continues to develop pressure, ventricular volume remains constant for a time, the period of "isovolumetric contraction" (Fig. 2.113). The pressure in the ventricle then reaches that in the aorta and, when the left ventricular systolic pressure exceeds the aortic pressure, the aortic valve opens and ejection of blood into the aorta begins (Fig. 2.113). Before ejection is complete, relaxation of the ventricle has already begun, and therefore the pressures in the ventricle and aorta begin to fall together, at first slowly and then more rapidly; then, as blood deceleration occurs and pressure approaches that in the aorta (recall that runoff of blood from the aorta is largely determined by the peripheral factors), the aortic valve begins to close and snaps shut as ventricular pressure falls below that in the aorta (Fig. 2.113). For practical purposes diastole begins with aortic valve closure.* Left ventricular pressure then continues to fall rapidly. This period of the cardiac cycle is called "isovolumetric relaxation," since there is no inflow to or outflow from the ventricle (Fig. 2.113).

As the pressure in the ventricle continues to fall it eventually becomes lower than that in the left atrium, the atrioventricular valve (mitral valve) opens, and blood flow across that valve commences (Fig. 2.113). With the onset of flow from atrium into the ventricle the period of *ventricular filling* commences. The flow from the atrium into the ventricle is at first rapid, then it slows somewhat (Fig. 2.113), only to speed up again as the left atrium contracts following the next P wave; the latter event leads to a rapid rise in the left ventricular end-diastolic volume and pressure (sometimes called the "atrial

---

*Some authors consider that ventricular diastole near midejection at the time of onset of ventricular relaxation, long before aortic valve closure (Brutsaert et al., 1984). Wiggers termed a briefer period during ventricular relaxation before aortic valve closure "protodiastole" (Wiggers, 1954).

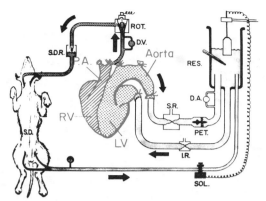

Figure 2.112.   Version of isolated heart preparation. The coronary venous return is passed through a donor dog to maintain the blood at a normal biochemical environment before it is infused into the coronary system of the isolated heart. In older versions of the Starling isolated heart or heart-lung preparation in which the coronary venous blood circulating through the myocardial wall was simply reoxygenated and returned to the coronary system, progressive deterioration in performance characteristics developed within 60 to 90 min. *S.D.,* support dog; *S.D.R.,* support dog reservoir; *SOL.,* solenoid valve electrically operated by microswitch at top of reservoir float; *S.R.,* air-filled Starling resistor; *PET.,* Potter electroturbinometer; *RES.,* reservoir; *I.R.,* water-filled inflow Starling resistance to regulate cardiac inflow; *D.V.,* venous densitometer; *ROT.,* rotameter. (From Sarnoff et al., 1958a.)

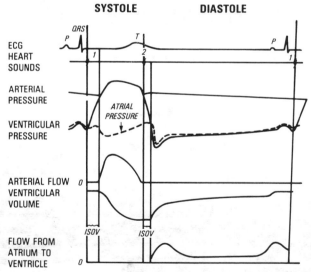

Figure 2.113.   Diagram of the relations between the electrocardiogram *(ECG,)* the heart sounds, the pressure in an artery and ventricle (either the right or left ventricle) and the corresponding atrium, flow in the artery leaving the ventricle, the volume of the ventricular chamber, and flow across the atrioventricular valve during diastole. The isovolumetric phases *(ISOV)* of ventricular contraction and relaxation are indicated. At the *far right,* at the end of diastole, the P wave is followed by a presytolic (a) wave in atrial and ventricular pressure pulses, which produces an increase in ventricular volume and an accelerated flow from atrium to ventricle just prior to the first heart sound, which occurs as the atrioventricular valve closes. For further discussion, see text; for atrial pressure, see Chapter 13.

kick" which contributes about 20% of filling volume) and reflects the "booster pump" function of the atrium (Fig. 2.113). Ventricular diastole ends with the arrival of the next QRS complex and the onset of ventricular isovolumetric contraction. Ventricular end-diastolic pressure is measured after the atrium has contracted and represents the final point of diastole, just at the onset of ventricular isovolumetric contraction; it corresponds with the largest volume or end-diastolic volume of the ventricle (Fig. 2.113).

Ventricular systole and diastole are also reflected by events in the aorta (shown as the arterial pressure pulse in Fig. 2.113). The maximum systolic pressure in the aorta, just above the aortic valve, is normally close to the peak left ventricular systolic pressure (slightly lower). The lowest pressure reached in the aorta before the next pressure wave arrives from the ventricle is called the aortic diastolic pressure. The pressure in the aorta as it crosses the ventricular pressure (at the second heart sound) (Fig. 2.113) corresponds closely with the pressure in the ventricle at the end of left ventricular ejection, which is termed the end-systolic ventricular pressure.

FUNCTIONAL GEOMETRY OF VENTRICULAR CONTRACTION

The left ventricle has a somewhat ellipsoidal shape and normally it ejects about two-thirds of its contents; that is, two-thirds of the volume present at end-diastole is ejected with each heart beat (Fig. 2.114) (Holt et al., 1968). Most of the inward movement of the left ventricular wall during ejection occurs near the minor or short axis of the ventricle, only a small reduction occurring in the long axis, giving the ventricle an even more elliptical shape at the end of systole (Fig. 2.114). As the ventricle fills during diastole it becomes somewhat more spherical, since the short axis enlarges more than the long axis.

The right ventricle has a much more irregular shape, and its free wall wraps around the interventricular septum (see Fig. 2.68, transverse plane).

PRESSURE-VOLUME LOOP FROM SINGLE CONTRACTIONS OF THE LEFT VENTRICLE

Referring to Figure 2.113 and plotting simultaneous points of the ventricular volume curve and the left ventricular pressure curve, it is possible to construct a loop, representing ventricular diastolic filling and contraction, which provides a highly useful way of describing the activity of the ventricle (Fig. 2.115), as discussed further below.

L.L. #G47952   Anteroposterior   Lateral

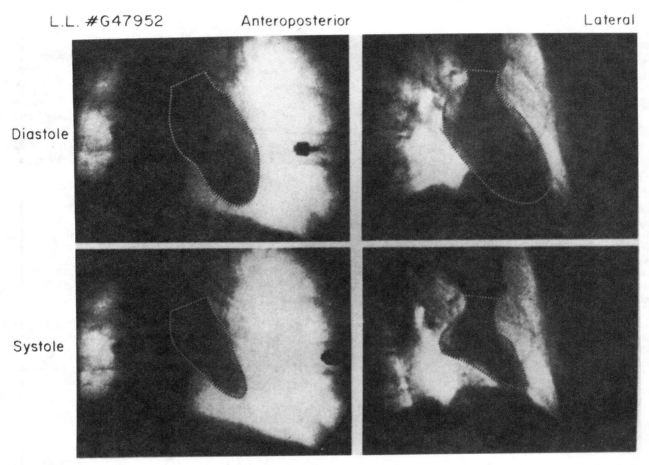

Diastole

Systole

Figure 2.114. Representative frames from cineangiograms with injection of contrast medium into the left ventricle in the frontal (antero-posterior) and lateral projections. The upper two panels show the end of diastole in these two projections, and the lower two panels show the end of systole (outlined for clarity). Note the ellipsoidal shape of the left ventricle in diastole (From Rackley et al., 1967.)

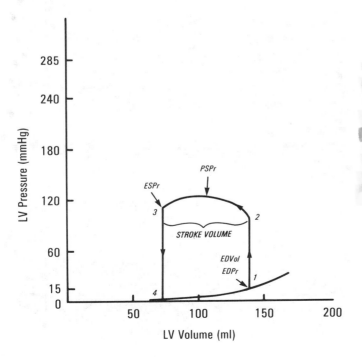

Figure 2.115. Left ventricular *(LV)* pressure-volume loop and the diastolic pressure-volume curve. During a normal contraction, *point 1* shows end-diastolic *(ED)* pressure *(Pr)* and volume *(Vol,) point 2* the onset of left ventricular ejection, which continues to *point 3* (the distance on the volume axis representing the stroke volume). This is followed by isovolumetric relaxation (*point 3* to *point 4*) and then by filling of the ventricle along its diastolic pressure-volume curve (*point 4* to *point 1*). The end-systolic *(ESPr)* pressure-volume point is indicated, and the peak systolic pressure *(PSPr)* is also shown.

Passive (Resting) Pressure-Volume Curve of the Ventricle

As with isolated cardiac muscle, the myofilaments do not appear to interact appreciably in the relaxed ventricle, and they slide freely between one another. Therefore, if blood is allowed to fill a relaxed ventricle, the shape of the resulting relationship between ventricular pressure and volume is determined primarily by the properties of the tissues supporting the myofilaments, in particular the connective tissue or collagen. As the normal left ventricle fills progressively (Fig. 2.115, *point 4* to *point 1*), a nonlinear relation between the rise in pressure resulting from each increment of volume is observed, the so-called passive pressure-volume curve. The shape of this curve is primarily due to the high compliance (low stiffness) of relaxed supporting tissues when heart volume is small and the reduced compliance (increased stiffness) of the heart wall when these tissues begin to reach their elastic limit as the volume of the ventricle continues to increase, leading to steepening of the pressure-volume relation. (It can also be influenced by other factors, including the pericardium. For review see Shabetai, 1981. See also Chapter 17 for pericardial restraining effects with intravascular volume overloading.)

As can be seen from Figure 2.116, when the diastolic or passive pressure-volume curve is normal, at low ventricular volumes large changes in volume are accompanied by only small changes in pressure, whereas at high ventricular volumes similar volume changes are accompanied by larger pressure alterations. The inverse slope of this relation at any point on the curve (dV/dP) can be called the chamber compliance, and the slope (dP/dV) represents the chamber stiffness (Fig. 2.116). The shape of this relationship has importance because it indicates that the heart is intrinsically difficult to overdistend; thus, on the steep portion of the pressure-volume curve, sarcomere lengths in the ventricular

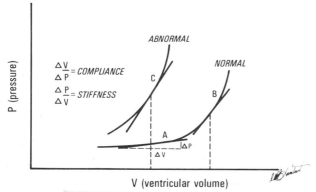

Figure 2.116. Diastolic pressure-volume relations of a normal and an abnormal ventricle. The slope drawn at *A* shows a relatively flat relation between changes of volume (Δ*V*) and the accompanying change in pressure (Δ*P*), a large change of the volume producing a small change in pressure. Since the ventricle normally operates at relatively low pressures, it is relatively compliant in this range (compliance is the ratio of volume change to pressure change). Stated another way, on that portion of the curve stiffness is low (stiffness = the ratio of pressure change to volume change, the inverse of compliance). The normal ventricle becomes less compliant (more stiff) at large ventricular volumes (slope at *B* dashed vertical line). In one type of abnormal ventricle (ventricular hypertrophy), the pressure-volume relation is steepened and shifted to the left, so that at the same ventricular volume *(dashed vertical line)* the compliance is reduced and stiffness is increased *(point C)*.

wall only slightly exceed 2.2 μm, and further efforts to increase volume result only in a very large rapid increase in pressure with little further overstretch of the sarcomeres. In the presence of heart disease, the shape and position of the diastolic pressure-volume relation can change considerably. For example, with a thickened left ventricular wall due to hypertrophy, the slope may be increased throughout its course and the pressure may be higher throughout diastole, indicating a less compliant (stiffer) ventricle (Fig. 2.116, *abnormal curve*).

In a single cardiac cycle, during the phase of ventricular filling the ventricle will tend to move up on a portion of its diastolic pressure-volume curve, as mentioned. Normally, the left ventricle operates on the lower, relatively flat portion of the curve (Fig. 2.115). (For application of this type of analysis to the human heart, see Grossman and McLaurin, 1976).

Systolic Left Ventricular Contraction

Beginning at *point 1* on Figure 2.115 (the point of end-diastolic pressure and volume), we can trace the remainder of the cardiac cycle by considering the moment-to-moment relation between left ventricular pressure and ventricular volume obtained from data shown in Figure 2.113. During isovolumetric contraction, pressure builds up in the ventricle with zero change in ventricular volume (*point 1* to *point 2*, Fig. 2.115). After the onset of left ventricular ejection into the aorta (*point 2*, Fig. 2.115) ventricular pressure rises and then falls (together with the aortic pressure), and the volume of the ventricle progressively diminishes, reaching its minimum at the point of end-systolic pressure and volume *(point 3)*. Ventricular pressure then falls at the smallest ventricular volume without further change in the volume (isovolumetric relaxation) until *point 4,* the lowest ventricular pressure, which marks the onset of ventricular filling. The sequence of events gives a counterclockwise loop (see *arrows,* Fig. 2.115), and the difference between the volume at end-diastole and that at the end of ejection is the stroke volume (Fig. 2.115).

We have previously considered the path of tension development and shortening in a piece of isolated cardiac muscle contracting isotonically. Although this type of contraction somewhat resembles the mode of contraction of the whole heart, it does not form a loop; that is, as shown in Figure 2.103, the muscle develops tension, shortens, and then begins to relax with the weight still borne by the muscle, so that it lengthens while bearing the same tension before relaxing isometrically. Of course, it would be possible to keep the muscle short while allowing tension to fall isometrically and then to stretch the muscle passively to its resting end-diastolic length, thereby even more closely imitating the contraction of the whole heart (in which ejection and aortic valve closure cause the ventricle to relax at a small ventricular volume). Recall that the shortest muscle length during active muscle shortening falls approximately on the isometric length-active tension curve of the muscle at any level of inotropic state (Fig. 2.102), a feature that also has an analogue in the whole heart (discussed below).

ISOVOLUMETRIC PRESSURE-VOLUME RELATION

If the isolated heart preparation is arranged with pressure at a very high level in the aorta so that the left ventricle cannot eject blood and if the end-diastolic volume of the ventricle is then progressively lowered and raised over a wide range, an essentially linear relation between ventricular end-diastolic volume and the maximum (peak) pressure developed by isovolumetric contractions of the left ventricle is described (Fig. 2.117, *left*) (Sagawa, 1967). (A similar linear relation can be plotted if ventric-

ular diastolic volume is plotted against calculated peak isovolumetric tension developed by the ventricle, resembling the isometric length-active tension relation in isolated muscle (Taylor et al., 1969).)† It has also been shown in the isolated heart that a positive inotropic stimulus, such as norepinephrine or digitalis, shifts this isovolumetric pressure-volume relation upward and to the left with an increase in its slope (with little change in the intercept on the volume axis), whereas a negative inotropic intervention can shift the relation downward and to the right with a reduced slope (Fig. 2.117, *left*).

Further, it has been demonstrated that the end-systolic pressure-volume point at the end of left ventricular ejection *(point 3,* Fig. 2.117, *right)* falls on or close to this isovolumetric pressure-volume relation (Fig. 2.117, *right*). This point can be measured at the aortic incisura on the aortic pulse wave marking the end of ejection (or by finding the maximum ratio of left ventricular pressure to volume). This behavior of the heart allows the passive pres-

---

†For detailed analysis of isolated hearts contracting under controlled conditions so that isovolumetric pressure-volume and tension-volume, as well as isotonic contractions, can be determined, see Covell et al., 1969; Burns et al., 1973; and Weber et al., 1976.

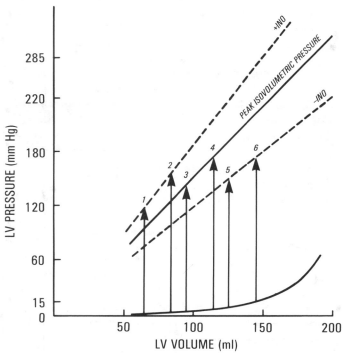 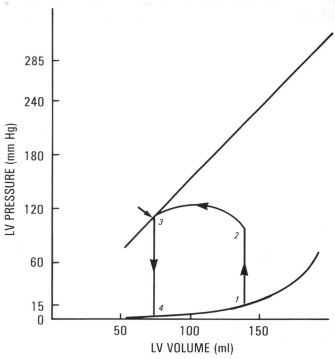

Figure 2.117. *Left panel,* Left ventricular *(LV)* pressure-volume relations, showing the relation between left ventricular diastolic volume and peak active isovolumetric pressure *(solid line,)* produced by a series of isovolumetric contractions at differing end-diastolic volumes such as *beats 3* and *4.* The active peak isovolumetric pressure-volume curve is shifted upward and to the left by a positive inotropic stimulus (+ *INO,* isovolumetric *beats 1* and *2)* and to the right by a negative inotropic stimulus (− *INO, beats 5* and *6). Right panel,* Normal passive and peak active isovolumetric pressure-volume relation combined with the pressure-volume loop. The pressure-volume point at the end of left ventricular ejection *(point 3, arrow)* falls near the isovolumetric pressure-volume relation.

sure-volume curve, the linear end-systolic pressure-volume relation, and the pressure-volume loop of the left ventricle to be combined in a single diagram (Fig. 2.117, *right*). (For review see Sagawa, 1978.)‡

FOUR MAJOR FACTORS THAT INFLUENCE VENTRICULAR PERFORMANCE

The same factors that affected the performance of isolated muscle can be shown to have a great influence on the performance of the whole ventricle (both the right and left ventricles), although only the left ventricle will be considered in this discussion. Although the mechanisms responsible for these factors are somewhat interrelated, as with isolated muscle it is useful to consider them separately; they include the *preload*, the *afterload*, the *inotropic state*, and the *heart rate* (for review see Braunwald and Ross, 1979; Weber et al., 1986) (Fig. 2.109).

Preload

In the whole heart, the preload should constitute the tension in the wall at the end of diastole (which determines the resting fiber length), but for practical purposes the ventricular end-diastolic volume or the ventricular end-diastolic pressure is used to indicate the preload. The preload affects the performance of the heart through "Starling's law of the heart." As stated by E. H. Starling (1915): "The mechanical energy set free on passage from the resting to the contracted state is a function of the

‡The maximum instantaneous ratios of pressure to volume occurring near end-ejection (elastance, E_{max}) over a range of loading conditions form a linear relation, which lies close to the end-systolic pressure-volume relation. Changes in the slope of this relation have also been found to be useful for defining a given level of inotropic state (Suga et al., 1973).

length of the muscle fiber, i.e., of the area of chemically active surfaces." This mechanism was also demonstrated in 1895 by Otto Frank in the isolated frog heart, and later Patterson, Piper, and Starling (1914) used the heart-lung preparation of the dog to show that, as the end-diastolic volume of the ventricle (inflow from reservoir, Fig. 2.112) was increased (aortic pressure being held relatively constant), the stroke volume of the ventricle increased. Hence, the law is sometimes referred to as the "Frank-Starling mechanism."

Based on the principles of contraction previously described for isolated cardiac muscle, the effects of this mechanism on the potential for active isovolumetric pressure development by the ventricle and for altered shortening of the ventricular wall are evident. As shown in Figure 2.118, if the end-diastolic volume of the ejecting left ventricle is increased and the aorta is then suddenly occluded for one beat, two phenomena are seen. First, in the ejecting beats the stroke volume and velocity of ejection are increased at the higher end-diastolic pressure and volume and, second, the peak pressure in the isovolumetric beat is augmented (Fig. 2.118, *right panel*).

This response can also be represented using the pressure-volume diagram, as shown in Figure 2.119. Both isovolumetric and ejecting contractions are shown at increasing end-diastolic volumes. At increasing end-diastolic volumes, the ejecting contractions (*beats 1, 3,* and 5) reach the same left ventricular end-systolic volume and pressure with the delivery of a larger stroke volume, as well as the potential for higher peak isovolumetric pressures (in *beats 2, 4,* and 6) (Fig. 2.119). It follows that, if ventricular end-diastolic volume is reduced, peak isovolumetric pressure and stroke volume will fall.

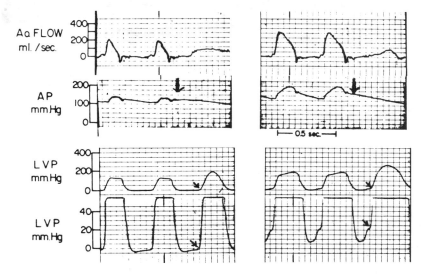

Figure 2.118. Tracings obtained in an experimental animal in which aortic flow *(Ao Flow)* was measured in the ascending aorta with an electromagnetic flowmeter, together with aortic pressure *(AP)*, left ventricular pressure *(LVP)*, and left ventricular pressure at high gain to allow visualization of the end-diastolic pressure *(small diagonal arrows, lower tracing)*. The preparation also allowed sudden occlusion of the aortic valve with a balloon *(large ventrical arrow)*, so that the third contraction in each panel is isovolumetric. In the *left-hand panel*, the left ventricular end-diastolic pressure is low (1 or 2 mm Hg) and, in the *right-hand panel*, a large transfusion has markedly elevated the left ventricular end-diastolic pressure to about 23 mm Hg. Note in the *right panel* that the two beats prior to aortic occlusion show an increased stroke volume and peak velocity of ejection despite the increased aortic pressure, compared to the *left panel*, and that the isovolumetric beat *(third beat, right panel)* develops a considerably higher peak pressure than the third beat in the *left panel*.

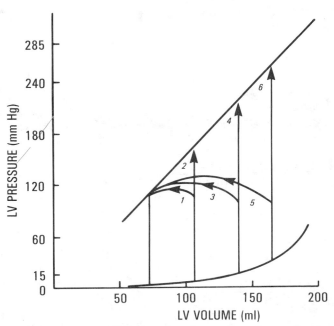

Figure 2.119.   Left ventricular *(LV)* pressure-volume diagram showing effects of progressively increasing end-diastolic volume. In isovolumetric contractions (*beats 2, 4,* and *6*) the peak pressure is increased at larger end-diastolic volumes, and at nearly the same left ventricular pressure during ejection the stroke volume progressively increases (*beats 1, 3,* and *5*).

Afterload

For practical purposes we will consider the systolic aortic pressure or systolic left ventricular pressure as the major determinant of the afterload, recognizing that the tension or force borne by the fibers of the ventricular wall (Fig. 2.111) provides a better measure of afterload (the aortic input impedance also affects the afterload; see Chapter 7).§ If an

---

§Under certain circumstances, full understanding of the factors that *oppose* ventricular ejection may require actual measurement of the force on the myocardial fibers (the true afterload) (Sonnenblick and Downing, 1963). The factors which influence wall force are complex and involve the inertial properties of the blood and the elastic properties of the aorta, both of which can change continuously during systole, as well as the size of the ventricle and the thickness of its walls. Left ventricular performance can be affected not only by changes in mean aortic pressure and resistance, but also by the compliance characteristics and impedance of the peripheral circulation, as discussed in Chapter 7 (Milnor, 1975; Noordergraaf, 1975). However, such changes do not negate the force-velocity-length framework, because changes in impedance are reflected by changes in instantaneous myocardial wall stress (afterload) which, in turn, affects the shortening of the muscle (Pouleur et al., 1980). The input impedance constitutes a means for determining those external factors which contribute to the instantaneous pressure-flow relation and thereby contributes importantly to the systolic loading conditions on the ventricle. Thus, factors affecting impedance, as well as factors relating to the geometry of the ventricle (heart size and wall thickness), determine the instantaneous afterload (force or ten-

experimental preparation is devised so that the end-diastolic volume of the ventricle is kept constant while the aortic pressure is suddenly varied for a single beat by means of a solenoid valve (Fig. 2.120), the effects of changing the afterload *alone* on left ventricular performance can be examined (Ross et al., 1966). The response of the ventricle remarkably resembles that in isolated cardiac muscle; when aortic pressure is lowered during diastole (while the aortic valve is shut), the ensuing beat encounters a lower aortic pressure, and the left ventricular stroke volume is larger and ejected at a more rapid velocity (Figs. 2.120 and 2.121). (Integration of the aortic flow velocity signal for each beat yields the stroke volume.) As the aortic pressure is sequentially raised (Fig. 2.120 and on the second beat of each two-beat sequence in Fig. 2.121), the stroke volume and peak velocity of ejection progressively fall (notice that preload, as reflected by the left ventricular end-diastolic pressure, is constant) until the final panel, in which aortic pressure is raised to a level higher than the amount of systolic pressure the ventricle can develop at that particular preload (which was set in this experiment at a relatively low level by means of a pump), and an isovolumetric contraction results (Fig. 2.121). Thus, an inverse relation exists between: (1) the level of afterload (systolic left ventricular and aortic pressure in this instance) and the extent of wall shortening (stroke volume) and (2) the maximum rate of velocity of ejection from a given level of preload (Fig. 2.121), just as in isolated muscle.

This type of response is illustrated in the pressure-volume framework in Figure 2.122. *Beat 1* illustrates a high stroke volume at a low level of left ventricular systolic pressure, *beats 2* and *3* show progressively reduced stroke volumes at higher levels of systolic pressure, and *beat 4* is an isovolumetric contraction, all beats originating from the same ventricular end-diastolic volume.

---

sion) on the muscle fibers. The calculation of the input impedance remains a research procedure, since it requires a number of assumptions as well as high-fidelity measurement of pressure and flow at nearly the same site in the aorta, but it has found application, for example, in study of the effects of changes in the peripheral circulation in heart failure (Pepine et al., 1978).

When afterload is measured as wall force or tension, we find that in the normal heart the wall force is maximum early in ejection and then rapidly falls, since ventricular size (radius) progressively diminishes and the wall thickness increases more rapidly than systolic pressure falls during later ejection. This importance of wall tension in heart failure and the application of the law of Laplace under these circumstances is discussed in Chapter 18.

AFTERLOAD

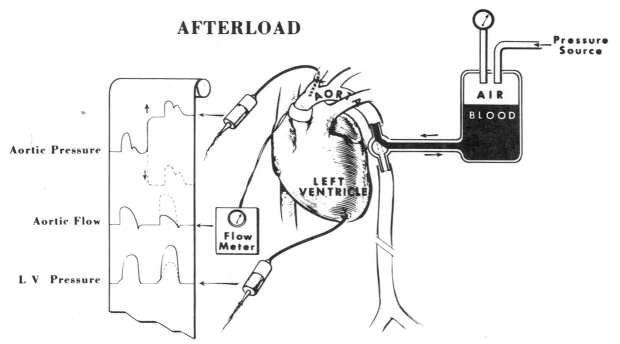

Figure 2.120. Diagram showing experiment in which the afterload in the left ventricle can be suddenly changed between beats by rapidly infusing or withdrawing blood from a reservoir attached to a compressed air source, the reservoir being connected to the aorta through a solenoid valve which is triggered from the electrocardiogram (Ross et al., 1966a). The effects of suddenly increasing or decreasing aortic pressure during diastole are shown diagrammatically in the *tracing:* increasing aortic and left ventricular *(LV)* pressure decreases aortic flow, and decreasing aortic pressure *(dashed lines)* increases aortic flow.

Inotropic State

Tracings obtained in an animal in which end-diastolic pressure (preload) and systolic pressure (afterload) were held relatively constant (by means of a pump or a reservoir and a Starling resistor, Fig. 2.112) while a negative inotropic agent was given, followed by the administration of a large dose of a positive inotropic agent (digitalis), are shown in Figure 2.123. Following the administration of the depressant drug, the stroke volume and velocity of ejection fall, and when the aorta is occluded (second beat in each panel) there is less peak pressure development during isovolumetric systole; following administration of digitalis, contractility rises above the control level with a rise in stroke volume and velocity of ejection, and there is a considerable increase in the peak rate of pressure development during isovolumetric contraction (peak + dP/dt) (Fig. 2.123).

These responses are diagrammed using the pressure-volume diagram in Figure 2.124. For clarity, the beats are drawn at slightly different left ventricular systolic pressures. *Beat 1* shows a normal stroke volume and *beat 2* an isovolumetric beat, both of which reach the normal linear relation between left ventricular end-systolic pressure and

volume. A negative inotropic intervention shifts the linear end-systolic pressure-volume relation down with reduction in its slope; *beat 5* exhibits a markedly reduced stroke volume, and there is a reduced maximum peak isovolumetric pressure (*beat 6*, Fig. 2.124). Finally, a positive inotropic stimulus shifts the end-systolic pressure-volume relation upward and increases its slope. *Beat 3* delivers a larger stroke volume than control, and *beat 4* reaches a higher peak isovolumetric pressure (Fig. 2.124).

Heart Rate

As in isolated cardiac muscle, changing the frequency of cardiac contraction has an effect on the myocardial inotropic state through the force-frequency (staircase) relation, an increase in heart rate increasing myocardial contractility. Although the effects of altered contractility through the force-frequency relation are relatively small in the whole heart, the effect of changing heart rate on overall cardiac performance per minute (cardiac output) can be very large. For example, in Figure 2.125, the filling of the heart is artificially maintained from a reservoir or by a pump to keep the end-diastolic volume constant while the heart rate is doubled from 70 to 140 beats/min. *Beat 1* represents the control condition, and in *beat 2* a slightly

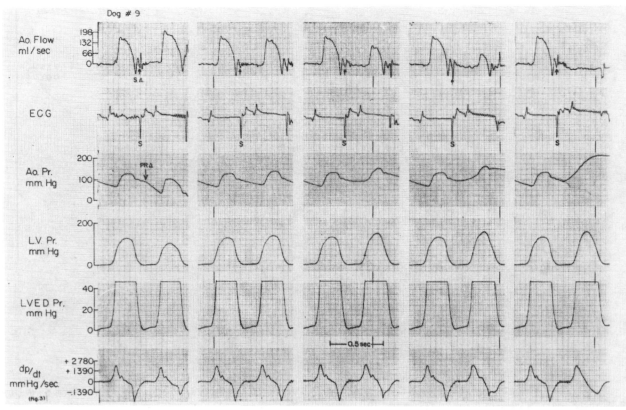

Figure 2.121. Actual tracings obtained in an experiment using the preparation shown in Figure 2.120. The first beat in each of the five panels represents a control contraction, and the second beat shows the effect of decreasing or increasing aortic pressure *Ao.*, aortic; *ECG*, electrocardiogram; *Pr.*, pressure; *L.V.*, left ventricular *LVED Pr.*, left ventricular end-diastolic pressure at high gain. The first derivative of the left ventricular pressure *(dP/dt)* is shown in the *lower tracing. S* on the electrocardigraphic tracing (and stimulus artifact *(S.A.)* on the flow tracing) shows the point of electrical triggering of the solenoid valve which causes withdrawal or infusion of blood into the aorta from the reservoir.

In the *left-hand panel*, the aortic pressure drop (*PRΔ, arrow*) resulted in a pronounced fall in diastolic aortic pressure, and the next contraction originates from the same left ventricular end-diastolic pressure but develops a lower left ventricular systolic pressure; hence, the stroke volume and peak aortic ejection velocity increased. Progressively higher aortic pressures in the *middle three panels* cause a progressive reduction of the aortic flow (stroke volume) and peak flow velocity, and in the *right-hand panel* the pressure is raised so high that the ventricular cannot eject from that end-diastolic pressure; an isovolumetric contraction results (flow rate is zero). (From Ross et al., 1966a.)

Figure 2.122. Pressure-volume diagram showing the effects of progressively increasing left ventricular systolic pressure from a constant left ventricular end-diastolic volume. There is progressive reduction of the stroke volume in *beats 1, 2,* and *3. Beat 4* represents an isovolumetric contraction (compare to Fig. 2.121). The *cross-hatched area* is used in calculating the pressure-volume area and represents the "end-systolic" potential energy," as discussed in Chapter 14.

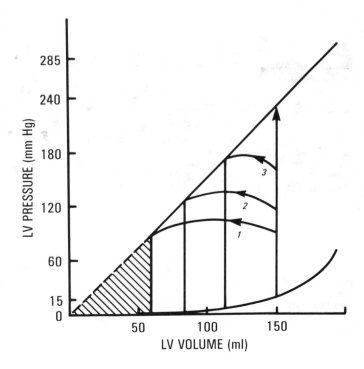

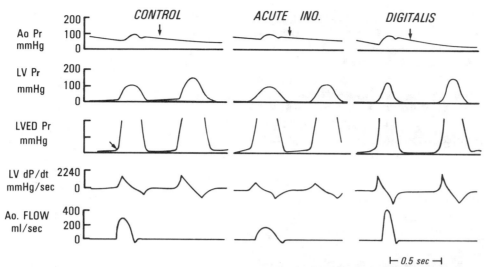

Figure 2.123. Tracings from an experiment similar to that shown in Figure 2.118 in which an isovolumetric beat could be produced by sudden inflation of the balloon in the ascending aorta, *(vertical arrows)*. The filling of the ventricle is controlled by means of a pump in this experiment, so that the ventricular end-diastolic pressure *(angled arrow)* and volume are held constant throughout. Heart rate is also constant. Control conditions are shown in the *left-hand panel,* and a negative inotropic intervention *(middle panel)* produces a decrease in the aortic flow velocity and stroke volume, a marked fall in the peak isovolumetric pressure, and a fall in the peak rate of pressure rise *(dP/dt)* of the left ventricular during isovolumetric systole. The positive inotropic effect of digitalis is shown on the *right* and results in an increase in aortic flow velocity and stroke volume, peak isovolumetric pressure, and dP/dt. *Ao Pr,* aortic pressure; *LV Pr,* left ventricular pressure; *LVED Pr,* left ventricular end-diastolic pressure.

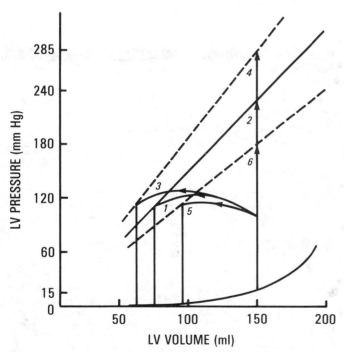

Figure 2.124. Pressure-volume diagram showing the effects of positive *(upper dashed line)* and negative *(lower dashed line)* inotropic stimuli compared to control *(solid line)*; peak pressure in isovolumetric contractions is increased and decreased, respectively *(beats 4 and 6 compared to beat 2)*, and the stroke volume is augmented with the positive inotropic stimulus *(beat 3)* and reduced with the negative inotropic intervention *(beat 5)*. *LV;* left ventricular.

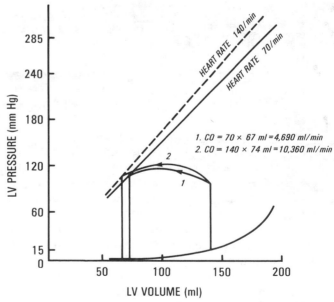

Figure 2.125. Pressure-volume diagram showing the effects of increasing heart rate while maintaining the end-diastolic pressure and volume of the ventricle constant with a pump or by transfusion. The end-systolic or isovolumetric pressure-volume relation is shifted upward and to the left by the force-frequency effect *(dashed line)*, and *beat 2* at the faster heart rate therefore delivers a slightly larger stroke volume. The effect of doubling the contraction frequency from 70 to 140/min at a costant end-diastolic volume is to more than double the cardiac output *(CO)*. *LV,* left ventricular.

larger stroke volume is delivered by each beat because of the leftward shift of the end-systolic pressure-volume relation (due to the positive inotropic effect of increased heart rate) (Fig. 2.125). Under these conditions, however, increasing the heart rate more than doubles the cardiac output (the product of stroke volume × heart rate, Fig. 2.125), indicating that the heart rate can be a powerful determinant of cardiac performance per minute under certain circumstances.

PHYSIOLOGICAL IMPORTANCE OF THE FOUR MAJOR DETERMINANTS OF PERFORMANCE

Under controlled conditions, we have varied each of the four major determinants of cardiac performance while holding the other three constant to demonstrate their separate effects. However, it is important to realize that all of the determinants are operating together in the intact circulation, and they may change in *different* directions. Also, the ventricle responds to changes in loading conditions during the course of *each* cardiac cycle. For example, if the afterload is varied *during* ejection in single beats by suddenly increasing aortic outflow resistance, the systolic wall force (afterload) during that beat progressively rises and the instantaneous fiber length and volume do not diminish as rapidly, thereby reducing the stroke volume (Fig. 2.126). Notice, however, that the ventricle still reaches the end-systolic volume-wall tension relation at the end of ejection in each beat (Fig. 2.126) (Taylor et al., 1969).

Thus, all four factors operate simultaneously in the intact animal, or in the human, to constantly modulate the heart's performance to meet different physiological conditions (changing posture, exercise, excitement) by altering the heart rate, the level of inotropic state (through the neurohumoral control of the heart and circulation), and the loading conditions (preload and afterload), as reflected in changing volumes of the right and left ventricles and the varying aortic pressure.

In Chapter 17 we consider the interaction of cardiac function and the peripheral circulation in integrated circulatory responses. However, it will be useful to consider here some additional features of the major determinants of ventricular performance and how they are expressed in the moment-to-moment regulation of the heart.

Frank-Starling Mechanism (Preload)

Starling's law operates continuously on a beat-to-beat basis to control the output of all four chambers of the heart. This mechanism is also occasionally termed "heterometric autoregulation" of the heart

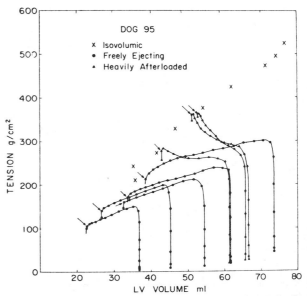

Figure 2.126.   Relations between left ventricular *(LV)* volume and calculated wall tension in an experimental animal. × peak tension developed by isovolumetric contractions at varying end-diastolic volumes. Portions of volume-tension loops are also shown, the *arrows* indicating the volume-tension point at the end of the left ventricular ejection. Notice that these points fall near the isovolumetric volume-tension relation both in freely ejecting beats from differing end-diastolic volumes, and in the heavily afterloaded beats produced by inflation of a balloon in the ascending aorta during ventricular ejection (see beats with rising wall tension during late ejection). (From Taylor et al., 1969.)

(intrinsic regulation due to changing muscle length). It is expressed importantly in the following ways.

1. Any change in the return of blood to the right or left ventricle is immediately (on that same beat) expressed as a corresponding increase or decrease in performance. For example, if one very suddenly stands up, blood will be pooled in the legs, venous return to the right ventricle will drop, and its stroke volume will immediately drop. Conversely, if one is lying in a supine posture and the legs are suddenly elevated, return of venous blood to the heart from the legs will immediately increase, and the stroke volume of the right ventricle will rise.

2. Moment-to-moment changes in the output of blood from the right ventricle, after a delay of several beats (to account for the transit time of blood through the lungs), are followed by similar changes in the output of blood from the left ventricle; that is, an increased volume of blood pumped into the lungs will subsequently reach the left heart, causing an increased left ventricular end-diastolic volume (enhanced stretch of the fibers) and thereby producing a correspondingly enhanced stroke volume of the left ventricle. Were this mechanism *not* to operate, blood would rapidly accumulate in or be

drained from the lungs, and the balance of the two ventricles could not be maintained.

This *balance* between the stroke volumes of the left and right ventricles over time is also manifested during normal respiration, and this type of response is illustrated in Figure 2.127. The velocity of flow in the superior vena cava (SVC) is recorded continuously together with the pressure in a human subject by means of a calibrated catheter-tip velocity meter and a pressure transducer; the arterial pressure is also recorded. During inspiration, increased negative intrathoracic pressure lowers the SVC pressure (by creating an increased gradient for drawing blood from the peripheral veins into the right heart), and the vena caval flow velocity therefore rises transiently (Fig. 2.127). There is then a delay as this extra volume of blood is pumped through the lungs by the right ventricle. Initially, there is a transient, small reduction in the aortic pressure and stroke volume during inspiration and an increase in heart rate (the reflex mechanisms for these responses are discussed in Chapter 16). Then, while expiration is occurring the increased output from the right heart reaches the left heart, and the left ventricular stroke volume and arterial pressure rise transiently (Fig. 2.127). Thus, the changes in stroke volume of the two ventricles are out of phase, but over a 1-minute period the integrated stroke volumes from the right and the left ventricles are equal, providing an illustration of the importance of the Frank-Starling mechanism.

3. Atrial muscle also is responsive to stretch, and as an increased volume of blood reaches the right heart or left heart it also stretches the atrium (or vice versa, if venous return decreases). This leads to variation in the force of atrial contraction with a consequent change in the "booster pump" function of the atrium (Mitchell et al., 1962; Mitchell and Shapiro, 1969). For example, when venous return to the right heart rises, increased atrial contraction helps to increase the end-diastolic volume of the right ventricle and hence its stroke volume, further assisting the heart in its roles as a "demand pump" (a pump which delivers whatever volume of blood it receives).

The atrial booster pump (Mitchell et al., 1962) can become highly important in exercise (as well as in disease). During normal strenuous exercise, when the heart rate is very high, the large number of systolic contractions per minute tends to abbreviate the duration of diastole markedly (Fig. 2.128) (time spent in systole is largely at the expense of diastolic time), despite the slight shortening of each beat produced by the positive inotropic effect. A forceful atrial contraction between each ventricular beat under these conditions greatly improves the transport of blood through the ventricles (Fig. 2.128).

4. Under abnormal conditions Starling's law may play an important compensatory role (see Chapter 18). For example, with complete heart block, when the cardiac rate is very slow (Chapter 9), filling of the ventricle increases during each prolonged diastole, and the increased end-diastolic volume therefore allows the ventricle to generate a much larger than normal stroke volume, so that the cardiac output per minute may be maintained at a level compatible with life.

5. Starling's law continues to operate even when the cardiac volume is reduced below normal (as in shock or during moderate exercise). Under mild exercise conditions or during excitement, for example, the inotropic state is enhanced, and even though overall ventricular size below the resting level, an increase or decrease of venous return to the ventricle will still result in a change in end-diastolic volume and hence in the stroke volume (Fig. 2.129).

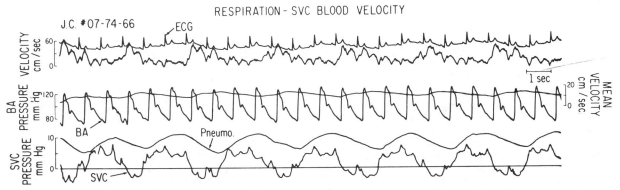

RESPIRATION - SVC BLOOD VELOCITY

Figure 2.127. Effects of respiration on flow velocity *(upper tracing)* and pressure *(lower panel)* in the superior vena cava *(SVC)* and on the pressure in the brachial artery *(BA, middle panel)* in a human subject. The pneumotachograph *(Pneumo.)* records respiration with downward movement indicating inspiration. For further discussion see text. *ECG,* electrocardiogram. (Tracings obtained by Gabe et al., 1969.)

Figure 2.128. Diagrammatic tracings showing the effect of the atrial booster pump *(arrow)* on ventricular end-diastolic pressure at rest *(left panel)* and during severe exercise *(right panel)*. During the short diastole of severe exercise, atrial transport becomes important in maintaining ventricular filling.

Afterload

As mentioned, the peak or mean systolic aortic pressure or pulmonary artery pressure during ejection provides a useful measure of the afterload. Although aortic pressure changes usually occur in the intact organism during activation of the autonomic nervous system, which also produces changes in myocardial contractility and heart rate (Chapter 16), the systolic aortic pressure alone can be varied by drug administration. Also, in some disease states rapid alterations in the afterload can suddenly occur; for example, pulmonary embolism (blood clot to the pulmonary artery) can cause a sudden increase in the afterload on the right ventricle.

1. A pure vasoconstrictor such as phenylephrine may be used to raise the blood pressure as, for example, when hypotension occurs during anesthesia, or is used purposely to produce an alteration of ventricular function in testing cardiac reserve (Chapter 17). Under the latter conditions, elevating the aortic pressure alone would *reduce* the stroke volume; however, as shown in Figure 2.130, in the intact circulation compensatory use of the Frank-Starling mechanism tends to restore the stroke volume of the normal heart to normal. If *beat 1* represents the control conditions (for simplicity, we consider a three-beat sequence) and *beat 2* encounters

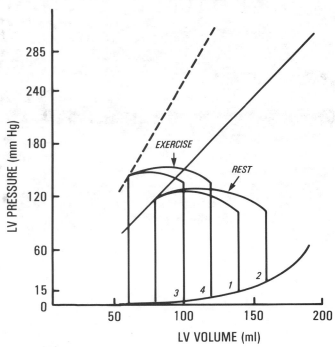

Figure 2.129. Left ventricular *(LV)* pressure-volume diagram showing the effect of increasing ventricular end-diastolic volume both at rest and during mild exercise or excitement. At rest, increasing the end-diastolic volume *(beat 1* to *beat 2)* produces an increase in the stroke volume. During exercise, the end-diastolic volume is reduced *(beat 3)* and the end-diastolic pressure-volume relation is shifted upward and to the left *(dashed line)*. If end-diastolic volume is increased *(beat 4)*, an increase in the stroke volume occurs even though the ventricle is smaller than under resting, control conditions *(beat 1)*, indicating continued operation of the Frank-Starling mechanism. Such changes might occur, for example, with deep respiration at rest and during exercise.

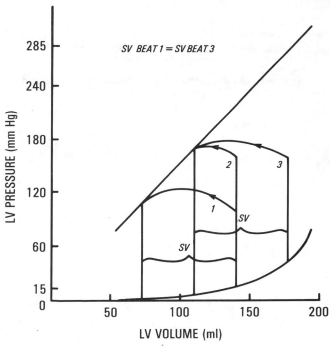

Figure 2.130. Left ventricular *(LV)* pressure-volume diagram showing the compensation for an increase in afterload. *Beat 1* shows the control condition, *beat 2* the initial response to an increase in afterload (the stroke volume falls), and *beat 3* the compensatory increase in end-diastolic volume at the higher left ventricular systolic pressure, in which the stroke volume *(SV)* is restored to normal (for further discussion, see text).

a higher systolic pressure and afterload as the drug is given, the stroke volume will drop; since the residual volume of *beat 2* is therefore increased (the volume at the end of ejection is larger than in *beat 1*) and since the right ventricle is unaffected by the vasopressor effect on the systemic circulation, the right ventricular stroke volume will remain normal and augment the diastolic filling of the next beat by a normal amount. Consequently, *beat 3* reaches a larger end-diastolic volume and the Frank-Starling mechanism restores the stroke volume to the control level, but at a higher systolic aortic pressure (Fig. 2.130).

2. "Homeometric autoregulation" is a term that has been applied to describe the response of the left ventricle to a very sudden increase in aortic pressure. It has also been called the "Anrep effect" (after the physiologist E. Anrep who first described it). In experiments in animals (under controlled conditions) in which a sudden marked increase in aortic pressure is produced, ventricular end-diastolic pressure initially rises and the stroke volume then recovers (described above) as the Frank-Starling mechanism compensates for the pressure increase. However, a gradual fall in left ventricular end-diastolic pressure than occurs in such experiments, while the stroke volume remains normal and the left ventricular and aortic systolic pressures remain elevated. This fall in end-diastolic pressure at a constant level of left ventricular systolic performance has been considered by some to represent a change in myocardial contractility.|| The term *homeometric autoregulation* (meaning at the same muscle length) has been used to distingush it from the "heterometric autoregulation" of the Frank-Starling mechanism. Whether homeometric autoregulation is of significance in the intact circulation remains controversial.

Inotropic State (Myocardial Contractility)

The level of sympathetic or adrenergic "tone" modulates the myocardial inotropic state and the heart rate through local norepinephrine release in the myocardium. Using peak left ventricular dP/dt

---

|| A length-dependent change in Ca^{2+} availability and myofilament Ca^{2+} sensitivity is discussed in Chapter 10. Such changes in isolated muscle are time-dependent. After sudden resting-length changes following the initial increase in developed tension, a gradual further augmentation in developed tension is caused by these Ca^{2+}-mediated mechanisms. Such a response has been implicated in "homeometric autoregulation" and is considered to represent a length-dependent change in contractility due to the initial increase in preload as aortic pressure is elevated (LaKatta, 1987). Others have suggested that the response represents recovery from transient insufficiency of blood flow to the inner wall of the heart caused by the sudden severe pressure change (Monroe et al., 1974).

(the maximum rate of pressure change during isovolumetric contraction) as a measure of contractility shortly *after* stimulation of the baroreceptor reflex by lowering blood pressure (when systolic and end-diastolic pressures had returned to normal), an increase in peak dP/dt is seen following the blood pressure change that is positively and linearly related to the degree of blood pressure fall (Fig. 2.131). Therefore, a decrease in the blood pressure produces a reflex positive inotropic effect on the heart muscle in the intact circulation (Chapter 16). This response was greatly attenuated by administration of the β-adrenergic blocking drug propranolol (Fig. 2.131.)

Stimulation of the vagus nerve has a marked negative inotropic effect on the atria (and slows heart rate), particularly when background sympathetic tone is elevated (as discussed Chapter 16). Under experimental conditions, a clear-cut negative inotropic effect on the ventricles of electrical stimulation of the vagus has been shown (DeGeest et al., 1965), although the effect is relatively small and its importance under normal physiological conditions is unclear, since there is little cholinergic innervation of the ventricles.

A variety of pharmacological agents, hormones, and metabolic factors also can profoundly affect myocardial performance by changing myocardial inotropic state. Also, the inotropic state of the myocardium is depressed in various types of heart failure. One of the goals of measuring performance of the heart in patients (Chapter 17) is to determine whether the function of the heart muscle is impaired compared to normal. For example, in patients with valvular heart disease, long-standing wear and tear on the muscle results in hypertrophy with some degree of scarring, and the severity of this myocardial impairment can affect whether the patient is a candidate for an operation to replace the defective heart valve. In some patients, the degree of myocardial depression (reduced inotropic state) or failure (Chapter 18) is so great that the risk of the operation becomes very high, and the benefits of valve replacement, even if the patient survives operation, are negligible.

Heart Rate

As mentioned earlier, changes in the heart rate can have profound importance in affecting the cardiac output, particularly under conditions, such as exercise, when the venous return to the heart is increased. However, when the heart rate is changed by artifical pacing (as by electrical stimulation of the right atrium with a pacemaker catheter), the cardiac output in normal subjects does not change

Figure 2.131. Experiments showing the important effect of reflex baroreceptor stimulation on the inotropic state of the left ventricle (LV), as measured by the change in the rate of isovolumetric LV pressure development (ΔLV dP/dt). The response shows the increase in dP/dt present a few seconds after correction of a *prior* decrease in systemic arterial pressure (ΔSAP). Note the positive correlation between decreasing SAP and increase of dP/dt. The increase is greatly attenuated by administration of the β adrenergic blocking agent propranolol *(dashed line)*. (From Yoran et al., 1981.)

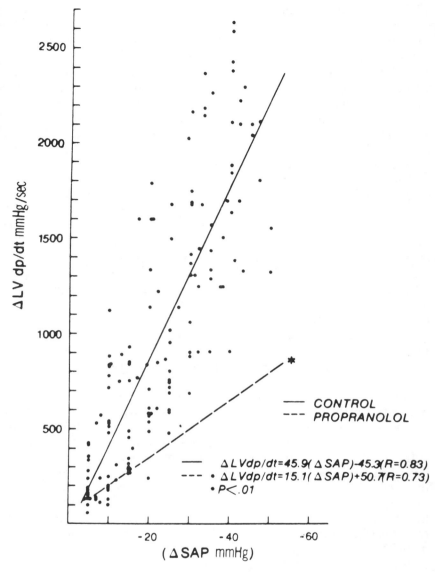

over a wide range of paced rates (Ross et al., 1965a). Under these conditions, the cardiac output is stabilized by metabolic and reflex factors and the venous return to the heart does not change, despite the altered rate and the enhanced myocardial contractility due to the force-frequency relation. (This type of response is also described as an interaction between the heart and the peripheral circulation in Chapter 17.) As shown in Figure 2.132, over a range of heart rates from about 50 to 180 beats/min the cardiac output remains relatively constant, and hence a progressive fall in the stroke volume occurs as rate is increased (cardiac output = heart rate × stroke volume) (Fig. 2.132). At higher paced heart rates, however, the cardiac output begins to fall, mainly because there is inadequate time available

for ventricular relaxation and complete cardiac filling (Fig. 2.132); also, at very low heart rates (below 50 beats/min) cardiac output falls since the ventricle reaches its maximum ability to deliver an increased stroke volume by use of the Frank-Starling mechanism (Fig. 2.132).

There is another factor that can influence cardiac contraction (and relaxation), particularly during exercise. Under resting conditions, changing the site of electrical stimulation of the heart from the normal right atrial location to a ventricular site using an electrical pacemaker has only a small effect on ventricular performance. However, with pacing of the ventricle during exercise, the marked increase in the rate of development of left ventricular pressure (dP/dt) which normally occurs during

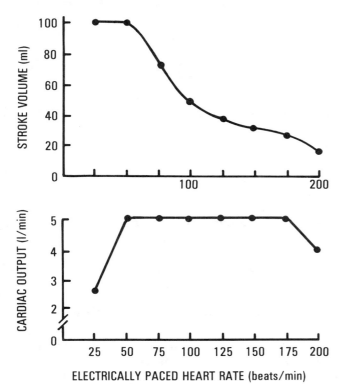

Figure 2.132. Diagram of the effects on the stroke volume and cardiac output of changing heart rate by electrical pacing of the right atrium or vagal stimulation. At very low heart rates (25–50 beats/min) the ventricle is maximally filled, and the cardiac output progressively drops as heart rate is slowed, without a change in the stroke volume. Over the range of 50–180/min the cardiac output does not change appreciably, so that the stroke volume progressively falls as heart rate is increased. Above about 180/min, the cardiac output falls with increasing rates, probably due to impaired ventricular filling. For further discussion, see text.

exercise is greatly reduced. Thus, the increased synchrony of contraction which accompanies inotropic stimulation during the sympathetic response of exercise cannot occur when there is abnormal, slowed electrical activation of the ventricles (Heyndrickx et al., 1985).

VENTRICULAR RELAXATION

Several factors influence the rate of relaxation of the ventricles and thereby can importantly influence early ventricular filling. Also, the time at which the onset of filling occurs can influence the degree of filling. Positive *inotropic stimulation* of the heart, as by epinephrine, both shortens systole and markedly increases the rate of relaxation. Augmentation of the heart rate also hastens relaxation and shortens systole. *Nonuniformity* of electrical activation of the heart (as by electrical pacing from the ventricle rather than the atrium) slows tension development and also reduces the rate of relaxation. So-called *"loaded-dependent" relaxation*, whereby the timing of changes in afterload importantly influence the rate of relaxation, has been demonstrated in isolated cardiac muscle and in the whole heart. With abrupt increases of ventricular volume during the first 30–50% of systole, as stretch occurs tension is increased and relaxation is slowed. On the other hand, when ventricular volume is abruptly increased later in systole, tension is lowered and relaxation is more rapid.¶ (for discussion of relaxation, see Brutsaert et al., 1984). The rate of early diastolic ventricular filling is importantly influenced not only by the speed of ventricular relaxation, reflected by the rate of fall of left ventricular pressure (sometimes expressed as peak $(-)dP/dt$, or as a time constant), but also by the level of *left atrial pressure* at the time of mitral valve opening. A higher left atrial pressure augments the early ventricular filling rate (Ishida et al., 1986). Late ventricular filling is dominantly influenced by atrial contraction.

¶Such effects most likely relate to availability of free Ca^{2+} to bind at muscle cross-bridges; early in the cardiac cycle Ca^{2+} is available and more force is developed as the muscle is stretched, whereas late in the cycle, when Ca^{2+} is being rapidly removed, cross-bridges are broken by the increased afterload and cannot reform, resulting in lowered tension and hastened relaxation.

CHAPTER 13

Intracardiac and Arterial Pressures and the Cardiac Output: Cardiac Catheterization

In experimental animals and in human subjects it is possible to measure the pressure wave forms within the heart and great vessels, and to determine the cardiac output, by passing flexible tubes (catheters) which are only 1–3 mm in diameter and opaque to X-rays, so that their travel can be guided using a fluoroscope. Although pressures were measured in the hearts of animals by means of catheters in the 19th century, the first person to pass a catheter into the human heart was a young German intern, Werner Forssmann, who, in 1929, passed a catheter through a vein in his own arm, pushed it further into the vena cava and then placed its tip in the right atrium where the position was verified by an X-ray. The technique did not come into general use, however, until the 1940s and 1950s when two physicians in the United States, André Cournand and Dickinson Richards, developed special small catheters for use in the heart, and in patients with heart disease they passed such catheters into the right ventricle and pulmonary artery to record pressure and measure the cardiac output. For this accomplishment, Forssmann, Cournand, and Richards received the Nobel Prize in 1956.

Through a catheter, it is possible not only to inject indicators for measurement of cardiac output by the indicator dilution principle but also to withdraw blood samples from various sites within the heart and circulation, to apply the Fick principle for measuring cardiac output, to record pressures through the fluid-filled lumen of the catheter, and to inject liquids that are opaque to X-rays and therefore permit visualization of the heart and blood vessels. Using the latter technique (called angiography), high-speed motion pictures can be taken as the contrast medium is injected into a ventricle or an artery, allowing visualization of the pumping heart (see Fig. 2.114), various congenital malformations of the heart, and identification of atherosclerotic lesions within blood vessels (such as in the coronary arteries, Fig. 2.154). (For details of cardiac catheterization methods of left and right sides of heart, see Grossman (1986).) Also, balloon-tipped catheters are now used to dilate arteries narrowed by atherosclerosis (coronary angioplasty) and to dilate narrowed cardiac valves (balloon valvuloplasty).

MEASUREMENT OF PRESSURE

When a catheter (or a needle, which is sometimes inserted directly into an artery) is filled with saline, the pressure wave form is transmitted from the catheter tip located within a heart chamber, pulmonary artery, or the aorta to the external end where the wave form can be recorded electronically by means of a transducer (Grossman, 1986). A typical transducer of the strain gauge type has a fluid-filled chamber which covers a thin metal diaphragm; the back of the diaphragm is attached to a metal plate which is sensitive to slight displacement of the diaphragm, with a change in resistance through the metal elements being proportional to the displacement of the diaphragm. The diaphragm follows the pressure wave forms transmitted to it through a fluid-filled catheter, and a voltage proportional to the pressure is generated, amplified, and recorded on a strip chart recorder. Calibration of the transducer with a mercury manometer allows measurement of absolute pressure, and it is important to set the reference level of the gauge to zero atmospheric pressure, at the level of the heart. Pressures recorded through such a catheter-transducer system are subject to some artifact from microbubbles in the catheter or transducer, and these must be carefully removed before use; in addition, movement of the catheter within the heart due to cardiac motion can cause artifacts by producing oscillation of the fluid column contained within the catheter. With care, the frequency response of such systems can be made adequate up to 10 or 15 cycles/second (Hz). Miniature pressure transducers have been developed (inductance, strain gauge, or piezoelectric types) that are small enough to be placed at the tip of a cardiac catheter, and these devices provide much higher fidelity pressure wave forms (up to 100 Hz).

The arterial pressure can also be measured *indirectly* by the auscultatory method using an inflatable cuff placed on an extremity, usually the arm, with the pressure inside the cuff measured either with a mercury or aneroid manometer; the above device is called a *sphygmomanometer*. Pressure in the cuff is first inflated above the level of systolic pressure in the aorta which collapses the artery, and as one listens over the artery in the arm (brachial artery) with a stethoscope, no sound is heard. The cuff is then slowly deflated (2–3 mm Hg/sec), and as soon as the systolic pressure is reached, a small spurt of blood passes beneath the cuff during each cardiac systole, producing a snapping sound due to turbulence and closure of the vessel (Fig. 2.133). As soon as this sound is audible, the systolic pressure is noted. As the cuff is then further deflated, the sounds change in character and intensity, and the flow becomes continuous, at which point the sound disappears and the diastolic pressure is recorded (Fig. 2.133). The changes in sound quality as the cuff pressure is lowered (Korotkoff sounds) are multiple,* but only the initial sound and the point of sound disappearance are now taken as systolic and diastolic pressures (a normal value would be recorded, for example, as 120/70 mm Hg).

---

*The Korotkoff sounds (Fig. 2.133) are: Phase I, faint tapping of increasing intensity (sound appearance up to 10–25 mm Hg lower); Phase II: softer bruit for next 15–20 mm Hg (sounds occasionally disappear near the "auscultatory gap," and systolic pressure should also be checked by palpating the pulse appearance); Phase III: sounds louder and clearer; Phase IV: muffling of sounds; Phase V: sounds disappear. The last phase corresponds clearly with the diastolic pressure at rest measured directly by intraarterial needle and transducer.

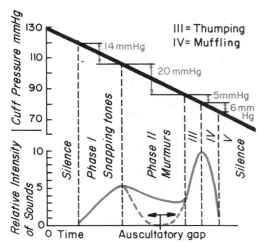

Figure 2.133. Characteristics of the auscultatory method of measuring blood pressure. (Redrawn from Geddes, 1970.)

Pressure Pulses in the Right and Left Sides of the Heart and the Heart Sounds

Knowledge of the sequence of electrical and mechanical events of both the right and left sides of the heart is central to understanding the shapes of the pressure waves recorded in the atria and ventricles, as well as the *timing* of heart sounds and murmurs (Wiggers, 1954). These events are shown diagrammatically in Figure 2.134, and a reference summary of terminology used to describe the ventricular cycle with the average duration of these periods at an average heart rate of 75/min is shown in Table 2.7.

First, the basic pattern of ventricular pressure development and blood flow into and out of the ventricle should be recalled; during ventricular systole, flow across the atrioventricular valve is zero, and there is a period at the onset of ventricular systole (isovolumetric systole) when no flow occurs across any heart valve (see Fig. 2.113). During ventricular ejection, flow into the aorta or pulmonary artery takes place through the open semilunar valves. Following ejection during the period of isovolumetric relaxation, which is the first phase of diastole, no flow takes place when all four valves are again closed. Then early rapid filling occurs followed by a phase of slow filling and then rapid filling again as atrial contraction occurs (see Fig. 2.113).

The entire sequence of pressure events in the right and left atria and ventricles is shown diagrammatically, together with the electrocardiogram and the heart sounds, in Figure 2.134. The P wave, spreading from the sinus node in the right atrium, first depolarizes the right atrium (the first mechanical event in the cardiac cycle) and shortly thereafter the left atrium with subsequent contraction of that chamber. The slightly asynchronous contraction of the right and then the left atrium occurs late in diastole and is reflected by a buildup of pressure within the atrial chambers and in the ventricles (the mitral and tricuspid valves are still open), and the resulting positive wave is termed the *a* wave (Fig. 2.134). The *a* wave in the right atrium slightly precedes and normally is at a lower pressure than that in the left atrium, and similar waves are produced in the ventricular pressure tracings as the pressure rises in those chambers due to the increase of volume provided by the atrial booster pump (see ventricular volume tracing, Fig. 2.113).

In disease conditions, when the ventricles are stiff, a thudding sound may be audible just following atrial contraction. This sound is termed a "gallop" and is called an "atrial diastolic gallop" or fourth heart sound (S_4) (Fig. 2.134). Gallop sounds

Figure 2.134. Diagram of pressure wave forms in the aorta and pulmonary artery, right ventricle and left ventricle (*RV* and *LV*), recorded together with the heart sounds on a phonocardiogram (*PHONO.*) and the electrocardiogram (*ECG.*). The *cross-hatched areas* show isovolumetric phases of right and left ventricular contraction and relaxation, and the *strippled areas* indicate the periods of right and left ventricular ejection. *OT* = opening of tricuspid valve, *OM* = opening of mitral valve. For further discussion, see text.

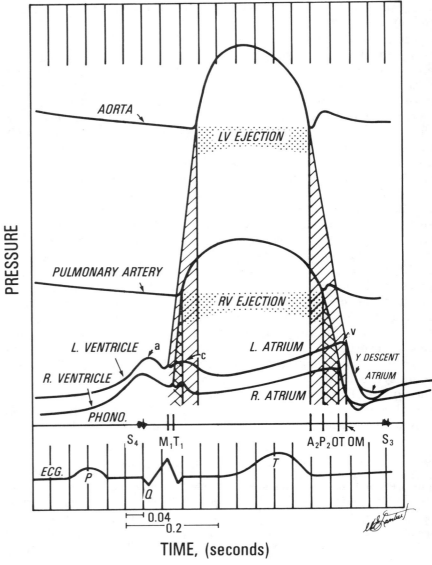

TIME, (seconds)

Table 2.7.
Average Times for Isovolumetric, Ejecting, and Filling Periods of the Two Ventricles at a Heart Rate of 75/min[a]

Systole, 260 ms (left ventricle)	Systole
	1. Onset of ventricular contraction
	2. Atrioventricular (inlet) valve closure
	3. *Isovolumetric ventricular contraction* (LV = 60 ms)
	(RV = 15 ms)
	4. Semilunar valve opening
	5. *Ventricular ejection* (LV = 200 ms)
	(RV = 270 ms)
	6. Semilunar (outlet) valve closure
Diastole, 600 ms	Diastole
	7. *Isovolumetric ventricular relaxation* (LV = 100 ms)
	8. Atrioventricular (inlet) valve opening
	9. *Ventricular filling* (about 0.5 s or 500 ms)
	a. Rapid filling
	b. Slow filling (diastasis)
	c. Atrial systole or contraction (rapid filling again)
Total: 860 ms (about 0.8 s)	

[a]The four most important periods of the cycle are italicized.

are usually related to rapid filling of a ventricle and may be likened to blowing up a paper bag which suddenly reaches its "elastic limit" and produces a popping sound.

After the a wave, the onset of isovolumetric ventricular contraction (first the left ventricle, then the right ventricle) follows the onset of the QRS complex by about 50 ms (Fig. 2.134). The left ventricular pressure then commences to rise rapidly, and as the mitral valve snaps shut the first heart sound (S_1) begins with closure of the mitral valve, followed soon thereafter by the tricuspid component of the first heart sound (M_1, T_1); Figs. 2.134 and 2.135).† Isovolumetric contraction of the two ventricles then occurs. The rapid rise of pressure in both chambers takes place at about the same rate; therefore, since the pulmonary artery pressure is only about one-fifth that in the aorta, the pulmonic valve opens sooner than the aortic valve, and the duration of isovolumetric contraction of the right ventricle is correspondingly shorter than that of the left ventricle (15 ms vs. 60 ms) (Table 2.7). No heart sounds accompany opening of the aortic and pulmonary valve. Thus, as shown in Figures 2.134 and 2.135, the entire duration of right ventricular isovolumetric contraction is enclosed within the isovolumetric period of the left ventricle. Isovolumetric ventricular contraction is accompanied by a positive wave in the atrial pressure tracings, which remain much lower than the ventricular pressure tracings because the tricuspid and mitral valves are shut. This small wave is termed the c wave. It is due in

---

†The first heart sound is relatively long and of low frequency. At least 4 components have been identified, beginning slightly before the rise in left ventricular pressure and ending during early ejection. The second sound is shorter and of higher frequency. It results primarily from vibrations set up in the blood and blood vessel walls as the pulmonic and aortic valves tense after their closure. (For detailed discussion of the origins of heart sounds, see Craige, 1988.)

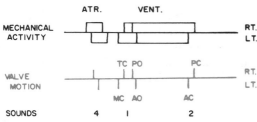

Figure 2.135. Sequence of valve motion as related to mechanical activity during the cardiac cycle. Right heart events above the line; left heart events below the line. *ATR.*, atrium; *VENT.*, ventricle; *RT.*, right; *LT.*, left; *MC.*, mitral closure; *TC.*, tricuspid closure. *PO.*, pulmonic opening; *AO.*, aortic opening; *AC.*, aortic closure; *PC.*, pulmonic closure; 4, 1, 2, fourth, first, and second heart sounds. Duration of isometric contraction for each ventricle is indicated in each ventricular rectangle by a verticle line. (Redrawn from Lewis, 1962.)

part to ballooning of the tricuspid and mitral valves backward into the atrium as higher pressure develops in the ventricular chambers (Fig. 2.134).

A negative wave in the atrial pressure pulse follows the c wave as the ventricles move away from the atria during early ejection (the so-called x descent, not labeled in Fig. 2.134). Throughout ventricular systole the tricuspid and mitral valves remain closed, but blood continues to flow into the atria from the venae cavae to the right atrium and from the pulmonary veins to the left atrium, producing a third positive wave in the atrial pressure tracings, the v wave (Fig. 2.134).

During ventricular ejection into the aorta and pulmonary artery, the pressure in the right ventricle is only 1 mm Hg or so higher than that in the pulmonary artery and the pressure in the left ventricle exceeds that in the ascending aorta by 1 or 2 mm Hg. It is during the period of ventricular ejection that the noise (or murmur) due to turbulent flow across an abnormally narrowed pulmonic or aortic valve would be heard. The ventricles then begin to relax, and the pressures also start to fall in the pulmonary artery and the aorta, as the rate of forward blood flow decreases sharply (this phase of the cardiac cycle has, in the past, sometimes been called "protodiastole"). The rapidly falling ventricular pressures then drop below those in the aorta and pulmonary artery. Deceleration of blood flow with reversal of the pressure gradient leads to closure of the aortic and pulmonary valves. The period of ventricular ejection, called the ejection time, is longer in the right ventricle than in the left (Fig. 2.134) (approximately 270 ms vs. 200 ms), so that pulmonic valve closure occurs *after* aortic valve closure. Therefore, two snapping sounds are audible as these valves close, and together they comprise the second heart sound (S_2), with its two components termed A_2 and P_2 for aortic and pulmonic valve closure, respectively (Figs. 2.134 and 2.135). As is seen in Table 2.7, at a normal heart rate diastole is considerably longer than systole. This gives the normal heart sounds a "1-2, pause" sequence.

Isovolumetric relaxation then occurs in the two ventricles, the right ventricular pressure rapidly falls below that in the right atrium and, when the tricuspid valve opens, the v wave rapidly falls (the so-called "y descent"), marking the onset of rapid filling of the ventricle. A few ms thereafter, the falling left ventricular pressure pulse crosses the left atrial pressure pulse, and when the mitral valve opens, the v wave in that chamber also rapidly falls during early diastole (Fig. 2.134). As rapid ventricular filling continues, the y descent ends because pressures begin to increase rapidly in the ventricles and atria simultaneously. Toward the end of this rapid filling wave, another sound is sometimes audible (often in the normal heart in younger individuals, or particularly when the ventricle becomes stiff in heart failure); this sound when abnormal is sometimes termed a "ventricular diastolic gallop," or third heart sound (S_3) (Fig. 2.134). Ventricular filling then slows, and the pressures in the right atrium and the right ventricle, and those in the left atrium and the left ventricle, rise slowly together (with the atrial pressure slightly exceeding the ventricular pressure) (Fig. 2.134). This period of slow ventricular filling, which is sometimes called diastasis, is lost at rapid heart rates. Toward the end of diastole, the next P wave occurs, initiating the next *a* wave, and the cardiac cycle recurs.

The sequence of events and flow rates across the mitral or tricuspid valve makes it quite easy to understand the motions of the mitral leaflet when an echocardiogram is recorded (Fig. 2.136). As the echo beam traverses the leaflets, rapid inflow initially moves the leaflets wide apart during diastole, and as the flow slows down, the leaflets move closer together (Fig. 2.137); they open again widely with atrial systole and then abruptly close and remain opposed throughout ventricular systole (Fig. 2.137). The echocardiogram is extremely useful in detecting abnormalities of valve motion and in detecting thickening of the leaflets (Feigenbaum, 1986).

Two or three of the normal atrial waves (*a, c,* and *v*) are usually visible in the neck veins. They can be recorded from the skin of the neck with an air-filled transducer and can provide considerable information about abnormal events on the *right side* of the heart. For example, if the tricuspid valve is leaking, blood fills the right atrium during ventricular sys-

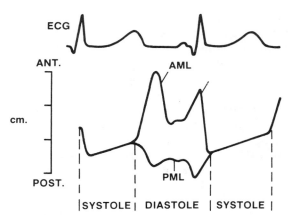

Figure 2.137. Diagram of an echocardiographic tracing obtained from the mitral valve leaflets (Fig. 2.136). During systole, both leaflets are closed, forming a single line, whereas during diastole, the anterior mitral leaflet *(AML)* moves anteriorly toward the front of the chest *(ANT.)* in a biphasic pattern; the posterior mitral leaflet *(PML)* moves similarly but in the opposite direction toward the posterior chest *(POST.)*. For further discussion, see text.

tole from the right ventricle, flowing backward through the incompetent valve as well as forward from the venae cavae. The *v* wave visible in the neck veins is then larger and earlier than normal.

The pressure wave forms in the left atrium may also become distinctly abnormal in various disease states. For example, when the mitral valve is narrowed (with "mitral stenosis" often due to the scarring of rheumatic fever), a higher pressure is required in the left atrium to force blood across the narrowed mitral valve orifice during diastole. As one might predict, the rate of emptying of the left atrium also is delayed, resulting in a slow fall of the downslope of the *v* wave (slowed *y* descent).

PULMONARY ARTERY WEDGE PRESSURE

The left atrial pressure can be measured indirectly by passing a small flexible catheter, which has a small inflatable balloon at its tip, into the pulmonary artery. This is accomplished by advancing the tip of the catheter into a small pulmonary artery while the balloon is deflated (Fig. 2.138*A*); the pressure recorded from the tip of the catheter through the fluid-filled lumen is that in the pulmonary artery (Fig. 2.138*A*). When the balloon is inflated, the catheter is swept by the blood flow further out into the pulmonary artery, where it "wedges" and therefore obstructs blood flow through the small artery. This results in a fall of pressure in the segment of artery beyond the inflated balloon, as equilibrium across the pulmonary capillary bed occurs with the pressure in the pulmonary veins and left atrium; the result is a somewhat "damped" but relatively accurate record-

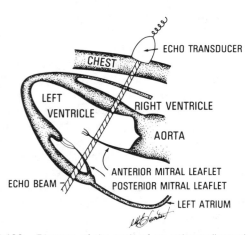

Figure 2.136. Diagram of the path of an echocardiographic beam passing through the chambers of the heart and the tips of the mitral valve leaflets.

A.

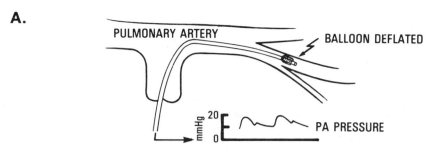

B.

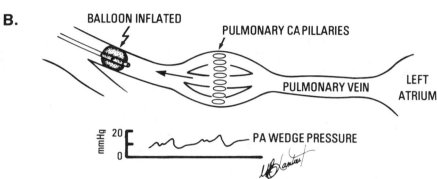

Figure 2.138. Technique for measurement of the pulmonary artery wedge pressure. *A,* Balloon at tip of catheter is deflated, and the pulmonary artery *(PA)* pressure is recorded. *B,* Magnified drawing showing the balloon inflated; pressure from the left atrium is transmitted across the pulmonary capillary bed into the obstructed segment of the pulmonary artery *(arrow)* to yield the pulmonary artery *(PA)* wedge pressure tracing.

ing of the left atrial pressure pulse (Fig. 2.138*B*), and the mean wedge pressure accurately reflects that in the left atrium. (This pressure is also sometimes called the pulmonary "capillary" wedge pressure.)

Such balloon "float" catheters are frequently used in critically ill patients at the bedside, since they can be inserted into a vein and passed without using X-ray guidance. The balloon is inflated, causing the catheter to be carried by the blood flow through the right heart chambers into the pulmonary artery. The position of the catheter tip can be determined simply by recording the different pressure waves as the catheter passes through the chambers of the right heart. Special balloon catheters are now manufactured for the purpose of measuring cardiac output. These have two lumens; one opens at the tip for measuring pressure, and the other more proximal opening is used to inject cold saline into the right atrium. There is also a thermistor at the catheter tip so that cardiac output can be measured by the "thermal dilution method" (see below).

Notice from Figure 2.134 and Table 2.8, in which the upper limits of normal for pressure values in the normal human heart are shown, that the pressure pulses in the left atrium are higher than those in the right atrium and that the left ventricular end-diastolic pressure is higher than that of the right ventricle. The differences are related primarily to two factors: first, the right ventricle has a much thinner wall than the left ventricle and is less stiff (more

compliant), as shown in Figure 2.139. Therefore, when contraction of the two atria delivers an equal volume of blood into the two ventricles, a larger rise in pressure occurs in the left ventricle because its diastolic pressure-volume curve is steeper (Fig. 2.139). The second major factor contributing to higher pressures in the left atrium than in the right is that the left atrium and pulmonary veins contain

Table 2.8.
Normal Intracardiac Pressures in Man[a]

		Adult Man[b]	
		Upper Limits of Normal, mm Hg	Average Level, mm Hg
Right atrium	a	7	5
	v	5	4
	Mean	6[a]	3
Right ventricle	S/D	30/5[a]	18/4
Pulmonary artery	S/D	30/15[a]	18/12
	Mean	20[a]	15
PA wedge	a	16	5
	v	21	10
	Mean	12[a]	8
Left atrium	a	16	10
	v	21	13
	Mean	12[a]	8
Left ventricle		140/12[a]	130/8
Systemic arterial	S/D	140/90[a]	130/75

S, systolic; D, diastolic (measured at end-diastole).
[a]Important values to remember at upper limit of normal. [b]*Other normal values* are as follows. Cardiac index: adults 2.5–3.5 liters/min/M² BSA; AVO₂ difference, upper limits of normal at rest 5.2 vol%; total body O₂ consumption, 110–150 ml/min/M² BSA; pulmonary vascular resistance (average 2 RU, upper limit 3 RU or 300 dynes·s/cm⁵); systemic vascular resistance (average, 20 RU, 700–1500 dynes·s/cm⁵); 1 RU = 80 dynes·s/cm⁵ (1333 is the constant for conversion of mm Hg to dynes/cm², and flow is expressed in ml/s for this calculation).

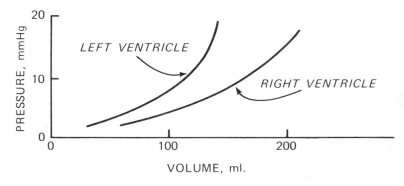

a smaller volume and are less compliant than the right atrium and the venae cavae. Thus, equal volumes of blood flowing into the left and right atria during ventricular systole (when the mitral and triscuspid valve are closed) begin at a higher pressure on the left side and result in a larger v wave in the left atrium than in the right atrium. Higher a and v waves on the left side lead to a higher mean pressure under normal conditions, and the upper limit of normal for the left ventricular end-diastolic pressure is about twice that for the right ventricle (Table 2.8).

MEASUREMENT OF CARDIAC OUTPUT

Many methods are used for measuring the cardiac output in experimental animals, including placement of an electromagnetic device directly around the aorta for determining blood flow velocity, insertion of mechanical recording flowmeters by cannulating the aorta and diverting blood through the flowmeter (for example, displacement of a float is proportional to velocity of blood flow), and use of indirect techniques, including the Fick principle and the indicator dilution method, as discussed below. The latter two methods are useful in human studies, and it is also possible to place an electromagnetic catheter tip velocity meter in the aorta or pulmonary artery. (The volume flow can then be calculated if the cross-sectional area of the vessel is known by angiography or echocardiography).

The total output of blood from the heart (the *cardiac output*) is equal to the product of the heart rate and the stroke volume. For example, in an adult human subject of medium size, the stroke volume might be 84 ml at a heart rate of 70/min, yielding a cardiac output of 5880 ml/min, or 5.88 liters/min. The cardiac output is often expressed as the *cardiac index*, which is the cardiac output per square meter of body surface area (BSA). This correction is helpful because, as might be expected, large individuals have a higher cardiac output at rest than small individuals. The total body energy requirements at

rest across species also correlate well with the cardiac output. In addition to body size, the amount of O_2 carried in the blood and the percent of O_2 extraction in the tissues importantly affect the cardiac output. For example, the blood of fish has a low O_2 carrying capacity, and cardiac output is proportionately higher. When anemia is present in the human subject, O_2 supply to the tissues is maintained by an increase in the cardiac output. The range of normal for the cardiac index is 2.4–3.5 liter/min/M^2 body surface area (Table 2.8); it is slightly higher in males than females and considerably higher in children.

Fick Method for Cardiac Output

The Fick principle (Adolph Fick, 1870) is an important concept, which can be applied for calculating the cardiac output, or the blood flow rate to any organ. It can also be used for calculating the oxygen consumption of the entire body or any organ, provided the flow rate and the Q_2 content of blood samples are known. It did not become possible to measure cardiac output accurately in man by means of the Fick method until a sample of mixed venous blood could be obtained from the pulmonary artery through a cardiac catheter (a requirement discussed further below).

The Fick principle actually is an indicator dilution method, the indicator in this case being a physiological one, oxygen. In the generalized form of the Fick equation the amount of any indicator accumulated or removed per minute (X) at a given site must be equal to the rate of flow per minute ($\dot{Q}$) of the substance carrying the indicator multiplied by the difference between the concentration of indicator in the carrier before and after it passes that site ($C_1 - C_2$). The analogy may be made to a coal train passing a dumping site. Each coal car is equally filled upon arriving at the site, and each dumps the same quantity at the dumping site; by knowing the amount of coal in the cars that are arriving and in the cars that are leaving, the amount dumped by each can be determined $(C_1 - C_2)$. It is readily apparent that the rate of accumulation of the coal per minute (X) will equal the product of the number of cars passing the site per minute $(\dot{Q})$ and the difference in content between a coal car arriving at and leaving the site: $X = \dot{Q}(C_1 - C_2)$ which, in order to calculate flow rate, can be arranged as:

$$\dot{Q} = \frac{X}{C_1 - C_2}$$

As usually applied to measure cardiac output in man, O_2 uptake is measured at the lungs by collecting a known volume of expired gas in a spirometer and analyzing it for residual O_2 content (the O_2 content of inspired air is known) (Fig. 2.140). The O_2 concentration in blood entering the lungs is measured from a sample of blood drawn through a catheter placed in the pulmonary artery, and the O_2 content of blood leaving the lungs is measured from a sample of blood collected from any peripheral artery, such as the brachial artery (Grossman,

THE FICK METHOD

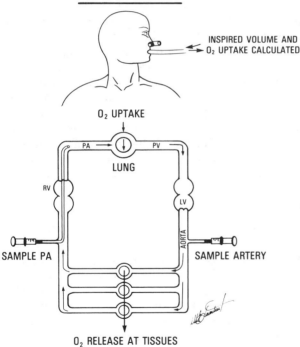

Figure 2.140. Diagram of the Fick method for measurement of cardiac output. The catheter is positioned in the pulmonary artery *(PA)*, and blood samples are withdrawn simultaneously from a peripheral artery and the *PA* as room air is inspired. The volume of expired air is measured, the inspired volume is calculated, and the difference between the O_2 content of inspired air and that measured in the expired air can then be used to calculate O_2 uptake at the lungs per minute. For further discussion, see text.

1986) (the small amount of blood that normally shunts through the lungs is ignored, and the oxygen content of the systemic arterial bed is similar everywhere in the circulation).

In practice (Fig. 2.140), O_2 uptake at the lungs per minute $(\dot{V}O_2)$ is measured for a 3-min period, and blood samples are drawn simultaneously from the pulmonary artery (V) and peripheral artery (A) during the middle minute of the gas collection period (Fig. 2.140) to obtain the arteriovenous O_2 difference (AVO_2):

$$\dot{Q} = \frac{\dot{V}O_2}{AVO_2}$$

Example: O_2 uptake measured at the lungs was found to equal 240 ml/min, arterial blood O_2 content = 18 vol% (18 ml O_2/100 ml blood = 0.18 ml O_2 per ml), and venous O_2 content = 14 vol%. Solving with the Fick equation, the blood flow or $\dot{Q}$ = 6000 ml/min. As measured, this value represents the pulmonary blood flow, but in the steady state it also equals the cardiac output to the body since, under basal resting conditions, the oxygen uptake of the lung is equal to the O_2 utilized by the tissues of the body (Fig. 2.140).

There are several prerequisites which must be fulfilled if the Fick equation is to be applied for measuring cardiac output.

1. A sample of *mixed* venous blood must be obtained from the pulmonary artery, since the O_2 content of the venous blood draining the various organs of the body varies widely, and some organs extract much more O_2 than others (Table 2.9). For example, the cerebral arteriovenous difference is about 6.5 vol% (and drains into the superior vena cava), while the difference in O_2 content across the heart is about 12 vol% (and drains into the right atrium); the renal circulation has an arteriovenous difference of about 2 vol% (and drains into the inferior vena cava). Therefore, the oxygen contents of blood in the venae cavae and right atrium can differ widely, usually being higher in the inferior vena cava because of the high renal contribution, whereas sampling in the right atrium near the exit of the coronary vein would result in a low O_2 content which is not representative of the entire body. It is apparent that in order to obtain a sample from the whole body it is necessary to draw the sample from a site where the venous blood is well mixed. This mixing is accomplished in the right ventricle, and a pulmonary artery sample there-

Table 2.9.
Distribution of Cardiac Output and Oxygen Usage

Region	Weight		Blood Flow			Oxygen Usage			O₂ Extraction	
	kg	% Total	liters/min	% Total	ml/100 g/min	ml/min	% Total	ml/100 g/min	Venous O₂ ml/100 ml Blood	AVO₂
Total	70	100	5.4	100		250	100		14.5	4.5
Brain	1.54	2.2	0.83	15	54	63	23	3.7	12.5	6.5
Heart	0.33	0.5	0.22	4	70	23	9	7.0	7	12
Liver, intestines	2.86	4.0	1.54	29	54	55	20	1.95	15	4.0
Kidney	0.33	0.5	1.43	27	430	20	7	6.0	17	2.0
Skeletal muscle	34.0	50.0	0.92	17	2.7	55	20	0.16	12.5	6.5
Exercise (total)			35.0	80	100	2700	90	7.9	3	16.0

fore is representative of the combined venous drainage. The difference between the mixed venous O₂ content and the arterial O₂ content for the whole body at rest is usually 3–4 vol%, the upper limit of normal at rest being 5.2 vol%, and it can rise much higher during heavy exercise.

When the Fick equation is applied to measure blood flow to any organ, the venous sample must be representative of the entire venous drainage from that organ.

2. The *total amount* of the substance taken up or removed must be accurately known (no indicator can be lost). The Fick equation can also be applied when foreign substances are infused or cleared from the circulation, and other physiological substances such as CO₂ production have also been used in the Fick equation to measure cardiac output. In the case of the application of the Fick method using O₂, it is necessary to make certain that all the expired gas is collected (there must be no leakage around the mouthpiece or noseclip).

3. A *steady-state* must exist. There must be no marked change in the rate of ventilation during the measurement period (for example, transiently increased ventilation might result in more oxygen uptake at the lungs than is used by the peripheral tissues).

A steady circulatory state must also exist during the measurement period (for example, if the subject were to become somewhat excited during the measurement period, the cardiac output might rise without a corresponding increase in ventilation, leading to inaccuracy in the measurement).

Given these precautions, the Fick principle can be widely applied in a variety of physiological settings. For example, if blood flow to an organ is measured by an independent method (such as a flow-meter) and the arteriovenous O₂ difference across

the organ is measured, the Fick equation in the form:

$$\dot{V}O_2 = \dot{Q}(AVO_2)$$

can be used to calculate the oxygen consumption of the organ.

Indicator Dilution Methods

A variety of indicators has been used for measuring cardiac output or organ blood flow. The earliest were dyes, the concentration of which could be measured in the blood by light transmission (densitometer) techniques. This approach is now rarely employed using indocyanine green dye, the concentration of which is measured as a sample of blood is withdrawn continuously through a recording densitometer. The thermal dilution method is now more widely used and involves the injection of chilled saline (of known volume and temperature) while the degree of cooling of the blood (dilution of the cold injectate) is measured by means of a calibrated thermistor placed directly in the bloodstream.

CONTINUOUS INFUSION METHOD (FIG. 2.141)

If an indicator (such as a dye or cold saline) is infused at a constant rate at one location (for instance, the vena cava) and a sample is drawn during a steady circulatory state further along in the circulatory system beyond a mixing chamber (for example, in the pulmonary artery), cardiac output can be calculated provided there is no recirculation (Fig. 2.141). The cardiac output will be directly proportional to rate of infusion of the indicator (which is known and constant) and inversely proportional to concentration at the sampling site. Thus:

$$\dot{Q} = X(\text{mg/min})/PA \text{ concentration (mg/liter)}$$

That is, the higher the blood flow, the more dilute the indicator (which is being delivered at a constant

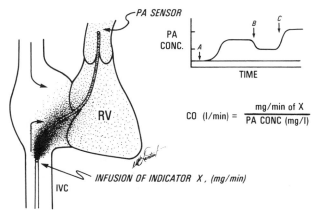

Figure 2.141. Continuous infusion method for measurement of cardiac output. The indicator is infused into the inferior vena cava *(IVC)* or right atrium, mixes in the right ventricle *(RV)*, and passes into the pulmonary artery *(PA)* where the PA concentration *(PA CONC.)* is recorded. The indicator injection is commenced at *A* and reaches a steady state. At *B* the cardiac output then increases, resulting in a lower PA concentration, and at *C* the cardiac output falls and PA concentration rises (less dilution).

rate); the lower the flow, the higher the indicator concentration at the sampling site (Fig. 2.141).

This approach would be highly useful were it not for the problem of recirculation which, of course, distorts the concentration in the pulmonary artery as soon as it occurs. A second sensor can be placed in the vena cava upstream from the injection site which samples the recirculation concentration to make a correction, but this approach is cumbersome. If the indicator is a highly insoluble gas (such as krypton) that is eliminated in one passage through the lungs, flow can be calculated by this technique, provided an appropriate sensor is available.

The continuous infusion approach has had some success for measuring blood flow in individual organs by continuously infusing cold saline into a vein draining the organ and by sampling with a thermistor further downstream. Adequate mixing appears to be obtained in some veins having high flow (such as the coronary sinus), since only a small amount of cold solution is injected to measure local blood flow, and the total amount reaching the body for recirculation is negligible, so that coronary blood flow to the left ventricle (mainly drained by the coronary sinus) can be recorded continuously by this method. Regional blood flow in the vein draining any organ also can be determined by the thermal technique using bolus injection (see below) provided good mixing of the injectate can be obtained.

RAPID (BOLUS) INJECTION METHOD

Cardiac output can also be determined by rapidly injecting a small amount of an indicator (green dye or a bolus of cold saline) into the circulation and sampling at a downstream site to produce an "indicator dilution curve," which represents a graphic recording of the indicator concentration vs. time. The approach commonly employed was derived experimentally by G. N. Stewart (1896) and subsequently modified by Hamilton (1930); hence, it is commonly called the Stewart-Hamilton method. (For review, see Dow, 1956).

In practice, 2 or 3 ml of the indicator is rapidly injected intravenously, with a needle for sampling being inserted into a systemic artery. The indicator injected into the vein mixes in the right heart and is diluted by the entire systemic blood flow. It is then further mixed and dispersed in the central circulation and left heart, and its concentration in the artery is sampled by withdrawing with a syringe at a constant rate through a densitometer or, most commonly, a thermistor is placed directly in the arterial stream. It is also possible to apply this method by injecting a bolus of cold saline into the right atrium and sampling with a thermistor placed in the pulmonary artery (Fig. 2.141), or dye or cold solution can be rapidly injected into and mixed by the left ventricle and sampled from a peripheral artery.

It is important to understand that once the indicator is injected and becomes mixed with the *entire* volume of blood passing a given site at a given time, only a *sample* of blood need be taken from *any* site further along in the circulation beyond the point of mixing. This principle is illustrated in a very simplified form in Figure 2.142. In the example, 6 mg of dye are injected into a container of fluid and thoroughly mixed. A stopcock is then opened, and the fluid is allowed to run out through a tube. When a sample of the flowing stream is continuously drawn past a densitometer, a concentration of 1 mg/liter is recorded (Fig. 2.142). This means that the indicator must have been *diluted* in 6 liters of fluid. If the duration of the recording is 2.0 min (the duration of the indicator-dilution "curve"), it can be intuitively seen that the flow rate (cardiac output) through the tube must have been 3 liters/min. The general form of the equation is:

$$\dot{Q} = \frac{X}{\bar{c}t}$$

where X = amount of indicator, $\bar{c}$ = mean concentration, and t = time (duration of the curve),

Figure 2.142. Simplified representation of an indicator dilution curve. The indicator is mixed in the container, and the stopcock is then opened to allow the container to empty as a sample is drawn by a syringe at a constant rate through a densitometer to yield a time concentration curve. For further discussion, see text.

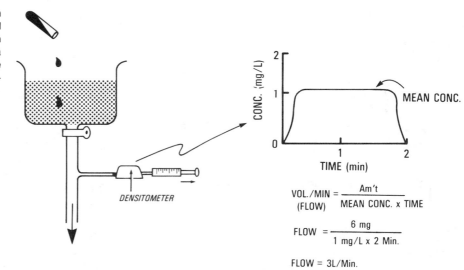

$$VOL./MIN = \frac{Am't}{MEAN\ CONC.\ x\ TIME}$$

$$FLOW = \frac{6\ mg}{1\ mg/L\ x\ 2\ Min.}$$

FLOW = 3L/Min.

and in the example above, $\dfrac{6\ mg}{1\ mg/liter \times 2\ min} =$ 3 liters/min. Time is often given in seconds, and the equation is written:

$$\dot{Q} = \frac{60 \cdot X}{\bar{c}t}$$

This equation resembles that shown for the continuous infusion method (note the resemblance to the Fick equation), and it can be intuitively seen that the lower the mean concentration (more dilution) and the shorter the duration of the curve, the higher the cardiac output, and vice versa (Fig. 2.143).

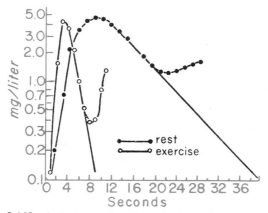

Figure 2.143. Indicator dilution determinations of cardiac output at rest and during exercise, plotted in semilog fashion for dye concentration, with time on the abscissa, and the downslope of the curve extrapolated linearly to zero. Recirculation is indicated by deviation from the exponential downslope. Flow equals the amount of dye injected divided by the average concentration of the dye. (From Asmussen and Nielsen, 1952.)

For the rapid injection method to be used reliably, several prerequisites must be fulfilled.

The effects of recirculation must be eliminated, that is, recirculation of indicator before the initial time-concentration curve is complete must be detected and removed. Recirculation poses a particular problem with this method, since early recirculation (as occurs through the coronary arteries, the first branches of the aorta) may not be apparent from the shape of the indicator dilution curve, although the bulk of the recirculating dye is usually detectable as a second wave on the trailing edge of the curve. Hamilton discovered that the downslope of an indicator dilution curve is a simple exponential function. This may be thought of in simplest terms as washout of the indicator from a ventricle of the heart. For example, if a dye is injected suddenly into the left ventricle and sampled in the aorta, since the ejection fraction remains constant during a sequence of heartbeats during a steady state, the concentration of the dye is diluted with each successive heartbeat by a constant fraction of undyed blood flowing in from the left atrium. This progressive reduction or concentration by a constant fraction results in an exponential fall in the concentration of the curve recorded in the aorta and is sometimes called a "washout curve."

In practice, recirculation is eliminated by plotting the curve on semilog paper (Fig. 2.143) as the log of dye concentration vs. time. Recirculation is then readily detected as a deviation from the straight line downslope, and the primary curve can be calculated by linearly extrapolating this downslope to zero. The *mean concentration* of the indicator during the curve can then be computed. Recirculation does not occur with the thermal method.

No indicator can be lost during injection; it must

be entirely mixed, and subsequently a representative sample must be obtained as the blood entirely mixed with indicator passes the sampling site.

The indicator must also exhibit the property called "stationarity," that is, it must be capable of being completely mixed and carried with the blood (a particulate dye must not settle out and thereby be lost in transit to the sampling site).

There must be a steady circulatory state during the entire measurement period of the curve, since sudden changes in blood flow will distort the characteristics of the curve and invalidate the principles of the method.

USE OF PRESSURE AND FLOW TO CALCULATE VASCULAR RESISTANCE

When the rate of blood flow and the mean pressure difference across a vascular bed are known, a measure of vascular resistance can be determined. As discussed in Chapter 6, calculation of true resistance involves application of the Poiseuille equation and knowledge of the length and radius of the vascular bed and the viscosity of blood, values which are not available. Nevertheless, measures of vascular resistance and the "vascular resistance unit" are highly useful in studying human disease.

For example, the pulmonary and systemic vascular resistances can be estimated in the following manner using the general resistance equation:

$$R = \frac{P_1 - P_2}{\dot{Q}}$$

where R = resistance, P_1 = mean arterial pressure, P_2 = mean venous pressure, and $\dot{Q}$ = flow. Thus:

$$PVR = \frac{P_{PA} - P_{LA}}{\dot{Q}}$$

where PVR = pulmonary vascular resistance, P_{PA} = pulmonary artery pressure, P_{LA} = left atrial pressure, and $\dot{Q}$ = pulmonary blood flow.

If the mean *PA* pressure is 20 mm Hg, the mean *LA* pressure is 10 mm Hg, and the pulmonary blood flow is 5 liters/min, the *PVR* = 2 *RU* (dimensionless resistance units). Note from Table 2.8 that the upper limit of normal is 3 resistance units.

Also:

$$TPVR = \frac{P_{SA} - P_{SV}}{\dot{Q}}$$

where $TPVR$ = total peripheral vascular resistance, P_{SA} = mean pressure in a systemic artery, P_{SV} = mean pressure in a systemic vein (or the right atrium), and $\dot{Q}$ = the cardiac output.

If the P_{SA} is 95 mm Hg, the venous pressure (P_{SV}) is 5 mm Hg, and the cardiac output 6 liters/min, the *TPVR* is 15 *RU*. This is a normal value for the systemic vascular resistance, which is nearly 10 times that of the normal pulmonary circulation. Vascular resistance is also sometimes expressed in dynes·s/cm⁵ instead of units, and this conversion is shown in footnote b of Table 2.8.

These resistance relations may be strikingly altered in disease states. For example, the systemic vascular resistance is elevated in hypertension. In patients who have a hole between two chambers of the heart (such as a ventricular septal defect), there is a left-right shunt of blood due to the higher pressures on the left side of the heart. This short circuit of blood causes an extra volume of blood to recirculate continuously through the right ventricle and lungs so that, in addition to the forward cardiac output, the left heart must pump this extra quantity of blood around and around, resulting in an elevated pulmonary blood flow (it may be several times the output of blood into the aorta). However, sometimes in this disorder the very high pulmonary blood flow produces a high pressure in the pulmonary artery, and eventually the arterioles in the lungs become thickened so that their resistance rises, causing the left-right shunt to disappear. Under these circumstances, it is no longer possible to correct the shunt by surgically repairing the hole between the heart chambers. However, the pulmonary artery pressure may be high simply because the blood flow is very high through the lungs (in which case an operation could be performed), and the pulmonary vascular resistance is normal; the distinction between these two conditions is obviously important in managing the patient.

Application of the equations given above can solve this diagnostic problem. For example, if the mean pulmonary artery pressure is elevated to 50 mm Hg, the mean left atrial pressure is 26 mm Hg, and the pulmonary blood flow is elevated at 12 liters/min (calculated by the Fick principle), the pulmonary vascular resistance is 2 RU, a normal value. Alternatively, if the pulmonary artery pressure is 50 mm Hg, the mean left atrial pressure is 2 mm Hg, and the pulmonary blood flow is reduced to only 2 liters/min, the pulmonary vascular resistance is markedly elevated to 24 RU, a value above the systemic level.

CHAPTER 14

Cardiac Energetics and Myocardial Oxygen Consumption

The heart is a constantly working organ, and therefore its energy use is high. Since it must continually develop a high pressure, the left ventricle in particular consumes a large amount of energy, and while it constitutes only about 0.3% of the body weight in man it uses about 7.0% of body's resting oxygen consumption. The heart must be continuously active, and important differences exist between the metabolism of cardiac and skeletal muscle; thus, there are many more mitochondria in heart muscle (about 30% by volume), it can utilize more substrates than skeletal muscle (which relies predominantly on glucose), and under normal conditions the energy production in heart muscle is entirely by aerobic means (skeletal muscle can readily use the anaerobic pathway). In fact, as we shall see, cardiac muscle does not develop an appreciable "oxygen debt" during each contraction, unlike skeletal muscle which develops an oxygen debt that is "repaid" during periods of rest.

There are several ways in which the energetics of muscle can be considered. The *total energy* (*TE*) output of the muscle can be equated with the mechanical work (*W*) it performs plus the heat (*H*) liberated:

$$TE = W + H$$

The *total chemical energy used* (*TE$_c$*) by the muscle can also be determined by measuring the loss of high energy phosphate ($\sim P$) so that

$$TE_c = \sim P \text{ depleted}$$

In an aerobic muscle, such as the myocardium, we can equate the oxygen utilized in resynthesizing $\sim P$ with the total chemical energy so that:

$$TE_c = \dot{V}O_2$$

Since the O_2 consumption of the heart, or the left ventricle, can be readily determined, whereas total heat release from the heart or its total high energy phosphate usage cannot, most of our knowledge concerning energy expenditure by the heart is based on determinations of myocardial oxygen consumption ($M\dot{V}O_2$) under varying conditions. However, it will be worthwhile to first consider briefly heat production and high energy phosphate use in isolated skeletal and cardiac muscle.

HEAT PRODUCTION BY MUSCLE

The experimental techniques and description of heat release measurements in isolated skeletal muscle (the frog sartorius muscle is often used) are described in detail in Section 1, Chapter 4. (See also Hill (1939) and Gibbs and Chapman (1979) for a review of cardiac muscle energetics.) When such a muscle is at rest and not contracting, if it is deprived of O_2 its resting heat production is reduced by about one-half, and lactic acid and other products of anaerobic metabolism accumulate. When oxygen is then resupplied a corresponding amount of extra heat production can be measured, which corresponds to repayment of the "oxygen debt." If the resting muscle is stimulated by frequent repetitive electrical depolarizations to contract isometrically (tetanic contraction), an immediate burst of additional heat production is measured. This heat release begins even before the onset of contraction (it is termed the *initial heat*), and during maintenance of the tetanic contraction a steady output of heat above the resting level is also observed *(maintenance heat)*. At the end of the tetanic contraction, heat production rapidly falls back toward, but not to, the resting level, and a small but prolonged output of additional heat occurs for several minutes *(the recovery heat)*, which is approximately equal to the initial heat. If the O_2 consumption is simultaneously measured the late recovery heat follows the time course of an additional consumption of O_2 above the basal level. Finally, if the contraction takes place under anaerobic conditions, the initial heat occurs without a measurable recovery heat.

Therefore, in skeletal muscle, the recovery heat after normal contractions is related to delayed oxidative "recovery" metabolism.

A component of the initial heat has been identified that is not related to the mechanical contraction of the muscle. The measurement is made by overstretching skeletal muscle to about 3.5 μm (so that no overlap of the myofilaments is present), and under these conditions electrical stimulation still produces some heat liberation (the *activation heat*). This heat is presumed to be related to the release and reuptake of activator calcium from the sarcoplasmic reticulum by energy-dependent processes.

Of course, contractions of the heart are normally not isometric, and measurements have also been made during isotonic contractions of skeletal muscle against varying afterloads. Under these conditions, W. O. Fenn (Fenn, 1923) showed that *more heat* is liberated by the isotonic contraction (which shortens actively) than by an isometric contraction developing the *same* tension. (The isotonic contraction is at a longer resting muscle length so that it can develop the same active tension as the isometric contraction (Fig. 2.144, *lower panel*).) It was found, moreover, that the extra heat associated with shortening was proportional to the work performed which was later correlated with oxygen and high energy phosphate use (Fig. 2.144), and this phenomenon is often called the "Fenn effect." A. V. Hill and others have developed equations for calculating the total energy output of a contracting muscle using certain constants, the various heat components discussed above, plus measurement of the mechanical work performed (tension $\times$ shortening).

Heat Release in Cardiac Muscle

Measurement of the heat production by isolated cardiac muscle has proved to be more difficult than in skeletal muscle, since single twitches must be analyzed rather than sustained tetanic contractions. However, measurements have been made in mammalian muscle during single twitches (Gibbs and Chapman, 1979), and significant differences have been observed in the patterns of heat release compared to skeletal muscle. The resting heat release is higher in cardiac muscle. Also, it has been found that following single twitches of cardiac muscle, the recovery heat is negligible. This finding suggests that oxidative recovery metabolism begins *during* contraction and is nearly or entirely complete by the end of the contraction, so that an "oxygen debt" does not occur. Such a conclusion is supported by studies in which NADH fluorescence was measured during cardiac muscle contractions

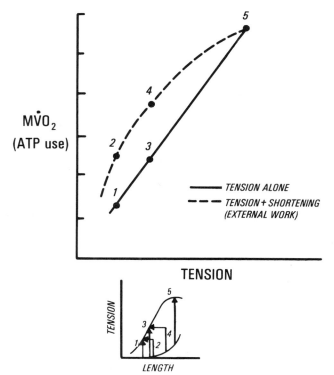

Figure 2.144. Relations between energy used by the myocardium ($M\dot{V}O_2$, *ATP*) and isometric tension development *(solid line)* or tension development associated with shortening *(dashed line)* to perform external work (the Fenn effect). The *dots* indicate isometric and isotonic contractions which are diagrammed in the lower panel. Three isometric contractions (beats *1, 3,* and *5*) are shown *(vertical arrows)*, and when the resting lengths in beats *1* and *3* are increased to those of beats *2* and *4,* isotonic contractions can be produced at the same level of tension that was produced isometrically in beats *1* and *3,* and extra energy is liberated. Beat *5* is at the peak of the isometric length-tension curve, no shortening occurs, and the *solid* and *dashed lines* in the *upper panel* converge (point *5*).

(Chapman, 1972). In these experiments, a transient reduction in fluorescence occurred during the onset of contraction (NADH $\rightarrow$ NAD in the respiratory chain).

In cardiac muscle that has been shortened markedly so that no tension can be developed, heat release that may be related to the activation heat has been measured. When the muscle is then stimulated electrically the ensuing series of action potentials is accompanied by a so-called "tension independent heat" (Gibbs and associates, 1969; 1979). The amount of this heat varies importantly in magnitude when the level of inotropic state of such noncontracting muscles is varied; for example, if catecholamines or a digitalis glycoside is administered, the tension-independent heat increases substantially (Fig. 2.145), suggesting that oxidative metabolism is importantly stimulated by alterations in contractility.

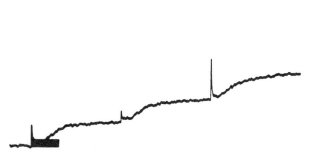

Figure 2.145. Tension-independent heat production in cardiac muscle which is markedly shortened so that no tension development can occur. The upright electrical spikes initiate three contractions in each panel which produces an associated burst of heat release, as shown. The *left tracing* is under control conditions, and the *right tracing* was obtained after addition of a digitalis preparation. (From Gibbs and Gibson, 1969.)

HIGH ENERGY PHOSPHATE USE BY HEART MUSCLE

As discussed in detail in Section 1, Chapter 4, the contractile event in both skeletal and cardiac muscle is closely linked to the hydrolysis of ATP by myosin ATPase. The ATP is rapidly replenished from the larger pool of creatine phosphate (CP) in equilibrium with ATP, and these high-energy phosphates are rapidly resynthesized by oxidative metabolism in heart muscle. If oxidative metabolism is blocked by removal of O_2, and if resynthesis of ATP by glycolysis is also blocked by use of iodoacetate, CP loss as the muscle contracts can be measured by rapid freezing and chemical analysis of the muscle after a given number of contractions. Under these conditions, the quantity of creatine phosphate hydrolyzed during isometric contractions is directly related to the amount of active tension developed and the number of contractions (Davies, 1966).

It has further been shown in cardiac muscle metabolically blocked in this manner (Pool and Sonnenblick, 1967) that if the muscle is stimulated to contract in the presence of a positive inotropic agent there is *increased* depletion of CP. Thus, under these conditions, for a given number of contractions that develop the same integral of tension as in the absence of the inotropic agent, more CP is hydrolyzed. Observations have also been made in isolated cardiac muscles while the O_2 consumption of the muscle was determined. Oxygen consumption is nearly linearly related to the level of peak tension developed in isometric contractions, and an extra O_2 cost of shortening against a load in isotonic contractions was determined (Fig. 2.144). Finally, it was shown that if a positive inotropic drug such as digitalis is administered, there is an increased oxygen consumption at all levels of isotonic tension (Fig. 2.146) (Coleman, 1967).

A COUPLED FRAMEWORK FOR CONTRACTION AND METABOLISM

The studies described above on heat release, high energy phosphate usage, and oxygen consumption in isolated muscle suggest that a close coupling exists between the initial chemical and mechanical events and the recovery metabolism. As shown in Figure 2.147, electrical activation produces calcium release, thereby triggering mechanical contraction which, in turn, determines how much ATP and CP are utilized. Thus, the initial chemical (ATP use) and mechanical (contractile) events do not involve oxidative metabolism; however, they *determine* the rate of oxygen consumption because of the close coupling with the recovery metabolism (Fig. 2.147), which in heart muscle appears to occur *during* the contraction. Uncoupling of oxidative phosphorylation could, of course, result in "slippage of the gears" (at coupling B), with more O_2 consumed than is used in high energy phosphate resynthesis. Also, there is evidence that different types of cardiac work require different energy expenditures, so that coupling A (Fig. 2.147) may be variable. Finally, as shown in Figure 2.147, events that do not directly relate to mechanical contraction, such as increased use of high energy phosphates by the calcium pump in the sarcoplasmic reticulum during inotropic stimulation of the myocardium (shown as "maintenance processes," many of which also use ATP; Fig. 2.147) could also increase O_2 consumption.

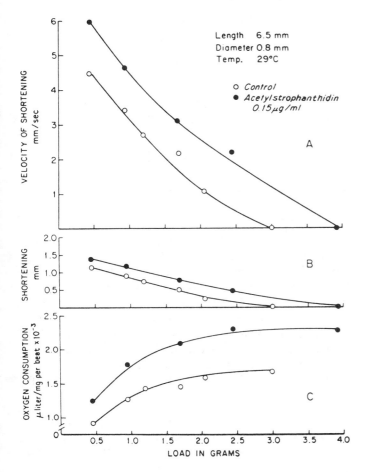

OXYGEN CONSUMPTION OF THE WHOLE HEART

Oxygen consumption ($M\dot{V}O_2$) of the whole heart can be determined using the Fick equation, provided the arteriovenous (AV) O_2 difference across the coronary bed is known, and the coronary blood flow can be measured:

$$M\dot{V}O_2 = CBF \times AVO_2.$$

Often, the O_2 consumption of the left ventricle alone is determined by measuring the O_2 content of blood in the large vein which drains that chamber (the coronary sinus) and in an artery, with measurements of coronary blood flow in experimental animals by means of a flowmeter on the left main coronary artery, or use of the microsphere technique, or in human subjects by using an indirect technique (Chapter 15).

The O_2 content of venous blood draining the heart is low relative to that of other organs, giving a wide arteriovenous O_2 difference. Although the percent-

Figure 2.146. Oxygen consumption in isolated cardiac muscle before *(open circles)* and after the administration of a digitalis preparation acetyl strophanthidin *(closed circles)* into the muscle bath. Notice that at the same levels of afterload, the velocity of shortening is increased with some increase in the extent of shortening, together with stimulation of the oxygen consumption. (From Coleman, 1967.)

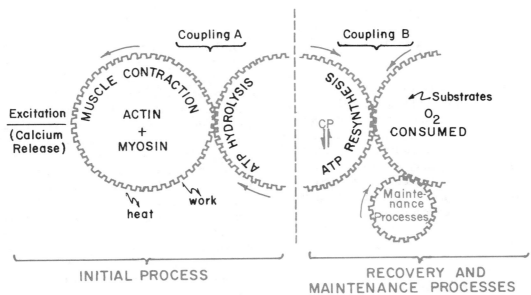

Figure 2.147. Diagram of mechanochemical coupling in cardiac muscle. The interaction of actin and myosin during contraction is initiated by Ca^{2+} release *(left-hand panel, initial process)* and utilizes ATP through *"Coupling A."* Oxidative phosphorylation provides for the formation of creatine phosphate *(CP)* and adenosine triphosphate (*ATP*),

CP serving as a reservoir for high energy phosphate bonds available for ATP resynthesis. Oxygen utilization is related to CP and ATP synthesis through *"Coupling B."* Oxygen is also consumed in various cellular maintenance processes. (Redrawn from Braunwald et al., 1967.)

age of O_2 extracted (AVO_2/AO_2) can increase some-what during severe exercise or other forms of stress, this reserve is small, and whenever cardiac O_2 demands rise or fall, it is primarily through adjustments in the coronary blood flow that oxygen delivery to the myocardium is varied.

DETERMINANTS OF MYOCARDIAL OXYGEN CONSUMPTION (MV̇O₂)

A number of factors contribute to the O_2 consumption of the beating heart. (For reviews see Braunwald et al., 1976; Gibbs and Chapman, 1979.) For simplicity, we will focus mainly on the factors that affect the energy requirements of the major pumping chamber, the left ventricle, which consumes about 8.0 ml O_2/100 g of myocardium/min in a normal human subject at rest. These determinants include:

Basal requirements
Activation energy
Heart rate
Tension or pressure development
Shortening against a load
Level of inotropic state

Basal O₂ Consumption

In the potassium-arrested mammalian left ventricle, the O_2 consumption is about 1.5 ml/100 g myocardium/min or almost 19% of the O_2 consumption of the normally contracting ventricle (Boerth et al., 1969). These basal energy requirements are related mainly to maintenance of chemical processes of the cell unrelated to active contraction.

Activation Energy

The amount of O_2 expended in activating contraction is unknown for the whole heart although, as discussed previously, the tension-independent heat and the related O_2 cost of changing inotropic state may be related to extra activation energy expended in operation of the sarcoplasmic reticular calcium pump. The energy required for *electrical* activation *alone* (action potential and repolarization) has been measured in the whole heart with calcium removed from the perfusing medium, and this O_2 cost is small (less than 0.05% of the total MV̇O₂) (Klocke et al., 1966).

Four Major Determinants

Those factors which remain primarily determine the MV̇O₂ of the intact, contracting heart as it changes its activity (Fig. 2.148). They are related to the performance of the myocardium and include the following.

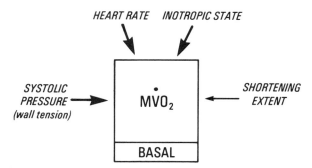

Figure 2.148.   Diagram of the major determinants of myocardial oxygen consumption, above the basal level.

HEART RATE

The frequency of cardiac contraction is a very important determinant of MV̇O₂. There is a nearly linear relation between increases in heart rate and increases in cardiac O_2 requirement. In fact, under controlled experimental conditions when the heart is contracting isovolumetrically (or when the stroke-volume is held constant in the ejecting heart), as the heart rate is doubled by electrical pacing, the MV̇O₂ approximately doubles to match the increased number of beats per minute (Fig. 2.149) (Boerth et al., 1969).

SYSTOLIC PRESSURE AND TENSION DEVELOPMENT

An almost linear increase in MV̇O₂ also results as either the peak systolic pressure or the calculated peak systolic tension developed by the left ventricle is increased (Fig. 2.150). This occurs either in ventricles made to contract isovolumetrically or in the ejecting heart.

Under many conditions, the product of systolic blood pressure × the heart rate been found to adequately reflect changes in MV̇O₂. For example, in the isolated supported heart preparation, the so-called "tension-time index" (TTI), which constitutes the integrated area under the left ventricular pressure curve per beat or per minute was found to correlate very highly with the O_2 consumption per beat or per minute. This index, developed by S. J. Sarnoff (Sarnoff et al., 1958b), was highly useful in an experimentally controlled setting, in which the heart is separated from autonomic connections, and spontaneous changes in inotropic state do not occur. However, as discussed subsequently, when changes take place in the level of inotropic state the TTI is no longer a useful predictor of MV̇O₂ (see below).

EJECTION OF BLOOD (WALL SHORTENING)

Shortening of the muscle in the ventricular wall as the stroke-volume is ejected is a less important, but nevertheless significant determinant of O_2 consumption (Fig. 2.148) and constitutes the equiva-

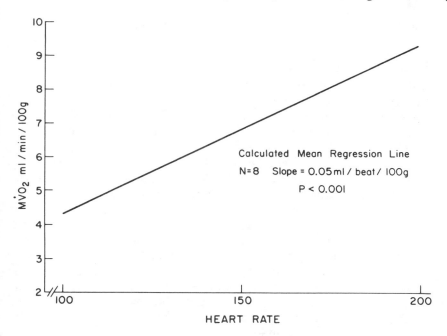

Figure 2.149. Relation between the heart rate and myocardial oxygen consumption in hearts in which the peak wall tension did not change. (From Boerth et al., 1969.)

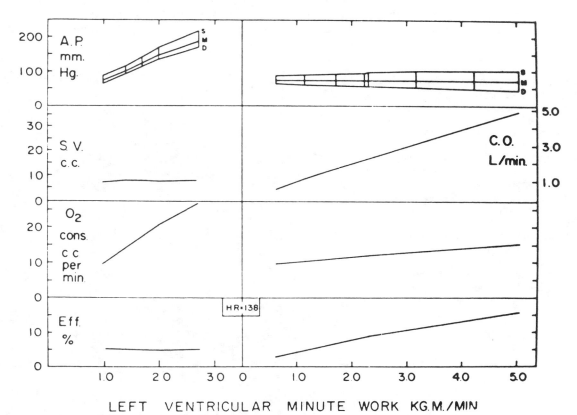

Figure 2.150. Data obtained in the isolated supported heart showing the hemodynamic determinants of cardiac O_2 consumption, left ventricular minute work, and efficiency *(Eff.)*. Heart rate was constant at 138 beats/min. In the *left-hand panel,* increasing the aortic pressure *(A.P.)* without a change in the stroke-volume markedly increases the O_2 consumption per minute, and the efficiency (the ratio of work to O_2 consumption) does not change. In the *right-hand panel* the cardiac output *(C.O.)* is markedly increased while the aortic pressure is held constant. The cardiac O_2 consumption increases only slightly, and therefore the cardiac efficiency increases markedly. (From Sarnoff et al., 1958b.)

lent of the Fenn effect (muscle shortening against a load) discussed earlier in isolated muscle. This extra O_2 cost is proportional to the extra work performed (Fig. 2.144). Recent studies indicate that the volume work of the ventricle ejecting under normal conditions contributes an extra 15% to the $M\dot{V}O_2$ above that which would have been developed with the ventricle contracting isovolumetrically at the same systolic pressure (Fig. 2.144) (Burns and Covell, 1972).

It has been known for many years that cardiac work (stroke volume × mean pressure) which is performed by ejecting a relatively large stroke-volume against a low aortic pressure is less costly in terms of O_2 consumption than the same amount of work (stroke-volume × mean pressure) performed against a high pressure. This is illustrated in Figure 2.150 in which, at constant heart rate, aortic pressure and cardiac work are raised while stroke-volume is held constant, or stroke-volume and work are raised while mean aortic pressure is held constant. $M\dot{V}O_2$ rises more steeply with increased "pressure work," compared to increased "volume work" (Fig. 2.150). This important point relates to the lower energy cost of shortening than of pressure

development, and will be discussed further under the section on cardiac efficiency.

LEVEL OF INOTROPIC STATE

When the above three factors (the heart rate, mean pressure or tension, and the stroke-volume) are held constant in an experimental setting another important determinant of $M\dot{V}O_2$ can be readily demonstrated (Sonnenblick et al., 1965; Ross et al., 1965b). This determinant was not recognized until studies were made in relatively intact animals during stresses such as exercise, types of interventions that were not examined in the heart-lung preparation. In such an experiment, when a substance having a positive inotropic effect on the myocardium such as calcium, digitalis, or norepinephrine is administered, an additional stimulation of $M\dot{V}O_2$ occurs (Sonnenblick et al., 1965). As shown in Figure 2.151 the administration of norepinephrine produced a marked increase in the velocity of contraction evidenced by an increased maximum rate of pressure development (peak dP/dt), and despite a constant heart rate and stroke-volume, and nearly unchanged arterial pressure, the $M\dot{V}O_2$ has almost doubled. At the same time, the positive

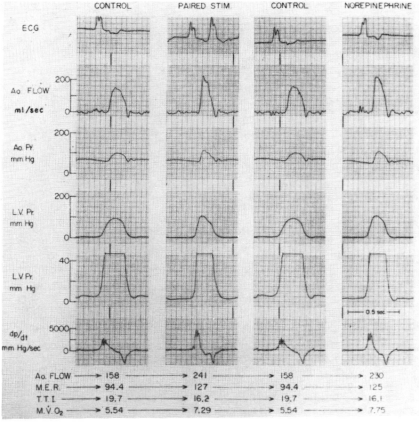

Figure 2.151. Effects of stimulating the inotropic state of the heart at a constant heart rate, stroke volume, and mean aortic pressure. In the *left two panels* is shown the positive inotropic effect of paired electrical stimulation (repeated postextrasystolic potentiation), and in the *right two panels* the effect of norepinephrine. *Ao. FLOW,* aortic flow measured with an electromagnetic flowmeter; *Ao. Pr.,* aortic pressure; *L.V. Pr.,* left ventricular pressure shown at full scale and at high gain to show the left ventricular end-diastolic pressure, and the first derivative of the left ventricular pressure (*dP/dt*). With norepinephrine, for example, the peak velocity of flow increases, the duration of contraction shortens, and the dP/dt rises. Little change occurs in the determinants of $M\dot{V}O_2$ other than inotropic state, although wall shortening must increase as well. $M\dot{V}O_2$ increases from 5.54 to 7.75 ml/min. (From Sonnenblick et al., 1965.)

inotropic stimulus shortens the duration of contraction, and the area under the left ventricular pressure pulse is diminished (Fig. 2.151); since the heart rate is constant in this experiment, the TTI falls while $M\dot{V}O_2$ increases, and it is clear that under such conditions directional changes in TTI or in the heart rate blood pressure product do *not* follow changes in $M\dot{V}O_2$. The mechanisms for changes in $M\dot{V}O_2$ with increased inotropic state, which sometimes have been termed an "oxygen-wasting" effect, are not fully elucidated. They may be related to energy costs associated with the calcium pump, as suggested earlier, or to increased velocity and extent of shortening (Sonnenblick et al., 1965), with additional kinetic energy expenditure (Chapter 7).

If a negative inotropic substance is given, such as a large dose of a β-blocking drug or a barbiturate, the $M\dot{V}O_2$ falls when the other three determinants of $M\dot{V}O_2$ are held constant in the experimental setting (Graham et al., 1967).

INTERRELATIONS BETWEEN THE FOUR DETERMINANTS OF $M\dot{V}O_2$

Changes in the relative importance of any of the determinants of $M\dot{V}O_2$ can occur, and under some circumstances either the heart rate, aortic pressure, or level of inotropic state can become the most important factor influencing energy requirements of the heart (Fig. 2.152). For example, if a marked increase in the blood pressure is produced by infusing a vasoconstrictor drug such as phenylephrine (which has little inotropic effect), the oxygen consumption of the heart will rise largely due to the energy cost of increased pressure development, even though some reflex slowing of heart rate occurs (Fig. 2.152A). If a positive inotropic agent such as dopamine is administered intravenously, stimulation of $M\dot{V}O_2$ will occur through increased myocardial contractility (Fig. 2.152B). Electrical pacing of the heart at a rate twice normal will double the $M\dot{V}O_2$, largely through the increased number of contractions per minute, with a small contribution from enhanced inotropic state (the staircase effect) (Fig. 2.152C). The response to exercise is shown in Figure 2.152D. At rest, sympathetic (adrenergic) tone is low, and there is little stimulation of the myocardial inotropic state. However, during exercise a number of factors operate together to markedly increase $M\dot{V}O_2$. These include activation of the sympathetic nervous system with an increase in the heart rate and myocardial inotropic state, and as exercise becomes severe the systolic blood pressure increases and stroke volume rises (Fig. 2.152D). Of course, if an increased stroke-volume accompanies the augmented cardiac output during exercise, it will also contribute to the increased $M\dot{V}O_2$.

ESTIMATING CARDIAC OXYGEN CONSUMPTION

A variety of methods has been developed to estimate the cardiac $M\dot{V}O_2$ from measurements of cardiac activity under various conditions. One is the product of mean arterial pressure and heart rate (the so-called "double product"); the "tension time index" has been discussed. While often useful for estimating directional changes of $M\dot{V}O_2$, they have limitations for more quantitative estimates, particularly when pronounced changes in myocardial inotropic state occur. A formula which uses the rate-pressure product $\times$ the left ventricular work, plus the O_2 consumption of the resting heart, has a relatively low error of prediction and also accounts for the effects of stroke-volume and therefore external work (Rooke and Feigl, 1982). This index seems less likely to account for effects of changes in the inotro-

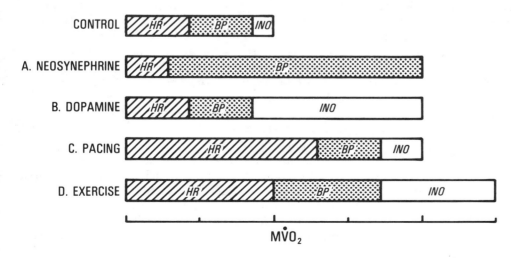

Figure 2.152. Diagram showing the manner in which various major determinants of cardiac energy utilization could influence the myocardial oxygen consumption. *HR*, heart rate; *BP*, blood pressure; *INO*, inotropic state. For further discussion, see text.

pic state on M$\dot{V}O_2$ as, for example, when heart rate, stroke-volume, and pressure are held constant while norepinephrine is administered (Fig. 2.151) (Sonnenblick et al., 1965).

Another correlate of M$\dot{V}O_2$, the pressure-volume area (PVA), which consists of the area within the pressure-volume loop of left ventricular contraction (stroke work) *plus* the area enclosed by the end-systolic pressure-volume relation (determined from multiple points at end-systole) extends beyond the end-systolic portion of the pressure-volume loop down to the zero volume intercept. The latter area lies between the extrapolated linear end-systolic pressure-volume relation and the extrapolated diastolic pressure-volume relation (see *cross-hatched area*, Fig. 2.122), and provides a measure of the "end-systolic potential energy" converted into heat instead of work (Suga, 1979). This approach accounts for the effects of stroke-volume and also for the greater M$\dot{V}O_2$ observed at the same stroke work when the work is augmented by increasing pressure compared to increasing stroke-volume (see Fig. 2.150); thus, the contribution of the "potential energy" area is larger when the stroke work is augmented by increasing systolic pressure than when increasing stroke-volume (Suga et al., 1982). The PVA method for estimating M$\dot{V}O_2$ also can be used to assess the influence of alterations in inotropic state. With positive inotropic stimulation, the linear end-systolic pressure-volume relationship is steepened and shifted upward by positive inotropic stimulation (see Fig. 2.124). The PVA is therefore increased, and over a range of work levels an added O_2 cost is predicted by the PVA compared to control conditions without inotropic stimulation (Nozawa et al., 1987) (Fig. 2.153).

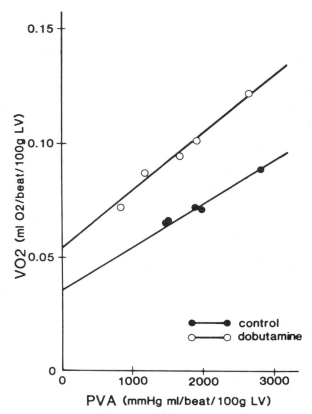

Figure 2.153. Relationship between the "pressure-volume area" (*PVA*) and myocardial oxygen consumption (*VO₂*) per 100 grams of left ventricle (*LV*) before *(closed circles)* and during dobutamine infusion to increase contractility *(open circles)*. Notice that at both levels of inotropic state, the relationship between these two variables is linear (the pressure-volume area being altered over a wide range under both conditions by changing aortic pressure and venous return to the heart). The entire linear relationship is shifted *upward* during dobutamine infusion *(open circles)*. Note that extrapolation of the relation to zero on the PVA axis indicates a higher oxygen consumption at zero cardiac work, suggesting an "oxygen wasting" effect of dobutamine. (Modified from Nozawa et al., 1987.)

CARDIAC EFFICIENCY

Efficiency represents the fraction of the total energy of contraction that appears as mechanical work. In the whole heart, the M$\dot{V}O_2$ represents the total energy expended, and the work is usually determined as the so-called "effective" or "external work" (equal to the product of the stroke-volume and the mean aortic pressure, or the mean systolic left ventricular pressure)*:

$$\text{Efficiency} = \frac{\text{cardiac work/min}}{\text{M}\dot{V}O_2}$$

---

*To solve the efficiency equation it is necessary to employ in the equation the "work equivalent" of M$\dot{V}O_2$. One milliliter of O_2 has a caloric equivalent of 4.86 calories, and the constant for conversion of calories to kilogram meters is 0.4268. Therefore, the energy equivalent of 1 ml of O_2 is 2.06 kg m, the units in which work is generally expressed.

The kinetic factor of cardiac work ordinarily comprises less than 5% of the total work, and this factor is usually neglected. Often, since the resting M$\dot{V}O_2$ does not contribute directly to energy expended in contraction, it is subtracted from the total M$\dot{V}O_2$, and under these conditions the external efficiency of the left ventricle in anesthetized animals has been calculated at about 15%.

In considering the efficiency of the whole heart, several factors are of importance. As mentioned, the heart uses relatively little extra oxygen to deliver an increased cardiac output (and substantially increase the minute work) at a normal aortic pressure (low oxygen cost of the Fenn effect), making the ratio of work to M$\dot{V}O_2$ high and hence increasing the efficiency (Fig. 2.150). With increasing aortic pressures at a constant cardiac output, however, $M\dot{V}O_2$ is elevated substantially as the minute work increases and efficiency may not change (Fig. 2.150); the stroke-volume also can fall as aortic pressure is elevated, and work may actually decrease as $M\dot{V}O_2$ rises, leading to decreased efficiency.

Enhanced inotropic state and heart rate stimulate $M\dot{V}O_2$, and it might be expected that under conditions of mild exercise, particularly in an untrained individual (in whom the heart rate response is large and the sympathetic nervous system response is substantial), the increase in blood pressure and cardiac output (work) will be small, so that cardiac efficiency may diminish substantially. The effects of β-adrenergic blockade may be useful under these circumstances in a patient who has angina pectoris (see Chapter 15), since the heart rate response and inotropic stimulation during exercise are markedly attenuated by β-blockade; the venous return and cardiac output increase under these circumstances, but the cardiac output now rises through an increase in the stroke-volume (use of the Frank-Starling mechanism), rather than by an increase in heart rate and inotropic state. This leads to an increase in cardiac efficiency, since the minute work is the same, but the $M\dot{V}O_2$ is substantially lower than prior to β blockade.

SUBSTRATE UTILIZATION BY THE MYOCARDIUM

The details of intermediary cellular metabolism, including aerobic and anaerobic glycolysis and oxidative metabolism, are presented in Section 1, Chapter 4. Important differences exist between the metabolism of cardiac and skeletal muscle, particularly in relation to the large number of substrates that heart muscle can utilize to generate high energy phosphates, predominantly by aerobic means. Fatty acids are utilized in preference to glucose when they are available, but also lactate, pyruvate, ketone bodies, and amino acids can be metabolized (Table 2.10). Normally 60–70% of oxidative metabolism comes from fatty acids, although intermediates formed during aerobic glycolysis are used in the citric acid cycle. During cardiac work overload, glycolysis becomes a more important source of high energy phosphate production through aerobic glycolysis, although anaerobic glycolysis occurs to some extent in heart muscle during ischemia as discussed briefly below. A few key features of fatty acid and carbohydrate metabolism in the heart should be noted. (For detailed review see Morgan and Neely, 1982.)

Fatty Acid Metabolism

Circulating plasma triglycerides are broken down to free fatty acids at capillary and cell membranes by lipoprotein lipases, and the subsequent uptake of fatty acids by myocardial cells is not energy dependent but depends on their plasma concentration. There is also some intracellular triglyceride storage, and these endogenous triglycerides can be mobilized to form free fatty acids when there is deprivation of glucose or of exogenous free fatty acids. Smaller activated fatty acid subunits are utilized within the mitochondria, and longer chain fatty acids employ the carnitine transport system to enter the mitochondrial matrix where β oxidation makes acetyl-CoA available for oxidation in the Kreb's cycle, with eventual production of high energy phosphates in the respiratory chain. Free fatty acid oxidation is controlled by the rate of removal of acetyl-CoA in the citric acid cycle, and a high NADH:NAD ratio inhibits β oxidation.

During increased cardiac work the rate of oxidative phosphorylation is increased; feedback regulation of the citric acid cycle occurs through the ratios of ATP:ADP and NADH:NAD. Thus, accumulation of ADP with a reduction of the ATP:ADP ratio and accumulation of NAD, with lowering of the NADH-NAD ratio, stimulate the cycle, and also accelerate high energy phosphate production in isolated mitochondria.

Glucose Metabolism

Glucose transport across myocardial cell membrane occurs by facilitated diffusion and is related to the blood glucose level. In the absence of insulin, there is a threshold for glucose transport of about 60 mg/100 ml, but in the presence of insulin this threshold is lowered to about 10 mg/100 ml. Increased cardiac work results in increased glucose transport into the cell, whereas oxidation of free fatty acids inhibits glucose transport into the cell.

Within the myocardial cell, glucose can then either be stored as glycogen or enter the aerobic glycolytic pathway, which supplies about 30% of the oxidative production of high energy phosphates, as

Table 2.10.
Percentage Contribution of Various Metabolic Substrates to Total Myocardial Oxygen Usage

Carbohydrate, %		Noncarbohydrate, %	
Glucose	17.90	Fatty acids	67.0
Pyruvate	0.54	Amino Acids	5.6
Lactate	16.46	Ketones	4.3
Total	34.90[a]	Total	76.9[a]

[a]The total is more than 100% because some of the substrates removed from the blood, in amounts measured by the arteriovenous differences, were stored by the heart or not completely metabolized. (From Bing, 1955.)

acetyl-CoA produced from pyruvate by the action of pyruvate dehydrogenase enters the mitochondria. Product inhibition largely regulates the rate of aerobic glycolysis, the conversion of fructose-6-phosphate to fructose-1-6-diphosphate by phosphofructokinase being inhibited by ATP, CP, and citrate, and pyruvate dehydrogenase being inhibited by ATP, NADH, and acetyl-CoA.

Hypoxemia and Ischemia

Inadequate O_2 content in the blood, or *hypoxemia*, as may occur on breathing low O_2 tension (or in congenital heart disease in which there is right to left shunting of blood) tends to result in high rates of blood flow through the organs of the body. Vasodilation therefore serves as the predominant means by which O_2 delivery to the tissues is maintained. *Ischemia*, on the other hand, is a condition in which inadequate O_2 delivery to the tissues occurs because of low blood flow, usually without hypoxemia of the arterial blood (see Rovetto et al., 1973). Not only is there inadequate O_2 delivery, but the low blood flow fails to wash out accumulating metabolic products and hydrogen ions, which can lead to further impairment of metabolism.

With hypoxemia, O_2 lack slows oxidative production of ATP but removes inhibition of phosphofructokinase and stimulates the glycolytic production of high energy phosphates. Accumulation of NADH also promotes the production of lactate from pyruvate. In heart muscle, such lactate production occurs *only* during hypoxemia or ischemia (Rovetto et al., 1973).

With ischemia as may occur with insufficient coronary blood flow due to coronary atherosclerosis (see Chapter 15), glycolysis initially is stimulated, but failure to wash out lactate and other acid products because of the reduced blood flow leads to inhibition of phosphofructokinase; accumulation of NADH and other products inhibits the action of pyruvate dehydrogenase, and marked slowing of glycolysis ensues. NADH accumulation also inhibits other steps in the glycolytic pathway.

When heart muscle is completely deprived of O_2 by coronary occlusion, glycogen stores are rapidly depleted, and there is insufficient ATP production by glycolysis to maintain normal ATP concentrations. Contraction rapidly diminishes in the involved region, and within about 30 s it ceases; insufficient high energy phosphate is produced to maintain even a basal level of cell metabolism, and irreversible cell damage occurs in about 40 min. Partial ischemia can be tolerated for a much longer period, but if low blood flow persists glycogen stores become exhausted, contraction is impaired, and tissue death eventually may result.

The Coronary Circulation

The coronary arteries are the first arterial branches to arise from the aorta just above the aortic valve. They are particularly vital arteries, since they supply blood directly to the muscle of the pumping chambers and to the conducting system of the heart, and control of their blood flow is intimately related to cardiac function. The heart must develop pressure and flow in the aorta to allow perfusion of all the other organs of the body, and it must also "autoperfuse" itself, in that it must develop sufficient pressure in the aorta to promote flow through its own vascular supply. Clearly, when the heart becomes too weak to accomplish this (as may occur, for example, during severe shock or the acute phase of a heart attack) a cycle of deterioration due to inadequate blood supply leading to further weakening of the heart ensues, and the blood pressure falls further with eventual brain death. (For reviews on the physiology of the coronary circulation see Klocke and Ellis, 1980; Feigl, 1983; Braunwald and Sobel, 1988.)

ANATOMY OF THE CORONARY CIRCULATION

The distribution of the coronary arteries varies among mammalian species. In most there are two main coronary arteries, the right and the left, which originate from the sinuses of Valsalva just above the aortic valve leaflets (see Fig. 2.6). In the dog, the right coronary artery is diminutive and provides blood only to the right ventricle and right atrium. In the pig and in primates, such as the baboon, the coronary anatomy tends to resemble that found in human subjects. In about 90% of human subjects, the right coronary artery is "dominant," being a large vessel that traverses down the groove between the right atrium and right ventricle, supplying branches to both of those chambers and then, upon reaching the posterior aspect of the heart, it turns toward the cardiac apex (the so-called posterior descending branch) and supplies blood to the lower (inferior) wall of the left ventricle (Fig. 2.6) (for details, see James et al., 1982).

This artery also usually supplies a branch to the atrioventricular node. The left coronary artery in man commences as a short left main branch and then divides into two branches, the anterior descending branch which supplies the septum between the right and left ventricles and the anterior surface of the left ventricle, and the circumflex branch which supplies the lateral aspect and part of the posterior surface of the left ventricle (see Figs. 2.6 and 2.168). There are many variations in coronary artery anatomy, and in some subjects a so-called "left dominant" pattern exists (the right coronary artery being small and more like that found in the dog) and the circumflex coronary artery supplies the inferior wall of the left ventricle. These patterns of distribution have importance, since whenever an artery becomes obstructed by an atherosclerotic lesion, specific areas of the heart that are involved by the lack of coronary blood flow can be identified (see Chapter 9, electrocardiographic changes in myocardial infarction).

The coronary arteries divide into smaller and smaller branches as they spread over the surface of the heart (the so-called epicardial arteries) and these give off tiny arteries that course at right angles through the heart muscle into the wall of the ventricle (intramural vessels). The heart has a rich blood supply, since the heart is continuously active and performs a large amount of work. Therefore, a large supply of oxygen per unit of muscle is necessary, and the potential capillary density in the myocardium is high. In the rat left ventricle, for example, when the coronary bed is maximally dilated the capillary number is as high as $3000/mm^2$ (Honig and Bourdeau-Martini, 1973). A constant supply of oxygen is necessary for this essentially aerobic organ, and the capillaries pass through the myocardium in close proximity to the working muscular elements and the mitochondria, yielding a relatively short diffusion distance for oxygen.

The venous blood from the heart is collected in venules which course with the small arteries and then converge on the left side of heart to form the coronary sinus, a large vein running in the groove between the left atrium and left ventricle and opening into the right atrium near the tricuspid valve. This vein is sufficiently large in humans that a plastic catheter or tube can be placed under fluoroscopic control directly within it to allow withdrawal of blood samples for determining oxygen content, or the content of other substances necessary for determining coronary blood flow. The veins from the right side of the heart (right atrium, right ventricle) do not form a common channel, but drain individually directly from the myocardium at multiple sites within the right atrium and right ventricle.

Coronary Collateral Vessels

Collateral channels are blood vessels (usually small) which allow blood to flow directly from one artery to another; hence, if an artery becomes obstructed, when collateral channels exist they can allow arterial blood to enter the blocked vessel beyond the site of obstruction (Fig. 2.154). In the dog, a large number of tiny collaterals exists even in the normal heart, and they can be readily seen under magnification as vessels somewhat larger than capillaries on the surface of the heart that provide direct anastamoses (connections) between adjacent arteries. Under normal conditions such collaterals appear to have little function, but they can undergo substantial enlargement if a coronary artery is occluded.

There is evidence that the pig has few collateral channels after coronary occlusion compared to the dog (Table 2.11); collateral flow can be collected from the distal end of an occluded vessel or measured with the microsphere technique (see below). In baboons and in most human subjects few collaterals are present under normal conditions. However, such channels may develop over time and become highly important when atherosclerotic disease involves the coronary arteries (Fig. 2.154) (For reviews on the coronary collateral circulation, see Schaper, 1971; Bloor, 1974.)

MEASUREMENT OF CORONARY BLOOD FLOW AND RESISTANCE

When a flow-sensing device (such as an electromagnetic flowprobe) is placed directly around one of the coronary arteries supplying the left ventricle, blood flow through the artery exhibits a unique pat-

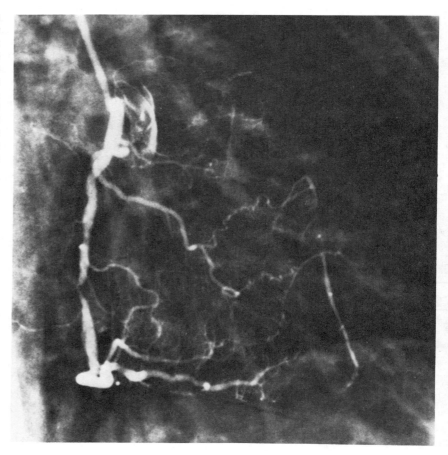

Figure 2.154. Coronary arteriogram with injection of contrast material into the right coronary artery (lateral view of the heart). The catheter is visible at the *upper left,* and several areas of narrowing, one nearly completely occluding the coronary artery, are evident. A small "bridging collateral" bypasses this very narrowed area, and other small collateral vessels can be seen reaching, and faintly opacifying, a vessel at the right which runs vertically. This is the left anterior descending coronary artery, which was found to be completely blocked at its origin from the left main coronary artery, and supplied only by these collaterals. (Courtesy of Dr. M. Judkins.)

Table 2.11.
Collateral Coronary Blood Flow after Acute Coronary Artery Occlusion as Determined by Three Different Methods in Three Animal Species

Species	Method			
	Retrograde Flow, %	TM Distribution, %	^{133}Xe ia, %	^{133}Xe im, %
Dog	8	16	33.5	11
Pig	0.5	5.3	23.5	
Sheep			35.0	

TM signifies tracer microsphere distribution. Xenon-133 was injected either into the coronary artery (ia) before occlusion, or directly into the heart muscle (im). Values are expressed as a percentage of the normal myocardial flow taken as 100%. Retrograde flow is collected directly from the cannulated distal end of a coronary artery after it is occluded. (From Schaper, 1971.)

tern. Blood flow through the coronary artery falls sharply during left ventricular systole, and with the onset of diastole it rises abruptly and then falls gradually during the remainder of the diastolic period (Fig. 2.155). A similar pattern of blood flow is seen in the right coronary artery when this vessel is "dominant" and supplies the posterior wall of the left ventricle as well as the right ventricle (as in human subjects). However, as might be anticipated, in the dog in which the right coronary artery supplies only the low-pressure right ventricle and atrium, the less forceful contraction of the right ventricle does not inhibit flow from the higher pressure aorta, and the pattern of flow therefore closely resembles the pressure pulse in the aorta, rising in systole and falling gradually during diastole. Such patterns indicate that the contraction of the *left ventricle* squeezes the small intramural arteries, so that the systolic aortic pressure cannot maintain

blood flow through the increased resistance offered by the compressed vessels, producing a marked fall in coronary flow during systole in that chamber (Fig. 2.155).

Coronary blood flow was first measured in man by the *inert gas washout technique* (Kety, 1960), a method which is still sometimes employed during cardiac catheterization. The procedure usually employed for determining coronary flow involves having the subject inhale a mixture of room air and an inert gas for a number of minutes, until the concentration of inert gas in the blood is in equilibrium with that in the heart muscle, in accordance with its partition coefficient. The inhalation of the gas is then abruptly stopped, and a series of blood samples is drawn simultaneously from a needle placed in a peripheral artery and from a catheter positioned in the coronary sinus (cardiac vein). Since the gas is selected to have a low solubility, it is rapidly eliminated (blown off) as blood passes through the lungs, and therefore the arterial concentration rapidly falls toward zero (Fig. 2.156). However, the

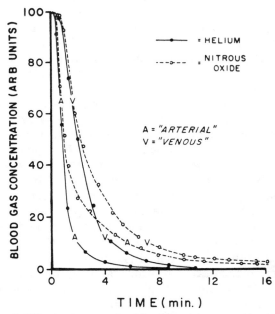

Figure 2.156. Inert gas washout for determining organ blood flow by inhaling the gas to equilibrium, followed by cessation of inhalation of the foreign gas and measurement of its concentration in serial samples of arterial and venous blood. (In the case of the heart, samples are taken from an artery and the coronary sinus). These data are from an in vitro model system to illustrate the principle. The more rapid elimination from the arterial blood and from the organ of the less soluble gas helium compared to nitrous oxide is shown; knowing the initial quantity of either gas taken up by the organ (such as the myocardium) per unit mass (from the partition coefficient of the gas relative to blood), and the integrated arteriovenous difference of the gas, the same blood flow per gram of myocardium is calculated using both gases. (From Klocke et al., 1969.)

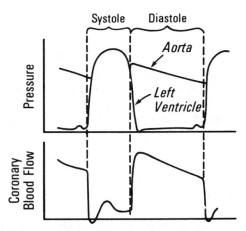

Figure 2.155. Diagram of aortic and left ventricular pressure pulses and blood flow through the left main coronary artery of a dog measured with a flowmeter. The abrupt drop in flow during systole is evident, and the diastolic flow tends to fall together with decreasing diastolic pressure in the aorta.

gas is more slowly eliminated from the organ under study, and the washout rate is proportional to the blood flow (Fig. 2.156). In accordance with a modified Fick equation, the arteriovenous difference can be calculated as the difference between the integrals of the arterial and coronary sinus concentrations (Fig. 2.156) and knowing the partition coefficient in ml of gas per gram of myocardium, coronary blood flow in ml/100 g left ventricle/min can be calculated. Gases such as hydrogen and helium have been employed, low concentrations of which can be readily measured by chromatographic methods (see Klocke et al., 1969). Using such inert gas techniques, the level of coronary blood flow in normal human subjects at rest is 60–90 ml/100 g left ventricle/min. During exercise and other forms of stress, coronary blood flow increases, and the coronary flow "reserve" is 4–5 times the basal level of flow.

Another technique for measuring coronary blood flow frequently used in experimental studies is the *radioactive microsphere technique*. The method is particularly useful in the heart because it allows determination of *regional myocardial blood flow*, including the *distribution* of blood flow across the ventricular wall. As applied for the measurement of coronary blood flow, a batch of radioactive microspheres (usually measuring 9–15 μm in diameter) is suspended in a saline-detergent solution and injected into the left atrium; it is mixed by blood flow within the left ventricle, ejected into the aorta, and then perfuses the coronary arteries. The microspheres lodge in only a few capillaries, so that there is no damage or effect on flow, but the number of spheres trapped per unit of myocardial tissue is directly proportional to the myocardial blood flow to that region. Following one or more injections, the experiment is terminated, the heart excised, and the radioactivity in samples of myocardium is directly measured in a gamma scintillation counter.

In practice, a number of isotopes are injected during experimental conditions at different coronary blood flows. Since each batch of microspheres is labeled with an isotope having a differing energy peak, the quantity of radioactivity at a given peak for a given isotope (i.e., an isotope of cesium or strontium) can be detected using a gamma spectrometer. In order to determine absolute myocardial blood flow, the so-called "reference withdrawal method" is used to determine a proportion between the rate of blood withdrawal from a sampling site and the counts in that sample, and the (unknown) coronary blood flow and the counts in the myocardium (Domenech et al., 1969). With this approach, blood is withdrawn by a motor-driven syringe from

the ascending aorta at a known rate during the injection of microspheres, and the following equation is applied:

$$Q_m/Q_r = C_m/C_r$$

where Q_m = blood flow per g myocardium, Q_r = blood flow into the syringe (reference sample, r), C_m = counts/min in myocardium, and C_r = counts/min in reference sample. Myocardial blood is then calculated as:

$$Q_m = \frac{C_m}{C_r} \cdot Q_r$$

In experimental animals, portions of the ventricular wall can be divided into a section from the endocardium, the midwall, and the epicardium, thereby allowing measurements of total transmural blood flow and its distribution across the wall.

Another method for determining coronary blood flow, which can be used in patients undergoing cardiac catheterization, is the *thermodilution technique* with coronary sinus sampling (Ganz et al., 1971). With this approach, a special catheter having an end hole is passed well up into the coronary vein (coronary sinus), a temperature sensor is placed further down on the catheter near the orifice of the coronary sinus, and the continuous injection method is used (see Chapter 13). Cold saline is injected continuously through the catheter and is diluted by the coronary sinus blood flow, the amount of dilution being proportional to the coronary sinus flow sampled at the thermister. Alternatively, rapid injection of the cold saline can be carried out using a spray tip catheter, and the bolus is sampled downstream in the coronary sinus from the same catheter.

Coronary Vascular Resistance

Determination of coronary vascular resistance (*CVR*) can be made from the ratio of mean pressure in the aorta (*P*), or in the coronary artery, to the mean coronary blood flow (*CBF*) per minute as:

$$\text{mean } CVR = P/CBF$$

If left ventricular pressure is measured, the ventricular diastolic pressure is subtracted from the aortic pressure, since this value provides the perfusion pressure difference across the coronary vascular bed. Coronary vascular resistance is high during systole, when the intramural arteries are compressed by the left ventricle, and when phasic blood flow is determined by the flowmeter technique, coronary vascular resistance is often measured at the end of diastole using the pressure and flow at that point in time. This method is free of mechanical effects on flow from left ventricular contraction and is therefore more representative of the tone of the coronary resistance vessels themselves.

MYOCARDIAL O₂ SUPPLY

Since extraction of O_2 by the coronary circulation is high, the major factor affecting O_2 delivery (O_2 supply) is the coronary blood flow (see next section). However, other factors also influence O_2 delivery. These include factors affecting O_2 availability: the P_{O_2} in arterial blood, the hemoglobin level, and the position of the oxyhemoglobin dissociation curve (see Section 5, Chapter 37). Also, several local factors including tissue P_{O_2} may directly or indirectly influence capillary density (see discussion of microcirculatory regulation in Chapter 6). Thus, while changes in blood flow through the major coronary arteries due to alterations in coronary arteriolar resistance are the most important determinant of O_2 delivery to the myocardium, other mechanisms may have an effect, particularly during stress or under abnormal conditions.

DETERMINANTS OF CORONARY BLOOD FLOW

The most important single factor determining the rate of coronary blood flow in the normal mammalian heart is the requirement of the heart for oxygen; thus, *metabolic* regulation of coronary vascular resistance is the major determinant. Systolic compression of the coronary arteries and coronary perfusion pressure affect coronary blood flow, particularly under abnormal conditions. Neurogenic control of coronary vascular resistance has also been demonstrated. These opposing determinants are summarized in Figure 2.157. Circulating neurohumoral factors and pharmacological agents also affect coronary vascular tone under some circumstances.

1. Metabolic control (O_2 demands)
2. Coronary perfusion pressure
3. Systolic compression
4. Autonomic nervous system
5. Circulating catecholamines and other vasoactive substances and drugs.

Metabolic Factors

The coronary blood flow can be shown to have a nearly linear correlation with the level of myocardial oxygen consumption ($M\dot{V}O_2$) in the normal heart (Fig. 2.158), coronary blood flow increasing or decreasing as the $M\dot{V}O_2$ increases or decreases (Eckenhof et al., 1947). As mentioned, the arteriovenous oxygen difference is wide in the heart under resting conditions; thus, O_2 extraction is nearly complete, the coronary sinus blood oxygen content being low (about 5 vol%) leading to an AVO_2 difference of about 12 vol% (the highest in the body). The requirements for increased oxygen by the heart are therefore met, not primarily by increased oxygen extraction, but by increased coronary blood flow.

Since the most important determinant of the $M\dot{V}O_2$ is the mechanical activity of the heart and its inotropic state, the determinants of $M\dot{V}O_2$ are also the determinants of coronary blood flow. They include the systolic ventricular pressure or wall tension, the heart rate, the amount of shortening against a load, and the level of inotropic state or contractility (see Chapter 14).

Precisely how the alterations in the $M\dot{V}O_2$ are translated into increases in coronary blood flow, and vice versa, remains controversial. Perfusion of the coronary circulation with blood having a low O_2 content is a potent stimulus to coronary vasodilatation, and increased K^+ concentration, decreased pH, and increased P_{CO_2} also have vasodilator effects. Vasodilator prostaglandins could play a role. Evidence has been accumulated to indicate that the major physiological vasodilator in the heart is the nucleoside adenosine, a metabolic product of ATP breakdown (Berne and Rubio, 1979). Thus, the concentration of adenosine (an extremely potent vasodilator) in the myocardium, interstitial space, and the coronary venous blood can be linked with level of coronary blood flow as follows:

$$ATP \rightarrow ADP \rightarrow AMP \rightarrow adenosine \rightarrow vasodilatation$$
$$\downarrow$$
$$inosine \text{ or } hypoxanthine$$

AMP is broken down by 5′-nucleotidase, located primarily in the sarcolemma and T tubules of car-

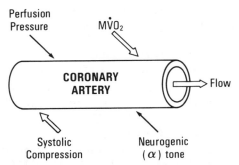

Figure 2.157 The determinants of coronary blood flow. The major determinant is metabolic, the myocardial oxygen requirement ($M\dot{V}O_2$); systolic compression limits forward flow, and a neurogenic increase in resistance is a minor determinant.

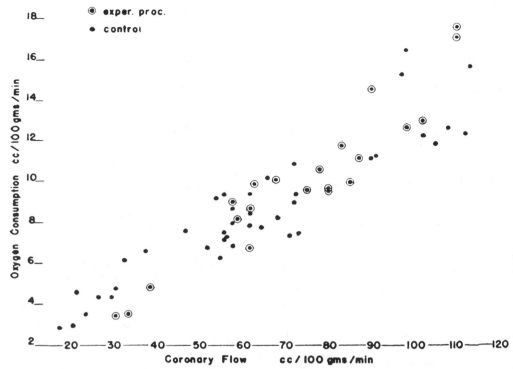

Figure 2.158. Relationship between MV̇O₂ and coronary blood flow. Various experimental procedures were used to alter cardiac work. (From Eckenhof et al., 1947.)

diac muscle cells, the dephosphorylation of AMP forming adenosine. Adenosine, in turn, is rapidly broken down by adenosine deaminase, an intracellular enzyme which is also found in the red blood cell, so that measurement of adenosine is complicated by its rapid disappearance. Adenosine may also reenter the myocardial cell.

The *adenosine hypothesis* is attractive, since it directly links the metabolic products of increased or decreased ATP usage to the regulation of coronary blood flow. Thus, as ATP usage increases, increased coronary blood flow would occur at a new steady state of adenosine production. Also, increased washout of this metabolite by augmented coronary blood flow, as with a sudden increase in coronary perfusion pressure alone, would lower adenosine concentration and produce vasoconstriction, thereby serving to bring coronary blood flow back toward control; conversely, if coronary blood flow transiently dropped because of hypotension, for example, the slowed rate of flow would result in accumulation of this metabolite in the myocardium, with coronary vasodilatation.

When a coronary artery is briefly occluded completely and the occlusion then released, a marked increase in coronary blood flow ensues for a period of time. This response is termed *reactive hyperemia,* and the length of the hyperemia increases as the duration of the coronary occlusion is prolonged (Fig. 2.159) (see Coffman and Gregg, 1960). Reactive hyperemia has been attributed to the accumulation of adenosine or other metabolites during the period of obstructed coronary blood flow, leading to vasodilatation in the distribution of the artery, although lowered perfusion pressure could also produce myogenic vasodilation. The vasodilation becomes evident when the occlusion is released, leading to the marked transitory increase in coronary blood flow; as metabolic recovery of the tissue takes place, the high blood flow washes out the accumulated metabolite, and blood flow is subsequently restored to normal. As shown in Figure 2.159, the "flow debt" during occlusion is overpaid in the coronary circulation by the reactive hyperemia response.

Coronary Perfusion Pressure

The pressure in the coronary arteries and the aorta importantly affects the level of coronary blood flow, particularly during the diastolic interval when the aortic pressure and coronary blood flow are directly related and fall together (Fig. 2.155). However, the level of mean and systolic aortic pressures also importantly influences the MV̇O₂, and when the aortic pressure increases or decreases, coronary blood flow will change as a consequence of the metabolically induced vasodilation or vasocon-

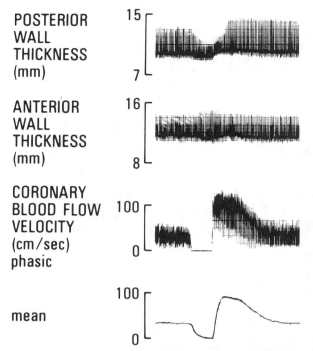

POSTERIOR WALL THICKNESS (mm)

15

7

ANTERIOR WALL THICKNESS (mm)

16

8

CORONARY BLOOD FLOW VELOCITY (cm/sec) phasic

100

0

mean

100

0

Figure 2.159. Reactive hyperemia. Tracings obtained in an experimental animal in which the degree of wall thickening during systole of the posterior wall and anterior walls *(upper tracings)* are recorded simultaneously with phasic and mean coronary blood flow velocity *(lower tracings)* in the circumflex coronary artery (which supplies the posterior left ventricular wall). The effects of a brief coronary occlusion (10 s) are shown. Blood flow promply drops to zero, while systolic wall thickening in the posterior wall promptly diminishes as ischemia occurs. There is slight, compensatory increase of systolic wall thickening in the anterior wall. With release of the coronary occlusion, mean blood flow promptly increases to nearly three times the preocclusion value, and then gradually returns to normal, demonstrating reactive hyperemia. Myocardial contraction promptly returns to normal after the brief coronary occlusion.

striction as the work of the left ventricle changes. Thus the relationship between aortic pressure and coronary blood flow is complex.

It is possible to design an experiment in which the aortic pressure and left ventricular pressure, the heart rate, and the inotropic state of the heart are held *constant* while coronary perfusion pressure *alone* is varied by means of a separate cannulation and perfusion of a coronary artery (Fig. 2.160) (Mosher et al., 1964). In such an experiment when the coronary perfusion pressure from the separate pressure-regulated blood reservoir is abruptly lowered and maintained at a new level, while the cardiac work is held constant, a transient rapid decrease in coronary blood flow ensures which is then followed over 10–12 s by return of the flow toward the previous level (Fig. 2.161); conversely, when the coronary perfusion pressure is abruptly elevated, coronary blood flow rapidly rises initially

but then returns toward the control level. Thus, the coronary vascular resistance varies *directly* with the coronary perfusion pressure under these conditions. When a series of such abrupt pressure changes is carried out over a perfusion pressure range of 10–170 mm Hg (left ventricular and aortic systolic pressures remaining constant at a normal level of approximately 120 mm Hg), a plot relating the perfusion pressure to coronary blood flow is obtained (Fig. 2.162), *solid circles*). This pressure-flow plot reveals a *plateau* between perfusion pressures of approximately 60 and 150 mm Hg, over which coronary blood flow is relatively independent of coronary perfusion pressure; this relationship is termed *autoregulation* of coronary blood flow (see additional discussion of this phenomenon in Chapter 6 and Fig. 2.14).

At the upper ranges of perfusion pressure, autoregulation *fails*, and coronary blood flow rises at these very high coronary perfusion pressures (Fig. 2.162). Likewise, at pressures below approximately 60 mm Hg, autoregulation *fails* and coronary blood flow falls as coronary perfusion pressure drops (Fig. 2.162). The latter portion of the autoregulation curve is of great importance, because it indicates the point at which the coronary bed is maximally

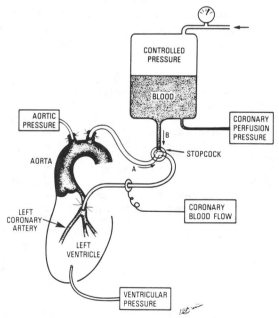

Figure 2.160. Experimental preparation in which the perfusion pressure to a coronary artery can be abruptly changed, while the aortic pressure and function of the left ventricle remain unaffected, allowing demonstration of autoregulation. A three-way stopcock allows blood to either perfuse the coronary artery from a cannula situated in the aorta (*A*) or from the controlled pressure reservoir (*B*). Coronary blood flow is measured with a flowmeter as shown. For further discussion, see text.

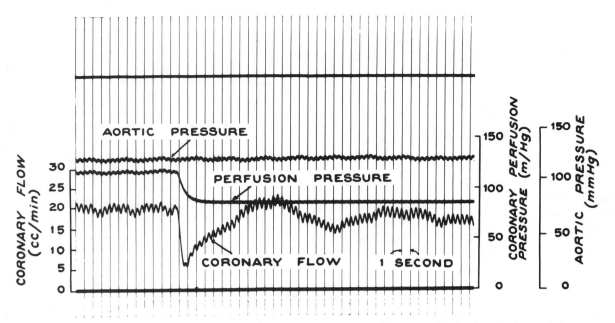

Figure 2.161. Experimental demonstrating coronary autoregulation obtained with experimental preparation shown in Figure 2.160. Note that the mean pressure in the aorta is constant, and when the mean coronary perfusion pressure is abruptly dropped, the coronary blood flow falls sharply and then recovers over 8–10 s to near the control value. (From Mosher et al., 1964.)

vasodilated, i.e., *autoregulatory reserve* has been fully utilized. For example, as coronary perfusion pressure is changed from a mean pressure of 100 to 70 mm Hg, the initial drop in coronary blood flow is followed by active vasodilation, with a drop in coronary resistance (resistance = perfusion pressure/ coronary flow; since pressure has fallen and flow is unchanged, resistance must have fallen). However, when perfusion pressure is dropped from 60 to 40 mm Hg, the blood flow falls, indicating the absence of further coronary vasodilation.

If the initial flow values *prior* to the autoregulatory responses (the lowest initial flow point after the pressure fall, see Fig. 2.161, or the initial flow

Figure 2.162. Pressure-flow curves in the left circumflex coronary artery of the dog demonstrating autoregulation of coronary blood flow *(solid circles with sigmoid dashed curve)*. The three curves shown by the *x*'s, *open circles,* and *triangles* indicate instantaneous pressure-flow curves, before coronary autoregulation had occurred, at three different levels of coronary perfusion pressure and coronary tone, or vascular resistance (50 mm Hg, 100 mm Hg, and 125 mm Hg, respectively). For further discussion, see text. (From Mosher et al., 1964.)

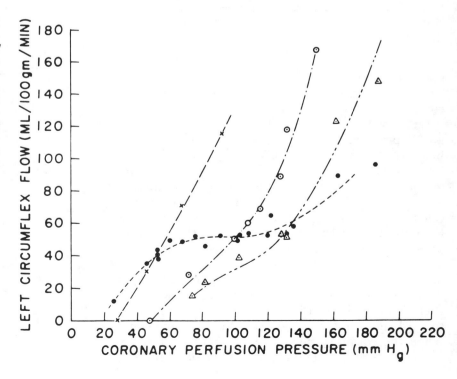

increase if perfusion pressure is raised) are plotted against coronary perfusion pressure the pressure-flow relation is characteristic of a vascular bed that does not autoregulate (*exponential dashed lines*, Figs. 2.162 and 2.163)* The instantaneous pressure sure-flow curve in Figure 2.163 *(triangles)* is the same as the pressure-flow plot obtained after administration of a coronary vasodilator such as nitroglycerin or adenosine; it is the relationship of *maximum coronary vasodilation*, in which flow is highly dependent on perfusion pressure, exhibiting a nearly exponential change in coronary blood flow as pressure is changed (characteristic of a passive vascular bed).

The mechanism of pressure-induced autoregulation (which is also seen in certain other vascular beds such as the cerebral and skeletal muscle circulations) is controversial (see Chapter 6). The so-called myogenic hypothesis holds that at least a portion of the response is due to altered stretch of the smooth muscle in the walls of the coronary arteries. An important component of the autoregulatory response may also relate to metabolic control of coronary vascular tone; as discussed earlier, with the increased coronary blood flow accompanying a

sudden increase in coronary perfusion pressure, washout of adenosine could rapidly occur, returning flow toward the normal level, whereas with a drop in perfusion pressure and decreased flow, diminished washout with accumulation of adenosine could result in coronary vasodilation.

The plateau of autoregulation is shifted upward by increased aortic pressure and $\dot{M}VO_2$ (Fig. 2.63) and vice versa. However, when the coronary artery is perfused from the aorta, with a fall in aortic pressure below a critical value, the vasodilatory reserve will be exceeded, and coronary blood flow will fall with unfavorable effects on the myocardium (see section on ischemia). It can also be seen from Figure 2.163 that when the $\dot{M}VO_2$ is increased (as during exercise), if the perfusion pressure remains at 60 mm Hg, coronary vascular resistance must fall so that flow can rise and meet the increased O_2 demands. If flow cannot rise during exercise (see below), the resulting insufficient O_2 supply to the myocardium in a region can cause chest pain during exertion, as discussed subsequently.

Systolic Compression

It has been pointed out that active contraction of the left ventricle causes mechanical compression with squeezing of the intramyocardial coronary blood vessels, a phenomenon that is responsible for the pattern of reduced *total* coronary blood flow observed during ventricular systole in coronary arteries supplying the left ventricle (see Fig. 2.155) (Kirk and Honig, 1964; Downy and Kirk, 1974). This

---

*Notice the similarity of these "instantaneous" pressure-flow curves at different levels of coronary vascular resistance in Figure 2.162 to those in a vascular bed that do not exhibit autoregulation at various degrees of sympathetic stimulation (Fig. 2.13); note also the change in critical closing pressure in both circumstances.

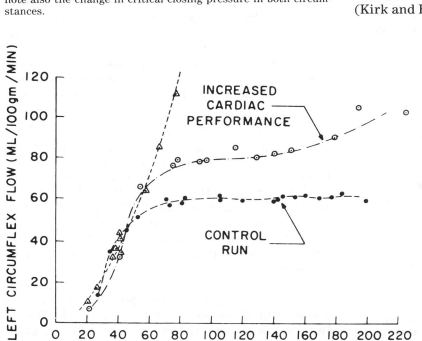

Figure 2.163. Pressure-flow curves demonstrating coronary autoregulation down to a perfusion pressure of approximately 60 mm Hg *(solid circles).* When cardiac work was increased by increasing the mean aortic pressure *(open circles),* coronary blood flow was auto-regulated at a higher level. The instantaneous pressure-flow relation with *maximum* coronary dilation is steep *(open triangles),* and indicates a marked dependency of coronary blood flow upon perfusion pressure below a critical pressure. For further discussion, see text. (From Mosher et al., 1964.)

compression increases the average resistance to coronary blood flow, and if the contraction of the heart is suddenly stopped momentarily in an experimental setting, coronary blood flow rises abruptly (Fig. 2.164).

When the *transmural* distribution of blood flow is analyzed systolic compression of the coronary circulation becomes particularly important when blood flow to the *inner wall* of the left ventricle is considered, as discussed below.

TRANSMURAL DISTRIBUTION OF BLOOD FLOW

Systolic contraction of the left ventricular myocardium is responsible for an important difference between the pattern of blood flow to the inner (subendocardial) regions of the left ventricular wall and the outer (subepicardial) regions. In fact, during ventricular systole, blood flow in the subendocardial region for practical purposes is zero and then rises rapidly during diastole, whereas in the subepicardial layers blood flow rises during systole and then remains relatively high during diastole (Fig. 2.165). Blood flow to the midwall region is intermediate between these two patterns (Fig. 2.165). As mentioned, the pattern of *total* coronary blood flow shows reduced flow during systole, with an abrupt rise and then a slow fall during diastole. These patterns of regional flow across the wall have been verified by injecting microspheres to perfuse the coronary arteries *only* during the phase of ventricular systole (Hess and Bache, 1976); such studies demonstrate that few microspheres reach the subendocardium under these conditions, whereas flow is roughly comparable during systole and diastole in

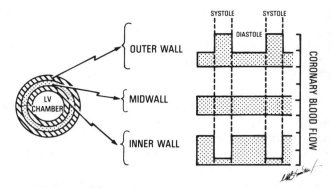

Figure 2.165. Diagrammatic representation of transmural coronary blood flow distribution in three layers of the left ventricular *(LV)* wall measured with radioactive microspheres during systole and diastole. Note that systolic flow predominates in the outer wall, whereas diastolic flow with minimal systolic flow is present in the inner wall. For further discussion, see text.

the midwall. This means that under normal conditions coronary vascular resistance is lower in the subendocardium, in order to compensate (during *diastole*) for this systolic flow limitation (Fig. 2.165). The *mean* blood flow (ml/min) to the subendocardium is slightly higher than to the subepicardium, yielding an endocardial:epicardial blood flow ratio of about 1.1:1 under normal conditions, and it has been postulated that the explanation for this observation may lie in a slightly higher O_2 consumption of the inner myocardial fibers since they must shorten further and perform more work.

The differing flow pattern in the inner and outer wall of the left ventricle is particularly important in certain disease states, and makes the subendocardium vulnerable to insufficient perfusion pressure. Thus, analysis of the *transmural* distribution of myocardial blood flow, with consideration of blood flow to the inner wall of the left ventricle, is important under some conditions (see subsequent section).

Autonomic Nervous System

Although metabolic and mechanical factors are normally the most important determinants of the level of coronary blood flow, the autonomic nervous system, via sympathetic nerve fibers directly supplying the coronary arteries, has some influence at rest and under other circumstances. Parasympathetic stimulation, as well as circulating catecholamines and certain other substances also can have an effect.

The sympathetic nerve fibers innervating the coronary arteries contain norepinephrine, and its release stimulates α-adrenergic receptors located in the smooth muscle of the coronary vessels. Local metabolic factors largely override the influence of

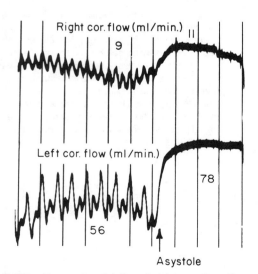

Figure 2.164. Response of left and right coronary flow to vagal induced asystole. Note the prompt increase with the induction of asystole. Time lines are 1 s apart. (From Sabiston and Gregg, 1957.)

these nerves, but vasoconstriction due to activation of the sympathetic nerves to the coronary arteries can be demonstrated under some circumstances (see Feigl, 1983, for review). Such an effect is clearly shown when sympathetic nerve stimulation is carried out (or norepinephrine is injected) after blockade of the myocardial β_1 receptors; under these conditions, the increase in $\dot{M}VO_2$ is blocked, and an *increase* in coronary vascular resistance can be demonstrated during nerve stimulation.

During exercise in experimental animals, increased sympathetic stimulation of the heart (which includes β_1-adrenergic stimulation of myocardial contractility and heart rate) causes marked metabolic coronary vasodilation, but with superimposed mild vasoconstriction due to increased α-adrenergic receptor stimulation; thus, after α-adrenergic blockade, coronary blood flow is somewhat higher during heavy exercise than in the absence of such blockade, and the coronary AVO_2 difference is narrower (Murray and Vatner, 1969). Such an effect does not impair the functional capacity of the normal heart, and there is some evidence that most of the mild vasoconstrictor effect is limited to the outer wall layers, which would favor increased subendocardial perfusion (Feigl, 1987).

There is also experimental evidence that some degree of reflex adrenergic constrictor tone may exist at rest (Vatner et al., 1970). Thus, electrical stimulation of the carotid sinus nerves (which *inhibits* sympathetic tone and produces reflex peripheral vasodilation, Chapter 16) causes a fall in the coronary vascular resistance, indicating reduced vasoconstriction, despite a decrease of $\dot{M}VO_2$ (resulting from a reflex slowing of heart rate and lowered arterial pressure, which should cause increased coronary vascular resistance) (Vatner, 1970). Nevertheless, the level of sympathetic "tone" in the coronary bed is low relative to that in many other vascular beds, and under most circumstances metabolic control predominates. A role for coronary sympathetic vasoconstriction in patients with atherosclerotic coronary artery lesions, or in patients with spasm of large coronary arteries (discussed further below), has been suggested and using coronary angiography, narrowing of large coronary arteries at the site of atherosclerotic lesions has been observed during exercise in some patients with coronary heart disease (Gage et al., 1986).

Direct stimulation of the vagal nerve supply to the heart causes mild coronary vasodilation when bradycardia is prevented by electrical pacing of the heart (Feigl, 1969), although this effect is not likely to be of much importance in the intact circulation. Intracoronary administration of acetylcholine produces pronounced coronary vasodilation.

Reflexes which appear to originate in the coronary arteries and produce reflex peripheral vasodilation have been described, but their role is unclear.

Circulating Catecholamines and Other Substances

β-Adrenergic (β_2) receptors exist in the smooth muscle of the coronary vascular bed, and their stimulation by blood-borne epinephrine or injected sympathomimetic drugs can produce coronary vasodilation. Such direct effects can be difficult to demonstrate, however, because these substances stimulate $\dot{M}VO_2$ and thereby produce secondary, metabolic coronary vasodilatation. Under experimental conditions, a direct coronary vasodilator effect due to the sympathomimetic drug isoproterenol can be shown in the potassium-arrested heart, where perfusion of isoproterenol into the coronary arteries produces a marked drop in coronary vascular resistance in the absence of a change in $\dot{M}VO_2$ (Klocke et al., 1965). If a "cardioselective" β blocker is administered to block only the myocardial β_1 receptors (but not β_2 receptors in the coronary arteries), administration of isoproterenol also produces direct coronary vasodilation, while $\dot{M}VO_2$ remains unaffected.

A number of other substances in the body as well as a variety of drugs can affect coronary blood flow indirectly by altering $\dot{M}VO_2$, or exhibit direct actions on the coronary vascular smooth muscle. Direct coronary vasodilators include nitroglycerin, prostacyclin, adenosine, papavarine, and calcium antagonist drugs such as verapamil and nifedipine.

Coronary vasoconstrictors include vasopressin, ergonovine, angiotensin II, and α-adrenergic agonist drugs such as methoxamine. Ergonovine is sometimes administered in the cardiac catheterization laboratory in order to induce coronary spasm in the diagnosis of patients suspected of having chest pain due to this phenomenon; its effects are readily reversed by the administration of nitroglycerin or a calcium antagonist.

Several other intrinsic mechanisms and endogenous substances may play a role in regulating coronary vascular resistance under some circumstances. Increased blood flow itself can cause further vasodilation (so-called flow-mediated vasodilation). This is mediated by *endothelial derived relaxing factor* (EDRF), a substance released by endothelial cells under some conditions (Furchgott, 1983). EDRF is released and causes vasodilation with acetylcholine or bradykinin injection and its absence may be a factor in coronary vasoconstriction or spasm occurring at the site of an atherosclerotic lesion. (For review of mechanisms affecting large coronary arteries, see Young and Vatner, 1986).

Naturally occurring prostaglandins are important in controlling platelet activity and can have a role in influencing vascular tone, as well as reflex autonomic responses (Kaley, 1987). In the heart, PGI_2 (prostacyclin) and PGE_2 (prostaglandin E_2) are coronary vasodilators, whereas PGH_2 which is converted to thromboxane (TXA_2) is a vasoconstrictor. These metabolites of arachidonic acid are formed in blood vessels and endothelial cells, and certain prostaglandins are released from platelets. They can be important in pathological states, in which endothelial injury can lead to complex effects including loss of PGI_2 (which diminishes platelet adhesion) and endothelial derived relaxing factor, and increased TXA_2 levels, which activate platelets and cause coronary vasoconstriction (Mullane and Pinto, 1987).

CORONARY VASODILATOR RESERVE

Myocardial O_2 demands are met primarily by moment to moment alterations in coronary blood flow, and a substantial vasodilator reserve exists in the normal heart. This can be demonstrated by experimentally performing brief coronary occlusions, which are followed by reactive hyperemia, as discussed earlier (Fig. 2.159). Also, during stresses such as exercise, a uniform large increase of coronary blood flow occurs across the normal left ventricular wall (as measured with microspheres) (Fig. 2.166, *closed circles* to *open circles*). In addition, *intracoronary* infusion of nitroglycerin or adenosine produces a marked increase in coronary blood flow (without significant change in $M\dot{V}O_2$). When the dose of adenosine is progressively increased up to the limit of coronary vasodilator reserve, the reserve is maximally utilized first in the subendocardial regions of the left ventricle. This occurs because the increase of reserve flow in the subendocardium takes place mainly during diastole, and in diastole there is already some encroachment on subendocardial vasodilator reserve (Fig. 2.165), as discussed earlier (Bache and Cobb, 1977). Some additional reserve then remains in the subepicardial regions when the subendocardium is maximally dilated by adenosine.

Under conditions of *maximal coronary vasodilation* by adenosine (Bache and Cobb, 1977), blood flow to the subendocardium becomes directly proportional to the so-called diastolic pressure-time index (DPTI), the integrated area under the aortic

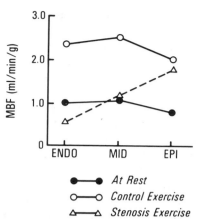

Figure 2.166. Myocardial blood flow *(MBF)* distribution across the wall of the left ventricle in three layers, *ENDO* (subendocardium), *MID* (midwall), and *EPI* (subepicardium) measured with radioactive microspheres. Data in an animal standing at rest on a treadmill *(closed circles)* show the slight predominance of subendocardial blood flow. During exercise under normal conditions, there is a uniform increase of blood flow across the wall *(open circles,* control exercise). When the left circumflex coronary artery was partially constricted, blood flow in the region supplied by that vessel showed a marked fall in subendocardial blood flow during exercise *(open triangles)*, although the subepicardial flow increased normally. Also, during exercise with coronary stenosis, the region of myocardium supplied by the left circumflex coronary artery exhibited severe impairment of contraction, whereas under normal conditions, contraction increased during exercise. (Courtesy of K. P. Gallagher.)

pressure pulse during diastole (Fig. 2.167) (Buckberg et al., 1972). Under conditions of maximal coronary vasodilation, a decrease in the time of diastolic perfusion per minute, as when the heart rate increases (an increased number of contractions is at the expense of diastolic time), would lead to a reduction in subendocardial coronary blood flow despite increasing $M\dot{V}O_2$ (Fig. 2.167); also, if the diastolic blood pressure falls at an unchanged heart rate, coronary blood flow will also fall (Fig. 2.167), whereas if diastolic blood pressure rises or heart rate slows, the DPTI (and coronary blood flow to the subendocardium) will increase (Fig. 2.167). These observations have led to the idea that the *ratio* of the systolic pressure-time index (SPTI) (the same as the "tension-time index" discussed in Chapter 14), as an indicator of $M\dot{V}O_2$, to the DPTI can provide an estimate of the adequacy of subendocardial perfusion under conditions of maximal coronary dilatation (Hoffman and Buckberg, 1978). This concept has importance to regions of the heart when stenosis or narrowing of a coronary artery due to coronary atherosclerosis causes the perfusion pressure to drop, resulting in maximum dilatation of the coronary bed beyond the area of stenosis (see below). The SPTI/DPTI also can have relevance to condi-

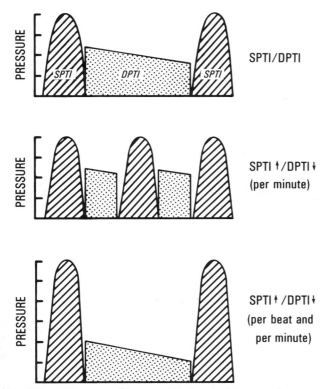

SPTI/DPTI

SPTI ↑/DPTI ↓
(per minute)

SPTI ↑ /DPTI ↓
(per beat and
per minute)

Figure 2.167. Diagram of the systolic pressure-time index *(SPTI)* and the diastolic pressure-time index *(DPTI)*, which provide a guide to the adequacy of myocardial oxygen demands and coronary blood flow to the subendocardium, respectively, during maximum coronary vasodilatation. (SPTI = the area under the systolic aortic pressure pulse and DPTI = the area under the diastolic aortic pressure pulse). Normal conditions are shown in the *top panel.* During exercise in the presence of maximum coronary vasodilatation *(middle panel),* the SPTI/minute increases, whereas the time for diastolic flow to the subendocardium (DPTI) decreases; coronary blood flow may become inadequate to meet myocardial oxygen demands under these conditions. In the *bottom panel,* which illustrates severe aortic regurgitation, the SPTI is increased because of high systolic pressure, whereas the DPTI is reduced because of the low aortic pressure due to the regurgitant leak. The ratio is therefore increased, and in the presence of maximum coronary vasodilatation in the subendocardium, insufficiency of coronary flow to that region may result.

tions in some valvular lesions such as severe aortic regurgitation, where a high systolic pressure (increased SPTI) and low diastolic pressure (reduced DPTI) coexist and could lead to subendocardial underperfusion (Fig. 2.162) and an imbalance in O_2 supply and demand under some circumstances.

CORONARY HEART DISEASE AND INADEQUATE CORONARY BLOOD FLOW

Understanding of the relationships between myocardial O_2 requirements and the coronary blood supply becomes of great importance in dealing with clinical disorders of the coronary arteries. An example of the interplay between coronary perfusion and $M\dot{V}O_2$ is the setting in which a coronary artery is narrowed but not completely occluded, and partially supplied by collateral vessels (Fig. 2.154). Under these conditions, coronary blood flow may be adequate to supply the resting metabolic needs of the myocardium, but when O_2 demands are increased as during physical exercise, the blood supply to that region becomes insufficient. Thus, in the zone of myocardium supplied by the diseased coronary artery, the increased heart rate in the presence of maximal subendocardial vasodilation can result in a fall in subendocardial blood flow during exercise (Fig. 2.166, *open triangles*), and regional contraction may rapidly deteriorate (Gallagher et al., 1982). In this setting, when the level of coronary blood supply relative to O_2 needs becomes insufficient to that region (Figs. 2.167 and 2.168), the condition is termed *ischemia* and it often leads to a clinical syndrome termed *angina pectoris,* or pain beneath the breastbone (Fig. 2.168). Occasionally, this syndrome is also produced by spasm of one or more large coronary arteries in the absence of an increase in $M\dot{V}O_2$ (Maseri et al., 1978).

Exercise-induced ischemia with angina pectoris due to the presence of an atherosclerotic coronary lesion can be treated by a variety of measures, which involve either decreasing $M\dot{V}O_2$ (reducing O_2 demands) or improving myocardial O_2 supply through enhanced coronary blood flow, or both (Fig. 2.169) (Ross, 1989). One important way of decreasing myocardial O_2 demands is the administration of nitroglycerin (a smooth muscle relaxant which generally dilates arteries and veins), thereby reducing the blood pressure and heart size and diminishing $M\dot{V}O_2$. Calcium channel blockers such as nifedipine or diltiazem also directly dilate coronary arteries, and may produce peripheral vasodilation, thereby reducing myocardial O_2 demand. In addition, nitroglycerin and other vasodilators can directly dilate large coronary arteries at the site of a narrowing due to atherosclerosis, thereby also directly increasing coronary blood flow and myocardial O_2 supply (Brown et al., 1984). Nitroglycerin and other vasodilators also may increase O_2 supply by enhancing coronary collateral blood flow. Also, nitroglycerin and calcium antagonists are highly useful in the treatment of angina pectoris due to coronary artery spasm. Another important approach for diminishing O_2 demands (Fig. 2.169) is use of β-adrenergic blocking drugs (such as propranolol) which, by slowing the heart rate and diminishing the increase in blood pressure and contractility pro-

Angina Pectoris
(chest pain)

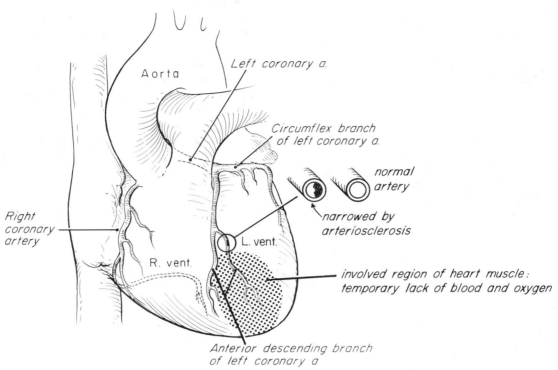

Aorta

Left coronary a.

Circumflex branch of left coronary a.

normal artery

narrowed by arteriosclerosis

Right coronary artery

R. vent.

L. vent.

involved region of heart muscle: temporary lack of blood and oxygen

Anterior descending branch of left coronary a.

Figure 2.168. Diagram showing the mechanism of angina pectoris. A coronary atherosclerotic lesion is present in a branch of the left anterior descending coronary artery. During exercise, blood flow through the narrowed area becomes inadequate and the region of muscle supplied by that branch *(stippled area)* becomes ischemic, causing chest pain or angina pectoris. (Drawing by G. Gloege. From Ross and O'Rourke, 1976c.)

Figure 2.169. Diagram representing oxygen supply and oxygen demand to the myocardium as a scale, which is normally in balance *(upper panel)* both at rest *(cross-hatched areas)* and during exercise, when the increased myocardial oxygen demands are met by increased oxygen supply (coronary flow rises). In the presence of coronary artery disease (CAD, *lower panel*), the scale may be in balance under basal conditions at rest *(cross-hatched areas, and dashed horizontal line)*, but during exercise the myocardial oxygen demand increases while the increase in coronary blood flow is inadequate. The scale is therefore unbalanced, producing angina pectoris.

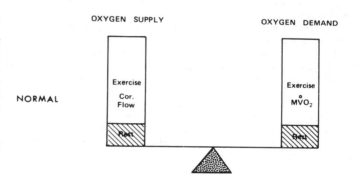

OXYGEN SUPPLY   OXYGEN DEMAND

NORMAL

Exercise
Cor. Flow
Rest

Exercise
$\overset{\circ}{M}VO_2$
Rest

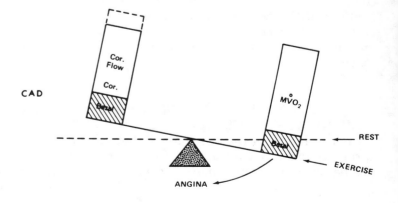

CAD

Cor. Flow
Cor.
Basal

$\overset{\circ}{M}VO_2$
Basal

REST

EXERCISE

ANGINA

duced by exercise, allow the same exercise workload on the body to be achieved at a lower level of cardiac work and $M\dot{V}O_2$; angina pectoris may thereby be prevented.† Limitation of the heart rate increase by β-blocking drugs also results in a longer time for diastolic coronary perfusion, and also should improve subendocardial blood flow during exercise by this mechanism (Ross, 1989). Another approach for treating angina pectoris is to increase O_2 supply (Fig. 2.169) by surgical means with the placement of grafts to bypass the obstructions in the coronary arteries. Alternatively, a balloon catheter placed in the coronary artery and positioned at the site of an atherosclerotic plaque may be used to dilate the lesion (so-called coronary angioplasty).

Abrupt obstruction of a coronary artery, usually caused by thrombosis (clotting) at the site of an atherosclerotic lesion, produces within 1–2 min loss of contraction in the involved region with local systolic expansion of the wall (Tennant and Wiggers, 1935; Theroux et al., 1974). If sustained beyond about 40 min, coronary occlusion produces an area of muscle damage or necrosis, which is termed an acute myocardial infarction, or "heart attack." The damage can be limited by improving the coronary blood supply through early use of thrombolytic (clot-lysing) drugs, and a variety of other approaches for reducing the damage due to coronary occlusion are under study (Braunwald and Sobel, 1988).

---

†If the heart rate cannot increase, cardiac output rises by use of the Frank-Starling mechanism when the venous return increases during exercise, so the β blockade does not impair the ability of the body to respond to modest levels of exertion.

CHAPTER 16

Neurohumoral Control of the Circulation

The circulatory system is regulated by neural and hormonal control systems, which are superimposed on local metabolic control of vascular resistance. This complex series of neural and hormonal systems and local regulation alter the heart and vascular system and blood volume to maintain a constant level of the arterial pressure and maintain the cardiac output to meet the body's demands. These systems are described in detail in this chapter.

ANATOMY OF THE NEURAL CONTROL SYSTEMS

The neural control mechanisms have inputs to the central nervous system (afferent nerves), which sense and relay various parameters of circulatory function, such as arterial pressure, and outputs, or efferent nerves which regulate cardiac function, vascular volume, vascular resistance, and flow through the peripheral circulation (Peis, 1965; Korner, 1979; Bishop, 1985). These are shown schematically in Figure 2.170, which diagrams afferent nerves on the left and efferent nerves on the right. There is, of course, a bilateral distribution of both inputs and outputs.

AFFERENT CONNECTIONS (INPUTS)

Afferent information (nerve traffic) from receptors sensitive to distortion or to chemical stimuli located throughout the body is carried by fibers in autonomic nerve trunks to the central nervous system. These afferent fibers are frequently classified by the speed of transmission of impulses along the nerve, which in turn, is directly proportional to the amount of myelin sheath present about the nerve. "C" fibers, for example, contain little myelin and have a slow transmission velocity (approximately 2 meters/s). "A" fibers are myelinated and have transmission velocities from 10–40 m/s.

Carotid Sinuses

The carotid sinus consists of a dilatation at the base of each internal carotid artery where it joins the external carotid artery. A branch of the glossopharyngeal nerve (IXth cranial nerve) supplies that region, and reflexes from the carotid sinus are abolished when the glossopharyngeal nerve is sectioned. There is a rich sensory innervation of the carotid sinus wall. With appropriate staining, diffuse arborizing nerve networks can be demonstrated, as well as lamella-like receptors. These endings are located in connective tissue (adventitia) of the vascular wall and are oriented parallel to the long axis of the vessel. Nerve fibers from these receptors course through a branch of the glossopharyngeal nerves (IXth cranial nerve) and impinge on the nucleus tractus solitarius in the medulla.

Carotid Bodies

These small organs, composed or rounded clumps of polyhedral cells near the carotid sinuses, are highly vascular, containing a network of vascular sinusoids in close connection with special sensory cells which respond to chemical stimuli and have afferent fibers passing to the IXth nerve. Blood flow through the carotid bodies is extremely high.

Aortic Baroreceptors

A network of branching nerves arborizes in the aortic wall in the region of the aortic arch, carotid, and subclavian arteries (Fig. 2.170). They are similar to the carotid sinus receptors, but their afferent fibers run through the vagus nerve (Xth cranial). There are both myelinated and nonmyelinated fibers in the carotid sinus, and the aortic arch.

Aortic Bodies

These chemoreceptors lie near the aortic arch and pulmonary artery, and recent studies have also identified such receptors near the coronary arteries. Structurally, they resemble the carotid bodies. Their afferent fibers pass toward the central nervous system in the vagus nerve.

Cardiopulmonary Receptors

Mechanosensitive and chemosensitive receptors are also found in the lungs, the walls of the atria

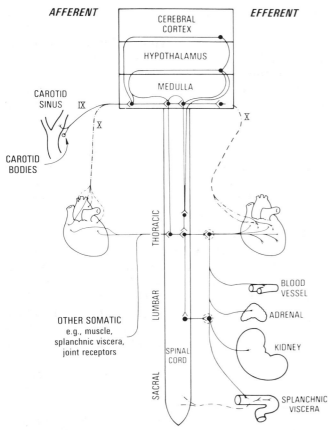

AFFERENT   *EFFERENT*

Figure 2.170. Schematic diagram of the neural connections important in circulation control. Although the system is bilaterally symmetrical, in this schematic drawing inputs (afferent fibers to the *left*) and outputs (sympathetic and parasympathetic efferent fibers, to the *right*) are shown only on one side. (Adapted from Korner, 1979.)

and ventricles (Fig. 2.170), and the junctions of the venae cavae and pulmonary veins with the atria; these afferents consist of both myelinated and nonmyelinated fibers that course to the central nervous system through sympathetic or parasympathetic (X) trunks.

EFFERENT CONNECTIONS (OUTPUTS)

Cardiac Efferent Connections

The autonomic nerve supply to the heart consists of many mixed fibers, containing both vagal and sympathetic nerves, which course over the aorta, pulmonary artery, and vena cava to reach the cardiac chambers, conducting system, and coronary arteries.

SYMPATHETIC FIBERS

Fibers from the central nervous system traverse the spinal cord and synapse in the sympathetic chain (there are multiple cervical ganglia and para-

vertebral ganglia, represented by two ganglia in Fig. 2.170). Efferent fibers from these ganglia provide a rich network of postganglionic nerve endings, whose neurotransmitter is norepinephrine, to the atria, ventricles, sinus node, and atrioventricular (AV) node (Fig. 2.8)

VAGAL (PARASYMPATHETIC) FIBERS

The vagal control of the heart is mediated by fibers which synapse in ganglia in the heart (Fig. 2.170, shown as–·–); these ganglia provide a network of fibers which contain acetylcholine and supply the atria, sinoatrial (SA) node, and AV node (see Fig. 2.7). A few fibers also supply the ventricular muscle.

Efferent Connections to the Peripheral Circulation

VASOCONSTRICTOR FIBERS

Distribution

These nerves were discovered in 1852 by Claude Bernard, who stimulated the cervical sympathetic nerve in the rabbit and observed constriction of the vessels of the ear. They belong to the thoracolumbar (sympathetic) division of the autonomic nervous system (Fig. 2.170). The constrictor fibers arise from groups of nerve cells situated in the intermediolateral columns of the spinal gray matter, extending in man from the first thoracic to the second or third lumbar segment, inclusive. Axons from these cells synapse with neurons lying in sympathetic ganglia outside the cerebrospinal axis. All the arterioles of the body, wherever situated, are supplied with these neurons which have their source in this relatively limited region of the central nervous system. Their adrenergic nerve endings contain norepinephrine and have been identified in all types of blood vessels, except the true capillaries (Chapter 6). Precapillary resistance vessels (small arteries and arterioles) have, in general, a rich innervation, although in the smallest precapillary vessels, the number of fibers is small. The venules have fewer adrenergic fibers than the larger veins which are less richly innervated than the precapillary vessels. It is solely through such sympathetic fibers traveling with somatic nerve trunks that constrictor impulses are conveyed to the minute vessels of the limbs. Vasoconstrictor fibers to the head and neck are conveyed from the sympathetic chain through plexuses investing the blood vessels, and via peripheral nerve trunks (cervical and certain cranial nerves). The vessels of the abdomen and pelvis are supplied with fibers which pass along the vascular walls from plexuses surrounding the aorta and its branches.

Site and Mode of Action

There is a resting discharge rate of the vasoconstrictor fibers of 1–2 impulses/s which acts to maintain normal vessel tone and reaches 10 impulses/s with maximal physiological excitation. The chemical transmitter released at the smooth muscle cell is norepinephrine, and norepinephrine is released into the bloodstream on intense sympathetic stimulation. Because these constrictor fibers exert tonic control over the resistance vessels, a marked increase in blood flow in a vasoconstricted limb immediately follows sympathetic or ganglionic blockade. During prolonged constriction accumulation of "vasodilator metabolites" can oppose neurogenic influence.

In addition, these fibers exert strong control over the capacitance vessels, mainly the veins. They can alter greatly the venous return to the heart and, thus, markedly influence heart size and cardiac output. The significance of vasomotor fibers in controlling vascular capacitance has been verified in functionally isolated parts of the superficial and deep venous system of dog and man by recording pressure changes evoked by various types of reflex stimulation. Veins in the intact forearm constrict during reflex sympathetic stimulation by cold, excitement, etc. Similar observations have been made on the capacity vessels of the splanchnic area. These effects are abolished by agents which block neurotransmission at any level (central, ganglionic, nerve terminals, or receptors).

Functional Significance

Vasoconstrictor fibers are of major importance in short-term homeostasis of blood pressure and blood flow, including those reflex adjustments that arise from baro- and chemoreceptors. Vasodilation is predominantly an inhibition of vasoconstriction. In fulfilling its role as the main neural modulator of the peripheral circulation, the vasoconstrictor system may exhibit generalized, segmental, or regional function, depending upon the type of stimulus (Simon and Riedel, 1975).

VASODILATOR FIBERS

Vasodilator impulses emerge from the central nervous system by (1) the thoracolumbar sympathetic outflow; (2) the cranial outflow of the parasympathetic division reaching the periphery by way of the chorda tympani nerve (supplying the salivary glands), glossopharyngeal, and vagus nerves; (3) sacral outflow of the pelvic nerve; and (4) the posterior spinal nerve roots.

Sympathetic Vasodilator Fibers

Although sympathetic control of vascular smooth muscle is accomplished primarily by regulation of tonic activity of sympathetic vasoconstrictor neurons, a separate sympathetic cholinergic (vasodilator) system has been shown to exist in the dog and cat and in certain nonhuman primates. These fibers arise in the cerebellum and course through the medulla before joining other sympathetic outflow. In the presence of adrenergic blockade, stimulation of sympathetic nerves produces vasodilation of skeletal and skin blood vessels.

There are also β-adrenergic receptors located in resistance vessels which may be demonstrated by the administration of isoproterenol or by sympathetic stimulation following atropine or α-adrenergic blockade (Shepard, 1983).

Parasympathetic Vasodilator Fibers

Parasympathetic vasodilator fibers run to restricted cranial and sacral areas such as the cerebral vessels, tongue, salivary glands, external genitalia, and to the bladder and rectum. These fibers are probably not concerned with baro- and chemoreceptor control of blood vessels, nor are they tonically active. It is generally believed that they are cholinergic, but their vasodilator effect is notably resistant to atropine. In skin and some exocrine glands, the active vasodilation associated with these fibers is mediated by bradykinin. Bradykinin produces vasodilation when given intraarterially in most beds (Roddie, 1983). Bradykinin is present in sweat and has been thought to be responsible for the dilation of cutaneous vessels during thermal stress. However, this finding is not universally accepted and the mediator of the vasodilation associated with thermal stress remains controversial.

Histaminergic Vasodilator Fibers

The presence of vasodilator fibers to the hind limb of the dog, which apparently liberates histamine, has been demonstrated (Oberg, 1976). The vasodilation is not prevented by muscarinic or β-adrenergic receptor blockade. It is abolished by antihistamines and is associated with release of histamine into the blood. The function of these fibers is unknown.

ADRENAL MEDULLA

In addition to vasomotor control by way of nerves ending on vascular smooth muscle there is another type of neurohumoral control that is autonomic in nature but that acts upon vascular smooth muscle through catecholamines circulating in blood. One of the oldest known neurohumoral relationships is secretion of epinephrine (with smaller quantities of norepinephrine) from the adrenal medulla following stimulation of (1) splanchnic nerves, (2) lateral columns of the spinal cord, (3) vasomotor center in medulla oblongata, or (4) the lateral hypothalamus.

In a direct comparison of the separate actions of the hormones and of vasomotor fibers on peripheral resistance in representative vascular beds, most blood vessels have been found to be dominated by their vasomotor fibers (Mellander and Johansson, 1968). The exception is vessels in skeletal muscle, where physiological concentrations of epinephrine cause almost maximal dilatation with a significant reduction in the ratio of precapillary to postcapillary resistance. Secreted epinephrine, therefore, is believed to produce vasodilatation in, e.g., exercise. If the amount of secreted epinephrine is large, as following hemorrhage, α-adrenergic effects predominate and reinforce neurally mediated α-receptor vasoconstrictor effects, especially in the renal vascular bed (Korner, 1974).

ADRENERGIC RECEPTORS IN VASCULAR SMOOTH MUSCLE

Receptors that respond to catecholamines are called adrenergic; at least two pharmacological types, α and β, and several subtypes α_1, α_2 and β_1, and β_2 have been identified. A third receptor for cholinergic effects, the muscarinic receptor, is postulated for sympathetic cholinergic vasodilation of vessels in skeletal muscle. A detailed description of the distribution and effects of these receptors is beyond the scope of this chapter and readers are referred to general texts of pharmacology. However, it is important to understand the effects of the adrenergic neurotransmitters. Norepinephrine acts primarily on α-receptors and is their natural activator when released from adrenergic sympathetic nerve endings. Following pharmacological α-receptor blockade intraarterially infused norepinephrine can produce vasodilatation, the result of β_2-receptor stimulation by norepinephrine. Epinephrine at low levels is predominantly a β-receptor agonist but if β-receptors are blocked, or if the concentration is increased, epinephrine can cause vasoconstriction. The major source of epinephrine is the adrenal medulla.

CENTRAL CONNECTIONS

In the past several years concepts concerning central nervous integration of cardiorespiratory control have been rapidly evolving. It is no longer appropriate to consider cardiorespiratory control as a series of independent reflex arcs with localized central nervous connections. There is now good evidence for modulation of afferent traffic at several different levels within the brain, and there are possibilities for interaction at various levels of efferent responses extending from the cortex to the spinal cord (Korner, 1979, Hilton and Spyer, 1980). Studies of reflex effects in man and in unanesthetized animals (Vatner and Boettcher, 1978; Eckberg, 1976) where the central nervous system is not depressed by anesthetics emphasize the important role of the central nervous system in integrating cardiovascular responses.

The anatomy of the central connections important in circulatory control is complex and properly the subject of neurophysiology. However, there are five major areas in the central nervous system that are of major importance to circulatory control: spinal cord, medulla oblongata, hypothalamus cerebellum, and cerebral cortex. The functional interconnections between these areas are shown schematically in Figure 2.170. Stimulation and regional ablation experiments have shown that specific regions of the brainstem may have predominant vasoconstrictor or cardioaccelerator activity. However, most recent work has shown extensive interconnections between ascending and descending pathways which emphasize the interaction between areas of the brainstem.

Spinal Cord

The preganglionic sympathetic neurons may, under some circumstances, exhibit spontaneous activity that is independent of an excitatory drive from afferent fibers or from higher levels of the nervous system. This phenomenon has been observed in animals with the cervical spinal cord sectioned so that it is isolated from central inputs. Under these circumstances local changes in the tension of oxygen or carbon dioxide are believed to be responsible for the "spontaneous" activity.

Various types of afferent stimulation are able to call forth reflex vasoconstriction via the spinal cord. For example, pain or cold stimulation of the skin induces a segmental constriction of splanchnic vessels in spinal animals. Cutaneous vasodilatation occurs when the skin is moderately warmed.

Medulla

Neurons subserving vasomotor function from the spinal cord are under the control of higher order neurons located particularly in the floor of the fourth ventricle of the medulla. Local electrical stimulation has revealed a lateral "pressor area" and a medial "depressor area" causing vasoconstriction and vasodilatation, respectively. The vasodilatation is caused by inhibition of vasoconstrictor tone, specific vasodilator fibers not being involved. In the intact animal or human these areas receive afferent information and other neural inputs from more central areas (hypothalamus and cerebral cortex) and from the nucleus tractus solitarius which are integrated into patterns of vasomotor activity.

Myelinated vasoconstrictor fibers descend through the spinal cord to thoracolumbar sympathetic ganglia and then unmyelinated postganglionic fibers carry vasoconstrictor activity to all vessels and many organs.

The vasoconstrictor areas exhibit inherent automaticity, i.e., they continue to discharge and to maintain arterial blood pressure through vasoconstriction even after elimination of all incoming nerve influences. Section of the brainstem above the medulla does not affect blood pressure; this indicates that upper levels do not dominate the medullary level, even though they can modify its state of activity. This tonic activity is frequently rhythmic in nature reflecting the phasic inputs from nucleus tractus solitarius and other medullary afferent activity. This phasic activity gives rise to phasic oscillations of blood pressure. Those occurring at the same frequency as respiration are termed Traube-Hering waves and those at lower frequencies are termed Mayer waves.

Hypothalamus

Both increases and decreases in heart rate and arterial blood pressure have been reported by investigators using electrical currents to stimulate the hypothalamus. The hypothalamus also plays an important role in the control of body temperature. The rostral hypothalamus and preoptic area contain neurons that protect the body against overheating; they also control the discharge to the vasoconstrictor fibers of the cutaneous blood vessels, and, thus, play an important role in adjusting blood pressure. Electrical stimulation or local cooling of this area brings about a rise in blood pressure (vasoconstriction), while direct warming of this region produces a fall in blood pressure (vasodilation). The cutaneous arterioles, and, especially, the arteriovenous anastomoses (shunts), are the vessels most sensitively engaged in control of heat loss.

Cerebral Cortex

A number of studies of cortical control of vasoconstriction are available, but relatively little is known about the importance of circulatory adjustments originating in the cortex. Stimulation of the motor and premotor cerebral cortex results in marked elevation of blood pressure with constriction of the cutaneous, splanchnic, and renal vessels, and, at the same time, a considerable vasodilation in the skeletal muscle. It is believed that these higher centers play significant roles in the blood pressure responses to pain and anxiety, and exercise, but may, as mentioned below, have their most important effects in modulating the response of "lower centers" in the brainstem to afferent input.

Interaction Between Central Areas

The major role of the central nervous system in cardiovascular control is to integrate different information ("sensory inputs") and adjust the tonic outflow of autonomic nerve traffic. In the regulation of autonomic outflow there is integration of descending traffic with inputs from all levels in the central nervous system impinging upon preganglionic vagal and sympathetic fibers. In the classic "defense" reaction which is elicited by stimulating small areas of the hypothalamus there is a rise in blood pressure, cardiac output, and heart rate, with vasoconstriction in renal, intestinal, and skin beds and marked cholinergic vasodilation in muscle. If one stimulates lower in the brainstem the responses tend to be more specific, e.g., only splanchnic or skin vasoconstriction may occur. Thus, the brainstem also allows for the possibility of differential regional responses which may be important in integrated activity such as exercise.

There are fluctuations in vagal efferent activity synchronous with respiration, which largely account for the normal respiratory variation in heart rate. These fluctuations are abolished by factors which inhibit the central areas responsible for the control of respiration and are not seen when respiration is suspended (apnea). Thus, cyclic variations in activity in areas of the brainstem not related to cardiovascular function may normally influence efferent autonomic activity to the heart and circulation. There are also phasic changes in sympathetic nerve activity that are strongly linked to the activity of respiratory motor neurons. These are most apparent as fluctuations in aortic blood pressure and are discussed in more detail later in this chapter.

Afferent input from the IXth and Xth (cranial) nerves enters the medulla and synapses in an area termed the nucleus tractus solitarius. From here there are "long latency" (presumably involving many synapses) connections to the pons, thalamus, and the cerebral hemispheres (Figs. 2.170). Other ascending traffic comes from "visceral" afferents (cardiac and pulmonary receptors, and somatic afferents).

REFLEX AND HORMONAL REGULATION OF THE HEART AND CIRCULATION

During local adjustments of blood flow to meet metabolic needs the arterial pressure driving the blood through the vessels must be maintained. According to the general resistance equation discussed in Chapter 7 (resistance = pressure/flow), if there is vasodilation with decreased resistance and increased flow, and yet the pressure gradient is found to be unchanged, some extrinsic regulatory system must have intervened to maintain arterial pressure in the vessels. In the case of a local decrease in the arterial resistance due to muscular exercise, for example, the resistance in other beds may increase, and regulatory mechanisms may also act to increase the cardiac output. These extrinsic regulatory mechanisms govern heart rate, cardiac contractility, blood vessel caliber, and the distribution of blood volume.

The control mechanisms responsible for maintaining arterial pressure may be divided into *short-term processes*, which are effective over a period of seconds to hours, and *long-term processes* that operate over days to weeks. The former are largely neural, utilizing receptors in the heart and blood vessels to sense blood pressure and the autonomic nervous system to regulate cardiac function and arteriolar diameter. The long-term controls are largely hormonal and renal. They regulate both arterial resistance and the blood volume. Both of these systems are negative feedback control systems and have many properties in common with other biological control systems which regulate a wide variety of body functions, ranging from intracellular metabolism to body posture.

The engineering theory which describes control systems was initially developed in connection with speed controls on steam engines, and now has progressed into a large and complex field utilizing modern computing equipment. The approach requires detailed knowledge of the function of the system being examined and the dynamic response of its various components. The application of this approach to cardiovascular control has been summarized in several texts, e.g., Milhorn (1966). The approach is extremely useful for investigation of the circulation, but it is beyond the scope of this discussion and frequently beyond the level of our current knowledge of the individual components of the cardiovascular systems. However, it is important to understand the basic principles of negative feedback control systems, since they form the basis for most cardiovascular control systems. A schematic diagram of a simplified negative feedback control system is shown in Figure 2.171. There are three essential components to this type of control system: a sensing device (sensor); a mechanism to change the controlled variable (effector) and a function which determines which way and how much to alter the variable (transfer function). Let us consider the example of the arterial baroreceptor system utilizing the generalized diagram shown above. Suppose that methoxamine, an agent which con-

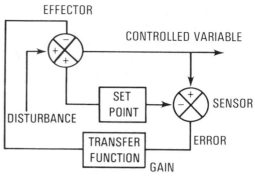

Figure 2.171.   Schematic diagram of a negative feedback control system. A disturbance applied to the system is sensed and compared to some reference point *(SET POINT)*. The difference between the set point and the controlled variable *(ERROR)* is used to return the controlled variable to its original value.

stricts the arteries, is given to increase blood pressure. In the control diagram, the agent would act at the *arrow* labeled disturbance to produce a 60-mm Hg increase in blood pressure. Thus a 60-mm Hg error would be determined at the sensors. The response of the transfer function, in this case the central nervous system, is designed to force blood pressure in the opposite direction (negative feedback). In this case a withdrawal of sympathetic arterial constrictor tone would be initiated by the central nervous system. If the initial mean arterial pressure is 100 mm Hg and is increased to 160 mm Hg by the drug, following "compensation" by the control system arterial blood pressure would return to perhaps 120 mm Hg. The overall gain of this system can be estimated as the ratio of the amount of correction to the remaining error. For the baroreceptors, the amount of correction would be 40 mm Hg, the remaining error, 20 mm Hg, for an overall gain of -2 (Sagawa, 1983). The greater the gain the less the final error. Most of the body's control systems and the arterial baroreceptor system are quite well damped (there is significant time delay in reacting to a particular stimulus), and the eventual steady state value is approached without substantial oscillation or overshoot but with a residual error.

BARORECEPTOR BLOOD PRESSURE CONTROL SYSTEM

Etienne Marey, a French physician, first noted in 1859 the inverse relationship between blood pressure and heart rate, and largely through the later pioneering work of Hering (1927), Heymans, and others, we have come to understand that the carotid sinus and aortic arch baroreceptor systems are responsible for a highly effective blood pressure control system (Heymans and Neil, 1958). This system consists of stretch receptors located in vessel walls which send information reflecting the level of blood pressure to the central nervous system which, in turn, sends efferent impulses to the cardiovascular system which then alter blood pressure. These sensitive stretch receptors are diagrammed in Figure 2.172. As mentioned earlier, they are located throughout the aortic arch along the major thoracic vessels, and at the bifurcation of the carotid arteries in the neck. Fibers from the latter area course through the carotid sinus nerves (these branches of the IXth nerve are sometimes termed "Hering's nerve") and thence to the medulla. Fibers from the aortic arch sensors course through the vagus nerve (these branches going to the aortic arch and heart are sometimes termed cardiac depressor nerves) (Fig. 2.172).

The receptors themselves respond only to distortion, and if the carotid sinus is encased in a plaster cast before raising the blood pressure, there will be no alteration in carotid sinus nerve traffic.

The precise mechanism by which distortion activates these receptors is unknown (Kunze and Brown, 1978) and their small size has prevented standard electrophysiological approaches. How-

ever, local alterations in sodium content of the vessel wall, as occurs during congestive heart failure, may be important in regulating receptor function (Brown, 1980).

In Figure 2.173 panels *A* and *B* schematically represent recordings from a single carotid sinus nerve fiber at several levels of mean carotid sinus blood pressure, each vertical spike representing a single depolarization from this fiber. At a very low average aortic blood (50 mm Hg) there is little activity. The activity is normally correlated in time to the systolic aortic pressure increase, indicating that these receptors are sensitive to rate of change of pressure as well as average level (Chapleau and Abboud, 1987). The average level of pressure at which a fiber first fires is termed the threshold. As mean pressure within the carotid sinus is progressively increased with the pulse pressure constant (Fig. 2.173, *panel A*) nerve traffic is increased until it becomes continuous (at 200 mm Hg in this example). With a fixed level of mean pressure, the number of impulses per second increases as the pulse pressure increases. (Fig. 2.173, panel *B*).

Both the threshold for firing and the frequency of impulses per unit change in the aortic pressure (gain) may vary substantially from fiber to fiber, but the average threshold in the carotid sinus at which firing begins is normally not less than 50 mm Hg, and the maximum output is achieved at approximately 170 mm Hg (Fig. 2.174). Although the sensitivity of myelinated and nonmyelinated fibers is similar the threshold is usually higher in nonmyelinated fibers and they may continue activity at higher pressures. The sigmoid shape of the relation-

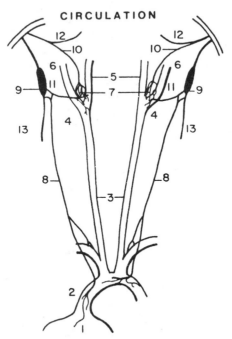

Figure 2.172. Innervation of the carotid sinus and arch of aorta: *1*, heart; *2*, arch of aorta; *3*, common carotid; *4*, carotid sinus; *5*, external carotid; *6*, internal carotid; *7*, carotid bodies; *8*, cardiac depressor nerve; *9*, ganglion of vagus; *10*, sinus nerve, branch of glossopharyngeal nerve; *11*, nerve branch connecting the carotid sinus with the vagus ganglion; *12*, glossopharyngeal nerve; and *13*, vagus nerve. (Adapted from Heymans et al., 1933.)

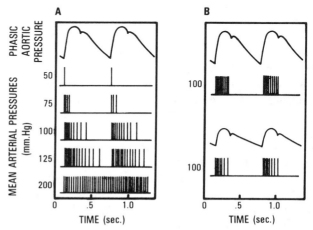

Figure 2.173. Relationship between phasic blood pressure in the carotid sinus and aorta and frequency of nerve traffic (represented by single vertical spikes). In panel *A* nerve traffic at several different mean levels of aortic pressure is shown, and in panel *B* activity at two levels of pulse pressure at the same mean pressure.

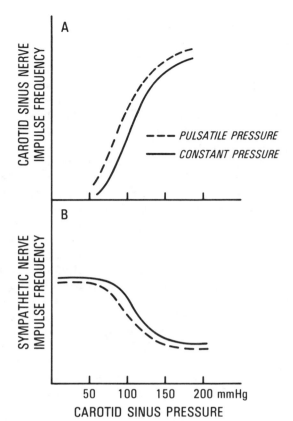

Figure 2.174. *A*, Relationship of frequency of firing of the carotid sinus nerve and carotid sinus pressure. *B*, Frequency of firing of postganglionic sympathetic fibers in response to changes in carotid sinus pressure. (Adapted from Kezdi and Geller, 1968.)

ship between nerve traffic and carotid sinus pressure leads to two important conclusions concerning the sensitivity of the baroreceptors. First, pressures below approximately 50 mm Hg are not sensed, and therefore alterations in blood pressure below this level will result in no change in baroreceptor output. Secondly, the slope (gain) of the relationship between nerve traffic (impulse frequency) and blood pressure is maximal at the normal level of mean arterial pressure (frequently referred to as the set point); thus, the system works most effectively here. It will return blood pressure to this level with the least error and the least time lag (see discussion of Fig. 2.171). At pressures above 170 mm Hg, where the gain becomes very low, changes in blood pressure will no longer result in significant changes in nerve activity, and hence the system will be ineffective in controlling blood pressure. Since these receptors are also sensitive to rate of change of pressure, an increase in pulse pressure at the same mean pressure increases the frequency of carotid sinus nerve traffic. Although the receptors in the aortic arch have a slightly higher threshold and a

somewhat reduced gain, the receptors in the carotid sinus and aortic arch are in most other respects quite similar.

The increase in afferent carotid sinus nerve or vagal (aortic) nerve traffic acts through the central nervous system to return blood pressure to its control value. This is achieved by a decrease in efferent sympathetic constrictor tone to the peripheral arteries, and an increase in parasympathetic cardiac tone. As shown in Figure 2.174, *panel B*, as carotid sinus traffic increases over the physiological range of arterial pressures, sympathetic nerve impulse frequency declines. Thus the net result of an increase in pressure in the carotid sinus will be to reduce sympathetic tone to the blood vessels and thereby to reduce blood pressure. Note that the *dashed line* in Figure 2.174 again indicates the response to increased pulse pressure at the same mean pressure. Several factors will contribute to the reduction of blood pressure. There will be withdrawal of sympathetic vasoconstriction tone to arterioles resulting in a fall of systolic and diastolic arterial pressure and a fall in vasoconstriction tone to capacitance vessels resulting in decreased cardiac filling; an increase in vagal tone will reduce heart rate.

The hemodynamic effects of lowering carotid sinus pressure are demonstrated in the experiment illustrated in Figure 2.175, which shows the effect of sudden occlusion of the carotid arteries in an open chest dog in which a force gauge is sewn to the ventricle. Following bilateral carotid occlusion the drop in pressure in the carotid sinuses causes a rise in blood pressure, by increasing sympathetic stimulation to the blood vessels. This occurs 10–15 s following carotid occlusion (the normal "time lag" for a sympathetic response). The effects of sympathetic nerve activation on myocardial inotropic state and the ventricular function curve are also illustrated in Figure 2.181. The increase in heart rate preceeds the increase in blood pressure (the time lag for vagal activity is 1–2 s). The tachycardia and hypertension, however, are not sustained. As blood pressure in the aortic arch increases aortic arch baroreceptor nerve traffic increases which reduces the response of medullary neurons to the carotid sinus activity. This results in a secondary fall in heart rate and blood pressure seen in Figure 2.175.

It is important to remember that the changes in efferent sympathetic and parasympathetic activity induced by the baroreceptor system have a variety of effects on the heart and circulation. For example, decreases in carotid sinus pressure increase sympathetic tone and will produce an increase in blood pressure; an increase in cardiac contractility, as

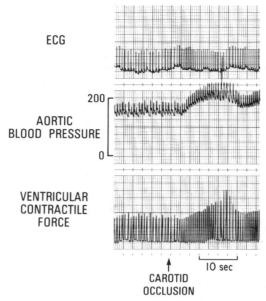

ECG

AORTIC
BLOOD PRESSURE

200 ⌐

0 ⌐

VENTRICULAR
CONTRACTILE
FORCE

↑
CAROTID
OCCLUSION

⊢ 10 sec ⊣

Figure 2.175. Recordings of aortic blood pressure, electrocardiogram, and right ventricular contractile force (measured with a gauge sewn to the right ventricular free wall) during abrupt sustained bilateral carotid occlusion. At the *arrow* both carotid arteries are occluded below the carotid sinus. The initial increase in blood pressure, heart rate, and contractile force is due to the activation of the baroreceptor reflex. Since the vagus nerves are still intact the secondary falls in heart rate and blood pressure are due to the firing of aortic arch baroreceptors in response to the *increased* blood pressure.

reflected in the change in the ventricular function curve; and decreased venous capacity with increased mean circulatory pressure and shift the venous return relationship (Chapter 17). The combination of these effects will produce a new equilibrium point resulting in an increase in cardiac output.

Modifications of the Baroreceptor Response

Alterations in the baroreceptor blood pressure control system occur in both physiological and pathophysiological circumstances. These alterations may occur at all levels of the system from the interaction of receptors with the vessel wall to arterial smooth muscle. The nerve endings in the carotid sinus normally adapt to a prolonged stimulus by reducing their firing rate despite a sustained increase in pressures in the carotid sinus. Carotid sinus receptors begin to adapt to prevailing pressure within minutes reaching a maximum adaptation within days (Brown, 1980). This adaptation does not normally influence gain, only threshold. The mechanism of this type of adaptation is unknown, but it may have to do with altered arrangement of the coupling between the vessel wall and the receptor itself. The interaction between receptors and the vessel wall is importantly altered in several disease states. One example

is the alteration occurring in essential hypertension (high blood pressure of unknown cause (Aars, 1975)). The vessel wall becomes stiffer in hypertension; thus, for a given increase in carotid sinus pressure there is less distortion of the receptor. Therefore, the gain of the system, the relationship between carotid sinus pressure and change in nerve traffic, is reduced (Kezdi, 1967). In addition, the threshold for firing has been reset upward (to the new and higher level). The effect would be to shift the entire relationship between carotid sinus nerve traffic and blood pressure (Fig. 2.174) to the right. The factors responsible for these changes are not well understood, but it is clear that they involve both changes in the vessel wall (stiffer) and in the receptor (Guo and Thames, 1983). Receptor sensitivity may even be increased, partially compensating for the vessel wall changes (Brown, 1981).

In congestive heart failure the gain of the baroreceptor control system is reduced without substantial change in threshold for firing (Higgins et al., 1972). Although the mechanism for this change is unknown, heart failure affects the baroreceptor control system at several levels. There is an increased sodium content of the arterial wall which may alter its physical properties or change the receptor response. Moreover, the efferent limb of the baroreceptor control system is altered with reduced stores of neurotransmitters in the myocardium and an altered response to sympathetic stimulation (Schmid, 1981).

Important changes in the baroreceptor response also can occur when information is integrated in the central nervous system. For example, recalling Figure 2.170, it is clear that afferent nerve traffic from other sympathetic inputs or centrally from "higher centers" may impinge upon nerve traffic coming from the baroreceptors and alter the net change of efferent sympathetic and parasympathetic traffic induced by a given change in blood pressure. The clearest example of this has been shown to occur in man during exercise, when increasing levels of exercise progressively reduce baroreceptor gain. This reduction is proportional to the extent of exercise (Korner, 1979).

OTHER REFLEXOGENIC AREAS

Two other major groups of receptors contribute importantly to circulatory regulation. One major group of receptors responds to alterations in their chemical environment (chemoreceptors). These receptors are located in specialized cells near the high pressure receptors (Heymans et al., 1931). The other group are stretch receptors located in the large vessels in the thorax, in the lung, in both atria, and in the ventricles. These receptors, frequently termed "low pressure receptors," are sensitive to distortion.

Chemoreceptors

Chemoreceptors may be divided conveniently into two groups, arterial and cardiopulmonary. The first group are located in the carotid and aortic bodies and respond to physiological alterations in blood gases and pH (Biscoe, 1971). The second group are located in the heart and lungs and are sensitive to minute amounts of substances such as veratrum alkaloids, phenyldiguanide, nicotine, bradykinin, and capsaicin. Such substances stimulate a variety of afferent vagal endings in the heart, lungs, and great vessels (Bevan, 1962). Recent studies have shown that these receptors may be sensitive to some substances formed and released in their vicinity such as bradykinin, prostaglandins, or perhaps metabolic products associated with ischemia.

Both the carotid and aortic chemoreceptors are sensitive to changes in the partial pressures of carbon dioxide, hydrogen ion, and oxygen in the blood. Blood flow through the carotid body per gram of tissue is by far the largest in the body. Direct measurements indicate flows of 2 liters/100 g/min through this organ (recall that the flow to the left ventricular muscle is approximately 100 ml/100 g/min).

The arterial chemoreceptors play an important role in the control of respiration (see Chapter 40) and are thought to be of relatively minor importance in the control of the circulation. The mechanism by which changes in blood chemical composition effect a response in the arterial chemoreceptors is not clearly understood. The threshold for changes in afferent traffic of the chemoreceptors from the carotid body and aortic arch is fairly high. Vascular effects in dogs require a severe degree of carotid body hypoxia; when arterial P_{O_2} progressively falls below 70 mm Hg afferent traffic increases and there is linear increase of sympathetic efferent traffic until a maximum is reached at a P_{O_2} in the mid 30s (Pelletier, 1972). Increases in nerve impulses from afferent nerves may be recorded with CO_2 tensions above 30 mm Hg and with small decreases in pH either together or independently (Neil and Howe, 1973).

With respiration controlled (eliminating the effects of the pulmonary stretch receptors on the circulation) the major cardiac effect of stimulation (increasing afferent traffic) of the arterial chemoreceptors is profound bradycardia, conduction defects (a common result of vagal efferent stimulation), and a diminution of cardiac contractility. Bradycardia is greatly diminished by the use of atropine but may not be abolished until sympathetic pathways to the heart are also blocked, suggesting a synergistic withdrawal of sympathetic tone (MacLeod and Scott, 1964). Carotid body reflexes not only slow the heart but also depress contractility. Downing et al. (1962), utilizing isolated heart preparations, showed that there was a depression of the ventricular function curve during carotid body hypoxia. Although the efferent neuromechanisms for this change are still controversial, they appear to involve both alterations in parasympathetic and sympathetic tone. The major effect on the peripheral circulation of hypoxic chemoreceptor stimulation is vasoconstriction (Pelletier, 1972; Biscoe, 1971; Heymans and Neil, 1958).

Recent evidence suggests that when stimulated with hypoxic blood aortic and carotid bodies elicit a directionally opposite response. Although the reflex effect of stimulation of the carotid bodies is primarily inhibitory (bradycardia), stimulation of aortic bodies causes an increase in heart rate and ventricular function. Perfusion of aortic bodies with hypoxic and hypercapnic blood produces the same degree of vasoconstriction as perfusion of the carotid bodies (Daly et al., 1975).

There are also direct effects of hypoxia on the brain and spinal cord, causing an increase in sympathetic efferent discharge with hypertension and tachycardia (Downing et al., 1963; DeGeest et al., 1965).

For most terrestrial mammals the peripheral arterial chemoreceptors have been shown to play a major role in the cardiovascular response to diving (Jones and Purves, 1970). On immersion of vertebrates, sensory receptors in the larynx are stimulated and trigger a prompt suppression of respiration. During the resultant hypoxia peripheral chemoreceptors are stimulated, and bradycardia and hypertension are the dominant cardiovascular response (Daly et al., 1975).

Cardiopulmonary Receptors

Receptors are located in the thorax and found along the large intrathoracic vessels in all chambers of the heart and the bronchi. Myelinated and nonmyelinated nerves from these receptors course through vagal or sympathetic trunks and connect to the nucleus tractus solitarius.

Cardiac Afferents

Receptors exist in all chambers of the heart and play important roles in the regulation of blood volume, heart rate, and in sensing ischemic pain. They can be classified into two general types, those sensitive to deformation, and nociceptive afferents sensitive to a variety of agents such as bradykinin and prostaglandins. Nociceptive receptors were first discovered by von Bezold in 1867. He showed that the intravenous injection of crude veratrum alkaloids caused bradycardia and hypotension. Later Jarisch and his associates in 1930–1948 showed that these cardiovascular effects were due to stimulation of afferent vagal endings in the heart and lungs (hence the Bezold-Jarisch reflex; see Heymans and Neil, 1958). These afferent endings exist in lungs, heart, and great vessels and the abdominal viscera (Dawes and Comroe, 1954; Coleridge et al., 1978; Coleridge and Coleridge, 1979). In the heart, nerve endings are present in the right ventricle and atria, but there are more nerve endings in the left ventricle. There tend to be more of these nonmyelinated vagal fibers on the posterior wall of the left ventricle (Thames et al., 1978). They normally have very sparse resting discharges, but the endings are extremely sensitive to chemicals. Stimulation of these fibers with veratridine produces a profound cardiac vagal efferent discharge, similar to that of the physiological chemoreceptors discussed above. The peripheral effects are opposite in direction, and stimulation of these receptors induces vasodilation and hypotension due to withdrawal of sympathetic tone.

Most current electroneurographic evidence has shown that ventricular fibers may be excited by a variety of agents, including veratradine, nicotine, phenyldiguanidine, acetylstrophanthidin, and capsaicin, all of which may evoke the chemoreceptor nerve traffic. There is increasing evidence that these types of fibers may also be sensitive to endogenously released substances, such as prostaglandins, serotonin, lactic acid, and bradykinin (Mark, 1987). These substances may be released by the myocardium during hypoxia and ischemia and may provide an important pathophysiological role for these receptors.

In recent years it has become apparent that there are also afferent fibers from chemosensitive receptors coursing through sympathetic pathways to the spinal cord. These afferents may be sensitive to nonphysiological substances and in vagotomized animals their activation with substances which also excite the chemoreceptors discussed above produces a tachycardia and vasoconstriction with an increase in contractility (Peterson and Brown, 1971). It is likely that these responses may be important in the excitatory reflexes occurring in myocardial ischemia in man.

Stretch Receptors

The most important of these receptors for circulatory control are located within the four cardiac chambers. These cardiac receptors participate in reflexes which are important in the regulation of heart rate, blood pressure, and blood volume. Nerve endings are found within the subendocardium and in the subepicardium distributed along the coronary vessels. Innervation of the myocardium is present but less dense. All receptors respond to distortion or stretch with increased nerve traffic. Unmyelinated vagal afferents are located in the atria and ventricles and increase in chamber volume and force of contraction activate these receptors. Through a significant resting discharge they exhibit a tonic inhibitory action on sympathetic nerve activity, vasopressin release, and indirectly (via reduction of sympathetic activity) on renin release. Unmyelinated fibers with receptors sensitive to distortion also course through sympathetic nerve trunks. Their function is unknown. However, it is quite likely that these receptors are also important in adaptations occurring under pathological conditions, including chronic cardiac dilatation and myocardial ischemia. The larger myelinated vagal fibers of this group have been studied extensively. The endings of these fibers are located primarily in the right and left venoatrial junction. These receptors probably monitor changes in cardiac volume. Activation is thought to produce a tachycardia and sympathetic activation. Renal sympathetic nerve activity is, however, inhibited.

INTEGRATED CONTROL OF CIRCULATORY FUNCTIONS

The receptor systems discussed above all contribute importantly to cardiovascular homeostasis. In the next section, we will discuss how the systems interact to regulate heart rate and blood volume, and will further illustrate these interactions when we discuss the effects of respiration on the circulation. Integrated responses to exercise, heart failure, and hypertension are discussed in the next two chapters.

Before considering these integrated responses it is helpful to recall the sites and mechanisms of vasomotion that are important in circulatory control (see Fig. 2.8). Arterial resistance and systemic blood pressure are primarily under the control of the resistance vessels, small arteries, and arterioles. The sympathetic (vasoconstrictor) nervous system plays an important role in regulating the diameter of these vessels through continuous levels of neural discharge (tone). The resistance across a given parallel vascular bed is primarily the result of the balance between this tone and the local metabolic and myogenic factors discussed in Chapter 6. This resistance, of course, also determines flow through the bed at a constant pressure and thus can regulate the circulating volume of a regional bed. The ability of a regional circulation to alter its resting flow varies markedly in different regional beds, with the myocardium, skeletal muscle, the splanchnic circulation, and skin all showing the capacity to increase flow by 3 to 4 fold under conditions of maximal vasodilation. The cerebral bed, however, is less sensitive to neural control (Rapela et al., 1967).

Referring to Figure 2.9 there are two further sites of neurally controlled vasomotion: First on the balance between pre- and postcapillary resistances and secondly on the capacitance vessels. From the Starling equilibrium recall that the filtration of fluids across the capillary is dependent on the oncotic and hydrostatic pressure gradients. The latter is determined by both neural and local factors. Thus changes in the ratio of pre- to postcapillary resistance can produce either net absorption or filtration of circulating volume across the capillary (Chapter 6). Regulation of tone to the capacitance beds also has important effects on circulating blood volume and venous return as discussed below.

CONTROL OF HEART RATE

The cells of the cardiac conduction system exhibit spontaneous depolarization; the intrinsic cardiac rate is normally dominated by the fastest pacemaker site (usually the SA node). In the denervated heart, the rate ranges from 100–110 beats/min, and continuous control by the autonomic nervous system (predominately the parasympathetic nervous system) is responsible for a suppression of the normal intrinsic rate of the SA node by 20–30 beats/min.

The SA node is richly innervated by both sympathetic and parasympathetic fibers. At rest there is continuous efferent vagal nerve traffic to this region and relatively little resting sympathetic activity. This results in suppression of the intrinsic rate and the low resting normal heart rate. However, changes in both sympathetic and parasympathetic tone may increase or decrease heart rate. In contrast to other regions of the myocardium, combined effects of changes in sympathetic and parasympathetic efferent activity at the sinus node do not result in entirely independent effects. Thus the parasympathetic effects on the SA node predominate and, when there is high parasympathetic activity, an increase in sympathetic activity will produce only a modest increase in heart rate. However, an alteration in parasympathetic activity will produce a large change in heart rate relatively independent of existing levels of sympathetic nervous activity (Levy and Zieske, 1969).

Due to the high concentration of cholinesterase in the SA node the effects of parasympathetic activity are very transient. The effects of a single burst of parasympathetic activity will depend to a large extent on the timing between atrial depolarization (P wave) and the arrival of the parasympathetic efferent activity. This transitory effect of parasympathetic activity was first shown in the classic studies of Brown and Eccles (1934). A burst of vagal activity which arrives just before a P wave is more effective in reducing heart rate and prolonging the RR interval than a burst of vagal activity which arrives just after a P wave. Later work has shown that repetitive vagal stimulation can lead either to large fluctuations in heart rate, depending on the stimulus to P wave interval, or even "pace" the heart by maintaining a constant timing relationship between vagal stimulation and the P wave (Levy et al., 1972).

The AV node is also richly innervated by both sympathetic and parasympathetic branches of the autonomic nervous system. In contrast to SA nodal tissue the effects of simultaneous stimulation of the sympathetic and parasympathetic systems at the AV node appear to be independent. Thus, an increase in sympathetic nerve traffic will decrease AV nodal conduction time independent of the level of vagal tone.

Although there is extensive overlap between the distribution of right and left sympathetic and parasympathetic nerves to the SA and AV nodes, stimulation of the left sympathetic stellate ganglion will predominantly stimulate the left side of the ventricle and the AV node. Indeed stimulation of the sympathetic nerve on the left may lead to acceleration of phase 4 depolarization in an area near the His bundles to such an extent that this area then becomes the predominant cardiac pacemaker.

In addition to the possibility of altering cardiac pacemaker sites the autonomic nervous system is thought to play an important role in the genesis of arrhythmias. As discussed in Chapter 8 norepinephrine, the sympathetic neurotransmitter, not only induces changes in phase 4 depolarization but also increases phase 0 dV/dt in the "slow fibers" of the AV node or in ischemic tissue. These changes can induce alterations in ventricular excitability and refractoriness and may be an important mechanism in the generation of arrhythmias when there is increased sympathetic nerve traffic.

In 1915 Bainbridge showed that rapidly infusing saline into the right atrium resulted in an increase in heart rate. This effect has been thought to be due to stimulation of mylenated atrial stretch receptors mentioned earlier. Later studies have shown that the nature of this reflex depends upon a variety of factors, including the extent of stimulation of baroreceptors which tend to produce an opposite effect. Moreover, the magnitude of tachycardia observed is extremely sensitive to the initial cardiac rate. With a high initial cardiac rate, even a bradycardic response to volume infusion may be observed. In the conscious animal a tachycardia may be induced by volume loading or by acute hemorrhage. Presumably the Bainbridge reflex dominated the baroreceptor response when blood volume was acutely raised, and the baroreceptor system dominated the response during acute reduction in blood volume (Vatner and Boettcher, 1978).

REGULATION OF CIRCULATING BLOOD VOLUME

The maintenance of adequate tissue perfusion is dependent on a sufficient level of cardiac output. The factors which influence the performance of the heart and therefore affect its output are discussed in Chapters 12 and 17. In the section below we will consider the factors which influence circulating blood volume and thereby determine venous return and cardiac output.

The kidney is the dominant organ associated with control of blood volume, and there are important reflex and hormonal systems which regulate salt and water excretion. These systems are discussed in detail in Section 4, Chapter 31 on the regulation of volume and osmolality of body fluids. Maintenance of normal volume and osmolality of the body fluids requires that the input and output of water and sodium are matched each day. Both the low pressure (atrial) stretch receptors and the arterial baroreceptors participate in this regulation. Receptors in the left atrium (and probably pulmonary veins) which respond to stretch have been implicated in the regulation of blood volume (Gauer et al., 1970; Goetz et al., 1975). Acute distention of the whole left atrium, as with a balloon, causes diuresis. As discussed below this diuresis is the result of several factors including: a decrease in antidiuretic hormone release due to increased atrial stretch receptor nerve traffic; reduced renal sympathetic nerve traffic (Bishop, 1987); and a reduction of renin release and release of atrial natriuretic factor (ANF) from the atria. These powerful natriuretic-vasoactive peptides (ANF) are stored in granules (Kisch, 1956). These peptides, when injected, produce a sodium diuresis, hypotension, and increase in hematocrit (deBould et al., 1981). Their renal effects are discussed in detail in Chapter 31. ANF tends to oppose the vasoconstrictor effects of angiotension and its vasodilation effects are more prominent in constricted vessels (Laragh, 1986). The fall in blood pressure is most likely due to a decrease in

venous return and may be due to movement of fluid out of capillaries (Kleinart et al., 1986).

To understand how these factors interact to control blood volume, we will examine the response to a loss of blood volume (hemorrhage). Figure 2.176 illustrates several of these effects in response to graded hemorrhage. With 10% loss of blood volume the firing of the atrial stretch receptors is reduced, and with further reduction in blood volume aortic pressure is reduced decreasing nerve traffic from the aortic and carotid sinus baroreceptors. Lowering of traffic from the atrial receptors increases the secretion of antidiuretic hormone which acts to

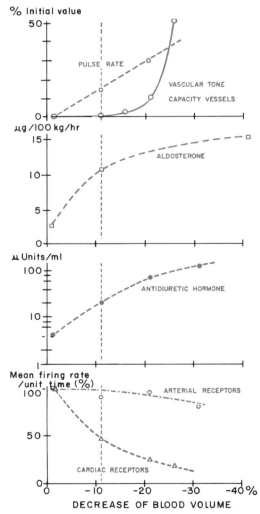

Figure 2.176. Responses to a graded hemorrhage. With loss of the first 10% of blood volume the firing rate of atrial *(cardiac)* receptors falls to one-half. As the loss exceeds 20% there is a fall in the firing rate of aortic *(arterial)* baroreceptors. The plasma aldosterone and antidiuretic hormone concentrations rise, causing sodium and water retention. The pulse rate increases steadily as blood volume falls. (Redrawn from Gauer et al., 1970.)

reduce water excretion by the kidney (see Chapter 31). Reduction of carotid sinus and aortic baroreceptor traffic increases renal sympathetic nerve traffic constricting the afferent arteriole and reducing filtration of fluids by the kidney. Moreover, increased sympathetic nerve traffic will increase the secretion of renin by the granular cells of the kidney. Renin is a proteolytic enzyme released by the kidney which acts on a substrate protein in plasma to liberate the decapeptide angiotensin I (Fig. 2.177). This inactive prohormone is in turn converted into a potent vasoconstrictor hormone angiotensin II by a converting enzyme. Angiotensin II performs three major related roles: first, increasing the arterial pressure by direct vasoconstriction; second, increasing the renal retention of salt and water through stimulation of aldosterone secretion; and third, angiotensin II is thought to stimulate thirst directly. All of these responses tend to correct the loss of circulating volume induced by hemorrhage. The relative importance of these mechanisms for control of circulatory volume is discussed in Chapter 32. However, the sensitive response of the hypothalamic osmoreceptors to decreases in plasma osmolality may be the dominant normal control mechanism. These receptors stimulate the release of antidiuretic hormone. The reflex control of capacitance vessels by the baroreceptor systems, an important part of the response to a decrease in blood volume, is illustrated by the following experiment in which the effects of changing blood pressure in the carotid sinus is examined (Shoukas and Sagawa, 1973). The aortic nerve is cut, eliminating the response of the aortic baroreceptors. In this preparation all the blood coming from the systemic veins is diverted to a reservoir and returned at a constant rate to the right heart. Thus, at a constant cardiac output regulated by the pump, changes in the reservoir volume reflect changes in venous capacitance. As shown in Figure 2.178, progressive decreases in the pressure within the isolated carotid sinus produce an increase in aortic pressure and an increase in volume in the reservoir due to increased sympathetic vasoconstrictor tone to the capacitance vessels.

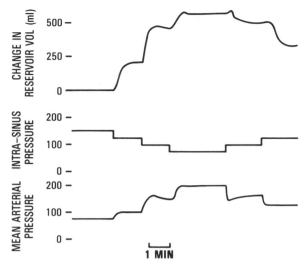

Figure 2.178. Results of an experiment in which blood pressure in the isolated carotid sinus is varied. All venous blood returning to the heart is drained into a reservoir and then pumped at a constant rate back to the pulmonary artery. Changes in reservoir volume thus represent changes in venous capacity. (Adapted from Shoukas et al., 1973.)

EFFECTS OF RESPIRATION

There are important effects of respiration on the heart and circulatory system. In the central nervous system there are interactions between respiratory "pacemakers" and efferent autonomic tone which directly affect heart rate and blood pressure. These pacemakers induce variations in systemic blood pressure at the same period as the frequency of respiration. These blood pressure waves, termed Traube-Hering waves, after their discoverers, are the result of increased efferent sympathetic nerve traffic occurring with the onset of respiration. There are also variations in vagal tone that are centrally mediated and related to respiration. These result in a decrease in vagal activity occurring during inspiration and an increase in tone during expiration. These latter two effects are normally thought to be independent of a variety of reflex effect mediated by either low or high pressure baroreceptors or pulmonary stretch receptors.

There are also mechanical effects of respiration on left ventricular output. With inspiration there is a decrease in intrathoracic pressure which results in an increase in right atrial inflow (and perhaps, a secondary increase in heart rate mediated by the "Bainbridge" effect), whereas left atrial inflow is transiently reduced consequent to the increase in pulmonary vascular capacity. This results in lowering of aortic blood pressure which occurs most predominantly in early inspiration and is normally

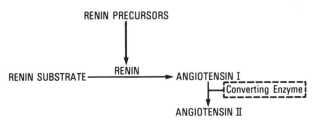

Figure 2.177. Production of angiotensin II from renin.

over by midinspiration. Secondly, when the increase in systemic venous return eventually reaches the left atrium, the increased left ventricular outflow results in an increase in systemic pressure returning blood pressure to its peak level, an effect which normally occurs in mid- or end expiration. This increase in arterial pressure frequently results in reflex cardiac slowing. Thus, the normal responses of cardiac frequency to inspiration are represented by an increase in the heart rate during inspiration due to central factors and possibly to a Bainbridge-like response, and a slowing of heart rate during late inspiration and early expiration when aortic pressure is increasing.

Frameworks for Analysis of Ventricular and Circulatory Function: Integrated Responses

Ventricular and peripheral circulatory function, and the interaction between the two, should be understood within frameworks that are useful for measuring responses to physiological stress, adaptations to disease, and pharmacological effects. Several approaches for assessing the function of the ventricle were introduced in Chapter 12, including analysis of pressure-volume loops, peak ventricular dP/dt, the ejection fraction, and the end-systolic pressure-volume relation (for review see Ross, 1990). In this chapter, we will summarize such approaches, together with the analysis of the ventricular function curve, and also introduce an additional method: the interaction between cardiac output and the venous return from the peripheral circulation. These frameworks will then be applied to analyze selected physiological perturbations.

ANALYSIS OF SINGLE CARDIAC CONTRACTIONS

The function of the right or left ventricle is often assessed in experimental animals, and in human subjects, by measuring its volume or dimensions. This can be accomplished by placing a catheter in the left or right ventricle and injecting radiopaque contrast medium while X-ray films are taken (angiocardiography). The films are calibrated, ventricular dimensions measured, and the volumes at different points in the cardiac cycle calculated, assuming an ellipsoidal shape of the left ventricle (see Fig. 2.114, Chapter 12) (Dodge et al., 1966). It is also possible to implant X-ray opaque markers on the walls of the heart in experimental animals in order to determine cardiac dimensions and calculate heart volumes, to use implanted ultrasonic dimension gauges to determine the long and short axes and wall thickness of the left ventricle continuously, or to use M-mode and two-dimensional echocardiography for the noninvasive determination of the ventricular dimensions and wall thickness, with calculation of ventricular volume (Feigenbaum, 1986). If pressure is measured simultaneously in the ventricle, it is pos-

sible to construct pressure-volume loops under a variety of conditions and to determine the linear end-systolic pressure-volume relation (Chapter 12).

In clinical studies, during cardiac catheterization the ventricular volume is determined by angiography, and the most useful and practical measure of ventricular function in the resting state with this technique is the ejection fraction. Sometimes, high fidelity pressure measurements also allow determination of the rate of change of left ventricular pressure (maximum dP/dt) and, as discussed further below, if the left ventricular end-diastolic pressure and the stroke volume (or stroke work) are known in two different circulatory states, it often is possible to determine whether or not a change in inotropic state has occurred from the analysis of only two contractions.

Ejection Fraction

The approach for determining the ejection fraction from an angiocardiogram using high-speed X-ray motion pictures (cineradiography) is shown in Figure 2.114 (Chapter 12). The volume of the ventricle is calculated both at end-diastole (the end-diastolic frame being selected just after the onset of the QRS complex) and at end-systole (the smallest ventricular volume observed in the motion picture series). The ejection fraction (*EF*), which is the ratio of the stroke volume to the end-diastolic volume, is then calculated from the end-diastolic volume (*EDV*) and the end-systolic volume (*ESV*) as follows:

$$EF = \frac{EDV - ESV}{EDV}$$

and expressed as the percentage or decimal fraction of the end-diastolic volume that is ejected.

The volumes of the ventricles and the ejection fraction can also be estimated by so-called "noninvasive" techniques. These include the use of echocardiography to determine the internal diameter of one axis of the left ventricle at end-diastole and end-systole from which the percent shortening of the diameter, or the volumes and the ejection fraction is estimated (using the formula for an ellipsoid or a sphere) or by two-dimensional echocardiography (Feigenbaum, 1986). The radionuclide technique involves injection of a radioisotope, such as technetium-99m-labeled red blood cells, which remains for a time within the vascular space. An image of either the left or right ventricle is then obtained using a γ scintillation camera, and the end-diastolic and end-systolic counts of the image are used to compute the ejection fraction (for review, see Ashburn et al., 1978).

The ejection fraction has been shown to detect depression of myocardial contractility (reduced inotropic state) sensitively under resting basal conditions. The normal ejection fraction lies between 55 and 80% (average 67%) and is similar across various mammalian species (Holt et al., 1968). When myocardial depression is present the left ventricular ejection fraction is reduced *below* 55% (Dodge et al., 1966).

The ejection fraction of the right ventricle can also be determined by angiography using somewhat more complicated analysis, since the shape of that chamber is neither spherical nor elliptical (see Chapter 12). The various models used to calculate right ventricular volume have remained somewhat controversial. The lower limit of normal for the ejection fraction of the right ventricle is about 45%, a value confirmed by the radionuclide method which relies on total counts within the area of the right ventricle, and therefore is not dependent upon geometric assumptions.

More sophisticated techniques for analysis of left ventricular function by angiocardiography have also been employed, combining pressure, volume, and wall thickness analysis to determine the force-velocity relation of the whole heart (but these approaches are beyond the scope of present discussion).

Peak dP/dt

The maximum rate of rise of left (or right) ventricular pressure measured with a high-fidelity catheter system has been used to assess the basal contractility of the heart, or to define a change in contractility produced by a drug or other intervention. As shown in Figure 2.179, the peak dP/dt occurs just before opening of the aortic valve during isovolumetric left ventricular contraction and the lower limit of normal is about 1200 mm Hg/s. The maximum slope can be determined manually by drawing the tangent to the steepest point of left ventricular pressure tracing during isovolumetric contraction (Fig. 2.179), but ordinarily peak dP/dt

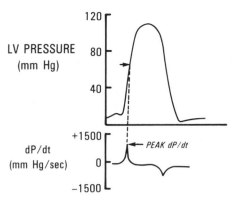

Figure 2.179. Relationship between left ventricular *(LV)* pressure and its first derivative, or rate of change of pressure *(dP/dt)* determined electronically in the lower tracing. Maximum dP/dt can also be determined manually by drawing a slope to the steepest portion of the left ventricular pressure tracing *(dashed line, arrow)*.

is determined from the first derivative of the pressure by means of an electronic differentiating circuit (Fig. 2.179). (See Fig. 2.123, Chapter 12, for changes in peak dP/dt induced by positive and negative inotropic interventions.)

VENTRICULAR FUNCTION CURVE

Ventricular function curves can be plotted for both the right and left ventricles, and in some disease states the function of one or the other ventricle (or both) can be impaired. The shape of ventricular function curves and how they are affected by inotropic influences are determined by the major factors which affect cardiac muscle performance (Chapter 11), that is, the preload (the positive relation between preload and tension development or muscle shortening); the afterload (the inverse relation between afterload and the extent and velocity of shortening); and the level of isotropic state (which shifts these curves upward or downward) (Fig. 2.109; Chapter 11). Direct measurements of wall force and shortening are not readily performed in laboratory studies, or in human subjects, and a simpler type of function curve is usually employed, in which the ventricular end-diastolic pressure (or sometimes the end-diastolic volume) is plotted against some measure of systolic ventricular performance, usually the stroke volume or the stroke work. Sometimes the stroke power (work per unit time) or peak dP/dt are used.

In the isolated heart preparation (Chapter 12), this type of curve is produced by progressively infusing fluid (or blood) into the circulation and obtaining a range of points relating the end-diastolic pressure or volume to heart performance. In such an experiment the mean arterial pressure is kept constant, but this is not possible in the intact circulation where the systolic pressure tends to rise as fluid is infused. In addition, it should be recognized that as the size of the heart increases, the wall becomes thinner and the radius increases, so that by the Laplace relation (Chapter 12) tension in the wall during ejection tends to increase. Therefore, such function curves do not represent a pure relation between preload and heart performance, but rather a composite effect of increased diastolic size and afterload and the curve tends to flatten as the ventricular volume becomes larger. This occurs in part because the effect of increased afterload tends to limit the increase in stroke volume produced by the augmented preload (Fig. 2.109, Chapter 11. An even more important factor in the flattening of the function curve, when end-diastolic pressure is used, is the shape of the diastolic pressure-volume relation (see next section). In other words, the stroke volume increase would be larger if the afterload or tension were held truly constant.) Nevertheless, the effect on the normal heart of increasing diastolic volume is dominant as shown in the *top panel*, Figure 2.109 (Chapter 11), and as the ventricular end-diastolic pressure or volume increases, the stroke volume and the stroke work increase. Typically, a function curve is determined during one state and compared with a function curve determined during another condition (such as after a drug has been administered), and the *position* of the curve then can define whether or not a positive or negative change in inotropic state has occurred. Thus, by definition, a function curve of the ventricle is shifted upward and to the left by a positive inotropic intervention, and downward and to the right by a negative inotropic influence resulting in a "family of curves" (Sarnoff et al., 1958a), provided significant changes in systolic pressure do not occur.*

Starling Curve

A Starling curve generally indicates the relation between the ventricular end-diastolic pressure or volume and the *stroke volume*. As mentioned, experimentally such curves should ideally be com-

pared at a constant level of mean arterial pressure. The Starling curve is the simplest to determine and, in practice, it is often plotted in intact animals or in man, even when the arterial pressure does not remain constant. Indeed, sometimes Starling curves are produced by administration of a vasopressor drug (that increases the afterload) by plotting of the relation between the rising end-diastolic pressure and the stroke volume over a range of doses of the drug (angiotensin or phenylephrine often are used); such curves are useful for describing differences, for example, between the normal and the failing heart (Chapter 18). Alternatively, the effect of a small transfusion of saline solution may be used to test ventricular functional reserve, as in patients following a heart attack (Fig. 2.180). If the heart is failing in such patients, a large rise in the left ventricular end-diastolic pressure or the left ventricular "filling pressure" (as determined by the mean pulmonary artery wedge pressure, see Chapter 13) may be associated with little change in the stroke volume (minimal ventricular reserve), in contrast to the response of the relatively normal ventricle (Fig. 2.180).

Ventricular end-diastolic pressure often is used instead of end-diastolic volume in describing cardiac function curves because it is much easier to measure. In this setting, the nearly exponential relation between ventricular end-diastolic volume and end-diastolic pressure becomes significant (see Fig. 2.182, *lower panel*). Thus, over the range of low end-diastolic pressures (below 12 mm Hg in the left ventricle) relatively large changes in ventricular

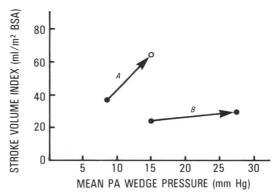

Figure 2.180. Examples of segments of ventricular function curves determined in the clinical setting as the relationship between the mean pulmonary artery *(PA)* wedge pressure (as a measure of the cardiac filling pressure or mean left atrial pressure) and the stroke volume index (stroke volume corrected for body surface area, *BSA*). Fluid is infused intravenously and *arrow A* shows the response in a subject with a relatively normal left ventricle, a large rise in stroke volume occurring with a modest increase in PA wedge pressure. *Arrow B* indicates the response in a patient who has had a large heart attack, in which a large rise in the PA wedge pressure occurs with a minimal change in the stroke volume.

---

*The importance of the aortic pressure (and afterload on the left ventricle) during volume loading is illustrated by studies on the relation between left ventricular end-diastolic volume and the stroke volume when aortic pressure is changed by infusion of a vasopressor drug. The relationship between left ventricular end-diastolic volume and the stroke volume is shifted downward, with an apparent descending limb of function, when aortic pressure is raised and vice versa (see Lee et al., 1986).

Recent studies in intact animals instrumented with ultrasonic devices so that ventricular volume can be continuously computed has shown that when ventricular end-diastolic volume is altered over a physiological range (with relatively small changes in aortic pressure) the positive relation between the *stroke work* and left ventricular end-diastolic volume is linear, as might be predicted from the pressure-volume loops of the pressure-volume diagram shown in Figure 2.119 (Glower et al., 1985).

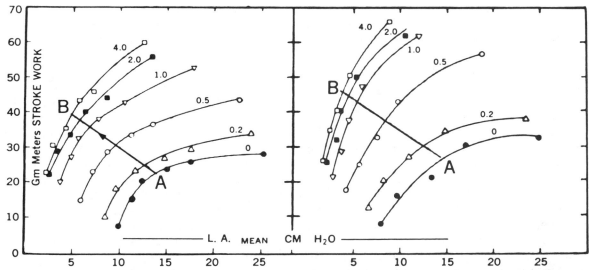

Figure 2.181. Examples of ventricular function curves obtained in the isolated heart in which the mean left atrial *(L.A.)* pressure is plotted against the stroke work of the left ventricle over a range of left atrial pressures. The progressive shift upward of the curves from *A* to *B* is due to a progressive increase of left ventricular inotropic state pro- duced by stimulation in the cardiac sympathetic nerves at the frequen- cies indicated (0–4 cycles/s). The *left-hand* and *right-hand panels* indi- cate responses in the same heart over 1 h later. (From Sarnoff et al., 1958a.)

diastolic volume occur with relatively small changes in diastolic pressure, whereas as more vol- ume expansion occurs at high left ventricular diastolic volumes small changes in end-diastolic vol- ume are accompanied by *large* changes in the end- diastolic pressure, causing a flattening of the upper portion of the function curve.

Ventricular Function Curve Using Stroke Work

The term ventricular function curve sometimes in the past has been applied specifically to indicate the relation between the ventricular end-diastolic pres- sure (or mean atrial pressure) and the ventricular *stroke work* (the product of the mean arterial pres- sure or the mean ventricular systolic pressure and the stroke volume)† (Sarnoff et al., 1958a). Again, a shift in the *position* of the function curve for either ventricle is *defined* as a change in inotropic state (Figs. 2.181 and 2.182), and increasing degrees of change of inotropic state will shift the curve pro- gressively upward (Fig. 2.181).

However, just as the work of isolated muscle is "load-dependent" (see Fig. 2.105, *panel D*, Chapter 11), the ventricular stroke work is also dependent upon the *level* of aortic pressure. As in isolated mus- cle, at very low levels of systolic pressure the prod- uct of mean arterial pressure and stroke volume will be low, and at very high levels of aortic pressure when the stroke volume is reduced, the stroke work will also be low. Thus, if large changes in aortic

pressure occur during the performance of such curves, alterations in the stroke work are not entirely dependent upon the level of inotropic state at any given end-diastolic pressure.

In using any type of function curve of the ventri- cle, it is important to understand clearly the differ- ence between *moving up and down* on a *single* func- tion curve due to altered preloading and changing the *position* of the entire function curve by altering the inotropic state of the ventricle.

Using the ventricular function curve concept, and having only two points of information about heart function, often it is possible to infer whether or not an intervention has caused a change in inotropic state. As shown in Figure 2.182, if the stroke vol- ume or stroke work *decreases* or does not change while the end-diastolic pressure shows no change or *increases* (control point to *points A, B* or *C*, Fig. 2.182), it can be inferred that the curve has shifted downward and that inotropic state has become depressed between the two determinations (assum- ing no major change in aortic pressure). On the other hand, if the stroke volume or stroke work shows no change or *increases* while the end-dia- stolic pressure shows no change or *decreases*, a pos- itive inotropic influence must have been operative (control point to *points D, E,* or *F*, Fig. 2.182). How- ever, if the stroke work or stroke volume increases *and* the end-diastolic pressure or volume also increases, or if they *both* decrease, one cannot infer that a change in inotropic state has occurred with- out knowing the shape of the *entire* curve, since these changes could simply represent moving up or

---

†The stroke work is usually expressed in gram meters (g M) where pressure in centimeters H_2O × stroke volume (in cm^3) = stroke work ÷ 100 (1 mm Hg = 1.36 cm H_2O).

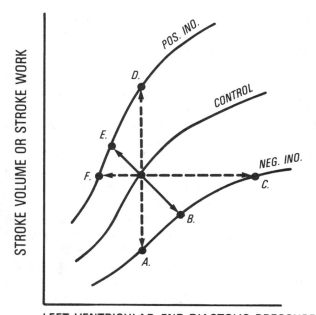

LEFT VENTRICULAR END-DIASTOLIC PRESSURE

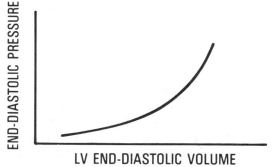

LV END-DIASTOLIC VOLUME

Figure 2.182. *Upper panel,* Ventricular function curves showing the relation between left ventricular stroke volume or stroke work to left ventricular end-diastolic pressure during control conditions *(center curve),* positive inotropic *(POS. INO.)* and negative inotropic *(NEG. INO.)* interventions. The *arrows* indicate how measurement only of a control and one other point can indicate a decrease in inotropic state *(points A, B,* or *C)* or an increase in inotropic state *(points D, E,* or *F).* For further discussion, see text. The *lower panel* shows the nonlinear relation between left ventricular *(LV)* end-diastolic volume and end-diastolic pressure.

down on a single function curve. A variety of normal physiological stimuli and pathological influences can affect the preload (end-diastolic volume of the ventricle) (Table 2.12), and a number of influences can alter the level of inotropic state and thereby shift the ventricular function curves (Table 2.13).

END-SYSTOLIC PRESSURE-VOLUME RELATION

The pressure-volume loop of the left ventricle was discussed in Chapter 12 and related to the performance of isolated cardiac muscle (in which the

Table 2.12.
Factors that Affect the Preload

Posture
Blood volume
Intrinsic venous tone
Sympathetic neural tone to veins and vasoactive drugs
Atrial contraction
Muscular activity
Intrapleural pressure
Intrapericardial pressure

isometric length-tension curve provides the limit of shortening for isotonic contractions). The use of the linear relation between the end-systolic volume and the end-systolic pressure of the ventricle has been studied extensively in the isolated heart preparation (Suga et al., 1973; Sagawa, 1978), and it has also been extended to study intact animals by infusing a range of doses of a vasoconstrictor drug (which does not itself have appreciable inotropic effects, such as the α-adrenergic agonist phenylephrine) (Mahler et al., 1975). In man, the end-systolic ventricular volume can be determined by performing two or more angiocardiograms during infusion of the vasoconstrictor. The end-systolic volumes are then related to the corresponding ventricular pressure at the end of ejection. This approach also has been applied using noninvasive techniques for measuring ventricular dimensions or volume (echocardiography and radionuclide methods). The linear relation has been found to shift downward and to the right in the presence of myocardial disease in human subjects (Fig. 2.183), and to shift upward and to the left (with steepening of its slope) during acute positive inotropic interventions in experimental animals and in man (Figs. 2.183 and 2.184) (for example, Mehmel et al., 1981).

Table 2.13.
Factors that Influence Myocardial Inotropic State

Reflex adrenergic neural stimulation of the heart
Vagal tone (primarily effects atrium)
Circulating catecholamines
Force-frequency relation (heart rate, postextrasystolic potentiation)
Myocardial failure
    Myocarditis
    Myocardial infarction and ischemia
    Myocardial failure in late hypertrophy
Cardioactive (positive inotropic) drugs
    Digitalis
    Amrinone
    Sympathomimetic amines
Cardiac depressant drugs
    Certain anesthetic agents
    Antiarrhythmic drugs
    Certain calcium antagonists
External ion concentrations
    Calcium
    Sodium

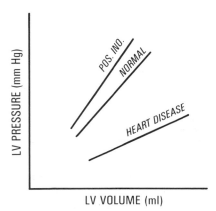

Figure 2.183. Diagrammatic linear relations between left ventricular *(LV)* end-systolic pressure and end-systolic volume in normal hearts and in patients with myocardial disease. The influence of an acute positive inotropic stimulus *(POS. INO.)* on this relation is also shown.

The relationship between this type of analysis and ventricular function curves before and after administration of a positive inotropic agent is shown in Figure 2.184. Ejecting beats *A, B,* and *C* before the inotropic intervention are shown as end-diastolic volume is increased *(left panel)* and as a ventricular function curve *(right panel)*. When beats arising from the same three end-diastolic volumes are plotted after the positive inotropic agent is given, the end-systolic pressure-volume relation is shifted and the stroke volumes increase (beats *D, E,* and *F, left panel)*, displacing the function curve upward *(right panel,* Fig. 2.184). Thus, the framework using pressure-volume loops and end-systolic pressure-volume relations can be readily interchanged with ventricular function curve responses.

The linear relation between ventricular end-systolic volume and end-systolic pressure is of particular importance, because it provides a method for detecting changes in the level of inotropic state under acutely changing conditions that is *independent* of the end-diastolic volume *(preload)* and the aortic pressure (as a measure of *afterload*). Thus, a given heartbeat will arrive at end-ejection and fall on this linear relation regardless of the starting point for end-diastolic volume and the level of aortic pressure that it encounters during ejection, and the entire relationship will be shifted acutely only by a change in inotropic state.

It should be recognized, however, that under conditions where there is *chronic* change in the shape and size of the ventricle, or in the thickness of its wall, the systolic pressure is *not* indicative of the level of afterload, and under those conditions the wall force must be calculated in order to define the linear relation between end-systolic volume and wall force (Sasayama et al., 1977) (see also section on hypertrophy, Chapter 18). Also the linear end-systolic pressure-volume relation may not provide a unique descriptor of inotropic state when the transmural pressure of the ventricle is altered (as in pericardial disease).

CONTROL OF CARDIAC OUTPUT

Cardiac output is determined by the interplay between the heart and the peripheral circulation. In discussions of how these two factors interact to determine cardiac output, it can be considered, rather arbitrarily, that the rate of blood flow returning to the heart, commonly called the venous return, is primarily influenced by the peripheral circulation and blood flow from the left ventricle primarily by cardiac function. It should be emphasized, however, that in the steady state cardiac output and venous return must be the same, and that the interdependence between cardiac and peripheral vascular function is not simple. Cardiac output is not regulated only by the heart, nor is the rate of blood flow returning to the heart regulated only by peripheral vascular factors. These factors are shown schematically in Figure 2.185.

Figure 2.184. The *left panel* shows relations between left ventricular *(LV)* pressure and volume and the linear end-systolic pressure-volume relations during control conditions and a positive inotropic stimulus. The pressure-volume loops are also shown with progressive increase in end-diastolic volume and stroke volume under control conditions *(beats A, B,* and *C)*, and during the positive inotropic stimulation at a slightly higher systolic pressure, for clarity *(beats D, E,* and *F)*. *Right panel* shows the same beats diagrammed in left panel plotted as ventricular function curves relating left ventricular *(LV)* end-diastolic volume to the stroke volume. *A, B,* and *C,* control conditions; *D, E,* and *F,* positive inotropic stimulus.

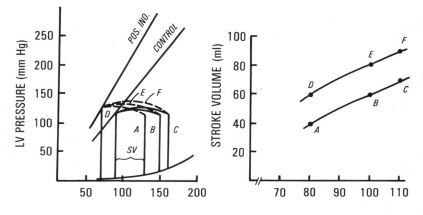

Four factors predominantly affect the performance of the heart: heart rate, inotropic state, preload, and afterload. As described in Chapter 12, heart rate and inotropic state can change the volume ejected on any particular beat and alter cardiac output by changing the position of the end-systolic pressure-volume relationship. Heart rate affects the cardiac output directly from the relationship of CO = SV × HR. Afterload and preload are determined both by the peripheral circulation and the heart. Thus, afterload is dependent on arterial vascular resistance, as well as factors which influence heart size and force of contraction. Preload, to a large extent, is influenced by the rate of return of blood to the heart as well as the heart's pumping ability. The rate of return of blood to the heart is determined by both cardiac and peripheral factors but is proportional to the pressure gradient across the venous bed. In the next section we will consider these factors separately.

Venous Return Curve

To explore the interplay between the peripheral circulation and the heart in determining the cardiac output, the approach devised by Guyton will be employed (Guyton et al., 1973). In this approach, the peripheral factors which influence the return of blood to the heart will be examined; secondly, using

a modified form of the ventricular function curve, the cardiac factors which influence the cardiac output will be reviewed.

Since the return of blood to the heart is regulated by the peripheral circulation, the concept to be used describes the pressure gradient for venous return to the heart and examines the factors that influence that pressure gradient. Guyton (1955) developed an experimental animal preparation in which the heart is removed and replaced by a pump oxygenator system which can perfuse the entire peripheral circulation at different flows (Fig. 2.185). If pumping is commenced at 5 liters/min with a systemic vascular resistance of approximately 20 resistance units, the mean arterial pressure will be 105 mm Hg and the venous pressure 5 mm Hg (Fig. 2.186), and if the pump is suddenly turned off, arterial pressure will fall and venous pressure will rise until an equilibrium pressure is reached throughout the circulation (Fig. 2.186). This equilibrium pressure, termed the "Mean Systemic Filling Pressure" by Guyton, represents the distending pressure of all vessels in the circulation at a particular blood volume and level of vascular tone. It is approximately equal to the capacitance of the peripheral circulation times the blood volume (Rothe, 1983). Since the capacitance of the veins is 18 to 20 times greater than that of the arteries, this pressure is largely a function of the venous system. The mean systemic filling pressure is normally in the range of 6–8 mm Hg, and it is importantly influenced by alterations in vascular volume and the tone of the veins. For example, if a blood transfusion is performed in the preparation

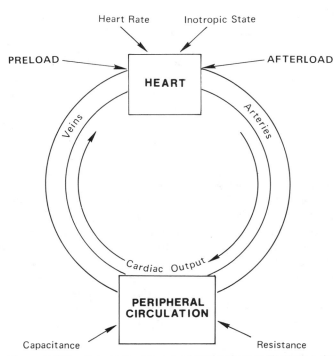

Figure 2.185. Schematic diagram of the circulation illustrating the relationship between peripheral and cardiac factors that determine cardiac output.

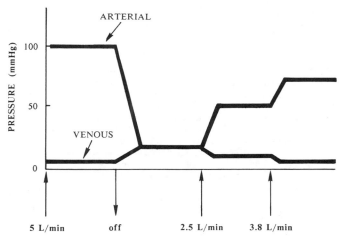

Figure 2.186. Schematic representation of the results of an experiment in which the heart is replaced with a pump. *Arrows* indicate changes in pump flow (cardiac output). When the pump is turned off venous and arterial pressures rapidly equilibrate at approximately 7 mm Hg (mean systemic filling pressure).

described above and the experiment of turning the pump off is repeated, mean systemic filling pressure would be increased in proportion to the amount of blood infused. Similarly, with an infusion of methoxamine (a pure α_1 agonist) or with sympathetic nerve stimulation, at constant blood volume (equivalent to an "isometric" contraction of the veins), the mean systemic filling pressure would also increase.

We have just considered the equilibrium at zero flow. What happens if flow around the system is altered? As shown in Figure 2.186, if pump flow is restarted at 2½ liters/min, a small amount of blood is transferred from the venous circuit to the arterial circuit resulting in an increase of arterial pressure and a decrease of right atrial pressure. A plot of the mean right atrial pressure as an independent variable against the steady state flows through the system (which can be termed either venous return or cardiac output) at several different flow rates is shown in Figures 2.187, *curve B* and 2.189, *curve A*.‡ It can be seen that as right atrial pressure falls (and the pressure gradient between peripheral veins and the right atrium increases) venous return increases. At flows between 0 and 5 liters/min there is a linear portion of the curve. A plateau portion of the curve occurs at a right atrial pressure below approximately 1 mm Hg; since at this point the large intrathoracic veins tend to collapse, no further

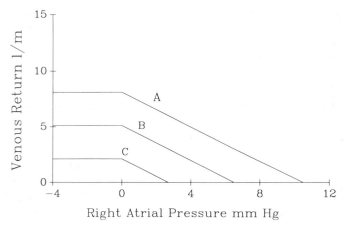

Figure 2.187. The effects of changes in blood volume on the venous return curve. Control = *B*. Increasing blood volume shifts the mean systemic pressure and plateau upward (curve *A*).

increase in venous return can occur as right atrial pressure is lowered.

With transfusion of the preparation or, as discussed earlier with contraction, i.e., stiffening the wall of the veins at constant volume, the entire venous return relationship is shifted upwards (*curve A* of Fig. 2.187). Correspondingly, an opposite shift downward occurs when blood volume or venous tone are reduced, with a lower mean circulatory filling pressure and reduced plateau (Fig. 2.187, *curve C*).

CHANGES IN THE RESISTANCE TO VENOUS RETURN

If a marked change in the *total* resistance to venous return is produced, such as by constricting the inferior vena cava, flow will be reduced at any right atrial pressure. At *zero flow*, however, there can be no effect of such a resistance change; thus, for example, when the circulation is at rest progressive contriction of a small area of the inferior vena cava cannot affect the mean systemic pressure. Hence, a change in vascular resistance will not affect the zero intercept of the venous return curve on the pressure axis (mean systemic filling pressure). However, as blood flow is progressively increased at an elevated venous resistance, cardiac output is decreased at all levels of flow and thus the venous return curve is shifted downward and to the left, with a pronounced change in the slope of the relationship (Fig. 2.188, *curve C*). Conversely, decreasing the venous resistance increases the venous return at all levels of right atrial pressure except zero, rotating the curve upward and to the right (Fig. 2.188, *curve A*).

Changes in arteriolar resistance have a much less pronounced effect on shifting the venous return

---

‡It has been argued (Levy, 1979) that it is not the right atrial pressure per se that influences the venous return to the heart by affecting the pressure gradient between the periphery and the right atrium, but that changes in right atrial pressure are rather a *consequence* of changes in cardiac output which cause redistribution of blood to the central (pulmonary) circulation and the arterial bed from the venous compartment. Thus, in a *closed* circulation (rather than using the right atrial reservoir employed by Guyton) it can be shown that as cardiac output is increased by means of a pump the right atrial pressure will fall, and vice versa. When curves are plotted relating right atrial (or venous) pressure to the cardiac output in such a closed circulation, they are affected by various interventions in a manner identical with the venous return curve (Levy, 1979). In analyzing curves it can be shown that at a given constant value for total resistance, and at a constant blood volume, the ratio of the capacitance of the venous bed $C_v (C = \Delta_v/\Delta_c)$ to that of the arterial bed, C_a, is 19:1, and assuming that these capacitances remain constant at a given blood volume, as flow is progressively increased through such a system the following equation applies: $\Delta P_a/\Delta P_v = C_v/C_a$. Thus, at any level of flow the rapid rise of pressure in the noncompliant arterial compartment can be calculated together with the fall in pressure in the more compliant venous compartment, as the blood is translocated into the arterial compartment at progressively higher flow rates (consideration of the lungs and central circulation is omitted in such a model). Thus, the *flow rate* is responsible for the pressure gradient (the difference between P_A and P_V at any level of flow) rather than the right atrial pressure per se (Levy, 1979).

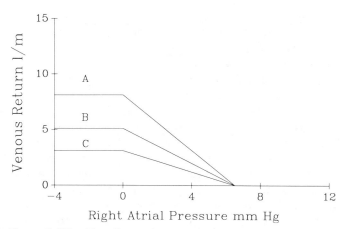

Figure 2.188. The effects of changes in venous resistance on the venous return curve increasing venous resistance does not change mean systemic filling pressure but reduces the slope of the venous return relationship.

curve than changes in the venous resistance (Guyton et al., 1959; 1973). This difference is related primarily to the much larger capacitance of venous bed compared to the arterial bed. The venous return during changing vascular resistance can be approximated from the following equation (Guyton et al., 1973):

$$\text{Venous return} = \frac{MSP - RAP}{R_v + \dfrac{R_A}{19}}$$

where MSP = mean systemic pressure, RAP = mean right atrial pressure, R_v = venous resistance, R_A = arteriolar resistance, and 19 = sum of the arteriolar and venous capacitances (1 + 18).

This equation indicates that if the total peripheral resistance increases by 20%, with all of the resistance increase occurring in the arterioles, venous return will be reduced by only about 6%. However, if this 20% resistance change occurs entirely on the venous side, venous return will be reduced by about 53% (a nine fold difference). This phenomenon has been noted for many years in that a striking drop in cardiac output occurs with mild compression of the inferior vena cava in the abdomen.

REGIONAL VARIATIONS IN VENOUS RETURN

The preparation shown in Figure 2.185 represents the circulation as a whole. As discussed in Chapters 6 and 7, there are profound differences in regional circulations. The systemic venous return curve approach represents the net effect of a great many regional distributions, regional changes in blood volume may occur without changes in overall

mean circulatory filling pressure, which represents the lumped contribution of several compartments. It is important to realize that the properties of the regional circulations are different and may be affected differently during a particular intervention. Thus, transfers of blood volume may occur within different regional circulations without changes in mean systemic filling pressure; for example, the studies of Caldini et al. (1974) demonstrated the differential effect of epinephrine on the splanchnic and systemic circulations resulted in transfers of blood volume between the two beds. The clearest example of this phenomenon is the treatment of congestive heart failure with vasodilator agents, discussed in Chapter 18, when blood volume is transferred between the pulmonary and systemic circulations.

Blood volume in an individual venous bed may be altered by several mechanisms, which can be classified into three different types: passive, active, and capillary transport. The volume of blood stored in a venous bed is determined by the pressure gradient across the bed and the tone (stiffness) of the vessels. If the flow through a regional bed is reduced by arteriolar constriction the pressure gradient across the venous bed will fall and result in loss of blood volume from that venous bed. Similarly, if outflow pressure to a particular venous bed is altered, say decreased, and inflow pressure is not altered, blood volume will be lost from the bed. Secondly, volume in a regional circulation may be altered by active processes, i.e., by isotonic contraction of the small venules. This reduction in caliber tends to slightly increase the resistance of the circuit but primarily to displace large amounts of blood from that venous circuit. An example of this effect is most clearly shown with reflex activation of the sympathetic nervous system shown in Figure 2.178 of Chapter 16. Lastly, changes in the ratio of pre- to postcapillary resistance can result in large changes in blood volume by loss of fluid across the capillary membranes. In many settings, all three of these factors may play important roles in shifts in blood volume during various circulatory changes. For example, it has been estimated (Rothe, 1983) that during hemorrhage, recruitment of volume into the circulation occurs by all three of these mechanisms, in approximately equal amounts.

Cardiac Output Curve

In order to understand the coupling between the venous return relationship discussed above and the heart, we must next derive a relationship that relates cardiac output to cardiac filling pressure.

The ventricular function curve, discussed earlier in this chapter, which can be derived from the pressure-volume loop and end-systolic pressure-volume relationship as shown in Figure 2.184, is a fundamental relationship which expresses the output of a single cardiac chamber, relative to its end-diastolic volume. Since, as shown in Figure 2.182, there is direct although *nonlinear* relationship between end-diastolic pressure and the ventricular end-diastolic volume, we can express the ventricular function relationship as the relationship between end-diastolic pressure and stroke volume. The left ventricular function curve is normally determined under conditions where arterial pressure and heart rate are constant and detects changes in inotropic state. In the more intact circulation, we must use a relationship that is different, one that is sensitive to changes in both heart rate and loading conditions, as well as inotropic state and other factors. To do this, we will utilize the relationship between cardiac output and filling pressure (Fig. 2.189, *panel A*), called *cardiac output curve* (Guyton et al., 1973). It is important to realize that not *only* changes in inotropic state, but also changes in afterload and heart rate, as well as factors such as wall thickness and heart size can influence this curve, in contrast to the standard ventricular function curves. The magnitude and direction of these shifts can be directly determined from the effects of afterload, inotropy, and heart rate changes on single contractions plotted in the end-systolic pressure-volume framework. Referring to Figure 2.122 in Chapter 12, contractions 1 and 2 show the effects of an increase in arterial pressure on stroke volume in contractions originating at the same end-diastolic volume or pressure and without a change in contractility. Stroke volume is reduced. Thus, we

would anticipate that with an increase in arterial pressure (or afterload) the relationship between filling pressure or right atrial pressure and cardiac output would be shifted downward and to the right, and upward with reduced afterload (Fig. 2.190). An increase in inotropic state in a single contraction (shifted end-systolic pressure-volume relation) arising from the same end-diastolic pressure would result in an increased stroke volume and thus a shift in the cardiac output right atrial pressure relationship upward and to the left with a negative inotropic influence having the opposite effect (Fig. 2.190). Effects of an increase in heart rate on this relationship are more complex as shown in Figure 2.125 (Chapter 12). A small increase in stroke volume at a fixed end-diastolic pressure would be anticipated, however, since the vertical axis in the right atrial pressure cardiac output relationship is cardiac output or the product of stroke volume and heart rate, the cardiac output relationship will shift upward (or downward) in proportion to the increase (or decrease) in heart rate (Fig. 2.190). These various influences are summarized in Table 2.14 and Figure 2.190.

Equilibrium Between Venous Return and Cardiac Output

To combine the cardiac output and venous return curves, some simplifying assumptions must be made. The left and the right ventricles, each have unique relations between cardiac output and filling pressure. However, under ordinary circumstances the right ventricular output relationship and the venous return curve for the pulmonary circulation are rarely altered by disease or by normal reflex and hormonal controls of the circulation. Thus, in order to simplify the analysis it will be assumed that the

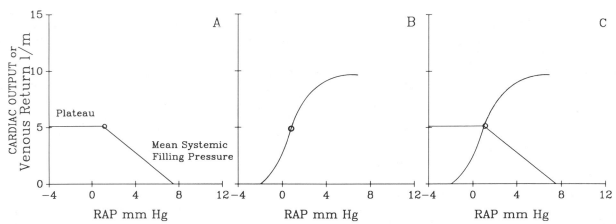

Figure 2.189. The relationship between the normal venous return curve and the cardiac output curve. *Panel A:* the normal venous return curve. *Panel B:* The normal cardiac output curve. *Panel C:* The relationship between cardiac output and venous return curves showing the resting level of cardiac output at their intersection. *RAP* is right atrial pressure.

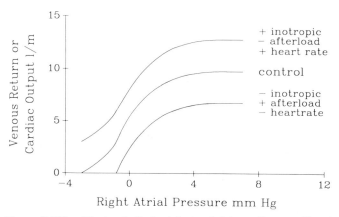

Figure 2.190. Effects of afterload, inotropic interventions, and heart rate on the relationship of cardiac output to right atrial pressure (cardiac output curve).

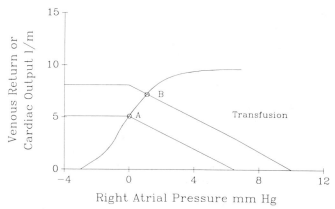

Figure 2.191. Effects of increases in blood volume (transfusion) on the relationship between cardiac output and venous return.

heart can be represented by a single chamber, the left ventricle, with the systemic venous return relationship (using right atrial pressure) providing its input. Thus, the relationship between left ventricular output and right atrial pressure, (as if we had replaced the right ventricle and pulmonary circulation with a simple conduit) will be examined.

Figure 2.190, *panel C* shows the combined relationship. The cardiac output curve (using right atrial pressure) is shown by the *dashed line,* and the venous return curve by the *solid line.* Note that the resting level of cardiac output occurs at the intersection of these two curves (the equilibrium point). Normally, this intersection occurs at low right atrial pressures at the "knee" of the venous return relationship, and hence unless the venous return curve is shifted the various factors causing a shift leftward or the cardiac output curve (Table 2.14) would not influence the level of cardiac output at the equilibrium point.

Curve *B* in Figure 2.191 indicates the effects of transfusion on resting cardiac output. Cardiac output under these circumstances would increase from *A* (5.1 liters/min) to *point B* (8 liters/min), the equilibrium point moving upward along the cardiac output curve as the venous return curve is shifted

upward. Figure 2.192 shows the effects of increasing sympathetic tone, e.g., by bilateral carotid artery occlusion. Under these circumstances, the venous return relationship will shift to the right and upward due to increased α-adrenergic tone to the peripheral veins with an increase in mean systemic filling pressure. Note that there would be a modest increase in both venous and arterial resistance resulting in a slight reduction in slope of the venous return relationship. The cardiac output relationship is shifted upward and to the left, due to the associated increase in inotropic state (Fig. 2.192), the magnitude of the shift being also affected by the extent to which arterial pressure (and afterload) increases during sympathetic stimulation. The net effect is an increased cardiac output at a slightly lower right atrial pressure.

SELECTED INTEGRATED RESPONSES

The purpose of the following descriptions is to illustrate that several different frameworks can be usefully employed for examining the cardiac and

Table 2.14.
Factors that Can Shift the Cardiac Output Curve

Shift Upward	Shift Downward
Decreased afterload (decreased vascular resistance)	Increased afterload (increased vascular resistance)
Increased inotropic state (Table 2.13)	Decreased inotropic state (Table 2.13)
Increased heart rate	Decreased heart rate
Cardiac hypertrophy	Pericardial fluid (tamponade)
Negative pressure breathing	Positive pressure breathing

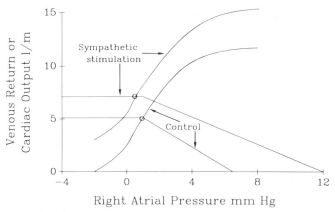

Figure 2.192. Effects of sympathetic stimulation on the cardiac output and venous return relationships.

peripheral circulatory responses to various forms of physiological stress.

Cardiac Pacing

When electrical stimulation of the right atrium is carried out in the normal human subject, as the heart rate is increased the mean left and right atrial pressures fall, but there is no change in the cardiac output despite the positive inotropic stimulation induced by the force-frequency relationship (Ross et al., 1965a). Based on the hemodynamic principles discussed above, cardiac output would not be expected to change. Thus, cardiac pacing shifts the cardiac output curve upward and to the left, but since the normal heart operates on the knee of the venous return curve, little effect or no effect on the cardiac output and venous return occurs (Fig. 2.193).

Supine to Erect Posture

A sudden change from the supine to the upright position on a tilt table produces a response somewhat resembling acute blood loss, due to pooling of blood in the veins of the lower extremities. The result is a decreased effective mean systemic pressure (relative to the level of the heart), which shifts the venous return curve downward and to the left, resulting in an immediate, severe fall in the cardiac output (Fig. 2.194, *point A* to *point B*). The fall in arterial pressure causes reflexes which increase ventricular contractility, heart rate, and the mean systemic pressure, and these compensations then

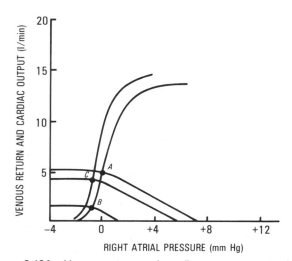

Figure 2.194. Venous return and cardiac output curves during upright body tilting. Control conditions are shown at point *A,* and during the initial phase of the tilt *(point B)* a marked fall of the cardiac output occurs; when reflex compensations occur the cardiac output improves *(point C).* For further discussion, see text.

partially restore the cardiac output by shifting both the venous return and cardiac output curves upward (Fig. 2.194, *point C*).

A similar response occurs after acute blood loss, and in addition a further upward shift of the venous return curve would occur over several hours; thus, the acute blood loss produces a drop in capillary pressure, and osmotic forces then draw fluid into the circulation, thereby tending to further increase the mean systemic pressure.

Valsalva Maneuver and Positive Pressure Breathing

The Valsalva maneuver, a forced expiration against a closed glottis, is an excellent demonstration of the effects of changes in intrathoracic pressure on heart rate and blood pressure. As shown in Figure 2.195 with the initiation of a forced expiration against a closed glottis intrathoracic pressure rises to extremely high levels. This abruptly limits venous return as reflected by the sharp decrease in pulmonary arterial flow velocity. Since this increase in intrathoracic pressure is transmitted directly to systemic blood vessels there is a slight increase in systemic blood pressure. Next, the decreased venous return reduces left ventricular output and decreases blood pressure (Fig. 2.195). Thus, the major chronotropic (heart rate) response to the Valsalva maneuver is tachycardia mediated by the baroreceptor response to hypotension. Since baroreflex sensitivity and ventricular function curves are both altered in congestive heart failure, the response to the Valsalva maneuver can be used to detect the presence of congestive failure. In congestive heart failure these responses are blunted, and

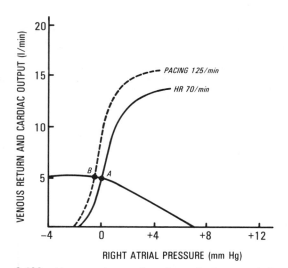

Figure 2.196. Venous return and cardiac output curves during various stages of the Valsalva maneuver. Control conditions are shown at *point A,* and the initial response to the Valsalva maneuver with a marked decrease in the cardiac output is at *point B;* reflex compensations partially restore the cardiac output at *point C.* For further discussion, see text.

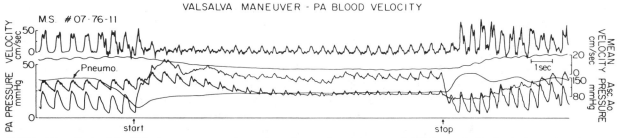

Figure 2.195. The effects of the Valsalva maneuver on hemodynamic variables in man. With the onset of increased intrathoracic pressure, pulmonary artery and aortic pressures are increased. There is an abrupt decrease in PA flow velocity due to the reduced venous return. This results in aortic hypotension and tachycardia secondary to baroreceptor activation. (Adapted from Mason et al., 1970.)

very frequently there will be no hypotension and tachycardia because the baroreceptor gain in reduced and the depressed cardiac function curve is flat.

The effects of the Valsalva maneuver on the cardiac output-venous return relation are shown in Figure 2.196. The elevation of the intrapleural pressure shifts the entire cardiac output curve to the right (cardiac filling pressures are now high relative to extrathoracic venous pressure), and a very marked initial drop in the cardiac output occurs (Fig. 2.196, *points A to B*), associated with a drop in the arterial pressure. However, the associated strong contraction of the abdominal muscles markedly increases the intraabdominal pressure, thereby increasing the mean systemic pressure; this, coupled with displacement of blood from the central circulation (lungs) to the periphery by the high intrathoracic pressure, shifts the venous return curve upward. Also, after 15–20 s a marked baroreceptor reflex results from the drop in arterial pressure, which further shifts the venous return curve upward and also displaces the cardiac output curve upward (increased heart rate and inotropic state), thereby partially restoring the cardiac output (Fig. 2.196, *point C*). Following release of the Valsalva maneuver, there is a marked overshoot of the blood pressure and cardiac output due to persistence of sympathetic tone to the arterioles and myocardium, and to the increased venous return as intrapleural pressure abruptly drops. Also, a sharp, reflex bradycardia occurs as the arterial pressure rises abruptly ("post-Valsalva overshoot").

Positive pressure breathing is frequently used during general anesthesia, and it can occur during scuba diving or while playing certain musical instruments. Similar effects to the Valsalva maneuver occur initially, with a drop in cardiac output during steady state positive pressure breathing (Guyton et al., 1973). By the same compensatory mechanisms shown in Figure 2.196 the cardiac output can be restored to nearly normal in this setting.

Circulatory Adjustments with Exercise

The circulatory response to exercise involves a complex series of changes resulting in a large increase (up to eight fold) in cardiac output which is proportional to the increased metabolic demands placed upon the body. These adjustments are designed to assure that the metabolic needs of the exercising muscles are met, that hyperthermia does not occur, and that blood flow to essential organ systems is protected. In strenuous exercise in trained subjects, when the cardiac output has increased six or seven fold it may attain values of 35 liters/min in man, while total oxygen consumption of the body is 10 to 12 times greater than at rest (Khouri et al., 1965).

There are two different types of exercise which have somewhat different cardiovascular effects. In static (isometric) exercise, force is generated with-

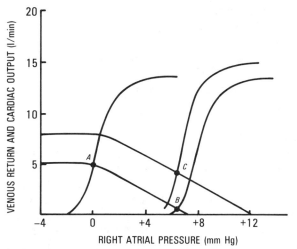

Figure 2.196. Venous return and cardiac output curves during various stages of the Valsalva maneuver. Control conditions are shown at *point A,* and the initial response to the Valsalva maneuver with a marked decrease in the cardiac output is at *point B;* reflex compensations partially restore the cardiac output at *point C.* For further discussion, see text.

out muscle shortening and thus no external "work" is performed (lifting and holding a heavy load, such as in weight lifting, is representative of an "isometric or static" type of exercise). In dynamic or "isotonic" exercise (such as running), external work (shortening against a load) is performed. There are some differences in the circulatory responses to static and dynamic exercise, with the former failing to show the decreased total peripheral vascular resistance and increased cardiac output characteristics of dynamic exercise (sustained isometric exercise decreases blood flow to skeletal muscle), while the mean systemic blood pressure responses are similar (Asmussen, 1981; Blomqvist et al., 1983).

During dynamic exercise the degree of increase in heart rate, stroke volume, cardiac output, and oxygen extraction depends upon the severity of the exercise and the amount of muscle mass involved. Total calculated peripheral resistance decreases significantly, because of vasodilation in active muscles, but systolic arterial pressure and the pulse pressure usually increase. The pulmonary artery wedge pressure rises slightly, while ventricular contractility increases during strenuous exercise (Higginbotham et al., 1986). Finally, it should be noted that the flow in the pulmonary circulation is just as great as in the systemic vessels; this flow is characterized by a mild increase in pulmonary arterial pressure during strenuous exercise, with a decrease in the resistance of the lung vasculature.

Regional Circulatory Responses

The changes in regional blood flow with dynamic exercise are summarized in Figure 2.197. As expected, the increased cardiac output supplies a greatly elevated blood flow to the exercising skeletal muscles. This local hyperemia is due primarily to the buildup of vasodilator metabolites in the exercising muscles. The increase in flow to skeletal muscle is directly proportional to the increase in

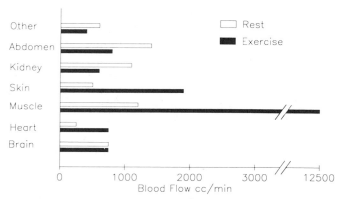

Figure 2.197. Distribution of blood flow to various tissues and organs during rest and strenuous exercise. The numbers refer to blood flow in ml per min. (From Vander et al., 1975.)

body oxygen consumption (Fig. 2.197). Skin flow initially increases until 50–60% of the maximal workload is reached; this increase favors body cooling. As exercise proceeds, body fluid is lost as sweat and through the lungs to produce some dehydration, and the hematocrit increases. Further increase in workload causes a progressive decrease in skin flow as cutaneous sympathetic vasoconstrictor tone continues to rise, apparently overcoming thermoregulatory vasodilator responses. Cerebral blood flow remains unchanged and splanchnic flow (to the abdominal viscera) decreases, and coronary flow increases in proportion to cardiac work (Fig. 2.197). These findings are, in general, based on indirect determinations of blood flow applied one at a time to various vascular beds. Renal blood flow and flow to the liver and stomach decrease with short-term exercise. On the other hand, in prolonged moderate work or sustained strenuous exercise blood flow distribution to the kidney, liver, gastrointestinal tract, and skin must be maintained at levels adequate for continued function of these organs.

Body Arteriovenous O₂ Difference

In addition to changes in cardiac output and local blood flow, there are other mechanisms that maintain oxygen supply to the tissues (see also Chapter 6). In exercise, the O_2 extraction rises, but considerably more slowly than the cardiac output, the maximum arteriovenous O_2 difference being of the order of 13–16 ml O_2/100 ml of blood (about three times the resting value). Since most of this increase in O_2 delivery presumably occurs in the exercising muscles, the oxygen extraction in muscle must be very nearly complete whenever the average AV difference for the entire body is as high as 16 ml/100 ml. (A similarly high extraction has been observed in heart muscle during maximal activity). For well-trained athletes, however, the AVO₂ difference does not attain these high values.

Cardiac Responses to Exercise

HEART RATE

At the transition from rest to very heavy work, the pulse frequency rises very rapidly, reaching levels of 160–180/min. During short bouts of near maximal exercise, heart rates as high as 240–270/min have been recorded in normal young persons. Lower heart rates are achieved with less strenuous exercise. The almost instant acceleration in cardiac rate is due to vagal withdrawal and not associated with an increase in sympathetic tone. The initial rapid increase suggests a central command, or a rapid reflex from mechanoreceptors in the active muscles (Hollander and Bouman, 1975). Later increases in heart rate stem from reflex activation of the pulmonary stretch receptors and reflexes from exercising muscles and are due to increases in sympathetic tone and vagal withdrawal as well as increases in circulating catecholamines.

VENTRICULAR FUNCTION

The positive inotropic effect of exercise due to effects of adrenergic nerve stimulation and circulating catecholamines shifts the end-systolic pressure-volume relation upward and to the left, and can deliver the same or increased stroke volume at a higher systemic arterial pressure from the same left ventricular end-diastolic volume and pressure; in this setting the associated marked acceleration of heart rate is the predominant factor increasing the cardiac output (Fig. 2.198). The positive inotropic stimulation also shortens ejection and increases the rate of relaxation which, together with increased atrial contraction due to augmented atrial contractility, enhances ventricular filling.

Exercise responses within the ventricular function curve framework are shown in Figure 2.198. The responses summarized are those in normal human subjects exercising upright on a bicycle odometer at various levels up to maximal exercise (exhaustion) (Plotnick et al., 1986; Higginbotham et al., 1986). Left ventricular volumes were measured by radionuclide ventriculography. The transition from sitting upright at rest to mild exercise shows an increase in ventricular end-diastolic volume and stroke volume, with a shift upward of the function curve (Fig. 2.198A, *points A to B*). With moderately severe exercise there is a further increase in the end-diastolic volume and stroke volume (Fig. 2.198A, *points B to C*). The corresponding changes in heart rates and cardiac outputs are shown in the insert. At maximum exercise, there is a further increase in heart rate (to 170/min) but, perhaps because of the limited time for cardiac filling, there is a small drop in the end-diastolic volume (which still remains higher than at rest); however, the stroke volume drops only slightly because of a further increase in contractility with decreased end-systolic volume (Fig. 2.198A, *points C to D*). Thus, during such exercise responses, the stroke volume increases by 30–40% from combined use of the Frank-Starling mechanism and increased myocardial contractility, which causes decreased ventricular end-systolic volume and increased ejection fraction.

VENOUS RETURN AND CARDIAC OUTPUT

The exercise response also can be analyzed in the venous return curve-cardiac output framework as shown in Fig. 2.199. The resting conditions are shown at *A*, and the responses during very strenu-

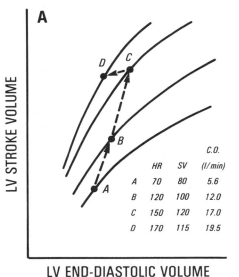

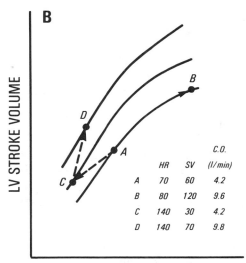

Figure 2.198. *Panel A,* Responses to upright exercise within the framework of left ventricular *(LV)* end-diastolic volume and the stroke volume. Cardiac outputs are calculated in the table. *Point A* shows the resting conditions; *point B,* mild exercise; *point C,* moderately severe exercise; *point D,* maximal exercise in normal subjects. For further discussion see text. *Panel B,* Relations between left ventricular end-diastolic volume and stroke volume, showing diagrammatically the responses to maximal exercise after cardiac denervation and adrenalectomy, or β-adrenergic blockade (*point A* to *point B*). *Point A* is also used as the control point for a normal heart to compare the effects of cardiac pacing alone, which shifts the ventricular function curve, and produces a fall in the stroke volume (*point A* to *point C*) (although cardiac output is unchanged). If moderate exercise is performed while the pacing is continued at a rate of 140 (see table), in the absence of β blockade a further shift upward to the ventricular function curve due to sympathetic stimulation occurs, but the cardiac output increase is achieved through an increase in stroke volume mediated by an increased end-diastolic volume and pressure and a decreased end-systolic volume (*points C to D*). For further discussion, see text.

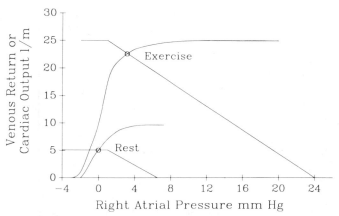

Figure 2.199. The effects of exercise on the equilibrium between cardiac output and venous return.

ous exercise in the untrained normal subject are indicated. During initial, milder exercise the mean systemic pressure is elevated by tensing of the abdominal muscles and the exercising muscles, accompanied by some increase in the resistance to venous return; however, as exercise continues, metabolic vasodilatation in the muscles coupled with a marked increase in sympathetic tone further elevates the mean systemic pressure and the venous return curve. Simultaneously, the cardiac output curve is shifted upward by sympathetic stimulation and by lowering of total peripheral vascular resistance, allowing the cardiac output to reach over 20 liters/min at a near maximum heart rate, with only a slight rise in the right atrial pressure above normal (Fig. 2.199).

In experimental animals, when the sympathetic adrenergic response of the heart is blocked by surgically denervating the heart (Donald and Shepherd, 1963) and the effects of circulating catecholamines are eliminated by adrenalectomy, the ability to exercise maximally becomes impaired (Fig. 2.198, *panel B*). Under these conditions, the ability of the heart to shift its function curve disappears, the heart rate response is only partially impaired (direct atrial stretch may increase the heart rate, see Chapter 16), but most of the cardiac output increase is now accomplished by an increase in stroke volume via the Frank-Starling mechanism and increased ventricular end-diastolic pressure (Donald, 1974) (Fig. 2.198B, *point A* to *point B*). Somewhat similar patterns of responses to exercise are seen in patients who are treated with β-adrenergic blocking drugs.

When the normal heart is paced by an electrical pacemaker under resting conditions to a rate simulating that during moderate exercise, as discussed earlier, the cardiac output does not change and the stroke volume therefore falls (Fig. 2.198B, *point A* to *point C*, also Fig. 2.189). If pacing is maintained and moderate exercise commenced, the heart rate fails to rise (since the heart is already paced to a rate which the heart would have reached at that level of exercise); however, there is sympathetic stimulation of the myocardium, shifting the left ventricular function curve upward and to the left. Under these conditions, the increase of cardiac output is achieved by an increase in the stroke volume, through increased venous return and operation of the Frank-Starling mechanism and by decreased end-systolic volume at the enhanced level of inotropic state (Fig. 2.198B, *point C* to *point D*) (Ross et al., 1965a).

Physical Training

In trained athletes, such as in long distance runners, the heart undergoes hypertrophy of the volume overload type (see Chapter 18) with enlargement of the end-diastolic volume of the chambers together with some thickening of the wall, and a normal level of basal inotropic state. This allows the cardiac output curve to shift even higher, and cardiac outputs over 35 liters/min may be reached during severe exercise with a right atrial pressure at or below normal. Weight lifters (static exercise) tend to exhibit increased left ventricular wall thickness, without associated chamber enlargement. (For review of the chronic adaptations to training see Wilmore, 1977; Blomqvist and Saltin, 1983.)

Physical conditioning or training affects the cardiopulmonary and skeletal muscular systems in a variety of ways which improve work performance (Clausen, 1969). Blood volume, total hemoglobin, the skeletal muscle capillary bed size, and mitochondrial content of muscle are all increased after training. Thus, all oxygen delivery and utilization systems involved in exercise are augmented. Muscle blood flow at a given workload is actually reduced (Clausen, 1969). This occurs because there is an increased oxygen extraction in trained muscle, made possible by the increase in oxidative mitochondrial enzymes and in number and size of mitochondria. Thus, "white" muscle fibers come to resemble "red" muscle fibers in histological appearance and enzymatic activity (for example, increased succinate dehydrogenase activity), an adaptation favoring aerobic metabolism and endurance (Gollnick et al., 1972). Endurance athletes have lower resting and exercise heart rates, which may in part relate to "downregulation" of cardiac β-adrenergic receptors secondary to repeated and prolonged episodes of sympathetic stimulation during exercise (Hammond et al., 1987).

CHAPTER 18

Heart Failure, Hypertrophy, and Other Abnormal Cardiocirculatory States

In this chapter, selected abnormal cardiac and circulatory states will be introduced briefly, in order to illustrate the application of physiological principles. These conditions include heart failure, certain aspects of pericardial function, cardiac hypertrophy, hypertension, anemia, and hemorrhagic shock.

HEART FAILURE

Failure of the whole heart can be defined in a rather simple manner if the heart is considered as a "black box" (Fig. 2.200, *upper panel*). When the heart fails it is unable to meet its obligations as a pump, and either the cardiac output falls, the venous pressures rise,* or both. In the early stages of heart failure, the mechanisms discussed below may compensate for the basic cardiac abnormality and serve to maintain the cardiac output and filling pressures within the normal range. At this stage, *mild heart failure* can be defined as: Abnormal elevation of the venous pressure (left or right) during exercise (or other stress), or failure of the cardiac output to rise normally during exercise, or both.

In later stages of heart failure compensatory mechanisms become inadequate, and *severe heart failure* can be defined as: Elevation of the left- or right-sided venous pressures, reduction of the cardiac output, or both, in the resting state.

Heart failure characterized by reduced cardiac output sometimes has been called "forward heart failure," whereas elevation of the venous pressure has been termed "backward heart failure." Often, however, the two phenomena occur simultaneously (Fig. 2.200, *upper panel*).

Causes of Heart Failure

Many factors can result in failure of the heart as a pump, and most of these can be related to the cardiac subsystems (Chapter 5). Heart failure can result from abnormal electrical conduction; for

---

*The ventricular "filling pressures" are elevated, either the left atrial (and pulmonary venous pressure), the right atrial (and systemic venous pressure), or both.

example, long-standing complete heart block with a very slow heart rate and large stroke volume can result in marked cardiac enlargement and eventual heart failure. Atherosclerotic disease of the coronary arteries can produce myocardial infarction with heart failure. Several kinds of valvular deformities can produce heart failure; for example, obstruction of the mitral valve (mitral stenosis) causes a high pressure in the left atrium and pulmonary circulation, a type of left heart failure, and can produce secondary failure of the right ventricle; in aortic stenosis the hypertrophied left ventricle (see section on hypertrophy) eventually can fail, and in aortic regurgitation severe left ventricular enlargement with coronary insufficiency can also lead to heart failure. Such failure of the heart muscle to contract normally results from the long-standing overload with hypertrophy, eventual loss of myofibrils from hypertrophied cells (Schwartz et al., 1981), and microscopic scarring. Mechanical constriction of the heart by a scarred and thickened pericardium can also produce all the signs of severe heart failure (Shabetai, 1981). Finally, primary diseases of heart muscle such as infections (viral myocarditis), toxic factors (certain antitumor agents, and possibly alcohol), and "cardiomyopathies" (heart muscle diseases of unknown cause) are associated with damage and loss of heart muscle cells. Such primary and secondary causes of heart muscle damage result in *myocardial failure* (that is, depression of the myocardial inotropic state).

The cellular causes for such myocardial failure remain controversial; multiple etiologies are likely, including the possibility of abnormal excitation-contraction coupling (restoration of intracellular calcium by digitalis, for example, improves the contraction of the failing heart); abnormalities in mitochondrial function, or in the structural proteins (myosin ATPase is low in some types of experimental heart failure); and, particularly in hypertrophy, inadequate capillary blood supply may lead to undersupply of substrates to the heart. (For

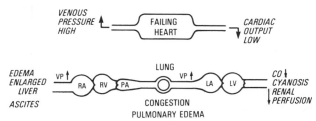

Figure 2.200. Heart failure. *Upper diagram* shows common features of heart failure, that is, either low cardiac output or high venous pressure (or both) at rest or during exercise. *Lower diagram* shows the physiological consequences of low forward cardiac output delivered by the left side cardiac chambers (*LV, LA*, left ventricle and left atrium) with decreased cardiac output *(CO)*, cyanosis, and decreased renal perfusion. The effects of increased venous pressure *(VP)* in the pulmonary veins due to an increased left atrial pressure causes pulmonary congestion and edema and can cause secondary elevation of the pulmonary artery *(PA)* pressure and right heart pressures (some lesions can cause right-sided heart failure alone). This, in turn, leads to an increased venous pressure *(VP)* in the systemic veins, which can cause tissue edema, enlargement of the liver, and fluid accumulation in the abdomen.

reviews on cellular and pathophysiological mechanisms in heart failure see Fanburg, 1974; Katz, 1976; Braunwald et al.,1988). It does not seem likely that simple overstretch of the sarcomeres (descending limb of the sarcomere length-tension relation) is responsible for myocardial failure, since it does not occur in acute or chronic volume overload (Ross et al., 1971).

Pathophysiological Effects of Heart Failure

The right heart may fail alone (for example, in long-standing right ventricular overload in chronic lung disease). More commonly, the left side of the heart fails, causing elevation of pressures in the pulmonary circulation with secondary right heart failure. The potential sites of physiological abnormalities are illustrated in Fig. 2.200, *lower panel.* Severe heart failure with a reduced cardiac output can produce fatigue, a weakened pulse, peripheral vasoconstriction with cyanosis of the extremities, and decreased flow through the organs of the body, including the kidneys. Elevation of the left atrial pressure and the pulmonary venous pressure behind the failing left heart produces perivascular edema and "congestion" of the lungs (Fig. 2.200), which increases their stiffness thereby causing shortness of breath; if this pressure elevation is severe, leakage of fluid directly into the lung air sacs causes respiratory distress and impairment of oxygen exchange ("pulmonary edema"). With either primary or secondary failure of the right ventricle (due to elevation of pulmonary artery pressure), the venous pressure in the systemic veins eventually becomes elevated (Fig. 2.200, *lower*

panel), and if the elevation is sufficiently severe leakage of fluid out of the capillaries can occur, causing fluid accumulation in the extravascular spaces (edema) in dependent locations such as the legs; enlargement of the liver with eventual leakage of fluid into the abdomen (ascites) may follow. This stage of circulatory failure is called "congestive heart failure."

Compensatory Mechanisms in Heart Failure

When the heart fails, a number of mechanisms come into play which tend to preserve the blood pressure and the cardiac output. The primary compensatory mechanisms are as follows.

1. Increased Cardiac Size. This takes place initially through acute dilation, allowing use of the Frank-Starling mechanism, but chronic heart failure then results in gradual further heart enlargement associated with some degree of hypertrophy. Hypertrophy is particularly marked when the heart failure is secondary to a long-standing increase in cardiac work (such as in hypertension or valvular heart disease). Thus, when heart failure is due to myocardial disease, the heart as seen on a chest X-ray often is enlarged.

This gradual, late ventricular enlargement helps to maintain the stroke volume by allowing greatly reduced shortening of each unit of myocardium to produce a normal or only mildly reduced stroke volume. This effect is illustrated in Figure 2.201. End-diastole and end-systole are shown in the normal heart, delivering a normal stroke volume. An 18% shortening of the circumference of the chamber produces a 30-ml stroke volume and a normal ejection fraction (60%). With chronic heart failure, the end-diastolic volume increases along the *dashed curved line* extending the normal curvilinear relationship between ventricular circumference and volume (for a sphere volume = $4/3\pi\ r^3$) (Fig. 2.201, *Failure*). Under these conditions, the end-diastolic volume is markedly increased (120 ml) and the systolic shortening of the diameter is only 4%, yet the stroke volume is maintained at 30 ml. The increased end-diastolic volume, even though the ejection fraction is decreased (to 25%), can result in a normal stroke volume.

The responses to acute left ventricular failure are shown as ventricular function (Starling) curves of the ventricle in Figure 2.202*A.* Acute heart failure results in displacement of the normal left ventricular function curve downward and to the right. Without compensation by the Frank-Starling mechanism, stroke volume would drop markedly (normal to *point A*), and the cardiac output would be greatly reduced. However, the stroke volume is partially

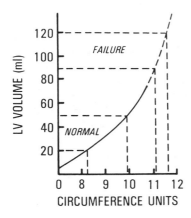

	△CIRC.	SV	EFx
NORMAL:	18%	30ml	60%
FAILURE:	4%	30ml	25%

Figure 2.201. Relation between left ventricular circumference and the volume of the normal and chronically failing ventricle. The *curve* shows the relation between the circumference and volume of a sphere, and its *dashed extension* shows the influence of chronic progressive left ventricular dilation in an experimental animal. The *straight dashed lines* intersecting the normal portion of the curve indicate end-diastole and end-systole, and in failure the much larger end-diastolic and end-systolic volumes and circumferences are also indicated by *straight dashed lines*. Notice that in the normal ventricle, 18% shortening of the internal chamber circumference can produce a stroke volume of 30 ml with a normal ejection fraction (60%). In heart failure, from a much larger end-diastolic volume (120 ml) the depressed myocardium can now shorten the internal circumference by only 4%; however, the dilated ventricle could theoretically still deliver a normal stroke volume of 30 ml at a much lower ejection fraction (25%).

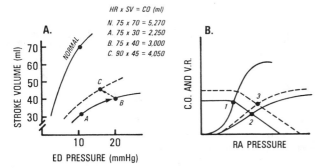

Figure 2.202. *Heart* failure as diagrammed in a framework of curves relating the stroke volume *(SV)* and left ventricular end-diastolic *(ED)* pressure *(panel A)*, and in a framework of cardiac output *(C.O.)* and venous return *(V.R.)* curves *(panel B)*. Reflex and other compensations are shown in *panel A* as *points B* and *C* with movements up a depressed function curve and a shift of the ventricular function curve together with reflex increase in heart rate (Table). In panel *B*, the normal state *(point 1)* and the failure state are shown *(point 2)* with compensations indicated by the *dashed lines* and *point 3*. For further discussion, see text.

restored by the higher left ventricular end-diastolic pressure (see Fig. 2.202 Table inset, *point B*), and the cardiac output therefore partially recovers; further compensations also occur (see below).

2. Autonomic and Renal Compensations. With decreased cardiac output and stroke volume, the systemic arterial compartment is relatively under-filled, a condition which activates the baroreceptors (Chapter 16). The reflex adrenergic responses elevate the peripheral vascular resistance, restoring the blood pressure and they also increase the heart rate (Zelis et al., 1973). In late phases of heart failure, circulating catecholamines are elevated. The result is a shift of the left ventricular function curve upward and to the left (*dashed line*, Fig. 2.202A). The positive inotropic effect of adrenergic stimulation tends to increase the stroke volume and lower the left ventricular end-diastolic pressure to some degree (point *C*). In addition, reflex elevation of the heart rate can further restore the cardiac output (*C.O.*) toward normal (Fig. 2.202A, *point C*, and Table inset).

Mechanisms also come into play which increase the circulating blood volume, thereby sustaining the increased cardiac volume. Thus, underperfusion of the kidneys (due to the decreased cardiac output, activation of the sympathetic nervous system, and perhaps elevated venous pressure acting on the renal veins) activates the renin-angiotensin system (Chapter 32) which, in turn, stimulates aldosterone secretion. The reduced renal blood flow per se, together with an increased plasma aldosterone level (Chapter 32), promote sodium and water reabsorption by the kidney, serving to increase the circulating blood volume and the volume of the heart. This effect is encompassed by the increased end-diastolic pressure of points *B* and *C*, Figure 2.202A.

Within the venous return curve framework (Fig. 2.202B), acute depression of the cardiac output curve by heart failure (point *1* to point *2*) results initially in a marked reduction of the cardiac output. Some shift upward of the cardiac output curve occurs by reflex autonomic stimulation and the increase in heart rate (Fig. 2.202B, *dashed line*), but even more importantly, activation of the sympathetic nervous system increases the venous tone (Zelis et al., 1973) and the mean systemic pressure together with increased blood volume brought about by renal effects, this shifts the venous return curve upward (Fig. 2.202B, *dashed venous return curve*). By these mechanisms the cardiac output can be restored toward normal, despite depression of the cardiac output curve (point *3*, Fig. 2.202B). In the later phases of heart failure, cardiac autonomic compensations become impaired (Braunwald et al.,

1988), and fluid retention by the kidney becomes excessive.

Cardiovascular reflex control is impaired in severe heart failure, with decreased sensitivity of the baroreceptors, and this may lead to diminished central inhibition of the autonomic nervous system, thereby contributing to the increase in sympathetic tone.

Other abnormalities in congestive heart failure include elevation of atrial natriuretic factor (ANF). ANF is a powerful peptide hormone, normally secreted primarily by the atria in response to stretch, which promotes renal sodium and water loss, relaxes vascular smooth muscle, and inhibits secretion of renin and aldosterone (Zimmerman et al., 1987) (see Chapter 31). In volume overload states and in congestive heart failure in man, increased circulating levels of immunoreactive ANF have been reported (Bates et al., 1986), but the significance of such observations is not yet clear. The potential therapeutic usefulness of ANF is also under investigation.

To summarize the *neurohumoral abnormalities* in heart failure: there are blunted autonomic reflexes, increased adrenergic stimulation with increased circulating catecholamines, and increased plasma renin and angiotensin II, the latter further contributing to the sympathetically mediated increase in peripheral vascular resistance. Atrial natriuretic factor, as well as plasma arginine vasopressin may also be elevated. (For reviews of neurohumoral effects, see Abboud and Thames, 1983; Zucker and Gilmore, 1985; Francis, 1985).

Treatment of Heart Failure

Traditional forms of treatment for heart failure have included use of the positive inotropic agent *digitalis*, which can shift the ventricular function curve and the cardiac output curve upward. *Diuretics* are added to rid the tissues of excessive interstitial fluid (peripheral edema and pulmonary edema), and to lower excessively elevated venous pressures. *Vasodilators* are also commonly used (see below). Of course, it may be possible to treat failure of the different subsystems of the heart by *specific therapies* aimed at the underlying cause of the heart failure. For example, heart block can be treated by implantation of a permanent electrical pacemaker; serious valvular lesions can be corrected by surgical replacement of the diseased valve with an artificial prosthesis; severe coronary artery narrowing can be treated with antianginal drugs, angioplasty, or by surgical bypass-grafting of the coronary arteries; and severe hypertension is treated by appropriate antihypertensive medications.

In late stage hypertension or aortic stenosis, and in severe left ventricular myocardial failure of any cause, the afterload on the left ventricle (reflected either by the peak systolic pressure or the wall tension) is usually elevated, leading to a state of *afterload mismatch* (Ross, 1976b). In this situation, the preload reserve of the ventricle is generally exhausted, and it is possible to show that any further increase of the afterload will result in a sharp drop in the stroke volume (Fig. 2.203). Thus, when the ventricle moves from a normal condition (Fig. 2.203, beat *1*) to severe heart failure with a high left ventricular end-diastolic volume and pressure (the heart is now operating on a steep portion of the diastolic pressure-volume curve), the limit of preload reserve may be reached (beat *2*, Fig. 2.203); under these conditions, if an additional pressure load is applied (such as by infusing the vasopressor agent phenylephrine), the ventricle behaves *as if* the preload were fixed (compare with Fig. 2.122, Chapter 12). Thus, the response to the vasopressor will be a sharp drop in the stroke volume (*beat 2* to *3*, Fig. 2.203), as demonstrated in patients with heart failure (Ross and Braunwald, 1964).

In the severely failing heart, when the peripheral vascular resistance is elevated by adrenergic compensations and the ventricle is operating at the limit of its preload reserve, the dilated ventricle can

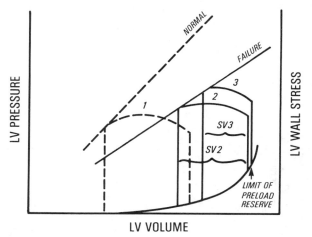

Figure 2.203. Left ventricular *(LV)* pressure-volume relations and pressure-volume loops under normal conditions *(dashed lines)* and during heart failure when the linear end-systolic pressure-volume relation is shifted downward and to the right (failure). *Beat 2* to *beat 3* shows the effect of acutely increasing the left ventricular systolic pressure with a vasoconstrictor when there is little or no preload reserve; the stroke volume drops (*SV2* to *SV3*). In chronic heart failure, the left ventricle may be operating under basal conditions similar to beat 3 and left ventricular wall stress *(right-hand ordinate)* may be elevated, despite a normal left ventricular systolic pressure. Under these circumstances use of a vasodilator drug may relieve this "afterload mismatch" and allow the ventricle to improve the stroke volume by lowering the wall stress (*beat 3* to *beat 2*). For further discussion, see text.

exhibit an elevated wall stress, even though the systolic pressure in the ventricle may be normal (Fig. 2.203, right-hand ordinate, *beat 3*). Remember, a large radius and a thin wall will yield a high wall stress even in the presence of a normal systolic pressure (see Fig. 2.111). Under these conditions, it can be reasoned that *reducing the afterload* on the failing ventricle will improve its stroke volume (moving the left ventricle from *beat 3* to *beat 2*, Fig. 2.203).

Afterload reduction (vasodilator therapy) has become an important form of treatment for acute and chronic heart failure. (For review, see Chatterjee and Parmley, 1977.) Experiments on venous return curves in the normal circulation and in acute severe left ventricular failure (produced by multiple coronary artery ligations), allow understanding of how peripheral and circulatory factors interact in vasodilator treatment using nitroprusside (an agent which dilates both arterioles and veins by relaxing vascular smooth muscle) (Pouleur et al., 1980).

In the *normal* circulation, intravenous infusion of nitroprusside leads to a *reduction* of cardiac output despite a lower systemic vascular resistance and more favorable systolic loading conditions on the normal left ventricle. This occurs because there is concomitant dilation of the venous bed, which is only partly compensated by a small shift of blood volume from the central to the peripheral circulation. Therefore, the venous return curve is displaced downward and to the left (Fig. 2.204) with a reduced mean systemic pressure and effective systemic blood volume. The reduction in the venous return to the right heart is responsible for a *reduction* in right ventricular output (and hence left) ventricular output), despite the more favorable systolic loading conditions on the normal left ventricle (Fig. 2.204).

In the presence of acute experimental left ventricular failure in which the left ventricular end-diastolic pressures were elevated to over 20 mm Hg, an opposite effect occurred, with an *increase* in cardiac output during nitroprusside infusion (Fig. 2.205). Again, venodilatation took place in the peripheral circulation; however, this time a large shift of blood volume from the distended central circulation to the peripheral circulation occurred with vasodilator therapy (the failing heart is now able to unload by ejecting against a lower vascular resistance, and blood dammed up behind it is released). This blood volume shift exactly counterbalanced the tendency for nitroprusside to lower the effective systemic blood volume and the mean systemic pressure in these experiments (Pouleur et al., 1980). Therefore,

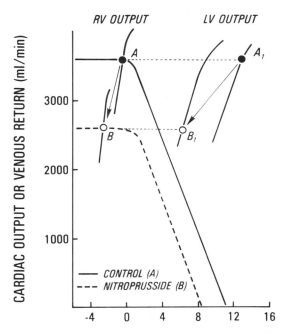

NORMAL

Figure 2.204. Relation between cardiac output and venous return in the normal heart (open chest, anesthetized dog). The inverse relation between right atrial pressure and venous return is shown under control conditions *(solid lines)* and during nitroprusside infusion *(dashed lines)*. Segments of cardiac output curves relating right ventricular output *(RV OUTPUT)* to right atrial pressure and left ventricular output *(LV OUTPUT)* to left atrial pressure are also shown in these two conditions. Under control conditions, the cardiac output is limited by the venous return, and the equilibrium *(point A)* where the venous return and cardiac output curves intersect is on the plateau of the venous return curve. In the steady-state, the right ventricular and left ventricular outputs are in equilibrium *(dashed horizontal line* and *point A₁)*. During nitroprusside infusion, venodilation produces a drop in mean systemic pressure and a shift downward of the venous return curve, the right ventricle reaching a new equilibrium point at a lower cardiac output *(point B)* which, in turn, is in equilibrium with a lower left ventricular output at a lower mean left atrial pressure *(point B₁)*. Thus, despite upward shifts of the right and left ventricular function curves due to reduced afterload with lowered impedance to ejection, the cardiac output is lower. (Adapted from Pouleur et al., 1980.)

the venous return curve was *not* shifted downward (Fig. 2.205), and the marked shift upward of the cardiac output curve, due to correction of the afterload mismatch on the failing ventricle, could now be expressed as an increase in the cardiac output (Fig. 2.205). Thus, is it clear that the effects of vasodilator drugs on the *peripheral circulation* as well as on the heart are highly important in determining the overall responses to acute vasodilator therapy (Pouleur et al., 1980).

A number of other drugs can decrease vascular resistance. For example, hydralazine (a vascular smooth muscle relaxant) has an effect largely lim-

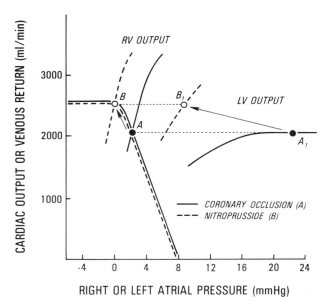

HEART FAILURE

Figure 2.205. Relations between cardiac output and venous return in acute heart failure produced by experimental coronary occlusion. Under control conditions *(solid lines)*, the intersect between right ventricular output and venous return *(point A)* is on the ascending portion of the venous return curve, cardiac output being limited by the failing left ventricle which is operating on a flat and depressed cardiac output curve *(point A₁)*. Following nitroprusside infusion *(dashed lines)* there is no shift of the venous return curve, a downward shift being prevented by a redistribution of the central blood volume to the periphery (see text). There is now a marked shift upward of the function curve of the left ventricle due to reduced afterload with correction of afterload mismatch, and a marked drop in left ventricular filling pressure occurs. The shift upward of the left and right ventricular output curves is now accompanied by an increase in the cardiac output *(point B)*, and at equilibrium the left ventricle *(point B₁)* is now operating at a lower filling pressure with an improved cardiac output. (Adapted from Pouleur et al., 1980.)

ited to the arterioles, and it increases cardiac output in large measure by shifting the cardiac output curve upward (through lowered afterload), with little effect on the venous bed (in contrast to nitroprusside). In normal subjects this results in little change in the cardiac output, since the function curve is shifted to the left on the flat portion of the venous return curve. In heart failure, however, when the failing ventricle is operating below the knee of the venous return curve (the cardiac output curve being shifted downward and to the right), such an agent substantially improves the cardiac output.

Frequently, nitroglycerin is added to hydralazine, and its venodilator properties serve to lower the elevated left and right atrial pressures. This drug combination can improve mortality in severe heart failure (Cohn et al., 1986).

More recently, angiotensin-converting enzyme

(ACE) inhibitors, such as captopril and enalapril, have been shown to improve the cardiac output, lower diastolic ventricular pressures, and to prolong life in patients with severe heart failure (Consensus trial, 1987). Beneficial effects of the ACE inhibitors (which block the conversion of angiotensin I to angiotensin II) result from decreasing circulating levels of angiotensin II and aldosterone, as well as through other mechanisms which lead to vasodilator effects and a decrease in circulating catecholamines.

Effects of the Pericardium in Overtransfusion: Possible Role in Heart Failure

The effect of the pericardium on the diastolic pressure-volume relation has been controversial. It appears to have little effect under normal conditions, but large downward shifts of the entire left ventricular diastolic pressure-volume curve have been described in patients with heart failure during treatment with nitroprusside; these shifts were associated with little change in the left ventricular diastolic volume but a large reduction of the left ventricular diastolic pressure. That such shifts in the pressure-volume curve may reflect an effect of the pericardium, rather than an alteration of left ventricular diastolic compliance, has been indicated by laboratory studies in acutely distended hearts of conscious dogs (Shirato et al., 1978; also Glantz and Parmley, 1978; Shabetai, 1981). Following acute circulatory distension by transfusion, the entire pressure-dimension curve was displaced upward (Fig. 2.206, *upper panel*). When nitroprusside was infused under these conditions with the pericardium intact there was downward shift of the *entire curve* (Fig. 2.206, *upper panel*). When the experiment was repeated after the pericardium was removed, however, the left ventricle now moved upward (with transfusion) and downward (nitroprusside) on a *single* curve (Fig. 2.206, *lower panel*). This finding suggests that during acute cardiac distension by overtransfusion, and perhaps in acute heart failure, the limit of pericardial distension can be reached, and elevated *intrapericardial pressure* can contribute to high ventricular filling pressures. Moreover, decreases in intrapericardial pressure may contribute substantially to the large reductions in left ventricular filling pressure observed during vasodilator therapy in some patients with acute or subacute heart failure.

CARDIAC HYPERTROPHY

Several types of cardiac hypertrophy (increased size of heart muscle cells causing cardiac enlargement) can be identified. One type occurs in trained

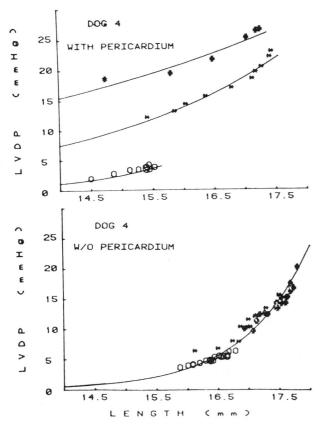

Figure 2.206. Relations between the length of a segment of the left ventricle (representing the volume of the chamber) and the left ventricular diastolic pressure (ordinate, in mm Hg) in a conscious dog. Points were obtained during slow cardiac filling (diastasis). The *upper panel* shows this relation with the pericardium intact before *(open symbols)* and after the intravenous infusion of dextran to produce acute cardiac dilatation *(asterisks, upper curve)*; the middle curve *(xs)* shows the effect of an intravenous infusion of nitroprusside in the presence of such acute cardiac dilatation. In the *lower panel,* the same dog is studied again without *(W/O)* the pericardium, following its surgical removal. The same interventions, volume loading and nitroprusside, are produced. The ventricle now appears to be operating on a single diastolic pressure-length relation. (From Shirato et al., 1978.)

athletes, and another is left ventricular hypertrophy of unknown cause; hypertrophy also occurs in response to valvular heart disease or to hypertension, and it also takes place in normal regions of the left ventricle after a heart attack, tending to compensate for scar formation in damaged regions.

Two basic types of hypertrophy that occur in response to stress generally are defined: that caused by pressure overload and that due to volume overload of the heart. For simplicity, we will confine the discussion primarily to the effects of these two types of overload on the left ventricle. (For review concerned with enhanced protein synthesis and mechanisms in hypertrophy, see Bugaisky and Zak, 1986).

Pressure Overload

Pressure overload on the left ventricle is most commonly produced by high blood pressure (hypertension), which raises the systolic and diastolic pressures in the aorta, and therefore increases the left ventricular systolic pressure. Narrowing of the aortic valve (aortic stenosis) also produces a high systolic pressure in the left ventricle, with a lower than normal pressure in the aorta. (The right ventricle can also undergo hypertrophy in response to pulmonary artery hypertension, or congenital pulmonic valve stenosis.)

The basic response of the left ventricle to sustained pressure overload is to develop *concentric hypertrophy.* This adaptation, which occurs over months or years, consists of increased *thickness* of the left ventricular wall (protein synthesis is stimulated, and the individual cells of the myocardium greatly enlarge with an increased number of myofilaments) but *without* an increase in the size of the left ventricular chamber (ventricular end-diastolic volume remains unchanged). Such a response in an experimental animal subjected to constriction of the ascending aorta for several weeks by means of an implanted inflatable cuff is shown in Figure 2.207 (Sasayama et al., 1976). Notice that the initial response to aortic constriction is that to an acute, severe increase in afterload: an increase in end-diastolic pressure with thinning of the left ventricular wall, and a fall in the extent of systolic wall shortening due to markedly increased systolic pressure (wall tension is increased during this acute response, see below). However, over the next few weeks the left ventricular wall thickens, and the left ventricular chamber size and the shortening of the wall return to normal (Fig. 2.207, *right-hand tracings*). This chronic adaptation is diagrammed in Figure 2.208, together with the changes in factors which determine mean wall tension by the Laplace relation. The normal ventricle during systole is in the *left panel,* and it can be seen that with chronic pressure overload the high systolic pressure, with no change in ventricular radius, is compensated for by the increased wall thickness, resulting in an unchanged tension in the left ventricular wall during systole (Fig. 2.208, *right panel*). Thus, the chronic adaptation to pressure overload returns the afterload (wall tension) on the muscle fibers toward normal, despite the elevated systolic pressure, and this near-normal level of afterload allows preservation of the extent of shortening of the wall and the delivery of a normal stroke volume.

This functional response is diagrammed within the end-systolic pressure-volume framework in Fig-

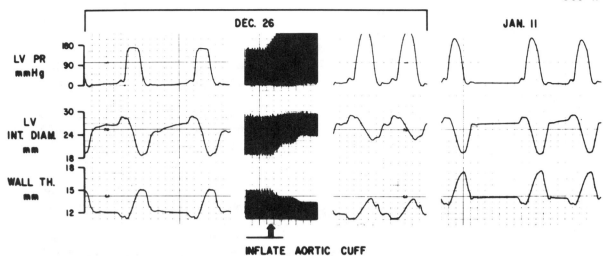

Figure 2.207. Responses of the left ventricle of a conscious dog to acute and chronic pressure overload produced by means of a cuff on the ascending aorta. Tracings on the *left* show left ventricular pressure *(LV PR)*, internal diameter *(INT. DIAM)*, and the thickness of the left ventricular wall *(WALL TH)*. measured with implanted ultrasonic dimension gauges. Notice the expansion of left ventricular diameter during atrial systole (synchronous with the presystolic wave in the left ventricular pressure tracing), the shortening of the diameter during left ventricular ejection followed by rapid and slow elongation during diastolic filling; the wall thickness shows a mirror image to the internal diameter, thinning with atrial systole, and thickening as the ventricle ejects blood.

Sudden inflation of the aortic cuff produces a high left ventricular systolic pressure (off scale approaching 240 mm Hg) and a decreased extent of systolic shortening of the internal diameter with decreased systolic wall thickening *(middle rapid tracing)*. Approximately 2 weeks later (January 11th) the left ventricular systolic pressure is still over 200 mm Hg, but now the wall thickness has markedly increased, whereas the chamber diameter is normal (concentric hypertrophy). This adaptation has allowed the ventricle to develop nearly normal systolic shortening at a high left ventricular systolic pressure. (Tracings obtained by Sasayama et al., 1976.)

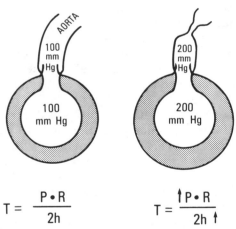

$$T = \frac{P \cdot R}{2h} \qquad T = \frac{\uparrow P \cdot R}{2h \uparrow}$$

Figure 2.208. Concentric hypertrophy. Diagram of a normal left ventricle developing a systolic pressure of 100 mm Hg during ventricular ejection. A simplified Laplace relation for wall tension (*T*) is shown where *P* = pressure, *R* = chamber radius, and *h* = wall thickness. In the *right* figure, constriction of the aorta above the aortic valve has been produced to raise the left ventricular systolic pressure to 200 mm Hg, and the chronic adaptation has occurred. Notice that the chamber has remained the same size, but the wall has thickened markedly due to hypertrophy. Thus, the increased pressure has been counterbalanced by the increased wall thickness, and wall tension (*T*) or stress remains normal.

ure 2.209*A*. *Beat A* shows the normal pressure-volume loop reaching the normal end-systolic pressure-volume relation, and *beat B* illustrates the chronic adaptation to pressure overload with concentric hypertrophy in which a normal stroke volume is delivered from the same end-diastolic volume but at a much higher level of left ventricular systolic pressure. Note that the linear end-systolic pressure-volume relation is shifted to the left and upward in this chronic adaptation (Fig. 2.209*A*), despite the fact that no change in the myocardial inotropic state is caused by the hypertrophy. Thus, under these chronic conditions, the left ventricular end-systolic pressure-volume relation is an unreliable indicator of myocardial inotropic state; however, it does indicate hyperfunction of the left ventricle (this would result in a shift upward of the cardiac output curve). This difficulty with the use of the end-systolic pressure-volume relation for defining inotropic state in chronic adaptations can be overcome by the use of left ventricular wall tension or wall stress (instead of left ventricular pressure) on the ordinate of the graph (Fig. 2.209*B*). Under these conditions, it can be seen that both the normal loop *(beat A)* and that after the chronic adaptation *(beat B)*

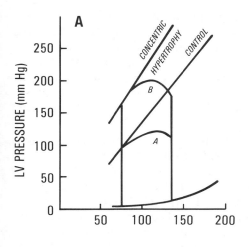

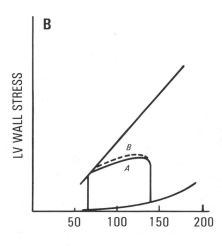

Figure 2.209. *Panel A,* Left ventricular *(LV)* pressure-volume relations and linear end-systolic pressure-volume relations before *(curve A)* and after *(curve B)* the development of concentric left ventricular hypertrophy. Left ventricular end-diastolic volume and the stroke volume are unchanged, but the left ventricular systolic pressure is much higher after hypertrophy so that the linear end-systolic pressure volume relation is shifted upward and to the left (for further discussion, see text). *Panel B,* The same responses as in *panel A* using left ventricular wall stress instead of pressure, the curves are now superimposable, since chronic thickening of the left ventricular wall lowers the systolic wall stress despite the elevated left ventricular systolic pressure.

are now superimposable (since the systolic wall stress during ejection is normal) and they fall on the same linear end-systolic volume-wall stress relation (see Sasayama et al., 1977, for further details). Eventually, with long-standing pressure overload, myocardial failure supervenes.

Volume Overload

In this setting, the left ventricle must adapt to the requirement for a larger than normal stroke volume. Such a need may arise in conditions which place a volume overload on the left (or right) ventricle such as chronic strenuous exercise in the athlete, anemia, mitral regurgitation, aortic regurgitation, systemic arteriovenous fistula, or a left-toright shunt due to ventricular septal defect. (The right ventricle may experience a volume overload in lesions such as tricuspid regurgitation, or a left-to-right shunt caused by an atrial septal defect).

The basic response of the left ventricle to these forms of sustained volume overload is termed *eccentric hypertrophy.* In this adaptation, the left ventricle undergoes hypertrophy in such a way that a large increase in the left ventricular volume occurs with relatively little increase in wall thickness (Ross, 1974). This results in a marked increase in overall muscle mass of the ventricle. Because of the curvilinear relation between the volume of a spherical chamber and its radius, if the myocardial contractility is normal a *normal* percentage shortening of the circumference when the end-diastolic volume is greatly increased can deliver a much larger than normal stroke volume (see Fig. 2.201). This adaptation to chronic volume overload is shown in Fig. 2.210; the normal left ventricle is on the *left.* With chronic ventricular dilation the stroke volume is tripled with a normal percent shortening of the circumference (and the sarcomeres) (Fig.

2.210, *right*). By this mechanism, in aortic regurgitation both the amount regurgitated and the forward stroke volume can thereby be delivered, and in other conditions of volume overload a high forward stroke volume and cardiac output can be maintained.

The response to chronic volume overload is represented within the end-systolic pressure-volume framework in Figure 2.211. The response of the normal heart to an acute volume increase is shown in *beats A* and *B.* In eccentric hypertrophy, the entire diastolic (resting) pressure-volume is shifted to the right, and with the ventricle operating at a much larger end-diastolic volume a much larger stroke volume can be delivered (Fig. 2.211, beat *C*).

HYPERTENSION

Chronic elevation of the blood pressure is a widespread condition in many populations, involving over 60 million people in the United States (Subcom-

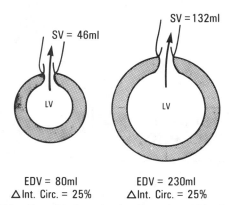

Figure 2.210. Eccentric hypertrophy. Normal ventricle shown on the *left* delivers a normal stroke volume with shortening of the chamber internal diameter of 25%. With left ventricular dilatation due to chronic volume overload, the same percentage shortening of the internal diameter now produces a nearly three fold increase in the stroke volume.

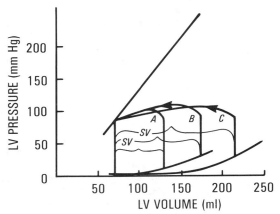

Figure 2.211. Left ventricular *(LV)* pressure-volume relations and the linear end-systolic pressure-volume relation during acute volume overloading (*beat A* to *beat B*) and after chronic eccentric hypertrophy has occurred. *SV*, stroke volume. The entire diastolic pressure-volume relation is shifted to the right, and under these circumstances the chronically dilated ventricle which has normal inotropic state can deliver a much larger stroke volume *(beat C)*.

mittee on Definition and Prevalence of the 1984 Joint Natl. Comm.: Hypertension prevalence and the status of awareness, treatment, and control in the United States, 1985). In hypertension, a diastolic arterial pressure above 90–95 mm Hg and a systolic pressure over 140–150 mm Hg are generally considered elevated (determined by repeated measurements using the cuff method, Chapter 13). In the great majority of patients, the cause of hypertension is unknown (essential hypertension), although in a small proportion (4–6%) secondary forms of hypertension due to adrenal tumors, renal disease, or other specific causes can be identified. Detailed reviews of the pathogenesis of hyperten-

sion are available (Kaplan, 1988) and their discussion is beyond the scope of this text. However, all mechanisms that control the blood pressure participate in the hypertensive response, and it will be useful to review briefly here the major findings in hypertension. It is usually unclear which control systems are responsible for the initial changes in blood pressure; once blood pressure becomes altered for any cause a constellation of control systems (Chapter 16) that are responsible for the normal regulation of blood pressure come into play.

The main factors regulating the arterial blood pressure, discussed in Chapter 16, are shown schematically in Figure 2.212. Blood pressure is determined by the product of cardiac output and the total peripheral vascular resistance. Cardiac output is controlled by factors such as the heart rate, inotropic state, and preload and afterload which, in turn, are dependent on both neural and humoral influences from the sympathetic and parasympathetic nervous systems and from vasoactive agents. Cardiac output is also dependent on the peripheral factors that influence venous return; these relate primarily to several renal mechanisms that influence blood volume such as the relation between blood pressure and sodium excretion (pressure-natriuresis) and the renin-angiotensin system. Peripheral resistance, on the other hand, is determined by neural factors, primarily the sympathetic nervous system's vasodilator and vasoconstrictor mechanisms, as well as by a variety of humoral factors such as prostaglandins, bradykinins, angiotensin, and catecholamines, and it is importantly influenced by autoregulation in local vascular beds.

In the majority of individuals with essential

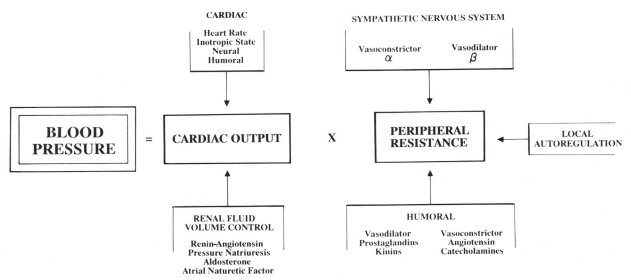

Figure 2.212. Factors influencing blood pressure.

hypertension, many of the systems involved in the control of blood pressure show abnormal responses. Lund-Johansen (1977) showed in a group of untreated young hypertensive subjects that the cardiac output was normal or slightly elevated whereas the peripheral resistance was normal, suggesting that early in the course of hypertension factors that increase the cardiac output may be important. Later in the course of hypertension, cardiac output is usually normal and the peripheral resistance is elevated. Thus, in the great majority of individuals with hypertension, the peripheral resistance is elevated, regardless of the initiating cause. Moreover, there is substantial recent evidence to support the hypothesis that the underlying factor leading to increased peripheral resistance is vasoconstriction due to augmented intracellular calcium (Postnov and Orlov, 1985). This increase of intracellular calcium may be an inherited membrane defect, a primary or inherited change in the sodium-potassium ATPases, although other hormonal and neural factors reviewed below may also play a role in the elevated peripheral resistance and increased intracellular Ca^{2+}.

The sympathetic nervous system can influence the blood pressure by producing changes in peripheral resistance and altering cardiac activity, as well as by changing the renal release of renin and influencing renal pressure-natriuresis. Evidence that sympathetic activity is involved in the production or maintenance of an elevated blood pressure derived from the findings that drugs which inhibit adrenergic nervous system activity lower the blood pressure when administered to hypertensive patients. In addition, some hypertensive individuals have high plasma catecholamine levels that correlate with the blood pressure; however, it may be that the major role of the sympathetic nervous system lies in perpetuating high blood pressure by altering the normal relationship between the blood pressure and sodium excretion via the kidney. However, the role of the sympathetic nervous system is still controversial and no current consensus exists as to its role in hypertension. The kidney has a central role in most hypertensive disorders. The relationship between blood pressure and sodium excretion (the pressure-natriuresis relationship) provides a control system with nearly infinite gain for the control of blood pressure (Guyton, 1972). In the normal circulation when arterial pressure is increased, the kidney rapidly excretes sodium and water, yielding the normal relationship between blood pressure and sodium excretion. Factors that alter this normal relationship, such as increased renal sensitivity to angiotensin or catecholamines,

could be responsible for steady-state alterations in blood pressure. There is evidence that many hypertensive subjects fail to respond normally to sodium loads, increasing blood pressure following intravenous administration of sodium, and some show high resting levels of renin secretion. Some essential hypertensives tend to have an abnormal pressure-natriuresis curve, with decreased sodium excretion relative to normals in response to same blood pressure rise. Both increased sympathetic nervous system activity and increased renin and angiotensin levels may lead to constriction of renal efferent arterioles, increasing the filtration faction and thus augmenting sodium reabsorption by the kidney (Chapter 32).

Elevated renin and angiotensin are directly responsible for hypertension associated with renovascular disease, and secondary forms of experimental hypertension may be induced by unilateral renal damage. Renin may be released by at least four different mechanisms: (1) changes in pressure within the renal afferent arteriole, (2) changes in the concentration of sodium chloride within the macula densa, (3) increased renal sympathetic nerve activity, and (4) alterations in plasma electrolyte concentration (e.g., potassium). Renin values vary substantially in patients with essential hypertension, being normal in over one-half of cases, high in approximately 10%, and low in 30% (Helmer, 1964). The presence of normal renin levels in patients with clearly elevated blood pressure is of interest, in that one would expect lowered levels of renin activity in these patients. Thus, these "normal" levels might be considered to be abnormally elevated and indicative of abnormalities in the renin-angiotensin control system in essential hypertension. Elevated renin leads directly to increases in angiotensin I through its actions on renin substrate. Angiotensin I is converted to angiotensin II by converting enzyme. Angiotensin II then acts to increase blood pressure by two powerful mechanisms: direct vasoconstriction and stimulation of aldosterone secretion from the adrenal medulla. Aldosterone promotes renal sodium retention. A number of other factors also may be involved in the initiation or maintenance of hypertension. These include hormones such as vasopressin, kinins, and various prostaglandins. The recently discovered atrial natriuretic peptides have a profound effect on sodium excretion by the kidney, but their specific role in hypertension remains uncertain at present.

In population studies, dietary sodium intake is proportional to the incidence of hypertension, and in some individuals (who are "salt sensitive") the effects of sodium intake may be linked to a genetic predisposition toward hypertension. Although the question of a genetic predisposition to hypertensive disease remains largely unanswered, in some populations inheritance may explain over half of the variability in blood pressure in the population (Longini, 1984). However, it is not at all clear exactly what defects in normal blood pressure control mechanisms are inherited. A variety of possibilities exist including membrane defects, defects in sodium or calcium transport, changes in the effects of stress on levels of autonomic nervous system activity, and the effects of excess dietary sodium.

In summary, in the great majority of individuals the initial cause of hypertension remains unknown. Most patients with long-standing hypertension demonstrate increased peripheral vascular resistance and normal or reduced levels of cardiac output, but changes in nearly every system involved in the control of blood pressure have been shown to exist and alterations in the renal handling of sodium can be demonstrated in the majority of patients; thus, alterations in the normal pressure-natriuresis relationship are likely to be central in the maintenance of hypertension.

ANEMIA

In anemia (Chapter 23), there is a reduction of circulating red blood cells with a normal intravascular blood volume, leading to a low hematocrit. The reduced hematocrit causes decreased viscosity of blood (Chapter 6), which lowers arterial and venous resistances, shifting the venous return curve (Fig. 2.213). Also, the need for increased oxygen delivery

to the tissues (to compensate for the low oxygen-carrying capacity of the blood) results in metabolic vasodilatation, with further lowering of peripheral vascular resistance. The lowered vascular resistance decreases the afterload on the left ventricle and shifts the cardiac ouput curve upward and to the left. Chronic hypertrophy of the volume overload type due to the high cardiac output, shifts the cardiac output curve further upward. Thus, a new equilibrium point is reached at a substantially increased cardiac output during anemia (Fig. 2.213, *point B*, compared to the normal state at *point A*).

HEMORRHAGIC SHOCK: AN ABNORMAL PERIPHERAL CIRCULATORY STATE

The hemodynamic features of one form of shock will be discussed briefly in order to illustrate certain abnormalities and compensatory mechanisms that can occur in various segments of the circulation. However, detailed discussion of mechanisms and neurohumoral responses in various forms of shock is beyond the scope of this discussion (for reviews, see Zweifach and Fronek, 1975; Weil et al., 1988).

In the human subject, intense bleeding due to trauma produces a marked fall in cardiac output and a secondary severe drop in the blood pressure, so-called hypotensive or "hemorrhagic shock." This condition can be mimicked in an experimental animal by attaching an artery to a blood reservoir where the pressure can be lowered to a given level (40 mm Hg, for example) and maintained there for many minutes before the blood is restored to the animal. When the blood pressure is maintained at this level for a sufficiently long period of time, so-called "irreversible shock" occurs; that is, even when the blood is retransfused, the blood pressure and cardiac output fail to return to normal, and the animal eventually dies.

In the early phase of shock, the loss of blood results in a decreased venous pressure and venous return to the heart, and a diminished cardiac output; in the intact circulation (if a pressure reservoir did not maintain the low arterial pressure) the resulting hypotension would result in baroreceptor reflexes (and later to chemoreceptor and cerebral reflexes resulting from diminished cerebral blood flow, see Chapter 16) which would lead to increased peripheral vascular resistance, reflex constriction of the veins (with central displacement of blood volume), and tachycardia. All of these mechanisms would tend to restore the cardiac output and arterial pressure toward normal, and if the hemorrhage were not too severe this compensation might be adequate. In the experimental shock model the blood pressure cannot be restored by these mechanisms

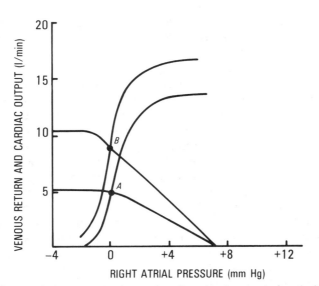

Figure 2.213.   Venous return and cardiac output curves before *(point A)* and after the adaptations to severe anemia *(point B)*. Notice the substantial increase in cardiac output. For further discussion, see text.

because of the pressure reservoir, and the blood mobilized by reflexes is initially transferred instead into the reservoir. Later in the shock state, as the pressure in the capillaries remains low, the unopposed colloid osmotic pressure tends to bring extravascular fluid into the circulation, and over time the hematocrit will fall (a manifestation of this dilution effect); this late compensatory mechanism may prevent a sustained shock state in the intact circulation when there is only a modest blood loss. However, again, in the experimental model it will only transfer more fluid into the pressure reservoir. The hypotension also causes increased release of circulating catecholamines from the adrenal glands, which serve to further stimulate the myocardium, augment heart rate, and raise the peripheral vascular resistance (Chien, 1967). Insufficient blood flow to the kidneys results in the release of angiotensin (a vasoconstrictor) as well as elevation of aldosterone levels which, along with direct effects of the low flow on the kidney, causes sodium and water retention with increased intravascular fluid volume.

The compensatory vasoconstriction initially involves primarily the skin, muscle, and splanchnic circulations (and to a much lesser degree the renal circulation), whereas the autoregulating coronary and cerebral circulations maintain relatively good flow.

After 1 or 2 h of severe hypotension, the shock state gradually becomes irreversible. If the shock remains untreated, sustained low blood flow with ischemia (insufficient O_2 and nutrient supply due to reduced flow) in the central nervous system results in decreased activity of the vasomotor centers and gradual loss of reflex compensations. Moreover, toxic substances absorbed when there is ischemia of the bowel, as well as progressive acidosis, can reduce the reactivity of vascular smooth muscle to catecholamines. This loss of vascular reactivity tends to result in sequestering of blood in the peripheral circulation (and depletion of blood from the reservoir in the experimental model). In addition, ischemia of the capillaries causes increased permeability, with reversal of fluid transfer (loss of intravascular fluid). The intense acidosis and metabolic products from ischemic tissues may lead to thrombosis of small blood vessels.

Damage to the internal organs results in the accumulation of other toxic factors, and one of these (the so-called "myocardial depressant factor") appears to reduce myocardial contractility in the late phase of shock (Lefer and Martin, 1970). Severe acidosis also tends to decrease myocardial contractility, and the ventricular function curve shifts downward in late shock (ischemia in the subendocardial regions may also be contributory). Finally, severely reduced flow through the bronchial arteries which supply the lung tissue eventually results in pulmonary damage, with edema and a decrease in the arterial oxygen saturation. This adds to generalized tissue hypoxia and ischemia and the downward spiral of circulatory function, as compensatory mechanisms fail.

BIBLIOGRAPHY

AARS, H. The baroreflex in arterial hypertension. *Scand. J. Clin. Lab Invest.* 35: 97–102, 1975.

ABBOUD, F. M., AND M. D. THAMES. Interaction of cardiovascular reflexes in circulatory control. In: *Handbook of Physiology, The Cardiovascular System Section 1, Vol. III*, edited by J. T. Shepherd and F. M. Abboud. Washington DC: American Physiological Society, p. 675–753, 1983.

ALLEN, D. G., AND J. C. KENTISH. The cellular basis of the length-tension relation in cardiac muscle. *J. Molec. Cell Cardiol.* 17: 821–840, 1985.

ALLESSIE, M. A., F. I. M. BONKE, AND F. J. G. SCHOPMAN. Circus movement in rabbit atrial muscle as a mechanism of tachycardia. *Circ. Res.* 33: 54–62, 1973.

ALPERT, N. R., AND M. S. GORDON. Myofibrillar adenosine triphosphatase activity in congestive heart failure. *Am. J. Physiol.* 202: 940–946, 1962.

ASHBURN, W. L., H. R. SCHELBERT, AND J. W. VERBA. Left ventricular ejection fraction—a review of several radionuclide angiographic approaches using the scintillation camera. *Prog. Cardiovasc. Dis.* 20: 267–284, 1978.

ASMUSSEN, E., AND M. NIELSEN. The cardiac output in rest and work determined simultaneously by the acetylene and the dye dilution methods. *Acta Physiol. Scand.* 27: 217–230, 1952.

ASMUSSEN, E. Similarities and dissimilarities between static and dynamic exercise. *Circ. Res.* 48: I3–I10 (Suppl.), 1981.

AYERS, C. R., J. O. DAVIES, F. LIBERMAN, C. C. J. CARPENTER, AND M. BERMAN. The effects of chronic hepatic venous congestion on the metabolism of d,1-aldosterone and daldosterone. *J. Clin. Invest.* 41: 884–895, 1962.

BACHE, R. J., AND F. R. COBB. Effect of maximal coronary vasodilation on transmural myocardial perfusion during tachycardia in the awake dog. *Circ. Res.* 41: 648–653, 1977.

BADEER, H. S. Contractile tension in the myocardium. *Am. Heart J.* 66: 432–434, 1963.

BAKER, C. H., W. L. NASTUK (eds.) *Microcirculatory Techniques.* Orlando: Academic Press, 1986.

BAKER, C. H., AND W. L. NASTUK (eds.) *Microvascular Technology.* New York: Academic Press, 1986.

BATES, E. R., Y. SHENKER, AND R. J. GREKIN. The relationship between plasma levels of immunoreactive atrial natriuretic factor hormone and hemodynamic function in man. *Circulation* 73: 1155–1161, 1986.

BAUEREISEN, E. Herz. In: *Kurzgefasstes Lehrbuch der Physiologie*, edited by W. D. Keidel. Stuttgart: Thieme Verlag, 1970, p. 71–99.

BERNARD, C. Sur les effects de la section de la portion cephalique de grand sympathique. *Compte Rendu des Seances de la Societe de Biologie*, November 1852, p. 168–170, 1964.

BERNE, R. M., AND R. RUBIO. Adenine nucleotide metabolism in the heart. *Circ. Res.* 35: 109, 1974.

BERNE, R. M., AND R. RUBIO. Coronary circulation. In: *Handbook of Physiology*, Sect. 2, edited by R. M. Berne. Washington DC: American Physiological Society, 1979, vol. 1, p. 873–952.

BEVAN, J. A. Action of lobeline and capsaicine on afferent endings in the pulmonary artery of the cat. *Circ. Res.* 10: 792–797, 1962.

BEVEGARD, B. S., AND J. D. SHEPHERD. Regulation of the circulation during exercise in man. *Physiol. Rev.* 47: 178–213, 1967.

BIRCH, G. E., AND T. WINSOR. *A Primer of Electrocardiography*, 6th ed. Philadelphia: Lea & Febiger, 1972.

BING, R. J. Myocardial metabolism. *Circulation* 12: 635–647, 1955.

BISCOE, T. J. Carotid body: Structure and function. *Physiol. Rev.* 51: 437–495, 1971.

BISHOP, V. S., AND E. M. HASSER. Arterial and cardiopulmonary reflexes in the regulation of the neurohumoral drive to the circulation. *Fed. Proc.* 44(8): 2377–2381, 1985.

BISHOP, V. S., AND H. L. STONE. Quantitative description of ventricular output curves in conscious dogs. *Circ. Res.* 20: 581–586, 1967.

BLINKS, J. R. Intracellular $[Ca^{2+}]$ measurements. In: *The Heart and Cardiovascular System*, edited by H. A. Fozzard, et al. New York: Raven Press, 1986, p. 671–702.

BLOMQVIST, C. G., S. F. LEWIS, W. F. TAYLOR, AND R. M. GRAHAM. Similarity of the hemodynamic responses to static and dynamic exercise of small muscle groups. *Circ. Res.* 48: I87–I92, 1981.

BLOMQVIST, C. G., AND B. SALTIN. Cardiovascular adaptations to physical training. *Annu. Rev. Physiol.* 45: 169–189, 1983.

BLOOR, C. M. Functional significance of the coronary collateral circulation. A review. *Am. J. Pathol.* 76: 561–586, 1974.

BOERTH, R. C., J. W. COVELL, P. E. POOL, AND J. ROSS, Jr. Increased myocardial oxygen consumption and contractile state associated with increased heart rate in dogs. *Circ. Res.* 24: 725–734, 1969.

BOHR, D. F. Vascular smooth muscle. In: *Peripheral Circulation*, edited by P. C. Johnson. New York: John Wiley & Sons, 1978.

BOINEAU, J. P., R. B. SCHUESSLER, C. R. MOONEY, C. B. MILLER, A. C. WYLDS, R. D. HUDSON, J. M. BORREMANS, AND C. W. BROCKUS. Natural and evoked atrial flutter due to circus movement in dogs. *Am. J. Cardiol.* 45: 1167–1181, 1980.

BOWDITCH, H. P. Ueber die Eigenthum-

lichkeiten der Reizbarkeit, welchedie Muskelfasern des Herzens zeigen. Leipzig: *Berichte Math. Phys.* 23: 652, 1871.

BRADY, A. J. Electrophysiology of cardiac muscle. In: *The Mammalian Myocardium*, edited by G. A. Langer, and A. J. Brady. New York: John Wiley & Sons, 1974, p. 135–161.

BRADY, A. J. Contractile and mechanical properties of the myocardium. In: *Physiology and Pathophysiology of the Heart*, edited by N. Sperelakis. Boston: Martinus Nijhoff Publ., 1984.

BRAUNWALD, E., E. H. SONNENBLICK, J. ROSS, Jr., G. GLICK, AND S. E. EPSTEIN. An analysis of the cardiac response to exercise. *Circ. Res.* 20–21 (Suppl. 1): 44–58, 1967.

BRAUNWALD, E., J. ROSS, Jr., AND E. H. SONNENBLICK. *Mechanisms of Contraction of the Normal and Failing Heart*, 2nd ed., Boston: Little Brown, 1976.

BRAUNWALD, E., AND J. ROSS, Jr. Control of cardiac performance. In: *Handbook of Physiology*, Sect. 2, edited by R. M. Berne. Washington DC: American Physiological Society, 1979, vol. 1, p. 533–580.

BRAUNWALD, E., K. J. ISSELBACHER, R. G. PETERSDORF, J. D. WILSON, J. B. MARTIN, AND A. S. FAUCI (eds.). *Harrison's Principles of Internal Medicine*, 11th ed. New York: McGraw Hill, 1987.

BRAUNWALD, E. Pathophysiology of heart failure. In: *Heart Disease*, edited by E. Braunwald. Philadelphia: W. B. Saunders Co., 1988, p. 426–448.

BRAUNWALD, E., AND B. E. SOBEL. Coronary blood flow and myocardial ischemia: In: *Heart Disease*, 3rd ed., edited by E. Braunwald. Philadelphia: W. B. Saunders, Co., 1988, p. 1191–1221.

BRAUNWALD, E., E. H. SONNENBLICK, AND J. ROSS, Jr. Mechanisms of cardiac contraction and relaxation. In: *Heart Disease*, edited by E. Braunwald, Philadelphia: W. B. Saunders Co., 1988, p. 383–425.

BROOKS, C. McC., B. F. HOFFMAN, E. E. SUCKLING, AND O. ORIAS. *Excitability of the Heart.* New York: Grune & Stratton, 1955.

BROWN, A. M. Motor innervation of the coronary arteries of the cat. *J. Physiol.* (London) 198: 311–328, 1968.

BROWN, A. M. Brief reviews. Receptors under pressure: an update on baroceptors. *Circ. Res.* 46: 1–10, 1980.

BROWN A. M., AND M. C. ANDERSEN. Plasticity of arterial baroreceptors in hypertension states. In: *Disturbances in Neurogenic Control of the Circulation*, edited by D. J. Reis. Bethesda: American Physiological Society, 1981.

BROWN, B. G., E. L. BOLSON, AND H. T. DODGE. Dynamic mechanisms in human coronary stenosis. *Circulation* 70: 917–922, 1984.

BROWN, G. L., AND J. C. ECCLES. The action of a single vagal volley on the rhythm of the heart beat. *J. Physiol.* (London) 82: 242–257, 1934.

BRUTSAERT, D. L., N. M. DeCLERCK, M.

A. GOETHALS, AND P. R. HOUSMANS. Relaxation of ventricular cardiac muscle. *J. Physiol. (London)* 283: 469–480, 1978.

BRUTSAERT, D. L., F. E. RADEMAKERS, AND S. U. SYS. Triple control of relaxation: Implications in cardiac disease. *Circulation* 69: 190–196, 1984.

BUCKBERG, G. D., D. E. FIXLER, J. P. ARCHIE, AND J. I. T. E. HOFFMAN. Experimental subendocardial ischemia in dogs with normal coronary arteries. *Circ. Res.* 30: 67–81, 1972.

BUGAISKY, L., AND R. ZAK. Biological mechanisms of hypertrophy. In: *The Heart and Cardiovascular System*, edited by H. A. Fozzard et al., New York: Raven Press, 1986, p. 1491–1506.

BULBRING, E., A. F. BRADING, A. W. JONES, AND T. TOMITA (eds.) *Smooth Muscle*. London: Edward Arnold, 1970.

BURCH, G. E., AND T. WINSOR. *A Primer of Electrocardiography*. Philadelphia: Lea & Febiger, 1968.

BURNS, J. W., AND J. W. COVELL. Myocardial oxygen consumption during isotonic and isovolumic contractions in the intact heart. *Am. J. Physiol.* 223: 1491–1497, 1972.

BURNS, J. W., J. W. COVELL, AND J. ROSS, Jr. Mechanics of isotonic left ventricular contractions. *Am. J. Physiol.* 224: 725–732, 1973.

BURTON, A. C. Relation of structure to function of the tissues of the wall of blood vessels. *Physiol. Rev.* 34: 619–642, 1954.

BURTON, A. C. Importance of shape and size of heart. *Am. Heart J.* 54: 801–810, 1957.

BURTON, A. C. *Physiology and Biophysics of the Circulation*, ed. 2. Chicago: Year Book, 1972.

CALDINI, P., S. PERNOTT, J. A. WADDELL, AND R. L. RILEY. The effects of epinephrine on pressure, flow and volume relationships in the systemic circulation of dogs. *Circ. Res.* 34: 606–623, 1974.

CARO, C. G., J. PEDLEY, R. C. SCHROTER, AND W. A. SEED. *The Mechanics of the Circulation*. New York: Oxford University Press, 1979.

CARONI, P., L. REINLIB, AND CARAFOLI. Change movements during the Na^+-Ca^{2+} exchange in heart sarcolemmal vescicles. *Proc. Natl. Acad. Sci. U.S.A.* 77: 6354, 1980.

CARONI, P., AND E. CARAFOLI. The Ca^{2+}-pumping ATPase of heart sarcolemma. Characterization, calmodulin, dependence and partial purification. *J. Biol.. Chem.* 256: 3263, 1981.

CASE, D. B., J. M. WALLACE, H. J. KLEIN, M. A. WEBER, J. E. SEALEY, AND J. H. LARAGH. Possible role of renin in hypertension as suggested by renin sodium profiling and inhibition of converting enzyme. *N. Engl. J. Med.* 296: 641–655, 1977.

CASLEY-SMITH, J. R. Lymph and lymphatics. In: *Microcirculation*, edited by G. Kaley and B. M. Altura. Baltimore: University Park Press, 1977, vol. 2, p. 423–502.

CASTELLANOS, A., AND R. J. MYERBURG. The resting electrocardiogram. In *The Heart*, edited by J. W. Hurst. New York: McGraw Hill, 1986, p. 206–229.

CHAPMAN, J. B. Fluorometric studies of oxidative metabolism in isolated papillary muscle of the rabbit. *J. Gen. Physiol.* 59: 135–154, 1972.

CHAPLEAU, M. W., AND F. M. ABBOUD. Contrasting effects of static and pulsatile pressure on carotid baroreceptor activity in dogs. *Circ. Res.* 61(5): 648–658, 1987.

CHATTERJEE, K., AND W. W. PARMLEY. The role of vasodilator therapy in heart failure. *Prog. Cardiovasc. Dis.* 19: 301–325, 1977.

CAULFIELD, J. B., AND T. K. BORG. The collagen network of the heart. *Lab Invest.* 40: 364–372, 1979.

CHIDSEY, C. A., E. BRAUNWALD, A. G. MORROW, AND D. T. MASON. Myocardial norepinephrine concentration in man: effects of reserpine and of congestive heart failure. *N. Engl. J. Med.* 269: 653–658, 1963.

CHIEN, S. Role of the sympathetic nervous system in hemorrhage. *Physiol. Rev.* 47: 214–288, 1967.

CHIEN, S., S. VSAMI, H. M. TAYLOR, J. L. LUNDBERG, AND M. I. GREGERSEN. Effects of hematocrit and plasma proteins on human blood rheology at low shear rates. *J. Appl. Physiol.* 21: 81–87, 1966.

CHIEN, S. Biophysical behavior of red cells in suspension. In: *The Red Blood Cells*, 2nd ed. New York: Academic Press, 1975, vol. 2, p. 1031–1133.

CHOU, T. C., R. A. HELM, AND S. KAPLAN. *Clinical Vector Cardiography*, 2nd ed. New York: Grune & Stratton, 1974.

CLAUSEN, J. P. The effects of physical conditioning. A hypothesis concerning circulatory adjustments to exercise. *Scand. J. Clin. Invest.* 24: 305–313, 1969.

CLAUSEN, J. P. Effect of physical training on cardiovascular adjustments to exercise in man. *Physiol. Rev.* 57: 779–815, 1977.

CLIFF, W. H., AND P. A. NICOLL. Structure and function of lymphatic vessels of the bat's wing. *Q. J. Exp. Physiol.* 55: 112–121, 1970.

COFFMAN, J. D. AND D. E. GREGG. Reactive hyperemia characteristics of the myocardium. *Am. J. Physiol.* 199: 1143–1149, 1960.

COHN, J. N., D. G. ARCHIBALD, S. ZIESCHE, J. A. FRANCIOSA, W. E. HARSTON, F. E. TRISTANI, W. B. DUNKMAN, W. JACOBS, G. S. FRANCIS, K. H. FLOHR, S. GOLDMAN, F. R. COBB, P. M. SHAH, R. SAUNDERS, R. D. FLETCHER, H. S. LOEB, V. C. HUGHES, AND B. BAKER. Efect of vasodilator therapy on mortality in chronic congestive heart failure: Results of a Veterans Administration Cooperative Study. *N. Engl. J. Med.* 314: 1547–1552, 1986.

COKELET, G. R., H. J. MEISELMAN, AND D. E. BROOKS. *Erythrocyte Mechanics and Blood Flow*. New York: A. R. Liss, 1980.

COLANTUONI, A., S. BERTUGLIA, AND M. INTAGLIETTA. Quantitation of rhythmic diameter changes in arterial microcirculation. *Am. J. Physiol.* 246: H508–H517, 1984.

COLEMAN, H. N. III. Rose of acetylstrophanthidin in augmenting myocardial oxygen consumption: relation of increased O_2 consumption to changes in velocity of contraction. *Circ. Res.* 21: 487–495, 1967.

COLEMAN, T. G., A. W. COWLEY, AND A. C. GUYTON. Experimental hypertension and long-term control of arterial pressure. In: *Cardiovascular Physiology*, edited by A. C. Guyton and C. E. Jones. Baltimore: University Park Press, 1974, p. 259–297.

COLERIDGE, H. M., AND J. C. G. COLERIDGE. Afferent C-fibers and cardiorespiratory chemorefelexes. *Am. Rev. Respirat. Diseases* 115: 251–260, 1977.

COLERIDGE, J. G., J. C. G. COLERIDGE, AND D. G. BAKER. In: *Cardiac Receptors*, edited by R. J. Linden. London: Cambridge University Press, 1978, p. 117–137.

COLERIDGE, J. C. G., AND H. M. COLERIDGE. Chemoreflex regulation of the heart. In: *Handbook of Physiology: The Cardiovascular System*, edited by S. R. Geiger. Bethesda: American Physiological Society, 1979, p. 653–676.

CONCENSUS TRIAL STUDY GROUP: Effects of enalapril on mortality in severe congestive heart failure. *N. Engl. J. Med.* 316: 1429–1435, 1987.

COPENHAVER, W. M. (ed.) *Bailey's Textbook of Histology*. Baltimore: Williams & Wilkins, 1967.

COURTICE, F. C. The lymphatic circulation. In: *Structure and Function of the Circulation*, edited by C. J. Schwartz, N. T. Werthessen and S. Wolf. New York: Plenum Press, 1981, vol. 2, p. 487–602.

COVELL, J. W., J. ROSS, Jr., E. H. SONNENBLICK, AND E. BRAUNWALD. Comparison of force-velocity relation and ventricular function curve as measures of contractile state of intact heart. *Circ. Res.* 19: 364–372, 1966.

COVELL, J. W., J. S. FUHRER, R. C. BOERTH, AND J. ROSS, Jr. Production of isotonic contractions in the intact canine left ventricle. *J. Appl. Physiol.* 27: 577–581, 1969.

COWLEY, A. M., J. F. LAIRD, AND A. C. GUYTON. Role of baroreceptor reflex in the daily control of arterial pressure and other variables in the dog. *Circ. Res.* 32: 564–576, 1973.

CRAIGE, E. Heart sounds: Phonocardiography; Carotid, apex and jugular venous pulse tracings; and systolic time intervals. In: *Heart Disease, A Textbook of Cardiovascular Medicine*, edited by E. Braunwald, Philadelphia: W. B. Saunders, 1988, p. 41–64.

CRANEFIELD, P. F. *The Conduction of the Cardiac Impulse*. New York: Futura Publishing Co., 1975.

CRANEFIELD, P. F. Action potentials, after potentials and arrhythmias. *Circ. Res.* 41: 415–423, 1977.

CRANEFIELD, P. F., AND A. L. WIT. Cardiac arrhythmias. *Annu. Rev. Physiol.* 41: 459–472, 1979.

CURRY, F. R. E. Determinants of capillary permeability: A review of mechanisms based on single capillary studies in the dog. *Circ. Res.* 59:367–380, 1986.

CURTIS, B. M. AND W. A. CATTERALL. Phosphorylation of the calcium antagonist receptor of the voltage-sensitive calcium channel by cAMP-dependent protein kinase. *Proc. Natl. Acad. Sci. U.S.A.* 82: 2528–2532, 1985.

DALY, M., J. E. JAMES, AND G. ANGELL. Chemoreflexes in the circulation. Role of the arterial chemoreceptors in the control of the cardiovascular responses to breath-hold diving. In: *The Peripheral Arterial Chemoreceptors*, edited by M. J. Purves. London: Cambridge University Press, 1975, p. 387–405.

DAVIES, R. E. Role of ATP in contraction. In: *The Myocardial Cell: Structure, Function and Modification by Cardiac Drugs*, edited by S. A. Briller and H. L. Conn. Philadelphia: University of Pennsylvania Press, 1966, p. 157–168.

DAWES, G. S. AND J. H. COMROE, Jr. Chemoreflexes from the heart and lungs. *Physiol. Rev.* 34: 167–201, 1954.

DeBOLD, A. J., H. B. BORENSTEIN, A. T. VERESS, AND H. SONENBERG. A rapid and potent natiuretic response to intravenous injection of atrial myocardial extract in rats. *Life Sci.* 28: 89–94, 1981.

DeGEEST, H., M. N. LEVY, H. ZIESKE, AND R. I. LIPMAN. Depression of ventricular contractility by stimulation of the vagus nerves. *Circ. Res.* 17: 222–235, 1965.

DINTENFASS, L. *Rheology of Blood in Diagnostic and Preventive Medicine*. London: Butterworths, 1976.

DOMENECH, R. J., J. I. E. HOFFMAN, M. I. H. NOBLE, K. B. SAUNDERS, J. R. HENSON, AND S. SUBIJANTO. Total and regional coronary blood flow measured by radioactive microspheres in conscious and unanesthetized dogs. *Circ. Res.* 25: 581–596, 1969.

DONALD, D. E., AND J. T. SHEPHERD. Response to exercise in dogs with cardiac denervation. *Am. J. Physiol.* 205: 393–400, 1963.

DONALD, D. E. Myocardial performance after excision of the extrinsic cardiac nerves in the dog. *Circ. Res.* 34: 417–424, 1974.

DOW, P. Estimation of cardiac output and central blood volume by dye dilution. *Physiol. Rev.* 36: 77–102, 1956.

DOWNEY, J. M., AND E. S. KIRK. Distribution of the coronary blood flow across the canine heart wall during systole. *Circ. Res.* 34: 251–257, 1974.

DOWNING, S. E., J. P. REMENSNYDER, AND J. H. MITCHELL. Cardiovascular responses to hypoxia stimulation of the carotid bodies. *Circ. Res.* 10: 676–685, 1962.

DOWNING, S. E., J. H. MITCHELL, AND A. G. WALLACE. Cardiovascular responses to ischemia, hypoxia and hypercapnia of the central nervous system. *Am. J. Physiol.* 204: 881–887, 1963.

DOWNING, S. E., AND E. H. SONNENBLICK. Cardiac muscle mechanics and ventricular performance force and time parameters. *Am. J. Physiol.* 207: 705–715, 1964.

DUDEL, J., AND W. TRAUTWEIN. Das Aktions potential unter Mechanogramm des Herz muskels unter den Einfluss der Duhnung. *Cardiologia* 25: 344–362, 1954.

DULING, B. R., AND B. KILTZMAN. Local control of microvascular function: Role of tissue oxygen supply. *Ann. Rev. Physiol.* 42: 373–382, 1980.

DULING, B. R. Coordination of microcirculatory function with oxygen demand in skeletal muscle. In: *Advances Physiological Science*, vol. 7: Cardiovascular Physiology. Microcirculation and Capillary Exchange, edited by A. G. B. Kovach, J. Hamar, L. Szabo, Budapest: Pergamon Press, 1980, p. 1–16.

DUSTAN, H. P. R. C. TARAZI, AND E. L. BRAVO. Physiological characteristics of hypertension. In: *Hypertension Manual*, edited by J. H. Laragh. New York: Yorke Medical Books, 1974, p. 227–256.

EBASHI, S. Excitation-contraction coupling. *Annu. Rev. Physiol.* 38: 293–313, 1976.

ECKBERG, D. L. Temporal response of the human sinus node to Brie's carotid baroreceptor stimuli. *J. Physiol. (London)* 258: 769–82, 1976.

ECKENHOF, J. E., J. H. HAFKENSHIEL, C. M. LANDMESSER, AND M. HARMEL. Cardiac oxygen metabolism and control of the coronary circulation. *Am. J. Physiol.* 149: 634–649, 1947.

EINTHOVEN, W. Weiteres uber das Elektrokardiogramm. *Pflugers Arch.* 122: 517–584, 1908.

EDMAN, K. A. The active state and the force velocity relationship in cardiac muscle. In: *Heart Failure*, edited by H. Reindell, J. Kevl, and E. Doll. Stuttgart: Thieme Verlag, 1968, p. 133–138.

ENDO, M. Calcium release from the sarcoplasmic reticulum. *Physiol. Rev.* 57: 71–108, 1977.

FABIATO, A., AND F. FABIATO. Calcium and cardiac excitation-contraction coupling. *Annu. Rev. Physiol.* 41: 473–484, 1979.

FAHRAEUS, R., AND T. LINDQVIST. The viscosity of blood in narrow capillary tubes. *Am. J. Physiol.* 96: 562, 1931.

FANBURG, B. L. Myocardial cell failure. In: *The Mammalian Myocardium*, edited by G. A. Langer and A. J. Brady. New York: John Wiley & Sons, 1974, p. 283–306.

FEIGENBAUM, H. *Echocardiography*, 4th ed. Philadelphia: Lea & Febiger, 1986.

FEIGL, E. O. Coronary physiology. *Physiol. Rev.* 63: 1–205, 1983.

FEIGL, E. O. Parasympathetic control of coronary blood flow in dogs. *Circ. Res.* 25: 509–579, 1969.

FEIGL, E. O. The paradox of adrenergic coronary vasoconstriction. *Circulation* 76: 737–745, 1987.

FELD, G. K., N. VENKATESH, AND B. N. SINGH. Pharmacologic conversion and suppression of experimental canine atrial flutter: Differing effects of d-sotalol, quinidine and lidocaine and the significance of changes in refractoriness and conduction. *Circulation* 74: 107–204, 1986.

FENN, W. O. A quantitative comparison between the energy liberated and the work performed by the isolated sartorius muscle of the frog. *J. Physiol. (London)* 58: 175–203, 1923.

FISHMAN, A. P., AND D. W. RICHARDS (eds.). *Circulation of the Blood, Men and Ideas*. New York: Oxford Univ. Press, 1964.

FOZZARD, H. A. Conduction of the action potential. In: *Handbook of Physiology*, Sect. 2, edited by R. M. Berne, et al. Baltimore: Williams & Wilkins, 1979, vol. 1, p. 335.

FOZZARD, H. A., AND M. F. ARNSDORF. Cardiac Electrophysiology. In: *The Heart and Cardiovascular System*. New York: Raven Press, 1986, p. 1–30.

FRAME, L. H., R. L. PAGE, AND B. F. HOFFMAN. Atrial re-entry around an anatomic barrier with a partially refractory excitable gap. A canine model of atrial flutter. *Circ. Res.* 58: 495–511, 1986.

FRANK, O. On the dynamics of cardiac muscle (translated by C. B. Chapman and E. Wasserman). *Am. Heart J.* 58: 282–317, 1959 (Original: Zur Dynamic des Herzmuskels. *Zeitschrift fur Biologie* 32: 370–447, 1895).

FRANCIS, G. S. Neurohumoral mechanisms involved in congestive heart failure. *Am. J. Cardiol.* 55: 15A–21A, 1985.

FRIDOVICH, I., P. O. HAGEN, AND J. J. MURRAY. Endothelium derived relaxing factor: In search of the endogenous nitroglycerin. *News in Phys. Sci.* 2: 61–64, 1987.

FRONEK, A. Noninvasive diagnostics of vascular diseases. *Physiologic and Bioengineering Principles*. New York: McGraw Hill, 1988.

FRONEK, U., AND Y. C. FUNG. Mechanical properties of arteries as a function of topography and age. *Bioheology* 17: 227–234, 1980.

FRY, D. L. Acute vascular endothelial changes associated with increased blood velocity gradients. *Circ. Res.* 22: 165–197, 1968.

FUNG, Y. C. *Biodynamics: Circulation.* New York: Springer-Verlag, 1984.

FUNG, Y. C., *Biomechanics, Mechanical Properties of Living Tissues.* New York: Springer-Verlag, 1981.

FUNG, Y. C. Introduction of biophysical aspects of microcirculation. In: *Microcirculation*, edited by G. Kaley and B. M. Altura. Baltimore: University Park Press, 1977, vol. 1, p. 253–254.

FUNG, Y. C. Rheology of blood in microvessels. In: *Microcirculation*, edited by G. Kaley and B. Altura. Baltimore: University Park Press, 1977, vol. 1, p. 279–298.

FURCHGOTT, R. F. Role of endothelium in responses of vascular smooth muscle. *Circ. Res.* 53: 557–573, 1983.

FURCHGOTT, R. F., AND Z. U. ZAWADZKI. The obligatory role of endothelial cells in the relaxation of arterial smooth muscle by acetylchaline. *Nature* 288: 377–376, 1980.

GABE, I. T., J. H. GAULT, J. ROSS, Jr, D. T. MASON, C. J. MILLS, J. P. SCHILLINGFORD, AND E. BRAUNWALD. Measurement of instantaneous blood flow velocity and pressure in conscious man with a catheter-tip velocity probe. *Circulation* 40: 603–614, 1969.

GAGE, J. E., O. M. HESS, T. MURAKAMI, M. RITTER, J. GRIMM., AND H. P. KRAYENBUEHL. Vasoconstrictions of stenotic coronary arteries during dynamic exercise in patients with classic angina pectoris: reversibility by nitroglycerin. *Circulation* 73: 865–876, 1986.

GALLAGHER, J. J., E. L. C. PRITCHETT, W. C. SEALY, J. KASELL, AND A. G. WALLACE. The pre-excitation syndromes. *Prog. Cardiovasc. Dis.* 20: 285–327, 1978.

GALLAGHER, K. P., G. OSAKADA, M., MATSUZAKI, W. S. KEMPER, AND J. ROSS, Jr. Myocardial blood flow and function with critical coronary stenosis in exercising dogs. *Am. J. Physiol.* 12: H698–H707, 1982.

GANZ, W., K. TAMURA, H. S. MARCUS, R. DONOSO, S. YOSHIDA, AND H. J. C. SWAN. Measurement of coronary sinus blood flow by continuous thermodilution in man. *Circulation* 44: 181–195, 1971.

GAUER, O. H., J. P. HENRY, AND C. BEHN. The regulation of extracellular fluid volume. *Ann. Rev.* 32: 547–595, 1970.

GAULT, J. H., J. ROSS, Jr., AND D. T. MASON. Patterns of brachial arterial blood flow in conscious human subjects with and without cardiac dysfunction. *Circulation* 34: 833–848, 1966.

GEDDES, L. A. *The Direct and Indirect Measurement of Blood Pressure.* Chicago: Year Book, 1970.

GEEST, H. De., M. N. LEVY, AND H. ZIESKE. Reflex effects of cephalic hypoxia, hypercapnia, and ischemia upon ventricular contractility. *Circ. Res.* 17: 349–357, 1965.

GEVERS, W. The mechanism of myocardial contraction. In: *The Heart;* edited by L. Opie, New York: Grune & Stratton, 1986, p. 98–107.

GIBBS, C. L. AND J. B. CHAPMAN. Cardiac energetics. In: *Handbook of Physiology*, Sect. 2, edited by R. M. Berne et al. Washington DC: American Physiological Society, 1979, vol 1, p. 775–804.

GIBBS, C. L., AND W. R. GIBSON. Effect of Quabain on the energy input of rabbit cardiac muscle. *Circ. Res.* 24: 951–967, 1969.

GIRLING, F. Vasomotor effects of electrical stimulation. *Am. J. Physiol.* 170: 131–135, 1952.

GLANTZ, S. A., AND W. W. PARMLEY. Factors which affect the diastolic pressure-volume curve. *Circ. Res.* 42: 171–180, 1978.

GLOWER, D. D., J. A. SPRATT, N. D. SNOW, J. S. KABAS, J. W. DAVIS, C. O. OLSEN, G. S. TYSON, D. C. SABISTON, AND J. S. RANKIN. Linearity of the Frank-Starling relationship in the intact heart: the concept of pre-load recruitable stroke work. *Circulation* 71: 994–1009, 1985.

GOLDBERGER, A. L., AND E. GOLDBERGER. *Clinical Electrocardiography: A Simplified Approach*, St. Louis: C. V. Mosby, 1986.

GOETZ, K. L., G. C. BOND, AND D. D. BLOXHAM. Atrial receptors and renal function. *Physiol. Rev.* 55: 157–205, 1975.

GOLLNICK, P. D., R. B. ARMSTRONG, C. W. SAUBERTIV, K. PIEHL, AND B. SALTIN. Enzyme activity and fiber composition in skeletal muscle of untrained and trained men. *J. Appl. Physiol.* 33: 312–319, 1972.

GOODMAN, A. H., A. C. GUYTON, R. DRAKE, AND J. H. LOFLIN. A television method for measuring capillary red cell velocities. *J. Appl. Physiol.* 37: 126–130, 1974.

GRAHAM, T. P., Jr., J. ROSS, Jr., AND J. W. COVELL. Myocardial oxygen consumption in acute experimental cardiac depression. *Circ. Res.* 21: 123–138, 1967.

GRANDE, P. O., P. BORGSTROM, AND S. MELLANDER. On the nature of basal vascular tone in cat skeletal muscle and its dependence on transmural pressure stimuli. *Acta Physiol. Scand.* 107: 365–376, 1979.

GRANGER, H., G. A. MEININGER, J. L. BORDERS, R. J. MORFF, AND A. H. GOODMAN. Microcirculation of skeletal muscle. *Phys. Pharm. Microcirc.* 2: 181–265, 1984.

GREGG, D. E., E. M. KHOURI, AND C. R. RAYFORD. Systemic and coronary energetics in the resting unanesthetized dog. *Circ. Res.* 16: 102–113, 1965.

GROSSMAN, W., AND L. P. McLAURIN. Diastolic properties of the left ventricle. *Ann. Int. Med.* 84: 316–326, 1976.

GROSSMAN, W. (ed.). *Cardiac Catheterization and Angiography*, 3rd ed. Philadelphia: Lea & Febiger, 1986.

GUO, G. B., AND M. D. THAMES. Abnormal baroreflex control in renal hypertension is due to abnormal baroreceptors. *Am. J. Physiol.* 245(3) H420–H428, 1983.

GUYTON, A. C. Determination of cardiac output by equating venous return curves with cardiac response curves. *Physiol. Rev.* 35: 123–129, 1955.

GUYTON, A. C., A. W. LINDSEY, J. B. ABERNATHY, AND T. RICHARDSON. Venous return at various right atrial pressures and the normal venous return curve. *Am. J. Physiol.* 189: 609–615, 1957.

GUYTON, A. C., B. ABERNATHY, J. B. LANGSTON, B. N. KAUFMAN, AND H. M. FAIRCHILD. Relative importance of venous and arterial resistance in controlling venous return and cardiac output. *Am. J. Physiol.* 196: 1008–1014, 1959.

GUYTON, A. C. A concept of negative interstitial pressure based on pressures in implanted perforated capsules. *Circ. Res.* 12: 399–414, 1963.

GUYTON, A. C., T. G. COLEMAN, A. W. COWLEY Jr., K. W. SCHELL, R. D. MANNING Jr. AND R. A. NORMAN Jr. Arterial pressure regulation: overriding dominance of the kidneys in long-term regulation and hypertension. *Am. J. Med.* 52: 584, 1972.

GUYTON, A. C., AND C. E. JONES. Central venous pressure: physiological significance and clinical implications. *Am. Heart J.* 86: 431–437, 1973.

GUYTON, A. C., C. E. JONES, AND T. G. COLEMAN. *Circulation Physiology: Cardiac Output and its Regulation*, 2nd ed. Philadelphia: W. B. Saunders, 1973.

GUYTON, A. C., T. G. COLEMAN, A. W. COWLEY, K. W. SCHEEL, R. D. MANNING, AND R. A. NORMAN. Arterial pressure regulation overriding dominance of the kidneys in long-term regulation and in hypertension. In: *Hypertension Manual*, edited by J. H. Laragh. New York: Yorke Medical Books, 1974, p. 111–134.

HADDY, F. J., AND J. B. SCOTT. Active hyperemia, reactive hyperemia and autoregulation of blood flow. In: *Microcirculation*, edited by G. Kaley and B. M. Altura. Baltimore: University Park Press, 1977, vol. 2.

HALES, S. Statical essays. *Haemastatics*, vol. 2. London: Innys and Manby, 1733.

HAMILL, O. P., A. MARTZ, E. NEHER, B. SAKMANN, AND F. J. SIGWORTH. Improved patchclamp techniques for high-resolution current recording from cells and cell free membrane patches. *Pfluegers Arch.* 391: 85–100, 1981.

HAMMOND, H. L., F. C. WHITE, L. L.

BRUNTON, AND J. C. LONGHURST. Association of decreased myocardial β-receptors and chronotropic response to isoproterenol and exercise in pigs following chronic dynamic exercise. *Circ. Res.* 60: 720–726, 1987.

HANSON, J., AND J. LOWY. Molecular basis of contractility in muscle. *Br. Med. Bull.* 21: 264–271, 1965.

HARLAN, J. C., E. E. SMITH, AND T. Q. RICHARDSON. Pressure-volume curves of systemic and pulmonary circuit. *Am. J. Physiol.* 213: 1499–1503, 1967.

HARRIS, P., AND L. H. OPIE (eds.) *Calcium and the Heart*. New York: Academic Press, 1971.

HARVEY, W. *Exercitatio Anatomica de Moto Cordis et Sanguinis in Animalibus* (translated by C. D. Leake). Springfield, IL: Charles C. Thomas, 1928.

HECHT, H. H. Electrical disorders of cardiac fibers. In: *Cardiovascular Disorders*, edited by A. N. Brest and S. J. Moyer. Philadelphia: F. A. Davis, 1968, p. 79–93.

HELMER, O. M. Renin activity in blood from patients with hypertension. *Can Med. Assoc. J.* 90: 221–225, 1964.

HELVIG, E. B., AND F. K. MOSTOFI (eds.). *The Skin*. New York: R. E. Kreiger, 1980.

HERING, H. E. *Die Karotidssinus Reflexe auf Herz und Fegasse*. Leipzig: D. Steinkopff, 1927.

HESS, D. S., AND R. J. BACHE. Transmural distribution of myocardial blood flow during systole in the awake dog. *Circ. Res.* 38: 5–15, 1976.

HEYMANS, C., AND E. NEIL. *Reflexogenic Areas of the Cardiovascular System*. London: Churchill, 1958.

HEYMANS, C. J., J. J. BOUCKAERT, AND P. REGNIERS. *Le Sinus Carotidien*. Paris: Doin, 1973.

HEYMANS, C., J. J. BOUCKAERT, AND L. DAUTREBANDE. Au sujet du mechanisme de la bradycardie provoquee par la nicotine la lobeline le cyanure le sulfure de sodium, les nitrites et al morphine et de la bradycardie asphyxique. *Arch. Intern. Pharmcodyn.* 41: 261–289, 1931.

HEYNDRICKX, G. R., J.-P. VILAINE, D. R. KNIGHT, AND S. F. VATNER. Effects of altered site of electrical activation on myocardial performance during inotropic stimulation. *Circulation* 71: 1010–1016, 1985.

HIGGINBOTHAM, M. B., K. G. MORRIS, R. S. WILLIAMS, P. A. McHALE, R. E. COLEMAN AND F. R. COBB. Regulation of stroke volume during submaximal and maximal uptight exercise in normal man. *Circ. Res.* 58: 281–291, 1986.

HIGGINS, C. B., S. E. VATNER, E. L. ECKBERG, AND E. BRAUNWALD. Alterations in the baroreceptor reflex in conscious dogs with heart failure. *J. Clin. Invest.* 51: 715–724, 1972.

HILL, A. V. Heat of shortening and dynamic constants of muscle. *Proc. R. Soc. London* (Biol.), Ser. B. 126: 136–195, 1939.

HILTON, S. M. AND K. M. SPYER. Central nervous regulation of vascular resistance. *Ann. Rev. Physiol.* 42: 399–411, 1980.

HINSHAW, L. B. Arterial and venous pressure-resistance relationships in perfused leg and intestine. *Am. J. Physiol.* 203: 271–274, 1962.

HODGKIN, A. L. The ionic basis of nervous conduction. *Science* 145: 1148–1154, 1964.

HOFFMAN, B. F., E. BINDLER, AND E. E. SUCKLING. Postextrasystolic potentiation of contraction in cardiac muscle. *Am. J. Physiol.* 185: 95–102, 1956.

HOFFMAN, B. F., P. F. CRANEFIELD, AND A. G. WALLACE. Physiologic basis of cardiac arrhythmias. *Mod. Concepts Cardiovas. Dis.* 35: 103–108, 1966.

HOFFMAN, B. F. Effects of digitalis on electrical activity of cardiac fibers. In: *Digitalis: Mechanisms of the Inotropic Effect of Digitalis*, edited by C. Fisch and B. Surawicz. New York: Grune & Stratton, 1969, p. 93–109.

HOFFMAN, J. T. E., AND G. D. BUCKBERG. The myocardial supply: demand ratio—a critical review. *Am. J. Cardiol.* 41: 327–332, 1978.

HOLT, J. P., E. A. RHODE, AND H. KINES. Ventricular volumes and body weight in mammals. *Am. J. Physiol.* 215: 704–715, 1968.

HONIG, C. R., AND J. BOURDEAU-MARTINI. Role of O_2 in control of the coronary capillary reserve. In: *Current Topics in Coronary Research*, edited by C. M. Bloor and R. A. Olsson. New York: Plenum Press, 1973, p. 55–71.

HORI, M., E. L. YELLIN, AND E. H. SONNENBLICK. Left ventricular diastolic suction as a mechanism of ventricular filling. *Circulation Journal*, 46: 124–129, 1982.

HUXLEY, A. F. Excitation and conduction in nerve: quantitative analysis. *Science* 145: 1154–1159, 1964.

HUXLEY, A. F. Local activation of muscle. *N.Y. Ann. Acad. Sci.* 81: 446–452, 1959.

HUXLEY, A. F. Muscular contraction. *J. Physiol. (London)* 243: 1–43, 1974.

IGNARRO, L. J. Endothelium-derived nitric oxide: action and properties. *FASEB J.* 3: 31–36, 1989.

IGNARRO, L. J., G. M. BUGA, K. S. WOOD, R. E. BYRUS, AND G. CHANDHURI. Endothelium derived relaxing factor produced and released from artery and vein in nitric oxide. *Proc. Natl. Acad. Sci. U.S.A.* 84: 9265–9269, 1987.

ISHIDA, Y., J. S. MEISNER, K. TSUJIOKA, J. I. GALLO, C. YORAN, W. M. FRATER, AND E. L. YELLIN. Left ventricular filling dynamics: influence of left ventricular relaxation and left atrial pressure. *Circulation* 74: 187–196, 1986.

JAMES, T. Morphology of the human atrioventricular node with remarks pertinent to its electrophysiology. *Am. Heart J.* 62: 756–771, 1961.

JAMES, T. N. Anatomy of the conduction system of the heart. In: *The Heart*, 2nd ed., edited by J. W. Hurst, and R. B. Logue, New York: McGraw-Hill, 1970.

JENSEN, C. R., AND G. FISHER. *Scientific Basis of Athletic Conditioning*. Philadelphia: Lea & Febiger, 1978.

JEWELL, B. R. A re-examination of the influence of muscle length on myocardial performance. *Circ. Res.* 40: 221–230, 1977.

JOHANNSON, B. Vascular smooth muscle reactivity. *Ann. Rev. Physiol.* 43: 355–370, 1981.

JOHNSON, E. A. Force-interval relationship of cardiac muscle. In: *Handbook of Physiology*, Sect. 2, edited by R. M. Berne, et al. Washington DC: American Physiological Society, 1979, vol 1, p. 475–496.

JOHNSON, P. C. Review of previous studies and current theories of autoregulation. *Circ. Res.* 15(suppl 2): 2–9, 1964.

JOHNSON, P. C. The role of intravascular pressure in regulation of the microcirculation. In: *Advances in Physiological Science. Vol 7: Cardiovascular Physiology, Microcirculation and Capillary Exchange*, edited by A. G. B. Kovach, J. Hamar, L. Szabo. Budapest: Pergamon Press, 1980, p. 17–34.

JOHNSON, P. C., AND V. SMIESKO. Flow-dependent and propagated dilation in arcade microvessels of the rat mesentery. *FASEB J* 3: A-943, 1988.

JONES, D. R., AND W. J. PURVES. The carotid body in the duck and the consequences of its denervation upon the cardiac responses to immersion. *J. Physiol. (London)* 211: 279–294, 1970.

JOSEPHSON, M. E., L. N. HOROWITZ, A. FARSHIDI, AND J. A. KASTOR. Recurrent sustained ventricular tacycardia. *Circulation* 57: 431–440, 1978.

JOSEPHSON, M. E., AND J. A. KASTOR. Supraventricular tachycardia: Mechanisms and management. *Ann. Int. Med.* 87: 346–358, 1977.

JOSEPHSON, M. E., AND S. F. SEIDES. *Clinical Cardiac Electrophysiology Techniques and Interpretations*. Philadelphia: Lea & Febiger, 1979.

JOYNER, W. L., AND T. J. DAVIS. Pressure profile along the microvascular network at its control. *Fed. Proc.* 46: 266–269, 1987.

JULIAN, F. J., AND M. R. SOLLING. Sarcomere length-tension relations in living rat papillary muscle. *Circ. Res.* 37: 299–308. 1975.

KALEY, G. Control of coronary circulation and myocardial function by eicosanoids. *Federation Proc.* 46: 46, 1987.

KAPLAN, N. M. Systemic hypertension: mechanisms and diagnosis. In: *Heart Disease*, edited by E. Braunwald. Philadelphia: W. B. Saunders Co., 1988, p. 879–861.

KARNOVSKY, M. J. The ultrastructural basis of transcapillary exchanges. *J. Gen. Physiol.* 52: 64s–95s, 1968.

KATZ, A. M. Contractile proteins of the heart. *Physiol. Rev.* 50: 63–158, 1970.

KATZ, A. M. Congestive heart failure: role of altered myocardial cellular control. *N. Engl. J. Med.* 293: 1184–1191, 1976.

KATZ, A. M. *Physiology of the Heart.* New York: Raven Press, 1977, p. 73–160.

KAUFMANN, R., T. BAYER, T. FURNISS, H. KRAUSE, AND H. TRITTHART. Calcium-movement controlling cardiac contractility. II. Analog computation of cardiac excitation-contraction coupling on the basis of calcium kinetics in a multicompartment model. *J. Mol. Cell. Cardiol.* 6: 543–559, 1974.

KENTISH, J. C., H. E. D. J. TERKEORS, M. I. M. NOBLE, L. RICCIARDI, AND V. J. A. SCHOOTEN. The relationships between force (Ca²⁺) and sarcomere length in skinned trabeculae from rat ventricle. *J. Physiol.* 251: 627–643, 1983.

KETY, S. Theory of blood-tissue exchange and its application to measurement of blood flow. *Methods Med. Res.* 8: 223–227, 1960.

KEZDI, P. Resetting of carotid sinus in experimental renal hypertension. In: *Baroreceptors and Hypertension,* edited by P. Kezdi, New York: Pergamon Press, 1967, p. 301–308.

KHOURI, E. M., D. E. GREGG, AND C. R. RAYFORD. Effect of exercise on cardiac output, left coronary flow and myocardial metabolism in the unanesthetized dog. *Circ. Res.* 17: 427–437, 1965.

KIRK, E. S., AND C. R. HONIG. An experimental and theoretical analysis of myocardial tissue pressure. *Am. J. Physiol.* 125: 361–367, 1964.

KISCH, B. Electron microscopy of the atrium of the heart. I. Guinea pig. *Exp. Med. Surg.* 14: 99–112, 1956.

KLEINERT, H. D., M. VOLPE, G. ODELL, D. MARION, S. A. ATLAS, M. J. F. CAMARGO, J. H. LARAGH, AND T. MAACK. Cardiovascular effects of atrial natriuretic factor in anaesthetized and conscious dogs. *Hypertension* 8: 312–316, 1986.

KLITZMAN, B., AND P. C. JOHNSON. Capillary network geometry and red cell distribution in hamster cremaster muscle. *Am. J. Physiol.* 11(2): H211–H219, 1982.

KLOCKE, F. J., G. A. KAISER, J. ROSS, Jr., AND E. BRAUNWALD. An intrinsic adrenergic vasodilator mechanism in the coronary vascular bed of the dog. *Circ. Res.* 16: 376–382, 1965.

KLOCKE, F. J., E. BRAUNWALD, AND J. ROSS Jr. Oxygen cost of electrical activation of the heart. *Circ. Res.* 18: 357–365, 1966.

KLOCKE, F. J., D. R. ROSING, AND D. E. PITTMAN. Inert gas measurements of coronary blood flow. *Am. J. Cardiol.* 23: 548–555, 1969.

KLOCKE, F. J., AND A. K. ELLIS. Control of coronary blood flow. *Annu. Rev. Med.* 31: 489–508, 1980.

KORNER, P. I. Control of blood flow to special vascular areas: brain, kidney, muscle, skin, liver, and intestine. In: *Cardiovascular Physiology,* edited by A. C. Guyton and C. E. Jones. Baltimore: University Park Press, 1974, p. 123–162.

KORNER, P. I. Central nervous control of autonomic cardiovascular function. In *Handbook of Physiology, Section 2, Vol. 1, The Cardiovascular System.* Washington, DC: American Physiologic Society, 1979, p. 691–739.

KRAYER, O. The history of the Bezold-Jarisch effects. *Arch. Exp. Pathol. Pharmakol.* 240: 361–368, 1961.

KROGH, A. The number and distribution of capillaries in muscles with calculations of the oxygen pressure head necessary for supplying the tissue. *J. Physiol. (London)* 52: 409–415, 1919.

KROGH, A. *The Anatomy and Physiology of the Capillaries,* reprint edition. New York: Habner, 1959.

KUNZE, D. L., AND A. M. BROWN. Sodium sensitivity of baroreceptors: reflex effects on blood pressure and fluid volumes in the cat. *Circ. Res.* 42: 714–720, 1978.

LaKATTA, E. G. Starling's law of the heart is explained by an intimate interaction of muscle length and myofilament calcium activation. *J. Am. Coll. Cardiol.* 10: 1157–1164, 1987.

LANDIS, E. M. The capillary circulation. In: *Circulation of Blood: Men and Ideas,* edited by A. P. Fishman and D. W. Richards, New York: Oxford University Press, 1964.

LANDIS, E. M., AND J. H. GIBBON, Jr. Effects of temperature and of tissue pressure on movement of fluid through human capillary wall. *J. Clin. Invest.* 12: 105–138, 1933.

LANDIS, E. M., AND J. R. PAPPENHEIMER. Exchange of substances through the capillary walls. In: *Handbook of Physiology,* Sect. 2: Circulation. Washington DC: American Physiological Society, 1963, vol 2, p. 961–1034.

LANGER, G. A., AND A. J. BRADY. *The Mammalian Myocardium.* New York: John Wiley & Sons, 1974.

LANGER, G. A. The intrinsic control of myocardial contraction-ionic factors. *N. Engl. J. Med.* 285: 1065–1071, 1971.

LANGER, G. A. The structure and function of the myocardial cell surface. *Am. J. Physiol.* 235: H461–H468, 1978.

LARAGH, J. H. The biochemical regulation of blood pressure: pathogenetic and clinical implications. In: *Biochemical Regulation of Blood Pressure,* edited by R. L. Soffer. New York: John Wiley & Sons, 1981, p. 393–410.

LARAGH, J. H. The endocrine control of blood volume, blood pressure and sodium balance: atrial hormone and renin system interactions. *J. Hypertension* 4(2): S143–S156, 1986.

LEAK, L. V., AND J. F. BURKE. Ultrastructural studies on the lymphatic anchoring filaments. *J. Cell Biol.* 36: 129–149, 1968.

LEAK, L. V. The structure of lymphatic capillaries in lymph formation. *Fed. Proc.* 35: 1863–1871, 1976.

LEE, J-D., T. TAJIMI, J. PATRITTI, AND J. ROSS, Jr. Preload reserve and mechanisms of afterload mismatch in normal conscious dog. *Am. J. Physiol.* 250: H464–H473, 1986.

LEFER, A. M., AND J. MARTIN. Relationship of plasma peptides to the myocardial depressant factor in hemorrhagic shock in cats. *Circ. Res.* 26: 59–69, 1970.

LEGATO, M. J. The correlation of ultrastructure and function in the mammalian myocardial cell. *Prog. Cardiovasc. Dis.* 11: 391–409, 1969.

LEGATO, M. J. *The Myocardial Cell for the Clinical Cardiologist.* Mount Kisco: Futura Publishing Co., 1973.

LEVY, M. N., N. G. MIL, AND H. ZIESKE. Functional distribution of the peripheral cardiac sympathetic pathways. *Circ. Res.* 19: 650–661, 1966.

LEVY, M. N., AND H. ZIESKE. Autonomic control of cardiac pacemaker activity and atrioventricular transmission. *J. Appl. Physiol.* 27: 465–470, 1969.

LEVY, M. N., T. IANO, AND H. ZIESKE. Effects of repetitive bursts of vagal activity on heart rate. *Circ. Res.* 30: 286–295, 1972.

LEVY, M. N. The cardiac and vascular factors that determine systemic blood flow. *Circ. Res.* 44: 739–746, 1979.

LEVY, M. N., AND P. MARTIN. Parasympathetic control of the heart. In: *Nervous Control of Cardiovascular Function,* edited by Walter C. Randall. New York: Oxford University Press, 1984, pp. 68–94.

LeWINTER, M. M., R. ENGLER, AND R. S. PAVELEC. Time-dependent shifts of the left ventricular diastolic filling relationship in conscious dogs. *Circ. Res.* 45: 641–653, 1979.

LIPMAN, B. S., AND E. MASSIE. *Clinical Scalar Electrocardiography,* 6th ed. Chicago: Year Book Medical Publications, 1972.

LIPOWSKY, H. H., S. USAMI, AND S. CHICU. In vivo measurements of "apparent viscosity" and microvessel hematocrit in the mesentery of the cat. *Microvasc. Res.* 19: 297–319, 1980.

LONGINI, I. M., M. W. HIGGINS, P. C. HINTON, P. P. MOLL, AND J. B. KELLER. Environmental and genetic sources of familial aggregation of blood pressure in Tecumseh, Michigan. *Am. J. Epidemiol.* 120: 131–144, 1984.

LUND-JOHANSEN, P. Hemodynamic alterations in hypertension—spontaneous changes and effect of drug therapy. *Acta Med. Scand.* (Suppl.) 603: 1, 1977.

LUNDGREN, O., AND M. JODAL. Regional blood flow. *Annu. Rev. Physiol.* 37: 395–414, 1975.

MacGREGOR, D. C., J. W. COVELL, F. MAHLER, R. B. DILLY, AND J. ROSS Jr. Relations between afterload, stroke volume and descending limb of Starling's curve. *Am. J. Physiol.* 227: 884–890, 1974.

MacLEOD, R. D. M., AND M. J. SCOTT. The heart rate responses to carotid body chemoreceptor stimulation in the cat. *J. Physiol. (London)* 175: 193–202, 1964.

McDONALD, D. A. *Blood Flow in Arteries.* Baltimore: Williams & Wilkins, 1974.

McNUTT, N. S., AND D. W. FAWCETT. Myocardial ultrastructure. In: *The Mammalian Myocardium*, edited by G. A. Langer and A. J. Brady. New York: John Wiley & Sons, 1974, p. 1–49.

McNUTT, N. S., AND R. S. WEINSTEIN. Membrane ultrastructure of mammalian intercellular functions. *Prog. Biophys. Mol. Biol.* 2: 45, 1973.

MAHLER, F., J. W. COVELL, AND J. ROSS, Jr. Systolic pressure-diameter relations in the normal conscious dog. *Cardiovasc. Res.* 9: 447–455, 1975.

MARK, A. L. Sensitization of cardiac vagal afferent reflexes at the sensory receptor level: an overview. *Fed. Proc.* 46(1) 36–40, 1987.

MARRIOTT, H. J. L., AND R. J. MYERBURG. Recognition of arrhythmias and conduction disturbances. In: *The Heart*, 6th ed., edited by J. W. Hurst. New York: McGraw-Hill, 1986, p. 431–475.

MARTINI, J., AND G. R. HONIG. Direct measurement of intercapillary distance in beating rat heart in situ under various conditions of O_2 supply. *Microvas. Res.* 1: 244–256, 1969.

MASERI, A., G. A. KLASSEN, AND W. LESCH, (eds.). *Primary and Secondary Angina Pectoris.* New York: Grune & Stratton, 1978.

MASON, D. T., I. T. GABE, J. H. GAULT, E. BRAUNWALD, J. ROSS, Jr., AND J. P. SHILLINGFORD. Applications of the catheter-tip electromagnetic velocity probe in the study of the central circulation in man. *Am. J. Med.* 49: 465–471, 1970.

MEHMEL, H. C., B. STOCKINS, K. RUFFMANN, K. von OLSHAUSEN, G. SCHULER, AND W. KUBLER. The linearity of the end-systolic pressure-volume relationship in man and its sensitivity for assessment of left ventricular function. *Circulation* 63: 1216–1222, 1981.

MEININGER, G. A., C. A. MACK, K. IEHU, AND H. G. BOHLEN. Myogenic vasoregulation overrides local metabolic control in resting rat skeletal muscle. *Circ. Res.* 60: 861–820, 1987.

MELLANDER, S., AND B. JOHANSSON. Control of resistance, exchanges and capacitance functions in the peripheral circulation. *Pharmacol. Rev.* 20: 117–196, 1968.

MILHORN, H. T. *Application of Control Theory to Physiological Systems.* Philadelphia: W. B. Saunders, 1966.

MILNOR, W. R. Pulmonary circulation. In: *Medical Physiology*, edited by V. B. Mountcastle. St. Louis, C. V. Mosby, 1968, vol. 1, p. 24–34, 209–220.

MILNOR, W. R. Arterial impedance as ventricular afterload. *Circ. Res.* 36: 565–570, 1975.

MINES, G. R. On circulating excitations in heart muscle and their possible relation to tachycardia and fibrillation. *Trans. Royal Soc. Canada (Biol)* 8: 43, 1914.

MIRSKY, I. Basic terminology and formulae for left ventricular wall stress. In: *Cardiac Mechanics: Physiological, Clinical and Mathematical Considerations*, edited by I. Mirsky, et al. New York: John Wiley & Sons, 1974, p. 3–10.

MITCHELL, J. H., AND W. SHAPIRO. Atrial function and the hemodynamic consequences of atrial fibrillation in man. *Am. J. Cardiol.* 23: 556–557, 1969.

MITCHELL, J. H., J. P. GILMORE, AND S. J. SARNOFF. The transport function of the atrium. Factors influencing the relation between mean left atrial pressure and left ventricular end-diastolic pressure. *Am. J. Cardiol.* 9:237–247, 1962.

MONROE, R. G., W. J. GAMBLE, C. G. LaFARGE, AND S. F. VATNER. Hemeometric autoregulation. In: *The Physiological Basis of Starling's Law of the Heart*, CIBA Foundation Symposium No. 27. Amsterdam: Elsevier/North Holland Biomedical Press, 1974, p. 257–277.

MORGAN, H. E., AND J. R. NEELY. Metabolic regulation and myocardial function. In: *The Heart*, edited by J. W. Hurst. New York: McGraw-Hill Book Co., 1982, p. 128–142.

MORGAN, J. P., AND K. G. MORGAN. Calcium and cardiovascular function. Intracellular calcium levels during contraction and relaxation of mammalian cardiac and vascular smooth muscle as detected with aequorin. *Am. J. Med.* November 5: 33–46, 1984.

MORGAN, J. P., J. K. GWATHMEY, T. T. DeFEO, AND K. G. MORGAN. The effects of amrinone and related drugs on the intracellular calcium in isolated mammalian cardiac and vascular smooth muscle. *Circulation* 73 (suppl. III), III 65–76, 1986.

MOSHER, P., J. ROSS, Jr., P. A. McFATE, AND R. F. SHAW. Control of coronary blood flow of an autoregulatory mechanism. *Circ. Res.* 14: 250–259, 1964.

MULLANE, K. M., AND A. PINTO. Endothelium, arachidonic acid, and coronary vascular tone. *Fed. Proc.* 46: 59–62, 1987.

MURPHY, Q. The influence of the accelerator nerves on the basal heart rate of the dog. *Am. J. Physiol.* 137: 727–730, 1942.

MURRAY, P. A., AND S. F. VATNER. α-Adrenoreceptor attenuation of the coronary vascular response to severe exercise in the conscious dog. *Circ. Res.* 45: 654–660, 1969.

NEIL, E., AND A. HOWE. Arterial chemoreceptors. In: *Handbook of Sensory Physiology*, edited by E. Neil, Vol. 3: *Enteroceptors*. Berlin and New York: Springer-Verlag, 1973, p. 47–80.

NICOLL, P. A., AND T. A. CORTESE, Jr. The physiology of skin. *Annu. Rev. Physiol.* 34: 177–203, 1972.

NOBLE, D. The surprising heart. A review of recent progress in cardiac electrophysiology. *J. Physiol.* 353: 1–50, 1984.

NOBLE, M. M., AND G. H. POLLACK. Controversies in cardiovascular research: molecular mechanisms of contraction. *Circ. Res.* 40: 333–342, 1977.

NOBLE, D. Application of Hodgkin-Huxley equations to excitable tissues. *Physiol. Rev.* 46: 1–50, 1966.

NOORDERGRAAF, A. *Circulatory System Dynamics.* New York: Academic Press, 1978.

NOZAWA, T., Y. YASAMURA, S. FUTAKI, N. TANAKA, Y. IGARASHI, Y. GOTO, AND H. SUSA. Relation between oxygen consumption and pressure-volume area of in situ dog heart. *Am. J. Physiol.* 253: H31–H40, 1987.

OBERG, B. Overall cardiovascular regulation. *Ann. Rev. Physiol.* 38: 537–570, 1976.

OLSSON, R. A. Local factors regulating cardiac and skeletal muscle blood flow. *Ann. Rev. Physiol.* 43: 385–395, 1981.

PATTERSON, S. W., H. PIPER, AND E. H. STARLING. The regulation of the heartbeat. *J. Physiol. (London)*, 48: 465–513, 1914.

PELLETIER, C. L. Circulatory responses to graded stimulation of carotid chemoreceptors in the dog. *Circ. Res.* 31: 431–443, 1972.

PEPINE, C. J., W. W. NICHOLS, AND C. R. CONTI. Aortic input impedance in heart failure. *Circulation* 58: 460–465, 1978.

PETERSON, D. F., AND A. M. BROWN. Pressor reflexes produced by stimulation of afferent fibers in the cardiac sympathetic nerves of the cat. *Circ. Res.* 28: 605–610, 1971.

PLOTNICK, G. D., L. C. BECKER, M. L. FISHER, G. GERSTENBLITH, D. G. RENLUND, J. L. FLEG, M. L. WEISFELDT, AND E. G. LAKATTA. Use of the Frank-Starling mechanism during submaximal versus maximal upright exercise. *Am. J. Physiol.* 251: H1101–H1105, 1986.

POISEUILLE, J. L. M. Recherches experimentals sur le mouvement des liquides dans les tubes de tres petits diametres, Paris: *Mem. Acad. Sci.* 9: 433–544, 1846.

POLME, R. M. J., R. G. FERRIGE, AND S. MONCADA. Nitric oxide release accounts for the biological activity of endothelium derived relaxing factor. *Nature* 327: 524–526, 1987.

POOL, P. E., AND E. H. SONNENBLICK. The mechanochemistry of cardiac muscle. *J. Gen. Physiol.* 50: 951–965, 1967.

POSTNOV, Y. V., AND S. N. ORLOV. Ion transport across plasma membrane in primary hypertension. *Physiol. Rev.* 65: 904–945, 1985.

POULEUR, H., J. W. COVELL, AND J. ROSS, Jr. Effects of alterations in aortic input impedance on the force-velocity-length relationships in the intact canine heart. *Circ. Res.* 45: 126–136, 1979.

POULEUR, H., J. W. COVELL, AND J. ROSS, Jr. Effects of nitroprusside on venous return and central blood volume in the absence and presence of acute heart failure. *Circulation,* 61: 328–337, 1980.

RACKLEY, C. E., V. S. BEHAR, R. E. WHALEN, AND H. D. McINTOSH. Biplane cineangiographic determinations of left ventricular function: pressure-volume relationships. *Am. Heart J.* 74: 766–799, 1967.

RAPELA, C. E., H. D. GREEN, AND A. B. DENNISON, Jr. Baroreceptor reflexes and autoregulation of cerebral blood flow in the dog. *Circ. Res.* 21: 559–568, 1967.

REEVES, J. P., AND J. L. SUTKO. Sodium-calcium exchange activity generates a current in cardiac membrane vescicles. *Science* 208: 1461, 1980.

RENKIN, E. M., AND C. C. MICHEL (eds.) Handbook of Physiology, Section 2, vol. IV, Parts 1 and 2, *Microcirculation.* American Physiological Society, Washington, D.C., 1984.

REUTER, H. Exchange of calcium ions in the mammalian myocardium: mechanisms and physiological significance. *Circ. Res.* 34: 599–605, 1974.

REUTER, H. Properties of two inward membrane currents in the heart. *Annu. Rev. Physiol.* 41: 413–414, 1979.

REUTER, H. Ion channel in cardiac cell membrane. *Annu. Rev. Physiol.* 46: 473, 1984.

REUTER, H. Modulation of ion channels by phosphorylation and second messengers. *News Physiol. Sci.* 2: 168–171, 1987.

ROBINSON, S. Physiology of muscular exercise. In: *Medical Physiology,* edited by V. Mountcastle. St. Louis: C. V. Mosby, 1974, vol. 2, p. 1273–1304.

RODDIE, I. C. Circulation to skin and adipose tissue. *Handbook of Physiology* 3: 285–317, 1983.

ROOKE, G. A., AND E. O. FEIGL. Work as a correlate of canine left ventricular oxygen consumption, and the problem of catecholamine oxygen wasting. *Circ. Res.* 50: 273–286, 1982.

ROSENBLEUTH, A., AND J. R. GARCIA. Studies on flutter and fibrillation. II. The influences of artificial obstacles on experimental auricular flutter. *Am. Heart J.* 33: 67, 1947.

ROSENBLUM, W. I., Viscosity: in vitro versus in vivo: In: *Microcirculation,* edited by G. Kaley and B. M. Altura. Baltimore: University Park Press, 1977, vol. 1, p. 325–334.

ROSS, J., Jr. Adaptations of the left ventricle to chronic volume overload. *Circ. Res..* 35 (Suppl II): 64–70, 1974.

ROSS, J., Jr. Afterload mismatch and preload reserve: a conceptual framework for the analysis of ventricular function. *Prog. Cardiovasc. Dis.* 18: 255–164, 1976b.

ROSS, J. Jr. Assessment of cardiac function and myocardial contractility. In: *The Heart,* 3rd ed., edited by J. W. Hurst. New York: McGraw-Hill Book Co., 1986, p. 265–298.

ROSS, J. Jr. Electrocardiographic ST-segment analysis in characterization of myocardial ischemia and infarction. *Circulation* 53(Suppl I): 73–81, 1976a.

ROSS, J. Jr. Mechanisms of cardiac contraction; what roles for preload, afterload and inotropic state in heart failure? *Europ. Heart J.* 4: 19–28, 1983.

ROSS, J., Jr. Mechanisms of regional ischemia and antianginal drug action during exercise. *Progr. in Cardiovasc. Dis.,* 31: 455–466, 1989.

ROSS, J., Jr., AND E. BRAUNWALD. The study of left ventricular function in man by increasing resistance to ventricular ejection with angiotensin. *Circulation* 29: 739–749, 1964.

ROSS, J., Jr., J. W. COVELL, E. H. SONNENBLICK, AND E. BRAUNWALD. Contractile state of heart characterized by force-velocity relations in variably afterloaded and isovolumic beats. *Circ. Res.* 18: 149–163, 1966a.

ROSS, J., Jr., C. J. FRAHM, AND E. BRAUNWALD. Influence of intracardiac baroreceptors on venous return, systemic vascular volume and peripheral resistance. *J. Clin. Invest.* 40: 563–572, 1961b.

ROSS, J., Jr., C. J. FRAHM, AND E. BRAUNWALD. The influence of the carotid baroreceptors and vasoactive drugs on systemic vascular volume and venous distensibility. *Circ. Res.* 9: 75–82, 1961a.

ROSS, J., Jr., J. H. GAULT, D. T. MASON, J. W. LINHART, AND E. BRAUNWALD. Left ventricular performance during muscular exercise in patients with and without cardiac dysfunction. *Circulation* 34: 597–608, 1966b.

ROSS, J., Jr., J. W. LINHART, AND E. BRAUNWALD. Effects of changing heart rate by electrical stimulation of the right atrium in man: studies at rest, during muscular exercise, and with isoproterenol. *Circulation* 32: 549–558, 1965a.

ROSS, J. Jr., AND R. A. O'ROURKE. *Understanding the Heart and its Diseases.* New York: McGraw-Hill Book Co., 1976c.

ROSS, J., Jr., E. H. SONNENBLICK, G. A. KAISER, P. L. FROMMER, AND E. BRAUNWALD. Electroaugmentation of ventricular performance and oxygen consumption by repetitive application of paired electrical stimuli. *Circ. Res.* 16: 332–342, 1965b.

ROSS, J., Jr., E. H. SONNENBLICK, R. R. TAYLOR, AND J. W. COVELL. Diastolic geometry and sarcomere lengths in the chronically dilated canine left ventricles. *Circ. Res.* 28: 49–61, 1971.

ROTHE, C. F. Venous System: Physiology of the Capacitance Vessels. In: *Handbook of Physiology, Section 2. The Cardiovascular System,* Vol. III. *Peripheral Circulation and Organ Blood Flow,* Part I. edited by J. T. Shepherd and F. M. Abboud. Bethesda, MD: American Physiological Society, 1983.

ROVETTO, M. J., J. J. WHITMER, AND J. R. NEELY. Comparison of the effects of anoxia and whole heart ischemia on carbohydrate utilization in isolated working rat hearts. *Circ. Res.* 32: 699, 1973.

SABISTON, D. C., AND D. E. GREGG. Effect of cardiac contraction on coronary blood flow. *Circulation* 15: 14–20, 1957.

SAGAWA, K. Analysis of the ventricular pumping capacity as a function of input and output pressure loads. In: *Physical Bases of Circulatory Transport: Regulation and Exchange,* edited by E. B. Reeve and A. C. Guyton. Philadelphia: W. B. Saunders, 1967, p. 141–149.

SAGAWA, K. The ventricular pressure-volume diagram revisited. *Circ. Res.* 43: 677–687, 1978.

SAGAWA, K. The end-systolic pressure-volume relation of the ventricle: definition, modifications, and clinical use. *Circulation* 63: 1223–1227, 1981.

SAGAWA, K. Baroreflex control of systemic arterial pressure and vascular bed. In: *Handbook of Physilogy, Section 2: The Cardiovascular System.* Washington DC: American Physiological Society, 1983, p. 453–496.

SARNOFF, S. J., R. B. CASE, G. H. WELCH, E. BRAUNWALD, AND W. N. STAINSBY. Performance characteristics and oxygen debt in a nonfailing metabolically supported, isolated heart preparation. *Am. J. Physiol.* 192: 141–147, 1958a.

SARNOFF, S. J., E. BRAUNWALD, G. H. WELCH, Jr., R. B. CASE, W. N. STAINSBY, AND R. MACRUZ. Hemodynamic determinants of oxygen consumption of the heart with special reference to the tension-time index. *Am. J. Physiol.* 192: 148–156, 1958b.

SASAYAMA, S., J. ROSS, Jr., D. FRANKLIN, C. BLOOR, S. BISHOP, AND R. B. DILLEY. Adaptations of the left ventricle to chronic pressure overload. *Circ. Res.* 38: 172–178, 1976.

SASAYAMA, S., D. FRANKLIN, AND J. ROSS, Jr. Hyperfunction with normal inotropic state of the hypertrophied left ventricle. *Am. J. Physiol.* 232: H418–H425, 1977.

SCHAPER, W. *The Collateral Circulation*

of the Heart. New York: Elsevier, 1971, p. 176.

SCHEUER, J., AND N. BRACHFELD. Coronary insufficiency: relations between hemodynamic, electrical and biochemical parameters. *Circ. Res.* 18: 178–189, 1966.

SCHLANT, R. C., AND M. E. SILVERMAN. Anatomy of the heart. In: *The Heart*, 6th ed., edited by J. W. Hurst, New York: McGraw-Hill, 1986, p. 16–37.

SCHMID-SCHÖNBEIN, G. W. Microheology of erythrocytes, blood viscosity and the distribution of blood flow in the microcirculation. *Int. Rev. Physiol.* 9: 1–62, 1976.

SCHMID-SCHÖNBEIN, G. W., R. SKALAK, S. USAUII, AND S. CHIEN. Cell distribution in capillary networks. *Microvasc. Res.* 19: 18–44, 1980.

SCHMID-SCHÖNBEIN, G. W., T. C. SHALAK, E. T. ENGELSON, AND B. W. ZWEIFACH. Microvascular Network Anatomy in Rat Skeletal Muscle. In *Microvascular Networks: Experimental and Theoretical Studies*, edited by S. Popel, P. C. Johnson. Basel: Karger, 1985, pp. 38–51.

SCHMID-SCHÖNBEIN, G. W., AND H. MURAKAMI. Blood flow in contracting arterioles. *Int. J. Microcirc. Clin. Exp.* 4: 311–328, 1985.

SCHMID-SCHÖNBEIN, G. W., AND R. L. ENGLER. Granulocytes as active participants in acute myocardial ischemia and infarction. *Am. J. Cardiovasc. Path.* 1: 15–30, 1987.

SCHMID, P. G., D. L. LUND, AND R. ROSKOSKI, Jr. Efferent autonomic dysfunction in heart failure. In: *Disturbances in Neurogenic Control of the Circulation*, edited by D. J. Reis. Bethesda: American Physiological Society, 1981.

SCHOLZ, H., AND W. MEYER. Phosphodiesterase-inhibiting properties of newer inotropic agents. *Circ.* 73 (suppl. III): III 99–106, 1986.

SCHWARTZ, A. Active transport in mammalian myocardium. In: *The Mammalian Myocardium*, edited by G. A. Langer and A. J. Brady. New York: John Wiley & Sons, 1974, p. 81–104.

SCHWARTZ, A. Is the cell membrane Na$^+$, K$^+$-ATPase enzyme system the pharmacological receptor for digitalis? *Circ. Res.* 39: 2–7, 1976.

SCHWARTZ, C. J., N. T. WERTHESSEN, AND S. WOLF (eds.) *Structure and Function of the Circulation.* New York: Plenum Press, 1981, vol. 3.

SCHWARZ, F., J. SCHAPER, D. KITTSTEIN, W. FLAMENG, P. WALTER, AND W. SCHAPER. Reduced volume fraction of myofibrils in myocardium of patients with decompensated pressure overload. *Circulation* 63: 1299–1304, 1981.

SHABETAI, R. *The Pericardium.* New York: Grune & Stratton, 1981.

SHALAK, T. C., G. W. SCHMID-SCHÖNBEIN, AND B. W. ZWEIFACH. New morphological evidence for a mechanism of lymph formation in skeletal muscle. *Microvas. Res.* 28: 95–112, 1984.

SHEPHERD, J. T. *Physiology of the Circulation in Human Limbs in Health and Disease.* Philadelphia: W. B. Saunders, 1963.

SHEPHERD, J. T., AND P. M. VANHOUTTE. *Veins and Their Control.* London: W. B. Saunders, 1975.

SHEPHERD, J. T., AND P. M. VANHOUTTE. *The Human Cardiovascular System: Facts and Concepts.* New York: Raven Press, 1979.

SHEPHERD, J. T. Circulation to skeletal muscle. *Handbook of Physiol.* 3(1): 319–370, Washington, DC: American Physiological Society, 1983.

SHIRATO, K., R. SHABETAI, V. BHARGAVA, D. FRANKLIN, AND J. ROSS,Jr. Alteration of the left ventricular diastolic pressure-segment length relation produced by the pericardium: effects of cardiac distension and afterload reduction in conscious dogs. *Circulation* 57: 1191–1198, 1978.

SHOUKAS, A. A., AND K. SAGAWA. Control of total systemic vascular capacity by the carotid sinus baroreceptor reflex. *Circ. Res.* 33: 22–33, 1973.

SIEBENS, A. A., B. F. HOFFMAN, P. F. CRANEFIELD, AND C. M. BROOKS. Regulation of contractile force during ventricular arrhythmias. *Am. J. Physiol.* 197: 971–977, 1959.

SIMON, E., AND W. RIEDEL. Diversity of regional sympathetic outflow in integrative cardiovascular control. Patterns and Mechanisms. *Brain Res.* 87: 232–333, 1975.

SOMLYO, A. P., AND A. V. SOMLYO. Excitation and contraction in vascular smooth muscle. In: *Structure and Function of the Circulation*, edited by C. J. Schwartz and N. T. Werthessen. New York: Plenum Press, 1981, vol 3, p. 239–286.

SOMMER, J. R., AND E. A. JOHNSON. Ultrastructure of cardiac muscle. In: *Handbook of Physiology*, Sect. 2, edited by R. M. Berne, et al. Washington DC: American Physiological Society, 1979, vol 1, p. 113–186.

SONNENBLICK, E. H. Implications of muscle mechanics in the heart. *Fed. Proc.* 21: 975–990, 1962.

SONNENBLICK, E. H., AND S. E. DOWNING. Afterload as a primary determinant of ventricular performance. *Am. J. Physiol.* 204: 604–610, 1963.

SONNENBLICK E. H., J. ROSS, Jr., J. W. COVELL, G. A. KAISER, AND E. BRAUNWALD. Velocity of contraction as a determinant of myocardial oxygen consumption. *Am. J. Physiol.* 209: 919–927, 1965.

SONNENBLICK, E. H. The mechanics of myocardial contraction. In: *The Myocardial Cell*, edited by S. A. Briller and H. L. Conn, Jr. Philadelphia: Univ. of Pennsylvania Press, 1966, p. 173–250.

SONNENBLICK, E. H., W. H. PARMLEY, C. W. URSHEL, AND D. L. BRUTSAERT. Ventricular function: evaluation of myocardial contractility in health and disease. *Prog. Cardiovasc. Dis.* 12: 449–456, 1970.

SPANN, J. F., Jr., R. A. BUCCINO, E. H. SONNENBLICK, AND E. BRAUNWALD. Contractile state of cardiac muscle obtained from cats with experimentally produced ventricular hypertrophy and heart failure. *Circ. Res.* 21: 341–354, 1967.

SPERELAKIS, N. Origin of the cardiac resting potential. In: *Handbook of Physiology*, Sect. 2, edited by R. M. Berne, et al. Washington DC: American Physiological Society, 1979, vol 1, p. 187–267.

SPIRO, D., AND E. H. SONNENBLICK. Comparison of contractile process in heart and skeletal muscle. *Circ. Res.* 15(suppl. 2): 14–37, 1964.

SPOTNITZ, H. M., E. H. SONNENBLICK, AND D. SPIRO. Relation of ultrastructure to function in intact heart: sarcomere structure relative to pressure volume curves of intact left ventricles of dog and cat. *Circ. Res.* 18: 49–66, 1966.

STAINSBY, W. N. Autoregulation of blood flow in skeletal muscle during increased metabolic activity. *Am. J. Physiol.* 202: 273–276, 1962.

STAINSBY, W. N., AND E. M. REKIN. Autoregulation of blood flow in resting skeletal muscle. *Am. J. Physiol.* 207: 117–122, 1961.

STARLING, E. H. *The Linacre Lecture on the Law of the Heart* (Given at Cambridge, 1915). London: Longmans, Green, 1918.

STEGALL, H. F. Muscle pumping in the dependent leg. *Circ. Res.* 19: 180–190, 1966.

STONE, H. L., V. S. BISHOP, AND E. DONG, Jr. Ventricular function in cardiac-denervated and cardiac-sympathectomized conscious dogs. *Circ. Res.* 20: 587–593, 1967.

STONE, H. O., H. K. THOMPSON, AND K. SCHMIDT-NIELSEN. Influence of erythrocytes on blood viscosity. *Am. J. Physiol.* 214: 913–918, 1968.

STREETER, D. D., Jr. Gross morphology and fiber geometry of the heart. In: *Handbook of Physiology*, Sect. 2, edited by R. M. Berne, et al., Washington DC: American Physiological Society, 1979, vol. 1, p. 61–112.

STREETER, D. D. Jr., H. M. SPOTNITZ, D. P. PATEL, J. ROSS Jr., AND E. H. SONNENBLICK. Fiber orientation in the canine left ventricle during diastole and systole. *Circ. Res.* 24: 339–347, 1969.

SUBCOMMITTEE ON DEFINITION AND PREVALENCE OF THE 1984 JOINT NATIONAL COMMITTEE. Hypertension prevalence and the status of awareness, treatment and control in the United States. *Hypertension* 7: 457–468, 1985.

SUGA, H., K. SAGAWA, AND A. A. SHOU-

KAS. Load independence of the instantaneous pressure-volume ratio of the canine left ventricle and effects of epinephrine and heart rate on the ratio. *Circ. Res.* 32: 314–322, 1973.

SUGA, H. Total mechanical energy of a ventricular model and cardiac oxygen consumption. *Am. J. Physiol.* 237: H498–H505, 1979.

SUGA, H., R. HISANO, S. HIRATA, T. HAYASHI, AND I. NINOMIYA. Mechanism of higher oxygen consumption rate: Pressure-loaded vs. volume-loaded heart. *Am. J. Physiol.* 242: H942–H948, 1982.

TAYLOR, M. G. An introduction to some recent developments in arterial hemodynamics. *Austral. Ann. Med.* 15: 71–86, 1966.

TAYLOR, R. R., J. W. COVELL, AND J. ROSS, Jr. Volume-tension diagrams of ejecting and isovolumic contractions in left ventricle. *Am. J. Physiol.* 216: 1097–1102, 1969.

TENNANT, R., AND C. J. WIGGERS. The effect of coronary occlusion on myocardial contraction. *Am. J. Physiol.* 112: 351–361, 1935.

THAMES, M. D., H. S. KLOPFENSTEIN, F. M. ABBOUD, A. L. MARK, AND J. L. WALKER. Preferential distribution of inhibitory cardiac receptors with vagal afferents to the inferoposterior wall of the left ventricle activated during coronary occlusion in the dog. *Circ. Res* 43: 512–514, 1978.

THEROUX, P., D. FRANKLIN, J. ROSS, Jr., AND W. S. KEMPER. Regional myocardial function during acute coronary occlusion and its modification by pharmacologic agents. *Circ. Res.* 35: 896–908, 1974.

TRAUTWEIN, W. Membrane currents in cardiac muscle fibers. *Physiol. Rev.* 53: 793–835, 1973.

TSE, W. W. Adrenergic potentiation on spontaneous activity of canine paranodal fibers. *Am. J. Physiol.* 16: H415–421, 1984.

VANDER, A. J., J. H. SHERMAN, D. S. LUCIANO. *Human Physiology In the Mechanisms of Body Function.* New York: McGraw-Hill, 1975.

VASSALLE, M. Cardiac automaticity and its control. *Am. J. Physiol.* 233: H625–H634, 1977.

VATNER, S. F., D. FRANKLIN, R. L. VAN CITTERS, AND E. BRAUNWALD. Effects of carotid sinus nerve stimulation on the coronary circulation of the conscious dog. *Circ. Res.* 27: 11–21, 1970.

VATNER, S. F., D. FRANKLIN, C. B. HIGGINS, T. PATRICK, AND E. BRAUNWALD. Left ventricular response to severe exertion in untethered dogs. *J. Clin. Invest.* 51: 3052–3060, 1972.

VATNER, S. F., D. FRANKLIN, R. L. VAN CITTERS, AND E. BRAUNWALD. Effects of carotid sinus nerve stimulation of blood flow distribution in conscious dogs at rest and during exercise. *Clin. Sci.* 15: 457–463, 1974.

VATNER, S. F., R. G. MONROE, AND R. J. McRITCHIE. Effects of anesthesia, tachycardia, and autonomic blockade of the Anrep effect in intact dogs. *Am. J. Physiol.* 226: 1450–1456, 1974.

VATNER, S. F., AND D. H. BOETTCHER. Regulation of cardiac output by stroke volume and heart rate in conscious dogs. *Circ. Res.* 42(4): 557–56, 1978.

VIVEROS, O. H., D. C. GARLICK, AND E. M. RENKIN. Sympathetic beta adrenergic vasodilatation in skeletal muscle of the dog. *Am. J. Physiol.* 215: 1218–1225, 1968.

VON BEYOLD, A., AND L. HIRT. Ueber die physiologischen Wirkungen des essigsauren Veratrins. *Physiol. Lab. Wurzburg. Untersuchungen* 1: 75–156, 1867.

WAGNER, M. L., R. LAZZARA, R. M. WEISS, AND B. F. HOFFMAN. Specialized conducting fibers in the interatrial band. *Circ. Res.* 18: 502–518, 1966.

WALLACE, A. G. Electrical activity of the heart. In: *The Heart*, 6th ed., edited by J. W. Hurst. New York: McGraw-Hill, 1986, p. 73–84.

WADE, O. L., B. L. COMBES, A. W. CHILDS, A. COURNAND, AND S. BRADLEY. The effect of exercise on the splanchnic blood flow and splanchnic blood volume in normal man. *Clin. Sci.* 15: 457–463, 1956.

WALDO, A. L. Mechanisms of atrial fibrillation, atrial flutter and ectopic atrial tachycardia—a brief review. *Circulation* 75: III37–III39, 1987.

WEBER, K. T., J. S. JANICKI, R. C. REEVES, AND L. L. HEFNER. Factors influencing left ventricular shortening in isolated canine heart. *Am. J. Physiol.* 230: 419–426, 1976.

WEBER, K. T., J. S. JANICKI, AND S. G. SHROFF. Measurement of ventricular function in the experimental laboratory. In: *The Heart and Cardiovascular System*, edited by H. A. Fozzard et al. New York: Raven Press, 1986, pp. 865–886.

WEIDMANN, S. Heart: electrophysiology. *Ann. Rev. Physiol.* 36: 155–169, 1974.

WEIDEMAN, M. P., R. F. TUMA, AND H. N. MAYROVITIZ. *An Introduction to Microcirculation.* New York: Academic Press, 1981.

WEIL, M. H., M. von PLANTA, AND E. C. RACKOW. Acute circulatory failure (shock). In: *Heart Disease*, edited by E. Braunwald. Philadelphia: W. B. Saunders Co., 1988, p. 561–580.

WHELAN, R. F. *Control of Peripheral Circulation in Man.* Springfield, IL: Charles C Thomas, 1967.

WHITMORE, R. I. *Rheology of the Circulation.* Oxford: Pergamon Press, 1968.

WIEDEMAN, M. P. Dimensions of blood vessels from distributing artery to collecting vein. *Circ. Res.* 12: 375–378, 1963.

WIEDERHIELM, C. A., AND B. V. WESTON. Microvascular, lymphatic and tissue pressures in the unanesthetized mammal. *Am. J. Physiol.* 225: 992–996, 1973.

WIGGERS, C. J. *Physiology in Health and Disease.* Philadelphia: Lea & Febiger, 1954.

WILMORE, J. H. Acute and chronic physiological responses to exercise. In: *Exercise in Cardiovascular Health and Disease*, edited by E. A. Amsterdam, J. H. Wilmore, and A. N. DeMaria. New York: Yorke Medical Books, 1977, p. 53–69.

WINEGRAD, S. Electromechanical coupling in heart muscle. In: *Handbook of Physiology*, Sect. 2, edited by R. M. Berne, et al. Washington DC: American Physiological Society, 1979, p. 393–428.

WIT, A. L., AND P. F. CRANEFIELD. Triggered and automatic activity in the canine coronary sinus. *Circ. Res.* 41: 435–445, 1977.

WIT, A. L., M. B. WEISS, W. D. BERKOWITZ, K. M. ROSEN, C. STEINER, AND A. N. DAMATO. Patterns of atrioventricular conduction in the human heart. *Circ. Res.* 27: 345–359, 1970.

WIT, A. L., M. R. ROSEN, AND B. F. HOFFMAN. Electrophysiology and pharmacology of cardiac arrhythmias. II. Relationship of normal and abnormal electrical activity of cardiac fibers to the genesis of arrhythmias. B. Reentry, section I. *Am. Heart J.* 88: 664–670, 1974.

WOOD, E. H. Cardiovascular and pulmonary dynamics by quantitative imaging. *Circ. Res.* 38: 131–139, 1976.

WOOD, E. H., R. L. HEPPNER, AND S. WEIDMAN. I. Positive and negative effects of constant electric currents of current pulses applied during cardiac action potentials. II. Hypotheses: calcium movements, excitation-contraction coupling and inotropic effects. *Circ. Res.* 24: 436–445, 1969.

WU, D. Demonstration of sustained sinus and atrial re-entry as a mechanism of paroxysmal supraventricular tachycardia. *Circulation* 51: 234–243, 1975.

YOFFEY, J. M., AND F. C. COURTICE. *Lymphatics, Lymph and Lymphoid Tissue*, 2nd ed., Cambridge, MA: Harvard Univ. Press, 1956.

YORAN, C., L. HIGGINSON, M. A. ROMERO, J. W. COVELL, AND J. ROSS, Jr. Reflex sympathetic augmentation of left ventricular inotropic state in the conscious dog. *Am. J. Physiol.* 10: H857–863, 1981.

YOUNG, M. A., AND S. F. VATNER. Regulation of large coronary arteries. *Circ. Res.* 59: 579–596, 1986.

ZELIS, R., J. LONGHURST, R. J. CAPONE, AND G. LEE. Peripheral circulatory control mechanisms in congestive heart failure. *Am. J. Cardiol.* 32: 481–490, 1973.

ZIMMERMAN, R. S., J. A. SCHIRGER, B. S. EDWARDS, T. R. SCHWAB, D. M. HEUBLEIN, AND J. C. BURNETT, Jr. Cardio-renal endocrine dynamics during stepwise infusion of physiologic and pharmacologic concentrations of atrial natriuretic factor in the dog. *Circ. Res.* 61: 63–69, 1987.

ZIPES, D. P. Genesis of Cardiac Arrthymias: Electrophysiologic Considerations. In *Heart Disease,* 3rd ed., edited by E. Braunwald. Philadelphia: W. R. Saunders, 1988. p. 581–620.

ZUCKER, I. H., AND J. P. GILMORE. Aspects of cardiovascular reflexes in pathologic states. *Fed. Proc.* 44: 2400–2407, 1985.

ZWEIFACH, B. W. Microcirculation. *Annu. Rev. Physiol.* 35: 117–150, 1973.

ZWEIFACH, B. W., AND A. FRONEK. The interplay of central and peripheral factors in irreversible hemorrhagic shock. *Prog. Cardiovasc. Dis.* 18: 147–180, 1975.

SECTION 3

Blood

CHAPTER 19

Blood and the Plasma Proteins: Functions and Composition of Blood

As blood moves through the capillaries of the organs and tissues, it performs its vital function of picking up and delivering a variety of materials whose transport through the circulation is required for the survival of complex multicellular organisms. First and foremost, it picks up O_2 from the lungs and delivers O_2 and glucose to all organs and tissues for the oxidative metabolic reactions essential for life. Of particular importance is an uninterrupted delivery of O_2 to the brain and heart. Blood also delivers amino acids, fatty acids, trace metals, and other substances to the cells for nutritive use or for incorporation into cellular components or secretory products. Blood brings to the cells the hormones and vitamins that modulate cellular metabolic processes. Through a constant exchange of molecules with the interstitial fluid, blood helps to maintain the pH and electrolyte concentrations of interstitial fluid within the ranges required for normal cell functions. Blood carries the waste products of metabolism to their organs of excretion: CO_2 to the lungs, bilirubin to the liver, nonprotein nitrogenous products to the kidneys. In warm-blooded animals, blood transports heat generated in deep organs to the skin and lungs for dissipation. Blood also serves essential body protective functions. Its white blood cells combat invading microorganisms, mediate inflammation, and initiate immune responses to foreign materials. Its antibodies and complement components play vital roles in these defense responses. Its platelets and blood coagulation proteins maintain hemostasis.

Blood consists of cells suspended in a fluid called *plasma*. If blood is mixed with an anticoagulant powder to prevent clotting, the cells can be separated from the plasma by centrifugation. After vigorous centrifugation, a unit volume of normal human blood will consist of a volume of packed red blood cells of about 45%, on top of which is a volume of between 0.5 and 1% of white blood cells and platelets. The supernatant plasma will occupy the remaining volume of about 54%. If an anticoagulant is not added and blood is allowed to clot, the supernatant fluid that can be expressed from the clot is called *serum*. Serum differs from plasma in lacking fibrinogen (which has been converted to the fibrin of the clot), prothrombin, and other coagulation factor activities consumed during clotting and in containing minute but physiologically important amounts of materials released from platelets during clotting, e.g., growth factors that can affect the function of cells of an injured vascular wall.

Plasma consists of an aqueous solution of proteins, electrolytes, and small organic molecules. Plasma (100 ml, or 1 dl) will contain about 7 g of plasma proteins; about 900 mg of electrolytes, primarily sodium, chloride, and bicarbonate ions, but also physiologically critical amounts of potassium, calcium, magnesium, and phosphate ions; and a few hundred milligrams of small organic molecules, including about 100 mg of glucose, about 25 mg of nonprotein nitrogenous waste products, of which urea is the major constituent, and several hundred milligrams of lipids in the form of triglycerides, phospholipid, and cholesterol. Plasma from a fasting individual is translucent and is a pale yellow color due to small amounts of the pigment bilirubin; plasma from an individual who has eaten a fatty meal is opaque because of dietary fat, primarily triglycerides, suspended in the plasma as particles called chylomicrons.

PLASMA PROTEINS

Studies of the plasma proteins have yielded much information about physiological processes and disease states. The accessibility of normal plasma, the development of methods for purifying proteins, and the powerful techniques of molecular biology have made possible the determination of the structural and functional properties of most of the more than 100 individual plasma proteins recognized to date. The precise structure of the different classes of immunoglobulin molecules has been delineated (Chapter 22). The known components of the complex plasma proteolytic systems participating in inflammation and hemostasis, including many present in only trace amounts, have been purified, and their functions have been identified. A growing number of plasma protease inhibitors have also been purified, and their mechanisms and spectra of protease inhibition have been clarified. An increasing number of apolipoproteins participating in lipid transport and metabolism have also been delineated. Their study has advanced understanding of the most prevalent of serious diseases of Western civilization, atherosclerosis. Carrier proteins have been identified in plasma for many small molecules such as vitamins, hormones, trace metals, and drugs. A number of proteins have also been separated from plasma, whose physiological functions remain unknown.

Studies of plasma proteins have also yielded important genetic information. Some plasma proteins exist in polymorphic forms differing in prevalence in different populations; they have proven a valuable tool in investigations of population genetics. At the molecular level, the determination of degrees of homology in amino acid sequences of purified proteins of related function, such as the vitamin K-dependent blood clotting factors, has broadened understanding of how multicomponent biological systems develop in evolution. Similar studies of proteins of different species have clarified phylogenetic relationships.

Separation and Measurement of the Plasma Proteins

Although the approximately 7 g/dl of human plasma proteins consist of many proteins; one protein, albumin, makes up over 4 g of the total. Initially, albumin was measured by salting out techniques that separate albumin from all the other plasma proteins, the globulins. Later, electrophoretic techniques were developed that separate the plasma proteins into albumin and several globulin fractions that migrate at different rates in an electrical field. These globulin fractions were given Greek letter names: α_1, α_2, β, and γ. The vast majority of the γ globulins, which make up another 1.0–1.5 g of total plasma proteins, turned out to be immunoglobulins. A number of other proteins, whose concentrations range between about 50 and 300 mg/dl, have been identified within or between the other electrophoretically separated globulin fractions (Fig. 3.1). These proteins, which are of widely diverse functions, include: fibrinogen (which is not seen when serum is used for electrophoresis); two major protease inhibitors, α_1-protease inhibitor (α_1-antitrypsin) and α_2-macroglobulin; the carrier proteins, haptoglobin and transferrin, two major

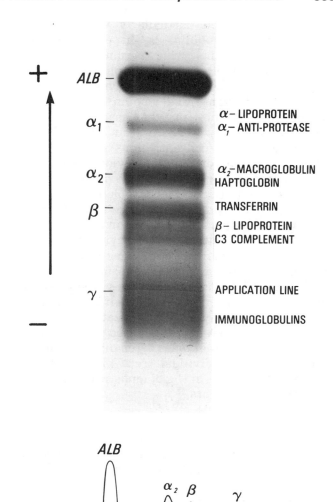

Figure 3.1.   The distribution of serum proteins obtained on electrophoresis of normal serum on cellulose acetate at pH 8.6. Albumin *(ALB)*, which is the most negatively charged of the major plasma proteins, moves farthest from the point of application toward the positive pole. The immunoglobulins remain close to the point of application. The intensity of the protein bands on the cellulose acetate strip can be measured with an instrument called a densitometer. The electrophoretic profile shown in the bottom half of the slide is the densitometer tracing obtained from the strip shown in the top half of the slide.

lipoproteins, and the C$_3$ component of complement. An immunoelectrophoretic technique was also developed in which the plasma proteins are separated by electrophoresis and then precipitated by immunodiffusion against an antiserum to whole human serum. This sensitive technique resolves the plasma proteins into multiple precipitin arcs, each representing a different protein.

Serum albumin levels are frequently measured (nowadays by automated dye-binding techniques) as part of the screening laboratory evaluation of a new patient. Serum protein electrophoresis is also commonly performed, particularly when a disorder affecting immunoglobulins is suspected. Immuno-electrophoresis against whole serum antiserum has proven of more limited clinical application.

Numerous assays have been developed to measure small or even trace amounts of individual plasma proteins. These may be based upon the specific enzymatic activity of a protein or may involve measuring minute concentrations of antigen by radioimmunoassay or ELISA (enzyme-linked immunoabsorbant assay) methodology. Some of these trace plasma proteins, such as blood coagulation and complement components, have key extracellular functions. Others have no known extracellular function; they are intracellular enzymes that escape into the blood in small amounts as a result of normal cell function or the minor tissue insults of daily living. Plasma levels of such enzymes frequently are increased in disorders affecting their cells of origin, e.g., elevation of amylase in pancreatic disease, alanine aminotransferase in hepatocellular disease, and creatine phosphokinase after injury to muscle. Measurement of individual trace plasma proteins has proven valuable in the diagnostic evaluation of many types of patients.

Synthesis of the Plasma Proteins

Albumin and the vast majority of the globulins are made in the liver. The major exception is the immunoglobulins, which are made by B-lymphocytes-plasma cells. In severe hepatocellular disease, the plasma level of albumin and many other plasma proteins falls. Cells other than hepatocytes or B-lymphocyte-plasma cells contribute significantly to the synthesis of a small number of plasma proteins. Thus, macrophages make substantial amounts of some components of the complement system; intestinal cells synthesize some apoproteins; and endothelial cells are the site of synthesis of several proteins involved in hemostasis (see Chapter 24).

Factors regulating the synthesis of albumin (Rothschild et al., 1977) are as yet poorly understood. The rise of plasma amino acids that follows ingestion of a meal containing protein stimulates the liver to synthesize albumin. A regulator of albumin synthesis sensitive to the colloid oncotic pressure of the plasma has been postulated. In inflammation, materials secreted by monocytes, particularly interleukin 1, stimulate the synthesis of a number of plasma proteins (see below) but diminish the synthesis of albumin (Moshage et al., 1987). For this reason plasma albumin levels fall in all chronic diseases, not just in those disorders directly affecting hepatocytes.

Exposure to antigen stimulates synthesis of immunoglobulin. Experimental animals kept in a germ-free environment have virtually absent immunoglobulins. Conversely, immunoglobulin levels rise in chronic infectious disorders causing persisting antigentic stimulation of B-lymphocytes-plasma cells.

Dual mechanisms regulate the hepatic synthesis of a number of plasma proteins known as the acute phase proteins. Unknown factors, possibly different for each protein, maintain basal levels of production of the individual proteins in this group, whereas interleukin 1 and tumor necrosis factor, materials released from activated macrophages and other cells, stimulate the increased synthesis of these proteins after tissue injury or inflammation.

Distribution and Catabolism of the Plasma Proteins

When a radiolabeled plasma protein, such as radiolabeled albumin, is injected intravenously and the radioactivity of serial plasma samples is plotted against time on semilogarithmic paper, a plasma radioactivity decay curve is obtained whose shape corresponds to the sum of two decreasing exponentials. Two rate constants can be calculated from such curves: an initial larger rate constant for the equilibration of the protein between intravascular and extravascular compartments and a second smaller rate constant for the biological decay of the protein. Analysis of such curves also provides an estimate of the distribution of the protein between the plasma and the interstitial fluid. From such data it is apparent that plasma proteins circulate not only intravascularly but also in a second slower pathway across capillary beds into the interstitial fluid and back into the plasma by way of lymphatic channels. About 5% of plasma albumin exchanges with albumin in interstitial fluid per hour. Moreover, such data indicate that about 60% of the total body mass of albumin is present within the extravascular fluid. Most other plasma proteins studied have similar high fractions of total mass within the extravascular fluid. Two plasma proteins are exceptions: IgM immunoglobulin and α_2-macroglobulin, which have difficulty in crossing capillary membranes because of their large size and are, therefore, largely confined to the intravascular compartment.

Plasma proteins appear to be broken down randomly, i.e., a constant proportion of molecules are lost per unit time (first order kinetics), independent of how long individual molecules may have aged in the circulation. Intravascular biological half-disappearance times vary widely for different proteins. At its normal plasma concentration, albumin has an intravascular half-disappearance time of 19 days, and IgG has an intravascular half-disappearance time of 21 days. These long half-times contrast with biological intravascular half-disappearance times of only about 4 days for haptoglobin and fibrinogen and contrast with the remarkably short half-times for some of the blood coagulation proteins, e.g., 5 h for factor VII.

The catabolism of albumin (Waldmann, 1977) may differ fundamentally from the catabolism of the other plasma proteins, since albumin is virtually the only plasma protein that is not a glycoprotein (see below). Albumin moves across capillary endotheluim into the extracellular fluid largely by the process of receptor-mediated transcytosis (Simionescu et al., 1987). Tissue cells take up albumin by pinocytosis, which is thought to be the major mechanism for albumin catabolism. The albumin is broken down within the lysosomes of tissue cells to amino acids, which are then available for protein synthesis within the cells. A small fraction of albumin, less than 10%, is lost into the digestive tract, where it is broken down into amino acids that can be reabsorbed.

The vast majority of the other plasma proteins contain carbohydrate attached at asparagine residues of the polypeptide chains. The carbohydrate residues form short branched chains that always terminate in sialic acid residues. When terminal sialic acid is removed in vitro by treatment of a plasma glycoprotein with an enzyme called neuramidase, the second residue from the end of the chain is exposed. It is always a galactose residue. In experiments first carried out with ceruloplasmin and then with other plasma glycoproteins, disialylated radiolabeled proteins were found to be removed rapidly from the circulation. Receptors on hepatocyte plasma membranes have been identified for galactose; these receptors bind the exposed galactose residues on desialylated plasma glycoproteins and remove the proteins from the circulation. Small amounts of desialylated proteins have recently been demonstrated in human serum, and it has been proposed that hepatic recognition of the exposed galactose on such molecules represents a major physiological mechanism for plasma glycoprotein catabolism (Ashwell and Steer, 1981).

The function of a number of plasma proteins is to combine with other materials in plasma. The resultant complexes are usually removed more rapidly from the circulation than are the free proteins. Thus, when immunoglobulin molecules combine with antigen to form immune complexes, they may be removed rapidly from the circulation by the mononuclear phagocyte system. Hemoglobin-haptoglobin complexes are cleared from the circulation by specific receptors on hepatocyte membranes for the hemoglobin-haptoglobin complex. Antithrombin III-thrombin complexes and α_2-plasmin inhibitor-plasmin complexes have substantially shorter half-disappearance times than do free antithrombin III and free α_2-plasmin inhibitor. Thus, the catabolism of many plasma proteins varies with the extent to which they are called upon to carry out their physiological functions.

Specific Plasma Proteins

ALBUMIN

As already mentioned, albumin makes up about 60% of the total plasma proteins, has a biological intravascular half-time of 19 days, and is distributed two-fifths intravascularly and three-fifths extravascularly (Peters, 1975). In a normal 70-kg individual in the steady state, 14–17 g of albumin are made daily. Because of its plasma concentration and relatively low MW of 66,000, albumin is the principal protein responsible for the colloid osmotic pressure of plasma. Albumin also performs important carrier functions. Binding to albumin is the only physiological mechanism for the transport of free fatty acids in plasma. Albumin transports bilirubin from mononuclear phagocytes, where it is formed from the heme of phagocytized red blood cells, to the liver for excretion. Albumin also serves as a secondary carrier for thyroxine, for cortisol, and for heme, when the capacities of their primary plasma transport proteins are exceeded. In addition, albumin binds many of the drugs given to patients. The tightness of binding and the competition for binding sites between drugs given simultaneously influences drug dosing.

Because of diminished synthesis, plasma albumin levels fall in any chronic illness. Very low levels are often found in patients with advanced hepatocellular disease and in patients with conditions in which diminished synthesis is accompanied by loss of large amounts of albumin into the urine (nephrosis) or into the gastrointestinal tract (protein-losing enteropathies).

PLASMA PROTEOLYTIC SYSTEMS

Four proteolytic systems are present in plasma: the complement system, the kinin system, the blood coagulation system, and the fibrinolytic system. Their reactions are discussed in Chapters 23 and 25.

PLASMA PROTEASE INHIBITORS

The properties of a number of plasma protease inhibitors are listed in Table 3.1. Most belong to the same protein family, called the serpins (Carrell and Boswell, 1986), and share common structural and functional properties. α_1-Protease inhibitor (α_1-antitrypsin) is the inhibitor normally present in plasma

Table 3.1.
Properties of Plasma Protease Inhibitors

Major Protease Inhibitors	Plasma Concentration		Physiological Inhibitor of	Manifestations of Deficiency
	mg/dl	μmol		
α_1-Protease inhibitor[a,b]	250	46	WBC elastase and other WBC proteases, factor XI$_a$, activated protein C	Emphysema
α_1-Antichymotrypsin[a]	50	7.0	WBC cathepsin G	Unknown
α_2-Macroglobulin	250	3.5[c]	Plasmin, thrombin, kallikrein	Unknown
Inter-α-trypsin inhibitor[a]	50	3.0	Unknown	Unknown
Antithrombin III	15	2.5	Thrombin, factor X$_a$, IX$_a$	Thrombosis
C1 inhibitor[a]	18	1.5	Activated C1r, C1s, kallikrein	Hereditary angioneurotic edema
α_2-Plasmin inhibitor[a,d]	7	1.0	Plasmin	Fibrinolytic bleeding

Ranges are wide, and the concentrations listed are approximations. [a]Concentration rises in tissue injury or inflammation. [b]Formerly called α_1-antitrypsin. [c]The low value results from a very high molecular weight of 725,000 daltons. [d]Also called α_2-antiplasmin.

in the highest concentration. Although able to inhibit a wide spectrum of proteases, α_1-protease inhibitor functions in vivo primarily as an inhibitor of elastase and cathepsin G released by stimulated neutrophil leukocytes in the inflammatory response (Carrell, 1986). However, α_1-protease inhibitor does appear to function as a major inhibitor of two of the blood coagulation enzymes (see Chapter 25), activated factor XI and activated protein C. Allelic structural genes for α_1-protease inhibitor exist (Pi genes). They determine an individual's basal plasma level of α_1-protease inhibitor. The vast majority of individuals are homozygous for the Pi^m gene (MM phenotype); they have basal levels of about 250 mg/dl, which are capable of rising substantially in inflammation. Homozygotes for the uncommon Pi^z gene (ZZ phenotype) have very low plasma α_1-protease inhibitor levels. Many of these individuals develop pulmonary emphysema, a condition resulting from destruction of the alveolar structure of the lungs. These individuals apparently have insufficient α_1-protease inhibitor in the alveolar walls and spaces to prevent elastases, continually liberated from white blood cells in response to repeated minor pulmonary inflammatory events, from gradually destroying the alveoli (Gadek et al., 1982).

α_2-Macroglobulin is another major protease inhibitor of plasma that inhibits a wide spectrum of proteases (Harpel and Rosenberg, 1976). α_2-Macroglobulin is unique among protease inhibitors in its large size and in its ability to inhibit the proteolytic activity of proteases without inhibiting their enzymatic activity against small substrates such as synthetic esters. The physiological functions of α_2-macroglobulin are not yet clear, although it has been shown to function as a back-up inhibitor for plasmin when the capacity of the primary inhibitor, α_2-plasmin inhibitor, is exceeded. α_2-Macroglobulin has been found to line the luminal surface of vascular endo-

thelial cells (Becker and Harpel, 1976), and it has been postulated that α_2-macroglobulin could serve the general physiological function of protecting cell surfaces from attack by serine protease enzymes.

CARRIER PROTEINS

A number of the carrier proteins of plasma and the substances that they bind are listed in Table 3.2. The carrier functions of albumin have already been described. The apolipoproteins are discussed in Chapter 49. Three carrier proteins, transferrin, haptoglobin, and hemopexin, are involved in processes affecting synthesis or catabolism of hemoglobin. The iron for new hemoglobin synthesis primarily comes from the iron of senescent red blood cells that are broken down within mononuclear phagocytes. *Transferrin* transports this iron from the phagocytes through the plasma to developing erythroid precursors (normoblasts) in the bone marrow (Huebers and Finch, 1987). Each molecule of transferrin can carry two atoms of iron, and plasma transferrin is normally about one-third saturated with iron. Normal plasma contains about 250 mg/dl

Table 3.2.
Important Carrier Proteins of Plasma

Protein	Materials Bound or Transported
Albumin	Primary carrier: fatty acids, bilirubin, many drugs Secondary carrier: heme, thyroxine, cortisol
Apolipoproteins	Triglycerides, phospholipids, cholesterol
Haptoglobin	Plasma hemoglobin from lysed erythrocytes
Hemopexin	Heme from plasma hemoglobin
Transferrin	Iron
Ceruloplasmin	Copper
Prealbumin	Thyroxine, vitamin A
Group specific (G) globulin	Vitamin D
Transcortin	Cortisol
Transcobalamins I and II	Cobalamin (vitamin B$_{12}$)

of transferrin. In clinical laboratories its concentration is not measured directly, but is measured indirectly by determining the amount of iron needed to saturate the iron-binding capacity of plasma. This value is called the total iron-binding capacity (TIBC). Mean normal values are: for serum iron, 100 μg/dl; for TIBC, 300 μg/dl. Transferrin is distributed in a 60:40 ratio between the extravascular and intravascular compartments. Although transferrin moves into and out of the normoblast during iron delivery, this process apparently does not affect its catabolism. Transferrin has an intravascular biological half-decay time of 10 days. Plasma transferrin concentration and, therefore, the TIBC, rises as a patient becomes iron deficient. Like albumin, transferrin synthesis is depressed in inflammation, and patients with chronic diseases usually have a low TIBC.

Haptoglobin forms complexes with dimers of hemoglobin. When red blood cells are lysed intravascularly, hemoglobin is released into the plasma and rapidly breaks up into $\alpha\beta$ dimers. A single haptoglobin molecule can combine with two dimers, i.e., with the equivalent of a single hemoglobin molecule. Hemoglobin-haptoglobin complexes are removed from plasma by receptors on hepatocytes, transported into hepatocyte lysosomes, and digested (Kino et al., 1980). Thus, haptoglobin functions to conserve body iron, conveying iron in plasma hemoglobin to the liver and preventing its loss in the urine through excretion of free dimers.

Normal plasma haptoglobin concentrations vary from about 40–180 mg/dl. In a 70-kg human, the daily breakdown of senescent erythrocytes results in a turnover of about 6 g of hemoglobin, of which about 0.6 g (representing 10% of erythrocyte breakdown) is released intravascularly. A haptoglobin concentration of 130 mg/dl in a 70-kg human can bind about 3 g of hemoglobin, or five times the amount liberated intravascularly each day. Plasma haptoglobin concentrations fall in patients with severe hepatocellular disease because the liver is the primary site of haptoglobin synthesis. Plasma haptoglobin levels also frequently fall in hemolytic disorders, because then most of the plasma haptoglobin is converted into hemoglobin-haptoglobin complexes and their rapid clearance does not stimulate compensatory increased haptoglobin production.

Hemoglobin in plasma not only breaks up into dimers but into free oxidized heme (metheme) and globin chains. Whereas haptoglobin binds dimers, a second protein, *hemopexin*, binds free metheme. The metheme-hemopexin complex is also rapidly removed from plasma by hepatocytes and catabo-

lized. Hemopexin is normally present in plasma in a concentration of 50–100 mg/dl. When intravascular hemolysis results in the formation of amounts of metheme exceeding the binding capacity of hemopexin, metheme then binds to albumin, forming methemalbumin. Methemalbumin functions as a circulating heme storage protein. As new hemopexin is produced, methemalbumin gives up its heme to hemopexin, and the process of hepatic clearance of metheme-hemopexin complexes continues. In this way, the body conserves iron in metheme that would otherwise be lost in the urine.

The copper in plasma circulates bound to the protein *ceruloplasmin*. The normal plasma ceruloplasmin concentration is about 30 mg/dl. Estrogens react with steroid receptors on hepatocytes to stimulate ceruloplasmin synthesis, and plasma levels rise substantially in pregnancy or when women take oral contraceptive agents containing estrogens. Ceruloplasmin levels also rise in disorders producing inflammation or tissue injury. Low ceruloplasmin levels are found in patients with a hereditary disturbance of copper metabolism (Wilson's disease) in which biliary excretion of copper is impaired, and large amounts of copper accumulate in the body, damaging the liver and brain. Ceruloplasmin is an oxidase and may perform important functions in addition to carrying copper. Ceruloplasmin has been postulated to serve as a major extracellular scavenger of superoxide radicals generated by white blood cells during the inflammatory response (Goldstein et al., 1982). Ceruloplasmin may also function physiologically as a ferroxidase, catalyzing the conversion of ferrous iron in tissues to ferric iron, a reaction that must occur before transferrin can pick up and transport iron.

ACUTE PHASE PROTEINS

When tissue injury or infection results in local inflammation, interleukin 1 and tumor necrosis factor are secreted from activated macrophages and induce a systemic acute phase response. This response usually includes fever, an increased release of certain hormones, the decreased synthesis of albumin, and the increased synthesis of acute phase proteins. These include proteins whose increased synthesis has recognizable survival value after tissue injury or infection: the hemostatic factors, fibrinogen, and von Willebrand factor; the C_3 and factor B components of complement; haptoglobin; ceruloplasmin; and the protease inhibitors, α_1-protease inhibitor, α_1-antichymotrypsin, and α_2-antiplasmin. The synthesis of two proteins that are normally present in only trace amounts in human plasma increases many hundred fold during the acute phase response. One protein is C-reactive protein, so named because it was first identified as a material in plasma that precipitates a pneumococcal polysaccharide fraction known as the C fraction. The second protein is SAA (serum amyloid A protein), which was first recognized as the major component of secondary amyloid and was later identified, in a slightly larger molecular weight form, in plasma (Gorevic et al., 1982). C-reactive protein has recently been shown to bind to altered cell membranes. Once bound, like antibody, it can then activate the classical complement pathway. Macrophages have been shown to ingest particles coated with C-reactive protein and complement. Thus, C-reactive protein has been postulated to function in host defense in reactions that supplement but are distinct from the immunologic response (Gewurz, 1982). SAA has recently been shown to circulate in plasma attached to high density lipoprotein. A host defense function for SAA has not yet been identifed.

Altered Plasma Proteins in Disease

Hereditary abnormalities of plasma proteins are an uncommon but well-recognized cause of disease. Thus, lifelong disorders of hemostasis, of which the hemophilias are the most important, result from hereditary abnormalities of the blood coagulation proteins. A hereditary deficiency of antithrombin III is associated with an increased risk for thromboembolic events. As already mentioned, a decreased plasma concentration of α_1-protease inhibitor in individuals homozygous for the Pi^z gene can result in the development of pulmonary emphysema. Failure to synthesize transferrin is a very rare cause of a refractory iron deficiency anemia.

Abnormalities of plasma proteins are common as manifestations of disease and, as discussed earlier, the plasma proteins are frequently examined in the diagnostic evaluation of patients.

CHAPTER 20

Hemopoiesis

Blood cells are made in early embryonic life in the liver and, to a lesser extent, in the spleen. Hemopoiesis begins in the bone marrow at about the 20th embryonic week. As the fetus matures, hemopoiesis increases in the bone marrow and decreases in the liver and spleen. After birth, all blood cells except lymphocytes are normally made only in the bone marrow. In young children, active hemopoietic marrow is found throughout both the axial skelton (cranium, ribs, sternum, vertebrae, and pelvis) and the bones of the extremities. In adults, hemopoietic marrow is confined to the axial skeleton and proximal ends of the femur and humerus.

Blood cells are divided into three categories: erythrocytes (red blood cells, RBC), white blood cells (WBC), and platelets. Erythrocytes carry oxygen from the lungs to the tissues. The WBC, of which there are five different types (neutrophilic granulocytes, lymphocytes, monocytes, eosinophils, and basophils) are involved in the body's defense against microorganisms and other foreign materials. Platelets play key roles in the reactions of hemostasis (see Chapter 24). The approximate concentrations of these cells in the blood are given in Table 3.3.

All blood cells, including lymphocytes, originate from a single class of primitive cells, *the pluripotent stem cell.* However, lymphocytes differ from the other blood cells, which normally arise after fetal life only within the bone marrow and therefore are referred to, inclusively, as myeloid cells. Lymphocytes, although originating from marrow precursor cells, mature and proliferate outside of the marrow in the thymus and peripheral lymphoid tissues (see Chapter 21).

Erythrocytes, neutrophilic granulocytes, eosinophils, monocytes, and platelets have finite lifespans, and, therefore, fractions of these cells must be replaced daily. Approximate daily rates of production are listed in Table 3.3. In addition, the bone marrow must respond intermittently to increased demands for specific cells, e.g., for granulocytes to fight a bacterial infection or for erythrocytes after acute blood loss. This requires regulation of the proliferation and differentiation of cells capable of diverse genetic expression, i.e., of stem cells and early progenitor cells, whose responses maintain appropriate blood cell production.

STEM CELLS AND PROGENITOR CELLS

The restoration of hemopoiesis in animals or humans given supralethal doses of radiation by infusion of donor bone marrow cells provides unequivocable evidence for the existence of hemopoietic stem cells. Hemopoietic stem cells (Quesenberry and Levitt, 1979) possess two fundamental properties: first, an ability by cell division to give rise to new stem cells, i.e., *self-renewal,* and second, an ability to differentiate into mature specialized blood cells (Fig. 3.2). The frequency of hemopoietic stem cells in human bone marrow has been estimated as 1–2 per 1000 nucleated cells. Small numbers also circulate in the blood.

When, after supralethal doses of radiation, normal mice or humans with leukemia are infused with donor bone marrow cells, hemopoiesis is re-established only in the marrow cavity and spleen of the mouse and in the marrow cavity of man. Specific supporting tissues in these organs permit growth and differentiation of stem cells and are referred to as the *hemopoietic microenvironment.* Stem cells "home" to the hemopoietic microenvironment by binding to recognition sites in the microenvironment for specific surface carbohydrates on the stem cells. In addition, glycosaminoglycans in the extracellular matrix of the hemopoietic microenvironment bind and localize hemopoietic growth factors. This restricts proliferation of stem cells and their immediate progeny to those cells in contact with the matrix.

Stem cells can be quantified by irradiating mice to destroy their own stem cells and then injecting the mice with a limited number of syngeneic bone marrow cells. After several days, cellular aggregates

Table 3.3.
Approximate Concentrations and Daily Production Rates of Peripheral Blood Cells

Cell Type	Mean Concentration, per microliter	Daily Production Rate, per kg body weight
Erythrocytes		
Males	5.4×10^6	3.0×10^9
	$(4.7-6.1 \times 10^6)$	
Females	4.8×10^6	
	$(4.2-5.4 \times 10^6)$	
White blood cells		
Granulocytes	4500	1.6×10^9
	(2600–7000)	
Monocytes	300	1.7×10^{8a}
Eosinophils	150	Variable
Basophils	40	Unknown
Lymphocytes	2500	Unknown
	(1500–4000)	
Platelets	2.5×10^6	2.8×10^9
	$(1.5-4.0 \times 10^5)$	

Values in parentheses are ranges. [a]Based upon assuming production rate equals blood turnover rate and using an intravascular half-disappearance time of 8.4 h.

(colonies) of differentiated blood cells are found in the spleen of the irradiated animals. Each spleen colony arises from a single cell and, therefore, counting visible colonies provides an estimate of the number of stem cells infused. Self-renewal of stem cells can be demonstrated by harvesting cells from a spleen colony, injecting these cells into a second irradiated mouse (serial passage), and observing formation of second generation spleen colonies.

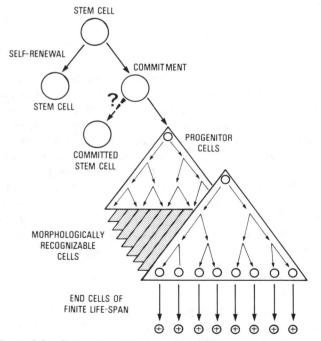

Figure 3.2. Schematic of clonal hemopoiesis beginning with a stem cell.

Since all cells in a spleen colony arise from a single stem cell, the second generation spleen colonies must have arisen from daughter stem cells present in the spleen colony from the first recipient mouse. Although all stem cells assayed by these spleen colony-forming assays have extensive proliferative capabilities, only a small number of such cells are actually capable of completely reconstituting an irradiated recipient mouse. Self-renewal prevents the marrow from running out of stem cells during a normal life-span.

When a multitude of cells arises from a single cell, as when a single stem cell gives rise to a spleen colony, that multitude of cells is called a clone. In normal hemopoiesis, mature blood cells originate from multiple stem cells, i.e., normal hemopoiesis is *polyclonal*. In hematologic malignancies, the progeny of a single abnormal stem cell may overgrow and suppress normal stem cells, and hemopoiesis may become *monoclonal*.

As a stem cell differentiates, it loses its ability to give rise to all blood cells and becomes committed to the production of one or more cell lines (Fig. 3.3). Different mechanisms have been postulated to regulate commitment (Till and McCulloch, 1980). One hypothesis holds that the probability of differentiation increases with each division of a stem cell. Another view is that extracellular inductive factors, perhaps hemopoietic growth factors or cell surface molecules in the hemopoietic microenvironment, determine the changes within a stem cell that lead to commitment. Others postulate a stochastic model in which a series of unknown factors influence the probabilities for self-renewal or commitment of the daughter cells of a dividing stem cell.

In key experiments in which chromosomal or retroviral markers were used to identify stem cell progeny serially passaged in mice (Abramson et al., 1977; Dick et al., 1985), some markers appeared in all blood cells, some only in myeloid cells, and a few only in lymphocytes. These observations suggest the existence of three functionally different stem cells: (1) a pluripotent stem cell; (2) a myeloid stem cell giving rise to erythrocytes, granulocytes of all types, monocytes, and platelets; and (3) a lymphocyte stem cell. Other investigators have proposed that the pluripotent stem cell loses its abilities to give rise to myeloid and to lymphoid cells in a sequence, and that different types of stem cells represent cells at varying points in this process.

As commitment progresses, the capability of a stem cell for self-renewal diminishes. When it is markedly limited or lost, the cell is no longer called a stem cell but is termed a *progenitor cell*. Progenitor cells are recognized by their ability to give rise

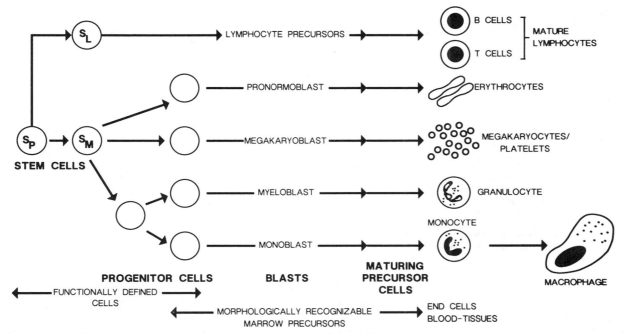

Figure 3.3. A line diagram illustrating the development and maturation of blood cells. Abbreviations are: S_P, pluripotent stem cells; S_M, committed myeloid stem cell; S_L, committed lymphoid stem cell.

to clonal aggregates (colonies) of differentiated cells on culture in vitro in the presence of growth factors. Such colonies may contain only a single cell line or combinations of multiple cell lines, an observation suggestive of probability-based or stochastic patterns of differentiation. Progenitor cells are more numerous than stem cells and, like stem cells, are present in both the bone marrow and the circulating blood.

Although not identifiable by conventional morphological techniques, stem cells and progenitor cells can be separated from other, more differentiated cells by cell-sorting techniques. Antibodies or lectins are used that react with molecules that become expressed on the surface membrane of stem cells. Stem or progenitor cells purified by such cell-sorting techniques look like small to medium-sized undifferentiated "blast" cells.

HEMOPOIETIC GROWTH FACTORS

Hemopoietic growth factors are *cytokines* (a general term for proteins released by cells that act as intercellular mediators) that control the growth, differentiation, and function of blood cells. They are physiochemically distinct acid glycoproteins defined by their abilities to alter the growth and function of progenitor cells, differentiated cells, or both types of cells of one or more blood cell lineages. The properties of several key hemopoietic growth factors are summarized in Table 3.4. A number of hemopoietic growth factors have now been cloned

and the proteins expressed using recombinant DNA techniques. Several (erythropoietin, GM-CSF, and G-CSF) have been administered to humans in studies of their potential clinical utility.

The nomenclature for hemopoietic growth factors is inconsistent, reflecting the different circumstances of their discovery. Some were identified as materials supporting the growth of colonies of blood cells in semi-solid media. They were therefore termed *colony-stimulating factors* or CSFs, with an initial or initials before the abbreviation CSF to identify the cell type or types whose growth the particular CSF supported. For example, a growth factor first recognized by its ability to support the in vitro growth of colonies of granulocytes and macrophages was named GM-CSF. Other growth factors were first recognized as materials affecting the proliferation or function of lymphocytes or monocytes in immunology. These materials were referred to as interleukins (IL); as new interleukins were identified, they were sequentially numbered (IL-1, IL-2, IL-3, and so forth). Some, e.g., IL-1, IL-3, are now known also to function as key growth factors for myeloid cell production.

Most hemopoietic growth factors are present in tissue fluids and blood in only minute concentrations (10^{-10} to 10^{-11} M). They affect their target cells through binding to specific cell-surface receptors, which may be present in only several hundred to several thousand copies per cell and may require only a 10% receptor occupancy for functional

Table 3.4.
Hemopoietic Cell Growth Factors

Growth Factor	Cellular Sources	Cells Affected[a]	
		Precursor Cells	Mature Cells
IL-1	Macrophages, also other cells	Early progenitors (?)	T lymphocyte, B lymphocyte, granulocyte, eosinophil, macrophage
IL-3	T cell	Granulocyte, macrophage, T lymphocyte, eosinophil, megakaryocyte, mast cell, basophil, erythrocyte	?T lymphocyte, eosinophil
GM-CSF	Fibroblast, endothelial cell, T cell	Granulocyte, macrophage, erythrocyte, eosinophil, megakaryocyte	Granulocyte, macrophage, eosinophil
G-CSF	Monocyte, fibroblast, endothelial cells	Granulocyte	Granulocyte
M-CSF	Macrophage, fibroblast, endothelial cell	Macrophage	Macrophage
EPO[c]	Kidney	Erythrocyte	

[a]Growth and differentiation of precursor cells; effector functions of mature cells. [b]Also modulator of secretion of IL-3 and GM-CSF by other cells. [c]Erythropoietin.

responses. Many of the growth factors have two types of action: the induction of cell growth and differentiation in immature cells and the modulation of effector functions in mature cells. Some of the growth factors are known to act synergistically, and multiple growth factors probably act simultaneously upon precursor cells in vivo, giving rise to complex patterns of cell production.

Macrophages play a central role in controlling hemopoiesis through the secretion of the cytokine, IL-1. Amongst its diverse effects on both hemopoietic and nonhemopoietic tissues (Nathan, 1987; Oppenheim et al., 1986), IL-1 stimulates growth of T lymphocytes, which elaborate IL-3, and stimulates fibroblasts and endothelial cells, including those of the marrow microenvironment, to secrete GM-CSF and G-CSF. IL-3 and GM-CSF, in turn, then stimulate the early cell divisions of progenitors of most if not all myeloid cell lineages. IL-1 also has effects upon mature blood cells, including the stimulation of release of neutrophilic granulocytes from the bone marrow and of macrophage cytocidal activities. A second cytokine from macrophages, tumor necrosis factor (TNF) or cachectin (Beutler and Cerami, 1988), has broad activities that overlap those of IL-1 and may also play a role in controlling marrow cell production.

Although both IL-3 and GM-CSF act upon myeloid cell progenitors, IL-3 may act upon an earlier progenitor cell subset, expanding and preparing the progenitor cells for subsequent exposure to GM-CSF and other growth factors (Donahue et al., 1988). Whereas GM-CSF was first recognized and named for its ability to stimulate in vitro colony growth of granulocytes, macrocytes, and eosinophils, GM-CSF is now known also to stimulate early division of erythrocyte progenitors and megakaryocyte pro-

genitors. GM-CSF also has prominent effects on mature granulocytes and macrophages, including enhancement of granulocyte responses to chemotactic stimuli and the ability of macrophages to kill intracellular organisms.

The full development in vitro of colonies containing mature blood cells requires not only a growth factor, such as IL-3 or GM-CSF, that stimulates early multilineage progenitor cell divisions but also the simultaneous or sequential presence of a second growth factor acting specifically upon more differentiated precursors. For neutrophilic granulocytes, this is G-CSF, a growth factor that stimulates terminal divisions of granulocyte precursors; for red blood cells, this second growth factor is erythropoietin.

One growth factor, macrophage CSF (M-CSF and also called CSF-1) differs from the other growth factors in being constitutively produced by macrophages, in being present in blood and urine in readily detectable levels, and in having many-fold higher numbers of cell surface receptors. These are expressed not only by macrophages but by human placenta and other tissues. The cell surface receptor for M-CSF has been identified as a tyrosine-specific protein kinase that is the product of the c-fms proto-oncogene (the cellular homologue of a feline sarcoma virus oncogene). M-CSF stimulates the in vitro growth of murine macrophage colonies. Its in vivo role in human hemopoiesis is not well defined, but when infused into humans, M-CSF increases the number and enhances the cytotoxic functions of circulating monocytes.

In addition to growth factors a variety of humoral agents have been identified that inhibit myeloid cell growth in vitro. These include TGF (transforming growth factor)-β, the granulocyte protein lactoferrin, certain intracellular iron storage proteins called acidic isoferritins, and E series prostaglandins. Definite roles for these materials in the regulation of hemopoiesis in vivo have yet to be established.

PRODUCTION OF MYELOID CELL LINES

Erythropoiesis

ERYTHROID PROGENITOR CELLS

Two types of erythroid progenitor cells, burst forming units-erythrocyte (BFU-E) and colony forming units-erythrocyte (CFU-E), have been recognized by the size of the colonies they form on culture (Testa, 1979). BFU-E form large, multilobulated colonies whereas CFU-E form much smaller colonies. In vivo, BFU-E gave rise through a continuum of intermediate forms to CFU-E, which represent the most differentiated erythroid progenitors.

Growth of BFU-E in culture requires the presence of one or more growth factors referred to collectively as "Burst Promoting Activity" or BPA. As mentioned earlier, both IL-3 and GM-CSF stimulate early erythroid progenitor growth and thus possess BPA. Other materials with BPA have also been described. In contrast, CFU-E are independent of BPA but require the hormone erythropoietin for proliferation in culture and maturation to hemoglobin-containing erythrocytes. BFU-E and CFU-E are also regulated differently in vivo. Thus, hypertransfusion of experimental animals suppresses elaboration of erythropoietin by the kidney and, consequently, proliferation of CFU-E, but has no effect upon the number or activity of BFU-E. Neither BFU-E nor CFU-E have morphological features permitting their distinction from each other or their recognition as members of the erythroid series.

MATURATION OF NORMOBLASTS

The earliest morphologically recognizable member of the erythroid series is the pronormoblast (erythroblast). On conventional methylene blue-eosin stains of the bone marrow, e.g., Wright's stain, the pronormoblast appears as a cell of moderate to large size, with a round nucleus occupying most of the cell and containing fine punctate chromatin and small blue nucleoli, and with a rim of basophilic cytoplasm. This cell matures into a small orange-pink-staining, non-nucleated red blood cell. First, the nucleoli disappear, and the cell is then called a normoblast. Subsequently, the size of the normoblast diminishes progressively; the nucleus shrinks into a purple-black pyknotic mass that is finally extruded; and the cytoplasm changes color from blue, through intermediate shades, to orange-pink as increasing amounts of hemoglobin are syn-

thesized. Three to five cell divisions take place during pronormoblast-normoblast maturation. Thus, one pronormoblast gives rise to from 8-32 red blood cells. Cell division stops at the late normoblast stage. The normal time required for an erythroid cell to mature from a pronormoblast to a newly released circulating red blood cell—which can be estimated from the time course of appearance of labeled red blood cells in the circulation after an intravenous injection of a tracer dose of radioactive iron—is 5-7 days.

Hemoglobin synthesis has three requisites. First, adequate amounts of the mRNAs for the polypeptide chains of globin must be transcribed and translated. Second, the cell must synthesize normal amounts of the tetrapyrrole protoporphyrin. Third, iron must be available for incorporation into protoporphyrin to form heme. Plasma transferrin brings the iron to the developing erythroid cells, which have surface membrane receptors for transferrin. When lack of body iron prevents transferrin from bringing sufficient iron to developing red blood cells for normal hemoglobin synthesis, erythropoiesis is impaired and anemia results. Insufficient body iron is the most common cause of anemia worldwide.

MATURATION OF THE RETICULOCYTE

When first formed by extrusion of the nucleus of the late normoblast, a non-nucleated red blood cell is larger than a mature red blood cell and does not yet have its full complement of hemoglobin. It still possesses some cytoplasmic RNA, mitochondria, and surface transferrin receptors, i.e., the synthetic machinery needed to continue to make hemoglobin. Staining with supravital dyes will precipitate the cell's residual RNA, giving rise to beads and strands of dark blue material that form a reticulin network within the cytoplasm. Because of this staining phenomenon, the newly made red blood cell is called a *reticulocyte.*

A reticulocyte normally matures for 1-2 days in the marrow before entering the circulation, during which time hemoglobin synthesis continues and the cell's size decreases. Enough cytoplasmic RNA persists for the first day of its circulating life-span for the newly released red blood cell to still be recognized by supravital staining as a reticulocyte. Thus, counting reticulocytes on a blood smear provides a convenient clinical method for estimating the rate of red blood cell production.

REGULATION AND ASSESSMENT OF ERYTHROPOIESIS

Although growth factors with BPA may influence the earliest stages of erythroid differentiation in vivo, erythropoietin functions as the major regulator of erythropoiesis. Erythropoietin is made by cells in the kidney that sense *tissue hypoxia*. The tissue hypoxia may stem from one of three general causes: a decreased blood hemoglobin content (anemia); a failure to oxygenate hemoglobin adequately in the lungs (lung diseases, certain congenital heart diseases, high altitude); or an impaired release of oxygen from hemoglobin at a normal tissue oxygen tension (as occurs when the blood carbon monoxide level is elevated).

Normal erythropoiesis requires a basal level of stimulation of erythroid progenitor cells by erythropoietin (Spivak, 1986). When, as in end-stage kidney disease or after bilateral nephrectomy, the basal secretion of erythopoietin is eliminated, patients become severely anemic unless they are infused with exogenous recombinant erythropoietin. In other chronic steady-state anemias, plasma erythropoietin concentration, as measured by radioimmunoassay, rises in inverse proportion to the degree of reduction of the blood hemoglobin concentration. Experimental hypertransfusion in animals or polycythemia (elevated blood hemoglobin concentration) in humans depresses, but does not eliminate, basal erythropoietin secretion.

When erythropoietin secretion is increased, CFU-E and their more mature progeny increase in number in the marrow. Total marrow cellularity and the relative proportion of erythroid precursors to granulocytic forms are augmented (erythroid hyperplasia). Reticulocytes are released without first maturing in the marrow and can be recognized on a Wright's stained blood smear as somewhat larger red blood cells with a purple-grayish hue (polychromatophilic macrocytes). The reticulocyte count will be elevated, reflecting both an elevated rate of red blood cell production and the persistence for a longer period in the circulating blood of reticulocytes released early from the marrow.

Total erythropoiesis can be estimated from the number of nucleated red blood cells seen in a bone marrow specimen. Effective erythropoiesis, i.e., erythropoiesis producing red blood cells that circulate, can be estimated from the absolute reticulocyte count corrected for the early release of marrow reticulocytes (Hillman and Finch, 1985). Ineffective erythropoiesis, i.e., erythropoiesis that gives rise to defective cells that do not circulate but are phagocytosed by macrophages within the marrow, can be evaluated by comparing the number of nucleated red blood cells seen in the marrow with the corrected, absolute reticulocyte count. (In healthy persons, only a negligibly small fraction of erythropoiesis is ineffective, but in certain disorders of erythropoiesis, such as pernicious anemia, ineffective erythropoiesis may predominate.)

When erythropoiesis is stimulated by erythropoietin and large amounts of iron for hemoglobin synthesis are readily available from the breakdown of red blood cells, as occurs in most hemolytic disorders, effective erythropoiesis may increase six to eight fold. When erythropoiesis is stimulated by erythropoietin, but only limited amounts of iron are readily available from normal body stores, as happens after acute blood loss, effective erythropoiesis increases only about two to three fold.

Production of Granulocytes and Monocytes-Macrophages

STAGES OF GRANULOCYTE MATURATION

Although neutrophilic granulocytes, eosinophils, and basophils all contain cytoplasmic granules, the term granulocyte is usually reserved for the neutrophilic granulocyte. Alternatively, the cell is referred to as a neutrophil.

The bone marrow and the blood contain progenitor cells that, depending upon in vitro culture conditions, give rise to colonies of granulocytes, macrophages, or both. The progenitor cell can not be distinguished morphologically from other undifferentiated marrow and blood cells. It displays antigens of Class II HLA-coded molecules (see Chapter 21) on the surface membrane. These disappear from the surface membrane in cells that differentiate into granulocytes but persist on the surface membrane of cells that differentiate into monocytes.

The earliest morphologically identifiable member of the granulocytic series is the myeloblast. On a bone marrow smear stained with Wright's stain, the myeloblast has a large round nucleus containing fine stippled chromatin and from 1–5 nucleoli. Its cytoplasm is blue and does not contain granules. As the cell matures its nucleus undergoes condensation and finally segmentation. Its cytoplasm acquires three types of granules: large azurophilic primary granules, which appear first, and which are later obscured by the appearance of two types of fine granules called specific and tertiary granules. A more detailed description of the cytological features of Wright's stained cells at different stages of maturation is available elsewhere (Rapaport, 1987).

During maturation the granulocyte acquires the properties it needs for chemotaxis, phagocytosis, and killing of bacteria. These are described in Chapter 22 and include: surface membrane receptors required for adherence, chemotaxis, and phagocytosis; proteins of a cytoplasmic contractile system needed for directed cell contraction; components of the NADPH oxidase catalyzed respiratory burst that generates toxic oxygen radicals after neutrophil stimulation; and the enzymes and other materials within the granules involved in the killing of phagocytosed bacteria.

KINETICS OF GRANULOCYTE PRODUCTION

When radiolabeled granulocytes are injected into a subject and a blood sample is drawn after equilibration, only about 50% of the injected radioactivity can be accounted for in the circulating blood. Only about one-half of the labeled granulocytes are circulating; the remainder are adherent to endothelial cells of postcapillary venules in a marginal pool that exchanges freely with the circulating pool.

If the radioactive counts obtained in serial blood samples are plotted against time on semilogarithmic paper, a straight line of disappearance of radioactivity is obtained. This means that granulocytes are lost randomly from the intravascular compartment into the tissues, i.e., a granulocyte just released from the bone marrow is as likely to leave the intravascular compartment as is a neutrophil that has been circulating for a number of hours. Gradients of materials called chemoattractants cause the granulocytes to move from the blood into tissue sites, a process termed *chemotaxis*. Once in the tissues the granulocyte never returns to the blood.

The half-time for the disappearance from blood of radiolabeled granulocytes is about 7 hours. A blood granulocyte turnover rate may be calculated from this value, the blood granulocyte count, the marginal pool size, and the blood volume. In a 70-kg man in a steady state, about 3 billion granulocytes enter the blood from the bone marrow and leave the blood for the tissues each hour (Golde, 1983).

Cells of granulocytic lineage exist in the marrow in two pools: a mitotic pool of cells that divide and a postmitotic or maturation pool of cells no longer capable of cell division (Cronkite, 1979). Early granulocytic precursors undergo four or five divisions in the mitotic compartment. Data obtained from counting proportions of cells in mitosis and from pulse labeling cells in DNA synthesis with ^{3}H-thymidine have been used to calculate the transit time of granulocyte precursors through the mitotic compartment. Values of from 3–5 days have been obtained.

The postmitotic compartment is made up of cells in the last stages of maturation (metamyelocytes and bands) and mature granulocytes. The transit time through the postmitotic compartment can be determined by pulse labeling with ^{3}H-thymidine and noting the elapsed time between the detection of labeled metamyelocytes in the marrow and the appearance of labeled granulocytes in the blood. This interval is 6–7 days. Adding this time to the transit time for the mitotic compartment gives a total time of about 10 days for normal granulopoiesis. This time shortens when the tissue demand for granulocytes increases. In infection and other inflammatory states, the postmitotic pool transit time may be reduced to 2 days.

A postmitotic pool size can be calculated from the postmitotic pool transit time and the blood granulocyte turnover rate. The number of mature granulocytes in the pool, which can be determined from differential marrow cell counts, equals about three times the number of blood granulocytes leaving the circulation each day. Thus, the normal marrow contains a 3-day reserve of mature granulocytes available for rapid release as needed.

The blood granulocyte count may rise because granulocytes within the intravascular compartment move from the marginal pool into the circulating pool (as occurs after heavy exercise or an injection of epinephrine); because increased numbers of granulocytes are released from the postmitotic marrow pool into the intravascular compartment (as occurs at the onset of most acute inflammatory states); or because of both processes. An increased count resulting solely from the shift of marginated cells does not cause incompletely mature granulocytes whose nucleus is not yet segmented (bands) to enter the circulation. An elevated blood granulocyte count, or even a normal or low blood granulocyte count, accompanied by bands on a peripheral blood smear alerts one to the possibility of a pathologic condition that has stimulated early release of marrow granulocytes.

PRODUCTION AND FATE OF MONOCYTES-MACROPHAGES

As discussed above, monocytes arise within the bone marrow from a common granulocyte-macrophage progenitor cell. Monoblasts, promonocytes, and monocytes make up from 1–3% of marrow nucleated cells. Labeling studies in mice suggest that a monoblast divides to give rise to two promonocytes and each promonocyte divides once to give rise to two monocytes. Mature marrow monocytes do not divide (van Furth et al., 1979). The marrow transit time is about 6 days. No evidence exists for a bone marrow reserve of monocytes in mice or, in a single study in which hematologically normal humans were given ^{3}H-thymidine (Whitelaw, 1972), in man. Monocytes released from the marrow into the blood migrate into the tissues with a reported intravascular half-time in humans of approximately 3 days (Whitelaw, 1972).

Whereas granulocytes arrive in the tissues fully differentiated and ready for phagocytosis and microbicidal killing, monocytes undergo further differentiation in tissues to give rise to the cells of the *mononuclear phagocyte system*. The multifunctional macrophages of this system (formerly referred to as the reticuloendothelial system) far exceed in number the combined size of the bone marrow and blood monocyte pools. They include: the von Kupffer cells lining the sinusoids of the liver; the macrophages within the substance and lining the sinusoids of the spleen and bone marrow; the macrophages lining the peripheral sinus of and distributed throughout lymph nodes; the alveolar macrophages of the lungs; the microglial cells of the central nervous system; the osteoclasts of bone; and the wandering macrophages free within the pleural and peritoneal cavities and scattered in variable numbers throughout most of the tissues.

Few data are available on the life-span of tissue macrophages but it is thought to be months. After bone marrow transplantation, replacement of host alveolar macrophages by donor alveolar macrophages requires approximately 80 days. Tissue macrophages may retain some ability to proliferate as mitoses have been observed in alveolar macrophages from patients with chronic inflammatory pulmonary diseases.

The multiple functions of macrophages (Johnston, 1988) include: (1) antigen processing and modulation of lymphocyte activity in immune responses (see Chapter 21); (2) phagocytosis of certain microorganisms, dead cells, and denatured connective tissue matrix materials; and (3) clearance from the blood (by fixed macrophages lining the sinusoids of the liver, spleen, and bone marrow) of denatured proteins, senescent red blood cells, and invading microorganisms. As noted above, activated macrophages also secrete the multifunctional cytokines, IL-1 and TNF, which play key roles both in the body's general response to infectious agents and in the regulation of hemopoiesis.

In acute inflammation, first granulocytes and then macrophages accumulate in the inflamed tissue. The macrophages are derived mainly from blood monocytes attracted by chemoattractants to the inflamed site and retained at the site by processes that cause the cells to adhere to surfaces and differentiate. One such process involves a striking increase in the surface area of the cell, a phenomenon called *spreading*. Spreading results from reactions at the monocyte surface involving two components of the complement system, factors B and C5 (see Chapter 22) and the fibrinolytic enzyme plasmin. A second mechanism for adhesion and retention involves the binding of monocytes-macrophages to a protein of plasma, connective tissue matrix, and cell surfaces called fibronectin (Bevilacquia et al., 1981). Binding to fibronectin enhances the activity of monocyte-

macrophage surface receptors involved in phagocytosis (Fc and C3b receptors) and induces secretion of proteases and plasminogen activator necessary for the scavenger function of macrophages.

A multinucleated giant cell, a terminal stage of monocyte-macrophage development, may be formed in areas of inflammation by the apparent fusion of macrophages. Multinucleated giant cells are particularly prominent in the coordinate inflammatory process known as a granuloma, where T lymphocytes, plasma cells, and monocytes are organized into a specific spatial relationship with multinucleated giant cells. A soluble factor released from T cells, possibly γ-inferferon, causes giant cell formation at the site of granulomas. Although multinucleated giant cells have phagocytic capabilities, their function in granulomatous inflammation remains unknown.

Eosinophils, Basophils, and Mast Cells

EOSINOPHILS

The growth factors IL-3 and GM-CSF can support the growth in vitro of colonies of both mixed myeloid cells and nearly pure eosinophils. A recently described and less well characterized growth factor, IL-5, also directly stimulates the proliferation and maturation of eosinophils. Cells of the eosinophilic series, which normally make up about 3% of nucleated marrow cells, mature in the bone marrow through stages similar to those of the granulocytic series. Most kinetic studies of eosinophil production have been carried out in rodents. These have revealed a mean marrow transit time of 5.5 days, an intravascular compartment made up of a marginal pool and a circulating pool of equal size, and an intravascular half-time of about 8 hours. For each circulating eosinophil, 100 are reportedly present in the tissues, where they are found primarily in the skin and in the submucosa of the respiratory, gastrointestinal, and genitourinary tracts.

The mature eosinophil contains a bilobed nucleus and distinctive large granules which stain orange-red using Wright's stain. A material called major basic protein (MBP) makes up about 50% of the mass of the large granules (Butterworth and David, 1981). MBP is thought to play a key role in the eosinophil's ability to damage the larva-tissue stage of helminthic parasites and is an extremely potent tissue toxin. A potent bactericidal and tissue toxin called eosinophilic cationic protein (ECP), a neurotoxic protein, and an eosinophil peroxidase with properties different from the myeloperoxidase of neutrophils, are also present in the large granules. Within its finer granules, the eosinophil contains a high concentration of aryl sulfatase B, an enzyme that hydrolyzes S-O ester linkages and which can inactivate the sulfur-containing leukotrienes that tissue mast cells liberate in immediate hypersensitivity reactions. Eosinophils also contain and secrete an unusual membrane-bound protein, lysophospholipase, which forms crystals called Charcot-Leyden crystals in the pulmonary secretions of patients with asthma.

Eosinophils increase in number when T cells are activated because of T cell-derived IL-5, although IL-3 and GM-CSF may play a role in some responses. Mast cells synthesize a number of eosinophil chemotactic factors including chemotactic tetrapeptides, higher molecular weight peptides, and platelet activating factor (PAF). For these reasons, the blood eosinophil count often rises in helminthic infestations and in allergic states. Even when blood eosinophil counts are normal, eosinophils are usually increased in number at tissue sites of allergic reactions, where, by degrading mast cell products, they may decrease the clinical manifestations of allergic responses.

Administration of adrenal glucocorticoids causes blood eosinophil levels to fall within about 2 hours, presumably because of margination or sequestration of circulating cells. Continued administration of adrenal glucocorticoids results in an eosinopenia associated with impaired release of eosinophils from the bone marrow. Eosinopenia is found in acute bacterial infections, and is associated with a diminished release of marrow eosinophils (Bass, 1975).

BASOPHILS AND TISSUE MAST CELLS

Basophils are multilobed cells that on Wright's stain contain large, purple-black metachromatic granules. They are the least numerous of human white blood cells (see Table 3.3), making up less than 0.1% of nucleated bone marrow cells. Basophils are produced from marrow stem cells and may follow a maturation sequence similar to that for neutrophils and eosinophils. The cytokines regulating basophil production are not fully defined, but infusion of IL-3 into primates has been found to increase circulating basophils (see below). Basophils are thought to remain in the blood only for hours; their fate in the tissues is uncertain.

The basophil surface membrane contains receptors for the Fc fragment of IgE molecules. In acute allergic reactions, a specific antigen reacts with IgE bound to basophils, and the basophils release their granule contents, which include the allergic mediator, histamine. Basophils also contain the proteoglycan, chondroitin sulfate, whose function within the basophil is not presently known.

A related cell found in the tissues, the mast cell, also possesses metachromatic granules containing histamine and surface membrane receptors for IgE. Mast cells differ morphologically from basophils in possessing a round rather than a multilobed nucleus and their granules contain a different proteoglycan, namely heparin. Thus, although mast cells are also derived from a marrow stem cell and can be stimulated to grow by IL-3, blood basophils are not the precursors of tissue mast cells. Mast cells are most abundant at host-environment interfaces (lung, skin, lymphoid tissues, and the submucosal layers of the digestive tract), where they may be present in concentrations as high as $1–2 \times 10^6$ cells per gram of tissue. Two types of mast cells, a connective tissue type and a mucosal type, have been identified by their different fixation properties and protease content. There may be interconversion of the two mast cell types under certain in vitro conditions.

Mast cells contain or can synthesize multiple mediators of immune and inflammatory responses: histamine, prostaglandins, leukotrienes, platelet-activating factor, proteases and other lysosomal enzymes, and materials affecting chemotaxis of eosinophils and neutrophils (Wasserman, 1989). Mast cells also contain the Charcot-Leyden crystal protein and small amounts of major basic protein. Through their activation and degranulation after the binding of antigen to IgE on the surface membrane, mast cells play a central role in triggering immediate hypersensitivity reactions. They also participate in inflammatory responses and tissue repair.

Thrombopoiesis

STAGES OF THROMBOPOIESIS

Platelets are derived from giant bone marrow cells called megakaryocytes, which are readily recognized by their large-size, multilobulated nucleus and abundant cytoplasm. Normal marrow contains about 1 megakaryocyte per 500 nucleated red blood cells. Megakaryocytes arise from a population of small to medium blast-like cells called megakaryoblasts. Analysis of their DNA content suggests that megakaryoblasts contain even number multiples of the normal human DNA content (2n, 4n, 8n, etc.). This results from nuclear duplication without cell division, a process called *endoreduplication* that is idiosyncratic to the megakaryocytic lineage. Megakaryoblasts cannot be identified by their morphology, but can be identified with antisera specific for antigens on a surface membrane glycoprotein complex, glycoprotein IIb-IIIa, which is present on all nucleated cells of megakaryocytic lineage as well as on mature platelets.

Endoreduplication leads to the formation of different ploidy classes of human megakaryocytes.

(Ploidy refers to the number of sets of chromosomes in a cell; almost all human cells contain two sets of chromosomes, i.e., are diploid or 2n in class). The modal ploidy distribution is 16n, i.e., most human megakaryocytes have a chromosome complement 8 times that of a diploid cell.

The ploidy class of the megakaryocyte determines the size and degree of lobulation of its nucleus; each nuclear lobule corresponds approximately to 2n chromosomal material. The ploidy class also determines the amount of cytoplasm; megakaryocytes with a higher ploidy have more cytoplasmic mass and, therefore, give rise to increased numbers of platelets, larger platelets, or both. At the ultrastructural level, cytoplasmic maturation is characterized by the development of granules and mitochondria, and by the appearance of an increasing amount of cytoplasmic plasma membrane in the form of tubules and branching cysternae (demarcation membrane system) whose lumen communicates with the outside of the cell. This demarcation membrane system becomes the plasma surface membrane and open canalicular system of the platelet (see Chapter 24).

Mature megakaryocytes are ameboid cells that extend cytoplasmic pseudopods through the endothelial lining cells into the lumen of the marrow sinusoids, where the pseudopods then fragment to form platelets. Platelets remain within the intravascular space for their lifetime, but at any moment about ⅓ are pooled within a unique component of the intravascular compartment, the red pulp of the spleen. (In certain diseases with splenomegaly, up to 85% of the platelet mass may be pooled in a greatly expanded red pulp, with a resultant decrease in the peripheral blood platelet count.)

Labeling studies with ^{3}H-thymidine in rats suggest a transit time of a little less than 3 days for a cell to progress from a 2n morphologically unrecognizable precursor cell in cycle to the bare nucleus of a megakaryocyte that has shed its cytoplasm as platelets. Similar data are not available for humans, but, about 5 days are required for the platelet count to return to normal in patients whose platelets have been depleted acutely.

REGULATION OF THROMBOPOIESIS

As mentioned earlier in this chapter, megakaryocytes are present in mixed colonies grown in vitro with IL-3 and GM-CSF. Under appropriate conditions, pure colonies of up to 50 megakaryocytes are also observed, which suggests that specific megakaryocyte progenitor cells may exist. A hemopoietic growth factor (CSF) specific for megakaryocytes has not yet been identified (Mazur, 1987).

Whether such a CSF exists or not, experimental

data are consistent with a 2-stage regulatory model for the control of platelet production. When megakaryocytes are depleted from the marrow by drugs or toxins, the marrow content of megakaryocyte colony-forming cells increases. However, when peripheral circulating platelets decrease, a material appears in plasma that stimulates megakaryocyte protein synthesis and the shedding of megakaryocyte cytoplasm as platelets but which does not support megakaryocyte colony growth. This material has been termed *thrombopoietin* and is only partially characterized (Levin, 1987). Thus, one material or set of materials appears to regulate early megakaryocyte progenitor proliferation and a second material (or materials) appears to regulate megakaryocytic maturation and platelet production (Williams and Levine, 1982). In addition, TGF-β, a material that is released from the α granules of platelets when they are activated, inhibits the growth in vitro of megakaryocytic colonies (Ishibashi et al., 1987) and could conceivably act as a negative feedback regulator of platelet production in vivo.

When an experimental animal receives a severe thrombocytopenic stimulus, a small number of platelets, equivalent to only about 1 day's expected platelet production, are rapidly released into the blood. (This limited marrow reserve of platelets contrasts with the larger marrow reserve of granulocytes.) Within 24 hours after induction of experimental thrombocytopenia in mice, both the number and the size of the megakaryocytes in the marrow are increased. The increase in size reflects extra mitoses during endoreduplication and a shift in megakaryocyte ploidy. The increase in megakaryocyte numbers probably reflects increased cell divisions by megakaryocyte progenitors already in cycle and their subsequent development into morphologically recognizable megakaryocytes. In acute thrombocytopenic states resulting from platelet destruction, platelet production may increase from two to four fold over several days; in chronic thrombocytopenic states resulting from platelet destruction, platelet production may be increased six to ten fold (Harker, 1974).

DISORDERS OF BLOOD STEM CELLS

Aplastic Anemia

Since blood cells have finite life-spans, hemopoietic stem cells must continually supply progenitor cells for production of new blood cells. Failure of this stem cell function results in *aplastic anemia*, a disorder in which the mature forms of all the myeloid cell lines are markedly decreased in the peripheral blood (pancytopenia). Within 2 weeks of an event that destroys stem cells or severely impairs their function (e.g., exposure to a very high dose of radiation), an individual will become highly susceptible to bacterial infection because of severe granulocytopenia and to serious bleeding because of severe thrombocytopenia. Within 2 months, the circulating red blood cell mass will fall and red blood cells must be transfused to maintain tissue oxygenation. On bone marrow examination, fat will be found to have replaced normal marrow cellular elements (marrow aplasia).

Damage to the stem cell pool by environmental toxins or radiation, suppression of stem cell proliferation and self-renewal by immune mechanisms, or some combination of these events are known mechanisms inducing permanent, severe depletion of self-renewing stem cells. However, in most patients with aplastic anemia a specific causal event is never identified. Young patients with aplastic anemia and a genetically compatible donor are often given a drug to suppress their immune system followed by an allogenic bone marrow transplantation to provide an exogenous source of self-renewing pluripotent stem cells.

Proliferative Stem Cell Disorders

The prototypical proliferative stem cell disorder is *chronic myelocytic leukemia (CML)*. In this dis-order, a monoclonal stem cell proliferates to produce excess numbers of all granulocytic forms (neutrophils, eosinophils, basophils), excess monocytes, and excess platelets. The erythrocytes also arise from the abnormal clone, but erythrocyte production is usually depressed by the time the patient is first seen. All myeloid cells contain a characteristic chromosomal abnormality, termed the Philadelphia chromosome after the site of its first description. Finding this abnormality established that CML is a disorder arising in a multilineage stem cell.

The Philadelphia chromosome results from a translocation in which genetic material is exchanged between chromosomes 9 and 22. As a result, a normal proto-oncogene on chromosome 9 (c-abl) is fused to a truncated normal cellular gene (called bcr for breakpoint cluster region) on chromosome 22. The translated product of this fusion gene is an abnormal protein with tyrosine kinase activity that is thought to play a key role in the pathogenesis of CML (Kurzrock et al., 1988).

Evidence exists that other proliferative blood disorders including polycythemia vera, myelofibrosis, and some but not all instances of acute myelogenous leukemia are disorders arising in multilineage stem cells. Descriptions of these disorders can be found elsewhere (Rapaport, 1987).

CHAPTER 21

Lymphocytes and Immune Responses

Lymphocytes are the effector cells for immune responses to immunogens, i.e., to materials not recognized as "self" and therefore eliciting reactions designed to neutralize or destroy "nonself." Immune responses combat invasion of the body by microorganisms, cause rejection of organ transplants between individuals with different histocompatibility antigens, and may also suppress the growth of malignant cells.

Immune responses are of two types, cell mediated and antibody mediated (humoral), and require the participation of both lymphocytes and macrophages (Nossal, 1987). The lymphocytes are of three functional classes: T lymphocytes, which both regulate and mediate cellular immune reactions and also regulate antibody synthesis; B lymphocytes, which make some antibodies themselves and are the precursors of plasma cells, the body's principal antibody-forming cells; and natural killer (NK) cells, which constitute a small fraction of lymphocytes that can lyse certain target cells without prior antigenic stimulation or exposure. Macrophages and related accessory cells (the dendritic cell in lymph nodes and the Langerhans cell in the skin) are required to process and present antigens, to secrete cytokines supporting immune responses, and, after macrophage activation by T lymphocytes, to kill ingested organisms in one form of cell-mediated immunity (Unanue and Allen, 1987).

MAJOR HISTOCOMPATIBILITY COMPLEX GENE PRODUCTS AND IMMUNE RECOGNITION

Immune recognition responses require that T lymphocytes interact with specific cell surface molecules, called class I and class II molecules (McDevitt, 1982; Benacceraf, 1981). These are products of genes in a cluster of loci on chromosome 6. This cluster has been named the major histocompatibility complex (MHC) because antigenic determinants on class I and class II molecules largely determine whether organ or tissue transplants are recognized as self or foreign and, therefore, are accepted or rejected (Bach and Sachs, 1988). In humans, the MHC is frequently referred to by a different name, the HLA (human leukocyte antigen) region, because it was first recognized by analyzing patterns of reactions of peripheral blood lymphocytes with sera containing antibodies to leukocytes.

The HLA system has been described most simply as consisting of four loci: HLA-A, HLA-B, HLA-C, and HLA-D. However, the HLA-D region is no longer recognized as a single locus but as at least 3 loci called DP, DQ, and DR. Genes at the HLA A, B, and C loci code for the heavy chain of class I molecules and genes at the DB, DQ, and DR loci code for class II molecules.

Most nucleated cells possess class I molecules on their surface membrane. These are noncovalently linked heterodimers (complexes of two nonidentical polypeptide chains) consisting of a smaller β_2-microglobulin chain and a larger heavy chain. The β_2-microglobulin chain, which is coded for by a gene on chromosome 15, is invariant, i.e., all β_2-microglobulin chains of class I molecules are the same. The heavy chains are coded for by genes at the HLA A, B, and C loci that are highly polymorphic, i.e., multiple alternative (allelic) genes exist for each locus. Since a person inherits 3 heavy chain genes from each parent, the class I molecules of any given individual can consist of a mixture of molecules containing up to 6 different heavy chains.

Class II molecules are expressed normally by only a few cell types: hemopoietic progenitor cells, B lymphocytes, and monocytes-macrophages including dendritic cells and Langerhans cells. However, T lymphocytes, when activated, also express class II molecules. In addition, activated T lymphocytes can secrete a lymphokine, γ-interferon, that amplifies the expression of class II molecules on macrophages and can cause class II molecules to be expressed on other cells, such as endothelial cells. Class II molecules are heterodimers of α and β chains, but differ from the heterodimers of class I molecules in that both polypeptide chains are coded for by genes of the HLA region, i.e., by genes at the DP, DQ, and DR loci. The β chain genes are highly polymorphic, as is the α chain gene at the DQ locus (but less so at the DR and DP loci). Because of this, a given individual may possess up to 20 different class II molecules.

In the regulation of antibody production and in the initiation of cell-mediated immune reactions of the delayed hypersensitivity type (see below), helper T lymphocytes must recognize immunogenic fragments of processed antigens embedded within clefts in class II molecules on the surface of macrophages or B lymphocytes. Since the ability to bind fragments varies with the topography of the clefts, and the latter varies for different allelic class II β chains, the helper T cells of different individuals may recognize different immunogenic determinants of a foreign antigen. This, in turn, means that different individuals may vary in their ability to mount an effective immune response to a given foreign antigen. Class I molecules are involved in the sensitization of a different type of T lymphocyte, a cytotoxic T lymphocyte (see below), to viral or cellular antigens that appear in association with class I molecules on the surface of cells infected with certain viruses. Moreover, not only the initial sensitizing event but the subsequent attack of the cytotoxic T cell upon the target cell depends upon the lymphocyte recognizing antigen in association with class I molecules on the cell surface.

It is not yet clear how T lymphocytes recognize the class I and II molecules of foreign cells as different and so trigger cell rejection. Since each individual expresses class I and II molecules coded for by a number of HLA region genes and each gene, in turn, has multiple alleles (e.g., more than 40 alternative genes exist for the β chain coded for at the HLA-B locus), there is almost no chance that the class I and class II molecules of unrelated individuals will all be identical. In contrast, a 25% chance exists that siblings will have identical class I and class II molecules. This is because the HLA genes are located so close to each other on chromosome 6 that they are inherited as a unit or *haplotype*. Each person inherits a particular combination of alleles for all HLA loci (haplotype) from one parent and a second different combination of alleles for all HLA loci (haplotype) from the other parent. HLA typing (which is carried out, for example, on someone needing a bone marrow transplant and all potential sibling donors) can be performed for the HLA A, B, and C loci by mixing an individual's peripheral blood lymphocytes with a panel of typing sera containing antibodies to different class I antigenic determinants. Antisera are also available for DR and DQ but

not for DP. In determining compatibility of class II molecules prior to certain organ transplants, donor and recipient lymphocytes are also cocultured (Mixed Lymphocyte Reaction). If their D antigens differ, lymphocytes are activated and begin to synthesize DNA, which is detected by their uptake of ^{3}H-thymidine.

PRODUCTION, DISTRIBUTION, AND FUNCTIONS OF LYMPHOCYTES

Organization of Lymphoid Tissue

In lymphopoiesis, marrow stem cells give rise to precursors of T and B cells that are processed into immunologically competent cells in sites termed the *central lymphoid tissues*. For T cells, this site is the thymus. For B cells, the site in mammals is thought to be the bone marrow. Processing, which is associated with multiple cell divisions, gives rise to a large repertoire of lymphocytes capable of distinguishing between self and nonself class I and II molecules and of recognizing different foreign immunogenic determinants.

Once processed, T and B cells leave the central lymphoid tissues and are distributed to the *peripheral lymphoid tissues*. The peripheral lymphoid tissues include: lymph nodes located throughout the body; the spleen; the ring of tonsillar tissues in the oropharynx (Waldeyer's ring); submucosal accumulations of lymphocytes in the respiratory tract, urinary tract, and in the gut (particularly in the terminal ileum as Peyer's patches); and a portion of the lymphocytes in the bone marrow. T cells, which make up about 70% of peripheral blood lymphocytes, do not usually settle within a single fixed location in the peripheral lymphoid tissues, but cycle back and forth between the tissues and the blood. B cells may also recirculate between the peripheral lymphoid tissues and the blood but to a limited extent.

Quiescent lymphocytes of the peripheral lymphoid tissues are dormant-appearing small lymphocytes, but are transformed into large, proliferating cells after exposure to antigen. When antigenic exposure increases in a particular body area, proliferative activity in the peripheral lymphoid tissue adjacent to that area increases. For example, if one develops a bacterial infection of the upper arm, the lymph nodes of the corresponding axilla will enlarge. Bacterial antigens are carried by afferent lymphatics to the lymph nodes and stimulate the clonal expansion of many different T and B lymphocytes, each programmed to recognize a single one of a number of specific immunogenic determinants present on a given bacterial antigen. Activated

T and B cells may also leave the nodes to seed more distant areas of the peripheral lymphoid tissue.

When stimulation by antigen ceases, proliferative activity regresses under the influence of cellular and humoral factors that suppress the immune response. However, a fraction of the T cells, educated by their contact with antigen, persist for many years as members of a body pool of memory T cells. Memory B cells of unknown but shorter lifespans are also formed, which, on subsequent exposure to the same immunological determinants, give rise to plasma cells that make high affinity antibodies of the IgG class (see below).

Neonatal thymectomy in mice produces a profound "wasting disease" due to a deficiency of T cells, whereas thymecotomy in older mice produces much less of an effect. This difference reflects the building up in the interval of a population of T cells in the peripheral lymphoid tissues capable of at least partially sustaining T cell functions in the absence of renewal of T cells from the thymus. In humans, the thymus begins to atrophy after puberty and is reduced to about 15% of its maximum size by age 50 years. Involution of the thymus, with decreased formation of new T cells and consequent increased dependence upon a shrinking pool of memory T cells for immune responses, may be causally related to senescence of the immune system in the aged (Weksler, 1981; van de Griend et al., 1982).

If not stimulated by antigens or nonspecific mitogens, newly processed B cells in the peripheral lymphoid tissues die within a few days (Cooper, 1987). Although an increasing pool of long-lived B memory cells is acquired during the first years of life that increase the efficiency of antibody production, normal B cell immune responses require the steady production of new B cells throughout an individual's life.

T Lymphocyte Ontogeny and Functions

THYMIC PROCESSING AND T CELL RECEPTOR SYNTHESIS

T cell precursors migrate from the bone marrow to the cortex of the thymus, where the cells undergo an extremely high rate of cell division. The vast majority of these early *thymocytes* die within the thymic cortex. A minority of lymphocytes from the dividing pool in the thymic cortex migrate to the medulla of the thymus, where they undergo further differentiation. T cells surviving this thymic processing will possess a cell surface T cell antigen receptor (Marrack and Kappler, 1987) with the following recognition properties:

(1) The receptor distinguishes between foreign MHC proteins (another name for class I and II molecules) and self MHC proteins, reacting with the former but tolerating the latter.
(2) Although the receptor does not react with self MHC proteins in the absence of antigen, the receptor recognizes and reacts with a complex made up of *an antigen-derived peptide bound to a self MHC protein.*
(3) The receptor of each T cell will be clonotypic, i.e., will recognize only a single specific immunogenic determinant (epitope) of an antigen-derived peptide bound to an MHC protein.

The T cell antigen receptor is a multisubunit complex (Fig. 3.4). On more than 95% of T cells, the receptor complex contains two disulfide-linked polypeptide chains, called α and β chains, that mediate antigen recognition. These chains are noncovalently linked to five invariant chains (α, δ, ϵ, and two ζ chains) that are involved in signal transduction, i.e., in sending a signal to the interior of the cell that the receptor is occupied (Royer and Reinherz, 1987). Monoclonal antibodies have been developed that recognize an antigen on the invariant chains of the complex. This antigen is called the CD3 antigen and the complex of the invariant chains is sometimes referred to simply as CD3 complex. (Since all mature T cells will possess CD3, this antigen is used as a surface marker to identify mature T cells in tissues and body fluids.)

The α and β polypeptide chains of each cell have

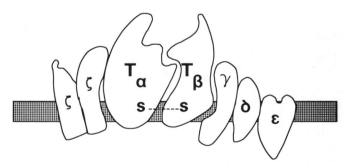

Figure 3.4. A schematic representation of the T cell antigen receptor. The α and β polypeptide chains contain unique amino acid sequences in their NH$_2$-terminal segments that form a binding pocket permitting the T cell antigen receptor of each cell to recognize a specific immunogenic determinant in association with an MHC-coded surface molecule. The γ, δ, ϵ, and ζ chains are referred to as the CD3 complex and are involved in signal transduction. (Modified by permission from Royer, H. D. and E. L. Reinherz. T lymphocytes: ontogeny, function and relevance to clinical disorders. *N. Engl. J. Med.* 317: 1136–1142, 1987.)

"hypervariable" segments containing unique amino acid sequences that permit the antigen receptors of that cell to recognize a specific epitope on an MHC protein-bound antigen fragment. This results from different physical rearrangements of the germ line DNA of each differentiating T cell during thymic maturation. The α gene on chromosome 14 undergoes rearrangements that bring together one of some 50 V (variable) region DNA segments with one of some 50 J (junctional) region DNA segments and a single C (constant) region gene. Similarly, the β chain gene on chromosome 7 undergoes rearrangements involving 50 V regions, 2 D (diversity) regions, and 13 J regions, which are then joined to 1 of 2 C genes. Moreover, nucleotides may be introduced or deleted during these rearrangements producing additional variations in the genes.

During thymic processing T cells also acquire surface glycoprotein molecules that, in as yet an unknown way, determine whether the antigen receptor on the T cell will react with an antigenic peptide bound to class I molecules or to an antigenic peptide bound to class II molecules. Two such surface glycoproteins can be identified by their reaction with monoclonal antibodies that recognize an antigen termed CD4 present on one of the glycoproteins and an antigen termed CD8 present on the other glycoprotein. Although thymic T cells exist that possess both the CD4 and the CD8 antigen, the T cells that are released from the thymus to enter the peripheral lymphoid tissue contain only one of the antigens, i.e., are either CD4+ (sometimes referred to as T4 cells) or CD8+ (sometimes referred to as T8 cells).

ANTIGEN RECOGNITION AND T CELL ACTIVATION

T cells released from the thymus circulate freely throughout the body, moving back and forth between the circulating blood and the peripheral lymphoid tissues. Each cell is programmed to recognize a specific antigenic determinant that has been processed within cells and expressed on the surface of cells in association with MHC proteins. Thus, the T cell must be in physical proximity to the antigen presenting cell in order to recognize and be activated by antigen. CD4+ cells recognize antigenic peptides expressed on the surface of cells in association with class II molecules, and, therefore, antigen that has been processed within and expressed by macrophages or B lymphocytes. CD8+ cells recognize surface antigenic determinants that become expressed in association with class I molecules on a variety of cells, including some cells infected with viruses. In addition, both

CD4+ and CD8+ cells interact with antigen-presenting cells through a newly recognized second mechanism involving an interaction between a molecule called CD2/T11 on the lymphocyte and a molecule called LFA (lymphocyte function-associated antigen) 3 on the antigen-presenting cells (Breitmeyer, 1987).

Activation by antigen through the above mechanisms plus the supporting effect of a cytokine secreted from macrophages (thought originally to be IL-1 but now thought possibly to be IL-6) causes antigen-activated cells to transform; to develop receptors for an autocrine growth factor, interleukin-2 (IL-2); and to secrete IL-2. As a consequence, what are initially a very few T cells recognizing a small number of epitopes of a processed antigen give rise to proliferating clones of daughter cells. The extent of the proliferation reflects the combined effect of the concentration of secreted IL-2, the duration of secretion, and the density of IL-2 receptors that develop on the surface of the activated T cells (Smith, 1988). Subpopulations with overlapping immunoregulatory and effector functions (Fig. 3.5) arise and may include:

(1) Helper T cells that help B cells develop into antibody-secreting plasma cells;
(2) T cells that mediate the delayed hypersensitivity type of cell-mediated immunity (see below);
(3) Cytotoxic T cells that kill target cells;
(4) T cells that induce the formation of suppressor cells;
(5) T suppressor cells that keep immune responses from escaping control by dampening the proliferation and differentiation of antibody-producing B cells and of cytotoxic T cells.

Activated T cells exert their immunoregulatory and effector functions through the secretion of proteins called lymphokines (Dinarello and Mier, 1987). These materials (e.g., IL-2, IL-4, γ interferon, and so forth) are active in very low concentrations and are neither antigenically nor genetically restricted in their biological effects. A detailed description of the lymphokines and their functions is beyond the scope of this book, but their actions are summarized in part in the schematic diagram shown in Figure 3.5. Although appearing complex, this figure represents a simplified version of the present understanding of the multiple reactions of the human immunoregulatory network.

Natural Killer Cells

A distinct population of lymphocytes involved in cell-mediated immune responses, the NK or natural

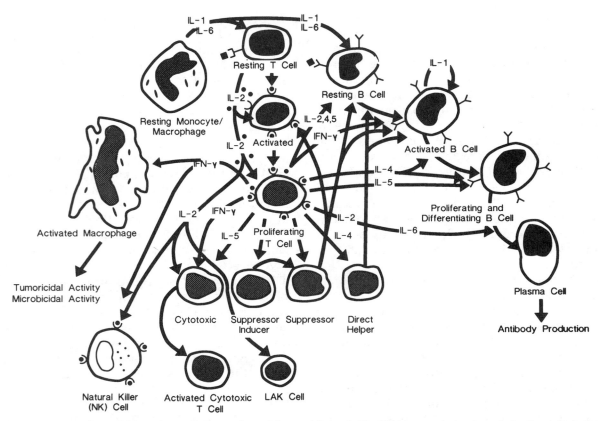

Figure 3.5. A schematic diagram illustrating multiple reactions of the human immunoregulatory network. See text for discussion. Abbreviations are: *IL,* interleukin; *IFN,* interferon; *LAK,* lymphokine-activated killer cell. (Modified by permission from A. S., Fauci, S. A. Rosenberg, S. A. Sherwin, C. A. Dinarello, D. L. Longo, and H. C. Lane. Immunomodulators in clinical medicine. *Ann. Intern. Med.* 1987; 106: 421–433.)

killer cell, is also shown in Figure 3.5. NK cells differ from cytotoxic T cells in that they lyse sensitive cell targets on first contact without a requirement for prior antigen sensitization. Their ontogeny is unclear, but marrow transplant studies indicate that they do not descend from the lymphoid stem cells giving rise to T and B lymphocytes. NK cells do not require thymic processing and do not express surface T cell antigen receptors. NK cells represent less than 10% of circulating blood lymphocytes. Their specific roles in immune responses are not yet clear.

B Cell Differentiation and Functions

B CELL ONTOGENY

B cell differentiation from a primitive precursor cell into a terminally differentiated plasma cell occurs in two phases (Cooper, 1987). The first phase, sometimes referred to as the nonantigen driven phase, takes place within the central lymphoid tissues (bone marrow of mammals). During this phase precursor cells undergo both very rapid cell division and a high rate of cell death. This wastage of cells, which is analogous to the wastage of

pre-T cells during T cell processing in the thymic cortex, probably represents a consequence of the extensive germline DNA rearrangements (see below) needed for the maintenance of a repertoire of B-cells in the peripheral lymphoid tissue capable of recognizing and responding to a virtually unlimited array of antigens. Cells with aberrant DNA rearrangements or with DNA rearrangements coding for antigenic determinants too closely resembling self presumably die during processing.

Several stages of the *antigen-independent phase* of B cell differentiation have been identified. It begins with the rearrangement of germline immunoglobulin gene segments, recognized by a molecular genetic technique as forming a functional gene for the heavy chain of a class of immunoglobulin molecules termed IgM. (The structure of immunoglobulin molecules and their different classes are described in the sections that follow.) This then allows the cell to begin synthesis of a heavy chain of IgM, termed a μ chain, which can be demonstrated by immunostaining within the cytoplasm of the cell. Following this, a gene for a light chain of immunoglobulin molecules is rearranged and light chain synthesis begins. The light chain combines with the μ chain, which can then move from the cytoplasm to be inserted into the cell membrane as part of a whole IgM molecule (Fig. 3.6). Next, by a mechanism described later, a second class of heavy chain, called a δ chain, is also made. This allows a second class of immunoglobulin molecule, an IgD molecule, also to become inserted into the cell membrane. The cell now possesses both IgM and IgD molecules embedded by their carboxy-terminus into the surface membrane and extending to the external environment their variable NH$_2$-terminus region (see below). These serve as the surface recognition sites for the specific antigenic determinants to which the cell, by its germline DNA rearrangements, has become programmed to respond. Thus, the B cell is ready to function as an immunologically competent cell.

In addition to immunoglobulin antigen receptors, the B cell possesses class II MHC molecules on its surface membrane. During its processing the B cell also acquires other surface membrane molecules, whose physiological functions are as yet incompletely understood. These include receptors for the Fc portion of IgG (see below and Fig. 3.7), receptors for an activated component of complement (see Chapter 22) called C3b, and for a degraded fragment of C3b termed C3d. (This last receptor is of particular clinical interest because it also serves as a receptor whereby Epstein-Barr virus (EBV) can enter and infect B cells.)

The second, or *antigen-dependent phase* of B cell differentiation results from activation in the peripheral lymphoid tissue of immunologically competent B cells by contact with antigens that they are programmed to recognize in the presence of cytokines from activated T cells and macrophages. This antigen-driven proliferation leads to the formation of two types of cells. One is an IgM-secreting cell with properties intermediate between a lymphocyte and a plasma cell (plasmacytoid lymphocyte). The other is a B cell whose surface IgM and IgD are replaced (as a result of immunoglobulin heavy chain constant gene switching as described later) with surface IgG, IgA, or IgE (see Fig. 3.6). With continuing exposure to antigen most of these B cells differentiate further to give rise to plasma cells that secrete IgG, IgA, or IgE, but a subset enter a B cell memory pool. The latter retain their surface IgG, IgA, or IgE and can give rise, on subsequent exposure to the antigen, to clones of plasma cells producing high affinity antibodies. As mentioned earlier, B cells

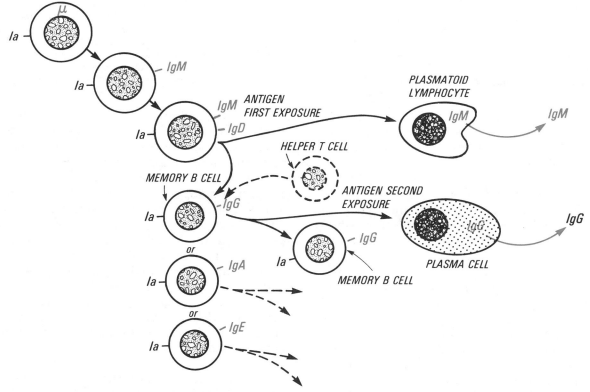

Figure 3.6. The development of B cells and their response to stimulation by antigen. Immunoglobulin B cell markers are printed in red. The symbol μ stands for the heavy chain of IgM. Ia is a synonym for class II major histocompatibility complex gene products.

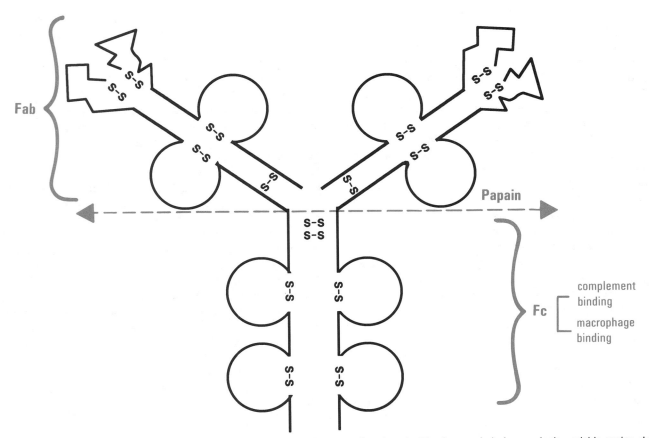

Figure 3.7. A schematic model of an IgG molecule. The molecule consists of two identical light chains and two identical heavy chains held together by disulfide bonds. Each chain contains an NH_2-terminal variable region, drawn with a *jagged line,* followed by one or more constant regions, drawn with a *straight line.* Interchain disulfide bonds cause the chains to be folded into globular regions called domains. The light chain has a single variable region domain and a single constant region domain. The heavy chain has a single variable region domain and three constant region domains. The proteolytic enzyme, papain, cleaves the heavy chains in an area known as the hinge, which splits the IgG molecule into two antibody-binding fragments, labeled Fab, and a single Fc fragment. Each Fab fragment is made up of a whole light chain and the NH_2-terminal portion of the heavy chain. The Fc fragment contains the carboxy-terminal portion of both heavy chains.

that fail to undergo an antigen-dependent phase of differentiation in the peripheral lymphoid tissue die quickly and must be continually replenished by newly processed B cells.

STRUCTURE OF IMMUNOGLOBULIN MOLECULES

The only known role of B lymphocytes and plasma cells is to secrete proteins with antibody activity, i.e., to secrete immunoglobulins. The immunoglobulins are remarkable proteins in that the NH_2-terminal polypeptide chains of an immunoglobulin molecule made by any given clone of B cells-plasma cells will differ from the polypeptide chains made by all other clones of B lymphocytes-plasma cells. This conveys upon immunoglobulin molecules their antigenic specificity.

Immunoglobulin molecules are made up of a basic unit consisting of two identical light polypeptide chains and two identical heavy polypeptide chains linked by interchain disulfide bonds (see Fig. 3.7).

Most immunoglobulin molecules are monomers of this basic unit. However, some immunoglobulins are polymers of two units or of five units. These polymeric immunoglobulins are composed of identical basic units plus a single additional chain, called a J chain, that holds the polymer together. J chains are also synthesized by B cells-plasma cells.

The amino acids of the light and heavy chains are not arranged linearly but form loops, called domains. The NH_2-terminal 110 amino acids of the chains contain a single domain known as the *variable region.* Within the variable region, are *hypervariable regions* made up of amino acid sequences unique for that polypeptide chain. The hypervariable regions of an adjacent light and heavy chain of an immunoglobulin molecule (see Fig. 3.7) form an antigen-binding pocket into which several antigenic determinants may fit partially (cross reactivity) but which will bind preferentially a single specific antigenic determinant.

The carboxy-terminal segment of the light and heavy polypeptide chains is called the *constant region*. Light chains have a constant region with a single domain, whereas heavy chains have a constant region of either three or four domains. Light chains have been divided into two types, κ and λ, on the basis of antigenic differences stemming from the different amino acid sequences of their constant region. Only a single germline gene exists for the κ chain constant region and therefore the κ chain constant region for a given individual is invariant. (However, κ chain constant regions may differ slightly between individuals because of the existence of allelic genes [allotypic differences].) In contrast, each individual possesses several germline genes for the constant region of the λ light chain and, therefore, each individual possesses several subtypes of λ light chains.

A series of constant region genes exists for heavy chains. Differences in the constant regions of the heavy chains give rise to the different classes and subclasses of immunoglobulin molecules described in the next section.

The proteolytic enzyme papain cleaves immunoglobulin molecules at the region of the heavy chain called the hinge. Two antibody-combining fragments (Fab) are formed, each made up of a whole light chain and the variable and first constant region domain of a heavy chain (see Fig. 3.7). A single fragment called the Fc fragment is also formed; it is made up of the remaining domains of the constant region of both heavy chains and determines the biological properties of an immunoglobulin molecule. For example, macrophages possess surface receptors that will recognize a binding site present on the Fc fragment of some but not all immunoglobulin molecules.

CLASSIFICATION AND BIOLOGICAL ACTIVITIES OF IMMUNOGLOBULINS

Immunoglobulins have been divided on the basis of differences in their heavy chains into five major classes; IgM (μ chain), IgG (γ chain), IgA (α chain), IgD (δ chain), and IgE (ϵ chain). Immunoglobulin classes can be divided further into subclasses based upon serological and physiochemical differences in their class-specific heavy chains. Thus, there are two subclasses of IgM, containing either μ_1 or μ_2 chains; two subclasses of IgA, containing either α_1 or α_2 chains; and four classes of IgG, containing γ_1, γ_2, γ_3, or γ_4 chains. Molecules with κ or λ light chains are distributed throughout each of the different classes and subclasses. Thus, every person's plasma contains a spectrum of immunoglobulin *isotypes*, made up of each possible combination of one type or subtype of light chain and one class or subclass of heavy chain.

Because each immunoglobulin synthesized by a different B lymphocyte-plasma cell clone differs in the amino acid sequences of its hypervariable regions, each immunoglobulin synthesized by a different clone is *antigenically* different. This antigenic difference confers upon each immunoglobulin molecule what is termed its own *idiotype*. Normally, no single immunoglobulin species is present in high enough concentration for its idiotype to act as an immunogen. However, when an antigen-driven immune response increases the production of specific immunoglobulins, their hypervariable regions can then function not only as antigen-binding sites, i.e., as antibodies, but also as idiotypic antigens capable of inducing their own immune responses. This anti-idiotype response can dampen antibody production to the primary antigens (see below).

Characteristics of the different classes of immunoglobulins are listed in Table 3.5. IgM antibodies are the first antibodies formed after exposure to antigen. IgM circulates as a pentamer with 10 antigen-binding sites on each molecule. Therefore, it functions as an efficient activator of the classical pathway of complement (see Chapter 22). Because of its large size (MW 900,000), plasma IgM is referred to as a macroglobulin.

Although a minority of antigens continue to evoke an IgM antibody response on continued expo-

Table 3.5.
Properties of Immunoglobulins

	IgG	IgA	IgM	IgD	IgE
MW (daltons)	150,000	170,000	900,000	180,000	200,000
Plasma concentration (mg/dl)	700–1500	250	100	3	0.03
Intravascular t½ (days)	21	6	5	3	2
Intravascular distribution (%)	45	40	75	75	50
Major Ig of plasma and extravascular fluid	+	−	−	−	−
Major Ig in secretions	−	+	−	−	−
First antibody made to antigen	−	−	+	−	−
Binds to mast cells	−	−	−	−	+
Crosses placental barrier	+	−	−	−	−

sure or re-exposure, for most antigens antibody production shifts to a different immunoglobulin class, usually IgG or IgA. IgG is the major antibody of plasma and extracellular fluid and the only class of antibody that can pass the human placental barrier. Phagocytes have receptors for a binding site on the Fc portion of IgG_1 (the subclass making up about 70% of IgG) and IgG_3. Phagocytes also have receptors for the complement component fragment iC3b (see Chapter 22). If antibodies of the IgG_1 or IgG_3 subclass bind to antigens on the surface of a microorganism or cell, and this binding, in turn, activates complement and brings down iC3b onto the surface, then a granulocyte or monocyte-macrophage can readily phagocytose the microorganism or cell.

IgA is the major antibody found in secretions of the respiratory, gastrointestinal, and genitourinary tracts, where it can bind foreign antigens and impair their entry into the body. In secretions, IgA exists as a polymer of two immunoglobulin molecules plus a J chain plus yet another protein called the secretory component. The secretory component is synthesized by mucosal epithelium and helps transport IgA molecules through epithelial cells into secretions. In plasma, IgA exists as either a monomer or a polymer containing a J chain but not a secretory component.

IgD and IgE are found in trace amounts in plasma. Whereas IgD on the surface of B cells serves as an antigen recognition site, the function of secreted IgD is unknown. IgE is secreted in increased amounts in patients with helminthic infections and in patients with atopic allergy. IgE binds by its Fc region to mast cells and can invoke mast cell-triggered immediate hypersensitivity reactions (described later).

GENETIC MECHANISMS OF IMMUNOGLOBULIN SYNTHESIS

In order to make an immunoglobulin molecule containing two identical light chains and two identical heavy chains, the diploid B cell-plasma cell must exclude one of the two parental genes it contains for making each of these polypeptide chains. Only one haploid set of genes are assembled and expressed in any given cell. Moreover, a B cell-plasma cell must possess genetic mechanisms permitting it to switch immunoglobulin class by attaching a different heavy chain constant region to its heavy chain variable region (Davis et al., 1980).

The rearrangements in the germline DNA required to produce immunoglobulin molecules that can bind a diverse array of antigenic determinants resemble the rearrangements described earlier to produce functional genes for the α and β polypeptides of the T cell antigen receptor. A functional gene for the variable region of a heavy chain results from assembly of three separate germline DNA fragments on chromosome 14 (Cooper, 1987). On both parental chromosomes, one of 20 D (diversity) germline gene segments and one of six J_H (*heavy* chain junctional) germline gene segments are brought together with deletion of the intervening DNA sequences. Next, on one chromosome, one of 50 V (variable) germline gene segments is joined to the DJ_H to give rise to a VDJ_H complex, which encodes the entire variable region domain of the heavy chain. Transcription then begins and gives rise to an mRNA transcript of the VDJ_H complex plus the DNA segments encoding the μ and δ constant regions (C_μ and C_δ). If the VDJ_H rearrangements have yielded a "useable" variable region mRNA, then the noncoding intervening mRNA segments are edited from the transcript, the mRNA is spliced to leave only the C_μ mRNA, and a μ heavy chain is transcribed. However, if the initial VDJ_H rearrangement has been nonproductive, then a V germline segment on the second chromosome will be translocated as the cell tries a second time to make a functional heavy chain variable region gene. This sequence ensures that each B lymphocyte makes a heavy chain with only a single variable region.

The assembly and transcription of light chains follow similar rules. The cell tries first to make a κ light chain by joining, on chromosome 2, a κ germline V (variable) segment to a κ J (junctional) segment to form a gene for the light chain's variable domain. Its RNA is transcribed and the transcript is spliced to the RNA transcript of the single κ constant region (C) gene to give rise to an mRNA for a complete κ light chain. However, in about ⅓ of B cells the rearrangements of the κ chain germline gene segments will be nonproductive. If so, the B cell will then attempt to assemble a functional λ chain gene by rearranging germline λ V, J, and C gene segments on chromosome 22. If successful, then a λ light chain is translated. Since κ rearrangements occur first, and are productive about ⅔ of the time, the ratio of κ- to λ-containing immunoglobulins for each antibody class is approximately 2:1.

The diversity of the antigen binding site of immunoglobulin molecules stems partly from the many combinations made possible by the assembly of different V, D, and J gene segments to form a functional heavy chain gene, by the assembly of different V and J genes to form a κ or λ light chain gene, and by the different combinations of heavy and light chain genes used to make the whole immunoglobulin molecule. A further source of diversity results from DNA frame shifts that may occur at the sites joining the individual gene segments. Such frame shifts greatly expand the possibilities for further amino acid variability of the variable region of the polypeptide chains.

The B cell-plasma cell uses two mechanisms for combining the same heavy chain variable region with different constant regions, i.e., to make more than one class of immunoglobulin molecule. Since the maturing B cell acquires both surface membrane-bound IgM and surface membrane-bound IgD, the cell must make both classes of immunoglobulin molecules during its processing. The constant region gene segments for IgM and IgD (C_u and C_δ) are located next to each other on the heavy chain gene locus. Both are transcribed, and the initial transcript is converted, by alternative sites of RNA splicing, into mRNA either for IgM or IgD. However, the sites for the series of constant region genes encoding the IgG and IgA subclasses are at a distance from the C_μ-C_δ sites on the germline heavy chain gene. Therefore, when a B cell switches from making IgM on initial exposure to an antigen to making IgG or IgA on continuing exposure to the antigen, germline DNA segments must be cut and rejoined at switch regions as described elsewhere (Cooper, 1987).

MECHANISMS OF IMMUNE RESPONSES

Humoral immune responses are particularly important for protection against acute bacterial infections. As already mentioned, bacteria coated by antibody and, as a consequence, by complement are phagocytosed with greatly increased efficiency. Diseases impairing antibody production increase a patient's risk for serious acute bacterial infections.

Cell-mediated immune responses suppress infections by microorganisms that reside and multiply within cells. These include many viral infections and infections with certain bacteria, such as mycobacteria, and fungi. As AIDS patients with depleted CD4+ cells so tragically illustrate, microorganisms that usually do not cause human disease may cause fatal disease when cell-mediated immune responses are lost. Cell-mediated immune responses may also provide immune surveillance against body cells that have undergone premalignant changes.

Synthesis of Antibodies and Regulation of Antibody Production

Antibody production usually requires interactions between macrophages, T cells, and B cells (Unanue and Allen, 1987). However a few antigens, known as *thymus-independent antigens*, induce B cells to form antibodies without T cell help. Such antigens are large molecules with repeating structural units, e.g., bacterial polysaccharides, in which the antigen apparently functions as a B lymphocyte mitogen. Thymus-independent antigens result in synthesis of only IgM antibody and induce little, if any, immunological memory. (The ABO blood group substances represent a clinically important example of thymus-independent antigens.)

Production of antibody to the vast majority of antigens requires that both antigen and helper T cells interact with B cells. Such antigens are therefore referred to as *thymus-dependent antigens*. Processed antigen fragments containing immunogenic determinants are presented in association with a cell surface class II molecule to CD4+ helper T cells. The presenting cell was originally thought to be either a macrophage or a macrophage-related cell, such as the dendritic cell or Langerhans cell. It is now known that the presenting cell may also be a B cell. The B cell can recognize antigen by use of its membrane-bound immunoglobulin receptor, internalize and process the antigen, and then display antigenic peptides on its cell surface in association with class II molecules. As mentioned earlier, in the presence of macrophage-secreted IL-1 or IL-6, CD4+ T cells activated by antigen presented by macrophages or B cells proliferate and secrete multiple cytokines (e.g., IL-2, IL-4, IL-5, γ-interferon). Under the influence of these lymphokines, of IL-6 from macrophages, and of a continuing exposure to antigen, clones of B cells expand and differentiate into plasma cells secreting large amounts of antibody (Fig. 3.6).

Activation of T cells by antigen also induces formation of suppressor cells that control the antibody response by dampening both the proliferation of helper T cells and the expansion of B cells (Fig. 3.5). A decreasing availability of the antigen further modulates the antibody response as the antibody products of reactive clones compete for diminishing amounts of antigen. Only the most effective clones, yielding antibodies with the highest affinity, persist as the immune response abates. Moreover, as discussed earlier, when sufficient immunoglobulin molecules of a given idiotype have accumulated, the idiotype may act as an antigen and anti-idiotypic antibodies may be formed. An anti-idiotypic antibody will have an antigen-binding pocket with the same spatial configuration as the antigenic determinant that stimulated formation of the original immunoglobulin molecule. Therefore, the anti-idiotypic antibody can occupy the surface immunoglobulin receptors of B cells making the original immunoglobulin molecule and thus interfere with continuing synthesis of the original antibody molecule (Jerne, 1975; Geha, 1981).

Cell-Mediated Immune Responses

Cell-mediated immune responses include *delayed hypersensitivity* reactions and *cytotoxic reactions* in which T cells and NK cells kill target cells. A positive tuberculin skin test is a well-studied example of a cutaneous delayed hypersensitivity reaction. The intradermal injection of tuberculin, an antigen extracted from the tubercle bacillus, results in the accumulation at the site of a small number of T lymphocytes sensitized by previous exposure to the antigen. These sensitized lymphocytes interact with the antigen on the surface of a macrophage or of a Langerhans cell. Just as for initiation of antibody production to a thymus-dependent antigen, the interaction is MHC restricted, requiring that the T cell antigen receptor of a CD4+ T cell recognize the antigen in association with a class II molecule.

The antigen-activated T lymphocyte then secretes cytokines that bring about the further steps of the reaction. They cause monocytes from the blood to accumulate at the site and differentiate into macrophages (monocyte chemotactic factor of MCF); they inhibit macrophage migration from the site (macrophage inhibitory factor or MIF); and they activate the macrophage, enhancing its microbicidal properties, its ability to secrete proteolytic enzymes and generate toxic oxygen metabolites, and its surface expression of class II molecules (γ-interferon and other macrophage-activating factors).

Other cytokines, called lymphotoxins, directly damage cells in the vicinity of the antigen. By 24–48 hours after the injection of the antigen, the reaction causes an area of erythema and induration at the injection site. If it were biopsied, one would see an infiltrate of mononuclear cells, primarily macrophages, edema, and deposits of fibrin. In some areas, the localized immune response consists of multiple macrophage (histiocytic) cells and lymphocytes organized into a pallisading configuration called a granuloma.

Delayed hypersensitivity reactions, which provide a mechanism whereby large numbers of macrophages become activated at a site of antigen, enhance the body's ability to fight infections by microogranisms that can survive and multiply within unactivated macrophages (e.g., mycobacteria, fungi). Moreover, once activated, macrophages not only possess an enhanced microbicidal activity against the organism initiating the delayed hypersensitivity reaction, but also an enhanced microbicidal activity against other organisms.

A second cell-mediated immune response results in the lysis by cytotoxic T cells of tissue cells infected with viruses. Unlike intracellular bacteria, which invade only monocyte-macrophages, viruses may invade a variety of tissue cells (e.g., hepatitis viruses invade hepatocytes and influenza viruses invade respiratory epithelial cells). When viruses infect tissue cells, viral antigens are translocated to the cell surface. Cytotoxic CD8+ T cells become sensitized to these viral antigens in conjunction not with class II molecules (which are absent from most tissue cells) but in conjunction with class I molecules. Once sensitized, the cytotoxic T cells can lyse infected cells in a reaction in which, once again, the cytotoxic T cell must recognize both the viral antigen and the class I molecule on the cell surface. A lytic agent resembling C9 of the membrane attack complex of the complement system (see Chapter 23) induces the cell lysis (Müller-Eberhard, 1988). Such cell lysis, by preventing intracellular viral replication, may play an important role in resistance to viral infections. However, the T cell-mediated death of virus-infected cells is also responsible for major manifestations of disease induced by some viruses in experimental animals and, probably, also in man (Zinkernagel, 1978).

ABNORMALITIES OF IMMUNE RESPONSES IN DISEASE

Pathological Manifestations of Humoral Immune Responses

Although humoral immune responses represent a key biological defense mechanism, under some circumstances antibodies damage or alter the function of cells and cause manifestations of disease. Important ways in which this may happen are described below.

1. *Immediate hypersensitivity reactions.* When an antigen stimulates B cells-plasma cells to make IgE, the antibodies bind by their Fc region to the surface of tissue mast cells. On subsequent exposure of the individual to the same antigen, it can then bind to the free antibody recognition sites of the IgE molecules on the mast cells. This triggers mast cell degranulation and release of mediators of immediate hypersensitivity (Serafin and Austen, 1987). This is the mechanism whereby inhaled antigens, such as pollens or molds, may precipitate an acute asthmatic attack in an allergic person. Rarely, individuals may develop IgE antibodies after exposure to a drug, such as penicillin, or to other allergens, such as bee venom. If these antigens subsequently gain access to the bloodstream, systemic mast cell degranulation may precipitate a potentially fatal anaphylactic reaction.

2. *Interactions of auto antibodies with cell antigens.* Although evidence now exists that on initial stimulation by some antigens B cells may briefly use selected, conserved V gene segments to make low affinity antibodies that can cross-react with certain self-antigens, multiple mechanisms prevent normal individuals from making clinically significant amounts of antibodies to self-antigens. Most maturing B cells with immunoglobulin gene arrangements that would program them to make "self-reacting" antibodies are thought to be silenced in the bone marrow. Similarly, thymic precursors are selected for their abilities to distinguish between self and nonself antigens. Finally, continuing interactions between B cells and suppressor T cells suppress unwanted immune responses.

However, under some conditions these monitoring mechanisms fail. A variety of clinical disorders result from antibodies that react with normal cells (autoantibodies). Some of these antibodies cause tissue damage, inflammation, or cell destruction. Examples are: the binding of autoantibodies to the surface of red cells that results in hemolytic anemia due to phagocytosis or lysis of the red cells; and the binding of autoantibodies to the glomerular basement membrane that damages the kidney by causing complement fixation, inflammation, and blood coagulation (Goodpasture's syndrome). Other autoantibodies can cause disease not by destroying cells but by altering their function. This occurs, for example, in Graves' disease, a disorder in which thyroid hormone is overproduced as a consequence of autoantibodies occupying the receptors on thyroid cells for thyroid-stimulating hormone (TSH).

3. *Precipitation of antigen-antibody complexes (immune complexes) in tissues.* Complexes of antibody and antigen(s) that form in the blood may become deposited in blood vessel walls. When antigen-antibody complexes precipitate in the glomerular capillary bed, as may occur in serum-sickness or in systemic lupus erythematosus, the resultant glomerular inflammation may lead to transient or permanent renal failure. When antigen-antibody complexes are deposited in the walls of small arteries, a systemic vasculitis may be incited that causes ischemic necrosis in many organs or tissues.

Immune complexes deposited in tissues produce inflammation and tissue damage by activating complement (see Chapter 22). Cleavage of complement components attracts granulocytes which phagocytose the immune complexes. In the process, proteolytic enzymes escape from the granulocyte and injure surrounding tissues. The activated complement components may also form a membrane attack complex that directly damages cell membranes.

Pathological Manifestations of Cell-Mediated Immune Responses

Cell-mediated immune responses are, by their nature, to some extent auto-aggressive. Lymphokines produced during delayed hypersensitivity reactions may damage cells directly or indirectly, through release of proteolytic enzymes from macrophages. The severity of the resulting inflammatory response varies with the extent and duration of the cell-mediated reaction. In pulmonary tuberculosis, uncontrolled infection results in a continuing immune response of the delayed hypersensitivity type that may destroy lung tissue. When cytotoxic T cells attack cell targets infected with viruses, not only the virus but the host cells may also be damaged or destroyed. This is thought to occur in certain forms of demyelinating central nervous system disorders. Cellular immune responses to B lymphocytes infected with the Epstein-Barr virus are a principal cause of the manifestations of infectious mononucleosis.

Immunodeficiency States

Hereditary disorders producing immunodeficiency are rare, but acquired immunodeficiency states are relatively common. One cause is the administration of immunosuppressive drugs to prevent rejection of organ transplants. The cause of primary concern today is the acquired immune deficiency syndrome (AIDS), a fatal disorder resulting from infection with a lymphotrophic retrovirus (HIV-1). The virus replicates within CD4+ T cells with ultimate depletion of these cells and resultant severe impairment of cell-mediated immune responses.

CHAPTER 22

Inflammation and Phagocytosis

When microorganisms breach local defenses at skin and mucosal surfaces, systemic reactions are set off to destroy the invading microorganisms. These reactions include the immune responses described in Chapter 21 and the responses described in the present chapter that lead to the ingestion and killing of pathogens by leukocytes. The high risk of life-threatening infection in the patient whose blood granulocyte count persists below $500/\mu l$ attests to the importance of the latter in maintaining normal natural resistance to infection.

The introduction into tissues of microorganisms (or materials such as antigen/antibody complexes or precipitates of crystals) provokes inflammation and local tissue damage. Although this may stem partly from direct effects of invading microorganisms, it stems mainly from the effects of mediators involved in the body's defense mechanisms. *The effectors of leukocyte activation, of phagocytosis, and of microbicidal killing are also the effectors of inflammation.* These include activated components of complement; enzymes released into tissues from the lysosomal granules of phagocytes; and two classes of materials, toxic oxygen products and lipid mediators, which are generated when leukocytes are activated.

The linkage of phagocytosis and microbicidal killing to the inflammatory response works both to the body's advantage and to its disadvantage. Release of materials from phagocytes into the surrounding tissues may facilitate the phagocytic attack upon microorganisms, particularly fungi and the larval forms of helminths that are too large for single phagocytes to ingest. However, inflammatory tissue damage induced by products of phagocytes also represents an important pathogenetic mechanism of disease (Malech and Gallin, 1987).

COMPLEMENT SYSTEM

The complement system consists of 20 plasma proteins plus additional cell surface proteins (Müller-Eberhard, 1988). The plasma proteins are either enzymes or binding proteins; the cell surface proteins are regulatory membrane proteins, cell surface receptors, or both. Activation of complement plays a key role in the killing of invading microorganisms and in the pathogenesis of the inflammatory response.

Nomenclature

Numbers, letters, and trivial names have all been used to identify components of the system. A series of native molecules of the classic pathway have been designated by the numbers C1 through C9. C1 was found to be a complex of three proteins: a binding protein, C1q, and two serine proteases, C1r and C1s. Components of the alternative pathway have been given capital letter names: an enzyme, factor D; a proenzyme, factor B; and a stabilizing protein, factor P (properdin). Some of the complement proteins are cleaved into fragments during activation, and fragments are identified by adding lower case letters from the beginning of the alphabet. For example, B is cleaved to its enzymatically active form, Bb, and C3 is cleaved during activation to C3a and C3b. Capital letter names have also been given to two proteins reacting with C3b: factor H, a binding protein, and factor I, an enzyme that further cleaves C3b into iC3b, C3c, and C3d. Complexes of components are formed during activation; these are identified by listing the components in order of their participation in the complex. Thus, the activity that cleaves C5 in the classic pathway is designated C4b,2a,3b. Some regulatory proteins have trivial names, e.g., the C1 inhibitor, a serine proteinase inhibitor that inhibits C1r and C1s, and the C4b binding protein. Finally, three cell surface receptors that bind fragments of C3 are termed CR1 (binds C3b), CR2 (binds C3d), and CR3 (binds iC3b).

Biosynthesis of the Complement Proteins

The liver is thought to be the primary site of synthesis of the majority of the plasma complement components. C3, C4, and C5 have some common biological properties and areas of similar amino acid sequences; they probably arose from a common ancestral gene. C3 and C4, but not C5, contain an unusual thioester, which undergoes hydrolysis after activation of C3 and C4 to create a transient site for covalent bonding of these molecules to surface polysaccharides or proteins. C2 and factor B are novel types of serine protease proenzymes. After activation, each cleaves the same bonds in C3 and C5 as does the other. The gene loci for C2 and factor B are close to each other in the HLA region of human chromosome 6, and these proteins probably arose from a common ancestral gene.

C3 may be viewed as the fulcrum protein of the complement system. Its plasma concentration is about 1.2 g/liter. Many of the other proteins are present in concentrations ranging from about 25 to 500 mg/liter. One protein, the enzyme factor D, is a trace protein, present in plasma in a concentration of 1 mg/liter.

General Description of Complement Activation

Two activation pathways for complement have been delineated: the *classic pathway* and the *alternative pathway*. They join at the step of activation of C3 and again at the step of activation of C5 (Fig. 3.8). Both pathways lead to the formation of a complex of terminal complement components, the membrane attack complex, that lyses biological membranes.

Synthesis of antibodies with resultant formation of antigen/antibody complexes on a membrane (for example, a bacterial surface membrane) is required to trigger the reactions of the classic pathway. C1q binds to such antigen/antibody complexes with

resultant activation of the serine proteases C1r and C1s. These enzymes cleave C4 to yield a large membrane-binding fragment, C4b, and they also cleave and activate the serine protease proenzyme C2. A C4b,2a complex is then formed on the membrane and functions as the classic pathway C3 convertase.

Antibody synthesis is not a requisite for activation of the alternative pathway. This pathway normally undergoes a slow, continuous activation in the fluid phase stemming from the spontaneous hydrolysis, in the presence of water, of the thioester of C3. This leads to the formation of a small number of C3b molecules. Whether the pathway proceeds beyond this depends upon the type of surface that the C3b molecules come down onto. If they come down onto certain foreign surfaces (e.g., the lipopolysaccharide molecules of Gram-negative bacterial wall endotoxin or certain artificial membranes used for hemodialysis), then factor B binds to the C3b and is cleaved by factor D to yield the active enzyme, Bb. Following this, factor P binds to and stabilizes the C3b,Bb complex. The resultant C3b,Bb,P complex functions as the alternative pathway C3 convertase (Fig. 3.8).

The membrane-bound C3 convertases cleave C3 into a small soluble fragment, C3a, and the large, previously mentioned fragment, C3b. With continuing C3 convertase activity, increasing numbers of C3b molecules come down onto the membrane, some in proximity to C3 convertases. Since the alternative pathway convertase contains C3b, additional C3b amplifies the activity of this convertase (see Fig. 3.8). Moreover, C3b has a recognition site for

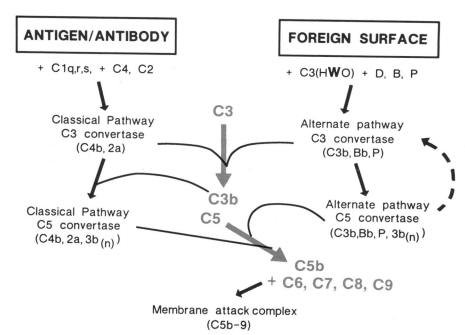

Figure 3.8. A simplified diagram of the reactions of complement activation by the classic pathway (*left side* of the diagram) and by the alternative pathway (*right side* of the diagram).

C5, and C5 molecules are also brought down onto the surface in proximity to the C3 convertases. This allows the C2a enzyme of the classic pathway convertase and the Bb enzyme of the alternative pathway convertase to begin to cleave C5. *The C3 convertases are converted into C5 convertases.*

Cleavage of C5 yields a small fragment, C5a, which functions as a major inflammatory mediator (see below), and C5b, which binds to membranes. Membrane binding of C5b initiates assembly on the membrane of the terminal complement components, which form a membrane attack complex, C5b,6,7,8,9, that lyses biological membranes.

A number of regulatory proteins influence different steps in the activation process. A plasma protein called C4b binding protein can bind to C4b, and a second plasma protein called factor H can bind to C3b. When so bound, C4b and C3b are susceptible to cleavage and inactivation by a circulating active protease called factor I. Binding of C3b to a membrane protein, the receptor CR1, which is found in human blood primarily on red blood cells, also facilitates the inactivation of C3b by factor I. A second surface membrane protein, DAF (delay-accelerating factor), present on red blood cells and other cells of myeloid lineage, can impair the membrane assembly of both the classic and alternative C3 convertases and so limit the amount of C3b found on these cells when complement is activated.

Two proteins regulate the molecular assembly of the membrane attack complex and its lytic activity. One, a plasma protein called vitronectin (formerly called S protein) competes with membrane lipid for the membrane binding site of a forming membrane attack complex. The other, a surface membrane protein called homologous restriction factor (HRF), which has been recognized on the surface membrane of all blood cells, can interact with C9 and diminish the lytic activity of the membrane attack complex.

The above represents a very much simplified description of the organization and activation of the complement system. For an authoritative discussion, the reader is referred to a recent comprehensive review (Müller-Eberhard, 1988).

Biological Effects of Complement Activation

Activation of complement may have a number of important biological consequences. If activation proceeds to completion, with formation of the membrane attack complex on a biological membrane, then that membrane may be lysed. Such lysis is important in combating certain bacterial infections, particularly meningococcal and gonococcal infections. In certain pathological conditions IgM antibodies bind to red blood cell antigens (as, for example, after a mismatched blood transfusion) and activate large amounts of complement on the cell surface. The resultant formation of the membrane attack complex may lyse the circulating red blood cells, producing intravascular hemolysis.

C3b bound to circulating antigen/antibody complexes by one binding site may bind by a second site to the C3b membrane receptor on red blood cells, CR1. The red blood cells may then transport the sequestered antigen/antibody complexes to the liver, where the antigen/antibody complexes are stripped from the red blood cells without damaging the red blood cells. This is now thought to be an important mechanism for clearing antigen/antibody complexes from the circulation (Frank, 1987).

Factor I cleaves C3b initially to iC3b and later to fragments called C3c and C3d. Granulocytes contain a receptor, CR3 (synonyms: Mac-1, CD11b/CD18), whose ligand is iC3b. When bound to this receptor, iC3b may serve as one of the "glues" by which neutrophils adhere to cells and tissue surfaces and to bacteria coated with antibody and complement (Malech and Gallin, 1987).

Finally, the small polypeptides cleaved from C3, C4, and C5 during their activation act as anaphylotoxins, which are mediators that cause mast cells to degranulate and smooth muscle cells to contract and that enhance vascular permeability. Moreover, C5a is a key, very potent mediator of inflammation that activates neutrophils.

RESPONSES OF NEUTROPHILIC GRANULOCYTES

In phagocytic body defense reactions and their accompanying inflammatory tissue response, neutrophils are activated, adhere to and pass through vascular endothelium, emigrate by chemotaxis to a tissue site, undergo partial degranulation, adhere to extravascular tissue surfaces and cells and to themselves to form aggregates, ingest microorganisms or other opsonized materials, and activate biochemical reactions leading to microbial killing (Lehrer et al., 1988). These responses are illustrated schematically in Figure 3.9.

Activation, Adherence, and Chemotaxis

Binding of ligands to specific surface membrane receptors initiates neutrophil activation. These ligands are often referred to as chemoattractants because neutrophils migrate in the direction of an increasing concentration of these materials while emigrating to a tissue site. Table 3.6 provides a list of chemoattractants considered important for the inflammatory response. These are derived from

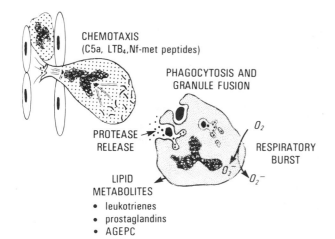

CHEMOTAXIS
(C5a, LTB$_4$, Nf-met peptides)

PHAGOCYTOSIS AND
GRANULE FUSION

PROTEASE
RELEASE

RESPIRATORY
BURST

LIPID
METABOLITES
• leukotrienes
• prostaglandins
• AGEPC

Figure 3.9. An illustration of the responses of granulocytes in inflammation. *LTB$_4$* is leukotriene B$_4$. *AGEPC* is acetyl glycerol ether phosphorylcholine.

diverse sources: from the breakdown of bacterial proteins and of mitochondrial proteins from damaged tissue cells (Carp, 1982); from the cleavage of the complement component, C5; from secreted constituents of the granules of platelets (Deuel et al., 1982); and from the release of lipid-derived mediators from stimulated white blood cells.

Occupancy of a specific receptor by its ligand induces signal transduction by a common mechanism now considered general to receptor-mediated

Table 3.6.
Chemoattractants of Probable Physiological Significance

Origin	Material
Bacterial protein breakdown	*N*-Formyl-methionyl-oligopeptides
Mitochondrial protein breakdown	*N*-Formyl-methionyl-oligopeptides
Complement cleavage product	C5a, C5a-des-Arg[a]
Platelet α granule secretion	Platelet factor 4 Platelet-derived growth factor
Neutrophilic granulocyte activation	Metabolites of lipoxygenation of arachidonic acid (leukotriene B$_4$, 5-HETE, etc.) AGEPC (acetyl glycerol ether phosphorylcholine) Crystal-induced chemotactic factor (CCF)[b]
Tissue mast cell degranulation	Eosinotactic tetrapeptides (eosinophil chemotactic factor of anaphylaxis)[c] HMW[d] neutrophil chemotactic factor of anaphylaxis
T-lymphocyte antigen or mitogen stimulation	Lymphokine chemoattractant for granulocytes

[a]C5a minus its carboxy-terminal arginine. Requires an anionic helper factor present in serum. [b]A chemotactic factor produced when granulocytes phagocytose monosodium urate crystals. [c]Chemotactic to a lesser degree for neutrophilic granulocytes. [d]HMW is an abbreviation for high molecular weight.

activation of many eukaryotic cells. It is described in detail elsewhere (Boxer and Smolen, 1988). In brief, occupancy of a receptor causes it to interact with a guanyl nucleotide binding protein (G protein), inducing a substitution of GTP for GDP on the protein. The G protein can now activate an enzyme, phospholipase C, which hydrolyzes the surface membrane phospholipid—phosphatidylinositol 4,5-biphosphate—to give rise to two second messengers, inositol 1,4,5-triphosphate (IP$_3$) and diacylglycerol. IP$_3$ increases cytosolic Ca^{2+} concentration through release of calcium from intracellular stores. Diacylglycerol activates protein kinase C with resultant phosphorylation of leukocyte proteins. These combined events set in motion the subsequent biochemical reactions coupling leukocyte stimulation to the events of leukocyte activation.

Neutrophil activation triggers a partial degranulation of neutrophils that results in the translocation of materials from specific and tertiary granules to the neutrophil surface membrane. This increases the surface membrane concentration of several materials important for the function of activated neutrophils: receptors for chemoattractants; the CR3 receptor (C11b/CD18), which is essential for normal neutrophil adherence (Malech and Gallin, 1987); and a unique cytochrome (cytochrome b_{558}) essential for the respiratory burst that causes activated neutrophils to generate microbicidal oxidants (Babior, 1988).

Under the influence of interleukin 1 and tumor necrosis factor, which are secreted in inflammatory states by mononuclear phagocytes, vascular endothelial cells develop surface membrane binding sites that enhance markedly the ability of neutrophils to adhere to endothelial cells. This adherence, which also requires the participation of CR3 on the neutrophil surface membrane, facilitates the chemotactic migration of neutrophils into tissue sites. The morphology of the granulocyte changes strikingly during chemotactic migration (Snyderman and Goetzl, 1981). The cell elongates and becomes polarized, developing a broad "head" called a lamellipodium, which is oriented toward the concentration gradient of the chemotactic factor, and a thin "tail" with terminal arborization called a uropod (Fig. 3.9). As the cell moves, the nucleus lags behind the center of the cell and is separated from its leading edge by the main mass of the cytoplasm. The cytoplasm of the leading edge of the lamellipodium appears free of granules, probably reflecting a degranulation of specific granules along the leading edge. Unoccupied chemotactic receptors are concentrated in the surface membrane of the leading edge, whereas occupied receptors accumulate in the uropod.

Generation of Lipid Mediators

As surface receptors are increasingly occupied by chemoattractants and by the ligands inducing phagocytosis (IgG and iC3b; see below), the signal transduction mechanism described above can lead to the activation of other phospholipases, called phospholipase A₂. These catalyze the release of arachidonic acid from membrane phospholipids. One consequence is the formation of lysophosphatidylcholine, a material with detergent-like properties that may be involved in the fusion of granule membranes with plasma membranes during degranulation.

Free arachidonic acid, which is liberated during cell activation in many different types of cells, is oxidized by two enzymes, cyclooxygenase and lipoxygenase. These oxidations yield end products—prostaglandins, thromboxanes, leukotrienes, and other derivatives of hydroperoxy-fatty acids—that differ in the different cells (Goetzl, 1981).

Oxidation of arachidonic acid by the cyclooxygenase pathway results in formation of the endoperoxide PGG₂, which is then reduced to PGH₂ with release of a reactive oxygen species. As shown in Figure 3.10, PGH₂ is converted in different cells into different prostaglandin and thromboxane end products. The inflammatory mediator effects of PGE₂, the major prostaglandin formed by granulocytes and monocytes/macrophages, is not yet clear. PGE₂ raises cAMP levels, which lowers cytosolic calcium concentration and so should dampen leukocyte functions in inflammation. Nevertheless, giving the drug aspirin, which blocks oxidation of arachidonic acid by cyclooxygenase and therefore leukocyte synthesis of PGE₂, diminishes the symptoms of inflammation.

The leukotrienes, products of the lipoxygenase pathway of arachidonic acid oxidation (Fig. 3.10), play dramatic roles in hypersensitivity and inflammatory responses (Dahlen et al., 1981). In leukocytes, lipoxygenase catalyzes the formation of 5-hydroperoxyeicosatetranoic acid (5-HPETE), which is a precursor of a material called leukotriene A₄ (the subscript 4 denotes its derivation from arachidonic acid, which has four double bonds). In mast cells and also in monocytes/macrophages, leukotriene A₄ (LTA₄) is conjugated with glutathione to give rise to leukotriene C₄ (LTC₄), which may then be converted to LTD₄, a molecule in which the glutamine residue is lost from glutathione and then to LTE₄ (Fig. 3.10). Degranulation of mast cells with release of LTC₄ and LTD₄ and other mediators of immediate hypersensitivity is thought to trigger the bronchospasm and the mucosal edema of bronchial asthma. In neutrophils, LTA₄ is not conjugated with glutathione but is instead enzymatically converted into a molecule called LTB₄. LTB₄ is one of the most potent inducers of leukocyte chemotaxis, aggregation, and degranulation yet discovered (Samuelsson, 1981).

Acetyl glycerol ether phosphorylcholine (Fig.

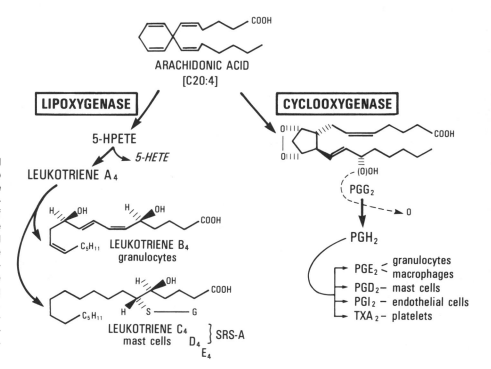

Figure 3.10. Oxidation of arachidonic acid by way of the cyclooxygenase pathway to form prostaglandins (PGs) and thromboxane (*TXA₂*) and by way of the lipoxygenase pathway to form leukotrienes. In the illustration of leukotriene C₄, the molecule attached at the 6 position that is represented by the symbol S——G is glutathione. In leukotriene D₄, the glutamine residue has been lost from glutathione. In leukotriene E₄, both the glutamine and glycine residues have been lost, leaving only the cysteinyl backbone of glutathione. *SRS-A*, slow-reacting substance of anaphylaxis; *5-HPETE*, 5-hydroperoxyeicosatetraenoic acid; *5-HETE*, 5-hydroxyeicosatetraenoic acid.

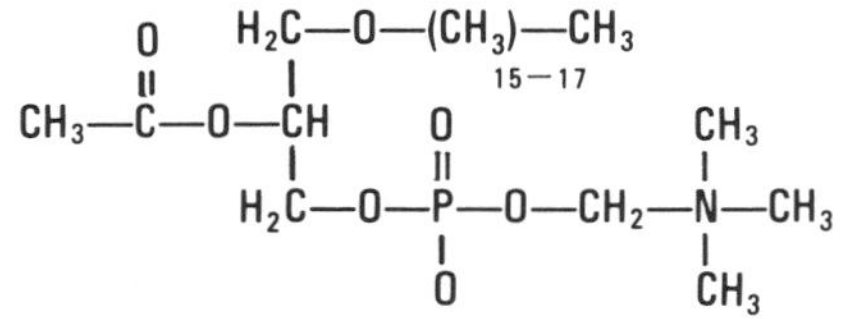

$$CH_3-\overset{\overset{\displaystyle O}{\|}}{C}-O-\overset{\overset{\displaystyle H_2C-O-(CH_3)-CH_3}{|}}{\underset{\underset{\displaystyle H_2C-O-\overset{\overset{\displaystyle O}{\|}}{P}-O-CH_2-\overset{\overset{\displaystyle CH_3}{|}}{\underset{\underset{\displaystyle CH_3}{|}}{N}}-CH_3}{|}}{CH}}$$

Figure 3.11. The structure of acetyl glycerol ether phosphorylcholine.

3.11) is another potent lipid inflammatory mediator (Pinckard, 1982). It is often referred to as platelet-activating factor (PAF) because it was first recognized as a material causing platelet aggregation that was released from IgE-sensitized rabbit basophils stimulated by antigen. However, it was later found to be liberated after different types of stimulation of many cells, including neutrophils, monocytes/macrophages, and platelets. In addition to causing platelet aggregation, PAF induces chemotaxis, aggregation, degranulation, and the respiratory burst in neutrophils. PAF does not exist preformed in resting neutrophils but forms within seconds of neutrophil activation, possibly by acetylation of a precursor molecule, lysoglycerol 3-phosphorylcholine.

Reactive Oxygen Species

Within seconds of stimulation, neutrophils sharply increase their oxygen uptake, a phenomenon known as the respiratory burst (Babior, 1988). The increased oxygen uptake is associated with two biochemical events. The first is a one-electron reduction of molecular oxygen to a form called superoxide ion (O_2^-). The second is the oxidation of glucose via the hexose monophosphate shunt, with resultant production of NADPH and reduced gluta-thione (GSH) (Fig. 3.12). It is established that neutrophils must generate O_2^- as a step in the process whereby they kill certain bacteria in oxygen-dependent reactions. Thus, in a group of rare hereditary disorders known collectively as chronic granulomatous disease (Curnette, 1988), an inability to generate O_2^- leads to recurrent, very troublesome bacterial infections.

The generation of O_2^- requires activation of a respiratory burst oxidase that catalyzes the transfer of electrons from NADPH, through an electron transfer chain thought to consist of a flavin and a unique cytochrome known as cytochrome b_{558}, to oxygen. The overall reaction may be written

$$2O_2 + NADPH \rightarrow 2O_2^- + NADP^+ + H^+$$

How stimulation of the neutrophil activates the oxidase is not yet clear, but the function of the oxidase requires the participation of several cytoplasmic factors as well as membrane-bound factors. The consumption of NADPH in the reaction stimulates the oxidation of glucose in the hexose monophosphate shunt (Fig. 3.12) to replenish NADPH and also to replenish GSH, which is consumed in the detoxification of H_2O_2 (see below).

Most of the O_2^- formed during the respiratory burst is rapidly converted into H_2O_2 through a reaction catalyzed by the enzyme superoxide dismutase. O_2^- itself has insignificant batericidal activity, and H_2O_2, acting alone, is only a weak microbicidal agent. However, in the presence of the primary granule enzyme myeloperoxidase and a halide ion, H_2O_2 exhibits potent microbicidal activity because myeloperoxidase utilizes H_2O_2 to oxidize halide ions,

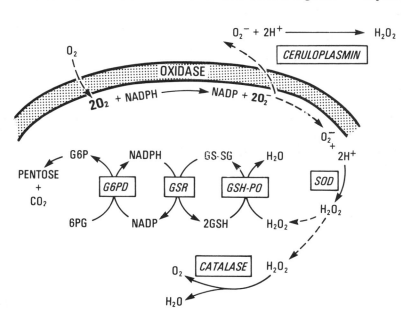

Figure 3.12. Reactions associated with the respiratory burst in which superoxide anion is generated and converted to H_2O_2. The reactions of the hexose monophosphate shunt that provide NADPH as a source of the electron for superoxide anion and provide reduced glutathionine as a substrate for the detoxification of H_2O_2 by the enzyme glutathione peroxidase are also shown. In addition, the detoxification of H_2O_2 by the intracytoplasmic enzyme catalase and the conversion of extracellular superoxide anion to H_2O_2 by the plasma protein ceruloplasmin are illustrated. O_2^-, superoxide anion; *G6P*, glucose 6-phosphate; *6PG*, 6-phosphogluconate; *GSH*, reduced glutathione; *GS-SG*, the oxidized form of glutathione; *G6PD*, the enzyme glucose 6-phosphate dehydrogenase; *GSR*, the enzyme glutathione reductase; *GHS-PO*, the enzyme glutathione peroxidase; *SOD*, the enzyme superoxide dismutase.

yielding potent microbicidal agents such as HOCl (Klebanoff, 1980).

Nevertheless, individuals with a hereditary deficiency of myeloperoxidase, although susceptible to infection with a particular fungus, *Candida albicans*, do not experience the severe infections of individuals with chronic granulomatous disease. One must assume, therefore, that toxic oxygen radicals are required for an additional mechanism of bacterial killing independent of myeloperoxidase. This could be mediated by the Haber-Weiss reaction, a reaction in which O_2^- and H_2O_2 react together in the presence of a metal such as iron to form another reduced form of oxygen called hydroxyl radical (OH·), which is a potent oxidizing agent.

In addition to inducing microbial killing, reactive oxygen metabolites that escape from the neutrophil can injure tissue cells and connective tissue matrix. Reduced oxygen species may damage cells and tissue both directly and indirectly, the latter through a chain reaction in which the peroxidation of lipids on cell membranes results in the generation of fatty acid radicals that can then react with other lipids, proteins, or free radicals in tissues (Fantone and Ward, 1982). O_2^- released extracellularly from neutrophils may be reduced to H_2O_2 by the oxidase activity of the acute phase protein ceruloplasmin (Goldstein et al., 1982). The role of this reaction in limiting oxidant damage to tissues requires further evaluation, but it supplies a teleological reason for the rise in concentration of this acute phase protein in inflammatory states.

The neutrophil utilizes two mechanisms to rid itself of unwanted H_2O_2 (Fig. 3.12). In one, H_2O_2 is reduced to water through the oxidation of GSH to GSSG in a reaction catalyzed by glutathione peroxidase. In the second, H_2O_2 is reduced to water by the enzyme catalase.

Phagocytosis

Neutrophils have on their surface membrane Fc receptors that may initiate phagocytosis when occupied by the Fc portion of IgG that has reacted with antigen on a microorganism or particle or when occupied by the Fc portion of IgG in immune complexes that have become deposited upon a particle. As mentioned earlier, activated neutrophils have an increased concentration of a second surface recognition unit, CR3, which facilitates the recognition and adherence to the neutrophil of microorganisms coated with iC3b and, thereby, increases the efficiency with which they are phagocytosed.

In the act of phagocytosis, the granulocyte surface membrane invaginates and extends pseudopods around the sides of the particle that fuse to enclose the particle within a vacuole, which is called a phagocytic vacuole. Primary (lysosomal), secondary (specific), and tertiary granules are attracted to the phagocytic vacuole. Their membranes fuse with the membrane of the phagocytic vacuole, and their contents are secreted into the phagocytic vacuole. Thus, potent materials accumulate in proximity to ingested microorganisms but isolated from the rest of the cell. These include lysozyme, myeloperoxidase, cathepsin G, elastase, lactoferrin, and cationic proteins called defensins (Selsted, 1988). In addition, superoxide anion and its metabolites diffuse into the phagocytic vacuole and initiate oxygen-dependent microbial killing.

In contrast to the specific and tertiary granules, whose contents are partially translocated to the surface membrane and secreted during the early response of neutrophils to inflammatory stimuli, the contents of the primary granules are primarily secreted into phagocytic vacuoles. Nevertheless, some primary granule enzymes may escape into surrounding tissues before a phagocytic vacuole is sealed off (Weissman et al., 1980). The extent of external secretion increases when the engulfment of particles is frustrated as, for example, by their large size. The escape of primary granule contents brings neutral proteases, particularly the enzyme elastase, into close contact with substrates in the surrounding tissue that they can cleave and damage. Such release of neutrophil enzymes during phagocytosis represents an important mechanism for producing tissue damage in many acute clinical disorders and also in some serious chronic disorders. The latter include disorders affecting the joints (rheumatoid arthritis) and the lungs (pulmonary emphysema).

CHAPTER 23

The Red Blood Cell

GENERAL DESCRIPTION, PRODUCTION, AND DESTRUCTION OF RED BLOOD CELLS

The red blood cell is the simplest cell of the human body. Formed as a nucleated cell in the bone marrow, it normally loses its nucleus before its release into the circulation. On entering the circulation, it still possesses residual ribosomes and mitochondria and a Golgi apparatus. These cytoplasmic organelles are lost after a day or so, and the red blood cell then assumes the shape of a flattened, indented sphere or biconcave disc (Fig. 3.13). Its dimensions are: diameter, 7.8 μm; thickness, 0.81 μm in the thin part and 2.6 μm in the thick part of the biconcave disc; surface area, 135 μm^2, and volume, 94 fl. The cell exhibits a remarkable plasticity, being able repeatedly to squeeze through capillaries only one-half of the diameter of the cell and then to return, undistorted, to its original biconcave shape. The high surface to volume ratio facilitates respiratory gas transfer, which is the red blood cell's sole known physiological function.

The mature red blood cell may be looked upon as a cell membrane enclosing proteins, electrolytes, and other components of energy systems. Ninety-five percent of the protein of the cell is hemoglobin. The remaining proteins are largely enzymes of its energy systems, many of which, without means of replacement, must maintain at least minimal catalytic activity for the approximately 4-month lifespan of the red cell.

Production of Red Blood Cells

Erythropoiesis requires the presence, in a normal marrow microenvironment, of a normal population of hemopoietic stem cells; their differentiation and maturation under the influence of burst promoting factors and erythropoietin (see Chapter 20); and the availability of three specific nutrients: the vitamins, folic acid and cobalamin (vitamin B$_{12}$), and iron. As also discussed in Chapter 20, counting reticulocytes in a sample of peripheral blood provides a simple way of assessing red cell production clinically.

FOLIC ACID AND COBALAMIN IN ERYTHROPOIESIS

Erythropoiesis requires synthesis of DNA for cell division. In DNA, the pyrimidine base thymine, which is 5-methyluracil, replaces its counterpart in RNA, uracil. Folate (a name used for folic acid or any of its derivatives) and cobalamin participate in coupled reactions that make available the methyl group needed for conversion of deoxyuridilate to deoxythymidilate in DNA synthesis.

Figure 3.14 is a diagram of the reactions thought to be involved. It shows that 5,10-methylenetetrahydrofolate may either be converted into 5-methyltetrahydrofolate or be utilized to methylate deoxyuridilate in a reaction that oxidizes the 5,10-methylenetetrahydrofolate to dihydrofolate. Kinetic conditions favor the former reaction, and 5-methyltetrahydrofolate must be recycled to tetrahydrofolate to maintain an adequate supply of 5,10-methylenetetrahydrofolate for normal synthesis of deoxythymidilate. The recycling is coupled to a second reaction, the methylation of homocysteine to methionine, which is catalyzed by the enzyme methionine synthase. However, this second reaction also requires a coenzyme, cobalamin, which receives the methyl group from 5-methyltetrahydrofolate and transfers it to homocysteine. The reactions shown in Figure 3.14 may be impaired in different ways: by an insufficient supply of folate within the developing red cells; by drugs, such as the chemotherapeutic agent, methotrexate, that compete for the enzyme reducing dihydrofolate to tetrahydrofolate (dihydrofolate reductase); or by insufficient cobalamin to support the recycling of 5-methyltetrahydrofolate, which then traps folate in a form unusable for DNA synthesis (Herbert and Zalusky, 1962).

Folate molecules contain either one terminal glutamyl residue (monoglutamate) or several terminal

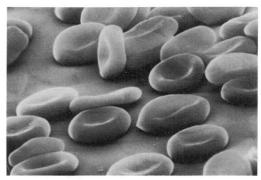

Figure 3.13. The biconcave shape of normal human red blood cells as seen by scanning electron microscopy (courtesy of Dr. Judith A. Berliner).

glutamyl residues (polyglutamate). Plasma folate in tissues is in the polyglutamate form, an apparent requirement for its intracellular storage. In cobalamin deficiency, levels of cellular polyglutamate folate fall while the plasma level of monoglutamate folate rises. The normal substrates for making polyglutamate tissue folate are formylfolates, which include the oxidized formylfolate, 5,10-methylenetetrahydrofolate, and two folate derivatives not shown in Figure 3.14, 5-formyltetrahydrofolate (folinic acid) and 10-formyltetrahydrofolate. Neither 5-methyltetrahydrofolate nor tetrahydrofolate can apparently serve as a substrate for making polyglutamate. The methyl group of methionine is thought to represent the major source of formate for formation of formylfolates (Chanarin et al., 1981).

Folic acid is provided in green vegetables, many fruits, beans, nuts, and liver; intake varies widely among individuals depending upon eating habits and methods of preparing food. Folic acid in food is present as polyglutamates, which must be deconjugated by intestinal enzymes called conjugases to form monoglutamic folate before folic acid can be absorbed. Normal tissue stores of folate are estimated as 5–10 mg. The daily folate requirement is 50–100 μg. Thus, failure of folate intake can lead to significant folate deficiency within about 3 months. Moreover, use of alcohol interferes with folate metabolism in as yet incompletely understood ways. Individuals drinking heavily and eating poorly for a prolonged period are particularly prone to develop serious folate deficiency.

Cobalamin is provided by animal protein in the diet; an average American diet provides 5–30 μg daily. A specific glycoprotein secreted by the parietal cells of the gastric mucosa, called intrinsic factor, must bind cobalamin before it can be absorbed from the gastrointestinal tract. Absorption takes place in the ileum as a result of the attachment of intrinsic factor to specific receptors in the brush border of mucosal cells. Cobalamin is transported to the tissues bound to a specific plasma transport protein called transcobalamin II. Cobalamin also binds to a second protein in plasma, transcobalamin I, which is synthesized within and secreted from granulocytes (see Chapter 20). The function of transcobalamin I is unknown; rare individuals with a hereditary absence of this protein apparently suffer no recognized ill effects. Cobalamin is found in tissues in two forms, the coenzymes methylcobalamin and adenosylcobalamin. Body tissue content is between 2 and 3 mg, most of which is present in

Figure 3.14. A diagram illustrating how the methylation of deoxyuridilate to thymidilate by 5,10-methylene FH$_4$ may be dependent upon the methylation of homocysteine to methionine. Cobalamin deficiency traps folate as 5-methyl FH$_4$ by impeding the methylation of homocysteine to methionine. A drug competing for the enzyme, FH$_2$ reductase, would also impair DNA synthesis. *FH$_2$*, dihydrofolate; *FH$_4$*, tetrahydrofolate. See text for further details. (From Rapaport, S.I.: *Introduction to Hematology*, 2nd Edition. Philadelphia: J. B. Lippincott, 1987.)

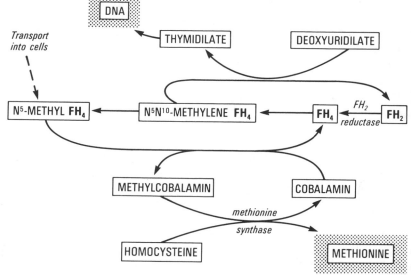

the liver. The daily requirement of cobalamin is 3 μg. Thus, several years of failure to absorb cobalamin are required to induce manifestations of cobalamin deficiency in a previously normal person. Atrophy of the gastric mucosa secondary to a suspected autoimmune reaction, with resultant lack of production of intrinsic factor, permanently impairs cobalamin absorption and results in the disease pernicious anemia.

Impaired DNA synthesis from either folate or cobalamin deficiency prevents the production of normal red blood cells and causes anemia. The red blood cells that are formed are larger than normal (macrocytic), with a mean volume exceeding 100 fl, and misshapen, with many oval and teardrop forms visible on a stained peripheral blood smear. The granulocytes are also larger than normal, and granulocytes containing an increased number of nuclear lobes (hypersegmentation) may be found. The nucleated red blood cell precursors in the bone marrow contain chromatin that appears abnormally fine and lacy with prominent parachromatin spaces (megaloblastic changes).

IRON METABOLISM

Iron is found within the body in heme, primarily in red blood cells and muscle cells, and as storage iron bound to ferritin or hemosiderin in mononuclear phagocytes and hepatic parenchymal cells. A normal 70-kg man has about 2.5 g of iron in hemoglobin, about 150 mg in muscle myoglobin, about 15 mg in trace heme tissue enzymes, and about 1 g in stores. A very small amount of iron (about 3 mg) is also present in plasma bound to transferrin on its way to cells. Women have lower iron stores than men, ranging from none to about 500 mg. This reflects loss of iron from menstrual bleeding and pregnancy. (The iron cost of a pregnancy is about 500 mg, which is why a woman should be given therapeutic iron during pregnancy.)

Daily body iron loss from the gut and in urine and sweat is minimal, less than 1 mg. During lactation a woman loses an additional 0.5 to 1 mg daily. In contrast to these minimal losses from other sources, bleeding causes loss of substantial amounts of body iron. One gram of hemoglobin contains 3.4 mg of iron. In a normal individual with 15 g of hemoglobin per dl, 100 ml of blood contains approximately 50 mg of iron. Thus, removal of only 2 ml of blood results in the loss of 1 mg of iron, i.e., an amount exceeding the normal daily iron loss from all other sources. A blood donation for transfusion (450 ml) depletes stores of 225 mg of iron. Normal menstrual bleeding has been estimated as 45 ml per month, but some normal women may lose up to 90 ml of blood per month.

The normal American diet contains an estimated 6 mg of iron per 1000 calories, which varies in its availability for absorption. Iron in meat is more readily absorbed than iron in foods such as wheat or eggs. Normally, only enough iron is absorbed from the diet to balance iron loss (0.6 to about 2.0 mg/day) and to allow the accumulation over a lifetime of about 1 g of storage iron. However, the dietary intake of a substantial fraction of otherwise normal American women apparently does not contain sufficient iron to allow them to build up any iron stores before the menopause (Cook et al., 1976).

Iron is absorbed in the upper small intestine. The amount absorbed depends not only upon the iron content of the diet but also upon a mucosal regulatory mechanism. Iron entering an intestinal mucosal cell either passes through the cell to enter the blood or is retained within the cell as ferritin. If the latter, the iron is eventually returned to the lumen of the gastrointestinal tract when the mucosal cell dies and is shed. What determines whether the cell lets iron through or holds it back is unknown. Increased amounts are let through in iron deficiency anemia but also in certain anemias characterized by markedly increased but ineffective erythropoiesis and plentiful body iron stores. Mucosal cells fail to hold back iron normally in a genetic disorder called idiopathic hemochromatosis. When the body absorbs iron from a normal diet as avidly as possible, between 3 and 4 mg of iron are taken in daily. This is insufficient to restore to normal the body iron content of an individual with even a mild iron deficiency anemia (who may need replacement of 750 mg of body iron), which is why such patients are treated with medicinal iron. Nevertheless, over many years the difference between taking in 3 or 4 mg of iron daily and losing 0.6 to 2.0 mg leads to the accumulation of grams of excess body iron, and the patient with idiopathic hemochromatosis usually develops evidence of organ failure from iron overload in mid-adult life.

Several tests can be used to evaluate body iron status. A bone marrow smear or biopsy can be examined for evidence of storage iron as granules of hemosiderin. The plasma transferrin concentration, which begins to rise as iron stores are depleted, can be measured (total iron-binding capacity or TIBC, see Chapter 19). Moreover, the trace amounts of ferritin in plasma can now be measured by radioimmunoassay. Although factors besides iron stores affect the serum ferritin level, particularly chronic infection and liver disease, this noninvasive and sensitive test is widely used for evaluating body iron status. A serum ferritin level of less than 12 μg/liter is diagnostic of iron deficiency (Cook, 1982).

About 20 mg of iron are needed daily to make new hemoglobin to replace the hemoglobin catabolized in the breakdown of senescent red blood cells in mononuclear phagocytes. This iron is recycled from the catabolized hemoglobin, carried back by plasma transferrin from the mononuclear phagocytes to the normoblasts in the marrow. When red blood cell production is increased, additional iron must be brought to the developing red cells. If the increased red blood cell production is in response to an increased breakdown of red blood cells (hemolysis), then the additional iron is usually readily available within the mononuclear phagocytes for transport back to the bone marrow. However, if the stimulus for increased red cell production is acute blood loss, e.g., a major gastrointestinal hemorrhage, then the additional iron must be mobilized from less readily available iron in stores. This usually limits the increase in red blood cell production to about two-fold. If the stores are empty, then despite increased stimulation by erythropoietin, the marrow may be able to respond with only a very limited increase in red blood cell production.

When body iron deficiency progresses beyond storage iron depletion, insufficient iron is available to support a normal level of erythropoiesis. An individual then develops iron deficiency anemia. Since normal daily iron loss is small, iron deficiency anemia in an adult should not be attributed to insufficient dietary iron intake. A source for abnormal blood loss should always be sought. In contrast, in an infant who requires dietary iron for an expanding hemoglobin mass, insufficient dietary intake is the usual cause for iron deficiency anemia.

In iron deficiency the inability to synthesize normal amounts of hemoglobin causes red cells to be made that are smaller than normal, with a mean volume of less than 80 fl (microcytic). On a stained blood smear the cells appear to contain a diminished complement of hemoglobin and an expanded central area of pallor (hypochromic). With mild to moderate iron deficiency the mean concentration of hemoglobin within the red cells (mean corpuscular hemoglobin concentration or MCHC), which is calculated from measured values for mean red cell volume, red blood cell count, and hemoglobin, may remain within the normal range of 32–34%. With severe iron deficiency anemia the value falls below 32%. In the uncomplicated patient the serum iron level will be below 30 μg/dl, the TIBC will be greater than 300

μg/dl, and the serum ferritin level will be less than 12 μg/liter.

Destruction of Red Blood Cells

Normal senescence and death of red blood cells occur as a function of their age, but the molecular change that determines the death of the normal red cell is not known. This change presumably alters the red cell membrane, since mononuclear phagocytes in the spleen, liver, and bone marrow recognize and remove the senescent cell. Hemoglobin, other proteins, and membrane lipids of the phagocytosed, senescent cell are catabolized within the mononuclear phagocyte. Heme is dissociated from the globin chains of hemoglobin, which are broken down into their amino acids. The heme is oxidized, in a reaction catalyzed by a microsomal enzyme, heme oxidase, with resultant opening up of the porphyrin ring structure and release of iron. As already mentioned, transferrin carries the iron back to the normoblast for incorporation into new hemoglobin. The scission of the porphyrin ring gives rise to a molecule of carbon monoxide, which is excreted in the lungs. The remainder of the molecule, called biliverdin, is then reduced to bilirubin and transported through the plasma bound to albumin for excretion by the liver. In a 70-kg adult, about 200 mg of bilirubin, derived from the breakdown of 6 g of hemoglobin, are excreted by the liver daily.

About 3% of the lipids of red cell membranes consist of a glycolipid called globoside, which contains a short carbohydrate chain. During the degradation of globoside in mononuclear phagocytes, different enzymes cleave successive sugar molecules from the carbohydrate chain. A series of rare hereditary disorders called lipid storage diseases have been recognized as resulting from hereditary deficiencies of different cleaving enzymes. The most common disorder, Gaucher's disease, results from a deficiency of the enzyme β-glucosidase, which cleaves the last remaining sugar molecule of the chain. In Gaucher's disease macrophages clogged with partially digested globoside accumulate in ever increasing numbers in the spleen, liver, bone marrow, and other organs. As a consequence, affected individuals develop enormously enlarged spleens and deformities of the bones.

About 10% of daily red cell breakdown takes place not within mononuclear phagocytes but intravascularly, with resultant release, in a 70-kg person, of about 0.6 g of hemoglobin each day into the plasma. Hemoglobin in plasma dissociates to α-β dimers; heme also dissociates from the globin chains. Two plasma proteins described in Chapter 19, haptoglobin which binds dimers and hemopexin which binds heme, prevent the daily urinary loss of approximately 2 mg of iron that would otherwise occur from a continuous glomerular filtration of dimers and heme.

In pathological states red blood cells may be destroyed at an accelerated rate. If production of new red blood cells can keep up with the accelerated rate of destruction, the red blood cell count will not fall (compensated hemolytic anemia). However, polychromatophilic macrocytes (see Chapter 20) will be seen on the peripheral blood smear, as evidence of an erythropoietin-stimulated premature release of marrow reticulocytes, and the reticulocyte count will be elevated. If the rate of destruction exceeds the capacity of the marrow to produce new red cells, then the red blood cell count will fall until a level is reached at which the number of cells being destroyed equals the number of cells being produced. In conditions with massive hemolysis, the red blood cell count may fall suddenly to very low levels. The body's compensatory mechanisms for delivering normal amounts of oxygen to the tissues in the face of a reduced hemoglobin level—an increase in cardiac output, redistribution of blood flow, and a shift to the right (see below) in the oxygen association curve of hemoglobin—may be inadequate to handle such a marked fall, and the patient may become very ill.

In many hemolytic processes altered red blood cells are removed primarily by mononuclear phagocytes (extravascular hemolysis). The unique circulation of the spleen allows this organ to play a particularly important role in extravascular hemolysis. As some of the small blood vessels enter the red pulp of the spleen they end in spaces called cords containing loosely meshed reticulin and mononuclear phagocytes. The red cells are emptied into the cords and must then percolate through the cords in intimate contact with mononuclear phagocytes and reenter venous sinuses in the red pulp by squeezing between endothelial cells. These conditions favor the removal of red blood cells with decreased deformability of the cell membrane. Because of an increased removal of red blood cells, the spleen usually enlarges in hemolytic disorders. In one disorder in which a hereditary red cell membrane defect causes cells to lose their biconcave shape and become microspheres with decreased deformability (hereditary spherocytosis), the spleen functions as the sole site of accelerated removal of the abnormal red cells. However, in most other extravascular hemolytic disorders, large numbers of damaged red blood cells are phagocytosed not only in the spleen but by mononuclear phagocytes in the liver and bone marrow.

In hemolytic states, the amount of bilirubin that the liver must excrete will usually increase severalfold. The liver can clear this increased load only after the bilirubin concentration of the plasma has risen moderately. Thus, in many patients with hemolytic anemia the total bilirubin level is elevated from a normal value of about 1 mg/dl to a value in the 2–4 mg/dl range.

When large numbers of red blood cells break down within the circulating blood (intravascular hemolysis), enough hemoglobin may enter the plasma to give it a pink to purple color. The capacity of haptoglobin to bind hemoglobin dimers is exceeded, and hemoglobin dimers are excreted in the urine (hemoglobinuria). The capacity of hemopexin to bind heme is also exceeded, and the excess heme binds to albumin, giving rise to a new plasma pigment, methemalbumin. Moreover, in chronic states of intravascular hemolysis, renal tubule cells become filled with hemosiderin due to the absorption of dimers by the tubule cells and the breakdown of hemoglobin within the cells. Shedding into the urine of tubule cells filled with hemosiderin (hemosiderinuria) can be demonstrated by centrifuging the urine and staining the pellet for iron. This is sometimes used as a screening test for chronic intravascular hemolysis.

RED BLOOD CELL MEMBRANE

As a nonnucleated cell without organelles, the red blood cell has a plasma membrane but lacks the internal membranes enclosing the nucleus and organelles found in other cells. Red cell plasma membranes can be obtained in relatively pure state in large quantities by hypotonic lysis of red cells, followed by washing to release the cytosolic proteins, mostly hemoglobin. The membranes so obtained are composed of a lipid bilayer with proteins that may enter or traverse the bilayer or adhere to its cytoplasmic surface. Like the membranes of other cells, the red cell plasma membrane is a selective permeability barrier with specific pumps, channels, and gates. It exhibits a remarkable deformability, a necessary property of a cell whose function is to transport respiratory gases.

Membrane Lipids

The membrane lipids are of three types: phospholipids, cholesterol, and small amounts of glycolipids. The phospholipids have hydrophilic and hydrophobic moieties that form a bimolecular sheet, with the polar hydrophilic groups oriented toward the exterior and cytoplasmic surfaces and the hydrophobic hydrocarbons oriented toward the interior of the bilayer. The choline-containing phospholipids, phosphatidylcholine (lecithin) and sphingomyelin, are primarily located in the outer layer of the bilayer, with their polar groups directed toward the exterior; the amino-containing phospholipids, phosphatidylethanolamine and phosphatidylserine, are primarily located in the inner layer, with their polar groups directed toward the cytoplasmic surface. The length and degree of saturation of the fatty acid residues of the phospholipids strongly influence the fluidity of the membrane. With increasing chain length and degree of saturation the fatty acids become less fluid, and the membrane becomes more viscous. Almost 55% of the membrane lipid is phospholipid and almost 45% is cholesterol, which is present in its free, nonesterified form. The cholesterol interacts with the phospholipid; if the ratio of cholesterol to phospholipid in the membrane rises, the membrane also becomes less fluid. The glycolipids of the membrane are made up of a lipid base called ceramide, consisting of sphingosine and a long chain fatty acid, to which, as mentioned earlier, hexose molecules are attached.

Membrane Proteins

The membrane proteins are classified into two types, depending upon their ease of separation from the lipid bilayer: *peripheral proteins*, which are separable by mild means, and *integral proteins*, whose separation requires organic solvents or detergents. The integral proteins are embedded within or traverse the lipid bilayer, wherein they interact extensively with the hydrocarbon chains of the membrane lipids. The peripheral proteins, which do not penetrate the hydrophobic core of the lipid layer, are located primarily at the cytoplasmic face of the membrane (Fig. 3.15). They are bound by electrostatic or hydrogen bonds to the inner polar surface of the lipid bilayer and also, through organization of the membrane skeleton (see below) to the integral protein, band 3, that protrudes through the cytoplasmic face of the membrane.

When membrane proteins extracted from red cell ghosts are subjected to polyacrylamide gel electrophoresis in sodium dodecyl sulfate (SDS-PAGE), the proteins migrate different distances into the gel depending upon their size. Major and minor protein bands are visualized (Fig. 3.15) after staining for protein with Coomassie blue stain. These bands have been assigned arabic numbers, with decimals being used to identify bands thought to be subfractions of a major band or minor bands in close proximity to major bands. A number of bands have been identified: band 5 is erythrocyte actin, and band 6 is the enzyme glyceraldehyde-3-phosphate dehydrogenase (G3PD), bands 1 and 2 are the two chains of spectrin, and a series of bands, 2.1 and its proteolytic products 2.2, 2.3, and 2.6, are the syndeins (also called ankyrin). When such SDS gels are stained for carbohydrate with periodic acid-Schiff stain, four additional bands are visualized which represent monomers or dimers of two glycoproteins called glycophorin A and glycophorin B.

The glycophorins and band 3 are the major integral proteins that pass from the exterior of the cell through the lipid bilayer and into the cell cyto-

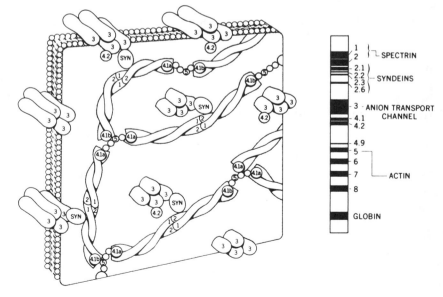

Figure 3.15. The organization of the membrane skeleton of the red blood cell. The integral protein, band 3, is shown protruding through the cytoplasmic face of the lipid bilayer and binding, through the syndeins, to the spectrin meshwork of the membrane skeleton. The heads of spectrin dimers associate to form spectrin tetramers at the sites marked *1* and *2* (for bands 1 and 2). The association of the 4.1 band proteins and of actin (labeled *5*) to the terminal portion of spectrin tetramers is also shown. The glycophorins, integral proteins that also penetrate the cytoplasmic surface of the lipid bilayer but which are not associated with the membrane skeleton, have not been illustrated. Also shown is an SDS-PAGE gel to illustrate the major membrane protein bands. (From Goodman, S. R., et al. Identification of the molecular defect in the erythrocyte membrane skeleton of some kindreds with hereditary spherocytosis. *Blood* 60: 772–784, 1982, by permission.)

plasm. Carbohydrate chains are attached to the external portion (NH_2-terminal segment) of the proteins. Glycophorin A, which makes up about 75% of red cell membrane glycoproteins, is particularly rich in carbohydrate, containing 16 oligosaccharide chains that represent 60% of the molecule. The complete amino acid sequence of glycophorin A has been determined. However, despite extensive knowledge of its structure, the function of glycophorin A remains unknown. Band 3 (Steck, 1978) contains 5–8% carbohydrate. Its amino acids are arranged in a tertiary structure that forms an anion channel, through which chloride ions enter and leave the red cell (chloride shift) as intracellular bicarbonate ion concentration changes with the CO_2 content of the blood (see below). The syndeins (ankyrin), two glycolytic enzymes and, under some circumstances, hemoglobin binds to the cytoplasmic portion of band 3.

When red cell ghosts are extracted with nonionic detergents, several of the membrane proteins are not extracted but remain as an insoluble proteinaceous remnant called the red cell membrane skeleton (Bennett, 1985). The membrane skeleton provides the structural integrity of the cell, determining at least partially the shape of the cell and its ability to change shape as it passes through the microcirculation. The principal protein of the membrane skeleton is spectrin; other proteins are erythrocyte actin and band 4.1 (a and b).

The spectrin molecule is made up of a 250,000-dalton polypeptide chain (band 1) and a 225,000-dalton polypeptide chain (band 2). The two chains, frequently referred to as the α and β chains, form a dimer consisting of two similar, flexible, fiber-like structures about 1000 Å in length, which lie parallel to each other and touch each other at multiple points. The β chain is phosphorylated on incubation in vitro with ATP; the functional significance of its phosphorylation in vivo is not clear. Recent evidence suggests that spectrin in the red cell membrane exists as a tetramer formed by the head to head association of two heterodimers. It has been suggested (Marchesi, 1983) that higher molecular weight oligomeric forms of spectrin are also present in the red cell membrane.

The spectrin tetramers (and possibly higher molecular weight oligomers) are linked to form the basic membrane skeleton network by erythrocyte actin filaments and the band 4.1 proteins, which are associated with the terminal ends of the spectrin tetramers. Through these bonds a cross-linked two-dimensional network of spectrin is formed just beneath the lipid bilayer (Fig. 3.15). Yet another protein, ankyrin (syndeins) associates with spec-

trin and also with the carboxy-terminal segment of band 3. This spectrin-ankyrin-band 3 association binds the spectrin meshwork to the inner surface of the membrane and also provides a potential link between events in the cytoplasm and the external surface of the red cell.

Abnormalities of association in the spectrin network have been found in some kindreds with elliptocytosis and pyropoikilocytosis. In a common hereditary hemolytic anemia, hereditary spherocytosis, the amount of spectrin is reduced (leading to conversion of normally biconcave discs to spherical shape), but the molecular defect or defects are generally not known. (In a few families with hereditary spherocytosis, defective binding of spectrin to band 4.1 has been found.)

A number of enzymes are present in red blood cell membrane preparations (Schrier, 1977). Some, like G3PD, the enzyme present in such large amounts that it forms band 6 on SDS gel electrophoresis, are found both in membrane preparations and in membrane-free cytosol. Others are enzymes whose activity is confined to the membrane. These include an ATPase that is part of the Na^+-K^+ pump which maintains a high intracellular K^+ and low Na^+ relative to the extracellular fluid; an ATPase that is part of a Ca^{2+} pump that extrudes Ca^{2+} from the cell against a 50-fold concentration gradient; and protein kinases stimulated by cyclic AMP. The majority of the enzymes are found on the cytoplasmic surface of the membrane. However, some enzymes, including an acetylcholine esterase and several enzymes generally thought of as lysosomal enzymes (glycosidases, acid phosphatase), are externally oriented membrane enzymes. Their physiological functions are not understood.

HEMOGLOBIN

Biosynthesis of Hemoglobin

The production of hemoglobin is a tightly regulated process requiring the coordinated synthesis of different polypeptide globin chains and of heme. Hemoglobin is synthesized primarily in the nucleated red cells of the marrow during the 6–8 days of erythroid cell maturation. Low levels of synthesis continue for a day or so in the reticulocyte; the mature erythrocyte cannot synthesize hemoglobin. The quantities of free globin subunits and of free heme in red cells are minute.

HEME SYNTHESIS

Heme, ferrous protoporphyrin IX, is synthesized in a series of steps that begin in the mitochondria with the condensation of glycine and succinyl coenzyme A to form aminolevulinic acid (ALA). This reaction, which is catalyzed by δ-aminolevulinate synthetase and requires pyridoxal phosphate, is a committed step. Once ALA is formed, reactions continue until heme is formed. Further steps take place in the cytoplasm where, catalyzed by a specific dehydrase, two moles of ALA condense to form porphobilinogen. Four porphobilinogens condense to form a linear tetrapyrrol, which then cyclizes. Side chains of the tetrapyrrol ring are successively reduced by additional enzymatic reactions to give rise to protoporphyrin IX. In the mitochondria, ferrous iron is inserted into the protoporphyrin ring, catalyzed by the enzyme ferrochelatase. The product, heme (Fig. 3.16) represses the synthesis of the first enzyme of the process, δ-aminolevulinate synthetase. Thus, heme regulates its own synthesis. Defects in enzymes of heme synthesis give rise to different clinical disorders known as the porphyrias.

Biosynthesis of Globin

Hemoglobin consists of two pairs of unlike chains: two α chains and two non-α chains, which may be β, γ, or δ chains. The major adult hemoglobin, hemoglobin A, is $\alpha_2\beta_2$; fetal hemoglobin is $\alpha_2\gamma_2$. The locus for the α chains is on chromosome 16 and is duplicated, i.e., a normal individual has four α genes, two on each chromosome. The products of the duplicated α genes are identical. The non-α genes all lie closely linked on chromosome 11 (Fig. 3.17). The γ locus is also duplicated; the gene products are identical except for residue 136, which is a glycine at one locus (G_γ) and an alanine at the other (A_γ). The order of the non-α genes on chromosome 11 is G_γ, A_γ, δ, β. The amino acid sequence of the chains is determined by the triplet groups of bases in the DNA (codons). The sequences in DNA that code for the globin gene polypeptides are split (Fig. 3.17); each globin gene has two intervening sequences

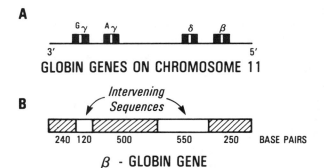

Figure 3.17. Globin genes. *A* illustrates the loci for the non-α globin genes on chromosome 11. *B* is a representation of the β-globin gene. The *hatched areas* indicate areas of base pairs that are translated. The *open areas* are intervening sequences that are transcribed and eliminated in processing.

that are not translated into the amino acid sequence of the polypeptide chain. In β globin the small intervening sequence (intron) corresponds to an untranslated portion between the codons for amino acid residues 30 and 31 and the large intervening sequence between the condons for amino acid residues 104 and 105. The total DNA sequence including the introns is transcribed into a large mRNA precursor which is then "processed" in the nucleus by the addition of modified nucleotides at each end of the molecule, excision of the introns, and ligation of the remaining mRNA sequences. The mRNA thus created is transported to the cytoplasm where it associates with the translational machinery. Translation is a highly coordinated process requiring the interaction of mRNA and tRNAs, proteins with specific enzymatic functions, and ribosomes. Chain growth proceeds from the amino to the carboxyl end of the polypeptide chain; release factors that recognize a termination codon terminate the process.

There appears to be a minute free pool of α chains in developing erythroid cells. Both free α chains and free β chains are unstable, however, and may precipitate in the cell and undergo proteolytic degradation.

During the last several years, techniques for establishing DNA sequences have provided powerful tools for the study of abnormalities of globin synthesis. Restriction endonucleases are enzymes that cleave DNA at specific sites. The resulting fragments can be separated and used to prepare maps of the DNA region of interest. Enzymatic replication techniques and chemical cleavage by compounds that react with specific bases are used to establish nucleotide sequences.

Genetic abnormalities may give rise to structural hemoglobin variants, may impair polypeptide chain

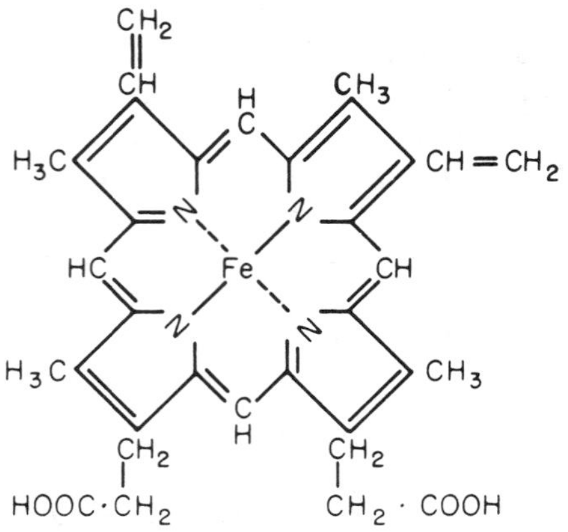

Figure 3.16.   The structure of heme.

synthesis, or may do both. Structural abnormalities most commonly are single amino acid substitutions that stem from substitution of a single base in a codon. However, in some structural variants, one or more amino acids have been deleted or a mutation in the normal termination codon has caused the elongation of a polypeptide chain.

Defects in the rates of synthesis of a specific polypeptide chain are expressed clinically as a group of disorders called thalassemia. Alpha thalassemia (impaired synthesis of α chains) usually results from a deletion of one to four of the α genes. (Deletion of all four genes is fatal, with death of the fetus in utero.) The non-α thalassemias may result either from deletions or from other gene abnormalities. In different β thalassemic disorders, part of the β gene, the entire β gene, or both the β and δ genes may fail to be transcribed. In some β thalassemias, β globin DNA is present but abnormal, giving rise to mRNAs that are defective, diminished, or absent. One of these abnormalities results from a mutation in a codon in an intron; another results from a mutation converting the codon for amino acid residue 17 into a termination codon. Other abnormalities result in defective splicing of mRNA during processing.

In yet other disorders a combination of decreased synthesis and structural abnormality of a polypeptide chain occurs together to yield a "thalassemic phenotype." In hemoglobin E, a substitution in a codon near the small intron of the β gene results in both an amino acid substitution at residue 26 and a decreased amount of mRNA (due to defective splicing). In the Lepore hemoglobin syndromes, a crossing over between δ and β genes in which the N-terminal residues are of δ chain and the C-terminal residues are of β chain origin results in a new chain of 146 amino acid residues.

The severest forms of thalassemia are characterized by anemia, hypochromic and microcytic red cells, and splenomegaly. The anemia, which often is severe enough to require transfusions, results both from inadequate red cell production and from damage to the red cells by the *normal* subunits which, lacking unlike subunits with which to form tetramers, precipitate in the red cells.

Functional Properties of Hemoglobin

Hemoglobin is found in an extraordinarily high concentration, 32 g/dl or a 5 mM solution, in the red blood cell. Within the cell, hemoglobin transports oxygen efficiently without exerting the large osmotic effect it would have as a plasma protein. Moreover, within the red blood cell a mechanism operates (discussed later) to reduce ferrihemoglo-

bin (methemoglobin) and so keep the iron of hemoglobin in its oxygen-carrying ferrous form.

The important properties of hemoglobin as a transporter of oxygen—understood before much was known about the structure of hemoglobin—were elegantly summarized by Barcroft in 1928. These are as follow (see also Chapter 37).

1. The *oxygen affinity* of hemoglobin is so arranged that hemoglobin becomes fully saturated with oxygen in the lungs exposed to ambient air and delivers oxygen at the partial pressure of oxygen encountered in tissues. The oxygen affinities of different hemoglobins or different red cells can be compared by determining the partial pressure of oxygen at which half the hemoglobin is oxygenated and half is deoxygenated, i.e., the P_{50}.

2. The initial binding of oxygen to hemoglobin facilitates the further binding of oxygen to hemoglobin. This characteristic of hemoglobin binding is called *heme-heme interaction* because the oxygen binding of one heme affects the binding properties of other hemes. The changing oxygen affinity of hemoglobin with oxygenation results in a sigmoid curve (Fig. 3.18) when the degree of oxygenation or percentage of saturation of hemoglobin with oxygen is plotted against the partial pressure of oxygen (P_{o2}). Such a plot is referred to as an oxygen dissociation

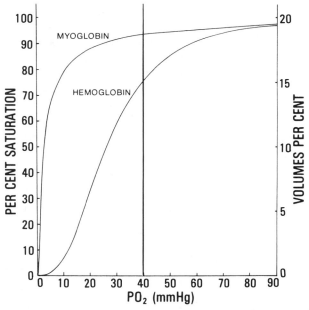

Figure 3.18. The difference between oxygen release from hemoglobin in whole blood and from myoglobin at 37°C and pH 7.4. The normal oxygen pressure of mixed venous blood is 40 mm Hg. (From Bunn, H. F., et al. *Human Hemoglobins.* Philadelphia: Saunders, 1977, by permission.)

curve of hemoglobin. The shape of its midportion is denoted by the value of n from the Hill equation:

$$\log (Y/(1 - Y)) = n \log (P_{o_2}/P_{50})$$

where Y indicates fractional saturation with oxygen and n is an empiric constant without physical basis. Values of n for oxygen equilibria of normal mammalian hemoglobins are 2.8 to 3. For myoglobin the value of n is 1, which reflects the hyperbolic shape of the oxygen dissociation curve of myoglobin (Fig. 3.18). (The high oxygen affinity of myoglobin at normal tissue oxygen pressure allows myoglobin to function as an oxygen storage protein of muscle that releases its oxygen at the very low intracellular partial pressures of oxygen resulting from exercise.)

3. The oxygen affinity of hemoglobin changes with intracellular pH (Fig. 3.19). This property of hemoglobin was first recognized in 1904 by Christian Bohr, who noted the effect of P_{CO_2} upon the oxygen dissociation curve (Fig. 3.19). In the capillaries of metabolizing tissues, CO_2 enters the plasma and red cells. Red cells contain carbonic anhydrase that rapidly converts CO_2 to H_2CO_3, a weak acid that ionizes to H^+ and HCO_3^-, lowering intracellular pH. This increase in hydrogen ion concentration decreases the oxygen affinity of the hemoglobin (Bohr effect) and facilitates oxygen delivery to the tissues. As deoxyhemoglobin, a weaker acid than oxyhemoglobin and so able to bind added protons, accumulates in the red cell, the deoxyhemoglobin binds the H^+ ions liberated from H_2CO_3. The increased HCO_3^- ions diffuse out of the red cell and are replaced by chloride ions in the "chloride shift." In the lungs the process is reversed; CO_2 is lost from the blood, pH rises, and the affinity of hemoglobin for oxygen increases. (The basis of relationships between

O_2, CO_2, and H^+ binding have been analyzed in a classic article by Wyman (1964).

Barcroft (1928) also suggested that an additional, then unidentified modulator of oxygen affinity of hemoglobin existed. Almost 40 years later this additional modulator was identified (Benesch and Benesch, 1967) as 2,3-diphosphoglycerate (2,3-DPG). Today, the *principal modulators* of oxygen affinity are recognized as hydrogen ion concentration (Bohr effect), temperature, and organic phosphates, particularly 2,3-DPG. ATP, the second most abundant organic phosphate in human red cells, is primarily bound to Mg^{2+} and the Mg^{2+}-ATP complex has little effect on oxygen affinity (Bunn et al., 1971). The effect of temperature upon oxygen affinity appears physiologically appropriate: with increasing temperature, oxygen affinity decreases and, with hypothermia, oxygen affinity increases. The effect of temperature change is significant. A 10° increase in temperature nearly doubles the P_{50} of hemoglobin. Direct binding of carbon dioxide as in a carbamino complex also lowers oxygen affinity, but this effect is minor in human hemoglobin, where 10% or less of the CO_2 is transported in this form. The concentration of hemoglobin in the red cell also seems to have little effect upon oxygen affinity except for abnormal hemoglobins, notably sickle hemoglobin.

The mechanism of synthesis of 2,3-DPG in the red cell is discussed later. In cells other than the erythrocytes, 2,3-DPG is present in minute concentrations; in the erythrocyte, the concentration of 2,3-DPG is about equimolar with hemoglobin, 5 mM. 2,3-DPG alters the oxygen affinity of hemoglobin by two mechanisms: by binding to deoxyhemoglobin (as also discussed later) and by its effect upon intracellular pH. Since 2,3-DPG is a highly charged impermeant anion with 5 titratable acid groups and 3.5 negative charges, it lowers intracellular pH rel-

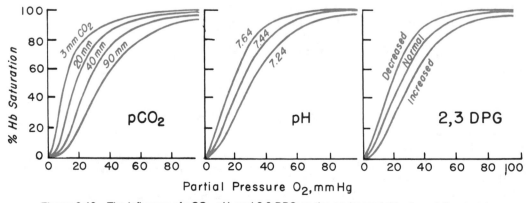

Figure 3.19   The influence of pCO₂, pH, and 2,3-DPG on the oxyhemoglobin dissociation curve.

ative to that of plasma. The further decrease of oxygen affinity of hemoglobin observed when the molar concentration of 2,3-DPG exceeds the molar concentration of hemoglobin reflects the contribution of 2,3-DPG to the Bohr effect.

Physiological factors that increase P_{50}, i.e., that shift the oxygen dissociation curve to the right, have an insignificant effect upon oxygen loading at the normal partial pressure of oxygen in arterial blood in the lungs (90 torr) but substantially increase oxygen release from hemoglobin at the mean end capillary partial pressure of oxygen in the tissues (40 torr). This is perhaps best illustrated when oxygen equilibria are plotted as O_2 content in volumes percent (rather than percentage of saturation) against P_{O_2}. As seen in Figure 3.20, for an oxygen dissociation curve with a normal P_{50} of 26.5 torr, 4.5 volumes of oxygen are delivered per 100 ml of blood. For a right-shifted curve with a P_{50} of 36.5 torr, the O_2 content of arterial blood is reduced slightly to 19 volumes percent, but each 100 ml of blood at an end capillary partial pressure of oxygen of 40 torr has delivered 7.3 volumes of oxygen.

Structural Properties of Hemoglobin

The molecular weight of human hemoglobin is 64,400 daltons. The molecule is a tetramer of two pairs of unlike polypeptide chains, which undergo conformational isomerization with oxygenation.

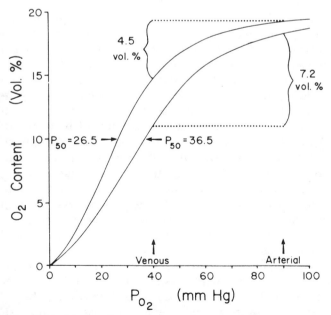

Figure 3.20 The enhanced unloading of oxygen at the oxygen pressure of mixed venous blood resulting from a right-shift in the oxyhemoglobin dissociation curve that increases the P_{50} from 26.5 to 36.5 mm Hg. (From Klocke, R. A. Oxygen transport and 2,3-diphosphoglycerate (DPG). *Chest* 62: 79–85S, 1972, by permission.)

Hemoglobin, therefore, has two different stable structures, oxy and deoxy. Much of our present knowledge of hemoglobin is based upon the X-ray crystallographic studies of Perutz and his associates (Perutz, 1970; 1978; 1979) and upon the work of many investigators who delineated amino acid sequences.

As already mentioned, hemoglobin A, which accounts for more than 95% of normal adult hemoglobin, has two α and two β chains. Human fetal hemoglobin (hemoglobin F) has two α chains and two non-α chains, called γ chains, that differ from β chains by 39 amino acid residues out of a total of 146 residues for each chain. A minor human hemoglobin, hemoglobin A_2, comprising only about 2–3% of total adult hemoglobin, is composed of two α and two δ chains; the δ chain differs in only 10 amino acid residues from the β chain.

The signal that determines the switch from fetal (γ chain) to adult (β or δ chain) hemoglobin synthesis is not known. It appears to be correlated with fetal maturity and occurs normally shortly before birth. The ability to turn off this switch mechanism would have major therapeutic implications in the common and serious hereditary anemias resulting from β chain defects, namely, the sickle cell disorders and the β thalassemias.

The individual chains of hemoglobin, α, β, and γ, have oxygen-carrying properties similar to those of the single chain of myoglobin; a tetramer of β chains (hemoglobin H), which is found in a form of α thalassemia, has a hyperbolic oxygen dissociation curve similar to that of myoglobin. Thus, the functional properties of hemoglobin that permit it to function as an effective oxygen transport protein require that the molecule be assembled as a tetramer of unlike chains, α with β, δ, or γ. The normal hemoglobins so formed, A, A_2, and F, differ little in their functional properties except for a lesser ability of hemoglobin F to bind organic phosphates. Fetal red cells therefore have a higher O_2 affinity than do adult red cells.

Hemoglobin in its deoxy or T (tense) conformation has many salt bridges between and within subunits. As the hemoglobin molecule takes up successive O_2 molecules, these salt bridges are successively broken and the molecule reaches a stable fully oxygenated or R (relaxed) conformation. During oxygenation the position of $\alpha\beta$ dimers within the tetramer shifts in a "ratchet" arrangement along the $\alpha_1\beta_2$ interface. The β chains move together by about 7 Å, narrowing the central cavity of the tetramer. The C-terminal portion of the β chains move from an anchored position between the F and H helices (Fig. 3.21) to a position in which their res-

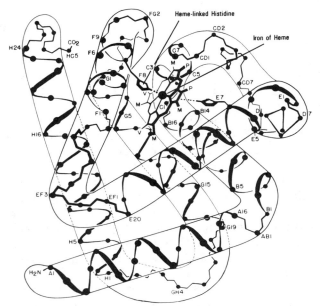

Figure 3.21. A three-dimensional representation of a hemoglobin subunit. The helical segments are assigned letters from A through H beginning at the NH₂-terminal segment of the molecule. Individual amino acids within segments are numbered consecutively. Interhelical segments bear the letters of the two adjacent helices (e.g., *EF*). The proximal heme-linked histidine and heme iron in a porphyrin ring are indicated. (From Dickerson: *The Proteins: Composition, Structure and Function,* 2nd ed., edited by H. Neurath, vol. 2. New York: Academic Press, 1963, by permission.)

idues are available to the solvent. These reactions occur sequentially as the reaction of a heme with oxygen affects the structure around the remaining hemes in sequence. Thus, the heme-heme interaction is an expression of the allosteric properties of hemoglobin.

2,3-DPG is bound in the central cavity of deoxyhemoglobin to specific chain residues (Arnone, 1972) where it, in effect, forms an additional salt bridge in deoxyhemoglobin. As the β chains move together in the oxy conformation, 2,3-DPG is expelled, and the opening becomes too narrow to permit binding at the specific deoxy site. The Bohr effect results from the high affinity of deoxyhemoglobin for protons. A change in the environment of three pairs of proton-binding groups mostly accounts for the effect. The largest contribution is from the C-terminal histidines of the β chains.

Heme is embedded in a hydrophobic pocket between two helical segments (E and F) of each polypeptide chain of hemoglobin (Fig. 3.21). The heme is so oriented that its hydrophobic side chains are in the interior of the globin subunit and its polar side chains are on the surface. It is anchored in the chain by many interatomic contacts. The iron of the heme is covalently bound to hemoglobin by the proximal F8 histidine (Fig. 3.21). In the deoxy form,

the heme iron is displaced 0.6 Å from the plane of the porphyrin ring, and the F8 heme-linked histidine is "tilted." On oxygenation, the iron moves into the plane of the ring, and the "tilt" changes. Oxygen is bound on the E7 (distal histidine) side of the heme in an area "guarded" by residue E11. The divalent heme iron—five coordinated in deoxyhemoglobin and six coordinated in oxyhemoglobin in which O₂ binds to the sixth position—binds oxygen without itself being oxidized.

Abnormalities of Hemoglobin as a Cause for Disease

GENETIC ABNORMALITIES OF HEMOGLOBIN

The thalassemias, a group of anemic disorders of varying severity dependent upon the extent of impaired synthesis of a polypeptide chain of hemoglobin, have been discussed earlier. Three common, clinically important structural hemoglobin variants result from single amino acid substitutions in the β chain: hemoglobin S (Pauling et al., 1949), in which a valine replaces the normal glutamic acid at the sixth residue of the chain; hemoglobin C, in which a lysine replaces glutamic acid at the same sixth residue site; and hemoglobin E, in which a lysine replaces a normal glutamic acid at the 26th residue of the chain. As mentioned earlier, the mutation for hemoglobin E also apparently decreases β^E messenger RNA, and affected individuals have hypochromic, microcytic red cells as seen in thalassemia. Hemoglobin E is found most frequently in persons of southeast Asian ancestry.

Hemoglobin S (sickle hemoglobin) which consists of two normal α chains and two β^s chains, forms intracellular polymers on deoxygenation. With polymer formation the red cells containing hemoglobin S lose their normal deformability and assume a sickled shape, which reverts to normal on oxygenation. In the heterozygote state, called sickle cell trait, the red cells contain a mixture of hemoglobin A and hemoglobin S, and affected individuals, with very rare exceptions, do not have any clinical manifestations. In contrast, in the homozygous state, sickle cell anemia, the patient has a life-long anemia due to the short survival time of sickled cells, and repeated painful episodes, called vasoocclusive crises, are brought on by sickled cells occluding small blood vessels.

Hemoglobin C also has a reduced solubility but not to the same degree as hemoglobin S. It does not form sickle cells. Homozygotes for hemoglobin C have a compensated hemolytic state, unusual-appearing red cells (target cells) on stained blood smears, and a minimally to moderately enlarged spleen. They do not have vasoocclusive crises. Both hemoglobin S and hemoglobin C are most often found in individuals of African or Mediterranean ancestry. The frequency of genes for thalassemia is also increased in these ethnic groups. Therefore, individuals who are heterozygotes for both hemoglobin S and hemoglobin C or for hemoglobin S and β thalassemia are encountered. Such individuals have hemoglobin S/C disease or S-thal disease. Their red blood cells sickle in vivo, and they have vasoocclusive crises that may be as severe as those in sickle cell disease. However, their anemia is usually less severe than the anemia of sickle cell disease.

Hemoglobins S and C produce disease because they decrease hemoglobin solubility. Other hereditary structural abnormalities of globin affect other properties of hemoglobin: its stability, oxygen affinity, or ability to maintain iron in heme in the ferrous form. These variant hemoglobins are of less clinical importance because of their rarity, but their study has provided valuable information on the relation between globin chain structure and both the structural and the functional integrity of hemoglobin.

Several *unstable hemoglobins* arising from single amino acid substitutions or deletions have been characterized. Their globin chains are abnormally susceptible to denaturation, forming precipitates of hemoglobin within the cell that are demonstrable on supravital staining and are called Heinz bodies. Unlike aggregates of sickle hemoglobin which form when oxygen is released and become soluble again when oxygen is taken up, Heinz bodies are neither ligand-dependent for their formation nor reversible. Heinz bodies damage the red cell membrane and lead to accelerated red cell destruction, with a resultant anemia of variable severity.

METHEMOGLOBINEMIA

The iron of heme normally is in the ferrous state; during reversible binding of oxygen the iron remains ferrous. When the heme iron is oxidized to the ferric state, the resulting methemoglobin is no longer capable of reacting with oxygen. In acid solutions, the oxygen binding site is occupied by water, in alkaline solutions by an OH^- group. Methemoglobin, which is brown in color, has a characteristic light absorption at 630 nm. Methemoglobinemia is encountered in three circumstances:

1. After exposure of normal red cells to toxic chemicals such as nitrites, aniline dyes, and certain oxidant drugs.
2. In heterozygotes (the homozygote state is lethal) for hemoglobin variants called M hemoglobins in which amino acid substitutions affecting the heme binding pocket cause the iron of heme to be oxidized.

3. In homozygotes for deficiency of NADH-dependent methemoglobin reductase, which, as discussed later, normally reduces the small amounts of methemoglobin that form in red blood cells.

Levels of methemoglobin between 10 and 25% result in cyanosis but usually do not require treatment. At levels of 35%, patients may develop dyspnea and headache; levels of 70% are lethal. The toxicity of methemoglobin stems not only from its own inability to carry oxygen but from its effect upon the oxygen equilibria of hemoglobin tetramers. In hemoglobin molecules some of the hemes of a tetramer will be oxidized, while the remainder will be in the ferrous form. In this circumstance, the oxygen affinity of the functional ferrous hemes is increased, with a resultant "left-shifted" oxygen dissociation curve and impaired delivery of oxygen to the tissues. Toxic levels of methemoglobin may be treated with methylene blue (see below).

CARBONMONOXY HEMOGLOBIN

Carbon monoxide (CO), like oxygen, binds to the sixth coordination position of the iron in heme, but the affinity of hemoglobin for CO is about 200 times that for oxygen. Carbon monoxide interferes with oxygen transport in two ways. In a portion of the hemoglobin molecule CO may occupy the iron of all of the hemes of the tetramer, and the molecule cannot bind oxygen. In other molecules, the hemoglobin is composed of mixed tetramers (e.g., α^{CO}, β^{CO}, α^O, β^O) with "left-shifted," high affinity, oxygen equilibria. Symptoms of carbonmonoxy hemoglobin toxicity appear at levels of 5–10% carbonmonoxy hemoglobin; at levels above 40% unconsciousness may proceed to death. At a given blood level, carbonmonoxy hemoglobin intoxication is more life-threatening than methemoglobinemia.

The high affinity of CO for hemoglobin results not from an increased rate of association but from a slow rate of dissociation of CO from heme. The half-disappearance time for CO hemoglobin in an individual with normal ventilatory function is about 4 hours; breathing pure oxygen will shorten the half-time to about 1 hour.

RED CELL METABOLISM

Although the anucleate red cell serving as a gas transporter has fewer metabolic needs than most other cells, it does require energy to maintain the shape and flexibility of the cell membrane, to maintain hemoglobin iron in its functional divalent form, and to preserve the high K^+, low Na^+, low Ca^{2+} ionic milieu of the red cell against the gradient of the different ionic composition of plasma. An anaerobic pathway (Embden-Meyerhof pathway) in which

ATP and NADH are generated supplies this energy. The pathway has two shunts: an aerobic shunt, the hexose monophosphate shunt, in which NADPH is generated, and a second shunt, in which 2,3-DPG is produced. The pathway and its shunts are shown in Figure 3.22.

Glucose Metabolism

The main energy source of the red cell is glucose, which is metabolized to pyruvate or lactate. Mature red cells do not have a citric acid cycle; only in reticulocytes which still possess mitochondria is pyruvate broken down to CO_2.

Glucose apparently enters the cell via a carrier protein; its transport is not insulin-dependent. Within the cell it is rapidly converted by hexokinase to glucose 6-phosphate. About 90% of the glucose 6-phosphate is processed via the anaerobic pathway in a series of enzymatic steps that require 2 moles of high energy phosphate (ATP) but generate up to 4 moles of ATP to provide a net potential yield of 2 moles of ATP per mole of glucose. Deficiencies, usually recessively inherited, of many of the enzymes of the anaerobic pathway have been recognized in association with manifestations of hereditary nonspherocytic hemolytic anemia (Valentine, 1979). These are very rare except for deficiency of pyruvate kinase, which, although uncommon, ranks behind glucose 6-phosphate dehydrogenase (see below) as the second most com-

mon inherited enzyme deficiency in human red blood cells. Deficiency of pyruvate kinase leads to hereditary anemia of variable severity that may be accompanied by jaundice and splenomegaly.

Hexose Monophosphate Shunt

In the hexose monophosphate shunt, glucose 6-phosphate is oxidized with reduction of NADP to NADPH. High energy phosphate bonds are not generated in this shunt; its function includes the production of NADPH for use in reducing glutathione disulfide (GSSG) to GSH (Beutler, 1983). As already mentioned, a deficiency of the enzyme glucose 6-phosphate dehydrogenase (G-6-PD), which catalyzes the oxidation of glucose 6-phosphate, is the most frequently encountered hereditary deficiency of a red cell enzyme.

Many different forms of G-6-PD have been recognized by electrophoretic and kinetic analyses; in only a few of these variants is the physiological enzymatic activity of the enzyme affected. G-6-PD activity normally declines as the red blood cell ages. In one variant, the A^- variant, enzymatic activity declines abnormally rapidly, and activity present in reticulocytes disappears in mature red cells. The A^- variant is inherited as a sex-linked trait with a high gene frequency in blacks; a deficiency of G-6-PD due to this variant is found in 11% of black American males. Affected individuals have no manifestations of the deficiency until exposed to certain stresses such as oxidant drugs or infections. Then, they may develop a hemolytic anemia.

The mechanism of hemolysis after exposure to drugs is only partially understood. Exposure of red cells to some of the offending drugs yields low levels of hydrogen peroxide or free radicals. Hydrogen peroxide is detoxified in a reaction in which GSH is oxidized to GSSG (Fig. 3.12, Chapter 22). Glutathione reductase, in a reaction requiring NADPH, catalyzes reduction of the GSSG to replenish GSH. Red cells deficient in G-6-PD cannot generate NADPH at a normal rate to support this reaction. As a consequence, the products of oxidant drugs may damage both hemoglobin and the cell membrane. Precipitates of damaged hemoglobin that appear attached to the membrane (Heinz bodies) are formed in the cells.

2,3-Diphosphoglycerate Shunt

A second shunt, the 2,3-DPG shunt (Rapoport-Luebering shunt) is present in the main glycolytic pathway at the step of processing of 1,3-diphosphoglycerate (1,3-DPG). In the straight anaerobic glycolytic pathway, 1,3-DPG is converted to 3-phosphoglycerate (3-PG) with a gain of one ATP (two

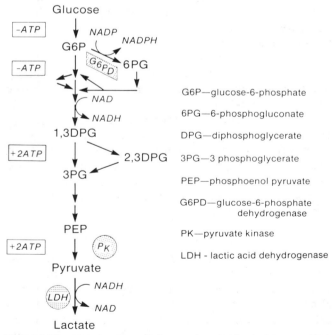

Figure 3.22. Embden-Meyerhof pathway, with its two shunts. (From Rapaport, S. I. *Introduction to Hematology,* 2nd edition. Philadelphia: J. B. Lippincott, 1987.)

G6P—glucose-6-phosphate

6PG—6-phosphogluconate

DPG—diphosphoglycerate

3PG—3 phosphoglycerate

PEP—phosphoenol pyruvate

G6PD—glucose-6-phosphate dehydrogenase

PK—pyruvate kinase

LDH - lactic acid dehydrogenase

ATPs for each starting glucose molecule, since one glucose molecule gives rise to two molecules of 1,3-DPG). In the shunt, an enzyme, diphosphoglycerate mutase, first catalyzes the formation of 2,3-DPG from 1,3-DPG and then catalyzes the loss of a phosphate from 2,3-DPG and its reentry into the glycolytic pathway as 3-PG (Fig. 3.22). ATP is not generated in these reactions. Therefore, when a molecule of glucose is processed to pyruvate by way of the 2,3-DPG shunt, two molecules of ATP are used up and two molecules of ADP are made. Glycolysis then generates NADH but no net gain in ATP.

2,3-DPG accounts for over half of the phosphorus in the erythrocyte and is about equimolar in quantity with hemoglobin. Several factors influence the amount of 2,3-DPG synthesized, including the ratio of cellular ADP and ATP levels and the overall rate of glycolysis. Alkalosis stimulates glycolysis because of the high pH optimum of the glycolytic enzyme, phosphofructokinase, whereas acidosis impairs glycolysis. Thus, the respiratory alkalosis induced by a rapid ascent to high altitudes, by stimulating glycolysis, increases red cell production of 2,3-DPG. Increased red cell levels of 2,3-DPG are also found in anemia, cardiac disease, and pulmonary disease. The stimulus for increased production of 2,3-DPG in these circumstances may arise from an increased binding of 2,3-DPG to deoxyhemoglobin, which shifts an increased proportion of 1,3-DPG into the shunt pathway.

The physiological usefulness of a glycolytic pathway shunt that prevents a gain in ATP was clarified when it was shown (Benesch and Benesch, 1967) that hemoglobin stripped of phosphate has a very high oxygen affinity and that addition of 2,3-DPG to such a hemoglobin preparation lowers its oxygen affinity to that observed in a hemolysate prepared from normal red blood cells. It was suggested that 2,3-DPG and other red cell organic phosphates modulate oxygen affinity by binding to deoxyhemoglobin (Benesch and Benesch, 1969), a binding that has since been demonstrated directly by X-ray crystallography (Arnone, 1972). 2,3-DPG binds less well to the γ chains of fetal hemoglobin than to the β chains of adult hemoglobin, which benefits the fetus, since fetal red cells take up oxygen not at the partial pressure of oxygen in the lungs but at the partial pressure of oxygen in maternal blood flowing through the placenta.

Reduction of Methemoglobin

The products of red cell metabolism also interact with hemoglobin to reduce to ferrous hemoglobin the small amounts of methemoglobin that are con-

stantly being produced in normal red cells. The red blood cell possesses both an NADH-dependent pathway and an NADPH-dependent pathway for reduction of methemoglobin. Only the NADH-dependent pathway functions physiologically. It utilizes the enzyme NADH methemoglobin reductase and cytochrome b_5 as an electron acceptor to transfer an electron from NADH to heme. The NADPH pathway does not function physiologically despite the presence of the enzyme NADPH reductase because it lacks an electron acceptor. Methylene blue, an exogenous electron acceptor, can mobilize the NADPH pathway and is given in the treatment of patients with severe methemoglobinemia after exposure to an oxidant agent.

ANEMIA AND ERYTHROCYTOSIS

A person is generally considered to be anemic when his or her hemoglobin level falls below 2 SDs of the mean for a normal population (below 14 g/dl for a man and 12 g/dl for a woman at sea level). When fully oxygenated, each gram of hemoglobin carries 1.39 ml of oxygen. Therefore, normal arterial blood with a hemoglobin concentration of 15 g/dl carries about 21 ml of oxygen, of which 4.5 ml are delivered to the tissues as the partial pressure of oxygen in blood flowing through a capillary bed falls to 40 torr (Fig. 3.20). In an anemic individual with a hemoglobin concentration of only half-normal, 7.5 g/dl 100 ml of blood can carry only about 10 ml of oxygen. The body compensates for this loss of oxygen-carrying capacity by three mechanisms: by increasing the 2,3-DPG concentration within the red blood cell, which, as already discussed, shifts the oxygen dissociation curve to the right and increases the amount of oxygen released from each gram of hemoglobin at normal capillary oxygen pressures; by an internal redistribution of blood flow that increases flow to tissues whose oxygen supply must be maintained; and by increasing cardiac output (Finch and Lenfant, 1973).

Anemia may stem from an inability to make red blood cells normally, from loss of red blood cells due to bleeding or hemolysis, or from a combination of impaired production and increased loss. Causes of impaired production include damage to or acquired proliferative abnormalities of stem cells or erythroid precursors; lack of erythropoietin due to destruction of kidney tissue; body iron deficiency; or deficiency of folic acid or cobalamin. Hemolysis may result from an intrinsic red blood cell defect, from an abnormality of the red cell environment, or both. Intrinsic defects include red cell membrane defects (as in hereditary spherocytosis); defects of globin chain structure or synthesis that increase cell deformity or cause hemoglobin to precipitate within the cell (sickle cell anemia, thalassemia, unstable hemoglobins); and defects in enzymes of the red blood cell's energy systems (as in pyruvate kinase deficiency, which impairs production of ATP). Environmental causes of hemolysis include the formation of plasma antibodies that react with antigens on the red cell membrane and mechanical damage to red cells as blood flows through multiple small blood vessels in which, for one reason or another, strands of fibrin have been deposited (microangiopathic hemolytic anemia). In the relatively common cause for anemia in black males, G-6-PD deficiency, anemia stems from a combination of the intrinsic defect, which impairs generation via the hexose monophosphate shunt of the hydride ion needed to form GSH, and environmental exposure to an oxidant drug that increases the need for GSH to protect hemoglobin from oxidant stress.

When someone becomes chronically ill, his or her hemoglobin level often falls 2 or 3 g/dl. This anemia of chronic disease has a complex pathogenesis. Apparently, mononuclear phagocytes are activated by the disease process and phagocytose red blood cells at a moderately increased rate. The marrow is unable to respond with an increase in red cell production and so the circulating mass of red cells falls until the number of cells being destroyed again equals the number being produced. The reasons why the marrow fails to increase production are not clear but include an impaired return of iron to normoblasts from the breakdown of hemoglobin in the mononuclear phagocytes and an inadequate erythropoietin response to the fall in hemoglobin concentration.

The terms *erythrocytosis* and *polycythemia* refer to a higher than normal red blood cell or hemoglobin concentration (greater than 18 g/dl in a man or than 16 g/dl in a woman at sea level). This may result from an increased circulating red blood cell mass, a contracted plasma volume, or both. An elevated hemoglobin concentration due solely to a contracted plasma volume is sometimes called *relative polycythemia*.

Erythrocytosis due to an increased circulating red blood cell mass usually stems from an increased production of erythropoietin. The increased production may be "appropriate," i.e., in response to some cause for impaired ability of hemoglobin to deliver oxygen to the tissues, or may be "inappropriate," arising from aberrant erythropoietin synthesis. Common causes of impaired arterial oxygen saturation include residence at a very high altitude, chronic pulmonary disorders impeding uptake of oxygen as blood flows through the lungs, and congenital heart disorders that cause a substantial fraction of the circulating blood to be shunted from the right to the left side of the heart without passing through the lungs. Another common cause for an impaired ability of hemoglobin to deliver oxygen is an elevated carboxyhemoglobin concentration due to heavy smoking. Rare patients may have erythrocytosis because of inherited structural globin chain abnormalities that increase the affinity of hemoglobin for oxygen at normal end capillary oxygen pressure. *Inappropriate erythropoietin production* is not common and usually results from synthesis of erythropoietin by a malignant tumor. Polycythemia not due to increased erythropoietin production occurs in a myeloproliferative disorder called primary polycythemia or polycythemia vera, in which erythroid precursors arising from an aberrant clone no longer respond appropriately to normal regulatory influences.

CHAPTER 24

Hemostasis

When a blood vessel is damaged, reactions are initiated to arrest bleeding, that is, to achieve hemostasis. The process involves at least four interrelated steps: contraction of the injured vessel; accumulation of platelets at the site of the lesion; activation of blood coagulation; and, as a secondary event, activation of fibrinolysis.

Vasoconstriction occurs immediately after injury, is usually transient, and stems primarily from a direct effect of the injury upon vascular smooth muscle cells. Disruption of the endothelial cells lining the vessel lumen brings platelets into contact with underlying subendothelial tissues and exposes trace plasma clotting proteins to materials in the vessel wall that initiate blood coagulation. Platelets adhere to the subendothelial tissue, are activated, and adhere to each other to form a growing, increasingly compacted mass, the platelet hemostatic plug. Blood coagulation proceeds as a series of amplifying enzymatic reactions in which plasma serine protease proenzymes serve first as substrates and then, after activation, as enzymes triggering further steps in the process. Nonenzymatic plasma factors and materials present on the surface membrane of activated platelets and tissue cells participate as cofactors. The final serine protease coagulant enzyme generated, thrombin, splits small fibrinopeptides from fibrinogen and activates a cross-linking cysteine protease enzyme, factor XIII. This results in the formation and stabilization of strands of fibrin that extend outward from the surface of the platelets and other cells. Thus a seal is formed that is made up of a fused mass of platelets reinforced by the meshwork of a fibrin clot.

Processes are also set in motion to limit growth of the platelet mass and fibrin clot. Endothelial cells bordering the injury site release prostacyclin, a prostaglandin that impedes platelet aggregation. Plasma proteinase inhibitors neutralize activated enzymes of blood coagulation. Moreover, thrombin, after binding to a surface recognition site on endothelium, acquires the ability to activate an anticoagulant serine protein proenzyme, protein C, that then inactivates key coagulation cofactors. Endothelial cells also release plasminogen activators that can activate plasminogen bound to the fibrin clot. The enzyme plasmin, which is thus formed within the clot, dissolves fibrin strands, liberating soluble degradation products that may reenter the circulation. Over a number of days fibrin continues to be both formed and dissolved in balanced reactions at the injury site. A hemostatic seal is thus maintained and remolded while the proliferation of smooth muscle cells and fibroblasts, the deposition of new connective tissue matrix, and the ingrowth of a new luminal lining of endothelial cells repair the vessel wall.

Normal hemostatic function prevents excessive bleeding after the minor tissue injuries of daily living. Platelet hemostatic plugs are particularly important in controlling bleeding from capillaries and small venules in erosions of mucosal surfaces. Abnormal bleeding from the gastrointestinal or genitourinary tract is a source of concern in a thrombocytopenic patient. When an effective fibrin clot cannot be formed, either because of impaired blood coagulation or excessive fibrinolysis, a trivial tissue injury may cause extensive bleeding. For example, a patient with hemophilia may bleed massively into the soft tissues of an extremity from minor trauma that, in a normal person, might cause a bruise no larger than a 50-cent piece.

If a large artery is severed by a lacerating injury, a tourniquet must be applied immediately to prevent the injured person from bleeding to death until the vessel can be surgically repaired. During surgical procedures many small arteries are severed; each is occluded with a surgical instrument (hemostat) and then sutured. Very large numbers of arterioles, capillaries, and venules are also severed; bleeding from these vessels ceases spontaneously as the result of the hemostatic reactions. However, if hemostasis is impaired from any cause, bleeding from the myriads of small vessels that are not sutured may result in serious blood loss during surgery. Moreover, bleeding may reoccur during the first 2 weeks of the postoperative period.

Although essential for survival, hemostatic reactions are harmful when they cause a clot to form within the lumen of a blood vessel (thrombosis). Patients at risk for thrombosis are often treated with drugs that impair hemostasis by interfering with platelet function or slowing blood coagulation. Patients with an acute thrombosis of a coronary artery are given a plasminogen activator in an attempt to reopen the occluded vessel.

FORMATION OF PLATELET HEMOSTATIC PLUGS

Platelets are nonnucleated cells, 2 to 4 μm in diameter, present in blood in a concentration of 150–400×10^9 per liter. Their production and distribution are discussed in Chapter 20. An excellent description of their ultrastructural anatomy is available (White, 1987).

Platelet Adhesion and von Willebrand Factor

When vascular endothelium is disrupted, platelets adhere to the exposed subendothelium, which is the initial step in the formation of platelet hemostatic plugs (Fig. 3.23). At the high rates of wall shear present in small blood vessels, this platelet adhesion requires the participation of von Willebrand factor (vWF) a protein synthesized in vascular endothelial cells and secreted both into the plasma and abluminally into the superficial layer of the subendothelium. Subendothelial vWF contributes to platelet adhesion (Turitto et al., 1985) but is insufficient, in itself, for normal adhesion. After endothelial cell disruption, plasma vWF must also come down onto the subendothelium through binding to an as yet unidentified subendothelial matrix material (Sixma, 1987). This binding alters the plasma vWF so that it can then also bind to a binding site present on the surface membrane of the unstimulated platelet and so link the platelet to the subendothelium. This platelet binding site is located on a major glycoprotein (GP) of the platelet membrane called GPIb (Nurden, 1987).

vWF is a member of a group of proteins, called the adhesive proteins, that contain a particular tripeptide sequence of amino acids, Arg-Gly-Asp (RGD). This sequence allows adhesive proteins to bind to one or more members of a family of cell surface heterodimeric proteins called integrins, which posses a recognition site for RGD (Ruoslahti and Pierschbacher, 1987). vWF can bind to a recognition site

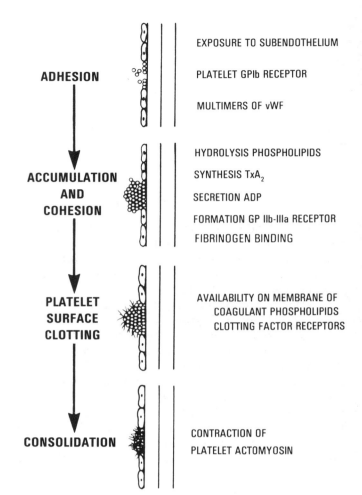

Figure 3.23. Steps in the formation of a stable hemostatic plug. *TxA₂*, thromboxane A₂; *vWF,* von Willebrand factor.

for RGD on a complex of two platelet surface membrane glycoproteins, GPIIb and GPIIIa, that becomes functional only after platelets are activated. The primary physiological function of this GPIIb-IIIa complex is to bind fibrinogen during platelet aggregation (see below). The physiological significance for platelet adhesion of the ability of vWF factor also to bind to GPIIb-IIIa is not yet clear.

Endothelial cells synthesize vWF as a 270,000-dalton molecule (pre-pro-vWF) that undergoes further complex processing within the cell (Wagner and Marder, 1984) to give rise to an approximately 500,000-dalton dimer. This is the basic unit of the mature vWF protein (Ruggeri and Zimmerman, 1987). The cell stores vWF within a large organelle called the Weibel-Palade body, wherein the dimers are assembled, through interchain disulfide bonding of their amino terminals, into vWF multimers of increasing size of up to 20,000,000 daltons. Normal plasma contains 5 to 10 μg of vWF per ml, made up of multimers of all sizes.

In a group of genetic disorders called collectively von Willebrand disease, patients bleed abnormally because of a deficiency of functional vWF. Usually this stems from a low plasma vWF concentration, but variant forms of the disease have been recognized in which total plasma vWF concentration is not significantly reduced, but the highest molecular weight multimers are markedly reduced or missing. The bleeding of these patients establishes that the high molecular weight multimers of plasma vWF are essential for platelets to adhere normally to subendothelium.

Megakaryocytes also synthesize vWF, which is a normal constituent of the platelet α granule and is secreted along with other granule constituents during platelet activation. A subtype of von Willebrand disease has been identified in which the high molecular weight multimers are missing from the plasma but present within the platelets. Since these patients also bleed abnormally, it appears that von Willebrand factor supplied by platelets cannot substitute fully for plasma von Willebrand factor in supporting normal platelet adhesion.

Vasoactive stimuli trigger release of stored vWF from vascular endothelium, which causes the plasma level to rise sharply within minutes. It then gradually falls back to basal levels over hours. Very large multimers are released that are particularly effective in supporting platelet adherence to subendothelium (Sporn et al., 1986). An analogue of vasopressin, 1-desamino-8D-arginine vasopressin (DDAVP), is often given just before surgery to induce release of endothelial cell stores of vWF in patients with the common form of von Willebrand disease. Moreover, vWF is an acute phase protein. Plasma levels rise in inflammatory states because of increased vWF synthesis and may remain elevated for as long as an inflammatory state persists. Plasma vWF levels also rise during the third trimester of pregnancy.

A final point about vWF needs mention. One of the blood coagulation factors, factor VIII, circulates in plasma bound to vWF. Plasma factor VIII levels cannot be maintained in the absence of vWF. Each of the protomers of the vWF dimer possesses a binding site for factor VIII, and the plasma factor VIII level is reduced in patients with von Willebrand disease in proportion to the reduction of total vWF antigen in the plasma.

Platelet Activation

As platelets adhere to exposed subendothelium they become activated. Platelets arriving subsequently at the injury site also begin to activate, adhering to the platelets already present. Thus, a platelet mass starts to grow. As platelets activate they undergo a series of progressive overlapping events: shape change; aggregation; liberation and oxidation of arachidonic acid; secretion of α granule and dense granule contents; reorganization of surface membrane phospholipid, which makes phosphatidylserine available on the outer surface of the bilayer, where it can participate in blood coagulation reactions; and an oriented centripetal contraction of actomyosin of the platelet cytoskeleton.

The early events of platelet activation—shape change and the primary phase of aggregation—are reversible, and loosely aggregated platelets may break away from the hemostatic plugs to reenter the circulation. However, as platelet activation progresses, an increasing contraction of the platelet cytoskeleton crushes the platelets together and causes them to appear to fuse. In addition, thrombin begins to clot fibrinogen on and around the platelets, giving rise to a scaffolding of fibrin to which the fused masses of platelets adhere. These events convert the plugs into stable, permanent plugs (Fig. 3.23).

STIMULUS/RESPONSE COUPLING IN PLATELET ACTIVATION

Two primary agonists trigger platelet activation: the initial thrombin formed at an injury site and sequences on collagen in the subendothelium. An increasing concentration of thrombin and agonists coming from the activated platelets themselves—ADP and arachidonic acid oxidation products—maintain and amplify the activation.

The agonists interact with specific, but as yet poorly characterized, membrane receptors to activate guanine nucleotide-binding proteins (G-proteins) in the platelet membrane (Haslam, 1987). This leads, in turn, to activation of phospholipase C, which catalyzes a hydrolysis of membrane inositol phospholipids with the formation of two second messengers, *diacylglycerol* and *inositol triphosphate*, $I(1,4,5)P_3$. These act synergistically (Nishizuka, 1984) as follows:

1. Diacylglycerol activates protein kinase C, an enzyme catalyzing transfer of phosphate from ATP to serine or threonine residues of proteins. The major substrate of platelet protein kinase C is a 47-kD protein. Although the function of this protein is as yet unknown, its phosphorylation is essential for the responses of platelet activation.
2. $I(1,4,5)P_3$ acts as a calcium ionophore that causes calcium to enter the cytosol from an internal

platelet reservoir (the dense tubular system) and from the platelet exterior. A rising cytosolic calcium concentration triggers:

a. Activation of myosin light chain kinase with resultant phosphorylation of myosin light chain. This is required for the reorientation of cytoskeletal proteins needed for platelet shape change, secretion, and contraction.

b. Activation of a calcium-dependent protease called calpain. Through proteolysis, calpain then activates other platelet enzymes.

c. Activation of phospholipase A_2.

Activated phospholipase A_2 liberates arachidonic acid from platelet membrane phospholipids. The arachidonic acid is then oxidized by two pathways (Fig. 3.24). One, catalyzed by lipoxygenase, yields eicosatetraneoic acid derivatives (hydroperoxyeicosatetraenoic acid (HPETE) and hydroxyeicosatetraenoic acid (HETE)) of unknown function. The other, catalyzed by cyclooxygenase, yields two products that play important amplifying roles in platelet activation: the prostaglandin, PGH_2, and its metabolite, thromboxane A_2. PGH_2 reportedly acts as a cofactor enhancing collagen's ability to function as a platelet agonist (Rittenhouse and Allen, 1982). Thromboxane A_2 binds to a specific platelet membrane receptor with resultant activation of phospholipase C and amplification of platelet activation through further generation of diacylglycerol and $I(1,4,5)P_3$.

FACTORS MODULATING PLATELET SENSITIVITY TO ACTIVATING STIMULI

Platelet cyclic AMP levels affect the responsiveness of platelets to activating stimuli through reg-

ulation of a calcium pump that lowers cytosolic calcium levels (Hawiger et al., 1987). Prostacyclin (PGI_2), a PGH_2 metabolite secreted by endothelial cells (Fig. 3.24), helps to keep circulating platelets in an unstimulated state through activation of platelet adenyl cyclase and a resultant rise in platelet cyclic AMP levels. PGD_2, a nonenzymatic breakdown product of PGH_2 formed in small amounts within platelets, can also activate platelet adenyl cyclase.

Alternations of platelet membrane lipids also affect platelet responsiveness. Ingestion of n-3 fatty acids found in fish oils decreases arachidonic acid release from platelet membrane phospholipids (Leaf and Weber, 1988) and thus can limit the synthesis of thromboxane A_2. As a consequence, the bleeding time—a test reflecting the rate of formation of the platelet hemostatic plugs that stop bleeding from the small vessels severed by a 1-mm cut in the skin—may lengthen.

Aspirin irreversibly inactivates platelet cyclooxygenase, which prevents synthesis of PGH_2 and thromboxane A_2 for the remaining life of the platelet. Taking a single aspirin tablet may increase the bleeding time by 1–5 minutes, and the prolongation may persist for several days. This impairs hemostatic function but in most persons only minimally. However, when aspirin is given to a patient whose ability to activate platelets through the generation of thrombin is also impaired (e.g., to a patient with severe hemophilia) platelet activation may be paralyzed. The bleeding time may lengthen markedly (e.g., to over 30 minutes) and the risk for bleeding is substantially enhanced. (Nevertheless, in patients with coronary artery disease, aspirin is given together with the anticoagulant heparin

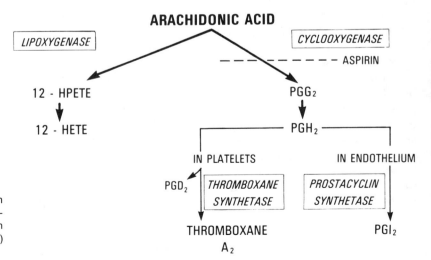

Figure 3.24. Pathways of arachidonic acid oxidation and products generated in the platelet. That the endoperoxide PGH_2 is a common substrate for the formation of thromboxane A_2 in the platelet and PGI_2 (prostacyclin) in the endothelial cell is also shown.

when it is crucial to keep the lumen of a coronary artery open after dissolving a thrombus with a fibrinolytic agent or after dilating a stenotic coronary artery by percutaneous angioplasty.)

PLATELET SHAPE CHANGE

Within seconds of exposure to activating agents and as a probable consequence of reactions dependent upon phosphorylation of both the 47-kD protein and myosin light chain, platelets change shape from flattened discs to spheres with multiple projecting pseudopods. Platelet actin is polymerized, and actin microfilaments, elongating from their barbed end, push the surface membrane out ahead of them. The microtubules that encircle the unstimulated platelet just below the platelet membrane dissolve and reform centrally, surrounding platelet granules that have moved to the center of the cell prior to platelet secretion. Despite these major structural alterations, platelet shape change is reversible. When inducing stimuli are weak and transient, the platelet may revert to its unstimulated appearance.

PLATELET AGGREGATION

When platelets are exposed to any of the agonists that may initiate platelet activation (thrombin, collagen, thromboxane A_2, ADP), fibrinogen receptors are expressed in a GPIIb-IIIa complex on the platelet surface membrane (Phillips et al., 1988). The GPIIb-IIIa complex is an integrin that recognizes two RGD sequences present on the α chain of fibrinogen. Moreover, the GPIIb-IIIa complex recognizes a second site on fibrinogen, the ultimate 12 amino acids of the carboxy-terminal segment of its γ chain (Kloczewiak et al., 1984).

Through binding to these recognition sites, fibrinogen, which is a dimeric molecule, can form a "bridge" between a GPIIb-IIIa complex on two opposing activated platelets. This causes the platelets to stick to each other in the primary phase of platelet aggregation. This primary phase is reversible but progresses to a secondary irreversible phase after platelet secretion begins. Thrombospondin, an α granule protein secreted during platelet activation, is thought to participate in the development of irreversibility (Nachman et al., 1987). Thrombospondin can bind to fibrinogen and also to a binding site on GPIV, another glycoprotein of the platelet surface membrane. By binding to both its platelet surface receptor and a fibrinogen molecule holding two platelets together, thrombospondin reinforces the fibrinogen "glue" of aggregation.

Patients with a rare hereditary disorder called thrombasthenia (also known as Glanzmann's disease) cannot form functional GPIIb-IIIa receptors for fibrinogen on the platelet surface membrane. Therefore, their platelets fail to aggregate after exposure to any of the agonists activating platelets. The serious bleeding of these patients from mild abrasions of mucosal surfaces attests to the fundamental importance of the binding of fibrinogen to GPIIb-IIIa for the normal formation of platelet hemostatic plugs during hemostasis.

The GPIIb-IIIa heterodimer traverses the platelet membrane. Occupancy by fibrinogen of binding sites on the external domain of GPIIb-IIIa alters the cytoplasmic domain so that it can then, through an intermediate protein, bind the actin filaments. Linking of the membrane to the platelet cytoskeleton by this mechanism is thought to orient the centripetal contraction that compacts and stabilizes the hemostatic plugs.

PLATELET SECRETION

Platelets contain three types of granules: α granules, dense granules, and lysosomal granules. Distributed randomly in the unstimulated platelet, they move to the center of the platelet after platelet activation. Their membranes then fuse with the membranes of an open canalicular system made up of invaginations of the platelet surface membrane. The contents of the granules are secreted through the open canalicular system—first from the α and dense granules and later, as the platelets appear to break down and fuse, from the lysosomal granules.

The dense granules contain ADP, ATP, calcium, and serotonin. Although a vasoconstrictor and a weak platelet agonist, serotonin has no recognized function in hemostasis. Depleting dense granules of serotonin with the drug reserpine does not impair hemostasis. In contrast, ADP secreted from the dense granules functions as a physiologically significant agonist amplifying platelet activation. Patients with hereditary disorders that prevent the storage of normal quantities of ADP in the dense granules (Weiss, 1987) have a mild to moderate bleeding diathesis resulting from impaired formation of hemostatic plugs.

The contents of the α granules may be divided into two groups: (1) proteins that are also plasma proteins and (2) proteins absent from plasma until secreted by the platelets. The former may be divided further into two groups. The first consists of proteins, of which albumin and IgG are the most prominent, that are taken up by platelets by a nonspecific mechanism. Their plasma concentration determines their platelet concentration. The proteins in this group have no known hemostatic functions. The second group consists of proteins that are present in platelets in a substantially higher concentration than can be accounted for by their plasma concentrations. These proteins may be synthesized or taken up from plasma by the maturing megakaryocyte. They include fibrinogen, vWF, and factor V, each of which, after its secretion from the activated platelet, binds to the platelet surface membrane. This makes them available in high concentration to participate in hemostatic reactions within the growing platelet hemostatic plugs.

The α granule proteins absent from plasma until secreted by activated platelets include thrombospondin, β-thromboglobulin, platelet factor 4 (PF4), platelet-derived growth factor (PDGF), and transforming growth factor-β (TGF-β). A possible role for thrombospondin in platelet aggregation has already been discussed. PF4, a small cationic polypeptide originally recognized because of its ability to neutralize the anticoagulant activity of heparin, may compete with antithrombin III for binding sites on heparan sulfate present on endothelial cells and in the subendothelium. This could help promote blood coagulation at a site of vessel wall injury. PF4, PGDF, and TGF-β are chemoattractants for white blood cells, smooth muscle cells, and fibroblasts;

activate these cells; and accelerate wound healing. They thus contribute to the processes of inflammation and repair.

BLOOD COAGULATION

Nomenclature

Numbers, letters, and trivial names are used to identify different components of the blood coagulation reactions (Table 3.7). Some years ago an international committee assigned Roman numerals to the then recognized clotting factors. The Roman numeral nomenclature was accepted for most but not all factors; fibrinogen, prothrombin, and tissue factor are rarely referred to as factors I, II, and III. When a Roman numeral clotting factor is in its activated form, the lower case letter a is added, e.g., the activated form of factor IX is written factor IXa or, simply IXa. Two proteins that participate both in the contact activation of blood coagulation and in the generation of kinin, prekallikrein and high molecular weight kininogen, and two more recently identified vitamin K-dependent proteins, protein C and protein S, have not been given Roman numeral names. The major protease inhibitor of blood coagulation, which neutralizes thrombin, factor Xa and factor IXa, is called antithrombin III, a name carried over from a time when the name antithrombin, followed by a Roman numeral, was used to distinguish between different thrombin-neutralizing activities

Table 3.7.
Properties of Moeities Involved in Blood Coagulation

Type	Name[a]	Molecular Weight, daltons	Plasma Concentration		Intravascular Half-time, days
			μg/ml	nM	
Contact system proenzymes	F XII	80,000	29	360	2
	Prekallikrein	88,000	45	510	—
	F XI	160,000	4[b]	25[b]	2.5
Vitamin K-dependent coagulant proenzymes	F VII	50,000	0.5	10	0.2
	F IX	57,000	4	70	1
	F X	57,000	8	140	1.5
	Prothrombin	70,000	150	2.1×10^3	3
Cofactors	Tissue factor	—	0	0	
	Platelet lipid	—	0	0	
	HMW kininogen	120,000	70	583	—
	Factor V	300,000	7	21	1.5
	Factor VIII	300,000	0.2	0.3	0.5
	Protein S	69,000	25	348	—
Factors of fibrin deposition	Fibrinogen	340,000	2500	7.0×10^3	4.5
	F XIII	320,000	8	25	7
Inhibitors	Protein C	62,000	4	66	0.3
	Antithrombin III	58,000	150	2.6×10^3	2.5
	EPI	~40,000	0.1	~3	—

[a]Abbreviations are as follows: F, factor; HMW, high molecular weight; EPI, extrinsic pathway inhibitor. EPI is also referred to as LACI (lipoprotein-associated coagulation inhibitor). [b]A dimer with two active enzymatic sites per molecule; therefore, the functional concentration is double the value given.

in clotting mixtures. A recently identified factor Xa-dependent inhibitor of the catalytic activity of the factor VIIa/tissue factor complex (Rapaport, 1989) does not yet have a generally accepted name. It has been referred to as the extrinsic pathway inhibitor (EPI) and as the lipoprotein-associated coagulation inhibitor (LACI).

Production and Catabolism of Clotting Proteins

The plasma concentration of all clotting proteins except factor VIII and EPI falls in patients with decompensated chronic hepatocellular disease. Therefore, the hepatocyte is considered the primary source of all clotting proteins in plasma except factor VIII and EPI. However, the hepatocyte is one of several cells that can synthesize factor VIII (Wion et al., 1985), and, after a patient with hereditary factor VIII deficiency (hemophilia A) and severe liver disease received a liver transplant, his plasma factor VIII level rose to normal. Both hepatocytes and endothelial cells can synthesize EPI and both are probably sources of plasma EPI. As mentioned earlier, endothelial cells are also the source of plasma vWF.

Vitamin K is required for the synthesis of functional molecules of six clotting factors: prothrombin; factors VII, IX, and X, which are zymogens of procoagulant serine proteases; protein C, which is a zymogen of an antiocoagulant serine protease; and protein S, which has no serine protease catalytic domain but functions as a cofactor for protein C. Several of these proteins possess a similar modular structure (Furie and Furie, 1988). All of the proteins contain 10–12 residues of a unique amino acid, γ-carboxyglutamic acid (Gla), in their NH_2-terminal segments. Vitamin K is needed for the posttranslational reaction in which these Gla residues are formed by addition of a second carboxy group at the γ-carbon of selected glutamic acid residues. Gla residues acting together form strong calcium-binding sites, and binding of calcium ions alters the conformation of these proteins so that they can function. In vitamin K deficiency or when vitamin K is inhibited by the oral anticoagulant warfarin, nonfunctional vitamin K-dependent proteins missing Gla residues can be demonstrated in the plasma by immunological techniques.

Little is yet known of the mechanisms regulating basal production of the clotting proteins. Fibrinogen and vWF are acute phase proteins whose plasma levels rise in inflammatory states because of increased synthesis stimulated by interleukin 1 and interleukin 6. Factor VIII levels also rise in inflammatory states, either because it, too, is an acute phase protein or because of the rise in vWF, the protein to which plasma factor VIII is bound. Plasma concentrations of fibrinogen, von Willebrand factor, and factor VIII also increase during the last weeks of pregnancy, as do factor VII levels, and, to a lesser extent, factor IX and X levels.

Fibrinogen has an intravascular half-time of 4–5 days, which the infusion of heparin to inhibit blood coagulation or of tranexamic acid to inhibit fibrinolysis does not alter. The intravascular half-time for factor XIII is about 7 days and for prothrombin is 3 days. The other clotting factors have unusually short intravascular half-times, e.g., 1.5 days for factor X, 10–12 hours for factor VIII, and only 5 hours for factor VII. The reasons for these rapid rates are unknown but do not reflect inherent instability of the molecules themselves. Because of the short intravascular half-times of the clotting factors, a patient with a clotting factor deficiency undergoing surgery usually requires repeated infusions of the missing clotting factor over 7–10 days to prevent excessive postoperative bleeding.

General Description of the Blood Coagulation Reactions

Blood coagulation may be viewed as occurring in three steps: a sequence of reactions resulting in the generation of an activator of prothrombin; a second step in which the prothrombin activator cleaves prothrombin to form thrombin; and a third step, in which the reactions of thrombin with fibrinogen and factor XIII lead to the deposition of cross-linked polymers of fibrin (Fig. 3.25). The first and second steps consist of successive reactions in which a zymogen precursor is converted, by proteolysis of one or two peptide bonds, into an active serine protease, which then activates the next zymogen in the sequence. With the exception of the contact activation reactions that lead to the activation of factor XI (see below), the reactions require the presence of calcium ions. (An anticoagulant that can chelate calcium ions is added to prevent blood from clotting when it is drawn for many types of laboratory tests.)

Three key enzymes of the sequence, factors VIIa, IXa, and Xa, are, in themselves, physiologically inert. They acquire powerful catalytic efficiency when bound to cofactors (see Fig. 3.25). The cofactors target the serine proteases to their specific substrates and also help to localize the reactions to the surfaces of tissue cells and activated platelets. Two of the cofactors, factors VIII and V, cannot function until they are themselves activated in key positive feedback reactions. Amplification occurs as the

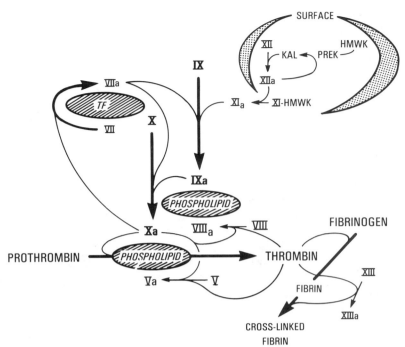

Figure 3.25. An overall diagram of the blood coagulation reactions. The reactions initiated when blood is exposed to a negatively charged surface in vitro are summarized in the *upper right.* These generate factor XIa. The reactions initiated when blood is clotted in the presence of tissue factor are shown in the *upper left.* Factor VII/tissue factor complexes are shown that are rapidly converted to factor VIIa/tissue factor complexes in a key feedback reaction catalyzed by factor Xa. (To avoid further complexity the presumed presence of trace amounts of factor VIIa in the blood, which allow formation of a very few factor VIIa/tissue factor complexes that initiate the generation of factor Xa, is not shown.) Factor IX is shown to be activated in two ways: by factor XIa and by factor VIIa/tissue factor. Factor X is also shown to be activated in two ways: by factor VIIa/tissue factor and by a factor IXa/factor VIIIa/phospholipid complex. Prothrombin is shown to be activated by a factor Xa/factor Va/phospholipid complex. Factor Xa and thrombin are shown to participate in key feedback reactions that activate factor VIII and factor V, which is necessary for their function as cofactors. Thrombin is also shown to convert fibrinogen to fibrin and to activate factor XIII. Also to avoid further complexity, the participation of calcium ions in the formation of all complexes, in the activation of factor IXa by factor XIa, and in the activation of factor XIII has not been shown. *HMWK,* high molecular weight kininogen; *KAL,* kallikrein; *PREK,* prekallikrein; *TF,* tissue factor.

blood coagulation reactions proceed. Activation of a few nanograms of the serine proteases initiating blood coagulation leads to the conversion of micrograms of prothrombin to thrombin and to the clotting of milligrams of fibrinogen.

Important regulatory mechanisms limit blood clotting to the site of injury. The protease inhibitor antithrombin III, in the presence of the glycosaminoglycans heparin and heparan sulfate, rapidly neutralizes the enzymatic activity of factors IXa, Xa, and thrombin. In addition, negative feedback reactions—involving factor Xa in the inhibition of the factor VIIa/tissue factor complex and involving thrombin complexed to thrombomodulin in the activation of protein C and consequent inactivation of factors VIIIa and Va—play important roles in dampening the blood coagulation reactions.

The above provides an overall orientation to the blood coagulation reactions. The individual reactions are described in the following sections.

Initiation of Blood Coagulation

The exposure of blood in vitro to a negatively charged surface such as a glass surface initiates reactions involving four plasma proteins sometimes referred to as the contact activation factors. Three of these proteins are zymogens of serine proteases—factor XII, prekallikrein, and factor XI—and the fourth is a nonenzymatic factor, high molecular weight kininogen (HMWK). Factor XII and HMWK come down onto the negatively charged surface. Prekallikrein and coagulation factor XI, which circulate bound to HMWK, are brought down onto the surface along with HMWK. A conformational change resulting from the coming down of factor XII onto a surface initiates its autoactivation (Kaplan and Silverberg, 1987). Reciprocal activation reactions then follow in which factor XIIa activates prekallikrein to kallikrein and kallikrein activates factor XII. Sufficient factor XIIa is thus generated for factor XIIa rapidly to activate factor XI.

Although essential for triggering blood coagulation in a glass test tube, factor XII, prekallikrein, and HMWK play no known role in normal hemostasis. Patients with isolated hereditary deficiencies of each of these factors do not bleed abnormally. Patients with hereditary factor XI deficiency have a bleeding disorder characterized by excessive bleeding after surgical procedures but not after tissue trauma. Their bleeding tendency, although mild, establishes that factor XIa is needed, at least in some circumstances, for normal hemostasis. Therefore, an as yet undiscovered reaction or reactions, independent of factor XII, prekallikrein, and HMWK, is postulated to activate factor XI in vivo.

However, the key event triggering blood coagulation during hemostasis is thought not to be the generation of factor XIa but the exposure of blood to tissue factor. Tissue factor is an approximately 30-kD transmembrane apoprotein that acquires coagulant activity when associated with phospholipid of the cell surface membrane (Nemerson, 1988). Although not normally present on cells to which the circulating blood is exposed, tissue factor is readily demonstrated by immunostaining on pericytes and fibroblasts in the adventitia of blood vessels, on fibroblasts in loose connective tissue, and on many other cells in different organs. Thus, when a blood vessel is disrupted, blood readily comes into contact with cells possessing surface membrane tissue factor activity.

Tissue factor is the cofactor for factor VII, and factor VII must bind to tissue factor to function in blood coagulation (Fig. 3.25). Although factor VII is a zymogen, a factor VII/tissue factor complex has been thought to possess a minimal enzymatic activity capable of initiating blood coagulation (Nemerson, 1988). However, in a recent study (Rao and Rapaport, 1988), factor VII/tissue factor was found incapable of a physiologically meaningful activation of either of its biological substrates, factor IX or factor X. The alternative possibility exists that normal circulating blood contains traces of factor VIIa, which would allow the formation of a few factor VIIa/tissue factor complexes at a site of vessel wall damage. These complexes could then activate a minute amount of factor X, with resultant initiation of a key positive feedback reaction in which factor Xa preferentially activates factor VII complexed to tissue factor (Fig. 3.25).

Indirect evidence supports the hypothesis that traces of factor VIIa may be normally present in circulating blood. Low levels of a prothrombin fragment generated during formation of thrombin have been demonstrated immunologically in normal plasma (Bauer and Rosenberg, 1987). This represents convincing evidence for a basal, minimal continuing activation of factor X in normal individuals. If this occurs intravascularly, then traces of factor Xa, particularly if its catalytic activity for factor VII is augmented by phospholipid on the luminal surface of endothelial cells (Rao et al., 1988), could activate traces of circulating factor VII. Furthermore, factor VIIa, once formed, will persist in the circulation with an intravascular half-time of over 2 hours (Seligsohn et al., 1979).

The serious and even fatal bleeding of patients with severe hereditary factor VII deficiency and essentially unmeasurable plasma factor VII activity testifies to the importance of the tissue factor pathway for initiating blood coagulation in hemostasis. However, patients with moderately severe hereditary factor VII deficiency and factor VII levels in the 5–20% range may experience little or no abnormal bleeding. Such reduced levels apparently still provide enough factor VII molecules to occupy the number of tissue factor sites to which the blood is usually exposed in normal hemostasis.

Intermediate Reactions Leading to the Generation of Thrombin

The two known ways of initiating blood coagulation—the "intrinsic" pathway in which the contact activation reactions result in formation of factor XIa and the "extrinsic" pathway in which exposure of blood to tissue factor leads to formation of factor VIIa/tissue factor complexes—come together at the step of activation of factor IX (Fig. 3.25). The mild bleeding tendency of patients with hereditary factor XI deficiency establishes that activation of factor IX by factor XIa is a physiologically significant reaction. The much more severe bleeding tendency of patients with hereditary factor IX deficiency (hemophilia B) leads one to believe that the activation of factor IX by factor VIIa/tissue factor must be of at least equal if not greater physiological importance.

Two enzymes also activate factor X: factor VIIa in a factor VIIa/tissue factor complex and factor IXa in a factor IXa/factor VIIIa/anionic phospholipid complex. A key role for factor VIIa/tissue factor activation of factor X in initiating blood coagulation during hemostasis has already been mentioned. However, activation of factor X by factor VIIa/tissue factor cannot suffice, in itself, for normal hemostasis. If it could, then neither hemophilia A (factor VIII deficiency) nor hemophilia B (factor IX deficiency) would be a bleeding disorder. The need for the factor IXa/factor VIIIa/phospholipid activator of factor X could stem from an increasing inhibition of factor VIIa/tissue factor catalytic activity as increasing amounts of factor Xa are formed. EPI, a plasma proteinase inhibitor belonging to the basic protease inhibitor gene superfamily (Wun et al., 1988), can bind factor Xa. The resultant EPI/factor Xa complex can then bind to factor VIIa/tissue factor, forming a quaternary EPI/factor Xa/factor VIIa/tissue factor complex lacking enzymatic activity (Rapaport, 1989).

Factor IXa cannot activate factor Xa effectively in vitro unless factor VIIIa, a source of anionic phospholipid, and calcium ions are also present in the reaction mixture. Native factor VIII will not support factor X activation in such mixtures. Native factor VIII must first be activated by limited proteolysis in a feedback reaction catalyzed most efficiently by traces of thrombin but also by factor Xa. Phospholipid vesicles containing phosphatidyl-serine or platelets activated with collagen and thrombin can be used to provide the phospholipid.

One assumes from these in vitro experiments that a factor IXa/factor VIIIa/phospholipid complex activates factor X on the platelet surface during normal hemostasis. Moreover, a factor IXa/factor VIIIa-induced activation of factor X has been demonstrated on the surface of vascular endothelial cells. A specific recognition site on the endothelial cell for factor IX and factor IXa (Rimon et al., 1987) is thought to participate in the reaction. Whether additional phospholipid constitutively present on the surface membrane of the endothelial cell also participates in the reaction is not yet clear. Although yet to be demonstrated, other cells are postulated to possess binding sites that would allow the factor IXa/factor VIIIa activation of factor X to take place on their surface when blood escapes into the tissues.

Factor Xa is the sole known physiological activator of prothrombin. Analogous to activation of factor X by factor IXa, activation of prothrombin by factor Xa requires the participation as cofactors of factor Va, phospholipid, and calcium ions. As with factor VIII, plasma factor V cannot function as a cofactor until activated by limited proteolysis (Kane and Davie, 1988) catalyzed by thrombin, factor Xa, or both. However, the factor V secreted from activated platelets is already partially activated by a platelet protease.

Assembly of a prothrombin activating complex on the surface of activated platelets has been studied in detail. Factor Va must bind first to the platelet surface, where it serves as a platelet surface binding site for factor Xa (Miletich et al., 1978). A factor Xa/factor Va/coagulant phospholipid complex is then formed in which phosphatidylserine, transposed after platelet activation from the inner to the outer leaflet of the platelet surface membrane, is thought to provide the phospholipid coagulant activity (Bevers et al., 1983).

A factor Xa/factor Va/membrane phospholipid (or functional equivalent) complex can also form on the surface of cells other than platelets (Tracy et al., 1985). Formation of the complex on these cells differs from formation of the complex on platelets in at least two ways. First, although factor Va is required for the complexes to function as effective prothrombin activators, factor Va, at least for endothelial cells (Rodgers and Shuman, 1985) and presumably for other cells, is not required for factor Xa to bind to the cell surface. Second, the moiety or moieties supplying the membrane equivalent of phospholipid vesicles in vitro does not appear to require cell activation for its expression.

Prothrombin binds to its activator on cell surfaces in two ways. It binds to phospholipid (or membrane equivalent) by a mechanism requiring participation of calcium ions and the Gla groups in the NH_2-terminal segment of the molecule. It also binds to factor Va through a factor Va recognition site present in the NH_2-terminal segment of the prothrombin molecule. Factor Xa cleaves two peptide bonds in prothrombin. On cell surfaces, an initial cleavage at a site within an interchain disulfide bond in the carboxy-terminal segment of the molecule gives rise to a molecule with thrombin activity called meizo-thrombin (Mann, 1987). A second cleavage then separates the NH_2-terminal segment of the molecule (called fragment 1–2) from the remainder of the molecule, which is then referred to as α thrombin or simply as thrombin. Minus fragment 1–2, thrombin lacks both Gla groups and the factor Va binding site. Therefore, thrombin can move off of cell surfaces and catalyze conversion of fibrinogen to fibrin in the surrounding plasma and extracellular fluid.

Formation and Stabilization of Fibrin

Fibrinogen is a dimeric protein made up of two pairs of three nonidentical polypeptide chains, called the α, β, and γ chains. The molecule possesses a trinodular structure. The central nodule is made up of the NH_2-terminal ends of all six chains held together by disulfide bonds. Each distal nodule is made up of the carboxy segments of one pair of the three chains. Thrombin cleaves small peptides from the NH_2-terminal of each α chain (fibrinopeptide A) and each β chain (fibrinopeptide B), giving rise to a molecule called fibrin monomer. Contact sites thus exposed in the central nodule then form noncovalent bonds with sites located in the distal nodules of other fibrin monomer molecules. This takes place in a partially overlapping pattern that results in both lateral and end-to-end polymerization of the molecules and the formation of insoluble fibrin strands (Doolittle, 1981).

Thrombin also activates factor XIII, an enzyme catalyzing the formation of covalent bonds between the polymerizing fibrin molecules (McDonagh, 1987). Plasma factor XIII is a tetrameric protein made up of two chains called *a* chains and two chains called *b* chains. The *a* chain contains the active enzymatic site; the *b* chain is needed for the release from hepatocytes of the *a* chain, which lacks a signal peptide. Thrombin cleaves a peptide from the *a* chain and, in the presence of calcium ions, the resultant thrombin-altered *a* chain separates from the *b* chain and develops enzymatic activity. The resultant enzyme is a transglutaminase that catalyzes the formation of amide bonds between a γ-carbonyl group of glutamine and an ε-amino group of lysine on γ chains of adjacent fibrin molecules oriented in an antiparallel direction (Fig. 3.26).

Additional cross-linking sites are present on the α chains. These α chain sites participate not only in the cross-linking of fibrin molecules to each other but in the cross-linking of fibrin to α₂-antiplasmin (discussed later) and to fibronectin, an adhesive protein of plasma and extracellular matrix (Mosher, 1987). Factor XIIIa can also cross-link fibronectin to collagen. Thus, through these reactions factor XIIIa makes possible the binding of fibrin to collagen.

The reactions catalyzed by plasma factor XIIIa stabilize fibrin and protect it from excessive fibrinolysis. They must take place for normal hemostasis; patients with a hereditary deficiency of factor XIII have a serious bleeding disorder.

Regulation of Blood Coagulation

Several reactions regulate blood coagulation in vivo. Adsorption onto fibrin and neutralization by plasma proteinase inhibitors, particularly antithrombin III, inhibit the activity of thrombin. Antithrombin III also inhibits the activity of the key intermediate enzymes, factors IXa and Xa. Activated protein C inactivates the cofactors for these enzymes, factors VIIIa and Va. A complex formed by the extrinsic pathway inhibitor (EPI) and factor Xa neutralizes the catalytic activity of the factor VIIa/tissue factor complex.

Antithrombin III is a proteinase inhibitor with an intravascular half-life of 2.5–3 days. Synthesized in the liver, it is present in plasma in a modest molar excess over prothrombin (2.5 μM for antithrombin III, 2.0 μM for prothrombin). Antithrombin III forms a 1:1 stoichiometric complex with thrombin in a reaction involving binding of the active site serine of thrombin to a reactive center arginine (Carrell et al., 1987). In solution, in vitro antithrombin III inactivates thrombin at a slow, physiologically ineffective rate. However, when antithrombin III binds to saccharide sequences present in many copies in the potent anticoagulant glycosaminoglycan, heparin, and in fewer copies in the weaker anticoagulant glycosaminoglycan, heparan sulfate, antithrombin III can rapidly neutralize thrombin. Heparan sulfate is

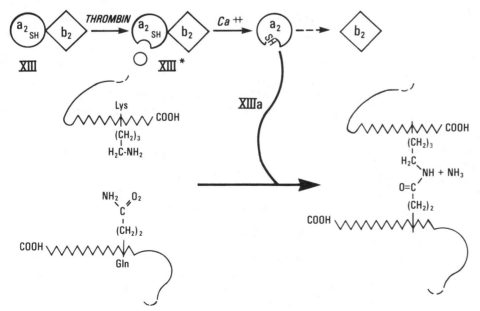

Figure 3.26. A schematic representation of activation of plasma factor XIII by thrombin and calcium ions. Thrombin cleaves a peptide from the *a* chains of the molecule, which contains the active site. In the presence of calcium ions the *b* chains separate from the *a* chains. Calcium ions also induce a conformational change in the *a* chains that exposes an active site cysteine *(SH)*, which activates the enzyme. Also shown is the catalysis by factor XIIIa of the covalent cross-linking of the carboxy ends of two γ chains of adjacent fibrin molecules, oriented in an antiparallel direction, by the formation of an amide bond between the ε-amino group of lysine on one γ chain and the γ-carbonyl group of glutamine on the other γ chain. XIII* represents thrombin-altered XIII. The symbol *SH* represents the exposed active site SH of the enzyme.

present on the luminal surface of vascular endothelial cells. Antithrombin III competes with platelet factor 4 secreted from activated platelets and with the plasma protein, histidine-rich glycoprotein, for saccharide-sequence binding sites on heparan sulfate (Rosenberg and Rosenberg, 1984). This competition for binding sites probably explains why heterozygotes for hereditary antithrombin III deficiency, whose plasma antithrombin III activity is about 50% of normal, have an increased risk for venous thrombotic disease. A homozygote for hereditary antithrombin III deficiency has never been identified, presumably because a total deficiency of antithrombin III is lethal in utero.

Other plasma serine proteinase inhibitors, notably α_2-macroglobulin and heparin cofactor II, can also neutralize thrombin. α_2-Macroglobulin binds thrombin and other serine proteases by a unique mechanism that prevents their cleaving large protein substrates but not small synthetic esters. The major physiological function of α_2-macroglobulin, which is concentrated along the luminal surface of endothelium, is not yet known. Heparin cofactor II resembles antithrombin III in that heparin and heparan sulfate enhance its ability to inhibit thrombin. However, heparin cofactor II differs functionally from antithrombin III in two ways. First, heparin cofactor II is not able to inhibit factors IXa and Xa. Second, an additional glycosaminoglycan, dermatan sulfate, which is present in extravascular tissues, can enhance heparin cofactor II's ability to neutralize thrombin. Therefore, a physiological role for heparin cofactor II in neutralizing thrombin generated extravascularly has been postulated (Tollefsen et al., 1983).

Serine proteinase inhibitors of plasma cannot neutralize free factor VIIa. As mentioned above, the enzymatic activity of the factor VIIa/tissue factor complex is inhibited by the formation of a quarternary EPI/factor Xa/factor VIIa/tissue factor complex. In contrast to the 2 μM concentration of antithrombin III, EPI is present in a 2–3 nM concentration in plasma.

Whereas factor VIIa and tissue factor must form a complex before factor VIIa can be neutralized, the components of the factor IXa/factor VIIIa/phospholipid complex and of the factor Xa/factor Va/phospholipid complex are more efficiently inhibited before rather than after they form complexes. As already mentioned, antithrombin III is the primary inhibitor of factors IXa and Xa. Limited proteolysis by activated protein C destroys the cofactor activity of factors VIIIa and Va.

The mechanism of activation of protein C and its subsequent inactivation of factors VIIIa and Va

have been an area of intense recent interest (Esmon, 1987). When thrombin is formed in vivo it can bind to a receptor on the luminal surface of vascular endothelial cells called thrombomodulin. Binding to thrombomodulin changes thrombin's enzymatic specificity, permitting it to activate protein C (Fig. 3.27). Activated protein C, with the help of its cofactor protein S, can form an activated protein C/phospholipid complex that may be thought of as competing with factor IXa/phospholipid for factor VIIIa and with factor Xa/phospholipid for factor Va. The cofactor role of protein S is as yet poorly understood but is related to bringing protein C down onto a surface membrane. Protein S exists in plasma partly free and partly bound to the C4b binding protein of the complement system. Only free protein S can serve as a cofactor for protein C.

Clinical observations have established the importance of protein C as a key natural anticoagulant. Infants born totally deficient in protein C die shortly after birth of fulminant thrombotic disease. Otherwise unexplained thrombotic events have been documented in a number of heterozygotes for protein C or protein S deficiency, who may have 40–50% of normal plasma activity of one or the other factor.

FIBRINOLYSIS

General Description

Whereas thrombin cleaves only small fibrinopeptides from fibrinogen, leaving the remainder of the molecule intact and ready to polymerize into fibrin strands, a second serine protease of plasma, plasmin, cleaves arginine residues and lysine residues at a number of sites in either fibrinogen or fibrin. These cleavages lyse fibrin, breaking it down into successively smaller soluble fragments called fibrin degradation products. As discussed in the beginning of this chapter, fibrinolysis is a normal secondary response in hemostasis. Moreover, plasmin-catalyzed proteolysis on or around cell surfaces in tissues plays an important role in the inflammatory

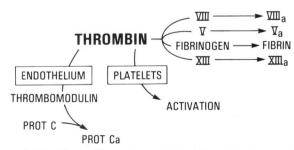

Figure 3.27. The multiple actions of thrombin in blood coagulation. *Prot C,* protein C; *Prot Ca,* activated protein C.

response, in tissue remodeling, and in the mechanisms whereby tumor cells invade tissues.

An inert circulating precursor of plasmin, plasminogen, is converted into plasmin through proteolysis of a single peptide bond. Serine protease enzymes capable of catalyzing this cleavage are called *plasminogen activators* (Fig. 3.28). Two functionally and immunologically distinct plasminogen activators are released from cells. One is called tissue-type plasminogen activator (t-PA) and the other, because it was first recognized in urine, is called urokinase-type plasminogen activator (u-PA). Endothelial cells secrete both t-PA and u-PA, and both participate in the secondary lysis of fibrin in hemostasis. u-PA is also secreted by epithelial cells, monocytes, fibroblasts, and decidual cells. These cells possess a surface membrane receptor for u-PA. Binding of secreted u-PA to its surface receptor results in a pericellular localization of plasmin-induced proteolysis in tissues.

Both t-PA and u-PA are secreted as single-chain, enzymatically active molecules, but traces of plasmin rapidly convert each to a two-chain molecule. Although conversion of single-chain t-PA (sct-PA) to two-chain t-PA (tct-PA) appears not to alter its plasminogen activator activity substantially, conversion of single-chain u-PA (scu-PA) to the two-chain molecule (tcu-PA) markedly alters its physiological function (see below).

The formation and deposition of fibrin within vessels give rise to stimuli triggering release of plasminogen activators from endothelial cells and initiation of secondary fibrinolysis. A plasminogen activator inhibitor that is released from endothelial cells and activated platelets (PAI-1) modulates the activity of the released activators. In addition, the rapid hepatic clearance of plasminogen activators from plasma limits their physiological activity (Fig. 3.28).

Several mechanisms confine the fibrinolytic reactions to the surface of fibrin in normal secondary fibrinolysis (Bachmann, 1987). The *first* is the binding of plasminogen to lysine binding sites on fibrin, which starts as fibrin is being laid down but is enhanced as the beginning cleavage of fibrin by plasmin exposes many new lysine binding sites. The *second* is the preferential activation of plasminogen bound to fibrin. Single-chain urokinase (scu-PA) can initiate plasminogen activation only on the surface of fibrin, and the binding of t-PA to plasminogen on fibrin markedly enhances t-PA's ability to activate plasminogen. The *third* is the remarkable ability of an inhibitor of plasmin, α_2-antiplasmin, to neutralize within milliseconds any plasmin escaping from a fibrin surface into the circulation. When plasminogen activators are given intravenously in very high concentrations to patients with an acute coronary artery thrombosis, these normal regulatory mechanisms are overwhelmed. Then, to an extent dependent upon the type and dosage of plasminogen activator given, plasmin is formed within the circulating blood and induces the proteolysis of circulating fibrinogen and other clotting factors.

Properties of the Components of the Fibrinolytic Reactions

PLASMINOGEN

Plasminogen is a single-chain 92-kD protein made in the liver and present in plasma in a 2 μM concentration (100 μg/ml). It has an intravascular half-time of 2–2.5 days. Cleavage of a single arginine-valine bond converts plasminogen into two-chain active plasmin. The A chain, which corresponds to the NH_2-terminal segment of plasminogen, contains structures called kringles that possess lysine binding sites. The B chain, corresponding to the carboxy-terminal segment of plasminogen, contains the active site—the triad of histidine, aspartic acid, and serine required for serine proteases to possess catalytic activity.

Plasminogen binds to fibrin through plasminogen's lysine binding sites. Other proteins in plasma can also bind to the lysine binding sites of plasminogen, and about 50% of plasma plasminogen circu-

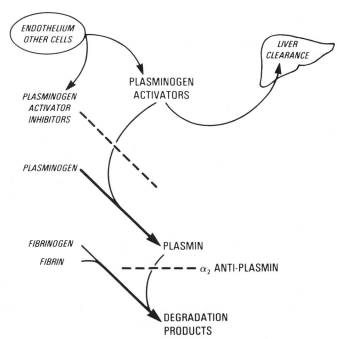

Figure 3.28. A simplified diagram of the overall reactions of fibrinolysis. (From Rapaport, S.I. Normal hemostasis. In: *Textbook of Internal Medicine*, edited by W. Kelley. Philadelphia: J.B. Lippincott, 1989.)

lates as a complex with a protein called histidine-rich glycoprotein. Occupancy of lysine binding sites by histidine-rich glycoprotein represents a natural regulatory mechanism limiting the binding of plasminogen to fibrin. Two analogues of lysine, ϵ-aminocaproic acid and tranexamic acid, are used therapeutically to suppress pathological fibrinolysis. They exert their antifibrinolytic effect through occupying the lysine binding sites of plasminogen.

Native plasminogen has an NH_2-terminal glutamic acid (Glu-plasminogen), and circulating plasma normally contains only Glu-plasminogen. The first traces of plasmin generated in fibrinolysis split an 8-kD peptide from the NH_2 segment of Glu-plasminogen, giving rise to a smaller molecule with an NH_2-terminal lysine (Lys-plasminogen). Under some in vitro experimental circumstances u-PA activates Lys-plasminogen much faster than it activates Glu-plasminogen. Whether the conversion of Glu-plasminogen to Lys-plasminogen significantly accelerates the activation of plasminogen during fibrinolysis in vivo is not yet clear.

PLASMINOGEN ACTIVATORS

The plasminogen activators are normally present in human plasma in only trace amounts, i.e., ~100 pM concentrations. These minimal levels reflect a balance between a limited, basal secretion of t-PA and u-PA from vascular endothelium and their rapid hepatic clearance. The intravascular half-time for these plasminogen activators is about 5 minutes.

Thrombin stimulates the release of plasminogen activators from endothelial cells in culture and could so function in vivo in the initiation of secondary fibrinolysis. Vasoactive stimuli—exercise, venous occlusion, catecholamines, vasopressin—also trigger the release of plasminogen activators from endothelium.

The single-chain t-PA and u-PA molecules secreted from endothelium possess enzymatic activity, but neither is normally physiologically active in the flowing blood. t-PA, although possessing a limited activity against plasminogen in solution, circulates almost entirely complexed with a plasminogen activator inhibitor (PAI-1, see below). For reasons as yet poorly understood, scu-PA has no activity except on a fibrin surface. However, when scu-PA is cleaved by plasmin (or by kallikrein forming during the contact activation reactions of blood coagulation (see Fig. 3.25)), the resultant two-chain molecule, tcu-PA, can readily activate circulating plasminogen.

Single-chain u-PA and t-PA may act synergistically in initiating lysis of intravascular deposits of fibrin. As fibrin is deposited, plasminogen initially binds only to an internal lysine on fibrin. Binding of plasminogen to this internal lysine apparently cannot enhance the ability of t-PA to activate plasminogen. However, scu-PA can activate this internally bound plasminogen. The resultant plasmin then cleaves peptides from the carboxy-terminals of the fibrin polypeptide chains, exposing new carboxy-terminal lysines. Binding of plasminogen and of t-PA, which also possesses a lysine binding site, to these newly exposed lysines creates the ternary complex responsible for the enhanced activation by t-PA of plasminogen bound to fibrin (Pannell et al., 1988).

PLASMINOGEN ACTIVATOR INHIBITORS

Two proteinase inhibitors with specificity for plasminogen activators, PAI-1 and PAI-2, are thought to play physiologically important roles in the regulation of fibrinolysis. (Sprengers and Kluft, 1987; Bachmann, 1987). PAI-1 is present in plasma in a 1 nM concentration, which is severalfold higher than the plasma concentration of t-PA. PAI-1 is secreted by vascular endothelium. Cytokines activating endothelial cells in inflammatory states increase the release from endothelium of PAI-1, and plasma PAI-1 levels rise after major surgery or other types of tissue trauma. PAI-1 is also present within platelet α granules; its release from activated platelets increases the local concentration of PAI-1 as platelet hemostatic plugs form at sites of vessel wall damage. PAI-1 inhibits both single-chain and two-chain t-PA and also two-chain u-PA. It does not inhibit single-chain u-PA.

PAI-2 is measurable in plasma only in the third trimester of pregnancy, when its concentration may rise to 2 nM. It is thought responsible for the decreased fibrinolytic activity of pregnancy. Cells in the placenta of unknown identity secrete PAI-2. They could be placental macrophages, since macrophages in cultures secrete PAI-2. PAI-2, which appears to be more active against u-PA than t-PA, may function primarily to modulate activation of plasminogen in the tissues. Binding of u-PA to a receptor present on the surface of tissue cells that secrete u-PA appears to protect u-PA against inactivation by PAI-2. This helps to confine activation of plasminogen in the tissues to the surface of cells.

PAI-1 and PAI-2 are members of the serine proteinase inhibitor family (serpins). In addition to their partial homology with each other, they have partial homology with antithrombin III, C1 inhibitor, α_1-protease inhibitor, and α_2-antiplasmin. Two other serpins with plasminogen activator inhibitor activity have been identified. One, present in urine, has been named PAI-3. The other, called proteus nexin, is an inhibitor released from a number of cultured anchorage-dependent cells (e.g., fibroblasts) that can inactivate many proteases, including plasminogen activators. PAI-3 and proteus nexin have as yet no established roles in the regulation of fibrinolysis.

PLASMIN INHIBITORS

The major plasma inhibitor of plasmin, α_2-antiplasmin, is made by the liver and is present in plasma in a 1 μM concentration. It inhibits plasmin stoichiometrically. Thus, it can neutralize only one-half of the plasmin that would be formed if all plasma plasminogen (2 μM concentration) were to be activated.

α_2-Antiplasmin inactivates plasmin by combining with both its lysine binding site and its active serine. These sites are unoccupied when plasmin is not degrading fibrin. Therefore, as long as α_2-antiplasmin is present in normal concentration in plasma, it will essentially instantaneously neutralize small amounts of plasmin escaping from a clot. However, if plasma α_2-antiplasmin levels are depleted, as occurs after therapeutic administration of certain plasminogen activators, then plasmin is free to attack circulating fibrinogen. A backup plasma protease inhibitor, α_2-macroglobulin, which has a substantial in vitro ability to inhibit plasmin, seems unable to stop this.

α_2-Antiplasmin also inactivates plasmin on fibrin but at a slower rate. Nevertheless, inactivation by α_2-antiplasmin of plasmin on fibrin appears to be key for the normal regulation of secondary fibrinolysis (Bachmann, 1987). About equal concentrations of plasminogen and α_2-antiplasmin apparently come down initially onto fibrin. The α_2-antiplasmin is bound covalently by factor XIIIa to a lysine residue of the α chains of fibrin (Kismura and Aoki, 1986). As the fibrin-bound plasminogen is activated, it is neutralized 1:1 by α_2-antiplasmin. Only when the fibrin-bound α_2-antiplasmin is consumed can plasminogen activated by plasminogen activators on fibrin begin to digest the fibrin clot. This protects the hemostatic seal from excessive lysis. Rare patients born with homozygous hereditary α_2-antiplasmin deficiency (Aoki et al., 1979) do not have evidence of an ongoing continuous fibrinolysis. However, they bleed extensively into the tissues following only minor tissue injury; normally initiated secondary fibrinolysis is not regulated by α_2-antiplasmin bound to fibrin and, consequently, hemostatic seals, although normally formed, cannot hold.

FIBRIN(OGEN) DEGRADATION PRODUCTS

Plasmin cleaves fibrinogen or fibrin in a sequence that gives rise to four degradation products, fragments X and Y, which are large degradation products, and fragments D and E, which are smaller degradation products resulting from further degradation of the molecules. When present in blood in substantial amounts, the larger degradation products can impair hemostasis by inhibiting normal fibrin polymerization. Fragments X, Y, and D can also compete with fibrinogen for binding to the GPIIb-IIIa fibrinogen receptor on platelets. Therefore they can impede the formation of hemostatic plugs and lengthen the bleeding time.

Fibrinogen/fibrin degradation products can be readily measured immunologically. Because fibrin molecules are cross-linked by factor XIII, fibrin degradation products contain fragments from more than a single fibrin molecule. Measurement of the plasma concentration of one such fragment, a dimer of two D domains of fibrin (D-D dimer), has come into increasing clinical use as a test for deposition of fibrin in vivo followed by secondary fibrinolysis. This is because the test specifically measures fibrin degradation products and not also fibrinogen degradation products.

Plasmin readily dissolves a freshly formed thrombus; however, after several days plasmin will no longer dissolve the thrombus. How fibrin, as it ages, acquires resistance to plasmin is not yet known.

HEMOSTASIS IN DISEASE

Screening Tests of Hemostasis

Hemostatic function is examined in clinical medicine to evaluate the possibility of abnormal bleeding and to obtain information for the diagnosis and mangement of diseases whose manifestations include disturbed hemostasis. Two tests are often carried out to screen for the adequacy of formation of hemostatic plugs—the platelet count and the bleeding time. The bleeding time measures the time it takes to form platelet plugs that stop bleeding (against a pressure of 40 mm Hg transmitted from an inflated blood pressure cuff on the upper arm) from tiny vessels severed by making a cut 1 mm deep in the skin of the forearm. It is used to screen for conditions other than thrombocytopenia that can impair the formation of hemostatic plugs.

Two tests are used to screen for the adequacy of blood coagulation. One test, the prothrombin time, measures the adequacy of the reactions that clot plasma when a very high concentration of tissue factor is present. The other test, the activated partial thromboplastin time (APTT), measures the adequacy of the clotting reactions that clot plasma when a reagent optimizing the contact activation reactions and providing procoagulant phospholipid is present.

The platelet count, bleeding time, prothrombin time, and APTT will usually provide a satisfactory screen of hemostatic function. In certain circumstances (Rapaport, 1983) other tests are added, e.g., a second bleeding time after giving a patient aspirin or tests of the stability of plasma clots incubated in saline and in a urea solution to look for evidence of excessive fibrinolysis and factor XIII deficiency. When screening tests are abnormal, the pattern of abnormality plus the patient's other clinical findings provide the information needed to proceed with specific tests of platelet function or specific coagulation factor assays.

Disorders Affecting Formation of Hemostatic Plugs

THROMBOCYTOPENIA

Whereas hereditary thrombocytopenia is very rare, acquired thrombocytopenia is the commonest cause of impaired hemostatic function. It may result from failure of platelet production or from accelerated platelet destruction. Moreover, moderate thrombocytopenia, in the 60,000–150,000/μl range, may result from the pooling of platelets in an enlarged spleen.

Thrombocytopenia due to decreased production usually occurs in patients with very serious bone marrow disease, for example, leukemia or marrow aplasia. Thrombocytopenia from accelerated peripheral destruction most often stems from the coating of platelets with IgG and the removal of the coated platelets by mononuclear phagocytes in the spleen, liver, and bone marrow. The IgG may be a true autoantibody to a platelet antigen, an antibody to an epitope formed when a drug binds to the surface of the platelets, or IgG in immune complexes that binds to Fc receptors on the platelets. Patients with severe gram-negative bacterial infections also frequently become thrombocytopenic. This apparently results from several processes: deposition of platelets on activated vascular endothelium, activation by endotoxin of complement on the platelet surface, binding of immune complexes to platelets, and consumption of platelets in intravascular coagulation.

DISORDERS WITH NORMAL PLATELET COUNTS BUT IMPAIRED FORMATION OF HEMOSTATIC PLUGS

Hereditary disorders affecting the formation of hemostatic plugs include von Willebrand's disease, in which the abnormality is in the plasma, and a group of intrinsic platelet disorders. von Wille-

brand's disease, an autosomal dominant disorder, is the most common hereditary hemostatic disorder. In most patients it stems from an inability to make normal amounts of von Willebrand factor, but in occasional patients it stems from synthesis of abnormal von Willebrand factor molecules. The usual patient with von Willebrand's disease has a mild bleeding disorder with a prolonged bleeding time and a concordant moderate reduction in the plasma levels of von Willebrand factor antigen and factor VIII clotting activity. The latter reflects the need for factor VIII to circulate in plasma bound to von Willebrand factor.

Hereditary intrinsic disorders of platelet function are uncommon but of historical importance. Many clues to how platelets normally form hemostatic plugs came from the study of patients with these disorders. For example, identifying the defect responsible for the serious bleeding of patients with thrombasthenia (Glanzmann's disease) led to the recognition that binding of fibrinogen to a recognition site on a GPIIb-IIIa complex that is expressed on the surface membrane of activated platelets is how platelets adhere to each other during platelet aggregation.

The formation of hemostatic plugs may be impaired despite a normal platelet count in a number of acquired conditions. In myelodysplastic or certain myeloproliferative diseases, abnormal megakaryocytes may make normal or increased numbers of platelets with defective hemostatic function. Certain therapeutic agents, particularly penicillin and penicillin derivatives when given in high doses, may coat platelets and interfere with their function. Uremic patients frequently have a prolonged bleeding time despite a normal or moderately depressed platelet count. The reason is not yet clear.

Disorders Affecting Blood Coagulation

Hereditary deficiency states have been identified for each of the known plasma coagulation factors. As mentioned earlier, factor XII deficiency, prekallikrein deficiency, and high molecular weight kininogen deficiency prolong coagulation in glass tubes but do not cause abnormal bleeding. The other disorders are all associated with abnormal bleeding.

Except for hemophilia A (factor VIII deficiency), hemophilia B (factor IX deficiency), and factor XI deficiency, the hereditary deficiency states are rare autosomal recessive disorders in which heterozygotes do not bleed abnormally. Factor XI deficiency is also an autosomal recessive disorder but has an unusually high gene frequency in Ashkenazi Jews. The structural genes for factor VIII and factor IX are located on the X chromosome. The hemizygote male who receives an abnormal gene from his mother is not protected by a normal X chromosome and develops the disease. Hemophilia, therefore, is the most common hereditary bleeding disorder due to a coagulation factor deficiency but is confined to males.

There are four major causes of acquired abnormalities of blood coagulation: vitamin K deficiency, liver disease, disseminated intravascular coagulation, and acquired antibodies against clotting factors. *Vitamin K deficiency* usually results from the combination of inadequate dietary intake plus the administration of broad spectrum antibiotics that suppress bacterial synthesis of vitamin K in the gut. Vitamin K is a fat-soluble vitamin, and patients with gastrointestinal disorders interfering with fat absorption are at particular risk for developing vitamin K deficiency. In *liver disease* seriously impairing the ability of hepatocytes to synthesize proteins, patients develop a bleeding tendency secondary to a fall in the plasma concentration of all clotting factors except factor VIII and to a fall in the plasma concentration of the plasmin inhibitor, α_2-

antiplasmin. Patients with advanced liver disease may bleed uncontrollably from relatively minor lesions in the gastrointestinal tract. *Disseminated intravascular coagulation* (DIC) may occur as a complication of pregnancy, of malignancy, or of infection, usually gram-negative bacterial infection. In DIC, tissue factor, activated clotting enzymes, or both gain access to the blood in sufficient concentration to cause fibrin to be formed within the flowing blood. Bleeding may result from the depletion of plasma fibrinogen, factor VIII, and factor V plus the antihemostatic effects of extensive secondary fibrinolysis. *Antibodies to clotting factors* may develop as a complication of a known autoimmune disorder or, without warning, in a patient with no known underlying disease. Antibodies that neutralize the coagulant activity of factor VIII and nonneutralizing antibodies that cause hypoprothrombinemia because of the rapid cellular clearance of prothrombin-antiprothrombin immune complexes are the most common causes of bleeding secondary to clotting factor antibodies.

BIBLIOGRAPHY

ABRAMSON, S., R. G. MILLER, AND R. A. PHILLIPS. The identification in adult bone marrow of pluripotent and restricted stem cells of the myeloid and lymphoid systems. *J. Exp. Med.* 145: 1567–1579, 1977.

ARNONE, A. Xray diffraction study of binding of 2,3 DPG to human deoxyhemoglobin. *Nature* 237: 146–149, 1972.

ASHWELL, G., AND C. J. STEER. Hepatic recognition and catabolism of serum glycoproteins. *J.A.M.A.* 246: 2358–2364, 1981.

AOKI, N., H. SAITO, T. KAMIYA, K. KOIE, Y. SAKATA, AND M. KOBAKURA. Congenital dificiency of α 2-plasmin inhibitor associated with severe hemorrhagic tendency. *J. Clin. Invest.* 63: 877–884, 1979.

BABIOR, B. M. The respiratory burst, p. 132–134. In: Lehrer, R. I., moderator. Neutrophils and Host Defense. *Ann. Intern. Med.* 109: 127–142, 1988.

BACH, F. H., AND D. H. SACHS. Transplantation Immunology. *N. Engl. J. Med.* 317:489–492, 1987.

BACHMANN, F. Fibrinolysis. In: *Thrombosis and Haemostasis 1987*, edited by M. Verstraete, J. Vermylen, R. Lijnen, and J. Arnout, Leuven: Leuven University Press, 1987, p. 227–265.

BARCROFT, J. *The Respiratory Function of the Blood. Part II. Haemoglobin.* London: Cambridge University Press, 1928.

BASS, D. A. Behavior of eosinophil leukocytes in acute inflammation. I. Lack of dependence on adrenal function. *J. Clin. Invest.* 55: 1229–1236, 1975.

BAUER, K. A., AND R. D. ROSENBERG. The pathophysiology of the prethrombotic state in humans: Insights gained from studies using markers of hemostatic system activation. *Blood* 70: 343–350, 1987.

BECKER, C. G., AND P. C. HARPEL. α_2-Macroglobulin on human vascular endothelium. *J. Exp. Med.* 144: 1–9, 1976.

BENACERRAF, B. Role of MHC gene products in immune regulation. *Science* 212: 1229–1238, 1981.

BENESCH, R., AND R. E. BENESCH. The effect of organic phosphates from the human erythrocyte on the allosteric properties of hemoglobin. *Biochem. Biophys. Res. Comm.* 26: 162–167, 1967.

BENESCH, R., AND R. E. BENESCH. Intracellular organic phosphates as regulators of oxygen release by hemoglobin. *Nature* 221: 618–622, 1969.

BENNETT, G. V. The membrane skeleton of human erythrocytes and its implications for more complex cells. *Ann. Rev. Biochem.* 54: 273–304, 1985.

BEUTLER B., AND A. CERAMI. Tumor necrosis, cachexia, shock, and inflammation: A common mediator. *Ann. Rev. Biochem.* 57: 505–518, 1988.

BEUTLER, E. Glucose-6-phosphate dehydrogenase deficiency. In: *The Metabolic Basis of Inherited Disease*, 5th ed. edited by J. B. Stanbury, J. B. Wyngaarden, D. S. Fredrickson, J. L. Goldstein and M. S. Brown, New York: McGraw Hill, 1983.

BEVERS, E. M., P. COMFURIUS, AND R. F. A. ZWAAL. Changes in membrane phospholipid distribution during platelet activation. *Biochim. Biophys. Acta* 736: 57–66, 1983.

BEVILACQUIA, M. P., D. AMRANI, M. W. MOSESSON, AND C. BIANCO. Receptors for cold-insoluble globulin (plasma fibronectin) on human monocytes. *J. Exp. Med.* 153: 42-60, 1981.

BOXER, L. A., AND J. E. SMOLEN. Neutrophil granule constituents and their release in health and disease. *Hem/Oncol. Clin. North Am.* 2: 101–134, 1988.

BREITMEYER, J. B. Lymphocyte activation: how T cells communicate. *Nature* 329: 760–761, 1987.

BUNN, H. F., B. J. RANSIL, AND A. CHAO. The interaction between erythrocyte organic phosphate, magnesium and hemoglobin. *J. Biol. Chem.* 246: 5273–5279, 1971.

BUTTERWORTH, A. E., AND J. R. DAVID. Eosinophil function. *N. Engl. J. Med.* 304: 154–156, 1981.

CARP, H. Mitochondrial N-formylmethionyl proteins as chemoattractants for neutrophils. *J. Exp. Med.* 155: 264–273, 1982.

CARRELL, R. W., AND D. R. BOSWELL. Serpins: the superfamily of plasma serine proteinase inhibitors. In: *Proteinase Inhibitors*, edited by A. J. Barrett and G. Salvesen. New York: Elsevier, 1986, p. 403–420.

CARRELL, R. W., P. B. CHRISTEY, AND D. R. BOSWELL. Serpins: antithrombin and other inhibitors of coagulation and fibrinolysis. Evidence from amino acid sequences. In: *Thrombosis and Hemostasis 1987*, edited by M. Verstraete, J. Vermylen, R. Lijnen, and J. Arnout. Leuven: Leuven University Press, 1987, p. 1–15.

CHANARIN, I., R. DEACON, J. PERRY, AND M. LUMB. How vitamin B12 acts. *Br. J. Haematol.* 46: 487–491, 1981.

COMP, P. C., R. R. NIXON, R. COOPER, AND C. T. ESMON. Familial protein S deficiency is associated with recurrent thrombosis. *J. Clin. Invest.* 74: 2082–2088, 1984.

COOK, J. D. Clinical evaluation of iron deficiency. *Semin. Hematol.* 19: 6–18, 1982.

COOK, J. D., C. A. FINCH, AND N. J. SMITH. Evaluation of the iron status of a population. *Blood* 48: 449–455, 1976.

COOPER, M. D. B-lymphocytes: normal development and function *N. Engl. J. Med.* 317: 1452–1456, 1987.

CRONKITE, E. P. Kinetics of granulocytopoiesis. *Clin. Haematol.* 8:351–370, 1979.

CURNETTE, J. T. Classification of chronic granulomatous disease. *Hem/Oncol. Clin. North Am.* 2: 241–252, 1988.

DAHLEN, S. E., J. BJORK, P. HEDQVIST, K. E. ARFORS, S. HAMMARSTROM, J. A. LINDGREN, AND B. SAMUELSSON. Leukotrienes promote plasma leakage and leukocyte adhesion in postcapillary venules: *in vivo* effects with relevance to the acute inflammatory response. *Proc. Natl. Acad. Sci. USA* 78: 3887–3891, 1981.

DAVIS, M. M., S. K. KIN, AND L. HOOD. Immunoglobulin class switching: developmentally regulated DNA rearrangements during differentiation. *Cell* 22: 1–2, 1980.

DEUEL, T. F., R. M. SENIOR, J. S. HUANG, AND G. L. GRIFFIN. Chemotaxis of monocytes and neutrophils to platelet-derived growth factor. *J. Clin. Invest.* 69: 1046–1049, 1982.

DICK, J. E., M. C. MAGLI, D. HUSZAR, R. A. PHILLIPS, AND A. BERNSTEIN. Introduction of a selectable gene into primitive stem cells capable of longterm reconstitution of the hemopoietic system of W/Wᵛ mice. *Cell* 42: 71–79, 1985.

DINARELLO, C. A., AND J. W. MIER. Lymphokines. *N. Engl. J. Med.* 317: 940–945, 1987.

DONAHUE, R. E., J. SEEHRA, M. METZGER, D. LEFEBRE, B. ROCK. et al. Human Il-3 and GM-CSF act synergistically in stimulating hematopoiesis in primates. *Science* 241: 1820–1823, 1988.

DOOLITTLE, R. F. Fibrinogen and fibrin. *Sci. Am.* 245: 126–135, 1981.

ESMON, C. T. The regulation of natural anticoagulant pathways. *Science* 235: 1348–1352, 1987.

FANTONE, J. C., AND P. A. WARD. Role of oxygen-derived free radicals and metabolites in leukocyte-dependent inflammatory reactions. *Am. J. Pathol.* 107: 397–418, 1982.

FINCH, C. A., AND C. LENFANT. Oxygen transport in man. *N. Engl. J. Med.* 206: 407–415, 1973.

FRANK, M. M. Complement in the pathophysiology of human disease. *N. Engl. J. Med.* 316: 1525–1530, 1987.

FURIE, B., AND B. C. FURIE. The molecular basis of blood coagulation. *Cell* 53: 505–518, 1988.

GADEK, J. E., G. A. FELLS, R. L. ZIMMERMAN, S. I. RENNARD, AND R. G. CRYSTAL. Antielastases of the human alveolar structures. Implications for the protease-antiprotease theory of emphysema. *J. Clin. Invest.* 68: 889–898, 1982.

GEHA, R. S. Current concepts in immunology. Regulation of the immune response by idiotypic-anti-idiotypic interactions. *N. Engl. J. Med.* 305: 25–28, 1981.

GEWURZ, H. Biology of C-reactive protein and the acute phase response. *Hosp. Prac.* 17: 67–81, 1982.

GOLDE, D. Production, distribution, and fate of neutrophils. In: *Hematology*, 3rd Ed., edited by W. J. Williams, E. Beutler, A. J. Erslev, and M. A. Lichtman. New York: McGraw-Hill, 1983.

GOLDSTEIN, I. M., H. B. KAPLAN, H. S. EDELSON, AND G. WEISSMANN. Ceruloplasmin: an acute phase reactant that scavenges oxygen-derived free radicals. *Ann. N.Y. Acad. Sci.* 389: 368–379, 1982.

GOETZL, E. J. Oxygenation products of arachidonic acid as mediators of hypersensitivity and inflammation. *Med. Clin. North Am.* 65: 809–828, 1981.

GOODMAN, S. R., K. A. SHIFFER, L. A. CASORIA, AND M. E. EYSTER. Identification of the molecular defect in the erythrocyte membrane skeleton of some kindreds with hereditary spherocytosis. *Blood* 60: 772–784, 1982.

GOREVIC, P. D., A. B. CLEVELAND, AND E. C. FRANKLIN. The biologic significance of amyloid. *Ann. N.Y. Acad. Sci.* 389: 380–394, 1982.

HARKER, L. A. Control of platelet production. *Ann. Rev. Med.* 25: 383–400, 1974.

HARPEL, P. C., AND R. D. ROSENBERG. $\alpha_2$macroglobulin and antithrombin-heparin cofactor: modulators of hemostatic and inflammatory reactions. In: *Progress in Hemostasis and Thrombosis*, edited by T. H. Spaet, New York: Grune & Stratton, 1976, vol. III, p. 145–189.

HASLAM, R. J. Signal transduction in platelet activation. In: *Thrombosis and Haemostasis 1987*, edited by M. Verstraete, J. Vermylen, R. Lijnen, and J. Arnout. Leuven: Leuven University Press, 1987, p. 147–174.

HAWIGER, J., M. L. STEER AND E. W. SALZMAN. Intracellular regulatory processes in platelets. In: *Hemostasis and Thrombosis: Basic Principles and Clinical Practice*, 2nd Ed., edited by R. W. Colman, J. Hirsch, V. J. Marder, and E. W. Salzman, Philadelphia: J. B. Lippincott, 1987, p. 710–725.

HERBERT, V., AND R. ZALUSKY. Interrelation of vitamin B_{12} and folic acid metabolism: folic acid clearance studies. *J. Clin. Invest.* 41: 1263–1276, 1962.

HILLMAN, R. S., AND C. A. FINCH. *Red Cell Manual*, 5th Ed. Philadelphia: F. A. Davis, 1985.

HUEBERS, H. A., AND C. A. FINCH. The physiology of transferrin and transferrin receptors. *Physiol. Rev.* 67: 520–582, 1987.

ISHIBASHI, T., S. L. MILLER AND S. A. BURSTEIN. Type beta transforming growth factor is a potent inhibitor of murine megakaryopoiesis in vitro. *Blood* 69: 1737–1741, 1987.

JERNE, N. The immune system: a web of V-domains. *Harvey Lecture Series* 70: 93–110, 1974–75.

JOHNSTON, R. B. Monocytes and macrophages. *N. Engl. J. Med.* 318: 747–751, 1988.

KANE, W. H., AND E. W. DAVIE. Blood coagulation factors V and VIII: structural and functional similarities and their relationship to hemorrhagic and thrombotic disorders. *Blood* 71: 539–555, 1988.

KAPLAN, A. P., AND M. SILVERBERG. The coagulation-kinin pathway of human plasma. *Blood* 70: 1–15, 1987.

KIMURA, S., AND N. AOKI. Cross-linking site in fibrinogen for α2-plasmin inhibitor. *J. Biol. Chem.* 261: 15591–15595, 1986.

KINO, K., H. TSUNOO, Y. HIGA, M. TAKAMI, H. HAMAGUCHI, AND H. NAKAJIMA. Hemoglobin-haptoglobin receptor in rat liver plasma membranes. *J. Biol. Chem.* 255: 9616–9620, 1980.

KLEBANOFF, S. J. Oxygen metabolism and the toxic properties of phagocytes. *Ann. Intern. Med.* 93: 480–489, 1980.

KLOCZEWIAK, M., S. TIMMONS, T. S. LUCAS, AND J. HAWIGER. Platelet receptor recognition site on human fibrinogen. Synthesis and structure-function relationship of peptides corresponding to the carboxy-terminal segment of the gamma chain. *Biochemistry* 23: 1767–1774, 1984.

KURZROCK, R., J. V. GUTTERMAN, AND M. TALPAZ. The molecular genetics of Philadelphia chromosome-positive leukemias. *N. Engl. J. Med.* 319: 990–998, 1988.

LEAF, A. AND P. C. WEBER. Cardiovascular effects of n-3 fatty acids. *N. Engl. J. Med.* 318: 549–557, 1988.

LEHRER, R. I., T. GANZ, M. E. SELSTED, B. M. BABIOR, AND J. T. CURNUTTE. Neutrophils and host defense. *Ann. Intern. Med.* 109: 127–142, 1988.

LEVIN, J. Thrombopoiesis. In: *Hemostasis and Thrombosis: Basic Principles and Clinical Practice*, 2nd Ed., edited by R. W. Colman, J. Hirsch, V. J. Marder, and E. W. Salzman. Philadelphia: J. B. Lippincott, 1987, p. 418–430.

McDEVITT, H. O. Current concepts in immunology. Regulation of the immune response by the major histocompatibility system. *N. Engl. J. Med.* 303: 1514–1517, 1980.

McDONAGH, J. Structure and function of factor XIII. In: *Hemostasis and Thrombosis: Basic Principles and Clinical Practice*, 2nd Ed., edited by R. W. Colman, J. Hirsch, V. J. Marder, and E. W. Salzman. Philadelphia: J. B. Lippincott, 1987, p. 289–300.

MALECH, H. L., AND J. H. GALLIN. Neutrophils in human diseases. *N. Engl. J. Med.* 317: 687–694, 1987.

MANN, K. G. The assembly of blood clotting complexes on membranes. *Trends Bioch. Sci.* 12: 229–233, 1987.

MARCHESI, V. T. The red cell membrane skeleton: recent progress. *Blood* 61: 1–11, 1983.

MARRACK, P., AND J. KAPPLER. The T cell receptor. *Science* 238: 1073–1079, 1987.

MAZUR, E. M. Megakaryocytopoiesis and platelet production: a review. *Exp. Hematol.* 15: 340–350, 1987.

MILETICH, J. P., C. M. JACKSON, AND P. W. MAJERUS. Properties of the factor Xa binding site on human platelets. *J. Biol. Chem.* 253: 6908–6916, 1978.

MOSHAGE, H. J., J. A. M. JANSSEN, J. H. FRANSSEN, J. C. M. HALKENSCHEID, AND S. H. YAP. Study of the molecular mechanism of decreased liver synthesis of albumin in inflammation. *J. Clin. Invest.* 79: 1635–1641, 1987.

MOSHER, D. F. Fibronectin-relevance to hemostasis and thrombosis. In: *Hemostasis and Thrombosis: Basic Principles and Clinical Practice*, 2nd Ed., edited by R. W. Colman, J. Hirsch, V. J. Marder, and E. W. Salzman. Philadelphia: J. B. Lippincott, 1987, p. 210–218.

MÜLLER-EBERHARD, H. J. Molecular organization and function of the complement system. *Ann. Rev. Biochem.* 57: 321–347, 1988.

NACHMAN, R. L., R. L. SILVERSTEIN, AND A. S. ASCH. Thrombospondin: cell biology of an adhesive glycoprotein. In: *Thrombosis and Haemostasis 1987*, edited by M. Verstraete, J. Vermylen, R. Lijnen, and J. Arnout. Leuven: Leuven University Press, 1987, p. 81–91.

NATHAN, C. F. Secretory products of macrophages. *J. Clin. Invest.* 79: 319–326, 1987.

NEMERSON, Y. Tissue factor and hemostasis. *Blood.* 71: 1–8, 1988.

NISHIZUKA, Y. Turnover of inositol phospholipids and signal transduction. *Science* 225: 1365–1370, 1984.

NOSSL, G. J. V. The basic components of the immune system. *N. Engl. J. Med.* 316: 1320–1325, 1987.

NURDEN, A. T. Platelet membrane glycoproteins and their clinical aspects. In: *Thrombosis and Haemostasis 1987*, edited by M. Verstraete, J. Vermylen, R. Lijnen, and J. Arnout. Leuven: Leuven University Press, 1987, p. 93–125.

OPPENHEIN, J. J., E. J. KOVACS, K. MATSUSHIMA, AND S. K. DURUM. There is more than one interleukin 1. *Immunol. Today* 7:45–56, 1986.

PANNELL, R., J. BLACK, AND V. GUREWICH. Complementary modes of action of tissue-type plasminogen activator and pro-urokinase by which their synergistic effect on clot lysis may be explained. *J. Clin. Invest.* 81: 853–859, 1988.

PAULING, L., H. A. ITANO, S. J. SINGER, AND J. C. WELLS. Sickle cell anemia: a molecular disease. *Science* 110: 543–548, 1949.

PERUTZ, M. F. Stereochemistry of cooperative effects in hemoglobin. *Nature* 228: 726–739, 1970.

PERUTZ, M. F. Hemoglobin structure and respiratory transport. *Sci. Am.* 239: 92–125, 1978.

PERUTZ, M. F. Regulation of oxygen affinity of hemoglobin: influence of structure of the globin on the heme iron. *Ann. Rev. Biochem.* 48: 327–386, 1979.

PHILLIPS, D. R., I. F. CHARO, L. V. PARISE, AND L. A. FITZGERALD. The platelet membrane glycoprotein IIb-IIIa complex. *Blood* 71: 831–843, 1988.

PINCKARD, R. N. The "new" chemical mediators of inflammation. In: *Current Topics in Inflammation and Infection*, edited by G. Majino, R. S. Cotran, and

N. Kaufman. Baltimore: Williams & Wilkins, 1982, p. 38–53.

QUESENBERRY, P., AND L. LEVITT. Hematopoietic stem cells. *N. Engl. J. Med.* 301: in 3 parts, 755–760, 819–823, 868–872, 1979.

RAO, L. V. M., S. I. RAPAPORT, AND M. LORENZI. Enchancement by human umbilical vein endothelial cells of factor Xa-catalyzed activation of factor VII. *Blood* 71: 791–796, 1988.

RAO, L. V. M., AND S. I. RAPAPORT. Activation of factor VII bound to tissue factor: A key early step in the tissue factor pathway of coagulation. *Proc. Natl. Acad. Sci. USA* 85: 6701–6705, 1988.

RAPAPORT, S. I. Preoperative hemostatic evaluation: which tests if any. *Blood* 61: 229–231, 1983.

RAPAPORT, S. I. *Introduction to Hematology*, 2nd Ed. Philadelphia: J. B. Lippincott, 1987.

RAPAPORT, S. I. Inhibition of factor VIIa/tissue factor-induced blood coagulation: with particular emphasis upon a factor Xa-dependent inhibitory mechanism. *Blood* 73: 359–365, 1989.

RIMON, S., R. MELAMED, T. SCOTT, P. P. NAWROTH, AND D. M. STERN. Identification of a factor IX/IXa binding protein on the endothelial cell surface. *J. Bio. Chem* 262: 6023–6031, 1987.

RITTENHOUSE, S. E., AND C. L. ALLEN. Synergistic activation by collagen and 15-hydroxy-9α, 11α-peroxidoprosta 5, 13-dienoic acid (PGH$_2$) of phosphatidylinositol metabolism and arachidonic acid release in human platelets. *J. Clin. Invest.* 70: 1216–1224, 1982.

RODGERS, G. M., AND M. H. SHUMAN. Characterization of the interactions between factor Xa and bovine aortic endothelial cells. *Biochim. Biophys. Acta* 844: 320–329, 1985.

ROSENBERG, R. D., AND J. S. ROSENBERG. Natural anticoagulant mechanisms. *J. Clin. Invest.* 74: 1–6, 1984.

ROYER, H. D., AND E. L. REINHERZ. T lymphocytes: ontogeny, function, and relevance to clinical disorders. *N. Engl. J. Med.* 317: 1136–1142, 1987.

RUGGERI, Z. M., AND T. S. ZIMMERMAN. Von Willebrand factor and von Willebrand disease. *Blood* 70: 895–904, 1987.

RUOSLAHTI, E., AND M. D. PIERSCHBACHER. New perspectives in cell adhesion. RGD and integrins. *Science* 238: 491–497, 1987.

SAMUELSSON, B. Leukotrienes: mediators of allergic reactions and inflammation. *Int. Arch. Allergy Appl. Immun.* 66 (Suppl. 1): 98–106, 1981.

SCHRIER, S. L. Human erythrocyte membrane enzymes: current status and clinical correlates. *Blood* 50: 227–237, 1977.

SELIGSOHN, U., C. K. KASPER, B. ØSTERUD, AND S. I. RAPAPORT. Activated factor VII: presence in factor IX concentrates and persistence in the circulation after infusion. *Blood* 53: 828–837, 1978.

SELSTED, M. E. Non-oxidative killing by neutrophils. p. 129–132. In: Lehrer, R. I. Moderator. Neutrophils and Host Defense. *Ann. Intern. Med.* 109: 127–142, 1988.

SERAFIN, W. E., AND K. F. AUSTEN. Mediators of immediate hypersensitivity reactions. *N. Engl. J. Med.* 317: 30–34, 1987.

SIMIONESCU, M., L. GHITESCU, A. FIXMAN, AND N. SIMIONESCU. How plasma macromolecules cross the endothelium. *News Physiol. Sci.* 2: 97–100, 1987.

SIXMA, J. J. Platelet adhesion in health and disease. In: *Thrombosis and Haemostasis 1987*, edited by M. Verstraete, J. Vermylen, R. Lijnen, and J. Arnout. Leuven: Leuven University Press, 1987, p. 127–146.

SMITH, K. A. Interleukin-2: inception, impact, and implications. *Science* 240: 1169–1176, 1988.

SNYDERMAN, R., AND E. J. GOETZL. Molecular and cellular mechanims of leukocyte chemotaxis. *Science* 213: 830–837, 1981.

SPIVAK, J. L. The mechanism of action of erythropoietin. *Int. J. Cell Cloning 4:* 139–166, 1986.

SPORN, L. A., V. J. MARDER, AND D. D. WAGNER. Inducible secretion of large, biologically potent von Willebrand factor multimers. *Cell* 46: 185–190, 1986.

SPRENGERS, E. D., AND C. KLUFT. Plasminogen activator inhibitors. *Blood* 69: 381–387, 1987.

STECK, T. L. The band 3 protein of the human red cell membrane: a review. *J. Supramolec. Struct.* 8: 311–324, 1978.

TESTA, N. D. Erythroid progenitor cells; their relevance for the study of haematological disease. *Clin. Haematol.* 8: 311–333, 1979.

TILL, J. E., AND E. A. McCULLOCH. Hematopoietic stem cell differentiation. *Biochem. Biophys. Acta* 605: 431–459, 1980.

TOLLEFSEN, D. M., C. A. PESTKA, AND W. J. MONAFO. Activation of heparin cofactor II by dermatan sulfate. *J. Biol. Chem.* 258: 6713–6716, 1983.

TRACY, P. B., L. L. EIDE, AND K. G. MANN. Human prothrombinase complex assembly and function on isolated peripheral blood cell populations. *J. Biol. Chem.* 260: 2119–2124, 1985.

TURITTO, V. T., H. J. WEISS, T. S. ZIMMERMAN, AND I. I. SUSSMAN. Factor VIII/von Willebrand factor in subendothelial mediates platelet adhesion. *Blood* 65: 823–831, 1985.

UNANUE, E. R., AND P. A. ALLEN. The basis for the immunoregulatory role of macrophages and other accessory cells. *Science* 236: 551–557, 1987.

VALENTINE, W. N. The Stratton lecture: hemolytic anemia and inborn errors of metabolism. *Blood* 54: 549–559, 1979.

VAN DE GRIEND, R. J., M. CARRENO, R. VAN DOORN, C. J. M. LEUPERS. A. VEN DEN ENDE, P. WIXERMANS, H. J. G. H. OOSTERHUIS, AND A. ASTALDI. Changes in human T lymphocytes after thymectomy and during senescence. *J. Clin. Immunol.* 2: 289–294, 1982.

VAN FURTH, R., J. A. RAEBURN, AND T. L. VAN ZWET. Characteristics of human mononuclear phagocytes. *Blood* 54: 485–500, 1979.

WAGNER, D. D., AND V. J. MARDER. Biosynthesis of von Willebrand protein by human endothelial cells: processing steps and their intracellular localization. *J. Cell. Biol.* 99: 2123–2130, 1984.

WALDMANN, T. A. Albumin catabolism. In: *Albumin Structure, Function and Uses*, edited by V. M. Rosenoer, M. Oratz, and M. A. Rothschild. New York: Permagon Press, 1977, p. 255–273.

WASSERMAN, S. I. Mast cells and eosinophils. In: *Textbook of Rheumatology*, edited by W. Kelley, S. Ruddy, and E. Harris. Philadelphia: W. B. Saunders, 1989.

WEISS, H. J. Inherited disorders of platelet secretion. In: *Haemostasis and Thrombosis: Basic Principles and Clinical Practice*, 2nd Ed., edited by R. W. Colman, J. Hirsch, V. J. Marder, and E. W. Salzman. Philadelphia: J. B. Lippincott, 1987, p. 741–749.

WEISSMAN, G., J. E. SMOLEN, AND H. M. KORCHAK. Release of inflammatory mediators from stimulated neutrophils. *N. Engl. J. Med.* 303: 27–34, 1980.

WEKSLER, M. E. The senescence of the immune system. *Hosp. Prac.* 16: 53–64, 1981.

WHITE, J. G. Anatomy and structural organization of the platelet. In: *Hemostasis and Thrombosis: Basic Principles and Clinical Practice, 2nd Ed.*, edited by R. W. Colman, J. Hirsch, V. J. Marder, and E. W. Salzman. Philadelphia: J. B. Lippincott, 1987, p. 537–554.

WHITELAW, D. M. Observations on human monocyte kinetics after pulse labeling. *Cell Tissue Kinet.* 5: 311–317, 1972.

WILLIAMS, N., AND R. F. LEVINE. The origin, development and regulation of megakaryocytes. *Br. J. Haematol.* 52: 173–180, 1982.

WION, K. L., D. KELLY, J. A. SUMMERFIELD, E. G. D. TUDDENHAM, AND R. M. LAWN. Distribution of factor VIII mRNA and antigen in human liver and other tissues. *Nature* 317: 726–729, 1985.

WUN, T-C, K. K. KRETZMER, T. J. GIRARD, J. P. MILETICH, AND G. J. BROZE, JR. Cloning and characterization of a c-DNA coding for the lipoprotein-associated coagulation inhibitor shows that it consists of three tandem Kunitz-type inhibitory domains. *J. Biol. Chem.* 263: 6001–6004, 1988.

WYMAN, J., JR. Linked functions and reciprocal effects in hemoglobin: a second look, *Adv. Prot. Chem.* 19: 223–286, 1964.

ZINKERNAGEL, R. M. Major transplantation antigens in host responses to infection. *Hosp. Prac.* 7: 83–92, 1978.

SECTION 4

Body Fluids and Renal Function

CHAPTER 25

Physiology of the Body Fluids

The primary function of the kidneys is the maintenance of the normal volume and composition of the body fluids. Thus, the kidneys are responsible for the excretion of excess water, ions, and waste products as well as for the conservation of solutes important to proper body function. In this chapter, an introduction to the physiology of the body fluids is presented to provide a background for the study of renal function.

BODY WATER AND ITS SUBDIVISIONS

Water is by far the most abundant component of the body, constituting 45–75% of the body weight. This large variation in water content is primarily a function of variations in the amount of *adipose tissue*. Whereas skeletal muscle is over 75% water, skin is over 70% water, and organs such as the heart, lungs, and kidneys are approximately 80% water, adipose tissue contains less than 10% water (Table 4.1) (Skelton, 1927). Thus, the percentage of body weight that is water will vary inversely with the body's fat content. The total body water (TBW) is about 60% of body weight in normal young adult males and 50% of body weight in normal young adult females, who have a somewhat larger amount of subcutaneous fat. TBW may, however, be a much lower percentage of body weight in obese individuals or a somewhat greater percentage in extremely thin persons. In both sexes, the percentage of body weight that is water decreases with age (Table 4.2), a trend that can be attributed primarily to an increasing percentage of adipose tissue (Hays, 1980). During the *1st year* of life, however, the marked decrease in the percentage of body weight that is water occurs principally because the cell mass (which contains at least 20% solids (see Table 4.1)) grows at a faster rate than the extracellular fluid volume (in which the percentage of solids is much less) (Friis-Hansen, 1957).

TBW is distributed into two major fluid compartments: *intracellular fluid* (ICF), which contains approximately 55% of the TBW, and *extracellular fluid* (ECF), which contains approximately 45% of the TBW (Table 4.3) (Edelman and Leibman, 1959). The ECF, in turn, is subdivided into several smaller compartments (Table 4.3). The most important ECF compartments are *plasma*, which contains about 7.5% of the TBW, and *interstitial fluid*, which contains about 20% of the TBW and includes the fluid between cells (i.e., in the interstitium) and in the lymphatics. Also classified as ECF compartments are the water crystallized in *bone* and the fluid in *dense connective tissues* such as cartilage, each containing about 7.5% of the TBW. Several minor ECF compartments, which together contain only about 2.5% of the TBW, include the fluid in the gastrointestinal, biliary, and urinary tracts, the intraocular and cerebrospinal fluids, and the fluid in the serosal spaces, such as the pleural, peritoneal, and pericardial fluids. The term *transcellular fluid* commonly is used to describe these minor ECF compartments because they are separated from the rest of the ECF by a layer of epithelial cells. While the transcellular fluid can be neglected in most experimental and clinical problems of fluid balance, its volume can be increased in certain disease states. For example, considerable fluid may accumulate in the *serosal spaces* in diseases involving the lung (pleural effusion), heart (pericardial effusion), or liver (ascites) or in the *gastrointestinal tract* in intestinal obstruction.

Although Table 4.3 accurately summarizes the distribution of body fluids among the various compartments, in clinical applications a somewhat simplified distribution often is used. For example, it is commonly stated that the ECF contains about one-third of the TBW (or 20% of the body weight of a normal adult man) and the ICF contains about two-thirds of the TBW (or 40% of the body weight of a normal adult man). This approximation is appropriate because the methods used to estimate ECF volume for clinical purposes exclude bone fluid and include variable amounts of dense connective tissue fluid and transcellular fluid (see below).

Table 4.1.
Water Content of Body Tissues

Tissue	% Water
Kidney	83
Heart	79
Lung	79
Skeletal muscle	76
Brain	75
Skin	72
Liver	68
Skeleton	22
Adipose tissue	10

From Skelton, 1927.

Measurement of Body Fluid Compartments

The volumes of the various body fluid compartments cannot be measured directly, but estimates useful for experimental and clinical purposes can be obtained with the aid of *dilution methods*. These methods utilize marker substances that distribute in a specific body fluid compartment. If a known quantity of such a marker X is administered and given time to distribute throughout the compartment, then the volume of the compartment can be determined from the concentration of the marker in a sample of fluid from that compartment:

$$\text{Volume of compartment} = \frac{\text{Mass of X administered}}{\text{Concentration of X in compartment}} \quad (1)$$

Radioactive markers or markers whose concentrations can be assayed colorimetrically are generally used. A more precise application of the dilution method includes a correction for the amount of the marker lost (e.g., in urine) during the period of distribution:

$$\text{Volume of compartment} = \frac{\text{Mass of X administered} - \text{mass of X lost}}{\text{Concentration of X in compartment}} \quad (2)$$

Dilution methods can be used to estimate the volumes of TBW, ECF, and plasma (Table 4.4).

Table 4.2.
Approximate Values for Total Body Water in Normal Humans as Percentage of Body Weight

Age, yr	Male, %	Female, %
Newborn	80	75
1–5	65	65
10–16	60	60
17–39	60	50
40–59	55	47
60+	50	45

From Hays, 1980.

Table 4.3.
Body Fluid Compartments

Compartment	Percent of Total Body Water[a]	Percent of Total Body Weight — Normal Adult Man	Percent of Total Body Weight — Normal Adult Woman
Intracellular fluid	55	33	27.5
Extracellular fluid	45	27	22.5
Interstitial	20	12	10
Plasma	7.5	4.5	3.75
Bone	7.5	4.5	3.75
Dense connective tissue	7.5	4.5	3.75
Transcellular	2.5	1.5	1.25
Total body water	100	60	50

[a]From Edelman and Leibman, 1959.

TOTAL BODY WATER

The volume of *TBW* is estimated using markers that distribute uniformly throughout *all* body fluids, such as *deuterated water* (D_2O) or *tritiated water* (HTO). The drug *antipyrine* also can be used, although it penetrates certain parts of the body water slowly and therefore tends to *underestimate* the TBW. Since plasma is part of the TBW, the concentration of the marker in the TBW compartment (for *Eqs. 1* and *2*) can be obtained from a *plasma sample*.

EXTRACELLULAR FLUID

The estimation of *ECF* volume with dilution methods requires markers that can freely cross the capillary endothelium but that are predominantly excluded from cells. Unfortunately, no ideal marker substance for ECF is available (Edelman and Leibman, 1959). *Radioisotopes of ions* such as Na^+, Cl^-, Br^-, SO_4^{2-}, and $S_2O_3^{2-}$ (thiosulfate) can be used, but these enter cells to variable extents and therefore tend to *overestimate* the ECF volume. *Nonmetabolizable saccharides* such as inulin, mannitol, and raffinose also can be used, but these do not readily distribute throughout the entire extracellular com-

Table 4.4.
Marker Substances Used to Measure Volumes of Body Fluid Compartments

Compartment	Marker Substance
TBW	D_2O HTO Antipyrine
ECF	Radioisotopes of selected ions (Na^+, Cl^-, Br^-, SO_4^{2-}, $S_2O_3^{2-}$) Nonmetabolizable saccharides (inulin, mannitol, raffinose)
Plasma	^{131}I-Albumin Evans blue dye (T-1824)

partment and therefore tend to *underestimate* the ECF volume. Furthermore, none of the ECF markers distributes into *bone water,* and many are excluded from, or penetrate very slowly, *dense connective tissue fluid* and *transcellular fluid.* Dilution methods therefore give ECF volumes that range from approximately 27% of the TBW (for markers such as inulin that distribute primarily into plasma and interstitial fluid) to as much as 45% of the TBW (for markers such as Na^+ that, while not distributing into bone water, enter cells to some extent). Because such different volumes are obtained with different markers, the approximation that ECF contains about one-third of TBW is widely used for clinical purposes, as indicated above. Like the TBW, ECF includes plasma, so that the concentration of a marker in ECF can be obtained from a *plasma sample.*

PLASMA

The most commonly used markers for *plasma* volume take advantage of the fact that plasma proteins are distributed almost exclusively in the vascular compartment. Thus, the volume of plasma can be estimated using *radioisotopes of albumin* (e.g., ^{131}I-albumin) or *Evans blue dye* (also called T-1824), which binds tightly to albumin (the small amount of albumin that enters interstitial fluid generally can be neglected in dilution studies). A somewhat different method for estimating plasma volume involves the use of labeled erythrocytes (e.g., ^{51}Cr-erythrocytes) to determine the *blood volume* by dilution. The *plasma volume* can then be calculated as follows:

$$\text{Plasma volume} = \text{Blood volume} (1 - \text{Hct}) \qquad (3)$$

where Hct is the *hematocrit.* Since the hematocrit measured from a peripheral vein slightly *overestimates* the actual hematocrit (primarily because the hematocrit in small capillaries is slightly less than that in larger vessels; see Chapter 7), the labeled erythrocyte method slightly *underestimates* the plasma volume.

ICF AND INTERSTITIAL FLUID

Dilution methods cannot be used to measure the volumes of ICF and interstitial fluid, since no markers that distribute exclusively in these compartments are available. However, once the volumes of the TBW and ECF have been estimated by dilution, the volume of *ICF* can be calculated as follows:

$$\text{ICF} = \text{TBW} - \text{ECF} \qquad (4)$$

Because different values for ECF volume are obtained with different markers, as indicated above,

the ICF volumes calculated using *Eq. 4* also will vary. The approximation that ICF contains about two-thirds of TBW is therefore widely used for clinical purposes (see above).

The volume of *interstitial fluid* can be calculated from the ECF and plasma volumes:

$$\text{Interstitial fluid} = \text{ECF} - \text{plasma} \qquad (5)$$

In general, the volume calculated using *Eq. 5* slightly *overestimates* the true interstitial fluid volume, since ECF as estimated by dilution methods includes not only interstitial fluid and plasma but also variable amounts of dense connective tissue fluid and transcellular fluid (see above).

COMPOSITION OF THE BODY FLUIDS

Units for Measuring Solute Concentrations

SI UNITS

While solute concentrations can be expressed in several different units, the *Système International (SI) Units* will be emphasized throughout this section. In SI units, concentrations are expressed in *moles per liter* (mol/liter), where a mole of a solute is defined as the molecular weight (or atomic weight) of the solute in grams. For example, 1 mol of Na^+ (atomic weight 23) contains 23 g of Na^+, 1 mol of $CaCl_2$ (molecular weight 111) contains 111 g of $CaCl_2$, 1 mol of glucose (molecular weight 180) contains 180 g of glucose, and 1 mol of albumin (molecular weight 69,000) contains 69,000 g of albumin. Since the body fluids are relatively dilute, the units *millimoles per liter* (mmol/liter) generally are more useful than mol/liter, where each mmol represents 10^{-3} mol (1/1000 mol).

In spite of the preference given to SI units in this section, some concepts in renal physiology, particularly those presented in this chapter regarding the body fluids, require a consideration of the number of *electrical charges* per unit volume or the number of *discrete solute particles* per unit volume. For this reason, units that express solute concentrations in terms of *electrical equivalents* or *osmoles* also will be used in certain circumstances.

ELECTRICAL EQUIVALENTS

For electrolytes, concentrations often are expressed in *equivalents per liter* (eq/liter). A mole of solute with valence v is equal to v equivalents of solute. For example, 1 mol of Na^+ (valence 1) is equal to 1 eq of Na^+, 1 mol of $CaCl_2$ (total valence 4: $Ca^{2+} = 2$, $2Cl^- = 2$) is equal to 4 eq of $CaCl_2$, and 1 mol of albumin (valence 18 at pH 7.4) is equal to 18 eq of albumin. Since the body fluids are relatively dilute, the units meq/liter are generally used, where each meq equals 10^{-3} eq. By the above definition of an equivalent, solute concentrations expressed in mmol/liter can be converted to meq/liter as follows:

$$meq/liter = mmol/liter \times v \qquad (6)$$

where v is the valence.

OSMOLES

As discussed in Chapter 2, the movement of water between different body fluid compartments is related to the concentration of discrete solute particles or *osmotically active particles*, regardless of their size or valence. Thus, for many problems involving the movement of water, solute concentrations are best expressed in *osmoles per liter* (osmol/liter) or *osmoles per kilogram water* (osmol/kg H_2O). A mole of solute that dissociates into n discrete particles in solution is equal to n osmoles of solute. For example, 1 mol of Na^+ is equal to 1 osmol of Na^+, 1 mol of $CaCl_2$ is equal to 3 osmol of $CaCl_2$ (since $CaCl_2$ dissociates into three discrete solute particles), 1 mol of glucose is equal to 1 osmol of glucose, and 1 mol of albumin is equal to 1 osmol of albumin. The units of osmol/liter and osmol/kg H_2O are respectively termed *osmolarity* and *osmolality*. In most physiological applications (and in this section), *osmolality* is the preferred unit since it is independent of the temperature of the solution and the volume occupied by the solutes in the solution. Again, since the body fluids are relatively dilute, the units mosmol/kg H_2O generally are used, where each mosmol equals 10^{-3} osmol. By the above definition of an osmole, solute concentrations expressed in mmol/liter can be converted to mosmol/kg H_2O as follows:

$$mosmol/kg \ H_2O = mmol/liter \times n \qquad (7)$$

where n is the number of discrete particles into which the solute dissociates. It should be noted that n cannot always be determined by examining the chemical formula of the solute. For example, in the body fluids n for NaCl actually equals 1.75 instead of 2 because NaCl does *not* completely dissociate. It also should be noted that *Eq. 7* will be precise only if the units mmol/liter refer to mmol/liter of *water* instead of mmol/liter of *solution*, since 1 liter of *water* does in fact equal 1 kg of water but 1 liter of *solution* may contain less than 1 kg of water if the solutes in the solution occupy a significant fraction of the solution volume (see below).

An alternative way of expressing the concentration of osmotically active solute particles in a solution is in terms of the *osmotic pressure* of the solution. At body temperature, the osmotic pressure

and osmolality are related as follows:

Osmotic pressure (mm Hg)
$$= 19.3 \times osmolality \ (mosmol/kg \ H_2O) \qquad (8)$$

A comparison of the various units for measuring solute concentrations is presented in Table 4.5.

Although the body fluids contain a large variety of solutes, the *electrolytes* will be emphasized in the following discussion of body fluid composition. This is because electrolytes are the predominant solutes in the body fluids and also are primarily responsible for determining the distribution of water among the various compartments.

Extracellular Fluid

PLASMA AND INTERSTITIAL FLUID

These two major compartments of the ECF have very similar compositions (Fig. 4.1), with Na^+ as the predominant cation and Cl^- and HCO_3^- as the predominant anions. However, an important difference between plasma and interstitial fluid is the larger concentration of *proteins* in plasma. This difference exists because the capillary endothelium is freely permeable to water and to small solutes (the so-called *crystalloids*), such as inorganic ions, glucose, and urea, but has limited permeability to larger solutes (*colloidal particles)*, such as large proteins and lipids (Chapters 2 and 27).

Given the high permeability of the capillary endothelium to small solutes, one might anticipate that the concentrations of such solutes in plasma and interstitial fluid would be *identical*. However, plasma concentrations typically are measured in clinical laboratories as meq/liter of *plasma volume* or mmol/liter of *plasma volume*. To compare plasma and interstitial fluid concentrations, one must *correct* the plasma concentrations to account for the significant fraction of the plasma volume occupied by proteins and lipids. The small solutes that can diffuse across the capillary endothelium are restricted to the *aqueous phase* of plasma, which

Table 4.5.
Comparison of Units for Measuring Solute Concentrations

Solute	g/mol	eq/mol	osmol/mol[a]
Na^+	23	1	1
Cl^-	35.5	1	1
NaCl	58.5	2[b]	2
Ca^{2+}	40	2	1
$CaCl_2$	111	4[c]	3
Glucose	180		1
Urea	60		1
Albumin	69,000	18[d]	1

[a]Values for osmoles per mole assume 100% dissociation and the properties of an ideal solution. [b]$Na^+ = 1$, $Cl^- = 1$. [c]$Ca^{2+} = 2$, $2Cl^- = 2$. [d]At pH 7.4.

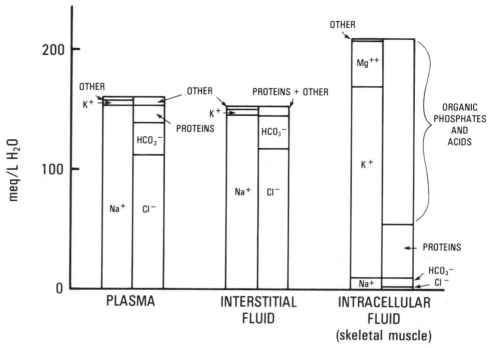

Figure 4.1. Electrolyte composition of the major body fluids.

normally occupies about 93% of the total plasma volume; the remaining 7% is occupied by plasma proteins and, to a lesser degree, lipids. Thus, plasma concentrations must be expressed as meq/liter of *plasma water* or mmol/liter of *plasma water* to be compared to interstitial fluid concentrations. Laboratory values expressed as meq/liter or mmol/liter of *plasma volume* can be converted to meq/liter or mmol/liter of *plasma water* by dividing by 0.93 (Table 4.6). Once such conversions are made, the plasma and interstitial fluid concentrations of small *nonelectrolytes* are in fact identical, but small *electrolytes* have slightly different concentrations in plasma and interstitial fluid. These differences can be attributed primarily to the *Gibbs-Donnan effect.*

Gibbs-Donnan Effect

Consider the hypothetical situation of equal size plasma and interstitial fluid compartments separated by a capillary endothelium barrier (Fig. 4.2). Initially, five Na^+ ions and five Cl^- ions are present in each compartment (Fig. 4.2*A*), and then a protein molecule that has five negative charges is added to the plasma compartment along with five additional Na^+ ions to function as counterions to the negative charges on the protein (Fig. 4.2*B*) (it is assumed that the volume of the protein molecule is small and

Table 4.6.
Approximate Concentrations of Solutes in Body Fluids[a]

	Plasma, meq/liter	Plasma Water, meq/liter H_2O	Interstitial Fluid, meq/liter H_2O	Intracellular Fluid (Skeletal Muscle), meq/liter H_2O
Cations				
Na^+	142	153	145	10
K^+	4	4.3	4.1	159
Ca^{2+b}	2.5	2.7	2.4	<1
Mg^{2+b}	1	1.1	1	40
Total	149.5	161.1	152.5	209
Anions				
Cl^-	104	112	117	3
HCO_3^-	24	25.8	27.1	7
Proteins	14	15.1	<0.1	45
Other	7.5	8.2	8.4	154
Total	149.5	161.1	152.5	209
	mmol/liter	mmol/liter H_2O	mmol/liter H_2O	mmol/liter H_2O
Nonelectrolytes				
Glucose	4.7	5.0	5.0	
Urea	5.6	6.0	6.0	6.0

[a]The plasma water, interstitial fluid, and intracellular fluid concentrations are expressed as meq/liter of H_2O. While proteins and protein-bound ions are strictly not part of the aqueous phase of these fluids, their charges have an important role in understanding body fluid composition (see text). Thus, the table presents the compositions of these fluids as if proteins and protein-bound ions were in fact part of the aqueous phase. [b]*Total* concentrations are given; free ionized concentrations are lower (Chapter 34).

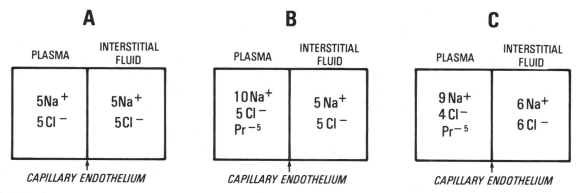

A

PLASMA | INTERSTITIAL FLUID

5 Na$^+$
5 Cl$^-$

5 Na$^+$
5 Cl$^-$

CAPILLARY ENDOTHELIUM

B

PLASMA | INTERSTITIAL FLUID

10 Na$^+$
5 Cl$^-$
Pr^{-5}

5 Na$^+$
5 Cl$^-$

CAPILLARY ENDOTHELIUM

C

PLASMA | INTERSTITIAL FLUID

9 Na$^+$
4 Cl$^-$
Pr^{-5}

6 Na$^+$
6 Cl$^-$

CAPILLARY ENDOTHELIUM

Figure 4.2. The Gibbs-Donnan effect in hypothetical plasma and interstitial fluid compartments. The capillary barrier is relatively impermeable to proteins and other large solutes. *A,* Initial condition of five Na$^+$ ions and five Cl$^-$ ions in each compartment. *B,* A protein molecule with five negative charges is added to the plasma compartment, with five additional Na$^+$ to function as counterions. *C,* Equilibrium state is characterized by (1) a higher concentration of diffusible cations (Na$^+$) in the plasma compartment and a higher concentration of diffusible anions (Cl$^-$) in the interstitial fluid compartment; (2) a greater total concentration of equivalents of charge in the plasma compartment; and (3) electrical neutrality within each compartment.

therefore does not significantly affect the volume of the aqueous phase of the plasma compartment). Since the capillary endothelium is permeable to Na$^+$, Na$^+$ diffuses from the plasma compartment to the interstitial fluid compartment along its concentration gradient, accompanied by Cl$^-$ to maintain electrical neutrality. However, this migration will increase the concentration of Cl$^-$ in the interstitial fluid compartment, thereby generating a concentration gradient that *opposes* the further diffusion of Cl$^-$ from the plasma compartment to the interstitial fluid compartment. According to thermodynamic principles that are beyond the scope of this text, at equilibrium the distribution of Na$^+$ and Cl$^-$ between the plasma and interstitial fluid compartments is given by the *Gibbs-Donnan relationship:*

$$P_{Na} \cdot P_{Cl} = ISF_{Na} \cdot ISF_{Cl} \qquad (9)$$

where P and ISF designate the concentrations of the ions in the plasma and interstitial fluid compartments, respectively. If x Na$^+$ ions and x Cl$^-$ ions migrate from the plasma compartment to the interstitial fluid compartment before equilibrium is established, then at equilibrium $(10 - x)(5 - x) = (5 + x)(5 + x)$ so that $x = 1$, i.e., at equilibrium nine Na$^+$ ions, four Cl$^-$ ions, and the protein molecule with five negative charges will be present in the plasma compartment, while six Na$^+$ ions and six Cl$^-$ ions will be present in the interstitial fluid compartment (Fig. 4.2*C*). In other words, because of the anionic protein in the plasma compartment, the equilibrium state between the two compartments is characterized by three important features: (1) the small diffusible ions (Na$^+$, Cl$^-$) do *not* have equal concentrations in the two compartments, the *cation*

(Na$^+$) concentration being slightly higher in the protein-containing *plasma compartment* and the *anion* (Cl$^-$) concentration being slightly higher in the *interstitial fluid compartment;* (2) the *total concentration* of equivalents of charge is greater in the protein-containing *plasma compartment;* and (3) in spite of the differences in ion concentrations and in the total concentration of equivalents of charge, *electrical neutrality* is maintained within each compartment, i.e., the total number of cationic charges equals the total number of anionic charges. Note that these same three features characterize the composition differences between plasma and interstitial fluid (Fig. 4.1, Table 4.6), although the situation is more complex than the example illustrated in Figure 4.2 because of the presence of additional small diffusible ions and the small amount of protein in interstitial fluid.

In spite of the differences in composition between plasma and interstitial fluid, these differences are sufficiently small that for most clinical purposes the electrolyte concentrations in plasma (which are readily measured) can be assumed to be representative of those in the ECF in general, and a similar assumption will be made in subsequent chapters. In addition, while the distinction between plasma concentrations expressed in terms of *plasma volume* and those expressed in terms of *plasma water* is important for a precise comparison of plasma and interstitial fluid compositions and for a consideration of the Gibbs-Donnan effect, in general the concentrations as measured by clinical laboratories (i.e., meq/liter or mmol/liter of *plasma volume*) are used in clinical medicine and will be used in subsequent chapters. However, in certain pathological conditions the failure to express solute concentrations in terms of plasma water can lead to the erroneous interpretation of laboratory data (Albrink et al., 1955). For example, consider the effects of severe hyperproteinemia (e.g., multiple myeloma) or hyperlipidemia (e.g., familial hyperlipidemia) on the plasma Na^+ concentration. In these conditions, water may represent considerably *less* than 93% of the plasma volume. As a result, even if the Na^+ concentration in *plasma water* is normal ($\simeq 153$ meq/liter of plasma water), the Na^+ concentration as measured in the clinical laboratory (meq/liter of plasma volume) may be reduced since each liter of plasma contains less water.

OTHER EXTRACELLULAR FLUID COMPARTMENTS

The electrolyte composition of the extracellular fluid in *bone* and *dense connective tissue* is believed to be similar to that in interstitial fluid. The various *transcellular fluids* have unique (and often varying) compositions that are considered in the sections dealing with the corresponding organ systems.

Intracellular Fluid

In contrast to plasma and interstitial fluid, ICF is not a continuous fluid phase, and the precise composition of ICF differs in different tissues. The data presented in Figure 4.1 and Table 4.6 are from mammalian skeletal muscle cells, but several characteristics of ICF in general can be identified (Manery and Hastings, 1939). In contrast to ECF, ICF contains relatively low concentrations of Na^+, Cl^-, and HCO_3^-. Instead, the predominant cation in ICF is K^+, while the predominant anions are *organic phosphates* (e.g., ATP) and *proteins* (Fig. 4.1). These striking composition differences between ICF and ECF can be attributed to several factors. First, the Na^+/K^+-*ATPase* in cell membranes actively transports Na^+ from and K^+ into cells, thereby accounting for the high Na^+ and low K^+ concentrations in ECF and the low Na^+ and high K^+ concentrations in ICF. Second, the cell membrane, which separates ICF from interstitial fluid, has very limited permeability to organic phosphates and proteins, resulting in the establishment of a *Gibbs-Donnan equilibrium* across the cell membrane. In analogy to the example presented in Figure 4.2, this Gibbs-Donnan equilibrium accounts for three important features that characterize the composition differences between ICF and interstitial fluid: (1) the concentration of small diffusible *cations* (Na^+, K^+) is greater in the protein-containing *ICF*, while the concentration of small diffusible *anions* (e.g., Cl^-) is greater in *interstitial fluid* (the distribution of cations is *not* governed solely by the Gibbs-Donnan equilibrium, however, because of the presence of the Na^+/

K^+-ATPase); (2) the *total concentration* of equivalents of charge is greater in the protein-containing *ICF*; and (3) in spite of the differences in ion concentrations and in the total concentration of equivalents of charge, *electrical neutrality* is maintained in each compartment. Of course, electrical potential differences exist between the ICF and interstitial fluid (Chapter 3), but these are created by excesses or deficits of such small numbers of ions (relative to the total number present) that a charge imbalance cannot be detected by measurements of electrolyte concentrations.

The ICF concentration values for a given tissue (such as those presented in Figure 4.1 and Table 4.6 for skeletal muscle cells) must be regarded as *mean* values for that tissue. This is because the various subcellular organelles (mitochondria, endoplasmic reticulum, nucleus, etc.) appear to differ somewhat in composition.

Measurement of Total Electrolyte Content of Body Fluids

Important information about the total content of a given ion in the body fluids can be obtained with dilution methods similar to those used to determine the volume of the body fluids, with a radioisotope of the ion employed as a marker substance. The total content of an ion as determined by such *isotope dilution* methods is termed the *exchangeable pool* of the ion in question (Edelman and Leibman, 1959).

For Na^+, K^+, and Cl^-, the major ions in the body fluids, the exchangeable pool represents the majority of the total body contents of these ions as determined by analyses of cadavers. For example, *exchangeable* Na^+ (Na_e^+) represents over 70% of the total body Na^+, while *exchangeable* K^+ (K_e^+) represents over 90% of the total body K^+. Most of the *nonexchangeable* pools of Na^+ and K^+ are found in *bone*. In contrast, a relatively small percentage of the total body contents of Ca^{2+}, Mg^{2+}, and phosphate is exchangeable. For example, *exchangeable* Ca^{2+} (Ca_e^{2+}) represents only about 1% of the total body Ca^{2+}, the remainder being part of the structure of bone (Chapter 34).

OSMOLALITY OF THE BODY FLUIDS

Many clinical problems involving fluid and electrolyte balance require an understanding of the osmolality of the body fluids and how this affects the distribution of water between the extracellular and intracellular body fluid compartments. The contributions of the various solutes to the osmolalities of plasma, interstitial fluid, and ICF are illustrated in Figure 4.3. Note that in spite of the differences in

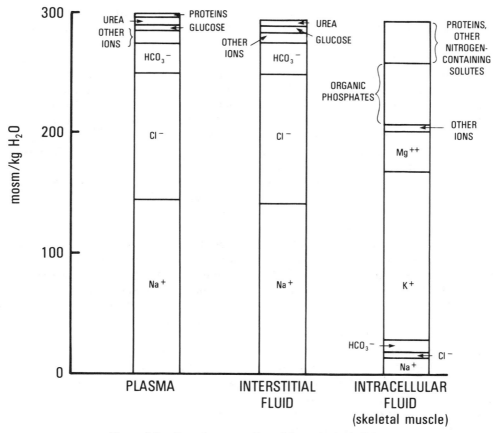

Figure 4.3. Osmotic composition of the major body fluids.

composition, these fluids have essentially *identical* total osmolalities ($\simeq$290 mosmol/kg H_2O). This is because the capillary endothelium and almost all cell membranes are freely permeable to water, allowing the plasma, interstitial fluid, and ICF to be isoosmotic. Hence, the primary determinant of the distribution of water between the extracellular and intracellular fluid compartments is the number of *osmotically active solute particles* in each compartment. The isoosmolality of the body fluids is *not* in conflict with the fact that plasma, interstitial fluid, and ICF contain different total concentrations of equivalents of charge (Fig. 4.1) because osmolality is a measure of the total concentration of discrete solute *particles* in solution. A divalent or polyvalent ion (e.g., Ca^{2+}, Mg^{2+}, SO_4^{2-}, albumin) may contribute two or more equivalents of charge, even though it represents a *single* solute particle.

It should be noted that the isoosmolality principle applies primarily to the *major* body fluids. A few minor volumes of fluid have osmolalities that differ significantly from 290 mosmol/kg H_2O, including the peritubular interstitial fluid in the *renal medulla* (which can have an osmolality as high as 1200 mosmol/kg H_2O; see Chapter 30) and certain *transcell-*

ular fluids (the most striking example being *urine*, whose osmolality can vary from 70 to 1200 mosmol/kg H_2O; see Chapter 30). It also should be noted that although Figure 4.3 illustrates the osmolalities of the major body fluids as identical, the osmolality of plasma actually exceeds that of interstitial fluid by approximately 1–2 mosmol/kg H_2O. This small osmolality difference can be attributed both to the osmotic contribution of the plasma proteins and to the larger concentration of diffusible ions in plasma as a result of the Gibbs-Donnan effect (Fig. 4.2 and Table 4.6). The deviation between the osmolality of plasma and that of interstitial fluid is so small that the major body fluids can be considered *isoosmotic* for most clinical problems involving the distribution of water between ECF and ICF. However, this deviation *does* have an important role in problems involving the distribution of water between plasma and interstitial fluid (Chapters 2 and 27). In such problems, the osmotic contribution of the plasma proteins and excess diffusible ions in plasma typically is expressed as an *osmotic pressure*, termed the *colloid osmotic pressure* or the *oncotic pressure* and designated by the symbol π. In normal plasma, the oncotic pressure is approximately 25 mm Hg, of which about two-thirds is caused by the plasma proteins and one-third by the excess diffusible ions. Problems involving the distribution of water between plasma and interstitial fluid also require a consideration of the oncotic pressure resulting from the small amount of protein in *interstitial fluid,* normally about 3 mm Hg. The relatively small magnitudes of these oncotic pressures is best illustrated by the fact that the *total* osmolality of plasma corresponds to an osmotic pressure of approximately 5600 mm Hg *(Eq. 8).* An osmolality difference does *not* exist between interstitial fluid and ICF in spite of the osmotic contribution of the intracellular proteins and the Gibbs-Donnan effect between interstitial fluid and ICF (see above). This is because the Na^+/K^+-ATPase actively transports *three* Na^+ ions from ICF to interstitial fluid for every *two* K^+ ions transported from interstitial fluid to ICF, thereby preventing the osmolality of ICF from exceeding that of interstitial fluid.

Analysis of Plasma Osmolality and Relationship of Plasma Sodium Concentration to Osmolality

An important consequence of the isoosmolality principle is that the osmolality of the major body fluids can be studied by analyzing the osmolality of *plasma*. The plasma osmolality can be estimated as a sum of contributions from *electrolytes* plus contributions from *glucose* and *urea:*

$$P_{osm} \simeq \text{Osmolality of electrolytes}$$
$$+ \text{osmolality of glucose} + \text{osmolality of urea} \quad (10)$$

ELECTROLYTE CONTRIBUTION

Since Na^+ and its attendant anions represent the major electrolytes in plasma, it can be assumed that

$$\text{Osmolality of electrolytes} \simeq \text{Osmolality of } (Na^+$$
$$+ \text{attendant anions)} \quad (11)$$

If it is further assumed that each Na^+ ion is paired with a *univalent* anion, then

$$\text{Osmolality of electrolytes (mosmol/kg } H_2O)$$
$$\simeq 2P_{Na} \text{ (mmol/liter)} \quad (12)$$

The assumptions made deriving *Eq. 12* actually are less serious than might be expected, since they introduce opposing errors of essentially equal magnitude. First of all, the assumption that Na^+ and its attendant anions are the only electrolytes *(Eq. 11)* tends to *underestimate* the true osmotic contribution of the electrolytes, since it excludes other cations such as K^+, Ca^{2+}, and Mg^{2+}. But the assumption that each Na^+ ion is paired with a univalent anion

(Eq. 12) tends to *overestimate* the true osmotic contribution of the electrolytes, since (1) Na^+ and its anions are *not* completely dissociated in the body fluids (even NaCl, for example, dissociates into only 1.75 particles, not 2 particles; see above) and (2) some Na^+ is paired with multivalent anions such as SO_4^{2-}, HPO_4^{2-}, and protein, resulting in a smaller number of solute particles than if all of the anions of Na^+ were univalent. Finally, *Eq. 12* tends to *underestimate* the true electrolyte contribution, since P_{Na} is typically measured by clinical laboratories as meq/liter or mmol/liter of plasma *volume* instead of meq/liter or mmol/liter of plasma *water* as would be required to convert P_{Na} *precisely* to mosmol/kg H_2O *(Eq. 7).* As previously noted, concentrations expressed as mmol/liter of plasma *volume* are *less* than those expressed as mmol/liter of plasma *water* (Table 4.6). Fortuitously, when these various errors are corrected, the osmolality of the electrolyte contribution is in fact closely approximated by $2P_{Na}$ *(Eq. 12).*

GLUCOSE AND UREA CONTRIBUTIONS

The contributions of glucose and urea to P_{osm} can be expressed as

$$\text{Osmolality of glucose (mosmol/kg } H_2O)$$
$$\simeq P_G \text{ (mmol/liter)} \quad (13)$$

and

$$\text{Osmolality of urea (mosmol/kg } H_2O) \simeq P_U \text{ (mmol/liter)} \quad (14)$$

where the subscripts G and U refer to glucose and urea, respectively. If the plasma concentrations are expressed in terms of mmol/liter of plasma *water* instead of mmol/liter of plasma *volume, Eqs. 13* and *14* become precise. Actually, many clinical laboratories measure plasma glucose and urea concentrations in the units milligrams of glucose per deciliter and milligrams of urea nitrogen per deciliter, respectively. Since the molecular weight of glucose is 180, the molecular weight of urea nitrogen is 28 (i.e., 28 g of the 60 g in 1 mol of urea consist of nitrogen), and 1 liter = 10 dl, the osmotic contributions of glucose and urea can be expressed as

$$\text{Osmolality of glucose (mosmol/kg } H_2O) \simeq P_G/18 \text{ (mg/dl)} \quad (15)$$

and

$$\text{Osmolality of urea (mosmol/kg } H_2O) \simeq P_{UN}/2.8 \text{ (mg/dl)} \quad (16)$$

where the subscript UN refers to urea nitrogen.

When the electrolyte, glucose, and urea contributions are substituted into *Eq. 10,* the following equations for P_{osm} are obtained:

$$P_{osm} \text{ (mosmol/kg } H_2O) \simeq 2P_{Na} \text{ (mmol/liter)}$$
$$+ P_G \text{ (mmol/liter)} + P_U \text{ (mmol/liter)} \quad (17)$$

and

$$P_{osm} \text{ (mosmol/kg } H_2O) \simeq 2P_{Na} \text{ (mmol/liter)}$$
$$+ P_G/18 \text{ (mg/dl)} + P_{UN}/2.8 \text{ (mg/dl)} \quad (18)$$

Since the glucose and urea contributions normally account for only about 10 mosmol/kg H_2O, *Eqs. 17* and *18* can be further simplified to

$$P_{osm}(\text{mosmol/kg } H_2O) \simeq 2P_{Na} \text{ (mmol/liter)} + 10 \quad (19)$$

i.e., in most cases a good estimate of P_{osm} can be obtained from P_{Na}, an observation consistent with the fact that Na^+ and its attendant anions are the principal osmotically active solutes in plasma (Fig. 4.3).

It must be kept in mind, however, that while P_{Na} can be a good index of P_{osm}, P_{Na} is *not* a good index of either the total amount of osmotically active solute in the body or the TBW (Edelman et al., 1958). This is best illustrated by a set of examples involving changes in body solute or TBW. Figure 4.4*A* illustrates the volumes and osmolality of the ECF and ICF in a normal young adult 60-kg man. TBW is 36 liters (60% of body weight) and, using the clinical approximation that ECF contains about one-third of TBW and ICF contains about two-thirds of TBW (see above), ECF and ICF are 12 and 24 liters, respectively. If P_{osm}, and hence the osmolality of the major body fluids, is 290 mosmol/kg H_2O, then the amount of osmotically active solute in the TBW, ECF, and ICF can be calculated as follows:

$$TBW_{osm} = \frac{\text{Total body osmotically active solute}}{\text{TBW}} \quad (20)$$

Total body osmotically active solute
$$= TBW_{osm} \times TBW$$
$$= 290 \text{ mosmol/kg } H_2O \times 36 \text{ liters}$$
$$= 10{,}440 \text{ mosmol}$$

$$ECF_{osm} = \frac{\text{Extracellular osmotically active solute}}{\text{ECF}} \quad (21)$$

Extracellular osmotically active solute
$$= ECF_{osm} \times ECF$$
$$= 290 \text{ mosmol/kg } H_2O \times 12 \text{ liters}$$
$$= 3480 \text{ mosmol}$$

$$ICF_{osm} = \frac{\text{Intracellular osmotically active solute}}{\text{ICF}} \quad (22)$$

Intracellular osmotically active solute
$$= ICF_{osm} \times ICF$$
$$= 290 \text{ mosmol/kg } H_2O \times 24 \text{ liters}$$
$$= 6960 \text{ mosmol}$$

Figure 4.4*B* illustrates the changes in the volumes and osmolality of ECF and ICF that would result from the sudden loss of 360 mosmol of Na^+ and 360 mosmol of anions *without* changing the TBW. Since Na^+ is predominantly an *extracellular* ion, it can be assumed that this 720 mosmol of solute is lost exclusively from the ECF. The sudden loss of solute from the ECF without any change in TBW reduces the osmolality of the ECF. However, this decrease in ECF osmolality can be only *transient,* since water will move from ECF to ICF until the osmolalities of all major body fluids are equal. After this redistribution of water is complete, the new osmolality of the body fluids can be calculated by substituting the *new* value for total body osmotically active solute into *Eq. 20:*

$$TBW_{osm} = \frac{10{,}440 \text{ mosmol} - 720 \text{ mosmol}}{36 \text{ liters}}$$
$$= 270 \text{ mosmol/kg } H_2O$$

i.e., the net result is a decrease in the osmolality of *all* major body fluids, even though the solute is lost primarily from the ECF. The new volume of the ECF after the redistribution of water can be calculated by substituting the *new* values for extracellular osmotically active solute and osmolality into *Eq. 21:*

$$ECF = \frac{3480 \text{ mosmol} - 720 \text{ mosmol}}{270 \text{ mosmol/kg } H_2O}$$
$$= 10.2 \text{ liters}$$

i.e., the loss of 720 mosmol of Na^+ and attendant anions from the ECF results in the shift of approximately 1.8 liters of the original 12 liters of ECF to the ICF. While this example may seem unrealistic, a similar decrease in P_{osm} and shift of water from ECF to ICF can occur in patients who lose large quantities of Na^+ and water (e.g., as a result of copious diarrhea) and then replace the volume lost by drinking water.

Figure 4.4*C* shows the changes in the volumes and osmolality of the ECF and ICF that would result from the sudden loss of 360 mosmol of K^+ and 360 mosmol of anions, again without changing the TBW. Since K^+ is primarily an *intracellular* ion, it can be assumed that this 720 mosmol of solute is lost exclusively from the ICF. The sudden loss of solute from the ICF without any change in TBW reduces the osmolality of the ICF. However, this decrease in ICF osmolality can only be *transient,* since water will move from ICF to ECF until the osmolalities of all major body fluids are equal. After this redistribution of water is complete, the new osmolality of the body fluids can be calculated by substituting the

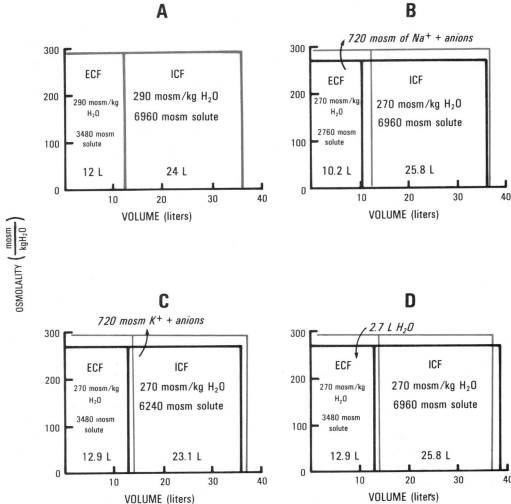

Figure 4.4. Effects of changes in body solute or TBW on osmolality of body fluids and on distribution of water between ECF and ICF in a 60-kg young adult man. *A,* Normal state, with a body fluid osmolality of 290 mosmol/kg H_2O and a TBW of 36 liters (=60% of total body weight), of which about one-third is in ECF (12 liters) and two-thirds is in ICF (24 liters). *B,* Following the sudden loss of 360 mosmol of Na^+ and 360 mosmol of anions from the ECF, with no change in TBW. *C,*

Following the sudden loss of 360 mosmol of K^+ and 360 mosmol of anions from the ICF, with no change in TBW. *D,* Following the sudden addition of 2.7 liters of distilled water to the ECF with no change in total body osmotically active solute. In *B, C,* and *D,* the original normal osmolality and distribution of water (from *A*) is outlined in *blue.* Additional details are provided in the text.

new value for total body osmotically active solute into *Eq. 20:*

$$TBW_{osm} = \frac{10,440 \text{ mosmol} - 720 \text{ mosmol}}{36 \text{ liters}}$$
$$= 270 \text{ mosmol/kg } H_2O$$

i.e., the net result is a decrease in the osmolality of *all* major body fluids, even though solute is lost primarily from the ICF. Note that the decrease in osmolality is the *same* as that resulting from the loss of 720 mosmol of solute from the ECF (Fig. 4.4*B*), a result that could be anticipated from *Eq. 20.* The new volume of the ICF after the redistribution of water can be calculated by substituting

the new values for intracellular osmotically active solute and osmolality into *Eq. 22:*

$$ICF = \frac{6960 \text{ mosmol} - 720 \text{ mosmol}}{270 \text{ mosmol/kg } H_2O}$$
$$= 23.1 \text{ liters}$$

i.e., the loss of 720 mosmol of K^+ and attendant anions from the ICF results in the shift of approximately 0.9 liter of the original 24 liters of ICF into the ECF.

Finally, Figure 4.4*D* illustrates the changes in the volumes and osmolality of the ECF and ICF that would result from the sudden addition of 2.7 liters of distilled water to the ECF with no change in total body osmotically active solute. This addition of water reduces the osmolality of the ECF. However, this decrease in ECF osmolality can only be *transient*, since water will move from ECF to ICF until the osmolalities of all major body fluids are equal. After the redistribution of water is complete (and assuming for simplicity that none of the water is excreted), the osmolality of the body fluids can be calculated by substituting the new value for TBW into *Eq. 20*:

$$TBW_{osm} = \frac{10,440 \text{ mosmol}}{38.7 \text{ liters}}$$
$$= 270 \text{ mosmol/kg } H_2O$$

The new volumes of the ECF and ICF can be calculated by substituting the new value for body fluid osmolality into *Eqs. 21* and *22* (Note that the quantity of osmotically active solute in each compartment does *not* change.)

$$ECF = \frac{3480 \text{ mosmol}}{270 \text{ mosmol/kg } H_2O}$$
$$= 12.9 \text{ liters}$$
$$ICF = \frac{6960 \text{ mosmol}}{270 \text{ mosmol/kg } H_2O}$$
$$= 25.8 \text{ liters}$$

i.e., approximately one-third of the added 2.7 liters of water distributes into the ECF, while two-thirds distributes into the ICF. Note that the added water distributes between ECF and ICF in the same proportions as the original TBW, because the number of osmotically active solute particles in each compartment does not change.

Thus, in each of the three examples of changes in body solute and water illustrated in Figure 4.4, the osmolality of the major body fluids decreases from 290 to 270 mosmol/kg H_2O. Furthermore, if it is assumed that the osmotic contribution of glucose and urea remains constant at approximately 10 mosmol/kg H_2O (see above), then P_{Na} decreases from 140 to 130 mmol/liter in each case *(Eq. 19)*. This result can also be predicted from the following considerations. Since the major body fluids are isoosmotic, *Eq. 20* can be used to calculate P_{osm} as well as TBW_{osm}:

$$P_{osm} = \frac{\text{Total body osmotically active solute}}{TBW} \quad (23)$$

The major osmotically active solutes in the body fluids are Na^+ and its attendant anions (primarily in ECF) and K^+ and its attendant anions (primarily in ICF). Furthermore, the osmotically active Na^+ and K^+ in the body fluids are closely approximated by

the *exchangeable Na^+* (Na_e^+) and *exchangeable K^+* (K_e^+), respectively.* Thus

$$P_{osm} \simeq \frac{2Na_e^+ + 2K_e^+}{TBW} \quad (24)$$

where the multiplier 2 is included to account for the osmotic contributions of the attendant anions of Na^+ and K^+ (these anions are assumed for simplicity to be *univalent;* see above). But P_{osm} also is approximated by $2P_{Na}$ if the osmotic contributions of glucose and urea are neglected *(Eq. 19)*. Substituting this approximation into *Eq. 24* gives

$$P_{Na} \simeq \frac{Na_e^+ + K_e^+}{TBW} \quad (25)$$

Thus, a decrease in body Na^+, a decrease in body K^+, *or* an increase in TBW could each result in a similar change in P_{Na}. *Eq. 25* and the examples illustrated in Figure 4.4 emphasize an important point about the use of P_{Na} as an index of P_{osm} and hence body fluid osmolality: P_{Na} does *not, by itself,* give any information about either the total amount of osmotically active solute in the body or the TBW.

Figure 4.5 illustrates that the relationship expressed by *Eq. 25* is valid over a wide range of P_{Na}

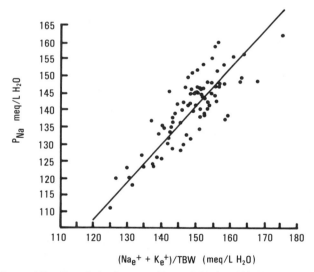

Figure 4.5. Correlation between P_{Na} and (Na_e^+ + K_e^+)/TBW, illustrating the validity of *Eq. 25*. (Reproduced from *The Journal of Clinical Investigation,* 1958, vol. 37, pp. 1236–1256, by copyright permission of The American Society for Clinical Investigation.)

---

*Actually, whereas virtually all of the K_e^+ is osmotically active, a small amount of Na_e^+ is absorbed to structures rich in mucopolysaccharides (e.g., tendons, cartilage) and is *not* osmotically active. However, the amount of this *residual Na^+* is so small that Na_e^+ represents a close approximation to the amount of osmotically active Na^+.

(Edelman et al., 1958). *Eq. 25* is particularly helpful in analyzing the changes in body solute or TBW that can occur in disease states, which can be considerably more complex than those illustrated in Figure 4.4. For example, sweating in a hot environment in the absence of water intake will decrease *both* Na_e^+ and TBW. Since the osmolality of sweat is *less* than that of plasma, however, the percentage decrease in TBW will be *greater* than the percentage decrease in Na_e^+. P_{Na} will therefore *increase* in spite of the decrease in Na_e^+ *(Eq. 25)*.

While P_{Na} can represent a good index of P_{osm} and hence the osmolality of the major body fluids, P_{Na} may *not* provide an acceptable estimate of P_{osm} in certain special circumstances:

1. In the presence of elevated plasma levels of osmotically active solutes such as glucose or urea, P_{Na}, and hence the *estimated* P_{osm}, will be *normal* when P_{osm} *actually* is *high*. For example, the osmotic contribution of glucose or urea can become so large in uncontrolled diabetes mellitus or renal failure, respectively, that P_{osm} would have to be estimated with *Eq. 17* instead of *Eq. 19*.
2. In conditions such as severe hyperproteinemia or hyperlipidemia, P_{Na}, and hence the *estimated* P_{osm}, may be *low* when P_{osm} *actually* is *normal*. This is because the laboratory values of P_{Na} typically used in *Eq. 19* are expressed as mmol/liter of plasma *volume* instead of mmol/liter of plasma *water* (see above). While the use of typical laboratory values normally does not introduce a significant error (and in fact helps to *compensate* for other approximations made in the derivation of *Eq. 19*, as previously noted), in severe hyperproteinemia or hyperlipidemia P_{Na} expressed as mmol/liter of plasma *volume* may be abnormally low, even though P_{Na} expressed as mmol/liter of plasma *water*, and hence P_{osm}, is *normal*.

Isoosmolality vs. Isotonicity

Throughout this chapter, the major body fluids have been described as *isoosmotic* because they have virtually identical osmolalities. However, when discussing osmolalities of the body fluids or the osmolalities of solutions used clinically to replace body fluids, the terms *isotonic, hypotonic,* and *hypertonic* often are used. An *isotonic solution* is one in which normal body cells (e.g., red blood cells) can be suspended without a change in cell volume occurring. In contrast, cells suspended in a *hypotonic fluid* will swell (or even rupture) due to the entry of water, while cells suspended in a *hypertonic fluid* will shrink due to the exit of water. For example, a solution containing NaCl at an osmolality of 290 mosmol/kg H_2O is an isotonic solution that is used clinically as a replacement fluid. This solution commonly is referred to as *0.9% saline* (since it contains 0.9 g of NaCl per dl) or ''normal saline.''

CHAPTER 26

Anatomy of the Kidneys

GROSS ANATOMY

The kidneys are bean-shaped organs located behind the peritoneal cavity, where they are protected from injury by a surrounding layer of fat. The two kidneys constitute about 0.5% of the total body weight, so that each kidney weighs approximately 150 g in a 60-kg man (Oliver, 1968). The concave surface of each kidney faces medially toward the vertebral column. Located in the middle of this surface is a longitudinal slit termed the *hilus*, through which a variety of structures, such as blood vessels, nerves, lymphatics, and the ureter, enter or exit the kidney (Fig. 4.6). The entire kidney is enclosed in a thin fibrous *capsule*.

Upon longitudinal section of the kidney (Asscher et al., 1982), two zones can be identified: an outer *cortex* and an inner *medulla* (Fig. 4.6). The medulla is composed of a number of pyramid-shaped structures, the *renal pyramids,* and can be divided into an *outer zone* (next to the cortex) and an *inner zone* (which forms the apexes of the pyramids, called *papillae*). The outer zone of the medulla, in turn, can be subdivided into an *outer stripe* and an *inner stripe* (Bulger, 1979). In some small animals, such as the rat or rabbit, each kidney has just one pyramid and is therefore termed *unipapillary*. In contrast, larger species such as the dog or human have *multipapillary* kidneys. Each human kidney, for example, contains 4 to 14 renal pyramids (Oliver, 1968). In multipapillary kidneys, the cortex not only forms a thick peripheral shell but also extends like a column toward the hilus between renal pyramids. These extensions of cortical tissue into the interior of the organ are the renal *columns of Bertin.*

Each papilla projects into a cup-shaped *minor calyx.* Several minor calyces join to form a *major calyx;* the major calyces unite into the funnel-shaped *renal pelvis* (Fig. 4.6). Urine continuously exits the tips of the papillae and is collected in the renal pelvis. From the renal pelvis, urine flows through the *ureter* to the *urinary bladder* for storage prior to intermittent voiding via the *urethra.* The structures from the minor calyces through the urethra commonly are termed the *urinary tract.*

The extensively regulated excretory functions of the kidneys are accomplished by individual functional units called *nephrons* (Fig. 4.7*A*). Approximately 1.0 to 1.5 million nephrons are packaged in the average human kidney (Oliver, 1968). Each nephron consists of two major structures, a *glomerular portion,* often termed *Bowman's capsule,* and a *tubule.* The tubule, in turn, can be divided into three major regions: the *proximal tubule,* the *loop of Henle,* and the *distal nephron.* Several adjacent distal nephrons share a common final segment, the *collecting duct.* In a longitudinal section of the kidney, the collecting ducts are evident as fan-like striations in the renal pyramids (Fig. 4.6).

Two basic types of nephrons can be distinguished. While all glomeruli are located in the cortex, *cortical nephrons* have glomeruli which lie in the *outer* region of the cortex, whereas *juxtamedullary nephrons* have glomeruli which lie in the *inner* cortex near the corticomedullary junction. Cortical nephrons tend to have short loops of Henle that remain exclusively in the cortex or that penetrate only the outer zone of the medulla. In contrast, juxtamedullary nephrons tend to have long loops of Henle that descend deep into the inner zone of the medulla, even as far as the tip of the papilla in some nephrons, before turning and ascending back to the cortex. However, the length of the loop of Henle cannot always be used to distinguish between cortical and juxtamedullary nephrons, as both nephron types may have short or long loops. In humans, approximately 85% of the nephrons are cortical, with the remaining 15% being juxtamedullary (Valtin, 1977).

In general terms, the function of the *glomerular portion* of the nephron is the filtration of fluid and its crystalloid constituents from plasma into the tubule, while the function of the *tubule* is to reduce the volume and modify the contents of the filtrate.

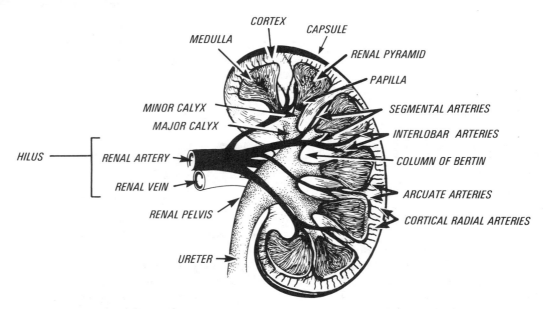

Figure 4.6.   Longitudinal section of the kidney, illustrating major anatomic features and blood vessels.

The tubule both *reabsorbs* substances from the tubular fluid and *secretes* substances into the tubular fluid, so that the final urine contains the constituents that must be excreted to preserve a normal body fluid volume and composition. The quantity of fluid and solute processed by the kidneys is *enormous:* the glomerular portions of the nephrons filter an average of 125 ml of plasma each minute or *180 liters* of plasma each day, while the tubular portions reabsorb up to 99+% of the filtered water and essential solutes.

BLOOD SUPPLY

The renal vasculature is highly specialized to bring plasma to the glomerular portions of the nephrons for filtration and then to the tubular portions to take up reabsorbed water and solutes and to deliver substances to be secreted (Figs. 4.6 and 4.7*B*). At the hilus, blood enters the kidney in the *renal artery* (in about 20% of human kidneys, more than one main renal artery is present). The main renal artery branches (Netter, 1973) a variable number of times into *segmental arteries,* which subdivide into *interlobar arteries.* Each interlobar artery penetrates the kidney through a column of Bertin, but before reaching the cortical surface, divides into arc-shaped *arcuate arteries,* which eventually run parallel to the surface. *Cortical radial arteries* (also termed *interlobular arteries*) arise from the arcuate arteries and penetrate the cortex perpendicularly toward the surface. Branching from each cortical radial artery are numerous small arterioles, the *afferent arterioles.* Each afferent arteriole leads blood toward a single nephron (the *a*fferent arteriole is *a*ttracted to the nephron).

The afferent arteriole intimately interacts with the glomerular portion of the nephron, where it breaks into a capillary network, the *glomerular capillary tuft.* About 20% of the water in plasma entering the afferent arteriole (along with ions and other crystalloids dissolved in plasma water) filters out of the glomerular capillary into the cup-shaped collection area inside Bowman's capsule, often termed *Bowman's space.* The remaining 80% of plasma, along with *all* larger solutes (e.g., proteins and lipids) and *all* cellular elements of blood, flows from the glomerular capillaries into an *efferent arteriole* (the *e*fferent arteriole *e*xits from the glomerular capillaries). The efferent arteriole disperses into a second capillary network that surrounds the tubular portions of the nephrons, delivering substances for tubules to secrete into the tubular fluid and taking up water and solutes reabsorbed by the tubules. Most of the capillaries in this second network surround the tubules in the *cortex* and are termed *peritubular capillaries* (Beeuwkes and Bonventre, 1975). However, the capillaries that are derived from the efferent arterioles of juxtamedullary nephrons descend to varying depths in the medulla (even as far as the tip of the papilla) before forming a capillary network. The capillaries reunite while turning around and ascend back to the cortex. These unique hairpin-shaped capillaries are called *vasa recta* ("straight vessels") and have an important role in the mechanism for the concentration of urine (Chapter 30).

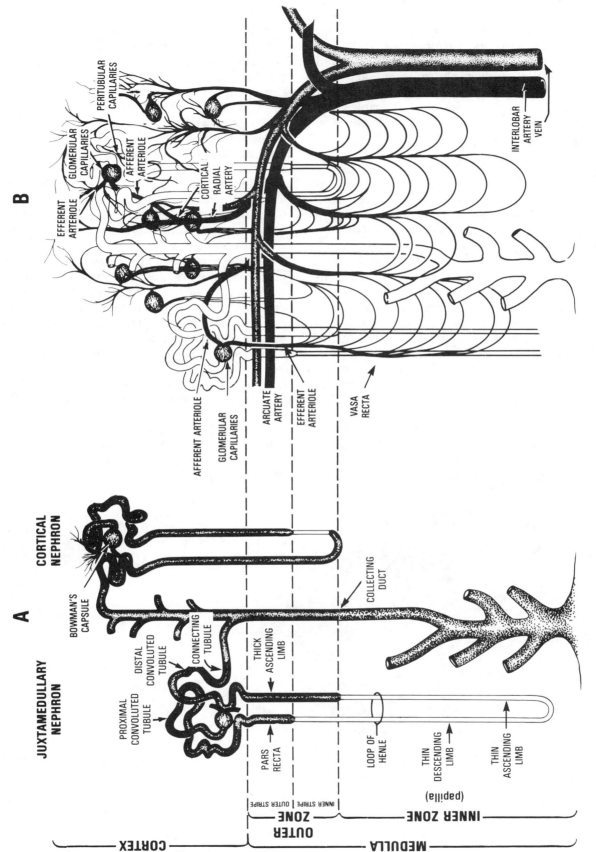

Figure 4.7. Cortical and juxtamedullary nephrons (*A*) and their vasculature (*B*). In *B*, the nephrons illustrated in *A* are shown in the background for reference.

From the peritubular capillaries and vasa recta, blood flows into *cortical radial veins* and then leaves the kidney in veins flowing counter to, and named in accordance with, the adjacent arteries. Note the rather unusual feature of the renal vasculature, wherein blood traverses *two* arterioles and *two* capillary networks before being collected into the venous system.

NEPHRON

The anatomy of the nephron will now be considered in greater detail, with reference to its relationship to the specialized renal vasculature and its general function. The glomerular portion of the nephron, or *Bowman's capsule*, is intimately associated with the glomerular capillary tuft. In fact, nephrologists (specialists in diseases of the kidneys) and physiologists often use the term *glomerulus* to include *both* the nephron and capillary components, a convention that will be followed in this textbook as well. The anatomic relationship between nephron and capillaries in the glomerulus is rather difficult to visualize, and the following analogy may prove helpful. First, imagine putting your hand in a rubber glove. Then, evert the portion of the glove covering the palm and back of your hand, so that it folds down over your fingers, while keeping your fingers in the fingers of the glove. In this analogy, your fingers represent the glomerular capillaries (with blood flowing into and out of your

fingers), while the glove represents the epithelial cells of Bowman's capsule. During filtration, then, fluid and solutes move from your fingers (the glomerular capillaries) across the fingers of the glove (the epithelial cells of Bowman's capsule) into a collecting area formed by the everted hand of the glove (Bowman's space).

The actual filtration barrier is somewhat more complex than indicated by the above analogy (Bulger, 1979), consisting of three layers: capillary endothelium, basement membrane, and epithelium (Fig. 4.8). The *capillary endothelium* is similar to that lining capillaries elsewhere in the body, except that the glomerular endothelial cells have numerous holes or *fenestrae* in their cytoplasm. The *basement membrane*, composed of collagen and proteoglycans (but *no* cells), surrounds the glomerular capillaries. Overlying the basement membrane is the *epithelial cell layer* of Bowman's capsule. The epithelial cells that surround the capillary tuft (i.e., the fingers of the glove in the analogy used above) differ from those forming the rest of Bowman's capsule (the everted hand of the glove), primarily because they have numerous *foot processes*, or *pedicels;* hence, these cells sometimes are called *podocytes.* The foot processes from adjacent podocytes interdigitate extensively with one another.

One final structural feature of the glomerulus remains to be described. Interspersed between adjacent capillaries, particularly in the center or *core* of

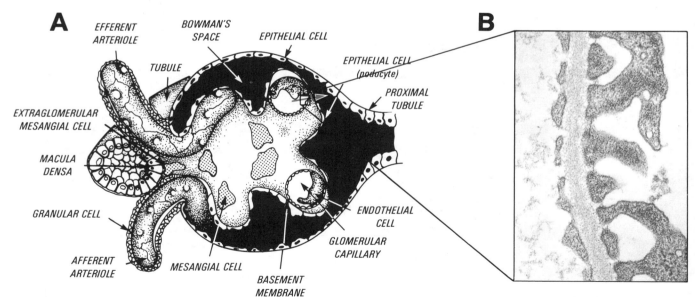

Figure 4.8. *A,* Cross-section of the glomerulus. The pathway followed by water and small solutes as they are filtered from the glomerular capillaries into Bowman's space is indicated by the *blue arrow.* The filtered molecules must cross the fenestrated *capillary endothelium,* the *basement membrane,* and the *epithelial cell layer* of Bowman's capsule. Also illustrated are the components of the specialized structure

found where the tubule contacts its originating glomerulus, the so-called *juxtaglomerular apparatus* (JG apparatus). *B,* Enlargement of the filtration barrier, showing (from *left* to *right*) the fenestrated capillary endothelium, the basement membrane, and the epithelial cell layer of Bowman's capsule. (Courtesy of Curtis B. Wilson.)

the glomerulus, are clusters of *mesangial cells.* These cells appear to provide support for the glomerular capillaries and may have additional functions as well, particularly in pathophysiological conditions.

The *tubular portion* of the nephron consists of an *epithelial cell layer* that is continuous with the epithelial cell layer forming Bowman's capsule. A *basement membrane,* continuous with the glomerular basement membrane (Fig. 4.8), encases the entire tubule. As previously indicated, the tubule can be divided into three major regions: *proximal tubule, loop of Henle,* and *distal nephron.*

The epithelial cells in various regions of the tubule differ in many respects, such as size, shape, number of mitochondria, and number of microvilli on the luminal surface (Fig. 4.9). However, certain structural features are common to all tubular epi-

thelial cells. For example, *cytoplasmic processes* extend from the lateral and basal surfaces of the cells to *interdigitate* with similar processes of adjacent cells, although the extent of this interdigitation varies in different regions. At the luminal surface, adjacent cells actually are *fused* together at a specialized junctional structure, the *zonula occludens* or *tight junction.* Below the tight junction, adjacent cells are separated by the so-called *lateral intercellular space,* which extends all the way to the bases of the cells. The lateral space is functionally continuous with the interstitial space surrounding the tubule and peritubular capillaries (the so-called *peritubular space*), as the basement membrane does not impede the diffusion of water or solutes. The tight junction therefore serves as a barrier to the movement of water and solutes between the lumen and peritubular space. However, the effec-

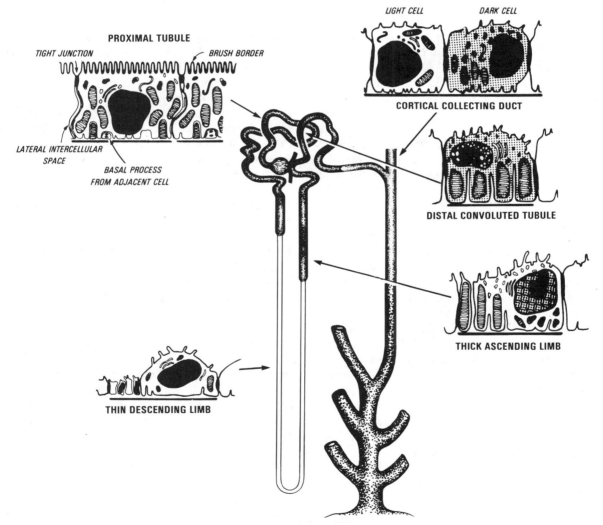

Figure 4.9. Comparison of major structural features of epithelial cells in different regions of the tubule. Certain features common to all regions (e.g., *tight junction, lateral intercellular space, basal process*) are identified on the proximal tubule illustration only.

tiveness of this barrier, or degree of "tightness," varies in different regions of the tubule (Fig. 4.9). In discussions of the tubular epithelium, the luminal surface often is termed the *mucosal* or *apical* surface, while the basal surface often is termed the *serosal* or *peritubular* surface.

Proximal Tubule

The proximal tubule (Venkatachalam, 1980) has traditionally been divided into the *proximal convoluted tubule*, which contorts itself in an apparently random fashion in the cortex, and the *proximal straight tubule* (or *pars recta*), which dives toward the medulla. More recently, the proximal tubule has been subdivided into three segments. *Segments 1* and *2* correspond to the beginning and end portions of the proximal convoluted tubule, while *segment 3* corresponds to the pars recta.

The epithelial cells of the proximal tubule have a *cuboidal* shape. A layer of closely packed *microvilli* covers the luminal surface, forming a *brush border*. The large *nuclei* are close to the basal surface. The cells have prominent *basal* and *lateral processes* and interdigitate extensively. *Mitochondria* are abundant, particularly in areas adjacent to the interdigitating processes.

The proximal tubule both *reabsorbs* substances from the tubular fluid and *secretes* substances into the tubular fluid. Thus, the proximal tubule is responsible for the initial processing of the glomerular filtrate. For example, approximately two-thirds of the filtered water and Na^+ and virtually all of the filtered glucose and amino acids are *reabsorbed* in this region. Organic acids and bases, such as drugs and drug metabolites, are *secreted*. The epithelial cells of the proximal tubule are well adapted for the processes of reabsorption and secretion, with their large luminal surface area (created by the brush border), large basal surface area (created by the basal processes), and abundant mitochondria (which undoubtedly have a role in supplying energy for the various transport processes). The numerous peritubular capillaries that surround the proximal tubule take up the reabsorbed water and solutes and deliver the substances to be secreted.

Loop of Henle

The loop of Henle consists of a *descending limb* and an *ascending limb* (Jamison and Kriz, 1982). While some authorities include the pars recta in the descending limb, many renal physiologists consider the loop of Henle to begin at the end of the pars recta, and a similar convention will be followed in this section. At this point, the cuboidal epithelial

cells of the proximal tubule are replaced by flat, squamous epithelial cells with a small number of short microvilli and few mitochondria. This segment of the tubule is termed the *thin descending limb* because the epithelial cell layer is so flat, i.e., "thin" does *not* mean that the tubular lumen is diminished in diameter.

Important differences between nephrons are present in the *ascending limb*. In *long-looped nephrons*, the ascending limb begins with a segment of flat, squamous epithelial cells, termed the *thin ascending limb*. Although these cells resemble the cells of the thin descending limb when viewed by light microscopy, structural differences are discernible with the electron microscope. At the junction between the inner and outer zones of the medulla, the *thick ascending limb* begins. In this segment, the epithelial cells are cuboidal and extensively interdigitated, and they contain numerous mitochondria like the epithelial cells of the proximal tubule. However, these cells lack a luminal brush border, although a small number of short microvilli are present. The thick ascending limb traverses the outer zone of the medulla (the *medullary portion* of the thick ascending limb) and then ascends through the cortex (the *cortical portion* of the thick ascending limb) to the level of its nephron's glomerulus. In *short-looped nephrons*, the ascending limb consists *entirely* of cuboidal epithelial cells, i.e., short-looped nephrons *lack* a thin ascending limb.

Like the proximal tubule, the loop of Henle both *reabsorbs* substances from the tubular fluid and *secretes* substances into the tubular fluid. For example, in long-looped nephrons as much as 25% of the filtered water and Na^+ is reabsorbed in the loop of Henle, while urea, a major waste product, is added to the tubular fluid. The loop of Henle also has an important role in the mechanisms for both urinary concentration and urinary dilution.

Distal Nephron

In this section, the term *distal nephron* will be used to refer to all portions of the tubule from the end of the cortical thick ascending limb through the tip of the papilla. The junction between the cortical thick ascending limb and the distal nephron is marked by an astonishing fact and a unique structure. The cortex of each human kidney contains the glomeruli of 1.0 to 1.5 million nephrons. Remarkably, in each nephron the thick ascending limb returns to the glomerulus of its own origin (Fig. 4.7*A*)! The tubule contacts its originating glomerulus at the *vascular pole*, the region where the afferent and efferent arterioles, respectively, enter and leave the glomerulus. A specialized structure, the *juxtaglomerular apparatus* (JG apparatus), is found at this contact point (see below).

The distal nephron includes several anatomically and functionally distinct segments (Kriz and Bankir, 1988). Three major segments can be identified: the *distal convoluted tubule*, the *connecting tubule*, and the *collecting duct*. The collecting duct, in turn, can be subdivided into the *cortical collecting duct*, the *medullary collecting duct*, and the *papillary collecting duct*. Actually, the term distal nephron is somewhat of a misnomer, because embryologically the distal convoluted tubule represents the last segment of the nephron proper. The collecting duct develops from the ureteral bud, and the connecting tubule is presumed to be the region where the embryonic nephron "connects" with the ureteral bud.

The *distal convoluted tubule* is a relatively short segment that extends outward from its contact point with the glomerulus toward the cortical surface. The *connecting tubule* has a somewhat convoluted shape and may join with adjacent connecting tubules before its junction with the cortical collecting duct. The *cortical collecting duct* begins as a short segment beyond the connecting tubule, continues as a slightly larger duct formed by the union of two or more such segments, and finally becomes a straight duct penetrating toward the medulla. The *medullary collecting duct* and *papillary collecting duct* represent the extensions of this straight duct into the outer medulla and papilla, respectively. Deep in the papilla, several papillary collecting ducts merge to form a common duct, often termed the *duct of Bellini*, that empties into a minor calyx. In the human kidney, a given duct of Bellini collects the urine from approximately 2800 nephrons (Oliver, 1968).

The epithelial cells of the distal convoluted tubule, like those of the proximal tubule, are cuboidal, exhibit extensive basal and lateral interdigitations, and contain numerous mitochondria. However, these cells lack a luminal brush border, although a small number of short microvilli are present as in the thick ascending limb. The epithelial cells of the connecting tubule resemble those of the distal convoluted tubule but are taller and have a more granular appearance.

In the collecting duct, the epithelial cells increase in height from a cuboidal shape in the cortical collecting duct to a columnar shape in the papillary collecting duct. Two basic cell types can be distinguished (Jamison and Kriz, 1982). The predominant cell type, the *principal* or *light cell*, stains lightly because of a relative paucity of cellular organelles. In contrast, the *intercalated* or *dark cell* stains heavily because of abundant mitochondria. The epithelial cells of the collecting duct lack a brush border, although short microvilli may be seen as in the distal convoluted tubule.

The distal nephron is responsible for the final transformation of the tubular fluid into urine. Its functions include the reabsorption of Na^+ and Cl^-, the secretion of H^+ and K^+, and an important role in the concentration and dilution of urine.

Juxtaglomerular Apparatus

As indicated above, the *juxtaglomerular apparatus* (JG apparatus) is a specialized structure found in the region where the tubule contacts its originating glomerulus (Barajas, 1979). The JG apparatus consists of three components: macula densa, extraglomerular mesangial cells, and granular cells (Fig. 4.8). The *macula densa* is a row of tightly packed cuboidal epithelial cells lining the tubule at the site of contact. This row of cells defines the beginning of the distal nephron. The *extraglomerular mesangial cells* can be regarded as an extension of the mesangial cells of the glomerulus into the triangular region bounded by the afferent arteriole, efferent arteriole, and macula densa. The extraglomerular mesangial cells are sometimes called *agranular cells* to distinguish them from the *granular cells* of the JG apparatus. The *granular cells* (also called JG cells) are located both in the region of the extraglomerular mesangial cells and in the walls of the adjacent afferent and efferent arterioles. In fact, the granular cells are believed to be specialized types of smooth muscle cells derived from the walls of these arterioles.

The granular cells receive their name from the existence of secretory granules that contain the precursor of *renin*, a proteolytic enzyme with a very important function. When secreted into the lumen of the arterioles, renin acts on a specific protein in plasma, *renin substrate* (also called angiotensinogen), to produce a decapeptide, *angiotensin I*. Angiotensin I, in turn, has two amino acids removed by *converting enzyme* to yield *angiotensin II*.

$$\text{Angiotensinogen} \xrightarrow{\text{renin}} \text{angiotensin I} \xrightarrow{\text{converting enzyme}} \text{angiotensin II}$$

Angiotensin II has at least two important actions: it stimulates the adrenal cortex to secrete aldosterone, and it is a powerful constrictor of arteriolar smooth muscle. Because of the latter action, angiotensin II may be able to modulate both the flow of blood through and the hydrostatic pressure in the glomerular capillaries (Chapter 27). In addition, the vasoconstrictor action of angiotensin II on *systemic* arterioles can increase the peripheral vascular resistance and thereby maintain or elevate the systemic arterial pressure. Indeed, abnormal production of renin, and therefore angiotensin II, is a cause of the high blood pressure in a small percentage of patients with hypertension.

The association of vessels, glomerulus, and macula densa from one nephron in the JG apparatus in conjunction with the vasoactive angiotensin system represents an ideal anatomic and functional arrangement whereby the fluid reaching the distal nephron might "signal" the arterioles to alter the blood flow to and rate of filtration in the glomerulus (Chapter 27). The term *tubuloglomerular feedback* has been used to describe this regulation of glomerular blood flow and filtration by the fluid delivered to the distal nephron. However, despite intensive investigation, the precise signal for such feedback remains to be established.

URINARY TRACT

As indicated above, the *urinary tract* consists of the minor and major calyces, renal pelvis, ureters, urinary bladder, and urethra (Netter, 1973). The entire urinary tract is lined by a transitional cell epithelium that does not further modify the composition of the urine leaving the distal nephron. Thus, the sole function of the urinary tract is the transmission and storage of urine.

In the calyces, renal pelvis, and ureters, the epithelial cell layer is surrounded by bundles of smooth muscle cells arranged in spiral configurations that contract when distended. Most importantly, distention of the renal pelvis and upper ureter initiates a *peristaltic contraction* that begins in the pelvis and spreads along the ureter to move urine toward the bladder. On the average, 1–5 such peristaltic contractions occur per minute, although the rate may increase to 20 or more per minute when distension is markedly increased due to high rates of urine production (Longrigg, 1982).

The ureters penetrate the bladder wall at an oblique angle and, although there are no anatomic sphincters at the junction between the ureter and the bladder (the *ureterovesicular junction*), this oblique penetration tends to keep the ureters closed except during ureteric peristalsis, thereby preventing reflux of urine from the bladder back into the ureters (Woodburne, 1964). A mechanism to prevent such reflux is important because the pressure in the bladder can exceed that in the ureter between periods of ureteric peristalsis and during voiding.

The urinary bladder, when distended, consists of a spherical portion (called the *body* or *fundus*) and an inferior cylindrical portion (called the *neck*). The epithelial cell layer lining the bladder is surrounded by the *detrusor muscle,* which consists of a randomly interwoven network of smooth muscle cell bundles. The smooth muscle in the bladder neck is mixed with a considerable amount of elastic tissue and functions as an *internal sphincter.* The urethra is surrounded by a band of skeletal (voluntary) muscle, the *external sphincter.*

During voiding *(micturition),* the detrusor muscle contracts, the neck of the bladder opens into a funnel shape, and the external sphincter is voluntarily relaxed. Once initiated, micturition normally continues until the bladder is entirely empty, although it can be interrupted by a powerful voluntary contraction of the external sphincter.

METHODS IN RENAL PHYSIOLOGY

An appreciation of the major experimental methods available for the study of renal function can contribute significantly to an understanding of the material presented in subsequent chapters. These methods can be divided into four general categories: clearance measurements, micropuncture methods, perfusion of microdissected nephron segments, and model tissues.

Clearance Measurements

Clearance is a central concept in renal physiology, as it provides a way of evaluating the elimination of a substance by the kidneys (Smith, 1951). To understand the meaning of clearance, consider first the *rate of excretion* of a substance X, i.e., the mass of X excreted per unit time. This can be equated to the mass of X per unit volume of urine (i.e., the concentration of X in urine, U_x) multiplied by the volume of urine excreted per unit time (i.e., the urine flow rate, $\dot{V}$)

$$\frac{\text{Mass of X excreted}}{\text{time}} = \frac{\text{Mass of X in urine}}{\text{urine volume}} \times \frac{\text{urine volume}}{\text{time}}$$

or

$$\frac{\text{Mass of X excreted}}{\text{time}} = U_x \dot{V}$$

Since an important function of the kidneys is the removal of substances from plasma, it is useful to describe the elimination of X in a somewhat different way. Instead of talking about the rate of excretion of X in *urine,* consider the rate of removal of X from *plasma.* This can be equated to the concentration of X in plasma (P_X) multiplied by the volume of plasma from which X is completely removed, or "cleared," per unit time

$$\frac{\text{Mass of X removed from plasma}}{\text{time}} = \text{Plasma concentration of X}$$
$$\times \frac{\text{volume of plasma "cleared" of X}}{\text{time}}$$

The volume of plasma from which X is completely "cleared" per unit time is termed the *clearance of X* and is designated C_X. Therefore

$$\frac{\text{Mass of X removed from plasma}}{\text{time}} = P_X C_X$$

It is important to note that the volume of plasma "cleared" of X is a *theoretical* volume rather than a volume that can be collected and directly measured. This is because no single milliliter of plasma has *all* of its X removed by the kidneys; instead, a certain *fraction* of the X in each milliliter of plasma is removed. But while C_X is only a theoretical volume, its value can be calculated from measurable quantities, since mass conservation requires that the rate of removal from plasma must equal the rate of excretion:

$$\frac{\text{Mass of X removed from plasma}}{\text{time}} = \frac{\text{Mass of X excreted in urine}}{\text{time}}$$

Therefore

$$P_X C_X = U_X \dot{V}$$

and

$$C_X = \frac{U_X \dot{V}}{P_X} \qquad (26)$$

The great advantage of clearance measurements over other methods for studying renal function is that they require only urine and peripheral blood samples, i.e., clearance measurements are virtually *noninvasive*. With rare exceptions, then, the clearance technique is the only method available for the study of renal physiology in humans. Clearance measurements are useful only for evaluating the *overall* elimination of a substance by the kidneys, however, *not* for studying the function of *individual segments* of the nephron. Almost all statements made in this text about the function of individual nephron segments, therefore, are based on studies conducted in experimental animals using methods in the remaining three categories.

Micropuncture Methods

The technique of micropuncture, in which a tiny micropipette is inserted into a nephron segment or adjacent blood vessel, has resulted in major advances in the understanding of renal physiology. Five important examples of micropuncture methods will be considered briefly here (Deetjen et al., 1975; Jamison and Kriz, 1982):

1. Micropuncture can be used to aspirate fluid from accessible nephron segments for analysis of composition. For example, the composition of the fluid filtered into Bowman's space can be determined by direct analysis of fluid collected by micropuncture. In a variation of this method, fluid of known composition is injected into an accessible nephron segment and held stationary by prior injection of a droplet of oil to block forward flow. The fluid is later aspirated and analyzed to determine how the tubule has modified its composition.

2. Micropipette-sized pressure transducers can be used to measure the hydrostatic pressures in glomerular and peritubular capillaries, Bowman's space, and accessible nephron segments, allowing the forces involved in glomerular filtration to be studied.

3. Micropipette-sized glass electrodes can be used to measure intracellular electrical potentials and ion activities as well as the transepithelial potential difference.

4. Micropuncture can be used to perfuse a single nephron distal to an oil block, a technique often termed *microperfusion*. For example, an oil droplet can be injected into segment 1 of the proximal tubule and perfusion fluid injected into segment 2. The perfusion fluid then flows through segment 3 of the proximal tubule, the descending limb of the loop of Henle, and the ascending limb of the loop of Henle. The fluid is aspirated from the distal convoluted tubule and analyzed, allowing the actions of the perfused segments on the volume and composition of the tubular fluid to be determined. A variation of this microperfusion method involves the simultaneous injection of perfusion fluid into the tubular lumen and surrounding peritubular capillaries. In this manner, the composition of the fluid on *both* sides of the tubular epithelium can be controlled.

5. A small catheter can be advanced upstream from the calyceal area through a duct of Bellini and into the lumen of a papillary collecting duct. By sampling fluid at various levels of the collecting duct, changes in the volume and composition of

the tubular fluid as it passes through the duct can be determined. This technique, which does not actually entail puncture of the tubule, is termed *microcatheterization.*

The study of renal physiology by these micropuncture techniques is limited to regions of the nephron that are accessible from the surfaces of the kidney. From the *cortical surface*, several regions of cortical nephrons can be micropunctured: the glomerular structures, segments 1 and 2 of the proximal tubule, and the distal convoluted tubule, connecting tubule, and cortical collecting duct. (However, much of the older micropuncture data obtained by distal puncture did not take into account the subdivision of the early distal nephron.) In contrast, *none* of the above segments in juxtamedullary nephrons is accessible by micropuncture. From the *papillary surface*, the thin descending and ascending limbs of the loop of Henle of long-looped nephrons can be punctured. Segment 3 of the proximal tubule (the pars recta), the thick ascending limb of the loop of Henle, and the medullary collecting duct are not accessible to micropuncture.

Perfusion of Microdissected Nephron Segments

The perfusion of microdissected nephron segments (Grantham and Burg, 1966) is a technique that can be used to study *all* nephron segments and has revealed significant functional specializations within the various segments of the nephron. Individual nephron segments are dissected by "teasing" and shaking small bits of tissue into smaller and smaller fragments while observing the process through a dissection microscope. The glomerulus and other landmarks such as the thin descending limb are used to identify the segments during the isolation procedure. The final isolated segment, which may be only 1–2 mm long and finer than a human hair, is placed in a bathing solution of known composition and then connected at both ends to micropipettes. Fluid is pumped through one pipette into the tubular lumen at a rate of a few nanoliters (10^{-9} liter) per minute and is collected in the second pipette as it exits the nephron segment. Changes in volume and composition as well as in the

flux (movement) of radioactive solutes into or out of the tubular fluid can therefore be determined. In addition, a small electrode can be advanced through one of the micropipettes so that the transepithelial potential difference can be monitored. The rabbit kidney has been most extensively studied by these methods because of its relative ease of dissection. In contrast, the normal rat kidney is virtually impossible to microdissect adequately. This is unfortunate, since most micropuncture data have been obtained in rats. Thus, a direct comparison between the findings obtained by micropuncture and those obtained in microdissected segments is hindered by potential species variations. A few tubule segments have been dissected from human kidneys removed surgically as a consequence of renal disease (usually cancer of the kidneys), and the results obtained to date are similar to those obtained from rabbit kidneys.

Model Tissues

Much of what is known about transport and the hormonal regulation of transport across renal tubular epithelial cells has been learned from studies of epithelial cell layers from lower vertebrates, such as amphibians or reptiles. In these species, regulation of the ionic composition of the body fluids is not the exclusive domain of the kidneys, but also involves the skin (amphibians) and urinary bladder. The latter may be considered analogous to the collecting duct in mammalian kidneys, to which it is embryologically and functionally similar. The large size of the urinary bladder (over 10 cm^2 per animal) compared to the size of a tubule segment allows the transport of water and solutes such as Na^+, Cl^-, H^+, and urea to be monitored in a rapid and convenient fashion. The large size of the bladder also facilitates biochemical investigations, such as the study of cyclic AMP metabolism in relation to the action of antidiuretic hormone (Chapter 30).

Another example of the use of model tissues in renal physiology has involved the study of fish. Some fish species, especially in the Antarctic Ocean, are *aglomerular*, i.e., they lack glomeruli. Such a species was used to demonstrate unequivocally that tubular cells could secrete solutes into the tubular lumen (Shannon, 1938).

CHAPTER 27

Filtration and Blood Flow

Although the kidneys represent only 0.5% of the total body weight, they receive 20–25% of the total cardiac output. Thus, in an average adult with a cardiac output of 6 liters/min, the *renal blood flow* exceeds 1.2 liters/min or 1700 liters/day. This means that the total blood volume, which averages 6 liters, passes through the renal vasculature nearly 300 times per day. The distribution of such a large share of the cardiac output to the kidneys is an important adaptation that enables the kidneys to regulate the normal quantity and composition of the body fluids.

Whereas the renal blood flow (RBF) represents the volume of *blood* flowing through the renal vasculature per unit time, the *renal plasma flow* (RPF) refers to the rate of *plasma* flow through the renal vasculature. If the hematocrit is 45%, RPF is 55% of RBF. The average RPF therefore exceeds 650 ml/min or 900 liters/day.

As discussed in Chapter 26, approximately 125 ml of plasma are filtered each minute at the glomerulus, i.e., the *glomerular filtration rate* (GFR) is 125 ml/min or 180 liters/day. Thus, approximately 20% of the RPF is filtered into Bowman's space. The ratio of GFR to RPF is termed the *filtration fraction* (FF).

PROPERTIES OF THE FILTRATION BARRIER

As discussed in Chapter 26, the filtration barrier consists of three layers: glomerular capillary endothelium, basement membrane, and epithelium. In spite of its complex structure, however, the filtration properties of the barrier are qualitatively similar to those of other capillaries in the body. Thus, the barrier is freely permeable to water and to small solutes *(crystalloids)*, such as ions, glucose, and urea. It has limited permeability to larger solutes *(colloidal particles)*, such as large proteins and lipids, and is almost completely impermeable to the cellular elements of blood. Except for the absence of proteins and lipids, then, the glomerular filtrate is virtually identical to *plasma*. The term *ultrafiltrate* often is used when describing the glomerular filtrate because of the exclusion of cellular elements *and* colloidal particles (a simple *filtrate* would exclude only the cellular elements of blood).

The size discrimination properties of the filtration barrier have been studied by comparing the concentration of a given substance in the filtrate to its concentration in plasma. Table 4.7 shows the ratio between the filtrate concentration and the plasma concentration for a variety of plasma constituents and infused substances. The [filtrate]/[plasma] ratio is 1.0 for substances with molecular weights up to approximately 5000, whose effective radii are less than 15 Å, but falls sharply for larger molecules. For example, less than 1% of serum albumin ($\simeq$ 69,000 daltons; effective radius, $\simeq$36 Å) is filtered. Data such as these suggest that the filtration barrier behaves as if it had channels or *pores* up to approximately 60 Å in diameter (Navar et al., 1986). The location of these pores in the filtration barrier has not been firmly established, however. The *holes* or *fenestrae* in the glomerular capillary endothelium (Chapter 26) are 500–1000 Å in diameter and are therefore much too large to prevent the filtration of large molecules. In the epithelial cell layer, the foot processes (pedicels) of adjacent podocytes appear to be separated by slits approximately 250 Å wide, also too large to account fully for the exclusion of large molecules. Present evidence suggests that the pores may be hydrated channels between the collagen and proteoglycan chains of the basement membrane (Kanwar, 1984). Such channels probably would be tortuous and might *not* be stable anatomic structures, which could explain why attempts to visualize pores by electron microscopy have not been successful.

While the [filtrate]/[plasma] ratios in Table 4.7 correlate fairly well with molecular weight and size, the extent to which a substance is filtered can depend on other factors as well. For example, the [filtrate]/[plasma] ratio for hemoglobin is 3 times larger than that for serum albumin, even though its molecular weight is just 3% less than that of albu-

Table 4.7.
Approximate [Filtrate]/[Plasma] Ratios for Selected Substances

Substance	Molecular Weight, daltons	Effective Radius,[a] Å	[Filtrate]/[Plasma]
Water	18	1.0	1.0
Urea	60	1.6	1.0
Glucose	180	3.6	1.0
Inulin	5000	14.8	0.98
Myoglobin	17,000	19.5	0.75
Egg albumin	43,500	28.5	0.22
Hemoglobin	68,000	32.5	0.03
Serum albumin	69,000	35.5	0.01

Adapted from Pitts, 1974. [a]Estimated from diffusion coefficient.

min and its effective radius is just 8% less than albumin's. One reason that hemoglobin is so much more readily filtered than serum albumin is *shape:* hemoglobin is a relatively compact, cylindrical molecule, whereas albumin is an asymmetrical ellipsoid. A second and more important reason involves *electrostatic factors.* Figure 4.10 illustrates how the [filtrate]/[plasma] ratio for *anionic* dextran is substantially less than that for *uncharged* dextran or *cationic* dextran of an equivalent size (Brenner et al., 1978). While both hemoglobin and albumin are anions and therefore have [filtrate]/[plasma] ratios less than the corresponding uncharged or cationic dextran molecules (Fig. 4.10), albumin is *more* negatively charged than hemoglobin, which significantly hinders its filtration.

Substances with a [filtrate]/[plasma] ratio of 1.0 are said to be *freely filtered.* The amount of such a substance that is filtered per unit time can be

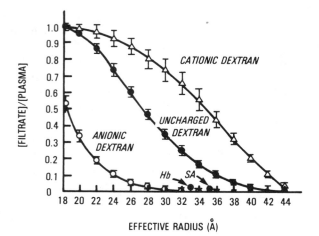

Figure 4.10. [Filtrate]/[plasma] ratios for uncharged dextran, anionic dextran, and cationic dextran as a function of effective radius. The [filtrate]/[plasma] ratios for hemoglobin *(Hb)* and serum albumin *(SA)* also are identified. (Reproduced from *The Journal of Clinical Investigation,* 1978, vol. 61, pp. 72–78, by copyright permission of the American Society for Clinical Investigation.)

readily calculated by multiplying its plasma concentration by the GFR. Thus, for a freely filtered substance X

$$\frac{\text{Mass of X filtered}}{\text{time}} = \text{Plasma concentration of X}$$
$$\times \frac{\text{volume of plasma filtered}}{\text{time}}$$

or

$$\frac{\text{Mass of X filtered}}{\text{time}} = P_X\, GFR \qquad (27)$$

The term *filtered load* often is used when referring to the quantity $P_X\, GFR$.

While freely filtered substances must have a molecular weight of ≤ 5000 daltons (Table 4.7), a further requirement is the absence of significant binding to plasma proteins such as albumin. The filtration of certain ions and small solutes (e.g., Ca^{2+}, Mg^{2+}, bilirubin) is markedly reduced as a result of plasma protein binding. The filtered load of a substance that exhibits significant plasma protein binding is given by

$$\frac{\text{Mass of X filtered}}{\text{time}} = P_X\, GFR \cdot F_X \qquad (28)$$

where F_X is the fraction of the substance in plasma that is *free* (i.e., unbound).

As indicated in Table 4.7, many of the small solutes in plasma have [filtrate]/[plasma] ratios of 1.0. However, if *very* precise concentration measurements are made, one finds that the filtrate and plasma concentrations of these solutes are *not* exactly the same. The reasons for such discrepancies were introduced in Chapter 25. First, plasma concentration values typically are measured in clinical laboratories as mmol/liter of *plasma volume.* To compare the plasma and filtrate concentrations of small solutes, one must *correct* the plasma concentrations to account for the fact that these solutes are dissolved only in the *aqueous phase* of plasma, which normally represents about 93% of the plasma volume. Since only the aqueous phase is filtered, plasma concentrations must be expressed as mmol/liter of *plasma water* in order to be compared to glomerular filtrate concentrations. Laboratory values expressed as mmol/liter of *plasma volume* can be converted to mmol/liter of *plasma water* by dividing by 0.93 (Chapter 25).

Once such conversions are made, the [filtrate]/[plasma] ratios of small *nonelectrolytes* do in fact equal 1.0, but small *electrolytes* still have slightly different concentrations in plasma and filtrate.

These remaining differences can be attributed to the *Gibbs-Donnan effect* (Chapter 25). The presence of proteins in plasma, but not in the filtrate, results in the establishment of a Gibbs-Donnan equilibrium between plasma and filtrate. Since the plasma proteins are negatively charged, *cations* such as Na^+ and K^+ have somewhat higher concentrations in *plasma water*, while *anions* such as Cl^- and HCO_3^- have somewhat higher concentrations in the *filtrate*. However, the concentration differences attributable to the Gibbs-Donnan effect are small and generally are neglected in renal physiology.

FORCES INVOLVED IN FILTRATION

The *Starling principle* provides a framework for analyzing fluid movement across capillaries. According to this principle, the rate and direction of fluid movement is determined by the balance of hydrostatic and oncotic pressures, as discussed in Chapter 7. The Starling principle can be expressed as follows:

$$\dot{q} = K_f [(P_c - P_i) - (\pi_c - \pi_i)] \qquad (29)$$

where $\dot{q}$ = rate of fluid movement across the capillary, K_f = filtration coefficient, P_c = capillary hydrostatic pressure, P_i = interstitial hydrostatic pressure, π_c = plasma oncotic pressure, and π_i = interstitial oncotic pressure.* *Eq. 29* is written such that a *positive* value of $\dot{q}$ signifies *net filtration*, whereas a *negative* $\dot{q}$ signifies *net reabsorption*. Thus, the expression $[(P_c - P_i) - (\pi_c - \pi_i)]$ has been termed the *net filtration pressure*.

It is instructive to review how the net filtration pressure varies along the length of a typical systemic *(extrarenal)* capillary (Table 4.8*A*; Fig. 4.11*A*). At the arterial end, the balance of pressures favors *filtration* (net filtration pressure $\simeq +16$ mm Hg). However, at the venous end *reabsorption*

*Note that in *Eq. 29* and in the discussion pertaining to it, P is used to represent *hydrostatic pressure, not* plasma concentration.

occurs (net filtration pressure $\simeq -14$ mm Hg), since capillary hydrostatic pressure (P_c) declines markedly along the length of the capillary. The result is that filtration and reabsorption approximately balance. Actually, if the net filtration pressure is averaged along the capillary, a *mean* net filtration pressure of 8 mm Hg is obtained. A slight excess of plasma therefore is filtered into the interstitium, although the net volume filtered in *all* extrarenal capillaries probably is less than 2 ml/min. This volume is returned to the systemic circulation via the lymphatics.

For *glomerular* capillaries, the Starling equation *(Eq. 29)* can be rewritten as

$$GFR = K_f [(P_G - P_B) - (\pi_G - \pi_B)] \qquad (30)$$

where the subscripts G and B refer to the glomerular capillaries and Bowman's space, respectively. Since the amount of protein filtered into Bowman's space is negligible, π_B approaches zero and *Eq. 30* simplifies to

$$GFR = K_f [P_G - P_B - \pi_G] \qquad (31)$$

The variation in the net filtration pressure along the length of a glomerular capillary (Brenner and Humes, 1977) differs somewhat from that in extrarenal capillaries (Table 4.8*B*; Fig. 4.11*B*). Most importantly:

1. Capillary hydrostatic pressure remains relatively constant along the glomerular capillary, in contrast to its marked decline in extrarenal capillaries. This may be due in part to the presence of a high resistance arteriole at the efferent end of glomerular capillaries but not extrarenal capillaries.
2. Because the volume of fluid filtered in glomerular capillaries is so large, the plasma oncotic pressure rises along the length of a glomerular capillary, whereas it remains relatively constant in extrarenal capillaries.

Table 4.8.
Net Filtration Pressure in Extrarenal vs. Glomerular Capillaries

	A, Extrarenal Capillary			*B*, Glomerular Capillary	
	Arterial End	Venous End		Afferent End	Efferent End
P_c	40mm Hg	10mm Hg	P_G	45mm Hg	45mm Hg
P_i	2	2	P_B	10	10
π_c	25	25	π_G	25	35
π_i	3	3	π_B	0	0
Net filtration pressure = $(P_c - P_i) - (\pi_c - \pi_i)$	+16mm Hg	−14mm Hg	Net filtration pressure = $(P_G - P_B) - (\pi_G - \pi_B)$	+10 mm Hg	$\simeq 0$

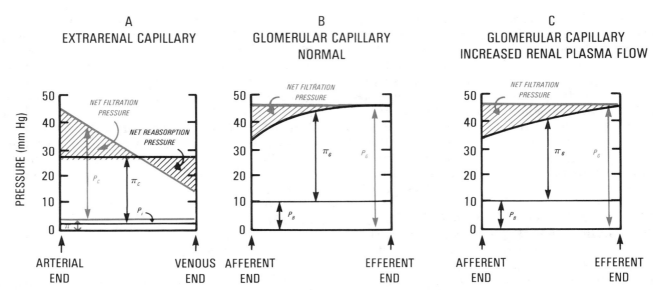

Figure 4.11. Variations in pressures affecting filtration along the length of extrarenal and glomerular capillaries. Pressures favoring *filtration* (the capillary hydrostatic pressure P_c and interstitial oncotic pressure π_i in *extrarenal* capillaries; the glomerular capillary hydrostatic pressure P_G in *glomerular* capillaries) are illustrated in *blue*, while those favoring *reabsorption* (the interstitial hydrostatic pressure P_i and plasma oncotic pressure π_c in *extrarenal* capillaries; the hydrostatic pressure in Bowman's space P_B and glomerular capillary oncotic pres-

sure π_G in *glomerular capillaries*) are illustrated in *black*. In *extrarenal* capillaries (*A*), *filtration* occurs at the arterial end and *reabsorption* occurs at the venous end. In *glomerular* capillaries (*B*), *filtration* occurs at the afferent end and *filtration equilibrium* is present at the efferent end. The point along the glomerular capillary at which filtration equilibrium is achieved is displaced toward the efferent end if RPF increases (*C*), since the glomerular capillary oncotic pressure π_G rises less rapidly.

Thus, *filtration* occurs at the afferent end (net filtration pressure $\simeq$ 10 mm Hg), as in extrarenal capillaries. However, at the efferent end fluid movement *ceases* (net filtration pressure $\simeq$ 0). *Filtration equilibrium* therefore is said to be present at the efferent end of glomerular capillaries.

The *mean* net filtration pressure in the glomerular capillaries cannot be determined precisely, since the exact point along the capillary at which filtration equilibrium is achieved is not known (Blantz and Pelayo, 1986). It has been estimated, however, that the mean net filtration pressure in glomerular capillaries is similar to that in extrarenal capillaries, $\simeq$8 mm Hg. The fact that the GFR averages 125 ml/min, whereas the total net filtration rate in *all* extrarenal capillaries is less than 2 ml/min, must therefore be attributed to differences in the *filtration coefficient* K_f. The larger filtration coefficient of glomerular capillaries may be partly due to the numerous fenestrae in the glomerular capillary endothelium (Chapter 26).

As indicated in Table 4.8*B*, the hydrostatic pressure in Bowman's space (P_B) is approximately 10 mm Hg. This pressure not only is an important determinant of the net filtration pressure, but also serves to propel the glomerular filtrate through the proximal tubule and the remainder of the nephron.

REGULATION OF RBF AND GFR

Renal blood flow, like the blood flow to any organ, is controlled by the arteriovenous pressure difference across the vascular bed (the *perfusion pressure*) and the vascular resistance, as expressed by the general resistance equation (Chapter 5):

$$RBF = \frac{\Delta P}{R} \qquad (32)$$

The perfusion pressure for the renal vasculature is identified in Figure 4.12, which illustrates the normal hydrostatic pressure profile of the renal circulation *(curve A)*. Note that the renal arterial pressure is essentially the same as that of other systemic arteries, with a mean normal value of 100 mm Hg. Also note that the glomerular capillary hydrostatic pressure (P_G) is approximately 45 mm Hg, as previously indicated (Table 4.8*B*). The large pressure drop that occurs in the afferent and efferent arterioles identifies these vessels as the major sites of renal vascular resistance (Blantz, 1980). Changes in the diameters of these vessels represent the primary mechanism for adjusting the renal vascular resistance.

Many of the factors that control *glomerular filtration rate* can be readily identified with the aid of the Starling equation *(Eq. 30)*. For example,

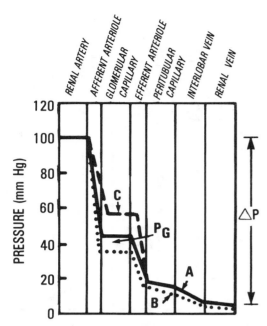

Figure 4.12. Hydrostatic pressure profile of the renal circulation. *A,* Normal profile. *B,* Following *afferent* arteriolar constriction and consequent *decrease* in glomerular capillary hydrostatic pressure (P$_G$). *C,* Following *efferent* arteriolar constriction and consequent *increase* in glomerular capillary hydrostatic pressure. ΔP represents the normal perfusion pressure of the renal circulation.

changes in *glomerular capillary hydrostatic pressure* can cause changes in GFR. Just as the hydrostatic pressure in a systemic capillary is controlled by the pressures and resistances in the adjacent arteriole and venule (Chapter 5), glomerular capillary hydrostatic pressure is controlled by the pressures and resistances in the adjacent afferent arteriole and efferent arteriole. Figure 4.12 illustrates examples of how changes in these pressures and resistances can affect P$_G$. Note that the effects of afferent arteriolar resistance changes on P$_G$, and hence GFR, are *opposite* to those of efferent arteriolar resistance changes. For example, an increase in *afferent* arteriolar resistance (e.g., due to afferent arteriolar constriction) *decreases* P$_G$ (Fig 4.12, *curve B*), whereas an increase in *efferent* arteriolar resistance *increases* P$_G$ (Fig. 4.12, *curve C*). In contrast, the effects of afferent and efferent arteriolar resistance changes on RBF are the *same*. For example, an increase in either afferent *or* efferent arteriolar resistance will decrease RBF.

While changes in P$_G$ can cause changes in GFR, the other terms in the Starling equation also can affect the GFR, particularly in disease states (Tucker and Blantz, 1977). For example, the *filtration coefficient* (K$_f$) can be decreased by diseases that cause a thickening of the filtration barrier or decrease its surface

area by destroying glomerular capillaries; in addition, many hormones and other endogenous substances may alter K$_f$, possibly by contracting or relaxing mesangial cells. The *hydrostatic pressure in Bowman's space* (P$_B$) increases in ureteral obstruction. Small changes in *glomerular capillary oncotic pressure* (π$_G$) can occur with dehydration (increased π$_G$) or hypoalbuminemia (decreased π$_G$). The *oncotic pressure in Bowman's space* (π$_B$) normally is negligible *(Eq. 31)*, but it can be increased by diseases that increase the permeability of the filtration barrier, thereby allowing plasma proteins to leak into the filtrate.

Although the Starling equation is helpful in identifying many of the factors that control GFR, it does *not* explicitly contain the *most important* regulator of GFR, namely, the *RPF*. However, the Starling equation *can* be used to understand the mechanism whereby RPF regulates GFR. This mechanism can be summarized as follows: The RPF determines the rate at which the plasma oncotic pressure (π$_G$) rises along the length of the glomerular capillary. For example, if filtration equilibrium normally occurs at a point two-thirds of the distance between the afferent and efferent ends of the glomerular capillary (Fig. 4.11*B*), an increase in RPF will increase GFR because π$_G$ rises less rapidly, thereby displacing the point at which the net filtration pressure becomes zero further toward the efferent end of the capillary (Fig. 4.11*C*). The importance of RPF as a regulator of GFR is illustrated by the observation that at least 50% of the change in GFR caused by many hormones and other endogenous substances (see below) occurs secondary to a change in RPF.

Autoregulation

From the above discussion, the changes in systemic arterial pressure that occur in different physiological situations would be expected to have significant effects on RPF and GFR. However, the maintenance of the normal quantity and composition of the body fluids by the kidneys is a continuous process that is performed most effectively if RPF and GFR remain relatively *constant* (Navar, 1978). Figure 4.13 illustrates that RPF and GFR do in fact remain nearly constant even in the presence of large changes in mean systemic arterial pressure. This maintenance of a constant flow in spite of changes in systemic arterial pressure is termed *autoregulation*. While autoregulation also occurs in the heart, brain, and other organs (Chapter 6), the constancy of flow is quite striking in the kidneys. For example, in the dog RPF and GFR change by less than 10% when mean systemic arterial pressure is

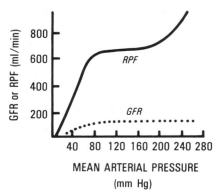

Figure 4.13. Effect of mean arterial pressure on glomerular filtration rate (GFR) and renal plasma flow (RPF) in an anesthetized dog, illustrating the phenomenon of autoregulation.

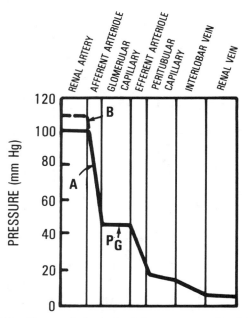

Figure 4.14. Changes in the hydrostatic pressure profile of the renal circulation during autoregulation. *A*, Normal profile. *B*, Following elevation of mean arterial pressure, afferent arteriolar constriction maintains constant glomerular capillary hydrostatic pressure (P_G).

varied between 80 and 180 mm Hg (Shipley and Study, 1951), and an autoregulatory range of approximately 60 to 130 mm Hg is believed to be present in humans. Because of autoregulation, the *filtration fraction* (GFR/RPF) also remains relatively constant in this pressure range.

From *Eq. 32*, it is evident that in the autoregulatory range an increase in arterial pressure must be accompanied by an increase in renal vascular resistance in order to maintain a constant RPF. Since GFR also is autoregulated, however, this increase in resistance must occur without significantly altering the glomerular capillary hydrostatic pressure. The *afferent arteriole* must therefore represent the major site of the autoregulatory resistance change, as illustrated in Figure 4.14.

The phenomenon of autoregulation can be demonstrated in transplanted, denervated kidneys as well as in isolated kidneys perfused in vitro, i.e., autoregulation definitely is *intrinsic* to the kidneys. Regarding the mechanism for autoregulation, current evidence favors the so-called *myogenic mechanism*, which attributes the increase in renal vascular resistance that accompanies an increase in systemic arterial pressure to contraction of the afferent arteriolar smooth muscle cells in response to stretch. It has been suggested that stretch leads to the opening of Ca^{2+} channels in the plasma membrane, resulting in a Ca^{2+} influx that causes contraction (Fray et al., 1986). The ability of arteriolar smooth muscle cells to contract when stretched has been demonstrated in vessels isolated from a variety of organs; in fact, the myogenic mechanism originally was proposed to explain autoregulation in organs other than the kidneys (Chapter 6). An alternative mechanism, the *juxtaglomerular hypothesis*, attributes autoregulation to changes in the rate of renin secretion from the granular cells, which in

turn alter the local concentration of angiotensin II, a powerful vasoconstrictor (see below).

As noted above, autoregulation represents an *intrinsic* mechanism for keeping RPF and GFR constant in the face of changing systemic arterial pressure. However, RPF and GFR also can be regulated by neural and hormonal influences. These *extrinsic factors* can *reset* the autoregulatory mechanism, altering RPF and GFR even in the autoregulatory range of systemic arterial pressure.

Neural Regulation

Like most blood vessels, the vessels of the kidneys are innervated exclusively by vasoconstrictor fibers from the sympathetic division of the autonomic nervous system. At rest, sympathetic tone to the renal vasculature appears to be minimal. The renal vasoconstriction resulting from sympathetic stimulation causes a reduction in both RPF and GFR. However, with moderate sympathetic stimulation, GFR does not decrease as much as RPF; the vasoconstriction therefore must involve both the afferent *and* efferent arterioles. The filtration fraction, then, actually *increases* at moderate levels of sympathetic stimulation. With further sympathetic stimulation, the decreases in GFR parallel those in RPF; apparently, afferent arteriolar constriction predominates at these higher levels of stimulation.

It is important to note that the sympathetically

mediated renal vasoconstriction represents part of the body's mechanism for controlling *systemic arterial pressure* rather than a mechanism for regulating RPF and GFR. Such vasoconstriction is initiated as a reflex response to a decrease in systemic arterial pressure, as sensed by the carotid sinus and aortic arch baroreceptors, e.g., during postural changes, hemorrhage, or syncope. A similar response can be initiated by decreases in pressure in the atria, as sensed by baroreceptors located in their walls. In all of these situations, the renal vasoconstriction contributes to a rise in total peripheral vascular resistance, which in turn helps to restore the arterial pressure to its normal value (Chapter 5). Sympathetic vasoconstriction also can be initiated in response to input from the central nervous system, for example during fright, pain, cold, exercise, and other stressful situations. The fact that GFR does not decrease as much as RPF at moderate levels of sympathetic stimulation can be regarded as an adaptation designed to maintain GFR as high as possible even if renal vascular resistance must increase to control systemic arterial pressure.

Hormonal Regulation

Many hormones and other endogenous substances can cause renal vasoconstriction or vasodilation. The substance most likely to have a physiological role in the regulation of RPF and GFR is *angiotensin II,* a potent vasoconstrictor that is synthesized in response to the secretion of renin from the granular cells (Chapter 26) (Wright and Briggs, 1979). Since renin is secreted in response to decreased renal perfusion and also in response to sympathetic stimulation of the granular cells (Chapter 31), at least part of the renal vasoconstriction occurring as a result of a decrease in systemic arterial pressure may be mediated by angiotensin II.

The *prostaglandins* also may have a physiological role in the regulation of RPF and GFR (Schnermann and Briggs, 1981). The kidneys can synthesize several members of the prostaglandin family, including prostaglandin E_2, prostacyclins, leukotrienes, and thromboxanes. In terms of effects on RPF and GFR, the *vasodilator* prostaglandins (such as prostaglandin E_2) are believed to be most important. These prostaglandins may modulate the constrictor effects of sympathetic stimulation and angiotensin II; in fact, the synthesis of vasodilator prostaglandins appears to be *increased* by sympathetic stimulation or increased angiotensin II levels. A role for the vasodilator prostaglandins also is supported by the fact that inhibitors of prostaglandin synthesis (e.g., indomethacin) may cause a dramatic decrease in RPF and GFR in patients with diseases that impair perfusion of the kidneys (e.g., congestive heart failure, cirrhosis of the liver). Since RPF and GFR appear to be altered little by inhibitors of prostaglandin synthesis in normal individuals, it has been suggested that the vasodilator prostaglandins have their most significant effects on RPF and GFR when renal perfusion is compromised by abnormal conditions or disease. *Bradykinin,* a potent vasodilator, may have a similar role in modulating RPF and GFR.

Antidiuretic hormone (ADH), which is of major importance in the urinary concentrating mechanism (Chapter 30), is a vasoconstrictor and can markedly decrease RPF and GFR. *Serotonin* also causes renal vasoconstriction, reducing RPF and GFR. However, since the effects of ADH and serotonin have primarily been demonstrated with large, pharmacological doses, the physiological role of these substances in the regulation of RPF and GFR is uncertain. *Dopamine* is a renal vasodilator and can increase RPF and GFR. In fact, dopamine and related drugs currently are being used to increase RPF and GFR in pathophysiological states. *Atrial natriuretic factor* (ANF), which has an important role in the regulation of Na^+ excretion (Chapter 31), also is a vasodilator and can increase GFR.

Pregnancy can increase RPF and GFR by as much as 50% (Atherton and Green, 1983). The mechanism for this marked effect is unknown, although the hormonal changes accompanying pregnancy may be involved. The *ingestion of a protein-rich meal* can increase RPF and GFR by as much as 30% (Krishna et al., 1988), an effect which can be mimicked by infusing amino acids intravenously and which may be mediated by prostaglandins.

INTRARENAL DIFFERENCES IN RBF AND GFR

Approximately 90% of the total RBF perfuses the *cortex;* only 10% perfuses the *medulla* (Barger and Herd, 1971). This marked discrepancy between cortical and medullary blood flow is not due to a difference in the number of blood vessels per unit of renal tissue, since the vascular volume is approximately 20% of the total tissue volume in both cortex and medulla. Instead, the reduced blood flow to the medulla apparently results from a relatively high vascular resistance in the vasa recta. Whether this high resistance is due to the small number of these vessels, the length of these vessels, an increased viscosity of medullary blood, or other factors is unknown. The existence of a low medullary blood flow has considerable significance in the urinary concentrating mechanism (Chapter 30) and probably explains, at least in part, the increased susceptibility of the medulla to hypoxic injury in disease states (Brezis et al., 1984).

RENAL OXYGEN CONSUMPTION

As noted at the beginning of this chapter, the high blood flow to the kidneys is related to their role in regulating the quantity and composition of the body fluids. Thus, the kidneys differ from most other organs, in which blood flow is related to the oxygen requirements of the organ. Since the flow to the kidneys is so high relative to their oxygen needs, the renal arteriovenous oxygen difference is quite low, approximately 1.7 ml O_2/100 ml blood, compared to 4–5 ml O_2/100 ml blood for the body as a whole. The relationship between RBF, the renal arteriovenous oxygen difference, and renal oxygen consumption is expressed by the Fick equation (Chapter 13):

$$\text{RBF} = \frac{\dot{Q}_{O_2}}{RA_{O_2} - RV_{O_2}} \qquad (33)$$

where $\dot{Q}_{O_2}$ is the rate of renal oxygen consumption and $RA_{O_2} - RV_{O_2}$ is the renal arteriovenous oxygen difference.

In most organs, if the blood flow is varied, the *arteriovenous oxygen difference* changes inversely to meet the oxygen needs of the organ (assuming, of course, that the metabolic activity of the organ, and hence its consumption of oxygen, remains constant). The kidneys are unique in that changes in blood flow are accompanied by parallel changes in *oxygen consumption*, with the ateriovenous oxygen difference remaining the same. The explanation for this unusual behavior is that a change in RBF generally is accompanied by a parallel change in GFR and hence in the quantity of ions and other solutes that must be reabsorbed (Kiil et al., 1961). Solute reabsorption, in turn, requires oxygen for energy, with as much as 75–85% of the renal oxygen consumption being used to support the active reabsorption of ions and other solutes, particularly Na^+. In fact, renal oxygen consumption is directly proportional to the amount of Na^+ reabsorbed (Fig. 4.15). Changes in *solute reabsorption*, then, are responsible for the changes in oxygen consumption that accompany changes in RBF.

Synthesis of Erythropoietin

Having considered renal *oxygen consumption*, it is appropriate to note that the kidneys also have a role in monitoring the adequacy of *oxygen delivery* to the tissues in general. In response to renal hypoxia, the endothelial cells of the peritubular capil-

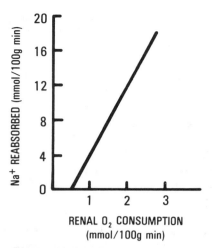

Figure 4.15. Relationship between renal oxygen consumption and amount of Na^+ reabsorbed by the kidneys. (Adapted from Deetjen and Kramer, *Pfluegers Arch* 273: 636–650, 1961.)

laries synthesize *erythropoietin*, a glycoprotein hormone that regulates the production of red blood cells (erythrocytes) from precursor cells in the bone marrow (Lacombe et al., 1988). For example, the elevated hematocrit (polycythemia) occurring upon exposure to high altitude or in chronic lung disease can be attributed to increased production of erythropoietin. Many diseases of the kidneys are associated with an erythropoietin deficiency, and hence anemia.

MEASUREMENT OF GFR

The GFR can be measured with the aid of certain substances that are freely filtered at the glomerulus but are *not* reabsorbed or secreted by the tubule. *Inulin*, a fructose polymer with a molecular weight of $\simeq 5000$ daltons, can be considered the "prototype" example of such a substance (Shannon and Smith, 1935). Since inulin is neither removed from the filtrate by reabsorption nor added to the filtrate by secretion, the entire amount filtered will be excreted in the urine. Hence, the rate of excretion must equal the filtered load:

$$\frac{\text{Mass of inulin excreted}}{\text{time}} = \frac{\text{Mass of inulin filtered}}{\text{time}}$$

or

$$U_{In}\dot{V} = P_{In}GFR \qquad (34)$$

If *Eq. 34* is substituted into the expression for the *clearance* of inulin *(Eq. 26)*, the following significant result is obtained:

$$C_{In} = \frac{U_{In}\dot{V}}{P_{In}} = \frac{P_{In}GFR}{P_{In}}$$

or

$$C_{In} = GFR \qquad (35)$$

C_{In}, of course, can be calculated from measurements made on plasma and urine samples. Thus, *Eq. 35* defines a *noninvasive* method for measuring GFR.

That GFR can be equated to C_{In} is actually quite logical. If some of the inulin in the filtrate were reabsorbed, the volume of plasma "cleared" of inulin would be *smaller* than the volume filtered. On the other hand, if some of the inulin that escaped filtration were added to the filtrate by secretion, the volume of plasma "cleared" would be *greater* than the volume filtered. Inulin is neither reabsorbed nor secreted, however, so the volume of plasma "cleared" of inulin per unit time (C_{In}) is *equal* to the volume filtered per unit time (GFR).

Because C_{In} is equal to GFR, C_{In} and GFR often are used interchangeably in discussions of renal function. It should be emphasized that the key characteristic of inulin necessary for the equality of C_{In} and GFR is the special way that the kidneys handle inulin: inulin is freely filtered but neither reabsorbed nor secreted. In addition, inulin is nontoxic, is not bound by plasma proteins, is neither metabolized nor synthesized by the kidneys, and has no effect on GFR.

Although inulin is considered the "prototype" substance for measuring GFR, other suitable substances have been identified, most notably *mannitol* and *iothalamate,* a contrast medium used in diagnostic radiology. However, all of these substances must be infused intravenously until a stable plasma concentration is achieved, an inconvenience for routine clinical determinations of GFR. A more common practice in the clinical setting is to use the endogenous substance *creatinine* to *estimate* GFR. Creatinine is an organic base formed during muscle protein metabolism as a degradation product of creatine phosphate. Like many other organic bases, creatinine is filtered at the glomerulus and secreted into the tubular lumen by the proximal tubular epithelial cells (Chapters 28 and 29). However, at normal plasma levels in humans the amount of creatinine secreted is only about 10–15% of the amount filtered. This means that

$$\frac{\text{Mass of creatinine excreted}}{\text{time}} \simeq \frac{\text{Mass of creatinine filtered}}{\text{time}}$$

or

$$U_{Cr}\dot{V} \simeq P_{Cr}GFR \qquad (36)$$

As an *approximation*, then, creatinine behaves like substances that are filtered only, such as *inulin (Eq.*

34).† The creatinine clearance can therefore be used clinically to obtain an approximate value for GFR:

$$C_{Cr} = \frac{U_{Cr}\dot{V}}{P_{Cr}} \simeq \frac{P_{Cr}GFR}{P_{Cr}}$$
$$\simeq GFR \qquad (37)$$

Because a small amount of creatinine is secreted in humans, C_{Cr} is of course slightly greater than C_{In}. The GFR value obtained from a C_{Cr} measurement is therefore slightly *greater* than the actual GFR, as determined from C_{In}. However, the size of the error is less than anticipated because the standard colorimetric methods for determining P_{Cr} measure other chromagens as well as true creatinine. The P_{Cr} values reported by clinical laboratories are therefore somewhat greater than the true P_{Cr} levels. This slight overestimate of P_{Cr} tends to reduce the C_{Cr} calculated with *Eq. 37* to a value closer to the actual GFR.

Clinically, when obtaining data for a creatinine clearance calculation, P_{Cr} typically is determined from a single plasma sample. However, a 24-hour urine collection generally is required for an accurate determination of $U_{Cr}\dot{V}$. Failure to collect a complete 24-hour urine sample is a common cause of erroneous C_{Cr} values. A simple method that can be used to check for a complete urine collection is to compare the C_{Cr} value determined from P_{Cr} and the 24-hour urine collection to the C_{Cr} value *estimated* from P_{Cr} *only*. Such an estimate is obtained from an empirical calculation that takes advantage of the dependence of creatinine production on muscle mass, which in turn is determined by age and body weight (Cockroft and Gault, 1976):

$$C_{Cr} \simeq \frac{(140 - \text{age}) \times \text{body weight}}{72 \times P_{Cr}} \qquad (38)$$

When using this experimentally derived equation, age is expressed in *years*, body weight in *kg*, and P_{Cr} in *mg/dl*. The C_{Cr} value calculated from *Eq. 38* has the units *ml/min* and its value will be within 35% of the C_{Cr} value determined from P_{Cr} and the 24-hour urine collection in 95% of patients. *Eq. 38* is *not* accurate in conditions in which the ratio of muscle mass to body weight is distorted, such as pregnancy, extreme obesity, or diseases that cause muscular wasting. In addition, *Eq. 38* is *not* accurate in children; through adolescence, better results are obtained with the following experimentally derived equation (Schwartz et al., 1976):

$$C_{Cr} \approx \frac{0.55 \times \text{length}}{P_{Cr}} \qquad (39)$$

---

†In the dog, creatinine is handled *exactly* like inulin, i.e., it is freely filtered but neither reabsorbed nor secreted.

When using this equation, length is expressed in *cm;* as in *Eq. 38,* P_{Cr} is expressed in *mg/dl* and C_{Cr} has the units *ml/min.*

While C_{Cr} represents a useful clinical approximation for GFR, important information about GFR actually can be obtained from a simple P_{Cr} measurement. This is because the rate of creatinine excretion must equal the rate of creatinine production and therefore tends to be constant from day to day in the steady state, i.e.,

$$U_{Cr}V = Constant$$

Since $U_{Cr}\dot{V} = P_{Cr}C_{Cr}$, the quantity $P_{Cr}C_{Cr}$ also tends to be constant:

$$P_{Cr}C_{Cr} = Constant$$

or

$$P_{Cr}GFR \simeq Constant \qquad (40)$$

P_{Cr} therefore is *inversely proportional* to the GFR, as illustrated in Figure 4.16.

An important feature of the P_{Cr} vs. GFR curve is its *shape.* Note that P_{Cr} changes little when GFR falls from its normal value of 125 ml/min to a value as low as 60 ml/min. However, P_{Cr} begins to increase markedly with further decreases in GFR. This means that P_{Cr} is a sensitive indicator of GFR only at fairly low GFRs. Thus, P_{Cr} is not useful for detecting small or even moderate decreases in GFR due to minor impairment of renal function but is extremely useful for detecting the large decreases in GFR seen in severe renal dysfunction. The upper limit of normal for P_{Cr} is approximately 110 μmol/liter (1.25 mg/dl) in men and 100 μmol/liter (1.1 mg/dl) in women.

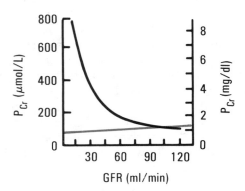

Figure 4.16. Effect of GFR on plasma creatinine concentration (P_{Cr}). Note that P_{Cr} is a sensitive indicator of GFR only at low GFRs. The upper limit of normal for P_{Cr} in men is indicated by the *blue line.*

MEASUREMENT OF RPF

The Fick equation, introduced in the discussion of renal oxygen consumption *(Eq. 33),* provides the basis for measuring RPF. For this application, the Fick equation is written as

$$RPF = \frac{\dot{Q}_X}{P_{RA_X} - P_{RV_X}} \qquad (41)$$

where $\dot{Q}_X$ is the rate at which a substance X is consumed by the kidneys and P_{RA_X} and P_{RV_X} represent the concentrations of X in renal arterial and renal venous plasma, respectively. For substances that are not metabolized or synthesized by the kidneys, the rate at which the substance is "consumed" by the kidneys can be equated to its rate of *excretion,* $U_X\dot{V}$. Furthermore, if the substance if not "consumed" by tissues other than the kidneys, P_{RA_X} can be equated to P_X, the concentration of X measured in *systemic* arterial or venous plasma (a venous sample generally is used for convenience). Thus

$$RPF = \frac{U_X V}{P_X - P_{RV_X}} \qquad (42)$$

A variety of substances could be used for the determination of RPF with *Eq. 42.* But whereas the rate of excretion $U_X\dot{V}$ and the systemic plasma concentration P_X are easily determined, renal venous plasma levels P_{RV_X} are *not* readily obtainable in routine clinical studies. The most ideal substance for a *noninvasive* determination of RPF with the Fick equation, then, would be one with a renal venous plasma concentration of *zero,* i.e., a substance that is *totally removed* from renal arterial plasma by the kidneys. For such a substance, the Fick equation *(Eq. 42)* reduces to the *clearance equation (Eq. 26):*

$$RPF = \frac{U_X\dot{V}}{P_X - 0}$$
$$= C_X \qquad (43)$$

This result is a logical one, since a renal venous plasma level of zero would indicate that the total RPF had been "cleared" of X.

What type of substance might have a renal venous plasma concentration of zero? Because only 20% of the RPF is filtered, the substance could *not* be completely "cleared" from renal plasma by filtration alone; it also would have to be added to the tubular fluid by secretion. Furthermore, this secretion would have to be efficient enough to remove virtually all of the substance that was not filtered. The substance that comes closest to meeting these criteria is *p-aminohippurate* (PAH). PAH is filtered at the glomerulus and, like other organic acids, is secreted by the proximal tubular epithelial cells, i.e., it is transported from the peritubular capillaries surrounding the proximal tubule into the tubular lumen (Chapters 28 and 29). Furthermore, at low plasma levels of PAH this secretion is so efficient that virtually all of the PAH in the peritubular capillaries is secreted into the proximal tubular lumen. But even PAH is not *completely* "cleared" from renal plasma because 10–15% of the total RPF supplies portions of the kidneys that *cannot* remove PAH (or any other substance), e.g., the renal capsule, renal pelvis, perirenal fat, medulla, and papilla. Thus, the renal venous plasma concentration of PAH (or any other substance) is *not* zero. The so-called *extraction ratio* (E_{PAH}) is used to represent the fraction of PAH that is "cleared," or *extracted*, from the renal plasma in a single passage through the kidneys:

$$E_{PAH} = \frac{P_{PAH} - P_{RV_{PAH}}}{P_{PAH}} \qquad (44)$$

Since 10–15% of the total RPF is not "cleared" of PAH, E_{PAH} has a value of 0.85–0.90.

While C_{PAH} therefore only *approximates* the total RPF, it *does* accurately measure the rate of plasma flow to regions of the kidneys that *can* remove PAH, the so-called *effective renal plasma flow* (ERPF):

$$\begin{aligned} C_{PAH} &= \frac{U_{PAH}\dot{V}}{P_{PAH}} \\ &= ERPF \end{aligned} \qquad (45)$$

By combining *Eqs. 42* and *44*, it is evident that

$$ERPF = RPF \cdot E_{PAH} \qquad (46)$$

Table 4.9.
Methods for Determining GFR and RPF

	Most Accurate Determination	Useful Approximation for Clinical Purposes
GFR	C_{In}	C_{Cr}
RPF	Fick equation	C_{PAH}

For clinical purposes, the ERPF values obtained from C_{PAH} measurements provide a useful approximation for the total RPF. *Iodohippurate* has an extraction ratio similar to that of PAH and therefore is another substance whose clearance can be used to approximate the total RPF.

$$RBF = \frac{RPF}{1 - Hct} \qquad (47)$$

where Hct is the *hematocrit*.

CALCULATION OF FILTRATION FRACTION

Table 4.9 summarizes the methods presented in this chapter for measuring the GFR and RPF. Once GFR and RPF have been determined, the *filtration fraction* can readily be calculated. The most accurate value of the filtration fraction would be obtained as follows:

$$FF = \frac{C_{In}}{RPF} \qquad (48)$$

where RPF is determined with the Fick equation. However, an approximation useful for clinical purposes is

$$FF \simeq \frac{C_{Cr}}{C_{PAH}} \qquad (49)$$

CHAPTER 28

Introduction to Tubular Function

As indicated in Chapter 26, the tubular portion of the nephron modifies the contents of the glomerular filtrate so that the final urine contains only those constituents that must be excreted to preserve the normal volume and composition of the body fluids. This modification involves both *tubular reabsorption*, the process whereby water and other essential substances in the glomerular filtrate are *recovered* and returned to the body via the peritubular capillaries, and *tubular secretion*, the process whereby substances in the peritubular capillaries are transported across the tubular epithelium and *added* to the tubular fluid. In this chapter, a general introduction to these important functions of the tubule will be presented.

TUBULAR REABSORPTION

Since approximately 180 liters of plasma are filtered each day in humans, the tremendous importance of tubular reabsorption in conserving the body stores of water and essential solutes should be obvious. For a substance X, the amount reabsorbed per unit time can be calculated as the difference between the filtered load and the rate of excretion:

$$\frac{\text{Mass of X reabsorbed}}{\text{time}}$$
$$= \frac{\text{Mass of X filtered}}{\text{time}} - \frac{\text{mass of X excreted}}{\text{time}}$$

or

$$T_X = P_X GFR - U_X \dot{V} \qquad (50)$$

where T_X represents the rate of tubular reabsorption of X. Table 4.10 presents typical reabsorption rates for several major plasma constituents, emphasizing the magnitude of the reabsorption process. Also listed are typical *fractional reabsorption* values for these constituents, i.e., the fraction of the filtered load that is reabsorbed. The fractional reabsorption can be calculated as follows:

$$\text{Fractional reabsorption} = \frac{T_X}{P_X GFR} \qquad (51)$$

Note that the reabsorption of certain essential constituents, such as glucose and amino acids, is virtually *complete*, i.e., the *entire* filtered load is reabsorbed.

Eq. 50 can be used to derive an expression for the *clearance* of a substance that is filtered and then reabsorbed. Rearranging this equation gives

$$U_X \dot{V} = P_X GFR - T_X \qquad (52)$$

Eq. 52 can be substituted into the clearance equation $C_X = U_X \dot{V}/P_X$ to give

$$C_X = \frac{P_X GFR - T_X}{P_X}$$

or

$$C_X = GFR - \frac{T_X}{P_X} \qquad (53)$$

Therefore, for a substance that is filtered and reabsorbed

$$C_X < GFR$$

or, since GFR can be equated to the *clearance of inulin (Eq. 35)*

$$C_X < C_{In} \qquad (54)$$

This result is expected, since a smaller volume of plasma will be "cleared" of a substance that is filtered and then reabsorbed than will be "cleared" of a substance, such as inulin, that is filtered only. It should be noted that *Eq. 53* was derived primarily to illustrate the relationship between C_X and C_{In} *(Eq. 54)*, not as a method of calculating C_X. C_X, of course, is most readily determined from the clearance equation *(Eq. 26)*.

The transport mechanisms involved in the tubular reabsorption of substances such as those listed in Table 4.10 can be broadly classified as active or passive. In *passive* reabsorption, a substance dif-

Table 4.10.
Typical Rates of Tubular Reabsorption in Normal Humans (GFR = 180 liters/day)

	P_X, mmol/liter	Filtered Load P_XGFR, mmol/day	Rate of Excretion $U_X\dot{V}$, mmol/day	Rate of Tubular Reabsorption $T_X = P_X$GFR $- U_X\dot{V}$, mmol/day	Fractional Reabsorption, T_X/P_XGFR $\times$ 100%
Na^+	140	25,200	100	25,100	>99%
Cl^-	105	18,900	100	18,800	>99%
K^+	4	720	50	670	93%
HCO_3^-	24	4,320	2	4,318	>99%
Glucose	5	900	0	900	100%
Alanine	0.35	63	0	63	100%
Urea	6	1080	432	648	60%
Water		180 liters	1.5 liters	178.5 liters	>99%

fuses from tubular lumen to peritubular capillary along an osmotic, electrical, and/or concentration gradient. Metabolic energy is not directly required, although energy may have been used to establish the initial gradient. In contrast, *active* reabsorption occurs against an electrical and/or concentration gradient and requires the expenditure of metabolic energy. The reabsorption of various components of the glomerular filtrate will be described in detail in later chapters. However, an important example of *active* reabsorption (T_m-limited reabsorption) and of *passive* reabsorption (urea) will be discussed here to illustrate certain general characteristics of active and passive reabsorption mechanisms

Active Reabsorption: T_m-limited Reabsorption

For many actively reabsorbed substances, the rate of reabsorption has a finite upper limit. The maximum rate that can be achieved is termed the *transport maximum* (T_m). Glucose, amino acids, phosphate, and sulfate are examples of substances that exhibit such *T_m-limited reabsorption*. Furthermore, the affinity of the transport system for many substances with T_m-limited reabsorption is so high that the *entire* filtered load is reabsorbed from the tubular fluid as long as the transport system is unsaturated. For example, the reabsorption of glucose and amino acids is virtually complete if the filtered load does not saturate the transport system.

It is instructive to consider the renal handling of a substance such as glucose from a quantitative standpoint. As noted above, when the filtered load of glucose does *not* saturate the transport system, the entire filtered load is reabsorbed, i.e., the rate of reabsorption equals the filtered load:

$$\frac{\text{Mass of glucose reabsorbed}}{\text{time}} = \frac{\text{Mass of glucose filtered}}{\text{time}}$$

or

$$T_G = P_G\text{GFR} \tag{55}$$

Clearly, under such conditions no glucose is excreted. However, when the transport system is *saturated,* the excess load will be excreted. The transport system will be operating at its maximum rate under such conditions, i.e.,

$$\frac{\text{Mass of glucose reabsorbed}}{\text{time}} = T_{m_G}$$

where T_{m_G} represents the *transport maximum* for glucose. The rate of excretion therefore will be (cf. *Eq. 52*)

$$U_G\dot{V} = P_G\text{GFR} - T_{m_G} \tag{56}$$

It is possible, then, to define a plasma concentration at which *glucosuria* first occurs. This critical plasma concentration is termed the *renal threshold* for glucose (P_{Th_G}). Theoretically, P_{Th_G} represents that plasma concentration at which the filtered load exactly saturates the transport system, i.e.,

$$P_{Th_G}\text{GFR} = T_{m_G}$$

or

$$P_{Th_G} = \frac{T_{m_G}}{\text{GFR}} \tag{57}$$

GLUCOSE TITRATION CURVES

The above discussion can be summarized by examining how the renal handling of glucose varies with plasma glucose concentration. Since the filtered load (P_GGFR), rate of excretion ($U_G\text{V}$), and rate of tubular reabsorption (T_G) all have the same units (e.g., mmol/min), plots of filtered load vs. P_G, rate of excretion vs. P_G, and rate of reabsorption vs. P_G can be drawn on a single graph (Fig. 4.17, *blue curves*). The important features of these so-called *glucose titration curves* are as follows:

1. *Filtered Load vs. P_G.* Since GFR is constant, the plot of filtered load (P_GGFR) vs. P_G is a straight line with slope GFR.

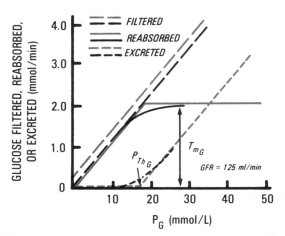

Figure 4.17.   Glucose titration curves, illustrating effect of plasma glucose concentration (P_G) on filtered load, rate of excretion, and rate of tubular reabsorption. The maximum rate of tubular reabsorption (T_{m_G}) is about 2 mmol/min. *Blue curves* are *idealized,* showing *abrupt* changes in the rate of tubular reabsorption and rate of excretion when the transport system is saturated, which is predicted to occur at a P_G of about 16 mmol/liter (P_{Th_G}; see text). *Black curves* resemble *actual data* and exhibit *gradual* changes in the rate of tubular reabsorption and rate of excretion *(splay).*

2. *Rate of Tubular Reabsorption vs. P_G.* At low P_G, the *entire* filtered load is reabsorbed. At higher P_G, the transport system becomes *saturated* and reabsorbs glucose at its maximum rate, T_{m_G}. In this example, T_{m_G} is approximately 2 mmol/min (375 mg/min), which is the average normal value in men (the average normal T_{m_G} in women is somewhat lower, about 1.7 mmol/min, or 303 mg/min).

3. *Rate of Excretion vs. P_G.* Since the entire filtered load is reabsorbed at low P_G, glucose is virtually absent from the urine. Once the transport system becomes saturated, however, the excess filtered load is excreted *(Eq. 56).* In this example, in which GFR is 125 ml/min, glucosuria first occurs at a P_G of about 16 mmol/liter (300 mg/dl), i.e., the *renal threshold* for glucose is 16 mmol/liter *(Eq. 57).* When P_G exceeds the renal threshold, the rate of excretion vs. P_G curve parallels the filtered load vs. P_G curve, i.e., it is a straight line with slope GFR, as predicted by *Eq. 56.*

It should be noted that the glucose titration curves drawn in blue in Figure 4.17 are somewhat *idealized,* as the plots of rate of reabsorption vs. P_G and rate of excretion vs. P_G show an *abrupt* change when the rate of reabsorption reaches its maximum level. *Actual* glucose titration curves show a more *gradual* change in this region, as illustrated by the *black curves* in Figure 4.17 (Mudge, 1958). This feature of renal titration curves is called *splay.* Because of splay, the renal threshold for glucose is

approximately 11 mmol/liter (200 mg/dl) instead of 16 mmol/liter (300 mg/dl), even though the transport system does not become saturated (i.e., the rate of tubular reabsorption does not reach its maximum level) until P_G is 16 mmol/liter. The threshold value of 16 mmol/liter shown by the *blue curves* in Figure 4.17 (and calculated by *Eq. 57*) therefore can only be regarded as the *theoretical* renal threshold for glucose. Although the measured renal threshold is closer to 11 than to 16 mmol/liter, it still is considerably larger than the normal P_G (about 5 mmol/liter). Thus, glucose normally is absent from the urine.

At least two factors account for the splay phenomenon. First, although the affinity of the transport system for glucose is very high, some glucose will appear in the urine before the transport system is fully saturated. An analysis of the kinetics of glucose transport shows why this is so. Using the symbol R to represent the carrier responsible for the reabsorption of glucose, the reaction between glucose and the carrier can be written as $GR \rightleftharpoons G + R$ and can be described by the following mass action equation

$$K = \frac{[G][R]}{[GR]}$$

where K is the dissociation constant for the glucose-carrier reaction. Even though K is very small (i.e., the affinity of the carrier for glucose is very large), a finite concentration of glucose must remain in the tubular fluid to saturate the transport system. Some glucose therefore will be excreted prior to saturation.

A second factor contributing to the presence of splay is nephron heterogeneity. Considerable variability in the filtration capacity of glomeruli and in the reabsorptive capacity of renal tubules is believed to exist. For example, some nephrons may have subnormal reabsorptive capacities and therefore are likely to excrete glucose at a lower P_G than the average nephron. In contrast, other nephrons may have high reabsorptive capacities and therefore are likely to excrete glucose at a higher P_G than the average nephron.

The effects of P_G on filtered load, rate of excretion, and rate of tubular reabsorption have now been considered. It also is of interest to examine the effect of P_G on the *clearance* of glucose (Fig. 4.18). At low P_G, the entire filtered load is reabsorbed, so that no plasma is "cleared" of glucose:

$$C_G = \frac{U_G \dot{V}}{P_G}$$
$$= 0 \qquad (58)$$

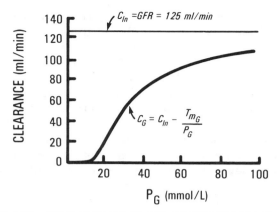

Figure 4.18. Effect of plasma glucose concentration (P_G) on clearance of glucose (C_G). C_G increases as P_G increases, asymptotically approaching the clearance of inulin (C_{In}).

Once the renal threshold for glucose is exceeded and glucosuria begins to occur, the clearance of glucose gradually increases. A formal expression for C_G can be derived by substituting *Eq. 56* into the clearance equation $C_G = U_G \dot{V}/P_G$:

$$C_G = \frac{P_G GFR - T_{mG}}{P_G}$$
$$= GFR - \frac{T_{mG}}{P_G}$$

or

$$C_G = C_{In} - \frac{T_{mG}}{P_G} \qquad (59)$$

Thus, C_G asymptotically approaches C_{In} at high P_G, as illustrated in Figure 4.18. But C_G *always* is less than C_{In}, since some glucose always is reabsorbed *(Eq. 54).*

Passive Reabsorption: Urea

The renal handling of urea represents an important example of *passive reabsorption.* In many regions of the nephron, urea will be passively reabsorbed whenever its concentration in the tubular fluid exceeds that in the surrounding peritubular fluid. Such a urea concentration gradient is created in these regions by the reabsorption of *water* (Chapters 29 and 30).

Because urea reabsorption occurs as a consequence of the reabsorption of water, urea reabsorption varies markedly with the urine flow rate (i.e., with the volume of water that is *not* reabsorbed). This dependence of reabsorption on $\dot{V}$ is a unique characteristic of passively reabsorbed substances like urea. It is therefore instructive to compare the effects of $\dot{V}$ on the renal handling of a substance

such as *inulin* to the effects of $\dot{V}$ on urea. Since inulin is neither reabsorbed nor secreted, the amount of inulin excreted always equals the amount filtered *(Eq. 34):*

$$U_{In}\dot{V} = P_{In}GFR$$

Thus, changes in $\dot{V}$ will result in reciprocal changes in U_{In}, so that the product $U_{In}\dot{V}$ always equals the filtered load $P_{In}GFR$. The clearance of inulin ($= U_{In}\dot{V}/P_{In}$) is therefore independent of $\dot{V}$ and, as previously discussed, is always equal to the GFR (Chapter 27). In contrast, with *urea* the amount excreted increases with increasing $\dot{V}$ (due to decreased reabsorption), so that the clearance of urea increases as $\dot{V}$ increases.

The effect of $\dot{V}$ on C_{In} and C_U is illustrated in Figure 4.19 (Smith, 1951). Note that C_U increases as $\dot{V}$ increases from approximately 0.5 ml/min to approximately 1.5 ml/min but then levels off as $\dot{V}$ exceeds 2 ml/min, approaching approximately two-thirds C_{In}.

TUBULAR SECRETION

Tubular secretion is the process whereby substances in the peritubular capillaries are transported across the tubular epithelium into the tubular lumen. Thus, whereas substances are *recovered* from the tubular fluid in tubular *reabsorption,* substances are *added* to the tubular fluid in tubular *secretion.* It is important to distinguish between secretion and *excretion,* the overall process by which the kidneys compose the final urine. For example, a substance could be filtered at the glomerulus, be partially reabsorbed in one or more regions of the nephron, never be secreted, and still be excreted.

For a substance X, the amount secreted per unit time can be calculated as the difference between the

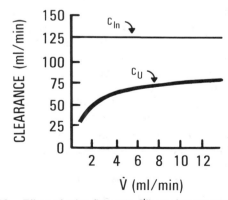

Figure 4.19. Effect of urine flow rate ($\dot{V}$) on clearances of urea (C_U) and inulin (C_{In}). As $\dot{V}$ exceeds 2 ml/min, C_U approaches two-thirds C_{In}. (Adapted from Chasis and Smith, 1938.)

rate of excretion and the filtered load:

$$\frac{\text{Mass of X secreted}}{\text{time}} = \frac{\text{Mass of X excreted}}{\text{time}} - \frac{\text{mass of X filtered}}{\text{time}}$$

The rate of tubular secretion, like the rate of tubular reabsorption, commonly is designated by the symbol T_X. Hence,

$$T_X = U_X\dot{V} - P_X GFR \qquad (60)$$

Eq. 60 can be used to derive an expression for the *clearance* of a substance that is filtered and then secreted. Rearranging this equation gives

$$U_X\dot{V} = P_X GFR + T_X \qquad (61)$$

Eq. 61 can be substituted into the clearance equation $C_X = U_X\dot{V}/P_X$ to give

$$C_X = \frac{P_X GFR + T_X}{P_X}$$

or

$$C_X = GFR + \frac{T_X}{P_X} \qquad (62)$$

Therefore, for a substance that is filtered and secreted,

$$C_X > GFR$$

or

$$C_X > C_{In} \qquad (63)$$

This result is expected, since a larger volume of plasma will be "cleared" of a substance that is filtered and then secreted than will be "cleared" of a substance, such as inulin, that is filtered only. It should be noted that *Eq. 62* was derived primarily to illustrate the relationship between C_X and C_{In} (*Eq. 63*), not as a method for calculating C_X. C_X, of course, is most readily determined from the clearance equation (*Eq. 26*).

Like the mechanisms for tubular reabsorption, the mechanisms involved in tubular secretion can be broadly characterized as active or passive. The secretion of specific substances will be described in detail in later chapters. However, an important example of *active secretion* (T_m-limited secretion) will be discussed here as an illustration of certain general characteristics of active secretion mecha-nisms and because of its close correspondence to the active reabsorption example discussed above.

Active Secretion T_m-limited Secretion

In analogy to T_m-limited reabsorption, certain substances that are actively secreted exhibit T_m-limited secretion, i.e., the rate of secretion has a finite upper limit, termed the *transport maximum* (T_m). *p*-Aminohippurate (PAH), iodohippurate, and penicillin are among the substances whose secretion is T_m-limited. Some of these substances also exhibit a phenomenon that is somewhat analogous to the threshold phenomenon for reabsorption. Specifically, the affinity of the transport system for some secreted substances is so high that essentially *all* of the substance delivered to the peritubular capillaries is secreted into the tubular fluid as long as the transport system is not saturated. For example, PAH and iodohippurate are almost completely removed from the peritubular capillaries when the transport system is unsaturated.

It is instructive to examine the renal handling of a substance such as PAH in a quantitative fashion. The rate of delivery of PAH to the peritubular capillaries is given by

$$\frac{\text{Mass of PAH delivered to peritubular capillaries}}{\text{time}}$$
$$= P_{PAH} \times \frac{\text{volume of plasma delivered to peritubular capillaries}}{\text{time}}$$
$$(64)$$

The rate of plasma delivery to the peritubular capillaries can be *approximated* by subtracting GFR from the total RPF, i.e.,

$$\frac{\text{Volume of plasma delivered to peritubular capillaries}}{\text{time}}$$
$$\simeq RPF - GFR \qquad (65)$$

Eq. 65 is an approximation because, as discussed in Chapter 27, about 10–15% of the total RPF supplies the renal capsule, renal pelvis, perirenal fat, medulla, and papilla, regions of the kidneys in which *no* filtration or secretion of PAH occurs. Thus, the rate of plasma delivery to the peritubular capillaries is more precisely given by

$$\frac{\text{Volume of plasma delivered to peritubular capillaries}}{\text{time}}$$
$$= ERPF - GFR \qquad (66)$$

where ERPF is the *effective renal plasma flow*, i.e., the rate of plasma flow to regions of the kidneys that *can* remove PAH (see *Eq. 45*). *Eq. 64* therefore can be expressed as

$$\frac{\text{Mass of PAH delivered to peritubular capillaries}}{\text{time}}$$
$$= P_{PAH}(ERPF - GFR) \quad (67)$$

As noted above, when the secretory mechanism is not saturated, essentially *all* of the PAH delivered to the peritubular capillaries is secreted:

$$\frac{\text{Mass of PAH secreted}}{\text{time}}$$
$$= \frac{\text{Mass of PAH delivered to peritubular capillaries}}{\text{time}}$$

i.e.,

$$T_{PAH} = P_{PAH}(ERPF - GFR) \quad (68)$$

Under such conditions, then, PAH is almost completely removed from the plasma by the kidneys. Approximately 20% of the PAH delivered to the kidneys enters the tubular fluid by glomerular filtration; the remaining 80% enters the peritubular capillaries and is secreted. This can be expressed formally by substituting *Eq. 68* into *Eq. 61:*

$$U_{PAH}\dot{V} = P_{PAH}GFR + P_{PAH}(ERPF - GFR) \quad (69)$$
$$= P_{PAH}ERPF$$

However, once the secretory mechanism is saturated, the rate of PAH secretion is fixed at its maximum value, $T_{m_{PAH}}$. The rate of PAH excretion then will be

$$U_{PAH}\dot{V} = P_{PAH}GFR + T_{m_{PAH}} \quad (70)$$

PAH TITRATION CURVES

The renal handling of PAH, as described above, can be summarized using PAH titration curves (Fig. 4.20). The important features of these curves are as follows:

1. *Filtered Load vs. P_{PAH}.* Since GFR is constant, the plot of filtered load ($P_{PAH}GFR$) vs. P_{PAH} is a straight line with slope GFR.
2. *Rate of Tubular Secretion vs. P_{PAH}.* At low P_{PAH}, where the transport system is able to secrete essentially all of the PAH delivered to the peritubular capillaries, the rate of secretion is considerably greater than the filtered load. At higher P_{PAH}, the transport system becomes *saturated* and secretes PAH at its maximum rate,

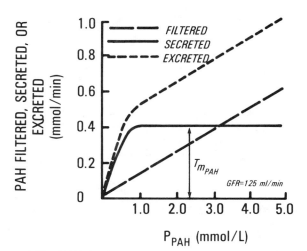

Figure 4.20. PAH titration curves, illustrating effect of plasma PAH concentration (P_{PAH}) on filtered load, rate of excretion, and rate of tubular secretion. The maximum rate of tubular secretion ($T_{m_{PAH}}$) is about 0.4 mmol/min. The rate of tubular secretion and rate of excretion change *gradually* as the transport system is saturated *(splay).*

$T_{m_{PAH}}$. In this example, $T_{m_{PAH}}$ has a value of 0.4 mmol/min (80 mg/min), which is the average normal value in men.
3. *Rate of Excretion vs. P_{PAH}.* The curve rises steeply at low P_{PAH}. Once the transport system becomes saturated, however, the rate of excretion vs. P_{PAH} curve parallels the filtered load vs. P_{PAH} curve, i.e., it is a straight line with slope GFR, as predicted by *Eq. 70.*

Note that the PAH titration curves show a *gradual* change as the rate of tubular secretion reaches its maximum level. Thus, like the titration curves for glucose (Fig. 4.17, *black curves*), the PAH titration curves exhibit *splay.* The factors accounting for the splay in the PAH curves are analogous to those presented above for glucose.

The effect of P_{PAH} on the *clearance* of PAH now will be examined. At *low* P_{PAH}, essentially all of the PAH is removed from the ERPF by the kidneys, as noted above, i.e., the entire ERPF is "cleared" of PAH. This can be expressed formally by substituting *Eq. 69* into the clearance equation $C_{PAH} = U_{PAH}\dot{V}/P_{PAH}$:

$$C_{PAH} = \frac{P_{PAH}ERPF}{P_{PAH}}$$

i.e.,

$$C_{PAH} = ERPF \quad (71)$$

Eq. 71 was introduced previously, when the clinical use of PAH clearance measurements to approximate

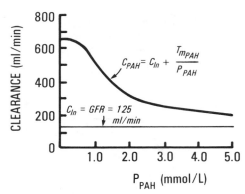

Figure 4.21. Effect of plasma PAH concentration (P_{PAH}) on the clearance of PAH (C_{PAH}). C_{PAH} decreases as P_{PAH} increases, asymptotically approaching the clearance of inulin (C_{In}).

RPF was considered (*Eq. 45*). At *high* P_{PAH}, when the transport system is saturated, some of the PAH in the peritubular capillaries will *not* be secreted, i.e., a smaller volume of plasma is "cleared" of PAH. An expression for C_{PAH} under these conditions can be obtained by substituting *Eq. 70* into the clearance equation $C_{PAH} = U_{PAH}\dot{V}/P_{PAH}$:

$$C_{PAH} = \frac{P_{PAH}GFR + T_{mPAH}}{P_{PAH}}$$

$$= GFR + \frac{T_{mPAH}}{P_{PAH}}$$

or

$$C_{PAH} = C_{In} + \frac{T_{mPAH}}{P_{PAH}} \qquad (72)$$

Thus, C_{PAH} asymptotically approaches C_{In} at high P_{PAH}, as illustrated in Figure 4.21. But C_{PAH} *always* is greater than C_{In}, since some PAH always is secreted (*Eq. 63*).

BIDIRECTIONAL TRANSPORT

Thus far in this chapter, consideration has been given to substances that are either filtered and then *reabsorbed* or filtered and then *secreted*. Certain substances, however, are transported across the tubular epithelium in *both* directions, i.e., they are filtered and then both reabsorbed *and* secreted. For quantitative purposes, a substance that undergoes such *bidirectional transport* is categorized according to whether there is *net* reabsorption or *net* secretion.

The renal handling of K^+ represents an important example of bidirectional transport. While K^+ generally undergoes *net reabsorption*, net secretion may occur under certain conditions, such as during the ingestion of a high K^+ diet (Chapter 33).

Another endogenous substance that exhibits bidirectional transport is *uric acid*. In contrast to K^+, uric acid *always* undergoes *net reabsorption* (Chapter 29).

Weak Organic Acids and Bases

The proximal tubular epithelium contains a special active transport system capable of secreting a variety of organic acids and another capable of secreting a variety of organic bases (Chapter 29). *p*-Aminohippurate (PAH), whose handling by the kidneys was discussed above, is an example of a substance that is secreted by the transport system for organic acids. It is important to emphasize that these transport systems only secrete organic molecules that are *ionized* in plasma. Whereas *strong* organic acids and bases are completely ionized in plasma (pH = 7.4), this is not necessarily true of *weak* organic acids and bases, which may exist partly or even primarily in the nonionized form at pH 7.4. Many weak organic acids and bases therefore undergo little active secretion. However, the nonionized form of most weak organic acids and bases is sufficiently lipid-soluble to diffuse *passively* across the tubular epithelium. Since many drugs, drug metabolites, and other exogenous organic molecules are weak acids and bases, the effect of this passive diffusion on the renal handling of such substances is a topic of considerable significance.

Like other plasma constituents, a weak organic acid or base that is freely filtered will be present in the filtrate at a concentration virtually identical with that in plasma. As water is reabsorbed, its concentration in the tubular fluid will exceed that in plasma, i.e., a concentration gradient favoring reabsorption develops. However, since only the *nonionized* form is sufficiently lipid-soluble to diffuse across the tubular epithelium, it is the concentration gradient of the *nonionized diffusible* form (*not* the *total* concentration gradient) that is significant. In any given region of the nephron, the concentration gradient of the nonionized form may favor either reabsorption *or* secretion, depending upon the relationship between the pH of the tubular fluid in that region and the pK_a of the acid or base. Weak organic acids and bases therefore can exhibit *bidirectional transport*.

As an example of such bidirectional transport, consider the renal handling of a *weak organic acid*, whose ionization reaction can be written as follows:

$$HA \rightleftharpoons H^+ + A^-$$

If the tubular fluid is *acidic*, the reaction will shift to the left, favoring the formation of the *nonion-*

ized, diffusible form HA (the exact relationship between the pH of the tubular fluid and the concentration of HA will depend on the pK_a of the acid). The concentration gradient for the reabsorption of the nonionized form created by the reabsorption of water (Fig. 4.22*A*) will be *augmented* by this shift of the equilibrium (Fig. 4.22*B*). Conversely, if the tubular fluid is *alkaline,* the formation of the *ionized, nondiffusible* form A⁻ is favored. The concentration gradient for the reabsorption of the nonionized form will therefore be small in spite of the reabsorption of water. In fact, depending on the relationship between the pH of the tubular fluid and the pK_a of the acid, the concentration of the nonionized form in the tubular fluid may become so small that a concentration gradient for the *secretion* of the nonionized form actually may develop (Fig. 4.22*C*). Hence, an *acidic* tubular fluid favors the *reabsorption* of weak organic acids, while an *alkaline* tubular fluid favors their *secretion* and hence eventual *excretion.* The phenomenon whereby the *nonionized* form of a molecule (e.g., HA) can readily diffuse across the tubular epithelium while the *ionized* form (e.g., A⁻) cannot is termed *nonionic diffusion* or *diffusion trapping.*

The renal handling of a *weak organic base* is governed by similar principles:

$$BH^+ \rightleftharpoons B + H^+$$

However, in this case an *alkaline* tubular fluid favors the formation of the *nonionized, diffusible* form (B), while an *acidic* tubular fluid favors the formation of the *ionized, nondiffusible* form (HB⁺). Thus, as illustrated in Figure 4.23 an *alkaline* tubular fluid favors the *reabsorption* of weak organic bases, while an *acidic* tubular fluid favors their *secretion* and hence eventual *excretion.*

As previously noted, many drugs and drug metabolites are weak organic acids and bases. The fact that the tubular fluid pH can significantly affect the renal handling of weak acids and bases therefore has important implications for *drug excretion.* Urine normally is acidic (average pH <6.0; see Chapter 32), which would favor the *reabsorption* of acidic drugs (Fig. 4.22*B*) and the *excretion* of *basic* drugs (Fig. 4.23*B*). Thus, in certain clinical situations the manipulation of urine pH to maximize or minimize drug excretion may be beneficial. For example, in cases of overdose involving certain *acidic* drugs (e.g., aspirin), *alkalinization* of the urine can promote drug excretion.

The *urine flow rate* also can have a significant effect on the excretion of weakly acidic and basic drugs. The reason is that the reabsorption of water has an important role in creating the concentration gradient for the passive reabsorption of the nonionized, diffusible form of the drug (Figs. 4.22*A* and 4.23*A*). With an increase in V̇, this concentration gradient is diminished, promoting drug excretion. The effect of V̇ on the passive reabsorption of weak organic acids and bases is completely analogous to the effect of V̇ on the passive reabsorption of *urea* (see above).

FRACTIONAL EXCRETION AND FRACTIONAL REABSORPTION

Important information about the renal handling of a substance can be obtained from measurements of its concentration in plasma, in urine, and in tubular fluid samples obtained by micropuncture (Chapter 26). In the following discussion, such concentration data will be used to calculate the *fractional excretion, fractional reabsorption,* and other significant quantities.

The *fractional excretion* of a substance X (FE_X)

A   **B**   **C**

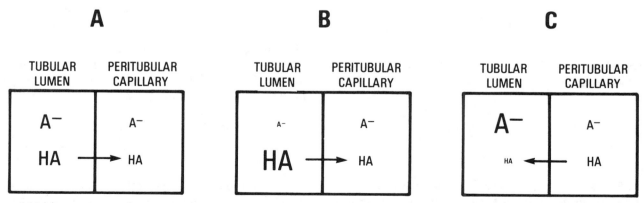

Figure 4.22. Renal handling of weak organic acid. The relative concentrations of HA and A⁻ are indicated by the size of the letters. *A,* Reabsorption of water creates a concentration gradient for the reabsorption of the nonionized, diffusible form HA. *B, Acidic* tubular fluid *augments* the concentration gradient for the reabsorption of HA. *C, Alkaline* tubular fluid may produce a concentration gradient for the *secretion* of HA.

A   **B**   **C**

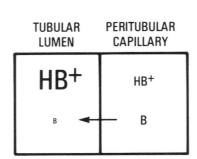
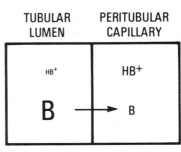

Figure 4.23. Renal handling of weak organic base. The relative concentrations of HB^+ and B are indicated by the size of the letters. *A,* Reabsorption of water creates a concentration gradient for the reabsorption of the nonionized, diffusible form B. *B, Acidic* tubular fluid may produce a concentration gradient for the *secretion* of B. *C, Alkaline* tubular fluid *augments* the concentration gradient for the reabsorption of B.

can be defined as the fraction of the total amount of X in the glomerular filtrate that appears in the final urine. Thus

$$FE_X = \frac{\text{Mass of X excreted}}{\text{mass of X filtered}}$$

Dividing both numerator and denominator by time gives

$$FE_X = \frac{\dfrac{\text{Mass of X excreted}}{\text{time}}}{\dfrac{\text{Mass of X filtered}}{\text{time}}}$$

or

$$FE_X = \frac{U_X \dot{V}}{P_X GFR} \qquad (73)$$

Several alternative expressions for FE_X can be derived from *Eq. 73.* For example, substituting C_X for $U_X\dot{V}/P_X$ and C_{In} for GFR gives

$$FE_X = \frac{C_X}{C_{In}} \qquad (74)$$

Alternatively, $U_{In}\dot{V}/P_{In}$ can be substituted for GFR in *Eq. 73* to give

$$FE_X = \frac{U_X\dot{V}/P_X}{U_{In}\dot{V}/P_{In}}$$
$$= \frac{U_X/P_X}{U_{In}/P_{In}} \qquad (75)$$

Eq. 75 is of particular interest because it allows FE_X to be determined *without* measuring the urine flow rate, $\dot{V}$.

The fractional excretion of X (FE_X) represents the fraction of the total amount of X in the filtrate that appears in *urine.* The fraction of the total amount of X in the filtrate that remains in the *tubular fluid* at some point in the nephron is termed the *fractional delivery* of X (FD_X), since it represents the fraction of X in the filtrate that is "delivered" to that point. By a derivation analogous to that for *Eq. 75*, the following expression for FD_X is obtained:

$$FD_X = \frac{TF_X/P_X}{TF_{In}/P_{In}} \qquad (76)$$

where TF refers to the concentration in the tubular fluid at the point in question. Such tubular fluid samples are obtained by micropuncture (Chapter 26).

Once the fractional excretion of a substance X is known *(Eq. 75),* the *fractional reabsorption* can be readily calculated, since

$$\text{Fractional reabsorption} = 1 - FE_X$$

i.e.,

$$\text{Fractional reabsorption} = 1 - \frac{U_X/P_X}{U_{In}/P_{In}} \qquad (77)$$

Similarly, once the fractional delivery of a substance X to a particular point in the nephron is known *(Eq. 76),* the fraction reabsorbed up to that point can be readily calculated:

Fraction reabsorbed up to point of sample

$$= 1 - \frac{TF_X/P_X}{TF_{In}/P_{In}} \qquad (78)$$

The fractional reabsorption also can be calculated with *Eq. 51,* but the parameters required for the calculation are less readily obtained.

Table 4.11 includes typical U/P ratios and fractional excretion data for several plasma constitu-

Table 4.11.
Typical Fractional Excretion Values Obtained from U/P Ratios

	U_X/P_X	Fractional Excretion $U_X/P_X \div U_{In}/P_{In} \times 100\%$
Inulin	150	100%
Na$^+$	0.6	0.4%
Cl$^-$	0.75	0.5%
K$^+$		
Normal diet	10.5	7%
Vegetarian diet	180	120%
HCO$_3^-$	0.075	0.05%
Urea	60	40%

ents from Table 4.10 as well as for inulin. To conform with the format for presenting such data in clinical medicine, the fractional excretion values have been converted to *percentages* by multiplying by 100%. Note that the fractional excretion of K$^+$ *exceeds* 100% when a vegetarian (high K$^+$) diet is ingested. This indicates that the amount of K$^+$ excreted is *greater* than the amount filtered, i.e., net *secretion* of K$^+$ has occurred. The fact that K$^+$ can undergo either net reabsorption or net secretion has been previously noted.

Fractional Excretion and Fractional Reabsorption of Water

The ratios U_{In}/P_{In} and TF_{In}/P_{In} are important because of their role in calculating the fractional excretion and fractional reabsorption of the various substances in the glomerular filtrate, as described above. However, these ratios *also* can be used to calculate the fractional excretion and fractional reabsorption of *water*.

The fractional excretion of water (FE_{H_2O}) is defined by the expression

$$FE_{H_2O} = \frac{\text{Volume of water excreted}}{\text{volume of water filtered}}$$

Dividing both numerator and denominator by time gives

$$FE_{H_2O} = \frac{\dfrac{\text{Volume of water excreted}}{\text{time}}}{\dfrac{\text{volume of water filtered}}{\text{time}}}$$

or

$$FE_{H_2O} = \frac{\dot{V}}{GFR} \qquad (79)$$

An expression for FE_{H_2O} that utilizes the U_{In}/P_{In} ratio is obtained by substituting $U_{In}\dot{V}/P_{In}$ for GFR:

$$FE_{H_2O} = \frac{\dot{V}}{U_{In}\dot{V}/P_{In}}$$

i.e.,

$$FE_{H_2O} = \frac{1}{U_{In}/P_{In}} \qquad (80)$$

By an analogous derivation, it can be shown that the fraction of the total volume of water filtered that remains in the tubule at some point in the nephron, the *fractional delivery* of water (FD_{H_2O}), is given by

$$FD_{H_2O} = \frac{1}{TF_{In}/P_{In}} \qquad (81)$$

The *fractional reabsorption* of water can be calculated from the fractional excretion (*Eq. 80*) as follows:

$$\text{Fractional reabsorption} = 1 - FE_{H_2O}$$

or

$$\text{Fractional reabsorption} = 1 - \frac{1}{U_{In}/P_{In}} \qquad (82)$$

Similarly, the fraction of the filtered water reabsorbed up to some point in the nephron can be calculated from the fractional delivery (*Eq. 81*):

$$\text{Fraction reabsorbed up to point of sample} = 1 - FD_{H_2O}$$

or

$$\text{Fraction reabsorbed up to point of sample}$$
$$= 1 - \frac{1}{TF_{In}/P_{In}} \qquad (83)$$

Table 4.12 summarizes the important equations introduced here that utilize U_{In}/P_{In} and TF_{In}/P_{In} ratios. U/P and TF/P ratios for *creatinine* can be substituted for the corresponding ratios for inulin if an approximate calculation is adequate, as in many clinical applications.

Table 4.12.
Summary of Equations for Calculating Fractional Excretion and Fractional Reabsorption

	Water	Other Substances
Fractional excretion (FE)	$\dfrac{1}{U_{In}/P_{In}}$	$\dfrac{U_X/P_X}{U_{In}/P_{In}}$
Fractional reabsorption	$1 - \dfrac{1}{U_{In}/P_{In}}$	$1 - \dfrac{U_X/P_X}{U_{In}/P_{In}}$
Fractional delivery to point of micropuncture sample (FD)	$\dfrac{1}{TF_{In}/P_{In}}$	$\dfrac{TF_X/P_X}{TF_{In}/P_{In}}$
Fraction reabsorbed up to point of micropuncture sample	$1 - \dfrac{1}{TF_{In}/P_{In}}$	$1 - \dfrac{TF_X/P_X}{TF_{In}/P_{In}}$

TF_{In}/P_{In} ratios will be used in later chapters, particularly when discussing the mechanisms for the concentration and dilution of urine (Chapter 30). Most importantly, TF_{In}/P_{In} will be used as an index of water reabsorption along the nephron. For example, at the end of the proximal tubule, TF_{In}/P_{In} is estimated to be approximately 3.0. Thus, approximately one-third of the filtered water is delivered to the end of the proximal tubule *(Eq. 81)*, i.e., approximately two-thirds of the filtered water is reabsorbed in the proximal tubule *(Eq. 83)*, as indicated in Chapter 26.

CHAPTER 29

The Proximal Tubule

As discussed in Chapter 26, the proximal tubule is responsible for the initial processing of the glomerular filtrate. Among its important functions are the reabsorption of approximately two-thirds of the filtered Na^+ and water, the reabsorption of virtually all of the filtered glucose and amino acids, and the secretion of organic acids and bases. These functions are considered in detail in this chapter.

WATER PERMEABILITY

An important determinant of proximal tubular function is the high permeability of the proximal tubule to water, which prevents the establishment of an appreciable osmotic gradient across the proximal tubular epithelium. Physiologically, this means that the reabsorption of *any* solute by the proximal tubule will result in the reabsorption of water so that the tubular fluid osmolality remains approximately equal to that of the peritubular fluid. Actually, the osmolality of the peritubular fluid has been shown to be about 1–5 mosmol/kg H_2O greater than that of the tubular fluid due to the accumulation of reabsorbed solute in the peritubular space (Green and Giebisch, 1984). The peritubular fluid surrounding the proximal tubule, i.e., the *cortical* peritubular fluid (see Fig. 4.7), has an osmolality essentially identical to that of plasma (assumed here for simplicity to be approximately 300 mosmol/kg H_2O). Thus, the proximal tubular fluid is slightly *hypotonic* (although for simplicity an *approximate* osmolality of 300 mosmol/kg H_2O can be assumed here as well).

SODIUM REABSORPTION

As previously noted, approximately two-thirds of the filtered Na^+ is reabsorbed by the proximal tubule. Na^+ must be accompanied by an anion to maintain electrical neutrality; approximately 75% is accompanied by Cl^-, while the remaining 25% is accompanied by HCO_3^-. Although the reabsorption of *any* solute by the proximal tubule will result in the reabsorption of water, Na^+ and accompanying anions are the *major* solutes reabsorbed by the proximal tubule and therefore are primarily responsible for generating the osmotic driving force for water reabsorption. Thus, it is the reabsorption of two-thirds of the filtered Na^+ and accompanying anions by the proximal tubule that is responsible for the reabsorption of two-thirds of the filtered water.

It is important to keep in mind this *general* picture of the handling of Na^+ by the proximal tubule (reabsorption of about two-thirds of the filtered Na^+; 75% accompanied by Cl^-, 25% by HCO_3^-) throughout the following discussion of the mechanisms for proximal tubular Na^+ reabsorption. Three major mechanisms have been identified: Na^+-solute symport, Na^+-H^+ exchange (antiport), and Cl^--driven Na^+ transport. These mechanisms are discussed in detail below. However, first certain basic features shared by the *Na^+-solute symport* and *Na^+-H^+ exchange* mechanisms are considered.

The Na^+-solute symport and Na^+-H^+ exchange mechanisms for proximal tubular Na^+ reabsorption have important characteristics in common because both are examples of *active Na^+ reabsorption*. In both mechanisms, Na^+ enters the tubular epithelial cell across the luminal (apical) or brush border surface and then is actively extruded across the basal-lateral surfaces of the cell (Fig. 4.24A and B). The *entry of Na^+* is carrier-mediated (although the carriers involved are different in each mechanism; see below) and is driven by an *electrochemical gradient*. The *active extrusion of Na^+* is accomplished by the *Na^+/K^+-ATPase* (Katz, 1982), which also generates the electrochemical gradient for Na^+ entry. Thus, the energy for both the Na^+-solute symport and Na^+-H^+ exchange mechanisms, including the electrochemical gradient for the entry of Na^+, comes from ATP via the Na^+/K^+-ATPase. It is important to note that the active Na^+ reabsorption mechanisms in later regions of the nephron (thick ascending limb; distal nephron) share the general characteristics presented here: apical entry of Na^+

SEGMENT 1

TRANSEPITHELIAL P.D. ≃ −2 mV (lumen negative)

TUBULAR LUMEN   **TRANSMEMBRANE P.D.≃ −60 to −80 mV**   **PERITUBULAR SPACE**

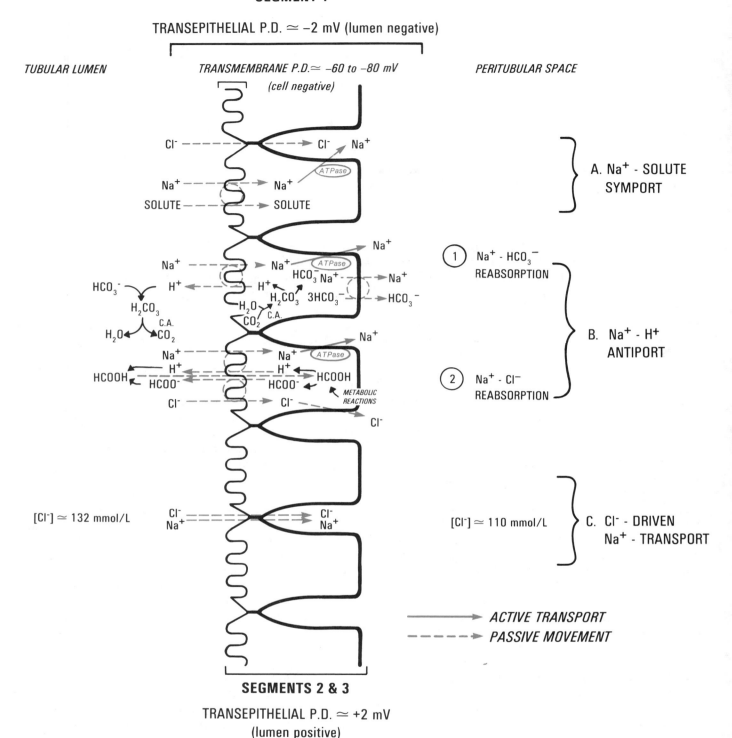

A. Na⁺ - SOLUTE SYMPORT

① Na⁺ - HCO₃⁻ REABSORPTION

B. Na⁺ - H⁺ ANTIPORT

② Na⁺ - Cl⁻ REABSORPTION

[Cl⁻] ≃ 132 mmol/L   [Cl⁻] ≃ 110 mmol/L   C. Cl⁻ - DRIVEN Na⁺ - TRANSPORT

→ ACTIVE TRANSPORT
---→ PASSIVE MOVEMENT

SEGMENTS 2 & 3

TRANSEPITHELIAL P.D. ≃ +2 mV
(lumen positive)

Figure 4.24. Mechanisms for Na⁺ reabsorption in the proximal tubule. *A*, Na⁺-solute symport. Apical entry of Na⁺ along its electrochemical gradient is coupled (as indicated by the *dashed circle*) to the transport of an organic solute (e.g., glucose, amino acids) or phosphate. The Na⁺ is then actively extruded across the basal-lateral surfaces by the Na⁺/K⁺-ATPase (illustrated here across the lateral surface). Since glucose, the major solute cotransported with Na⁺, is uncharged, this reabsorption of Na⁺ generates a lumen-negative transepithelial.potential difference (P.D.), but only of about −2 mV because Cl⁻ accompanies Na⁺ by diffusing across the "leaky" tight junctions. *B*, Na⁺-H⁺ exchange (antiport). The apical entry of Na⁺ is coupled (as indicated by the *dashed circle*) to the secretion of H⁺ into the tubular fluid. The Na⁺ is then actively extruded across the basal-lateral surfaces by the Na⁺/K⁺-ATPase. If the secreted H⁺ is derived from H_2CO_3, the *net result* of the Na⁺-H⁺ antiport is *Na⁺-HCO₃⁻ reabsorption* (*B*①); if the secreted H⁺ is derived from *formic acid*, the *net result* of the Na⁺-H⁺ antiport is *Na⁺-Cl⁻ reabsorption* (*B*②) (see text). *C*, Cl⁻-driven Na⁺ transport. The Cl⁻ concentration in the tubular fluid increases to about 132 mmol/liter in segments 2 and 3, resulting in a concentration gradient for the diffusion of Cl⁻ across the "leaky" tight junctions into the peritubular fluid, where the Cl⁻ concentration is about 110 mmol/liter. This reabsorption of Cl⁻ generates a lumen-positive transepithelial P.D. in segments 2 and 3, but only of about +2 mV because Na⁺ accompanies Cl⁻ across the "leaky" tight junctions.

along its electrochemical gradient, followed by active extrusion across the basal-lateral surfaces by the Na^+/K^+-ATPase. However, the carriers mediating the apical entry are different in each region (Chapter 30).

Na⁺-Solute Symport

As indicated above, the Na^+-solute symport mechanism is an active Na^+ reabsorption process that is distinguished from other active Na^+ reabsorption processes in the nephron by the carriers mediating the apical entry of Na^+. Specifically, apical entry in the Na^+-solute symport mechanism is coupled to the transport of one of several important organic molecules (glucose, amino acids, lactate, etc.; see below) or phosphate (Chapter 34) (Fig. 4.24*A*). The Na^+-solute symport mechanism is primarily important at the beginning of *segment 1* of the proximal tubule. In fact, the reabsorption of organic molecules such as glucose and amino acids normally is virtually *complete* before the end of segment 1 (see below).

Since glucose, the major solute cotransported with Na^+ via the Na^+-solute symport mechanism, is *uncharged,* the transport of Na^+ from tubular lumen to peritubular space by this mechanism will produce a transepithelial electrical potential difference (lumen negative with respect to peritubular space) whose magnitude will depend on the permeability of the tubular epithelium to one or more anions, which then could accompany Na^+. In the proximal tubule, the high permeability to Cl^- effectively prevents the establishment of a large potential difference; in fact, the transepithelial potential difference in segment 1 is only about -2 mV (lumen negative) (Fig. 4.24). The pathway for the reabsorption of Cl^- is presumed to include the tight junctions between the epithelial cells (Fig. 4.24*A*). However, the amount of Cl^- reabsorbed by this mechanism is relatively small compared to the total amount of Cl^- reabsorbed by the proximal tubule (see below). Because the high Cl^- permeability of the proximal tubule prevents the establishment of a large potential difference, the proximal tubule is said to have a *low electrical resistance*. The proximal tubule, then, can reabsorb Na^+, Cl^-, and water *without* establishing *either* a large transepithelial electrical gradient *or* an appreciable osmotic gradient. The proximal tubule is therefore said to have *"leaky" tight junctions.*

Na⁺-H⁺ Exchange

Like the Na^+-solute symport mechanism, the Na^+-H^+ exchange mechanism is an active Na^+ reabsorption process that is distinguished from other active Na^+ reabsorption processes in the nephron by the carriers involved in the apical entry of Na^+. Specifically, Na^+ entry in the Na^+-H^+ exchange mechanism is coupled to the extrusion or *secretion* of H^+ into the tubular fluid (Fig. 4.24*B*) (Ullrich et al., 1977). Thus, the Na^+-H^+ exchange mechanism represents an *antiport* or *countertransport system* whereby the movement of a substance in one direction provides the energy for the coupled movement of a second substance in the opposite direction.

Because the Na^+-H^+ antiport involves the exchange of two cations, it may initially appear that this mechanism for Na^+ reabsorption would *not* produce a transfer of anions. However, a detailed examination of the Na^+-H^+ antiport reveals that the reabsorbed Na^+ actually is accompanied by either HCO_3^- or Cl^- and that the *net result* of the Na^+-H^+ antiport is either *Na^+-HCO_3^- reabsorption* or *Na^+-Cl^- reabsorption*, respectively.

1. *Na^+-H^+ Antiport as a Mechanism for Na^+-HCO_3^- Reabsorption.* The Na^+-H^+ antiport can accomplish *Na^+-HCO_3^- reabsorption* if the secreted H^+ is derived from *H_2CO_3* (carbonic acid) (Fig. 4.24*B*①) (Ullrich et al., 1977). As discussed in detail in Chapter 32, the enzyme carbonic anhydrase (C.A.), which is located in the epithelial cells of the proximal tubule, catalyzes the formation of H_2CO_3 from CO_2 and water in the epithelial cells; the H_2CO_3 then dissociates into H^+ and HCO_3^-:

$$CO_2 + H_2O \underset{C.A.}{\rightleftarrows} H_2CO_3 \rightleftarrows H^+ + HCO_3^- \qquad (84)$$

a. The H^+ formed in reaction *84* is *secreted* into the tubular fluid in exchange for Na^+, where most of it reacts with filtered HCO_3^- to form H_2CO_3; the H_2CO_3, in turn, dissociates into CO_2 and water in a reaction catalyzed by carbonic anhydrase, which is found in large quantities in the brush border of proximal tubule as well as in the proximal tubular epithelial cells. Thus, reaction *84* occurs in *reverse* in the tubular fluid (Fig. 4.24*B*①).

b. The HCO_3^- formed in reaction *84* can exit the cell, crossing the *basal-lateral surfaces* by a carrier-mediated process in which the exit of three HCO_3^- ions is coupled to the exit of one Na^+ ion (Fig. 4.24*B*①) (Soleimani et al., 1987). The HCO_3^- then enters the peritubular fluid and peritubular capillaries.

As illustrated in Figure 4.24*B*①, this means that when the secreted H^+ is derived from H_2CO_3, the *net result* of the Na^+-H^+ antiport is the removal of Na^+ and HCO_3^- from the tubular fluid and the addition of Na^+ and HCO_3^- to the peritubular fluid, i.e., *Na⁺-HCO₃⁻ reabsorption*. It is important to note that the atoms in the HCO_3^- removed from the tubular fluid are *not* the same as those in the HCO_3^- returned to the peritubular fluid, but the net result is *equivalent* to HCO_3^- reabsorption.

2. *Na⁺-H⁺ Antiport as a Mechanism for Na⁺-Cl⁻ Reabsorption.* The Na^+-H^+ antiport can accomplish *Na⁺-Cl⁻ reabsorption* if the secreted H^+ is derived from *formic acid* (Fig. 4.24B②) (Karniski and Aronson, 1987). Formic acid, probably from metabolic reactions, can dissociate into H^+ and formate in the proximal tubular epithelial cells:

$$HCOOH \rightleftharpoons H^+ + HCOO^- \qquad (85)$$

a. The H^+ formed in reaction *85* is *secreted* into the tubular fluid in exchange for Na^+, where most of it reacts with formate to form formic acid. Thus, reaction *85* occurs in *reverse* in the tubular fluid (Fig. 4.24B②). Although formate is not normally present in large amounts in the glomerular filtrate, reaction *85* represents a source of formate, as discussed in the following paragraph.

b. The *formate* formed in reaction *85* is *secreted* into the proximal tubular fluid, crossing the luminal surface via a *Cl⁻-formate exchange mechanism* (antiport) whereby filtered Cl^- is exchanged for formate in the proximal tubular epithelial cells (Alpern, 1987). The Cl^- that enters the epithelial cell in this exchange can then diffuse across the basal-lateral surfaces into the peritubular fluid and peritubular capillaries.

As illustrated in Figure 4.24B②, this means that when the secreted H^+ is derived from formic acid, the *net result* of the Na^+-H^+ antiport is the removal of Na^+ and Cl^- from the tubular fluid and the addition of Na^+ and Cl^- to the peritubular fluid, i.e., *Na⁺-Cl⁻ reabsorption*.

Only a small amount of formic acid is required to reabsorb significant quantities of Na^+ and Cl^- by the above mechanism. This is because the formic acid formed in the tubular fluid is *nonionized* and can passively diffuse into the epithelial cell, where it can *once again* serve as a source of H^+ for secretion and *formate* for the Cl^--formate antiport (Fig. 4.24B②), a process that can be termed *formic acid recycling*. In fact, it has been estimated that as much as 50% of the Na^+ reabsorbed in the proximal tubule is reabsorbed by the Na^+-H^+ antiport operating in parallel with the Cl^--formate antiport (see below).

Cl⁻-driven Na⁺ Transport

To understand Cl^--driven Na^+ transport, the third mechanism for proximal tubular Na^+ reabsorption, one must consider the consequences of the first two mechanisms—Na^+-solute symport and Na^+-H^+ antiport—upon the *composition* of the tubular fluid. The concentrations of HCO_3^- and Cl^- in the glomerular filtrate are approximately equal to their respective concentrations in plasma, 24 and 110 mmol/liter. If HCO_3^- and Cl^- were to accompany the Na^+ reabsorbed by the Na^+-solute symport and Na^+-H^+ antiport mechanisms in the ratio of 24 HCO_3^-:110 Cl^-, the concentrations of HCO_3^- and Cl^- in the tubular fluid would not change. However, the rate of *Na⁺-HCO₃⁻ reabsorption* by the Na^+-H^+ antiport (Fig. 4.24B①), particularly at the beginning of segment 1 of the proximal tubule, is such that the ratio of HCO_3^- reabsorbed to Cl^- reabsorbed is *greater* than 24:110. The concentration of HCO_3^- in the tubular fluid therefore decreases, to approximately 8 mmol/liter, whereas the concentration of Cl^- increases, to approximately 132 mmol/liter. The primary determinant of the rate of HCO_3^- reabsorption, and hence of the lowering of the tubular fluid HCO_3^- concentration, is the rate of H_2CO_3-derived H^+ secretion by the Na^+-H^+ antiport system, although some H^+ also may be secreted into the tubular fluid via an active proton pump in the brush border.

These reciprocal changes in tubular fluid HCO_3^- and Cl^- concentrations occur by the end of segment 1 of the proximal tubule (Gottschalk et al., 1960). For the remainder of the proximal tubule, it appears that HCO_3^- and Cl^- are reabsorbed in the ratio of 8 HCO_3^-:132 Cl^-. The concentrations of HCO_3^- and Cl^- therefore do not change from their respective levels of 8 and 132 mmol/liter in segments 2 and 3 of the proximal tubule. Since the Cl^- concentration in the peritubular fluid is approximately equal to that in plasma (110 mmol/liter), this means that in segments 2 and 3 a concentration gradient for Cl^- exists between the tubular fluid and peritubular fluid. Because of the high permeability of the proximal tubule to Cl^- (see above), Cl^- will diffuse passively from tubular lumen to peritubular space in segments 2 and 3, creating a transepithelial potential difference (lumen positive with respect to peritubular space). This transepithelial potential difference is only about +2 mV (lumen positive), however, because the proximal tubule is sufficiently permeable to Na^+ to allow Na^+ to accompany Cl^-. The Cl^- concentration gradient in segments 2 and 3 therefore results in a third mechanism for proximal tubular Na^+ reabsorption, *Cl⁻-driven Na⁺ transport* (Fig. 4.24C). The pathway for Na^+ reabsorption probably includes the "leaky" tight junctions between the tubular cells (Boulpaep and Sackin, 1977). It is important to emphasize that although Cl^--driven Na^+ transport generally is referred to as *passive* rather than *active* Na^+ reabsorption, the energy for this mechanism, like that for the Na^+-solute symport and Na^+-H^+ antiport, actually is provided by the Na^+/K^+-ATPase. This is because the Cl^- concentration gradient that causes Cl^--driven Na^+ transport is generated by the Na^+-H^+ antiport (see above), which in turn requires the Na^+/K^+-ATPase to generate an electrochemical gradient for Na^+.

The relative proportions of Na^+ reabsorbed by the three major mechanisms for Na^+ transport in the proximal tubule have not yet been established. However, it can be estimated that approximately 10% is reabsorbed via the Na^+-solute symport, approximately 20–25% via the Na^+-H^+ antiport with Na^+ being accompanied by HCO_3^-, approximately 40–50% via the Na^+-H^+ antiport with Na^+ being accompanied by Cl^-, and approximately 20–30% via the Cl^--driven Na^+ transport mechanism.

A general feature of proximal tubular Na^+ reabsorption that should be mentioned at this time is that approximately two-thirds of the filtered Na^+ is reabsorbed *regardless* of the filtered load of Na^+. For example, if the filtered load of Na^+ increases, an increased amount of Na^+ will be reabsorbed so that the *fraction* reabsorbed remains nearly constant (Burg and Orloff, 1968). This phenomenon, termed *load-dependent Na^+ reabsorption*, also is present in later regions of the nephron (Chapter 30) and helps to minimize any changes in Na^+ excretion caused by changes in GFR (Chapter 31). In addition, if the filtered load of Na^+ is constant, Na^+ reabsorption in the proximal tubule can be increased by angiotensin II and by sympathetic stimulation, and also can be affected by the hydrostatic and oncotic pressures in the peritubular capillaries (part of the so-called *third factor* effect) and possibly by as yet unidentified factors (Chapter 31).

The concentrations of Na^+, Cl^-, HCO_3^-, and other solutes entering and leaving the proximal tubule in a normal person are shown in Table 4.13. TF_{In}/P_{In} ratios also are included. As indicated in Chapter 28, TF_{In}/P_{In} is particularly useful as an index of water reabsorption in the nephron, with the TF_{In}/P_{In} ratio of 3 at the end of the proximal tubule indicating that approximately one-third of the filtered water remains in the tubule at this point *(Eq. 81)*.

WATER REABSORPTION

Thus far, the pathway for the movement of water across the proximal tubular epithelium has not been discussed. Indeed, there is uncertainty about the

Table 4.13.
Approximate Concentrations of Substances Entering and Leaving the Proximal Tubule in Normal Humans

	Entering the Proximal Tubule (via Glomerular Filtration)	Leaving the Proximal Tubule (to the Loop of Henle)
[Na^+], mmol/liter	140	140
[Cl^-], mmol/liter	110	132
[HCO_3^-], mmol/liter	24	8
[Urea], mmol/liter	6	20
[Glucose, amino acids, other solutes], mmol/liter	20	$\simeq 0$
Osmolality, mosmol/kg H_2O	300	300
TF_{In}/P_{In}	1	3

relative importance of two pathways: the *transcellular route* and the *paracellular route* (Fig. 4.25). As previously noted, the Na^+/K^+-ATPase is distributed along both the basal and lateral surfaces of the cell. However, since the lateral space between adjacent cells (the *lateral intercellular space*) is restricted in size (Welling and Welling, 1979), the osmolality of the fluid in the lateral space will be increased slightly by the extrusion of Na^+ from the cell into this space by the Na^+/K^+-ATPase. Any HCO_3^- or Cl^- that accompanies the extruded Na^+ also will contribute to this osmolality increase, as will any Na^+ or Cl^- that enters the lateral space by crossing the leaky tight junctions. The osmolality increase results in the movement of water from the tubular lumen into the lateral space. Some of the water goes from lumen to lateral space via the cell interior (the *transcellular* route), while the remainder moves from the lumen across the leaky tight junctions into the lateral space (the *paracellular* route) (Andreoli et al., 1979).

The increase in the volume of fluid in the confines of the lateral space produces an increase in the hydrostatic pressure in the lateral space which, in turn, drives water and dissolved solutes from the lateral space to the peritubular space, from which they are returned to the body via the peritubular

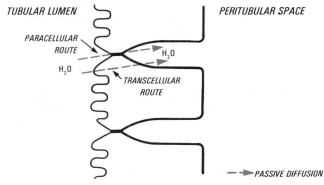

Figure 4.25. Pathways for water reabsorption in the proximal tubule.

capillaries. The rate of water and solute movement from peritubular space to peritubular capillaries can be markedly affected by the hydrostatic and oncotic pressures in the peritubular capillaries. This influence of peritubular capillary hydrostatic and oncotic pressure on proximal tubular reabsorption represents part of the third factor effect on Na^+ reabsorption (see above) and is discussed in greater detail when the regulation of Na^+ excretion is considered (Chapter 31).

REABSORPTION OF GLUCOSE AND AMINO ACIDS

Over 99% of the filtered glucose normally is reabsorbed by the proximal tubule. Thus, at a GFR of 180 liters/day and a normal plasma glucose of 5 mmol/liter (90 mg/dl), about 1 mol of glucose is reabsorbed daily from the proximal tubular fluid. As noted above, the reabsorption of glucose is coupled to the reabsorption of Na^+ via the Na^+-solute symport mechanism. Specifically, the carrier-mediated transport of glucose across the brush border is coupled to the entry of Na^+ across the brush border (Fig. 4.24*A*). The energy for the transport of glucose across the brush border therefore is provided by the Na^+/K^+-ATPase, which establishes the electrochemical gradient for Na^+ entry.

The carrier for the proximal tubular transport of glucose can be selectively inhibited by *phlorizin* (Turner and Silverman, 1981). The administration of phlorizin to an animal will therefore decrease reabsorption of glucose by the proximal tubule and cause glucosuria, even though the plasma glucose concentration is *normal*. As discussed in Chapter 28, glucosuria may also occur when the plasma glucose concentration is *elevated*, thereby saturating the Na^+-glucose transport system.

The proximal tubule also reabsorbs over 99% of the filtered amino acids. The normal plasma concentration of free amino acids (i.e., amino acids *not* linked to one another to form peptides or proteins) is about 3 mmol/liter. Thus, at a GFR of 180 liters/day, nearly 540 mmol of amino acids are reabsorbed daily by the proximal tubule. Like glucose reabsorption, the reabsorption of amino acids by the proximal tubule is coupled to Na^+ reabsorption via the Na^+-solute symport mechanism (Murer and Kinne, 1980). As with glucose, then, the energy for the transport of amino acids across the brush border is provided by the Na^+/K^+-ATPase. The reabsorption of different classes of amino acids (neutral, basic, etc.) apparently occurs via different carriers, although the exact number of carriers is not known.

While the occurrence of glucose-Na^+ and amino acid-Na^+ cotransport at the *brush border* is fairly

well established, there is less certainty about the mechanisms for the exit of glucose and amino acids across the *basal-lateral surfaces* of the proximal tubular cells. Presumably, some carriers are present in the basal-lateral surfaces that allow the glucose and amino acids to diffuse (probably down their concentration gradients) out of the cells into the peritubular fluid and peritubular capillaries (Murer and Kinne, 1980).

REABSORPTION OF PEPTIDES AND PROTEINS

As discussed in Chapter 27, peptides and proteins smaller than serum albumin are filtered to a variable extent (Table 4.7). Normally, these molecules are almost completely reabsorbed by the proximal tubule so that the urinary excretion of peptides and proteins is small. The reabsorption of *small* peptides (especially dipeptides and tripeptides) from the proximal tubular fluid may involve a carrier mechanism (Silbernagl et al., 1987). In contrast, the reabsorption of *larger* peptides and proteins occurs predominantly by pinocytosis (Carone et al., 1982). It is not certain whether all of these larger peptides and proteins bind to specific sites on the brush border surface prior to the pinocytosis. However, once inside the cell, they enter lysosomes, where they are degraded to their constituent amino acids. In other words, peptides and proteins are *not* returned intact to the peritubular capillaries. The degradation of proteins by the proximal tubular epithelial cells can be important in patients with diabetes mellitus. When an insulin-requiring diabetic develops renal disease, the dose of insulin necessary to control his plasma glucose may decrease since the diseased kidneys degrade less insulin.

SECRETION OF ORGANIC ACIDS AND BASES

The secretion of the organic acid *p*-aminohippurate (PAH) was discussed in Chapters 27 and 28. This secretion occurs exclusively in the *proximal tubule*. Specifically, PAH can be actively transported from the peritubular fluid across the basal-lateral surfaces of the tubular cells. Once inside the cell, the PAH can passively diffuse across the brush border into the proximal tubular fluid, probably by a carrier-mediated process (Somogyi, 1987).

The system that secretes PAH is *nonselective*, i.e., it can secrete a variety of different organic acids (Rennick, 1972). Table 4.14 lists some important examples of such compounds. Of particular interest is the large number of *antibiotics* and other drugs secreted by the organic acid system (Weiner and Mudge, 1964). A clinically useful *competitive inhibitor* of the organic acid secretory system, *probenecid,* can be administered in conjunction with rapidly

Table 4.14.
Organic Acids Secreted by the Proximal Tubule

Endogenous Substances	Drugs and Other Exogenous Substances
Bile acids	Cephalothin
cAMP	Chlorothiazide
Hydroxyindoleacetic acid	Furosemide
Oxalic acid	Iodohippuric acid
Uric acid	*p*-Aminohippuric acid (PAH)
	Penicillin
	Salicylic acid

secreted antibiotics such as penicillin to prolong their duration of action. In the presence of probenecid, penicillin is secreted (and therefore excreted) less rapidly.

A secretory system for organic bases also is present in proximal tubular cells and appears to function in a manner analogous to the organic acid system (Somogyi, 1987). Table 4.15 lists some of the important compounds secreted by the organic base system.

It is not known whether the same cells secrete both organic acids and organic bases or whether cellular specialization exists. In the case of the *organic acid system*, some *regional* specialization has been demonstrated in certain species.

REABSORPTION AND SECRETION OF URIC ACID

Uric acid is the end product of purine metabolism and is eliminated exclusively by the kidneys. The fractional excretion of uric acid in humans is approximately 10%, i.e., the amount excreted is about 10% of the amount filtered. Because of the unavailability of a suitable animal model for studying uric acid excretion (and other experimental problems, such as the difficulty in analyzing small quantities of uric acid), considerable uncertainty exists about the exact mechanisms involved in the renal handling of uric acid in humans. However, the proximal tubule is believed to both reabsorb and secrete uric acid, such that the filtered uric acid is

Table 4.15.
Organic Bases Secreted by the Proximal Tubule

Endogenous Substances	Drugs and Other Exogenous Substances
Acetylcholine	Amiloride
Creatinine	Atropine
Dopamine	Cimetidine
Epinephrine	Isoproterenol
Histamine	Morphine
N-Methylnicotinamide	Neostigmine
Norepinephrine	Procaine
Serotonin	Quinine
Thiamine	Ranitidine
	Trimethoprim

gradually reabsorbed and the excreted uric acid is almost completely derived from secretion (Diamond and Paolino, 1973).

The reabsorption of uric acid is inhibited by *probenecid* (see below). Although probenecid also inhibits the secretion of organic acids, as previously discussed, the uric acid reabsorption system probably is not related to the organic acid secretory system. An experimental agent, *pyrazinoic acid*, inhibits the secretion of uric acid.

Although the fractional excretion of uric acid in humans averages about 10%, there is considerable individual variation. In persons with a lower fractional excretion, the plasma uric acid becomes elevated, thereby allowing the normal quantity of uric acid to be eliminated. This can be illustrated by considering an example of a person who must eliminate 3 mmol of uric acid daily ($U_{UA}\dot{V} = 3$ mmol/day). If the fractional excretion of uric acid (FE_{UA}) is 10% and the GFR is 180 liters/day, the *clearance* of uric acid (C_{UA}) can be calculated from *Eq. 74*:

$$FE_{UA} = \frac{C_{UA}}{C_{In}}$$
$$= \frac{C_{UA}}{GFR}$$

Thus

$$C_{UA} = 18 \text{ liters/day}$$

and

$$P_{UA} = \frac{U_{UA}\dot{V}}{C_{UA}} = 0.17 \text{ mmol/liter}$$

If, however, FE_{UA} is only 3% but the GFR is still 180 liters/day, by similar calculations $C_{UA} = 5.4$ liters/day and $P_{UA} = 0.56$ mmol/liter. A P_{UA} level of 0.56 mmol/liter is sufficient to cause attacks of *gouty arthritis* in susceptible individuals. In such persons, *probenecid* can be used to decrease uric acid reabsorption, thereby increasing FE_{UA} and lowering P_{UA}.

Another potential problem with uric acid relates to its insolubility at acid pH (i.e., at the normal urine pH in humans). In some individuals, uric acid *crystallizes* in the upper urinary tract to form *kidney stones*. Uric acid is involved in other diseases of the kidneys that are beyond the scope of this text.

UREA

Urea is the major end product of protein metabolism and, like uric acid, is eliminated exclusively by the kidneys. Neither the proximal tubule nor other regions of the nephron can actively reabsorb urea. However, as water is reabsorbed by the proximal tubule, the concentration of urea rises modestly, thereby establishing a concentration gradient for the passive diffusion of urea from tubular lumen to peritubular space. By the end of segment 2 of the proximal tubule (the last part of the proximal tubule accessible to micropuncture), about 40% of the filtered urea has been reabsorbed by this mechanism.

In segment 3 of the proximal tubule, urea may be *secreted* into the tubular fluid (Knepper, 1983). It has been estimated that the urea concentration increases from about 6 mmol/liter in plasma and glomerular filtrate to as high as 20 mmol/liter by the end of the proximal tubule (Table 4.13).

OTHER FUNCTIONS OF THE PROXIMAL TUBULE

The proximal tubule has an important role in the handling of other ions in the glomerular filtrate. It reabsorbs K^+ (in segments 1 and 2), secretes K^+ (in segment 3), and reabsorbs Ca^{2+}, Mg^{2+}, and phosphate. The renal handling of these ions is considered in greater detail in Chapters 33 and 34.

The proximal tubule also reabsorbs several miscellaneous organic molecules that are present in the glomerular filtrate, including lactate, citrate and other Krebs cycle intermediates (Wright et al., 1982), and water-soluble vitamins. In addition, proximal tubular cells synthesize ammonia, a compound that is important in H^+ excretion, as discussed in Chapter 32.

CHAPTER 30

The Loop of Henle and the Distal Nephron

LOOP OF HENLE

In the loop of Henle, the fluid that leaves the proximal tubule, which is rich in Na^+Cl^- and approximately isotonic, is reduced in volume and transformed into a hypotonic fluid in which a major osmotically active solute is urea. The loop of Henle accomplishes this by reabsorbing approximately 25% of the filtered Na^+ and 20% of the filtered water from the tubular fluid while adding substantial amounts of urea to the tubular fluid. In addition, as discussed at the end of this chapter, the loop of Henle has an important role in the mechanisms for both urinary concentration and urinary dilution (Knepper and Stephenson, 1986).

Medullary Gradient

The function of the loop of Henle must be considered in light of the somewhat unusual characteristics of the peritubular fluid in the medulla. First of all, whereas the osmolality of the peritubular fluid in the renal *cortex* is essentially identical with that of plasma (assumed here for simplicity to be approximately 300 mosmol/kg H_2O), the osmolality of the peritubular fluid in the *medulla* increases from about 300 mosmol/kg H_2O at the corticomedullary junction to approximately 1200 mosmol/kg H_2O at the tip of the papilla. Second, whereas Na^+ and Cl^- are the predominant osmotically active solutes in the *cortical* peritubular fluid, a substantial fraction of the osmotically active solutes in the *medullary* peritubular fluid is urea. In fact, the concentration of urea in the medullary peritubular fluid increases from the corticomedullary junction, where its concentration is similar to that in plasma, to the tip of the papilla, where it accounts for about half of the osmotically active solutes (Fig. 4.26) (Ullrich et al., 1961). The term *medullary gradient* has been used to describe the osmotic gradient that exists in the peritubular fluid between the corticomedullary junction and the tip of the papilla. The mechanism for the establishment of this gradient is considered at the end of this chapter.

Sodium, Chloride, and Water Reabsorption

The reabsorption of Na^+, Cl^-, and water in the loop of Henle is markedly dependent upon the *length of the loop* (Schmidt-Nielsen and O'Dell, 1961). Thus, solute and water reabsorption in *long-looped nephrons* differs from that in *short-looped nephrons*, primarily as a result of differences in *ascending limb structure*, and hence function. As noted in Chapter 26, the ascending limb of long-looped nephrons can be divided into two regions: the thin ascending limb, which begins at the hairpin turn, and the thick ascending limb, which begins at or near the junction between the inner and outer medulla. In contrast, short-looped nephrons lack a thin ascending limb; the thick ascending limb therefore begins near the hairpin turn, i.e., at or above the junction between the inner and outer medulla (see Fig. 4.7).

In this discussion, a "prototype" long-looped juxtamedullary nephron, whose loop reaches the tip of the papilla, and a "prototype" short-looped cortical nephron, whose loop reaches the junction between the outer and inner medulla, are used to illustrate how the loop of Henle reabsorbs Na^+, Cl^-, and water (Fig. 4.26). "Prototype" solute concentrations and osmolalities in the tubular fluid and surrounding peritubular fluid as well as "prototype" TF_{In}/P_{In} ratios also are given. However, it should be emphasized that these prototype concentrations, osmolalities, and TF_{In}/P_{In} ratios are *estimates* presented to illustrate the function of the loop rather than precise values that should be mastered. Such estimation is necessary because micropuncture sampling of the tubular fluid and peritubular fluid is not possible in humans. Furthermore, even in animal studies the only sites accessible to micropuncture are the cortical surface and near the tip of the papilla (Chapter 26).

THIN DESCENDING LIMB

The thin descending limb of the loop of Henle has a relatively low permeability to solutes and lacks

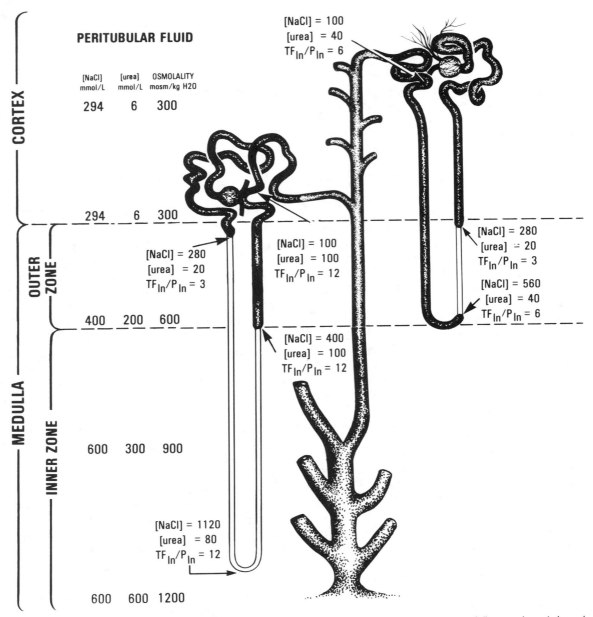

Figure 4.26. Approximate solute concentrations, osmolalities, and TF_{In}/P_{In} ratios in loop of Henle tubular fluid and the surrounding peritubular fluid for a "prototype" short-looped cortical nephron (whose loop reaches the junction between the inner and outer medulla) and a "prototype" long-looped juxtamedullary nephron (whose loop reaches the tip of the papilla). The [NaCl] values given include small contributions from other electrolytes and nonurea solutes.

mechanisms for active solute transport but is highly permeable to water (Kokko, 1970). Thus, as the tubular fluid entering the thin descending limb flows downward through regions of increasingly hypertonic peritubular fluid (the medullary gradient), water is reabsorbed. In fact, the thin descending limb is so permeable to water that the osmolality of the tubular fluid becomes equal to that of the surrounding peritubular fluid. The mechanism whereby the water reabsorbed from the thin

descending limb is removed from the medulla without diluting or washing out the medullary gradient is considered below.

Because of the reabsorption of water, the tubular fluid osmolality increases as fluid flows through the thin descending limb. In long-looped nephrons whose loops reach the tip of the papilla (e.g., the prototype long-looped nephron of Fig. 4.26), the tubular fluid osmolality increases from about 300 mosmol/kg H_2O at the start of the thin descending limb to as high as 1200 mosmol/kg H_2O at its termination at the hairpin turn. In short-looped nephrons whose loops reach the junction between the outer and inner medulla (e.g., the prototype short-looped nephron of Fig. 4.26), the tubular fluid osmolality is estimated to increase from about 300 to about 600 mosmol/kg H_2O at the hairpin turn. The available micropuncture data indicate that these substantial increases in osmolality can in fact be accomplished primarily by the *reabsorption of water* from the tubular fluid rather than by the secretion of solutes into the tubular fluid. Specifically, TF_{In}/P_{In} is estimated to increase from about 3 at the end of the proximal tubule to approximately 12 at the hairpin turn in long-looped nephrons whose loops reach the tip of the papilla, like the prototype long-looped nephron considered here. Hence, approximately three-fourths of the water entering the thin descending limb of such long-looped nephrons is reabsorbed there, i.e., the fourfold increase in osmolality (from 300 to 1200 mosmol/kg H_2O) can be attributed almost entirely to the reabsorption of water. Although the hairpin turn of short-looped nephrons is not accessible to micropuncture, it is estimated that TF_{In}/P_{In} increases from about 3 at the end of the proximal tubule to approximately 6 at the hairpin turn in the prototype short-looped nephron considered here. According to this estimate, then, approximately one-half of the water entering the thin descending limb of this prototype short-looped nephron is reabsorbed there, i.e., the twofold increase in osmolality (from 300 to 600 mosmol/kg H_2O) can be attributed almost entirely to the reabsorption of water.

While the *osmolality* of the tubular fluid in the thin descending limb is equal to that of the surrounding peritubular fluid, there are important *composition* differences between the tubular fluid and the peritubular fluid, particularly in long-looped nephrons. The tubular fluid that enters the thin descending limb contains approximately 280 mmol/liter of electrolytes (predominantly Na^+ and Cl^-) and other solutes and approximately 20 mmol/liter of urea (Tables 4.13 and 4.16). These solutes are *concentrated* in the thin descending limb as water is reabsorbed. In the long-looped nephron example presented here, the solutes are concentrated approximately fourfold in the thin descending limb, as noted above. The tubular fluid at the hairpin turn therefore contains about 1120 mmol/liter of electrolytes (again, predominantly Na^+ and Cl^-) and 80 mmol/liter of urea. The surrounding peritubular fluid, on the other hand, contains about 600 mmol/liter of electrolytes and 600 mmol/liter of urea (Fig. 4.26). In the short-looped nephron example, the solutes are concentrated approximately twofold in the thin descending limb. The tubular fluid at the hairpin turn therefore is estimated to contain about 560 mmol/liter of electrolytes and 40 mmol/liter of urea, which compares with approximately 400 mmol/liter of electrolytes and 200 mmol/liter of urea in the surrounding peritubular fluid (Fig. 4.26). Thus, in both long-looped and short-looped nephrons the predominant solutes in the *tubular fluid* are Na^+ and Cl^-, but a substantial fraction of the solutes in the surrounding *peritubular fluid* is urea.

THIN ASCENDING LIMB

After passing the hairpin turn, the tubular fluid enters the *ascending limb*. In *long-looped nephrons*, the ascending limb begins with the *thin ascending limb*. In spite of its structural similarity to the thin descending limb, the thin ascending limb has com-

Table 4.16.
Approximate Concentrations of Substances Entering and Leaving the Loop of Henle in Normal Humans

	Entering the Thin Descending Limb (from the Proximal Tubule[a])		Leaving the Thin Descending Limb	Leaving the Thin Ascending Limb	Leaving the Thick Ascending Limb (to the Distal Nephron)
[NaCl],[b] mmol/liter	280	short-looped	560		100
		long-looped	1120	400	100
[Urea], mmol/liter	20	short-looped	40		40
		long-looped	80	100	100
Osmolality, mosmol/kg H_2O	300	short-looped	600		140
		long-looped	1200	500	200
TF_{In}/P_{In}	3	short-looped	6		6
		long-looped	12	12	12

[a]See Table 4.13. [b]Includes small contributions from other electrolytes and nonurea solutes.

pletely different transport and permeability characteristics, being virtually impermeable to water but highly permeable to Na^+ and Cl^- and moderately permeable to urea (Imai and Kokko, 1974). Thus, as a result of the substantial concentration gradients for Na^+, Cl^-, and urea between the tubular fluid entering the thin ascending limb and the peritubular fluid (Fig. 4.26), Na^+ and Cl^- diffuse passively from lumen to peritubular space, while urea diffuses passively from peritubular space to lumen. Of course, the greater permeability of the thin ascending limb to Na^+ and Cl^- than to urea means that the number of moles of Na^+ plus Cl^- *leaving* the lumen *exceeds* the number of moles of urea *entering*. In spite of this net solute exit and in spite of the fact that the tubular fluid flows upward through a region of progressively decreasing peritubular fluid osmolality, the *volume* of the tubular fluid does not significantly change in the thin ascending limb because of the low water permeability. With net solute exit and no appreciable volume change, the osmolality of the tubular fluid in the thin ascending limb falls slightly below that of the surrounding peritubular fluid (Jamison et al., 1967). In the long-looped nephron example presented here, the tubular fluid at the end of the thin ascending limb is estimated to contain about 400 mmol/liter of electrolytes and 100 mmol/liter of urea for a total osmolality of approximately 500 mosmol/kg H_2O. In contrast, the surrounding peritubular fluid is estimated to contain approximately 400 mmol/liter of electrolytes and 200 mmol/liter of urea for a total osmolality of 600 mosmol/kg H_2O (Fig. 4.26). Note that TF_{In}/P_{In} remains constant throughout this region, since the volume of the tubular fluid does not change appreciably.

THICK ASCENDING LIMB

The tubular fluid enters the thick ascending limb from the thin ascending limb in long-looped nephrons and from the thin descending limb in short-looped nephrons (Fig. 4.26). Like the thin ascending limb, the water permeability of the thick ascending limb is negligible (Rocha and Kokko, 1973). However, the other functional characteristics of the thick ascending limb differ markedly from those of the thin ascending limb. First of all, the urea permeability of the thick ascending limb is quite low. Second and most important, the thick ascending limb *actively* transports Na^+ and Cl^- from lumen to peritubular space (Greger and Schlatter, 1981). The combination of *low water permeability* and *active reabsorption of Na^+ and Cl^-* means that the thick ascending limb *lowers* both the osmolality of the tubular fluid and the concentration of Na^+ and Cl^- in the tubular fluid to levels below those in the surrounding peritubular fluid. The combination of *low water permeability* and *low urea permeability* means that the urea concentration is *unaltered* by the thick ascending limb. In the long-looped nephron example presented here, the tubular fluid at the end of the thick ascending limb is estimated to contain about 100 mmol/liter of electrolytes and 100 mmol/liter of urea for a total osmolality of approximately 200 mosmol/kg H_2O. In the short-looped nephron example, the tubular fluid at the end of the thick ascending limb is estimated to contain approximately 100 mmol/liter of electrolytes and 40 mmol/liter of urea for a total osmolality of 140 mosmol/kg H_2O (Fig. 4.26). Of course, since the thick ascending limb terminates in the *cortex* in both long-looped and short-looped nephrons, the solute concentrations and osmolality of the surrounding peritubular fluid are similar to those of plasma, i.e., approximately 294 mmol/liter of electrolytes and other nonurea solutes and 6 mmol/liter of urea for a total osmolality of approximately 300 mosmol/kg H_2O. In *both* nephron types, then, the tubular fluid that leaves the loop of Henle is hypotonic and (particularly in long-looped nephrons) has a urea concentration that is considerably greater than that of plasma and the surrounding peritubular fluid. It should be noted that TF_{In}/P_{In} remains constant throughout the thick ascending limb in both long-looped and short-looped nephrons, since the volume of the tubular fluid does not change significantly due to the low water permeability.

Like other active Na^+ reabsorption processes in the nephron (Chapter 29), the mechanism for the active reabsorption of Na^+ and Cl^- in the thick ascending limb is characterized by the apical entry of Na^+ along its electrochemical gradient followed by the active extrusion of Na^+ across the basal-lateral surfaces by the Na^+/K^+-ATPase. However, the active reabsorption mechanism in the thick ascending limb is distinguished from these other active reabsorption processes by the carriers mediating the apical entry of Na^+. Specifically, the entry of a Na^+ ion in the thick ascending limb is coupled to the entry of a K^+ ion and two Cl^- ions (Fig. 4.27) (Greger and Schlatter, 1981); the mechanism for the active reabsorption of Na^+ and Cl^- in the thick ascending limb therefore is termed the *Na^+/K^+-2Cl$^-$ symport*.

Although the Na^+/K^+-2Cl$^-$ symport involves two cations and two anions and therefore produces no net transfer of electrical charge, a transepithelial potential difference of approximately $+6$ to $+10$ mV (lumen positive) is generated in the thick ascending limb. Apparently, some of the K^+ that

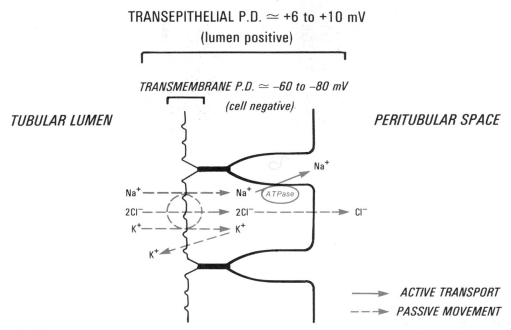

TRANSEPITHELIAL P.D. ≃ +6 to +10 mV

(lumen positive)

TRANSMEMBRANE P.D. ≃ −60 to −80 mV

(cell negative)

TUBULAR LUMEN   *PERITUBULAR SPACE*

Na⁺

Na⁺ → → → → Na⁺ (ATPase)

2Cl⁻ → → → → → 2Cl⁻ → → → → Cl⁻

K⁺ → → → → → K⁺

K⁺

→→→ *ACTIVE TRANSPORT*
---→ *PASSIVE MOVEMENT*

Figure 4.27. Na⁺/K⁺-2Cl⁻ symport in the thick ascending limb. The apical entry of Na⁺ is coupled to the entry of K⁺ and Cl⁻ (as indicated by the *dashed circle*). The Na⁺ is then actively extruded across the basal-lateral surfaces by the Na⁺/K⁺-ATPase, with Cl⁻ following pas- sively to maintain electrical neutrality. Some of the K⁺ that enters the cell leaks back across the apical membrane into the tubular lumen, gen- erating a lumen-positive transepithelial potential difference (P.D.).

enters the cell by the Na⁺/K⁺-2Cl⁻ symport leaks back across the apical membrane into the tubular lumen (Fig. 4.27), thereby producing the lumen-pos- itive transepithelial potential difference.

An important characteristic of the Na⁺/K⁺-2Cl⁻ symport is that an increased amount of Na⁺ will be reabsorbed by the thick ascending limb if an increased load of Na⁺ is delivered to this region, a phenomenon termed *load dependence*. For example, if the reabsorption of Na⁺ in the *proximal tubule* is decreased as a result of plasma volume expansion (part of the *third factor* effect; see Chapter 31), more Na⁺ will be delivered to the thick ascending limb and an *increased* amount of Na⁺ will be reab- sorbed (Cortney et al., 1965). Such load-dependent Na⁺ reabsorption also is present in the *proximal tubule* (Chapter 29). The active reabsorption of Na⁺ and Cl⁻ by the thick ascending limb is *inhibited* by certain *prostaglandins* (see below) and by an important class of diuretic drugs, the so-called *loop diuretics*, whose members include *furosemide* and *bumetanide*.

Table 4.16 summarizes the modifications in the composition of the tubular fluid that occur in the loop of Henle of the prototype short-looped and long-looped nephrons considered here. The perme- ability and transport characteristics of the various segments of the loop of Henle are summarized in Table 4.17.

Other Functions of the Loop of Henle

The loop of Henle not only reabsorbs K⁺ in the thick ascending limb (via the Na⁺/K⁺-2Cl⁻ sym- port), but also secretes K⁺ in the thin descending limb. In addition, the thick ascending limb reab- sorbs Ca²⁺. These functions are considered in greater detail in Chapters 33 and 34.

DISTAL NEPHRON

The distal nephron produces the modifications in tubular fluid composition and volume necessary to complete the transformation of the tubular fluid into urine. These modifications include the reab- sorption of as much as 5–10% of the filtered Na⁺ (thereby making possible the recovery of virtually all of the filtered Na⁺), the reabsorption of urea, the secretion of H⁺ and K⁺, and the reabsorption of up to 15% of the filtered water. Since the distal neph- ron includes several anatomically distinct segments (Chapter 26), it should not be surprising that the distal nephron is not a functionally homogeneous structure. Thus, the distal convoluted tubule, the connecting tubule, and the various segments of the collecting duct (cortical, medullary, and papillary collecting ducts) have specialized as well as com- mon functions.

To appreciate the modifications in volume and composition that occur in the distal nephron, "pro-

Table 4.17.
Permeability and Transport Characteristics of Loop of Henle and Distal Nephron[a]

	Loop of Henle			Distal Nephron		
					Collecting Duct	
	Thin Descending Limb	Thin Ascending Limb	Thick Ascending Limb	Distal Convoluted Tubule and Connecting Tubule	Cortical Collecting Duct and Medullary Collecting Duct	Papillary Collecting Duct
Na^+						
Permeability	0	+++	+	0	0	0
Active reabsorption	0	0	Yes	Yes	Yes	Yes
Urea permeability	0	+	0	0	0	+ADH +++ / −ADH 0
Water permeability	++++	0	0	0	+ADH ++++ / −ADH 0	++++ / 0

[a]Relative permeabilities indicated by number of + symbols; 0 indicates low or negligible permeability.

totype" solute concentrations and osmolalities in the tubular fluid and surrounding peritubular fluid as well as TF_{In}/P_{In} ratios are given in the discussion that follows. Since the composition of the tubular fluid leaving the thick ascending limb is dependent upon the length of the loop of Henle (Fig. 4.26; Table 4.16), the composition of the tubular fluid entering the distal nephron will be different in different nephrons. However, because several adjacent nephrons share a common collecting duct, throughout much of the distal nephron the tubular fluid will be a *mixture* of contributions from different nephrons. It is therefore sufficient to consider *average* tubular fluid compositions in the distal nephron. The average tubular fluid entering the distal nephron is estimated to contain approximately 100 mmol/liter of electrolytes and other nonurea solutes and 50 mmol/liter of urea for a total osmolality of approximately 150 mosmol/kg H_2O, and to have a TF_{In}/P_{In} ratio of approximately 7 (Fig. 4.28). The average tubular fluid entering the distal nephron is therefore estimated to be quite similar to the tubular fluid leaving the thick ascending limb of the prototype short-looped nephron discussed above (approximately 85% of the nephrons in humans are short-looped cortical nephrons). Like the "prototype" concentrations and osmolalities for the loop of Henle, the distal nephron prototype concentrations and osmolalities must be regarded as *estimates* that illustrate the function of the kidneys rather than precise values that should be mastered.

It should be noted that whereas Na^+ and Cl^- are the primary nonurea solutes *entering* the distal nephron, other nonurea solutes *(nus)* become increasingly important *in* the distal nephron. This is because tubular reabsorption in the distal nephron can reduce the Na^+ and Cl^- concentrations to such low levels (see below) that the concentrations of solutes such as creatinine, uric acid, K^+, NH_4^+, Ca^{2+}, Mg^{2+}, and phosphate can no longer be neglected.

Sodium, Chloride, and Water Reabsorption

One transport function common to all segments of the distal nephron is the active reabsorption of Na^+. Like other active Na^+ reabsorption processes in the nephron, the active reabsorption of Na^+ in the distal nephron is characterized by the apical entry of Na^+ along its electrochemical gradient followed by the active extrusion of Na^+ across the basal-lateral surfaces by the Na^+/K^+-ATPase. However, the active Na^+ reabsorption in the distal nephron is distinguished from these other active reabsorption processes by the carriers mediating the apical entry of Na^+. In fact, apical Na^+ entry in the *distal convoluted tubule* and *connecting tubule* differs from that in the *collecting duct*.*

DISTAL CONVOLUTED TUBULE AND CONNECTING TUBULE

In the first two segments of the distal nephron, the *distal convoluted tubule* and *connecting tubule*, the apical entry of Na^+ is coupled to the entry of Cl^-; the mechanism for active Na^+ reabsorption in these segments therefore can be termed the *Na^+-Cl^- symport* (Fig. 4.29*A*) (Ellison et al., 1987). An important characteristic of the Na^+-Cl^- symport is that an increased amount of Na^+ will be reabsorbed by the distal convoluted tubule and connecting tubule if an increased load of Na^+ is delivered to

---

*The two carriers mediating apical Na^+ entry in the distal nephron may be located in different segments in different species.

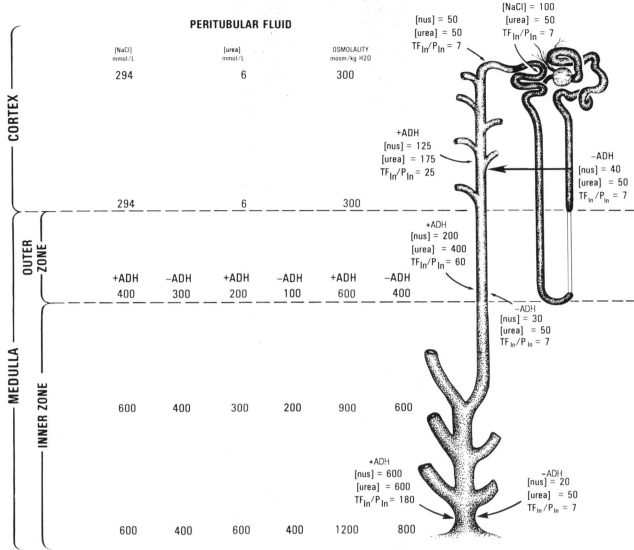

Figure 4.28. Approximate solute concentrations, osmolalities, and TF$_{In}$/P$_{In}$ ratios in distal nephron tubular fluid and the surrounding peritubular fluid in the presence of maximal ADH (+ADH) and in the absence of ADH (−ADH). Because the tubular fluid throughout much of the distal nephron is a mixture of contributions from different nephrons, only *average* tubular fluid compositions are shown. The [NaCl] values given include small contributions from other electrolytes and nonurea solutes; this term is replaced by [nus] *(nonurea solutes)* in the later parts of the distal nephron because the Na$^+$ and Cl$^-$ concentrations can be reduced to such low levels that the concentrations of other electrolytes and nonurea solutes can no longer be neglected.

these segments. Like the proximal tubule and thick ascending limb, then, the distal convoluted tubule and connecting tubule exhibit *load-dependent Na$^+$ reabsorption.* The Na$^+$-Cl$^-$ symport is *inhibited* by an important class of diuretic drugs, the *thiazide diuretics.*

To understand the modifications in tubular fluid volume and composition that occur in the distal convoluted tubule and connecting tubule, it is important to note that certain characteristics of the distal convoluted tubule and connecting tubule closely resemble those of the thick ascending limb, including low permeability to water and urea (Imai, 1979). The combination of *low water permeability* and *active Na$^+$ reabsorption* means that the distal convoluted tubule and connecting tubule *lower* both the osmolality of the tubular fluid and its electrolyte concentration. The combination of *low water permeability* and *low urea permeability* means that the urea concentration remains essentially *unchanged* in these segments. At the end of the connecting tubule, the tubular fluid is estimated to contain approximately 50 mmol/liter of urea and 50 mmol/liter of nonurea solutes (including electro-

DISTAL CONVOLUTED TUBULE AND CONNECTING TUBULE

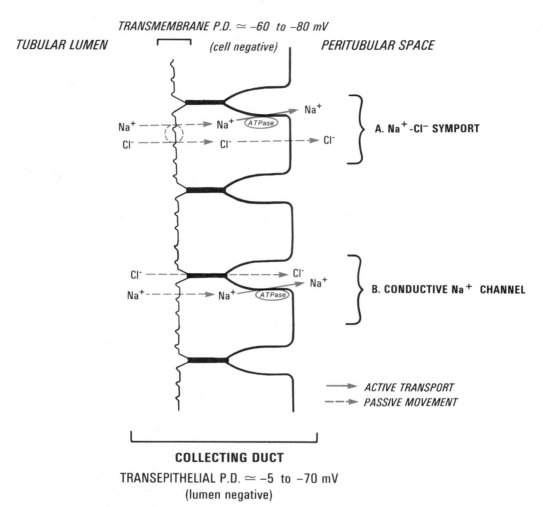

Figure 4.29. Mechanisms for Na$^+$ reabsorption in the distal nephron. *A*, Na$^+$-Cl$^-$ symport in the distal convoluted tubule and connecting tubule. The apical entry of Na$^+$ along its electrochemical gradient is coupled (as indicated by the *dashed circle*) to the entry of Cl$^-$. The Na$^+$ is then actively extruded across the basal-lateral surfaces by the Na$^+$/K$^+$-ATPase. *B*, Conductive Na$^+$ channel in the collecting duct. The apical entry of Na$^+$ along its electrochemical gradient is not directly coupled to the entry or secretion of another ion or solute. The Na$^+$ is then actively extruded across the basal-lateral surfaces by the Na$^+$/K$^+$-ATPase. This Na$^+$ transport mechanism generates a lumen-negative transepithelial potential difference (P.D.), which can be as large as -70 mV because the "tight" tight junctions of the collecting duct limit the ability of Cl$^-$ to accompany Na$^+$. Cl$^-$ can be reabsorbed via a transcellular route as well as across the "tight" tight junctions as shown (Chapter 32).

lytes), for a total osmolality of 100 mosmol/kg H$_2$O (Fig. 4.28). Of course, the solute concentrations and osmolality in the surrounding cortical peritubular fluid are similar to those of plasma. Thus, the hypotonic tubular fluid entering the distal nephron has its osmolality further reduced in the distal convoluted tubule and connecting tubule, but the urea concentration remains considerably greater than that in plasma. TF$_{In}$/P$_{In}$ remains constant in these regions, with an approximate value of 7, since the volume of the tubular fluid does not change ap-

preciably due to the low water permeability (Fig. 4.28).

COLLECTING DUCT

In the various segments of the collecting duct (*cortical, medullary,* and *papillary collecting ducts*), the apical entry of Na$^+$ is *not* directly coupled to the entry or secretion of other ions or solutes; active Na$^+$ reabsorption in these segments therefore is said to occur via the *conductive Na$^+$ channel* (Fig. 4.29*B*) (Stokes, 1982). This mechanism for Na^+ transport will produce a transepithelial electrical potential difference (lumen negative) whose magnitude will depend on the permeability of the tubular epithelium to one or more anions, which then could accompany Na^+. Although the tubular epithelium allows much of the Na^+ to be accompanied by Cl^- (Fig. 4.29B), the Cl^- permeability of the collecting duct is sufficiently low that a substantial transepithelial potential difference can be generated. In fact, the transepithelial potential difference can be as large as -70 mV (lumen negative) in the collecting duct (Fig. 4.29B), depending upon the amount of Na^+ that is reabsorbed. Because the relatively low Cl^- permeability allows the establishment of such a large potential difference, the collecting duct is said to have a *high electrical resistance*, in contrast to the low electrical resistance of the proximal tubule (Chapter 29). The high resistance properties of the collecting duct suggest that the tight junctions between epithelial cells are "tighter" than the "leaky" tight junctions of the proximal tubule. These "tight" tight junctions not only limit the ability of Cl^- to accompany Na^+, but also prevent the Na^+ pumped into the lateral intercellular space from leaking back into the tubular lumen to any significant extent.

An important characteristic of the conductive Na^+ channel is that an increased amount of Na^+ will be reabsorbed by the collecting duct if an increased load of Na^+ is delivered to this region (Khuri et al., 1975). Like the proximal tubule, thick ascending limb, and distal convoluted tubule and connecting tubule, then, the collecting duct exhibits *load-dependent Na^+ reabsorption*. Because of this load dependence of Na^+ reabsorption in the collecting duct, a larger transepithelial potential difference will be generated if Na^+ delivery to the collecting duct is increased. Na^+ reabsorption via the conductive Na^+ channel in the collecting duct is *stimulated* by the hormone *aldosterone* (Chapter 31). Thus, aldosterone represents an additional cause of a larger transepithelial potential difference. Na^+ reabsorption via the conductive Na^+ channel is *inhibited* by *atrial natriuretic factor* (Chapter 31), certain *prostaglandins*, and several important diuretic drugs, including *triamterene* and *amiloride*.

To understand the modifications in tubular fluid volume and composition that occur in the collecting duct, it is important to note that the collecting duct differs from other segments of the nephron (as well as from all other tissues in the body) in that its water and urea permeability properties are regulated by a *hormone* (Imai, 1979)[†]. This hormone, *antidiuretic hormone* (abbreviated ADH; also called vasopressin), is a nonapeptide that is synthesized in the hypothalamus and stored in the posterior pituitary (Chapter 53). The factors that stimulate the secretion of ADH from the posterior pituitary are discussed in Chapter 31, but the effects of ADH on the water and urea permeability properties of the collecting duct are considered here (Hebert and Andreoli, 1986). In the *absence* of ADH, the collecting duct is relatively impermeable to water and urea. However, in the *presence* of ADH, the water permeability of the entire collecting duct and the urea permeability of the papillary collecting duct increase significantly. The changes in tubular fluid composition and osmolality that occur in the collecting duct will therefore be a function of the ADH level. In the discussion that follows, composition and osmolality changes will be examined first in the presence of *maximal* ADH and then in the *absence* of ADH (Kokko and Rector, 1972).

Maximal ADH

In the presence of maximal ADH, the high water permeability of the collecting duct prevents the establishment of an osmotic gradient across the tubular epithelium. Recall that the osmolality of the tubular fluid entering the collecting duct is only 100 mosmol/kg H_2O. Since the osmolality of the surrounding peritubular fluid ranges from approximately 300 mosmol/kg H_2O in the cortex to 1200 mosmol/kg H_2O at the tip of the papilla, this means that water is reabsorbed as the tubular fluid flows through the collecting duct. The continued reabsorption of Na^+ and Cl^- as the fluid flows through the collecting duct further promotes the reabsorption of water. In the *cortical collecting duct*, over 70% of the water entering the collecting duct is reabsorbed, as evidenced by the fact that TF_{In}/P_{In} increases by a factor of approximately 3.5, from 7 to about 25. Since the high water permeability of the cortical collecting duct is combined with low urea permeability, the urea concentration also increases by a factor of approximately 3.5, from about 50 to about 175 mmol/liter. The concentration of nonurea solutes increases less markedly, from about 50 to about 125 mmol/liter, because Na^+ and Cl^- are reabsorbed. At the end of the cortical collecting duct, then, the osmolality of the tubular fluid is approximately 300 mosmol/kg H_2O, in osmotic equilibrium with the peritubular fluid at the corti-

---

[†]The urea permeability of the papillary collecting duct may not be hormonally regulated in all species.

comedullary junction (Fig. 4.28). The fact that the increase in osmolality (about threefold, from 100 to 300 mosmol/kg H$_2$O) is smaller than the increase in TF$_{In}$/P$_{In}$ (3.5-fold) or urea concentration (3.5-fold) can be attributed primarily to the reabsorption of Na$^+$ and Cl$^-$.

In the *medullary collecting duct,* water continues to be reabsorbed as the tubular fluid flows downward through regions of increasingly hypertonic peritubular fluid (the medullary gradient) and as additional Na$^+$ and Cl$^-$ are reabsorbed. It is estimated that over 50% of the water entering the medullary collecting duct is reabsorbed in this region, as evidenced by the fact that TF$_{In}$/P$_{In}$ increases over twofold, from 25 to about 60. Since the high water permeability of the medullary collecting duct is combined with low urea permeability, as in the cortical collecting duct, the urea concentration also increases over twofold, from about 175 to about 400 mmol/liter. The concentration of nonurea solutes increases less markedly, from about 125 to about 200 mmol/liter, because Na$^+$ and Cl$^-$ are reabsorbed. The osmolality of the tubular fluid at the end of the medullary collecting duct is therefore approximately 600 mosmol/kg H$_2$O, in osmotic equilibrium with the peritubular fluid at the junction between the inner and the outer medulla (Fig. 4.28). The fact that the osmolality increases less than TF$_{In}$/P$_{In}$ or the urea concentration can be attributed primarily to the reabsorption of Na$^+$ and Cl$^-$.

Additional water is reabsorbed in the *papillary collecting duct* as the tubular fluid continues flowing downward through the medullary gradient. However, the papillary collecting duct differs from the cortical and medullary collecting ducts in that ADH increases its permeability to *urea* as well as to water. Since the concentration of urea in the tubular fluid entering the papillary collecting duct (about 400 mmol/liter) exceeds that in the surrounding peritubular fluid (about 200 mmol/liter at the junction between the inner and outer medulla), urea will be reabsorbed by the papillary collecting duct. This reabsorption of urea has an important role in the generation of the medullary gradient (see below) and also promotes the further reabsorption of water. It is estimated that approximately two-thirds of the water entering the papillary collecting duct is reabsorbed there, as TF$_{In}$/P$_{In}$ increases about threefold, from approximately 60 to about 180. The concentration of nonurea solutes also increases approximately threefold, from about 200 to 600 mmol/liter because, in spite of some Na$^+$ and Cl$^-$ reabsorption, the amount of nonurea solutes in the tubular fluid does not change appreciably. Since

urea is reabsorbed, its concentration increases less markedly, from approximately 400 to 600 mmol/liter, i.e., at the end of the papillary collecting duct a transepithelial concentration gradient for urea no longer exists. The osmolality of the tubular fluid at the end of the papillary collecting duct is therefore approximately 1200 mosmol/kg H$_2$O, in osmotic equilibrium with the peritubular fluid at the tip of the papilla (Fig. 4.28). The fact that the osmolality increases less than TF$_{In}$/P$_{In}$ or the concentration of nonurea solutes can be attributed primarily to the reabsorption of urea.

The tubular fluid from several papillary collecting ducts empties into a common duct of Bellini and from there flows through the minor calyx, major calyx, renal pelvis, and ureter into the urinary bladder (Chapter 26). No modifications in the composition or volume of the tubular fluid are believed to occur in any of these structures, i.e., the final urine is identical with the tubular fluid leaving the papillary collecting duct. In the presence of maximal ADH, then, the final urine has an osmolality of approximately 1200 mosmol/kg H$_2$O, of which about 600 mmol/liter is urea and about 600 mmol/liter represents nonurea solutes such as Na$^+$, Cl$^-$, creatinine, uric acid, K$^+$, NH$_4^+$, Ca^{2+}, Mg^{2+}, and phosphate. The U$_{In}$/P$_{In}$ ratio is approximately 180. Recall from Chapter 28 that U$_{In}$/P$_{In}$ can be used to calculate the fractional excretion of water (FE$_{H_2O}$) *(Eq. 80):*

$$FE_{H_2O} = \frac{1}{U_{In}/P_{In}}$$

In the presence of maximal ADH, then, only about 0.5% of the filtered water is excreted (1/180 $\simeq$ 0.005), 99.5% is reabsorbed. If the GFR is 125 ml/min (180 liters/day), the urine flow rate ($\dot{V}$) will be only 0.6 ml/min (0.9 liter/day) in the presence of maximal ADH.

The effect of ADH on the water permeability of the collecting duct involves the binding of ADH to receptors on the basal-lateral surfaces of the epithelial cells, the activation of adenylate cyclase, and the generation of cyclic AMP (Dousa and Valtin, 1976), which in turn leads to the insertion of protein-containing aggregates into the luminal membrane (Wade et al., 1977). These aggregates, which presumably function as channels for water movement, are believed to be components of tubular vesicles in the cells that fuse with the luminal membrane in the presence of ADH. The effect of ADH on the urea permeability of the papillary collecting duct appears to be mediated by a somewhat different mechanism, possibly involving a carrier for urea that is independent of the above-mentioned aggregates.

Absence of ADH

In the absence of ADH, the entire collecting duct is relatively *impermeable* to water. Since the thin ascending limb, thick ascending limb, distal convoluted tubule, and connecting tubule *always* are relatively impermeable to water, this means that in the absence of ADH the nephron is water-impermeable from the hairpin turn at the tip of the loop of Henle all the way to the end of the papillary collecting duct. As a result, the volume of the tubular fluid remains virtually unchanged not only in the ascending limb, distal convoluted tubule, and connecting tubule (see above), but in the collecting duct as well. In the absence of ADH, then, TF_{In}/P_{In} remains approximately constant in the collecting duct, with a value of about 7. The urea concentration also remains approximately constant, with a value of about 50 mmol/liter, since the entire collecting duct has a low permeability to urea in the absence of ADH. However, the reabsorption of Na^+ and Cl^- continues, thereby reducing the concentration of nonurea solutes from about 50 mmol/liter at the beginning of the cortical collecting duct to as low as 20 mmol/liter at the end of the papillary collecting duct. The osmolality of the tubular fluid at the end of the papillary collecting duct is therefore as low as 70 mosmol/kg H_2O in the absence of ADH. Note that the transepithelial osmotic, Na^+-Cl^-, and urea gradients become quite substantial in the collecting duct in the absence of ADH (Fig. 4.28), even though the medullary gradient is somewhat diminished (see below). The "tight" tight junctions of the collecting duct epithelium allow these gradients to be maintained.

Since the final urine is identical with the tubular fluid leaving the papillary collecting duct, this means that in the absence of ADH the final urine has an osmolality of approximately 70 mosmol/kg H_2O, of which about 50 mmol/liter is urea and about 20 mmol/liter represents nonurea solutes. The U_{In}/P_{In} ratio is approximately 7, i.e., in the absence of ADH nearly 15% of the filtered water is excreted ($1/7 \simeq 0.14$); 85% is reabsorbed *(Eq. 80)*. With a GFR of 125 ml/min (180 liters/day), the urine flow rate ($\dot{V}$) will be greater than 15 ml/min (26 liters/day). Because the volume of the tubular fluid remains virtually constant throughout the entire ascending limb and distal nephron in the absence of ADH, this urine flow rate also should represent the flow rate of tubular fluid at the tip of the loop of Henle. However, even in the absence of ADH the water impermeability of the collecting duct is *relative, not* absolute, so that a small fraction of the water flowing through the collecting duct will be reabsorbed. The actual urine flow rate may therefore be slightly less than the flow rate of tubular fluid at the tip of the loop.

Table 4.18 summarizes the modifications in the composition of the tubular fluid that occur in the distal nephron in the presence of maximal ADH and in absence of ADH. The permeability and transport characteristics of the various segments of the distal nephron are summarized in Table 4.17.

Other Functions of the Distal Nephron

The distal nephron has an essential role in the regulation of K^+ excretion. In fact, virtually *all* regulation of K^+ excretion occurs in the distal nephron, which can both reabsorb and secrete K^+. The distal nephron also has an essential role in regulating the body's acid-base balance via secretion of H^+ and HCO_3^-. In addition, the distal nephron has a role in the regulation of Ca^{2+} excretion. These important functions of the distal nephron are discussed in greater detail in Chapters 32–34.

Table 4.18.
Approximate Concentrations of Substances Entering and Leaving the Distal Nephron in Normal Humans

	Entering the Distal Nephron (from the Loop of Henle)	Leaving the Connecting Tubule		Leaving the Cortical Collecting Duct	Leaving the Medullary Collecting Duct	Leaving the Papillary Collecting Duct (Final Urine)
[nus],[a] mmol/liter	100	50	+ADH	125	200	600
			−ADH	40	30	20
[Urea], mmol/liter	50	50	+ADH	175	400	600
			−ADH	50	50	50
Osmolality, mosmol/kg H_2O	150	100	+ADH	300	600	1200
			−ADH	90	80	70
TF_{In}/P_{In}	7	7	+ADH	25	60	180
			−ADH	7	7	7

[a]nus, nonurea solutes.

URINARY CONCENTRATION AND DILUTION

The above discussion of Na$^+$, Cl$^-$, and water reabsorption in the distal nephron includes a description of the events leading to the excretion of a concentrated urine and a dilute urine (Kokko and Rector, 1972). To summarize briefly, because the ascending limb reabsorbs Na$^+$ and Cl$^-$ but is impermeable to water, the tubular fluid entering the distal nephron is *hypotonic* to plasma (average osmolality $\simeq$ 150 mosmol/kg H$_2$O). The osmolality of the tubular fluid is further reduced (to $\simeq$ 100 mosmol/kg H$_2$O) in the distal convoluted tubule and connecting tubule which, like the ascending limb, reabsorb Na$^+$ and Cl$^-$ but are water-impermeable. Events in the collecting duct determine whether a *concentrated urine* or a *dilute urine* will be excreted. In the *presence of ADH*, the entire collecting duct is relatively permeable to water. Water is therefore reabsorbed, since the osmolality of the peritubular fluid surrounding the collecting duct ranges from approximately 300 mosmol/kg H$_2$O in the cortex to 1200 mosmol/kg H$_2$O at the tip of the papilla. With *maximal* ADH, the final urine has the same osmolality as the tubular fluid at the tip of the papilla (about 1200 mosmol/kg H$_2$O) and represents only about 0.5% of the filtered volume. In the *absence of ADH*, the entire collecting duct is relatively impermeable to water. The volume of the tubular fluid therefore does not change appreciably, but since solute is reabsorbed, the osmolality of the tubular fluid is further reduced (below 100 mosmol/kg H$_2$O). The final urine has an osmolality as low as 70 mosmol/kg H$_2$O and represents up to 15% of the filtered volume.

In all discussions of urinary concentration and dilution presented thus far in this chapter, the presence of the *medullary gradient* was assumed. It is now appropriate to consider how this gradient is formed and maintained, as well as how the tubular epithelial cells in the medulla maintain their intracellular volume in the presence of the high peritubular fluid osmolalities of the gradient.

Formation of the Medullary Gradient: Countercurrent Multiplication

A key factor in the formation of the medullary gradient is the unusual anatomic configuration of the loop of Henle, which allows fluid to flow in opposite, or *countercurrent*, directions in its two limbs. Because of the specific transport and permeability properties of the ascending and descending limbs (Table 4.17), countercurrent flow allows a large osmotic gradient to be established in the peritubular fluid between the corticomedullary junction and the tip of the papilla. This is illustrated schemati-

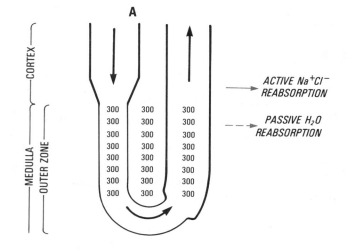

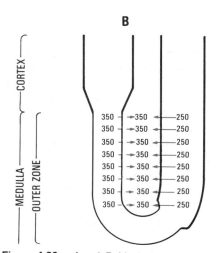

Figure 4.30. *A* and *B*, Mechanisms for the formation of an osmotic gradient in the medullary peritubular fluid surrounding the loop of Henle of a short-looped nephron, starting at a hypothetical time zero. The tubular fluid and peritubular fluid osmolalities are given in mosmol/kg H$_2$O. Details are provided in the text.

cally in Figure 4.30, which demonstrates how an osmotic gradient could develop in the medullary peritubular fluid surrounding the loop of a *short-looped nephron*. Recall that the loop of a short-looped nephron consists of a *thin descending limb*, which is relatively impermeable to solutes but is highly permeable to water, and a *thick ascending limb*, which actively reabsorbs Na$^+$ and Cl$^-$ but is relatively impermeable to water (short-looped nephrons lack a thin ascending limb). Initially (i.e., at a hypothetical time zero), the osmolality of the fluid in both limbs of the loop and in the surrounding peritubular fluid would be similar to plasma; an approximate value of 300 mosmol/kg H$_2$O is used here for simplicity (Fig. 4.30*A*). Assume now that

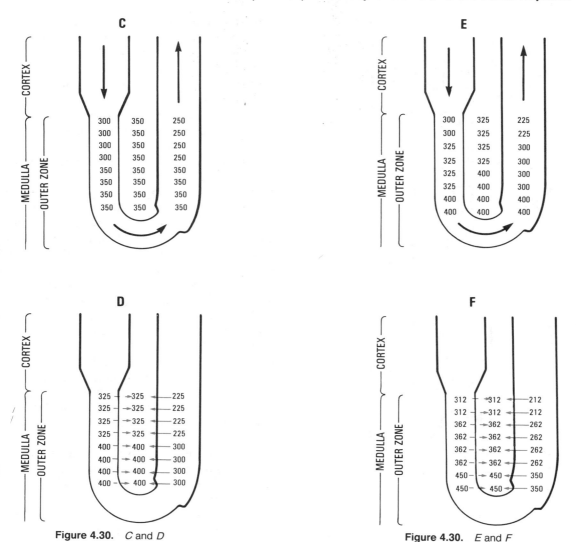

Figure 4.30. *C* and *D*

Figure 4.30. *E* and *F*

flow through the loop is momentarily stopped. Since the ascending limb actively reabsorbs Na^+ and Cl^- but is impermeable to water, the peritubular fluid osmolality increases and a small osmotic gradient is established between the ascending limb and the surrounding peritubular fluid at each horizontal level in the loop; the increased peritubular fluid osmolality, in turn, causes water to be reabsorbed from the water-permeable descending limb (Fig. 4.30B). For illustrative purposes, it is arbitrarily assumed that a gradient of 100 mosmol/kg H_2O can be developed between the tubular fluid in the ascending limb and the fluid in the peritubular space and descending limb. In analogy to certain countercurrent flow applications in physical chemistry and engineering, this horizontal gradient often is referred to as the "single effect."

When flow continues (Fig. 4.30C), additional fluid with an osmolality of 300 mosmol/kg H_2O is intro-

duced from the proximal tubule into the descending limb. Flow again is momentarily stopped, and a gradient of 100 mosmol/kg H_2O is again established between the fluid in the ascending limb and the fluid in the peritubular space and descending limb (Fig. 4.30D). As the sequence repeats (Figs. 4.30E to 4.30H), because of countercurrent flow the *small* osmotic gradient between the ascending limb and the surrounding peritubular fluid at each *horizontal* level becomes "multiplied" considerably in the *vertical* direction, resulting in a *large* osmotic gradient between the peritubular fluid at the corticomedullary junction and the peritubular fluid surrounding the hairpin turn. This mechanism for the formation of a medullary osmotic gradient therefore is termed *countercurrent multiplication*. It should be evident that the basic requirements for the operation of the countercurrent multiplication mechanism are (1) countercurrent flow, (2) an active transport process

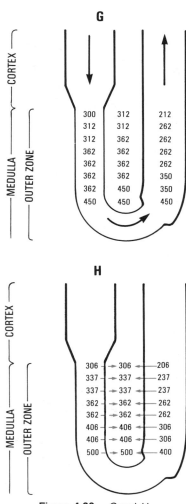

Figure 4.30. *G* and *H*

allowing the establishment of the horizontal osmotic gradient, and (3) the different water permeability properties of the descending and ascending limbs. Although these basic requirements for countercurrent multiplication also are satisfied by the *cortical* part of the ascending limb and segment 3 of the proximal tubule (the pars recta), the vertical osmotic gradient develops in the *medullary* peritubular fluid *only* (Fig. 4.30). This is because the relatively high blood flow to the renal cortex washes away the reabsorbed Na⁺ and Cl⁻, thereby preventing the establishment of an osmotic gradient in the cortical peritubular fluid.

In the short-looped nephron example illustrated in Figure 4.30, the osmolality of the peritubular fluid surrounding the hairpin turn was increased to about 500 mosmol/kg H₂O by countercurrent multiplication. Recall that the hairpin turn in short-looped nephrons occurs at or above the junction between the outer and inner medulla (Fig. 4.26). Since the peritubular fluid osmolality at this junc-

tion is estimated to be approximately 600 mosmol/ kg H₂O (Fig. 4.26), Figure 4.30 represents a realistic example. It should be evident that the maximum peritubular fluid osmolality achieved by the countercurrent mechanism will depend upon the *length* of the loop. Thus, countercurrent multiplication in *long-looped nephrons* is responsible for developing the very high osmolalities measured in the inner medulla, with values up to 1200 mosmol/kg H₂O at the tip of the papilla (Fig. 4.26). However, the countercurrent mechanism must operate somewhat differently in long-looped nephrons. This is because a basic requirement of the mechanism is an active transport process along the entire length of the ascending limb to establish the horizontal osmotic gradient. In the thin ascending limb of long-looped nephrons, a horizontal osmotic gradient can be generated as a result of *passive* reabsorption of Na⁺ and Cl⁻, as discussed previously, but this passive reabsorption process depends upon the *prior* existence of the medullary gradient to reabsorb water from the thin *descending* limb, thereby increasing the concentration of Na⁺ and Cl⁻ in the tubular fluid delivered to the thin ascending limb. To understand how passive Na⁺ and Cl⁻ reabsorption in the thin ascending limb might *first* be initiated, one must consider how *urea* becomes a component of the inner medullary gradient.

The steps leading to the addition of urea to the inner medulla are illustrated schematically in Figure 4.31. At a hypothetical time zero, the tubular fluid entering the thick ascending limb is similar to that leaving the proximal tubule, with a urea concentration of 20 mmol/liter and a total osmolality of 300 mosmol/kg H₂O (Fig. 4.31, ①). As this fluid flows through the thick ascending limb, distal convoluted tubule, and connecting tubule (Fig. 4.31, ② and ③), the urea concentration remains *unchanged* because of the low water permeability and low urea permeability of these regions. However, Na⁺ and Cl⁻ are actively reabsorbed, thereby reducing the osmolality of the tubular fluid. The cortical and medullary collecting ducts also have a low urea permeability. Thus, if ADH is present, the concentration of urea will *increase* in these regions because of the reabsorption of water (Fig. 4.31, ④ and ⑤). Tubular fluid with a high urea concentration is therefore delivered to the papillary collecting duct, which is relatively *permeable* to urea in the presence of ADH. As a result, urea is passively reabsorbed in the papillary collecting duct (Fig. 4.31, ⑥) and enters the peritubular fluid in the inner medulla. This addition of urea to the inner medulla is sufficient to initiate the passive reabsorption of Na⁺ and Cl⁻ from the thin ascending limb. First, the urea causes water to

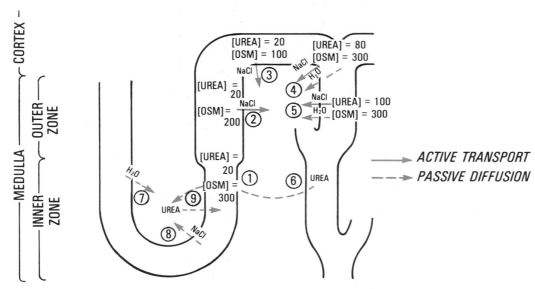

Figure 4.31. Mechanism for the addition of urea to the inner medullary gradient, starting at a hypothetical time zero. Tubular fluid urea concentrations and osmolalities are given in mmol/liter and mosmol/kg H₂O, respectively. The events at the *circled numbers* are described in the text.

be reabsorbed from the water-permeable thin descending limb (Fig. 4.31, ⑦), thereby increasing the concentration of Na⁺ and Cl⁻ in the tubular fluid. Then, when this Na⁺Cl⁻-rich fluid flows through the water-impermeable, Na⁺Cl⁻-permeable thin ascending limb, Na⁺ and Cl⁻ are *passively reabsorbed* (Fig. 4.31, ⑧). Although some of the urea added to the peritubular fluid in the inner medulla enters the thin ascending limb (which is somewhat permeable to urea) (Fig. 4.31, ⑨), the *net* result is solute *exit*. In other words, because of the urea added to the medullary peritubular fluid from the papillary collecting duct, a horizontal osmotic gradient can be created in the thin ascending limb *in spite* of the lack of an active transport process in this region. The countercurrent multiplication mechanism can therefore operate in *long-looped nephrons* as well as in short-looped nephrons.

Of course, even though the horizontal osmotic gradient in the inner medulla is created by a *passive* process, energy *is* expended in delivering a urea-rich tubular fluid to the papillary collecting duct (Fig. 4.31), which in turn allows urea to enter the inner medulla. Most of this energy is expended by the *thick ascending limb*. Recall that the formation of a urea-rich tubular fluid depends on the reabsorption of water from the urea-impermeable cortical and medullary collecting ducts in the presence of ADH. By actively reabsorbing Na⁺ and Cl⁻, the thick ascending limb allows water to be reabsorbed (1) from the *cortical collecting duct* in the presence of ADH by delivering tubular fluid with a reduced osmolality to the distal nephron and (2) from the

medullary collecting duct in the presence of ADH by establishing an osmotic gradient in the peritubular fluid of the outer medulla. In fact, the *entire* medullary gradient will completely *vanish* within minutes after the administration of a *loop diuretic* such as furosemide, which inhibits the active reabsorption of Na⁺ and Cl⁻ by the thick ascending limb (see above).

It also should be noted that the delivery of a urea-rich tubular fluid to the papillary collecting duct is dependent upon the presence of *ADH* as well as upon the active reabsorption of Na⁺ and Cl⁻ by the thick ascending limb. This in part explains why the medullary gradient is *diminished* in the *absence* of ADH, as previously indicated (see Fig. 4.28); the failure of urea to enter the medullary interstitium in the absence of ADH also diminishes the gradient. Furthermore, it should be noted that once the medullary gradient is established, the urea concentration in the tubular fluid delivered to the papillary collecting duct will be even *higher* than that shown in Figure 4.31 (see Fig. 4.28). This is because (1) additional water will be reabsorbed from the medullary collecting duct when the medullary gradient is present and (2) the urea that enters the thin ascending limb augments the urea concentration in the tubular fluid delivered to the thick ascending limb.

Maintenance of the Medullary Gradient: Countercurrent Exchange

Whereas countercurrent multiplication is the mechanism whereby the medullary gradient is *formed,* a second mechanism is required to *maintain* the medullary gradient. The importance of a mechanism for maintaining the medullary gradient is best illustrated by the fact that such a gradient could be diluted or washed out by *blood flow* to the medulla. For example, recall that even though the conditions for countercurrent multiplication are satisfied by the *cortical* part of the thick ascending limb and segment 3 of the proximal tubule, an osmotic gradient does *not* form in the cortical peritubular fluid because of the relatively high blood flow to the cortex, which washes away the reabsorbed solute. *Medullary* blood flow is relatively low (only about 10% of the total RBF), but even this amount could significantly dilute the medullary gradient. That the medullary blood flow does *not* dilute the gradient can be attributed to the unusual anatomic configuration of the medullary blood supply.

Recall that the medullary blood supply is derived from the *vasa recta,* special hairpin-shaped capillaries that exit from the efferent arteriole of juxtamedullary nephrons (Fig. 4.7). Thus, *countercurrent flow* occurs not only in the loop of Henle, but in the capillaries supplying the medullary peritubular space as well. Figure 4.32 illustrates how this countercurrent blood flow helps to *maintain* (rather than dilute or wash out) the medullary gradient. It is assumed that the vasa recta, like other capillaries, are highly permeable to water and small solutes

and that the blood entering the vasa recta has an osmolality similar to that of systemic plasma; an approximate value of 300 mosmol/kg H_2O is used here for simplicity (Fig. 4.32, ①). Thus, as the blood flows downward through regions of increasingly hypertonic peritubular fluid (Fig. 4.32, ②), water passively diffuses out of the blood and solute diffuses in. The osmolality of the blood in the vasa recta therefore increases as the hairpin turn is approached. However, until the hairpin turn is reached, the osmolality of the blood always is slightly *below* that of the surrounding peritubular fluid. The reason is that the blood flow is sufficiently rapid to cause a small lag in equilibration between the blood and the adjacent peritubular fluid. If the vasa recta did *not* have a hairpin configuration, the capillary would exit the kidney at the tip of the papilla (Fig. 4.32, ③) and would have *diluted* the gradient by adding water to and removing solute from the peritubular space. Because of the hairpin turn, however, the blood ascends through regions of decreasing peritubular fluid hypertonicity. Water and solute fluxes *opposite* to those in the descending limb therefore occur: water passively diffuses into the blood and solute diffuses out (Fig. 4.32, ④). Note that in the ascending limb, the osmolality of the blood is always slightly *greater* than that in the adjacent peritubular fluid because of a small lag in equilibration. Since the *vol-*

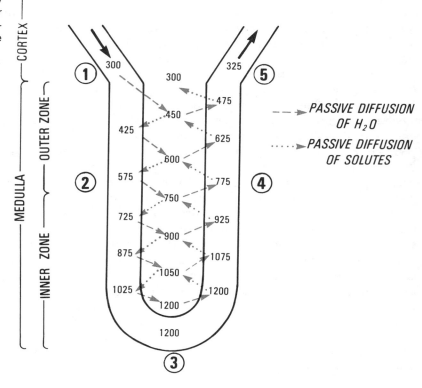

Figure 4.32.   Maintenance of the medullary gradient by countercurrent exchange in the vasa recta. The tubular fluid and peritubular fluid osmolalities are given in mosmol/kg H_2O. The events at the *circled numbers* are described in the text.

ume of blood leaving the medulla via the vasa recta is slightly greater than that entering (see below), some solute is removed from the peritubular space. However, the depletion of peritubular solutes is minimized because of countercurrent flow (*without* countercurrent flow, the blood would leave the medulla at ③). This mechanism for the preservation of the gradient is termed *countercurrent exchange.*

As indicated above, the volume of blood leaving the medulla via the vasa recta is slightly greater than that entering. This increase in volume can be attributed primarily to the above-mentioned lag in equilibration between the vasa recta blood and the adjacent peritubular fluid. In the *descending* limb, the somewhat higher osmolality of the adjacent peritubular fluid tends to extract fluid from the capillary. However, oncotic forces exerted by plasma proteins *oppose* this fluid exit. In the *ascending* limb, the somewhat higher osmolality of the blood favors the uptake of fluid into the capillary. The oncotic forces exerted by plasma proteins now act *together* with the effect of the osmolality difference created by the lag in equilibration, to favor fluid uptake. As a result, fluid uptake by the ascending vasa recta exceeds fluid loss from the descending vasa recta. This serves to remove the water reabsorbed from the thin descending limb and medullary and papillary collecting ducts, which otherwise would dilute the gradient.

It should be emphasized that countercurrent exchange is a *passive* process, involving only the diffusion ("exchange") of solutes and water across the permeable walls of the vasa recta capillaries. The vasa recta have *no* role in creating the osmotic gradient and in fact would rapidly dissipate the gradient if countercurrent multiplication were to cease, e.g., following the administration of a loop diuretic (see above).

Maintenance of Intracellular Volume in the Medulla

As discussed in Chapter 25, cells suspended in a hypertonic fluid generally will *shrink* because of the exit of water. Given the high osmolalities present in the medullary peritubular fluid, the tubular epithelial cells in the medulla certainly would be expected to shrink. Such shrinkage is *avoided,* however, because the medullary cells transport and synthesize certain *organic solutes* that increase their intracellular osmolality, thereby balancing the high peritubular fluid osmolalities (Bagnasco et al., 1986). Important examples of these solutes are sorbitol, inositol, betaine, and glycerophosphorylcholine, which are found in high concentrations in the medullary cells.

Factors Affecting the Concentrating and Diluting Mechanisms

In the normal kidney, the tubular fluid delivered to the collecting duct always is *hypotonic.* Then, in the collecting duct ADH regulates the final urine osmolality through its effects on water permeability. The osmolality of the urine is not solely dependent upon the ADH level, however. Most importantly, both the concentrating and diluting mechanisms are markedly dependent on the proper function of the *thick ascending limb.* Even if plasma ADH is *high,* a concentrated (hypertonic) urine can be excreted *only* in the presence of the *medullary gradient* to provide the osmotic driving force for the reabsorption of water from the medullary and papillary collecting ducts. As discussed above, the active reabsorption of Na^+ and Cl^- by the thick ascending limb is the key factor in the formation of the medullary gradient. If ADH is *absent,* the osmolality of the tubular fluid is further reduced in the distal nephron and a dilute (hypotonic) urine is excreted. However, the urine will *not* be *maximally* dilute unless the tubular fluid entering the distal nephron is hypotonic, i.e., the production of a maximally dilute urine requires the active reabsorption of Na^+ and Cl^- from the water-impermeable thick ascending limb.

The concentrating and diluting mechanisms can be affected by other factors as well (Jamison et al., 1979). For example, although the proper function of the thick ascending limb is the most crucial factor in *establishing* the medullary gradient, the *maximum osmolality* of the gradient has several additional determinants (Knepper and Stephenson, 1986):

1. *Length of the Loop of Henle and Percentage of Nephrons with Long Loops.* In humans, approximately 15% of the nephrons are long-looped juxtamedullary nephrons, and the maximum peritubular fluid osmolality is approximately 1200 mosmol/kg H_2O. In the Saharan desert rat *Psammomys,* however, nearly 35% of the nephrons have very long loops, and the maximum peritubular fluid osmolality is as high as 5000 mosmol/kg H_2O (this means that the *urine* osmolality also can be as high as 5000 mosmol/kg H_2O, a very important water-conserving adaptation).

2. *Availability of Urea.* When a protein-deficient diet is ingested, the maximum peritubular fluid osmolality will be diminished (recall that urea is a protein breakdown product).

3. *Rate of Flow through the Loop of Henle.* A rapid flow through the loop of Henle tends to wash out the gradient because additional water is then

reabsorbed from the thin descending limb, diluting the solutes in the medulla.

4. *Rate of Flow through the Collecting Duct.* A rapid flow through the collecting duct tends to wash out the gradient, primarily by diluting the urea in the tubular fluid delivered to the papillary collecting duct and thereby reducing the concentration gradient for the reabsorption of urea into the inner medulla.

5. *Rate of Flow through the Vasa Recta.* With increased flow, the equilibration between peritubular fluid and vasa recta blood is impaired, and additional solute is removed from the medulla (see above). Thus, the maximum osmolality of the gradient is highest when flow through the loop of Henle, collecting duct, *and* vasa recta is low.

6. *Presence of Prostaglandins.* Certain prostaglandins may increase blood flow through the vasa recta, thereby reducing the maximum osmolality of the gradient (cf. paragraph 5 above). Prostaglandins also can reduce the maximum osmolality of the gradient by inhibiting the active reabsorption of Na^+ and Cl^- by the thick ascending limb (see above).

Prostaglandins also impair the formation of a concentrated urine by inhibiting adenylate cyclase (Stokes, 1981), thereby inhibiting the effect of ADH on the water permeability of the collecting duct and on the urea permeability of the papillary collecting duct. Since ADH *stimulates prostaglandin synthesis* by the kidneys, this inhibition can be regarded as a form of *negative feedback.* The various mechanisms by which the prostaglandins can impair the formation of a concentrated urine are summarized in Figure 4.33.

Quantitation of Urinary Concentration and Dilution

To assess quantitatively the operation of the concentrating and diluting mechanisms, an index of the ability of the kidneys to excrete a concentrated urine or a dilute urine is required. One simple index is the *urine osmolality,* U_{osm}. When measured at *high ADH* or in the *absence of ADH,* U_{osm} assesses the ability of the kidneys to excrete a *maximally concentrated* urine or a *maximally dilute* urine, respectively. As indicated above, in humans U_{osm} approaches 1200 mosmol/kg H_2O at maximum ADH and can be as low as 70 mosmol/kg H_2O in the absence of ADH.

Another index for assessing the operation of the concentrating and diluting mechanisms is the so-called *free water clearance.* The free water clearance differs from other clearances discussed in this

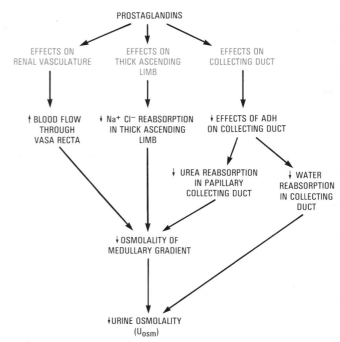

Figure 4.33. Mechanisms by which prostaglandins can impair the formation of a concentrated urine. (Adapted from Beck and Dunn, 1981.)

text since it is *not* defined by the standard clearance relationship, $C_X = U_X \dot{V}/P_X$ *(Eq. 26)*. Instead, the free water clearance (C_{H_2O}) is defined as the difference between the urine flow rate ($\dot{V}$) and the osmolar clearance (C_{osm}):

$$C_{H_2O} = \dot{V} - C_{osm} \qquad (86)$$

To understand the significance of free water clearance, it is helpful to examine the *osmolar clearance* more closely. The osmolar clearance is a standard clearance except that it refers to the *total* number of osmotically active solute particles instead of to a *single* substance. Thus, C_{osm} represents the volume of plasma completely cleared of *all* osmotically active solute particles per unit time and is calculated as $U_{osm}\dot{V}/P_{osm}$. However, C_{osm} also represents the *hypothetical* urine flow rate that *would* be measured if urine were *isotonic* to plasma, i.e., if U_{osm} were equal to P_{osm}. This can be shown by starting with the clearance equation $C_{osm} = U_{osm}\dot{V}/P_{osm}$. If $U_{osm} = P_{osm}$, then

$$C_{osm} = \dot{V} \qquad (87)$$

The free water clearance ($= \dot{V} - C_{osm}$) therefore represents the difference between the *actual* urine flow rate ($\dot{V}$) and this *hypothetical isotonic* urine flow rate (C_{osm}). When a *dilute* (hypotonic) urine is excreted, the *actual* urine flow rate is *greater* than

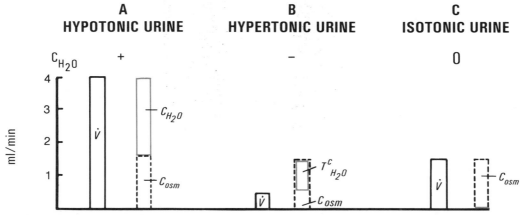

Figure 4.34. Use of *free water clearance* to quantitate urinary concentration and dilution. The free water clearance (C_{H_2O}, *solid blue lines*) can be calculated by comparing the *actual* urine flow rate ($\dot{V}$, *solid black lines*) to the *hypothetical* flow rate that would be present if urine were isotonic (C_{osm}, *dashed black lines*). *A*, In *hypotonic* urine, C_{H_2O} is *positive* and represents the solute-free water that must be *added* to the hypothetical isotonic urine to make the actual urine. *B*, In *hypertonic* urine, C_{H_2O} is *negative* and represents the solute-free water that must be *removed* from the hypothetical isotonic urine to make the actual urine. *C*, In *isotonic* urine, C_{H_2O} is *zero* since the actual flow is the same as the hypothetical isotonic urine flow. Note that the negative free water clearance in *B* is designated $T^c_{H_2O}$ $(=(-C_{H_2O}))$. C_{osm} is *independent* of urine tonicity, with an average normal value of 1.5 ml/min.

the *hypothetical isotonic* urine flow rate. C_{H_2O} therefore is a *positive* number and represents the volume of distilled water, or *"free water"* (i.e., *solute*-free water), that would have to be added to the hypothetical isotonic urine to make the actual urine (Fig. 4.34*A*). Physiologically, this solute-free water is formed by reabsorbing solute in a water-impermeable region, such as the thick ascending limb, distal convoluted tubule, connecting tubule, or (in the absence of ADH) collecting duct. When a *concentrated* (hypertonic) urine is excreted, the *actual* urine flow rate is *less* than the *hypothetical isotonic* urine flow rate. C_{H_2O} therefore is a *negative* number and represents the volume of "free water" (i.e., *solute*-free water) that would have to be removed from the hypothetical isotonic urine to make the actual urine (Fig. 4.34*B*). Physiologically, this solute-free water is removed by reabsorbing water in excess of solute, as in the cortical and medullary collecting ducts when ADH is present. Of course, if the urine is *isotonic*, the actual urine flow rate is *equal* to the

hypothetical isotonic urine flow rate (Fig. 4.34*C*), and the free water clearance is *zero*.

When a *concentrated* urine is excreted, reference often is made to the quantity $C_{osm} - \dot{V}$, a *positive* number, instead of to $\dot{V} - C_{osm}$, a *negative* number for a concentrated urine. This quantity is designated $T^c_{H_2O}$ (Fig. 4.34*B*) and is commonly termed the *negative free water clearance*, since $T^c_{H_2O}$ equals $(-C_{H_2O})$.

It should be noted in Figure 4.34 that C_{osm} is *independent* of the urine tonicity. This is because an average of 650 mosmol of electrolytes and other solutes must be excreted per day in a normal human to eliminate waste products such as creatinine and urea and to maintain a normal electrolyte balance, i.e., the average $U_{osm}\dot{V}$ equals 650 mosmol/day or 0.45 mosmol/min (typical normal range: 400–900 mosmol/day or 0.3–0.6 mosmol/min). C_{osm} therefore has an average normal value of 1.5 ml/min *(Eq. 26)*, as illustrated in Figure 4.34 (typical normal range: 1–2 ml/min).

CHAPTER 31

Regulation of Volume and Osmolality of the Body Fluids

Maintenance of the normal volume and osmolality of the body fluids requires that the *input* of solvent and solutes to the body equals the *output* of solvent and solutes from the body each day. This principle, input = output, is referred to as *balance* and applies to the normal, nongrowing, nonpregnant adult. In the normal person, of course, water is the only solvent of concern. In contrast, many different solutes contribute to the osmolality of the body fluids (Chapter 25). Since the major extracellular solute is Na^+ (along with its attendant anions), the regulation of the volume and osmolality of the *extracellular* fluid is dependent almost exclusively upon the regulation of the balances of water and Na^+. Moreover, since cell membranes are permeable to water, the volume and osmolality of the *intracellular* fluid also will be influenced by the balances of water and Na^+ in the extracellular fluid (Chapter 25). Therefore, this chapter on the regulation of volume and osmolality of the body fluids is directed at understanding the inputs and outputs of water and Na^+ and their regulation.

WATER

Input of Water

Water is added to the body fluids from three sources: the water content of food, water generated during the oxidation of food, and water consumed as a liquid (Table 4.19). The amount of water *in food* can vary widely, but the average diet of adult Americans contains about 800–1000 ml/day. About 300–400 ml of water are generated per day during the *oxidation of food*. For example, the oxidation of 1 mole of glucose (180 g) generates 6 moles of water (108 ml):

$$C_6H_{12}O_6 + 6\ O_2 \rightarrow 6\ CO_2 + 6\ H_2O$$

Neither the water obtained from food nor the water generated during the oxidation of food is an important variable in the regulation of water input. Thus,

water input is regulated primarily by changes in the volume of water consumed *as a liquid*, which averages 1–2 liters/day but can vary from less than 1 liter/day to more than 20 liters/day. For example, when large volumes of fluid are lost from the body, liquid consumption increases in response to the sensation of *thirst* (see below). However, it should be noted that when environmental conditions are normal and constant, liquid consumption is primarily determined by an individual's habits rather than by thirst. Man appears to be able to anticipate future requirements of water and to drink appropriate amounts of liquid (Fitzsimons, 1976).

Output of Water

Water is lost from the body via four routes (Table 4.19): insensible loss, sweat, feces, and urine. *Insensible loss* averages about 800–1000 ml/day, of which about half is lost as moisture in expired air and half is lost as water that evaporates through the skin *(water of transpiration)*. The water of transpiration differs from *sweat*, which is produced by the sweat glands in response to thermal stress. The volume of sweat can vary from less than 200 ml/day in an individual at rest in a cool environment to as much as 8–10 liters/day if the environmental temperature, humidity, and/or level of physical activity are increased. *Fecal loss* of water in normal man is only 100–200 ml/day, but it can exceed 5 liters/day in diarrheal diseases such as cholera. Insensible loss, sweat, and fecal water do not represent important variables in the physiological regulation of water output (although sweat is of extreme importance in the regulation of *body temperature;* see Chapter 51). Thus, water output is regulated primarily by changes in the volume of *urine*, which averages 1–2 liters/day but can vary from less than 1 liter/day to more than 20 liters/day. As discussed in Chapter 30, the urine volume is determined primarily by the plasma level of *ADH*.

Because the body is unable to prevent insensible water loss, sweat, or fecal loss, these outputs of

Table 4.19.
Major Inputs and Outputs of Water[a]

Inputs	Average Volume per Day, ml/day	Outputs	Average Volume per Day, ml/day
Water content of food	800–1000	Insensible loss	800–1000
Water generated during oxidation of food	300–400	Sweat	200
Water consumed as a liquid	1000–2000	Feces	100–200
		Urine	1000–2000
Total	2100–3400	Total	2100–3400

Adapted from Weitzman and Kleeman, 1980. [a]Inputs and outputs represent average basal values for an adult in a cool environment.

water commonly are termed *obligatory losses*. In addition, part of the *urine* output can be regarded as obligatory. As indicated in Chapter 30, an average of 650 mosmol of electrolytes and other solutes must be excreted per day to eliminate waste products such as creatinine and urea and to maintain a normal electrolyte balance. Thus, even if the urine is *maximally concentrated* ($U_{osm} \simeq 1200$ mosmol/kg H_2O), the *minimum* urine volume is 500–600 ml, i.e., approximately 500–600 ml of the daily urine volume represents an obligatory water loss.

Control of Water Balance

The input and output of water, then, are regulated primarily by changes in the volume of *liquid ingested* and the *urine volume,* which, in turn, are controlled by *thirst* and the plasma level of *ADH,* respectively. Both thirst and the secretion of ADH from the posterior pituitary are controlled by centers located in the hypothalamus that are stimulated primarily by two physiological conditions, increases in plasma osmolality and decreases in plasma volume:

1. Increases in *plasma osmolality* are sensed by cells within the hypothalamus termed *osmoreceptors* that are stimulated if water is removed from the cells, causing them to shrink. Hence, these cells respond to increases in plasma osmolality produced by (a) deficits of total body water; or (b) solutes that are predominantly excluded from the intracellular fluid, such as Na^+ and Cl^-, which cause water to shift from the intracellular compartment to the extracellular compartment. Solutes that enter cells readily, such as urea, do *not* change the volume of the osmoreceptors and therefore do not stimulate thirst or ADH secretion (Robertson et al., 1976).

2. Decreases in *plasma volume* are sensed by baroreceptors ("pressure receptors") located in both the low-pressure regions (atria) and the high-pressure regions (carotid sinus, aortic arch) of the circulation (Zerbe and Robertson, 1987). A decrease in plasma volume inhibits the firing of these receptors, which, in turn, causes a reflex stimulation of thirst and ADH secretion. The baroreceptors can sense such changes in plasma volume because they actually respond to *stretch* rather than to pressure. In fact, they frequently are termed "volume receptors" rather than "pressure receptors."

Under normal conditions, thirst and ADH secretion are primarily under the control of the *osmoreceptors*. This is because the osmoreceptors are extremely sensitive to *small* changes in plasma osmolality. In fact, significant changes in ADH secretion occur when the plasma osmolality changes by less than 1%. In contrast, as much as a 10% change in plasma volume may be necessary before significant changes in ADH secretion are observed (Robertson et al., 1976). Thus, changes in plasma volume primarily affect thirst and ADH secretion in extreme circumstances such as severe dehydration, hemorrhage, or redistribution of body water away from the central circulation (e.g., edema of an extremity due to occlusion of venous outflow or loss of fluid into an infected area or body cavity).

Although the osmoreceptors are more *sensitive* than the volume receptors, the volume receptors can result in a *stronger* response. For example, a very low plasma volume will result in thirst and ADH secretion even in the presence of plasma *hypotonicity*. The fact that ADH secretion is *stimulated* at a *low* plasma osmolality if plasma volume is very low and, conversely, that ADH secretion is *suppressed* at a *high* plasma osmolality if plasma volume is very high is illustrated in Figure 4.35. The data presented in this figure are commonly summarized by the statement that *volume overrides tonicity*.

It should be noted that while thirst and ADH secretion are primarily determined by the plasma osmolality and volume, they can be affected by several additional conditions and pharmacological agents (Zerbe and Robertson, 1987), as summarized in Table 4.20.

To summarize the regulation of water input and output by thirst and ADH secretion, the pathway by which plasma osmolality and volume are returned to normal following an acute episode of dehydration is illustrated in Figure 4.36.

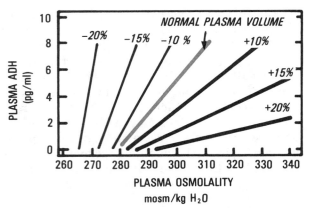

Figure 4.35. Relationship between plasma osmolality and plasma concentration of ADH at different plasma volumes. The *blue line* represents the relationship at a normal plasma volume. *Thin black lines* represent the relationship at plasma volumes that are *below* the normal plasma volume by the indicated percentages, illustrating that ADH secretion can be stimulated at *low* plasma osmolalities if the plasma volume is very low. *Thick black lines* represent the relationship at plasma volumes that *exceed* the normal plasma volume by the indicated percentages, illustrating that ADH secretion can be suppressed at *high* plasma osmolalities if the plasma volume is very high. (Adapted from *Kidney International* 10: 25–37, 1976, with permission.)

SODIUM

Input of Sodium

The input of Na^+ depends entirely upon the Na^+ content of food and water consumed (Table 4.21). While specially constructed diets may contain less than 10 mmol of Na^+ per day, individuals who use a salt shaker liberally may ingest more than 600 mmol/day. Thus, the daily input of Na^+ can vary markedly, depending upon an individual's diet and habits. Geographical factors also can influence Na^+ input by affecting the Na^+ content of water and agricultural products. The average diet of adult Americans contains 100–400 mmol Na^+/day. Although, an appetite for Na^+ can be demonstrated in animals, Na^+ input does not appear to be physiologically regulated in man.

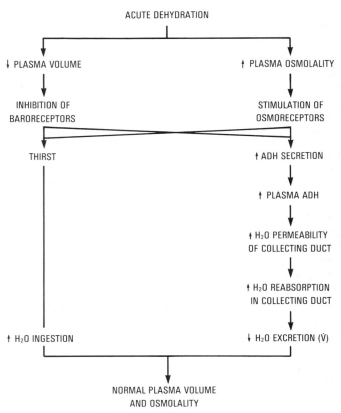

Figure 4.36. Pathway by which normal plasma volume and osmolality are restored following an acute episode of dehydration.

Output of Sodium

Na^+ is lost from the body via three routes: sweat, feces, and urine (Table 4.21). The amount of Na^+ lost in *sweat* depends upon both the volume of sweat and the degree of adaptation to a hot environment. The concentration of Na^+ in sweat decreases as adaptation to heat occurs over a period of several days. Thus, Na^+ loss can vary from negligible in a person at rest in a cool environment to several hundred mmol/day in a nonheat-adapted person exercising in a hot enviroment. Under nor-

Table 4.20.
Conditions That Affect Thirst and ADH Secretion

Stimuli	Inhibitors
↑ Plasma osmolality	↓ Plasma osmolality
↓ Plasma volume	↑ Plasma volume
Angiotensin II	Ethanol[a]
Nausea[a]	Phenytoin[a]
Pain, emotion, stress[a]	
Cholinergic agents	
Barbiturates	
Nicotine[a]	

[a]Affects ADH secretion only.

Table 4.21.
Major Inputs and Outputs of Sodium[a]

Inputs	Average Amount per Day, mmol/day	Outputs	Average Amount per Day, mmol/day
Food and water	100–400	Sweat	[b]
		Feces	[b]
		Urine	100–400
Total	100–400	Total	100–400

[a]Inputs and outputs represent average basal values for an adult ingesting a conventional American diet in a cool environment. [b]Negligible in basal state, but can become appreciable under certain conditions (see text).

mal conditions, *fecal* Na^+ is negligible, although in diarrheal diseases such as cholera it may exceed 1000 mmol/day. Neither sweat nor fecal loss is an important physiological variable in the regulation of Na^+ output. Like water output, then, Na^+ output is regulated primarily by changes in the amount of Na^+ excreted in *urine*. Urinary Na^+ excretion averages 100–400 mmol/day in adult Americans, matching the average input, but can vary from almost negligible amounts to over 600 mmol/day.

Regulation of Sodium Balance

With no significant physiological regulation of Na^+ input in man, Na^+ balance must be achieved by changing Na^+ output to match Na^+ input. As indicated above, Na^+ output is regulated primarily by changes in the amount of Na^+ excreted in urine. In Chapters 29 and 30, the renal handling of Na^+ was discussed in detail. It is now appropriate to consider the mechanisms for regulating Na^+ excretion. Classically, three major regulatory mechanisms have been identified: changes in GFR, aldosterone, and the so-called "third factor" effect (DeWardener, 1978).

CHANGES IN GFR

When Na^+ intake changes, compensatory changes in GFR occur that can affect Na^+ excretion. Recall from Chapter 27 that the most important regulator of GFR is the *RPF*, although GFR also can be affected by changes in the *net filtration pressure (Eq. 31)*. Changes in Na^+ intake can change both the RPF and the net filtration pressure and, hence, GFR. Consider, for example, the changes in RPF and net filtration pressure that could result from an *increase* in Na^+ intake (Fig. 4.37). Because Na^+ is primarily an *extracellular* solute, an increase in Na^+ intake causes an increase in plasma osmolality that stimulates the osmoreceptors (see above). The resulting stimulation of thirst and ADH secretion causes an expansion of the plasma volume, which (by the pathways illustrated in Fig. 4.37) causes an increase in both RPF and net filtration pressure, and hence an increase in GFR. The increase in GFR increases the filtered load of Na^+ and promotes Na^+ excretion (Fig. 4.37).

However, it should be noted that such changes in GFR probably have a *minor* role in the regulation of Na^+ excretion, for at least two reasons. First of all, *autoregulation* and *tubuloglomerular feedback* (Chapters 26 and 27) attenuate or even prevent any significant changes in GFR that might result from changes in Na^+ intake. Secondly, the amount of Na^+ reabsorbed by the proximal tubule, loop of Henle,

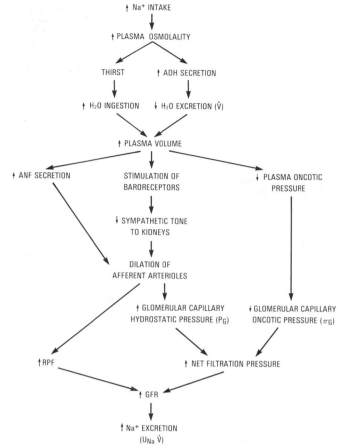

Figure 4.37. Pathway by which an increase in GFR could restore Na^+ balance following an increase in Na^+ intake.

and distal nephron varies directly with the load of Na^+ delivered to that region (Chapters 29 and 30). Because of this *load-dependent* reabsorption of Na^+, a constant percentage of the filtered load of Na^+ continues to be reabsorbed as GFR varies in the physiological range, a phenomenon that is termed *glomerulotubular balance*. The exact mechanism for load-dependent reabsorption and glomerulotubular balance is not completely understood but probably can be attributed in part to the very large capacity of the tubular reabsorptive mechanisms for Na^+.

Thus, changes in GFR are of minor importance in the regulation of Na^+ excretion. Such regulation must therefore be primarily accomplished by the remaining two mechanisms, aldosterone and the "third factor" effect.

ALDOSTERONE

A second factor regulating Na^+ excretion is *aldosterone*, a steroid hormone produced by the zona glomerulosa of the adrenal cortex. Increased plasma levels of aldosterone stimulate Na^+ reabsorption in the collecting duct by increasing Na^+ entry through the conductive Na^+ channel (Chapter 30) (Petty et al., 1981). Increased plasma aldosterone also stimulates H^+ and K^+ secretion in the collecting duct, as discussed in Chapters 32 and 33.

The secretion of aldosterone by the adrenal cortex is stimulated by five factors (Carey, 1986): (1) increased plasma angiotensin II, (2) decreased plasma atrial natriuretic factor (ANF) (Elliott and Goodfriend, 1986), (3) increased P_K, (4) increased plasma adrenocorticotrophic hormone (ACTH), and (5) decreased P_{Na}. *Increases in* P_K represent an important stimulus for aldosterone secretion (Chapter 33), but do not represent a mechanism for changing aldosterone secretion when Na^+ intake changes. *ACTH* has a relatively minor effect on aldosterone secretion, although it does have an important role in regulating the secretion of adrenal glucocorticoids and androgens. The increased aldosterone secretion seen with *decreased* P_{Na} represents an appropriate response for maintaining Na^+ balance:

$\downarrow P_{Na} \rightarrow \uparrow$ aldosterone
$\qquad \rightarrow \uparrow Na^+$ reabsorption by collecting duct
$\qquad\qquad \rightarrow \downarrow Na^+$ excretion

However, the effect of P_{Na} on aldosterone secretion is of minor importance in the regulation of Na^+ excretion for two reasons. First of all, decreases in P_{Na} have a relatively *weak* stimulatory effect on aldosterone secretion. Secondly, changes in Na^+ intake have minimal effects on P_{Na}. For example, while an increased Na^+ intake adds Na^+ to the extracellular fluid and produces a *transient* increase in P_{Na}, the *plasma osmolality* also increases, stimulating the osmoreceptors. The resulting stimulation of thirst and ADH secretion leads to expansion of the plasma volume (Fig. 4.37) and dilution of the ingested Na^+, so that the overall change in P_{Na} is small. Thus, the changes in aldosterone secretion that accompany changes in Na^+ intake must be primarily mediated by *angiotensin II* and *ANF*.

As indicated in Chapter 26, *angiotensin II* is an octapeptide that is formed from angiotensinogen in the following 2-step reaction:

$$\text{Angiotensinogen} \xrightarrow{\text{renin}} \text{angiotensin I} \xrightarrow{\text{converting enzyme}} \text{angiotensin II}$$

Renin, the proteolytic enzyme that catalyzes the first step in this reaction, is secreted into plasma by the granular cells of the JG apparatus (Chapter 26). The secretion of renin appears to be stimulated by:

(1) decreased renal perfusion pressure, as sensed by baroreceptors (stretch receptors) in the afferent arteriole; (2) changes in the volume or composition of the tubular fluid reaching the macula densa; (3) stimulation of the renal sympathetic nerves; and (4) a variety of other factors (Stella and Zanchetti, 1987; Fray et al., 1986). For example, since inhibitors of prostaglandin synthesis (e.g., indomethacin) impair renin secretion, it has been suggested that prostaglandins modulate renin secretion. *ANF* is a peptide hormone synthesized in atrial muscle cells and secreted from granules in these cells in response to stretch. To illustrate how angiotensin II and ANF could change aldosterone secretion in response to changes in Na^+ intake, consider the changes in angiotensin II production and ANF secretion that could result from an *increase* in Na^+ intake (Fig. 4.38). By mechanisms analogous to those illustrated in Figure 4.37, an increase in Na^+ intake results in an increase in plasma volume,

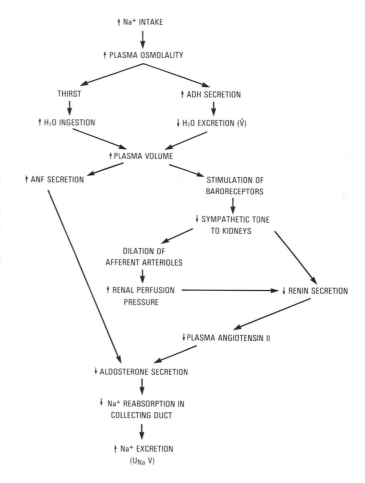

Figure 4.38. Pathway by which a decrease in aldosterone secretion could restore Na^+ balance following an increase in Na^+ intake.

which, in turn, causes (1) a reflex decrease in sympathetic tone to the kidneys, decreasing renin secretion, and hence angiotensin II production, both *directly* (by an effect on the granular cells) and *indirectly* (by increasing renal perfusion pressure); and (2) an increase in ANF secretion. Both the decrease in angiotensin II and the increase in ANF would decrease aldosterone secretion, thereby promoting Na$^+$ excretion. However, it should be noted that the physiological role of ANF in the regulation of aldosterone secretion is not as clearly established as that of angiotensin II, although ANF does appear to have an important *direct* role in regulating Na$^+$ excretion (see below).

Like the regulation of Na$^+$ excretion by changes in GFR, then, the regulation of Na$^+$ excretion by aldosterone occurs as a result of changes in the *plasma volume* caused by changes in Na$^+$ input. However, as previously indicated, aldosterone has a much more important role in regulating Na$^+$ excretion than changes in GFR.

THIRD FACTOR EFFECT

When the first two factors that regulate Na$^+$ excretion, GFR and aldosterone, are controlled experimentally, an animal can *still* regulate Na$^+$ excretion to match Na$^+$ input. For example, in an experimental animal in whom a *constant GFR* is maintained by controlling blood flow to the kidneys and a high plasma concentration of *aldosterone* is maintained by administering large doses of the hormone, intravenous infusion of isotonic saline *still* will be followed by a decrease in Na$^+$ reabsorption and, hence, an increase in Na$^+$ excretion. The phenomenon whereby an increase in Na$^+$ input can result in an increase in Na$^+$ excretion *independent* of any significant increase in GFR or decrease in aldosterone level is termed the *third factor effect.* Conversely, a decrease in Na$^+$ input can result in a decrease in Na$^+$ excretion *independent* of any significant change in GFR or aldosterone level, a phenomenon that can be referred to as the *absence of* third factor.

Despite intensive investigation, the mechanism for the third factor effect remains incompletely understood, probably because several interrelated mechanisms appear to be involved, including the sympathetic nervous system, angiotensin II, ANF, and changes in the hydrostatic and oncotic pressures in the peritubular capillaries. To illustrate how these mechanisms could alter Na$^+$ excretion *independently* of changes in GFR or aldosterone, consider again the response to an *increase* in Na$^+$ intake (Fig. 4.39). By mechanisms analogous to those illustrated in Figures 4.37 and 4.38, an

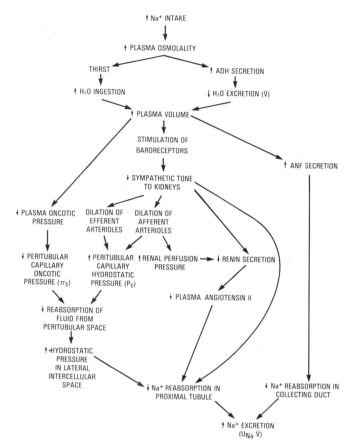

Figure 4.39. Postulated mechanisms for the third factor effect. The interrelated pathways by which the sympathetic nervous system, angiotensin II, ANF, and changes in the hydrostatic and oncotic pressures in the peritubular capillaries could increase Na$^+$ excretion following an increase in Na$^+$ intake are shown.

increase in Na$^+$ intake results in an increase in plasma volume, which, in turn, influences the postulated mechanisms for the third factor effect as follows:

1. *Sympathetic stimulation* of the kidneys directly stimulates Na$^+$ reabsorption in the proximal tubule (Chapter 29) (Gottschalk, 1979) and also stimulates, both directly and indirectly, renin secretion (see above). An increase in plasma volume would stimulate the baroreceptors and thereby *decrease* sympathetic tone to the kidneys, inhibiting proximal tubular Na$^+$ reabsorption and renin secretion and increasing Na$^+$ excretion (Fig. 4.39). Recall that this decrease in sympathetic tone also would promote Na$^+$ excretion by increasing GFR (Fig. 4.37) and by decreasing aldosterone secretion (Fig. 4.38).

2. *Angiotensin II* directly stimulates Na$^+$ reabsorption in the proximal tubule (Chapter 29) (Liu and Cogan, 1987). An increase in plasma volume

would *decrease* angiotensin II production by inhibiting the secretion of *renin* (see paragraph 1 above), thereby inhibiting proximal tubular Na$^+$ reabsorption and increasing Na$^+$ excretion (Fig. 4.39). Recall that this decrease in angiotensin II production also would promote Na$^+$ excretion by decreasing aldosterone secretion (Fig. 4.38).

3. *ANF* inhibits Na$^+$ reabsorption in the collecting duct (Chapter 30) (Van de Stolpe and Jamison, 1988). An increase in plasma volume would *increase* ANF secretion, thereby inhibiting Na$^+$ reabsorption in the collecting duct and increasing Na$^+$ excretion (Fig. 4.39). This increase in ANF secretion also would promote Na$^+$ excretion by increasing GFR (Fig. 4.37) and decreasing aldosterone secretion (Fig. 4.38).

4. The *hydrostatic and oncotic pressures in the peritubular capillaries* can affect Na$^+$ reabsorption in the proximal tubule, particularly following *large* changes in Na$^+$ intake. According to the Starling principle, the rate of fluid movement from *capillaries* to *interstitial space* (i.e., the rate of *filtration*) is proportional to the difference between the hydrostatic and oncotic pressure gradients across the capillary wall, the so-called *net filtration pressure (Eq.29):*

$$\text{Rate of filtration} \propto [(P_c - P_i) - (\pi_c - \pi_i)]$$

The rate of fluid movement from *interstitial space* to *capillaries* (i.e., the rate of *reabsorption*) is therefore proportional to

$$\text{Rate of reabsorption} \propto [(P_i - P_c) - (\pi_i - \pi_c)] \quad (88)$$

If *Eq. 88* is applied to the reabsorption of fluid from the *peritubular* interstitial space into the *peritubular* capillaries, it is evident that an increase in peritubular capillary hydrostatic pressure (P_c) or a decrease in peritubular capillary oncotic pressure (π_c) will retard the reabsorption of fluid into the capillaries. The movement of fluid from the lateral intercellular space to the peritubular space will therefore be retarded, and the hydrostatic pressure in the lateral space will increase. This increased hydrostatic pressure, in turn, will impair the reabsorption of water and solutes by the proximal tubule, perhaps by allowing water and solutes that *already* have been transported into the lateral space to leak back *("pump leak")* into the tubular lumen. Figure 4.39 illustrates how an increase in plasma volume resulting from an increase in Na$^+$ intake could increase peritubular

capillary hydrostatic pressure and decrease peritubular capillary oncotic pressure, thereby decreasing proximal tubular Na$^+$ reabsorption and increasing Na$^+$ excretion. Because such peritubular capillary pressure changes would decrease the proximal tubular reabsorption not only of Na$^+$, but of water and other solutes as well, this mechanism can explain the observation that *all* proximal tubular reabsorption is decreased following the ingestion of a large quantity of Na$^+$ or another cause of plasma volume expansion.

The relative importance of each of the above mechanisms is not known. Furthermore, it should be noted that prostaglandins and additional hormones also may contribute to the regulation of Na$^+$ excretion that occurs independently of changes in GFR and aldosterone.

Overall Response to Changes in Sodium Intake

Figure 4.40 summarizes the three factors that regulate Na$^+$ excretion in response to changes in Na$^+$ intake (changes in GFR, aldosterone, third factor effect) and shows the interrelationships among them. It is important to emphasize that these three factors are activated *not* by a direct effect of the changes in Na$^+$ input, but instead by the resulting changes in *plasma volume* (Figs. 4.37–4.40). Figure 4.41 illustrates that the increase in Na$^+$ excretion following an abrupt increase in daily Na$^+$ intake in

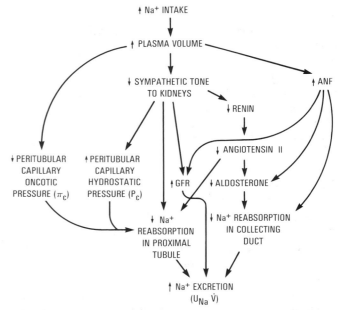

Figure 4.40. Summary of and interrelationships among the three factors that regulate Na$^+$ excretion in response to changes in Na$^+$ intake (changes in GFR, aldosterone, third factor effect).

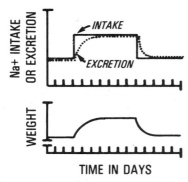

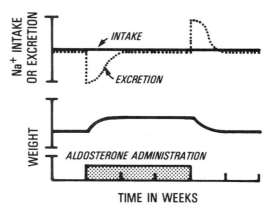

Figure 4.41. Increases in Na⁺ excretion and body weight following an abrupt increase in daily Na⁺ intake. The increase in Na⁺ excretion parallels the increase in body weight, which, in turn, parallels an increase in plasma volume. In this example, the restoration of Na⁺ balance (Na⁺ input = Na⁺ output) requires approximately 3 days. The changes are reversed when daily Na⁺ intake returns to its original level. (Adapted from Seely and Levy, 1981.)

Figure 4.42. The mineralocorticoid escape phenomenon. When aldosterone is administered to an experimental animal with a constant daily Na⁺ intake, Na⁺ excretion initially decreases, resulting in Na⁺ retention and an increase in body weight. However, after several days of Na⁺ retention, Na⁺ balance is again achieved (Na⁺ input = Na⁺ output) despite continued aldosterone administration. These changes are reversed when aldosterone is terminated. (Adapted from Seely and Levy, 1981.)

fact occurs parallel to an increase in body weight, which, in turn, parallels an increase in plasma volume. Note that the complete restoration of Na⁺ balance (Na⁺ input = Na⁺ output) requires several days (an average of 2–3 days in humans) (Walser, 1985). These changes are reversed when Na⁺ intake returns to its original level.

A similar relationship between changes in Na⁺ excretion and changes in plasma volume is seen in the phenomenon of *mineralocorticoid escape* (August et al., 1958). If a mineralocorticoid such as *aldosterone* is administered in an experimental setting, Na⁺ excretion will, of course, initially decrease. However, after several days of Na⁺ retention (and concomitant increase in plasma volume and weight), Na⁺ excretion increases to its original

level, matching Na⁺ intake, i.e., Na⁺ balance is again achieved *despite* the continued administration of the hormone (Fig. 4.42). Studies of experimental animals with mineralocorticoid escape suggest that ANF may play a role in the escape phenomenon (Granger et al., 1987). Mineralocorticoid escape is clinically significant because it explains why patients with aldosterone-secreting tumors or hyperplasia of the adrenal cortex experience only moderate Na⁺ retention and weight gain. However, such patients generally develop metabolic alkalosis and hypokalemia, since aldosterone continues to stimulate H⁺ and K⁺ secretion in the collecting duct (Chapters 32 and 33).

CHAPTER 32

Acid-Base Balance and Regulation of H^+ Excretion

The kidneys have a major role in regulating the H^+ ion concentration, or pH, of the body fluids. In a healthy individual, the pH of the ECF generally is maintained within a rather narrow range, with a mean normal value of 7.40 ± 0.02 in arterial plasma and 7.38 ± 0.02 in mixed venous plasma. The clinical evaluation of a patient's acid-base status generally is based on laboratory studies of *arterial* samples. Thus, an individual is considered to have an *acidosis* when the arterial pH falls below 7.38 and an *alkalosis* when the arterial pH rises above 7.42. Such precise control of pH is necessary because of the marked effects of pH changes on protein conformation, enzymatic reactions, and central nervous system function.

It should be noted that the use of pH rather than H^+ concentration is widespread in physiology and clinical medicine because clinical laboratories typically measure pH. However, since pH $= -\log [H^+]$, the H^+ concentration can be readily calculated if necessary, using the relation $[H^+] = 10^{-pH}$. Furthermore, in the pH range of 7.30–7.50 a *rapid* method for obtaining an *approximate* H^+ concentration is available. This method takes advantage of the fact that $[H^+]$ is *40* nmol/liter at pH *7.40* and that each *0.01-unit* change in pH from 7.40 corresponds to an *inverse* change of approximately *1 nmol/liter* in $[H^+]$ in the 7.30–7.50 pH range (Table 4.22). Thus, the normal arterial pH range of 7.38–7.42 corresponds to a $[H^+]$ range of 42–38 nmol/liter.

THREATS TO pH

Although small quantities of acid or base are present in certain foods and medications, the major threats to the pH of the body fluids are *acids* formed in *metabolic processes*. These metabolic acids can be conveniently divided into three categories:

1. *Volatile Acids: Carbon Dioxide.* CO_2, a major end product in the oxidation of carbohydrates, fats, and amino acids, can be regarded as an acid by virtue of its ability to react with water to form *carbonic acid* (H_2CO_3), which, in turn, can dissociate to form H^+ and HCO_3^-:

$$CO_2 + H_2O \rightleftharpoons H_2CO_3 \rightleftharpoons H^+ + HCO_3^- \qquad (89)$$

Because it is a gas (and can be eliminated by the lungs, as discussed below), CO_2 often is termed a *volatile acid*. Enough CO_2 is produced each day to add approximately 10,000 mmol of H^+ to the body fluids, and even greater amounts of CO_2 can be produced in exercise and hypermetabolic states such as thyrotoxicosis. Without compensating mechanisms, the addition of 10,000 mmol of H^+ would have a *catastrophic* effect on the pH of the body fluids. For example, in a 60-kg man with 36 liters of total body water, without compensating mechanisms 10,000 mmol of H^+ would increase the H^+ concentration of the body fluids by nearly 300 mmol/liter. This would represent more than a *7 million fold* increase in the H^+ concentration, since the normal H^+ concentration at pH 7.4 is only 40 nmol/liter or 0.00004 mmol/liter.

2. *Fixed Acids: Sulfuric Acid and Phosphoric Acid.* Sulfuric acid is an end product of the oxidation of the sulfur-containing amino acids methionine and cysteine, while phosphoric acid is formed in the metabolism of phospholipids, nucleic acids, phosphoproteins, and phosphoglycerides. In contrast to CO_2, sulfuric acid and phosphoric acid are *nonvolatile* and have therefore been termed *fixed acids*. The production of fixed acids varies with the diet, but typically results in the formation of 50–100 mmol of H^+ per day. While much less threatening to the body than the 10,000 mmol of H^+ from CO_2, even the effect of 50–100 mmol of H^+ would be disastrous without compensating mechanisms. For example, without compensating mechanisms the addition of just 50 mmol of H^+ to the body fluids of a 60-kg man would increase the H^+ concentration of the body fluids by over 1 mmol/liter, which would

Table 4.22.
Correlation of pH with Estimated and Actual Levels of [H$^+$]

pH	Estimated [H$^+$], nmol/liter	Actual [H$^+$], nmol/liter
7.30	50	50.1
7.35	45	44.7
7.40	40	40
7.45	35	35.5
7.50	30	31.6

represent more than a *25,000 fold increase* in the H$^+$ concentration.

3. *Organic Acids.* Organic acids such as lactic acid, acetoacetic acid, and β-OH butyric acid are formed during the metabolism of carbohydrates and fats. Normally, these acids are further oxidized to CO$_2$ and water and therefore do not directly affect the pH of the body fluids. However, in certain abnormal circumstances these organic acids may accumulate, causing an *acidosis.* For example, in hypovolemic and other forms of circulatory shock, *lactic acid* levels may increase markedly due to inadequate perfusion of tissues and the resulting increase in anaerobic glycolysis *(lactic acidosis).* In uncontrolled diabetes mellitus, *acetoacetic acid* and *β-OH butyric acid* may accumulate due to increased lipid catabolism and consequent overloading of the body's ability to metabolize acetyl CoA *(diabetic ketoacidosis).*

ACID-BASE BUFFER SYSTEMS

The kidneys and the lungs together share the responsibility for regulating the pH of the body fluids. However, it is the *buffer systems* of the body fluids that actually provide the most *immediate* defense against changes in pH. These buffers are *weak acids* that exist as a mixture of a protonated form and an unprotonated form in the physiological pH range. The dissociation reaction for such a weak acid, HA, is as follows:

$$HA \rightleftharpoons H^+ + A^- \qquad (90)$$

In order to understand how such an acid could attenuate, or "buffer," the pH changes caused by the addition or loss of H$^+$ or OH$^-$, it is necessary to examine reaction *90* more closely. The law of mass action for reaction *90* can be written as follows:

$$K_a = \frac{[H^+][A^-]}{[HA]} \qquad (91)$$

where K$_a$ is the apparent dissociation constant for the acid. Taking the logarithm of *Eq. 91* results in

the following expression:

$$\log K_a = \log \frac{[H^+][A^-]}{[HA]}$$

or

$$\log K_a = \log [H^+] + \log \frac{[A^-]}{[HA]} \qquad (92)$$

Eq. 92 can be rearranged to give

$$-\log [H^+] = -\log K_a + \log \frac{[A^-]}{[HA]} \qquad (93)$$

Substituting pH for $-\log [H^+]$ and defining pK$_a$ = $-\log K_a$, the so-called *Henderson-Hasselbalch equation* is obtained:

$$pH = pK_a + \log \frac{[A^-]}{[HA]} \qquad (94)$$

This equation can be used to illustrate how a weak acid functions as a buffer. Consider a weak acid with a pK$_a$ of 7.0. If in 1 liter of solution the unprotonated form A$^-$ and the protonated form HA are each present in a concentration of 10 mmol/liter, then

$$pH = 7.0 + \log \frac{10}{10}$$
$$= 7.0 \qquad (95)$$

Assume now that 1 mmol of a *strong* acid HX is added to the solution. The strong acid will be completely dissociated:

$$HX \rightarrow H^+ + X^-$$

However, its H$^+$ ion can combine with the unprotonated form of the weak acid A$^-$, as follows:

$$H^+ + X^- + A^- \rightarrow HA + X^-$$

Since the weak acid HA is much less dissociated than the strong acid HX, it can be assumed that virtually all of the added H$^+$ reacts with unprotonated A$^-$. The concentration of A$^-$ therefore will decrease to 9 mmol/liter and the concentration of HA will increase to 11 mmol/liter. According to the Henderson-Hasselbalch equation *(Eq. 94)*, the pH of the solution then will be

$$pH = 7.0 + \log \frac{9}{11}$$
$$\simeq 6.9 \qquad (96)$$

i.e., the 1 mmol of strong acid has decreased the pH of the buffered solution by approximately *0.1 units.* This can be contrasted to a *"control"* in which the same amount of strong acid HX (1 mmol) is added to 1 liter of an *unbuffered* solution such as *water.* Like the buffered solution considered above, water has a pH of 7.0. However, due to the absence of buffer the addition of 1 mmol of HX results in 1 mmol of H^+ being added to the solution, which originally contained only 0.0001 mmol of H^+ (pH 7.0). The final pH is therefore 3.0 (= $-\log 10^{-3}$ mol/liter), i.e., the 1 mmol of strong acid has decreased the pH of the *unbuffered* solution by *4.0 units.*

A more general demonstration of the ability of the weak acid HA to function as a buffer is presented in Figure 4.43, which illustrates the changes in pH that would result from the addition of *varying* amounts of a strong acid *or* a strong base to the buffer solution discussed above. *Point A* represents the pH of the buffer solution *and* of water *before* the addition of acid or base (pH 7.0). The pH falls as increasing amounts of strong acid are added to the buffer solution *(solid curve to the left of A)*, but not nearly as much as it would fall if the strong acid were added to water *(dashed curve to the left of A)*. Similarly, the pH rises as increasing amounts of strong base are added to the buffer solution *(solid curve to the right of A)*, but not nearly as much as it would rise if the strong base were added to water *(dashed curve to the right of A)*. Note that the slope of the *solid curve* is flattest near point *A*, i.e., 1 mmol of strong acid or strong base causes the *smallest* change in pH when added to the buffer in the vicinity of pH 7.0. But recall that 7.0 also represents the pK_a of the weak acid used in these examples.

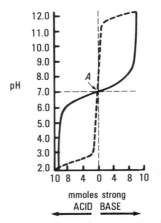

Figure 4.43. Changes in pH resulting from the addition of a strong acid or strong base to the weak acid buffer solution described in the text *(solid curve)* and to water *(dashed curve)*. Point A represents the pH of the buffer solution and of water before the addition of acid or base (pH 7.0). The weak acid is most effective as a buffer in the vicinity of its pK_a (7.0).

This means that a weak acid is *most effective* as a buffer when the pH of the solution remains close to the pK_a of the acid.

A second factor that determines the buffering effectiveness of a weak acid is the total *concentration* of the weak acid in the solution. For example, if the HA buffer solution discussed above contained only 2 mmol/liter of A^- and 2 mmol/liter of HA (instead of 10 mmol/liter of A^- and 10 mmol/liter of HA), its pH still would be 7.0 (compare to *Eq. 95*):

$$pH = 7.0 + \log \frac{2}{2}$$
$$= 7.0$$

However, the addition of 1 mmol of strong acid now would decrease the pH of the buffer solution by nearly *0.5 units* (instead of by only 0.1 units; see *Eq. 96*):

$$pH = 7.0 + \log \frac{1}{3}$$
$$\simeq 6.5$$

Buffer Systems of the Body Fluids

From the above discussion, it should be evident that the *optimal* buffer system for keeping the pH of the body fluids close to 7.40 would (1) have a pK_a close to 7.4; and (2) be present in a high concentration. Although there is no such optimal buffer in the body, the major body buffer systems, to be introduced here, satisfy these criteria to different extents.

HEMOGLOBIN AND OTHER PROTEINS

Proteins contain several ionizable groups that are weak acids, including C-terminal carboxyl groups, N-terminal amino groups, side chain carboxyl groups of glutamic and aspartic acid, side chain amino groups of lysine, and imidazole groups of histidine. While the pK_a of a given ionizable group will be different in different proteins, depending upon the local environment, in many proteins the *imidazole groups of histidine* and the *N-terminal amino groups* have pK_as sufficiently close to 7.4 to enable these proteins to function as effective buffers. Proteins are present in relatively high concentrations in cells and plasma, which further contributes to their effectiveness as buffers.

Hemoglobin is a particularly effective protein buffer and is a major buffer in blood. In fact, hemoglobin has approximately six times the buffering capacity of the plasma proteins in blood due to its high concentration (nearly four times greater than that of plasma proteins) and its 38 histidine residues (over twice as many as albumin, even though the molecular weights of albumin and hemoglobin are similar). Also contributing to the effectiveness of hemoglobin as a blood buffer is the fact that *deoxygenated* hemoglobin has imidazole groups with a somewhat higher pK_a (smaller K_a) than *oxygenated* hemoglobin (Van Slyke et al., 1922). As a result, once hemoglobin becomes deoxygenated in the capillaries, it is better able to bind the H^+ ions formed when CO_2 enters capillary blood from the tissues (Chapter 37). Thus, hemoglobin helps to minimize the fall in pH caused by the loading of CO_2 into capillary blood.

PHOSPHATE

The phosphate buffer system involves the weak acid $H_2PO_4^-$, which undergoes the following dissociation reaction

$$H_2PO_4^- \rightleftharpoons H^+ + HPO_4^= \qquad (97)$$

Since the pK_a for this reaction is 6.8, the phosphate buffer system is a suitable candidate for a body buffer. However, not all body fluids contain sufficient concentrations of phosphate for this buffer system to be effective. For example, the phosphate buffer system is much more effective in *ICF* than in *ECF*, since (1) the total concentration of phosphate is much greater in ICF than in ECF (Fig. 4.3); and (2) the intracellular pH generally is somewhat lower than the extracellular pH and therefore is closer to the pK_a of the phosphate buffer.

Phosphate also is effective in buffering the tubular fluid in the *kidneys,* as discussed below. Even though the tubular fluid is derived from plasma (i.e., from *ECF*), phosphate can be an effective buffer in the tubular fluid because (1) phosphate becomes greatly *concentrated* in the tubular fluid due to the reabsorption of water in excess of phosphate; and (2) the pH of the tubular fluid generally becomes more *acidic* than the pH of ECF and therefore is closer to the pK_a of the phosphate buffer.

BICARBONATE

The bicarbonate buffer system involves the weak acid H_2CO_3, which undergoes the following dissociation reaction

$$H_2CO_3 \rightleftharpoons H^+ + HCO_3^- \qquad (98)$$

The Henderson-Hasselbalch equation for this reaction is as follows:

$$pH = pK_a + \log \frac{[HCO_3^-]}{[H_2CO_3]}$$

Since the pK_a for reaction *98* is 3.7, the bicarbonate buffer system appears to represent a very poor candidate for a body buffer. However, in body fluids H_2CO_3 is in equilibrium with dissolved CO_2 and water *(Eq. 89)*. Thus, the Henderson-Hasselbalch equation for the bicarbonate buffer system is most appropriately written as follows:

$$pH = pK_a' + \log \frac{[HCO_3^-]}{[CO_2]} \qquad (99)$$

where pK_a' includes the equilibrium constant for the formation of H_2CO_3 from CO_2 and water. But the value of pK_a' is only 6.1, i.e., even when the equilibrium between H_2CO_3 and dissolved CO_2 is taken into account, the bicarbonate buffer system does not appear to represent an optimal body buffer. Nevertheless, the bicarbonate buffer system probably can be regarded as the most *important* buffer system in the body. This is because the concentrations of its components can be *independently regulated,* the CO_2 concentration by the lungs and the HCO_3^- concentration by the kidneys (see below). The significance of such regulation can best be appreciated by noting that according to the Henderson-Hasselbalch equation, the pH of a weak acid buffer solution is determined by the ratio of the concentrations of the two forms of the buffer. Thus, by regulating the CO_2 and HCO_3^- concentrations, the lungs and kidneys also regulate the pH of the body fluids *(Eq. 99)*.

Although the Henderson-Hasselbalch equation applies to the other buffer systems of the body as well, in *none* of the other systems can the concentrations of the components be independently regulated. In other words, although the nonbicarbonate buffer systems can defend the body fluids against changes in pH, they cannot be used as a primary mechanism for adjusting the pH of the body fluids. Of course, any change in the $[HCO_3^-]/[CO_2]$ ratio will affect the concentration ratios for the *nonbicarbonate* buffers in the body fluids as well, since

$$\begin{aligned} pH &= pK_{a_1} + \log \frac{[HCO_3^-]}{[CO_2]} \\ &= pK_{a_2} + \log \frac{[HPO_4^=]}{[H_2PO_4^-]} = pK_{a_3} + \log \frac{[Hb^-]}{[HHb]} \\ &= pK_{a_4} + \log \frac{[Prot^-]}{[HProt]} = \cdots \cdots \quad (100) \end{aligned}$$

The interrelationship between the various buffer systems of the body, as expressed in *Eq. 100*, is termed the *isohydric principle.*

Before proceeding to describe the respiratory and renal regulation of the CO_2 and HCO_3^- concentrations, it should be noted that the Henderson-Hasselbalch equation for the bicarbonate buffer system most commonly is written in a form that allows the *concentration* of carbon dioxide, [CO_2], to be replaced by the *partial pressure* of carbon dioxide, P_{CO_2}, a more conveniently measured quantity. According to Henry's law, the concentration of a gas dissolved in solution is directly proportional to its partial pressure. In plasma at 37°C, the relationship between the concentration of dissolved CO_2 (in mmol/liter) and the partial pressure of CO_2 (in mm Hg) is as follows (Chapter 37):

$$[CO_2] = 0.03 \times P_{CO_2}$$

Substituting into *Eq. 99*, the Henderson-Hasselbalch equation for the bicarbonate buffer system becomes

$$pH = pK'_a + \log \frac{[HCO_3^-]}{0.03\ P_{CO_2}} \qquad (101)$$

Applying this equation to arterial plasma, the normal arterial pH of 7.40 corresponds to a [HCO_3^-] of approximately 24 mmol/liter and a P_{CO_2} of approximately 40 mm Hg

$$\begin{aligned} pH &= 6.1 + \log \frac{24}{0.03 \times 40} \\ &= 7.40 \end{aligned} \qquad (102)$$

In the remainder of this chapter, the plasma HCO_3^- concentration will be designated by $[HCO_3^-]_p$ (instead of by P_{HCO_3}, as in other chapters) to be consistent with the widespread use of square brackets to designate concentrations in the Henderson-Hasselbalch equation.

RESPIRATORY REGULATION OF pH

The lungs participate in the regulation of pH by regulating the partial pressure of CO_2 in arterial blood. As previously indicated, the CO_2 produced in metabolic processes can combine with water to form H_2CO_3, which, in turn, can dissociate to form H^+ and HCO_3^- *(Eq. 89)*. However, this entire reaction is *reversed* in the lungs when CO_2 is eliminated, or "blown off," by ventilation, i.e., CO_2 is a *volatile acid*. In fact, in a person with normal lungs and a fixed rate of CO_2 production, the arterial P_{CO_2} is determined solely by, and is inversely proportional to, the alveolar ventilation, $\dot{V}_A$ (Chapter 35):

$$P_{CO_2} \propto \frac{1}{\dot{V}_A} \qquad (103)$$

Since enough CO_2 is produced each day to add approximately 10,000 mmol of H^+ to the body fluids, the elimination of CO_2 by the lungs has an *essential* role in regulating the pH of the body fluids. Of course, H^+ *is* added to the blood when CO_2 enters the capillaries for transport from the tissues to the lungs. However, virtually *all* of this H^+ reacts with blood buffers, primarily hemoglobin (see above), as evidenced by the fact that the venous pH is only slightly lower than the arterial pH (see above). The fact that the body's buffer systems provide the most *immediate* defense against changes in pH has been noted previously.

Because of the important relation between alveolar ventilation and arterial P_{CO_2} *(Eq. 103)*, it should not be surprising that many of the mechanisms that allow the body to maintain a normal arterial pH are mediated by changes in alveolar ventilation. For example, chemoreceptors in the medulla oblongata, aortic bodies, and carotid bodies respond to increases in arterial P_{CO_2} by stimulating alveolar ventilation. Chemoreceptors in the carotid bodies also stimulate alveolar ventilation in response to decreases in arterial pH that occur *independently* of increases in arterial P_{CO_2}. These regulatory mechanisms, which are discussed in greater detail in Chapter 40, are important not only in the day-to-day maintenance of a normal arterial pH but also in the body's response to certain acid-base disturbances (see below).

RENAL REGULATION OF pH

The kidneys participate in the regulation of pH by regulating the concentration of HCO_3^- in plasma, designated here by $[HCO_3^-]_p$ (see above). This regulation of $[HCO_3^-]_p$ involves three tasks. First, the kidneys regulate the amount of HCO_3^- *recovered* or *reabsorbed* from the glomerular filtrate. With a $[HCO_3^-]_p$ of 24 mmol/liter and a GFR of 125 ml/min (180 liters/day), approximately 3 mmol of HCO_3^- are filtered per minute (4320 mmol/day), i.e., without a mechanism for HCO_3^- recovery, the daily loss of HCO_3^- would be equivalent to that resulting from the addition of over 4000 mmol of a strong acid to the body fluids. Second, the kidneys *generate* HCO_3^- to replace the HCO_3^- lost in buffering the various strong acids formed in the body. Recall that as much as 100 mmol of H^+ are added to the body fluids each day from the *fixed* (i.e., *nonvolatile*) *acids* produced during metabolism. The body's buffer systems, particularly the *bicarbonate buffer*, provide the most *immediate* defense against this H^+. For example, the following reaction illustrates how *sulfuric acid* produced during the metabolism of sulfur-containing amino acids is buffered by HCO_3^-:

$$2H^+ + SO_4^= + 2HCO_3^- \rightleftharpoons SO_4^= + 2H_2CO_3 \rightleftharpoons SO_4^=$$
$$+ 2CO_2 + 2H_2O \qquad (104)$$

Since the CO_2 formed in this reaction is eliminated by the lungs, the net result of the reaction is the loss of two HCO_3^- ions from the body, which must be replaced by the kidneys. Even if the H^+ ions from the sulfuric acid were initially to react with a *different* body buffer, such as phosphate, by the isohydric principle *(Eq. 100)* a decrease in the concentration of the unprotonated form of *any* buffer will cause a decrease in the concentration of HCO_3^- as well. *Organic acids* such as lactic acid, acetoacetic acid, and β-OH butyric acid can similarly deplete the body's store of HCO_3^- whenever significant accumulation of these acids occurs (e.g., in lactic acidosis, diabetic ketoacidosis; see above). A third way that the kidneys regulate $[HCO_3^-]_p$ is by *secreting* HCO_3^- under conditions of chronic alkalosis.

It is important to note that two of the tasks involved in the renal regulation of $[HCO_3^-]_p$, the reabsorption of filtered HCO_3^- and the generation of new HCO_3^-, are accomplished by a *single* process, the *secretion of H_2CO_3-derived H^+* by the tubular epithelial cells (Fig. 4.44) (Warnock and Rector, 1979). This H^+ secretion occurs in the *proximal tubule* and *distal nephron*. Although there are certain differences between these regions in the mechanism and regulation of the secretion process, in both regions H_2CO_3-derived H^+ secretion begins in the epithelial cells as CO_2 reacts with water to form H_2CO_3. The CO_2 for this reaction is either produced in the epithelial cells by metabolic processes or diffuses into the cells from the peritubular capillaries or tubular lumen. While the formation of H_2CO_3 can proceed spontaneously, the epithelial cells of both the proximal tubule and the distal nephron contain the enzyme *carbonic anhydrase*, which catalyzes the reaction. The H_2CO_3 *dissociates* into H^+ and HCO_3^-; the H^+ is then secreted into the tubular lumen, while the HCO_3^- exits the cell via a carrier-mediated process, crossing the basal-lateral surfaces and entering the peritubular fluid and peritubular capillaries. Thus, *for every secreted H^+ that is derived from H_2CO_3, a HCO_3^- ion is returned to the systemic circulation*. In the *proximal tubule, H^+ secretion* is primarily coupled to the movement of Na^+ from the tubular lumen into the epithelial cells (Na^+-H^+ exchange, or *antiport*) (Murer et al., 1976); carrier-mediated HCO_3^- exit involves the coupled transport of three HCO_3^- ions and one Na^+ ion (Chapter 29) (Fig. 4.44*A*) (Soleimani et al., 1987). In contrast, in the *distal nephron, H^+ secretion* is primarily an active process, while carrier-mediated HCO_3^- exit involves the exchange of one HCO_3^- ion for one Cl^- ion (Fig. 4.44*B*) (Schuster and Stokes, 1987). It should be noted that in spite of the reference to "distal nephron H^+ secretion," active H^+ secretion occurs primarily in the intercalated

cells of the *collecting duct* (Schuster and Stokes, 1987).

Recovery of Filtered Bicarbonate

The kidneys must be able to reabsorb virtually *all* of the 4300+ mmol of HCO_3^- filtered each day. As indicated above, this reabsorption of filtered HCO_3^- is accomplished by the *secretion of H_2CO_3-derived H^+*. To understand the relationship between this H^+ secretion and HCO_3^- reabsorption, it is necessary to examine the fate of the secreted H^+ ions.

Most of the H_2CO_3-derived H^+ that is secreted into the tubular fluid will react with HCO_3^- in the tubular fluid to form H_2CO_3, which in turn dissociates into CO_2 and water (Fig. 4.45). Because all cellular membranes are freely permeable to CO_2, the CO_2 formed in this reaction can diffuse (1) into the *epithelial cells* to generate additional H^+ for secretion; or (2) through the epithelial cells into the *peritubular capillaries* for eventual elimination by the lungs. The water formed in the reaction mixes with the water in the tubular fluid, but its contribution to the total volume of the tubular fluid is negligible. It should be noted that in the *proximal* tubular fluid, the dissociation of H_2CO_3 into CO_2 and water is catalyzed by *carbonic anhydrase*, which is found in large quantities in the brush border of the proximal tubule (Lönnerholm, 1971). Thus, in the proximal tubule, carbonic anhydrase catalyzes not only the *formation* of H_2CO_3 in the epithelial cells, but also the *dissociation* of H_2CO_3 in the lumen.

As illustrated in Figure 4.45, each time a secreted H^+ reacts with a HCO_3^- in the tubular fluid, a HCO_3^- ion is *lost* from the tubular fluid. But this secretion process also results in the *addition* of a HCO_3^- ion to the peritubular fluid and peritubular capillaries. Thus, when H^+ reacts with HCO_3^- in the tubular fluid, the *net result* of H_2CO_3-derived H^+ secretion is the *reabsorption of HCO_3^-* (Fig. 4.45). Of course, the atoms in the HCO_3^- removed from the tubular fluid are *not* the same as those in the HCO_3^- returned to the peritubular fluid and capillaries, but the net result is *equivalent* to HCO_3^- reabsorption. The secretion of H_2CO_3-derived H^+ and its subsequent reaction with HCO_3^- in the tubular fluid therefore represents the mechanism whereby the kidneys reabsorb filtered HCO_3^-.

The fraction of the filtered HCO_3^- that is reabsorbed can be calculated using the methods described in Chapter 28. In normal man, approximately 90% of the filtered HCO_3^- is reabsorbed in the *proximal tubule*. Most of the remaining HCO_3^- is reabsorbed in the *distal nephron*. In fact, the *fractional excretion* of HCO_3^- typically is less than 0.1%, i.e., over 99.9% of the filtered HCO_3^- is normally reabsorbed. Since over 4300 mmol of HCO_3^-

A. PROXIMAL TUBULE

TUBULAR LUMEN   *PERITUBULAR SPACE*

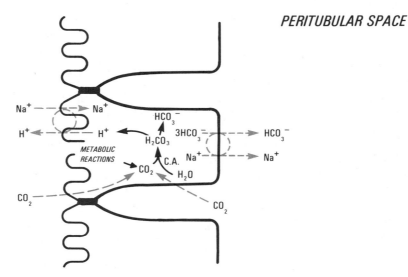

B. DISTAL NEPHRON

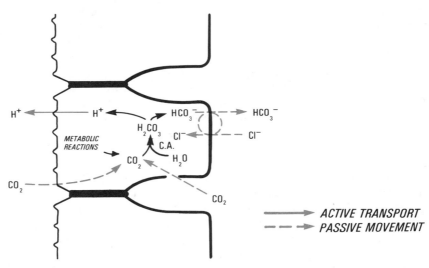

→ *ACTIVE TRANSPORT*
--→ *PASSIVE MOVEMENT*

Figure 4.44. Mechanisms for H_2CO_3-derived H^+ secretion in the proximal tubule (*A*) and distal nephron (*B*). In both regions, CO_2 from metabolic processes, the peritubular capillaries, or the tubular lumen reacts with water to form H_2CO_3, a reaction catalyzed by carbonic anhydrase (C.A.) in the epithelial cells. The H_2CO_3 dissociates to form H^+ and HCO_3^-. The H^+ is secreted via the Na^+-H^+ exchange process (antiport) in the proximal tubule (indicated by the *dashed circle*) and via an active process in the distal nephron; the HCO_3^- exits the cell via a carrier-mediated process, crossing the basal-lateral surfaces and entering the peritubular fluid and peritubular capillaries, so that for every secreted H^+ that is derived from H_2CO_3, a HCO_3^- is returned to the systemic circulation.

are filtered per day, this means that over 4300 mmol of H^+ must be secreted per day to reabsorb the filtered HCO_3^-.

Generation of New Bicarbonate

The kidneys must be able to generate 50–100 mmol of new HCO_3^- per day to replace the HCO_3^- lost in titrating the strong acids produced by the body (see above). Like the reabsorption of filtered HCO_3^-,

the generation of new HCO_3^- is accomplished by the *H_2CO_3-derived H^+ secretion* process illustrated in Figure 4.44, in which a HCO_3^- ion is returned to the systemic circulation for each H^+ ion secreted. If the secreted H^+ reacts with HCO_3^- in the tubular fluid, the net result is *HCO_3^- reabsorption*, since the HCO_3^- returned to the systemic circulation simply replaces a HCO_3^- lost from the tubular fluid in the reaction with H^+ (Fig. 4.45). However, if *excess* H^+

PROXIMAL TUBULE

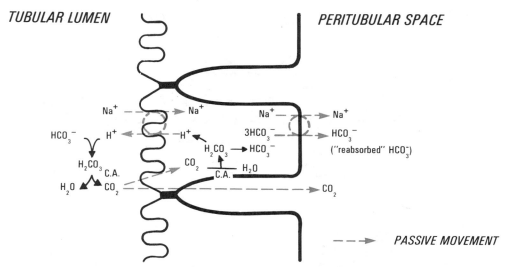

Figure 4.45. Reabsorption of filtered HCO_3^- via H_2CO_3-derived H^+ secretion. The secreted H^+ reacts with HCO_3^- in the tubular fluid to form H_2CO_3, which then dissociates into CO_2 and water, i.e., a HCO_3^- ion is *lost* from the tubular fluid. But since this H^+ secretion process *adds* a HCO_3^- ion to the peritubular fluid and peritubular capillaries, the *net result* is HCO_3^- reabsorption. The proximal tubule is illustrated here, since approximately 90% of the filtered HCO_3^- is reabsorbed in the proximal tubule. However, similar reactions occur in the distal nephron (except for the absence of carbonic anhydrase in the luminal surface) and are responsible for the reabsorption of most of the remaining HCO_3^- in the tubular fluid.

is secreted, the HCO_3^- returned to the systemic circulation represents *new* HCO_3^-. The kidneys can therefore generate 50–100 mmol of new HCO_3^- per day simply by secreting 50–100 mmol of H_2CO_3-derived H^+ *in excess* of the 4300+ mmol of H^+ that must be secreted to reabsorb the filtered HCO_3^-.

The excess H^+ that is secreted in order to generate new HCO_3^- *cannot*, however, be *excreted* from the body as *free* H^+ ion. This is because the magnitude of the H^+ concentration gradient that can be maintained across the tubular epithelium is *limited*. The *proximal tubule*, with its "leaky" tight junctions, can maintain a pH gradient of only 0.5 pH units, i.e., the pH of the proximal tubular fluid cannot be reduced below ≃6.9. The *distal nephron*, with its "tight" tight junctions, can maintain a pH gradient of almost 3 pH units, i.e., the pH of the tubular fluid in the distal nephron can be as low as 4.5. The minimum *urine* pH is therefore also 4.5. But even at this minimum urine pH, the concentration of free H^+ in urine is only 0.03 mmol/liter. Thus, if the 50–100 mmol of excess H^+ secreted each day were to be excreted *solely* as free H^+ ion, over *1000 liters* of urine would be required. The excess H^+ must therefore be excreted in combination with *buffers*.

The predominant buffer in the tubular fluid is, of course, HCO_3^-. However, since the reaction between H^+ and HCO_3^- results in the *reabsorption* of HCO_3^- (see above), *excess* H^+ cannot be excreted by combining with HCO_3^-. The excess H^+ must therefore

combine with *nonbicarbonate* buffers in the tubular fluid, the most important of which are *ammonia* and *phosphate*.

AMMONIA

The specific role of the ammonia buffer system in acid-base balance and the existence of renal transport systems for NH_4^+ are current areas of considerable study. The traditional understanding of the role of the ammonia buffer system is emphasized here. The dissociation reaction for the ammonia buffer system is as follows:

$$NH_4^+ \rightleftharpoons NH_3 + H^+ \qquad (105)$$

Since this reaction has a pK_a of 9.3, in *any* of the body fluids virtually *all* of the ammonia is present as the *protonated* ammonium ion (NH_4^+). For example, in plasma and the glomerular filtrate (pH = 7.4), the $[NH_3]/[NH_4^+]$ ratio is approximately 1:100. It might appear, then, that the amount of *unprotonated* NH_3 available to react with secreted H^+ would be insignificant. However, NH_3 is actually a *highly* effective buffer for secreted H^+ as a result of two important factors (Fig. 4.46A):

1. NH_3 is synthesized in the epithelial cells of the proximal tubule and distal nephron and secreted into the tubular fluid. In other words, the amount of NH_3 available to function as a buffer is *not* lim-

DISTAL NEPHRON

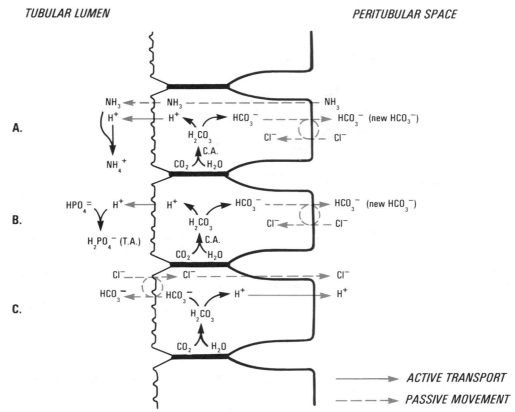

TUBULAR LUMEN   *PERITUBULAR SPACE*

Figure 4.46. *A, B,* Generation of new HCO_3^- via H_2CO_3-derived H^+ secretion. The HCO_3^- generated in the H^+ secretion process represents a *new* HCO_3^- if the secreted H^+ ion reacts with NH_3 in the tubular fluid to form NH_4^+ (*A*) or with $HPO_4^=$ in the tubular fluid to form $H_2PO_4^-$, which represents the major component of the urinary titratable acids (T.A.) (*B*). Although the distal nephron is illustrated here, similar reactions can occur in the proximal tubule. However, most of the H_2CO_3-derived H^+ that is secreted in the proximal tubule reacts with HCO_3^- in the tubular fluid and therefore accomplishes the reabsorption of filtered HCO_3^- (Fig. 4.45) rather than the generation of new HCO_3^-. *C,* Secretion of HCO_3^- in chronic alkalosis. This HCO_3^- secretion process, which occurs in a subpopulation of intercalated cells in the cortical collecting duct, is essentially a mirror image of H_2CO_3-derived H^+ secretion in the distal nephron (*A, B*) and represents a route for the transcellular reabsorption of Cl^-.

ited to the extremely small amount that enters the tubular fluid by glomerular filtration. It is estimated that approximately 60% of the NH_3 synthesized in the epithelial cells is derived from *glutamine* in the following reactions:

$$\text{Glutamine} \xrightarrow[\text{glutaminase}]{\overset{NH_3}{\nearrow}} \text{glutamic acid}$$

$$\underset{\substack{\text{glutamic} \\ \text{dehydrogenase}}}{\overset{NH_3}{\nearrow}} \alpha\text{-ketoglutarate}$$

(*106*)

The enzyme *glutaminase* is abundant in the mitochondria of tubular epithelial cells. The remaining 40% of the NH_3 synthesized in the epithelial cells is derived from other amino acids, particularly glycine and alanine (Pitts, 1971).

2. NH_3 and NH_4^+ have markedly different *solubility characteristics.* NH_3 is highly lipid-soluble and can passively diffuse across cellular membranes. In contrast, NH_4^+ is highly polar and crosses membranes poorly.

Thus, as NH_3 is synthesized and its concentration in the epithelial cells increases, NH_3 diffuses out of the cells. Although NH_3 can diffuse into *either* the tubular lumen *or* the peritubular capillaries, NH_3 diffusion into the *lumen* (i.e., NH_3 secretion) is favored because the secreted NH_3 immediately reacts with previously secreted H^+ to form NH_4^+. The concentration of NH_3 in the tubular fluid therefore remains extremely low, maintaining a large concentration gradient for the further secretion of NH_3 into the lumen. Because of its polar character, the NH_4^+ stays in the tubular fluid and can be excreted, although some of the NH_4^+ formed in the *proximal* tubular fluid is reabsorbed in the thick ascending limb of the loop of Henle, substituting for K^+ in the Na^+/K^+-

$2Cl^-$ symport. The NH_3 formed from this reabsorbed NH_4^+ *(Eq. 105)* can diffuse into the epithelial cells of the adjacent collecting duct and is then available for secretion, i.e., most of the NH_3 secreted in the *distal nephron* actually is synthesized in the *proximal tubule*. The passive secretion of *nonionized* NH_3 into the lumen with subsequent trapping of *ionized* NH_4^+ in the tubular fluid represents an example of *nonionic diffusion* or *diffusion trapping*, a phenomenon that also is important in the renal handling of weak organic acids and bases (Chapter 28).

It is important to note that because the maintenance of a large concentration gradient for the secretion of NH_3 into the lumen is dependent upon the reaction between NH_3 and H^+, the rate of NH_3 secretion into the lumen is to some extent proportional to the amount of H^+ that is secreted. Sufficient NH_3 is therefore secreted to react with virtually *all* of the secreted H^+ that cannot be buffered by the low capacity phosphate buffer system (see below), allowing the 50–100 mmol of H^+ that must be *secreted* per day in order to generate new HCO_3^- to then be *excreted* with a minimum fall in tubular fluid pH. Furthermore, it should be noted that the formation of NH_3 from glutamine in the epithelial cells is pH-dependent, increasing in acidosis and decreasing in alkalosis. For example, although only 30–70 mmol of H^+ normally is excreted as NH_4^+ per day, over 300 mmol of H^+ can be excreted as NH_4^+ in *chronic acidosis*. This *adaptation* of NH_3 production, which takes 3–5 days to develop fully, may be due to increased glutaminase activity *(Eq. 106)* and/or facilitation of glutamine transport into the mitochondria, where the conversion to glutamic acid occurs. The ability of the kidneys to augment NH_3 production and NH_4^+ excretion in acidotic conditions is of major importance in the body's response to acid-base disturbances (see below) (Simpson, 1971).

PHOSPHATE

The phosphate buffer system, described previously *(Eq. 97)*, also represents an important buffer for secreted H^+ ions. Specifically, secreted H^+ can react with $HPO_4^=$ in the tubular fluid to form $H_2PO_4^-$, which is then excreted (Fig. 4.46*B*). Because the pK_a for the phosphate buffer system is 6.8, the ratio of $[HPO_4^=]/[H_2PO_4^-]$ is 4:1 in the glomerular filtrate (pH 7.4), i.e., most of the buffer is in the form that can react with H^+. However, in the *proximal tubule*, as in plasma (see above), the effectiveness of the phosphate buffer system is limited by its low concentration. In fact, in spite of the reabsorption of water in the proximal tubule, the phosphate concentration at the end of the proximal tubule may be even *lower* than in plasma and the glomerular fil-

trate, since 75–85% of the filtered phosphate normally is reabsorbed in the proximal tubule (Chapter 34), compared to approximately two-thirds of the filtered water. But in the *distal nephron*, the phosphate concentration in the tubular fluid can be significantly increased due to the reabsorption of water in excess of phosphate. Thus, phosphate represents an important tubular fluid buffer primarily in the *distal nephron*. In a maximally acidic urine (pH 4.5), the $[HPO_4^=]/[H_2PO_4^-]$ ratio is less than 1:100, i.e., almost *all* of the $HPO_4^=$ has been converted to $H_2PO_4^-$. But in spite of such maximum utilization of the phosphate buffer system, in normal man the amount of phosphate delivered to the distal nephron allows only 12–40 mmol of H^+ to be excreted as $H_2PO_4^-$ each day. Furthermore, the amount of available phosphate remains relatively *constant*, even in *acidotic* conditions, when the kidneys must excrete additional H^+ ions (see below). As previously indicated, the *ammonia* buffer system lacks these limitations of the phosphate system and therefore has a more important role in the excretion of excess H^+.

It should be noted that the excreted $H_2PO_4^-$ represents the major component of the so-called urinary *titratable acids* (TA), defined as those weak acids in urine that can be titrated when a strong base such as NaOH is used to bring the pH of an acidic urine back up to the pH of the glomerular filtrate (normally, pH 7.4). In addition to $H_2PO_4^-$, the titratable acid fraction of urine includes small amounts of several other weak acids that can function as urinary buffers, such as uric acid (pK_a = 5.75), creatinine (pK_a = 4.97), β-OH butyric acid (pK_a = 4.8), and acetoacetic acid (pK_a = 4.8). The titratable acid fraction does *not* include NH_4^+: since the pK_a of the ammonia buffer system is 9.3, insignificant amounts of NH_4^+ will be titrated by bringing the urine pH up to 7.4. It should be noted that β-OH butyric acid and acetoacetic acid are normally present in such small amounts that their role as urinary buffers is *negligible*. However, in *diabetic ketoacidosis* these acids are excreted in the urine in increased quantities and therefore can assume greater importance as urinary buffers (Schiess et al., 1948). Diabetic ketoacidosis, then, represents one example of an acidosis in which *both* the titratable acid fraction and the ammonia buffer system exhibit *adaptation* to allow increased H^+ excretion.

Secretion of Bicarbonate

Under conditions of chronic alkalosis, a subpopulation of intercalated cells in the cortical collecting duct can *secrete* HCO_3^- (Fig. 4.46*C*) (Schuster and Stokes, 1987). The HCO_3^- secretion process is essentially a mirror image of H_2CO_3-derived H^+ secretion in the distal nephron (Fig. 4.46*A, B*) and begins with the formation of H_2CO_3 in the epithelial cells; the H_2CO_3 then dissociates to form HCO_3^-, which is secreted into the tubular lumen in exchange for Cl^-, and H^+, which is actively transported across the basal-lateral surfaces of the cells into the peritubular fluid and peritubular capillaries. Because of the luminal HCO_3^--Cl^- exchange, this HCO_3^- secretion process also represents a transcellular route for the reabsorption of Cl^-. Certain intercalated cells may exhibit *plasticity*, secreting H^+ in acidosis and HCO_3^- in alkalosis (Schwartz et al., 1985).

Quantitation of Acid Excretion

As previously indicated, whenever a H^+ ion that is derived from H_2CO_3 is excreted as either NH_4^+ or titratable acid such as $H_2PO_4^-$, a new HCO_3^- ion is generated for the body. Thus

$$\begin{matrix} \text{New } HCO_3^- \\ \text{generated} \end{matrix} = U_{NH_4}\dot{V} + U_{TA}\dot{V} \qquad (107)$$

However, since the kidneys may fail to recover all of the filtered HCO_3^- and may even secrete HCO_3^- in chronic alkalosis, a more important quantity is the *net* amount of new HCO_3^- generated for the body. Since every HCO_3^- that is not recovered or is secreted will be excreted in the urine, the net amount of new HCO_3^- generated can be calculated as follows:

$$\begin{matrix} \text{Net new } HCO_3^- \\ \text{generated} \end{matrix} = U_{NH_4}\dot{V} + U_{TA}\dot{V} - U_{HCO_3}\dot{V}$$

But because the *addition* of a new HCO_3^- ion to the body has an effect equivalent to the *elimination* of a *H^+ ion* from the body, the net amount of HCO_3^- generated is more commonly termed the *net acid excretion*, i.e.,

$$\begin{matrix} \text{Net acid} \\ \text{excretion} \end{matrix} = U_{NH_4}\dot{V} + U_{TA}\dot{V} - U_{HCO_3}\dot{V} \qquad (108)$$

In normal humans, HCO_3^- excretion is negligible, so that the net acid excretion is approximately equal to $U_{NH_4}\dot{V} + U_{TA}\dot{V}$, i.e., the amount of H^+ secreted in generating new HCO_3^- for the body. Thus, net acid excretion in normal human is approximately 50–100 mmol/day.

Control of H₂CO₃-Derived H⁺ Secretion

Since both the recovery of filtered HCO_3^- and the generation of new HCO_3^- are accomplished by *H_2CO_3-derived H^+ secretion*, the kidneys can regulate the pH of the body fluids just by controlling this one process. An important determinant of H_2CO_3-derived H^+ secretion by the epithelial cells of both the proximal tubule and distal nephron is the *intracellular pH*, H^+ secretion increasing as the intracellular pH falls and decreasing as the intracellular pH rises. The intracellular pH, in turn, is primarily determined by the *arterial pH* and the *plasma K^+ concentration*.

Arterial pH

Since changes in arterial pH produce corresponding changes in intracellular pH, H_2CO_3-derived H^+ secretion increases in acidosis and decreases in alkalosis. This relationship between H^+ secretion and arterial pH is of major importance in the renal regulation of the pH of the body fluids. For example, the primary cause of an *acidosis* could be either a decrease in $[HCO_3^-]_p$ or an increase in arterial P_{CO_2}. In both cases, the increased renal tubular H^+ secretion caused by the acidosis generates additional HCO_3^- for the body, which, in turn, tends to return the arterial pH toward normal (see below). This *adaptation* of H_2CO_3- derived H^+ secretion to changes in arterial pH requires 4–5 days to develop fully. It should be noted that the intracellular pH, and hence H_2CO_3-derived H^+ secretion, is most *immediately* sensitive to arterial pH changes caused by changes in arterial P_{CO_2}. This is because CO_2 can cross cellular membranes much more readily than H^+ or HCO_3^-. In fact, it can be experimentally demonstrated that increases in arterial P_{CO_2} that occur while the arterial pH is kept *constant*, or even *increased*, by the simultaneous addition of HCO_3^- will still augment H_2CO_3-derived H^+ secretion.

Plasma K⁺ Concentration

Intracellular pH is directly proportional to the plasma K^+ concentration. As a result, H_2CO_3-derived H^+ secretion increases in hypokalemia and decreases in hyperkalemia. The relationship between intracellular pH and plasma K^+ concentration derives from the fact that cells contain many large *anions*, such as proteins and organic phosphates (see Fig. 4.1), and therefore must contain a sufficient number of cations to maintain electrical neutrality. K^+, being the *major* intracellular cation, is primarily responsible for this maintenance of electrical neutrality (see Fig. 4.1), but small changes in the role of the less abundant intracellular cations such as Na^+ or H^+ can be significant. Of particular relevance here is the fact that the roles of K^+ and H^+ are *reciprocally related*. For example, in *hypokalemia* K^+ leaves cells along its concentration gradient, and H^+ enters cells to maintain electrical neutrality. As a result, the intracellular pH falls, and H_2CO_3-derived H^+ secretion increases. Conversely, in *hyperkalemia* K⁺ enters cells, and H⁺ leaves to maintain electrical neutrality. The intracellular pH therefore rises, and H_2CO_3- derived H⁺ secretion decreases.

It should be noted that because H⁺ enters cells in *hypokalemia*, the *plasma* actually becomes slightly *alkalotic*, even though the *intracellular* pH *falls*. Conversely, in *hyperkalemia* the *plasma* becomes slightly *acidotic*, even though the *intracellular* pH *rises*. The relationship between plasma K⁺ and arterial pH is discussed further in Chapter 33.

While intracellular pH is an important determinant of H_2CO_3-derived H⁺ secretion, another factor that can become significant in pathological states is the mass of *functioning renal tissue*. Patients with advanced renal disease frequently become acidotic due to a reduced capacity for H⁺ secretion. H_2CO_3-derived H⁺ secretion also can be significantly reduced by pharmacological agents that inhibit the enzyme *carbonic anhydrase*, which has an important role in the H⁺ secretion process (Fig. 4.44). *Acetazolamide* represents the most important example of such an agent.

All of the preceding discussion of the control of H_2CO_3-derived H⁺ secretion applies to *both* the proximal tubule and the distal nephron. However, certain mechanisms that affect H⁺ secretion in just *one* of these regions also can be identified.

CONTROL OF H_2CO_3-DERIVED H⁺ SECRETION IN THE PROXIMAL TUBULE

Since H_2CO_3-derived H⁺ secretion in the proximal tubule is primarily coupled to the movement of Na⁺ from the tubular lumen into the epithelial cells, it should not be surprising that H⁺ secretion in the proximal tubule can be affected by changes in proximal tubular *Na⁺ reabsorption* (Malnic and Giebisch, 1979). Recall that Na⁺ reabsorption in the proximal tubule can be influenced by changes in plasma volume, with Na⁺ reabsorption decreasing in volume expansion (part of the so-called *third factor* effect) and increasing in volume contraction (Chapter 31). Given the coupling between Na⁺ reabsorption and H⁺ secretion, this means that H_2CO_3-derived H⁺ secretion also decreases in plasma volume expansion and increases in plasma volume contraction. The increased H⁺ secretion in plasma volume contraction has an important effect on the renal response to certain acid-base disturbances (see below).

CONTROL OF H_2CO_3-DERIVED H⁺ SECRETION IN THE DISTAL NEPHRON

Although H_2CO_3-derived H⁺ secretion in the distal nephron is an active process that does *not* appear to be coupled to Na⁺ reabsorption (see above), distal nephron H⁺ secretion, like proximal tubular H⁺ secretion, is markedly stimulated by increased Na⁺ reabsorption (Warnock and Rector, 1979). This is because a major factor affecting H_2CO_3-derived H⁺ secretion in the distal nephron is the magnitude of the *transepithelial potential difference*, with H⁺ secretion increasing as the lumen becomes more negative relative to the peritubular space (Ziegler et al., 1976). As indicated in Chapter 30, the transepithelial potential difference in the distal nephron is primarily determined by the amount of Na⁺ reabsorbed, with the lumen becoming more negative as Na⁺ reabsorption increases, e.g., due to aldosterone (Chapter 31) or increased Na⁺ delivery to the distal nephron (Chapter 30). Thus, H_2CO_3-derived H⁺ secretion in the distal nephron is markedly stimulated by aldosterone and increased Na⁺ delivery to the distal nephron. For example, the *aldosterone* secreted in response to a decrease in Na⁺ intake and plasma volume depletion will stimulate not only Na⁺ reabsorption but also H⁺ secretion (it should be noted, however, that aldosterone may have a *direct* stimulatory effect on distal nephron H⁺ secretion (Ludens and Fanestil, 1972) in addition to this indirect stimulatory effect that occurs via stimulation of Na⁺ reabsorption). An important illustration of how *increased Na⁺ delivery* to the distal nephron can stimulate H⁺ secretion is provided by the *diuretic drugs*. Diuretics that inhibit Na⁺ reabsorption in the proximal tubule *(mannitol)* or in the thick ascending limb of the loop of Henle *(furosemide, bumetanide)* will increase the delivery of Na⁺ to the distal nephron, resulting in increased Na⁺ reabsorption, an increased transepithelial potential difference, and increased H⁺ secretion. *Thiazide* diuretics, which inhibit Na⁺ reabsorption in the distal convoluted tubule and connecting tubule, have a similar effect on H⁺ secretion, by increasing Na⁺ delivery to the collecting duct. Thus, all of these drugs increase H⁺ excretion, which contributes to the *metabolic alkalosis* seen with many diuretics (see below). In contrast, diuretics that impair Na⁺ reabsorption in the collecting duct *(spironolactone, triamterene, amiloride)* decrease the transepithelial potential difference in the distal nephron and therefore decrease H⁺ secretion.

H_2CO_3-derived H⁺ secretion in the distal nephron also is increased if the tubular fluid delivered to the distal nephron contains a substantial concentration of anions such as sulfate and nitrate, to which the distal nephron epithelium has limited permeability. Since these so-called *impermeant anions* cannot accompany reabsorbed Na⁺ as readily as Cl⁻, the magnitude of the transepithelial potential difference generated by Na^+ reabsorption increases, thereby promoting H^+ secretion.

INTRODUCTION TO ACID-BASE DISTURBANCES

As previously indicated, the bicarbonate buffer system is the most important one in the body because the concentrations of its components can be independently regulated, P_{CO_2} by the lungs and $[HCO_3^-]_p$ by the kidneys. Thus, via the bicarbonate buffer system the lungs and kidneys not only have an important role in the day-to-day maintenance of a *normal* pH, but also can help to restore acid-base homeostasis when the pH becomes *abnormal*. The lungs and kidneys also can *cause* an abnormal pH by changing P_{CO_2} and $[HCO_3^-]_p$, respectively, as in certain pathological conditions (see below).

The *Henderson-Hasselbalch equation* for the bicarbonate buffer system provides a convenient starting point for the study of acid-base disturbances *(Eq. 101)*. According to this equation, an *acidosis* can be produced by an increase in the arterial P_{CO_2} or a decrease in $[HCO_3^-]_p$, while an *alkalosis* can be produced by a decrease in the arterial P_{CO_2} or an increase in $[HCO_3^-]_p$. Since the arterial P_{CO_2} is regulated by the lungs, the acid-base disturbances resulting from a change in arterial P_{CO_2} are termed *respiratory disturbances*. In contrast, the acid-base disturbances resulting from a change in $[HCO_3^-]_p$ are termed *metabolic disturbances*, even though such changes in $[HCO_3^-]_p$ can be caused by abnormal renal or gastrointestinal function as well as by abnormal metabolic function (see below). The four so-called *primary acid-base disturbances*, then, are respiratory acidosis (increased arterial P_{CO_2}), respiratory alkalosis (decreased arterial P_{CO_2}), metabolic acidosis (decreased $[HCO_3^-]_p$), and metabolic alkalosis (increased $[HCO_3^-]_p$) (Table 4.23).

Several graphical methods have been developed to illustrate the changes in pH, P_{CO_2}, and $[HCO_3^-]_p$ that occur during various acid-base disturbances. One of the most useful is the so-called *pH-bicarbon-*

ate diagram (Fig. 4.47) (Davenport 1974). Two types of curves are drawn on this graph:

1. P_{CO_2} *isobars,* which illustrate the relation between pH and $[HCO_3^-]_p$ at a *constant* P_{CO_2} (hence, the term *isobar*). Each isobar is obtained by selecting the P_{CO_2} and then using the Henderson-Hasselbalch equation to calculate how the pH changes when $[HCO_3^-]_p$ changes at that P_{CO_2}. Since the normal arterial P_{CO_2} is 40 mm Hg and normal $[HCO_3^-]_p$ is 24 mmol/liter, *point A* in Figure 4.47 represents the *normal arterial point*, with a pH of 7.4.

2. *Blood-buffer lines,* which illustrate the changes in pH and $[HCO_3^-]_p$ that occur when P_{CO_2} varies. Although pH and $[HCO_3^-]_p$ must change in such a way that the Henderson-Hasselbalch equation is satisfied, the Henderson-Hasselbalch equation cannot be used to calculate these changes since only one of the three variables in the equation (P_{CO_2}) is known. Thus, blood-buffer lines must be *experimentally determined*. The term *blood-buffer line* is used because the slope of the line can be regarded as an index of the buffering capacity of blood.* Consider, for example, two blood samples with blood-buffer lines X and Y, as shown in Figure 4.48. If P_{CO_2} increases from 40 mm Hg to 80 mm Hg, the pH falls 0.25 units in blood sample Y (to pH 7.15) compared with a fall of only 0.2 units in blood sample X (to pH 7.2).

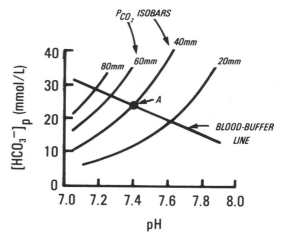

Figure 4.47. The pH-bicarbonate diagram. The P_{CO_2} isobars illustrate the relationship between pH and $[HCO_3^-]_p$ at the indicated P_{CO_2} values. The *blood-buffer line* illustrates the changes in pH and $[HCO_3^-]_p$ of normal blood that occur when P_{CO_2} is varied. *Point A* represents the *normal arterial point* (P_{CO_2} = 40 mm Hg, $[HCO_3^-]_p$ = 24 mmol/liter, pH = 7.4).

Table 4.23.
Primary Acid-Base Disturbances

Disturbance	Acute			Chronic		
	pH	P_{CO_2}	$[HCO_3^-]_p$	pH	P_{CO_2}	$[HCO_3^-]_p$
Respiratory acidosis	↓↓	↑↑[a]	↑	↓	↑↑	↑↑
Respiratory alkalosis	↑↑	↓↓[a]	↓	↑	↓↓	↓↓
Metabolic acidosis	↓↓	N[b]	↓↓[a]	↓	↓	↓↓↓
Metabolic alkalosis	↑↑	N[b]	↑↑[a]	↑	↑	↑↑↑

[a]Primary abnormality. [b]N, no significant change in hypothetical acute state before compensation has occurred (see text).

*It should be noted that the slope of the line measured on whole blood in vitro usually differs somewhat from that measured in vivo because of the buffering action of the interstitial fluid and other body tissues.

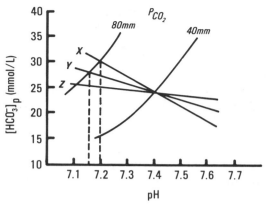

Figure 4.48. Blood-buffer lines as an index of the buffering capacity of blood. In blood with a normal hemoglobin concentration of 15 g/dl (*X*), the pH falls about 0.2 units (to 7.2) when P_{CO_2} increases from 40 to 80 mm Hg. In contrast, blood with a hemoglobin concentration of 5 g/dl (*Y*) exhibits a 0.25-unit decrease in pH (to 7.15) when P_{CO_2} increases from 40 to 80 mm Hg. The buffer line of plasma also is shown (*Z*). (Adapted from Davenport: *The ABC of Acid-Base Chemistry,* 6 ed. Chicago: University of Chicago Press, 1974. © 1947, 1949, 1950, 1958, 1969, 1974 by the University of Chicago.)

Thus, sample X is a *better buffer.* The primary determinant of the buffering capacity of the blood is the concentration of *hemoglobin.* In fact, line X represents the buffer line of blood with a normal hemoglobin concentration of 15 g/dl, while line Y represents the buffer line of blood with a hemoglobin concentration of only 5 g/dl. For comparison, line Z represents the buffer line of *plasma.* It should be noted that on all three buffer lines, $[HCO_3^-]_p$ increases slightly as P_{CO_2} increases because some additional H_2CO_3 is formed, which then dissociates. Of course, the pH falls in spite of this increase in $[HCO_3^-]_p$ because the *ratio* $[HCO_3^-]_p/P_{CO_2}$ falls.

In the following brief discussions of the four primary acid-base disturbances, the pH-bicarbonate diagram will be used to illustrate the changes in pH, P_{CO_2}, and $[HCO_3^-]_p$ that occur in each disturbance (Narins and Emmett, 1980; Masoro, 1982). It is necessary to consider what happens to pH, P_{CO_2}, and $[HCO_3^-]_p$ not only as a result of the primary disturbance itself, but also as a result of the secondary or *compensatory* responses that occur when the body attempts to restore the pH to normal. Thus, each acid-base disturbance is characterized by an *acute* or "uncompensated" phase followed by a *chronic* or "compensated" phase. It should be noted that in spite of the widespread use of the terms "uncompensated" and "compensated," the terms *acute* and *chronic* are preferable because the compensatory responses do *not* fully restore the pH to normal.

Respiratory Acidosis

The primary abnormality in respiratory acidosis is an increase in arterial P_{CO_2}. Given the important inverse relationship between arterial P_{CO_2} and alveolar ventilation *(Eq. 103)*, a major cause of respiratory acidosis is *hypoventilation.* For example, barbiturates and other drugs that depress ventilation are a common cause of respiratory acidosis, particularly in the emergency setting. Respiratory acidosis also can be caused by lung diseases that impair gas exchange.

The changes in pH and $[HCO_3^-]_p$ that result from an *acute* increase in arterial P_{CO_2} can be shown readily on the pH-bicarbonate diagram, since the blood-buffer line illustrates how pH and $[HCO_3^-]_p$ change when the arterial P_{CO_2} changes (see above). For example, Figure 4.49 illustrates how pH and $[HCO_3^-]_p$ change when the arterial P_{CO_2} increases *abruptly* from its normal value of 40 mm Hg *(point A)* to 60 mm Hg *(point B)*. Or more generally, the *entire* blood-buffer line to the *left* of the normal arterial point depicts the pH and $[HCO_3^-]_p$ changes that occur in different cases of *acute respiratory acidosis.*

If the arterial P_{CO_2} *remains* elevated (i.e., if the cause of the respiratory acidosis *persists*), the body attempts to restore the pH to normal. This *compensatory response* is accomplished by the *kidneys.* Recall that an important determinant of H_2CO_3-derived H$^+$ secretion by the proximal tubule and

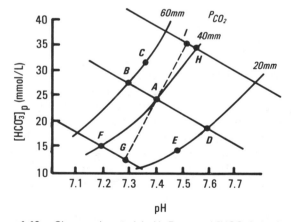

Figure 4.49. Changes in arterial pH, P_{CO_2}, and $[HCO_3^-]_p$ in the four primary acid-base disturbances and in the compensatory responses to these disturbances. *A,* Normal arterial point. *B,* Acute respiratory acidosis. *C,* Chronic respiratory acidosis. *D,* Acute respiratory alkalosis. *E,* Chronic respiratory alkalosis. *F,* Hypothetical acute metabolic acidosis. *G,* Chronic metabolic acidosis. *H,* Hypothetical acute metabolic alkalosis. *I,* Chronic metabolic alkalosis. Since the metabolic disturbances seldom develop acutely, the states of chronic metabolic acidosis and chronic metabolic alkalosis shown here are most likely to develop along the *dashed lines* connecting *A* to *G* and *A* to *I*, respectively. Details are provided in the text.

distal nephron is the *intracellular pH*. With increased arterial P_{CO_2}, the intracellular pH falls and H^+ secretion increases. The increased H^+ secretion, in turn, generates additional HCO_3^- for the body, i.e., $[HCO_3^-]_p$ increases significantly. Thus, although the arterial P_{CO_2} remains elevated, the $[HCO_3^-]_p/P_{CO_2}$ *ratio*, and hence pH, increases toward normal. An example of such a compensatory response is depicted in Figure 4.49, which illustrates the renal compensation to the acute respiratory acidosis represented by *point B*. If the arterial P_{CO_2} remains elevated at 60 mm Hg, the generation of new HCO_3^- by the kidneys causes $[HCO_3^-]_p$ to increase along the isobar for $P_{CO_2} = 60$ mm Hg, and the pH increases toward normal *(point C)*. *Point C*, then, represents a state of *chronic respiratory acidosis*. Note that the compensatory response is *not* complete, i.e., a mild acidosis remains.

Respiratory Alkalosis

In respiratory alkalosis, the primary abnormality is a decrease in arterial P_{CO_2}. Virtually all cases of respiratory alkalosis result from *hyperventilation*. Thus, major causes of respiratory alkalosis include hypoxic conditions such as high altitude, certain CNS disorders, and psychological disturbances such as anxiety.

The changes in pH and $[HCO_3^-]_p$ that result from an *acute* decrease in arterial P_{CO_2} are depicted by the blood-buffer line to the *right* of the normal arterial point. For example, Figure 4.49 illustrates how pH and $[HCO_3^-]_p$ change when the arterial P_{CO_2} falls *abruptly* from its normal value of 40 mm Hg *(point A)* to 20 mm Hg *(point D)*. If the arterial P_{CO_2} *remains* low, the body attempts to restore the pH to normal. As in respiratory acidosis, the compensatory response is accomplished by the *kidneys*. Specifically, the low arterial P_{CO_2} causes the intracellular pH to rise, thereby diminishing H_2CO_3-derived H^+ secretion. As a result, the kidneys not only fail to generate any new HCO_3^- to replace the HCO_3^- lost in titrating the strong acids produced by the body, but also fail to reabsorb all of the filtered HCO_3^-. In addition, the kidneys can secrete HCO_3^- under conditions of chronic alkalosis. $[HCO_3^-]_p$ can therefore fall significantly. Thus, although the arterial P_{CO_2} remains low, the $[HCO_3^-]_p/P_{CO_2}$ *ratio*, and hence pH, decreases toward normal. Figure 4.49 illustrates the renal compensation to the acute respiratory alkalosis represented by *point D*. If the arterial P_{CO_2} remains low at 20 mm Hg, the failure of the kidneys to conserve HCO_3^- causes $[HCO_3^-]_p$ to fall along the isobar for $P_{CO_2} = 20$ mm Hg, and the pH decreases toward normal *(point E)*. *Point E*, then, represents a state of *chronic respiratory alkalosis*. As in res-

piratory acidosis, the compensatory response is *not* complete, i.e., a mild alkalosis remains.

Metabolic Acidosis

In metabolic acidosis, the primary abnormality is a decrease in $[HCO_3^-]_p$. Most cases of metabolic acidosis result from the abnormal accumulation of organic acids. Thus, the common causes of metabolic acidosis include hypovolemic and other forms of circulatory shock, in which lactic acid may accumulate *(lactic acidosis)*, and uncontrolled diabetes mellitus, in which acetoacetic acid and β-OH butyric acid may accumulate *(diabetic ketoacidosis)* (see above). Metabolic acidosis also can occur in severe renal failure, since the kidneys are unable to secrete sufficient H^+ to generate new HCO_3^- to replace the HCO_3^- lost in titrating strong acids produced by the body or even to reabsorb all of the filtered HCO_3^-.

Although metabolic acidosis seldom develops *acutely*, it is instructive to consider a hypothetical case of an acute decrease in $[HCO_3^-]_p$. Figure 4.49 illustrates the fall in pH that would occur if $[HCO_3^-]_p$ dropped *abruptly* from its normal value of 24 mmol/liter *(point A)* to 15 mmol/liter *(point F)*. Note that $[HCO_3^-]_p$ decreases along the isobar for $P_{CO_2} = 40$ mm Hg, since respiratory function remains unchanged in this hypothetical acute state. The compensatory response to the decrease in pH is accomplished by the *lungs*. Specifically, the decreased arterial pH stimulates chemoreceptors in the carotid bodies, which in turn initiate a reflex stimulation of alveolar ventilation (Chapter 40). Because of the inverse relationship between alveolar ventilation and arterial P_{CO_2} *(Eq. 103)*, this increase in alveolar ventilation results in a decrease in arterial P_{CO_2}. Thus, although $[HCO_3^-]_p$ remains low (and in fact becomes even *lower* as P_{CO_2} decreases), the $[HCO_3^-]_p/P_{CO_2}$ *ratio*, and hence arterial pH, increases toward normal. Figure 4.49 illustrates the respiratory compensation to the hypothetical acute metabolic acidosis represented by *point F*. The increased alveolar ventilation causes the arterial P_{CO_2} to fall, and the pH increases toward normal *(point G)*. *Point G*, then, represents a state of *chronic metabolic acidosis*. Note that between *point F* and *point G* the arterial P_{CO_2} decreases along a blood-buffer line that *differs* from the buffer line of normal arterial blood, because the buffering capacity of arterial blood is *changed* in metabolic acidosis due to the decreased $[HCO_3^-]_p$. As in the respiratory acid-base disturbances, the compensatory response to a metabolic acidosis is *not* complete, i.e., a state of mild acidosis remains. The incompleteness of the compensatory response can be attributed to the decrease in arterial P_{CO_2} that occurs when alveolar ventilation is reflexly stimulated by the low arterial pH. This decrease in arterial P_{CO_2} is sensed by chemoreceptors in the medulla oblongata, carotid bodies, and aortic bodies, thereby attenuating the reflex stimulation of alveolar ventilation.

As indicated above, metabolic acidosis seldom develops acutely. Thus, respiratory compensation begins as soon as $[HCO_3^-]_p$ begins to fall. The state of chronic metabolic acidosis is therefore most likely to develop along the *dashed line* connecting *points A* and *G* in Figure 4.49, instead of from *point A* to *point F* to *point G*.

When metabolic acidosis is caused by the accumulation of organic acids (or other *nonrenal* factors), it should be noted that the *kidneys* as well as the lungs can have a role in the body's compensatory response. The mechanism for this renal correction of arterial pH is *not* an increase in H_2CO_3-derived H^+ secretion secondary to the reduced arterial, and hence intracellular, pH. In fact, H_2CO_3-derived H^+ secretion in metabolic acidosis generally does *not* increase significantly, or may actually *decrease*, because of the reduction in arterial P_{CO_2}. However, because of the low $[HCO_3^-]_p$ and consequent reduction in the filtered load of HCO_3^-, even a *diminished* rate of H^+ secretion is sufficient not only to reabsorb all of the filtered HCO_3^- but also to generate additional HCO_3^- to replace the HCO_3^- lost in titrating the organic acids. The resulting increase in $[HCO_3^-]_p$ helps to increase the arterial pH toward normal.

Metabolic Alkalosis

In metabolic alkalosis, the primary abnormality is an increase in $[HCO_3^-]_p$. Most cases of metabolic alkalosis result from the loss of fluid from the body that contains little or no HCO_3^-: the body's store of HCO_3^- is therefore contained in a smaller volume, and $[HCO_3^-]_p$ increases. Thus, common causes of metabolic alkalosis include (1) vomiting and nasogastric suction, in which H^+-rich gastric fluid is lost; and (2) diuretic drugs that result in the excretion of a large volume of acidic urine, such as the loop diuretics and thiazides (see above).

Like metabolic acidosis, metabolic alkalosis seldom develops *acutely*. However, it is still instructive to consider the hypothetical case of an acute increase in $[HCO_3^-]_p$. Figure 4.49 illustrates the increase in pH that would occur if $[HCO_3^-]_p$ increased *abruptly* from its normal value of 24 mmol/liter *(point A)* to 35 mmol/liter *(point H)*; the increase occurs along the isobar for $P_{CO_2} = 40$ mm Hg, since respiratory function remains unchanged in this hypothetical acute state. A compensatory response to the increased pH can be accomplished by the

lungs, since the increased arterial pH is sensed by chemoreceptors in the carotid bodies, which, in turn, initiate a reflex decrease in alveolar ventilation. However, the magnitude of the respiratory compensation normally is quite *small*, since the response of the carotid body chemoreceptors to an *increase* in arterial pH is much less than their response to a *fall* in arterial pH, as occurs in metabolic acidosis. But even with a small increase in arterial P_{CO_2}, the $[HCO_3^-]_p/P_{CO_2}$ *ratio*, and hence pH, will decrease toward normal. Figure 4.49 illustrates a small respiratory compensation to the hypothetical acute metabolic alkalosis represented by *point H*. The decrease in alveolar ventilation causes the arterial P_{CO_2} to increase slightly and the pH to decrease toward normal *(point I)*. *Point I*, then, represents a state of *chronic metabolic alkalosis*. As in metabolic acidosis, between *points H* and *I* the arterial P_{CO_2} changes along a blood-buffer line that differs from the buffer line of normal arterial blood. Furthermore, as in metabolic acidosis any respiratory compensation would begin as soon as $[HCO_3^-]_p$ begins to change. The state of chronic metabolic alkalosis is therefore most likely to develop along the *dashed line* between *points A* and *I* in Figure 4.49, instead of from *point A* to *point H* to *point I*.

Given the limited *respiratory* response to metabolic alkalosis, the role of the *kidneys* in correcting the arterial pH must be explored. A possible mechanism for such a renal response is completely analogous to the mechanism for the renal correction of metabolic acidosis (see above). To summarize, H_2CO_3-derived H^+ secretion may be normal or slightly elevated (due to the increased arterial P_{CO_2}), but because of the increased filtered load of HCO_3^- even a somewhat *increased* rate of H^+ secretion is *insufficient* to recover all of the filtered HCO_3^-. In addition, the kidneys can secrete HCO_3^- under conditions of chronic alkalosis. HCO_3^- excretion therefore increases, $[HCO_3^-]_p$ falls, and the arterial pH decreases toward normal. However, it should be noted that this mechanism for the renal correction of metabolic alkalosis is of importance in a *limited* number of cases. This is because the most common causes of metabolic alkalosis involve a fluid loss (see above), i.e., most cases of metabolic alkalosis are accompanied by a *plasma volume contraction*. Recall that H_2CO_3-derived H^+ secretion in both the proximal tubule and distal nephron is related to the amount of Na^+ reabsorbed. To review, in the *proximal tubule* H^+ secretion is directly coupled to Na^+ reabsorption, while in the *distal nephron* H^+ secretion is regulated in part by the transepithelial potential difference, which, in turn, is primarily determined by the amount of Na^+ reabsorbed. Since plasma volume contraction is a potent stimulus to Na^+ reabsorption in *both* the proximal tubule (via negation of the third factor effect) and distal nephron (via aldosterone), this means that plasma volume contraction causes a significant increase in H_2CO_3-derived H^+ secretion as well. In fact, H_2CO_3-derived H^+ secretion can increase so much that not only is all of the filtered HCO_3^- reabsorbed, but additional HCO_3^- is generated for the body, and $[HCO_3^-]_p$ actually may *increase*. Thus, when metabolic alkalosis is accompanied by plasma volume contraction, the renal response serves to *perpetuate* the alkalosis rather than to correct the arterial pH. In such cases of metabolic alkalosis, the kidneys will respond to the alkalosis by excreting HCO_3^- *only* after the plasma volume deficiency has been corrected by administering suitable fluids.

The compensatory responses that occur in each of the four primary acid-base disturbances are summarized in Table 4.23.

CHAPTER 33

Potassium Balance and the Regulation of Potassium Excretion

K^+ has an important role in the excitability of nerve and muscle, cell metabolism, and other physiological processes. Although K^+ is primarily an *intracellular* ion, as indicated in Chapter 25, the discussion of K^+ balance in this chapter will focus on the maintenance of a normal K^+ concentration in *extracellular fluid* (ECF), i.e., a normal *plasma* K^+ concentration (P_K). This is because the value of P_K is so low (average normal $P_K \approx 4$ mmol/liter) that even a small change in P_K can have significant adverse effects, particularly on the transmembrane potential of cardiac and skeletal muscle cells. For example, if an amount of K^+ equivalent to only 1% of the total body K^+ were added to the extracellular fluid, P_K would nearly double, and cardiac and skeletal muscle cells would be depolarized by 15 mV or more. Such depolarization is particularly serious in *cardiac* cells, leading to abnormal impulse conduction and, at a P_K above approximately 7 mmol/liter, the possibility of life-threatening or even fatal arrhythmias (Chapter 9). Conversely, a decrease in P_K would hyperpolarize cells. Such hyperpolarization is particularly serious in *skeletal muscle*, leading to muscle weakness and, at a P_K below about 1 mmol/liter, to paralysis.

The maintenance of a normal P_K requires, of course, that the input of K^+ equal the output of K^+. However, P_K also can be affected by the *distribution* of K^+ between the ECF and intracellular fluid (ICF). This chapter will therefore begin by examining the factors that affect the distribution of K^+ between ECF and ICF (Bia and DeFronzo, 1981). Then, the inputs and outputs of K^+ and their regulation will be considered.

DISTRIBUTION OF POTASSIUM

The high concentration of K^+ in ICF is generated by the Na^+/K^+-ATPase, which is present in the plasma membrane of all cells, including red and white blood cells, and platelets. Since resting cells are highly permeable to K^+, under normal conditions a steady state develops in which the quantity of K^+ *pumped into* cells by the Na^+/K^+-ATPase is equal to that passively *diffusing out* of cells. A change in P_K can therefore result from changes in either the active uptake of K^+ by cells or the passive diffusion of K^+ from cells, or both.

Changes in Active K^+ Uptake

Active K^+ uptake is accelerated by *insulin* (Andres et al., 1962), probably via stimulation of the Na^+/K^+-ATPase. In fact, insulin can be used clinically to lower P_K (glucose must be administered simultaneously, however, to prevent hypoglycemia). Uptake of K^+ by some cells is also increased by β_2-*adrenergic agonists* (Rosa et al., 1980). The mechanism for this effect is not certain, but like the insulin effect may involve stimulation of the Na^+/K^+-ATPase. The importance of the Na^+/K^+-ATPase in maintaining a normal P_K is further demonstrated by the finding that severe hyperkalemia may develop in patients who ingest large quantities of *digitalis* (Elkins and Watanabe, 1978), which inhibits the Na^+/K^+-ATPase.

Changes in Passive K^+ Diffusion

The passive diffusion of K^+ out of cells may be accelerated as a result of *cellular injury* or *death*. For example, tissues damaged as a result of severe burns or trauma can release sufficient K^+ to produce hyperkalemia. Rupture of red blood cells *(hemolysis)* also will increase P_K. If hemolysis occurs *after* a blood sample has been withdrawn from a patient with a *normal* P_K, the *measured* P_K will be factitiously elevated, i.e., P_K is normal but the in vitro hemolysis produces an abnormal laboratory value for P_K. The measured value for P_K also may be factitiously elevated due to in vitro rupture of platelets or white blood cells but only in patients with extremely high platelet counts *(thrombocytosis)* or extremely high white blood cell counts *(leukocytosis,* as in leukemia), respectively. Failure to recognize these potential factitious causes of hyperkalemia could lead to inappropriate treatment of a patient whose P_K actually is normal.

The passive diffusion of K^+ out of cells can also be influenced by the *arterial pH* (Leibman and Edelman, 1959), increasing as the arterial pH falls. In fact, in some types of metabolic acidosis every 0.1 unit decrease in arterial pH causes sufficient K^+ efflux to increase P_K by an average of 0.5–1.0 mmol/liter, i.e., even a moderate acidosis can produce hyperkalemia. This relationship between arterial pH and P_K can be primarily attributed to the reciprocal relation between the roles of K^+ and H^+ in maintaining electrical neutrality inside cells (Adler and Fraley, 1977), introduced in Chapter 32. Recall that cells contain many large *anions* such as proteins and organic phosphates and therefore must contain a sufficient number of *cations* to maintain electrical neutrality. While K^+, being the *major* intracellular cation, is primarily responsible for this maintenance of electrical neutrality, the reciprocal relation between the roles of K^+ and H^+ can have important consequences. For example, in *acidosis* H^+ enters cells and protonates negatively charged groups on the large anions. K^+ therefore leaves cells to maintain electrical neutrality, augmenting P_K. Conversely, in *alkalosis* H^+ leaves cells, and K^+ enters to maintain electrical neutrality, lowering P_K. In fact, $NaHCO_3$ can be administered clinically to lower P_K in hyperkalemia (it should be noted, however, that an increase in P_{HCO_3} may *by itself* cause K^+ to enter cells, *independent* of any effect on arterial pH or protonation of intracellular anions) (Fraley and Adler, 1977).

INPUT OF POTASSIUM

The input of K^+, like the input of Na^+ (Chapter 31), depends entirely upon the K^+ content of food and water consumed (Table 4.24). The average diet of adult Americans contains 50–100 mmol of K^+ per day, but the K^+ input can increase to over 500 mmol/day in persons whose diets contain large quantities of K^+-rich fruits and vegetables. At the other extreme, dietary K^+ input may be limited by

starvation or disease (e.g., anorexia nervosa) to less than 10 mmol/day. Like Na^+ input, K^+ input does not appear to be physiologically regulated in man.

OUTPUT OF POTASSIUM

K^+, like Na^+ (Chapter 31), can be lost from the body via three routes: sweat, feces, and urine (Table 4.24). The average concentration of K^+ in *sweat* is similar to, or just slightly greater than, that in plasma. Thus, except in extreme conditions (e.g., an individual exercising in a hot environment), the amount of K^+ lost in sweat is negligible. *Fecal loss* of K^+ averages only 5–10 mmol/day but may exceed 100 mmol/day in diarrheal diseases such as cholera. Neither sweat nor fecal loss is an important variable in the physiological regulation of K^+ output. Like Na^+ output, then, K^+ output is regulated primarily by changes in the amount of K^+ excreted in *urine*. Urinary K^+ excretion averages 45–90 mmol/day in adult Americans, thereby maintaining K^+ balance (Table 4.24), but it can vary from less than 10 mmol/day to over 500 mmol/day.

REGULATION OF POTASSIUM BALANCE

With no physiological regulation of K^+ input, K^+ balance must be achieved by changing K^+ output to match K^+ input. As indicated above, K^+ output is regulated primarily by changing the amount of K^+ excreted in urine. The handling of K^+ in the various nephron segments has been noted briefly in previous chapters. To review, K^+ is reabsorbed in segments 1 and 2 of the proximal tubule, secreted in segment 3 of the proximal tubule and the thin descending limb of the loop of Henle, reabsorbed in the thick ascending limb of the loop of Henle, and both reabsorbed and secreted in the distal nephron (Jamison, 1987). Thus, as indicated in Chapter 28, K^+ is one of the few substances that is reabsorbed *and* secreted.

The reabsorption of K^+ in segments 1 and 2 of the proximal tubule occurs in approximate proportion to the volume of water reabsorbed (LeGrimellec, 1975). Although the mechanism for this K^+ reabsorption is poorly understood, it is important to note that proximal tubular K^+ reabsorption does *not* vary with changes in K^+ balance, i.e., the quantity of K^+ reabsorbed in segments 1 and 2 of the proximal tubule is *independent* of K^+ input and output. The secretion of K^+ in segment 3 of the proximal tubule and the thin descending limb of the loop of Henle is substantial (Jamison, 1987). In fact, micropuncture samples of the tubular fluid at the hairpin turn of long-looped nephrons indicate that the

Table 4.24.
Major Inputs and Outputs of Potassium[a]

Inputs	Average Amount per Day, mmol/day	Outputs	Average Amount per Day, mmol/day
Food and water	50–100	Sweat	[b]
		Feces	5–10
		Urine	45–90
Total	50–100	Total	50–100

[a]Inputs and outputs represent average basal values for an adult ingesting a conventional American diet in a cool environment. [b]Negligible in basal state, but can become appreciable under certain conditions (see text).

turn of long-looped nephrons indicate that the quantity of K^+ reaching the hairpin turn can *exceed* the filtered load of K^+ (Battilana et al., 1978). K^+ secretion in segment 3 and the thin descending limb increases when K^+ intake increases and therefore has a potential role in matching K^+ output to K^+ input. However, the reabsorption of K^+ in the thick ascending limb of the loop of Henle is so extensive that less than 10% of the filtered K^+ reaches the distal nephron *regardless* of K^+ input and output (Wright and Giebisch, 1978).

Since less than 10% of the filtered K^+ reaches the distal nephron *regardless* of K^+ input and output, the regulation of K^+ output to match K^+ input must occur almost exclusively in the *distal nephron,* which both reabsorbs and secretes K^+ (Jamison, 1987). Specifically, K^+ *reabsorption* occurs in the *intercalated cells* of the collecting duct (Doucet and Marcy, 1987), while K^+ *secretion* occurs in the *principal cells* of the collecting duct (Morel and Doucet, 1986). Although K^+ reabsorption increases under conditions of extremely low dietary K^+ intake, the regulation of K^+ output to match K^+ input appears to occur primarily by changing the rate of K^+ *secretion.* Under conditions of extremely low dietary K^+ intake, K^+ secretion is negligible and *net K^+ reabsorption* occurs in the distal nephron. In contrast, with a high dietary K^+ intake, K^+ secretion increases markedly, and *net K^+ secretion* occurs. In adult Americans with an average dietary intake of 50–100 mmol/day, 45–90 mmol of K^+ must be excreted per day (Table 4.24). Assuming a filtered load of 720 mmol of K^+ per day (P_K = 4 mmol/liter,

GFR = 180 liters/day), about 72 mmol of K^+ (= 10% of 720 mmol) reaches the distal nephron per day, i.e., an average dietary intake could result in *either* net K^+ reabsorption *or* net K^+ secretion in the distal nephron.

To understand how distal nephron K^+ secretion changes to regulate K^+ output, it is important to consider the secretory mechanism (Giebisch and Stanton, 1979). As illustrated in Figure 4.50, distal nephron K^+ secretion involves two steps: (1) active uptake of K^+ from the peritubular fluid into the cell by the Na^+/K^+-ATPase; and (2) diffusion of K^+ from the cell across the luminal surface into the tubular fluid (some K^+ may be *actively* transported across the luminal surface by a K^+ pump, as indicated in Fig. 4.50). The K^+ secretion process can therefore be controlled in a variety of ways. For example, K^+ secretion can be altered by changes in K^+ *uptake* by the tubular cells (e.g., due to changes in the activity of the Na^+/K^+-ATPase or in the number of Na^+/K^+-ATPase molecules), the *permeability* of the luminal membrane to K^+, the *intracellular K^+ concentration* (which affects the magnitude of the *concentration* gradient for diffusion), or the *transepithelial potential difference* (which affects the magnitude of the *electrical* gradient for diffusion). It should be noted that the electrical gradient across the luminal surface always *opposes* diffusion into the tubular fluid, but with a more negative transepithelial potential difference this opposing effect is minimized. For example, with a transepithelial potential difference of −5 mV (lumen negative), an electrical gradient of 85 mV opposes K^+ diffusion (since the

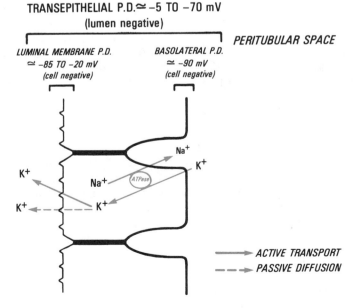

Figure 4.50. Mechanism for K^+ secretion in the distal nephron. K^+ secretion is a two-step process, involving (1) active uptake of K^+ by the Na^+/K^+-ATPase; and (2) passive diffusion of K^+ across the luminal surface into the tubular fluid (some K^+ also may be *actively* transported across the luminal surface). Note that K^+ *reabsorption* in the distal nephron is not illustrated here.

intracellular electrical potential is 85 mV negative with respect to the lumen; see Fig. 4.50). In contrast, with a transepithelial potential difference of −70 mV (lumen negative), the electrical gradient opposing K^+ diffusion from the cell into the tubular fluid is only 20 mV.

Variations in dietary K^+ input cause several changes in the K^+ secretion process (Silva et al., 1977), some of which are mediated by *aldosterone*. Consider, for example, the changes in distal nephron K^+ secretion that result from an *increase* in dietary K^+ intake (Fig. 4.51). The increased K^+ input causes a slight increase in P_K, which, in turn, stimulates K^+ secretion by increasing *K^+ uptake* by the tubular cells. With chronic increases in K^+ input, this increased uptake is due, at least in part, to an increased number of Na^+/K^+-ATPase molecules. The increase in P_K also represents one of the major physiological stimuli for the secretion of aldosterone, which *further* stimulates K^+ secretion by (1) increasing the activity of the Na^+/K^+-ATPase, thereby increasing *K^+ uptake;* (2) stimulating Na^+ reabsorption in the distal nephron, thereby increasing the *transepithelial potential difference* (see below); and (3) increasing the *permeability* of the luminal membrane to K^+. Note that all of these changes in K^+ secretion can be attributed, either directly or indirectly, to the initial increase in P_K (Fig. 4.51).

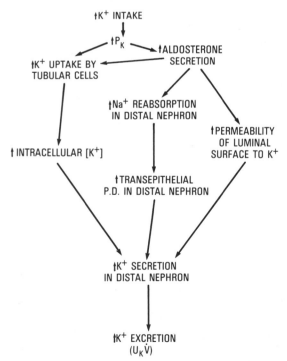

Figure 4.51. Pathway by which K^+ balance is restored following an increase in K^+ intake. P.D., potential difference.

Distal nephron K^+ secretion also can be influenced by factors that are *unrelated* to changes in K^+ input, including distal nephron Na^+ reabsorption, the presence of impermeant anions in the tubular fluid, the rate of tubular fluid flow, and acid-base status (Giebisch and Stanton, 1979).

Na⁺ Reabsorption in the Distal Nephron

As indicated above, the rate of K^+ secretion can be altered by changes in the *transepithelial potential difference* in the distal nephron. Since the transepithelial potential difference is primarily a function of the amount of Na^+ reabsorbed (Chapter 30), this means that K^+ secretion, like H^+ secretion, is markedly stimulated by increased Na^+ reabsorption in the distal nephron, e.g., due to aldosterone or increased Na^+ delivery to the distal nephron (Chapter 32). The fact that increased Na^+ reabsorption is responsible for part of the stimulatory effect of *aldosterone* on K^+ secretion has been noted above (Fig. 4.51). However, it should be emphasized that aldosterone can affect K^+ secretion not only in response to changes in K^+ intake (Fig. 4.51) but also in response to changes in *Na^+ intake* (Chapter 31). For example, the aldosterone secreted in response to a decrease in Na^+ intake and plasma volume depletion will stimulate K^+ secretion. An important illustration of how *increased Na^+ delivery* to the distal nephron can stimulate K^+ secretion is provided by the *diuretic drugs*. Diuretics that inhibit Na^+ reabsorption in the proximal tubule *(acetazolamide, mannitol)* or in the thick ascending limb of the loop of Henle *(furosemide, bumetanide)* will increase the delivery of Na^+ to the distal nephron, resulting in increased Na^+ reabsorption, an increased transepithelial potential difference, and increased K^+ secretion. *Thiazide* diuretics, which inhibit Na^+ reabsorption in the distal convoluted tubule and connecting tubule, have a similar effect on K^+ secretion, by increasing Na^+ delivery to the collecting duct. Thus, all of these drugs increase K^+ excretion and commonly cause *hypokalemia* as a side effect. In contrast, diuretics that impair Na^+ reabsorption in the collecting duct *(spironolactone, triamterene, amiloride)* decrease the transepithelial potential difference in the distal nephron and therefore decrease K^+ secretion. Consequently, such drugs decrease K^+ excretion and are referred to as *K^+-sparing diuretics*. Note that the effects of the diuretics on K^+ secretion are almost completely analogous to their effects on H^+ secretion (Chapter 32).

Impermeant Anions in the Tubular Fluid

K^+ secretion, like H^+ secretion, is increased if the tubular fluid delivered to the distal nephron contains a substantial concentration of anions such as sulfate and nitrate, which cannot accompany reabsorbed Na^+ as readily as Cl^-. In the presence of these *impermeant anions*, the magnitude of the transepithelial potential difference generated by Na^+ reabsorption is increased, thereby stimulating K^+ secretion (Chapter 32).

Rate of Tubular Fluid Flow

K^+ secretion can be affected by the rate of *tubular fluid flow* through the distal nephron. For example, an *increased* rate of tubular fluid flow creates a more favorable concentration gradient for K^+ diffusion across the luminal surface by "washing out" the secreted K^+ (Good and Wright, 1979). In fact, the increased K^+ secretion seen with many diuretic drugs (see above) can be attributed in part to an increased rate of tubular fluid flow.

Acid-Base Status

Changes in acid-base balance can have significant effects on K^+ secretion (Gennari and Cohen, 1975). For example, an *acute alkalosis*, whether of respi-

Table 4.25.
Major Causes of Increased K^+ Secretion in Distal Nephron

↑ Dietary intake
Aldosterone
↑ Na^+ reabsorption in distal nephron
↑ Impermeant anions in tubular fluid delivered to distal nephron
↑ Tubular fluid flow
Acute alkalosis

ratory or metabolic origin, stimulates K^+ secretion by causing K^+ to enter tubular cells, like other cells of the body (Fig. 4.52). As indicated above, this increased K^+ entry can be attributed to the reciprocal relationship between the roles of K^+ and H^+ in maintaining electrical neutrality inside cells. Acute alkalosis also stimulates K^+ secretion by increasing the *transepithelial potential difference* and the *permeability* of the luminal membrane to K^+, although somewhat indirectly (Fig. 4.52). By reducing H^+ secretion in the proximal tubule and distal nephron and thereby decreasing the reabsorption of HCO_3^- from the tubular fluid (Chapter 32), acute alkalosis causes the tubular fluid in the distal nephron to have an elevated HCO_3^- concentration and an elevated pH. The *excess HCO_3^-* functions as an *impermeant anion* and therefore stimulates K^+ secretion by increasing the transepithelial potential difference (see above). The *elevated pH* increases the K^+ permeability of the luminal membrane (Boudry et al., 1976), an effect similar to that of aldosterone. *Acute acidosis* causes a decrease in K^+ secretion by pathways opposite to those illustrated in Figure 4.52.

The effects of *chronic* acid-base disturbances on K^+ secretion can be quite complex. For example, since alkalosis stimulates K^+ secretion and, hence, K^+ excretion, *chronic alkalosis* (days or weeks) may produce depletion of total body K^+ and, hence, a decrease in the concentration of K^+ in cells, including those of the distal nephron. Thus, chronic alkalosis may be accompanied by a *low* rate of K^+ secretion and K^+ excretion. Moreover, as K^+ is lost from cells, H^+ tends to enter to maintain electrical neutrality, which may result in the paradoxical situation of *extracellular alkalosis* and *intracellular acidosis*. Such an intracellular acidosis would stimulate H^+ secretion in the proximal tubule and distal nephron (Chapter 32), thereby *perpetuating* the extracellular alkalosis.

The major causes of increased distal nephron K^+ secretion are summarized in Table 4.25.

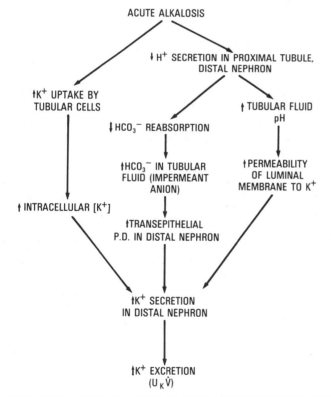

Figure 4.52. Pathway by which an acute alkalosis stimulates distal nephron K^+ secretion and, hence, K^+ excretion. P.D., potential difference.

Regulation of Calcium, Magnesium, and Phosphate Excretion

Ca^{2+}, Mg^{2+}, and phosphate are considered together in this chapter primarily because the same conditions and hormones tend to control the excretion of all three ions, a fact that is related to their role as the major constituents of bone mineral. The regulation of Ca^{2+}, Mg^{2+}, and phosphate balance differs from that of Na^+ and K^+ balance (Chapters 31 and 33) in several respects. First of all, the inputs of Na^+ and K^+ are equal to the amounts of these ions in food and water consumed and are not subject to physiological regulation. In contrast, the inputs of Ca^{2+}, Mg^{2+}, and phosphate do *not* equal the amounts of these ions in food and water consumed because (particularly in the case of Ca^{2+}) gastrointestinal absorption is incomplete and is hormonally regulated. A second important difference between Na^+ and K^+ and the ions considered here is that Na^+ and K^+ balance are achieved primarily by adjusting urinary output to match dietary input. In contrast, Ca^{2+}, Mg^{2+}, and phosphate balance are determined not only by the inputs and outputs of these ions, but also by the factors that affect their deposition into *bone*. The importance of the bone deposits of Ca^{2+}, Mg^{2+}, and phosphate is emphasized by the large percentages of the total body stores of these ions that are found in bone (Table 4.26).

In contrast to Na^+ and K^+, then, the regulation of the body balance of Ca^{2+}, Mg^{2+}, and phosphate involves the *gastrointestinal tract* and *bone* as well as the kidneys. In this chapter, emphasis is placed on the *renal component* of the regulation of the total body balance of these ions. The reader is referred to the gastrointestinal and endocrine sections of this text for further discussion of Ca^{2+}, Mg^{2+}, and phosphate balance.

CALCIUM

Although over 99% of the total body Ca^{2+} is in bone (Table 4.26), the remaining Ca^{2+}, which is primarily extracellular, has an important role in the excitation of nerve and muscle, muscle contraction, blood coagulation, and other physiological pro-

cesses. The average normal concentration of Ca^{2+} in plasma is 2.5 mmol/liter (5 meq/liter, 10 mg/dl). About 40–50% of the Ca^{2+} in plasma is bound to plasma proteins; of the remaining 50–60% (1.25–1.5 mmol/liter), about 90% is ionized and 10% is complexed with anions such as phosphate, sulfate, and citrate (Table 4.27) (Walser, 1973). It is the concentration of *ionized Ca^{2+}* that affects physiological processes such as nerve and muscle function. However, at present clinical laboratories typically measure the *total* concentration of Ca^{2+} in plasma rather than the concentration of *ionized* Ca^{2+}. Accurate evaluation of P_{Ca} values reported by clinical laboratories therefore requires knowledge of the plasma protein concentration. For example, lowering of the plasma protein concentration by disease (e.g., cirrhosis of the liver) can lower the *total* concentration of Ca^{2+} in plasma without necessarily altering the physiologically relevant concentration of *ionized* Ca^{2+}.

Renal Regulation of Calcium Balance

The 40–50% of the plasma Ca^{2+} that is protein bound cannot be filtered. Thus, at a GFR of 125 ml/min (180 liters/day) and a total P_{Ca} of 2.5 mmol/liter (i.e., a *filterable* P_{Ca} of 1.25–1.5 mmol/liter), the filtered load of Ca^{2+} is approximately 225–270 mmol/day. In the normal adult, the maintenance of Ca^{2+} balance involves the reabsorption of about 97–99% of the filtered Ca^{2+}, i.e., about 2.5–7.5 mmol of Ca^{2+} are excreted per day (Kesteloot and Geboers, 1982).

The renal handling of Ca^{2+} is similar in many ways to the renal handling of Na^+ (Chapters 29–31). For example, about two-thirds of the filtered load of each ion is reabsorbed in the proximal tubule, 20–25% in the loop of Henle, and 5–10% in the distal nephron (Lassiter et al., 1963). The mechanisms for Ca^{2+} reabsorption in the *proximal tubule* and *distal nephron* are not well understood (Suki, 1979). In the *loop of Henle*, the major portion of the Ca^{2+} reabsorption occurs in the *thick ascending limb*, where the lumen-positive transepithelial

Table 4.26.
Distribution of Calcium, Magnesium, and Phosphate

	Calcium, %	Magnesium, %	Phosphate, %
Bone	99	50–60	85
Nonbone			
Intracellular	<0.1	39–49	$\simeq 14.9$
Extracellular	$\simeq 1$	$\simeq 1$	$\simeq 0.1$

Adapted from Potts and Deftos, 1974.

potential difference generated by the Na^+/K^+-2 Cl^- symport (Chapter 30) favors Ca^{2+} reabsorption (Boudreau and Burg, 1979). The similarities between the renal handling of Na^+ and Ca^{2+} in the proximal tubule and loop of Henle are underscored by the finding that the two ions have virtually identical TF/P ratios in the early distal nephron, i.e., identical fractions of the filtered Na^+ and Ca^{2+} are reabsorbed prior to the distal nephron *(Eq. 78)*. Furthermore, in most cases a very good correlation exists between the *fractional excretions* of Na^+ and Ca^{2+} (Walser, 1961b). For example, an increase in Na^+ intake will increase not only Na^+ excretion, but Ca^{2+} excretion as well, an outcome that can be used clinically to treat hypercalcemia. The mechanism for this effect of Na^+ intake on Ca^{2+} excretion probably includes the *third factor* effect in the proximal tubule, whereby *all* proximal tubular reabsorption is decreased following a large increase in Na^+ intake or another cause of plasma volume expansion (Chapter 31). Another important example of the correlation between Na^+ and Ca^{2+} excretion is provided by the *loop diuretics*, which inhibit the active reabsorption of Na^+ and Cl^- by the thick ascending limb (Chapter 30). This inhibition increases not only Na^+ excretion but also Ca^{2+} excretion by impairing the formation of the lumen-positive transepithelial potential difference and therefore inhibiting Ca^{2+} reabsorption (see above). An *exception* to the general correlation between Na^+ and Ca^{2+} excretion involves the *thiazide diuretics*, which increase the excretion of Na^+ but *decrease* Ca^{2+} excretion, an effect used to treat some patients who form Ca^{2+}-containing renal stones.

The regulation of Ca^{2+} excretion to maintain Ca^{2+} balance is accomplished primarily by *parathormone* (PTH), a polypeptide hormone secreted by the parathyroid glands in response to a decrease in the concentration of ionized Ca^{2+} in plasma. PTH stimulates Ca^{2+} reabsorption in the thick ascending limb (Suki et al., 1980) and distal nephron (Shareghi and Stoner, 1978), thereby decreasing Ca^{2+} excretion and helping to restore plasma Ca^{2+} to normal levels. Ca^{2+} excretion also can be affected by factors not directly related to Ca^{2+} balance (Massry and Coburn, 1973). For example, Ca^{2+} excretion is increased by chronic adrenocorticosteroid therapy, chronic vitamin D excess, growth hormone, hypermagnesemia, and chronic metabolic acidosis as well as by increased Na^+ intake, plasma volume expansion, and the loop diuretics as noted above. Some of these factors may *not* have a *direct* effect on the renal handling of Ca^{2+}. For example, an agent that stimulates Ca^{2+} absorption from the gastrointestinal tract will elevate the concentration of ionized Ca^{2+} in plasma and decrease the secretion of PTH, thereby increasing the renal excretion of Ca^{2+}.

MAGNESIUM

About 50–60% of the total body Mg^{2+} is in bone (Table 4.26). Most of the remaining Mg^{2+} is found in ICF, where it represents an essential cofactor for enzymatic reactions and also has a role in nerve and muscle function (Wacker and Parisi, 1968). The normal concentration of Mg^{2+} in plasma ranges from about 0.75–1.25 mmol/liter (1.5–2.5 meq/liter); why the normal plasma level of Mg^{2+} varies by a greater percentage than the normal plasma levels of most other ions is unknown. About 30% of the Mg^{2+}

Table 4.27.
Forms of Calcium, Magnesium, and Phosphate in Normal Plasma

	Calcium	Magnesium	Phosphate	
Protein bound	40–50%	30%	Phospholipids Protein-bound	} 67%
Free (filterable)	50–60%	70%	Acid-soluble (filterable)	33%
Ionized	$\simeq 90\%$ of free	$\simeq 70\%$ of free	Inorganic Ionized	$\simeq 90\%$ of acid-soluble $\simeq 45\%$ of acid-soluble
Complexed	$\simeq 10\%$ of free	$\simeq 30\%$ of free	Complexed Organic	$\simeq 45\%$ of acid-soluble $\simeq 10\%$ of acid-soluble
Concentration measured by clinical laboratories	*Total* plasma calcium	*Total* plasma magnesium	Acid-soluble phosphate (termed "plasma phosphate")	
Normal value	2.5 mmol/liter (10 mg/dl)	0.75–1.25 mmol/liter	1.0–1.6 mmol/liter (3–5 mg phosphorus/dl)	

Includes data from Walser, 1961a; Walser, 1973; and Krane, 1970.

in plasma is bound to plasma proteins; of the remaining 70% (0.5–0.9 mmol/liter), about 70% is ionized and 30% is complexed with the same anions (e.g., phosphate, sulfate, and citrate) that complex with Ca^{2+} (Table 4.27) (Walser, 1973). As with Ca^{2+}, clinical laboratories typically measure the *total* concentration of Mg^{2+} in plasma, not just the concentration of ionized Mg^{2+}.

Renal Regulation of Magnesium Balance

The 30% of the plasma Mg^{2+} that is protein bound cannot be filtered. Thus, at a GFR of 125 ml/min (180 liters/day) and a total P_{Mg} of 1 mmol/liter (i.e., a *filterable* P_{Mg} of 0.7 mmol/liter), the filtered load of Mg^{2+} is approximately 125 mmol/day. In the normal adult, the maintenance of Mg^{2+} balance involves the reabsorption of 90–99% of the filtered Mg^{2+} (Quamme, 1986), i.e., about 1–10 mmol of Mg^{2+} are excreted per day.

While the renal handling of Mg^{2+} is poorly understood, changes in Mg^{2+} excretion generally parallel changes in Ca^{2+} excretion (Massry and Coburn, 1973). For example, Mg^{2+} excretion is decreased by *PTH*, although PTH levels are not as closely related to Mg^{2+} excretion as they are to Ca^{2+} excretion (Quamme, 1986).

It should be noted that although the renal excretion of Mg^{2+} can be reduced to less than 1 mmol/day, it *cannot* be reduced to zero. As a result, individuals who ingest a low Mg^{2+} diet (e.g., a chronic "diet" of ethanol) can become Mg^{2+} depleted. In contrast, such a large quantity of Mg^{2+} can be excreted by individuals with normal renal function that even the ingestion of excessive amounts of Mg^{2+} (e.g., during chronic use of antacids) does not cause Mg^{2+} overload.

PHOSPHATE

Approximately 85% of the total body phosphate is in bone (Table 4.26). Of the remainder, over 99% is intracellular and consists primarily of *organic phosphates* such as phospholipids, nucleic acids, nucleotides, phosphoproteins, and intermediates in carbohydrate metabolism. Emphasis in this chapter is placed on the small amount of *inorganic phosphate* in the body fluids, which exists primarily as the anions $HPO_4^=$ and $H_2PO_4^-$. Inorganic phosphate is required for the synthesis of organic phosphates (e.g., ATP) and also is important as a body buffer (Chapter 32).

In plasma, about two-thirds of the total phosphate is in phospholipids or bound to plasma proteins. The remaining one-third is termed *acid-soluble phosphate* since it represents the phosphate that remains after plasma is treated with an acid such as trichloroacetic acid to precipitate plasma proteins and phospholipids. About 90% of the acid-soluble phosphate is inorganic phosphate, of which approximately half is ionized and half is complexed with ions such as Ca^{2+} and Mg^{2+} (see above); the remaining 10% represents organic phosphates such as ATP (Table 4.27). It is the *acid-soluble* phosphate, *not* the *total* phosphate, that is measured by clinical laboratories and that is referred to as *plasma phosphate*. Actually, clinical laboratories typically report plasma phosphate as *plasma phosphorus*. A plasma *phosphorus* concentration expressed in mmol/liter is the same as a plasma *phosphate* concentration expressed in mmol/liter, since one atom of phosphorus is found in each molecule of phosphate. However, plasma phosphorus concentrations often are expressed in mg/dl, in which case a plasma *phosphorus* concentration of 3.1 mg/dl is equal to a plasma *phosphate* concentration of 1 mmol/liter, since each mmol of phosphate contains 31 mg of phosphorus. The normal plasma phosphate concentration (i.e., the *acid-soluble phosphate* concentration measured by clinical laboratories) ranges from about 1.0–1.6 mmol/liter (3–5 mg phosphorus/dl). Like the plasma Mg^{2+} concentration, then, the normal plasma phosphate varies by a greater percentage than the normal plasma levels of most other ions.

Renal Regulation of Phosphate Balance

Phospholipids and protein-bound phosphate cannot be filtered, i.e., only the *acid-soluble* phosphate is filtered. Thus, at a GFR of 125 ml/min (180 liters/day) and a plasma phosphate of 1.3 mmol/liter, the filtered load of phosphate is about 235 mmol/day (recall that what is termed *plasma phosphate* actually represents *acid-soluble phosphate*, i.e., filterable phosphate). In the normal adult, the maintenance of phosphate balance involves the reabsorption of 85–95% of the filtered phosphate, i.e., about 12–40 mmol of phosphate are excreted per day.

Normally, about 75–85% of the filtered phosphate is reabsorbed in the *proximal tubule* (Lang, 1980). Since the fractional excretion of phosphate is about 5–15%, it is generally assumed that about 10% of the filtered load is reabsorbed in the loop of Henle and/or distal nephron. However, it has also been proposed that a proximal tubular phosphate reabsorption of 75–85% is found only in *cortical nephrons* and that *juxtamedullary nephrons* reabsorb a much larger percentage of the filtered phosphate in their proximal tubules (Knox et al., 1977). By this theory, the combined effect of cortical nephrons and juxtamedullary nephrons could yield an overall fractional reabsorption of 85–95%.

Like glucose and amino acid reabsorption in the proximal tubule (Chapter 29), the reabsorption of phosphate in the proximal tubule is coupled to the reabsorption of Na^+ *(Na$^+$-solute symport)* (Hoffman et al., 1976). Furthermore, like glucose and amino acids, phosphate exhibits T_m-*limited reabsorption* (Chapter 28). Thus, variations in plasma phosphate can result in changes in phosphate excretion analogous to those illustrated in Figure 4.17 for glucose.

The regulation of phosphate excretion to maintain phosphate balance is accomplished primarily by *PTH,* which inhibits phosphate reabsorption in the proximal tubule such that 40% or more of the filtered phosphate can be excreted at high PTH levels and less than 5% in the absence of PTH (Knox et al., 1977). To understand how PTH can have a role in maintaining phosphate balance when the secretion of PTH is regulated primarily by the concentration of ionized Ca^{2+} in plasma (see above), consider, for example, how phosphate balance is restored following an increase in plasma phosphate concentration (Fig. 4.53). Some of the excess phosphate will complex with ionized Ca^{2+} in plasma, thereby lowering the plasma concentration of ionized Ca^{2+}. The decrease in ionized Ca^{2+}, in turn, stimulates the secretion of PTH, which then inhibits proximal tubular phosphate reabsorption, enabling the kidneys to excrete the excess phosphate. PTH also stimulates Ca^{2+} reabsorption in the thick ascending limb and distal nephron, thereby enabling the

plasma concentration of ionized Ca^{2+} to return toward normal. Phosphate excretion also appears to be regulated by *dietary phosphate,* as subjects on an experimental low phosphate diet reabsorb nearly 100% of the filtered load of phosphate. The mediator(s) of this important homeostatic mechanism, which does *not* require PTH, is (are) unknown.

Phosphate excretion also can be affected by factors not directly related to phosphate balance (Lang, 1980). For example, phosphate excretion, like Ca^{2+} excretion, is increased by plasma volume expansion, probably as a result of the *third factor effect* that depresses *all* proximal tubular reabsorption in the presence of a plasma volume expansion. If the relationship between plasma phosphate and phosphate excretion is analyzed in a manner analogous to that illustrated in Figure 4.17, one can demonstrate that the *transport maximum* for phosphate ($T_{m_{PO_4}}$) is decreased by factors that inhibit phosphate reabsorption (e.g., PTH, plasma volume expansion) and increased by factors that increase phosphate reabsorption (e.g., ingestion of a low phosphate diet).

It should be noted that plasma phosphate can be affected by the *distribution* of phosphate between the ECF and ICF as well as by the renal handling of phosphate. For example, alkalosis, insulin, glucose, and β_2-adrenergic agonists cause phosphate to move from ECF to ICF (Body et al., 1983; Brautbar et al., 1983).

Actions of Parathormone

Since PTH can have effects on the renal handling of Ca^{2+}, Mg^{2+}, *and* phosphate, a brief description of its mechanism of action is appropriate. PTH acts via specific receptors on the basal-lateral surfaces of cells in the responsive segments of the nephron, including the proximal tubule, thick ascending limb, and distal nephron (Morel, 1981). The PTH-receptor complex activates adenylate cyclase, stimulating the production of cyclic AMP. Cyclic AMP, in turn, mediates the inhibition of phosphate reabsorption in the proximal tubule and the stimulation of Ca^{2+} (and Mg^{2+}) reabsorption in the thick ascending limb and distal nephron.

PTH also promotes an important metabolic reaction in the kidneys: the hydroxylation of 25-hydroxy vitamin D_3 to the more active 1,25-dihydroxy vitamin D_3 *(calcitriol).* Calcitriol stimulates the absorption of Ca^{2+} and phosphate from the gastrointestinal tract and has an important role in regulating the total body balance of Ca^{2+} and phosphate. In addition, calcitriol can decrease the renal excretion of Ca^{2+} and phosphate, but the physiological significance of this effect is unknown.

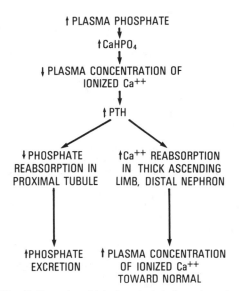

Figure 4.53. Pathway by which phosphate balance is restored following an increase in plasma phosphate concentration.

BIBLIOGRAPHY

ADLER, S., AND D. S. FRALEY. Potassium and intracellular pH. *Kidney Int.* 11: 433–442, 1977.

ALBRINK, M. J., P. M. HALD, E. B. MAN, AND J. P. PETERS. The displacement of serum water by the lipids of hyperlipemic serum. A new method for the rapid determination of serum water. *J. Clin. Invest.* 34: 1483–1488, 1955.

ALPERN, R. J. Apical membrane chloride/base exchange in the rat proximal convoluted tubule. *J. Clin. Invest.* 79: 1026–1030, 1987.

ANDREOLI, T. E., AND J. A. SCHAFER. Effective luminal hypotonicity: the driving force for isotonic proximal tubular fluid absorption. *Am. J. Physiol.* 236: F89–F96, 1979.

ANDREOLI, T. E., J. A. SCHAFER, S. L. TROUTMAN, AND M. L. WATKINS. Solvent drag component of Cl$^-$ flux in superficial proximal straight tubules: evidence for a paracellular component of isotonic fluid absorption. *Am. J. Physiol.* 237: F455–462, 1979.

ANDRES, R., M. A. BALTZAN, G. CADER, AND K. L. ZIERLER. Effect of insulin on carbohydrate metabolism and on potassium in the forearm of man. *J. Clin. Invest.* 41: 108–114, 1962.

ASSCHER, A. W., D. B. MOFFAT, AND E. SANDERS. *Nephrology Illustrated.* Philadelphia: W. B. Saunders, 1982.

ATHERTON, J. C., AND R. GREEN. Renal function in pregnancy. *Clin. Sci.* 65: 449–455, 1983.

AUGUST, J. T., D. H. NELSON, AND G. W. THORN. Response of normal subjects to large amounts of aldosterone. *J. Clin. Invest.* 37: 1549–1555, 1958.

BAGNASCO, S., R. BALABAN, H. M. FALES, Y.-M. YANG, AND M. BURG. Predominant osmotically active organic solutes in rat and rabbit renal medullas. *J. Biol. Chem.* 261: 5872–5877, 1986.

BARAJAS, L. Anatomy of the juxtaglomerular apparatus. *Am. J. Physiol.* 237: F333–F343, 1979.

BARGER, A. C., AND J. H. HERD. The renal circulation. *N. Engl. J. Med.* 284: 482–490, 1971.

BATTILANA, C. A., D. C. DOBYAN, F. B. LACY, J. BHATTACHARYA, P. A. JOHNSTON, AND R. L. JAMISON. Effect of chronic potassium loading on potassium secretion by the pars recta or descending limb of the juxtamedullary nephron in the rat. *J. Clin. Invest.* 62: 1093–1103, 1978.

BECK, T. R., AND M. J. DUNN. The relationship of antidiuretic hormone and renal prostaglandins. *Mineral Electrolyte Metab.* 6: 46–59, 1981.

BEEUWKES, R., III, AND J. V. BONVENTRE. Tubular organization and vascular-tubular relations in the dog kidney. *Am. J. Physiol.* 229: 695–713, 1975.

BIA, M. J., AND R. A. DeFRONZO. Extrarenal potassium homeostasis. *Am. J. Physiol.* 240: F257–F268, 1981.

BLANTZ, R. C. Segmental renal vascular resistance: single nephron. *Ann. Rev. Physiol.* 42: 573–588, 1980.

BLANTZ, R. C., AND J. C. PELAYO. Disorders of glomerular filtration. In: *Physiology of Membrane Disorders*, 2nd ed., edited by T. E. Andreoli, J. F. Hoffman, D. D. Fanestil, and S. G. Schultz. New York: Plenum Medical, 1986, p. 919–938.

BODY, J.-J., P. E. CRYER, K. P. OFFORD, AND H. HEATH, III. Epinephrine is a hypophosphatemic hormone in man. *J. Clin. Invest.* 71: 572–578, 1983.

BOUDREAU, J. E., AND M. B. BURG. Voltage dependence of calcium transport in the thick ascending limb of Henle's loop. *Am. J. Physiol.* 236: F357–F364, 1979.

BOUDRY, J., L. C. STONER, AND M. B. BURG. Effect of acid lumen pH on potassium transport in renal cortical collecting tubules. *Am. J. Physiol.* 230: 239–244, 1976.

BOULPAEP, E. L., AND H. SACKIN. Role of the paracellular pathway in isotonic fluid movement across the renal tubule. *Yale J. Biol. Med.* 50: 115–131, 1977.

BRAUTBAR, N., H. LEIBOVICI, AND S. MASSRY. On the mechanism of hypophosphatemia during acute hyperventilation: evidence for increased muscle glycolysis. *Mineral Electrolyte Metab.* 9: 45–50, 1983.

BRENNER, B. M., AND H. D. HUMES. Mechanics of glomerular ultrafiltration. *N. Engl. J. Med.* 297: 148–154, 1977.

BRENNER, B. M., T. H. HOSTETTER, AND H. D. HUMES. Glomerular permselectivity: barrier function based on discrimination of molecular size and charge. *Am. J. Physiol.* 234: F455–F460, 1978.

BREZIS, M., S. ROSEN, P. SILVA, AND F. H. EPSTEIN. Renal ischemia: a new perspective. *Kidney Int.* 26: 375–383, 1984.

BULGER, R. E. Kidney morphology. In: *Diseases of the Kidney*, edited by L. E. Earley and C. W. Gottschalk. Boston: Little, Brown and Company, 1979, p. 3–39.

BURG, M. B., AND J. ORLOFF. Control of fluid absorption in the renal proximal tubule. *J. Clin. Invest.* 47: 2016–2024, 1968.

CAREY, R. M. Recent progress in the control of aldosterone secretion. *Recent Prog. Horm. Res.* 42: 251–296, 1986.

CARONE, F. A., D. R. PETERSON, AND G. FLOURET. Renal tubular processing of small peptide hormones. *J. Lab. Clin. Med.* 100: 1–14, 1982.

CHASIS, H., AND H. W. SMITH. The excretion of urea in normal man and in subjects with glomerulonephritis. *J. Clin. Invest.* 17: 347–358, 1938.

COCKCROFT, D. W., AND M. H. GAULT. Prediction of creatinine clearance from serum creatinine. *Nephron* 16: 31–41, 1976.

CORTNEY, M. A., M. MYLLE, W. E. LASSITER, AND C. W. GOTTSCHALK. Renal tubular transport of water, solute and PAH in rats loaded with isotonic saline. *Am. J. Physiol.* 209: 1199–1205, 1965.

DAVENPORT, H. W. *The ABC of Acid-Base Chemistry*, 6th ed. Chicago: University of Chicago Press, 1974.

DEETJEN, P., J. W. BOYLAN, AND K. KRAMER. *Physiology of the Kidney and of Water Balance.* New York: Springer, 1975.

DEETJEN, P., AND K. KRAMER. Die Abhängigkeit des O_2-Verbrauchs der Niere von der Na-Rückresorption. *Pflugers Arch.* 273: 636–650, 1961.

DeWARDENER, H. E. The control of sodium excretion. *Am. J. Physiol.* 235: F163–F173, 1978.

DIAMOND, H. S., AND J. S. PAOLINO. Evidence for a postsecretory reabsorptive site for uric acid in man. *J. Clin. Invest.* 52: 1491–1499, 1973.

DOUCET, A., AND S. MARCY. Characterization of K-ATPase activity in distal nephron: stimulation by potassium depletion. *Am. J. Physiol.* 253: F418–423, 1987.

DOUSA, T. P., AND H. VALTIN. Cellular actions of vasopressin in the mammalian kidney. *Kidney Int.* 10: 46–63, 1976.

EDELMAN, I. S., AND J. LEIBMAN. Anatomy of body water and electrolytes. *Am. J. Med.* 27: 256–277, 1959.

EDELMAN, I. S., J. LEIBMAN, M. P. O'MEARA, AND L. W. BIRKENFELD. Interrelations between serum sodium concentration, serum osmolarity and total exchangeable sodium, total exchangeable potassium and total body water. *J. Clin. Invest.* 37: 1236–1256, 1958.

ELKINS, B. R., AND A. S. WATANABE. Acute digoxin poisonings: review of therapy. *Am. J. Hosp. Pharm.* 35: 268–277, 1978.

ELLIOTT, M. E., AND T. L. GOODFRIEND. Inhibition of aldosterone synthesis by atrial natriuretic factor. *Fed. Proc.* 45: 2376–2381, 1986.

ELLISON, D. H., H. VELAZQUEZ, AND F. S. WRIGHT. Thiazide-sensitive sodium chloride cotransport in early distal tubule. *Am. J. Physiol.* 253: F546–F555, 1987.

FITZSIMONS, J. T. The physiological basis of thirst. *Kidney Int.* 10: 3–18, 1976.

FRALEY, D. S., AND S. ADLER. Correction of hyperkalemia by bicarbonate despite constant blood pH. *Kidney Int.* 12: 354–360, 1977.

FRAY, J. C. S., D. J. LUSH, AND C. S. PARK. Interrelationship of blood flow, juxtaglomerular cells, and hypertension: role of equilibrium and Ca^{++}. *Am. J. Physiol.* 251: R643–R662, 1986.

FRIIS-HANSEN, B. Changes in body water compartments during growth. *Acta Paediatr.* 46(Suppl. 110): 1–68, 1957.

GENNARI, F. J., AND J. J. COHEN. Role of the kidney in potassium homeostasis:

lessons from acid-base disturbances. *Kidney Int.* 8: 1–5, 1975.

GIEBISCH, G., AND B. STANTON. Potassium transport in the nephron. *Ann. Rev. Physiol.* 41: 241–256, 1979.

GOOD, D. W., AND F. S. WRIGHT. Luminal influences on potassium secretion: sodium concentration and fluid flow rate. *Am. J. Physiol.* 236: F192–F205, 1979.

GOTTSCHALK, C. W. Renal nerves and sodium excretion. *Ann. Rev. Phys.* 41: 229–240, 1979.

GOTTSCHALK, C. W., W. LASSITER, AND M. MYLLE. Localization of urine acidification in the mammalian kidney. *Am. J. Physiol.* 198: 581–585, 1960.

GRANGER, J.-P., J. C. BURNETT, J. C. ROMERO, T. J. OPGENORTH, J. SALAZAR, AND M. JOYCE. Elevated levels of atrial natriuretic peptide during aldosterone escape. *Am. J. Physiol.* 252: R878–R882, 1987.

GRANTHAM, J. J., AND M. B. BURG. Effect of vasopressin and cyclic AMP on permeability of isolated collecting tubules. *Am. J. Physiol.* 211: 255–259, 1966.

GREEN, R., AND G. GIEBISCH. Luminal hypotonicity: a driving force for fluid absorption from the proximal tubule. *Am. J. Physiol.* 246: F167–F174, 1984.

GREGER, R., AND E. SCHLATTER. Presence of luminal K$^+$, a prerequisite for active NaCl transport in the cortical thick ascending limb of Henle's loop of rabbit kidney. *Pflugers Arch.* 392: 92–94, 1981.

HAYS, R. M. Dynamics of body water and electrolytes. In: *Clinical Disorders of Fluid and Electrolyte Metabolism*, edited by M. H. Maxwell and C. R. Kleeman. New York: McGraw-Hill, 1980, p. 1–36.

HEBERT, S. C., AND T. E. ANDREOLI. The effects of ADH on salt and water transport in the mammalian nephron. In: *Physiology of Membrane Disorders*, 2nd ed., edited by T. E. Andreoli, J. F. Hoffman, D. D. Fanestil, and S. G. Schultz. New York: Plenum Medical, 1986, p. 701–711.

HOFFMANN, N., M. THEES, AND R. KINNE. Phosphate transport by isolated renal brush border vesicles. *Pflugers Arch.* 362: 147–156, 1976.

IMAI, M. The connecting tubule: a functional subdivision of the rabbit distal nephron segments. *Kidney Int.* 15: 346–356, 1979.

IMAI, M., AND J. KOKKO. Sodium-chloride, urea, and water transport in the thin ascending limb of Henle. Generation of osmotic gradients by passive diffusion of solutes. *J. Clin. Invest.* 53: 393–402, 1974.

JAMISON, R. L. Potassium recycling. *Kidney Int.* 31: 695–703, 1987.

JAMISON, R. L., C. M. BENNETT, AND R. W. BERLINER. Countercurrent multiplication by the thin loops of Henle. *Am. J. Physiol.* 212: 357–366, 1967.

JAMISON, R. L., AND W. KRIZ. *Urinary Concentrating Mechanism: Structure and Function.* Oxford: Oxford University Press, 1982.

JAMISON, R. L., H. SONNENBERG, AND J. H. STEIN. Questions and replies: role of the collecting tubule in fluid, sodium, and potassium balance. *Am. J. Physiol.* 237: F247–F261, 1979.

KANWAR, Y. S. Biophysiology of glomerular filtration and proteinuria. *Lab Invest.* 51: 7–21, 1984.

KARNISKI, L. P., AND P. S. ARONSON. Anion exchange pathways for Cl$^-$ transport in rabbit renal microvillus membranes. *Am. J. Physiol.* 253: F513–F521, 1987.

KATZ, A. I. Renal Na-K-ATPase: its role in tubular sodium and potassium transport. *Am. J. Physiol.* 242: F207–F219, 1982.

KESTELOOT, H., AND J. GEBOERS. Calcium and blood pressure. *Lancet* 1: 813–815, 1982.

KHURI, R. N., M. WIEDERHOLT, N. STRIEDER, AND G. GIEBISCH. Effects of graded solute diuresis on renal tubular sodium transport in the rat. *Am. J. Physiol.* 228: 1262–1268, 1975.

KIIL, F., K. AUKLAND, AND H. E. REFSUM. Renal sodium transport and oxygen consumption. *Am. J. Physiol.* 201: 511–516, 1961.

KNEPPER, M. A. Urea transport in nephron segments from medullary rays of rabbits. *Am. J. Physiol.* 244: F622–F627, 1983.

KNEPPER, M. A., AND J. L. STEPHENSON. Urinary concentrating and diluting processes. In: *Physiology of Membrane Disorders*, 2nd ed., edited by T. E. Andreoli, J. F. Hoffman, D. D. Fanestil, and S. G. Schultz. New York: Plenum Medical, 1986, p. 713–726.

KNOX, F. G., H. OSSWALD, G. R. MARCHAND, W. S. SPIELMAN, J. A. HAAS, T. BERNDT, AND S. P. YOUNGBERG. Phosphate transport along the nephron. *Am. J. Physiol.* 233: F261–F268, 1977.

KOKKO, J. Sodium chloride and water transport in the descending limb of Henle. *J. Clin. Invest.* 49: 1838–1846, 1970.

KOKKO, J. P., AND F. C. RECTOR, Jr. Countercurrent multiplication system without active transport in inner medulla. *Kidney Int.* 2: 214–233, 1972.

KRANE, S. M. Calcium, phosphate and magnesium. In: *International Encyclopedia of Pharmacology and Therapeutics*, edited by H. Rasmussen. London: Pergamon Press, 1970, p. 19–59.

KRISHNA, G. G., G. NEWELL, E. MILLER, P. HEEGER, R. SMITH, M. POLANSKY, S. KAPOOR, AND R. HOELDTKE. Protein-induced glomerular hyperfiltration: Role of hormonal factors. *Kidney Int.* 33: 578–583, 1988.

KRIZ, W., AND L. BANKIR. A standard nomenclature for structures of the kidney. *Kidney Int.* 33: 1–7, 1988.

LACOMBE, C., J.-L. DA SILVA, P. BRUNEVAL, J.-G. FOURNIER, F. WENDLING, N. CASADEVALL, J.-P. CAMILLERI, J. BARIETY, B. VARET, AND P. TAMBOURIN. Peritubular cells are the site of erythropoietin synthesis in the murine hypoxic kidney. *J. Clin. Invest.* 81: 620–623, 1988.

LANG, F. Renal handling of calcium and phosphate. *Klin. Wochenschr.* 58: 985–1003, 1980.

LASSITER, W. C., C. W. GOTTSCHALK, AND M. MYLLE. Micropuncture study of renal tubular reabsorption of calcium in normal rodents. *Am. J. Physiol.* 204: 771–775, 1963.

LeGRIMELLEC, C. Micropuncture study along the proximal convoluted tubule. *Pflugers Arch.* 354: 133–150, 1975.

LEIBMAN, J., AND I. S. EDELMAN. Interrelations of plasma potassium concentration, plasma sodium concentration, arterial pH and total exchangeable potassium. *J. Clin. Invest.* 38: 2176–2188, 1959.

LIU, F.-Y., AND M. S. COGAN. Angiotensin II: a potent regulator of acidification in the rat early proximal convoluted tubule. *J. Clin. Invest.* 80: 272–275, 1987.

LONGRIGG, J. N. The upper urinary tract. *Br. J. Clin. Pharmacol.* 13: 461–468, 1982.

LÖNNERHOLM, G. Histochemical demonstration of carbonic anhydrase activity in rat kidney. *Acta Physiol. Scand.* 81: 433–439, 1971.

LUDENS, J. H., AND D. D. FANESTIL. Aldosterone stimulation of acidification of urine by isolated urinary bladder of the Colombian toad. *Am. J. Physiol.* 226: 1321–1326, 1972.

MALNIC, G., AND G. GIEBISCH. Cellular aspects of renal tubular acidification. In: *Membrane Transport in Biology. IVA. Transport Organs*, edited by G. Giebisch, D. C. Tosteson, and H. H. Ussing. Berlin: Springer, 1979, p. 299–355.

MANERY, J. F., AND A. B. HASTINGS. The distribution of electrolytes in mammalian tissues. *J. Biol. Chem.* 127: 657–676, 1939.

MASORO, E. J. An overview of hydrogen ion regulation. *Arch. Intern. Med.* 142: 1019–1023, 1982.

MASSRY, S. G., AND J. W. COBURN. The hormonal and nonhormonal control of renal excretion of calcium and magnesium. *Nephron* 10: 66–112, 1973.

MOREL, F. Sites of hormonal action in the mammalian nephron. *Am. J. Physiol.* 240: F159–F164, 1981.

MOREL, F., AND A. DOUCET. Hormonal control of kidney functions at the cell level. *Physiol. Rev.* 66: 377–468, 1986.

MUDGE, G. H. Clinical patterns of tubular dysfunction. *Am. J. Med.* 24: 785–804, 1958.

MURER, H., AND R. KINNE. The use of isolated membrane vesicles to study epithelial transport processes. *J. Memb. Biol.* 55: 81–95, 1980.

MURER, H., U. HOPFER, AND R. KINNE. Sodium/proton antiport in brush-border-membrane vesicles isolated from rat small intestine and kidney. *Biochem. J.* 154: 597–604, 1976.

NARINS, R. G., AND M. EMMETT. Simple and mixed acid-base disorders: a practical approach. *Medicine* 59: 161–187, 1980.

NAVAR, L. G. Renal autoregulation: perspectives from whole kidney and single nephron studies. *Am. J. Physiol.* 234: F357–F370, 1978.

NAVAR, L. G., P. D. BELL, AND A. P. EVAN. The regulation of glomerular filtration rate in mammalian kidneys. In: *Physiology of Membrane Disorders*, 2nd ed., edited by T. E. Andreoli, J. F. Hoffman, D. D. Fanestil, and S. G. Schultz. New York: Plenum Medical, 1986, p. 637–667.

NETTER, F. H. *Kidneys, Ureters, and Urinary Bladder. CIBA Collection of Medical Illustrations.* Summit, NJ: CIBA Pharmaceutical Company, 1973.

OLIVER, J. *Nephrons and Kidneys.* New York: Harper and Row, 1968.

PETTY, K. J., J. P. KOKKO, AND D. MARVER. Secondary effect of aldosterone on Na-K ATPase activity in the rabbit cortical collecting tubule. *J. Clin. Invest.* 68: 1514–1521, 1981.

PITTS, R. F. The role of ammonia production and excretion in regulation of acid-base balance. *N. Engl. J. Med.* 284: 32–38, 1971.

PITTS, R. F. *Physiology of the Kidney and Body Fluids*, 3rd ed. Chicago: Year Book Medical Publishers, 1974.

POTTS, J. T., AND L. J. DEFTOS. Parathyroid hormone, calcitonin, vitamin D, bone and bone mineral metabolism. In: *Diseases of Metabolism*, edited by P. K. Bondy and L. E. Rosenberg. Philadelphia: W. B. Saunders, 1974, p. 1225–1430.

QUAMME, G. A. Renal handling of magnesium: drug and hormone interactions. *Magnesium* 5: 248–272, 1986.

RENNICK, B. R. Renal excretion of drugs: tubular transport and metabolism. *Ann. Rev. Pharmacol.* 12: 141–156, 1972.

ROBERTSON, G. L., R. L. SHELTON, AND S. ATHAR. The osmoregulation of vasopressin. *Kidney Int.* 10: 25–37, 1976.

ROCHA, A. S., AND J. P. KOKKO. Sodium chloride and water transport in the medullary thick ascending limb of Henle. Evidence for active chloride transport. *J. Clin. Invest.* 52: 612–623, 1973.

ROSA, R. M., P. SILVA, J. B. YOUNG, L. LANDSBERG, R. S. BROWN, J. W. ROWE, AND F. H. EPSTEIN. Adrenergic modulation of extrarenal potassium disposal. *N. Engl. J. Med.* 302: 431–434, 1980.

SCHIESS, W. A., J. L. AYER, W. D. LOTSPEICH, AND R. F. PITTS. The renal regulation of acid-base balance in man. II. Factors affecting the excretion of titratable acid by the normal human subject. *J. Clin. Invest.* 27: 57–64, 1948.

SCHMIDT-NIELSEN, B., AND R. O'DELL. Structure and concentrating mechanism in the mammalian kidney. *Am. J. Physiol.* 200: 1119–1124, 1961.

SCHNERMANN, J., AND J. P. BRIGGS. Participation of renal cortical prostaglandins in the regulation of glomerular filtration rate. *Kidney Int.* 19: 802–815, 1981.

SCHUSTER, V. L., AND J. B. STOKES. Chloride transport by the cortical and outer medullary collecting duct. *Am. J. Physiol.* 253: F203–F212, 1987.

SCHWARTZ, G. J., J. BARASCH, AND Q. AL-AWQATI. Plasticity of functional epithelial polarity. *Nature (Lond.)* 318: 368–371, 1985.

SCHWARTZ, G. J., G. B. HAYCOCK, C. M. EDELMAN, AND A. SPITZER. A simple estimate of glomerular filtration rate in children derived from body length and plasma creatinine. *Pediatrics* 58: 259–263, 1976.

SEELY, J. F., AND M. LEVY. Control of extracellular fluid volume. In: *The Kidney*, edited by B. M. Brenner and F. C. Rector, Jr. Philadelphia: W. B. Saunders, 1981, p. 371–407.

SHANNON, J. A. The renal excretion of phenol red by the aglomerular fishes *Opsanus tau* and *Lophius piscatorius. J. Cell Comp. Physiol.* 11: 315–323, 1938.

SHANNON, J. A., AND H. W. SMITH. The excretion of inulin, xylose and urea by normal and phlorizinized man. *J. Clin. Invest.* 14: 393–401, 1935.

SHAREGHI, G. R., AND L. C. STONER. Calcium transport across segments of the rabbit distal nephron *in vitro. Am. J. Physiol.* 235: F367–F375, 1978.

SHIPLEY, R. E., AND R. S. STUDY. Changes in renal blood flow, extraction of inulin, GFR, tissue pressure, and urine flow with acute alterations of renal artery pressure. *Am. J. Physiol.* 167: 676–688, 1951.

SILBERNAGL, S., V. GANAPATHY, AND F. H. LEIBACH. H^+ gradient-driven dipeptide reabsorption in proximal tubule of rat kidney. Studies in vivo and in vitro. *Am. J. Physiol.* 253: F448–F457, 1987.

SILVA, P., R. S. BROWN, AND F. H. EPSTEIN. Adaptation to potassium. *Kidney Int.* 11: 466–475, 1977.

SIMPSON, D. P. Control of hydrogen ion homeostasis and renal acidosis. *Medicine* 50: 503–541, 1971.

SKELTON, H. The storage of water by various tissues of the body. *Arch. Int Med.* 40: 140–152, 1927.

SMITH, H. W. *The Kidney.* New York: Oxford University Press, 1951.

SOLEIMANI, M., S. M. GRASSI, AND P. S. ARONSON. Stoichiometry of Na^+-HCO_3^- cotransport in basolateral membrane vesicles isolated from rabbit renal cortex. *J. Clin. Invest.* 79: 1276–1280, 1987.

SOMOGYI, A. New insights into the renal secretion of drugs. *Trends in Pharmacological Sciences* 8: 354–357, 1987.

STELLA, A., AND A. ZANCHETTI. Control of renin release. *Kidney Int.* 20: 889–894, 1987.

STOKES, J. B. Integrated actions of renal medullary prostaglandins in the control of water excretion. *Am. J. Physiol.* 240: F471–F480, 1981.

STOKES, J. B. Ion transport by the cortical and outer medullary collecting tubule. *Kidney Int.* 22: 473–484, 1982.

SUKI, W. N. Calcium transport in the nephron. *Am. J. Physiol.* 237: F1–F6, 1979.

SUKI, W. N., D. ROUSE, R. C. K. NG, AND J. P. KOKKO. Calcium transport in the thick ascending limb of Henle. *J. Clin. Invest.* 66: 1004–1009, 1980.

TUCKER, B. J., AND R. C. BLANTZ. An analysis of the determinants of nephron filtration rate. *Am. J. Physiol.* 232: F477–F483, 1977.

TURNER, R. J., AND M. SILVERMAN. Interaction of phlorizin and sodium with the renal brush-border membrane D-glucose transporter: stoichiometry and order of binding. *J. Memb. Biol.* 58: 43–55, 1981.

ULLRICH, K. J., G. CAPASSO, F. RUMRICH, F. PAPVASSILIOUS, AND S. KLÖSS. Coupling between proximal tubular transport processes. *Pflugers Arch.* 368: 245–252, 1977.

ULLRICH, K. J., K. KRAMER, AND J. W. BOYLAN. Present knowledge of the countercurrent system in the mammalian kidney. *Prog. Cardiovasc. Dis.* 3: 395–431, 1961.

VALTIN, H. Structural and functional heterogeneity of mammalian nephrons. *Am. J. Physiol.* 233: F491–F501, 1977.

VAN DE STOLPE, A., AND R. L. JAMISON. Micropuncture study of the effect of ANP on the papillary collecting duct in the rat. *Am. J. Physiol.* 254: F477–F483, 1988.

VAN SLYKE, D. D., A. B. HASTINGS, M. HEIDELBERGER, AND J. M. NEILL. Studies of gas and electrolyte equilibria in the blood. III. The alkali-binding and buffer values of oxyhemoglobin and reduced hemoglobin. *J. Biol. Chem.* 54: 481–506, 1922.

VENKATACHALAM, M. A. Anatomy and histology of the kidney. In: *Nephrology*, edited by J. H. Stein. New York: Grune & Stratton, 1980, p. 1–14.

WACKER, W. E. C., AND A. F. PARISI. Magnesium metabolism. *N. Engl. J. Med.* 278: 658–663, 1968.

WADE, J. B., W. A. KACHADORIAN, AND V. A. DISCALA. Freeze-fracture electron microscopy: relationship of membrane structural features to transport physiology. *Am. J. Physiol.* 232: F77–F83, 1977.

WALSER, M. Ion association. VI. Interactions between calcium, magnesium, inorganic phosphate, citrate, and protein in normal human plasma. *J. Clin. Invest.* 40: 723–730, 1961a.

WALSER, M. Calcium clearance as a function of sodium clearance in the dog. *Am. J. Physiol.* 200: 1099–1104, 1961b.

WALSER, M. Divalent cations: physicochemical state in glomerular filtrate and urine and renal excretion. In: *Handbook of Physiology, Sect. 8: Renal Physiology*, edited by J. Orloff and R. W. Berliner. Bethesda, MD: American Physiological Society, 1973, p. 555–586.

WALSER, M. Phenomenological analysis of renal regulation of sodium and potassium balance. *Kidney Int.* 27: 837–841, 1985.

WARNOCK, D. G., AND F. C. RECTOR, JR. Proton secretion by the kidney. *Ann. Rev. Physiol.* 41: 197–210, 1979.

WEINER, I. M., AND G. H. MUDGE. Renal tubular mechanisms for excretion of organic acids and bases. *Am. J. Med.* 36: 743–762, 1964.

WEITZMAN, R., AND C. R. KLEEMAN. Water metabolism and the neurohypophyseal hormones. In: *Clinical Disorders of Fluid and Electrolyte Metabolism*, edited by M. H. Maxwell and C. R. Kleeman. New York: McGraw-Hill, 1980, p. 531–645.

WELLING, D. J., AND L. W. WELLING. Cell shape as an indicator of volume reabsorption in proximal nephron. *Fed. Proc.* 38: 121–127, 1979.

WOODBURNE, R. T. Anatomy of the ureterovesical junction. *J. Urol.* 92: 431–435, 1964.

WRIGHT, F. S., AND G. GIEBISCH. Renal potassium transport: contributions of individual nephron segments and populations. *Am. J. Physiol.* 235: F515–F527, 1978.

WRIGHT, F. S., AND J. S. BRIGGS. Feedback control of glomerular blood flow, pressure, and filtration rate. *Physiol. Rev.* 59: 958–1006, 1979.

WRIGHT, S. H., I. KIPPEN, AND E. M. WRIGHT. Stoichiometry of Na^+-succinate cotransport in renal brush-border membranes. *J. Biol. Chem.* 257: 1773–1778, 1982.

ZERBE, R. L., AND G. L. ROBERTSON. Osmotic and nonosmotic regulation of thirst and vasopressin secretion. In: *Clinical Disorders of Fluid and Electrolyte Metabolism*, 4th ed., edited by M. H. Maxwell, C. R. Kleeman, and R. G. Narins. New York: McGraw-Hill, 1987, p. 61–78.

ZIEGLER, T. W., D. D. FANESTIL, AND J. H. LUDENS. Influence of transepithelial potential difference on acidification in the toad urinary bladder. *Kidney Int.* 10: 279–286, 1976.

SECTION 5

Respiration

CHAPTER 35

Structure-Function Relationships of the Lung; Ventilation

SCOPE OF RESPIRATORY PHYSIOLOGY

Human beings and other higher animals remove oxygen from the air and add carbon dioxide to it in the course of satisfying the metabolic demands of their tissues. The business of getting oxygen from the atmosphere into the cells and carbon dioxide out, that is, *gas exchange*, is the essence of respiratory physiology.

We can identify several links in the chain of processes involved in gas exchange (Fig. 5.1):

1. Ventilation—the process of moving oxygen from the air into the alveoli in the depths of the lung (and carbon dioxide in the opposite direction)
2. Diffusion—movement of gases across the gas-blood barrier
3. Matching of ventilation and blood flow—not easily shown on the diagram but critical for efficient gas exchange
4. Pulmonary blood flow—to move the gases out of the lung
5. Blood gas transport—carriage of oxygen and carbon dioxide in the blood
6. Transfer of gases between the peripheral capillaries and the cells
7. Utilization of oxygen (and production of carbon dioxide) in the cells

Six other topics should also be included in the domain of respiratory physiology:

8. Structure-function relationships of the lung
9. Lung mechanics—that is, the forces involved in supporting and moving the lung and chest wall
10. Control of ventilation—the mechanism which regulates the gas exchange function of the lung
11. Metabolic functions of the lung
12. Respiration in unusual environments—including high altitude, diving, space, and other situations were special problems are encountered
13. Tests of lung function—these are important in assessing lung disease

The organization of Section 5 is as follows: Chapter 35 begins with an introductory section on structure-function relationships of the lung (no. 8, above), the physical gas laws, and the language of respiratory physiology. It then deals with ventilation (no. 1). Chapter 36 covers pulmonary blood flow and metabolism (nos. 4 and 11), while Chapter 37 is devoted to gas transport to the periphery (nos. 5 and 6). Chapter 38 addresses the central topic of pulmonary gas exchange including diffusion (no. 2) and ventilation-perfusion relationships (no. 3). Chapters 39, 40, and 41 discuss the mechanics of breathing (no. 9), control of ventilation (no. 10), and respiration in unusual environments (no. 12), respectively. Intracellular oxidation (no. 7) is referred to in Chapters 37 and 1 but for details the reader should consult a textbook of biochemistry. Tests of lung function are alluded to in various places, but for a full account see other texts such as Cotes (1979) or West (1987).

STRUCTURE-FUNCTION RELATIONSHIPS OF THE LUNG

Throughout the body, the function of an organ is reflected in its structure; this is particularly true of the lung. Since the essential function of the lung is gas exchange, let us start with the *blood-gas barrier*, which separates the blood in the pulmonary capillary from the alveolar gas (Figs. 5.1 and 5.2). The barrier is composed of an alveolar epithelial cell, interstitial layer, and capillary endothelial cell. In addition, there is a layer of phospholipid (surfactant) on the epithelial cell not shown in this preparation (see Chapter 39). The barrier is less than half a micrometer thick in places. Calculations indicate that the surface area is between 50 and 100 m^2 so that its structure is well suited to the business of gas exchange by diffusion through it.

Air is brought to one side of the barrier by ventilation, and blood to the other by the pulmonary circulation (Fig. 5.1). The *airways* or bronchi start at the trachea and form an intricate branching system

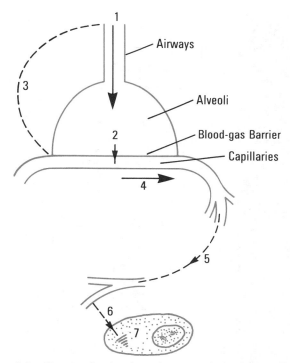

Figure 5.1. Diagram showing the various steps in the delivery of oxygen from air to tissues. *1,* ventilation; *2,* diffusion across the blood-gas barrier; *3,* matching of ventilation and blood flow; *4,* pulmonary blood flow; *5,* transport of gas in the blood; *6,* diffusion from capillary to cell; *7,* utilization of O_2 by mitochondria.

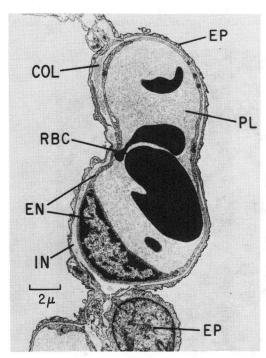

Figure 5.2. Electron micrograph showing a pulmonary capillary in the alveolar wall. Note the extremely thin blood-gas barrier less than 0.5 μm thick. *EP,* alveolar epithelium; *IN,* interstitial tissue; *EN,* capillary endothelium; *PL,* plasma; *RBC,* red blood cell; *COL,* collagen. (From Weibel, 1973.)

(Fig. 5.3*A*) which leads gas into the depths of the lung. The number and size of the airways in a typical human lung were counted by the Swiss anatomist, Weibel, and his idealization is shown in Figure 5.3*B* (Weibel, 1963). Although it is an oversimplification, this model has been of great value in respiratory physiology.

Figure 5.3*B* shows that, on the average, the airways branch approximately 16 times before any alveoli with their thin blood-gas barrier (Fig. 5.2) are seen. Since no gas exchange can occur in these generations, they compose the *anatomic dead space.* Its volume is approximately 150 ml.* The smallest airways of this conducting zone are known as the *terminal bronchioles.* Beyond these lie airways with a few alveoli in their walls, the *respiratory bronchioles.* These lead to *alveolar ducts,* whose walls are comprised entirely of alveoli, and finally to the blindly ending *alveolar sacs.* All gas exchange occurs in the alveolated portion of lung, which is known as the respiratory zone. Its volume is about 3 liters.

Blood is brought to the other side of the blood-gas barrier (Fig. 5.1) from the right heart by the pulmonary arteries, which lead to the pulmonary capillaries (Fig. 5.4*A*). The capillaries lie in the walls of the alveoli and, seen from the vantage point of the alveolar space, form a dense network of interconnecting vessels (Fig. 5.4*B*). So dense is the network that the blood forms an almost continuous sheet in the alveolar wall. At normal capillary pressures, not all the bed is open, but recruitment of closed capillaries can occur if the pressure rises (for example, on exercise). When all of the capillaries are open, over 80% of the area of the alveolar wall is apparently available for gas exchange.

The lung has a second blood supply, the bronchial circulation via the bronchial arteries, which arise from the aorta. This flow is only about a hundredth of that in the pulmonary circulation, and its main purpose is to supply the walls of the large airways. There is also a small lymph flow from the lung. The lymphatics run chiefly around the larger airways and blood vessels.

The alveolar tissue (parenchyma) contains fibers of elastin and collagen, and the fluid lining the alveoli has surface tension. As a result the lung is elastic and is held expanded by keeping the pressure around it (intrapleural pressure) lower than alveolar pressure. During inspiration the diaphragm contracts and moves downward and the intercostal

---

*These figures for lung volumes, ventilation, etc., are intended only to give the reader a general idea. The values depend on age, sex, height, and weight, and appropriate tables have been published (Cotes, 1979).

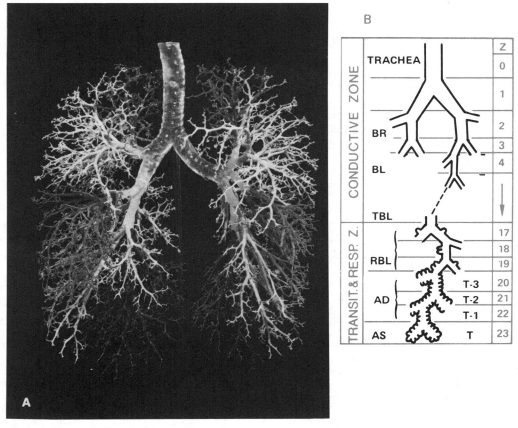

Figure 5.3. *A,* Cast of the airways of a human lung. *B,* Idealization of the human airways according to Weibel: *Z,* generation number; *BR,* bronchus; *BL,* bronchiole; *TBL,* terminal bronchiole; *RBL,* respiratory bronchiole; *AD,* alveolar duct; *AS,* alveolar sac. (From Weibel, 1963.)

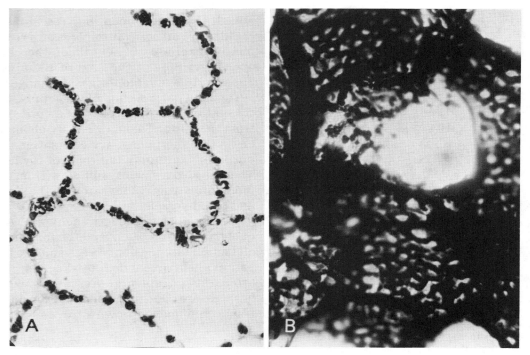

Figure 5.4. Microscopic appearances of capillaries in the alveolar walls of canine lung. *A* shows a thin cross-section; *B* shows the appearance of the network of capillaries in a thick section when the alveolar wall is seen face-on. (*A* from Glazier et al., 1969.)

520

muscles increase the diameter of the rib cage. As a result the intrapleural pressure falls further and the lung expands. Expiration is normally passive, and the lung returns to its resting position because of its elasticity. This topic is dealt with in more detail in Chapter 39.

BEHAVIOR OF GASES

Before we turn to the movement of gas into the lung (ventilation), a brief review of the physical laws of gases may be useful.

Kinetic Theory of Gases

The molecules of a gas are in continuous random motion and are deflected from their course only by collision with other molecules or with the walls of a container. When they strike the walls and rebound, the resulting bombardment results in a pressure. The magnitude of the pressure depends on the number of molecules present, their mass, and their speed.

BOYLE'S LAW

At constant temperature, the pressure (P) of a given mass of gas is inversely proportional to its volume (V), or

$$P_1 V_1 = P_2 V_2 \text{ (temperature constant)}$$

This can be explained by the fact that, as the molecules are brought closer together (smaller volume), the rate of bombardment on a unit surface increases (greater pressure).

CHARLES'S LAW

At constant pressure, the volume is proportional to the absolute temperature (T), or

$$\frac{V_1}{V_2} = \frac{T_1}{T_2} \text{ (pressure constant)}$$

The explanation is that a rise in temperature increases the speed and momentum of the molecules, thus increasing the force of bombardment on the container. Another form of Charles's law states that, at constant volume, the pressure is proportional to absolute temperature. (Note that absolute (Kelvin) temperature is obtained by adding 273 to the Celsius temperature. Thus 37°C = 310° absolute.)

AVOGADRO'S LAW

Equal volumes of different gases at the same temperature and pressure contain the same number of molecules. Thus a gram molecule (for example, 32 g of oxygen) occupies 22.4 liters at STPD (standard temperature (0°C) and pressure (760 mm Hg), dry).

IDEAL GAS LAW

This combines the above three laws thus:

$$PV = nRT$$

where n is the number of gram molecules of the gas and R is the "gas constant." When the units employed are millimeters of mercury, liters, and degrees absolute, then $R = 62.4$. Real gases deviate from ideal gas behavior under certain conditions.

DALTON'S LAW

Each gas in a mixture exerts a pressure according to its own concentration, independently of the other gases present. That is, each component behaves as though it were present alone. The pressure of each gas is referred to as its *partial pressure* or tension. The total pressure is the sum of the partial pressures of all gases present. In symbols,

$$P_x = P \cdot F_x$$

where P_x is the partial pressure of gas x, P is the total pressure, and F_x is the fractional concentration of gas x (for example, if half of the gas is oxygen, $F_{O_2} = 0.5$). Conventionally the fractional concentration always refers to dry gas. Thus, in gas which is saturated with vapor at 37°C where the water vapor pressure is 47 mm Hg,

$$P_x = (P_B - 47) \cdot F_x$$

where P_B is the total (barometric) pressure and $(P_B - 47)$ is the dry gas pressure.

HENRY'S LAW

The volume of gas dissolved in a liquid is proportional to its partial pressure. Thus:

$$C_x = K \cdot P_x$$

where C is the concentration of gas in the liquid, for example, in milliliters of gas per 100 ml of blood, and K is a solubility constant whose value depends on the units of C and P. Tables of solubility constants are available (Altman, 1961). The partial pressure of a gas in solution is best defined as its partial pressure in a gas which is in equilibrium with that solution.

DIFFUSION IN THE GAS PHASE

Because of their random motion, gas molecules tend to distribute themselves uniformly throughout any available space until the partial pressure is everywhere the same. The process by which gas moves from a region of high to low partial pressure is known as diffusion. Light gases diffuse faster than heavy gases because the mean velocity of the molecules is higher. In fact the kinetic theory of gases states that the kinetic energy ($0.5 \, mv^2$, where m is mass and v is velocity) of all gases is the same (at a given temperature and pressure). From this it follows that the rate of diffusion of a gas is inversely proportional to the square root of its density *(Graham's Law)*. Thus hydrogen (MW = 2) diffuses 4 times as rapidly as oxygen (MW = 32).

DIFFUSION OF DISSOLVED GASES

Gas moves from a region of high to low partial pressure in a liquid by diffusion, just as it does in the gas phase. Again the rate is inversely proportional to the square root of the molecular weight of the gas. In addition, the amount of gas that moves through a film of liquid (or a membrane) is proportional to the solubility of the gas (see Chapter 38 and Fig. 5.37).

LANGUAGE OF RESPIRATORY PHYSIOLOGY

Symbols and equations will be used sparingly in this section because many students are discouraged by them. However, some kind of shorthand is essential if the quantities are to be manipulated algebraically. Fortunately there is general agreement on most of the symbols used, and these are shown in Table 5.1.

Table 5.1.
Some Symbols Used in Respiratory Physiology

Primary symbols
    C Concentration of gas in blood
    F Fractional concentration in dry gas
    P Pressure or partial pressure
    Q Volume of blood
    $\dot{Q}$ Volume of blood per unit time (flow)
    R Respiratory exchange ratio
    S Saturation of hemoglobin with O_2
    V Volume of gas
    $\dot{V}$ Volume of gas per unit time (flow)
Secondary symbols for gas phase
    A Alveolar
    B Barometric
    D Dead space
    E Expired
    I Inspired
    L Lung
    T Tidal
Secondary symbols for blood phase
    a Arterial
    c Capillary
    c' End-capillary
    i Ideal
    v Venous
    $\bar{v}$ Mixed venous
Examples
    O_2 concentration in arterial blood Ca_{O_2}
    Fractional concentration of N_2 in expired gas $F_{E_{N2}}$
    Partial pressure of O_2 in mixed venous blood $P\bar{v}_{O_2}$

VENTILATION

Ventilation is the business of getting gas to and from the alveoli. Figure 5.5 is a highly simplified diagram of the lung to show the volumes and flows involved. The various bronchi which make up the conducting airways (Fig. 5.3) are now represented by a single tube labeled anatomic dead space. This leads into the gas exchanging region of the lung which is bounded by the blood-gas interface and the pulmonary capillary blood. With each inspiration, about 500 ml of air enter the lung (tidal volume). Note how small a proportion of the total lung volume is represented by the anatomic dead space. Also note the very small volume of capillary blood compared with alveolar gas (compare Fig. 5.4*A*). In this chapter we shall not consider the forces exerted by the lung and chest wall during ventilation; this subject is taken up in Chapter 39.

Lung Volumes

Before looking in detail at the movement of gas in the lung, it is useful to consider the static volumes of the lung. At one time a great deal of emphasis was placed on these measurements because they were thought to be valuable aids to the diagnosis of lung disease. Less attention is paid to them now, but they are still frequently determined.

SIMPLE SPIROMETRY

A spirometer is a light bell-shaped container which is immersed in a water tank to form a seal (Fig. 5.6). Although it was invented as early as 1846 by Hutchinson to measure lung volumes in various groups of people in London (including 121 sailors, 24 "pugilists and wrestlers," and 4 "giants and dwarfs"!) it is still used today. Modern modifications include dry spirometers made with a bellows or a piston in a large cylinder. These often have an electrical output.

Figure 5.6 shows that, as the bell moves up during

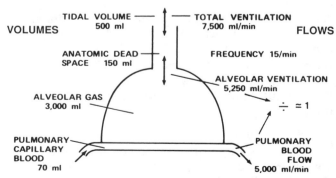

Figure 5.5. Diagram of a lung showing typical volumes and flows. There is considerable variation around these values. (From West, 1985b.)

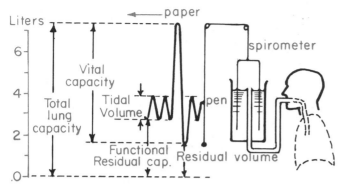

Figure 5.6. Lung volumes. Note that the functional residual capacity and residual volume cannot be measured with the spirometer.

exhalation, the pen moves down marking the chart. In the tracing shown, the subject first breathed normally and the excursion of the pen gave the *tidal volume.* Next he made a maximal inspiration followed by a maximal expiration. The exhaled volume is called the *vital capacity.* Normal breathing was then resumed.

Some gas remained in the lung after the maximal expiration; this is the *residual volume.* The volume of gas in the lung after a normal expiration is the *functional residual capacity* (FRC). Neither this nor the residual volume can be obtained by simple spirometry.

Additional terms are sometimes used. The maximum volume which can be inhaled from FRC is the *inspiratory capacity.* Subtracting the tidal volume from this gives the *inspiratory reserve volume.* The difference between the FRC and the residual volume is sometimes called the *expiratory reserve volume.*

MEASUREMENT OF FUNCTIONAL RESIDUAL CAPACITY

As stated above, the FRC and residual volume cannot be measured by simple spirometry because the lung cannot be emptied completely by a maximal exhalation. One method of determining the FRC is by helium dilution in a "closed circuit," that is, a system which does not allow gas to escape. The subject is connected to a spirometer containing a known concentration of helium, which is almost insoluble in blood (Fig. 5.7A). After some breaths, the helium concentration in the spirometer and that in the lung become the same (Fig. 5.7B). Since a negligible amount of helim has been lost, the amount of helium present before equilibration (concentration × volume) is $C_1 \times V_1$ and equals the amount after equilibration, $C_2 \times (V_1 + V_2)$. Thus:

$$C_1 \times V_1 = C_2(V_1 + V_2)$$
$$V_2 = \frac{V_1(C_1 - C_2)}{C_2}$$

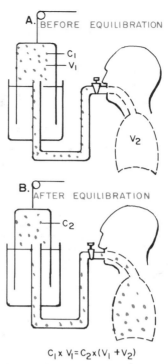

$$C_1 \times V_1 = C_2 \times (V_1 + V_2)$$

Figure 5.7. Measurement of the functional residual capacity by helium dilution.

In practice, the CO_2 produced by the subject is absorbed during equilibration, and oxygen is added to keep the spirometer volume constant.

The functional residual capacity can also be obtained by "open circuit" nitrogen washout. The subject breathes pure oxygen from a valve box which separates inspired and expired gas, and the expired gas is collected in a large spirometer. After about 7 min almost all of the nitrogen has been washed out of a normal lung. The nitrogen concentration in the spirometer is then determined, and the volume of exhaled nitrogen is calculated. Knowing that the concentration in the lung before the washout was 80%, the lung volume can be computed. This method has the disadvantage that a very accurate measurement of nitrogen in the spirometer is required. In addition, some nitrogen is washed out of body tissues, and patients with diseased lungs may not wash out all of the nitrogen from their lungs. Thus errors are introduced.

A very different method of measuring FRC is with a body plethysmograph (Dubois et al., 1956). This is a large airtight box like a telephone booth in which the subject sits (Fig. 5.8). The pressure inside the box can be measured very accurately. At the end of a normal expiration, a shutter closes the mouthpiece and the subject is asked to make respiratory efforts. As he tries to inhale, he expands the gas in his lungs, lung volume increases, and the box pres-

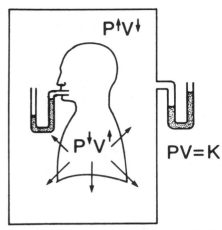

Figure 5.8. Measurement of lung volume with a body plethysmograph. When the subject makes an inspiratory effort against a closed airway, he slightly increases the volume of his lung, airway pressure decreases, and box pressure rises. From Boyle's law, lung volume is obtained (see text).

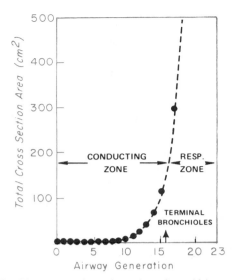

Figure 5.9. Diagram to show the extremely rapid increase in total cross-sectional area of the airways in the respiratory zone (compare Fig. 5.3). As a result, the forward velocity of the gas during inspiration becomes very small in the region of the respiratory bronchioles, and gaseous diffusion becomes the dominant mechanism of ventilation.

sure rises since its gas volume decreases. Boyle's law states that pressure × volume is constant (at constant temperature). Therefore, if the pressures in the box before and after the inspiratory effort are P_1 and P_2, respectively; V_1 is the preinspiratory box volume; and ΔV is the change in volume of the box (or lung), we can write $P_1 V_1 = P_2(V_1 - \Delta V)$. Thus ΔV can be obtained.

Next, Boyle's law is applied to the gas in the lung. Now $P_3 V_2 = P_4(V_2 + \Delta V)$, where P_3 and P_4 are the mouth pressures before and after the inspiratory effort and V_2 is the FRC. Thus, FRC can be obtained.

The body plethysmograph and the gas dilution (or washout) method may measure different volumes. The body plethysmograph measures the total volume of gas in the lungs, including any which is trapped behind closed airways (for example, Fig. 5.73) and which therefore does not communicate with the mouth. By contrast, the helium dilution and nitrogen washout methods measure only communicating gas or ventilated lung volume. In young normal subjects these volumes are virtually the same, but in patients with lung disease the ventilated volume may be considerably less than the total volume because of gas trapped behind obstructed airways.

Bulk Flow and Diffusion in the Airways

Figure 5.3 shows that the system of airways through which ventilation occurs branches successively at each generation. Using Weibel's data it is possible to calculate the cross-sectional area of each airway generation, and this is shown in Figure 5.9. Generation 0 is the trachea, with a cross-sectional area of about 2.5 cm². Next comes generation 1, com-

prised of the right and left main bronchi which have a total cross-section of 2.3 cm², and so on. The area changes relatively slowly up to the region of the terminal bronchioles (the end of the conducting airways) but then increases very rapidly. In other words, the combined airways have a shape something like a trumpet which is flared at the end.

This geometry has an important influence on the mode of airflow (Cumming et al., 1966). In the airways down to the terminal bronchioles, gas moves predominantly by bulk flow or convection, like water through a garden hose. However, since the same volume of gas traverses each generation, Figure 5.9 implies that the forward velocity of the inspired gas rapidly decreases as the gas enters the respiratory zone. In fact the velocity becomes so small that another mode of flow takes over as the dominant one. This is gaseous diffusion due to the random movement of the molecules. The rate of diffusion of the molecules is so rapid and the distance to be covered is so short (only a few millimeters) that differences in concentration along the terminal airways are virtually abolished within a second. However, inhaled dust particles diffuse very slowly because of their relatively large size, and they therefore tend not to reach the alveoli, but to settle out in the region of the respiratory bronchioles.

Total and Alveolar Ventilation

Suppose the volume exhaled with each breath is 500 ml (Fig. 5.5) and there are 15 breaths/min. Then the total volume leaving the lung each minute is 500

× 15 = 7500 ml/min. This is known as the *total ventilation* or *minute volume*. The volume of air entering the lung is slightly greater because more oxygen is taken in than carbon dioxide is given out.

However, not all the air that passes the lips reaches the alveolar gas compartment where gas exchange occurs. Of each 500 ml inhaled in Figure 5.5, 150 ml remains behind in the anatomic dead space. Thus the volume of fresh gas entering the respiratory zone each minute is (500 − 150) × 15 or 5250 ml/min. This is called the *alveolar ventilation* and is of key importance because it represents the amount of fresh inspired air available for gas exchange. (Strictly, the alveolar ventilation is also measured on expiration, but the volume is almost the same.)

The total ventilation can easily be measured by having the subject breathe through a valve box and collecting all of the expired gas in a large bag (sometimes called a Douglas bag). The alveolar ventilation, however, is more difficult to determine.

MEASUREMENT OF ALVEOLAR VENTILATION

One way to determine the alveolar ventilation is to measure the volume of the anatomic dead space (see below) and calculate the dead space ventilation (volume × respiratory frequency). This is then subtracted from the total ventilation.

This process can be summarized using the symbols of Table 5.1 (also compare Fig. 5.10). If V denotes volume and the subscripts T, D, and A refer to tidal, dead space, and alveolar, respectively,† then

$$V_T = V_D + V_A$$

Multiplying each term by n, the respiratory frequency, gives

$$V_T \cdot n = V_D \cdot n + V_A \cdot n$$

whence

$$\dot{V}_E = \dot{V}_D + \dot{V}_A$$

where $\dot{V}$ means volume per unit time, $\dot{V}_E$ is total expired ventilation, and $\dot{V}_D$ and $\dot{V}_A$ are the dead space and alveolar ventilations, respectively. Thus

$$\dot{V}_A = \dot{V}_E - \dot{V}_D$$

A difficulty with this method is that the anatomic dead space is not easy to measure, although a value

†Note that V_A here means the volume of alveolar gas in the tidal volume, not the total volume of alveolar gas in the lung.

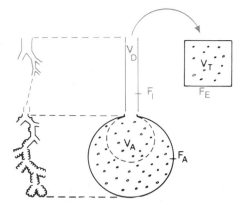

Figure 5.10. The tidal volume V_T is a mixture of gas from the anatomic dead space V_D and a contribution from the alveolar gas V_A. The concentrations of CO_2 are shown by the *dots; F*, fractional concentration; *I*, inspired; *E*, expired. Compare Figure 5.3B. (Adapted from Piiper.)

for it can be assumed with little error. A simple approximation for seated subjects is that the dead space in milliliters is equal to the subject's weight in pounds.

Another way of measuring alveolar ventilation in normal subjects is from the concentration of CO_2 in expired gas (Fig. 5.10). Since no gas exchange occurs in the anatomic dead space, there is no CO_2 there at the end of inspiration. (The extremely small amount of CO_2 in the atmosphere is neglected.) Thus, since all expired CO_2 comes from the alveolar gas,

$$\dot{V}_{CO_2} = \dot{V}_A \cdot F_{A_{CO_2}}$$

where $\dot{V}_{CO_2}$ means the volume of CO_2 exhaled per unit time and $F_{A_{CO_2}}$ is the fractional concentration of CO_2 in alveolar gas. (In other words if the concentration of CO_2 is 5%, the fractional concentration is 0.05). Therefore,

$$\dot{V}_A = \frac{\dot{V}_{CO_2}}{F_{A_{CO_2}}}$$

Thus the alveolar ventilation can be obtained by dividing the CO_2 output by the alveolar concentration of this gas. The CO_2 output is derived by collecting expired gas in a bag and analyzing it for CO_2. The alveolar concentration can be determined by means of a rapid CO_2 analyzer which samples gas in a mouthpiece through which the subject breathes. The last portion of a single expiration is representative of alveolar gas (see discussion of Fig. 5.11).

The partial pressure of CO_2 (denoted P_{CO_2}) is proportional to the concentration of the gas in the alveoli, or

$$P_{A_{CO_2}} = F_{A_{CO_2}} \cdot K$$

where K is a constant. Therefore,

$$\dot{V}_A = \frac{\dot{V}_{CO_2}}{P_{A_{CO_2}}} \cdot K$$

Since the P_{CO_2} of alveolar gas and arterial blood are almost identical in normal subjects,

$$\dot{V}_A = \frac{\dot{V}_{CO_2}}{Pa_{CO_2}} \cdot K$$

and thus the arterial P_{CO_2} can be used to determine alveolar ventilation. If the alveolar ventilation is halved (and the CO_2 production remains unchanged, as it will do under steady state conditions), the alveolar and arterial P_{CO_2} will double.

Anatomic Dead Space

We have seen that the anatomic dead space is the volume of the conducting airways (Figs. 5.3, 5.5, and 5.10). The normal value is in the region of 150 ml, and it increases with large inspirations because of the traction exerted on the bronchi by the surrounding lung parencyhma. The dead space also depends on the size and posture of the subject.

The volume of the dead space can be measured by *Fowler's method* (Fowler, 1948). The subject breathes through a valve box, and the sampling tube of a rapid nitrogen analyzer continuously samples gas at the lips (Fig. 5.11). During air breathing, the concentration of nitrogen changes little between inspiration and expiration. To perform the test, the

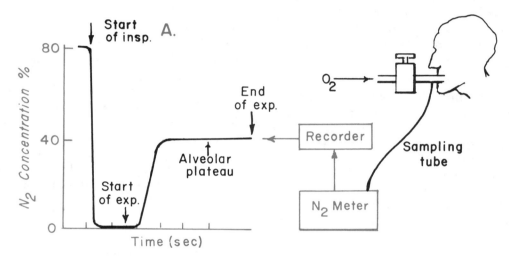

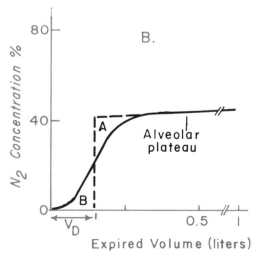

Figure 5.11. Fowler's method of measuring the anatomic dead space with a rapid N_2 analyzer. *A* shows that, following a test inspiration of 100% oxygen, the nitrogen concentration rises during expiration to an almost level "plateau" representing pure alveolar gas. *B* shows a plot of nitrogen concentration against expired volume. The dead space is the expired volume up to the *vertical broken line* which makes the areas *A* and *B* equal.

subject takes a single inspiration of 100% oxygen and then exhales fully. The nitrogen concentration changes that occur are shown in Figure 5.11A. During inspiration, the nitrogen concentration falls to zero. On expiration, first pure dead space (100% oxygen) is sampled, and then the nitrogen concentration rises as the dead space is increasingly washed out by alveolar gas. Finally an almost uniform gas concentration is seen, representing pure alveolar gas. This phase is often called the "alveolar plateau," although in normal subjects it is not quite flat and in patients with lung disease it may rise steeply.

Expired volume is recorded at the same time, and then the nitrogen concentration is plotted against this. (Alternatively nitrogen concentration and volume can be recorded together on an XY recorder). The dead space is found by drawing a vertical line such that A is equal to area B on Figure 5.11B. The dead space is then the volume exhaled up to the vertical line. In effect this method measures the volume of the conducting airways down to the midpoint of the transition from dead space to alveolar gas. The volume of the anatomic dead space is not frequently measured, partly because the dead space changes little in disease and partly because measurement requires a very rapid N_2 analyzer.

Physiological Dead Space

Another method of measuring dead space was described by Bohr (1891). Figure 5.10 shows that all expired CO_2 comes from the alveolar gas and none from the dead space. Therefore,

$$V_T \cdot F_{E_{CO_2}} = V_A \cdot F_{A_{CO_2}}$$

where V_A indicates the component of the tidal volume which comes from the dead space.

But, as stated earlier,

$$V_T = V_A + V_D$$

Therefore

$$V_A = V_T - V_D$$

and by substitution

$$V_T \cdot F_{E_{CO_2}} = (V_T - V_D) \cdot F_{A_{CO_2}}$$

whence

$$\frac{V_D}{V_T} = \frac{F_{A_{CO_2}} - F_{E_{CO_2}}}{F_{A_{CO_2}}}$$

Because the partial pressure of a gas is proportional to its concentration,

$$\frac{V_D}{V_T} = \frac{P_{A_{CO_2}} - P_{E_{CO_2}}}{P_{A_{CO_2}}} \text{ (Bohr equation)}$$

In normal subjects the P_{CO_2} in alveolar gas and arterial blood are almost identical so that the equation is often written

$$\frac{V_D}{V_T} = \frac{Pa_{CO_2} - P_{E_{CO_2}}}{Pa_{CO_2}}$$

The normal ratio of dead space to tidal volume is in the range of 0.2–0.35 during resting breathing. The ratio decreases on exercise but increases with age.

Fowler's and Bohr's methods measure somewhat different things. Fowler's method measures the volume of the conducting airways down to the level where the rapid dilution of inspired gas occurs with gas already in the lung. This volume is determined by the geometry of the rapidly expanding airways (Figs. 5.3 and 5.9) and, because it reflects the morphology of the lung, it is called the anatomic dead space. Bohr's method measures the volume of the lung which does not eliminate CO_2. Because this is a functional measurement, the volume is called the physiological dead space. In normal subjects the volumes are very nearly the same. In patients with lung disease, however, the physiological dead space may be considerably larger because of inequality of blood flow and ventilation within the lung (Chapter 38).

Analysis of Respiratory Gases

CHEMICAL ABSORPTION METHODS

The original methods of measuring the composition of respiratory gases depended on absorption of the carbon dioxide and oxygen and the corresponding changes in volume of the gas (at constant pressure). In *Haldane's method* (Peters and Van Slyke, 1932), 10 ml of gas is drawn into a gas burette, and the CO_2 is absorbed by adding potassium hydroxide and shaking the mixture. After recording the change in volume, potassium pyrogallate or some other oxygen absorber is added, and the change in volume is again determined. The remainder of the gas is then nitrogen (plus argon which is included with nitrogen in respiratory studies). In *Scholander's micromethod* (1947), only 0.5 ml of gas is required, and the volume changes are measured in a burette fitted with a micrometer gauge.

PHYSICAL METHODS

The above methods are still sometimes used for calibration but have been replaced by physical methods for most purposes. The *respiratory mass spectrometer* separates the components of a gas mixture according to their mass by accelerating the ions in electric and magnetic fields. The results of the analysis are immediately available on a pen recorder, and the time of analysis is typically only 0.1 s. These analyzers are expensive but very convenient.

Infrared analyzers depend on the absorption of infrared radiation from the test gas. They are extensively employed for CO_2 and CO. *Thermal conductivity analyzers* measure the rate at which heat is lost from a hot wire immersed in the test gas. This principle is particularly suitable for helium. *Radiation emission* can be used to measure nitrogen when the gas is electrically excited (Fig. 5.11). Oxygen is sometimes measured by a *paramagnetic analyzer* or a *polaragraphic electrode*. Finally *gas chromatography* is being increasingly used for the analysis of individual gas samples.

CHAPTER 36

Pulmonary Blood Flow and Metabolism

PULMONARY CIRCULATION

Just as ventilation brings oxygen to the blood-gas barrier where gas exchange occurs, so the pulmonary circulation picks up the oxygen in the alveoli and delivers it to the left heart from where it is distributed to the rest of the body.

The pulmonary circulation begins at the main pulmonary artery, which receives the mixed venous blood pumped by the right ventricle. This artery then branches successively like the system of airways (Fig. 5.3), and indeed the pulmonary arteries accompany the bronchi down the centers of the secondary lobules as far as the terminal bronchioles. Beyond that they break up to supply the capillary bed which lies in the walls of the alveoli (Fig. 5.4). The pulmonary capillaries form a dense network in the alveolar wall which makes an exceedingly efficient arrangement for gas exchange (Figs. 5.2 and 5.4). So rich is the mesh that some physiologists feel that it is misleading to talk of a network of individual capillary segments and prefer to regard the capillary bed as a sheet of flowing blood interrupted in places by posts, rather like an underground parking garage. The oxygenated blood is then collected from the capillary bed by the small pulmonary veins which run between the lobules and eventually unite to form the four large veins (in humans) which drain into the left atrium.

At first sight, this circulation appears to be simply a small version of the systemic circulation which begins at the aorta and ends in the right atrium. Indeed, the pulmonary circulation is often called the "lesser circulation." However, there are important differences between the two circulations, and confusion frequently results from attempts to emphasize similarities between them.

Pressures within the Pulmonary Blood Vessels

The pressures in the pulmonary circulation are remarkably low. The mean pressure in the main pulmonary artery is only about 15 mm Hg; the systolic and diastolic pressures are about 25 and 8 mm Hg,

respectively (Fig. 5.12). The pressure is therefore very pulsatile. By contrast, the mean pressure in the aorta is about 100 mm Hg—about 6 times more than in the pulmonary artery. The pressures in the right and left atria are not very dissimilar—about 2 and 5 mm Hg, respectively. Thus the pressure differences from inlet to outlet of the pulmonary and systemic systems are about $(15 - 5) = 10$ and $(100 - 2) = 98$ mm Hg, respectively—differing by a factor of 10.

In keeping with these low pressures, the walls of the pulmonary artery and its branches are remarkably thin and contain relatively little smooth muscle (they are easily mistaken for veins). This is in striking contrast to the systemic circulation, where the arteries generally have thick walls and the arterioles in particular have abundant smooth muscle.

The reasons for these differences become clear when the functions of the two circulations are compared. The systemic circulation regulates the supply of blood to various organs and body parts, including those which may be far above the level of the heart (the upstretched arm, for example). By contrast, the lung is required to accept the whole of the cardiac output at all times. It is rarely concerned with directing blood from one region to another (an exception is localized alveolar hypoxia, see below), and its arterial pressure is therefore as low as is consistent with lifting blood to the top of the lung. This keeps the work of the right heart as small as is feasible for efficient gas exchange to occur in the lung.

The pressure within the pulmonary capillaries is uncertain. Several pieces of evidence suggest that it lies about halfway between pulmonary arterial and venous pressure, and some work indicates that much of the pressure drop occurs within the capillary bed itself. Certainly the distribution of pressures along the pulmonary circulation is far more symmetrical than in its systemic counterpart, where most of the pressure drop is just upstream of the capillaries (Fig. 5.12). In addition, the pressure

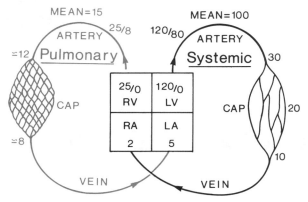

Figure 5.12.   Comparison of pressures (mm Hg) in the pulmonary and systemic circulations. Hydrostatic differences modify these. *CAP,* capillaries; *RV,* right ventricle; *LV,* left ventricle; *RA,* right atrium; *LA,* left atrium.

within the pulmonary capillaries varies considerably throughout the lung because of hydrostatic effects (see below).

Pressures Around the Pulmonary Blood Vessels

The pulmonary capillaries are unique in that they are virtually surrounded by gas (Figs. 5.2 and 5.4). It is true that there is a very thin layer of epithelial cells lining the alveoli, but the capillaries receive little support from this and, consequently, they are liable to collapse or distend, depending on the pressures within and around them. The pressure outside the capillaries is alveolar pressure, and during normal breathing this is close to atmospheric pressure;

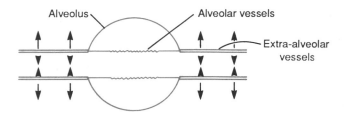

Figure 5.13.   Alveolar and extra-alveolar vessels. The first are mainly the capillaries and are exposed to alveolar pressure. The second are pulled open by the radial traction of the surrounding lung parenchyma, and the effective pressure around them is therefore lower than alveolar pressure. (Modified from Hughes et al., 1968.)

indeed, during breath holding with the glottis open, the two pressures are identical. Under some special conditions, the effective pressure around the capillaries is reduced by the surface tension of the fluid lining the alveoli. Usually, however, the effective pressure is alveolar pressure, and, when this rises above the pressure inside the capillaries, they collapse. The pressure difference between the inside and outside of the vessels is often called the *transmural pressure.*

What is the pressure around the pulmonary arteries and veins? This can be considerably less than alveolar pressure. As the lung expands, these larger blood vessels are pulled open by the radial traction of the elastic lung parenchyma which surrounds them (Figs. 5.13 and 5.14). Consequently, the effective pressure around them is low; in fact, there is

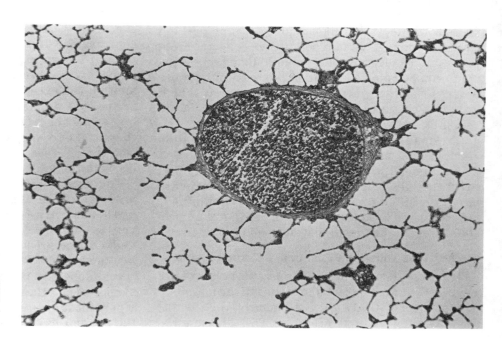

Figure 5.14.   Section of lung showing many alveoli and an extra-alveolar vessel (in this case a small vein) with its perivascular sheath. (From West, 1985a.)

some evidence that this pressure is even less than the pressure around the whole lung (intrapleural pressure). This paradox can be explained by the mechanical advantage which develops when a relatively rigid structure like a blood vessel or bronchus is surrounded by a rapidly expanding elastic material such as lung parenchyma. In any event, both the arteries and the veins increase their caliber as the lung expands because of the radial traction of the surrounding alveolar walls.

The behavior of the capillaries and the larger blood vessels is so different that they are often referred to as alveolar and extra-alveolar vessels (Fig. 5.13). Alveolar vessels are exposed to alveolar pressure and include the capillaries and the slightly larger vessels in the corners of the alveolar walls. Their caliber is determined by the relationship between alveolar pressure and the pressure within them. Extra-alveolar vessels include all arteries and veins which run through the lung parenchyma. Their caliber is greatly affected by the lung volume, since this determines the expanding pull of the parenchyma on their walls. The very large vessels near the hilum are outside the lung substance and are exposed to intrapleural pressure.

Pulmonary Vascular Resistance

It is useful to describe the resistance of a system of blood vessels as

$$\text{Vascular resistance} = \frac{\text{Input pressure} - \text{output pressure}}{\text{Blood flow}}$$

(see Chapter 5). This number is certainly not a complete description of the pressure-flow properties of the system. For example, the number usually depends on the magnitude of the blood flow. Nevertheless, it often allows a helpful comparison of different circulations or of the same circulation under different conditions.

We have seen that the total pressure drop from pulmonary artery to left atrium in the pulmonary circulation is only some 10 mm Hg against about 100 mm Hg for the systemic circulation. Since the blood flows through the two circulations are virtually identical, it follows that the pulmonary vascular resistance is only one-tenth of that of the systemic circulation. The pulmonary blood flow is about 6 liters/min so that, in numbers, the pulmonary vascular resistance = (15 − 5)/6 or about 1.7 mm Hg/liter/min. The high resistance of the systemic circulation is largely caused by the muscular arterioles which allow the regulation of blood flow to various organs of the body. The pulmonary circulation has no such vessels and appears to have as low a resist-

ance as is compatible with distributing the blood in a thin film over a vast area in the alveolar walls.

Although the normal pulmonary vascular resistance is extraordinarily small, it has a remarkable facility for becoming even smaller as the pressure within it rises. Figure 5.15 shows that an increase in either pulmonary arterial or venous pressure causes pulmonary vascular resistance to fall. This figure may appear confusing at first. The key to understanding it is to realize that, when one pressure (either arterial or venous) was changed, the other was held constant. The figure shows that, when either pulmonary arterial or pulmonary venous pressure was raised, thus raising capillary pressure, vascular resistance fell.

Two mechanisms are responsible for the fall in pulmonary vascular resistance as capillary pressure is raised. Under normal conditions, some capillaries are either closed or open with no blood flow. As the pressure rises, these vessels begin to conduct blood, thus lowering the overall resistance. This is termed *recruitment* (Fig. 5.16) and is apparently the chief mechanism for the fall in pulmonary vascular resistance which occurs as the pulmonary artery pressure is raised from low levels. The reason some vessels are unperfused at low perfusing pressures is not fully understood, but perhaps this is caused by random differences in the geometry of the complex network (Fig. 5.4*B*) which result in preferential channels for flow.

At higher vascular pressures widening of individual capillary segments occurs. This increase in caliber or *distension* is hardly surprising in view of the

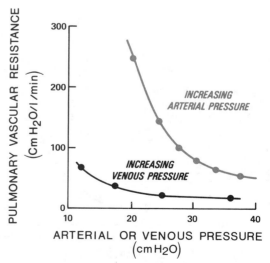

Figure 5.15. Fall in pulmonary vascular resistance as the pulmonary arterial or venous pressure is raised. When the arterial pressure was changed, the venous pressure was held constant at 12 cm H_2O, and when the venous pressure was changed, the arterial pressure was held at 37 cm H_2O. (Data from an excised canine lung preparation.)

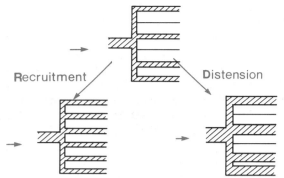

Figure 5.16. Recruitment (opening up of previously closed vessels) and distension (increase in caliber of vessels). These are the two mechanisms for the decrease in pulmonary vascular resistance which occurs as vascular pressures are raised.

very thin membrane which separates the capillary from the alveolar space (Fig. 5.2). Distension is apparently the predominant mechanism causing the fall in pulmonary vascular resistance at relatively high vascular pressures.

Another important determinant of pulmonary vascular resistance is *lung volume*. The caliber of the extra-alveolar vessels (Fig. 5.13) is determined by a balance between various forces. As we have seen, these vessels are pulled open as the lung expands. As a result their vascular resistance is low at large lung volumes. On the other hand, their walls contain smooth muscle and elastic tissue which resist distension and tend to reduce their caliber. Consequently they have a high resistance when lung volume is low (Fig. 5.17). Indeed, if the lung is completely collapsed, the smooth muscle

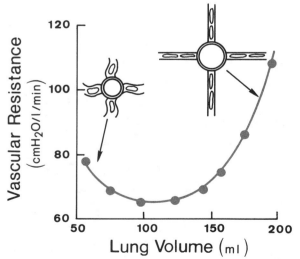

Figure 5.17. Effect of lung volume on pulmonary vascular resistance when the transmural pressure of the capillaries is held constant. At low lung volumes, resistance is high because the extra-alveolar vessels become narrow. At high volumes, the capillaries are stretched and their caliber is reduced. (Data from a canine lobe preparation.)

tone of these vessels is so effective that the pulmonary artery pressure has to be raised several centimeters of water above downstream pressure before any flow at all occurs. This is called a *critical opening pressure* (Chapter 6).

Is the vascular resistance of the capillaries influenced by lung volume? This depends on whether alveolar pressure changes with respect to the pressure inside the capillaries, that is, whether their transmural pressure alters. If alveolar pressure rises with respect to capillary pressure, the vessels tend to be squashed and their resistance rises. This usually occurs when a normal subject takes a deep inspiration because the pulmonary vascular pressures fall. The reason for this is that the heart is surrounded by intrapleural pressure which falls on inspiration (Chapter 39). However, the pressures in the pulmonary circulation do not remain steady after such a maneuver, and this complicates ensuing events.

An additional factor is that the caliber of the capillaries is reduced at large lung volumes because of stretching of the alveolar walls. The same thing happens in a thin-walled rubber tube if it is stretched laterally across its diameter. Thus, even if the transmural pressure of the capillaries is not changed with large lung inflations their vascular resistance increases (Fig. 5.17).

Because of the role of smooth muscle in determining the caliber of the extra-alveolar vessels, drugs which cause contraction of the muscle increase pulmonary vascular resistance. These include serotonin, histamine, and norepinephrine. These drugs are particularly effective vasoconstrictors when the lung volume is low and the expanding forces on the vessels are weak. Drugs which can relax smooth muscle in the pulmonary circulation include acetylcholine and isoproterenol. However, the vascular tone of the normal pulmonary circulation is so small that these drugs have little or no effect on vascular resistance.

Measurement of Pulmonary Blood Flow

Since the whole of the cardiac output normally goes through the lungs, pulmonary blood flow can be obtained from the Fick principle or indicator dilution methods as described in Chapter 13. The Fick and dye methods give the average flow over a number of heart cycles. It is also possible to measure *instantaneous pulmonary blood flow* using the body plethysmograph (Fig. 5.18). For this application the subject inhales a gas mixture containing 79% nitrous oxide and 21% O_2 from a rubber bag inside the box. Nitrous oxide is a very soluble gas and, as it is taken up by the blood, the box pressure

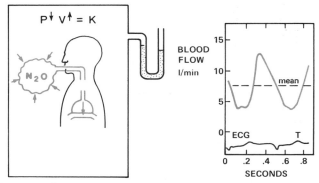

Figure 5.18. Measurement of instantaneous pulmonary capillary blood flow by recording nitrous oxide uptake in the body plethysmograph. Calculated blood flow is shown on the *right* with the electrocardiogram (*ECG*).

falls in a series of small steps which are synchronous with the heart beat. Since the uptake of nitrous oxide is flow-limited (Fig. 5.3*B*), instantaneous blood flow can be calculated. In normal subjects, there is considerable pulsatility of pulmonary capillary blood flow, and this is altered by disease.

Distribution of Blood Flow

Thus far we have been assuming that all parts of the pulmonary circulation behave identically. However, considerable inequality of blood flow exists within the human lung. This can be shown by using the technique shown in Figure 5.19. The subject is seated with a bank of counters or a radiation camera mounted behind the chest. Radioactive xenon gas is dissolved in saline and injected into a peripheral vein. When the xenon reaches the pulmonary capillaries it is evolved into alveolar gas because of

its low solubility, and the distribution of radioactivity can be measured by the counters during breath holding.

In the upright human lung, blood flow decreases almost linearly from bottom to top, reaching very low values at the apex (Fig. 5.19). This distribution is affected by change of posture and exercise. When the subject lies supine, the apical zone blood flow increases, but the basal zone flow remains virtually unchanged, with the result that the distribution from apex to base becomes almost uniform. However, in this posture, blood flow in the posterior (dependent) regions of the lung exceeds flow in the anterior parts. Measurements on men suspended upside down show that apical blood flow may exceed basal flow in this position. On mild exercise, both upper and lower zone blood flows increase, and the regional differences become less.

The uneven distribution of blood flow can be explained by the hydrostatic pressure differences within the blood vessels. If we consider the pulmonary arterial system as a continuous column of blood, the difference in pressure between the top and bottom of a lung 30 cm high will be about 30 cm H_2O, or 23 mm Hg. This is a large pressure difference for such a low pressure system as the pulmonary circulation (Fig. 5.12) and its effects on regional blood flow are shown in Figure 5.20.

There may be a region at the top of the lung *(zone 1)* where pulmonary arterial pressure falls below alveolar pressure (normally close to atmospheric pressure). If this occurs, the capillaries are squashed flat and no flow is possible. This zone 1 does *not* occur under normal conditions, because the

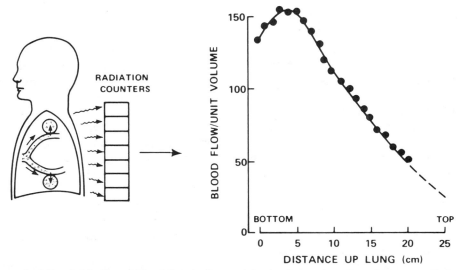

Figure 5.19. Measurement of the distribution of blood flow in the upright human lung using radioactive xenon. Xenon gas dissolved in saline is injected into a vein, and when it reaches the lung it is evolved into alveolar gas from the pulmonary capillaries. The units of blood flow are such that, if flow were uniform, all values would be 100. Note the small flow at the apex. (Redrawn from Hughes et al., 1968.)

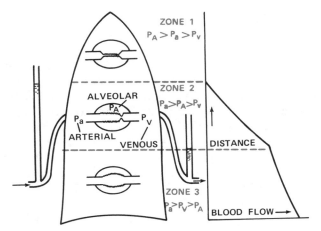

Figure 5.20. Model to explain the uneven distribution of blood flow in the lung based on the pressures affecting the capillaries. (From West et al., 1964.)

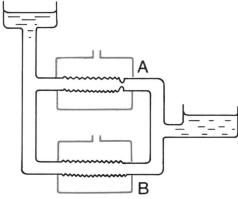

Figure 5.21. Two Starling resistors each consisting of a thin rubber tube inside a container. When chamber pressure exceeds downstream pressure as in *A*, flow is independent of downstream pressure. However, when downstream pressure exceeds chamber pressure as in *B*, flow is determined by the upstream-downstream difference.

pulmonary arterial pressure is just sufficient to raise blood to the top of the lung, but may be present if the arterial pressure is reduced (following severe hemorrhage, for example) or if alveolar pressure is raised (during positive pressure ventilation). This ventilated but unperfused lung is useless for gas exchange and is called *alveolar dead space*.

Further down the lung *(zone 2)*, pulmonary arterial pressure increases because of the hydrostatic effect and now exceeds alveolar pressure. However, venous pressure is still very low and is less than alveolar pressure, and this leads to remarkable pressure-flow characteristics. Under these conditions blood flow is determined by the difference between arterial and alveolar pressures (not the usual arterial-venous pressure difference). Indeed venous pressure has no influence on flow unless it exceeds alveolar pressure.

This situation can be modeled with a flexible rubber tube inside a glass chamber (Fig. 5.21). When chamber pressure is greater than downstream pressure, the rubber tube collapses at its downstream end, and the pressure in the tube at this point limits flow. The pulmonary capillary bed is clearly very different from a rubber tube, but nevertheless the overall behavior is similar and is often called the Starling resistor, sluice, or waterfall effect. Since arterial pressure is increasing down the zone but alveolar pressure is the same throughout the lung, the pressure difference responsible for flow increases. In addition, increasing recruitment of capillaries occurs down this zone.

In *zone 3*, venous pressure now exceeds alveolar pressure, and flow is determined in the usual way by the arterial-venous pressure difference. The increase in blood flow down this region of the lung is apparently caused chiefly by distension of the

capillaries. The pressure within them (lying between arterial and venous) increases down the zone, while the pressure outside (alveolar) remains constant. Thus their transmural pressure rises, and indeed measurements show that their mean width increases. Recruitment of previously closed vessels may also play some part in the increase in blood flow down this zone.

The scheme shown in Figure 5.20 summarizes the role played by the capillaries in determining the distribution of blood flow. At low lung volumes, the resistance of the extra-alveolar vessels becomes important, and a reduction of regional blood flow is seen starting first at the base of the lung where the parenchyma is least expanded (see Fig. 5.72). This reduced blood flow can be explained by the narrowing of the extra-alveolar vessels which occurs when the lung around them is poorly inflated (Fig. 5.17).

Hypoxic Pulmonary Vasoconstriction

We have seen that passive factors dominate the vascular resistance and the distribution of flow in the pulmonary circulation under normal conditions. However, a remarkable active response occurs when the P_{O_2} of alveolar gas is reduced. This consists of contraction of smooth muscle in the walls of the small arterioles in the hypoxic region. The precise mechanism of this response is obscure, but it occurs in excised isolated lung and so does not depend on central nervous connections. Excised segments of pulmonary artery can be shown to constrict if their environment is made hypoxic, so that it may be a local action of the hypoxia on the artery itself. One hypothesis is that cells in the perivascular tissue release some vasoconstrictor substance in response to hypoxia, but an intensive search for the mediator has not been successful. Interestingly, it is the P_{O_2} of the alveolar gas, not the pulmonary arterial blood, which chiefly determines the response. This can be proved by perfusing a lung with blood of a high P_{O_2} while keeping the alveolar P_{O_2} low. Under these conditions the response occurs.

The vessel wall presumably becomes hypoxic through diffusion of oxygen over the very short distance from the wall to the surrounding alveoli. Recall that a small pulmonary artery is very closely surrounded by alveoli (compare the proximity of alveoli to the small pulmonary vein in Fig. 5.14). The stimulus-response curve of this constriction is very nonlinear. When the alveolar P_{O_2} is altered in the region above 100 mm Hg, little change in vascular resistance is seen. However, when the alveolar P_{O_2} is reduced below approximately 70 mm Hg, marked vasoconstriction may occur, and at a very low P_{O_2} the local blood flow may be almost abolished (Fig. 5.22).

The vasoconstriction has the effect of directing blood flow away from hypoxic regions of lung. These regions may result from bronchial obstruction, and the diversion of blood flow reduces the deleterious effects on gas exchange. At high altitude, generalized pulmonary vasoconstriction may occur, leading to a large rise in pulmonary arterial pressure and a substantial increase in work for the right heart. Probably the most important situation where this mechanism operates is at birth. During fetal life, the pulmonary vascular resistance is very high, partly because of hypoxic vasoconstriction, and only some 15% of the cardiac output goes through the lungs (see Chapter 41). When the first breath oxygenates the alveoli, the vascular resistance falls dramatically because of relaxation of vascular smooth muscle, and the pulmonary blood flow enormously increases.

Other active responses of the pulmonary circulation have been described. A low blood pH causes vasoconstriction, especially when alveolar hypoxia is present. There is also evidence that the autonomic nervous system exerts a weak control, with an increase in sympathetic outflow causing stiffening of the walls of the pulmonary arteries and vasoconstriction.

Water Balance in the Lung

Since only 0.5 μm of tissue separates the capillary blood from the air in the lung (Fig. 5.2), the problem of keeping the alveoli free of fluid is critical. Fluid exchange across the capillary endothelial wall is believed to obey Starling's law (Chapter 6). The force tending to push fluid *out* of the capillary is the capillary hydrostatic pressure minus the hydrostatic pressure in the interstitial fluid, or $P_c - P_i$. The force tending to pull fluid *in* is the colloid osmotic pressure of the proteins of the blood minus that of the proteins of the interstitial fluid, or $\pi_c - \pi_i$. The magnitude of this force depends on the reflection coefficient σ which indicates the effectiveness of the capillary wall in preventing the passage of proteins across it (Chapter 2). Thus

$$\text{Net fluid out} = K[(P_c - P_i) - \sigma(\pi_c - \pi_i)]$$

where K is a constant called the filtration coefficient.

Unfortunately, the practical use of this equation is limited because of our ignorance of many of the values. The colloid osmotic pressure within the capillary is about 28 mm Hg. The capillary hydrostatic pressure is probably about halfway between arterial and venous pressure but is much higher at the bottom of the lung than at the top (Fig. 5.20). The colloid osmotic pressure of the interstitial fluid is not known but is about 20 mm Hg in lung lymph. However, this value may be higher than that in the interstitial fluid around the capillaries. The interstitial hydrostatic pressure is unknown but is thought by some physiologists to be substantially below atmospheric pressure. It is probable that the net pressure of the Starling equation is outward, causing a small lymph flow of perhaps 20 ml/hr in humans under normal conditions.

Where does fluid go when it leaves the capillaries? Figure 5.23 shows that fluid which leaks out into the interstitium of the alveolar wall tracks through the interstitial space to the perivascular and peribronchial spaces within the lung. Numerous lymphatics run in the peribronchial and perivascular spaces, and these help to transport the fluid to the

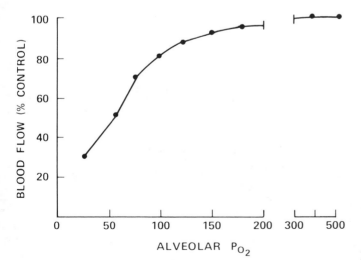

Figure 5.22. Effect of reducing alveolar P_{O_2} on pulmonary blood flow. Data from anesthetized cat. (From Barer et al., 1970.)

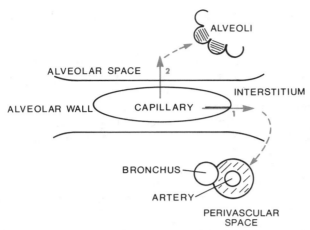

Figure 5.23. Two routes for loss of fluid from pulmonary capillaries. Fluid which enters the interstitium initially finds its way into the perivascular and peribronchial spaces. Later, fluid may cross the alveolar wall, filling alveolar spaces.

hilar lymph nodes. In addition, the pressure in these perivascular spaces is low, thus forming a natural sump for the drainage of fluid (compare Fig. 5.13). The earliest form of pulmonary edema is characterized by engorgement of these peribronchial and perivascular spaces and is known as *interstitial edema.*

In a later stage of pulmonary edema, fluid crosses the alveolar epithelium into the alveolar spaces (Fig. 5.23). When this occurs the alveoli fill with fluid one by one, and since they are then unventilated, no oxygenation of the blood passing through them is possible. This stage is known as *alveolar edema.* What causes fluid to start moving across the alveolar spaces is not known, but it may be that this occurs when the maximal drainage rate through the interstitial space is exceeded and the pressure there rises too high. The normal rate of lymph flow from the lung is only a few milliliters per hour, but it can be shown to increase greatly if the capillary pressure is raised over a long period. Alveolar edema is much more serious than interstitial edema because of the interference with pulmonary gas exchange.

Other Functions of the Pulmonary Circulation

The chief function of the pulmonary circulation is to move blood to and from the blood-gas barrier so that gas exchange can occur. However, the pulmonary circulation has other important functions. One is to act as a reservoir for blood. We saw that the lung has a remarkable ability to reduce its pulmonary vascular resistance as its vascular pressures are raised through the mechanisms of capillary recruitment and distension (Fig. 5.16). The same mechanisms allow the lung to increase its blood volume with relatively small rises in pulmonary arte-

rial or venous pressure. This occurs, for example, when a subject lies down after standing. Blood then drains from the legs into the lung.

Another function of the lung is to filter blood. Small blood thrombi are removed from the circulation before they can reach the brain or other vital organs. There is also evidence that many white blood cells are trapped by the lung, although the value of this is not clear.

METABOLIC FUNCTIONS OF THE LUNG

The lung has important metabolic functions in addition to gas exchange, and this is a convenient place to discuss them briefly. This topic overlaps with pulmonary pharmacology, and further details can be found in such textbooks. One of the most important metabolic functions is the synthesis of phospholipids such as dipalmitoyl phosphatidylcholine, which is a component of pulmonary surfactant. This is considered further in Chapter 39. Protein synthesis is also clearly important since collagen and elastin form the structural framework of the lung. Under abnormal conditions, proteases are apparently liberated from leukocytes or macrophages in the lung, causing breakdown of proteins and thus emphysema. Another significant area is carbohydrate metabolism, especially the elaboration of mucopolysaccharides of bronchial mucus.

A number of vasoactive substances are metabolized by the lung, as shown in Figure 5.24. Since the lung is the only organ except the heart which receives the whole circulation, it is uniquely suited to modifying blood-borne substances. A substantial fraction of all the vascular endothelial cells of the body are located in the lung.

The only known example of biological activation

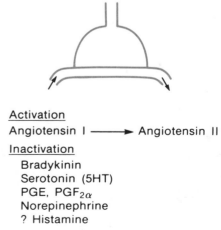

Activation
Angiotensin I ⟶ Angiotensin II

Inactivation
  Bradykinin
  Serotonin (5HT)
  PGE, $PGF_{2\alpha}$
  Norepinephrine
  ? Histamine

Figure 5.24. Pulmonary metabolism of vasoactive substances. The inactivation of bradykinin, serotonin, PGE, and $PGF_{2\alpha}$ is highly effective but only partial for norepinephrine and histamine.

by passage through the pulmonary circulation is the conversion of the relatively inactive polypeptide, angiotensin I, to the potent vasoconstrictor, angiotensin II. The latter, which is up to 50 times more active than its precursor, is unaffected by passage through the lung. The conversion of angiotensin I is catalyzed by an enzyme, angiotensin-converting enzyme or ACE, which is located in small pits *(caveolae intracellulares)* in the surface of the capillary endothelial cells.

Many vasoactive substances are completely or partially inactivated during passage through the lung. Bradykinin is largely inactivated (up to 80%), and the enzyme responsible is ACE. The lung is the major site of inactivation of serotonin (5-hydroxytryptamine), but this is not by enzymatic degradation, but by an uptake and storage process. Some of the serotonin may be transferred to platelets in the lung or stored in some other way and released during anaphylaxis. The prostaglandins E_1, E_2, and $F_{2\alpha}$ are also inactivated in the lung, which is a rich source of the responsible enzymes. Norepinephrine is also taken up by the lung to some extent (up to 30%). Histamine appears not to be affected by the intact lung but is readily inactivated by lung slices.

Some vasoactive materials pass through the lung without significant gain or loss of activity. These include epinephrine, prostaglandins A_1 and A_2, angiotensin II, and vasopressin (ADH).

Several vasoactive and bronchoactive substances are metabolized in the lung and may be released into the circulation under certain conditions. Important among these are the arachidonic acid metabolites (Fig. 5.25). Arachidonic acid is formed through the action of the enzyme phospholipase A_2 on phospholipid bound to cell membranes. There are two major synthetic pathways, with the initial reactions being catalyzed by the enzymes lipoxygenase and cyclooxygenase, respectively. The first produces the leu-

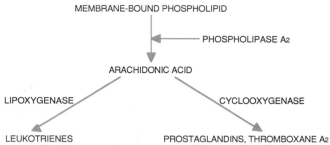

Figure 5.25. Two pathways of arachidonic acid metabolism. The leukotrienes are generated via the lipoxygenase pathway, while the prostaglandins and thromboxane A_2 come from the cyclooxygenase pathway.

kotrienes, which include the mediator originally described as slow-reacting substance of anaphylaxis (SRS-A). These compounds cause airway constriction and may have an important role in asthma. Other leukotrienes are involved in inflammatory responses.

The prostaglandins are potent vasoconstrictors or vasodilators. PGE_2 plays an important role during the perinatal period because it helps to constrict the patent ductus arteriosus. Prostaglandins also affect platelet aggregation and are active in other systems such as the kallikrein-kinin clotting cascade. They also may have a role in the bronchoconstriction of asthma.

There is also evidence that the lung plays a role in the clotting mechanism of blood under normal and abnormal conditions. For example, there are large numbers of mast cells containing heparin in the interstitium. Finally, the lung is able to secrete special immunoglobulins, particularly IgA, in the bronchial mucus; these contribute to its defenses against infection. The significance of some of these metabolic functions is still obscure, but it is clear that the lung has important functions in addition to its main role of gas exchange.

CHAPTER 37

Gas Transport to the Periphery

CARRIAGE OF OXYGEN AND CARBON DIOXIDE BY THE BLOOD

In some ways it would be more logical to consider gas exchange within the lung before dealing with the transport of oxygen and carbon dioxide by the blood to the periphery of the body. However, knowledge of the oxygen and carbon dioxide blood dissociation curves is essential for an understanding of pulmonary gas exchange, so we shall deal with this topic now.

Oxygen

Oxygen is carried in the blood in two forms—dissolved and in combination with hemoglobin.

DISSOLVED OXYGEN

This obeys Henry's law, that is, the amount dissolved is proportional to the partial pressure (see *bottom broken line* in Fig. 5.26). For each mm Hg of P_{O_2} there is 0.003 ml of O_2/100 ml of blood (sometimes written 0.003 vol%). Thus normal arterial blood with a P_{O_2} of 100 mm Hg contains 0.3 ml of oxygen/100 ml.

It is easy to see that this way of transporting oxygen must be inadequate for man. Suppose that on exercise the maximal cardiac output is 25 liters/min (25,000 ml/min). Then the total amount of oxygen which could be delivered to the muscles would be only 25,000 × 0.3/100 or 75 ml of oxygen/min, assuming that all the oxygen could be unloaded. But the maximal oxygen consumption on exercise is of the order of 3,000 ml/min, so clearly an additional method of transporting oxygen is required.

OXYGEN DISSOCIATION CURVE

At this point the reader may wish to review those aspects of the biochemistry of hemoglobin which were covered in Chapter 23. Oxygen forms an easily reversible combination with hemoglobin (Hb) to give oxyhemoglobin:

$$O_2 + Hb \rightleftharpoons HbO_2$$

Suppose we take a number of glass containers (tonometers), each containing a small volume of blood, and add gas with various concentrations of oxygen. After allowing time for the gas and blood to reach equilibrium, we measure the P_{O_2} of the gas and the oxygen content of the blood. Knowing that 0.003 ml of oxygen will be dissolved in each 100 ml of blood/mm Hg of P_{O_2}, we can calculate the oxygen combined with Hb (Fig. 5.26). Note that the amount of oxygen carried by the Hb increases rapidly up to a P_{O_2} of about 50 mm Hg, but at a higher P_{O_2} the curve becomes much flatter. The maximum amount of oxygen that can be combined with Hb is called the *oxygen capacity*. It can be measured by exposing the blood to a very high P_{O_2} (say 600 mm Hg) and subtracting the dissolved oxygen. One gram of pure Hb can combine with 1.39 ml of oxygen* and since normal blood has about 15 g of Hb/100 ml the oxygen capacity is about 20.8 ml of O_2/100 ml of blood.

The *oxygen saturation* of Hb is given by:

$$\frac{O_2 \text{ combined with Hb}}{O_2 \text{ capacity}} \times 100$$

The oxygen saturation of arterial blood with a P_{O_2} of 100 mm Hg is about 97.5%, while that of mixed venous blood with a P_{O_2} of 40 mm Hg is about 75%. It is important to grasp the relationships between P_{O_2}, O_2 saturation, and oxygen content (or concentration). For example, suppose a severely anemic patient with a Hb concentration of only 7.5 g/100 ml of blood has normal lungs and an arterial P_{O_2} of 100 mm Hg. His oxygen saturation will be 97.5% (at normal pH, P_{O_2}, and temperature), but the oxygen combined with Hb will be only 10.4 ml/100 ml. Dissolved oxygen will contribute 0.3 ml giving a total oxygen content of 10.7 ml/100 ml of blood. In gen-

---

*Some authorities give 1.34 or 1.36 ml/100 ml. The reason is that under the normal conditions of the body, some of the hemoglobin is in forms such as methemoglobin which cannot combine with oxygen.

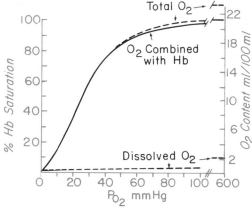

Figure 5.26. Oxygen dissociation curve *(solid line)* for pH 7.4, P_{CO_2} 40 mm Hg and temperature 37°C. The total blood oxygen content is also shown for a hemoglobin concentration of 15 g/100 ml of blood.

eral, the oxygen content (or concentration) of blood (in ml O_2/100 ml blood) is given by:

$$\left(1.39 \times Hb \times \frac{Sat}{100}\right) + 0.003\, P_{O_2}$$

where Hb is the hemoglobin concentration in g/100 ml, *Sat* is the percentage saturation of the hemoglobin, and P_{O_2} is in mm Hg.

The curved shape of the oxygen dissociation curve has several physiological advantages. The nearly flat upper portion assists in the diffusion of oxygen across the blood-gas barrier in the lung and thus the loading of oxygen by the blood; this will become clearer when we consider pulmonary gas exchange in Chapter 38. An additional advantage is that small falls in the P_{O_2} of alveolar gas will hardly affect the oxygen content of arterial blood and thus the amount of oxygen available to the tissues. The steep lower part of the dissociation curve means that the peripheral tissues can withdraw large amounts of oxygen for only a small drop in capillary P_{O_2}. This maintenance of blood P_{O_2} assists the diffusion of oxygen into the tissue cells.

Oxygenated Hb is bright red but reduced Hb is purple, so that a low arterial oxygen saturation causes *cyanosis*. This is not a reliable clinical sign of mild desaturation, however, because its recognition depends on so many variables, such as lighting conditions and skin pigmentation. Since it is the amount of reduced Hb that is important, cyanosis is often marked when polycythemia is present, but it may be difficult to detect in anemic patients.

Several factors may shift the position of the oxygen dissociation curve, including changes in the pH, P_{CO_2}, and temperature of the blood, and the concentration of organic phosphates within the red blood

cell. Figure 5.27 shows that a fall in pH, rise in P_{CO_2}, and rise in temperature all shift the curve to the right, that is they decrease the affinity of hemoglobin for oxygen. Opposite changes shift it to the left. Most of the effect of P_{CO_2}, which is known as the *Bohr effect* (1904), can be attributed to its action on pH. As discussed in Chapter 23, an increase in H^+ concentration slightly alters the configuration of the Hb molecule and thus reduces the accessibility of oxygen to the heme groups. The consequent rightward shift (decreased affinity) enhances the unloading of oxygen for a given P_{O_2} in a tissue capillary. A simple way of remembering these shifts is that an exercising muscle is acid, hot, and has a high P_{CO_2}. Consequently, it benefits from the increased unloading of oxygen in its capillaries.

An increase in organic phosphates, particularly 2,3-diphosphoglycerate (2,3-DPG), within the red blood cell also shifts the oxygen dissociation to the right and thus assists unloading of oxygen (Benesch and Benesch, 1969; Chanutin and Curnish, 1967). Some biochemical aspects of this subject are discussed in Chapter 23. An increase in red cell 2,3-DPG occurs in chronic hypoxia, for example, after 2 days of ascent to an altitude of 4,500 m (about 15,000 feet) primarily because of the respiratory alkalosis (Lenfant et al., 1970). Calculations suggest that the resultant shift of the dissociation curve increases the amount of oxygen released from the blood by about 10%. This may be a useful feature of acclimatization to moderately high altitudes, although it is much less important than other factors such as hyperventilation (see Chapter 41). At

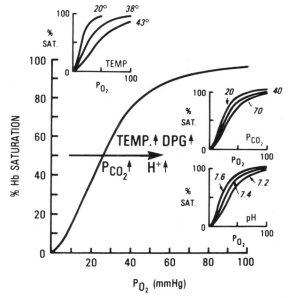

Figure 5.27. Shift of the oxygen dissociation curve by pH, P_{CO_2}, temperature *(temp.)*; and 2,3-diphosphoglycerate *(DPG)*; *Sat.*, saturation.

much higher altitudes the advantages of the increase in 2,3-DPG disappear because the loading of oxygen in the pulmonary capillaries is adversely affected by the rightward shift. Other hypoxic states in which a useful increase in 2,3-DPG are seen include chronic lung disease, cyanotic heart disease, and severe anemias. The resultant enhanced unloading of oxygen in the peripheral tissues can be regarded as an intrinsic adaptive mechanism in response to hypoxemia.

In blood stored for transfusions, a slow decrease in 2,3-DPG brings about an increased oxygen affinity, with the result that oxygen to the tissues is subnormal. The change in the blood is influenced by the preservative used and can be retarded by the addition of inosine (Bunn et al., 1969).

A useful measure of the position of the oxygen dissociation curve is the P_{O_2} for 50% oxygen saturation. This is known as the P_{50}. The normal value for human blood is about 26 mm Hg at P_{CO_2} 40 mm Hg, pH 7.4, and temperature 37°C. The P_{50} differs between species of animals, often being appreciably lower in animals with large body weight (for example the small rhesus monkey and the larger gorilla have P_{50} values of approximately 32 and 25 mm Hg, respectively). The changes in P_{50} can presumably be explained by variations in composition of the globin portion of the molecule. The oxygen affinity of the hemoglobin is also often increased in animals that live at high altitudes (such as the South American llama), or other hypoxic environments (fishes in stagnant water). Various strategies for increasing the affinity include changes in the concentration of organic phosphates, and alterations in the Bohr effect.

Carbon Monoxide

Carbon monoxide interferes with the oxygen transport function of blood by combining with Hb to form carboxyhemoglobin (COHb). The dissociation curves of oxy- and carboxyhemoglobin have basically similar shapes. However, CO has about 250 times the affinity of oxygen for Hb; this means that CO will combine with the same amount of Hb as oxygen when the CO partial pressure is 250 times lower. For example, at a P_{CO} of 0.16 mm Hg, 75% of the Hb is combined with CO as COHb. Thus when plotted on the same scale (Fig. 5.28) the CO dissociation curve has almost a right-angled bend near a P_{CO} of zero.

The higher affinity of CO for Hb means that people inadvertently exposed to small concentrations of CO in the air (for example, during a fire in a building) may have a large proportion of their Hb tied up as COHb, and thus unavailable for oxygen carriage.

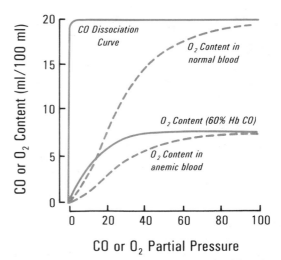

Figure 5.28. Comparison of dissociation curves for CO and oxygen. Note that the hemoglobin (Hb) is almost completely saturated with CO at a very low partial pressure. Note also that the CO dissociation curve for 60% HbCO is displaced to the *left* compared with the oxygen dissociation curve for anemic blood with the Hb concentration reduced by 60%.

If this happens, the Hb concentration and P_{O_2} of the blood may be normal, but its oxygen content is grossly reduced. Small amounts of COHb also shift the oxygen dissociation curve to the left, thus making it difficult for the blood to unload the oxygen that it does carry. Figure 5.28 also shows the oxygen dissociation curve of blood containing 60% COHb contrasted with the curve for anemic blood with the Hb concentration reduced by 60%.

A patient with CO poisoning is treated by being given pure oxygen or 95% O_2 with 5% CO_2 to breathe. The addition of 5% CO_2 to the inspired gas increases the rate of washout of CO from the blood. In specialized medical centers, hyperbaric oxygen therapy is often used because by raising the inspired P_{O_2} to 2000 mm Hg, all the oxygen required by the body tissues can be carried by the blood in the dissolved form (see Chapter 41).

Carbon Dioxide

CARBON DIOXIDE CARRIAGE

Carbon dioxide is carried in the blood in three forms—dissolved, as bicarbonate, and in combination with proteins as carbamino compounds (Fig. 5.29).

Dissolved CO_2

Like oxygen, dissolved CO_2 obeys Henry's law, but because CO_2 is about 20 times more soluble than O_2, the dissolved form plays a more significant role in the normal carriage in the blood. In fact about 10% of the gas which is evolved into the lung from

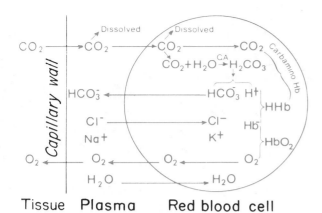

Figure 5.29. Scheme of the uptake of O_2 and liberation of carbon dioxide in systemic capillaries. Exactly opposite events occur in the pulmonary capillaries.

the mixed venous blood is in the dissolved form (Fig. 5.30).

Bicarbonate

This is formed in the blood by the following sequence:

$$\overset{C.A.}{CO_2 + H_2O \rightleftharpoons H_2CO_3} \rightleftharpoons H^+ + HCO_3^-$$

The first reaction is very slow in plasma but fast within the red blood cell because of the presence there of the enzyme *carbonic anhydrase (C.A.)*. This is a zinc-containing protein which is present in

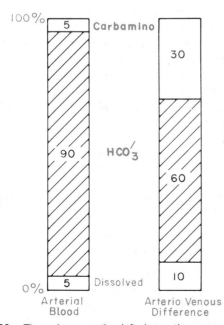

Figure 5.30. The column on the *left* shows the proportions of the total CO_2 content in arterial blood. The column on the *right* shows the proportions which make up the arterial-venous difference.

a high concentration in the red cell but not in the plasma. It is also found in other tissues, including gastric and intestinal mucosa, renal cortex, and muscle. It can be inhibited by the administration of acetazolamide (Diamox) which causes diuresis in man through its action on the kidney (Chapter 31). In experimental animals, very large doses of acetazolamide interfere with CO_2 transport by slowing the hydration reaction (Roughton, 1964).

Ionization of the carbonic acid within the red cell occurs rapidly and does not need an enzyme. When the concentration of hydrogen and bicarbonate ions in the cell rises, HCO_3^- diffuses out, but H^+ cannot move out easily because the cell membrane is relatively impermeable to cations. Thus in order to maintain electrical neutrality, Cl^- ions diffuse into the cell from the plasma, the so-called *chloride shift* (Fig. 5.29). The movement of chloride is in accordance with the Gibbs-Donnan equilibrium (Chapter 2).

Some of H^+ ions which are liberated are bound to hemoglobin:

$$H^+ + HbO_2 \rightleftharpoons H \cdot Hb + O_2$$

This reaction occurs because reduced Hb is a better proton acceptor (less "acid") than the oxygenated form. Thus the presence of reduced Hb in the peripheral blood helps with the loading of CO_2, while the oxygenation which occurs in the pulmonary capillary assists in the unloading. The fact that the deoxygenation of the blood increases its ability to carry CO_2 is often known as the *Haldane effect* (Christiansen et al., 1914).

These events associated with the uptake of CO_2 by the blood increase the osmolar content of the red cell, and consequently water enters the cell, thus increasing its volume. When the cells pass through the lung, they shrink a little.

Carbamino Compounds

These are formed by the combination of CO_2 with terminal amine groups of blood proteins. The most important protein is the globin of hemoglobin, and the reaction can be represented:

$$CO_2 + Hb \cdot NH_2 \rightleftharpoons Hb \cdot NH \cdot COOH \rightleftharpoons Hb \cdot NH \cdot COO + H^+$$

This occurs very rapidly without an enzyme, and most of the carbamic acid is in the ionized form. Reduced Hb can bind much more CO_2 than HbO_2. Thus again unloading of oxygen in peripheral capillaries facilitates the loading of CO_2 while oxygenation enhances the unloading of CO_2 in the lung (Haldane effect).

The relative contributions of the various forms of CO_2 in blood to the total CO_2 content are shown in Figure 5.30. Note that the great bulk of the CO_2 is in the form of bicarbonate. The amount in the dissolved form is small as is that as carbaminohemoglobin. However, these proportions do not reflect the changes that take place when CO_2 is loaded or unloaded by the blood. Of the total venous-arterial difference, about 60% is attributable to HCO_3^-, 30% to carbamino compounds, and 10% to dissolved CO_2.

CO₂ DISSOCIATION CURVE

The relationship between the P_{CO_2} and the total CO_2 content of blood is shown in Figure 5.31. Note that the CO_2 dissociation curve is much more linear than the oxygen dissociation curve (Fig. 5.26). Note also that the lower the saturation of Hb with oxygen the larger the CO_2 content for a given P_{CO_2}. As mentioned earlier, this Haldane effect can be explained by the better ability of reduced Hb to mop up the H^+ ions produced when carbonic acid dissociates, and the greater facility of reduced Hb to form carbaminohemoglobin.

Figure 5.32*A* shows that the CO_2 dissociation curve is considerably steeper than that for oxygen. For example, in the range 40–50 mm Hg, the CO_2 content changes by about 4.7 ml/100 ml compared with a change of oxygen content of only about 1.7 ml/100 ml.

A useful way of displaying the interactions between the oxygen and CO_2 dissociation curves is by means of the O_2-CO_2 diagram (Rahn and Fenn, 1955) (compare Fig. 5.47). In this diagram (Fig.

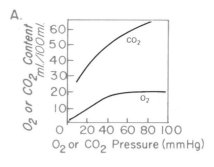

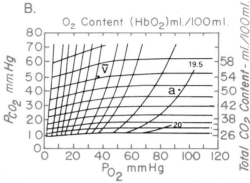

Figure 5.32. *A.*, Typical O_2 and CO_2 dissociation curves plotted with the same scales. Note that the CO_2 curve is much steeper. *B.*, O_2-CO_2 diagram showing lines of equal oxygen and CO_2 content. The lines are not parallel to the X and Y axes because of the Bohr and Haldane effects (see text).

5.32*B*) the X and Y axes show the partial pressures of O_2 and CO_2, respectively, in samples of blood. Note that the lines of O_2 and CO_2 content are not straight and parallel to the axes as they would be if the contents were simply proportional to the partial pressures. Choose a P_{O_2} on the X axis (say 50 mm Hg). Follow this P_{O_2} vertically (increasing P_{CO_2}) and lines of decreasing oxygen content will be encountered (Bohr effect). The same procedure following a given P_{CO_2} to the right (increasing P_{O_2}) will give decreasing CO_2 contents (Haldane effect). This diagram will be used in the next chapter for analysis of pulmonary gas exchange.

It should be emphasized that the events involved in the transport of CO_2 by the blood have a profound effect on the acid-base status of the body. This important subject is discussed in detail in Chapter 32.

ANALYSIS OF BLOOD GASES

Removal of Blood Samples

Arterial or venous blood can be drawn with a syringe and needle. Arterial blood is commonly obtained by puncture of the brachial or radial artery after infiltrating the skin with local anesthetic. The dead space of the syringe is filled with

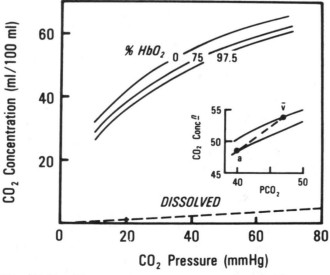

Figure 5.31. CO_2 dissociation curves for blood of different oxygen saturations. Note that well-oxygenated blood carries less CO_2 for the same P_{CO_2}. The "physiological" curve between arterial and mixed venous blood is also shown in the inset.

dilute heparin solution to prevent clotting of the blood. At the end of the procedure, pressure is applied to the site to prevent extravasation of blood. Venous blood can be obtained from an arm vein for determination of the hematocrit and electrolyte concentrations. Such a sample is of little value for blood gases, however, because it reflects only the local conditions.

To obtain mixed venous blood from the pulmonary artery, a long catheter must be passed through the right heart. Fine plastic catheters, sometimes with a small balloon on the end (Swan-Ganz), can be floated in from a peripheral vein along the blood stream without much difficulty. However, very occasionally an abnormal heart rhythm develops, and the procedure is not as innocuous as an arterial puncture.

Blood P_{O_2}, P_{CO_2}, and pH

These can be measured in a few minutes with blood gas electrodes.

Oxygen partial pressure is measured with a polarographic oxygen electrode (Fig. 5.33). The principle is that if a small voltage (0.6 V) is applied to a platinum electrode immersed in a buffer solution, the current which flows is proportional to the P_{O_2}. In practice the buffer is separated from the blood by a semipermeable membrane through which oxygen diffuses. Oxygen is consumed by the electrode; consequently, the P_{O_2} falls with time. The electrode is calibrated with gas or with a solution of known P_{O_2}.

Carbon dioxide partial pressure is measured with a P_{CO_2} electrode (Fig. 5.33). This is essentially a glass pH electrode surrounded by a bicarbonate buffer which is separated from blood by a thin membrane. CO_2 diffuses through the membrane and alters the pH of the buffer. This is measured by the electrode which reads out P_{CO_2} directly. Calibration is by means of CO_2 gas mixtures.

Blood pH can be measured with a glass pH electrode immersed in the blood. The reading gives the plasma pH, the value inside the cells being 0.1–0.2 pH units lower. In practice the P_{O_2}, P_{CO_2}, and pH electrodes are part of the same piece of equipment, and all three measurements can be made on 1 ml of blood.

Oxygen Saturation

Blood oxygen saturation can be determined by spectrophotometry because of the different absorption of light of certain wavelengths by oxygenated and reduced hemoglobin. The sample of blood is placed in a cuvette through which light is passed, and the oxygen saturation is read out directly on a meter. Other devices measure the light which is reflected from the surface of a blood sample. The amount of carboxyhemoglobin can sometimes be measured in the same instrument because of its special absorption characteristics.

The *pulse oximeter* is a specialized small spectrophotometer which measures the oxygen saturation of blood flowing through the finger. Another device, the *ear oximeter* measures the oxygen saturation of "arterialized" blood in the heated ear lobe. These may not be accurate measurements, especially when the oxygen saturation is very low, but they are often useful for monitoring changes in saturation.

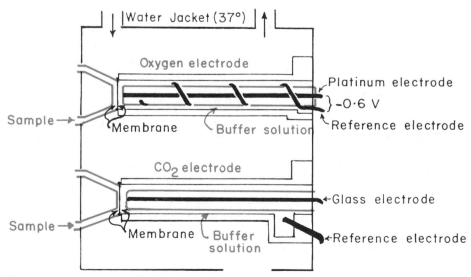

Figure 5.33. O_2 and CO_2 electrodes for measuring the partial pressures of these gases in blood. For the measurement of O_2, the gas diffuses through the semipermeable membrane and causes a flow of current which is proportional to P_{O_2}. For CO_2, the gas diffuses through the semipermeable membrane and changes the pH of the buffer solution. This is measured by the glass electrode.

O$_2$ and CO$_2$ Contents

These can be determined with the Van Slyke manometric apparatus (Peters and Van Slyke, 1932). The gases are liberated from the blood by adding lactic acid, potassium ferricyanide, and saponin, and vigorously shaking the mixture in a partial vacuum. The evolved gas is then analyzed by absorption with sodium hydroxide and an anthraquinone mixture. Oxygen content can also be derived from devices employing polarographic electrodes or fuel cells to consume the oxygen.

GAS TRANSPORT TO THE TISSUES

Diffusion from Capillary to Cell

Oxygen and carbon dioxide move between the systemic capillary blood and the tissue cells by diffusion from regions of high to low partial pressures. The principles governing diffusion (Fick's law) will be discussed in detail in the next chapter. Here it should be pointed out that the distance to be covered by diffusion in the peripheral tissues is considerably greater than in the lung. For example, the distance between open capillaries in resting muscle is of the order of 50 μm, whereas the thickness of the blood-gas barrier in the lung is only 1/100 of this. On the other hand, during exercise when the oxygen consumption of the muscle increases, additional capillaries open up, thus reducing the diffusion distance and increasing the cross-sectional area available for diffusion. Because CO$_2$ diffuses about 20 times faster than O$_2$ through tissue (see Fig. 5.37) elimination of CO$_2$ poses less of a problem than O$_2$ delivery.

Some measurements suggest that the movement of oxygen through some tissues in vitro is too fast to be accounted for by simple passive diffusion. It is possible that *facilitated diffusion* occurs under some circumstances (see Chapter 2), a possible carrier within muscle cells being myoglobin. Another possibility is that convective processes ("stirring") occur on a small scale, perhaps within cells.

Tissue Partial Pressures

Various models have been suggested to account for the way in which the P$_{O_2}$ falls in tissue between adjacent open capillaries. Figure 5.34*A* shows a hypothetical cylinder of tissue between capillaries in which blood comes in with a P$_{O_2}$ of 100 mm Hg. As the oxygen diffuses from the peripheral capillary into the center of the cylinder, the oxygen is consumed and the P$_{O_2}$ falls. In *1*. the balance between oxygen consumption and delivery (determined by the capillary P$_{O_2}$ and the intercapillary distance) results in an adequate P$_{O_2}$ in all the tissue.

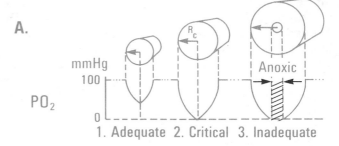

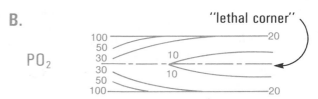

Figure 5.34. Scheme showing the fall in P$_{O_2}$ between adjacent open capillaries. In *A*. three hypothetical cylinders of tissue are shown. In *2*., the cylinder has a critical radius R_c while in *3*., the radius of the cylinder is so large that there is an anoxic zone in the middle of the cylinder. *B*. shows a section along the hypothetical cylinder of tissue. The P$_{O_2}$ in the blood adjacent to the tissue is assumed to fall from 100 to 20 mm Hg along the capillary. Lines of equal P$_{O_2}$ are shown. Note the possibility of a "lethal corner" in the middle of the cylinder at the venous end.

In *2*., the intercapillary distance or the oxygen consumption has been increased until the P$_{O_2}$ at one point in the tissue falls to zero. This is referred to as a *critical* situation. In *3*., there is an anoxic region where aerobic (that is, oxygen-utilizing) metabolism is impossible. Under these conditions, the tissue turns to anaerobic glycolysis with the formation of lactic acid.

The situation *along* the cylinder is shown in Figure 5.34*B*. Here it is assumed that the P$_{O_2}$ in the capillaries at the periphery of the tissue cylinder falls from 100 to 20 mm Hg. As a consequence the P$_{O_2}$ in the center of the cylinder falls toward the venous end of the capillary. It is clear that, on the basis of this model, the most vulnerable tissue is that furthest from the capillary at its downstream end. This was referred to as the "lethal corner" by Krogh. Direct measurements of the P$_{O_2}$ in tissues of experimental animals (for example, the surface of the cerebral cortex), have been made by means of oxygen microelectrodes. These have confirmed the steep descent of P$_{O_2}$ between open blood vessels.

How low can the P$_{O_2}$ fall before oxygen utilization ceases? Probably the value depends on the type of tissue involved. For example, there is evidence that the cells of the central nervous system cease functioning at a higher P$_{O_2}$ than liver cells. In measurements on suspensions of liver mitochondria in vitro, oxygen consumption continues at the same rate until the P$_{O_2}$ falls to the region of 1–3 mm Hg. In general it appears that the much higher P_{O_2} in capillary blood is to ensure an adequate pressure for diffusion of oxygen to the mitochondria and that at the sites of oxygen utilization the P_{O_2} may be very low.

Oxygen Delivery and Utilization

The amount of oxygen theoretically available to a tissue per minute is given by:

$$\dot{Q} \times Ca_{O_2}$$

where $\dot{Q}$ is the blood flow to the tissue and Ca_{O_2} is the arterial oxygen content (see Table 5.1).

The amount of oxygen consumed is given by:

$$\dot{Q} \times (Ca_{O_2} - Cv_{O_2})$$

where Cv_{O_2} is the oxygen content of the blood draining from the tissue. The ratio of the amount consumed to the amount available is sometimes called the oxygen utilization. This is given by:

$$\frac{\dot{Q}(Ca_{O_2} - Cv_{O_2})}{\dot{Q} \cdot Ca_{O_2}} = \frac{Ca_{O_2} - Cv_{O_2}}{Ca_{O_2}}$$

The utilization varies widely from organ to organ, being as small as 10% in the kidney, 60% in the coronary circulation, and over 90% in exercising muscle. For the body as a whole at rest, the value is about 25%, rising to 75% on severe exercise.

In most types of tissue hypoxia, the oxygen utilization is high. This is true of (1) "hypoxic hypoxia" when the P_{O_2} of the arterial blood is low, for example, in pulmonary disease; (2) "anemic hypoxia" when the oxygen-carrying capacity of the blood is reduced as in anemia or CO poisoning; and (3) "circulatory hypoxia" when tissue blood flow is reduced, as in shock or local obstruction. In (4) "histotoxic hypoxia," however, such as cyanide poisoning, the arterial-venous oxygen difference and the oxygen utilization are very low. This is because the cyanide prevents the use of O_2 by cytochrome oxidase within the tissues.

A different situation is seen with poisoning by dinitrophenol. Here oxygen is consumed giving energy as heat, but adenosine triphosphate production is prevented because of uncoupling of phosphorylation. As a consequence, useful work by the muscles is not possible.

Tissue Survival

The body has only small stores of oxygen which can be called upon during complete anoxia or asphyxia. The total store is only about 1500 ml of oxygen—enough to maintain life for only 6 min, if it were appropriately distributed. Tissues vary considerably in their ability to withstand oxygen deprivation, depending on how easily they can utilize anaerobic glycolysis. The cerebral cortex and the myocardium are particularly vulnerable to anoxia; in man, cessation of blood flow to the cerebral cortex results in loss of function within 4–6 s, loss of consciousness in 10–20 s, and irreversible changes in 3–5 min. The newborn of many species are less vulnerable to complete ischemia than the adults. Also there is great variation from one species to another.

CHAPTER 38

Pulmonary Gas Exchange

OXYGEN TRANSPORT FROM AIR TO TISSUES

We saw in the introductory chapter (Chapter 35) that the primary function of the lung is to allow oxygen to move into the blood, and carbon dioxide to move out. Figure 5.35 shows a simple scheme of the partial pressures of oxygen as it moves from the air in which we live to the mitochondria of the tissue cells where it is utilized. The scheme is shown for a hypothetical perfect lung. This does not exist but we shall use it as a model to show how the actual lung falls short of this perfect scheme, both in health and disease. In practice there are four causes of hypoxemia, that is, an abnormally low P_{O_2} in arterial blood, and each will be considered in turn.

First look at the P_{O_2} in the air that is warmed and moistened as it is inhaled into the upper airways. If the body temperature is 37°C, the partial pressure of water vapor will be 47 mm Hg. Thus, assuming a barometric pressure of 760 mm Hg, the total dry gas pressure is $760 - 47 = 713$ mm Hg. Since the fractional concentration of oxygen in dry air is 0.2093, the P_{O_2} of moist inspired air is 0.2093×713 or 149 mm Hg (say, 150 mm Hg).

The scheme of Figure 5.35 is similar to the problem of designing a hydroelectric power station in a town which is 150 m below the level of a water reservoir. There is nothing we can do about the figure of 150, and we are therefore concerned about losses of pressure in the pipes leading from the reservoir to the turbines where the power is generated. In the same way, a physiologist (or physician) is concerned about losses in the partial pressure of oxygen from the atmosphere to the mitochondria.

The first surprising feature of Figure 5.35 is that by the time the air has reached the alveolar gas, it has already lost about one-third of the available oxygen pressure. What is the reason for this apparent extravagance? The level of P_{O_2} in the alveolar gas is set by a balance between two processes. First, there is the addition of oxygen by alveolar ventilation which was examined in Chapter 35. This occurs

with each breath and is therefore a discontinuous process. However, it can be looked upon as continuous with little error because the fluctuations in alveolar P_{O_2} during the normal breathing cycle are only 2 or 3 mm Hg. This is because the tidal volume is only about ½ liter, which is added to some 3 liters of gas already in the lung. Moreover, only about one-third of the oxygen which is added with each breath is taken up by the blood. The second process is the removal of oxygen from the lung by the pulmonary capillary blood. This again can be regarded as a continuous process with little error, though in fact it is pulsatile because of fluctuations in the capillary flow (Fig. 5.18). It is the balance between these two opposing processes that sets the normal level of P_{O_2} at 100 mm Hg.

In this hypothetical perfect lung, the blood draining from the lung will have the same partial pressure as the alveolar gas. When the blood reaches the systemic capillaries, however, oxygen diffuses out to the tissues where the P_{O_2} is much lower. Actually it is an oversimplification to talk of a single value for tissue P_{O_2} because it varies considerably within a particular tissue (Fig. 5.34) and between tissues in different parts of the body. In addition, there is evidence that the P_{O_2} at the mitochondria is extremely low, perhaps about 1 mm Hg. Nevertheless, it is useful to include this last link in the chain because it reminds us that, other things being equal, any losses of partial pressure along the chain must inevitably depress the P_{O_2} in the tissues where it is used.

Hypoventilation

The first cause of hypoxemia to be discussed is hypoventilation (Fig. 5.36). Recall that the level of P_{O_2} in alveolar gas is determined by a balance between two processes. On the one hand, oxygen is added by alveolar ventilation, and on the other, oxygen is removed by the blood flow. Generally the oxygen uptake by the body at rest is nearly constant. It is true that under exceptional conditions, such as hypothermia during cardiac surgery, for

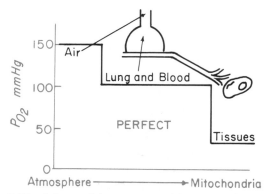

Figure 5.35.   Scheme of the oxygen partial pressures from air to tissues. This depicts a hypothetical perfect situation which does not exist in practice.

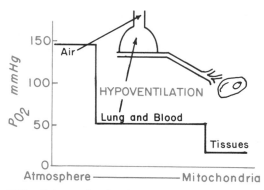

Figure 5.36.   Scheme showing the result of hypoventilation (compare Fig. 5.35). Note the depression of P_{O_2} in alveolar gas and, therefore, in the arterial blood and tissues.

example, it is possible to reduce resting oxygen consumption, but by and large there is little variation in the level of oxygen uptake at rest. Therefore, if for any reason the level of alveolar ventilation falls, so does the P_{O_2} in the alveolar gas and therefore in the arterial blood (Fig. 5.36). Essentially, the same argument shows that the P_{CO_2} must rise, since this is determined by a balance between the rate at which CO_2 is added to alveolar gas by the blood flow, and the rate at which it is washed out by alveolar ventilation. Thus hypoventilation always causes a fall in alveolar and arterial P_{O_2} and a rise in P_{CO_2}.

How far does the P_{CO_2} rise? In Chapter 35 we saw that there is a very simple relationship between the alveolar ventilation $\dot{V}_A$ and the alveolar P_{CO_2} which is known as the *alveolar ventilation equation:*

$$\dot{V}_A = \frac{\dot{V}_{CO_2}}{P_{A_{CO2}}} \cdot K$$

where K is a constant. This expression can be rearranged:

$$P_{A_{CO2}} = \frac{\dot{V}_{CO_2}}{\dot{V}_A} \cdot K$$

and since, in normal lungs, the alveolar and arterial P_{CO_2} are almost identical we can write:

$$Pa_{CO2} = \frac{\dot{V}_{CO_2}}{\dot{V}_A} \cdot K$$

This very important equation indicates that the level of P_{CO_2} in alveolar gas or arterial blood is inversely related to the alveolar ventilation. For example, if the alveolar ventilation is halved, the P_{CO_2} doubles. This is true only after a steady state is reestablished so that the CO_2 production rate is the same as before. In practice the P_{CO_2} will rise over

10–20 min, rapidly at first and then more slowly, as the body stores of CO_2 are gradually filled.

How far will the P_{O_2} fall during hypoventilation? One way to look at this is by means of an equation which is analogous to the alveolar ventilation equation for CO_2 given above,

$$(\dot{V}_{A_I} \cdot P_{I_{O2}} - \dot{V}_A \cdot P_{A_{O2}}) = \dot{V}_{O_2} \cdot K$$

where $\dot{V}_{A_I}$ and $\dot{V}_A$ are the inspired and expired alveolar ventilation, respectively. If the inspired and expired alveolar ventilations are assumed to be the same, this equation reduces to:

$$P_{I_{O2}} - P_{A_{O2}}{}^* = \frac{\dot{V}_{O_2}}{\dot{V}_A} \cdot K$$

(Compare the alveolar ventilation equation for CO_2). This means that the difference between the inspired and alveolar P_{O_2} is inversely related to the alveolar ventilation. However, this equation is not strictly correct because the inspired and expired alveolar ventilations are generally slightly different because less CO_2 is eliminated than oxygen is taken up (respiratory exchange ratio is less than 1). The asterisk in the equation means that it is only an approximation.

Another way of calculating the alveolar P_{O_2} during hypoventilation is from the *alveolar gas equation:*

$$P_{A_{O2}} = P_{I_{O2}} - \frac{P_{A_{CO2}}}{R} + F$$

where R is the respiratory exchange ratio ($\dot{V}_{CO_2}/\dot{V}_{O_2}$) and F is the correction factor which is generally small during air breathing (1–3 mm Hg) and can be ignored. Suppose a person with normal lungs takes an overdose of a barbiturate drug which depresses the alveolar ventilation. Typically the alveolar P_{CO_2} might rise from 40 to 60 mm Hg (the actual value will be determined by the alveolar ventilation equation). Assuming that the respiratory exchange ratio is one, the alveolar P_{O_2} will be given by:

$$P_{A_{O2}} = P_{I_{O2}} - \frac{P_{A_{CO2}}}{R} + F$$
$$= 149 - \frac{60}{1}$$
$$= 89 \text{ mm Hg}$$

Note that the alveolar P_{O_2} falls by 20 mm Hg which is the same amount that the P_{CO_2} rises. If $R = 0.8$, which is a more typical resting value:

$$P_{A_{O2}} = 149 - \frac{60}{0.8} + F$$
$$= 74 \text{ mm Hg}$$

In this case, the P_{O_2} falls more than the P_{CO_2} rises. Note, however, that in practical terms the hypoxemia is generally of minor importance compared with the hypercapnia (CO_2 retention) and consequent respiratory acidosis.

A patient with alveolar hypoventilation (and normal CO_2 production) always has an increased arterial P_{O_2}. However, the arterial P_{O_2} may be normal or high if he is receiving supplementary oxygen. Suppose the person with barbiturate intoxication referred to above is given 30% oxygen to breathe. Assuming that the ventilation remains unchanged, the alveolar P_{O_2} will rise from 74 to about 139 mm Hg (try the sum yourself). Thus a small increase in inspired P_{O_2} is very effective in eliminating the hypoxemia of hypoventilation.

Causes of hypoventilation include drugs such as morphine and barbiturates which depress the respiratory center, paralysis of the respiratory muscles, trauma to the chest wall, and any situation which greatly increases the resistance to breathing such as might occur when a diver breathes very dense gas at a great depth. If there is any doubt as to whether a patient is hypoventilating, measure the arterial P_{CO_2}. If it is not raised he is not hypoventilating.

Diffusion

Figure 5.35 and 5.36 imply that the partial pressures of oxygen in the alveolar gas and in the blood that drains from the lung are identical. However, in practice the P_{O_2} in the blood is always slightly lower, and the difference may become large in disease. To understand how this occurs we must consider the movement of gases across the blood-gas barrier.

It is now generally believed that O_2 and CO_2 cross the blood-gas barrier by simple diffusion. In the early part of this century, however, there were physiologists such as Bohr and Haldane who believed that the alveolar epithelium could actively secrete oxygen and CO_2 against a partial pressure gradient (Bohr, 1909; Haldane, 1922). Indeed, in the 1935 edition of Haldane and Priestley's book *Respiration* they devoted a whole chapter to the evidence for oxygen secretion. Such a mechanism was thought to be responsible for the very high P_{O_2} that occurs in the swim bladder of some fish, although it is now thought that this is the result of a counter-current exchange. However, early in this century a number of experiments were carried out at high altitudes in an attempt to prove the theory of oxygen secretion by the lung.

LAW OF DIFFUSION

The evidence now available indicates that the diffusion of O_2 and CO_2 through the blood-gas barrier obeys *Fick's law* as shown in Figure 5.37. This states that the volume of gas per unit time moving across a tissue sheet is directly proportional to the area of the sheet and the difference in partial pressures between the two sides but inversely proportional to the tissue thickness:

$$\dot{V}_{gas} = \frac{A}{T} \cdot D \cdot (P_1 - P_2)$$

This law was originally formulated in terms of concentrations but partial pressures are more convenient in this context; the two are directly related by Henry's law.

It is clear that the properties of a tissue sheet that would enhance diffusion are a large surface area and a small thickness. We saw in Chapter 35 that

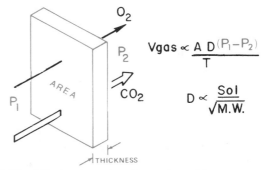

Figure 5.37. Diffusion through a tissue sheet. The amount of gas transferred is proportional to the area (*A*), a diffusion constant (*D*), and the difference in partial pressures (*P₁ − P₂*), and is inversely proportional to the thickness (*T*). The constant is proportional to the gas solubility *(Sol)*, but inversely proportional to the square root of its molecular weight *(M.W.)*.

the surface area of the blood-gas barrier of the human lung is between 50 and 100 m², with a thickness of less than 0.5 μm in many places. Its geometry is therefore ideally suited to rapid diffusion. The Fick equation also contains a diffusion constant (*D*) often called the diffusivity. This depends upon the structure of the sheet and the species of gas. For a given tissue, the diffusivity is proportional to the solubility (*S*) of the gas in the sheet and inversely proportional to the square root of the molecular weight *(MW)* of the gas:

$$D \propto \frac{S}{\sqrt{MW}}$$

If we compare the diffusion of O_2 and CO_2, we find that the latter has a great advantage. The solubility of CO_2 in normal saline at 37°C is approximately 24 times greater than that of O_2. Their MWs are not far apart, at 44 for CO_2 and 32 for O_2, so that when the square root of the quotient is taken, the result is 1.2. Thus, CO_2 diffuses about 20% slower by virtue of its high MW but some 24 times faster by virtue of its high solubility. The net result is that the diffusion rate of CO_2 through a tissue sheet is about 20 times that of O_2.

DIFFUSION AND PERFUSION LIMITATIONS

Consider what happens when blood enters the pulmonary capillary of an alveolus that contains a foreign gas such as carbon monoxide (CO) or nitrous oxide. How rapidly will the partial pressure in the blood rise? Figure 5.38 shows the time courses as the blood moves through the capillary, a process which takes about 0.75 s. Look first at CO. When the red cells enter the capillary, CO moves rapidly across the extremely thin blood-gas barrier from the alveolar gas into the red cell. As a result, the partial pressure of CO in the cell rises. However, as we saw in the last chapter, blood can combine with a large amount of CO for a very small rise in partial pressure (Fig. 5.28). This is because of a tight bond that CO forms with hemoglobin. Thus, as the blood moves along the capillary, the P_{CO} in the blood hardly changes; consequently, no appreciable back pressure develops, and the gas continues to move rapidly across the alveolar wall. It is clear, therefore, that the amount of CO that gets into the blood is limited by the diffusion properties of the blood-gas barrier, not by the amount of blood available.*

---

*This introductory description of CO transfer is not completely accurate because of the limit imposed by the rate of reaction of CO with hemoglobin (see later). Also, the example assumes that the partial pressure of CO in the alveolar gas is very small, typically about 1 mm Hg.

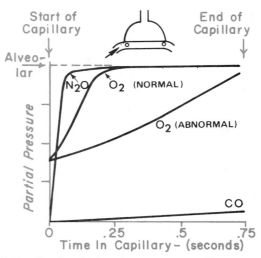

Figure 5.38. Uptake of CO, N_2O, and O_2 along the pulmonary capillary. Note that the blood partial pressure of N_2O virtually reaches that of alveolar gas very early in the capillary so that the transfer of this gas is perfusion limited. By contrast, the partial pressure of CO in the blood is almost unchanged so that its transfer is diffusion limited. O_2 can be perfusion or partly diffusion limited, depending on the conditions.

The rate of transfer of carbon monoxide is therefore said to be *diffusion limited*.

Contrast this with the time course of nitrous oxide. When this gas moves across the alveolar wall into the blood, no combination with hemoglobin takes place. As a result the blood has nothing like the avidity for nitrous oxide that it had for CO, and the partial pressure rises rapidly. Indeed Figure 5.38 shows that the partial pressure of nitrous oxide in the blood virtually reaches that of the alveolar gas by the time the red cell is only one-fourth of the way along the capillary. After this point almost no nitrous oxide is transferred. Thus, the amount of this gas taken up by the blood depends entirely on the amount of available blood flow and not at all on the diffusion properties of the blood-gas barrier. The transfer of nitrous oxide is therefore *perfusion limited*.

Now consider oxygen. Its time course lies between that of CO and nitrous oxide. Oxygen combines with hemoglobin (unlike nitrous oxide) but with nothing like the avidity of CO. In other words, the rise in partial pressure when oxygen enters a red blood cell is much greater than is the case for the same volume of CO. Figure 5.38 shows that the P_{O_2} of the red cell as it enters the capillary is already about four-tenths of the alveolar value because of the oxygen in mixed venous blood. Under resting conditions the capillary P_{O_2} virtually reaches that of alveolar gas when the red cell is about one-third of the way along the capillary. Therefore, under these conditions oxygen transfer is perfusion limited like nitrous oxide. In some abnormal circumstances, however, when the diffusion properties of the lung are impaired (for example, because of thickening of the alveolar wall by disease), the blood P_{O_2} does not reach the alveolar value by the end of the capillary, and now there is some diffusion limitation as well.

OXYGEN UPTAKE ALONG THE PULMONARY CAPILLARY

Figure 5.39 shows time courses for P_{O_2} in the pulmonary capillary of the normal lung and in a lung in which the blood-gas barrier is thickened by disease. Note that the P_{O_2} in the red cell as it enters the capillary is normally about 40 mm Hg. Across the blood-gas barrier, less than 0.5 μm away, is the alveolar P_{O_2} of 100 mm Hg (compare Figs. 5.2 and 5.35). Oxygen floods down this large pressure gradient, and the P_{O_2} in the red cell rises rapidly; indeed, we have seen that it very nearly reaches the P_{O_2} of the alveolar gas by the time the red cell is only one-third of its way along the capillary. Thus, under normal circumstances the difference in P_{O_2} between alveolar gas and end-capillary blood is immeasurably small—a mere fraction of a mm Hg (see also Fig.

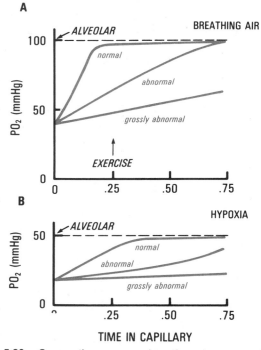

Figure 5.39. Oxygen time courses along the pulmonary capillary. *A* shows the situation when the alveolar P_{O_2} is 100 mm Hg and the diffusion properties of the blood-gas barrier are normal, abnormal, or grossly abnormal because of thickening of the alveolar wall. Note that exercise reduces the time available for oxygenation. *B* shows the situation when the alveolar P_{O_2} is abnormally low. Note the slower time courses. A combination of alveolar hypoxia and exercise can result in hypoxemia.

5.42). In other words, the diffusion reserves of the normal lung are enormous.

During severe exercise the pulmonary blood flow is greatly increased, and the time spent by the red cell in the capillary, normally about 0.75 s, may be reduced to as little as one-third of this. This reduced time for oxygenation greatly stresses the diffusion reserves of the lung, but in normal subjects breathing air, there is generally still no measurable reduction in end-capillary P_{O_2}. If the blood-gas barrier is markedly thickened, however, so that oxygen diffusion is impeded, the rate of rise of P_{O_2} in the red blood cell is correspondingly slower, and the P_{O_2} may not reach that of alveolar gas before the time available for oxygenation in the capillary has run out. In this case a measurable P_{O_2} difference between alveolar gas and end-capillary blood may occur.

Another way of stressing the diffusion properties of the lung is to lower the alveolar P_{O_2} (Fig. 5.39*B*). Suppose that the P_{O_2} has been reduced to 50 mm Hg, either by the subject going to high altitude, or breathing a low oxygen mixture. Now the rise in P_{O_2} along the capillary is relatively slow, and failure to reach the alveolar P_{O_2} is more likely. The slow rate of rise of P_{O_2} is associated with the fact that we are operating on a much steeper part of the O_2 dissociation curve (Fig. 5.26) and the hemoglobin is therefore more avid for the O_2 (compare the CO curve in Fig. 5.38). It can be shown that the rate of rise of partial pressure is determined by the ratio of the solubility of the gas in the blood-gas barrier, to the effective "solubility" of the gas in blood, that is, the slope of the dissociation curve (Wagner, 1977). Thus when the P_{O_2} is low and the O_2 dissociation curve is steep, there is a slower rate of rise of P_{O_2} in the capillary. Heavy exercise at very high altitude is one of the few situations where diffusion impairment of oxygen in normal subjects can be convincingly demonstrated. By the same token, a patient with a thickened blood-gas barrier will be most likely to show evidence of diffusion impairment if he breathes a low oxygen mixture, especially if he exercises as well.

MEASUREMENT OF DIFFUSING CAPACITY

In order to measure the diffusion properties of the lung, we must clearly choose a gas whose uptake is diffusion limited. Although this is true of oxygen under some very special conditions, and although at one time this gas was therefore used to measure the diffusing capacity of the lung, CO is the obvious choice.

Fick's law (Fig. 5.37) states that:

$$\dot{V}_{gas} = \frac{A}{T} D (P_1 - P_2)$$

However, for a complex structure like the blood-gas barrier of the lung, it is not possible to measure the area and thickness during life. Instead, the equation is rewritten:

$$\dot{V}_{gas} = D_L \cdot (P_1 - P_2)$$

where D_L is called the diffusing capacity of the lung, a term which includes the area, thickness, and diffusion properties of the tissue sheet and the gas concerned. Thus the diffusing capacity for carbon monoxide is given by:

$$D_L = \frac{\dot{V}_{CO}}{P_1 - P_2}$$

where P_1 and P_2 are the partial pressures of alveolar gas and capillary blood, respectively. But as we saw in Figure 5.38, the partial pressure of CO in capillary blood is so small that it usually can be neglected, although exceptions to this sometimes occur in smokers who have significant levels of CO in their blood. In general, however:

$$D_L = \frac{\dot{V}_{CO}}{P_{A_{CO}}}$$

Or, in other words, the diffusing capacity of the lung for CO is the volume of CO transferred in ml per min per mm Hg alveolar partial pressure.

Several techniques for making this measurement are available. In the *single breath method,* a single inspiration of a dilute mixture of CO is made, and the rate of disappearance of CO from the alveolar gas during a breath holding period of 10 s is calculated. This is usually done by measuring the inspired and expired concentrations of CO with an infrared analyzer. The alveolar concentration of CO is not constant during the breath holding period but allowance can be made for that. Helium is also added to the inspired gas to give a measurement of lung volume by dilution. This method is extensively used in clinical lung function laboratories.

In the *steady-state method,* the subject breathes a low concentartion of CO (about 0.1%) for ½ min or so until a steady state has been reached. The constant rate of disappearance of CO from alveolar gas is then measured for a further short period along with the alveolar concentration. The latter can be obtained by sampling gas at the end of each expiration, or by making allowance for the dead space using the Bohr method (see Chapter 35).

The normal value for the diffusing capacity for CO at rest is about 25 ml/min/mm Hg, and it increases to two or three times this value on exer-

cise. It is reduced by diseases which thicken the blood-gas barrier, such as interstitial lung diseases, including sarcoidosis and asbestosis.

REACTION RATES WITH HEMOGLOBIN

So far we have assumed that all the resistance to the movement of O_2 and CO resides in the barrier between blood and gas. However, Figure 5.2 shows that the path length from the alveolar wall to the center of a red blood cell exceeds that in the wall itself so that some of the diffusion resistance is located *within* the capillary. In addition, there is another type of resistance to gas transfer which is most conveniently discussed at this point, that is, the resistance caused by the finite rate of reaction of oxygen or CO with hemoglobin inside the red blood cell.

When oxygen is added to blood, its combination with hemoglobin is quite fast, being well on the way to completion in 0.2 s. However, oxygenation occurs so rapidly in the pulmonary capillary (Fig. 5.39) that even this rapid reaction significantly delays the loading of oxygen by the red cell. Thus, the uptake of oxygen can be regarded as occurring in two stages: (a) diffusion of oxygen through the blood-gas barrier (including the plasma and red cell interior); and (b) reaction of the oxygen with hemoglobin (Fig. 5.40). Each stage can be regarded as contributing a resistance to the movement of oxygen, and it is possible to sum the two resultant resistances to produce an overall "diffusion" resistance (Roughton and Forster, 1957).

The diffusing capacity of the lung can be written as:

$$D_L = \frac{\dot{V}_{gas}}{P_A - P_c} \quad \text{or} \quad P_A - P_c = \frac{\dot{V}_{gas}}{D_L}$$

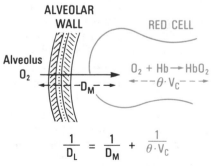

$$\frac{1}{D_L} = \frac{1}{D_M} + \frac{1}{\theta \cdot V_C}$$

Figure 5.40. The diffusing capacity of the lung D_L is made up of two components, that due to the diffusion process itself and that attributable to the time taken for O_2 (or CO) to react with hemoglobin *(Hb).* θ, rate of reaction of O_2 (or CO) with hemoglobin; V_c, volume of capillary blood; D_m, diffusing capacity of the membrane.

where P_C is the partial pressure inside the red cell. By analogy, the diffusing capacity of the membrane can be written:

$$D_m = \frac{\dot{V}_{gas}}{P_A - P_{pl}} \quad \text{or} \quad P_A - P_{pl} = \frac{\dot{V}_{gas}}{D_M}$$

where P_{pl} means the plasma partial pressure. Furthermore, we can write a similar equation to represent the rate of reaction of oxygen (or CO) with hemoglobin. This rate can be described by a variable θ whose units are ml of oxygen per min per mm Hg per ml of blood. This is analogous to the "diffusing capacity" of 1 ml blood, and when multiplied by the volume of capillary blood (V_c) gives the effective "diffusing capacity" of the rate of reaction of oxygen with hemoglobin:

$$\theta \cdot V_c = \frac{\dot{V}_{gas}}{P_{pl} - P_c} \quad \text{or} \quad P_{pl} - P_c = \frac{\dot{V}_{gas}}{\theta \cdot V_c}$$

now

$$P_A - P_c = (P_A - P_{pl}) + (P_{pl} - P_c)$$

$$\frac{\dot{V}_{gas}}{D_L} = \frac{\dot{V}_{gas}}{D_M} + \frac{\dot{V}_{gas}}{\theta \cdot V_c}$$

and therefore

$$\frac{1}{D_L} = \frac{1}{D_M} + \frac{1}{\theta \cdot V_c}$$

This holds for both O_2 and CO.

When the diffusing capacity for CO uptake is measured, the value of θ depends on the prevailing P_{O_2} in the alveolar gas because of the competition of O_2 and CO for hemoglobin. Therefore, by performing the diffusing capacity measurement at different levels of alveolar P_{O_2}, two equations for D_L are obtained with two unknowns, D_M and V_C. In this way D_M and V_c can be separately evaluated.

In practice, the resistances offered by the membrane and blood components to the uptake of CO are approximately equal. Thus, diseases which cause the capillary blood volume to fall can reduce the diffusing capacity of the lung. This discussion also applies to oxygen uptake along the pulmonary capillary, although the values of θ for this gas are different.

The measurement of θ is difficult because the unstirred layer of plasma immediately adjacent to the red cell offers its own resistance in most techniques. For this reason there is uncertainty about the values of θ in the living lung.

CARBON DIOXIDE TRANSFER ALONG THE PULMONARY CAPILLARY

We have seen that the rate of diffusion of CO_2 through tissue is about 20 times faster than oxygen because of the much higher solubility of CO_2 (Fig. 5.37). At first sight, therefore, it seems unlikely that CO_2 elimination could be adversely affected by diffusion difficulties, and, indeed, for many years that was held to be so. The reaction of CO_2 with blood, however, is complex, involving the hydration of CO_2, the chloride shift across the red cell membrane, and the formation of carbamino compounds (see Fig. 5.29). Some of these reactions are relatively slow. Moreover, the CO_2 dissociation curve is quite steep (Fig. 5.32A) and, as we have seen, this tends to slow the rate of change of partial pressure of a gas in the pulmonary capillary (Fig. 5.39B).

Recent work suggests that a partial pressure difference for CO_2 can develop between end-capillary blood and alveolar gas if the blood-gas barrier is diseased. Figure 5.41 shows the calculated normal time course for CO_2 and how this might be affected by thickening of the blood-gas barrier. Note that the P_{CO_2} of the blood as it enters the capillary is about 45–47 mm Hg and that the normal P_{CO_2} of alveolar

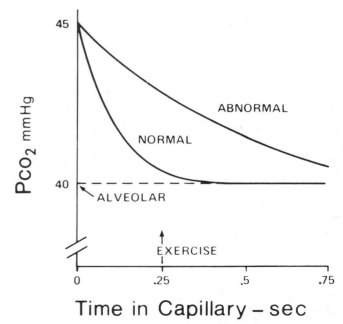

Figure 5.41. Change in P_{CO_2} along the pulmonary capillary when the diffusion properties are normal and abnormal (compare the change in P_{O_2} in Fig. 5.39). (From Wagner and West, 1972.)

gas is about 40 mm Hg. The diagram shows that the time taken for the blood to reach virtually the same partial pressure as alveolar gas is similar to that for oxygen under normal conditions (Fig. 5.39A) so that a considerable time reserve is available. However, when the diffusing capacity of the membrane is reduced to one-fourth of its normal value, a small difference between end-capillary blood and alveolar gas develops. To what extent the elimination of CO_2 can be limited by such a mechanism is unknown at the present time.

Shunt

This is the third cause of hypoxemia. A shunt is defined as any mechanism by which blood that has not been through ventilated areas of lung is added to the systemic arteries. This is shown schematically in Figure 5.42. Even in the normal lung there is some blood in this category. For example, some of the bronchial arterial blood which has supplied the walls of the large airways finds its way into the pulmonary veins downstream of the pulmonary capillaries. Since this blood has been partly depleted of oxygen, its addition to the pulmonary venous blood reduces the P_{O_2} there. Again, there is a small amount of blood which drains from the myocardium of the left ventricle directly into the cavity of the ventricle via Thebesian veins. This blood contributes to a lowering of systemic arterial P_{O_2}.

In normal subjects the reduction of arterial P_{O_2} from these sources amounts to only 5 mm Hg or so. However, in some patients, especially those with congenital heart disease who have blood moving from the right to the left sides of the heart through septal defects, the depression of arterial P_{O_2} by shunt may be very severe indeed.

When the shunt is caused by the addition of mixed venous blood (pulmonary arterial) to blood draining from the capillaries (pulmonary venous), it is possible to calculate the amount of shunt flow, as shown in Figure 5.43. The total amount of oxygen leaving the system is the total blood flow *($\dot{Q}_T$)* multiplied by the oxygen concentration in the arterial blood (Ca_{O_2}):

$$\dot{Q}_T \times Ca_{O_2}$$

This must equal the sum of the amounts of oxygen in the shunted blood:

$$\dot{Q}_s \times C_{\bar{v}O_2}$$

and end-capillary blood:

$$(\dot{Q}_T - \dot{Q}_S) \times Cc'_{O_2}$$

Thus,

$$\dot{Q}_T \times Ca_{O_2} = (\dot{Q}_S \times C_{\bar{v}O_2}) + (\dot{Q}_T - \dot{Q}_S) \times Cc'_{O_2}$$

Rearranging, we have:

$$\frac{\dot{Q}_S}{\dot{Q}_T} = \frac{Cc'_{O_2} - Ca_{O_2}}{Cc'_{O_2} - C_{\bar{v}O_2}}$$

This is known as the *shunt equation.*

The oxygen concentration in arterial blood can be obtained by arterial puncture. End-capillary blood cannot be sampled directly, but it can be calculated if the P_{O_2} of alveolar gas is known, assuming no alveolar to end-capillary difference caused by diffusion impairment. Mixed venous blood can be

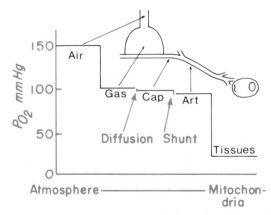

Figure 5.42. Scheme of oxygen transfer from air to tissues showing the depression of arterial P_{O_2} caused by diffusion and shunt. (From West, 1985b.)

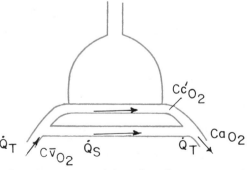

Figure 5.43. Measurement of shunt flow. The oxygen carried in the arterial blood equals the sum of the oxygen carried in the capillary blood and that in the shunted blood (see text). $\dot{Q}_T$, total blood flow; Ca_{O_2}, oxygen concentration in arterial blood; $\dot{Q}_S$, shunted blood flow; Cc'_{O_2} and $C\bar{v}_{O_2}$ are oxygen concentrations in end-capillary and mixed venous blood, respectively.

obtained only from a catheter in the pulmonary artery. In practice the denominator in the equation is sometimes assumed to be 5 ml oxygen/100 ml blood, but this assumption can lead to substantial errors if the cardiac output is abnormally high or low.

Notice that this equation is only strictly true if the shunted blood has the same composition as mixed venous blood. This is not always the case, for example, when the shunted blood comes from the bronchial veins. In such a case it may be useful to calculate an "as if" shunt, that is, what the shunt *would* be if all the observed depression of arterial oxygen content were caused by the addition of true mixed venous blood.

An important feature of a shunt is that if the subject breathes 100% oxygen, the arterial P_{O_2} does not rise to the level it does in a normal subject. This is because the shunted blood which bypasses ventilated alveoli is never exposed to the high alveolar P_{O_2} resulting from oxygen breathing, so that this blood continues to depress the arterial P_{O_2}. Some elevation of the arterial P_{O_2} occurs, however, because of the oxygen added to the capillary blood. Most of the added oxygen is in the dissolved form rather than attached to hemoglobin because the blood which is perfusing ventilated alveoli is nearly fully saturated (see Fig. 5.26).

Figure 5.44 shows why a relatively small shunt results in a large depression of arterial P_{O_2} during 100% oxygen breathing. A small shunt depresses the oxygen *content* of arterial blood (see the shunt equation given above), and when the alveolar P_{O_2} is very high, the curve relating oxygen content and partial pressure is nearly flat. In fact, the slope reflects only the addition of dissolved oxygen (compare Fig. 5.26). Therefore, a given reduction in oxygen content causes a much larger fall in arterial P_{O_2} when the alveolar P_{O_2} is high than when the alveolar P_{O_2} is normal or low. Figure 5.44 shows that the fall in O_2 content of venous blood is the same, but the depression of P_{O_2} is much less.

Does a shunt result in an elevated arterial P_{CO_2}? This might be expected since the mixed venous blood is relatively rich in CO_2. In practice, however, the arterial P_{CO_2} is usually not raised because any tendency for it to increase stimulates ventilation (Chapter 40). The result is that the P_{CO_2} in the end-capillary blood is reduced until the arterial P_{CO_2} is normal. In fact, some patients with a large shunt have an abnormally low arterial P_{CO_2} because of the excessive stimulation of ventilation by the hypoxemia.

Ventilation-Perfusion Inequality

The fourth cause of hypoxemia is both the most common in practice and also the most difficult to understand. It is the mismatching of ventilation and blood flow within the lung; the key to understanding how this affects gas exchange is the ventilation-perfusion ratio.

VENTILATION-PERFUSION RATIO

Consider a model of a lung unit (Fig. 5.5) in which the uptake of oxygen is being mimicked using dye and water (Fig. 5.45). Powdered dye is continuously poured into the unit to represent the addition of oxygen by alveolar ventilation. (In practice, ventilation is discontinuous but it makes little difference to the result.) Water is pumped continuously through the model to represent the blood flow which removes the oxygen. A stirrer mixes the alveolar contents, a process normally accomplished by gaseous diffusion. The key question is: What deter-

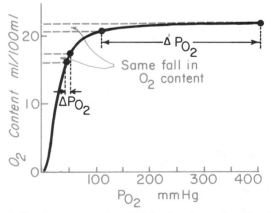

Figure 5.44. Depression of arterial P_{O_2} by shunt when the alveolar P_{O_2} is high. The addition of a small amount of shunted blood lowers the arterial O_2 content (shunt equation), and this greatly reduces the arterial P_{O_2} because the O_2 dissociation curve is so nearly flat. The depression of P_{O_2} in venous blood is much less.

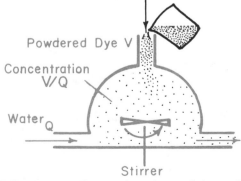

Figure 5.45. Model to illustrate how the ventilation-perfusion ratio determines the P_{O_2} in a lung unit. Powdered dye is added by ventilation of the rate V and removed by blood flow Q to represent the factors controlling alveolar P_{O_2}. The concentration of dye is given by V/Q. (From West, 1985b.)

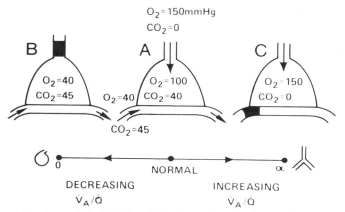

Figure 5.46. Effect of alterations of the ventilation-perfusion ratio on the P_{O_2} and P_{CO_2} in a lung unit. $\dot{V}_A/\dot{Q}$, ventilation-perfusion ratio. (From West, 1985b.)

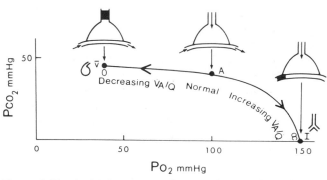

Figure 5.47. O_2-CO_2 diagram showing a ventilation-perfusion ratio line. The P_{O_2} and P_{CO_2} of a lung unit move along this line from the mixed venous point v to the inspired gas point I as its ventilation-perfusion ratio is increased (compare Fig. 5.46). (From West, 1985b.)

mines the concentration of dye (or oxygen) in the alveolar compartment and therefore in the effluent water (or blood)?

It is clear that both the rate at which the dye is added (ventilation) and the rate at which water is pumped (blood flow) will affect the concentration of dye in the model. What may not be intuitively clear is that the concentration of dye is determined by the ratio of these rates. In other words, if dye is added at the rate of V g/min and water is pumped through at Q liters/min, the concentration of dye in the alveolar compartment and effluent water is V/Q g/liter.

In exactly the same way, the concentration of oxygen (or, better, P_{O_2}) in any lung unit is determined by the ratio of ventilation to blood flow, and this is true not only for oxygen but also for CO_2, nitrogen, or any other gas which is present under steady-state conditions. This is the reason why the ventilation-perfusion ratio plays such a key role in pulmonary gas exchange.

EFFECT OF ALTERING THE VENTILATION-PERFUSION RATIO OF A LUNG UNIT

Consider first a lung unit with a normal ventilation-perfusion ratio of about 1 (compare Fig. 5.5). This is shown in Figure 5.46A. The inspired air has a P_{O_2} of 150 mm Hg (Fig. 5.35) and a P_{CO_2} of zero. The mixed venous blood entering the unit has a P_{O_2} of 40 mm Hg and P_{CO_2} of 45 mm Hg. The alveolar P_{O_2} of 100 mm Hg is determined by a balance between the addition of oxygen by ventilation and its removal by blood flow. The normal alveolar P_{CO_2} of 40 mm Hg is set similarly.

Now suppose that the ventilation-perfusion ratio of the unit is gradually reduced by obstructing its ventilation, leaving its blood flow unchanged (Fig. 5.46B). It is clear that the O_2 in the unit will fall and

the CO_2 will rise, although the relative changes of these two are not immediately obvious.† It is easy to predict, however, what will eventually happen when the ventilation is completely abolished (ventilation-perfusion ratio of zero). Now the O_2 and CO_2 of alveolar gas and of end-capillary blood must be the same as those of mixed venous blood. (In practice, completely obstructed units eventually collapse, but such long-term effects can be neglected at the moment.) Note that in this model we are assuming that what happens in one lung unit out of a very large number does not affect the composition of the mixed venous blood.

Suppose instead that the ventilation-perfusion ratio is increased by gradually obstructing blood flow (Fig. 5.46C). Now the alveolar O_2 rises and the CO_2 falls, eventually reaching the composition of inspired gas when blood flow is abolished (ventilation-perfusion ratio of infinity). Thus, as the ventilation-perfusion ratio of the unit is altered, its gas composition approaches either that of mixed venous blood or that of inspired gas.

A convenient way of depicting these changes is to use the O_2-CO_2 diagram (Rahn and Fenn, 1955; Fig. 5.47). Here the P_{O_2} is plotted on the X axis and P_{CO_2} on the Y axis. First, locate the normal alveolar gas composition, point A ($P_{O_2} = 100$, $P_{CO_2} = 40$). If blood has in fact equilibrated with alveolar gas along the capillary (Fig. 5.39 and 5.41), this point can equally well represent the end-capillary blood. Next, find the mixed venous point $\bar{v}$ ($P_{O_2} = 40$, $P_{CO_2} = 45$). The

---

†The alveolar gas equation is not completely applicable here because the respiratory exchange ratio is not constant. The appropriate equation is

$$\frac{\dot{V}_A}{\dot{Q}} = \frac{8.63 \cdot R \cdot (Ca_{O_2} - C\bar{v}_{O_2})}{P_{A_{CO_2}}}$$

This is called the ventilation-perfusion ratio equation.

bar above v means "mixed" or "mean." Finally, find the inspired point I ($P_{O_2} = 150$, $P_{CO_2} = 0$). Also note the similarities between Figs. 5.46 and 5.47.

The line joining $\bar{v}$ to I passing through A shows the changes in alveolar gas (and end-capillary blood) composition that can occur when the ventilation-perfusion ratio is either decreased below normal ($A \rightarrow \bar{v}$) or increased above normal ($A \rightarrow I$). Indeed, this line indicates *all* the possible alveolar gas compositions in a lung which is supplied with gas of composition I and blood of composition $\bar{v}$. For example, such a lung could not contain an alveolus with a P_{O_2} of 70 and P_{CO_2} of 30 mm Hg, since this point does not lie on the ventilation-perfusion ratio line. This alveolar composition could exist, however, if the mixed venous blood or inspired gas were changed so that the line then passed through this point.

REGIONAL GAS EXCHANGE IN THE LUNG

In many diseases, such as the very common chronic obstructive lung diseases (including chronic bronchitis and emphysema), there is evidence of great mismatching of ventilation and blood flow throughout the lung. As a consequence, lung units will be dispersed most of the way along a ventilation-perfusion ratio line as shown in Figure 5.47. It is instructive to look at the pattern of regional gas exchange in the normal upright lung. Here the distribution of ventilation-perfusion ratios follows a simple topographical pattern, and the effects on regional gas exchange are easily appreciated. In addition there is some evidence that these regional differences affect the localization of some types of disease, for example, adult tuberculosis.

It was pointed out earlier that blood flow is unevenly distributed in the normal upright human lung (Fig. 5.19). At normal lung volumes, blood flow increases markedly from apex to base. Ventilation also increases down the lung (see Chapter 39), but the regional differences are less marked than those for blood flow. The approximate distributions of blood flow and ventilation are depicted in Figure 5.48, where both are taken to change linearly when plotted against rib number. Note that the ventilation-perfusion ratio increases from a low value near the base of the lung to a very high value near the apex.

Since the ventilation-perfusion ratio determines the gas exchange in any lung unit, the resulting regional differences in gas exchange can be calculated from the principles shown in Figures 5.47 and 5.48. The details of such calculations need not be given here; suffice it to say that they are greatly complicated by the O_2 and CO_2 dissociation curves

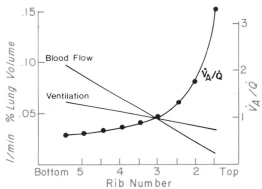

Figure 5.48. Distribution of ventilation and blood flow down the upright lung (compare to Fig. 5.71). Note that the ventilation-perfusion ratio ($\dot{V}_A/\dot{Q}$) decreases down the lung. (From West, 1985b.)

which are nonlinear and interdependent (Figs. 5.26 and 5.31). Originally the solutions to the ventilation:perfusion ratio equation were obtained graphically (Rahn and Fenn, 1955), but they are now more easily found by numerical analysis with a computer (West and Wagner, 1977).

Figure 5.49 shows typical values calculated for a lung model divided into nine imaginary horizontal slices. (Naturally there will be variations between subjects; the chief aim of this approach is to describe the principles underlying gas exchange). Note first that the volume of the lung in the slices is less near the apex than the base. Ventilation is less at the top than the bottom, but the differences of blood flow are much more marked. Consequently, the ventilation-perfusion ratio decreases down the lung, and all the differences in gas exchange follow from this. Note that the P_{O_2} changes by over 40 mm

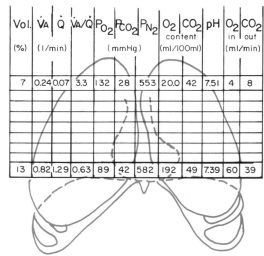

Vol.	$\dot{V}_A$	$\dot{Q}$	$\dot{V}_A/\dot{Q}$	P_{O_2}	P_{CO_2}	P_{N_2}	O_2 content	CO_2	pH	O_2 in	CO_2 out
(%)	(l/min)			(mmHg)			(ml/100ml)			(ml/min)	
7	0.24	0.07	3.3	132	28	553	20.0	42	7.51	4	8
13	0.82	1.29	0.63	89	42	582	19.2	49	7.39	60	39

Figure 5.49. Regional differences in gas exchange down the normal lung. Only the apical and basal values are shown for clarity. $\dot{V}_A$, alveolar ventilation; $\dot{Q}$, blood flow; P_{O_2}, partial pressure of oxygen; P_{CO_2}, partial pressure of CO_2; P_{N_2}, partial pressure of nitrogen.

Hg, while the difference in P_{CO_2} between apex and base is much less. (Incidentally, the high P_{O_2} at the apex probably accounts for the predilection of adult tuberculosis for this region since it provides a more favorable environment for this organism.) The variation in P_{N_2} is, in effect, by default since the total pressure in the alveolar gas is the same throughout the lung.,

The regional differences of P_{O_2} and P_{CO_2} imply differences in the end-capillary contents of these gases which can be obtained from the appropriate dissociation curves (Figs. 5.26 and 5.31). Note the surprisingly large difference in pH down the lung which reflects the considerable variation in P_{CO_2} of the blood in the presence of a constant base excess. The minimal contribution to overall O_2 uptake made by the apex can be mainly attributed to the very low blood flow there. The differences in CO_2 output are much less since these can be shown to be more closely related to ventilation. As a result, the respiratory exchange ratio (CO_2 output:O_2 uptake) is higher at the apex than the base. During exercise, when the distribution of blood flow becomes more uniform, the apex assumes a more appropriate share of oxygen uptake.

VENTILATION-PERFUSION INEQUALITY AS A CAUSE OF HYPOXEMIA

While the regional differences in gas exchange discussed above are of interest, more important to the body as a whole is whether uneven ventilation and blood flow affect the overall gas exchange of the lung, that is, its ability to take up O_2 and put out CO_2. It transpires that a lung with ventilation-perfusion inequality is not able to transfer as much O_2 and CO_2 as a lung which is uniformly ventilated and perfused, other things being equal. Or, if the same amounts of gas are being transferred (because these are set by the metabolic demands of the body), the lung with ventilation-perfusion inequality cannot maintain as high an arterial P_{O_2} or as low an arterial P_{CO_2} as a homogeneous lung, again other things being equal.

The reason why a lung with uneven ventilation and blood flow has difficulty oxygenating arterial blood can be illustrated by looking at the differences down the upright lung (Fig. 5.50). Note that the P_{O_2} at the apex is some 40 mm Hg higher than at the base of the lung. However, the major share of the blood leaving the lung comes from the lower zones where the P_{O_2} is low. This has the result of depressing the arterial P_{O_2}. By contrast, the expired alveolar gas comes more uniformly from apex and base because the differences of ventilation are much less than those for blood flow (Fig. 5.48). By the same

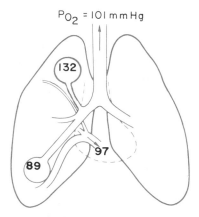

Figure 5.50. Depression of the arterial P_{O_2} by ventilation-perfusion inequality. In this diagram of the upright lung, only alveoli at the apex and base are shown. The relative sizes of the airways and blood vessels indicate their relative ventilations and blood flows. Because most of the blood comes from the poorly oxygenated base, depression of the blood P_{O_2} is inevitable. (From West, 1963.)

reasoning, the arterial P_{CO_2} will be elevated since this is higher at the base of the lung than at the apex (Fig. 5.49).

An additional reason why uneven ventilation and blood flow depress the arterial P_{O_2} is shown in Figure 5.51. This depicts three groups of alveoli with low, normal, and high ventilation-perfusion ratios. The oxygen contents of the effluent blood are 16, 19.5, and 20 ml/100 ml, respectively, and thus the units with the high ventilation-perfusion ratio add relatively little oxygen to the blood compared with the decrement caused by the alveoli with the low ventilation-perfusion ratio. Thus, the mixed capillary blood has a lower oxygen content than blood from units with a normal ventilation-perfusion ratio. This can be explained by the non-linear shape of the oxygen dissociation curve, which means that although units with a high ventilation-perfusion ratio have a relatively high P_{O_2}, this does not

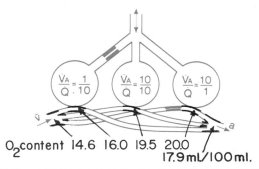

Figure 5.51. Additional reason for the depression of arterial P_{O_2} by mismatching of ventilation and blood flow. The lung units with a high ventilation-perfusion ratio add relatively little oxygen to the blood compared with the decrement caused by alveoli with a low ventilation-perfusion ratio. $\dot{V}_A$, alveolar ventilation; $\dot{Q}$, blood flow. (From West, 1985b.)

increase the oxygen content of their blood very much. This additional reason for the depression of P_{O_2} does not apply to the elevation of the P_{CO_2} because the CO_2 dissociation curve is almost linear in the working range.

The net result of these mechanisms is a depression of the arterial P_{O_2} below that of the mixed alveolar P_{O_2}—the so-called alveolar-arterial oxygen difference. In the normal upright lung this is of trivial magnitude, being only about 4 mm Hg as a result of ventilation-perfusion inequality. Its development is described here only to illustrate how uneven ventilation and blood flow must result in depression of the arterial P_{O_2}. In lung disease, the lowering of arterial P_{O_2} by this mechanism can be extreme.

Oxygen breathing is very effective in raising the arterial P_{O_2} of a patient whose hypoxemia is caused by ventilation-perfusion inequality. The reason is that lung units which are ventilated (if only poorly) will eventually have all their nitrogen washed out; their alveolar P_{O_2} is then given by the expression:

$$P_{O_2} = P_B - P_{H_2O} - P_{CO_2}$$

The basis for this equation is that the sum of all the partial pressures must equal the barometric pressure. Since the P_{CO_2} is limited by the value in mixed venous blood (Fig. 5.47), the alveolar P_{O_2} of all units generally will now exceed 600 mm Hg. In practice, however, the arterial P_{O_2} may take many minutes to complete its rise because the nitrogen is washed out of poorly ventilated regions so slowly.

VENTILATION-PERFUSION INEQUALITY AS A CAUSE OF CARBON DIOXIDE RETENTION

Imagine a lung which is uniformly ventilated and perfused and which is transferring normal amounts of O_2 and CO_2. Suppose that in some magical way the matching of ventilation and blood flow is suddenly disturbed while everything else remains unchanged. What happens to gas exchange? It transpires that the effect of this "pure" ventilation-perfusion inequality (that is, everything else held constant) is to reduce *both* the O_2 uptake and CO_2 output of the lung. In other words, the lung becomes less efficient as a gas exchanger for both gases. Thus, mismatching ventilation and blood flow must cause both hypoxemia and hypercapnia (CO_2 retention), other things being equal.

In practice, however, patients with undoubted ventilation-perfusion inequality often have a normal arterial P_{CO_2}. The reason for this is that whenever the chemoreceptors sense a rising P_{CO_2}, there is an increase in ventilatory drive (Chapter 40), with a consequent increase in ventilation to the alveoli.

This usually effectively returns the arterial P_{CO_2} to normal. Such patients, however, can only maintain a normal P_{CO_2} at the expense of this increased ventilation to their alveoli; the ventilation in excess of what they would normally require is sometimes referred to as *wasted ventilation* and is necessary because the lung units with abnormally high ventilation-perfusion ratios are inefficient at eliminating CO_2. Such units are said to constitute an *alveolar dead space.*

While the increase in ventilation to a lung with ventilation-perfusion inequality is usually effective at reducing the arterial P_{CO_2}, it is much less effective at increasing the arterial P_{O_2}. The reason for the different behavior of the two gases lies in the shapes of the CO_2 and O_2 dissociation curves (Figs. 5.26 and 5.31). The CO_2 dissociation curve is almost straight in the physiological range, with the result that an increase in ventilation will raise the CO_2 output of lung units with both high and low ventilation-perfusion ratios. By contrast, the almost flat top of the O_2 dissociation curve means that only units with moderately low ventilation-perfusion ratios will benefit appreciably from the increased ventilation. Those units which are very high on the dissociation curve (high ventilation-perfusion ratio) increase the O_2 content of their effluent blood very little because the hemoglobin is almost fully saturated with O_2 anyway (Fig. 5.51). Those units which have a very low ventilation-perfusion ratio continue to put out blood which has an O_2 content close to that of mixed venous blood. The net result is that the mixed arterial P_{O_2} rises only modestly, and some hypoxemia always remains.

MEASUREMENT OF VENTILATION-PERFUSION INEQUALITY

The topographical distribution of ventilation-perfusion ratios can be obtained with radioactive gases, for example, xenon-133. The distribution of ventilation is obtained by having the subject inhale a single breath of the gas and then recording the counting rates over different regions of the chest (see Fig. 5.71). The distribution of blood flow can be determined in a similar way if a solution of xenon-133 is injected into a vein (Fig. 5.19). The patterns of ventilation and blood flow shown in Figure 5.48 are based on this type of measurement. However, in most diseased lungs, so much inequality of ventilation and blood flow exists between closely adjacent lung units that methods based on external counting of radioactivity have inadequate resolution.

Continuous distributions of ventilation-perfusion ratios can be obtained from the pattern of elimination by the lung of a series of foreign gases infused into a vein in solution (Wagner et al., 1974). The gases have a large range of such solubilities and this allows information on a large range of ventilation-perfusion ratios to be obtained. This method shows narrow distributions of ventilation-perfusion ratios in normal subjects. In other words, most of the ventilation and blood flow goes to lung units with nearly normal ventilation-perfusion ratios and, therefore, efficient gas exchange. However, patients with lung disease have broader dispersions and sometimes bimodal patterns with the result that some ventilation and blood flow are wasted on alveoli with very low or very high ventilation-perfusion ratios. Much information about new and diseased lungs has been obtained from this relatively new method but it is too complicated for use in the routine clinical laboratory.

Another method of assessing the amount of ventilation-perfusion inequality is based on measurements of the P_{O_2} and P_{CO_2} of arterial blood and expired gas (Riley and Cournand, 1949). First, the *alveolar-arterial oxygen difference* is measured. This is obtained by subtracting the arterial P_{O_2} from the so-called "ideal" alveolar P_{O_2}. The latter is the P_{O_2} which the lung *would* have if there were no ventilation-perfusion inequality and gases were exchanging at the same respiratory exchange ratio as in the real lung. It is derived from the alveolar gas equation in its full form:

$$P_{A_{O2}} = P_{I_{O2}} - \frac{P_{A_{CO2}}}{R} + P_{A_{CO2}} \cdot F_{I_{O2}} \cdot \frac{1-R}{R}$$

The arterial P_{CO_2} is used for the alveolar value.

An increased alveolar-arterial oxygen difference is caused both by abnormally low and abnormally high ventilation-perfusion ratios within the lung, though chiefly by the former. It is possible to assess separately the approximate contribution of these two groups to the impairment of gas exchange. The effect of the units with low ventilation-perfusion ratios is determined by calculating the *physiological shunt*. Here we pretend that all the hypoxemia is caused by blood passing through unventilated alveoli (although this is a gross oversimplification). The shunt equation is then used in the form:

$$\frac{\dot{Q}_{PS}}{\dot{Q}_T} = \frac{C_{i_{O2}} - C_{a_{O2}}}{C_{i_{O2}} - C_{\bar{v}_{O2}}}$$

where $\dot{Q}_{PS}$ refers to physiological shunt and $C_{i_{O2}}$ denotes the O_2 content of blood draining from "ideal" alveoli. The latter is found from the ideal alveolar P_{O_2} and the oxygen dissociation curve.

The effect of lung units with abnormally high ventilation-perfusion ratios is assessed by calculating the *physiological dead space*. Here we pretend that all the lowering of P_{CO_2} in expired gas is caused by unperfused alveoli together with the anatomic dead space. The Bohr dead space equation is used in the form:

$$\frac{V_{D_{phys}}}{V_T} = \frac{P_{a_{CO2}} - P_{E_{CO2}}}{P_{a_{CO2}}}$$

where $V_{D_{phys}}$ refers to physiological dead space. Most patients with chronic obstructive lung disease, for example, have increases in both the physiological shunt and dead space.

CHAPTER 39

Mechanics of Breathing

The subject of the mechanics of breathing includes the forces that support and move the lung and chest wall, together with the resistances they overcome and the resulting flows. The topic is sometimes considered along with ventilation (Chapter 35), which is the process by which gas gets to and from the alveoli. The field of mechanics is so large and important, however, that it is convenient to discuss it in a separate chapter.

MUSCLES OF RESPIRATION

Inspiration

In normal quiet breathing, inspiration is active, but expiration passive. In other words, the inspiratory muscles distort the lung and chest wall from their equilibrium position, and the elastic properties of the system return them to their resting position during expiration. The most important muscle of inspiration is the *diaphragm*. This is a thin sheet of muscle in the shape of a dome which is attached to the lower ribs, sternum, and vertebral column. When the diaphragm contracts two things happen (Fig. 5.52). First, the abdominal contents are forced downward, thus enlarging the vertical dimension of the chest. Second, the rib margins are moved upward and outward. It may seem paradoxical that shortening of the diaphragm should lead to an increase in the transverse diameter of the rib cage. However, when the rib cage is lifted by the leverage action of the diaphragm on the abdominal contents, the ribs also move out because of the way they are hinged to the spine.

The diaphragm is supplied by two phrenic nerves, one to each lateral half. The fibers come from the spinal cord at cervical levels 3 and 4. During normal tidal breathing the dome moves about a centimeter or so, but on forced inspiration and expiration, a total excursion of up to 10 cm may be seen. If one-half of the diaphragm is paralyzed because the phrenic nerve supplying it is damaged, this portion moves up rather than down with inspiration when the intrathoracic pressure falls. This is called *paradoxical movement* and can be demonstrated at fluoroscopy by asking the patient to sniff.

The effectiveness of the diaphragm for changing the dimensions of the chest is related to the strength of its contraction and to its shape when relaxed. Normal descent of the diaphragm is impeded by advanced pregnancy, extreme obesity, or tight abdominal garments. In addition, patients who have had upper abdominal surgery often have limited diaphragmatic movement because of pain. Patients with advanced pulmonary emphysema often have very large lung volumes with almost flat diaphragms. Under these circumstances the diaphragm works very inefficiently and the work of breathing is increased. Some work suggests that fatigue of the diaphragm may also occur in some patients with severe lung disease.

The next most important group of muscles of inspiration are the *external intercostals*. Figure 5.53 shows that these run downward and forward between adjacent ribs. When they contract the shortening of the muscle causes the ribs to rise. As a result the anteroposterior diameter of the chest is increased because of the downward slanting angle of the ribs. In addition, the ribs can be thought of as forming a bucket-handle shape, hinged both at the spine posteriorly and at the sternum anteriorly. For this reason the lateral diameter of the chest also increases as the ribs are raised.

The extent to which a given rib moves when the external intercostal muscles contract depends on the relative stability or fixation of adjacent ribs. The uppermost ribs are supported by the shoulder girdle and, as a result, contraction of the external intercostal muscles tends to raise the whole of the rib cage.

The intercostal muscles are supplied by intercostal nerves which come off the spinal cord at the same level. Paralysis of the intercostal muscles alone by transection of the spinal cord in the lower cervical region below the origin of the fibers supply-

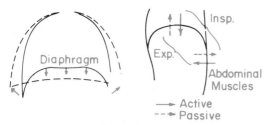

FIGURE 5.52. On inspiration *(Insp.),* the dome-shaped diaphragm contracts, the abdominal contents are forced down and forward, and the rib cage is lifted. Both increase the volume of the thorax. On forced expiration *(Exp.),* the abdominal muscles contract and push the diaphragm up.

ing the phrenic nerves does not seriously affect breathing because the diaphragm is so effective.

The last group of inspiratory muscles is called the *accessory muscles* of inspiration. They have this name because they make little if any contribution during normal quiet breathing, but on exercise or forced respiratory maneuvers they may contract vigorously. They include the scalene muscles in the neck, which elevate the first two ribs, and the sternocleidomastoids, which insert into the top of the sternum. Patients with breathlessness at rest because of severe lung disease can be seen to be using these muscles. Other muscles which sometimes come into play include the alae nasi, which cause flaring of the nostrils (particularly striking in the horse), and the small muscles of the head and neck.

Expiration

As indicated above, this is normally passive during quiet breathing. On exercise and during forced respiratory maneuvers, however, the expiratory muscles contract. The most important are those of the *abdominal wall* (Fig. 5.52). When they contract the intraabdominal pressure is raised and the diaphragm is pushed upward into the chest, thus reducing its volume. The abdominal muscles include

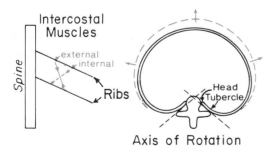

Figure 5.53. When the external intercostal muscles contract, the ribs are pulled upward and forward, and they rotate on an axis joining the tubercle and head of the rib. As a result both the lateral and anteroposterior diameters of the thorax increase. The internal intercostals have the opposite action.

the rectus abdominus, internal and external oblique muscles, and transversus abdominus. These muscles also contract forcefully during coughing, vomiting, and defecation.

The other expiratory muscles are the *internal intercostals.* Their action is opposite to that of the external intercostal muscles (Fig. 5.53) so that, as they shorten, the ribs are pulled downward, backward, and inward, thus decreasing both the anteroposterior and the lateral dimensions of the thoracic cage. Electromyographic studies show that these muscles also contract during straining, thus stiffening the intercostal spaces and preventing them from bulging outward.

There is evidence that the actions of the respiratory muscles are more complicated than indicated in the relatively simple explanation given here. The subject continues to be an active area of research.

ELASTIC PROPERTIES OF THE LUNG

Pressure-Volume Curve

Figure 5.54 shows how the pressure-volume curve of an excised canine lung can be measured. The bronchus is cannulated, and the lung is placed inside a jar in which the pressure is gradually reduced in steps of say 5 cm H_2O by means of a pump. Notice first that, when there is no pressure distending the lung, there is a small volume of gas in it. Then, as the pressure is gradually reduced in steps, the volume increases. Initially this occurs rapidly, but after the expanding pressure exceeds about 20 cm H_2O the volume changes are much less. In other words, the lung is much stiffer when it is well expanded. During the reverse procedure, when the pressure in the jar is allowed to return to atmospheric pressure in similar steps the volume decreases. At any given pressure, however, say −10 cm H_2O, the volume on the descending limb of the pressure-volume curve exceeds that obtained while the lung was being expanded. This failure of the lung to follow the same course during deflation as it did during inflation is called *hysteresis.*

It is important to be clear about what is meant by these negative values of pressure. The pressures here are recorded with respect to atmospheric pressure. In other words, zero is the pressure recorded when both limbs of the water manometer are at the same level. In absolute units zero pressure is about 760 mm Hg so that what is called −10 cm H_2O here is about 753 mm Hg (10 cm H_2O is equivalent to 10/1.36 or 7.4 mm Hg). This use of atmospheric pressure as a zero reference is a convenient and universal practice. Imagine the confusion that would be

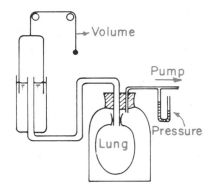

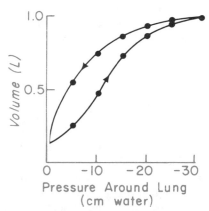

Figure 5.54. Measurement of the pressure-volume curve of excised lung. The lung is held at each pressure for a few seconds while its volume is measured. The curve is nonlinear and becomes flatter at high expanding pressures. Note that the inflation and deflation curves are not the same; this is called *hysteresis*.

caused if a patient's blood pressure were reported as 880 mm Hg instead of 120!

Suppose we expand the lung of Figure 5.54 with a pressure of say -20 cm H_2O and then clamp the line leading from the lung to the spirometer. We then turn off the pump so that the pressure in the jar rises to atmospheric pressure and measure the pressure inside the lung by attaching the saline manometer to the airway. What pressure would be recorded? The answer is 20 cm H_2O above atmospheric pressure. This is because the pressure that matters is the difference between the inside and outside of the lung. In fact, an identical pressure-volume curve can be obtained by inflating the lung with positive pressure while the pressure outside the lung is atmospheric. The pressure difference across the lung is often called the *transpulmonary pressure* (compare the term *transmural pressure* which is often used to indicate the pressure difference across a blood vessel wall).

In Figure 5.54 the pressure outside the lung was reduced by exhausting the gas in the jar with a pump. However, the same result could have been obtained if the jar itself had been expanded like a concertina. This is what happens in the intact animal. As the volume of the thoracic cage is increased through the action of the diaphragm and the intercostal muscles, the pressure within it is reduced and the lung is expanded. The fact that the space between the lung and the chest wall is extremely small does not matter. The normal intrapleural space contains no gas and only a few milliliters of liquid. Nevertheless, it is possible to measure the pressure in this space by inserting a small balloon catheter into the space in experimental animals. If this is done, a pressure-volume curve can be drawn just like the one shown in Figure 5.54 for an excised canine lung. In practice a measurement of pressure inside the esophagus is a useful approximation to the intrapleural pressure and can fairly easily be obtained by having a subject swallow a small balloon on the end of a catheter. This is how pressure-volume curves are measured in the clinical pulmonary function laboratory.

Compliance

Figure 5.54 shows that the pressure-volume curve of the lung is nonlinear. However, it is useful in practice to refer to the slope of the pressure-volume curve over a particular range of interest. For example, we might measure the change in pressure over the liter above functional residual capacity on the descending limb of the pressure-volume curve. (The deflation curve is used because it is a more repeatable measurement.) The volume change per unit pressure change is known as the *compliance* of the lung. It is clearly not a complete description of the pressure-volume properties of the lung, but nevertheless it is useful in practice as a measure of the comparative stiffness of lungs. The stiffer the lung the less the compliance.

The compliance of the lung is *reduced* by diseases which cause an accumulation of fibrous tissue in the lung or by edema in the alveolar spaces, which prevents the inflation of some lung units. The compliance of the lung is *increased* in pulmonary emphysema and also with age, probably because of alterations in the elastic tissue in both instances. The normal value of the compliance in humans is about 200 ml/cm H_2O. In other words, to inspire a normal tidal volume of about 500 ml, intrapleural pressure must fall by 2–3 cm H_2O.

The compliance of a lung clearly depends on its size. For example, a mouse lung will have a much smaller volume change per unit change in transpulmonary pressure than an elephant lung. In the same way, a patient who has had one lung removed surgically will have a reduced compliance. Thus, if compliance is used to get information about the elastic properties of the lung, one must take account of size. Sometimes this is done by calculating the *specific compliance*, which is the compliance per unit volume. Another index of elastic properties that is sometimes useful in the pulmonary function laboratory is the transpulmonary pressure at total lung capacity. This is abnormally low in emphysema in spite of the fact that the lung volume is usually abnormally high.

What is responsible for the elastic behavior of the lung, that is, its tendency to return to its original position after distortion? There are two factors. One is the elastic components of the tissue of the lung, which are visible in microscopic sections. Figure 5.55 shows a thick section of normal adult lung stained to show the fibers of elastin in the pulmonary parenchyma. In addition, not shown in this section, there are numerous collagen fibers distributed throughout the lung tissue. Elastin fibers are easily stretched, while collagen fibers are not. However, the elastic behavior of the lung probably does not depend on simple elongation of elastic fibers but more on their geometric arrangement. This has been called "nylon stocking" elasticity. The analogy is apt because in a stocking the nylon threads are very difficult to stretch, and the elastic behavior of the stocking is the result of distortion of the geometri-

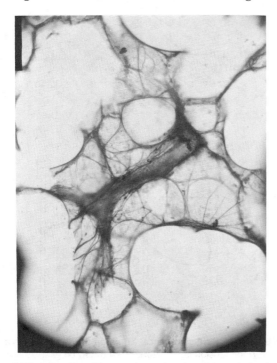

Figure 5.55. Section of human lung showing fibers of elastin in the alveolar walls and around blood vessels. These contribute to the elastic behavior of the lung. (Courtesy of Dr. R. R. Wright.)

cal arrangement of the knitted fibers. The changes in the pressure-volume properties of lungs that occur with age may have more to do with changes in the geometrical arrangement of the fibers because the actual amounts of collagen and elastin apparently do not change.

Surface Tension

The second factor responsible for the elastic behavior of the lung is the surface tension of the liquid film lining the alveoli. Figure 5.56*A* indicates that the surface tension is defined as the force (in dynes, for example) acting across an imaginary line 1 cm long in a liquid surface. This tension develops because the cohesive forces between adjacent liquid molecules are greater than the forces between the molecules of liquid and gas outside the surface. One way to think of surface tension is to imagine that the surface consists of a thin rubber membrane under stretch. If an incision 1 cm long were made in this membrane it would gape, and if sutures were put in to bring the two cut sides together the total force of the sutures would be equal to the surface tension.

An important property of surface tension is that it generates a pressure in a bubble (Fig. 5.56*B*). Again imagine the surface as a thin rubber membrane under stretch like a balloon. The tension in the wall develops a pressure just as a hoop encircling a barrel tends to crush it inward. The relationship between the surface tension in the wall and the pressure developed in a soap bubble is given by Laplace's Law (Fig. 5.56*C*). This states that, for each surface of the bubble, the pressure (*P*) is equal to twice the tension (*T*) divided by the radius (*r*) or, for both surfaces,

$$P = \frac{4T}{r}$$

In two bubbles with the same surface tension, the pressure developed by the smaller bubble will

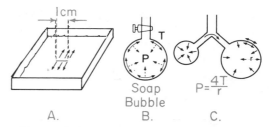

Figure 5.56. *A.*, Surface tension (*T*) is the force in dynes acting across an imaginary line 1 cm long in a liquid surface. *B.*, Surface forces in a soap bubble tend to reduce the area of the surface and generate a pressure (*P*) within the bubble. *C.*, Because the smaller bubble generates a larger pressure, it blows up the larger bubble. *r*, radius.

exceed that of the larger bubble; if they are connected the smaller bubble will therefore blow up the larger bubble.

This raises the question of why the lung is not unstable. At first sight it can be regarded as a series of 300 million bubbles all connected by airways and all containing a gas-liquid interface with surface tension. It is true that each "bubble" or alveolus is surrounded by many other alveoli, which tend to confer some stability (see below). Nevertheless, why is such a mass of "bubbles" not unstable like a soap foam?

The answer to this question leads to a fascinating story in the development of pulmonary mechanics. As long ago as 1929, von Neergaard showed that lungs inflated with saline are easier to distend than are air-filled lungs. Figure 5.57 shows the results of a typical experiment using excised feline lungs; it can be seen that, at a given distending volume, the pressure developed by the air-inflated lungs is considerably greater than that for saline-filled lungs. Presumably the only difference between the two lungs is the abolition of the gas-liquid interface by saline filling. Therefore, this experiment demonstrates that a large part of the retractive force comes from the surface tension. Some years later Radford (1954) made some calculations of the pressure-volume properties that a lung would have if the surface tension of the fluid lining the alveoli was that of plasma, namely about 70 dynes/cm. He found that the transpulmonary pressures that would be needed to inflate the lung were far too high and concluded that the assumptions he had made about the geometry of the air spaces were incorrect.

At about this time Pattle (1955) was working in a government defense laboratory on the effects of chemical gases on the lung. In studying the edema foam that comes from the lungs of animals exposed to noxious gases, he noticed that the bubbles of this foam were extremely stable. He recognized that they therefore had an extremely low surface tension, since small bubbles with a normal surface tension shrink rapidly (this is because the large pressure developed inside them is sufficient to force gas out of them by diffusion). Pattle's observation led to the discovery of pulmonary *surfactant*, which is a phospholipid secreted by type 2 alveolar cells and which has the property of reducing the surface tension of the liquid lining layer.

Clements (1962) showed that lung extracts lowered surface tension on a surface balance as shown in Figure 5.58*A*. In this device the material to be investigated is added to the surface and this is alternately compressed and expanded by a barrier across the trough. The tension of the surface is measured by dipping a platinum strip into it and attaching this to a force transducer. Figure 5.58*B* shows that, if the trough contains pure water, the surface tension obtained is independent of the area of the surface and has a value of about 70 dynes/cm. If

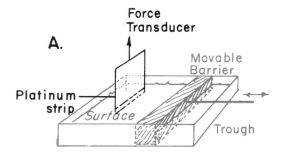

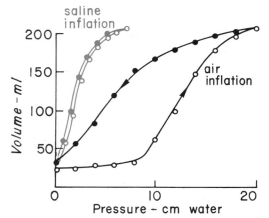

Figure 5.57. Comparison of pressure-volume curves of air-filled and saline-filled lungs (cat). *Open circles*, inflation; *closed circles*, deflation. The saline-filled lung has a higher compliance and also much less hysteresis than the airfilled lung. (From Radford, 1957.)

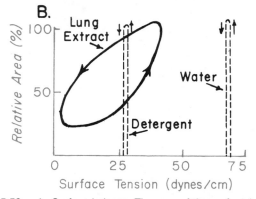

Figure 5.58. *A.*, Surface balance. The area of the surface is altered and the surface tension is measured from the force exerted on a platinum strip dipped into the surface. *B.*, Plots of surface tension and area obtained with a surface balance. Lung washings show a change in surface tension with area, and the minimum tension is very small.

detergent is added to the water, the surface tension falls considerably, but it is still independent of the area of the surface. However, when a lung extract is obtained (by rinsing out the airways with saline) and this is added to the trough, a curve is drawn as the area of the surface is altered.

Two features of this curve should be noted. One is that, in general, the smaller the area the lower the value of the surface tension. Thus, if we extrapolate this behavior to the lung and assume that all alveoli secrete the same amount of surfactant, the smaller alveoli will presumably have a lower surface tension than the larger alveoli, and therefore, the tendency for the smaller alveoli to inflate the larger alveoli (Fig. 5.56C) will be lessened. Also, the curve shows the same kind of hysteresis as the pressure-volume curve of the lung; that is, for a given surface area the surface tension during expansion is much greater than during compression. It seems likely that most of the hysteresis of the intact lung (Fig. 5.54) can be explained by the behavior of the surface tension. Additional evidence for this

statement is that the pressure-volume curve for saline-filled lungs shows little hysteresis. (Fig. 5.57).

The composition of pulmonary surfactant is not completely known, but it contains the phospholipid, dipalmitoyl phosphatidylcholine (DPPC). This material shows the same kind of surface tension behavior on the surface tension balance as does lung extract. The surfactant is secreted by the type 2 alveolar cells. Osmiophilic lamellated bodies can be seen in electron micrographs of type 2 cells in the alveolar walls (Fig. 5.59). These lamellated bodies are extruded to form surfactant. The type 2 cells synthesize the phospholipid from fatty acids that either are extracted from the blood or are themselves synthesized in the lung. Synthesis is fast, and there is a rapid turnover of surfactant. Different chemical pathways are available for the incorporation of fatty acids into DPPC, and there apparently are differences between the pathways used by the adult and by the fetal lung. Because surfactant is formed relatively late in fetal life, babies born with-

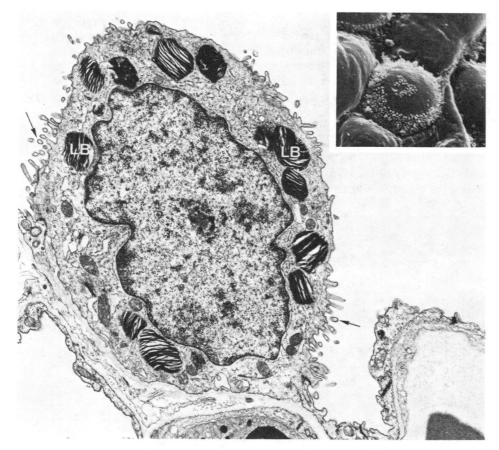

Figure 5.59. Electron micrograph of type 2 epithelial cell (×10,000). Note the osmiophilic lamellated bodies *(LB)*, large nucleus, and microvilli (indicated by *arrows*). The *inset* at *top right* is a scanning electron micrograph showing the surface view of a type 2 cell with its characteristic distribution of microvilli (×3400). (From Weibel and Gil, 1977.)

out adequate amounts develop respiratory distress and may die.

What are the physiological advantages of pulmonary surfactant? First, a low surface tension in the alveoli increases the compliance of the lung and reduces the work of expanding it with each breath. Second, stability of the alveoli is promoted. Third, surfactant helps to keep the alveoli dry. This action is more difficult to grasp, but the inward-acting forces that tend to collapse the alveoli also tend to pull water into the alveolar spaces from the pulmonary capillaries in the alveolar wall. This tendency is reduced by pulmonary surfactant. Indeed, some physiologists believe that this may be the major role of pulmonary surfactant.

In some pathological conditions there is an absence of pulmonary surfactant. A good example is the respiratory distress syndrome of the newborn, which is particularly likely to develop in premature infants whose surfactant system is poorly developed. As a result the lungs are still and difficult to inflate, and they contain areas of alveolar atelectasis and also areas of edema.

Interdependence

Another mechanism that apparently contributes to the stability of the alveoli in the lung has been described (Mead et al., 1970). Figure 5.60 illustrates how all the alveoli (except those immediately adjacent to the pleural surface) are surrounded by other alveoli and are therefore supported by each other. Furthermore, in a structure with many connecting links, any tendency for one group of units to reduce or increase its volume relative to the rest of the structure is opposed. For example, if a group of alveoli has a tendency to collapse, large expanding

forces will be developed on them by the surrounding parenchyma, which then tends to be overexpanded. In fact, calculations and measurements show that surprisingly large expanding forces can be developed on a portion of collapsed lung as the lung around it is expanded.

This support offered to lung units by those surrounding them is termed *interdependence*. The same mechanism is responsible for the development of low pressure around large blood vessels (Fig. 5.13) and airways as the lung expands; in this case the fact that the relatively stiff blood vessels do not expand to the same extent as the parenchyma around them is responsible for the development of low perivascular pressures. Interdependence also may well play an important role in the prevention of atelectasis and in the opening up of areas that have collapsed for some reason. In fact, some physiologists believe that this mechanism may be more important than surfactant in maintaining the stability of the small air spaces.

ELASTIC PROPERTIES OF THE CHEST WALL

Just as the lung is elastic, so is the thoracic cage. Although it is not so immediately obvious, the chest wall is pulled in by the elastic retractive forces of the lung just as the lung is expanded by the elastic force developed by the chest wall. This becomes clearer if we consider what happens when air is introduced into the intrapleural space and a pneumothorax is formed, as shown in Figure 5.61. When the lung collapses and it no longer pulls the chest wall in, the wall springs out.

In clinical practice it is often possible to detect differences in expansion of the chest wall by careful

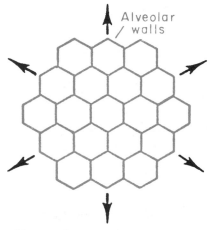

Figure 5.60. Diagram of an array of alveoli emphasizing how each unit is supported by surrounding units. This is referred to as *interdependence* and is important in maintaining alveolar stability.

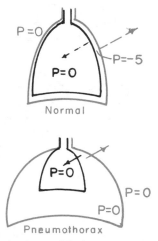

Figure 5.61. The tendency of the lung to recoil to its deflated volume is balanced by the tendency of the chest cage to bow out. As a result, the intrapleural pressure is subatmospheric. Pneumothorax allows the lung to collapse and the thorax to spring out. *P*, pressure.

examination of the chest or by looking at a chest radiograph and noticing whether the ribs are crowded. Typically, a patient with a reduced compliance caused by fibrosis of one lung will have the chest wall pulled in to some extent on that side. By contrast a patient with a unilateral pneumothorax often has an overexpanded chest wall on the affected side.

The interaction between the elasticity of the lung and the chest wall can be summarized in what is known as a *relaxation pressure-volume diagram* (Rahn et al., 1946). This is shown in Figure 5.62, which contains a large amount of information that may be rather confusing at first sight. The relaxation pressure is the airway pressure obtained when the subject is completely relaxed, that is, when he is not attempting either to inflate or to deflate his lung and chest wall by muscle activity. It can be measured as shown on the *left half* of the figure by having the subject inhale or exhale to a particular volume as measured by the spirometer, closing the tap, and then measuring the airway pressure with the subject completely relaxed. Incidentally, this is difficult for untrained subjects to do.

In the graph on the *right,* look first at the *continuous line* for the lung and chest wall. Notice that, at functional residual capacity (FRC), the relaxation pressure is zero. That is to say, this is the equilibrium position of the lung and chest wall when airway pressure is the same as atmospheric pressure. As the volume is increased, the relaxation pressure becomes positive, since the lung and chest wall tend to return to the equilibrium FRC position. At total lung capacity, the relaxation pressure is about 30

cm H_2O. Below FRC, the residual volume is gradually approached with a very low relaxation pressure. Under these conditions the lung and chest wall tend to spring out when the expiratory muscles are relaxed, thus generating the negative relaxation pressure.

Now look at the *broken line* for the lung alone. This is the same curve on the human lung as can be obtained on canine lung when it is inflated by positive pressure (compare Fig. 5.54). At FRC, a positive pressure of some 5 cm H_2O is developed as the lung tries to collapse. At total lung capacity this pressure has increased to about 25 cm H_2O. Below FRC the relaxation pressure falls to zero at what is known as the minimal volume of the lung, that is, the volume that the lung takes up when there is no expanding pressure on it. However, as we saw previously (Fig. 5.54), some gas ramains in the lung under these conditions.

Finally, look at the *broken line* for the chest wall alone. We can imagine this being measured on a man with a normal chest wall and no lung! Notice that at FRC the chest wall develops a negative relaxation pressure. In other words it is tending to spring out at FRC just as the lung is tending to collapse inward. In fact, the negative relaxation pressure of the chest wall and the positive relaxation pressure of the lung are identical, thus making this the equilibrium position for the lung and chest wall together. Indeed, at any volume, the curve for the combined lung and chest wall can be explained by the addition of the individual lung and chest wall curves. It is not until the volume of the chest wall is increased to about 70% of vital capacity that it no longer tends to spring out. In fact, this is the resting posi-

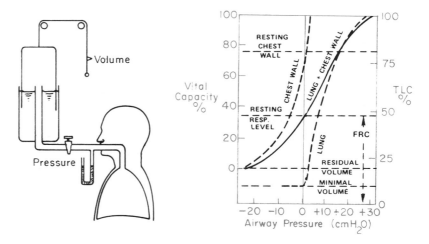

Figure 5.62. Relaxation pressure-volume curve of the lung and chest wall. The subject inspires (or expires) to a certain volume from the spirometer, the tap is closed, and he then relaxes his respiratory muscles. The curve for lung + chest wall can be explained by the addition of the individual lung and chest wall curves. See text for details. *FRC,* functional residual capacity; *TLC,* total lung capacity. (From Rahn et al., 1946.)

tion of the chest wall, and at volumes above this the tendency of the chest wall to collapse back results in positive relaxation pressures.

We saw above that, at every volume, the relaxation pressure of the lung plus chest wall is the sum of the pressures for the lung and chest wall separately. Since the pressure (at a given lung volume) is inversely proportional to compliance, this implies that the total compliance of the lung and chest wall is the sum of the reciprocals of the lung and chest wall compliances measured separately, or $1/C_T = 1/C_L + 1/C_{CW}$.

AIRWAY RESISTANCE

Airflow through Tubes

If air flows through a tube (Fig. 5.63), a difference of pressure exists between the ends. The pressure difference depends on the rate and pattern of flow. At low flows rates, the streamlines may everywhere be parallel to the sides of the tube (*A*). This is known as *laminar flow*. As the flow rate is increased, unsteadiness develops, especially at branches. Here separation of the streamlines from the wall may occur, with the formation of local eddies (*B*). At still higher flow rates, complete disorganization of the streamline is seen; this is *turbulence* (*C*).

The pressure-flow characteristics for *laminar flow* were first described by the French physician Poiseuille. In smooth, straight circular tubes, the volume flow rate $\dot{V}$ is given by

$$\dot{V} = \frac{\Delta P \pi r^4}{8nl}$$

where ΔP is the driving pressure, r radius, n viscosity, and l length. It can be seen that driving pressure is proportional to flow rate, or $\Delta P = K\dot{V}$. Since flow resistance R is driving pressure divided by flow, we have

$$R = \frac{8nl}{\pi r^4}$$

Note the critical importance of tube radius; if the radius is halved, the resistance increases 16 fold! However, doubling the length only doubles resistance. Note also that the viscosity of the gas but not its density affects the pressure-flow relationship.

Another feature of laminar flow when this is fully developed is that the gas in the center of the tube moves twice as fast as the average velocity. Thus, a spike of rapidly moving gas travels down the axis

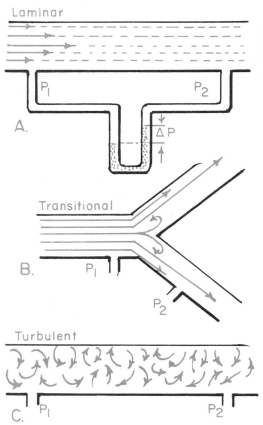

FIGURE 5.63.   Patterns of airflow in tubes. In *A*, the flow is laminar, in *B*, transitional with eddy formation at branches, and in *C* turbulent. Resistance is equal to $(P_1 - P_2)$/flow.

of the tube (Fig. 5.63*A*). This changing velocity across the diameter of the tube is known as the *velocity profile*, and its shape is a parabola. The axial velocity, that is, the velocity in the center of the stream, is twice the mean flow velocity, but the gas immediately adjacent to the wall of the tube is not moving at all. This latter is known as the *boundary layer*. This parabolic velocity profile is only seen if the tube is long enough for it to develop, and the pattern is then called "fully developed laminar flow." If gas enters the tube from a larger chamber, for example, the length of the tube must be many times its diameter before this pattern becomes completely established.

Turbulent flow has different properties. Here pressure is not proportional to flow rate but, approximately, to its square: $P = K\dot{V}^2$. In addition, the viscosity of the gas becomes less important, but an increase in gas density increases the pressure drop for a given flow. Turbulent flow does not have the very high axial flow velocity that is characteristic of laminar flow.

Whether flow will be laminar or turbulent

depends to a large extent on the Reynolds number, Re. This is given by

$$\mathrm{Re} = \frac{2rvd}{n}$$

where d is density, v average velocity, r radius, and n viscosity. In straight, smooth tubes, turbulence is probable when the Reynolds number exceeds 2,000. The expression shows that turbulence is most likely to occur when the velocity of flow is high and the tube diameter large (for a given velocity). Note also that a low density gas like helium tends to produce less tubulence.

In such a complicated system of tubes as the bronchial tree with its many branches, changes in caliber during ventilation, and irregular wall surfaces, application of the above principles is difficult. In practice, for laminar flow to occur the entrance conditions of the tube are critical. If eddy formation occurs upstream at a branch point, this disturbance is carried downstream some distance before it disappears. Thus, in a rapidly branching system like the lung, fully developed laminar flow (Fig. 5.63*A*) probably occurs only in the very small airways where the Reynolds numbers are very low (approximately 1 in terminal bronchioles). In most of the bronchial tree, flow is transitional (*B*), although true turbulence may occur in the trachea, especially on exercise when flow velocities are high. In general, driving pressure is determined by both the flow rate and its square:

$$\Delta P = K_1 \dot{V} + K_2 \dot{V}^2$$

For a review of this area the reader is referred to Macklem and Mead (1986).

Pressures During the Breathing Cycle

Figure 5.64 shows the intrapleural and alveolar pressures during a normal breathing cycle. Intrapleural pressure can be measured approximately by means of a balloon catheter placed in the esophagus. Alveolar pressure is more difficult to obtain but can be deduced from measurements made in a body plethysmograph. The *top panel* on the *right* shows that spirometer volume decreases on inspiration and then increases during expiration. The *third panel* shows that inspiratory flow rate (downward deflection) increases and then decreases during inspiration; expiratory flow rate follows. Alveolar pressure *(bottom panel)* is measured with respect to atmospheric pressure.

Before inspiration begins there is no flow and

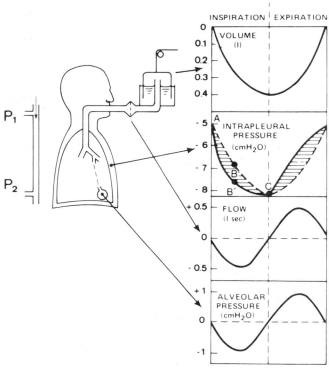

Figure 5.64. Pressures during the breathing cycle. If there was no airway resistance, alveolar pressure would remain at zero and intrapleural pressure would follow the broken line *ABC,* which is determined by the elastic recoil of the lung. Airway (and tissue) resistance contributes the cross-hatched portion of intrapleural pressure (see text).

therefore alveolar pressure is zero. For inspiratory flow to occur, alveolar pressure falls and then returns to zero at the end of inspiration. On expiration, alveolar pressure rises above atmospheric pressure to overcome the resistance of the airways during expiration. If airway resistance remained constant during the breathing cycle, the shapes of the curves for flow and alveolar pressure *(bottom two panels)* would be identical.

Now look at intrapleural pressure (the *second panel*). Before inspiration, this is -5 cm of H_2O with respect to atmospheric pressure because of the elastic recoil forces of the lung. Intrapleural pressure falls during inspiration for two reasons. The first is that lung volume increases and the elastic recoil forces become greater. If these were the only forces operating, intrapleural pressure would fall long the broken line *ABC,* and if the compliance of the lung were constant the shape of this line would be identical to that of the volume change *(top panel).* An additional fall of pressure is required, however, to move gas along the airways; the result is that intrapleural pressure falls along the *AB'C.* Thus, the *cross-hatched portion* of the intrapleural pressure curves represents the pressure necessary to overcome airway resistance. In addition, there is a contribution made by the viscous resistance of the lung tissues, that is, the forces required to move one layer of tissue over another; this will be discussed below.

The total pressure drop between the mouth and the intrapleural space is thus made up of two components: (1) that caused by the pressure drop along the airways due to their resistance to air flow and (2) the pressure drop across the lung caused chiefly by the static recoil forces. In addition, there is a small component of tissue viscous resistance. As an equation of pressures:

(Mouth − intrapleural)
      = (Mouth − alveolar) + (alveolar − intrapleural)

On expiration, opposite changes occur such that intrapleural pressure is now less negative than it would be in the absence of flow because alveolar pressure rises to overcome airway resistance. Indeed, with a forced expiration, intrapleural pressure may easily go positive, that is, above atmospheric pressure.

Measurement of Airway Resistance

For the complete system of airways, resistance (R_{aw}) is defined as

$$R_{aw} = \frac{P_{mouth} - P_{alv}}{\dot{V}}$$

where P_{mouth} and P_{alv} are the pressure in the mouth and alveoli, respectively. Mouth pressure is easy to measure, but alveolar pressure can be obtained only indirectly. One method of doing this is with a body plethysmograph (Fig. 5.65).

The subject sits in an air-tight box and breathes rapidly and shallowly. Before inspiration (*A*), the box pressure is atmospheric. At the onset of inspiration, the pressure in the alveoli falls, and the alveolar gas expands by a volume ΔV. This compresses the gas in the box, and from its change in pressure ΔV can be calculated (compare Fig. 5.8). If lung volume is known, ΔV can be converted into alveolar pressure using Boyle's Law. Flow is measured simultaneously and thus airway resistance is obtained. A measurement can be made during expiration in the same way. Lung volume is determined with the plethysmograph as described in Figure 5.8.

Airway resistance can also be measured during normal breathing from an intrapleural pressure record as obtained with an esophageal balloon (Fig. 5.64). However, in this case tissue viscous resistance is included as well, and the result is often

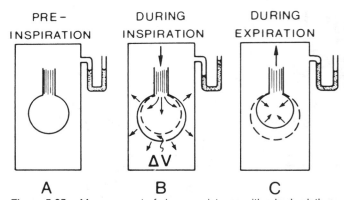

Figure 5.65. Measurement of airway resistance with a body plethysmograph. During inspiration, the alveolar gas expands and box pressure therefore rises. From this, alveolar pressure can be calculated. When divided by the flow, this gives airway resistance (see text). ΔV, change in volume. (Adapted from Comroe et al, 1962.)

referred to as pulmonary resistance. As we saw above, intrapleural pressure reflects two sets of forces, those opposing the elastic recoil of the lung and those overcoming the resistance to air and tissue flow. It is possible to subtract the pressure caused by the recoil forces during quiet breathing because this is proportional to lung volume (if compliance is constant). The subtraction is done with an electrical circuit so as to leave a plot of pressure against flow which gives (airway + tissue) resistance. This method is not satisfactory in lungs with severe airway disease because the uneven time constants prevent all regions from moving together (Fig. 5.75).

Chief Site of Airway Resistance

As the airways penetrate toward the periphery of the lung, they become more numerous but much narrower (Figs. 5.3 and 5.9). Based on Poiseuille's equation with its (radius)[4] term, it would be natural to think that the major part of the resistance lies in the very narrow airways. Indeed, for many years this was thought to be the case. However, it has been shown by direct measurements of the pressure drop along the bronchial tree that the major site of resistance is the medium-sized bronchi and that the very small bronchioles contribute relatively little resistance (Macklem and Mead, 1967). Figure 5.66 shows that most of the pressure drop occurs in the airways up to approximately the seventh generation. Less than 20% can be attributed to airways less than 2 mm in diameter. The reason for this apparent paradox is the prodigious number of small airways. In other words, the resistance of each airway is relatively large, but there are so many airways arranged in parallel that the combined resistance is small.

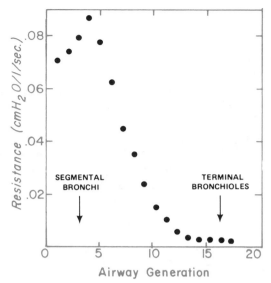

Figure 5.66. Location of the chief site of airway resistance. Note that the intermediate sized bronchi contribute most of the resistance and that relatively little is located in the very small airways. (From Pedley et al., 1970.)

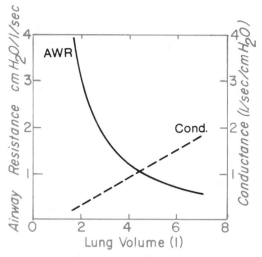

Figure 5.67. Variation of airway resistance with lung volume. If the reciprocal of airway resistance *(AWR)*, that is, conductance *(Cond.)*, is plotted, the graph is a straight line. (From Briscoe and DuBois, 1958.)

The fact that the peripheral airways contribute so little resistance is important in the detection of early airway disease. Because they constitute a "silent zone" it is probable that a considerable degree of small airway disease can be present before the usual measurements of airway resistance can pick up an abnormality. As a consequence, special methods aimed at detecting the resistance of small airways have been explored.

Factors Determining Airway Resistance

Lung volume has an important effect on airway resistance. Because the bronchi which run within the lung parenchyma are supported by the guy-wire action of the surrounding lung tissue, their caliber is increased as the lung expands. This can be appreciated at bronchoscopy, a form of examination of the bronchial tree in which a lighted tube is passed down into the large bronchi via the mouth or nose. With each inspiration, the bronchi can be seen to enlarge. Figure 5.67 shows that, as lung volume is reduced, airway resistance rises rapidly. This relationship is very nonlinear. However, if the reciprocal of resistance, known as *conductance*, is calculated, this increases approximately linearly with lung volume.

It is clearly important to standardize lung volume in reporting a measurement of airway resistance. Many patients who have disease of the airways and therefore a high airway resistance (for example, patients with bronchial asthma) often breathe at a high lung volume, and this helps to reduce their resistance in the direction of normal. At very low lung volumes, the small airways in the region of the terminal and respiratory bronchioles may actually close completely, thus trapping gas in the alveoli. As is explained below, this is especially likely to occur at the bottom of the lung where the lung is least well expanded.

Another important determinant of the caliber of the airways is the *tone of the bronchial smooth muscle*. This is under autonomic control. Sympathetic stimulation causes dilation, whereas parasympathetic activity causes bronchial constriction. Drugs such as isoproterenol and epinephrine, which stimulate the β-adrenergic receptors in the airways, cause bronchial dilation and are used to treat the bronchial constriction which occurs in asthma. A fall in alveolar P_{CO_2} as a result of hyperventilation or a local reduction of pulmonary blood flow also causes bronchoconstriction, apparently by a direct action on the smooth muscle of the airways. Bronchial constriciton also occurs reflexly through stimulation of receptors in the trachea and large bronchi by irritants such as cigarette smoke, inhaled dusts, cold air and noxious gases (Chapter 40). Motor innervation is by the vagus nerve. The injection of histamine into a pulmonary artery causes constriction of smooth muscle located in the mouths of the alveolar ducts. A similar response occurs when some types of microemboli are injected into the pulmonary circulation. Swelling of the bronchial mucosa caused by inflammation will also increase airway resistance.

The *density and viscosity* of the gas that is breathed affect the pressure drop along the airways because flow is transitional in most regions of the bronchial tree (Fig. 5.63*B*). In diving at great depths under the sea where the density of the gas is enormously increased, large pressures are required to move the gas but, if a helium-oxygen mixture is substituted, the pressures are considerably smaller. In the past, patients with severe airway disease were occasionally treated with helium-oxygen mixtures to reduce their work of breathing. The fact that changes in density rather than viscosity have such an influence of resistance is evidence that flow is not purely laminar in the medium-sized airways, the main site of resistance (Fig. 5.66).

Dynamic Compression of Airways

Suppose we ask a subject to inspire to his total lung capacity and then breath out as hard and as fast as he can down to residual volume. We can then plot expiratory flow rate against lung volume as shown in Figure 5.68, the so-called *flow-volume curve*. The result will be a curve like *A*, showing that flow rises very rapidly to a high value and then declines over most of expiration.

A curious feature of this flow-volume envelope is that it is virtually impossible to penetrate it. For example, suppose that the subject starts breathing out slowly from total lung capacity as shown in *curve B* and then halfway through expiration accelerates as hard as he can. We would find that the latter part of the flow-volume curve is almost superimposed on the first curve inscribed. Finally, suppose he exhales from total lung capacity but this time makes much less effort overall. We would obtain a *curve C* where the maximum flow rate is somewhat reduced, but over much of expiration the flow-volume curve will be almost superimposed on that obtained with the forced expiration.

Two conclusions follow from these surprising observations. The first is that something powerful is limiting the flow rate that can be achieved at a given lung volume, since it is impossible to get outside the flow-volume envelope no matter what tricks are tried. Second, over most of expiration the flow rate is virtually independent of effort, in that whether we try very hard or very much less the flow rates are the same.

More information can be obtained about this curious state of affairs by plotting the data in another way, as shown in Figure 5.69. To obtain this figure the subject inhales to total lung capacity and then exhales fully, with varying degrees of effort. We then plot flow rate and intrapleural pressure at the *same* lung volume for each different inspiration and expiration and thus obtain these so-called *isovolume pressure-flow curves*. The figure shows that at high lung volumes the flow rate continues to increase as the intrapleural pressure is raised, as we might expect. At midvolume, however, there is surprisingly no further increase in flow rate once an intrapleural pressure of some 5–10 cm H_2O has been exceeded. This phenomenon is even more marked at low lung volumes. Thus, at these low and intermediate lung volumes, flow is effort-independent.

The reason for these remarkable findings is that compression of the airways effectively limits the flow rate. Figure 5.70 shows the mechanism. In *A*, before inspiration, intrapleural pressure is −5 cm H_2O and since there is no air flow the pressure throughout the airways is atmospheric. If we assume that the pressure outside the large airways is intrapleural, the transmural pressure across these airways is 5 cm H_2O tending to hold them open. In *B*, at the start of inspiration, intrapleural

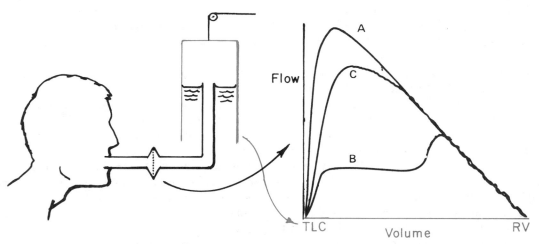

Figure 5.68. Flow-volume curves. In *A*, a maximal inspiration was followed by a forced expiration. In *B*, expiration was initially slow and then forced. In *C*, expiratory effort was submaximal. In all three, the descending portions of the curves are almost superimposed. *TLC*, total lung capacity; *RV*, residual volume.

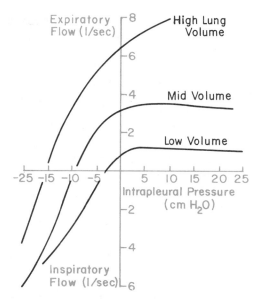

Figure 5.69. Isovolume pressure-flow curves drawn for three lung volumes. Each of these was obtained from a series of forced expirations or inspirations (see text). At the high lung volume, a rise in intrapleural pressure (obtained by increasing expiratory effort) resulted in a greater expiratory flow. However, at midvolume and low volume, flow became independent of effort after a certain intrapleural pressure had been exceeded. (From Fry and Hyatt, 1960.)

pressure falls to -7 cm H_2O and alveolar pressure falls to -2 cm H_2O. We are assuming at ths point that there has been a negligible change in lung volume, so that the pressure difference between intrapleural and alveolar spaces remains at 5 cm H_2O. There will be a pressure drop along the airways because of resistance to air flow. Suppose that at a particular point the pressure is -1 cm H_2O. At this place we see that the transmural pressure holding

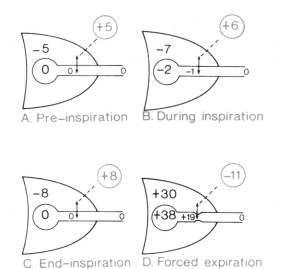

Figure 5.70. Scheme showing why airways are compressed during a forced expiration. Note that the pressure across the airway is holding it open except during the forced expiration. See text for details.

the airway open has now increased to 6 cm H_2O. *C* shows the situation at the end of inspiration. Intrapleural pressure is -8 cm H_2O (larger lung volume) and, since the air has stopped moving, the pressure inside the airways at every point is atmospheric. There is now a pressure of 8 cm H_2O holding the airway open.

A striking change occurs at the beginning of forced expiration, as shown in *D*. Intrapleural pressure rises abruptly, perhaps to 30 cm H_2O. The pressure difference between intrapleural and alveolar spaces is still 8 because lung volume has changed by a negligible amount so early in the expiration. Thus alveolar pressure is now 38 cm H_2O. Again there will be a pressure drop along the airways because of their flow resistance, and at the point in which we are interested the pressure is 19 cm H_2O. It can be seen, however, that now there is a pressure of 11 cm H_2O across the airways tending to *close* it. The airway therefore is compressed and partially collapses.

We now have a situation in which flow occurs through a collapsible tube when the outside pressure exceeds downstream pressure. This was discussed earlier as the "vascular waterfall" which determines blood flow in zone 2 of the lung (Fig. 5.21). Under such conditions, flow is independent of downstream pressure, being determined only by the difference between upstream pressure and the pressure outside the collapsible tube. In the case of the lung, this becomes alveolar pressure minus intrapleural pressure.

Two important conclusions follow from this. One is that, no matter how forceful the expiration, the flow rate cannot be increased because, as intrapleural pressure goes up, it takes alveolar pressure with it. Thus, the driving pressure (alveolar-intrapleural pressure) remains constant. This explains why flow rate becomes effort-independent. Second, maximum flow rate will be determined partly by the elastic recoil forces of the lung because these are what generate the difference between alveolar and intrapleural pressure. These elastic recoil forces will decrease as lung volume becomes smaller, and this is one reason why the maximum flow rate decreases as lung volume falls (Fig. 5.68). Another reason is that the resistance of the peripheral airways increases as lung volume-reduces (Fig. 5.67).

Several factors will exaggerate this flow-limiting mechanism. One is any increase in the resistance of the airways peripheral to the collapse point, since this will magnify the pressure drop along them and thus decrease the intrabronchial pressure during expiration. Once collapse occurs, the resistance of the airways proximal (mouthward) to the point of collapse becomes irrelevant. Flow rate is determined by the lung elastic recoil pressure divided by the resistance of the peripheral airways up to the collapse point. This point is sometimes called the *equal pressure point*. As expiration progresses, the equal pressure point moves distally, deeper into the lung. This occurs because the resistance of the airways rises as lung volume falls, and therefore the pressure within the airways falls more rapidly with distance from the alveoli.

Another factor which exaggerates dynamic compression is a reduction in radical traction on the airways offered by surrounding lung tisue. Such a reduction occurs in emphysema, which is characterized by destruction of alveolar walls. Also in the emphysematous lung the elastic recoil forces are reduced (increased compliance). As a result the driving pressure under conditions of dynamic compression (alveolar minus intrapleural pressure) is reduced. Indeed, while this type of flow limitation is seen only during forced expiration in normal subjects, it may well occur during normal expiration in patients with severe emphysema. By contrast, in restrictive lung disease, e.g., caused by fibrosis, the maximal flow rate for a given lung volume may be somewhat higher than normal because of the increased retractive forces.

INEQUALITY OF VENTILATION

Topographical Differences Within the Lung

Suppose a seated normal subject inhales a breath of radioactive xenon gas (Fig. 5.71). When the xenon-133 enters the counting field, its radiation penetrates the chest wall and can be recorded by a bank of counters or a radiation camera. In this way the volume of the inhaled xenon going to various regions can be determined (Ball et al., 1962).

Figure 5.71 shows results obtained in a series of normal volunteers using this method. It can be seen that ventilation per unit volume is greater near the bottom of the lung and becomes progressively smaller toward the top. Other measurements made when the subject is in the supine position show that the difference disappears, with the result that apical and basal ventilations become the same. In that posture, however, the ventilation of the lowermost (posterior) part of the lung exceeds that of the uppermost (anterior) part. Again in the lateral position (subject on his side) the dependent lung is the better ventilated.

The cause of these topographical differences in ventilation lies in the distortion that occurs in the lung as a consequence of its weight (Milic-Emili et al., 1966). Figure 5.72 shows that the intrapleural pressure is less negative at the bottom than at the top of the upright lung. These differences are apparently caused by the weight of the lung. Anything that is supported requires a larger pressure below it than above it to balance the downward-acting weight forces, and the lung, which is supported by the chest wall and diaphragm, is no exception. Thus the pressure near the base is higher (less negative) than at the apex.

Figure 5.72 shows the way in which the volume of a portion of lung (for example, a lobe) expands as the pressure around it is decreased (compare Fig. 5.54). The pressure inside the lung is the same as atmospheric pressure. The lung is easier to inflate at low volumes than at high volumes, where it becomes stiffer. Because the expanding pressure at the base of the lung is small, this region has a small resting volume. However, because it is situated on a steep part of the pressure-volume curve, it expands well on inspiration. By contrast, the apex of the lung has a large expanding pressure, a big resting volume, and small change in volume on inspiration.

When we talk of regional differences in ventila-

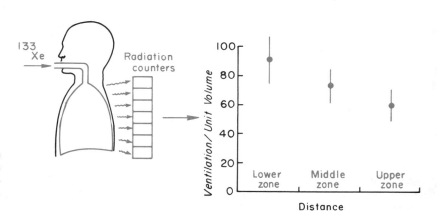

Figure 5.71.   Measurement of regional differences in ventilation with radioactive xenon. When the gas is inhaled, its radiation can be detected by counters outside the chest. Note that the ventilation decreases from the lower to upper regions of the upright lung.

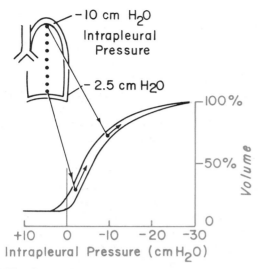

Figure 5.72. Cause of the regional differences of ventilation down the lung. Because of the weight of the lung, the intrapleural pressure is less negative at the base than at the apex. As a consequence, the basal lung is relatively compressed in its resting state but expands more on inspiration than the apex. (From West, 1985b.)

tion, we mean the change in volume per unit resting volume. It is clear from Figure 5.72 that the base of the lung has both a larger *change* in volume and smaller *resting* volume than the apex. Thus, its ventilation is greater. Note the paradox that, although the base of the lung is relatively poorly expanded compared with the apex, it is better ventilated. The same explanation can be given for the large ventilation of dependent lung in both the supine and lateral positions.

A remarkable change in the distribution of ventilation occurs at low lung volumes. Figure 5.73 is

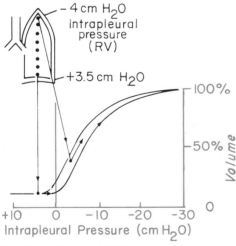

Figure 5.73. Distribution of ventilation at very low lung volumes. Now intrapleural pressures are generally less negative, and the pressure at the base actually exceeds airway (atmospheric) pressure. As a consequence airway closure occurs in this region, and no gas enters with small inspirations. *RV*, residual volume. (From West, 1985b.)

similar to Figure 5.72 except that it represents the situation at residual volume (that is, after a maximal expiration). Now the intrapleural pressures are less negative because the lung is not so well expanded and the elastic recoil forces are consequently smaller. Nevertheless, pressure differences between apex and base are still present because of the weight of the lung. The intrapleural pressure at the base now actually exceeds airway (atmospheric) pressure. Under these conditions the lung at the base is not being expanded but compressed, and ventilation is impossible until the local intrapleural pressure falls below atmospheric pressure. By contrast the apex of the lung is on a favorable part of the pressure-volume curve and ventilates well. Thus the normal distribution of ventilation is inverted, the upper regions ventilating better than the lower zones.

Airway Closure

Figure 5.73 shows that the compressed region of lung at the base does not have all its gas squeezed out. In practice, small airways, probably in the region of respiratory bronchioles (Fig. 5.3), close first, thus trapping gas in the distal alveoli (compare Fig. 5.54). This airway closure occurs only at very low lung volumes in young, normal subjects. In elderly, apparently normal people, however, airway closure in the lowermost regions of the lung occurs at somewhat higher volumes and may be present at functional residual capacity. The reason for this is that the aging lung loses some of its elastic recoil and the intrapleural pressures therefore become less negative, thus approaching the situation shown in Figure 5.73. In these circumstances, dependent regions of the lung may be only intermittently ventilated, and this leads to defective gas exchange. A similar situation frequently develops in patients with chronic lung disease.

Closing Volume

We have seen that during a full expiration a volume is reached at which the airways at the base of the lung begin to close. This is known as the *closing volume*. It can be measured as follows. The subject first breathes out to residual volume, and then takes a vital capacity breath of pure oxygen. During the subsequent slow, full expiration, the nitrogen concentration at the lips is measured with a rapid nitrogen meter. Figure 5.74 shows the pattern obtained. Four phases can be recognized (McCarthy et al., 1972). First pure dead space gas is exhaled (phase *1*), followed by a mixture of dead space and alveolar gas (phase *2*), and then by pure alveolar gas (phase *3*). The same sequence is shown in Figure

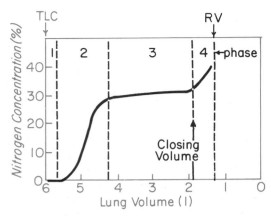

Figure 5.74. Measurement of closing volume, that is, the lung volume at which the airways at the bottom of the lung close. The subject makes an inspiration of 100% oxygen and then exhales to residual volume. Four phases can be recognized (see text). The last is caused by preferential emptying of the apex of the lung after the lower zone airways have closed. The apex has a higher nitrogen concentration because it receives less oxygen (compare Fig. 5.11). *TLC,* total lung capacity; *RV,* residual volume.

5.11. Toward the end of expiration, an abrupt increase in nitrogen concentration is seen (phase *4*). This signals closure of airways at the base of the lung. The nitrogen concentration rises at this point because, when the breath of oxygen was taken, the alveoli at lung apex were initially larger and therefore increased their volume less during the inspiration (Fig. 5.73). Therefore the apex received less oxygen. As a consequence, the nitrogen in the apex was less diluted with oxygen, and the preferential emptying of the apex in phase 4 caused a rise in nitrogen concentration to be seen.

In young, normal subjects, the closing volume is about 10% of the vital capacity. It increases steadily with age and is equal to about 40% of the vital capacity, that is, the functional residual capacity, at about the age of 65 years. There is some evidence that small amounts of disease of the small airways increase the closing volume, although the mechanism of this is not fully understood. Cigarette smokers may have increased closing volumes when other tests of lung function are still normal. Unfortunately the repeatability of the test is often poor. Nevertheless the test is sometimes used to detect early disease of the small airways.

Time Constants and Dynamic Compliance

In addition to the topographical inequality referred to above, the diseased lung often shows uneven ventilation as a result of regional differences in airway resistance and compliance. These

factors also probably operate to a small extent in the normal lung.

Suppose we regard a lung unit (Fig. 5.5) as an elastic chamber connected to the atmosphere by a tube. Then the amount of ventilation will depend on the compliance of the chamber and the resistance of the tube. In Figure 5.75 unit *A* has a normal distensibility and airway resistance. Its volume change on inspiration is large and rapid so that it is complete before expiration for the whole lung begins *(vertical broken line)*. By contrast, unit *B* has a low compliance, and its change in volume is rapid but small. Finally, unit *C* has a large airway resistance so that inspiration is slow and not complete before the lung begins to exhale. The shorter the time available for inspiration (fast breathing rate), the smaller the inspired volume. Such a unit is said to have a long *time constant,* the value of which is given by the product of the compliance and the resistance. Thus, inequality of ventilation can result either from alterations in local distensibility or in airway resistance.

Since the tidal volume of a lung with uneven time constants decreases as the time available for inspiration is reduced, that is, as the breathing frequency increases, it is sometimes said that its *dynamic compliance* is reduced. This is a somewhat loose term because the reduced tidal volume is caused in part by the resistance of the airways rather than by the elastic properties of the lung. This frequency dependence of dynamic compliance can be used, however, as a sensitive test of increased airway resistance (Woolcock et al., 1969).

We saw earlier (Fig. 5.66) that very little of the

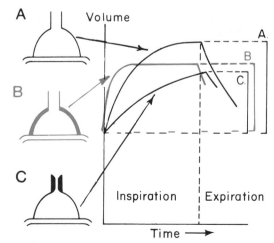

Figure 5.75. Effects of decreased compliance (*B*) and increased airway resistance (*C*) on ventilation of lung units compared with normal (*A*). In both *B* and *C* the inspired volume is abnormally low. (From West, 1985b.)

resistance of the normal bronchial tree is located in the very peripheral airways. This region, therefore, constitutes a "silent zone" in which considerable disease probably can occur without being detectable by measurements of total airway resistance. Nevertheless, changes of resistance in peripheral airways do cause uneven time constants and therefore a fall in dynamic compliance with increased breathing frequency. In practice, the dynamic compliance is measured at a series of frequencies from about 10 up to 120 breaths/min.

Stratified Inhomogeneity

Another possible mechanism of uneven ventilation is incomplete diffusion within the airways of the respiratory zone. We saw in Chapter 35 that the dominant mechanism of ventilation of the lung beyond the terminal bronchioles is diffusion. Normally this occurs so rapidly that differences in gas concentration within the acinus (lung unit supplied by a terminal bronchiole) are virtually abolished within a fraction of a second (Cumming et al., 1966). However, if there is dilation of the airways in the region of the respiratory bronchioles as in some disease such as centrilobular emphysema, the distance to be covered by diffusion may be enormously increased. In these circumstances, inspired gas is not distributed uniformly within the respiratory zone because of uneven ventilation *along* the lung units. This is known as stratified inhomogeneity.

Measurement of Ventilatory Inequality

The single-breath nitrogen test (Figs. 5.11 and 5.74) can be used to measure the degree of uneven ventilation in the lung. In normal subjects, the alveolar plateau (phase 3) is almost flat because the nitrogen in the lung is almost uniformly diluted by the breath of oxygen. In patients with uneven ventilation, however, the nitrogen concentration rises in phase 3. This can be explained by the fact that poorly ventilated regions (which receive relatively little oxygen and therefore have a relatively high nitrogen concentration) empty last. This late emptying is caused, in part, by the increased time constants (Fig. 5.75C) of the poorly ventilated regions.

Uneven ventilation can also be detected by a multibreath nitrogen washout. The subject breathes pure oxygen via a one-way valve, and the expired nitrogen concentration is measured at the lips. When the nitrogen concentration is plotted against the number of breaths on semilogarithmic paper, an almost straight line is found in normal subjects. This is because the successive dilution of the nitrogen in the lung by each breath of oxygen causes an exponential decay in nitrogen concentration. In the presence of uneven ventilation, however, the line becomes successively flatter because different regions of the lung lose their nitrogen at different rates until finally only the nitrogen in the worse-ventilated spaces is being washed out. Various indices of uneven ventilation can be derived from the washout pattern.

Tissue Resistance

When the lung and chest wall are moved, some pressure is required to overcome the viscous forces within the tissues as they slide over each other. Thus, part of the cross-hatched portion Figure 5.64 should be attributed to these tissue forces. This tissue resistance is only about 20% of the total (tissue + airway) resistance in young normal subjects, although it increases considerably in some diseases. This total resistance is sometimes called *pulmonary resistance* to distinguish it from airway resistance.

WORK OF BREATHING

Work is required to move the lung and chest wall. In this context, it is most convenient to measure work as pressure × volume.

Work Done on the Lung

This can be illustrated on a pressure-volume curve (Fig. 5.76). During inspiration the intrapleural pressure follows the curve *ABC* and the work done on the lung is given by the area *0ABCD*. Of this, the trapezoid *0AECD* represents the work required to overcome the elastic forces, and the cross-hatched area *ABCEA* represents the work overcoming viscous (airway and tissue) resistance (compare Fig. 5.64). The higher the airway resistance or the inspiratory flow rate, the more negative (rightward) would be the intrapleural pressure excursion between *A* and *C* and the larger the area.

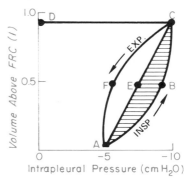

Figure 5.76. Pressure-volume curve of the lung showing the inspiratory work done overcoming elastic forces *(area 0AECD0)* and viscous forces *(hatched area ABCEA)*. INSP, inspiration; EXP, expiration.

On expiration, the area *AECFA* is the work required to overcome airway (+ tissue) resistance. Normally this falls within the trapezoid *0AECD*, and thus this work can be accomplished by the energy stored in the expanded elastic structures and released during a passive expiration. The difference between the areas *AECFA* and *0AECD* represents the work dissipated as heat.

The higher the breathing rate, the faster the flow rates and the larger the viscous work area *ABCEA*. On the other hand, the larger the tidal volume, the larger the elastic work area *0AECD*. Patients who have a reduced compliance (stiff lungs) tend to take small rapid breaths, while patients with severe airway obstruction often breathe more slowly. These patterns tend to reduce the work done by the lungs.

Total Work of Breathing

The total work done moving the lung and chest wall is difficult to measure, although estimates have been obtained by artificially ventilating paralyzed patients (or "completely relaxed" volunteers) in an iron lung type of respirator. Alternatively, the total work can be calculated by measuring the oxygen cost of breathing and assuming a figure for the *efficiency* as given by:

$$\text{Efficiency in \%} = \frac{\text{Useful work}}{\text{Total energy expended (or } O_2 \text{ cost)}} \times 100$$

The efficiency is believed to be about 5–10%.

The oxygen cost of quiet breathing is extremely small, being less than 5% of the total resting oxygen consumption. With voluntary hyperventilation it is possible to increase this to 30%. In patients with obstructive lung disease, the oxygen cost of breathing may limit their exercise ability.

CHAPTER 40

Control of Ventilation

We have seen that the chief function of the lung is to exchange oxygen and carbon dioxide between blood and gas and thus maintain normal levels of P_{O_2} and P_{CO_2} in arterial blood. In this chapter we shall see that, despite widely differing demands for oxygen uptake and carbon dioxide output made by the body, the arterial P_{O_2} and P_{CO_2} are normally kept within close limits. This remarkable regulation of gas exchange is possible because the level of ventilation is so carefully controlled.

The three basic elements of the respiratory control system (Fig. 5.77) are

1. *sensors*, which gather information and feed it to the
2. *central controller* in the brain, which coordinates the information and, in turn, sends impulses to the
3. *effectors* (respiratory muscles), which cause ventilation.

We shall see that increased activity of the effectors generally ultimately decreases the sensory input to the brain, for example, by decreasing the arterial P_{CO_2}. This is an example of negative feedback.

CENTRAL CONTROLLER

The normal automatic process of breathing originates in impulses which come from the brainstem. The cortex can override these centers if voluntary control is desired. Additional input from other parts of the brain occurs under certain conditions.

Brainstem

The periodic nature of inspiration and expiration is controlled by neurons located in the pons and medulla. These have been designated the *respiratory centers*. However, these should not be thought of as composing a discrete nucleus but rather as a somewhat poorly defined collection of neurons with various components.

Three main groups of neurons are recognized.

1. Medullary respiratory center in the reticular formation of the medulla. This comprises two identifiable areas. One group of cells in the dorsal region of the medulla (dorsal respiratory group) is associated chiefly with inspiration; the other in the ventral area (ventral respiratory group) is mainly for expiration. One popular (though not universally accepted) view is that the cells of the *inspiratory area* have the property of intrinsic periodic firing and are responsible for the basic rhythm of ventilation. When all known afferent stimuli have been abolished, these inspiratory cells generate repetitive bursts of action potentials which result in nervous impulses going to the diaphragm and other inspiratory muscles.

The intrinsic rhythm pattern of the inspiratory area starts with a latent period of several seconds during which there is no activity. Action potentials then begin to appear, increasing in a crescendo over the next few seconds. During this time inspiratory muscle activity becomes stronger in a "ramp"-type pattern. Finally, the inspiratory action potentials cease, and inspiratory muscle tone falls to its preinspiratory level.

This inspiratory ramp can be "turned off" prematurely by inhibiting impulses from the *pneumotaxic center* (see below). In this way inspiration is shortened and, as a consequence, the breathing rate increases. The output of the inspiratory cells is further modulated by impulses from the vagal and glossopharyngeal nerves. Indeed, these terminate in the tractus solitarius, which is situated close to the inspiratory area.

The *expiratory area* is quiescent during normal quiet breathing because ventilation is then achieved by active contraction of the inspiratory muscles (chiefly the diaphragm), followed by passive relaxation of the chest wall to its equilibrium position (Chapter 39). However, in more forceful breathing, for example, on exercise, expiration becomes active as a result of the activity of the expiratory cells. There is still not universal agreement on how the

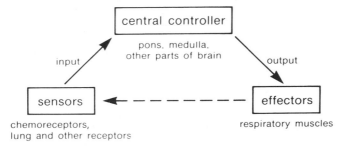

Figure 5.77. Basic elements of the respiratory control system. Information from various sensors is fed to the central controller, the output of which goes to the respiratory muscles. By changing ventilation, the respiratory muscles reduce perturbations of the sensors (negative feedback).

intrinsic rhythmicity of respiration is brought out by the medullary centers.

2. Apneustic center in the lower pons. This area is so named because, if the brain of an experimental animal is sectioned just above this site, prolonged inspiratory gasps *(apneuses)* interrupted by transient expiratory efforts are seen. Apparently, the impulses from the center have an excitatory effect on the inspiratory area of the medulla, tending to prolong the ramp action potentials. Whether this apneustic center plays a role in normal human respiration is not known, although in some types of severe brain injury this type of abnormal breathing is seen.

3. Pneumotaxic center in the upper pons. As indicated above, this area appears to "switch off" or inhibit inspiration and thus regulate inspiratory volume and, secondarily, respiratory rate. This has been demonstrated experimentally in animals by direct electrical stimulation of the pneumotaxic center. Some investigators believe that the role of this center is "fine tuning" of respiratory rhythm because a normal rhythm can exist in the absence of this center.

Cortex

Breathing is under voluntary control to a considerable extent, and the cortex can override the function of the brainstem within limits. It is not difficult to halve the arterial P_{CO_2} by voluntary hyperventilation, although the consequent alkalosis may cause tetany with contraction of the muscles of the hand and foot (carpopedal spasm). Halving the P_{CO_2} increases the pH by about 0.2 units (Chapter 32).

Voluntary hypoventilation is more difficult. The duration of breath-holding is limited by several factors, including the arterial P_{CO_2} and P_{O_2}. A preliminary period of hyperventilation increases breath-holding time, especially if oxygen is breathed. However, factors other than chemical are involved.

This is shown by the observation that if, at the breaking point of breath-holding, a gas mixture is inhaled which *raises* the arterial P_{CO_2} and *lowers* the P_{O_2}, a further period of breath-holding is possible.

Other Parts of the Brain

Other parts of the brain, such as the limbic system and hypothalamus, can affect the pattern of breathing, for example, in affective states such as rage and fear.

EFFECTORS

The muscles of respiration include the diaphragm, intercostal muscles, abdominal muscles, and accessory muscles such as the sternomastoids. The actions of these muscles were described at the beginning of Chapter 39. In the context of the control of ventilation, it is crucially important that these various muscle groups work in a coordinated manner, and this is the responsibility of the central controller. There is evidence that some newborn children, particularly those who are premature, have uncoordinated respiratory muscle activity, especially during sleep. For example, the thoracic muscles may try to inspire while the abdominal muscles expire. This may be a factor in the "sudden infant death syndrome."

SENSORS

Central Chemoreceptors

A chemoreceptor is a receptor which responds to a change in the chemical composition of the blood or some other fluid around it. The most important receptors involved in the minute-by-minute control of ventilation are those situated near the ventral surface of the medulla in the vicinity of the exit of the IXth and Xth nerves. In animals, local application of H^+ or dissolved carbon dioxide to this area stimulates breathing within a few seconds. For some time it was thought that the medullary respiratory center itself was the site of action of CO_2, but most physiologists now believe that the chemoreceptors are anatomically separate. Some evidence suggests that they lie about 200–400 μm below the ventral surface of the medulla (Fig. 5.78).

The central chemoreceptors are surrounded by brain extracellular fluid and respond to changes in its H^+ concentration. An increase in H^+ concentration stimulates ventilation while a decrease inhibits it. The composition of the extracellular fluid around the receptors is governed by the cerebrospinal fluid (CSF), local blood flow, and local metabolism.

Of these, the CSF is apparently the most important. It is separated from the blood by the blood-

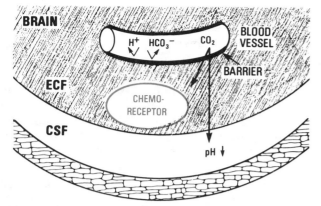

Figure 5.78. Scheme of the environment of the central chemoreceptors. They are bathed in brain extracellular fluid (ECF) through which CO_2 easily diffuses from blood vessels to CSF. The CO_2 reduces the CSF pH, thus stimulating the chemoreceptor. H^+ and HCO_3^- ions cannot easily cross the blood-brain barrier.

which takes several days. The partial resetting of the CSF pH in relation to blood pH results in its predominating influence on ventilation and arterial P_{CO_2}.

An example showing the importance of CSF pH in the control of ventilation is a patient with chronic lung disease and CO_2 retention of long standing. Such a patient may have a nearly normal CSF pH and, therefore, an abnormally low ventilation for his arterial P_{CO_2}. A similar situation is seen in normal subjects who are exposed to an atmosphere containing 3% CO_2 for some days.

Peripheral Chemoreceptors

These are located in the carotid bodies at the bifurcation of the common carotid arteries and in the aortic bodies above and below the aortic arch. The carotid bodies are the most important in humans. They contain glomus cells of two or more types which show an intense fluorescence staining because of their large content of dopamine. The mechanism of chemoreception is not yet understood. A popular view has been that glomus cells themselves are chemoreceptors. Another more recent hypothesis is that they are inhibitory interneurons and that the impulses are generated in the afferent terminals of the carotid sinus nerve. This theory proposes that impulses generated at the nerve fiber terminal release an excitatory transmitter that causes the glomus cell to release dopamine. In turn, dopamine acts on the nerve fiber terminal to inhibit impulse generation. The nerve in which impulses travel from the carotid body is known as Hering's nerve; it is a branch of the glossopharyngeal nerve.

The peripheral chemoreceptors respond to decreases in arterial P_{O_2} and pH and increases in arterial P_{CO_2}. They are unique among tissues of the body in that their sensitivity to changes in arterial P_{O_2} begins around 500 mm Hg. Figure 5.79 shows that the relationship between firing rate and arterial P_{O_2} is very nonlinear; relatively little response

brain barrier, which is relatively impermeable to H^+ and HCO_3^- ions, although molecular CO_2 diffuses across it easily. When the blood P_{CO_2} rises, CO_2 diffuses into the CSF from the cerebral blood vessels, liberating H^+ ions, which stimulate the chemoreceptors. Thus, the CO_2 level in blood regulates ventilation chiefly by its effect on the pH of the CSF. The resulting hyperventilation reduces the P_{CO_2} in the blood and, therefore, in the CSF. The cerebral vasodilation which accompanies an increased arterial P_{CO_2} enhances diffusion of CO_2 into the CSF and the brain extracellular fluid.

The normal pH of the CSF is 7.32 and, since the CSF contains much less protein than blood, it has a much lower buffering capacity. As a result the change in CSF pH for a given change in P_{CO_2} is greater than that in blood. If the CSF pH is displaced over a prolonged period, a compensatory change in HCO_3^- occurs as a result of transport across the blood-brain barrier. However the CSF pH does not usually return to 7.32. The change in CSF pH occurs more promptly than that of the pH of arterial blood by renal compensation, a process

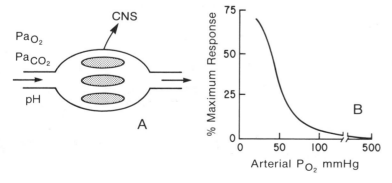

Figure 5.79. *A,* Carotid body which responds to changes of P_{O_2}, P_{CO_2}, and pH in arterial blood. Impulses travel to the central nervous system *(CNS)* through Hering's nerve. *B,* The nonlinear response to arterial P_{O_2}. Note that the maximum response occurs below a P_{O_2} of 50 mm Hg.

occurs until the arterial P_{O_2} is reduced below 100 mm Hg, but then the rate rapidly increases. The carotid bodies have a very high blood flow for their size (20 ml/min/g tissue) and consequently a very small arterial-venous O_2 difference despite a very high metabolic rate. As a consequence, they respond to arterial rather than venous P_{O_2}. The response of these receptors can be very fast; indeed, their discharge rate can alter during the respiratory cycle as a result of the small cyclic changes in blood gases. The peripheral chemoreceptors are responsible for all of the increase of ventilation which occurs in humans in response to arterial hypoxemia. Indeed, in the absence of these receptors, severe hypoxemia depresses respiration, presumably through a direct effect on the respiratory centers. Complete loss of hypoxic ventilatory drive has been shown in patients with bilateral carotid body resection.

The response of the peripheral chemoreceptors to arterial P_{CO_2} is much less important than that of the central chemoreceptors. For example, when a normal subject is given a mixture of CO_2 in O_2 to breathe, less than 20% of the ventilatory response can be attributed to the peripheral chemoreceptors. However, their response is more rapid, and they may be useful in matching ventilation to abrupt changes in P_{CO_2}.

In humans, the carotid but not the aortic bodies respond to a fall in arterial pH. This occurs regardless of whether the cause is respiratory or metabolic. Interaction of the various stimuli occurs. Thus, increases in chemoreceptor activity in response to decreases in arterial P_{O_2} are potentiated by increases in P_{CO_2} and, in the carotid bodies, decreases in pH.

Lung Receptors

PULMONARY STRETCH RECEPTORS

These are believed to lie within airway smooth muscle. They discharge in response to distension of the lung, and their activity is sustained with lung inflation; that is, they show little adaptation. The impulses travel in the vagus nerve via large myelinated fibers.

The main reflex effect of stimulating these receptors is a slowing of respiratory frequency due to an increase in expiratory time. This is known as the Hering-Breuer inflation reflex. It can be well demonstrated in a rabbit preparation where the diaphragm has a slip of muscle from which recordings can be made without interfering with the other respiratory muscles. Classic experiments showed that inflation of the lungs tended to inhibit further inspi-

ratory muscle activity (Fig. 5.80). The opposite response is also seen, that is, deflation of the lungs tends to initiate inspiratory activity (deflation reflex). Thus these reflexes can provide a self-regulatory mechanism or negative feedback.

The Hering-Breuer reflexes were once thought to play a major role in ventilation by determining the rate and depth of breathing. This could be done by using the information from these stretch receptors to modulate the "switching off" mechanism in the medulla. Thus, bilateral vagotomy which removes the input of these receptors causes slow, deep breathing in most animals. However, more recent work indicates that the reflexes are largely inactive in adult humans unless the tidal volume exceeds one liter, as in exercise. Transient bilateral blockade of the vagi by local anesthesia in awake humans does not change either breathing rate or volume. There is some evidence that these reflexes may be more important in newborn babies.

IRRITANT RECEPTORS

These are thought to lie between airway epithelial cells, and they are stimulated by noxious gases, cigarette smoke, inhaled dusts, and cold air. The impulses travel up the vagus in myelinated fibers, and the reflex effects include bronchoconstriction and hyperpnea. If the stimulus is maintained, the activity of the receptors rapidly lessens, and they are therefore classified as "rapidly adapting." In fact, some physiologists prefer to call these receptors "rapidly adapting receptors" rather than "irritant receptors" because they are apparently involved in additional mechanoreceptor functions as well as responding to noxious stimulants on the airway walls. It is possible that irritant receptors play a role in the bronchoconstriction of asthma attacks as a result of their response to released histamine.

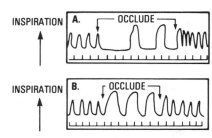

Figure 5.80. Hering-Breuer reflexes. The records show contraction of an isolated strip of rabbit diaphragm. In *A*, the trachea was occluded at the end of inspiration. Note the inhibition of further inspiratory contraction (inflation reflex). In *B*, the trachea was occluded at the end of expiration; this occlusion is followed by prolonged inspiratory contractions (deflation reflex).

J RECEPTORS

The term "juxtacapillary," or J, is used because the receptors are believed to be in the alveolar walls close to the capillaries (Paintal, 1970). The evidence for this location is that they respond very quickly to chemicals such as capsaicin injected into the pulmonary circulation. Identification of the receptors is uncertain, but sparse nonmyelinated fibers have been demonstrated in alveolar walls in histological sections. The impulses pass up the vagus nerve in slowly conducting nonmyelinated fibers and can result in rapid shallow breathing, although intense stimulation causes apnea. These receptors do not show a firing pattern that can be associated with normal respiration.

There is evidence that engorgement of pulmonary capillaries and increases in the interstitial fluid volume of the alveolar wall activate these receptors. They may play a role in the dyspnea (see below) associated with left heart failure, interstitial lung disease, pneumonia, and microembolism. It has been speculated that stimulation of the J receptors may contribute to the increase in ventilation which occurs on exercise.

A fourth group of receptors has recently been described. These are lodged in the airways and are innervated by nonmyelinated fibers, but their role is unclear.

Other Receptors

NOSE AND UPPER AIRWAY RECEPTORS

The nose, nasopharynx, larynx, and trachea contain receptors that respond to mechanical and chemical stimulation. Various reflex responses have been described, including sneezing, coughing, and bronchoconstriction. Laryngeal spasm may occur if the larynx is irritated mechanically, for example, during insertion of an endotracheal tube with insufficient local anesthesia.

JOINT AND MUSCLE RECEPTORS

Impulses from moving limbs are believed to be part of the stimulus to ventilation during exercise, especially in the early stages.

GAMMA SYSTEM

Many muscles, including the intercostal muscles and diaphragm, contain muscle spindles which sense elongation of the muscle. This information is used to reflexly control the strength of contraction. These receptors may be involved in the sensation of dyspnea which occurs when unusually large respiratory efforts are required to move the lung and chest wall, for example, because of airway obstruction.

ARTERIAL BARORECEPTORS

Increase in arterial blood pressure can cause reflex hypoventilation or apnea through stimulation of the aortic and carotid sinus baroreceptors (Chapter 16). Conversely, a decrease in blood pressure may result in hyperventilation. A possible physiological advantage of this reflex is enhancement of venous return following severe hemorrhage. However, the duration of these reflexes is typically short.

PAIN AND TEMPERATURE

Stimulation of many afferent nerves can bring about a change in respiration. Pain often causes a period of apnea followed by hyperventilation. Heating the skin may result in hyperventilation. The increased ventilation seen in fever is thought to be due in part to stimulation of hypothalamic thermoreceptors.

INTEGRATED RESPONSES

Now that we have looked at the various units that make up the respiratory control system (Fig. 5.77), it is useful to consider the overall responses of the system to changes in the arterial CO_2, O_2, and pH and to exercise.

Response to Carbon Dioxide

The most important factor in the control of ventilation under normal conditions is the P_{CO_2} of the arterial blood. The sensitivity of this control is remarkable. In the course of daily activity with periods of rest and exercise, the arterial P_{CO_2} is probably held to within 3 mm Hg. During sleep it may rise a little more.

The ventilatory response to CO_2 is normally measured by having the subject inhale CO_2 mixtures or rebreathe from a bag so that the inspired P_{CO_2} gradually rises. In one technique which can easily be used in the pulmonary function laboratory (Read, 1967), the subject rebreathes from a bag which is prefilled with 7% CO_2 and 93% O_2. As he rebreathes, he adds metabolic CO_2 to the bag, but the O_2 concentration remains relatively high. In such a procedure the P_{CO_2} of the bag gas increases at the rate of about 4 mm Hg/min.

Figure 5.81 shows the results of experiments in which the inspired mixture was adjusted to yield a constant alveolar P_{O_2}. (In this type of experiment on normal subjects, alveolar end-tidal P_{O_2} and P_{CO_2} are generally taken to reflect the arterial levels.) With a normal P_{O_2} the ventilation increases by about 2–3 liters/min for each 1 mm Hg rise in P_{CO_2}. Lowering the P_{O_2} produces two effects: there is a higher ven-

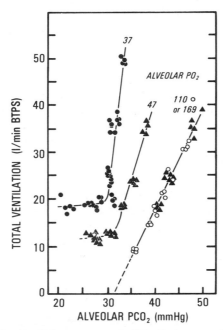

Figure 5.81. Ventilatory response to CO_2. Each curve of total ventilation against alveolar P_{CO_2} is for a different P_{O_2}. In this study, no difference was found between an alveolar P_{O_2} of 110 and one of 169 mm Hg, although some investigators have found that the slope of the line is slightly less at the higher P_{O_2}. *BTPS*, body temperature, pressure, and saturation. (From Nielsen and Smith, 1952.)

tilation for a given P_{CO_2} and, also, the slope of the line becomes steeper. It should be emphasized that there is considerable variation between subjects.

Another way of measuring respiratory drive is to record the inspiratory pressure during a brief period of airway occlusion (Whitelaw et al., 1975). The subject breathes through a mouthpiece attached to a valve box, and the inspiratory port is provided with a shutter. This is closed during an expiration (the subject being unaware), so that the first part of his next inspiration is against an occluded airway. The shutter is opened after about 0.5 s. The pressure generated during the first 0.1 s of attempted inspiration (known as $P_{0.1}$) is taken as a measure of respiratory center output. This is largely unaffected by the mechanical properties of the respiratory system, although it may be influenced by lung volume. This method can be used to study the respiratory sensitivity to CO_2, hypoxia, and other variables as well.

A reduction in arterial P_{CO_2} is very effective in reducing the stimulus to ventilation. For example, if the reader hyperventilates voluntarily for a few seconds, he or she will have no urge to breathe for a short period. An anesthetized patient will frequently stop breathing for a minute or so if the patient is first overventilated by the anesthesiologist.

The ventilatory response to CO_2 is reduced by sleep, increasing age, and genetic, racial, and personality factors. Trained athletes and divers tend to have a low CO_2 sensitivity. Various drugs depress the respiratory center, including morphine and barbiturates. Patients who have taken an overdose of one of these drugs often have marked hypoventilation. The ventilatory response to CO_2 is also reduced if the work of breathing is increased. This can be demonstrated by having normal subjects breathe through a narrow tube. The neural output of the respiratory center is not reduced, but it is not so effective in producing ventilation. The abnormally small ventilatory response to CO_2 and the CO_2 retention in some patients with lung disease can be partly explained by the same mechanism. In such patients, reducing the airway resistance with bronchodilators often increases their ventilatory response. There is also some evidence that the sensitivity of the respiratory center is reduced in these patients.

As we have seen, the main stimulus to increase ventilation when the arterial P_{CO_2} rises comes from the central chemoreceptors which respond to the increased H^+ concentration of the brain extracellular fluid near the receptors. An additional stimulus comes from the peripheral chemoreceptors, both because of the rise in arterial P_{CO_2} and the fall in pH.

Response to Oxygen

The way in which a reduction of P_{O_2} in arterial blood stimulates ventilation can be studied by having a subject breathe hypoxic gas mixtures. The end-tidal P_{O_2} and P_{CO_2} are used as measures of the arterial values. Figure 5.82 shows that when the alveolar P_{CO_2} is kept at about 36 mm Hg the alveolar P_{O_2} can be reduced to the vicinity of 50 mm Hg before any appreciable increase in ventilation occurs. Raising the P_{CO_2} (by altering the inspired mixture) increases the ventilation at any P_{O_2} (compare Fig. 5.81). When the P_{CO_2} is increased, a reduction in P_{O_2} below 100 mm Hg causes some stimulation of ventilation, unlike the situation when the P_{CO_2} is normal. Thus, the combined effects of both stimuli exceed the sum of each stimulus given separately; this is referred to as interaction between the high CO_2 and low O_2 stimuli.

Various indices of hypoxic sensitivity are in use. One is the increment in ventilation when the arterial (or alveolar) P_{O_2} is reduced from 150 to 40 mm Hg (so-called $\dot{V}_{40}$). The mean value in normal subjects is about 35 liters/min, but the normal range is very large. Another index that is sometimes used in the pulmonary function laboratory is based on the

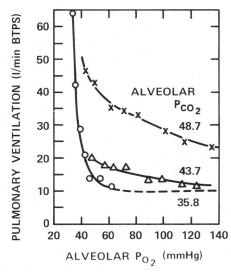

Figure 5.82. Hypoxic response curves. When the P_{CO_2} is 36 mm Hg, almost no increase in ventilation occurs until the P_{O_2} is reduced to about 50 mm Hg. *BTPS*, body temperature, pressure, and saturation. (Adapted from Loeschke and Gertz, 1958.)

finding that there is a nearly linear relationship between arterial O_2 saturation and ventilation as the inspired P_{O_2} is reduced. Since the O_2 saturation can conveniently be measured with a pulse or ear oximeter, the hypoxic sensitivity can be expressed as the change in ventilation per 1% fall in O_2 saturation.

Large differences in hypoxic ventilatory response occur between individuals, and the sensitivity is, in part, genetically determined. Interestingly, the sensitivity is considerably reduced or "blunted" in persons who have been hypoxemic since birth, such as those born at high altitude, or patients with cyanotic congenital heart disease ("blue babies"). When someone who was born and has lived for years at high altitude descends to sea level, the blunted response persists for many years.

We have seen (Fig. 5.82) that the P_{O_2} can normally be reduced a long way without evoking any increase in ventilation. Therefore, the role of this hypoxic drive in the day-by-day control of ventilation is small. However, on ascent to high altitude, a large and persistent increase in ventilation occurs in response to the hypoxia. This increase is further considered in Chapter 41.

In some patients with severe lung disease, the hypoxic drive to ventilation becomes very important. These patients have chronic CO_2 retention, and the pH of their brain extracellular fluid may have returned to near normal despite a raised P_{CO_2}. Thus they have lost most of their increase in the CO_2 stimulus to ventilation. In addition, the initial depression of blood pH has been nearly abolished by

renal compensation so that there is little pH stimulation of the peripheral chemoreceptors (see below). Thus, the arterial hypoxemia becomes the chief stimulus to ventilation. If such a patient is given a high O_2 mixture to breathe to relieve his hypoxemia, his ventilation may become grossly depressed. His ventilatory state is best monitored by measuring his arterial P_{CO_2}.

As we have seen, hypoxemia reflexly stimulates ventilation by its action on the carotid and aortic body chemoreceptors. It has no action on the central chemoreceptors; indeed, in the absence of peripheral chemoreceptors, hypoxemia depresses respiration. However, prolonged hypoxemia can cause mild cerebral acidosis, which in turn can stimulate ventilation.

Response to pH

A reduction in arterial blood pH stimulates ventilation. In practice, it is often difficult to separate the ventilatory response resulting from a fall in pH from that caused by an accompanying rise in P_{CO_2}. However, in experimental animals in whom it is possible to reduce the pH at a constant P_{CO_2}, the stimulus to ventilation can be convincingly demonstrated. Patients with a partly compensated metabolic acidosis (such as in diabetes mellitus) who have a low pH and low P_{CO_2} show an increased ventilation. Indeed, this is responsible for the reduced P_{CO_2}.

As we have seen, the chief site of action of a reduced arterial pH is the peripheral chemoreceptors, especially the carotid bodies in humans. It is also possible that the central chemoreceptors or the respiratory center itself is affected by a change in blood pH if it is large enough. In this case, the blood-brain barrier becomes partly permeable to H^+ ions.

Response to Exercise

On exercise, ventilation increases promptly and, during strenuous exertion, it may reach very high levels. A fit young person who attains a maximum O_2 consumption of 4 liters/min may have a total ventilation of 120 liters/min, that is, about 15 times his resting level. This increase in ventilation closely matches the increase in O_2 uptake and CO_2 output. It is remarkable that the cause of the increased ventilation on exercise remains largely unknown.

The arterial P_{CO_2} does not increase during most forms of exercise; indeed, during severe exercise it typically falls slightly. The arterial P_{O_2} usually increases slightly, although it may fall at very high work levels. The arterial pH remains nearly constant for moderate exercise, although during heavy exercise it falls because of the liberation of lactic acid through anaerobic metabolism. If this occurs, ventilation is further stimulated by the increased H^+ concentration. However, it is clear that none of the mechanisms discussed thus far can account for the large increase in ventilation observed during light to moderate exercise.

Other stimuli have been suggested. *Passive movement of the limbs* stimulates ventilation in both anesthetized animals and awake humans. This is a reflex with receptors presumably located in joints or muscles. It is probably responsible for the abrupt increase in ventilation which occurs during the first few seconds of exercise. Another hypothesis is that *oscillations in arterial P_{O_2} and P_{CO_2}* may stimulate the peripheral chemoreceptors even though the mean level remains unaltered. These fluctuations are caused by the periodic nature of ventilation and increase when the tidal volume rises, as on exercise. Another theory is that the central chemoreceptors increase ventilation to hold the *arterial P_{CO_2} constant* by some kind of servomechanism, just as a thermostat can control a furnace with little change in temperature. The objection that the arterial P_{CO_2} often *falls* on exercise is countered by the assertion that the preferred level of P_{CO_2} is reset in some way. Proponents of this theory believe that the ventilatory response to inhaled CO_2 may not be a reliable guide to what happens on exercise.

Yet another hypothesis is that ventilation is linked in some way to the additional *CO_2 load* presented to the lungs in the mixed venous blood during exercise. In animal experiments, an increase in this load produced either by infusing CO_2 into the venous blood or by increasing venous return has been shown to correlate well with ventilation. However, a problem with this hypothesis is that no suitable venous receptor has been found.

Additional factors which have been suggested include the *increase in body temperature* during exercise which stimulates ventilation. Finally, it has been proposed that "irradiation" of the respiratory centers by *impulses from the motor cortex or hypothalamus* is responsible for linking ventilation to muscle activity during exercise. However, none of the theories proposed thus far is completely satisfactory.

ABNORMAL PATTERNS OF BREATHING

Subjects with severe hypoxemia often exhibit a striking pattern of periodic breathing known as *Cheyne-Stokes respiration*. This pattern is characterized by periods of apnea of 5–20 s separated by approximately equal periods of hyperventilation when the tidal volume gradually waxes and then wanes. This pattern is frequently seen at high alti-

tude, especially at night during sleep. It may be associated with marked cyclic changes in the arterial oxygen saturation caused by variations in the alveolar P_{O_2} as the ventilation decreases and increases again (Fig. 5.83). The pattern is also found in some patients with severe heart disease or brain damage.

The cause of periodic breathing is not fully understood. One hypothesis is that, during sleep at high altitude, the large stimulus to ventilation caused by the severe hypoxemia causes instability of the ventilatory control system. It is well known that control systems can become unstable if the gain of the system is greatly increased. An example is seen when a microphone is placed near the loudspeaker of a public address system; the unpleasant howl is caused by instability resulting from the high gain of the system.

It is likely that not all examples of periodic breathing have the same explanation. The pattern can be reproduced in experimental animals by lengthening the distance through which the blood travels on its way to the brain from the lung. Under these conditions, there is a long delay before the central chemoreceptors sense the alteration in P_{CO_2} caused by the change of ventilation. As a result, the respiratory center hunts for the equilibrium condition, always overshooting it. This may be the explanation in some patients with severe heart disease and increased circulation times.

DYSPNEA

A feeling of shortness of breath is one of the most important symptoms of lung disease. Dyspnea refers to the sensation of difficulty with breathing

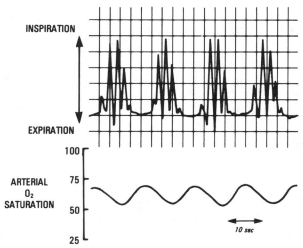

Figure 5.83. Cheyne-Stokes breathing during sleep at an altitude of 6300 m. Note the waxing and waning of breathing and the consequent fluctuation of arterial O_2 saturation.

and should be distinguished from simple tachypnea (rapid breathing) or hypernea (increased ventilation). Because dyspnea is a subjective phenomenon, it is difficult to measure, and the factors responsible for it are poorly understood. Broadly speaking, dyspnea occurs when the *demand for ventilation* is out of proportion to the patient's *ability to respond* to that demand. As a result, breathing becomes difficult, uncomfortable, or labored.

An *increased demand for ventilation* is often caused by changes in the blood gases and pH. High ventilations on exercise are commonly seen in patients with inefficient pulmonary gas exchange, especially those with large physiological dead spaces who tend to develop CO_2 retention and acidosis unless they achieve high ventilations. Another important factor is stimulation of intrapulmonary stretch receptors or J receptors. These factors may explain, in part, the increased ventilation of patients with pulmonary fibrosis and congestive heart failure, respectively.

A *reduced ability to respond* to ventilatory needs is generally caused by abnormal mechanics of the lung or chest wall. Frequently, increased airway resistance is the culprit, as in asthma, but other causes include a stiff chest wall, as in the patient with a hunchback (kyphoscoliosis).

The assessment of dyspnea is difficult. Usually, exercise tolerance is determined from a standard questionnaire which grades breathlessness according to how far the patient can walk on the level or upstairs without pausing for breath. Occasionally, ventilation is measured at a standard level of exercise and then related to the patient's maximum voluntary ventilation in an attempt to obtain an index of dyspnea. However, it should be remembered that dyspnea is something that only the patient feels, and as such it cannot be accurately measured.

CHAPTER 41

Respiration in Unusual Environments

The surface area of the lung is approximately 50–100 m^2, compared with only some 2 m^2 for the skin. Therefore, in one sense the lung serves as our principal physiological link with the environment. Respiratory physiologists have always been intrigued by the challenges of unusual environments, as illustrated, for example, by a long and colorful history of physiological expeditions to high altitudes. Recently the first measurements of respiratory function were made at the highest point on earth, the summit of Mt. Everest, where the inspired P_{O_2} is only about 43 mm Hg. At the other end of the pressure spectrum, there is now intense research activity in hyperbaric physiology. Divers connected with the exploration and recovery of oil in the North Sea area, for example, are exposed to great hazards, and physiological accidents are disturbingly common.

Nearer home, everyone is exposed to atmospheric pollution, which continues to increase in most parts of the world. Moreover, the composition of the pollutants changes as attempts are made to clean the environment, and new problems therefore arise all the time. Even those who live at normal pressures breathing clean air have, at one stage, coped with the most momentous challenge of all—that of changing from placental to lung respiration in a few seconds at the time of birth.

HIGH ALTITUDE

Hypoxia of High Altitude

The barometric pressure decreases with distance above the earth's surface in an approximately exponential manner (Fig. 5.84). The pressure at 5,500 m (18,000 feet) is only about one-half the normal 760 mm Hg, so the P_{O_2} of moist inspired gas is $(380 - 47) \times 0.2093 = 70$ mm Hg (Table 5.2). (Recall that 47 mm Hg is the partial pressure of water vapor at body temperature—see Chapter 35). At the summit of Mt. Everest (altitude 8,848 m (29,028 feet)), where the barometric pressure is approximately 253 mm Hg, the inspired P_{O_2} is $(253 - 47) \times 0.2093$

$= 43$ mm Hg. For a person who has the misfortune to bail out of an aircraft flying at 19,200 m (63,000 feet) at a barometric pressure of 47 mm Hg, the P_{O_2} of inspired gas in the trachea will be zero! Indeed, the tissue fluids vaporize ("boil") near this pressure.

Even a modest ascent, for example, in a modern skiing resort, can cause appreciable hypoxemia. Thus, at a cable car station at 3,400 m (11,000 feet), the inspired P_{O_2} is about 95 mm Hg. If one assumes an alveolar P_{CO_2} of 35 mm Hg, respiratory exchange ratio of 0.8, and alveolar-arterial oxygen difference of 3 mm Hg, the alveolar gas equation (Chapter 38) gives an arterial P_{O_2} of only about 53 mm Hg. No wonder it is difficult to adjust one's skis! In terms of oxygen transport, however, it should be remembered that, because of the shape of the oxygen dissociation curve, the arterial oxygen saturation even then will be approximately 89% (assuming a base excess of zero).

In spite of the hypoxia associated with high altitude, some 15 million people live at elevations over 3,050 m (10,000 feet), and permanent residents live higher than 4,900 m (16,000 feet) in the Peruvian Andes. A remarkable degree of acclimatization occurs when humans ascend to these altitudes; indeed, climbers have lived for several days at altitudes that would cause unconsciousness within a minute or two in the absence of acclimatization. For example, mountaineers have spent over a week above 7,600 m (25,000 feet) at an oxygen pressure which usually produces unconsciousness within about 2 min in persons abruptly exposed to this low pressure.

In all discussions of the effects of hypoxia caused by low barometric pressure, a clear distinction should be made between (1) sudden exposure to low pressure, such as occurs when an aircraft loses cabin pressure; (2) exposure over several weeks, as when a lowlander mountaineer acclimatizes during a high climb; and (3) the lifelong exposure of a permanent resident at high altitude. In general, the tol-

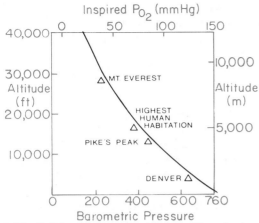

Figure 5.84. Relationship between altitude and barometric pressure. At 1500m (5000 ft) (Denver), the P_{O_2} of moist inspired gas is about 130 mm Hg, but it is only 43 mm Hg on the summit of Everest.

erance to high altitude, as reflected (for example) in the amount of physical activity than can be accomplished, increases with the length of stay as a result of the process of acclimatization.

ACUTE MOUNTAIN SICKNESS

Newcomers to high altitude frequently complain of headache, fatigue, dizziness, palpitations, nausea, loss of appetite, and insomnia. This is known as *acute mountain sickness*. The cause is not certain, although a combination of hypoxemia and respiratory alkalosis is likely. Some investigators believe that the condition is due to mild cerebral edema. There is evidence that cerebral blood flow may be considerably increased as a result of severe hypoxemia, although the reduced arterial P_{CO_2} causes cerebral vasoconstriction. Measurements of retinal blood flow show that this can double in lowlanders moved to an altitude of 5,300 m (17,500 feet) (Frayser et al., 1974). Retinal hemorrhages apparently

frequently occur, although they usually heal leaving no visual defects. Periodic or Cheyne-Stokes breathing is common at high altitude, especially during sleep (Chapter 40).

Long-term residents at high altitude sometimes develop an ill-defined syndrome characterized by fatigue, reduced exercise tolerance, severe hypoxemia, and excessive polycythemia. This is called *chronic mountain sickness* or Monge's disease. Removal to lower altitudes usually results in improvement.

Hyperventilation

This is one of the most beneficial physiological responses to exposure to a low oxygen partial pressure. Its value can be seen by considering the alveolar gas equation (see Chapter 38) for a climber breathing air on the summit of Mt. Everest. If his ventilation were normal giving an alveolar P_{CO_2} of 40 mm Hg, his alveolar P_{O_2} would be

$$P_{A_{O_2}} = P_{I_{O_2}} - \frac{P_{A_{CO_2}}}{R} + F$$
$$= 43 - \frac{40}{0.8} + 2$$
$$= -5 \text{ mm Hg!}$$

In this calculation the respiratory exchange ratio is assumed to be 0.8; under these conditions the small correction factor is 2 mm Hg.

Clearly something must be done about this! In practice the climber increases his alveolar ventilation approximately fivefold and thus reduces his alveolar P_{CO_2} to about 7.5 mm Hg. The equation then becomes

$$P_{A_{O_2}} = 43 - \frac{7.5}{0.8} + 0.4$$
$$= 34 \text{ mm Hg}$$

The arterial P_{O_2} will be several mm Hg below this. In fact, there is severe diffusion limitation of oxygen transfer under these conditions of extreme hypoxia (compare Fig. 5.39*B*), and the arterial P_{O_2} is believed to be about 28–30 mm Hg. Moreover, because of the diffusion limitation, it will fall further during exercise. Nevertheless, climbers have now reached the summit of Mt. Everest without supplementary oxygen, though their accounts (e.g., Messner, 1979) show that they are extremely close to the limit of human tolerance.

Less striking degrees of hyperventilation occur at intermediate altitudes. Permanent residents who live at 4,600 m (15,000 feet) in the Peruvian Andes have an arterial P_{CO_2} of about 33 mm Hg (Hurtado, 1964). A group of physiologists who wintered for some 4 mo in the Himalayas at 5,800 m (19,000 feet) had an average alveolar P_{CO_2} of about 23 mm Hg

Table 5.2.
Barometric Pressure (P_B) at Various Altitudes

Altitude		Location	Barometric Pressure, mm Hg	Inspired P_{O_2}, mm Hg[a]
m	ft			
0	0	Sea level	760	149
2,240	7,350	Mexico City (Olympic games, 1968)	580	112
3,400	11,000	Skiing resort	500	95
5,300	17,500	Highest permanent habitation	390	72
8,848	29,028	Summit of Mt. Everest	253	43
19,000	63,000	Bail out from high performance aircraft	47	0

[a]The inspired P_{O_2} is for moist inspired gas and is calculated from $0.2093 \times (P_B - 47)$.

(West et al., 1962). Figure 5.85 shows the changes in alveolar P_{O_2} and P_{CO_2} that occur in acclimatized subjects and also in subjects acutely exposed to a low barometric pressure. The latter typically do not reduce their P_{CO_2} (that is, hyperventilate) until the pressure is reduced below that corresponding to an altitude of about 3,000 m (about 10,000 feet).

The cause of the hyperventilation is hypoxic stimulation of the peripheral chemoreceptors (see Chapter 40). The resulting low arterial P_{CO_2} and alkalosis of the cerebrospinal fluid (CSF) and blood tend to inhibit this increase in ventilation initially. After a day or so, however, the pH of the CSF is reduced to some extent by outward movement of bicarbonate, and after 2 or 3 days the pH of the arterial blood is returned to near normal by the renal excretion of bicarbonate. These brakes on ventilation then diminish and ventilation tends to increase further. However, the cause of the sustained hyperventilation at high altitude is still not fully understood.

Interestingly, people born at high altitude have a diminished ventilatory response to hypoxia which is apparently only slowly increased by subsequent residence at sea level. Conversely, those born at sea level who move to high altitudes for a prolonged period retain their hypoxic response intact. Thus, this sensitivity is determined very early in life (Lahiri et al., 1969). Patients with congenital heart disease and large right-to-left shunts causing severe hypoxemia also have a blunted ventilatory response to hypoxia.

POLYCYTHEMIA

Another feature of acclimatization to high altitude is an increase in the red blood cell concentration of the blood. The resulting rise in hemoglobin concentration, and therefore oxygen-carrying capacity, means that, although the arterial P_{O_2} and oxygen saturation are diminished, the oxygen *content* of the arterial blood may be normal or even above normal. For example, in permanent residents at 4,540 m (15,000 feet) in the Peruvian Andes, the arterial P_{O_2} is only 45 mm Hg and the corresponding arterial oxygen saturation is only 81% (Hurtado, 1964). Ordinarily this would considerably decrease the arterial oxygen content but, because of the polycythemia, the hemoglobin concentration is increased from 15–19.8 g/100 ml, giving an arterial oxygen content of 22.4 ml/100 ml, which is above the normal sea level value.

The polycythemia also tends to maintain the P_{O_2} of mixed venous blood; typically in Andean permanent residents at 4,540 m (15,000 feet) this P_{O_2} is

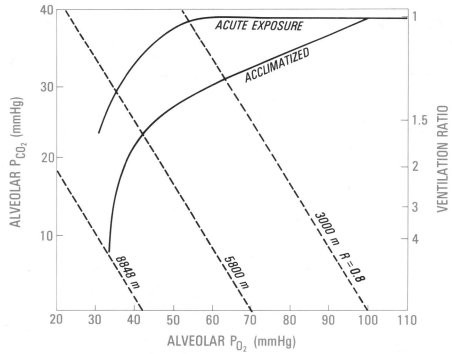

Figure 5.85. O_2-CO_2 diagram showing the effects of acutely exposing subjects to low pressures and the results of acclimatization to high altitude. Acutely exposed subjects do not increase their ventilation until the pressure is reduced below the equivalent of about 3000 m. The P_{CO_2} on the summit of Everest (8848 m) is about 7.5 mm Hg. This corresponds to an approximately fivefold increase in ventilation *(right-hand scale)*. The *diagonal lines* indicate the possible values for P_{O_2} and P_{CO_2} at a given altitude if the subject maintains a respiratory exchange ratio (*R*) of 0.8 (From Rahn and Otis 1949, with additional data from West et al., 1983.)

only 7 mm Hg below normal, as shown in Figure 5.86. Note how little the P_{O_2} falls from the arterial to the mixed venous blood. This is partly a reflection of the polycythemia, which means that the blood P_{O_2} falls less for a given reduction in oxygen content in the peripheral capillaries, and partly due to the steepness of the oxygen dissociation curve in this region.

The stimulus for the increased production of red blood cells is tissue hypoxia, which releases erythropoietin from the kidney, which in turn stimulates the bone marrow. Polycythemia is also seen in many patients with chronic hypoxemia caused by lung or heart disease.

Recent work suggests that the polycythemia of high altitude may not be wholly advantageous. The resulting increase in viscosity of the blood (Chapter 7) increases the work of the heart and may cause uneven blood flow in systemic capillaries. Indeed some physicians have advocated hemodilution (that is, removal of some red cells with replacement by cell-free fluid) in people at high altitude. Certainly this procedure does not appear to reduce exercise tolerance. It may be that high altitude polycythemia is an inappropriate response to tissue hypoxia, the genetic pressure for the response being developed over thousands of years at sea level.

OTHER FEATURES OF ACCLIMATIZATION

These include a *shift to the right of the oxygen dissociation curve* (at pH 7.4), which results in a better unloading of oxygen in venous blood at a given P_{O_2}.

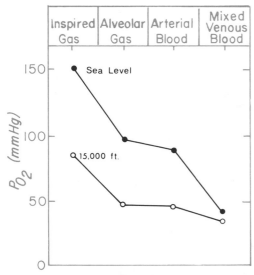

Figure 5.86. P_{O_2} from inspired air to mixed venous blood at sea level and in residents at an altitude of 4,540 m (15,000 ft). In spite of the much lower inspired P_{O_2} at altitude, the P_{O_2} of the mixed venous blood is only 7 mm Hg lower. (From Hurtado, 1964.)

The cause of the shift is an increase of the concentration of 2,3-diphosphoglycerate within the red cells, which develops primarily because of the alkalosis. This shift has been observed to occur after 2 days of residence at 4,600 m (15,000 feet) and is calculated to increase the amount of oxygen released from the blood by about 10% (Lenfant et al., 1970). At higher altitudes the benefit is less because of interference with the loading of oxygen in the pulmonary capillaries. Indeed, during exercise at extreme altitudes where oxygen transfer across the pulmonary capillary is diffusion-limited, a left-shifted curve is advantageous. In practice, a left-shifted curve is often seen under in vivo conditions because of the respiratory alkalosis. This is especially true at extreme altitudes but has also been described in permanent residents at 4600 m altitude.

There is also evidence that the *concentration of capillaries* in peripheral muscles increases because the muscle fibers become smaller. There are also increases in some *oxidative enzymes* inside the cells. The *maximum breathing capacity* increases because the air is less dense (see Chapter 39) and this makes possible the very high ventilations (up to 200 liters/min) which occur on exercise. The maximum oxygen uptake, however, declines rapidly above 4500 m. This is improved to some extent if 100% oxygen is breathed but does not return to sea level values for reasons which are poorly understood.

Pulmonary vasoconstriction frequently occurs at high altitudes as a result of the alveolar hypoxia (see Chapter 36). As a consequence of the hypoxic vasoconstriction the pulmonary arterial pressure rises, as does the work done by the right heart. This is exaggerated by the polycythemia, which increases the viscosity of the blood. Hypertrophy of the right heart is seen with characteristic changes in the electrocardiogram. There seems to be no physiological advantage to this response except that the topographical distribution of blood flow becomes more uniform (see Fig. 5.20) and the degree of ventilation-perfusion inequality is therefore slightly reduced.

The pulmonary hypertension is occasionally associated with *high altitude pulmonary edema*. Typically a climber or skier who has ascended to high altitude, perhaps without an adequate period of acclimatization, develops shortness of breath and may cough up pink, frothy sputum. The attack often comes on at night. The cause of this condition is not understood, but the fact that measurements of pulmonary artery wedge pressure are normal implies that the pulmonary venous pressure is not raised. One hypothesis is that the hypoxic vasoconstriction is uneven, and leakage of fluid therefore occurs in those regions of the capillary bed which are not protected from the high pulmonary arterial pressure. The treatment is to move the patient to a lower altitude and also to give oxygen if this is available.

EXERCISE TOLERANCE AT HIGH ALTITUDE

Permanent residents at an altitude of 4,600 m (15,000 feet) in the Peruvian Andes have a remarkably high work capacity in spite of their hypoxemia. Barcroft (1925) in a colorful early study of these people expressed astonishment at their athletic ability, for example, during a game of football. More extensive studies by Hurtado (1964) have confirmed the remarkable exercise tolerance of these permanent residents. It is of interest that the South American Incas maintained two armies—one that was kept at high altitude and therefore remained acclimatized and another for fighting on the plains.

Lowlanders who acclimatize to an altitude of 4,600 m (15,000 feet) can achieve only about 75% of the maximal oxygen consumption that they had at sea level. Above this altitude their exercise tolerance falls rapidly, being about 60% of the sea level value at 5,800 m (19,000 feet) and less than 40% at 7,400 m (24,400 feet). When well-acclimatized subjects breathe a mixture with the same inspired P_{O_2} as exists on the summit of Mt. Everest, their maximal oxygen consumption is just over 1 liter/min, that is, less than one-quarter of their sea level value. Under these extremely hypoxic conditions, the maximal oxygen uptake is exquisitely sensitive to changes in inspired P_{O_2}. It has even been suggested that whether a climber can scale Mt. Everest breathing ambient air may depend on the barometric pressure for that day! Certainly it is a remarkable coincidence that the P_{O_2} at the highest point on earth is just sufficient for humans to survive.

SPACE FLIGHT

Compared to the high altitude climber exposed to low oxygen, cold, and wind, the astronaut is indulged because the atmosphere and climatic conditions are carefully controlled. In the Apollo flights to the moon, the environment was 100% oxygen with a cabin pressure of one-third of an atmosphere; this gave an inspired P_{O_2} of about 260 mm Hg. The atmosphere of the Space Shuttle and the Mir Space Station is air at 760 mm Hg.

A number of physiological problems have been encountered by the astronauts. Motion sickness because of the disorientation is often disabling early in the flight. Diuresis with loss of circulating blood volume is seen early in space flight. This is caused by movement of blood from the lower body into the thorax in the weightless state and consequent stimulation of the atrial volume receptors (Chapter 16). Cardiovascular "deconditioning" is liable to occur; as a result of this, postural hypotension is frequently seen when the astronaut returns to earth. Decalcification of bone and muscle atrophy occur through disuse. There may also be a small reduction in red cell mass. Presumably the distribution of ventilation and blood flow in the lung will become more uniform (Figs. 5.19 and 5.71) in the gravity-free environment, and as a result there may be small improvement in pulmonary gas exchange. However, other changes such as a reduction of lung volume caused by upward movement of the diaphragm and possible interstitial edema as a result of increased pulmonary vascular pressures may impair gas exchange. No measurements have yet been made.

Oxygen Toxicity

Breathing pure oxygen is associated with several hazards; this is a convenient place to discuss them. The usual problem is getting enough oxygen into the body, but it is possible to have too much. When high concentrations of oxygen are breathed for many hours, damage to the lung may occur (Clark and Lambertsen, 1971). If guinea pigs are placed in 100% oxygen at atmospheric pressure for 48 hr they develop pulmonary edema. Some of the first pathological changes are seen in the endothelial cells of the pulmonary capillaries (Fig. 5.2). It is not easy to administer very high concentrations of oxygen to patients. However, this is sometimes done in intensive care units where the patients are mechanically ventilated with a respirator via an endotracheal tube. Under these conditions evidence of impaired gas exchange has been demonstrated after 30 hr of inhalation of 100% oxygen (Fig. 5.87). Normal volunteers who breathe 100% oxygen at atmospheric pressure for 24 hr complain of substernal distress which is aggravated by deep breathing, and they develop a diminution of vital capacity of 500–800 ml. This is probably caused by absorption atelectasis (see below).

Another hazard of breathing 100% oxygen is seen in premature infants who develop blindness because of retrolental fibroplasia, that is, fibrous tissue formation behind the lens. Here the mechanism is local vasoconstriction caused by the high P_{O_2} in the incubator. This fibroplasia can be avoided

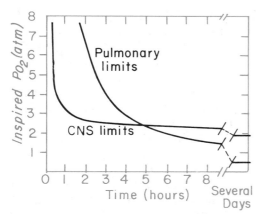

Figure 5.87. Relationship between P_{O_2} and exposure time responsible for oxygen toxicity. *CNS*, central nervous system. (From Lambertsen, 1974.)

if the oxygen concentration is less than 40% or if the arterial P_{O_2} is kept below 140 mm Hg.

Absorption Atelectasis

This is another danger of breathing 100% oxygen. Suppose that an airway is completely obstructed by mucus (Fig. 5.88A). The total pressure in the trapped gas is close to 760 mm Hg (it may be a few mm Hg less as it is absorbed because of elastic forces in the lung), but the sum of the partial pressures in the mixed venous blood is far less than 760 mm Hg. This is because the P_{O_2} of the mixed venous blood remains relatively low even when oxygen is breathed. In fact, the rise in oxygen *content* of arterial and mixed venous blood when oxygen is

breathed will be the same if the cardiac output and oxygen consumption remain unchanged. This follows from the Fick principle:

$$\dot{V}_{O_2} = \dot{Q}(Ca_{O_2} - C\bar{v}_{O_2})$$

as discussed in Chapter 36. The arterial oxygen content will rise by only some 1.5 ml/100 ml due to the increased dissolved oxygen since the hemoglobin is normally nearly fully saturated (Fig. 5.26). As a result, the increase in venous P_{O_2} is only about 10–15 mm Hg because of the shape of the oxygen dissociation curve, in spite of the fact that the arterial P_{O_2} increases several hundred mm Hg (compare Figure 5.44). Thus, since the sum of the partial pressures in the alveolar gas greatly exceeds that in the venous blood, gas diffuses into the blood, and rapid collapse of the alveoli occurs. Reopening such an atelectatic area may be difficult because of surface tension forces in such small units (Fig. 5.56).

Absorption collapse also occurs in a blocked region, even when air is breathed, although here the process is slower. Figure 5.88B shows that again the sum of the partial pressures in venous blood is less than 760 mm Hg because the fall in P_{O_2} from arterial to venous blood is much greater than the rise in P_{CO_2} (this is a reflection of the steeper slope of the CO_2 dissociation curve compared with that of O_2—see Fig. 5.32A). Since the total gas pressure in the alveoli is nearly 760 mm Hg, absorption is inevitable. Actually the changes in the alveolar partial pressures during absorption are somewhat compli-

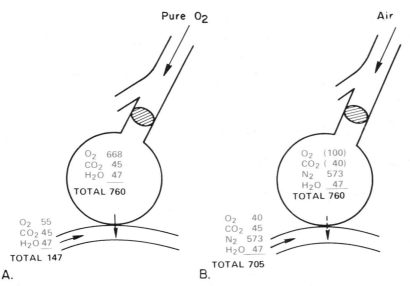

Figure 5.88. Reasons for atelectasis of alveoli beyond blocked airways when oxygen (*A*) and when air (*B*) is breathed. In both cases, the sum of the gas partial pressure in the mixed venous blood is less than

in the alveoli. In *B*, the P_{O_2} and P_{CO_2} are shown in parentheses because these values change with time. However, the total alveolar pressure remains within a few mm Hg of 760.

cated, but it can be shown that the rate of collapse is limited by the rate of absorption of nitrogen. Since this gas has a low solubility, its presence acts as a "splint" which, as it were, supports the alveoli and delays collapse. Even relatively small concentrations of nitrogen or other poorly soluble inert gases in alveolar gas have a useful splinting effect. Nevertheless, postoperative atelectasis is a common problem in patients who are treated with high oxygen mixtures. Collapse is particularly likely to occur at the bottom of the lung where the parenchyma is least well expanded (Fig. 5.72) or the airways are actually closed (Fig. 5.73). The same basic mechanism of absorption is responsible for the gradual disappearance of a pneumothorax, or a gas pocket introduced under the skin.

INCREASED PRESSURE

During diving under water, the pressure increases by 1 atmosphere for every 10 m (33 feet) of descent. Pressure by itself is relatively innocuous down to 300–450 m (1000–1500 feet) of depth so long as it is balanced. However, if a gas cavity such as the lung, middle ear, or intracranial sinus fails to communicate with the outside, the pressure difference may cause compression on descent or overexpansion on ascent. For example, it is very important for a scuba diver to exhale as he ascends to prevent overinflation and possible rupture of alveoli of the lungs. The increased density of the gas at depth increases the work of breathing (see Chapter 39). This may result in CO_2 retention, especially on exercise.

Decompression Sickness (Dysbarism)

During diving or working in caissons at high pressure, the high partial pressures of nitrogen force this poorly soluble gas into solution in body tissues. This occurs particularly in fat, which has a relatively high nitrogen solubility. The blood supply of adipose tissue is meager, however, and the blood can carry little nitrogen. In addition, the gas diffuses slowly because of its low solubility (Fig. 5.37). As a result, equilibration of nitrogen between the tissues and the environment takes hours. Nevertheless, while the total volume of nitrogen in the body after equilibration is only about 1 liter at sea level, this increases to about 10 liters (measured at sea level pressure) in a diver at a depth of 100 m (330 feet).

During ascent, nitrogen is slowly removed from the tissues since the partial pressure there is higher than that in the alveolar gas and arterial blood. If decompression is unduly rapid, bubbles of gaseous nitrogen are released, just as CO_2 comes out of solu-

tion when a bottle of champagne is opened. The first observation of this phenomenon was made by Robert Boyle in 1670 when he decompressed a viper and noticed that a bubble formed in the eye. Some small bubbles can occur without physiological disturbances, but large numbers of bubbles cause pain, especially in the region of joints ("bends"). Bubbles of gas in the pulmonary circulation may cause the "chokes"—a feeling of shortness of breath often accompanied by coughing. In severe cases there may be neurological impairment such as deafness, impaired vision, vestibular disturbances, and even paralysis caused by bubbles which obstruct blood flow in vessels in the central nervous system (CNS). In overwhelming decompression sickness, the patient may collapse and die. Professional divers or caisson workers sometimes develop avascular necrosis of the head of the femur or other long bones late in life (that is, the bone dies because of obstruction to its blood supply). This is possibly due to intermittent obstruction by nitrogen bubbles over many years.

Decompression sickness can be avoided if the diver ascends slowly enough to prevent bubbles from becoming unduly large. In practice divers use tables which indicate the depths at which they should pause for periods of time during ascent, a process known as *staging*. Table 5.3 shows an example of a decompression table as used by the U.S. Navy when the diver breathes compressed air. For a 40-min dive at a depth of 200 feet, the total decompression time is almost 2 hr. If a dive at 200

Table 5.3.
Typical Decompression Table Showing the Depth and Time of the Stops Required to Prevent Decompression Sickness after Dives to Various Depths

Depth of Dive, ft	Time on Bottom, min	Time at Stops, min					Total Time, min[b]
		50 ft	40 ft	30 ft	20 ft	10 ft	
40	250					11	12
50	200					35	36
60	160					48	49
70	120				4	47	52
80	120				17	56	74
90	100				21	54	77
100	90			3	23	57	85
110	80			7	23	57	89
120	70			9	23	55	89
130	60			9	23	52	86
140	60			16	23	56	97
150	50			12	23	51	89
160	50		2	16	23	55	99
170	40		1	10	23	45	82
180	40		3	14	23	50	93
200	40	2	8	17	23	59	112

[a]This table is for illustrative purposes only and should not be used for recreational diving. Ascent rate assumed to be 60 ft/min. [b]Approximate total decompression time.

feet is extended to 90 min, the decompression time is over 5 hr! By contrast, a 4-hr dive at 40 feet requires only 12 min of decompression time.

Tables of the type shown in Table 5.3 were originally developed by Haldane and his colleagues (Boycott et al., 1908) and were based partly on observations of decompression sickness in experimental animals and partly on calculations of how rapidly nitrogen is washed out of body compartments. Figure 5.89 shows the rate of washout of three typical tissues. There is considerable variation, and these numbers should be considered illustrative only. The half-time of blood is about 2 min (that is, half the nitrogen is removed in this time), that of muscle is about 20 min, while fat has a half-time of about 1 hr. The rates of washin of nitrogen when the diver is at depth are similar. As a result the events which take place in the various tissues when a diver descends to 300 feet for 20 min and then starts the ascent (Table 5.3) are complicated and depend on the relative washin and washout rates of the various tissues. While Figure 5.89 indicates the behavior of typical tissues, the body actually consists of a spectrum of half-times determined by the nitrogen solubility and blood supply of the various components.

For many years it was thought that the value of slow decompression was in preventing the formation of nitrogen bubbles. Recent work, however (for example, with ultrasonic bubble detectors), shows that small bubbles usually occur in the blood during normal decompression even in the absence of symptoms. Thus the modern objective is to prevent these bubbles, which are inevitably present, from growing large enough to cause decompression sickness

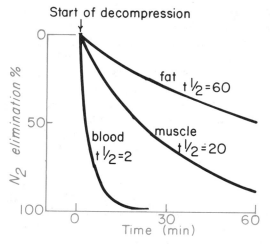

Figure 5.89. Rate of nitrogen elimination during decompression following a dive for three typical tissues. The half times (*t½*) shown depend on the blood flow, which may vary. These are therefore illustrative curves only.

(Hills, 1969). This change of emphasis has affected the design of decompression tables so that divers now tend to spend more time decompressing at greater depths but actually reach the surface faster.

The treatment of decompression sickness is by recompression in a pressure chamber. This reduces the volume of the bubbles, forces them back into solution, and often results in a dramatic reduction of symptoms. Industrial operations involving deep dives, for example, deep-sea drilling platforms, have compression chambers available. The amateur scuba diver who develops bends is not so fortunate, however, and there may be long delay in transporting him to an appropriate facility. Oxygen inhalation during decompression can be used to accelerate the washout of nitrogen.

The risk of decompression sickness following very deep dives can be reduced if a helium-oxygen mixture is breathed during the dive. Helium is about one-half as soluble as nitrogen so that less is dissolved in tissues. In addition, it has one-seventh of the molecular weight of nitrogen and therefore diffuses more rapidly through tissue (Fig. 5.37). Both of these factors reduce the risk of bends after long dives. Another advantage of a helium-oxygen mixture for divers is its low density, which reduces the work of breathing. Pure oxygen or enriched oxygen mixtures cannot be used at depth because of the dangers of oxygen toxicity (see below).

Aviators also may develop decompression sickness due to nitrogen bubble formation if they are suddenly decompressed at very high altitudes as a result of cabin failure. This can be prevented if they breathe 100% oxygen prior to the flight and thus wash out the nitrogen. Astronauts do the same prior to extra-vehicular activity (outside the spacecraft) because the pressure in the suit is so low.

SATURATION DIVING

To avoid the long periods of decompression associated with deep dives, some divers have lived in underwater habitats for periods of 2 wk or so. Under these conditions, the tissues equilibrate with nitrogen at that depth, often about 50 feet. It is then possible to make excursion dives to, say, 150 feet and return to the habitat with little or no decompression time (compare Table 5.3). Of course when the divers eventually return to the surface, they must decompress for a long period. In present day commercial diving in the North Sea, divers sometimes live for several days in a pressure chamber on the tender at a simulated depth of 300 feet. This allows them to make dives to depths of 600 feet.

Inert Gas Narcosis

Although nitrogen is usually regarded as a physiologically inert gas, at high partial pressures it affects the CNS. At a depth under water of about 50 m (160 feet) there is a feeling of euphoria (not unlike that following a martini or two), and divers have been known to offer their mouthpiece to a fish! At higher partial pressures, loss of coordination and eventually coma may develop. The mechanism of action is not understood but may be related to the high fat/water solubility ratio of nitrogen, which is a general property of anesthetic agents. Other gases such as helium and hydrogen can be used at much greater depths without narcotic effects.

Oxygen Toxicity

Earlier in this chapter it was noted that inhalation of 100% oxygen at 1 atmosphere can damage the lung. Another form of oxygen toxicity affects the CNS, leading to convulsions when the inspired P_{O_2} considerably exceeds 760 mm Hg. The convulsions may be preceded by premonitory symptoms such as nausea, ringing in the ears, and twitching of the face.

The likelihood of convulsions depends on the inspired P_{O_2} and the duration of exposure (Fig. 5.87), and it is increased if the subject is exercising. At a P_{O_2} of 4 atmospheres, convulsions frequently occur within 30 min. For increasingly deep dives, the oxygen concentration must be progressively reduced to avoid toxic effects and may eventually be less than 1% for a normal inspired P_{O_2}! The amateur scuba diver should *never* fill his tanks with oxygen because of the danger of a convulsion underwater. However, pure oxygen is sometimes used by the military for shallow dives because a closed breathing circuit with a CO_2 absorber leaves no telltale bubbles. The biochemical basis for the deleterious effects of a high P_{O_2} on the CNS is not fully understood but is probably the inactivation of certain enzymes, especially dehydrogenases containing sulfhydryl groups.

Scuba Diving

Self-contained underwater breathing apparatus (scuba) is now used extensively by sport divers as well as professionals. The hazards of decompression sickness and inert gas narcosis have been discussed above. Additional dangers include lung rupture and consequent *air embolism,* which may occur if a diver ascends without exhaling. Under these conditions the gas in the lungs expands so much that the alveolar walls tear, air escapes into the small pulmonary veins, and some of it travels to the cerebral and coronary circulations. Unconsciousness and collapse may rapidly supervene. Treatment is by recompression; oxygen breathing hastens the absorption of bubbles in the blood. Scuba divers should always be accompanied (buddy system) and be familiar with the physiological principles of diving (see, for example, Council for National Cooperation in Aquatics, 1974).

Shallow Water Blackout

There have been a number of accidents in relatively shallow water (for example, a swimming pool), where the victim loses consciousness during ascent from a breath-hold dive. One reason for this is that a diver may hyperventilate before a dive to reduce the ventilatory drive caused by CO_2 accumulation. The diver can then stay submerged for a long period until the arterial P_{O_2} falls to a low level and stimulates breathing. During ascent, the ambient pressure falls and the alveolar, and therefore arterial, P_{O_2} becomes so low that the diver loses consciousness. Hyperventilation prior to a long breath-hold dive should therefore be discouraged, and another adult should ideally be present.

Drowning

Drowning is suffocation by submersion, usually in water. Studies of drowned or nearly drowned people and of experimental animals show that the most important blood gas changes are severe hypoxemia combined with hypercapnia and respiratory acidosis (Modell, 1971). After recovery, a metabolic acidosis may persist. While the initial cause of the hypoxemia is blood flow through unventilated lung, aspiration of fresh water apparently interferes with the function of pulmonary surfactant and leads to areas of alveolar atelectasis in people who recover from near drowning. This atelectasis may cause persistent hypoxemia. Aspiration of sea water moves additional fluid into the lung from the blood through osmotic forces. This fluid increases the proportion of unventilated lung.

A few victims of drowning do not inhale water into their lungs because of reflex contraction of their laryngeal muscles; they die from acute asphyxia. Transient changes in blood volume commonly occur in drowning. The blood volume increases in fresh water drowning due to the rapid transfer of hypotonic fluid into the circulation. Opposite changes occur in sea water drowning. At one time it was thought that changes in blood electrolytes were of great importance in drowning, but more recent work shows that in most cases the electrolytes are within the normal range.

Inert Gas Counterdiffusion

An interesting complication of exposure to high pressures has been described by Lambertsen and his colleagues (Graves et al., 1973). Men in compression chambers who breathe one inert gas (for example, nitrogen) while surrounded by another (for example, helium) sometimes develop itching and swollen patches in their skin. This is due to the formation of bubbles, which occur without any change in ambient pressure. The cause of the bubbles is localized supersaturation; that is, the sum of the partial pressures of the gases exceeds the pressure in the surrounding tissues and consequently the gases come out of solution. This occurs at lipid-water interfaces because one gas diffuses particularly rapidly through the lipid, while the other gas diffuses fast through the water. The result is that the sum of the partial pressures at the interface becomes very high. In experimental animals, the bubbles can grow so large that gas embolism can occur, resulting in death. Although this phenomenon was first noticed during pressure breathing, it can also occur at normal pressure.

Hyperbaric Oxygen Therapy

Increasing the arterial P_{O_2} to a very high level is useful in some clinical situations. One is severe carbon monoxide poisoning, where most of the hemoglobin is bound to CO and is therefore unavailable to carry oxygen. By raising the inspired P_{O_2} to 3 atmospheres in special chambers, the amount of dissolved oxygen in arterial blood can be increased to about 6 ml/100 ml (Fig. 5.26), and thus the needs of the tissues can be met without functioning hemoglobin. Occasionally, an anemic crisis is managed in this way. Hyperbaric oxygen is also useful for treating gas gangrene because the organism cannot live in a high P_{O_2} environment. A hyperbaric chamber is also useful for treating decompression sickness.

Fire and explosions are serious hazards of a 100% oxygen atmosphere, especially at increased pressure. For this reason, oxygen in a hyperbaric chamber is given by mask, and the chamber itself is filled with air.

Artificial Gills

Various ingenious devices (Paganelli et al., 1967) have been proposed to enable a diver to remove dissolved oxygen from water, a process which is normally accomplished by the fish gill. In Figure 5.90 the diver is connected to a bag with a large gas-exchanging surface made of a silicon polymer which is highly permeable to O_2 and CO_2. O_2 from the water diffuses into the bag and CO_2 diffuses out.

Figure 5.90. Proposed artificial gill consisting of a bag made of a material highly permeable for oxygen and CO_2. Exchange occurs with the gases in the surrounding water; however, eventually the bag empties because nitrogen diffuses out (see text). *Hz*, Hertz.

At first sight the system appears to offer an unlimited period of submersion, but it eventually fails for an interesting reason. Because the volume of CO_2 exhaled is generally less than the volume of O_2 consumed, the nitrogen in the bag is gradually concentrated and its partial pressure rises. As a result, nitrogen gradually diffuses into the water and the bag progressively empties.

Oddly enough nature uses the same principle to allow the water boatman *Corixa* to remain submerged for long periods. This beetle dives with a tiny bubble of air adhering to its ventral surface and connected with the respiratory air tubes (spiracles) which lead to the tissues. O_2 diffuses into the bubble from the water and CO_2 diffuses out, but eventually the bubble collapses through loss of nitrogen.

A way around this difficulty was suggested by Rahn when he designed a chamber for fish-watching in the Niagara River (Paganelli et al., 1967). This has a frame which supports the permeable membrane and thus allows the chamber pressure to fall by a small amount (Fig. 5.91). The result is that the nitrogen partial pressure is in equilibrium with that of the water, and the system is in a true steady state of gas exchange.

Again nature has priority with a similar solution. Some insects exchange gas under water by means of a "plastron," that is, a bubble supported on stiff hairs which protrude from the insect's body (Crisp, 1964; Rahn and Paganelli, 1968). By means of surface tension, the framework of hairs prevents the bubble from collapsing even when the pressure inside it falls. Essentially the same steady state of gas exchange as that shown in Figure. 5.91 is thus developed.

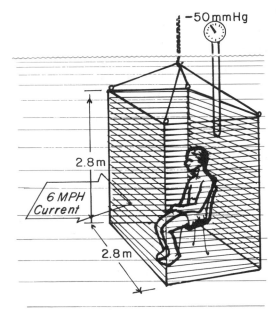

Figure 5.91. Artificial gill for fish watching in the Niagara River. The permeable membrane is supported on a frame, and under steady state conditions the pressure in the box is less than that in the surrounding water, so that nitrogen exchange is in a steady state. (From Paganelli et al., 1967.)

Liquid Breathing

As explained earlier, in very deep diving the carrier gas (usually nitrogen) is responsible for most of the problems, including decompression sickness and inert gas narcosis. An exotic solution, therefore, is to do away with this vehicle gas entirely and breathe the oxygen dissolved in liquid!

Kylstra and his colleagues (1962) were the first to show that it is possible for mammals to survive for some hours breathing liquid instead of air. They immersed mice in saline in which the oxygen concentration was increased by exposure to 100% oxygen at 8 atmospheres pressure. Subsequently, mice, rats, and dogs have survived a period of breathing fluorocarbon exposed to pure oxygen at 1 atmosphere. This liquid has a high solubility for both O_2 and CO_2. The animals successfully returned to air breathing.

Because liquids have a much higher density and viscosity than air, the work of breathing is enormously increased. Adequate oxygenation of the arterial blood can be obtained, however, if the inspired concentration is raised sufficiently. Interestingly, a serious problem is eliminating CO_2. In Chapter 35 we saw that diffusion within the airways is chiefly responsible for the gas exchange which occurs between the alveoli and the terminal or respiratory bronchioles where bulk flow takes over. Because the diffusion rates of gases in liquid are several orders of magnitude slower than in the gas phase, a large partial pressure difference for CO_2 between the alveoli and terminal bronchioles must be maintained if CO_2 is to be eliminated adequately. Animals breathing liquid, therefore, commonly develop CO_2 retention and acidosis. The diffusion pressure for O_2 can always be raised by increasing the inspired P_{O_2}, but the converse option is not available to assist in the elimination of CO_2.

Apart from its possible use in deep diving, liquid breathing has other potential applications. In Chapter 39 we saw that premature babies sometimes develop a respiratory distress syndrome characterized by patchy atelectasis, pulmonary edema, and stiff lungs. This is caused by the absence of pulmonary surfactant. A suggested method of treatment is to ventilate the lungs of these infants with fluorocarbon until the surfactant system matures. This would abolish the liquid-air interface and thus the high surface tension forces which are chiefly responsible for the abnormalities. Preliminary experiments in premature experimental animals have given promising results (Schaffer et al., 1976).

Another possible application of liquid breathing is in future space flight, where very large accelerations may be necessary. The lung is vulnerable to high accelerations because of the different densities of air and blood within it. Filling the lung with saline confers a great degree of protection against injury by acceleration.

Finally, it is of interest that washing out one lung with normal saline to remove accumulated secretions is now an accepted form of therapy in some types of disease, for example, alveolar proteinosis. The saline is introduced by means of a special catheter (Carlens) which separates the two lungs, and the patient uses the dry lung for gas exchange while the other is being rinsed.

OTHER UNUSUAL ENVIRONMENTS

Perinatal Respiration

All mammalian lungs undergo the transition from a liquid-filled to an air-filled condition at birth. During fetal life, gas exchange takes place through the placenta. Maternal blood enters from the uterine arteries and surges into small spaces called the intervillous sinusoids. Fetal blood is supplied through the umbilical arteries to capillary loops which protrude into the intervillous spaces. Gas exchange occurs across the blood-blood barrier, which is approximately 3.5 μm thick.

This arrangement is much less efficient for gas exchange than is the adult lung. Maternal blood apparently swirls around the sinusoids somewhat haphazardly, and there are probably large differences in P_{O_2} within these blood spaces. Contrast this situation with the air-filled alveoli, where rapid gaseous diffusion maintains an almost uniform composition in each lung unit. The result is that the blood leaving the placenta in the uterine vein has a P_{O_2} of only about 30 mm Hg (Fig. 5.92). Note the very low P_{O_2} of about 22 mm Hg in the descending aorta. However, the blood going to the head has a somewhat higher P_{O_2} of 25 mm Hg. These differences of P_{O_2} are the result of streaming within the fetal heart (see Chapter 61). For example, much of the relatively well oxygenated blood from the inferior vena cava crosses the patent foramen ovale into the left atrium and is available for distribution to the fetal head.

FIRST BREATH

The emergence of the baby into the outside world is perhaps the most cataclysmic event of his or her life. The baby is suddenly bombarded with a variety of external stimuli. In addition, the process of birth interferes with placental gas exchange, with result-

ing hypoxemia and hypercapnia which stimulate the chemoreceptors. As a combined result of these stimuli, the baby makes a first gasp.

The fetal lung is not collapsed but is inflated with liquid to about 40% of total lung capacity. This fluid is continuously secreted by alveolar cells during fetal life and has a low pH. Some of it is squeezed out as the infant moves through the birth canal, but the remainder has an important role in the subsequent inflation of the lung. As air enters the lung, large surface tension forces have to be overcome. We know that the larger the radius of curvature, the lower the forces (Fig. 5.56); therefore, this preinflation reduces the pressures required. Nevertheless, the intrapleural pressure during the first breath may fall to −40 cm H_2O before any air enters the lung (Fig. 5.93), and peak pressures as low as −100 cm H_2O during the first few breaths have been recorded. These very large transient pressures are partly caused by the high viscosity of the lung liquid compared with air. The inspiratory gasps may be augmented reflexly, the afferent impulses coming from stretch receptors in the stiff lung. The fetus makes small, rapid breathing movements in the uterus over a considerable period before birth.

Expansion of the lung is very uneven at first. However, pulmonary surfactant, which is formed relatively late in fetal life, is available to stabilize open alveoli, and the lung liquid is removed by the lymphatics and capillaries. Within a few moments the functional residual capacity has reached almost its normal value, and an adequate gas-exchanging surface has been established. It is several days, however, before uniform ventilation is achieved.

The circulatory changes which occur at birth are discussed in Chapter 61.

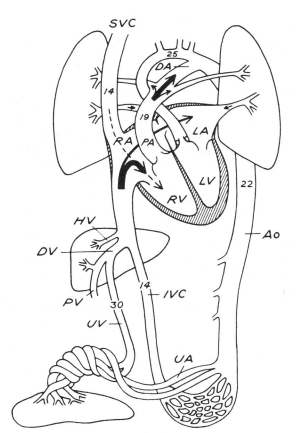

Figure 5.92. P_{O_2} values in the fetal circulation. *UV,* umbilical vein; *PV,* portal vein; *DV,* ductus venosus; *HV,* hepatic vein; *RA,* right atrium; *SVC,* superior vena cava; *RV,* right ventricle; *PA,* pulmonary artery; *DA,* ductus arteriosus; *LA,* left atrium; *LV,* left ventricle; *Ao,* aorta; *UA,* uterine artery. (From Comroe, 1974.)

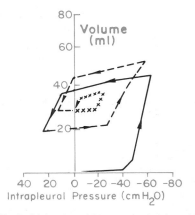

Figure 5.93. Typical intrapleural pressures developed in a newborn human infant during the first three breaths after birth. (From Avery, 1968.)

Polluted Atmospheres

A polluted atmosphere is, unfortunately, hardly an unusual environment in modern urban areas. One of the chief sources of pollutants is the automobile, which accounts for some 40% of all pollutants by weight in the United States. It produces major portions not only of CO but also of hydrocarbons and nitrogen oxides. Electrical utilities and industrial plants are major sources of sulfur oxides because the coal and oil fuels contain sulfur as an impurity. Table 5.4 lists the sources of major air pollutants in this country in 1974.

TYPES OF POLLUTANTS

Carbon monoxide is the largest pollutant by weight; it is produced by the incomplete combustion of carbon in fuels, principally in automobile engines. The propensity of CO to tie up hemoglobin was discussed in Chapter 37. A commuter using a Los Angeles freeway may have up to 10% of his hemoglobin bound to CO, especially if he is a cigarette smoker, and there is evidence that this can impair his mental skills (Coburn et al., 1976). The emission of CO and other pollutants by automobile engines can be reduced by installing a catalytic converter, which processes the exhaust gases.

Sulfur oxides are corrosive, poisonous gases produced when sulfur-containing fuels are burnt, chiefly by power stations. Sulfur oxides irritate the upper respiratory tract and the eyes, and long-term exposure causes chronic bronchitis. The best way to reduce these emissions is to use low sulfur fuels, but these are more expensive.

Hydrocarbons, like carbon monoxide, represent unburned wasted fuel. They are not toxic at concentrations normally found in the atmosphere. Their hazardous nature stems from their role in forming photochemical oxidants under the influence of sunlight.

Particulate matter includes particles with a wide range of sizes up to visible smoke and soot. Major sources are power stations and industrial plants. Often their emission can be reduced by processing the waste air stream by filtering or scrubbing, although removing the smallest particles is often expensive.

Nitrogen oxides are produced when fossil fuel is burned at very high temperatures in power stations and automobiles. These gases irritate the eyes and upper respiratory tract and are responsible for the yellow haze of smog.

Photochemical oxidants include ozone and other substances such as peroxyacyl nitrates, aldehydes, and acrolein. They are not primary emissions but are produced by the action of sunlight on hydrocarbons and nitrogen oxides. They cause eye and lung irritation, damage to vegetation, and offensive odors, and they contribute to the thick haze of smog.

The concentration of atmospheric pollutants is often greatly increased by a temperature inversion, especially in a low-lying area surrounded by hills such as the Los Angeles basin. The inversion prevents the normal escape of warm surface air to the upper atmosphere.

BEHAVIOR OF AEROSOLS

Many pollutants exist as *aerosols*, that is, very small particles which remain suspended in the air. When an aerosol is inhaled, its fate depends on the size of the particles (Fig. 5.94). Large particles are removed by *impaction* in the nose and pharynx. This means that the particles are unable to turn corners rapidly because of their inertia, and when they impinge on the wet mucosa they are trapped. Many medium-sized particles deposit in small airways because of their weight. This is called *sedimentation* and tends to occur particularly where the flow velocity is suddenly reduced because of the enormous increase in combined airway cross-section (Fig. 5.9). For this reason deposition is heavy in the terminal and respiratory bronchioles, and this region of a coal miner's lung shows a heavy dust concentration. The smallest particles (less than 0.1 μm in diameter) reach the alveoli, where some deposition occurs through *diffusion* to the walls. Many

Table 5.4.

Estimated Emissions of Major Air Pollutants (in Millions of Tons per Year) in U.S.A., 1974, as Reported by the Environmental Protection Agency

Source	CO	Sulfur Oxides	Hydrocarbons	Particulates	Nitrogen Oxides	Total
Transportation sources (especially automobiles)	73.5	0.8	12.8	1.3	10.7	99.1
Stationary sources (especially power stations)	0.9	24.3	1.7	5.9	11.0	43.8
Industrial processes	12.7	6.2	3.1	11.0	0.6	33.6
Other, including solid waste disposal, forest fires	7.5	0.1	12.8	1.3	0.2	21.9
Total	94.6	31.4	30.4	19.5	22.5	198.4

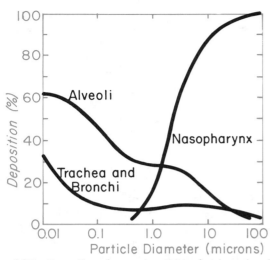

Figure 5.94. Deposition of aerosol particles of various sizes in different regions of the respiratory tract. The largest particles impact in the nasopharynx, while the smallest particles can reach the alveoli.

small particles are not deposited at all but are exhaled with the next breath.

CLEARANCE OF DEPOSITED PARTICLES

Fortunately the normal lung is very efficient at removing particles deposited within it. Particles that fall on the bronchial walls are swept up by a moving staircase of mucus propelled by cilia. The mucus is secreted by specialized cells of the bronchi, including cells of mucous glands which lie deep in the bronchial wall and secrete the mucus through ducts and goblet cells, which make up part of the bronchial mucosa. There is evidence that the normal mucus film has two layers, a superficial gel layer that is relatively viscous and is efficient at trapping deposited particles and a deeper layer that is less viscous and in which the cilia can beat easily. It may be that the retention of secretions that occurs in disease is caused in part by changes in the composition of the mucus so that it cannot be propelled by the cilia.

The cilia are 5–7 μm in length and they beat in a synchronized fashion at between 1000 and 1500 times a minute. This moves the mucus carpet upward at a speed of 1–3 cm/min in the large airways. Eventually the particles reach the level of the pharynx, where they are swallowed. In very dusty environments, mucus secretion may be increased so much that cough and expectoration assist in the clearance. The normal operation of the mucociliary system can be adversely affected by atmospheric pollution. For example, tobacco smoke, sulfur oxides, and nitrogen oxides may paralyze the cilia and also change the character of the mucus.

Particles that are deposited in the alveoli are not

cleared by the mucociliary system since it does not extent down to that level. Instead the particles are engulfed by phagocytic macrophages that roam around the alveoli by ameboid action. Laden macrophages leave the lung via the lymphatics or blood, while some migrate to the mucociliary escalator. Normal macrophage activity can be impaired by cigarette smoke, oxidant gases, radiation, and the ingestion of alcohol.

Oxygen Therapy

In the treatment of patients with severe lung disease, the environment is often changed by the physician, who increases the concentration of inspired oxygen or controls ventilation by means of a mechanical ventilator. Oxygen therapy is usually very effective in relieving hypoxemia, except when this is caused by a shunt (blood flow through unventilated alveoli). In this case the added oxygen does not have access to the shunted blood. The oxygen content of the nonshunted blood will nevertheless rise somewhat, chiefly because of the addition of dissolved oxygen, and the arterial P_{O_2} may therefore rise by a few mm Hg. By contrast, the hypoxemia of hypoventilation, diffusion impairment, or ventilation-perfusion inequality is readily corrected by oxygen therapy, although in the last instance the arterial P_{O_2} may not reach a steady level for many minutes because poorly ventilated regions of lung wash out their nitrogen so slowly.

Oxygen is now normally administered by small cannulas inserted into the nostrils or by means of a plastic oronasal mask. This may be of the Venturi type, in which the oxygen flow entrains air and thus allows a preset concentration of oxygen to be inspired. Newborn babies can be nursed in incubators in which the P_{O_2} can be controlled.

There are dangers in giving too much oxygen. Reference has already been made to the pulmonary edema which occurs in lungs exposed to 100% oxygen for more than 2 or 3 days. Other hazards include a propensity for atelectasis and retrolental fibroplasia in neonates.

Some patients with respiratory failure develop serious CO_2 retention when given high concentrations of oxygen to breathe. These are patients with compensated respiratory acidosis whose ventilatory drive comes chiefly from the effect of the severe hypoxemia on their peripheral chemoreceptors (see Chapter 40). When this hypoxemia is relieved, the patient's ventilation falls, and dangerous CO_2 retention and acidosis may develop. The proper management is to give a judicious concentration of oxygen, usually 24–28% initially, and to monitor the arterial P_{O_2}, P_{CO_2}, and pH frequently.

Mechanically Assisted Ventilation

While hypoxemia can usually be relieved by administering oxygen, CO_2 retention can be corrected only by increasing the ventilation. Mechanical ventilators are now frequently used in management of patients in respiratory failure. These are of two types: constant pressure and constant volume respirators. The former generate a preset pressure for each inspiration and are relatively simple to use. A gas tank is employed as the source of the pressure. If the lungs or chest wall is unusually still (low compliance), however, constant volume ventilators are more reliable because they deliver a preset ventilation even if there are changes in the mechanical properties of the lung.

Before a ventilator can be used, an airtight connection to the lungs must be made. This can be done by means of a plastic tube inserted through the mouth or nose into the trachea (endotracheal tube). An inflated cuff on the end gives an airtight seal. Alternatively a tracheostomy tube may be inserted through the neck and anterior wall of the trachea. The inspired gas should be humidified to prevent drying of the upper airways. Sometimes a small pressure of about 5 cm H_2O or more is applied to the airways at the end of expiration (positive end expiratory pressure or PEEP). This tends to prevent atelectasis and often increases the arterial P_{O_2} in patients with respiratory failure. High positive pressures, however, may interfere with venous return to the chest.

There are numerous problems in ventilator therapy. These include the accumulation of secretions in the lung, damage to the trachea by the endotracheal cuff, and malfunction of the equipment. Effective therapy requires a skilled team of personnel in a specialized intensive care unit.

Resuscitation

Resuscitation may be life-saving for victims of near drowning, electric shock, or a heart attack (myocardial infarct). Resuscitation should be begun immediately when it is determined that the victim is truly unresponsive and not just asleep. The victim should be placed on his or her back and an open airway established by lifting the neck and pulling the jaw bone (mandible) forward. Any foreign materials should be removed from the mouth. If the victim is not breathing, artificial ventilation should be started by pinching the nose and performing mouth-to-mouth breathing. If there is no carotid or femoral artery pulsation, external compression should be given by applying the heel of the hand to the lower third of the sternum and applying firm vertical pressure about once a second. If only one rescuer is present, two quick deep lung inflations should be given after every fifteen sternal compressions. Details can be found elsewhere (*Journal of the American Medical Association*, 1980).

BIBLIOGRAPHY

ALTMAN, P. L. *Blood and Other Body Fluids.* Washington, D.C.: Federation of American Societies for Experimental Biology, 1961, p. 12.

AVERY, M. E. *The Lung and Its Disorders in the Newborn Infant.* Philadelphia: Saunders, 1968.

BALL, W. C., JR., P. B. STEWART, L. G. S. NEWSHAM, AND D. V. BATES. Regional pulmonary function studied with xenon[133]. *J. Clin. Invest.* 41: 519–531, 1962.

BARCROFT, J. *The Respiratory Function of the Blood. Part 1. Lessons From High Altitudes.* London: Cambridge Univ. Press, 1925.

BARER, G. R., P. HOWARD, AND J. W SHAW. Stimulus-response curves for the pulmonary vascular bed to hypoxia and hypercapnia. *J. Physiol. (Lond.)* 211: 139–155, 1970.

BENESCH, R., AND R. E. BENESCH. Intracellular organic phosphates as regulators of oxygen release by hemoglobin. *Nature (Lond.)* 221: 618–622, 1969.

BOHR, C. Über die Lungenathmung. *Skand. Arch. Physiol.* 2: 236–268, 1891.*

BOHR, C. Über die spezifische Tatigkeit der Lungen bei der respiratorischen Gasaufnahme und ihr Verhalten zu der durch die Alveolarwand stattfindenden Gasdiffusion. *Skand Arch. Physiol.* 22: 221–280, 1909.*

BOHR, C., K. A. HASSELBALCH, AND A. KROGH. Ueber einen in biologischer Beziehung wichtigen Einfluss, den die Kohlensaurespannung des Blutes auf dessen. Sauerstoffbinding übt. *Skand. Arch. Physiol.* 16: 402–412, 1904.*

BOYCOTT, A. E., G. C. C. DAMANT, AND J. S. HALDANE. The prevention of compressed air illness. *J. Hyg. (Camb.)* 8: 342–443, 1908.

BOYLE, R. Continuation of the experiments concerning respiration. *Philos. Trans. R. Soc. Lond.* 5: 2035–2056, 1670.

BRISCOE, W. A., AND A. B. DUBOIS. The relationship between airway resistance, airway conductance, and lung volume in subjects of different age and body size. *J. Clin. Invest.* 37: 1279–1285, 1958.

BUNN, H. F., M. H. MAY, W. F. KOCHOLATY, AND C. E. SHIELDS. Hemoglobin function in stored blood. *J. Clin. Invest.* 48: 311–321, 1969.

CHANUTIN, A., AND R. R. CURNISH. Effect of organic and inorganic phosphates on the oxygen equilibrium of human erythrocytes. *Arch. Biochem.* 121: 96–102, 1967.

---

*English translations of these papers can be found in: West J. B. (ed.). *Translations in Respiratory Physiology.* Stroudsburg: Dowden, Hutchinson and Ross, 1975.

CHRISTIANSEN, J., C. G. DOUGLAS, AND J. S. HALDANE. The absorption and dissociation of carbon dioxide by human blood. *J. Physiol. (Lond.)* 48: 244–271, 1914.

CLARK, J. M., AND C. J. LAMBERTSEN. Pulmonary oxygen toxicity: a review. *Pharmacol. Rev.* 23: 37–133, 1971.

CLEMENTS, J. A. Surface phenomenon in relation to pulmonary function. *Physiologist* 5: 11–28, 1962.

COBURN, R. E., E. R. ALLEN, A. AYERS, D. BARTLETT, JR., S. M. HORWATH. L. H. KULLER, V. G. LATIES, L. D. LONGO, AND E. P. RADFORD, JR. *The Biologic Effects of Carbon Monoxide.* Washington, D.C.: National Academy of Sciences, National Research Council, 1976.

COMROE, J. H. *Physiology of Respiration,* 2nd ed. Chicago: Year Book, 1974, p. 234–271.

COMROE, J. H., R. E. FORSTER, A. B. DUBOIS, W. A. BRISCOE, AND E. CARLSEN. *The Lung—Clinical Physiology and Pulmonary Function Tests,* 2nd ed. Chicago: Year Book, 1962.

COTES, J. E. *Lung Function,* 4th ed. Oxford: Blackwell, 1979.

COUNCIL FOR NATIONAL COOPERATION IN AQUATICS. *The New Science of Skin and Scuba Diving,* 4th ed. New York: Association Press, 1974.

CRISP, D. J. Plastron respiration. In: *Recent Progress in Surface Science,* edited by J. F. Danielli, K. G. A. Pankhurst, and A. C. Riddiford. New York: Academic Press, 1964, vol. 2, p. 377–425.

CUMMING, G., J. CRANK, K. HORSFIELD, AND I. PARKER. Gaseous diffusion in the airways of the human lung. *Respir. Physiol.* 1: 58–74, 1966.

DUBOIS, A. B., S. Y. BOTELHO, G. N. BEDELL, R. MARSHALL, AND J. H. COMROE, JR. A rapid plethysmographic method for measuring thoracic gas volume. *J. Clin. Invest.* 35: 322–326, 1956.

FOWLER, W. S. Lung function studies. II. The respiratory dead space. *Am. J. Physiol.* 154: 405–416, 1948.

FRAYSER, R., G. W. GRAY, AND C. S. HOUSTON. Control of the retinal circulation at altitude. *J. Appl. Physiol.* 37: 302–304, 1974.

FRY, D. L., AND R. E. HYATT. Pulmonary mechanics: a unified analysis of the relationship between pressure, volume and gas flow in the lungs of normal and diseased human subjects. *Am. J. Med.* 29: 672–689, 1960.

GLAZIER, J. B., J. M. B. HUGHES, J. E. MALONEY, AND J. B. WEST. Measurements of capillary dimensions and blood volume in rapidly frozen lungs. *J. Appl. Physiol.* 26: 65–76, 1969.

GRAVES, D. J., J. IDICULA, C. J. LAMBERTSEN, AND J. A. QUINN. Bubble formation resulting from counterdiffusion supersaturation: a possible explanation for inert gas "urticaria" and vertigo. *Phys. Med. Biol.* 18: 256, 1973.

HALDANE, J. S. *Respiration.* New Haven: Yale Univ. Press, 1922.

HALDANE, J. S., AND J. G. PRIESTLEY. *Respiration.* New Haven: Yale Univ. Press, 1935.

HILLS, B. A. Thermodynamic decompression. In: *The Physiology and Medicine of Diving and Compressed Air Work,* edited by P. B. Bennett and D. H. Elliot. London: Baillière, Tindall and Cassell, 1969.

HUGHES, J. M. B., J. B. GLAZIER, J. E. MALONEY, AND J. B. WEST. Effect of lung volume on the distribution of pulmonary blood flow in man. *Respir. Physiol.* 4: 58–72, 1968.

HURTADO, A. Animals in high altitudes: resident man. In: *Handbook of Physiology, Adaptation to the Environment,* Sect. 4, edited by D. B. Dill. Washington, D.C.: American Physiological Society, 1964, p. 843–860.

HUTCHINSON, J. On the capacity of the lungs, and on the respiratory functions, with a view of establishing a precise and easy method of detecting disease by the spirometer. *Med. Chir. Trans. (Lond.)* 29: 137, 1846.

JOURNAL OF THE AMERICAN MEDICAL ASSOCIATION. Standards and guidelines for Cardiopulmonary Resuscitation (CPR) and Emergency Cardiac Care (ECC). 244: 453–509, 1980.

KYLSTRA, J. A., M. O. TISSING, AND A. VAN DER MAEN. Of mice as fish. *Trans. Am. Soc. Artif. Intern. Organs* 8: 378–383, 1962.

LAHIRI, S., F. F. KAO, T. VELASQUEZ, C. MARTINEZ, AND W. PEZZIA. Irreversible blunted respiratory sensitivity to hypoxia in high altitude natives. *Respir. Physiol.* 6: 360–374, 1969.

LAMBERTSEN, C. Respiration. In: V. Mountcastle, *Medical Physiology.* St. Louis: C. V. Mosby, 1974, p. 1361–1597.

LAMBERTSEN, C. J. Therapeutic gases: Oxygen, carbon dioxide, and helium. In: *Drill's Pharmacology in Medicine.* 3rd ed., edited by J. R. DiPalma. New York: McGraw Hill, 1965.

LENFANT, C., J. D. TORRANCE, R. D. WOODSON, P. JACOBS, AND C. A. FINCH. Role of organic phosphates in the adaptation of man to hypoxia. *Fed. Proc.* 29: 1115–1117, 1970.

LOESCHCKE, H. H., AND K. H. GERTZ. Einfluss des O_2-Druckes in der Einatmungsluft auf die Atemtätigkeit des Menschen, geprüft unter Konstanthaltung des alveolaren CO_2-Druckes. *Pflugers Arch. Ges. Physiol.* 267: 460–477, 1958.

MACKLEM, P. T., AND J. MEAD. Resistance of central and peripheral airways measured by a retrograde catheter. *J. Appl. Physiol.* 22: 395–401, 1967.

MACKLEM P. T., AND J. MEAD. *Handbook of Physiology.* Sect. 3. The Respiratory System. Bethesda; American Physiological Society, 1986.

McCARTHY, D. S., R. SPENCER, R. GREEN, AND J. MILICEMILL. Measurement of "closing volume" as a simple and sensitive test for early detection of small airway disease. *Am. J. Med.* 52: 747–753, 1972.

MEAD, J., T. TAKISHIMA, AND D. LEITH. Stress distribution in lungs: a model of pulmonary elasticity. *J. Appl. Physiol.* 28: 596–608, 1970.

MESSNER, R. The Mountain. In: *Everest: Expedition to the Ultimate.* London: Kay and Ward, 1979.

MILIC-EMILI, J., J. A. M. HENDERSON, M. B. DOLOVICH, D. TROP, AND K. KANEKO. Regional distribution of inspired gas in the lung. *J. Appl. Physiol.* 21: 749–759, 1966.

MODELL, J. H. *The Pathophysiology and Treatment of Drowning and Near-Drowning.* Springfield, IL: Charles C Thomas, 1971.

NIELSEN, M., AND H. SMITH. Studies on the regulation of respiration in acute hypoxia. (With an appendix on respiratory control during prolonged hypoxia.) *Acta Physiol. Scand.* 24: 293–313, 1952.

PAGANELLI, C. V., N. BATEMAN, AND H. RAHN. Artificial gills for gas exchange in water. In: *Proceedings of the Third Symposium on Underwater Physiology*, edited by C. J. Lambertsen. Baltimore: Williams and Wilkins, 1967, ch. 38.

PAINTAL, A. S. The mechanism of excitation of type J receptors, and the J reflex. In: *Breathing: Hering-Breuer Centenary Symposium*, edited by R. Porter. London: Churchill, 1970, p. 59–71.

PATTLE, R. E. Properties, function and origin of the alveolar lining layer. *Nature* 175: 1125–1126, 1955.

PEDLEY, T. J., R. C. SCHROTER, AND M. F. SUDLOW. The prediction of pressure drop and variation of resistance within the human bronchial airways. *Respir. Physiol.* 9: 387–405, 1970.

PETERS, J. P., AND D. D. VAN SLYKE. *Quantitative Clinical Chemistry 2, Methods.* Baltimore: Williams & Wilkins, 1932.

RADFORD, E. P., JR. Method for estimating respiratory surface area of mammalian lungs from their physical characteristics. *Proc. Soc. Exp. Biol. Med.* 87: 58–61, 1954.

RADFORD, E. P., JR. Recent studies of mechanical properties of mammalian lungs. In: *Tissue Elasticity*, edited by J. W. Remington, Washington, D.C.: American Physiological Society, 1957, p. 177–190.

RAHN, H., AND W. O. FENN. *A Graphical Analysis of the Respiratory Gas Exchange.* Washington, D.C.: American Physiological Society, 1955.

RAHN, H., AND A. B. OTIS. Man's respiratory response during and after acclimatization to high altitude. *Am J. Physiol.* 157: 445–462, 1949.

RAHN, H., A. B. OTIS, L. E. CHADWICK, AND W. O. FENN. The pressure-volume diagram of the thorax and lung. *Am. J. Physiol.* 146: 161–178, 1946.

RAHN, H., AND V. PAGANELLI. Gas exchange in gas gills of diving insects. *Respir. Physiol.* 5: 145–164, 1968.

READ, D. J. C. A clinical method for assessing the ventilatory response to carbon dioxide. *Austral. Ann. Med.* 16: 20–32, 1967.

RILEY, R. L., AND A. COURNAND. "Ideal" alveolar air and the analysis of ventilation-perfusion relationships in the lungs. *J. Appl. Physiol.* 1: 825–847, 1949.

ROUGHTON, F. J. W., AND R. E. FORSTER. Relative importance of diffusion and chemical reaction rates in determining rate of exchange of gases in the human lung, with special reference to true diffusing capacity of pulmonary membrane and volume of blood in the lung capillaries. *J. Appl. Physiol.* 11: 290–302, 1957.

SCHAFFER, T. H., D. RUBENSTEIN, G. D. MOSKOWITZ, AND M. DELIVORIA-PAPADOPOULOS. Gaseous exchange and acid balance in premature lambs during liquid ventilation since birth. *Pediatr. Res.* 10: 227–231, 1976.

SCHOLANDER, P. F. Analyzer for accurate estimation of respiratory gases in one half cubic centimeter samples. *J. Biol. Chem.* 167: 235–250, 1947.

VON NEERGAARD, K. Neue Auffassungen über einen Grundbegriff der Atemmechanik: Die Retraktionskraft der Lunge, abhängig von der Oberflächenspannung in den Alveolen. *Zeit. ges. exper. Med.* 66: 373–394, 1929.*

WAGNER, P. D. Diffusion and chemical reaction in pulmonary gas exchange. *Physiol. Rev.* 57: 257–312, 1977.

WAGNER, P. D., R. B. LARAVUSO, R. R. UHL, AND J. B. WEST. Continuous distributions of ventilation-perfusion ratios in normal subjects breathing air and 100% O_2. *J. Clin. Invest.* 54: 54–68, 1974.

WAGNER, P. D., AND J. B. WEST. Effects of diffusion impairment on O_2 and CO_2 time course in pulmonary capillaries. *J. Appl. Physiol.* 33: 62–71, 1972.

WEIBEL, E. R. *Morphometry of the Human Lung.* New York: Academic Press. 1963.

WEIBEL, E. R. Morphological basis of alveolar-capillary gas exchange, *Physiol. Rev.* 53: 419–495, 1973.

WEIBEL, E. R., AND J. GIL. Structure-function relationships at the alveolar level. In: *Bioengineering Aspects of the Lung*, edited by J. B. West. New York: Marcel Dekker, 1977.

WEST, J. B. Blood-flow, ventilation, and gas exchange in the lung. *Lancet* 2: 1055–1058, 1963.

WEST, J. B. *Respiratory Physiology: The Essentials*, 3rd ed. Baltimore: Williams & Wilkins, 1985a.

WEST, J. B. *Pulmonary Pathophysiology: The Essentials*, 3rd ed. Baltimore: Williams & Wilkins, 1987.

WEST, J. B. *Ventilation/Bloodflow and Gas Exchange*, 4th ed. Oxford: Blackwell, 1985b.

WEST, J. B., C. T. DOLLERY, AND A. NAIMARK. Distribution of blood flow in isolated lung: relation to vascular and alveolar pressures. *J. Appl. Physiol.* 19: 713–724, 1964.

WEST, J. B., S. LAHIRI, M. B. GILL, J. S. MILLEDGE, L. G. C. E. PUGH, AND M. P. WARD. Arterial oxygen saturation during exercise at high altitude. *J. Appl. Physiol.* 17: 617–621, 1962.

WEST, J. B., AND P. D. WAGNER. Pulmonary gas exchange. In: *Bioengineering Aspects of the Lung*, edited by J. B. West. New York: Marcel Dekker, 1977.

WEST, J. B., P. H. HACKETT, K. H. MARET, J. S. MILLEDGE, R. M. PETERS, JR., C. H. PIZZO, AND R. M. WINSLOW. Pulmonary gas exchange on the summit of Mt. Everest. *J. Appl. Physiol.: Respirat. Environ. Exercise Physiol.* 55: 678–687, 1983.

WHITELAW, W. A., J-P DERENNE, AND J. MILIC-EMILI. Occlusion pressure as a measure of respiratory center output in conscious man. *Respir. Physiol.* 23: 181–199, 1975.

WOOLCOCK, A. J., N. J. VINCENT, AND P. T. MACKLEM. Frequency dependence of compliance as a test for obstruction in the small airways. *J. Clin. Invest.* 1097–1106, 1969.

SECTION 6

Gastrointestinal System

CHAPTER 42

Introduction to the Function and Control of the Gastrointestinal System

The gastrointestinal system consists of the alimentary canal from the mouth to the anus and those organs that empty their contents into the alimentary canal, such as the salivary glands, the liver and biliary tract, and the pancreas. Aspects of liver physiology concerned with the secretion of bile are considered gastrointestinal physiology, whereas the aspects concerned with protein synthesis, lipid and carbohydrate metabolism, and synthesis of urea are considered under intermediary metabolism (Chapters 48 and 49). Similarly, pancreatic physiology relating to its exocrine secretions are considered gastrointestinal physiology, whereas the endocrine secretions of the pancreas are considered in the physiology of the endocrine system (Chapter 50).

The overall functions of the gastrointestinal system are to take in nutrients and to eliminate wastes. The major physiological processes occurring in the gastrointestinal system to achieve these functions are movements of the gut (motility), secretion, excretion, digestion, and absorption. Although presented separately in this section, these processes are interdependent. For example, absorption of nutrients by the intestinal mucosa cannot occur unless there is digestion of food constituents to smaller structures that can be absorbed. Digestion cannot occur unless pancreatic digestive enzymes are secreted into the alimentary canal. Finally, none of these processes can occur unless the movements of the gut bring the meal to the proper locations for mixing with digestive secretions, for absorption, or for eliminating wastes. Thus, the overall function of the gastrointestinal system is an integration of these separate physiological processes.

Individuals can now survive for indefinite periods with intravenous feedings providing water, electrolytes, and nutrients including vitamins and trace elements. Thus, the absorption surface of the alimentary canal is not absolutely necessary for life. In addition, the loss of the endocrine and exocrine functions of the pancreas can be tolerated with the exogenous administration of insulin and pancreatic enzymes. However, removal of the liver is fatal unless replaced by a transplanted liver because presently its metabolic and excretory functions cannot be replaced.

Although considered in detail elsewhere (Chapter 21), the gastrointestinal tract is an important part of the body's immune system. Both humoral antibodies secreted by the mucosa of the gut and the cellular immune system within the gut mucosa play important roles in protecting the body against microorganisms in the lumen of the gut and foreign proteins capable of acting as antigens.

CONTROL SYSTEMS

Controls in the gastrointestinal tract can be divided into three categories: endocrine, neural, and paracrine (Fig. 6.1). *Endocrine controls* are those in which a stimulus causes secretion of a hormone that travels by way of the bloodstream to interact with a target to cause a response. The gastrointestinal tract is lined with numerous and varied endocrine cell populations containing hormones such as gastrin, secretin, and cholecystokinin. Specific stimuli from the luminal contents cause release of such hormones into the bloodstream. Each hormone then causes specific responses by interacting with tissue-specific receptors. An example of such an endocrine-mediated response is the mechanism by which fatty acids in the duodenal lumen cause pancreatic enzyme secretion. The fatty acids stimulate cholecystokinin (CCK) release from CCK-containing endocrine cells in the duodenal mucosa. The CCK enters the bloodstream and interacts with tissues containing specific CCK receptors such as the exocrine pancreas. This interaction leads to digestive enzyme secretion.

Neurally mediated responses are those where the stimulus-mediated response is mediated by nerves

---

Nora Laiken, Ph.D., reviewed the chapters for this section and made many important contributions.

ENDOCRINE

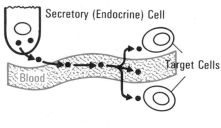

Secretory (Endocrine) Cell

Blood

Target Cells

PARACRINE

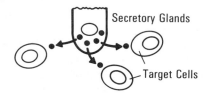

Secretory Glands

Target Cells

NEURAL

Target Cell

Nerve Cell

Figure 6.1. Types of extracellular chemical transmission in the gastrointestinal tract.

(Fig. 6.1). In the gastrointestinal tract there are a variety of neural receptors that detect both chemical and mechanical stimuli. The transmission to the target tissue is rapid and may involve only the nervous system of the gastrointestinal tract (enteric nervous system, ENS) or the sensory information may be processed in the central nervous system (CNS). Efferents from the CNS are parasympathetic, sympathetic, and nonadrenergic noncholinergic. These efferent systems synapse with the ENS to effect tissue-specific responses. The neurotransmitters in the enteric nervous system mediating the tissue responses include a variety of agents in addition to acetylcholine and adrenergic agonists.

An example of a neurally mediated response is that of peristalsis in the body of the esophagus. Esophageal peristalsis is a wave of circular muscle relaxation followed by contraction transversing the length of the esophagus. The peristalsis moves a bolus of food from the pharynx to the stomach. The process in the esophageal body is controlled by the CNS and ENS. There are many examples of neurally mediated secretions as well. In fact, the control of salivary secretion is completely neurally mediated.

Paracrine-mediated responses are those where the chemical signal released by a cell acts as a local mediator only on cells in its immediate environment (Fig. 6.1). This happens because the transmitter is either rapidly degraded, taken up, or diluted out so that its action is confined to the vicinity of its release. The best example of a paracrine-mediated response is that of histamine-stimulated HCl secretion from the parietal cell. The histamine is stored and released from histamine-containing cells that are in close proximity to the parietal cells in the gastric mucosa.

Proof that a humoral agent, paracrine agent, or neurotransmitter is the physiological mediator of a stimulus-induced response requires several criteria. First, for a hormone, the agent must be shown to increase in concentration in the circulation with the proposed stimulus. For a neurotransmitter, there must be evidence for the presence of the specific transmitter in the neural endings and release at the target tissue with neural stimulation. Second, there should be a temporal correlation between the increase of the hormone in the blood or neurotransmitter at the target tissue and the proposed tissue response. Third, exogenous administration of the hormone, paracrine agent, or neurotransmitter should reproduce the response. Fourth, the response should be reproduced with concentrations of mediator achieved during the physiological stimulus. Finally, the response to the stimulus should be blocked by specific receptor antagonists or eliminated with specific antibodies against the proposed agent.

Endocrine and Paracrine Control Systems

ENDOCRINE-PARACRINE CELLS

Scattered in the epithelium of the alimentary canal are the endocrine-paracrine cells (Fig. 6.2). The existence of endocrine cells in the gastrointestinal mucosa was described as early as 1870. The first group to be recognized were the argentaffin or enterochromafin cells, so called because of their histological staining with silver or chromic salt solutions. These cells store and secrete serotonin (5-hydroxytryptamine). In recent years, several other classes of endocrine cells have been distinguished from enterochromaffin cells by both electron microscopy and immunohistochemistry. Each distinct class of these cells stores and secretes one of a large series of endocrine agents. Such agents include 5-hydroxytryptamine, gastrin, cholecystokinin, secretin, gastric inhibitory polypeptide, motilin, neurotensin, and somatostatin. Of interest is the finding that these agents are also found in the neural endings of the ENS. The secretory granules of each class of cells contains one or more than one humoral agent. In many cases, the secretory granules are located at the basal aspect of the cytoplasm while the apical surface contains microvilli facing the

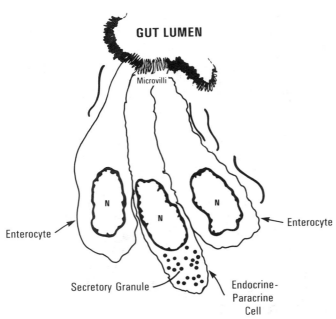

GUT LUMEN

Microvilli

Enterocyte

Enterocyte

N

N

N

Secretory Granule

Endocrine-Paracrine Cell

Figure 6.2. Endocrine-paracrine cell of the alimentary canal. Microvilli face the gut lumen to receive chemical signals from luminal contents. Secretory granules are released from the basal cytoplasm into the interstitial space and then into the bloodstream. The secretory agent may act on surrounding cells such as enterocytes or distant tissues by way of the bloodstream.

lumen of the alimentary canal. With this arrangement, the cell responds directly to specific chemical stimuli in the luminal contents resulting in basal release of the secretory granules. The local diffusion of the secreted agent in the intracellular spaces results in paracrine effects on nearby target cells, whereas diffusion into the bloodstream results in endocrine effects on distal target organs. The mechanisms by which these cells "sense" luminal stimuli and transmit this information in the cell to cause secretion are not known. Each class of endocrine-paracrine cells responds to specific stimuli, suggesting specific "receptors" for the stimuli. The various endocrine-paracrine cells will be discussed with respect to their humoral agent as follows.

For a review of endocrine-paracrine cells see Solcia et al., 1987.

GASTRIN

Gastrin is contained in gastrin cells (G-cells) that are located mainly in the glands of the gastric antrum but also in the mucosa of the upper small bowel (Greider et al., 1972). The most abundant forms are a 17 amino acid (G-17) and 34 amino acid (G-34) peptide. Gastrin is also present as a 14 amino acid peptide (minigastrin). The COOH-terminal amino acid sequences of the gastrins are identical (Table 6.1). Thus, they differ in their NH$_2$-terminal

extensions. Full biological activity of gastrin is retained in the COOH-terminal tetrapeptide. The tyrosine in the sixth position from the COOH-terminal may be present as either a sulfated or nonsulfated residue. The sulfated and nonsulfated forms are approximately equal in potency for their biological effects.

Gastrin is released by small peptides, amino acids (especially phenylalanine and tryptophan) and calcium in the gastric lumen; by neural stimulation; and by catecholamines and gastrin-releasing peptide. Small peptides, amino acids, and calcium directly stimulate the G-cell to release gastrin. The neural-stimulated release of gastrin is mediated by both direct vagal stimulation (cephalic stimulation) and distension of the stomach. Distension of the stomach activates reflexes in the ENS that mediate gastrin release. The distension also activates vagal sensory afferents that mediate gastrin release by way of the CNS and vagal efferents (vagovagal reflex). The putative neurotransmitter for gastrin release is a peptide named gastrin-releasing peptide (GRP). Gastrin release is inhibited by a low gastric luminal pH (below 3) and somatostatin. Recent evidence suggests that paracrine release of somatostatin in the gastric antrum may be a physiological mediator of the low luminal pH on gastrin release (Yamada, 1987).

Gastrin is an established physiological regulator of gastric acid secretion. Gastrin mediates secretion stimulated by both distension of the stomach and intraluminal peptides and amino acids. Specific high affinity gastrin receptors on the parietal cell mediate this response. Gastrin also is important for growth of the acid-secreting mucosa of the stomach. Gastrin may also mediate intrinsic factor and pepsinogen secretion and it may regulate antral motor activity.

For a review of gastrin, see Walsh and Grossman, 1975; Walsh, 1987.

CHOLECYSTOKININ (CCK)

CCK-containing cells are located in the mucosa of the duodenum and jejunum. Like gastrin, CCK occurs in multiple forms sharing the COOH-terminus but differing in the length of the NH$_2$-terminal extension (Table 6.1). Identified forms include CCK-58 (i.e., 58 amino acid peptide), CCK-39, CCK-33, CCK-22, and CCK-8. The COOH-terminal pentapeptide of CCK and gastrin are identical and, like gastrin, the COOH-terminal tetrapeptide of CCK retains full biological activity. Thus, the specificity of CCK or gastrin responses are provided by the remainder of the molecule. One principal difference between gastrin and CCK is that all forms of CCK are sul-

Table 6.1.
Amino Acid Sequences of Some Endocrine Gastrointestinal Peptides[a]

	Gastrin G-34	Gastrin G-17	Gastrin G-14	CCK-39	CCK-33	CCK-8		Secretin	Somatostatin-14
NH2-terminus									
39				TYR					
38				ILE					
37				GLN					
36				GLN					
35				ALA					
34	GLP			ARG					
33	LEU			LYS	LYS				
32	GYL			ALA	ALA				
31	PRO			PRO	PRO				
30	GLN			SER	SER				
29	GLN			GYL	GYL				
28	HIS			ARC	ARC		NH2-terminus		
27	PRO			VAL	VAL		1	HIS	ALA
26	SER			SER	SER		2	SER	GLY
25	LEU			MET	MET		3	ASP	CYS
24	VAL			ILE	ILE		4	GYL	LYS
23	ALA			LYS	LYS		5	THR	ASN
22	ASP			ASN	ASN		6	PHE	PHE
21	PRO			LEU	LEU		7	THR	PHE
20	PRO			GLN	GLN		8	SER	TRP
19	LYS			SER	SER		9	GLU	LYS
18	LYS			LEU	LEU		10	LEU	THR
17	GLN	GLP		ASP	ASP		11	SER	PHE
16	GLY	GLY		PRO	PRO		12	ARG	THR
15	PRO	PRO		SER	SER		13	LEU	SER
14	TRP	TRP	TRP	HIS	HIS		14	ARG	CYS
13	LEU	LEU	LEU	ARG	ARG		15	ASP	
12	GLU	GLU	GLU	ILE	ILE		16	SER	
11	GLU	GLU	GLU	SER	SER		17	ALA	
10	GLU	GLU	GLU	ASP	ASP	ASP	18	ARG	
9	GLU	GLU	GLU	ARG	ARG	ARG	19	LEU	
8	GLU	GLU	GLU	ASP	ASP	ASP	20	GLN	
7	ALA	ALA	ALA	TYR-SOH	TYR-SOH	TYR-SOH	21	ARG	
6[b]	TYR-R*	TYR-R*	TYR-R*	MET	MET	MET	22	LEU	
5	GLY	GLY	GLY	GLY	GLY	GLY	23	LEU	
4	TRP	TRP	TRP	TRP	TRP	TRP	24	GLN	
3	MET	MET	MET	MET	MET	MET	25	GLY	
2	ASP	ASP	ASP	ASP	ASP	ASP	26	LEU	
1	PHE-NH2	PHE-NH2	PHE-NH2	PHE-NH2	PHE-NH2	PHE-NH2	27	VAL-NH2	
COOH-terminus							COOH-terminus		

[a]Note that several of the peptides are amidated at the COOH-terminus.
[b]R*—gastrin may be sulfated in this position.

fated at the tyrosine residue in the seventh position from the COOH-terminus.

CCK is released by intraluminal fatty acids of chain length of nine or more carbons and their corresponding monoglycerides; by intraluminal protein digestion products; by amino acids such as phenylalanine and tryptophan; and slightly by intraluminal glucose. GRP and its amphibian analogue, bombesin, also release CCK suggesting that GRP may be involved in the physiological regulation of CCK release.

The established physiological actions of CCK are pancreatic enzyme secretion (Stubbs and Stabile, 1985), gallbladder contraction (Liddle et al., 1985), and relaxation of the sphincter of Oddi. CCK has been proposed as a satiety factor. CCK has effects on gastric and intestinal motility and a trophic effect on the pancreas that have not yet been proved to be physiological.

For a review of CCK, see Walsh, 1987.

SECRETIN

Secretin was the first hormone identified in 1901 by Bayliss and Starling. Secretin-containing cells are located in the duodenum and jejunum. In contrast to CCK and gastrin, there is one circulating form of secretin—a 27 amino acid peptide (Table 6.1). Secretin is structurally related to glucagon, vasoactive intestinal polypeptide (VIP), PHI (*Pep*tide with *H*istidine at the NH2-terminus and *I*soleucine at the COOH-terminus), gastric inhibitory peptide, and growth hormone-releasing factor.

The major intraluminal stimulant of secretin release is hydrogen ion. Increases in plasma secretin occur when intraluminal pH decreases to less than 4.5. Intraluminal fatty acids may also release secretin. The established physiological effect of secretin is the stimulation of ductal secretion from the pancreas (Chey et al., 1979). This is a high volume secretion that is alkaline because of its high bicarbonate concentration. Secretin also causes a bicarbonate and water secretion from the biliary system. Secretin causes inhibition of gastric emptying and gastric acid secretion but these effects are probably not physiological. Finally, secretin causes pepsin secretion in the stomach. Whether this effect of secretin is physiological has not been tested.

For a review of secretin, see Walsh, 1987.

SOMATOSTATIN

This peptide is contained in D-cells in the islets of Langerhans and in the mucosa throughout the gastrointestinal tract. Somatostatin also is present in neuron cell bodies and in nerve fibers in both the CNS and the ENS. Somatostatin occurs as a 14 (SS-14) and 28 (SS-28) amino acid peptide. Circulating concentrations of somatostatin increase with fat and protein in the intestine and increase slightly with acidification of the gastric antrum and duodenum. The physiological functions of somatostatin include the inhibition of growth hormone release from the anterior pituitary and insulin and glucagon release from the islets of Langerhans. Somatostatin probably also has a physiological role in inhibiting gastrin release from the G-cell; acid secretion from the parietal cell; and pancreatic enzyme secretion. Somatostatin has several biological effects that have not been established as physiological. These effects include inhibition of release of hormones such as secretin and CCK from endocrine-paracrine cells; inhibition of absorption of amino acids in the small intestine; various effects on electrolyte and water absorption and secretion; and inhibition of gastrointestinal motility.

For a review of somatostatin, see Walsh, 1987; Yamada, 1987.

GASTRIC INHIBITORY PEPTIDE (GIP)

This is a 42 amino acid peptide in the glucagon-secretin family of peptides. It is present in highest concentrations in endocrine-paracrine cells in the duodenum and jejunum. It is present in lower concentrations in the gastric antrum and ileum. GIP is released by several components of a meal—intraluminal glucose, amino acids, and hydrolyzed triglycerides. It is also released by intraduodenal acidification. Although this peptide was named for its ability to inhibit gastric acid secretion, this effect of GIP has not been demonstrated to be physiologic. GIP does have a physiologic role in augmenting glucose-stimulated insulin release. This has lead to the suggestion that "glucose-dependent insulinotropic peptide" is a better name for this peptide.

For a review of GIP, see Walsh, 1987.

MOTILIN

Cells containing motilin are present in the mucosa of the upper small intestine. Motilin is a 22 amino acid peptide. Motilin is released during the fasting state and at the initiation of the interdigestive migrating myoelectric complex (MMC) in the duodenum. Because of this relationship, it has been suggested that motilin regulates the contractions of the MMC, especially in the stomach and duodenum.

For a review of motilin, see Walsh, 1987.

OTHER PEPTIDE MEDIATORS

Neurotensin, pancreatic polypeptide, and peptide YY are peptide transmitters also found in specific endocrine-paracrine cells in the alimentary tract (Walsh, 1987). For these peptides, physiological roles have not been established. Neurotensin is a 13 amino acid peptide present in endocrine cells in the ileal mucosa and released by fat in the meal. A possible physiological role for neurotensin is inhibition of gastric acid secretion caused by fat in the intestinal lumen.

Pancreatic polypeptide is a 36 amino acid peptide contained in endocrine cells in the islets of Langerhans and in the substance of the exocrine pancreas. It increases with protein-rich meals and vagal stimulation. Pancreatic polypeptide inhibits pancreatic exocrine secretion; however, the physiological role of this inhibition is in question because antiserum to pancreatic polypeptide does not alter meal-stimulated pancreatic secretion.

Peptide YY is a 36 amino acid peptide related to pancreatic polypeptide that is present in endocrine-paracrine cells in the mucosa of the ileum and colon. Fat is the most potent stimulant for release. Peptide YY may mediate inhibition of exocrine pancreatic and gastric acid secretion.

HISTAMINE

Histamine is a nonpeptide paracrine mediator that stimulates HCl secretion from the gastric parietal cell by interacting with histamine-H_2 receptors on the cell (Soll and Berglindh, 1987). Histamine is synthesized from histidine by the action of histidine decarboxylase and stored in secretory granules in histamine-containing cells in the gastric mucosa. The physiological mechanisms of release of histamine from the gastric mast cell are not established but the role of histamine in acid secretion is. Specific histamine-H_2-receptor antagonists inhibit parietal cell secretion stimulated by meals, gastrin, acetylcholine analogues, and histamine analogues. Thus, histamine stimulation is central to the physiological mechanism of acid secretion.

Neural Control Systems

The nervous control of the gastrointestinal system is mediated by the ENS and the CNS. The ENS is an independent integrative system with structural and functional properties that are similar to those in the CNS (Wood, 1987). The ENS contains circuitry for processing information supplied to it from various kinds of sensory receptors in the gut (Fig. 6.3). The sensory receptors are osmoreceptors, thermoreceptors, mechanicoreceptors, and chemoreceptors that constantly supply information to the ENS from the wall of the gut and its intraluminal contents.

The sensory neurons synapse with interneurons in the ENS which, in turn, are connected to networks that process the information and control the activity of motor neurons. The motor neurons effect responses in the gut musculature, endocrine-paracrine cells, secretory epithelium, absorptive epithelium, and blood vessels. The integrative circuitry with the ENS also mediates programs for the automatic control of repetitive cyclical motor responses or stereotyped sequences of motor responses such as peristalsis and segmentation. These characteristics of the ENS have led some to refer to it as the "brain of the gut" because it can integrate sensory information and effect complex motor response independent of the CNS. The advantage of this arrangement is efficiency of control of a complex set of effector functions eliminating the need for many long-distance conducting pathways to the CNS. A

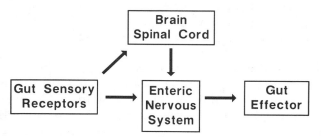

Figure 6.3. Functional model of the enteric nervous system (ENS) and its relationship to the central nervous system (CNS). Sensory information from the gut can be processed completely within the integrative circuitry of the ENS to effect a response or can be processed through the CNS. There are programs within the ENS that effect complex motor responses.

detailed anatomy of the ENS is presented in Chapter 43.

Extrinsic gut efferent innervation is by way of the parasympathetic and sympathetic nerves. The nerves carrying the parasympathetic efferents (vagal and sacral nerves) and sympathetic efferents (splanchnic nerves) also contain sensory neurons from the gut. The anatomy of this innervation is discussed elsewhere in the text (Chapter 69). Parasympathetic stimulation generally causes muscle contraction and secretion. This is thought to occur by synapse of the parasympathetic efferents with the circuitry of the ENS. Sensory neurons in the vagus allow processing of sensory information from the gut in the CNS leading to an alteration of the output of vagal efferents to the circuitry of the ENS. This CNS processing constitutes a reflex referred to as a vagovagal or long reflex. Vagovagal reflexes mediate part of the acid secretion caused by distention of the stomach (see above) and pancreatic enzyme secretion stimulated by fatty acids in the duodenum.

Sympathetic efferents to the gastrointestinal tract function to inhibit blood flow and motility. The mechanism of this inhibition results from the ability of the sympathetic input acting on presynaptic terminals to prevent release of transmitter substances.

NEUROTRANSMITTERS IN THE ENTERIC NERVOUS SYSTEM

The importance of the classic neurotransmitters, acetylcholine and norephinephrine, as neurotransmitters in the ENS has been accepted for many years. Acetylcholine mediates contraction of smooth muscle in the gut and secretions from the salivary glands, stomach, pancreas, and small intestine. Norepinephrine inhibits contraction, secretion, and blood flow. It has also been recognized for several years that there are noncholingergic, nonadrenergic neurotransmitters in the gut. Electron microscopic studies demonstrate that certain neural endings in the gut have structural characteristics that are distinct from adrenergic and cholinergic neurons. Several ultrastructurally distinct classes of nerve endings have been described. These classes may contain one or more than one potential neurotransmitter. The putative transmitters fall into two categories—peptide and nonpeptide. Putative nonpeptide transmitters include ATP, adenosine, serotonin (5-hydroxytryptamine), dopamine, γ-aminobutyric acid (GABA), histamine, glycine, and prostaglandins. Putative peptide transmitters include substance P, VIP, PHI, GRP, enkephalin, somatostatin, CCK, secretin, neurotensin, motilin, gastrin, thyrotropin-releasing hormone, angiotensin, and neuropeptide Y.

The proposal that the various peptide and nonpeptide agents listed above are neurotransmitters comes from studies that have localized the agents to neural endings using radioimmunoassay, histochemistry, and immunohistochemistry. However, our understanding of whether these agents have physiological roles is incomplete. For several of these agents, it has not been established that they are released by physiolgical stimuli. Furthermore, difficulties in establishing physiological roles for these agents as neurotransmitters occur because in many cases the agents are also released by endocrine-paracrine cells. In addition, specific receptor antagonists are available for only a few of these agents. For some transmitters, immunoneutralization experiments have suggested physiological roles.

Examples of nonadrenergic noncholinergic agents where the evidence favors a physiological role as a neurotransmitter are VIP and GRP (Table 6.2) Dockray, 1987). VIP is a 28 amino acid peptide in the glucagon-secretin family of peptides. VIP-containing neurons occurs throughout the alimentary canal, salivary glands, and pancreas. Neural stimulation to a variety of organs such as the salivary glands, pancreas, and small and large intestine results in VIP release. Also, distension of the esophagus, small intestine, and large intestine results in VIP release. VIP causes relaxation of the lower esophageal sphincter, the proximal stomach, and the internal anal sphincter. Immunoneutralization studies with specific VIP antibodies suggest that VIP physiologically mediates these processes. Similar types of experiments suggest a role for VIP in mediating exocrine pancreatic secretion. VIP has long been known to be a potent stimulant of intestinal secretion. However, a role for VIP as a mediator of intestinal secretion has not been established.

GRP is a 27 amino acid peptide that is released form the antrum, fundus, and pancreas with neural stimulation (Table 6.2). GRP (and its amphibian analogue, bombesin) stimulates release of gastrin, CCK, pancreatic polypeptide, insulin, glucagon, GIP, and somatostatin from endocrine-paracrine cells. Bombesin-specific antibodies inhibit the gastrin-releasing effect of electrical field stimulation on isolated antrum suggesting that GRP mediates vagal stimulation of gastrin release. The physiological role of GRP in releasing the other endocrine agents has not been sufficiently established. GRP stimulates gastric acid and exocrine pancreatic secretion. Its ability to stimulate acid secretion results from its effect on gastrin release. GRP's physiological role in exocrine pancreatic secretion has not been established. GRP and bombesin have a variety of effects on gut motility but these may be mediated by release of other gut hormones or neurotransmitters.

Endocrine Tumors

Clinical syndromes resulting from tumors that synthesize and secrete excess peptide transmitters have been described for some of the transmitters described in the previous sections. Well-characterized clinical syndromes occur with tumors that synthesize and secrete either gastrin, VIP, or somatostatin. These tumors are most commonly located in the pancreas. Patients with *gastrin-secreting* tumors (gastrinoma or Zollinger-Ellison syndrome) develop massive gastric secretion due the effect of the gastrin on the parietal cell to cause HCl secretion and the trophic effects of gastrin on the acid-secreting mucosa of the stomach. These patients develop extensive ulcerations in their stomachs and proximal intestine because of the acid secretion. They also develop diarrhea because of the large volume of gastric secretion and the fact that the resulting low pH in the duodenum has deleterious effects on pancreatic enzyme—especially lipase. The

Table 6.2.
Amino Acid Sequences of VIP, GRP, and Bombesin[a]

	VIP	GRP	Bombesin
NH$_2$-terminus			
1	HIS	VAL	
2	SER	PRO	
3	ASP	LEU	
4	ALA	PRO	
5	VAL	ALA	
6	PHE	GLY	
7	THR	GLY	
8	ASP	GLY	
9	ASN	THR	
10	TYR	VAL	
11	THR	LEU	
12	ARG	THR	
13	LEU	LYS	
14	ARG	MET	pGLU
15	LYS	TYR	GLN
16	GLN	PRO	ARG
17	MET	ARG	LEU
18	ALA	GLY	GLY
19	VAL	ASN	ASN
20	LYS	HIS	GLN
21	LYS	TRP	TRP
22	TYR	ALA	ALA
23	LEU	VAL	VAL
24	ASN	GLY	GLY
25	SER	HIS	HIS
26	ILE	LEU	LEU
27	LEU	MET-NH$_2$	MET-NH$_2$
28	ASN-NH$_2$		
COOH-terminus			

[a]The COOH-termini of these peptides are amidated.

pathophysiological effects of the tumor can be eliminated by complete removal of the tumor, by surgical removal of the stomach, by pharmacological inhibition of the acid secretion caused by the gastrin, or by pharmocological inhibition of gastrin release from the tumor by somatostatin analogues.

The major effect of *VIP-secreting tumors* (VIPoma or pancreatic cholera) is massive diarrhea. This results because VIP is a potent stimulant of electrolyte and water secretion from the intestine. The increased circulating VIP also causes inhibition of gastrointestinal motility. The effects of the tumor can be treated by removal of the tumor, pharmacological inhibition of VIP release from the tumor with somatostatin analogues, or by various pharmacological agents that either inhibit intestinal secretion or increase absorption.

Somatostatin-secreting tumors (somatostatinoma) result in a constellation of clinical features that include diabetes mellitus, diarrhea, achlorhydria (absent HCl secretion from the stomach), steatorrhea (fat in the stools), and gallstones. The diabetes mellitus results from the inhibitory effects of somatostatin on insulin release. The achlorhydria and steatorrhea result from the inhibitory actions of somatostatin on release of the hormones gastrin, secretin, and CCK; and on somatostatin-induced inhibition of gastric HCl and exocrine pancreatic secretion. The gallstones result from the inhibitory action of somatostatin on gallbladder contraction. Effective treatment requires removal of the tumor. Exogenous replacement of insulin and pancreatic enzymes can ameliorate some of the features of the syndrome.

CHAPTER 43

Gastrointestinal Motility

NATURE OF GASTROINTESTINAL MOTILITY

Motions of the gut walls govern flow. Controlling systems regulate these wall motions so that flow patterns vary with circumstances. Thus, gastrointestinal motility really encompasses three causally related matters, the *control system, wall motions,* and *fluid flow* (Fig. 6.4).

The controls constitute (a) the influence of the autonomic nerves, (b) the influence of some hormones, and (c) the features of the smooth muscle itself. The wall motions include contractions both of the circular and of the longitudinal muscle layers. The flows produced include both net flows along the axis of the tubular conduit and local flows that mix the contents of the gut.

Control systems, wall motions, and fluid flows are not easily studied simultaneously. In clinical practice, one usually studies flow and infers the nature of the wall motions. To a lesser extent, one can examine the wall motions and infer the patterns of flow they produce. The control systems are much less accessible to direct study in the clinic. Most of what is known about the control systems comes from the study of animals where the gut is more accessible than it is in man.

One should try to maintain a unitary view of gastrointestinal physiology. That is, motility should not be considered as separate from the other gastrointestinal functions: absorption, secretion, and digestion. Also, the different organs cannot be considered independent of one another in their motor operations. The neural and hormonal control systems that regulate gastrointestinal functions, including motility, are linked among gastrointestinal organs to make such coordinated operation possible.

METHODS TO STUDY GASTROINTESTINAL MOTILITY

Many different methods, both in the clinic and in the laboratory, are used to examine gastrointestinal motility. All of them have limitations, but these restrictions are overcome by the fact that the methods are complementary.

Flow

The direct visualization of fluid flow is possible through the use of radiopaque fluids that can be observed radiographically, of solutions containing radioactive isotopes that can be detected by scintigraphic cameras, and of markers (like dyes or nonabsorbed polymers) that can be injected at a known site and detected when they reach a determined point, like the tip of an aspirating catheter. Such methods have the advantage that they can be used in humans.

The common radiopaque fluid used in *radiography,* a barium sulfate suspension, is much heavier than water, and so it may not flow exactly as does the normal luminal content, which has about the specific gravity of water. Barium sulfate clings to the gut mucosa well enough to allow some observation of wall motions, but this is not a satisfactory method for this purpose because safety considerations in man limit the time of observation, and because the potential for quantitation is limited.

Scintigraphic methods use aqueous solutions of isotopes so that the flow observed is that of water. Safety considerations do not allow studies of long duration in man. The fact that the solution does not cling to the wall and the poor resolution of the γ-camera prevent the observation of wall motion.

Timing the passage of dyes or other markers from one determined point to another gives a measure of *transit time* from one point to another. Since the gut is elastic or variable in its length, distance is not a constant along the gut. It can and does show variations in length, or "sleeving," and this makes transit time an imprecise determination. The degree of dispersion of a bolus during its passage between the points can be used to infer something of the amount of dilution by secretion that occurred during passage, but this cannot easily be made quantitative. Such methods are of no use in examining wall motions.

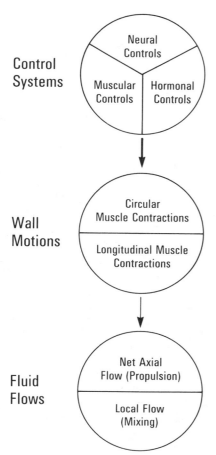

Figure 6.4. The three elements of gastrointestinal motility and their component parts.

Wall Motions

Wall motions can be observed directly for short times at *surgical operations*, but the method has severe limitations. The gut so observed is not in a normal state, being affected by preoperative fasting, by anesthetics and other drugs, by reflexes induced by the opening of the peritoneum, by abnormal temperature, and by the effects of manipulation of the gut. Also, the method does not allow easy quantitation or objective analysis.

Wall motions can be observed indirectly by *manometric technology*, in which pressures are recorded from distended balloons or open-ended catheters placed in the gut lumen. Balloon manometry, or *kymography*, finds little contemporary use. The balloon covers, usually, several centimeters of the length of the gut, so that it gives poor resolution. The compliance of balloons makes it impossible to use them to measure pressures accurately. Also, balloons, being distended, tend to stimulate contractions and to be carried along the gut by them. The measurement of pressures by *open-ended catheter manometry* is much more useful. In this technique,

a bundle of several catheters is cemented together. Each has its open tip at a location that is fixed in relation to those of the others. The bundle is introduced into the organ (mainly the esophagus and small intestine), and tethered at the mouth so that it cannot move. All the catheters are perfused very slowly with water from a low-compliance pump, and the pressure in each catheter is recorded separately. This makes it possible to record pressures with some accuracy. Ring-contractions of the circular muscle layer transiently occlude the open tips of the perfused catheters so that the pressure in the system rises. Since the locations of the catheter tips are known, the method displays the spatial localization of ring-contractions of the circular muscle layer, but "sleeving" of the gut limits the accuracy of this. The amplitude of the pressure peak reflects the force of the contraction that produces it. The time-course of the pressure peak reflects that of the contraction. The method is able to detect only those contractions of the circular muscle layer that completely occlude the gut lumen, for only these occlude the tips of the small catheters. Most contractions do so in uniform tubular viscera like the esophagus (where catheter manometry is especially useful clinically) and the small intestine. Manometry finds little use in capacious organs, like the stomach and colon, where many contractions may occur that do not occlude the gut lumen completely.

Electromyography is useful to detect wall movements in animals (and to a limited extent in man), because most gut muscle generates a *spike burst*, a burst of rapid electrical transients, when it contracts. These signals can be detected through the mucosa with luminal electrodes, as well as with electrodes that can be implanted in the gut wall at surgical operation. Electromyography is also used to record the pacesetting slow electrical transients, called *slow waves*, of the gastric antrum, intestine, and colon. These signals recur constantly while the circular muscle layer responds to only a fraction of them by generating a spike burst and contracting (Fig. 6.5). Thus, electromyography can be used either to examine the localization of contractions, as the spike burst, or the characteristics of the pacesetting control mechanism, the slow waves.

Control Systems

For the most part, neither the invasive nor the noninvasive methods that are used to study flow and wall motions allow the extended exploration of the control systems. The control systems constitute both neural and hormonal mechanisms that operate simultaneously in the whole animal. The isolation of one or another of these mechanisms for study

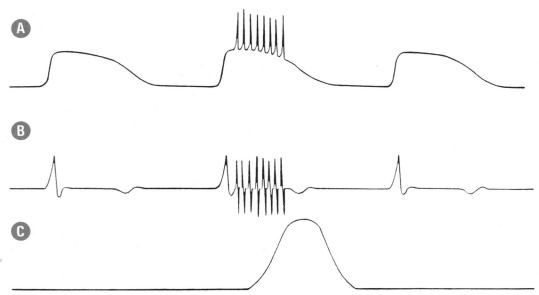

Figure 6.5. Diagram of the simultaneous recording of electrical activity and tension from a gut muscle that exhibits rhythmic activity. The trace at the top (*A*) shows three sequential *slow waves,* the second of which contains a *spike burst,* as recorded from a single muscle cell with an intracellular microelectrode. The trace in the middle (*B*) shows the same features as recorded from an extracellular electrode in contact with hundreds or thousands of cells. The signal, as diagrammed, is the second derivative of that in the top trace as occurs with the usual recording methods. The bottom trace (*C*) of muscle tension indicates that contraction occurs with the spike burst. (From Christensen, 1971.)

requires the study of *isolated tissues* or cells by laboratory methods. In such methods, isolated segments of gut, or strips of smooth muscle, are mounted in organ or tissue baths and attached to force transducers, linear displacement transducers, or pressure transducers to allow the monitoring of contractions and relaxations. Extracellular electrodes applied to such muscle segments or to muscle strips can detect the pacesetting slow waves and the spike bursts that occur with rhythmic contractions (Fig. 6.5). Intracellular microelectrodes in such preparations can be used for the detailed study of electrical events in the muscle.

The effects of the several different kinds of nerves on muscle function can be examined by the application of agents known to stimulate, antagonize, or mimic the actions of the various sorts of nerves. Also, the intrinsic nerves of the gut wall can be stimulated in isolated muscle strips directly by *electrical field stimulation,* to stimulate the nerves selectively.

The direct study of the operation of the intramural nerves themselves is also possible in isolated preparations of the gut wall. Separation of the longitudinal and circular muscle layers in vitro exposes the myenteric plexus whose nerve cells then become accessible to extracellular and intracellular electrodes.

Tetrodotoxin, an extract of the puffer fish, selectively and reversibly blocks action potentials in most nerves in isolated preparations of the gut wall, leaving muscle function intact. This effect allows

the elimination of neural influences, so that the properties of the muscle, and the influences of hormones directly on the muscle, can be examined in vitro.

GENERAL ANATOMY OF MUSCLES AND NERVES OF THE GUT

Anatomy of Gut Muscles

Gut muscle lies in three layers throughout most of the gut, the muscularis mucosae and the inner (circular) and outer (longitudinal) layers of the muscularis propria. The gastric fundus contains a fourth layer, the oblique muscle layer, internal to the circular muscle layer.

Differences exist among organs and regions in the thickness of these layers. For example, the muscularis mucosae is much thicker in the esophagus than it is in the other organs. Also, thickenings in the circular muscle layer form sphincters at the esophagogastric and gastroduodenal junctions and at the caudal end of the colon.

Smooth muscle constitutes most of the muscle involved in gastrointestinal motility, but not quite all of it. *Striated muscle* makes up the walls of the pharynx and the proximal one-third of the esophagus in primates and marsupials. It extends much farther down the esophagus in carnivores and rodents. Also, striated muscle makes up the external and sphincter and pelvic floor muscles, which function in coordination with colonic motor events in defecation. This striated muscle is somatic muscle, and no structural features have been described that distinguish it from that of all other somatic muscles.

Gut *smooth muscle* is generally very similar in structure to that of blood vessels (Fig. 6.6). The fusiform smooth muscle cells are packed together in bundles separated by collagen. Gap junctions connect adjacent smooth muscle cells, but the number of gap junctions varies considerably among organs and layers. Numerous caveolae indent the muscle cell membrane, and a sarcoplasmic reticulum ramifies among these caveolae. The elliptical nucleus is surrounded by a condensation of the sarcoplasmic reticulum and by mitochondria that extend from the perinuclear zone into the cytoplasm. The Golgi apparatus and the rough endoplasmic reticulum are sparse. The cytoplasm contains the thick and thin filaments of the contractile proteins, the dense bands and dense bodies into which they insert, and the intermediate filaments of the cytoskeleton.

These structures serve specific functions. The gap junctions establish an electrical linkage between cell membranes so that all the muscle cells over great distances work electrically and mechanically in synchrony. This makes the smooth muscle a functional syncytium. The caveolae probably serve to carry cell membrane events into close proximity to the sarcoplasmic reticulum: the system is thought to

be like the T-tubule system of striated muscle. The dense bodies, composed of α-actinin, are homologous to the Z lines of striated muscle. The thin filaments are actin filaments and the thick filaments are myosin.

For reviews of gut muscle anatomy, see Gabella, 1984, 1987; Murphy, 1979; and Somlyo, 1980.

Anatomy of the Intramural Nerves in the Gut

The cell bodies of the intrinsic nerves of the gut lie in ganglia that constitute the two *ganglionated plexuses* of the gut, one in the submucosa (the submucous plexus) and the other in the intermuscular plane between the two layers of the muscularis propria (the myenteric plexus). There are also other plexuses of nerves, *aganglionic plexuses*, that lie in other planes in the gut wall. These aganglionic plexuses are ramifications of nerve processes arising from the ganglion cells of the ganglionated plexuses. They seem to represent planes where a concentration of nerve processes is required to serve particular functions. The location of these aganglionic neural plexuses in the wall differs among the different organs (Fig. 6.7). An aganglionic plexus lies between the circular and oblique muscle layers in the gastric fundus. Another lies deep within the circular muscle layer in the small intestine. Still another lies on the submucosal surface of the cir-

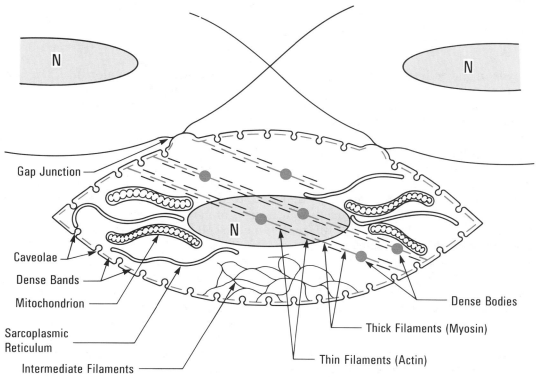

Figure 6.6. A diagram of a gut smooth muscle cell to show the general structural features. N represents the nucleus. Two cells are shown

only in partial outline. The obliquity of the contractile filaments is controversial and is exaggerated in this sketch.

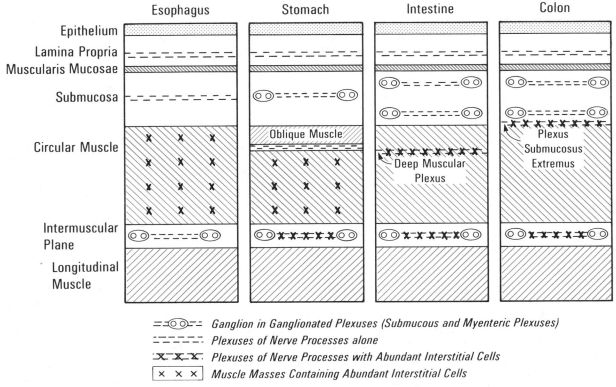

=⊙⊙=- *Ganglion in Ganglionated Plexuses (Submucous and Myenteric Plexuses)*

------- *Plexuses of Nerve Processes alone*

X̄-X̄-X̄ *Plexuses of Nerve Processes with Abundant Interstitial Cells*

| × × × | *Muscle Masses Containing Abundant Interstitial Cells*

Figure 6.7. A diagram to contrast the parts of the gut in respect to the intramural nerve plexuses. The *esophagus* lacks ganglia in the submucous plexus, lacks many interstitial cells of Cajal in both major plexuses, and contains many interstitial cells in the circular muscle layer. The *stomach* has a ganglionated submucous plexus, an aganglionic plexus between oblique and circular layers (in the fundus), contains many interstitial cells in the circular muscle layer (in the antrum), and has a ganglionic intermuscular (myenteric) plexus with many interstitial cells (in the antrum). The *intestine* has a bilayered ganglionic submucous plexus, an aganglionic plexus with many interstitial cells deep in the circular muscle layer, and a ganglionic intermuscular (myenteric) plexus with many interstitial cells. The *colon* has a bilayered ganglionic submucous plexus, an aganglionic plexus with many interstitial cells on the submucosal surface of the circular muscle layer, and a ganglionic intermuscular (myenteric) plexus with many interstitial cells.

cular muscle layer in the colon. The lamina propria contains an aganglionic plexus in all organs. Terminal axons extend from the ganglionic and aganglionic plexuses to supply effector structures in adjacent planes.

A *ganglion* of the submucous and myenteric plexuses is a compact body covered with a collagen sheath. It contains nerve cell bodies and processes embedded in a dense stroma of neurites and Schwann cells. The *ganglion cells* are variable in form, the various forms probably having different functions. Major interganglionic fascicles of nerve processes connect adjacent ganglia to form an irregular pattern of polygons. In both submucous and myenteric plexuses, the density of distribution of ganglia varies among organs, and so also does the distribution density of ganglion cells (Fig. 6.8).

In those organs where the *myenteric plexus* is most dense, the plexus can be subdivided into three components, all in the same intermuscular plane.

The ganglia and the major interganglionic fascicles constitute the *primary plexus*. Small bundles of nerve processes that branch from the interganglionic fascicles form a second-order network called the *secondary plexus*. Still smaller branches from the secondary plexus form the *tertiary plexus*. These three elements form a continuum.

In many regions, both the ganglionic (submucous and myenteric) plexuses and the aganglionic plexuses of nerves contain other distinctive cells besides nerve cells and their processes, the *interstitial cells of Cajal* (or interstitial cells) (Fig. 6.7). These are very small pleomorphic cells with a single ovoid nucleus surrounded by a sparse cytoplasm. The cytoplasm gives rise to broad branching processes that intersect with those of adjacent interstitial cells. Neurites run among these cells. Interstitial cells are especially dense in the myenteric plexus of the gastric antrum, small intestine, and colon, but they are sparse in the myenteric plexus of the esophagus and gastric fundus. They are abundant

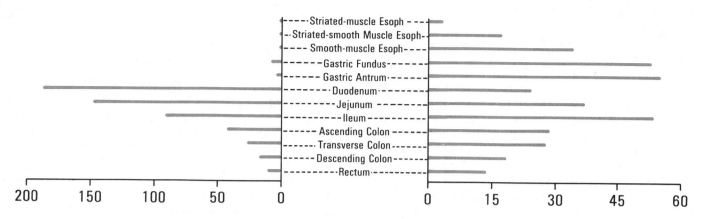

Ganglia Per Square Centimeter

Figure 6.8. Ganglionic densities in the submucous and myenteric plexuses all along the gut. The data on the submucous plexus come from a cat. Those on the myenteric plexus come from an opossum. Note the much greater density of the submucous plexus in the intestine than in other organs. Myenteric plexus density declines toward the two ends of the gut. Note the differences in scales. The submucous plexus ganglia are much smaller than those of the myenteric plexus, and so this graph does not demonstrate the differences in *neuronal* density. (Data compiled from Christensen et al., 1983; and Christensen and Rick, 1985.)

in the aganglionic plexuses in the circular muscle layer of the small intestine and on the submucosal surface of the circular muscle layer in the colon, but they are sparse in the aganglionic plexuses of the lamina propria and in the aganglionic plexus that lies between the circular and oblique layers in the stomach.

For reviews of the anatomy of gut intramural nerves, see Gabella, 1979, 1987; and Thuneberg, 1982.

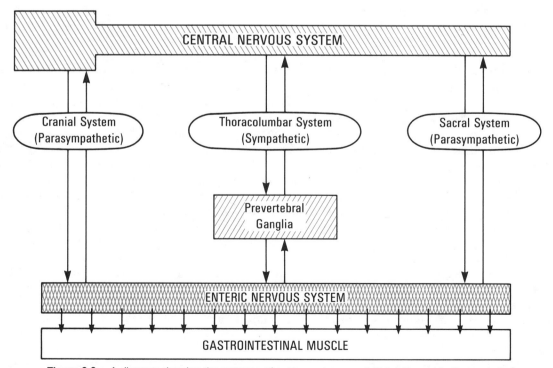

Figure 6.9. A diagram showing the systems of extramural nerves that link the gut to the central nervous system.

Anatomy of the Extrinsic Nerves to the Gut

The intramural nerves of gut are not entirely autonomous. They are linked to the central nervous system by pathways of two sorts, the craniosacral (or parasympathetic) tracts and the thoracolumbar (or sympathetic) tracts (Fig. 6.9).

The *cranial part* of the *craniosacral supply* to the gut constitutes the IXth cranial nerve, the glossopharyngeus, and Xth cranial nerve, the vagus. The domain of the IXth nerve ends at the upper esophageal sphincter and that of the Xth nerve covers the stomach, small intestine and, probably, proximal colon. The nerve fibers of the vagus arise from the nucleus ambiguus and the dorsal motor nucleus in the brainstem. The vagi divide in the lower mediastinum to form a loose network about the distal esophagus, the esophageal plexus, whose bundles come together again above the diaphragm to form the vagal trunks that enter the abdomen. The vagal trunks in the abdomen give branches especially to the stomach, liver, and pancreas. Branches of the vagal trunks cannot be readily traced beyond the duodenum, but the influence of vagal stimulation can be seen as far as the level of the ascending colon. The vagi are mixed nerves, about 90% of their fibers being afferents. The efferent fibers of the vagi constitute three classes, somatic motor nerves to the striated muscle of the proximal esophagus, postganglionic sympathetic fibers that enter the vagus from the cervical sympathetic ganglia, and preganglionic parasympathetic fibers that supply the ganglion cells of the intramural plexus.

The *sacral part* of the *craniosacral* supply constitutes the outflows from sacral segments 2–4, the pelvic nerves, which supply the rectum and distal colon.

The *thoracolumbar supply* to the gut constitutes the sympathetic innervation. Pathways to the gut mainly traverse the splanchnic nerves to the prevertebral ganglia, from which paravascular nerve bundles are distributed to the gut.

In the *central nervous system,* various structures influence the function of the autonomic nerves that supply the gut. Motor responses in the gut can be elicited by electrical stimulation of the cortex, thalamus, amygdaloid complex nuclei, hypothalamus, pons, cerebellum, and other regions. Certain integrated functions like vomiting, swallowing, and defecation clearly require "centers" in the central nervous system that can coordinate complex motor behavior of the gut and of the somatic musculature. These centers are not completely defined anatomically. They provide for the coordinated activation of many identified structures in the central nervous system.

For a review of the anatomy of the extrinsic nerves to the gut, see Gabella, 1979.

NEUROGENIC CONTROL OF GUT SMOOTH MUSCLE FUNCTION

Extrinsic Nerves in the Control of Motility

The nature and degree of central nervous participation in the regulation of motility remains to be fully explored. The functions seem to be largely subcortical. The ability of people consciously to change the functions of some autonomically innervated structures has been claimed, not without controversy, but this ability has not been demonstrated in respect to gastrointestinal motility.

Both sensory and motor fibers make up the major autonomic nerve trunks to the gut. In the vagus, the sensory fibers are axons from bipolar cells in the nodose ganglion. These fibers serve reflex pathways through which mechanical and chemical stimulation of neuroreceptors in the gut modify the operation of the gut. The integration of sensory and motor functions occurs, in part, in the vagal nuclei of the brainstem and in related structures. In the thoracolumbar supply, much of the sensory-motor integration occurs in the prevertebral ganglia, the celiac, superior mesenteric, and inferior mesenteric ganglia. They are linked together so that sympathetic reflexes can integrate activities of the widely separated parts of the gut.

For a review of the extrinsic nerves in the control of gut motility, see Gabella, 1979; Thuneberg, 1982; and Davison, 1983.

Intrinsic Nerves in the Control of Motility

A variety of reflex motor responses can be demonstrated in the extrinsically denervated gut. The classical one is the reflex response of the intestine to localized distension. The stimulus elicits contraction above and inhibition below the level of stimulation. Thus, the intramural plexuses also must contain sensory and motor nerve cells that are linked together directly, or linked by interposed internuncial neurons.

Direct evidence to support this idea comes from *electrophysiological studies* of myenteric plexus neurons in the small intestine exposed by the separation of the muscle layers. Extracellular electrodes applied to such cells give records of electrical events that exhibit different patterns of spontaneous and evoked activity (Fig. 6.10). Some cells discharge spontaneously with bursts of action potentials. Others discharge continuously. Still others are silent, but discharge in response to mechanical deformation of the gut (Fig. 6.11). One type of nerve cell discharges in a burst at the onset (and some-

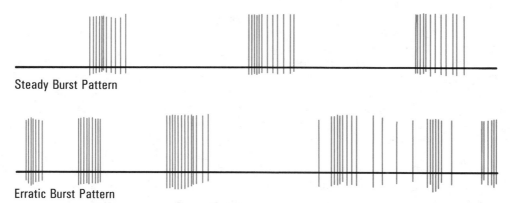

Steady Burst Pattern

Erratic Burst Pattern

Figure 6.10. Two patterns of spontaneous discharge of action potentials from enteric neurons, as recorded by extracellular electrodes. Some cells continuously discharge in periodic bursts *(top trace)*. Others discharge in bursts from time to time, reverting at times to a continuous discharge *(bottom trace)*. Others discharge continuously with no pattern (not shown). (From Davison, 1983.)

times at the offset) of a mechanical stimulus. Others discharge continuously at the onset and continue to do so until the offset of the stimulus. Still others begin to discharge with a long delay after the onset of the stimulus, and they continue to do so for a long period after the offset.

These different patterns of discharge imply different functions. The neurons with a spontaneous and constant discharge of bursts of action potentials are thought to be pacemaking cells. Those that discharge at the onset and offset of a mechanical stimulus are thought to be sensory cells that are linked to a rapidly adapting mechanoreceptors. Those that discharge continuously during a mechanical stimulus are probably sensory cells that are linked to slowly adapting mechanoreceptors. Those whose

discharge continues long after the end of the mechanical stimulus may be internuncial cells, linking sensory and motor neurons.

For reviews on the intrinsic nerves in the control of gut motility, see Davison, 1983; Burnstock, 1972; Costa et al., 1987; and Wood, 1987.

Transmission by Nerves

Transmission by nerves to other nerves, to smooth muscle or to other effector structures is effected mainly by the release of neurohormones, or *neurotransmitters*, from axon terminals. Electrotonic transmission is also a possible mechanism, but it has not been demonstrated in relation to gastrointestinal motility.

Neurotransmitters, synthesized in the nerve, are

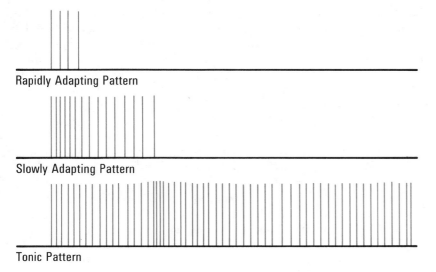

Rapidly Adapting Pattern

Slowly Adapting Pattern

Tonic Pattern

Stimulus

Figure 6.11. Three patterns of discharge of action potentials by enteric neurons in response to mechanical deformation of the intestinal wall. The period of mechanical stimulation appears on the *bottom trace*. Some cells discharge only at the onset of the stimulus, others discharge throughout the stimulus, and still others discharge long after the end of the stimulus. (From Davison, 1983.)

stored in vesicles that are contained in the varicosities of axons. The neurotransmitters escape upon discharge of the nerve, diffuse over a space of up to about 200 mμ, and reach the receptive sites on the membrane of the postsynaptic nerve or muscle. Enzymatic destruction of the neurotransmitter, diffusion of the neurotransmitter, or its reuptake into the axons can terminate the action of the neurotransmitter at its receptive sites on the postsynaptic membrane.

Many substances are candidate neurotransmitters in relation to gastrointestinal motility, but few can be considered to be firmly established as such. Established neurotransmitters are substances that are localized in the enteric neurons in high concentrations, and that are potent in affecting enteric nerve or muscle function. The substance must be shown to be released upon nerve stimulation. Also evidence of specific mechanisms for its synthesis, degradation, or reuptake into nerves must be demonstrated. In relation to gastrointestinal motility, the best established neurotransmitters are acetylcholine, norepinephrine, dopamine, and 5-hydroxytryptamine. The many candidate neurotransmitters include more than 20 nerve-related peptides, the chief one being vasoactive intestinal polypeptide.

Nerves can both excite and inhibit gut muscle. For most excitatory nerves, evidence indicates that acetylcholine is the responsible excitatory neurotransmitter, but in some cases noncholinergic neural excitation of muscle also seems to exist.

Neurogenic inhibition is a more obscure matter. Norepinephrine may either excite or inhibit gut smooth muscle, depending upon the location of that muscle and the relative proportions of α-receptors (which are mostly excitatory) and β-receptors (which are inhibitory), but few adrenergic nerves are found within the muscle layers of the gut. Furthermore, the effects of the major inhibitory nerves to gut muscle are not generally opposed by selective antagonists of adrenergic neurotransmission. The major inhibitory innervation of gut muscle, therefore, seems to act by the release of a neurotransmitter that is still unidentified. Vasoactive intestinal polypeptide and the adenine nucleotides are the most likely of the candidate inhibitory neurotransmitters, but they are not yet firmly established as such. At present, these inhibitory nerves are best called the nonadrenergic noncholinergic inhibitory nerves (often abbreviated NANC nerves). It may well be that there are several nonadrenergic inhibitory neurotransmitters.

HORMONAL CONTROL OF GUT SMOOTH MUSCLE FUNCTION

Until about 10 years ago, the nerves were considered to be the major control system for gastrointestinal motility, but one hormonal control was accepted long before that. The peptide hormone, cholecystokinin, was accepted as a hormone that induced gallbladder contraction. Also, the fact that various extracts of the gut wall could induce excitation or inhibition of gut muscle kept alive the idea that other substances present in the gut wall could affect motility as hormones. The advent of radioimmunoassay and other techniques in peptide chemistry have revealed a long and growing list of peptides present in the gut wall that are potent in affecting gut smooth muscle function, and many nonpeptides have been identified as well. Some of these substances are localized in enteric nerves, and so they are candidate neurotransmitters, but others are located in epithelial endocrine-like cells. These substances are collectively called the enteric hormones.

The physiological function of most of these hormones remains obscure. There is little evidence yet that they are major factors in the control of normal gastrointestinal motility (aside from that of the gallbladder), yet some evidence suggests that they may be. In many cases, it may be that they act only in pathological situations, in motor dysfunction.

Cholecystokinin, the only enteric peptide that is accepted as an important enteric hormone in motility, is secreted by the intestinal mucosa in response to a meal. It circulates systematically to induce the contraction of the gallbladder, the relaxation of the sphincter of Oddi, and the secretion of enzymes by the pancreas. For many of the other peptides, there is evidence for their presence in the gut wall, their release from the gut into the systemic circulation and their potency in affecting gut muscle. Their physiological significance, however, remains obscure.

Nonenteric hormones might also be expected to affect motility, and some do. For example, glucagon is a potent inhibitor of intestinal motility, but this effect does not seem to be very important physiologically. Also progesterone inhibits gastrointestinal motility. This effect may well be physiologically significant, for evidence indicates that it is responsible for the weakening of lower esophageal sphincter closure force and the slowed gastric emptying that occur in pregnancy.

The gut wall contains other non-neural myoactive substances in abundance, not peptides, that could affect motility. These include histamine, adenosine, the prostaglandins, and the leukotrienes. Such substances could conceivably be involved in motility changes in relation to inflammatory states, but as yet they have no established place in the control of normal motor function.

For reviews on the hormonal control of gut smooth muscle function, see Johnson, 1977; Dockray, 1987; Walsh, 1987; and Yamada, 1987.

MYOGENIC CONTROL OF GUT SMOOTH MUSCLE FUNCTION

Muscle Membrane Events—The Source of Rhythmicity

The resting membrane potential in gut smooth muscle has the same origin and magnitude as that in other smooth muscle. There seem to be some differences in its magnitude from one region or muscle layer to another, an observation consistent with the differences in the motor functions among organs and muscle layers.

The instability of resting membrane potential is a further indication of the inhomogeneity of gut smooth muscle. Regular, periodic, and relatively slow partial depolarizations of the membranes characterize the muscles of the gastric antrum, the small intestine, the colon, and the part of the common bile duct that lies within the duodenal wall. These rhythmic depolarizations are not present in the smooth muscle of the esophagus, gastric fundus or gallbladder.

These slow partial depolarizations recur constantly in most of those regions where they occur, regardless of the motor activity of the region. When the region contracts rhythmically, the rhythmic contractions are phase-locked to the rhythmic depolarizations. For this reason, these constant rhythmic depolarizations have been called "pacemaker potentials" or "electrical control activity." Several other terms have been used for them as well, including "the basic electrical rhythm" and "electrical slow waves." *Electrical slow waves* and *slow waves* are the terms most widely used.

The slow waves, as recorded from a single muscle cell with an intracellular microelectrode, have a fairly consistent form in all regions. From a resting membrane potential of approximately -40 to -70 mV, a rapid depolarization of approximately 10–20 mV occurs, followed by a plateau depolarization that is slightly less in magnitude, lasting several seconds, with a slower repolarization of the membrane to its resting value (Fig. 6.5). Electrotonic linkages between adjacent muscle cells in a tissue, presumably through gap junctions between muscle cells, assure that all cells in the immediate area generate slow waves in synchrony. For this reason, slow waves can also be detected as discrete events by extracellular electrodes that simultaneously contact hundreds of cells (Fig. 6.5).

When the muscle responds to a slow wave with a phasic contraction, the slow wave acquires a different configuration. In the small intestine, the colon, and the gastric antrum, one or more rapid depolarizations of the membrane appear on the plateau of the slow wave, spike potentials that approach the isopotential line. Since these rapid electrical transients occur with a contraction that is a response to the slow wave, they are called "action potentials," the "electrical response activity," "spike potentials," or "the spike burst." *Spike burst* is the most widely used term. In some parts of the gastric antrum, the spikes of the spike burst are fused so that the spike burst appears as an exaggeration of the plateau potential. In the colon the rhythmic contraction may occur without the generation of a spike burst. Thus, the spike burst cannot be taken as a required event for contraction in all organs, but its occurrence always signals a contraction.

Spike bursts and rhythmic contractions are always phase-locked to slow waves when these events are examined simultaneously at a single point. When slow waves are recorded simultaneously from several closely spaced points in the organ, the slow waves at these points are not synchronous but appear in a sequence such that they seem to spread away from one point in a constant pattern and with a constant velocity (Fig. 6.12). The phase-lock that exists between slow waves and contractions imposes this pattern and this velocity of the slow waves upon the contractions themselves, so that the slow waves establish the features of peristalsis—frequency, direction, and velocity of migration. The slow waves thus constitute a clock, fixing discrete points in time and space to which the occurrence of contraction is restricted.

With each slow wave cycle the muscle has a choice either to contract or not to contract. The likelihood of contraction is raised by excitatory elements like the excitatory nerves (perhaps hormones), and depressed by inhibitory nerves and hormones. Neural and hormonal factors may also influence the frequency, velocity, and patterns of spread of the slow waves themselves. This matter remains incompletely explored.

The source of the slow waves was long thought to be the muscle itself, and such possibilities as rhythmic oscillations in the function of the sodium pump in the muscle cell membrane, for example, were sought to explain their occurrence. In fact, the cell

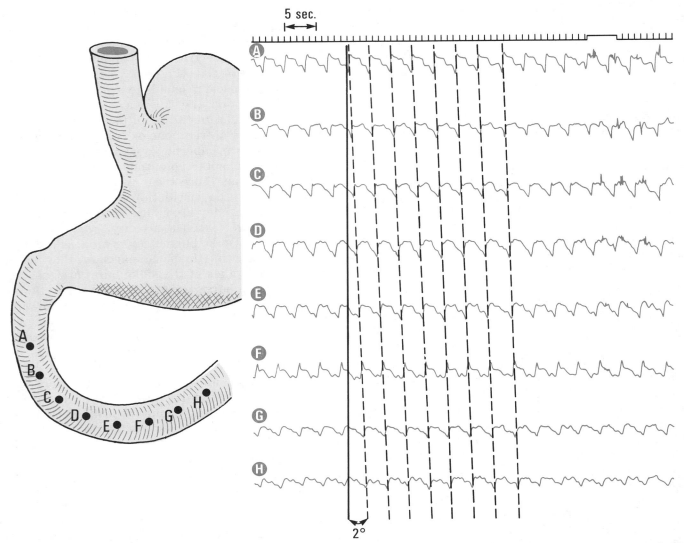

Figure 6.12. A record of slow waves taken from the duodenum of the cat. Eight electrodes (*A-H*) are spaced uniformly along the duodenum. The tracings show the uniform slow signals that are slow waves. The *dashed lines* join sequential slow waves. These lines deviate from the isotemporal line by 2°, which is a function of the slow wave velocity and of the paper speed. The slow waves spread at a uniform velocity from the first electrode site to the last. Slow waves to the right in channels *A-E* show spike bursts. (From Christensen, 1971.)

processes underlying slow waves remain obscure, and it was never established that they arise in the muscle itself. Recent evidence supports the view that slow waves are responses of the muscle to signals arising in relation to plexuses of axons containing interstitial cells of Cajal (Thuneberg, 1982; Szurszewski, 1987). The interstitial cells of Cajal are abundant in the plexuses in those regions where slow waves occur, the interstitial cells form gap junctions with the smooth muscle, and they are always closely related to axons. Also, slow waves seem to have the greatest magnitude at those muscle surfaces that are adjacent to the planes in the gut wall where the nerve plexuses containing interstitial cells of Cajal lie. The evidence remains cir-

cumstantial, however, and the hypothesis, if correct, does not answer the question of the cellular events that give rise to slow waves, since the biology of the interstitial cells of Cajal is not known.

For reviews on muscle membrane events, see Szurszewski, 1987; Christensen et al., 1964, 1966; Bartoff, 1976; and Prosser and Mangels, 1982.

Source of Tone in Tonic Muscles of the Gut

A constant and stable low level of contraction, called *tone*, characterizes virtually all of the gut musculature. The principal exception is the circular layer of smooth muscle of the esophageal body which exhibits no tone in vitro. Tone is especially prominent in the gastrointestinal sphincters, where it accomplishes the sustained closure of the gut lumen that is the principal function of the sphincters.

It was once thought that tone in gastrointestinal smooth muscle is neurogenic, the consequence of the maintained release of an excitatory neurotransmitter, like acetylcholine, from intramural nerves. But the exposure of sphincteric muscle to the neurotoxin, tetrodotoxin, or to other more selective antagonists of neuromuscular transmission, causes little or no reduction in tone. This indicates that much of the tone of such tonic muscles originates in some specialization of the muscle. Some evidence suggests that the muscle of the sphincters differs morphologically from that of other regions in the number and distribution of mitochondria, and that intracellular calcium concentrations may be greater in sphincters than in nonsphincteric muscles. These specializations in tonic muscle may be such as to make the muscle more sensitive than nontonic muscles to the effects of circulating excitatory hormones.

Mechanism of Contraction in Gut Smooth Muscle

Contraction in gut smooth muscle seems to make use of the same general processes as contraction in other kinds of muscle. The sliding of thin (actin) and thick (myosin) filaments exerts tension on the dense bodies (composed of α-actinin) to shorten the cell. This sliding involves cross-bridge cycling that is driven by ATP hydrolysis. The slowness in contraction and the economy of energy utilization of smooth muscle, as compared to striated muscle, is attributed to the slowness of detachment of ADP from the actomyosin complex. As a result, actin and myosin remain attached longer, in the force-generating state, so that ATP hydrolysis per unit time is less.

The process of contraction is regulated by the intracellular calcium concentration. Regulation depends upon a calcium-calmodulin-myosin light chain complex. Thus, the regulation of the smooth muscle cross-bridge cycle is "myosin-linked regulation."

The cytosolic concentration of calcium can rise both through the entry of extracellular calcium and from the release of calcium from intracellular stores. The main intracellular storage site is the sarcoplasmic reticulum. The effect of membrane depolarization to raise intracellular calcium concentration is called *electromechanical coupling*. Membrane depolarization can raise intracellular calcium concentration either by increasing the diffusion of calcium through voltage-gated calcium channels, or by inducing calcium release from the sarcoplasmic

reticulum in response to the depolarization. Also, calcium is released from the sarcoplasmic reticulum in direct response to the raised calcium concentration that occurs by the diffusion of calcium into the cell (calcium-induced calcium release). The relative magnitudes of the contributions of these mechanisms to the induction of contraction probably vary among different kinds of gut smooth muscle, and among different stimuli.

Pharmacomechanical coupling refers to the means by which neurohormonal agents can raise intracellular calcium concentration without depolarizing the membrane. Neurotransmitters, hormones, and drugs can interact with specific plasma membrane receptors in the muscle either to modulate ion channels directly or to stimulate intracellular mediators, called *second messengers*, that can modify the activity of ion channels, membrane ion pumps, or the contractile proteins themselves. The two major second messengers are inositol 1,4,5-triphosphate and the cyclic nucleotides.

For example, acetylcholine binds to its plasma membrane receptor to activate phospholipase C which hydrolyzes an inositol lipid of the membrane, phosphatidylinositol 4,5-biphosphate, to produce inositol 1,4,5-triphosphate and diacylglycerol, both of which act as second messengers. The former substance acts directly on the sarcoplasmic reticulum to cause release of intracellular calcium. The latter, diacylglycerol, activates a calcium- and phospholipid-dependent protein kinase C, capable of phosphorylating a variety of intracellular proteins.

For reviews on the mechanism of contraction in gut smooth muscle, see Hartshorne, 1987; Hurwitz, 1986; Rasmussen and Barrett, 1984.

DEGLUTITION AND ESOPHAGEAL MOTOR FUNCTION

Functional Anatomy of the Pharynx and Esophagus (Fig. 6.13)

Three overlapping sheets of striated muscle, the superior, middle and inferior pharyngeal constrictors, make up the muscular wall of the pharynx. The inferior constrictor thickens at the level of the cricoid cartilage to form a muscle band, the *cricopharyngeus*, that joins the ends of the cricoid cartilage. The cricopharyngeus constitutes most if not all of the pharyngoesophageal or upper esophageal sphincter.

Striated muscle continues into the esophagus, in both the outer longitudinal and inner circular muscle layers, approximately to the level of the aortic arch, about one-third of the way down the esophageal body. Here, striated muscle gives way to

Figure 6.13. A posterior oblique view of the pharynx and the cephalic end of the esophagus show the arrangement of the muscle layers.

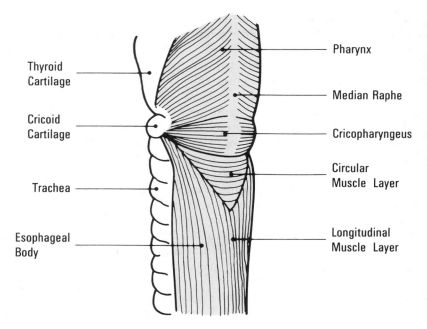

Thyroid Cartilage

Cricoid Cartilage

Trachea

Esophageal Body

Pharynx

Median Raphe

Cricopharyngeus

Circular Muscle Layer

Longitudinal Muscle Layer

smooth muscle, bundles of the two types being mingled together over a variable distance, up to several centimeters. Smooth muscle makes up both layers all the way to the level of the diaphragm where a thickening of the circular muscle layer forms the *esophagogastric* or *lower esophageal sphincter*. The muscle of this sphincter, 2 cm or so wide, is continuous not only with the circular muscle layer of the esophageal body above and the stomach below, but also with the oblique muscle of the gastric fundus, whose margins, the gastric sling fibers, radiate from the sphincter along both sides of the lesser gastric curvature. The sphincter lies partly within the esophageal hiatus of the diaphragm, being encased within the leaflets of the phrenoesophageal ligament.

The muscularis mucosae of the esophagus, smooth muscle throughout, extends from a centimeter or so below the cricopharyngeus to the stomach, where it is continuous with that of the stomach. Its bundles are generally oriented in the longitudinal axis.

The pharyngeal branches of the IXth nerve supply somatic motor nerves to the striated muscle of the pharynx, ending in motor end-plates. Vagal branches, including the recurrent laryngeal nerves, carry somatic motor fibers that end in motor end-plates in the striated muscle of the cricopharyngeus and rostral part of the esophageal body. The esophageal plexus surrounds the smooth-muscle part of the esophageal body, supplying it through numerous fine branches.

The *myenteric plexus* of the esophagus extends from a centimeter or so below the cricopharyngeus all the way to the stomach where it is continuous with that of the stomach (Christensen et al., 1983; 1987). It consists of coarse nerve fascicles, oriented mainly in the longitudinal axis, which extend into the gastric myenteric plexus. These fascicles are connected by smaller oblique lateral branches. Ganglia lie at some, but not all, branch points. Some ganglia are located away from branch points along the course of major fascicles, and others lie at the ends of short stalks off the main fascicles. The distribution density of ganglion cells in the myenteric plexus is very low relative to that in other organs, and it declines along the esophagus to reach a nadir at the esophagogastric junction.

About half of the ganglia of the esophageal myenteric plexus contain distinctive structures, called *intraganglionic laminar endings* (Rodrigo et al., 1982), which are unique to the esophagus. These pleomorphic leafy and stellate flat structures are presumed to be mechanoreceptive sensory nerve endings, since they disappear upon destruction of the sensory ganglion of the vagus, the nodose ganglion.

The *submucous plexus* of the esophagus (Christensen and Rick, 1985) contains few or no ganglia (Fig. 6.7). It is a loose and sparse network of fascicles whose component axons, by exclusion, seem to arise from the ganglion cells of the myenteric plexus.

The innervation of the muscularis mucosae is assumed to arise from the submucous plexus, and that of the circular and longitudinal layers of the smooth muscle from the myenteric plexus. The varicose axons extend along and within the bundles of smooth muscle cells.

In the circular layer of smooth muscle, but not in the others, terminal axons pass in close proximity to interstitial cells of Cajal that are abundant throughout the whole thickness of the circular layer. These interstitial cells of Cajal form gap junctions with the smooth muscle cells.

Muscle Function at Rest and in Deglutition (Fig. 6.14)

At rest, the pharyngeal muscle is inactive, exhibiting only a low level of tone. The cricopharyngeus muscle, in contrast, is strongly contracted to close the esophageal inlet by compressing it against the cricoid cartilage, forming a crescentic slit. The resting luminal pressure measured by manometry at the pharyngoesophageal junction is high, about 50–150 mm Hg above atmospheric pressure. The striated and smooth muscle of the esophageal body is flaccid, without appreciable tone. At the gastroesophageal junction, a tonic contraction closes the lumen in a rosette, creating a manometrically measured pressure of about 15–50 mm Hg above atmospheric pressure.

The degree of luminal closure at the gastroesophageal junction is not quite constant. Pressure at this region seems to vary to some degree. The muscle at the junction relaxes briefly and spontaneously at intervals, particularly during sleep, when such relaxations can result in brief periods of acidification of the esophageal lumen, detected by a pH probe put into the distal esophageal body, as gastric contents enter the esophagus. Such reflux of gastric content is usually quickly cleared by a reflex esophageal peristaltic contraction, called *secondary peristalsis.*

Swallowing involves a series of motions, some voluntary and others involuntary, that occur in an unvarying sequence. Initiation of a swallow involves three voluntary actions, the clenching of the jaws, the elevation of the tongue against the hard palate, and the elevation of the soft palate to separate the nasopharynx from the oropharynx.

These motions are followed by the voluntary and then involuntary contraction of the pharyngeal muscles as a very rapidly moving event that sweeps toward the esophageal inlet. At about the same time, the strap muscles of the neck move the pharyngoesophageal junction so that the lumen of the inlet to the esophagus is aligned with that of the pharynx. The upper esophageal sphincter relaxes at some point in this process so that the bolus, forced from the mouth into the pharynx by the tongue and through the pharynx by its peristaltic contraction, encounters no delay in entry into the esophagus. The upper esophageal sphincter remains relaxed for about 1 ½ s and then contracts quickly, very powerfully for about 1 s, before returning to its basal level of tone, which then remains stable.

In the muscle of the esophageal body, a contraction begins just below the cricopharyngeus. Seemingly, the contraction that marks the closure of the

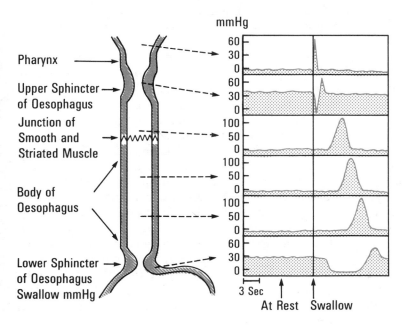

Figure 6.14. Diagram to show pharyngeal and esophageal intraluminal pressures at rest and after a swallow. Pressure scales show only approximate values. (From Christensen, 1983b.)

upper esophageal sphincter simply continues on into the esophageal body. This contraction moves caudad as a front along the esophageal body, passing through the junction between striated and smooth muscle without delay, and reaching the level of the lower esophageal sphincter about 6–9 s after it has first appeared at the rostral limit of the esophagus. The relaxation after this contraction also moves as a wave-front, the length of the contracted segment at any one point in time being some 8–10 cm.

The lower esophageal sphincter relaxes at about the same time that the peristaltic contractions of the esophageal body begins at the esophageal inlet, and it remains relaxed until the moving front of the peristaltic contraction reaches it. As the lower esophageal sphincter contracts again, the force of its contraction exceeds resting values for about 1 s or so before the sphincter returns to its stable level. Sphincter closure at the end of peristalsis appears to be an extension of the peristaltic contraction. The duration of sphincteric relaxation is about the same as the time required for esophageal peristalsis to be completed, 5–9 s.

There is rather little variability in these events in normal swallowing. The most common variation is the failure of esophageal peristalsis to be propagated. In such *incomplete peristalsis*, the peristaltic contraction passes only part of the way along the esophageal body and then fades before it reaches the sphincter. A less common occurrence is a *repeated peristaltic contraction*, in which a second contraction quickly follows the first. This occurs mainly in the distal part of the esophagus. Occasional swallows will also fail to trigger esophageal peristalsis after closure of the upper esophageal sphincter. In normal people, such abnormal peristaltic sequences occur in less than about 10% of efforts, at least in the brief study periods that are involved in manometric and radiographic observations.

Flow in the Esophagus

The esophageal lumen, lying mostly within the thorax, is exposed to the fluctuating negative intrathoracic pressure that accompanies respiration. The sphincters at the esophageal inlet and outlet oppose the ingress of air from the pharynx and of gastric content from the stomach which would occur in their absence. The flow of air through the esophageal inlet is of no great consequence, but gastroesophageal reflux is, the gastric contents being noxious to the esophageal epithelium. Such gastroesophageal reflux is the cause of reflux esophagitis and its consequences, ulceration and stricture.

Flow in the esophagus is also facilitated by gravity. The effect of gravity on flow is greater for liquids than it is for solids or viscous liquids. For such materials, the peristaltic phenomenon is more important in accomplishing peristaltic flow. Peristalsis is essential in animals that do not eat in the upright position. In man, though, gravity often carries liquid boluses ahead of the peristaltic contraction, so that liquids may reach the lower esophageal sphincter before it opens completely.

The spontaneous relaxations of the lower esophageal sphincter that normally occur from time to time allow gastric contents to enter the esophagus. Normally, these are cleared quickly from the esophagus by a reflex in which the stimulus (which may be the excitation of either or both chemoreceptors and mechanoreceptors) excites a peristaltic contraction of the esophageal body. Such a contraction can be induced experimentally by localized esophageal distension. It is called *secondary peristalsis*, to distinguish it from the *primary peristalsis* induced by swallowing.

Control Systems That Govern Motility

STRIATED MUSCLE REGIONS

The tonic closure of the upper esophageal sphincter is the consequence of the constant high level of discharge of the somatic motor nerves that supply this striated muscle. The level of tonic discharge of these excitatory nerves is much less to the remainder of the striated muscles innervated by the IXth nerve, including the pharyngeal musculature and many of the strap muscles of the neck. This difference between the pharynx and the cricopharyngeus in tone can be explained if one assumes that these different striated muscles constitute separate motor units representing the outputs of different neurons or groups of neurons within the IXth cranial nerve nuclei. The absence of tone in the striated muscle of the rostral part of the esophageal body can be similarly explained in terms of motor units that are not tonically discharging. The swallowing-induced contractions of the striated muscle of the pharynx and of the strap muscles of the neck are always preceded by a brief disappearance of the resting tone, indicating that transient inhibition of the motor neurons occurs in the IXth nerve nuclei before the neurons are stimulated to maximal discharge (Fig. 6.15). This inhibition coincides in time with the inhibition of the cricopharyngeus muscle. The progressive nature of the swallowing-induced peristaltic contractions of the pharynx and proximal esophagus can be explained as the consequence of the excitation of adjacent motor units in a craniocaudad sequence.

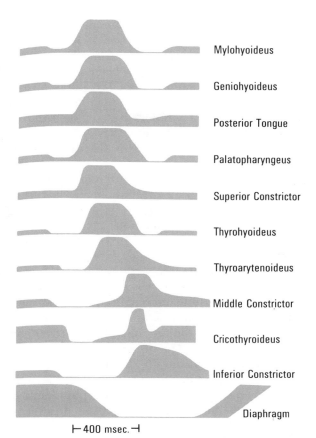

Mylohyoideus

Geniohyoideus

Posterior Tongue

Palatopharyngeus

Superior Constrictor

Thyrohyoideus

Thyroarytenoideus

Middle Constrictor

Cricothyroideus

Inferior Constrictor

Diaphragm

⊢400 msec.⊣

Figure 6.15. A summary of the outflow of the swallowing center in the dog, demonstrated in terms of the electromyographic recording from various striated muscles. The upper five muscles, related to the pharynx, contract nearly simultaneously, while the lower five muscles, related to the hypopharynx, contract in craniocaudad sequence. Note the inhibition of resting tonic discharge that precedes contractions in all muscle but the tongue and superior constrictor. (Doty and Bosma, 1956.)

Thus, the coordinating mechanism for the striated muscle regions is in the brainstem rather than in the gut. The coordinating mechanism is the *swallowing center,* an anatomically poorly defined part of the limbic system in the region of the fourth ventricle where the IXth and Xth cranial nerve nuclei lie. The swallowing center must have sensory inputs, for swallowing is difficult (but not impossible) to initiate if the jaws are not apposed and the tongue elevated against the hard palate. The center must also have connections to some other autonomic centers. Respiration, for example, is normally suspended momentarily at the initiation of swallowing.

SMOOTH MUSCLE REGION OF THE ESOPHAGEAL BODY

The operation of the smooth muscle part of the esophagus cannot be the same as that of the striated muscle part, because smooth muscle is innervated by autonomic nerves, and so a motor unit organiza-

tion is not possible. The origin of peristalsis in the smooth muscle part has been a matter for a great deal of investigation. The initiation of contraction of the esophageal smooth muscle in swallowing is clearly neurogenic, governed in part by the swallowing center through its control on cells in the Xth nerve nuclei. This is evident from the fact that the peristaltic contractions of the striated muscle and the smooth muscle regions are so closely coordinated. The progressive nature of peristalsis in the smooth muscle segment of the esophageal body was once thought to be programmed in the central nervous system as well. But this is not so. In the smooth muscled segment, peristalsis and sphincteric relaxation can be induced in vitro by nerve stimulation. This establishes that intramural, rather than central, mechanisms can provide for the progressive nature of the esophageal peristaltic contraction in this region.

The peristaltic contraction of the smooth muscle part of the esophagus represents the transient discharge of nerve cell bodies within the myenteric plexus. When these nerves are stimulated artificially, the circular muscle layer responds by contracting once briefly after the end of the period of nerve stimulation. This response, called the "*off response,*" is characteristic only of the circular muscle layer (Fig. 6.16). If it is induced simultaneously in strips of muscle cut from all parts of the esophageal body, the latency of the off-response (the time elapsing between the end of the period of

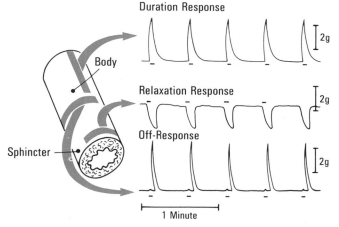

Duration Response

Body

Relaxation Response

Off-Response

Sphincter

⊢ 1 Minute ⊣

Figure 6.16. The three kinds of responses of the esophageal smooth muscle to electrical field stimulation of intramural nerves. The longitudinal muscle responds to stimulation *(bars)* by contracting *during* the stimulus (the "*duration response*") which is cholinergically mediated. The circular muscle of the esophageal body contracts *after* the stimulus (the "*off-response*"). The muscle of the lower esophageal sphincter has a large tone and relaxes during the stimulus (*"relaxation response"*). The *off-response* and the *relaxation response* are due to the action of nerves whose transmitter(s) is (are) unknown. (From Christensen, 1983b.)

nerve stimulation and the occurrence of the off-response) can be seen to form a gradient along the esophageal body. Latency is shortest at the top of the smooth muscled segment and longest at the bottom (Fig. 6.17). The off-response is not opposed by any antagonists of known neurotransmitters acting selectively at defined receptor sites, but it is sensitive to tetrodotoxin, and so it must be initiated by nerves. When the circular muscle layer is made to contract by superfusion with an excitatory agent (like acetylcholine) before nerve stimulation, a relaxation is revealed during the period of intra-

mural nerve stimulation, preceding the off-response.

From these and other observations, the following mechanism of peristalsis has been postulated for the smooth-muscle segment. Peristalsis is proposed to represent the excitation of a noncholinergic nonadrenergic inhibitory innervation. The inhibitory effect of the nerves in the smooth muscle is not ordinarily apparent because of the absence of tone in this muscle. The occurrence of the rebound contraction, the off-response, is an intrinsic part of the inhibitory transmission mechanism. It is the only part of the response to nerve stimulation that is seen because this muscle has no tone. In this way, an inhibitory innervation is, in fact, excitatory in its major effect.

Controversy exists and is related to the nature of the relationship between the off-response and the antecedent inhibition that occurs during the period in which the responsible inhibitory nerves are stimulated. One theory holds that the off-response is essentially passive, a contraction produced by a rebound depolarization of the muscle cell membrane, following the hyperpolarization, that is related to the intrinsic properties of that membrane. If so, the muscle itself should exhibit gradients of some kind that correspond to the gradient in the duration of the latency period of the off-response. Indeed, gradients in resting membrane potential of the circular muscle layer have been described, such that the magnitude of the resting membrane potential is less distally along the smooth muscle segment. These gradients in resting membrane potential result from gradients in intracellular K^+ and plasma membrane permeability to K^+.

An alternative theory holds that the off-response is due to the discharge of a separate excitatory innervation, distinct from the inhibitory innervation, through a direct linkage between excitatory and inhibitory nerves in the myenteric plexus, or perhaps through a local reflex whose pathway resembles that of the peristaltic reflex of the intestine. The relative simplicity of structure of the esophageal myenteric plexus argues against this. Also, no one has been able to dissociate the off-response from the inhibition that precedes it, which should be possible if two separate classes of nerves are responsible. The interstitial cells of Cajal in the circular layer makes the matter even more complex. The intimate relationship of these cells to the axons in that layer and to the muscle cells, their abundance, and their unique occurrence in this layer in this organ, all suggest strongly that these cells are involved in the genesis of the off-response, but it is

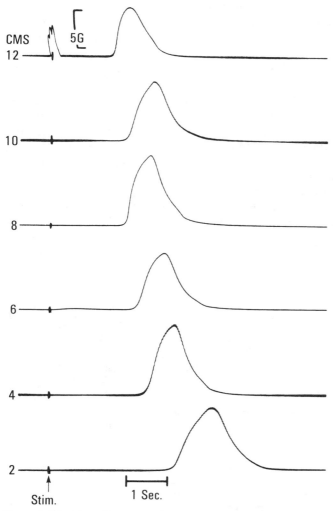

Figure 6.17. The off-response elicited in isolated strips of muscle from various levels of the smooth muscle segment of the opossum esophagus. The numbers at the *left* indicate distance above the esophagogastric junction. The stimulus is a 250-ms train of I-ms pulses at 15 Hz. Note that the latency of the off-response (time from the stimulus to the onset of contraction) increases along the segment. The *topmost strip* contained some striated muscle which contracted immediately with the stimulus. (From Weisbrodt and Christensen, 1972.)

not known how they participate. Thus, a full understanding of the origin of peristalsis in the smooth muscle part of the esophagus remains to be found.

The esophagus shortens considerably in swallowing. This is presumably due to contraction of the longitudinal muscle layer and perhaps of the muscularis mucosae. These muscles are innervated by excitatory cholinergic nerves. The contraction of the longitudinal muscle with cholinergic stimulation is described as the *duration response*. The effect of this shortening on esophageal transit is unknown.

LOWER ESOPHAGEAL SPHINCTER

The tonic contraction that characterizes the muscle of the lower esophageal sphincter is mainly myogenic, for it is only a little depressed by tetrodotoxin in vivo and in vitro. In vivo, tone is variably reduced in some species by atropine, but never abolished, so it seems likely that the cholinergic innervation of the sphincter can be regulated neurogenically in vivo, to some degree. The specialization in this muscle that accounts for its tonic contraction is not known.

The relaxation of the lower esophageal sphincter that occurs in swallowing is due to the action of nonadrenergic noncholinergic inhibitory nerves, indistinguishable from those that accomplish peristalsis in the esophageal body. The time of sphincteric relaxation in relation to peristalsis in the esophageal body indicates that nonadrenergic inhibitory nerves are excited simultaneously all along the esophagus to relax the sphincter before peristalsis has started in the smooth-muscle segment. The occasional spontaneous relaxations of the sphincter that occur without swallowing probably represent local activation of the inhibitory nerves to the sphincter, perhaps from gastric stimulation, through the connections of the myenteric plexus between the two organs.

Clinical Disorders of Esophageal Motor Function

The clinical disorders of the swallowing process demonstrate the difference between the controls of the striated muscle and the smooth muscle regions of the pharyngoesophageal conduit. *Occlusion of the vascular supply to the brainstem* damages the brainstem nuclei and this impairs the function of the striated muscle regions with little effect on the smooth muscle segment.

Some diseases selectively affect the smooth-muscled segment. The fact that these disorders affect only the smooth-muscled part emphasizes the difference between the innervation of this segment and that of the rest of the esophagus. In one such disorder, *achalasia* or *cardiospasm*, the striated muscle operates normally in swallowing, but the smooth-muscled esophageal body fails to exhibit peristalsis, and the lower esophageal sphincter fails to relax. Swallowed material accumulates in the dilated esophagus, escaping into the stomach only when hydrostatic pressure rises enough to breach the resistance offered by the constantly contracted sphincter. Histological and physiological studies suggest that this disorder is due to death or dysfunction of intrinsic nerves. The neuropathy is selective for this region, for patients with achalasia do not seem to have deranged motility in other parts of the gut. It is not clear why the neuropathy is so selective for the esophagus. The damage could be only to one class of nerves, the nonadrenergic inhibitory nerves that seem to be the major nerves governing esophageal peristalsis in the smooth muscle segment and lower esophageal sphincteric relaxation.

In another disorder, *esophageal spasm*, the lower esophageal sphincter operates normally, while peristalsis in the smooth muscle segment of the esophageal body is disordered. Several different kinds of dysfunction occur. The contraction may be simultaneous throughout the whole segment rather than progressive, it may be greatly increased in magnitude, and repetitive contractions may occur after a single swallow. Such abnormalities may reflect dysfunction of the smooth muscle itself rather than a neuropathy, but there is little firm evidence on this point.

In *reflux esophagitis*, the impaired function of the lower esophageal sphincter allows the reflux of gastric contents that damage the esophageal mucosa, leading to diffuse inflammation, ulceration, and stricture formation. This may be a consequence of the primary loss of sphincteric tone, the reason for which is not clear.

In *diffuse systemic sclerosis*, reflux esophagitis is a major problem. This disease is associated with degeneration and fibrosis of smooth muscle throughout the gut, but the esophagus is particularly affected. The consequent loss of tone in the lower esophageal sphincter and the weakness of the peristaltic contraction in the smooth muscle of the esophageal body lead to the enhanced reflux of gastric contents and the incomplete clearance of the refluxed fluid from the esophageal body by secondary peristalsis.

For reviews of deglutition and esophageal motor function, see Christensen, 1983, 1987; Weisbrodt and Christensen, 1972b.

GASTRIC MOTOR FUNCTIONS: RECEPTIVE RELAXATION, GRINDING, SIEVING, AND EMPTYING

Functional Anatomy of the Stomach

The stomach consists of three anatomically continuous but functionally distinct units, the proximal stomach, the distal stomach, and the pylorus. These functional units are not equivalent to the anatomic divisions of the stomach (the cardia, fundus, body, antrum, and pylorus), since the body of the stomach encompasses part of the functional proximal stomach and a part of the functional distal stomach. Also, the cardia lies wholly within the functional proximal stomach. In discussing gastric motor function, it is most convenient to use the functional terms, *proximal stomach, distal stomach,* and *pylorus.*

There are muscular distinctions among the three functional parts of the stomach (Fig. 6.18). In the proximal stomach, the oblique muscle layer invests the fundus and part of the gastric body, lying between the circular muscle layer and the submucosa, and fusing with the circular muscle layer at the greater curvature and with the lower esophageal sphincter. The circular muscle layer thickens distally along the distal stomach. The pylorus constitutes a distinct thickening of the circular layer.

Differences in the myenteric plexus also distinguish the functional units. In the proximal stomach the ganglia are smaller and more sparse than they are in the distal stomach, and the secondary and tertiary components of the myenteric plexus are poorly developed. Thick nerve fascicles enter the proximal stomach from the esophagus and radiate from the cardia toward the greater curvature, bypassing ganglia but giving branches to them, to end in branches that disappear among the ganglia of the distal stomach. These structures are called *shunt fascicles.* In the distal stomach, the ganglia are larger and more abundant, the interganglionic fascicles are thicker, the secondary and tertiary plexuses are more exuberantly developed, and

interstitial cells of Cajal are abundant in the tertiary component of the myenteric plexus and throughout the thick circular muscle layer. These anatomic differences are consistent with the different motor functions of the two parts of the stomach.

Motor Function of the Proximal Stomach and Its Control

The muscle of the proximal stomach is a tonic muscle, exhibiting slow changes in tone but little in the way of regular spontaneous rhythmic contractions. As the stomach fills, the tone falls to allow the massive expansion that must occur to accommodate the ingested volume. The "empty" stomach contains, usually, less than 100 ml of fluid, while the "filled" stomach may contain 2 liters, and even more in pathological states. The reservoir function of the proximal stomach is possible because of the ability of the proximal stomach to relax in receiving the volume that is eaten, and the volume of fluid that is secreted in response to eating. This expansion occurs through the process called *receptive relaxation* or *gastric accommodation.* After eating has ceased, tone is gradually restored to press the contents of the region into the distal stomach.

These changes in tone may reside especially in the oblique layer of muscle that is unique to this region, and in the circular layer, for the longitudinal muscle layer is very thin in the fundus. The tone of the proximal stomach muscle is presumably myogenic but it may be due to a cholinergic innervation as well. The relaxation of tone with gastric filling is neurogenic, representing the action of nonadrenergic inhibitory nerves whose cell bodies apparently lie within the myenteric plexus. The activation of these nerves in gastric filling is a reflex response to gastric distension, and both central (vagovagal) and local intramural pathways exist that can serve the function. It is possible also that the nerve fascicles joining the myenteric plexus of the esophagus to that of the stomach convey the message to the stomach that swallowing is occurring.

Figure 6.18. A drawing to show the three muscle layers of the stomach. (From Dubois, 1983.)

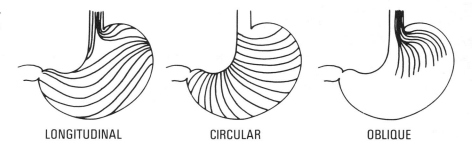

LONGITUDINAL   CIRCULAR   OBLIQUE

Motor Function of the Distal Stomach and Its Control

The muscle of the distal stomach exhibits rhythmic peristalsis. Ring-contractions of the circular layer develop near the middle of the gastric body and sweep toward the pylorus. These occur at a maximum frequency of 3/min. They accelerate and deepen as they move so that the length of the contracted segment increases from a centimeter or so at their origin to perhaps 6–8 cm in the distal antrum. The complete and simultaneous occlusion in the pre-pyloric antrum and pylorus is called the *terminal antral contraction.*

Slow waves underlie the peristalsis in the distal stomach. The slow waves begin at a point in the greater curvature in the middle of the gastric body as small signals that increase in amplitude and velocity as they sweep in a ring to the pylorus (Fig. 6.19). The gastric slow wave frequency is quite constant (3/min), but the likelihood of a contractile response to the slow wave is highly variable. All slow waves produce contractions when the stomach is emptying at its maximal rate, and none do when emptying is maximally inhibited. The relative abundance of interstitial cells of Cajal in the antrum, where slow waves are found, suggests that here, as in other regions, these cells are involved in slow wave generation.

The peristaltic contractions of the distal stomach

accomplish the grinding and sieving functions of the stomach. As these contractions move with increasing force and acceleration, they force the contents toward the pylorus. The antral contents that lie at the perimeter of the lumen are carried forward with the contractions, but the matter that lies at the core is forced back, in compensation (Fig. 6.20). This retropulsion carries solid particles back into the gastric body. They are picked up again by peristalsis to be carried toward the pylorus, only to undergo retrograde flow once more. The grinding and retropulsion of large solid masses, which is quite violent, generates considerable energy for the mechanical disintegration of solid masses as they are softened by intragastric digestion. The solid masses are retropelled before the antral peristaltic contraction has reached the terminal antrum as the completely occluding terminal antral contraction. As a result, the terminal antral contraction carries before it mainly liquids. This is the mechanism of gastric *sieving*, which assures that the solid particles which are allowed to escape the stomach are generally less than about 2 mm in diameter.

Motor Function of the Pylorus

The pyloric muscle functions in concert with that of the distal stomach. As the peristaltic contraction begins in the middle of the gastric body, the myo-

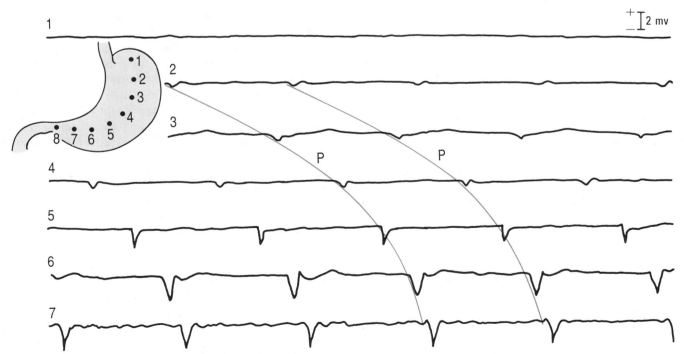

Figure 6.19. A diagram to show the migration pattern with amplification and acceleration of gastric slow waves. Seven tracings are shown from the seven sites indicated. *P* indicates a peristaltic wave. (From Kelly et al., 1969.)

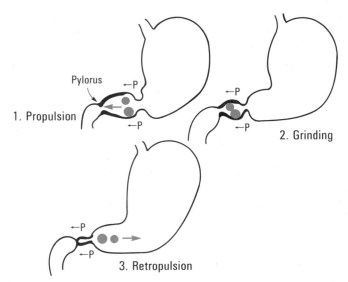

Figure 6.20. Antral Functions. These are drawings made from cineradiographs of the dog stomach. (From Kelly, 1981.)

genic tone in the pyloric musculature falls, and pyloric resistance seems to remain low until the moving front of the terminal antral contraction reaches it. The pylorus then contracts, and it seems to remain at least partially contracted until the next peristaltic wave begins. The actual magnitude and timing of pyloric contraction and relaxation have not been critically examined. Relaxation of the pylorus is brought about by nonadrenergic inhibitory nerves, but it is not clear how the discharge of these nerves is tied to the occurrence of antral peristalsis. The operation of the pylorus regulates gastric emptying in cooperation with antral peristalsis. It may have an equally important function in opposing duodenogastric reflux.

Factors Involved in the Control of Gastric Emptying

The rate at which the filled stomach empties depends both upon the rate at which tone is restored to the relaxed muscle of the proximal stomach, and upon the responsiveness of the distal gastric muscle to the pacing slow waves. These functions are not readily examined separately in vivo, and so the emptying of the stomach is commonly examined as a single process, although it is the net consequence of the two processes.

Gastric emptying is generally studied by the use of markers that allow the monitoring of the fractional decay of that volume which remains in the stomach. The occurrence of gastric secretion requires that correction be made for the dilution of marker solutions. Emptying behaves as a single exponential process. Markers placed in solution ("liquid markers") empty more rapidly than markers that are integrated into solid foods ("solid markers"), as would be predicted from the sieving function of the peristaltic contraction. But even among liquid markers, the rates of emptying vary greatly with the chemical composition of the meal. Studies carried out in conscious man as well as in animals indicate that gastric emptying of liquids is slowed in proportion to the concentrations of fatty acids, of the hydrogen ion, and of osmolytes in the gastric content. Also, tryptophan is unique among amino acids as a potent inhibitor of gastric emptying. Chemoreceptors (responsive to fats, acids, osmotic pressure, and tryptophan) in the duodenal mucosa mediate the regulation of the rate of gastric emptying in response to specific chemical features of the gastric effluent. The chemoreceptors are very sensitive, and the response to their stimulation is prompt. This suggests that the receptors act by a neural reflex mechanism rather than by the release of a hormone. If so, the reflex pathways and the nervous mechanisms are unknown.

Gastric emptying is also inhibited by central mechanisms, by vagovagal reflexes, and as a reflex response to visceral and somatic pain. Probably, both vagal and splanchnic pathways can be involved. The inhibition can involve both suppression of antral contractions and disorganizations of the antral slow waves. These central systems have had little study.

Clinical Disorders of Gastric Motor Function

Delayed gastric emptying is a common problem, usually attributable to a mechanical obstruction like pyloric stenosis from pyloric or duodenal ulceration. But it occurs as well in the absence of obstructing lesions, especially in complicated diabetes mellitus. Delayed gastric emptying in complicated diabetes, called *gastroparesis diabeticorum*, is a major problem not only in nutrition but also in the control of the diabetes, for management programs with insulin and with oral hypoglycemic agents both depend upon a predictably regulated delivery of carbohydrates to the small intestine. Both solid and liquid emptying are delayed in gastroparesis. The delay is due to a neuropathy, the autonomic nerves probably being affected by the same process as that which effects the somatic nerves in complicated diabetes.

Altered gastric emptying invariably occurs with the conventional gastric operations used to treat gastric and duodenal ulceration. These operations involve truncal, selective, or highly selective *vagotomy* done in the hope of reducing gastric acid secretion. Truncal vagotomy alone causes a gastroparesis like that seen in complicated diabetes, but the more selective vagotomies affect motility much less. This is the case because the selective vagotomies are designed to spare the nerve of Laterjet which seems to be more important in governing gastric motility than are the other gastric branches of the vagus. Surgeons never perform vagotomy alone but always combine it with a partial resection of the distal stomach, or at least a pyloroplasty (in which the pyloric ring is cut across and reconstructed in such a way as to prevent pyloric closure). Pyloroplasty is done to reduce resistance to flow at the gastric outlet. The disturbances in emptying may be quite profound, even with pyloroplasty and vagotomy alone. Liquid emptying is normal or accelerated, and there may be a defect in the braking effect of the duodenal chemoreceptor mechanism. As a result, the sudden delivery of a large volume of hypertonic fluid to the duodenum may induce vascular hypovolemia because of osmotic equilibration across the intestinal mucosa. This can cause nausea, dizziness, and other symptoms, collectively called the *dumping syndrome.* The symptoms of the dumping syndrome have also been attributed to the release of hormones and vasoactive amines from the duodenum, so that it is not hypovolemia alone that is responsible. Paradoxically, for unknown reasons, the stomach after vagotomy often retains solids, sometimes with such avidity that they are cemented together by proteins to form a mass of fibrous material, called a *gastric bezoar.* In any operation in which the pylorus is breached, either by removal or by pyloroplasty, duodenal content refluxes freely into the stomach. The refluxed fluid contains, among other things, bile salts and pancreatic enzymes which may work together with gastric acid to damage the epithelium of the gastric body to lead to *gastritis* or *gastric ulceration.*

"Tachygastria" is blamed for some cases of otherwise-unexplained cases of delayed gastric emptying. In a few such cases, the electrical slow waves of the antrum have been observed to occur with a very rapid frequency and a chaotic pattern of spread. There is no doubt that this can occur and produce symptoms, but it is so difficult to demonstrate that we have no clear idea of its incidence.

For reviews of gastric motor function, see Dubois, 1983; Meyer et al., 1979; Meyer, 1987; Hunt and Knox, 1968; Kelly et al., 1969; and Kelly, 1981.

SMALL INTESTINAL MOTOR FUNCTIONS: MIXING AND PROPULSION

Functional Anatomy of the Small Intestine

There is no important structural specialization that distinguishes one level of the small intestine from another. The muscle bundles of the two major layers, the circular being thicker than the longitudinal, are oriented perpendicularly to one another. The muscularis mucosa is thin, its bundles lacking a major orientation in one direction or the other. It gives off small bundles that extend into the villi.

The myenteric plexus is very highly developed, with large regularly spaced ganglia, thick interganglionic fascicles, and prominent secondary and tertiary plexuses with abundant interstitial cells of Cajal.

An aganglionic plexus of small fiber bundles with single fibers lies at a plane deep within the substance of the circular muscle layer. This plexus, the *deep muscular plexus,* is present throughout the small intestine. It lies somewhat closer to the submucosal surface of the circular layer than it does to the intermuscular plane. Its nervous elements arise from the ganglion cells of the myenteric plexus. It contains abundant interstitial cells of Cajal.

The submucous plexus of the intestine is very much denser than is that of any other part of the gut (Fig. 6.8). The ganglia lie in two planes, one close to the mucosa and the other close to the circular muscle layer. It seems possible, but it is not established, that the one layer of ganglia supplies mainly the innervation of the mucosa and the other the circular muscle layer. There are interstitial cells in relation to the submucous plexus, but they are not so abundant as they are in the myenteric plexus and the deep muscular plexus.

Motor Function of the Intestine

The intestinal muscle is a rhythmic muscle, generating rhythmic peristalsis. Contraction rings of the circular muscle may develop at any point and move caudad over highly variable distances, at velocities of approximately 2–3 cm/s. This is *peristalsis.* Another motor pattern is *segmentation,* in which tonic contraction rings develop at uniform intervals to divide the lumen into segments. The longitudinal muscle also can contract rhythmically with peristalsis. The flow pattern produced by these contractions constitutes both mixing flows and net caudad flow. There is evidence, from mathematical modeling, that the moving ring contractions (peristalsis) mainly provide for net caudad flow, and that the longitudinal muscle contractions mainly enhance mixing.

The rhythmic contractions occur at a maximum frequency of 12/min in the duodenum, about 9/min in the midjejunum, and about 7/min in the ileum. When they occur at a lower frequency, intercontractile intervals for contractions measured at a single point have, on the average, a period that is an integral multiple of the fundamental period. For example, intercontractile intervals in the human duodenum cluster around multiples of 5 s (Fig. 6.21). This suggests that a fundamental "clock" regulates their distribution, as discussed below.

The function of the intestinal mucosal muscle is often neglected, for it is a difficult matter to study. The intestinal mucosa moves back and forth across the circular muscle layer, and the intestinal villi themselves move in quick pumping or shortening movements. These movements appear to increase with eating. The mechanism of this effect is not known. Villous and mucosal motions, independent of the motions of the main muscular coats, could serve to enhance mucosal blood and lymph flow, or to stir luminal contents against the epithelial surface, where a thick unstirred layer would provide a major limiting step in the rate of nutrient absorption.

Controls of Intestinal Motility

As in the gastric antrum, rhythmic contractions in the intestine are governed by slow waves. The slow wave frequency declines along the intestine in steps. The frequency "plateaus" extend over some long distances, to a level where an abrupt decline occurs to a new lower value, and so on (Fig. 6.22). These plateaus seem to be inconstant in length, the points of sudden decline in frequency shifting back and forth along the intestine from time to time. The frequency of the most proximal plateau is firmly established at 12/min; that of the most distal plateau is about 7/min. Slow wave frequencies at any point in a frequency plateau are remarkably constant. Slow waves propagate along the plateau at a fixed velocity. Spike bursts appear on slow waves to signal contractions. Thus, as in other regions, slow waves are pacemaking potentials, establishing the frequency, direction, and velocity of rhythmic peristalsis. In the fed state, contractions accompany about 30–40% of slow wave cycles, and contractions are randomly distributed among slow wave cycles. A contraction may begin at any location, and follow the slow wave for any distance. The distance a peristaltic contraction moves seems to be extremely variable.

The characteristics of the intestinal slow waves—frequency, direction, and velocity of propagation—are very constant, and they are not apparently subject to major controls by neurogenic or hormonal influences. The inclination of the muscle to respond to a slow wave cycle by contracting, however, is subject to such controls. Excitatory neural and hormonal influences raise the likelihood of muscle response, and inhibitory control mechanisms reduce it.

The same slow waves also pace rhythmic contractions of the longitudinal muscle layer. It appears that the neural controls of the two layers are separate, that some neural or hormonal influences can selectively enhance or depress the responsiveness of one layer or the other.

Much has been made of the "peristaltic reflex" of the intestine, in which localized distension of the intestine induces peristaltic contraction above the point of stretch and inhibition below it. It seems unlikely that this reflex is of major importance in the control of normal function because the intestine is rarely distended to the degree required to induce the reflex and because distension is rarely sharply localized. The reflex is best considered to be a means to examine polarity in the connections of the myenteric plexus experimentally.

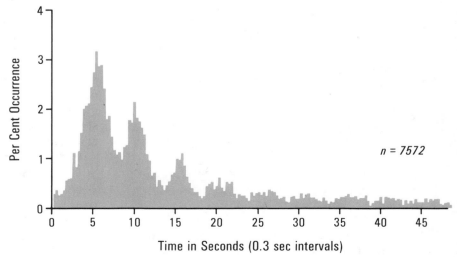

Figure 6.21. Frequency distribution of intercontractile intervals of rhythmic contractions from one locus in human duodenum in the fed state. This is from a manometric recording with an open-tip tube. The *horizontal axis* shows time. Observe that peaks in the distribution occur at multiples of 5 s, reflecting the influence of the 5-s period of the slow wave in the human duodenum. Blurring of these peaks may reflect either variations in the 6-s fundamental period or shifts in the position of the manometric catheter. (From Christensen et al., 1971.)

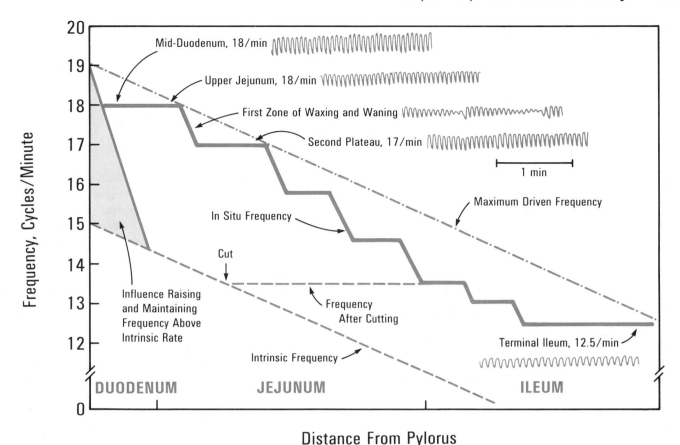

Figure 6.22. Frequency of slow waves in the small intestine of anesthetized cats. The *stepwise heavy line* shows the observed frequency. Each plateau is separated by a zone of waxing and waning slow waves. The *lower slanting dashed line* shows the intrinsic frequency of slow waves in short isolated segments of the intestine. The *upper slanted dotted line* indicates the maximal driven frequency of slow waves for the various points along the intestine. If the intestine is transected at the point labeled *CuT*, the frequency distal to the cut falls to that of the intrinsic frequency at the point of transection. Frequencies shown are those in the dog. In man, frequency declines from 12 cycles/min in the duodenum to about 7 cycles/min in the ileum, as stated in the test. (From Davenport, 1982; and Diamant and Bortoff, 1969.)

Flow in the Intestine

Peristaltic contractions are the main mechanisms for caudad propulsion of luminal content. As would be predicted from the slow wave frequency gradient, propulsion rates in the proximal intestine exceed those in the ileum, a fact demonstrated by studies of transit of nonabsorbed markers. The function of the longitudinal contractions is less clear. Mathematical modeling experiments suggest that they mainly produce mixing between the luminal contents at the core and at the periphery of the intestinal lumen.

Motility in Fasting—The Migrating Myoelectric Complex

Intestinal motility changes radically about 4–6 hours after a meal. Actually, the change involves the motility of other gastrointestinal organs as well, but the controls of the process seem to be the same in all participating organs, and the phenomenon has been best described in the intestine.

As indicated above, contractions in the fed state are randomly distributed among slow wave cycles. In the fasted state, this pattern changes. Intestinal motility enters into a pattern in which the contraction incidence changes cyclically from none—no contractions at all with slow waves, to the maximum—contractions with every slow wave. The cyclic phenomenon migrates along the intestine, and so it is called the *migrating myoelectric complex* (MMC).

The MMC cycle consists of three phases. In phase I, which lasts about 70 min in man, there are no contractions. Phase II follows, with contractions occurring with each slow wave in groups of 1–5, but with intervening periods of 1–5 slow wave cycles without contractions. This phase lasts 10–20 min. In phase III, contractions occur with every slow wave for a period of 1–5 min. Then, the incidence of contractions quickly declines to zero, marking the onset of a new phase I. At any one point in the intestine, this whole cycle of the MMC lasts approximately 90 min, and the length of the cycle is quite constant.

The MMC in the intestine begins at the level of the pylorus and migrates through the whole intestine (Fig. 6.23). Its progress is most easily detected as phase III, called the *activity front,* which takes about 90 min to traverse the intestine. The MMC effectively evacuates the intestine, for which reason it has been called "the interdigestive housekeeper." Periodic fluctuations in lower esophageal sphincter pressure, changes in the incidence of gastric antral peristalsis, and variations in gallbladder contraction occur in sequence with the migrating myoelectric complex in the intestine.

Despite much study, the origin of the MMC pattern remains unknown. The idea that it may be hormonally mediated prompted investigation of a variety of the enteric peptides which might mediate the cycle, especially motilin, which is abundant in the intestine. Peaks in the plasma levels of motilin correspond in time to the initiation of the migrating myoelectric complex in the duodenum. Various observations, however, suggest that these peaks may be the consequence of the MMC rather than the cause. The MMC in conscious human subjects is disrupted by situations that induce psychological stress. This finding indicates that cortical functions can affect it, and it strengthens the idea that the cycle is governed through the extrinsic innervation of the gut. The cycle ends very promptly with feeding, the promptness also supporting the idea that a neurogenic control mechanism is responsible.

For a review of the MMC, see Wingate, 1981.

Motility in Vomiting

Small intestinal motility changes strikingly in vomiting as studied experimentally in the cat. Shortly before vomiting is to occur, contractions and slow waves disappear over much, if not all, of the intestine. Then a single prolonged spike burst appears somewhere in the distal intestine and sweeps rapidly cephalad. It seems to represent a prolonged (approximately 5–7 s) occlusive retrograde peristaltic contraction. When it reaches the stomach, vomiting ensues within less then a minute or so. Then the intestinal electromyogram returns to normal. It is presumed that the stomach is emptied by the raised intraabdominal pressure produced by the characteristic somatic movements. The profound change in intestinal motility and the characteristic somatic movements indicate that the *vom-*

Figure 6.23. The temporal distribution of contractions at three points along the small intestine in the interdigestive period and after a meal. Periods of no contractions (phase I) are followed by periods of less than maximal contraction (phase II), *open columns.* Phase III, periods of maximal activity *(solid blue bars),* follow. Feeding causes the pattern to revert to a constant level of activity. Note how the interdigestive complex (phases I–III) migrates along the bowel. (From Granger et al., 1985.)

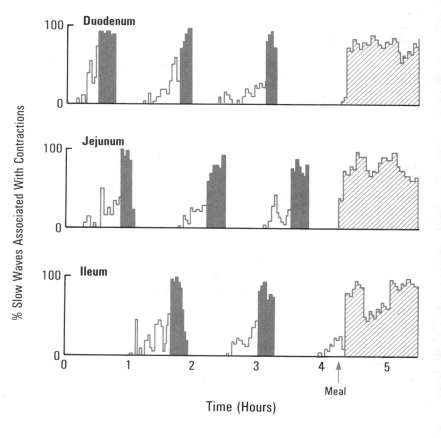

iting center, in the region of the 4th ventricle, controls both autonomic and somatic motor centers.

For review of motility in vomiting, see Weisbrodt and Christensen, 1972a.

Clinical Disorders of Intestinal Motor Function

The motility of the small bowel is difficult to study in man, so clinical dysfunction is difficult to discover. It is best examined by duodenal and jejunal manometry. Dysfunction can sometimes be confined to the intestine, but most examples represent a generalized gastrointestinal motor disorder in which the intestinal dysfunction is the most conspicuous.

The main dysfunction found in the intestine is the failure of the MMC to develop fully. This has been found especially in patients with *diabetic enteropathy,* where the MMC may not develop at all or may exhibit various abnormalities, such as a prolonged cycle or incomplete migration. The force of contractions may also be diminished. The abnormal motility leads to bacterial overgrowth in the intestine which can result in diarrhea. Diabetic diarrhea, like diabetic gastroparesis, is generally a complication in diabetic patients with extensive vascular and neurological damage. It seems, therefore, to be a consequence of autonomic neuropathy.

The small intestine is also affected in systemic disorders that lead to muscle atrophy and fibrosis, notably *diffuse systemic sclerosis* and the *primary visceral myopathies.* The former is uncommon, the latter are rare. In both cases the force of contractions is diminished. The weakness can lead to diarrhea and to a clinical picture resembling that of bowel obstruction. These myopathies are among the causes of the syndrome of *intestinal pseudo-obstruction,* in which the abnormal motility creates a clinical picture resembling that of mechanical intestinal obstruction.

For reviews of small intestinal motor function, see Weisbrodt, 1981, 1987; Davenport, 1982; Granger et al., 1985; Weisbrodt and Christensen, 1972; Weems, 1981; Christensen, 1971; Christensen et al., 1971; Wingate, 1981.

COLONIC MOTOR FUNCTIONS: MIXING AND PROPULSION

Functional Anatomy of the Colon

The gross structure of the colon varies more among mammals than does that of any other gastrointestinal organ. In man and many primates, the outer longitudinal muscle layer is not uniformly thick in the circumference of the organ but is condensed into three thick bands, the taeniae coli, with a very thin coat of longitudinal muscle between the taeniae. This arrangement characterizes most of the human colon, except for the rectum and part of the sigmoid colon, where the taeniae broaden and fuse to produce a longitudinal muscle coat of uniform thickness about the colon. In most animals used in laboratory research—rodents and carnivores—no such taeniae occur, the longitudinal coat being uniform in thickness around the circumference. In many herbivores, taeniated regions characterize a part of the proximal colon, and in some there are two taeniated segments separated by nontaeniated segments. The relative lengths of the various parts of the colon vary among species as well.

This anatomic variability may reflect physiological variability among species. This makes it difficult to apply observations made in animals to humans with confidence.

The thickness of the circular muscle layer in humans increases in the rectum, and a distinct thickening of this smooth muscle in the anal canal forms the internal anal sphincter. The muscles of the pelvic floor, all striated muscle, include the external anal sphincter which is outside and a little caudad to the internal anal sphincter.

The myenteric plexus of most of the colon contains large uniformly spaced ganglia that are connected by thick interganglionic fascicles, and the secondary and tertiary plexuses are well-developed. The tertiary plexus contains abundant interstitial cells of Cajal. The density of the myenteric plexus diminishes in the rectum, the ganglia becoming both smaller and less abundant. Here, there are many thick fascicles of nerves, branching irregularly with many branch points devoid of ganglia.

Coarse nerve bundles, having the structure of peripheral nerves and containing many myelinated fibers, run cephalocaudad within the myenteric plexus in the distal colon in many species. These, the *ascending nerves of the colon* or "shunt fascicles," represent intramural extensions of the colonic branches from the pelvic plexus. These colonic branches perforate the longitudinal muscle layer at the rostral end of the rectum and branch to extend both cephalad and caudad. They extend a variable distance up the colon, up to two-thirds of the length of the colon in some species. They constitute the main route for the distribution of the pelvic innervation to the distal colon.

The submucous plexus consists of a network of small ganglia arranged in two planes, as in the small intestine. The submucosa of the colon also contains a third plane of nerves, a dense aganglionic plexus of neurites with abundant interstitial cells of Cajal spread over the submucosal surface of the circular muscle layer. This plexus, called the *plexus submucosus extremus,* is probably derived from the deeper plane of ganglia in the submucosa, Henle's or Schabadasch's plexus. Current evidence indicates that it is involved in the generation of slow waves in the colon.

The parasympathetic nerves to the colon are supplied through the vagus to variable parts of the proximal colon, and by the sacral outflow to the rest. Sacral segments 2–4 give rise to the pelvic and pudendal nerves. The pelvic nerves receive contributions from the prevertebral ganglia (sympathetic outflow) and branch to form the pelvic plexus within the depths of the pelvis, from which branches extend to the colon (to form the intramural colonic nerves), to the urinary bladder, and other autonomically innervated structures. The pudendal nerves follow a course separate from the pelvic nerves to supply the somatic innervation of the striated musculature of the pelvic floor. They also supply some fibers to the ascending nerves of the colon.

Motor Function of the Colon

The inaccessibility of the colon limits our understanding of the nature of its wall movements. There are both peristaltic and tonic contractions in the colon.

Rhythmic peristalsis, moving short distances, occurs in the colon, as in the small intestine. This peristalsis has been seen best in experimental animals. The peristaltic contractions differ in character from one part of the colon to another. In the proximal colon (of the cat) these contractions occur at a maximal rate of about 6/min. They begin at about the level of the hepatic flexure and migrate cephalad toward the cecum. This seems to be the only place in the gut where such retrograde ("antiperistaltic") contractions normally occur. At many points between the hepatic flexure and the sigmoid colon, rhythmic peristaltic contractions develop at a similar rate to move caudad. A different form of peristalsis characterizes the most caudad part of the colon, corresponding to the sigmoid colon and rectum. Here, spontaneous rhythmic peristalsis is a less common event, but single powerful peristaltic contractions, moving caudad over a long distance can be provoked by pelvic nerve stimulation in experimental animals. Similar powerful peristaltic contractions, moving over great distances, can occur very infrequently in other parts of the colon as well.

Tonic contraction is best demonstrated in the internal anal sphincter, but some degree of tonic contraction also occurs in the body of the colon. This is inferred from the fact that haustral markings, which are often seen to divide the colon into pockets (or haustra) in humans, appear and disappear at long intervals in the radiographically examined colon. Thus, the haustral markings appear to be long-lasting, but not permanent, narrow ring-contractions of the circular muscle layer. Nothing is known of the local mechanisms that produce these contractions.

Thus, the colon, from the standpoint of the nature of its wall motions, constitutes three somewhat different regions. The proximal colon (from the cecum to about the hepatic flexure) exhibits rhythmic peristaltic contractions that mainly move short distances cephalad. The middle part of the colon exhibits rhythmic peristaltic contractions that move short distances caudad, with occasional single powerful peristaltic contractions that move over long distances caudad. The third part of the colon, corresponding to the sigmoid colon and rectum, exhibits little rhythmic peristalsis, but has rare single powerful long-moving peristaltic contractions, which are apparently controlled by the pelvic nerves.

Flow in the Colon

Flow in the colon is more readily examined than are wall movements, and flow studies reveal patterns that seem to reflect the varieties of wall motions described. Transit studies show a prolonged retention and mixing of markers in the proximal colon, which would be expected if rhythmic contractions moving cephalad dominate the activity of this segment. Slow caudad flow of markers occurs throughout the midcolon, as would be expected if rhythmic peristalsis is characteristic of that part. Colonic flow beyond the proximal colon is especially characterized by its intermittent nature. Radiographic studies show that a fecal mass can reside for a very long period in the transverse and descending colon, being molded there by the haustral markings. The haustral markings can both induce and impede colonic flow, in both directions (Fig. 6.24). From time to time, the haustral markings disappear and the fecal mass is moved into a new position, where haustral markings then reform to indent the fecal mass. Such sudden bulk flows of the fecal mass are called *mass movements.* They are infrequent, occurring at intervals of up to several hours. It seems likely that a mass movement reflects the occurrence of a single occlusive peristaltic contraction. The rectum and sigmoid colon also retain a fecal mass for many hours, evacuating

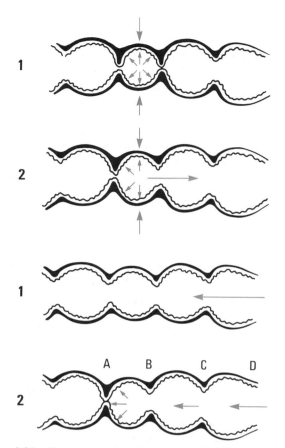

Figure 6.24. Diagrams to show how pressure localized to one haustrum may either initiate movement of colonic content *(upper diagram)* or halt movement either way through haustra. The *short arrows* show the directions of forces caused by the haustral ring-contractions. The *long arrows* show directions of flows induced. (From Paitner et al., 1965.)

that mass only in defecation. This, too, appears to be a consequence of a single mass movement through these regions.

Controls of Colonic Motility

The controls of colonic motility, like the wall motions themselves, are not very clearly understood. The controls include neurogenic, myogenic, and hormonal systems.

Electrical slow waves in the colonic musculature govern rhythmic peristaltic contractions, just as they do in the stomach and intestine. These have been studied most extensively in animals. In the cat (the usual model), they correspond to rhythmic peristalsis in frequency and in the direction and velocity of migration, including the predilection of those of the proximal colon to migrate cephalad toward the cecum. Colonic slow waves differ in some respects from those of the gastric antrum and

intestine. Their frequency is somewhat more labile, and the direction of this migration is not so constant. That is, the source can shift, so that slow waves (and rhythmic peristalsis) may not always be retrograde in the proximal colon, but antegrade from time to time. Also, colonic contractions with slow waves may occur in the absence of spike bursts superimposed on the slow wave plateau.

In the human colon, slow waves have not been wholly satisfactorily demonstrated, in part because recording techniques make it difficult to separate the slow wave signals from noise. Nevertheless, signals that probably constitute the slow waves can be extracted from noise. In humans, several widely different frequencies have been described, in contrast to the findings in the dog and cat.

These slow waves in the colon also differ from those of the gastric antrum and intestine in that they are best recorded from the submucosal surface of the circular muscle layer, seemingly originating there, perhaps in relation to the plexus submucosus extremus, the mat of neurites and interstitial cells of Cajal that covers this surface in the colon.

The colon exhibits another element of the electromyogram that may be related to the occasional single occlusive peristaltic contractions that seem to account for mass movements. These are oscillatory signals at a rather high frequency, about 1/s in the cat, which occur in short bursts of approximately 5–15 s, the bursts migrating cephalad. These sinusoidal waves may give rise to rapid electrical transients (spikes) and they are associated with powerful and prolonged contractions. The source of these signals seems to be near the plane of the myenteric plexus where the tertiary plexus with interstitial cells of Cajal is highly developed. This plexus may well be the source of this control system. The burst of sinusoidal oscillations, often with spikes superimposed, has been called the *"migrating spike burst,"* and *"myenteric oscillatory potentials."* The factors that regulate it are unknown. It may be significant that the phenomenon is most prominently displayed, in the cat colon at least, in the more distal regions, the domain of the pelvic innervation. It may well be the electrical correlate of the peristaltic contraction that produces mass movement.

The influence of nerves on colonic motility is apparently considerable. The fact that the powerful single peristaltic contraction (that which apparently accomplishes mass movement) can be induced by pelvic nerve stimulation has already been mentioned. There is also a prominent inhibitory innervation of the colon, best demonstrable by electrical field stimulation of intrinsic nerves. These inhibitory nerves are nonadrenergic. The exposure of colonic muscle in vitro to tetrodotoxin in some species excites powerful contractions, suggesting that the colon may be chronically dominated by a tonic neurogenic inhibition. Clearly, variations in the degree of such tonic neurogenic inhibition could be responsible for variations in the level of colonic motor activity. There could also be variations in the activity of excitatory nerves, probably mainly cholinergic nerves.

Hormonal influences on colonic motor function seem not to be very important except, perhaps, in the "gastrocolic reflex." This term refers to the fact that colonic motility transiently increases soon after eating in animals that eat intermittently. This is readily apparent in the fact that defecation often occurs soon after eating in many animals, and it has been observed by the experimental study of colonic motility as well. The increased colonic motor activity is triggered by the arrival of nutrients in the small intestine, which induces the release of peptide enteric hormones. Of these, cholecystokinin is the most-studied candidate for the excitation of the colonic response to eating, but it is not yet established that the effect is, indeed, hormonal. It could as well be a neurogenic phenomenon.

Defecation

Like swallowing, defecation involves a mixture of voluntary and involuntary actions, but they have not been sorted out so well as they have been in swallowing.

The excitation of anorectal mechanoreceptors, sensitive to movement and distension of the rectum, initiates the process. These mechanoreceptors are located in the mucosa in or just above the anal canal. The mucosa of the anal canal contains abundant free epithelial nerve endings and organized endings of recognized forms, such as Golgi-Mazzoni, Krause, Meissner, Pacinian, and genital corpuscles.

Whatever the source of the defecation reflex, it gives rise to a sensation that triggers both the voluntary somatic movements that raise intraabdominal pressure and the involuntary movements that lower resistance in the anal canal. The internal anal sphincter relaxes reflexly before defecation in response to the action of nonadrenergic noncholinergic inhibitory nerves. The pathway of this *rectoanal inhibitory reflex* is not known. As the fecal mass begins to move, a powerful occluding peristaltic contraction sweeps the distal sigmoid colon and rectum to evacuate the region. The striated muscles of the pelvic floor relax, and the puborectalis muscle

relaxes to align the rectum with the anal canal. The precise timing of these events is not known.

A mechanism exists for the voluntary resistance of the involuntary urge to defecate. This involves the conscious contraction of the external anal sphincter and pelvic floor musculature, and suppression of the urge to initiate the somatic movements involved in defecation. From these facts, it is clear that a "center" is functioning in the central nervous system. It must be linked to many voluntary motor centers and to the cerebral cortex, as well as to the autonomic and somatic outflows from the sacral segments through the pelvic and pudendal nerves. The location of this "defecation center" and its exact connections are unknown.

Clinical Disorders of Colonic Motor Function

Diverticulosis of the colon is the consequence of an incompletely defined colonic motor disorder. The diverticula are "pockets," herniations of the colonic mucosa through the circular muscle layer at points of weakness, where vessels perforate the muscle. They are most numerous in the descending and sigmoid colon, although they can occur throughout the whole organ. They are the consequence of abnormally high pressures in the colonic lumen. Hypertrophy of the circular muscle layer long precedes the development of the herniations, and this hypertrophy produces the abnormally high pressures in the colonic lumen. Epidemiological studies suggest that the pathological process is related to the prolonged adherence to a low-residue diet, the standard diet of Western countries. Diverticula are uncommon in people from undeveloped countries and in those who consume a high-residue diet. It is not clear how a low-residue diet would lead to hypertrophy of the circular muscle layer of the sigmoid colon.

Aganglionosis of the colon, or Hirschsprung disease, is a congenital neuropathic disorder in which ganglion cells of the rectum fail to achieve maturation after they migrate from the neural crest, and so die. The smooth muscle of the noninnervated segment, which involves the anal canal and a variable length of the rectum, is tonically contracted. It does not relax in response to rectal distension. The diagnosis of aganglionosis rests upon demonstration of the absence of the rectoanal inhibitory reflex and upon histological demonstration of the absence of ganglion cells in the ganglionated plexuses of the affected region.

The *irritable bowel syndrome,* a commonly used diagnostic term, is a clinically and physiologically undefined syndrome characterized by symptoms thought to represent abnormalities in colonic motor function. The symptoms may reflect emotional disturbances, and the emotions are widely recognized to influence defecation, at least, if not the motor function of the whole colon. This disorder is probably not a single nosological entity, and it cannot be discussed in physiological or anatomical terms.

For reviews of colonic motor function, see Christensen, 1983a; 1987; Christensen et al., 1984; 1987; and Painter et al., 1965.

GALLBLADDER AND BILIARY TRACT

Functional Anatomy of the Biliary System

A single layer of muscle makes up the muscular wall of the gallbladder, a loose reticulum of muscle bundles extending in all directions in the plane. The widely separated muscle bundles are separated by connective tissue. The muscle layer is a little thicker and denser in the cystic duct than it is in the gallbladder. The walls of the hepatic and common ducts contain only scattered longitudinal and circular muscle fibers in the connective tissue stroma. A condensation of circular muscle fibers encircles the distal part of the common duct that lies in the duodenal wall above the sphincter of Oddi. These fibers wrap around both the pancreatic and the common bile ducts.

Ganglionated plexuses lie throughout the walls of the gallbladder and bile ducts, and on the outer surface of the circular muscle layer in the sphincter of Oddi. The ganglia are very small and the plexuses are sparse, as compared to those of the intestine. Extrinsic nerves supply these organs, both from the hepatic plexus formed by vagal branches and from the prevertebral ganglia.

Biliary Motor Function and the Bile Flow

The 1 liter of bile produced daily from the liver enters a low-pressure (<10 cm H_2O), low-flow system. In fasting humans, about ¾ of the bile produced enters the gallbladder while the rest flows to the duodenum directly. The rate of bile delivery to the intestine fluctuates with the fasting interdigestive cycling of gastrointestinal motility (the MMC), peaking late in duodenal phase II and phase III. During and after a meal, the motor actions of the biliary system express gallbladder bile into the duodenum.

Brief contractions of the gallbladder occur in fasting, in rhythm with the interdigestive MMC. With eating, the gallbladder contracts for a longer time, the volume diminishing approximately 50% by 30–60 min after the meal. The rate of emptying is controlled more by the rate and force of contraction of the gallbladder than it is by any variable resistance in the biliary ducts. The common duct constitutes mainly a passive conduit for bile flow.

In the fasting state, the sphincter of Oddi is tonically contracted to produce a resistance to flow that is small, but sufficient to divert most of the secreted bile into the gallbladder. During gallbladder emptying, it exhibits a reduced tone and rhythmic contractions that are peristaltic, sweeping antegrade through the sphincter segment. Retrograde peristaltic contractions can also occur. Thus, the sphincter can change bile flow into the duodenum both through changes in the level of its tone and through changes in its rhythmic peristalsis which serve to pump bile into the duodenum.

Controls of Motility in the Biliary System

Cholecystokinin, released from the mucosa of the duodenum in response to nutrients passing over it, is considered to be the main stimulus to gallbladder contraction. Other enteric hormones could also participate in the effect. Cholecystokinin activates gallbladder muscle to contract partly directly and partly through excitation of intrinsic cholinergic nerves. An excitatory effect of the nerves, independent of cholecystokinin, is likely as another controlling factor, for stimulation of both parasympathetic and sympathetic nerves can both contract and relax the gallbladder, respectively.

The operation of the sphincter of Oddi also depends, probably, mainly upon the action of cholecystokinin. The hormone affects the sphincter directly (to excite contraction) and indirectly (to inhibit contraction) through the stimulation of nonadrenergic inhibitory nerves. The indirect effect is normally dominant. Control by other peptide hormones and by nerves has not been excluded.

The peristaltic contractions of the sphincter are governed by electrical signals resembling the slow waves of the intestine and other organs, but these signals differ in two respects. The slow potentials in the sphincter of Oddi are not continuously running, but they appear only with the onset of peristaltic contractions. They carry rapid electrical transients with some contractions, but not with all. The source of these signals, and the neural or homonal factors that regulate their frequency and direction of migration, are unknown.

Clinical Disorders of Motility of the Biliary System

The difficulties in examining the motions of the biliary system limit our understanding of possible disordered motor function. Still, many physicians believe that motor abnormalities of the biliary system exist, and that they give rise to symptoms.

Abnormal gallbladder emptying may contribute to *gallstone* formation through stasis, and this is an explanation advanced for the high incidence of gallstones in patients with diabetes, in patients who have had a vagotomy, and in pregnancy. In the first two situations, the defect is postulated to be in the neural regulation of contraction of the organ. In pregnancy, the mechanism is presumed to be due to the general depression of smooth muscle function that is produced by progesterone.

Biliary dyskinesia is a clinical term used to refer to a fairly well-defined pattern of pain, characteristic of that of biliary tract origin, that is unaccompanied by morphological evidence of biliary tract disease. Sphincter "spasm," paradoxical responses of the sphincter to cholecystokinin, and abnormalities in rhythmic peristalsis in the sphincter have all been detected by manometric study in some such patients. The origin of such physiological dysfunctions remains unknown.

For a review of gallbladder and biliary motility, see Hogan et al., 1983.

CHAPTER 44

Salivary, Gastric, Duodenal, and Pancreatic Secretions

GENERAL CONSIDERATIONS

The secretions of the salivary glands, stomach, duodenum, and pancreas are central to the delivery of a digested meal to the small intestine for absorption. These secretions include water, electrolytes, digestive enzymes, proteins, and humoral agents. The secretory responses are mediated by multiple stimuli derived from the meal. For example, the sight and smell of an appetizing meal cause salivary gland, gastric, and pancreatic secretions through central nervous system (CNS)-mediated cholinergic and nonadrenergic noncholinergic efferents to these organs. Mechanical and chemical stimuli in the meal stimulate secretions from each of these organs via both neural and humoral mechanisms.

In general, the water secreted by these organs aids in both the transport of the meal through the gut and the digestion and absorption of the meal. Without water secretion from the salivary glands, for example, it would be difficult with a dry meal to form a bolus for swallowing. In like manner, the secretions from the stomach and pancreas, as well as the intestine, act as a medium for the transport of the meal through the gut. The watery secretions also act as a medium for the action of water-soluble digestive enzymes and solubilization of nutrients in the meal so that they can be absorbed.

The ionic composition of the secretions varies from organ to organ. The organ-specific ion secretion gives rise to the optimal environment for the digestive functions of the organ. The clearest examples are the hydrogen ion secretion by the stomach and bicarbonate secretion by the duodenum and pancreas. The hydrogen ion secretion activates pepsins that initiate protein digestion in stomach. In contrast, the bicarbonate secretion by the duodenum and pancreas neutralizes the hydrogen ion delivered to the duodenum. This neutralization is essential for digestion because digestive enzyme action (especially lipase) is optimal at a neutral pH. Bicarbonate secretions from the gastric and duodenal mucosa are also important for the protection of the epithelium from the damaging effects of intraluminal hydrogen ion on the mucosa.

A variety of proteins are synthesized, stored, and secreted on stimulation from the salivary glands, stomach, duodenum, and pancreas. One class of proteins is the digestive enzymes that are secreted from each organ. The pancreas secretes a full complement of digestive enzymes, while the salivary glands and the stomach also secrete digestive enzymes, such as salivary amylase, and gastric pepsin. The roles of the nonpancreatic enzymes in digestion is probably minor. Mucus is a specialized glycoprotein secreted by each of the organs discussed in this chapter. It has major protective functions for the gut mucosa. For example, mucus secreted by the salivary gland is mixed with the bolus of food and lubricates it to minimize trauma to the mucosa of the mouth, pharynx, and esophagus. Mucus secreted by the stomach protects the gastric mucosa from trauma and also acts as a chemical buffering zone overlying the mucosa to protect it from damaging effects of intraluminal gastric juice.

In addition to the exocrine secretions, the stomach, duodenum, and pancreas also secrete humoral agents such as gastrin, cholecystokinin, secretin, insulin, glucagon, somatostatin, and pancreatic polypeptide. The physiology of these secretions are discussed both in the present chapter and elsewhere in the text (Chapters 42 and 50).

Protein Synthesis, Storage, and Secretion

Each of the organs discussed in this chapter synthesizes and stores proteins that are secreted with neurohumoral stimuli. The mechanisms for these processes have been best investigated in pancreatic acinar cells. The synthesis of exportable proteins takes place at the rough endoplasmic reticulum (RER) (Palade, 1975; Gorelick and Jamieson, 1987). Several lines of evidence indicate that these proteins are synthesized and segregated within the cisternal space of the RER. Electron microscopic autoradiography of radiolabeled proteins demonstrates that early in the course of synthesis, the proteins are found in the cisternal space of the RER. The newly synthesized protein is resistant to the action of proteolytic enzymes added to preparations of RER and the protein can be released from the cisternal space of the RER if the membrane is solubilized. In vitro studies of protein synthesis using rough microsomes (equivalent to the RER) also demonstrate that newly synthesized proteins are segregated into the internal space of the microsomes.

The mechanism of ribosome attachment to the endoplasmic reticulum (ER) and segregation of newly synthesized protein into the cisternal space is explained by the signal hypothesis (Fig. 6.25). This hypothesis states that on the mRNA for exportable protein there is a signal sequence following the AUG initiation codon. The sequence codes for an amino terminal extension of the protein termed the signal peptide. Thus, after attachment of the initiation codon of mRNA to ribosomes free in the cytosol, the signal peptide is synthesized. This peptide allows for the attachment of the ribosome to the ER and segregation of the protein being synthesized into the cisternal space of the ER. The association of this complex with the ER requires a recognition site on the ER referred to as the docking protein. After synthesis of the signal peptide, translation remains arrested in the cytoplasm until the complex interacts with the docking protein. Although signal peptides vary in their composition of amino acids from protein to protein, they all have a central domain of hydrophobic residues and charged residues on their amino terminus. It is thought that the hydrophobicity allows the peptide to cross the lipid membrane of the ER. After the signal peptide enters the internal space of the ER, it is cleaved from the protein and the remainder of the

protein is translated and sequestered in the ER. The special feature of this segregation of exportable protein is that it keeps potentially damaging substances such as digestive enzymes from the cytosol of the cell.

Proteins in the RER can be modified after removal of the signal sequence and completion of synthesis. These modifications include processes such as disulfide bridge formation, phosphorylation, sulfation, and glycosylation. Conformational changes resulting in tertiary and quaternary structure of the protein also take place in the RER.

Cell fractionation and autoradiographic studies indicate that processed proteins within the RER move to the Golgi complex. The transfer may involve vesicles pinched off from the RER that act as transport containers to the Golgi complex. In the Golgi complex, further posttranslational modification occurs and the proteins are concentrated. One major posttranslational modification function of the Golgi complex is glycosylation and modeling of glycoproteins. The concentrating function of the Golgi complex results in mature secretory granules that move to the apical portion of the cell where they are stored until the appropriate neurohumoral stimulus results in exocytosis.

Exocytosis includes movement of the secretory granule to the apical surface, the recognition of a plasma membrane site for fusion, and fission of the granule membrane/plasma membrane fusion site. The mechanisms of movement, recognition, fusion, and fission are not established nor is it established how intracellular signalling systems promote these processes. It is very likely that microtubule and microfilament systems are involved in these processes.

Although the scheme presented above has been largely accepted for excretory proteins, an alterna-

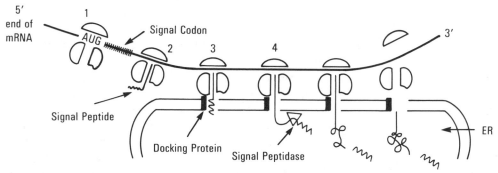

Figure 6.25. Signal hypothesis for protein synthesis. A single mRNA is attached to six ribosomes (polysome). The first two ribosomes near the 5′ end of the mRNA are not yet bound to the endoplasmic reticulum (ER). The nascent protein elongates through a tunnel in the large ribosomal subunit (second ribosome) until the signal peptide is synthe- sized. Then the ribosome with signal peptide attaches to the docking protein (third ribosome). Translation resumes and the signal peptide is removed by a peptidase (the fourth ribosome). Translation continues until the synthesis of the protein is complete.

tive mechanism of secretion has been presented by Rothman, especially with respect to the pancreas (Rothman, 1977). This hypothesis states that secretory proteins can be preferentially released. This hypothesis is referred to as the nonparallel secretion hypothesis and proposes that there is a soluble cytoplasmic pool of secretory proteins that is in equilibrium with the proteins in the zymogen granule and that a stimulus can promote secretion of specific enzymes out of the cell.

SALIVARY GLAND SECRETION

The regulation of salivary gland secretions has been studied extensively and as early as 1833 by Mitscherlich. The reason for the interest lies in the accessibility of the secretions for epithelial physiologists to study the transport of electrolytes and water. Salivary secretions are composed of water, electrolytes, enzymes, glycoproteins, and growth factors. Saliva is important for lubrication and moistening of food for swallowing; for solubilizing material so it can be tasted; for cleansing the mouth and teeth to prevent caries; for initiation of starch digestion; for helping to clear the esophagus of refluxed gastric secretions; and for speech.

Functional Anatomy

There are three pairs of major salivary glands: the parotid, submandibular, and sublingual glands. The glands are tubuloalveolar structures as illustrated in Figure 6.26. The acinar aspect of the structure is referred to as the secretory end-piece. The secretory end-pieces of salivary glands are classified into three main groups depending on the cell type contained in the end-piece: serous, mucous, and mixed (Fig. 6.27). Mixed secretory end-pieces contain both serous and mucous elements. Salivary

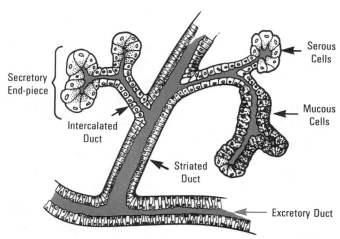

Figure 6.26. Diagrammatic representation of the secretory end-pieces and duct system of the human submandibular gland. This is a mixed gland containing both serous and mucous cells in the secretory end-pieces. (Modified from Young et al., 1987.)

glands containing predominantly serous or mucous cells in the end-pieces are referred to as serous or mucous glands, respectively. Glands containing end-pieces with both elements are referred to as mixed glands. The parotid gland is purely serous. The sublingual gland is largely mucous and the submandibular is a mixed gland.

Cells of the secretory end-piece are characterized by a basally located nucleus and RER (Fig. 6.27). The RER is developed to a greater extent in serous cells. Both mucous and serous cells contain apically located secretory granules. The apical secretory granules in serous cells are smaller and contain amylase and variable amounts of glycoprotein. Mucous cell granules are larger and filled with the specialized glycoprotein, mucin.

A branching system of ducts leading from the

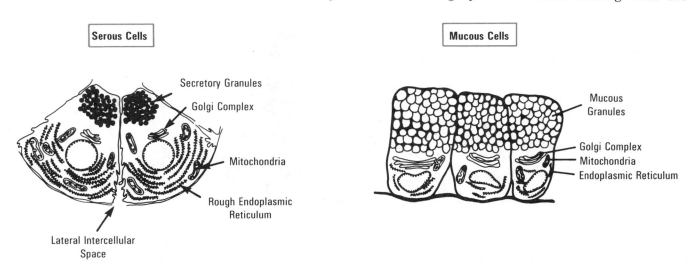

Figure 6.27. Diagrammatic illustration of salivary serous and mucous cells. (Modified from Young and van Lennep, 1978.)

secretory end-pieces to the oral cavity are referred to as intercalated, striated, and excretory ducts (Fig. 6.26). The intercalated ducts are lined by small cuboidal cells containing a prominent nucleus and scant cytoplasm. The intercalated ducts are interposed between the secretory end-piece and the striated ducts. The epithelial cells lining striated ducts take their name from the appearance of basal striations that represent extensive infoldings of the basal cell membrane around mitochondria and arranged in a series of parallel columns. This membrane infolding may be important for increasing the surface area for transport of electrolytes across the epithelium. The striated ducts empty into the excretory ducts that lead into the oral cavity. The epithelium of excretory ducts in most glands is composed of two layers—a superficial layer of columnar cells and a basal layer of flattened cells.

Secretory end-pieces and intercalated ducts are surrounded by myoepithelial cells. These cells are capable of compressing the underlying structures, forcing saliva into the main ducts.

Salivary secretion is exclusively under neural control. The salivary glands are innervated by both parasympathetic cholinergic nerves and sympathetic adrenergic nerves. In addition, immunohistochemical techniques have demonstrated the presence of VIP and substance P in nerve fibers of the salivary glands. Parasympathetic postganglionic cholinergic nerve fibers supply cells of both the secretory end-piece and ducts and parasympathetic stimulation results in the highest rates of salivary secretion. Sympathetic stimulation of the salivary glands causes a small and variable increase in secretion. The myoepithelial cells also are sympatheti-

cally innervated and the secretory response may be, at least in part, mediated by the ability of myoepithelial contractions to express secretions from the gland.

For a review of salivary gland junctional anatomy, see Leeson, 1967.

Composition of Salivary Gland Secretions

INORGANIC COMPONENTS

The volume of saliva excretion in humans is 1–1.5 liters/day. The composition of saliva depends on the particular salivary gland, the stimulus, and the rate of flow. The major inorganic components are sodium, potassium, chloride, bicarbonate, calcium, magnesium, and phosphate.

At rest, saliva is hypotonic with a sodium concentration of less than 5 meq/liter and potassium concentration of about 30 meq/liter (Fig. 6.28). With stimulation, the concentration of potassium decreases (but remains greater than the plasma concentration) while that of sodium increases toward the plasma concentration. With stimulation, the concentrations of both bicarbonate and chloride increase. The pH of resting salivary secretions is less than the blood pH, whereas, with stimulated secretory flow rates, the pH may be slightly greater than blood pH (7.8).

Thaysen postulated in 1954 that the secretion from secretory end-pieces (primary secretion) is isotonic and approximately plasma-like in electrolyte content (Thaysen et al., 1954). He further postulated that the primary secretion is modified during its subsequent passage along the gland ducts. Recent micropuncture studies have confirmed that the primary fluid leaving the secretory end-piece

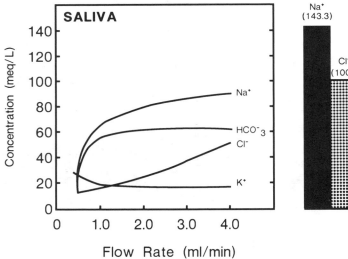
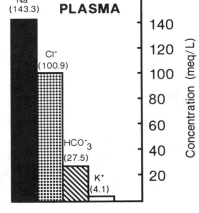

Figure 6.28. Relationship between the salivary concentration of sodium, potassium chloride, and bicarbonate; and salivary flow rate.

Plasma concentrations of the ions are shown on the *right*. (Redrawn from Thaysen et al., 1954.)

approaches electrolyte composition of plasma and that the concentrations change little with increasing flow rates.

Investigations of the mechanism of electrolyte transport from end-pieces (Young et al., 1987) provide evidence that during stimulation there is an increase in basolateral membrane K^+ conductance and apical membrane Cl^- conductance. The basolateral membrane also contains Na^+/K^+-ATPase and a furosemide-sensitive $Na^+/K^+/2Cl^-$ symport. A hypothetical model incorporating these findings is illustrated in Figure 6.29A. The driving force for the ion flow is the Na^+/K^+-ATPase which expels 3 Na^+ ions from the cell for 2 K^+ ions taken up by the cell creating an intracellular driving force for Cl^- exit. Also, the increased K^+ conductance of basolateral membranes during stimulation results in K^+ efflux. The resulting chemical gradients resulting from the movements of Na^+, K^+, and Cl^- drive the basolateral $Na^+/K^+/2Cl^-$ symport. Electroneutrality is maintained because the 6 Cl^- ions moving

across the apical membrane are balanced by a net 3 K^+ ions and 3 Na^+ ions moving out of the cell across the basolateral membrane by the K^+ channel and Na^+/K^+-ATPase. The circuit is completed by a Na^+ current passing paracellularly across semipermeable junctional complexes between cells. The mechanism by which agonists activate secretion is not currently understood. However, the model in Figure 6.29A suggests that secretion could occur with activation at one of several transports (i.e., Na^+/K^+-ATPase, Cl^- channel, or K^+ channel).

Agonists that mediate fluid, electrolyte, and protein secretion from secretory end-pieces include cholinergic (muscarinic) agents, adrenergic agents, and possibly substance P. These agents act to cause secretion presumably by their ability to stimulate metabolism of membrane phosphoinositides and to increase the concentration of free calcium in the cytosol (see "Regulation of Exocrine Pancreatic Secretion, Stimulus-Secretion Coupling" later in this chapter). β-Adrenergic agonists increase cyclic

A.   END-PIECE SECRETORY CELL

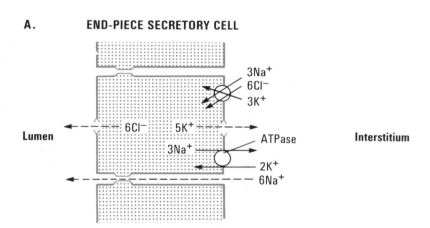

Figure 6.29. Hypothetic models of electrolyte transport in end-piece secretory cells (*A*) and duct absorptive cells (*B*). (Redrawn from Young et al., 1987).

B.   DUCT ABSORPTIVE CELL

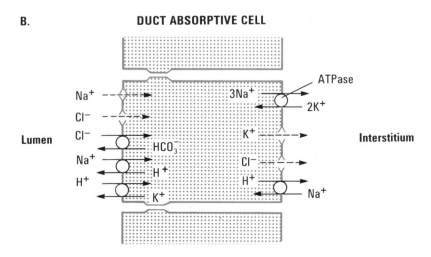

AMP resulting in amylase secretion but only small changes in fluid and electrolyte secretion.

As indicated earlier, the primary secretion from the end-piece is modified by its passage in the duct. The model in Figure 6.29B incorporates the current hypothesis of ductal action on the primary secretion. The duct acts to reabsorb sodium and chloride so that at low flow states the major portion of the NaCl is reabsorbed, resulting in a hypotonic fluid. Na^+ can enter the cell from the lumen by either a Na^+ channel or a Na^+/H^+ antiport. Cl^- can enter by either a Cl^- channel or a Cl^-/HCO^-_3 antiport. The driving force for Na^+ absorption comes from the basolateral Na^+/K^+-ATPase that maintains a low intracellular concentration of Na^+ and a high concentration of K^+. Cl^- leaves the cell by way of a basolateral Cl^- channel. K^+ that enters the cell by way of the Na^+/K^+-ATPase can leave the cell by way of a basolateral K^+ channel or an apical K^+/H^+ antiport. At low intraluminal Na^+ concentration, the K^+/H^+ antiport is more active than the Na^+/H^+ antiport. This leads to secretion of K^+ and an augmentation of HCO_3^- concentration in the lumen. There also appears to be a Na^+/H^+ antiport on the basolateral membrane that allows HCO_3^- accumulation in the cytosol.

The net result of ductal transport is NaCl absorption and $KHCO_3$ secretion. The absorption of NaCl is more rapid than the secretion of $KHCO_3$. Because the duct epithelium is relatively impermeable to water, the resulting intraluminal fluid is hypotonic. Both parasympathetic and α-adrenergic stimulation increase salivary flow rate, in part, by acting on ductal epithelium to inhibit Na^+ transport by both the Na^+/H^+ antiport and Na^+ channel leading to less Na^+ absorption.

ORGANIC COMPONENTS

The organic components are mainly proteins and glycoproteins that are synthesized, stored, and secreted by end-piece cells (Mandel and Wotman, 1976). These components include amylase, glycoproteins, secretory piece, kallikrein, lactoperoxidase, and lactoferrin. There are a few proteins that are synthesized, stored, and secreted by duct cells. These are nerve growth factor, epidermal growth factor, lysozyme, and ribonuclease. Amylase and glycoproteins are the major organic components of end-piece secretion. The glycoprotein of parotid salivary secretion is a calcium-binding, proline-rich protein. The backbone of the protein carries approximately six oligosaccharides with three mannose residues in their core linked to asparagine by N-glycosylation. The specialized glycoprotein, mucin, from sublingual glands is an o-glycosylated

protein with a large number of short oligosaccharides covalently linked to serine or threonine residues.

The secretion of proteins has been studied by measuring amylase release (Young et al., 1987). Studies using dispersed parotid acinar cells demonstrate that β-adrenergic agonists, α-adrenergic agonists, cholinergic agents, and substance P stimulate release of amylase. The action of β-adrenergic agonists is mediated by increases in cellular cyclic AMP, whereas, the actions of α-adrenergic agonists, cholinergic agents, and substance P are mediated by increased phosphoinositide metabolism and increases in the concentration of free cytosolic calcium.

Regulation of Salivary Gland Secretion

The control of salivary secretion is exclusively neural (Young et al., 1987). The flow rate of saliva during sleep is small. This spontaneous secretion keeps the mucous membranes moist. Stimulated secretion occurs by way of nervous reflexes. Neural mechanoreceptors and chemoreceptors in the oral cavity respond to dryness of the mucosa, chewing, chemicals in the food, and texture of the food. The afferent impulses are integrated in the medullary area of the CNS (salivation center). This center is in proximity to centers regulating respiration and vomiting. The salivation center also receives inputs from the cortex, amygdala, and hypothalamus, accounting for the conditioned salivation response described by Pavlov where dogs were conditioned to salivate at the sound of a bell and the response in humans to an appetizing meal before it is ingested.

Pathophysiology of Salivary Gland Secretion

Salivary gland secretion may be inhibited temporarily with infections or drugs such as anticholinergic agents. More permanent inhibition of secretion occurs with damage that occurs when head or neck tumors are irradiated. Permanent damage of salivary glands also occurs with Sjögren's syndrome, an immune-mediated disorder. These disorders lead to dryness of the mouth (xerostomia), difficulty with speech and swallowing, extensive dental caries, and disturbances of taste.

GASTRIC SECRETION

General Considerations

The stomach is not required to sustain life, yet it has a number of important physiological functions. First, it serves as a temporary reservoir following the ingestion of meals. The musculature of the proximal stomach (corpus) has the capacity to relax with the entry of nutrients. This is referred to as receptive relaxation, and is mediated by the vagus nerve. Therefore, as food enters the stomach it expands with a minimal increase in luminal pressure. This capacity to relax permits the ingestion of large meals over relatively brief time periods followed by the gradual delivery of nutrients into the small intestine over the subsequent few hours. Moreover, the distal portion of the stomach, the gastric antrum, triturates large solid particles into smaller particles. This markedly increases the surface area of solid nutrients, and thereby facilitates their absorption from the small intestine. In humans, solids are divided into particles of 2 mm in diameter or less before passage through the pylorus and entry into the duodenum. The process of gastric emptying is carefully regulated to avoid the rapid delivery of highly acidic or hyperosmolar substances into the small intestine.

The cells that compose the gastric mucosa secrete a number of important substances (Table 6.3). In humans, hydrochloric acid and intrinsic factor are secreted by the parietal, or oxyntic (acid-producing), cells. Pepsinogens are secreted principally by the chief, or zymogen, cells. Mucus and bicarbonate are produced by the mucous cells. Water and electrolytes are transported by the cells lining the gastric glands. This constellation of secretory products has a number of important physiological functions. For example, gastric acid kills ingested microorganisms and converts inactive pepsinogen into its active proteolytic enzymatic form, pepsin. Pepsins are proteolytic enzymes (peptidases) that initiate the hydrolysis of protein into smaller peptides and amino acid. Intrinsic factor is necessary to bind with vitamin B_{12}, also known as extrinsic factor, to

facilitate the absorption of vitamin B_{12} in the terminal ileum. Mucus secreted by the gastric mucous cells combined with surface epithelial bicarbonate secretion contributes to a barrier (the mucus-bicarbonate barrier) that prevents gastric acid and pepsin from damaging the gastric mucosa. Finally, water and electrolytes (sodium, potassium, calcium, chloride, and bicarbonate) lubricate and dilute food particles to produce a liquid suspension of nutrients prior to their entry into the duodenum. In spite of the stomach's multiple functions, it is not necessary for survival provided that vitamin B_{12} is administered parentally. However, in the absence of the stomach, the quantity and content of ingested nutrients requires judicious care.

Functional Anatomy

In humans, the stomach is divided into four portions: the cardia, fundus, corpus (body), and antrum (Fig. 6.30). The cardiac area is located just distal to the entry of the esophagus, is a few millimeters in length, and contains mucous cells, but not parietal or chief cells. The functional significance of the cardiac area is unknown. The fundus is the dome-shaped portion of the stomach that extends above the cardia. The body of the stomach is the largest portion representing about 80% of the entire stomach. The gastric antrum, or antral gland area, occupies about 20% of the stomach area (Ito, 1987).

The entire stomach is lined by a columnar epithelium of surface mucous cells. These cells also extend into the gastric pits, or foveola (Fig. 6.31). At the

Table 6.3.
Cells Within Gastric Glands and Their Secretory Product(s)

Gland Area	Cell(s)	Secretory Product(s)
Cardiac	Mucous	Mucus, HCO_3^-, pepsinogens (group II)
	Endocrine[a]	
Oxyntic	Parietal (oxyntic)	HCl, intrinsic factor
	Chief	Pepsinogens (groups I and II)
	Mucous neck	Mucus, HCO_3^-, pepsinogens (groups I and II)
	Enterochromaffin	Serotonin
	Endocrine[a]	
Pyloric	Mucous	Mucus, HCO_3^-, pepsinogens (group II)
	G cell	Gastrin
	Enterochromaffin	Serotonin
	Other endocrine[a]	

[a]Cardiac, oxyntic, and pyloric glandular mucosas contain at least nine different types of endocrine cells. In some of these cells the hormonal product has been identified. Examples include D cells (somatostatin) and A cells (gut glucagon). (From Feldman, 1983).

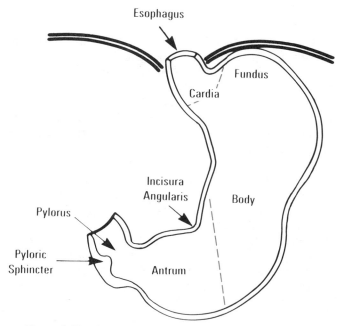

Figure 6.30. Anatomic regions of the stomach of carnivores.

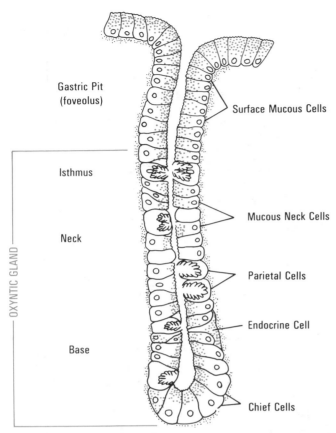

Gastric Pit
(foveolus)

Surface Mucous Cells

OXYNTIC GLAND

Isthmus

Neck

Base

Mucous Neck Cells

Parietal Cells

Endocrine Cell

Chief Cells

Figure 6.31. Diagrammatic representation of a gastric fundic gland. The stomach is lined with surface mucous cells that entend into the upper portions of the gastric pit. Mucous neck and parietal cells are located in the isthmus and neck regions, while chief cells are at the base. This diagram shows one gland attached to a gastric pit. In actuality, more than one gland are attached to each pit. (Redrawn from Ito, 1987, with permission.)

base of each gastric pit a few simple or tubular glands join together so that their combined secretory products exit together from the single gastric pit.

The glands in the mucosa of the fundus and body are referred to as parietal or oxyntic glands. These glands are divided into three areas: the isthmus, which contains surface mucous and parietal cells; the neck, which contains mainly parietal and mucous neck cells; and the base, which contains chief cells as well as some parietal cells (Fig. 6.31). Endocrine cells are also present scattered throughout the gland. Parietal cells make up approximately one-third of the cells lining the gland; chief cells account for about 20–25% of the cells lining the gland, and the remaining cells consist of about 20% surface mucous cells, 10% mucous neck cells, and a few endocrine cells.

The pyloric, or antral, gland area contains deep gastric pits lined by surface mucous cells. Deeper in

the pits are located the gastrin, or G, cells (see Fig. 6.34). As in the oxyntic gland area, scattered cells that contain biogenic amines, somatostatin, or other endocrine agents are present. Of note, the somatostatin-containing endocrine-paracrine cells have long cytoplasmic extensions that contain somatostatin and terminate in close proximity to potential effector cells such as G cells, chief cells, and parietal cells. Preliminary studies suggest that somatostatin exerts a local paracrine inhibitory effect via these cellular extensions. The glands of the cardiac area contain mucous lining cells and scattered endocrine cells.

The stomach has a rich blood supply that has been the subject of extensive investigation (Guth and Leung, 1987). Recent studies using radiolabeled microspheres indicate that mucosal blood flow increases with a meal and that blood flow is regulated independently of the secretory regulation (Perry et al., 1983). Although autonomic stimulation, biogenic amines, gastrointestinal hormones, prostaglandins, and leukotrienes affect blood flow, the physiological regulation of mucosal blood flow is unknown.

The stomach contains both sympathetic and parasympathetic nerves from the autonomic nervous system. The sympathetic fibers arise from the celiac plexus, while the parasympathetic fibers are derived from the right and left vagal trunks. As the vagal trunks enter the abdomen the left vagus divides: one branch innervates the liver (the hepatic branch), and a second supplies the anterior wall of the stomach (the gastric branch). The right vagus also divides after entering the abdomen, sending one branch (the celiac branch) to the small intestine and to the right colon, and the second branch to the posterior wall of the stomach. The preganglionic vagal fibers synapse with the enteric nervous system.

CELLULAR STRUCTURE

Mucous Cell

Mucous cells are columnar in shape. They line the entire stomach wall as well as extending down into the upper one-third of the gastric pits. These cells avidly absorb carbohydrates from the bloodstream and process them into special glycoproteins called mucins. When hydrated, mucins form mucus. The intracellular transport steps involved in glycoprotein synthesis are similar to glycoprotein synthesis in other organs (Neutra and Forstner, 1987). Gastric mucus is extruded from the surface mucous cells into the lumen by three mechanisms: exocytosis, a slow and continuous process; apical expulsion, a rapid release of stored mucus; and cell exfoliation, a slow and continous process that occurs in the intrafoveolar area and along the gastric surface.

Gastric mucus consists principally of carbohydrates (fucose, galactose, and N-acetylglucose amine). These carbohydrates are arranged as oligosaccharides with chain lengths ranging from 2–20 residues. Mucins have a molecular weight of between 2 million and 44 million. Allen has proposed a "windmill type" of structure with a protein core and carbohydrate side-chains (Allen, 1981), while others have suggested that gastric mucins consist of coiled threads. In either case, the mucins become filled with large volumes of water, some lipid, and protein after extrusion from the mucous cells. Mucus is viscous and has elastic properties.

The functions of gastric and duodenal mucus are first, to serve as a protective barrier from acid and peptic damage, as well as ingested potentially toxic nutrients (e.g., alcohol, aspirin, other drugs; see below). Secondly, mucus acts as a lubricant for food particles in their passage throughout the gut. The mucous layer provides a rather continuous barrier throughout the stomach and remainder of the intestine. In humans, it is approximately 200 μm in thickness. Mucus impedes the movement of pepsin (see below) as well as HCl from the lumen to the surface epithelial cells. Moreover, the surface epithelial cells secrete bicarbonate (see below). Therefore, the pH immediately adjacent to the surface epithelial cell is neutral while the pH within the gastric lumen may be less than 1.0 (Flemstrom and Turnberg, 1984). This barrier has been referred to as the mucus-bicarbonate barrier (Fig. 6.32).

Parietal Cell

The parietal (or oxyntic-acid secreting) cell has been the subject of intense study by anatomists,

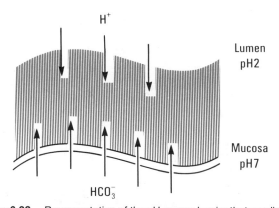

Figure 6.32.   Representation of the pH-mucus barrier that overlies the gastroduodenal mucosa. While the luminal pH is acidic, the pH adjacent to the epithelial cell membrane is near neutral due to epithelial bicarbonate production and the presence of the mucus layer. (Redrawn from Selling et al., 1987.)

physiologists, and pharmacologists (Ito, 1987). The parietal cell is of great interest because of its remarkable structure-function relationships (Fig. 6.33). The parietal cell is pyramidal, or triangular, in shape, relatively large (approximately 25 μm in diameter), and the basal side bulges into the lamina propria. The parietal cell is densely acidophilic due to the large number of densely packed mitochondria.

Electron microscopic examination of the parietal cell reveal unique structures. These are the tubulovesicular apparatus and the intracellular secretory canaliculi (Fig. 6.33). The microvilli of the secretory canaliculi are long and numerous and face the lumen of the stomach. They serve to increase the luminal cell surface area approximately four- to fivefold. H^+/K^+ ATPase (the enzyme involved in the final step of H^+ secretion) is located within the cell membrane of the intracellular canaliculi. Intrinsic factor has also been localized on this membrane. Immediately beneath the microvilli multiple tubulovesicles are located. When gastric acid secretion is stimulated, the microvillar surface increases rapidly. Moreover, the number of vesicles decreases as the microvillus membrane expands. Current evidence indicates that the tubulovesicles fuse with the cell membrane and thereby expand the intracellular microvillus membrane (Schofield et al., 1979). This process is reversible. Therefore, when gastric acid secretion decreases, the microvillus membrane decreases in size while the number of tubulovesicles increases.

The parietal cell also contains an unusually large number of mitochondria occupying approximately 30–40% of the cytoplasm. These are necessary for the high level of oxidative function involved in gastric acid secretion. The parietal cell also contains small amounts of endoplasmic reticulum (likely involved with the synthesis of intrinsic factor) as well as a small Golgi complex.

Chief or Peptic Cell

The chief cell is morphologically quite similar to the pancreatic acinar cell (see below). Chief cells are primarily located at the base of the oxyntic gland in the gastric corpus, absent in the cardiac gland area, and exceedingly rare in the gastric antrum. The chief cells stain with basic dyes indicating an abundant ER. The unique structures of the chief cell include the apically located zymogen granules and the abundant ER that is involved with pepsinogen synthesis. Pepsinogens are released from the apical cytoplasm principally by exocytosis. Therefore, following stimulation of pepsin secretion there is a decrease in the number of zymogen gran-

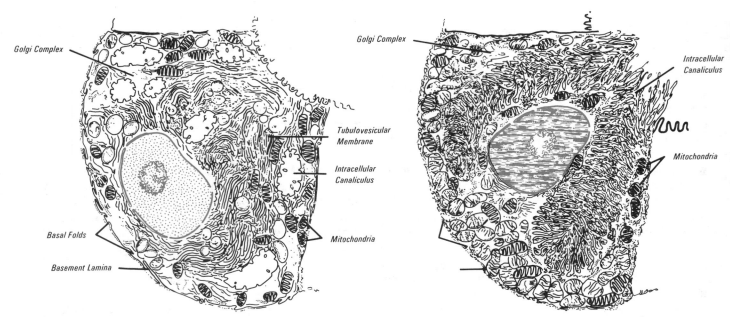

Figure 6.33. *Left,* a resting parietal cell is depicted. Note that the intracellular canaliculus is internalized as the tubulovesicular apparatus, and the cell has a limited microvillus membrane. Also note the extensive tubulovesicular membrane. Also note the extensive mitochondrial network. *Right,* the acid-secreting parietal cell is depicted. Note the abundant long microvilli and the decrease in the tubulovesicles. As HCl secretion occurs the tubulovesicles fuse to enlarge the intracellular microvilli. (Modified from Ito, 1987.)

ules in the apical cytoplasm. Recent work by Samloff using immunofluorescent techniques has demonstrated the presence of pepsinogen within the chief cells as well as the mucous neck cells (Samloff, 1971).

The intracellular events involved in the synthesis, storage, and secretion of pepsinogen are almost identical to those involved with pancreatic enzymes (see above).

Gastrin or G Cells

The G cells, located within the gastric antrum, are triangular in shape with a narrow apical cell border containing long microvilli (Fig. 6.34). It is postulated that these microvilli are able to sense the microenvironment in the gastric lumen. Granules within the G cell that contain gastrin are approximately 200 nm in diameter and are principally located in the basal cytoplasm. Gastrin is released principally across the basal cell membrane and is regulated in part by changes in the microenvironment of the microvilli.

Other Cells

There are also endocrine-paracrine cells at the base of the gastric glands containing histamine, somatostatin, serotonin, and other peptide hormones.

Gastric Electrolyte Secretion

There are two principal theories to account for the composition of gastric juice. One, referred to as the two-component hypothesis (Makhlouf, 1981),

postulates that gastric juice represents an admixture of nonparietal secretions that are secreted at a fixed rate plus a parietal cell component that varies according to the intensity of the stimulation of acid secretion. The second theory, the diffusion hypothesis, proposes that HCl is the primary secretion from the parietal cell and that hydrogen ion then

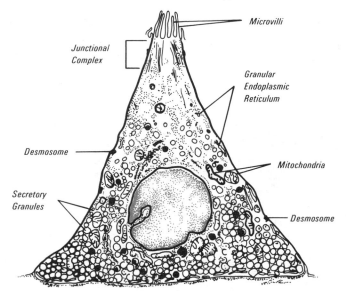

Figure 6.34. Representation of the gastrin cell (G cell) from the antral gland area. Note that the apical surface contains microvilli that likely sense the microenvironment of the gastric antrum. Also note the gastrin-containing granules at the base of the cell. (Redrawn from Ito, 1987.)

exchanges for sodium as the secretion passes from the glands through the gastric pits. Thus, when the parietal cell secretion of HCl is at a low rate, complete exchange results in NaCl secretion into the stomach lumen. In either case, when parietal cells are at rest, a neutral solution containing principally sodium and chloride is recovered (Fig. 6.35). However, when the parietal cells are stimulated and hydrogen ion secretion increases there is a reciprocal decrease in the sodium concentration. As the secretory flow rate increases there is also a modest increase in chloride concentration from approximately 120–160 meq/liter (Fig. 6.35). In humans, pure parietal cell secretion is estimated to contain approximately 149 meq/liter HCl.

Potassium is present in both parietal and nonparietal components. At high flow rates gastric juice consists almost entirely of HCl and KCl. Moreover the potassium concentration, which is similar to the plasma concentration at low rates of secretion, increases approximately twofold. Potassium moves into the lumen either by transcellular or paracellular pathways.

Gastric calcium concentration is similar to the concentration of ionized calcium in plasma at low flow rates. At high flow rates in humans gastric calcium concentration increases modestly. Chloride is the sole anion in the parietal secretion as well as the main anion in the nonparietal secretion. Water transport across the gastric mucosa occurs in

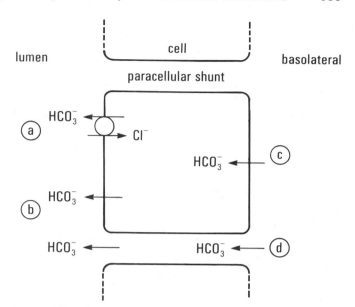

Figure 6.36. Proposed mechanisms for duodenal surface epithelial secretion of HCO_3^- in stomach and duodenum. HCO_3^- transport involves (*a*) electroneutral Cl^-/HCO_3^- exchange stimulated by some gastrointestinal hormones and inhibited by furosemide; (*b*) active electrogenic transport, independent of luminal Cl^-, stimulated by prostaglandins and insensitive to furosemide; anion carrier which under some conditions also displays an affinity for Cl^-; (*c*) and (*d*) movement of HCO_3^- through paracellular pathways which are sensitive to variations in transmucosal hydrostatic pressure. (Redrawn from Flemstrom, 1987.)

response to osmotic gradients that are created by the active transport of solutes, primarily HCl.

The gastric mucosa also secretes small quantities of HCO_3^- (Flemstrom, 1987). This was observed in animals with isolated antral gland pouches, free of parietal cells, and also in patients with pernicious anemia who do not secrete HCl. HCO_3^- is likely to be secreted both by the mucous cells within the gastric pits as well as those lining the surface epithelium. In humans, cholinergic stimulation by sham feeding or the cholinergic agonist, bethanechol, increases HCO_3^- secretion. However, the physiological regulation of HCO_3^- secretion is not yet fully understood. Possible transports involved in gastric HCO_3^- secretion are illustrated in Figure 6.36. At the luminal membrane there is a Cl^-/HCO_3^- exchange stimulated by glucagon as well as an active HCO_3^- transporter that is chloride independent and stimulated by prostaglandins and cyclic AMP. At the basolateral surface there is an anion carrier for HCO_3^-. Also, HCO_3^- diffuse into the lumen by paracellular movement.

ION TRANSPORT OF THE PARIETAL CELL

The pathways involved in ion transport across the parietal cell have been the subject of intense study (Forte and Wolosin, 1987; Sachs, 1987). Hydrogen ions are secreted into the gastric lumen

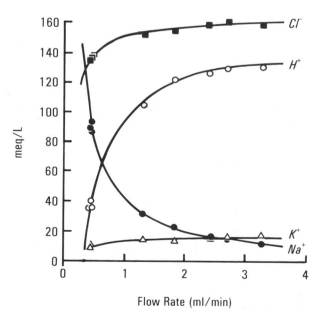

Figure 6.35. Relationship between gastric juice flow rate and gastric juice hydrogen, chloride, sodium, and potassium concentrations in humans. Note that as hydrogen ion concentration increases there is a reciprocal decrease in the sodium concentration. Also, with increased flow both potassium and chloride concentrations increase modestly. (Redrawn from Makhlouf, 1981.)

where the pH can be less than 1 while the pH in the oxyntic cell cystosol remains about 7.1, thereby generating a membrane concentration gradient of 6.4 log units or about 2½ million. This secretion is mediated by an active transport pump, the H^+/K^+-ATPase (Fig. 6.37). H^+ ion for the pump is made available by carbonic anhydrase which enhances the formation of H_2CO_3 from CO_2 and H_2O. H_2CO_3 dissociates to H^+ and HCO_3^-. To maintain intracellular neutrality as H^+ is secreted into the canaliculus, HCO_3^- is exchanged for Cl^- (HCO_3^-/Cl^- exchanger) across the basolateral membrane. The resulting increased intracellular Cl^- drives Cl^- through conductive pathways across the luminal surface of the cell. High intracellular K^+ and low Na^+ concentrations are maintained by the Na^+/K^+-ATPase at the basolateral membrane, while K^+ also moves through conductive pathways into the lumen (thereby providing K^+ for the H^+/K^+-ATPase) across the canalicular membrane. The overall process generates HCl secretion that is followed by passive H_2O flow (Fig. 6.37).

Parietal Cell Receptors

Since the gastric mucosa is a heterogeneous cell population, assessment of parietal cell function in vivo is complicated by the presence of a mixed cell population as well as interacting neural and humoral factors. To avoid these difficulties, isolated parietal cells or isolated gastric glands have been examined in vitro (Soll and Berglindh, 1987). Isolated gastric glands contain endocrine and chief cells along with parietal cells, but maintain their cellular polarity. Glands are obtained by stripping the stomach mucosa from the underlining layers followed by exposure to proteolytic enzymes. To obtain almost pure parietal cell preparations, the cells of the glands are dissociated. The parietal cells are then further separated by differential centrifugation.

Since the parietal cell secretes hydrogen ion from the intracellular canaliculus and an equal amount of bicarbonate ion from the basal surface of the cell, it is not possible to measure directly hydrogen ion secretion in either isolated glands or parietal cells. Indirect measurements of parietal cell function include oxygen consumption, glucose oxidation, and aminopyrine accumulation (Soll and Berglindh, 1987). Oxygen consumption occurs with acid secretion and can be directly measured. Glucose oxidation can be directly measured by determining the $^{14}CO_2$ generation from ^{14}C-labeled glucose. Aminopyrine is a weak base that accumulates within the intracellular canaliculus of stimulated parietal cells. It is lipophilic at neutral pH but becomes ionized at low pH. Thus, aminopyrine becomes trapped in the ionized state in the secretory canaliculus.

Three separate receptor classes have been identified on the parietal cells that mediate HCl secretion. These receptor classes interact with histamine, gastrin, or acetylcholine (Fig. 6.38).

Histamine Receptor. In isolated parietal cells and gastric glands, histamine stimulates aminopyrine accumulation, oxygen consumption, and glucose oxidation. Histamine H_2-receptor antagonists but not histamine H_1-receptor antagonists competitively inhibit the effect of histamine. Histamine stimulates parietal cell secretion in vitro by activating adenylate cyclase and increasing cyclic AMP (Soll and Wollin, 1979). The secretory response is enhanced markedly by the phosphodiesterase inhibitor, isobutylmethylxanthine. The response to histamine can also be augmented by cholinergic agonists and gastrin (Fig. 6.39). The intracellular mechanism that mediates the augmented responses between histamine and gastrin or histamine and cholinergic agents is not understood. However, when parietal cells are exposed to a combination of secretagogues, the blockade of the histamine response with H_2-receptor antagonists results in blockade of the augmented responses as well. Histamine H_2-receptor antagonists are potent

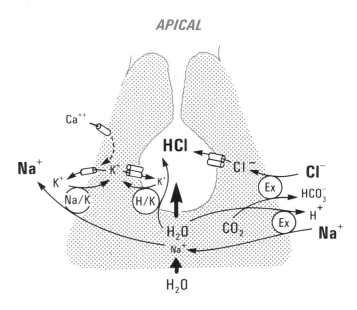

APICAL

BASOLATERAL

Figure 6.37.   Graphic representation of ion transport in the secreting parietal cell. On the *left hand* side the homeostatic mechanisms of K^+ are depicted while on the *right hand* side the homeostatic mechanisms involved with Cl^- are shown. Potassium is transported through conductive pathways at both the basolateral and apical membranes as well as by a Na^+/K^+ ATPase. Cl^- enters the cell via a neutral Cl^-/HCO_3^- exchange and then passes across the apical membrane via conductive channels. The final step involved with H^+ secretion involves the H^+/K^+ ATPase along the intracellular canaliculus. See text for details. (Redrawn from Forte and Wolosin, 1987.)

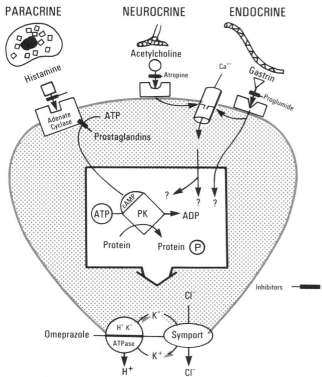

Figure 6.38. A model of the parietal cell representing the pathways for stimulation of acid secretion. Note that along the basolateral membrane there are three classes of receptors, one each for histamine, acetylcholine, and gastrin. Histamine stimulation is mediated through cyclic AMP while acetylcholine and possibly gastrin stimulation are mediated by the influx of calcium. Also note the final step in H$^+$ secretion is the H$^+$/K$^+$-ATPase proton pump located on the microvillus and tubulovesicular membranes. Prostaglandins inhibit acid secretion by inhibiting adenylate cyclase activation. (Redrawn from Olson and Soll, 1984.)

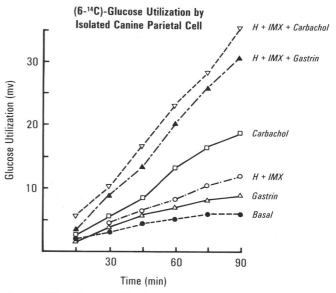

Figure 6.39. Glucose utilization by isolated canine parietal cells. Glucose utilization is a measure of H$^+$ secretion. Note that gastrin is a modest stimulant of acid secretion as is histamine (*H*) and isobutylmethylxanthine (*IMX*). The combination of agonists results in a potentiated response. (Redrawn from Soll and Berglindh, 1987.)

inhibitors of basal and stimulated gastric acid secretion.

Cholinergic Receptor. Cholinergic agents stimulate both isolated parietal cells and gastric glands to secrete HCl. They are generally the most effective natural secretagogues *in vitro*. Antimuscarinic agents inhibit the effect of carbachol. The M$_1$ and M$_2$ muscarinic antagonist, atropine, and the specific M$_1$ antagonist, pirenzepine, inhibit cholinergically stimulated aminopyrine accumulation.

Gastrin Receptor. The strongest evidence for the presence of a specific gastrin receptor are based on binding studies with ^{125}I-gastrin (Soll et al., 1984) that has been partially characterized as a 75 K dalton single protein. Gastrin produces only a small increase in oxygen consumption, aminopyrine accumulation, and glucose oxidation in isolated canine parietal cells (Fig. 6.39). This effect is not blocked with either H$_2$-receptor antagonists or anticholinergic agents indicating a specific parietal cell receptor for gastrin. Furthermore, proglumide, a receptor antagonist for both cholecystokinin and gastrin

receptors, inhibits the effect of gastrin but not histamine.

Activation of HCl Secretion

Stimulation of parietal cell function is linked to either calcium-dependent or cyclic AMP-dependent mechanisms (Soll and Berglindh, 1987). Histamine stimulates cyclic AMP production in isolated parietal cells while neither gastrin nor carbachol produce this effect. Although cyclic AMP mediates HCl secretion, the intracellular pathways mediating its effect are not known.

Studies in isolated parietal cells indicate that cholinergic and gastrin stimulation are associated with increases in intracellular calcium. These changes result from both the ability of these agents to mobilize intracellular calcium stores and to increase influx of extracellular calcium which, at least in part, mediates HCl secretion.

Inhibitors of Parietal Cell Function

There are specific receptor antagonists for histamine, acetylcholine, and gastrin. As mentioned above, proglumide, which blocks the gastrin receptor, has antisecretory properties in vitro, but its in vivo potency is low. The histamine H$_2$-receptor antagonists specifically block the histamine receptor. However, these agents are also potent inhibitors of all stimulants of gastric acid secretion. This suggests that histamine is involved in the secretory response either by serving as the single common final mediator of acid secretion on the parietal cell, or that histamine augments the H^+ secretory effect of gastrin or acetylcholine in the parietal cell. The model illustrated in Figure 6.38 is consistent with the second mechanism. Finally, anticholinergic drugs are modest inhibitors of acid secretion in vivo.

Inhibitors have been developed recently that block the H^+/K^+-ATPase (Sachs, 1987). This class of agents is referred to as the substituted benzimidazoles. An example, omeprazole, was designed to have a pKa of less than 5 and thereby becomes trapped in acidic spaces such as the intracellular canaliculus and tubulovesicular apparatus of the parietal cell, but not at the sites of other H^+/K^+-ATPases. In vivo omeprazole produces irreversible inhibition of the H^+/K^+-ATPase, leading to prolonged suppression of gastric acid secretion, approximately 24 hours in duration in humans.

MUCOSAL PERMEABILITY

The gastric mucosa is a tight epithelium (Diamond, 1978). This is reflected in its high transmucosal potential difference, high electrical resistance, low osmotic permeability, and steep solute gradients that occur in response to the active transport of ions. The permeability of the antral mucosa is somewhat greater than that of the corpus mucosa. Both are far less permeable than the duodenal and jejunal mucosa, which are leaky membranes to facilitate the absorption of nutrients, electrolytes, and water. Davenport conducted a classic series of experiments in dogs with denervated pouches designed to examine the movement of H^+ from the lumen and entry of Na^+ into the lumen (Davenport et al., 1964). Hydrogen loss is minimal and tends to equal sodium gain. Water flow across the intact gastric mucosa from the lumen is also negligible. Thus, the gastric mucosa is impermeable to HCl and water movement from lumen-to-blood. Gastric mucosal permeability to HCl, electrolytes, and water is increased by weak acids such as acetic acid, acetylsalicylic acid, and taurocholic acid. Moreover, this functional barrier is also damaged by alcohol, hyperosmolar substances such as urea, ethanol, and concentrated salt solutions. During, and shortly after epithelial damage, the mucosa becomes leaky permitting hydrogen ion to diffuse from the lumen into the submucosal area and sodium to enter the gastric lumen. The gastric mucosa promptly (within 30–60 min after most injury) undergoes reconstitution by the migration of epithelial cells from the gastric crypts to the surface lumen.

Phases of Gastric Acid Secretion

The gastric acid secretory response to a meal can arbitrarily be divided into three phases: cephalic, gastric, and intestinal. These, however, interact in concert (Fig. 6.40; Table 6.4).

BASAL SECRETION

Basal or interdigestive secretion occurs in the absence of environmental and gastrointestinal stimulation. It is measured usually 12 hours after a meal to ensure that complete digestion and absorption of the previous meal has occurred (Grossman, 1981). There is considerable variation between species in the amount of basal acid secretion. For example, in dogs, basal acid secretion is usually zero while in rats, basal secretion is approximately 30% of maximal acid secretion, whereas in humans, it is approximately 10% of maximal acid secretion (Fig. 6.41). In humans, a circadian rhythm of basal acid secretion exists, being lowest between 5 a.m. and 11 a.m. and highest in the evening between approximately 7 p.m. and 1 a.m.

The factors that regulate basal acid secretion are not fully understood. However, it is likely that tonic vagal, parasympathetic stimulation as well as small amounts of circulating gastrin contribute to basal secretion. This is based on the observation in animals and humans that following vagotomy (interruption of the vagal innervation to the stomach) or antrectomy (removal of the major source of gastrin) basal acid secretion is usually abolished. Basal acid secretion increases moderately during the activity

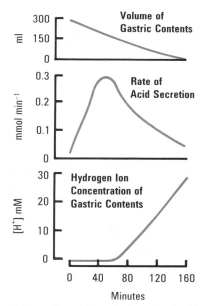

Figure 6.40. Volume of gastric contents, rate of acid secretion, and concentration of hydrogen ion and gastric contents at intervals after instilling a 300-ml peptone meal into the stomach. Note that the gastric contents only become acidified after the buffering capacity of the protein has been overcome. Luminal acidification occurs after the maximal rate of acid secretion is past and much of the meal has already emptied from the stomach. (Redrawn from Grossman, 1981.)

Table 6.4.
Phases, Initiators, Paths to Parietal Cell, and Mediators at Parietal Cell

Phase	Initiator	Path	Mediator at Parietal Cell
Cephalic	Feeding, sham or real[a]	Direct vagal	Acetylcholine
		Vagal-antral gastrin-blood	Gastrin
	Impairment of supply utilization of glucose by brain		
Gastric	Distention[a]	Vagovagal reflexes (same two postganglionic paths as for cephalic phase)	Acetylcholine
			Gastrin
		Local intramural reflexes (same two postganglionic paths as for cephalic phase)	
	Calcium, amino acids, peptides	Release of gastrin into blood by intraluminal action on G cells	Gastrin
Intestinal	Distension	At least in part by release into blood unidentified hormone	Enterooxyntin (postulated)
	Amino acids and peptides	At least in part by amino acids absorbed into blood	Amino acids

[a]In mechanisms involving both direct neural action on the parietal cell and neural release of gastrin, the relative importance of these two components varies with species, initiators, and conditions of testing. (From Flemstrom, 1987.)

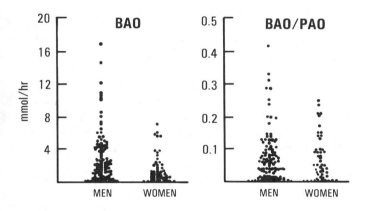

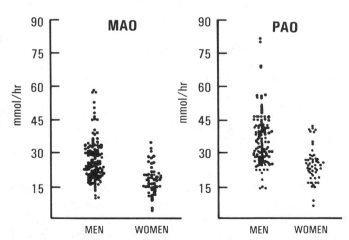

Figure 6.41. Basal acid output *(BAO)*, maximal acid output *(MAO)*, peak acid output *(PAO)*, and basal acid output/peak acid output *(BAO/PAO)* ratios, in healthy men and women. Note that both the *BAO* and *PAO* are less in women than in men. Also note that the *BAO/PAO* ratio is approximately 10%. (Redrawn from Feldman, 1983.)

front of the interdigestive myoelectric motility (Chapter 42).

CEPHALIC PHASE

Stimulation of the cephalic, or vagal, phase of gastric acid secretion is initiated by the thought, sight, smell, and taste of appetizing food. This stimulates afferent neural signals to the brainstem which, in turn, stimulate vagal efferent fibers to the oxyntic and antral gland portions of the stomach. Studies in dogs with cervical esophagostomies, which permit the animals to ingest the meal continuously while the meal exits from the esophagus and does not enter the stomach, reveal that the acid secretory response and its duration are related to the duration of feeding (Tache, 1987). In humans, the cephalic phase of gastric acid secretion is stimulated by having subjects visualize and smell an appetizing meal as well as chew the meal and then spit out the oral contents (Fig. 6.42). Another stimulus for the vagally mediated gastric secretion is the interference with the supply of glucose to the hypothalmus, medulla, and possibly areas of the diencephalon and limbic system of the brain. Therefore, insulin-induced hypoglycemia, or analogues of glucose that compete for glucose transport within the brain such as 2-deoxy-D-glucose (2-DG), stimulate vagal efferents. Because of their toxicity, insulin and 2-DG are not used in humans for diagnostic testing.

The brain contains numerous oligopeptides (Tache, 1987). Intraventricular injection of a number of peptides alters gastric secretion; for example, thyrotropin-releasing hormone (TRH) and synthetic analogues of TRH, cholecystokinin, and gastrin increase gastric secretion while bombesin, calcitonin gene-related peptide (CGRP), β-endorphin, and enkephalins inhibit acid secretion. The physiological regulatory roles of these agents require further investigation.

The vagal efferents from the brainstem to the stomach increase gastric acid and pepsinogen secretion by at least two mechanisms: first, by directly stimulating the oxyntic glands, and second, indirectly by stimulating the release of gastrin from the antral gland area. Preganglionic vagal efferents synapse with postganglionic neurons in close proximity to the oxyntic glands. Acetylcholine, released by postganglionic fibers, stimulates the parietal and chief cells directly. Secondly, the postganglionic vagal efferents also terminate near the gastrin cells resulting in the secretion of small quantities of gastrin. The peak vagally stimulated gastrin response precedes the peak gastric secretory resonse by about 15 min. Gastrin secreted in response to the cephalic stimulation augments the direct acetylcholine stimulation of the parietal cells as discussed previously.

The serum gastrin response to sham feeding is approximately one-half that produced by a protein-containing meal (Feldman, 1983). Acidification of the gastric contents to pH 1.5, thereby essentially abolishing gastrin secretion, decreases the acid secretory response to sham feeding by approximately 50% Furthermore, resection of the gastric antrum, the principal source of gastrin, reduces the acid secretion to sham feeding by approximately 50%. Therefore, a portion of the sham-induced acid secretory response is gastrin mediated while a part is gastrin independent.

GASTRIC PHASE

The gastric phase of acid secretion (Fig. 6.42) is mediated by gastric distension and factors that increase antral gastrin release. Distension of the stomach stimulates acid secretion by the activation of both "long" vagovagal reflexes as well as "short" cholinergic intragastric reflexes (Feldman, 1983). These latter "short" reflexes represent a complex network of cholinergic fibers within the wall of the stomach that travel to and from the corpus and antrum. Efferents in both types of reflexes stimulate acid secretion by both directly stimulating the oxyntic cells and indirectly by stimulating gastrin release from the antral gland area.

The acid secretory response to a mixed meal (protein, carbohydrate, and fat) is significantly greater than that to an equal volume of saline, and approximately 70% of the maximal acid output respose to

the gastrin analogue, pentagastrin. The major constituents of food that stimulate acid secretion are small peptides and amino acids. The amino acids that are the most potent stimulants of acid secretion include phenylalanine and tryptophan, while aspartic acid and leucine have a lesser effect (Walsh, 1987). Furthermore, phenylalanine and tryptophan, when infused into the stomach, release antral gastrin. Instillation of undigested albumin does not stimulate acid secretion more than its distension effect alone. Yet light peptic digestion of albumin into smaller peptide fragments results in a marked stimulatory effect. Therefore, large protein fragments require hydrolysis in order to induce acid secretion.

Water-soluble extracts of meat, as well as gravies, bouillon, and peptone, are strong stimulants of acid secretion and frequently are used as test meals in both animals and humans. Some, but not all, studies report a positive correlation between the increase in serum gastrin and the increase in acid secretion. It is likely that gastrin secretion is a major mechansim by which chemical agents stimulate gastric acid secretion (Feldman, 1983). It is also possible that amino acids may directly stimulate the parietal cells, a mechanism demonstrated in animals but not yet in humans. Gastrin is released by intragastric peptides, amino acids (particularly phenylalanine and tryptophan), and calcium (Walsh, 1987). Vagal stimulation also releases modest amounts of gastrin. Interestingly, after vagotomy serum gastrin increases promptly indicating the presence of vagal inhibitory fibers as well. Gastrin secretion is inhibited by antral acidification below pH 2.5, somatostatin, and some prostaglandins. The nature of the acid-induced inhibition of gastrin secretion is not fully understood. It may be secondary to the release of an inhibitor (e.g., somatostatin) or prevention of the interaction of amino acids with the receptors on the G-cell. About two-thirds of fasting circulating gastrin is in the form of gastrin-34 (G-34, big gastrin) while the remainder is primarily G-17 (little gastrin, heptadecapeptide gastrin). The secretory potency of G-17 and G-34 are similar, but the circulating half-time for G-34 is approximately four to five times longer (3 min vs. about 12 min, respectively).

Other chemical substances that stimulate acid secretion include caffeine and calcium salts (Feldman, 1983). In addition, a number of popular beverages such as Coca-Cola, Tab, 7-Up, coffee, milk, beer, and wine stimulate acid secretion. The stimulant activity of these beverages is not all due to caffeine, calcium, or protein since drinks without these

GASTRIC ACID SECRETION: PERCENTAGE OF PEAK G-17 ACID OUTPUT (MEAN ± SE)

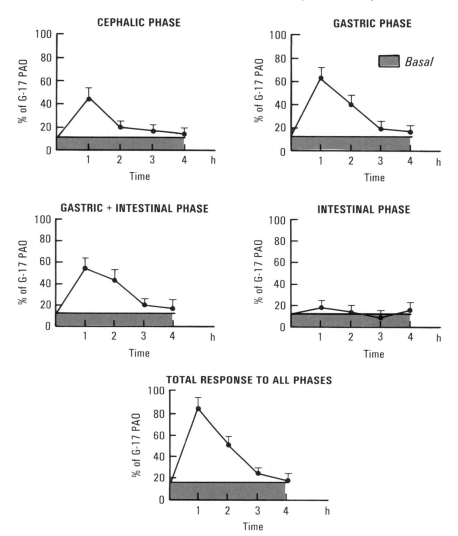

Figure 6.42. Responses to the cephalic, gastric, intestinal, and gastric plus intestinal phases as well as the total secretory response to a mixed (protein, carbohydrate, and fat) meal in humans. Note that the cephalic phase occurs within 1 hour and terminates shortly thereafter. Also note that the gastric phase is more prolonged, lasting into the second postprandial hour, while the total response to the meal is greater than the individual components and lasts somewhat longer.

substances have major stimulant effects on acid secretin.

INTESTINAL PHASE

The acid secretory response to a meal consisting of protein, fat, and carbohydrates in humans lasts for about 4 hours (Feldman, 1983). Therefore, since meal-stimulated acid secretion continues after the meal has emptied from the stomach, it was postulated that stimuli originating from within the small intestine stimulate gastric acid secretion. The secretagogues of the intestinal phase of gastric secretion in dog and humans are similar to gastric stimulants; that is, peptides and amino acids. In dogs, perfusion of the duodenum with peptides and amino acids increases gastric acid secretion to approximately 30% of the maximal acid response produced by histamine. Moreover, intestinal phase stimulation potentiates the response to stimuli such as histamine and gastrin.

In humans, perfusion of the proximal duodenum with either peptone, a mixture of small peptides and amino acids, or a mixture of L-amino acids also increases acid secretion to approximately one-third of the maximal gastric secretory rate with the gastrin analogue, pentagastrin. In humans, duodenal perfusion was not accompanied by a significant increase in serum gastrin suggesting that the intestinal phase operates independently of gastrin release. Subsequent studies revealed that intravenous infusion of a mixture of L-amino acids increased gastric acid secretion similar to intraduodenal amino acid infusion. Therefore, a substantial proportion of the response to intestinal amino acids is likely secondary to their absorption and entry into the circulation. Additional studies revealed that the specific amino acids that were most potent in stimulating acid secretion when infused intravenously include the aromatic amino acids, phenylalanine, and tryptophan.

Table 6.4 summarizes the phases of gastric acid secretion.

Inhibitors of Acid Secretion

CEPHALIC PHASE INHIBITORS

The cephalic phase of gastric acid and pepsin secretion can be abolished by vagotomy. In dogs with vagally denervated oxyntic gland (Heidenhain) pouches, sham feeding decreases acid secretion from the pouch during a low-dose background infusion of pentagastrin. These findings suggest that at least in dogs cephalic stimulation releases a circulating inhibitor of gastric acid secretion. The precise substance or substances that are released have not been isolated. Furthermore, these observations have not been evaluated in humans.

Injection of neuropeptides such as gastrin-releasing peptide, bombesin, neurotensin, calcitonin gene-related peptide (CGRP), and corticotropin releasing factor into either the cisterna magna or cerebral ventricles causes inhibition of acid secretion (Tache, 1987). These peptides act by either stimulating sympathetic nerve activity or inhibiting parasympathetic activity. The physiological role of the peptides in the brain in regulating acid secretion is unknown.

GASTRIC PHASE INHIBITORS

The major inhibitor of the gastric phase of acid secretion is antral acidification to a pH of 2 or less. This abolishes gastrin secretion. Furthermore, although antral distension has principally a stimulatory effect on gastric acid secretion, balloon distension in humans produces a modest inhibitory effect of pentagastrin-stimulated acid secretion. Therefore, antral distension produces both stimulatory and inhibitory effects.

INTESTINAL PHASE INHIBITORS

Three substances inhibit gastric acid secretion when instilled into the proximal small intestine:

acid, fat, and hyperosmolar solutions (Debas, 1987). Moreover, these agents also delay gastric emptying.

When unbuffered acid is introduced into the duodenum, gastric acid secretion stimulated by a meal or gastrin is inhibited markedly. This effect is possibly mediated by secretin secretion from the duodenum. In dogs secretin plays a physiological role in the inhibition of gastric acid secretion (Debas, 1987). The physiological role of secretin in humans as an inhibitor of acid secretion is still not fully resolved although preliminary data suggest a physiological inhibitory function.

Luminal fatty acids or monoglycerides are also potent inhibitors of acid secretion (Fig. 6.43). Those fatty acids with 10 or more carbons are most effective. The humoral agent responsible for the fat-induced inhibition of acid secretion remains unknown. When fat is perfused into the proximal small intestine several gut peptides are released. These include gastric inhibitory peptide (GIP), neurotensin, glucagon, vasoactive intestinal polypeptide (VIP), and cholecystokinin (CCK). The humoral agent, or agents, that inhibit gastric acid secretion are referred to as enterogastrones. Peptides that inhibit gastric secretion include GIP, neurotensin, and peptide YY (PYY). GIP is released by fat. However, in humans, GIP is only a weak inhibitor of acid secretion (Debas, 1987). This would suggest that at least in humans GIP is not a major enterogastrone. Another candidate enterogastrone is PYY. This hormone is released by fat in the distal small intestinal and proximal colon. Furthermore, PYY inhibits meal-stimulated acid secretion. However, PYY does not inhibit acid secretion stimulated by pentagastrin or histamine in the dog. Therefore, it is unlikely that PYY plays a major role as an enterogastrone. The most likely current candidate as an enterogastrone is neurotensin. It too is released by fat in the

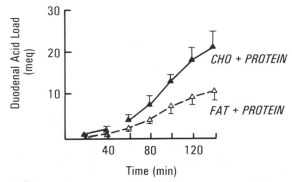

Figure 6.43. Duodenal acid load (acid delivered to the duodenum) and gastric emptying rates following an equicaloric meal containing protein plus carbohydrate *(CHO)* compared with protein plus fat. Note that the fat-containing meal significantly suppresses the secretory response.

distal small intestine and inhibits gastric acid secretion. It is a very potent inhibitor in the innervated stomach. Following vagotomy, the inhibitory effect of neurotensin is abolished. In humans there is a direct correlation between the plasma response to neurotensin and the inhibition of pentagastrin-stimulated acid secretion.

In summary, there are a number of humoral agents released from the small intestine as well as the colon that inhibit gastric acid secretion. However, their physiological roles require further study.

Pepsinogens and Intrinsic Factor

PEPSINOGENS

Pepsinogens are a series of inactive proenzymes that upon conversion to pepsins can hydrolyze peptide bonds, particularly those formed by aromatic amino acids (Samloff and Townes, 1970). Gastric pepsins initiate hydrolysis of ingested proteins; however, they are not required for amino acid absorption due to the excess of pancreatic peptidases. As indicated in Table 6.3 pepsinogens are located principally in the chief and mucous cells. There are three major classes of pepsinogens: pepsinogen I (group 1), pepsinogen II (group II), and an electrophoretically slow-moving protease (SMP) (Samloff and Townes, 1970). In the presence of acid, pepsinogens are converted to active pepsins by the loss of variable NH_2-terminus sequences. This occurs rapidly at pH values of less than 2, is autocatalytic, and is slower as the pH approaches 5. Pepsins are denatured irreversibly at pH values of 7 or greater. The pH optimum value of the group I pepsins is about 1.5–2, while it is slightly higher for the group II pepsins. Groups I and II are present in the gastric cardia and corpus, but only the group II pepsinogens are located in the gastric antrum and duodenum (Samloff and Townes, 1970). The group I pepsinogens can be further distinguished by their electrophoretic mobility into five fractions, while the group II pepsinogens contain two fractions. Pepsinogens I and II can also be measured in blood by radioimmunoassay. Pepsinogen secretion occurs in response to those agents that stimulate gastric acid secretion and is inhibited by those agents that decrease acid secretion. For the most part, pepsinogen secretion is qualitatively similar to acid secretion. Cephalic-vagal stimulation is the most potent stimulus of pepsinogen secretion. In addition, luminal HCl stimulates pepsinogen secretion.

INTRINSIC FACTOR SECRETION

Intrinsic factor is a glycoprotein with a molecular weight of 45,000. Intrinsic factor is synthesized and secreted by the parietal cells and required for the active absorption of vitamin B_{12}, cyanocobalamine, or extrinsic factor (Donaldson, 1987). Vitamin B_{12} is ingested principally with dietary protein and released by intragastric peptic digestion. Intrinsic factor secretion occurs by membrane translocation and is stimulated by gastric acid secretagogues (e.g., histamine, gastrin, cholinergic agents) (Donaldson, 1987). However, in contrast to gastric acids, intrinsic factor secretion reaches a peak response shortly after stimulation suggesting a washout of stored intrinsic factor. In the gastric lumen intrinsic factor competes with R proteins for the binding of vitamin B_{12}. R proteins are a family of glycoproteins that migrate rapidly ("R") during electrophoresis and are secreted into the gastric lumen. However, in the small intestine, the R protein-vitamin B_{12} complex is cleaved by pancreatic peptidases thereby permitting intrinsic factor to bind the free B_{12} and facilitate ileal absorption (Donaldson, 1987).

DUODENAL MUCOSAL BICARBONATE SECRETION

The proximal duodenum is the crucible in which gastric acid is neutralized. In order to prevent acid-peptic damage to the duodenal mucosa prompt neutralization of H^+ and inactivation of pepsin are required. The mechanisms of neutralization of H^+ involve surface epithelial, pancreatic, and hepatobiliary bicarbonate production. Recent studies indicate that the duodenal surface epithelial cells secrete bicarbonate at rest as well as in response to a number of agonists (Flemstrom, 1987; Hogan and Isenberg, 1988). Bicarbonate production is greater in the proximal than the distal duodenum. Although it is possible that the Brunner's glands, located in the proximal duodenum, produce bicarbonate, epithelial bicarbonate secretion is present in amphibians, a species that does not contain Brunner's glands. The two substances that likely have a physiological role in regulating duodenal bicarbonate secretion include hydrochloric acid and prostaglandins of the E class. Duodenal bicarbonate secretion increases promptly in response to luminal acidification (Fig. 6.44) as well as topical prostaglandin E_2 or its analogues. Furthermore, suppression of cyclooxygenase, a key enzyme in prostaglandin synthesis, decreases duodenal bicarbonate production. Also, vagal stimulation either electrically in animals or by sham feeding in humans is a potent, and likely physiological, agonist of duodenal bicarbonate secretion.

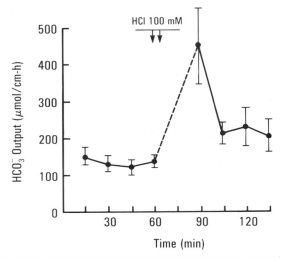

Figure 6.44. Proximal duodenal mucosal bicarbonate secretion in humans. Note that at rest the proximal duodenum, the duodenal bulb, secretes about 600 μmol/hour; also, brief acidification of the mucosa results in a prompt increase in mucosal bicarbonate output. (From Isenberg, 1987.)

Pathophysiological Correlates

There are numerous examples of pathophysiological relationships between abnormalities in gastric acid secretion and human disease (Grossman et al., 1982). Only two will be commented upon: (1) gastrinoma (Zollinger-Ellison syndrome); and (2) duodenal ulcer disease.

Gastrinoma is a gastrin-producing neoplasm originating in non-β cells of the islets of Langerhans in the pancreas. These tumors synthesize and secret gastrin. The clinical effects are largely due to the increased circulating gastrin. Gastrin stimulates both acid secretion and growth of the oxyntic gland mucosa including the parietal cells. The hypergastrinemia results in an increased basal (or resting) gastric acid secretion, and the increased parietal cell mass results in an increased response to gastric secretagogues (e.g., pentagastrin, food). The elevated basal and stimulated gastric acid secretion result in acidification of the duodenum (normally at a near neutral pH). This, in turn, frequently results in severe ulceration of the duodenum, damage to the jejunal mucosa, and inactivation of pancreatic enzymes. Peptic ulcers can produce severe complications such as hemorrhage, penetration into an adjacent viscus, perforation into the peritoneum, or obstruction to the passage of nutrients. Jejunal mucosal damage often results in diarrhea from impaired absorption of water and nutrients. Inactivation of pancreatic enzymes, particularly lipase, results in impaired absorption of fat, producing steatorrhea and weight loss. Prior to the development of the histamine H_2-receptor antagonists and the H^+/K^+-ATPase inhibitors, treatment was primarily surgical removal of the stomach. Many patients with gastrinoma are now treated quite satisfactorily by these potent antisecretory drugs.

A second example of a disease due to abnormalities of gastric secretion is duodenal ulcer disease. Ulcer disease develops when the normal balance between aggressive and defensive factors is altered. Although genetic and other factors (e.g., cigarette smoking) are involved with ulcer pathogenesis, patients with duodenal ulcer disease as a group have an increased parietal cell mass; elevated basal and stimulated gastric acid and pepsin secretion; impaired suppression of gastrin release when the gastric contents are acidic; rapid gastric emptying; and a more acidic duodenal bulbar pH when contrasted to normal subjects. Therefore, duodenal ulcer disease is due to the interaction of a number of factors (multifactorial). Also, surface epithelial bicarbonate secretion is impaired at rest and following acidification in patients with duodenal ulcer disease, indicating a defect in duodenal defense mechanisms. Therapy is currently directed at diminishing the aggressive factors by decreasing acid and pepsin secretion. However, newer forms of therapy are being tested to increase the defensive factors.

PANCREATIC SECRETION

The pancreas is both an exocrine and endocrine secretory organ. The exocrine secretions are composed of digestive enzymes that are essential for the digestion of lipids, carbohydrate, and protein in the gut; and a bicarbonate rich fluid that is an important factor in neutralizing the gastric hydrochloric acid emptied into the duodenum. The endocrine secretions are from the islets of Langerhans. The islets contain cells that synthesize and secrete glucagon, insulin, somatostatin, and pancreatic polypeptide into the blood. The endocrine pancreas is discussed elsewhere in this text. However, recent studies demonstrate that venous blood draining the islets passes to the acinar cells before returning to the systemic circulation (Williams and Goldfine, 1987). This drainage arrangement should expose acinar cells surrounding the islets to relatively high concentrations of islet hormones. A case for the physiological role for endocrine secretions in pancreatic exocrine function is strongest for insulin. In the exocrine pancreas insulin has long-term effects on the regulation of the synthesis of digestive enzymes.

Functional Pancreatic Anatomy

ACINUS AND DUCT

The exocrine pancreas is composed functionally of an acinus and its draining ductule (Fig. 6.45). These functional units are arranged into pancreatic lobules. The organization of acini and intralobular ducts within the lobule may be quite complex as illustrated in Fig. 6.46. The acini can be spherical, tubular, or irregular in shape. An acinus may be the terminal structure of the duct or it may be interpolated between ducts. In addition to intralobular ducts, the duct system of the pancreas is composed of interlobular ducts and the main pancreatic ducts. The interlobular ducts drain the secretion of the lobule into the main pancreatic duct while the main pancreatic duct drains exocrine secretions into the duodenum. In addition to transporting secretions, the ducts secrete an electrolyte solution rich in bicarbonate.

The exocrine pancreas also contains centroacinar cells. These cells constitute the final subdivision of the pancreatic duct and abut the cells of the acinus (Fig. 6.45). Centroacinar cells are probably involved in electrolyte transport.

The cells of the acinus (acinar cells) synthesize, store, and secrete digestive enzymes. They are polarized with zymogen granules containing stores of digestive enzymes restricted to the apical cytoplasm (Fig. 6.47). The nucleus is basally located and the paranuclear and basal areas of the cytoplasm are rich in RER. The area between the nucleus and zymogen granules is rich in Golgi complex. The apical surface of the acinar cell possesses microvilli. In the cytoplasm underlying the apical plasma membrane there is a filamentous meshwork consisting of actin that effectively excludes cell organelles such

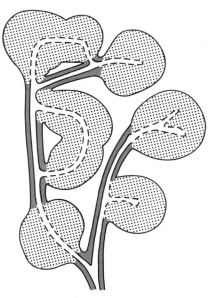

Figure 6.46. Organization of acini and intralobular ducts. The *shaded areas* represent acinar cells and the *blue areas* represent ductules. (Adapted from Akao et al., 1986.)

as the zymogen granules from this subapical zone in the resting cell. During stimulation of enzyme secretion this apical meshwork appears reduced in areas of approximation of zymogen granule and apical plasma membrane suggesting that it may have a role in the secretory process. Microtubules are distributed randomly in the apical cytoplasm.

The acinar lumen is separated from the interstitial space by specialized connections between acinar cells, the junctional complexes (Fig. 6.47). These complexes are located near the apical surface and act as a barrier to passage of large molecules. The junctional complexes also act as permeable barriers to the passage of water and ions.

Another intercellular connection between acinar cells is the gap junction. This specialized area of the plasma membrane between adjacent cells acts as a pore to allow small molecules to pass between cells. The gap junction allows chemical and electrical communication between cells.

The duct epithelium contains cuboidal to pyramidal cells that are devoid of zymogen granules and have little RER. The duct cells as well as the centroacinar cells contain carbonic anhydrase, which is important for their ability to secrete bicarbonate. The main pancreatic duct and interlobular ducts also contain mucus-secreting cells.

INNERVATION

The pancreas receives innervation from the vagal and splanchnic nerves (Holst, 1987). Vagal pregan-

DUCTULE   ACINUS

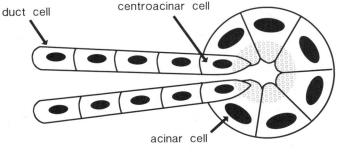

duct cell

centroacinar cell

acinar cell

Figure 6.45. Functional unit of the exocrine pancreas. (Adapted from the Undergraduate Teaching Project in Gastroenterology and Liver Disease, American Gastroenterological Association, Unit 16, Exocrine Pancreas: Pancreatitis, 1984.)

Figure 6.47. Diagram of the ultrastructure of the pancreatic acinar cell. Microfilaments and gap junctions are not shown but drawn in the text. (Adapted from Case, 1978.)

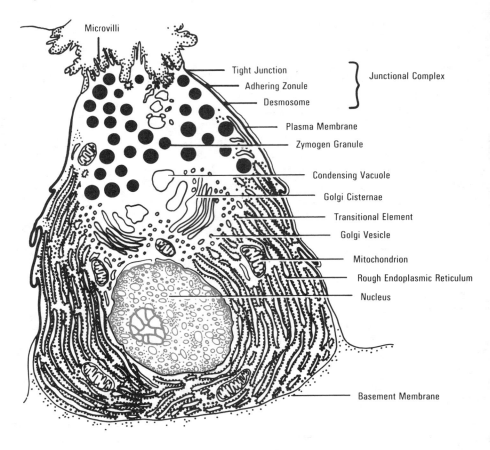

Microvilli

Tight Junction
Adhering Zonule — Junctional Complex
Desmosome

Plasma Membrane

Zymogen Granule

Condensing Vacuole

Golgi Cisternae

Transitional Element

Golgi Vesicle

Mitochondrion

Rough Endoplasmic Reticulum

Nucleus

Basement Membrane

glionic neurons synapse with collections of postganglionic neurons in the substance of the pancreas. Postganglionic neurons are distributed to acini and ducts. Vagal stimulation causes stimulation of pancreatic secretion by a parasympathetic mechanism by release of acetylcholine from postganglionic neurons. The acinar cell has muscarinic receptors for acetylcholine that mediate digestive enzyme secretion. Vagal stimulation also results in release of VIP from peptidergic nonadrenergic noncholinergic postganglionic neurons. The acinar cell has receptors for VIP that mediate enzyme secretion suggesting a physiological role for VIP in secretion. Recent studies have identified fibers and cell bodies for other peptides, including gastrin-releasing peptide (mammalian analogue of bombesin), substance P, cholecystokinin, and enkephalins. Gastrin-releasing peptide, substance P, and cholecystokinin each stimulate enzyme secretion from preparations of exocrine pancreas from various animals so that their presence in intrapancreatic neurons may have physiological importance.

Sympathetic innervation of the pancreas is by way of postganglionic neurons from the celiac ganglia. Sympathetic output is inhibitory to stimulate secretion from the exocrine pancreas.

Composition of Exocrine Pancreatic Secretion

Exocrine secretion accounts for the physiological function of the pancreas to digest components of the meal and neutralize the hydrogen ion emptied from the stomach into the duodenum. For the purposes of this discussion the exocrine secretion will be divided into inorganic and organic components.

INORGANIC COMPONENTS

The inorganic components of the exocrine secretion are water, sodium, potassium, chloride, bicarbonate, calcium, and magnesium.

Pancreatic juice secreted during stimulation is clear, colorless, alkaline, and isotonic with plasma (Zimmerman and Janowitz, 1987). The flow rate increases from 0.2–0.3 ml/min in the resting state to 3.0 ml/min during stimulation. Total volume of secretion per day is about 2.5 liters. The osmolality does not change with flow rate. During secretin stimulation the bicarbonate and chloride concentrations vary inversely (Fig. 6.48). The bicarbonate concentration varies between 25 meq/liter (low flow state) and 150 meq/L (high flow state). From the results of the micropuncture and microperfusion technique in animal exocrine pancreas, the simplest explanation for the variation in chloride and bicar-

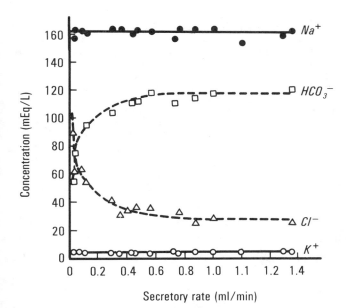

Figure 6.48. Relationships between pancreatic juice ion concentrations and secretory flow rate. (Adapted from Bro-Rasmussen et al., 1956.)

bonate concentrations is that with secretin stimulation there is a high volume secretion with a high concentration of bicarbonate originating from the duct system (Case and Argent, 1987). Because the secretory flow rate from the acini is relatively small (even if stimulated to secrete) the electrolyte composition of the combined secretion from acini and ducts will approach that of ductal secretion when ductal secretion is stimulated.

Micropuncture and microperfusion techniques in pancreas from several animals indicate that fluid secreted from the acinus contains chloride concentrations that are significantly greater than that of the fluid secreted from the ducts (Case and Argent, 1987). However, the concentration of chloride in acinar secretion varies between species. The concentrations of chloride and bicarbonate secreted by acinar cells probably change little with stimulation of acinar cell secretion. Specific stimulation of secretion from acini results in a low volume secretion. In animals, where the chloride concentration secreted by the acinus is less than that of plasma, the acinar fluid can be modified by the ducts so that the concentration of chloride increases as the fluid traverses in the ductal system. This is thought to occur via a HCO_3^-/Cl^- antiport.

The mechanisms of ion transport in ductal epithelium are poorly understood (Case and Argent, 1987). Current hypotheses favor an active transport of hydrogen ion from the epithelial cell into the interstitial space (i.e., away from the lumen) because the pancreas will actively secrete buffer

anions other than bicarbonate. Since it is highly unlikely that a single ion pump could handle anions of differing structure, the most likely explanation for their active transport is the movement of protons. Another finding, favoring an active proton pump, is that secretin stimulation causes a fall in pH of the pancreatic venous drainage. Several models of ductal ion transport have been proposed, one of which is illustrated in Figure 6.49. In this model, bicarbonate is derived from metabolic CO_2. CO_2 is converted to bicarbonate by carbonic anhydrase in the presence of water. Bicarbonate is transported into the duct lumen by a HCO_3^-/Cl^- antiport. Other ions and water enter the lumen via the paracellular space. The cellular content of bicarbonate is dependent on a Na^+/H^+ antiport. The H^+ leaving the cell enhances the production of HCO_3^-. The Na^+/H^+ antiport is driven by the basolateral Na^+/K^+-ATPase, which removes cell Na^+ that enters through the Na^+/H^+ antiport.

The cellular mechanism by which secretin stimulates bicarbonate secretion is poorly understood. Secretin activates adenylate cyclase and increases cyclic AMP in rat pancreatic duct cells indicating that cyclic AMP probably mediates the active secretion of bicarbonate (Schulz, 1980). However, the cellular mechanism of action of cyclic AMP is unknown.

Calcium and magnesium are present in pancreatic juice at about 25–35% of their concentrations in the plasma. With stimulation of acinar secretion, both calcium and magnesium enter the pancreatic secretions.

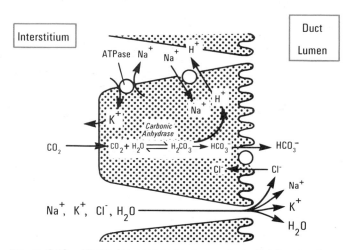

Figure 6.49. Model for pancreatic duct cell electrolyte secretion. (Adapted From The Undergraduate Teaching Project in Gastroenterology and Liver Disease, American Gastroenterological Association, Unit 16, Exocrine Pancreas: Pancreatitis, 1984.)

ORGANIC COMPONENTS

The human pancreas has an extraordinary capacity to synthesize protein. It produces and secretes more protein (mostly digestive enzymes) per gram of tissue than any other organ. The concentration of protein in pancreatic juice is dependent on the rate of secretion of the pancreatic acinar cells. Classes of digestive enzyme that hydrolyze proteins, carbohydrate, lipids, and nuclei acid are listed in Table 6.5. Several of the enzymes are stored in the acinar cell and secreted into the pancreatic juice as inactive precursors. These are shown in Figure 6.50. Their storage as inactive precursors prevents them from digesting the pancreas. Activation of these enzymes takes place in the duodenum where the brush border enzyme, enterokinase, activates trypsinogen by removing (by hydrolysis) an N-terminal portion of the molecule (Fig. 6.51). Then active trypsin catalyzes the activation of more trypsinogen and activation of all of the other inactive proenzymes.

In addition to digestive enzymes, the acinar cell secretes a trypsin inhibitor and procolipase into pancreatic juice. Trypsin inhibitor inactivates trypsin by forming a relatively stable complex with the enzyme. The function of this inhibitor is probably to inactivate any trypsins formed autocatalytically in the pancreas or pancreatic juice. This would block the effect of the small amounts of active trypsin formed in the pancreas from activating the cascade of pancreatic enzymes.

The colipase is secreted by the acinar cell and is a cofactor for lipase. It acts by preventing the inhibitory effect that bile acids have on lipase activity.

Other proteins present in small amounts in pancreatic juice include immunoglobilins, kallikrein,

Table 6.5.
Digestive Enzymes in the Pancreatic Acinar Cell

Proteolytic Enzymes	Lipolytic Enzymes
Trypsinogen	Lipase
Chymotrypsinogen	Prophospholipase A_1, A_2
Proelastase	Nonspecific esterase
Procarboxypeptidase A	
Procarboxypeptidase B	Nucleases
	Deoxyribonuclease (DNAse)
Amylolytic Enzyme	Ribonuclease (RNAse)
α-Amylase	
	Others
	Procolipase
	Trypsin inhibitor

lysosomal enzymes, alkaline phosphatase, and small amounts of albumin.

For a review of the organic components, see Rinderknecht, 1987.

DIGESTIVE ENZYME SYNTHESIS AND TRANSPORT

As discussed early in this chapter, the pancreatic synthesis of exportable protein takes place in the RER. The synthesized proteins are transported in an orderly fashion to the Golgi complex where modification and concentration take place. Secretory or zymogen granules are formed from Golgi complex and move to the apical portion of the cytoplasm where they are stored until a neurohumoral stimulus activates exocytosis and secretion.

The mechanisms involved in pancreatic tissue-specific expression of genes for digestive enzyme have been partially elucidated recently (Swift et al., 1984; Walker et al., 1983). Studies demonstrate that genes for the enzymes chymotrypsin and elastase contain enhancer regions in the 5′ flanking nucleotide sequence that regulate the transcription of their messenger RNAs. Recombinant DNA and

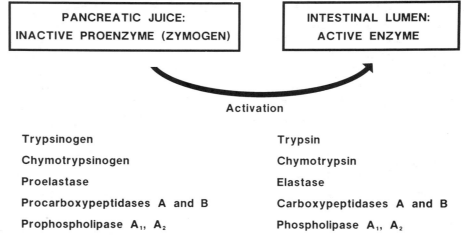

Figure 6.50. Proenzymes activated in the intestinal lumen. (Adapted from The Undergraduate Teaching Project in Gastroenterology and Liver Disease, American Gastroenterological Association, Unit 16, Exocrine Pancreas: Pancreatitis, 1984.)

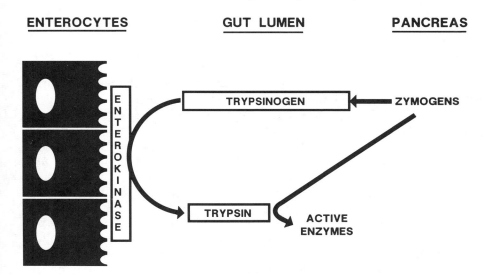

ENTEROCYTES   **GUT LUMEN**   **PANCREAS**

Figure 6.51. Enterokinase activation of trypsinogen leads to activation of other pancreatic proenzymes (Adapted from The Undergraduate Teaching Project in Gastroenterology and Liver Disease, American Gastroenterological Association, Unit 16, Exocrine Pancreas: Pancreatitis, 1984.)

transgenic techniques demonstrate that these enhancer regions allow specific expression in exocrine pancreatic tissue and not other tissues. Such studies suggest that acinar cells contain tissue-specific enhancer elements that regulate transcription of exocrine pancreatic digestive enzyme genes. The nature of these elements is not known.

Regulation of Exocrine Pancreatic Secretion

The regulation of pancreatic exocrine secretion is considered both at the cellular level and the organ level. The regulation at the cellular level is often referred to as stimulus-secretion coupling, while that of the whole exocrine pancreas as organ physiology.

STIMULUS-SECRETION COUPLING

Pancreatic Acinar Cell

Studies using in vitro preparations of dispersed acinar cells and acini from a variety of species over the past decade have demonstrated several classes of receptors on the acinar cell (Fig. 6.52). Studies using tissue from human pancreas are limited. Using radiolabeled ligands and specific receptor antagonists, receptors for cholecystokinin, acetylcholine, bombesin, substance P, VIP, and secretin have been identified in preparations from species such as guinea pig, rat, mouse (Gardner and Jensen, 1987). In cases where it has been evaluated, the receptors are on the basolateral surface of the acinar cell.

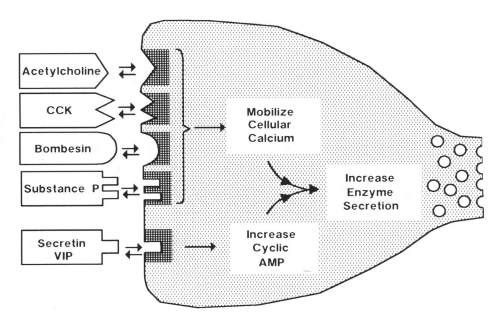

Figure 6.52. Receptors and stimulus-secretion coupling mechanisms for pancreatic secretagogues.

These receptors have been divided into two categories depending on their mode of stimulus-secretion coupling. In one category are VIP and secretin. The interaction of these agents with acinar cells leads to activation of adenylate cyclase and an increase in cellular cyclic AMP. Detailed studies of receptor classes interacting with VIP and secretin in guinea pig and rat reveal that there are three or four receptor classes that these agents interact with. Interaction with only some of the receptor classes is linked to both an increase in cyclic AMP and secretion. Although the picture is complex, those receptor classes that have high affinities for these agents and are coupled to a response will have physiological significance.

Several lines of evidence indicate that the increase in cellular cyclic AMP stimulated by either VIP or secretin mediates the secretory response. First, analogues of cyclic AMP such as dibutryryl-cyclic AMP and 8-bromo-cyclic AMP stimulate secretion. Second, agents that directly stimulate adenylate cyclase, such as cholera toxin and forskolin, stimulate secretion. Lastly, phosphodiesterase inhibitors augment both the increase in cellular cyclic AMP and the secretory response. Although it is clear that cyclic AMP mediates the secretory response, the actual intracellular mechanisms of its action are unknown.

The pancreatic acinar cell contains specific classes of receptors for the agonists, CCK, acetylcholine analogues, bombesin, and substance P. These agents do not act to stimulate enzyme secretions through changes in cellular cyclic AMP, but by stimulating cellular metabolism of membrane phosphoinositides and calcium. In contrast to the complex nature of the VIP and secretin receptor interactions, each of these agents interacts only with its specific receptors.

The CCK receptor class has a high affinity for CCK-OP (the COOH-terminal octapeptide of CCK), CCK-33, CCK-39, and caerulein (a CCK analogue originally isolated from the skin of Hyla caerulea). This receptor also interacts with gastrin, but with a low affinity. Each of the agents interacts only with the CCK receptor to stimulate enzyme secretion.

The bombesin receptor also interacts with gastrin-releasing peptide (the mammalian analogue of bombesin) with a high affinity resulting in enzyme secretion. The acetylcholine receptor is muscarinic and will respond to muscarinic agonists such as carbachol and bethanachol. Substance P receptors may be activated by substance P-like peptides released from intrapancreatic nerves.

One of the major effects that CCK, acetylcholine, bombesin, and substance P cause is the mobilization and release of intracellular stores of calcium (Williams and Hootman, 1987). Recent studies suggest that the ability of these agents to cause cellular calcium mobilization results from their effect on membrane phosphoinositides. Specifically, the agonist-receptor interaction leads to an apparent phospholipase C-mediated hydrolysis of phosphatidylinositol 4,5-bisphosphate to 1,2 diacylglycerol and inositol 1,4,5-trisphosphate (IP_3). IP_3, in turn, releases calcium from a nonmitochondrial intracellular store that probably includes the RER. The calcium is released into the cytosol resulting in a rapid rise in the concentration of free calcium in the cytosol (Pandol et al., 1985). This rise in $[Ca^{2+}]$ in the cytosol is, in large part, transient, lasting for the first 2–5 min of stimulation, but it mediates enzyme secretion. The mechanism by which increases in cytosolic $[Ca^{2+}]$ mediate secretion is not clear, but may include calmodulin-dependent processes. In contrast to the first few minutes of stimulation, the continued stimulation of enzyme secretion by these agents occurs at near resting concentrations of cytosolic $[Ca^{2+}]$ and is dependent on extracellular calcium. The intracellular mechanism of the sustained response is not established but may be regulated by both calcium and 1,2 diacylglycerol (the other phosphoinositide hydrolysis product) by their ability to activate a phospholipid-dependent, calcium-activated enzyme (protein kinase C).

The agonists that act by changes in cell calcium also stimulate an increase in cellular cyclic GMP, an increase in arachidonate release from membrane phospholipids, and electrical membrane potential changes (Williams and Hootman, 1987). The increase in cellular cyclic GMP results from the ability of these agents to release intracellular calcium. The roles of both cyclic GMP and arachidonate release in the cell response are not known. The effects of these agonists on membrane potential is complex and species dependent. Acinar cells from some species depolarize with stimulation while others hyperpolarize. In some, the membrane potential changes are biphasic with an initial depolarization followed by a sustained hyperpolarization. The electrical changes are probably involved in the ion secretion by the acinar cell.

As indicated above, although the initial steps in stimulus-secretion coupling have been partially elucidated, the mechanisms by which these steps lead to the secretory process are not well understood. Specific phosphorylations and dephosphorylations of cellular proteins occur with both cyclic AMP agonists and calcium-phosphoinositide agonists. These events are probably mediated by specific kinases and phosphatases activated by the specific messenger system (i.e., Ca^{2+}, cyclic AMP). However, neither the function of these proteins nor their roles in the secretory process have been elucidated.

The enzyme secretory response of the acinar cell to a combination of an agonist that acts by cyclic AMP and an agonist that acts by changes in calcium is greater than the additive response. Such a combination would be VIP or secretin and CCK or acetylcholine. The mechanism of this potentiated response is not known but it probably functions physiologically so that significant quantities of secretion occur with a combination of small increases in individual agonists.

Pancreatic Duct Cell

The mechanisms of pancreatic duct cell secretion were described above. The major mediator of ductal secretion is secretin. Like the acinar cell, the duct cell responds to a combination of secretin and CCK or acetylcholine with a potentiated response.

ORGAN PHYSIOLOGY

Interdigestive Secretion

Human exocrine pancreatic secretion occurs both during the fasting state (interdigestive) and after ingestion of a meal (digestive) (DiMagno, 1987). The interdigestive pattern of secretin begins when the upper gastrointestinal tract is cleared of food. In the normal individual who eats three meals per day, the digestive pattern begins after breakfast and continues until late in the evening after the evening meal is cleared from the upper gastrointestinal tract.

The interdigestive pancreatic secretory pattern is cyclical and follows the pattern of the migrating myoelectric complex (MMC) (Chapter 43). The first of the three phases (phase I) of the MMC is characterized by both a lack of motor activity and secretion of enzymes. Enzyme secretion increases during both phases II and III as does intestinal motor activity. With the return of phase I of the MMC, both the motor activity and enzyme secretory rate decrease. In addition to pancreatic enzyme secretion, there is increased bicarbonate and bile secretion into the duodenum during phases II and III of the MMC. There is also partial gallbladder contraction. The role of these secretions during the MMC is not clear, but may be related to the "housekeeping" function of the MMC.

As is the case for the motility patterns during the interdigestive period, the neurohumoral mechanisms regulating secretion during the interdigestive period are unknown.

Digestive Secretion

Like gastric secretion, exocrine pancreatic secretion with ingestion of a meal is divided into three phases—cephalic, gastric, and intestinal (DiMagno, 1987; Holst, 1987; Chey, 1987; Singer, 1987).

The vagal nerves mediate the *cephalic phase* of exocrine secretion. The extent of cephalic stimulation of exocrine pancreatic secretion in humans has been evaluated by measuring exocrine secretions stimulated by sham feeding (chewing and then spitting out the food). A recent study indicates that sham feeding stimulates pancreatic enzyme secretion up to about 50% of the maximal secretory rate. If acid secreted by the stomach is allowed to enter the duodenum, the rate of enzyme secretion increases to about 90% of maximal, and bicarbonate is also secreted. If gastric acid is prevented from entering the duodenum, there is no bicarbonate secretion. These results suggest that cephalic stimulation specifically stimulates acinar secretion and that a low pH in the duodenum augments this secretion. In addition, the low pH in the duodenum stimulates ductal bicarbonate secretion. The effects of acid in the duodenum are probably mediated by secretin.

Investigations of the mechanism of neurotransmission during cephalic stimulation indicate that acetylcholine is not the only neurotransmitter involved. For example, the effect of sham feeding is only partially blocked by atropine in humans. Neural endings containing peptides, VIP, gastrin-releasing peptide (GRP), CCK, and enkephalins, have been identified in the pancreas. Data supporting the role of these peptides in the cephalic phase of secretion are strongest for VIP and GRP. Both are released into the venous effluent with vagal stimulation in animals. Furthermore, as discussed prevously, acinar cells have receptors for GRP and VIP that mediate enzyme secretion. The duct epithelium cells also respond to VIP by secretion of water and bicarbonate. In experiments using isolated perfused pig pancreas, antiserum against VIP inhibited the secretory effects (water and bicarbonate) of electrical vagal stimulation.

Another postulated mediator of sham feeding-induced pancreatic secretion is gastrin. Vagal stimulation results in gastrin secretion in the dog. However, in the human, sham feeding causes little or no increase in gastrin. Also the potency for gastrin to stimulate pancreatic secretion is significantly less than that for CCK. Thus, gastrin is an unlikely mediator of pancreatic secretion.

The *gastric phase* of pancreatic secretion results from stimuli acting in the stomach. The gastric stimuli cause stimulation predominantly of enzymes with little stimulation of water and bicarbonate secretion. In humans and dogs, balloon distension of either the gastric fundus or the antrum results in a low-volume, enzyme-rich secretion by way of a vagovagal reflex. Gastrin does not mediate the pancreatic secretory effect of antral distension because antral acidification prevents gastrin release but not pancreatic secretion. Thus, the gastric phase is neurally mediated. The magnitude of the gastric phase secretion in humans is not established.

When gastric juice and contents of a meal enter the duodenum, a variety of intraluminal stimulants can act on the intestinal mucosa to stimulate pancreatic secretion through both neural and humoral mechanisms. Thus, the physiology of gastric acid secretion and gastric emptying are tightly coupled to the mechanisms of the intestinal phase of pancreatic secretion.

The quantitatively most important phase of pancreatic enzyme secretion is the *intestinal phase*. It begins when chyme first enters the small intestine from the stomach. It is mediated by both hormones and enteropancreatic vagovagal reflexes that interact resulting in augmentation of the secretory response.

First, we will consider the mechanisms involved in activating *ductal water and bicarbonate secretion* (Fig. 6.53).The major intestinal stimulant of ductal secretion is hydrogen ion in the duodenal lumen. The mediator of hydrogen ion-stimulated bicarbonate and water secretion is secretin. Secretin measured by radioimmunoassay is released from the duodenal mucosa with a threshold pH of 4.5. The quantity of secretin released is dependent on the load of titratable acid delivered to the duodenum. That secretin mediates pancreatic exocrine secretion is established by several findings. First, exogenous secretin infused to give a plasma concentration equal to that observed with a meal, stimulates water and bicarbonate secretion. Second, immunoneutralization of secretin with specific antisecretin antibody decreases pancreatic volume and bicarbonate secretion by as much as 80% to a mixed meal in the dog. The antisecretin antibody also inhibited enzyme secretion stimulated by the mixed meal by as much as 50%, suggesting that secretin has a role in enzyme secretion, possibly by potentiating the action of agonists such as CCK and acetylcholine.

Although acid load in the duodenum is the best established stimulant for releasing secretin, fatty acids greater than eight carbons in length and bile acids have been found to increase circulating secretin.

The pancreatic bicarbonate output with exoge-

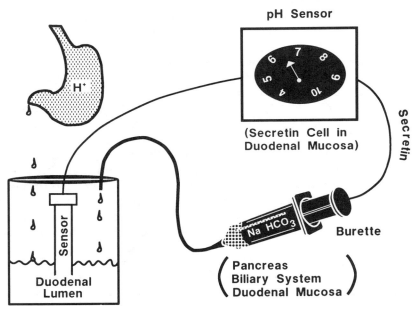

Figure 6.53. Secretin mediates bicarbonate secretion. The acid load delivered to the duodenum from the stomach stimulates secretin release from the secretin-containing endocrine-paracrine cell in the duodenal mucosa. Secretin then acts as a humoral agent to stimulate bicarbonate secretion from the pancreas, biliary system, and duodenal mucosa. (Adapted from The Undergraduate Teaching Project in Gastroenterology and Liver Disease, American Gastroenterological Association, Unit 16, Endocrine Pancreas: Pancreatitis, 1984.)

nous secretin infused to reproduce the plasma concentrations of secretin during a meal is less than the bicarbonate output observed with a meal. Recent studies demonstrate that secretin-induced bicarbonate secretion is augmented by CCK when both agents are infused to reproduce concentrations observed during a meal. CCK alone caused no bicarbonate secretion. The bicarbonate response to secretin is also dependent on cholinergic output because atropine partially inhibits the response stimulated by exogenous secretin. The intestinal stimulants mediating CCK release and cholinergic output are discussed below. Thus, the full meal-stimulated response results from a combination of mediators (secretin plus either CCK or acetylcholine) that lead to a potentiated response.

Human *secretion of digestive enzymes* is mediated by intraluminal (intestinal) fatty acids more than eight carbons in chain length, monoglycerides of these fatty acids, peptides, and amino acids. Intestinal infusion of calcium in humans also causes a significant enzyme secretory response, whereas starch and glucose have only small effects. The most potent amino acids for stimulating secretion in man

are phenylalanine, valine, methionine, and tryptophan. The response to peptides and amino acids is related to the total load perfused into the intestine rather than the concentration.

The mediators of the enzyme secretory response from intestinal stimuli are both neural and humoral (Fig. 6.54). Truncal vagotomy and atropine markedly inhibit the enzyme (and bicarbonate) responses to low intestinal loads of amino acids and fatty acids. These results suggest a vagovagal enteropancreatic reflex that mediates enzyme secretion and augments bicarbonate secretion stimulated by secretin.

CCK is the major humoral mediator of meal-stimulated enzyme secretion. Using both radioimmunoassay and bioassay techniques, the circulating concentration is found to increase with a meal. Furthermore, infusion of an amount of exogenous CCK that results in a circulating concentration mimicking that measured during a meal causes a secretion that is rich in enzymes. CCK is released from the upper small intestinal mucosa by digestion products of fat, protein, and to a small extent with starch digestion products. In contrast to low loads

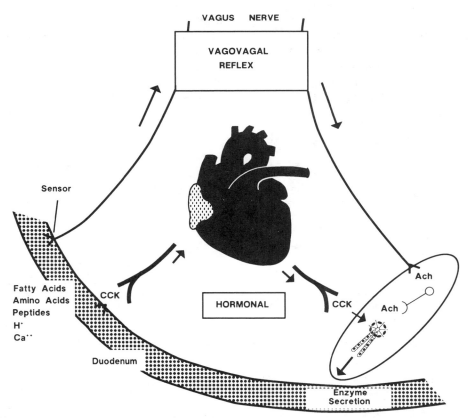

Figure 6.54. Intestinal phase of pancreatic enzyme secretion is mediated by both neural and humoral mechanisms. (Adapted from The Undergraduate Teaching Project in Gastroenterology and Liver Disease, American Gastroenterological Association, Unit 16, Endocrine Pancreas: Pancreatitis, 1984.)

of these stimulants, large loads of the stimulants cause secretion that is little affected by vagotomy or atropine. Thus, low loads of intestinal stimulants mediate enzyme secretion predominantly by an enteropancreatic neural reflex, while high loads mediate secretion predominantly by CCK.

INHIBITION OF SECRETION

Only a few physiological mechanisms involved in the inhibition of pancreatic exocrine secretion have been well defined. An inhibitory feedback mechanism exists between intraluminal pH and pancreatic bicarbonate secretion. With a low pH in the duodenum, secretin is released and pancreatic bicarbonate secreted. The intraluminal bicarbonate increases duodenal pH resulting in a cessation of secretin release and bicarbonate secretion.

In animals, diversions of pancreatic juice from the intestine result in augmented secretion (DiMagno, 1987). Infusion of trypsin into the intestine suppresses the increased secretion. The increased secretion is probably due to an increase in circulating CCK. These results suggest a mechanism whereby the concentration of pancreatic enzymes in the intestine regulates CCK release. The existence of this mechanism in humans is not established.

Sympathetic innervation and several peptide hormones inhibit stimulated secretion. These include pancreatic polypeptide, glucagon, somatostatin, α- and β-adrenergic agents, and enkephalins. However, whether these agents act physiologically has not been established.

Pathological States

Two common processes leading to decreased exocrine pancreatic secretion are pancreatic carcinoma and chronic pancreatitis. With pancreatic carcinoma, insufficiency occurs because the carcinoma usually arises from the pancreatic duct and obstructs it. In chronic pancreatitis, insufficiency occurs because a chronic inflammatory process destroys both acinar and ductal tissue. The exocrine pancreas has a remarkable reserve so that enzyme output must fall below 10% of normal before maldigestion and malabsorption occur. The most common symptom of patients with pancreatic insufficiency is steatorrhea (fat or oil in the stool). Also with pancreatic insufficiency, bicarbonate secretion is decreased and intraduodenal pH decreased. Lipase is inactivated at low pH values. Thus, with marginal secretion of lipase, maldigestion of triglyceride is worsened because of a decrease in bicarbonate secretion. For the same reason, patients with massive gastric acid hypersecretion as in the Zollinger-Ellison syndrome may have malabsorption of fat. The quantity of hydrogen ion delivered to the duodenum exceeds the neutralizing capacity of the pancreas, biliary system, and duodenal mucosa so that the pH in the duodenum is low and lipase is inactivated.

Secondary pancreatic insufficiency may occur in diseases of the small intestinal mucosa such that there is an inadequate activation of enteropancreatic reflexes and/or release of CCK and secretin. This is the case for the intestinal mucosal disease, sprue.

CHAPTER 45

Biliary Secretion and Excretion

SIGNIFICANCE

Biliary secretion by the liver delivers bile acids and immunoglobulins to the small intestine. Bile acids are important in nutrition, because they play a key role in solubilizing dietary lipids and thereby promoting their absorption (see Chapter 46). Immunoglobulins (mostly IgA) are important in immune defense, since they are secreted in response to the antigens present in intestinal bacteria and viruses and act to inhibit the growth of these organisms.

In addition to being a digestive secretion, bile is also a key fluid for excretion of metabolic end-products. Secretion of bile is essential for excretion of cholesterol and its metabolic products, bile acids. Bile is also essential for the excretion of bilirubin, derived from heme, and is the major route of elimination of some heavy metals such as copper. In addition, conjugated derivatives of drugs, hormonal steroids, prostaglandins, and fat-soluble vitamins are excreted in bile, as well as some amphipathic drugs (Smith, 1973).

Obstruction of the biliary tract causes two defects. The first is a deficiency of bile acids in the small intestine, which causes lipid malabsorption. The second is the retention in liver followed by regurgitation into blood of biliary constituents—bile acids, cholesterol, and bilirubin. Neither of these abnormalities is life-threatening on a short-term basis. With time, continued obstruction to bile flow causes accumulation of bile acids in the hepatocytes which kills them. Scarring and fibrosis ensue, ultimately leading to hepatic insufficiency, portal hypertension, and death.

A recent review of biliary secretion can be found in Hofmann, 1989.

FUNCTIONAL ANATOMY OF THE BILIARY SYSTEM

The fundamental organization of the liver is the lobule which is the system of rows of plates of hepatocytes arranged around the terminal portal vein and arteriole (Rappaport, 1975; Jones, 1982). Sinusoidal blood flows from the terminal portal vein and arteriole between the plates of hepatocytes to the terminal hepatic vein (Fig. 6.55). The biliary system begins in the canaliculi (*Latin*, canalis plus the diminutive suffix iculi, meaning "little"), which are a network of spaces between the hepatocytes. The blind end of the canaliculus begins at the central region of the hepatic lobule (in the region of the hepatic vein). Its lumen extends toward the portal triad. The canaliculus has no true epithelium, as its lining is merely a specialized region of hepatocyte membrane. This membrane is termed the canalicular membrane and because of its enzyme composition and function is also termed an "apical" membrane. The canaliculi are separated from the space of Disse by the paracellular junctions between the hepatocytes (Fig. 6.56). Bile flows from the central, blind end of the canaliculi toward the portal triad, and thus moves in the opposite direction of blood flow in the sinusoids. Thus, bile flow is countercurrent.

The canaliculi connect to the biliary ductules via specialized ducts (termed canals of Hering) which are lined by primitive epithelial cells. The canals of Hering end in the biliary ductules which join to form bile ducts. The general arrangement is arborescent, hence the term "biliary tree" (see Fig. 6.58). The biliary ductules are surrounded by a periductular capillary plexus which originates from branches of the hepatic artery (Fig. 6.57). The periductular capillary plexus drains into the sinusoid at its portal end. Thus, substances which are produced by or absorbed by the biliary ductular cells immediately enter sinusoidal blood and not the systemic circulation.

The right and left hepatic ducts merge to form the common hepatic duct. The common hepatic duct is joined by a smaller duct from the gallbladder termed the cystic duct. The cystic duct and common hepatic duct join to form the common bile duct which passes under the second portion of the duo-

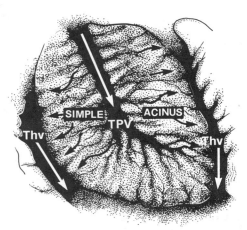

Figure 6.55. Schematic representation of the structure of a liver lobule, the basic unit of the parenchymal liver. Blood enters by the terminal portal vein *(TPV)* and terminal hepatic arteriole (not shown). Blood flows between the hepatic plates in the sinusoids to the terminal hepatic vein *(Thv)*. Bile flows in the opposite direction from sinusoidal blood flow to enter the bile duct (not shown). The terminal portal vein, terminal hepatic arteriole (also not shown), and bile duct form the "portal triad." (From Rappaport, 1975.)

denum and unites with the pancreatic duct close to the sphincter of Oddi (Fig. 6.58). Bile flows from the liver in the common hepatic duct. If the sphincter of Oddi is contracted, bile is diverted into the gallbladder. With ingestion of a meal, the gallbladder contracts and the sphincter of Oddi relaxes, so that bile flows from the gallbladder into the duodenum.

Bile secreted by the liver is termed hepatic bile; after it is concentrated and modified during storage in the gallbladder, it is termed gallbladder bile.

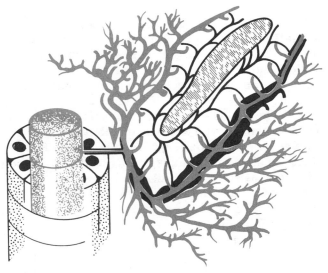

Figure 6.56. Schematic drawing of the biliary canaliculi, canal of Hering *(arrow)*, and bile duct. The drawing on the *right* shows the canaliculi, the canal of Hering, and the bile duct. The sinusoidal space is shown between the hepatic plates. (From Goresky et al., 1983.)

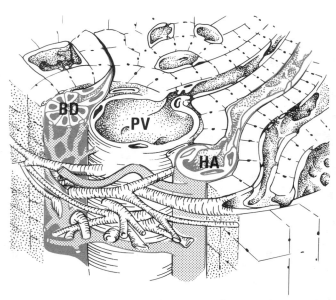

Figure 6.57. Anatomy of the portal triad showing the branches of the terminal portal vein *(PV)*, hepatic artery *(HA)*, and the periductular biliary plexus *(BD* = bile duct). The majority of hepatic arteriolar blood goes directly to the sinusoid; a minority is diverted to the periductular biliary plexus and flows around the bile duct and then back to the portal vein. Presumably each sinusoid receives the majority of its blood from the terminal portal vein and a minority from the terminal hepatic arteriole. (From Jones, 1983.)

SECRETION OF BILE

Formation of Canalicular Bile

BILE ACID-DEPENDENT FLOW

Canalicular bile is formed in response to the osmotic effect of bile acid anions (and their accompanying cation) (Jones, 1982; Boyer, 1980). The bile acid anions present in sinusoidal plasma pass through the fenestrations of the endothelial cells, diffuse through the space of Disse, and are actively taken up across the sinusoidal membrane of the hepatocyte by a Na^+ coupled cotransport system. During their intracellular transport, they are bound to specific cytosolic proteins. They are then secreted actively across the canalicular membrane into the canaliculi. There is a membrane potential across the canalicular membrane (about 30 mV) with the canalicular lumen being positive. This potential difference also contributes to the transport of bile acid anions. The anatomy is shown in Figure 6.59.

The transport is remarkably concentrative—20–200 fold. The concentration of bile acids in the hepatocyte is not known, but estimates place it in the range of 10–50 μM. The concentration of monomeric bile acid anions in the canaliculus is usually about 2000 μM, but experimentally, may be increased 10 fold higher.

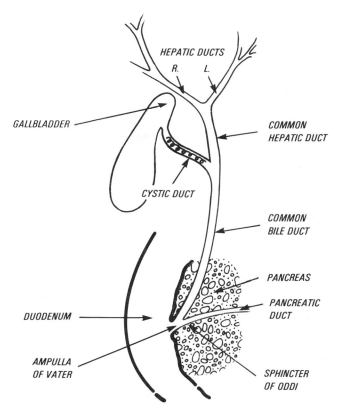

Figure 6.58. Anatomy of the external biliary tract. In some mammals such as the rat and horse, a gallbladder is not present. In man, the common bile duct and major pancreatic duct enter the ampulla of Vater together. In some animals such as the rabbit and guinea pig, the common bile duct enters the second portion of the duodenum, whereas the pancreatic duct enters the third portion of the duodenum.

The paracellular junctions (tight junctions) which separate the canalicular lumen from the space of Disse are too small to permit permeation of the large (charged) bile acid molecules (Fig. 6.60). Plasma water and accompanying solutes flow through the tight junctions to restore isotonicity in the canaliculus. Canalicular bile is thus formed. Bile acids also appear to induce canalicular contraction. The end result is the active secretion of canalicular bile. In animals a secretory "pressure" can be measured by using a simple L-shaped tube. The value is about 30 cm H_2O.

Canalicular bile consists of primary solutes—those which induce bile formation—and secondary solutes—those which enter the canalicular lumen in response to the osmotic effect of the primary solutes. The chemical composition of the secondary solutes will be determined by the reflection coefficients of individual plasma solutes, which are influenced by the size and charge of solutes in relation to the pore size (and charge) of the paracellular junctions. The major secondary solutes are plasma electrolytes, monosaccharides, amino acids, and organic acids. Since bile acids are impermeable, the cutoff point for permeation in terms of molecular size is quite low; sucrose, for example, permeates poorly.

BILE ACID-INDEPENDENT BILE FLOW

If one infuses a solution of bile acids into an animal, the acids are immediately extracted by the liver and secreted into bile, inducing bile flow. Such bile flow is the bile acid-dependent bile flow (Boyer, 1980; Scharschmidt, 1982). The slope of the line relating bile flow to bile acid output in bile indicates the choleretic activity of a bile acid molecule (Fig. 6.61). In humans, this value is about 7–10 μl/μmol bile acid or 3.5–5 μl/μmol osmotically active anion, but in other animals, this value may be considerably higher for unknown reasons.

The intercept of the line relating bile flow to bile acid recovery with the vertical axis (i.e., no bile acid secretion) is termed "bile acid-independent bile flow" and is a major component of bile flow in some animals. The solutes which are secreted into the canaliculus and are responsible for bile acid-independent flow are not known. In humans, in contrast to many other mammals, bile acid-independent flow is much lower than bile acid-dependent bile flow. Thus, canalicular bile flow in man is almost entirely driven by active bile acid secretion.

For measurement of canalicular bile flow, one wishes to use a marker substance which is present only in canalicular bile, whose biliary secretion is proportional to canalicular bile flow and which is neither secreted nor absorbed by the biliary ductules. The nonmetabolizable carbohydrates such as erythritol or mannitol are useful for this purpose. The use of these substances is analogous to the use of inulin for the measurement of glomerular filtration.

In addition to inducing bile flow, bile acids also induce the secretion of the two major biliary lipids—phospholipid and cholesterol. Although biliary phospholipid may play a minor role in promoting the emulsification of dietary lipid, there is no evidence that it is essential for efficient fat absorption. Cholesterol is present in bile solely as an excretory product. Accordingly, biliary lipid secretion is discussed later in the context of hepatic excretory function.

Ductular Modification

Bile is modified as it flows down the biliary tree. The present view is that these modifications are not very important, at least in man because they do not influence the two key functions of bile—delivery of bile acids to the small intestine, and excretion of cholesterol, bile acids, and bilirubin.

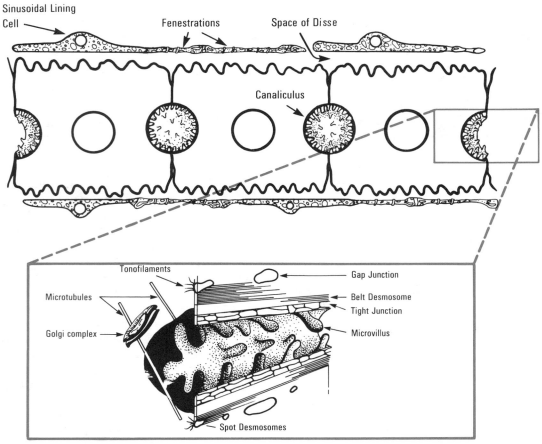

Figure 6.59. Structure of the liver cell plate showing the sinusoid, fenestrations, space of Disse, and canaliculus (the lumen of which is perpendicular to the sinusoid). The anatomy of the canaliculus is enlarged below. Gap junctions allow the passage of small molecules between adjacent hepatocytes. Belt desmosomes and tight junctions comprise the junctional complex. (From Scharschmidt, 1983.)

Ductular absorption of filtered plasma solutes may be considered as salvage. Glucose and amino acids, which enter bile passively as canalicular bile is formed, are efficiently absorbed. It is not clear why such absorption is useful to the organism, since all of these substances should be efficiently absorbed by the small intestine. One speculation is that absorption of these substances contributes to the resistance of gallbladder bile to serve as a culture medium for bacteria.

DUCTULAR SECRETION

The infusion of hormones such as secretin result in increased water and bicarbonate secretion in bile. Since the biliary clearance of inert sugars such as erythritol or mannitol does not increase, it is believed assumed that such bicarbonate is of ductular origin. Somatostatin causes inhibition of ductular secretion of bicarbonate and water, and may cause active absorption of bicarbonate (Rene et al., 1984).

Modification of Hepatic Bile in the Gallbladder

The gallbladder concentrates, acidifies, stores, and discharges bile. Continuous concentration during overnight fasting permits the storage of an ever-increasing mass of bile acids in a constant volume of bile. The longer one fasts, the greater the quantity of bile acids available to promote digestion (Holzbach, 1984).

The paracellular junctions of the gallbladder epithelium are permeable to the inorganic electrolytes of bile. Thus, bile is concentrated by removal of sodium and chloride. The lipids of bile are present in aggregates (micelles and vesicles) and gallbladder bile remains isotonic, despite the concentration of sodium ions increasing to as high as 300 mM after overnight fasting. The limiting case is the removal of all electrolytes in bile other than bile acid anions and their accompanying cations (Fig. 6.62).

The biliary epithelium removes Na^+ from bile in exchange for H^+ ions. The H^+ ions combine with HCO_3^- to form CO_2 (Rege and Moore, 1986). The

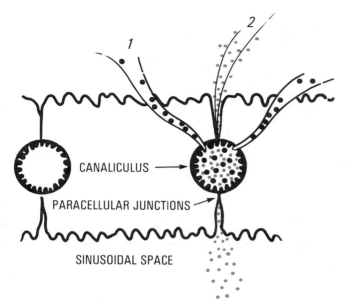

Figure 6.60. Schematic depiction of canalicular bile flow. Bile acids are actively secreted across the canalicular membrane (*1*). The transmembrane potential contributes to their concentrative transport. Osmotic activity of the bile acids draws water and filterable plasma solutes across the paracellular junctions (*2*) thus forming canalicular bile; some water may also pass across the canalicular membrane. In addition to bile acids, other molecules may be secreted across the canalicular membrane, and generate "bile acid-independent" bile flow. (From Scharschmidt, 1983.)

mechanism is identical to that proposed to occur in the proximal renal tubule. Experimentally, the pH of bile decreases during concentration, and P_{CO_2} is

found to be elevated close to the epithelium. An active Na^+/K^+-ATPase on the basolateral membrane of the epithelium pumps Na^+ from the biliary epithelial cell into the blood. Chloride ions flow paracellularly into the blood to maintain electrical neutrality. The net result is NaCl absorption. Some Ca^{2+} ions are absorbed, but the concentration of Ca^{2+} in gallbladder bile is higher than plasma-ionized Ca^{2+} because the distribution of Ca^{2+} ions is determined by the rules of Gibbs-Donnan equilibrium.

Storage and Discharge of Gallbladder Bile into the Intestine

During overnight fasting, about half of the hepatic bile enters the gallbladder. The remainder passes into the small intestine. Bile acids are swept down the small intestine, absorbed from the terminal ileum, resecreted into bile; again, half of the secreted bile is stored in the gallbladder. Depending on the length of the overnight fast, a progressively greater fraction of the bile acids present in the body are stored in the gallbladder. The time course of hepatic secretion of bile acids, their gallbladder storage, and their secretion into the duodenum is summarized in Figure 6.63.

With ingestion of a meal, vagal discharges initiate gallbladder contraction. Release of cholecystokinin (CCK) by intraluminal stimuli such as fats and amino acids acting on the CCK-containing cells in the duodenum cause sustained gallbladder contraction and relaxation of the sphincter of Oddi (Fig.

MAN

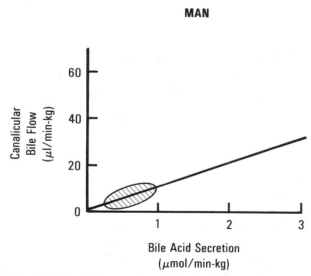

SOME OTHER MAMMALS

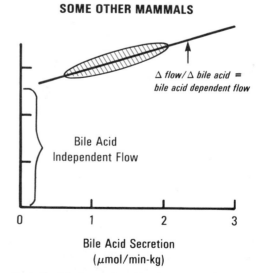

Figure 6.61. Relationship between canalicular bile flow and bile acid secretion in humans *(left)* and in some other mammals *(right)*. Representative values are shown for the digestive state. Bile flow in humans is characterized by little or no bile acid-independent bile flow, whereas in animals this may have a value from 10–180 μl/min-kg. The slope of

the line reflects the bile flow induced by the secretion of bile acids and their accompanying cations. Most animals have a considerably greater flux of bile acids through the liver than humans do, when expressed in relation to body weight.

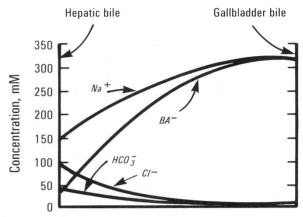

Figure 6.62. Change in concentration of major ions in hepatic bile *(left)* as it is concentrated in the gallbladder during storage to form gallbladder bile *(right)*. Micelle formation permits extremely high concentrations of bile acids and cations to be reached, because the number of osmotically active particles is far less than would be indicated by the chemical concentration; yet bile remains isotonic at all times. The volume of the gallbladder will be determined by the net sum of water loss from absorption (or contraction) and water gain (from entry of hepatic bile). (Modified from Dietschy, 1964.)

6.64) (Ryan, 1987). The gallbladder contracts slowly, discharging about two thirds of its content during the first hour of digestion. It remains in a contracted state, so that hepatic bile is not stored in the gallbladder during digestion.

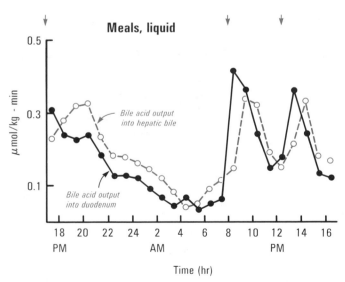

Figure 6.63. Relationship between time of meal ingestion *(arrows at top)*, secretion of bile acids into hepatic bile *(dashed line, open circles)*, and secretion of bile acids into the duodenum *(solid line, solid circles)*. During overnight fasting, bile acid secretion into the duodenum is less than hepatic bile acid secretion, indicating storage of bile acids in the gallbladder. When a meal is ingested, duodenal bile acid secretion exceeds hepatic bile acid secretion indicating discharge of bile acids stored in the gallbladder. (From van Berge Henegouwen and Hofmann, 1978.)

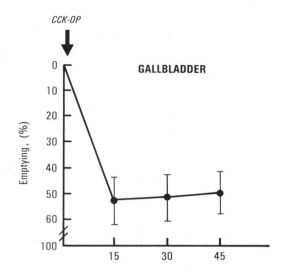

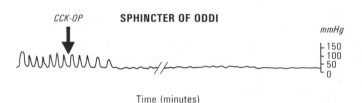

Figure 6.64. Change in gallbladder volume, expressed as % of original volume *(above)* and sphincter of Oddi contractions *(below)* in response to an intravenous injection of cholecystokinin (20 ng/kg) in man. As the gallbladder contracts, the sphincter of Oddi releases, permitting discharge of gallbladder contents. (From Ryan, 1987.)

HEPATIC EXCRETORY FUNCTION

Overview

Bile is an essential excretory route for a number of substances which because of their physicochemical properties cannot be eliminated in urine. In general, such substances are organic, are large (MW > 300), and are hydrophobic resulting in binding to plasma albumin (Smith, 1973). As a consequence, such substances do not enter the glomerular filtrate. The most important of such endogenous substance is unconjugated bilirubin. In addition, extremely hydrophobic lipids such as cholesterol are transported in lipoproteins, again precluding elimination in urine.

The fenestrations of the endothelial cells of the hepatic sinusoids are sufficiently large to allow ready passage of albumin and lipoproteins smaller than chylomicrons across the space of Disse and reach the surface of the hepatocyte. Uptake of albumin-bound molecules involves direct collision of the albumin molecule with transporters in the hepatocyte membrane. Somehow this collision enhances dissociation of molecules from albumin and uptake is more rapid than predicted from the unbound concentration.

Intact low-density lipoprotein molecules enter the hepatocyte by receptor-mediated endocytosis. For high-density lipoproteins, individual lipid constituents are selectively taken up by the hepatocyte in a manner that is poorly understood.

Bile Acid-Dependent Biliary Lipid Excretion

CHOLESTEROL-PHOSPHOLIPID VESICLES

Bile acids induce the secretion of lipid vesicles from the hepatocyte across the canalicular membrane. The vesicles contain mostly (unesterified) cholesterol and phospholipid in a molar ratio of about 1:3. Cholesterol is a crystalline solid at body temperature and to be dispersed in vesicular form, it must be present together with another lipid. In the case of biliary lipid vesicles, the phospholipid is mostly phosphatidylcholine with a fatty acid composition that differs from that of other hepatocyte membranes. Details as to how bile acids induce the movement of vesicles from an unknown site (possibly the Golgi apparatus) to the canaliculus are not understood, how the vesicles are exocytosed across the canalicular membrane is also not understood, but current speculations are shown in Figure 6.65 (Carey and Cahalane, 1988; Admirand and Small, 1968; Carey and Small, 1978).

Bile acids are believed to be secreted uphill across the canalicular membrane, presumably by a secondary cotransport mechanism. Vesicles contain mostly phospholipid and cholesterol, and as they are exocytosed, they encounter a high concentration of bile

acids present in the form of simple micelles (see below). The bile acid molecules and micelles interact with the lipid vesicles, transforming them to mixed micelles containing phospholipid, cholesterol, and bile acids. This process of vesicle to micelle transformation continues to occur while bile is stored in the gallbladder. The current view of the molecular arrangement of the molecules composing the mixed micelle present in gallbladder bile is shown in Figure 6.66.

The composition of lipid vesicles is influenced by the bile acid flow rate. At low bile acid secretion rates such as occur during the fasting state the vesicles tend to have a high cholesterol phospholipid ratio. At higher bile acid secretion rates, this ratio falls (Fig. 6.67).

In the micelle there are about seven bile acid mol-

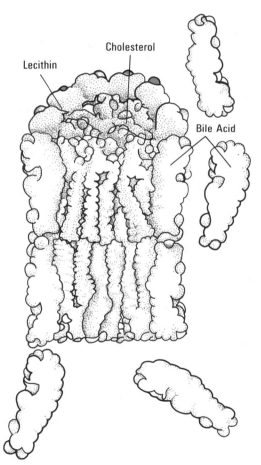

Figure 6.66. Current view of the molecular arrangement of the mixed micelle present in gallbladder bile. (Although this model is consistent with a variety of physicochemical data, other models such as a helical arrangement have not been excluded.) The inside of the micelle is composed mostly of hydrocarbon chains of phospholipid (in a fluid state) and other lipid molecules can dissolve in the center of the micelle or can replace bile acid molecules which surround the outside of the micelle. (From Hofmann, 1978–1979.)

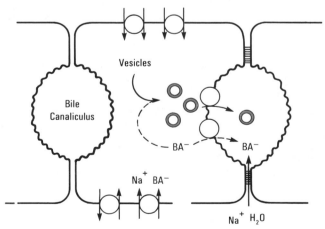

Figure 6.65. Current view of bile acid-dependent biliary lipid secretion by the hepatocyte. During their transport, bile acids stimulate the movement of exocytosis of phospholipid-cholesterol vesicles across the canalicular membrane by an unknown mechanism. Bile acid molecules are transported independently across the canalicular membrane. In the canaliculus, the vesicles are transformed to micelles by collision with bile acids at a concentration above their CMC.

EFFECT OF BILE ACID SECRETION RATE ON
BILIARY LIPID SECRETION AND SATURATION WITH CHOLESTEROL

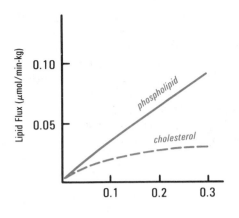
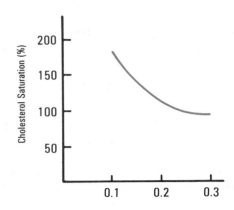

Bile Acid Flux (μmol/min-kg)

Figure 6.67.   Relationship between bile acid secretion rate and biliary lipid secretion *(left).* At low bile acid secretion rates, the vesicles contain relatively more cholesterol. The explanation for this is not known, although bile acid-independent vesicle secretion might be an explanation. Because the vesicles contain a higher cholesterol/phospholipid ratio at low bile acid secretion rates, the calculated saturation with cholesterol is higher at low bile acid secretion rates *(right.)* The significance of this relationship is that conditions which tend to cause low bile acid secretion rates may make an individual at greater risk for precipitation of cholesterol in bile.

ecules for every two to three phospholipid molecules, and one-half to one cholesterol. One can prepare (aqueous) mixtures of the three major biliary lipids and thus simulate gallbladder bile. The composition of any mixture is defined by the molar proportions of the three lipids and is conveniently shown using triangular coordinates, a method commonly used in physical chemistry to depict composition of mixtures containing three substances. [(Such a mixture, as a triangle, has only two degrees of freedom, hence, if the proportion of two components is known, the third is obtained by difference) (Fig. 6.68).] A line can be drawn which separates samples that are unsaturated in cholesterol from those which are supersaturated in cholesterol. If a bile sample falls within the micellar zone, it is said to be unsaturated (Carey and Small, 1978).

For any bile sample, cholesterol saturation (in percent) may be calculated from the equaton:

$$\text{cholesterol saturation} = \frac{[\text{cholesterol}]_{\text{sample}}}{[\text{cholesterol}]_{\substack{\text{at saturation in the model system} \\ \text{having the identical bile acid and} \\ \text{phospholipid concentration}}}}$$

Tables are available giving the cholesterol value at saturation in the model system.

When this approach to describing biliary lipid composition was developed, it was hoped that defining the cholesterol saturation of gallbladder bile samples would have clinical value. This has not proved to be the case, since supersaturation of bile with cholesterol is common in adults whether or not they have cholesterol gallstones. Nonetheless, discovery of the concept of the cholesterol saturation of bile was important since (*a*) cholesterol gallstones cannot occur unless bile is supersaturated, and (*b*) bile desaturation by oral bile acid therapy can induce gallstone dissolution.

Bile Acid-Independent Biliary Excretion

BILIRUBIN CONJUGATION AND BILIARY EXCRETION

Heme, a closed tetrapyrrole, is converted to bilirubin, a linear tetrapyrrole with two carboxylic acid groups, in the reticuloendothelial system (Kupfer cells and spleen) (Fig. 6.69) (Ostrow, 1986). The major source of heme is from dying red cells, although heme in muscle protein and cytochrome enzymes also contributes. After its formation, bilirubin (in unconjugated form) leaves the reticuloendothelial cells, enters the portal circulation, and binds to serum albumin. Hepatic uptake is not efficient—first pass clearance is about 20%—so unconjugated bilirubin is always present in plasma.

Bilirubin is then transported through the hepatocyte bound to a hydrophobic cytosolic protein named ligandin. During its transit, bilirubin is esterified with one molecule of glucuronic acid on each of its acidic groups to form bilirubin diglucuronide. This glucuronidation is catalyzed by a microsomal enzyme termed bilirubin glucuronyl transferase (Fig. 6.70).

Glucuronidation is required for the efficient can-

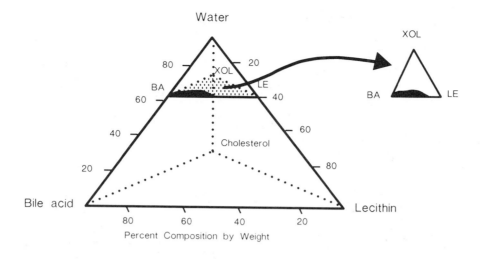

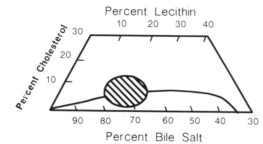

Figure 6.68. Triangular coordinate system used to indicate the limits of cholesterol solubility in the model system composed of biliary lipids to simulate human bile. The *line* indicates the limit of the micellar zone. Samples lying outside of this zone are supersaturated and will tend to precipitate cholesterol. Bile whose composition lies within the micellar zone is unsaturated in cholesterol and cannot form cholesterol gallstones. The *hatched area* indicates average gallbladder composition in healthy adults. (In such triangular coordinates, the sum of the three constituents is always constant, that is, 100%.) (From Holzbach, 1984.)

Figure 6.69. *Above,* chemical structure of heme, bilirubin, and its diglucuronide conjugate. In the conversion of heme to bilirubin, the iron atom is lost and the cyclic tetrapyrrole molecule is opened. Below, bilirubin in its "closed" internally hydrogen bonded form *(left)* (the form in which bilirubin is believed to circulate in plasma), or in its open form *(right)*. In conjugated bilirubin, the carboxylic acid groups of bilirubin are each esterified to the 1-hydroxyl group of glucuronic acid. Unconjugated bilirubin in closed form is quite hydrophobic; it is unreactive with the diazo-reagent commonly used for bilirubin measurement; accordingly, such bilirubin is termed "indirect." When bilirubin is conjugated, it becomes hydrophilic. It reacts instantaneously with the diazo-reagent, and is termed "direct." (From Schmid, 1978.)

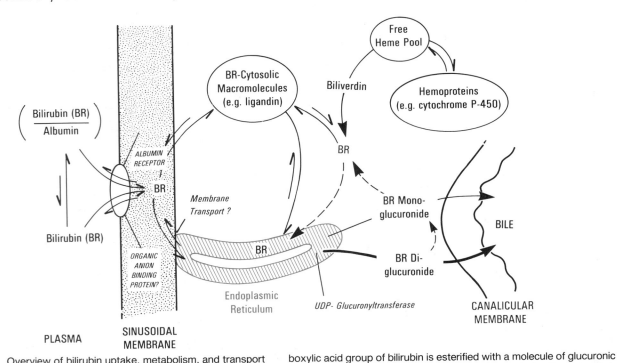

Figure 6.70. Overview of bilirubin uptake, metabolism, and transport in the hepatocyte. In plasma, newly formed (unconjugated) bilirubin is en route from the spleen and other biosynthesis sites to the hepatocyte. Uptake is believed to be carrier mediated, with a bilirubin-binding protein termed "ligandin" involved in intracellular transport. Each car- boxylic acid group of bilirubin is esterified with a molecule of glucuronic acid, a reaction catalyzed by bilirubin glucuronyl transferase, which uses UDP-glucuronic acid. The diglucuronide is secreted actively into bile and not reabsorbed during transit in the biliary tract or small intestine. (From Gollan and Schmid, 1982.)

alicular secretion of bilirubin. In the absence of glucuronidation, as occurs in experimental animals or patients lacking bilirubin glucuronyl transferase, only trace amounts of unconjugated bilirubin are secreted into canalicular bile. The mechanism of this secretion of unconjugated bilirubin is not known. The canalicular carrier involved in excretion into canalicular bile of bilirubin diglucuronide differs from that involved in the transport of bile acids. Thus, there are at least two organic anion transporters in the canalicular membrane.

The canalicular secretion of bilirubin conjugates is efficient, and normally the concentration of conjugated bilirubin in plasma is extremely low. Increased conjugated bilirubin in plasma is considered a definite sign of liver or biliary disease.

OTHER ORGANIC ANIONS

The major organic anions transported across the canalicular membrane are bile acids. During a meal, the rate of bile acid secretion is about 300 nmol/kg-min; during the fasting state, it is about one-third of this value. Bilirubin glucuronide secretion is constant throughout the day—at a rate of about 5 nmol/kg-min.

Other organic anions are also excreted in bile in trace amounts. These include sulfate or glucuronide conjugates of sex steroids and fat-soluble vitamins such as vitamin A and D. These have no effect on bile flow or biliary composition. Drugs are also secreted in bile as such or as sulfate, glucuronide, or glutathione conjugates.

There are some instances when the increased biliary secretion of organic anions becomes clinically significant. Hemolytic conditions increase the production of bilirubin glucuronide many fold. The bilirubin glucuronide may be deconjugated in the gallbladder and precipitate with calcium and form gallstones. Some drugs that are excreted largely in bile may also form precipitates of insoluble calcium salts termed "sludge."

The efficient hepatic extraction and biliary secretion of large (lipophilic) organic anions has stimulated a large body of investigation aimed at using the plasma disappearance rate of an injected bolus to indicate hepatic excretory function. The traditional dyes that have been used are bromosulfophthalein or indocyanine green. These are not widely used today because it is recognized that retarded plasma disappearance may reflect impaired hepatic blood flow, portal systemic shunting, as well as impaired hepatocyte uptake. Indocyanine green is also used to estimate "functional" hepatic blood flow by the Fick principle. A constant infusion of indocyanine green is given, and the arterial venous concentration difference across the liver is measured.

The ability of the liver and biliary system to

extract and excrete large lipophilic anions has been used to design tests of gallbladder function. Substituted imino acetic acid derivatives bind ^{99}Tc avidly and can be used to image the gallbladder (Fig. 6.71); failure of the gallbladder to image indicates gallbladder disease.

Oral cholecystographic agents are iodinated benzoic acid derivatives which are selectively secreted into bile. During storage in the gallbladder, the concentration of the ''dye'' becomes sufficiently great to make the gallbladder radiopaque to x-rays. Oral cholecystography is therefore used to measure gallbladder function, that is, filling *and* concentration. Failure of the gallbladder to opacify is a sensitive and specific indicator of gallbladder disease, usually caused by cholelithiasis.

FATE OF BILIARY CONSTITUENTS IN THE BOWEL

Biliary phospholipids mix with dietary phospholipids and are digested and absorbed. Biliary cholesterol mixes with dietary cholesterol and less than one-third of each is absorbed. Thus, the majority of cholesterol that enters the intestine in bile is excreted, and bile serves as the major excretory route for cholesterol.

Sulfate and glucuronide conjugates are not hydrolyzed in the small intestine, because sulfatase and glucuronidase activity is not present in pancreatic juice or the brush border enzymes. Hydrolysis may occur in the colon, but the absorptive surface of the colon is small and little absorption of the unconjugated moiety occurs. As a consequence, glucuronide and sulfate conjugates that are secreted into bile are not absorbed from the intestine but are eliminated in feces (Fig. 6.72).

Conjugated bile acids, the major constituents of bile, are not absorbed in the proximal small intestine because they are too polar (because of their charge) to partition into the lipid domains of the enterocyte brush border membranes and because they are too large to pass through the paracellular junctions. After promoting fat absorption, bile acids are efficiently reabsorbed from the distal ileum (largely in conjugated form) by an active transport system. The bile acids which are absorbed from the intestine are transported to the liver in portal blood, are reextracted by the liver and secreted into bile, and then back into the intestine. This circular route—from intestinal lumen to liver to bile and back into the intestine—is termed the enterohepatic circulation (Fig. 6.72) (Hofmann, 1989b).

Chemistry and Enterohepatic Circulation of Bile Acids

Of all the lipid constituents in bile, only bile acids have a useful function and only bile acids undergo an enterohepatic circulation. The enterohepatic circulation of bile acids is useful because it results in a large mass of previously synthesized bile acids available to pass through the hepatocyte, where they generate bile flow and induce the secretion of cholesterol-phospholipid vesicles. (It also results in

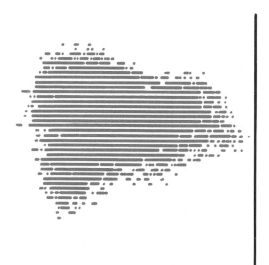

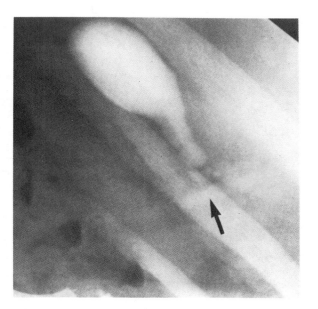

Figure 6.71. *Left,* use of a ^{99}Tc-labeled iminodiacetic acid derivative which is rapidly taken up by the liver to image the liver using a gamma camera. *Right,* opacification of a gallbladder using an oral cholecystographic agent; the compound, which is an iodinated benzoic acid deriv-ative, achieves sufficiently high concentration in bile to become radi-opaque after gallbladder concentration. The *arrow* indicates the location of the junction of the cystic duct with the common duct. (From Iio et al., 1973; and Zeman and Burrell, 1987.)

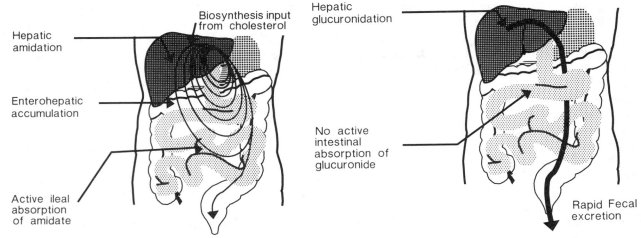

Figure 6.72. Influence of mode of conjugation on fate of organic anions secreted into the small intestine. Natural bile acids conjugated with glycine or taurine are efficiently absorbed from the terminal ileum.

Drugs which are conjugated with glucuronic acid (or sulfate) and secreted into bile are not reabsorbed from the small intestine. (From Cohen et al., 1986.)

a large mass of bile acids being available to facilitate fat digestion, as summarized in Chapter 46.)

Bile acids are a group of water-soluble acidic steroids with powerful detergent properties; bile acids are formed in the liver from cholesterol. Formation of bile acids from cholesterol involves a number of chemical changes on both the nucleus and the side chain. On the nucleus, additional hydroxyl groups are added, the double bond is saturated stereospecifically, as the A/B ring juncture is *cis* in bile acids.

The side chain is oxidatively cleaved; bile acids have a C5 branched isopentanoic side chain compared to the C8 iso-octane side chain of cholesterol (Fig. 6.73). The end result is a highly water-soluble amphipathic anion. The conversion of cholesterol to bile acids is inefficient in man—and more cholesterol is eliminated as such than in the chemical form of bile acids. In man, two bile acids are formed—cholic acid and chenodeoxycholic acid; these two bile acids are termed *primary* bile acids (Fig. 6.74).

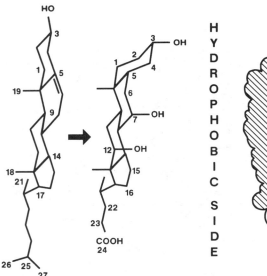

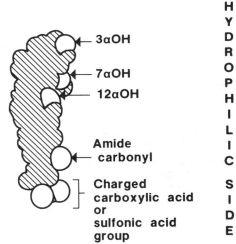

Figure 6.73. Formation of cholic acid, a typical bile acid, from cholesterol *(left)*. Bile acid biosynthesis involves stereospecific saturation of the double bond to form a 5β A/B cis steroid, stereospecific nuclear hydroxylations, and oxidative cleavage of the C_8 side chain to form a C_5 side chain with the structure of isopentanoic acid. Cholesterol, a flat insoluble molecule is converted to bile acids, kinked, highly water-soluble yet amphipathic molecules.

On the *right* is shown a space-filling model of the bile acid molecule indicating it is an amphipathic molecule with a hydrophobic side and a hydrophilic side; it is these amphipathic properties that are responsible for the ability of bile acids to form mixed micelles with other biliary lipids.

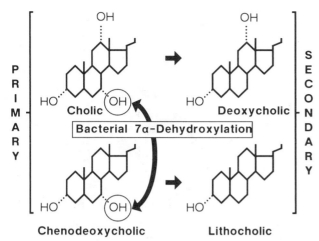

Figure 6.74. Chemical structure of the major primary and secondary bile acids in man. The major change involved in biotransformation of a primary bile acid to a secondary bile acid is 7-dehydroxylation or 7-epimerization (in which a 7α hydroxy group is changed to a 7β hydroxy group). Many other secondary bile acids are formed in the colonic lumen, but such bacterial biotransformations are not considered to be of physiological significance. The 7β hydroxy epimer of chenodeoxy-cholic and is named ursodeoxycholic acid and is used widely as an orally administered agent to induce cholesterol gallstone dissolution.

During the enterohepatic cycling, the hydroxyl groups of the bile acids may be removed, oxidized, or epimerized. Such new bile acids formed by bacterial enzymes are termed secondary bile acids.

Bile acid biosynthesis from cholesterol is under negative feedback control. Ordinarily it is repressed. When the concentration of bile acids falls in the hepatocyte as occurs in instances of bile acid malabsorption, bile acid biosynthesis increases by up to 20 fold. Cholesterol biosynthesis (and/or recruitment of plasma cholesterol) must also increase in a parallel fashion, since cholesterol molecules are the substrate for bile acid biosynthesis. The marked increase in cholesterol and bile acid biosynthesis is associated with upregulation of the low-density lipoprotein receptors of the hepatocyte.

After their biosynthesis from cholesterol, bile acids are conjugated with glycine or taurine to form "amidates" (Fig. 6.75). The four conjugated bile acids are then secreted into bile, where they join with the far larger flux of bile acids continuously flowing through the hepatocyte.

In the healthy human, a "pool" of about 8 mmol of bile acids circulates continuously in the enterohepatic circulation, as determined by the isotope dilution technique. The circulation slows during overnight fasting, and increases during meals; the number of cycles is related to meal size and frequency. With average meals, the pool circulates about twice during digestion. The half-life of the molecules in the enterohepatic circulation is 2–3 days. About twice as much cholic acid as chenode-

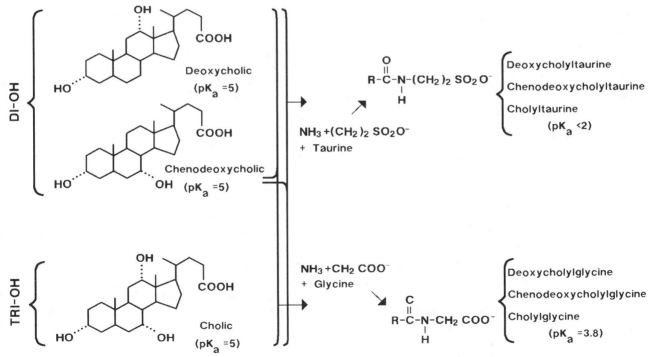

Figure 6.75. Chemical structures of the major conjugated bile acids present in human bile. The amide bond linking the glycine or taurine is resistant to pancreatic carboxypeptidases. The glycine conjugates have a pK_a of about 3.8; the taurine conjugates, < 1. The pK_a values of all unconjugated bile acids are identical to those of isopentanoic acid, that is, about 5.0.

oxycholic acid is biosynthesized from cholesterol; but intestinal absorption of the chenodeoxycholic conjugates is more efficient than that of cholyl conjugates and the pool size of these two primary bile acids is about the same.

Bile acids are efficiently absorbed by active transport from the terminal ileum; absorption is remarkably efficient—about 90%/meal. Bile acids are transported to the liver in portal blood—about 60–80% of the conjugates of cholic acid are bound to albumin and nearly 99% of the conjugates of chenodeoxycholic acid are bound to albumin.

First pass extraction by the liver is remarkably efficient—70–90%. Those bile acids not extracted spill over into the systemic circulation. Since the first pass hepatic extraction is not identical for all bile acids, the pattern of plasma bile acids differs from that of gallbladder bile. Plasma is enriched in those bile acids with less efficient hepatic extraction. In the healthy person, hepatic fractional extraction remains constant during the fasting state and during digestion. As a result, the level of bile acids in the peripheral plasma is linearly proportional to the rate at which bile acids are returning from the ileum and increases after meals. The small fraction of bile acids not bound to plasma albumin enters the glomerular filtrate; however, conjugated bile acids are actively absorbed by the renal tubule so that urinary loss of bile acids in the healthy human is negligible.

Ileal transport of bile acids is saturated during digestion. The T_{max} for bile acid transport is likely to be about 0.3–0.6 μmol/kg-min, and if bile acids are presented at a greater rate, bile acids escape ileal absorption and pass into the colon. Here the bile acids are modified by bacterial enzymes. The major changes are deconjugation and dehydroxylation at the 7 position to form secondary bile acids (Fig. 6.74). Cholic acid is converted by 7-dehydroxylation to deoxycholic acid; and chenodeoxycholic acid is converted by 7-dehydroxylation to lithocholic acid.

About one-third to one-half of the deoxycholic acid which is newly formed is absorbed from the colon. After conjugation with glycine or taurine by the liver, the deoxycholic acid becomes a part of the circulating bile acid pool (Fig. 6.76). Indeed, its ileal transport may compete with that of the primary bile acid conjugates and it may become the most common biliary bile acid. The significance of such accumulation is that deoxycholic acid appears to induce the secretion of lipid vesicles whose cholesterol content is higher than those induced by other bile acids.

Lithocholic acid is an insoluble bile acid which is

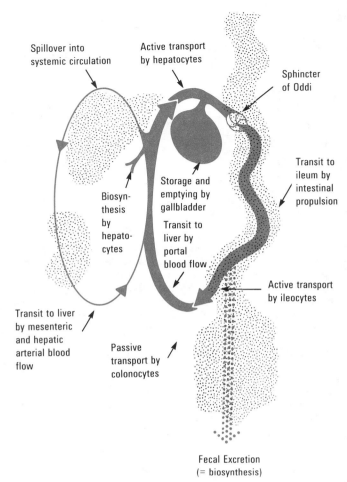

Figure 6.76. Enterohepatic circulation of bile acids. As a consequence of efficient intestinal absorption, a large "pool" of bile acids accumulates and recycles; a small fraction spills over into the systemic circulation. There is no urinary loss. Some secondary bile acids enter the enterohepatic circulation by absorption from the colon. The mass of the circulating bile acids is termed the "bile acid pool." It circulates about twice per meal, on the average, or six times a day. (From Hofmann, 1988.)

toxic when administered to animals. In man, it is poorly absorbed from the colon. When lithocholic acid reaches the liver, it is "detoxified" by not only amidation with glycine or taurine but also by sulfation at the 3 position. The resultant "double conjugates" are secreted in bile but undergo little ileal absorption. As a consequence, lithocholic acid is present in bile in only trace proportions (Fig. 6.77).

During colonic transit, most primary bile acids are 7-dehydroxylated so that the predominant fecal bile acids are deoxycholic acid (from cholic acid) and lithocholic acid (from chenodeoxycholic acid) (Fig. 6.78).

During their enterohepatic circulation, bile acids are deconjugated in part before ileal absorption, even though most bile acids are absorbed from the

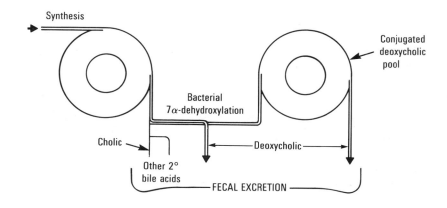

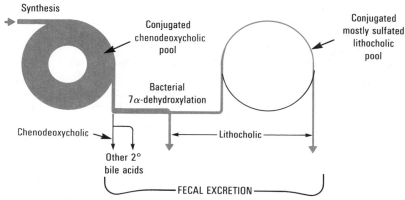

Figure 6.77. *Upper,* Enterohepatic circulation of cholic acid and deoxycholic acid, its major metabolic product in humans. About one-third to one-half of newly formed deoxycholic acid is absorbed from the colon; it is conjugated with glycine or taurine in the liver and circulates with the primary bile acids. In some individuals, it may become the major biliary bile acid. Note that cholic acid input refers to hepatic synthesis of cholic acid from cholesterol, while deoxycholic acid input refers to the absorption of deoxycholic acid from the colon following its formation from cholic acid by bacterial modification (7α-hydroxylation). The amount of cholic acid excreted in feces is negligible in normal individuals. (From Hofmann, 1978–1979.) *Lower,* Enterohepatic circulation of chenodeoxycholic acid and lithocholic acid, its bacterial 7-dehydroxylation product, in man. About one-fifth of lithocholic acid which is formed is absorbed; it is not only conjugated with glycine or taurine, but it is also sulfated. The resultant "double" conjugates are poorly absorbed from the terminal ileum; as a consequence, conjugates of lithocholate do not accumulate in the enterohepatic circulation and lithocholic acid is only a trace bile acid in biliary bile acids in man. Note that spillover into plasma is not shown. (From Hofmann, 1984.)

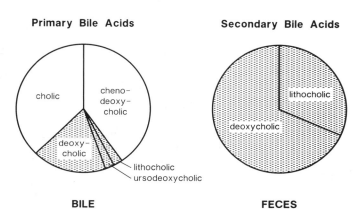

BILE

FECES

Figure 6.78. Composition of biliary bile acids is contrasted with that of fecal bile acids. Biliary bile acids contain both primary and secondary bile acids; in contrast, fecal bile acids contain solely secondary bile acids. (From Hofmann, 1983.)

ileum in conjugated form. Unconjugated bile acids which reach the liver from the small or large intestine are reconjugated during hepatocyte transport.

PHYSIOCHEMICAL PROPERTIES OF BILE ACIDS

Bile acids are water-soluble amphipathic acidic compounds (Hofmann and Roda, 1984). The unconjugated bile acids, which have a pK_a of 5, are insoluble below a pH of about 7, depending on the bile acid. Conjugation with glycine or taurine lowers the pK_a to 3.8 for glycine conjugates and <1 for taurine conjugates. The effect of such conjugation is to decrease passive absorption in the biliary tract and small intestine as well as to lower the pH at which bile acids precipitate from solution.

Bile acids are amphipathic, having a hydrophobic side and a hydrophilic side. Over a narrow range of concentration, bile acids self-associate to form polymolecular clusters called micelles. This concentration is termed the critical micellization concentration (CMC), and varies considerably according to bile acid structure. The natural primary bile acids all have CMC values from 2–5 mM. The simple micelles formed by bile acids range from tetramers (formed by trihydroxy bile acids) to 20-mers formed by dihydroxy bile acids. The micelles formed by bile acids interact with cholesterol-phospholipid vesicles, transforming them to mixed micelles. The simple and mixed bile acid micelles also bind Ca^{2+} ions in hepatic bile.

COMMON PATHOLOGICAL DISTURBANCES

Biliary Obstruction

As was noted in the beginning of this chapter, biliary obstruction causes retention of biliary constituents and absence of bile acids in the small intestine. The soluble constituents of bile such as bile acids and bilirubin conjugates increase greatly in concentration in plasma causing itching. They are excreted in urine in greatly increased amounts and as a result, urine becomes brown and foams. The insoluble constituents of bile—phospholipid and cholesterol—accumulate in plasma as disc-shaped aggregates. Because of their insolubility, urinary elimination is impossible. High blood cholesterol causes the formation of yellow lipid deposits in the skin termed xanthomata.

A second common disturbance in bile flow is external drainage of bile to the outside via a catheter or a T-tube. When this occurs, the enterohepatic circulation of bile acids is interrupted. A marked increase in bile acid biosynthesis occurs, and eventually all of the secreted bile acids originate from de novo synthesis. The biliary excretion of other substances is unchanged.

An absence of bile acids in the small intestine causes malabsorption of dietary lipids and a deficiency of fat-soluble vitamins. The deficiency of vitamin K causes hypoprothrombinemia and hemorrhage; of E, neuropathies; of D, bone disease, and of A, impaired retinal function.

Bile Acid Metabolism

Inborn errors of bile acid biosynthesis are extremely rare (Bjorkhem, 1985), but involve defective side chain hydroxylation or defective nuclear hydroxylation and/or double bond saturation. Defective ileal transport of bile acids causing chronic diarrhea has been reported but is also extremely rare. A far more common disturbance in bile acid metabolism is bile acid malabsorption caused by ileal dysfunction resulting from ileal inflammatory disease or ileal resection. This is discussed in Chapter 46.

Bilirubin Metabolism

Neonatal jaundice is common because of immaturity of the bilirubin conjugating system of the canalicular transport system at birth. The bilirubin molecule is so hydrophobic it can cross cell membranes passively. Bilirubin crosses the blood-brain barrier and may deposit in the nuclei of central nervous systems causing cell death. By light microscopy, the nuclei are stained yellow (termed kernicterus). Therapy involves ultraviolet radiation which isomerizes the ring junctions of the unconjugated bilirubin molecule increasing its hydrophilicity, thus permitting biliary and urinary excretion (Fig. 6.79).

A common genetic defect in bilirubin metabolism is decreased activity of the glucuronyl transferase. In this defect, termed Gilbert's syndrome, the plasma level of unconjugated bilirubin is elevated, and a major fraction of bilirubin in bile is present as the monoglucuronide. An extremely rare genetic defect is the Crigler-Najjar syndrome in which bilirubin glucuronyl transferase is totally absent; such patients have markedly elevated levels of unconjugated bilirubin in plasma, and in time die from the consequences of excessive deposition of this bilirubin in brain cells.

Organic Anion Transport

Rare defects in canalicular organic anion transport occur and are diagnosed by infusing dyes which enter the liver cell, are conjugated, and then reflux back into the serum in conjugated form instead of being secreted into bile. The most common of these defects (Dubin Johnson syndrome) involves bilirubin, but not bile acid transport (Arias et al., 1988).

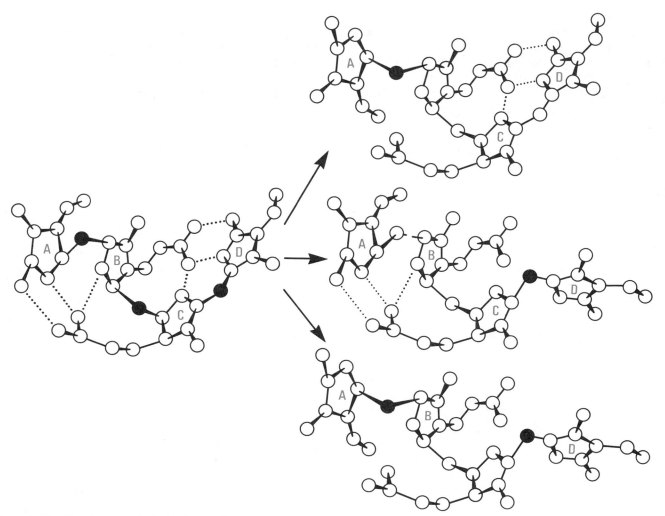

Figure 6.79. Photoisomers of bilirubin formed during ultraviolet radiation of infants with unconjugated hyperbilirubinemia. The isomers are more hydrophilic and can be excreted into bile without glucuronidation. The letters *A, B, C,* and *D* are the labels assigned to the designated tetrapyrrole rings.

Cholelithiasis: Gallstone Disease

Gallstone disease (gallstones in the gallbladder) is endemic in Western populations afflicting one-fourth of the population older than 60 years of age. Approximately one-half of these patients are symptom free and are not treated; most symptomatic patients undergo cholecystectomy, a safe and curative operation.

Cholesterol gallstones form because cholesterol-rich vesicles are secreted into bile and nucleate cholesterol crystals in the gallbladder. In general, patients with cholesterol gallstones have vesicles which contain a higher cholesterol/phospholipid ratio than healthy subjects. In addition, for a given vesicle composition, nucleation occurs more rapidly in gallstone patients than in healthy subjects. Cholesterol gallstone patients also have defective gallbladder contraction.

The oral administration of bile acids such as chenodeoxycholic acid (or its 7 beta epimer, ursodeoxycholic acid) induces the secretion of vesicles containing a much lower cholesterol/phospholipid ratio. As a consequence, bile becomes unsaturated in cholesterol, and cholesterol gallstones slowly dissolve (Schoenfield et al., 1981).

Other gallstones are composed of insoluble calcium salts. These include unconjugated bilirubin, carbonate, and phosphate. Since unconjugated bilirubin is not secreted into bile, its presence in gallstones must indicate deconjugation after secretion. Calcium bilirubinate in gallstones may polymerize forming "black" pigment gallstones.

Both cholesterol and black calcium bilirubinate stones are metabolic in origin being formed in the gallbladder in the absence of inflammation or infection.

Hepatic bile is supersaturated in calcium carbonate because both calcium and bicarbonate enter canalicular bile at their concentrations present in plasma. Acidification of gallbladder bile by Na^+/H^+ exchange causes decreased bicarbonate concentration, and gallbladder bile is unsaturated in $CaCO_3$. Patients with gallstones appear to have a defect in epithelial acidification and gallstones often contain calcium carbonate.

Cholestasis is a generic term for impaired bile flow without mechanical obstruction to bile flow. Cholestasis arises from impaired canalicular transport of bile acids into the canaliculus or loss of canalicular integrity such that bile acid anions leak out of the canaliculus into plasma.

CHAPTER 46

Digestion and Absorption

OVERVIEW

The absorption of nutrients (protein, fat, and carbohydrate), minerals, trace elements, vitamins, and water is the key role of the small intestine. Loss of the absorptive function of the small intestine is fatal, unless nutrients are supplied parenterally. The intestine also absorbs endogenous secretions such as bile acids, digestive enzymes (after their hydrolysis), and electrolytes. In addition, the intestine has an important excretory function, since it acts as a conduit for biliary substances such as conjugates of bilirubin, sex hormones, and drugs, which are not absorbed, as well as unabsorbable substances (such as fiber) in the diet.

Some dietary constituents are not absorbable as such but must be modified to permit absorption. The most common modification—for proteins and carbohydrates—is enzymatic hydrolysis; fats undergo not only enzymatic hydrolysis but also micellar solubilization. Together, these chemical and physical modifications which render dietary constituents absorbable are termed digestion. Digestion is mediated by hydrolytic enzymes in the major digestive secretions—salivary, gastric, and pancreatic as well as hydrolytic enzymes on the brush border of the epithelial cells (these are termed enterocytes). Digestion is also mediated by bile acids, which solubilize the dietary lipids or their digestive products. The digestive process is buffered by bicarbonate which is secreted by the pancreas and duodenal epithelium.

Intestinal absorption is characterized by great rapidity: a complex meal may be fully digested and absorbed in 3 hours. Intestinal absorption is also characterized by great efficiency—the absorptive efficiency of both fat and protein exceeds 95%. For the major nutrients, absorptive efficiency is not regulated—the greater the amount ingested, the greater the amount absorbed. The capacity of the small intestine to absorb nutrients greatly exceeds that necessary for caloric balance. In the healthy human, absorption of the major nutrients is largely complete by the distal jejunum (Borgstrom et al., 1957).

The intestine is also a permeability barrier since it is almost impermeable to molecules larger than glucose unless they are lipophilic or a carrier mechanism is present. The permeability barrier function of the intestine must remain intact despite the presence of powerful digestive enzymes in the lumen. The intestine also contains an immense and powerful immune system, as discussed elsewhere.

In humans, absorption occurs under nearly sterile conditions, inasmuch as the bacterial flora of the stomach and small intestine are sparse. The major constituents of colonic content are bacteria, but in healthy humans these do not play an important role in nutrient digestion and absorption, as the only important site of entry of nutrients and other important dietary constituents into the circulation is the small intestine. In other animals, the situation may be quite different. For example, the cow, which is a ruminant, has a complex prestomach where bacterial hydrolysis of ingested carbohydrates such as cellulose generates short chain fatty acids which are absorbed there. In the horse, bacterial degradation of dietary carbohydrates (grass) in the large intestine generates short chain fatty acids which are absorbed there and are an important caloric source.

The mechanisms of transport across the enterocyte are both active and passive. Small molecules such as water, ethanol, and even some monosaccharides and amino acids not only pass through the enterocyte (a route termed transcellular) but also pass between the enterocytes across the junctional complexes between the cells (termed the paracellular route). The most absorption occurs in the proximal third of the small intestine; however, conjugated bile acids and vitamin B_{12} are absorbed in the distal ileum. For bile acids, there are distinct advantages to maintaining a high intraluminal concentration throughout the length of the small intestine, and the localization of the absorptive site to the distal small intestine can be considered to offer a survival advantage (Hofmann and Mysels, 1988).

Efficient digestion and absorption requires the coordination of gastric emptying, small intestinal mixing and propulsion, and secretion of bile and pancreatic juice. Secretion must increase when a meal is ingested and decrease when digestion is completed.

The fate of substances which traverse the intestinal epithelium is determined by their physical properties. Monosaccharides, amino acids, electrolytes, minerals, and water-soluble vitamins enter portal blood and are thus transported to the liver. Dietary lipids including fat-soluble vitamins and cholesterol esters as well as extremely lipophilic drugs such as cyclosporin are packaged by the enterocyte into small lipid droplets termed chylomicrons. These pass into the intestinal lymphatics and are discharged via the thoracic duct into the subclavian vein; chylomicrons thus enter the systemic circulation (Barrowman, 1978).

FUNCTIONAL ANATOMY OF THE SMALL INTESTINE

The small intestine in humans is a tube about 3 cm in diameter and about 300 cm long. Its volume increases to about 500–1000 ml during digestion. The presence of folds in the mucosa (termed valves of Kerkring, plicae circulares, or valvulae conniventes), and villi (on which the columnar epithelial cells rest) increase the surface area of the small intestine. The surface area is still further amplified by the presence of microvilli on the individual enterocytes.

The intestine is suspended on a mesentery of connective tissue which contains its blood supply and draining lymphatics.

The first portion of the small intestine—30 cm in length and shaped like the letter C—is termed the duodenum (from its length of 12 inches); it has three portions, equal in length. The first portion begins with a short area of wide diameter termed the duodenal bulb. In the second portion of the duodenum, the ampulla of Vater indicates the site of the entry of the common bile duct and pancreatic duct. The jejunum (Latin, *jejun-*, of no interest, presumably denoting its rather monotonous anatomy) is about one-half of the remaining length and is tethered to the diaphragm by a muscular band known as the ligament of Treitz. The latter half of the small intestine is termed the ileum (from its location in the ileal region of the abdomen). The small intestine begins with the pyloric sphincter and ends with the ileocecal valve.

The epithelial surface of the small intestine is termed the mucosa which is arranged in villi between which are crypts containing the cells which are precursors of enterocytes as well as the endocrine cells. The submucosa consists of the nutrient vessels and the central lymphatic, as well as nerve fibers and collagen fibers in the lamina propria. The submucosa is rich in white cells—lymphocytes, macrophages, and mast cells. Between the villi, especially in the ileum, there are collections of lymphocytes termed Peyer's patches; the luminal surface of the lymph node is coated with a specialized macrophage termed an "M" cell (see Chapter 21). The submucosa is bounded by two layers of smooth muscle—an inner circular layer and an outer, longitudinal layer.

The villus, whose upper portion is coated with mature epithelial cells, is the functional absorptive unit of the small intestine. Its major structural figures are shown in Figure 6.80. Absorbed solutes enter the body by two routes. The first is *through* the enterocytes, that is, via a transcellular path. The second, which is limited to smaller solutes such as water, monosaccharides, and amino acids is between the enterocytes, that is, via the paracellular pathway. Enterocytes are considered to be formed from dividing precursor cells in the crypts and migrate upward to the tip of the vessel. Migration and maturation are coordinated, with cells maturing partway up the villus. The mature cells are eventually sloughed from the tip of the villus into the intestinal lumen; the average enterocyte has a lifetime of several days (Lipkin, 1987). The rapid turnover of enterocytes may be useful, since it prevents the occurrence of storage diseases.

The center of the villus contains a nutrient arteriole which gives rise to a capillary plexus and a draining venule. In the center of the villus is a terminal (closed) lymphatic channel termed a lacteal (since it becomes milky during fat digestion). The villus contains smooth muscle elements and contracts slowly during digestion; such contraction should serve to pump the contents of the lacteals into the lymphatic vessels draining the small intestine. The arrangement of the blood flow in the villus appears to be countercurrent, and the interstitial fluid of the villus appears to contain sodium ion at extremely hyperosmotic concentrations. This arrangement is likely to facilitate water absorption.

Intestinal blood flow is characterized by richness (25% of cardiac output), redundancy, and autoregulation (Parks and Jacobson, 1987). Intestinal blood flow increases during meals and during digestion, reaching values as high as 100 ml/min. Only the brain and kidneys receive as much blood flow per

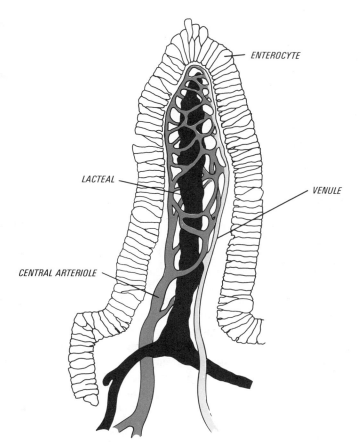

Figure 6.80. Structural organization of the villus showing enterocytes on villus surface, blood vessels, and central lacteal. Connective tissue elements and leukocytes present in the lamina propria are not shown. Enterocytes are formed in the crypts at the base of the villi, and mature as they migrate to the tip of the villus. There is a capillary network originating from a central, nutrient arteriole. A central lacteal drains fluid and chylomicrons from the interstitial space at the base of the enterocytes.

unit weight as the intestine. The intestine is supplied by arcades of arterioles, each of which has multiple origins. Interruption of one arteriole leads to collateral blood flow. Occlusion of any single nutrient artery, if gradual, causes little diminution in blood flow; if acute, however, ischemic necrosis may occur.

DIGESTION AND ABSORPTION OF MAJOR NUTRIENTS

Methods of Study

Luminal digestion is usually studied in vitro since it is an extracellular process, and enzymes and substrates are usually available for study. Digestion by brush border enzymes (surface digestion) requires intestinal perfusion or use of isolated cells or membrane fractions. Historically, intestinal absorption studies used the balance approach (output in rela-

tion to intake), but in recent years, intestinal perfusion techniques have been used. The advantage of such a technique is that the absorption rate can be measured for a given intraluminal concentration (Soergel, 1968). In such intestinal perfusion studies, a nonabsorbable marker such as an inert water-soluble polymer for example, polyethylene glycol, is often used to facilitate calculations of the rates of absorption.

Carbohydrate Digestion and Absorption

CARBOHYDRATE DIGESTION

Dietary carbohydrate consists mostly of starch (polyglucose in α-glycosidic linkage), sucrose (a disaccharide, glucose-fructose in α-glycosidic linkage, table sugar), lactose (a disaccharide glucose-galactose in β linkage; milk sugar), and fructose (a monosaccharide). Carbohydrate digestion involves hydrolysis of starch by salivary and pancreatic amylase as well as surface digestion of oligosaccharides. Absorption involves carrier-mediated transcellular transport as well as some paracellular absorption.

Starch digestion is initiated by salivary amylase, which attacks the inner α 1,4 glycosidic bonds to form maltose (glucose dimers), maltotriose (glucose trimers), and higher oligomers (Alpers, 1987). Starch hydrolysis continues in the stomach until the bolus is macerated and disintegrated into small particles. However, salivary digestion of starch is not essential, since carbohydrate digestion is unimpaired in the absence of salivary secretion; indeed, measurements indicate there is a large excess of pancreatic amylase. (The exception is the newborn in which salivary amylase is important because of decreased pancreatic amylase secretion.) Pancreatic amylase is secreted in active form. The enzyme hydrolyzes the inner 1,4 glycosidic linkages of starch generating tri- and disaccharides, which are termed maltotriose and maltose, respectively. Some of the starch chains are cross-linked by 1,6 linkages which are resistant to the action of amylase; units of maltose or maltotriose connected by these resistant 1,6 linkages are termed α-limit dextrins.

The remaining carbohydrate digestion is mediated by glycosidases present on the brush border of the enterocytes. Six disaccharidases have been identified, of which lactase, sucrase, and isomaltase are most important. The properties of the six glycosidases known to be present on the human enterocyte are summarized in Table 6.6.

The monosaccharides formed by the action of disaccharidases are glucose, fructose, and galactose. These are actively transported into the enterocyte by sodium-coupled secondary active trans-

Table 6.6.
Disaccharidases Present on the Apical (Luminal) Membrane of the Enterocyte

Enzyme	Substrate	Products	K_m, mM
Lactase	Lactose	Glucose, galactose	18
Sucrase	Sucrose	Glucose, fructose	20
	α-dextrins (1,4 branched)	Oligosaccharides	
	Linear glucosyl oligosaccharides (G_2–G_9)	Glucose	
Isomaltase	α-dextrins (1,6 linear)	Glucose	5–11
Trehalase	Trehalose	Glucose	2–5
Gluco-amylase	Linear glucosyl oligosaccharides (G_2–G_9)	Glucose	1–5
α-Limit dextrinase	α-dextrins (1,6 links; G_5–G_6)		1

port systems located in the brush border; these transport systems are similar in function to those of the renal proximal tubule. The carbohydrates are not metabolized to any major extent in the enterocyte, but are transported out of the enterocyte into the interstitial space by sodium-independent carriers present in the basolateral membrane. Energy for the uptake step is provided by the lumen to cytosol sodium gradient, and, ultimately, the basolateral Na^+-K^+-ATPase, as in the renal tubule. Paracellular absorption also contributes to the overall rapidity of carbohydrate absorption, because the paracellular junctions of the jejunum have some permeability to monosaccharides. Carbohydrate digestion and absorption is summarized in Figure 6.81.

The evolutionary significance of partly luminal/partly surface digestion of carbohydrates is not understood. It may be that the absence of monosaccharides in the intestinal lumen makes bacterial proliferation there more difficult. There may be a

kinetic advantage to diffusion of oligosaccharides rather than monosaccharides through the unstirred water layer coating the mucosa. The glycocalyx fronds which project from the brush border should impede access of luminal bacteria to the site of monosaccharide production. Brush border hydrolysis generates a local high concentration of monosaccharides which promotes transport across the brush border membrane. The paracellular junctions of the small intestine are readily permeable to water, and small intestinal content is invariably isotonic except in the upper duodenum after a meal. Hydrolysis of poly- or oligosaccharide to form monosaccharide must occur at the same rate as monosaccharide absorption to prevent excessive osmotic movement of water into the small intestinal lumen.

Some types of starch, for example, rice starch, are digested and absorbed more easily than others. Cooking alters the structure of starch and acceler-

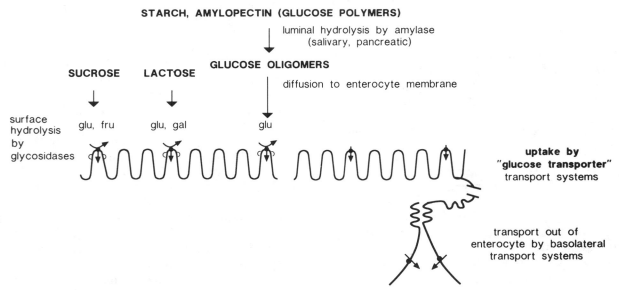

Figure 6.81. Major chemical events in the digestion and absorption of carbohydrates. Properties of the glycosidases present on the micro-villus membrane are summarized in Table 6.6.

ates its hydrolysis by amylase. Potato starch is not completely absorbed—only about 90%, and that which is not absorbed passes into the colon where it is degraded by colonic bacteria.

DEFECTIVE CARBOHYDRATE DIGESTION AND ABSORPTION

In patients with advanced exocrine pancreatic insufficiency, starch maldigestion will be severe. Nonetheless, this is not an important clinical problem, because salivary amylase activity compensates in part and because dietary starch can be substituted by dietary sucrose. Moreover, brush border enzymes can hydrolyze disaccharides.

The small intestine of the infant is rich in lactase, but in humans other than Caucasians, lactase disappears during childhood. As a consequence, many Blacks and Orientals, for example, are lactase-deficient (Paige and Bayless, 1981). Such individuals will develop signs of malabsorption when challenged with oral lactose loads. Lactase deficiency may be acquired transiently during the course of a viral enteritis. A congenital deficiency of sucrase-isomaltase is a rare, but important cause of diarrhea in the newborn. As would be expected, loss of intestinal surface which occurs, for example, in celiac sprue (a clinical condition characterized by loss of villi) can cause carbohydrate malabsorption.

FATE OF MALABSORBED CARBOHYDRATE

Carbohydrate which is not absorbed passes into the colon which is rich in anaerobic bacteria. Most malabsorbed carbohydrate is reduced to short chain fatty acids which are absorbed, but in children with severe carbohydrate malabsorption, monosaccharides may actually be present in diarrheal stools. Severe carbohydrate malabsorption causes diarrhea and cramps, presumably from the generation of an osmotic load in the colon.

In most individuals, hydrogen gas is also formed by the action of anaerobic colonic flora on malabsorbed carbohydrate. The hydrogen gas is absorbed and expired in breath, and increased breath H_2 after a lactose load is used clinically to diagnose lactose malabsorption (usually because of lactase deficiency) (Levitt and Donaldson, 1970). A minority of individuals form methane gas from malabsorbed carbohydrate, and such individuals have flatus that burns with a blue flame!

A few years ago, agents which inhibited amylase activity were touted as effective weight loss drugs. Such "starch blockers" were shown to induce little carbohydrate malabsorption, presumably because of the great excess of pancreatic amylase. Carbo-

hydrate malabsorption if sufficiently great to reduce nutrient absorption is likely to cause diarrhea and cramps.

Digestion and Absorption of Protein

Dietary protein digestion begins in the stomach where the powerful action of pepsins form peptides and amino acids (Alpers, 1987). Digestion continues in the small intestine where the combined action of the endopeptidases (trypsin, chymotrypsin, elastase) and exopeptidases (carboxypeptidases) rapidly convert the peptides to oligopeptides, dipeptides, and amino acids. Hydrolysis is so rapid that diffusion to the enterocyte is likely to be rate-limiting. Paracellular absorption of some amino acids seems likely.

At the surface of the epithelium, aminopeptidases which are present on the surface of the microvilli hydrolyze most but not all of the remaining oligo- and dipeptides to amino acids. The amino acids are rapidly absorbed by a group of Na^+-coupled cotransport systems, similar to those present in the proximal tubular cell of the kidney. In addition, there are specific hydrolysis-transport proteins for dipeptides, which simultaneously hydrolyze dipeptides and transport the amino acids which are formed; absorption of some dipeptides is more rapid than absorption of the constituent amino acids. Amino acids pass through the enterocyte rapidly, with only a small fraction being used for protein synthesis within the enterocyte. The small fraction of peptides which enter the enterocyte are mostly hydrolyzed during transport through the enterocyte, so that amino acids are considered the dominant chemical form in which dietary protein is absorbed. Carrier proteins for amino acid exit are likely to be present on the basolateral membranes, but have not been characterized. Protein digestion and absorption are summarized in Figure 6.82.

DEFECTIVE PROTEIN DIGESTION AND ABSORPTION

Protein malabsorption is indicated by increased fecal nitrogen, termed creatorrhea. This occurs when pancreatic exocrine function is decreased by more than 90% (DiMagno, 1987). Defective protein absorption because of impaired dipeptide or amino acid transport by the enterocyte is detected extremely rarely, perhaps because of the redundancy in substrate specificity. A generalized defect in protein absorption would be fatal. A deficiency of enterokinase causing defective activation of trypsinogen has been described, and is characterized by creatorrhea.

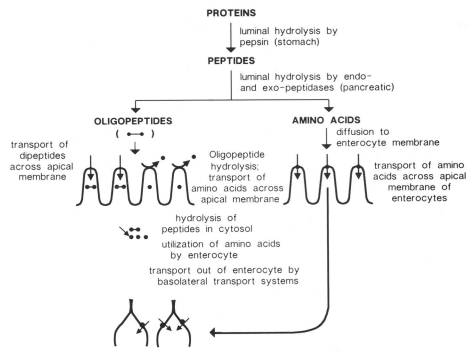

Figure 6.82. Major chemical events in protein digestion and absorption.

FATE OF MALABSORBED PROTEIN

Protein which enters the large intestine is degraded by enteric flora to a variety of substances including ammonia and products of amino acid decarboxylation such as cadaverine. Protein malabsorption causes diarrhea which has an offensive odor—termed putrefactive diarrhea.

Digestion and Absorption of Fat

Fat is the common name for the adipose tissue of animals, whereas oils (vegetable) refers to the non-aqueous fluid obtained by pressing vegetables, seeds, or nuts. Chemically, both fats and oils are mostly triglyceride, and this class of dietary constituent is termed "fat." Fat is an important caloric source because it generates more calories per gram than carbohydrate or protein. In addition, natural fats such as butter or vegetable oils contain little or no water. Thus, a potato contains about one-fiftieth as many calories as an equivalent sized lump of butter!

Triglyceride molecules are insoluble in water and do not form mixed micelles with bile acids. For efficient absorption of triglyceride, it must be hydrolyzed enzymatically and the resulting products solubilized in mixed micelles. The bile acid micelles have a remarkable ability to solubilize the digestive products of dietary triglyceride. This process of fat digestion involves the coordination of gastric emptying, pancreatic and biliary secretion, and a com-

plex sequence of physicochemical events. Some lipid molecules such as fat-soluble vitamins do not require hydrolysis, but do require micellar solubilization to be absorbed because of their low aqueous solubility (Table 6.7) (Borgstrom, 1986).

HYDROLYSIS OF TRIGLYCERIDE MOLECULES

Gastric lipase activity is weak in humans, and only a little fatty acid is formed in the stomach. When a fatty meal begins to pass into the small intestine, it stimulates biliary and pancreatic secretion neurohumorally. Zymogen granules containing the pancreatic enzymes are discharged. Trypsinogen undergoes proteolytic cleavage by brush border enterokinase; the trypsin thus formed acts on prophospholipase, procarboxyl ester hydrolyase, as well as chymotrypsinogen, proelastase, and the procarboxypeptidases to form the active enzymes; trypsin also acts on procolipase to form the active cofactor. Bile contains mixed bile acid-phospho-

Table 6.7.
Physicochemical Events in Lipid Digestion

Process	Substrate
Enzymatic hydrolysis and micellar solubilization of products	Triglycerides of long chain fatty acids
	Sterol esters
	Phospholipids
Enzymatic hydrolysis alone	Esters of short chain fatty acids
Micellar solubilization alone	Fat-soluble vitamins (D, E, K)
	Cholesterol

lipid-cholesterol micelles. When bile mixes with dietary fat droplets, the phospholipids adsorb to the oil/water interface, promoting emulsification of the fatty acids which were formed in the stomach and are present on the surface of the fat droplets. At the interface, the phosphatidylcholine molecules are hydrolyzed by pancreatic phospholipase forming lysophosphatidylcholine and fatty acid. The lysophosphatidylcholine is hydrophilic, leaves the interface, and is rapidly absorbed.

At the same time, pancreatic lipase adsorbs to the interface and generates fatty acid and monoglyceride. Desorption of pancreatic lipase by the surface-active bile acid molecules is prevented by the action of pancreatic colipase, a cationic protein which aids the anchoring of lipase to the interface. The lipase hydrolyzes the outer ester bonds of the triglyceride molecules forming 2-monoglyceride and fatty acid molecules. The fatty acid molecules which are formed ionize, in part, so that a hydrated layer of monoglyceride and partly ionized fatty acid is formed at the oil/water interface. Somehow, bile acids interact with this layer, probably initially forming vesicles composed of these lipolytic products and a small proportion of bile acids; as the proportion of bile acids increases, the vesicle is transformed into a mixed micelle. The fatty acids in the mixed micelle are neutralized by pancreatic bicarbonate; the pK_a of fatty acids solubilized in the mixed micelle is about 6.5, so that about one-half of the fatty acids in the mixed micelle are ionized at the pH present in jejunal content during digestion (pH 6–7). The physicochemical events in fat digestion, micellar solubilization of lipolytic products, and uptake by the microvillus membrane are summarized in Figure 6.83.

Both vesicles and mixed micelles diffuse through the unstirred layer coating the surface of the small intestine. The current view is that dispersion of lipolytic products into vesicles and to mixed micelles promotes their diffusion toward the enterocyte membrane; however, uptake of lipolytic products can only occur when they are present in monomeric form. For the monomeric form to reach an appreciable concentration, mixed micelles must be present (see Table 6.8). The micelles are "flickering clusters" and the fatty acids exchange rapidly between other micelles and a low concentration of monomeric fatty acids in the aqueous phase. Some of the fatty acid molecules partition (as the lipid-soluble protonated form) into the lipid domains of the microvillous membranes, and in addition, fatty acid anions enter by a Na^+ coupled cotransport system.

The fatty acids are swept down the microvillus by

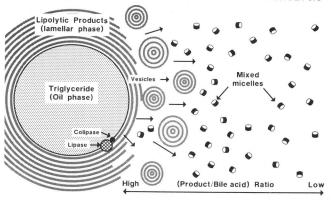

BILE ACIDS DISPERSE THE PRODUCTS OF FAT HYDROLYSIS

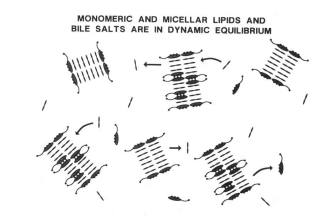

MONOMERIC AND MICELLAR LIPIDS AND BILE SALTS ARE IN DYNAMIC EQUILIBRIUM

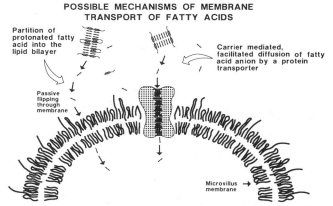

POSSIBLE MECHANISMS OF MEMBRANE TRANSPORT OF FATTY ACIDS

Figure 6.83. Schematic depiction of major physicochemical events in triglyceride digestion and absorption. *Top,* Formation of fatty acids and 2-monoglycerides (termed lipolytic products) at the oil/water interface by the action of pancreatic lipase. Colipase assists in anchoring the lipase to the interface, preventing its desorption by bile acids. *Center,* Mixed micelles containing fatty acid and monoglycerides, as well as other lipid solutes. Fatty acids and monoglycerides exchange between the micelles at a rapid rate. Note that the molecular arrangement of the micelle is similar to that present in bile, except that the solubilized lipids differ in chemical composition. *Bottom,* Mechanisms of uptake of fatty acids and monoglycerides by the microvillus membrane. (Adapted from the Undergraduate Teaching Project in Gastroenterology and Liver Disease, American Gastroenterological Association, Unit 19, Lipid Digestion, 1986.)

Table 6.8.
Effect of Micellar Solubilization of Fatty Acid on Diffusive Flux through Unstirred Layer[a]

Concentration × Diffusion Constant = Flux

Form	Fatty Acid Concentration (μmoles/ml)	Diffusion Constant (cm^2/s)	Flux (μmoles/cm-s)	Relative Flux
Molecules	0.01	7×10^6	0.07×10^6	1
Micelles	10	1×10^6	10×10^6	142

[a]Approximate values.

an undefined mechanism and pass into the cytosol, where they are rapidly converted to their CoA derivatives in the endoplasmic reticulum of the enterocyte. The fatty acid CoA derivatives esterify the 1- and 3- positions of the absorbed 2-monoglyceride, thus forming a triglyceride molecule. The triglyceride molecules aggregate to form oil droplets, which are readily seen during fat absorption, using light microscopy. At the same time, absorbed lysophosphatidylcholine and fatty acid (either of dietary or endogenous sources) combine to form phosphatidylcholine.

The triglyceride droplets thus formed are coated with a stabilizing layer of polar lipids (phosphatidylcholine and unesterified cholesterol) and, in turn, with a specific apoprotein. The final droplet with an oily interior and a surface of polar lipids and apoprotein is termed a chylomicron. The chemical composition of a chylomicron is shown schematically in Figure 6.84.

Chylomicrons are exocytosed across the basolateral membrane of the enterocyte, and enter the central lacteal of the villus. From there they pass into the larger lymphatic channels draining the intestine, into the thoracic duct, and ultimately into the subclavian vein. Chylomicrons have a diameter averaging 0.05–1 μm and are readily visible in postprandial plasma by darkfield microscopy; their increase in plasma is responsible for postprandial lipemia (Gage and Fish, 1924). The many steps in triglyceride absorption are summarized in Figure 6.85.

Although hydrolysis of a triglyceride molecule generates two fatty acids for each monoglyceride, measurements of the lipid composition of digestive intestinal content indicate there is a much higher ratio of fatty acid to monoglyceride. Lipids that can contribute to this fatty acid excess are other dietary esters such as cholesterol esters and phospholipids. Since all absorbed fatty acids are esterified to triglyceride during absorption, an additional source of glycerol is necessary. Indeed, the enterocyte has an alternative triglyceride synthesis pathway (via phosphatidic acid) in which the glycerol is derived from cellular glucose (Johnston, 1968). In principle, one can survive on a diet in which the dietary lipid is entirely in the form of fatty acid; this explains the use of candles as a dietery source in times of starvation.

Absorption of Other Lipids

Cholesterol, in intestinal content, originates from dietary cholesterol and from biliary cholesterol. In most adults, biliary cholesterol exceeds dietary cholesterol (biliary cholesterol = 1 g/day; dietary cholesterol, 0.3–0.6 g/day). Cholesterol is solubilized poorly by the fatty acid-monoglyceride mixed micelles, and during digestion, a considerable fraction of cholesterol is found in particulate form. The overall efficiency of cholesterol absorption is 20–40%. The mechanism of cholesterol absorption is not understood, but much evidence indicates that bile acids are required, not only for micellar solubilization but also for mucosal uptake. During enterocyte transport the cholesterol is, in part, esterified with long chain fatty acids. Esterified cholesterol is

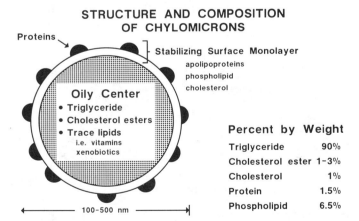

Figure 6.84. Structure and composition of chylomicrons. As soon as chylomicrons enter the blood, their central core of triglyceride is hydrolyzed rapidly by lipoprotein lipase, which is released from the capillary endothelium. The remainder of the chylomicron, now enriched in its polar surface lipids (cholesterol and phospholipid) is termed a chylomicron remnant; it is sufficiently small to pass through the fenestrations of the endothelial cells of the hepatic sinusoid. The remnants are taken up by the hepatocyte. (Adapted from the Undergraduate Teaching Project in Gastroenterology and Liver Disease, American Gastroenterological Association, Unit 19, Lipid Digestion, 1986.)

STEPS IN TRIGLYCERIDE ABSORPTION

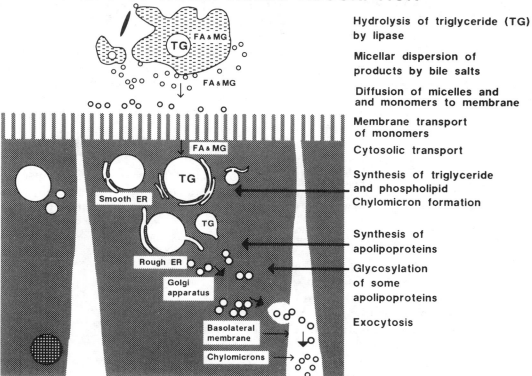

Hydrolysis of triglyceride (TG) by lipase

Micellar dispersion of products by bile salts

Diffusion of micelles and and monomers to membrane

Membrane transport of monomers

Cytosolic transport

Synthesis of triglyceride and phospholipid Chylomicron formation

Synthesis of apolipoproteins

Glycosylation of some apolipoproteins

Exocytosis

Figure 6.85. Summary of the multiple steps in triglyceride digestion and absorption. Intraluminal events are detailed in Figure 6.82. Fatty acids and 2-monoglycerides move to the smooth endoplasmic reticulum *(ER)* where the fatty acid is esterified with CoA, and then esterified to 2-monoglyceride forming triglyceride. Lysolecithin (lysophosphatidylcholine) is esterified with fatty acid to form phosphatidylcholine.

Aproproteins are added to the surface of the triglyceride droplets to form chylomicrons which are exocytosed across the basolateral membrane of the enterocyte to enter the central lacteal. (Adapted from the Undergraduate Teaching Project in Gastroenterology and Liver Disease, American Gastroenterological Association, Unit 19, Lipid Digestion, 1986.)

transported in the hydrophobic center of the chylomicron.

SHORT CHAIN (C_2-C_6) OR MEDIUM CHAIN (C_8-C_{12}) FATTY ACIDS

These fatty acids are present in dietary fat in proportions far smaller than long chain fatty acids. Short and medium chain fatty acids differ in physical properties from long chain fatty acids, inasmuch as they are water soluble and do not require micellar solubilization for efficient absorption. Consequently, synthetic triglycerides greatly enriched in such uncommon fatty acids are used in the treatment of children with an intraluminal deficiency of bile acids, as might occur in biliary atresia. These fatty acids are sufficiently small to pass paracellularly as well as transcellularly. A second difference between short and medium chain fatty acids is their metabolism in the enterocyte. During enterocyte transport, they do not form a CoA derivative, and they are not incorporated into chylomicrons. Absorption into the body is via the portal circulation. Consequently, synthetic triglycerides enriched

in short and medium chain triglycerides are used to treat patients with defective chylomicron formation or obstructed lymphatic systems (Senior, 1968). Differences between long chain and short chain fatty acid metabolism are summarized in Table 6.9.

Fat-Soluble Vitamins

The fat-soluble vitamins—vitamins A, D, E, and K⁻ have such a limited aqueous solubility that micellar solubilization by bile acids is required for intestinal absorption. Deficiency states which occur because of bile acid deficiency or mucosal dysfunction are important to recognize because they are readily treated by synthetic vitamins given parenterally.

Vitamin A is ingested in two chemicals. In plant sources, vitamin A is present as carotene, a C36 hydrocarbon, whereas in animal sources, vitamin A is present as fatty acid esters of retinol. Carotene is partly hydrolyzed and partly absorbed intact, provided it is solubilized by micelles; in the enterocyte, intact carotene is cleaved and reduced by mucosal

Table 6.9.
Difference between Long Chain ($>C_{12}$) and Medium Chain (C_5–C_{11}) or Short Chain ($<C_5$) Fatty Acids

	Long Chain	Medium and Short Chain
Aqueous solubility	Low	High
Micellar solubilization required	Yes	No
Paracellular absorption	Minor role	Major role likely
Biotransformation in enterocytes	CoA ester formation; re-esterification	None
Form of absorption	Chylomicron triglyceride	Portal plasma fatty acid
Fate	Hydrolysis in capillaries by lipoprotein lipase; Storage after re-esterification	Efficient hepatic extraction; Oxidation to CO_2

enzymes to two molecules of retinol. In contrast, dietary retinol esters are hydrolyzed in the intestinal lumen by pancreatic carboxyl esterase, solubilized in the mixed micelles, and absorbed. The retinol which is either formed from carotene in the enterocyte or absorbed from the lumen is then esterified with palmitic acid to form retinol esters which, like cholesterol esters are absorbed by transport in the hydrophobic core of the chylomicron. Retinol is stored in esterified form in specialized cells lining the sinusoids. Transport to tissues is in the form of retinol which is complexed to a specific plasma protein.

Vitamin E, tocopherol, is an antioxidant and occurs in the diet as such in leafy vegetables. Tocopherol requires micellar solubilization for absorption, and deficiency which causes a variety of neurological problems is common in children with biliary obstruction.

Vitamin K is a long chain (poly-isoprenoid) hydroquinone, which, as other fat-soluble vitamins, requires solubilization by micelles to be absorbed. Some vitamin K is formed by bacteria in the colon; if colonic content refluxes into the terminal ileum, or if micelles are transiently present in the colon, this vitamin K can be absorbed, providing an endogenous source of this vitamin which is necessary for posttranslational modification of prothrombin.

Vitamin D refers to two similar sterols formed by the action of light on a 7-ene precursor; irradiation ruptures the B ring. Vitamin D is absorbed as such, and is incorporated without esterification into chylomicrons. Presumably it is delivered to the liver in chylomicron remnants, where it undergoes 25-hydroxylation (for further details on vitamin D metabolism, see Chapter 56). Its subsequent metab-

olism involves renal 1-hydroxylation to form the 1,25 dihydroxy derivative; this metabolite is a potent regulator of Ca^{2+} metabolism.

DEFECTIVE LIPID DIGESTION AND ABSORPTION

Severe pancreatic disease causes impaired lipase secretion. When secretion is less than 10% of normal, steatorrhea occurs because of deficient lipolysis (Fig. 6.86), *top*). Some lipolysis persists because of the presence of prepancreatic (gastric and perhaps lingual) lipases (Fredrikson and Blackberg, 1980).

Bile acid deficiency occurs because of obstruction or diversion of biliary secretion, or when severe bile acid malabsorption prevents intestinal conserva-

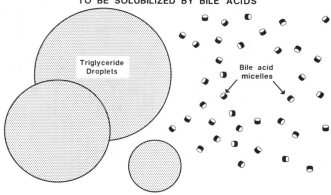

IN LIPASE DEFICIENCY THERE ARE FEW LIPOLYTIC PRODUCTS TO BE SOLUBILIZED BY BILE ACIDS

Triglyceride Droplets

Bile acid micelles

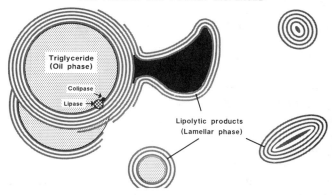

IN BILE ACID DEFICIENCY LIPOLYTIC PRODUCTS ARE FORMED BUT POORLY DISPERSED

Triglyceride (Oil phase)

Colipase

Lipase

Lipolytic products (Lamellar phase)

Figure 6.86. Schematic depiction of the disturbances in fat digestion occurring in two major types of disturbances. *Top,* In pancreatic insufficiency causing lipase deficiency, fatty acids and monoglycerides are not formed. Micelles contain only biliary lipids. *Bottom,* In bile acid (salt) deficiency as occurs in hepatobiliary obstruction or bile acid malabsorption, lipolytic products are formed, but cannot be solubilized into mixed micelles. (Adapted from the Undergraduate Teaching Project in Gastroenterology and Liver Disease, American Gastroenterological Association, Unit 19, Lipid Digestion, 1986.)

tion of bile acids. Under such circumstances, lipolysis proceeds normally, but micellar solubilization does not occur (Fig. 6.86, *bottom*). The diffusive flux of monomers is much less than that of micelles (see Table 6.8), and as a consequence, fat absorption is slowed. Overall absorption is incomplete, and occurs throughout the small intestine. The formation of calcium soaps may be increased. Some dietary constituents such as oxalate, which normally form insoluble Ca^{2+} salts in the intestinal lumen, may remain in solution because so much luminal calcium is involved in soap formation. Deficiency is common in patients with biliary obstruction and is evidenced by a prolonged prothrombin time. Correction of this clotting defect by parenteral vitamin K administration suggests good hepatocyte function and defect absorption. Failure of the prothrombin time to be corrected suggests that low prothrombin levels have resulted from defective protein synthesis because of impaired hepatocyte function.

Mucosal atrophy as occurs in celiac sprue also causes deficient fat absorption. Genetic conditions of impaired chylomicron synthesis are extremely rare; they are characterized by storage of triglyceride droplets in the enterocyte.

FATE OF MALABSORBED LIPID

Fatty acids which enter the large intestine are neither absorbed (since micelles are not present) nor appreciably degraded, and are nearly all excreted in stools. Accordingly, increased fecal fat (termed steatorrhea) is considered a reliable sign of defective fat absorption. In pancreatic deficiency, di- and triglycerides enter the large intestine. These undergo partial hydrolysis during colonic transit mediated by the continuing action of nonpancreatic lipases and/or the presence of bacterial lipases.

Fatty acids also have pharmacological effects in the colon. They stimulate colonic motility and inhibit electrolyte absorption by the colon. Hydroxy fatty acids which are formed by bacterial "hydration" of unsaturated fatty acids are still more potent than their unsaturated precursors. In some patients with steatorrhea, reduction in dietary triglyceride causes decreased fecal weight, presumably by decreasing the secretory effect of malabsorbed fatty acids.

Fatty acids may also increase paracellular permeability in the colon when they are present in excess and this, in turn, may permit other substances to be absorbed. In some patients with steatorrhea, as noted above, dietary oxalate may persist in solution in the small intestine. When it reaches the large intestine, it is absorbed from the colon causing hyperoxaluria and contributing to nephrolithiasis which is a common consequence of fat malabsorption (Chadwick et al., 1973).

Bile Acid Absorption

Bile acids are not absorbed together with other micellar constituents because they are charged at physiological pH and, thus, do not partition into the lipid domains of the brush border membranes of the jejunal enterocytes. The bile acid molecule is too large to pass paracellularly. As a consequence, bile acids remain in relatively high concentration throughout the small intestine. The concentration at which bile acids form mixed micelles with lipolytic products is about 2 mM, and bile acids are present at a concentration three to five times higher throughout the proximal small intestine.

Bile acid absorption commences in the distal half of the ileum. Distal ileocytes actively transport bile acid anions across the brush border membrane using a Na^+ coupled secondary active cotransport system. Exit from the cell involves an ion exchange protein on the basolateral membrane (Wilson, 1989).

Bile acid absorption is highly efficient. About 90% of the bile acids secreted with each meal are absorbed, so that about one-third of the bile acids are lost each day. The bile acids that are absorbed are returned to the liver in portal blood, efficiently extracted, and resecreted into canalicular bile. As a consequence of this efficient intestinal conservation, a pool of 8–10 mmol of bile acids circulates about 6 times a day giving a total daily bile acid secretion of 48–60 mmol/day despite a hepatic biosynthesis rate averaging only 1 mmol/day. Details of the enterohepatic circulation of bile acids are discussed in the preceding chapter.

BILE ACID MALABSORPTION

Bile acid malabsorption occurs after surgical excision of the terminal ileum or in severe inflammatory bowel disease (ileitis). The consequences of bile acid malabsorption depend on its severity and are invariably associated with increased hepatic bile acid biosynthesis. When bile acid malabsorption is mild, the increased synthesis can restore the hepatic secretion of bile acids to normal. In such a situation, the only consequence of bile acid malabsorption is an increased passage of bile acids into the colon, which may induce diarrhea because of the secretory effect of some dihydroxy bile acids (Fig. 6.87, *center*).

When bile acid malabsorption is severe, increased

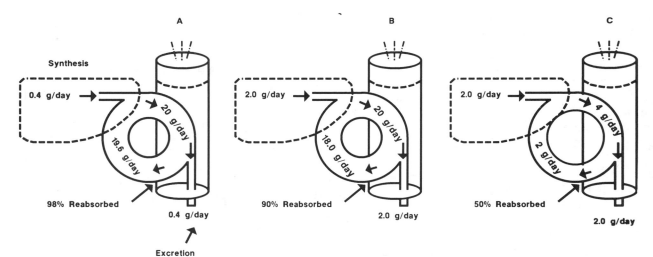

Figure 6.87. Effect of bile acid malabsorption on the enterohepatic circulation of bile acids and bile acid secretion. *Left* (A), The situation in health, when bile absorption is 98% efficient. *Center* (B), The situation in mild bile acid absorption when bile acid absorption falls to 90%. In these circumstances, increased hepatic biosynthesis from cholesterol can compensate for decreased ileal absorption, and bile acid secretion into the proximal intestine is maintained. Patients may have diarrhea resulting from the bile acid excess in the colon, which induces secretion of water and electrolytes. *Right,* (C) The situation in severe bile acid malabsorption, when maximal hepatic biosynthesis of bile acids cannot compensate for decreased absorption. Hepatic secretion of bile acids into the proximal intestine is impaired and fat maldigestion occurs; such patients may have steatorrhea. The increased concentration of fatty acids in the colon may cause secretion of water and electrolytes.

hepatic synthesis cannot compensate for increased loss. Maximal bile acid synthesis is 10 mmol/day which is far below the usual bile acid secretion rate of 48–60 mmol/day. Under such circumstances, a bile acid deficiency occurs and causes defective micellar solubilization of lipolytic products and fat-soluble vitamins (Fig. 6.87, *right*). Bile acid replacement is not satisfactory because the large quantities of bile acids which are required to correct intraluminal concentrations cause diarrhea. Thus, such patients have triglyceride and fat-soluble vitamin malabsorption because of a bile acid deficiency in the small intestine, as well as diarrhea because of the pharmacological effects of malabsorbed fatty acids in the colon. The concentration of malabsorbed bile acids in the colon in such patients is too low to induce secretion (Hofmann and Poley, 1972).

FATE OF MALABSORBED BILE ACIDS

Bile acids are biotransformed by colonic bacteria (see previous chapter) and undergo limited (passive) colonic absorption. However, like fatty acids, bile acids have pharmacological properties when present in high aqueous concentrations—above 1–2 mM. (In healthy persons, the concentrations of bile acids in the aqueous phase of colonic content is less than 0.1 mM). When present above an aqueous concentration of 1–2 mM, the two dihydroxy bile acids—deoxycholic acid and chenodeoxycholic acid—stimulate colonic motility, increase electro-

lyte and water secretion, and cause increased paracellular permeability.

Bile acid malabsorption may or may not be associated with increased aqueous concentrations. The concentration of bile acids in solution—which is the important determinant of their activity—depends on interaction with colonic bacteria, intraluminal pH, and other factors. The normal action of colonic bacteria is to remove bile acids from solution, since deoxycholic acid and lithocholic acid are less soluble than their precursors. Clinically, abnormally increased colonic bile acid concentrations caused by bile acid malabsorption may be manifest as a watery diarrhea, which often will respond to the administration of a bile acid-binding resin.

Absorption of Vitamin B_{12}

Vitamin B_{12} (now called cobalamin) refers to a group of large water-soluble molecules containing cobalt bound in a corrin ring. Like conjugated bile acids, cobalamin is absorbed in the distal ileum. It is not known whether this distal site of absorption is "better" for the organism than a more proximal site. Cobalamin is absorbed by receptor-mediated endocytosis of a complex formed with a protein termed intrinsic factor. In the ileocyte, the intrinsic factor is degraded in an acidic endosomal compartment, and the B_{12} exits the cell across the basolateral membrane.

Cobalamin is present in the diet bound to

enzymes. Cobalamin is released from dietary proteins by gastric peptidases, but then combines with salivary proteins termed R proteins. Pancreatic proteinases subsequently degrade the R proteins, releasing cobalamin which combines with intrinsic factor, a protein secreted by the gastric parietal cell. The cobalamin-intrinsic factor complex is resistant to pancreatic digestion.

Cobalamin is present in bile bound to R proteins, so that endogenous and dietary cobalamin are absorbed together in the distal ileum. Loss of ileal function leads to the gradual development of cobalamin deficiency which can cause megaloblastic anemia and neurological disease. Cobalamin deficiency is treated by parenteral administration of synthetic cobalamin.

Folic Acid Absorption

Folic acids (also termed folacin or folate) are a group of water-soluble vitamins containing pteroic acid which is essential for normal red cell maturation and cell division since it is required for nucleotide synthesis. It occurs in the diet conjugated with oligomers of glutamate (often heptamers), which are cleaved by a brush border enzyme to the monoglutamate; folate enters the enterocyte as the monoglutamate. It is stored in tissues as a polyglutamate, and circulated as the monoglutamate. Folate, in contrast to cobalamin, is absorbed in the jejunum. A folate deficiency is common in diseases of impaired mucosal function.

Other Water-Soluble Vitamins

Thiamine and riboflavin are large polar molecules and also require carrier systems for their absorption. Pyridoxine is small enough to pass paracellularly, but a carrier system is also likely to be present (Alpers, 1988). These are absorbed in the proximal small intestine.

Absorption of Divalent Cations

CALCIUM

Dietary calcium is ingested in milk and other dietary products and ionic Ca^{2+} is released during gastric digestion. Calcium is a small cation and should be freely permeable across the paracellular junctions of the proximal small intestine, provided it is in solution. Its passive absorption is likely to occur throughout the small intestine, especially in the distal small intestine where most water absorption occurs. The physical state of the calcium ion in small intestinal content is not well defined, but at least some of the Ca^{2+} ions should be present as

counterions to the mixed micelles present during fat digestion. The enterocytes of the proximal intestine contain a Ca^{2+}-binding protein, vitamin D inducible, which enhances Ca^{2+} entry into the enterocytes of the duodenum. Export of Ca^{2+} across the basolateral membrane is likely to involve a Ca^{2+}-ATPase. Luminal Ca^{2+} is cleaved from dietary Ca^{2+} (1000 mg/day) and Ca^{2+} present in biliary and pancreatic secretions (300 mg/day). Absorption is inefficient—about 33%. In Ca^{2+} deficiency, efficiency of intestinal absorption increases, this effect being modulated by 1, 25 dihydroxy vitamin D and parathormone. Body calcium metabolism is discussed in Chapter 56.

IRON

Iron has a limited solubility at small intestinal pH and is believed to be absorbed mostly in the duodenum where pH is relatively low and bile acid concentrations are high, both factors which promote solubilization of Fe^{2+} ion. The presence of iron-binding anions such as citrate in the diet should diminish the activity of the iron ion and decrease its absorption. Iron which enters the enterocyte is only well absorbed into the systemic circulation if the individual is iron deficient; otherwise, the iron is stored as ferritin (Conrad, 1987); if this iron is not mobilized, the ferritin will be lost when the enterocyte dies. The daily requirement for iron is far greater in the menstruating woman than in the adult man. In hemochromatosis, the enterocyte control mechanism is lost, and iron absorption continues despite a body burden so high that it causes destruction of sensitive tissues such as the liver and pancreas.

OTHER CATIONS

The student is referred to specialized texts for information on the absorption of other important cations.

Magnesium metabolism resembles that of Ca^{2+} and a Ca^{2+} deficiency is often associated with a magnesium deficiency. Zinc is a key constituent of many enzymes and because body stores of zinc are small, zinc deficiency is common in patients with small intestinal disease or in alcoholics with diets deficient in protein. Absorption of the common electrolytes is discussed in Chapter 47.

Absorption of Drugs

Drug absorption involves intraluminal and mucosal events. The most important intraluminal event is usually dissolution of the solid phase. The drug molecules then dissolve in monomeric form or associate with the mixed micelles. Intestinal transport is considered to involve passive permeation by partition into the lipid domains of the brush border membrane. Active transport mechanisms are rarely utilized. For low molecular weight drugs, paracellular absorption can be important. Little is known about the transport mechanisms involved in the transport of drugs from the enterocyte into portal plasma. Drugs formulated to have prolonged dissolution times (sustained release preparations) will often pass into the colon before absorption is complete. In the colon, absorption is less efficient and bacterial degradation and/or adsorption to solid residue also occurs.

Intestinal Water and Electrolyte Secretion and Absorption

Water and electrolyte secretion and absorption processes in the intestine are required for normal digestion and absorption of nutrients and elimination of wastes. Digestion and absorption in the intestine require water to maintain the fluidity of the luminal contents, to serve as a medium bringing digestive enzymes in contact with food particles, and to allow the diffusion of digested nutrients to the epithelial cells where absorption occurs. A large amount of water is secreted by various gastrointestinal organs to facilitate these processes. The daily fluid load varies with meals but includes approximately 2000 ml of oral intake, 1500 ml of saliva, 2500 ml of gastric juice, 500 ml of bile, 1500 ml of pancreatic juice, and 1000 ml from the intestine itself. Most of the secreted fluid is absorbed downstream. By the time a day's intestinal content reaches the colon it is reduced to only 1500–2000 ml by absorptive processes. Most of the fluid entering the colon is absorbed leaving only 100 ml or so to be excreted in the stool. Thus, the intestine clearly has a trememdous capacity for water absorption. Interestingly, the capcity of the intestine to secrete water and electrolytes is as great or greater that its capacity to absorb. The fluid secreted under normal circumstances (1000 ml/day) can be augmented many times in the presence of toxins or endogenous secretagogues. The most dramatic examples are patients with cholera who can secrete more than 30 liters of stool water a day.

Electrolytes play a central role in the regulation of water absorption and secretion. Electrolytes are absorbed or secreted by active transport mechanisms. The actively transported electrolyte then causes water to move so that isoosmolality between compartments is maintained. In other words, water transport is regulated indirectly by the regulation of electrolyte transport. Active electrolyte transport mechanisms usually drive the absorption (or secretion) of only one electrolyte across the epithelial cells. Other ions, with an opposite charge, and water follow the actively transported ion(s) passively and paracellularly through the tight junction.

Electrolyte transport processes, both absorptive and secretory, are regulated by a variety of endogenous and exogenous factors. Endogenous regulatory factors are peptide hormones and neurotransmitters secreted by regulatory cells. Luminally active agents and bacterial toxins are major exogenous factors affecting fluid and electrolyte secretion. Besides these endogenous and exogenous factors which directly affect the epithelial cell, food particles and nonabsorbable substances are also important. They contribute to the osmolality of the luminal contents and thus affect passive water absorption or secretion via the tight junction. When these molecules remain in the intestinal lumen, (e.g., lactose in lactase-deficient subjects), they increase water secretion. The opposite is also true, when molecules are absorbed and leave the lumen, they increase water absorption.

ANATOMICAL CONSIDERATIONS

Epithelial Cells

Epithelial cells cover the surface of the gastrointestinal tract. Besides serving as barriers, these cells serve to absorb nutrients, water, and electrolytes from the gut lumen. They also secrete water and electrolytes. The intestine has a tremendous reserve capacity for both absorptive and secretory functions, having a surface area larger than a doubles tennis court (> 200 square meters). The large surface area of the small intestine is achieved by intestinal folds, villi, and microvilli as shown in Figure 6.88. These structural features amplify the surface area tremendously. The large intestine has a smaller amplification factor as compared to the small intestine due to the absence of villi. In addition to the anatomical amplification of surface area, the motility function of the intestine also provides physiological amplification. Contractions of intestinal smooth muscle increase or decrease the flow of the luminal contents to assure optimal contact time. Therefore, deficiency of mucosal absorptive function is rarely observed except in (1) patients with

CHAPTER 47

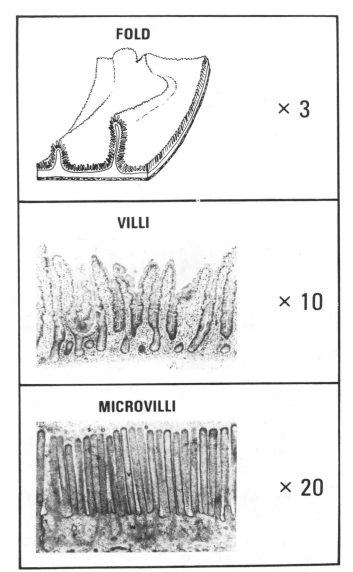

Figure 6.88. Amplification of the intestinal surface area by intestinal folds (plicae conniventes), villi, and microvilli. The numbers indicate the factor by which the surface area is amplified over a flat surface and which together amplify the surface area by approximately 600 fold. (From Trier et al., 1983. Used by permission.)

massive intestinal resection; (2) patients with diseases that destroy the absorptive cells; or (3) patients with diseases that form a submucosal barrier to the absorptive process. On the other hand, the vast reserve capacity allows exogenous toxins to cause excessive secretion readily. For this reason, almost everyone has some personal experience with diarrhea. It should be emphasized here that the absorptive and secretory functions of the intestine are unrelated and have different regulatory mechanisms. Therefore, the absorptive function may remain intact while the secretory function is excessively stimulated by toxins or other secreta-

gogues. The belief that the crypt cells are responsible for secretion (Welsh et al., 1982) while the villous and surface cells are responsible for absorption provides a potential anatomical basis for these distinct functions.

Tight Junctions

Structurally, the tight junctions which seal the luminal side of the epithelial cells can be compared to the plastic holder that holds a six-pack of beer (Fig. 6.89). Water movement across the gastrointestinal epithelium occurs passively through the tight junction depending on the osmotic gradient across the epithelium. Beside water, water-soluble molecules of small molecular size (< 300 MW) and electrolytes can also move across the tight junction. It should be pointed out that the plasma membrane of the epithelial cell, being a lipid bilayer, is quite impermeable to water and water-soluble substances unless specialized transport pathways exist for them. To absorb or secrete water, the epithelial cells actively transport certain electrolytes transcellularly via the specialized transport pathways discussed below. The tight junction then allows water and ions with an opposite electrical charge to follow passively and paracellularly between the epithelial cells. Thus, the gastrointestinal epithelium has the properties of a semipermeable membrane with the tight junctions serving as selective pores. The permeability of tight junctions allows passive movement of water and small molecules from one side to the other according to the direction of the osmotic or chemical gradients. Passive water movement across the tight junction is a key physiological feature that permits water absorption or secretion to be controlled via regulation of electrolyte transport. Ions of charge opposite to those being actively transported follow them passively across the tight junctions preventing the accumulation of electrical charges. In general, the permeability of tight junctions decreases as one moves distally through the intestine. The upper gut is therefore significantly more permeable than the colon to passive movement of fluids and electrolytes. Variations of tight junction permeability also exist locally in the same segment of intestine, with the villous region being less permeable than the crypts. It appears that most of the passive paracellular movement of water and electrolytes occurs in the crypt (Madara and Trier, 1987; Claude and Goodenough, 1973).

The tight junction is composed of tight junction strands. Increasing the number of strands increases resistance and decreases epithelial permeability. Recent studies have demonstrated that the cytoskeleton of the cells inserts itself directly into the

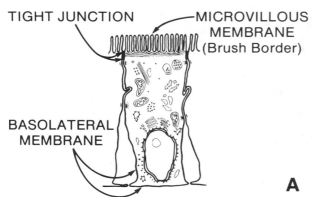

TIGHT JUNCTION ——MICROVILLOUS
MEMBRANE
(Brush Border)

BASOLATERAL
MEMBRANE

A

B

Figure 6.89. Tight junctions (occluding junctions). These structures seal the top of each epithelial cell to adjacent cells. *A,* Diagram shows the tight junction separating the apical (microvillous membrane) and basolateral membrane. *B,* Freeze fracture replicas of tight junctions on the lateral surface of a villous cell, demonstrating the network of tight junction strands. (From Madara et al., 1980.)

tight junction strands; the number of strands increases with hypertonicity and is decreased by agents that disrupt the cytoskeleton. It is likely, therefore, that the tight junction permeability can be regulated, but the precise regulatory factors involved are currently unknown.

Regulatory Cells

Endocrine cells in the crypts, nerve cells in the submucosal and myenteric plexuses, and the vagus nerve provide coordinated regulations of epithelial functions. The absorptive and secretory functions of the epithelial cells are regulated by a large number of peptide hormones and neurotransmitters released from endocrine cells and nerve endings, as well as a number of luminally active agents (Binder and Sandle, 1987; Cooke, 1987; Donowitz and Welsh, 1987). These regulatory mechanisms can be of paracrine, neurocrine, or endocrine nature as illustrated in Figure 6.90. The endocrine cells located in the crypt region provide most, if not all, of the paracrine regulations in addition to the classical endocrine regulations. The messengers (peptides or other substances) contained in these cells are released in response to the luminal environment and provide local regulation of the epithelial cells nearby. Many of these compounds stimulate or inhibit electrolyte secretion, while others regulate electrolyte absorption. Neural elements provide neurocrine regulation. As in the skin, nerve cells and nerve endings in the gut are in close approximation to the surface lining cells. The neurons also innervate the intestinal smooth muscles and may have an important role in coordinating epithelial transport with gut motility. For example, endogenous opiates released from gut neurons constrict circular smooth muscle, delaying intestinal transit while simultaneously increasing mucosal absorp-

tion. In addition, the central nervous system plays a role in the regulation of fluid and electrolyte transport via the vagus nerve.

Besides paracrine and neurocrine regulation, endocrine regulation provides distant coordination between different portions of the gastrointestinal tract, or between the gut and other organs. This effect results from the release of peptides or neurotransmitters from endocrine cells and possibly nerve cells into the bloodstream. As stated above, the luminal content including bile salts, fatty acids, and microbial toxins may also affect electrolyte transport function. With such a large number of potential regulatory mechanisms for ion transport, the precise physiological contribution of each factor is difficult to determine and remains largely unclear.

ELECTROLYTE TRANSPORT PATHWAYS

The term "transport mechanism" will be used in this chapter to represent an active transport process that enables ions to move across the intestinal epithelium *transcellularly.* The term "transport pathway" represents a large membrane protein which acts either as a carrier, a channel, or a pump in the cell membrane; "transport pathways" enable ions to move *across the plasma membranes* of an epithelial cell. In epithelial cells, usually each of these transport pathways is localized only to either the apical or basolateral membrane.

In general, each active transport mechanism requires the participation of at least two or three transport pathways: an uptake step across one plasma membrane, an exit step across the other plasma membrane, and a pump which provides the energy. The pump may also serve as an uptake or exit step; thus, a minimum of two transport path-

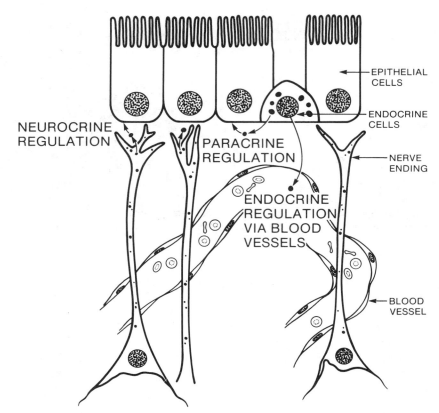

Figure 6.90. Neurohumoral regulation of intestinal epithelium. Endocrine cells in the crypt region release peptides or other substances across their basolateral membranes which regulate cells nearby (paracrine regulation) or enter the bloodstream to regulate distant organs (endocrine regulation), while nerve cells release peptides or neurotransmitters from the nerve endings to regulate the epithelial cells or muscle cells (neurocrine regulation).

ways are required for each absorptive or secretory mechanism (Figs. 6.91–6.93). One may consider an absorptive or secretory mechanism as a specialized function possessed by the epithelial cells. To exhibit a specialized transport mechanism, an epithelial cell must possess every electrolyte transport pathway required in the transport process. Because most electrolyte transport pathways have been identified only recently, information concerning them is less complete than physiological knowledge regarding the overall transport mechanism itself. At present, the participation of transport pathways in a large

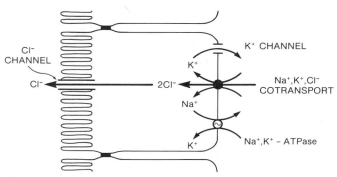

Figure 6.91. Chloride secretory mechanism. The Na^+, K^+, Cl^- cotransport on the basolateral membrane serves as the Cl^- uptake step with the Na^+/K^+-ATPase pump providing the driving force and recycling the Na^+. The excess K^+ is recycled via the K^+ channels on the basolateral membrane. Chloride exits via a Cl^- channel on the apical membrane. Regulatory sites for the Cl^- secretory process are either at the Cl^- channels and/or K^+ channels.

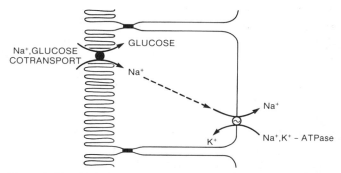

Figure 6.92. Na^+-glucose cotransport. A Na^+-glucose cotransport carrier on the apical membrane serves to bring glucose and Na^+ into the cell. Sodium is then pumped out via the Na^+/K^+-ATPase; glucose proceeds across the basolateral membrane via another facilitated transport carrier for glucose. Similar Na^+ cotransport mechanisms also exist for amino acids, dipeptides, tripeptides, many B vitamins, and bile salts.

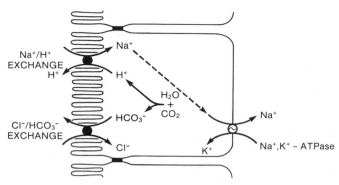

Figure 6.93. Electroneutral NaCl cotransport. At the present, it appears that the electroneutral NaCl cotransport mechanism in the intestine results from a dual exchange system, comprised of a Na^+-H^+ exchange and a Cl^--HCO_3^- exchange, working in concert. These exchange carriers serve to bring Na^+ and Cl^- across the apical membrane into the cell. Sodium is then pumped out via the Na^+/K^+-ATPase, with Cl^- following via another Cl^- transport pathway.

which serve as transport pathways. Biochemical identification is important because it provides direct verification of the existence of a transport pathway, while verification by physiological techniques is indirect. The electrolyte transport pathways known to exist in gastrointestinal epithelial cells are summarized in Table 6.10, together with their possible participation in an absorptive or a secretory mechanism. Evidence that these transport pathways are regulated by peptides or neurotransmitters is accumulating rapidly and is providing important insight into the regulation of various transport mechanisms. Short-term regulation can be directed either at pumps, carriers, or channels that participate in a given transport mechanism. In the longer term, cells can also adjust their transport capacity by increasing or decreasing the number of pumps, carriers, or channels as well as regulatory receptors on the plasma membrane. Regulation of channels, due to the large capacity of channels as discussed below, is usually required for immediate or short-term adjustment. The discussion to follow will first deal with the properties or different types of transport pathways. Subsequently, this chapter will review how the arrangement of these transport pathways enables different portions of the gastrointestinal tract to carry out their specialized functions, and how such pathways are regulated.

number of transport mechanisms is still being investigated. However, advances in experimental techniques, particularly the use of membrane vesicles, cultured cell lines and patch clamp techniques, has allowed rapid progress in the area. An increasing number of pumps, carriers, and channels are being identified by these physiological techniques. Meanwhile, significant progress has also been made in the biochemical identification of the membrane proteins

Table 6.10.
Electrolyte Transport Pathways, their Participation in Transport Mechanisms, and their Regulation

Transport Pathway	Transport Mechanism	Endogenous Regulation
Transcellular Pathways		
Pumps		
Na^+/K^+-ATPase	Active transport mechanism	Aldosterone or corticosteroid by affecting synthesis and insertion of the pump
Ca^{2+} and Mg^{2+} ATPases	Cellular handling of Ca^{2+} and Mg^{2+}	Unknown
Carriers		
Na^+-H^+ exchange	Electroneutral NaCl absorption	Intracellular substances which increase Ca^{2+}-calmodulin
Cl^--HCO_3^- exchange	Electroneutral NaCl absorption HCO_3^- secretion	Ca^{2+}-calmodulin
Na^+, K^+, Cl^- cotransport	Cl^- secretion	Unknown
Na^+, glucose cotransport	Glucose absorption	Unknown
Other Na^+ cotransport pathways	Absorption of other nutrients	Unknown
Channels		
Na^+ channel	Electrogenic Na^+ absorption (colon only)	Aldosterone by increasing synthesis and insertion
Cl^- channel	Cl^- secretion	Intracellular substances which increase cyclic AMP and ? Ca^{2+}
K^+ channel	Cl^- secretion ? K^+ absorption and secretion	Intracellular substances which increase cyclic AMP and Ca^{2+}
Paracellular Pathways		
Tight junction	Movement of water and oppositely charged ions	Unknown, involves regulation of cytoskeleton

Properties of Pumps, Carriers, and Channels

ATPase Pumps

For active transport to occur, a pump is required to provide the necessary driving force for ion movement. In intestinal epithelial cells, the Na^+/K^+-ATPase pumps provide this driving force and are localized basolaterally (Dibona and Mills, 1979). In the presence of Mg^{2+}, the Na^+/K^+-ATPase pump catalyzes outward movement of three Na^+ ions and inward movement of two K^+ ions per cycle across the basolateral membrane using one ATP molecule each time. Thus, this ATPase pump creates a negative intracellular electrical potential and a relatively low intracellular Na^+ concentration. This electrochemical gradient favors Na^+ influx across the plasma membrane via the Na^+ channel and the cotransport and exchange transport for which Na^+ is required (Figs. 6.91–6.93). The role of these Na^+/K^+-ATPase pumps in active transport of nutrients and electrolytes has been well characterized. Most, if not all, active transport processes require the activity of this pump. This is supported by the fact that cardiac glycosides (e.g., ouabain) which inactivate the pump also inhibit every active transport mechanism currently known to exist in the intestine.

CARRIERS

Carriers are either cotransport carriers, also called symports, or exchange carriers, also called antiports (Aronson, 1981). In the case of a cotransport carrier, the carrier protein binds to ions or other molecules on one side of the plasma membrane and shuttles them across to another. In the case of an exchange carrier, an ion is transported in exchange for another ion on the opposite side of the plasma membrane with the same electrical charge. A diversity of active transport mechanisms is possible because a large number of carriers serve to convert the energy provided by the Na^+/K^+-ATPase pumps into specialized functions. Since the carriers are integral parts of active transport mechanisms, they are called secondary active transport pathways. By themselves, these carriers are best described as facilitated transport pathways which allow certain ions to move across the plasma membrane faster than could be expected by diffusion along the existing electrochemical gradient. In sharp contrast with the pumps, which drive ion movement across the plasma membrane in only one direction, the carriers catalyze ion movement in both directions. The direction of movement depends on the existing electrochemical gradient. For a cotransport mechanism, the net chemical gradient is established by the concentrations of the molecules involved on each side of the membrane. For example, in the case of Na^+-glucose cotransport, the effective gradient can be calculated as difference between $[Na^+] \times [glucose]$ on either side of the membrane. Therefore, by coupling the transport of the needed molecules such as glucose with Na^+, it is possible for the cell to move such a molecule even against its own chemical gradient and effectively use the energy provided by the Na^+/K^+-ATPase pump for this purpose. Accordingly, there are a large number of cotransport mechanisms which include Na^+, taking advantage of the favorable Na^+ gradient provided by the Na^+/K^+-ATPase pump.

The cotransport carriers or symports are major transport pathways for absorption of a wide variety of nutrients as well as electrolytes by the intestine. Of the nutrient-related carriers, the Na^+-glucose cotransport carrier has been studied more than others (Fig. 6.92). In fact, the Na^+-glucose cotransport carrier is one of the few transport pathways that have been characterized biochemically. The primary role of the Na^+-glucose cotransporter is nutrient absorption, but the process also brings in Na^+ and water, and promotes positive electrolyte and water balance. Besides glucose, many other food-derived products, e.g., amino acids, di- and tripeptides, vitamins, bile salts, etc., are taken up by cotransport mechanisms with Na^+. Each of them involves a different specific carrier protein. Intestinal cotransport carriers which deal only with electrolytes also exist. Among these are the Na^+, K^+, Cl^- cotransport carrier which is required for the Cl^- secretory mechanism (Fig. 6.91) (O'Grady et al., 1987). Other cotransport pathways such as a K^+, Cl^- cotransporter have been postulated, but at the present time, it is unclear whether they truly exist and what role they may play in the gut.

The exchange carriers or antiports serve to carry a needed ion into the cell in exchange for a readily available intracellular ion. There are two major exchange carriers in the intestine, the Na^+-H^+ exchange and Cl^--HCO_3^- exchange pathways. They are coupled in a so-called "electroneutral NaCl absorptive process." The exchanges make good sense, since these pathways effectively exchange the waste product of respiration, CO_2 (which in the presence of carbonic anhydrase forms H^+ and HCO_3^-) for Na^+ and Cl^- (Fig. 6.93).

CHANNELS

Ionic channels serve a similar but distinct function from carriers. While both systems serve as integral parts of transport mechanisms, the channel is a high-flow system that is capable of moving a large number of specific ions rapidly. When open, it allows a large number of selected ions to flow downstream according to the existing electrochemical gradient. The carrier system, on the other hand, moves smaller numbers of molecules, but it can do so against a chemical gradient as discussed above. Each channel on the plasma membrane, when open, allows 10^6–10^8 ions to move across per second. Because of its large capacity, the physiological characteristics of a channel approach those of passive diffusion systems which are not saturable. Also, because of their capacity, ionic channels are probably the most important transport pathways for electrolytes. It is likely that the immediate actions of many peptides or neurotransmitters are the result of the opening or closing of ionic channels. Specific ionic channels for Na^+, Cl^-, K^+, and Ca^{2+} have been described and will be reviewed briefly.

Sodium channels have been recognized longer and hence investigated more extensively than the other channels mentioned. There are at least two types of Na^+ channels, one of which functions in colonic Na^+ absorption and another which serves as the acetylcholine receptor. The former Na^+ channel can be regulated by aldosterone and blocked by amiloride. Many studies of this Na^+ channel prior to the availability of the patch clamp technique were indirect using amiloride sensitivity as an indicator (Will et al., 1981). The biochemistry of both types of Na^+ channels has recently been elucidated (Chien and Gargus, 1987).

Chloride channels have not yet been as well investigated. Their existence has been suggested by the observation that Cl^- ion can be secreted alone in response to many peptides or neurotransmitters. Recent patch clamp studies confirmed their presence on the apical membrane of intestinal epithelial cells (Frizzell et al., 1986). These studies also suggest the presence of more than one type of Cl^- channel, one of which can be regulated by cellular cyclic AMP. The search for potent and specific blockers for Cl^- channels is currently in progress. Once available, these Cl^- channel blockers may facilitate our understanding of electrolyte transport physiology. More importantly, they may have useful clinical applications, as discussed below.

Enough evidence is available to confirm the existence of multiple K^+ channels in the gastrointestinal tract, and their involvement in the Cl^- secretory process (Fig. 6.91). At least two types of K^+ channels, one regulated by intracellular cAMP and another by intracellular Ca^{2+}, are thought to be localized in the basolateral membrane of mucosal epithelial cells. Verification of their existence by patch clamp techniques are pending. K^+ channels are likely to be involved in active K^+ absorption or secretion.

Lastly, the ability of Ca^{2+} channel blockers to inhibit certain secretory mechanisms in the intestine has provided evidence for the existence of Ca^{2+} channels in intestinal epithelia. Ca^{2+} transport in the intestine is relatively complex and certain mechanisms attributed to Ca^{2+} channels may actually involve the Na^+/Ca^{2+} exchange carriers. As is true for other ionic channels, definition of the precise role of Ca^{2+} channels in the intestine awaits further investigation with the patch clamp technique.

OTHER TRANSPORT PATHWAYS

Transport pathways for other ions, e.g., bicarbonate, sulfate, phosphate, organic ions, and divalent cations, are poorly defined although on theoretical grounds they undoubtedly exist. There may exist a HCO_3^- channel in addition to the Cl^-/HCO_3^- exchange carrier, and specific transport pathways for sulfate and phosphate have been postulated.

ELECTROLYTE TRANSPORT MECHANISMS IN THE GASTROINTESTINAL TRACT

The preceding section dealt with the electrolyte transport pathways which serve as basic building blocks for electrolyte transport mechanisms. This section will discuss each major transport mechanism in the intestinal tract individually. Table 6.11 provides an overall view of electrolyte transport mechanisms in the small and large intestines. Transport functions exhibited by the intestinal tract are diverse. The diversity of transport mechanisms in the gut accommodate absorption of a broad range of nutrients as well as maintaining proper amounts of fluid in the gut lumen. The small intestine is responsible for most of the absorption of nutrients and water. When needed, the intestine can absorb more water by delaying intestinal transit.

In contrast to the small intestine, the large intestine of man does not absorb nutrients, but plays a role in the conservation of fluid and electrolytes (Devroede and Phillips, 1969). The colon actively absorbs most of the fluid and electrolyte presented to it. Similar to the small intestine, the colon has a large secretory capacity that can be stimulated by luminal toxins or endogenous hormones. Furthermore, the colon can also absorb short chain fatty acids that are produced by bacterial catabolism of unabsorbed carbohydrates which may account for up to 10% of the ingested calories (Bond et al., 1980). Since the small and large intestines share a

Table 6.11.
Electrolyte Transport Mechanisms along the Intestinal Tract

Intestinal Segment	Major Transport Mechanisms	Pathological Condition(s)	Disease
Small intestine	HCO_3^- secretion (mainly in proximal duodenum)	Decreased HCO_3^- secretion	Duodenal ulcer
	Absorption of glucose and other nutrients in symport with Na^+	Decreased nutrient absorption	Osmotic diarrhea
	Electroneutral NaCl absorption	Decreased NaCl absorption	Infectious diarrhea or other secretory diarrhea
	Cl^- secretion	Excessive Cl^- secretion	Infectious diarrhea or other secretory diarrhea
	Absorption of bile in symport with Na^+ (terminal ileum only)	Malabsorption of bile salts	Bile salt-induced diarrhea
Large intestine	Electrogenic Na^+ absorption	Decreased Na^+ absorption	Secretory diarrhea
	Electroneutral NaCl absorption	Decrease NaCl absorption	Infectious diarrhea or other secretory diarrhea
	Cl^- (and HCO_3^-) secretion	Excessive secretion	Infectious diarrhea or other secretory diarrhea

number of transport mechanisms, to minimize repetition the discussion to follow is divided into secretory and absorptive mechanisms.

Secretory Mechanisms

Secretory mechanisms throughout the gastrointestinal tract center around Cl^- ion. Hydrochloric acid is the major secretory product of the stomach. In other parts of the intestinal tract, the predominant ion secreted is either Cl^- or HCO_3^- (Davis et al., 1980; 1981; 1983). The latter may be related to and require the active secretion of Cl^- as discussed below.

INTESTINAL Cl^- SECRETION (ELECTROGENIC Cl^- SECRETION)

The components of the Cl^- secretory mechanism have been reasonably well established (Fig. 6.91). The transcellular process is called electrogenic Cl^- secretion because the anion Cl^- is secreted by the intestinal epithelium, without an accompanying cation or without involving exchange with another anion. In the Cl^- secretory process, the cell takes up Cl^- from the bloodstream across the basolateral membrane via the Na^+, K^+, Cl^- cotransport pathway. The Na^+/K^+-ATPase provides energy for this mechanism and recycles Na^+; while K^+ channels on the basolateral membrane allow for K^+ recycling. Cl^- accumulates intracellularly above its electrochemical equilibrium and exits through the apical membrane via Cl^- channels. In this secretory mechanism, the Cl^- ion is secreted actively and transcellularly with Na^+, K^+, and water following passively

via the tight junction. The sites of regulation by the intracellular messengers, cyclic AMP, cyclic GMP, and Ca^{2+}, are at the Cl^- and/or K^+ channels.

INTESTINAL HCO_3^- SECRETION

HCO_3^- is the predominant ion secreted by the biliary and pancreatic ducts. Recently, the importance of a HCO_3^- secretory mechanism in the stomach and proximal duodenum has been recognized (Isenberg et al., 1987). HCO_3^- secretion in these regions of the gut is important because of its possible role in mucosal defense against peptic ulcer information. HCO_3^- secretion is also important in other parts of the intestine. The upper small intestine secretes more HCO_3^- than the lower portion. However, due to the acid load from the stomach, HCO_3^- is neutralized and HCO_3^- concentration in the upper intestinal lumen is relatively low. HCO_3^- gradually becomes the predominant anion of the luminal content in the lower gut as a means to conserve Cl^- via the Cl^- HCO_3^- exchange carrier. At the present time, HCO_3^- secretory mechanisms have not been fully elucidated and more than one mechanism may exist. HCO_3^- secretion may occur via the Cl^-/HCO_3^- exchange carrier with the Cl^- secretory mechanism providing the intraluminal Cl^- for exchange. HCO_3^- could either be produced intracellularly by the action of carbonic anhydrase or possibly transported from the bloodstream. Other mechanisms for HCO_3^- secretion by the gastrointestinal epithelium which do not involve the Cl^-/HCO_3^- exchange pathway may exist, but they have yet to be explored. As in the case of Cl^- secretion, HCO_3^- secretion also promotes increased secretion of cations and water by paracellular pathways.

Absorptive Mechanisms

Na^+ is the primary ion that drives water absorption in the intestine. In the small intestine, cotransport of Na^+ with other food-derived products, together with the electroneutral NaCl absorptive mechanism, is responsible for most, if not all, of the water and electrolyte absorption. The large intestine absorbs Na^+ avidly via the Na^+ channel and by the electroneutral NaCl absorptive mechanism shared with the small intestine.

SODIUM-GLUCOSE COTRANSPORT AND SIMILAR COTRANSPORT MECHANISMS IN THE SMALL INTESTINE

The absorption of carbohydrates, amino acids, dipeptides, tripeptides, fat, minerals, vitamins, and bile salts is the key function of the small intestine. Many of these substances are absorbed by a cotransport pathway together with Na^+. Glucose is absorbed via a Na^+-glucose cotransport carrier on the apical membrane of the small intestinal epithelium (Goldner et al., 1969) (Fig. 6.92). Na^+ and glucose bind to the carrier which then shuttles from the outer surface of the apical membrane to the inner surface where both of them are released before the carrier returns to the outer surface. Because transport carriers function in both directions, an appropriate electrochemical gradient must exist to provide a proper direction for the vectorial transport. The Na^+ gradient created by the Na^+/K^+-ATPase pump provides the needed energy for the absorptive direction. The glucose that accumulates in the cell diffuses across the basolateral membrane by a facilitated transport pathway, and the Na^+ is pumped out basolaterally by the Na^+/K^+-ATPase. The charge of the absorbed Na^+ promotes absorption of an anion via the paracellular pathway; Cl^- primarily serves this function. Water passively follows to keep the intercellular space isoosmolar. Therefore, this intestinal glucose transport mechanism also drives water and electrolyte absorption (Fordtran et al., 1965). The ability of this transport mechanism to promote positive electrolyte and water balance remains unaffected by most disease processes. Therefore, glucose salt solution, which increases the activity of this transport mechanism, proves to be very useful clinically for the management of diarrhea and dehydration to replace electrolytes and water. Sugar and starch, which yield glucose after digestion, provide similar results (Mehta and Subramaniam, 1986).

Similarly, many cotransport mechanisms for nutrients depend on Na^+ gradients across the apical membrane created by the Na^+-K^+-ATPase pump. Of the large number of cotransport mechanisms which involve Na^+, most exist only in the small intestine. Cotransport of Na^+ and amino acids, dipeptides, or tripeptides provides for the absorption of needed protein components. Cotransport of Na^+ and many B vitamins has been recognized. Finally, ileal uptake of bile acids is also provided for by a Na^+ cotransport mechanism.

ELECTRONEUTRAL Na^+ AND Cl^- ABSORPTION IN THE INTESTINE

A Na^+-Cl^- cotransport mechanism has been proposed because in the intestine a significant portion of Na^+ absorption has been shown to require the presence of Cl^-, and vice versa. It is called electroneutral Na^+-Cl^- absorption to distinguish this mechanism from an electrogenic Na^+ absorptive mechanism discussed below. However, the interdependence of Na^+ and Cl^- transport could not be readily demonstrated when apical membrane vesicles from intestinal mucosa were studied, raising questions about the nature of this cotransport phenomenon. The debate about the nature of this mechanism is likely to continue. Many investigators now believe that electroneutral Na^+-Cl^- cotransport is actually comprised of a Na^+-H^+ exchange working in concert with a Cl^--HCO_3^- exchange mechanism. This mechanism allows both Na^+ and Cl^- to enter the cell in exchange for H^+ and HCO_3^- (Fig. 6.93). Protons and HCO_3^- are produced intracellularly by the action of carbonic anhydrase on CO_2. Sodium, which enters the cell, is then pumped out by the Na^+/K^+-ATPase with Cl^- following via another Cl^- transport pathway yet to be identifed. The HCO_3^- and H^+ transported into the lumen are converted to CO_2 and H_2O. CO_2 is simply reabsorbed and excreted in the expired air without the need for a special transport mechanism. The result is net NaCl absorption. Water then follows the absorbed ions via the tight junction. This Na^+-Cl^- cotransport mechanism appears to be widely distributed in the epithelium of the intestinal tract, including the small intestine, the proximal part of the large intestine, and the gallbladder.

ELECTROGENIC Na^+ ABSORPTION IN THE COLON

Electrogenic Na^+ absorption is predominant in the colon. Electrogenic Na^+ absorption is so named because the absorbed Na^+ ion, with a positive electrical charge, is unaccompanied by cation exchange or anion cotransport. The Na^+ ion enters the cell via a Na^+ channel on the apical membrane. It is pumped out across the basolateral membrane by the Na^+/K^+-ATPase pump. Water and anion follow the absorbed Na^+ from the luminal side to the bloodstream side via paracellular routes.

Absorption and Secretion of Potassium, Calcium, and Other Ions

There are specific transport mechanisms to handle other electrolytes, including potassium, calcium and other divalent cations, sulfate, phosphate, and organic anions. In general, the body does not totally permit the absorption of important substances to occur passively. Rather, mechanisms exist to regulate these processes and permit adaptation to changes in the environment. It is reasonable to assume that active transport processes exist for all of the important ions. Unfortunately, the transport mechanisms for many of these ions are poorly understood at the present time.

Only recently have active transport mechanisms in the intestine for potassium been recognized and investigated. In the past, it was wrongly assumed that potassium simply leaked passively through the epithelial tight junction according to existing gradients. Most investigators now believe that active transcellular mechanisms are the predominant means by which potassium is absorbed or secreted. Furthermore, active participation of the colonic mucosa in potassium homeostasis has also been emphasized. However, the transport pathways involved in the potassium absorptive or secretory mechanisms have only been postulated. For potassium absorption, it is possible that a K^+-H^+ exchange pathway serves to exchange intracellular H^+ for potassium. The accumulated potassium then diffuses across the basolateral membrane via a potassium channel or other potassium carriers. For potassium secretion, either an increase in Na^+/K^+-ATPase activity that pumps K^+ across the basolateral membrane into the cell, or an increased K^+ movement across the apical membrane, may serve this function.

Luminal absorption of Ca^{2+} is regulated by the active metabolites of vitamin D. Parathyroid hormone and calcitonin also influence Ca^{2+} transport. Ca^{2+} in different intracellular pools serves different functions. This localization is possible because the pools are associated with different calcium-binding proteins. A number of such calcium-binding proteins exist in cells. Some of these calcium-binding proteins bind to absorbed Ca^{2+} and thus, regulate Ca^{2+} absorption, while others bind free cytosolic Ca^{2+} and mediate cellular function. Therefore, absorbed Ca^{2+} can be transported across the epithelial cell without affecting any Ca^{2+}-mediated cellu-

lar events. It does not appear that Ca^{2+} absorbed from the lumen mixes with the pool that mediates epithelial cell functions. Consequently, the source of Ca^{2+} for Ca^{2+}-mediated cellular events is derived from the bloodstream rather than from the gut lumen. This Ca^{2+} must therefore enter the epithelial cells through the basolateral membrane. The transport processes involved at this site are thought to be a Na^+-Ca^{2+} exchange carrier as well as a Ca^{2+} channel. In addition, a number of Ca^{2+}-ATPases may be very important in maintaining different intracellular compartments of Ca^{2+}.

REGULATION OF ELECTROLYTE TRANSPORT

Our understanding of the regulation of electrolyte transport mechanisms has the potential to be very useful clinically. Therefore, a great deal of attention has been given to research in the area of regulatory mechanisms. At the epithelial cell level, receptor-mediated regulation has been widely investigated. Therefore, the discussion emphasizes this aspect. Other factors that influence intestinal water and electrolyte transport are less well understood. They are briefly discussed together at the end.

Receptor-Mediated Regulation of the Epithelial Cells

For the purpose of discussing peptide or neurotransmitter regulation, we can consider the intestine to be a self-modulated organ. The endocrine cells and nerve cells serve as sensors, while the hormones or neurotransmitters serve as messengers. Hormones or neurotransmitters are released in response to stimuli, and these substances serve to modulate and coordinate the functions of the intestine and other digestive organs.

Significant advances in our understanding of the mechanisms of action of peptides or neurotransmitters that regulate intestinal secretion or absorption have been made in recent years. In the past, only cyclic AMP and cyclic guanosine monophosphate (GMP) could be measured. At present, calmodulin, free cytosolic calcium, products of phosphatidylinositol metabolism, G-proteins, products of arachidonic acid metabolism, and various protein kinases can be measured directly. These new developments allow better and more complete investigation into the roles of secondary messengers in receptor-mediated regulation. The isolation of cellular organelles (e.g., endoplasmic reticulum, Golgi bodies, etc.) also facilitates our knowledge regarding the transport of ions between different intracellular pools. Better understanding of the regulatory controls in conjunction with the growth of knowledge regarding the participation of basic transport pathways in a given transport mechanism should allow us to map the cellular regulatory events from the receptor site to the transport pathway. Obviously, our knowledge in this regard which has important clinical applications is still quite limited.

For practical purposes, receptor-mediated regulation can be divided into three steps: (1) receptor binding and activation by the regulatory substances; (2) activation of intracellular mediators (secondary messengers); and (3) activation and regulation of the cellular transport pathway.

RECEPTOR BINDING AND ACTIVATION BY REGULATORY SUBSTANCES

Receptor binding and activation by endogenous regulatory peptides or neurotransmitters occur on the basolateral membrane while actions by bacterial toxins usually occur on the apical membrane. The receptor recognizes the signal (hormone or neurotransmitter) released by sensor cells (endocrine cells or nerve cells) and initiates the biological event. There are a large number of hormones and neurotransmitters which affect or modulate the transport function of the epithelial cells. A list of hormones and neurotransmitters that increase fluid

and electrolyte secretion or absorption, and the intracellular mediators involved in their action, is given in Tables 6.12 and 6.13, respectively. The physiological role of each peptide or neurotransmitter cannot be clearly delineated at the present time because phrmacological doses are needed for most of the biological studies and very few specific receptor antagonists are available.

Activation of some receptor-mediated regulatory mechanisms has been applied clinically. Drugs whose tertiary structures resemble hormones or neurotransmitters bind to their receptors and mimic their actions. Those with antisecretory properties (e.g., opiates) have been used successfully in the treatment of diarrhea. Others with secretory properties have been used for constipation. Some of the drugs were widely used before their mechanisms of action were well understood. A classic example is the use of synthetic opiates to treat diarrhea.

SECONDARY MESSENGERS

After binding to receptors, peptides or neurotransmitters activate a cascade of intracellular mediators (secondary messengers). An increase of these secondary messengers regulates various

Table 6.12.
Intestinal Secretory Stimuli

Type of Substance	Substance	Intracellular Mediator
Endogenous stimuli (peptides, neurotransmitters, and products of inflammation)	Acetylcholine[a]	Ca^{2+}
	Histamine[a]	Ca^{2+}
	Serotonin	Ca^{2+}
	Substance P	Ca^{2+}
	Neurotensin	Ca^{2+}
	Calcitonin	? Ca^{2+}
	Cholecystokinin	? Ca^{2+}
	Prostaglandins[a]	Cyclic AMP
	Bradykinin	Cyclic AMP[b]
	VIP[a]	Cyclic AMP
	Secretin	Cyclic AMP
	Glucagon	Cyclic AMP
	PHI	Cyclic AMP
	Adenosine[a]	? Cyclic AMP
	Atrial natriuretic factor	Cyclic GMP
	Gastrin	Unknown
	GIP	Unknown
	Bombesin	Unknown
	5-HPETE and 5-HETE	Unknown
Exogenous stimuli (microbial enterotoxins and luminally active agents)	Bacterial enterotoxins:	
	Vibrio cholerae toxin	Cyclic AMP
	Escherichia coli (heat-lable) toxin	Cyclic AMP
	Escherichia coli (heat-stable) toxin	Cyclic AMP
	Yersinia enterocolitica toxin	Cyclic GMP
	Clostridium difficile toxin	? Ca^{2+}
	Other microbial enterotoxins	Unknown
	Bile salts and fatty acids	Ca^{2+} and ? Cyclic AMP
Pharmaceutical agents	Laxatives	Mostly unknown

[a]Among endogenous agents, acetylcholine, prostglandins, VIP, histamine, and adenosine have been shown to directly regulate epithelial cells and probably are of physiological importance. [b]Bradykinin increases submucosal prostaglandin production which then causes an increase in epithelial cyclic AMP.

Table 6.13.
Intestinal Absorptive Stimuli

Type of Substance	Substance[a]
Endogenous stimuli (peptides and neurotransmitters)	α-Adrenergic agonists Enkephalins and other opoids Somatostatin Glucocorticoids Angiotensin Peptide YY and neuropeptide Y GABA
Exogenous stimuli (luminally active agents)	Nutrients
Pharmaceutical agents	Cyclooxygenase inhibitors, e.g., indomethacin, aspirin Phenothiazines Propranolol Nicotinic acid and analogs Local anesthetics Chloroquine Lithium Verapamil Berberine

[a]Except for α-adrenergic agonists, which activate epithelial cells directly, the effects of other agents are mostly indirect via neural regulatory controls either locally or centrally. α-Adrenergic agonists are probably of physiological importance.
Intracellular mediators involved in the absorptive stimuli of the intestine are largely unknown, except for cyclooxygenase inhibitors which inhibit prostaglandin synthesis. Others may affect the secretory mechanism anywhere in the cascade of regulation by cyclic nucleotides or calcium.

absorptive or secretory processes in the intestinal epithelial cells. In contrast to receptors which are of numerous types, the types of intracellular mediators are relatively few. Many hormones and neurotransmitters thus share similar amplifying events in the epithelial cells. A few major patterns of secondary messenger responses have been recognized in epithelia. They can be divided into two main groups, one mediated by cyclic nucleotides and another mediated by calcium.

For cyclic nucleotide-mediated mechanisms, cyclic AMP is more commonly called upon than cyclic GMP. Both, however, serve important roles in electrolyte transport. Increased cellular cyclic AMP or cyclic GMP stimulates Cl^- secretion by crypt cells and inhibits the neutral NaCl absorptive mechanism of surface or villous cells. The secretory response results directly from the action of cyclic nucleotides in the secretory cells. The ability of cyclic nucleotides to inhibit NaCl absorption directly in absorptive cells is less clearly understood but may be secondary to their ability to increase cellular calcium.

Two important endogenous regulators, prostaglandins and vasoactive intestinal peptide (VIP), mediate their actions by activating adenylate cyclase and increasing cAMP production. Exogenous cholera toxin causes intestinal secretion by

ADP-ribosylation of the G_s protein (the regulatory subunit of the adenylate cyclase system). ADP-ribosylation sustains the G_s protein in an active form capable of stimulating the adenylate cyclase to increase cellular cyclic AMP. *Escherichia coli* heat-stable enterotoxin (ST_a), which increases guanylate cyclase activity and thus increases cellular cyclic GMP, also mediates a secretory response. In this case, cyclic GMP rather than cyclic AMP stimulates chloride secretion and inhibits NaCl absorption.

Some receptor-mediated responses occur via changes in phosphatidylinositol metabolism and changes in free cytosolic calcium. Calmodulin may regulate absorptive mechanisms in villous or surface cells, while phosphatidylinositol metabolism and changes in free cytosolic calcium regulate secretory mechanisms in crypt cells. The calcium calmodulin-dependent mechanism is known to be important because trifluoperazine and chlorpromazine, which inactivate the calcium-calmodulin complex, have antidiarrheal properties. In the surface absorptive cells or villous cells, the calcium-calmodulin complex inhibits apical membrane transport pathways, the Na^+-H^+ and the Cl^-/HCO_3^- exchange carriers, resulting in an inhibition of NaCl absorption. In crypt cells, phosphatidylinositol metabolism and free cytosolic calcium appear to be important regulatory mechanisms for Cl^- secretion. Acetylcholine and histamine cause Cl^- secretion via this mechanism probably by causing breakdown of phosphatidylinositol into inositoltrisphosphate and diacylglycerol. Inositoltrisphosphate then mobilizes calcium from intracellular calcium pools resulting in an increase in free cytosolic calcium. This information is relatively new and it is not known whether the calcium-mediated responses listed in Tables 6.12 and 6.13 involve calmodulin.

Besides the cyclic nucleotide and calcium-related mechanisms discussed above, the importance of arachidonic acid metabolites has recently been recognized. Arachidonic acid and products of its metabolism via cyclooxygenase or lipoxygenase pathways (the latter normally considered a result of inflammation) are also produced during phosphatidylinositol breakdown. Such potent mediators can profoundly affect electrolyte transport and should be considered secondary messengers. Finally, it is likely that most if not all of the secondary messengers described above mediate their action by activating a protein kinase or a phosphatase. For example, cyclic AMP may activate the cyclic AMP-dependent protein kinase, cyclic GMP may activate the cyclic GMP-dependent kinase and diacylglycerol may activate the C-kinase, etc. Due to the complexity of intestinal tissues, the role of protein kinases and phosphatases should be investigated in isolated cell systems.

Pharmacological interference with the cascade of secondary messengers has been applied successfully to treat secretory diarrhea. The use of trifluoperazine, a calmodulin inhibitor, is an example. However, clinical application of knowledge in this area is still limited.

REGULATION OF CELLULAR TRANSPORT PATHWAYS

The final event regulated by receptor activation is at the transport pathway level, either on the basolateral or apical membrane. For short-term regulation, hormones and neurotransmitters can activate or inactivate an existing transport pathway via one of the secondary messengers. For example, VIP and PGE_1 increase cyclic AMP which activates a cyclic AMP-dependent Cl^- channel on the apical membrane, causing Cl^- secretion. Acetylcholine increases cytosolic calcium which activates a calcium-dependent K^+ channel on the basolateral membrane; the resultant hyperpolarization caused by K^+ exit creates a favorable electrical gradient for Cl^- secretion across the apical membrane. The effect of this calcium-dependent mechanism is to augment the Cl^- secretion caused by the action of cyclic nucleotides (Cartwright et al., 1985). It is likely that each receptor-mediated event results from the activation of one or two transport pathways via a cascade of secondary messengers. For long-term regulation, a number of hormones, e.g., steroids, can increase the number of pumps, carriers, or channels. For example, aldosterone increases the number of Na^+ channels on the colonic apical membrane and $Na^+,$-K^+-ATPase pump on the basolateral membrane to increase Na^+ absorption. Epithelial cells may also increase or decrease the number of receptors on the plasma membrane causing up- or downregulation of a receptor-mediated event.

Clinical applications directed at the transport pathways are relatively new and novel. The effectiveness of omeprazole to inhibit gastric acid secretion has stimulated interest in this area. In the intestine, blockage of the Cl^- exit pathway should inhibit intestinal secretion and diarrhea. With successful development of potent and specific Cl^- channel blockers, this approach may be possible in the future.

Other Factors that Influence Intestinal Water and Electrolyte Transport

There are a few other factors which may indirectly influence water and electrolyte transport by intestinal epithelial cells. Among these factors are intestinal motility, intestinal permeability, oncotic pressure of the blood, arterial pressure, venous pressure, and luminal pressure. The important roles of the central nervous system and immune system are being recognized. For clinical purposes, we need to consider only the important role of intestinal motility in water and electrolyte transport. Coordinated contraction of gastrointestinal smooth muscles may allow more contact time for absorption and decrease stool volume significantly. The effectiveness of many antidiarrheal drugs, including synthetic opiates, results mainly from their effect on gut motility. Other factors besides gut motility are not considered important clinically at the present time, but as we learn more about them, they may also become important. For example, intestinal permeability increases a great deal in diseases that cause mucosal injury such as inflammatory bowel diseases. Colonic permeability also increases when bile salts or fatty acids are present in the colon. Theoretically, an increased intestinal permeability may allow the oncotic and hydrostatic pressures to have a more prominent effect on electrolyte secretion. Increased permeability may also allow a number of larger molecules, which normally are not absorbed, to diffuse across the intestinal mucosa more easily.

One relatively difficult area to study is immune-related intestinal secretion, which involves the participation of lymphocytes, mast cells, and other immune cells. Undoubtedly, the immune cells in the submucosa release messengers (e.g., lymphokines and mast cell products) which may activate intestinal secretion and cause diarrhea in a manner similar to peptides or neurotransmitters. The availability of cell culture models derived from the intestine may allow better elucidation of this important mechanism in immune-related diseases.

IMPLICATION IN DISEASES

The area of intestinal electrolyte transport physiology is applied mainly to the management of diarrhea. Pharmacological bypass of the secretory defect in cystic fibrosis is another current topic of investigation.

Diarrheas usually result from toxins produced by microorganisms or from excessive release of endogenous regulatory substances. The pathological processes usually mimic the physiological mechanisms. The primary actions of cholera toxin and heat-stable toxin of *E. coli* are at the epithelial cells where they stimulate Cl^- secretion and inhibit NaCl absorption by activating the adenylate cyclase and guanylate cyclase, respectively. Some toxins such as *Clostridum* and *Shigella* toxins are cytotoxic to the epithelial cells. Once these toxins reach the submucosa, they may cause diarrhea by stimulating gut motility or by activating a neural or immune response which then causes intestinal secretion. Endogenous sources of hormone may also cause diarrhea when excessive amounts are released. Examples are endocrine tumors such as carcinoid tumors, medullary carcinoma of the thyroid, and VIP-secreting tumors.

When discussing the treatment of diarrhea, two basic approaches should be considered: (1) stimulation of the absorptive mechanism, which typically remains intact; and (2) inhibition of the excessive secretory function. For the former, glucose-NaCl solutions have been successfully applied clinically to stimulate water absorption without affecting the secretory process. The latter approach may be applied by manipulating the receptor-mediated process, or by blocking Cl^- exit. Pharmacological activation of receptors by drugs such as opiates, α_2-adrenergic agents, and somatostatin, have been applied clinically and proved effective in inhibiting secretion and stimulating absorption. Trifluoperazine is the only drug available that interferes with secondary messengers. In the future, direct inhibition of a transport pathway, such as the Cl^- secretory pathway, may be possible.

METHODOLOGIES FOR INTESTINAL ELECTROLYTE TRANSPORT STUDIES

In Vivo Perfusion

In vivo perfusion techniques are used to study water and electrolyte transport in healthy humans. These techniques allow measurement of the rate of disappearance or appearance of water, and electrolyte, or other substance in the intestinal lumen. With these techniques, it is assumed that what disappears has been absorbed, and what appears has been secreted. In general, this assumption is valid and the technique is quite useful for identification and study of absorptive or secretory phenomena. These techniques allow for the influence of most contributing factors, including gut motility and blood flow. Perfusion studies have significant limitations when employed in hopes of elucidating the mechanisms involved in absorptive or secretory processes. Studies of these mechanisms can be better carried out with the in vitro techniques described below.

In Vitro Techniques

In vitro techniques use either isolated segments of the intestine, isolated enterocytes, or plasma membranes. The advantage of each method arises mainly from its ability to exclude certain confounding factors. Isolated intestine stripped of the muscular layers excludes the effects of blood flow and motility. Isolated enterocytes exclude the influence of preexisting peptides or neurotransmitters. Purified apical or basolateral membrane allows identification of transport pathways on either side of the plasma membrane. The disadvantages of each method come mainly as a result of the preparation's isolation from normal regulatory factors and normal environment. The phenomena observed may therefore not be pertinent to whole organs.

EVERTED GUT SAC

A segment of the intestine is everted, tied at both ends, and suspended in the same solution that fills the sac. The everted gut sac enlarges upon absorption and shrinks with secretion. The preparation retains most intestinal structure including local regulatory cells. Except for blood flow and CNS control, the phenomena observed should be pertinent to the whole intestine.

USSING CHAMBER

The Ussing chamber (Ussing and Zerahn, 1951) is designed to eliminate all passive forces, and therefore detects only active transport mechanisms. In essence, it consists of two fluid-filled reservoirs with a piece of the isolated intestine or epithelium separating the two reservoirs (Fig. 6.94). A voltage-clamp apparatus constantly generates sufficient electrical current to nullify the electrical potential difference. The current needed is called the short-circuit current. This short-circuit current reflects the net amount of electrical charges carried by various transported ions and can be used to monitor electrolyte transport activity once the ion transported is identified. By itself, the current does not identify the exact types of ion transported. The Ussing chamber allows the active transport phenomena across the epithelial sheet to be investigated without the confounding effects of gut motility and blood flow.

SPECIFIC ION TRANSPORTS

Uptake or efflux of a radionuclide into or out of cells or across epithelial sheets can be used to investigate specific channels, symporters, antiporters, and pumps.

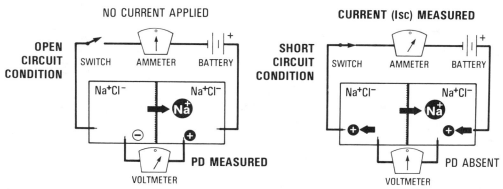

Figure 6.94.   Ussing chamber. A piece of intestine or epithelial sheet is clamped between two water baths. Chemical gradients are eliminated by the use of similar bathing solutions on each side; the electro-chemical gradient, which may result from active transport, is nullified by an electrical current. *Open circuit condition:* the voltmeter measures the potential difference *(PD)* resulting from the active transport. *Short circuit condition:* the battery with ammeter adjusts the potential differ-ence to zero by applying an electrical current; under this condition, there is no electrochemical gradient across the isolated intestine or epithelial sheet. The amount of current needed to nullify the potential difference quantifies the net active transport of charged ion. Radioactive tracers can be used to follow quantiatively the movement of ion from one side to another. (Adapted from Moore, 1976. Used by permission.)

Several techniques are used to measure the intracellular concentration of ions such as Na^+, K^+, H^+, Cl^-, and Ca^{2+}. These techniques include ion-specific microelectrodes, electron probe analysis, and intracellular chemical probes. The chemical probes are dyes with fluorescent properties that are specific-ion dependent.

The patch clamp is a technique that can be used to measure specific ion channels. In this technique the tip of a microelectrode is sealed with a tiny patch of plasma membrane from the epithelial cell. This tiny patch of plasma membrane, therefore, acts like an ion-selective microelectrode containing one or a few ionic channels of interest. When exposed to proper ion gradients, a small burst of electrical current can be detected if the ionic channel is present

MEMBRANE VESICLES AND CELL CULTURE

Membrane vesicles can be prepared from apical or basolateral plasma membrane. Specific transport processes can then be localized to a specific region of the plasma membrane and can be measured using radionuclide fluxes or intravesicular fluorescent dyes.

Cultured epithelial cells exclude the confounding factors introduced by endocrine cells and nerve endings. A cultured cell line contains only one cell type and thus serves as a better model for study of specific transport mechanisms.

BIBLIOGRAPHY

ADMIRAND, W. H. AND SMALL, S. M. The physicochemical basis of cholesterol gallstone formation in man. *J. Clin. Invest.* 47: 1043–1052, 1968.

AKAO, S., D. E. BOCKMAN, P. LECHENE DE LA PORTE, AND H. SARLES. Three-dimensional pattern of ductuloacinar associations in normal and pathological human pancreas. *Gastroenterology* 90: 661–668, 1986.

ALLEN, A. Structure and function of gastrointestinal mucus. In: *Physiology of the Gastrointestinal Tract*, vol. 1, edited by L. R. Johnson, J. Christensen, M. I. Grossman, E. D. Jacobson, S. G. Schultz. New York: Raven Press, 1981, p. 617–640.

ALLEN, R. H., B. SEETHARAM, E. PODELL, AND D. H. ALPERS. Effect of proteolytic enzymes on the binding of cobalamin to R protein and intrinsic factor. In vitro evidence that a failure to partially degrade R protein is responsible for cobalamin malabsorption in pancreatic insufficiency. *J Clin Invest* 61: 47–54, 1978.

ALPERS, D. H. Digestion and absorption of carbohydrates and proteins. In: *Physiology of the Gastrointestinal Tract*, 2nd ed., vol. 2, edited by L. R. Johnson, J. Christensen, M. J. Jackson, E. D. Jacobson, J. H. Walsh. New York: Raven Press, 1987; p.1469–1487.

ARIAS, I. M., W. B. JACOBY, H. POPPER, D. SCHACHTER, AND D. A. SHAFRITZ. *The Liver: Biology and Pathobiology*, 2nd ed. New York: Raven Press, 1988.

ARONSON, P. S. Identifying secondary active solute transport in epithelia. *Am. J. Physiol.* 240: F1–F11, 1981.

BARROWMAN, J. A. *Physiology of the Gastrointestinal Lymphatic System.* Monographs of the Physiological Society, No. 33. Cambridge: Cambridge University Press, 1978.

BINDER, H. J., AND G. I. SANDLE. Electrolyte absorption and secretion in the mammalian colon. In: *Physiology of the Gastrointestinal Tract*, 2nd ed., vol. 2, edited by L. R. Johnson, J. Christensen, M. J. Jackson, E. D. Jacobson, J. H. Walsh. New York: Raven Press, 1987, p. 1389–1418.

BJORKHEM, I. Mechanism of bile acid biosynthesis in mammalian liver. In: *Sterols and Bile Acids*, edited by H. Danielsson and J. S. Sjovall. New York: Elsevier, 1985; p. 231–278.

BOND, J. H., B. CURRIER, H. BUCHWALD, AND M. LEVITT. Colonic conservation of malabsorbed carbohydrate. *Gastroenterology* 78: 444–447, 1980.

BORGSTROM, B. Luminal digestion of fats. In: *The Exocrine Pancreas: Biology, Pathobiology and Diseases*, edited by V. L. W. Go, J. D. Gardner, F. P. Brooks, E. Lebenthal, E. P. DiMagno, G. A. Scheele. New York: Raven Press, 1986, p. 361–373.

BORGSTROM, B., A. DAHLQVIST, G. LUNDH, AND J. SJOVALL, Studies of intestinal digestion and absorption in the human. *J. Clin. Invest.* 36: 1521–1536, 1957.

BORTOFF, A. Myogenic control of intestinal motility. *Physiol. Rev.* 56: 418–434, 1976.

BOYER, J. L. New concepts of mechanisms of hetpatocyte bile formation. *Physiol. Rev.* 60: 303–320, 1980.

BRO-RASMUSSEN, F., S. A. KELLMAN, AND J. H. THAYSEN. The composition of pancreatic juice as compared to sweat, parotid saliva and tears. *Acta Physiol. Scand.* 37: 97–113, 1956.

BURNSTOCK, G. Purinergic nerves. *Pharmacol. Rev.* 1972; 54: 418–440.

CAREY, M. C., AND M. J. CAHALANE. Enterohepatic circulation. In: *The Liver: Biology and Pathobiology*, 2nd ed. edited by I. M. Arias, W. B. Jakoby, H. Popper, D. Schachter, D. A. Shafritz. New York: Raven Press, 1988; p. 573–616.

CAREY, M. C. AND D. M. SMALL. The physical chemistry of cholesterol solubility in bile. *J. Clin. Invest.* 61: 998–1026, 1978.

CARTWRIGHT, C. A., J. A. McROBERTS, K. G. MANDEL, AND K. DHARMSATHAPHORN. Synergistic action of cyclic AMP and calcium mediated chloride secretion in a colonic epithelial cell line. *J. Clin. Invest.* 76: 1837–1842, 1985.

CASE, R. M. Synthesis, intracellular transport and discharge of exportable proteins in the pancreatic acinar cell and other cells. *Biol. Rev.* 53: 211–354, 1978.

CASE, R. M., AND B. E. ARGENT. Bicarbonate secretion by pancreatic duct cells: mechanism and control. In: *The Exocrine Pancreas. Biology, Pathobiology, and Diseases.* edited by V. L. W. Go, J. D. Gardner, F. P. Brooks, E. Lebenthal, E. P. Dimagno, G. A. Scheele. New York: Raven Press, 1987; p. 213–243.

CHADWICK, V. A., K. MODHA, AND R. H. DOWLING. Mechanism for hyperoxaluria in patients with ileal dysfunction. *N. Engl. J. Med.* 289: 172–176, 1973.

CHEY, W., M. KIM, K. LEE, AND T-M. CHANG. Effect of rabbit antisecretin serum on postprandial pancreatic secretion in dogs. *Gastroenterology* 77: 1268–1275, 1979.

CHEY, W. Y. Hormonal control of pancreatic exocrine secretion. In: *The Exocrine Pancreas. Biology, Pathobiology and Diseases*, edited by V. L. W. Go, J. D. Gardner, F. P. Brooks, E. Lebenthal, E. P. DiMagno, G. A. Scheele. New York: Raven Press, 1987, p. 301–211.

CHIEN, S., AND J. J. GARGUS. Molecular biology in physiology. *FASEB. J.* 1: 97–102, 1987.

CHRISTENSEN, J. Motility of the colon. In: *Physiology of the Gastrointestinal Tract.* 2nd ed., vol. 1, edited by L. R. Johnson, J. Christensen, M. J. Jackson, E. D. Jacobson, J. H. Walsh. New York: Raven Press, 1987, p. 665–693.

CHRISTENSEN, J. Motor functions of the pharynx and esophagus. In: *Physiology of the Gastrointestinal Tract.* 2nd ed., vol. 1, edited by L. R. Johnson, J. Christensen, M. J. Jackson, E. D. Jacobson, J. H. Walsh. New York: Raven Press, 1987, p. 595–612.

CHRISTENSEN, J. The colon. In: *A Guide to Gastrointestinal Motility*, edited by J. Christensen, D. L. Wingate. Bristol, England: Wright-PSG, 1983a, p. 198–214.

CHRISTENSEN, J. The controls of gastrointestinal movements: some old and new views. *N. Engl. J. Med.* 285: 85–98, 1971.

CHRISTENSEN, J. The oesophagus. In: *A Guide to Gastrointestinal Motility*, edited by J. Christensen, D. L. Wingate. Bristol, England: John Wright and Sons, 1983b, p. 75–100.

CHRISTENSEN, J., J. GLOVER, E. O. MACAGNO, R. SINGERMAN, AND N. W. WEISBRODT. Statistics of concentration at a point in the human duodenum. *Am. J. Physiol.* 221: 1818–1823, 1971.

CHRISTENSEN, J., AND G. A. RICK. Nerve cell density in submucous plexus throughout the gut of cat and opossum. *Gastroenterology* 89: 1064–1069, 1985.

CHRISTENSEN, J., AND G. A. RICK. The distribution of myelinated nerves in the ascending nerves and myenteric plexus of the cat colon. *Am. J. Anat.* 178: 250–258, 1987.

CHRISTENSEN, J., G. A. RICK, B. A. ROBISON, M. J. STILES, AND M. A. WIX. The arrangement of the myenteric plexus throughout the gastrointestinal tract of the opossum. *Gastroenterology* 85: 890–899, 1983.

CHRISTENSEN, J., G. A. RICK, AND D. J. SOLL. Intramural nerves and interstitial cells revealed by the Champy-Maillet stain in the opossum esophagus. *J. Auton. Nerv. Syst.* 19: 137–151, 1987.

CHRISTENSEN, J., AND B. A. ROBINSON. Anatomy of the myenteric plexus of the opossum esophagus. *Gastroenterology* 83: 1033–1042, 1982.

CHRISTENSEN, J., H. P. SCHEDL, AND J. A. CLIFTON. The basic electrical rhythm of the duodenum in normal human subjects and in patients with thyroid disease. *J. Clin. Invest.* 43: 1659–1667, 1964.

CHRISTENSEN, J., H. P. SCHEDL, AND J. A. CLIFTON. The small intestinal basic electrical rhythm (slow wave) frequency gradient in normal men and in patients with a variety of diseases. *Gastroenterology* 50: 309–315, 1966.

CHRISTENSEN, J., M. J. STILES, G. A. RICK, AND J. SUTHERLAND. Comparative anatomy of the myenteric plexus of the distal colon in eight mammals. *Gastroenterology* 86: 706–713, 1984.

CLAUDE, P., AND D. A. GOODENOUGH. Fracture faces of zonulae of occludentes from "tight" and "leaky" epithelia. *J. Cell. Biol.* 58: 390–400, 1973.

COHEN, B. I., A. F. HOFMANN, E. H. MOSBACH, R. J. STENGER, M. H. ROTHSCHILD, L. R. HAGEY, AND Y. B. YOON. During effects of non-ursodeoxycholic or ursodeoxycholic acid on hepatic histology and bile acid metabolism in the rabbit. *Gastroenterology* 91: 189–197, 1986.

CONRAD, M. E. Iron absorption. In: *Physiology of the Gastrointestinal Tract,* 2nd ed., vol. 2, edited by L. R. Johnson, J. Christensen, M. J. Jackson, E. D. Jacobson, J. H. Walsh. New York: Raven Press, 1987, p. 1437–1453.

COOKE, H. J. Neural and humoral regulation of small intestinal electrolyte transport. In: *Physiology of the Gastrointestinal Tract,* 2nd ed., vol. 2, edited by L. R. Johnson, J. Christensen, M. J. Jackson, S. D. Jacobson, J. H. Walsh. New York: Raven Press, 1987, p. 1307–1350.

COOPER, H., R. LEVITAN, J. S. FORDTRAN, AND J. F. INGELFINGER. A method for studying absorption of water and solute from the human small intestine. *Gastroenterology* 50: 1–7, 1966.

COSTA, M. J. B. FURNESS, AND I. J. LLEWELLYN-SMITH. Histochemistry of the enteric nervous system. In: *Physiology of the Gastrointestinal Tract.* 2nd ed., vol. 1, edited by L. R. Johnson, J. Christensen, M. J. Jackson, E. D. Jacobson, J. H. Walsh. New York: Raven Press, 1987, p. 1–40.

DAVENPORT, H. W. *Physiology of the Digestive Tract,* 5th ed. Chicago, London: Year Book Medical Publishers, 1982.

DAVENPORT, H. W., H. A. WARNER, AND C. F. CODE. Functional significance of gastric mucosal barrier to sodium. *Gastroenterology* 47: 142–152, 1964.

DAVIS, G., C. SANTA ANA, S. MORAWSKI, AND J. FORDTRAN. Active chloride secretion in the normal human jejunum. *J. Clin. Invest.* 66: 1326–1333, 1980.

DAVIS, G. R., C. A. SANT ANA, S. MORAWSKI, AND J. S. FORDTRAN. Effect of vasoactive intestinal polypeptide on active and passive transport in the human jejunum. *J. Clin. Invest.* 67: 1687–1694, 1981.

DAVIS, G., S. MORAWSKI, C. SANTA ANA, AND J. FORDTRAN. Evaluation of chloride/bicarbonate exchange in the human colon in vivo. *J. Clin. Invest.* 71: 201–207, 1983.

DAVISON, J. S. Innervation of the gastrointestinal tract. In: *A Guide to Gastrointestinal Motility,* edited by J. Christensen, D. L. Wingate. Bristol, London, Boston: Wright-PSG, 1983, p. 1–47.

DEBAS, H. T. Peripheral regulation of gastric acid secretion. In: *Physiology of the Gastrointestinal Tract,* 2nd ed., vol. 2, edited by L. R. Johnson, J. Christensen, M. J. Jackson, E. D. Jacobson, J. H. Walsh. New York: Raven Press, 1987, p. 931–946.

DEVROEDE, G., AND S. PHILLIPS. Conservation of sodium, chloride and water by the human colon. *Gastroenterology* 56: 134–143, 1969.

DIAMANT, N. E., AND A. BORTOFF. Effects of transection on the intestinal slow wave frequency gradient. *Am. J. Physiol.* 216: 734–743, 1969.

DIAMOND, J. M. Channels in epithelial cell membranes and junctions. *Fed. Proc.* 37: 2639–2644, 1978.

DiBONA, D. R., AND J. W. MILLS. Distribution of Na-pump sites in transporting epithelia. *Fed. Proc.* 38: 134–143, 1979.

DIETSCHY, J. M. Water amd solute movement across the wall of the everted rabbit gallbladder. *Gastroenterology* 47: 395–408, 1964.

DiMAGNO, E. P. Human exocrine pancreatic enzyme secretion. In: *The Exocrine Pancreas. Biology, Pathobiology and Diseases,* edited by V. L. W. Go, J. D. Gardner, F. P. Brooks, E. Lebenthal, E. P. DiMagno, G. A. Scheele. New York: Raven Press, 1987, p. 193–211.

DOCKRAY, G. J. Physiology of enteric neuropeptides. In: *Physiology of the Gastrointestinal Tract,* 2nd ed., vol. 2, edited by L. R. Johnson, J. Christensen, M. J. Jackson, E. D. Jacobson. New York: Raven Press, 1987, p. 41–66.

DONALDSON, R. M. Intrinsic factor and the transport of cobal In: *Physiology of the Gastrointestinal Tract,* 2nd ed., vol. 1., edited by L. R. Johnson, J. Christensen, M. J. Jackson, E. D. Jacobson, J. H. Walsh. New York: Raven Press, 1987, p. 959–974.

DONOWITZ, M., AND M. J. WELSH. Regulation of mammalian small intestinal electrolyte secretion. In: *Physiology of the Gastrointestinal Tract,* 2nd ed., vol. 2, edited by L. R. Johnson, J. Christensen, M. J. Jackson, E. D. Jacobson, J. H. Walsh. New York: Raven Press, 1987, p. 1351–1388.

DOTY, R. W., AND J. F. BOSMA. An electromyographic analysis of reflex deglutition. *J. Neurophysiol.* 19: 44–60, 1956.

DUBOIS, A. The stomach. In: *A Guide to Gastrointestinal Motility,* edited by J. Christensen, D. L. Wingate. Bristol, England: Wright-PSG, 1983, p. 101–127.

FELDMAN, M. Gastric secretion. In: *Gastrointestinal Disease. Pathophysiology, Diagnosis, Management.* 3rd ed, edited by M. H. Sleisenger, J. S. Fordtran. Philadelphia: W. B. Saunders, 1983, p. 541–558.

FLEMSTROM, G. Gastric and duodenal mucosal bicarbonate secretion. In: *Physiology of the Gastrointestinal Tract,* 2nd ed. vol. 2. L. R. Johnson, J. Christensen, M. J. Jackson, E. D. Jacobson, J. H. Walsh. New York: Raven Press, 1987, p. 1011–1030.

FLEMSTROM, G., AND L. A. TURNBERG. Gastroduodenal defense mechanisms. *Clin. Gastroenterol.* 13: 327–354, 1984.

FORDTRAN, J., F. RECTOR, M. EWTON, N. SOTER, AND J. KINNEY. Permeability characteristics of the human small intestine. *J. Clin. Invest.* 44: 1935–1944, 1965.

FORTE, J. G., AND J. M. WOLOSIN. HCl secretion by the gastric oxyntic cell. In: *Physiology of the Gastrointestinal Tract,* 2nd ed., vol. 1, edited by L. R. Johnson, J. Christensen, M. J. Jackson, E. D. Jacobson, J. H. Walsh. New York: Raven Press, 1987, p. 853–863.

FREDRIKSON, B., AND L. BLACKBERG. Lingual lipase. An important lipase in the digestion of dietary lipids in cystic fibrosis. *Pediatr. Res.* 14: 1387–1390, 1980.

FRIZZELL, R. A., D. R. HALM, G. RECHKEMMER, AND R. L. SHOEMAKER. Chloride channel regulation in secretory epithelia. *Fed. Proc.* 45: 2727–2731, 1986.

GABELLA, G. Innervation of the gastrointestinal tract. *Int. Rev. Cytol.* 59: 129–193, 1979.

GABELLA, G. Structural apparatus for force transmission in smooth muscle. *Physiol. Rev.* 64: 455–477, 1984.

GABELLA, G. Structure of muscles and nerves in the gastrointestinal tract. In: *Physiology of the Gastrointestinal Tract,* 2nd ed., vol. 1, edited by L. R. Johnson, J. Christensen, M. J. Jackson, E. D. Jacobson, J. H. Walsh. New York: Raven Press, 1987, p. 335–382.

GAGE, S. H., AND P. A. FISH. Fat digestion. Absorption and assimilation in man and animals as determined by the dark-field microscope and a fat-soluble dye. *Am. J. Anat.* 34: 1–77, 1924.

GARDNER, J. D., AND R. T. JENSEN. Receptors mediating the actions of secretagogues on pancreatic acinar cells. In: *The Exocrine Pancreas. Biology, Pathobiology and Diseases,* edited by V. L. W. Go, J. D. Gardner, F. P. Brooks, E. Lebenthal, E. P. DiMagno, G. A. Scheele. New York: Raven Press, 1987, 109–122.

GEENEN, J. E., W. J. HOGAN, W. J. DODDS, E. T. STEWARD, AND R. C. ARNDORFER. Intraluminal pressure recording from the human sphincter of Oddi. *Gastroenterology* 78: 317–324, 1980.

GOLDNER, A. M., S. G. SCHULTZ, AND P. F. CURRAN. Sodium and sugar fluxes across the mucosal border of rabbit ileum. *J. Gen. Physiol.* 53: 362–383, 1969.

GOLLAN, J. L., AND R. SCHMID. Bilirubin update: formation, transport, and metabolism. In: *Progress in Liver Diseases,* vol. VII, edited by H. Popper, F. Schaffner. New York: Grune and Stratton, 1982, p. 261–283.

GORELICK, F. S., AND J. D. JAMIESON. Structure-function relationship of the pancreas. In: *Physiology of the Gastrointestinal Tract,* 2nd ed., vol. 2, edited by L. R. Johnson, J. Christensen, M. J. Jackson, E. D. Jacobson, J. H. Walsh. New York: Raven Press, 1987, p. 1089–1108.

GORESKY, C. A., P. M. HUET, AND J. P. VILLENEUVE. Blood-tissue exchange and blood flow in the liver. In: Hepatology. *A Textbook of Liver Disease,* edited by D. Zakim, T. A. Boyer. Philadelphia: W. B. Saunders, 1982, p. 32–63.

GRANGER, D. N., J. A. BARROWMAN, AND P. R. KUIETYS. *Clinical Gastrointestinal Physiology.* Philadelphia: W. B. Saunders, 1985.

GREIDER, M. H., V. STEINBERG, J. E. McGUIGAN. Electron microscopic identification of the gastrin cell of the human antral mucosa by means of immunocytochemistry. *Gastroenterology* 63: 572–582, 1972.

GROSS, R. A., HOGAN, D., AND J. I. ISENBERG. The effect of fat on meal-stimulated duodenal acid load, duodenal pepsin load, and serum gastrin in duodenal ulcer and normal subjects. *Gastroenterology* 75: 357–362, 1978.

GROSSMAN, M. I. Control of gastric secretion. In: *Gastrointestinal Disease. Pathophysiology, Diagnosis, Management,* 2nd ed., edited by M. H. Sleisenger, J. S. Fordtran. Philadelphia: W.B. Saunders, 1978, p. 651.

GROSSMAN, M. I. Regulation of gastric acid secretion. In: *Physiology of the Gastrointestinal Tract,* vol. 1, edited by L. R. Johnson, J. Christensen, M. I. Grossman, E. D. Jacobson, S. G. Schultz. New York: Raven Press, 1981, p. 659–672.

GROSSMAN, M. I., J. H. WALSH, J. I. ISENBERG, AND J. H. MEYER. Peptic ulcer. In: *Cecil Textbook of Medicine,* 16th ed., edited by J. B. Wyngaarden, L. H. Smith, Jr. Philadelphia: W.B. Saunders, 1982, p. 635–654.

GUTH, P. H., AND F. W. LEUNG. Physiology of the gastric circulation. In: *Physiology of the Gastrointestinal Tract,* 2nd ed., vol. 2, L. R. Johnson, J. Christensen, M. J. Jackson, E. D. Jacobson, J. H. Walsh. New York: Raven Press, 1987, p. 1031–1054.

HARTSHORNE, D. J. Biochemistry of the contractile process in smooth muscle. In: *Physiology of the Gastrointestinal Tract,* 2nd ed., vol. 1, edited by L. R. Johnson, J. Christensen, M. J. Jackson, E. D. Jacobson, J. H. Walsh. New York: Raven Press, 1987, p. 423–482.

HOFMANN, A. F., AND J. R. POLEY, Role of bile acid malabsorption in pathogenesis of diarrhea and steatorrhea in patients with ileal resection. I. Response to cholestyramine or replacement of dietary long chain triglyceride by medium chain triglyceride. *Gastroenterology* 62: 918–934, 1972.

HOFMANN, A. F. The medical treatment of gallstones: a clinical application of the new biology of bile acids. *The Harvey Lectures.* Series 74: 1978–1979, p. 23–48.

HOFMANN, A. F. The enterohepatic circulation of bile acids in health and disease. In: *Gastrointestinal Disease. Pathophysiology, Diagnosis, Management,* 3rd ed., vol. 1, edited by M. H. Sleisenger, J. S. Fordtran. Philadelphia: W. B. Saunders, 1983, p. 115–131.

HOFMANN, A. F. Chemistry and enterohepatic circulation of bile acids. *Hepatology* Suppl. 4: 4S–14S, 1984.

HOFMANN, A. F., AND A. RODA. Physicochemical properties of bile acids and their relationship to biological properties: an overview of the problem. *J. Lipid. Res.* 25: 1477–1489, 1984.

HOFMANN, A. F., AND K. J. MYSELS. Bile salts as biological surfactants. *Colloids and Surfaces* 30: 145–173, 1988.

HOFMANN, A. F. Overview: Enterohepatic circulation of bile acids—a topic in molecular physiology. In: *Bile Acids in Health and Disease.* Update on Cholesterol Gallstones and Bile Acid Diarrhea, edited by T. Northfield, R. Jazrawi, Zentler-Munro, P. Dordrecht, The Netherlands: Kluwer Academic Publishers BV, 1988, p. 1–18.

HOFMANN, A. F. Overview of bile secretion. In: *Handbook of Physiology. Section on the Gastrointestinal System,* edited by S. G. Schultz. Bethesda, MD: American Physiological Society, 1989a, p. 549–566.

HOFMANN, A. F. The enterohepatic circulation of bile acids. In: *Handbook of Physiology. Section on the Gastrointestinal System,* edited by S. G. Schultz. Bethesda, MD: American Physiological Society, 1989b, p. 567–596.

HOGAN, W. J., W. J. DODDS, AND J. E. GEENEN. The biliary tract. In: *A Guide to Gastrointestinal Motility,* edited by J. Christensen, D. L. Wingate. Bristol, England: Wright-PSG, 1983, p.157–197.

HOGAN, D. L., J. I. ISENBERG. Gastroduodenal bicarbonate production. In: *Advances in Internal Medicine,* vol. 33, edited by G. H. Stollerman. Chicago: Year Book Medical Publishers, Inc., 1988, p. 385–408.

HOLST, J. J. Neural regulation of pancreatic exocrine function. In: *The Exocrine Pancreas. Biology, Pathobiology, and Diseases,* edited by V. L. W. Go, J. D. Gardner, F. P. Brooks, E. Lebenthal, E. P. DiMagno, G. A. Scheele. New York: Raven Press, 1987, p. 287–300.

HOLZBACH, R. T. Effects of gallbladder function of human bile: compositional and structural changes. *Hepatology* (Suppl.) 4: 57S-60S, 1984.

HOLZBACH, R. T. Metastability behavior of supersaturated bile. *Hepatology* (Suppl.) 4: 155S–158S, 1984.

HUNT J. N. AND M. T. KNOX. Regulation of gastric emptying. In: *Handbook of Physiology,* section 6, vol. 4: Alimentary Canal, edited by C. F. Code, H. Heidel. Washington, DC: American Physiological Society, 1968, p. 1917–1935.

HURWITZ, L. Pharmacology of calcium channels and smooth muscle. *Ann. Rev. Pharmacol. Toxicol.* 1986; 26: 225–258.

IIO, M., H. YAMADA, K. KITANI, AND Y. SASAKI. Anatomy of the liver in nuclear hepatology. In: *Nuclear Hepatology. Clinical and Physiological Aspects of Liver Disease by Radioisotopes,* edited by M. Iio, H. Yamada, K. Kitani, Y. Sasaki. Tokyo: Igaku Shoin LTD, 1973, p. 8–17.

ISENBERG J. I., J. A. SELLING, D. L. HOGAN AND M. A. KOSS. Impaired proximal duodenal mucosal bicarbonate secretion in patients with duodenal ulcer. *N. Engl. J. Med.* 316: 374–379, 1987.

ITO, S. Functional gastric morphology. In: *Physiology of the Gastrointestinal Tract,* 2nd ed., vol. 1, edited by L. R. Johnson, J. Christensen, M. J. Jackson, E. D. Jacobson, J. H. Wash. New York: Raven Press, 1987, p. 817–852.

JOHNSON, L. R. Gastrointestinal hormones and their functions. *Annu. Rev. Physiol.* 39: 135–158, 1977.

JOHNSTON, J. M. Mechanism of fat absorption. In: *Handbook of Physiology.* Section 6: Alimentary Canal. Volume III. Intestinal Absorption, edited by C. F. Code. Washington, DC: American Physiological Society, 1968, p. 1353–1375.

JONES, A. L. Anatomy of the normal liver. In: *Hepatology: A Textbook of Liver Disease,* edited by D. Zakim, T. D. Boyer. Philadelphia: W. B. Saunders, 1982, p. 3–31.

KELLY, K. A., C. F. CODE, AND L. R. ELVEBACK. Patterns of canine electrical activity. *Am. J. Physiol.* 217: 461–470, 1969.

KELLY, K. A. Motility of the stomach and gastroduodenal junction. In: *Physiology of the Gastrointestinal Tract,* 1st ed., vol. 1, edited by L. R. Johnson, J. Christensen, M. I. Grossman, E. D. Jacobson, S. G. Schulz. New York: Raven Press, 1981, p. 393–410.

LEESON, C. R. Structure of the salivary glands. In: *Handbook of Physiology.* Section 6: Alimentary Canal. vol. II, Secretion, edited by C. F. Code. Washington, DC: Am Physiol Soc, 1967, p. 463–495.

LEVITT, M. D., AND R. M. DONALDSON. Use of respiratory hydrogen (H_2) excretion to detect carbohydrate malabsorption. *J. Lab. Clin. Med.* 75: 937–945, 1970.

LIDDLE, R., I. GOLDFINE, M. ROSEN, R. TAPLITZ AND J. WILLIAMS. Cholecystokinin bioactivity in human plasma: molecular forms, responses to feeding, and relationship to gallbladder contraction. *J. Clin. Invest.* 75: 1144–1152, 1985.

LIPKIN, M. Proliferation and differentiation of normal and diseased gastrointestinal cells. In: *Physiology of the Gastrointestinal Tract,* 2nd ed., edited by L. R. Johnson. New York: Raven Press, 1987, p. 255–284.

MADARA, J. L., AND J. S. TRIER. Functional morphology of the mucosa of the small intestine. In: *Physiology of the Gastrointestinal Tract,* 2nd ed., vol. 2. edited by L. R. Johnson, J. Christensen, M. J. Jackson, E. D. Jacobson, and J. H. Walsh. New York: Raven Press, 1987, p. 1209–1249.

MADARA, J. L., J. S. TRIER, AND M. R. NEUTRA. Structural changes in the plasma membrane accompanying differentiation of epithelial cells in human and monkey small intestine. *Gastroenterology* 78: 963–975, 1980.

MAKHLOUF, G. M. Electrolyte composition of gastric secretion. In: *Physiology of the Gastrointestinal Tract*, vol. 1, edited by L. R. Johnson, J. Christensen, M. I. Grossman, E. D. Jacobson, and S. G. Schultz. New York: Raven Press, 1981, p. 551–566.

MANDEL, I. D., AND S. WOTMAN. The salivary secretions in health and disease. *Oral Sci. Rev.* 8: 25–47, 1976.

MEHTA, M. N., AND S. SUBRAMANIAM. Comparison of rice water, rice electrolyte solution, and glucose electrolyte solution in the management of infantile diarrhea. *Lancet* 1: 843–845, 1986.

MEYER, J. H. Motility of the stomach and gastroduodenal junction. In: *Physiology of the Gastrointestinal Tract*, 2nd ed., vol. 1, edited by L. R. Johnson, J. Christensen, M. J. Jackson, E. D. Jacobson, J. H. Walsh. New York: Raven Press, 1987, p. 613–630.

MEYER, J. H., J. B. THOMSON, M. B. COHEN, A. SHADCHEHR, AND S. A. MANDIOLA. Sieving of solid food by the canine stomach and sieving after gastric surgery. *Gastroenterology* 76: 804–813, 1979.

MOORE, E. W. Gastroenterology—Liver Disease Teaching Material, American Gastroenterological Association, Unit 7, Physiology of Intestinal Water and Electrolyte Absorption, Milner-Fenwick, Inc., 1976.

MURPHY, R. A. Filament organization and contractile function in vertebrate smooth muscle. *Annu. Rev. Physiol.* 41: 737–748, 1979.

NEUTRA, M. R., AND J. F. FORSTNER. Gastrointestinal mucus: synthesis, secretion, and function. In: *Physiology of the Gastrointestinal Tract*, 2nd ed., vol. 2, edited by L. R. Johnson, J. Christensen, M. J. Jackson, E. D. Jacobson, J. H. Walsh. New York: Raven Press, 1987, p. 975–1010.

O'GRADY, S. M., H. C. PALFREY, AND M. FIELD. Characteristics and functions of Na-K-Cl cotransport in epithelial tissues. *Am. J. Physiol.* 253:C177–C192, 1987.

OLSON, C. AND A. H. SOLL. The parietal cell and regulation of gastric acid secretion. *Viewpoints on Digestive Diseases* 16: 1–4, 1984.

OSTROW, J. D. *Bile Pigments and Jaundice. Molecular, Metabolic, and Medical Aspects.* New York: Marcel Dekker, 1986.

PAIGE, D. M., AND T. M. BAYLESS. *Lactose Digestion: Clinical and Nutritional Implications.* Baltimore: Johns Hopkins University Press, 1981.

PAINTER, N. S., S. C. TRUELOVE, G. M. ARDAN, AND M. TUCKEY. Segmentation and the localization of intraluminal pressures in the human colon, with special reference to the pathogenesis of colonic diverticula. *Gastroenterology* 49: 169–177, 1965.

PALADE, G. E. Intracellular aspects of the process of protein secretion. *Science.* 189: 347–358, 1975.

PANDOL, S. J., M. S. SCHOEFFIELD, G. SACHS, AND S. MUALLEM. Role of free cytosolic calcium in secretagogue-stimulated amylase release from dispersed acini from guinea pig pancreas. *J. Biol. Chem.* 260: 10081–10086, 1985.

PARKS, D. A., AND E. D. JACOBSON. Mesenteric circulation. In: *Physiology of the Gastrointestinal Tract*, 2nd ed., vol. 2, L. R. Johnson, J. Christensen, M. J. Jackson, E. D. Jacobson, J. H. Walsh. New York: Raven Press, 1987, p. 1649–1670.

PERRY, M. A., G. J. HAEDICKE, G. B. BULKLEY, P. R. KVIETYS, D. N. GRANGER. Relationship between acid secretion and blood flow in the canine stomach: Role of oxygen consumption. *Gastroenterology* 85: 529–534, 1983.

PROSSER, C. L. AND A. W. MANGEL. Mechanisms of spike and slow wave pacemaker activity in smooth muscle cells. In: *Cellular Pacemakers*, edited by D. O. Carpenter. New York: Wiley-Interscience, 1982, p. 273–301.

RAPPAPORT, A. M. Anatomic considerations. In: *Diseases of the Liver*, 4th ed., edited by L. Schiff. Philadelphia: J. B. Lippincott, 1975, p. 1–50.

RASMUSSEN, H. AND P. Q. BARRETT. Calcium messenger system: an integrated view. *Physiol. Rev.* 64: 939–984, 1984.

REGE, R. V., AND E. W. MOORE. Pathogenesis of calcium-containing gallstones. Canine ductular bile, but not gallbladder bile, is supersaturated with calcium carbonate. *J. Clin. Invest.* 77: 21–26, 1986.

RENE, E., R. G. DANZINGER, A. F. HOFMANN, AND M. NAKAGAKI. Pharmacologic effect of somatostatin on bile formation in the dog. *Gastroenterology* 84: 120–129, 1984.

RODRIGO, J., J. DE FILIPE, E. M. ROBLES-CHILLIDA, J. A. PEREZ ANTON, I. MAYO, AND A. GOMEZ. Sensory vagal nature and anatomical access paths to esophageal laminar nerve endings in myenteric ganglia. Determination by surgical degeneration methods. *Acta Anat.* 112: 47–57, 1982.

ROSE, R. C. Intestinal absorption of water-soluble vitamins. In: *Physiology of the Gastrointestinal Tract*, 2nd ed., vol. 2, edited by L. R. Johnson, J. Christensen, M. J. Jackson, E. D. Jacobson, J. H. Walsh. New York: Raven Press, 1987, p. 1581–1596.

ROTHMAN, S. S. The digestive enzymes of the pancreas: a mixture of inconstant proportions. *Ann. Rev. Physiol.* 39: 373–389, 1977.

RYAN, J. P. Motility of the gallbladder and biliary tree. In: *Physiology of the Gastrointestinal Tract*, 2nd ed., vol. 1, edited by L. R. Johnson, J. Christensen, M. J. Jackson, E. D. Jacobson, J. H. Walsh. New York: Raven Press, 1987, p. 695–722.

SACHS, G. The gastric proton pump: The H^+, K^+-ATPase. In: *Physiology of the Gastrointestinal Tract*, 2nd ed., vol. 2, edited by L. R. Johnson, J. Christensen, M. J. Jackson, E. D. Jacobson, J. H. Walsh. New York: Raven Press, 1987, p. 865–882.

SAMLOFF, I. M. AND P. L. TOWNES. Electrophoretic heterogeneity and relationships of pepsinogens in human urine, serum and gastric mucosa. *Gastroenterology* 58: 462–469, 1970.

SAMLOFF, M. Cellular localization of group pepsinogens in human gastric mucosa of immunofluorecence. *Gastroenterology* 61: 185–188, 1971.

SCHARSCHMIDT, B. F. Bile formation and cholestasis, metabolism and enterohepatic circulation of bile acids, and gallstone formation. In: *Hepatology: A Textbook of Liver Disease*, edited by D. Zakim, T. D. Boyer. Philadelphia: W. B. Saunders, 1982, p. 297–351.

SCHENK, D. B., J. J. HUBERT, AND H. L. LEFFERT. Use of a monoclonal antibody to quantify (Na^+, K^+)-ATPase activity and sites in normal and regenerating rat liver. *J. Biol. Chem.* 259: 14941–14951, 1984.

SCHMID, R. Bilirubin metabolism: state of the art. *Gastroenterology* 74: 1307–1312, 1978.

SCHOENFIELD, G. C., S. ITO, AND R. P. BOLANDER. Changes in membrane surface areas in mouse parietal cells in relation to high levels of acid secretion. *J. Anat.* 128: 669–692, 1979.

SCHOENFIELD, L. J., J. M. LACHIN. The Steering Committee, the National Cooperative Gallstone Study Group. Chenidiol (chenodeoxycholic acid) for dissolution of gallstones: The National Cooperative Gallstone Study. *Ann. Intern. Med.* 95: 257–282, 1981.

SCHULZ, I. Bicarbonate transport in the exocrine pancreas. *Ann. N.Y. Acad. Sci.* 341: 191–209, 1980.

SELLING, J. A., D. L. HOGAN, A. ALY, M. A. KOSS, AND J. I. ISENBERG. Indomethacin inhibits duodenal mucosal bicarbonate secretion and endogenous prostaglandin E_2 output in human subjects. *Ann. Intern. Med.* 106: 368–371, 1987.

SENIOR, J. R. *Medium Chain Triglycerides.* Philadelphia: University of Pennsylvania Press, 1968.

SINGER, M. V. Neurohumoral control of pancreatic enzyme secretion in animals. In: *The Exocrine Pancreas. Biology, Pathobiology and Diseases*, edited by V. L. W. Go, J. D. Gardner, F. P. Brooks, E. Lebenthal, E. P. DiMagno, G. A. Scheele. New York: Raven Press, 1987, p. 315–331.

SMITH, R. L. *The Excretory Function of Bile: The Elimination of Drugs and*

Toxic Substances in Biles. London: Chapman and Hall, 1973.

SOERGEL, K. H. Inert markers. *Gastroenterology* 54: 449–452, 1968.

SOLCIA, F., C. CAPELLA, R. BUFFA, L. USELLINI, R. FIOCCA, AND F. SESSA. Endocrine cells of the digestive system. In: *Physiology of the Gastrointestinal Tract*, 2nd ed., vol. 2, edited by L. R. Johnson, J. Christensen, M. J. Jackson, E. D. Jacobson, J. H. Walsh. New York: Raven Press, 1987, p. 111–130.

SOLL, A. H. AND A. WOLLIN. Histamine and cyclic AMP in isolated canine parietal cells. *Am. J. Physiol.* 237: E444-E450, 1979.

SOLL, A. H., D. A. AMIRIAN, L. P. THOMAS, T. J. REEDY AND J. D. ELASHOFF. Gastrin receptors on isolated canine parietal cells. *J. Clin. Invest.* 73: 1434–1447, 1984.

SOLL, A. H., AND T. BERGLINDH. Physiology of isolated glands and parietal cells: receptors and effectors regulating function. In: *Physiology of the Gastrointestinal Tract*, vol. 2, 2nd ed., vol. 1. edited by L. R. Johnson, J. Christensen, M. J. Jackson, E. D. Jacobson, J. H. Walsh. New York: Raven Press, 1987, p. 883–909.

SOMLYO, A. V. Ultrastructure of vascular smooth muscle. In: *Handbook of Physiology*, Section 2, vol. 2, edited by D. F. Bohr, A. P. Somlyo, H. V. Sparks. Washington, DC: Am Physiol Soc, 1980, p. 33–67.

STUBBS, R., AND B. STABILE. Role of cholecystokinin in pancreatic exocrine response to intraluminal amino acids and fat. *Am. J. Physiol.* 248: G347–G352, 1985.

SWIFT, G., R. HAMMER, R. MacDONALD, AND R. BRINSTER. Tissue specific expression of the rat pancreatic elastase 1 gene in transgenic mice. *Cell* 38: 639–646, 1984.

SZURSZEWSKI, J. H. Electrical basis for gastrointestinal motility. In: *Physiology of the Gastrointestinal Tract*, 2nd ed., vol. 1, edited by L. R. Johnson, J. Christensen, M. J. Jackson, E. d. Jacobson, J. H. Walsh. New York: Raven Press, 1987, p. 383–422.

TACHE, Y. Central nervous system regulation of gastric acid secretion. In: *Physiology of the Gastrointestinal Tract*, 2nd ed., vol.2, edited by L. R. Johnson, J. Christensen, M. J. Jackson, E. D. Jacobson, J. H. Walsh. New York: Raven Press, 1987, p. 911–930.

THAYSEN, J. H., N. A. THORN, AND I. L. SCHWARTZ. Excretion of sodium, potassium, chloride and carbon dioxide in human parotid saliva. *Am. J. Physiol.* 178: 155–159, 1954.

THUNEBERG, L. Interstitial cells of Cajal: intestinal pacemaker cells. In: *Advances in Anatomy, Embryology and Cell Biology*, vol. 71, edited by F. Beck, W. Hild, J. van Limbrorgh, R. Ortmann, J. E. Pauly, T. H. Schiebler. Berlin, Heidelberg, New York: Springer-Verlag, 1–130, 1982.

TRIER, J. S., C. L. KRONE, AND M. H. SLEISENGER. Anatomy, embryology, and developmental abnormalities of the small intestine and colon. In: *Gastrointestinal Disease: Pathophysiology, Diagnosis, Management*, 3rd ed., edited by M. H. Sleisenger, J. S. Fordtran. Philadelphia: W. B. Saunders, 1983, p. 780–810.

UNDERGRADUATE TEACHING PROJECT IN GASTROENTEROLOGY AND LIVER DISEASE. American Gastroenterological Association, Unit 16, Exocrine Pancreas: Pancreatitis, 1984.

UNDERGRADUATE TEACHING PROJECT IN GASTROENTEROLOGY AND LIVER DISEASE. American Gastroenterological Association, Unit 19, Lipid Digestion, 1986.

USSING, H. H., AND K. ZERAHN. Active transport of sodium as the source of electric current in the short circuited frog skin. *Acta Physiol. Scand.* 214: 110–127, 1951.

VAN BERGE HENEGOUWEN, G. P., AND A. F. HOFMANN. Nocturnal gallbladder storage and emptying in gallstone patients and healthy subjects. *Gastroenterology* 75: 879–885, 1978.

WALKER, M., T. EDLUND, A. BOULET, AND W. RATTER. Cell specific expression controlled by the 5′ flanking region of insulin and chymotrypsin genes. *Nature* 306: 557–561, 1983.

WALSH, J., AND M. GROSSMAN. Medical progress: Gastrin. *N. Engl. J. Med.* 292: 1324–1332, 1975.

WALSH, J. H. Gastrointestinal hormones. In: *Physiology of the Gastrointestinal Tract*, 2nd ed., vol. 1, edited by L. R. Johnson, J. Christensen, M. J. Jackson, E. D. Jacobson, J. H. Walsh. New York: Raven Press, 1987, p. 181–254.

WEEMS, W. A. The intestine as a fluid propelling system. *Annu. Rev. Physiol.* 43: 9–19, 1981.

WEISBRODT, N. W. Patterns of intestinal motility. *Annu. Rev. Physiol.* 43: 21–31, 1981.

WEISBRODT, N. W., AND J. CHRISTENSEN. Electrical activity of the cat duodenum in fasting and vomiting. *Gastroenterology* 63: 1004–1010, 1972a.

WEISBRODT, N. W., AND J. CHRISTENSEN. Gradients of contraction in opossum esophagus. *Gastroenterology* 62: 1159–1166, 1972b.

WELSH, M., P. SMITH, M. FROMM, AND R. FRIZELL. Crypts are the site of intestinal fluid and electrolyte secretion. *Science* 218: 1219–1221, 1982.

WILL, P. C., F. L. LEBOVITZ, AND U. HOPFER. Induction of amiloride-sensitive sodium transport in the rat colon by mineral ocorticoids. *Am. J. Physiol.* 238: F261–F268, 1981.

WILLIAMS, J. A. AND I. D. GOLDFINE. The insulin-acinar relationship. In: *The Exocrine Pancreas. Biology, Pathobiology, and Diseases*, edited by V. L. W. Go, J. D. Gardner, F. P. Brooks, E. Lebenthal, E. P. DiMagno, G. A. Scheele. New York: Raven Press, 1987, p. 347–360.

WILLIAMS, J. A., AND S. R. HOOTMAN Stimulus-secretion coupling in pancreatic acinar cells. In: *The Exocrine Pancreas. Biology, Pathobiology and Diseases*, edited by V. L. W. Go, J. D. Gardner, F. P. Brooks, E. Lebenthal, E. P. DiMagno, G. A. Scheele. New York: Raven Press, 1987, p. 123–139.

WILSON, F. A. The intestinal transport of bile acids. In: *Handbook of Physiology Section on the Gastrointestinal System*, edited by S. G. Schultz. Bethesda, MD: American Physiological Society, 1989.

WINGATE, D. L. Backwards and forwards with the migrating complex. *Dig. Dis. Sci.* 26: 641–664, 1981.

WOOD, J. D. Physiology of the enteric nervous system. In: *Physiology of the Gastrointestinal Tract*, 2nd ed., vol. 1, edited by L. R. Johnson, J. Christensen, M. J. Jackson, E. D. Jacobson, J. H. Walsh. New York: Raven Press, 1987, p. 67–110.

YAMADA, T. Local regulatory actions of gastrointestinal peptides. In: *Physiology of the Gastrointestinal Tract*, 2nd ed., vol. 1, edited by L. R. Johnson, J. Christensen, M. J. Jackson, E. D. Jacobson, J. H. Walsh. New York: Raven Press, 1987, p. 131–142.

YOUNG, J. A., D. I. COOK, E. W. VAN LENNEP, AND M. ROBERTS. Secretion by the major salivary glands. In: *Physiology of the Gastrointestinal Tract*, 2nd ed., vol. 1, edited by L. R. Johnson, J. Christensen, M. J. Jackson, E. D. Jacobson, J. H. Walsh. New York: Raven Press, 1987, p. 773–815.

YOUNG, J. A., AND E. W. VAN LENNEP. *The Morphology of Salivary Glands.* London: Academic Press, 1978.

ZEMAN, R. K., AND M. I. BURRELL. Biliary imaging techniques. In: *Gallbladder and Bile Duct Imaging. A Clinical Radiologic Approach*, edited by R. K. Zeman, M. I. Burrell. New York: Churchill Livingstone, 1987, p. 47–104.

ZIMMERMAN, M. J., AND H. D. JANOWITZ. Water and electrolyte secretion in the human pancreas. In: *The Exocrine Pancreas. Biology, Pathobiology, and Diseases*, edited by V. L. W. Go, J. D. Gardner, F. P. Brooks, E. Lebenthal, E. P. DiMagno, G. A. Scheele. New York: Raven Press, 1987, p. 275–281.

SECTION 7

Metabolism

CHAPTER 48

Regulation of Carbohydrate Metabolism

VITAL IMPORTANCE OF PLASMA GLUCOSE HOMEOSTASIS

Maintenance of plasma glucose levels within fairly narrow limits is of vital importance in the mammalian organism. If the plasma glucose rapidly falls to low levels (below 40–50 mg/dl) and stays low, even for 5 or 10 min, the consequences can be dramatic and drastic. This is because under ordinary circumstances the central nervous system depends absolutely upon a continuing minute-to-minute supply of glucose. Whereas most tissues can readily utilize free fatty acids or other blood-transported substrates when glucose becomes unavailable, nerve tissue depends absolutely on glucose, the only energy substrate it can utilize at a significant rate. Consequently, sustained hypoglycemia (e.g., after an overdose of insulin) can lead to coma and, if uncorrected, death. Prompt intervention to correct the hypoglycemia will save the patient's life, but if the hypoglycemia has been profound and prolonged, there will be irreversible brain damage that can be extensive and totally disabling.

Abnormal elevation of plasma glucose levels *(hyperglycemia)* does not pose an analogous acute threat, yet prolonged hyperglycemia is also ultimately life-threatening. If levels above 300 or 400 mg/dl are sustained for days, the patient will lose large amounts of glucose in the urine *(glucosuria)*. This will entail an obligatory loss of water and electrolytes, leading to progressive dehydration, decrease in blood volume *(hypovolemia)*, hypotension, shock, and coma. Prolonged hyperglycemia also, then, can ultimately lead to death.

These dramatic consequences of extreme departures from the norm illustrate the importance of glucose homeostasis. It should be stressed, however, that even relatively small departures from the norm may be deleterious if sustained chronically. For example, a growing body of evidence indicates that even modest hyperglycemia over a period of years may account for the dysfunction of the nervous system, blood vessels, kidneys, and other tissues associated with diabetes mellitus (the so-called late complications of diabetes mellitus). Repeated episodes of hypoglycemia, no one of which approaches the life-threatening severity discussed above, may nevertheless have a cumulative effect and lead to nervous system damage.

Because glucose homeostasis is so obviously vital, this chapter will deal primarily with the mechanisms that operate to maintain normal plasma glucose levels. Of course, there are many additional facets of carbohydrate metabolism of importance in the body economy. For example, synthesis of ribose and deoxyribose is sine qua non for RNA and DNA synthesis; glycolipids play essential roles in membranes generally; complex carbohydrates are fundamental components of connective tissue matrices. These, however, are more the province of biochemistry and pathophysiology. Here we confine ourselves to a consideration of the "physiological chemistry" of glucose homeostasis. This is best done by an input-output analysis in which we consider in turn the various *sources* of the pool of plasma glucose and the various *sinks*, i.e., the exit routes for glucose from the plasma compartment.

It is useful to deal with the problem at two levels. First, we simply ask where the glucose comes from and where it goes without regard to the intracellular enzymatic machinery that operates to govern transport and utilization—a "black box" approach. Second, we look into the black boxes (organs and cells) to see what can be said about enzymes and their regulation. One reason for keeping approaches at these two levels separate is that the levels of our certainty about them differ. At the higher level of organization—at the level of physiological chemistry—matters are more nearly settled, and it is this level that we deal with in clinical situations. For example, there is absolutely no doubt that insulin favors deposition of liver glycogen; on the other hand, there is still uncertainty about the precise molecular mechanisms by which insulin increases glycogen synthase activity. Again, there is no doubt that insulin favors the transport of glucose across plasma membranes and into the bulk of body tissues, namely, muscle and adipose tissue; the molecular mechanisms involved are still incompletely defined. Consequently, we shall discuss glucose homeostasis primarily first at the physiological level and secondarily at the level of cell biology and intracellular mechanisms.

SOURCES OF THE PLASMA GLUCOSE POOL

There are just three sources of the plasma glucose (see Fig. 7.1). The two major sources are: (1) intestinal absorption of dietary glucose and its precursors and (2) release of glucose from the liver. Under ordinary circumstances, the kidney is a relatively minor third source; however, in prolonged starvation it becomes significant (Owen et al., 1969).

Dietary Sources

The dietary sources of glucose are listed in Table 7.1. Very few foods contain significant amounts of free glucose. Significant quantities of glucose are presented in the form of disaccharides, especially sucrose and lactose, and in the form of polysaccharides (mainly starch in plant foods and some glycogen in animal foods). Disaccharides are rapidly hydrolyzed; the fructose and galactose moieties are rapidly absorbed and converted to glucose. Consequently, monosaccharides and disaccharides cause a prompt increase in plasma glucose concentration whereas glucose presented in the form of polysaccharides generally enters the bloodstream more slowly and causes less of a spike in plasma glucose concentration. However, there are many exceptions to this general rule and foodstuffs need to be evaluated individually. These differences become important in the dietary management of patients with

Table 7.1.
Dietary Sources of Plasma Glucose

Glucose per se (minor source)
Glucose-containing disaccharides
Sucrose (fructosyl-glucose)
Lactose (galactosyl-glucose)
Maltose (glucosyl-glucose)
Glucose-containing polysaccharides (major source)
Starch
Glycogen
Sugars readily converted to glucose
Fructose
Galactose
Gluconeogenic amino acids
Glycerol moiety of triglycerides

diabetes mellitus, in whom spikes in plasma glucose concentration are to be avoided.

Amino acids derived from dietary protein (with the exception of leucine) can all contribute to de novo glucose formation. Some amino acids can contribute all of their carbon atoms to the formation of glucose. For example, the carbons of alanine, after transamination to yield pyruvate, can be converted quantitatively to glucose under the right conditions. This generation of glucose from protein or any other nonglucose precursors is designated *gluconeogenesis*. Other amino acids can contribute some proportion of their carbon atoms for gluconeogenesis. For example, during the degradation of tyrosine, four carbon atoms become available as fumaric acid, which can enter the Krebs cycle and be converted to oxaloacetic acid. The latter, in turn, can be converted to phosphoenolpyruvate on the pathway toward glucose-6-phosphate formation. Leucine, on the other hand, cannot contribute carbons for gluconeogenesis; its degradation yields CO_2, acetyl CoA, and acetoacetic acid. In the mammalian organism, acetate cannot make any *net* contribution to glucose formation. Acetate can enter the Krebs cycle, but the Krebs cycle can be thought of as simply a complicated way of converting acetate carbons to 2 moles of CO_2. After one turn of the cycle, no carbon atoms are left from which to generate precursors for net glucose formation. Thus fatty acids, like leucine, cannot make any *net* contribution to gluconeogenesis via the Krebs cycle (although radioactive carbons from isotopically labeled fatty acids or labeled leucine will eventually find their way into glucose and glycogen). In prolonged fasting or in diabetic ketoacidosis, acetone, generated by decarboxylation of acetoacetate, has the potential to contribute carbons to glucose formation, but this is probably a minor pathway. Thus, the triglycerides in the diet are a rather insignificant source of carbon atoms for gluconeogenesis, since only the carbon atoms in the glycerol moiety

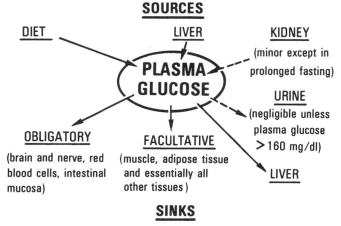

Figure 7.1. Sources and sinks for plasma glucose.

can contribute. The glycerol released from the triglycerides can be readily converted to glucose in the liver. Thus about 10% of the carbon atoms in a triglyceride constitute a source for glucose formation. While fatty acids do not directly contribute carbon atoms for gluconeogenesis, they nevertheless strongly stimulate gluconeogenesis from *other* precursors (see Chapter 50 and Fig. 7.19).

In summary, then, input of glucose or glucose precursors from the diet is one way to maintain plasma glucose levels. This happens only during a small portion of the day, i.e., for the few hours after each meal. How do we introduce glucose into the plasma pool during the rest of the day?

Hepatic Input

If we measure the concentration of plasma glucose in various vascular beds during the fasting state, we find that the ateriovenous (A-V) difference is positive everywhere (i.e., arterial concentration exceeds venous concentration), except across the liver and the kidney. From the magnitude of the A-V differences and simultaneous measurements of blood flow, we can calculate the rate of delivery of glucose into the plasma and show that the contribution of the liver far exceeds that of the kidney. The latter only makes a quantitatively important contribution during prolonged fasting. The greatest bulk of glucose input during the periods between meals and during an overnight fast must come from the liver. This is dramatically demonstrated when a total hepatectomy is performed. Very quickly, the plasma glucose level begins to drop, and the animal goes into hypoglycemic convulsions unless glucose is infused intravenously (Mann and Magath, 1924). The kidney can and does make a small contribution, but its contribution is normally too small to prevent fatal hypoglycemia when hepatic output is abruptly cut off. During an extended period of fasting, on the other hand, the kidney can account for as much as one-third of the total glucose production, but total production is much reduced. If we look into the "black boxes," we find that the liver and the kidney are the only two organs that contain significant levels of glucose-6-phosphatase, the enzyme needed to convert glycogen degradation products or glycolytic intermediates to free glucose so that they can be channeled out to the plasma.

The liver can contribute glucose to the plasma by two general mechanisms (Fig. 7.2): (1) by breakdown of stored glycogen (glycogenolysis) and (2) by the formation of glucose from nonglucose precursors (gluconeogenesis). When glucose is readily available from dietary sources, the liver increases its store of glycogen, but this store can only reach a

HEPATIC CONTRIBUTION TO PLASMA GLUCOSE

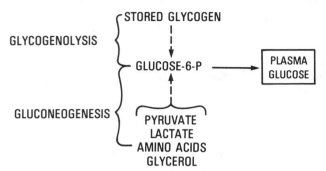

Figure 7.2. Hepatic contributions to plasma glucose via glycogenolysis and gluconeogenesis.

maximum of about 75 g. Between meals and during the early hours of fasting, this stored glycogen is broken down to provide glucose to the plasma, but the supply is very limited. The absolute minimum total body requirement for glucose during a fast is approximately 125–150 g/day in order to supply the brain, for which glucose is an obligatory substrate. Smaller additional amounts (30–40 g) are needed for peripheral nerves, red and white blood cells, and the renal medulla. Thus, the glucose available in stored hepatic glycogen is only sufficient to provide glucose requirements for less than 12 hours. If fasting continues, other sources must be invoked, and the mechanisms of hepatic gluconeogenesis become all important (Fig. 7.3).

The most important source of carbons for gluconeogenesis during fasting is the amino acids released from tissue protein, predominantly muscle proteins. During starvation, gluconeogenesis from amino acids contributes about 75 g of glucose daily. Most of this derives from breakdown of muscle protein, but there is some protein degradation in virtually every organ. All of the amino acids (except leucine) are available to some extent to help provide the needed glucose. Plasma concentrations of amino acids increase acutely during starvation, but the relative magnitude of the increase varies, and the efficiency of extraction of these amino acids by the liver varies. From studies of A-V differences, it is clear that alanine and glutamine are quantitatively the most important amino acids in the gluconeogenesis of starvation. The contribution of alanine is out of proportion to the amount of alanine in muscle proteins. A significant amount of the alanine brought to the liver during starvation probably represents alanine formed from pyruvate by transamination using amino groups from other amino acids.

A second source of carbons for gluconeogenesis is the lactate release from muscle as a result of anaer-

CARBON SOURCES FOR GLUCONEOGENESIS

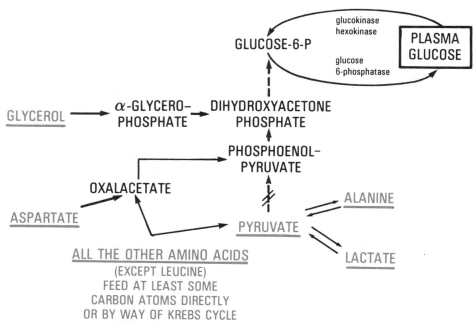

Figure 7.3. The major sources of carbon atoms contributing to gluconeogenesis.

obic glycolysis. As shown in Figure 7.4, this lactate can be carried to the liver and there converted to glucose. If this glucose returns to the muscle and undergoes glycolysis there, it has clearly made no *net* contribution to the glucose stores of the body. However, the muscle has been provided with a further source of energy for anaerobic glycolysis; that energy was provided by catabolism of other substrates in the liver. This cycling of carbons from lactate in the muscle to glucose in the liver and back to

muscle is designated the *Cori cycle* (Fig. 7.4). Some of the glucose generated in the liver from muscle lactate can be channeled up to the brain and thus can help provide the much needed glucose to maintain central nervous system function during fasting.

Finally, glycerol released from adipose tissue during the mobilization of stored triglycerides can be almost quantitatively converted to glucose in the liver. Even during total starvation, however, this triglyceride glycerol can only account for about 20 g

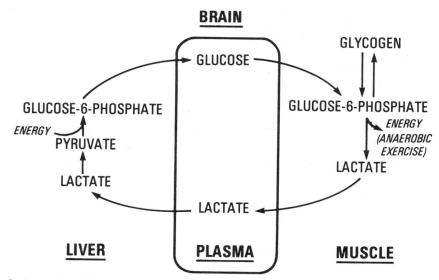

Figure 7.4. The Cori cycle, by which lactate produced in muscle is converted to glucose by the liver and then returned to muscle.

of new glucose generated daily. As discussed above, the free fatty acids released from stored triglycerides cannot make any net contribution toward gluconeogenesis. Nevertheless, the free fatty acids delivered to the liver play a crucial role in gluconeogenesis because they act as a trigger to stimulate gluconeogenesis from *other carbon sources*.

The mammalian organism simply does not have mechanisms for storing important quantities of glucose. As indicated above, the amount stored in liver glycogen can only maintain plasma glucose levels for 12 hours at most. Muscle and adipose tissue contain glycogen, but that glycogen cannot be released directly to the plasma as glucose (these tissues lack glucose-6-phosphatase), and only a small fraction of it can contribute via the Cori cycle. During fasting, then, we depend primarily on gluconeogenesis from the amino acids released from muscle protein. Knowing that the available stores of glucose are almost completely exhausted after less than 12 hours of starvation, we can easily infer the tremendous potential for gluconeogenesis from simple studies of patients with uncontrolled diabetes mellitus. These patients will maintain plasma glucose levels of over 400 mg/dl for many days, even though they have nausea and vomiting, and may take in no food at all. Furthermore, they excrete as much as 200–300 g of glucose daily for many days on end. These findings vividly demonstrate that gluconeogenesis *must* be occurring at the rate of several hundred grams daily in these patients.

SINKS FOR PLASMA GLUCOSE

All body tissues can and do utilize plasma glucose. As shown in Figure 7.1, some are *obligatory* users, i.e., they cannot use alternative substrates when glucose is unavailable. Nerve tissue, for example, cannot use free fatty acids (FFA), the major alternative circulating fuel, at a significant rate; hence the serious neurological consequences of acute hypoglycemia. The nervous system requires about 125–150 g of glucose daily under most conditions. However, in prolonged starvation (5–6 wk) the brain undergoes an interesting metabolic switch that allows it to utilize ketone bodies (β-hydroxybutyrate and acetoacetate) in place of over 50% of its usual glucose requirement (Owen et al., 1967). During such starvation, fatty acids are continually mobilized from the huge stores of adipose tissue triglycerides, and a portion of them is continually converted to ketone bodies in the liver. This adaptation in metabolism of the brain ensures that it can survive without requiring drastic depletion of muscle protein to provide substrate for gluconeogenesis.

A few other tissues utilize glucose almost exclusively. These include red blood cells, the intestinal mucosa, and the renal medulla. Their use of glucose is largely or exclusively via anaerobic glycolysis. Most of the body tissues, however, are *facultative* users of glucose. When FFA levels are high and glucose and insulin levels are low, as in the fasting state, these tissues can and do switch to use FFA as their primary metabolic fuel.

The liver is both a source and a sink for plasma glucose. In fact, both uptake and release are going on at all times. The *net balance* is under multiple controls that determine whether at any given time it represents a source (net input into plasma) or a sink (net uptake from plasma). In the absence of supervening hormonal or neural signals, the liver shows a net output of glucose when the plasma level is low (below 120–150 mg/dl) and a net uptake when it is high (above 120–150 mg/dl). This "autoregulation" can be overridden by changes in hormonal balance.

Under normal conditions no glucose is lost by renal excretion. Glucose is rapidly filtered into the glomerular fluid, but it is normally reabsorbed very efficiently by the renal tubules. The T_m is about 340 mg/min. If the glomerular filtration rate is normal, spillage will only occur when the plasma glucose exceeds 160–190 mg/dl (renal threshold for glucose). Generally the amount of glucose excreted in the urine will be a reasonably good indicator of the degree of hyperglycemia. However, there are significant and instructive exceptions. For example, during pregnancy, when the mother's glomerular filtration rate (GFR) is normally elevated, glucosuria may occur even at normal or only slightly elevated plasma glucose levels. Conversely, with renal damage, such as can occur as a late complication of diabetes mellitus, the GFR may be abnormally low. Under these circumstances, the urine may remain free of glucose even though plasma glucose levels are high. Under these conditions, quantification of the glucose in the urine will underestimate the severity of the hyperglycemia. Finally, mention should be made of certain rare inherited disorders in which renal tubular transport mechanisms are defective, including tubular glucose reabsorption (e.g., Fanconi's syndrome). Because of the low renal threshold, glucosuria occurs without hyperglycemia.

NORMAL PLASMA GLUCOSE LEVELS AND GLUCOSE TOLERANCE

Normal plasma glucose levels after an overnight fast range from 70–110 mg/dl. Values in the older literature, representing whole blood glucose concentrations, are about 15% lower than plasma concentrations. In evaluating the older literature, one must also be aware that some values are falsely high because they were based on nonspecific methods that included reducing substances other than glucose. Today, almost without exception, *plasma* glucose is measured and measured specifically, e.g., using enzymatic methods based on glucose oxidase.

Plasma glucose increases with the ingestion of each meal, rising to a peak at 30–60 min and returning to basal values at about 120 min. The magnitude of the response and its duration depends on the size and composition of the meal. Between meals the glucose level is remarkably constant with a slight downward drift during the night. Precise measurements show minor phasic changes of a few milligrams per deciliter, reflecting pulsatile release of regulatory hormones, but these changes are too small to be of immediate practical significance.

The changes in plasma glucose levels in response to ingestion of a glucose load have long been used clinically and in research to evaluate the effectiveness of glucose homeostatic mechanisms. This is the oral glucose tolerance test (OGTT). As currently employed, a 75- to 100-g dose of glucose is given after an overnight fast. Plasma glucose levels are determined on venous blood samples taken just before the glucose load, at 1 hour and at 2 hours. In normal subjects the fasting level is below 115, the 1-hour value is below 200, and the 2-hour value is below 140 mg/dl. If both the 1- and 2-hour values are more than 200, the test is definitely abnormal, indicating primary or secondary diabetes mellitus. Intermediate test results indicate "impaired glucose tolerance," which may or may not have clinical implications (National Diabetes Data Group, 1979).

The sequence of events that determine the shape of the OGTT curve illustrates some of the basic elements involved in glucose homeostasis (Fig. 7.5). Plasma glucose begins to rise within a few minutes of glucose ingestion, reflecting the rapid, efficient absorption of glucose from the upper duodenum. Administration of glucose by mouth is thus very effective in quickly correcting hypoglycemia. Glucose is absorbed by active transport at a rate that is a function of the concentration at the mucosal surface. When the concentration is high, the transport mechanism becomes saturated, and the rate of absorption does not increase further with increasing concentration in the intestine. The large dose of glucose used in the OGTT is enough to maximize the initial rate of absorption in the proximal duodenum. The rate at which the plasma glucose level rises and the peak value reached will be determined, in part, by this input rate. For example, patients with

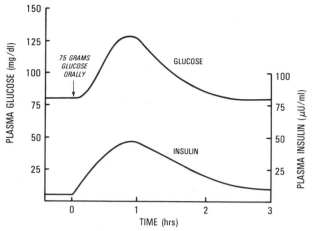

Figure 7.5. Idealized oral glucose tolerance test (OGTT) showing a typical normal response in plasma glucose level and plasma insulin level after a 75-g oral dose of glucose.

hyperthyroidism absorb glucose more rapidly, and this plays a role in the abnormal OGTT seen in a significant number of them (although other factors are also involved). If the full glucose load is delivered into the intestine too rapidly, as in patients with partial gastrectomy or gastrojejunostomy, the OGTT curve rises rapidly to an abnormally high level that may peak at 30 min but generally returns to normal by 2 hours or even overshoots to hypoglycemic values.

Returning to the normal OGTT curve, we see that a peak value is reached at 30–60 min, yet intestinal glucose absorption is still going on very actively at that time. The levels begin to fall despite continuing absorption because the rate of glucose removal from plasma has progressively increased and now exceeds the rate of absorption. This is mainly due to enhanced secretion of insulin, which begins almost immediately on ingestion of the glucose (Fig. 7.5). Even before plasma glucose levels rise appreciably, the output of insulin into the portal vein has already started to increase. This initial response is triggered by hormones released from the gastrointestinal mucosa in response to food ingestion. Several gastrointestinal hormones have been shown to have the *potential* of stimulating insulin secretion (including secretin, pancreozymin, gastrin, glucagon, and gastric inhibitory polypeptide). Most of these act only at very high concentrations; gastric inhibitory peptide (GIP), however, is effective at concentrations actually reached in vivo and is probably importantly involved (Dupre et al., 1973; Ebert and Creutzfeldt, 1987). The importance of the intestinal hormones in augmenting the insulin response was nicely demonstrated by showing that the integrated response to intravenous glucose is less than that to the equivalent dose given orally, even though arterial glucose levels are lower in the latter case (Perley and Kipnis, 1967). Once the glucose level begins to rise, by all odds the major stimulus for the continuing increase in insulin output from the pancreas is the direct action of glucose on the B-cells. However, there are many additional factors that can modulate the magnitude of the insulin response and the shape of the OGTT curve.

Until recently it was thought that most of the 100 g of glucose given for the OGTT was removed by the liver and converted directly to glycogen via glucose-6-phosphate and glucose-1-phosphate. This conclusion was largely based on extrapolation from studies in perfused tissues and in isolated hepatocytes or on studies using much larger doses of glucose given as a bolus to increase plasma glucose concentrations dramatically. More recent studies clearly show that much of the glucose ultimately incorporated into liver glycogen is first broken down to 3-carbon fragments (see review by McGarry et al., 1987). Paradoxically, however, A-V difference studies provide no evidence for net uptake of lactate by the liver during the disposal of a glucose load. This makes it necessary to postulate that the breakdown to 3-carbon fragments must occur within the liver itself after which these 3-carbon fragments are resynthesized to the level of glycogen and stored. If all of this occurred in the same set of hepatocytes, there would be a good deal of futile recycling. Thus it has been postulated that different subsets of hepatocytes may be responsible for the breakdown of the glucose, on the one hand, and its reincorporation up to the level of glycogen, on the other (Pilkis et al., 1985).

A smaller portion of the glucose absorbed is taken up in peripheral tissues and stored as glycogen, either in adipose tissue or in muscle; another portion is converted to fatty acids and stored in adipose tissue as triglycerides (see Chapter 49). Finally, some of the extra glucose taken up in the periphery is immediately oxidized as a source of calories in place of free fatty acids.

A wide variety of metabolic and hormonal perturbations can, in principle, cause an abnormal OGTT by interfering with the efficient disposition of the glucose load. In insulin-dependent diabetes mellitus (IDDM) there is a primary defect in the ability of the pancreatic islet tissue to respond to glucose with a release of insulin; in noninsulin-dependent diabetes mellitus (NIDDM), the most prominent defect is in a relative insensitivity of the tissues to plasma insulin. Suffice it to say here that the underlying specific abnormalities in the various primary forms of diabetes mellitus are in the process of being elucidated (see Chapter 50). There are also

many disease states that secondarily cause an abnormality in the OGTT. For example, liver disease severe enough to limit the rate of hepatic glycogenesis can cause a "diabetic" glucose tolerance curve. Anything that secondarily affects the sensitivity of the tissues to insulin can also yield an abnormal OGTT. For example, adrenal glucocorticoids decrease insulin sensitivity; therefore, in many acute stress situations associated with excess glucocorticoid output (e.g., febrile illnesses, trauma, surgery) the OGTT may be abnormal. Increased sympathetic nervous system activity and adrenal medullary secretion of catecholamines also contribute by inhibiting insulin output.

An alternative method for evaluating glucose homeostasis, primarily for research purposes, is the intravenous glucose tolerance test (IGTT). A 25-g dose is given rapidly by vein, plasma glucose levels are measured at 10-min intervals for 1 hour, and the results are plotted on semilogarithmic graph paper. The slope of the disappearance curve gives a rate constant, k, proportional to the efficiency of glucose disposition. In normal individuals the k value is 1.2 or greater. The IGTT is obviously independent of one set of the variables affecting the OGTT, namely, those influencing glucose absorption. Thus it may be useful in evaluating patients with gastrointestinal disorders.

This brief discussion of the events determining the response to a glucose load will provide a context for the following discussion of mechanisms regulating glucose uptake and utilization. One last point should be made in regard to the remarkable efficiency of the process. If 75 g of glucose, the dose used most commonly in the standard oral test, were suddenly introduced into the extracellular fluid volume of a 70-kg man (14 liters), the glucose level would rise by over 500 mg/dl. In normal individuals, the observed rise is often less than 50 mg/dl. Partly this is due to the fact that the absorption is stretched over an extended time period. However, in diabetics, most of whom have normal rates of glucose absorption, glucose levels after an OGTT do rise by as much as 300 mg/dl because of inefficient uptake and utilization. (Diagnostic OGTTs are not necessary in frank diabetes mellitus with fasting hyperglycemia and are not advisable because of the marked hyperglycemia induced.)

REGULATION OF GLUCOSE UPTAKE FROM PLASMA

Glucose Transport Across Cell Membranes

Different mechanisms for glucose transport are found in different tissues. *Active transport*, i.e., an energy-dependent process that can transport against a concentration gradient, occurs in the intestinal epithelium and in the renal tubular epithelium. Neither intestinal absorption of glucose nor tubular reabsorption of glucose by this active transport mechanism is insulin dependent.

In muscle, adipose tissue and other insulin-dependent tissues, glucose uptake occurs by a *carrier-mediated transport* mechanism. Glucose cannot be transported against a concentration gradient by this mechanism. It is a form of passive transport but more is involved than simple diffusion. This is evident from four characteristic properties of such a *facilitated diffusion* process:

(a) *Saturation kinetics.* Transport increases with external glucose concentration but only to a certain maximum rate.

(b) *Stereospecificity.* The unnatural isomer (L-glucose) is not transported.

(c) *Competition.* Sugars of similar structure competitively inhibit glucose transport.

(d) *Countertransport.* By taking advantage of the fact that a nonmetabolizable glucose analogue (3-*O*-methylglucose) is transported by the same system it can be shown (using an isolated rat diaphragm as an example of muscle tissue) that the carrier system is operating bidirectionally (Morgan et al., 1964). This principle is illustrated in Figure 7.6. If only 3-*O*-methylglucose is present in the medium, its intracellular concentration builds up and stabilizes when the concentration in the cytoplasm equals that outside. At this point, the rate of movement of the analogue into the cell equals its rate of movement out. If now glucose itself is added to the medium, the intracellular concentration of the analogue suddenly begins to fall. This appears to be paradoxical since the analogue is moving out against a concen-

tration gradient. The explanation lies in the fact that on the outside glucose is now competing with the analogue for a limited number of carrier molecules in the membrane and, thus, the inward movement of the analogue is slowed down. On the inside of the membrane there is no comparable competition because the glucose molecules are phosphorylated almost as soon as they enter the cell, and glucose-6-phosphate cannot bind to the carrier. Consequently, the *outward* movement of the analogue continues for a while at about the same rate that prevailed when the glucose was added. As the concentration of the analogue on the inner side of the membrane falls, its rate of transport out falls until a new steady-state is established (Fig. 7.6).

All of these properties can be accounted for if one postulates a set of carriers within the plasma membrane that shuttle from outer to inner surfaces and back, able to bind glucose reversibly at either surface. If the number of carriers is finite, there will be a maximum rate of transport when all are occupied *(saturation kinetics).* If the configuration of the carrier is sharply defined, it will combine only with molecules of certain configurations *(stereospecificity).* If the configuration is not absolutely specific, some molecules of very closely related structure will also bind *(competition).* Even though the system operates symmetrically, differential metabolism of one of two competing ligands can lead to an apparent paradoxical transport against a gradient *(countertransport).*

In muscle and adipose tissue, the rate-limiting step in the uptake of glucose is its transport across the cell membrane by the carrier-mediated mechanism. The rate of glucose entry increases with the concentration of glucose presented to the cell. However, even at very high levels the rate at which entering glucose is phosphorylated to form glucose-6-phosphate is so rapid that there is an almost unmeasurably low concentration of free glucose inside the cell (Fig. 7.7). Insulin, the most important regulator of glucose transport in these tissues, can accelerate the inflow to the point that intracellular free glucose can be demonstrated (which supports

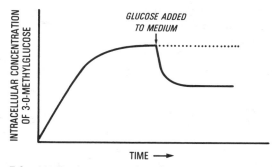

Figure 7.6. Idealized representation of an experiment demonstrating the phenomenon of countertransport (Morgan et al., 1964). The glucose analog, 3-*O*-methylglucose, enters the cell via the same carrier-mediated transport system involved in glucose uptake. In the absence of added glucose, the intracellular concentration of the analog rises to reach a plateau value which is maintained. The analog is not a substrate for phosphorylation and is not metabolized. At the time indicated by the *arrow,* glucose is added to the medium, and the intracellular concentration of the analog falls to reach a new lower plateau level (see text for further discussion).

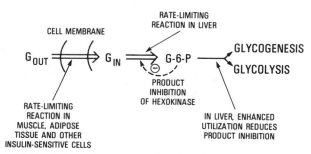

Figure 7.7. Schematic representation of glucose transport and phosphorylation.

the proposition that free glucose and not a derivative of it is delivered at the inner surface of the cell membrane).

In contrast to muscle and adipose tissue membranes, liver cell membranes appear to be permeable to glucose, and the rate-limiting step for uptake is the rate at which the free glucose delivered into the cytoplasm can be phosphorylated (Fig. 7.7). In the liver, glucose is phosphorylated by two major types of kinases: hexokinase, a relatively nonspecific enzyme saturated at very low glucose concentrations (K_m about 10^{-5} M); glucokinase, a specific enzyme with a K_m value near the physiological range of glucose concentrations (K_m about 10^{-2} M). The latter is only partially saturated at normal glucose levels so that flux can increase when glucose levels increase and fall when glucose levels fall. Glucokinase is not inhibited by glucose-6-phosphate; thus the latter can build up and drive glycogen synthesis without inhibiting the continuing hepatic uptake of glucose when plasma levels are high. The activity of glucokinase decreases in starvation and in diabetes mellitus (Weinhouse, 1976).

Insulin does not *directly* enhance membrane transport in liver, but it definitely does so *indirectly*. This effect depends on induced glucokinase and the fact that glucose-6-phosphate *is* a feedback inhibitor of *hexokinase*. Reducing glucose-6-phosphate concentration accelerates glucose phosphorylation and, thus, glucose transport. Insulin reduces glucose-6-phosphate levels by stimulating its conversion to glygocen (by favoring activation of glycogen synthase) and by stimulating glycolysis. Insulin definitely stimulates hepatic glucose uptake and reduces hepatic glucose output, even though the hepatic plasma *membrane* is not insulin responsive.

Factors Influencing Rates of Glucose Uptake

The systems regulating glucose uptake (and also glucose production) are highly interactive. For purposes of discussion, we dissect them but recognize that rarely does a *single* factor act independently. Indeed, the successful operation of the homeostatic system depends on a balance of simultaneously interacting stimuli and counterstimuli.

PLASMA GLUCOSE LEVEL. AUTOREGULATION

In both insulin-sensitive and insulin-insensitive tissues, the glucose concentration available is a key factor determining rate of uptake. As discussed above, the transport mechanism is saturable but only at extremely high glucose concentrations. The maximum rate can be increased under the influence of insulin, and insulin will enhance uptake at any glucose concentration. However, even in the com-

plete absence of insulin, glucose uptake continues in all tissues. In the diabetic deprived of all insulin, the plasma glucose level rises to very high values. At those high levels, and driven by them, the *absolute* rate of glucose uptake and oxidation (milligrams of glucose per hour) can be as high as it is in a normal subject; however, the *fractional* rate of glucose uptake (percentage of plasma pool per hour) is much lower because of the lesser efficiency of the transport mechanism.

Autoregulation in the liver has been demonstrated by perfusing the isolated organ with media containing different concentrations of glucose. Below about 150 mg/dl the liver shows a net release of glucose; above this level it shows a net uptake. In vivo, however, changes in glucose levels will also induce changes in key regulatory hormones, especially insulin and glucagon, and these superimposed hormonal stimuli are of overriding importance. For example, in the insulin-deprived diabetic the liver continues to show net release of glucose even in the face of plasma glucose concentrations above 500 mg/dl.

FREE FATTY ACIDS

In the presence of high plasma levels of FFA, the rate of glucose uptake by skeletal and cardiac muscle at any given level of insulin is reduced (*Randle effect*). In the fasting state, FFA levels are elevated because of rapid mobilization of adipose tissue triglycerides. Thus, the utilization of glucose by peripheral tissues is inhibited, helping to conserve the limited supply of glucose, which is needed to supply obligatory substrate for the brain. In the fed state, A-V difference studies across the heart show that glucose oxidation accounts for the bulk of myocardial oxygen consumption; in the fasting state, glucose utilization drops, and FFA oxidation accounts for most of the oxygen consumed. Again, hormone effects are superimposed: insulin levels are lower in the fasted state, reducing glucose utilization in muscle but not affecting it in the brain.

MUSCULAR WORK

At any given glucose concentration, the rate of glucose uptake into muscle (skeletal and cardiac) is enhanced by muscular contraction. This effect does not depend on an increase in insulin concentration. Electrical stimulation of isolated skeletal muscle strips in vitro enhances glucose uptake; increasing the work of the isolated perfused heart by increasing the filling pressure does the same thing. The effects of low levels of insulin and low levels of exercise are additive. This phenomenon is of practical importance in the management of diabetic patients. Their insulin requirement is less on days when they engage in vigorous muscular exercise; failure to adjust the dose downward may result in a hypoglycemic reaction.

Prolonged anaerobiosis will also increase glucose uptake at a given level of glucose and insulin. This effect can be dissociated from the effect of exercise per se and is of lesser magnitude.

HORMONAL EFFECTS

Insulin is without question the key hormone controlling rates of glucose removal from plasma. In the bulk of the body tissues—muscle and adipose tissue—it directly enhances the rate of glucose transport into the cells. In the liver, it both enhances glucose uptake and inhibits glucose release. The dominant role of insulin is clearly shown by the profound hypoglycemia produced by injection of a large dose of this hormone, despite the recruitment of a battery of counterregulatory responses. The cellular mechanisms involved in the action of insulin are discussed in Chapter 50.

Glucocorticoids decrease glucose uptake in peripheral tissues, both basal and insulin-stimulated. Part of the effect is associated with a decrease in the number of insulin receptors, an effect that appears, however, to be transient; the more important and longer term effect relates to insulin resistance at a postreceptor level. In addition, cortisol plays a permissive role in hormone-stimulated release of FFAs from adipose tissue which, as pointed out above, would also tend to reduce glucose transport (Randle effect). As discussed below, a major additional basis for the hyperglycemic effect of glucocorticoids is enhancement of gluconeogenesis.

Catecholamines tend to reduce glucose uptake. A direct effect on glucose uptake into diaphragm in vitro has been demonstrated, which may reflect increased glycogenolysis and inhibition of hexokinase by glucose-6-phosphate. In addition, catecholamines have an indirect effect by increasing FFA levels (Randle effect). However, the major effect, as with cortisol, is not on glucose uptake but on hepatic glucose production (see below).

Growth hormone has biphasic effects on glucose uptake. Acutely it has an insulin-like effect, increasing glucose uptake, but this lasts only an hour or two. Long-term elevation of growth hormone levels, as in acromegaly, decreases glucose uptake by muscle and adipose tissue and induces a form of secondary diabetes mellitus. Sensitivity to insulin is reduced at a postreceptor level and is only partially compensated for by increased production of insulin. An additional inhibition of glucose uptake is refer-

able to increased FFA levels secondary to more rapid lipolysis in adipose tissue.

Factors Influencing Rates of Hepatic Release of Glucose

There are four critical points at which metabolic and hormonal control of hepatic glucose production can be exercised. These control points, as in almost every regulated metabolic system, are points in the pathway at which the forward and reverse reactions are catalyzed by *different*, independently controlled enzymes. The four sites of control are indicated in the simplified scheme shown in Figure 7.8. Two general categories of control are involved: (1) regulation of glycogen synthesis and degradation; (2) regulation of glycolysis and gluconeogenesis.

GLYCOGEN SYNTHESIS AND DEGRADATION

The rate-limiting step for synthesis is the incorporation of glucose from uridine diphosphoglucose (UDPG) into glycogen, catalyzed by glycogen synthase; degradation is catalyzed by glycogen phosphorylase. This is the classic example of a "push-pull" system in which there is coordinate regulation of the two enzymes involved. For example, glucagon simultaneously activates phosphorylase and deactivates glycogen synthase, as briefly reviewed in Figure 7.9. The same cyclic AMP-dependent protein kinase leads to the phosphorylation of both glycogen synthase, which it acts on directly, and phosphorylase, through an intermediate step (i.e., phos-

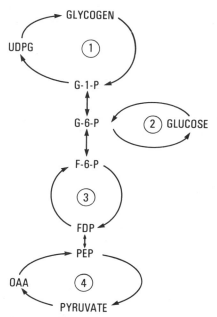

Figure 7.8. Skeleton outline of the pathways for glucose metabolism indicating the four major control points at which the forward and reverse reactions are catalyzed by separate enzymes.

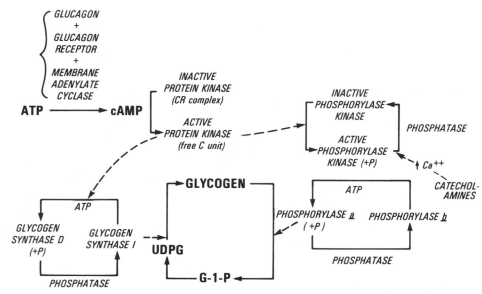

Figure 7.9. Schematic representation of glycogen metabolism in the liver and the enzymes involved in its regulation.

phorylation and activation of phosphorylase kinase). The latter, in turn, phosphorylates phosphorylase *b*, converting it to its *active* form, phosphorylase *a*. The phosphorylation of glycogen synthase, on the other hand, converts it to a *less* active form or forms. Thus, the hormone effect is maximized by accelerating breakdown while simultaneously inhibiting synthesis of glycogen.

Glucagon is the most important physiological stimulus for glycogen degradation and release of glucose from the liver. It acts through its specific membrane receptor to increase adenylate cyclase activity and, through cyclic AMP, protein kinase activity (see Fig. 7.9).

Catecholamines, like glucagon, enhance glycogen degradation but the mechanism is now recognized to be different (Exton, 1979). The *major* glycogenolytic action of epinephrine is due to its α-adrenergic effects. They appear to be mediated not by changes in cAMP but by control of calcium ion concentration which regulates the activity of phosphorylase kinase. The β-adrenergic effects of catecholamines are exercised, like those of glucagon, through the adenylate cyclase system. Concurrently, the catecholamines inhibit insulin release, reducing the action of the major antagonist of catecholamine effects on glycogen metabolism.

Insulin favors glycogen deposition and inhibits glycogen degradation. Thus insulin counteracts the effects of glucagon and the catecholamines, favoring glucose uptake into and inhibiting glucose release from the liver. The mechanism is not yet firmly established. Under some circumstances, insulin can counter the effects of glucagon and catecholamines on cAMP formation, probably by enhancing phosphodiesterase activity, but under other circumstances this does not adequately account for the effect, either in the liver or in adipose tissue. Recent studies suggest that insulin interaction with its receptor may generate a second messenger, not yet definitively characterized, or that the intrinsic tyrosine kinase activity of the insulin receptor acts on cytosolic enzymes.

GLUCONEOGENESIS AND GLYCOLYSIS

Gluconeogenesis can be stimulated either directly or indirectly. Direct effects are brought about by affecting the activities of the enzymes at control points *2*, *3*, or *4* shown in Figure 7.8. Indirect effects are brought about by affecting the delivery of gluconeogenic substrates to the liver. Under ordinary circumstances, the rate of delivery of gluconeogenic substrates to the liver (amino acids, lactate, glycerol) does not saturate the capacity of the liver to convert them to glucose. Consequently, changes in the extrahepatic tissues that increase the concentrations of these substrates reaching the liver can increase gluconeogenesis even without changes in hepatic enzymes (e.g., increased protein degradation, increased anaerobic glycolysis in muscle, and increased lipolysis in adipose tissue increasing release of glycerol). The effect of enhanced uptake of FFA to increase gluconeogenesis is in a special category. The FFA are not themselves substrates for gluconeogenesis. As detailed below, increased uptake of FFA shifts the metabolic stance of the liver and channels *other* substrates toward glucose-6-phosphate (glycerol, lactate, amino acids).

Glucagon has a major direct effect through its regulation at control points *3* and *4* (Fig. 7.8). Pyruvate kinase, catalyzing the essentially irreversible conversion of phosphoenolpyruvate (PEP) to pyruvate at point *4*, is converted to an inactive form by cAMP-dependent protein kinase, the activity of which is increased by glucagon. This event inhibits glycolysis and of itself tends to drive the reverse (gluconeogenic) pathway. At the same time glucagon stimulates conversion of pyruvate to PEP, probably by increasing the activity of PEP carboxykinase (conversion of oxaloacetate to PEP). Traffic at point *3* is directed by the available levels of fructose 2,6-bisphosphate (Pilkis and El-Maghrabi, 1988). Those levels are, in turn, controlled by cAMP-dependent protein kinase. Glucagon, by increasing cAMP levels, decreases levels of fructose 2,6-bisphosphate, resulting in stimulation of gluconeogenesis and inhibition of glycolysis. In addition, at least in the diabetic state, glucagon may play a significant role in FFA mobilization and thus indirectly enhance gluconeogenesis.

Catecholamines, by stimulating β_2-adrenergic receptors, can mimic the effects of glucagon but they are less potent. Moreover, catecholamines have the potential actually to decrease cAMP levels through their interaction with α_2-adrenergic receptors. In any case, the effects of catecholamines on gluconeogenesis are not tightly linked to changes in tissue cAMP levels. The possibility that they act, in part, by regulating cytosolic levels of calcium ion is under evaluation. In the intact animal, the powerful effect of epinephrine on FFA mobilization provides an indirect stimulus to gluconeogenesis and also provides more substrate in the form of glycerol.

Insulin suppresses gluconeogenesis, and several mechanisms are operative. Studies in perfused livers show that there is an acute direct effect that may relate to decreases in cAMP levels. Long-term insulin deprivation induces adaptive changes in the levels of gluconeogenic enzymes that favor glucose production. In addition to these direct effects, insulin reduces gluconeogenesis by suppressing the flow of substrates to the liver. It inhibits protein degradation in muscle and lipolysis in adipose tissue, reducing the flow of FFA to the liver, thus removing a "trigger" for gluconeogenesis. A deficiency of insulin, reciprocally, favors gluconeogenesis.

Cortisol increases gluconeogenesis by increasing protein degradation and thus the flow of amino acids to the liver (indirect effect). It also induces synthesis of gluconeogenic enzymes.

Free Fatty Acids

While not themselves a substrate for net new glucose formation, high FFA levels channel gluconeogenic substrates to glucose-6-phosphate. Several effects are involved:

1. Generation of high intrahepatic concentrations of acetyl CoA, which inhibits oxidation of pyruvate to acetyl CoA via pyruvate dehydrogenase and in this way channels pyruvate to oxalacetate via pyruvate carboxylase (see Fig. 7.19).

2. Generation of high intrahepatic concentrations of fatty acyl CoA, which inhibits pyruvate kinase (PEP to pyruvate) and, at control point *3* (Fig. 7.8), inhibition of phosphofructokinase due to high levels of citrate and ATP generated during FFA oxidation (Garland et al., 1963). The rapid oxidation of FFA also provides reduced pyridine nucleotide needed for gluconeogenesis.

Long-term Hormonal Control

In addition to the acute, short-term regulation discussed above, there is long-term regulation of gluconeogenesis and glycolysis by regulation of the genes controlling the relevant enzymes (reviewed by Pilkis and El-Maghrabi, 1988). For example, administration of either glucagon or cAMP causes a marked decrease in the level of mRNA for pyruvate kinase. Glucagon, again acting through cAMP, enhances the expression of mRNA for phosphoenolpyruvate carboxykinase. Thus, in prolonged fasting the levels and patterns of enzymes involved in gluconeogenesis and glycolysis can be reset appropriately to continue the supply of glucose to the central nervous system.

SUMMARY

Plasma glucose levels are affected by a wide variety of factors—metabolic, neural, and hormonal. The multiple influences and their mechanisms of action are summarized in Table 7.2. How these hormones are secreted and how that secretion is regulated is discussed in Chapters 50, 52, and 54.

The interplay involved can be illustrated taking the response to insulin-induced hypoglycemia as an example. If the plasma glucose begins to drop rapidly, the patient feels cold and anxious. He is pale and shows a tremor. These responses signal a sympathetic discharge with release of catecholamines, a major counterregulatory response. The catecholamines tend to correct the hypoglycemia by the several mechanisms listed in Table 7.2. The hypoglycemia per se suppresses further endogenous insulin secretion and favors glucagon secretion. Insulin

Table 7.2.
Summary of Hormonal Effects on Plasma Glucose Levels

	Major Mechanisms	Minor Mechanisms
Hormones Tending to Raise Levels		
Glucagon	Increases glycogenolysis (via cAMP) Increases gluconeogenesis (via cAMP).	Increases FFA levels (which inhibits glucose uptake via the Randle effect and stimulates gluconeogenesis)
Catecholamines	Increase glycogenolysis (α-adrenergic effect via Ca^{2+}; β-adrenergic effect via cAMP Inhibit insulin release (α-adrenergic effect)	Increase FFA levels Some direct effect on glucose uptake into muscle
Glucocorticoids	Increase gluconeogenesis (by increasing flow of substrate amino acids from muscle protein degradation; by inducing hepatic synthesis of gluconeogenic enzymes (e.g., transaminases, pyruvate carboxylase, glucose-6-phosphatase) Decrease glucose uptake by decreasing insulin sensitivity	Increase FFA levels (by permissive effect favoring lipolytic action of catecholamines)
Growth hormone	Decreases glucose uptake by decreasing insulin sensitivity (downregulation of receptors?; competition between somatomedins and insulin?) (N.B.: for an hour or so after administration, *lowers* glucose levels but dominant, long-term effect is to raise them)	Increases FFA levels
Hormones Tending to Lower Levels		
Insulin	Increases glucose uptake in all insulin-sensitive tissues (direct effect on transport; increase of pyruvate dehydrogenase activity) and also in liver (secondary to effects on glycogen metabolism and glycolysis) Decreases glucose release by liver (decreases activity of phosphorylase and glucose-6-phosphatase) Decreases gluconeogenesis (by reducing flow of amino acid substrate from muscle; reducing FFA levels; decreasing activity of glucose-6-phosphatase and other gluconeogenic enzymes)	Decreases FFA levels (counteracts lipolytic effects of catecholamines and glucagon)
Somatostatin	Inhibits glucagon release (and also insulin release) Inhibits intestinal absorption of glucose	

secretion is also suppressed by the direct α-adrenergic effect of catecholamines on the pancreas. Output of growth hormone and of glucocorticoids is enhanced, both of which tend to increase plasma glucose levels (Table 7.2). Finally, the patient feels hungry, ingests food, and thus corrects the hypoglycemia—a complex behavioral mechanism, still poorly understood at the molecular level, but very effective!

The reader is referred to several excellent reviews for further elaboration and documentation (Cahill and Owen, 1968; Felig, 1980; Hers, 1976; Larner et al., 1968; Phillips and Vassilopoulou-Sellin, 1980; Pilkis and El-Maghrabi, 1988).

CHAPTER 49

Regulation of Lipid and Lipoprotein Metabolism

In the preceding chapter, it was pointed out that the total available stores of glucose are strictly limited; except during the several hours following each meal, most body tissues depend upon oxidation of fatty acids, the other major caloric substrate, to provide energy. Those fatty acids are mobilized from the huge stores of triglyceride in adipose tissue depots and transported through the bloodstream. The first part of this chapter deals with the regulation of fatty acid mobilization from adipose tissue, the transport mechanisms involved, and the metabolic fate of the mobilized *free fatty acids* (FFA). In addition, there is an elaborate system for the transport of triglycerides, phospholipids, cholesterol, and some less common lipids in several classes of *plasma lipoproteins.* The second part of this chapter deals with the nature of this complex lipoprotein system, the ways in which it is regulated, and the relationship between it and atherosclerosis.

ADIPOSE TISSUE "ORGAN"

Adipose tissue accounts for about 20% of the total body weight of a normal young adult—about 15 kg in the average person. There is (unfortunately for some) almost no limit to the extent to which this store of adipose tissue can be increased. Almost all of the increase in body mass of the extremely obese represents enlargement of the stores of adipose tissue. Thus, in an obese individual weighing 140 kg, more than 50% of body weight is represented by adipose tissue. About 90% of the mass of the adipose tissue represents stored triglycerides (over 13 kg in normal adult). Thus there is enough triglyceride available to provide fuel for 2 months or more at average rates of calorie consumption. This is in striking contrast with the limited amounts of stored glucose available, only enough to provide energy for less than a day. The extensive store of adipose tissue triglycerides is distributed widely throughout the mammalian organism, the pattern differing from species to species. Migrating birds deposit huge quantities of fat in the abdomen (omental fat) in the days just before their arduous southward flight, which may involve as much as 36 hours continuously in the air. Their body weight may increase by over 50% while "stuffing" for the trip, and every bit of the extra fat is burned up en route (Odum, 1965).

In humans, adipose tissue is found subcutaneously over the entire body, with some extra deposits in the areas of the buttocks and breasts. There are large deposits of adipose tissue in the mesentery, some around the kidneys, and some around the heart. While careful studies have revealed some quantitative differences in metabolism of adipose tissue from various anatomic sites, these differences are subtle, and it is almost correct to say that the adipose tissue can be regarded as a homogeneous "organ" that happens to be anatomically widely distributed (like the skin) (Wasserman, 1965). there are some regional differences as shown, for example, by the selective deposition of fat in hips, buttocks, and breasts under the influence of estrogenic hormone. In addition to its physiological role as a store of calories, adipose tissue plays a structural role, for example, in the cushioning of viscera (such as the kidney) and as an insulating layer reducing the rates of loss of body heat.

The mature adipose tissue cell *(adipocyte)* consists of a large structureless droplet of lipid surrounded by a very thin rim of cytoplasm. The area containing the nucleus is somewhat thickened, giving the cell the appearance of a signet ring in microscopic section. In the embryo and early fetus, the adipocytes contain very little stored triglyceride, and they do not have this characteristic appearance. Nevertheless, careful histological examination can define nests of cells that are in fact differentiated and destined to become mature triglyceride-filled adipocytes in the adult. The adipocyte is a unique cell in its ability to continue to take up and store large quantities of fatty acids in the form of triglycerides. Other cells have at most a very limited ability to increase their triglyceride stores and that increased storage is never in the form of a uni-locular central fat droplet.

The bulk of the adipose tissue in man is *white fat* as just described; in addition, particularly in the infant, there is a small amount of so-called *brown fat.* This form of adipose tissue is literally brown because of its high concentration of mitochondria which contain cytochromes and other pigments. Brown adipose tissue can also store large quantities of triglycerides, but the storage occurs in multiple droplets in the cytoplasm rather than in the form of a single, central fat droplet. Brown adipose tissue is particularly prominent in hibernating mammals and is believed to play a role in the generation of heat when the animal is roused from its hibernating state (Joel, 1965). Metabolism in brown fat may be relevant to maintenance of energy balance in humans (Jung et al., 1979; Himms-Hagen, 1979).

Enlargement of the mass of the adipose tissue organ can occur: (1) because the total number of fat cells of a given size is increased, or (2) because the size of the fat cells is increased, or (3) a combination of these two. In most obese adults, an increase in cell size without an increase in cell number is found. Especially in individuals becoming obese during childhood, there tends to be both an increase in cell size and an increase in cell number. The total number of potential fat cells is affected by early nutrition, with overnutrition tending to increase and undernutrition tending to decrease the potential number of adipocytes in the mature individual (Hirsch and Batchelor, 1976; Sallans, 1980).

METABOLISM OF ADIPOSE TISSUE

For many years adipose tissue was considered to be a relatively inert tissue. This was based, in part, on the very low rate of oxygen consumption of adipose tissue when expressed per gram of total wet weight. What was overlooked was the fact that 90% of the weight of the adipose tissue represents inert stored triglycerides. When adipose tissue oxygen consumption is expressed per milligram of cell protein, the value turns out to be surprisingly high—comparable to that of the liver and other metabolically highly active organs. Another long-standing misconception about adipose tissue was that the triglycerides stored there were metabolically inert. The classical studies of Schoenheimer (1942) using stable isotopes dispelled that notion many years ago by demonstrating the continuing turnover of the fatty acids in adipose tissue. We now know that the deposition and mobilization of fatty acids goes on all the time, even when the mass of adipose tissue is not changing.

The importance of the adipose tissue in sustain-ing the body during starvation is evident even to the casual observer. Yet it was quite mysterious for many decades as to just how the adipose tissue triglycerides were mobilized for utilization in the various body tissues. Most of the lipids in the plasma are present in the form of *lipoproteins,* and about 90% of the lipoprotein triglycerides are found in the *very low density lipoproteins* (VLDL). Consequently, efforts were made to demonstrate the release of lipoproteins and/or triglycerides from adipose tissue, but all such efforts gave negative results. In the mid-1950s, the presence of a small but, as it turned out, highly significant concentration of *free fatty acids* (FFA) was demonstrated in plama, i.e., fatty acids not covalently bonded to any other compound (Gordon and Cherkes, 1956; Dole, 1956). The rate at which these free fatty acids turned over was enormously fast, the half-life being only 2 or 3 min. Thus, even though the concentration was very low—less than 10% of the concentration of very low density lipoprotein triglycerides—the turnover of this fraction in the fasting state could account for transport of enough fatty acids to equal or exceed the daily energy utilization. Subsequent studies of adipose tissue in vitro confirmed that indeed stored triglycerides were mobilized as FFA and free glycerol, i.e., prior hydrolysis was required. Release of FFA represents the only significant output from the adipose tissue triglyceride fatty acid stores (Steinberg and Vaughan, 1965).

A highly simplified input-output scheme of adipocyte metabolism is shown in Figure 7.10. Hydrolysis of the stored triglycerides is under the regulation of a unique lipase, *hormone-sensitive lipase.* Some of the FFA, after being activated to fatty acetyl coenzyme A, can be reincorporated into depot triglycerides by reaction with glycerol-3-phosphate. Part of the FFA released is always thus rein-

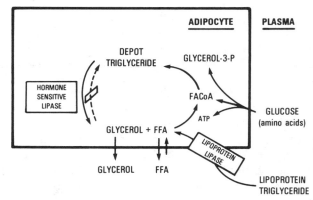

Figure 7.10. Schematic representation of the major inputs to the tri-glyceride stores of the adipocyte, the continuing lipolysis and reesterification of fatty acids, and the output exclusively as FFA and glycerol.

corporated; the remainder is released to the plasma. The glycerol released from hydrolyzed triglycerides is lost from the cell almost quantitatively because the adipocyte contains little or no glycerophosphokinase. In the liver, which is rich in glycerophosphokinase, free glycerol can be very rapidly incorporated into glucose or into triglycerides after first being converted to glycerophosphate. Because adipose tissue cannot reutilize the liberated glycerol, measurement of the rate of glycerol release constitutes an index of the true rate of triglyceride hydrolysis; the rate of release of FFA will always underestimate it to some extent because of concurrent reesterification (Steinberg and Vaughan, 1965).

The sources of adipose tissue triglycerides are several (Fig. 7.10). Glucose contributes carbons both to the formation of glycerol-3-phosphate and, after glycolysis, to the synthesis of fatty acids. Glucose, in addition, provides energy in the form of ATP generated from its own degradation, which can be used to synthesize triglycerides and add them to the depot stores. Amino acids can also be taken up by the adipose tissue and metabolized to yield both fatty acids (after generation of acetyl coenzyme A) and, in the case of certain amino acids, glycerophosphate (glyceroneogenesis).

Another important input is from chylomicron triglycerides and from the triglyceride-rich very low density lipoproteins. The chylomicrons represent exogenous fat, newly absorbed from the intestine; the very low density lipoproteins represent endogenous fat, released from the liver. These triglyceride-rich lipoproteins are acted on by a second lipase in which the adipose tissue is rich—*lipoprotein lipase.* This enzyme is located predominantly on the capillary endothelium of the adipose tissue (and of other tissues), but it is synthesized by the adipocyte and then transported to its location on the endothelial cells. There it can act on the triglycerides contained in triglyceride-rich lipoproteins, initially generating FFA and 2-monoglycerides because of the positional specificity of the enzyme. Spontaneous isomerization, however, can make the remaining fatty acid available. A large fraction of the triglycerides in chylomicrons is taken up into adipose tissue by this mechanism while the remainder of the chylomicron, including a residual small fraction of the triglyceride but almost all of the cholesterol, is taken up predominantly in the liver (see Havel et al., 1980, for further discussion).

The characteristics of plasma FFA are summarized in Table 7.3. The term "free" indicates that these fatty acids are not covalently bonded (in contrast to the fatty acids esterified to cholesterol or to glycerol, in cholesterol esters and triglycerides,

Table 7.3.
Characteristics of Plasma Free Fatty Acids

Chemical Form: Not *covalently* bonded; hence, designated free fatty acids (FFA), unesterified fatty acids (UFA), or nonesterified fatty acids (NEFA).
Physical Form: Tightly bound, but noncovalently, to serum albumin; association constants 10^6–10^7; <1% unbound, i.e., in free solution.
Concentration: Normally about 0.5 μEq/ml (12.5 mg/dl)
Turnover: $t_{1/2}$ 2–3 min; plasma pool size 1750 μEq; total turnover in fasting state, 550 μEq/min or over 200 g/day

respectively). While they are free in the strict chemical sense, they are actually very tightly bound, albeit *noncovalently,* to serum albumin and transported as albumin-FFA complexes. The affinity of FFA for binding sites on albumin is so great that less than 0.5% of the total FFA in the plasma is present in unbound form. Each albumin molecule can bind two molecules of FFA very tightly, with an association constant of approximately 10^7; an additional four molecules can be bound with a somewhat lower affinity (Goodman, 1958). Thus in a normal individual with an albumin concentration of about 7 g/dl, corresponding to approximately 1 μeq of albumin per ml, 2 μeq of FFA can be tightly bound per milliliter of plasma. As the total concentration of FFA rises, the percentage present in unbound form (at any given albumin level) increases. Even at the very highest concentrations reached under any ordinary circumstances, however, less than 1% of the total FFA is present in unbound form. Yet, the physical form of FFA that actually leaves the plasma must be the unbound FFA; the rate at which albumin leaves the plasma compartment is orders of magnitude slower than the rate at which FFA leaves. Despite the relatively modest concentration of FFA normally present, its turnover is so rapid that an enormous amount of substrate is potentially channeled through this fraction in the fasting individual. Indeed, as indicated in Table 7.3, the daily transport of FFA in the fasting state is more than sufficient to account for total daily caloric expenditure. Some fraction of transported FFA is redeposited in lipid esters (in the liver, for example) and not oxidized. In the fed state, when glucose provides a major portion of caloric substrate, FFA levels are low, and turnover is correspondingly reduced. As discussed below, the major if not exclusive control point for regulating FFA turnover is through regulation of mobilization. The half-life of FFA appears to be relatively independent of its concentration, and the fractional uptake into various tissues appears to be relatively constant.

The metabolic fate of mobilized FFA is depicted schematically in Figure 7.11 (Steinberg, 1966). The

PLASMA COMPARTMENT

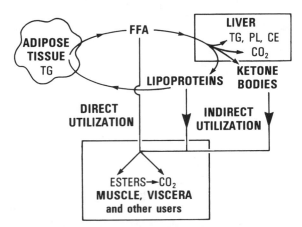

Figure 7.11.   The fatty acid transport cycle, illustrating the fate of FFA released from adipose tissue, their metabolism in liver and extrahepatic tissues, and their reentry into the adipose tissue stores.

liver is the organ most active in removal of FFA from the plasma. About one-third of the total FFA in the plasma passing through the liver is extracted during a single passage. The FFA taken up is, in part, reesterified and stored temporarily in the form of triglycerides, phospholipids, or cholesterol esters. Another portion is oxidized completely to CO_2 and water. Another fraction is converted to ketone bodies (incomplete oxidation), which are released from the liver and then oxidized completely to CO_2 and water by the extrahepatic tissues. (The liver lacks the enzymes necessary for oxidation of ketone bodies.) Finally, either immediately or subsequent to the initial deposition, fatty acids that have been incorporated into lipid esters can be secreted from the liver again in the form of lipoprotein-borne lipids. As discussed above, the lipoprotein-borne lipids, especially the triglycerides in VLDL, can be redeposited into adipose tissue through the action of lipoprotein lipase. This completes a "fatty acid transport" cycle. When mobilization of FFA from adipose tissue occurs in excess of that needed immediately for fuel, the liver will increase its stores of lipid esters. Those stored esters can at a later time be oxidized, or they can be transported back to the adipose tissue in lipoproteins, restoring the balance. Finally, some FFA is taken up by essentially every tissue (with the notable exception of brain and nerve tissue), providing them with substrate for immediate combustion or, under some circumstances, for storage in ester form until needed at a later time. The amount of this temporary storage is, however, very limited.

From the above discussion, one can predict certain consequences of overly rapid and excessive mobilization of FFA from adipose tissue. These include the following: (1) deposition of triglycerides in the liver and a tendency to develop a *fatty liver;* (2) increased rates of production of ketone bodies and a tendency to develop *ketoacidosis;* (3) an increased output of lipoproteins from the liver, especially triglyceride-rich lipoproteins, and a tendency to develop *hyperlipidemia;* (4) inhibition of the rate of glucose uptake in peripheral tissues (Randle effect) and a tendency toward *hyperglycemia.* Two other consequences should be mentioned, although their importance is less firmly established. First, there may be some increase in total body oxygen consumption, since excessively high FFA levels may uncouple oxidative phosphorylation in mitochondria; second, there may be an increased tendency for intravascular thrombosis because of the effects of FFA and/or the consequent hyperlipoproteinemia on platelet function and other aspects of the clotting mechanism.

REGULATION OF FFA MOBILIZATION

The triglycerides in adipose tissue are always in a dynamic state, as mentioned above, even when the total mass of triglyceride is not changing. The balance can be shifted by altering the rate of lipolysis, i.e., by changing the activity of hormone-sensitive lipase. It can also be shifted by altering the rate of reesterification of released FFA back into triglycerides. The latter can be increased by anything that favors activation of FFA (e.g., providing additional ATP) or increases the availability of glycerol-3-phosphate as acceptor. Glucose provides both ATP and glycerophosphate and therefore strongly stimulates the reesterification arm of the cyclic process. Insulin, by favoring glucose uptake at any given glucose concentration, will act to inhibit FFA release, in part, by this mechanism. However, the more important effect of insulin is directly on the state of activation of hormone-sensitive lipase and thus on the degradation of stored triglycerides (see below).

There are four major modes of regulation of fat mobilization:

Metabolic Control. As already indicated, glucose will suppress mobilization by favoring reesterification. Amino acids can do the same, although less potently.

Neural Control. Adipose tissue is richly innervated, and stimulation of its nerve supply causes release of norepinephrine at sympathetic endings. This, in turn, activates hormone-sensitive lipase and promotes FFA mobilization.

Direct Hormonal Control. A surprisingly large number of different hormones can act on adipose tissue in vitro to enhance FFA release. Two of these are clearly established to play a role in regulation in vivo: norepinephrine and epinephrine. Intravenous infusion of these catecholamines elicits a prompt increase in plasma FFA levels. Glucagon is a potent stimulus to FFA release in vitro. Infusion of glucagon in vivo, however, does not normally increase FFA levels. This is probably because of the accompanying glycogenolytic action of glucagon, leading to an increase in plasma glucose levels and a consequent increase in plasma insulin levels. Glucagon also has a direct effect in stimulating insulin release. These effects in normal subjects mask any tendency for glucagon to increase fatty acid release. On the other hand, in Type I diabetic patients who cannot respond with an increase in insulin levels, or in experimental subjects in whom insulin response has been deliberately suppressed by infusion of somatostatin, a glucagon effect on FFA mobilization has been demonstrated. Thus, glucagon regulation of fat mobilization may only be significant in the fasting state and in diabetic ketoacidosis.

The other hormones shown to stimulate FFA release in vitro include ACTH, vasopressin, melanocyte-stimulating hormone, and a number of others. However, there is no conclusive evidence as yet that changes in these hormones within the physiological range have a significant impact on FFA release in vivo.

Indirect Hormonal Control. Cortisol and other glucocorticoids have little or no direct effect in stimulating FFA release from adipose tissue in vitro. Yet, an intact adrenal cortex is necessary in order to obtain an optimal response to lipolytic hormones such as catecholamines in vivo. This is another example of the so-called "permissive effect" of the glucocorticoids. Presumably, the continuing action of the glucocorticoid is necessary to maintain appropriate levels of receptors or one or more of the enzymes involved in the acute lipolytic response to catecholamines and other lipid-mobilizing hormones.

Thyroid hormone, again, has little direct activity in mobilizing FFA but is essential to maintain responsiveness to rapidly acting fat-mobilizing hormones. Again the action can be described as predominantly a permissive one. Thyroid hormone modulates a number of other cyclic AMP-dependent hormonal responses in a similar way (see Chapter 53).

HORMONE-SENSITIVE LIPASE

Hormone-sensitive lipase is a neutral lipase mediating direct hormonal effects on the rates of lipid mobilization from adipose tissue (Vaughan et al.,

1964). As shown schematically in Figure 7.12, it is activated by a phosphorylation reaction catalyzed by cyclic AMP-dependent protein kinase (Butcher et al., 1965; Huttunen et al., 1970; Corbin et al., 1970; Stralfors et al., 1984). The active phosphorylated form is deactivated by one or more protein phosphatases (Severson et al., 1977; Olsson and Belfrage, 1987). Hormones that directly stimulate FFA release (β-adrenergic agents, glucagon, ACTH) do so primarily by increasing the activity of adenylate cyclase; changes in the activity of lipase phosphatases could also play a role but that possibility remains to be fully evaluated. Rat hormone-sensitive lipase has been cloned and sequenced (Holm et al., 1988). It is a 757-amino acid protein with an activity-controlling phosphorylation site at Ser^{563}.

During the mobilization of triglycerides, there is very little accumulation of diglycerides or monoglycerides in adipose tissue, indicating that the rate-limiting reaction is the hydrolysis of the first ester bond in the triglyceride. In addition to hydrolyzing triglycerides, the same hormone-sensitive lipase also hydrolyzes cholesterol esters. The significance of this in adipose tissue is not certain, but it could prevent progressive buildup in cholesterol ester content during the mobilization of adipose tissue triglycerides. Steroidogenic tissues also contain hormone-sensitive lipase (Holm et al., 1987; 1988) and there the function of the cholesterol esterase activity is evident—to hydrolyze stored cholesterol esters and thus provide free cholesterol as steroidogenic precursor.

Adipose tissue contains glycogen and the full complement of glycogen-metabolizing enzymes, analogous to those found in the liver. In animals that are fasted and refed, the glycogen content of adipose tissue can rise to very high levels. Presum-

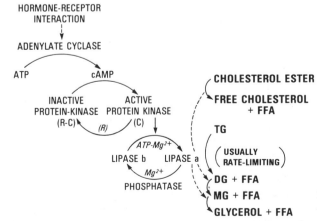

Figure 7.12. Mechanisms involved in the activation and deactivation of hormone-sensitive lipase in adipose tissue.

ably this provides a mechanism to help dispose of large quantities of ingested glucose which can later be converted to fat. The glycogen synthase and phosphorylase of adipose tissue are under hormonal regulation analogous to that in the liver. Thus, catecholamines activate phosphorylase but convert glycogen synthase to its inactive form; conversely, insulin increases the activity of glycogen phosphorylase. Thus, when glucose and insulin are available in abundance, the adipose tissue will tend to take up glucose and deposit it as glycogen for later utilization.

Insulin is the only major hormone that tends to *suppress* the mobilization of FFA. Injection of antibody against insulin, creating a sudden insulin deficiency, is associated with an almost immediate rise in plasma FFA levels. In chronic insulin deficiency, such as occurs as in young diabetic subjects, FFA levels reach values higher than those seen under any other circumstances and are associated with many adverse metabolic consequences, discussed later. The mechanism by which insulin blocks FFA mobilization includes its effects to enhance fatty acid reesterification through promoting glucose uptake. Much more important, however, is its direct effect on the state of activation of the enzyme: insulin prevents the epinephrine-induced conversion of lipase *b* to lipase *a* (Steinberg, et al., 1975). Under some conditions, this effect can be explained by insulin regulation of cyclic AMP levels (Butcher et al., 1966; Loten and Sneyd, 1970; Manganiello and Vaughan, 1973). However, under other circumstances it is clear that the insulin must have an additional effect since it can reduce the activity of hormone-sensitive lipase without much effect on the tissue concentrations of cyclic AMP (Khoo et al., 1973).

The effects of cortisol, as mentioned above, seem to be permissive. In the isolated adipose tissue, cortisol can enhance FFA release when presented in combination with growth hormone. The effects are blocked by puromycin or by actinomycin D, indicating that the effects relate to induction of new enzyme synthesis (Fain et al., 1965). The actions of the rapidly acting hormones (catecholamines and glucagon) are not blocked by inhibitors of protein synthesis or inhibitors of RNA synthesis.

LIPOPROTEIN LIPASE

Adipose tissue uptake of triglycerides from chylomicrons and VLDL is catalyzed by lipoprotein lipase, and the activity of this enzyme is evidently rate-limiting. Its activity is increased by insulin and glucose. Thus its activity is high in the fed and low in the fasted state—the inverse of the changes seen

in hormone-sensitive lipase. Again, catecholamines, which *increase* hormone-sensitive lipase activity, *decrease* lipoprotein lipase activity. This reciprocal control led to speculation that the two major lipases in adipose tissue might represent mirror images of the same enzyme protein, having different substrate specificities and different functions at different times. However, both enzymes have now been isolated, fully characterized, and cloned (Wion et al., 1987; Holm et al., 1988).

Lipoprotein lipase, present in almost every tissue, appears to act at the capillary surface. Thus, triglyceride-rich lipoproteins entering the microcirculation bind to the capillary-bound enzyme, for which they have a high affinity, and there they undergo partial hydrolysis. The lipolytic products can then enter the parenchymal cells of the organ involved. For example, the uptake of chylomicron and VLDL triglycerides by cardiac muscle is mediated by the action of lipoprotein lipase. Whereas fasting causes a *decrease* in the lipoprotein lipase level in adipose tissue, it causes an *increase* in the levels of the enzyme in the heart and in many other tissues. Thus, in the fasting state there is a shunting of newly ingested triglyceride to the tissues that require substrate and away from storage in the adipose tissue, an appropriate response. In the fed state, the reciprocal changes occur, and the triglycerides tend to be deposited in the adipose tissue (reviewed by Garfinkel and Schotz, 1987).

After injection of heparin, which causes a marked increase in the levels of lipoprotein lipase in the plasma, there is an increase also in a second lipase with a number of properties similar to those of lipoprotein lipase. This lipase has its origin primarily in the liver and has been named hepatic lipase. It is found primarily on the sinusoidal surfaces of the liver and acts to hydrolyze phospholipids and glycerides on lipoproteins (Jackson, 1983; Kinnunen, 1984). Its exact role remains to be established but it is presumed to play a role in the metabolism of high density lipoproteins (HDL); it may also participate in the clearance of VLDL and chylomicron remnants. The cDNA for the rat hepatic lipase has recently been cloned and this should facilitate establishing its biological function (Komaromy and Schotz, 1987).

PLASMA LIPOPROTEINS

Lipids are found in the plasma at concentrations far above their solubility in aqueous systems. This is only possible because of the way in which lipids are associated with specialized apoproteins to form large spherical lipoprotein complexes. The highly nonpolar lipids (cholesterol esters and triglycerides) are concentrated almost exclusively in the central core while the more polar lipids, mainly phospholipids, and the protein moieties make up a polar shell. The physical properties of the package are those of proteins, not lipids, and the lipoproteins can be resolved by precipitation, electrophoresis, or differential centrifugation using methods like those used with ordinary proteins. The interaction of lipoproteins with cell membrane receptors, and thus their metabolic fate, is also dictated largely by the nature of their protein moieties.

Lipoprotein Classes

Five major classes of plasma lipoproteins are currently recognized (Table 7.4). However, these classes do not constitute stoichiometrically precise complexes. Each can be subdivided into subclasses or even regarded as a spectrum of molecules. Nevertheless, the crude classification is sufficient for many purposes, including broad definition of metabolic functions and relationship to disease states.

DENSITY CLASSES

Lipids have densities much lower than those of proteins. For example, triglycerides have densities less than 1.00 g/ml, and cholesterol has a density of 1.05; proteins have densities of 1.25–1.35. The densities of the lipoproteins thus vary according to the proportions of lipid and protein. As a general class they all have densities less than that of all the other plasma proteins and can be separated from them on that basis. If enough salt is added to plasma to raise its background density (density of the nonprotein fluid phase) to 1.21 g/ml and if the sample is centrifuged overnight in the ultracentrifuge at >100,000 × *g*, all the lipoproteins rise to the surface, well separated from all the other plasma proteins, which sediment toward the bottom of the tube. If plasma is centrifuged without changing its background density (which is 1.006), any chylomicrons present will float after just 30 min at 100,000 × *g*. A subsequent overnight centrifugation brings the VLDL class to the top. Successive centrifugation steps at salt densities adjusted up to 1.019, then to 1.063, and finally to 1.21 float up the intermediate density (IDL), low density (LDL), and high density lipoprotein (HDL) classes, respectively. This separation by preparative ultracentrifugation is widely used in research since the individual fractions can be obtained in quantity for further characterization (Havel et al., 1955).

FLOTATION CLASSES

In the analytic ultracentrifuge one normally measures the rate at which protein boundaries sediment under specified conditions and designates a Svedberg sedimentation coefficient (S). Using a higher medium density, lipoproteins are made to float (i.e., move centripetally), and the rate of flotation is specified in analogous fashion as a Svedberg flotation coefficient (S_f). The rate of flotation increases according to the difference between the density of the particle and that of the medium; the large, lipid-rich particles float fastest. The S_f values shown in Table 7.4 apply when the medium has a density of 1.063

ELECTROPHORETIC MOBILITY CLASSES

The several classes can also be resolved by electrophoresis. By staining specifically for lipid, the lipoproteins can be located readily, even in the presence of all the other serum proteins. The mobilities of the several classes of lipoproteins, shown in Table 7.4, are determined primarily by the apoproteins associated with them. Size also plays a role, especially in the case of the chylomicrons, which are too large to penetrate most gels and do not leave the origin. Electrophoretic separation, being the simplest, is the most widely used method for clinical purposes.

SIZE AND CHEMICAL COMPOSITION

Chylomicrons consist predominantly of triglycerides (85–95% by weight) surrounded by a thin shell of protein (1–2% by weight) and phospholipid. They also contain newly absorbed cholesterol, mainly in esterified form. The particles are greater than 80 nm in diameter, visible in a dark field in the light microscope.

VLDL, except after a fat-containing meal, account for most of the triglycerides in plasma. They vary tremendously in size (30–80 nm), mainly because of different triglyceride content (20–80% by weight).

LDL (about 20 nm) carry most (about two-thirds) of the cholesterol in normal plasma. The LDL is

Table 7.4.
The Major Classes of Plasma Lipoproteins

Class of Lipoprotein	Density (g/ml)	Flotation Constant (S_f)	Electrophoretic Mobility (Paper or Agarose)
Chylomicrons	<0.96	>400	Remain at origin
Very low density (VDL)	<1.006	20–400	pre-β (α)
Intermediate density (IDL)	1.006–1.019	12–20	β to pre-β
Low density (LDL)	1.019–1.063	0–12	β
High density (HDL)	1.063–1.21	(Analyzed separately)	α

about 50% by weight cholesterol and 20% by weight protein.

HDL molecules (about 8 nm) are about 50% protein by mass and only 50% lipid—hence their high density.

The clinical laboratory usually analyzes only for total cholesterol content and total triglyceride content of the plasma. Only for relatively sophisticated studies is a further breakdown into relative concentrations of different lipoprotein classes necessary. However, knowing the general composition of the lipoproteins, one can draw useful inferences about lipoprotein patterns knowing only total cholesterol and triglyceride values. For example, a marked elevation of cholesterol level with a normal triglyceride level cannot be due to an increase in VLDL alone; since VLDL is rich in triglycerides, an increased VLDL sufficient to raise cholesterol level would necessarily increase triglyceride levels even more. One can conclude if cholesterol is elevated and triglycerides are normal that the patient has an increase in LDL concentration. (Elevation of HDL sufficient to cause a marked increase in cholesterol level can occur but is unusual.)

APOPROTEINS

The protein moieties associated with the lipoproteins play not only a structural role, stabilizing the lipid particles and conferring protein-like properties on them, but also a variety of functional roles. The nomenclature is not particularly helpful because letter designations were applied chronologically as the apoproteins were identified. In several instances, heterogeneity has been discovered subsequently, and several different apoproteins with no apparent genetic or functional relationship are grouped into a given "family." For example, the apoprotein associated with HDL was originally designated "apoprotein A." Subsequent studies established that there are two major apoproteins, now designated apo A-I and apo A-II, that differ considerably in structure and function. Both have been completely sequenced. Apo A-I, with a molecular weight of about 27,000, is a cofactor for the enzyme lecithin-cholesterol acyltransferase; apo A-II, with a molecular weight of about 16,000, consists of two identical polypeptides joined by a disulfide bridge and has no established functional role. Apoprotein E, found in both VLDL and HDL, has turned out to include three different major isoforms separable by isoelectric focusing. This is more nearly what one would designate a "family" since they derive from allelic genes at a given locus. The apo C group, on the other hand, consists of proteins of quite different structure and different function. For example,

apoprotein C-II is an obligatory cofactor for the action of lipoprotein lipase; the other C apoproteins, apo C-I and apo C-III, cannot substitute for it. Thus, several kindreds have been discovered in which apo C-II is specifically deleted, and these families suffer from familial hyperchylomicronemia.

In addition to the enzyme and cofactor functions mentioned above, the most important role of the apoproteins is to target lipoproteins to the appropriate cells and tissues by virtue of receptor-apoprotein interactions. The best established of these is the receptor recognizing LDL, which binds apoprotein B with high affinity (Goldstein and Brown, 1977). Deletion of this receptor results in extremely high plasma concentrations of LDL and therefore of cholesterol (familial hypercholesterolemia). This same receptor also recognizes and binds apoprotein E with even higher high affinity (Innerarity and Mahley, 1978). A separate receptor for chylomicrons has been described in the liver (Sherrill et al., 1980). Apoprotein-receptor interactions are being actively investigated, and a clearer picture should emerge over the next few years.

Lipoprotein Metabolism

At the outset it should be recognized that there are important species differences in lipoprotein patterns. For example, most animals have extremely low plasma concentrations of LDL; man has the highest levels found in any animal species. In the following discussion, whenever possible, we will present patterns of lipoprotein metabolism as they occur in man. Several detailed reviews are available (Goldstein and Brown, 1977; Steinberg, 1979; Havel et al., 1980; Stanbury et al., 1983).

VLDL, IDL, AND LDL

The liver is the major source of VLDL, with the intestine making a relatively small contribution (about 15% or less). The large, triglyceride-rich VLDL particles emerging from the liver are acted upon by lipoprotein lipase as they pass through the capillary beds. This enzyme breaks down VLDL triglycerides, and the products are taken up into the various extrahepatic tissues, as shown in Figure 7.13. At the same time, some of the apoproteins of the VLDL are lost. Over a relatively short time (the half-life of VLDL in the plasma is only about 60 min), a large fraction of the triglyceride has been removed, and the size of the particle has decreased to that characteristic of an IDL. Some of the IDL molecules are removed intact from the circulation, but in man most are further degraded either by the action of hepatic lipase and/or further action of lipoprotein lipase to generate LDL. This final IDL-to-

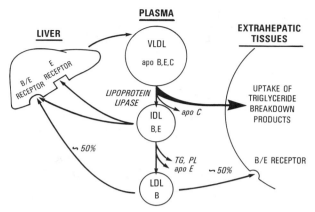

Figure 7.13. Schematic representation of the metabolic interconversions of very low density lipoprotein *(VLDL)* to intermediate density lipoprotein *(IDL)* and on to low density lipoprotein *(LDL)* (see text for elaboration). The hepatic *"E receptor"* has yet to be fully characterized. It must be involved in removal of chylomicron remnants but it may or may not play a major role in removal of VLDL remnants.

LDL transformation is associated with the loss of any remaining apoproteins other than apoprotein B and with the loss of some additional residual triglyceride from IDL. During this elaborate transformation, the amount of apoprotein B per lipoprotein particle remains constant. In other words, the apoprotein B of the VLDL particle is the linchpin—the constant that defines the VLDL-IDL-LDL pathway. During this interconversion, most of the VLDL cholesterol is also retained.

LDL is taken up and degraded by tissues throughout the body. The liver is the major site of removal, accounting for about one-half of all LDL degradation; the remainder is distributed among a variety of tissues, no one of which approaches the liver in individual importance. Weight for weight, however, the adrenal cortex and the gonads are extremely active in LDL degradation, which provides cholesterol precursor for the biosynthesis of the steroid hormones they produce. These tissues also synthesize cholesterol endogenously and can produce their steroid hormones in the absence of LDL, but maximal steroid hormone production may depend upon LDL being provided. About two-thirds of LDL degradation in vivo is mediated by the LDL receptor. That receptor recognizes both apo B and apo E and is also designated the B/E receptor. In the rat, it appears that HDL plays an important role analogous to this role of LDL, i.e., HDL, recognized by its apoprotein E moiety, delivers a significant amount of cholesterol to adrenal and gonads.

The uptake of VLDL remnants or IDL into the liver appears to depend on recognition of their apoprotein E moiety rather than the apoprotein B. This conclusion is based on the inherited abnormality of apoprotein E in families with dysbetalipoproteinemia (type III hyperlipoproteinemia), in which there can be an accumulation of lipoproteins with characteristics similar to those of IDL. These patients are characterized by homozygosity for one of the isoforms of apoprotein E not well recognized by receptors (apo E_2 or other mutant forms). These patients, like those with familial hypercholesterolemia, suffer premature atherosclerosis leading to myocardial infarctions and strokes.

CHYLOMICRONS

Chylomicrons generated during the digestion and absorption of fat have a very short lifetime in the plasma (half-life about 10 min). Their metabolism, like that of VLDL, depends upon the action of lipoprotein lipase and apoprotein C-II is an obligatory cofactor. Chylomicrons contain a form of apoprotein B that is synthesized only by the intestine in man (but synthesized also by the liver in the rat). This form of apoprotein B has a lower molecular weight than that synthesized in the liver, but its physical properties and immunochemical properties are closely related. The intestinal form has been designated B48 and the hepatic form B100, designations roughly indicating their relative molecular weights. Since no B48 is found in the LDL fraction in man, it is evident that in man essentially all of the chylomicron apoprotein B is removed from the plasma compartment before degradation to the molecular size of LDL can take place.

HDL

HDL is synthesized in the liver and in the intestine. It emerges into plasma in the form of flattened discs, essentially bilayers of phospholipid containing some cholesterol, mainly free, and some apoproteins. In the plasma compartment, the nascent HDL is acted upon by lecithin-cholesterol acyltransferase (LCAT), which converts free cholesterol to ester cholesterol by transfer of a fatty acid from lecithin (Fig. 7.14). Most of the lipoproteins secreted into the plasma contain only limited amounts of cholesterol ester; most of the cholesterol ester found in the plasma at steady state is the result of the action of this enzyme. Cholesterol esters can be transferred among lipoprotein classes, a process mediated by cholesterol ester transfer protein, present in normal human plasma and in the plasma of most animals (absent in the rat). The result is that the lipoproteins undergo a great deal of "remodeling" after their secretion.

As a result of the action of LCAT, HDL gradually assumes a spherical configuration with cholesterol esters in the central core of the molecule. In the pro-

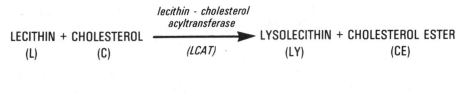

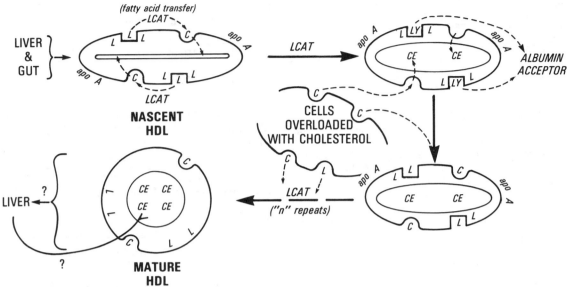

Figure 7.14. Schematic representation of nascent HDL from the liver, its conversion to mature HDL by the action of lecithin-cholesterol acyl- transferase *(LCAT)*, and the postulated role of HDL in reverse choles- terol transport (see text).

cess, it becomes capable of accepting free choles- terol from tissues, as indicated in Figure 7.14, which may be of central importance in preventing overload of cholesterol in various peripheral tis- sues. All tissues participate to some extent in the uptake and degradation of LDL as discussed above, but none of them except for the gonads and adrenal cortex can further catabolize cholesterol. Further- more, all tissues synthesize cholesterol de novo at some rate. Therefore, there must be some mecha- nism for returning cholesterol to the liver for excre- tion in order to avoid progressive buildup of tissue cholesterol levels. HDL probably plays a key role in this "reverse cholesterol transport."

During the action of lipoprotein lipase on chylo- microns or on VLDL, sections of their phospholipid shell become redundant as the core of triglycerides gets smaller. These surface components make an additional contribution to HDL formation. In man the HDL fraction includes at least two subfractions, HDL_2 and HDL_3, the latter being smaller and more dense. The addition of lipids from chylomicrons or VLDL can convert HDL_3 to HDL_2. Epidemiological studies leave no doubt that subjects with low con- centrations of HDL_2 are much more likely to develop premature atherosclerosis than subjects with high concentrations of HDL_2. Whether this "protective

effect" is fully accounted for by the mechanism of reverse cholesterol transport we have just dis- cussed is not fully established.

Lipoprotein Uptake at the Cellular Level

Lipoprotein uptake into the cell can occur by one of three classes of mechanism: (1) receptor-medi- ated endocytosis; (2) adsorptive endocytosis not mediated by a specific receptor; and (3) fluid endo- cytosis (not involving adsorption to the cell mem- brane).

The best established receptor-mediated mecha- nism is that involved in the uptake of LDL. The work of Brown and Goldstein establishing the nature and function of this receptor was a milestone in the development of our understanding of lipopro- tein metabolism. The receptor is widely distributed among all body tissues, including the liver, and about two-thirds of total body LDL degradation is mediated by this receptor. Its function is schemati- cally indicated in Figure 7.15. LDL binds with high affinity to the receptor, which tends to cluster in specialized areas of the plasma membrane known as "coated pits." The LDL concentrated in the coated pit is taken into the cell by invagination of the mem- brane (endocytosis), and a vesicle containing LDL pinches off from the plasma membrane. That vesicle

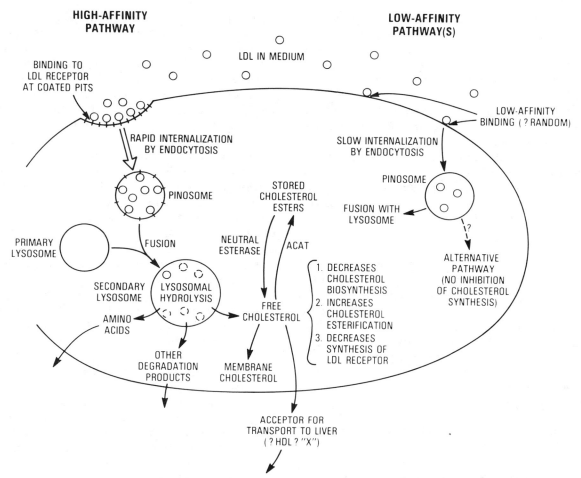

Figure 7.15. Schematic representation of the mechanism for uptake of *LDL* by endocytosis, including both the specific receptor-mediated mechanism and low-affinity nonspecific mechanisms (see text).

ultimately fuses with a lysosome. There the lysosomal hydrolyases degrade the various components of the LDL, including the protein and cholesterol esters. The products of lysosomal hydrolysis are released, and the free cholesterol formed regulates cholesterol metabolism of the cell. It inhibits cholesterol biosynthesis by suppressing the production of HMG CoA reductase; it stimulates the rate of cholesterol esterification so that any excess cholesterol is temporarily stored in the form of cholesterol esters, primarily cholesterol oleate; finally, it in some way inhibits the further production of LDL receptors. The latter effect provides a feedback control so that the cell can limit the amount of cholesterol delivered to it in LDL.

Adsorptive endocytosis not involving the LDL receptor occurs at all times. In patients who totally lack the LDL receptor (homozygous familial hypercholesterolemia), there is no receptor-mediated uptake, yet the total uptake and degradation of LDL per day actually exceeds that in a normal individual. This must all obviously occur by pathways independent of the LDL receptor, but these pathways are incompletely characterized.

Finally, some incorporation and degradation takes place by the random imbibition of LDL molecules when the cell takes up the surrounding medium by fluid or bulk endocytosis. The amounts taken up by this mechanism are limited and are strictly in proportion to the concentration of LDL in the surrounding medium.

Regulation of the LDL receptor is an important mechanism for controlling plasma LDL concentrations. For example, one of the most widely used drugs for treating hypercholesterolemia is cholestyramine, which sequesters bile acids in the intestine. It has been shown that the drug is effective, in part, because it leads to an induction of hepatic LDL receptors and thus an increase in the rate of hepatic degradation of LDL. Other drugs have also been shown to affect LDL receptor number.

The LDL receptor recognizes not only apo B, the

only significant apoprotein of LDL, but also apo E. The latter apoprotein is found in some subfractions of HDL, and those subfractions are avidly adsorbed and taken up by the so-called LDL receptor. It might better be referred to as the apo B/E receptor since it recognizes both.

The liver, in addition to the apo B/E receptor, expresses a receptor involved in the binding and uptake of chylomicron remnants and VLDL remnants. Strong evidence suggests that the binding to this receptor like the binding of VLDL remnants to the LDL receptor, depends upon apoprotein E. In fact, these receptors appear to bind only two of the three major isoforms of apoprotein E—apo E_3 and apo E_4—with high affinity. Apo E_2 binds very poorly, and patients who are homozygous for apo E_2 may have accumulation of remnants (type III hyperlipoproteinemia). However, hyperlipidemia is *not* always associated with this pattern of apo E isoforms, although all of the patients who do have familial dysbetalipoproteinemia are homozygous for apo E_2. Additional genetic or environmental factors are needed to evoke the hyperlipidemia.

HDL appears to be taken up and degraded, in part, by saturable mechanisms, but the receptors have not been fully characterized.

Physiological and Pathophysiological Significance of Lipoprotein Transport

The role of the lipoproteins in triglyceride transport and cholesterol transport is readily apparent. Chylomicrons are responsible for transporting whatever quantity of fat one ingests daily. In the average American diet, this amounts to about 100 g daily. Triglyceride transport in VLDL is more modest by far, amounting to about 10–15 g daily. However, this retransport of fatty acids from liver to extrahepatic tissues in the form of VLDL triglycerides is essential to avoid excessive accumulation of triglycerides and development of a fatty liver (see Fig. 7.11).

The importance of delivery of cholesterol in LDL to tissues producing steroid hormones was discussed above. In addition, delivery of cholesterol may be essential in other tissues under certain circumstances. For example, during growth or tissue repair the requirements for cholesterol may exceed the capacity of the cells involved to synthesize cholesterol de novo. Thus, during growth LDL may play a more important role as a vehicle for transport of cholesterol than it does in the adult at steady-state.

From a pathophysiological point of view, the lipoproteins are extremely important. There is no question about the strong positive correlation between plasma cholesterol levels and, more specif-

ically, LDL levels and susceptibility to atherosclerosis. The myocardial infarction rate and stroke rate in patients with very high LDL and high total plasma cholesterol levels is much greater than that in other patients. The best evidence that elevated LDL levels are in themselves atherogenic is provided by the families with inherited deficiency of the LDL receptor (Goldstein and Brown, 1977). This disorder is monogenic. Consequently, if one accepts the one gene-one protein hypothesis, all of the phenotypic expressions must be related ultimately to the single underlying genetic error. In this case, the genetic error is the deletion of the LDL receptor, and its immediate consequence is an elevation of the plasma concentrations of LDL. The development of atherosclerosis in these patients, then, must be linked to this elevation of LDL, although the precise mechanisms of that linkage remain to be established. The causative LDL-atherosclerosis connection is further strengthened by the recently described rabbit model of LDL receptor deficiency. These animals have plasma cholesterol levels over 400 mg/dl on their usual diet and develop atherosclerotic changes as early as age 3 months (Watanabe, 1980; Tanzawa et al., 1980).

The notion of what constitutes a normal value for total plasma cholesterol level has undergone some drastic revisions in recent years. Traditionally, a laboratory value below the 95th percentile or above the 5th percentile has been considered "normal," i.e., within 2 standard deviations of the mean. For cholesterol levels in the United States, this definition fails us. Epidemiological studies show clearly that in countries where the mean cholesterol level is much lower than it is in the United States the incidence of coronary heart disease is also much lower. Moreover, studies within our country show clearly that reducing plasma cholesterol levels from what used to be considered "normal" reduces the risk of coronary heart disease. For many reasons, the whole issue was reevaluated in 1984 (Consensus Conference, 1985) and a new set of guidelines for interpreting plasma cholesterol levels and dealing with them was issued through the National Cholesterol Education Program (Expert Panel, 1988). Most of the difference between Americans and Japanese with respect to cholesterol levels probably relates to the differences in their diets (Marmot et al., 1975). As we reduce our intake of saturated fats and cholesterol in this country, the mean plasma cholesterol levels will predictably fall and begin to approach that of the Japanese in the 1960s and 1970s. (The plasma cholesterol levels in Japanese are tending to rise as their intake of saturated fat increases in recent years!)

Other lipoprotein fractions have also been implicated as possibly contributing to atherosclerosis, although the epidemiological correlations are less striking. Patients with hypertriglyceridemia only (i.e., having elevation of chylomicrons and/or VLDL) have been found in some studies to show a higher than average incidence of atherosclerosis and its complications, but other epidemiological studies find no such increased risk. If one assesses the risk associated with elevated cholesterol levels and "corrects" for it, then there is little or no additional risk associated with being hypertriglyceridemic. On the other hand, from a mechanistic point of view, the presence of lipoproteins enriched in triglycerides may not be without consequences and implications for atherogenesis.

Finally, as mentioned previously, elevated levels of HDL are associated with a *decreased* atherosclerotic tendency. This negative correlation with HDL is just as strong as the positive correlation between atherogenesis and *increased* LDL levels. One hypothesis has been mentioned above, namely, that HDL may play a role in reverse cholesterol transport. By accepting cholesterol from cholesterol-laden cells in the arteries and carrying it back to the liver for excretion, HDL may defer the development of atherosclerosis. Another hypothesis relates to the competition between certain subclasses of HDL and LDL for uptake and degradation. Apoprotein E-containing HDL competes effectively with LDL for binding to the apo B/E receptor. Thus, a high concentration of HDL might reduce the rate of uptake of LDL. There may be competition also for nonsaturable binding and uptake. Finally, there is evidence that HDL may protect the endothelial lining against damage induced by LDL.

Space does not allow a complete discussion of the complexities of the atherogenic process. Several comprehensive reviews are available (Goldstein and Brown, 1977; Benditt, 1978; Wissler, 1978; Ross, 1981; Steinberg, 1988). Recent studies suggest that the atherogenicity of LDL is increased if it undergoes oxidative modification in vivo (Steinberg et al., 1989). Abnormal lipoprotein patterns are by no means the only cause, of course. Many individuals develop serious atherosclerosis in the presence of normal lipoprotein patterns. Other factors that are known to be associated with increased risk include: (1) hypertension; (2) hyperglycemia and diabetes mellitus; (3) obesity; (4) cigarette smoking; (5) being male rather than female; and (6) aging. Just how these so-called risk factors relate to the disease process at the cellular level remains to be established. Damage to the endothelium, such as might occur in association with hypertension, is believed to be an important factor. The tendency toward platelet aggregation with consequent release of factors stimulating cell growth and favoring local thrombosis is believed to be a factor. Elevated lipoprotein levels and some of these other mechanisms may relate closely to one another. For example, high concentrations of LDL that have undergone oxidative modification may themselves be damaging to endothelial cells and can initiate the consequences of endothelial denudation.

In summary, the involvement of LDL in atherogenesis is clearly established by experimental studies, epidemiological studies and, most importantly, genetic studies in patients and animals lacking the LDL receptor. The apo B/apo E-containing remnants of incomplete VLDL degradation are also rather clearly implicated. Other lipoproteins are probably involved as well, but the data are less clear. Effective intervention to decrease the toll of myocardial infarction and stroke requires attention to a series of risk factors, most notable of which are hyperlipoproteinemia, hypertension, and cigarette smoking (Levy, 1981).

CHAPTER 50

The Endocrine Pancreas

The normal human pancreas, weighing about 80 g, is a flattened, elongate organ lying against the posterior wall of the upper abdomen (Fig. 7.16). The vast bulk of the gland consists of acinar cells that synthesize and secrete digestive enzymes that enter the duodenum via the pancreatic duct system. This *exocrine* function of the pancreas is discussed in Chapter 44.

Scattered more or less randomly through the pancreas and accounting for only 1–2% of its weight are several hundred thousand microscopic nests of cells, the islets of Langerhans. These cells do not connect with the system of pancreatic ducts. The islets collectively constitute the *endocrine pancreas,* secreting several critically important hormones directly into the bloodstream. This arrangement of an endocrine organ buried within an exocrine organ has no established functional significance. In fact, in some lower species the islet tissue occurs anatomically separated from the exocrine pancreas. For example, the islet tissue of the angler fish (and other bony fish) is concentrated in one or two discrete tiny organs ("principal islets") that can be as large as a pea. Many early studies of islet physiology were done using these fish so that the endocrine cells could be investigated separately. Later, techniques were developed for isolating islets from mammalian pancreas. The independence of the islets and the exocrine pancreas is nicely illustrated by the fact that ligation of the pancreatic duct causes atrophy of all the acinar tissue (because of back pressure), yet the islets survive and continue to function. Banting and Best (1922) exploited this differential atrophy in order to isolate insulin for the first time. Previous attempts had failed because the enormous amounts of digestive enzymes in the homogenates of the pancreas rapidly degraded the small amounts of insulin; in homogenates of the atrophied pancreas this was much less of a problem. Also Banting and Best used acidic ethanol to extract the insulin, and this inactivated any residual proteolytic enzymes.

Each islet in the mammalian pancreas contains several cell types that differ in morphology, staining properties, and functions (Orci, 1976; Munger, 1981; Orci, 1982). The use of specific antibodies and immunofluorescent labeling techniques has identified four major cell types and the hormones each secretes:

1. A cells, secreting glucagon (also designated α or α_2 cells)
2. B cells, secreting insulin (β cells)
3. D cells, secreting somatostatin (α_1 cells)
4. F cells, secreting pancreatic polypeptide (PP cells)

The "missing" C and E cells have been tentatively identified on morphological grounds in some animal species but the limited data available do not establish them as unique cell types.

The four major cell types are not randomly distributed within the islet. The B cells, the predominant cell type, make up the large central mass of the islet, as diagrammatically illustrated in Figure 7.16. The A cells, the second most common cell type (about 20% of the total), form a rim around this central mass in most of the islets. However, in the inferior portion of the head of the pancreas islets contain very few A cells and many more F cells, i.e., the islets are morphologically and possibly functionally different in this portion of the pancreas. D cells are also found almost exclusively near the outer rim, in close proximity to both A cells and the inner core of B cells. As we shall see, the hormones secreted by these different cell types can affect the rate of hormone secretion by their neighbors. Thus *local* release of somatostatin from D cells, for example, could inhibit secretion of glucagon and insulin by A and B cells, respectively. The physiological importance of this *paracrine* regulation is under current investigation. In addition, the different islet cell types have been shown to communicate directly through "gap junctions," providing the potential for a second, more direct form of mutual control.

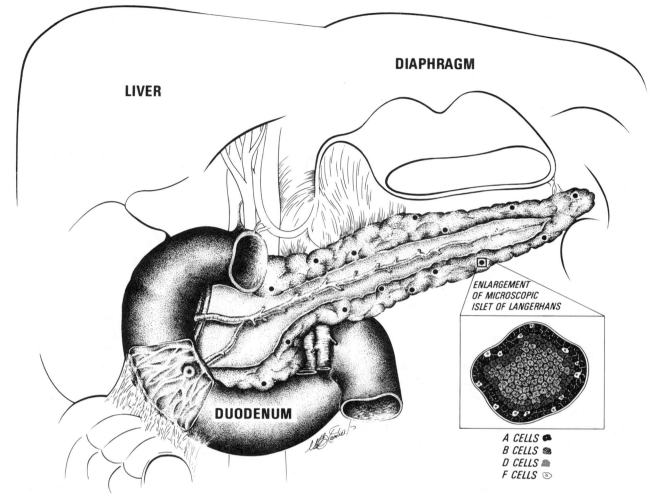

Figure 7.16.   Schematic representation of the anatomy of the pancreas. *Inset,*
Arrangement of the various cell types in a typical islet of Langerhans.

INSULIN

Synthesis and Secretion

The biosynthesis and secretion of insulin follow the general pattern that applies to polypeptide hormones generally (see Chapter 52). The mature messenger RNA moves from the nucleus to take up its position on ribosomes dotting the external or cytoplasmic surfaces of the rough endoplasmic reticulum. There it is translated to yield a single, long polypeptide chain, designated *preproinsulin*. The N-terminal sequence of preproinsulin (signal peptide) is cleaved as the nascent protein is secreted into the cisternae of the endoplasmic reticulum and the cleaved product, proinsulin, moves along the cisternae down to the Golgi apparatus. There it is packaged into secretory vesicles that bud from the Golgi membrane. The secretory vesicles of the B cells contain mainly insulin but also a small percentage of proinsulin. The vesicle contains a high

concentration of zinc, which is known to form tight complexes with insulin, but the physiological significance of zinc is not yet established. The vesicles are moved to the inner aspect of the plasma membrane by a process that involves the microtubular system, and they then fuse with the membrane, discharging their content into the extracellular space (exocytosis or emiocytosis).

The proinsulin molecule (Fig. 7.17) is coiled and cross-linked internally by two specific disulfide bonds. In the Golgi apparatus, and continuing after packaging into secretory vesicles, proinsulin undergoes proteolytic cleavage by enzymes with a high degree of specificity. They attack arginine and lysine residues that are paired at either end of a 31-residue "C-peptide" in human proinsulin; two basic amino acids at either end are released in free amino acid form. Thus, a total of 35 residues, constituting the "connecting peptide," are excised from the middle of the proinsulin molecule (see Fig. 7.17). The

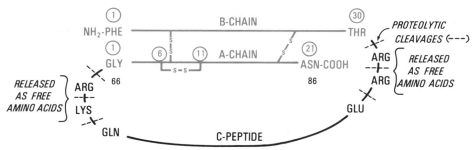

Figure 7.17. General structure of proinsulin and of the mature insulin derived from it *(black)*. The B-chain of insulin represents the first 30 residues of the proinsulin molecule and the A-chain the last 21. The intervening 35 amino acid-residue peptide, the connecting peptide, is split out posttranslationally. Additional peptide cleavages take place with the loss of four free amino acids (three arginines and one lysine) leaving the so-called C-peptide, 31 residues long. (See text for further discussion.)

disulfide bridges are not affected. This generates the mature hormone, which consists of two polypeptide chains linked only by the disulfide bridges. There is, in addition, an intrachain disulfide bridge connecting residues six and eleven in the A chain. The C-peptide accumulates with insulin in the secretory granules and is released into the plasma with the insulin in approximately equimolar amounts (Horwitz et al., 1976). Normally the conversion of proinsulin to insulin is almost completed prior to emiocytosis, but proinsulin may account for as much as 15% of the secreted product.

The biosynthesis of insulin as a single polypeptide chain is essential in order that the mature hormone have the appropriately placed disulfide bridges. If the disulfide bridges in insulin are chemically reduced and then reoxidized in vitro, random coupling of half-cystine residues takes place, and only a small percentage of the biological activity is restored. Reduction and reoxidation of proinsulin, on the other hand, restores almost all of the biological activity. The specific configuration of the proinsulin molecule, dictated by its primary amino acid sequence, presumably brings into close apposition the correct pairs of half-cystine residues that are to be linked through disulfide bridges in the mature insulin molecule.

Human insulin has been produced in bacteria using recombinant DNA techniques (Crea et al., 1978; Goeddel et al., 1979). The final product did not differ in chemical or biological properties from pure pancreatic human insulin (Chance et al., 1981).

The genomic DNA sequences coding for preproinsulin in man and in the rat have been completely determined (Bell et al., 1979; Cordell et al., 1979). Complementary DNA has been prepared from messenger RNA (from a human insulinoma and from isolated rat islets), cloned in bacteria, and the base sequences of the cloned cDNA determined. The amino acid sequences predicted for proinsulin agreed precisely with those previously established for the purified prohormone. The 72-base sequence preceding the sequence coding for proinsulin established for the first time the full amino acid sequence of the "signal" peptide or *pre* region of preproinsulin. Finally, the actual genomic DNA for rat preproinsulin has been isolated in amounts sufficient for sequencing. Two slightly different forms of rat insulin (differing at just two positions on the B chain) have been identified, and two nonallelic genes have been identified. (Lomedico et al., 1979). The genes for preproinsulin I and preproinsulin II both contain introns, i.e., sequences transcribed into primary messenger RNA but then spliced out before mRNA translation and therefore not represented in the protein product. Both genes contain an intron of 119 residues upstream from the *pre* region; only the preproinsulin II gene contains an additional long intron (499 residues) within the connecting peptide region. The base sequences of the two genes upstream from the 5′-capping site, where mRNA transcription starts, show striking homology. These areas presumably participate in regulation of gene expression, and thus synthesis of the two forms of rat insulin may be under equivalent control.

A number of genetic variants of human insulin have been described, and more undoubtedly will be discovered. In one case the linkage between the C-peptide and the B chain resisted cleavage, leading to hyperproinsulinemia but with normal carbohydrate metabolism, presumably because of the biological activity contributed by high proinsulin levels (Gabbay et al., 1976). In the other example, the patient produced both normal insulin and insulin in which leucine was substituted for phenylalanine at residue 24 or 25 in the B chain (Olefsky et al., 1980). The biological activity of the mutant insulin was very low and, in addition, it competed with the normal insulin for binding to receptors. Thus the patient was diabetic and "hyperinsulinemic" (since the abnormal insulin was recognized by standard radioimmunoassay). For reviews of insulin synthesis and secretion see Steiner, 1977; Permutt, 1981; and Tager et al., 1981.

Insulin Assay

Hormone concentrations are most accurately and most sensitively determined by radioimmunoassay (RIA) as first developed for insulin assay by Berson and Yalow (1959). An unknown sample of unlabeled insulin is allowed to compete with a known amount of radioactive insulin for binding to an antibody against insulin. The degree of displacement of labeled insulin from the antibody by the unlabeled unknown is measured by separating bound from free insulin and assaying the radioactivity in one or both fractions. Several variant methods are in use, including enzyme-linked assays, but the principle is the same.

Since the chemical structure of insulin is fully established, concentrations can be expressed in nanomolar terms or in nanograms per milliliter. By convention, however, concentrations are still most commonly stated in terms of in vivo biological activity. This was defined many years ago for the U.S. Pharmacopoeia (U.S.P.) in terms of the potency in lowering the blood glucose level of rabbits! The U.S.P. standard has a potency of about 22 U/mg, but it is not 100% pure. Biological activity has also been assayed in vitro by measuring stimulation of glucose uptake by isolated rat diaphragm or rat adipose tissue (epididymal fat pad). The fat pad bioassay for serum insulin yields a considerably higher value than RIA because of the presence in plasma of insulin-like hormones. Antibodies against insulin itself suppress the biological activity of serum insulin but not that of the insulin-like hormones, which were thus designated *nonsuppressible insulin-like activity* (NSILA). It is now known that some or all of NSILA can be attributed to somatomedins (insulin-like growth factors; IGF) which are produced under growth hormone regulation and are recognized by insulin receptors, although with low affinity (see Chapter 53).

The RIA for insulin is accurate and specific. However, because the structure of insulin in mammals is highly conserved, antibodies prepared against one mammalian insulin cross-react with insulin from other species. Proinsulin also cross-reacts—understandably since its structure embraces both the chains of the mature insulin molecule—but with lower affinity than insulin itself. In standard assays, about 25–50% of proinsulin will be detected. C-peptide, on the other hand, does not cross-react with antiinsulin antibodies. C-peptide levels can be assayed by RIA using specific anti-C-peptide antibodies. This can be helpful clinically under some circumstances. For example, it is useful in assessing insulin secretion in patients whose serum contains high levels of antibodies against insulin, which precludes assay by conventional RIA. Since C-peptide and insulin are secreted in approximately equimolar amounts, C-peptide levels reflect rates of insulin secretion (although C-peptide levels are somewhat higher because it is more slowly removed from the plasma) (Horwitz et al., 1976). RIA for C-peptide can also distinguish hyperinsulinemia due to endogenous secretion (from an insulin-secreting tumor, for example) from that due to exogenous insulin administration (from surreptitious dosage). In the former, C-peptide levels will be elevated; in the latter they will be low.

Factors Regulating Insulin Release

These are shown in Table 7.5. The routing of ingested nutrients and of metabolic intermediates derived from them is generally stated to be under the control of the endocrine system. For example, insulin facilitates the disposition of ingested glucose by stimulating its uptake into liver, muscle, and adipose tissue. However, it would be equally true to say, reciprocally, that the endocrine system is under the control of foodstuffs and metabolic intermediates. Thus, ingestion of glucose stimulates the output of insulin and suppresses the output of glucagon. This interlocking system works well because during evolution there has been selection for an optimal hormone response to metabolites and an optimal reciprocal response of metabolites to hormones. Insulin is intimately involved in the regulation not only of glucose metabolism but also of

Table 7.5.
Factors Influencing Insulin Release

Stimulation	Inhibition
Physiological	
Glucose	
Amino acids	
Gastrointestinal peptide hormones (esp. GIP)	
Ketone bodies (esp. in starvation)	
Glucagon	Somatostatin
Parasympathetic stimulation	Sympathetic stimulation (splanchnic nerve)
β-Adrenergic stimulation	α-Adrenergic stimulation
Pharmacological and experimental	
Cyclic AMP	α-Deoxyglucose
Theophylline	Mannoheptulose
Sulfonylureas	Diazoxide
Salicylates	Prostaglandins
	Diphenylhydantoin
	β cell poisons: Alloxan, streptozotocin

protein and fat metabolism. One might predict, then, that any and all of the major foodstuffs might play some role in regulating insulin release, and that is indeed the case.

After each meal the rate of insulin secretion increases, and the plasma insulin levels rise. During the overnight fasting period insulin levels tend to drift downward and are usually below 10–20 μU/ml by morning; after a meal they may reach peak values as high as 100 μU/ml. These levels, measured in peripheral plasma, are much lower than those in the portal circulation for two reasons. First, the newly secreted insulin enters the portal circulation initially, and only later is it diluted into the larger volume of the systemic circulation. Second, the liver is highly efficient in the uptake of insulin, taking up in one circulation about 50% of what is delivered to it. Thus the liver is normally exposed to insulin concentrations in portal vein blood 3 to 10 fold higher than those to which other tissues are exposed. When the diabetic patient is treated with subcutaneous injections of insulin, however, there is no such differential exposure. Consequently, the response to exogenous insulin may be different from that to endogenously secreted insulin.

The normal basal rate of insulin secretion in man is estimated to be 0.5–1.0 U/hour. Because of the postprandial bursts of secretion, the total daily secretion may be as much as 40 U/day. During fasting, less is secreted, and plasma levels drift down progressively. For a review of factors regulating insulin release see Gerich et al., 1976.

GLUCOSE

Without question the plasma level of glucose is the most important determinant of the rate of insulin release. Both synthesis and secretion are stimulated when plasma glucose levels rise and inhibited when they fall. Glucose has a direct effect that is readily demonstrated using the perfused pancreas or isolated islet tissue. In addition, glucose exerts indirect stimulation when given orally, by stimulating the release of peptide hormones from the gastrointestinal tract (see below).

Two major mechanisms have been proposed to explain the direct stimulatory effect of glucose on insulin secretion. The *glucose receptor theory* proposes that the β cells of the islet express specific receptors that recognize glucose and respond to it by increasing insulin synthesis and secretion, perhaps by generating a "second messenger." The second hypothesis holds that the metabolism of glucose in the β cell is necessary and that one or more of the metabolites of glucose represent the direct stimulus to an increased synthesis and secretion of

insulin. Evidence for both hypotheses is available, and it may be that both are operative.

The response of the islet cell to glucose stimulation occurs in two phases. With very little lag time there is an immediate response, and the rate of insulin release reaches a peak within a minute or two. This first phase is transient, and the rate of release drops sharply back toward normal levels over the next 5 min or so. Then there follows a second phase, the rate of release increasing again over the following hour or so. This second phase can be blocked by inhibitors of protein synthesis and presumably reflects a late stimulation of new insulin formation. The first phase represents primarily the release of preformed insulin present in the secretory granules.

AMINO ACIDS

Plasma insulin levels rise after a meal consisting exclusively of protein. This response is partly attributable to a direct effect of higher plasma amino acid levels on the β cell. Insulin release is increased by intravenously administered amino acids, and isolated islet tissue responds directly. In addition, a protein meal stimulates insulin release indirectly by causing secretion of hormones derived from the gastrointestinal tract, as does glucose. The potency of the different amino acids in stimulating insulin secretion varies. The most potent are arginine, leucine, and lysine. Intravenous doses of these amino acids have been used to evaluate the ability of the pancreas to secrete insulin. Other amino acids, such as valine and histidine, for example, are much less potent.

GASTROINTESTINAL PEPTIDE HORMONES

The insulin response to orally administered glucose is considerably greater than the response to the same amount of glucose given intravenously at a rate to match the plasma glucose levels reached after ingestion of glucose. The same disparity holds for amino acids given orally vs. those given intravenously. The greater effectiveness of the oral dose is due to the release from the gastrointestinal tract of hormones that reinforce the stimulus of the metabolite and boost the insulin response of the pancreas. Many different gastrointestinal hormones have been shown to have the *potential* of stimulating insulin release, but it is not yet clear which of these are important physiologically. Secretin, gastrin, and pancreozymin are all effective, but the amounts released in response to meals do not appear to be sufficient to account for the incremental insulin response to oral glucose. Gastric inhibitory polypeptide (GIP), a 43-amino acid residue peptide, is a very potent stimulus, and it is effective at plasma concentrations comparable to those reached under physiological conditions (Dupre et al., 1973; Ebert and Creutzfeldt, 1987). Its release is stimulated not only by glucose but also by fats and amino acids. Thus, the ingestion of a meal sends an "anticipatory" signal to the pancreas to increase insulin release even before substrate levels have risen very much, and this amplifies the response.

KETONE BODIES AND FATTY ACIDS

Ketone bodies (acetoacetate and β-hydroxybutyrate) and free fatty acids administered intravenously enhance insulin release in experimental animals, but evidence for their effectiveness in man is limited. In starvation, when free fatty acid (FFA) and ketone body levels are elevated, insulin levels are low, indicating that these stimuli are probably of secondary importance.

GLUCAGON AND SOMATOSTATIN

Glucagon stimulates and somatostatin inhibits insulin release. These effects are readily demonstrated by administering the hormones intravenously. Thus systemic levels of these hormones may be relevant to regulation of insulin release. However, regulation within the islet itself is probably more significant (*paracrine* control). As summarized in Figure 7.18, the three major hormones produced by islet cells affect each other's synthesis and secretion in a complex, interactive manner. Local concentrations of the hormones within the islet are probably very high and could therefore exercise control locally more effectively than they do after release into the general circulation with the consequent dilution. In addition, some islet cells have been shown to have *gap junctions*, i.e., open connections that allow transfer of materials from cell to cell without escape into the extracellular milieu. How much of the interaction that occurs relates to such cell-cell connections is not known. Somatostatin inhibits the release of both insulin and glucagon. Advantage has been taken of this somatostatin suppression to study independently the

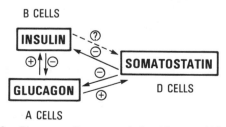

Figure 7.18. Diagrammatic representation of currently known regulatory interactions that may be important in paracrine regulation of islet hormone production and release.

metabolic effects of insulin and of glucagon. Constant intravenous infusion of somatostatin reduces release of both insulin and glucagon to very low levels. Then the infusion of insulin alone or of glucagon alone allows the investigator to study the effects of one of these hormones without the complicating, counteracting effects of the other. The extent to which somatostatin operates under physiological conditions as a regulator remains uncertain.

Insulin suppresses glucagon release. Consequently, an increase in insulin release will tend to be accompanied by a decrease in glucagon release. The increase in insulin release in response to glucose may be an important factor in decreasing glucagon levels. Conversely, when insulin levels fall, as during fasting, glucagon levels will tend to increase.

Glucagon stimulates the release both of somatostatin and of insulin. The former will tend to *decrease* insulin release, and it becomes apparent that there is no simple way to predict quantitatively the responses to a given stimulus. The interactions are obviously complex, and much remains to be learned about the detailed kinetic interactions and their relative importance in determining patterns of hormonal and metabolic responses. For reviews of the roles of glucagon and somatostatin see Unger and Orci, 1981; and Arimura, 1981.

AUTONOMIC NERVOUS SYSTEM REGULATION

Parasympathetic stimulation via the vagus nerve increases insulin release, whereas sympathetic stimulation via the splanchnic nerve inhibits insulin release. Administration of epinephrine or norepinephrine systemically inhibits insulin release. This effect is exercised via interaction with α-receptors. Agonists that affect primarily the β-adrenergic receptors actually stimulate release of insulin, but the α effect of the naturally occurring catecholamines dominates under ordinary circumstances. The decrease in insulin release associated with stress (such as severe infections, exercise, or hypothermia) may be attributable in part to an increase in sympathetic discharge. Administration of phentolamine, an α-adrenergic blocking agent, increases insulin release, suggesting that a tonic suppressive effect of the sympathetic nervous system is constantly operative.

The central nervous system also plays a role in regulating insulin secretion, and the outflow probably occurs via the hypothalamus and the autonomic nervous system (Louis-Sylvestre, 1976; Powley, 1977; Berthaud et al., 1981). Stimulation of the ventromedial nuclei of the hypothalamus suppresses insulin release, whereas destruction of these nuclei causes hyperinsulinism. While difficult to document and quantify, it is likely that the emotional state of the patient may importantly alter insulin release by way of the hypothalamus and its autonomic outflow tract.

The mechanism of action of the catecholamines on β cell secretion is not fully established. Cyclic AMP will stimulate insulin release, and the effects of β-adrenergic stimuli may be exercised, in part, by increases in cyclic AMP. There is also evidence that changes in cytosolic concentrations of calcium ion may play a role and that calmodulin, a specific calcium-binding protein, may be involved. The exact relationships among cyclic AMP, calcium ion, and calmodulin remain to be worked out.

PHARMACOLOGICAL AND EXPERIMENTAL FACTORS

Theophylline stimulates insulin release, presumably through its ability to inhibit phosphodiesterase and increase cyclic AMP concentrations. The sulfonylureas, orally effective hypoglycemic drugs, are effective primarily because of their ability to stimulate insulin secretion directly. Finally, salicylates are effective hypoglycemic agents. These drugs inhibit the conversion of arachidonic acid to prostaglandins. The prostaglandins have been demonstrated to inhibit insulin release.

Insulin release is inhibited by mannoheptulose and 2-deoxyglucose, substances that interfere with the utilization of glucose by the islet cells. Diazoxide is a potent inhibitor of insulin release, sufficiently potent to be sometimes helpful in managing patients with insulin-secreting tumors prior to resection or in patients with metastatic insulin-secreting tumors. Diphenylhydantoin, a drug in widespread use for control of epileptic seizures, inhibits insulin release and significantly increases the insulin requirement of diabetic patients. Finally, certain cytotoxic agents with a high degree of specificity for the β cell can be used to produce experimental diabetes in animals (alloxan and streptozotocin). The latter compound has been used to control intractable hypoglycemia in patients with insulin-secreting tumors.

Biological Effects of Insulin

Because of its dramatic effects on glucose metabolism and because these were historically the first to be described, there has been a tendency to categorize insulin as a hormone regulating carbohydrate metabolism. Actually insulin exercises important controls over the metabolism of all the major foodstuffs—carbohydrates, fats, and proteins. These effects are reasonably well summarized by saying that insulin favors anabolism and storage. It favors the synthesis and deposition of glycogen in the liver, the synthesis of fatty acids in liver and adipose tissue and their deposition and retention in adipose tissue as stored triglycerides, and the uptake of amino acids and their incorporation into proteins in muscle and other tissues. When insulin levels are high, as in the hours following the ingestion of a meal, nutrients are thus directed to appropriate storage sites; when insulin levels are low, as in the fasting state, all of these processes are reversed and metabolic substrates are mobilized in the form of glucose, FFA, and amino acids. In addition insulin stimulates synthesis of RNA and DNA. These effects are somehow selective although the exact mechanisms that are operative remain unknown. Thus some of the same enzymes that are acutely regulated (e.g., by changes in the state of phosphorylation of the enzyme protein) are also regulated in a sustained, long-term fashion by increasing (or decreasing) the *amount* of enzyme protein available.

In Table 7.6 and in the following sections, the metabolic effects of insulin are dealt with individually. While useful as a first approach, this is quite artificial. In vivo all of these effects are highly interactive—with one another and with the simultaneous influence of many other metabolic, hormonal, and neural factors. For example, a number of so-called counterregulatory hormones (including the catecholamines, glucagon, glucocorticoids, and growth hormone) modulate the effects of insulin. Furthermore, the net impact of insulin will depend on the number of receptors available, on the levels

Table 7.6.
Biological Effects of Insulin

A. On carbohydrate metabolism
  1. Reduces rate of release of glucose from liver
    a. by inhibiting glycogenolysis
    b. by stimulating glycogen synthesis
    c. by stimulating glucose uptake
    d. by stimulating glycolysis
    e. by indirectly inhibiting gluconeogenesis via inhibition of fatty acid mobilization from adipose tissue
  2. Increases rate of uptake of glucose into all insulin-sensitive tissues, notably muscle and adipose tissue
    a. directly, by stimulating glucose transport across the plasma membrane
    b. indirectly, by reducing plasma-free fatty acid levels
B. On lipid metabolism
  1. Reduces rate of release of free fatty acids from adipose tissue
  2. Stimulates de novo fatty acid synthesis and also conversion of fatty acids to triglycerides in liver
C. On protein metabolism
  1. Stimulates transport of free amino acids across the plasma membrane in liver and muscle
  2. Stimulates protein biosynthesis and reduces release of amino acid from muscle
D. On ion transport
E. On growth and development

of substrate available, and on a number of substrate-determined autoregulatory controls.

INSULIN-RECEPTOR INTERACTION

The metabolic effects of insulin are initiated by the interaction of the insulin molecule with a highly specific receptor on the plasma membrane (Stadie et al., 1953; Roth et al., 1975; Kahn et al., 1981; Czech, 1985; Kahn, 1985). This binding is saturable, of very high affinity, and highly specific. Most if not all of the biological effects are probably initiated by insulin-receptor interaction independent of subsequent internalization of the receptor-bound insulin, although it remains possible that hormone action continues (or may even be expressed differently) after the hormone-receptor complex is internalized. Evidence favoring this view is the finding that antibodies against the insulin receptor mimic the biological effects of insulin itself. Monovalent F_{ab} fragments are ineffective, suggesting that clustering of the receptors is necessary to elicit biological activity.

By use of affinity chromatography and by photoaffinity labeling, the insulin receptor has been purified and fully characterized (reviewed by Czech, 1985; and Kahn, 1985) and the cDNA for the receptor has been cloned (Ullrich et al., 1985; Ebina et al., 1985). It is a membrane protein of high molecular weight made up of four disulphide-linked subunits: two identical α chains (apparent M_r about 135,000) each capable of binding one mole of insulin, and two β chains (apparent M_r about 95,000). The fact that each receptor molecule can bind two moles of insulin per mole may relate to the unusual "negative cooperativity" observed for insulin effects as a function of concentration (i.e., the binding of the first mole of insulin to the receptor makes it more difficult to bind the next).

The binding of insulin induces autophosphorylation of the receptor and it is a tyrosine residue that becomes phosphorylated (Kasuga et al., 1982; Roth and Cassell, 1983). Regulation of enzyme function by phosphorylation of enzyme proteins is common, but it is generally a serine residue that is phosphorylated. The first example of tyrosine phosphorylation was discovered in relation to the oncogenic viruses that insert the *src* gene into the host's genome (see review by Houslay, 1981). The gene product is a membrane protein that phosphorylates tyrosine residues. Furthermore, epidermal growth factor (EGF) also causes phosphorylation of tyrosine residues and itself undergoes tyrosine autophosphorylation. The exact mechanisms linking tyrosine phosphorylation to growth control remain to be worked out, but these findings strongly suggest that it must be of general importance. Effects of insulin on growth are reviewed in Chapter 57 and the general issue of growth-regulating genes by Hunter (1984).

Precisely how insulin-receptor interaction leads to the wide spectrum of biological effects produced by insulin is not known. The stimulation of glucose transport, amino acid transport, and ion transport are easily visualized as more or less immediate consequences of changes in membrane configuration induced by interactions with the membrane receptor. If the effects on intracellular enzymes are to be accounted for also on the basis of interaction with the membrane-bound receptor, however, it becomes necessary to postulate the generation of a second messenger that carries information into the cell or phosphorylation of cytoplasmic proteins by the tyrosine kinase activity of the β subunit. For example, an immediate effect of insulin is the activation of pyruvate dehydrogenase, a mitochondrial enzyme. Pyruvate dehydrogenase is an interconvertible enzyme, i.e., it can be covalently modified to affect its activity. Phosphorylation of the enzyme protein deactivates, and removal of the phosphate activates. How, then, can insulin-receptor interaction bring about the dephosphorylation of this enzyme? One possibility is that insulin interaction with the receptor generates a product that can be transported to the mitochondria and act at that site. The effects of insulin on glycogen synthase and phosphorylase in the liver might be similarly explained. Suggestive evidence for generation of low molecular weight second messengers of this type has been presented but not yet confirmed (Larner et al., 1979; Seals and Jarett, 1980).

The functional importance of the insulin receptor in vivo has been established beyond doubt in a number of ways. A number of diabetic patients with marked resistance to the action of insulin have been shown to be deficient in the number of insulin receptors; others have shown a normal complement of receptors but fail to respond biologically because of some defect in the tyrosine kinase activity of the receptor; still others have had a normal number of receptors but fail to respond because of circulating autoantibodies against their own insulin receptors (Kahn et al., 1976; Kahn, 1985; Taylor, 1987).

Like many other receptors, the insulin receptor is subject to downregulation. If for any reason insulin levels in the plasma remain elevated for an extended period of time (hours), the number of insulin receptors expressed at the cell membrane decreases. A given dose of insulin in the downregulated state will have a smaller effect than it has in the control or "upregulated" state (Roth et al., 1975). This concept that response depends not only on the dose of hormone administered but also on the number of receptors currently expressed is, of course, of general applicability in endocrinology. Changes in receptor number on circulating blood cells (lymphocytes or red blood cells) appear to reflect changes in receptor number occurring generally.

EFFECTS ON CARBOHYDRATE METABOLISM

Insulin reduces plasma glucose levels both by stimulating uptake of glucose into tissues and by inhibiting the production and release of glucose from the liver. Glucose uptake by muscle and adipose tissue is stimulated directly, by enhancing the carrier-mediated transport (Morgan et al., 1964; Park et al., 1968), and indirectly, by inhibiting FFA release and reducing plasma FFA levels. Plasma FFA tends to inhibit glucose uptake (Randle effect), and thus a decrease in FFA levels favors glucose uptake.

Production and release of glucose by the liver is reduced in a concerted manner involving several points of attack. Insulin stimulates glycogen synthesis by enhancing the activity of glycogen synthase and simultaneously inhibiting glycogen breakdown by decreasing the activity of glycogen phosphorylase (see Fig. 7.9 and accompanying discussion in Chapter 48). Insulin simultaneously promotes glycolysis and inhibits gluconeogenesis. Directly or indirectly, insulin influences each of the four key control points at which forward and reverse reactions are catalyzed by different enzymes (Fig. 7.8). Conversion of glucose to glucose-6-phosphate (G-6-P), fructose-6-phosphate (F-6-P) to fructose diphosphate (FDP) and phosphoenolpyruvate (PEP) to pyruvate (PYR) are all favored. In addition, insulin strongly activates pyruvate dehydrogenase, shunting PYR on to acetyl CoA (AcCoA) which can be converted to fatty acids and stored or enter the Krebs cycle and be oxidized.

A centrally important indirect mechanism by which insulin influences hepatic gluconeogenesis is through its effects on FFA mobilization. When insulin levels are low and FFA are being taken up avidly by the liver, gluconeogenesis is stimulated by the AcCoA derived from fatty acid oxidation. As shown in Figure 7.19, AcCoA tends to shunt pyruvate back up the gluconeogenic pathway by simultaneously exerting a feedback inhibition of pyruvate dehydrogenase and powerfully stimulating pyruvate carboxylase. At the same time utilization of the AcCoA for resynthesis into fatty acyl CoA is inhibited because of feedback inhibition. Additionally, flow of substrate is directed up to glucose because FDPase is enhanced and phosphofructokinase (PFK) is inhibited. Thus, pyruvate and any substrates that contribute to the pyruvate pool are directed up to G-6-P. Under these conditions (starvation or diabetic ketoacidosis), alanine and other gluconeogenic amino acids released from muscle become key sources of glucose.

EFFECTS ON LIPID METABOLISM

Insulin is remarkably potent in suppressing the release of FFA from adipose tissue. In fact, this effect occurs at insulin concentrations below those

Figure 7.19. Key control points at which rapid metabolism of FFA in the liver leads to stimulation of gluconeogenesis (see text for discussion).

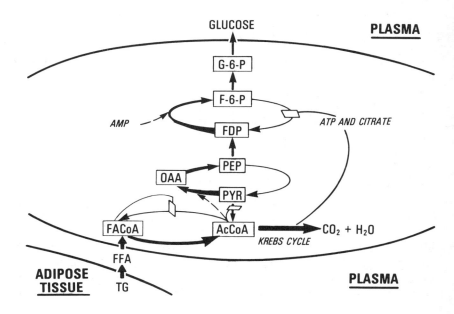

needed to stimulate glucose uptake in most tissues. Furthermore, the effect on FFA mobilization is immediate—even faster than the effect on plasma glucose levels. Under normal conditions insulin is constantly exerting a "braking" effect on FFA mobilization. Thus, when antiinsulin antiserum is injected intravenously into an animal the plasma FFA levels begin to rise immediately. Since there do not appear to be any important controls over the rate of utilization of FFA (other than FFA levels per se), this insulin effect is centrally important. Insulin is the key hormone monitoring the availability of glucose—suppressing FFA release when glucose is available and permitting rapid mobilization when glucose levels fall and lipids are needed to provide metabolic fuel. Insulin acts at the level of the fatty acid-mobilizing enzyme, hormone-sensitive lipase. In vivo this enzyme is under tonic control by catecholamines such that it is always at least partly in the active (phosphorylated) state. Insulin counteracts these effects, inhibiting conversion of lipase *b* to lipase *a*. Under some conditions this insulin effect is exercised at the level of cyclic AMP, but other mechanisms must also be involved, perhaps through control of lipase phosphatase activity.

Insulin stimulates fatty acid biosynthesis and also the incorporation of fatty acids into triglycerides in the liver and in adipose tissue. In addition, by increasing lipoprotein lipase activity it favors the uptake and deposition of triglycerides from very low density lipoproteins (VLDL) and chylomicrons into adipose tissue. Thus insulin channels substrate into lipid stores in the fed state; in the fasted state, when insulin levels are low, fat mobilization is favored.

EFFECTS ON PROTEIN METABOLISM

Insulin has a direct stimulatory effect on the transport of amino acids across the plasma membrane, an effect that is primary, not secondary to its effect on glucose transport. In addition, it directly stimulates protein biosynthesis by other mechanisms. In part, this effect is at the translational level, involving initiation or the state of aggregation of polysomes. In addition, insulin stimulates RNA and DNA synthesis, effects presumably related to its growth-promoting properties. The interrelationships between insulin, growth hormone, and the insulin-like growth factors (somatomedins) are discussed in Chapter 53.

EFFECTS ON ION TRANSPORT

One of the immediate effects of insulin is to cause hyperpolarization of the plasma membrane, an effect that must reflect changes in ion fluxes. In vivo it alters the distribution of sodium and potassium between extracellular and intracellular spaces. It favors movement of potassium into cells, and this is of great practical significance when treating patients with insulin deficiency. Vigorous treatment with insulin, without appropriate replacement of potassium, can cause serious or even fatal hypokalemia. There is some evidence that insulin acts directly on membrane-bound Na^+/K^+-ATPase (Hougen et al., 1978).

Insulin increases cytosolic calcium ion levels by effecting release of calcium from bound form within the cell and/or favoring entrance from the extracellular fluid. Since calcium ion, via calmodulin, exercises control over a number of the enzyme systems regulated by insulin, this mechanism may have more general significance.

GLUCAGON

Synthesis, Secretion, and Assay

Circulating glucagon is a simple polypeptide of 29 amino acid residues (Mr 3,485) (Bromer et al., 1957). It has no cross-linkages and exists in solution as a random coil. It is synthesized in the A cells as a prohormone of much higher molecular weight (about 9,000) which is then cleaved prior to secretion.

The general pattern for biosynthesis and secretion is like that for insulin. Preformed glucagon is stored in secretory granules that can disgorge their content into the surrounding medium rapidly under appropriate stimulation.

The earlier literature on glucagon is confusing because it was not at first recognized that the intestine secretes polypeptides that cross-react with many or most antisera against pancreatic glucagon but do not replace glucagon functionally. Highly specific antisera are now available that react only with true "pancreatic glucagon." In some species there are cells in the gastrointestinal tract that actually do secrete the same polypeptide secreted by the α cells, but in man the pancreas appears to be the sole source of true glucagon. None is found in the plasma after total pancreatectomy; glucagon-like immunoreactivity, however, persists, as it does in most mammalian species. The physiological significance of the latter is not clear.

Like insulin, glucagon enters the portal circulation and is first presented to the liver. However, the efficiency of hepatic extraction of glucagon is considerably less than that for insulin, and so the portal:systemic ratio is much less (about 1.5:1 instead of almost 10:1). Normal peripheral concentrations in the basal state are about 100–150 pg/ml; basal rates of secretion are about 100–150 μg/day.

Factors Regulating Glucagon Release

The effects of glucose on glucagon secretion are reciprocal to those on insulin secretion, i.e., hypoglycemia stimulates and hyperglycemia suppresses glucagon release. Hyperglycemia suppresses glucagon release mainly through its stimulation of insulin release (Table 7.7). Insulin is a potent inhibitor of glucagon release and acts within the islet (paracrine control; cf. Fig. 7.18). The high levels of glucagon in patients with insulin deficiency are mainly due to release of the A cells from the suppressive action of insulin. However, hypoglycemia appears to be able to stimulate glucagon release by acting directly on the A cells (Unger, 1985).

Glucagon secretion, like that of insulin, is stimulated by amino acids. The rise in glucagon level together with the rise in insulin stimulated by amino acids may act to prevent hypoglycemia when food intake is primarily protein. Again like insulin, glucagon is secreted in response to a number of gastrointestinal polypeptides, including pancreozymin.

A number of factors that *inhibit* release of insulin and/or antagonize its biological activities *stimulate* glucagon release. These include catecholamines, growth hormone, and glucocorticoids. Glucagon levels tend to rise during exercise, whereas insulin levels fall; both may be due to increased catecholamine release secondary to increased sympathetic nervous system activity during exercise.

The half-life of plasma glucagon is about 10 min—2 to 3 times as long as that of insulin. Still the duration of action is quite short and acute responses can be aborted quickly.

Biological Effects of Glucagon

In most respects the biological effects of glucagon are opposed to those of insulin. Thus, glucagon tends to raise rather than lower plasma glucose levels, mostly by its potent stimulation of hepatic glycogenolysis. This effect is mediated via stimulation of adenylate cyclase and activation of cAMP-dependent protein kinase. The latter simultaneously activates phosphorylase (via phosphorylase kinase) and inactivates glycogen synthase.

Table 7.7.
Factors Influencing Glucagon Release

Stimulation	Inhibition
Amino acids	Glucose
Gastrointestinal polypeptide hormones	Insulin
Catecholamines (exercise)	Free fatty acids (FFA)
Growth hormone	
Glucocorticoids	

Glucagon increases hepatic gluconeogenesis, again opposing the effect of insulin. The glucagon effect is exerted, in part, at the level of pyruvate kinase, which it deactivates. In addition, it stimulates the conversion of PYR to PEP. Finally, it may increase the uptake of amino acids into the liver to provide substrate for gluconeogenesis.

Glucagon has the potential of enhancing lipolysis in adipose tissue (whereas insulin suppresses it). While the glucagon effects are readily demonstrated with isolated adipose tissue, the response of plasma FFA levels to glucagon administered in vivo is usually minimal or even, paradoxically, a fall rather than a rise. This is readily explained on the basis of the sharp increase in glucose levels that accompanies administration of glucagon and the consequent rise in insulin levels. These override any tendency of glucagon to mobilize FFA. However, in diabetic subjects or in subjects that have fasted for some time, the glucagon effects on FFA may become important.

Other effects of glucagon have been described (including effects on the gastrointestinal tract, on calcium metabolism, on ion transport, and on myocardial function), but their physiological significance remains to be fully established.

An assessment of the relative importance of glucagon and insulin in respect to control of carbohydrate metabolism and, in particular, of their relative importance in diabetes mellitus is difficult to make. To some extent they may be regarded as paired hormones, and some argue that the ratio of insulin to glucagon is more relevant in evaluating the diabetic tendency than the concentration of either one alone. However, we know that pancreatectomy, which removes both, leads to gross deterioration of glucose homeostasis. This would seem to argue for the predominant importance of insulin. Patients with glucagon-secreting tumors do have glucose intolerance, but it tends to be moderate compared to that of patients with insulin deficiency. Furthermore, the diurnal fluctuations in glucagon level are small, and the glucagon response to a glucose meal is relatively small; small increases in plasma glucose may increase insulin levels without much effect on glucagon levels. On the other hand, as discussed above, when there is insulin deficiency (absolute or relative) the severity of the disturbance in glucose homeostasis may be importantly determined by glucagon levels.

SOMATOSTATIN

Somatostatin is a tetradecapeptide hormone first discovered and characterized as a potent hypothalamic inhibitor of the release of growth hormone from the pituitary (Brazeau et al., 1973). Studies of its biological effects when administered intravenously revealed, unexpectedly, that it also potently inhibited secretion of both insulin and glucagon (Koerker et al., 1974; Yen et al., 1974). It is now clear that somatostatin has a much broader biological significance (Arimura, 1981). It inhibits release of both growth hormone and thyroid-stimulating hormone (TSH) from the pituitary and under some circumstances also inhibits release of prolactin and ACTH. It has been shown to occur throughout the gastrointestinal tract and to act as an inhibitor of many gastrointestinal functions: the secretion of gastric acid and pepsin; the secretion of pancreatic digestive enzymes; intestinal motility and intestinal absorption (including the absorption of glucose); secretion of many gastrointestinal hormones (secretin, pancreozymin, vasoactive intestinal peptide, GIP). Somatostatin is widely distributed throughout the central and peripheral nervous system and probably plays a role in neurotransmission. In the present context we shall limit consideration to the role of somatostatin in the regulation of the function of the pancreatic islets where it is synthesized and secreted by the D cells.

Structure and Synthesis

Somatostatin is a 14-amino acid peptide with a carboxy-terminal cysteine residue in disulfide linkage to another cysteine residue at position 3. More recently, a 28-amino acid residue peptide has been purified from hypothalamus and intestine in which the carboxy-terminal 14 residues are identical to somatostatin. This larger peptide is much more potent than somatostatin (as much as 10 times more potent in inhibiting growth hormone and insulin release) and may represent the form active within cells. Other, still larger forms have been reported. Recently, the complete sequence of the preprohormone has been deduced by cloning the cDNA for human somatostatin (using mRNA from a human somatostatin-secreting tumor) (Shen et al., 1982). The coding region predicts a 116 residue precursor protein with the sequence of somatostatin-28 at the carboxy-terminus. A highly nonpolar region at the amino-terminus presumably represents a signal peptide; whether or not the prohormone has biological functions different from those of somatostatin remains to be determined. The cDNA for somatostatin has also been cloned from the angler fish (Hobart et al., 1980). These fish have discrete islets separate from the pancreas, and some of these are relatively enriched in D cells. Two cDNA forms have been identified. In one, the carboxy-terminal 14-amino acid somatostatin has the same sequence as the mammalian peptide, but the other differs at

two residues and there is some evidence that they may differ functionally. It is possible that local effects of "somatostatin" may relate to portions of the prohormone structure or other peptides than the 14-amino acid or even the 28-amino acid forms studied intensively in the past. It is also possible that more than one gene is involved.

Biological Effects (in Relation to Islet Function)

Somatostatin probably has no intrinsic direct effects on glucose metabolism analogous to the effects of insulin and glucagon. It does, however, inhibit the intestinal absorption of glucose, and this can result in apparent effects on glucose tolerance. Its major effects are probably mediated through its inhibition of insulin and glucagon secretion. Because local concentrations within the islet are so much higher than systemic concentrations it is likely that local, paracrine regulation is most important.

Intravenous infusion of somatostatin into man causes a fall in plasma levels of both insulin and glucagon. Glucose levels also fall, and this, in the face of lower insulin levels, provides evidence that glucagon does normally play a role maintaining plasma glucose levels. Continuous somatostatin infusion has been useful as a research tool for evaluating the roles of glucagon and insulin separately, maintaining the levels of one or the other of them by infusion.

That somatostatin exerts tonic control over insulin and glucagon secretion within the islet has been demonstrated using antisomatostatin antibodies. Added to isolated islets such antibodies can cause an increase in the rate of secretion of insulin and glucagon. It remains to be determined whether and how somatostatin can exert *selective* control. The anatomical relationships and/or intercellular connections via gap junctions could play a role. Since it acts to inhibit secretion of two antagonistic hormones, its impact on metabolism cannot be predicted until more is known about selective mechanisms.

Somatostatin secretion in vitro is regulated by many of the same factors that regulate insulin and glucagon secretion, but the patterns are difficult to interpret, as just mentioned. Glucose, amino acids, and gastrointestinal peptides stimulate somatostatin secretion. Glucagon also stimulates, but no effect of insulin has been established. Acetylcholine stimulates, and epinephrine inhibits. The latter effect is via α-adrenergic mechanisms. Pure β-adrenergic agonists stimulate release via cyclic AMP. The physiological relevance of these various potential regulatory factors remains to be established (Gerich, 1981).

PANCREATIC POLYPEPTIDE

Specialized cells in the islets, tentatively designated F cells, secrete a basic polypeptide of 36 amino acid residues—pancreatic polypeptide (PP). This peptide, first discovered as a minor contaminant of avian insulin, has now been found in many avian and mammalian species, including man. Complete sequences are available for human, bovine, ovine, and porcine PP, and they are very similar. Plasma levels of PP in man (60–100 pg/ml) are comparable to those of glucagon, and they respond to a number of metabolic stimuli, especially protein-containing meals. Responses to intravenous amino acids or glucose are much smaller and variable, suggesting the mediation of gastrointestinal polypeptide hormones. Despite a large and growing literature on the chemistry, secretion, and metabolic effects of PP, it is still uncertain just what its physiological or pathophysiological significance may be (Hazelwood, 1981) (see Chapter 44).

DIABETES MELLITUS

Perhaps the best way to bring together and to interrelate the multiple, complex effects of insulin and glucagon on carbohydrate, lipid, and protein metabolism is to consider the disturbances in the animal or patient with diabetes mellitus. Patients with Type I diabetes mellitus (insulin-dependent diabetes mellitus; IDDM) have an absolute insulin deficiency. Their basal insulin levels are very low, and they respond poorly or not at all to stimuli that normally increase plasma insulin levels. Animal models for the disease can be produced by selective destruction of the β cells using alloxan or streptozotocin, which are toxic to the β cells (but at higher dosages have some effect on other tissues as well). The metabolic derangements in diabetes mellitus are multiple, and many are secondary rather than immediate consequences of insulin deficiency (or glucagon excess). Yet almost without exception these disturbances can be traced back to the primary inability to secrete insulin. A similar but not identical array of metabolic disturbances can be encountered in another category of diabetes mellitus, Type II diabetes mellitus (noninsulin-dependent diabetes mellitus; NIDDM). Here insulin levels are normal or, commonly, elevated, but the responses of the tissues to insulin is for some reason below normal, i.e., there is a *relative* rather than an *absolute* insulin deficiency. There are undoubtedly many different mechanisms underlying an apparent resistance or insensitivity to insulin, some involving receptor-insulin interactions and some involving subsequent steps (postreceptor defects) (Olefsky and Kolterman, 1981). The following discussion of mechanisms is limited to the patient with Type I diabetes mellitus.

Major Signs and Symptoms in Patients with Absolute Insulin Deficiency

A typical history for children with diabetes mellitus often begins with *polyuria*. They urinate frequently and copiously, often having to get up several times during the night. They drink large quantities of water *(polydipsia)* and also increase food intake strikingly *(polyphagia)*. Despite this increase in food intake, they lose weight, a paradoxical situation seen in diabetes and in hyperthyroidism but rarely in other clinical situations.

If untreated the children become listless and drowsy. They may complain of nausea. Appetite now dwindles, and weight loss becomes more marked. Later there may be vomiting and severe abdominal pain. Eventually they slip into a coma, unresponsive to ordinary stimuli. Breathing is deep and regular (Kussmaul breathing), a characteristic finding in metabolic acidosis.

In the emergency room it is seen that they are severely dehydrated. Blood pressure is dangerously low. The urine is strongly positive for glucose *(glucosuria)* and acetoacetic acid *(ketonuria)*; plasma levels are also markedly elevated *(hyperglycemia* and *ketonemia)*. The plasma is milky in appearance, and the trigylceride level is enormously elevated due to accumulation of VLDL and chylomicrons *(hyperlipemia)*. Plasma pH may be as low as 7.0, and plasma bicarbonate is also very low *(metabolic acidosis)*. Plasma FFA levels are as much as 5 fold elevated. Plasma sodium and potassium are low *(hyponatremia* and *hypokalemia)*. Insulin levels, finally, are very low or unmeasurable.

In the following paragraphs, with the help of the flow diagrams in Figure 7.20*A–C*, we show how all of these disturbances can be accounted for as ultimate consequences of insulin deficiency. It should be pointed out, however, that when the deficiency has persisted for any length of time it may not be enough to treat with insulin alone. For example, it may be absolutely essential to restore as rapidly as possible the large volumes of fluid and the large amounts of electrolytes that have been lost over a period of days.

Consequences of Disturbed Carbohydrate Metabolism

Polyuria, polydipsia, and polyphagia (the three Ps of diabetes) are the consequences of sustained hyperglycemia with the accompanying loss of large quantities of glucose in the urine (because the renal

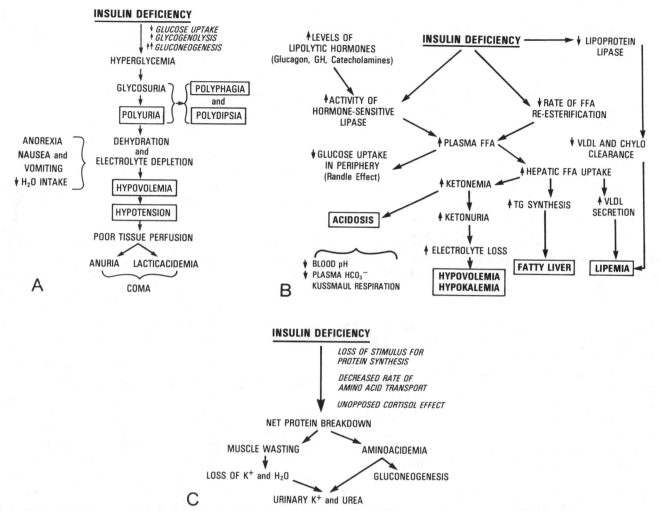

Figure 7.20. *A*, Flow diagram indicating the sequence of events by which impaired carbohydrate metabolism in diabetes mellitus is expressed pathophysiologically. *B*, Flow diagram indicating the sequence of events by which disturbances in lipid metabolism express themselves pathophysiologically in diabetes mellitus. *C*, Flow diagram indicating how disturbances in protein metabolism in diabetes mellitus express themselves pathophysiologically.

threshold is exceeded) (see Fig. 7.20*A*). Since not all of the glucose can be reabsorbed there is an accompanying obligatory loss of water into the collecting tubules (osmotic diuresis). This loss of water also entails some accompanying loss of electrolytes as well. The quantities of glucose lost can be enormous—hundreds of grams per day. To maintain energy balance the patient takes in larger quantities of food. When nausea and vomiting supervene there is negative caloric balance and weight loss. Not only adipose tissues but also muscle protein becomes depleted. In the absence of insulin, hormone-sensitive lipase is fully active and FFA release is maximal. Protein breakdown predominates over protein synthesis. The amino acids mobilized to the liver provide most of the substrate for gluconeogenesis. As indicated in Figure 7.20*A*, with the onset of nausea and vomiting dehydration is accentuated and additional electrolyte losses are incurred. Total blood volume falls, and eventually compensatory mechanisms fail to sustain blood pressure. In the extreme, tissue perfusion falls below critical values; there is tissue anoxia and production of excess lactic acid. Finally, renal function is shut down partially or completely.

Consequences of Disturbed Lipid Metabolism

The acidosis stems from an overproduction of ketone bodies (β-hydroxybutyric acid, acetoacetic acid, and acetone) (see Fig. 7.20*B*). These have their origin in the liver, to which FFA are being delivered at an enormous rate. As shown in Figure 7.21, acetyl coenzyme A derived from fatty acid oxidation is generated at a rate that exceeds the ability of the liver to oxidize it completely. This leads to synthesis of acetoacetyl coenzyme A and from it free ace-

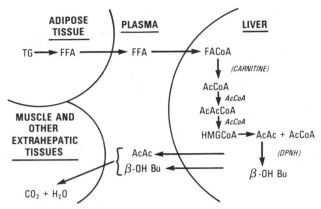

Figure 7.21. Diagrammatic representation of the pathways involved in generation of ketone bodies in the liver and their catabolism by extrahepatic tissues. *TG,* triglyceride; *FFA,* free fatty acid; *FACoA,* fatty acyl CoA; *AcCoA,* acetyl CoA; *AcAcCoA,* acetoacetyl CoA; *HMGCoA,* β-hydroxy-β-methylglutaryl CoA; *β-OH Bu,* β-hydroxybutyric acid; *AcAc,* acetoacetic acid.

toacetic acid. Glucagon enhances ketogenesis by increasing hepatic carnitine levels, making it possible for fatty acids to be transported more rapidly across the mitochondrial membrane barrier (as fatty acylcarnitine). Thus, the maximal rate of formation of acetoacetate is increased (McGarry and Foster, 1977). The latter cannot be oxidized by liver, because the enzymes needed for its activation are not present. It leaves the liver as acetoacetate and also in the reduced form, β-hydroxybutyrate. Muscle and other extrahepatic tissues *can* activate acetoacetate to its coenzyme A derivative, convert it to acetyl CoA by thiolytic cleavage, and oxidize it via the Krebs cycle. Some acetoacetate is nonenzymatically oxidized to acetone and CO_2; the acetone is excreted, in part, via the lungs and can sometimes be smelled on the patient's breath. It is primarily the accumulation of acetoacetic acid and β-hydroxybutyric acid that causes the blood pH to fall. With the onset of acidosis there is stimulation of the respiratory center, and the increased loss of CO_2 tends to prevent the pH from falling, but this compensatory mechanism eventually fails to keep up. Meanwhile the ketone bodies are being shed in the urine. In fact, the renal threshold for them is quite low (and so ketonuria of mild degree does not necessarily signal severe acidosis). The renal loss of these acids is a major basis for the accompanying loss of sodium and potassium.

The unbridled mobilization of FFA has other unfortunate consequences. Through the mechanisms illustrated in Figure 7.19, it stimulates gluconeogenesis and in this way indirectly contributes to the consequences of hyperglycemia. In addition,

it is the basis for the hyperlipemia. When FFA are delivered to the liver at a high rate, fatty acid oxidation and triglyceride synthesis are stimulated. Secretion of the newly synthesized triglycerides in the form of VLDL is one mechanism called into play to prevent development of a fatty liver. Thus an increased rate of VLDL secretion is one basis for lipemia. The clearance of VLDL, and also of chylomicrons, depends on lipoprotein lipase. The synthesis and secretion of this enzyme is stimulated by glucose and insulin. In the diabetic there is a relative lipoprotein deficiency, and so the rate of removal of triglyceride-rich lipoproteins is abnormally slow, accentuating the lipemia.

Consequences of Disturbed Protein Metabolism

These are summarized in Figure 7.20C. The net breakdown of muscle protein causes blood amino acid levels to rise and, as discussed above, hepatic uptake of these feeds gluconeogenesis. The loss of muscle mass also entails losses of cellular potassium; urinary nitrogen and potassium are increased.

Consequences of Chronic Insulin Deficiency

With modern methods of management, diabetic patients do not frequently slip into severe diabetic ketoacidosis like that discussed above. If they do, they can almost always be successfully treated. Disability and death in diabetes much more often relate to atherosclerosis, which occurs prematurely, or to the so-called late complications of diabetes. These include damage to the kidneys, eyes, nervous system, and skin. The precise cause of these devastating chronic changes, which may only appear after decades, is not understood. However, there is a good deal of experimental evidence and some clinical evidence to suggest that they may be the result of hyperglycemia per se sustained over many years. It is known that hemoglobin can combine nonenzymatically with glucose, and the hemoglobin conjugate, hemoglobin A_{1c}, is present at elevated levels in diabetics (Rahbar, 1968; Rahbar et al., 1969). It increases in concentration roughly in proportion to the degree of sustained hyperglycemia and is used clinically as a guide to the degree of control of the patient's hyperglycemia (Koenig et al., 1976; Bunn, 1981). Other plasma proteins and a number of tissue proteins are also more heavily glycated in diabetics. Whether such glycation of proteins is involved in causing complications is still not known. In any case, the weight of evidence linking complications to hyperglycemia is such that current best management calls for maintenance of normoglycemia at all times as long as this does not entail risks to the patient (e.g., from intermittent hypoglycemia).

Consequences of Insulin Excess

Overdosage with insulin can produce symptomatic hypoglycemia which, if severe and persistent, can go on to coma and death. If the plasma glucose level falls rapidly, as it does after a dose of regular insulin (rapidly absorbed from the injection site), the symptoms include anxiety, hunger, tremulousness, and cold sweat. These symptoms may be recognized as those accompanying an injection of epinephrine, and indeed they reflect a sympathetic discharge, a key response to hypoglycemia. This will tend to prevent the developing hypoglycemia by stimulating release of glucose from the liver (activation of phosphorylase) and reducing utilization in the periphery (conserving glucose for the central nervous system). Glucagon levels rise, again in an attempt to correct glucose levels. Other so-called counterregulatory hormones include glucocorticords and growth hormone, but their effects are more important in long-term control. If the acute fall in plasma glucose continues, the patient may become comatose and suffer irreversible brain damage. The same chain of events can occur in patients with insulin-secreting tumors (insulinoma).

If the plasma glucose falls only gradually, the symptoms referable to sympathetic discharge may not occur. Instead the patient becomes listless, mentally confused, slurred of speech, and slips gradually into coma. These symptoms all reflect the progressive deterioration of central nervous system functions, since glucose is the only substrate the brain can utilize at a significant rate (except during prolonged starvation).

CHAPTER 51

Energy Balance

THERMODYNAMIC CONSIDERATIONS

The advertisements for miracle weight loss programs ("Eat All You Want and Lose Weight") raise false hopes that the first law of thermodynamics has been repealed or at least amended. Unfortunately, the human body is subject to the same fundamental laws of thermodynamics that govern any other physicochemical system. The first law states that energy can neither be created nor destroyed. The various forms of energy (thermal, chemical, electrical, mechanical) are interconvertible, but the *total* energy of any closed system is immutable. Thus, the total energy of the body will only decrease if it does work on the external environment or transfers heat or waste materials to the external environment. This basic notion of energy balance can be illustrated by a theoretical experiment. We place a man in a box that isolates him completely from his environment. To simplify matters we will not permit any external work to be done. Now we need to concern ourselves only with changes in the physicochemical state of the man, the box, and heat exchanges with the environment. Let us define the physicochemical state at the beginning (state A) in terms of the temperature, pressure, chemical composition, phases, etc. of the total system, as customary in physical chemistry. Then we introduce into the box an amount of glucose which, when completely combusted, will yield exactly 2000 kcal, and we have our man in the box ingest this amount over the ensuing 24 h. During this time the only energy loss from the system is heat. At the end of the 24 hours we find that the physicochemical state of the system (state B) is identical to the initial state (state A), i.e., the man has neither gained nor lost weight, his body temperature is the same, and the composition of his tissues is unchanged. From precise measurements of the heat transfers taking place we find that the net heat loss was exactly 2000 kcal! Had our subject been able to do work on the external environment, such as lifting bricks and setting them on a table outside his box, we would then have had to subtract a work term in our calculations, and our subject would weigh less. Had we introduced more than 2000 cal and had the subject produced the same amount of heat and done no work, then we would have found that the physicochemical state B was *not* identical to state A. The difference would lie in the deposition of some uncombusted glucose carbons, mostly in the form of fat. Note that energy used to do work *within* the body over our 24-hour experiment will have been reconverted to heat energy if state A equals state B. For example, the heart does do mechanical work in pumping blood, but that reverts to heat as the kinetic energy of the blood, through frictional forces, is degraded to heat. Similarly, energy is used to maintain ionic gradients—ATP is consumed— but if the gradients do not change, all is expressed ultimately as heat produced.

In the case of a substrate like glucose, which is completely absorbed and completely converted to carbon dioxide and water, matters are relatively simple. Had we done our experiment with foodstuffs not completely absorbed or not completely combusted, we would have had to measure the residual caloric equivalents appearing in feces and urine. Actually, the absorption of most foodstuffs is highly efficient, and the residual caloric equivalents in feces is generally small (less than 5% of calorie intake). Combustion of proteins, however, *is* incomplete. The nitrogen is excreted in the form of urea, and some of the complex proteins leave other uncatabolized residues which must be taken into account when striking an energy balance.

The metabolic pathways by which glucose and other substrates are catabolized can be very complex. Glucose, for example, can be oxidized by way of the Embden-Meyerhof glycolytic pathway or by way of the pentose shunt; it can be oxidized directly or can first enter glycogen and only later enter the glycolytic pathway. However, as long as the final products are the same, the amount of energy released will be the same *independent of the pathway followed*. This fundamental law, the law of Hess, greatly simplifies how we deal with energy balance problems.

The applicability of the laws of thermodynamics to the living system does not have to be taken on faith. It has been experimentally demonstrated and all of the results confirm the applicability of the first law. For example, Rubner (1894) directly measured heat production of dogs in a specially designed large calorimeter and calculated the amounts of substrates burned from measurements of gaseous exchange (see section on Indirect Calorimetry). The results agreed with an error of less than 3%.

Man is not exempt. The applicability of the first law to man was first demonstrated at the turn of the century by Atwater and Benedict (1905). These investigators constructed a large metabolic chamber in which a man could live for days and in which total heat production could be measured accurately along with rates of oxygen consumption and carbon dioxide production. They carried out meticulous balance studies in which the caloric values of foods were determined directly (or calculated from gaseous exchange measurements) as the input; the heat produced was measured precisely, and the caloric values of urine and feces were measured. Heat production was measured by monitoring the increase in temperature of water circulating in pipes that jacketed the chamber and arranging matters so that no other heat losses occurred. A very important element in assessing total heat production is the measurement of the heat used for vaporization of water lost through the skin and lungs. This process of "insensible water loss" (i.e., conversion of water from liquid phase to gas phase in the absence of sweating) is a major mechanism for loss of heat from the body. The heat capacity of water is very large (0.58 kcal/g), and approximately 700 ml of water are converted to vapor by the above-mentioned process every day. As much as 25% of basal calorie production daily occurs by this mechanism. When all of these various inputs and outputs were precisely measured by Atwater and coworkers they found that they could achieve a total energy balance to within less than 2%, a remarkable engineering feat for the time and a direct confirmation that the human system obeys the laws of thermodynamics (Atwater and Benedict, 1905).

CALORIC VALUES OF FOODSTUFFS

The energy released during the catabolism of carbohydrates and fats in the body is identical to that released by their ordinary combustion because the catabolism is complete, i.e., the products are carbon dioxide and water. The caloric values for various dietary foodstuffs are given in Table 7.8. Values are expressed in kilocalories (kcal or Cal = 1000 cal = the amount of heat necessary to raise the temperature of 1000 g of water by 1°, from 14.5 to 15.5°C). While the different forms of carbohydrate have somewhat different caloric values, the mixture of carbohydrates in most diets is such that an approximation using 4.1 kcal/g is satisfactory for most purposes. The high concentration of calories in ethanol is sometimes overlooked in assessing energy balance. Like refined sugar, ethanol in the diet adds calories without essential nutrients (amino acids, vitamins, and minerals). These "empty calories" can be a basis for poor nutrition.

The only quantitatively important lipid in the diet comes in the form of triglycerides, although there are small quantities of sterols, phospholipids, and other lipids in some foodstuffs. Fatty acids account for about 90% of the mass of the triglycerides. Their caloric equivalent is about 9.5 kcal/g. The other 10% of the triglyceride molecule is the glycerol backbone. Glycerol is rapidly converted to glucose in the body, and its metabolism is thus that of a carbohydrate. The caloric equivalent of glycerol is 4.32 kcal/g, similar to that of other carbohydrates. The ingestion of "fats" in the diet therefore contributes a small amount of carbohydrate precursor as well. As discussed in Chapter 49, when triglycerides are mobilized from adipose tissue during starvation the glycerol released along with the fatty acids makes a small but real contribution to gluconeogenesis.

For nutritional purposes a value of 9.3 kcal/g can be used to estimate the caloric contribution of dietary fats. This much higher value per gram for fats, as compared to that for carbohydrates, relates to the lower state of oxidation of the carbon atoms in fatty acids contrasted with the partially oxidized state of the carbon atoms in carbohydrates. By the same token, however, the amount of oxygen needed to combust a gram of fat is considerably larger than

Table 7.8.
Calorie Values of Foodstuffs (Cal/g)

Carbohydrates (avg 4.1)	
Glucose	3.7
Starch	4.2
Ethanol	7.1
Fats (avg 9.3)	
Stearic acid	9.5
Glycerol	4.3
Acetic acid	3.5
Proteins (avg 4.1)	
Glycine	2.1
Leucine	5.9

that needed to combust a gram of carbohydrate completely. Because of these relationships, the amount of energy released *per liter of oxygen utilized* is very nearly the same, whether the substrate is fat or carbohydrate:

$$C_6H_{12}O_6 + 6O_2 \rightarrow 6CO_2 + 6H_2O$$
(glucose; 180 g)
Energy yield: 670 kcal; 112 kcal/mole O_2 or 5 kcal/liter   *(1)*
$$C_{16}H_{32}O_2 + 23O_2 \rightarrow 16CO_2 + 16H_2O$$
(palmitic acid, 256 g)
Energy yield: 2430 kcal; 105 kcal/mole O_2 or 4.7 kcal/liter

Proteins are not completely catabolized in vivo, and therefore their caloric value as nutrients is less than the energy released by ordinary combustion. For example, the complete combustion of glycine yields 3.12 kcal/g whereas the caloric value of glycine catabolized in the body is only 2.12 kcal/g. The difference represents the caloric value of the urea formed during metabolism and excreted in the urine. Proteins also contain other substituents that are incompletely combusted. Furthermore, caloric values of different amino acids range widely—from 2.12 kcal/g for glycine to 5.9 kcal/g for leucine. However, the average amino acid composition of a diet that includes a wide variety of protein sources tends toward an average that can be used for making reasonable estimates of caloric intake. The nutritional average for protein is exactly like that for carbohydrate, 4.1 kcal/g.

RESPIRATORY QUOTIENT

The respiratory quotient (RQ) is defined as the rate of production of carbon dioxide divided by the rate of utilization of oxygen:

$$RQ = \frac{\dot{V}_{CO_2}}{\dot{V}_{O_2}}$$

This value will be 1.0 when carbohydrate is the only substrate being oxidized; it will be 0.7 when only fatty acids are being oxidized. A valid measurement requires that these gases appear in the expired air at the same rate at which they are being produced by metabolic processes. If the subject overbreathes, then the rate of CO_2 output will be accelerated, and a false high value obtained; if the patient breathes too shallowly, there will be some CO_2 retention and a false low value will be obtained.

If only carbohydrates and fats are being catabolized, an estimate of the relative contribution of these two metabolic fuels can be made from a measurement of the respiratory quotient. This will cover a range of values from 0.7–1.0, depending upon whether fat or carbohydrate is the exclusive fuel and with a spectrum of values between for different ratios of fat to carbohydrate. If protein is also being metabolized, this simple approach cannot be applied. However, it is possible to make a correction for the contribution of protein in the diet by measuring nitrogen excretion in the urine and calculating from it the amount of protein being catabolized. The amounts of oxygen and of carbon dioxide attributable to catabolism of that amount of protein can be subtracted from the total measured oxygen consumption and carbon dioxide production. This reduces the data so that a "nonprotein respiratory quotient" can be calculated and from that the contribution of fat and carbohydrate can be determined (see Richardson, 1929, for review).

CALORIMETRY

Direct Calorimetry

Total body heat production in humans and animals has been measured using a variety of techniques. The Atwater-Benedict metabolic chamber has already been described above. Less cumbersome devices using various heat sensor systems have been utilized, but these are largely confined to sophisticated research studies. For most purposes the indirect method is adequate.

Indirect Calorimetry

The heat produced from the combustion of 1 g of carbohydrate is considerably less than that produced from the combustion of 1 g of fat. However, the amounts of oxygen needed for combustion of these two substrates are in an inverse relationship, as shown by *Eq. 7.1.* Consequently, the amount of heat produced when 1 liter of oxygen is utilized is similar whether that oxygen is used to combust carbohydrate (5.0 kcal/liter) or to combust fat (4.7 kcal/liter). The estimated value when protein is the exclusive substrate is not too different (4.3 kcal/liter). If precise values are needed for research purposes, it is necessary to take account of these differences. That can be done by measuring urinary nitrogen and calculating the amount of protein undergoing catabolism. Knowing that, one can correct respiratory gas data to derive the nonprotein respiratory quotient, as discussed above. In this way, it is possible to determine just what mixture of metabolic foodstuffs is being combusted and calculate the heat production attributable to each class. However, for many purposes the values do not need to be calculated so exactly. In the postabsorptive state, after an overnight fast, the mixture of metabolic fuels is rather constant from individual to individual. The postabsorptive respiratory quotient is about 0.82, indicating that fat is the predominant fuel being oxidized.

The respiratory quotient for the complete combustion of various substrates is readily determined from the chemical equation for their oxidation. Thus it can be readily shown that the respiratory quotient for complete combustion of lactate is 1.0 and for the complete combustion of glycerol 0.86. The respiratory quotient for protein oxidation is not so readily determined because of its incomplete combustion, but an empirical average respiratory quotient of about 0.80 will generally apply. Some anomalies arise when oxidation is incomplete. For example, if fatty acids are partially oxidized, only to the level of acetoacetic acid, the respiratory quotient will be less than 0.7. If carbohydrate is converted to fat and stored, the respiratory quotient will be greater than 1.0. Ordinarily, it is assumed that one is dealing with a steady-state, i.e., without any storage or accumulation of intermediates, and then the values discussed above apply. For detailed discussion of physiological methods of measurement and calculation of gaseous exchange, see Consolazio et al. (1963).

BASAL METABOLIC RATE (BMR)

The basasl metabolic rate is defined as the rate of calorie consumption in the postabsorptive state (after an overnight fast), in the absence of any muscular activity, at a comfortable environmental temperature, and with the patient resting comfortably (but not sleeping). The measurement is a useful research tool in physiology. It was at one time widely used in clinical medicine for the assessment of thyroid function, but it has been displaced by more direct measurements based on immunoassay of hormone levels in the plasma.

The BMR is determined using indirect calorimetry, measuring the rate of oxygen consumption. This can be measured in a closed system, such as the Benedict-Roth apparatus, in which the patient breathes from an oxygen reservoir and expired CO_2 is trapped. The decrease in volume of the reservoir reflects utilization of oxygen. Alternatively, the measurement can be made in an open system using the Pauling oxygen analyzer, which directly measures the partial pressure of oxygen in a stream of air supplied to the patient's mask (or into a chamber) and also measures the concentration in the expired air. Coupled with volume-flow measurements this allows a calculation of rates of oxygen utilization. In all cases the measurements are corrected to standard temperature and pressure (0°C and 760 mm of pressure). In the basal state it is

assumed that the respiratory quotient is 0.82 and that the consumption of 1 liter of oxygen corresponds to the expenditure of 4.82 kcal.

Obviously, calorie consumption will be greater per unit time in a large person than in a small person, and one way of normalizing is to express the results in terms of kcal/kg/hour. Even better normalization is obtained if the results are expressed in relationship to the estimated surface area of the body (kcal/m^2/hour). Surface areas can be estimated from height and weight using the empirical formula developed by DuBois and DuBois (1916): (area in cm^2 = 71.8 $\times$ weight$^{0.425}$ $\times$ height$^{0.725}$), or by using a nomogram or appropriate tables. As an example, an 80-kg man 180 cm tall has a body surface area of exactly 2 m^2. The difference between using weight and surface area as the reference denominator is striking when one compares animals of very different sizes. For example, the BMR of the mouse is 654 kcal/kg/day and that of man about 35 kcal/kg/day. When related to surface area, however, the values are much closer, i.e., 1888 vs. 1042 kcal/m^2/day (see Kleiber, 1947, for review).

A convenient rough value to remember for BMR in normal man is about 1 kcal/kg/hour or 35–40 kcal/m^2/hour. Some feeling for amounts of heat involved comes from comparison with a familiar source. The rate of heat production of an 80-watt light bulb is about the same as that of a 70-kg man in the basal state.

A number of factors influence BMR. It decreases with age, showing approximately a 10% decrease between the ages of 20 and 60 years. The BMR is slightly higher in males than in females (about 10%). If the body temperature rises (in a very hot climate or as a result of a fever), the BMR can increase considerably. On exposure to cold, rats and other animals increase their metabolic rate even without muscle tensing or frank shivering. Humans, however, do not appear to be capable of this *nonshivering thermogenesis*, and only increase metabolic rate significantly by virtue of muscle tensing or shivering. The level of thyroid hormone activity is a key determinant of metabolic rate, which rises in hyperthyroidism and falls in hypothyroidism (see Chapter 54). Values as much as 50% above or below the norm can be observed. In part this thyroid effect is mediated via increases in ion flux associated with increased activity of membrane-bound Na/K-ATPase (Smith and Edelman, 1979). The catecholamines increase BMR, hence the importance of avoiding stress when attempting to measure BMR. Caffeine and theophylline also will increase BMR.

BMR is always measured after an overnight fast because there is an increment in energy production in response to ingestion of foods (diet-induced thermogenesis or "specific dynamic action"). This added heat production, which can amount to as much as 20% of the nominal caloric value of the foods ingested, is not related exclusively to gastrointestinal activity because it can be elicited by intravenous injection of nutrients. The possibility that there may be variations in the magnitude of food-induced thermogenesis from individual to individual is currently being closely examined to see if it can explain differences in tendencies toward obesity.

METABOLIC "COSTS" OF PHYSICAL ACTIVITY

The total daily caloric expenditure can be considered as the sum of basal caloric expenditure plus that attributable to physical activity. Actually, the rate of caloric expenditure during sleep is somewhat less than that observed during a formal measurement of BMR (by about 10%), but during most of the day it is well above BMR. During sleep, caloric expenditure may be about 1 kcal/min; running hard, one can expend as much as 30–40 kcal/min. Obviously, the physical costs of activity range very widely (Table 7.9). Precise measurements of these energy costs are difficult. Most of the data have been obtained by having the subject breathe from a reservoir of oxygen while engaging in various activities (measuring the actual oxygen consumption) and then calculating energy expenditure by indirect calorimetry. Eight hours of hard work by a miner may cost up to 3000 kcal whereas a clerk may only expend an additional 300 kcal in connection with his/her 8 hours of work. While we can measure BMR rather exactly and while we can measure caloric intake exactly by careful accounting for food intake, it is very difficult to assess exactly the total caloric expenditure of an individual over any given 24-hour period, let alone the expenditure over a more extended period of time. The importance of this fact in connection with studies of energy balance and studies of obesity becomes apparent when

we look at the narrow margin of error that is allowable.

Consider an individual who is in caloric balance, i.e., weight is stable over an extended period of time with usual food intake and usual energy expenditure. Suppose that individual increases his/her caloric intake by only 200 kcal/day, i.e., an increase of only about 10%. Over a 1-yr period the individual will take in 73,000 extra kcal. Now, a pound of adipose tissue (450 g), which is 90% triglyceride by mass, has a caloric equivalent of about 3500 kcal. By the first law of thermodynamics the extra calories taken in over 1 yr will result in the deposition of about 20 lbs, and if continued for 5 yr, 100 lbs! Conversely, seemingly trivial differences in physical activity would also result in large changes in body weight if not accompanied by appropriate changes in calorie intake. For example, the difference in calorie expenditure when simply standing quietly and when standing and engaging in animated conversation is about 18 kcal/hour. If an individual became talkative for 4 hours/day instead of being taciturn, he might be expending 70 extra kcal daily. Over the period of a year, again, he might *lose* as much as 17 lbs if he did not increase calorie intake. Normal individuals manage to maintain their body weight within rather narrow limits, even though they change their physical activity very drastically, far beyond the narrow limits we have outlined above. For example, a male college student may work intensively as a lumberjack during the summer and return to the relatively sessile existence of the college campus during the rest of the year. However, he adjusts his food intake appropriately and usually huge changes in weight are not shown. The basis for this fine tuning that adjusts calorie intake to calorie expenditure is poorly understood. Regulation at the level of hypothalamus is certainly involved, as discussed in Chapter 69, but the precise mechanisms remain to be established. It is essential to keep in mind that small discrepancies between caloric intake and caloric expenditure, discrepancies that are at the very limits of our ability to quantify them, can account for the development of obesity. It is not necessary to break the first law. For a review of the metabolic costs of physical activity, see Passmore and Durnin, 1955.

OBESITY

Obesity represents a major medical problem in the United States and in most of the developed countries. The obese as a group are more likely to develop a number of diseases (e.g., hypertension, gallstones, and diabetes mellitus), and their life

Table 7.9.
Approximate Energy Costs of Various Physical Activities[a]

Activity	Rate of Energy Expenditure (Cal/min)
Sleeping	ca 1.0
Dressing, washing, and shaving	3.8
Walking at 2 mph (70-kg man)	3.2
Walking at 4 mph (70-kg man)	5.8
Dancing the foxtrot	5.2
Dancing the rumba	7.0
Skiing	10–18
Woodchopping	11–13

[a]Data from Passmore and Durnin, 1955.

span is significantly shortened. Obesity is defined in terms of an increase in the amount of adipose tissue, which normally accounts for about 15% of total body weight. A football player, however, can be *overweight* (i.e., above the tabulated norm for his height) yet not be *obese* because most of the excess weight represents muscle, not fat. Generally, however, increases in weight reflect increases in adipose tissue, i.e., body fat. The exact relationship between mild degrees of obesity and threat to health is still not fully established, but there is no doubt about the life-threatening implications of moderately severe obesity.

Anything that leads to a continuing imbalance between food intake and energy consumption (positive energy balance) will lead to obesity. However, to state that obesity is due to excess intake of food is no more profound or enlightening than to say that alcoholism is due to excess intake of alcohol! *Why* do some individuals manage to maintain a nice balance throughout a lifetime, despite wide swings from time to time in their energy expenditure, while others become seriously obese? Food-seeking behavior is a complex, highly integrated process. Consider the multiple steps involved in deciding to rise from the desk, walk to the kitchen, take down the peanut butter jar, make a sandwich, chew, swallow, etc., etc. What determines the frequency with which such a behavioral pattern recurs? The amounts eaten each time? Somehow, it must normally be linked to energy expenditure, but the precise linkage mechanisms are still incompletely understood. Clearly, central nervous system control is involved and, as discussed in Chapter 69, the hypothalamic centers are known to play a key role.

Bilateral destructive lesions in the ventrolateral nuclei of the hypothalamus can totally arrest food-seeking behavior, causing the animal actually to starve to death with food right there in the cage with it! Bilateral destructive lesions in the ventromedial nuclei, on the other hand, induce voracious eating and gross obesity. Direct hypothalamic damage secondary to tumors or trauma has on occasion caused marked obesity in patients, but this is extremely rare. Certainly most obese patients have no evidence of central nervous system dysfunction. Whether more subtle dysfunction occurs in hypothalamic sensitivity to various input signals or in effectiveness in sending appropriate output control signals is not known.

The hypothalamic centers are obviously subject to modulating influences originating in the cortex. The mental image of a peanut butter jar on the shelf or the smell of hamburgers grilling has behavioral consequences! Many elegant psychological studies reveal potentially significant behavior differences between the obese and the lean with respect to perception of and the response to appetite.

A number of metabolic and endocrine abnormalities have been observed in obese subjects. For example, they have abnormally high plasma insulin levels, and their insulin response to the ingestion of glucose is greater than normal. However, normal individuals who deliberately gain weight for experimental purposes by forcing themselves to overeat also develop hyperinsulinemia; when they revert to their normal eating habits and return to their normal weight, their hyperinsulinemia disappears. Thus, hyperinsulinemia and most of the "metabolic abnormalities" described in the obese represent *results* of obesity rather than causes (Sims et al., 1973).

Obesity is almost certainly not a single disorder but a common manifestation of a wide variety of disorders that affect energy balance in some way. Whatever the cause or causes, behavior modification to decrease food intake and moderate exercise to increase caloric expenditure represent effective management. A useful overview of many basic and clinical aspects of obesity can be found in the NIH conference *Obesity in Perspective* (Bray, 1973).

BIBLIOGRAPHY

ARIMURA, A. Recent progress in somato-statin research. *Biomed. Res.* 2: 233–257, 1981.

ATWATER, W. O. Neue Versuche uber Stoff- und Kraftwechsel im menschlichen Korper. *Ergebnisse der Physiol.* 3: 497, 1904.

ATWATER, W. O. AND F. G. BENEDICT. A respiration calorimeter. Publ. No. 42, Carnegie Institution of Washington, 1905.

BANTING, F. G., AND BEST, C. H. Internal secretion of pancreas. *J. Lab. Clin. Med.* 7: 251, 1922.

BELL, G. I., W. F. SWAIN, R. PICTET, B. CORDELL, H. M. GOODMAN, AND W. J. RUTTER. Nucleotide sequence of a cDNA clone encoding for human preproinsulin. *Nature* 282: 525–527, 1979.

BENDITT, E. P. The monoclonal theory of atherogenesis. *Atheroscler. Rev.* 3: 77–86, 1978.

BERSON, S. A., AND R. S. YALOW. Quantitative aspects of reaction between insulin and insulin-binding antibody. *J. Clin. Invest.* 38: 1996–2016, 1959.

BERTHAUD, H. R., D. A. BEREITER, E. R. TRIMBLE, E. C. SIEGEL, AND B. JEAN-RENAUD. Cephalic phase reflex insulin secretion: neuroanatomical and physiological characterization. *Diabetologia* 20 (Suppl.): 393–401, 1981.

BRAY, G. A. (ed.). Proceedings of a Conference Sponsored by the John E. Fogarty International Center for Advanced Study in the Health Science. DHEW Publication No. (NIH) 75-708, 1973.

BRAZEAU, P., W. VALE, R. BURGUS, N. LING, M. BUTCHER, J. RIVIER, AND R. GUILLEMIN. Hypothalamic polypeptide that inhibits secretion of immunoreactive pituitary growth hormone. *Science* 179: 77–79, 1973.

BROMER, W. W., L. G. SINN, A. STAUB, AND O. K. BEHRENS. The amino acid sequence of glucagon. *Diabetes* 6: 234, 1957.

BUNN, H. F. Nonenzymatic glycosylation of protein: relevance to diabetes. In: *Diabetes Mellitus*, edited by J. S. Skyler and G. F. Cahill, Jr. New York: Yorke Medical Books, 1981, p. 173–178.

BUTCHER, R. W., R. J. HO, H. C. MENG, AND E. W. SUTHERLAND. Adenosine 3′,5′-monophosphate in biological materials. *J. Biol. Chem.* 240: 4515–4523, 1965.

BUTCHER, R. W., J. G. T. SNEYD, C. R. PARK, AND E. W. SUTHERLAND, JR. Effect of insulin on adenosine 3′,5′-monophosphate in the rat epididymal fat pad. *J. Biol. Chem.* 241: 1651–1653, 1966.

CAHILL, G. F., JR., AND O. E. OWEN. Some observations on carbohydrate metabolism in man. In: *Carbohydrate Metabolism*, edited by F. Dickens, P. S. Randle, and W. J. Whelan. London: Academic Press, 1968, vol. I, p. 457–522.

CHANCE, R. E., E. P. KROEFF, J. A. HOFFMAN, AND B. H. FRANK. Chemical, physical and biological properties of biosynthetic human insulin. *Diabetes Care* 4: 147–154, 1981.

CONSENSUS CONFERENCE. Lowering blood cholesterol to prevent heart disease. *J.A.M.A.* 253: 2080–2090, 1985.

CONSOLAZIO, C. F., R. E. JOHNSON, AND L. J. PECORA. *Physiologic Measurements of Metabolic Function in Man.* New York: McGraw-Hill Book Co., 1963.

CORBIN, J. D., E. M. REIMANN, D. A. WALSH, AND E. G. KREBS. Activation of adipose tissue lipase by skeletal muscle cyclic adenosine 3′,5′-mono-phosphate-stimulated protein kinase. *J. Biol. Chem.* 245: 4849–4857, 1970.

CORDELL, B., G. BELL, E. TISCHER, F. M. DeNOTO, A. ULLRICH, R. PICTET, W. J. RUTTER, AND H. M. GOODMAN. Isolation and characterization of a cloned rat insulin gene. *Cell* 18: 533–543, 1979.

CREA, R., A. KRASZEWSKI, H. TAD-ASKI, AND K. ITAKURA. Chemical synthesis of genes of human insulin. *Proc. Natl. Acad. Sci. (USA)* 75: 5765–5769, 1978.

CZECH, M. P. The nature and regulation of the insulin receptor: Structure and function. *Ann. Rev. Physiol.* 47: 357–381, 1985.

DENTON, R. M., R. W. BROWNSEY, AND G. J. BELSHAM. A partial view of the mechanism of insulin action. *Diabetologia* 21: 347–362, 1981.

DOLE, V. P. A relation between non-esterified fatty acids in plasma and the metabolism of glucose. *J. Clin. Invest.* 35: 150, 1956.

DuBOIS, D., AND E. F. DuBOIS. Clinical calorimetry. X. A formula to estimate the approximate surface area if height and weight be known. *Arch. Intern. Med.* 17: 863, 1916.

DUPRE, J., S. A. ROSS, D. WATSON, AND J. D. BROWN. Stimulation of insulin secretion by a gastric inhibitory polypeptide in man. *J. Clin. Endocrinol. Metab.* 37: 826–828, 1973.

EBERT, R., AND CREUTZFELDT, W. Gastrointestinal peptides and insulin secretion. *Diabetes/Metabolism Rev.* 3: 1–26, 1987.

EBINA, Y., L. ELLIS, K. JARNAGIN, M. EDERY, L. GRAF, E. CLAUSER, J-H. OU, F. MASIARZ, Y. W. KAN, I. D. GOLDFINE, R. A. ROTH, AND W. J. RUTTER. The human insulin receptor cDNA: The structural basis for hormone-activated transmembrane signalling. *Cell* 40: 747–758, 1985.

EXPERT PANEL. Report of the National Cholesterol Education Program Expert Panel on Detection, Evaluation, and Treatment of High Blood Cholesterol in Adults. *Arch. Intern. Med.* 148: 36–69, 1988.

EXTON, J. H. Mechanisms involved in effects of catecholamines on liver carbohydrate metabolism. *Biochem. Pharmacol.* 28: 2237–2240, 1979.

FAIN, J. N., V. P. KOVACEV, AND R. O. SCOW. Effects of growth hormone and dexamethasone on lipolysis and metabolism in isolated fat cells of the rat. *J. Biol. Chem.* 240: 3522–3529, 1965.

FELIG, P. Disorders of carbohydrate metabolism. In: Metabolic Control and Disease, edited by P. K. Bondy and L. E. Rosenberg, Philadelphia: W. B. Saunders, 1980, pp. 276–392.

FOA, P. P., J. S. BAJAJ, AND N. L. FOA (eds.). *Glucagon: Its Role in Physiology and Clinical Medicine.* New York: Springer-Verlag, 1977.

FREDRIKSON, G., P. STRALFORS, N.-O. NILSSON, AND P. BELFRAGE. Hormone-sensitive lipase of rat adipose tissue. *J. Biol. Chem.* 256: 6311–6320, 1981.

GABBAY, K. H., K. DeLUCA, J. N. FISHER, JR., M. E. MAKO, AND A. H. RUBENSTEIN. Familial hyperproinsulinemia. *N. Engl. J. Med.* 294: 911–915, 1976.

GARFINKEL, A. S., AND M. C. SCHOTZ. Lipoprotein lipase. In: *Plasma Lipoproteins*, edited by A. M. Gotto, Jr. Amsterdam: Elsevier Science Publishers, 1987, p. 335–357.

GARLAND, P. B., P. J. RANDLE, AND E. A. NEWSHOLME. Citrate as an intermediary in the inhibition of phosphofructokinase in rat heart muscle by fatty acids, ketone bodies, pyruvate, diabetes and starvation. *Nature (London)* 200: 167–170, 1963.

GERICH, J. E., M. A. CHARLES, AND G. GRODSKY. Regulation of pancreatic insulin and glucagon secretion. *Annu. Rev. Physiol.* 38: 353, 1976.

GERICH, J. E. Somatostatin and diabetes. In: *Diabetes Mellitus*, edited by J. S. Skyler and G. F. Cahill, Jr. New York: Yorke Medical Books, 1981, p. 48–55.

GOEDDEL, D. V., D. G. KLEID, F. BOLI-VAR, H. L. HEYNEKER, D. G. YAN-SURA, R. CREA, T. HIROSE, A. KRASZ-EWSKI, K. ITAKURA, AND A. RIGGS. Expression in *Escherichia coli* of chemically synthesized genes for human insulin. *Proc. Natl. Acad. Sci. (USA)* 76: 106–110, 1979.

GOLDSTEIN, J. L., AND M. S. BROWN. The low-density lipoprotein pathway and its relation to atherosclerosis. *Annu. Rev. Biochem.* 46: 897–930, 1977.

GOODMAN, D. S. The interaction of human serum albumin with long-chain fatty acid anions. *J. Am. Chem. Sci.* 80: 3892, 1958.

GORDON, R. S., JR., AND A. CHERKES. Unesterified fatty acid in human plasma. I. *J. Clin. Invest.* 35: 206, 1956.

HAVEL, R. J., J. L. GOLDSTEIN, AND M. S. BROWN. Lipoproteins and lipid transport. In: *Metabolic Control and Disease*, edited by P. K. Bondy and L. E. Rosenberg. Philadelphia: W. B. Saunders, 1980, p. 393–494.

HAVEL, R. J., H. A. EDER, AND J. H. BRAGDON. The distribution and chemical composition of ultracentrifugally separated lipoproteins in human serum. *J. Clin. Invest.* 34: 1345–1353, 1955.

HAZELWOOD, R. L. Synthesis, storage, secretion, and significance of pancreatic polypeptide in vertebrates. In: *The Islets of Langerhans*, edited by S. J. Cooperstein and D. Watkins. New York: Academic Press, 1981.

HERS, H. G. The control of glycogen metabolism in the liver. *Annu. Rev. Biochem.* 45: 167–189, 1976.

HIMMS-HAGEN, J. Obesity may be due to a malfunctioning of brown fat. *J. Canad. Med. Assoc.* 121: 1361–1364, 1979.

HIRSCH, J., AND B. BATCHELOR. Adipose tissue cellularity in human obesity. *Clin. Endocrinol. Metab.* 5: 299–311, 1976.

HOBART, P., R. CRAWFORD, L.-P. SHEN, R. PICTET, AND W. J. RUTTER. Cloning and sequence analysis of cDNAs encoding two distinct somatostatin precursors found in the endocrine pancreas of anglerfish. *Nature* 288: 137, 1980.

HOLM, C., P. BELFRAGE, AND G. FREDRIKSON. Immunological evidence for the presence of hormone-sensitive lipase in rat tissues other than adipose tissue. *Biochem. Biophys. Res. Comm.* 148: 99–105, 1987.

HOLM, C., T. G. KIRCHGESSNER, K. L. SVENSON, G. FREDRIKSON, S. NILSSON, C. G. MILLER, J. E. SHIVELY, C. HEINZMANN, R. S. SPARKES, T. MOHANDAS, A. J. LUSIS, P. BELFRAGE, AND M. C. SCHOTZ. Hormone-sensitive lipase: sequence, expression and chromosomal localization to 19 cent—q13.3. *Science*, 241: 1503–1506, 1988.

HORWITZ, D. L., H. KUZUYA, AND A. H. RUBENSTEIN. Circulating serum C-peptide. *N. Engl. J. Med.* 295: 207, 1976.

HOUGEN, T. J., B. E. HOPKINS, AND T. W. SMITH. Insulin effects on monovalent cation transport and Na-K-ATPase activity. *Am. J. Physiol.* 234: C59, 1978.

HOUSLAY, M. D. Membrane phosphorylation: a crucial role in the action of insulin, EGF, and pp.60^{arc}? *Bioscience Reports* 1: 19–34, 1981.

HUNTER, T. The proteins of oncogenes. *Scientif. Am.* 251: 70–79, 1989.

HUTTUNEN, J. K., D. STEINBERG, AND S. E. MAYER. Protein kinase activation and phosphorylation of purified hormone-sensitive lipase. *Biochem. Biophys. Res. Commun.* 41: 1350, 1970.

INNERARITY, T. L., AND R. W. MAHLEY. Enhanced binding by cultured human fibroblasts of apo E-containing lipoproteins as compared with low density lipoproteins. *Biochemistry* 17: 1440–1447, 1978.

JACKSON, R. L. Lipoprotein lipase and hepatic lipase. In: *The Enzymes*, edited by P. Boyer, Vol. 16, New York: Academic Press, 1983, p. 141–181.

JOEL, C. D. The physiological role of brown adipose tissue. In: *Adipose Tissue. Handbook of Physiology*, edited by A. E. Renold and G. F. Cahill, Jr. Washington, D.C.: American Physiological Society, 1965, sect. 5, p. 87–100.

JUNG, R. T., P. S. SHETTY, W. P. T. JAMES, M. A. BARRAND, AND B. A. CALLINGHAM. Reduced thermogenesis in obesity. *Nature* 279: 322–323, 1979.

KAHN, C. R., J. S. FLIER, R. S. BAR, J. A. ARCHER, P. GORDEN, M. A. MARTIN, AND J. ROTH. The syndromes of insulin resistance and acanthosis nigricans: insulin-receptor disorders in man. *N. Engl. J. Med.* 294: 739, 1976.

KAHN, C. R., K. L. BAIRD, J. S. FLIER, C. GRUMFELD, J. T. HARMON, L. C. HARRISON, F. A. KARLSSON, M. KASUGA, G. L. KING, U. C. LANG, J. M. PODSKALNY, AND E. VAN OBBERGEHEN. Insulin receptors, receptor antibodies, and the mechanism of insulin action. *Recent Prog. Horm. Res.* 37: 477–538, 1981.

KAHN, C. R. The molecular mechanism of insulin action. *Annu. Rev. Med.* 36: 429–451, 1985.

KASUGA, M., Y. ZICK, W. L. BLITHE, M. CRETTAZ, AND C. R. KAHN. Insulin stimulates tyrosine phosphorylation of the insulin receptor in a cell-free system. *Nature (London)* 298: 667, 1982.

KHOO, J. C., D. STEINBERG, B. THOMPSON, AND S. E. MAYER. Hormonal regulation of adipocyte enzymes: the effects of epinephrine and insulin on the control of lipase, phosphorylase kinase, phosphorylase, and glycogen synthase. *J. Biol. Chem.* 248: 3823–3830, 1973.

KINNUNEN, P. K. J. Hepatic endothelial lipase. In *Lipases*, edited by B. Borgström and H. L. Brockman. Amsterdam: Elsevier, 1984, p. 307–328.

KLEIBER, M. Body size and metabolic rate. *Physiol. Rev.* 27: 511, 1947.

KOENIG, R. J., C. M. PETERSON, R. L. JONES, C. SABDEK, M. LEHRMAN, AND A. CERAMI. Correlation of glucose regulation and hemoglobin A$_{lc}$ in diabetes mellitus. *N. Engl. J. Med.* 295: 417, 1976.

KOERKER, D. J., W. RUCH, E. CHIDECKEL, J. PALMER, C. J. GOODNER, J. ENSINCK, AND C. C. GALE. Somatostatin: hypothalamic inhibitor of the endocrine pancreas. *Science.* 184: 482–484, 1974.

KOMAROMY, M. C., AND M. C. SCHOTZ. Cloning of rat hepatic lipase cDNA: Evidence for a lipase gene family. *Proc. Natl. Acad. Sci. (USA)* 84: 1526–1530, 1987.

LARNER, J., C. VILLAR-PALASI, N. D. GOLDBERG, S. J. BISHOP, F. HUIJING, J. I. WENGER, H. SASKO, AND N. P. BROWN. Hormonal and non-hormonal control of glycogen synthesis—control of transferase phosphatase and trans-

ferase I kinase. *Adv. Enzyme Regul.* 6: 409–423, 1968.

LARNER, J., G. GALASKO, K. CHENG, A. A. DEPALOIROACH, L. HUANG, P. DAGGY, AND J. KELLOGG. Generation by insulin of a chemical mediator that controls protein phosphorylation and dephosphorylation. *Science* 206: 1408, 1979.

LEVY, R. I. Declining mortality in coronary heart disease. *Arteriosclerosis* I: 312–325, 1981.

LOMEDICO, P., N. ROSENTHAL, A. EFSTRATIADIS, W. GILBERT, R. KOLODNER, AND R. TIZARD. The structure and evolution of the two nonallelic rat preproinsulin genes. *Cell* 18: 545–558, 1979.

LOTEN, E. G., AND J. G. SNEYD. An effect of insulin on adipose tissue adenosine 3′,5′-monophosphate diesterase. *Biochem. J.* 120: 187, 1970.

LOUIS-SYLVESTRE, J. Preabsorptive insulin release and hypoglycemia in rats. *Am. J. Physiol.* 230: 56–60, 1976.

MANGANIELLO, V., AND M. VAUGHAN. An effect of insulin on cyclic adenosine 3′,5′-monophosphate phosphodiesterase activity in fat cells. *J. Biol. Chem.* 248: 7164–7170, 1973.

MANN, F. C., AND T. B. MAGATH. Die Werkung der totalen Leberextirpation. *Ergeb. Physiol.* 23: 212–262, 1924.

MARMOT, S. G., S. L. SYME, A. KAGAN, H. KATO, J. B. COHEN, J. BELSKY. Epidemiologic studies of coronary heart disease and stroke in Japanese men living in Japan, Hawaii and California: prevalence of coronary and hypertensive heart disease and associated risk factors. *Am. J. Epidemiol.* 102: 514–525, 1975.

McGARRY, J. D., AND D. W. FOSTER. Hormonal control of ketogenesis. *Arch. Intern. Med.* 137: 495, 1977.

McGARRY, J. D., M. KUWAJIMA, C. B. NEWGARD, AND D. W. FOSTER. From dietary glucose to liver glycogen. *Ann. Rev. Nutr.* 7: 51–73, 1987.

MENDELSOHN, E. *Heat and Life.* Cambridge, MA: Harvard Univ. Press, 1964, p. 148–153.

MORGAN, H. E., D. M. REGEN, AND C. R. PARK. Identification of a mobile carrier-mediated sugar transport system in muscle. *J. Biol. Chem.* 239: 369, 1964.

MUNGER, B. L. Morphological characterization of islet cell diversity. In: *The Islets of Langerhans*, edited by S. J. Cooperstein and D. Watkins. New York: Academic Press, 1981, p. 1–49.

NATIONAL DIABETES DATA GROUP. Classification and diagnosis of diabetes mellitus and other categories of glucose intolerance. *Diabetes* 28: 1039, 1979.

ODUM, E. P. Adipose tissue in migrating birds. In: *Adipose Tissue. Handbook of Physiology*, edited by A. E. Renold and G. F. Cahill, Jr. Washington, D.C.:

American Physiological Society, 1965, sect. 5, p. 37–43.

OLEFSKY, J. M., AND O. G. KOLTERMAN. Mechanisms of insulin resistance in obesity and noninsulin-dependent (type II) diabetes. In: *Diabetes Mellitus*, edited by J. S. Skyler and G. F. Cahill, Jr. New York: Yorke Medical Books, 1981, p. 73–90.

OLEFSKY, J. M., M. SAEKOW, H. TAGER, AND A. H. RUBENSTEIN. Characterization of a mutant human insulin species. *J. Biol. Chem.* 255: 6098–6105, 1980.

OLSSON, H., AND P. BELFRAGE. The regulatory and basal phosphorylation sites of hormone-sensitive lipase are dephosphorylated by protein phosphatase-1, 2A and 2C but not by protein phosphatase-2B. *Eur. J. Biochem.* 168: 399–405, 1987.

ORCI, L. The microanatomy of the islets of Langerhans. *Metabolism* 25 (Suppl. 1): 1303, 1976.

ORCI, L. Macro- and micro-domains in the endocrine pancreas. *Diabetes* 31: 538–565, 1982.

OWEN, O. E., A. P. MORGAN, H. G. KEMP, J. M. SULLIVAN, M. G. HERRERA, AND G. F. CAHILL, JR. Brain metabolism during fasting. *J. Clin. Invest.* 46: 1589–1595, 1967.

OWEN, O. E., P. FELIG, A. P. MORGAN, J. WAHREN, AND G. F. CAHILL, JR. Liver and kidney metabolism during prolonged starvation. *J. Clin. Invest.* 48: 574–583, 1969.

PARK, C. R., O. B. CROFFORD, AND T. KONO. Mediated (nonactive) transport of glucose in mammalian cells and its regulation. *J. Gen. Physiol.* 52: 296, 1968.

PASSMORE, R., AND J. V. G. A. DURNIN. Human energy expenditure. *Physiol. Rev.* 35: 801–840, 1955.

PERLEY, M. J., AND D. M. KIPNIS. Plasma insulin responses to oral and intravenous glucose: studies in normal and diabetic subjects. *J. Clin. Invest.* 26: 1954–1962, 1967.

PERMUTT, M. A. Biosynthesis of insulin. In: *The Islets of Langerhans*, edited by S. J. Cooperstein and D. Watkins. New York: Academic Press, 1981, p. 75–95.

PHILLIPS, L. S., AND R. VASSILOPOU-LOU-SELLIN. Somatomedins. *N. Engl. J. Med.* 302: 371–438, 1980.

PILKIS, S. J., AND M. R. EL-MAGHRABI. Hormonal regulation of hepatic gluconeugenesis and glycolysis. *Ann. Rev. Biochem.* 57: 755–783, 1988.

PILKIS, S. J., D. M. REGEN, T. H. CLAUS, AND A. D. CHERRINGTON. Role of hepatic glycolysis and gluconeogenesis in glycogensynthesis. *BioEssays* 2: 273–276, 1985.

POWLEY, T. The ventromedial hypothalamic syndrome, satiety and a cephalic phase hypothesis. *Psychol. Rev.* 84: 89–126, 1977.

RAHBAR, S. An abnormal haemoglobin in red cells of diabetes. *Clin. Chem. Acta* 22: 296, 1968.

RAHBAR, S., D. BLUMENFELD, AND H. M. RANNEY. Studies of the unusual hemoglobin in patients with diabetes mellitus. *Biophys. Res. Commun.* 36: 838, 1969.

RICHARDSON, H. B. The respiratory gradient. *Physiol. Rev.* 9: 61, 1929.

ROSS, R. Atherosclerosis: a problem of the biology of arterial wall cells and their interaction with blood components. *Arteriosclerosis* 1: 293–311, 1981.

ROTH, R. A., AND D. J. CASSELL. Insulin receptor: evidence that it is a protein kinase. *Science* 219: 299–301, 1983.

ROTH, J., C. R. KAHN, M. A. LESNIAK, P. GORDEN, P. DeMEYTS, K. KEGGESI, D. M. NEVILLE, JR. J. R. GARVIN III, A. H. SOLL, P. FREYCHET, I. D. GOLDFINE, R. S. BAR, AND J. A. ARCHER. Receptors for insulin, NSILA-s and growth hormone: application to disease states in man. *Recent Prog. Horm. Res.* 31: 95, 1975.

RUBNER, M. Die Quelle der thierischen Warme [the source of animal heat]. *Ztschn. Biol.* 30: 73, 1894.

SALANS, L. B. Obesity and the adipose cell. In: *Metabolic Control and Disease*, edited by P. K. Bondy and L. E. Rosenberg. Philadelphia: W. B. Saunders, 1980, p. 495–521.

SCHOENHEIMER, R. The dynamic state of body constituents. Cambridge, MA: Harvard University Press, 1942.

SEALS, J. R., AND L. JARETT. Activation of pyruvate dehydrogenase by direct addition of insulin to an isolated plasma membrane/mitochondria mixture: evidence for generation of insulin's second messenger in a subcellular system. *Proc. Natl. Acad. Sci. (USA)* 77: 77, 1980.

SEVERSON, D. L., J. C. KHOO, AND D. STEINBERG. The role of phosphoprotein phosphatases in the reversible deactivation of chicken adipose tissues hormone-sensitive lipase. *J. Biol. Chem.* 252: 1484–1489, 1977.

SHEN, L-P., R. L. PICTET, AND W. J. RUTTER. Human somatostatin. I. Sequence of the cDNA. *Proc. Natl. Acad. Sci. (USA)* 79: 4575, 1982.

SHERRILL, B. C., T. L. INNERARITY, AND R. W. MAHLEY. Rapid hepatic clearance of the canine lipoproteins containing only the E apoprotein by a high affinity receptor. *J. Biol. Chem.* 255: 1804–1807, 1980.

SIMS, E. A. H., E. DANFORTH, JR., E. S. HORTON, G. A. BRAY, J. GLENNON, AND L. A. SALANS. Effects of experimental obesity in man. *Recent Prog. Horm. Res.* 29: 457–496, 1973.

SKYLER, J. S., AND G. F. CAHILL, JR. (eds). *Diabetes Mellitus.* New York: Yorke Medical Books, 1981.

SMITH, T. J., AND I. S. EDELMAN. The role of sodium transport in thyroid thermogenesis. *Fed. Proc.* 38: 2150–2153, 1979.

STADIE, W. C., N. HAUGAARD, AND M. VAUGHAN. The quantitative relation between insulin and its biological activity. *J. Biol. Chem.* 200: 745–781, 1953.

STANBURY, J. B., J. B. WYNGAARDEN, D. S. FREDRICKSON, J. L. GOLDSTEIN, AND M. S. BROWN (eds). *The Metabolic Basis of Inherited Disease. Part 4. Disorders of Lipoprotein and Lipid Metabolism.* New York: McGraw-Hill Book Co., 1983, p. 589–710.

STEINBERG, D. Catecholamine stimulation of fat mobilization and its metabolic consequences. *Pharmacol. Rev.* 18: 217–235, 1966.

STEINBERG, D. Interconvertible enzymes in adipose tissue regulated by cyclic AMP-dependent protein kinase. *Adv. Cyclic Nucleotides Res.* 7: 157–198, 1976.

STEINBERG, D. Origin, turnover and fate of plasma low density lipoprotein. *Prog. Biochem. Pharmacol.* 15: 166–199, 1979.

STEINBERG, D. Metabolism of lipoproteins and their role in the pathogenesis of atherosclerosis. In: *Atherosclerosis Reviews*, edited by J. Stokes III and M. Mancini. New York: Raven Press, 1988, vol 18, p. 1–23.

STEINBERG, D., S. PARTHASARATHY, T. E. CAREW, J. C. KITOO, AND J. L. WITZTUM. Beyond cholesterol: Modifications of low-density lipoprotein that increase its atherogenicity. *N. Engl. J. Med.* 320: 915–924, 1989.

STEINBERG, D., AND M. VAUGHAN. Release of free fatty acids from adipose tissues *in vitro* in relation to rates of triglyceride synthesis and degradation. In: *Adipose Tissue. Handbook of Physiology*, edited by A. E. Renold and G. F. Cahill. Washington, D.C.: American Physiological Society, 1965, sect. 5, p. 335.

STEINBERG, D., S. E. MAYER, J. C. KHOO, E. A. MILLER, R. E. MILLER, B. FREDHOLM, AND R. EICHNER. Hormonal regulation of lipase, phosphorylase, and glycogen synthase in adipose tissue. *Adv. Cyclic Nucleotide Res.* 5: 459–568, 1975.

STEINER, D. F. Insulin today (Banting lecture). *Diabetes* 26: 322, 1977.

STEINER, D. F., AND N. FREINKEL (eds.). *Endocrine Pancreas Handbook of Physiology.* Washington, D.C.: American Physiological Society, 1972, sect. 7, vol. 1.

STRALFORS, P., P. BJÖRGELL, AND P. BELFRAGE. Hormonal regulation of hormone-sensitive lipase in intact adipocytes: Identification of phosphorylated sites and effects on the phosphorylation by lipolytic hormones and insulin. *Proc. Natl. Acad. Sci. (USA)* 81: 3317–3321, 1984.

TAGER, H. S., D. F. STEINER, AND C. PLATZELT. Biosynthesis of insulin and glucagon. In: *Methods in Cell Biol-*

ogy. New York: Academic Press, 1981, p. 73.

TANZAWA, K., Y. SHIMADA, M. KURODA, Y. TSUJITA, M. ARAI, AND Y. WATANABE. WHHL-rabbit: a low density lipoprotein receptor-deficient animal model for familial hypercholesterolemia. *FEBS Lett.* 118: 81–84, 1980.

TAYLOR, S. I. Insulin action and inaction. *Clin. Res.* 35: 459–472, 1987.

ULLRICH, A., J. R. BELL, E. Y. CHEN, R. HERRERA, L. M. PETRUZZELLI, T. J. DULL, A GRAY, L. COUSSENS, Y.-C. LIAO, M. TSUBOKAWA, A. MASON, P. H. SEEBURG, C. GRUNFELD, O. M. ROSEN, AND J. RAMACHANDRAN. Human insulin receptor and its relationship to the tyrosine kinase family of oncogenes. *Nature* 313: 756–761, 1985.

UNGER, R. H., AND L. ORCI. glucagon and the A cell. *N. Engl. J. Med.* 304: 1518, 1575, 1981.

UNGER, R. H. Glucagon physiology and pathophysiology in the light of new advances. *Diabetologia* 28: 574–578, 1985.

VAUGHAN, M., J. E. BERGER, AND D. STEINBERG. Hormone-sensitive lipase and monoglyceride lipase activities in adipose tissue. *J. Biol. Chem.* 239: 401–409, 1964.

WASSERMAN, F. The development of adipose tissue. In: *Adipose Tissue Handbook of Physiology,* edited by A. E. Renold and G. F. Cahill, Jr. Washington, D.C.: American Physiological Society, 1965, sect. 5, p. 87–100.

WATANABE, Y. Serial inbreeding of rabbits with hereditary hyperlipidemia (WHHL-rabbits). *Atherosclerosis* 36: 261–268, 1980.

WEINHOUSE, S. Regulation of glucokinase in liver. *Curr. Top. Cell. Regul.* 11: 1–50, 1976.

WION, K. L., T. G. KIRCHGESSNER, A. J. LUSUS, M. C. SCHOTZ, AND R. M. LAWN. Human lipoprotein lipase complementary DNA sequence. *Science* 235: 1638–1641, 1987.

WISSLER, R. W. Current status of regression studies. *Atherosclerosis Rev.* 3: 213–230, 1978.

YEN, S. S. C., T. M. SILER, AND G. W. DeVANE. Effects of somatostatin in patients with acromegaly: suppression of growth hormone, prolactin, insulin and glucose levels. *N. Engl. J. Med.* 290: 935–938, 1974.

SECTION 8

Endocrine System

CHAPTER 52

Principles of Hormone Action and Endocrine Control

Even the most primitive forms of life regulate the expression of their genetic material in a sophisticated manner. With increasing complexity and differentiation of structure and function, regulatory mechanisms have assumed increasing importance in the survival of the organism, in its adaptation to the environment, and in its reproduction. The two primary communication systems, the nervous system and the endocrine system, serve as a biological communication network for integration of the organism's response to a changing environment. Many nervous system functions are mediated by hormones, and the endocrine system is regulated to a large degree by the nervous system. Together the nervous and endocrine systems institute alterations in metabolism, behavior, and development to conform to internal and external environmental demands.

The endocrine glands synthesize and secrete hormones, discrete chemical substances which transfer information from one set of cells to another. The term "hormone" was first used by E. H. Starling in 1905 to describe the gastrointestinal peptide secretin. The derivation is from the Greek word meaning "set in motion" or "excite." Hormones have either of two major chemical structures: peptide or steroid. The molecular mechanisms of action of biogenic amines such as epinephrine, which is derived from the amino acid tyrosine, resemble the peptide group of hormones, whereas thyroid hormone, which is also derived from tyrosine, more closely resembles the steroid group of hormones.

In contrast to the nervous system, where neurotransmitters are released from axons in direct proximity to their target, the endocrine glands secrete hormones into the bloodstream, which carries them to distal targets. The distance between hormone producer cell and hormone responder cell may be large, as from pituitary to gonads; moderate, as from hypothalamus to pituitary; or small, as from δ or β cell in pancreatic islets. The latter is designated as a paracrine system to distinguish it from the first

two, which are endocrine systems. Cells also produce hormones that are secreted and act on the producer cell itself, an autocrine system. At their site of action, hormones serve as regulators and integrators of the target cell response with the demands of the organism as whole.

HORMONES AS ALLOSTERIC EFFECTORS

Monod et al. (1963) defined regulatory control of protein structures and consequent function by small molecule effectors. The small molecules are not metabolized by interaction with specific proteins, but by binding to one site on the protein, they alter the tertiary configuration and subsequent function of a distal site. Proteins can thus exist in various structural conformations and states of activity; their activity can be modified by regulatory molecules which are not direct substrates of the protein. Proteins subject to this form of regulation are termed "allosteric proteins," and the binding molecules are termed "allosteric effectors." All hormones are allosteric effectors. The allosteric proteins to which hormones bind are termed hormone receptors. Although the kinetics of interaction between hormone and receptor may be complex, the interaction can be represented overall as a bimolecular reaction:

$$\text{Hormone} + \text{Receptor} \underset{k_2}{\overset{k_1}{\rightleftharpoons}} \text{Hormone} \cdot \text{Receptor} \qquad (8.1)$$
$$[\text{H}] \qquad [\text{R}] \qquad [\text{HR}]$$

The binding of hormone to receptor protein involves hydrophobic interactions, hydrogen bonds, Van der Waals forces, and salt bridges but not covalent linkages. Hence allosteric regulation can be quickly terminated once the allosteric effector is reduced in concentration. In the mass action equation shown, formation of the active [HR] complex depends on the concentration of both the hormone and the receptor, as well as on the intrinsic affinity of the receptor for the hormone. Classical endocrinology dealt with variations in the concentration of hormone as the primary factor determining formation of the active [HR] complex and, thus, biological activity. When an endocrine gland is ablated and hormone concentration decreases, [HR] decreases, and an endocrine deficiency state ensues; conversely, when excess hormone is present, increased [HR] is formed, and an endocrine excess syndrome results. Extensive control systems regulate the synthesis, secretion, and transport of hormones. These systems control the concentration of hormone in body fluids and, thus, hormone-dependent biological responses. The receptor is equally important in formation of the active [HR] complex. Like hormones, receptors are subject to extensive regulation. Both genetic and acquired disease states involving hormone receptors have been described. Patients with a decrease in a specific hormone receptor manifest a deficiency state despite normal or elevated concentrations of hormones. When receptor concentrations increase, an endocrine excess state may occur despite normal circulating concentrations of hormone.

Receptor proteins contain at least two principal domains: a hormone binding site and an activity site. The hormone binding site demonstrates high specificity and high affinity for the proper hormone ligand. Specificity is sufficient to distinguish between two hormone molecules with only minor structural modifications such as the presence or absence of a hydroxyl group on a steroid molecule or an amino acid change in a peptide. Affinity is high with equilibrium dissociation constants for hormone-receptor interactions of 10^{-8} to 10^{-12} M. Specificity and affinity of receptors are sufficient for recognition of the nanogram to picogram quantities of hormones present per ml of blood (1 part in a million to 1 part in a billion of total protein present). The activity site on receptor proteins couples them directly or indirectly to enzyme systems and biological responses. The activity site serves to distinguish receptors from other proteins, such as the hormone transport proteins of plasma, which have a binding site for recognition but no known biological activity. Hormone receptors are therefore identified by their ability to bind natural and synthetic hormones with high specificity and affinity in the same order and concentration at which these hormones activate target cell biological responses.

The distribution of hormone receptors determines the ability of cells to respond to a particular hormone. Receptors for certain hormones are localized to few cell types, whereas receptors for other hormones are widely distributed. For example, ACTH receptors are localized to the adrenal cortex and, to a lesser extent, to fat cells and circulating mononuclear cells; only these target cells respond to ACTH. Estrogen receptors are high in uterus, ovary, and breast; these are principal sites of estrogen action. Receptors for cortisol and thyroid hormone are widely distributed, and most cell types respond to these hormones.

Reaction with a specific receptor is the initial event in hormone action. The induced conformational change and altered activity of the receptor then causes a series of biochemical changes which produces the physiological response. At a biochemical level these changes include regulation of enzyme activity, regulation of the biosynthesis and degradation of nucleic acids and proteins, and regulation of the ability of target cells to grow and to replicate. Hormones thus control the differentiated function and growth potential of their target tissues. In differentiated tissues, hormones regulate the degree of expression of the genetic material but do not change the type of genetic information expressed. For example, hormones such as epinephrine, insulin, and cortisol profoundly alter the metabolic activity of liver cells; liver cell-specific functions are, however, always maintained. During the process of differentiation temporal expression of hormones and receptors control sequential events (Cate et al., 1986; Hafen et al., 1987).

There are two principal classes of hormone receptors. Receptors for peptide hormones are located in the cell membrane, with the hormone binding site expressed on the exterior cell surface (Fig. 8.1). Binding of circulating peptide hormones to these cell surface receptors alters the activity of membrane-bound enzymes and transport processes. The altered enzyme activity may act directly on target sites or may change the cellular concentration of regulatory molecules such as cAMP, diacylglycerol (DAG) and Ca^{2+}, which then function as second messengers of hormone action. Enzyme activity is then altered either directly by modifying existing enzyme molecules or by altering the synthesis and concentration of enzyme molecules. Receptors for steroid hormones are located within the cell. Hydrophobic, lipid-soluble steroid hormone molecules diffuse readily into cells from the circulation to bind to specific intracellular receptors. The steroid hormone-receptor complex acts at a nuclear level to initiate the biochemical events specifying the biological effect. Thyroid hormone resembles steroid hormones in its molecular mechanism of action; both undergo allosteric alterations on binding hormone. In their bound, activated state these receptors interact with specific DNA sequences to regulate the transcription of specific genes. Action at a

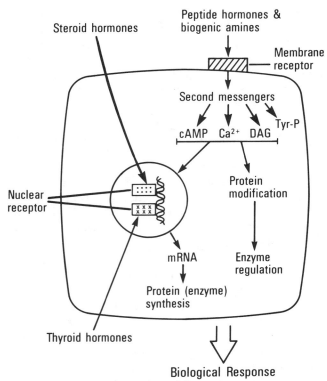

Figure 8.1.   Pathways through which hormones function.

nuclear level alters the biosynthesis and concentration of enzyme molecules. Alterations in enzyme activity via changes in concentration or by modification underlie the biological response of each cell, which is integrated with that of other cell types in the organism, into the physiological response.

CELL SURFACE RECEPTORS FOR PEPTIDE HORMONES

A large number of peptide hormones have been identified, including classical pituitary, pancreatic, and parathyroid hormones, neurohormones, gastrointestinal hormones, cell growth factors, and biogenic amines. Each peptide hormone is recognized by a specific and distinct cell surface receptor which translates hormone recognition into a biological response. Although each hormone receptor is a unique molecular species, there are certain features which are similar among peptide hormone receptors. Peptide hormone receptor proteins have a hormone-binding region exposed on the external cell surface, a hydrophobic region which anchors them within the lipid bilayer of the cell membrane, and an activity site which interacts either with other membrane proteins or with the interior of the cell. Peptide hormone receptors are glycosylated, often contain more than one subunit, and require phospholipid or a membrane environment for biological activity.

Two general structures for cell surface hormone receptors have been identified. The first structural motif consists of seven hydrophobic membrane-spanning helices. Surface loops between membrane-spanning segments form the ligand-binding domain; cytoplasmic interloop regions are involved in effector coupling. Receptors for rhodopsin, muscarinic cholinergic, α and β adrenergic agonists, and ligand-gated ion channels have this structure (Nathans and Hogness, 1984; Kubo et al., 1986; Kobilka et al., 1987 a, b). The second structural motif consists of a single membrane-spanning domain which separates the extracellular ligand binding domain from the cytoplasmic domain which often contains intrinsic enzymatic activity. Receptors for insulin, insulin-like growth factor 1 (IGF-1), epidermal growth factor (EGF), and platelet-derived growth factor (PDGF) have this structure (Ullrich et al., 1985; Ebina et al., 1985; Ullrich et al., 1984; Yarden et al., 1986).

These receptors move within the plane of the cell membrane. Mobility was initially demonstrated for catecholamine receptors by fusing a cell containing β-receptors but no adenylate cyclase with a cell containing adenylate cyclase but no β-receptors (Schramm et al., 1977). In membranes from the resulting hybrid cell, isoproterenol readily stimulated formation of cAMP, indicating that β-receptors and catalytic cyclase units from the two different parents had moved together into a functional complex. Mobility of receptors was directly demonstrated using hormones conjugated to fluorescent indicators (Schlessinger et al., 1978). Cytoskeletal elements which are anchored at the cell membrane may contribute to receptor mobility.

There are approximatley 10^4–10^5 receptors on a target cell. These receptors are constantly turning over. The insulin receptor, for example, has a half-life of ~7 h (Krupp and Lane, 1981). After synthesis, receptors are diffusely inserted into the cell membrane. When hormones are bound, they cluster and localize in membrane invaginations (Fig. 8.2). Receptors for nutrients such as low density lipoproteins (LDL), which carry cholesterol, are largely localized to these coated pits (Brown et al., 1983). Receptors with bound hormone are then internalized by adsorptive endocytosis. Within the acidic environment of the endocytic vesicle, ligand and receptor are uncoupled (Tycko and Maxfield, 1982). Receptors which deliver nutrients such as the LDL and transferrin receptors (Willingham and Pastan, 1980) recycle back to the cell surface while the endosome containing the ligand fuses with lysosomes. Many signaling receptors such as those for insulin and EGF remain with their ligand in endocytic vesicles which fuse with lysosomes where

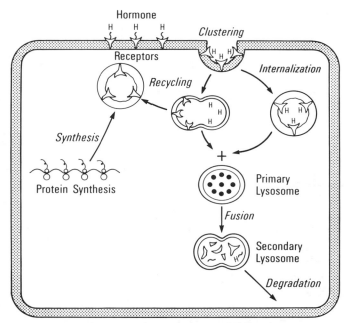

Figure 8.2. Metabolism of peptide hormone receptors.

both peptide hormone and receptor are degraded (Carpenter, 1987). Most available evidence indicates that the hormonal signal is transmitted at the cell membrane, although hormone-induced clustering of receptors may in some cases be required for signal generation (Kahn et al., 1978; Schreiber et al., 1983). Internalization is clearly involved in degradation of peptide hormones and peptide hormone receptors. Although it is possible that hormone signals are transmitted after internalization, there is little evidence for this. Mutations which abolish internalization result in ligand-activated cell surface receptors which strongly signal but fail to attenuate by down regulation (Chen et al., 1989)

Peptide hormone receptors are dynamic molecules which are extensively regulated. Regulation of receptors by their homologous hormone results most commonly in downregulation, a decrease in the number of receptors on each cell. Chronic exposure to hormone causes a decrease in receptor concentration which results in decreased sensitivity to circulating hormone (desensitization, tachyphylaxis). Downregulation involves conformational changes in receptors and internalization. Insulin resistance in Type II diabetes mellitus and desensitization to repeated administration of epinephrine are examples where this occurs. Conversely, receptor concentrations rise when homologous hormone levels are low, resulting in increased cell sensitivity. Regulation of receptors by heterologous hormones is a major mechanism for integration of responses by various cell types. Examples include follicle-stimulating hormone (FSH) induction of

luteinizing hormone (LH) receptors in ovarian granulosa cells so that LH now stimulates steroid hormone production, thyroid hormone elevation of β-adrenergic receptors so that increased sympathetic activity occurs in thyrotoxicosis without alterations in catecholamine concentrations, and estrogen induction of progesterone receptors in uterus and breast tissues so that enhanced responses occur during pregnancy. Both homologous and heterologous hormones primarily control the concentration of receptors present in target tissues. Less commonly, the affinity of the receptor for the hormone may be altered. Insulin receptors exhibit negative cooperativity, which decreases receptor affinity as occupancy by hormone increases (DeMeyts et al., 1976); membrane active agents such as phorbol esters affect the affinity of epidermal growth factor receptors for their ligand (Shayab et al., 1979; Lin et al., 1986).

The concentration of hormone required to induce the biological response half-maximally is frequently less than the concentration required for half-maximal receptor occupancy (Fig. 8.3*A*). Because full biological responses occur with occupancy of only a fraction of available receptors, the additional receptors are termed "spare." This dissociation between biological response and receptor occupancy is the predicted one when a second mediator is generated which itself binds reversibly to an intracellular receptor (effector) protein (Strickland and Loeb, 1981; Goldbeter and Koshland, 1984). Receptors are spare, not for generation of the second messenger (cAMP), but for the biological response elicited by this second allosteric effector. Second messengers thus provide amplification of the initial hormone signal. Such a system with spare receptors provides high sensitivity to changes in circulating hormone concentrations. In systems with spare receptors, a reduction in receptor concentration results in a requirement for more hormone to elicit the same response, i.e., decreased sensitivity but no decrease in the maximum biological effect possible (Fig. 8.3*B*). In systems where receptors are directly coupled to the biological response, a decrease in receptors results in a decrease in the maximal biological response attainable.

INTRACELLULAR MEDIATORS OF HORMONE ACTION

Biological information contained in the active peptide hormone-receptor complex is transmitted inside the cell via second messengers. The concept of second messengers was first proposed by E. W. Sutherland following his landmark discovery of cAMP in 1957 (Sutherland and Rall, 1958). cAMP is the intracellular mediator for a number of hor-

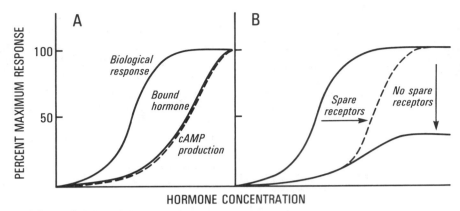

Figure 8.3. *A,* Relationship between receptor occupancy, second messenger production, and biological response. *B,* Effect of decreasing receptor concentration in "spare" and "nonspare" conditions.

mones, including adrenocorticotropin (ACTH), antidiuretic hormone (ADH), β-adrenergic catecholamines, calcitonin, FSH, glucagon, human chorionic gonadotropin (HCG), LH, melanocyte-stimulating hormone (MSH), parathyroid hormone (PTH), and TSH. An equally large number of peptide hormones use other intracellular second messenger systems such as cGMP, Ca^{2+}, and DAG; these include angiotensin II, α-adrenergic catecholamines, gonadotropin releasing-hormone (GnRH), thyrotropin-releasing hormone (TRH), oxytocin, and somatostatin.

Another group of peptide hormones bind to receptors which contain intrinsic protein tyrosine kinase activity to initiate a cascade of biochemical reactions directly; these include insulin, insulin-like growth factor, EGF, and PDGF.

The well-defined second messenger systems regulate the activity of protein kinase enzymes which catalyze transfer of the γ phosphate of ATP to substrate proteins. Because phosphorylation of proteins alters their activity, this covalent modification is a major mechanism through which enzyme activity is modified. The diversity of protein kinases present within a cell provides for complex regulatory circuits which control numerous metabolic and growth processes (Hunter, 1987). Peptide hormones via binding to cell surface receptors activate second messenger systems which then control the activity of distinct protein kinases which alter sets of substrates and thus metabolic pathways. Protein kinases are sophisticated intracellular control units subject to both positive and negative inputs and having multiple outputs to control cellular responses coordinately.

cAMP and cGMP

cAMP is formed from ATP by the membrane-bound enzyme adenylate cyclase. Hormone-sensitive adenylate cyclase is made up of three compo-

nents: the peptide hormone receptor, a coupling unit (G), and catalytic cyclase (C) (Ross and Gilman, 1980). The hormone receptor does not interact directly with the catalytic cyclase but interacts through the coupling unit (Rodbell, 1980; Casperson and Bourne, 1987). There are two major G proteins which are transducers of receptor-generated signals in the adenylate cyclase systems; other G proteins function in vision (rhodopsin) and other receptor systems (Gilman, 1987). G proteins are composed of two major subunits, α and β as well as a smaller subunit, γ. The α subunit binds guanine nucleotides and is the activator of the catalytic cyclase subunit. When hormone binds to receptor, there is rapid interaction of H·R with G to form H·R·G (Fig. 8.4). In this form, G becomes activated and preferentially binds GTP. G is active only when GTP is bound; it is inactive when GDP is bound. Formation of H·R·G increases removal of inhibitory GDP and facilitates GTP binding (Cassel and Selinger, 1978). When Gα binds GTP, the inhibitory subunit Gβ dissociates so that active Gα·GTP binds to the catalytic cyclase protein to activate it to catalyze formation of cAMP. There is also an inhibitory G protein complex which is activated via inhibitory hormone receptors to inhibit adenylate cyclase. Gα subunits differ; the Gβ subunits appear identical.

Reversals of activation of adenylate cyclase results from GTP hydrolysis by Gα and reversal of the reactions shown in Figure 8.4. GTP is hydrolyzed to GDP, which induces a confirmation of G unable to bind to C but able to bind Gβ. Cholera toxin, a ubiquitous stimulator of cAMP formation, acts via ADP ribosylation of Gα to inhibit GTP hydrolysis. By preventing GTP hydrolysis, cholera toxin causes prolonged activation of adenylate cyclase. Once GTP is bound to Gα the affinity of the receptor for the hormone is reduced to also favor reversal of the active state.

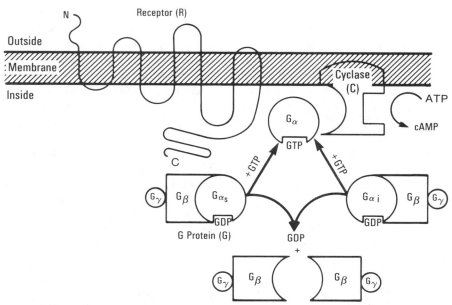

Figure 8.4. Hormone-sensitive adenylate cyclase. This model is based on the β-adrenergic receptor system because it has been most extensively characterized. A stimulatory receptor is shown; not shown is an inhibitory receptor which would act via the inhibitory G protein complex.

Hormone responses thus depend not only on the hormone and its receptor but also on the coupling protein G and the cellular concentrations of GTP and GDP. G protein and catalytic cyclase are the same in all hormone-sensitive adenylate cyclase systems. Molecular differences reside in receptor proteins, but these too may contain a common activity region which binds to G. In pseudohypoparathyroidism, the genetic defect in G protein results in inadequate responses to a number of peptide hormones, including PTH, glucagon, and ADH (Farfel et al., 1980).

By coupling hormone receptors to an enzyme, the initial hormonal signal is greatly amplified. Each hormone molecule stimulates formation of many molecules of cAMP. cAMP, in turn, interacts with a specific intracellular alloteric receptor protein which is coupled to an enzyme to give further amplication. The receptor for cAMP is the regulatory subunit of cAMP-dependent protein kinase (Gill and Garren, 1971) (Fig. 8.5). Binding of cAMP induces a conformation change in the regulatory subunit so that restraints on the catalytic kinase subunit are removed. The active catalytic subunit then catalyzes the transfer of the γ phosphate of ATP to substrate proteins.

There are several isoenzyme forms of cAMP-dependent protein kinase (Edelman et al., 1987). Specific biological effects depend on the genetic expression of a particular cell type, i.e, the potential substrates available for phosphorylation. The first discriminant is the expression of cell surface pep-tide receptors. Only target cells with specific receptors are activated. The second discriminant is the cell type. Elevations in cAMP in hepatic cells result in glycogenolysis and gluconeogenesis, whereas elevations in cAMP in adrenal cells result in increased formation of steroid hormones. In cells such as fat cells, where multiple hormones such as β-adrenergic agonists, glucagon, and ACTH all increase cAMP formation, stimulation of lipolysis is the common resultant. The long series of biochemical steps between the hormonal first messenger and the ulti-

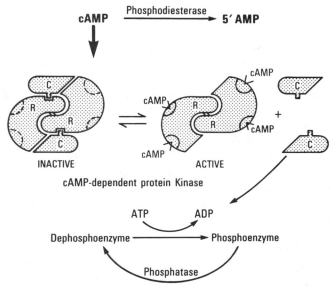

Figure 8.5. Mechanism of action of cAMP.

mate biological effect provide for quantitative and graded responses, amplification, diversification, and interaction with other regulatory signals.

The hormone-induced increase in cAMP is terminated by degradation of cAMP to 5′-AMP. Although cAMP concentrations are determined primarily by the rate of synthesis, degradative phosphodiesterase enzymes are also regulated. Drugs such as methyxanthines, which are inhibitors of phosphodiesterase, are frequently used clinically to augment cAMP responses. Phosphate groups are removed from proteins by phosphatase enzymes which are themselves subject to regulation.

cGMP rises within target cells stimulated by muscarinic cholinergic agents, oxytocin, atrial natriuretic hormone (ANH) and several other hormones. A metabolic system similar to that for cAMP exists, consisting of guanylate cyclase, cGMP-dependent protein kinase, and phosphodiesterase. The exact biological role of cGMP, however, remains uncertain because few specific biochemical processes subject to regulation by this system are known (Gill and McCune, 1979).

Calcium and Diacylglycerol

Calcium plays a regulatory role in muscle contraction, neuromuscular transmission, secretion, and hormone action. Hormone-receptor interactions alter intracellular calcium concentrations by facilitating calcium uptake from extracellular fluids and redistribution from cellular storage sites. Intracellular calcium concentrations are low so that large changes in cellular concentration occur when calcium is transported from either extracellular or sequestered intracellular pools. One intracellular receptor for calcium is the protein calmodulin, which is closely related to troponin C, the calcium binding protein involved in muscle contraction (Means and Dedman, 1980). Occupancy of calcium binding sites results in a conformational change in calmodulin so that this regulatory protein interacts with a number of cellular enzymes to alter their activity (Cheung, 1980). Calcium-calmodulin activates myosin light chain kinase to stimulate contraction in smooth muscle and other cells. Calcium-calmodulin stimulates phosphorylase kinase to increase glycogen breakdown. Calcium-calmodulin activates a protein kinase which is highly expressed in the nervous system in synaptic vesicles (Kelly et al., 1984). This calmodulin-dependent protein kinase undergoes sequential regulatory self-phosphorylations to modify its activity even after calcium is removed (Schworer et al., 1986). Calcium-calmodulin activates cyclic nucleotide phosphodiesterase and guanylate cyclase; it also acti-

vates adenylate cyclase in some cell types. Calcium-calmodulin activates membrane-bound enzymes, such as phospholipase A_2, and is an important component of the mitotic apparatus, where it increases microtubule disassembly. Calcium-calmodulin affects additional cellular processes, suggesting that a number of enzymes have evolved with calmodulin recognition sites. As with cAMP-dependent protein kinase, which catalyzes phosphorylation of a number of substrates, calmodulin interacts with a number of targets to induce diverse cellular effects dependent on the genetic expression of each cell type. Calcium-calmodulin interactions with cyclic nucleotide systems provides a mechanism for intracellular communication.

Many hormones deliver information to cells via interaction with receptors which regulate the phosphatidylinositol (PtdIns) cycle. This metabolic cycle produces two intracellular second messengers: DAG and calcium (Fig. 8.6). Like hormone-sensitive adenylate cyclase, hormone receptors appear coupled to the enzyme phosphoinositidase (a phopholipase) via a G protein complex. PtdIns 4,5-P_2, the substrate for activated phosphoinositidase, is produced in the membrane by sequential phosphorylations of PtdIns (Berridge, 1987). On hydrolysis, PtdIns 4,5-P_2 yields inositol 1,4,5-trisphosphate (Ins 1,4,5-P_3) and DAG. Ins 1,4,5-P_3 mobilizes intracellular calcium from storage sites (Berridge and Irvine, 1984); this and other metabolites also facilitates calcium entry into the cell. Calcium acts via calmodulin and other calcium binding proteins; it also binds to protein kinase C to participate in enzyme activation. DAG binding to protein kinase C activates this important regulatory enzyme (Nishizuka, 1984). Because protein kinase C is the receptor for tumor promoters which function as long-acting DAG analogues, this enzyme mediates the diverse effects of tumor promoters as well. Sphingosine, a building block of sphingomyelin and glycosphingolipids, may function as an inhibitor of protein kinase C to provide dual control of this important regulatory protein (Hannun et al., 1986).

Protein Tyrosine Kinases

Several hormone receptors contain intrinsic protein tyrosine kinase activity. Ligand binding to the extracellular domain of the receptor results in allosteric activation of the intracellular protein tyrosine kinase domain. The intrinsic protein tyrosine kinase activity mediates all measured biological effects (Chen et al., 1987; Rosen, 1987). Because the transforming proteins of several RNA tumor viruses are also tyrosine-specific protein kinases and because several tumor-produced growth factors also stimulate tyrosine-specific kinase activity, this

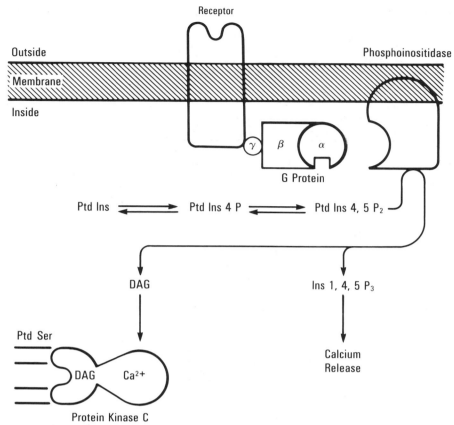

Figure 8.6. Production and action of the two intracellular mediators: DAG and Ins 1,4,5 P_3. DAG activates protein kinase C via coordination with calcium and phosphatidylserine (PtdSer) which anchors the acti- vated enzyme in the cell membrane. Ins 1,4,5 P_3 causes release of sequestered intracellular calcium.

class of enzymes may mediate cell growth (Bishop, 1985; Todaro et al., 1980). The cell surface receptors for insulin, IGF-1, EGF, PDGF, and macrophage colony-stimulating factor are examples of this mechanism of hormone-induced signal transduction. Hormone-receptor interactions likely directly affect additional membrane-associated enzymes, as well as exerting effects on cellular transport processes.

INTRACELLULAR RECEPTORS FOR STEROID AND THYROID HORMONES

Steroid Hormones

Steroid hormones are made in adrenal and gonadal tissues from cholesterol. The adrenal cortex produces cortisol, aldosterone, and androgen precursors; the ovary produces estrogen and progesterone; and the testis produces testosterone. Vitamin D is also synthesized from precursor cholesterol and is a steroid hormone which regulates mineral metabolism. Although each steroid hormone has a basic cholesterol nucleus to which different modifications and substitutions have been

made, the receptor proteins for each possess sufficient specificity and affinity to provide recognition for a single class of steroid hormones.

In contrast to peptide hormone receptors which are located in cell membranes, receptors for steroid hormones are found within the cell. Steroid hormone receptors are members of an extended gene family which includes the receptors for thyroid hormone, vitamin D, and retinoic acid (Hollenberg et al., 1985; Green et al., 1986; Loosfelt et al., 1986; Sap et al., 1986; Weinberger et al., 1986; Giguere et al., 1987; McDonnell et al., 1987) (Fig. 8.7). Their structures are highly conserved with a central DNA-binding domain containing Zn^{2+}-binding "fingers" characteristic of regulatory proteins which bind to DNA (Miller et al., 1985). The steroid binding domain is located at the C' terminus; the N' terminus, which is more variable, is readily distinguished by antibodies.

Hydrophobic steroid hormones are soluble in the lipid environment of the cell membrane and likely enter cells by passive diffusion (Fig. 8.8). They are retained within cells by binding to specific receptor proteins; cells which contain these receptors are the

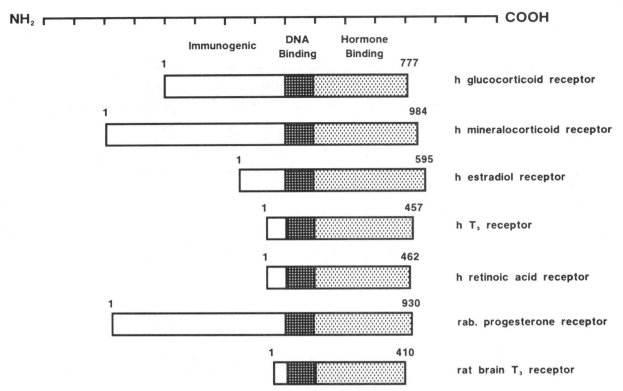

Figure 8.7. Structures of steroid hormone receptors.

target for that particular steroid hormone, changing their function in response to the hormone. Biological responses are proportional to the number of receptors with bound steroid hormone, suggesting that there are no intermediate second messengers, as with polypeptide hormones.

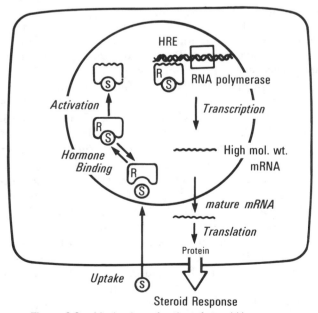

Figure 8.8. Mechanism of action of steroid hormones.

After binding the steroid hormone with high specificity and affinity, the receptor protein is activated. Activation, which requires heat in addition to substances present in the cell, results in a physical alteration in the receptor so that it binds tightly in the nucleus to chromatin, which is the site of action of steriod hormones. Binding of the hormone and activation expose the DNA-binding domain so it can bind tightly to specific DNA sequences (Hollenberg, et al., 1987). The DNA sequences or hormone response elements are enhancers which by binding the proteins at specific sites permit them to regulate transcription of specific genes. There is strong conservation in structure of the hormone receptor and of the response elements, but these differ sufficiently to allow specificity in ligand binding and specificity in DNA binding. This is manifest in biological responses specific to each hormone.

Steroid hormone receptors are one family of a larger class of transcriptional factors which control gene expression (Dynan and Tijian, 1985). These proteins provide cells with mechanisms to adjust gene transcription in response to environmental and developmental signals, many of which are hormonal. The DNA-binding regions of these protein factors target them to specific sites on the gene. The protein then either enhances or suppresses transcription of a particular gene by RNA polymerase. The exact molecular mechanism of action of transcription factors is incompletely defined but may involve looping of the DNA so these proteins can interact with RNA polymerase to control its activity (Ptashne, 1986).

The major effect of steroid hormones is to increase the synthesis of messenger RNA. This has been well documented in many cases, for example, in cortisol induction of tyrosine aminotransferase, tryptophan oxygenase, and growth hormone; estrogen induction of ovalbumin, vitellogenin, and prolactin; and progesterone induction of avidin, α_2-globulin, and aldolase. Even when the biological effect is an inhibitory one, as in cortisol-induced inhibition of the inflammatory response, specific proteins are induced (Blackwell et al., 1980). Steroid hormones may decrease mRNA synthesis as well. The inhibitory effect of cortisol on production of the messenger RNA for the ACTH precursor is a well-studied example (Nakanishi et al., 1977).

Although receptors exhibit a high degree of specificity for their particular steroid, crossover may occur. For example, aldosterone receptors may be occupied by cortisol so that higher concentrations of the latter hormone cause sodium retention similar to that induced by aldosterone. Steroid hormone receptors may also be occupied by antagonists as well as by agonists. Antagonist occupancy of binding sites induces an inactive receptor conformation. For example, spironolactone, a mineralocorticoid antagonist, forms a spironolactone-mineral ocorticoid receptor complex which does not bind to nuclear components (Marver et al., 1974). Tamoxifen, an estrogen antagonist, allows nuclear binding, but not actuation of gene transcription.

Thyroid Hormone

Like steroid hormones, thyroid hormone acts by binding to specific intracellular receptors. The intracellular receptors for thyroid hormone are located in the nucleus bound to chromatin (Fig. 8.1). Nuclear thyroid hormone receptors bind T_3, the metabolically active form of thyroid hormone, with an affinity greater than that for T_4, the precursor of T_3. Thyroid hormone occupancy of nuclear receptor sites increases expression of genetic information by increasing the synthesis of specific messenger RNAs which code for enzyme proteins specifying the biological response. The pattern of thyroid hormone-induced proteins depends not only on the particular target cell but also on the nutritional status of the animal (Oppenheimer and Dillmann, 1979). This provides integrated appropriate responses among various hormonal signals.

BIOSYNTHESIS AND METABOLISM OF HORMONES

Although regulation of hormone synthesis and metabolism will be considered in chapters dealing with individual endocrine glands, general features of these processes, which are common to all hormones of the peptide and steroid classes, will be considered here. Thyroid hormones and biogenic amines will be specifically covered in Chapters 54 and 55.

Peptide Hormones

Because peptide hormones are secretory proteins, their biosynthesis and secretion follow pathways of protein synthesis and packaging common to secretory proteins. The genetic information in DNA which codes for peptide hormones is discontinuous with intervening sequences between regions which code for protein (Fig. 8.9). The gene is initially transcribed into a large molecular weight messenger RNA precursor. Intervening sequences (introns) are progressively removed, and the coding portions (exons) are spliced together into the mature messenger RNA, which is capped and tailed. This genetic arrangement is thought to have facilitated evolution through recombination. Differential splicing also provides for generation of more than one hormone from a single gene region. The calcitonin gene

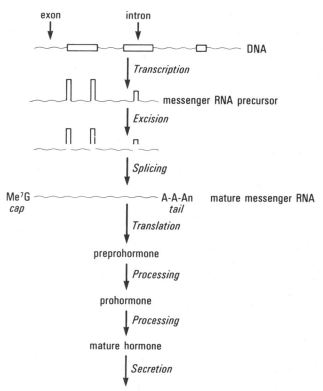

Figure 8.9. Biosynthesis of peptide hormones.

product is differentially spliced in the C cells to yield calcitonin and in the hypothalamus to yield a different peptide product (Amara et al., 1982). The major regulation of protein hormone biosynthesis by steroid hormones, polypeptide hormones, and other environmental signals is through increases or decreases in the rate of transcription of the gene into messenger RNA.

Messenger RNA is transcribed into protein which is in some cases considerably larger than the final mature circulating hormone. There is a "pre" sequence, coded from the 5' end of the messenger RNA, which contains many hydrophobic amino acids. This leader or signal sequence is characteristic of secretory proteins and specifies vectoral synthesis of the protein into the inside of the membrane of the endoplasmic reticulum (Lingappa and Blobel, 1980). The signal sequence is removed within the membrane space. The remaining prohormone is transferred within the cell inside the membranes of the endoplasmic reticulum to the Golgi complex, which packages the protein into secretory granules. The prohormone is converted by further proteolytic activity to the mature hormone within the Golgi and secretory granules. The extent and location of the proprotein removed varies among different peptide hormones. In proinsulin the connecting, or "C", peptide which joins the A and B chains and facilitates proper disulfide bonding is removed from the middle of the prohormone, and mature insulin containing two peptide chains is secreted. ACTH is processed from a large pro-form which contains information for lipotropin, melanocyte-stimulating hormone, and endorphins on the carboxy-terminal and information which may affect steroid synthesis on the amino-terminal side of ACTH (Mains et al., 1977). The pro-form of smaller peptide hormones may thus contain sequences for more than one biologically active peptid hormone which may be generated during proteolytic processing of the prohormone. Certain larger polypeptide hormones such as prolactin and growth hormone have little or no pro-sequences.

Secretory granules fuse with the plasma membrane to discharge the mature peptide hormone into the circulation. This process is under hormonal regulation and requires calcium. When a producer cell is highly stimulated, secretory granules which store hormone are largely depleted; conversely, when the cell is quiescent, secretory granules accumulate.

Peptide hormones are short lived within the circulation. Small neurohormones are so rapidly degraded that significant concentrations are reached only in the hypothalamic-pituitary portal circulation; insulin has a half-life of about 7 min and ACTH has a half-life of 3 min; the glycoprotein hormones such as TSH, LH, and FSH have the longest half-lives, ranging up to 180 min for FSH. Peptide hormones are destroyed by circulating proteases, via lysosomal proteases after receptor binding, and for the glycoproteins after binding to the hepatic asialoglycoprotein receptor. It is evident that continuous synthesis and secretion must occur to maintain circulating concentrations of these short-lived species. It is also evident that a form of continuous parenteral administration is required when peptide hormones are given to replace glandular secretion.

Steroid Hormones

All steroid hormones are derived from cholesterol. Many of the modifications of the cholesterol nucleus are common between classes of steroid hormones, although sufficient differences exist for specific recognition by receptors and consequent specific biological action. Cholesterol is obtained either from the circulation via LDL uptake (HDL in some species) or via biosynthesis from acetate. Steroidogenic tissues require large amounts of cholesterol and have increased numbers of LDL receptors and possess many lipid droplets containing cholesterol esterified to fatty acids. The principal modification of cholesterol precursor involves hydroxylations which require NADPH and molecular oxygen and which are catalyzed by specific cytochrome P-450 enzymes. Dehydrogenase, isomerase, and aromatase enzymes also modify the cholesterol nucleus. In all steroidogenic tissues (adrenal cortex, ovary, testis), the rate-limiting step in steroid hormone biosynthesis which is under trophic hormone (ACTH, LH, FSH) control lies in the conversion of cholesterol to pregnenolone (Fig. 8.10). This step involves the mitochondrial cytochrome P-450 enzyme which hydroxylates and cleaves the side chain of cholesterol. Once this slow step is speeded up, the flow of substrate through the biosynthetic enzyme pathway is rapid. Steroidogenic tissues contain large numbers of specialized mitochondria and an extensive network of endoplasmic reticulum. Substrates flow from mitochondria to endoplasmic reticulum and back to mitochondria as the molecule is progressively modified. Modifications depend on the particular biosynthetic enzymes which are expressed in each gland. For example, 11β-hydroxylase is largely localized to the adrenal cortex while aromatase is largely localized to the ovary. In contrast to peptide hormones, steroid hormones are not stored but are secreted as synthesized. Increased

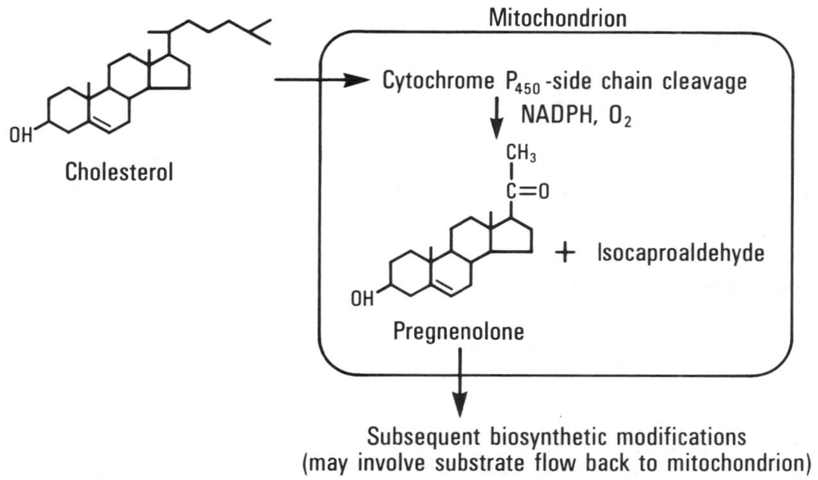

Figure 8.10.   The rate-limiting step of steroid hormone biosynthesis.

secretion therefore directly represents increased synthesis.

After secretion into the circulation, steroid hormones are bound to transport proteins. The transport glycoproteins are synthesized in the liver and bind most circulating steroid hormones. Corticosteroid-binding globulin (CBG or transcortin) binds more than 90% of circulating cortisol, and sex steroid-binding globulin binds more than 90% of circulating testosterone and lesser amounts of estradiol. Aldosterone is weakly bound to albumin rather than to a specific transport protein. The free unbound fraction of circulating steroid is the active species which enters cells to bind to specific receptor proteins. Plasma binding provides a reservoir of hormone protected from metabolism and renal clearance which can be released to cells. The circulating half-life of steroid hormones is thus significantly longer than that of peptide hormones. Transport proteins which exhibit specific binding are not, however, required for the action of steroid hormones. Genetic defects in transport proteins are compatible with normal endocrine function, and a number of active synthetic analogs bind poorly to transport proteins.

Metabolism of steroid hormones occurs principally in the liver and results in molecules which are biologically inert. These are conjugated with sulfate or glucuronide to render them water soluble for excretion. The principal exception to the general rule that metabolism leads to inactivation is testosterone. In target tissues testosterone is converted to dihydrotestosterone by the enzyme 5α-reductase. This conversion is required in many target tissues in which dihydrotestosterone is the active androgen species preferentially bound by receptors.

ORGANIZATION OF ENDOCRINE CONTROL SYSTEMS

One principal function of the endocrine system is to maintain internal homeostasis despite changes in the external environment. The second principal function is reproduction. In both of these processes, multiple hormones cooperate to bring about appropriate biochemical and physiological responses. For example, the primitive signal of glucose (substrate) lack has expanded to the broader signal of stress or fright and evokes a coordinated neural and endocrine response (Tomkins, 1975). Increases in glucagon and catecholamines stimulate glycogenolysis to release glucose from the liver into the circulation for immediate use by critical organs such as the brain. Cortisol is produced in greater amounts by the adrenal cortex and induces gluconeogenesis so that hepatic glucose production is maintained from muscle and other body proteins. Lipolysis is stimulated by a number of hormones, including epinephrine, glucagon, ACTH, and growth hormone, and FFA are released into the circulation to be used as an alternate metabolic fuel. Insulin, which signals glucose utilization, is suppressed. This metabolic adaptation is coordinated with a number of physiological and behavioral changes, such as increased sympathetic nervous system activity causing peripheral vasoconstriction and increased blood pressure.

Reproduction requires coordinated endocrine changes. During the menstrual cycle the ovarian follicle develops in response to the pituitary hormones, LH and FSH, so that ovulation can occur during the midcycle surge of LH. The uterus undergoes coordinated changes in response to ovarian estrogen and after ovulation to ovarian progesterone produced by the corpus luteum. This allows preparation of a uterine implantation site for a fertilized egg. Sexual behavior is strongly affected by changes in endocrine function.

Mechanisms which underlie these coordinated responses are many and include neural input to the endocrine system as well as hormonal modulation of nervous function, regulation of hormone biosynthesis and secretion, alterations in hormone receptors, and changes in metabolism of hormones.

During evolution many endocrine glands became distant from the nervous sytem. A hierarchy of neuroendocrine control developed, with the hypothalamus serving as the proximal neural signaling station to produce neurohormones which reach the pituitary gland via a short portal circulation (Fig. 8.11). The anterior pituitary produces a number of metabolic and reproductive peptide hormones which are trophic, i.e., which control the production of other hormones by distal endocrine glands. This arrangement provides for integration with the nervous system and for an expansion and modulation of the effects of the initial signal.

Once a hormonal response has been initiated and an appropriate metabolic or reproductive response has occurred, the signal must be terminated. Feedback control is one of the principal mechanisms through which this occurs. As shown in Figure 8.11, the hormonal product of the peripheral endocrine gland, such as the thyroid, adrenal cortex, testis, or ovary, exerts negative feedback control over the production and secretion of the stimulatory pituitary hormone. Feedback occurs at the level of the pituitary producer cell and likely at the level of the central nervous system and hypothalamus as well. Conversely, when concentrations of the peripheral endocrine gland hormone are low, feedback inhibition is lessened so that increased production of pituitary hormone occurs to stimulate production of the lowered hormone—and thus reestablish homeostasis. Feedback control occurs as well in endocrine glands not under pituitary control. For example, an

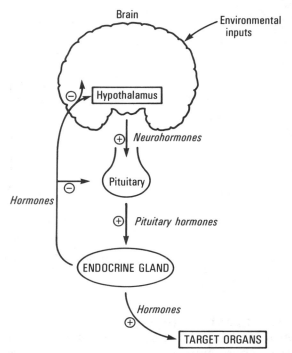

Figure 8.11. Feedback control of hypothalamic-pituitary-peripheral endocrine gland function.

elevation in serum Ca^{2+} feedback inhibits production of parathyroid hormone. A decrease in serum Ca^{2+} concentration results in elevated PTH which acts on bone, kidney, and gastrointestinal tract to reestablish a normal serum Ca^{2+} concentration.

The sophisticated endocrine control system which will be considered in subsequent chapters evolved from more primitive systems whose funciton was similar—regulation. cAMP is found in bacteria and serves as a symbol of glucose lack. Polypeptide hormones such as insulin and ACTH are made by paramecia. These primitive signals have evolved into the more complex systems necessitated by organ specialization. This communication system coordinates behavior with metabolism and reproduction in a manner most conducive to survival of the individual and species.

CHAPTER 53

The Hypothalamic-Pituitary Control System

The pituitary is the proximal endocrine gland link to the central nervous system for neuroendocrine control of both metabolism and reproduction. Neurohormones synthesized in the hypothalamus reach the anterior lobe of the pituitary via a specialized portal vascular system. These neurohormones control synthesis and secretion of the six major peptide hormones of the anterior pituitary; these anterior pituitary hormones regulate peripheral endocrine glands (thyroid, adrenal, gonads), as well as growth and lactation. The posterior lobe of the pituitary is composed of axons derived from neuronal cells of the hypothalamus and serves as a storage site for two peptide hormones which are synthesized in the hypothalamus but which act in the periphery to regulate water balance, uterine contraction, and milk ejection.

Isolation, characterization, and synthesis of neurohormones has been a formidable task because these peptides are synthesized in only a small number of cells which are distributed in several areas of the hypothalamus. Only small amounts of specific neurohormones are produced by these cells, and significant concentrations occur only in local areas such as the hypophyseal portal vascular system. Collection of over 1 million animal hypothalami and a Herculean labor of protein isolation were required to yield the presently known neurohormones (Table 8.1) (Burgus et al., 1969; Schally et al., 1971; Brazeau et al., 1973; Vale et al., 1981; Guillemin et al., 1982; Rivier et al., 1982). Because these are small peptides, once the structure was known, chemical synthesis yielded large amounts of both natural and modified neurohormones, and by using immunological and pharmacological techniques, it was found that these hypothalamic neurohormones are produced in the periphery as well as in the hypothalamus and function in local paracrine systems, especially in the gastrointestinal tract; they also function as specialized neurotransmitters in the central nervous system. Neurohormones are being increasingly identified by molecular biology techniques.

Identified neurohormones and their target cells in the anterior pituitary are shown in Figure 8.12. Two patterns of action are evident; first, control of more than one target cell by a single neurohormone and, second, both positive and negative regulation of particular anterior pituitary cell types. For example, thyrotropin-releasing hormone (TRH) stimulates production and secretion of both thyroid-stimulating hormone (TSH) and prolactin. Under pathological conditions TRH may also stimulate production and secretion of growth hormone. Somatostatin (SS) also affects more than one cell type, exerting negative control over growth hormone and TSH synthesis and secretion. Growth hormone-producing cells provide an example of dual control: these cells are stimulated by growth hormone-releasing hormone (GRH) and are inhibited by SS. The rate of production of growth hormone depends on the relative strength of these two stimuli.

Neurohormones are peptides, except for the biogenic amine dopamine, which is a potent prolactin-inhibiting hormone. Regulation of most anterior pituitary function depends primarily on positive stimulatory signals from the hypothalamus. When the anterior pituitary is removed from its hypothalamic connections and transplanted in a distal site such as the kidney capsule, production of all anterior pituitary hormones except prolactin ceases. Increased prolactin production indicates that dopamine and prolactin-inhibiting hormone (PIH) are the principal regulations of lactotrophs because when the negative influences are removed, prolactin production increases.

ANATOMY

Neurohormones are synthesized in specialized neurosecretory cells which are concentrated within the hypothalamus. Numerous nerve fibers terminate around these cells to bring information from other parts of the nervous system important for regulating production of neurohormones. The hypothalamus, which is located below the thalamus, lies

Table 8.1.
Hypothalamic-Releasing Hormones

Neurohormone	Structure	Pituitary Site of Action	Effect	Extrahypothalamic Synthesis	Extrapituitary Effects
Thyrotropin releasing hormone (TRH)	PyroEHPNH$_2$	Thyrotroph Prolactotroph	Stimulate Stimulate	Brain and GI tract	Yes
Growth hormone-releasing hormone (GRH)	YADAIFTKSYRKVLGQLSARKLLQDI-MSROOGESNQERGARARL	Somatotroph	Stimulate	Pancreas	Yes
Somatostatin (SS)	AGCKNFFWKTFTSC	Somatotroph Thyrotroph	Inhibit Inhibit	Brain, pancreas, and gastrointestinal tract	Yes
Gonadotropin-releasing hormone (GnRH)	PyroEHYSYGLRPGNH$_2$	Gonadotroph	Stimulate	Gonads	Yes
Corticotropin-releasing hormone (CRH)	SQEPPISLDLTFHLLREVLEMTKAD-QLAQQAHSNRKLLDIANH$_2$	Corticotroph	Stimulate	Pancreas	Yes
Prolactin-inhibiting hormone (PIH)	DAENLIDSFQEIVK EVGQLAETQRFEC TTHQPRSPLRDLK GALESLIEEETGQ KKI	Prolactotroph	Inhibit	Gonads	
Dopamine	HO—⬡—CH$_2$CH$_2$NH$_2$ (HO)	Prolactotroph	Inhibit	Adrenal medulla and nervous system	Yes

*Single letter symbols for amino acids are used.

behind the optic chiasm, between the optic tracts and anterior to the mammillary bodies. Anatomic localization of neurosecretory cells which produce a specific neurohormone is not precise, except for the large cells of the supraoptic and paraventricular areas, which synthesize antidiuretic hormone (ADH) and oxytocin, which are stored in the posterior lobe of the pituitary gland.

Neurohormones reach the anterior pituitary gland via a specialized portal vascular system. Nerve fibers originating in neurosecretory cells terminate in the median eminence of the hypothalamus located below the third ventricle, and the highest concentrations of hypothalamic neurohormones are found there. Blood supply to the anterior pituitary is through branches of the internal carotid artery, principally the superior hypophyseal artery (Fig. 8.13). Branches of the superior hypophyseal

artery form a capillary plexus in the median eminence and upper portion of the pituitary stalk. The region of the median eminence has a poorly developed blood-brain barrier so that blood which reaches the anterior pituitary via the portal vessels has first been in contact with the nervous tissue of the median eminence. The portal vessels, which arise from this capillary network and carry neurohormones, terminate in sinusoidal capillaries in the anterior lobe of the pituitary. This specialized cir-

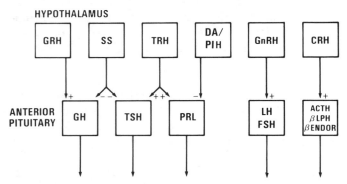

Figure 8.12. Regulation of anterior pituitary cell function by hypothalamic neurohormones.

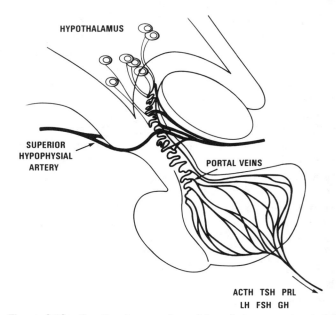

Figure 8.13. Functional connection of hypothalamus and anterior pituitary via the hypophyseal portal vascular system.

culation with two capillary networks delivers physiologically significant concentrations of neurohormones to their target cells in the anterior pituitary, but the concentrations of neurohormones which reach the peripheral circulation are insignificant. Retrograde flow may also occur via the hypophyseal portal system so that anterior pituitary hormones can also reach the hypothalamus.

In contrast to the anterior lobe of the pituitary, which is connected to the hypothalamus via the hypophyseal portal vascular system, the posterior lobe of the pituitary is an elongated extension of the ventral hypothalamus. Neurons from neurosecretory cells in the supraoptic and paraventricular nuclei of the hypothalamus terminate partially on the median eminence and primarily in the posterior lobe of the pituitary, where secretory granules containing ADH or oxytocin are in proximity to adjacent capillaries. Secretion of posterior pituitary hormones occurs in response to nerve impulses originating in hypothalamic cell bodies.

The pituitary gland which weighs approximately 500 mg is centrally located in the sella turcica, a bony cavity within the sphenoid (Fig. 8.14). The anterior lobe, which comprises about 80% of glandular weight, is derived from oral ectoderm, whereas the posterior lobe is derived from an outpouching of neural tissue from the region of the third ventricle. The pituitary stalk extends to the

hypothalamus through a dural reflection. An intermediate lobe located between anterior and posterior lobes is present in certain species and during fetal development, but it is vestigial in adult man. Because surrounding structures are vital ones, expansion of the pituitary due to tumor formation can result in superior extension with compression of the optic chiasm and visual loss or lateral extension with compression of cavernous sinuses containing cranial nerves.

ANTERIOR PITUITARY

Adrenocorticotropin (ACTH) and Related Peptides

BIOSYNTHESIS AND STRUCTURE

ACTH, the principal regulator of the adrenal cortex is a 39-amino acid single chain polypeptide hormone which is derived by proteolytic processing from a larger precursor protein made in corticotrophic cells of the anterior pituitary (Eipper and Mains, 1980). The structure of the precursor protein deduced from the nucleotide sequence of cloned DNA complementary to messenger RNA of the bovine species is shown in Figure 8.15 (Nakanishi et al., 1979). A 26-amino acid signal sequence containing a middle region rich in hydrophobic residues directs synthesis of the precursor protein into membrane structures for ultimate secretion. The core sequence for melanocyte-stimulating hormone

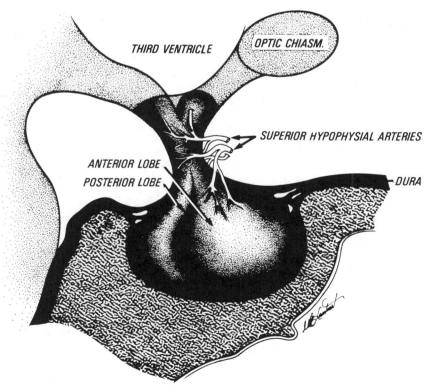

Figure 8.14.   Anatomy of the sella turcica and surrounding structures (sagittal view).

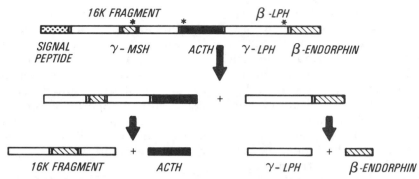

Figure 8.15. Structure of the ACTH precursor molecule. MSH core sequences are indicated by *. γ-MSH is located within the 16,000 M$_r$ amino terminal fragment; α-MSH is located at the amino terminus of ACTH; β-MSH is located at the carboxy terminus of γ-LPH. Dibasic amino acid residues which serve as the recognition site for proteolytic processing enzymes are indicated by *open spaces* in the *bars*. (Adapted from Nakanishi et al., 1979; Eipper and Mains, 1980.)

(MSH) is reiterated within the structure of the precursor, suggesting that it arose during evolution by gene duplication. A 16,000 M$_r$ fragment is located on the amino-terminal side, and β-lipotropin is located on the carboxyl-terminal side of the ACTH sequence. Dibasic amino acid residues serve as recognition signals for proteolytic enzymes which cleave the precursor protein into the final hormonal products. The first step is separation of the β-LPH containing carboxy-terminus. The amino-terminus then undergoes further processing to yield mature ACTH, and the 16,000 M$_r$ fragment, whereas β-LPH is processed into γ-LPH and β-endorphin. Although β-endorphin contains the sequence for met-enkephalin, β-endorphin is the final product which accounts for opiate agonist activity derived from this pathway.

Biological activity resides in the first 20 amino acids of the ACTH sequence, which are identical in all mammalian species studied (Fig. 8.16) (Riniker et al., 1972). Species variations occur in the carboxy-terminal 15 amino acids which are not requried for biological activity. The amino-terminal 13 amino acids are essential for biological activity, and the sequence of basic amino acids lys[15]-lys[16]-arg[17]-arg[18] is important for binding to adrenocortical receptors (Gill, 1979). Residues 1–13 of ACTH correspond to the sequence of α-MSH and account for the intrinsic skin darkening effects of the ACTH molecule.

Synthesis and secretion of ACTH are regulated by corticotropin-releasing hormone (CRH), a 41-amino acid peptide produced in hypothalamic neurosecretory cells located near the supraoptic and paraventricular nuclei (see Fig. 8.33, Chapter 55) (Vale et al., 1981). CRH has significant sequence homology with Sauvagene, a peptide isolated from frog skin which was previously found to have CRH activity (Erspamer and Melchiorri, 1980). ACTH secretion is also stimulated by ADH, and under certain conditions CRH and ADH synergize to give maximal ACTH production and secretion (Yasuda et al., 1982). CRH binds to receptors on ACTH-producing cells to stimulate formation of cAMP and, as expected, CRH stimulates coordinate production and secretion of both ACTH and β-endorphin. Feedback inhibition of ACTH and β-endorphin production by cortisol occurs at the level of the corticotroph cell of the anterior pituitary where CRH and cortisol have opposing effects (Allen et al., 1978). Glucocorticoids decrease CRH-stimulated cAMP formation (Bilezikjian and Vale, 1982), transcription of the ACTH gene (Charron and Provin, 1986) and production of messenger RNA for the ACTH precursor protein (Nakanishi et al., 1977). Inhibition of hypothalamic CRH production by cortisol has been implied from indirect studies (Dallman, 1979) and immunocytochemistry co-localizes the glucocorticoid receptor to CRH-producing neurons in addition to other brain areas (Gustafsson et al., 1987).

The intermediate lobe of the pituitary is composed almost entirely of cells which synthesize the ACTH precursor protein. In the intermediate lobe, ACTH is processed to α-MSH in which the amino-terminal serine is acetylated and the amino-terminal valine is amidated from the adjacent glycine at residue 14 and to corticotropin-like intermediate

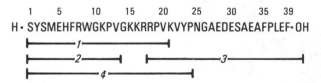

Figure 8.16. The amino acid sequence of human ACTH. Numbered segments indicate: *1*, the first 20 amino acids which possess full biological activity; *2*, sequence of α-MSH; *3*, sequence of CLIP; *4*, the major synthetic ACTH available for clinical use. Single letter symbols for amino acids are used.

peptide (CLIP, $ACTH_{18-39}$). Processing of the precursor protein thus varies between anterior and intermediate lobes, yielding ACTH in the anterior pituitary and α-MSH and CLIP in the intermediate lobe. Regulation also differs. The intermediate lobe is controlled by neural inputs, including dopamine as an inhibitory neurohormone, but responds poorly to CRH and glucocorticoids. Although the intermediate lobe is vestigial in man after fetal life, some cases of Cushing's disease with overproduction of ACTH are resistant to suppression with glucocorticoids but suppress with dopamine, a pattern similar to that which occurs in intermediate lobe cells (Lamberts et al., 1982).

THE CRH-ACTH-CORTISOL AXIS

The response to stress which involves multiple hormones, including ACTH, catecholamines, ADH, angiotensin, glucagon, and growth hormone, is integrated by the nervous system. The CRH-ACTH-cortisol axis is central to these integrated responses to stress (Rivier and Plotsky, 1986). The most primitive signal was substrate lack, and the gluconeogenic actions of cortisol are essential for adaptation to fasting. The signal has become more complex, and this axis has become an important part of a general neuroendocrine response to a variety of noxious environmental stimuli such as pain, trauma, and hypovolemia. CRH neurosecretory cells are regulated by a number of different afferent nervous impulses which arise in various regions of the nervous system and go through several synaptic intermediates prior to connection with CRH neurons. Several neurotransmitters, including norepinephrine, serotonin, acetylcholine, dopamine, and GABA, act on intermediates in this complex pathway. When CRH is released, it activates the sympathetic nervous system, in addition to stimulating ACTH biosynthesis and secretion (Brown et al., 1982). Centrally administered CRH increases blood pressure, heart rate, and behavioral responses characteristic of stress (Fisher et al., 1982). Activation of the sympathetic nervous system includes the adrenal medulla and its catecholamines. In response to CRH, both ACTH and β-endorphin are released in equimolar quantities (β-endorphin is present in secreted β-LPH and as mature β-endorphin). ACTH acts principally on the adrenal cortex to stimulate cortisol production, although ACTH also stimulates lipolysis in fat cells. Cortisol, through its actions on target organs, is the major regulator of adaptive responses to stress. The coordinate actions of cortisol and catecholamines mobilize substrate to maintain blood glucose and energy metabolism during the period of stress. Cardiovascular and behavioral responses are integrated with these metabolic adaptations in response to the CRH signal. Fluid volume and circulation are maintained in concert. ADH potentiates CRH actions on ACTH secretion and causes renal water conservation. ACTH and angiotensin increase adrenocortical aldosterone synthesis to enhance salt and water retention. These responses mediated through the hypothalamic-pituitary-adrenal-sympathetic nervous system allow survival under numerous adverse conditions.

The physiological role of the regions of the precursor protein, other than that of ACTH, are unclear. Skin pigmentation which accompanies states of ACTH excess is due to the intrinsic MSH activity of the ACTH molecule ($ACTH_{1-13}$) and to β-LPH and γ-LPH because β-MSH does not appear as a product of the ACTH precursor molecule in man. β-Endorphin does not exert analgesia in the periphery but may exert central effects, and ACTH-β-endorphin-containing cells are located within the nervous system. Products of the ACTH precursor protein may also effect nervous system responses. Both the amino- and carboxy-terminal regions of the ACTH precursor affect adrenal steroidogenesis, but whether this occurs under physiological conditions is uncertain.

Glycoprotein Hormones

The pituitary glycoprotein hormones, thyroid-stimulating hormone, luteinizing hormone (LH), follicle-stimulating hormone (FSH), and the placental hormone, chorionic gonadotropin (CG), are each composed of two subunits (α and β). The α subunits of all four glycoprotein hormones have the same amino acid sequence, whereas the sequences of the β subunits differ. Isolated subunits are inactive but can be combined to give fully active molecules which have the biological specificity of the β subunit (Pierce et al., 1971). Branched carbohydrate side chains are attached to α and β subunits (Pierce and Parsons, 1981). The α subunits each contain two oligosaccharides; β subunits contain one or two oligosaccharides. Human CG (HCG) of placental origin resembles LH with a 30-amino acid extension at the carboxy-terminus where additional carbohydrate is attached. The carbohydrate substitutions contribute to both the folding of α and β subunits into mature hormone (Weintraub et al., 1980) and to the relatively long circulating half-lives of glycoprotein hormones, compared to those of peptide hormones. Half-lives range from 30 min for LH to many hours for HCG.

Synthesis of TSH occurs in thyrotrophs which contain small basophilic storage granules and which constitute about 5% of anterior pituitary cells. The α and β subunits of TSH are synthesized on separate messenger RNAs and glycosylated prior to folding into the active dimer in secretory granules. More α and β subunits are made, and an excess of free α subunits is found in plasma under certain conditions. Mature TSH has a molecular weight of 28,000 and contains about 16% carbohydrate.

The two pituitary gonadotropins, LH and FSH are synthesized in the same cells. α and β subunits are synthesized on separate messenger RNAs, glycosylated in endoplasmic reticulum, and folded into the mature $\alpha\beta$ fully glycosylated structure prior to secretion. The interaction between subunits, while noncovalent, is very strong. As indicated for TSH, the intact $\alpha\beta$ dimer is required for biological activity.

TRH-TSH-THYROID HORMONE AXIS

TSH regulates the structure and function of the throid gland. TSH synthesis and secretion are controlled by neuroendocrine signals from the hypothalamus and by circulating thyroid hormones from the periphery (Table 8.2; see also Fig. 8.27 in Chapter 54). Thyrotropin-releasing hormone (TRH) is the major hypothalamic hormone which stimulates synthesis and secretion of TSH (Morley, 1981). TRH, a tri-peptide with the structure pyroglutamyl-histidyl-prolineamide, binds to receptors not only on thyrotrophs to stimulate TSH but also on lactotrophs to stimulate synthesis and secretion of prolactin (Fig. 8.12). Although cAMP reproduces some of the effects of TRH, it does not appear to be the intracellular mediator of TRH action, and there is better evidence for an important role of the Ptd Ins cycle and its two intracellular messengers: DAG and calcium in TRH action (Gershengorn, 1985). From studies of the effects of TRH on prolactin synthesis, it is clear that the tripeptide rapidly induces major changes in the rate of gene transcription. TRH thus has major regulatory effects not only on secretion ("releasing hormone") but on biosynthesis. Synthesis and secretion of TSH are inhibited by somatostatin, dopamine, and cholecystokinin, which reach the anterior pituitary via the hypophyseal portal system to act on the anterior pituitary; these also exert regulatory effects within the hypothalamus.

Table 8.2.
Control of TSH Synthesis and Secretion

Classification	Stimulated by	Inhibited by
Major	TRH	T_4, T_3
Minor	Estrogen	Somatostatin
		Cholecystokinin
		Glucocorticoids
		Androgens

The feedback effects of thyroid hormone occur at the level of the thyrotroph cells of the anterior pituitary. Increased concentrations of thyroid hormone decrease biosynthesis and secretion of TSH in part by decreasing the number of TRH receptors (Hinkel et al., 1981). Thyroid hormone, via binding to its receptor, also decreases transcription of the TSH-β chain gene through action on a strong promoter proximal to that gene (Carr et al., 1987). When thyroid hormone concentrations are high, TRH is ineffective in stimulating TSH secretion. When thyroid hormone concentrations decrease, this feedback inhibition is removed, and TSH synthesis and secretion increase in response to unopposed actions of TRH. Feedback effects of thyroid hormone also occur in hypothalamic TRH-producer cells. Thyroid hormone decreases TRH mRNA in these cells, implying negative regulation of transcription of the TRH gene (Segerson et al., 1987). Glucocorticoids and androgens have minor inhibitory, whereas estrogen has minor stimulatory, effects on TSH secretion.

Although under most conditions changes in circulating thyroid hormone concentrations are the most important determinants of plasma TSH concentrations (see Chapter 54), changes in TRH concentrations may also be important. In newborn humans and in many animals, acute exposure to cold increases plasma TSH; subsequent increases in thyroid hormone are important for thermogenesis. Because anti-TRH antibodies block cold-induced increases in TSH, increased TRH probably mediates this response (Szabo and Frohman, 1977). Only minor TSH responses to cold occur in adult humans. There are examples of hypothalamic hypothyroidism in which low TSH levels are reversed by administration of TRH. Plasma TSH concentrations exhibit a circadian rhythm with peak concentrations around midnight preceding the onset of deep sleep; this circadian rhythm in TSH concentrations may be due to changes in TRH.

HYPOTHALAMIC-PITUITARY-GONADAL AXIS

This axis, which is essential to the reproductive process in both males and females, is discussed fully in Chapters 58 and 59. Synthesis and secretion of both LH and FSH are stimulated by a single hypothalamic neurohormone, gonadotropin-releasing hormone (GnRH), a decapeptide (Fig. 8.12, Table 8.1). In females, elevations in GnRH initiate the midcycle LH surge which causes ovulation. Feedback regulation by ovarian estrogens and testicular androgens is complex. During the follicular phase of the menstrual cycle, estrogen exerts positive feedback effects, augmenting gonadotroph responses to GnRH (Yen et al., 1975). Estrogen also exerts negative feedback control over gonadotropin synthesis. This effect, which is also essential to normal progression of the menstrual cycle, is utilized in oral and injectable steroidal contraceptives. Androgens also exert negative feedback effects essential to normal spermatogenesis. In addition to feedback regulation of pituitary gonadotropin production by steroids, the gonads produce regulatory peptides which influence production, primarily of FSH. These peptides, containing α and β subunits, are termed inhibin and activin, reflecting their negative and positive feedback effects (Ling et al., 1985; 1986). Like müllerian inhibiting substance (MIS), which is involved in sexual differentiation, inhibin and activin have homology to the tumor growth factor β (TGF-β) gene family (Cate et al., 1986).

In males, LH acts on Leydig cells of the testis to stimulate testosterone biosynthesis. FSH acts on Sertoli cells and is essential for seminiferous tubule function and spermatogenesis. In females, FSH causes follicular growth in a process which depends on both LH and estrogens. Ovarian granulosa cells are the principal target for FSH. The LH surge at midcycle initiates ovulation, and LH regulates corpus luteum formation and function following ovulation.

Growth Hormone

STRUCTURE AND BIOSYNTHESIS

Human growth hormone is a 191-amino acid single chain polypeptide containing two intrachain disulfide bridges. The three-dimensional structure of porcine growth hormone consists of four α-helices arranged in a tightly packed antiparallel bundle (Abdel-Meguid et al., 1987). It has major sequence homology with the placental hormone chorionic somatomammotropin (CS) (83%) and lesser homology with prolactin (16%) (Niall et al., 1973). These three peptides arose from a common ancestral gene and although distinct, each causes some of the biological effects of the other. Prolactin evolved with growth hormone and somatomammotropin from a common ancestor by gene duplication about 400 million years ago (Cooke et al., 1980). The GH and somatomammotropin genes are located on chromosome 17, whereas the prolactin gene is located on chromosome 6 (Owerbach et al., 1980; 1981). Human GH has a high degree of sequence homology with growth hormones from other species, but only human or other primate GH is active in humans. Therefore, in the past, GH for human use was obtained by extraction from human pituitaries removed at autopsy. Using molecular cloning technology, the human gene has been expressed in bac-

teria, and this is now the major source of human GH.

Like other mammalian genes, that for GH consists of exons interrupted by introns. Several forms of GH, in addition to the 191-amino acid native peptide, are present in material extracted from human pituitaries (Lewis et al., 1980). One form which contains a deletion of 15 amino acids between residues 32 and 46 corresponds to removal of one exon; this form stimulates growth but lacks other actions of GH. In somatotrophs of the anterior pituitary, GH is made as a prohormone and after removal of the signal peptide is stored in large secretory granules. Somatotrophs are abundant in the anterior pituitary, and GH constitutes up to 10% of the dry weight of the gland. After secretion, GH has a circulating half-life of about 20 min.

REGULATION OF SYNTHESIS AND SECRETION

Growth hormone-releasing hormone (GRH) is the major stimulator, and SS is the major inhibitor of the synthesis and secretion of growth hormone. GRH is a 40- to 44-amino acid peptide which was originally isolated from a human pancreatic tumor (Table 8.1) (Guillemin et al., 1982; Rivier et al., 1982). SS is a 14-amino acid peptide which inhibits the synthesis and secretion of both GH and TSH (Vale et al., 1975). SS is found not only in the hypothalamus but also in other areas of the brain, as well as outside the central nervous system. It is synthesized in δ cells of the pancreas and inhibits secretion of both insulin and glucagon. SS also inhibits secretion of gastrin, secretin, and renin, and it decreases gastrointestinal motility. It functions therefore not only as a hypothalamic neurohormone but also in paracrine systems in the periphery and as a neuromodulator within the central nervous system.

GH synthesis and secretion are regulated by a number of factors (Table 8.3). Metabolic substrates, neuropharmacological agents, and peripheral hormones all affect GH secretion; many act via hypothalamic regulation of GRH and SS, but some act directly on anterior pituitary somatotrophs. Several of the stimuli are used clinically to assess GH secretory capacity in children with short stature or adults with suspected hypopituitarism, and several of the inhibitors are used to determine suppressibility of GH secretion in suspected acromegaly or gigantism.

Plasma levels of GH are high at the end of fetal life and during the first weeks after birth but then fall to normal resting values of 2–5 ng/ml. Under physiological conditions, GH secretion is suppressed by elevated plasma glucose concentrations

Table 8.3.
Control of Growth Hormone Synthesis and Secretion

Stimulated by	Inhibited by
Neurohormone	
GRH	Somatostatin
Central effects	
Hypoglycemia	Hyperglycemia
Amino acids	Free fatty acids
(arginine, lysine, leucine)	
Onset of deep sleep	
α-Adrenergic agonists	β-Adrenergic agonists
Dopamine	
Serotonin	
Enkephalins	
Pituitary effects	
Thyroid hormone	Progesterone
Cortisol (physiological concentrations)	Cortisol (supraphysiological concentrations)
Estrogen	

and is therefore low early after food intake (Fig. 8.17). When plasma glucose falls in response to postprandial increases in insulin, GH rises and functions to prevent excessively low plasma glucose concentrations. Insulin-induced hypoglycemia is a more major stimulus to GH secretion, but this procedure may cause excessive hypoglycemia when hypopituitarism is present. GH secretion is also stimulated by several amino acids, and infusion of L-arginine is a safe procedure for evaluating GH secretory capacity. Fatty acids inhibit GH secretion.

The metabolic factors act via the hypothalamus, as do most neuropharmacological agents. During a 24-hour period, elevations in GH occur after each meal, but the major peak of GH, accounting for up to 70% of total daily secretion, occurs during sleep. GH secretion occurs primarily during deep sleep (stages III and IV) and is triggered by the same neural mechanisms which initiate deep sleep (Weitzman et al., 1975). Nocturnal secretion of GH is particularly high in children at puberty, when nocturnal rises in LH, FSH, and TSH are also especially marked. α-Adrenergic agonists, L-dopa, serotonin, and enkephalins stimulate GH secretion, whereas β-adrenergic agonists inhibit secretion.

Thyroid hormone is required for GH synthesis. It acts directly on pituitary cells to enhance the rate of production of GH messenger RNA (Martial et al., 1977). In hypothyroidism the release of GH in response to hypoglycemic and arginine stimuli is diminished; treatment with thyroid hormone restores normal responsiveness. Thyroid hormone is also necessary for stimulation of GH synthesis by glucocorticoids (Samuels et al., 1977). Effects of glucocorticoids are biphasic. At physiological concentrations, cortisol stimulates GH synthesis and secre-

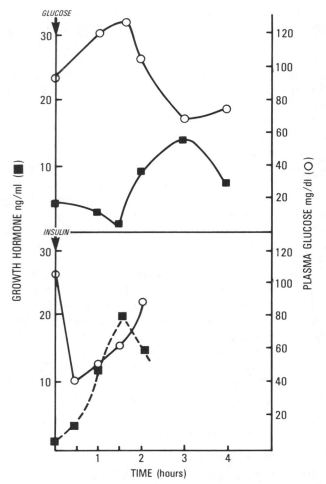

Figure 8.17. Response of plasma GH to a meal and to insulin-induced hypoglycemia.

tion, but at high concentrations cortisol inhibits GH production. Estrogens facilitate, whereas progesterone inhibits, growth hormone secretion.

BIOLOGICAL EFFECTS

GH has two major actions: stimulation of somatic growth and regulation of metabolism. Excesses or deficits in GH in man lead to striking changes in somatic growth. GH deficiency in childhood results in lack of growth and proportional dwarfism (ateliotic dwarfism). Treatment with GH results in marked acceleration of growth so that height, bone age, and body size become more correct for chronological age. Conversely, excesses of GH as occur with certain tumors result in marked acceleration of growth. If excess GH is present before puberty, gigantism will result. Often the GH-secreting pituitary tumor will compress and destroy surrounding pituitary tissue or interfere with the hypophyseal portal circulation and delivery of releasing hormones so that other pituitary hormones will be deficient. Without adequate gonadotropins, sex steroids will be deficient, epiphyses will not close, and linear growth will continue (Chapter 57). If excessive GH is present in adult life after epiphyseal closure, height will not increase but acral bond width, soft tissue, and organ growth will occur, producing the clinical syndrome of acromegaly.

The action of GH is initiated by binding to a specific receptor protein which contains a single transmembrane sequence separating its extracellular ligand-binding domain from a 350-amino acid cytoplasmic domain (Leung et al., 1987). The GH receptor does not resemble the protein tyrosine kinase gene family of growth factor receptors, but its extracellular domain appears identical to a plasma GH-binding transporter protein.

Growth is mediated in part by somatomedins, insulin-like growth factors, whose synthesis is controlled by growth hormone (Philips and Vassilopoulou-Sellin, 1980). An astute observation led to the discovery of somatomedins. It was found that the addition of GH to plasma obtained from hypophysectomized animals failed to stimulate cartilage explants, but if GH were administered to hypophysectomized animals, their plasma would stimulate cartilage explants (Salmon and Daughaday, 1957). This implied that a substance was generated in response to GH and that this substance, rather than GH itself, directly stimulated growth. Two somatomedins (A and C) have been characterized as two insulin-like growth factors (IGF-2 and IGF-1) (Svoboda et al., 1980). The primary structure of the IGF is similar to that of insulin and includes a portion equivalent to the connecting peptide of proinsulin (Fig. 8.18) (Blundell et al., 1978). There are distinct receptors for insulin and for each of the insulin-like growth factors. The structure of the receptor for IGF-1 resembles the receptor for insulin; both contain intrinsic protein tyrosine kinase activity (Ullrich et al., 1985). IGF-1 binds to insulin receptors

with a lower affinity than insulin and exerts insulin-like effects. Conversely, insulin binds to IGF-1 receptors with lower affinity than IGF-1 and exerts growth effects.

The receptor for IGF-2 is distinct. It is similar, if not identical, to the receptor for mannose 6-phosphate (Morgan et al., 1987). IGF-2 has actions similar to IGF-1, but its growth-promoting activities involve different signaling and biochemical pathways.

IGF-1 concentrations depend on GH (Fig. 8.19). In acromegaly, where GH levels are high, somatomedin levels are high, and in hypopituitarism, where GH levels are low, somatomedin levels are low. In one form of dwarfism, the Laron syndrome, GH concentrations are normal to elevated, but IGF-1 concentrations are low, suggesting that dwarfism results from IGF-1 deficiency (Daughaday et al., 1969). The observation that Laron dwarfs have low concentrations of plasma GH transport protein suggests that the primary genetic defect may lie in the GH receptor whose extracellular binding domain appears identical to the transport protein. Somatomedin production is also dependent on insulin and on nutritional status. In poorly controlled diabetic patients and in those with protein malnutrition, GH concentrations are elevated, but somatomedin concentrations are low. Growth is poor under both con-

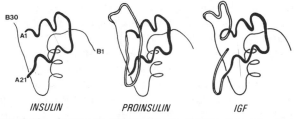

Figure 8.18. Schematic representation of three-dimensional structures based on x-ray analysis of insulin and on model building of proinsulin and IGF-1. The A chain of insulin is shown by a *thickened line*, the B chain by a *solid line* and the connecting peptide by a *dashed line*. (From Blundell et al, 1978.)

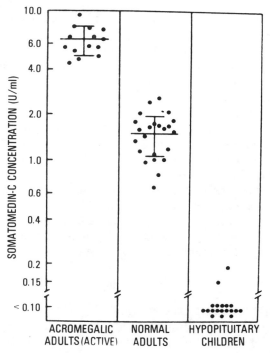

Figure 8.19. Somatomedin C concentrations in GH excess and deficiency states. (From Van Wyk and Underwood, 1978.)

ditions. Correction of the defects by either insulin administration to diabetics or by protein addition to deficient diets results in rises in somatomedin, even though GH concentrations decrease. As discussed in Chapter 57, growth is an integrated process requiring multiple hormonal signals in addition to adequate nutrition; one level of integration is at the level of somatomedin generation. Somatomedins are produced in the liver and in fibroblasts. They circulate bound to carrier proteins whose synthesis is also under GH control.

Metabolic effects of GH overlap with growth effects but may result directly from GH interactions with its receptor in addition to effects which are due to somatomedins. Metabolic effects of GH are biphasic. Acutely, GH exerts insulin-like effects. It increases glucose uptake in muscle and adipose tissue, stimulates amino acid uptake and protein synthesis in liver and muscle, and inhibits lipolysis in adipose tissue. Several hours after administration of GH, these initial effects disappear, and the more profound metabolic effects of GH occur. These effects which persist with prolonged elevations in plasma GH are antiinsulin-like. Glucose uptake and utilization are inhibited so that blood glucose rises, and lipolysis is increased so that plasma-free fatty acids (FFA) rise. GH, which rises during fasting, is thus important in the body's adaptation to lack of food. Along with cortisol, epinephrine, and glucagon, it maintains blood glucose for central nervous system use and mobilizes fat depots to provide alternate metabolic fuel.

Effects of GH are appropriate to the needs of the whole body. When adequate food in ingested and insulin is active, growth is appropriate, and the major effects of GH are anabolic and growth promoting. Under these conditions, GH stimulates macromolecular synthesis and growth in a number of organs. Insulin and adequate nutrition facilitate GH induction of somatomedins. During fasting, when insulin is low, GH increases blood glucose and FFA as part of the integrated hormonal response to substrate lack. Generation of somatomedins is impaired when insulin is low and nutrient intake is poor, so that growth is not enhanced. In insulin deficiency these GH responses may worsen the diabetic state.

Prolactin

STRUCTURE AND BIOSYNTHESIS

Prolactin is a single chain polypeptide hormone of 198 amino acids with three intrachain disulfide bridges (Shome and Parlow, 1977). Lactotrophs constitute approximately 30% of the cells of the ante-

rior pituitary, and an increase in lactotrophs accounts for most of the enlargement of the pituitary which occurs during pregnancy. Prolactin turns over more rapidly than GH so that pituitary stores are significantly less. The circulating half-life of prolactin is 30–50 min.

REGULATION OF SYNTHESIS AND SECRETION

In humans the major function of prolactin is regulation of milk production. During pregnancy, prolactin rises progressively to concentrations 10 fold higher than those in the nonpregnant state (Fig. 8.20) (Riggs and Yen, 1977). Increased estrogen concentrations of pregnancy are the major stimulator of this increase in prolactin production and lactotroph growth. After parturition, prolactin levels fall, but periodic rises in prolactin concentrations occur with each period of suckling. These increases in prolactin maintain milk production for subsequent feeding periods. Prolactin concentrations are high in fetuses during the last 10 wk of intrauterine life, high in amniotic fluid, and high in umbilical vein blood (Aubert et al., 1975). After parturition, infant prolactin levels fall to adult values over a 1- to 2-month period.

Prolactin synthesis is controlled principally by increasing or decreasing the rate of transcription of the prolactin gene. The major regulators are PIH (Nikolics et al., 1985) and dopamine, which inhibit prolactin synthesis (Fig. 8.12; Table 8.4). When the pituitary stalk is severed, as may occur with acceleration-deceleration head injuries, or when the anterior pituitary is transplanted away from its hypothalamic connections, prolactin synthesis increases while synthesis of all other anterior pituitary hormones decreases. Lactotrophs are thus unique among anterior pituitary cells in being under major hypothalamic restraint. Dopamine agonists such as bromoergocryptine are widely used to inhibit prolactin synthesis. Bromoergocryptine also frequently inhibits growth of prolactin-secreting pituitary tumors, suggesting that dopamine provides a physiological brake on lactotroph growth as well as prolactin synthesis (Thorner et al., 1980). Because a number of drugs either interfere with the synthesis of dopamine or act as dopamine antagonists, increases in prolactin commonly occur, leading to alterations in the menstrual cycle and to abnormal lactation. The anti-hypertensives reserpine and α-methyldopa deplete catecholamines, including dopamine, to increase prolactin. Phenothiazines and butyrophenones, potent antipsychotic drugs, are dopamine antagonists and increase prolactin synthesis by blocking the restraining

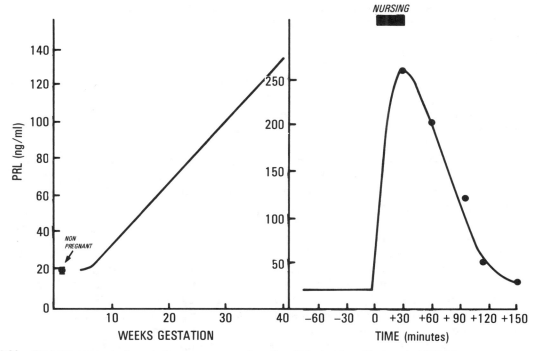

Figure 8.20. Prolactin concentrations during pregnancy and nursing. (Adapted from Noel et al., 1974; Riggs and Yen, 1977.)

influence of dopamine. Metoclopramide is another dopamine antagonist which increases prolactin synthesis and secretion. Glucocorticoids and thyroxine in higher concentrations inhibit prolactin synthesis by direct action on the lactotroph.

Although prolactin synthesis is under the major restraining influence of dopamine, hormone production responds to a number of stimulatory factors.

Table 8.4.
Control of Prolactin Synthesis and Secretion

Inhibitors	Stimulators
Dopamine	Dopamine antagonists
Dopamine agonists	Phenothiazines
Bromoergocryptine and other ergot alkaloids	Butyrophenones
	Metoclopramide
Prolactin-inhibiting hormone	Catecholamine depletors
Glucocorticoids	Reserpine
Thyroid hormone	α-Methyldopa
	TRH
	Estrogen
	Epidermal growth factor
	Sleep
	Pregnancy
	Suckling
	Sexual intercourse (females only)
	Exercise
	Hypoglycemia
	Arginine
	Serotonin
	Histamine H_2 antagonists
	Opiate agonists

TRH binds to lactotroph membrane receptors to stimulate prolactin gene transcription. Estrogen is the major stimulator of prolactin synthesis during pregnancy, acting directly on lactotrophs. Epidermal growth factor also increases prolactin synthesis although its role in normal physiology is uncertain (Schonbrunn et al., 1980).

Prolactin secretion is maximal during sleep, with peak levels occurring 5–8 hours after the onset of sleep. Even during pregnancy when prolactin concentrations are increased, diurnal peaks of secretion of prolactin continue. Nipple stimulation, in addition to suckling, causes increased prolactin production; increases in prolactin also occur with some chest wall lesions involving the nipple areas. Prolactin secretion also increases with exercise, with sexual intercourse (in females only), and with hypoglycemia and arginine administration. Serotonin, histamine H_2 antagonists, and opiate agonists are additional stimulators of prolactin secretion.

Prolactin is the most frequent hormone which is produced in excess by pituitary tumors. Because prolactin synthesis and secretion are influenced by many agents, evaluation of patients with hyperprolactinemia requires a careful history and physical examination to differentiate between factors included in Table 8.4 and a pituitary tumor. Clinical evaluation is especially important because stimualtion or suppression tests have been of little value in the differential diagnosis of hyperprolactinemia. Extremely high levels of prolactin (>200 ng/ml) indicate a pituitary tumor, but with levels this high the tumor is usually evident from radiographic procedures.

BIOLOGICAL EFFECTS

The breast is the principal site of action of prolactin in humans. Development of the breast requires estrogen, progesterone, glucocorticoids, and insulin in addition to prolactin. Chorionic somatomammotropin, which is synthesized in high concentrations by the placenta, also stimulates mammary gland development. During pregnancy there are both ductal and alveolar-lobular growth. Even though prolactin concentrations increase greatly, lactation does not occur because of inhibitory effects of progesterone and estrogen. After parturition progesterone and estrogen levels fall, and prolactin-dependent lactogenesis and milk secretion occur. Lactation is maintained by periodic suckling which, through a neural arc, stimulates prolactin, which acts on mammary epithelium to increase milk production. In the prepared differentiated mammary epithelial cell, prolactin exerts major regulatory effects so that over 50% of the total protein synthesized by the cell is milk protein.

Throughout human evolution lactation has provided the principal nourishment during the first 1–2 years of life. In addition to stimulating lactation, prolactin inhibits synthesis and secretion of LH and FSH. This results in amenorrhea which may persist throughout the period of nursing. Although not a reliable method of contraception, historically nursing has played a major role in limiting additional pregnancy so that adequate nutrition and resources are available to the living child and mother. Nonnursing elevations in prolactin, such as occur with pituitary tumors and in response to drugs listed in Table 8.4, also inhibit LH and FSH, resulting in the clinical syndromes of amenorrhea and galactorrhea. In males, high concentrations of prolactin result in impotence via this same process of gonadotropin inhibition. Galactorrhea in males with hyperprolactinemia is uncommon because estrogen-dependent breast growth and differentiation have not occurred.

In species other than man, prolactin has major effects on organs other than the breast. In pigeons, prolactin stimulates the crop sac which produces glandular secretion for feeding the young. In rats and mice, prolactin has direct actions on the ovary in conjunction with LH and FSH to prolong life of the functioning corpus luteum. In fish, especially those which have a life cycle involving both fresh water and salt water, prolactin is a major osmoregulatory hormone.

Specific high affinity prolactin receptors are located in breast, liver, kidney, adrenals, and gonads (Posner et al., 1974). However, in humans it is uncertain whether prolactin significantly effects organs other than the breast. In mammary epithelial cells, prolactin stimulates milk protein synthesis not only by increasing the rate of gene transcription but also by significantly prolonging the half-life of the milk protein messenger RNAs (Rosen et al., 1980). The latter effect may be due to increased polyadenylaton which stabilizes messenger RNA.

POSTERIOR PITUITARY

Structure and Biosynthesis of ADH and Oxytocin

The posterior pituitary secretes antidiuretic hormone (ADH, vasopressin) and oxytocin, two peptide hormones which are synthesized in the supraoptic and paraventricular nuclei of the hypothalamus. Each hormone is made up of nine amino acids with a six-amino acid ring formed by a disulfide bond between cysteine residues at positions 1 and 6 and a three-amino acid tail containing a carboxy-terminal glycineamide (Fig. 8.21). ADH in all mammals except pigs contains arginine at residue 8; pig ADH contains lysine at this position. Oxytocin differs from ADH by having isoleucine at position 3 and leucine at position 8. Nonmammalian vertebrates synthesize vasotocin which has an isoleucine at position 3. Because vasotocin is found in the nervous system of early aquatic species, it is thought to be the ancestral hormone from which both ADH and oxytocin evolved (Sawyer and Pang, 1977).

The disulfide bridge is required for biological activity. A basic residue in position 8 is necessary for antidiuretic activity while a leucine at position 8 promotes oxytocic activity. Among the many analogues synthesized, 1-desamino-8-D-arginine vasopressin (DDAVP) is used clinically to treat ADH deficiency states. Removal of the amino group from cystein at position 1 and substitution of D-arginine at position 8 favor antidiuretic activity over pres-

AMINO ACID   1 2 3 4 5 6 7 8 9

ADH   NH₂-CYS-TYR-PHE-GLU-ASN-CYS-PRO-ARG-GLYNH₂

ADH PORCINE   NH₂-CYS-TYR-PHE-GLU-ASN-CYS-PRO-LYS-GLYNH₂

OXYTOCIN   NH₂-CYS-TYR-ILE-GLU-ASN-CYS-PRO-LEU-GLYNH₂

VASOTOCIN   NH₂-CYS-TYR-ILE-GLU-ASN-CYS-PRO-ARG-GLYNH₂

DDAVP   CYS-TYR-PHE-GLU-ASN-CYS-PRO-D-ARG-GLYNH₂

Figure 8.21.   Structure of posterior pituitary hormones.

sor activity and result in a molecule which is less susceptible to proteolytic degradation.

Although ADH and oxytocin are synthesized in discrete cells, producer cells for each hormone are located in both the supraoptic and paraventricular nuclei areas. Each peptide is synthesized as part of a larger precursor protein and after processing remains noncovalently bound to a portion of the precursor termed neurophysin. The 166-amino acid precursor contains a 19-amino acid signal sequence at the amino-terminus followed by the sequence for ADH, which is connected to the sequence of neurophysin by gly-lys-arg (Land et al., 1982). This glycine provides for amidation of glycine 9 of ADH, and the dibasic region provides a site for cleavage of ADH from neurophysin. A 39-amino acid glycopeptide present in the precursor is located on the carboxy-terminal side of neurophysin. After synthesis in the cell body, ADH and oxytocin bound to their specific neurophysins are transported down the axons to be stored in secretory granules in nerve terminals in the posterior pituitary. In response to nerve impulses there is concordant secretion of both ADH and its neurophysin or oxytocin and its neurophysin. The hormones rapidly dissociate from their neurophysin after secretion; subsequent functions of the neurophysins are unknown. Like other small peptide hormones, ADH and oxytocin have short half-lives in the circulation of approximately 10 min. ADH is degraded by a postproline cleaving enzyme (position 7), trypic activity (position 8), aminopeptidase (position 1), and reduction of the disulfide bond (Walter and Simmons, 1977). Metabolism occurs principally in the kidney and liver.

Regulation of Synthesis and Secretion of ADH

The major action of ADH is to promote water conservation by the collecting duct of the kidney. In higher concentrations, it also causes vasoconstriction.* ADH, like aldosterone and atrial naturietic hormone, importantly maintains normal fluid homeostasis and vascular and cellular hydration. The osmotic pressure of body water is the principal variable controlling secretion of ADH. Osmotic pressure is sensed by osmoreceptor cells in the hypothalamus which transmit signals to ADH-synthesizing cells in the supraoptic and paraventricular areas. Small increases in plasma osmolality of approximately 1–2% stimulate increased secretion of ADH and subsequent water conservation to reestablish normal plasma osmolality which in humans is approximately 287 mosmol/kg (Fig. 8.22). Volume depletion is the second major stimulus to ADH secretion. As shown in Figure 8.22, more major changes in plasma volume and mean arterial pressure are required to stimulate ADH secretion, but the magnitude of the rise in plasma ADH concentration is larger. For example, loss of 500 ml of blood, or approximately 7% of total blood volume, does not alter plasma ADH in recumbent subjects, but if these subjects stand and reduce their central blood volume by an additional 10%, ADH rises (Robertson, 1977). Changes in circulatory volume and mean arterial pressure are sensed by baroreceptors in the left atrium, pulmonary veins, carotid sinus, and aortic arch and are transmitted to the central nervous system through vagal and glossopharyngeal nerves. Changes in circulatory volume will increase plasma ADH regardless of the plasma osmolality. However, ADH secretion responds to osmolality over a wide range of circulatory volumes, with the

---

*Water retention equals antidiuresis, and a more appropriate nomenclature than antidiuretic hormone might be water-conserving hormone. Vasopressin is also an unfortunate choice of a name because water conservation is the major function of the hormone rather than vasoconstriction.

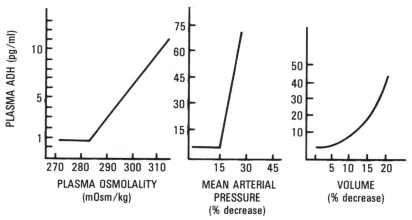

Figure 8.22. Control of plasma ADH concentration by osmolality *(left panel)*, mean arterial pressure *(center panel)*, and circulatory volume *(right panel)*. Note changes in scale on the ordinate. (Adapted from Robertson et al., 1973; Dunn et al., 1973.)

rise in ADH steeper when circulatory volume is low.

ADH is a first line of defense of water balance. When water intake is decreased, osmolality rises, and ADH is secreted to promote renal water conservation and reestablishment of normal osmolality. With major losses of intravascular volume, as in hemorrhage, ADH, which is secreted in large amounts, promotes water conservation to reestablish intravascular volume and causes vasoconstriction to maintain blood pressure. When water intake is increased, osmolality is decreased, and ADH is suppressed so that the kidney excretes free water to reestablish normal osmolality.

A number of factors in addition to osmolality and circulatory volume affect ADH secretion. Nausea is a potent stimulus, as are pain, stress, hypoglycemia, and exercise. ADH secretion is stimulated by cholinergic agonists, β-adrenergic agonists, angiotensin, prostaglandins, and several other pharmacological agents. ADH secretion is inhibited by alcohol, α-adrenergic agonists, glucocorticoids, and several other pharmacological agents.

Thirst is importantly entrained to ADH control of water balance (Fitzsimons, 1972). Hypothalamic osmoreceptors are connected to hypothalamic thirst centers as well as to ADH synthesizing neurons. When water intake decreases, a rise in osmolality of about 1% results in an increase in ADH to stimulate renal water conservation. A 2–3% rise in osmolality results in significant thirst and water-seeking behavior. The osmotic threshold for thirst of about 295 mosmol/kg is higher than that for ADH secretion so that thirst and water intake provide a second line of defense after renal water conservation. A decrease in circulatory volume and blood pressure strongly stimulates thirst, but this stimulus operates only under severe conditions. Angiotensin is a potent stimulus to thirst, but its role under normal physiological conditions is incompletely defined.

Biological Effects of ADH

In the normal kidney 85–90% of the 200 liters/day of glomerular filtrate are reabsorbed isosmotically in the proximal tubule. There is selective reabsorption of Na^+ in the ascending limb of Henle's loop so that the approximately 10% (20 liters) of glomerular filtrate which reaches the distal nephron is hypotonic. The amount of hypotonic fluid reaching the distal nephron varies, depending on the glomerular filtration rate and sodium intake. When sodium intake and GFR are high, the volume reaching the distal nephron is higher than 20 liters, whereas when sodium intake and GFR are low, the

volume reaching the distal nephron is lower. ADH regulates the amount of water which is further reabsorbed from the hypotonic fluid which reaches the collecting duct. Without ADH the collecting duct is impermeable to water so that large volumes of urine (~ 16 ml/min) as dilute as 50 mosmol/kg are excreted (positive free water clearance). With maximum concentrations of ADH the collecting duct is freely permeable so that water moves along an osmotic gradient from the hypotonic lumen to the hypertonic medullary interstitium. Small volumes of urine (0.5 ml/min) maximally concentrated to about 1200 mosmol/kg are excreted (negative free water clearance). Variations in urine osmolality over the full range of 50–1200 mosmol/kg occur in response to changes in plasma ADH from 1–10 pg/ml.

High affinity receptors for ADH which are located in epithelial cells of the distal nephron stimulate cAMP production. cAMP-mediated changes in membrane proteins and in cytoskeletal elements anchored to the membrane control water permeability channels (Dousa and Valtin, 1976). ADH also increases prostaglandin production, and prostaglandins exert a local negative feedback effect on ADH action (Dunn and Hood, 1977). Inhibitors of prostaglandin synthesis thus augment ADH action.

ADH also stimulates smooth muscle contraction and causes generalized vasoconstriction. Concentrations of ADH higher than those causing maximum renal water conservation are required so that vasopressor actions are important primarily when volume depletion stimulates secretion of high concentrations of ADH (Fig. 8.22). ADH has been used as an adjunct to control bleeding esophageal varices by causing splanchnic vasoconstriction. Because ADH may cause coronary arteries to constrict, high doses must be used with caution. ADH also increases ACTH secretion and has been used as a test of pituitary reserve.

Deficits and Excesses of ADH

Diabetes insipidus results from either lack of ADH or from inability of the kidney to respond to normal amounts of ADH (nephrogenic diabetes insipidus). Hypophysectomy does not usually result in permanent diabetes insipidus because, although significant retrograde degeneration of hypothalamic producer cells occurs, a number of neurons terminate on the median eminence, and those cells continue to function. With ADH deficiency, the thirst mechanism assumes a primary role in maintenance of water balance, and most patients with diabetes insipidus maintain their serum osmolality at the thirst threshold of about 295 mosmol/kg by ingestion of large amounts of water each day. When free access to water is interrupted, these patients develop severe dehydration because of continued urinary water losses. DDAVP and vasopressin tannate in oil are effective replacement therapy for ADH deficiency. Nephrogenic diabetes insipidus is more difficult to control, although thiazide diuretics have been useful in reducing urine volume by decreasing the volume of fluid reaching the distal nephron.

The response of 24 normal adults to a water load of 20 ml/kg given over 30 min is shown in Figure 8.23. The dilution of body solutes reduces osmolality, decreases ADH, decreases urine osmolality, and results in excretion of more than 90% of the water load within 4 hours. An excess of ADH, defined as an inappropriately high concentration relative to plasma osmolality, would prevent the fall in urine osmolality and excretion of free water so that progressive hypo-osmolality would develop. Because water is freely permeable across most cell membranes, plasma hypo-osmolality reflects cellular overhydration. When osmolality falls below 250 mosmol/kg, brain cell swelling and disturbances of mental function occur. When the expansion of body water reaches 10%, the kidney rejects sodium in the proximal tubule so that urinary sodium excretion increases. Suppression of the renin-angiotensin-aldosterone system by volume expansion also contributes to urinary sodium excretion. The paradox-ical excretion of sodium in the presence of a low serum sodium provides an important clue that the hyponatremia is caused by dilution of normal amounts of sodium rather than a body deficit of sodium. The syndrome of inappropriate antidiuretic hormone secretion (SIADH) occurs in a variety of clinical circumstances and is recognized by the triad of low plasma osmolality, inappropriately high urine osmolality, and significant amounts of sodium in the urine. When this occurs, it is necessary to restrict water intake. Declomycin, which causes a reversible form of nephrogenic diabetes insipidus, may also be useful.

Biological Effects of Oxytocin

The two principal targets for oxytocin are myoepithelial cells of the breast and smooth muscle cells of the uterus. Myoepithelial cells are specialized smooth muscle cells surrounding the alveoli of the mammary gland. In response to oxytocin, these cells contract and move milk from the alveoli to the large sinuses in the process of milk ejection. Oxytocin release occurs in response to sensory impulses arising from the nipple during suckling but may also occur when women play with their infants without suckling. ADH does not change, but prolactin also rises in response to suckling to maintain milk production.

Oxytocin stimulates contraction of uterine smooth muscle cells. The contractile response to oxytocin is dependent on estrogen and is antagonized by progesterone. The role of oxytocin in initiation of labor is uncertain because plasma oxytocin concentrations do not clearly increase with parturition. During the first two trimesters of pregnancy, the uterus is relatively resistant to oxytocin, but during the latter part of the third trimester, uterine sensitivity to oxytocin increases significantly so that changes in target tissue responsiveness or in local oxytocin concentrations, rather than changes in circulating concentrations, may be important. During pregnancy a proteolytic oxytocinase which cleaves the amino-terminal cysteine appears in the uterus and the placenta, and its concentration rises in plasma. This may prevent oxytocin from stimulating uterine contractions until parturition. Oxytocin is used clinically to increase uterine contraction in certain carefully defined obstetrical conditions.

NEUROHORMONES OUTSIDE THE HYPOTHALAMUS

Neurohormones initially identified in the hypothalamus by their effects on anterior pituitary function are also synthesized in several locations

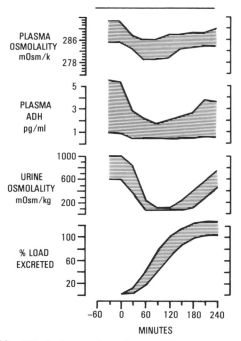

Figure 8.23. Effect of water ingestion on ADH secretion and water balance. A water load of 20 ml/kg was ingested between 0 and 30 min. (Adapted from Robertson, 1984.)

throughout the body, especially within the gastro-intestinal tract. Conversely, peptide hormones initially identified in the gastrointestinal tract are also synthesized in the hypothalamus and other central nervous system areas. These peptide hormones therefore have many more actions than those initially identified. Generally, these small peptide hormones act near their site of synthesis, rather than being released into the general circulation. It has been more difficult to define local paracrine actions because classical endocrine gland ablation followed by hormone replacement cannot be carried out. The hormones have been localized by immunochemical methods and by the ability of isolated cells to make the hormone in vitro. A functional role has been implied by either local application or by neutralizing endogenous hormones with injected antisera. Because local actions often modulate effects of other hormones and neurotransmitters, experimental approaches used have only partially defined the effects of these peptide hormones.

SS, initially identified as an inhibitor of GH release, is made in δ cells of the pancreatic islets and in the gut. The δ cells are in close proximity to β cells which synthesize insulin and α cells which synthesize glucagon. SS inhibits secretion of both insulin and glucagon and reduces a variety of digestive functions to slow the rate of nutrient entry. SS may thus serve as a paracrine signal to coordinate the rate of absorption of nutrients from the gut with pancreatic islet function (Unger et al., 1978).

GnRH, initially identified as the stimulator of LH and FSH synthesis and secretion, has direct effects on the ovary and testis. Acute effects are stimulatory, whereas chronic effects are inhibitory (Hseuh and Jones, 1981). The prolonged desensitization which occurs with potent GnRH agonists has proven useful in inhibiting the pituitary-gonadal axis in precocious puberty and in palliation of prostate concer via decreased androgen production. Material with GnRH activity (gonadocrinin) has been isolated from gonadal tissue and is thought to play a local modulatory role within the gonads (Ying et al., 1981).

TRH is widely distributed throughout the gastrointestinal tract, and the gastrointestinal active hormones cholecystokinin, secretin, vasoactive intestinal peptide, bombesin, and substance P are found in the hypothalamus and central nervous system. SS and GnRH are also found in extrahypothalamic areas of the nervous system. These peptides appear to function as specialized neurotransmitters acting at a limited number of sites. A large number of behavioral effects have been observed with central nervous system administration. Changes in body temperature, sexual behavior, pain perception, mood, learning, and memory all occur (Moss, 1979). It is likely that these central nervous system effects are important in coordinating nervous system function and behavior with metabolic and reproductive effects of the endocrine system.

The Thyroid Gland

Thyroid hormone regulates metabolic processes in most organs and is essential for normal nervous system development. In contrast to many hormones whose concentrations fluctuate rapidly in response to environmental signals, thyroid hormone exhibits remarkable stability. It is synthesized in the largest endocrine gland, stored in large quantities within that gland, secreted as a less active prohormone which is tightly bound to plasma proteins, and is converted into its most active form primarily in peripheral tissues.

STRUCTURE

The normal adult thyroid gland, which weighs approximately 20 g, is composed of two lobes joined by a midline isthmus (Fig. 8.24). The gland is located in the lower part of the neck, anterior to the trachea, between the sternocleidomastoid muscles. It receives an extensive arterial blood supply from superior and inferior thyroid arteries on each side. The four parathyroid glands are located posterior to the thyroid gland, as are the recurrent laryngeal nerves. Microscopically, the thyroid gland is composed of closely packed follicles. The wall of each follicle is composed of thyroid cells which become taller as their metabolic activity increases. The interior of the follicle is filled with colloid, a proteinaceous material containing principally thyroglobulin. In addition to supporting cells, the thyroid also contains C cells of neuroectodermal origin which synthesize calcitonin (Chapter 56).

BIOSYNTHESIS OF THYROID HORMONES

Thyroid hormones are iodinated derivatives of the amino acid tyrosine. As shown in Figure 8.25, thyroxine (T_4), the major secretory product of the thyroid gland, consists of two phenyl rings linked via an ether bridge with an alanine side chain on the inner ring. It contains four iodine atoms attached at carbons 3 and 5 of the inner ring and 3' and 5' of the outer ring. These substitutions impose a three-dimensional structure on the molecule in which the planes of the aromatic rings are perpendicular to each other (Cody, 1978). 3,5,3'-Triiodothyronine, or T_3, the active form of thyroid hormone, is principally derived by peripheral removal of one iodine from the outer ring. Removal of iodine from carbon 5 of the inner ring yields reverse T_3, an inactive metabolite.

Hormone synthesis occurs in thyroid follicular cells which exhibit marked functional polarity from their basal to apical sides. The biosynthetic process shown in Figure 8.26 has been divided into a number of steps which have been demonstrated either from genetic defects or by the use of inhibitors. The first step is active transport of iodide into the thyroid cell. This process occurs at the basal surface and requires ATP and Na^+/K^+-ATPase-mediated sodium transport (DeGroot and Niepomniszcze, 1977). By this active transport process, the thyroid gland efficiently extracts iodide, even when plasma concentrations are low; through action of this transport system, ratios of thyroid to plasma iodide may exceed 100. Other anions, such as perchlorate and pertechnetate, are also transported. Perchlorate has been used as a functional test to displace inorganic iodide from the cell, and pertechnetate as $^{99m}TcO_4^-$ is used to image the thyroid. Iodide is also actively concentrated by salivary glands, gastric mucosa, small intestine, skin, breast, and placenta but to a lesser extent; it is not significantly incorporated into protein within those tissues.

Within the thyroid cell, iodide is rapidly oxidized and incorporated into tyrosine residues in thyroglobulin molecules. This results in monoiodotyrosine (MIT) and diiodotyrosine (DIT) residues within the protein. There is subsequent coupling of these iodinated tyrosines in the thyroglobulin molecule to yield T_4 and T_3. Both organification and coupling reactions are catalyzed by thyroid peroxidase and require H_2O_2 (Taurog, 1970). These reactions, which occur at the apical surface of the thyroid cell, take place on the thyroglobulin molecule. Thyroglobulin, a large glycoprotein of 660,000 M_r, constitutes

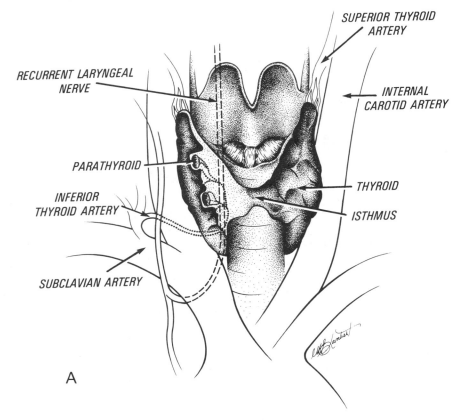

SUPERIOR THYROID ARTERY

RECURRENT LARYNGEAL NERVE

INTERNAL CAROTID ARTERY

PARATHYROID

INFERIOR THYROID ARTERY

THYROID

ISTHMUS

SUBCLAVIAN ARTERY

A

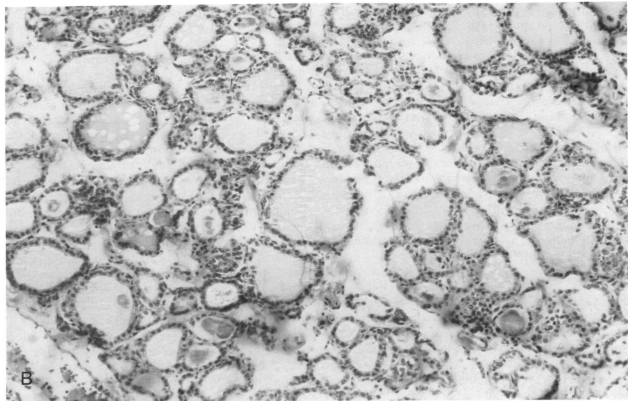

B

Figure 8.24. Gross and microscopic anatomy of the thyroid gland.

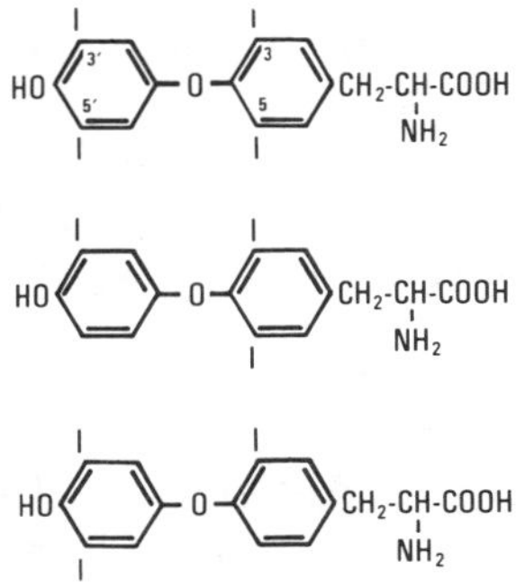

HO ⬡ O ⬡ CH₂-CH-COOH THYROXINE (T₄)
                    NH₂

HO ⬡ O ⬡ CH₂-CH-COOH 3,5,3' TRIIODOTHYRONINE (T₃)
                    NH₂

HO ⬡ O ⬡ CH₂-CH-COOH 3,3',5'- TRIIODOTHYRONINE
                    NH₂   (reverse T₃)

Figure 8.25. Structures of the major circulating thyroid hormones.

about 50% of the total protein synthesized by the thryoid gland. It is not unusually rich in tyrosine residues nor is it rich in thyroid hormone. Normally iodinated thyroglobulin contains about 26 atoms of iodine with 3–4 T_4 and 0.2 T_3 residues per molecule (Izumi and Larsen, 1977). The unique feature of thyroglobulin appears to be its tertiary structure which exposes tyrosine residues for iodination and favors the coupling reaction.

Mature thyroglobulin is stored as colloid in the lumen of the thyroid follicles. Formation of active thyroid hormones requires reabsorption of colloid by endocytosis, fusion of endocytic vesicles with lysosomes, and degradation of thyroglobulin via lysosomal proteases. T_4 and T_3 are secreted into the circulation in a ratio of about 10:1. This is some-

what less than the ratio in thyroglobulin and may reflect some conversion of T_4 to T_3 within the thyroid gland. The iodine in MIT and DIT is reclaimed for hormone biosynthesis by a thyroid deiodinase enzyme. This deiodination pathway reclaims approximately 50% of the iodine of thyroglobulin and is quantitatively important in thyroid gland economy, as demonstrated by the fact that congenital defects in deiodinase result in goitrous hypothyroidism.

REGULATION OF THE THYROID GLAND

Hypothalamic-Pituitary-Thyroid Axis

Thyroid-stimulating hormone (TSH) is the major regulator of thyroid gland function (Fig. 8.27). Central nervous system control is exerted via the hypothalamic neurohormone thyrotropin-releasing hormone (TRH), which binds to specific cell surface receptors on pituitary thyrotrophs to stimulate TSH synthesis and secretion. Somatostatin reduces TSH secretion and may transmit inhibitory signals from the hypothalamus. The glycoprotein TSH consists of α and β subunits, with biological specificity residing in the β subunit (Pierce and Parsons, 1981). TSH increases thyroid gland production of thyroid hormones which exert negative feedback control at the level of pituitary thyrotrophs and of hypothalamic TRH-producer cells. Variations in circulating con-

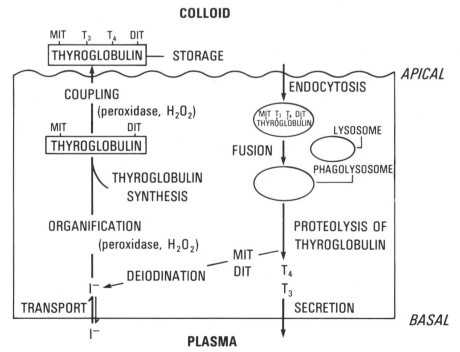

Figure 8.26. Pathway of biosynthesis of thyroid hormones.

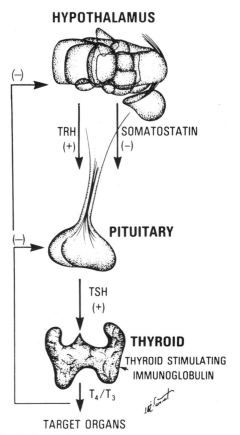

HYPOTHALAMUS

TRH (+)

SOMATOSTATIN (−)

PITUITARY

TSH (+)

THYROID

THYROID STIMULATING IMMUNOGLOBULIN

T_4/T_3

TARGET ORGANS

Figure 8.27.   The hypothalamic-pituitary-thyroid axis.

centrations of thyroid hormone are the principal regulator of TSH through this feedback loop. The pituitary gland and nervous system contain an unusually active 5′-deiodinase system which converts T_4 to T_3 so that circulating T_4, in addition to T_3, is an important feedback inhibitor of TSH secretion. For example, during fasting when peripheral conversion of T_4 to T_3 is reduced, TSH and T_4 remain normal because of maintenance of intrapituitary conversion of T_4 to T_3 (Larsen et al., 1981). Inhibitory effects of thyroid hormone include reduction of TRH receptors, which reduces the biological effectiveness of a given concentration of TRH (Perrone and Hinkle, 1978), inhibition of transcription of the TSH-β chain gene (Carr et al., 1987), and reduced hypothalamic cell TRH mRNA (Segerson et al., 1987).

Precise quantitation of TSH, T_4, and T_3 in plasma provides the basis for accurate diagnosis of many thyroid diseases. In hypothyroidism due to thyroid gland disease, TSH concentrations are elevated, whereas in hypothyroidism due to hypothalamic or pituitary disease, TSH concentrations are low (Fig. 8.28). Because elevations in TSH rarely cause hyperthyroidism, elevated concentrations of T_4 and

T_3 due to either thyroid-stimulating immunoglobulin or glandular autonomy result in low TSH concentrations. Reestablishment of normal TSH has also been useful in defining physiological replacement doses of thyroid hormone for treatment of hypothyroidism. Because thyroid hormone feedback is at the level of the pituitary, administration of TRH causes little or no rise in TSH in hyperthyroid states, whereas it causes an exaggerated rise in TSH in hypothyroidism due to diseases of the thyroid gland (Fig. 8.28). Measurement of the response to TRH may thus be helpful in identifying mild abnormalities of thyroid gland function.

An intact hypothalamic-pituitary-thyroid axis is well established by 12–16 weeks of gestation. Fetal thryoid function depends on this axis and is independent of maternal thyroid function because neither thyroid hormones nor TSH cross the placenta in significant amounts (Fisher et al., 1977). Attempts to supply the fetus via maternal ingestion of large amounts of thyroid hormone have been unsuccessful. Both iodide and antithyroid drugs do cross the placenta and maternal ingestion of these may affect fetal thyroid function. T_3 concentrations in the fetus and cord blood at term are low, whereas reverse T_3 is high (Chopra et al., 1975), suggesting that 5′-deiodinase activity is low in utero. At birth there is a sharp rise in TSH, probably due to exposure to the cold outside world, followed by an increase in T_4 and T_3, which are maximum at 24 hours. Normal metabolism of T_4 and T_3 is established soon after birth.

TSH binds to specific high affinity cell surface receptors on thyroid cells to activate adenylate cyclase and increase cellular concentrations of cAMP (Dumont et al., 1978). TSH increases the entire pathway of hormone biosynthesis, including iodide transport, organification, thyroglobulin synthesis, coupling, endocytosis, and thyroglobulin proteolysis. In Graves' disease, thyroid-stimulating immunoglobulins recognize the TSH receptor and increase glandular function via cAMP and the same metabolic pathways utilized by TSH (Volpe, 1981). In vivo elevated concentrations of TSH are associated with progressive growth of the thyroid gland unless destruction of thyroid cells has occurred. Intermittent increases of TSH in nontoxic nodular goiters and continuous increases in TSH in iodide deficiency and in enzymatic defects in the biosynthetic pathway result in large glands which may reach 5–10 times normal size and obstruct the trachea, neck veins, and esophagus. Following hypophysectomy, atrophy of the thyroid gland occurs. It is not certain as to whether TSH directly stimulates thyroid gland growth or whether a con-

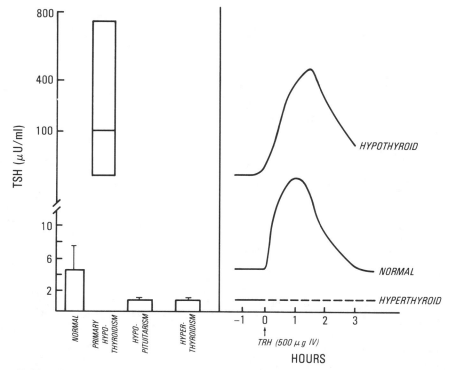

Figure 8.28. Plasma TSH concentrations in various forms of thyroid disease. In the *right panel,* TSH responses to 500 μg of intravenous TRH are shown.

comitant increase in delivery of growth factors is required. Studies in cell culture have confirmed the trophic effect of TSH on differentiated function, but data on a direct growth stimulatory role are conflicting.

Autoregulation

Autoregulation of the thyroid gland is essential because dietary intake of iodine may vary greatly. Iodine is obtained principally from seafood in the diet. In inland areas of the world, especially mountainous ones where, over millions of years, iodine has been washed from the soil, dietary iodine deficiency occurs. Although the thyroid iodide trapping mechanism becomes remarkably efficient, in areas where iodine intake is less than 60 μg/day, goiter is common. Growth of the thyroid gland, which is dependent on TSH, augments the ability of the gland to trap iodide and synthesize sufficient hormone to maintain a euthyroid state. In areas where iodine intake is less than 20 μg/day, very large glands are common; compensation may be incomplete, and endemic cretinism results. In iodine deficiency, glandular production of hormone also changes so that T_3 becomes the major secretory product (Greer et al., 1968). This adaptation provides a more active hormone containing less iodine.

Normal iodine intake in Western cultures is approximately 250 μg/day but may be greater because of use of iodized salt and use of iodates in making bread. Approximately 50 μg is used for thyroid hormone biosynthesis, and 200 μg is excreted by the kidneys (Fig. 8.29). Additional iodide for hormone biosynthesis is derived from metabolism of thyroid hormones. Autoregulatory mechanisms which enhance TSH-mediated adaptations to reduced iodine intake also protect against excessive iodine ingestion so that plasma thyroid hormone concentrations remain constant. In response to large quantities of iodine (2 mg or more) there is a sharp decrease in organification (Wolff and Chiakoff, 1948). This decrease in organification depends on a high concentration of iodide within the gland and results in decreased formation of thyroid hor-

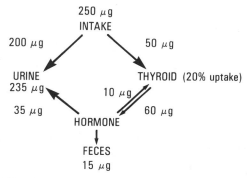

Figure 8.29. Daily iodine balance in the USA. (From Robbins, 1980.)

mone. Decreased organification is an acute response to high cellular iodide concentrations. When excessive iodide ingestion continues, the thyroid gland adapts by decreasing active transport of iodide into the thyroid cell (Braverman and Ingbar, 1963). Increased cellular concentrations of iodide are reduced and the block to organification is removed. In normal people, ingestion of large amounts of iodine thus does not adversely affect thyroid function. Because large iodide pools dilute any radioactive iodine ingested, iodine supplementation has been recommended as protection against radioactive iodine. In abnormal thyroids, autoregulatory adaptations may not occur. If inhibition of organification does not occur, iodine ingestion will result in hyperthyroidism (Jod-Basedow effect). If organification is inhibited, but transport does not adaptively decrease, high cellular concentrations of iodide will persist and result in hypothyroidism (iodide goiter and hypothyroidism) (Wolff, 1969).

Very large doses of iodine also inhibit release of thyroglobulin-bound thyroid hormones from the gland. This distinct inhibitory effect was used in the past to treat thyrotoxicosis and is still useful in preparing patients for thyroid surgery.

TRANSPORT AND METABOLISM

In a normal adult the thyroid gland secretes approximately 80 μg of T_4 per day. Approximately 6 μg of T_3 are secreted by the thyroid gland each day, with 80% of total daily production of T_3 arising from peripheral deiodination of T_4 (Schimmel and Utiger, 1977; Chopra et al., 1978). Conversion of T_4 to T_3, which occurs primarily in liver and kidney, is catalyzed by 5'-deiodinase, a microsomal enzyme which requires reduced sulfhydryl groups from glutathionine and other compounds for activity (Fig. 8.30). In addition to conversion of T_4 to the active hormone T_3, T_4 also undergoes deiodination in the inner ring to yield the inactive metabolite reverse T_3. Inactivation of the hormone occurs by further deiodinations, deamination-decarboxylation of the alanine side chain, and conjugation with glucuronide (Engler and Burger, 1984).

Circulating concentrations of T_3 are determined not only by glandular production of T_4 and T_3 but also by the activity of 5'-deoidinase. As shown in Figure 8.30, a number of conditions are associated with decreased 5'-deiodinase activity which result not only in decreased production of T_3 but also in increased concentrations of reverse T_3 due, in part, to the inability of reverse T_3 to be further metabolized. The best studied paradigm is fasting; changes occurring during illness may relate to similar metabolic factors. During fasting there is a 20% decrease

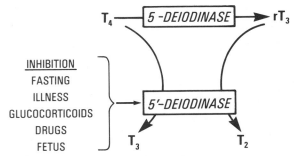

Figure 8.30. Peripheral metabolism of thyroid hormones. (From Schimmel and Utiger, 1977.)

in plasma T_3 by the end of the first day and a 50% decrease at the end of 3 days with a reciprocal increase in reverse T_3 (Wartofsky and Burman, 1982). With prolonged fasting of more than 3 wk, reverse T_3 tends to normalize, but T_3 remains low. Refeeding rapidly reverses these changes. The decrease in T_3 production with fasting is an important adaptive response contributing to the observed decreased metabolic rate and conservation of body tissues because, clearly, maintenance of normal or elevated rates of metabolism would be inappropriate when substrate is not available. Decreased production of T_3 is accompanied by adaptive changes in the pattern of enzymes induced in the liver so that T_3 induction of α-glycerol phosphate dehydrogenase is maintained but T_3 induction of malic enzyme is not (Oppenheimer et al., 1979). Hepatic T_3 nuclear receptors are reduced (Burman et al., 1977). Even though T_3 concentrations are reduced during fasting, T_4, free T_4, and TSH concentrations remain normal. This serves to maintain normal glandular production and prevents the thyroid from inappropriately compensating for the physiologically important reduction in T_3.

Upon secretion into the circulation, T_4 and T_3 are tightly bound to serum proteins so that only a small fraction of the total amount circulates in the free form to enter cells and to exert metabolic control (Table 8.5). Binding proteins serve as a protected reservoir to prevent renal clearance and provide free hormone as needed. T_4 is bound to three proteins: thyroid-binding globulin (TBG), thyroid-bind-

Table 8.5.
Quantitative Aspects of Thyroid Hormone Metabolism

	Production, μg/day		Mean Concentration in Plasma, ng/ml		Half-life, days
	Gland	Periphery	Total	Free	
T_4	78		80	0.03	6.5
T_3	6	24	1.2	0.004	1.3
rT_3	3	27	0.4	0.00098	0.1

Data kindly provided by Dr. Inder Chopra.

ing prealbumin (TBPA), and human serum albumin (HSA), all of which are synthesized in the liver (Robbins et al., 1978). TBG, a glycoprotein of 63,000 M_r, has the highest affinity for thyroid hormones; it binds approximately 75% of T_4 and 50% of T_3 in the circulation. Affinity constants are 2×10^{10} M^{-1} for T_4 and 2×10^8 M^{-1} for T_3. TBPA has a lower affinity for T_4, being 100 fold less than that of TBG but has a much greater capacity to bind primarily T_4. HSA has an even weaker affinity but is present in high concentrations. Protein binding, principally to TBG, results in the long half-life of circulating thyroid hormones (Table 8.5). T_4 has a half-life of about 1 week while T_3, which is bound less tightly to proteins, has a half-life of 1.3 days.

The concentration of the free active fraction is determined by the mass action equation shown in Chapter 52, Equation 8.1. Present methodology is sufficient to quantitate total T_4 and T_3 concentrations in plasma accurately, but measurement of the free fraction is tedious and difficult. Usual clinical measurements of total hormone concentrations must, therefore, be interpreted with knowledge of factors likely to affect hepatic synthesis of TBG and other binding proteins. During pregnancy, TBG concentrations increase because of estrogen stimulation of TBG production. An increase is evident by 1 month and is maximal by the 3rd month of pregnancy. When TBG increases, the initial effect will be to shift hormone from the free to the bound state. The total concentration will increase, first, because of decreased metabolic clearance and, second, because of increased glandular production due to the hypothalamic-pituitary-thyroid axis, which responds to the initial decreased concentration of free hormones. The free concentration will be restored to normal, with equilibrium established at higher concentrations of bound hormone. The pregnant woman thus has normal free hormone levels and normal thyroid hormone function but an elevated total plasma concentration of hormones because of an increase in the bound fraction. Conversely, androgens and certain illnesses such as hepatic cirrhosis may decrease TBG and TBPA production. Again, patients will maintain normal free T_4 concentrations, although the total will be decreased. Measurement of TBG or available TBG binding sites may occasionally be necessary to interpret measured total thyroid hormone concentrations accurately.

BIOLOGICAL EFFECTS OF THYROID HORMONE

The multiple effects of thyroid hormone result from occupancy of nuclear receptors with subsequent effects on gene expression. Two forms of thyroid hormone receptor have been identified via cDNA cloning (Sap et al., 1986; Weinberger et al., 1986; Thompson et al., 1987). One type appears more localized in the nervous system. Differing biological effects of thyroid hormone in different tissues result from the pattern of gene expression in each cell type. The actions of thyroid hormone appear mechanistically similar with the activated receptor binding to specific DNA sequences to control transcription of genes containing those sequences in their promoter regions. Receptors specifically bind T_3. Under physiological conditions 90% of receptor bound hormone is T_3; the affinity of nuclear receptors for T_3 of 2.9×10^{-11} M exceeds that for T_4 10 fold (Samuels and Tsai, 1973); and 5–7 fold lower concentrations of T_3 than T_4 are required to induce biological effects. T_3 receptor complexes increase messenger RNA and protein production which mediate observed biological responses (Baxter et al., 1979).

Growth and Development

Thyroid hormone is required for normal nervous system development and linear growth. In the absence of thyroid hormone, axonal and dendritic development and myelination of the nervous system are defective. In humans, thyroid hormone appears especially critical during the immediate postnatal period. With early hormone replacement, hypothyroid newborns attain normal intellectual development; if therapy is delayed by 6–12 months, mental retardation is permanent. Routine laboratory screening of newborn infants is now carried out to allow treatment of this preventable form of mental retardation, which occurs in about 1 in 4400 live births. Once nervous system development is complete, hypothyroidism causes mental slowing, but this is completely reversible with thyroid hormone therapy.

When thyroid hormone is deficient, linear growth is slowed, maturation of growing epiphyseal end plates is retarded, and tooth eruption is delayed. In cell culture, T_3 stimulates proliferation of several cell types, reflecting the direct growth-promoting effects of T_3 on many organs. In addition, T_3 is required for production of both growth hormone and insulin-like growth factors. It is also required for action of insulin-like growth factors on epiphyseal cartilage (Froesch et al., 1976).

Energy Metabolism

Thyroid hormone stimulates the basal rate of metabolism, oxygen consumption, and heat production. Biochemical mechanisms underlying this fundamental property of thyroid hormone, which was first recognized in 1895 by Magnus-Levy, is incompletely understood. At least a part of the T_3-induced increase in oxygen consumption in tissues such as liver and diaphragm muscle results from increased activity of Na^+/K^+-ATPase (Ismail-Beigi and Edelman, 1970). T_3 induces increased synthesis of Na^+/K^+-ATPase, which consumes ATP during its activity to transport Na^+ and K^+ across the cell membrane (Smith and Edelman, 1979). Generation of ATP to support the increased activity of this enzyme depends on oxidative metabolism. The stimulatory effects of thyroid hormone on general protein synthesis and on several metabolic pathways also contribute to oxygen consumption.

Thyroid hormone is important for thermogenesis in homeotherms. In humans, thyroid hormone deficiency results in a reduction in basal metabolic rate and in core temperature. When excessive thyroid hormone is present, the basal metabolic rate is increased, and core temperature is elevated. Sympathetic nervous system activity contributes importantly to temperature control as discussed below.

Organ Systems

Thyroid hormone importantly regulates sympathetic nervous system activity. Although T_3 does not appear to alter production or concentration of catecholamines, it does induce synthesis of β-adrenergic receptors. As indicated by the mass action equation shown in Chapter 52, *Equation 8.1,* an increase in receptors will increase formation of active hormone-receptor complexes and biological activity. Although many of the effects of T_3 on adrenergic receptors have been identified in animal studies, it is likely that this is also an important action of thyroid hormone in man. Many of the signs and symptoms of hyperthyroidism reflect increased β-adrenergic activity. Patients exhibit tachycardia, increased cardiac output, wide pulse pressure, peripheral vasodilatation, frequent movement, diffuse anxiety, and lid retraction. There is heat dissipation with warm skin and increased sweating. Many of these signs and symptoms of excessive thyroid hormone can be abolished by β-adrenergic blockade, although increased doses are often required to saturate the increased number of receptors. Conversely, in hypothyroidism, where β-adrenergic receptor synthesis is impaired, α-adrenergic activity may predominate, resulting in peripheral vasoconstriction and increased blood pressure.

T_3 affects most organs of the body. Although the effects of thyroid hormone are most easily appreciated by analysis of deficiency and excess states, it is important to remember that physiologically T_3 acts as a metabolic regulator to maintain homeosta-

sis between these two extremes. In addition to effects on the heart via adrenergic receptors, T_3 maintains normal myocardial contractility, in part through stimulating the expression of the most active isoenzyme form of myosin ATPase (Hjalmarsen et al., 1970). In hypothyroidism less active isoenzyme predominates; administration of T_3 restores the normal isoenzyme (Hoy et al., 1977). T_3 also restores normal myocardial contractility. Normal skeletal muscle function also requires thyroid hormone. Muscle weakness occurs both in hypothyroidism and in hyperthyroidism; in the latter it may result from excessive catabolism of muscle proteins.

In hypothyroidism, weight gain is common because of decreased metabolic rate. Hypomotility of the colon and constipation are common. Conversely, in hyperthyroidism there is increased motility and hyperdefecation. The increased metabolic rate results in weight loss despite a significant increase in caloric intake. In hypothyroidism, hypoventilation occurs because of muscle weakness, decreased respiratory effort, and accumulation of pleural fluid. In hypothyroidism the skin is not only cool but also coarse and dry, and subcutaneous accumulation of mucopolysaccharides may result in characteristic myxedema. In hyperthyroidism the skin is warm, smooth, and moist. Excesses or deficiencies of thyroid hormone also lead to alteration in function of other endocrine systems. This results in infertility in both hypo- and hyperthyroid states.

The altered metabolic states resulting from either an excess or a deficiency of thyroid hormone affect the metabolism of a number of other hormones, vitamins, and drugs. T_3 increases the metabolic clearance of cortisol; it increases sex steroid-binding globulin to decrease the metabolic clearance of estrogen and testosterone; it increases both the utilization and clearance of vitamins; and it increases the clearance of a number of drugs, including digitalis. Consequently, patients with hypothyroidism are exquisitely sensitive to a number of drugs and may exhibit toxic effects at usual therapeutic doses. Patients with hyperthyroidism require a larger vitamin intake and often require increased doses of a drug to achieve a therapeutic response.

PHARMACOLOGY

Because most T_3 is derived under normal physiological conditions from deiodination of T_4 in peripheral tissues, T_4 is the preferred hormone for treatment of hypothyroidism, regardless of etiology. The long circulatory half-life of T_4 provides remarkable stability once a therapeutic dose is established. Approximately 150 μg/day (100–200 μg) provides full replacement with restoration of normal physiology, including normalization of TSH concentrations. The shorter half-life of T_3 makes it less appropriate for long-term use; it has an additional disadvantage of bypassing regulatory mechanisms controlling T_4 to T_3 conversion. Regardless of the degree of hypothyroidism—mild, moderate, or severe—the goal is ultimate provision of a full replacement dose of T_4. In thyroid disease, residual production of T_4 may decrease and may require excessive TSH and thyroid enlargement to sustain a euthyroid state. Maintenance of a euthyroid state with suboptimal replacement dosages is uncertain, and full replacement is recommended once a decision to use thyroid hormone is made. It is clear from consideration of the normal hypothalamic-pituitary-thyroid axis that a "little" thyroid cannot be given to increase metabolism, cause weight loss, or regularize menses. With normal thyroid function, replacement of 50% of T_4 production by exogenous hormone will lead to a 50% reduction of endogenous production so that free T_4 and T_3 concentrations will remain normal. Full replacement will fully shut off endogenous thyroid synthesis. Thyroid function can only be increased long-term by administering supraphysiological doses which create a hyperthyroid and pathological state. T_4 is optimally used in full physiological replacement doses only for hypothyroid states or to suppress TSH-dependent growth of thyroid tissue.

Antithyroid drugs presently used to treat hyperthyroidism are thioamides. Propylthiouracil and methimazole are used in the United States, and carbimazole is available in England. These drugs inhibit the biosynthesis of thyroid hormones by blocking the organification reaction (Marchant et al., 1978). Full return to a euthyroid state requires 4–6 weeks because of the long circulating half-life of T_4 and the large glandular stores of thyroglobulin in colloid which must be depleted. Propylthiouracil has an additional peripheral action to block the deiodination of T_4 and T_3 and may thus have a somewhat more rapid onset of action. Antithyroid drugs are effective in restoring a euthyroid state to patients with both Graves' disease and toxic nodular goiters. In Graves' disease, approximately 50% of patients who are treated for 1 year will remain euthyroid after discontinuing the drug. The other 50% will relapse and thus require ultimate treatment with radioactive iodine or surgery. Toxic nodular goiters do not usually remit, so antithyroid drugs are used only to control hyperthyroidism and prepare a patient for ultimate therapy with radioactive iodine or surgery.

Radioactive isotopes of iodine are metabolized identically to stable ^{127}I. They are concentrated by the thyroid iodide transport system and organified in thyroid cells. Radioactive iodine thus specifically delivers radioactive energy primarily to one cell type and is useful for imaging and for radiation therapy. ^{131}I, which has a half-life of 8 days and emits both x-rays and β particles, is the principal isotope used to treat hyperthyroidism. ^{131}I is administered orally and, in doses sufficient to correct hyperthyroidism, delivers little radiation to any other cells. The major disadvantage of radioactive iodine therapy is hypothyroidism, which occurs in about 20% of treated patients in the first year with an incidence of about 1% per year thereafter (Becker et al., 1971). By comparison the incidence of hypothyroidism in surgically treated patients is 20–30%. Because many papillary and follicular thyroid cancers also concentrate iodide, radioactive iodine can be used to treat metastatic disease with minimal scatter of radiation to surrounding tissues (Beierwaltes, 1978). ^{123}I with a half-life of 13 hours and emission of x-rays is a good isotope for imaging the thyroid gland because little radiation is delivered.

CHAPTER 55

The Adrenal Gland

ADRENAL CORTEX

The adrenal glands are composed of two organs of separate embryological derivation: an outer cortex of mesodermal origin and an inner medulla of neuroectodermal origin. In a broad sense, both the steroid hormones synthesized by the cortex and the catecholamines synthesized by the medulla mediate adaptive responses of the organism to a changing environment. The model of "stress" which evokes secretion of both steroid hormones and catecholamines may have evolved as a response to the primitive signal of substrate lack (Baxter and Rosseau, 1979). When food and water intake are not possible, cortisol stimulates gluconeogenesis to maintain blood glucose for nervous system use; aldosterone stimulates sodium retention to maintain intravascular volume and cellular hydration; and epinephrine stimulates both mobilization of energy-providing substances and cardiovascular and neuromuscular adaptation. Because the environment is constantly changing for free living animals, including man, such adaptive hormones have evolved as major regulators of homeostasis and control a number of metabolic processes in a wide variety of cell types.

Structure

The adrenal glands are paired, pyramid-shaped organs each weighing approximately 5 g in man, located above the upper pole of each kidney (Fig. 8.31). The adult adrenal cortex, which makes up approximately 90% of the gland, is composed of three zones. The outer subcapsular glomerulosa zone synthesizes aldosterone while the middle fasciculata and inner reticularis zones synthesize cortisol and androgen precursors. Blood supply is from several arteries which enter the outer portion of the cortex, and blood flows centrally through fenestrated capillaries to the inner medulla. Venous drainage is through a single vein into the vena cava on the right and into the renal vein on the left. Integrated functioning of the adrenal gland depends on this anatomical arrangement. The adrenal cortex regenerates with normal zonation when only a thin rim of glomerulosa cells remain attached to the capsule or when cells are transplanted to other sites (Ingle and Higgins, 1938). Zonation appears to depend on centripetal blood flow, which generates a gradient of steroid hormone concentration which increases from subcapsular cells to those more centrally located (Hornsby and Crivello, 1983). These steroids inhibit selected steroidogenic enzymes so that biosynthesis of aldosterone is limited to outer glomerulosa cells, and precursor flow into androgens is facilitated in inner reticularis cells. The medullary enzyme phenylethanolamine N-methyl transferase, which catalyzes conversion of norepinephrine to epineprhine, is regulated by cortisol (Wurtman and Axelrod, 1966), which is delivered in high concentrations in centrally flowing sinusoidal blood.

Biosynthesis and Metabolism of Adrenocortical Hormones

Cholesterol which is utilized for steroid hormone synthesis is derived from cholesterol esters which are stored in abundant cytoplasmic lipid droplets in adrenocortical cells and from circulating cholesterol carried in low density lipoprotein (LDL) particles. In response to ACTH, increased precursor cholesterol is made available by activation of cholesterol esterase which releases free cholesterol from ester storage depots (Beckett and Boyd, 1977) and by facilitated uptake of LDL cholesterol via LDL receptors which are especially abundant in adrenocortical tissue (Brown et al., 1979). With acute stimulation of steroidogenesis, precursor cholesterol is derived largely from lipid droplet ester storage depots; on prolonged stimulation, the major source of cholesterol precursor is LDL. Free cholesterol is transported to mitochondria, where it is metabolized to pregnenolone and isocaproaldehyde by the cytochrome P-450 side chain cleavage enzyme located in the inner mitochondrial membrane (see

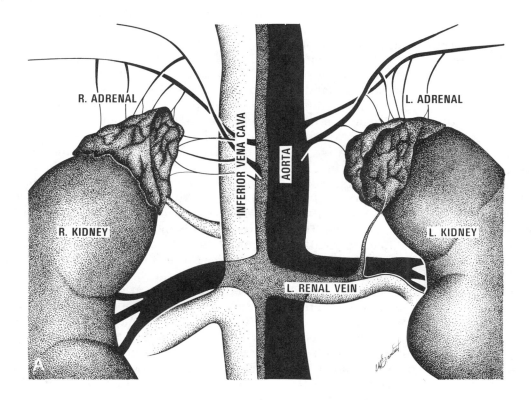

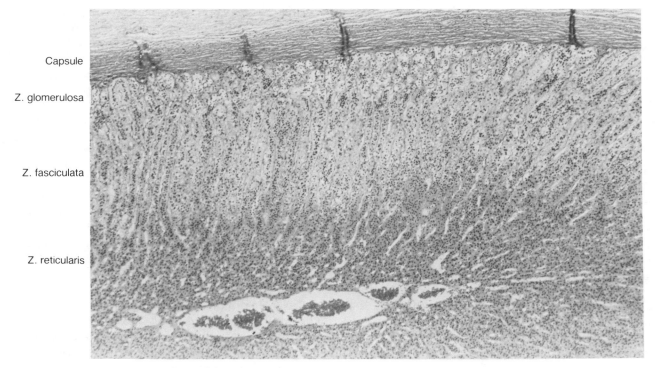

Capsule

Z. glomerulosa

Z. fasciculata

Z. reticularis

Figure 8.31.   Gross and microscopic anatomy of the adrenal gland.

Fig. 8.10, Chapter 52) (Burstein and Gut, 1971). The rate-limiting step in steroid hormone biosynthesis is the interaction of free cholesterol with cytochrome P-450 side chain cleavage enzyme (Simpson et al., 1978). This step requires continuing protein synthesis and may involve formation of polyphosphoinositol to facilitate this interaction. The limiting step in steroid hormone biosynthesis is thus availability of cholesterol substrate to the mitochondrial enzyme; ACTH and all steroidogenic stimuli increase this interaction of substrate with enzyme to acutely increase steroid hormone synthesis.

The cholesterol transformation (which involves hydroxylations at carbons 20 and 22) and subsequent steroid hydroxylations require NADPH, molecular oxygen, and specific cytochrome P-450 enzymes. Pregnenolone is rapidly removed from mitochondria and sequentially modified to yield the three major classes of adrenocortical steroids (Fig. 8.32). In addition to the major transformations shown in Figure 8.32, other intermediates, including sulfated ones, may exist. In endoplasmic reticulum, pregnenolone is hydroxylated at the 17 position to yield 17α-OH pregnenolone or is converted to progesterone by a 3β-hydroxysteroid dehydrogenase, $\Delta^{4,5}$-isomerase enzyme complex, which also converts 17α-OH pregnenolone to 17α-OH progesterone and dehydroepiandrosterone to androstenedione. 17α-OH progesterone is then hydroxylated at carbon 21 to yield 11-deoxycortisol. The final 11β-hydroxylation of cortisol is catalyzed by mitochondrial 11β-hydroxylase. Reactants flow through this highly organized biosynthetic pathway from cytosol to mitochondria to endoplasmic reticulum to mitochondria.

Formation of aldosterone in the zona glomerulosa proceeds from progesterone by hydroxylations at carbons 21, 11, and 18. 17α-Hydroxylation is not required, and the activity of this enzyme in glomerulosa cells is very low. The 18-hydroxyl group is oxidized to an aldehyde in mitochondria to yield aldosterone.

Dehydroepiandrosterone (DHEA), which is secreted primarily as the sulfated derivative, and androstenedione are the principal androgenic steroids produced by the adrenal cortex. 17α-Hydroxylation of either pregnenolone or progesterone permits 17, 20-lyase activity to remove the side chain from carbon 17. The adrenal androgens have little intrinsic biological activity and are primarily active after conversion in peripheral tissues to testosterone and estrogen.

Daily production rates and plasma concentrations of the principal secretory products of the adrenal cortex are shown in Table 8.6. Cortisol is the feedback regulator of the hypothalamic-pituitary-adrenocortical system and determines the rate of production of all steroids except aldosterone via this feedback axis. In plasma, more than 90% of cortisol is bound, principally to corticosteroid-binding globulin (CBG), a glycoprotein synthesized in the liver (Burton and Westphal, 1972). CBG is not required for biological effects of cortisol but serves as a reservoir to protect cortisol from renal clearance and degradation. CBG, which contains one high affinity binding site per molecule, is normally present in concentrations sufficient to bind about 250 ng of cortisol/ml. When cortisol production is elevated and the above capacity is exceeded, cortisol is bound weakly to serum albumin, and the urinary concentration of free cortisol rises sharply. CBG production is increased by estrogens so that by the third trimester of pregnancy CBG concentrations are twice those of the nonpregnant state. When CBG is increased, the biologically active free fraction will be maintained at normal levels via the hypothalamic-pituitary-adrenocortical feedback axis. Total plasma cortisol as measured in clinical situations will be elevated due to the increase in the bound fraction; free cortisol concentrations will, however, remain normal. CBG is decreased in certain liver diseases, hypothyroidism, and nephrosis, but again free cortisol concentrations are normally maintained. Interpretation of measured total plasma cortisol concentrations therefore requires knowledge of the metabolic state of the patient as well as of the time of sampling in relation to the circadian rhythm.

Aldosterone is only weakly bound to plasma proteins so that most aldosterone is metabolically inactivated during a single passage through the liver. DHEA and androstenedione are only weakly bound to plasma proteins, but testosterone and estrogen are tightly bound to a specific sex steroid-binding globulin.

Cortisol has a circulating half-life of about 90 min, whereas aldosterone, which is weakly protein bound, has a half-life of about 15 min. The principal site of metabolism is the liver, where inactivation of steroid hormones occurs and conjugation with glucuronic acid and sulfate render metabolites water soluble for excretion by the kidney. The principal alteration of cortisol is reduction of the double bond between carbons 4 and 5 and reduction of the ketone at carbon 3. The resulting inactive tetrahydrocortisol is conjugated with glucuronic acid at carbon 3 for renal excretion. Reduction of the ketone at carbon 20 yields the cortol series of metabolites. Aldosterone is metabolized to tetrahydroaldosterone and conjugated with glucuronic acid

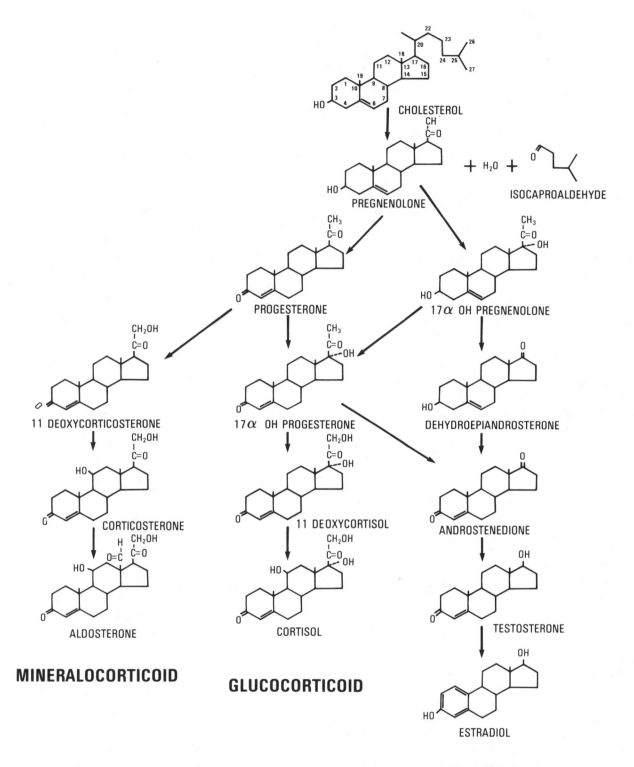

Figure 8.32. Pathway for biosynthesis of adrenocortical hormones.

Table 8.6.
Production Rates and Plasma Concentrations of Major Adrenocortical Hormones[a]

Class	Steroid	Production Rate, mg/day	Plasma Concentration, ng/ml
Glucocorticoid	Cortisol	8–25	40–180
	Corticosterone	1–4	2–4
Mineralocorticoid	Aldosterone	0.05–0.15	0.15
	Deoxycorticosterone	0.6	0.15
Androgenic	DHEA	7–15	5
	DHEA-S		1200
	Androstenedione	2–3	1.8

[a]Concentrations vary with time of day, sex, and stage of menstrual cycle for particular steroids.

at the 3 position or is conjugated with glucuronic acid at carbon 18. DHEA and androstenedione can be converted to the more active androgen testosterone in the periphery. However, most of these steroids are metabolized by reduction of the double bond between carbons 4 and 5 and by hydroxylations at carbons 7 and 16.

Clinical assessment of adrenocortical function is principally done by quantitation of circulating steroids using competitive protein-binding methods such as radioimmunoassay. Traditionally, production has been inferred by quantitation of urinary metabolites by chemical methods. The Porter-Silber measurement of the 17,21-dihydroxy-20-ketone group (17-hydroxysteroids) quantitates about one-third of cortisol metabolites and is an accurate reflection of cortisol production in both excess and deficiency states. The Zimmermann reaction, which measures 17-ketosteroids, provides an assessment of adrenal androgen production. Testosterone, the principal active androgen is quantitated by competitive protein-binding methods.

Regulation

HYPOTHALAMIC-PITUITARY-ADRENOCORTICAL AXIS

Production of cortisol and androgen precursors is controlled by ACTH, whereas aldosterone production is additionally regulated by angiotensin and potassium. Corticotropin-releasing hormone (CRH), a neuropeptide of 41 amino acids (Vale et al., 1981), is synthesized in the hypothalamus and reaches ACTH-producing cells of the anterior pituitary via the hypophyseal portal system (Fig. 8.33). In response to CRH, corticotroph cells of the pituitary synthesize and secrete ACTH, which circulates and specifically binds to high affinity receptors on the surface of adrenocortical cells to stimulate synthesis and secretion of cortisol (Buckley and Ramachandran, 1981). ACTH also stimulates synthesis of aldosterone and of adrenal androgens. Cortisol, but not other adrenal steroids, exerts negative feedback control on ACTH synthesis by suppressing transcription of the ACTH gene in the pituitary and by

suppressing formation of CRH in the hypothalamus. Like other endocrine systems, the hypothalamic-pituitary-adrenocortical axis tends to maintain its own homeostatic set point unless strongly driven by continuous environmental signals received by the nervous system. Net synthesis of ACTH is a result of the relative strength of stimulatory (CRH) and inhibitory (cortisol) signals. When cortisol or a synthetic derivative such as dexamethasone is given in pharmacological amounts, ACTH synthesis is decreased. However, when patients are given suppressive amounts of dexamethasone prior to major surgery, the stress of surgery is sufficient to over-

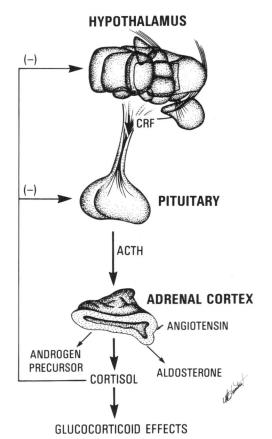

Figure 8.33. Hypothalamic-pituitary-adrenocortical axis.

ride the suppression and to increase ACTH synthesis.

Secretion of ACTH and of cortisol exhibits a circadian rhythm, with the highest rate occurring around the time of morning awakening (Fig. 8.34). Both ACTH and cortisol are secreted in episodic bursts, with the integrated sum of these bursts giving the characteristic circadian rhythm where mean morning plasma ACTH and cortisol concentrations exceed evening plasma concentrations approximately 2 fold. This circadian rhythm appears importantly tied to sleep-wake cycles and to food intake. In rats, which are nocturnal feeders, peak concentrations of ACTH and corticosterone occur prior to initiation of the feeding period (Miyabo et al., 1980). The circadian rhythm thus favors higher cortisol concentrations during the major period of fasting each 24 hours, whether that occurs during the day or night.

RENIN-ANGIOTENSIN-ALDOSTERONE AXIS

Aldosterone, the principal sodium-retaining steroid hormone in humans maintains normal fluid balance and circulatory volume. Aldosterone production is regulated by the renin-angiotensin system, as well as by ACTH and potassium. The enzyme, renin, is produced principally in juxtaglomerular cells located in the media of afferent arterioles at their entry into the renal glomerulus (Fig. 8.35). Renin, a proteolytic enzyme, cleaves the amino-terminal decapeptide from angiotensinogen or renin substrate, a glycoprotein which is synthesized in the liver. The decapeptide angiotensin I is further processed by converting enzyme which removes the two carboxy-terminal amino acids to give the active octapeptide, angiotensin II. Angiotensin-converting enzyme is widely distributed in the endothelium of vascular beds, especially in the lung. Although angiotensin II is the principal active peptide hormone, angiotensin III (des Asp[1] angiotensin II) is also active on zona glomerulosa adrenocortical cells.

The concentration of renin in blood is the principal regulator of this system. Juxtaglomerular cells synthesize and secrete renin in response to several signals: baroreceptor, β-adrenergic, prostaglandin, and the fluid composition of the distal nephron at the macula densa. The juxtaglomerular cells, located in the wall of the afferent glomerular arteriole, respond as baroreceptors and secrete renin in response to changes in renal perfusion pressure: decreased perfusion pressure stimulates renin secretion, whereas increased perfusion pressure inhibits renin secretion. Renal sympathetic nerves, which terminate at the juxtaglomerular and afferent arteriolar smooth muscle cells, increase renin secretion via β-adrenergic receptors. Prostaglandins also increase renin secretion. The macula densa, which is a specialized area of the distal nephron in proximity to juxtaglomerular cells, monitors tubular fluid composition and mediates renin release. There is controversy as to which signal is sensed (Na^+, Cl^-, Ca^{2+}) and how it is transmitted to the juxtaglomerulosa cell.

When intravascular volume is decreased, as occurs with dehydration or hemorrhage, renal perfusion pressure is decreased, the sympathetic nervous system is activated, and renin synthesis and secretion are increased (Fig. 8.35). Under most conditions adequate concentrations of angiotensinogen are present in blood, and increased renin produces increased angiotensin I, which is rapidly converted to angiotensin II by the large amount of converting enzyme available. Angiotensin II is an extremely potent vasoconstrictor through its action on arterial smooth muscle cells. It acts not only systemically to raise blood pressure but also locally to decrease filtration at the glomerulus. Angiotensin II and III bind to specific high affinity receptors in the adrenal

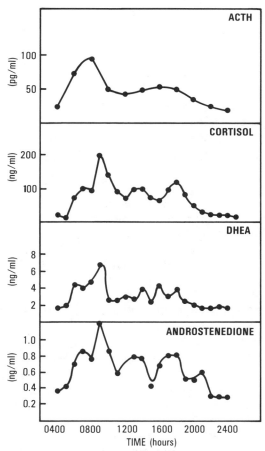

Figure 8.34. Diurnal variation in ACTH and adrenocortical steroid hormones. Data derived by hourly sampling in a 24-year-old woman. DHEA-S levels (not shown) do not vary because of slow clearance.

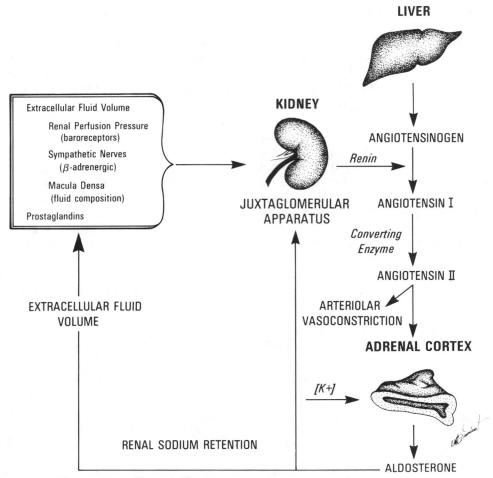

Figure 8.35. Renin-angiotensin-aldosterone axis.

cortex to stimulate synthesis and secretion of aldosterone. Aldosterone acts on its targets, primarily the distal nephron, to cause sodium readsorption and consequent expansion of intravascular and extracellular fluid volumes. When intravascular volume and renal perfusion pressure are increased, renin secretion is suppressed.

Potassium is also a potent regulator of aldosterone synthesis with small changes in plasma [K$^+$] within the physiological range altering aldosterone synthesis. Increases in plasma [K$^+$] elevate aldosterone synthesis, whereas decreases in plasma [K$^+$] lower aldosterone synthesis. Because aldosterone promotes Na$^+$ readsorption by facilitating K$^+$ and H$^+$ secretion, this regulatory system is a homeostatic one to maintain normal plasma [K$^+$].

Several additional regulators of zona glomerulosa and zona fasciculata steroid hormone synthesis have been identified, but the physiological role of these is questionable. β-Lipotropin and β-melanotropin (a fragment of β-lipotropin), which are derived from the ACTH precursor, preferentially stimulate aldosterone synthesis in glomerulosa cells (Matsuoka et al., 1981). The concentration required exceeds that usually found in plasma. The 16,000 M$_r$ fragment and the γ-MSH portion of that derived from the ACTH precursor potentiate low concentrations of ACTH to stimulate corticosterone synthesis (Pedersen and Brownie, 1980). Although the concentration required is within the physiological range, the total effect is small.

Effects of ACTH and Angiotensin on the Adrenal Cortex

ACTH works via cAMP. The binding curve for ACTH is superimposable on the concentration curve for cAMP formation; however, only a small fraction of receptors need to be occupied to give a maximal steroidogenic response (Hornsby and Gill, 1978) (see Chapter 52). Proof of the hypothesis that cAMP mediates the effects of ACTH was provided by showing that mutant adrenocortical cells with altered cAMP-dependent protein kinase have responses to ACTH that are altered in parallel (Rae et al., 1979). Acutely, ACTH stimulates steroid hormone biosynthesis by facilitating interaction of free cholesterol with cytochrome P-450 side chain cleavage enzyme. All steroidogenic stimuli acutely increase pregnenolone formation; the conversion of pregnenolone to final steroid products depends on the enzyme complement of the cell. Chronically, ACTH induces the enzymes of the steroidogenic pathway to increase biosynthetic capacity. ACTH increases the quantity of 17α-, 11β-, and 21-hydroxylase and 3β-hydroxysteroid dehydrogenase $\Delta^{4,5}$isomerase enzymes so that cortisol is the principal product of the biosynthetic pathway. Following hypophysectomy there is atrophy of the adrenal cortex, with a corresponding decrease in the enzymes of the steroidogenic pathway; administration of ACTH restores adrenocortical structure and steroidogenic enzymes to normal. When ACTH concentrations are elevated, there is enlargement of the cortex and increased steroidogenic responsiveness. ACTH and cAMP increase the concentrations of mRNA and their encoded enzymes via enhanced gene transcription (Waterman and Simpson, 1985). cAMP response elements have been identified in the 5′ end of several genes (Montminy et al., 1986; Silver et al., 1987), and similar sequences exist in the control regions of genes encoding steroidogenic enzymes. Induction of steroid biosynthetic enzymes by ACTH requires concomitant cellular hypertrophy which depends on the availability of growth factors and nutrients (Gill et al., 1982). ACTH and cAMP inhibit cell replication (Masui and Garren, 1971) so that in conjunction with growth factors the adrenocortical cell is hypertrophied with increased functional capacity (Gill et al., 1979). Inhibition of cell division is of advantage for a system which is continuously changing its function, because hypertrophy is a readily reversible process whereas cell replication is reversible only by cell death exceeding cell division. Cell replication occurs only with prolonged ACTH treatment and significant receptor desensitization (Hornsby and Gill, 1977). The link coordinating adrenal delivery of growth factors with ACTH is not known. Following unilateral adrenalectomy, growth of the remaining gland is mediated via neural pathways, suggesting that neural signals are important growth promoters for the adrenal cortex (Engeland and Dallman, 1975).

When cortisol or synthetic derivatives are used clinically, ACTH production is diminished, and atrophy of the adrenal cortex occurs. When high-dose glucocorticoids are administered for about 4 weeks or longer, recovery of the normal feedback axis is delayed. A similar pattern is observed following removal of cortisol-secreting adrenal and ACTH-secreting pituitary tumors. As shown in Figure 8.36 recovery of ACTH is delayed. When ACTH production resumes, it exhibits a diurnal secretory pattern, even though adrenal production of cortisol is low. Recovery of the adrenal cortex follows pituitary recovery so that the entire process may require 6–9 months (Graber et al., 1965). Because the zona glomerulosa is also regulated by angiotensin, exogenous or endogenous elevations of cortisol do not suppress aldosterone production. Patients receiving pharmacological amounts of glucocorticoids, therefore, maintain normal fluid and electrolyte balance via the renin-angiotensin-aldosterone axis.

Angiotensin principally affects the zona glomerulosa to increase aldosterone synthesis. Angiotensin does not appear to work via cAMP but may use DAG and calcium as second messengers. Both the early step of pregnenolone formation from cholesterol and the later step of aldosterone formation from corticosterone are increased. Angiotensin exerts a trophic effect as well as an acute stimulatory effect on the zona glomerulosa. The glomerulosa zone widens during sodium depletion, and angiotensin II stimulates growth of adrenal cells in culture (Gill et al., 1977). The boundary between the zona glomerulosa and zona fasciculata moves inward or outward, depending on the long-term need for mineralocorticoid or glucocorticoid. Inward movement with enlargement of the zona glomerulosa occurs with sodium depletion and angiotensin elevation via cell growth and inhibition of conversion of glomerulosa to fasciculata cells. Stimulation with ACTH, which acutely increases aldosterone synthesis, moves the boundary outward by facilitating conversion of glomerulosa to fasciculata cells. As noted, zonation depends on the blood flow and may

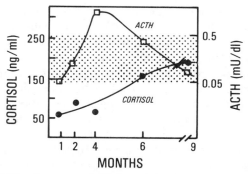

Figure 8.36. Recovery of the hypothalamic-pituitary-adrenocortical axis after exposure to long-term elevated concentrations of glucocorticoids. All measurements were made at 0600 hours and *closed circles* represent median values from 14 patients. The *dotted area* represents the range for normal subjects. (From Tyrrell and Baxter, 1981; based on original data of Graber et al., 1965.)

involve gradients of steroid hormones and free radical effects on enzymes.

Life History of the Adrenal Cortex

In man and other primates the fetal adrenal gland is very large relative to body size; it approaches the size of the fetal kidney at term. The gland is composed of an outer subcapsular definitive or adult zone, which is the precursor of the three zones of the postnatal cortex, and a large inner fetal zone, which constitutes about 80% of the fetal adrenal cortex. The cells of the fetal zone lack 3β-hydroxysteroid dehydrogenase, $\Delta^{4,5}$-isomerase (Fig. 8.32) so that precursors are converted to dehydroepiandrosterone, which is sulfated and secreted as the principal steroid product (Diczfalusy, 1964). The fetal zone synthesizes large amounts of steroids (100–200 mg/day at term) which serve as precursors for synthesis of estrogens by the placenta (Sitteri and McDonald, 1966). This fetoplacental unit is terminated at parturition, and after birth there is rapid involution of the fetal zone and establishment of the three zones of the definitive cortex from subcapsular cells. Fetal ACTH is required for maintenance of both fetal and definitive zones (Bernischke, 1956). When fetal zone cells are placed into cell culture, ACTH induces the full steroid biosynthetic pathway of the definitive cortex (Simonian and Gill, 1981), suggesting that, in utero, local conditions specifically suppress 3β-hydroxysteroid dehydrogenase.

Cortisol production remains constant throughout life, as measured under both basal and ACTH-stimulated conditions (Pintor et al., 1980). During early childhood to age 5 years, adrenal androgen secretion is low. Production of adrenal androgens, which contribute to the development of axillary and pubic hair (adrenarche), then increases progressively after age 5 to reach a maximum around age 13 years. This rise coincides with progressive development of the inner reticularis zone (Dhom, 1973), which produces more androgens relative to cortisol than does the fasciculata zone (O'Hare et al., 1980). Although a specific adrenal androgen-stimulating hormone has been proposed, it has not been identified. Alternatively, autoregulatory mechanisms within the adrenal cortex may cause increased adrenal androgen production. The enlarging reticularis zone is exposed via centripetal blood flow to increased concentrations of steroid hormones produced by outer zones; these steroids may preferentially decrease steroidogenic enzymes such as cytochrome P-450$_{11\beta}$ and 21 to divert precursors into androgen end products (Hornsby, 1980). After

puberty there is a progressive decline in adrenal androgen secretion, with DHEA declining about 5 fold from ages 30–80 years.

Biological Effects of Glucocorticoids

Cortisol, the principal glucocorticoid in humans, has widespread effects on most organs of the body to regulate metabolism of protein, nucleic acid, and fat, as well as carbohydrate. Physiologically, cortisol mediates adaptive responses to stress and fasting; pharmacologically, cortisol is used to suppress the inflammatory response. In fasting, which is the best-studied paradigm, cortisol maintains blood glucose by stimulating gluconeogenesis. Glucose is the required substrate for nervous system metabolism until chronic adaptation to fasting allows ketone bodies to be utilized. In the absence of food intake, the approximately 75 g of glycogen stored in the liver are sufficient to maintain blood glucose for only 12–24 hours. The large amounts of muscle and other body proteins, as well as fat, provide a much greater potential substrate reserve to maintain blood glucose and nervous system function. As shown in Figure 8.37, the liver is the principal site of gluconeogenesis, although in prolonged fasting the kidneys also contribute. Cortisol has anabolic effects on the liver, inducing the synthesis of a number of enzymes involved in transamination of amino acids and in gluconeogenesis (Baxter, 1979). Cortisol also increases glycogen synthesis and accumulation in the liver. The effects of cortisol on most other organs are catabolic and provide substrate for hepatic glucose production. In muscle, which is quantitatively the most important source, cortisol inhibits protein and nucleic acid synthesis and enhances protein breakdown to provide amino acids, principally alanine, for use by the liver (Felig, 1975). Cortisol blocks both glucose and amino acid uptake in the periphery to further enhance gluconeogenesis and blood glucose levels. Cortisol increases lipolysis and enhances the effects of other lipolytic stimuli such as catecholamines. Free fatty acids are an alternative fuel source via conversion to ketone bodies and provide reducing equivalents to sustain gluconeogenesis.

The effects of cortisol on individual organ systems are based on this pattern of reactions. Clinically, the role of cortisol is most evident when it is present in excessive amounts. Its role in reestablishing homeostasis after transient disturbances must often be inferred from analysis of deficiency or excess states. Because adrenal insufficiency involves loss of additional hormones, the effects of cortisol are most easily seen when cortisol is produced in exces-

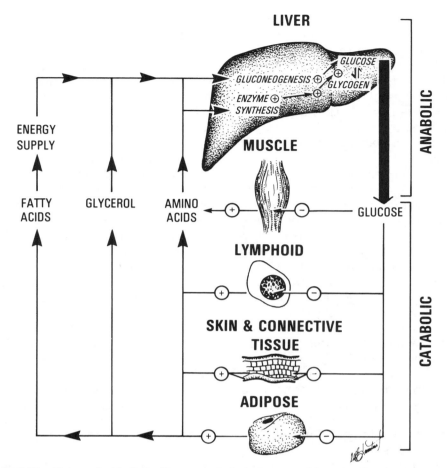

Figure 8.37. Glucocorticoid effects. The *arrows* indicate the general flow of substrate in response to glucocorticoids. The + and − signs indicate stimulation or inhibition, respectively. (From Baxter, 1979.)

sive amounts or when it is administered in pharmacological doses.

The principal clinical use of cortisol and synthetic glucocorticoids is to suppress inflammatory responses. Cortisol inhibits migration of polymorphonuclear leukocytes, monocyte-macrophages, and lymphocytes at the site of inflammation; it inhibits release of vasoactive, proteolytic, and other peptides; it inhibits fibroblast growth and wound healing. Although the mechanisms through which cortisol exerts its anti-inflammatory effects are not well understood, it may work, in part, through increased synthesis of "macrocortin" (Blackwell et al., 1980; Hirata et al., 1980). This polypeptide inhibits phospholipase A_2, thereby decreasing membrane release of arachidonic acid, the precursor of both prostaglandins and leukotrienes, two mediators of inflammation. Cortisol suppresses cell-mediated immunity with greater effects on T and B lymphocytes. A lympholytic effect is used in treatment of leukemia and lymphoma but is probably not a

prominent effect under normal conditions. Cortisol decreases circulating lymphocytes, monocytes, eosinophils, and basophils. It increases circulating polymorphonuclear leukocytes, primarily by release from storage in vessel walls. Antibody production is not usually affected by glucocorticoids.

The catabolic effects of high concentrations of cortisol are evident on muscle where there is loss of muscle mass and development of weakness, on skin where there is thinning and separation of subcutaneous tissues to result in striae, and on bone where there is osteoporosis and a proclivity to bone fractures. Glucocorticoids inhibit bone cell function and deposition of the collagen matrix; they also inhibit gastrointestinal calcium absorption so that calcification of the matrix is impaired. Glucocorticoids may interfere with gastric synthesis of prostaglandins, which are required for maintenance of the normal protective barrier against gastric acid and pepsin. Capillary fragility is increased, and bruising with minor trauma is common. High concentrations of glucocorticoids inhibit linear growth and skeletal maturation in children. Growth hormone is suppressed, but the major factor in reduced growth is inhibition of protein synthesis in many cell types.

In fed humans the effects of high concentrations of glucocorticoids may result in hyperglycemia. Cortisol also stimulates appetite so that weight gain is common. Insulin rises in response to hyperglycemia and may accentuate a peculiar central deposition of fat in the face, neck, and abdomen.

In response to severe stress, adrenal production of cortisol may rise as much as 10 fold. This is an adaptive response which favors survival. The mineralocorticoid, deoxycorticosterone, was introduced into clinical medicine before cortisol. This advance improved survival of patients with adrenal insufficiency 3–5 fold. The introduction of cortisol had a much greater impact, increasing survival more than 10 fold (Dunlop, 1963). Knowledge of basal production rates and of the capacity of the gland to increase production in response to stress allows ready calculation of replacement doses of cortisol for patients with adrenal insufficiency.

In addition to the effects of cortisol mentioned above, it is required ("permissive") for the action of a number of hormones (e.g., catecholamines) and for synthesis of others (e.g., growth hormones). Cortisol is also important during development. Cortisol increases surfactant synthesis in fetal lung, glutamine synthetase in neural retina, and hepatic enzymes in fetal liver, and is required for breast development. It is required for normal growth. When cortisol concentrations are elevated over prolonged periods as in Cushing's disease or with pharmacological use, catabolic effects are sustained. Therapy may yield desired antiinflammatory effects, but a price is paid by all organ systems.

Biological Effects of Mineralocorticoids

Aldosterone, the principal mineralocorticoid, stimulates reabsorption of Na^+ in the collecting tubules of the kidney. When Na^+ is reabsorbed from tubular fluid, electrical neutrality must be maintained by secretion of K^+ or H^+ or by concomitant reabsorption of an anion such as Cl^-. The exchange of K^+ and H^+ for Na^+ will be greater when the quantity of Na^+ reaching the collecting tubule is increased. When extracellular fluid volume is decreased as, for example, by gastrointestinal losses or by hemorrhage, activation of the renin-angiotensin system increases aldosterone to promote Na^+ conservation. Because Na^+ is the principal extracellular substance active as an osmole, water is also retained to expand the depleted extracellular fluid volume (Fig. 8.35). With adrenal insufficiency, Na^+ conservation in the distal tubule is defective so that excessive Na^+ is lost in the urine, with resultant hyponatremia and contraction of the extracellular fluid volume. Although most Na^+ is reabsorbed in the proximal tubule, with aldosterone deficiency up to 240 meq of Na^+ may be lost per day because of failure of Na^+ conservation by the collecting tubules. Because cation exchange is not occurring, hyperkalemia and acidosis result. The diminished extracellular fluid volume results in decreased blood pressure and reduced cardiac output. Cortisol also exerts important effects on the circulation and is required both for normal cardiac output and for vasoconstrictor effects of catecholamines. If adrenal insufficiency is uncorrected, circulatory collapse and death occur. Mineralocorticoid deficiency is a strong contributor to the muscle weakness characteristic of adrenal insufficiency and also contributes to poor growth in children. In adrenal insufficiency ADH concentrations are increased to facilitate water conservation and cellular hydration so that hyponatremia represents both Na^+ deficiency and dilution of extracellular $[Na^+]$. Patients with adrenal insufficiency are characteristically unable to excrete an administered load of free water because of elevated concentrations of ADH and lack of effects of cortisol on the collecting duct. Aldosterone also regulates ion transport in sweat glands, salivary glands, and the gastrointestinal tract. When exposed to increased ambient temperatures, patients with adrenal insufficiency continue to produce sweat with a high Na^+ content and are unable to adapt to climate changes.

When aldosterone is produced or administered in excessive amounts, there is increased Na^+ reabsorption and facilitated K^+ and H^+ exchange, resulting in hypokalemia and alkalosis. Hypokalemia and alkalosis are more pronounced when increased Na^+ reaches the collecting tubule for aldosterone driven exchange. Initially, isotonic volume expansion of 2–4 liters will occur over 2–3 days. After this, increased atrial natriuretic hormone and intrarenal compensatory mechanisms are activated to result in reduced Na^+ reabsorption in the proximal tubule. The body thus "escapes" from the effects of aldosterone, and Na^+ excretion then equals Na^+ intake. Facilitated cation exchange in the collecting tubule continues, and hypokalemic alkalosis is sustained. The modest volume expansion also continues. Although edema does not occur, hypertension results when excess aldosterone is present for a long time. Magnesium and calcium excretion are increased when aldosterone is elevated because of decreased proximal tubular reabsorption in parallel with decreased proximal tubular reabsorption of

Na$^+$ in the compensated state. When aldosterone is elevated because of autonomous production by adenomas or hyperplasia of the glomerulosa zone, the renin-angiotensin system will be suppressed because of volume expansion. As occurs more commonly with processes such as congestive cardiac failure and liver cirrhosis, the renin-angiotensin system is activated, and secondary hyperaldosteronism results. In these conditions, edema results from Na$^+$ retention and the primary disease process.

As for all steroid hormones, aldosterone acts via a specific receptor protein to regulate specific gene expression. In renal collecting tubules, Na$^+$ enters at the luminal membrane and is extruded at the serosal surface into the interstitial fluid by action of the Na$^+$/K$^+$-ATPase enzyme, which moves a Na$^+$ ion out of the cell and a K$^+$ ion into the cell. Aldosterone induces synthesis of at least four mitochondrial enzymes involved in energy generation to drive the Na$^+$/K$^+$-ATPase (Edelman and Marver, 1980). It may also increase Na$^+$ channels, with the ion gradient being a major driving force for activity of Na$^+$/K$^+$-ATPase (Ludens and Fanestil, 1979). Aldosterone also appears to increase the H$^+$ antiporter and may increase K$^+$ secretion into tubular fluid by either increasing K$^+$ channels or by means of a potassium pump.

Pharmacology

In adrenal insufficiency, replacement of glucocorticoid and mineralocorticoid hormones sustains normal life. However, the principal clinical use of adrenal steroids is to suppress undesirable inflammatory reactions, and more than 10 million people in the United States receive them each year. Because glucocorticoid receptors appear identical in all cell types, it has been impossible to synthesize a steroid molecule which has antiinflammatory but not other glucocorticoid effects. Because mineralocorticoid receptors are distinct from glucocorticoid receptors, it has been possible to synthesize steroid molecules with relatively pure glucocorticoid or mineralocorticoid effects. As shown in Figure 8.38, a number of substitutions can be made on the cortisol molecule. Introduction of a double bond between carbons 1 and 2 increases glucocorticoid potency (prednisone and prednisolone). A halogen substitution at carbon 9 increases both glucocorticoid and mineralocorticoid activities. Because this substitution increases mineralocorticoid activity especially, it results in an orally active mineralocorticoid (9α-fluorocortisol) which is used clinically because the degradation of aldosterone in the liver precludes its use as an oral agent. Methylation at carbon 16 greatly reduces

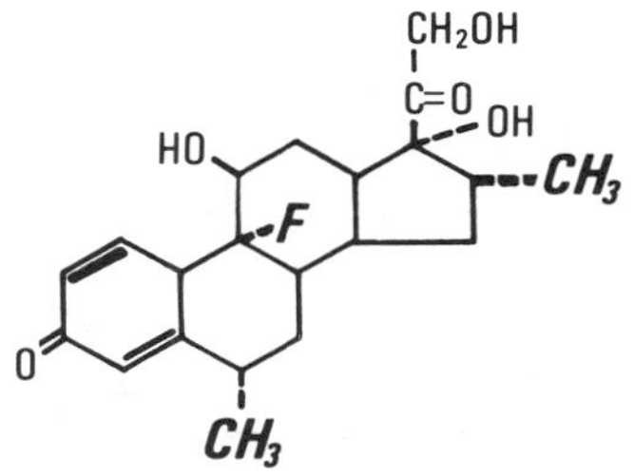

Figure 8.38. Pharmacological modifications of the cortisol molecule.

mineralocorticoid activity. 9α-Fluoro-16α-methylprednisolone (dexamethasone) is thus a potent glucocorticoid with essentially no mineralocorticoid activity. Methylation at carbon 6 increases water solubility. A variety of combinations of substitutions at these indicated positions results in the many available steroid products.

Glucocorticoid antagonists are not available, but the mineralocorticoid antagonist, spironolactone, which binds to the mineralocorticoid receptor to induce an inactive conformation, is used in conditions of aldosterone excess.

It is evident that long-term high-dose glucocorticoid therapy results in side effects and prolonged suppression of the hypothalamic-pituitary-adrenocortical axis. A number of strategies have been useful in minimizing these two unwanted effects. Strategies include use of minimal doses which are effective in treating the disease process, timing of administration to morning or every other day schedules to allow function of the normal circadian rhythm, and local application. Because the renin-angiotensin system is not affected by administered glucocorticoids, mineralocorticoid function remains normal.

ADRENAL MEDULLA

The adrenal medulla is both a part of the autonomic nervous system and the endocrine system. The sympathetic chromaffin granule-containing cells of the adrenal medulla are equivalent to postganglionic neurons. The adrenal medulla is innervated by cholinergic preganglionic fibers from the greater splanchnic nerve. Acetylcholine released from these preganglionic neurons activates cells of the adrenal medulla to synthesize and secrete catecholamines. In contrast to other postganglionic sympathetic neurons, which release norepinephrine from axon termini at their site of action, adrenal medullary cells release epinephrine into the circulation. Adrenal medulla cells are thus neuronal cells which function as an endocrine gland.

Because the adrenal medulla is properly considered as an integral part of the sympathetic nervous system, full discussion of catecholamine synthesis, metabolism, and action is included in Chapter 69. Development of neuronal cells into specialized adrenal medullary cells is dictated by cortisol (Anderson and Axel, 1986). Nerve growth factor directs precursor cells into sympathetic neurons; cortisol directs the precursors into the alternative choice pathway of adrenal medullary cells. The anatomic localization of the adrenal medulla in the center of the adrenal cortex reflects this early developmental choice. The adrenal medulla is also unique in containing phenylethanolamine-*N*-methyltransferase, an enzyme which transfers a methyl group to the amino-terminus of norepinephrine to form epinephrine. Epinephrine is the principal hormonal product of the adrenal medulla, although smaller amounts of norepinephrine are also released. In addition to catecholamines, the adrenal medulla also produces opioid peptides, including met-enkephalin, leu-enkephalin, and related heptapeptide and octapeptide sequences (Noda et al., 1982).

Epinephrine functions as an integral part of the sympathetic nervous system, stress, and metabolic responses. Secretion is coordinated with cortisol production via the CRH axis described in Chapter 53. Despite the important physiological role of the adrenal medulla, it is not essential for survival. Following adrenalectomy, glucocorticoid and mineralocorticoid replacement are essential, but epinephrine is not required because the sympathetic nervous system and norepinephrine are sufficient.

CHAPTER 56

Hormonal Regulation of Mineral Metabolism

Three hormones, vitamin D, parathyroid hormone (PTH), and calcitonin are the principal regulators of mineral metabolism. The endocrine system composed of these three hormones serves to maintain a remarkably constant concentration of ionized calcium in extracellular fluids. This constancy, which is necessary for proper bone mineralization, neuromuscular excitability, membrane integrity, cellular biochemical reactions, stimulus-secretion coupling, and blood coagulation is maintained in spite of wide variations in dietary intake.

The role of this endocrine control system can be appreciated by considering normal calcium balance. A normal adult body contains about 1 kg of calcium. Approximately 99% is present in the skeleton as hydroxyapatite $[Ca_{10}(PO_4)_6(OH)_2]$ and 1% in soft tissues and extracellular fluids. About 1% of skeletal calcium is exchangeable with extracellular fluid, but this 10 g is large relative to the 900 mg present in extracellular fluid. A normal diet provides from 200–2000 mg with an average of 1000 mg of calcium per day (Fig. 8.39). Of this, approximately 300 mg is absorbed, the majority in the ileum because of its large absorptive surface. Calcium is also secreted into the intestinal tract in bile, pancreatic juice, and intestinal secretions so that net absorption of calcium equals about 175 mg/day. The total extracellular fluid space contains about 900 mg of calcium which is in dynamic equilibrium with other compartments. Approximately 500 mg of calcium is deposited into and reabsorbed from bone in the ongoing process of bone remodeling. Sixty percent of serum calcium is ultrafilterable so that approximately 10,000 mg is filtered at the renal glomerulus daily. Renal reabsorption of calcium is extremely efficient, so that only a small amount of calcium equal to that absorbed in the gut is excreted in the urine daily. As shown in Figure 8.39, calcium balance is zero, with excretion in feces and urine exactly equaling dietary intake.

In spite of these large movements of calcium between body compartments, serum calcium is maintained constant at about 10 mg/dl (2.5 mM). The essential fraction is the ionized one which equals 50% of the total serum calcium (1.3 mM). Forty percent of total serum calcium is protein-bound, principally to albumin, and 10% is complexed with diffusible anions. Because total serum calcium is usually measured, knowledge of the concentration of serum albumin and of pH may be required to estimate the ionized fraction. A decrease in serum albumin of 1 g/dl (normal = 4 g/dl) results in a decrease in total serum calcium of 0.8 mg/dl without affecting the ionized fraction significantly. Because calcium is bound to carboxyl groups on albumin, pH changes affect ionized calcium. Acidosis decreases binding and increases ionized calcium, whereas alkalosis increases binding and decreases ionized calcium. Acidosis thus protects against hypocalcemia by shifting albumin-bound calcium to the ionized form, and alkalosis such as that which occurs with acute hyperventilation decreases ionized serum calcium by increasing binding to serum albumin.

The constancy of extracellular fluid concentrations of calcium is maintained by hormonal regulation of the absorption of calcium from the gastrointestinal tract, the mobilization of skeletal calcium, and the reabsorption of filtered calcium in kidney tubules. Vitamin D, PTH, and calcitonin are the regulators of these processes.

VITAMIN D

Structure and Biosynthesis

Although initially discovered as a fat-soluble vitamin which would prevent the bone disease rickets, vitamin D is a classical steroid hormone having both dietary and endogenous precursors. Historically, rickets was a major health problem accentuated by poor nutrition and by limited exposure to sunlight; it increased with the urban migration and the environment created by the industrial revolution. Poor nutrition and limited exposure to sunlight

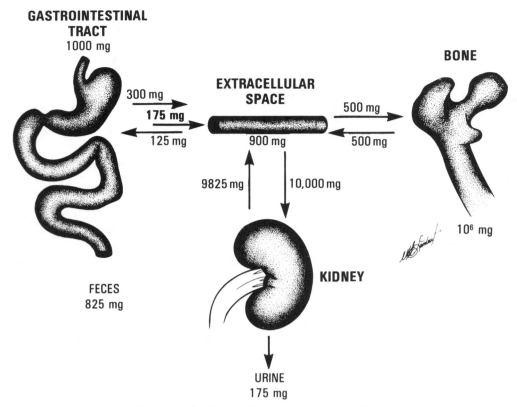

Figure 8.39.   Calcium balance on an average diet.

which were worsened during World War I led to a corresponding increase in clinically evident rickets—and increased study of this disease. It was found that rickets could be treated either by exposure to sunlight or by use of the folk remedy cod liver oil. Vitamin D was isolated in 1930 and is now widely used as a dietary supplement, so that simple vitamin D-deficient rickets is now distinctly unusual.

Vitamin D_3, cholecalciferol, is formed by ultraviolet irradiation of precursor 7-dehydrocholesterol present in skin (Fig. 8.40). This reaction can provide adequate vitamin D so a dietary source is not essential. Adequate exposure to UV light is necessary, with more exposure being required for darker skinned races. Dark-skinned children in northern urban environments where there is less sunlight exposure therefore have an especial requirement for dietary sources of vitamin D. Ergocalciferol, vitamin D_2, is the compound formed in plants from precursor ergosterol and differs from vitamin D_3 by the presence of a double bond between carbon 22 and 23 and a methyl group at carbon 24. Vitamin D_2 and D_3 are not abundant in foods (with the exception of fish liver), and irradiation to convert precursors to vitamin D_2 and D_3 are required. Both vitamin D_2 and D_3 are prohormones and require subsequent

modification to yield active hormone. Because hormones derived from either vitamin D_2 or D_3 are equally active in man, the generic term vitamin D will be used.

Formation of vitamin D from 7-dehydrocholesterol precedes via a 6,7-cis isomer intermediate, previtamin D (Holick and Clark, 1978). Vitamin D is transported in plasma bound to a specific vitamin D-binding protein which binds vitamin D with a greater affinity than previtamin D. Previtamin D thus preferentially remains in skin which serves as a storage depot after irradiation, while vitamin D is removed by the binding protein. Vitamin D-binding protein, an α-globulin synthesized in the liver, functions as do other steroid-binding proteins to provide a reservoir of hormone protected from renal clearance (Haddad and Walgate, 1976). It binds vitamin D, 25-(OH)D, 24,25-(OH)$_2$D, and 1,25-(OH)$_2$D. Vitamin D is metabolized in liver by hydroxylation at carbon 25 to yield 25-(OH)D. This is the major circulating form of vitamin D with a serum concentration of about 25 ng/ml and a half-life of 15 days.

The final metabolic conversion of vitamin D occurs in the kidney where hydroxylation at carbon 1 yields the active hormone 1,25-(OH)$_2$D (Deluca and Schnoes, 1976; Fraser, 1980). The 1α-hydroxylation is catalyzed by a mitochondrial mixed func-

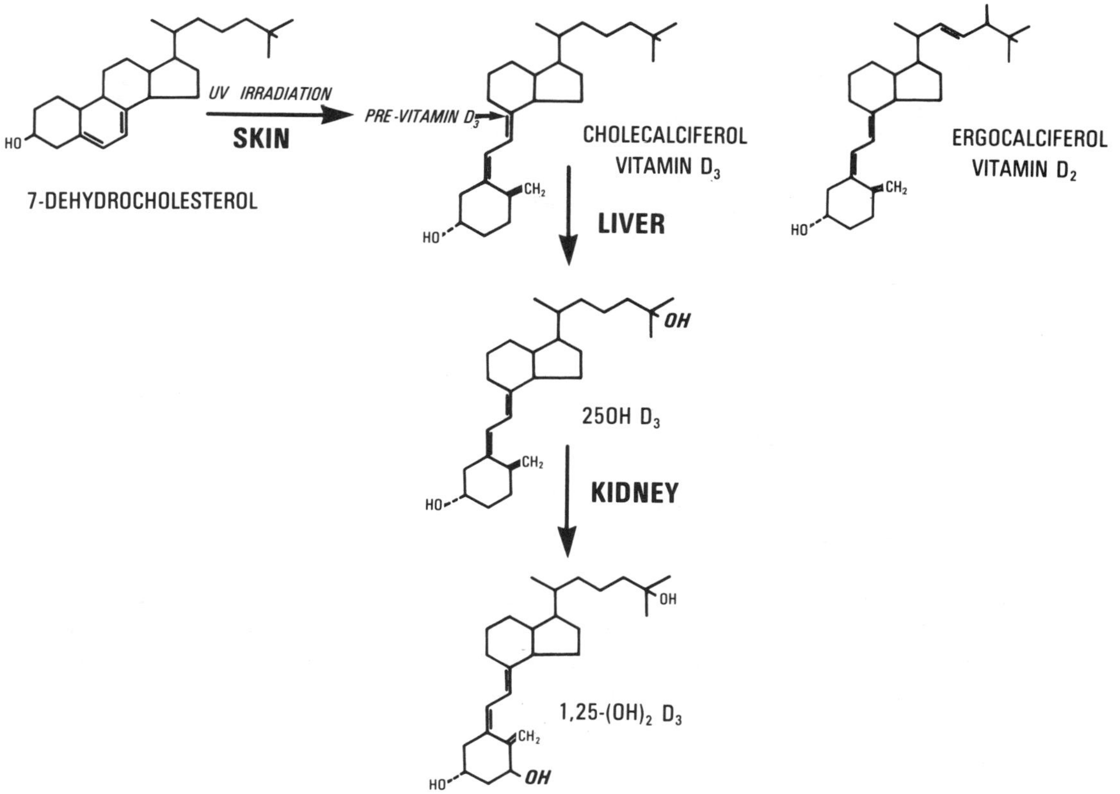

Figure 8.40. Biosynthesis of metabolically active 1,25-(OH)$_2$D.

tion oxidase similar to adrenocortical steroid hydroxylases. This last enzymatic step in the formation of biologically active 1,25-(OH)$_2$D is the principal site of regulation. 1,25-(OH)$_2$D has a circulating half-life of about 15 hours and a plasma concentration of 20–50 pg/ml.

Several additional modifications of the vitamin D steroid nucleus occur. Hydroxylation at carbon 24 is favored when hydroxylation at carbon 1 is low, and conversely hydroxylation at carbon 24 is low when hydroxylation at carbon 1 is high, suggesting that formation of 24,25-(OH)$_2$D represents an inactivation pathway. It is possible, however, that 24,25-(OH)$_2$D has some biological effects on bone (Bordier et al., 1978). Hydroxylation at carbon 26 appears to result in an inactive compound.

Regulation of 1,25-(OH)$_2$D Formation

To maintain normal calcium homeostasis, the portion of dietary calcium which is absorbed varies. When calcium intake is low the absorbed fraction may be as high as 90%, whereas, when calcium intake is high, much lower fractional absorption occurs. Intestinal absorption of calcium and proper adaptation to dietary intake are mediated by vitamin D. Regulation of 1,25-(OH)$_2$D formation is the

principal control point for this homeostatic system (Haussler and McCain, 1977).

When oral intake of calcium, phosphate, and vitamin D are decreased, serum levels of calcium and phosphate decrease. Parathyroid hormone secretion is stimulated in response to lowering of ionized serum calcium concentrations and acts to reestablish normal concentrations of serum-ionized calcium by stimulating bone resorption and renal tubular calcium reabsorption. PTH also increases renal phosphate excretion and lowers serum phosphate concentrations. These two signals, elevated PTH and low serum phosphate, act on renal tubular cells to stimulate 1α-hydroxylase activity which increases formation of 1,25-(OH)$_2$D (Fig. 8.41). This active form of the vitamin increases intestinal absorption of calcium and phosphate so that serum calcium concentrations return to normal, the hypocalcemic stimulus for PTH secretion is removed, and homeostasis is maintained. 1,25-(OH)$_2$D may also directly decrease PTH formation via a classical feedback loop.

Dietary intake and UV exposure determine the availability of precursor vitamin D$_2$ and D$_3$. Intestinal absorption occurs primarily in the ileum and because vitamin D is fat-soluble requires bile salts.

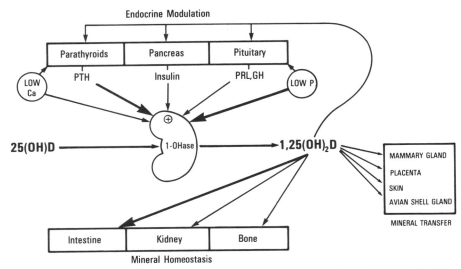

Figure 8.41. Function and regulation of 1,25-(OH)$_2$D. (From Haussler and McCain, 1977.)

Daily requirements are estimated to be 400 U/day (10 μg). Vitamin D deficiency may occur with chronic biliary obstruction and steatorrhea and leads to osteomalacia. Formation of 25-(OH)D depends largely on the concentration of precursor vitamin D$_2$ and D$_3$ and is not a site of major regulation of plasma concentrations of 1,25-(OH)$_2$D. The activity of renal 1α-hydroxylase controls the concentration of 1,25-(OH)$_2$D. Because the kidney is the major source of 1α-hydroxylase, inadequate formation of 1,25-(OH)$_2$D occurs in renal failure (Stanbury, 1978). Not only is the mass of kidney tissue and therefore enzyme decreased, but also with renal failure, phosphate excretion is reduced and serum phosphate rises. Increased phosphate inhibits 1α-hydroxylation so that little 1,25-(OH)$_2$D is formed. Acidosis, a frequent result of renal failure, also impairs 1α-hydroxylase activity. Deficiency of the active form of vitamin D causes osteomalacia, a prominent feature of renal osteodystrophy. Therapy is directed toward use of 1,25-(OH)$_2$D, reduction of serum phosphate, and correction of acidosis, so that residual 1α-hydroxlyase can be expressed.

During growth, pregnancy, and lactation, the body's requirement for calcium increases. Circulating concentrations of 1,25-(OH)$_2$D increase to facilitate increased calcium absorption. During pregnancy and lactation, the increase in plasma concentrations is due, in part, to estrogen-mediated increases in the vitamin D-binding protein and to effects of prolactin on 1α-hydroxylase activity (Haussler and McCain, 1977). Insulin may also facilitate 1,25-(OH)$_2$D formation. Mononuclear phagocytic cells also metabolize precursor 25-(OH)D to 1,25-(OH)$_2$D (Cohen and Gray, 1984). This may be important in vitamin D effects on the immune system. The hypercalcemia associated with chronic granulomatous diseases such as sarcoid appears to result from formation of 1,25-(OH)$_2$D by the large number of mononuclear cells in these lesions (Barbour et al., 1981).

Biological Effects

The intestine is the principal target for 1,25-(OH)$_2$D, although the hormone also affects bone, kidney, and the immune system. Like other steroid hormones, 1,25-(OH)$_2$D binds with high specificity and affinity to a receptor protein which functions to increase gene expression (McDonnell et al., 1987). The net effect of 1,25-(OH)$_2$D is to increase intestinal absorption of calcium.

At a normal calcium intake of 1000 mg about 15% of calcium absorption occurs by passive diffusion (Fig. 8.39). The amount of calcium absorbed via this mechanism is not sufficient to sustain calcium balance, even when calcium intake is significantly increased. The major component of calcium absorption occurs by an active transport process which is regulated by 1,25-(OH)$_2$D. The rate-limiting step is uptake of calcium at the mucosal surface; 1,25-(OH)$_2$D increases this process. In intestinal as in other cells the cytoplasmic concentration of calcium is low (μM) relative to that in the extracellular fluid (mM). Calcium which enters the cell must therefore be sequestered as it moves from the mucosal surface to the luminal surface where it is transported out of the cell into the extracellular compartment. Calcium is sequestered inside the cell in mitochondria and by a 15,000 M_r calcium-binding protein whose concentration is regulated by 1,25-(OH)$_2$D (Wasserman et al., 1976). Vitamin D also induces alkaline phosphatase and calcium-dependent ATPase. Although all of these identified mechanisms contribute to 1,25-$(OH)_2$D-induced intestinal absorption of calcium, none are sufficient to explain the increased mucosal transport and additional biochemical events are undoubtedly induced by 1,25-$(OH)_2$.

1,25-$(OH)_2$D also increases intestinal phosphorus absorption by increasing the active transport of phosphorus. Phosphorus is abundant in the diet, and its absorption is less tightly regulated than that of calcium. A normal diet provides about 1400 mg of phosphorus per day with a net phosphorus absorption of about 900 mg/day. Five hundred milligrams is excreted in feces and 900 mg in urine to maintain phosphorus balance. Phosphorus balance is positive during growth.

Bone is the second major target for vitamin D. Mineralization of bone is a complex process involving both simple and complex molecules in addition to calcium and phosphate (Boskey, 1981). The collagen matrix at the growing end of long bones is synthesized and secreted by chondrocytes. Influx of calcium and phosphate as well as other substances to the extracellular matrix is regulated via these cells. Mineralization is initiated on the extracellular matrix, and crystal growth then fills the matrix with mineral. Blood vessels invade the calcified cartilage matrix, and bone remodels to achieve optimal strength. Bone remodeling is a process which continues throughout life, long after epiphyseal fusion and cessation of linear growth of bone. Bone remodeling, which consists of bone formation and bone resorption, occurs continually, and a significant fraction of bone is replaced each year. Osteoblasts are the primary cells concerned with synthesis of new bone. These cells, which cover bone-forming surfaces, produce bone osteoid which will subsequently undergo calcification. Osteocytes are mature bone cells which are less active than osteoblasts. Osteoclasts are multinucleated cells derived from macrophages which function to resorb bone.

Vitamin D deficiency is characterized by defective mineralization of bone osteoid. By increasing intestinal absorption of calcium and phosphate, vitamin D provides sufficient calcium and phosphate to initiate the crystallization process at bone surfaces. 1,25-$(OH)_2$D also directly affects bone cells to facilitate mineralization even without measurable changes in the plasma concentrations of calcium and phosphate (Boris et al., 1978). 1,25-$(OH)_2$D increases alkaline phosphatase enzyme activity in osteoblast-like cells. In vitro, 1,25-$(OH)_2$D also acts on bone to increase resorption, an effect potentiated by parathyroid hormone. Although this resorptive effect of vitamin D appears antithetical to its major action to increase bone mineralization, it may be

important in the dynamic process of bone remodeling.

Vitamin D facilitates calcium reabsorption in the distal nephron. There are several other tissues which translocate calcium which have documented 1,25-$(OH)_2$D receptors, including skin, mammary gland, placenta, and the avian shell gland. Parathyroid, pituitary, and pancreas also contain 1,25-$(OH)_2$D receptors. Although the precise role of 1,25-$(OH)_2$D in these tissues is uncertain, effects on calcium and phosphate transport and feedback effects on overall calcium homeostasis are likely.

1,25-$(OH)_2$D also exerts important regulatory effects on the immune system (Manolagas et al., 1985). The hormone promotes differentiation of monocyte precursors to monocytes and macrophages. Bone resorbing osteoclasts arise from this differentiation pathway. 1,25-$(OH)_2$D also affects both T and B cells, inhibiting interleukin-2 production and other effector functions.

PARATHYROID HORMONE

Structure and Biosynthesis

Parathyroid hormone (PTH) is an 84-amino acid peptide synthesized within chief cells of parathyroid glands. There are four parathyroid glands in man, each weighing about 40 mg, located behind the thyroid gland (Fig. 8.24, Chapter 54). Occasionally, inferior parathyroid glands may migrate into the mediastinum during development, and occasionally accessory glands are present. In parathyroid chief cells, PTH is synthesized as part of a larger precursor protein containing a 25-amino acid hydrophobic residue-rich signal sequence and a 6-amino acid prosequence preceding the amino-terminus of PTH (Habener and Potts, 1978). These portions of the precursor are sequentially removed within the cell, and mature 84-amino acid PTH is secreted from storage granules. A large parathyroid secretory protein which is cosecreted with PTH may function as a PTH carrier during maturation of the PTH precursor in chief cells (Majzoub et al., 1979). Biosynthesis is closely coupled with secretion: the amount of stored hormone is small and would be exhausted under stimulated conditions within 1–2 hours without a concomitant increased rate of hormone synthesis. After secretion, PTH is cleaved between residues 33 and 34. The amino-terminal fragment possesses the full biological activity of the intact molecule, whereas the carboxy-terminal fragment is biologically inactive. Because clearance of the carboxy-terminal fragment from the circulation is relatively slow, it provides a convenient marker for PTH secretion. Radioimmunoassay quantitation of the carboxy-terminal fragment correlates well with PTH overproduction states provided renal function is normal. In renal failure the fragment accumulates because of diminished clearance as well as overproduction.

Regulation of Synthesis and Secretion

Synthesis and secretion of PTH are regulated by the concentration of ionized calcium in serum. When ionized serum calcium decreases below the physiological set point of 1.3 mM, PTH synthesis and secretion increase to reestablish homeostasis. When hypocalcemia is prolonged, significant parathyroid growth occurs to augment biosynthetic capacity. When ionized serum calcium increases above 1.3 mM, PTH synthesis is suppressed and serum calcium diminishes. A low rate of residual PTH synthesis and secretion persist even with very high levels of serum calcium.

By simultaneously measuring serum calcium and PTH, the feedback axis can be accurately evaluated. When hypocalcemia results from PTH deficiency, PTH concentrations are low, whereas when hypocalcemia results from vitamin D deficiency, PTH concentrations are elevated as a compensatory response to hypocalcemia (Fig. 8.42). In hypercalcemia due to malignancy and vitamin D intoxica-

tion, PTH concentrations are low. In contrast in primary hyperparathyroidism where PTH secretion from adenomatous or hyperplastic parathyroid glands is autonomous, PTH concentrations are elevated even though hypercalcemia is present.

PTH synthesis is also stimulated by β-adrenergic agonists, but this is a minor effect physiologically compared to that of ionized calcium (Brown et al., 1977). Phosphate does not affect PTH synthesis. Magnesium is required for PTH synthesis, and in severe hypomagnesemia PTH synthesis is impaired.

Biological Effects

Bone and kidney are the two principal target organs affected by PTH. PTH cell surface receptors are coupled to adenyl cyclase, and cAMP is the intracellular mediator of PTH action (Chase and Aurbach, 1967). In one form of pseudohypoparathyroidism characterized by tissue resistance to the biological effects of PTH, a genetic defect exists in the GTP-binding protein which couples hormone receptors to catalytic cyclase subunits (Farfel et al., 1980). Ineffective generation of cAMP results in apparent resistance to the biological effects of PTH.

The major effect of PTH is to maintain normal ionized serum calcium concentrations. PTH stimulates

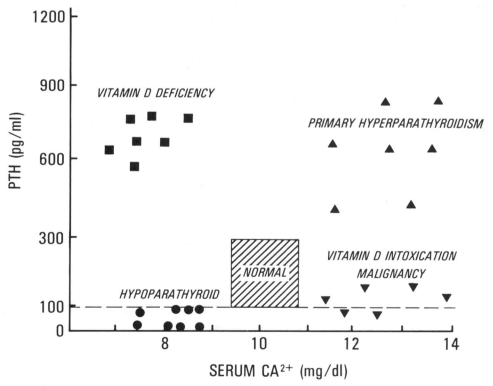

Figure 8.42. Simultaneous PTH and calcium concentrations in patients with various forms of hypo- and hypercalcemia with normal renal function.

bone resorption, releasing calcium into extracellular fluid. The principal target of PTH is the osteoblast. PTH inhibits osteoblast synthesis of new bone, but increases osteoblast-initiated recruitment of osteocytes and osteoclasts which participate in bone resorption. The overall effect is increased bone resorption and decreased new bone formation. When PTH concentrations are persistently elevated, significant bone loss occurs with subperiosteal resorption, formation of bone cysts, and spontaneous fractures.

PTH has two major effects on the kidney: to increase calcium reabsorption in the distal nephron and to decrease proximal tubular reabsorption of phosphate. Both responses function to maintain a normal ionized serum calcium. Of the approximately 10,000 mg of calcium filtered at the glomerulus each day, more than 90% is reabsorbed in the proximal tubule and loop of Henle in a process strongly linked to sodium. Reabsorption of the approximately 1000 mg which reaches the distal tubule is regulated by PTH. When PTH is increased, more calcium is reabsorbed in the distal nephron, whereas, when PTH is decreased, less calcium is reabsorbed, and urinary calcium excretion rises. Calcium excretion thus depends both on the amount of calcium which reaches the distal tubule and on the concentration of circulating PTH. The amount of calcium which reaches the distal nephron, in turn, depends on the filtered load of calcium and on sodium intake. In hypercalcemia the filtered load of calcium increases so that more calcium is delivered to the distal tubule. In hyperparathyroidism PTH increases fractional reabsorption of calcium, but because of the large calcium load delivered, hypercalcuria results. Calcium reabsorption in the proximal tubule is linked to sodium reabsorption. When sodium intake is high, more sodium escapes reabsorption in the proximal tubule to reach the distal nephron. Under these conditions more calcium also reaches the distal tubule and is excreted. Increased sodium intake is therefore used in many forms of hypercalcemia to increase calcium excretion and lower serum calcium.

The reabsorption of phosphate which occurs in the proximal tubule is controlled by PTH. PTH decreases proximal tubular reabsorption of phosphate, so that increased urinary excretion of phosphate occurs. Sustained elevations in PTH thus result in hypophosphatemia in addition to hypercalcemia. Both PTH and hypophosphatemia stimulate 1α-hydroxylase activity to increase formation of $1,25-(OH)_2D$ (Fig. 8.41). PTH effects on gastrointestinal absorption of calcium and phosphate are mediated via $1,25-(OH)_2D$.

CALCITONIN

Structure and Biosynthesis

Calcitonin is a 32-amino acid polypeptide with an amino-terminal disulfide bridge linking positions 1 and 7 and an amidated carboxy terminus (Fig. 8.43). Its two major biological effects are to lower serum calcium and phosphate. Calcitonin is synthesized in C cells of neuroendocrine origin which are located primarily within the thyroid gland but to a lesser extent in thymus. In lower vertebrates calcitonin is synthesized in a separate gland, the ultimobranchial body. Like other secretory proteins, calcitonin is synthesized as a portion of a larger precursor protein (Amara et al., 1980). The function of the large 76-amino acid amino-terminal fragment and of the smaller 16-amino acid carboxy-terminal fragments processed from the calcitonin precursor is not known. In other tissues, such as hypothalamus, an alternate exon is retained in the precursor, giving rise to a calcitonin-gene-related peptide (CGRP) (Amara et al., 1982). The alternate processing of a single precursor mRNA product to yield calcitonin in C cells and CGRP in neuronal cells appears determined by cell-specific factors (Crenshaw et al., 1987). CGRP is a neurohormone of incompletely defined function which appears involved in regulation of cholinergic receptors, ingestive and nociceptive responses (Rosenfeld et al., 1982). Circulating concentrations of immunoreactive calcitonin are less than 100 pg/ml; it has a circulatory half-life of about 10 min (Austin and Heath, 1981).

The sequence of calcitonin varies significantly between humans, pigs, sheep, rats, and salmon. All have a 7-membered amino-terminal disulfide-bonded ring and a carboxy-terminal prolineamide, but differ greatly in the central three-fourths of the molecule. Salmon calcitonin, which has increased biological potency in humans, is the form available for human use.

Regulation of Synthesis and Secretion

Synthesis and secretion of calcitonin are controlled by the concentration of ionized serum calcium. When serum calcium rises above about 9 mg/dl, calcitonin secretion increases in a linear fashion. The stimulus for calcitonin synthesis is thus opposite that for PTH secretion. When ionized serum calcium decreases, PTH secretion rises to restore serum calcium, whereas calcitonin falls to remove a hypocalcemic stimulus. When ionized serum calcium rises, PTH synthesis is suppressed, and calcitonin synthesis is increased to provide a hypocalcemic signal. As noted below, the physiological effects of calcitonin on serum calcium in humans are

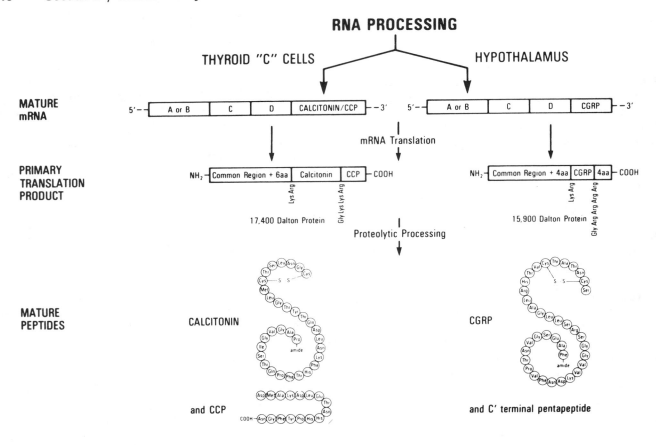

Figure 8.43. Differential processing of the calcitonin gene into calcitonin in thyroid C cells and into calcitonin gene-related product in hypothalamus. *CCP*, calcitonin carboxy-terminal peptide; *CGRP*, calcitonin gene-related peptide. (From Amara et al., 1982.)

minor, and the change in PTH synthesis in response to variations in ionized serum calcium is the major homeostatic adjustment.

Calcitonin synthesis and secretion are stimulated by gastrin, cholecystokinin, glucagon, and β-adrenergic agonists (Cooper et al., 1978). The effects of these hormones on calcitonin synthesis may represent a mechanism for integration of disposition of dietary calcium or may be a reflection of the neuroectodermal origin of C cells.

Biological Effects

Calcitonin reduces bone resorption by inhibiting osteoclast function. This results in a decrease in the concentration of both calcium and phosphate in serum. cAMP serves as the second messenger for calcitonin action. In spite of the reproducible effects of calcitonin on bone and serum mineral content, the hormone does not appear to be a major physiological regulator. After removal of the thyroid-containing C cells and documented calcitonin deficiency or after development of medullary thyroid carcinoma, a neoplasm of C cell origin which results in

markedly elevated calcitonin production, serum calcium, and phosphate concentrations remain normal. There is decreased osteoclast function in medullary thyroid carcinoma, indicating that calcitonin is biologically active, but it does not affect serum calcium concentrations. Effects of calcitonin are most marked when the rates of bone turnover and osteoclast function are highest, as in the young or in diseases such as Paget's disease. The major recognized use of calcitonin is thus in treatment of Paget's disease, where the hormone reduces the accelerated rate of bone turnover.

Calcitonin increases the urinary excretion of calcium, phosphate, sodium, potassium, and magnesium. This renal effect of calcitonin contributes to the hypocalcemic effects of calcitonin but persists only as long as calcitonin concentrations are elevated.

HYPO- AND HYPERCALCEMIC STATES

The integrated axis involving vitamin D, PTH, and, to a lesser extent, calcitonin maintains a constant ionized serum calcium concentration in response to either hypocalcemic or hypercalcemic challenge. When these challenges are severe and persistent or when the endocrine control system is deranged, clinically evident disorders of mineral homeostasis result.

When dietary calcium intake is decreased or when fecal or urinary calcium excretion is increased, the tendency to systemic hypocalcemia is counteracted by increased PTH and 1,25-(OH)$_2$D. In response to transient hypocalcemia, PTH synthesis and secretion increase with resultant mobilization of calcium and phosphate from bone and increased renal tubular reabsorption of calcium. PTH and lowered phosphate increase synthesis of 1,25-(OH)$_2$D which increases calcium and phosphate absorption from the gastrointestinal tract. Initial deficits in calcium are thus ultimately replaced from dietary sources, and overall bone mineral content is protected. When the endocrine defense system is impaired, as with either vitamin D or PTH deficiency, hypocalcemia results. As noted, vitamin D deficiency, whether due to inadequate intake, inadequate absorption, lack of UV irradiation, or failure to convert precursor to 1,25-(OH)$_2$D, results in hypocalcemia and inadequate mineralization of bone osteoid. Sustained elevations of PTH occur in response to hypocalcemia. PTH partially corrects extracellular fluid hypocalcemia by mobilization of bone calcium and increased renal tubular reabsorption of calcium. Maintenance of serum calcium near normal values occurs only at the expense of bone, so that demineralization is severe due both to lack of 1,25-(OH)$_2$D and to compensatory hyperparathyroidism. PTH deficiency results from either removal of or damage to parathyroid glands (hypoparathyroidism) or from resistance to the biological effects of PTH (pseudohypoparathyroidism). With PTH deficiency, hypocalcemia results from lack of calcium mobilization from bone and failure of renal calcium conservation. In addition, both the PTH and hypophosphatemic stimuli for formation of 1,25-(OH)$_2$D are missing, and secondary vitamin D deficiency results (Sinha et al., 1977). Like most other peptide hormones, PTH is ineffective when given orally because of degradation in the gastrointestinal tract and only transiently effective when given parentally because of the short circulating half-life. To prevent signs and symptoms of hypocalcemia, including neuromuscular irritability, muscle cramps, paresthesias, tetany, laryngospasm, and seizures, vitamin D therapy is utilized. Because formation of 1,25-(OH)$_2$D is impaired, massive doses of vitamin D$_3$ must be given (up to 1000 times normal daily intake) or active 1,25-(OH)$_2$D must be used.

Calcium supplementation is useful but ineffective unless adequate active 1,25-(OH)$_2$D is provided.

Hypercalcemia is avoided under physiological conditions when calcium intake rises by a fall in serum PTH and a resulting decrease in formation of 1,25-(OH)$_2$D. This reduces both gastrointestinal absorption of calcium and bone resorption, and increases urinary calcium excretion. Hypercalcemia will result from excessive PTH secretion or from excessive intake of vitamin D. It may result also from malignancy when bone dissolution delivers so much calcium to the extracellular space that renal excretory capacity is exceeded. A number of factors mediate bone resorption in malignancy, including an osteoclast-activating factor (Mundy et al., 1974), prostaglandins (Seyberth et al., 1976), and a parathyroid hormone-like peptide (Mangin et al., 1988). The PTH-like peptide is highly homologous to PTH in its amino-terminal 13 amino acids, but differs in the remainder of the molecule. This PTH-like peptide, which is a secretory product of keratinocytes, likely acts via PTH receptors.

In hyperparathyroidism, where autonomous excessive synthesis and secretion of PTH occur, hypercalcemia and hypophosphatemia result. Bone resorption is increased leading to decreased mineral and osteoid, and the increased filtered load of calcium predisposes to renal stones and renal calcification. Hypercalcemia may result in muscle weakness, lethargy, coma, constipation, nausea, and resistance to ADH, with inability to concentrate the urine. The latter may lead to dehydration and acceleration of the hypercalcemic state. Therapy is directed toward surgical removal of abnormal parathyroid tissue. Short-term control of hypercalcemia can be achieved with saline diuresis and oral phosphate which tends to favor calcium-phosphate deposition in bone. Calcitonin has occasionally been used to reverse hypercalcemia.

Excessive vitamin D results from ingestion or from increased sensitivity in sarcoidosis. Hypercalcemia results from increased absorption from the gastrointestinal tract and from bone resorption. In addition to cessation of vitamin D, glucocorticoids counteract hypercalcemia by opposing the effects of vitamin D.

In hypercalcemia resulting from malignancy involving bone, both PTH and 1,25-(OH)$_2$D are appropriately decreased; therapy must therefore be directed toward increasing renal calcium excretion and toward treatment of the malignant process. Oral phosphates and mitramycin, a cytotoxic antibiotic, are useful in addition.

CHAPTER 57

Hormonal Regulation of Growth and Development

Ordered, controlled growth is essential for development and maintenance of normal humans. The mechanisms which control growth are selective for different types of cells, for example, wound healing, growth of the contralateral kidney or adrenal cortex after unilateral nephrectomy or adrenalectomy, and hepatic regeneration following partial hepatic resection. Growth is orderly, and there are strict controls on the extent of proliferation of cells in any organ. Loss of strict controls on cell proliferation are manifest in cancer.

Growth of individual cells, termed hypertrophy, is a process resulting in reversible amplification of the cells' differentiated function, whereas cell division, termed hyperplasia, results in an increased number of functional cells. Both hypertrophy and hyperplasia are controlled by hormones and are dependent on availability of adequate nutrients and ions. To study growth, simpler tissue cultures of varying cell types have been frequently used, and results are extrapolated to more complex in vivo situations. Using cell culture systems, a frequent experimental approach has been to arrest cell growth by removing essential compounds from culture media and by measuring the effects of added compounds on growth of resting cells. Under most conditions of deprivation, mammalian cells reversibly cease growing with a diploid complement of DNA (G_1 growth arrest). Addition of stimulatory compounds results in cellular hypertrophy followed by DNA replication and cell division. Using this approach, a number of new hormones termed growth factors have been identified. Because normal cells exhibit orderly growth in culture and cease growing at saturation (density-dependent inhibition of cell proliferation), endogenous oncogenes have been identified by their ability to impose abnormalities on cell growth control and cause piling up of cells beyond contact inhibition (Cooper et al., 1980; Shih, 1981).

REQUIREMENTS FOR CELLULAR GROWTH

Nutrients and Ions

Adequate nutrients and ions are required for growth of any population of cells whether bacterial or mammalian. By creating deficiency states in vitro, a requirement for a number of small molecular weight nutrients can be demonstrated. These include ions such as calcium, phosphate, and magnesium; amino acids; glucose, fatty acids; vitamins; transferrin and bound iron; ceruloplasmin and bound copper; and nucleic acids. Because of extensive metabolic transformations, normal man requires only 20 essential organic compounds in addition to a source of calories and water. Eight amino acids, 11 vitamins, and linoleic acid are essential. Deficiencies of any single element are unusual. The most generalized deficiencies in humans are caloric malnutrition (marasmus) and protein malnutrition (kwashiorkor). In caloric malnutrition, protein may be utilized as metabolic fuel, resulting in combined protein-caloric malnutrition; similarly, in protein malnutrition, interference with the ingestion of the usually bulky carbohydrate diet often results in a combined deficiency state. These deficiencies are widespread in underdeveloped countries and are more common than usually recognized among disadvantaged, elderly, chronically ill, and alcoholic persons in developed countries. In children with protein-caloric malnutrition, growth is markedly retarded and sexual maturity is delayed. In people of all ages, processes involving cell growth are impaired, including wound healing and function of the immune system, so that death from infection is common.

In nutrient deficiency a number of hormonal adaptations occur. Adaptations to fasting have been discussed in other chapters and include a reduction in insulin secretion, decreased peripheral conversion of T_4 to T_3, elevations in cortisol, epinephrine, and glucagon, and increased growth hormone but impaired somatomedin production. With nutrient lack, growth is obviously inappropriate, and hormonal changes mediate mobilization of stored energy and suppress cellular growth. Reproduction is also inappropriate, and low gonadotropins and suppression of ovarian and testicular function occur.

Adequate nutrient intake is required for growth, but excess intake does not cause growth of tissues other than fat. Utilization of nutrients for growth of cells and organs requires hormones, and these regulate the process of growth, provided adequate nutrients are available. Growth factors may facilitate cellular uptake of nutrients required for hypertrophy and division of one cell into two (Holley, 1975).

Growth Factors

Animal serum has been used for many years to support growth of cells in culture. The essential macromolecular components in serum are hormones, and, by using an appropriate mixture of hormones, cell growth can be optimized without serum (Barnes and Sato, 1980). Even though cell growth in vivo is specific, i.e., after partial hepatectomy only liver grows, no organ-specific growth factors have been identified. Rather a combination of hormones regulate optimal growth, and delivery of this combination to tissues along with other factors such as nutrient availability, the extracellular matrix on which cells grow, the concentration of hormone receptors, and the concentration of antigrowth factors control organ growth. In addition to well characterized hormones, a number of new polypeptide hormone growth factors have been identified which stimulate cell growth and division (Table 8.7).

Epidermal growth factor (EGF), a 53-amino acid polypeptide, was initially identified by its ability to accelerate eye opening in newborn mice (Cohen, 1959). It was independently isolated later from human urine, based on its ability to inhibit gastric acid secretion (Gregory, 1975). EGF was originally isolated from submaxillary glands where its production is increased by androgens and adrenergic agonists. EGF is synthesized in many other tissues, including platelets, which are the major circulating source of EGF in humans (Oka and Orth, 1983). EGF is processed from a much larger precursor protein whose structure resembles that of transmembrane receptors, especially the LDL receptor (Russell et al., 1984). EGF-like sequences are a commonly

Table 8.7.
Growth Factors for Epithelial, Endothelial, and Mesenchymal Cells

  I. Epidermal growth factor (EGF)
    A. Single chain polypeptide of 53 AA containing 3 intrachain disulfide bonds
    B. Synthesized as a larger precursor protein homologous to transmembrane receptors; similar structure found in a wide variety of gene products
    C. Stimulates growth of cells of endodermal, ectodermal, and mesodermal origin; involved in development as well as regenerative growth
    D. Acts via a transmembrane receptor with intrinsic protein tyrosine kinase activity
 II. Platelet-derived growth factor (PDGF)
    A. A dimeric protein of 30,000 M_r consisting of disulfide-linked A and B chains
    B. Synthesized in megakaryocytes and carried in α granules of platelets; also synthesized by endothelial, glial, and fibroblast cells
    C. Stimulates growth primarily of mesenchymal-derived cells, including smooth muscle, glial, and fibroblast
    D. Acts via a transmembrane receptor with intrinsic protein tyrosine kinase activity
III. Fibroblast growth factor (FGF)
    A. Two single chain polypeptides of 146 amino acids (basic form) or 140 amino acids (acidic form)
    B. Basic FGF is synthesized in many cells; acidic FGF is synthesized primarily in brain and retina
    C. Stimulates growth of a variety of cells derived from mesoderm and neuroectoderm; is the major identified stimulator of angiogenesis
    D. Both forms of FGF act via a single receptor
 IV. Insulin-like growth factors (IGF)
    A. Structural homology to proinsulin
    B. Synthesis is dependent on growth hormone; synthesis is also regulated by thyroid hormone, insulin, and nutritional state
    C. Stimulate growth of mesodermally derived cells; important regulators of growth of epiphyseal end plates of long bone
    D. IGF_1 acts via a receptor which is homologous to that for insulin; this receptor contains intrinsic protein tyrosine kinase activity; IGF_2 acts via the receptor for mannose-6-phosphate
  V. Nerve growth factor
    A. Dimer composed of two identical 13,000 M_r subunits; distant homology to insulin
    B. Synthesized at synaptic junctions
    C. Stimulates replication of embryonic neurons; stimulates neurite outgrowth and hypertrophy of mature neurons; required for maintenance and survival of neurons
    D. Binds to a 399-amino acid transmembrane receptor identified on both sensory and sympathetic neurons; the receptor is also found on other cells of neural crest origin such as melanocytes and Schwann's cells

recurring structural motif found in clotting factors, tumor growth factor α (TGF-α), a vaccinia viral protein, and in genes regulating development in *Drosophila* and nematodes (Doolittle et al., 1984; Wharton et al., 1985; Greenwald, 1985). EGF stimulates proliferation of a large number of cells, including epithelial, fibroblastic, and mesodermally derived cells (Carpenter and Cohen, 1979). EGF binds to an 1186-amino acid transmembrane glycoprotein which has intrinsic protein kinase activity to phosphorylate tyrosine residues in proteins (Ullrich et al., 1984; Gill et al., 1987). The EGF receptor is a member of a family of transmembrane hormone receptors which have intrinsic protein tyrosine kinase activity (Fig. 8.44. Ligand binding to the extracellular domain activates the intrinsic protein tyrosine kinase activity to phosphorylate cellular proteins (Cohen et al., 1982; Chen et al., 1987). This results in a variety of acute and chronic changes in cells, including $[Ca^{2+}]$ increases, an increase in pH, stimulation of gene transcription, and ultimate cell replication.

A 30,000 M_r cationic polypeptide, platelet-derived growth factor (PDGF), is the major mitogen found in serum (Ross et al., 1979). It is stored in α-granules in platelets and released during the platelet release reaction. PDGF consists of two peptide chains (A and B) which are linked by disulfide bonds. Neither chain is active alone. The major form of PDGF is an AB heterodimer, but BB and AA homodimers are also active (Ross et al., 1986). PDGF stimulates proliferation of several cell types, including glial, smooth muscle, and fibroblast. Because PDGF is transported and released from platelets, it appears to have a major role in wound healing and may also contribute to growth of arterial plaques in arteriosclerosis. Although PDGF was initially discovered as a megakaryocyte product, it is synthesized in several cell types, including endothelial, glial, and fibroblast. Synthesis of PDGF is stimulated by several factors, including thrombin, tumor growth factor β (TGF-β) and hypoxia. PDGF produced by stimulated endothelial cells may stimulate growth of adjacent smooth muscle cells. The receptor for PDGF is a transmembrane protein with intrinsic protein tyrosine kinase activity (Fig. 8.44) (Yarden et al., 1986). The extracellular domain of the receptor binds PDGF with high affinity to activate the kinase activity intrinsic to its intracellular portion (Ek et al., 1982).

Fibroblast growth factor (FGF), a peptide of 13,400 M_r, was first purified from the pituitary and brain based on its ability to stimulate proliferation of cultured fibroblasts (Gospodarowicz, 1975). There are two types of FGF, acidic and basic, which have 55% amino acid homology and a similar range of biological activities (Abraham et al., 1986). Basic FGF is produced in many mesoderm- and neuroectoderm-derived tissues, whereas acidic FGF is produced primarily in the brain and retina (Gospodarowicz et al., 1987). FGF stimulates proliferation of a variety of cell types, including endothelial, myoblast, chondrocyte, adrenal cortex, ovarian granulosa, mesothelial, and fibroblast cells. Its effects on angiogenesis are particularly notable. Growth of new capillary blood vessels is important in normal

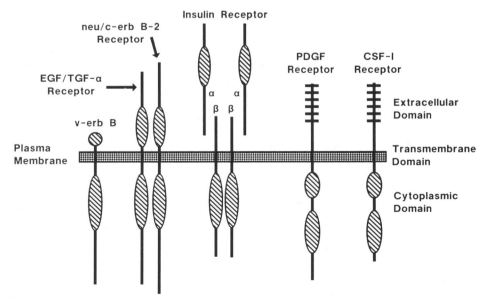

Figure 8.44. Structures of some growth factor receptors with intrinsic protein tyrosine kinase activity. Cysteine-rich regions in the extracellular ligand-binding domain are indicated by *enclosed areas* or *dots;* the conserved cytoplasmic kinase domain is also enclosed.

processes such as development of the embryo, formation of the corpus luteum, and wound healing, as well as in pathological processes such as chronic inflammation and tumor growth. Because nutrient and growth factors diffuse out of blood vessels to adjacent tissue cells over only a limited distance, organ or tumor growth occurs only when blood supply is maintained by capillary ingrowth. Soluble, diffusible factors which stimulate capillary proliferation were initially identified in extracts from tumor cells (Folkman et al., 1971). The major tumor angiogenesis factor appears to be FGF. Production of this factor is not unique to tumors, but is a property of normal cells and is required for organ regeneration and wound healing. For example, cyclic changes in the ovary involve vascular changes mediated in part by angiogenic factors produced by the corpus luteum (Gospodarowicz and Thaknal, 1978).

The receptor for FGF is widely distributed. Its mechanism of action is via protein tyrosine kinase, adding it to the protein tyrosine kinase gene family. FGF binds tightly to heparin; it may not only stimulate production of components of the extracellular matrix but may also be bound there to enhance interaction with its receptor (Gospodarowicz et al., 1986).

Nerve growth factor (NGF) is an insulin-like protein which induces morphologic and metabolic differentiation of sympathetic and sensory neurons (Levi-Montalcini and Angeletti, 1968; Bradshaw et al., 1974). Sensory neurons are stimulated by NGF only during a short period of embryonic development, whereas sympathetic cells retain sensitivity even in adult animals. NGF stimulates neurite outgrowth with accompaning cellular hypertrophy and neurotubule polymerization. It also stimulates noradrenergic neurotransmitter synthesis and regeneration of adrenergic fibers in brain. Because mature nerve cells do not divide, the principal effect of NGF is stimulation of cellular hypertrophy and differentiated function. Neurite outgrowth appears guided directionally by NGF. Production of NGF at synaptic junctions provides a concentration gradient toward which neurites grow. NGF is required also for maintenance and survival of neurons. NGF is a noncovalent dimer composed of two identical 13,000 M_r chains. It binds to a cell surface receptor containing a single transmembrane domain (Johnson et al., 1986). The second messenger system utilized by activated NGF receptors is not known, but endocytosed receptors may travel long distances from the axon periphery to the nerve cell body (Andres et al., 1977).

The somatomedin:insulin-like growth factors (IGF) are under the control of growth hormone

(Chapter 53). Several more specific growth factors regulate proliferation of hematopoietic lineages which give rise to erythrocytes (erythropoietin), granulocytes (granulocyte colony-stimulating factor and granulocyte/macrophage colony-stimulating factor), platelets (thrombopoietin), macrophages (macrophage colony-stimulating factor-1), and various types of T and B lymphocytes (interleukins and T cell growth factors) (see Chapter 21).

Growth Inhibitors

Growth of cells depends on a number of factors in addition to growth factors and nutrients. The presence of neighboring cells, the extracellular matrix, blood supply, and antigrowth factors are all important. Antigrowth factors, termed chalones, may be produced locally to limit tissue growth (Lozzio et al., 1975). For example, cartilage-produced factors inhibit growth of both cartilage and capillaries, and epithelial cells produce a factor which inhibits epithelial cell growth (Holley et al., 1978). Several well-characterized growth inhibitors serve to control cell function as well as negatively regulate growth (Table 8.8). Transforming growth factor β, also termed polyergin ("many functions"), inhibits proliferation of epithelial, mesenchymal, and hematopoietic cells (Sporn et al., 1986; Hanks et al., 1988). TGF-β is composed of two 112-amino acid chains linked by disulfide bonds. It is synthesized from a larger precursor and must be activated by release from a latent complex. The major source of TGF-β is platelets although it is also synthesized by other cells. Many epithelial and mesenchymal cells respond to TGF-β by increasing synthesis of extracellular matrix proteins, including collagen, fibronectin, and cell adhesion proteins (Ignotz and Massague, 1986). The production and organization

Table 8.8.
Growth Inhibitors

I. Tumor growth factor β (TGF-β, polyergin)
  A. Dimer of two 112 AA chains linked by disulfide bonds; several forms are identified (β_1, β_2, β_3)
  B. Synthesized as a larger precursor; exists as an inactive precursor which must be activated to function
  C. Inhibits growth of epithelial, mesenchymal, and immunological cells; involved in development and in synthesis of extracellular matrix proteins
  D. Acts via transmembrane receptor which is widely distributed on many cell types
II. Tumor necrosis factor (TNF, cachectin)
  A. Multimer of 157-amino acid peptide chains
  B. Produced by macrophages in response to endotoxin; structurally related to lymphotoxin
  C. Causes necrosis of tumors and a cachectic or wasting state; pleiotropic effects include inhibition of lipoprotein lipase, induction of β_2 interferon, and stimulation of fibroblast growth
  D. Receptor similar, if not identical, to that for lymphotoxin

---

of the extracellular matrix is essential to wound healing so TGF-β augments the growth promoting effects of PDGF, EGF, and FGF in healing wounds. TGF-β is structurally related to factors controlling mesenchymal differentiation, sexual differentiation, and gonadotropin secretion.

Tumor necrosis factor (TNF, cachectin) is a 157-amino acid polypeptide, arranged in multiple forms, which is produced by macrophages (Beutler and Cerami, 1986). Production of TNF is stimulated by endotoxin; it acts to cause hemorrhagic necrosis in tumors and to induce a wasting or cachectic state in the host. The actions of TNF, like those of TGF-β, are pleiotropic. It suppresses lipoprotein lipase activity and, although it induces a state of generalized wasting, activates polymorphonuclear leukocytes, and stimulates growth of fibroblasts. It also induces β_2-interferon (Kohase et al., 1986). Interferons, which were discovered and characterized by their antiviral activity, inhibit growth of a variety of cells (Pestka et al., 1987). TNF stimulation of interferon production may play a role in the host protective effects of TNF; interferon β_2 may also by its growth inhibitory effects limit TNF-stimulated growth. The tumor cell destructive and cachectic effects of TNF may represent an extreme of what is likely an evolutionarily conserved host defense.

Hormones such as ACTH control both structure and function of the adrenal cortex. ACTH inhibits DNA replication in adrenocortical cells, but, in the presence of growth factors, increases the differentiated function of steroid hormone production in hypertrophied cells. Because the requirement for increased steroid hormone production is intermittent, this mechanism allows amplification by reversible cellular hypertrophy. Hyperplasia, which is reversible only when cell death exceeds cell generation, occurs only when concentrations of ACTH are elevated over long periods and receptors decrease (Hornsby and Gill, 1977).

GROWTH AND DEVELOPMENT

Embryonic development depends on cell to cell interactions. Development is directed by growth regulatory hormones which later in life act on differentiated cells. Although little is known about early human development, some of these developmental hormones, which have been identified in lower organisms, will likely play similar roles in humans. Studies of early amphibian embryos indicate that signals from the vegetal pole, which gives rise to endoderm, signal cells from the animal pole, which gives rise to ectoderm, to cause development of the mesoderm (Nieuwkoop et al., 1985). Amphibian FGF and TGF-β homologs appear to represent

the signals which direct this very early tissue differentiation (Slack et al., 1987; Kimelman and Kirschner, 1987; Weeks and Melton, 1987). Another member of the TGF-β gene family, müllerian inhibiting substance (MIS) causes regression of the female sex organ progenitor müllerian duct in males (Cate et al., 1986) (see Chapter 58). EGF-like sequences are present in multiple copies in a *Drosophila* gene-controlling cell switching between epidermis and nervous system (Wharton et al., 1985), and in a nematode homeotic gene (Greenwald, 1985). In the compound eye of *Drosophila* a specific cell senses ultraviolet light. A genetic mutation in this developmental pathway identifies a transmembrane protein tyrosine kinase as essential for development of ultraviolet light perception (Hafen et al., 1987). It is postulated that this receptor, which has structural homology to the gene family shown in Figure 8.44, is expressed and activated by a specific hormonal ligand from an adjacent cell to direct this developmental pathway (Tomlinson et al., 1987). Retinoic acid, which acts via a nuclear receptor that is a member of the steroid hormone receptor gene family, is an important morphogen, directing chick limb bud formation and development of stem cells into parietal endoderm (Giguere et al., 1987). Thyroid hormone is required for tadpole morphogenesis to frogs. Thus similar, and in some cases identical, growth factors, hormones, and their receptors regulate the processes of differentiation as well as proliferation of differentiated cells.

In human intrauterine life, these hormonal signals are important for growth, development, and maturation of organs (for example, see Chapters 58–60). The orderly process of postnatal growth is also critically regulated by hormones as well as genetic makeup and nutrition. During the first year of postnatal life, the rate of growth is rapid, though less than in utero. By 1 year, birth length doubles and birth weight triples. Thyroid hormone is essential for myelinization and brain growth during this time. By 6–8 months of postnatal life, cell multiplication in the brain is completed, and, if thyroid hormone is deficient during this time, irreversible mental retardation occurs. Deficiency of thyroid hormone beginning later in life results in growth retardation and mental slowing, but these processes are reversible with thyroid hormone.

As shown in Figure 8.45, the rate of growth then slows during childhood years until puberty. During childhood, height increases 5–7.5 cm/year, and weight increases 2–2.5 kg/year. Charting of measured increases in height and weight on graphs which include 95% confidence limits for a large number of normal children is a valuable method for

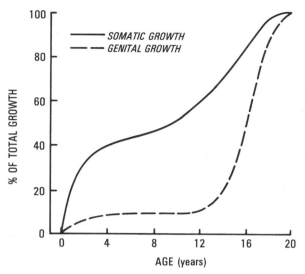

Figure 8.45. Relative rate of somatic and genital growth during normal human development.

continued assessment of a child's progress. Failure to grow at a normal rate provides an important clue to underlying abnormalities. Impaired nutrition, including diseases such as chronic diarrheal states, psychological disturbances, and hormonal derangements must be considered. The most common endocrine deficiencies are of growth hormone, thyroid hormone, and insulin.

Puberty and maturation of gonadal systems results in an increase in growth rate. In females, growth rate increases to 9 cm/year and in males to 10 cm/year. The increase in growth rate is due to the gonadal steroid hormones, estrogen, and testosterone, which accelerate the rate of growth but also cause ultimate fusion of the epiphyseal end plate growth center to the end of bones so that growth is terminated. Radiographic assessment of epiphyseal end plates provides an accurate indication of bone age for comparison with chronological age and allows prediction of the potential for further height increases (Greulich and Pyle, 1959).

In females puberty begins at approximately age 11 years with breast budding followed by menarche at age 13 years. Secondary sexual characteristics develop, and psychological and behavioral changes ensue. Regular ovulatory menstrual cycles are established 2–3 years after menarche, and linear growth ceases at age 16–17 years. In males testicular size begins to increase at age 12 followed 1 year later by penis enlargement and increased rates of linear growth. Secondary sexual characteristics occur at approximately age 14–15, and growth ceases at approximately age 18.

Abnormalities in gonadotropin or gonadal steroid production will alter this orderly program of growth and sexual development. When sex steroid deficiencies occur, growth of long bones may continue, so that a typical eunuchoid stature consisting of an arm span greater than height and of a lower body segment from symphysis pusis to heel greater than the upper body length occurs. Adequate genital growth and development of secondary sexual characteristics are lacking. With precocious puberty or with excessive production of adrenal androgen precursors resulting from enzyme defects in cortisol synthesis, early growth will be accelerated, but premature fusion of epiphyseal end plates will result in ultimate short stature.

In adulthood growth is limited to repair and renewal processes. Change in a number of hormones, i.e., excess cortisol, deficient thyroid hormone, will impair processes such as wound healing.

ABNORMALITIES OF GROWTH

Deficiency States

Detailed analysis of inadequate rates of growth and abnormal patterns of sexual maturity requires careful assessment of nutritional intake, endocrine gland function, psychosocial background, the presence of chronic disease, and genetic makeup. Any or all of these may be responsible for abnormalities in postnatal growth and development. Because appropriate hormonal therapy is available, identification of endocrine abnormalities is especially valuable. Similarly, identified endocrine abnormalities contributing to impaired adult growth processes such as wound healing are correctable.

Growth Factors and Oncogenes

Cancer is a disorder of growth control. Restraints on proliferation are abrogated and cells not only grow in an unrestrained fashion but also orchestrate accompanying growth of supportive cells and blood vessels. Cancer cells also break down normal tissue boundaries to spread via the circulation and lymphatics and implant in distant sites (metastasis). Cancer arises from previously normal cells via genetic mutations which are both dominant and recessive in type (Bishop, 1987). Although the etiologies and metabolic derangements in cancer cells are incompletely understood, several mutations have been identified in genes which normally regulate cell growth and differentiation. These mutant genes, termed oncogenes, were originally identified in RNA tumor viruses and subsequently in human tumors (Weinberg, 1985). They arise via mutation of cellular genes, termed proto-oncogenes, involved in controlling normal cell growth and differentiation. Two general types of mutations occur. Dominant mutations affect genes by either removing normal structural control elements which regulate their function or by enhancing expression of the genes by gene amplification or enhanced transcription. Recessive oncogenes result in deletion of genetic loci which encode growth-restraining proteins (Friend et al., 1988).

Oncogenes representing each of the known elements in the general mechanism of growth factor action have been identified. Single oncogenes do not suffice to convert normal cells into malignant ones. Instead, two or more independent mutations in growth-controlling genes are necessary for the final loss of growth control recognized as cancer (Land et al., 1983).

Table 8.9 shows constitutive expression of a growth factor that has been recognized as the viral oncogene derived from cellular proto-oncogenes (Hunter, 1985). The *sis* oncogene encodes the B chain of PDGF and the *int-2* oncogene encodes an FGF analogue. Growth factor receptors also give rise to oncogenes. The EGF receptor gives rise, via deletion of regulatory regions, to *erb* B (Downward et al., 1984); a point mutation in the transmembrane domain of c-*erb* B-2 results in constitutive activation to the *neu* oncogene (Bargman et al., 1986). In human brain tumors the EGF receptor gene is amplified (Libermann et al., 1985) and in human breast cancer amplification of the c-*erb* B-2/neu receptor gene is associated with rapidly fatal disease (Slamon et al., 1987).

Table 8.9.
Growth Control Steps and Oncogenes

	Oncogenes
I. Growth factors	
PDGF	*sis*
FGF	*int-2*
II. Growth factor receptors	
EGF/TGF-α receptor	*erb* B
c-*erb* B-2 receptor	*neu*
Macrophage colony-stimulating factor 1 receptor	*fms*
III. G-proteins	
Normal *ras* genes	n-*ras*
	Harvey *ras*
	Kirsten *ras*
IV. Nuclear DNA-binding proteins	
Normal cellular proto-oncogenes	*myc*
	myb
	fos
	ski
	jun
	rb

The *ras* protein family of GTP-binding proteins which have structural homology to the G-proteins involved in control of adenylate cyclase and vision are among the most commonly detected oncogenes in human cancer (Bos et al., 1987). Point mutations convert the signal transducing proto-oncogene into the active oncogene which is constitutively active (Gibbs et al., 1984).

Nuclear DNA-binding proteins were initially identified as RNA tumor viral oncogenes. The *myc* gene is rearranged in human Burkitt's lymphoma and may be amplified in other human tumors such as neuroblastoma (Taub et al., 1982; Seeger et al., 1985). The *jun* oncogene is derived from the transcription regulatory factor AP-1, a protein mediating some of the nuclear responses to protein kinase C and is related to the yeast transcription factor GCN4 (Imagawa et al., 1987; Struhl, 1987). These DNA-binding proteins ultimately mediate many of the transcriptional regulatory effects of growth factors acting initially at the cell surface. Mutations in these proteins or in their control elements may remove tight controls on gene transcription to drive cell replication. The best characterized recessive oncogene, that causing hereditary retinoblastoma, may code for a DNA-binding protein which is a negative regulator of cell growth (Lee et al., 1987).

Tumors arising in some organs continue to respond to hormones which stimulate growth of the normal cells of the tissue of origin. For example, many breast cancers grow in response to estrogen as does the normal breast; reduction of estrogen concentration and use of antiestrogens slow growth and cause regression of metastases in many breast cancers which contain estrogen receptors (Heuson and Coure, 1987). Prostate cancer, like normal prostate tissue, may grow in response to androgens. Reduction in androgens and estrogen therapy may retard the progress of this disease.

The mechanisms through which growth factors and hormones stimulate cell proliferation will provide essential information about the mechanisms through which malignant processes subvert normal growth control. Conversely, studies of oncogenes in retroviruses and in human tumors will provide important information concerning the essential networks through which normal growth controls function.

CHAPTER 58

The Testis

The testis, like the ovary, has both endocrine and reproductive functions because it produces both hormones and mature sperm cells. These testis functions are fulfilled by three major cell types: Leydig cells that secrete testosterone, Sertoli cells that support spermatogenesis, and male germ cells. Testicular cells are compartmentalized by seminiferous tubules. Sertoli cells and germ cells are located inside the tubules, whereas Leydig cells are situated in the interstitial area. The close interaction between the tubular and interstitial compartments forms the basis of intratesticular control mechanisms to ensure optimal testis functions.

The anterior pituitary gland regulates testis functions through secretion of two gonadotropins; follicle-stimulating hormone (FSH) and luteinizing hormone (LH), also known as interstitial cell-stimulating hormone. The secretion of gonadotropins is, in turn, under the control of a hypothalamic decapeptide, gonadotropin-releasing hormone (GnRH), and several testis hormones including gonadal steroids and inhibin. Through feedback control of the hypothalamic-pituitary-testis axis, a dynamic equilibrium of male reproductive functions is maintained.

ANATOMY OF THE TESTIS

The adult human testis weighs 30–45 g and has an approximate diameter of 2.5–4.5 cm. It is surrounded by a fibrous capsule, the tunica albuginea, and is subdivided by septa into many lobules (Fig. 8.46). Each lobule is composed of one to three seminiferous tubules. The coiled seminiferous tubules may extend as long as 60 cm in length and are drained at both ends into the rete testis, from which the efferent ducts empty into the epididymis. The epididymis, located adjacent to the testis, consists of the caput (head), corpus (body), and cauda (tail) and is connected to the vas deferens. The distal segment of the vas deferens is called the ejaculatory duct, and in this area the seminal vesicle and the prostate gland are connected (Fig. 8.47). The ejaculatory duct drains into the urethra near the prostate gland.

The interstitial tissue occupies one-third of the total testicular volume and contains the adrogen-producing Leydig cells, blood vessels, lymphatics, and nerves. In addition, collagen, elastic fibers, and a large number of macrophages are found in the interstitial space (Fabbrini and Hafez, 1980). All blood supply to the seminiferous tubules must pass through the interstitial compartment.

The Sertoli cells are 'nursing' cells with their base attached to the basement membrane of the tubule and their apex extending towards the lumen. They surround all germ cells except the stem cells (spermatogonia) (Fig. 8.48). The plasma membranes of adjacent Sertoli cells form tight junctions which constitute the blood-testis barrier. This barrier excludes most materials from passing through and divides the germinal epithelium into basal and adluminal compartments. The basal compartment is adjacent to the basement membrane and contains the spermatogonia and the early spermatocytes, whereas the adluminal compartment is closer to the tubular lumen and contains spermatocytes, spermatids, and spermatozoa (Fig. 8.48). All substances from the blood and the interstitium must be transported through the Sertoli cell cytoplasm to reach germ cells in the adluminal compartment. This unique functional compartmentalization minimizes the exposure of germ cells to harmful substances in the general circulation and ensures optimal support and maturation of the dividing germ cells. For example, intratubular spermatogenic cells are protected from antibodies circulating in the blood, and breakdown of the blood-testis barrier may lead to autoimmune responses.

BLOOD SUPPLY AND THERMOREGULATION

The blood supply of the testis originates from the internal spermatic artery which arises from the aorta and drains into the testis and epididymis. The spermatic vein passes along the vas deferens in a

Figure 8.46. Anatomy of the testis showing the structures involved in spermatogenesis and sperm transport.

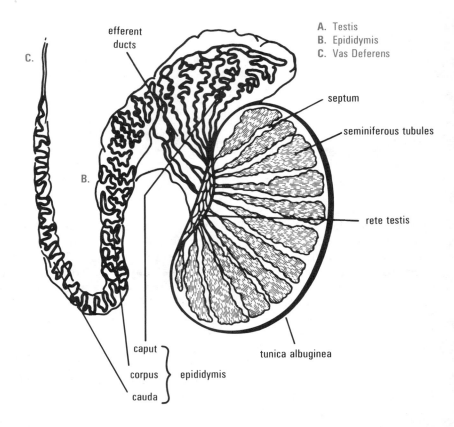

A. Testis
B. Epididymis
C. Vas Deferens

efferent ducts

C.

B.

septum

seminiferous tubules

rete testis

caput
corpus } epididymis
cauda

tunica albuginea

very tortuous course through the pampiniform plexus which surrounds the convoluted spermatic artery. This anatomical feature facilitates counter-current heat and androgen exchange between the venous and arterial systems. The testes are suspended outside the body in most mammals and are maintained at a temperature (34° C) below that of the body. This lower temperature is necessary for optimal spermatogenesis. Contraction and relaxation of the cremasteric muscle alter the distance between the testis and the body core, thus changing the temperature of the testis. Moreover, the ana-

Figure 8.47. Representation of the male reproductive tract, accessory sex organs, and neighboring structures in the lower abdomen and pelvis.

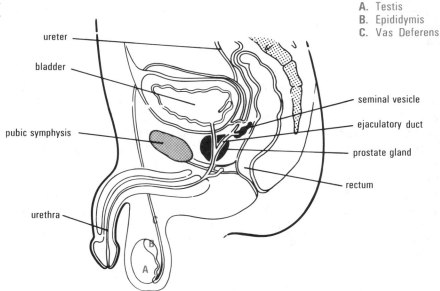

A. Testis
B. Epididymis
C. Vas Deferens

ureter

bladder

seminal vesicle

ejaculatory duct

pubic symphysis

prostate gland

rectum

urethra

A

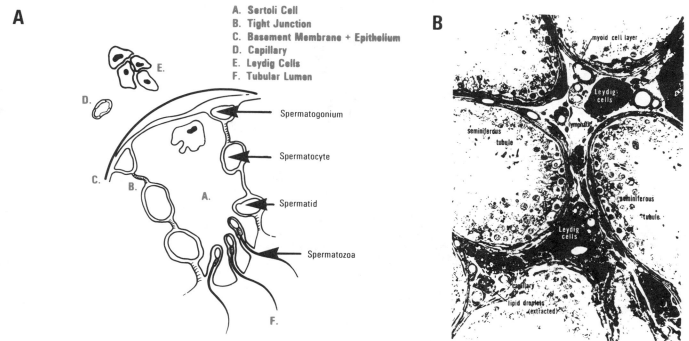

A. Sertoli Cell
B. Tight Junction
C. Basement Membrane + Epithelium
D. Capillary
E. Leydig Cells
F. Tubular Lumen

E.
D.
C.
B.
A.
F.

Spermatogonium
Spermatocyte
Spermatid
Spermatozoa

B

myoid cell layer
Leydig cells
seminiferous tubule
lymphatic
seminiferous tubule
Leydig cells
capillary
lipid droplets (extracted)

Figure 8.48. *A*, Structural arrangement of Sertoli cells and neighboring cell types in the testis. Sertoli cells and their tight junctions divide the germinal epithelium into the basal compartment containing the spermatogonia and the early spermatocytes, as well as the adluminal compartment containing spermatocytes, spermatids, and spermato- zoa. *B*, Cross-section of human testis showing the same structures plus lymphatics and lipid droplets in lower magnification (From Christensen. *Handbook of Physiology,* Section 7, Endocrinology, Vol V, edited by Hamilton & Greep, Washington, D.C., American Physiological Society, 1975, p. 60. With permission).

tomical arrangement of the pampiniform plexus provides an effective mechanism for cooling the arterial blood entering the testis. Impaired spermatogenesis is found in patients with undescended testes (cryptorchidism). The testes of these patients are exposed to the higher body temperature, resulting in decreased or absent spermatogenesis depending on the number of years the testes remain outside the scrotum. Besides maintaining an optimal temperature, the pampiniform plexus facilitates the exchange of androgens between venous and arterial blood thus assuring elevated intratesticular androgen levels required for optimal spermatogenesis.

A varicocele is the dilatation of veins of the pampiniform plexus associated with incompetence of the venous valves, thus resulting in retrograde blood flow. The varicocele condition often causes impaired sperm production.

SPERMATOGENIC FUNCTIONS OF THE TESTIS

Sertoli Cells

Sertoli cells, situated inside the seminiferous tubules, provide physical support for the germ cells and are considered to be the primary regulators of spermatogenesis (Parvinen, 1982). Germ cells are surrounded by, and in close contact with, the Sertoli cells. Moreover, germ cells are bathed in tubular fluid, which is mainly produced by Sertoli cells. The formation of the blood-testis barrier by neighboring Sertoli cells also minimizes the exposure of germ cells to harmful circulating substances. Sertoli cells have distinct endocrine and secretory functions by producing various hormones and proteins. Circulating FSH and intratesticular androgens stimulate Sertoli cell functions leading to optimal spermatogenesis.

Spermatogenesis

Spermatogenesis takes place throughout adult life and a continuous supply of germ cell precursors is found in the seminiferous tubules. In a typical cross-section of the tubule, germ cells at advancing stages of spermatogenesis are lined up in concentric rings (Fig. 8.48). The precursor germ cells are located near the basement membrane and move toward the tubular lumen as they mature. The process of spermatogenesis is highly programmed and each stage is fixed in duration with a total length of 65 days in man.

Spermatogenesis can be divided into three phases; mitotic multiplication of the stem cells, meiosis, and spermiogenesis (Fig. 8.49). The division and maturation of immature stem cells (spermatogonia) takes

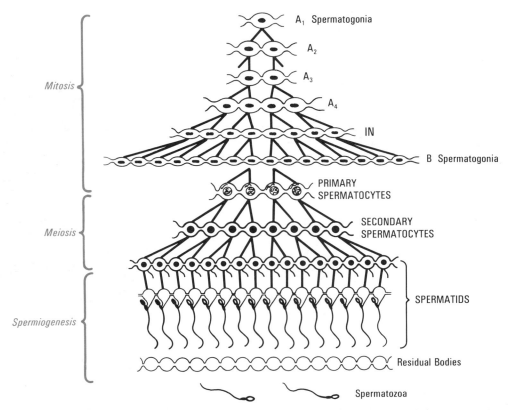

A₁ Spermatogonia

A₂

A₃

A₄

IN

B Spermatogonia

PRIMARY SPERMATOCYTES

SECONDARY SPERMATOCYTES

SPERMATIDS

Residual Bodies

Spermatozoa

Mitosis

Meiosis

Spermiogenesis

Figure 8.49. Defined stages of spermatogenesis showing progression of spermatogonia into spermatozoa.

place in the adluminal compartment of the tubules next to the basement membrane. Mitotic division of spermatogonia results in the formation of additional stem cells which remain resting until the next spermatogenic cycle. As a consequence of successive mitoses, the spermatogonia progress from $A_1 \rightarrow A_2 \rightarrow A_3 \rightarrow A_4 \rightarrow$ intermediate (IN) types to type B (Fig. 8.49). Due to incomplete cytokinesis, all cells derived from a single spermatogonium are connected through cytoplasmic bridges and are synchronized in subsequent cell divisions. This type of cell association is maintained throughout spermatogenesis until the germ cells differentiate into spermatids which then separate into individual spermatozoa. The second phase of spermatogenesis involves meiotic division. The type B spermatogonia progress into primary spermatocytes through one round of mitosis. The primary spermatocytes then enter meiotic division. Meiosis consists of two consecutive cell divisions ($4N \rightarrow 2N \rightarrow 1N$ chromosomes) accompanied by only one duplication of the chromosomes, so that the resulting cells (spermatids) contain only half the number of chromosomes (haploid; 1N).

The third phase of spermatogenesis is termed spermiogenesis during which the complex transfor-

mation of spermatids into specialized spermatozoa takes place. The Golgi apparatus, containing hyaluronidase and other proteases, is transformed to the acrosome. The acrosome is a caplike structure covering the anterior two-thirds of the sperm head. By releasing its enzyme-rich contents during fertilization, the acrosome is believed to play a major role in the penetration of the egg. The centrioles and mitochondria of the spermatids are transformed to the flagellae (sperm tails). Mitochondria generate energy and are therefore important for movement of the spermatozoa. In addition, the nuclear protein histone is replaced by protamine which surrounds the inactive and highly condensed chromatin of the spermatozoa. The nuclear condensation in the spermatozoa protects the genome from the deleterious effect of mutagens in the foreign environment. Furthermore, unnecessary cellular organelles are cast off from the spermatozoa as residual bodies. The Sertoli cells absorb these bodies which contain ribosomes, lipids, and degenerating mitochondria and Golgi apparatus. Spermatozoa are highly specialized cells that do not grow or divide. A spermatozoon consists of a head, containing the heredity material, a middle piece, and a tail which provides a means of motility.

Testis biopsy with subsequent histological examination may be helpful in the diagnosis of oligo- and azoospermia. One of the following histological patterns are commonly found in these patients: (1) normal testis histology, suggesting the existence of an obstruction in the efferent pathway; (2) spermatogenic arrest, where the spermatogenic process fails to progress beyond one of the early stages; (3) decreased or absent germ cells inside the tubules; (4) tubular atrophy.

Sperm Transport

Only fully formed spermatozoa are released by the Sertoli cells into the lumen of the seminiferous tubules. Spermatozoa are transported from the testis to the vas deferens through the epididymis. During their stay in the epididymis (approximately 1–2 weeks), the sperm undergo maturation, thus achieving full motility and the capacity to fertilize. Small amounts of fluid accompany spermatozoa during their passage through the epididymis. In the ejaculatory duct, sperm are mixed with the secretory products of the seminal vesicle, prostate gland, and the bulbourethral gland (the remnant of the müllerian duct). Spermatozoa normally constitute no more than 5% of the volume of the ejaculate.

Ejaculation involves the transport of semen into the urethra, the propulsion of semen out of the urethra during orgasm, and concomitant closure of the bladder neck. Injury of the sympathetic nervous system, due to trauma, surgery, or metabolic diseases such as diabetes, may result in retrograde ejaculation. Inability of the bladder neck to close during ejaculation results in an antegrade transport of semen into the bladder.

Semen Analysis

Both qualitative and quantitative techniques have been established to characterize semen quality. Semen consists of spermatozoa and seminal fluid. Criteria for functional tests of spermatozoa include: concentration of spermatozoa (under normal conditions more than 20×10^6 ml), percentage of spermatozoa with normal progressive forward motility (more than 60%), and the percentage of spermatozoa having abnormal heads or tails (less than 40%). In addition, the motility of spermatozoa in cervical mucus in vivo (postcoitum test) and in vitro (Kremer test) can be assessed. Routine semen analysis also includes the determination of the volume of the ejaculate (between 1 and 5 ml), investigation for the presence of leukocytes as a sign of inflammation, as well as the determination of acid phosphatase (index of prostate gland function), and fructose levels (index of seminal vesicle function).

The presence of autoantibodies on the spermatozoa surface can also be tested based on their agglutination abilities.

Sperm qualities vary significantly within one individual. Acute illness may temporarily depress spermatogenesis. To obtain adequate information about the quality of spermatogenesis, repeated semen analyses are necessary. Because of the prolonged duration and complexity of spermatogenesis, possible changes in sperm quality can only be judged after 3 months. Sperm production can be affected by various factors such as stress, medication, malnutrition, and chronic renal failure. Approximately 50% of infertility cases are caused by male factors and decreased sperm production (oligospermia) in men is a common finding in infertile couples. A reduction in the number of sperm is often associated with diminished motility and an increased percentage of abnormal sperm morphology. In men suffering from oligospermia, the underlying cause can only be detected in a minority of cases and clear endocrine abnormalities are rarely involved.

PITUITARY-TESTIS AXIS

Hypothalamic Regulation of Pituitary Functions

To obtain sustained pituitary gonadotropin release in subjects without endogenous GnRH, this decapeptide has to be administered in pulses (Wildt et al., 1981; Crowley et al., 1985). In animal models, simultaneous measurement of GnRH concentrations in the hypophyseal portal circulation and LH levels in peripheral blood suggests that pituitary LH is released in pulses as a consequence of intermittent GnRH stimulation. In humans, indirect estimates of fluctuating endogenous GnRH levels in the portal circulation can be derived by studying the pattern of stimulated LH secretion in peripheral blood (Huseman and Kelch, 1978; Fauser et al., 1987).

Patients with Kallmann's syndrome—a genetic disorder characterized by defective hypothalamic GnRH production—have been effectively treated with GnRH replacement therapy (Crowley et al., 1985). Administration of GnRH in discrete pulses at intervals of 90–150 min completely induces testis functions in these patients. It is believed that the onset of puberty and sexual maturation is caused by changes in the central nervous system, leading to an increased production of hypothalamic GnRH, and subsequent pituitary LH and FSH release. In contrast, true precocious puberty is caused by premature activation of the hypothalamic-pituitary axis. In some cases, this activation is caused by abnormalities in the central nervous system such as tumors or hydrocephalus. However, in the majority of cases, the underlying mechanisms are unknown.

The half-life of circulating GnRH is very short due to rapid degradation of the hormone by pituitary and brain enzymes (Griffiths and McDermot, 1983). Selective structural modifications of this decapeptide have led to the synthesis of superactive agonist analogues with a prolonged half-life and enhanced target cell action. In addition, GnRH antagonist analogues with potent receptor binding ability but without biological activity have also been synthesized. Treatment with GnRH antagonists prevents the action of endogenous GnRH whereas the administration of high doses of GnRH agonists results in an initial burst of gonadotropin release followed by desensitization of gonadotrophs and decreased production of LH and FSH. These compounds are effective in causing a reversible 'chemical castration' and are useful in the treatment of precocious puberty and reproductive organ tumors that are dependent on sex steroids (Vickery, 1986). Attempts to utilize GnRH analogues for male contraceptive purposes have not been successful, because a substantial reduction in spermatogenesis is accompanied by consistent suppression of circulating testosterone levels resulting in impotence and hot flashes. Testosterone supplementation therapy may reduce the side effects, but may also attenuate the antifertility effect of the GnRH analogues.

Secretion and Release of Gonadotropins

LH and FSH are synthesized by gonadotrophs present in the anterior lobe of the pituitary. Immunohistochemical studies have shown that both LH and FSH are present in most gonadotrophs, whereas a minority of cells contain only one form of the gonadotropin (Phifer et al., 1973). LH and FSH, stored in secretory granules, are discharged rapidly following GnRH stimulation. While the effect of GnRH on de novo gonadotropin synthesis occurs only after several hours of exposure, continuous infusion of GnRH results in two phases of LH responses; an initial acute release of stored hormones, followed by a greater secondary increase of LH due to enhanced synthesis (Bremner and Paulsen, 1977). Pituitary responsiveness to GnRH also varies markedly due to the negative feedback effect of changing steroid levels (Santen, 1981).

LH and FSH are released into the general circulation in a pulsatile manner. FSH pulses, however, are less pronounced than LH pulses, in part due to differences in their metabolic clearance rates. The serum half-life for FSH is about 2–3 hours as compared to approximately 30 min for LH. Because the biosynthesis of FSH seems to occur faster than that of LH (Chappel et al., 1983), the rate of FSH secretion is more related to its biosynthesis whereas LH secretion is more dependent on the release of stored forms. In men, episodic increases in plasma LH occur at intervals of 90–120 min with a wide range of pulse amplitudes (Santen and Bardin, 1973). Precise estimation of pulse-amplitude and pulse-frequency has been performed using frequent blood sampling and detailed data analysis. Serum gonadotropins are degraded and cleared by the liver and the kidney. In the urine, intact immunoreactive and bioactive gonadotropins have been identified.

Chemistry of Gonadotropins

LH and FSH are protein dimers consisting of two glycosylated polypeptide chains, noncovalently linked together. The three-dimensional structure of each subunit is maintained by internally cross-linked disulfide bonds. The α-subunits of all three pituitary glycoprotein hormones (LH, FSH, and thyroid-stimulating hormone [TSH]) and the placental human chorionic gonadotropin (hCG) are identical in amino acid sequence, whereas the β-subunit is hormone specific (Pierce and Parsons, 1981; Chappel et al., 1983). A hybrid molecule containing purified β-subunit of LH and α-subunit isolated from TSH have full LH bioactivity.

Although all subunits are believed to be derived from a common ancestral gene, the common α- and different β-subunits are encoded by different genes. The primary amino acid sequence of the gonadotropins has been identified and the genes for the subunits have been isolated (Fiddes and Goodman, 1981; Boorstein et al., 1982), allowing the production of human LH and FSH using recombinant DNA technology. Because the structure of hCG-β is highly homologous to LH-β except for an additional 24 amino acids on the carboxyl-terminus, hCG can be used for LH replacement therapy. Comparison of nucleotide sequences of the genes for LH-β and hCG-β subunits indicates that the hCG-β genes have evolved from an ancestral LH-β gene and the carboxyl-terminal extension of the hCG-β subunits appears to have arisen by a single base deletion resulting in a reading frame shift of the gene. The evolutionary origin of various glycoproteins of the gonadotropin and TSH family is shown in Figure 8.50. The β-chains have developed separately for TSH and gonadotropins, which in turn evolved separately into β-chains for LH, FSH, hCG, and pregnant mare serum gonadotropin (PMSG).

The α-subunit appears to be species-specific whereas β-subunits are not. α-Subunits are found in excess in the anterior pituitary and the regulation of β-subunit gene expression is rate-limiting. In gonadotrophs, synthesis of the β-subunit and the combination of both subunits are important. Circu-

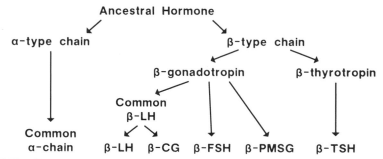

Figure 8.50. Evolution tree of α- and β-subunits of pituitary and placental glycoprotein hormones.

lating free-α and free-β subunits are not recognized by gonadotropin receptors and are therefore not biologically active.

Post- or cotranslational processing of the gonadotropins involves glycosylation or addition of carbohydrate side chains by microsomal enzymes in the endoplasmic reticulum (Catt and Pierce, 1986). Removal of terminal sialic acid residues from the carbohydrate side chains of gonadotropins decreases the half-life of these hormones in circulation because these asialo molecules are rapidly metabolized by hepatic cells (van Hall et al., 1971). Although the asialo-gonadotropins still retain bioactivity in vitro, their action in vivo is substantially diminished. Binding of gonadotropins to target cell receptors is primarily determined by the protein domain of the hormone, whereas the carbohydrate moieties are essential for optimal target cell responses. Deglycosylated forms of LH and FSH bind to the receptors but do not increase steroidogenic activity, and therefore act as competitive antagonists of the intact hormone (Sairam et al., 1983; Dahl et al., 1988). More information on the biological activity of gonadotropin hormones under different physiological and pathophysiological conditions has become available using sensitive in vitro bioassays for LH (Dufau et al., 1974), and FSH (Jia et al., 1985).

Pituitary Regulation of Testis Functions

The two pituitary gonadotropins act exclusively on specific testis cell types. Binding of FSH to receptors located on the plasma membrane of Sertoli cells activates one of the G proteins with a subsequent increase in adenyl cyclase activity. Through the cAMP-dependent protein kinase pathway, specific Sertoli cell genes for various proteins are activated. Likewise, Leydig cells have specific membrane receptors for LH. Following a similar activation of the cAMP-dependent protein kinase pathway, Leydig cell androgen secretion is stimulated by LH. Only a small proportion of the LH receptors needs to be occupied to initiate optimal steroidogenic activity. However, long-term exposure to LH or

administration of high doses of hCG causes a decrease in LH receptor content resulting in decreased Leydig cell responsiveness to LH. Recent cloning studies indicate that the LH receptor is structurally related to other G-protein-coupled receptors with seven transmembrane domains but it has an extended extracellular ligand-binding domain.

Lactotrophs present in the anterior pituitary secret prolactin, the regulation of which is controlled by multiple factors (Yen and Jaffe, 1986). Dopamine, sometimes referred to as prolactin-inhibiting factor, inhibits, whereas thyrotropin-releasing factor stimulates prolactin release. Unlike its well known role in the control of lactation in the female, prolactin does not have a well defined physiological role in the male. Secretion of adequate amounts of prolactin seems to be necessary for normal testosterone production, whereas elevated prolactin levels are associated with depressed testosterone biosynthesis. Patients with severe hyperprolactinemia are impotent as a result of decreased androgen production. The negative effect of prolactin on testosterone production may be due to indirect inhibition of testis functions secondary to changes at the hypothalamic-pituitary axis, or through a direct inhibitory action of prolactin mediated by Leydig cell prolactin receptors.

Steroid Feedback Mechanisms

The hypothalamus-pituitary-testis axis maintains a dynamic equilibrium of serum levels of reproductive hormones through a closed-loop feedback mechanism (Fig. 8.51). Synthesis and secretion of pituitary LH, and to a lesser extent FSH, is stimulated by hypothalamic GnRH released into the portal circulation. Pituitary gonadotropins, in turn, stimulate testis secretion of androgens and inhibin. The production of gonadotropins is negatively controlled through an inhibitory action of testicular androgens both at the central nervous system and the pituitary. Thus, LH-producing pituitary cells and androgen-producing Leydig cells form a closed-loop axis. Testosterone is also converted to estro-

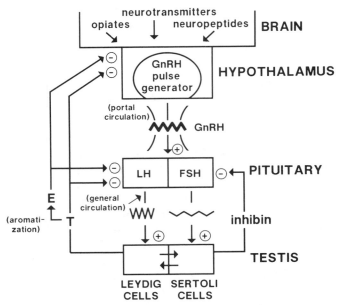

Figure 8.51. Hormones involved in the feedback regulation of the hypothalamic-pituitary-testis axis.

gens by the aromatase enzymes present in peripheral tissues and in brain cells. Part of the negative feedback of sex steroids is mediated through estrogens. Indeed, receptors for androgens and estrogens are found in the hypothalamus (Moore, 1986). Studies in men suggest that the suppressive effect of testosterone on LH secretion mainly takes place at the hypothalamic level, whereas estrogens primarily affect pituitary sensitivity to GnRH stimulation (Santen, 1981).

Although gonadal sex steroids have minimal effects on FSH secretion, an additional gonadal hormone—inhibin (Mason et al., 1985)—has been shown to suppress pituitary FSH secretion preferentially. Inhibin is produced by the Sertoli cells (Steinberger and Steinberger, 1976; Bicsak, et al., 1987) and its secretion is stimulated by pituitary FSH. The presence of this protein hormone forms the basis of an additional closed-loop feedback axis between pituitary FSH-producing cells and testis Sertoli cells. Inhibin is composed of two subunits covalently linked by S-S bonds. The higher molecular weight α-subunit combines with one of two β-subunits designated as β_A and β_B to form $\alpha\beta_A$ and $\alpha\beta_B$ dimers of inhibin with similar biological activity. In some patients with seminiferous tubule damage or with impaired spermatogenesis, selective elevations of serum FSH but not LH are detected, presumably due to decreased Sertoli cell production of inhibin.

The β-chains of inhibin show striking sequence homology with transforming growth factor-β (TGF-β) (Mason et al., 1985) and müllerian duct-inhibiting factor (MIF), another Sertoli cell product only found

during fetal development. In contrast to the inhibitory effect of the two $\alpha\beta$ dimer inhibins on FSH secretion, recent studies indicate that the $\beta\beta$ dimers of the inhibin subunits ($\beta_A\beta_B$ or $\beta_A\beta_A$) are potent stimulators of FSH synthesis and secretion. The stimulatory action of these homodimers on FSH release is different from that of GnRH in that they have no effect on LH secretion.

ENDOCRINE FUNCTIONS OF THE TESTIS

Leydig Cell Androgen Production

The biosynthesis of steroid hormones from cholesterol takes place in several steroidogenic tissues and the final product is dependent on specific enzymes present in each tissue (Waterman and Simpson, 1985). Many of the reactions involved in steroid biosynthesis are of the mixed-function oxidase type mediated by various forms of cytochrome P-450 enzymes.

Stimulation of Leydig cells by LH results in a cascade of intracellular events which eventually lead to the formation of testosterone (Preslock, 1980). LH can generate free cholesterol inside the cell through different mechanisms. Cellular cholesterol is stored in the form of cholesterol esters in the lipid granules. After LH stimulation of the cholesterol esterase enzyme, free cholesterol is formed. Free cholesterol, in turn, is transported to the mitochondria for conversion to pregnenolone. The C27 steroid cholesterol, which is the substrate for androgen biosynthesis, may also be synthesized in the Leydig cells from acetate or derived from the circulation in the form of lipoproteins. The rate-limiting enzyme involved in cholesterol biosynthesis is the cytoplasmic 3-hydroxy-3-methylglutaryl coenzyme A reductase (HMG-CoA reductase). The 6-carbon mevalonate is further converted to the 27-carbon cholesterol through phosphorylation, condensation of isomers, and cyclization. Lipoproteins are complex macromolecules that carry plasma lipids, and low-density lipoprotein (LDL) is the predominant form in man. After binding to Leydig cell plasma membrane receptors, LDL is internalized together with the lipoprotein receptor. The internalized vesicles fuse with lysosomes; the protein component of LDL is cleaved to amino acids while the cholesterol ester component is hydrolyzed to free fatty acids and unesterized cholesterol.

All steroidogenic tissues, namely the adrenals, testis, ovaries, and placenta, contain enzymes necessary for cleaving the cholesterol side chain to remove the 6-carbon isocaproic acid and thus converting cholesterol to pregnenolone. Conversion of cholesterol to pregnenolone is catalyzed by the mitochondrial side chain cleavage cytochrome P-450 (P-450$_{scc}$) complexes, consisting of three protein components: a hemoprotein cytochrome P-450$_{scc}$, a flavoprotein NADPH adrenodoxin reductase, and an iron-sulfur protein known as adrenodoxin or testodoxin (Simpson 1979). The synthesis of cytochrome P-450$_{scc}$ is stimulated by LH and this enzymatic conversion is irreversible and rate-limiting. These enzymes catalyze the hydroxylation and oxidation between the C20 and C22 position of the steroid backbone to reduce the C27 to a C21 steroid. Clinically useful drugs such as aminoglutethimide inhibit the side chain cleavage reaction.

After the release of pregnenolone from the mitochondria, the predominant steroidogenic pathway in the human is the Δ^5-route (steroid with double bound at C5 position) (Fig. 8.52). The formation of the C19 steroid testosterone from the C21 pregnenolone requires the activity of the cytochrome P-450$_{c17}$ (Payne et al., 1985) present in the endoplasmic reticulum (Fig. 8.52). The LH dependent P-450$_{c17}$ enzyme is also of the cytochrome P-450 protein family and catalyzes two separate reactions: hydroxylation of pregnenolone to 17α-hydroxy-pregnenolone (referred to as 17α-hydroxylase activity), and cleavage between C17 and C20 (17,20 lyase activity) to form the C19 steroid dehydroepiandrosterone (DHEA) (Fig. 8.52). Both enzyme activities can be demonstrated in nonsteroidogenic cells transfected with a cDNA clone for the P-450$_{c17}$ enzyme. This enzyme requires the participation of a flavoprotein, NADPH cytochrome P-450 reductase (Fig. 8.52).

The final conversion of DHEA to testosterone is mediated by two enzymes not related to the cytochrome P-450 protein family. The 3β-hydroxysteroid dehydrogenase (3βHSD) enzyme converts DHEA to androstenedione which is further converted by a reversible enzyme 17β-hydroxysteroid dehydrogenase (17βHSD) to testosterone. Alternatively, DHEA can be altered to androstenediol by the 3βHSD enzyme followed by transformation to testosterone mediated by 3βHSD. In addition to the Δ5 pathway, a parallel Δ4-pathway (through the formation of progesterone, 17α-hydroxyprogesterone and androstenedione as intermediate products) is predominantly operative in steroidogenic tissue of other species to convert C21 to C19 steroids (Fig.

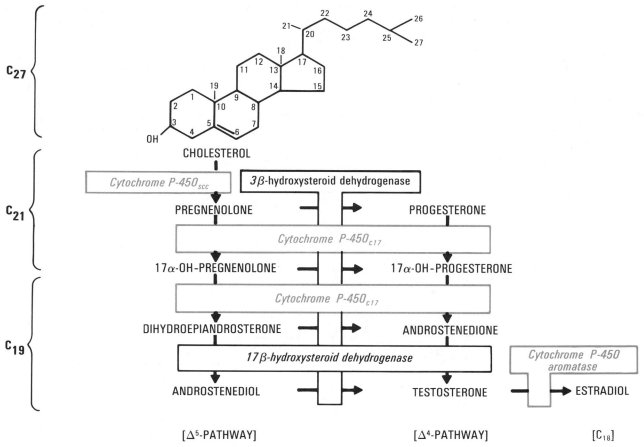

Figure 8.52. Representation of the two androgen biosynthetic pathways (Δ^4 and Δ^5 steroids).

8.52). Testosterone, produced by Leydig cells, is either secreted into the general circulation, or concentrated in the tubular lumen. In addition, testosterone can be converted to estradiol in Sertoli cells by another cytochrome P-450 enzyme complex (P-450$_{aromatase}$ enzyme).

Deficiencies of the enzymes involved in testosterone formation may be manifested in the adrenal gland as well as in the testis. Although most disorders are rare, individuals with defective production of each separate enzyme have been described. Impaired capacity to produce androgens results in abnormalities in the development of androgen-dependent structures such as internal and external genitalia. Patients with deficient production of the 17βHSD enzyme have elevated androstenedione and low testosterone levels resulting in partial virilization during puberty.

Endocrine Roles of Sertoli Cells

Sertoli cells have steroidogenic functions in that they convert androgens, produced by the Leydig cells, to estrogens (Fig. 8.53). This irreversible reaction takes place in the smooth endoplasmic reticulum and requires the participation of the cytochrome P-450$_{aromatase}$ enzyme and the ubiquitous NADPH cytochrome P-450 reductase. Estrogen production by the testis has only minimal contribution to circulating estradiol levels, but this steroid may play intratesticular roles. Sertoli cells also produce inhibin, a protein involved in the feedback regulation of pituitary FSH secretion.

Many other proteins are produced by Sertoli cells, including androgen binding protein (ABP) (Ritzen et al., 1977), transferrin, ceruloplasmin, and plasminogen activators. Androgen-binding protein has high affinity for testosterone and is responsible for the maintenance of high intratubular testosterone levels by concentrating testosterone in the tubular lumen. The structure of androgen-binding protein is homologous to the hepatic sex-hormone binding globulin (SHBG) (Joseph et al., 1987), which is responsible for sex steroid transport in the general circulation. Secretion of transferrin and ceruloplasmin is believed to be important for the transport of iron and copper, respectively, to tubular cells whereas the plasminogen activators may mediate proteolytic reactions important for the migration of maturing germ cells from the basal compartment of the seminiferous tubules to the adluminal compartment.

Intratesticular Regulatory Mechanisms

Leydig cell androgen biosynthesis is primarily regulated by LH. In addition, hormones present in the general circulation, such as insulin, prolactin, and thyroxin, may also have direct effects on Leydig cell functions as testicular receptors for these hormones have been identified. Leydig cell responses to LH may also be modified by intratesticular hormones. Estrogens, mainly synthesized in Sertoli cells, bind to specific receptors in the Leydig cells and inhibit androgen production (Hsueh et al., 1978; Smals et al., 1980). In addition, androgens produced by Leydig cells may exert autocrine actions by regulating testosterone biosynthesis. Recent studies also suggest that multiple protein and peptide hormones of testis origin modulate LH-stimulated androgen biosynthesis. Because inhibin, transforming growth factor-β and insulin-like growth factor-I are not only produced by Sertoli cells, but also exert direct effects on Leydig cell steroidogenesis (Hsueh

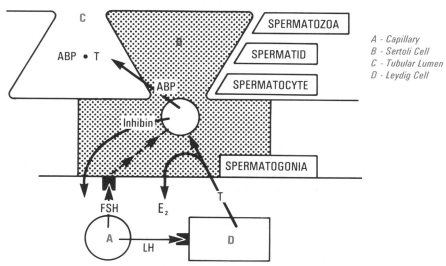

Figure 8.53. Hormonal regulation and function of Sertoli cells.

et al., 1987), these hormones can be considered as paracrine factors. The observed decreases of intratesticular testosterone levels in some patients with impaired sperm production (Rodriquez-Rigau et al., 1980) is consistent with the hypothesis that intratesticular communication between the intratubular and the interstitial compartment is important.

Hormonal Regulation of Spermatogenesis

High intratubular testosterone levels are essential for normal spermatogenesis. Testosterone, produced by Leydig cells, is concentrated in the tubular fluid because of the presence of high levels of androgen-binding protein secreted by Sertoli cells. Although androgens appear to regulate steps in meiotic division, it is unclear whether germ cells are direct targets of androgens (Fritz, 1978). Androgens may mediate their actions through interaction with Sertoli cell androgen receptors. In men, FSH appears to be necessary for the completion of spermatogenesis. Animal studies suggest that both FSH and LH are required for the initiation of spermatogenesis, whereas LH or testosterone alone is capable of maintaining spermatogenesis.

Transport and Action of Androgens

Serum testosterone levels show diurnal and seasonal variations. More than 95% of serum testosterone is transported in the general circulation after binding to serum proteins and only the unbound form of the androgen is biologically active. These carrier proteins include sex hormone-binding globulin and albumin. They are synthesized in the liver and serve to prevent diffusion of testosterone into tissues. The production rate of sex hormone-binding globulin is enhanced by estrogens and suppressed by androgens. In addition, elevated levels can be found in obese individuals via raised estrogen levels and patients with liver cirrhosis. Decreased levels are detected in patients with hypothyroidism. Because the serum concentrations of binding proteins may vary under different conditions, differences in target cell responses may occur without distinct changes in total testosterone levels. Measurement of free (unbound) testosterone levels therefore provides more adequate information regarding target cell responses and measurement of testosterone levels in saliva (only the free fraction passes the salivary gland) may prove to be clinically useful.

Testosterone acts on multiple tissues and is essential for the development of reproductive organs, sexual behavior, and secondary sex characteristics (Mooradian et al., 1987). Moreover, the regulation of some hypothalamic and pituitary functions is also mediated through testosterone. Although testosterone acts directly on androgen receptors in many tissues, it is converted to dihydrotestosterone in most reproductive tissues because of the presence of the 5α-reductase enzyme. Androgen receptors bind dihydrotestosterone with much higher affinity resulting in higher target cell responses. In contrast, muscle cells utilize testosterone exclusively, whereas adipose tissue and brain are also capable of converting testosterone to estrogens.

Like other steroid hormones, testosterone binds to specific high-affinity receptor proteins following passive diffusion into target cells. The hormone-receptor complex undergoes conformational changes and binds to specific DNA sequences in the androgen-induced genes (Grody et al., 1982), resulting in new mRNA accumulation and specific protein synthesis.

The testis also secretes small amounts of estradiol and several androgen precursors, including androstenedione, DHEA, and several 17α-hydroxysteroids. Most of these precursor steroids are weak androgens which are also produced by the adrenal cortex. In addition, small amounts of testosterone can be formed in the peripheral tissue from hormonally inactive metabolites. Secreted androgens are metabolized in the liver by oxidation and reduction. The reduced metabolites are conjugated to glucuronides and sulfates and are excreted in urine.

SEXUAL DEVELOPMENT AND DIFFERENTIATION

Genetic Sex

The sex chromosomes are separated during meiotic division, when spermatocytes differentiate into spermatids, and therefore the spermatozoa contain either a Y or X chromosome. In contrast, ova contain only an X chromosome. When an ovum is fertilized by a spermatozoon bearing the X chromosome, the resulting XX zygote develops into a female. Alternatively, when the fertilizing spermatozoon contains a Y chromosome, the XY combination ensures male development (Fig. 8.54). In rare cases of sex reversal, XX males and XY females have been documented. Although earlier work suggests that a specific cell surface antigen (H-Y antigen) is involved in testicular development, preliminary observations indicate that a gene located in the short arm of the Y chromosome is responsible for fetal testicular development (Page et al., 1987). This gene may be one of the putative testis-determining factors (TDF) and is translocated to the X chromosome of the XX males, but is deleted from

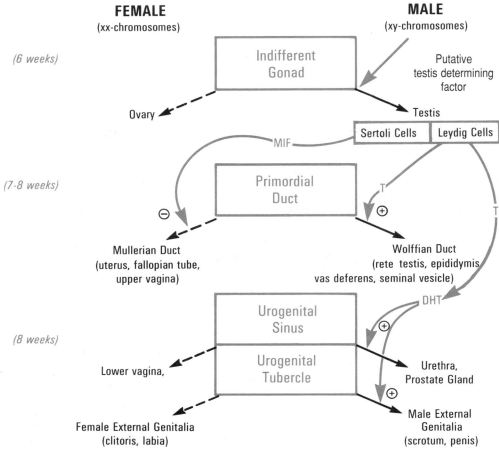

FEMALE
(xx-chromosomes)

MALE
(xy-chromosomes)

(6 weeks)

Indifferent
Gonad

Putative
testis determining
factor

Ovary

Testis

Sertoli Cells | Leydig Cells

MIF

(7-8 weeks)

Primordial
Duct

T

⊖

⊕

Mullerian Duct
(uterus, fallopian tube,
upper vagina)

Wolffian Duct
(rete testis, epididymis,
vas deferens, seminal vesicle)

T

DHT

(8 weeks)

Urogenital
Sinus

⊕

Urogenital
Tubercle

Lower vagina,

Urethra,
Prostate Gland

⊕

Female External Genitalia
(clitoris, labia)

Male External
Genitalia
(scrotum, penis)

Figure 8.54. Schematic representation of male sexual development, emphasizing that gonadal sex is determined by the Y-chromosome whereas the genital sex is hormonally regulated. *T* = testosterone; *DHT* = dihydrotestosterone; *MIF* = müllerian duct-inhibiting factor. At the left-hand side, fetal age is indicated.

the Y chromosome of the XY females. The genes encoding for the H-Y antigen and the putative TDF are located at different parts of the Y chromosome (Simpson et al., 1987).

Klinefelter's syndrome is the classical disorder of male gonadal development and is characterized by the presence of an extra X chromosome (XXY karyotype). Clinical features of this disease include small testes, gynecomastia (breast development), tall stature, and azoospermia.

Gonadal Sex

During the first 6 weeks of fetal life, the primitive gonad is found in both sexes and this structure is connected with the primitive kidney. In the absence of TDF, the primitive gonad develops into the ovary, whereas testicular differentiation occurs during the 7th week of fetal life in the presence of this factor. Testis parenchyme is organized into aggregates of germ cells and Sertoli cells, whereas Leydig cells are found in the interstitial area. Germ cells undergo a series of mitotic divisions, but the initiation of meiosis does not take place until sper-

matogenesis starts during the onset of puberty. The presence of a meiosis-inhibiting substance produced by Sertoli cells has been hypothesized (Gondos, 1980). Fetal Leydig cells have a similar structure as their adult counterparts, and the onset of testosterone production correlates closely with the differentiation of Leydig cells. The way by which fetal Leydig cell steroid production is regulated is not yet understood, but placental chorionic gonadotropin and pituitary LH are probably involved.

Genital Sex

Although gonadal sex is determined by genes in the Y chromosome, the development of genital sex is dependent upon hormones secreted by the developing gonads. In general, female genital organs develop spontaneously, whereas the action of various male hormones secreted by the fetal testis are responsible for the development of male genital structures. Two sets of primordial ducts are responsible for the development of the reproductive tracts, and are present until the 7th week of fetal life. The müllerian duct gives rise to the female structures, whereas the male structures are derived from the Wolffian duct. Even without the support of ovarian hormones, the müllerian duct develops into uterus, fallopian tube, and the upper part of the vagina. In the male, regression of the müllerian duct occurs under the influence of the müllerian duct-inhibiting factor MIF, a protein with sequence homology to TGF-β and the inhibin β-subunits, produced by fetal Sertoli cells (Donahoe et al., 1982; Josso, 1986). In contrast, testosterone produced by fetal Leydig cells stimulates Wolffian duct differentiation with the formation of the rete testis, epididymis, vas deferens, and seminal vesicle. Disturbance of fetal testis functions leads to various types of intersex and genital abnormalities.

Up until the 8th week of fetal life, no differentiation of the external genitalia takes place. Androgen receptors that mediate the action of testosterone and dihydrotestosterone are present in embryos of both sexes, so that the differences in phenotypical development are dependent only on differences in hormone production. In the absence of androgens, the urogenital sinus spontaneously develops into the lower part of the vagina, and the urogenital tubercle into clitoris and the labia. In contrast, testosterone stimulates the virilization of the Wolffian duct, whereas dihydrotestosterone formed at its target cells from testosterone, causes development of the male external genitalia. The urogenital sinus gives rise to the urethra and the prostate gland, whereas the urogenital tubercle differentiates into scrotum and penis.

The development of the male urogenital tract is largely completed by the end of the first trimester. Androgens also induce the descent of the testis by the end of the 7th month of pregnancy. This descent is guided by the degeneration of the gubernaculum, a fibromuscular band connecting the testis to the bottom of the scrotum.

Prepubertal and Pubertal Development

Leydig cells undergo involutionary changes during the late fetal period and shortly after birth, but a new generation of differentiated Leydig cells is found at 5–6 weeks after birth. During the 2nd week of life the pituitary starts to secrete gonadotropins as a consequence of absent negative feedback by placental steroids. Serum testosterone levels are low during the neonatal period, rise after 2 weeks, and reach adult levels within 2–3 months of age. After 6 months of age, testosterone declines to very low levels, and remains suppressed until the onset of puberty.

Puberty is initiated by alterations in the setpoint of the central nervous system to the negative feedback action of gonadal steroids and by changes in the neurohormonal regulation of GnRH and gonadotropin secretion. Initially, nocturnal LH increments in serum ensure a gradual, but progressive increase in testosterone levels (Boyar et al., 1974). The onset of physical signs of sexual maturation (testicular size increment, pubic hair development, and body growth) which generally takes place around the age of 14 years, is preceded by hormonal changes starting 2–3 years earlier.

Abnormal Sexual Development

Disorders of male sexual development (Griffin et al., 1982) are related to defective androgen production or target cell insensitivity, and may cause incomplete virilization of male individuals (male pseudohermaphroditism). Clinical expression can vary greatly from complete female phenotype to normal male phenotype but with infertility as the sole defect. Patients with defective androgen production include inborn errors of testosterone synthesis and the inability to convert testosterone to dihydrotestosterone (5α-reductase deficiency). The 5α-reductase deficiency is an inherited autosomal recessive defect. These patients show normal Wolffian duct differentiation (dependent on testosterone), whereas the development of the external genitalia and the prostate gland (dependent on dihydrotestosterone) fails. Consequently, these patients have a female phenotype at birth. Depending on the severity of the enzyme defect, the external genitalia of some patients may convert into the male type during puberty onset, due to the increased production of testosterone (Peterson et al., 1977). Disorders of male sexual development may also be related to androgen receptor defects. Testicular feminization (complete or incomplete) and related syndromes (Reinfenstein syndrome, and the infertile male syndrome) are X chromosome-linked, recessive disorders with a decrease in number or activity of androgen receptors. In extreme cases, these patients are phenotypically normal females, with a short, blind-ending vagina but with undescended testes located in the abdomen.

Precocious puberty is generally caused by a premature activation of the hypothalamic-pituitary-testis axis due to brain tumors or hydrocephalus. A primary excess in androgen secretion in young children, caused by testosterone-producing tumors or other causes, may have the same clinical appearance (pseudoprecocious puberty).

CHAPTER 59

The Female Reproductive System

Reproduction in women is dependent on endocrine interactions between different components of the female reproductive system. The basic components of this system are: (1) the hypothalamic-pituitary unit, (2) the ovaries, and (3) the uterine-endometrial compartment. On the basis of research over the past two decades (Knobil, 1980), it is now evident that reproductive activity is regulated in a cyclic fashion by an intricate neuroendocrine control system.

In women, the reproductive or menstrual cycle is approximately 28 days in duration and can be divided into three phases: the follicular phase (days 1–13), ovulation, and the luteal phase (days 14–28). The initiation of each menstrual cycle is regulated by pulsatile secretion of gonadotropin-releasing hormone (GnRH), a neurohormone secreted from the hypothalamus. GnRH is transported directly to the pituitary gland via the portal circulation where it stimulates the pituitary gland to release luteinizing hormone (LH) and follicle-stimulating hormone (FSH) in a pulsatile fashion. At the ovarian level, these two hormones coordinate to stimulate the maturation of ovarian follicles in the follicular phase. During follicular maturation, estradiol is produced, increasing until actual ovulation. Following ovulation, both progesterone and estrogen are secreted by the ovarian corpus luteum during the luteal phase. Estrogen and progesterone act at the level of the uterus and endometrium in sequential fashion to stimulate the growth and differentiation of the endometrium in preparation for embryo implantation. If fertilization of the oocyte and embryo implantation fail to occur, the corpus luteum spontaneously regresses. This is followed by a dramatic decline in estrogen and progesterone secretion resulting in endometrial breakdown and shedding. The function and regulation of the menstrual cycle as well as the dynamic relationships between each component of the reproductive system will be discussed in this chapter.

HYPOTHALAMIC-PITUITARY UNIT

Hypothalamus

The hypothalamus is located at the base of the brain adjacent to the third ventricle. Localized within the hypothalamus is a group of neurons (arcuate nucleus) which synthesize and secrete GnRH, a 10-amino acid peptide [pyro-Glu_1 His_2-Trp_3-Ser_4-Tyr_5-Gly_6-Leu_7-Arg_8-Pro_9-Gly_{10}-NH_2] that stimulates the pituitary to release LH and FSH. GnRH, like several other brain peptides, is synthesized as part of a much larger 92-amino acid precursor peptide (Seeburg and Adelman, 1984). This precursor peptide is enzymatically processed to the more biologically active decapeptide prior to release. Once secreted into the circulation, degradation of GnRH is rapid with a half-life of approximately 2–4 mins.

The neurosecretion of GnRH is not continuous, but is characterized by an episodic release of short bursts which coincide with the depolarization of arcuate neurons. Measurements of electrical activity within this group of neurons in monkeys show an almost coincidental release of LH along with changes in electrical potential (Fig. 8.55; Kesner et al., 1986). The average frequency of GnRH release based on peripheral measurements of LH secretion in humans varies from 60 min to 4 hours. Because GnRH neurons are localized within the hypothalamus adjacent to other neuronal systems, the release of GnRH can be modified by central nervous system input. For example, the opioidergic system, through the secretion of β-endorphin can decrease the secretion of GnRH whereas norepinephrine neurons can stimulate the release of GnRH (Rasmussen et al., 1983).

Pituitary-Portal Circulation

A specialized capillary network, the pituitary-portal system, located between the median eminence of the hypothalamus and the pituitary serves to transport neuropeptides between these two

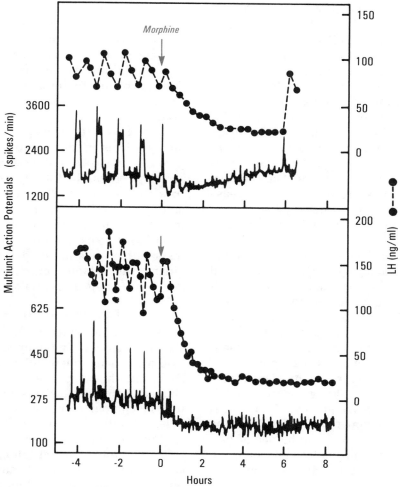

Figure 8.55. Pulsatile LH secretion in the circulation is functionally coupled to multiunit action potentials from GnRH neurons recorded with electrodes implanted in the arcuate nucleus of the rhesus monkey. (Modified from Kesner et al., 1986.)

structures. This circulatory system can not only deliver neuropeptides from hypothalamus to pituitary, but through retrograde blood flow allows endocrine communication from the anterior pituitary to the hypothalamus and between anterior and posterior pituitary (Bergland and Page, 1978).

Pituitary Gland

The pituitary gland is responsible for the synthesis and release of the two glycoprotein hormones, LH and FSH which regulate ovarian function. Located within the pituitary are also cells which can synthesize adrenocorticotrophic hormone (ACTH), growth hormone (GH), thyroid-stimulating hormone (TSH), and prolactin (see Chapter 53). Both LH and FSH are synthesized in specialized cells called gonadotropes. Within the population of gonadotropes are cells which can synthesize and secrete both LH and FSH while other gonadotropes produce and secrete only one glycoprotein hormone.

Specific receptors for GnRH are localized on the cell membrane of the gonadotropes. Receptor activation by GnRH results in a rapid influx of calcium into the cell through specialized channels in the cell membrane and subsequent release of LH and FSH (Clayton and Catt, 1981).

Structurally, LH and FSH molecules are composed of two subunits, α and β which are noncovalently bound. Each subunit is encoded by a separate gene. Although the α-subunit is common to the glycoprotein hormones LH, FSH, TSH, and human chorionic gonadotropin (hCG), the β-subunit structure is unique for each hormone and determines specificity for the different receptors. Membrane receptor activation requires binding by the combined α- and β-subunits. The half-lives of LH and FSH in the peripheral circulation are approximately 20 and 180 min, respectively.

Two unique properties of gonadotrophs in response to GnRH stimulation have been described.

When GnRH is administered continuously, circulating LH and FSH concentrations will increase. After several days of exposure to GnRH, secretion of LH and FSH gradually diminishes. Thus, with constant exposure to GnRH, GnRH binding to receptors on the gonadotrope decreases. This reversible process is termed down-regulation (Clayton and Catt, 1981). Down-regulation and diminished secretion of LH and FSH can also be induced by the administration of long-acting synthetic agonist analogues of GnRH. These GnRH agonists can reduce LH and FSH levels essentially to abolish gonadal function. On a clinical basis these GnRH agonists are utilized to decrease ovarian function in the treatment of estrogen-dependent neoplasms, precocious puberty, and endometriosis.

When GnRH is administered in a pulsatile fashion at a physiological frequency (60–90 min) increased LH and FSH concentrations are maintained, GnRH receptor binding on the gonadotrope increases, and eventually ovulation is induced. This process is called upregulation. The overall pituitary gonadotropin output is regulated by both the ambient estradiol concentrations as well as the frequency of GnRH release because both estradiol and GnRH promote gonadotropin synthesis (Wang and Yen, 1975).

In the female, the mean frequency of GnRH-LH secretion during the follicular phase is 60–120 min. During the luteal phase, under the influence of progesterone produced by the corpus luteum, the mean frequency of GnRH-LH secretion shifts to every 4 hours.

OVARY

Embryology

The adult ovaries are oval-shaped paired structures measuring $4 \times 3 \times 4$ cm situated in the pelvis. During embryonic development, the gonadal structures are organized between the 3rd through 4th week of gestation when the primordial germ cells migrate from the yolk sac to the genital ridge. Over the next few weeks, these germ cells undergo successive mitotic divisions to form oogonia such that by 20 weeks' gestation there are 6–7 million germ cells. These germ cells are enveloped by coelomic surface epithelial cells and medullary mesenchymal tissue. In the absence of a Y chromosome and testis-determining factor (Page et al., 1987), the epithelial cells which surround these oocytes differentiate into the granulosa cell layer forming the primordial follicle. This follicle complex is separated by a basement membrane from the less organized mesenchymal cells which form the theca and interstitial compartment (Ohno et al., 1962). In some oogonia, the first meiotic division is initiated but becomes arrested at the prophase stage. There is evidence that the granulosa cell secretes a substance responsible for inhibiting the progression of meiosis (meiotic inhibiting factor). Once arrested, the meiotic process does not resume until just prior to ovulation.

Throughout the life-span of the female, primordial follicles undergo spontaneous degeneration through a process called atresia. Through this atretic process, the total follicle number will decline from 6–7 million at 20 weeks' gestation to approximately 1–2 million at birth. At the time of puberty, follicle numbers will have further decreased to 400,000. It is estimated that the atretic process is responsible for a 99.9% decrease in oocyte number during the female reproductive life-span (Mandl and Zuckerman, 1951; Baker and Sum, 1976).

Follicular Growth and Ovulation

As shown in Figure 8.56, the follicle complex consists of the oocyte enveloped by a mucopolysaccharide translucent shell called the zona pellucida. Surrounding the zona are several layers of spindle-shaped granulosa cells separated from the theca interna compartment by a basement membrane. Follicles are situated close to the surface of the ovary within the ovarian cortex.

When the primordial follicle has formed, the process of follicular growth and atresia continues in fetal life and in childhood. However, prior to puberty, follicular development is always incomplete, resulting in atresia. During follicular maturation, the oocyte accumulates cytoplasm and increases in diameter from 25–80 μm (Fig. 8.56). The surrounding granulosa cells undergo rapid mitotic division forming multiple cell layers while the theca cells organize from the surrounding ovarian stroma and matrix. As the follicle matures, there is approximately a 1000 fold increase in follicle diameter primarily due to accumulation of follicular fluid. This fluid compartment contains substances which regulate follicular development including estradiol, progesterone, androgens, LH, FSH, prolactin, inhibin, follicular aromatase-inhibiting protein, and oocyte maturation inhibitor.

Just prior to ovulation, the preovulatory follicle enlarges to 2.0–2.5 cm in diameter. Ovulation is triggered by a rapid release or surge of LH from the pituitary coincident with a smaller surge of FSH. This event precedes follicle rupture and oocyte release by 24–36 hours (Fig. 8.57A). The LH surge triggers an increase in the production of progester-

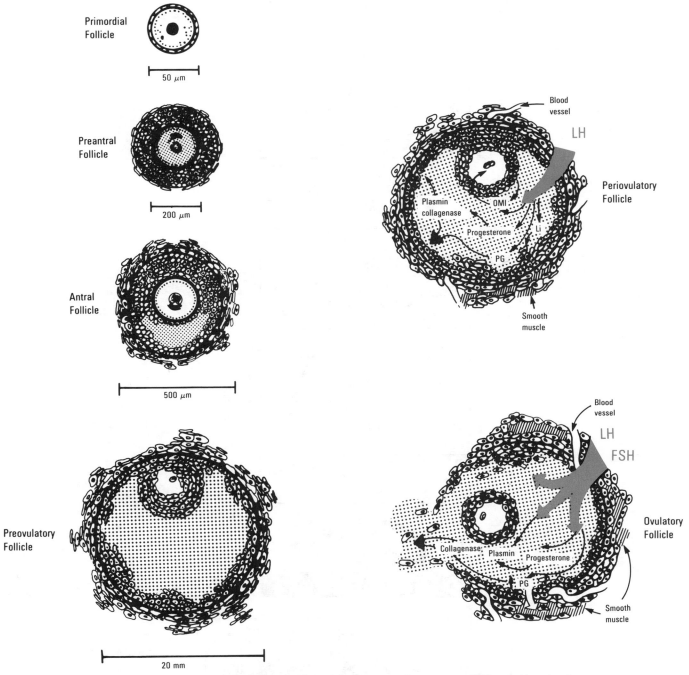

Figure 8.56. Diagrammatic illustration of the follicular maturation process. *OMI*, oocyte maturation inhibitor; *Li*, luteinization inhibitor; *PG*, prostaglandin.

one, prostaglandins, and proteolytic enzymes such as collagenase and plasmin, as well as resumption in meiosis (Fig. 8.57*B* and *C*). These enzymes and prostaglandins are responsible for increasing follicle wall distensibility causing formation of a "stigma," a conical elevation of the follicle wall where rupture of the follicle and release of the ovum is destined to occur (Fig. 8.57*E*) (Espey, 1974; Beers and Strickland, 1976).

Corpus Luteum

The process of luteinization and corpus luteum formation is triggered by the midcycle LH surge. After ovulation, hemorrhage occurs within the fol-

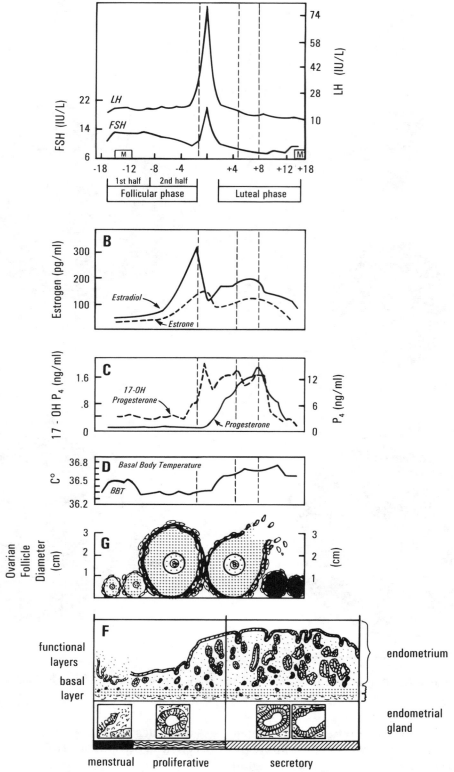

Figure 8.57. Composite illustration of the dynamic changes in gonadotropins, estrogen, progesterone, basal body temperature, ovarian follicle growth, and endometrial growth during a typical menstrual cycle.

licle remnant (Fig. 8.58). Repair and formation of the corpus luteum is initiated by capillary proliferation from the theca-stromal matrix through the basement membrane to supply the luteinized granulosa cells. The term corpus luteum is derived from the yellowish (lutein) appearance of these cells. Under LH stimulation, the corpus luteum is responsible for the production of progesterone, estradiol, and androgens during the postovulatory phase of the menstrual cycle (Mikhail, 1970). Steroidogenesis proceeds predominantly via the Δ^4 pathway

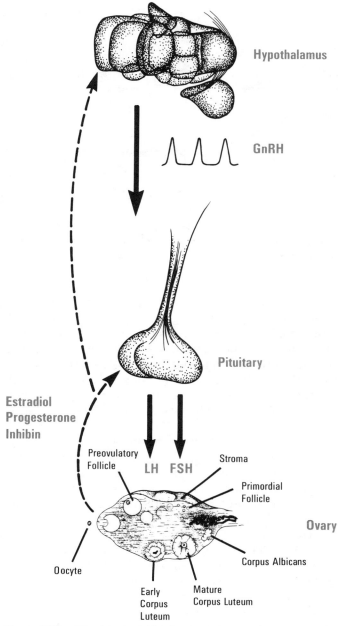

Figure 8.58. Stimulatory and feedback relationships between different compartments of the hypothalamic-pituitary-ovarian axis during the menstrual cycle.

(Fig. 8.59). If implantation of the embryo is successful, production of hCG by the embryo is sufficient to "rescue" and maintain corpus luteum function for approximately 8–10 weeks until placental steroidogenesis is established. In the absence of pregnancy and hormonal stimulation by hCG, corpus luteum function declines spontaneously 9–10 days after ovulation. In lower mammals, this luteolytic process is prostaglandin mediated.

UTERINE-ENDOMETRIAL COMPARTMENT

During early embryonic development of the internal genital tract, both the wolffian and müllerian ducts are present. In the male, the testes secrete müllerian-inhibiting factor which induces regression of the müllerian ducts. In the absence of müllerian-inhibiting factor, the müllerian duct system differentiates to form the uterus and the fallopian tubes.

Both the uterine cervix and the endometrium serve as target tissues for estrogen and progesterone. Under estrogen stimulation the glandular epithelium of the cervix secretes an abundant clear, stretchy, mucus. This cervical mucus can offer protection of sperm in the vagina, act as a biological filter for abnormal sperm, and serve as a reservoir for sperm release into the upper genital tract. Under the influence of progesterone following ovulation, mucus secretion decreases and becomes thick and whitish. These characteristic changes of cervical mucus can be utilized clinically to predict the "fertile period" and is the basis for the "natural" or "rhythm" methods of contraception.

Structurally, the endometrium is composed of superficial epithelium which lines the uterine cavity. The two principal components of this epithelium are the glandular epithelium and supporting stromal cells (Fig. 8.57*F*). During the menstrual cycle the epithelium differentiates to form three functional zones: the basalis, spongiosum, and stratum compactum. The beginning of each endometrial cycle is characterized by complete shedding of the spongiosum and stratum compactum layers during menstruation (day 1). By the second day of menses, the surface of the endometrium is reepithelialized through increased mitotic activity of both glands and stroma from the basalis layer. Through the stimulatory effects of estrogen, there is an increase in estrogen receptors, vascularity of the stroma, and in endometrial thickness from 0.5–5 mm. Following ovulation, increasing progesterone production antagonizes the growth-promoting effect of estrogen by decreasing estrogen receptors in the endometrium. Under the influence of progesterone, cylindrical vacuoles form within the base of the

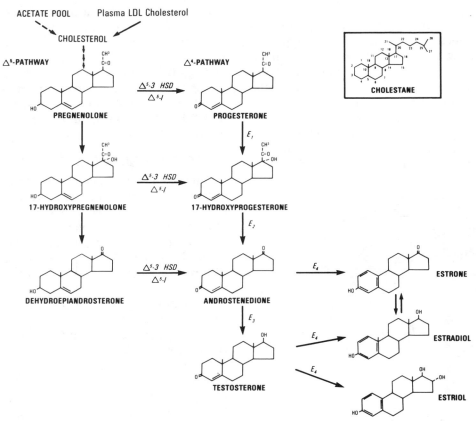

Figure 8.59. An overall summary of the ovarian steroidogenic pathways. Δ^5-3β HSD, Δ^5 3β-hydroxysteroid dehydrogenase; Δ^5-I, isomerase; E_1, 17 α-hydroxylase; E_2, 17-20 desmolase; E_3, 17 β-hydroxysteroid dehydrogenase; E_4, aromatase, *Inset:* 27-carbon skeletal structure for steroids.

glandular epithelium. These vacuoles gradually migrate toward the lumen of the glands such that by the day of expected implantation (day 21) there is active secretion into the glandular lumen. If implantation is successful, the corpus luteum production of progesterone is maintained by hCG while the endometrium undergoes decidua formation. Stromal cells become plump, polygonal, with a pavement-like arrangement. In the absence of hCG secretion, luteolysis and loss of progesterone support cause an increase in prostaglandin content in the endometrium. This is followed by prostaglandin-induced vasomotor spasms of the spiral arterioles that supply the endometrium, resulting in tissue hypoxia, fragmentation of the stromal-glandular matrix, and shedding of the endometrium (Markee, 1948; Rock et al., 1959).

NEUROENDOCRINE CONTROL OF THE MENSTRUAL CYCLE

From the foregoing discussion, it is apparent that both the ovary and endometrium undergo cyclic changes during the menstrual cycle and serve as the ultimate target organs under the control of pulsatile

gonadotropin secretion generated by the hypothalamic-pituitary unit (Fig. 8.58). The initiation of pulsatile LH and FSH secretion becomes most evident during the immediate peripubertal period when there is a marked increase in LH pulsatile secretion at night. After puberty, pulsatile LH secretion throughout the day becomes a constant feature of the reproductive adult female.

Negative Feedback

Like most endocrine systems, secretion of the stimulating hormones LH and FSH is regulated by a feedback mechanism. Increased estrogen secretion by the developing ovarian follicle inhibits gonadotropin secretion from the anterior pituitary and GnRH secretion from the hypothalamus (Ferin et al., 1984). The most important hormone regulating gonadotropin secretion during the follicular phase of the cycle is 17-β estradiol. Only small increases in estradiol levels are required to initiate negative-feedback action. When estradiol levels are very low, such as following castration, sustained increases in LH and FSH concentrations occur. After ovulation, the combination of increasing progesterone levels and estradiol synergistically suppress GnRH and gonadotropin secretion.

Inhibin, a nonsteroidal protein which selectively suppresses FSH secretion at the anterior pituitary level has been isolated and characterized by genetic cloning techniques (Mason et al., 1985). This protein is synthesized by the ovarian granulosa cells (Bicsak et al., 1986) and is composed of two dissimilar subunits, an α-subunit (MW 18,000) and a β-subunit (MW 14,000) cross-linked by disulfide bridges. Two forms of inhibin have been identified, inhibin A and inhibin B. Both forms share the same α-subunit but have differences in the peptide sequence of the β-subunit. There appears to be little difference in the biological activity between these two forms of inhibin. Preliminary studies have shown that the synthesis of inhibin is regulated, in part, by FSH. Measurements of circulating concentrations of inhibin during the menstrual cycle show that the highest concentrations occur during the late follicular and luteal phase when FSH secretion is suppressed.

Positive Feedback

A preovulatory surge of LH and FSH is required to induce ovulation. This midcycle gonadotropin surge is a consequence of rising estradiol concentrations produced by the developing follicle and is called positive feedback. The sustained increased estradiol production can be viewed as an "ovarian signal" to the hypothalamic-pituitary unit. In addition to estradiol, small increments of progesterone produced by the preovulatory follicle can also serve to amplify the magnitude and duration of the gonadotropin surge (Yen and Lein, 1976; Liu et al., 1983).

Regulation of Follicular Development

In addition to the negative and positive feedback control mechanisms, there are intraovarian processes which also regulate ovarian follicular development. At the beginning of each menstrual cycle, a pool or cohort of primordial follicles is stimulated to begin development. This process is termed *recruitment* (di Zerega and Hodgen, 1981). Animal experiments suggest that FSH and LH may be responsible for initiating the recruitment process in the fetus, but when the initial wave of recruitment begins, the ovaries have the ability to recruit primordial follicles independent of gonadotropins.

Because only a single ovum is normally released during ovulation in the human, the pool of developing primordial follicles must decrease by *selection* and *atresia* such that only one follicle proceeds to ovulation. This selection process is coupled with

ovarian steroidogenesis and requires interaction between LH, FSH, granulosa cells, and theca-interstitial cells. Figure 8.60 shows that initiation of steroidogenesis in the ovary requires LH stimulation of the theca-interstitial cell compartment. Through a cyclic AMP-mediated mechanism, low density lipoprotein (LDL) uptake through its own receptors is increased. This lipoprotein is then cleaved to yield cholesterol, the major precursor for steroidogenesis. Cholesterol is then sequentially processed to androstenedione, a weak androgen predominantly via the Δ^5 pathway (Fig. 8.59).

Under FSH stimulation, aromatase enzyme is induced in the granulosa cells. Androstenedione diffuses freely from the interstitial compartment through the basal lamina into the follicular compartment, enters the granulosa cell where it is further processed by aromatase to yield 17 β-estradiol (Ryan et al., 1968). Based on studies in the rat, estradiol and FSH act synergistically to increase the number of FSH receptors thereby further enhancing induction of aromatase enzyme activity and increasing estradiol production (Hsueh et al., 1984). Estradiol diffuses into the peripheral circulation to decrease FSH secretion.

In the presence of decreased FSH support, most of the recruited follicles are unable to sustain adequate levels of estrogen biosynthesis and undergo atresia. These atretric follicles contain low intrafollicular FSH concentrations, decreased aromatase enzyme activity, and low intrafollicular estradiol levels. Eventually only one follicle survives to achieve *dominance*. Thus, selection of the dominant follicle is dependent on the ability of the follicle unit to maintain estrogen biosynthesis in the presence of decreasing FSH support.

Hormonal Changes During the Menstrual Cycle

A summary of the hormonal changes during the menstrual cycle is shown in Fig. 8.57*A-C*. During the late luteal phase of the preceding cycle, LH and FSH levels begin to rise concurrent with declining progesterone and estradiol production from the degenerating corpus luteum. Rising FSH levels stimulate further development of the primary follicles, initiate estrogen production from the granulosa cells, and acquisition of LH receptors. By day 7 of the cycle, a dominant follicle has emerged and estrogen levels increase steadily. The production rates and serum concentrations of ovarian steroid hormones are shown in Table 8.10. Through negative feedback imposed by both estradiol and inhibin secretion, there is suppression of FSH secretion.

With acquisition of LH receptors on the granulosa

Figure 8.60.   Steroid biosynthesis in the ovary is compartmentalized between the theca and granulosa cell unit and is separately regulated by LH and FSH.

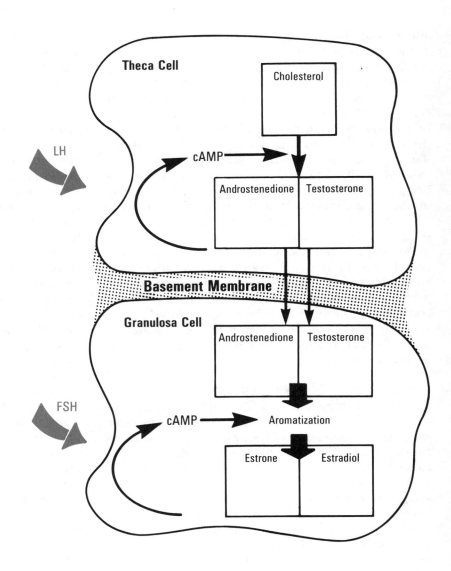

Table 8.10.
Production Rates and Serum Concentrations of Major Ovarian Steroid Hormones

Steroid	Menstrual Cycle Phase	Serum Concentration (ng/ml)	Production Rate (mg/day)
Estradiol	Follicular	0.037–0.315	0.081–0.945
	Midluteal	0.086–0.294	0.270
Estrone	Follicular	0.014–0.073	0.110–0.662
	Midluteal	0.029–0.133	0.243
Progesterone	Follicular	0.15–0.56	2.1
	Midluteal	5.0–27.0	25
Testosterone		0.469–0.55	0.26
Androstenedione		0.418–2.259	3.2

cells, the follicle is now primed to respond to the gonadotropin surge. After increasing exposure to rising levels of estrogen, the threshold for estrogen positive feedback is reached, resulting in a surge of LH and FSH lasting 24–48 hours. During the LH surge, luteinization of the granulosa cells initiates a rapid shift from estrogen production to progesterone production resulting in a dramatic decline in estradiol. The LH surge also activates proteolytic enzymes causing release of the ovum 25–36 hours later. The remaining granulosa and theca-interstitial cells reorganize to form the corpus luteum. The activity of FSH and LH on the follicle complex and corpus luteum is summarized in Table 8.11.

During the luteal phase, increasing production of progesterone and estrogen from the corpus luteum suppress LH and FSH levels through negative feedback. Progesterone also influences the hypothalamic thermoregulatory center to raise the thermal set point such that the basal body temperature (BBT) increases by approximately 0.5° C following ovulation (Fig. 8.57*D*). In the absence of fertilization and implantation, estradiol and progesterone levels begin to decrease by the 9th or 10th day following ovulation.

MECHANISM OF STEROID ACTION, TRANSPORT, AND METABOLISM

To exert physiological effects, steroids must first gain entry into the cell by diffusion. The steroid must then bind to a specific receptor protein in the cytoplasm to form an activated complex, be translocated into the nucleus, and bind to acceptor sites located on the DNA chromatin. This then stimulates transcription of new mRNA which directs protein synthesis. To regulate the biological action of of various steroids, the distribution of specific steroid receptors is limited to the target tissues. For example, estrogen receptors are concentrated in the uterus, endometrium, pituitary, and arcuate neu-

rons. In some target tissues such as endometrium, different steroid receptors must be activated in sequence. For example, estrogen must activate its own receptors in order to promote production of progesterone receptors.

The biological activity at target tissues is dependent not only on the circulating concentrations of the steroid, but also on the plasma concentration of the carrier protein (Partridge, 1981). Like many other steroids, the majority of circulating estradiol is bound to albumin and a specific carrier protein called sex hormone-binding globulin (SHBG, MW 80,000–90,000). Both proteins are synthesized by the liver (Rosner and Deakins, 1968). Similarly, the majority of circulating progesterone is bound to the specific carrier protein cortisol-binding globulin (CBG) and albumin. These carrier proteins provide a reservoir of hormones, protected from renal clearance, which release hormone to tissues based on mass action equilibrium.

The physiological action of sex steroids can be modified at the target tissue level. For example, in endometrium, estradiol is directly converted to a less potent estrogen, estrone, via the enzyme estradiol hydroxysteroid dehydrogenase while progesterone is metabolized to a less potent metabolite, pregnanediol. Other peripheral tissues may also be capable of metabolic conversion of steroids. Both fat and muscle tissues are able to convert androstenedione to estrone via the aromatase enzyme (Siiteri and MacDonald, 1973). In general, circulating steroids and their active metabolites are deactivated by conjugation with sulfate or glucuronic acid in the liver and intestinal mucosa and are then excreted into urine and bile.

PUBERTY

Puberty can be viewed as a transition period between childhood and adulthood. It is characterized by progressive sexual maturation and, with attainment of physical, mental, and emotional maturity. Pubertal changes are triggered in large part by the maturation of the hypothalamic-pituitary unit which stimulates increased production of ovarian steroids, inducing growth and maturation of the reproductive organs. On an endocrine basis, human puberty in the female is characterized by the establishment of monthly menstrual cycles.

Physical Changes During Puberty

Female pubertal development can be divided into four stages and occurs in the following sequence: (1) accelerated growth spurt; (2) thelarche or breast development (median age 9.8 years); (3) adrenarche

Table 8.11.
Summary of FSH and LH Action on the Follicle Unit and Corpus Luteum.

FSH	Initiation of follicular growth
	Induction of aromatase enzyme activity
	Induction of FSH and LH receptors
	Acts synergistically with estradiol to increase FSH receptors and granulosa cells
	Stimulation of inhibin production
LH	Initiates luteinization and progesterone production by granulosa cells
	LH surge stimulates completion of meiotic ooycte division
	Maintains corpus luteum progesterone production

or the appearance of pubic hair (median age 10.5 years); and (4) menarche or onset of menstrual cycles (median age 12.8 years). These physical changes generally take place over a span of 4.5 years with a range of 1.5–6 years. In males, pubertal maturation is delayed by 2 years.

Hormonal Changes During Puberty

Beginning as early as age 7 or 8 years, there is an increase in the secretion of circulating adrenal androgens such as dehydroepiandrosterone, dehydroepiandrosterone sulfate, and androstenedione. This process is called adrenarche. This increase in adrenocortical androgen production is dependent on the growth of the inner zone of the adrenal cortex, the zona reticularis (Grumbach et al., 1977).

From early infancy to the prepubertal period, gonadotropin secretion is suppressed to very low levels. One of the first signs of pubertal activation of ovarian function is the appearance of a sleep-entrained pulsatile release of GnRH-LH (Fig. 8.61; Boyar et al., 1972). The increase in GnRH secretion is responsible for reactivation of gonadotropin synthesis and secretion and is accompanied by a shift in gonadotropin secretion from a relatively apulsatile pattern to the normal adult pulsatile pattern. The timing for the increase in GnRH secretion during pubertal maturation is not known but is believed to be regulated by the central nervous system.

MENOPAUSE

As discussed previously, the pool of functional oocytes in the ovary is constantly decreasing such that by the 5th decade of life, very few ovarian follicles are capable of estrogen production. Of the remaining follicles, most are relatively refractory to gonadotropin stimulation, and menstrual cycles are characterized by inadequate estrogen production, failure of ovulation, deficient corpus luteum func-

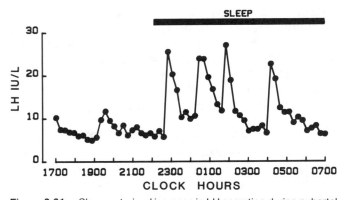

Figure 8.61. Sleep-entrained increase in LH secretion during pubertal sexual maturation.

tion, and irregular menstrual cycle intervals (Sherman et al., 1976) This physiological period during which regression of ovarian function occurs is termed the climacteric. Menopause can be arbitrarily defined as the final menstrual period. In the United States, the average age of menopause is 51.4 years (range 48–55 years). Because the average woman could live another 25–30 years, it is estimated that women will live one-third of their lives in the postmenopausal state.

Signs and Symptoms of the Menopause

Most of the signs and symptoms of the menopause reflect a deficiency in estrogen production. In general, over 80% of women will experience hot flushes. In a typical hot flush, the face becomes red and warm for a few seconds to 2 min, followed by cold chills and sweating. Each flush episode is accompanied by a drop in core body temperature, vasodilatation of head and neck, a rapid increase in heart rate, and a simultaneous release of LH (Casper et al., 1979). The flush mechanism is not dependent on LH release because following hypophysectomy (removal of the pituitary gland) in women hot flushes persist.

The decline in estrogen production also results in a decrease in vaginal secretions, atrophy and thinning of vaginal epithelium, and a gradual loss in bone density (osteoporosis). In addition, menopausal women may also manifest psychological symptoms such as increased irritability, tension, anxiety, and depression.

Hormonal Changes in the Menopause

In the absence of significant estradiol production from the ovary, the negative feedback action of estradiol decreases and circulating levels of LH and FSH increase to greater than 40 IU/liter, with the ratio of FSH to LH being greater than 1. Although there is little estradiol production by the ovary, the postmenopausal ovary is still capable of producing significant amounts of androgen, primarily androstenedione and testosterone (Judd et al., 1974). These androgens can be further metabolized to estrogens in peripheral tissue. This conversion accounts for most of the circulating estrogen in the postmenopausal women.

HORMONAL REGULATION OF FERTILITY

The most effective form of contraception, the birth control pill (BCP) has been commercially available since 1960. Three types of oral contraceptives are utilized in the United States: (1) a combination pill containing a fixed ratio of estrogen/progestin, (2) a biphasic or triphasic combination pill containing a variable ratio of estrogen/progestin, and (3) a continuous progestin "minipill."

Combined Birth Control Pill

The discovery that the addition of an ethinyl group at position 17 of estradiol made the compound ethinyl estradiol orally active was critical in the development of oral contraceptives. Ethinyl estradiol and its analogue, mestranol, are the principal estrogens utilized in combination birth control pills. Current formulations of combination BCPs generally contain 20–50 µg of ethinyl estradiol.

In 1951, Djerassi et al. discovered that elimination of the 19 carbon of ethisterone (a derivative of testosterone) enhanced its progestational activity. This discovery led to the synthesis of a new class of oral progestational compounds, the 19-nortestosterones. These progestins are utilized extensively in oral contraceptives with most BCP containing approximately 0.4–10 mg/day of norethindrone.

Combination BCPs are administered daily for 21 days beginning approximately 5 days after onset of menses. The estrogenic and progestational compo-

nents of the pill act synergistically to suppress gonadotropin secretion and ovulation at both the hypothalamic and pituitary levels. The estrogenic component also serves to stabilize the endometrium while the progestin component induces a decidual change in the endometrial glands which prevents successful embryo implantation. The pregnancy rates of oral contraceptives in actual use is approximately 0.44/100 woman-years.

Progestin Pill

The progestin pill is administered on a continuous basis and contains either norethindrone or norgesterel. Because this pill does not suppress gonadotropin secretion adequately, there is continued production of estrogen and ovulation in 40% of cycles. Pregnancy is, in part, prevented by the induction of atrophic endometrium and decreased cervical mucus. Pregnancy rates with this approach have been reported to be as high as 6.4/100 woman-years. A major side effect of this method is irregular vaginal bleeding due to the atrophic effects of continuous progestin stimulation of the endometrium.

CHAPTER 60

Fertilization, Pregnancy, and Lactation

Perpetuation of a species requires reproduction. In this chapter the processes of fertilization and implantation and the necessary changes which occur during pregnancy, at parturition, and in the puerperium are considered with special emphasis on the human.

FERTILIZATION

At the time of ovulation, the ovum is extruded and retrieved by the fimbria of the fallopian tube. The ovum, which had been arrested in development in the diplotene stage of prophase of the first meiotic division since early in fetal life, completes meiosis I just prior to ovulation. The layer of granulosa cells in the follicle surrounding the oocyte, termed the cumulus oophorus, is released with the oocyte and binds to the epithelial cells of the fallopian tube to aid in the initiation of tubal transport. Transport of the egg to the fertilization site, in the ampullary portion of the fallopian tube, appears to be a coordinated effect of the ciliary movement of the epithelium and muscular contractions of the myosalpinx. The second meiotic division is only completed if this ovum is fertilized by the spermatozoa. If sperm penetration is not completed within a 15- to 18-hour period after ovulation, the ovum degenerates.

During coitus, approximately 2–5 ml is ejaculated by men with each milliliter containing normally 20–200 million sperm. Fewer than 200 of the ejaculated sperm actually reach the ampulla of the fallopian tube, where fertilization can occur. Sperm have been noted in the tube within 5 minutes of ejaculation, but, on the average, it seems that 4–6 hours is the usual transit time from the vagina.

Movement of the spermatozoa is caused by flagellar action. The sperm must make their way through the cervix and uterotubal junction, which are thought to be natural barriers. Once in the tube, the sperm must move against the tubal cilia, which beat in the direction of the uterus.

When sperm are ejaculated into the vagina, they are not initially capable of fertilization. A further change, capacitation, occurs in the female reproductive tract to allow spermatozoa to be competent to fertilize an ovum. The exact process of capacitation is ill-defined, but at least two functional changes must occur: (1) the sperm must become hypermotile at the proper time; and (2) membrane changes, termed the acrosome reaction, must occur in sperm to permit union with the oocyte at fertilization.

The activated motility provides the optimal thrust that is required to penetrate the two outer layers surrounding the ovum, the cumulus oophorus and the zona pellucida. The acrosome reaction occurs when the limiting membrane of the sperm head, the acrosomal membrane, which has specific receptors for the attachment to the zona, becomes fenestrated in part due to influx of calcium and uses its proteolytic enzymes to navigate through the cumulus oophorus. By this point the outer acrosome membrane of the sperm has been lost, but the membrane of the midportion then fuses with the oolemma, which is the limiting barrier of the unfertilized egg. Soon after this union of the egg and sperm membrane, microvilli of the ovum encompass the sperm head. The sperm head is then engulfed into the cytoplasm of the egg. Once this occurs, the sperm head swells rapidly and forms the male pronucleus. The maternal genome is activated with passage of a second polar body and formation of a female pronucleus. As the cells enter mitosis, the nuclear membranes of both pronuclei break down. Therefore, at the first cleavage there is a symmetrical division of the fertilized egg, and the two daughter blastomeres that have formed now have fused nuclei containing both the maternal and paternal genome.

IMPLANTATION

The preimplantation period of development encompasses the time from conception to the time of implantation and is about 7 days in humans. The newly fertilized egg (Fig. 8.62) is similar in size to

PREIMPLANTATION
Days 5-7

IMPLANTATION
Days 7-8

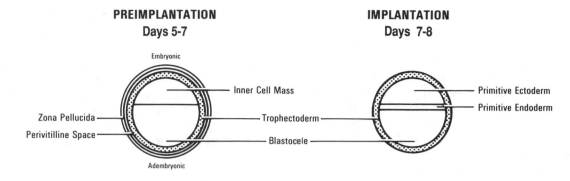

POSTIMPLANTATION

Days 10-11

Days 13-15

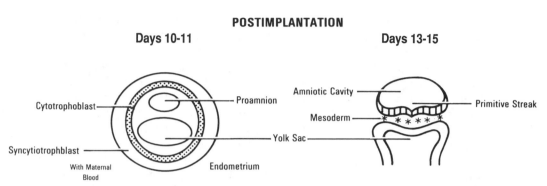

Figure 8.62. Diagrammatic representation of the early stages of mammalian development. The preimplantation stages shown are blastocysts with and without their zonae pellucidae. The embryonic end of the embryo contains the inner cell mass and will form the fetus. Trophectodermal cells are fated to form the extraembryonic tissues included in the placenta. The primitive endoderm forms at the time of implantation and will eventually form yolk sac structures. The primitive endoderm will form the definitive ectoderm, endoderm, and mesoderm following the stages shown in the lower half of the diagram (postim-plantation). The proamnion forms within the substance of the primitive ectoderm, and the trophoblast begins to differentiate into definitive placental structures (cytotrophoblast and syncytiotrophoblast). By day 13 the primitive ectoderm has formed a single layer of columnar cells and the craniocaudal groove (primitive streak) begins to form. Mesoderm differentiates from the primitive ectoderm at the point of the primitive streak in most primates. (Modified from P. M. Iannacone, in *Principles and Practice of Endocrinology,* edited by K. Becker et al. Philadelphia: JB Lippincott, 1990.)

the mature unfertilized egg. This embryo has two polar bodies that remain from the previous meiotic division and is 100 μm in diameter. The embryo remains surrounded by the zona pellucida, which is composed of complex glycoproteins. The zona is important to early development because (1) it responds instantaneously at the time of sperm penetration and renders the egg impervious to additional sperm (i.e., prevents polyspermy); (2) it provides a constraint to cleavage and ensures that as the blastomeres divide they remain together and in proper orientation; and (3) it helps prevent the sticky embryo from adhering to the wall of the oviduct as it makes its way to uterus. The early embryo divides by mitosis to form equal cells, also known as blastomeres.

The embryo at the late 8-cell stage undergoes a process known as compaction after which each individual blastomere is less prominent. Compaction is associated with polar migration of microvilli on the surface of the outside blastomeres.

At the 16-cell stage of development, the embryo has sufficient cells to have an inside and outside. The cells on the inside form the inner cell mass which ultimately becomes the fetus, while the cells on the outside (at this stage called the morula) will form the trophoblast and will go on to develop the extraembryonic tissues, which include the placenta and membranes.

The embryo, in addition to dividing, traverses the fallopian tube as the result of the combination of directed ciliary motion and muscular contraction. As the conceptus makes its way into the uterine cavity, it develops from the 2-cell stage into a morula which consists of up to 60 cells. After 6 days the morula enters the uterus and is transformed into a blastocyst with a large fluid-filled cavity (blastocele). During this early stage, the embryo is still surrounded by the zona pellucida. Seven days after ovulation, the zona is dissolved by enzymatic digestion and the blastocyst is now ready for implantation.

Implantation begins with the attachment of the blastocyst into the uterine tissue. The embryo typically attaches to the lining of the fundus of the uterus. The endometrial cells of the uterus have microvilli on their luminal surface, and these begin to combine with the microvilli of the trophoblast cells. The endocrine signals to implantation are not yet understood. It is clear that even the 4- to 8-cell embryo secretes human chorionic gonadotropin (Fishel et al., 1984), but what role, if any, this hormone plays in implantation is unknown.

Development following implantation is rapid and complex. The embryo quickly establishes both the placental component and its fetal structures. The trophoblast differentiates into the syncytiotrophoblast and cytotrophoblast. Lacunar spaces form within the syncytiotrophoblast, which eventually becomes continuous with maternal capillary circulation and into chorionic villi. The placenta is hemochorial (i.e., intimate contact with maternal bloodstream) and is noted to contain three fetal tissues (fetal endothelium, fetal connective tissue, and chorionic epithelium), which are bathed in maternal blood.

Immediately following implantation, two layers of cells appear at the blastocele margin on the inner cell mass side. The first layer of cells is called the embryonic endoderm. The endoderm proliferates rapidly and eventually surrounds the blastocele. The remaining inner cell mass is now called the embryonic (or primitive) ectoderm. These cells are arranged in a columnar fashion. By the 8th day of development, small spaces appear between the inner cell mass and the invading trophoblast, eventually coalescing to form the amniotic cavity.

By 13 days, three germ layers are present. The primitive ectoderm gives rise to the definitive ectoderm, endoderm, and mesoderm. The mesoderm is noted between the endoderm and ectoderm. The endoderm gives rise to a number of extraembryonic tissues (Fig. 8.63). The first indication of a craniocaudal axis and bilateral symmetry in the embryo occurs when a longitudinal depression in the columnar primitive ectoderm appears. This depression is called the primitive streak and seems to be the site of origin of mesodermal cells.

Early postimplantation stages are responsible for establishing the structures which will ultimately allow organogenesis to proceed. If there is failure of embryonic and fetal development, spontaneous abortion will usually occur. Most commonly abortion occurs because of gross chromosomal abnormalities.

ENDOCRINE FUNCTIONS OF THE PLACENTA

The maintenance of pregnancy requires the complex interaction of the fetus, placenta, and mother, together forming the so-called fetoplacental-maternal unit. It is now clear that the placenta itself is a complex endocrine organ with the capacity of producing neuropeptides similar or identical to those synthesized by the hypothalamus, pituitary-like peptide hormones, and a number of steroid hormones (Table 8.12).

Placental Steroid Hormone Formation, Metabolism, and Function

With regard to steroid production, the placenta has been termed an incomplete endocrine organ because it is unable to synthesize steroids de novo from cholesterol but is able to transform steroid precursors into other biologically active steroids. Given that other steroid-producing organs almost invariably begin synthesis with cholesterol from the circulation rather than with two carbon fragments, this distinction for the placenta is probably of little importance.

The biosynthetic pathways for the synthesis of steroids (progestogens, androgens, and estrogens) do not change during pregnancy. However, the anatomic sites of steroid synthesis are altered by the fact that both the placenta and the fetus, as well as the mother, are capable of steroidogenesis.

In humans the maternal ovary is required for pregnancy maintenance during the first 7–8 weeks of pregnancy dating from the last menstrual period (or during the first 5–6 weeks dating from ovulation and fertilization). After that time, the placenta can maintain pregnancy. The need for the maternal ovaries and the ability of the placenta to synthesize and secrete the steroid hormones necessary for the maintenance of pregnancy, however, varies among species.

The placenta never becomes an autonomous endocrine organ. The steroids secreted by the placenta are derived from precursors of both maternal and fetal origin. Relative enzyme deficiencies in the fetus and placenta determine the sites at which various steps in steroidogenesis take place.

ESTROGENS

The major estrogen formed in pregnancy is estriol (E_3) (Fig. 8.64A). Estriol is not secreted by the ovary of nonpregnant women, yet it comprises over 90% of the known estrogen excreted in the urine of pregnant women, being excreted primarily as sulfate and glucuronide conjugates. Estriol is not the only estrogen secreted by the placenta (Fig. 8.64B).

A

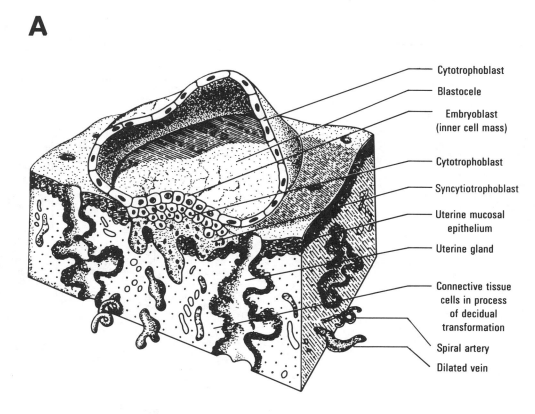

Cytotrophoblast

Blastocele

Embryoblast
(inner cell mass)

Cytotrophoblast

Syncytiotrophoblast

Uterine mucosal
epithelium

Uterine gland

Connective tissue
cells in process
of decidual
transformation

Spiral artery

Dilated vein

B

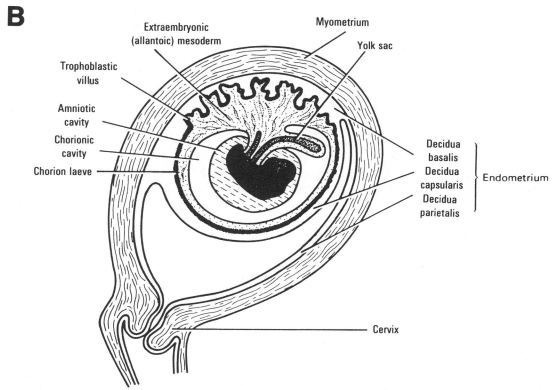

Extraembryonic
(allantoic) mesoderm

Myometrium

Yolk sac

Trophoblastic
villus

Amniotic
cavity

Chorionic
cavity

Chorion laeve

Decidua
basalis

Decidua
capsularis

Decidua
parietalis

Endometrium

Cervix

Figure 8.63. *A*, Diagram of the human implantation site. Trophoblast invasion of uterine epithelium at the time of attachment. (Modified from H. Tuchmann-Duplessis et al., 1972). *B*, Implanted embryo. (Adapted from Begley et al., 1980.)

Table 8.12.
Hormones Synthesized by the Placenta

Steroids
    Estriol
    Estradiol 17-β
    Estrone
    Progesterone
    Pregnenolone

Peptides (pituitary-like)
    Human chorionic gonadotropin (hCG)
    Human placental somatomammotropin (also known as human
        placental lactogen)
    β-glycoprotein, (β-lipotropic hormone(β-LPH),β-endor-
        phin,alpha-Melanocyte stimulating hormone (alpha-MSH)
    (?) Human chorionic thyrotropin (hCT)
    Human chorionic corticotropin (hCC)
    Growth hormone (GH)

Hypothalamic-like—produced in the cytotrophoblast
    Corticotropin-releasing hormone (CRH)
    Thyrotropin-releasing hormone (TRH)
    Gonadotropin-releasing hormone (GnRH)
    Somatostatin
    (?) Dopamine

Other Peptides
    Epidermal growth factor (EGF)
    Inhibin
    (?) Activin
    (?) Follistatin
    Relaxin
    (?) Gastrin
    (?) VIP

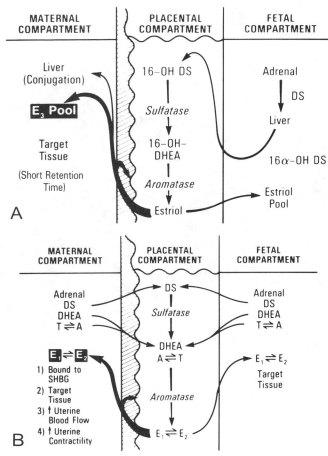

Figure 8.64. *A*, Formation of estriol (E_3) by the placenta is dependent initially on fetal adrenal secretion of dehydroepiandrosterone sulfate (*DS*). The *DS* is hydroxylated in the 16α position by the fetal liver before being transported to the placenta. Most E_3 is transported to the maternal circulation. *B*, Estrone (E_1) and estradiol (E_2) are also formed in large quantities by the placenta. Androgenic precursors include *DS*, dehydroepiandrosterone (*DHEA*), testosterone (*T*), or androstenedione (*A*) of maternal or fetal origin. (Courtesy of S. S. C. Yen, University of California, San Diego).

In general, concentrations of all estrogens in the maternal circulation increase with advancing gestation (Fig. 8.65). In contrast to the other common estrogens, estrone (E_1) and estradiol (E_2), which also increase markedly during pregnancy, E_3 has a very low affinity for sex hormone-binding globulin (SHBG) and thus a very rapid rate of clearance from maternal plasma.

To form estrogens, the placenta, which has active aromatizing capacity for the A ring of steroids, utilizes circulating androgens, primarily from the fetus, but also from the mother. The major androgenic precursor to placental estrogen formation is dehydroepiandrosterone sulfate (DHEAS), the major androgen produced by the fetal adrenal cortex. DHEAS is transported to the placenta and cleaved by the sulfatase (sulfate-cleaving) enzyme, which the normal placenta has in abundance, to form free (unconjugated) DHEA. The DHEA is converted to androstenedione (A) and testosterone (T), and then aromatized to estrone and estradiol.

Very little E_1 and E_2 are converted to estriol by the placenta. Rather most DHEAS undergoes 16α-hydroxylation within the fetus, primarily in the liver, although in part in the fetal adrenal. The resulting 16α-hydroxy-DHEAS then is cleaved of its sulfate by sulfatase in the placenta to form free

16OH-DHEA, which is aromatized to estriol. Estriol is then secreted into the maternal circulation and conjugated in the maternal liver to form estriol sulfate, estriol glucosiduronate, and mixed conjugates and excreted via the maternal urine.

Hydroxylation at the C_2 position of the phenolic A ring results in the formation of catecholestrogens (2-hydroxy-estrone, -estradiol, or -estriol) and is a major process in estrogen formation during pregnancy. Although the physiological significance of catecholestrogens are unknown, these substances can alter catecholamine synthesis and metabolism. They are potent inhibitors of catecholamine inactivation by competing for carboxyl-O-methyl transferase (COMT) and reduce catecholamine synthesis by inhibiting tyrosine hydrolase. Catecholestrogens also function as antiestrogens, having no intrinsic

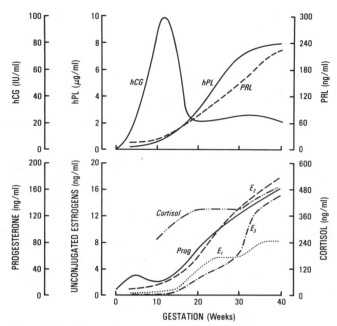

Figure 8.65. Serum concentrations of *PRL, hCG, hPL, cortisol, progesterone,* and unconjugated estrogens during pregnancy. The values have been obtained from several sources in the literature. (Reproduced with permission from Rebar, 1984.)

estrogenic activity, but competing with estrogens for their receptors.

PROGESTOGENS

The major progestogen progesterone is formed in the placenta mainly from circulating maternal cholesterol (Fig. 8.66). Most progesterone is ultimately metabolized to pregnanediol and excreted in the maternal urine. Before the 7th–8th week of pregnancy, progesterone produced by the corpus luteum maintains the pregnancy. Ovariectomy early in pregnancy will lead to abortion (Csapo et al., 1973).

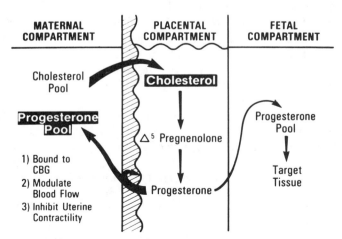

Figure 8.66. Formation of progesterone by the placenta is derived largely from maternal cholesterol. (Courtesy of S. S. C. Yen, University of California, San Diego.)

What is termed the luteal-placental shift then occurs, with virtually all progesterone being produced subsequently by the placenta. By the end of pregnancy, the placental production of progesterone approximates 250 mg/day with circulating levels of about 130 ng/ml. By comparison follicular phase production is about 2.5 mg/day and in the luteal phase is about 25 mg/day.

During the first 6 weeks of pregnancy, 17α-hydroxyprogesterone (17 OH-P) is also produced by the corpus luteum. Levels in the maternal circulation are increased, paralleling those of progesterone. After 6 weeks, levels of 17 OH-P decline progressively, becoming undetectable during the middle third of pregnancy, while progesterone exhibits only a transient fall at about the 9th week followed by a progressive increase thereafter (Fig. 8.65). The decrease in 17 OH-P and the dip in maternal progesterone concentrations reflects the transition of progesterone secretion from the corpus luteum to the placenta.

FUNCTIONS OF ESTROGENS AND PROGESTINS IN PREGNANCY

Estrogens and progestins are essential for pregnancy, but not all of their functions are known. These steroids must induce the secretory endometrium necessary for implantation. Once a pregnancy is established, the ratio of estrogen to progestin appears important for maintaining uterine quiescence until term. Together, these two steroids modualte uterine blood flow. In this regard estriol is a very weak estrogen except in its ability to increase uteroplacental blood flow, in which it appears to be as effective as other estrogens (Resnik, 1981). It has been suggested that estriol is thus ideal to increase blood flow to the fetus with a minimum of other estrogenic effects. Progesterone can inhibit estrogen-induced increases in uterine blood flow. Estrogens also stimulate liver enzymes to increase binding proteins such as those for sex steroids (sex hormone-binding globulin, SHBG), corticoids (corticoid-binding globulin, CBG), and thyroid hormones (thyroxine-binding globulin, TBG) in pregnancy. These steroids also participate in the complex neuroendocrine events involved in the initiation of parturition.

ANDROGENS

Several androgens increase in the maternal circulation during pregnancy as well. In large measure, these increases reflect fetal production with modification by the placenta. Of note is the observation that concentrations of androgens both in the amniotic fluid and in the maternal circulation are increased in pregnancies in which the fetus is male compared to those which are female. Because of the vast capacity of the placenta to aromatize androgens to estrogens, only markedly increased concentrations of naturally secreted androgens in the mother have the potential for masculinizing a female fetus. Because synthetic androgens cannot be aromatized, these agents do have the potential for such masculinization if ingested during pregnancy.

Placental Peptide Hormones

Several peptide hormones are synthesized de novo by the placenta. These hormones are similar but not identical in their chemical, biological, and immunological properties to the trophic hormones of pituitary origin. In general, these pregnancy-specific hormones assume important roles in the regulation of endocrine and metabolic processes in the maternal compartment to meet the unique requirements of the fetus.

HUMAN CHORIONIC GONADOTROPIN

Human chorionic gonadotropin (hCG) was the first placental protein identifed and was first detected in the urine of pregnant women by Ascheim and Zondek in 1927. Like the structurally similar glycoprotein hormones of the pituitary (i.e., LH, FSH, and TSH), hCG is composed of an α and a β subunit. Because it contains more carbohydrate, the half-life of hCG is longer than that of either LH or FSH, with the first component approximating 6–8 hours. Measurement of the specific β subunit of hCG is useful clinically to diagnose pregnancy and to monitor trophoblastic disease.

hCG is elaborated by all types of trophoblastic tissue, including that from hydatidiform moles, chorioadenoma destruens, and choriocarcinoma. It is also produced by choriocarcinoma of the testes and ovaries, numerous other neoplasms, and apparently in very small quantities by normal nonpregnant individuals, both male and female. For several years it has been debated as to whether hCG is produced by the syncytiotrophoblast or cytotrophoblast of the placenta. Recent studies indicate that messenger RNA for the subunits of hCG is produced by both.

It is now clear that hCG is produced by the blastocyst even before implantation (Fishel et al., 1984). It is generally detectable by radioimmunoassay 9 days after the midcycle LH peak, or 8 days after ovulation, and about 1 day after implantation. hCG rises to peak levels by 60–90 days of gestation and then decreases to a plateau maintained through the remainder of pregnancy (Fig. 8.65). In contrast, maternal LH and FSH levels are virtually undetectable throughout pregnancy.

The functions of hCG are incompletely understood. It is clear that hCG has a luteotrophic effect on the corpus luteum, stimulating its continued production of progesterone and estrogen until the placenta begins producing progesterone at 8–9 weeks of gestation. Evidence also has accumulated to suggest that hCG may regulate steroid production in the fetus early in pregnancy. It may stimulate secretion of both DHEAS by the fetal zone of the adrenal gland and testosterone by the testes (Seròn-Ferre et al., 1978). Stimulation of testicular testosterone secretion may be important in differentiation of male genitalia. Lastly, hCG has intrinsic thyroid-stimulating activity (Nisula et al., 1974), largely because of its structural similarity to TSH, and appears to be the only thyrotropic hormone produced by the placenta (Pekonen et al., 1988).

HUMAN CHORIONIC SOMATOMAMMOTROPIN

Human chorionic somatomammotropin (hCS), also known as human placental lactogen (hPL) and human chorionic growth hormone-prolactin (hCGP), is biologically and immunologically similar to both pituitary growth hormone and prolactin. It is a single chain polypeptide of 191 amino acids with two disulfide bridges and with a 96% homology with growth hormone (GH). It is found in maternal serum and urine in both normal and molar pregnancies and has also been found in the urine of men with choriocarcinoma of the testis. HCS rapidly disappears from the maternal circulation after removal of the placenta and has an initial half-life of 9–15 min.

The concentration of hCS in maternal serum increases through pregnancy, with peak levels achieved during the last 4 weeks of gestation (Fig. 8.65). Low levels of hCS (7–9 ng/ml) are present by 20–40 days of gestation. By delivery maternal levels of about 8 μg/ml are attained.

Although the somatotropic activity of hCS is ≤3% that of human pituitary GH, both animal and human studies indicate hCS has definite GH-like activity. In contrast, although hCS has potent mammotropic and lactogenic activity similar to that of GH in animal bioassays, no prolactin-like activity has been demonstrated in humans. Furthermore, hCS has no luteotropic activity in humans.

The effects of hCS are quite similar to those of growth hormone: (1) It mobilizes free fatty acids from depot fat. (2) It inhibits the peripheral actions of insulin on glucose uptake. (3) It stimulates insulin release because of this relative insulin "resistance." (4) It promotes nitrogen retention. Although a number of factors known to alter pituitary GH secretion are ineffective in altering maternal hCS concentrations, prolonged fasting at midgestation and insulin-induced hypoglycemia raise and intraamniotic instillation of prostaglandin $F_2\alpha$ reduces markedly hCS levels. The relatively autonomous secretion of this hormone allows hCS to exert its major metabolic effect on the mother to ensure the nutritional demands of the fetus, thus functioning as the "growth hormone" of pregnancy.

During pregnancy, blood glucose levels are decreased, plasma free fatty acids are increased, and insulin secretion is increased with resistance to endogenous insulin as a consequence of the GH-like and contrainsulin effects of hCS. Transplacental passage of free fatty acids is slow, but glucose crosses the placenta freely, and amino acids are actively transported to the fetus against a concentration gradient, while the placenta is impermeable to insulin and other protein hormones. The end result of these processes is to conserve maternal glucose and gluconeogenic precursors and allow for utilization of fat in the mother. The enhanced catabolism, particularly lipolysis, resulting from increased hCS thus constitutes an unique system of sparing nonlipid fuel for the fetus whenever food deprivation (i.e., fasting) occurs in the mother. In contrast, during feasting, increased insulin secretion effectively nullifies the effects of hCS (Felig et al., 1972; Spellacy et al., 1971).

OTHER "PREGNANCY-SPECIFIC" PEPTIDES

A series of other glycoproteins synthesized by the placenta have been identified, but their significance remains to be determined. The best characterized is "pregnancy-specific" β_1-glycoprotein (SP$_1$). It appears to be synthesized by syncytiotrophoblast and has a molecular weight of approximately 90,000 (Kaminska et al., 1979). Its concentration in maternal serum increases through pregnancy. It also has been found in low concentrations in the blood of healthy men and women and in higher levels in blood from individuals with nontrophoblastic malignant neoplasms. Its measurement both as a tumor marker and as an indicator of placental function has been suggested.

Relaxin has now been purified from human placenta and is also synthesized and secreted by the corpus luteum of pregnancy (Weiss et al., 1978; Fields and Larkin, 1981). Its possible role in preventing uterine contractions and in inducing cervical "ripening" are now under investigation.

It is also becoming clear that the placenta synthesizes a number of peptide growth factors such as epidermal growth factor (EGF) (Mauro and Mochi-

zuki, 1987) as well as the "ovarian" peptides inhibin (Petraglia et al., 1987), activin, and follistatin (Bicsak et al., 1987; Mason et al., 1985). The functions of these peptides, too, remains undetermined.

NEUROPEPTIDES IN THE PLACENTA

A number of peptides which are similar if not identical to hypothalamic inhibiting and releasing peptides are synthesized by the placenta. These include gonadotropin-releasing hormone (GnRH) (Khodr and Siler-Khodr, 1978), thyrotropin-releasing hormone (TRH) (Gibbons et al., 1975), corticotropin-releasing hormone (CRH) (Sasaki et al., 1988; Grino et al., 1987), adrenocorticotrophic hormone (ACTH) (Rees et al., 1975), somatostatin (Siler-Khodr and Khodr, 1981), and perhaps dopamine. The presence of such peptides suggests that the fetoplacental unit may function as a "complete" hypothalamic-pituitary-target organ unit.

While the release of anterior pituitary hormones is under the influence of hypothalamic neuropeptides, the mechanisms controlling release of hCG and hCS by the placenta have not been determined. That dopamine may be involved in hCS secretion (as it is in that of pituitary prolactin) is now suggested by studies demonstrating that administration of dopamine antagonists results in increased release of hCS (Macaron et al., 1978; Belleville et al., 1978).

It has also been suggested that GnRH (or a related peptide) is involved in regulation of hCG and that somatostatin is involved in inhibition of hCS release. Such a theory is predicated on the belief that these neuropeptides are synthesized primarily in cytotrophoblast. Because cytotrophoblast is most prominent in the first trimester, GnRH and somatostatin secretion are also greatest during this period, accounting for the peak in hCG secretion and reduced hCS early in pregnancy. Inhibition of hCS and stimulation of hCG secretion then decrease through the remainder of pregnancy.

Several reports have documented the presence of the pituitary intermediate lobe-like peptides β-lipotropin (β-LPH), ACTH, and β-endorphin in human placenta (Odagiri et al., 1979). The possible roles of these peptides in the maintenance or termination of human pregnancy remains to be determined. It is known that β-endorphin levels rise in maternal serum throughout pregnancy. At term, higher levels of β-endorphin in fetal as opposed to maternal blood are present, and amniotic fluid levels are increased in stressful situations (Wardlaw et al., 1979). It is possible that endogenous opiates contribute to a maternal sense of well-being during pregnancy (Ginstler, 1980) and also serve to calm the fetus in the amniotic fluid.

MATERNAL PITUITARY GLAND IN PREGNANCY

The maternal pituitary gland is not necessary for successful pregnancy and labor. With appropriate stimulation to permit ovulation, hypophysectomized women can conceive and carry infants normally to term.

Of the anterior pituitary hormones, LH, FSH, and GH are secreted in greatly reduced quantities during pregnancy. Moreover, exogenous GnRH cannot elicit release of maternal gonadotropins by 5 weeks after the last menstrual period (Reyes et al., 1976). Basal GH levels are decreased throughout pregnancy and GH responses to provocative stimuli such as arginine and hypoglycemia decrease in direct relation to the duration of pregnancy (Spellacy et al., 1970).

Maternal TSH secretion during pregnancy is largely unchanged. Total thyroxine (T_4) and triiodothyronine (T_3) in the circulation are increased because thyroxine-binding globulin (TBG) is increased during pregnancy. However, both free T_4 and free T_3 remain in the normal nonpregnant range (Furth, 1983).

It appears that maternal ACTH is increased during pregnancy (Carr et al., 1981). The concentrations of total and free plasma and urinary-free cortisol are also increased during pregnancy. Many of these changes are due to increased levels of CBG, the binding protein for cortisol, during pregnancy (Doe et al., 1964) and the free fraction of cortisol within the circulation is unchanged. It is presumed that an adjustment in the set point for ACTH occurs during pregnancy, because ACTH levels are increased two to three fold in most studies. Of note is the observation that exogenous estrogens mimic the changes on the maternal ACTH-cortisol axis observed during pregnancy.

Maternal prolactin secretion is also increased during pregnancy. However, secretion of hypothalamic-neurohypophysial oxytocin remains unchanged, at least until just prior to the onset of parturition.

PROLACTIN PHYSIOLOGY IN PREGNANCY

Maternal Prolactin Secretion

Serum prolactin (PRL) levels begin to rise in the first trimester of human pregnancy and increase progressively in a linear fashion to approximately 10 times the levels found in nonpregnant women at term (Fig. 8.65) (Rigg et al., 1977; Jaffe et al., 1973). Increased PRL secretion appears causally related to marked estrogen stimulation (Yen et al., 1974) and reflects the hyperplasia and hypertrophy of the PRL-secreting cells of the pituitary. The pituitary gland typically doubles in size during pregnancy,

largely because of the increased numbers and activity of the lactotropes. Just as in the nonpregnant state, PRL secretion is pulsatile during pregnancy with greater release during sleep also persisting (Boyar et al., 1975). In contrast to nonpregnant women, maternal PRL levels in late pregnancy do not appear to be influenced by surgical stress or anesthesia (Quigley et al., 1982). Maternal pituitary prolactin is important in promoting mammary growth (i.e., mammogenic), in initiating milk secretion after delivery (i.e., lactogenic), and in maintaining established milk secretion (i.e., galactopoietic).

Fetal Prolactin Secretion

The pituitary gland of the human fetus is able to synthesize, store, and secrete PRL in early gestation, with an accelerated increase occurring during the last several weeks of pregnancy (Aubert et al., 1975). At term, the mean umbilical vein concentration is higher than mean maternal plasma levels of PRL. In normal newborns, PRL levels then decrease to the low levels present in normal children by the end of the first week of postnatal life. The increase in fetal PRL secretion is probably due to direct estrogenic stimulation of the pituitary cells secreting prolactin in the fetus. Although the functional role of fetal PRL is unknown, animal studies suggest that PRL in the fetal compartment may be important in inducing fetal lung maturation (Hamosh and Hamosh, 1977). Fetal PRL may also play a role in osmotic regulation by the fetal kidney.

Prolactin Levels in Amniotic Fluid

PRL levels in amniotic fluid are 5 to 10 times as high as in maternal plasma (Schenker et al., 1975). The high concentrations in amniotic fluid are present even during the first trimester, when both fetal and maternal serum concentrations are relatively low. Decidua and fetal membranes, but not placenta, can synthesize PRL which is immunologically and chemically identical to PRL of pituitary origin (Riddick and Kusmik, 1977; Bigazzi et al., 1979). The function of PRL in the amniotic fluid is unknown, but it may be important in regulating the osmotic exchange of amniotic fluid (Josimovich et al., 1977). It may also be important in effecting maturation of the fetal lung and may help modulate uterine contractility (Bigazzi and Nardi, 1981).

LACTATION AND LACTOGENESIS

Development of the Human Breast

EMBRYONIC AND PREPUBERTAL DEVELOPMENT

The mammary gland is derived from ectoderm and is first apparent in the embryo 4 mm in length as a mammary band (Raynaud, 1969). By the time the embryo is 7 mm in length, the mammary band and surrounding tissue have thickened to form a ridge (known as the mammary crest or milk line) extending along the ventrolateral body wall from the axillary to the inguinal region on each side. The epithelium along the caudal portions of this ridge regresses and the crest in the thoracic region further thickens to form a primordial mammary bud by the time the embryo is 10–12 mm in length. After the 5th month of intrauterine life, the primitive mammary bud begins to form 15–25 secondary buds which will provide the basis for the ductal system in the mature gland. By the time of parturition, these epithelial ducts have undergone limited proliferation, presumably due to stimulation by a number of hormones present in the fetal circulation in the last trimester of pregnancy. At birth the mammary structures are still rudimentary, but in some newborns some secretion (i.e., "witch's milk") may be noted within the first few days after birth, no doubt as a result of the high PRL levels present in the newborn (Friesen, 1973).

Subsequent to birth the breast enters a quiescent stage. After a brief period of growth during the last several days of intrauterine life, the gland regresses and then grows proportionately with the remainder of the body until puberty. Some enlargement of the ducts does occur, but there is no lobuloalveolar development (Turner, 1952).

PUBERTAL AND POSTPUBERTAL DEVELOPMENT

With the onset of puberty in the girl, further growth of the breasts occurs and the areolae enlarge and become more pigmented. The growth involves an increase in connective tissue, adipose tissue, and vascular channels, but proliferation of the epithelial ducts and lobuloalveolar development are most prominent. Most of the molding of the breast, however, is due to accumulation of adipose tissue. Numerous studies have demonstrated that these changes are related to the increased secretion of estrogen and progesterone by the ovary (Folley, 1948; Meites, 1965; Huseby and Thomas, 1954; Mayer and Klein, 1961).

Considerable variation in the development of the mammary gland exists in different species and is dependent on the characteristics of the particular estrous or menstrual cycle (Anderson, 1974). In species with short estrous cycles, such as the rat and the mouse, in which the follicular phase is predominant and the luteal phase very brief, only the duct system is developed, with complete development occurring during pregnancy. In species with long luteal phases, such as primates, development during each menstrual cycle is considerable and is similar to the changes occurring during pregnancy.

BREAST CHANGES DURING THE MENSTRUAL CYCLE

Cyclic changes, with proliferation and regression of ductal breast tissue, occur during each menstrual cycle (Aragona and Friesen, 1979). The proliferative changes reach a maximum late in each luteal phase. Histological examination of normal breast tissue reveals that after puberty mammary development consists of a slow, cyclic, progressive increase in glandular tissue with further division of the epithelial ducts to form rudimentary lobules and regression with each menses. No further growth occurs in nonpregnant patients and true alveoli are generally not present until the 3rd month of pregnancy.

BREAST CHANGES DURING PREGNANCY AND POSTPARTUM

At the beginning of pregnancy, there is rapid growth and branching from terminal portions of the gland associated with some loss of interstitial adipose tissue (Cowie and Tindall, 1971). Visible enlargement of the breasts is first noticeable after the 2nd month. The nipples also increase in size and the areolae become larger and more pigmented.

The greatest portion of the increase in the ductal system occurs during the first two trimesters. True glandular acini, necessary for milk secretion, appear early in the 3rd month. During the last trimester the changes involve the differentiation of the acini leading toward their secretory activity. In the final month of pregnancy, the enlargement of the breast results largely from hypertrophy of parenchymal cells of the alveoli with a hyaline, eosinophilic, proteinaceous secretion termed colostrum. Approximately 3 days postpartum, the fat content of this secretion increases abruptly, and it becomes typical milk.

Grossly, the functional anatomy of the lactating breast is apparent upon inspection. The nipple projects as much as 2 cm beyond the areola in response to tactile or psychological stimuli because it is supplied with smooth muscle fibers. Beneath the areola are 15–20 sinuses that become engorged with milk when pressure is applied to the peripheral lactating glandular tissues. Milk is expressed by a relative compression of the subareolar sinuses by the lips and gums of the infant while suckling. Many ductules connect the peripheral alveoli with the subareolar sinuses. These ductules have smooth muscle in their walls, and myofibrils also surround the individual alveoli. Surrounding the alveoli, which are lined with lactating cells, is a rich capillary bed and lymphatic supply. In fact, blood flow through the lactating mammary gland is 400–500 times the volume of milk secreted, with the rate of blood flow influencing the volume (Linzell, 1974).

HORMONAL EFFECTS ON MAMMARY DEVELOPMENT

Growth and appropriate functioning of the mammary gland depend on the integrated actions of pituitary, ovarian, thyroid, and adrenal hormones (Fig. 8.67) (Lyons, 1958). Growth of mammary ducts is dependent on estrogen synergized by GH, PRL, and adrenal corticosteroids. The subsequent development of the lobuloalveolar glandular system requires both estrogen and progesterone in the presence of PRL. Lactogenesis and milk secretion are regulated principally by PRL and facilitated by corticoids.

That the ovarian steroids 17β-estradiol and progesterone are necessary for mammary development in women is based on clinical observations: no breast changes occur at puberty in girls with gonadal dysgenesis. Furthermore, enlargement of the breasts is frequently noted by women using oral contraceptive agents. However, once breast development is complete, the effects of removal of the ovaries are less obvious. Numerous studies have established that estrogens generally stimulate growth of the duct system, whereas lobuloalveolar

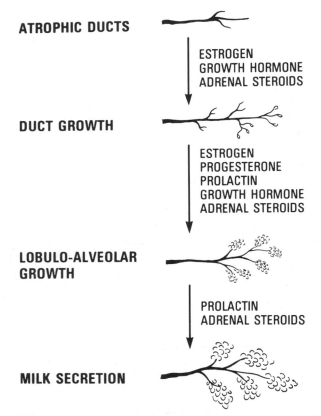

ATROPHIC DUCTS

ESTROGEN
GROWTH HORMONE
ADRENAL STEROIDS

DUCT GROWTH

ESTROGEN
PROGESTERONE
PROLACTIN
GROWTH HORMONE
ADRENAL STEROIDS

LOBULO-ALVEOLAR GROWTH

PROLACTIN
ADRENAL STEROIDS

MILK SECRETION

Figure 8.67. Multihormonal interaction in the development of the mammary gland and in the initiation of lactogenesis and lactation. Based on studies in the hypophysectomized, ovariectomized, adrenalectomized rat. (Reproduced with permission from Rebar, 1984; adapated from Lyons et al., 1958.)

development is dependent on progesterone (Folley, 1948; Meites, 1965; Topper and Freeman, 1980).

During pregnancy, the increasing levels of PRL, hCS, estrogens, and progesterone combine to stimulate the development of the secretory apparatus of the breast, but lactogenesis is minimal and lactation absent. Although estrogen and progesterone stimulate mammary development, they inhibit actual formation of milk during pregnancy. PRL, necessary for milk production, increases the number of receptor sites for both estrogen and PRL. However, estrogen in large amounts inhibits PRL-binding to mammary tissues, and, under such circumstances, lactation does not occur despite large numbers of PRL receptors (Bohnet et al., 1977; Djiane and Durand, 1977). Following delivery, estrogen and progesterone levels fall rapidly, and the lactogenic action of prolactin is then unopposed. Androgens also appear to inhibit the action of PRL, thus accounting for the successful use of preparations combining an estrogen and an androgen for suppression of lactation, provided such agents are administered immediately after delivery. Although women may note breast secretions in the last trimester of pregnancy, the secretions do not constitute milk and contain little fat. If true lactation begins before delivery, some condition associated with a fall in or very low levels of circulating estrogen in the mother should be suspected. These include fetal demise, an anencephalic fetus, a molar pregnancy, or placental sulfatase deficiency. PRL levels which are increased by a maternal pituitary tumor must also be considered.

Neuroendocrinology of the Suckling Reflex

During suckling, sensory signals originating in nerve terminals innervating the nipple are conveyed through an afferent fiber system to the hypothalamus where neurosecretory neurons institute a complex response (Fig. 8.68). Denervation of the nipple, transection of the spinal cord, and brain stem lesions can abolish the normal response to suckling (Voloschin and Dottaviano, 1976).

Three interrelated neuroendocrine events result from suckling: (1) Oxytocin (an octapeptide synthesized in cell bodies in the paraventricular and supraoptic nuclei of the hypothalamus) is released from axons in the posterior pituitary without concomitant release of vasopressin. Oxytocin release is induced by afferent signals to both paraventricular and supraoptic nuclei. The oxytocin stimulates contraction of myoepithelial cells in the mammary alveoli and ducts to result in milk ejection. Spontaneous oxytocin release will occur in a lactating woman allowed to play with her infant or upon

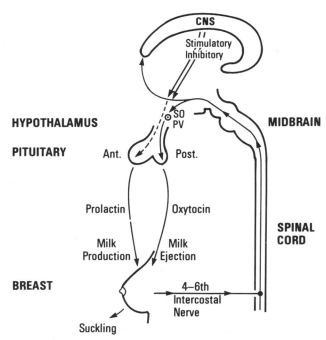

Figure 8.68. Neuroendocrine reflexes initiated by suckling. *PV* = paraventricular nucleus; *SO* = supraoptic nucleus; *Ant.* = anterior pituitary; *Post.* = posterior pituitary. (Reproduced with permission from Rebar, 1984.)

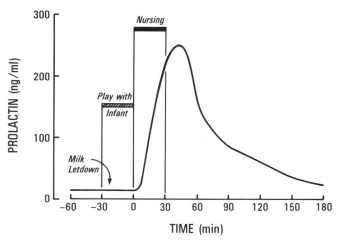

Figure 8.69. Schematic representation of serum prolactin concentrations during anticipation of nursing and during nursing in the first 40 days postpartum. The response decreases with time after delivery. No increase in growth hormone above basal levels was observed. (Reproduced with permission from Rebar, 1984; adapted from Noel et al., 1974.)

hearing the infant cry (so-called "milk letdown") (McNeilly et al., 1983; Noel et al., 1974). (2) With the onset of nursing, a short-lasting release of PRL occurs. This response must be separate from the release of oxytocin because no release of PRL occurs with playing with or hearing the infant cry. Lactation is maintained by periodic surges or PRL stimulated by each episode of suckling (Fig. 8.69). These surges in PRL prime the breast for the next feeding. In fact, lactation can continue indefinitely as long as the stimulus of suckling is continued (i.e., the "wet nurse"). Lactation has even been induced by repeated suckling in nonpregnant and postmenopausal women. (3) Gonadotropin secretion is inhibited during lactation, possibly because of increased dopamine and/or endorphin activity. These substances are believed to inhibit hypothalamic GnRH secretion (Rasmussen et al., 1983).

Composition of Colostrum and Milk

The composition of human milk has been extensively analyzed (Jenness, 1974; 1979). Physicochemically milk is an emulsion of fat in water which is isotonic with plasma, with water being quantitatively the most prominent constituent. Mature human milk contains 3–5% fat, 0.8–1.2% protein, 6.8–7.2% carbohydrate (calculated as lactose), and 0.2% mineral constituents. Human milk contains over 100 known constituents, and additional com-

ponents are continually being identified. Its energy content is 60–75 kcal/100 ml. The protein content is markedly higher and the carbohydrate content lower in colostrum (i.e., initial breast fluid secreted by the mother for 2–3 days after delivery) than in mature milk. In general, the fat content of milk does not vary consistently during the course of lactation but exhibits large diurnal variations, being highest in midmorning and lowest during the night, and increases during the course of each nursing. Race, age, parity, and diet do not have much effect on milk composition, and there is no difference in the composition of milk from each of the two breasts. Inadequate protein intake does not seem to affect the fat, protein, or lactose content of milk, but the volume of milk produced may be reduced.

The principal proteins in human milk are caseins, α-lactalbumin, lactoferrin, immunoglobulin A, lysozyme, and serum albumin (Bezkorovainy, 1977). α-Lactalbumin is a distinctive milk protein, being a component of the enzyme complex lactose synthetase, with its synthesis controlled by PRL. Lactoferrin and IgA are found in greater quantities in colostrum than in later milk. Of note is the observation that the essential amino acid content in human milk appears to parallel what is optimal for human infants.

The principal carbohydrate of human milk is lactose but 30 or more oligosaccharides, all containing a reducing group, terminal Gal-(B1,4)-G1c, and ranging from 3–14 saccharide units per molecule are also present. These may compose as much as 1 g/100 ml in mature milk and 2.5 g/100 ml in colostrum. Since many of the oligosaccharides are growth factors for certain strains of *Lactobacilli*, they may have an effect on intestinal flora in infants. Some nucleotide sugars and lipid-bound and protein-bound carbohydrates are present as well.

The principal class of human milk lipids is triglycerides. The most common are palmitic and oleic acids. The fatty acid composition of milk fat is somewhat influenced by diet. When the diet is deficient in calories, the milk fat tends to resemble adipose tissue fat in fatty acid composition, no doubt as a result of lipolysis of maternal stores. When dietary intake is adequate, the main influence is the type of diet (Insull et al., 1959).

The major mineral constituents of human milk are Na, K, Ca, Mg, P, and Cl. Iron, copper, and zinc contents of human milk vary considerably. Several other trace elements have been detected as well.

About 25%, or 40 mg/100 ml, of the total nitrogen in human milk represents nonprotein compounds including urea, uric acid, creatine, creatinine, and free amino acids. The total nonprotein nitrogen content of colostrum is much higher than that of later milk. Of the free amino acids, glutamic acid and taurine are the most abundant. It has been suggested that taurine may be required for early states of brain development in some species (Sturman et al., 1977). All of the vitamins, except for vitamin K, are found in human milk in nutritionally significant quantities.

At the end of the first week postpartum, 550 ml of milk per day are produced; by 2–3 weeks postpartum, production has increased to approximately 800 ml/day, peaking at 1.5–2 liters/day.

Ovarian Function in the Puerperium

Following delivery the hormones produced by the fetal-placental unit disappear rapidly (Fig. 8.70). Whether the patient is nursing or not seems to be the major determinant influencing the timing of the first ovulation. Postpartum women rarely ovulate before the 30th day after delivery (Vorherr, 1973). Furthermore, in most non-nursing women menstruation occurs by the 3rd postpartum month and the 1st postpartum period is anovulatory. In comparison, both ovulation and menstruation are significantly delayed in nursing relative to non-nursing mothers.

Presumably as a result of pituitary suppression persisting from pregnancy, low levels of estradiol postpartum do not result in an increase in serum FSH concentrations until approximately 2 weeks postpartum (Rolland et al., 1975). In fact, the pituitary has been shown to be unresponsive to GnRH in the early puerperium (Miyake et al., 1978). Sub-

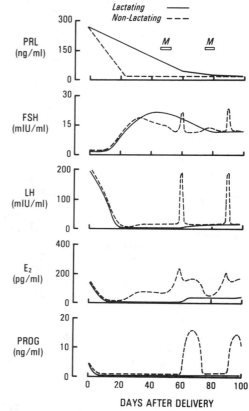

Figure 8.70. Schematic representation of serum concentrations of pituitary and gonadal hormones in lactating and nonlactating women in the puerperium. In the top graph, the boxed *M* refers to menses in the nonlactating woman. The lactating woman is continuing to lactate at the end of the depicted time interval. Two LH surges are seen in the nonlactating woman. The elevated LH level immediately postpartum is no doubt due to cross-reactions of assays with hCG, and the estradiol (E_2) and progesterone (*PROG*) are of placental origin. (Reproduced with permission from Rebar, 1984; based on data from Rolland et al., 1975.)

sequently, FSH but not LH rises above the normal range observed in normal eumenorrheic women. However, no increase in estradiol is noted until about the 9th week postpartum. Only after PRL levels have fallen to normal does ovarian secretion of estradiol begin to rise.

The importance of PRL and lactation to the return of gonadal function is perhaps even more dramatically demonstrated by studies of !Kung hunter-gatherers in Africa (Konner and Worthman, 1980). Despite initiation of coitus soon after delivery and the lack of contraception, the average interval between births in this population is 44.1 months. It now appears that this long birth spacing is directly attributable to the pattern of nursing. Mothers nurse briefly and frequently (with a mean interval between nursing episodes of approximately 13 min), and infants are always in immediate physical proximity to their mothers until age 2 years or older. Infants even customarily sleep on the same mat with their mothers until they are weaned and are allowed to nurse at night as desired. Low levels of 17β-estradiol and progesterone in the serum of the mothers correlated with the age of the infants and intervals between nursing but not with total nursing time. Because mean PRL levels and nursing frequency are also correlated (Delvoye et al., 1977; Rolland et al., 1975), the conclusion is that maternal ovarian function is suppressed by a timing-dependent, PRL-mediated effect of breast stimulation (Konner and Worthman, 1980).

DEVELOPMENT OF FETAL ENDOCRINE GLANDS

Fetal Hypothalamic-Pituitary Axis

The fetal hypothalamus begins to differentiate from the forebrain during the first few weeks of life and is fairly well developed by 12 weeks of gestation. Many hypothalamic hormones including GnRH, TRH, dopamine, norepinephrine, and somatostatin have been identified as early as 6–8 weeks of fetal life (Gluckman et al., 1980). By 14 weeks the supraoptic and paraventricular nuclei are fully developed.

Fetal Thyroid Gland

The fetal thyroid is capable of concentrating iodine and synthesizing iodothyronines by the end of the first trimester. The levels of TSH and thyroid hormone in fetal blood are relatively low until midgestation (Fisher and Klein, 1981). At 24–28 weeks of gestation serum TSH levels rise sharply but then decrease slightly thereafter until delivery. In response to the surge of TSH, T_4 levels rise progressively after midgestation until term. At birth abrupt release of TSH, T_4, and T_3 occurs. Circulating concentrations of these hormones then decrease during the first few weeks after birth. The fetus is believed to be in a relative hyperthyroid state during the last half of pregnancy in order to prepare the fetus for the thermoregulatory adjustments of extrauterine life. The abrupt hormonal changes occurring at birth possibly are stimulated by the cooling which occurs after delivery (Fisher and Odell, 1969).

Fetal Gonads

By 6 weeks' gestation, the structure of the fetal testis can be recognized, and functional Leydig cells, which secrete testosterone, are present. Because the patterns of plasma hCG concentrations, testicular hCG receptors, and plasma testosterone concentrations in the fetus are closely related, it has been pos-

tulated that hCG may regulate fetal testicular testosterone production and play an important role in male differentiation (Fig. 8.71) (Molsberry et al., 1982).

Much less is known about the differentiation of the fetal ovary. Morphologically the ovary becomes recognizable at 7–8 weeks of intrauterine life. It is unclear to what extent the fetal ovary can synthesize steroids.

Plasma concentrations of both LH and FSH slowly rise to a peak near the 25th week of fetal life and fall to lower levels by term. In general, levels of gonadotropin are higher in females than in males, possibly due to greater negative feedback in males from the high levels of fetal testosterone (Kaplan and Grumbach, 1976). It appears that fetal gonadotropins are required for complete differentiation of the fetal testis and ovary. For example, anencephalic fetuses with low levels of circulating LH and FSH have appropriate secretion of testosterone at 15–20 weeks due to normal levels of hCG but have decreased numbers of Leydig cells, hypoplastic external genitalia, and frequent undescended testes at birth (Bearn, 1968). Similarly, male fetuses with congenital hypopituitarism often have an associated micropenis. Although follicular development is seemingly relatively independent of gonadotropins, the anencephalic female fetus has small ovaries and decreased numbers of ovarian follicles.

Fetal Adrenal Gland

By midpregnancy the size of the fetal adrenal exceeds that of the fetal kidney (Serón-Ferré and Jaffe, 1981). Histologically, the fetal adrenal cortex is composed of two zones: (1) an unique, inner *fetal zone*, which comprises 80% of the gland and accounts for its large fetal size, and (2) an outer *definitive*, or adult, *zone*, destined to form the adult adrenal cortex. Dehydroepiandrosterone sulfate (DHEAS) is produced principally by the fetal zone, and cortisol is synthesized primarily by the definitive zone. Both DHEAS and cortisol increase progressively in fetal plasma through pregnancy. Fetal ACTH appears to regulate cortisol production by the definitive zone. Both ACTH and hCG, which is produced in greatest quantities in the first half of pregnancy, can stimulate DHEAS secretion by the fetal zone. Following delivery the fetal zone rapidly undergoes involution, with much of the zone gone by the age of 2–3 months, and with it completely absent at 1 year of age. After birth, DHEAS levels decline, paralleling the regression of the fetal zone.

Fetal DHEAS is the key precursor in the synthesis of estrogens by the placenta. The roles for fetal corticosteroids are many but less clearly defined. Cor-

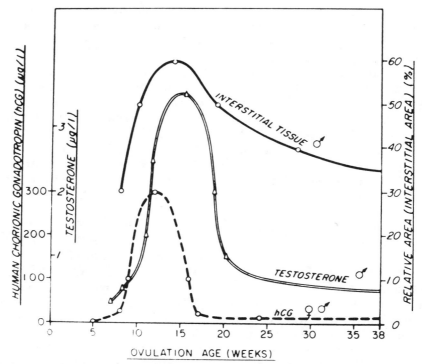

Figure 8.71. Coincidence of mass of interstitial tissue (i.e., Leydig cells), concentrations of testosterone in testicular tissue, and concentrations of hCG in maternal plasma early in pregnancy in the male fetus supporting a role for hCG in the sexual differentiation of the male. (Reproduced with permission from Ryan, 1980; based on data from Niemi et al., 1967.)

tisol may (1) induce development and maturation of a variety of enzyme systems in the liver, (2) aid in maturation of pancreatic β cells, (3) play a role in the change from fetal to adult hemoglobin, (4) induce cytodifferentiation of the type II alveolar cell and stimulate synthesis of surfactant, important for lung maturity, (5) play a role in parturition, and (6) stimulate formation of the adrenal medulla, and synthesis of catecholamines.

The adrenal medulla is formed by 10 weeks of gestation but is relatively immature at term (Fisher, 1979). Both epinephrine and norepinephrine can be detected in the human fetal medulla by 10–15 weeks of gestation. Except for the stimulatory effect of cortisol on catecholamines, the regulation of catecholamine secretion in the human fetus is not understood.

IMMUNOLOGICAL FUNCTIONS OF THE PLACENTA

The most important function of the immune system is to distinguish between "self" and "nonself." The human placenta performs two functions in this respect: it protects the fetus which is histoincompatible with the mother against maternal immunological attack and it provides the fetus transiently with immunological defenses by the transfer of selected maternal immunoglobulins which protect the newborn infant until its own immunological defense mechanisms have been developed. The newborns' antibody complement is also supplemented by the transfer of maternal immunoglobulins during breastfeeding.

The implanted placenta has been likened to a graft, but despite the intimacy of fetal and maternal tissues in the placenta they do not share a common circulation (Pijnenborg et al., 1981). This separation of vasculature is maintained by a continuous layer of tissue composed of fetal trophoblasts and a sheet of fibrinoid at the boundary between fetal and maternal tissue. This physical separation of mother and fetus is probably essential for the survival of the fetus, which is histoincompatible with the mother because of the paternal component of its genetic constitution. How the fetal trophoblast which is in direct contact with the maternal circulation avoids immunological rejection is not yet clear.

The gene system at the center of recognition of "self" from "nonself" is the major histocompatibility complex (MHC) on chromosome 6, which has several major subregions or loci known as HLA-A, -B, -C, and -D, with the D subregion being comprised of at least three loci. Each locus codes for the production of a glycoprotein which has important functions in the initiation and control of immune mechanisms. HLA-A, -B, and -C produce glycoproteins called Class I major histocompatibility antigens: those produced by HLA-D are Class II. The Class I antigens are expressed on the surface of most cells. Class II antigens are in general restricted to certain immunologically active cells such as macrophages.

Paternally derived antigens are present in the 8-cell zygote (Webb et al., 1977) and MHC antigens are expressed on the embryo and fetus (Heyner, 1980). Invasion of maternal tissues by trophoblast and their juxtaposition then pose important immunological questions. However, neither cyto- nor syncytiotrophoblast express MHC antigens at the time of implantation (Faulk and Temple, 1976; Kawata et al., 1984; Sunderland et al., 1981). Following implantation, the situation changes but villous trophoblast fails to express Class I or II MHC antigens at any time, whereas other types of trophoblast express Class I MHC antigens but in a nonuniform manner. No trophoblast population has been shown to express Class II MHC antigens (Redman, 1985).

This protects the trophoblast and the pregnancy from immune attack because the lack of Class II MHC antigens prevents induction of an immune response and the absence of Class I MHC antigens on trophoblast in contact with maternal blood prevents it being a target for cell-mediated attack. Other antigens besides MHC must be considered. A set of trophoblast antigens, some shared by lymphocytes, designated TLX (trophoblast-lymphocyte cross-reactivity) have been proposed. Expression of these antigens on trophoblast may directly or indirectly modulate lymphocyte function (McIntyre and Faulk, 1982).

The current understanding of the main protective mechanism between mother and the immunologically disparate fetus is the absence of MHC antigens from trophoblast; however, other mechanisms may be operative.

PARTURITION

General Information

The sequence of hormonal events occurring during pregnancy and immediately preceding the onset of labor varies among animal species. Even though no one animal model suffices to explain parturition in the human, most of the early work on this subject was done in sheep by Liggins and co-workers (1973). Regardless of the species, however, it appears that (1) the fetus plays a role in the initiation of parturition (Bassett and Thornburn, 1969), and (2) prostaglandins are important in the final steps in the initiation of myometrial contractility (Fig. 8.72).

In the sheep, the fetus, through the activation of the pituitary-adrenal axis, plays a major role in determining the initiation of labor. There is a slow

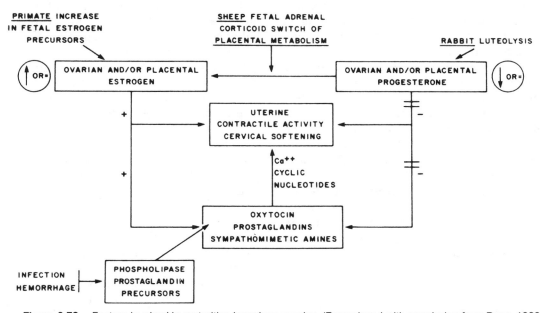

PREGNANCY TERMINATION

Figure 8.72. Factors involved in parturition in various species. (Reproduced with permission from Ryan, 1980.)

and steady increase in cortisol concentrations in the fetal lamb during the last 15 days of pregnancy, with a dramatic increase occurring in the last 3–4 days (Bassett and Thornburn, 1969). An increase in adrenal gland weight is also seen. This increase in cortisol is due primarily to increased sensitivity of the adrenal gland to ACTH, but the mechanism is not understood. This elevation in glucocorticoids also acts on the placenta to induce or increase the activity of 17 α-hydroxylase, which results in a fall in progesterone production by increasing metabolism to estrogen. The fall in progesterone (together with the corresponding change in the estrogen:progesterone ratio) stimulates synthesis and release of prostaglandin from the uterus, measured initially as an increase in the concentration of prostaglandin $F_2\alpha$ in utero-ovarian blood (Thornburn and Challis, 1979). Hypophysectomy or adrenalectomy of the sheep fetus results in prolonged pregnancy, but administration of either ACTH or glucocorticoids to the fetus stimulates the sequence of hormonal events that results in labor prematurely (Liggins, 1969). These changing hormonal events result in increases in the sensitivity of the myometrium to prostaglandins and oxytocin. These latter two substances increase myometrial contractility.

Although the relationship between the sheep fetal adrenal gland and the initiation of labor has been adequately established, a similar mechanism in humans is not as well elucidated. In women, studies have shown no good evidence of a decrease in plasma progesterone, nor is there a sharp increase in unconjugated estrogen concentrations in plasma in the period preceding birth. It is currently believed that the control of birth in women depends on steroid hormones and on prostaglandin production within the fetal intrauterine tissues—the membranes and the decidua. Within the membranes, the amnion primarily has been shown to be the major site of stimulatory prostaglandin production (PGE, PGF) at the time of labor. Prostaglandins are also produced to a lesser degree by the chorion, which also has a higher capacity to metabolize these substances (Bleasdale and Johnson, 1984). The fetal membranes, particularly the chorion and the decidua, also have the capacity to produce steroid hormones. These membrane-produced steroids seem to influence prostaglandin production and to increase myometrial sensitivity. Recent data suggest that a growth factor in membranes also may stimulate prostaglandin production to initiate labor.

Prostaglandins

Prostaglandins clearly play a major role in the onset of labor in humans. An increase in prostaglandins concentrations (PGE, PGF) has been found in amniotic fluid, and their metabolites have also been found in maternal plasma and urine in the latter part of human pregnancy. Giving prostaglandin synthesis inhibitors, such as aspirin and indomethacin, suppresses uterine activity and prolongs the length of pregnancy. In addition, the human uterus has been shown to be extremely sensitive to the stimulatory effects of exogenously applied prostaglandins throughout pregnancy.

The endometrium, whether in the nonpregnant or pregnant uterus, appears to be the major site of uterine prostaglandin synthesis and contains the multienzyme complex prostaglandin synthetase. The availability of free arachidonic acid (the obligatory precursor of $PGF_2\alpha$, PGE_2, prostacyclin, and thromboxane) is thought to be the determining factor in prostaglandin synthesis, but the esterified form of arachidonic acid cannot be utilized for prostaglandin synthesis.

As a woman approaches term, human decidua and membranes incorporate unusually large amounts of arachidonic acid into their glycerophospholipid stores. It is thought that spontaneous labor is secondary to increased action of phospholipase A_2, a lysosomal enzyme, that liberates arachidonic acid from its esterified form in the decidua and membranes to increase the production of the stimulatory prostaglandins PGE_2 and $PGF_2\alpha$ (Bleasdale and Johnson, 1984). $PGF_2\alpha$ has been found to be the more potent contractile stimulus (Novy and Liggins, 1980).

Free calcium availability also is a critical factor in influencing prostaglandin synthesis by the fetal membranes (Bleasdale et al., 1983). Calcium is believed to stimulate phospholipases A_2 and C, but this remains controversial (Okazaki et al., 1981). It is also possible that calcium may have its effects on other enzymes, or on a protein kinase which phosphorylates, and thereby inactivates, the macromolecular protein lipomodulin, which has been shown to inhibit phospholipase activity (Bleasdale et al., 1983).

Even though prostaglandins play a key role in the onset of labor, other hormones also may be important, particularly oxytocin and relaxin.

Oxytocin

Initially it was thought that oxytocin alone was the primary stimulus for the onset of labor, but studies have shown that women with prior pituitary ablations do not have prolonged pregnancy. Moreover, oxytocin is not significantly increased in maternal plasma until after labor is well advanced. Thus, despite its role in "milk letdown," oxytocin's role in human parturition is undefined.

Oxytocin exerts its contractile effects by stimulating prostaglandin release (Garrioch, 1978). Liggins has also shown that term induction of labor may depend, in part, on oxytocin effects on smooth muscle and in part on a stimulus to decidual prostaglandin synthesis (Liggins, 1985).

Oxytocin responsiveness has been shown to be partially enhanced by prostaglandin-induced formation of gap junctions in late pregnancy (Fuchs et al., 1982). These gap junctions are found in the myometrium and thought to provide low-resistance electrical pathways between cells, contributing to labor's coordinated muscle activity. This responsiveness also is partially enhanced by prostaglandin control of calcium mobilization.

Relaxin

Relaxin is synthesized by the corpus luteum and decidua with the highest concentrations occurring in the first trimester (Weiss, 1985). Relaxin acts on both the smooth muscles and connective tissue, with sites of action that are thought to be similar to those of prostaglandins.

Most of the effects of relaxin have been shown in studies using a rat model, where its physiological role in cervical dilatation has been noted (Steinetz and O'Byrne, 1983). Human studies are still unclear, but it is thought that relaxin produced by the decidua contributes to the maintenance of uterine quiescence and cervical softening in late pregnancy (Huszar and Naftolin, 1984).

CHAPTER 61

Physiology of Pregnancy

The physiological changes which occur during pregnancy are among the most remarkable and dramatic events in biology, and serve as adaptive mechanisms to sustain the mother while supporting the growth and development of her fetus. This chapter will explore maternal physiology in contrast to that of the nonpregnant state, summarize the fundamental mechanisms which control nutrient transfer and respiratory gas exchange across the placenta, and provide an overview of basic fetal physiology. By necessity, the chapter will separate the physiological alterations by organ system. It should be noted that these systems are highly integrated by the interplay of a variety of mechanisms acting in concert to ensure optimal pregnancy outcome.

MATERNAL PHYSIOLOGY

Blood Volume and Hemodynamic Changes

Pregnancy is characterized by a rapid increase in *total blood volume,* reaching levels approximately 40% over the nonpregnant mean by the middle of the third trimester. Total body water increases progressively, reaching 6–8 liters, most of which is distributed in the extracellular space. The increase in the *plasma volume* component occurs very early postconception, with increases of 400 ml and 900 ml observed by 8 and 16 weeks of gestational age, respectively (Clapp, 1988). Despite the fact that there is a 20–30% real increment in *red cell mass,* the plasma volume component rises at a more rapid rate, resulting in what is known as a "physiological anemia" of pregnancy (Hytten and Painten, 1964).

This striking change in plasma volume is a consequence of sodium and water retention, but the mechanisms responsible for their net accumulation are not clearly understood. Plasma renin activity increases during gestation, an alteration attributed to estrogenic stimulation of renin substrate production by the liver (Lipsett et al., 1971). Further, increased blood aldosterone concentrations promote sodium retention, and it is likely that these hormonal changes are, in part, responsible for the increase in blood volume (Seitchik, 1967). The augmentation in red cell mass is a consequence of an increase in erythropoietin activity.

The hemodynamic alteration of greatest significance during pregnancy is the increase in *cardiac output.* This change has been extensively studied, and the observed results are variable depending upon the techniques of evaluation and maternal positioning during measurement. Regardless of the methodology, it is quite clear that cardiac output increases significantly by as early as the end of the first trimester. This is of particular interest, since the conceptus is very small, and the increase in cardiac output occurs long before any substantial change in oxygen consumption by the maternal-fetal unit. By midpregnancy, cardiac output peaks at levels which are 30–40% greater than the nonpregnant state (Metcalfe et al., 1981).

Invasive techniques to measure cardiac output involve central arterial and venous catheterization, and have utilized indicators such as indocyanine green dye (dye dilution) and chilled saline (thermodilution). Echocardiography has been used as a noninvasive technique. Proper maternal positioning throughout the study is of the utmost importance because, in the supine position, the pregnant uterus compresses the inferior vena cava, decreasing venous return to the right heart, and consequently, decreasing cardiac output. Although some studies suggest that cardiac output begins to decrease from peak levels late in pregnancy, measurements obtained in the lateral recumbent position indicate that these levels are sustained throughout late gestation. These discrepancies serve to emphasize the lability of maternal cardiovascular hemodynamics.

Cardiac output changes are a function of incremental alterations in *stroke volume* and *heart rate.* Stroke volume increases by approximately 30%, reaching a peak at midpregnancy, and remaining at that level or decreasing slightly as term approaches.

Meanwhile, heart rate increases by 10–15%, and may reach 85–90 beats per minute (bpm) at term.

Systolic and *diastolic pressures* are also altered during pregnancy. By the late first or early second trimester, a decrease is observed in both parameters. When evaluated in the sitting or standing position, minimal changes are observed in systolic pressure dynamics. However, diastolic pressure decreases by 10 torr or more, reaching a nadir at around 28 weeks of gestation followed by a return to first trimester levels. As with measurements of cardiac output, maternal positioning is of central importance, and observed changes are more striking in lateral recumbency.

A decrease in the *left ventricular ejection time*, as well as its pre-ejection period, is also observed beginning in the second trimester. The end-diastolic volume increases without a concomitant increase in filling pressure, leading to an increase in stroke volume (Katz et al., 1978). Many of these changes are summarized in Figure 8.73.

The mechanisms responsible for these profound and fundamental physiological alterations are not entirely clear, but may be related to early pregnancy changes in the circulating concentrations of estrogenic steroids. As shown in Figure 8.74, estrogens are known to alter vascular resistance in the uterus, resulting in dramatic increases in uterine blood flow (Resnik et al., 1974). In addition, many of the changes in the distribution of cardiac output observed in humans, such as increases in uterine, renal, mammary and skin blood flow, can be repro-

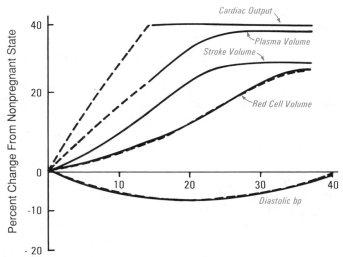

Figure 8.73. Changes in cardiovascular hemodynamics throughout the course of gestation.

duced by estrogen administration in animal models (Rosenfeld et al., 1976). Uterine blood flow is approximately 40 ml/min in the nonpregnant state, and may reach levels from 600 to >1000 ml/min at term, values which are close to 20% of the total cardiac output (Metcalfe et al., 1955). Early in pregnancy, the increase in estrogen production parallels the early increase in uterine blood flow. Finally, oral administration of substances containing estrogens and progestational compounds causes an increase in cardiac output secondary to an increase in stroke volume (Walters et al., 1969). Since the

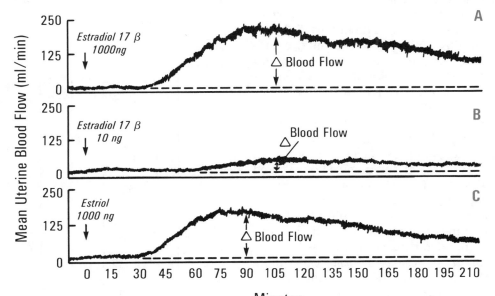

Figure 8.74. The effect of estradiol-17 β and estriol on uterine blood flow. Estrogen was administered as a pulse dose directly into the uterine artery of a nonpregnant sheep with electromagnetic flow probes around the uterine arteries. Within 90–110 min, blood flow reaches levels observed at midpregnancy in the sheep. (Adapted from Resnik et al., 1974.)

increase in cardiac output during pregnancy occurs long before marked increases in uterine blood flow are observed, it is likely that estrogens exert an influence directly on the myocardium and elsewhere within the vascular system.

No significant changes are noted in *venous pressure* in the upper extremities. However, femoral venous pressure increases markedly, largely due to the external pressure exerted by the pregnant uterus on the iliac veins and the inferior vena cava.

Right ventricular and *pulmonary artery pressures* are unchanged during pregnancy, a consequence of the high degree of pulmonary vascular compliance. This enables pulmonary blood flow to increase in parallel with the expanded cardiac output, a characteristic of considerable clinical importance.

Respiratory System Changes

The most striking alteration in respiratory physiology is the increase in *minute ventilation*, which increases by 36% by the 8th week of pregnancy, and ultimately reaches levels which are 50% above the nonpregnant mean (Clapp, 1988; Pernoll et al., 1975; Alaily et al., 1978). This adjustment is required to satisfy the increase in oxygen consumption (30–35%) by the growing fetus. However, the timing and magnitude of the increase in minute ventilation is in excess of the oxygen consumption requirements for fetal development, and may be due to the stimulatory effect of progesterone on respiratory centers (Skatrud et al., 1978). This expansion in minute ventilation leads to a slight decrease in alveolar P_{CO_2}, and a lower Pa_{CO_2} from 38 torr to approximately 30 torr at term. The kidney compensates metabolically by an increase in tubular excretion of bicarbonate, partially offsetting the changes in blood pH. Nevertheless, there is a mild respiratory alkalosis, and the pH rises from 7.35 to 7.4 near term (Lucius et al., 1970).

It is important to emphasize that the increase in ventilation occurs without an increase in respiratory rate. This is accomplished largely by a rise in the tidal volume (volume of gas inspired or expired with each ventilation) from 500–700 ml/min, and explains the frequent sighing observed in pregnant women. *Residual volume* (the volume of gas remaining in the lungs at the end of maximal expiration, not including dead space) decreases by 20%, as does the *expiratory reserve volume* (maximum amount of air that can be expired at the end of normal expiration). Consequently, the *functional residual capacity* (residual volume plus expiratory reserve volume) is diminished, reducing the amount by which tidal volume is diluted during each breath.

No significant change in diffusing capacity has been observed. These combined alterations serve to increase the effective alveolar ventilation, and are summarized in Figure 8.75.

The enlarging uterus alters the resting position of the diaphragm. Despite elevation of the diaphragm, there is no evidence that its dynamic motion is impaired.

Alterations in Renal Function

Renal plasma flow rises early and sharply during pregnancy, to levels 35% above the nonpregnant mean (Sims and Krantz, 1958; Dunlop, 1981). The increase parallels and is a consequence of the rise in blood volume and cardiac output, combined with a reduction in renal vascular resistance. Studies utilizing paraaminohippurate reveal flows of 800 ml/min by the end of the first trimester, 700 ml/min later in pregnancy, and 480–500 ml/min postpartum. The cause of the decrease in renal vascular resistance has not been entirely elucidated, but augmented renal production of the vasodilating prostaglandins, prostacyclin (PGI_2), and prostaglandin E_2 (PGE_2) during pregnancy has been implicated (Bay and Ferris, 1979). Pregnancy hormones such as human placental lactogen (HPL) and progesterone have been noted to produce modest increases in renal blood flow when administered to nonpregnant subjects, but other evidence to support their role in the changes observed during pregnancy is lacking.

Concomitant with the rise in renal plasma flow is a remarkable increase in *glomerular filtration rate* (GFR) as measured by inulin clearance. This change

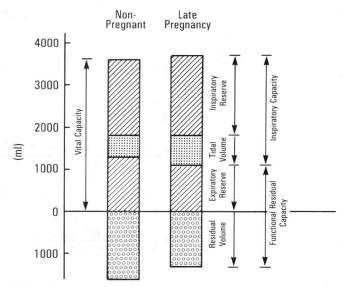

Figure 8.75. Comparison of lung volumes in the nonpregnant and late-pregnant female. (Adapted from Hytten F et al., *Clinical Physiology and Obstetrics.* Oxford: Blackwell Scientific Publications, 1980.)

occurs rapidly following conception, with significant alterations observed as early as 2 weeks following conception (Davison and Dunlop, 1984a). The magnitude of this change is approximately 40%, rising from a nonpregnant mean of 100–140 ml/min by midpregnancy. Although data are conflicting, the preponderance of evidence suggests that this level is maintained until term (Pippig, 1969). The changes in renal plasma flow and GFR, as well as the impact of maternal positioning on measurements, are displayed in Figure 8.76.

The increase in glomerular flow, without noticeable change in glomerular pressure, results in greater quantities of plasma solute traversing the glomerulus per unit time, and may serve to explain, in part, the appearance of glucose and amino acids in the urine. Despite the increased glucose load filtered, the capacity of the tubules to reabsorb glucose (T_{MAXG}) is unaltered in the pregnant state, and the plasma glucose concentration at which glucose "spills" into the urine falls from approximately 190 mg/dl to 155 mg/dl. Consequently, glucosuria is common among pregnant women (Davis and Hytten, 1975).

There is a tendency toward *aminoaciduria* during pregnancy, with substantial losses of glycine, histidine, threonine, serine, and alanine. Because administration of glucocorticoids in the nonpregnant state results in a similar pattern of losses, it is possible that pregnancy increases the sensitivity of the renal tubules to cortisol. The significance of this altered excretion pattern is unknown.

Despite the profound changes in GFR and the large plasma solute load filtered, the kidney retains its ability to maintain *sodium balance* and *tonicity*. Pregnant women are able to maintain a normal serum sodium concentration over a wide range of low to high sodium diets, and can concentrate or dilute their urine appropriately depending on the volume of water intake. Attempts to explain these compensatory mechanisms on the basis of homeostatic adjustments induced by aldosterone and progesterone alone have been unsuccessful, and it is likely that several complex integrated physiological and hormonal events are responsible. Despite renal adjustments, there is a modest decrease in serum sodium and tonicity (5 meq/liter and 10 mosm/kg water, respectively). This is apparently a consequence of a "resetting" of the osmostat for ADH secretion (Davison et al., 1984b).

Coincidental with the physiological changes are anatomic adjustments of great significance. These include dilation of the renal calyces and pelvis, as well as ureters, leading to a state of physiological hydronephrosis and hydroureter. This generalized expansion of the urinary tract, combined with decreased smooth muscle peristalsis, combine to increase the overall volume capacity of the urinary tract. The volume of urine within the system at any given time during pregnancy, above the level of the urinary bladder, may reach as much as 200 ml.

Hematological Changes

In addition to the previously mentioned augmentation in plasma volume, notable changes occur in the cellular components and clotting factors. The *erythron*, or red blood cell mass, increases to levels between 250 and 450 ml over nonpregnant values, a change of 20–30%. This is accomplished at a slower rate than the increase in plasma volume, leading to a normal decrease in the packed cell volume (hematocrit). The net increment in iron required over the entire course of pregnancy is approximately 800 mg (500 mg for increases in maternal red cell mass and 300 mg for the fetus).

Simultaneously, there is a modest increase in the *leukocyte* count, limited to neutrophils. Normal counts may reach 15×10^9 cells/liter. Furthermore, up to 20% of women may have a shift to more immature white blood cells in their peripheral blood smear. Although reports vary, the *platelet* count is not significantly altered during normal pregnancy.

One of the most striking changes occurring during pregnancy is the increase in procoagulant activity, likely an effect of the high concentrations of circu-

GESTATIONAL CHANGES IN GLOMERULAR FILTRATION RATE MEASURED IN THE SUPINE AND LATERAL RECUMBENT POSITIONS

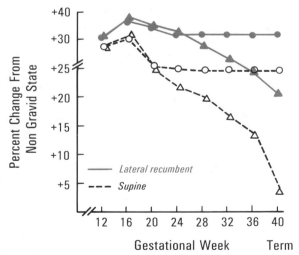

Figure 8.76. Gestational changes in glomerular filtration rate measured in the supine and lateral recumbent positions. (Adapted from Lindheimer MD, Katz AI. *Kidney Function and Disease in Pregnancy,* Philadelphia: Lea and Febiger, 1977, p. 20.)

lating estrogens on the hepatic production of proteins. Fibrinogen levels may increase two to four fold, and Factors VII, VIII (antihemophilic factor), IX and X are also found in higher concentrations.

Liver, Biliary, and Gastrointestinal Tract

LIVER

Despite the marked increase in blood volume and cardiac output during pregnancy, blood flow to the liver is unchanged. Consequently, that proportion of the cardiac output distributed to the liver decreases by approximately 35% (Haemmerli, 1967). Meanwhile, most of the serum proteins of hepatic origin change during pregnancy, a response which can be reproduced by estrogen administration to nonpregnant women.

Total serum protein declines approximately 20% by the middle of pregnancy, primarily due to a decrease in serum *albumin*. This is largely a result of dilution due to the increase in plasma volume, without any concomitant increase in albumin synthesis.

Alterations also occur in the *globulin* fraction, and the mechanisms responsible for their variable changes are not known. Slight increases are observed in the α and β fractions, while γ-globulin concentrations decrease. Levels of binding proteins are generally increased, including thyroid, sex steroid-hormone, and corticosteroid-binding globulins, ceruloplasmin, and transferrin. Serum enzymes of hepatic origin, such as glutamic-oxaloacetic transaminase (SGOT) and glutamic-pyruvic transaminase (SGPT) do not change, and any increase observed should be considered a deviation from normal. Total serum alkaline phosphatase concentration increases two to four times over nonpregnant normal levels by late pregnancy, but only as a result of the large amounts of this enzyme produced by the placenta.

Cholesterol and serum *lipids* rise sharply in gestation to levels at least two times higher than those levels observed in nonpregnant women. Similar increases are observed in nonpregnant women receiving estrogen, perhaps due to inhibition of lipoprotein lipase activity (Fabian et al., 1968). Values return to normal rapidly after delivery.

GALLBLADDER

Progesterone appears to inhibit smooth muscle contractility produced by cholecystokinin, thus producing a relaxing effect on the entire biliary tract (Braverman et al., 1980). This leads to an increase in gallbladder capacity, increased residual volume, and impaired contractility. Estrogen decreases the amount of chenodeoxycholic acid in the bile, a substance which increases solubility of cholesterol in the bile. This effect, combined with the increase in secretion of cholesterol into bile, predisposes to cholesterol stone formation.

GASTROINTESTINAL TRACT

The predominant changes in gastrointestinal tract function observed during pregnancy are the result of generalized decreases in smooth muscle motility. Gastric emptying time is increased, and intestinal tract motility is decreased, with a resultant prolongation in transit time (Davison et al., 1970). This leads to increased water reabsorption and constipation. In the upper gastrointestinal tract, the differential pressures which exist between the lower esophagus and stomach are diminished, due in part to the expanding uterus producing an increase in intragastric pressure (Ulmsten et al., 1978). The result is reflux of acidic gastric contents and disturbing esophageal symptoms (heartburn).

Endocrine and Metabolic Alterations

THYROID

The most notable change in thyroid function during gestation is a consequence of the increase in the GFR which raises the renal clearance of iodine (Aboul-Khair et al., 1964). This leads to an accelerated renal excretion of iodine and depletion of the iodine pool, unless it is replaced by an adequate dietary source. To compensate for this altered excretion pattern, the gland enlarges modestly and there is an increase in iodine uptake, which under normal dietary circumstances, is sufficient to maintain synthesis of adequate amounts of thyroid hormone and consequently a euthyroid state. There is no definitive evidence of a significant alteration in the hypothalamic-pituitary-thyroid negative feedback mechanism (Burrow et al., 1975). However, the high levels of circulating estrogen stimulate thyroid-binding globulin (TBG). Inasmuch as many of the thyroid function tests used in a clinical setting are dependent upon levels of TBG as well as the uptake of iodine by the thyroid gland, knowledge of the normal physiological changes in pregnancy is central to a correct interpretation of thyroid function tests. The reader is referred to Chapter 54 for a detailed discussion of thyroid physiology.

PARATHYROID

A dynamic alteration occurs in calcium metabolism during pregnancy, the purpose of which is to provide 25–30 g of net calcium accumulation required for fetal bone development, and to maintain the integrity of the maternal bone structure.

This is accomplished by the interaction between parathyroid hormone (PTH), calcitonin (CT), and 1,25-dihydroxy vitamin D_3 1,25(OH)$_2$D. In addition, gastrointestinal absorption of calcium is facilitated by increased levels of calcium-binding protein. *Total serum calcium* declines, reaching a nadir in the midthird trimester, rising again toward term. However, *ionized serum calcium* remains unchanged, and the decrease in total calcium is due to the fall in the concentration of serum albumin. To maintain normal ionized serum calcium concentrations in the presence of active transport of calcium across the placenta, PTH secretion is augmented. Furthermore, plasma concentrations of 1,25(OH)$_2$D increase, thus promoting calcium absorption from the gastrointestinal tract. Calcitonin is presumed to act in concert with other calcitropic hormones, but its role is not entirely clear. Various reports suggest that calcitonin concentrations in the blood are either unchanged or rise slightly (Pitkin, 1985). These changes are summarized in Figure 8.77.

PITUITARY

The anterior lobe of the pituitary gland increases two to three times during pregnancy, primarily as a result of an increase in prolactin-secreting cells (Goluboff et al., 1969), and circulating *prolactin* concentrations increase substantially. The other notable changes include a decrease in *gonadotropins* (FSH and LH), presumably a negative feedback consequence of the high circulating plasma estrogens of fetal-placental origin. Release of *growth hormone* in response to stimulus with arginine, or secondary to hypoglycemia, appears blunted (Tyson et al., 1969). Finally, although evidence is conflicting, the response of TSH to TRH may be augmented, also mediated by estrogens.

ADRENAL

Although remaining in the normal range, ACTH levels increase between the first and third trimester (Carr et al., 1976). Total plasma cortisol concentrations rise during pregnancy, largely due to the rise in corticosteroid-binding globulin induced by estrogens (Slaunwhite et al., 1959). The identical response is observed in nonpregnant women receiving estrogen therapy. Several observations indicate that free cortisol concentrations are increased during pregnancy as well. This may be related to the increase in cortisol production rate and its prolonged half-life in the circulation.

Metabolic Adjustments

Numerous profound changes occur in maternal metabolism, all serving to provide adequate and appropriate substrates for fetal growth and devel-

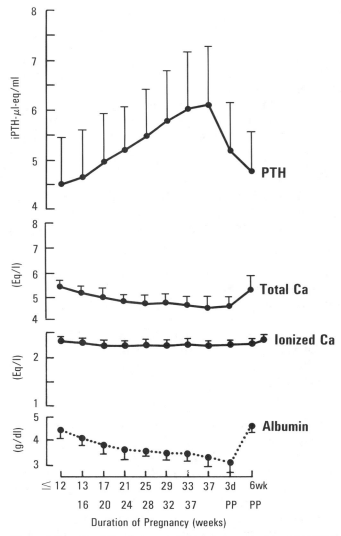

Figure 8.77. Gestational changes in blood concentrations of parathyroid hormone, total and ionized calcium, and albumin. (Adapted from Pitkin, 1985.)

opment, while ensuring that the mother continues normal physiological function. The most notable alterations are a reduction in maternal blood concentrations of glucose and amino acids, diminished responsiveness to exogenous insulin, and a simultaneous increase in the levels of free fatty acids, ketones, and triglycerides. There is also an augmented secretion of insulin in response to a normal meal or otherwise administered glucose load. The role of glucose and amino acids as metabolic substrates is discussed later in this chapter. Presumably, the higher maternal concentrations of cholesterol and triglycerides serve as alternative metabolic fuels for maternal metabolism.

Early in the second trimester, maternal blood glucose concentration falls following an overnight fast, resulting in lower fasting insulin concentrations.

This combination of events leads to an increase in the plasma levels of acetoacetic and β-hydroxybutyric acids, the net effect of which is a picture similar to the response to starvation. Indeed, pregnancy has been characterized as a state of "accelerated starvation" (Felig, 1977). Following feeding, the maternal response is one of hyperglycemia, hyperinsulinemia, and increased resistance to insulin in the liver, muscle, and adipose tissue (Freinkel, 1980).

The mechanisms responsible for these dramatic alterations are likely mediated by hormones of placental origin, including HPL, progesterone, estrogen, and well as prolactin of pituitary origin. HPL produces a slight alteration in glucose tolerance and enhances insulin resistance while stimulating lipolysis (Kalkhoff et al., 1969). Pituitary prolactin is also known to decrease insulin binding within adipose tissue sites (Jarrett et al., 1984), and estrogen has been shown to impair glucose tolerance and to antagonize the effects of insulin. Although each of these hormones exerts a modest effect only, it is possible that they act synergistically to produce the complex metabolic alterations observed in the pregnant state.

PLACENTA

No adult organ serves as many exchange and transport needs as does the placenta. This becomes apparent when it is recognized that the placenta provides to the fetus the exchange, absorption, and excretion functions of the lungs, kidneys, and gastrointestinal tract of the adult. In what follows, we briefly consider the basic transport mechanisms, anatomical and physiological factors which affect transplacental transport, the transport of individual substances, and the transport requirements to meet fetal needs. More detailed reviews are available (Faber and Thornburg, 1983; Longo and Bartels, 1972; Morriss and Boyd, 1988).

Basic Transport Mechanisms

All five of the basic transport mechanisms are utilized by the placenta. (1) *Passive diffusion* occurs in the presence of a concentration difference between the maternal and fetal blood bathing both sides of the placenta and is unaffected by electrical gradients since a transplacental electrical potential difference does not exist in humans; (2) *facilitated diffusion* occurs down a concentration gradient, utilizing a carrier mechanism usually in the presence of a cotransported substance; (3) *active transport* occurs against a concentration gradient and requires an energy expenditure; (4) *convection* (i.e., bulk flow or solvent drag) occurs when a volume of fluid flows through pores in the placental membrane and drags dissolved solutes along with the volume flow; (5) *vesicular transport* (i.e., pinocytosis) is a receptor-mediated process which transports specific molecules across the placenta.

Anatomic and Physiological Factors Which Affect Transport

For substances which cross the placenta slowly and from which only a small fraction is extracted from the maternal blood as it passes through the placenta, a change in fetal or maternal blood flow to the placenta has little effect on net transfer rates. The delivery of these substances is termed *diffusion limited* because a change in the transport mechanisms within the placental membrane is required in order to alter net transfer rates. For substances such as oxygen, which rapidly cross the placenta and which have a high fractional extraction, an increase in maternal blood flow rate to the placenta will result in an increase in delivery to the fetus, whereas an increase in oxygen permeability of the placenta would have little effect. Thus, the delivery of substances which are rapidly consumed by the fetus are termed *blood flow-limited*. For substances which are blood flow-limited, two additional factors have important effects on exchange: one is shunting of blood past the exchange area within the placenta and the other is the flow pattern of fetal blood relative to maternal blood. Maternal and fetal blood flows to the placenta average 200 ml/min/kg of fetal weight throughout the last two-thirds of gestation. Animal studies have shown that between 5 and 15% of the blood on both sides of the placenta flows to nonexchange areas, thereby reducing the exchange efficiency of the placenta. There are many possible blood flow patterns within the placenta, and Figure 8.78 shows a few idealized ones. The countercurrent exchange system has a maximum theoretical exchange efficiency of 100%, whereas all of the others have a maximum efficiency of 50%. In the human with maternal blood flowing through the intervillous space, the pool flow model is thought to approximate the maternal and fetal blood flow patterns. However, the exchange characteristics of this model are essentially the same as the concurrent system so that the terms concurrent exchange and pool flow exchange are both used.

Placental Transfer of Respiratory Gases

Maternal-fetal respiratory gas exchange occurs by simple diffusion, driven by the partial pressure differences of O_2 and CO_2 between the respective circulations (Battaglia and Meschia, 1986a). The barrier between the uterine and umbilical circulations,

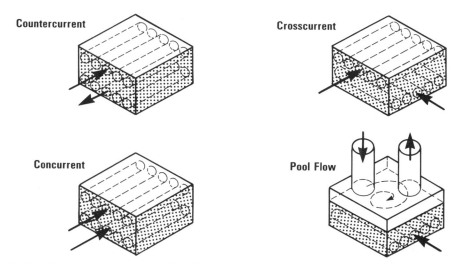

Figure 8.78.   Idealized flow patterns for maternal and fetal blood within the placenta. (Modified from Faber and Hart, 1966.)

the placental membrane, is highly permeable to both gases. Schematically, the exchange unit is represented in Figure 8.79. Oxygenated maternal blood perfuses the uterus at rates of 600–1200 ml/min at term, 85% of which is distributed to the placenta. At the same time, the fetus perfuses the umbilical circulation at flow rates of 600 ml/min at term, approximately 90% of which goes to the *chorionic villi*, which carry the fetal capillaries. Blood spurts into the *intervillous space* and circulates among the chorionic villi, much like water among reeds in a swamp. The maternal blood then returns to the circulation through the uterine veins. Respiratory

gases diffuse between the intervillous space and the terminal fetal capillaries, a distance of approximately 3.5 μ (compared to the distance between the alveolus and pulmonary capillary of less than 0.5 μ). Oxygen, which has now diffused from the intervillous space into the fetal capillary, is then delivered to the fetal tissues via the umbilical vein.

When one samples aliquots of blood from the vessels supplying the placenta, the partial pressures of oxygen observed at steady-state and at sea level are shown in Figure 8.80. Of particular importance is that the P_{O_2} of umbilical venous blood, the most arterialized blood in the fetal circulation, averages 35 torr. Indeed, the fact that the fetus undergoes such a complex growth and maturational process within a low P_{O_2} environment is one of the most interesting and central issues in fetal physiology. Although the P_{O_2} gradient between maternal and fetal blood is responsible for the mechanism of oxygen transfer across the placenta, it does not explain the low P_{O_2} in the umbilical vein. Placental structural characteristics, as well as placental metabolism of oxygen, are responsible for the large differ-

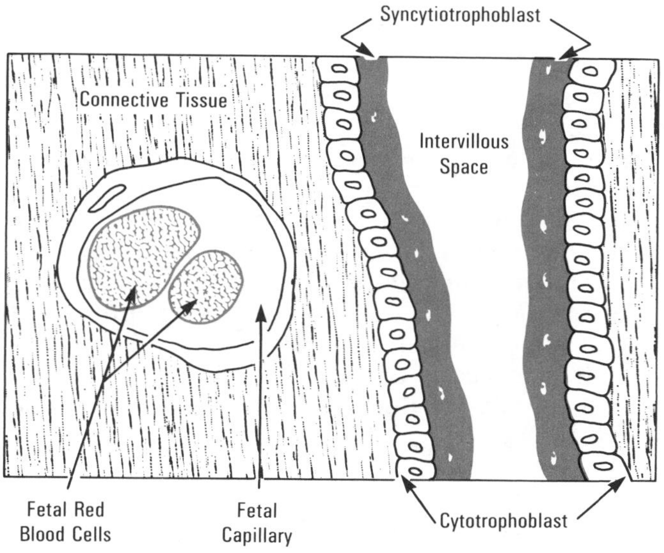

Figure 8.79.   Schematic representation of the respiratory gas exchange surface between the maternal and fetal circulations. The approximate distance between maternal blood in the intervillous space and fetal red blood cells within chorionic villi is 3.5 μ.

Oxygen Pressures in Maternal and Umbilical Blood Vessels

	pO$_2$ torr
Uterine artery	100
Uterine vein	45
Umbilical vein	35
Umbilical artery	20

Figure 8.80.   The partial pressures of oxygen in the four vessels supplying the placenta.

ences between maternal arterial P_{O_2} and umbilical venous P_{O_2}.

Much of our information regarding fetal oxygenation is derived from experiments utilizing sheep models with indwelling, chronically implanted catheters in the four vessels supplying the placenta. Manipulation of maternal inspired gases reveals that the P_{O_2} of umbilical venous blood tends always to equilibrate with that of uterine venous blood, suggesting the concept of a venous equilibration system, which one would expect from a *concurrent exchange* mechanism (see Fig. 8.78). However, the histology of the sheep and human placentas is quite different, leading some to conclude that the human placenta, based upon its microanatomy, is a countercurrent exchange system. There are no data to support this conclusion. Data derived from the rhesus monkey, which has a placenta similar to that of the human, reveals P_{O_2} values which are virtually identical to those observed in the sheep (Wulf et al., 1972). Thus, despite the histological appearance of the human placenta, all currently available evidence suggests that it behaves, from a physiological point of view, as a concurrent exchanger.

The umbilical venous P_{O_2}, as previously noted, equilibrates with but never achieves the same level as uterine venous P_{O_2}. This is due to uneven flow at the placental interface, as well as the high oxygen consumption rate by the placenta itself. There are two important mechanisms by which the fetus adapts to, and indeed thrives, at a low P_{O_2} environment. First, the fetus produces red blood cells which contain hemoglobin with a high affinity for oxygen. The oxyhemoglobin dissociation curve of fetal blood is to the left of maternal, allowing it to bind larger amounts of oxygen at a lower P_{O_2}, ensuring transfer of oxygen from the placenta to fetal blood (Fig. 8.81). Second, fetal cardiac output is much higher than the adult relative to body weight. For example, comparison of perfusion rates of the adult and fetal brain and limbs demonstrates that fetal tissues receive 2.5 times more blood/ml oxygen consumed (Jones et al., 1975; Singh et al., 1984). These two adaptive mechanisms allow the fetus to transfer adequate amounts of oxygen from the placenta to its tissues despite the low P_{O_2}.

The mechanisms regulating elimination of CO_2 by transfer from the fetal to the maternal circulation appear to be identical to those of oxygen transfer, but in the opposite direction. The P_{CO_2} gradient between the umbilical and uterine circulations is less than that of P_{O_2}, but is compensated for by the much higher placental permeability to CO_2. Fetal CO_2 production is equal, mole for mole, to the fetal oxygen consumption rate (James et al., 1972).

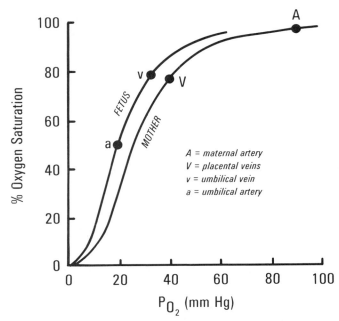

Figure 8.81. Oxyhemoglobin dissociation curves of maternal and fetal blood. The position of the fetal curve allows for greater binding of oxygen at a lower P_{O_2}. (Adapted from Meschia 1979.)

Placental Transfer of Other Substances

ELECTROLYTES

For *sodium* and *chloride* ions, movement across the placenta may be entirely passive (Morriss and Boyd, 1988) in that increases or decreases in maternal concentrations are followed by parallel changes in fetal concentrations. Potassium is actively transported as demonstrated not only by the observation that basal fetal levels are slightly but significantly higher than maternal levels, but also because a reduction in maternal potassium is not accompanied by a parallel decrease in fetal potassium ion concentration.

The fetal plasma concentration of total as well as ionized *calcium* is higher than in maternal plasma. Thus, calcium must be actively transported into the fetus. In addition, there are substantial unidirectional fluxes from mother to fetus as well as from fetus to mother, indicating that passive diffusion of calcium is significant. *Phosphate* is similar to calcium in that active transfer occurs but different in that little passive diffusion occurs (Channing and Boyd, 1984).

IRON

The transplacental transfer of iron delivers 300 mg to the human fetus. This transfer is largely unidirectional as well as against a concentration gradient. However, iron is very tightly bound to transferrin at physiological pH. In addition, only small amounts of transferrin cross the placenta. To overcome this, the placenta has special uptake mechanisms which absorb transferrin, remove the iron, and pass only the iron on to the fetus.

GLUCOSE

Fetal glucose concentrations are slightly lower than maternal values, indicating that glucose transfer may be by passive diffusion. Experiments have shown that a majority of glucose transfer is by facilitated rather than passive diffusion. The concept of facilitated diffusion was first introduced to explain the transfer of glucose in the sheep placenta (Morriss and Boyd, 1988). This transfer process saturates a supraphysiological maternal concentration and there is specificity for D-glucose (the naturally occurring isomer) as demonstrated in studies where L-glucose is transferred at significantly lower rates (Faber and Thornburg, 1983).

The placenta is a major metabolic organ and consumes large quantities of glucose and other substrates. Late in gestation the placenta consumes 60% of the glucose removed from the maternal circulation and passes less than 40% to the fetus. Most of the glucose consumed by the placenta is converted to lactate which passes into both the fetal and maternal circulations, with a preference to the fetus.

AMINO ACIDS

Although diffusion occurs in both directions, the concentrations of most amino acids are higher in fetal than maternal plasma with even higher concentrations within the placental tissues. On average, fetal concentrations are twice those of maternal. In addition, stereospecificity for maternal to fetal transfer exists in that the unnatural D-isomer traverses the placenta more slowly than the natural L-isomer. This specificity for the L- over the D-isomers does not exist in the fetal to maternal direction and movement out of the fetus is at the same rate for both isomers.

IMMUNOGLOBULINS

In general, large molecules such as proteins cross the placenta only slowly if at all. The primary exception is that there are special transfer mechanisms for the transport of immunoglobulins which import passive immunity to the fetus prior to birth. In humans, this occurs by vesicular transport and involves specific receptors on the maternal surface of the placenta. Selection is very specific in that only IgG molecules readily cross the placenta whereas immunoglobulins A, M, D, and E either do not cross the placenta or cross only in small quantities (Adinolfi and Stern, 1984).

A potentially serious consequence of this IgG transfer occurs when Rh-positive mothers have anti-Rh antibodies which are transferred into the circulation of Rh-negative fetuses. Maternal anti-Rh antibodies have been found in fetuses as early as 10 weeks' gestation. They bind to and cause rapid hemolysis of fetal red cells, producing the potentially fatal disease, erythroblastosis fetalis.

Fetal Nutrient Requirements

The fetus uses nutrients not only to support its rapid growth but also to provide energy to run its metabolic machinery. Experiments in animals have shown that the fetus utilizes carbohydrates and amino acids almost exclusively with little utilization of free fatty acids. In contrast, there is a dramatic shift in the source of calories to fat after birth (Fig. 8.82).

The amount of calories needed by the fetus for normal growth depends on the composition of the new tissues as they are formed as well as on the rate of growth. At the end of the first trimester, the human fetus is approximately 95% water with little fat. The water content of the fetus gradually decreases throughout gestation and averages 85% at 32 weeks' gestation. Beginning at this time and continuing until birth, the fetus deposits significant quantities of fat so that the normal term fetus is composed of 65–70% water at birth. Because the calorie requirement for the accretion of fat (9.5 kcal/g) is much greater than that of nonfat tissue (0.75 kcal/g), the caloric requirements of the fetus (per unit body weight) may be much greater late in gestation when fat is being deposited. Whether this is true has yet to be established but it should also be recognized that the rate of growth of the fetus (%/day) is decreasing at this time.

Overall, the factors which regulate the transplacental delivery of nutrients to the fetus are not well understood. One frequently asked question is whether the inability of the placenta to transfer adequate amounts of nutrients is the cause of fetal growth retardation. The term "placental insufficiency" is widely used to describe such a condition but our understanding of this is incomplete.

FETUS

Physiological function of most organ systems begins either in the embryonic (postfertilization weeks 2–7) or early fetal period. Even though individual organ systems are functioning in the fetus, their mode of function is often dramatically different from the normal physiology of the adult. This occurs as a consequence of anatomic and structural

THE DIET OF THE FETUS AND THE NEWBORN

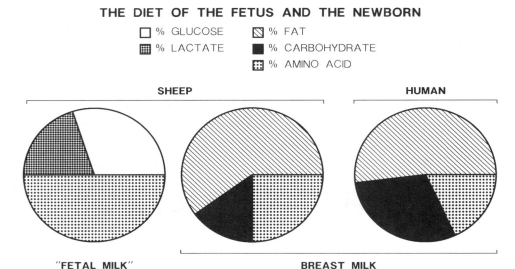

☐ % GLUCOSE   ▨ % FAT
▦ % LACTATE   ■ % CARBOHYDRATE
▦ % AMINO ACID

SHEEP   HUMAN

"FETAL MILK"   BREAST MILK

Figure 8.82. Calories delivered to the fetus and neonate. (Adapted from Battaglia and Meschia, 1986.)

immaturity as well as immaturity of the regulatory systems.

Fetal Growth and Metabolism

The fertilized ovum implants within the endometrium 4–5 days after reaching the uterine cavity. Following progressive cell division and growth, the embryonic stage begins 2 weeks after fertilization and the fetal stage begins at the end of the 7th week when the outline of most organs has been established.

Fetal weight changes during human gestation are shown in Figure 8.83, with average weight at delivery being 3300 g. The amount of weight gained by the fetus increases progressively to a maximum of 230 g/week at 34 weeks' gestation and declines thereafter (Fig. 8.84). Note that the weekly fetal growth rate is a maximum of 45% of body weight per week at the beginning of the fetal period and gradually declines to 3% per week at term (40 weeks from the last menstrual period).

Animal studies performed largely in fetal lambs in the steady-state have shown that the fetus utilizes glucose and lactate as its major metabolic substrates, with amino acids supplying 40–50% of the calories delivered to the fetus. Lactate normally is present in fetal plasma in concentrations higher than in the mother, and is derived from glucose metabolism by the placenta as well as from fetal conversion of glucose to lactate. Roughly half of the glucose, lactate, and amino acids is used for accretion of new tissues with the remainder being used to provide energy. Free fatty acids seem to be little utilized by the fetus and this may be due to a small capacity for transport across the placenta.

How the delivery of metabolic substrates affects fetal growth has not been established although there are several observations which suggest a direct link. Pregnant women with noninsulin-dependent diabetes mellitus have elevated plasma concentrations of glucose, fatty acids, and insulin (Hollingsworth and Cousins, 1984). The glucose and fatty acids cross the placenta in larger than normal amounts and may, together with elevated fetal insulin levels, promote fetal growth. These metabolic alterations may contribute to the macrosomic

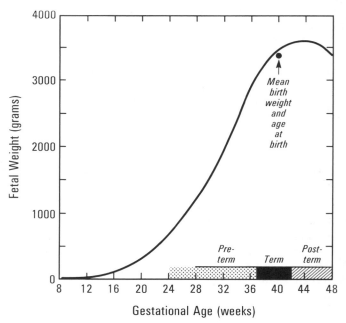

Figure 8.83. Variation in fetal weight with gestational age. (Data from Williams et al., 1982; Javert, 1957.)

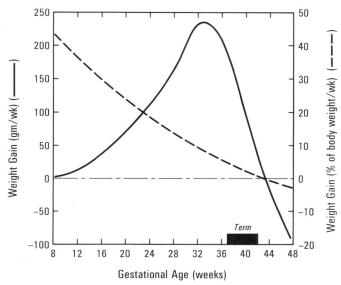

Figure 8.84. Gestational changes in the rate at which fetuses gain weight. (Data calculated from Fig. 8.83.)

Table 8.13.
Influences on Variations in Birth Weight[a]

	% Contribution
Genetic	
Maternal genotype	20%
Fetal genotype	15%
Fetal sex	3%
	38%
Environmental	
General maternal environment	18%
Immediate maternal environment	6%
Maternal age and parity	8%
	32%
Unknown	30%
Total	100%

[a]Source: Gluckman and Liggins, 1984.

(abnormally large) babies born to women with poorly controlled diabetes mellitus. Conversely, growth-retarded human fetuses usually are hypoxic and acidotic and have reduced amounts of amino acids in their blood (Cetin et al., 1988; Nicolaides et al., 1987).

Whether these variations in fetal growth are attributable directly to the amounts of metabolic substrates delivered to the fetus or are mediated secondarily through the effects of various growth factors has not yet been established. It is known that growth hormone has little effect on fetal growth because anencephalic fetuses without pituitary tissue grow to essentially normal size. Fetuses with renal agenesis are usually growth retarded, suggesting that a growth factor may be derived from the fetal kidneys. Insulin and insulin-like growth factors are thought to be major regulators of fetal growth (Gluckman and Liggins, 1984). In addition, both genetic and environmental factors are significant determinants of birth weight as shown in Table 8.13.

Fetal Circulation

DEVELOPMENT OF HEART AND BLOOD VESSELS

In the human fetus, the heart tube and primitive blood vessels begin to develop toward the end of the 3rd week after fertilization. The heart begins to beat and the fetal circulation is established at the end of the 3rd week or beginning of the 4th week (postfertilization day 21 or 22) when the fetus is approximately 0.4 mm in diameter. Thus, the cardiovascular system is the first system to begin functioning in the fetus. Structurally, the heart and vasculature continue to develop, with the addition of atria, septum, and blood vessels over the next few weeks so that the major structural features are completed by the end of the 7th week. After this time, the heart enlarges and the blood vessels proliferate as new tissues develop but the basic cardiovascular structures change little.

SPECIAL FETAL CARDIOVASCULAR STRUCTURES

There are four unique structures in the fetus which have major effects on the fetal circulation and these are depicted in Figure 8.85.

1. *Ductus arteriosus*—a large opening between the pulmonary artery and aortic arch which shunts most of the blood flowing out of the right ventricle into the descending aorta rather than into the lungs.
2. *Placenta*—a low resistance pathway which receives 35–40% of the total fetal cardiac output.
3. *Ductus venosus*—a connection between the umbilical vein and inferior vena cava which allows the oxygen-rich blood returning from the placenta to bypass the liver and flow into the inferior vena cava.
4. *Foramen ovale*—an opening in the septum between the left and right atria which allows blood to flow from the inferior vena cava into the left atrium. The foramen ovale is covered by a flap which readily permits a right to left atrial flow but prevents flow in the opposite direction.

There is an additional feature of the fetal circulation which contributes significantly to the pattern of oxygen delivery. As the poorly oxygenated blood from the lower inferior vena cava and the richly oxygenated blood from the ductus venosus meet and flow toward the heart, there is little mixing. The result is that the richly oxygenated blood flows preferentially through the foramen ovale, into the

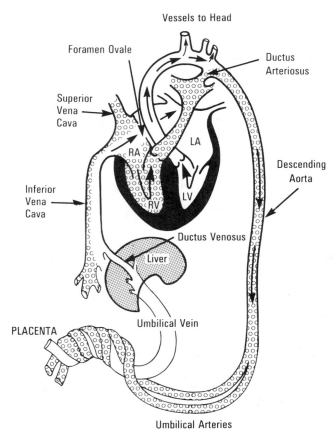

Vessels to Head

Foramen Ovale

Ductus Arteriosus

Superior Vena Cava

Descending Aorta

Inferior Vena Cava

LA

RA

LV

RV

Ductus Venosus

Liver

Umbilical Vein

PLACENTA

Umbilical Arteries

Figure 8.85. Diagram of the normal fetal circulation.

left heart, and into the ascending aorta, thereby supplying blood with a higher oxygen content to the heart and brain.

FORMATION OF RED BLOOD CELLS

During the early stages of development of the embryo, red cells are formed in the yolk sac. The liver, and to a lesser extent the spleen, soon take over this erythropoietic function and continue to produce red cells until late in gestation. Starting with the beginning of the third trimester, bone marrow becomes a major source of the red cells which circulate in the fetus.

The hemoglobin formed early in gestation is referred to as embryonic hemoglobin. A switch to fetal hemoglobin occurs at the transition from the embryonic to the fetal period and another switch from fetal to adult hemoglobin occurs at approximately 34 weeks of gestation. These different forms of hemoglobin differ in the nature of the globin but not the α chain. Functionally, a major difference between fetal and adult hemoglobin is that the fetal hemoglobin has a higher affinity for oxygen. Because of this, fetal hemoglobin has a higher saturation at the same oxygen tension than does adult hemoglobin. This effects the transplacental delivery

of oxygen to the fetus. Even though the fetus produces adult hemoglobin for the last 5–6 weeks of gestation, fetal hemoglobin persists into the neonatal period due to the long life of red cells in the circulation.

Erythropoietin, the hormone which stimulates the formation of red cells, is present with the fetal liver being the major source. Fetal plasma erythropoietin concentration increases with gestational age and there is a parallel increase in fetal hemoglobin concentration and hematocrit (Brace, 1988).

DISTRIBUTION OF FETAL BLOOD

The amount of fetal blood within the cord and fetal side of the placenta has been estimated to be approximately 30% of the total circulating volume during the latter half of gestation. Acute hypoxia causes a redistribution of fetal blood out of the placenta and into the fetal body (Oh et al., 1975). Fetuses which are chronically hypoxic also have reduced amounts of blood within the placenta but the mechanisms which cause these changes are unknown.

FETAL ARTERIAL PRESSURE AND VENOUS PRESSURES

Little is known about normal values for arterial or venous pressure in the human fetus. From animal studies in a variety of species, it is clear that fetal arterial pressure is near zero shortly after the heart first begins to beat (Nakazawa et al., 1988). Fetal arterial pressure increases gradually with advancing gestational age and then increases more abruptly at the time of labor and delivery. Arterial pressure in the term fetus correlates positively with fetal maturity at the time of birth and averages roughly half of adult values in fetal monkeys and sheep.

From animal studies, the average fetal venous pressure is 2–4 torr and is the same as in the adult if appropriate zero pressure reference is taken. There is no indication of gestational changes.

FETAL HEART RATE

At 20 weeks of gestation, the average heart rate is 155 bpm, decreasing to 144 bpm at 30 weeks, and 140 bpm at term prior to labor. Variations of 20 bpm above or below these values are taken as normal whereas variations of 40 bpm or greater may be indicative of fetal problems (Parer, 1984).

Both the parasympathetic and sympathetic nervous systems have major influences on fetal heart rate. Atropine (parasympathetic blocker with dose-dependent ganglionic blockade) administration to human fetuses at term elevates heart rate by 20 bpm and decreases beat-to-beat variability. Propranolol (β-receptor blocker) administration to monkey fetuses suppresses mean heart rate by only 10 bpm and decreases variability. Thus both the sympathetics and parasympathetics influence basal fetal heart rate as well as heart rate variability but the parasympathetic system dominates under resting conditions.

Fetal heart rate is reduced during acute episodes of hypoxia. Although both sympathetic and parasympathetic stimulation to the heart is increased, the parasympathetic system dominates so that heart rate decreases, sometimes to as low as 60 bpm. Conversely, chronic long-term fetal hypoxia is sometimes associated with an elevated fetal heart rate. The mechanisms underlying this paradox have yet to be fully clarified but appear to involve a decrease in parasympathetic tone and a general increase in myocardial sensitivity to catecholamines.

During severe hypoxia, the concentration of arginine vasopressin (AVP) in the fetal plasma also increases. AVP directly reduces heart rate independent of nervous stimulation to the heart. Thus, chronic fetal hypoxia can be associated with either maintained increases or maintained decreases in heart rate, depending in part on the actual changes in plasma catecholamines relative to the changes in AVP.

FETAL CARDIAC OUTPUT AND ITS DISTRIBUTION

In the adult, blood flows out of the left ventricle, through the systemic circulation, the right heart, the lungs, and back to the left heart. Thus, the left and right halves of the adult heart work in series and measurement of either right or left heart output provides an equivalent measurement of cardiac output. In the fetus, most of the blood pumped by the right heart bypasses the lungs and enters the aorta through the ductus arteriosus. Thus the two ventricles of the fetal heart pump in parallel and the term "combined ventricular output" is used to describe cardiac output in the fetus.

Fetal cardiac output has been most extensively studied in the chronically catheterized sheep. Under resting conditions, cardiac output in this animal model averages 500 ml/min/kg of body weight (Rudolph and Heymann, 1967; Walker, 1984). This number is enormous in comparison to the adult with a cardiac output of only 80 ml/min/kg. In addition, the distribution of this blood to the various organs in the body is quite different from that of the adult as seen in Table 8.14.

Estimates of cardiac output in human fetuses using ultrasonographic techniques are somewhat lower and average 350 ml/min/kg of fetal body weight. However, the hematocrit of the near-term human fetus averages 50% greater than in the

Table 8.14.
Distribution of Cardiac Output (% of Total)

	Fetus	Adult
Lungs	6	100
Heart	5	5
Kidneys	2	20
Brain	20	20
Skeletal muscle/bone	20	20
Splanchnic bed	7	30
Placenta	40	

sheep fetus so that oxygen delivery may be comparable.

REGULATION OF FETAL CARDIAC OUTPUT

Unlike the adult, cardiac output in the fetus is at its maximum under resting conditions. This occurs because the amount of work which the fetal heart can perform is maximal under basal conditions (Gilbert, 1980). Thus, it is not possible for the fetus to augment its cardiac output acutely in order to deliver more nutrients such as oxygen to its tissues. From the perspective of work performed, if either arterial pressure or blood viscosity were to increase, fetal cardiac output would decrease proportionately, and vice versa.

During stress conditions, such as acute hypoxia, there is a redistribution of cardiac output in the fetus in favor of the vital organs, i.e., the heart, brain, adrenals, and placenta and away from skin, skeletal muscle, and bone (Cohn et al., 1974). This occurs due to regional differences in vascular responses to increases in plasma concentrations of vasoconstrictor hormones and to regional differences in sympathetic innervation as well as sympathetic tone.

Fetal cardiac output decreases during severe hypoxia due to an increase in afterload caused by a concomitant arterial hypertension. In addition, cardiac output decreases when heart rate is elevated beyond normal limits due to a reduction in filling time. At one time it was widely believed that stroke volume in the fetus was relatively constant so that cardiac output varied in direct proportion to heart rate. It is now known that changes in cardiac output in the fetus are brought about by changes in the cardiac function just as occurs in the adult. However, it must be recognized that the fetus normally operates on the plateau of its cardiac function curve.

RHEOLOGICAL FACTORS WHICH EFFECT FETAL HEMODYNAMICS

The viscosity of blood potentially has major effects on fetal cardiovascular function. The two primary determinants of blood viscosity are the plasma protein concentration and hematocrit. Fetal plasma protein concentration is very low early in gestation and gradually increases with gestational age to a little more than one-half of the adult levels at term. Fetal hematocrit increases from 30% at midgestation to 44% at term prior to labor, with further increases during labor and delivery (Teramo et al., 1987).

Fetal Renal Function

Presently, little is known about kidney function in the human fetus. Even with major technological developments, including ultrasonography, understanding fetal renal function remains an elusive task. Consequently, a large majority of the details known about kidney function in the fetus has been derived from animal studies late in gestation.

Anatomically, the fetal kidneys in humans begin to develop during the 4th week of gestation, i.e., in the early embryonic period, with the formation of the pronephros, mesonephros, and then the metanephros, which is the permanent embryonic kidney. Urine is first formed and enters the urinary bladder around 9–10 weeks' gestation with the fetus weighing less than 0.01 kg. After this time, the kidneys continue to grow with an increase in the number of loops of Henle which penetrate the renal medulla, but the fetal kidney never establishes an intrarenal osmolality gradient with hyperosmolality in the medulla which is characteristic of the adult kidney.

MEASUREMENT OF FETAL URINARY OUTPUT

By combining high resolution ultrasonography with the assumption that the fetal bladder is ellipsoidal in shape, the rate of urine formation has been calculated in the normal human fetus from changes in dimensions of the bladder observed over time. In animal models, urine flow has been measured directly by catheterizing the fetal urinary bladder and performing timed collections. A significant observation was that the stress of fetal surgery produces major changes in renal function, presumably due to elevations in vasopressin concentration. Urine flow was reduced and osmolality elevated and several days were required for renal function to return to normal (Gresham et al., 1972).

FETAL URINE FLOW RATES

At 18 weeks of gestation, the human fetus produces 7–14 ml/day of urine. This increases to 50 ml/day at 22 weeks, 250 ml/day at 30 weeks, and 600 ml/day at term, i.e., 40 weeks (Kurjak et al., 1981). The significance of these numbers is made clear by normalizing with respect to fetal weight. Thus, throughout the last half of gestation, human fetuses excrete a volume of urine equivalent to 20–25% of

their body weight per day. A comparable number in the normal adult is 2–3% of body weight per day. In addition, it appears that the rate of fetal urine production decreases prior to labor and delivery and in postterm gestations.

COMPOSITION OF FETAL URINE

Although the basic solutes in fetal urine, i.e., sodium, chloride, urea, are the same as in the adult, there are major differences in the total concentration of these solutes. In contrast to the normal hypertonic urine that the adult produces, fetal urine is markedly hypotonic, with normal osmolality ranging from 60–220 mosm/kg during the third trimester. Fetal urine osmolality may be near isotonic with plasma when the kidneys first begin to function. In addition, fetal urine osmolality increases slightly very late in gestation for unknown reasons but remains hypotonic relative to plasma. Further increases in urine osmolality occur during labor and delivery, presumably due to increased arginine vasopressin concentrations.

BLOOD FLOW TO THE KIDNEYS

Renal blood flow has been measured in late-gestation fetal sheep and found to average only 2% of combined ventricular output compared with 20% of cardiac output in the adult. However, the fetus has a much higher cardiac output per unit body weight so that fetal renal blood flow normalized to body weight averages 75% of adult values.

GLOMERULAR FILTRATION RATE (GFR), FILTRATION FRACTION, AND FRACTIONAL SODIUM EXCRETION

Fetal GFR is stable and averages 1.0 ml/min/kg of body weight throughout the last half of gestation in fetal sheep (Robillard, 1988). Over this same time period, the filtration fraction increases from a value of 6% at the end of the second trimester to reach a value of 10% at term. The fractional excretion of sodium averages 8% and is stable with time throughout the last trimester.

ENDOCRINE REGULATION OF RENAL FUNCTION

The hormones arginine vasopressin (AVP), atrial natriuretic factor (ANF), angiotensin, and aldosterone are present in the fetus either before or when the kidneys first begin to function. Animal studies during the last trimester have shown that each of these hormones affects renal function in that they alter urinary output and/or electrolyte excretion. Small amounts of intravascularly infused AVP decrease the rate of urine formation and increase urine osmolality due to a decrease in free water clearance whereas larger amounts can cause a diuresis because of a pressor effect of the hormone (Lingwood et al., 1978). The immature kidney is less sensitive to AVP in that urine osmolality under maximal AVP stimulation is less in the immature fetus than in the mature fetus. Neither the immature nor mature fetus, however, can produce urine which is significantly hypertonic to plasma. Administration of exogenous aldosterone changes excretion of sodium and potassium in the fetus similar to that which occurs in the adult. ANF also appears to modulate renal function in the fetus. Under resting conditions, there is a positive correlation between fetal urine flow rate and plasma ANF concentration. In addition, intravenous infusion of ANF produces a comparable diuresis and natriuresis in fetus and adult. Finally, cortisol has effects on renal function which are unique to the fetus in that intravenous infusions of sufficient cortisol to increase plasma concentration four fold cause urine flow to double while electrolyte excretion quadruples (Wintour et al., 1985). Thus, it appears that all of the endocrine regulatory pathways are present and functional in the fetus at, or perhaps even before, the time when the kidneys themselves begin to function.

Fetal Pulmonary Physiology

At one time it was thought that the fetal lungs were collapsed in utero and would expand with the first breath of air following birth. Instead of being collapsed, it is now known that the fetal lungs are expanded and have a normal volume prior to birth. However, in contrast to the adult lungs, the fetal lungs are filled with liquid rather than gas. For many years it was thought this liquid was derived from amniotic fluid and was absorbed across the alveolar surface into the fetal circulation. The opposite is now known to occur in that the liquid in the fetal lungs is formed within the lungs (Johnston and Gluckman, 1986). This occurs as a consequence of and active secretion of chloride ions across the alveolar and tracheal surfaces into the future airways. Fetal lung fluid is isotonic to plasma and has a sodium ion concentration similar to plasma but a much higher chloride ion concentration (150 versus 110 meq/liter). Animal studies have shown that in late gestation an average of 200–500 ml/day (10–15% of body weight/day) of lung fluid flows out of the trachea and either enters the amniotic fluid or is swallowed as it exits the trachea (Harding et al., 1984). It is believed that this secretion of fluid into the fetal lungs plays an important role in expanding the fetal lungs as they grow and plays a major role in preventing the debris in the amniotic fluid from entering the lungs. The rate of formation of fetal lung fluid decreases just prior to birth due primarily

to catecholamine stimulation of pulmonary β-adrenergic receptors, although other factors also contribute.

As described in the fetal circulation section above, most of the blood pumped by the right ventricle into the pulmonary artery of the fetus passes through the ductus arteriosus rather than through the circulation of the lungs. This occurs because pulmonary artery pressure is 5 mm Hg greater than aortic pressure, as a consequence of the high resistance of the pulmonary vascular bed in combination with a low resistance of the ductus arteriosus. Although the details of the biochemical pathways which maintain the constriction of the pulmonary circulation and dilation of the ductus arteriosus have not been fully defined, it is clear that local oxygen tension plays a direct role as do local concentrations of prostaglandins. Prostaglandin E_2 plays a dominant role in maintaining the normal dilated state of the ductus arteriosus and data are beginning to suggest that the leukotrienes may be involved in producing the tonic constriction of the fetal pulmonary vasculature.

FETAL BREATHING MOVEMENTS

Although claimed for several centuries, it has only recently become firmly established that the normal fetus undergoes episodes of rhythmic breathing-like activity in utero and these are termed fetal breathing movements (Patrick, 1984). During fetal breathing movements, there are rhythmic contractions of the diaphragm and intercostal muscles similar to those which occur during normal air breathing in the adult, except that the frequency is higher and averages 50/min. These occur as early as 11 weeks of gestation and continue until delivery. In humans, fetal breathing movements are present 30% of the time during the third trimester. Although the changes in intrathoracic pressure are comparable to those which occur during air breathing in the adult, little volume moves in or out of the trachea during fetal breathing movements due to the high viscosity of the tracheal fluid. It is thought that these breathing-like activities play an important role in the development of the respiratory muscles, and prepare them for their continuous rhythmic activity following birth.

The factors which control the extent to which the fetus undergoes fetal breathing movements are beginning to be described. The incidence of fetal breathing movements increases after meals, in response to a glucose load, and during the night. Conversely, the incidence decreases during fetal hypoxia and there is a progressive decrease in the last few days prior to normal labor and delivery.

There are also indications that the absence of fetal breathing movements is a strong indicator of impending delivery, including during the preterm period (Agustsson and Patel, 1987).

In anatomically normal fetuses which are born prematurely, gas exchange capacity of the lungs is the major factor which determines whether the neonate will survive. This is largely determined by the amount of surfactant present in the lung at the time of birth because the alveoli cannot be patent without adequate amounts of surfactant. Presently, regulation of surfactant production by the fetus is poorly understood. A few neonates born as early as 25 weeks of gestation and weighing less than 1 kg have adequate surfactant for ventilation as well as anatomic lung maturity and survive. On the other hand, a few full-term fetuses have an inadequate amount of surfactant in their lungs and develop respiratory distress syndrome.

Fetal Gastrointestinal Physiology

Each day the fetus swallows large volumes of amniotic fluid which are processed by the gastrointestinal system. Estimates of the volume of amniotic fluid swallowed by the fetus late in gestation vary from 200–1000 ml/day and these appear to correlate with fetal urinary output. These measurements underestimate the total volume swallowed by the fetus because much of the fluid flowing out of the fetal lungs is swallowed prior to entering the amniotic fluid and because there appear to be additional fluids from the mouth and nasal cavities which are swallowed by the fetus.

This large volume of fluid, the electrolytes, and other solutes are absorbed by the fetal gastrointestinal system into the fetal circulation. Although the digestive enzymes and digestive processes are functional, the fetus has a limited ability to absorb glucose and amino acids as determined by injection of these substances into the amniotic fluid. It has been estimated from direct intragastric infusions that the fetus can absorb a maximum of 40% of its caloric requirement needed to maintain growth (Charlton, 1984). This is not a problem for the fetus because the amniotic fluid normally contains only small amounts of hydrocarbons and amino acids.

Even though the fetal gastrointestinal system processes large volumes of fluid each day, the fetus does not usually defecate in utero. Instead, a small amount of solids (meconium) accumulates in the lower segment of the intestinal tract. Fetal defecation may occur under conditions of severe hypoxia and appears related to greatly elevated AVP concentrations. This may cause severe problems for the newborn because aspiration of amniotic fluid containing meconium into the lungs at the time of birth obstructs the airways and damages the lungs due to lung tissue reactions to the meconium.

Amniotic Fluid Dynamics

As the fetus develops, it is surrounded by a pool of amniotic fluid. Amniotic fluid volume averages 125 ml at 15 weeks of gestation, increasing to a maximum of 900 ml at 35 weeks and decreasing thereafter (Fig. 8.86). Fetal urine is the major source of amniotic fluid during the latter half of pregnancy because little amniotic fluid is present (oligohydramnios) after midgestation if the fetus does not form and excrete urine. Less is known about the source of amniotic fluid early in gestation although it is clear that amniotic fluid is present long before the fetal kidneys produce urine. The most likely source of this early amniotic fluid is the amnion and fetal skin, with active transport of electrolytes into the amniotic cavity causing water to follow passively. At 24–26 weeks' gestation, the fetal skin undergoes keratinization so that little water or solute crosses the skin after this time.

Although aberrations in amniotic fluid volume either above or below normal are important clinically because they are associated with poor outcome (Chamberlain et al., 1984), our knowledge of the mechanisms which regulate amniotic fluid volume is incomplete (Brace, 1986; 1988). In addition to urine during the latter half of gestation, the outflow of fetal lung fluid contributes to amniotic volume. Fetal swallowing is the major route for removal of amniotic fluid and the potential for

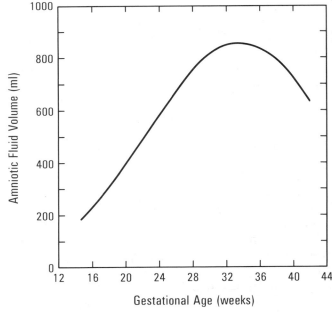

Figure 8.86.   Normal amniotic fluid volume during pregnancy.

movement of fluid and solute across the membranes (amnion and chorion) is great because of the large surface area. However, an understanding of the mechanisms which regulate fluid transfer through each of these routes, or amniotic fluid volume as a whole, is not presently available.

Transitions at Birth and Neonatal Physiology

In the last few days prior to labor and delivery, the fetus undergoes a great number of changes which are indicative of its preparation for birth. These include an increase in arterial pressure, an elevation in hematocrit, decreases in the amount of fluid secreted by the lungs and excreted by the kidney, and a decrease in frequency of fetal breathing movements. These and other changes are all brought about by changes in the hormonal milieu. In addition, during the process of birth, there is a massive release of catecholamines and other vasoactive hormones which further stimulate the neonate and assist in the transition from fetal to neonatal life.

Clamping of the umbilical cord and the first breath have dramatic effects on the circulatory system. With removal of the placenta, systemic vascular resistance almost doubles so that arterial pressure increases as does afterload on the heart. With the first breath, pulmonary vascular resistance is reduced by a factor of 5. The consequence is that pulmonary arterial pressure falls and the increase in pulmonary venous blood flow elevates left atrial pressure by a few mm Hg, a change which is sufficient to close the foramen ovale. The ductus arteriosus closes over a period of a few hours, thereby establishing the adult blood flow pattern in the circulation.

The stimuli for the first breath are multiple: tactile stimulation from the skin, decreased temperature due to evaporation, and chemoreceptor drive arising from the asphyxia following cord clamping. The resulting increase in pulmonary oxygen tension is largely responsible for the decrease in pulmonary vascular resistance. This same rise in P_{O_2} has oppo-

site effects elsewhere in that it causes the smooth muscle cells within the ductus arteriosus and umbilical vessels to constrict, although stretching of the cord is also a primary stimulus for constriction.

If the umbilical cord is not clamped immediately at the time of delivery, up to 75 ml (i.e., 20–25 ml/kg of newborn weight) of blood are transferred out of the placenta and into the circulation of the newborn during the process of placental transfusion which naturally occurs at the time of birth. In humans, the extent of this placental transfusion increases with postdelivery time of cord clamping and reaches a maximum at roughly 3 min. This placental transfusion is now thought to be detrimental, resulting in a vascular volume overload. The hematocrit can undergo large increases after delivery if the placental transfusion is extensive because the volume overload elevates vascular pressures, with a resultant transcapillary filtration of fluid while leaving the red cells in the circulation. Neonates with hematocrits greater than 65% are exchange transfused with saline or colloids in an attempt to reduce blood viscosity.

There is an early postnatal shift from the liver to the kidney as the major source of erythropoietin in the newborn. This occurs gradually over a period of a few months and the neonate undergoes a period of relative anemia during this time.

Although some of the physiological differences between the late-gestation fetus and the adult have been emphasized, physiological functions continue to change throughout life. The early-gestation fetus is different from that in late gestation, which, in turn, is different from the neonate, the juvenile, the adult, and the older individual. Thus, changes in physiological functions with age are a continuum. It is important to recognize that the fetus and mother are in constant communication mostly through hormonal messages so that there is an integration of maternal and fetal physiology to ensure normal fetal development, while maintaining physiological integrity of the maternal organism.

BIBLIOGRAPHY

ABDEL-MEGUID, S. S., H. S. SHIEH, W. W. SMITH, H. E. DAYRINGER, B. N. VIOLAND, AND L. A. BENTLE. Three-dimensional structure of a genetically engineered variant of porcine growth hormone. *Proc. Natl. Acad. Sci. U.S.A.* 84: 6434–6437, 1987.

ABOUL-KHAIR, S. A., J. CROOKS, A. C. TURNBULL, ET AL. The physiologic changes in thyroid function during pregnancy. *Clin. Sci.* 27: 195, 1964.

ABRAHAM, J. A., A. MERGIA, J. L. WHANG, A. TUMOLO, J. FRIEDMAN, K. A. HJERRILD, D. GOSPODAROW-ICZ, AND J. C. FIDDES. Nucleotide sequence of a bovine clone encoding the angiogenic protein, basic fibroblast growth factor. *Science* 233: 545–548, 1986.

ADINOLFI, M., AND C. STERN. Ontogeny of acquired immunity and maternofetal immunological interactions. In: *Fetal Physiology and Medicine*, edited by R. W. Beard and P. W. Nathanielsz. New York: Marcel Dekker, Inc., 1984.

AGUSTSSON, P., AND N. B. PATEL. The predictive value of fetal breathing movements in the diagnosis of preterm labor. *Br. J. Obstet. Gynecol.* 94: 860–863, 1987.

ALAILY, A. B., AND K. B. CARROL. Pulmonary ventilation in pregnancy. *Br. J. Obstet. Gynecol.* 85: 518, 1978.

ALLEN, R. G., R. HERBERT, M. HINMAN, H. SHIBUYA, AND C. B. PERT. Coordinate control of corticotropin, β-lipotropin and β-endorphin release in mouse pituitary cell cultures. *Proc. Natl. Acad. Sci. U.S.A.* 75: 4972–4976, 1978.

AMARA, S. G., V. JONAS, M. G. ROSENFELD, E. S. ONG, AND R. M. EVANS. Alternate RNA processing in calcitonin gene expression generates mRNAs encoding different polypeptide gene products. *Nature 298:* 240–244, 1982.

AMARA, S. G., M. G. ROSENFELD, R. S. BIRNBAUM, AND B. A. ROOS. Identification of the putative cell-free translation product of rat calcitonin mRNA. *J. Biol. Chem.* 255: 2645–2648, 1980.

ANDERSON, D. J., AND R. AXEL. A bipotential neuroendocrine precursor whose choice of cell fate is determined by NGF and glucocorticoids. *Cell* 47: 1079–1090, 1986.

ANDERSON, R. R. Endocrinological control. In: *Lactation. A Complete Treatise*, Vol. 1, edited by B. L. Larson and V. R. Smith. New York: Academic Press, 1974, p. 97.

ANDRES, R. Y., I. JENG, AND R. A. BRADSHAW. Nerve growth factor receptors: identification of distinct classes in plasma membranes and nuclei of embryonic dorsal root neurons. *Proc. Natl. Acad. Sci. U.S.A.* 74: 2785–2789, 1977.

ARAGONA, C., AND H. G. FRIESEN. Lactation and galactorrhea. In: *Endocrinology*, Vol III, edited by L. J. DeGroot, G. F. Cahill Jr., L. Martini, D. H. Nelson, W. D. Odell, J. T. Potts, E. Steinberger, A. I. Winegrad. New York: Grune & Stratton, Inc., 1979, p. 1613.

ASCHHEIM, S., AND B. ZONDEK. Anterior pituitary hormone and ovarian hormone in the urine of pregnant women. *Klin. Wochenschr.* 61: 1322, 1927.

AUBERT, M. L., M. M. GRUMBACH, AND S. L. KAPLAN. The ontogenesis of human fetal hormones. III. Prolactin. *J. Clin. Invest.* 56: 155–164, 1975.

AUSTIN, L. A., AND H. HEATH, III. Calcitonin. Physiology and pathophysiology. *N. Engl. J. Med.* 304: 269–278, 1981.

BAKER, T. G., AND SUM, O. W. Development of the ovary and oogenesis. *Clin. Obstet. Gynecol.* 3: 3–26, 1976.

BARBOUR, G. L., J. W. COBURN, E. SLATOPOLSKY, A. W. NORMAN, AND R. L. HORST. Hypercalcemia in an anephric patient with sarcoidosis: Evidence of extrarenal generation of 1,25-dihydroxyvitamin D. *N. Engl. J. Med.* 305: 440–443, 1981.

BARGMANN, C. I., M-C. HUNG, AND R. A. WEINBERG. Multiple independent activations of the *neu* oncogene by a point mutation altering the transmembrane domain of p185. *Cell* 45: 649–657, 1986.

BARNES, D., AND G. SATO. Serum-free cell culture: a unifying approach. *Cell* 22: 649–655, 1980.

BASSETT, J. M., G. D. THORNBURN. Foetal plasma corticosteroids and the initiation of parturition in the sheep. *J. Endocrinol.* 44: 285–286, 1969.

BATTAGLIA, F. C., AND G. MESCHIA. Fetal respiratory physiology. In: *An Introduction of Fetal Physiology*. Orlando: Academic Press, Inc., 1986a, p. 154.

BATTAGLIA, F. C., AND G. MESCHIA. *An Introduction to Fetal Physiology*. Orlando: Academic Press, 1986b.

BAXTER, J. D. Glucocorticoid hormone action. In: *Pharmacology of Adrenal Cortical Hormones*, edited by G. N. Gill. Oxford: Pergamon Press, 1979, p. 67–121.

BAXTER, J. D., N. L. EBERHARDT, J. W. APRILETTI, L. K. JOHNSON, R. D. IVARIE, B. S. SCHACHTER, J. A. MORRIS, P. H. SEEBURG, H. M. GOODMAN, K. R. LATHAM, J. R. POLANSKY, AND J. A. MARTIAL. Thyroid hormone receptors and responses. *Recent Prog. Horm. Res.* 35: 97–153, 1979.

BAXTER, J. D., AND C. G. ROSSEAU. In *Glucocorticoids and the Metabolic Code in Glucocorticoid Hormone Action*, edited by J. D. Baxter and C. G. Rousseau. Heidelberg: Springer-Verlag, 1979, p. 613–627.

BAY W. H., AND T. F. FERRIS. Factor controlling plasma renin and aldosterone during pregnancy. *Hypertension* 1: 410, 1979.

BEARN, J. G. Anencephaly and the development of the male genital tract. *Acta Paediatr Acad Sci (Hungary)* 9: 159–180, 1968.

BECKER, D. V., W. M. MCCONAHEY, AND B. M. DOBYNS. In: *Further Advances in Thyroid Research*, edited by K. Fellinger and R. Hofer. Vienna: Verlag der Miener Medizinischen academie, 1971, vol. 1, p. 6y03–609.

BECKETT, G. J., AND G. S. BOYD. Purification and control of bovine adrenal cortical cholesterol ester hydrolase and evidence for the activation of the enzyme by a phosphorylation. *Eur. J. Biochem.* 72: 223–233, 1977.

BEERS, W. H., AND S. STRICKLAND. Studies on the role of plasminogen activator in ovulation. In vitro response of granulosa cells to gonadotropins, cyclic nucleotide, and prostaglandins. *J. Biol. Chem.* 251: 5694–5702, 1976.

BEGLEY, D. J., J. A. FIRTH, AND J. R. S. HOULT. *Human Reproduction and Developmental Biology*. New York: MacMillan Press Ltd., 1980.

BEIERWALTES, W. H. The treatment of thyroid cancer with radioactive iodine. Semin. *Nucl. Med.* 8: 79–94, 1978.

BILEZIKJIAN, L. M., AND W. W. VALE. Stimulation of adenosine 3′,5′-monophosphate production by growth hormone-releasing factor and its inhibition by somatostatin in anterior pituitary cells, in vitro. *Endocrinology* 113: 657–662, 1983.

BELLEVILLE, F., A. LAS BEENNES, P. NABET, AND P. PAYSANT. HLS-HCG regulation in cultured placenta. *Acta Endocrinol* 88: 169–181, 1978.

BENIRSCHKE, K. Adrenals in anencephaly and hydrocephaly. *Obstet. Gynecol.* 8: 412–425, 1956.

BERGLAND, R. M., AND R. B. PAGE. Can the pituitary secrete directly to the brain? (affirmative anatomical evidence). *Endocrinology* 102: 1325–1338, 1978.

BERGLAND, R. M., S. L. DAVIS, AND R. B. PAGE. Pituitary secretes to brain. *Lancet* 2: 276–277, 1977.

BERRIDGE, M. J., AND R. F. IRVINE. Inositol trisphosphate, a novel second messenger in cellular signal transduction. *Nature* 312: 315–321, 1984.

BERRIDGE, M. J. Inositol trisphosphate and diacyglycerol: Two interacting second messengers. *Annu. Rev. Biochem.* 56: 159–194, 1987.

BEUTLER, B., AND A. CERAMI. Cachectin and tumour necrosis factor as two sides of the same biological coin. *Nature* 320: 584–588, 1986.

BICSAK, T. A., E. M. TUCKER, S. CAPPEL, J. VAUGHAN, J. RIVIER, W. VALE, AND A. J. W. HSUEH. Hormonal regulation of granulosa cell inhibin biosynthesis. *Endocrinology* 119: 2711–2719, 1986.

BICSAK, T. A., W. VALE, J. VAUGHAN, E. M. TUCKER, S. CAPPEL, AND A. J. W. HSUEH. Hormonal regulation of inhibin production by cultured Sertoli cells. *Mol. Cell Endocrinol.* 49: 211–217, 1987.

BIGAZZI, M., AND E. NARDI. Prolactin and relaxin: Antagonism on the spontaneous motility of the uterus. *J. Clin. Endocrinol. Metab.* 53: 665–667, 1981.

BIGAZAI, M., G. POLLICINO, AND E. NARDI. Is human decidua a specialized endocrine organ? *J. Clin. Endocrinol. Metab.* 49: 847–850, 1979.

BISHOP, J. M. The nolecular genetics of cancer. *Science* 235: 305–311, 1987.

BISHOP, J. M. Viral oncogenes. *Cell* 42: 23–38, 1985.

BLACKWELL, G. J., R. CARNUCCIO, M. DI ROSA, R. J. FLOWER, L. PARENTE, AND P. PERSICO. Macrocortin: a polypeptide causing the anti-phospholipase effect of glucocorticoids. *Nature* 287: 147–149, 1980.

BLEASDALE, J. E., AND J. M. JOHNSON. Prostaglandins and human parturition: regulation of arachidonic acid mobilization. *Rev. Perinat. Med.* 5: 157, 1984.

BLEASDALE, J. E., T. OKAZAKI, N. SAGAWA, G. C. DI RENZO, J. R. OKITA, P. C. MACDONALD, AND J. M. JOHNSTON. The mobilization of arachidonic acid for prostaglandin production during parturition. In: *Initiation of Parturition: Prevention of Prematurity*, edited by P. C. MacDonald and J. Porter. Fourth Ross Conference on Obstetric Research, Ross Laboratories, Columbus, OH. 1983, p. 129.

BLUNDELL, T. L., S. BEDARKAR, E. RINDERKNECHT, AND R. E. HUMBEL. Insulin-like growth factor: a model for tertiary structure accounting for immunoreactivity and receptor binding. *Proc. Natl. Acad. U.S.A.* 75: 180–184, 1978.

BOHNET, H., F. GOMEZ, AND H. G. FRIESEN. Prolactin and estrogen binding sites in the mammary gland of the lactating and non-lactating rat. *Endocrinology* 101: 1111–1121, 1977.

BOORSTEIN, W. R., N. C. VAMVAKOPOULOS, AND J. C. FIDDES. Human chorionic gonadotropin β-subunit is encoded by at least eight genes arranged in tandem and repeated pairs. *Nature* 419–42, 1982.

BORDIER, P., H. RASMUSSEN, P. MARIE, L. MIRAVET, J. GUERIS, AND A. RYCKWAERT. Vitamin D metabolites and bone mineralization in man. *J. Clin. Endocrinol. Metab.* 46: 284–294, 1978.

BORIS, A., J. F. HURLEY, T. TRMAL, J. P. MALLON, AND D. S. MATUSZEWSKI. Evidence for the promotion of bone mineralization by 1α,25-dihydroxycholecalciferol in the rat unrelated to the correction of deficiencies in serum calcium and phosphorus. *J. Nutr.* 108: 1899–1906, 1978.

BOS, J. L., E. R. FEARON, S. R. HAMILTON, M. VERLAAN-de VRIES, J. H. van BOOM, A. J. van der EB, AND B. VOGELSTEIN. Prevalence of *ras* gene mutations in human colorectal cancers. *Nature* 327: 293–298, 1987.

BOSKEY, A. L. Current concepts of the physiology and biochemistry of calcification. *Clin. Orthop. Relat. Res.* 157: 225–257, 1981.

BOYAR, R., J. FINKELSTEIN, H. ROFFWARG, S. KAPEN, E. D. WEITZMAN, AND L. HELLMAN. Synchronization of augmented luteinizing hormone with sleep during puberty. *N. Engl. J. Med.* 287: 582–586, 1972.

BOYER, R. M., J. W. FINKELSTEIN, S. KAPEN, AND L. HELLMAN. Twenty-four hour prolactin secretory pattern during pregnancy. *J. Clin. Endocrinol. Metab.* 40: 1117–1120, 1975.

BOYAR, R. M., R. S. ROSENFELD, S. KAPEN, J. W. FINKELSTEIN, H. P. ROFFWARG, E. D. WEITZMAN, AND L. HELLMAN. Human puberty. *J. Clin. Invest.* 54: 609–618, 1974.

BRACE, R. A. Amniotic fluid dynamics. In: *Maternal-Fetal Medicine: Principles and Practice*, edited by R. K. Creasy and R. Resnik, Philadelphia: W. B. Saunders Company, 1988.

BRACE, R. A. Amniotic fluid volume and its relationship to fetal fluid balance: Review of experimental data. *Sem. Perinatol.* 10: 103–122, 1986.

BRACE, R. A. Fetal blood volume, extracellular fluid, and lymphatic function. In: *Fetal and Neonatal Body Fluids: The Scientific Basis for Clinical Practice*. Ithaca: Perinatology Press, 1988.

BRACE, R. A., M. G. ROSS, AND J. E. ROBILLARD. *Fetal and Neonatal Body Fluids: The Scientific Basis for Clinical Practice*. Ithaca: Perinatology Press, 1988.

BRADSHAW, R. A., R. A. HOGUE-ANGELETTI, AND W. A. FRAZIER. Nerve growth factor and insulin: evidence of similarities in structure, function, and mechanism of action. *Recent Prog. Horm. Res.* 30: 575–596, 1974.

BRAVERMAN, L. E., AND S. H. INGBAR. Changes in thyroidal function during adaptation to large doses of iodide. *J. Clin. Invest.* 42: 1216–1231, 1963.

BRAVERMAN, D. Z., M. L. JOHNSON, AND F. KERN. Effects of pregnancy and contraceptive steroids on gall bladder function. *N. Engl. J. Med.* 302: 363, 1980.

BRAZEAU, P., W. VALE, R. BURGUS, N. LING, M. BUTCHER, J. RIVIER, AND R. GUILLEMIN. Hypothalamic polypeptide that inhibits the secretion of immunoreactive pituitary growth hormone. *Science* 179: 77–79, 1973.

BREMNER W. J., AND C. A. PAULSEN. Prolonged intravenous infusion of LH-releasing hormone into normal men. *Horm. Metab. Res.* 9: 13–16, 1977.

BROWN, E. M., S. HURWITZ, C. J. WOODARD, AND G. D. AURBACH. Direct identification of beta-adrenergic receptors on isolated bovine parathyroid cells. *Endocrinology* 100: 1703–1709, 1977.

BROWN, M. S., R. G. W. ANDERSON, AND J. L. GOLDSTEIN. Recycling receptors: The round trip itinerary of migrant membrane proteins. *Cell* 32: 663–667, 1983.

BROWN, M. R., L. A. FISHER, J. RIVIER, J. SPIESS, C. RIVIER, AND W. VALE. Corticotropin-releasing factor: effects on the sympathetic nervous system and oxygen consumption. *Life Sci.* 30: 207–210, 1982.

BROWN, M. S., P. T. KOVANEN, AND J. L. GOLDSTEIN. Receptor-mediated uptake of lipoprotein-cholesterol and its utilization for steroid synthesis in the adrenal cortex. *Recent Prog. Horm. Res.* 35: 215–257, 1979.

BUCKLEY, D. I., AND J. RAMACHANDRAN. Characterization of corticotropin receptors on adrenocortical cells. *Proc. Natl. Acad. Sci. U.S.A.* 78: 7431–7435, 1981.

BURGUS, R., T. F. DUNN, D. DESIDERIO, W. VALE, AND R. GUILLEMIN. Derives polypeptidiques de synthaese doues d'activite hypophysiotrope TRH: nouvelles observation. *C. R. Acad. Sci. [D] Paris* 269: 154, 1969.

BURMAN, K. D., Y. LUKES, F. D. WRIGHT, AND L. WARTOFSKY. Reduction in hepatic triiodothyronine binding capacity induced by fasting. *Endocrinology* 101: 1331–1334, 1977.

BURROW, G. N., R. POLACKWICH, AND R. DONABEDIAN. The hypothalamic-pituitary-thyroid axis in normal pregnancy. In: *Perinatal Thyroid Physiology and Disease*, edited by D. A. Fisher and G. N. Burrow. New York: Raven Press, 1975, p. 1–10.

BURSTEIN, S., AND M. GUT. Biosynthesis of pregnenolone. *Recent Prog. Horm. Res.* 27: 303–349, 1971.

BURTON, R. M., AND U. WESTPHAL. Steroid hormone-binding proteins in blood plasma. *Metabolism* 21: 253–276, 1972.

CATT, K. J., AND J. G. PIERCE. Gonadotropic hormones and the adenohypophysis. In: *Reproductive Endocrinology*, edited by S. S. C. Yen, and R. B. Jaffe. Philadelphia: W. B. Saunders, 1986, p. 75–114.

CARPENTER, G. Receptors for epidermal growth factor and other polypeptide mitogens. *Annu. Rev. Biochem.* 56: 881–914, 1987.

CARPENTER, G., AND S. COHEN. Epidermal growth factor. *Annu. Rev. Biochem.* 48: 193–216, 1979.

CARR, F. E., L. R. NEED, AND W. W. CHIN. Isolation and characterization of the rat thyrotropin α-subunit gene: Differential regulation of two transcriptional start sites by thyroid hormone. *J. Biol. Chem.* 262: 981–987, 1987.

CARR, B. R., C. R. PARKER, JR., AND J. D. MADDEN. Maternal plasma adrenocorticotropin and cortisol relations throughout human pregnancy. *Am. J. Obstet. Gynecol.* 125: 1136, 1976.

CARR, B. R., C. R. PARKER, JR., J. D. MADDEN, P. C. MACDONALD, AND J. C. PORTER. Maternal plasma adrenocorticotropin (ACTH) and cortisol relationships throughout human pregnancy. *Am. J. Obstet. Gynecol.* 139: 416–422, 1981.

CARSTEN, M. E. Prostaglandins and cellular calcium transport in the pregnant uterus. *Am. J. Obstet. Gynecol.* 117: 824–832, 1973.

CASPER, R. F., S. S. C. YEN, AND M. M. WILKES. Menopausal flushes: A neuroendocrine link with pulsatile luteinizing hormone secretion. *Science* 205: 823–825, 1979.

CASPERSON, G. F., AND H. R. BOURNE. Biochemical and molecular genetic analysis of hormone-sensitive adenylyl cyclase. *Annu. Rev. Pharmacol. Toxicol.* 27: 371–384, 1987.

CASSEL, D., AND Z. SELINGER. Mechanism of adenylate cyclase activation through the β-adrenergic receptor: catecholamine-induced displacement of bound GDP by GTP. *Proc. Natl. Acad. Sci. U.S.A.* 75: 4155–4159, 1978.

CATE, R. L., R. J. MATTALIANO, C. HESSION, R. TIZARD, N. M. FARBER, A. CHEUNG, E. G. NINFA, A. Z. FREY, D. J. GASH, E. P. CHOW, R. A. FISHER, J. M. BERTONIS, G. TORRES, B. P. WALLNER, K. L. RAMACHANDRAN, R. C. RAGIN, T. F. MANGANARO, D. T. MacLAUGHLIN, AND P. K. DONAHOE. Isolation of the bovine and human genes for Mullerian inhibiting substance and expression of the human gene in animal cells. *Cell* 45: 685–698, 1986.

CETIN, I., A. M. MARCONI, P. BOZZETTI, L. B. SERENI, C. CORBETTA, G. PARDI, AND F. C. BATTAGLIA. Umbilical amino acid concentrations in appropriate and small for gestational age infants: a biochemical difference present in utero. *Am. J. Obstet. Gynecol.* 158: 120–126, 1988.

CHAMBERLAIN, P. F., F. A. MANNING, I. MORRISON, C. R. MARMAN, AND I. R. LANGE. Ultrasound evaluation of amniotic fluid volume. I and II. *Am. J. Obstet. Gynecol.* 150: 245–254, 1984.

CHANNING, J. F., AND R. D. H. BOYD. Mineral and water exchange between mother and fetus. In: *Fetal Physiology and Medicine*, edited by R. W. Beard and P. W. Nathanielsz. New York: Marcel Dekker, Inc., 1984.

CHAPPEL, S. C., A. ULLOA-AGUIRRE, AND C. COUTIFARIS. Biosynthesis and secretion of follicle-stimulating hormone. *Endocr. Rev.* 4: 179–211, 1983.

CHARLTON, V. Fetal nutritional supplementation. *Sem. Perinatol.*, 8: 25–30, 1984.

CHARRON, J., AND J. DROUIN. Glucocorticoid inhibition of transcription from episomal proopiomelanocortin gene promoter. *Proc. Natl. Acad. Sci. U.S.A.* 83: 8903–8907, 1986.

CHASE, L. R., AND G. D. AURBACH. Parathyroid function and renal excretion of 3'5'-adenylic acid. *Proc. Natl. Acad. Sci. U.S.A.* 58: 518–525, 1967.

CHEN, W. S., C. S. LAZAR, K. A. LUND, J. B. WELSH, C.-P. CHANG, G. M. WALTON, C. J. DER, S. H. WILEY, G. N. GILL, AND M. G. ROSENFELD. Functional dependence of the epiderman growth factor receptor from a domain required for ligand-induced internalization and calcium regulation. *Cell* 59: 33–43, 1989.

CHEUNG, W. Y. Calmodulin plays a pivotal role in cellular regulation. *Science* 207: 19–27, 1980.

CHRISTENSEN, H. *Handbook of Physiology*, Section 7, *Endocrinology*. Vol V. Edited by R. HAMILTON, AND R. G. GICEP. Washington, D.C. American Physiological Society, 1975; 1960.

CHOPRA, I. J., J. SACK, AND D. A. FISHER. Circulating 3,3',5'-triiodothyronine (reverse T_3) in the human newborn. *J. Clin. Invest.* 55: 1137–1141, 1975.

CHOPRA, I. J., D. H. SOLOMON, U. CHOPRA, S. Y. WU, D. A. FISHER, AND Y. NAKAMURA. Pathways of metabolism of thyroid hormones. *Recent Prog. Horm. Res.* 34: 521–567, 1978.

CLAPP, J. F. Maternal physiological adaptation during early human pregnancy. *Am. J. Obstet. Gynecol.* 1988, in press.

CLAYTON, R. N., AND K. J. CATT. Gonadotropin-releasing hormone receptors. Characterization, physiological regulation, and relationship to reproductive function. *Endocr. Rev.* 2: 186–209, 1981.

COBO, E. Neuroendocrine control of milk ejection in women. In *Lactogenic Hormones, Fetal Nutrition, and Lactation*, edited by J. B. Josimovich, M. Reynolds, E. Cobo. New York: John Wiley & Sons, 1974, p. 433.

CODY, V. Thyroid hormones: crystal structure, molecular conformation, binding, and structure-function relationship. *Recent Prog. Horm. Res.* 34: 437–475, 1978.

COHEN, M. S., AND T. K. GREY. Phagocytic cells metabolize 25-hydroxyvitamin D_3 in vitro. *Proc. Natl. Acad. Sci. U.S.A.* 81: 931–984, 1984.

COHEN, S. Purification and metabolic effects of a nerve growth-promoting protein from snake venom. *J. Biol. Chem.* 234: 1129–1137, 1959.

COHEN, S., H. USHIRO, C. STOSCHECK, AND M. CHINKERS. A native 170,000 epidermal growth factor receptor-kinase complex from shed plasma membrane vesicles. *J. Biol. Chem.* 257: 1523–1531, 1982.

COHN, H. E., E. J. SACKS, M. A. HEYMANN, AND A. M. RUDOLPH. Cardiovascular responses to hypoxemia and acidemia in fetal lambs. *Am. J. Obstet. Gynecol.* 120: 817–824, 1974.

COOKE, N. E., D. COIT, R. I. WEINER, J. D. BAXTER, AND J. A. MARTIAL. Structure of cloned DNA complementary to rat prolactin messenger RNA. *J. Biol. Chem.* 255: 6502–6510, 1980.

COOPER, C. W., R. M. BOLMAN, III, W. M. LINEHAN, AND S. A. WELLS, JR. Interrelationships between calcium, calcemic hormones and gastrointestinal hormones. *Recent Prog. Horm. Res.* 34: 259–283, 1978.

COOPER, G. M., S. OKENQUIST, AND L. SILVERMAN. Transforming activity of DNA of chemically transformed and normal cells. *Nature* 284: 418–421, 1980.

COWIE, A. T., AND TINDAL, J. S. *The Physiology of Lactation*. London: Edward Arnold, 1971.

CRENSHAW, III, E. B., A. F. RUSSO, L. W. SWANSON, AND M. G. ROSENFELD. Neuron-specific alternative RNA processing in transgenic mice expressing a metallothionein-calcitonin fusion gene. 49: 389–398, 1987.

CROWLEY, W. F., M. FILICORI, D. I. SPRATT, AND N. F. SANTORO. The physiology of gonadotropin-releasing hormone (GnRH) secretion in men and women. *Rec. Prog. Horm. Res.* 41: 473–531, 1985.

CSAPO, A. L., M. O. PULKKINEN, AND W. G. WIEST. Effects of lutectomy and progesterone replacement in early pregnant patients. *Am. J. Obstet. Gynecol.* 115: 759–765, 1973.

DAHL, K. D., T. A. BICSAK, AND A. J. W. HSUEH. Naturally occurring antihormones: Secretion of FSH antagonist by women treated with a GnRH analog. *Science* 239: 72–74, 1988.

DALLMAN, M. F. Adrenal feedback on stress-induced corticoliberin (CRF) and corticotropin (ACTH) secretion. In: *Interactions Within the Brain-Pituitary-Adrenocortical System*, edited by M. T. Jones, B. Gillham, M. F. Dallman, and S. Chattopadahyay. New York: Academic Press, 1979, p. 149–162.

DAUGHADAY, W. H., Z. LARON, A. PERTZELAN, AND J. N. HEINS. Defective sulfation factor generation: a possible etiological link in dwarfism. *Trans. Assoc. Am. Physician* 82: 129–140, 1969.

DAVISON, J. S., M. C. DAVISON, AND D. M. HAY. Gastric emptying time in late pregnancy and labor. *J. Obstet. Gynaecol. Br. Commonw.* 77: 37, 1970.

DAVISON, J. M., AND F. E. HYTTEN. The effect of pregnancy on the renal handling of glucose. *J. Obstet. Gynaecol. Brit. Commonw.* 82: 374, 1975.

DAVISON, J. M., AND W. DUNLOP. Changes in renal haemodynamics and tubular function induced by normal human pregnancy. *Semin. Nephrol.* 4: 198, 1984a.

DAVISON, J. M., E. A. GILMORE, D. DURR, ET AL. Altered osmotic thresholds for vasopressin secretion and thirst in human pregnancy. *Am. J. Physiol.* 246: F105, 1984b.

DICZFALUSY, E. The feto-placental unit. *Excerpta Med. Int. Congr. Ser.* 183: 65–109, 1964.

DEGROOT, L. J., AND H. NIEPOMNISZCZE. Biosynthesis of thyroid hormone: basic and clinical aspects. *Metabolism* 26: 665–718, 1977.

DELUCA, H. F., AND H. K. SCHNOES. Metabolism and mechanism of action of vitamin D. *Annu. Rev. Biochem.* 45: 631–666, 1976.

DELVOYE, P., P. DEMAEGD, J. DELOGNE-DESNOECK, AND C. ROBYN.

The influence of the frequency of nursing and of previous lactation experience on serum prolactin in lactating women. *J. Biosoc. Sci.* 9: 447–451, 1977.

DE MEYTS, P., A. R. BIANCO, AND J. ROTH. Site-site interactions among insulin receptors. *J. Biol. Chem.* 251: 1877–1888, 1976.

DHOM, G. The prepuberal and puberal growth of the adrenal (adrenarche). *Beitr. Pathol.* 150: 357–377, 1973.

DI ZEREGA, G. S., AND G. D. HODGEN. Folliculogenesis in the primate ovarian cycle. *Endocri. Rev.* 2: 27–49, 1981.

DJERASSI, C. L., L. MIRAMONTES, AND G. ROSENKRANZ. The synthesis of 19-norprogesterone. Steroids XXII. *J. Am. Chem. Soc.* 73: 3540–3545, 1951.

DJIANE, J., AND P. DURAND. Prolactin-progesterone antagonism in self-regulation of prolactin receptors in the mammary gland. *Nature* 266: 641–643, 1977.

DOE, R. P., R. FERNANDEZ, AND V. S. SEAL. Measurement of corticosteroid-binding globulin in man. *J. Clin. Endocrinol. Metab.* 24: 1029–1039, 1964.

DONAHOE, P. K., G. P. BUDZIK, R. TRELSTAD, M. MUDGETT-HUNTER, A. FULLER, JR., J. M. HUTSON, H. IKAWA, A. HAYASHI, AND D. McLAUGHLIN. Mullerian-inhibiting substance: An update. *Rec. Prog. Horm. Res.* 38: 279–330, 1982.

DOOLITTLE, R. F., D. F. FENG, AND M. S. JOHNSON. Computer-based characterization of epidermal growth factor precursor. *Nature* 307: 558–560, 1984.

DOUSA, T. P., AND H. VALTIN. Cellular actions of vasopressin in the mammalian kidney. *Kidney Int.* 10: 46–63, 1976.

DOWNWARD, J., Y. YARDEN, E. MAYES, G. SCRACE, N. TOTTY, P. STOCKWELL, A. ULLRICH, J. SCHLESSINGER, AND M. D. WATERFIELD. Close similarity of epidermal growth factor receptor and v-*erb-B* oncogene protein sequences. *Nature* 307: 521–527, 1984.

DUFAU, M. L., C. R. MENDELSON, AND K. J. CATT. A highly sensitive in vitro bioassay for luteinizing hormone and chorionic gonadotropin: Testosterone production by dispersed Leydig cells. *J. Clin. Endocrinol. Metab.* 39: 610–613, 1974.

DUMONT, J. E., J. M. BOEYNAEMS, C. DECOSTER, C. ERNEAU, F. LAMY, R. LECOCQ, J. MOCKEL, J. UNGER, AND J. VAN SANDE. Biochemical mechanisms in the control of thyroid function and growth. *Adv. Cyclic Nucleotide Res.* 9: 723–734, 1978.

DUNLOP, D. Eighty-six cases of Addison's disease. *Br. Med. J.* 2: 887–891, 1963.

DUNLOP, W. Serial changes in renal haemodynamics during normal human pregnancy. *Br. J. Obstet. Gynaecol.* 88: 1, 1981.

DUNN, F. L., T. J. BRENNAN, A. E. NELSON, AND G. L. ROBERTSON. The role of blood osmolality and volume in regulating vasopressin secretion in the rat. *J. Clin. Invest.* 52: 3212–3219, 1973.

DUNN, M. J., AND V. L. HOOD. Prostaglandins and the kidney. *Am. J. Physiol.* 233: F169–F184, 1977.

DYNAN, W. S., AND R. TJIAN. Control of eukaryotic messenger RNA synthesis by sequence-specific DNA-binding proteins. *Nature* 316: 774–778, 1985.

EBINA, Y., L. ELLIS, K. JARNAGIN, M. EDERY, L. GRAF, E. CLAUSER, J-H. OU, F. MASIARZ, Y. W. KAN, I. D. GOLDFINE, R. A. ROTH, AND W. J. RUTTER. The human insulin receptor cDNA: The structural basis for hormone-activated transmembrane signalling. *Cell* 40: 747–758, 1985.

EDELMAN, A. M., D. K. BLUMENTAL, AND KREBS, E. G. Protein serine/threonine kinases. *Annu. Rev. Biochem.* 56: 567–614, 1987.

EDELMAN, I. S., AND D. MARVER. Mediating events in the action of aldosterone. *J. Steroid Biochem.* 12: 219–224, 1980.

EIPPER, B. A., AND R. E. MAINS. Structure and biosynthesis of proadrenocorticotropin/endorphin and related peptides. *Endocr. Rev.* 1: 1–27, 1980.

EK, B., B. WESTERMARK, A. WASTESON, AND C-H. HELDIN. Stimulation of tyrosine-specific phosphorylation by platelet-derived growth factor. *Nature* 295: 419–420, 1982.

ENGELAND, W. C., AND M. F. DALLMAN. Compensatory adrenal growth is neurally mediated. *Neuroendocrinology* 19: 352–362, 1975.

ENGLER, D., AND A. G. BURGER. The deiodination of the iodothysonines and of their derivatives in man. *Endocr. Rev.* 5: 151–184, 1984.

ERSPAMER, V., AND P. MELCHIORRI. Active polypeptides: from amphibian skin to gastrointestinal tract and brain of mammals. *Trends Pharmacol. Sci.* 1: 391–395, 1980.

ESPEY, L. L. Ovarian proteolytic enzymes and ovulation. *Biol. Reproduc.* 10: 216–235, 1974.

FABER, J. J. AND F. M. HART. The rabbit placenta as an organ of diffusional exchange. *Circ. Res.* 19: 816, 1966.

FABER, J. J., AND K. L. THORNBURG. Placental physiology. New York: Raven Press, 1983.

FABIAN, E., A. STORK, AND L. KUCEROVA. Plasma levels of free fatty acids, lipoprotein lipase, and postheparin esterase in pregnancy. *Am. J. Obstet. Gynecol.* 100: 904, 1968.

FABRINI, A., AND E. S. E. HAFEZ. Testis and epididymis. In: *Human Reproduction: Concception and Contraception*, edited by E. S. E. Hafez. New York: Harper and Row, 1980, p. 35–59.

FARFEL, Z., A. S. BRICKMAN, H. R. KASLOW, V. M. BROTHERS, AND H. R. BOURNE. Defect of receptor-cyclase coupling protein in pseudohypoparathyroidism. *N. Engl. J. Med.* 303: 237–242, 1980.

FAULK, W. P. AND TEMPLE, A. Distribution of β^2 microglobulin and HLA in chorionic villi of human placentae. *Nature* 262: 799–802, 1976.

FAUSER, B. C. J. M., R. ROLLAND, J. A. M. KREMER, W. H. DOESBURG, AND C. M. G. THOMAS. Pharmacokinetics of intravenous luteinizing hormone-releasing hormone administration in men: effects of various dosages. *Fertil. Steril.* 47: 144–149, 1987.

FELIG, P. Amino acid metabolism in man. *Annu. Rev. Biochem.* 44: 933–955, 1975.

FELIG, P. Body fuel metabolism and diabetes mellitus in pregnancy. *Med. Clin. North Am.* 61: 43, 1977.

FELIG, P. KIM, Y. J., LYNCH, V., HENDLER, R. Amino acid metabolism during starvation in human pregnancy. *J. Clin. Invest.* 51: 1195–1202, 1972.

FERIN, M. D. VAN VUGT, AND S. WARDLAW. The hypothalamic control of the menstrual cycle and the role of endogenous opioid peptides. *Rec. Prog. Horm. Res.* 40: 441–485, 1984.

FIDDES, J. C., AND H. M. GOODMAN. The gene encoding the common alpha subunit of the four human glycoprotein hormones. *J. Molec. App. Genet.* 1: 3–18, 1981.

FIELDS, P. A., AND LARKIN, L. H. Purification and immunohistochemical localization of relaxin in the human term placenta. *J. Clin. Endocrinol. Metab.* 52: 79–85, 1981.

FILICORI, M., N. SANTORO, G. R. MERRIAM, AND W. F. CROWLEY. Characterization of the physiological pattern of episodic gonadotropin secretion throughout the human menstrual cycle. *J. Clin. Endocrinol. Metab.* 62: 1136–1144, 1986.

FISHEL, S. B., EDWARDS, R. G., AND EVANS, C. J. Human chorionic gonadotropin secretion by preimplantation embryos cultured in vitro. *Science* 223: 816–818, 1984.

FISHER, D. A. Fetal endocrinology: Endocrine disease and pregnancy. In *Endocrinology*, Vol. 3, edited by J. De Groot. New York: Grune & Stratton, 1979, p. 1649.

FISHER, D. A., AND A. H. KLEIN. Thyroid development and disorders of thyroid function in the newborn. *N. Engl. J. Med.* 304: 702–712, 1981.

FISHER, D. A., AND W. D. ODELL. Acute release of thyrotropin in the newborn. *J. Clin. Invest.* 48: 1670–1677, 1969.

FISHER, D. A., J. H. DUSSAULT, J. SACK, AND I. J. CHOPRA. Ontogenesis of hypothalamic-pituitary-thyroid function and metabolism in man, sheep, and rat. *Recent Prog. Horm. Res.* 33: 59–116, 1977.

FISHER, L. A., J. RIVIER, C. RIVIER, J. SPIESS, W. VALE, AND M. R. BROWN.

Corticotropin-releasing factor (CRF): central effects on mean arterial pressure and heart rate in rats. *Endocrinology* 110: 2222–2224, 1982.

FITZSIMONS, J. T. Thirst. *Physiol. Rev.* 52: 468, 1972.

FOLKMAN, J., E. MERLER, C. ABERNATHY, AND G. WILLIAMS. Isolation of a tumor factor responsible for angiotenesis. *J. Exp. Med.* 133: 275–288, 1971.

FOLLEY, S. J. Endocrine control of the mammary gland. II. Lactation. *Br. Med. Bull.* 5: 135, 1948.

FRASER, D. R. Regulation of the metabolism of vitamin D. *Physiol. Rev.* 60: 551–613, 1980.

FREINKEL, N. Of pregnancy and progeny. *Diabetes* 29: 102, 1980.

FRIEND, S. H., T. P. DRYJA, AND R. A. WEINBERG. Oncogenes and tumor-suppressing genes. *N. Engl. J. Med.* 318: 618–622, 1988.

FRIESEN, H. G. Human prolactin in clinical endocrinology: The impact of radioimmunoassays. *Metabolism* 22: 1039–1045, 1973.

FRITZ, I. B. Sites of action of androgens and follicle stimulating hormone on cells of the seminiferous tubule. In: *Biochemical Actions of Hormones*, Vol V, edited by G. Litwack. New York: Academic Press, 1978, p. 249–281.

FROESCH, E. R., J. ZAPF, T. K. AUDHYA, E. BEN-PORATH, B. J. SEGEN, AND K. D. GIBSON. Nonsuppressible insulin-like activity and thyroid hormones: Major pituitary-dependent sulfation factors for check embryo cartilage. *Proc. Natl. Sci. U.S.A.* 73: 2904–2908, 1976.

FUCHS, A-R, F. FUCHS, P. HUSSELEIN, M. S. SOLOFF, AND M. J. FERNSTROM. Oxytocin receptors and human parturition: a dual role for oxytocin in the initiation of labor. *Science* 215: 1396–1398, 1982.

FURTH, E. D. Thyroid and parathyroid hormone function in pregnancy. In *Endocrinology of Pregnancy*, edited by F. Fuchs and A. Klopper. Philadelphia: Harper and Row Publishers, 1983, p. 176.

GARRIOCH, D. B. The effect of indomethacin on spontaneous activity in the isolated human myometrium and on the response to oxytocin and prostaglandin. *Br. J. Obstet. Gynaecol.* 85: 47–52, 1978.

GERSHENGORN, M. C. Mechanism of thyrotropin-releasing hormone stimulation of pituitary hormone secretion. *Annu. Rev. Physiol.* 48: 515–526, 1986.

GIBBONS, J. M. M. MITNICK, AND V. CHIEFFO. In vitro biosynthesis of TSH- and LH-releasing factors by the human placenta. *Am. J. Obstet. Gynaecol.* 121: 127–131, 1975.

GIBBS, J. C., I. S. SIGAL, M. POE, AND E. M. SCOLNICK. Intrinsic GTPase activity distinguishes normal and oncogenic *ras* p21 molecules. *Proc. Natl. Acad. Sci. U.S.A.* 81: 5704–5708, 1984.

GIGUERE, V., E. S. ONG, P. SEGUI, AND R.

M. EVANS. Identification of a receptor for the morphogen retinoic acid. *Nature* 330: 624–629, 1987.

GILBERT, R. D. Control of fetal cardiac output during changes in blood volume. *Am. J. Physiol.* 238: H80–H86, 1980.

GILL, G. N. ACTH regulation of the adrenal cortex. In *Pharmacology of Adrenal Cortical Hormones*, edited by G. N. Gill. Oxford: Pergamon Press, 1979.

GILL, G. N., P. J. BERTICS, AND J. G. SANTON. Epidermal growth factor and its receptor. *Mol. Cell. Endocr.* 51: 169–186, 1987.

GILL, G. N., J. F. CRIVELLO, AND P. J. HORNSBY. Growth, function, and development of the adrenal cortex: insights from cell culture. *Cold Spring Harbor Conf. Cell Proliferation* 9: 461–482, 1982.

GILL, G. N., AND L. D. GARREN. Role of the receptor in the mechanisms of action of adenosine 3′:5′-cyclic monophosphate. *Proc. Natl. Acad. Sci. U.S.A.* 68: 786–790, 1971.

GILL, G. N., P. J. HORNSBY, AND M. H. SIMONIAN. Regulation of growth and differentiated function of cultured bovine adrenocortical cells. *Cold Spring Harbor Conf. Cell Proliferations* 6: 701–715, 1979.

GILL, G. N., C. R. ILL, AND M. H. SIMONIAN. Angiotensin stimulation of bovine adrenocortical cell growth. *Proc. Natl. Acad. Sci. U.S.A.* 74: 5569–5573, 1977.

GILL, G. N. AND R. W. MCCUNE. Guanosine 3′,5′-monophosphate-dependent protein kinase. *Curr. Top. Cell. Regul.* 15: 1–45, 1979.

GILMAN, A. G. G proteins: Transducers of receptor-generated signals. *Annu. Rev. Biochem.* 56: 615–650, 1987.

GINSTLER, A. R. Endorphin-mediated increases in pain threshold during pregnancy. *Science* 210: 193–195, 1980.

GLUCKMAN, P. D., M. M. GRUMBACH, AND S. L. KAPLAN. The human fetal hypothalamus and pituitary gland. In *Maternal-Fetal Endocrinology*, edited by D. Tulchinsky & K. J. Ryan. Philadelphia: W. B. Saunders Co., 1980, p. 196.

GLUCKMAN, P. D., AND G. C. LIGGINS. Regulation of fetal growth, In: *Fetal Physiology and Medicine*, edited by R. W. Beard and P. W. Nathanielsz. New York: Marcel Dekker, Inc., 1984, p. 511–559.

GOLDBETER, A., AND D. E. KOSHLAND, JR. Ultrasensitivity in biochemical systems controlled by covalent modification. *J. Biol. Chem.* 259: 14441–14447, 1984.

GOLUBOFF, L. G., AND C. EZRIN. Effect of pregnancy on the somatotrophy and the prolactin cell of the human adenohypophysis. *J. Clin. Endocrinol. Metab.* 29: 1533, 1969.

GONDOS, B. Development and differentiation of the testis and male reproductive tract. In: *Testicular Development,*

Structure, and Function, edited by A. Steinberger, and E. Steinberger. New York: Raven Press, 1980, p. 3–20.

GOODFELLOW, P. N., C. J. BARSTABLE, W. F. BODMER, ET AL. Expression of HLA antigens on placenta. *Transplantation* 22: 595–603, 1976.

GOSPODAROWICZ, D. Purification of a fibroblast growth factor from bovine pituitary. *J. Biol. Chem.* 250: 2515–2520, 1975.

GOSPODAROWICZ, D., AND J-P. TAUBER. Growth factors and the extracellular matrix. *Endocr. Rev.* 1: 201–227, 1980.

GOSPODAROWICZ, D., N. FERRARA, L. SCHWEIGEIER, AND G. NEUFELD. Structural characterization and biological functions of fibroblast growth factor. *Endocr. Rev.* 8: 95–114, 1987.

GOSPODAROWICZ, D., G. NEUFELD, AND L. SCHWEIGERER. Fibroblast growth factor. *Mol. Cell Endocr.* 46: 187–204, 1986.

GOSPODAROWICZ, D., G. GREENBURG, H. BIALECKI, AND B. R. ZETTER. Factors involved in the modulation of cell proliferation in vivo and in vitro: the role of fibroblast and epidermal growth factors in the proliferative response of mammalian cells. *In Vitro* 14: 85–118, 1978.

GOSPODAROWICZ, D., AND K. K. THAKRAL. Production of a corpus luteum angiogenic factor responsible for proliferations of capillaries and neovascularization of the corpus luteum. *Proc. Natl. Acad. Sci. U.S.A.* 75: 847–851, 1978.

GRABER, A. L., R. L. NEY, W. E. NICHOLSON, D. P. ISLAND, AND G. W. LIDDLE. Natural history of pituitary-adrenal recovery following long-term suppression with corticosteroids. *J. Clin. Endocrinol. Metab.* 25: 11–16, 1965.

GREEN, S., P. WALTER, V. KUMAR, A. KRUST, J-M. BORNERT, P. ARGOS, AND P. CHAMBON. Human oestrogen receptor cDNA: Sequence, expression and homology to v-erb-A. *Nature* 320: 134–139, 1986.

GREENWALD, I. *lin-12*, A nematode homeotic gene, is homologous to a set of mammalian proteins that includes epidermal growth factor. *Cell* 43: 583–590, 1985.

GREER, M. A., Y. GRIMM, AND H. STUDER. Qualitative changes in the secretion of thyroid hormones induced by iodine deficiency. *Endocrinology* 83: 1193–1198, 1968.

GREGORY, H. Isolation and structure of urogastrone and its relationship to epidermal growth factor. *Nature* 257: 325–327, 1975.

GRESHAM, E. L., J. H. G. RANKIN, E. L. MAKOWSKI, G. MESCHIA, AND F. C. BATTAGLIA. An evaluation of fetal renal function in a chronic sheep preparation. *J. Clin. Invest.* 51: 149–155, 1972.

GREULICH, W. W., AND S. E. PYLE. *Radiographic Atlas of Skeletal Devel-*

opment of the Hand and Wrist. Stanford, CA: Stanford University Press, 1959.

GRIFFIN, J. E., M. LESHIN, AND J. D. WILSON. Androgen resistance syndromes. *Am. J. Physiol.* 243: E81–E87, 1982.

GRIFFITHS, E. C. AND J. R. McDERMOT. Enzymatic inactivation of hypothalamic hormones. *Molec. Cell Endocrinol.* 33: 1–25, 1983.

GRINO, M., G. P. CHROUSOS, AND A. N. MARGIORIS. The corticotropin releasing hormone gene is expressed in human placenta. *Biochem. Biophy. Res. Commun.* 148: 1208–1214, 1987.

GRODY, W. W., W. T. SCHRADER, AND B. W. O'MALLEY. Activation, transformation, and subunit structure of steroid hormone receptors. *Endocr. Rev.* 3: 141–163, 1982.

GRUMBACH, M. M., H. E. RICHARDS, F. A. CONTE, AND S. L. KAPLAN. Clinical disorders of adrenal function and puberty: An assessment of the role of the adrenal cortex in normal and abnormal puberty in man and evidence for an ACTH-like pituitary adrenal androgen stimulating hormone. In: *The Endocrine Function of the Human Adrenal Cortex,* edited by M. Serio. Serono Symposium, New York: Academic Press, 1977.

GUILLEMIN, R., P. BRAZEAU, P. BOHLEN, F. ESCH, N. LING, AND W. B. WEHRENBERG. Growth hormone-releasing factor from a human pancreatic tumor that caused acromegaly. *Science* 218: 585–587, 1982.

GUSTAFSSON, J-A., J. CARLSTEDT-DUKE, L. POELLINGER, S. OKRET, A-C. WIKSTROM, M. BRONNEGARD, M. GILLNER, Y. DONG, K. FUXE, A. CINTRA, A. HARFSTRAND, AND L. AGNATI. Biochemistry, molecular biology and physiology of the glucocorticoid-receptor. *Endocr. Rev.* 8:185–197, 1987.

HABENER, J. R., AND J. T. POTTS, JR. Biosynthesis of parathyroid hormone. *N. Engl. J. Med.* 299: 580–585, 635–644, 1978.

HADDAD, J. G., JR., AND J. WALGATE. 25-Hydroxyvitamin D transport in human plasma. Isolation and partial characterization of calcifidiol-binding protein. *J. Biol. Chem.* 251: 4803–4809, 1976.

HAEMMERLI, V. P. Jaundice during pregnancy with special emphasis of recurrent jaundice during pregnancy and its differential diagnosis. *Acta. Med. Scand. (Suppl)* 44: 1, 1967.

HAFEN, E., K. BASLER, J-E. EDSTROEM, AND G. M. RUBIN. *Sevenless,* a cell-specific homeotic gene of Drosophilia, encodes a putative transmembrane receptor with a tyrosine Kinase domain. *Science* 236: 55–63, 1987.

VAN HALL, C. E., J. L. VAITUKAITIS, G. T. ROSS, J. W. HICKMAN, AND G. ASHWELL. Effects of progressive desialylation on the rate of disappearance of immunoreactive HCG from plasma in rats. *Endocrinology* 89: 11–15, 1971.

HAMOSH, M., AND P. HAMOSH. The effect of prolactin on the lecithin content of fetal rabbit lung. *J. Clin. Invest.* 59: 1002–1005, 1977.

HANKS, S. K., R. ARMOUR, J. H. BALDWIN, F. MALDONADO, J. SPIESS, AND R. W. HOLLEY. Amino acid sequence of the BSC-1 cell growth inhibitor (polyergin) deduced from the nucleotide sequence of the cDNA. *Proc. Natl. Acad. Sci.* 85: 79–82, 1988.

HANNUN, Y. A., C. R. LOOMIS, A. H. MERRILL, JR., AND R. M. BELL. Sphingosine inhibition of protein kinase C activity and phorbol dibutyrate binding in vitro and in human platelets. *J. Biol. Chem.* 261:12604–12609, 1986.

HARDING, R., A. D. BOKING, J. N. SIGGER, AND P. J. D. Composition and volume of fluid swallowed by fetal sheep. *Quart. J. Exp. Physiol.* 69: 487–495, 1984.

HAUSSLER, M. R., AND A. McCAIN. Basic and clinical concepts related to vitamin D metabolism and action. (Two parts) *N. Engl. J. Med.* 297: 974–983, 1041–1050, 1977.

HAUSSLER, M. R., J. W. PIKE, J. S. CHANDLER, S. C. MANOLAGAS, AND L. J. DEFTOS. Molecular action of 1,25-dihydroxyvitamin D_3: new cultured cell models. *Ann N.Y. Acad. Sci.* 372: 501–517, 1981.

HEUSON, J. C. AND A. COURE. Hormone-responsive tumors. In: *Endocrinology and Metabolism,* edited by P. Felig, J. D. Baxter, A. E. Broadus, AND L. A. Frohman. New York: McGraw-Hill, 1987, p. 1736–1767.

HEYNER, S. Antigens of trophoblast and early embryo. In *Immunological Aspects of Infertility and Fertility Regulation,* edited by D. S. Dhindsa, R. S. Schumacher. New York: North Holland Biomedical Press, 1980, p. 183.

HINKLE, P. M., M. H. PERRONE, AND A. SCHONBRUNN. Mechanism of thyroid hormone inhibition of thyrotropinreleasing hormone action. *Endocrinology* 108: 199–205, 1981.

HIRATA, F., E. SCHIFFMANN, K. VENKATASUBRAMANIAN, D. SALOMON, AND J. AXELROD. A phospholipase A_2 inhibitory protein in rabbit neutrophils induced by glucocorticoids. *Proc. Natl. Acad. Sci. U.S.A.* 77: 2533–2536, 1980.

HJALMARSON, A. C., C. F. WHITFIELD, AND H. E. MORGAN. Hormonal control of heart function and myosin ATPase activity. *Biochem. Biophys. Res. Commun.* 41: 1584–1589, 1970.

HOLICK, M. F., AND M. B. CLARK. The photobiogenesis and metabolism of vitamin D. *Fed. Proc.* 37: 2567–2574, 1978.

HOLLENBERG, S. M., C. WEINBERGER, E. S. ONG, G. CERELLI, A. ORO, R. LEBO, E. B. THOMPSON, M. G. ROSENFELD, AND R. M. EVANS. Primary structure and expression of a functional human glucocorticoid receptor cDNA. *Nature* 318: 635–641, 1985.

HOLLEY, R. W. Control of growth of mammalian cells in cell culture. *Nature* 258: 487–490, 1975.

HOLLEY, R. W., R. ARMOUR, AND J. H. BALDWIN. Density-dependent regulation of growth of BSC-1 cells in cell culture: growth inhibitors formed by the cells. *Proc. Natl. Acad. Sci. U.S.A.* 75: 1864–1866, 1978.

HOLLINGSWORTH, D. R., AND L. COUSINS. Endocrine and metabolic disorders: disorders of carbohydrate metabolism. In: *Maternal Fetal Medicine: Principles and Practice,* edited by R. K. Creasy and R. Resnik. Philadelphia: W. B. Saunders Company, 1984.

HORNSBY, P. J. Regulation of cytochrome P-450-supported 11β-hydroxylation of deoxycortisol by steroids, oxygen, and antioxidants in adrenocortical cell cultures. *J. Biol. Chem.* 255: 4020–4027, 1980.

HORNSBY, P. J. AND J. F. CRIVELLO. Role of lipid peroxidation and biological antioxidants in the function of the adrenal cortex. *Mol. Cell Endocr.* 30: 1–20, 30: 123–147, 1983.

HORNSBY, P. J. AND G. N. GILL. Hormonal control of adrenocortical cell proliferation. Desensitization to ACTH and interaction between ACTH and fibroblast growth factor in bovine adrenocortical cell cultures. *J. Clin. Invest.* 60: 342–352, 1977.

HORNSBY, P. J., AND G. N. GILL. Characterization of adult bovine adrenocortical cells throughout their life span in tissue culture. *Endocrinology* 102: 926–936, 1978.

HOY, J. F. Y., P. A. McGRATH, AND P. T. HALE. Electrophoretic analysis of multiple forms of rat cardiac myosin: effects of hypophysectomy and thyroxine replacement. *J. Mol. Cardiol.* 10: 1053–1076, 1977.

HSUEH, A. J. W., E. Y. ADAHSI, P. B. C. JONES, AND T. H. WELSH. Hormonal regulation of the differentiation of cultured ovarian granulosa cells. *Endocr. Rev.* 5: 76–127, 1984.

HSUEH, A. J. W., K. D. DAHL, J. VAUGHAN, E. M. TUCKER, J. RIVIER, C. W. BARDIN, AND J. VALE. Heterodimers and homodimers of inhibin subunits have different paracrine action in the modulation of luteinizing hormone-stimulated androgen biosynthesis. *Proc. Natl. Acad. Sci. USA* 84: 5082–5086, 1987.

HSUEH, A. J. W., M. L. DUFAU, AND K. J. CATT. Direct inhibitory effect of estrogen on Leydig cell function in hypophysectomized rats. *Endocrinology* 103: 1096–1102, 1978.

HSUEH, A. J. W., AND P. B. C. JONES. Extrapituitary actions of gonadotropin-releasing hormone. *Endocr. Rev.* 2: 437–461, 1981.

HUNTER, T. Oncogenes and growth control. *TIBS* July: 275–280, 1985.

HUNTER, T. A thousand and one protein kinases. *Cell* 50: 823–829, 1987.

HUSEBY, R. A., AND THOMAS, L. B. Histological and histochemical alterations in the normal breast tissues of patients with advanced breast cancer being treated with estrogenic hormones. *Cancer* 7: 54–74, 1954.

HUSEMAN, C. A., AND R. P. KELCH. Gonadotropin responses and metabolism of synthetic gonadotropin-releasing hormone (GnRH) during constant infusion of GnRH in men and boys with delayed adolescence. *J. Clin.* Endocrinol. Metab. 47: 1325–1331, 1978.

HUSZAR, G., AND G. NAFTOLIN. The myometrium and uterine cervix in normal and preterm labor. *N. Engl. J. Med.* 311: 571–581, 1984.

HYTTEN, F. E. AND D. B. PAINTIN. *The Physiology of Human Pregnancy.* Philadelphia: F. A. Davis Company, 1964.

IANNACONE, P. M. Conception, implantation and early development. In: *Principles and Practice of Endocrinology,* edited by K. Becker et al. Lipincott: In Press.

IGNOTZ, R. A., AND J. MASSAGUE. Transforming growth factor-β stimulates the expression of fibronectin and collagen and their incorporation into the extracellular matrix. *J. Biol. Chem.* 261: 4337–4345, 1986.

IMAGAWA, M., R. CHIU, AND M. KARIN. Transcription factor AP-2 mediates induction by two different signal-transduction pathways: Protein kinase C and cAMP. *Cell* 51: 251–260, 1987.

IMPERATO-McGINLEY, J., AND R. E. PETERSON. Male pseudohermaphroditism: the complexities of male phenotypic development. *Am. J. Med.* 61: 251, 1976.

INGLE, D. K., AND HIGGINS, G. M. Regeneration of adrenal gland following enucleation. *Am. J. Med. Sci.* 196: 232–240, 1938.

INSULL, W. J., J. HIRSCH, T. JAMES, AND E. H. AHRENS. The fatty acids of human milk. II. Alterations produced by manipulation of caloric balance and exchange of dietary fats. *J. Clin. Invest.* 38: 443–450, 1959.

ISMAIL-BEIGI, F. AND I. S. EDELMAN. The mechanism of thyroid calorigenesis: role of active sodium transport. *Proc. Natl. Acad. Sci. U.S.A.* 67: 1071–1078, 1970.

IZUMI, M., AND P. R. LARSEN. Triiodothyronine, thyrozine, and iodine in purified thyroglobulin from patients with Graves' disease. *J. Clin. Invest.* 59: 1105–1112, 1977.

JAFFE, R. B., B. H. YUEN, W. R. KEYE JR., AND A. R. MIDGLEY, JR. Physiologic and pathologic profiles of circulating human prolactin. *Am. J. Obstet. Gynecol.* 117: 757–773, 1973.

JAMES, E. J., J. R. RAYE, E. L. GRESHAM, E. L. MAKOWSKI, G. MESCHIA, AND F. C. BATTAGLIA. Fetal oxygen consumption, carbon dioxide production and glucose uptake in a chronic sheep preparation. *Pediatrics* 50: 361, 1972.

JARRETT, J. C. III, G. BALLEJO, T. H. SALEEM, ET AL. The effect of prolactin and relaxin on insulin binding by adipocytes from pregnant women. *Am. J. Obstet. Gynecol.* 149: 250, 1984.

JENNESS, R. The composition of human milk. *Semin. Perinat.* 3: 225–239, 1979.

JENNESS, R. The composition of milk. In: *Lactation, A Comprehensive Treatise,* Vol III, edited by B. L. Larson and V. R. Smith. New York: Academic Press, 1974, p. 3.

JIA, X. C., B. KESSEL, S. S. C. YEN, E. M. TUCKER, AND A. J. W. HSUEH. Serum bioactive follicle-stimulating hormone during the human menstrual cycle and in hyper-and hupogonadotropic states: Application of a sensitive granulosa cell aromatase bioassay (GAB). *J. Clin. Endocrinol.* Metab. 62: 1243–1249, 1985.

JOHNSON, D., A. LANAHAN, C. R. BUCK, A. SEHGAL, C. MORGAN, E. MERCER, M. BOTHWELL, AND M. CHAO. Expression and structure of the human NGF receptor. *Cell* 47: 545–554, 1986.

JOHNSTON, B. M., AND P. D. GLUCKMAN. Respiratory control and lung development in the fetus and newborn. In: *The Scientific Basis for Clinical Practice.* Ithaca: Perinatology Press, 1986.

JONES JR., M. D., L. I. BURD, E. L. MAKOWSKI, G. MESCHIA, AND F. C. BATTAGLIA. Cerebral metabolism and sheep: A comparative study of the adult, the lamb and the fetus. *Am. J. Physiol.* 229: 235, 1975.

JOSEPH, D. R., S. H. HALL, AND F. S. FRANCH. Rat androgen-binding protein; evidence for identical subunits and amino acid sequence homology with human sex hormone-binding globulin. *Proc. Natl. Acad. Sci. USA* 84: 339–343, 1987.

JOSIMOVICH, J. B., K. MERISKO, AND L. BOCELLA, Amniotic prolactin control over amniotic and fetal extracellular fluid water and electrolytes in the rhesus monkey. *Endocrinology* 100: 564–570, 1977.

JOSSO N. AntiMullerian hormone: New perspectives for a sexist molecule. *Endocrinol. Rev.* 7: 421–433, 1986.

JUDD, H. L., G. E. JUDD, W. G. LUCAS, S. S. C. YEN. Endocrine function of the postmenopausal ovary; concentrations of androgens and estrogens in ovarian and peripheral vein blood. *J. Clin. Endocrinol.* Metab. 39: 1020–1024, 1974.

KAHN, C. R., K. L. BAIRD, D. B. JARRETT, AND J. S. FLIER. Direct demonstration that receptor crosslinking or aggregation is important in insulin action. *Proc. Natl. Acad. Sci. U.S.A.* 75: 4209–4213, 1978.

KALKOFF, R. K., B. L. RICHARDSON, AND P. BECK. Relative effects of pregnancy, human placental lactogen and prednisone on carbohydrate tolerance in normal and subclinical diabetic subjects. *Diabetes* 18: 153, 1969.

KAMINSKA, J., I. CALVERT, AND S. W. ROJEN. Radioimmunoassay of "pregnancy-specific"-β-glycoprotein (SP$_1$). *Clin. Chem.* 25: 557–580, 1979.

KAPLAN, S. L,, AND M. M. GRUMBACH. The ontogenesis of human foetal hormones. II. Luteinizing hormone (LH) and follicle stimulating hormone (FSH). *Acta. Endocrinol.* 81: 808–829, 1976.

KATZ, R., J. S. KARLINER, AND R. RESNIK. Effects of a natural volume overload state (pregnancy) on left ventricular performance in normal human subjects. *Circulation* 58: 434, 1978.

KAWATA, M., J. R. PARNES, AND L. A. HERZENBERG. Transcriptional control of HLA-A, B,C, antigens in human placental cytotrophoblast isolated using trophoblast and HLA-specific monoclonal antibodies and the fluorescence-activated cell sorter. *J. Exp. Med.* 160: 633–651, 1984.

KELLY, P. T., T. L. McGUINNESS, AND P. GREENGARD. Evidence that the major postsynaptic density protein is a component of a Ca^{2+}/calmodulin-dependent protein kinase. *Proc. Natl. Acad. Sci. U.S.A.* 81: 945–949.

KESSNER, J. S., J. KAUFMAN, R. C. WILSON, G. KURODA, AND E. KNOBIL. On the short-loop feedback regulation of the hypothalamic luteinizing hormone releasing hormone "pulse generator" in the rhesus monkey. *Neuroendocrinology* 42: 109–111, 1986.

KHODR, G. S., AND T. SILER-KHODR. Localization of luteinizing hormone-releasing factor in the human placenta. *Fertil. Steril.* 29: 523–526, 1978.

KIMELMAN, D., AND M. KIRSCHNER. Synergistic induction of mesoderm by FGF and TGF-β and the identification of an mRNA coding for FGF in the early Xenopus embryo. *Cell* 51: 869–877, 1987.

KNOBIL, F. The neuroendocrine control of the menstrual cycle. *Rec. Prog. Horm. Res.* 36: 53–88, 1980.

KOBILKA, G. K., H. MATSUI, T. S. KOBILKA, T. L. YANG-FENG, U. FRANCKE, M. G. CARON, R. J. LEFKOWITZ, AND J. W. REGAN. Cloning, sequencing and expression of the gene coding for the human platelet α_2-adrenergic receptor. *Science* 238: 650–656, 1987.

KOBILKA, B. K., R. A. F. DIXON, T. FRIELLE, H. G. DOHLMAN, M. A. BOLANOWSKI, I. S. T. L. YANG-FENG, U. FRANCKE, M. G. CARON, AND R. J. LEFKOWITZ. cDNA for the human adrenergic receptor: A protein with multiple membrane-spanning domains encoded by a gene whose chromosomal location is shared with that of the receptor for platelet-derived growth factor. *Proc. Natl. Acad. Sci. U.S.A.* 84: 46–50, 1987.

KOHASE, M., D. HENRIKSEN-DeSTE-

FANO, L. T. MAY, J. VILCEK, AND P. B. SEHGAL. Induction of β_2-interferon by tumor necrosis factor: A homeostatic mechanism in the control of cell proliferation. *Cell* 45: 659–666, 1986.

KONNER, M., AND C. WORTHMAN. Nursing frequency, gonadal function, and birth spacing among !Kung hunter-gatherers. *Science* 207: 778–791, 1980.

KRUPP, M., AND M. D. LANE. On the mechanism of ligand-induced down-regulation of insulin receptor level in the liver cell. *J. Biol. Chem.* 256: 1689–1694, 1981.

KUBO, T., K. FUKUDA, A. MIKAMI, A. MAEDA, H. TAKAHASHI, M. MISHINA, T. HAGA, ICHIYAMA, K. KANGAWA, M. KOJIMA, H. MATSUO, T. HIROSE, AND S. NUMA. Cloning, sequencing and expression of complementary DNA encoding the muscarinic acetylcholine receptor. *Nature* 323: 411–416, 1986.

KURJAK, A., P. KIRKINEN, V. LATIN, AND D. IVANKOVIC. Ultrasonic assessment of fetal kidney function in normal and complicated pregnancies. *Am. J. Obstet. Gynecol.* 141: 266–270, 1981.

LAMBERTS, S. W. J., A. deLANGE, AND S. Z. STEFANKO. Adrenocorticotropin-secreting pituitary adenomas originate from the anterior or the intermediate lobe in Cushing's disease: Differences in the regulation of hormone secretion. *J. Clin. Endocrinol. Metab.* 54: 286–291, 1982.

LAND, H., L. F. PARADA, AND R. A. WEINBERG. Tumorigenic conversion of primary enbryo fibroblasts requires at least two cooperating oncogenes. *Nature* 304: 596–602, 1983.

LAND, H., G. SCHUTZ, H. SCHMALE, AND D. RICHTER. Nucleotide sequence of cloned cDNA encoding bovine arginine vasopressin-neurophysin II precursor. *Nature* 295: 299–303, 1982.

LARSEN, P. R., J. E. SILVA, AND M. M. KAPLAN. Relationships between circulating and intracellular thyroid hormones: Physiological and clinical implications. *Endocr. Rev.* 2: 87–102, 1981.

LEE, W-H., J-Y. SHEW, F. D. HONG, T. W. SERY, L. A. DONOSO, L-J. YOUNG, R. BOOKSTEIN, AND E. Y-H. LEE. The retinoblastoma susceptibility gene encodes a nuclear phosphoprotein associated with DNA binding activity. *Nature* 329: 642–645, 1987.

LEUNG, D. W., S. A. SPENCER, G. CACHIANES, R. G. HAMMONDS, C. COLLINS, W. J. HEN BARBARD, M. J. WATERS, AND W. I. WOOD. Growth hormone receptor and serum binding protein. Purification, cloning and expression. *Nature* 330: 537–543, 1987.

LEVI-MONTALCINI, R., AND U. ANGELETTI. Nerve growth factor. *Physiol. Rev.* 48: 53–569, 1968.

LEWIS, U. J., R. N. P. SINGH, G. F. TUTWILER, M. B. SIGEL, E. F. VANDERLAAN, AND W. P. VANDERLAAN.

Human growth hormone: a complex of proteins. *Recent Prog. Horm. Res.* 36: 477–508, 1980.

LIBERMANN, T. A., H. R. NUSBAUM, N. RAZON, R. KRIS, I. LAX, H. SOREQ, N. WHITTLE, WATERFIELD, A. ULLRICH, AND J. SCHLESSINGER. Amplification enhanced expression and possible rearrangement of EGF receptor gene in primary human brain tumors of glial origin. *Nature* 313: 144–147, 1985.

LIGGINS, G. Premature delivery of fetal lambs infused with glucocorticoids. *J. Endocrinol.* 45: 515–523, 1969.

LIGGINS, G. The paracrine system controlling human parturition. In: *The Endocrine Physiology of Pregnancy and the Peripartal Period*, edited by R. B. Jaffe, S. Dell'Acqua. New York: Raven Press, 1985, p. 205.

LIGGINS, G. C., R. J. FAIRCLOUGH, AND S. A. GRIEVES. The mechanism of initiation of parturition in the ewe. *Rec. Prog. Horm. Res.* 29: 111–159, 1973.

LIN, C. R., W. S. CHEN, C. S. LAZAR, C. D. CARPENTER, G. N. GILL, R. M. EVANS, AND M. G. ROSENFELD. Protein kinase C phosphorylation of Thr 654 of the unoccupied EGF receptor and EGF binding regulate functional receptor loss by idependent mechanisms. *Cell* 44: 839–848, 1986.

LINDHEIMER, M. D., KATZ, A. I. Kidney Function and Disease in Pregnancy. Philadelphia, Lea & Febiger, 1977, p. 20.

LING, N., S-Y. YING, N. UENO, S. SHIMASAKI, F. ESCH, M. HOTTA, AND R. GUILLEMIN. Pituitary FSH is released by a heterodimer of the &-subunits from the two forms of inhibin. *Nature* 321: 779–782, 1986.

LING, N., S-Y. YING, N. UENO, F. ESCH, L. DENOROY, AND R. GUILLEMIN. Isolation and partial characterization of a M_r 32,000 protein with inhibin activity from porcine follicular fluid. *Proc. Natl. Acad. Sci. U.S.A.* 82: 7217–7221, 1985.

LINGAPPA, V. R., AND G. BLOBEL. Early events in the biosynthesis of secretory and membrane proteins: the signal hypothesis. *Recent Prog. Horm. Res.* 36: 451–475, 1980.

LINGWOOD, B., K. J. HARDY, I. HORACEK, M. L. MCPHEE, B. A. SCOGGINS, AND E. M. WINTOUR. The effects of antidiuretic hormone on urine flow and composition in the chronically-cannulated ovine fetus. *Quart. J. Exp. Physiol.* 63: 315–330, 1978.

LINZELL, J. F. Mammary blood flow and methods of identifying and measuring precursors of milk. In: *Lactation, A comprehensive Treatise*, Vol I, edited by B. L. Larson and V. R. Smith. New York: Academic Press, 1974, p. 143.

LIPSETT, M. G., J. W. COMBS, AND D. G. SIEGEL. Problems in contraception. *Am. Intern. Med.* 74: 251, 1971.

LIU, J. H., AND S. S. C. YEN. Induction of a midcycle gonadotropin surge by

ovarian steroids in women: a critical evaluation. *J. Clin. Endocrinol. Metab.L* 57: 797–802, 1983.

LONGO, L. D., AND H. BARTELS. Respiratory gas exchange and blood flow in the placenta. U.S. Dept. of Health, Education and Welfare, Bethesda, 1972.

LOOSFELT, H., M. ATGER, M. MISRAHI, A. GUIOCHON-MANTEL, C. MERIERL, F. LOGEAT, R. BENAROUS, AND E. MILGROM. Cloning and sequence analysis of rabbit progesterone-receptor complementary DNA. *Proc. Natl. Acad. Sci. U.S.A.* 83:9045–9049, 1986.

LOZZIO, B. B., C. B. LOZZIO, E. G. BAMBERGER, AND S. V. LAIR. Regulators of cell division: endogenous mitotic inhibitors of mammalian cells. *Int. Rev. Cytol.* 42: 1–47, 1975.

LUCIUS, H., H. GAHLENBECK, AND H-O. KLEIN, ET AL. Respiratory functions, buffer system and electrolyte concentrations of blood during human pregnancy. *Respir. Physiol.* 9: 311, 1970.

LUDENS, J. H., AND D. D. FANESTIL. The mechanism of aldosterone function. In: *Pharmacology of Adrenal Cortical Hormones*, edited by G. N. Gill. Oxford: Pergamon Press, 1979, p. 143–185.

LYONS, W. R., C. H. LI, AND R. E. JOHNSON. The hormonal control of mammary growth and lactation. *Rec. Prog. Horm. Res.* 14: 219–254, 1985.

MACARON, C., O. FAMUYIWA, AND S. P. SINGH. In vitro effect of dopamine and pimozide on human chorionic samatomammotropin (hCS) secretion. *J. Clin. Endocrinol. Metab.* 47: 168–170, 1978.

MAINS, R. E., B. A. EIPPER, AND N. LING. Common precursor to corticotropins and endorphins. *Proc. Natl. Acad. Sci. U.S.A.* 74: 3014–3018, 1077.

MAJZOUB, J. A., H. M. KRONENBERG, J. T. POTTS, JR., A. RICH, AND J. F. HABENER. Identification and cell-free translation of mRNA coding for a precursor of parathyroid secretory protein. *J. Biol. Chem.* 254: 7449–7455, 1979.

MALPAS, P. Postmaturity and malformation of the foetus. *J. Obstet. Gynecol. Br. Empire* 40: 1046–1053, 1933.

MANDL, A., AND A. ZUCKERMAN. The relation of age to the number of oocytes. *J. Endocrinol.* 7: 190–193, 1951.

MANGIN, M., A. C. WEBB, B. E. DREYER, J. T. POSILLICO, K. IKEDA, E. C. WEIR, A. F. STEWART, N. H. BANDER, L. MILSTONE, D. E. BARTON, U. FRANCKE, AND A. E. BROADUS. Identification of a cDNA encoding a parathyroid hormone-like peptide from a human tumor associated with humoral hypercalcemia of malignancy. *Proc. Natl. Acad. Sci. U.S.A.* 85: 597–601, 1988.

MANOLAGAS, S. C., D. M. PROVEDINI, AND C. D. TSOUKAS. Interactions of 1,25-dihydroxyvitamin D_3 and the immune system. *Mol. Cell. Endocr.* 43: 113–122.

MARCHANT, B., J. F. H. LEES, AND W. D.

ALEXANDER. Antithyroid drugs. *Pharmacol. Ther. [B]* 3: 305–348, 1978.

MARKEE, J. E. Menstruation in intraocular endometrial transplants in the Rhesus monkey. *Contrib. Embrol.* 24: 223, 1940.

MARKEE, J. E. Morphological basis for menstrual bleeding. *Bull. N.Y. Acad. Med.* 24: 253–268, 1948.

MARTIAL, J. A., J. D. BAXTER, H. M. GOODMAN, AND P. H. SEEBURG. Regulation of growth hormone messenger RNA by thyroid and glucocorticoid hormones. *Proc. Natl. Acad. Sci. U.S.A.* 74: 1816–1820, 1977.

MARUO, T., M. MOCHIZUKI. Immunohistochemical localization of epidermal growth factor receptor and oncogene products in human placenta: Implication for trophoblast proliferation and differentiation. *Am. J. Obstet. Gynecol.* 156: 721–727, 1987.

MARVER, D., J. STEWART, J. W. FUNDER, D. FELDMAN, AND I. S. EDELMAN. Renal aldosterone receptors: studies with [³H]aldosterone and the antimineralocorticoid [³H]spirolactone (SC-26304). *Proc. Natl. Acad. Sci. U.S.A.* 71: 1431–1435, 1974.

MASON, A. J., J. S. HAYFLICK, N. LING, F. ESCH, N. UENO, S.-Y. YING, R. GUILLEMIN, N. NIALL, AND P. H. SEEBURG. Complementary DNA sequences of ovarian follicular fluid inhibin show precursor structure end homology with transforming growth factor-beta. *Nature* 318: 659–663, 1985.

MASON, A. J., J. S. HAYFLICK, N. LING, F. ESCH, N. UENO, S. Y. YING, R. GUILLEMIN, H. NIALL, AND P. H. SEEBURG. Complementary DNA sequence of ovarian follicular fluid inhibin show precursor structure and homology with transforming growth factor-beta. *Nature* 318: 659–663, 1985.

MASON, A. J., H. D. NIALL, AND P. H. SEEBURG. Structure of human ovarian inhibins. *Biochem. Biophys. Res. Commun.* 135: 957–964, 1986.

MASUI, H., AND L. D. GARREN. Inhibition of replication in functional mouse adrenal tumor cells by adrenocorticotropic hormone mediated by adenosine 3′:5′-cyclic monophosphate. *Proc. Natl. Acad. Sci. U.S.A.* 68: 3208–3210, 1971.

MATSUO, H., Y. BABA, R. M. G. NAIR, A. ARIMURA, AND A. V. SCHALLY. Structure of the porcine LH- and FSH-releasing hormone. I. The proposed amino acid sequence. *Biochem. Biophys. Res. Commun.* 43: 1334–1339.

MATSUOKA, H., P. J. MULROW, R. FRANCO-SAENZ, AND C. H. LI. Effects of β-lipotropin and β-lipotropin-derived peptides on aldosterone production in the rat adrenal gland. *J. Clin. Invest.* 68: 752–759, 1981.

MAYER, G. AND M. KLEIN. Histology and cytology of the mammary gland. In *Milk: The Mammary Gland and Its Secretion.* Vol. I, edited by S. K. Kon and A. T. Cowie. New York: Academic Press, 1961, p. 47.

McDONNELL, D. P., D. J. MANGELSDORF, J. W. PIKE, M. R. HAUSSLER, AND B. W. O'MALLEY. Molecular cloning of complementary DNA encoding the avian receptor for vitamin D. Science 235: 1214–1217, 1987.

McINTYRE, J. A., AND W. P. FAULK. Allotypic trophoblastlymphocyte cross-reactive (TLX) cell surface antigens. *Hum. Immunol.* 4: 27–35, 1982.

McLACHLAN, R. I., D. M. ROBERTSON, D. L. HEALY, H. G. BURGER, AND D. M. DE KRETSER. Circulating immunoreactive inhibin levels during the normal human menstrual cycle. *J. Clin. Endocrinol. Metab.* 65: 954–961, 1987.

McNEILLY, A. S., C. A. F. ROBINSON, M. J. HOUSTON, AND P. W. HOWIE. Release of oxytocin and PRL in response to suckling. *Br. Med. J.* 286: 257–259, 1983.

MEANS, A. R., AND J. R. DEDMAN. Calmodulin—an intacellular calcium receptor. *Nature* 285: 73–77, 1980.

MEITES, J. Maintenance of the mammary lobulo-alveolar system in rats after adreno-orchidectomy by prolactin and growth hormone. *Endocrinology* 76: 1220–1223, 1965.

MESCHIA, G. Supply of oxygen to the fetus. *J. Reproduc. Med.* 23: 160, 1979.

METCALFE, J., S. L. ROMNEY, AND L. H. RAMESY, ET AL. Estimation of uterine blood flow to women at term. *J. Clin. Invest.* 34: 1632, 1955.

METCALFE, J., J. H. MCANULTY, AND K. UELAND. Cardiovascular physiology. *Clin. Obstet. Gynecol.* 24: 693, 1981.

Mikhail, F. Hormone secretion by the human ovary. *Gynecol. Invest.* 1: 5–20, 1970.

MILLER, J., A. D. MCLACHLAN, AND A. KLUG. Repetitive-zinc-binding domains in the protein transcription factor IIIA from *Xenopus* oocytes. *EmBO J.* 4: 1609–1614, 1985.

MITTWOCH, U. *Genetics of Sex Differentiation.* New York: Academic Press, 1973, p. 1–253.

MIYABO, S., K.-I. YANAGISAWA, E. OOYA, T. HISADA, AND S. KISHIDA. Ontogeny of circadian corticosterone rhythm in female rats: effects of periodic maternal deprivation and food restriction. *Endocrinology* 106: 636–642, 1980.

MIYAKE, A., O. TANIZAWA, T. AONO, AND K. KURACHI. Pituitary LH response to LHRH during puerperium. *Obstet. Gynecol.* 51: 37–40, 1978.

MOLSBERRY, R. L., B. R. CARR, C. R. MENDELSON, AND E. R. SIMPSON, Human chorionic gonadotropin binding to human fetal testes as a function of gestational age. *J. Clin. Endocrinol. Metab.* 55: 791–794, 1982.

MONAD, J., J. P. CHANGEUX, AND F. JACOB. Allosteric proteins and cellular control systems. *J. Mol. Biol.* 6: 306–329, 1963.

MONTMINY, M. R., K. A. SEVARINO, J. A. WAGNER, G. MANDEL, AND R. H. GOODMAN. Identification of a cyclic-

AMP-responsive element within the rat somatostatin gene. *Proc. Natl. Acad. Sci. U.S.A.* 83: 6682–6686, 1986.

MOORADIAN, A. D., J. E. MORLEY, AND S. G. KORENMAN. Biological actions of androgens. *Endocr. Rev.* 8: 1–28, 1987.

MOORE, R. Y. Neuroendocrine mechanisms. In: *Reproductive Endocrinology*, edited by S. S. C. Yen, and R. B. Jaffe. Philadelphia: W. B. Saunders, 1986, p. 3–32.

MORGAN, D. O., J. C. EDMAN, D. N. STANDRING, V. A. FRIED, M. C. SMITH, R. A. ROTH, AND J. RUTTER. Insulin-like growth factor II receptor as a multifunctional binding protein. *Nature* 329: 301–307, 1987.

MORLEY, J. E. Neuroendocrine Control of Thyrotropin Secretion. *Endocr. Rev.* 2: 396–436, 1981.

MORRISS, F. H. JR., AND R. D. H. BOYD. Placental transport. In: *The Physiology of Reproduction*, edited by E. Knobil and J. Neill. New York: Raven Press, 1988, pp. 203–208.

MOSS, R. L. Actions of hypothalamic-hypophysiotropic hormones on the brain. *Annu. Rev. Physiol.* 41: 617, 1979.

MUNDY, G. R., L. G. RAISZ, R. A. COOPER, G. P. SCHECHTER, AND S. E. SALMON. Evidence for the secretion of an osteoclast stimulating factor in myeloma. *N. Engl. J. Med.* 291: 1041–1046, 1974.

NAKANISHI, S., A. INOUE, T. KITA, M. NAKAMURA, A. C. Y. CHANG, S. N. COHEN, AND S. NUMA. Nucleotide sequence of cloned cDNA for the bovine corticotropin-β-lipotropin precursor. *Nature* 278: 423–427, 1979.

NAKANISHI, S., T. KITA, S. TAII, H. IMURA, AND S. NUMA. Glucocorticoid effect on the level of corticotropin messenger RNA activity in rat pituitary. *Proc. Natl. Acad. Sci. U.S.A.* 74: 3283–3286, 1977.

NAKAZAWA, M., S. MIYAGAWA, T. OHNO, S. MIURA, AND A. TAKAO. Developmental hemodynamic changes in rat embryos at 11 to 15 days of gestation: normal data on blood pressure and the effect of caffeine compared to data from chick embryo. *Pediatr. Res.* 23: 200–205, 1988.

NATHANS, J., AND D. S. HOGNESS. Isolation and nucleotide sequence of the gene encoding human rhodopsin. *Proc. Natl. Acad. Sci. U.S.A.* 81: 4851–4855, 1984.

NIALL, H. D., M. L. HOGAN, G. W. TREGEAR, G. V. SEGRE, P. HWANG, AND H. FRIESEN. The chemistry of growth hormone and the lactogenic hormones. *Recent Prog. Horm. Res.* 29: 387–416, 1973.

NICOLAIDES, K. H., S. CAMPBELL, R. J. BRADLEY, C. M. BILARDO, P. W. SOOTHILL, AND D. GIBB. Maternal oxygen therapy for intrauterine growth retardation. *Lancet* 1: 942–945, 1987.

NIEMI, M., M. IKONEN, AND A. HERVONEN. Histochemistry and fine structure of the interstial tissue in the foetal testis. In *CIBA Foundation Colloquia on Endocrinology: Endocrinology of the Testis*, Vol. 16, edited by G. E. W. Wolstenholme, M. O'Connor. London: J. & A Churchill, 1967, p. 31.

NIEUWKOOP, P. D., A. G. JOHNEN, AND B. ALBERS. *The Epigenetic Nature of Early Chordate Development. Inductive Interaction and Competence.* Cambridge University Press, Cambridge, England, 1985.

NIKOLICS, K., A. J. MASON, E. SZONYI, J. RAMACHANDRAN, AND P. H. SEEBURG. A prolactin-inhibiting factor within the precursor for human gonadotropin-releasing hormone. *Nature* 316: 511–517, 1985.

NISHIZUKA, Y. The role protein kinase C in cell surface signal transduction and tumor promotion. *Nature* 308: 693–698, 1984.

NISULA, B. C., F. J. MORGAN, AND R. E. CANFIELD, Evidence that chorionic gonadotropin has intrinsic thyrotropic activity. *Biochem. Biophys. Res. Commun.* 59: 86–91, 1974.

NODA, M., Y. FURUTANI, H. TAKAHASHI, M. TOYOSATO, T. HIROSE, S. INAYAMA, S. NAKANISHI, AND S. NUMA. Cloning and sequence analysis of cDNA for bovine adrenal preproenkephalin. *Nature* 295: 202–206, 1982.

NOEL, G. L., H. K. SUH, AND A. G. FRANTZ. Prolactin release during nursing and breast stimulation in postpartum and nonpostpartum subjects. *J. Clin. Endocrinol. Metab.* 38: 413–423, 1974.

NOVY, M. J., AND G. LIGGINS. Role of prostaglandins, prostacyclin, and thromboxane in the physiologic control of the uterus and in parturition. *Semin. Perinatol.* 4: 45–66, 1980.

ODAGIRI, E., B. J. SHERRELL, C. D. MOUNT, W. E. MICHOLSON, AND D. N. ORTH. Human placental immunoreactive corticotropin, lipotropin and B-endorphin: Evidence for a common precursor. *Proc. Natl. Acad. Sci. USA* 76: 2027–2031, 1979.

OH, W., K. OMORI, G. C. EMMANOULIDES, AND D. L. PHELPS. Placental to lamb fetus transfusion in utero during acute hypoxia. *Am. J. Obstet. Gynecol.* 122: 316–322, 1975.

O'HARE, M. J., E. C. NICE, AND M. NEVILLE. Regulation of androgen secretion and sulfoconjugation in the adult human adrenal cortex: studies with primary monolayer cell cultures. In: *Adrenal Androgens*, edited by A. R. Genazzani, J. H. H. Thijssen, and P. K. Siiteri. New York: Raven Press, 1980, p. 7–25.

OHNO, S., H. P. KLINGER, AND N. B. ATKIN. Human oogenesis. *Cytogenetics* 1: 42–51, 1962.

OKA, Y., AND D. N. ORTH. Human plasma epidermal growth factor/&-urogastrone is associated with blood platelets. *J. Clin. Invest.* 72: 249–259, 1983.

OKAZAKI, T., M. L., CASEY, J. R. OKITA, P. C. MAC DONALD, AND J. M. JOHNSTON. Initiation of human parturition, XII. Biosynthesis and metabolism of prostaglandins in human fetal membranes and uterine decidua. *Am. J. Obstet. Gynecol.* 139: 373–381, 1981.

OPPENHEIMER, J. H., W. H. DILLMANN, H. L. SCHWARTZ, AND H. C. TOWLE. Nuclear receptors and thyroid hormone action: a progress report. *Fed. Proc.* 38: 2154–2160, 1979.

OWERBACH, D., W. J. RUTTER, N. E. COOKE, J. A. MARTIAL, AND T. B. SHOWS. The prolactin gene is located on chromosome 6 in humans. *Science* 212: 815–816, 1981.

OWERBACH, D., W. J. RUTTER, J. A. MARTIAL. J. D. BAXTER, AND T. B. SHOWS. Genes for growth hormones, chorionic somatomammotropin, and growth hormone-like gene on chromosome 17 in humans. *Science* 209: 289–292, 1980.

PAGE, D. G., R. MOSHER, E. M. SIMPSON, E. M. C. FISHER, G. MARDON, J. POLLACK, B. McGILLIVRAY, A. DE LA CHAPELLE, AND L. G. BROWN. The sex-determining region of the human Y chromosome encodes a finger protein. *Science* 51: 1091–1104, 1987.

PARER, J. T. Fetal heart rate. In: *Maternal-Fetal Medicine: Principles and Practice*, edited by R. K. Creasy and R. Resnik, Philadelphia: W. B. Saunders Company, 1984, pp. 285–319.

PARTRIDGE, W. M. Transport of protein-bound hormones into tissues *in vivo*. *Endocr. Rev.* 2: 103–132, 1981.

PARVINEN, M. Regulation of the seminiferous epithelium. *Endocr. Rev.* 3: 404–417, 1982.

PATRICK, J. Fetal breathing and body movements. In: *Maternal-Fetal Medicine: Principles and Practice*, edited by R. K. Creasy and R, Resnik, Philadelphia: W. B. Saunders Company, 1984, pp. 239–257.

PAYNE, A. H., P. G. QUINN, AND S. S. RANI. Regulation of microsomal cytochrome P-450 enzymes and testosterone production in Leydig cells. *Rec. Prog. Horm. Res.* 41: 153–195, 1985.

PEDERSEN, R. C., AND A. C. BROWNIE. Adrenocortical response to corticotropin is potentiated by part of the aminoterminal region of procorticotropin/endorphin. *Proc. Natl. Acad. Sci. U.S.A.* 77: 2239–2243, 1980.

PEKONEN, F., H. ALFTHAN, U-H. STENMAN, AND O. YLIKORKALA. Human chorionic gonadotropin (hCG) and thyroid function in early human pregnancy: Circadian variation and evidence for intrinsic thyrotropic activity of hCG. *J. Clin. Endocrinol. Metab.* 66: 853–856, 1988.

PELLINIEMI, L. J., AND M. DYM. The fetal gonad and sexual differentiation. In *Maternal-Fetal Endocrinology*, edited by D. Tulchinsky, K. J. Ryan. Philadelphia: W. B. Saunders, 1980, p. 252.

PERNOLL, M. L., J. METCALFE, P. A. KOVACH, ET AL. Ventilation during rest and exercise in pregnancy and postpartum. *Respir. Physiol.* 25: 295, 1975.

PERRONE, M. H. AND P. M. HINKLE. Regulation of pituitary receptors for thyrotropin-releasing hormone by thyroid hormones. *J. Biol. Chem.* 253: 5168–5173, 1978.

PESTKA, S., J. A. LANGER, K. C. ZOON, AND C. E. SAMUEL. Interferons and their actions. *Annu. Rev. Biochem.* 56: 727–777, 1987.

PETERSON, R. E., J. IMPERATO-McGINLEY, T. GAUTIER, AND E. STURLA. Male pseudohermaphroditism due to steroid 5α-reductase deficiency. *Am. J. Med.* 62: 170–191, 1977.

PETRAGLIA, F., P. SAWCHENKO, A. T. LIM, J. RIVIER, AND W. VALE. Localization, secretion, and action of inhibin in human placenta. *Science* 237: 187–189, 1987.

PHIFER, R. F., A. R. MIDGLEY, AND S. S. SPICER. Immunologic and histologic evidence that follicle-stimulating hormone and luteinizing hormone are present in the same cell type in the human pars distales. *J. Clin. Endocrinol. Metab.* 36: 125–141, 1973.

PHILLIPS, L. S., AND R. VASSILOPOULOU-SELLIN. Somatomedins. (Two parts). *N. Engl. J. Med.* 302: 371–380, 438–446, 1980.

PICKLES, V. R. Prostaglandins in human endometrium. *Int. J. Fertil.* 12: 335, 1967.

PIERCE, J. G., T-H. LIAO, S. M. HOWARD, B. SHOME, AND J. S. CORNELL. Studies on the structure of thyrotropin: its relationship to luteinizing hormone. *Recent Prog. Horm. Res.* 27: 165–212, 1971.

PIERCE, J. G., AND T. F. PARSONS. Glycoprotein hormones: structure and function. *Annu. Rev. Biochem.* 50: 465–495, 1981.

PIJNENBORG, R., W. B. ROBERTSON, I. BROSENS, AND G. DIXON. Trophoblast invasion and the establishment of hemochorial placentation in man and laboratory animals. *Placenta* 2: 71–92, 1981.

PINTOR, C., A. R. GENAZZANI, G. CARBONI, T. FANNI, S. ORANI, F. FACCLINETTI, AND R. CORDA. Adrenal androgens and pubertal development in physiological and pathological conditions. In: *Adrenal Androgens*, edited by A. R. Genazzani, J. H. H. Thijssen, and P. K. Sitteri. New York: Raven Press, 1980, p. 173–182.

PIPPIG, L. Clinical aspects of renal disease during pregnancy. *Med. Hyg.* 27: 181, 1969.

PITKIN, R. M. Calcium metabolism in pregnancy and the perinatal period: A review. *Am. J. Obstet. Gynecol.* 151: 99, 1985.

POSNER, B. I., P. A. KELLY, R. P. C. SHIU, AND H. G. FRIESEN. Studies of insulin, growth hormone and prolactin binding: tissue distribution, species variation and characterization. *Endocrinology* 95: 521–531, 1974.

PRESLOCK, J. P. Steroidogenesis in the mammalian testis. *Endocr. Rev.* 1: 132–139, 1980.

PTASHNE, M. Gene regulation by proteins acting nearby and at a distance. *Nature* 322: 697–701, 1986.

QUIGLEY, M. E., B. ISHIZUKA, J. F. ROPERT, AND S. S. C. YEN. The food-entrained prolactin and cortisol release in late pregnancy and prolactinoma patients. *J. Clin. Endocrinol. Metab.* 54: 1109–1112, 1982.

RAE, P. A., N. S. GUTMANN, J. TSAO, AND B. P. SCHIMMER. Mutations in cyclic AMP-dependent protein kinase and corticotropin (ACTH)-sensitive adenylate cyclase affect adrenal steroidogenesis. *Proc. Natl. Acad. Sci. U.S.A.* 76: 1896–1900, 1979.

RASMUSSEN, D. D., J. H. LIU, P. L. WOLF, AND S. S. C. YEN. Endogenous opioid regulation of GnRH release from human fetal hypothalamus in vitro. *J. Clin. Endocrinol. Metab.* 57: 881–884, 1983.

RASMUSSEN, D. D., J. H. LIU, AND S. S. C. YEN. Endogenous opioid regulation of GnRH release from the human mediobasal hypothalamus (MBH) *in vitro. J. Clin. Endocrinol. Metab.* 57: 881–884, 1983.

RAYNAUD, A. Morphogenesis of the mammary gland. In: *Milk: The Mammary Gland and Its Secretion*, edited by S. K. Kon and A. T. Cowie. Vol I New York: Academic Press, 1969, p. 3.

REBAR, R. W. The breast and physiology of lactation. In: *Maternal-Fetal Medicine. Principles and Practice*, edited by R. K. Creasy and R. Resnik. Philadelphia: W. B. Saunders Co., 1984, p. 159.

REDMAN, C. W. G. HLA-DR antigen on human trophoblast: A review. *Am. J. Reproduc. Immunol.* 3: 175–177, 1985.

REES, L. H., C. W. BUARKE, T. CHARD, S. W. EVANS, AND A. T. LETCHORTH. Possible placental origin of ACTH in normal human pregnancy. *Nature* 254: 620–622, 1975.

RESNIK, R. The endocrine regulation of uterine blood flow in the non-pregnant uterus: A review. *Am. J. Obstet. Gynecol.* 140: 151–156, 1981.

RESNIK, R., A. P. KILLAM, AND F. C. BATTAGLIA, ET AL. The stimulation of uterine blood flow by various estrogens. *Endocrinology* 94: 1192, 1974.

REYES, F. I., J. S. D. AND C. FAIMAN. Pituitary gonadotropin function during human pregnancy: Serum FSH and LH Levels before and after LHRH administration. *J.Clin. Endocrinol. Metab.* 42: 590–592, 1976.

RIDDICK, D. H. AND W. F. KUSMIK, Decidua: A possible source of amniotic fluid PRL. *Am. J. Obstet. Gynecol.* 127: 187–190, 1977.

RIGG, L. A. A. LEIN, AND S. S. C. YEN. The pattern of increase in circulating prolactin levels during human gesta-

tion. *Am. J. Obstet. Gynecol.* 129: 454–456, 1977.

RINIKER, B., P. SIEBER, W. RITTEL, AND H. ZUBER. Revised amino-acid sequences for porcine and human adrenocorticotrophic hormone. *Nature New Biol.* 235: 114–115, 1972.

RITZEN, E. M., L. HAGENAS, L. PLOEN, F. S. FRENCH, AND V. HANSSON. In vitro synthesis of rat testicular androgen-binding protein. (ABP). *Molec. Cell Endocrinol.* 8: 335–346, 1977.

RIVIER, C. L., AND P. M. PLOTSKY. Mediation by corticotropin releasing factor (CRF) of adenohypophysial hormone secretion. *Annu. Rev. Physiol.* 48: 475–494, 1986.

RIVIER, J., J. SPIESS, M. THORNER, AND W. VALE. Characterization of a growth hormone-releasing factor from a human pancreatic islet tumour. *Nature* 300: 276–278, 1982.

ROBBINS, J. Iodine deficiency, iodine excess and the use of iodine for protection against radioactive iodine. *Thyroid Today* 3, 8: 1–5, 1980.

ROBBINS, J., S-Y. CHENG, M. C. GERSHENGORN, D. GLINOER, H. J. CAHNMANN, AND H. EDELNOCK. Thyroxine transport proteins of plasma: molecular properties and biosynthesis. *Recent Prog. Horm. Res.* 34: 477–519, 1978.

ROBERTSON, G. L. The regulation of vasopressin function in health and disease. *Recent Prog. Horm. Res.* 33: 333–385, 1977.

ROBERTSON, G. L. Disease of the posterior pituitary. In: *Endocrinology and Metabolism*, edited by P. Felig, J. D. Baxter, A. E. Broadus, and L. A. Frohman. New York: McGraw-Hill, 1984.

ROBERTSON, G. L., E. A. MAHR, S. ATHAR, AND T. SINHA. Development and clinical application of a new method for the radioimmunoassay of arginine vasopressin in human plasma. *J. Clin. Invest.* 52: 2340–2352, 1973.

ROBILLARD, J. E. Fetal renal function and regulation of water and electrolyte excretion. In: *Fetal and Neonatal Body Fluids: The Scientific Basis for Clinical Practice.* Ithaca: Perinatology Press, 1988.

ROCK, J., C. R. GARCIA, AND M. MENKIN. A theory of menstruation. *Ann. N.Y. Acad. Sci.* 75: 830–839, 1959.

RODBELL, M. The role of hormone receptors and GTP-regulatory proteins in membrane transduction. *Nature* 284: 17–22, 1980.

RODRIQUEZ-RIGAU L. J., Z. ZUKERMAN, D. B. WEISS, K. D. SMITH, AND E. STEINBERGER. In: *Testicular Development, Structure, and Function*, edited by A. Steinberger, and E. Steinberger. New York: Raven Press, 1980, p. 139–146.

ROLLAND, R., R. M. LEGUIN, L. A. SCHELLEKENS, AND F. H. DEJONG. The role of prolactin in the restoration of ovarian function during the early postpartum period in the human female. I. A study during physiological

lactation. *Clin. Endocrinol.* 4: 15–25, 1975.

ROSEN, J. M., R. J. MATUSIK, D. A. RICHARDS, P. GUPTA, AND J. R. RODGERS. Multihormonal regulation of casein gene expression at the R. RODGERS. Multihormonal regulation of casein gene expression at the transcriptional and posttranscriptional levels in the mammary gland. *Recent Prog. Horm. Res.* 36: 157–193, 1980.

ROSEN, O. M. After insulin binds. *Science* 237: 1452–1458, 1987.

ROSENFELD, C. R., F. H. MORRIS, JR., AND F. C. BATTAGLIA, ET AL. Effects of estradiol 17-β on blood flow to reproductive and nonreproductive tissues in pregnant ewes. *Am. J. Obstet. Gynecol.* 124: 618, 1976.

ROSENFELD, M. G., J.-J MERMOD, S. G. AMARA, L. W. SWANSON, P. E. SAWCHENKO, J. RIVIER, W. W. VALE, AND R. M. EVANS. Production of a novel neuropeptide encoded by the calcitonin gene via tissue-specific RNA processing. *Nature* 304: 129–135, 1983.

ROSNER, W. AND S. M. DEAKINS. Testosterone-binding globulin human plasma: Studies on sex distribution and specificity. *J. Clin. Invest.* 47: 2109–2116, 1968.

ROSS, E. M., AND A. G. GILMAN. Biochemical properties of hormone-sesitive adenylate cyclase. *Annu. Rev. Biochem.* 49: 533–564, 1980.

ROSS, R., A. VOGEL, P. DAVIES, E. RAINES, B. KARIYA, M. J. RIVEST, C. GUSTAFSON, AND J. GLOMSET. The platelet-derived growth factor and plasma control cell proliferation. *Cold Spring Harbor Conf. Cell Proliferation* 6: 3–16, 1979.

ROSS, R., E. W. RAINES, AND D. F. BOWEN-POPE. The biology of platelet-derived growth factor. *Cell* 46: 155–169, 1986.

RUDOLPH, A. M. AND M. A. HEYMANN. The circulation of the fetus in utero: methods for studying distribution of blood flow, cardiac output and organ blood flow. *Circ. Res.* 21: 163–184, 1967.

RUSSELL, D. W., W. J. SCHNEIDER, T. YAMAMOTO, K. L. LUSKEY, M. S. BROWN, AND J. L. GOLDSTEIN. Domain map of the LDL receptor: Sequence homology with the epidermal growth factor precursor. *Cell* 37: 577–585, 1984.

RYAN, K. J. Maintenance of pregnancy and the initiation of labor. In *Maternal-Fetal Endocrinology*, edited by Philadelphia: W. B. Saunders Co., 1980, p. 297.

RYAN, K. J., Z. PETRO, AND J. KAISER. Steroid formation by isolated and recombined granulosa and thecal cells. *J. Clin. Endocrinol. Metab.* 28: 355–358, 1968.

SAIRAM, M. R. Gonadotropin hormones: relationship between structure and function with emphasis on antagonists. In: *Hormonal Proteins and Pep-*

tides: Vol. XI. *Gonadotropic Hormones,* edited by C. H. Li. New York: Academic Press, 1983, p. 1–79.

SALMON, W. D., JR., AND W. H. DAUGHADAY. A hormonally controlled serum factor which stimulates sulfate incorporation by cartilege in vitro. *J. Lab. Clin. Med.* 49: 825, 1957.

SAMAAN, N., S. C. C. YEN, D. GONZALEZ, AND O. H. PEARSON. Metabolic effects of placental lactogen (HPL) in man. *J. Clin. Endocrinol. Metab.* 28: 485–491, 1968.

SAMUELS, H. H., Z. D. HORWITZ, F. STANLEY, J. CASANOVA, AND L. E. SHAPIRO. Thyroid hormone controls glucocorticoid action in cultured GH₁ cells. *Nature* 268: 254–257, 1977.

SAMUELS, H. H., AND J. S. TSAI. Thyroid hormone action in cell culture: demonstration of nuclear receptors in intact cells and isolated nuclei. *Proc. Natl. Acad. Sci. U.S.A.* 70: 3488–3492, 1973.

SANTEN, R. J. Independent control of luteinizing hormone secretion by testosterone and estradiol in males. In: *Hormones in Normal and Abnormal Human Tissues,* Vol. 1. edited by K. Fotherby, and S. B. Pahl. Berlin and New York: Walter de Gruyter, 1981, p. 459–490.

SANTEN, R. J., AND C. W. BARDIN. Episodic luteinizing hormone secretion in men. *J. Clin. Invest.* 52: 2617–2628, 1973.

SANTEN, R. S. The Testis. In: *Endocrinology and Metabolism,* edited by P. Felig, J. D. Baxter, A. E. Broadus, and L. A. Frohman. New York: McGraw-Hill, 1987, p. 821–905.

SAP, J., A. MUNOZ, K. DAMM, Y. GOLDBERG, J. GHYSDAEL, A. LEUTZ, H. BEUG, AND B. JANNSTROM. The c-erb-A protein is a high-affinity receptor for thyroid hormone. *Nature* 324: 635–640, 1986.

SASAKI, A., P. TEMPST, A. S. LIOTTA, A. N. MARGIORIS, L. E. HOOD, S. B. H. KENT, S. SATO, O. SHINKAWA, Y. YOSHINGA, AND D. T. KRIEGER. Isolation and characterization of a corticotropin-releasing hormone-like peptide from human placenta. *J. Clin. Endocrinol. Metab.* 67: 768–773, 1988.

SAWYER, W. H., AND P. K. T. PANG. Evolution of neurohypophyseal hormones and their function. In: *Neurohypophysis: International Conference on the Neurohypophysis,* edited by A. M. Moses and L. Share. Basel: Karger, 1977.

SCHALLY, A. V., A. ARIMURA, A. J. KASTIN, H. MATSUO, Y. BABA, T. W. REDDING, R. M. G. NAIR, L. DEBELJUK, AND W. F. WHITE. Gonadotropin-releasing hormone: one polypeptide regulates secretion of luteinizing and follicle-stimulating hormones. *Science* 173: 1036–1038, 1971.

SCHENKER, J. G., M. BEN-DAVID, AND W. Z. POLISHUK. Prolactin in normal pregnancy: Relationship of maternal,

fetal, and amniotic fluid levels. *Am. J. Obstet. Gynecol.* 123: 834–838, 1975.

SCHIMMEL, M., AND R. D. UTIGER. Thyroid and peripheral production of thyroid hormones. *Ann. Intern. Med.* 87: 760–768, 1977.

SCHLESSINGER, J., Y. SHECHTER, M. C. WILLINGHAM, AND I. PASTAN. Direct visualization of binding, aggregation, and internalization of insulin and epidermal growth factor on living fibroblastic cells. *Proc. Natl. Acad. Sci. U.S.A.* 75: 2659–2663, 1978.

SCHONBRUNN, A., M. KRASNOFF, J. M. WESTENDORF, AND A. H. TASHJIAN, JR. Epidermal growth factor and thyrotropin-releasing hormone act similarly on a clonal pituitary cell strain. *J. Cell. Biol.* 85: 785, 1980.

SCHRAMM, M., J. ORLY, S. EIMERL., AND M. KORNER. Coupling of hormone receptors to adenylate cyclase of different cells by cell fusion. *Nature* 268: 310–313, 1977.

SCHREIBER, A. B., T. A. LIBERMANN, I. LAX, Y. YARDIN, AND J. SCHLESSINGER. Biological role of epidermal growth factor receptor clustering. *J. Biol. Chem.* 258:846–853, 1983.

SCHWORER, C. M., R. J. COLBRAN, AND T. R. SODERLING. Reversible generation of a Ca²⁺-independent form of Ca²⁺(calmodulin)-dependent protein kinase II by an autophosphorylation mechanism. *J. Biol. Chem.* 261: 8581–8584, 1966.

SEEBURG, P. H., AND J. P. ADELMAN. Characterization of cDNA for the precursor of human luteinizing hormone releasing hormone. *Nature* 311: 666–668, 1984.

SEEGER, R. C., G. M. BRODEUR, H. SATHER, A. DALTON, S. E. SIEGEL, K. Y. WONG, AND D. HAMMOND. Association of multiple copies of the N-*myc* oncogene with rapid progression of neuroblastomas. *N. Engl. J. Med.* 313: 1111-1116, 1985.

SEGERSON, T. P., J. KAVER, H. C. WOLFE, H. MOBTAKER, P. WU, I. M. D. JACKSON, AND R. M. LECHAN. Thyroid hormone regulates TRH biosynthesis in paraventricular nucleus of the rat hypothalamus. *Science* 238: 78–80, 1987.

SEITCHIK, J. Total body water and total body density of pregnant women. *Obstet. Gynecol.* 29: 155, 1967.

SERÒN-FERRE, M., AND R. B. JAFFE. The fetal adrenal gland. *Ann. Rev. Physiol.* 43: 141–162, 1981.

SERÒN-FERRE, M., C. C. LAWRENCE, AND R. B. JAFFE. Role of hCG in regulation of the fetal zone of the human fetal adrenal gland. *J. Clin. Endocrinol. Metab.* 46: 834–837, 1978.

SEYBERTH, H. W., G. V. SEGRE, P. HAMET, B. J. SWEETMAN, J. T. POTTS, AND J. A. OATES. Characterization of the group of patients with the hypercalcemia of malignancy who respond to treatment with prostaglan-

din synthesis inhibitors. *Trans. Assoc. Am. Physicians* 89: 92, 1976.

SHERMAN, B. M. J. H. WEST, AND S. G. KORENMAN. The menopausal transition: analysis of LH, FSH, estradiol, and progesterone concentration during menstrual cycles of older women. *J. Clin. Endocrinol. Metab.* 42: 629–636, 1976.

SHIH, C., L. C. PADHY, M. MURRAY, AND R. A. WEINBERG. Transforming genes of carcinomas and neuroblastomas introduced in mouse fibroblasts. *Nature* 290: 261–264, 1981.

SHOME, B., AND A. F. PARLOW. Human pituitary prolactin (hPRL): the entire linear amino acid sequence. *J. Clin. Endocrinol. Metab.* 45: 1112–1115, 1977.

SHOYAB, M., J. E. DeLARCO, AND G. J. TODARO. Biologically active phorbol esters specifically alter affinity of epidermal growth factor membrane receptors. *Nature* 279: 387–391, 1979.

SIITERI, P. K., AND P. C. MacDONALD. Placental estrogen biosynthesis during human pregnancy. *J. Clin. Endocrinol. Metab.* 26: 751–761, 1966.

SIITERI, P. K. AND P. C. MacDONALD. Role of extraglandular estrogen in human endocrinology. In: *Handbook of Physiology,* Section 7, *Endocrinology,* edited by S. R. Geiger, E. B. Astwood, R. O. Greep. Washington D.C.: American Physiology Society, 1973, pp. 615–629.

SILER-KHODR, T. M., AND G. S. KHODR. Production and activity of placental releasing hormones. In: *Fetal Endocrinology,* edited M. J. Novy and J. A. Resko, New York: Academic Press, 1981, p. 183.

SILVER, B. J., J. A. BOKAR, J. B. VIRGIN, E. A. VALLEN, A. MILSTED, AND J. H. NILSON. Cyclic AMP regulation of the human glycoprotein hormone α subunit gene is mediated by an 18 base pan element. *Proc. Nat. Acad. Sci. U.S.A.* 84: 2198–2202, 1987.

SIMONIAN, M. H., AND G. N. GILL. Regulation of the fetal human adrenal cortex: effects of adrenocorticotropin on growth and function of monolayer cultures of fetal and definitive zone cells. *Endocrinology* 108: 1769–1779, 1981.

SIMPSON, E., P. CHANDLER, P. GOULMY, E. DISTECHE, C. M. FERGUSON, M. A. SMITH, AND D. C. PAGE. Separation of the genetic loci for H-Y antigen and for testis determining factor on human Y chromosome. *Nature* 326: 876–878, 1987.

SIMPSON, E. R. Cholesterol side-chain cleavage, cytochrome P-450, and the control of steroidogenesis. *Molec. Cell Endocrinol.* 13: 213–227, 1979.

SIMPSON, E. R., J. L. McCARTHY, AND J. A. PETERSON. Evidence that the cycloheximide-sensitive site of adrenocorticotropic hormone action is in the mitochondrion. *J. Biol. Chem.* 253: 3135–3139, 1978.

SIMS, E. A. H. AND K. E. KRANTZ. Serial

studies of renal function during pregnancy and the puerperium in normal women. *J. Clin. Invest.* 37: 1764, 1958.

SINGH, S., J. W. SPARKS, J. MESCHIA, F. C. BATTAGLIA, AND E. L. MAKOWSKI. Comparison of fetal and maternal hind limb metabolic rates in sheep. *Am. J. Obstet. Gynecol.* 149: 441, 1984.

SINHA, T. K., H. F. DELUCA, AND N. H. BELL. Evidence for a defect in the formation of $1\alpha,25$-dihydroxy vitamin D in pseudohypoparathyroidism. *Metabolism* 26: 731–738, 1977.

SKATRUD, J. B., J. A. DEMPSEY, AND J. G. KAISER. Ventilatory response to medroxyprogesterone acetate in normal subjects: Time course and mechanism. *J. Appl. Physiol.: Respir. Environ. Exercise Physiol.* 44: 939, 1978.

SLACK, J. M. W., B. G. DARLINGTON, J. K. HEATH, AND S. F. GODSAVE. Mesoderm induction in early Xenopus embryos by heparin-binding growth factors. *Nature* 326: 197–200, 1987.

SLAMON, D. J., G. M. CLARK, S. G. WONG, W. J. LEVIN, A. ULLRICH, AND W. L. McGUIRE. Human breast cancer: Correlation of relapse and survival with amplification of the HER-2/*neu* oncogene. *Science* 235: 177–182, 1987.

SLAUNWHITE, W. R. JR., AND A. A. SANDBERG. Transcortin: A corticosteroid binding protein of plasma. *J. Clin. Invest.* 38: 384, 1959.

SMALS, A. G. H., G. F. F. M. PIETERS, D. C. LOZEKOOT, TH.J. BENRAAD, AND P. W. C. KLOPPENBORG. Dissociated responses of plasma testosterone and 17-hydroxyprogesterone to repeated human chorionic gonadotropin administration in normal men. *J. Clin. Endocrinol. Metab.* 50: 190–193, 1980.

SMITH, T. J., AND I. S. EDELMAN. The role of sodium transport in thyroid thermogenesis. *Fed. Proc.* 38: 2150–2153, 1979.

SPELLACY, W. N. W. C. BUHI, AND S. A. BIRK, Human growth hormone and placental lactogen levels in midpregnancy and late postpartum. *Obstet. Gynecol.* 36: 238–243, 1970.

SPELLACY, W. N. W. C. BUHI, J. C. SCHRAM, S. A. BIRK, AND S. A. McCREARY. Control of human chorionic somatommortropin levels during pregnancy. *Obstet. Gynecol.* 37: 567–573, 1971.

SPORN, M. B., A. B. ROBERTS, L. M. WAKEFIELD, AND R. K. ASSOIAN. Transforming growth ROBERTS, L. M. WAKEFIELD, AND R. K. ASSOIAN. Transforming growth factor-&: Biological function and chemical structure. *Science* 233: 532–534, 1986.

STANBURY, S. W. Vitamin D and the syndromes of azotaemic osteodystrophy. *Contrib Nephrol.* 13: 132–146, 1978.

STEINBERGER A., AND E. STEINBERGER. Secretion of an FSH-inhibiting factor by cultured Sertoli cells. *Endocrinology* 99: 918–921, 1976.

STEINBERGER, E. Hormonal control of mammalian spermatogenesis. *Physiol. Rev.* 51: 1, 1971.

STEINETZ, B. G., AND E. M. O'BYRNE. Speculation on the probable role of relaxin in cervical dilatation and parturition in rats. *Semin. Reproduc. Endocrinol.* 1: 355–342, 1983.

STRICKLAND, S., AND J. N. LOEB. Obligatory separation of hormone binding and biological response curves in systems dependent upon secondary mediators of hormone action. *Proc. Natl. Acad. Sci. U.S.A.* 78: 1366–1370, 1981.

STRUHL, L. The DNA-binding domains of the jun oncoprotein and the yeast GCN4 transcriptional activator protein are functionally homologous. *Cell* 50: 841–846, 1987.

STURMAN, J. A. D. K. RASSIN, K. C. HAYES, AND G. E. GAULL. Taurine in the developing kitten: Nutritional importance. *Prediatr. Res.* 11: 450, 1977.

SUNDERLAND, C. A. M. NAIEM, D. Y. MASON, C. W. G. REDMAN, AND G. M. STIRRAT. The expression of major histocompatibility antigens by human chorionic villi. *J. Reprod. Immunol.* 3: 323–331, 1981.

SUTHERLAND, E. W. AND T. W. RALL. Fractionation and characterization of a cyclic adenine ribonucleotide formed by tissue particles. *J. Biol. Chem.* 232: 1077–1091, 1958.

SVOBODA, M. E., J. J. VAN WYK, D. G. KNAPPER, R. E. FELLOWS, F. E. GRISSOM, AND R. J. SCHLUETER. Purification of somatomedin C from human plasma: chemical and biological properties, partial sequence analysis and relationship to other somatomedins. *Biochemistry* 19: 790, 1980.

SZABO, M., AND L. A. FROHMAN. Suppression of cold-stimulated thyrotropin secretion by antiserum to thyrotropinreleasing hormone. *Endocrinology* 101: 1023–1033, 1977.

TAUB, R., I. KIRSCH, C. MORTON, G. LENOIR, D. SWAIN, S. TRONICK, S. AARONSON, AND P. LEDER. Translocation of the c-*myc* gene into the immunoglobulin heavy chain locus in human Burkitt lymphoma and murine plasmacytoma cells. *Proc. Natl. Acad. Sci. U.S.A.* 79, 7837–7841, 1982.

TAUROG, A. Thyroid peroxidase and thyroxine biosynthesis. *Recent Prog. Horm. Res.* 26: 189–247, 1970.

TERAMO, K. A., J. A. WIDNESS, G. C. CLEMONS, P. VOUTILAINEN, S. McKINLAY, AND R. SWARTZ. Amniotic fluid erythropoietin correlates with umbilical plasma erythropoietin in normal and abnormal pregnancy. *Obstet. Gynecol.* 69: 710–716, 1987.

THOMPSON, C. C., C. WEINBERGER, R. LEBO, AND R. M. EVANS. Identification of a novel thyroid hormone receptor expressed in the mammalian central nervous system. *Science* 237: 1610–1614, 1987.

THORBURN, G. D., AND J. R. G. CHALLIS. Control of parturition. *Physiol. Rev.* 59: 863–918, 1979.

THORNER, M. O., W. H. MARTIN, A. D. ROGOL, J. L. MORRIS, R. L. PERRY-MAN, B. P. CONWAY, S. S. HOWARDS, M. G. WOLFMAN, AND R. M. MacLEOD. Rapid regression of pituitary prolactinomas during bromocriptine treatment. *J. Clin. Endocrinol. Metab.* 51: 438–445, 1980.

TODARO, G. J., C. FRYLING, AND J. E. De LARCO. Transforming growth factors produced by certain human tumor cells: polypeptides that interact with epidermal growth factor receptors. *Proc. Natl. Acad. Sci. U.S.A.* 77: 5258–5262, 1980.

TOMKINS, G. M. The metabolic code. *Science* 189: 760–763, 1975.

TOMLINSON, A., D. D. L. BOWTELL, E. HAFEN, AND G. M. RUBIN. Localization of the *sevenless* protein, a putative receptor for positional information, in the eye imaginal disc of Drosophila. *Cell* 51: 143–150, 1987.

TOPPER, Y. Multiple hormone interactions in the development of mammary gland in vitro. Recent *Prog. Horm. Res.* 26: 287–308, 1970.

TOPPER, Y. J., FREEMAN, C. S. Multiple hormone interactions in the developmental biology of the mammary gland. *Physiol. Rev.* 60: 1049–1106, 1980.

TUCHMANN-DUPLESSIS, H., G. DAVID, AND P. HAEGEL. *Illustrated Human Embryology, Vol. 1, Embryogenesis,* editeure, Masson et Cie, Springer-Verlag, New York: 1972.

TURKINGTON, R. W. Molecular biological aspects of prolactin. In *Lactogenic Hormones,* edited by G. E. W. Wolstenholme, J. Knight. Edinburgh: Livingstone, 1972, p. 111.

TURNER, C. W. *The Mammary Gland,* Vol. I. Columbia, MO: Lucas Brothers, 1952.

TYCKO, B., AND F. R. MAXFIELD. Rapid acidification of endocytic vesicles containing α_2-macroglobulin. *Cell* 28: 643–651, 1982.

TYRREL, J. B., AND J. C. BAXTER. Glucocorticoid therapy. In: *Endocrinology and Metabolism,* edited by P. Felig, J. D. Baxter, A. E. Broadus, and L. A. Frohman, New York: McGraw-Hill, 1981, p. 620.

TYSON, J. E., M. KHOJANDI, J. HUTH, AND B. ANDREASSEN. The influence of prolactin secretion on human lactation. *J. Clin. Endocrinol. Metab.* 40: 764–773, 1975.

TYSON, J. E., D. RABINOWITZ, AND J. T. MERIMEE, ET AL. Response of plasma insulin and human growth hormone to arginine in pregnant and postpartum females. *Am. J. Obstet. Gynecol.* 103: 313, 1969.

ULLRICH, A., L. COUSSENS, J. S. HAYFLICK, T. J. DULL, A. GRAY, A. W. TAM, J. LEE, Y. YARDEN, T. A. LIBERMANN, J. SCHLESSINGER, J.

DOWNWARD, E. L. V. MAYES, N. WHITTLE, M. D. WATERFIELD, AND P. H. SEEBURG. Human epidermal growth factor receptor cDNA sequence and aberrant expression of the amplified gene in A431 epidermoid carcinoma cells. *Nature* 309: 418–425, 1984.

ULLRICH, A., J. R. BELL, E. Y. CHEN, R. HERRERA, L. M. PETRUZZELLI, T. J. DULL, A. L. COUSSENS, Y-C. LIAO, M. TSUBOKAWA, A. MASON, P. H. SEEBURG, C. GRUNFELD, O. M. ROSAN, AND J. RAMACHANDRAN. Human insulin receptor and its relationship to the tyrosine kinase family of oncogenes. *Nature* 313: 756–761, 1985.

ULMSTEN, U., AND G. SUNDSTROM. Esophageal manometry in pregnant and nonpregnant women. *Am. J. Obstet. Gynecol.* 132: 260, 1978.

UNGER, R. H., R. E. DOBBS, AND L. ORCI. Insulin, glucagon, and somatostatin secretion in the regulation of metabolism. *Annu. Rev. Physiol.* 40: 307, 1978.

UOTILA, U. U. The early embryological development of the fetal and permanent adrenal cortex in man. *Anat. Rec.* 76: 183–203, 1940.

VALE, W., P. BRAZEAU, C. RIVIER, M. BROWN, B. BOSS, J. RIVIER, R. BURGUS, N. LING, AND R. GUILLEMIN. Somatostatin. *Recent Prog. Horm. Res.* 31: 365–397, 1975.

VALE, W., R. BURGUS, AND R. GUILLEMIN. On the mechanism of action of TRH: effects of cycloheximide and actinomycin on the release of TSH stimulated in vitro by TRH and its inhibition by thyrosin. *Neuroendocrinology* 3: 34–46, 1968.

VALE, W., J. SPIESS, C. RIVIER, AND J. RIVIER. Characterization of a 41-residue ovine hypothalamic peptide that stimulates secretion of corticotropin and β-endorphin. *Science* 213: 1394–1397, 1981.

VAN WYK, J. J., AND L. E. UNDERWOOD. The somatomedins and their action. In: *Biochemical Actions of Hormones*, edited by G. Litwack. New York: Academic Press, 1978. vol. V.

VICKERY, B. H. Comparison of the potential for therapeutic utilities with gonadotropin-releasing hormone agonists and antagonists. *Endocr. Rev.* 7: 115–124, 1986.

VOLOSCHIN, L. M. AND E. J. DOTTAVIANO. The channeling of natural stimuli that evoke the ejection of milk in the rat: Effect of transections in the midbrain and hypothalamus. *Endocrinology* 99: 49–58, 1976.

VOLPE, R. Autoimmunity in the endocrine system. In: *Monographs in Endocrinology*, No. 2. Heidelberg: Springer-Verlag 1981.

VORHERR, H. Contraception after abortion and postpartum. *Am. J. Obstet. Gynecol.* 117: 1002–1025, 1973.

WALKER, A. M. Physiological control of the fetal cardiovascular system. In: Fetal Physiology and Medicine, edited by R. W. Beard and P. W. Nathanielsz, New York: Mercel Dekker, 1984, p. 287–316.

WALTER, R., AND W. H. SIMMONS. Metabolism of neurohypophyseal hormones: considerations from a molecular viewpoint. In: *Neurohypophysis: International Conference on the Neurohypophysis*, edited by A. M. Moses and L. Share. Basel: Karger, 1977.

WALTERS, W. A. W., AND Y. L. LIM. Cardiovascular dynamics in women receiving oral contraceptive therapy. *Lancet* 2: 879, 1969.

WANG, C. F., AND S. S. C. YEN. Direct evidence of estrogen modulation of pituitary sensitivity to luteinizing hormone releasing factor during the menstrual cycle. *J. Clin. Invest.* 55: 201–204, 1975.

WARDLAW, S. L. R. I. STARK, L. BAXI, AND A. G. FRANTZ. Plasma B-endorphin and B-lipotropin in the human fetus at delivery: Correlation with arterial pH and pO$_2$. *J. Clin. Endocrinol. Metab.* 49: 888–891, 1979.

WARTOFSKY, L., AND K. D. BURMAN. Alterations in thyroid function in patients with systemic illness: the "euthyroid sick syndrome." *Endocr. Rev.* 3: 164–217, 1982.

WASSERMAN, R. H., R. A. CORRADINO, AND C. S. FULLMER. Some aspects of vitamin D action, calcium absorption and vitamin D-dependent calcium-binding protein. *Vitam. Horm.* 32: 299–324, 1976.

WATERMAN, M. R., AND E. R. SIMPSON. Regulation of the biosynthesis of cytochromes P-450 involved in steroid hormone synthesis. *Molec. Cell Endocrinol.* 39: 81–89, 1985.

WEBB, C. G., W. E. GALL, AND G. M. EDELMANN. Synthesis and distribution of H$_2$ antigens in preimplantation mouse embryos. *J. Exp. Med.* 146: 923–932, 1977.

WEEKS, D. L., AND D. A. MELTON. A maternal mRNA localized to the vegetal hemisphere in Xenopus eggs codes for a growth factor related to TGF-B. *Cell* 51: 861–867, 1987.

WEINBERG, R. A. The action of oncogenes in the cytoplasm and nucleus. *Science* 230: 770–776, 1985.

WEINBERGER, C., C. C. THOMPSON, E. S. ONG, LEBO, R., D. J. GROUL, AND R. M. EVANS. c-erb-A gene encodes a thyroid hormone receptor. *Nature* 324: 641–646, 1986.

WEINTRAUB, B. D., B. S. STANNARD, D. LINNEKIN, AND M. MARSHALL. Relationship of glycosylation to *de novo* thyroid-stimulating hormone biosynthesis and secretion by mouse pituitary tumor cells. *J. Biol. Chem.* 255: 5715–5723, 1980.

WEISS, G. Relaxin in human pregnancy. In *The Endocrine Physiology of Pregnancy and the Peripartal Period*, edited by R. B. Jaffe, S. Dell'Acqua. New York: Raven Press, 1985, p. 241.

WEISS, G., E. FACOG, E. M. O'BYRNE, J. HOCHMAN, B. G. STEINETZ, L. GOLDSMITH, AND J. G. FLITCRAFT. Distribution of relaxin in women during pregnancy. *Obstet. Gynecol.* 52: 569–570, 1978.

WEITZMAN, E. D., R. M. BOYAR, S. KAPEN, AND L. HELLMANN. The relationship of sleep and sleep stages to neuroendocrine secretion and biological rhythms in man. *Recent Prog. Horm. Res.* 31: 399–446, 1975.

WHARTON, K. A., K. M. JOHANSEN, T. XU, AND S. ARTAVANIS-TSAKONAS. Nucleotide sequence from the neurogenic locus *notch* implies a gene product that shares homology with proteins containing EGF-like repeats. *Cell* 43: 567–581, 1985.

WILDT, L., A. HAUSLER, G. MARSHALL, J. S. HUTCHINSON, T. M. PLANT, P. E. BELCHETZ, AND E. KNOBIL. Frequency and amplitude of gonadotropin-releasing hormone stimulation and gonadotropin secretion in the rhesus monkey. *Endocrinology* 109: 376–385, 1981.

WILLIAMS, R. L., R. K. CREASY, G. C. CUNNINGHAM, W. E. HAWES, F. D. NORRIS, AND M. TASHIRO. Fetal growth and perinatal viability in California. *Obstet. Gynecol.* 59: 624–632, 1982.

WILLINGHAM, M. C., AND I. PASTAN. The receptosome: An intermediate organelle of receptor-mediated endocytosis in cultured fibroblasts. *Cell* 21: 67–77, 1980.

WINTOUR, E. M., J. P. COGHLAN, AND M. TOWSTOLESS. Cortisol is natriuretic in the immature ovine fetus. *J. Endocr.* 106: R13–R15, 1985.

WINTROBE, M. M. *Clinical Hematology*, 8th Ed., edited by Philadelphia: Lea & Febiger, 1981.

WOLFF, J. Iodide goiter and the pharmacologic effects of excess iodide. *Am. J. Med.* 47: 101–124, 1969.

WOLFF, J., AND I. L. CHAIKOFF. The inhibitory action of iodide upon organic binding of iodine by the normal thyroid gland. *J. Biol. Chem.* 172: 855–856, 1948.

WULF, K. H. W. KUNZEL, AND V. LEHMANN. Clinical aspects of placental gas exchange. In: *Respiratory Gas Exchange and Blood Flow in the Placenta*, edited by L. D. Longo and H. Bartels, DHEW Publication (NIH) No. 73–361, Washington, D.C. 1972.

WURTMAN, R. J., AND J. AXELROD. Control of enzymatic synthesis of adrenaline in the adrenal medulla by adrenal cortical steroids. *J. Biol. Chem.* 241: 2301–2305, 1966.

YARDEN, Y., J. A. ESCOBEDO, W-J. KUANG, T. L. TANG-FENG, T. O. DANIEL, P. M. TREMBLE E. Y. CHEN, M. E. ANDO, R. N. HARKINS, U. FRANCKE, V. A. FRIED, A. ULLRICH, AND L. T. WILLIAMS. Structure of the receptor for platelet-derived growth factor helps define a family of closely related growth factor receptors. *Nature* 323: 226–232, 1986.

YASUDA, N., M. A. GREER, AND T. AIZAWA. Corticotropinreleasing factor. *Endocr. Rev.* 3: 123–140, 1982.

YEN, S. S. C. Prolactin and human reproduction. In: *Reproductive Endocrinology*, edited by S. S. C. Yen and R. B. Jaffe. Philadelphia: W. B. Saunders, 1986, p. 237–261.

YEN, S. S. C., Y. EHARLA, AND T. M. SILER. Augmentation of prolactin by estrogen in hypogonadal women. *J. Clin. Invest.* 53: 652–655, 1974.

YEN, S. S. C., AND R. B. JAFFE. (eds.). *Reproductive Endocrinology, Physiology, Pathophysiology and Clinical Management.* Philadelphia: W. B. Saunders, 1986.

YEN, S. S. C., B. L. LASLEY, C. F. WANG, H. LEBLANC, AND T. M. SILER. The operating characteristics of the hypothalamic-pituitary system during the menstrual cycle and observations of biological action of somatostatin. *Recent Prog. Horm. Res.* 31: 321–363, 1975.

YEN, S. S. C., AND A. LEIN. The apparent paradox of the negative and positive feedback control system on gonadotropin secretion. *Am. J. Obstet. Gynecol.* 126: 942–954, 1976.

YING, S-Y., N. LING, P. BOHLEN, AND R. GUILLEMIN. Gonadocrinins: peptides in ovarian follicular fluid stimulating the secretion of pituitary gonadotropins. *Endocrinology* 108: 1206–1215, 1981.

Neurophysiology

CHAPTER 62

Sensory Processing

SYSTEMS NEUROPHYSIOLOGY

Chapter 3 presents neurophysiology in terms of single cell mechanism—neuron physiology. It depicts specialized structural and functional features of neurons that contribute to excitability and conductivity. It considers action potentials, special metabolic requirements, and synaptic transmission: excitatory, inhibitory, electrical, and chemical. The following chapters capitalize on such mechanisms to portray a *systems neurophysiological basis of medical practice.*

SIZING UP THE HUMAN BRAIN

By rough approximation, the human nervous system functions with about 10 million afferent (input or sensory) neurons, 50 billion central neurons, and ½ million efferent (output or motor) neurons. This provides a ratio of about 20:1 between the aggregate input and output channels, with many more channels available for inspecting the environment than for acting in it. There are several thousand central neurons for every input or output neuron, thus providing abundant circuits for perceptual processing and the organization of behavior.

Importance of Evolution and Development

Everything concerning the nervous system can usefully be considered in light of the consequences of evolution and development. Three mechanisms, each high in information content, link these processes in orderly biological sequences: (1) *linear molecules* (DNA, RNA, and protein) provide the initial information necessary for building an individual representative of the species; (2) *specific cell markers and intercellular signaling systems* execute plans—including useful epigenetic variations—for construction of a unique three-dimensional brain; (3) *brain-governed behavior* contributes to successes and failures in adapting to a given dynamic environment so that the original linear information plus epigenetic variations can be

adaptively selected in accordance with the living requirements for a unique, complex, dynamic environment; (4) *the processes of selection and consolidation of brain circuits* include the renewal of novel variations upon which further ongoing selection for adaptive perception, learning, judgment, memory, and behavior is organized.

DEVELOPMENT

Three classic germinal layers are laid down during embryogenesis: *entoderm*, which faces the internal environment of the embryo and generates lungs, gut, liver, and other visceral organs; *mesoderm*, the middle germinal layer which generates skeletal and muscular tissues; and, *ectoderm*, which faces outward from the embryo and generates skin and all its derivatives—including the neural tube, the neural crest, and their generation of the entire nervous system.

During embryogenesis, the wall of the neural tube exhibits three structurally and functionally distinctive layers: an inner *matrix layer;* an intermediate *mantle layer;* and an outer *marginal layer.* These form the neurological foundations for three distinctive aspects of motor activity. The matrix layer provides nervous system connections to visceral organs. This includes autonomic (sympathetic and parasympathetic systems), limbic mechanisms, and neuroendocrine relations. It organizes and integrates neurological mechanisms that control *visceration.* The mantle layer, in the middle of the wall of the neural tube, generates nuclear masses that control skeletomuscular actions involved in bodily attitude, posture, and locomotion. These include the basal ganglia and share the limbic system in motor control of visceration. These systems organize and integrate motor activities that express an individual's internal state—feelings, moods, and emotions (e-motions, ex-motions, outward expressions of the internal state), in short, motor *expression.* The marginal layer, which is mainly elaborated as neocortex, is principally related to movement and communication in respect to the external world of objects and people. Motor performances of the marginal layer mainly serve *effectuation*, that is, doing something interactively with objects and with people. It is largely through effectuation that civilization and the institutions and artifacts of humankind have been generated (Yakovlev, 1948).

Nervous System Variations Through Normal Inheritance

We need to consider brain variability according to the elementary processes of gene sorting and other factors that contribute to individuality. Varieties of offspring of a single human pair stem from random sorting of gametes which result in more than 8.39 million varieties of sex cells available to each parent (2^{23}). Fertilization multiplies the convergence of the varieties of sex cells, yielding 70.37 trillion different possibilities for each child of the mated pair ($2^{23} \times 2^{23}$). This potentiality for genetically distinctive individuals, out of a single human mating, is probably greater than the total number of individuals in the hominid line who reached reproductive age from the time of the origins of *Homo sapiens!*

Many further individuating events occur during development and throughout life: genetically normal equal random crossovers occur at a high rate; unequal crossovers representing massive mutations, the equivalent of antibody diversity, occur frequently. Somatic mutations, 10^{-7} per cell division (with the vast number of cell divisions in the brain, there will be thousands of such events), plus alterations by virus particles, including "jumping genes," which abound, and which operate on thousands of occasions in a lifetime, expressing the effects of thousands of different viruses, by insertion sequences and gene conversions. Also, other mutations are induced by thermal agitation, chemical mutagens, ionizing radiation, infectious diseases, etc. Each of these innumerable mutational effects multiplies the 70.37 trillion number.

Anatomy laboratory and neurosurgical experiences illustrate that gross neuroanatomy varies greatly among presumably normal human brains. It is frequently difficult to locate the central sulcus that separates primary sensory and motor cortices. A neurosurgeon usually needs to turn a big flap to obtain adequate visual exposure, in order to stimulate the cortex of a waking patient so that he or she can locate functional boundaries if it should prove necessary to resect brain tissue in any critical region such as a speech area. Stereotactic atlases for estimating the location of subcortical and brainstem structures are usually based on only a few brains; they are correct in cautioning with respect to gross morphological variability. We have produced cinemorphology films of some 70 postmortem human brains and find no two alike even according to gross morphological criteria; observable variations include gross structural differences in all regions of the brain, cortical and subcortical.

CHARACTERISTICS OF GLOBAL NERVOUS SYSTEM FUNCTIONS

What structural-functional characteristics distinguish the nervous system as an organ system? First, the nervous system has developmental priority. It is the first organ to differentiate during embryogenesis and it remains the fastest growing; hence it is the largest organ at delivery. The nervous system gathers information from and coordinates activities for all other organ systems. It begins this comprehensive integrative control over the rest of the body early in gestation, already well before movement can be detected by the mother. At the termination of pregnancy, the baby's hypothalamus signals its pituitary that the fetus is ready for birth; thus, an endocrine signal from the baby ordinarily triggers labor. The central nervous system serves the *functional integrity* of that individual for life, and provides a variety of combined visceral and somatic adaptations necessary for survival and reproductive activities, including contributions to the survival of siblings, mate and offspring, and others (MacLean, 1982).

Second, the nervous system is composed of an extraordinary variety of nerve cells which are remarkably specialized in form, chemistry, physiology, pharmacology, immunology, and other properties, including unique responses to injury. It is conservatively estimated that *the human nervous system contains more than 50 million different kinds of nerve cells*, meaning clearly distinguishable nerve cell types (Bullock, 1980).

Third, a principal function of the nervous system is *to plan and organize behavior*: this is a continuous space-time transaction that plays a governing role in subjective experience. What conditions inside or outside the body present threats or opportunities? What is the present and anticipated fitness of the body? What physiological and behavioral events are contributing to states of satisfaction and dissatisfaction? How can bodily events, and events in the environment be organized to optimize internal satisfactions? What priorities can be applied to competing behavioral options in order to optimize future satisfactions?

Fourth, the nervous system is engaged in neuronal mapping of dynamic strategies that are concerned with experiences, plans, and behavior regarding parts of the world as well as parts of the body (Mountcastle, 1975). These strategies are partly based on genetic information, partly on the outcome of structural and developmental processes, and partly on information which the nervous system itself generates, gathers from other sources, evaluates, and integrates to improve pursuits of internal satisfactions. Mapping strategies involve numerous maps for each sensory modality and for each motor control system. Maps relate back and forth re-entrantly with one another, and between cerebrum and cerebellum, brainstem and thalamus, thalamus and cortex, thalamus, cortex, and the limbic system, etc.

SIGNAL PROCESSING IN THE NERVOUS SYSTEM

The nervous system is a self-organizing, self-regulating signal processing system which continues to operate for a lifetime. It accomplishes this within a partially specified architecture that allows for selection during development and further selection during early-life experiences. And it continues to generate programs, utilizing neuronal group resources selected in memory, for deliberate adaptation of goals to secure future internal satisfactions (Edelman, 1987). These internal satisfactions attach also to mate, offspring, and a wide range of others. These are part of the dowry of our species. Mammalian existence could not have evolved without powerful intergenerational and interindividual faith, mutual trust, and altruism. The compass of individual empathy and compassion can be further augmented or eroded according to social and cultural guidance.

Thus, individuality in the nervous system is expressed by genetic uniqueness, structural and functional self-organizing characteristics, and cumulative neuronal group selection processing throughout development and postnatal experiences. Neuronal group selection takes place along all sensory as well as along central and motor pathways, in accordance with ongoing experiences. This means that the brain develops individualistic, culture-bound "sensory lenses" that are invisible to us. This is because they are located along sensory input pathways prior to conscious perceptual experience. This indicates that exactly as all brains are unique, there undoubtedly are as many perceptual worlds as there are individuals. Interindividual differences are more or less great or small depending on the extent of differences in original neuronal architecture, neuronal selection during development, and continuing selection during subsequent experiences.

This feature of neurophysiology creates problems, not only for the neurophysiological and psychological sciences that deal with perception, but also for disciplines concerned with epistemology and with interindividual and cross-cultural communication and understanding. It has fundamentally to do with the problems of conflict resolution.

Neuroscience provides an advancing scientific revelation of nervous system functions, particulary of higher neural process previously thought beyond reach. Current endeavors are helping to provide a more adequate understanding of the risks and potentialities of human experience. Requisite knowledge of brain processes relating to social perceptions, judgments, and behavior can come none too soon, for as neuroscientist Paul MacLean remarked, "It is too bad the brain wasn't cracked before the atom!"

SENSORY PROCESSING

The adaptation of man to the environment and to society requires constant processing of information relating to these sources. All behavioral activity—visceration, expression, and effectuation—constitutes a combination of immensely complicated central neuronal processing. Detection of changes in the environment, including the social environment, can be detected by sensory feedback from distance receptors and body wall viscera; this includes information fed back from the environment as affected by ongoing behavior. Sensory monitoring of the consequences of behavior permits the regulation and control of subsequent behavior.

Much of consciousness, even in the absence of significant sensory inputs, seems to involve sensory imagery. We tend to think according to visual and auditory imagery and imagery involving other senses. Although in the analytic investigation of sensory mechanisms there is a tendency to consider sensory systems as independent, our perception of the environment and our relations with it depend in fact on many interdependent transactions among individual sensory modalities (Gibson, 1966). It is our purpose in the following chapters to consider various important sensory processes individually, acknowledging the obvious fact that they do not function independently but rather provide a collective contribution to consciousness and the control of behavior (see Edelman, 1989).

Receptor Organs

Methods for studying different sensory systems are considered subsequently. Different sensory modalities are investigated according to the location and characteristics of specialized receptors which mediate responses to stimuli appropriate for the transmission of information in each sensory domain. The physical forms of stimulation that are most natural or appropriate for a given sensory receptor are called *adequate* stimuli. The physical dimensions of the stimuli, and the range of energies to which the receptor responds, characterize the transduction process.

Receptor cells set in motion chains of sensory processing events that can subsequently yield conscious perceptions. Most cells are affected by certain wave lengths of light, specific ions, protons, CO_2, and various forms of mechanical disturbance. Receptors have specialized through evolutionary selection to be especially sensitive to particular aspects of the stimulating world. This is called *transduction* (Changeux and Dennis, 1982). They retain their ability to respond to more prosaic influences, but, being specialized, they function by converting particular physical events into membrane responses that can set in train sensory processing signals.

Sensory transducers may be spontaneously active, as are photoreceptors, even in the dark. This makes possible differential activation by very subtle stimuli. Photoreceptors in the dark-adapted eye are credited with being able to respond to a single photon. In the case of hair cell receptors in the cochlea the threshold for best tone detection is obtained with vibration of the tympanic membrane with excursions less than half the diameter of the hydrogen atom. Many sense receptors are so sensitive that they verge on responding to thermal noise.

Sensory receptors may be classified in three categories: *exteroceptors, interoceptors,* and *proprioceptors.* Exteroceptors respond to stimuli impinging from outside, e.g., light, sound, touch, and chemical agents which stimulate olfactory receptors and taste buds. Interoceptors lie within the mucous linings and smooth muscle walls of the respiratory, digestive, and urinary tracts. They respond to materials inhaled, ingested, or passed (e.g., stones in the biliary duct or ureter), and to changes in chemical surroundings, mechanical pressure or shearing force. Proprioceptors are sensory receptor endings that respond to stimuli generated by muscular movement of the body or passive displacement of body parts. They reside in skeletal muscles, tendons, joints, and in the walls of the heart, blood vessels, and gastrointestinal tract. Vestibular receptors in the labyrinth respond to linear and angular accelrations generated by motor actions or motions otherwise imposed on the head. Their responses con-

tribute importantly to perception of the body in space. The vestibular system works in close harmony with proprioceptors and in the control of eye movements.

CONSCIOUS CONCOMITANTS OF SENSATION

Of course, the prominence or intensity of sensation in consciousness varies with circumstances. For someone listening to a musical work or engrossed in conversation, auditory perceptions may be more prominent than visual. For someone engaged in athletic competition, the vestibular sense may make a critical contribution yet be without concomitant consciousness. Olfactory and gustatory senses may be prominent under given circumstances or may intrude themselves on consciousness especially if one is hungry or if the odors or tastes are particularly meaningful. Other classes of receptors responding to chemical stimuli, e.g., those in the carotid sinus and aortic bodies, contribute importantly to the control of physiological processes yet do not enter consciousness. Receptors for sensations of pain, whether in skin or deep tissues, can elicit a compelling conscious experience, but under normal circumstances they are unobtrusive. Their role is usually protective but they may become disordered and dominate consciousness.

Although convenient to think of the senses separately, our perceptions depend on multisensory inputs, interactions among several senses, often combined with motor systems. Thus, interpretations of visual stimuli may be profoundly influenced by vestibular, proprioceptive, and auditory stimuli. "Global mapping" systems developed early in life provide synthesizing correlations among different stimulus modalities, categorizing "objects" and "events" that may be composed of several sensory and motor patterns. Situations that result in noncorrespondences of different stimuli that have ordinarily combined relationships can be disconcerting, intrude on consciousness and disrupt behavior. Noncorrespondences may be encountered between visual and vestibular inputs, e.g., on exposure to motion of a ship in rough water. Noncorrespondences between visual and tactual or proprioceptive senses interfere with attempts to carry out simple maneuvers in a "distorted room."

NEURAL CODING OF STIMULUS SIGNALS

Some units in sensory systems are spontaneously active. This allows detection of exceedingly subtle differences, the onset of a very weak stimulus or a shift in stimulus intensity or quality, differences in background or state that might not be detected if the receptor needed to be excited to some threshold value before it could discharge. Figure 9.1 shows neuronal responses at the thalamic relay (lateral geniculate nucleus) along the visual pathway. The geniculate neuron is receiving signals at the third synaptic relay along the most direct channel of visual information throughput. Recording is from a fourth-order neuron, following activities in the receptors, bipolar cells, and ganglion cells. Both excitatory and inhibitory impulses converge on this neuron along a path that is spontaneously active. This unit is closely linked in response pattern to that of the ganglion cell serving that precise part of the visual field. The upper part of the figure shows analog responses in sequence along the time axis for 10 successive trials of a spot of light directed at the center of that unit's receptive field. Each time the spot of light is flashed, for 50 ms, the unit responds with a strong burst of impulses. When the same spot is flashed in the same location, but is coupled with

an annular surrounding light (see inset, diagrammatic representation on Fig. 9.1), and the annular stimulus is continued for 500 ms, the unit responds with less vigor. It does not resume its usual background activity until the annular stimulus is turned off, when there is a moderate "off" response. When the annular stimulus is shown alone, the unit is dramatically inhibited, gradually beginning to escape but remaining inhibited until the annular light is turned off, when there is a more marked "off" response. One single unit, then, has monitored (and relayed) considerable information: *where* on the retina, *when* there has been a change in lighting, and *whether* the light consisted of a local spot, a surrounding ring of light, or the combination.

It is evident that a particular physical event in the visual world can trigger changes in photoreceptors and activate a chain of successive neurons to whatever end station may be involved in the central nervous system. Once the physical event has been transduced to membrane transmittable signals,

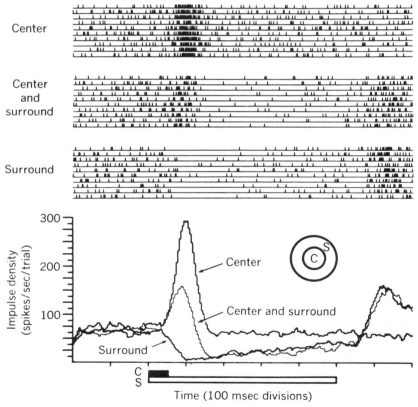

Figure 9.1. An important feature of integration involves convergence of excitatory and inhibitory influences on neurons. A unit in the lateral geniculate nucleus relay between the retina and cortex displays background activity prior to simulus onset. When a small spot of light strikes the center of its receptive field, excitation is followed by return to background activity. When light is applied to center and surround together, excitation is less marked. It is followed by inhibition and a rebound "off response." When surround light is applied alone, pure inhibition is followed by an "off response." Because the unit is spontaneously active it can be raised or reduced in rate of activity and thereby can register more subtle stimulus differences. *C,* center spot, 0.6° in diameter with flash duration of 50 ms; *S,* surround 1–2.7° in inner-outer diameter with flash duration of 500 ms. (From Poggio et al., 1969.)

there is no longer anything left of the light; everything is now transformed into the common currency of graded neuronal coding. Perkel and Bullock have written an authoritative analysis of this subject (1968).

We can see that part of the code is represented by *which neurons* are activated: the "labeled line." The *timing* of the physical event is signaled (with a certain latency) by onset of a *change* in background signaling, a "doorbell" code. Additional information is coded by, whether the unit is made more, or less, active and *how* the unit responds to the intensity and duration of the stimulus. Further, the unit conveys information relating to *differential physical effects* within the unit's field. Not shown is the fact that when any other region of the retina is illuminated, outside the receptive field of this lateral geniculate unit, its "spontaneous" activity is not altered. It thereby conveys information that there has been *no (local) change*.

COMPOUND ACTION POTENTIAL

When a bundle of nerve fibers transmits information in response to sensory stimulation, a compound action potential can be recorded from the bundle by means of a gross electrode (Fig. 9.2). This shows changes in voltage with time, reflecting the summed electrical activity of the active component fibers within the bundle. When a large number of fibers fires simultaneously or at nearly the same time, the amplitude of the action potential is high. Compound action potentials typically show a succession of peaks in response to a single stimulus volley. Each peak is associated with a different class of

neurons, characterized by having a distinctive range of conduction velocities. Thus, a wave of action potentials for a given class of fibers reaches the electrode at nearly the same time for most of the fibers in that class. Activity among these fibers is then reflected by a specific component of the compound action potential. Velocity categories are associated with different axonal diameters as shown in the inset of Figure 9.2 showing numbers of fibers by fiber diameter, and below the compound action wave form in brackets identified with those diameters and velocities.

Some Problems of Central Integration

Receptors introduce a spectrum of initial delays. Nerve conduction varies approximately with axon diameter, thickness of myelin sheath (if myelinated), temperature, and other factors. Conduction velocity slows at points of axonal branching, where there is typically some diminution of fiber diameter. Also, nerve impulses slow down in regions of axonal termination because of discontinuation of the myelin sheath and because of terminal branching and terminal narrowing or broadening of fiber diameters. It is obvious that conduction times for any neuronal message, from its origin to its destination, may exhibit a wide dispersal of action potentials, even in response to a single, abrupt initiating stimulus. The compound action potential will therefore show increasing dispersion of activities between classes of neurons and among individual fibers within any class.

At each synaptic junction there is a modest (0.1–0.5 ms), somewhat variable *synaptic delay*. Most pathways and circuits display a spectrum of temporal variables in addition because there may be different numbers of synaptic relays along the course of what is ostensibly a single pathway, as in the example of the dorsal column path in the spinal cord. Second-order neurons contribute additional complications because they can change the phase lag according to their reaction to signals delivered to them by first-order neurons. When sensory and central signals deliver their messages to motor units, there are additional delays in response, because of delay at the neuromuscular (or neuroglandular) junction and because of mechanical delays in muscular contraction, tendon elasticity, limb inertia, and whatever may be involved in the manifestation of glandular secretion.

We can see that there must be extremely variable time-distributed dispersals of nerve impulses occurring along any neuronal path and circuit, and that this variability is also susceptible to change according to the recent history of activity along that route.

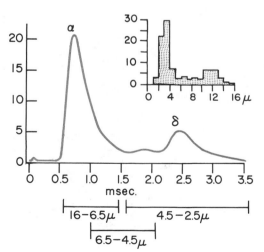

Figure 9.2. Compound action potential in a human sensory nerve. Counts of fiber diameters encountered in that nerve are shown. Voltage (arbitrary units) as a function of time illustrates how impulses spread out in time in accordance with fiber diameter (From Gasser, 1960.)

Then, how can the nervous system so accurately interpret simultaneity in sensory experience and command simultaneity in motor performance? How can we blink, clench jaw, tap finger, and tap foot simultaneously, linked gestures that involve very distributed numbers of neurons and synapses, exceedingly different conduction distances, and different inertial and other mechanical variables as well. Yet such gestures are easily demonstrated by untrained subjects to within about ½ ms, roughly the interval required for conduction across a single synapse.

In the outside world, events impinge on different sense receptors and on different parts of our body, yet we can estimate remarkably accurately whether such events are simultaneous. Perceptual processes must be able to "read" time-dispersed central events very precisely to recognize that they represent environmental events simultaneously. Conduction along the spinal cord between upper and lower limbs requires central command to be issued well in advance by about 8–12 ms to allow for spinal interlimb conduction. And the command, which must travel along variably dispersed neurons must be initiated in such a manner that the slower and faster signals are calculatedly dispersed, spatially and temporally. *Both in perception and command, and generally with respect to all central nervous transactions, the nervous system must be able to take into account the variables of transmission occurring along peripheral and central pathways.* How is this possible? An easy answer is that evolution would not have given advantage to a nervous system that could not solve such problems. It had to be solved by primitive nervous systems, and we enjoy the legacy.

In thinking neurophysiologically, we need to recognize that any and all functions are distributed in space and time; there is no focal point where either perception or motor command are provided a common datum. There is no central station where all attributes involved in seeing, listening, reading, or speaking are concentrated or where all the attributes of some perceived object ("Mom," for example) are located. The nervous system traffics in ionic and molecular events that are remote from the outside world. The spatial, temporal, and characteristic features of the outside world, as represented in the central nervous system, are vastly and radically different from what we think we understand of the physical universe (MacLean, 1990).

ANATOMICAL CORRELATES OF SENSORY INTEGRATION

Obviously structure is important for all sensory systems, from the nature of the sensory field, a ret-

ina, a sheet of olfactory receptors, a muscle tendon organ, for examples. These dictate how the physical or chemical stimulus can reach the receptor. Transduction may occur in a specialized cell such as a photoreceptor, or in an unmyelinated ("naked") nerve ending, or some specialized organ associated with the nerve terminal (e.g., Merkel's disk). Directly or indirectly, the process of transduction is converted into nerve impulses that are variously coded: instant of firing, phase locking to the stimulus, frequency of discharge (instantaneous frequency, increment above background, rate of change, weighted average) temporal pattern of impulses, number of impulses, duration of burst

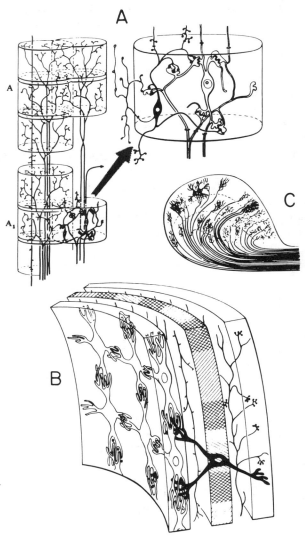

Figure 9.3.   A comparison of the preterminal arborizations of specific sensory afferents that gives some idea of the possibilities for convergence and divergence. *A,* The lateral geniculate nucleus; *B,* the medial geniculate nucleus; *C,* the ventroposterolateral nucleus. (From Szentagothai, 1975.)

of impulses, and so forth (Perkel and Bullock, 1968).

At the initial junction between receptor and nerve terminal and at the first synaptic relay of a neuron that has been directly activated, a host of integrative effects may be applied. There can be immediate feedback from the second-order unit to the first, reciprocal "push-me-pull-you" synaptic relations, or straight throughput of signals, with or without effects from neighboring or laterally integrating interneurons. In any event, modulation occurs at each sensory relay station.

Figure 9.3 illustrates how afferent neuron preterminal arborizations distribute the spent results of action potentials in the form of graded responses. This involves slowing of the rate of spread of membrane responses at each branching point and results in multiple synaptic events affecting a variety of cells in accordance with the time-space distribution of terminal signals. Figure 9.3A shows the long penetration and multiple arborizations of ganglion cell axons entering the lateral geniculate nucleus. The inset drawing of Figure 9.3A shows an enlarged version of one of the planes of section. Close inspection of the connections reveals that there is both *divergence* and *convergence* of signaling at this relay. In Figure 9.3B layers of cells in the medial geniculate nucleus are shown, the major thalamic relay for auditory signals en route to auditory cortex. Incoming axons enter alternating layers and distribute signals to many cells in these layers with effects that are relayed to other layers once removed, by interneurons, one of which is shown in black. Figure 9.3C shows the arrangement of afferent fibers in the somatic sensory system entering the ventroposterolateral nucleus which relays that modality to primary sensory cortex.

The most obvious and important point is that these three thalamic sensory relays are organized very differently in terms of access of incoming signals to intrinsic neurons, both interneurons and neurons that will relay messages to cortex. Clearly, different geometries are conducive to different patterns of integration. Analogous, extremely simplified diagrams are presented in Figure 9.4, which allows rough comparisons between several sensory modalities and their counterpart structural organization. These diagrams are deliberately scaled differently to make such comparisons easier between first-order, second-order, nth-order neurons along each vertical column. At the bottom are the three major thalamic relay nuclei, lateral geniculate nucleus (LGB), which provides visual relay; medial geniculate nucleus (MGB), auditory relay; and ventroposterolateral (VPL), somatic sensory relay.

Initial relays in olfactory and visual systems are remarkably similar, with strong lateral integrative controls. Olfaction, however, does not relay in the

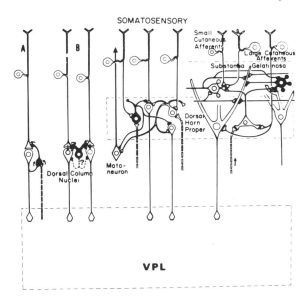

Figure 9.4. Simplified diagram to compare elementary neuron couplings in the four main sensory pathways. Inhibitory interneurons are indicated by *thick solid lines*. Only "main line" couplings can be indicated, and convergences and divergences are largely neglected. Descending control is indicated by connections drawn in *dashed* neurons. In the primary afferent-dorsal column pathway, diagram of the somatosensory system *(middle right)*, two possible interneuron couplings are indicated. *A* demonstrates the original concept of an elementary recurrent feedback coupling, while *B* shows an arrangement using dendritic synapses with triads. *LGB,* lateral geniculate body; *MGB,* medial geniculate body: *VPL,* ventroposterolateral nucleus (Szenthagothai and Shepherd, 1970). (From Szentagothai and Arbib, 1975.)

thalamus before going to cortex. This system—olfactory neuroepithelium, bulb, pathways and olfactory cortex—is phylogenetically older than the thalamus. At a later stage of olfactory discrimination, secondary routings involve the thalamus, but this is analogous to secondary association relays through thalamus in other sensory systems.

The LGB relay includes descending projections (interrupted axons entering the thalamus from the side opposite the entrace of ganglion cell axons from the retina). The auditory system has complex integration going on close to the sense organ, but conspicuously less lateral integration at lower levels. The auditory system also includes strong descending controls, only the lower levels of which are depicted here. For the auditory pathway, there are more relays along the ascending trajectory between cochlea and auditory cortex. There are also more connections crossing from one side of the brainstem to the other, meaning more bilaterality of representation than with other modalities.

The somatic sensory diagram reveals three different patterns, from left to right, the phylogenetically most recent—dorsal column system, with long first-order neurons (the longest neurons in the body), with tightly concentrated integration at the first central relay in lower brainstem. In the center is a phylogenetically intermediately old somatic sensory system with large-scale integration occurring in the sensory nuclei at the segmental level in the dorsal horn of the spinal cord. The diagram on the right shows the phylogenetically oldest somatic sensory group which conveys a wide range of somatic sensations in addition to pain and temperature. Integration in this oldest system is closer to the periphery than in either of the other somatic sensory systems. This illustrates a general principle: phylogenetically newer sensory systems reach further toward the brain to begin integrative processing; they include larger diameter fibers, and thus indicate priority for getting their messages reliably and quickly to higher levels. In the same strategic context, the phylogenetically newer motor pathways have the long axons, which extend from motor cortex directly, monosynaptically, to spinal motoneurons.

CHAPTER 63

Hearing

SIGNIFICANCE OF HEARING

Much of our understanding of the physical, biological and social universe is gained through hearing. Hearing rivals vision as a system for gathering and utilizing highly complex environmental information. Hearing depends on "sounds," described in physical terms as condensation-rarefaction waves of certain amplitudes and frequencies, borne in sound-conducting media: air, water, bone, wood, metal, etc. Sounds continually inform us about environmental activities, even around corners and during sleep. An intern in San Francisco, wakening regularly at 2 a.m. without knowing why, realized it was the absence of otherwise continuous background noise when the cable car cables stopped running that wakened him.

Voices are ordinarily the most important sounds people hear; individual capacities for hearing and understanding speech and language are inevitably medically important. Information conveyed by sounds, i.e., the *meaning of sounds,* is determined not so much by their physical characteristics as by the cultural experiences, expectations, and purposes of the listener. Perhaps the most important decisions a physician makes in clinical practice depend on careful listening to what the patient says, and to a lesser extent, on interpretation of certain sounds generated within the patient's body.

Loss of hearing, whether congenital or acquired, imposes severe social and communicative handicaps. Deafness should be compensated when hearing is lost, and as soon as intervention is feasible in hearing-deprived children. Hearing is more important than vision with respect to language and its development. Much of our thinking is auditory. We are surrounded by social and cognitive signals that involve gesture vocalization, speech and hearing, and, in deaf persons, equivalent cognitive substitutes for hearing, such as sign language.

It is said that within a few days of birth, a baby shows attentive preference for listening to a tape recording of his/her own mother's reading voice, as compared with the attention given another mother's voice reading the same material (DeCasper and Fifer, 1980). Within 6 months, an infant generates babbling sounds that are characteristic of local speech. The ascent of man is reaffirmed every time an infant creates de novo a personal consciousness that social significance attaches to certain sounds and begins to vocalize as a means of communication. The child gains rapid access to social and intellectual development on the magic carpet of language.

A deaf child with deaf parents develops normally because the parents begin presenting sign language early, and soon gesture-visual comprehension becomes a competent basis for family communication and language development. The child quickly learns to comprehend signing and progresses along a normal path and pace for developing cognitive skills and language learning (Klima and Bellugi, 1979). Deaf children born of hearing parents are in relative jeopardy. Both they and their parents need help in communicating. Unless the parents or other caregivers provide sign language, early and adequately, and begin communicating through that rich, dynamic medium, both generations have difficulty establishing mutually confident social relations. The child soon encounters difficulties developing cognitive and learning skills as well as communicating, and the frustrations of isolation may become overwhelming.

PHYSICAL CHARACTERISTICS OF SOUND AND SOUND RECEPTION

Simply put, hearing is the subjective experience of exposure to vibrations in the range from about 20 cycles per second (cps, Hertz, or Hz), to about 20,000 Hz, or 20 kHz (although some children can hear frequencies as high as 25 kHz). During normal aging, sensitivity to the higher frequencies is gradually lost so that by midlife, hearing is often limited to less than 5 kHz. Notwithstanding, since frequencies produced during speech lie mainly in the range between 200 Hz to 4–5 kHz, such losses do not substantially interfere with oral communication.

Receptors for both auditory and vestibular sensory input are cuboidal epithelial cells, characterized by stereocilia that protrude from their exposed surface. These so-called "hair cells" are served by first-order afferent *and* efferent nerve fibers in the VIIIth cranial nerve. The acoustic neuroepithelium—bearing hair cells—is arranged in the spiral *cochlea*, which, together with the attached vestibular apparatus (Fig. 9.5), is completely encased by dense temporal bone.

Sounds obtain access to the cochlea and its neuroepithelium by way of the oval window in the wall of the vestibule. The oval window appears as a large oval outline on the base of the labyrinth in Figure 9.5. This window is covered by a membrane to which is attached the foot plate of the *stapes*

(stirrup), the last in a chain of three middle ear ossicles (Fig. 9.6). The stapes is physically linked to the world of sound via these middle ear ossicles and the tympanic membrane. Displacement of the oval window against fluid in the vestibule is compensated by reciprocal displacement of a membrane-covered round window in the bony wall of the cochlea. Vibrations delivered to the oval window directly impact fluid in the scala vestibuli. To reach the round window, this kinetic energy must pass along and across two intervening membranes—a sound-transparent vestibular membrane and a drum-taut basilar membrane (see Fig. 9.9). It is on the basilar membrane that hair cells and nerve terminals ride the sound waves.

Figure 9.6 shows the tympanic membrane and middle ear ossicles. The ossicular chain provides a 1.3 mechanical advantage. The difference between dimensions of the most responsive area of tympanic

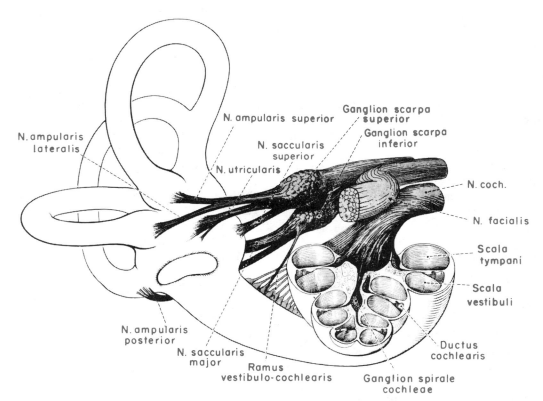

Figure 9.5. General schematic depiction of the inner ear showing three-dimensional relations of vestibular apparatus and cochlea with their respective branches of the VIIIth cranial nerve. Each ampulla of the three vestibular semicircular canals has its own nerve bundle, and large bundles also serve the utricles and the saccule. All vestibular afferent neurons have their cell bodies in the two Scarpa's ganglia, and they convey a sense of balance and motion of the head in three-dimensional space. The cochlea has been sectioned through the modiolus, its central bony axis, to show three turns of the cochlea, composed of three spiraling canals, scala vestibuli, which connects with the vestibular apparatus, scala tympani, and between, the cochlear partition—

the scala media or cochlear duct. Afferent innervation of the cochlea comes from the spiral ganglion which winds through the spiraling core of the modiolus. The oval window to which middle ear ossicles deliver vibrations interpreted by the nervous system as hearing can be seen as an indented oval outline on the base of the labyrinth. The oval window communicates with the scala vestibuli. Note the close proximity of the important facial nerve which may be implicated by inflammatory or neoplastic diseases affecting the inner ear, or by cold exposure to wind against the ear, leading to paralysis of facial muscles on one side, Bell's palsy. (From Melloni, 1957.)

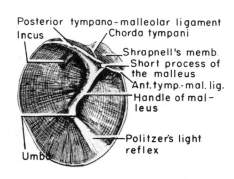

A.

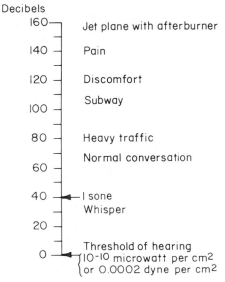

B.

Figure 9.6. *A*, View of outside of tympanic membrane as seen with an otoscope (Portman, 1951). Position of handle of malleus is visible. *B*, Cutaway section of tympanic membrane to show attachment of malleus for maximum excursion of membrane. Malleus articulates with incus, so the combination of two leverages applies to piston-like motion of stapes against flexing membrane which covers oval window of labyrinth. (From Melloni, 1957.)

membrane (about 40 mm²), and the area of the oval window (about 3 mm²), gives an additional 13:1 mechanical advantage. This improves impedance matching between air and fluid and readily transmits acoustic power from air to neuroepithelium. Impedance matching efficiency varies between 50 and 75% across the whole range of frequencies used in human speech.

If the oval window becomes ossified, as it does in *otosclerosis*, the tympanic membrane (TM) and the ossicular chain are obviously vitiated, and the acoustic neuroepithelium has to depend solely on vibrations transmitted to the temporal bone. In that event, hearing loss is even more severe, because there is a very poor impedance match between air and bone. This was apparently Beethoven's problem (Goodhill, 1983). His hearing became limited to bone conduction. To be able to hear some of his own music, he used to hold a stick between his teeth and press it against the sounding board of his piano. Modern hearing aids apply electromagnetically driven vibrations, impedance-matched to bone, directly over bone near the skin surface, usually over the mastoid process, and this can restore considerable hearing to persons similarly afflicted.

Specification of Acoustic Power

Referring to Figure 9.7, we see that the dynamic range of sound intensities to which the ear can respond is exceedingly broad. This makes a log scale advantageous for specifying sound intensities. An arbitrary standard minimum power to which the human ear can respond is taken to be 10^{-16} watt/cm², or 0.0002 dyne/cm² at 20°C. This standard, when designated in *newtons*, refers to a sound pressure level (SPL) of 2×10^{-5} N/m². The Bel is the ratio, increasing by powers of 10, between a given sound and this reference power; 12 Bels covers the 10^{12} human hearing intensity range, at 1 kHz. The decibel (dB), 10 to the Bel, is convenient for reference to measurement of normal and disordered hearing. Since subjective loudness is not directly related to sound intensity, the *sone*, a unit of loudness based on a subjective ratio scale, is used for psychophysical comparisons (Fig. 9.7).

Decibels
160 — Jet plane with afterburner

140 — Pain

120 — Discomfort

Subway
100 —

80 — Heavy traffic

Normal conversation
60 —

40 — I sone
Whisper

20 —

Threshold of hearing
0 — 10⁻¹⁰ microwatt per cm²
or 0.0002 dyne per cm²

Figure 9.7. A decibel (dB) scale showing approximate acoustic power encountered in the presence of various sources. The sone is a unit of perceived loudness based on ratio estimates for which the loudness of a 1000-Hz tone at 40 dB provides reference. (From Stevens, 1958.)

ANATOMY AND PHYSIOLOGY OF THE EAR

The organ of hearing consists of an outer, middle, and inner ear. The pinna helps to localize sound. The external auditory canal channels sound pressures to the TM. The TM, which divides outer from middle ear, has a reflective sheen that provides a useful diagnostic index. A too pale, a red, or a bulging TM indicates middle ear trouble. The middle ear cavity, normally air-filled, is connected to the throat by a narrow passage, the eustachian tube, through which air can escape from the middle ear cavity more easily than it opens to air. A rapid rise in altitude (with accompanying reduction in ambient pressure) allows air to escape from the middle ear. Rapid descent without equilibration induces an unpleasant inward pressure of air against the TM, associated with diminished hearing. This can usually be relieved by swallowing and by thrusting the larynx and trachea downward and forward. Nasal and throat congestion make such clearance difficult. Babies have difficulty admitting air to the middle ear; letting them drink milk during descent in an airplane is preventive.

The auditory nerve responds to frequencies between about 20 and 20 kHz, but it is not equally responsive over this range. Optimal response occurs at about 2 kHz. The frequency response curve relates primarily to physical characteristics of the ear: external, middle, and inner, and especially of the basilar membrane.

The three middle ear ossicles are: the *malleus* (hammer), *incus* (anvil), and *stapes* (stirrup), which articulate with one another in that order, from TM to oval window (Fig. 9.6). When the TM is vibrated, the ossicles are set in motion whereby the stapes transmits the vibrations to the oval window and the fluid-filled scala vestibuli.

Tailoring of Sounds by Contraction of Middle Ear Muscles

There are two middle ear muscles (MEMs): the *tensor tympani*, innervated by the trigeminal (Vth cranial nerve), which attaches to the handle of the malleus; and the *stapedius*, innervated by the facial (VIIth cranial nerve), which attaches to the stapes. Sound delivered to one ear elicits contralateral as well as ipsilateral reflex contraction of the MEMs. Binaural sound stimulation is more effective than monaural. MEM acoustic reflexes involve four neurons: cochlear afferent, ventral cochlear nucleus, medial superior olive, thence to motor nuclei V and VII.

Contraction of MEMs in hearing may be considered analagous to contraction of pupils in vision, but the consequences of MEM actions are more elaborate; they tailor the shape of sound prior to its reaching the cochlea. There is enhancement as well as attenuation of sound intensities and there are frequency-selective effects. MEMs have long been known to provide *reflex protection* of the cochlea from damaging loud sounds. Since latency for reflex contraction of MEMs is about 10–15 ms, they cannot protect against damage from abrupt, unexpected sounds such as a pistol shot; they do, however, provide some protection by contracting in anticipation of sudden loud sounds and against enduring loud sounds.

Protection is not all that MEMs do, however. They respond reflexly even to faint acoustic stimuli. Starr (1969) demonstrated that MEM contractions enhance sound transmission to the cochlea in the range of 1–2 kHz while attenuating responses to frequencies above and below that range. This improves sound transmission in the range of voice communication while reducing interference from other frequencies. MEMs contract prior to and during vocalization, thus attenuating acoustic responses to one's own voice before as well as during vocalization. Note that MEM contractions associated with vocalization and movement are not reflex responses since they contract in anticipation of these activities.

MEMs also contract in anticipation of loud sounds, jarring movements, etc. Every step in walking, for example, is associated with corresponding MEM contractions. This may protect middle ear ossicles from potential damage by jarring. MEM reflex contractions are also elicited by stimulating skin of the external ear, face, and neck, by yawning, chewing, swallowing, and by gross bodily movement. MEM activity can be modified by classical conditioning (Simmons et al., 1959): in respect to a sound's significance; according to prior acoustic experiences; and in response to attention being directed to another sensory system (Carmel and Starr, 1963). MEMs can also be contracted voluntarily. Nonacoustic reflex and nonreflex MEM contractions are equally active in totally deafened animals. Because MEM contractions are accompanied subjectively by a faint roar or flutter, it is advantageous when listening intently, to stand "stock still," to minimize MEM contractions.

In humans, with acoustic movement-induced and voluntary MEM contractions, there is up to a 15-dB elevation of hearing thresholds (Salomon and Starr, 1963). Patients lacking MEMs experience a deterioration in speech intelligibility during increasing sound intensities: intelligibility is about 20 dB below what is experienced by normal subjects (Borg and Jackrisson, 1973). This confirms that MEMs improve discriminative hearing in the presence of loud sounds. If the middle ear ossicles are lost, the system loses its advantage in impedance matching. Sound waves can still reach the oval and round windows via air in the middle ear, but they impact both windows nearly simultaneously and are scarcely effective in vibrating the basilar membrane. There is a consequent reduction in sensitivity of hearing of about 30 dB, equivalent to reducing the sound of a loud voice to a barely audible whisper.

The Inner Ear

As noted above, the functional interface between middle ear and inner ear is provided by two small openings in the temporal bone, the oval window, through which vibratory motion is delivered to fluid in the scala vestibuli, and the round window, which lies in the wall of the scala tympani. Wedged between these two spiraling canals is the scala media or cochlear duct (see Figs. 9.5 and 9.9). The basilar membrane constitutes the partition between scala media and scala tympani. The vestibular membrane divides scala vestibuli from scala media and is so thin it provides no bound barrier (Fig. 9.9). The scala vestibuli and scala tympani connect freely with one another via a wide opening at the apex of the cochlear spiral, the *helicotrema.* These two chambers contain *perilymph,* which resembles spinal fluid in composition. The scala media, however, contains *endolymph* which bathes the acoustic neuroepithelium. Endolymph has a distinctive chemical composition, being relatively higher in potassium, and lower in sodium, chloride, and protein. This difference in electrolyte and protein composition is established and maintained by the *stria vascularis* which lies along the outer wall of the endolymph-containing canal, the scala media (Fig. 9.9). The cochlea, including the stria, receives important autonomic innervation. Autonomic sympathetic fibers release norepinephrine to control the vascular supply (Densert, 1974). The difference in chemical composition between endolymph and perilymph creates a steady potential across the basilar membrane, approximately double that across ordinary cell membranes (Tasaki, 1954). This exalted potential is sustained by active metabolic processes and is rapidly lost if circulation to the stria is compromised. The net effect of strial metabolic activity is to boost the electrical potential across the basilar membrane.

Vibratory motion begins near the base of the cochlear spiral, at the opposite end of the canals from the helicotrema. When pressure is applied against the scala vestibuli by inward motion of the footplate of the stapes, fluid in scala vestibuli and scala media displace the basilar membrane which,

in turn, displaces fluid in scala tympani, pushing the round window outward. The sensory apparatus, the acoustic neuroepithelium, contained in the *organ of Corti,* rests on the basilar membrane, facing into the scala media, the endolymph-containing chamber.

The basilar membrane is a sheet of radially oriented collagen fibers stretched between a bony partition that extends partway across the cochlea from its origin on the bony modiolus, the central axis of the helical spiral of the cochlea, to the outer wall of the cochlea. Motion of the basilar membrane follows from its physical characteristics. The membrane becomes broader toward the apex although the cochlea as a whole becomes narrower there. The membrane becomes progressively less stiff toward the apex. Virtually all sound energy transmitted to the cochlea is absorbed by the basilar membrane near its stiff basal region.

Waves travel along the membrane as if one were shaking a rug from one end. Propagation of traveling waves from base to apex of the cochlea takes nearly 3 ms, thus successive points along the organ of Corti are activated with increasing delay. Motion of the cochlear partition is entirely damped within that time. Amplitudes are attenuated as the waves progress. As illustrated in Figure 9.8, the envelope of the traveling wave shows that damping results in falling off of the amplitude of motion to zero beyond a certain region of membrane, depending on the given frequency. The rate of attenuation increases with sound frequency; as waves travel toward the apex and the membrane becomes less stiff, higher frequencies are increasingly damped.

The pattern of basilar membrane motion is determined by both the stimulating frequency and its energy. The lower the sound frequency, the greater the proportion of membrane set in motion. The distribution and amplitude of physical displacement of hair cells lined up along the membrane is uniquely determined by the nature of the sound stimulus.

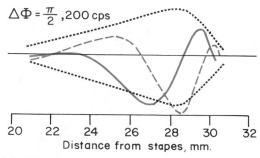

Figure 9.8. Traveling wave along the cochlear partition for a 200-Hz tone. The *solid line* indicates the deformation pattern at a given instant. The line with the *short dashes* shows the same traveling wave one-fourth of a cycle later. The *envelope* shows the maximum displacement at each point. (From von Békésy, 1947.)

Khanna and Leonard (1982) succeeded in making laser measurements of mechanical responses of the basilar membrane. The more they avoided damage to the basilar membrane, the more their curves resembled the curve of lowest amplitudes of responses of first-order auditory afferents. This implies that the main transducer of acoustic stimulus patterns is the basilar membrane itself.

Motion of the basilar membrane is translated into neuronal firing by action of hair cell receptors in the organ of Corti. There are about 3500 inner hair cells and somewhat more than 20,000 outer hair cells (Figs. 9.9 and 9.10). Inner hair cells form a single file. Outer hair cells form a bank of three or more rows on the outer margin of the tunnel of Corti. Inner hair cells have only about 50 stereocilia whereas outer hair cells have about 100.

Viscous coupling of hair cells with the tectorial membrane in the cochlea plays a significant role by patterning stereocilia displacement. The stereocilia of inner hair cells seem barely to graze the tectorial membrane whereas the stereocilia of outer hair cells are partly embedded in that structure. Since the tectorial membrane is hinged independently from the basilar membrane and has a free border, the relative motions introduce shearing forces which bend the stereocilia, providing adequate stimulation of the hair cells. Because of the unusually high standing potential between endolymph and the interior of the hair cell, changes in membrane conductance induced by the shearing forces are accompanied by a nearly double-rapid flux of ions. This generates a receptor potential that initiates release of neurotransmitter at synaptic junctions between hair cells and the afferent nerve terminals.

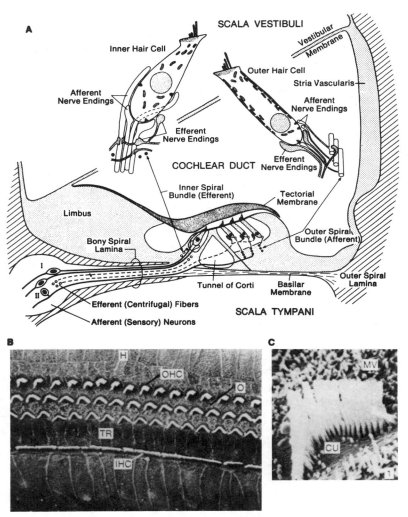

Figure 9.9. *A*, Cellular organization of organ of Corti in the guinea pig. Enlarged diagrams, located by *arrows,* show fine structure of inner and outer hair cells. *B*, Scanning electron micrograph looking down on hair cells, showing differences in arrangement of inner (IHC) and outer (OHC) hair cells and their stereocilia. *H*, Henson's cell; *TR,* tunnel rod. *C*, Scanning electron micrograph of stereocilia of an outer hair cell. *MV,* microvilli (stereocilia); *CU,* cuticular plate. (*A*, redrawn from Smith in Brodal, 1981; Smith, in Eagles, 1975; Shephard, 1983.)

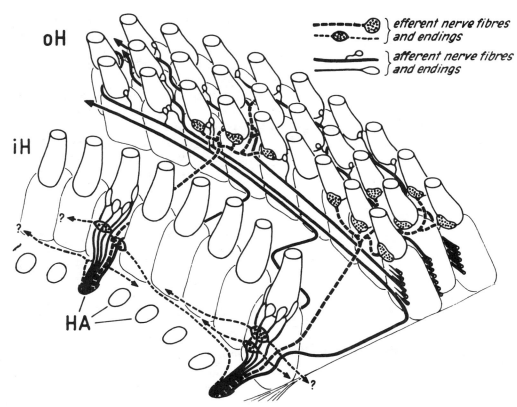

oH

iH

HA

) *efferent nerve fibres*
) *and endings*
) *afferent nerve fibres*
) *and endings*

Figure 9.10. Schema of general innervation pattern of the organ of Corti. Afferent innervation of outer hair cells goes spirally a long distance in one direction only, basalward. Efferent synaptic endings are abundant, inhibiting afferents shortly after they leave their attachment to the inner hair cells and directly ending on outer hair cells. Efferents come from both ipsilateral and contralateral superior olives. *iH,* inner hair cell; *oH,* outer hair cell; *HA,* habenula. (From Spoendlin, 1970.)

The general innervation pattern of the organ of Corti is shown in Figures 9.9 and 9.10. Practically all afferent fibers innervate single inner hair cells with a single synapse each. Each inner hair cell is innervated in this way by 8 or 10 different afferent neurons. Branches of these afferent axons also cross the floor of the tunnel of Corti and spiral toward the base of the cochlea, a distance of about 1 mm, using their last 0.2 mm to innervate several outer hair cells. A subpopulation of spiral ganglion cells (5%), consisting of large myelinated axons, do not branch to cross the tunnel but bifurcate and course toward both apex and base of the cochlea, innervating about 10 inner hair cells each.

Cochlear efferents, which number about 1200, are distributed more diffusely. Efferents to inner hair cells, shown in Figure 9.10 as large synaptic endings, with stippling to represent vesicles, apply several synapses to each of several *afferent fibers* near their point of synaptic attachments to the inner hair cells. Other efferents have multiple, large diameter terminations which synapse directly on a large number of *outer hair cells*. Both inner and outer hair cells contribute to integrated responses of auditory afferents. Outer hair cells enhance low thresh-

old responses and sharpen the tuning curves of auditory afferents. Following destruction of outer hair cells by the drug kanamycin, cochlear microphonics (CM, see below) are reduced to ⅟₃₀th of their normal amplitude.

COCHLEAR POTENTIALS

Beside the standing potential across the organ of Corti, there are three principal *dynamic potentials* which accompany sound stimuli delivered to the cochlea: *cochlear microphonics* (CMs), positive and negative *summation potentials* (+SP and −SP), and *nerve action potentials* (APs), in the auditory nerve.

Cochlear microphonics (CMs) are generated instantaneously by the pattern of vibration of the cochlear partition (Fig. 9.11). Analagous to the function of a microphone, CMs are faithful to acoustic stimuli. Summation potentials relate separately to the bending of the stereocilia (+SP) and the generation of nerve activity (−SP). The APs are ordinarily seen as nearly synchronous discharges of first-order afferents in response to brief transient acoustic stimuli. APs represent the first nervous system sampling of the world of sound. CMs

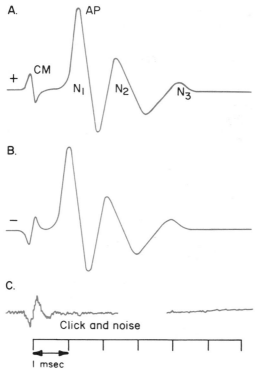

Figure 9.11. The electrical response to clicks as recorded from the round window of the cochlea of the cat. *CM*, cochlear microphonic potential; *AP*, action potential of the auditory nerve. The three components of the action potential are represented by N_1, N_2, and N_3. A reversal of polarity of the stimulus between *A*, at the top of the figure, and *B* results in an inversion of the cochlear microphonic but no change in the action potential except for a slight shift in latency. At *C*, the action potential is completely masked by white noise, while the cochlear microphonic remains.

induced by clicks or tone pips are followed by large amplitude negative-positive-negative sequences of APs. Negative AP peaks are referred to as N_1, N_2, and N_3. If the acoustic stimulus polarity is reversed from condensation-rarefaction to rarefaction-condensation, the CMs reverse polarity, as does a microphone, but the APs do not (Fig. 9.11).

COCHLEAR MICROPHONICS (CMs)

CMs represent the summed activity of many hair cells. The phase shift of CMs generated along the basilar membrane coincides with basilar membrane motion, and the CM longitudinal velocity is independent of stimulus frequency (Honrubia, 1970). Electric currents of the CMs must flow through the bodies of hair cells and they probably enhance receptor responses. Thus, CMs represent an electrical effect generated by cellular activity—the sum of which (the CM) has a feedback effect on the same individual cells that contributed to the sum in the first place—an example of receptor population integration.

CMs cannot be more than *indirectly* responsible for nerve firing, however, for the following reasons: CMs have virtually no latency, auditory nerve impulses do; CMs have no refractory period, auditory nerves do; CMs do not fatigue, neurons do; CMs follow stimulus rates in excess of those that can be followed by auditory neurons; there is a linearity in CMs following shifts of sound intensity which is not matched by neural responses. CMs, therefore, are preliminary to nerve responses, but probably boost receptor excitation.

SUMMATING POTENTIALS (+SP AND −SP).

Additional dynamic cochlear potentials are known as positive and negative summating potentials (+SP and −SP). +SP are thought to be due to bending of hair cell cilia (Davis, 1958). −SP may have a neural origin (Kupperman, 1966; 1970); moreover, −SP can be masked, as neural response can, whereas neither +SP nor CMs can be masked.

NEURONAL ACTION POTENTIALS (APS)

There are about 30,000 primary afferent neurons with cell bodies in the spiral ganglion lying within the bony modiolus along the axis of the cochlea. This gives a ratio of nearly 10:1 primary afferents to the number of inner hair cells. The afferent neurons are bipolar cells with axonal branches that extend peripherally to innervate the cochlea and centrally to innervate the cochlear nuclei. Peripheral branches lose their myelin on entry into the organ of Corti. Since the unmyelinated region is relatively long, there are opportunities for integration among graded responses to occur within a single axon and interneuronal integration among bare axons traveling together.

Most if not all primary auditory fibers of mammals are spontaneously active, exhibiting from a few to one hundred or more spikes/s (Liberman, 1978). Each auditory afferent has a critical frequency (CF) which subtends a narrow frequency band of greatly reduced threshold. This lowest CF threshold is a way of characterizing a given neuron although it can respond to a broad range of frequencies at higher stimulus intensities. The CF for a given neuron has a functional importance like that of the sensory field for a first-order tactile unit and the visual receptive area for a unit in the visual system. Like those classifications, CF is used to identify units which occupy higher stations throughout central auditory pathways.

The overall frequency response curve of a given neuron may be narrow or broad, and symmetrical or asymmetrical. The frequency response curve for a typical first-order auditory neuron is abruptly steep on the high frequency side, with a generally broader shoulder on the low frequency side. Even though the overall frequency response curves may be broad, the CF is important because the aggregate of CF profiles defines the threshold for hearing over the entire acoustic range, a neural equivalent of the audiogram (Fig. 9.12). The sensitivity of these units is not only dependent on basilar membrane and outer hair cell sharpening, but also on an adequate supply of oxygen (Evans, 1974). A modest reduction of oxygen tension markedly elevates CF thresholds.

When two tones rather than one are used for stimulation, there appear discrete patterns of lateral inhibition between afferent neurons. The consequence is that individual unit frequency response curves become even narrower. All primary auditory neurons show such two-tone inhibition. There is also apparent recruitment of one fiber by another, influencing others to fire when they would not otherwise have done so, or to increase their rate of firing.

It has been estimated that when the ear is stimulated by a 40 dB tone at 1000 Hz, approximately 3000 hair cells are activated. With a change in frequency by a just discriminable amount, it is likely that less than 2% of these hair cells are no longer activated and are supplanted by others which were formerly inactive. Evidence indicates that a population pattern of activation of many receptors rather than the activation of just a few is involved in even relatively simple discriminations of one pure tone frequency from another.

At onset of noise or tonal stimulation that includes appropriate frequencies, the spontaneous rate of firing of individual afferent neurons increases. If the tone is continued, units will shortly adapt to a lower level of activity which resembles their spontaneous activity, but at an elevated rate. If a pure tone is presented, there may be *phase-locking*, that is, units will tend to fire at a certain phase of the repeated stimulus pattern but not at every cycle. Whether a unit fires or not remains probabilistic, as it was during spontaneous activity.

ACCOUNTING FOR A DYNAMIC HEARING RANGE

We are able to distinguish speech and other sounds in the middle range of audible frequencies over a dynamic range of intensities from the faintest whisper to the loudest shout, a span of about 100 dB. How do afferent nerve fibers react so as to cover this extraordinary dynamic range? First, while most individual first-order neurons have a dynamic range of 30–50 dB, Evans (1981) found exceptional cochlear fibers with dynamic ranges in excess or 60 or 70 dB, and some even greater. Second, because of lateral inhibition and mechanical interferences affecting the organ of Corti, not all cochlear fibers become maximally activated. Third, the dynamic range may be coded by some combined index of axonal activity, i.e., the rate and degree of phase-locking. Fourth, at higher intensities, individual units shift their CF from both higher and lower frequencies toward an acoustic midrange, around 1000 Hz. These intensity-imposed shifts may be up to ½ octave (Evans, 1981).

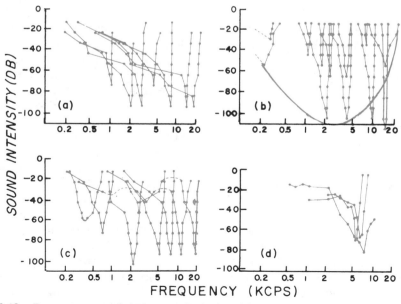

Figure 9.12. Response areas of single neurons obtained from *a*, cochlear nerve; *b*, inferior colliculus; *c*, trapezoid body; and *d*, medial geniculate body of the cat. (From Katsuki, 1961.)

The natural environment is relatively noisy. Prolonged exposure to loud noise at a given frequency elevates the auditory threshold for some minutes. Long-duration background noises bias units to shift their CFs toward higher frequencies. Threshold shifts may be from ½ to 1 octave toward frequencies higher than the frequency of the sound exposure. This is not fatigue; it is an adaptive mechanism, an automatic gain control in the graded response region of the afferent axon terminal prior to the probabilistic spike generator.

FACTORS THAT DISTORT COCHLEAR INPUT

There are, of course, many disease processes which can affect the external, middle, and inner ear, and thereby affect hearing capabilities. And there are commonplace mechanical interferences in the outer canal, and pressure disequilibria in the middle ear, which distort delivery of environmental sounds to the cochlea. There are additional built-in distortions:

First, the human ear responds to only a limited range of vibrations. Thresholds over this range conform to the limits of physical transmission via the peripheral auditory apparatus. For example, the tympanic membrane-ossicular linkage to the oval window is not fully impedance matched.

Second, the MEMs are dynamically active during quiet or noisy conditions, and they modify incoming signals on an individually idiosyncratic basis. They are dynamically active in anticipation as well as in reflex operations; they act differentially on the basis of acoustic experience and are subject to voluntary control as well as Pavlovian conditioning.

Third, the basilar membrane has imperfect response characteristics which can be especially nonlinear in the context of abruptly shifting, complex sounds. Both the basilar membrane and the outer hair cells contribute to sharpening critical frequency (CF) profiles among first-order neurons. Moreover, mechanical events in the cochlear partition reveal interference phenomena which resemble and interact with neuronal lateral inhibition.

Fourth, there are important dynamic as well as standing potentials which affect receptors and nerve endings as well as interact with one another. The composition of endolymph and the responsivity of hair cells are both dependent on an adequate supply of oxygen. If an individual becomes moderately hypoxic, hearing is affected in unpredictable ways.

Fifth, auditory afferent neurons are spontaneously active; they exhibit a variety of stimulus-response curves and phase-locking propensities under pure-tone conditions; they are more complexly affected by two-tone stimulation, and even more radically by complex sounds. In response to loud background noises there may be large CF shifts. Thresholds, CFs, and other general aspects of a unit's acoustic responses are altered by tonal interactions, shifts in sound intensity, background noise, and units may be shifted as much as one-half to whole octaves—according to recent acoustic history.

This admittedly incomplete inventory of factors that interfere with veridical auditory responses indicates that modifiability of first-order afferents, as units and as ensembles, is not adequately characterized by analyzing responses to pure tones at near-threshold levels of energy. The ear as a hearing organ is dynamically and radically modified in its performance by noisy backgrounds and loud sounds to which our ears are commonly exposed. And yet, notwithstanding, the auditory system is remarkably resilient, and under standard conditions, it is remarkably reliable. Still, these considerations do not take into account the influences of *efferent* nerves to the cochlea which add to the dynamics of receptors and first-order afferent activities and responses.

Efferent Control of the Cochlea

A compact bundle of *efferent* fibers leaves the brainstem and runs parallel to the incoming VIIIth cranial nerve auditory and vestibular afferents. The efferent fibers arise bilaterally from the superior olivary complex and are called the *olivocochlear bundle (OCB)*. The OCB divides to innervate both vestibular and cochlear neuroepithelia. At both locations efferents influence hair cell reception and impulse generation in first-order neurons. Some efferent fibers to the cochlea branch before the bundle leaves the brainstem and contribute terminals to the cochlear nuclei where they influence first-order to second-order auditory neuron relays. Further, all central auditory ascending pathways include *descending auditory projections* which influence the ascending auditory pathways nucleus-by-nucleus, throughout the auditory system. This central auditory descending cascade also contributes to control of the OCB.

Approximately 1800 axons comprising the OCB originate from both ipsilateral and contralateral neurons in the superior olivary complex (Rasmussen, 1960). About one-third of the cochlear efferents, mostly from the contralateral side, branch extensively and provide dense synaptic terminals on all 20,000 outer hair cells (Figs. 9.9 and 9.10). Activation of the OCB attenuates outer hair cell responses, reducing thresholds for weak acoustic stimuli, and improving sharpness of the critical frequency (CF) characteristics of afferent neurons. These efferent influences represent *presynaptic inhibition.*

The pattern of efferent innervation relating to inner hair cells is different. There, efferent fibers rarely synapse directly on the hair cells but instead synapse en passant on afferent neurons close to their synaptic junctions at the base of the inner hair cells. At these junctions, efferent signals attenuate or block afferent nerve impulse generation. This represents *postsynaptic inhibition.* Corresponding to this inhibition of auditory first-order neurons, there is a marked reduction in amplitude of evoked responses throughout all central auditory projections (Desmedt, 1962).

If the cochlear action potential is being masked by another sound, activation of the OCB improves signal-to-noise ratios in the cochlea (Dewson, 1967). Dewson trained monkeys to discriminate between human speech sounds presented in the presence of noise of different intensities; there was significant impairment of discrimination following sectioning of the crossed OCB fibers (Dewson, 1968).

By means of the OCB, the central nervous system can exert control over peripheral acoustic receptors and first-order central relays. The net effect is relative exclusion of some sounds and improvement of signal analysis among others. The OCB controls auditory functions in some ways analogously to middle ear muscles, one step closer to the central nervous system. The physiological design of both these centrifugal control systems allows the nervous system both to be more selective about what it hears and to hear better what it wants to hear.

INTRODUCTION TO CENTRAL AUDITORY PROJECTIONS

Central auditory pathways are described in a sequence of ascending and descending projections. But it should be understood that both directions of projections have feedforward and feedback loops which selectively influence neuronal impulse traffic in both ascending and descending directions simultaneously.

Four distinctive but interrelated "ascending" auditory systems are described: (1) The classical lateral lemniscal (LL) projection system, which relays via "core" components of the inferior colliculus (IC) and medial geniculate body (MGB) to innervate a "core" field of auditory cortex. The classical LL system has rather exclusively to do with acoustic signals. This projection system in the human is illustrated in Figure 9.13. (2) The "lemniscal adjunct" or "belt" projection system; (3) the nonspecific projection system (involving generalized arousal) through the diffusely projecting nuclei of the thalamus; and (4) specific neurotransmitter projection systems which have widespread cortical projections to all sensory and motor representations in cortex and which circumvent the thalamus.

From the point-of-view of successful behavior, which is what pays off in evolution, locating decision-making at levels closest to input/output is important, especially when rapid turnaround responses and excercise of practiced initiatives are required for the sake of speedy performance and safety. A great many neurophysiological decisions take place at brainstem levels. The auditory system is heavily engaged in brainstem and spinal acoustic reflexes, brainstem mechanisms relating to audio-visual coordination, to generalized arousal and focus of attention, and to exchange of signals between cerebral and cerebellar hemispheres.

Imagine the wide panoply of potentialities: Conscious auditory experiences and consciously directed, acoustically related performances are available in the same comprehensive system that provides unconscious anticipatory contraction of the middle ear muscles (MEMs), where acoustic startle reactions can trigger involuntary brainstem and spinal reflexes prior to conscious awareness, all functioning in a system that supports a magnificent auditory memory bank, a system that can learn exceedingly rapid and accurate acoustic-motor reactions—such as those required for expert musical performance.

There must be decisive switching between operations that require shortest possible feedforward feedback looping circuits at lower levels and those operations that require longer-looping, more reflective analysis. This switching is accomplished in the brainstem, mostly automatically, and also on the basis of successful overtraining. Overlearning and expert skill development contribute to earlier and unusually rapid and accurate, lower-level controls for decision-making and sensorimotor coordination. Skilled performance is accomplished with little deliberate conscious intervention. Indeed, conscious reflection on such processes interferes with performance.

There are abundant ascending auditory pathways that participate in longer-looping circuits and, ultimately, in panoramic telencephalic cortical and subcortical circuits for more contemplative, subjective, auditory experiences and creative pursuits. Such shorter- and longer-looping circuits operate in con-

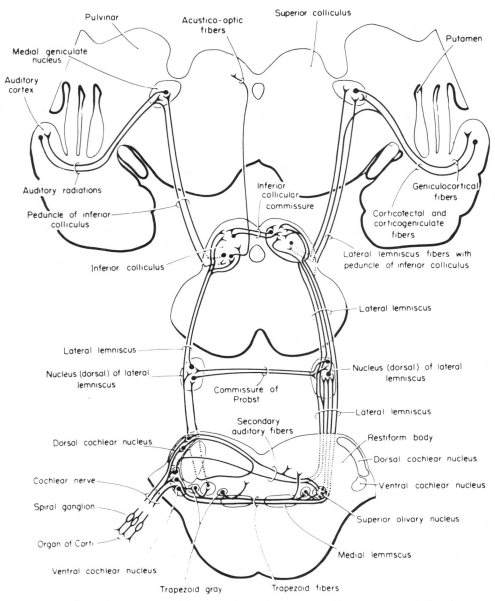

Figure 9.13.   Major central pathways of the auditory system. (From Crosby et al, 1962.)

tinuous parallel fashion, simultaneously accessible for integration in accordance with shifting global purposes.

Plasticity of Acoustic Responses

Permanently indwelling thin electrodes have been placed in the brains of experimental animals under anesthesia in many different auditory and nonauditory brain regions. Bipolar electrodes with fine tips close together ensure that recorded responses reflect local activity. In fully awake animals it is commonplace to record large amplitude responses to acoustic stimuli in areas of the brain not ordinarily considered to be auditory. It depends on whether

the experimenter has conditioned the animal by making the acoustic stimuli "biologically significant." Simple Pavlovian conditioning, with weakly positive or negative reinforcements, can induce click-evoked responses in practically any region of the brain.

When a simple acoustic stimulus is repeated many times without reinforcement, responses fluctuate, but slowly decline to low-amplitude steady-state responses which are said to be "habituated." An "unconditioned stimulus," mildly rewarding or punishing, can be reintroduced just before each click. The result is even more widespread and larger amplitude responses to the combination. Even when the reinforcing stimulus is discontinued, the click-evoked responses remain elevated for some time. After numerous extinction trials it is difficult to see any evoked potentials to clicks, even along the classical auditory pathway, i.e., the medial geniculate body. However, if only one biologically meaningful association is recoupled to the clicks, the click-evoked responses immediately become widespread, enlarged in amplitude, and are thereafter more difficult to rehabituate (Galambos et al., 1956, cited in: Galambos, 1961; Hernández-Peón et al., 1956; Hernández-Peón, 1960).

Such experiments suggest that brain circuits are more rapidly and extensively modifiable than we ordinarily assume, and that routing of signals depends on which signals have been given importance to a particular individual at a particular time in a particular context. They also show that neurophysiological consequences depend on past experiences along a given pathway and are labile to the introduction of only mildly "biologically meaningful" signals affecting other sensory or motor systems.

Ascending Auditory Pathways: Cochlea to Cortex

Ascending auditory pathways (Fig. 9.13) comprise a succession of neuronal relays for transmission of acoustic information. Cochlear transduction of sound into neuronal events gives rise in the central nervous system to detailed, dynamic spatiotemporal representations and re-representations of complex motions of the cochlear partition. Frequency, duration, time interval, and other time-dependent distributions provide the basis for individuals with normal hearing to discriminate pitch, loudness, localization of sound in space, and to interpret the time-, intensity- and frequency-modulations necessary for subjective interpretations of complex sounds, such as are involved in spoken language.

Even a low intensity pure tone activates a considerable length of the cochlear partition, hence there can never be extremely localized neuronal activation as there can be in vision and touch. Two *just-discriminable tones* excite two extensive and nearly completely overlapping neuronal populations. Such subtle differences are distinguished against background spontaneous activity occurring among almost all auditory neurons. The problem is: how does the nervous system differentiate patterns of activity that differ by so little? Similarly, how can we distinguish two *just-discriminable intensities* that have the same tonal pattern? How can we locate a sound in space that must be distinguished by *just-discriminable differences in intensity and latency* (intensity, because of the sound shadow of the head, and latency, because of differences in time of transit between arrivals of the sound at near and far ears, with sound traveling 340 m/s in air)? Further, how do perceptual processes project the significances of these representations outside the ear into the surrounding three-dimensional space-world of sound. These are all matters of population discrimination in which the group selection theory of Edelman (1987 and 1989) provides a comprehensive theoretical foundation.

COCHLEAR NUCLEI (CN)

First-order afferents from cell bodies in the spiral ganglion of the cochlea bifurcate on their way to the brainstem and deliver impulses to both dorsal and ventral cochlear nuclei (CN) (Fig. 9.13). First-order and second-order neurons maintain a strict cochleotopic (tonotopic) pattern of representations, creating dynamic three-dimensional nuclear "maps" of the cochlear partition. It is characteristic of all sensory relays that they are arranged for extraction and analysis of incoming information. Sensory nuclei are therefore arranged in layers, like cortex. Layering is characteristic for major components of the cochlear nuclei, inferior colliculus, and medial geniculate body, the main ascending auditory relay stations. This layered architecture of sensory nuclei is to be contrasted with motor nuclei which have an architecture like reticular formation and are organized for convergence of information from a variety of sources, and for *synthesis*, as contrasted to the sensory *analysis*, of signals.

In the dorsal CN, tuning curves are very complex, the relationship between intensity and rate of firing is complex, and there is strong lateral inhibition. Lateral inhibition along the entire primary ascending auditory pathway serves to extract signals from noise over a wide range of stimulus intensities. Responses in the two ventral CN maps are closely representative of primary input.

The dorsal CN contributes a "throughway" projection that passes directly and indirectly to the inferior colliculus (IC) (Fig. 9.13). This projection continues, but not exclusively to "core" representation in auditory cortex. The anteroventral and posteroventral CN contribute "byway" projections which pass mainly bilaterally to the trapezoid body, superior olivary complex, and reticular formation. This becomes a polysynaptically relayed ascending auditory pathway that carries along with it much else in addition to acoustic information. This "byway" projection contributes mainly to "belt" cortical representation surrounding the "core" representation in auditory cortex.

The olivary complex is organized tonotopically, somewhat differently in different regions of the complex, as is appropriate for locally distinctive kinds of information analysis (Fig. 9.13). Most olivary units resemble first- and second-order neurons in their response characteristics. Some, however, introduce a new type of signal, an *"off"* response; neurons becoming active or more active only when sound is discontinued. In this, they resemble "off" units in the visual system.

The superior olivary complex integrates binaural stimuli and provides a major ascending auditory relay. It also gives rise to the efferent projections to the cochlea, the olivocochlear bundle (OCB), described above. The SOC transmits acoustic messages bilaterally, preponderantly contralaterally, to nuclei of the lateral lemniscus (LL) and to the inferior colliculus (IC, see Fig. 9.13). It receives second-order inputs from the ventral cochlear nuclei on both sides of the brainstem, and from the contralateral dorsal cochlear nucleus. It also receives important information for comparative timing of events in the two ears, from axons of neurons in the trapezoid body and from collaterals of the CN neurons that serve the trapezoid body.

Since the OCB strongly influences primary and second-order auditory neurons, the SOC participates in unilateral and bilateral braided peripheral and central feedback loops. It accepts trapezoid contributions at the brainstem level, while it is involved with major ascending and descending auditory projections. SOC feedback loops exert both horizontal control (comparable to spinal segmental reflexes) and vertical control (comparable to spinal longitudinal reflexes).

OLIVARY ANALYSIS OF SOUND LOCALIZATION

Everyday experience indicates that the direction of a sound source can be fairly accurately localized in three-dimensional space. The human auditory system can detect very small binaural differences, delays as short as 10^{-5} s under optimal conditions, corresponding to an angle of about 3° between sound source and the midsagittal plane of the head (Schmidt, 1978). The SOC is the first station along the ascending auditory pathways to receive bilateral input. It is not known how the nervous system localizes sound in three-dimensional space, but the SOC provides the first neurophysiological location where information is available for making such determinations. Binaural-beat potentials can be recorded from the SOC but their characteristics do not correspond to perceptual experience, hence it is

inferred that we probably perceive binaural beats at some higher level (Wernick and Starr, 1968). The medial superior olive (part of the olivary complex) has an architecture suitable for localization of sound in space (Morest, 1973).

Cells showing bilateral responses have thresholds and CFs that are almost identical for the two ears. This helps to compare maps from the two ears for differential latencies and intensities. Specialized cells in the lateral part of the SOC respond specifically to latency differences between impulses initiated in the two ears. This provides information not only for subjective localization of sound but also for indexing brainstem reflexes involved in moving the head and eyes toward sound sources. These cells are inhibited by neuronal signals from the contralateral ear. Delay of the contralateral input is minimized by its being relayed through cells in the CN which lie closest to cochlear afferent input. These cells have large axons that cross the midline and relay via neurons in the trapezoid gray which are among the largest in the brain (Fig. 9.13). The CN-trapezoid synaptic junction is also impressive: each CN axon forms a cup-like calyx ending that almost completely surrounds the soma of a principal neuron in the trapezoid body, one-on-one. The total conduction time from the contralateral ear is therefore almost identical with that from the ipsilateral ear, despite the greater distance and the additional synapse.

Among binaurally responsive cells in the SOC, some are excited by stimulation of each ear (excitatory-excitatory units). Other SOC neurons are excited by sound in one ear and inhibited by sound in the opposite ear (excitatory-inhibitory units). Changes in interaural intensity elicit large response changes because each side exaggerates patterns in the other side reciprocally.

Goldberg and Brown (1969) found that many SOC neurons are responsive to differences in time of arrival of tones at the two ears. With low tones, when there is likely to be phase-locking among primary afferents, these SOC units discharge maximally when binaural coincidence is closest, and at progressively lower response rates when binaural latencies are increasingly separated, as when a sound source moves away from the listener's midline. Some delay units contribute to encoding information concerning the location of signal sources that emit several different low-frequency sounds. This is evidence for representation in the central nervous system of a topologically valid "map" of cochlear neuroepithelium within which high-fidelity preservation of exact timing combines to generate new, specific localizing information.

NUCLEUS OF THE TRAPEZOID BODY

This projection system is also tonotopically organized. Each principal neuron receives a CN axonal terminal which provides a single, large calyx terminal. The calyx embraces the trapezoid neuron with axonal expansions like tulip petals covering half of the surface of the recipient cell with synapses. Trapezoid cells also receive additional synapses from collaterals of the same CN neurons that supplied the calyx. Trapezoid nuclei project to olivary ascending relay cells and to cells from which the olivocochlear bundle (OCB) originates.

NUCLEI OF THE LATERAL LEMNISCUS (LL)

The lateral lemniscus (lemniscus = ribbon) is a major dense bundle of ascending auditory projections passing through pons and midbrain (Fig. 9.13). It is analogous to the medial lemniscus which is made up of somesthetic projections ascending from bulb to thalamus. A number of nuclei are scattered along the path of the LL just below the level of the inferior colliculus (IC). Cells in the LL nuclei ascend to the IC or cross the midline to ascend to the IC on the opposite side. They also send terminals to counterpart contralateral LL nuclei. These crossing fibers constitute a commissure (of Probst) for the auditory system (Fig. 9.13).

Inferior Colliculus

The inferior colliculus (IC) provides the main functional link between lower and higher auditory circuits (Fig. 9.13). The IC is an obligatory relay for auditory impulses traveling both up and down, except that auditory cortex can bypass IC in projecting to midbrain and pons, presumably to be able to exercise speedier, nonconsultative control.

The superior and inferior colliculi interchange visual and auditory information and they both project to midbrain mechanisms for arousal—the reticular activating sytem. Both contribute to cerebral-cerebellar servo-control of bodily movements, integrated to be effective in an environment with dynamic sights and sounds. A major auditory commissure crossing between the two ICs enables each to contribute to contralateral auditory cortex. Since the IC already consists largely of crossed projections, this means that, *although there is bilateral representation, auditory functions are predominantly crossed.*

The IC divides conveniently into two main functional regions: a "core," laminated, central nucleus, and a "belt" region that includes pericentral and external nuclei. The central nuclear lamellae are tonotopically "tuned" from low to high frequencies, going from superficial to deep layers. The central nucleus sends core information to a laminated part of the medial geniculate body (MGB) which, in turn, relays to core auditory cortex, bilaterally, with the greater number of fibers ascending on the same side.

Correlations begin to be made with other sensory modalities early in the central projections of all sensory systems. In the case of hearing, the pericentral and external nuclei of IC receive somesthetic information from collaterals of the spinothalamic tracts and medial lemniscus. These same nuclei receive visual signals from the superior colliculus. In turn, they send auditory information to deep layers of the superior colliculus. Information involved in such intersensory correlations is kept separate from core auditory information. The belt projection system of IC combines visual and somesthetic with auditory information and projects to nonlaminated parts of the medial geniculate body which, it turn, project to the belt areas of auditory cortex. This provides separate parallel ascending "auditory" systems for intramodality and intermodality sensory processing.

IC neurons respond to monoaural and binaural stimulation. Cells in the central IC of the cat are spatially segregated according to categories of binaural response characteristics, i.e., excited by both ears, excited by one ear and inhibited by the other, excited by one ear and not by the other, etc. (Semple and Aitkin, 1979). This means that neural information for location of sound in space is further specified at the IC level. The IC receives from auditory cortex bilaterally, from the medial geniculate body, contralateral IC, bilateral LL, and CN projections. IC sends fibers along with the superior colliculus to midbrain reticular formation and periaqueductal gray thereby contributing to acoustic and visual mechanisms of arousal and attention. It also sends projections that relay to the vermis of the cerebellum. These circuits are probably involved in auditorily guided motor performance, from startle responses to responses relating to complex sounds, such as speech and music.

The IC projects to all caudal auditory nuclei, with the exception of cells engaged in the calyx junctions in the trapezoid body. Every part of IC projects to the medial geniculate body. The inferior colliculus is not only rich in distributive and integrative responsibilities, it is exceptionally busy. The inferior colliculus has the highest metabolic rate per unit weight of any region in the nervous system (Sokoloff, 1961).

HUMAN AUDITORY BRAINSTEM-EVOKED POTENTIALS

"Far field" recordings (Jewett and Williston, 1971), permit remote electrode pick-up of very weak potentials which, by averaging, become recognizable signals. Figure 9.14 depicts auditory projections from cochlea to cortex, illustrates auditory evoked potentials, and assigns anatomical sources for these. Brainstem relayed activity takes place within the first 10 ms of the stimuli. Wave I represents auditory nerve activity. Waves of opposite polarity at 2–4 ms are generated in the CN and SOC (Galambos et al., 1959; Erulkar, 1972). Both LL (Webster, 1971) and IC (Kitahata et al., 1969) show later responses (wave V–VI).

Auditory evoked responses (see Callaway et al., 1978; Moore, 1983) help to distinguish and localize peripheral and brain stem interferences along the auditory pathway and to aid in the interpretation of various kinds of coma, drug effects, trauma, and neurological diseases (Starr, 1976). They are also used in assessing the neurological status of premature infants (Schulman-Galambos and Galambos, 1975).

Medial Geniculate Body (MGB)

Core projections are relayed via the laminated portion of the MGB which has a detailed tonotopical cochlear representation. Connections between IC and MGB units are essentially point-to-point between the same best frequencies at each relay (see Fig. 9.13). In belt regions of MGB there is convergence of somatic, vestibular, and auditory input onto single neurons (Blum et al., 1979). A medial division of the MGB provides integrative contributions to all areas of auditory cortex, including core and belt areas, and in addition, to somesthetic (second sensory) cortex.

Auditory Cortex

A primary auditory cortex can be identified in all mammals. Figure 9.15 provides diagrammatic representation of some major nuclear and cortical stations in the cat; the figure illustrates the descending projections, including the efferent projections to the cochlea itself which will be discussed later.

Mammalian auditory cortex receives from the laminated portion of MGB a complete and orderly representation of audible frequencies as monaural and binaural isofrequency strip maps. Isofrequency strips of primary auditory cortex consist of successions of *isofrequency columns,* each of which encodes the same frequency for all units encountered during electrode penetration through the full thickness of cortex (Merzenich and Kaas, 1975). *Binaural interaction columns* exhibit binaural summation or inhibition which occupy alternating bands oriented across the isofrequency strips (Imig

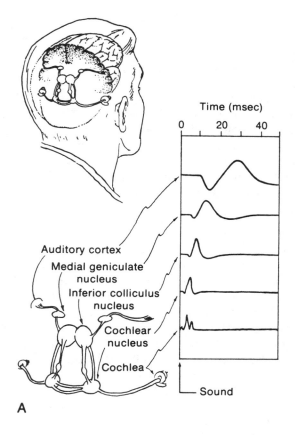

A

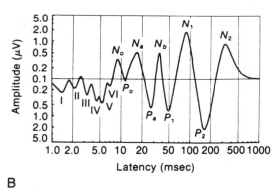

B

Figure 9.14. Schematic representation of central ascending auditory pathways and their event-related potentials (ERPs) from cochlea to cortex in humans. Approximated three-dimensional representation of main auditory pathway in a cutaway diagram of the brain and head. This schema is enlarged beneath, where *arrows* indicate the approximate form of evoked potentials at successively higher anatomical levels. Time scale on *right* indicates roughly the latencies for the early waves reflecting mass activity at the cochlea, cochlear nucleus, inferior colliculus, medial geniculate nucleus, and auditory cortex. *Below* is relative amplitude of summed brainstem waves designated I–VI, followed by a series of wavelets designated by *N* (negative peaks) and *P* (positive deflections). These auditory ERPs have clinical importance because the earlier ones are affected by brainstem lesions and the later ones by brain lesions, anesthesia, and even ongoing psychological processes (Picton et al., 1974). (From Bullock, 1977.)

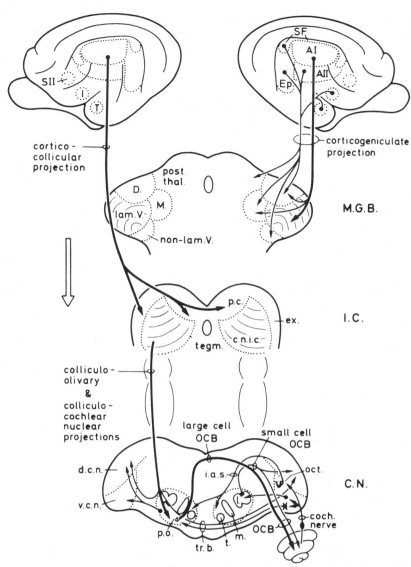

Figure 9.15. Diagram of the descending auditory pathway in the cat. Primary auditory cortex *(AI)*, secondary auditory area *(AII,)* posterior ectosylvian area *(Ep)*, suprasylvian fringe *(SF)*, medial geniculate body *(MGB)*, inferior colliculus *(IC)*, cochlear nuclear complex *(CN)*, olivocochlear bundle *(OCB)*, periolivary body *(p.o.)*, and trapezoid body *(tr.b.)*. (From Brodal, 1981.)

and Adrian, 1977). Within binaural interaction columns, there is a mosaic of *ipsilateral and contralateral dominance columns.* Distributed throughout this area of core representation are corticocortical projections, ipsilateral and contralateral, which represent distinctive analytic fields within primary auditory cortex (Imig and Reale, 1981). The peripheral auditory best cortex includes patterns of ipsilateral and contralateral projections relating to binaural summation and contralateral-dominant suppression.

In the monkey, auditory cortex consists of a central koniocortex surrounded by at least four different belt areas. These belt areas constitute a progres-

sion of connections from primary auditory cortex, each step with reciprocal connections, to cortex of the frontal lobes. This includes the frontal eye fields for the voluntary direction of gaze, and areas further anterior on the convexity that are concerned with forward planning of behavior. This march of progressive auditory associations and information elaboration contributes additional intersensory and sensorimotor patterns that can be increasingly readily converted to commands for behavior.

According to the comprehensive cytoarchitectonic study of Economo and Horn (1930), the auditory field of the human cerebral cortex occupies the *transverse temporal gyrus* (Heschl's gyrus) and a variable amount of the caudally situated *planum temporale,* fields corresponding approximately to Brodmann's areas 41 and 42. Auditory sensations can be produced by electrical stimulation of this region in waking patients (Penfield and Jasper, 1954). Sensations are described as ringing, humming, clicking, buzzing, and are referred mostly to the contralateral ear.

HUMAN HEMISPHERIC DIFFERENCES

Heschl's gyrus and immediately posterior planum temporale are generally larger in the left hemisphere than in the right (Witelson and Pallie, 1973). These gross morphological differences may be seen in computed axial tomography (Galaburda et al., 1978) and magnetic resonance imaging. A functional difference between the two hemispheres is apparent with dichotic listening. Scores for the two ears indicate superior competence for interpreting information from the contralateral ear (tests delivered to both ears through earphones coupled to a dual-channel tape player). Usually the right ear (left hemisphere) reveals a better score for verbal tests, while the left ear (right hemisphere) tests better for recognition of music (Kimura, 1961a; 1964); Broadbent finds that the two hemispheres contribute to different analytic components of complicated tasks that require integrative analysis, rather than functioning independently (Broadbent, 1974).

Descending Auditory Pathways

Paralleling the ascending auditory pathway is a stepwise, descending auditory projection system, from auditory cortex to the cochlear hair cells (Fig. 9.15). Auditory cortex sends three descending tracts: directly to thalamus (MGB), midbrain (IC), and, with a single synaptic relay, to pons (to project to cerebellar vermis). Auditory cortex also send fibers to SC, mainly from auditory belt areas.

From IC (Fig. 9.15), descending fibers project to the superior olivary complex and the cochlear nuclei, bilaterally. Colliculo-olivary projections ensure that the efferent olivocochlear bundle is strongly influenced by descending as well as ascending auditory signals.

CEREBRAL AUDITORY POTENTIALS

Figure 9.14 shows that following wave V-VI, which relates to activity induced in the LL and/or IC, there is a succession of middle latency components, N_o, P_o, N_a, P_a, N_b, which Picton et al. (1974) assigned to thalamus and cortex. Later components, P_1, N_1, P_2, N_2, are thought to be generated by cortex. Auditory-evoked far field potentials allow sampling of neural mechanisms involved in acoustic trans-

mission all the way from cochlea to cortex. They provide a noninvasive means for measuring the integrity of the human auditory path, and for neurological evaluation and diagnosis using objective signals relating to neurophysiolgical processes involved in hearing. Since the late waves are strongly influenced by processes of attention, perception and language, evoked potentials open a neurophysiological window to mental states and mental strategies.

COGNITIVE AUDITORY POTENTIALS

Figure 9.16 shows so-called cognitive potentials, recorded from a human subject listening to a train of clicks. The subject was counting each occasion when one click was omitted from the train. The subject's brain emits a large-amplitude vertex-positive wave peaking at about 300 ms following the time of the expected click, which is a nonevent in the "outside world." Similar potentials called P300 waves, appear when the stimulus train is made up of brief tones at two different frequencies, one presented

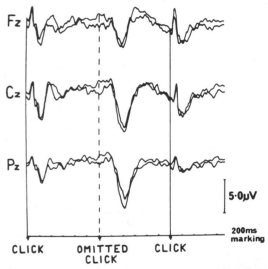

Figure 9.16. Cerebral electrical responses to an expected but non-occurring event. Event-related (in this case nonevent) potentials recorded from vertex and mastoid and averaged over 64 responses in human subjects presented a train of clicks or tones at regular intervals (1.1 s). An occasional member of the train is omitted. Subject anticipates a stimulus at each appropriate interval and is asked to count the number of times the expected stimulus does *not* occur. A large positive wave at 300 ms (P300) occurs regularly in the place of the expected signals. This is obviously a brain event, perhaps a recognition of novelty, or associated with matching expectation while looking for a nonevent, perhaps the resolution of prior uncertainty. Similar electrical events at this 300-ms interval have been recorded in relation to various detection and problem-solving auditory, visual, and tactile experiments. Hence the P300 appears to be related to central activity aroused by the task, regardless of the particular sensory pathway. (From Picton et al., 1974.)

frequently, the other rarely, and the listener counts the number of rare ones. The P300 wave has been associated in this way with a variety of problem-solving experiences by human and animal subjects. Included in this list are nonevents in all three modalities, auditory, visual and tactile, recognition of novelty, non-match to expectation, and resolution of uncertainty.

When a subject encounters a word that "does not make sense" in a sentence, a different cognitive wave, the N400 appears. Discovered by Kutas and Hillyard in 1980, this phenomenon is illustrated in Figure 9.17 where the *top line* shows scalp-recorded responses to the successive words in the four sentences inserted below. The words were flashed on a screen and the event-related potentials they produced are shown superimposed for comparison purposes. The N400 wave, produced by the completely incongruous "dog" terminating one of the sentences appears in the *dotted line* trace. N400s associated with such linguistic (semantic) noncongruities also appear when the words are presented to the ear, or, in deaf subjects, through American Sign Language. In contrast, simple grammatical errors, deviant notes in musical sequences, and substitution of art slides in place of words, and the absence of an anticipated signal all evoke P300s, not N400 waves (Galambos, 1974).

GENERAL AUDITORY FUNCTIONS AND HEARING IMPAIRMENT

A variety of properties of the auditory system have been investigated utilizing psychophysical procedures. These include studies of the way in which the characteristics of a stimulus must be varied for a listener to be able to discriminate between two frequencies (roughly speaking, pitch discrimination) or intensities (roughly speaking, loudness) or the comparative loudness of tones of different frequencies. This method can also be used to show changes in the auditory system during development and aging. These approaches test the auditory system as a whole and depend on the provision of simple stimuli, standard conditions, and standard instructions. Despite the simplicity and uniformity of test stimuli, the responses are highly individualistic.

Long-sustained motivation and training can make a difference. For example, there is evidence that blind people can acquire superior hearing (Niemyer and Starlinger, 1981). The authors compared 18 chronically blind subjects and 18 normally sighted subjects with normal hearing. The blind showed disinhibition of auditory signal interpretations when exposed simultaneously to broad-band noise. They also showed decrease inhibition with the same intervening stimulus delivered to the contralateral ear. Blind subjects had better speech discrimination, especially with sentence comprehension tests, both with and without competing noise. Especially interesting is the fact that the blind subjects had a significantly shorter latency for N_1: 140 ms, instead of 160 ms (p < 0.0002). Acoustic training can speed up as well as improve discrimination.

THRESHOLDS

The least sound power that the ear can detect at any frequency is necessarily a statistical problem. Variable results are observed in different subjects, and from time to time in the same subject. Nonetheless, minimum sound pressure levels have been

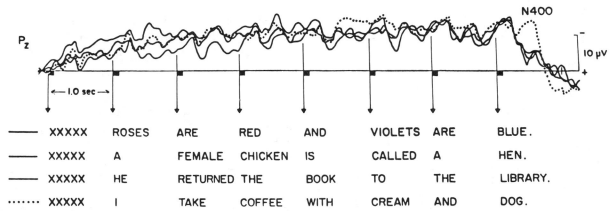

——	XXXXX	ROSES	ARE	RED	AND	VIOLETS	ARE	BLUE.
——	XXXXX	A	FEMALE	CHICKEN	IS	CALLED	A	HEN.
——	XXXXX	HE	RETURNED	THE	BOOK	TO	THE	LIBRARY.
⋯⋯⋯	XXXXX	I	TAKE	COFFEE	WITH	CREAM	AND	DOG.

Figure 9.17. Event-related potentials (ERPs) recorded throughout the presentation of 7-word sentences, with words flashed individually at 1/s. Each tracing is the averaged ERP across 40 sentences in which the last word was either appropriate to the context *(solid lines)* or semantically inappropriate *(dotted lines)*. Samples of the 160 different sentences that were presented to this subject are shown below. The "semantic mismatches" occurred in 25% of the sentences and elicited a prominent N400 component (Kutas and Hillyard, 1980). (From Galambos and Hillyard, 1981.)

selected to serve as standards representative of a population of normal listeners. Two functions which represent the minimum thresholds are illustrated in Figure 9.18, including the international audiometric zone curve which serves as the standard.

The nature of frequency discrimination can be examined from the level of threshold for detection of sound at a given frequency up to the maximum comfortable level of stimulation by matching the loudness at different levels over the entire frequency spectrum. There is notable flattening of the frequency response with increases in level of intensity (loudness), particularly at the low frequency end of the spectrum.

MASKING

Background sound serves to mask signals. In general masking is more effective the closer the frequency of the masking tone is to that of the signal, although this relation can be complicated by coincidences of phase, producing "beats." Low-frequency tones mask high-frequency tones more effectively than the converse. There is relatively little masking effect when the masking tone is presented to one ear and a test tone to the other. Thus, the two channels are not critically interfering with one another. A fairly broad spectrum noise serves as a useful mask for a wide range of test frequencies. This fact can be utilized by suppressing the better ear when one tests an ear in which hearing impairment is suspected.

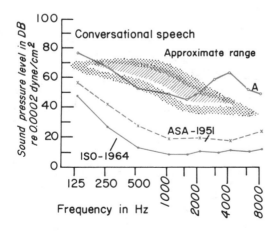

Figure 9.18. The International Audiometric Zone *(ISO-1964)* curve compared with the American Standards Association *(ASA)* 1951 curve. The ISO curve represents a better approximation of the thresholds of normal young adults than does the ASA 1951 curve. The *stippled area* indicates the pressure-frequency region in which most conversational speech sounds occur. Curve *A* represents thresholds typical of a case of moderate hearing impairment. (From Davis, 1965b.)

Types of Hearing Impairment

The causes of deafness can be considered in five groups (Davis, 1951). Hearing difficulties result from interference with conduction of sound to the neuroepithelium, *conduction deafness.* Another group is associated with damage to cochlear mechanisms or with the auditory nerve, *nerve deafness.* Another three groups include *central deafness; diplacusis,* a false sense of pitch; and, *tinnitus,* a ringing or noise "in the ears," thought to be caused by hypersensitivity of the hair cells or the peripheral (afferent or efferent) nerves.

CONDUCTION DEAFNESS

Conduction deafness may be caused by anything that interferes with getting the stimulus to the neuroepithelium, water or wax in the external ear canal, perforation or hardening of the tympanic membrane, or loss of mobility or destruction of the ossicular chain. In *otosclerosis,* described earlier, abnormal growth of bone anchors the footplate of the stapes so that it cannot move the oval window. Initial effects are progressive losses in hearing for low frequencies.

Acute illnesses may result in hearing impairment. If the eustachian tube is blocked for any reason, the balance of pressure between the external meatus and the middle ear is upset. Rupture of the tympanic membrane does not occur until the pressure differential reaches the range of 100–500 mm Hg. Rupture may occur in a diver executing rapid descent into deep water. If the eustachian tube remains closed, air trapped in the middle ear is gradually absorbed, with consequent lowering of partial pressure. If the pressure falls below hydrostatic pressure in the capillary bed, secretion of fluid into the middle ear occurs. Accumulated secretions or fluid resulting from infection, bleeding, or other problems increases ossicular friction which interferes with impedance matching and sound conduction.

NERVE DEAFNESS

The most common cause of sensorineural impairment is exposure to high intensity sound. After exposure to loud sound or noise, there is a temporary shift in hearing threshold. This can affect neural coding and it may result in perceptual disabilities following even short intervals of loud acoustic exposure, even though sounds can still be readily heard. Spontaneously active units may be inactivated and units that normally respond only to the "onset" of stimuli may become continuously active (Willott and Lu, 1982). If sound is loud enough, exposure long enough, or repeated often, the elevation of threshold may be permanent. Overexposure can lead to local damage of the cochlear partition and permanent hearing loss in a limited frequency range. There is special vulnerability in the vicinity of 4000 Hz. When the elevation of hearing threshold is on the order of 40–60 dB, *tinnitus* attributed to spontaneous discharge of hair cells complicates partial nerve deafness. Following injurious exposure to loud sound or noise, deafness is usually bilateral. When sensorineural impairment results from infection, deafness is usually unilateral. Ménière's syndrome, resulting from irritation of neuroepithelia in both labyrinth and cochlea, is accompanied by hearing loss in the low frequencies and sometimes by violent episodes of vertigo.

Hearing shows progressive degradation with increasing age (Fig. 9.19), particularly for the higher frequencies. Fortunately, discrimination of speech depends primarily on frequencies between

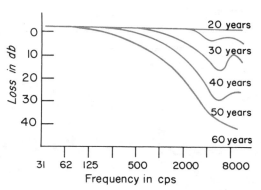

Figure 9.19. Progressive loss of sensitivity at high frequencies with increasing age. The normal audiogram at 20 years of age serves as a reference for the losses represented at greater ages. (From Licklider, 1951.)

about 500 and 3–5 kHz, and the characteristic loss of hearing at high frequencies with increasing age causes minimal difficulty in most circumstances (French and Steinberg, 1947).

CHAPTER 64

Vestibular Functions

Vestibular functions are provided by a generally unobtrusive system of sensorimotor controls which operates reflexly to stabilize vision and coordinate balance and motion of head, trunk, and limbs. Stability of the visual world requires rapid reflexive integration of information from the vestibular system to control the position of the eyes. The vestibular system provides continual inertial guidance to resist gravity and avoid falling down during risky maneuvers (e.g., jumping with ice skates). Vestibular information is essential for moving in three-dimensional space (e.g., scuba diving) to maintain an intended course and orientation. The vestibular system works effectively to coordinate such behavior even in sustained zero-gravity. The biological advantages of vestibular function are affirmed by the fact that vestibular systems are similar in design throughout vertebrate history and that there are analogous inertial guidance systems in most invertebrates.

PERIPHERAL COMPONENTS OF THE VESTIBULAR SYSTEM

Anatomy and Physiology of the Vestibuloacoustic Apparatus

The vestibuloacoustic apparatus consists of semicircular canals, a vestibule, and a spiral cochlea (Fig. 9.20). Two such apparatuses on each side of the head, encased in temporal bone, are mirror images of one another (Fig. 9.21). Each is equipped with six independent patches of neuroepithelium, and each is served by afferent and efferent axons in the VIIIth cranial nerve. Each neuroepithelium is located at a strategic mechanohydraulic vantage point. Central representations of neural activities elicited by hair cell receptors in these neuroepithelia detect and measure linear and angular accelerations in three-dimensional space and analyze the acoustic environment.

The semicircular canals consist of three toroidal tubular arcs fixed at right angles to one another in three orthogonal planes. They are attached to the posterior end of the vestibule (Fig. 9.20). Their three neuroepithelia, in combination, can detect and measure angular accelerations around any axis in three-dimensional space. Two additional neuroepithelia, in the utricle and the saccule, respond to various linear accelerations and vibrations. The acoustic part, the spiral cochlea, attached to the anterior end of the vestibule, has neuroepithelium on the basilar membrane to detect sound.

Each of the six neuroepithelia in the vestibuloacoustic apparatus depends on mechanoelectric transduction by *hair cells*. Each hair cell has synaptic contact with nerve endings of *afferent axons* which go to the brainstem (Fig. 9.22, NC and NE_1). Many of these afferent fiber endings, and hair cell receptors are influenced by *efferent axon terminals* (NE_2).

Sensory data from the vestibuloacoustic apparatus are separately processed for vestibular and acoustic functions. Signals from the semicircular canals and maculae are integrated reciprocally in a functionally reinforcing way from the two sides of the head. They provide a powerful reflex system for maintaining equilibrium, posture, and visual stability. Hearing is dealt with in Chapter 62.

SEMICIRCULAR CANALS

Each semicircular canal is maximally stimulated when the plane of its circumference is at right angles to the plane of rotation. Each canal has an enlargement, an *ampulla*, which contains a semielastic pendular hillock of specialized neuroepithelium, a *crista*, covered by a gelatinous cupula (Fig. 9.23). By inertial displacement of fluid in the canals this deformable barricade in the ampulla moves relative to the wall of the ampulla. This distorts hair cell processes in the crista and determines neuronal signaling to the brainstem. The ampullae of the horizontal and superior (anterior) canals are close together anteriorly where they attach to the vestibule. The ampulla of the posterior canal attaches to

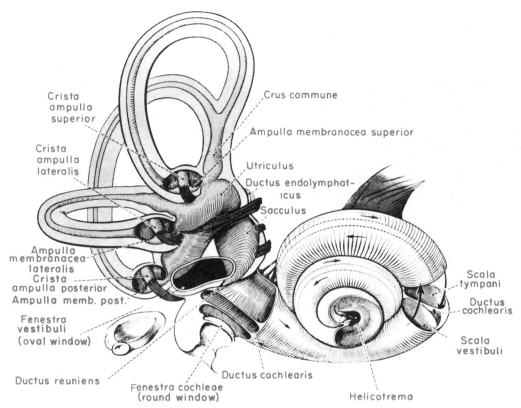

Figure 9.20. A schematic illustration of relations among the semicircular canals, utricle, saccule, and the cochlear duct on the right side. The membranous semicircular canals are shown within their surrounding bony structure. Nerve connections into the ampullae, utricle, and saccule are shown. (From Melloni, 1957.)

the vestibule posteriorly. The two vertical canals, superior and posterior, share a common limb (crus) which also attaches to the vestibule (Fig. 9.20).

THE UTRICLE AND SACCULE

The vestibule contains two additional neuroepithelia: in the utricle, and, on a stalk from the utricle, in the saccule. Each neuroepithelium is sensitive to linear displacements in determined directions (Fig. 9.24). Stimulation is provided by *otoliths*, "ear stones," which have a higher specific gravity than neighboring fluids and tissues and which, therefore, exhibit inertial delay during motion of the head.

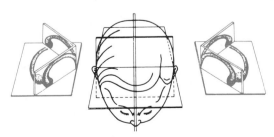

Figure 9.21. Relations of the semicircular canals to the median sagittal, transverse frontal, and horizontal planes of the head.

The utricle is placed approximately horizontally and the saccule approximately sagittally, so that the four neuroepithelia in the two ears monitor gravity and detect and measure other linear accelerations.

Physiology of the Semicircular Canals

The semicircular canals are filled with *endolymph*, rich in potassium and low in sodium. Endolymph is confined by a continuous membrane which isolates endolymph from surrounding *perilymph*. Perilymph is bounded by periosteum and connective tissue lining the tubular labyrinth inside the temporal bone. Perilymph is in communication with cerebrospinal fluid via a narrow duct. Its ionic composition is essentially the same as that of CSF. Analogous to cerebrospinal fluid in the central nervous system, perilymph provides the extracellular fluid for tissues of the inner ear. The only exception is the *stria vascularis*, in the cochlea, which is bathed in endolymph. The stria provides a steady electrical potential which greatly enhances the sensitivity of hair cells along the basilar membrane (see Chapter 62, Hearing). Because of differences in ionic composition, in endolymph in the canals which bathes

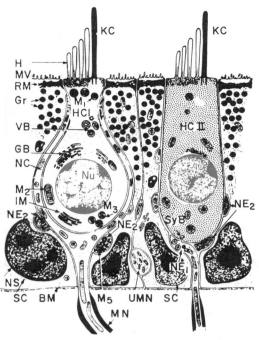

Figure 9.22. Two types of vestibular sensory cells. The flask-shaped type I cell *(HCI)* is surrounded by a nerve calyx *(NC)*. Nerve endings presumed to be efferent *(NE₂)* make contact with the nerve calyx and directly with the type II cylindrical sense cell *(HC II)*. Type II cells are supplied by both NE₁ and NE₂ types of nerve endings. Several types of mitochondria (M₁ to M₅) are found in the sensory cells and neural elements. Additional elements identified in the figure include kinocilia *(KC)*, stereocillia or hairs *(H)*, microvilli *(MV)*, granules *(Gr)*, cell nuclei *(Nu)*, supporting cells *(SC)*, unmyelinated nerve fibers *(UMN)*, myelinated fibers *(MN)*, basement membrane *(BM)*, Golgi complex *(GB)*, synaptic bar *(SyB)*, rerticular membranes *(RM)*, intracellular membrane *(IM)*, vesiculated body *(VB)*, and nucleus of supporting cell *(NS)*. (From Engström et al., 1965b.)

the exposed surface of hair cells, and in perilymph bathing the unexposed surfaces of hair cells, the hair cell responsivity is nearly doubled.

Each semicircular canal constitutes a circular "mill race" in one plane, and the three planes are at right angles to one another (Fig. 9.21). The crista and cupula yield, passively and dynamically, to slight rotational displacements of fluid in the canal. They extend from side to side and floor to ceiling of the ampulla, presenting a deformable barricade that permits endolymph to shift around the arc of the canal, with bending of the crista and cupula. Dye introduced into the canal on one side of this barricade does not escape to the other side.

Each hair cell has from 60–100 stereocilia which emerge from its exposed surface. A single *kinocilium* is the longest of the cilia. Bending cilia in the direction of the kinocilium, by as little as 100 picometers (trillionths of a meter, equivalent to atomic diameters), results in *depolarization* of the

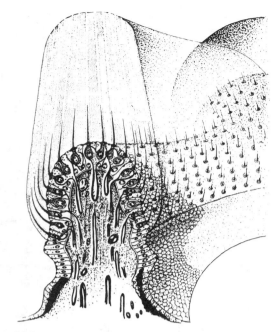

Figure 9.23. Illustration of the crista ampullaris. (From Wersäll, 1956).

hair cell. Depolarization of the hair cell leads to release of neurotransmitter and thereby to increased activity in the corresponding afferent nerve terminal. Bending of the stereocilia in the opposite direction results in *hyperpolarization* which, by reducing the spontaneous rate of release of neurotransmitter, leads to reduction of afferent nerve activity. The kinocilium is not essential to this receptor response, the stereocilia being sufficient (Hudspeth and Jacobs, 1979).

Whenever head movements are such that endolymph, by virtue of its inertial mass, moves relative to the membranous wall of the canals, there is excitation in one of a pair of canals, and inhibition in the other. Rotation of the head to the right around a vertical axis, with the head tilted forward about 30°, causes bending of the cupula in the right horizontal canal toward the kinocilium, resulting in excitation. The horizontal canal on the left side is inhibited because its cupula is bent oppositely. The resting level of activity of afferent elements in the system increases due to receptor excitation and decreases due to receptor inhibition, conveying afferent nerve patterns of precise spatial information. The pattern of reciprocal excitation and inhibition among the six canals is unique for any given axis of rotation.

If for any reason the function of one canal is lost, the respective paired canal apparently furnishes enough information for central integrative processes in the brainstem to compensate for that loss.

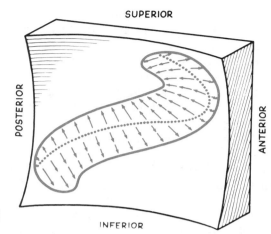

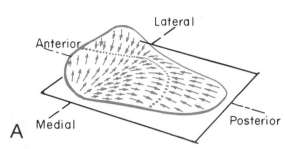

Figure 9.24 *A*, The pattern of polarization of sensory cells in the utricular macula of the guinea pig. *Arrows* indicate the direction of bending of stereocilia for excitation. *B*, The polarization of sensory cells in the saccular macula of the guinea pig. Note that although the direction for depolarization is reversed in the region of the middle of the structure as was true for the utricle, the direction of bending for excitation is reversed as contrasted with that shown in *A*. (From Spoendlin, 1965.)

The superior canal on one side and the posterior canal on the other side function as a matched pair; whenever one is excited, the other is inhibited. In both pairs of vertical canals, rotations that result in bending the cupula away from the utricle (i.e., toward the kinocilium) are excitatory.

Excitation and inhibition of hair cells begins with onset of angular acceleration. If a constant angular velocity is maintained, responses from the canals may continue for 20 seconds or more following start-up acceleration and then gradually return to resting level activity. This is because there is elastic recovery from deformation of the crista and cupula which thereby return to their original position. When deceleration occurs, the inertial mass of the endolymph imposes a reverse deformation on the crista and cupula whereupon hair cell deformation is reversed and neural responses are reduced below the resting level of activity for about 20 seconds.

HAIR CELL RECEPTOR PHYSIOLOGY

Furukawa and Ishii recorded from single first-order afferent fibers which innervate the macula of the saccule (Furukawa and Ishii, 1967a, 1967b). These primary sensory axons showed spontaneous activity. The otolith membrane was trimmed to overlie a small region of the macula and was moved by means of a piezoelectric probe. Hair cells in the saccule project their stereocilia into the gelatinous undersurface of the otolith membrane. When this was displaced in one direction, the nerve fiber discharged; displacement in the opposite direction was without effect.

The flask-shaped hair cells are nearly completely surrounded by a calyx-type afferent nerve ending, presumably for better reliability in receptor-neuron transmission. The *apical surfaces* of both flask-shaped and cuboidal receptor cells contain stereocilia which face into the membranous labyrinth, where they are bathed by *endolymph*. Plasma membrane covers each of the stereocilia, like multifingered gloves, providing a greatly expanded membrane for ion exchange. The main bulk of the hair cells, the *basal surfaces*, are surrounded by supporting cells which are bathed by *perilymph*, in the same way that central nervous system cells are bathed by cerebrospinal fluid (Hudspeth, 1983). Tight junctions form a collar around the apical end of the hair cells, thereby preventing the two fluids from mingling.

Displacement of the kinocilium and stereocilia in the direction of the asymmetrically placed kinocilium is believed to result in opening ion channels in the apical plasma membrane. The high concentration of potassium in endolymph obliges a rapid entry of potassium ions across that membrane. Thus, the membrane is more readily *depolarized* (e.g., going from -60 mV to -40 mV).

Depolarization alters the receptor so that it immediately admits calcium ions in the basal region of the cell, in the same way that a terminal axon admits calcium ions on arrival of an action potential. The effect of calcium ions in the presynaptic region of the receptor is to release neurotransmitter molecules into the synaptic cleft. This activates the afferent nerve ending.

When receptor cell stereocilia are displaced in the opposite direction, away from the direction of the kinocilium, some of the spontaneously (steady-state) open ionic channels are closed. As a result of continuing action of the cell membrane to pump potassium ions out of the cell, the membrane potential becomes *hyperpolarized* (e.g., going from −60 mV to −65 mV). Hyperpolarization reduces calcium ion influx into the base of the hair cell and thereby reduces the amount of steady-state neurotransmitter released. This reduces spontaneous activity among primary afferent neurons.

Several aspects of hair cell receptor function deserve emphasis: The amount of bending of the stereocilia to be effective is very small, about 1 μm distance, bending the stereocilia about 3°. The more distal parts of the cilia are more responsive. A small shift in membrane potential has large amplifying effects. Activation of the receptor is *probabilistic* rather than requiring overcoming some kind of threshold. This enables discrimination of extremely small displacements. The process is very rapid; cell membrane potentials begin to respond to movement of stereocilia within a few tens of microseconds. Further, the stereocilia adapt very quickly. If they are displaced by one micrometer, within a few tenths of a second the membrane potential has been almost completely restored to its status prior to displacement. This allows the same cell to be nearly maximally excitable when subsequent displacements occur, even though the previous displacement may be continued.

FLUID DYNAMICS IN ANGULAR ACCELERATION

Endolymph provides inertial mass during angular acceleration of the head. Because each canal lies in a single plane, endolymph moves relative to the walls of a given canal only when motion takes place around an axis that is nearly at right angles to the plane of the canal. When rotation of the semicircular canal ceases, the endolymph continues rotation, resulting in a reversal of the receptor membrane potential and hence in a corresponding shift in rate of release of neurotransmitter. There are physiological consequences during start-up but not during continued acceleration, and reversed physiological consequences during deceleration.

Neurophysiology of the Vestibuloacoustic Apparatus

INNERVATION OF THE VESTIBULOACOUSTIC APPARATUS

Rasmussen (1960) found an average of 18,500 nerve fibers in human vestibular nerves, including efferent as well as afferent axons. Judging from the ratios of efferent to afferent fibers in other species, a conservative guess is that betwen 1 and 2% of the total, say 300 fibers, may be efferent. The effects of vestibular efferent activity are likely to be diffuse and relatively nonspecific.

AFFERENT INNERVATION OF THE VESTIBULAR APPARATUS

Figure 9.22 illustrates schematically the flask-shaped and cuboidal types of hair cells that populate the neuroepithelium of the vestibuloacoustic apparatus. Individual receptor cells provide a wide range of mechanical sensitivities throughout both cristae and maculae. The two-choice direction of movement of endolymphatic fluid in each canal, with respect to orientation of the receptors in each crista (Fig. 9.22), dictates whether a given angular acceleration will be excitatory or inhibitory. The systematic distribution of differently oriented receptors in the macular neuroepithelium (see Fig. 9.24) provides a basis for mapping the direction of effective mechanical forces, static, linear, and rotary, which act across the two macular surfaces, each of which approximates a single plane (horizontal in the utricle and sagittal in the saccule).

Figure 9.22 shows afferent (NE_1) and efferent (NE_2) axon terminals which innervate both flask-shaped and cuboidal receptor hair cells. The terminal *afferent* axons form either calyx-type endings (NC) which embrace the flask-shaped hair cell bodies (to the left), or relatively bulbous bouton endings which are applied to the basal regions of the cuboidal receptors (NE_1). Receptor-axonal synapses provide central nervous system input (afferent) from and output (efferent) for each of the five neuroepithelia in the vestibular portion of the inner ear. Both afferent and efferent fibers travel in the VIIIth cranial nerve. Vestibular afferent nerve cell bodies are located in the *vestibular ganglion*. Their peripherally directed branches innervate hair cells, their centrally directed branches innervate second-order neurons in the brainstem.

The spontaneous activity of primary afferent neurons in the semicircular canals averages around 90 impulses/s in the monkey. This is another instance where a sense organ modulates ongoing activity rather than depending on thresholding. This provides very subtle time and motion information to the central nervous system. Afferents from the utricular macula show distinctive patterns of firing (tonic, phasic-tonic, phasic) associated with distinctive head positions and motions (Precht, 1979). Different units convey magnitude and rate of gravitational force change, linear acceleration of head position, and direction of head motion.

EFFERENT INNERVATION OF THE VESTIBULAR APPARATUS

Vestibular efferent components of the vestibular nerve (Fig. 9.23, [NE_2]), arise from three cell groups. One group originates in the ipsilateral lateral vestibular nucleus. The other two arise bilaterally from cells in the reticular formation near the VIth cranial nerve, in a "center for lateral gaze" (Gacek and Lyon, 1974; Warr, 1975). Vestibular efferents innervate all six neuroepithelia of the vestibuloacoustic apparatus (Smith and Rasmussen, 1967).

Efferent axons terminate in boutons that contain abundant synaptic vesicles (NE_2 in Fig. 9.22). These show morphological characteristics of presynaptic terminals and degenerate when the vestibular nerve is cut close to the brainstem. (Primary afferents do not degenerate following such transections because their cell bodies lie distal to the nerve section which therefore interrupts only the central portion of their connections.) Some efferents terminate on the outside of the afferent calices surrounding the flask-shaped receptors. There they are in a position to modify primary afferent responses. Since the primary afferents are spontaneously active, efferent effects can be very subtle. Other efferent axons attach directly to the base and sides of the cuboidal receptors, thus presumably directly inhibiting the receptors. This is a form of *presynaptic inhibition.* The receptors show appropriate postsynaptic morphology at the sites of such efferent junctions. Vestibular efferents release acetylcholine, and its effects are evidently exclusively inhibitory.

Vestibular afferents from each of the cristae of the semicircular canals deliver information that is comparable to that of a "bidirectional angular accelerometer." Afferents from the utricle and saccule contribute information comparable to that of a "multidirectional linear accelerator" operating in each of two major planes. The combination of information from angular and linear accelerators is equivalent to an "inertial guidance system." Each ear thus has its own complete inertial guidance system. The two systems integratively reinforce one another in the central nervous system. The reliability of information from this combination is excellent.

CENTRAL PROJECTIONS OF THE VESTIBULAR SYSTEM

Central processes of the bipolar sensory cells of the VIIIth cranial nerve, with cell bodies located in the vestibular ganglion, divide into ascending and descending branches as they enter the brainstem. They innervate the *four main vestibular nuclei,* the *floccular lobe of the cerebellum,* and the *reticular formation* in the "center for lateral gaze." The four main vestibular nuclei are: *superior, lateral, medial,* and *descending (Fig. 9.25).* Ascending branches of the incoming afferent fibers pass to the superior vestibular nucleus and also directly to the

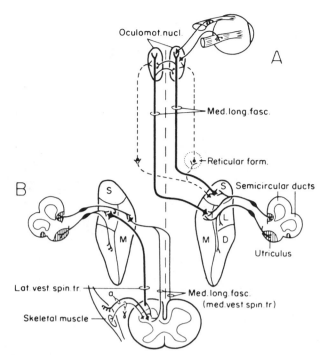

Figure 9.25. Simplified diagrams of main features in the organization of primary vestibular circuits to the vestibular nuclei and from there the ascending (*A*) and descending (*B*) projections to eye motor and spinal motor controls. *D, L, M,* and *S:* Descending, lateral, medial, and superior vestibular nuclei. (From Brodal, 1981.)

cerebellum, the only first-order sensory neurons to have direct access to the cerebellum. Descending branches of the incoming afferent axons pass to the medial and descending vestibular nuclei.

The vestibular nuclei project to the *spinal cord, cerebellum, reticular formation, and superior colliculus* (see Figs. 9.25–9.28). There is a relatively private system of projections from the lateral vestibular nucleus to the lateral reticular nucleus which, in turn, projects to the cerebellum. Roughly half of the neurons in the vestibular nuclei generate reflex control of the eye movements in relation to motion of the head in three-dimensional space in respect to the gravitational field. The other half generates reflex controls for movements of the head, trunk, and limbs on a dynamic center of gravity of the whole body moving in three-dimensional space. The result is a very powerful vestibulo-centric balance system which provides controlled coordination of eyes-in-head, head-on-neck, neck-on-trunk, and limbs-on-trunk, all in reference to accelerations in a gravitational field.

MEDIAL LONGITUDINAL FASCICULUS (MLF)

The MLF contributes to vestibular-directed head and eye motor coordination. This, in turn, is essential for an individual's perception of a stable world

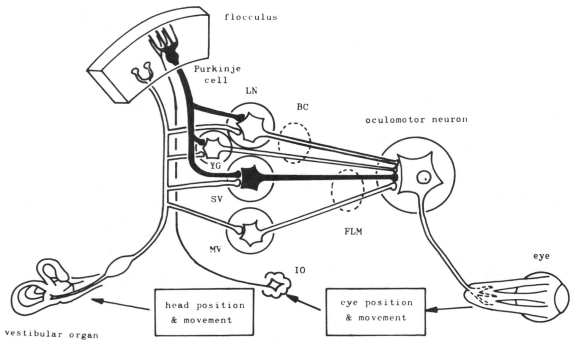

Figure 9.26. Neuronal connections between the vestibuloocular reflex and cerebellar cortex. Inhibitory neurons are filled in *black;* excitatory neurons are indicated by hollow structures. The vestibular branch of the VIIIth nerve has direct projections via mossy fibers to vestibulocerebellum, the flocculus, as well as to vestibular nuclei. Vestibuloocular three-neuron reflex (VOR) begins with signals from the semicircular canals and otolith organs, which are relayed by second order neurons in the vestibular nuclei, and project to the eye motor nuclei. Inhibitory action is largely relayed by the superior vestibular nucleus *(SV)* and excitatory action by the medial vestibular nucleus *(MV)*. Both of these nuclear groups are under inhibitory influences of the cerebellum. Y group of vestibular nuclear complex *(YG)* and lateral vestibular nucleus *(LN)* receive from both VIIIth nerve excitation and cerebellar inhibition. The cerebellum appears to improve a feedforward control of eye movements in response to vestibular stimulation. This establishes a two-way relationship between the VOR and the flocculus. *IO,* inferior olive; *BC,* brachium conjunctivum; *FLM,* medial longitudinal fasciculus. (From Ito, 1974.)

outside the dynamic self moving in three-dimensional space. Head and eye coordination is indispensable to stabilization of vision with respect to passive and active head movements and movements of the environment, and to visual grasping of targets of interest. Ascending axons from medial and superior vestibular nuclei (Fig. 9.25) pass upward in the *medial longitudinal fasciculus (MLF)* to the oculomotor nuclei (IIIrd, IVth, and VIth cranial nerve nuclei) in the midbrain. The MLF extends from the level of nuclei of the IIIrd cranial nerve to upper thoracic spinal levels. The majority of fibers in the medial longitudinal fasciculus come from the vestibular nuclei. The MLF interconnects the three pairs of eye motor nuclei which innervate the six extraocular muscles that direct each globe. The MLF also interconnects the vestibular-ocular system with motor nuclei that serve cervical and upper thoracic motor nuclei, thereby controlling turning of the head in coordination with the direction of gaze.

Additional ascending and descending vestibular contributions, outside the MLF, serve skeletal muscle responses controlled by the vestibular system all along the entire longitudinal column of brainstem and spinal motor nuclei. The more caudal parts of the vestibular nuclei projecting to the spinal cord are shown to the left (*B*) in Brodal's diagram (Fig. 9.25). These adjust the body to gravity and other static and acceleratory forces in accordance with vestibular integrative analyses (Gernandt et al., 1957; 1959). Lateral (L), descending (D), and medial (M) vestibular nuclei (Fig. 9.25) contribute descending axons via the lateral vestibulospinal tract, medial vestibulospinal tract, as well as the medial longitudinal fasciculus (MLF).

The vestibular nuclei send axons and axon collaterals to widespread regions in the brainstem reticular formation. The reticular formation reciprocates by sending axons to each of the vestibular nuclei. The reticular formation provides input to vestibular balance and motor control systems in reference to the direction of attention and level of arousal plus integrated information relating to important events taking place elsewhere in the nervous system.

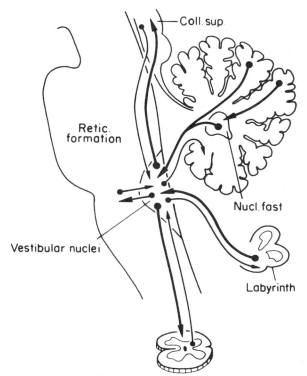

Figure 9.27. Note that the vestibular nuclei have reciprocal connections with the spinal cord, the labyrinth, the cerebellum, the reticular formation, and the superior colliculus. The connections indicated by *heavy lines* are more massive than those shown as *thin lines*. It can be seen that the major afferent input to the vestibular nuclei comes from the labyrinth and cerebellum, while their main output goes to the spinal cord and mesencephalon (mainly the eye motor nuclei). (From Brodal, 1981.)

Vestibular Reflexes

The eyes float in a moving head. Since the eyes are cushioned in soft tissues within the orbit that are of approximately the same specific gravity, rapid eye movements can be generated with little force. The visual system by itself has no way of detecting the basis for shifts in retinal patterns. If the eyes were fixed in a moving head there would be dizzying sensations of motion of the visual scene. Thus, head and body movements introduce complicated mechanical problems for perceiving a stable environment. These problems are solved by providing reflex eye movements that are opposite in direction, equal in displacement, and equivalent in rate for every active or passive motion of the head. This intricate physiological problem is solved by the vestibulo-ocular reflex.

VESTIBULO-OCULAR REFLEX (VOR)

The VOR is a high-speed, accurate, consistent reflex, almost machine-like in its performance.

There is only a 12-ms delay from head movement to eye motor response. Over a frequency range of 0.1–1.0 Hz, the gain of the monkey's VOR, peak eye velocity divided by peak head velocity, is between 0.9 and 1.0. Stable retinal images are provided almost exclusively by the VOR; other mechanisms such as the optokinetic reflex and target pursuit reflexes contribute minimally (Miles and Lisberger, 1981; Lisberger and Pavelko, 1988). Because execution of the VOR does not depend on visual feedback, it operates equally well with low illumination and in the dark.

When the upright head is rotated to the *left*, the VOR compensates by promptly rotating the eyes to the *right* at a corresponding rate. This compensatory eye movement stabilizes the retinal image. The VOR system thus operates to prevent head movements from disturbing retinal image stability. This compensation allows the gaze to fix on the nonmoving surround even though the head is moving, a compensation that succeeds for about 60° of head rotation. If head rotation is greater than that, the eye motor control system introduces a saccadic eye movement so that the eyes move rapidly to a new point in the visual scene. This gives rise to a rhythmic sequence of alternating slow and rapid eye movements known as *nystagmus*. The direction of nystagmus movements is designated according to the direction of the slow movement. Thus, rotation of the upright head to the left leads to nystagmus to the right.

Input to the vestibulo-ocular reflex, induced by head movements, starts in neuroepithelium of the vestibular apparatus. Primary vestibular impulses provide information that defines all motions of the head. The vestibular nuclei, with assistance from the cerebellum and reticular formation, generate signals to the oculomotor apparatus which activate the extraocular muscles and provide the compensatory eye movements (Fig. 9.26). The shortest VOR path involves two central synapses. There are receptors and effectors and three neurons. The first-order neuron delivers vestibular sense data to vestibular, reticular, and cerebellar neurons. Second-order neurons in the vestibular nuclei and reticular formation send messages to the oculomotor nuclei (IIIrd, IVth, and VIth cranial nerve nuclei). Third-order neurons deliver commands to the extraocular muscles. Relative directness of the VOR pathway accounts for the short latency for the response (12 ms) and there is no time for longer loops in the circuit to have access to higher-order neurons in this reflex action.

During head movements, the VOR must be respon-

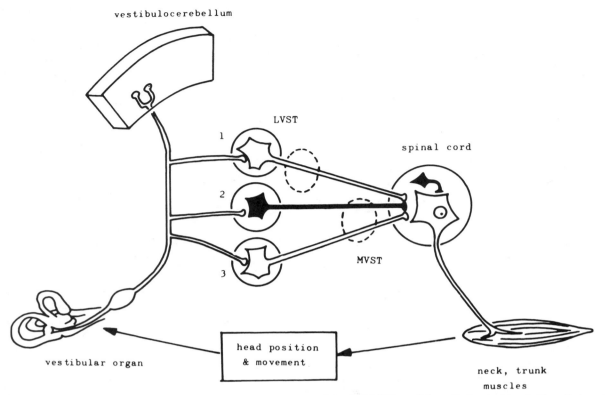

Figure 9.28. Neuronal organization of the three-neuron vestibulospinal reflex arc composed of primary vestibular afferents, secondary vestibular neurons, and spinal motor neurons. *LVST,* lateral vestibulospinal tract; *MVST,* medial vestibulospinal tract. The whole column of motor nuclei the length of the brainstem and spinal cord is under powerful control by vestibulospinal reflexes. (From Ito, 1974.)

sive, accurate, and rapid to maintain a stable perception of the visual scene. The VOR is primarily responsible for our being able to organize visual perceptions in a useful way while we are in motion. Even when both target pursuit and target-following optokinetic systems operate in concert, latency for their responses is of the order of 80 ms, hence reliance on visual feedback would greatly delay visual stability in the course of head movements (see Miles and Lisberger, 1981; Lisberger and Pavelko, 1988).

The VOR operates, as do head and eye movements, mainly in the horizontal plane. It is an *open-loop control system* because neither the eye motor events and nor the retinal images themselves influence the vestibular receptors which initiate the actions. The *optokinetic reflex* system operates on the basis of visual feedback from objects moving across the visual field. It requires about 80 ms between movement of objects and the compensatory "catch-up" saccadic eye movement. The *visual pursuit* system tracks small objects across a textured background. This requires over-riding the optokinetic reflex.

Dynamic Adaptations of the Vestibulo-Ocular Reflex (VOR)

If the VOR regularly fails to stabilize retinal images during head movements, the VOR reflex undergoes prolonged adaptive changes that restore visual stability. The VOR system has been experimentally adapted to radically different visual demands, such as: (a) image magnifications and minifications—which require complicated *motion-velocity adaptations* to achieve visual stability; (b) left-right, up-down, and left-right combined with up-down image reversals which required complicated *motion-direction adaptations.* Information that defines the operational requirements for this reflex adaptation originate from vision. These adaptations require several days for completion. They succeed in overcoming a variety of radically different problems that can be imposed on the system. The VOR is an excellent example of goal-seeking adaptation. In each different sort of imposed task the adaptations are made-to-order reorganizations that restore a rapid, precise, machine-like VOR. How is this accomplished? How can information that must originate in vision apply its discrepancies to shape new, accurate responses along a three-neuron arc in the brainstem? Sophisticated experiments have been carried out with spectacle-mounted lenses and prisms in humans, monkeys, rabbits, rats, birds, and fishes. It is clear that changes in the VOR following the introduction of displacing or magnifying lenses are due to modifications of the basic three-neuron reflex pathway and *not* due to some learning strategy that is improvised to adjust the reflex. Miles and Lisberger (1981) summarize the evidence:

1. Neither rate of adaptation to lenses nor rate of reversal when the lenses are removed suggest a learning strategy. Changes in gain are almost machine-like in acquisition and recovery, showing a similar time course in different individuals and in the same individual with repeated exposures. (There are no "savings" as might be expected with learning.)
2. After the VOR has adapted to a given experimentally imposed regimen, caloric tests of the VOR match the visually induced adaptations.
3. After the VOR has become fully adapted, the adaptation is retained at least several days without requiring further vestibular reinforcement.
4. Abrupt, unexpected head movements reflect the adapted state. Moreover, following adaptation to magnifying spectacles, trials in which zero gain would be preferable indicate that the VOR, in the adapted state, is fixed and immutable at least for days (see Lisberger and Pavelko, 1988).

MECHANISMS OF ADAPTATION OF VESTIBULO-OCULAR REFLEXES

One possibility for adaptation of the VOR is revealed by brainstem reorganization in the flatfish. During metamorphosis, this fish changes from a normally oriented, dorsal-ventral, up-down fish to being tilted 90° to one side when they become bottom-adapted adults. The vestibular apparatus, fixed in cartilage, must rotate by 90° relative to the direction of the gravitational field. Former horizontal semicircular canals become vertical canals. Brainstem neurons sprout new axonal arborizations which terminate on different extraocular motor neurons including some across the midline. If such circuit remodeling takes place in human VOR adaptations, that would account for several facts: (1) prolonged time is required for adaptation; (2) when adaptation is approaching completion, excellent control is rapidly instituted; and (3) there appear to

be no "savings" with adaptation and readaptation to different, including normal, VOR conditions.

The important lesson is that *the nervous system has a remarkable capacity to adapt to radically different environmental conditions*. When adaptation has been effected in the VOR, the reflex is once again high speed, 12 ms, accurate and consistent, machine-like in its performance. Such a capacity for adaptation seems to be "designed" (by evolution and development) to solve a variety of outwardly imposed problems, a set of examples that favors the theory of neuronal group selection (Edelman, 1987). One wonders whether a bat or a dolphin, operating in an acoustically defined environment must depend for its echoic stability on vestibular feedforward information as in the VOR (a VEAR, vestibulo-echo-acoustic reflex!)

Persistent breakdown of visual stability requires that this adaptation take place. Perhaps everyday experiences involving recurrent perturbations invoke custom-made adaptations that contribute to improvements in performance. The musician or ice skater who does not practice for a few days notices a deterioration in some aspects of performance. This implies that skill acquisition in athletics, music, and other high speed and accurate technical skills may involve adaptations that take place along similarly "low level" circuits in brainstem and spinal cord, perhaps including corrections applied to the primary sensory input (via efferent controls) and detailed precision of control applied to shaping the final motor execution.

OTHER VESTIBULAR SYSTEMS OF CONTROL

Figures 9.25 and 9.27 illustrate the main projections and functional circuits of the vestibular system. The peripheral vestibular apparatus projects mainly to the vestibular nuclei and these interact reciprocally with the brainstem reticular formation. There is a particular focus of vestibular projections to the reticular formation in the vicinity of the VIth cranial nerve nucleus a "center for lateral gaze." The VIth cranial nerve innervates the lateral rectus muscle which turns the eye laterally. Excitation of the reticular formation in this vicinity elicits not just a single muscle pull but a coordinated, conjugate deviation of both eyes to the ipsilateral side, with reciprocal innervation of the medial rectus on the same side and double reciprocal innervation of the pair of horizontal tractors (medial and lateral recti) on the contralateral side. Thus, conjugate deviation of the eyes is under bilateral control of the vestibular nuclei on each side—a double bilateral vestibular conjugate eye motor control system.

The vestibular nuclei also project strongly to the lateral part of the reticular formation which, in turn, projects to the cerebellar cortex. As shown in Figure 9.27, the vestibular nuclei send direct fibers to the flocculus. Note that the vestibular nuclei are under powerful inhibitory influences from cerebellar cortex.

VESTIBULOSPINAL REFLEXES (VSR)

First-order vestibular afferents, as we have seen, project to vestibular nuclei, reticular formation, and the floccular lobe in the cerebellum. A primary vestibular neuron sends mossy fiber terminals (knob-endings) to the cerebellum, with collaterals (or other axons) ending on three types of second-order sensorimotor relay cells. These have been categorized by Ito (1974): (1) fast-velocity lateral vestibulospinal tract neurons (LVST) which excite motor neurons; (2) slow-velocity medial vestibulospinal tract neurons (MVST) which inhibit spinal motor neurons; and (3) fast-velocity excitatory medial vestibulospinal tract neurons (Fig. 9.28).

The primary afferent vestibular neuron is intended to represent vestibular input from all five vestibular neuroepithelia. This provides information equivalent to an inertial guidance system. It is promptly distributed to second-order neurons in the vestibular nuclei and reticular formation which are engaged in organizing neuromuscular control relating to all skeletal musculature. They make additional integrative contributions depending on what they receive from other sources. Second-order vestibular neurons receive from the cerebellum, from midbrain superior collicular neurons involved in visually controlled head-eye coordination (see Chapter 64, Vision), the brainstem VOR system, and sensory representations in cortex, vision, somesthesis, and hearing. Thus, second-order vestibular neurons constitute a point of convergence from many different sources of information relating to the control of head, body, and limbs in space. Vestibular input, contributing to this busy pool of integrative neurons, delivers up-to-date information relating to the position and motion of the head. Output of these second-order neurons in the vestibular nuclei and reticular formation goes to all motor neurons for skeletal muscles, along the whole length of the neuraxis, from eye muscles to the external anal sphincter.

The vestibulospinal reflex (VSR), like the VOR, is a three-neuron system which provides rapid responses, with rapidly conducting fibers in both lateral and medial vestibulospinal tracts. There is

urgency to this command system: with either active or passive motion of the body in a gravitational field there are risks of falling. The vestibular apparatus has swift access to skeletal muscles in antigravity groups that can restore equilibrium. This is perhaps even more urgent and demanding than stability of visual images under control of the VOR. To meet this requirement some vestibulospinal neurons and corresponding reticulospinal neurons are among the largest diameter and most rapidly conducting of any vertebrate neurons.

SPINO-VESTIBULOSPINAL REFLEXES

We need to examine how somatic sensory information from the body informs the VSR system in relation to balance and equilibrium. This requires a spino-vestibulospinal reflex (SVS) system. *The somatic input (SVS reflex system) provides both feedback and feedforward contributions to VSR circuits.* Thus, somatic information provides corrective information relating to which body and limb activities are occurring, to compare with the vestibular command for their control. This system also feeds forward information about body movements actively or passively occurring prior to motion of the head, hence in advance of vestibular activation. The SVS thus contributes to appropriate execution of the VSR.

Tonic neck reflexes relate to the sequence of getting up from a reclining position. First, the head is raised. This activates appropriate patterns among vestibular afferents. Cervical somatic sensory input, activated simultaneously, especially afferent input from vertebral joint receptors further informs VSR circuits concerning requirements and progress in elevating the head and neck. Following this initiative, somatic (SVS) messages to VSR contribute a cervical to caudal ordering of successive axial skeletal muscle and limb contractions to fulfill the act of arising. One normally gets up head first, and somatosensory contributions to VSR combine to provide the organization of successive axial and limb muscle contractions in proper sequence.

The cerebellum does not initiate motor commands but it contributes a "servo-mechanism" boost to the VSR for execution of smoothly coordinated control of balance and equilibrium. On the left are shown cutaneous receptors and receptors from tendon organs and muscle spindles (see Chapter 70, Segmental Controls). First-order somatic afferents convey information on positions of axial and limb musculature, setpoint of skeletal muscle readiness, degree of tension exerted, and cutaneous component pressure or stress responding to body and limb position and strain.

These circuits are adjusted for inertial mass of body parts, and they are temporarily adjusted for extracorporeal loads applied to body parts (e.g., skater spinning a partner; skier carrying a heavy backpack). These somatic afferent signals are relayed by second-order neurons in the dorsal horn of the spinal cord whose axons ascend in the lateral funiculus of the spinal cord. They project to the vermis of the cerebellum and give off collateral terminals to each of the three types of vestibulospinal projections described above, plus a fourth type of neuron which projects by way of relatively slowly conducting inhibitory neurons via the medial vestibulospinal tract (MVST) to motor neurons serving skeletal muscles all along the neuraxis (Fig. 9.28). Here is a grand convergence of pertinent information on third-order vestibular, reticular, and cerebellar units responsible for vestibulospinal outflow.

Simple Clinical Testing of Vestibular Function

A knowledge of orientation of the semicircular canals is helpful for simple clinical testing of their function. When the head is held erect, the horizontal canals are tilted down to the rear about 30°. When the horizontal canal is horizontal, it is maximally sensitive to rotations about an axis at right angles to that plane, i.e., rotation by turning a chair with the subject sitting upright. The superior canal on one side is in a plane parallel to that of the posterior canal on the opposite side. Therefore, the contralateral pair of superior and posterior canals can be maximally stimulated by rotation around common axes (Fig. 9.21).

The horizontal semicircular canal can also be stimulated by convectional movement of fluid in the canal because that canal is close enough to the external auditory meatus to be artificially cooled or warmed. If the horizontal canal on the right side is cooled, the canal fluid becomes relatively dense and streams away from the ampulla, accompanied by an illusion of the head turning to the left. Nystagmus, past-pointing (drift of hand pointing analogous to nystagmus), and a tendency to fall occur to the right.

VESTIBULAR THRESHOLD DETERMINATIONS

There is a minimum rate of angular acceleration that can be detected by human subjects. Under optimum conditions this rate is on the order of $0.10°/s^2$ (Clark, 1967; Clark and Stewart, 1968). Threshold for normal subjects for horizontal motion sensation, measured in a parallel swing is about $2–6$ cm/s^2. Thresholds are considerably higher in patients in whom both labyrinths are deficient, yet normal if one labyrinth remains functional.

The otolith organs, particularly the utricles on which detection of head tilt depends, can be studied by tilting the body very slowly from an erect posture in a chair device. Subjects are especially sensitive to slight tilting away from the vertical. Analogous to nystagmus, on stimulation of the semicircular canals is counter-rolling of the eyes which occurs when the head is tilted away from the vertical. Counter-rolling in the earth's gravitational field is approximately 10° for 60° of tilt away from vertical. Labyrinth defective subjects show little or no counter-rolling under any circumstances. Normal subjects show no counter-rolling of the eyes with tilt of the body when they are in zero gravity conditions (Miller et al., 1966).

TEMPORAL DURATION OF EFFECTS OF VESTIBULAR STIMULATION

When subjects are exposed to constant angular acceleration about a given axis, a sensation of rotation is induced along with nystagmus. There appears to be an increase in the perceived angular velocity for about 30 s, followed by an impression of gradual decrease in angular velocity. With continued angular acceleration, there is a loss of both consciousness of acceleration and nystagmus. This corresponds to the damped torsion pendulum function of the crista and cupula which gradually resume their normal undeflected position when angular velocity remains constant. When angular acceleration stops, a normal subject perceives a reversed sense of rotation and shows reversed nystagmus for 20–30 s.

AVIATION

Prior to the advent of high speed vehicles, particularly aircraft which can maneuver relatively freely in three-dimensional space, man was not ordinarily subjected to prolonged strong angular rotation. The vestibular system is ideally suited to coordination of bodily movement and fixation of vision for brief, low amplitude rotations, but not for prolonged, high amplitude rotations. Following prolonged rotation at constant high angular velocity, deceleration results in sensory signals from the canals equivalent to rotations in the opposite direction but they do not correspond to actual flight conditions. This causes difficulties in subjective interpretation of the aircraft path and may be responsible for high-speed aircraft accidents. Additional circulatory problems may be associated with prolonged high amplitude rotations around a horizontal axis: blood return to the heart may be impaired so much that cerebral hypoxia leads to loss of consciousness and grand mal seizures.

SPACE FLIGHT

The zero gravity environment of space is inconvenient from the standpoint of manipulating liquids and certain aspects of daily living (kitchen and toilet activities, loose objects, powder, lint, etc. are troublesome). Zero gravity also presents chronic physiological risks with respect to loss of autonomic nervous system "tone" responsible for venous return of blood to the heart and loss of compression and torsion stimuli that contribute to calcium deposits in bones. Artificial gravity by application of a constant angular velocity of a space station with its occupants living at an appropriate radius has been recommended as being useful for space stations and space travel. Physiological problems arise, however, because for engineering reasons that radius must be kept relatively short. Moving about in such an environment, individuals may be obliged to tilt their heads around various axes which may be orthogonal to the axis of rotation. This yields equivalent Coriolis accelerations which can induce an oculogravic illusion of tilting of the visual world (Clark, 1967). There appears to be little adaptation in the otolith organs when these are tilted to unusual positions in gravitational fields, and these effects may contribute to motion sickness.

NYSTAGMUS

Nystagmus movements of the eyes associated with angular acceleration consist of a relatively slow component (fixation relative to the environment), in the direction opposite to the acceleration and a fast component in the direction of acceleration. Nystagmus ceases if the acceleration is continued in the initial direction. Stopping acceleration is followed by a secondary nystagmus with the slow phase in the opposite direction, which may continue for 20–30 s after angular acceleration has stopped. For clinical testing, using a rotating chair, the usual procedure is to rotate the subject at a rate of one revolution every 2 s for a total of 10 revolutions. The chair is then stopped abruptly. Nystagmus is induced by the deceleration. The slow phase is thus in the same direction as chair rotation.

If during rotation the head is nearly erect but tilted forward about 30°, the horizontal canals are maximally stimulated. Induced nystagmus is horizontal. If the head is tilted down to one side, nearly 90° onto one shoulder, with rotation about the vertical axis, then the vertical semicircular canals are maximally stimulated. The direction of induced nystagmus is nearly vertical with respect to the head. If the head is tilted back 60° or down 120° from the erect position, the nystagmus is rotary, i.e., there will be strong tortional movements. Tortion of the globes can be easily detected by referring to patterns characteristic of the iris or vessels in the sclera. By such maneuvers, a wide variety of nystagmus movements can be elicted. It is clear that the three pairs of semicircular canals work in concert to drive the appropriate extraocular muscles to compensate for motion of the head around any axis of rotation.

Nystagmus can also be induced by moving objects in a consistent direction before the eyes (e.g., motion picture of acrobatics in aircraft or riding a roller coaster). If an individual is rotated about the vertical axis, eyes open, the corresponding optokinetic nystagmus will be in the same direction as the vestibular nystagmus and the two will reinforce one another. Postrotational nystagmus is reduced if the eyes remain open during rotation. With eyes open during postrotatory nystagmus, optokinetic nystagmus tends to cancel out the vestibular induced nystagmus. In testing vestibular function, therefore, subjects are instructed to keep their eyes closed until rotation has stopped.

Vestibular stimulation that results in nystagmus movements of the eyes influences other musculature as well. There is a tendency for increased tone in the extensors on the side of the body that is leading in the direction of rotation, and increased tone in flexors on the side away from the direction of rotation. These effects are reversed following deceleration. Thus, if a subject is instructed to stand and walk as soon as the rotation of the chair is stopped, he will tend to stagger or fall in the direction of rotation and must be protected from injury.

Effects of Training

Tumblers and springboard divers maintain a sense of orientation in space while performing elaborate rotational movements that are quite disorienting to the unaccustomed. Figure skaters are capable of high-speed rotations, as high as 280 rpm for several seconds, immediately following which they can go into another complicated skating routine or stand immobile. To do this they must suppress nystagmus by intermittently fixing their gaze on stationary objects (Collins, 1966). Skaters' adaptation to these maneuvers rapidly diminishes and can be detected by a skater after 2 or 3 days of lack of rotational practice. Habituation is highly selective and may be lost if the head position is not also habituated, or if the direction of rotation is changed (Guedry, 1965).

LABYRINTH DYSFUNCTION

Damage to the vestibular apparatus is accompanied by disturbing symptoms of vertigo and dizziness along with ataxia and other signs of incoordination. This condition may be induced by inflammation of the vestibular branch of the VIIIth nerve, local infection in the middle ear, and abnormal ionic balance. Treatment is specific for the condition and the symptoms. Drugs useful to prevent motion sickness may help.

Gross vestibular impairment accompanied by paroxysmal attacks of vertigo, subjective noise (tinnitus), and a progressive impairment of hearing characterize a syndrome known as Mèniére's disease. It is generally attributed to disturbed ionic balance in tissues of the inner ear. Symptoms may be relieved by reducing sodium intake. Intractable cases have been treated by surgical destruction of the labyrinth, intracranial section of the vestibular branch of the VIIIth nerve, or ultrasonic destruction of vestibular end organs.

Motion Sickness

Prevailing notions emphasize overstimulation or hypersensitivity of vestibular neuroepithelium, or in the case of motion sickness in the weightless state, lack of sufficient stimulation of the vestibular neuroepithelium. Motion sickness may originate in part from neurophysiological conflicts among messages that cannot be integrated into meaningful experiences (Mayne, 1974).

Symptoms of motion sickness reflect widespread and sometimes violent reactions of the autonomic nervous system. Signs and symptoms of motion sickness may include pallor, cold sweating, yawning, salivation, vomiting, diarrhea, vertiginous disorientation, and a lurking fear of, or wish for, death. There may be severe, progressive dehydration, electrolyte imbalance, reduced circulating blood volume, reduced blood pressure, cardiac arrhythmias from electrolyte imbalance, and other effects. Motion sickness may be life-threatening because of dehydration, reduced circulating blood volume, reduced blood pressure, cardiac arrhythmias, and other problems.

Intractable, incapacitating motion sickness where continuing exposure to the cause may be necessary, may be eliminated by extremely drastic measures such as removal of the vestibular apparatus bilaterally, sectioning the vestibular branch of the VIIIth cranial nerve bilaterally (Johnson et al., 1951), or removal of the flocculonodular lobe of the cerebellum (Tyler and Bard, 1949).

CHAPTER 65

Vision

LIGHT AND OPTICS OF THE EYE

Light

Without photoreceptors and nervous systems there would be neither light nor color in the universe. Within the broad range of electromagnetic radiations, the human eye provides access to wave lengths that lie between two boundaries, both of which present risk to tissue damage from radiation. Thus, only a narrow band of wave lengths is utilized for the transduction of electromagnetic energy to the molecular and neuronal events experienced as "light" (Fig. 9.29). Within that range, specialized photoreceptors differentially respond to still narrower bands of radiation which are experienced as "colors." Resulting "vision" provides what are perhaps the richest and most varied phenomena of all perceptual experience.

The first step in vision is gained by light-gathering powers of transparent tissues of the eye. These tissues are vulnerable to damage by electromagnetic energies—infrared and ultraviolet—which lie just outside, respectively, the low and high frequency margins of the band of visible light (Fig. 9.29). Images are initiated by radiation beams brought to focus on a bowl-shaped layer of millions of photoreceptors lining the interior of the back of the eye.

The experience of light depends on transformations which begin with the transduction of photic energy by photoreceptors. Surprisingly, spontaneously active photoreceptors are made less active during transduction of electromagnetic radiation! Neural events are initiated by *decreases* in photoreceptor activities. This requires continuing expenditure of photoreceptor energy—but it allows neurophysiological signals to be elicited by an ultimate minimum of energy. With ideal conditions, human subjects can perceive exposures to radiation at a level of single photons!

Photoreceptors trigger patterns of neuronal activity that begin visual processing during their trajectory through the retina. Vision is "created" through the analysis of signals from the initial patterns of light and color and through subsequent central reintegrations that take place among various widespread neuronal projections. Central neuronal projections from the retina provide correspondences, "mappings," between the retina and various widespread central nervous system rerepresentations. These projections involve distinctive, extensive, parallel processing of visual signals.

Integration of information relating to detailed boundaries, geometry, and movement, brightness, color, surface texture, etc., analyzed in separately distributed cortical and subcortical subsystems, results in our subjective categorization or "visualization" of "objects-in-space." General characteristics and detailed features are synthesized into perceptually coherent representations which categorize objects and events that are further reified by memories, and by employing other sensory inputs: hearing, olfaction, taste, especially by the measurement of objects through touch and bodily movement. We "project" inferred objects and events outside ourselves. We "visualize" objects and events as happening "out there" in a space-time world that we create through sensorimotor activities.

Since visual pathways and those for eye motor controls traverse widespread regions of the brain, careful analysis of visual and oculomotor dysfunctions contributes important clues for localization and functional characterization of neurological and neuropsychological problems.

We are not passive observers. Vision is a dynamic cognitive process organized to suit motor behavior. In the process, individual past experiences, expectations, and purposes greatly influence perception. It is only within idealized laboratory conditions and by employing very simple stimuli and rigidly standardized instructions that roughly comparable sensory experiences can be documented. The study of vision thus helps us to understand some of the rich-

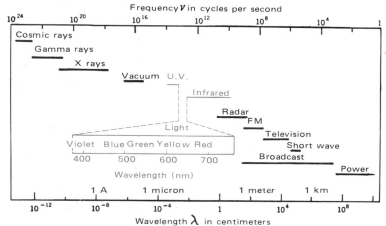

Figure 9.29. The electromagnetic spectrum. The region visible to the human eye is expanded. (From McKinley, 1947.)

ness of our subjective experiences and some of the limitations of our knowledge of the universe. All senses can be confounded by illusions, and visual illusions abound. Although we may *know* that we are experiencing an illusion, we nevertheless *see* the illusion. Separateness between knowledge and perception applies to illusions, and we may be more subject to errors because perception must operate rapidly.

How is perception corrected and improved? Perception can be improved by information acquired from other sensory modalities, but perception is fundamentally corrected only by engaging in purposeful actions which depend on perception. Passive exposure to perceptual errors is not corrective. Actions that reveal noncorrespondences between the perception and events as experienced open the way to correcting perception. Behavioral mistakes uncover perceptual errors and contribute to the improvement of perception. Perception guides behavior and behavior corrects perception.

REFRACTION OF LIGHT

Light passing from one medium to another changes direction at the interface. If the surface between two media is curved, parallel rays of light in a beam of large cross-section are refracted by different amounts and propagate through the new medium along correspondingly different paths. Based on principles of light refraction, lenses are designed to cause parallel beams of light to converge or diverge (Fig. 9.30) in various combinations.

White light consists of a distribution of energies at all visible wave lengths. Light of shorter wavelength is refracted more than light of longer wavelength. Thus, when a parallel beam of white light enters glass at an angle other than perpendicular, the different wavelengths travel along different paths, separating into different spectral components.

Light-Gathering Properties of the Eye

CORNEA

Most of the approximately 60 diopter refractive power of the eye is provided by curvature of the cornea (Fig. 9.31). This avascular structure is about 1-mm thick at its margins, gradually tapering to about 0.5 mm at its center. The cornea consists of a stroma of collagen fibrils, proteoglycans, and electrolytes which transmit light selectively from about 350 nm, slightly into ultraviolet, to about 700 nm, slightly into infrared.

Partial dehydration of the cornea is essential for corneal transparency. Some movement of water takes place across the transparent Bowman's membrane on the outer surface of the cornea as a result of evaporation through corneal epithelium to air. The epithelium is highly proliferative; it undergoes

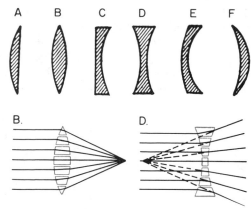

Figure 9.30. Cross-sections of lenses. *A*, Planoconvex; *B*, Bi-convex; *C*, Planoconcave; *D*, Binconcave; *E*, Convexoconcave; *F*, Concavoconvex. The way in which refraction of a beam at curved surfaces may result in convergence or divergence is illustrated for *B* and *D*.

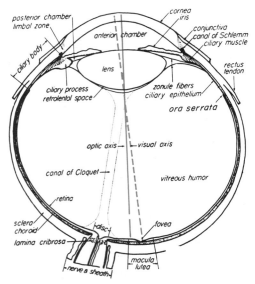

Figure 9.31. Horizontal section of the right human eye. (From Walls, 1942.)

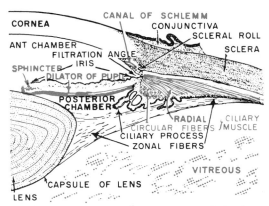

Figure 9.32. Detail of the anterior segment of the human eye. (Redrawn from Weymouth, 1955.)

a complete turnover in about a week and heals correspondingly rapidly when damaged. The healing process does not ordinarily interfere with the optics. But if Bowman's membrane is damaged, the cornea develops a scar that can distort the passage of light. The inner surface of the cornea is covered with a single layer of endothelial cells. This endothelium is leaky and performs the important task of dehydrating the corneal stroma against strong oncotic pressure. As long as this endothelial pump is effective, the cornea remains transparent. If this endothelial function fails, however, the cornea becomes chalky white, like the sclera.

AQUEOUS HUMOR

The transparent fluid which fills the anterior and posterior chambers on either side of the iris diaphragm is formed from blood plasma (Figs. 9.31 and 9.32). Production of aqueous humor takes place continuously. A complete turnover of aqueous humor occurs approximately hourly. Aqueous humor closely approximates a dialysate of blood plasma and provides the principal source of metabolites for both lens and cornea. Its blood supply is via the ciliary body. The main route for aqueous drainage is the canal of Schlemm (see Fig. 9.32). Other outflow occurs via the posterior chamber, retina, and choroid. Normally aqueous pressure is about 15–18 mm Hg higher than intracranial pressure. This pressure differential helps maintain the shape of the eyeball and appropriate spacing and alignment of its optical contents. If for any reason aqueous outflow is blocked, or aqueous production is increased without matching outflow, intraocular pressure increases.

Carbonic anhydrase inhibitors reduce the rate of aqueous production and may be of assistance in reducing intraocular pressure. Clinical measurement of this pressure is an important part of physical examinations, particularly in the elderly.

PUPIL

The optical path can be shuttered by closure of the eyelids which, depending on pigmentation, admit less than 1% of incident light. The iris diaphragm or pupil varies in diameter according to background illumination from about 2–8 mm. This 1:4 change in diameter corresponds to a 1:16 change in pupillary cross-section, controlling that range of retinal illumination. An optically optimum pupil diameter in bright illumination is between 2 and 3 mm. Parasympathetic innervation of a sphincter muscle provides active constriction of the pupil; sympathetic innervation of radial dilator muscles provides active dilation of the pupil. An abrupt increase in external illumination of one eye yields a brisk consensual reflex contraction. Decrease in external illumination yields a slower pupillary dilation.

When the pupil constricts, spherical and chromatic aberrations of the optical system are reduced and the depth of focus is increased. During observation of near objects, the lens increases its curvature (the accommodation reflex), and the pupil constricts. Both effects contribute to a sharper retinal image.

LENS

Additional refractive power, up to 9 or 10 diopters, is provided by the crystalline lens (Fig. 9.31), which is made up of similarly transparent collagen fibrils. The shape of the lens is dynamically altered by contraction of radial fibers which pull the ciliary body forward, reducing tension on the suspensory zonal fibers (Fig. 9.32). This relaxes the lens capsule and allows its surface to become more convex. Dynamic lens control permits sharp image resolution over a range from a remote point of about 6 meters and beyond, to a near point at about 10 cm. This range varies with age.

Variation in the refractive index with the wavelength of light results in chromatic aberration. There is remarkably little spherical aberration because the lens is parabolic in curvature; therefore peripheral and central rays are brought to focus at close to the same depth. Chromatic and spherical aberration are illustrated in Figure 9.33.

Lens Changes with Aging

In early childhood, the lens is capable of altering the resolving power of the eye over a range of about 20 diopters, providing clear vision of objects from very near to very far. The lens grows with the child, adding layers like an onion. It reaches adult size by about age 13–14 years. Thereafter, the lens begins to lose its high water content and becomes increasingly hard; it thereby becomes less capable of assuming full curvature when the ciliary muscles relax tension on the lens capsule. The range of contribution by the lens to the resolving power of the eye is therefore gradually reduced. The resolving power of the eye diminishes throughout childhood and adolescence until, at about age 20 years, the range is limited to about 9–10 diopters. The emmetropic eye continues to serve distance vision well but is not adequate for reading. Beginning in the fifth decade many people require additional refractive power to compensate for reduced maximal curvature of the lens. This problem may be corrected by convex lens reading glasses.

There is also a gradual loss of lens transparency which is exaggerated by trauma, radiation (ultraviolet light and high energy photons such as gamma-rays and x-rays), and by metabolic problems such as uncontrolled diabetes. An eye with opacity in the lens is said to have a cataract. An increasingly opaque lens may need to be removed to preserve sight, and be replaced by a corresponding power of plastic lens implantation, or by convex lens glasses.

Vitreous Humor

Vitreous humor occupies the posterior chamber of the eye. It consists of a gel-like substance that contains a network of thin fibers of a highly hygroscopic vitrein protein analogous to gelatin.

OPTICAL DEFECTS

In a normal eye, with relaxed accommodation, objects at an optical distance of infinity come to focus on the retina (Fig. 9.34). Such an eye is said to be emmetropic. If the point of focus falls in front of the retina, the target must be moved toward the eye, displacing the focal point posteriorly until it falls on the retina. This describes a myopic eye. When the image of an object at optical infinity falls behind the retina, increased curvature of the lens in necessary to bring the focal point forward to the retina. This describes a hyperopic eye (Fig. 9.34). Either myopia or hyperopia may be due to inappropriate size or optical axial length of the globe, or to abnormal curvature of its optical elements. Partial compensation occurs naturally because eyes with short axes tend to have more rounded corneas and long axis eyes tend to have more flattened corneas.

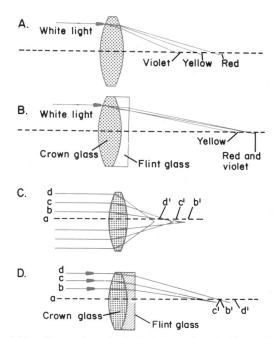

Figure 9.33. Chromatic and spherical aberration. *A.*, Chromatic aberration in simple lens; *B.*, Achromatic lens partially correcting chromatic aberration; *C.*, Spherical aberration in a simple lens; *D.*, Achromatic lens partially correcting spherical aberration. (From Riggs, 1965a.)

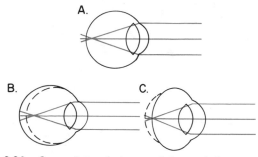

Figure 9.34. Comparison of shapes of the eyeball which may be associated with *A.*, Emmetropia; *B.*, Myopia; *C.*, Hypermetropia.

Myopia is corrected by means of negative or diverging lenses to empower distance vision. Hyperopia is corrected by positive lenses for close visual work.

Anatomy of the Globe

EXTRAOCULAR TISSUES

Eyes are physically shielded by the nose, bony orbit, eyelids, and the blink reflex. A relatively slack optic nerve tethers the globe posteriorly. Anteriorly, the eye is loosely elastically attached to the base of the eyelids by conjunctivae and by medial and lateral canthal ligaments. An anterior orbital septum forms a hammock-like partition which suspends the globe in a "catcher's mitt" of orbital fat. Tenon's capsule attaches to the sclera in a ring anterior to the insertion of the tendons of the extraocular muscles and provides separate capsular sheaths for each extraocular muscle. The eyeball has a specific gravity that approximates that of the soft tissues of the orbit. This gives them nearly negligible inertial mass, an advantage in efficient muscular direction of gaze. The extraocular muscles are correspondingly light and reflexly swiftly responsive.

Eye fixation can be maintained on moving objects even during head movements. When the distance of an object changes, the axes of direction of gaze are converged or made parallel so as to maintain conjugate fixation. Four rectus muscles control movements around two axes that are at right-angles to each other and are, in turn, nearly perpendicular to the line of sight when the eyes are directed straight ahead (see Figs. 9.35 and 9.36). The lateral (external) rectus is located on the temporal side of the globe; the medial (internal) rectus is attached to the nasal side. The superior and inferior recti are attached superiorly and inferiorly. Two additional muscles, superior and inferior obliques, in combi-

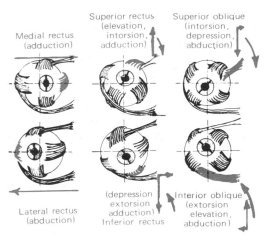

Figure 9.36. Action of the three pairs of antagonistic eye muscles. (From Adler 1965.)

nation with the other eye muscles, control tortional rotations of the eye around the visual axes (see Fig. 9.36). These rotations are also accompanied by depression or elevation of gaze. Their context-specific functional roles are described and diagrammed in Figures 9.35 and 9.36.

GLOBE

The globe is enclosed by three membranes, an outer opaque sclera which is continuous with the cornea. The major blood supply of the retina lies in the heavily pigmented choroid coat directly inside the sclera. This pigmented layer absorbs light within the eye and reduces light scatter. Anteriorly, the choroid coat is continuous with the ciliary body and iris diaphragm (Figs. 9.31 and 9.32). The innermost layer is the retina, containing visual receptors, the retinal nerve circuits and accompanying blood vessels. The light path passes through the anterior chamber, lens, and vitreous humor.

CORNEA

The cornea is highly transparent and devoid of blood vessels. It is covered by epithelium which is supplied by unmyelinated, bare nerve endings from the ophthalmic branch of the trigeminal nerve. Appropriate stimulation of the corneal surface triggers a blink reflex. Although it is ordinarily thought that bare nerve endings subserve only pain sensation, free nerve endings elsewhere in the body subserve a variety of sensations. The cornea responds to light touch; experienced subjects are able in the dark to discriminate differently shaped objects lightly applied to the cornea. It is a commonplace experience that appropriately fitted corneal lenses are reasonably well tolerated, but because contact lenses diminish respiration of the cornea it is advis-

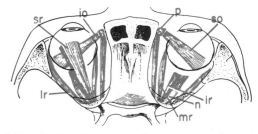

Figure 9.35. Oculomotor muscle of man, as seen from above in a dissected head. On the *left,* a portion of the superior oblique has been cut away to reveal the inferior oblique; on the *right,* the superior rectus has been removed to permit a view of the inferior rectus; *io,* inferior oblique; *ir,* inferior rectus; *lr,* lateral rectus; *mr,* medial rectus; *n,* optic nerve; *p,* pulley for tendon of superior oblique; *so,* superior oblique; *sr,* superior rectus. (From Walls, 1942.)

able that they be removed for several hours each day.

Transparency of the cornea depends on proper osmotic and hydrostatic pressures in the aqueous humor and contiguous tissues. Tears keep the outer surface of the cornea lubricated and moist and this permits some diffusion of nutrients and gasses into and out of the cornea. Cells in the outer layer of epithelium share tight junctions so that the rate of outward transport of fluid is limited. Accumulation of intracellular acidic metabolites, systemic acidosis, cell damage by contact lenses, drugs, etc., result in corneal swelling as a consequence of inhibition of electrolyte transport. Corneal epithelial cells provide sodium and chloride transport that moves fluid from the stroma to the tear side and is capable of dehydrating a swollen cornea (Wiederholt, 1980). Active sodium transport is inhibited by β-blocking drugs. There are serotonin and dopamine receptors, and sympathetic innervation can modulate corneal electrolyte transport. Electrolyte and water transport is stimulated by cAMP at pH > 7.5 and inhibited at pH < 7.5.

The main transport of corneal water takes place across endothelial cells on the inner surface of the cornea (see below). The pH of the thin tear film averages 7.6 between blinks. With eyelids closed, pH of the tear film becomes less alkaline, with accompanying inhibition of electrolyte transport. The normal human cornea swells by about 3–6% during sleep because of reduced diffusion of CO_2 through the tear film when the eyelids are closed. Transport of CO_2 slows when a contact lens is in place. This contributes to blurring of vision with prolonged wearing of contact lenses. Edema of the corneal epithelium results in the appearance of halos around bright lights.

Corneal transparency is threatened by physical abrasion, too great an elevation of intraocular pressure, osmotic imbalance, severe temperatures, gamma-rays, x-rays, ultraviolet and infrared radiation. Fortunately, the corneal epithelium and trigeminal innervation repair themselves within a few days following superficial trauma or mild ultraviolet light injury (e.g., sunburn, snow blindness). But penetrating wounds leave scars that interfere with refraction and transmission of light.

Although the cornea lacks blood supply, oxygen is supplied and CO_2 leaves by exchange diffusion with the surrounding tissues and outside air. The cornea consumes oxygen at a high rate as a result of endothelial activity essential for dehydration of the cornea. Since oncotic pressure of the stroma is about -40 mm Hg and the aqueous pressure is about 15 mm Hg, water and electrolytes tend to pass into the stroma; this induces corneal swelling and loss of transparency. The endothelium provides sufficient active transport to compensate for passive leakage of water and ions into the tissue and to maintain the fluid pressure differential. Competence of the endothelium decreases with age, and with insults such as trauma, blood in the anterior chamber, radiation injury, inflammation, and glaucoma.

PUPIL

The iris diaphragm controls a dynamic pupillary aperture (Figs. 9.31 and 9.32). Pupillary constriction to light is accomplished by contraction of sphincter muscles which encircle the inner margin of the iris. Retinal excitation by incident light travels centralward by way of ganglion cell axons. Some of these leave the optic tract to enter the cephalic brainstem whence they travel rostrally to a pretectal pupilloconstrictor zone. This region, near the junction of the aqueduct and third ventricle is especially vulnerable to damage from elevated intracranial pressure. *Sluggishness or disappearance of the pupilloconstrictor response to light is one of the most important vital signs to monitor whenever there is a possibility of excessive or increasing intracranial pressure.*

From this pupilloconstrictor zone, impulses are relayed to the Edinger-Westphal nuclei bilaterally, whence each pupil is served by ipsilateral preganglionic parasympathetic fibers. Axons course with the third cranial nerve to the ciliary ganglion, whence postganglionic parasympathetic fibers pass to smooth muscle fibers in the iris sphincter (Fig. 9.32). Inhibition of tonic activity in the Edinger-Westphal nucleus contributes to pupillary dilation.

If the gaze is shifted from a distant object to a near one, accommodation, convergence, and pupillary constriction occur. This *accommodation reflex* involves both smooth and skeletal muscle responses. Impulses from the retina reach visual cortex whence additional relays pass down to an "accommodation-convergence region" in the vicinity of the oculomotor nuclei. The oculomotor response brings the optical axes to bear on the near object. This reflex loop through visual cortex continues to "search" for binocular disparities among visual details of that object. Neighboring Edinger-Westphal cells are activated to constrict the pupils.

Thus, there are two distinct initiatory pathways for pupilloconstrictor reflexes: responses to light, and accommodation. These are widely separated in their retinodiencephalic and retinocortical-diencephalic relays, and may be separately affected by disease. Central nervous system syphilis, for example, can cause a narrow pupil that does not constrict to light but does constrict during accommodation and convergence (Argyll-Robertson pupil).

Radially oriented pupillodilator muscles in the iris are served by an upper thoracic spinal center which relays through the superior cervical sympathetic ganglion. In addition to inhibition of tonic activity in the Edinger-Westphal nuclei, active pupillary dilation occurs reflexly on shading the eye and as an accompaniment of pain and strong emotions such as anger, fear, and excitement. Pupillary dilation is also associated with positive affectional interest and serves as an involuntary indicator of directed affection. This introduces a "love-detector"—in contrast to a "lie detector." The efferent path proceeds from frontal cortex to posterior hypothalamus to neurons in the reticular formation of the lower brainstem which, in turn, project to the intermediolateral (visceral) column of motor neurons in upper thoracic spinal levels. Impulses from this region reach radial dilator muscles in the iris by relay through the superior cervical sympathetic ganglion.

LENS

The crystalline lens is a biconvex, avascular body composed of collagen fibers enclosed within an elastic lens capsule (Fig. 9.32). The reduced refractive power of the lens as compared with the cornea partly corrects for chromatic and spherical aberrations along the optical path. Energy for lens metabolism is derived from glucose oxidation: oxygen, ascorbic acid, and gluthathione are supplied by diffusion from the aqueous humor. As noted, new lens stroma is laid down very slowly after age 13–14 years, and as water is slowly withdrawn, the size of the lens remains relatively constant thereafter. Dehydration leads eventually to increased lens rigidity. This accounts for the reduced ability for visual accommodation with increasing age.

The lens capsule is held in a state of tension and the lens curvature is flattened by the pull of elastic suspensory ligaments anchored to the ciliary body (Fig. 9.32). The ciliary body is attached to the sclera, which is prevented from collapse by sustained intraocular pressure and partly also by the turgidly hydrated vitreous humor which fills the posterior chamber and supports the lens from behind.

Without active contraction of the ciliary muscle, the normal eye is focused at infinity. For near vision, accommodation is achieved by contracting the ciliary muscle which relaxes the suspensory ligaments and allows the lens to become more convex, thus increasing its refractive power. The degree of contraction of the ciliary muscle is dictated by

information relating to the amount of noncorrespondence obtaining between binocular foveal representations of objects in striate cortex. Accommodation thus depends on integrity of the primary visual cortex. Disparate images from the two eyes, compared in primary visual cortex, drive the oculomotor system so that the visual axes of the two globes cross one another at an angle that secures the best correspondence of images of objects at the distance of visual attention. Blurring of images from the two retinas thus serves simultaneously to bring the two globes to bear on the same point in three-dimensional space and to control the accommodation reflex whereby the lenses are driven to find the sharpest focus.

DISTRIBUTION OF RECEPTORS

The distribution of retinal receptors is nonuniform, having greatest concentration of cones in the neighborhood of the visual center of the eye, the *fovea* (Fig. 9.31). The concentration of rods increases from a point just outside the rod-free fovea to a maximum at a distance of about 20° from the foveal center (Fig. 9.37). There is a region in the normal eye which is devoid of any receptors, the optic disc (Fig. 9.31). It is here that ganglion cell axons leave the eye to form the optic nerve and where about half of the retinal circulation enters and leaves the eye. The remaining blood supply to the retina comes from a ring of about 15–20 posterior ciliary arteries and veins which provide circulation to the peripheral retina and ciliary body. The blood supply is further ensured by anastomoses with seven anterior ciliary arteries which supply circulation to the iris and ciliary body.

The optic disc occupies an area about 3° in diameter centered at a point about 15–17° from the foveal center and slightly below the nasal side of the horizontal meridian. The area devoid of receptors constitutes a "blind" spot. The presense of the blind spot is ordinarily not noticed subjectively, even with monocular vision. Yet you can easily demonstrate your own blind spot (see Fig. 9.38). Perceptual processing ordinarily "fills in" this area as though it were continuous with its surround.

ADAPTATIONS FOR VISUAL ACUITY

Cones for bright illumination and color vision dominate foveal vision (Fig. 9.39 *B* and *C*). There are about 35,000 cones in the rod-free region at the center of the fovea, subtending an angle of about 1°. The narrowest cones, in the center of the fovea, have a diameter of 1–1.5 μm, and the light responsive outer segment is roughly double in length thus maximizing visual acuity and sensitivity (Fig. 9.39

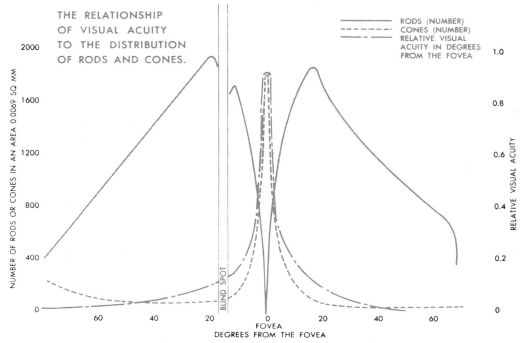

Figure 9.37. Distribution of rods and cones along a horizontal meridian. *Parellel vertical lines* show the limits of the blind spot. Visual acuity for a high luminance as a function of retinal location is included for comparison. (From Woodson, 1954; data from Osterberg, 1935 and Wertheim, 1894.)

C). Foveal cones communicate by a relatively direct "private line" consisting of only two synaptic relays between single cones in the central fovea via single bipolar cells to single ganglion cells. This ensures fine foveal discrimination. There are no rods in the fovea.

Arteries and veins are distributed over the inner surface of the retina and therefore lie in the light path, contributing some optical distortion. Vessels serve the fovea from its margins rather than overlying that region. Parts of retinal neurons and fibers not engaged in photoreception are shifted as far as possible outside the fovea, and ganglion cell axons in transit to the optic nerve detour around the fovea.

The Retina

EMBRYOLOGY AND ORGANIZATION OF THE RETINA

The retina is a true sensory outreach of the brain; its cells come directly from an outpouching of the

Figure 9.38. Demonstration of the blind spot. With the left eye closed, fixate the small cross. Vary the distance of the page until the black spot disappears. The horizontal extent of the blind spot is indicated indirectly by the range of depth in which the black spot is invisible. Some indication of the vertical extent of the blind spot may be obtained by scanning the page along an imaginary vertical line through the fixation cross.

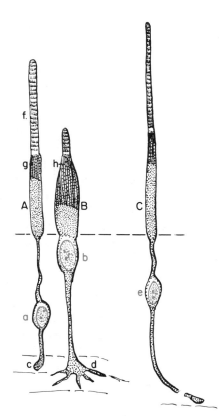

Figure 9.39. Drawings of human rods and cones as seen under light microscopy. *A*, Rod; *a*, Nucleus; *c*, Central connections; *g*, Ellipsoid; *f*, Outer segment, *B*, Cone; *b*, Nucleus; *d*, Central pedicle; *h*, Ellipsoid; *C*, Cone in the fovea centralis, *e*, Nucleus. (From Cajal, 1904.)

embryonic neural tube. During embryogenesis, the eyes begin with emergence of an optic vesicle on either side of the primitive forebrain. Arising as it does from the wall of the neural tube, the optic vesicle is lined with ependymal cells. The leading wall of the ballooning optic vesicle falls back into the stem (like a pushed-in hollow rubber ball) so that the end of the emerging stalk becomes cup-shaped).

As the optic cup approaches the surface epithelium of the face, it induces formation by epithelium of a translucent cornea and lens. The inner wall of the cup becomes the retinal nerve net which sends axons back along the optic stalk, while the outer wall becomes the pigmented layer of the retina. The retina, like the olfactory bulb, is thus a "migrated ganglion of the brain." This enables the brain to initiate sensory analyses as early as possible following receptor activation.

An orderly arrangement of retinal circuits initiates a cascade of information processing which proceeds from *photoreceptors* to *bipolar cells* to *ganglion cells* (shown diagrammatically in Fig. 9.40). Each human eye contains about 120 million receptors, with a ratio of about 20:1 of rods to cones. The count of bipolar cells is much smaller, and ganglion cells number slightly more than 1 million. Within the retina, therefore, there is a functional convergence of more than 100:1.

The main "throughput" of neurons (plus their feedback via local microcircuits)—from photoreceptors to bipolar cells to ganglion cells—is modulated by two layers of distinctive cells which make lateral connections within the retina (Fig. 9.40). These are (a) *horizontal cells,* which lie within the outer nuclear layer, at the level of the junction between photoreceptors and bipolar cells, and (b) *amacrine cells,* which lie within the inner nuclear layer at the junction between bipolar cells and ganglion cells (Fig. 9.40).

All horizontal cells have synaptic junctions with many photoreceptors and they influence both receptors (by feedback) and bipolar cells (by feedforward). Horizontal cells have receptive fields that extend well beyond their anatomical boundaries. This is achieved by gap junctions which interconnect many horizontal cells serially with increasingly remote horizontal cells. One class of horizontal cells, which lack axons, has connections preponderantly with rods and has to do mainly with dim light vision. Another class of horizontal cells, which possess axons, has connections preponderantly with cones and has to do mainly with bright illumination and color vision.

Amacrine cells receive from bipolar cells and feed

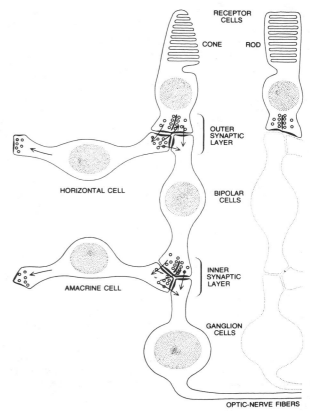

Figure 9.40. Circuitry of the retina was deduced from electron micrographs by Dowling, depicted schematically to show the direction of neurotransmitter release and hence the direction of excitatory or inhibitory influence from receptor to ganglion cells. Bipolar and ganglion cells pass along signals based on interaction of vertical and lateral cells; lateral interneurons transmit back to cells that drive them, across to one another and on to the succeeding mainline transmission cells. (From Werblin, 1973.)

back modulating signals to them as well as feed forward signals to ganglion cells. Horizontal and amacrine cells are involved in lateral integration within the retina thus contributing to sensory processing by integration of both local and remote signals occurring throughout the retina. When impulses leave the retina by way of ganglion cell axons in the optic nerve, they are complexly representative of time- and space-related image organization.

PHYSIOLOGY OF THE RETINA

We have considered how the pupil regulates the amount of light entering the eye. Additional controls of light flux operate within the retina. These preserve the full range of retinal responses in accordance with local retinal illumination and local luminance contrasts, for any given level of overall intensity of incident light, from noon at the beach to dusk under the trees. First, photochemical processes in individual receptors adjust the efficacy of photochemical transduction over a span of seconds to minutes so that the receptors operate discriminatively within a wide range of intensities, depending on any given time-averaged light intensity at that cell location. Further, horizontal and amacrine cells rapidly modulate retinal throughput in accordance with the average of light falling on the retina as a whole as well as its local incidence.

Obviously it is not feasible for each bipolar and each ganglion cell to respond to all nuances of all the photoreceptors with which they are linked. The consequences of horizontal and amacrine cell contributions are that the retina can form high-contrast, sharply defined neural images throughout the extremely wide range of intensities of incident light. This integrated image takes into account a multitude of factors occurring across the retinal nerve net while conveying specific and detailed information relating to each locale.

The *receptive field* for any neuron in the visual system is defined as that area of the retina or that corresponding part of the visual field from which the discharge of that neuron is most effectively influenced. Within a receptive field, illumination of some areas of the retina produce increased firing, and illumination of others produce inhibition of a given cell's activity. In lightly anesthetized cats and monkeys, particular parts of the retina can be stimulated by shining spots or patterns of light onto a screen the animal is facing, at a distance for which the eye is appropriately refracted. By this means the entire visual system can be investigated for individual neuronal responses to various patterns of light and color.

Ganglion cells have two basic kinds of receptive fields and are characterized as being *on-center* or *off-center* neurons. Because receptive fields overlap extensively, even small (0.1-mm) retinal spots of light strike many receptive fields simultaneously and generate antagonistic patterns of response in many different ganglion cells. Receptive fields near the central retina are generally smaller than those in the periphery, the range being from 0.5° diameter for the center zone of central as compared to 8° in peripheral regions of the retina.

Characteristically, light-receptive cells and other units in the retina, including most ganglion cells, and most cells throughout the rest of the visual system, are spontaneously active, exhibiting graded responses and spike discharges, even in the absence of illumination. Such background activity enables the nervous system to respond to subtle shifts in visual stimulus characteristics, and subtler changes in the organization of visual information than would be the case if some fixed threshold of stimulus intensity were required to induce activity on the part of each relay along the visual pathway.

RETINAL RECEPTORS

Retinal receptors are of two kinds: *rods*, for gray level discrimination and dark-adapted (scotopic) vision, and *cones*, for color discrimination and light-adapted (photopic) vision. Both kinds of receptors are segmented. The outer segment is a modified cilium which is filled with photoreceptive double-membrane discs. A typical rod (Fig. 9.39A) has a long, slender outer segment. A typical cone (B) has a short, squat outer segment. Cones in the fovea however, have long, slender outer segments with thousands of photoreceptive discs (C), an advantage for light-gathering purposes. Further, foveal cones are only 1–2 μm in diameter, thus enabling fine spatial discriminations in central vision. Outer segments of both kinds of receptors contain hundreds to thousands of discs with an aggregate disc surface of hundreds of thousands of square microns. Discs undergo rapid turnover, being replaced from specialized plasma membrane at a rate of about 30 discs per day.

Rhodopsin, a globular molecule, is embedded in the disc membrane. It undergoes a *cis* to *trans* isomerization when exposed to light. Rod outer segments contain biochemical machinery for light adaptation as well as photic transduction. The initial influence of light is a 10 fold activation of a cyclic GMP phosphodiesterase (PDE). Bleaching of one rhodopsin molecule leads to the disappearance of about 5×10^4 molecules of cyclic GMP, a large amplification of the light signal. The dark level of GMP concentration is restored in less than a minute after light is extinguished.

In the dark, photoreceptor cells undergo continuous active *depolarization*, resulting in a continuous emission of neurotransmitter. This provides extraordinary sensitivity for weak light stimulation. Continuous depolarization is accomplished by an ion pump, energy for which is provided by mitochondria. Sodium permeability of the plasma membrane of the outer segment is maintained in the dark by GMP-mediated protein phosphorylation. The inner segment of the photoreceptor releases neurotransmitter at a near maximum rate in the dark whereas during bright illumination, transmitter release is almost completely arrested. A consequence of photons striking a rhodopsin molecule in a disc membrane is that GMP is reduced. This induces a transient suppression of sodium permeability in the rod plasma membrane whereby the photoreceptor cell membrane *hyperpolarizes* (increases internal negativity).

Gap junctions link neighboring cones of the same color-receptor type. This provides an electrical coupling which enables spatial summation within a radius of about 40 μm for functionally similar cells. Spatial summation among rods covers a much wider area, with a radius of some hundreds of μm. Cone receptor potentials rise and decay more rapidly than rod receptor potentials. This relates to how well two closely spaced light flashes can be distinguished from a single flash. The lowest frequency when repeated flashes appear to fuse together into a single stimulus is called the critical fusion frequency (CFF). Optimal CFF for the fovea is about 50 Hz, reflecting response recovery characteristics of cones. CFF for flashes 20° outside the fovea is about 20 Hz. At low levels of flash intensity, the CFF falls off abruptly to levels of 10 Hz and below. Motion picture perception depends upon CFF; images are satisfactorily fused, out to about 10° outside the center of vision, at normal projection rates of 24 frames per second. At levels of high intensity illumination, the rates of recovery of both rods and cones are retarded. This is the basis for positive after images, as when the image of a flash persists for several seconds.

ELECTRORETINOGRAM (ERG)

The electroretinogram (ERG) is generated by electrical activity of the retina responding in synchrony to large-scale illumination (Fig. 9.41). The ERG is used to investigate retinal physiology and its disorders. An initial, brief *a-wave*, negative on the vitreal side of the retina, reflects *receptor cell responses to light*. A slightly later vitreal-positive *b-wave* presumably arises from *ganglion cell activity*.

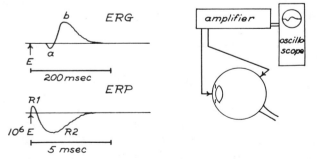

Figure 9.41. Diagram shows method of recording an electroretinogram (*ERG*) and early receptor potential (*ERP*) in a dark-adapted eye. Each potential fluctuation is biphasic, cornea-positive (upward), and cornea-negative (downward). The ERP has no measurable latency but requires much more intense illumination. (From Wald, 1972.)

A large, slow *c-wave* is triggered by receptors and sustained by potassium leakage which prolongedly hyperpolarizes the pigmented epithelium.

Neural Connections of the Visual System

RETINA

Entering light rays converge at a point between the lens and the retina and rediverge in reverse order so that topographic relations between the visual field and its image on the retina, like a camera's image on photographic film, are reversed left-right and up-down. Topological relations established in the retina are preserved in projections throughout the several different parallel central visual pathways. Retinal projections are thus mapped and remapped while serving a variety of analytic and associative purposes in several different regions of the brain.

Retinal throughput involves receptor, bipolar, and ganglion cells from "outer" to "inner" layers of the retina, referring to the interior of the eye (Fig. 9.40). Bipolar cell bodies constitute a middle nuclear layer. One proto-plasmic process of bipolar cells reaches outward to make direct synaptic contact with both receptors and horizontal cells in the outer nuclear layer. The other process reaches inward to make direct synaptic contact with both amacrine and ganglion cells in the inner nuclear layer. All cells of the retina, except ganglion cells, react with graded responses rather than with fully regenerative action potentials. Ganglion cells are the only cells in the retina that have typical axons, and their axons convey all visual information from retina to brain (Fig. 9.40).

Horizontal cells make synaptic contacts with large numbers of receptors and receive indirect influences from many other receptors via gap junctions which they share with other horizontal cells. Synapses between receptors and horizontal cells are reciprocal, hence, horizontal cells can provide large-scale, dynamic spatial interactions among many receptors. Light falling on the retina a few degrees distant from a given patch of receptors reduces the amplitude of the receptors' graded responses because of lateral inhibition by horizontal cells. Lateral inhibition of cones by horizontal cells is color specific.

Large bipolar cells receive from both rods and cones; small bipolars receive exclusively from cones. In much of the fovea, "midget bipolar cells" have synapses with only one cone and activate corresponding "midget ganglion cells." This provides a direct "private line" to the brain from single cones in the centralmost visual areas.

Responses of both large and small bipolar cells are of two kinds. On-center bipolars are depolarized by a spot of light in the center of their receptive area and hyperpolarized by illumination of an annular surround. Off-center bipolars are hyperpolarized by center spot illumination and depolarized by annular surround illumination. These two types of bipolar cells connect directly and via amacrine cells to two classes of ganglion cells which are specifically activated either by spot on-center lighting or by off-center lighting (dark center) events. The corresponding ganglion cells are named *"on-center neurons"* and *"off-center neurons,"* as are subsequent neurons throughout the visual system which respond to these distinctive center-surround signals.

Outer plexiform ("outer synaptic layer" in Fig. 9.40) retinal microcircuits, among receptors and horizontal cells, contribute the first integrative steps by establishing *antagonistic center-surround and color distinctions*. Inner plexiform ("inner synaptic layer" in Fig. 9.40) retinal microcircuits, among bipolar, amacrine and ganglion cells, establish the *orientations of light-dark and color boundaries*. These integrated visual signals characterize ganglion cell, thalamic, and initial cortical neuronal responses.

Efferent Influences on the Human Retina?

It is well-established that efferent control is exerted on the retina in birds. Neurons originating in the thalamus send axons to terminate on amacrine cells. These projections modify retinal responses to visual stimuli and improve visual discrimination in dim light (Miles, 1972). It is moot whether such efferent controls exist in humans, although fibers from diencephalon to retina have been demonstrated in fishes, reptiles, amphibia, birds, and nonprimate mammals. There is, however, some indirect evidence for efferent control of the retina in humans: (1) postmortem, years after enucleation of an eye, the optic nerve contains about 10% of the normal population of axons (Wolter, 1965; 1968); (2) it has been observed in humans that visual attention alters electroretinographic (ERG) responses to visual stimuli (Eason, 1981; Eason et al., 1983). These observations suggest that functionally influential centrifugal projections do exist in humans. If efferent fibers go to amacrine cells, as they do in birds, they could influence the final consequences of retinal integration and filter retinal signaling to the brain. It is noteworthy that analogous efferent projections are found in other sensory systems and likewise influence early stages of sensory processing (Livingston, 1978).

Categories of Ganglion Cells

A powerful method for following neuronal pathways involves the injection of selected populations of cells with horseradish peroxidase. This "retrograde-uptake" method of staining whole cells has been used to define input/output relations for several different categories of retinal ganglion cells in monkeys. All types of ganglion cells send projections, mostly by way of collaterals, to the pretectal area where they participate in pupilloconstrictor reflexes, and to the superior colliculus (SC) where they contribute to the control of eye movements.

"A" cells have large dendritic fields and large cell bodies in the retina which give rise to coarse axons projecting to the magnocellular portion of the lateral geniculate nucleus (LGNd). "B" cells have small cell bodies with small dendritic fields and medium-diameter axons. Within the central retina, especially small "B" cells make up a population of "midget" ganglion cells. These provide nearly "direct line" connections to central relay stations from individual cones and bipolar cells. "C" and "E" cells constitute two more heterogeneous groups with small to medium-sized cell bodies, large dendritic fields and small-diameter axons. Both "C" and "E" cells project to the superior colliculus and the immediately adjacent pretectal region of the diencephalon.

A small population of ganglion cells shows a miscellany of attributes: some show directional selectivity, responding to a certain direction of light-contrast movement, but not to movement in the opposite direction; some maintain continous activity which is suppressed by the presence of contrast within their receptive field; others are excited by the presence of contrast within their receptive field.

OPTIC NERVE, OPTIC CHIASMA, RETINAL PROJECTIONS

Ganglion cell axons sweep across the retina and converge at the optic disc, exiting the globe as a bundle of fibers—the *optic nerve* (Figs. 9.31 and 9.42). Optic nerve fibers represent the whole visual field as projected onto each retina. They continue centralward with half of the fibers crossing in the *optic chiasm*. Fibers from the paired nasal and temporal retinal halves of each eye separate and recombine so that representations of the contralateral visual fields in both eyes are combined on each side (Fig. 9.42). Thus, dual optic nerve representations of the whole visual field are converted by the partial crossing of the optic chiasma into bilateral representations that align information from the contralateral visual fields for both eyes in each of the optic tracts (Fig. 9.42).

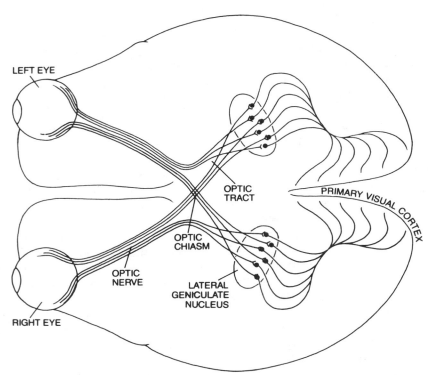

Figure 9.42. Visual pathway represented schematically in the human brain, seen from below. Output from the retina is conveyed by ganglion cell axons via the optic nerve and, after partial crossing, the optic tract to the lateral geniculate nuclei. Each half of the visual scene is projected to the geniculate of the opposite hemisphere. Neurons in the geniculates send their axons to the primary visual cortex. Enucleation of one eye or section of the optic nerve obviously yields an ipsilateral blind eye. Section of fibers crossing in the optic chiasm leads to "tunnel vision" by blinding both nasal retinal halves. Section of the optic tract or destruction of the geniculate or visual cortex causes loss of vision for the contralateral half visual field. The pathways are so orderly that quadratic and even smaller deficits can localize lesions to specific parts of the optic radiations and cortex. (From Hubel and Wiesel, 1979.)

The optic chiasm and optic tract send collaterals to at least six brain regions: (1) the hypothalamus (*suprachiasmatic nucleus—SN*), (2) the pretectum (*accessory optic system—AOS*), (3) the pretectum (*pupillary reflex system—PRS*), (4) the optic tectum (*superior colliculus—SC*), (5) the ventral thalamus (*ventral lateral geniculate nucleus—LGNv*), and (6) the lateral thalamus (*dorsal lateral geniculate nucleus—LGNd*).

Each of these destinations of ganglion cell axons and their onward projections plays an important role in various sensory and motor mechanisms relating to vision. For example, Sprague et al. (1981) showed that removal of cortex analogous to striate and peristriate cortex in humans, in cats trained and tested before and after cortical ablation, there is excellent retention of pattern and form discrimination and only minimal deficiencies in visual attention, orientation, depth discrimination, and visuomotor activity, postoperatively. The authors interpret this to mean that the suprasylvian cortex which receives predominantly from pulvinar in the thalamus must be providing rich information for visual perception. Note that visual perceptual and other cognitive functions are widely distributed, and that many traditional assumptions about what is primary and what is secondary in perceptual processing are subject to continuing evaluation.

The inverted image in each retina is relayed in a straightforward way through the LGN to visual cortex so that the upper portion of visual cortex in each cerebral hemisphere, above the calcarine fissure, represents for both eyes the lower quadrant of the contralateral visual field. Analogously, the lower portion of visual cortex, below the calcarine fissure, represent for both eyes the upper quadrant of the contralateral visual field. The anatomical topology of each retina is thus preserved with the original optical left-right, up-down reversals continuing through cortical representations. Thus, corresponding halves of the visual field in both eyes are represented upside down on each hemisphere.

Suprachiasmatic Circadian Control

The suprachiasmatic nuclei lie dorsal to the optic chiasm and immediately lateral to the third ventricle. Each nucleus receives bilateral projections from retinal fibers, collateral branches of axons in the optic tract. The SN exerts neural control over several circadian rhythms: eating and drinking, locomotor activity, sleep/wakefulness, and levels of pineal secretion of melatonin and adrenal cortisol (Moore, 1974). SN generates these rhythms and, with some hysteresis, regulates their schedule according to the intensity and duration of environmental light. The highest glucose consumption of the SN is in daytime, even without light cues. Bilateral ablation of the SN eliminates these integrative circadian controls and interferes with reproductive behavior.

PUPILLARY REGULATION AND OTHER OCULAR REFLEXES

The pretectal region, immediately anterior to the superior colliculus, receives fibers from many sources: retina, superior colliculus, the ventral part of the lateral geniculate nucleus, striate and extrastriate visual cortex, and frontal eye fields. These inputs are preponderantly retino-topically organized. Retinal input is bilateral, involving all ganglion cell types. They contribute to the *pupilloconstrictor reflex*, already described; *ocular fixation reflexes*; and *optokinetic nystagmus*. The latter two depend on the integrity of striate cortex.

Accessory Optic System (AOS) for Integration of Visual-Vestibular Reflexes

Terminal nuclei in the pretectal region receive axon collaterals from the optic tract which respond selectively to slow vertical movements of targets and visual surround. Neighboring *nuclei of the optic tract* receive collaterals which respond to slow movements in the horizontal plane. These two nuclear groups relay distinctive visual messages via an accessory optic system (AOS) which integrates visual, vestibular, and gaze control information for perception and analysis of movements in the environment in relation to the position and motion of head and eyes in three-dimensional space.

Optic fibers which contribute to the AOS originate from large, directionally selective retinal ganglion cells which possess very large receptive fields, up to 90°. Cells in the terminal and optic nuclei respond optimally to slow movements of broad, textured patterns. They send axons across the midline to enter the flocculonodular lobe of the cerebellum as mossy fibers. Other axons go uncrossed to the ipsilateral inferior olive which relays impulses across the midline to enter the same cerebellar cortical areas via climbing fibers. This area of cerebellar cortex receives directly from the vestibular apparatus and indirectly from the vestibular nuclei. The cerebellar cortex provides oculomotor centers with velocity command signals that serve ocular pursuit movements.

Retinal impulses influence cerebellar cortex with respect to movements that sweep across large areas of the visual field, especially movements along the horizontal and vertical axes of the retina. The flocculonodular lobe integrates horizontal and vertical motions of visual targets and surround with head and eye movements indexed to three-dimensional space.

AOS activation by visual surround and target motion contributes to stabilizing the gaze at frequencies too low for the vestibular system to be effective. The AOS improves vestibular contributions relating to slow head movements. The resulting integrated signal, representing head velocity in three-dimensional space, is transmitted to thalamus and cortex where it contributes to subjective perception of motion in three-dimensional space and to the interpretation and control of relative head and eye movements.

Access of visual signals to vestibular nuclei is through the AOS via the cerebellum. The AOS thereby contributes to vestibulo-ocular reflexes (VOR) which function to stabilize the gaze during active and passive head movements. Intention to fix the gaze modifies neuronal activity in the vestibular nuclei. This is one way by which motion sickness may be partially ameliorated, by employing deliberate visual activity.

As related in Chapter 63 (Vestibular Function), the vestibulo-ocular reflex (VOR), enables eyes in a moving head to stabilize on a nonmoving environment. *The AOS appropriately overrides the VOR when the eyes are tracking a target that is moving through the visual field.* The cerebellum, based on AOS and vestibular information, adjusts gains for perception and motor control to match the relative velocities of head and eyes to the environment, and to the relative motion of targets moving through the visual field. Plasticity in the AOS system is demonstrated when a human or animal subject wears prisms or reversing lenses. Surgical removal of the flocculonodular lobe eliminates this remarkable integrative plasticity (Melvill Jones, 1977) and reduces or eliminates motion sickness (Bard, 1947).

Superior Colliculus and Visuosomatic Integration

The superior colliculus (SC) is organized with retinotopically accurate representation of target-directed eye movements. The SC resembles cortex by being a richly layered structure. Collateral fibers from the optic tract, including axons from certain retinal ganglion cells, distribute terminals within the surface layers of the SC. The SC receives additional topologically organized projections from both striate and extrastriate visual cortex. These project to layers of the SC immediately beneath the topologically corresponding retinal projections. The deeper layers of the superior colliculus are involved in the generation of saccadic (rapid) eye movements. A large population of superior collicular neurons is active before each saccade. The direction of gaze appears to be determined by population averaging whereby the direction, amplitude, and velocity of saccadic eye movements is generated by a vector summation of the movement tendencies within a large population of active neurons. Variability in the discharge frequency of a given neuron ("noise") is minimized because the contribution of any given neuron to the direction and amplitude of the movement is slight. The large population of neurons active during a specific saccade ensures probabilistic accuracy of the movement. This appears to be an exemplary system for neuronal group selection (Edelman, 1987).

The circuitry of the SC is also organized for the integration of visual with nonvisual information, and for the generation of appropriate head and eye movements and other bodily movements directed toward objects of visual interest. Deeper-lying SC cells direct body, head, and conjugate eye movements toward that part of the visual field which corresponds to the particular collicular site activated. This provides a *visual orienting response* which can be initiated by visual, acoustic, or somesthetic excitation and by voluntary direction of attention. The SC is also aroused by stimulation of the reticular activating system. Using its informational sources, the SC generates a *visual grasp reflex*, an obvious contribution to evolutionary advantage.

Deep-lying collicular layers receive a variety of other visual and related contributions, including (a) projections from the frontal eye fields (area 8 of Brodmann) by which voluntary eye movements are elicited, (b) topologically organized projections from sensorimotor and auditory cortices, and (c) projections from cells in the substantia nigra which enable integration of visual exploration, body imagery, and visuoacoustic relations throughout the spatial environment.

The SC receives additional inputs from the hypothalamus (motivation and neuroendocrine status), the lateral geniculate nucleus (spatially organized visual scene), diffusely projecting thalamic nuclei (state of arousal and direction of attention) the inferior colliculus (spatially organized acoustic scene), the periaqueductal grey (pain and other somatic information), the cerebellum (sensorimotor servo-controls), and from trigeminal and spinal sources, all of which preserve somatotopic representations of the body.

Visual cortex and frontal eye fields are needed to *search* for visual targets when there are no localizing clues, and the SC is undoubtedly also utilized in this search process. Once localizing clues, accoustic, somesthetic, or visual, are available, the colliculus can direct the gaze binocularly toward such sources and *follow* moving targets, using convergent centrifugal eye, head, and body command and control information.

The SC contributes to two major thalamocortical visual systems: first, projections to the classical visual pathway via both ventral and dorsal components of the lateral geniculate nucleus. Second, projections to the lateral-posterior part of the pulvinar from which next-order projections are relayed, preserving topological correspondences, to striate and extrastriate visual cortex. This latter parallels the classical retinogeniculocortical pathway, with but one additional synaptic relay.

The superior colliculus gives rise to a crossed tectospinal tract which terminates in grey matter of the dorsal horn in the cervical spinal cord. By means of this pathway the SC contributes to head-turning and head-eye coordination. Both crossed and uncrossed tectal fibers project to brainstem interneurons which relate to the eye motor nuclei (cranial nerves III, IV, and VI). These tecto-oculomotor projections contribute importantly to *binocular fixation*. Similar projections reach the cerebellum by relay through both the pontine nuclei and inferior olives.

LATERAL GENICULATE NUCLEUS (LGN)

By the time visual information leaves the retina, significant visual pattern analyses have taken place. Signals have passed through two to four direct-path synaptic relays, and have been influenced along the way by selective laterally influenced microcircuits. Impulses generated by distinctive retinal events have combined at each relay into new patterns which take collateral patterned influences into account.

Organization of the Lateral Geniculate Nucleus

Whereas each retina involves convergence from about 120 photoreceptors to each ganglion cell, the thalamic relay is approximately one-to-one, i.e., there are about a million cells in each optic tract and a million cells to relay via each lateral geniculate nucleus (LGN). Individual optic tract fibers enter the LGN and divide into several branches, each of which may terminate on a different geniculate neuron. LGN organization provides for divergence, convergence, and overlap.

Cells of the LGN are organized into six layers: the

first, fourth, and sixth receive fibers from the contralateral eye; the second, third, and fifth receive fibers from the ipsilateral eye (Fig. 9.43). The upper image shows a normal LGN. The two reduced images show the gross anatomical consequences in both LGNs of removal of the right eye.

Even more precise retinotopic mapping in the LGN follows from the fact that small retinal lesions result in transneuronal degeneration of discrete clusters of geniculate neurons, confined to the appropriate laminae. A precise correspondence is maintained throughout the LGN between the two

RIGHT EYE

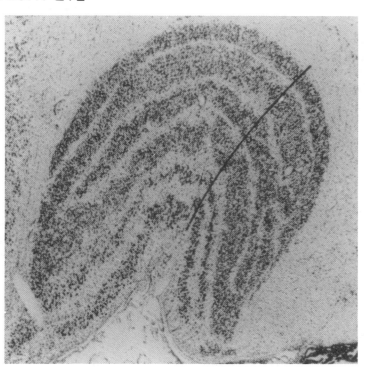

LEFT EYE

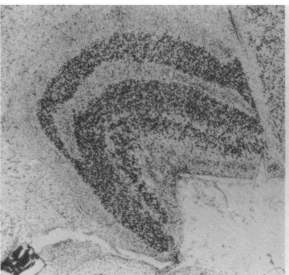

RIGHT EYE

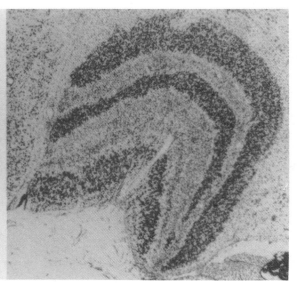

Figure 9.43. Lateral geniculate nucleus of a normal monkey *(top)* is a layered structure in which cells in layers I, IV, and VI (numbered from *bottom* to *top* receive their input from the contralateral eye, and those in layers II, III, and V receive information from the eye on the same side. The retinogeniculate maps are in register, so that neurons along any radius *(black line)* receive signals from the same part of the visual scene. The two lateral geniculates in the slightly reduced micrographs to the *below* are from an animal with vision limited to the left eye only. Note that in each of these geniculates the three layers with inputs from the right eye have atrophied. (From Hubel and Wiesel, 1979.)

corresponding retinal projections as they engage cells in their assigned laminae. Corresponding points from the two retinas line up over one another across the stack of laminae. The *black arc* in the normal LGN in Figure 9.43 indicates the alignment of cells that receive signals from the same part of the visual scene in both eyes. There is no evidence, however, that comparisons are made in the LGN between signals coming from the two eyes. Interocular comparisons begin only after visual information reaches striate cortex.

Responses of cells in the LGN to light patterns cast on the retina are remarkably similar to those of retinal ganglion cells. Almost all LGN units respond best to small light spots with "on-center" or "off-center" illumination. Opposite effects are elicited by changes of illumination in the annular surround. There is spontaneous activity against which excitation, inhibition, adaptation, and rebound effects are seen. Single LGN units, however, convey considerable information: precise stimulus location, approximate shape and size, intensity, and various temporal features including rate of onset, duration, and offset of illumination, whether the stimulus is steady, and whether it is moving. Most LGN cells include information about color. They respond well to light or dark line segments in any orientation which projects onto the center and across the surround of their retinal field. They integrate local light intensity levels in comparison with the average illumination in neighboring regions, and they are generally unaffected by changes in illumination across the whole retina.

Many individual LGN cells respond by increasing their activity on exposure to all wavelengths of light. Others, called color-opponent cells, respond with "red-on, green-off"; "green-on, red-off"; "yellow-on, blue-off"; or "blue-on, yellow-off".

Importance of Visual Experiences Early in Life

An interesting consequence of experience during development of the visual system is manifested in the LGN. If one eye is simply occluded for some months early in life, the corresponding laminae of the LGN do not disappear but are notably retarded in development. There is a limited window of time within which considerable functional restoration following deprivation is possible, provided that appropriate visual experiences become available. These phenomena have been demonstrated in a variety of species, including humans (for reviews, see Movshon and Van Slyters, 1981; Sherman and Spear, 1982).

The visual system for competent monocular and binocular vision is already available in the new-born, but its enduring functional integrity depends on there being appropriate visual experiences early in life. Clinically, it is well known that interference anywhere along the light path of the eye in infants needs to be corrected in early childhood in order to obtain adequate vision. Useful binocular vision depends on early binocular visual experiences. If a child has defective conjugate eye motor control, uncorrected during early life, even though each eye may have normal vision separately, binocular vision may be permanently impaired.

When animals are deprived of patterned vision (reared in the dark or with sutured eyelids), the circuit competence that is present at birth deteriorates. Even a relatively brief exposure (6 hours) to patterned vision after 6 weeks of dark rearing can largely restore competence. Early limitation or distortion of visual patterns lead to corresponding adaptations of visual responses (Held and Bauer, 1967; Blakemore, 1974). Neurons are apparently selected according to their effectiveness in coping with a given environment. Adaptive value is obvious; an animal develops a visual system to match its visual environment. Plastic adaptations are swift and powerful during early life, and some plasticity endures lifelong and is reinvigorated when there is brain injury or radically different environmental experiences are introduced (e.g., if a person wears disorting lenses).

Lateral Geniculate Projections to Visual Cortex

Axons of relay neurons in the LGN contribute optic radiations which project exclusively to striate cortex. This pathway maintains strict retinotopic representation so that even small defects in geniculostriate projections can be detected clinically by visual field testing. Moreover, small lesions in striate cortex result in retrograde degeneration of narrow sectors extending through all six layers of the LGN, indicating that the geniculostriate projections maintain precise retinotopic relations with respect to both retinal halves as the projections engage cortex. Close juxtaposition of the two retinal representations now involves comparisons of visual information coming from the two eyes. Binocular image matching is essential for convergence and accommodation.

STRIATE CORTEX

Striate cortex in humans (area 17) is located largely in the posterior half of the medial surface of the occipital lobe (Fig. 9.42). Retinal receptive fields appear smaller as visual targets are projected farther into the retinal periphery; the same millimeter along striate cortex represents an increasingly larger segment of visual field; this corresponds roughly to differences in density of ganglion cells between retinal paracentral and peripheral regions. An equivalent of a few thousand retinal ganglion cells are represented in each square millimeter of cortex. With over a million ganglion cells, the retina is represented across a few thousand square millimeters of primary visual cortex.

The number of neurons observed along a line perpendicular to the surface of the cortex, that is, through its thickness in a 30-μm-diameter column, is remarkably constant, about 110 cells in most areas of cortex and in most mammals (Rockel et al., 1974). The count for striate cortex in primates, however, is approximately double that, 260, and thickness of striate cortex is also uniquely increased. Furthermore, striate cortex is distinguished by sharply demarcated boundaries within which cortical layering is distinctly visible, hence the name "striate."

Afferent Input to Striate Cortex

The main afferent input to striate cortex comes from the LGN which distributes abundant terminals within layer IV. Geniculate axon terminals establish synaptic endings on spines of dendrites and dendritic surfaces of pyramidal and stellate cells in layer IV. Afferents from the pulvinar nucleus, representing the major alternative access to visual cortex, in contrast, end in layer I. Since this is the most superficial layer of cortex, offering synaptic terminals on the farthest branches of apical dendrites of striate pyramidal cells, the influence of pulvinar input on those cells is presumed to be modulatory rather than controlling.

The only interhemispheric cortical projections within primary visual cortex lie in a narrow vertical zone of area 17 along its border with area 18. These corpus callosum crossings contribute to a seamless subjective matching of the vertical midline which separates the two retinal visual half-fields. The two half-fields were divided into separate half-field hemispheric maps by the partial crossing of optic nerve fibers in the optic chiasma.

Other Visual Projections to Striate Cortex

As noted previously, there is a retinotopic longer-circuiting visual pathway that goes by way of the superior colliculus to the LGN and then projects to striate and extrastriate visual cortex. This slightly roundabout path contributes to classical visual cortical information gained from the superior colliculus, a convergent point for information from a variety of sensory and motor sources. Another visual projection via the superior colliculus is relayed to the lateral posterior pulvinar and thence to striate and extrastriate visual cortex. Both colliculus and pulvinar receive reciprocal feedback projections from each of their respective cortical projection fields. There is a less abundant projection from retina to the pretectal area, which also relays to the lateral pulvinar.

Visual Mechanisms Relating to Arousal

Another system of projections important to visual processing comes from the midbrain reticular formation. This region receives visual information mainly from the superior colliculus. The reticular formation is readily activated by changes in visual input, as by a flash of light or sudden darkness, or by significant changes in visual pattern. The reticular formation, in turn, activates thalamic diffusely projecting nuclei which relay cephalically to arouse the entire forebrain, including striate and extrastriate visual cortex. Effective forebrain arousal can also be achieved by visual attention that is generated in relation to cognitive processing of visual signals (Galambos and Hillyard, 1981).

Efferent Output from Striate Cortex

Neurons in different layers of striate cortex show different response characteristics and different degrees of complexity. After an unknown number of intracortical synaptic transformations, layer VI of striate cortex projects to LGN, providing a direct sensory feedback. Layer V projects to the superior colliculus, contributing to eye motor control, particularly to keeping both eyes oriented and focused on the target of interest. Layers II and III provide projections to other cortical regions.

Functional Characteristics of Intrinsic Striate Neurons

Hubel and Wiesel divided visual cortical cells into categories of *simple*, *complex*, and *hypercomplex*. They found that cells possessing similar characteristics: receptive field location, line segment orientation, and ocular dominance, tend to be grouped together. Many "on-center" and "off-center" cells respond like geniculate cells, with circularly symmetrical receptive fields. Another large population responds preferentially not to spots of light but to line segments that have specific orientations (Fig. 9.44).

Oriented Line Segments

Line segment discrimination specifies a particular orientation. Recall that angular rotation of the hour hand on a clock for 1 hour is 30° and appreciate the precision of cortical cells for line segment orientations. By identifying the orientation to which a neu-

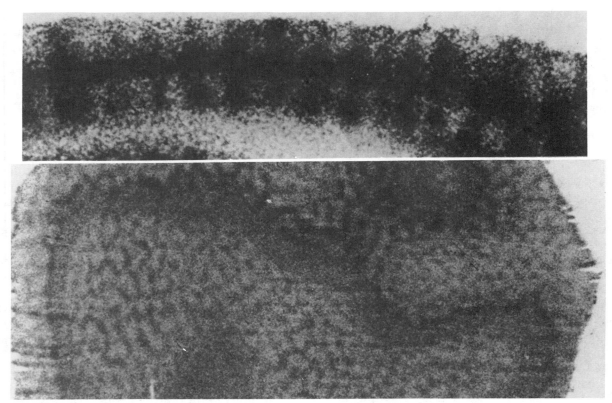

Figure 9.44. Orientation columns in the primary visual cortex of the monkey, visualized by deoxyglucose autoradiography, seen in cross-section *(top)* and tangential section *(bottom)*. Immediately after injection of the radioactively labeled deoxyglucose the animal was stimulated by a pattern of vertical stripes so that the cells responding to vertical lines were most active and accumulated greater stores of the incompletely metabolized glucose. Active cell regions constitute nar-row bands about 0.5 mm apart. Layer IV, with no orientation preference, is uniformly radioactive. Seen in the tangential section *(bottom)* the large oval darker region represents layer IV. In the other layers the vertical orientation columns constitute intricately curved bands, like the walls of a maze seen from above, and the distance from one band to next is uniform. (From Hubel and Wiesel, 1979.)

ron responds best, one can establish that it discriminates differences about 10–20°, i.e., a clockwise or counterclockwise shift of orientation of only about 5–10° to either side of best-response orientation will reduce or abolish the response.

Within a local region of cortex all orientations are separately represented so that each region of the visual field can generate localized cortical activities according to given orientations of line segments and hence specify the outlines of any objects in view (Fig. 9.44).

Neurons which respond to an oriented line in a restricted location are called *simple cells.* Figure 9.45 illustrates schematically the way in which it is supposed that a linear array of retinal center-surround ganglion cells could integrate information that would define a line segment and activate simple cells in striate cortex via the lateral geniculate nucleus. Such cells are found in abundance in the vicinity of layer IV.

Complex cells respond best to lines of a particular orientation moving across their receptive field.

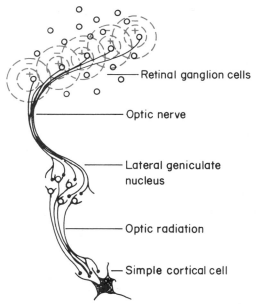

Figure 9.45. Connections from retinal ganglion cells through cells of the lateral geniculate body of the simple cells of the cortex which respond to line element stimulation of the retina.

Somewhat more complex neurons respond specifically to oriented lines moving in one direction only, and are inhibited by lines of similar orientation moving in the opposite direction. Some cells show a strong preference for line segments of limited length and thus contribute to corner as well as edge definition.

Hypercomplex cells respond best to two or more borders presented anywhere within a large retinal area, regardless of the size of the stimulus. Cells having these characteristics provide a basis for complex boundary detection and detection of movement of contrasting figures without necessarily being restricted to particular regions of the visual field, a significant degree of increased abstraction.

Figure 9.46 shows how a mixture of ganglion cells in a localized region of the retina, connected with corresponding cellular elements in the LGN, can provide not only a series of local line segment detec-

tors which have the same orientation, but how such cells could combine information necessary for complex cells to respond to directional motion of line segments. A combination of information from two such complex cells could account for the performance characteristics of hypercomplex cells responding to two or more borders in motion.

All simple cells in striate cortex appear to respond only monocularly. About half of all complex cortical cells respond to stimulation of corresponding points in each half-retina. Cells which respond binocularly have identical positions in respect to the visual field, respond to similar complexity of target illumination, and have the same orientation and preference for direction of movement.

Table 9.1 summarizes the incremental gain of differentiated capabilities of neurons at successive levels of the visual system, from receptors to striate cortex. All cells in striate cortex are presumed to

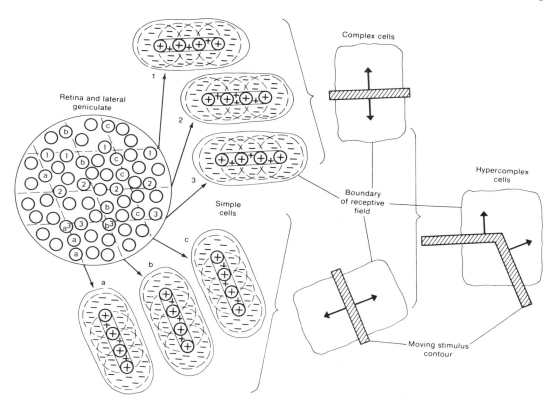

Figure 9.46. Response patterns and extraction of signal patterns in visual pathway. Many units are shown at the level of lateral geniculate body (at which responses are essentially like those of the retina). Each cell has a small concentric field with an on or off spot center and opposing surround; here all are assumed to be on-center units. If all cells marked *1* converge on a single *simple cell* in the primary visual cortex, this cell will have a larger, linear receptive field, specifically responsive to a thin bar of light or a border at orientation shown, with light falling on the centers of the field of first order units and darkness covering only part of the surround. Stimuli at other orientations are relatively ineffective or inhibitory. Other simple cells are illustrated, *2, 3, a, b,* and *c.* The same lateral geniculate cells could be contributing at the same

time, in different orientations. If several simple cells, each responsive to stimuli at the same orientation, converge on a *complex cell,* the resultant receptive field will be larger and specific for stimulus orientation and, often, movement or direction of movement, but the stimulus would not have to fill any particular part of the receptive field to be excitatory. Two or more complex cells may provide information for a *hypercomplex cell* in which stimulation is most effective when there are two or more borders whose orientation is determined by inputs. With continuing convergence of this sort, still higher order cells can be programmed to respond only to highly specific and complex shapes falling anywhere within a large retinal area, irrespective of the absolute size of the image. (Fom Bullock, 1977.)

Table 9.1.
Characteristics of Receptive Fields at Successive Levels of the Visual System*

Type of Cell	What is Best Stimulus?	How Good Is Diffuse Light as a Stimulus?	Is Orientation of Stimulus Important?	Is Position of Stimulus Important?	Are There Distinct "On" and "Off" Areas within Receptor Fields?	Are Cells Driven by Both Eyes?	Can Cells Respond Selectively to Movement in One Direction?
Receptor	Light	Good	No	Yes	No	No	No
Ganglion	Small spot or narrow bar over center	Moderate	No	Yes	Yes	No	No
Geniculate	Small spot or narrow bar over center	Poor	No	Yes	Yes	No	No
Simple	Narrow bar or edge	Ineffective	Yes	Yes	Yes	Yes (except in monkey layer IV)	Some can
Complex	Bar or edge	Ineffective	Yes	No	No	Yes	Some can
Hypercomplex	Line or edge that stops; corner or angle	Ineffective	Yes	Yes	Yes	Yes	Some can

(From Kuffler and Nicholls, 1976).

*Information is collected from retinal areas of varying size, emphasizing contrast between center and surround, and at higher levels, defining edges, corners, and directional motion of contrast boundaries.

derive their visual response characteristics from similar kinds of specific connections received from initial input signals resembling those of retinal ganglion cells, modulated by LGN as influenced by corticogeniculate projections, and further modulated by superior collicular contributions. These latter contain information concerning the direction of gaze, information from other sense modalities, and level of arousal and direction of attention.

Ocular-Dominance Patterns

Where geniculate axons terminate in visual cortex, inputs from the right and left eyes are still separated. Axons stemming from LGN layers I, IV and VI (contralateral eye) and from layers II, III, and V (ipsilateral eye) terminate in alternating slabs. Figure 9.47 shows ocular dominance columns in two views at right angles to each other, perpendicular to the cortical surface and tangential to the cortex. Adjacent parts of the ocular dominance slabs relating to the two eyes are closely matched so that corresponding regions of the two retinae lie side by side. Thus a general map of one-half of the visual field, suitably magnified, spreads across each occipital cortex. Ocular dominance for each eye is clearly evident in layer IV. Figure 9.48 shows a photomontage reconstruction of a portion of monkey visual cortex, with bright parts of the "map" representing the contralateral eye and dark parts representing the ipsilateral eye.

Orientation-specific neurons are disposed throughout the cortex in an orderly progression of orientation preferences. These are arranged in strips that course more or less orthogonally to the ocular dominance zones so that each member of a local ocular dominace pair (representing corresponding loci in each retina) contains a complete 180° sequence of line segment orientations. For given neurons in such a pair, the left-right dominance is determined by afferent input while the degree of dominance and the line segment orientation are determined intracortically. The combination of local systems analysis for binocularity plus orientation occupies an area of about 800×800 μm and constitutes a *macrocolumn* which is recurrent throughout primary cortical representation.

Both behavioral and neurophysiological evidence indicates that increasing the difficulty of a visual discrimination task correlates with enhanced responses and sharpened selectivity among those cortical neurons involved in processing the critical stimuli in striate cortex. Increased attention to a specific task induces changes that are limited to neurons activated by the attended stimuli. This is not a generalized effect such as level of arousal but is task specific and related to the task difficulty.

Poggio and Fischer (1977; and Fischer and Poggio, 1979) identified neurons in area 17 of waking, behaving monkeys that respond to receptive field disparities. They found excitatory and inhibitory neurons responding to depth cues in the visual field. This facilitates stereoscopic vision and contributes accurate information for visual fixation. Such neurons are presumably involved in the continual "search" for best binocular matching of images pro-

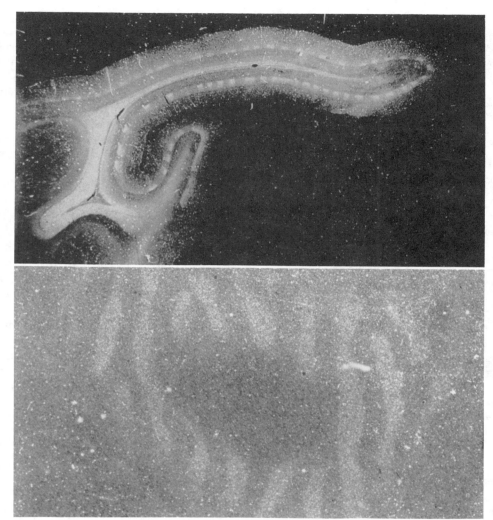

Figure 9.47. Ocular dominance columns in the primary visual cortex of the monkey seen in cross-section *(top)* and tangential section *(bottom)*. Bright bands are columns dominated by cells conveying right eye information. A radioactively labeled amino acid was injected into the right eye, picked up and transported by ganglion cells of the retina to the lateral geniculate nucleus whence it was picked up and transported to ocular dominance cells in the visual cortex where the radioactive particles accumulated and are seen as slab-like ocular dominance columns (about 0.4 mm diameter) from side *(top)* and facing *(bottom)* views. The dark intervals separating the bright regions represent ocular dominance columns relating to the left (uninjected) eye. (From Hubel and Wiesel, 1970.)

jected from the two eyes. Such depth-tuned neurons are widely distributed throughout the thickness of striate cortex, but most densely in layers V and VI which project to the superior colliculus and LGN. Neurons sensitive to larger degrees of binocular disparity, in front of the plane of fixation ("near neurons") or behind it ("far neurons") are more numerous in layers II and III, known to project to nearby cortical fields (areas 18 and 19) where stereoscopic perception appears to take place.

Columnar Organization of Striate Cortex

Mountcastle (1957) characterized the physiological significance of columnar organization of neocortex in his classical analysis of somatosensory func-

tions. The cortical column is a modular input-output circuit that processes and distributes information. Particularly complicated analyses apparently require expanded modular dimensions—as in striate cortex. The basic module is a vertically arranged group of cells heavily interconnected along that vertical axis, sparsely so horizontally (Edelman and Mountcastle, 1978). Individual columns may be further isolated horizontally by means of pericolumnar lateral inhibition. The human cortex is composed of an estimated 50 billion neurons, with a surface area of about 4000 cm^2. The number of columns is estimated at 600 million, each with a diameter of about 30 μm. These columns belong to larger processing units, estimated at 600,000 in number, which vary

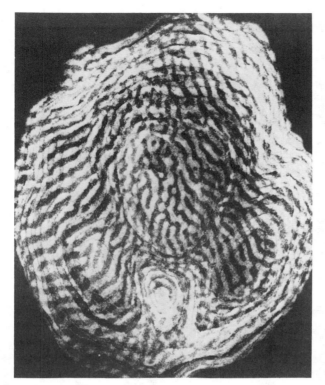

Figure 9.48. Ocular dominance pattern in about a fourth of the visual cortex of a macque monkey, a montage produced by LeVay from tissue slices. One eye was injected with tritium-labeled proline. The radioactive amino acid was transported from eye to lateral geniculate nucleus and thence to visual cortex. Bright stripes due to exposure of photographic emulsion reveal cortical zones that represent the contralateral eye. The dark stripes represent the uninjected eye. Each stripe is about 350 μm in width (From Constantine-Paton and Law, 1982.)

from 500–1000 μm in diameter in different cortical areas.

Columnar organization enables simultaneous mapping of several variables. Divergent connections among cells within a column enable selective information processing, e.g., "feature extraction" from among the input signals, and routing of new information to particular output destinations. Feature extraction in striate cortex has been established for binocular correspondence, line-segment orientation, as noted above, and for *distance* (Hubel and Wiesel, 1970), *direction* (Blakemore and Pettigrew, 1970), *movement* (Zeki, 1974), and *color* (Zeki, 1973; 1977).

By such means, separate subdivisions of the primate visual system analyze different aspects of the same retinal image. Major subdivisions provide for color selection, contrast sensitivity, and temporal and spatial qualities. Pathways selecting for form and color pass mainly via the dorsal (parvocellular) layers of the LGN, while those selecting for depth and movement pass mainly via the ventral (magnocellular) layers. Perceptual experiments indicate that visual abilities such as figure/ground discrimination, perception of depth from perspective and relative motion are differently segregated in these two subdivisions. This is consistent with observations that patients with strokes may experience loss of color discrimination without impairment of form perception, or loss of motion perception without loss of color or form perception, or loss of face recognition without loss of the ability to perceive color, motion, and depth.

Visual Representation in the Claustrum

A thin shell of subcortical grey matter lying beneath most of the cortex of the hemisphere is the *claustrum*. The posterior part of the claustrum has neurons that are primarily concerned with vision (LeVay and Sherk, 1981), neurons that receive from and send to striate cortex. Retinotopy is preserved in both directions. The claustrum appears to have no subcortical projections and only a small contralateral projection. Claustrocortical axons terminate in all cortical layers, but most heavily in layers IV and VI, whereby the claustrum can influence geniculocortical input (IV) and corticogeniculate output (VI) throughout striate cortex. Thus, the claustrum can modulate or impose gates at these two important cortical input and output channels.

EXTRASTRIATE CORTICAL AREAS CONTRIBUTING TO VISION

A partial map of striate and extrastriate cortical areas concerned with vision in the monkey is presented in Figure 9.49 (from Crick, 1979; originally from Allman and Kaas, 1972). Visual projection systems in the monkey probably resemble what we may expect in the human, although there is likely to be greater complexity in the latter. Nine separate areas are depicted, each of which has a distinctive magnification, orientation, and specialization of connections and functions. Although topology is preserved in nine of these areas, various magnifications and orientations are rerepresented, sometimes with mirror-images abutting one another. Each map has a somewhat different "trick" for analyzing visual data and each is integrated into the system as a whole by means of distinctive input-output connections. *Perception emerges from selective integrative processes among such distributed multiple reentrant transformations.*

The prestriate area contributes to retinal disparity detection and to integration of visual processing in the two hemispheres. Lesions here interfere with object and pattern discrimination, visual spatial ability, small-scale brightness discrimination, and visual learning and relearning. The prestriate area

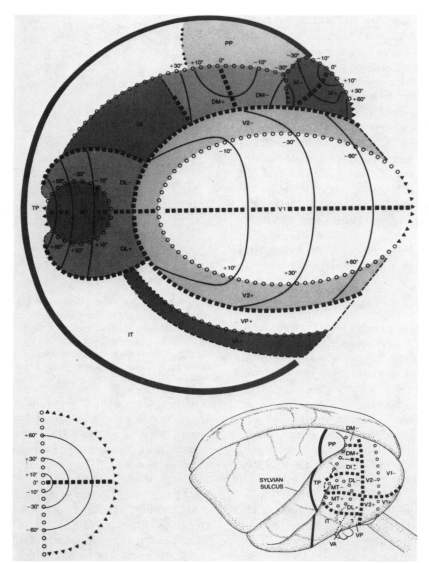

Figure 9.49. Visual cortex of the owl monkey exemplifies the tendency of the cerebral cortex to be "mapped into areas topologically related to their function. Nine areas are depicted, based on studies of Allman and Kaas, which are orderly maps of the monkey's visual field (and those areas that respond to stimuli from the visual field but do not seem to represent it in an orderly way). The *top map* depicts schematically the posterior third of left hemisphere cortex unfolded and oriented so that it can be viewed from above. The nine orderly areas are: first visual *(V1)*, second visual *(V2)*, dorsolateral crescent *(DL)*, the middle temporal *(MT)*, dorsointermediate *(DI)*, dorsomedial parietal *(PP)*, temporoparietal *(TP)*, and inferotemporal *(IT)*. The chart at the *bottom left* shows the right half of the visual field. *Solid squares* mark the horizontal meridian, *open circles* the vertical meridian, and *solid triangles* the outer boundaries of the visual field. These symbols are superimposed on corresponding parts of the sketch of the left hemisphere and on the schematic cortical map. *Plus signs* indicate the upper part of the visual field, *minus signs* the lower part. (From Crick, 1979.)

projects to an inferotemporal cortical region which lacks topographic representation of the field of vision. When this latter area is removed bilaterally in the monkey, behavior appears normal. There are no deficits in auditory or somesthetic functions, and no detectable defect in visual fields. Nonetheless, functional losses following these lesions relate to vision. They consist of faulty selection and maintenance of attention to visual cues, faulty objection and color discrimination, and impaired memory for learned visual discriminations. The syndrome reflects impaired categorization of visual memories (Weiskrantz, 1974). In awake monkeys, neurons in the parietal cortex activated by visual stimuli integrate visual cues that coordinate visuomotor mechanisms relating to the direction of visual attention (Yin and Mountcastle, 1977; 1978).

Extrastriate visual cortical areas are found anatomically and physiologically to belong to "family clusters" which are more intimately intracortically connected among one another with their special thalamic connections than with striate visual cortex and the classical visual "association" areas (Crick, 1979). These extrastriate groupings appear to have more to do with conscious visual experience than does striate cortex. Instead of there being a dominating principal visual pathway from retina via LGN to striate cortex, from which corticocortical projections distribute to "visual association cortex," there are three or more relatively separate but parallel projection systems between retina and cortex: retinothalamic (LGN relay), retinotectothalamic (superior colliculus-LGN relay) and retinopretectothalamic (pulvinar relay) pathways to cortex. Included with these are the still more roundabout transmissions of visual signals relating to arousal and attention that pass by way of the midbrain reticular formation.

Frontal Eye Fields

In addition to more than a score of cortical areas which relate to sensory processing for visual perception, there is a region in the middle of the convex surface of the frontal lobes, corresponding to area 8 of Brodmann, which is known as the frontal eye field. *The frontal eye fields are involved in voluntary eye movements.* Pupillary dilatation and conjugate movement of the eyes to the contralateral side are elicited by electrical stimulation in this region in monkeys, apes, and humans. Eye deviation may be accompanied by turning of the head in the same direction.

In unanesthetized monkeys, single neurons in this area discharge systematically during voluntary eye movements (Bizzi and Schiller, 1970; Bizzi, 1974). Like other voluntary actions, the direction of gaze can be interfered with by reflex mechanisms, in this context by head and neck reflexes, vestibulo-ocular (VOR) and related reflexes (e.g., AOS) involving the cerebellum. The frontal eye fields, among other sensory and motor fields relating to vision, send retinotopically organized fibers to the superior colliculus and pretectal area.

With respect to eye movements, *each frontal lobe contributes tonically to moving both eyes toward the contralateral side* (see Fig. 9.50, area 8). A destructive lesion in area 8, on one side, is followed by paralysis of conjugate gaze toward the contralateral side. The tonic balance is upset so that the eyes at rest are conjugately deviated toward the side of the lesion. Such paralytic deviation of the eyes occurs frequently with vascular lesions in the internal capsule, because of interference with projections between frontal eye fields and brainstem. When asked to look to the side opposite such

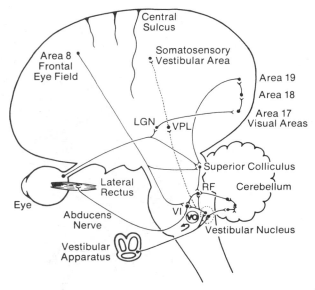

Figure 9.50. Circuits involved in control of eye movements. For simplicity, only control of the lateral rectus muscle is illustrated. Similar connections are involved in control of other extraocular muscles. Abducens nucleus *(VI)*, reticular formation *(RF)*, lateral geniculate nucleus *(LGN)*, thalamic ventroposterior lateral nucleus *(VPL)*, vestibulooocular reflex *(VO)*. (From Shepherd, 1983, adapted from many sources.)

a lesion, the patient is unable to do so. If, however, the patient is asked to look at an object that is slowly moved across the visual fields, the eyes can pursue a normal range of movements in either direction. Moreover, if the patient is asked to fixate on a point directly ahead, and the head is then turned slowly toward the side of the lesion, the gaze can be maintained straight ahead even though this involves directing the eyes toward the side to which gaze cannot be voluntarily directed.

Eye movements can also be evoked by stimulation of striate cortex (Fig. 9.50, areas 17 and 18). The resulting movements direct and fix binocular gaze onto that region of the visual field which is represented in that part of the cortical "map." Since the hemisphere stimulated relates to the contralateral visual field the eyes are directed contralaterally, but not so abruptly and not predominantly in the horizontal plane as with stimulation of the frontal eye fields (Fig. 9.50, area 8). Occipitally induced eye movements represent activation of a cortical reflex mechanism rather than a mechanism for voluntary direction of gaze. *Searching for visual targets depends on the frontal eye fields,* (area 8), whereas *grasping and following visual targets depends on striate cortex* (Fig. 9.50, areas 17 and 18).

The *saccadic system* involves abrupt, conjugate, darting of the eyes in movements that jump-advance fixations. Saccadic movements can be observed by watching someone reading. It is advantageous to give them something with a full page line so that their eyes come to rest twice or three times in the course of reading left to right. The eyes then dart back to start the next line, and may make fine saccadic adjustments before performing larger saccades while reading that line.

The speed of a saccade is swift and visual perception is momentarily blanked during the eye motion. Pursuit movements, following a moving target through the visual field, are generally slow because distant objects cannot move very rapidly across the visual field. If the target is close and moving fast, the eyes are capable of rapid following.

The eyes converge as distance of fixation is shifted from far to near, and diverge as distance of fixation is shifted from near to far. The *vergence system* is very much slower compared with either saccades or pursuit movements. The vestibular system (Chapter 63) affects mainly conjugate movements that have a slow phase followed by rapid recovery, i.e., *nystagmus*.

Saccades are probably largely controlled by the frontal eye fields. Pursuit and vergence movements are controlled by the occipital cortex. All three systems, saccades, pursuit and vergence exert their control mainly through the superior colliculus. The vestibular system has its own access to the eye motor nuclei and can utilize longer-circuiting through cerebellum and superior colliculus. All four systems ultimately converge on motor neurons in the IIIrd, IVth, and VIth cranial nerve nuclei.

Topological Distributions and Perceptual Completeness

The preservation of functional integrity in neurophysiological "maps" of sensory and motor functions does not require that representations be equal in area, but does require that certain topological relations be preserved. This permits differential magnification of central retinal representations while maintaining point-to-point topological relations. Thus, the foveal region of the retina has a primary cortical representation which is 35 times expanded as compared with the extreme periphery of the retina. Topology permits a succession of retinal, geniculate, and cortical representations to undergo elastic distortions without gaps, shearing displacements, or intrusions.

But wait! The optic chiasm establishes a split through an otherwise continuous map of the visual world. Interhemispheric visual division contains the germ of a more general question: how do we construct a coherent perceptual entirety of our body image which is similarly split down the middle in its representation between the two hemispheres; and how do we make a composite of a fluctuating envi-

ronment which contains many dynamic objects and in which we move about? *Projection* provides the foundation to account for this enigma.

By the time sensory signals begin to acquire awareness, they have already been "projected" onto a world which we infer to exist outside our nervous system—projected onto our body and outside ourselves onto an extracorporeal environment. Everything in sensory processing that serves to guide behavior is projected. Perception not only guides behavior into that externally projected world but acting out in that world is the means for perceptual organization and correction. Perceptual organization, projection, behavior, and perceptual correction can all be understood in relation to the processes of neuronal group selection and consciousness (Edelman, 1987, 1989).

Behavior inevitably moves into an inferred, "created" world. Very much as our blind spot, which represents a tiny hole in our visual field, is filled in, so the seams and separations between brain-divided visual fields and the two half-maps of sensory inputs from the surface of our body, which enter opposite sides of the spinal cord and brainstem and deliver signals to two separate hemispheres, are joined perceptually for purposes of behavior. Such synthesis of perceptual experience is clearly needed for successful motor performance. And if successful behavior selects neurons accordingly, all is brought into appropriate functional contexts and joined by reentrant pathways even though representations may be widely distributed. Neuronal group selection, among distributed populations, with feedback and feedforward reentry, would obviously make possible more comprehensive and more flexible integration than is likely to be attained by more linearly organized and coupled mechanisms (Edelman, 1987, 1989).

Limb movements are coordinated within each of the two hemispheres and on both sides of the body to move into an inferred behavioral space that is projected and perceived as being seamless. Even monosynaptic reflexes reflect outward projection. It is obvious that integrated sensory projections are implicitly involved in every aspect of behavior—from the humblest reflex to the most carefully elaborated conscious performance.

An obvious riposte is: "there must be some location where the dispersed information is all assembled into a coherent whole." But no such location is apparently available. Self-organizing systems operate as-a-whole and have widely distributed rather than localized functions that provide overall control (Prigogine, 1980; Jantsch, 1980; Edelman, 1987).

The reticular formation plays an important role

in integrative convergence of signals from multiple sources during the course of motor performance. Brainstem and spinal reticular formation operate to integrate converging motor patterns as they come to bear on efferent neurons. The reticular formation apparently plays a similar integrative role during the early stages of sensory processing. The reticular formation can influence receptors and sensory neuroepithelial surfaces, and sensory relays, as well as higher perceptual processing, and it plays a special role in mechanisms of arousal and attention.

Neuronal population selection ensures that corrections and adjustments are possible in accord with changes in state of the body as well as changes in the environment. There is a balance between analysis and synthesis. Regardless of the geometries of central representation, both perception and behavior maintain an obligatory outward projection of a dynamic body as-a-whole into a dynamic environment as-a-whole (Edelman, 1989).

Basic Visual Functions

Vision is an information-creating and organizing process. What basic functions are required for formation of a visual image? Imaging requires discrimination of *brightness, wavelength, movement, and distance*. How does visual system manage these basics and what are some of the clues it employs for higher level interpretation?

BRIGHTNESS DISCRIMINATION

Brightness discrimination in fine detail is spectacular. Foveal visual acuity can discriminate vernier lines down to a fraction of the diameter of the narrowest of cone photoreceptors. Subjects integrate information over short-line segments of two limbs of a vernier even when the lines are curved rather than straight, even though they are moving or the subject's eyes are moving, even though three dots are substituted for the vernier, and the dots are not presented simultaneously (up to 20 ms temporal dispersion), and even though the duration of exposure is brief (1.5 ms). The eye maintains comparably excellent vision through light intensity changes of a billion-fold. Adaptation is more rapid (2–3 minutes) for changes in light flux from low to high intensities than it is (30 plus minutes) for changes from high to low intensities.

DARK ADAPTATION

Dark adaptation requires half an hour or longer before "scotopic" vision becomes optimal (Fig. 9.51). In that diagram, the minimum amount of light

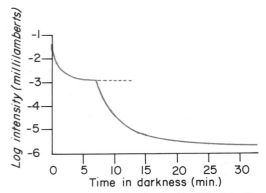

Figure 9.51. The time course of reduction in a light detection threshold during dark adaptation. The initial brach of the curve is associated with photopic vision and is attributed to cone function. The later branch is often called the scotopic branch and is associated with rod function.

which can just be detected is represented logarithmically on the ordinate, and time in darkness linearly on the abscissa. The threshold falls rapidly and gradually reaches a minimum at about 10 minutes. A second-order threshold reduction approaches its minimum only after about 30 minutes in the dark (Fig. 9.51). The initial threshold reduction is associated with cone adaptation and the prolonged one with rod adaptation.

A subject, asked to identify the color of test flashes, loses that ability after the first 8–10 minutes of dark adaptation. Because red light does not as rapidly exhaust rod responses as do other wavelengths, dark adaptation can be preserved by wearing red goggles so that only red light enters the eye.

The extraordinary range of visual adaptation to brightness/darkness is accounted for by alterations in retinal networking that control summation of activities over retinal area and time. There are enormous increases in responsiveness, in "gain," over the whole retinal system at very low levels of illumination. Even so, when brightly illuminated targets are presented to the dark-adapted eye, it is equal to the light-adapted eye in spatial and temporal discriminations.

LIGHT ADAPTATION

Just as the eye adjusts when the visual environment is darkened, so it also adjusts when the environment is made brighter. The nature of the brightness response involves a great reduction in "gain" in the retina. Light adaptation is ordinarily complete in from 3–5 minutes.

The activation of cones has an inhibitory effect on rods. Suppression of rod activity reduces rod interference with signals occurring along the cone path. It also helps to reduce dazzle during bright illumination exposure for a dark-adapted individual. In cases of total color blindness, without cones, the inhibitory effect on rods is absent. The color-blind individual may therefore need dark glasses in situations of moderate illumination.

WAVELENGTH DISCRIMINATION

The eye is not nearly as good a discriminator of wavelength or frequency of light as the ear is of sound frequency. Exploration using targets of different wavelengths projected onto different regions of the retina show that "red" and "green" are identifiable out to the same distance from the fovea. Targets which elicit "yellow" or "blue" are discriminable for equal distances but far beyond those for red and green. There are inhibitory interactions between these same pairs of colors at stages central to the primary receptors where they appear to operate as opponents and cancel one another. Electrical responses at the level of the lateral geniculate body show that in some units turning off a red stimulus is similar to turning on a green stimulus. For other cells the complements of yellow and blue seem to be correspondingly linked (DeValois, 1973).

Chromatic Properties

Rhodopsin provides the molecular basis for light sensitivity in rods. Light absorbance in vitro and light reflectance in vivo operate on the same molecular process (Rushton, 1975). There is evidence that three pigments may each be associated with different cones (see Fig. 9.52).

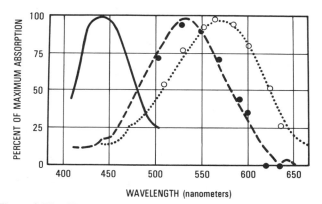

Figure 9.52. The cone pigments of normal color vision absorb lights of different wavelengths. The *curves* show average spectral absorbance from single cones in excised eyes of humans or monkeys scaled to equivalent maxima. *Open circles* represent the reflectivity of pigments in red-sensitive cones, *closed circles,* the green-sensitive cones in the living human eye. Coincidence of the measurements demonstrates that single cones contain single pigments. Curves, from left to right indicate blue-, blue-green-, and red-sensitive receptor populations. (From Rushton, 1975.)

Color Blindness

Most individuals who are deficient in color discrimination are able to discriminate some colors, although not all. Rarely is color blindness complete. Where three primaries are required by the normal observer, many color-blind observers require only two. Hence they are called dichromats to distinguish them from trichromats—normal observers, and monochromats, who apparently lack color vision entirely.

Hereditary Aspects

Color vision deficits are known to be genetically linked and some of the code is borne on the x-chromosome. More men are color blind then women and there are different incidences in different races, and different proportions for different kinds of color blindness. Blue-blindness, however, occurs in about equal frequency in the two sexes and is evidently autosomal.

SPATIAL RESOLUTION

The light-gathering power of the globe casts a greatly reduced image of an enormous visual field onto the tiny retina where some 120 million receptor cells can be exposed. The retinal and central nervous events associated with visual processing are projected outward so we perceive objects large scale. If all the aberrations of the optical system, the density of receptors, the size of receptive fields, etc. are considered, the quality of the visual image for the normal observer is better than one would expect theoretically. Perception is smoothed over much as in the case of our obliviousness to our blind spot.

The boundary between a bright field and a dark field is emphasized by edge enhancement. Representation of the light and dark fields are only slightly shifted above and below the level of spontaneous activity. Activity defining brightness, largely contrast, is confined to the boundary which also defines the direction of the brightness difference. Edge enhancement is accomplished by lateral inhibition in the retina and acting similarly at successive central relays. This provides an enormous economy for central nervous representation. Neurons do not have to "paint all the surfaces," but only recognize the borders and indicate the direction of differences on the two sides. Signals relating to other important aspects of the scene are also similarly specially enhanced. These effects improve perceptual discrimination of orientation, depth, size, color and movement (Cowey, 1982).

Visual Acuity

Visual acuity depends on the recent illumination history and on the way by which visual acuity is measured. The simplest is discrimination of a small dark spot against a lighter background, or a dark line against a contrasting background. Under ideal conditions, the finest line that can be resolved subtends a visual angle of less than 1 s of arc. Various easily administered clinical tests for acuity are usually expressed in relation to the distance at which a target can be correctly detected by a given observer in relation to a normal population. Thus, an acuity of 20/200 implies that an individual can resolve a test figure at a distance of 20 ft which the normal observer can resolve at 200 ft; this represents a severe deficit. A visual acuity of 20/10 indicates that an observer can resolve a test pattern at 20 ft whereas the general population requires that the target be within 10 ft to discriminate correctly.

TEMPORAL DISCRIMINATION

Perceptual Duration

At fairly high levels of light adaptation, a black outlined figure on a white background has a perceptual duration of about 250 ms. The interval of fixation of the eye on a printed page is about 250 ms, and it appears that an interval of about 200–250 ms is required for central processing of perceptual information during each visual fixation (Haber and Nathanson, 1969).

DEPTH PERCEPTION

There are a variety of cues that might provide information as to differences in depth; some work with monocular vision and two with binocular vision. Neither accommodation nor convergence provides reliable information for depth perception. Retinal image size is not very effective for estimation of distance, even for familiar objects.

Motion of an observer and/or of objects located at different distances from the observer results in changes in perspectives. This provides really effective clues to discriminate relative distances of objects. Even slight motion of an observer's head introduces monocular motion parallax effects that provide important distance cues.

Binocular Retinal Disparity

Pupillary centers of the two eyes are separated in different individuals over a range of about 60–75 mm. Their location provides two horizontally displaced different perspectives on the environment. Except for the point of fixation, the rest of the images of most objects in the visual field stimulate noncorresponding points on the two retinas. Barlow et al. (1967) and Hubel and Wiesel (1970) demonstrated the existence of cortical cells that respond differentially to disparity, particularly in the horizontal dimension. This is the main neurophysiological basis for depth perception.

CHAPTER 66

Olfaction and Taste

CHEMICAL SENSES IN GENERAL

Our preferences for tastes and odors are not easily described, but physiologically important decisions are continuously being made by our nervous systems according to the ways things taste and smell. Powerfully "good" or "bad" tastes and odors abruptly arouse, orient, and compel behaviors. In subtle but far-reaching ways nervous system discriminations based on taste and olfaction guide us toward vital survival and reproductive behavior.

Taste and olfactory discriminations are among the oldest and most influential of all sensory mechanisms. Every cell in our bodies performs selective chemo-acceptance, -rejection, and -elimination. Cells survive by taking in specific substances needed to satisfy fluctuating metabolic requirements and by selectively eliminating wastes. They actively exchange with their environment a throughput of specific chemicals upon which cell survival depends. Cells in multicelled organisms retain this capacity even though the internal environment is protected and favorably conditioned, as compared to the "outside world."

Multicelled organisms require particular chemoselective and regulatory mechanisms to satisfy chemical throughput for the body as a whole. These consist of local and regional internal chemosensitive mechanisms—interoceptors and associated circuits—which facilitate tissue and organ delivery of oxygen and nutrients, and the disposal of carbon dioxide (CO_2) and wastes. Specialized chemoceptive systems are organized for the control of digestion, circulation, respiration, temperature regulation, excretion, etc. Hypothalamic mechanisms guide overall behavior according to chemoceptively defined states of nutritional and sexual appetites and satiety. Chemosensitive systems in the hypothalamus measure appetite and satiety and guide the individual toward suitable water and food sources and reproductive opportunities.

In all but the most primitive vertebrates, in addition to interoceptive chemosensory systems with associated regulatory mechanisms and influences on behavior, there are two specialized exteroceptive chemosensory systems—olfaction and taste. Each has its own chemoselective neuroepithelia, central mechanisms for sensory processing and representation, and access to behavioral control mechanisms. Together, they govern approach and avoidance behavior relating to environmental chemistry. *Olfaction is adapted for distance-to-neighboring chemoselection* (smelling a freshly-baked pie) while *taste is adapted for direct contact chemoselection* (tasting the pie for ingestion purposes). They work together to organize behavior relating to smelling (tracking), eating, drinking, and reproduction. A remarkable feature of olfaction and taste is that their sensory relay systems become selectively sensitized with respect to chemicals that are in biological need.

Figure 9.53 shows the relationship between the olfactory neuroepithelium located in the roof of the nasal cavity and the overlying cribriform plate through which olfactory nerves reach the olfactory bulb. Figure 9.54*A* illustrates the relationship between tongue, palate, and glottis fields of taste receptors and their respective innervations.

When olfaction and taste are investigated using single cell recordings, it is obvious that both systems are extremely subtle and versatile. Single olfactory receptor neurons respond differentially to a wide variety and mixture of odorants; single taste receptors respond differentially to a wide variety and mixture of taste stimuli. In the course of smelling and tasting, complex dynamic patterns of receptor activation are distributed over neuroepithelial space and are dynamic over time. Receptors in both systems are affected according to recent past receptor history, e.g., sensory adaptation, slowing the firing rate in response to persistent stimuli.

Chemoreception and Molecular Signaling

Insights into the molecular mechanisms of chemoselection are aided by studying bacterial chemo-

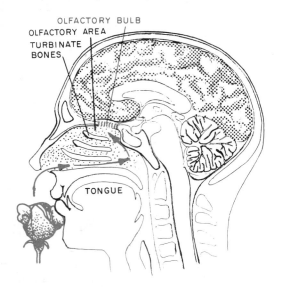

Figure 9.53 Parasagittal cross-section of the head showing of the three turbinate bones, the olfactory area, and the termination of the endings of nerves from the olfactory bulb in the olfactory epithelium. The passage of air-bearing odorous substances is shown by *arrows*. (From Amoore et al., 1964.)

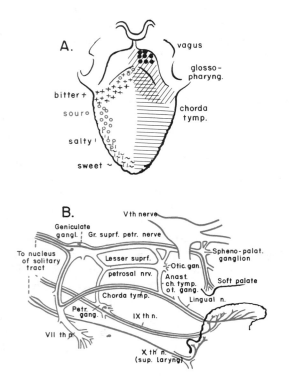

Figure 9.54. *A*, Sensitivity of the tongue in different regions for four primary taste qualities. *Diagonal cross-hatching* represents the region innervated by the glossopharyngeal nerve. *Solid dots* show the region innervated by the vagus. *Horizontal lines* represent the region innervated by the chorda tympani (von Békésy, 1966). *B*, The pathways of the nerve supplies are shown in greater detail and a sagittal cross-sectional view of the tongue. (Adapted from Brodal, 1959.)

taxis. Bacteria demonstrate how molecular machinery can provide sensory discriminations like those in olfaction and taste. Koshland et al., 1982 (see Koshland review 1980; Russo and Koshland, 1983) found that single bacteria have about 30 receptor molecules, each of which responds specifically to a limited number of chemical compounds that are important for metabolism and survival. Each receptor molecule presents its chemoselective branch outside the cell while the bulk of the molecule extends through the cell membrane to signal the state of outside reception to the cytoplasm. It also signals flagellar locomotion to propel the bacterium toward rising gradients of needed chemicals and away from rising gradients of noxious chemicals. Bacteria also illustrate how neurons can link neurotransmitter reception to intracellular responses. Receptors on neurons respond to specific neurotransmitters and activate consequential changes in membrane conductance and cytoplasmic mechanisms.

OLFACTION

Nasal Passages and Olfactory Innervation

Five separate neuroepithelia in the interior of the human nose are affected by odorants; the last four of these are bona fide chemoselective sensory systems: (1) trigeminal innervation of general nasal mucous membranes; (2) terminal nerve innervation of upper nasal epithelium; (3) the septal organ (of Masera); (4) the vomeronasal organ (of Jacobson); and (5) classical olfactory neuroepithelium.

TRIGEMINAL INNERVATION

The trigeminal nerve innervates the respiratory mucous membranes along nasal and oropharyngeal passageways, including the olfactory neuroepithelium itself. Trigeminal innervation responds to touch, temperature, and pain stimuli, and reacts strongly to many noxious odorants. It makes partial contributions to perceptions of smell and taste. For example, food described as peppery, piquant, hot, etc. is detected more effectively by trigeminal than by taste innervation.

TERMINAL NERVE INNERVATION

Evidence indicates that the terminal nerve (TN) plays a role in mediating sensory and behavioral responses to sexual pheromones. (A pheromone is a substance secreted by one individual and responded to in a particular manner by another individual; the effect may be on endocrine development and readiness and/or more immediately on behavior.) Bipolar TN cells, which lie alongside the olfactory bulb, innervate neuroepithelia near the roof of the nasal passages. Centrally directed axons project to the supracommissural nucleus and send collaterals (oddly enough!) to the inner plexiform layer of the retina.

SEPTAL ORGAN

This is an additional patch of neuroepithelium which conveys olfactory information directly to the olfactory bulb (Masera, 1943) and supplements the main olfactory input.

VOMERONASAL ORGAN

This forms a cul de sac with a narrow tubular entrance that empties into the nasal cavity. The walls of the tube are lined with neuroepithelium which has microvilli but lacks cilia. These first-order neurons respond to similar odorants as does the main olfactory neuroepithelium—although at higher thresholds. This indicates that microvilli are adequate for chemoreception and suggests that the cilia provide advantage by exposing a greater surface area. The vomeronasal organ contributes a separate chemoselective system, specifically concerned with pheromone discriminations (Powers et al., 1979; Meredith et al., 1980).

Vomeronasal neuroreceptors project to the accessory olfactory bulb which, in turn, projects to particular areas of olfactory cortex and the amygdala. These projections are independent of central projections of the main olfactory neuroepithelium and olfactory bulb. They relay to the ventromedial nucleus of the hypothalamus where they play an important role in neuroendocrine and behavioral functions.

Sexually related behavior elicited by olfaction depends on the vomeronasal system and not on the main olfactory system. The main olfactory system can guide an individual to approach odorant sources where, by means of the vomeronasal system, a specific pheromone source is categorized whereby cooperative and sexual behaviors may follow.

Principal Olfactory Neuroepithelium

Olfactory neuroepithelium appears as a yellowish patch on the lateral wall of the upper posterior nasal passages. It partly covers the superior turbinate and extends medially onto the nasal septum. In humans this sensory surface occupies an area of approximately 250 mm² in each nostril. Its location beneath the cribriform plate and the overlying olfactory bulb is illustrated in Figure 9.53. From the basal surface of each olfactory neuron, a fine unmyelinated axon (about 0.2 μm in diameter) leads up through the cribriform plate to the olfactory bulb.

Normal breathing generates currents that sweep air past the olfactory epithelium from posterior to anterior. More effective neuroepithelial exposure to odorants is obtained by sniffing. Stimulation of the olfactory neuroepithelium occurs during normal expiration and contributes to our sense of "taste" in food. If olfaction is blocked, eating an apple cannot be distinguished from eating a potato or an onion.

Spatial information has clear significance in other sensory systems and is important for olfaction. Information from differences in concentration of odorants in the two nostrils contributes to olfactory tracking. Discrete topographic organization is strictly maintained in projections from olfactory neuroepithelium to the bulb and throughout subsequent central olfactory projections.

Olfactory neuroepithelium consists of primary olfactory neurons, stem cells, and supporting cells. The surface consists of the apical borders of slender olfactory neurons, surrounded by supporting cells with which they share tight junctions. These junctions seal the neuroepithelium against penetration of substances through the surface and maintain a cross-epithelial resistance of a few thousand ohms/cm². Supporting cells are linked to one another by gap junctions which exchange ions and thereby stabilize epithelial electrical potentials against local perturbations. Primary olfactory neurons are frequently recycled. For replacement, new olfactory neurons, daughters of stem cells, push up between supporting cells, and new tight junctions are formed, maintaining structural and electrical integrity.

The nasal surface of the supporting cells bears microvilli which cushion long, sweeping olfactory cilia. Each olfactory neuron extends 10-15 long, whip-like motile cilia into the mucus. Olfactory nerve membrane covers each of these cilia like a many-fingered glove, supporting an enormously expanded nerve membrane. This membrane bears numerous intramembranous particles which constitute molecular chemoselective receptors.

PRIMITIVE CHARACTER OF OLFACTORY EPITHELIUM

The sensitive ends of olfactory neurons are exposed to the outside world from which they are protected only by a thin layer of mucus. Chemoreceptive neurons of this type have been widely retained throughout invertebrate and vertebrate evolution. All olfactory neurons develop, mature, and die in a matter of weeks and are replaced by stem cells which also replace olfactory neurons after injury. Neuronal death and regeneration requires routing of regenerating axons and synaptic junctions onto neurons in the olfactory bulb (and from the vomeronasal organ) onto neurons in the accessory olfactory bulb (Graziadei, 1976). The olfactory system thus presents continuous synaptic regeneration. It is thought that no other nerve cells in the adult human nervous system regenerate, although a great deal of central axonal sprouting and synaptic remodeling occurs.

Olfactory Neuron Responses

Olfactory neurons exhibit spontaneous impulse discharges; this provides for extremely sensitive responsiveness. Odorants ordinarily elicit increased spiking, but strong odorants may suppress activity. Following activation, there may be postexcitatory inhibition. On introduction of an odorant near behavioral threshold concentration, there is widespread prolonged neuroepithelial excitation.

It has been found possible to record intracellularly from the primary olfactory neurons, as shown in Figure 9.55. Note the spontaneous activity. Near absolute threshold for anisole, there is a barely discernable pause, activity, and slowing. At successively higher concentrations, there is a more definitive and prompt response followed by more prolonged postexcitatory inhibition. According to Mathews (1972), individual olfactory neurons are selectively sensitive in response to a specifiable list of odorants, and there does not seem to be any obvious chemical signature to the way odorants cluster in their ability to activate individual neurons. There are evidently no single pure-odorant neurons. Odorants with very different chemical properties may activate the same cell, and chemicals with similar properties may not do so.

There is great variation in the extent of olfactory neuroepithelium in different mammals and this seems to correspond to objective differences in their ability to detect and discriminate odorants. The threshold for detecting certain fatty acids by German shepherd dogs is at concentrations about 100 times lower than that for humans. This advantage probably lies in the dog's 56X greater surface area of olfactory neuroepithelium. At the single-neuron level, the dog does not have much advantage because one molecule of a suitable odorant is sufficient to excite individual human olfactory neurons (Moulton, 1976). The olfactory system is organized to encode an extremely large number of different odorants and mixtures of odorants, as indicated by an enormous variety of different activity profiles elicited in individual primary afferents (Gesteland et al., 1965).

During olfactory exploration, the neurological consequences depend on the concentration of an odorant (and whatever complementary and competing odorants may be in the same atmosphere), on sniffing patterns, on solubility of the odorant and physical interactions of the odorant with receptor surfaces of the neuroepithelium (Fig. 9.56, Macrides, 1976).

Functions of the Olfactory Bulb

FUNCTIONAL DEVELOPMENT IN THE OLFACTORY BULB

Olfaction is important in relation to suckling behavior and bonding between mother and infant. Restricted parts of the olfactory bulb mature earlier than does the remainder; in contrast to relatively immature glomeruli in most of the bulb, early maturing glomeruli are histologically well defined at birth. These specialized glomeruli lie along the border between the main bulb and the accessory olfactory bulb. They are activated by pheromones associated with breast feeding. Greer et al. (1982) showed that receptor cell terminals and their postsynaptic targets in this limited bulbar region are active within a few hours of birth. Thus, the newborn has special olfactory precocity which facilitates biologically essential behavior.

Additional evidence for localization of function within the olfactory bulb is reported by Sharp et al. (1975), who analyzed 2-deoxyglucose accumulation in rats exposed for 45 minutes to strong odorant

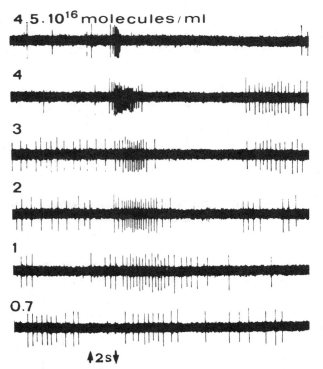

4.5 · 10^16 molecules/ml

4

3

2

1

0.7

↑2s↓

Figure 9.55. Responses of primary olfactory neurons to various concentrations of anisole. (From Holley and Døving, 1977.)

RESPONSE SURFACES

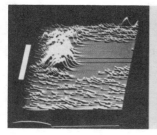 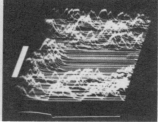

FREQUENCY RESPONSE HISTOGRAMS

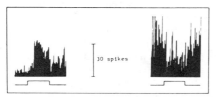

30 spikes

Figure 9.56. Frequency reponse histograms for hamster single unit olfactory neuron (*top*) in response to amyl nitrate puffs during artificial sniffing. Tracing below middle histograms indicates nasal air flow. Height of *slanted bar* indicates duration of introduction of amyl acetate. Variations in bursting pattern within inhalation cycles are batched into representative "overall" responses as recorded from response surfaces. Two units depicted below show differential responses to amyl acetate. (From Macrides, 1976.)

stimulation. Restricted regions of uptake in the olfactory bulb correlated with specific odorants. Further, exposure of rats to strong odors for periods of a week or more led to restricted regions of transneuronal degeneration of tufted and mitral cells, the output cells of the olfactory bulb (Døving and Pinching, 1973).

There is a reduction of roughly 500:1 from the number of neurons in the olfactory epithelium to the number of mitral and tufted cells in the bulb. There are about 2000 glomeruli in each bulb. Each glomerulus receives axons from more than 25,000 primary olfactory neurons. There are about 50 mitral and tufted cells in each glomerulus. First-order olfactory neurons have synapses on dendrites of mitral and tufted cells which, in turn, generate bulbar output signals. Sensory processing within the bulb involves two additional types of cells, periglomerular and granule cells, whose processes remain within the bulb.

Shepherd (1970) recognized useful analogies between the strategic organization in the olfactory bulb and retina (Fig. 9.57). Olfactory short-axon periglomerular cells are analagous to retinal short-axon horizontal cells; olfactory axonless granule cells are analagous to retinal axonless amacrine cells. Bulbar integrative circuits, as in the retina, are organized in two tiers: periglomerular cells modulate bulbar input while granule cells modulate bulbar output. Information processing in the bulb involves reciprocal synaptic (microcircuit) connections, as it does in the retina (see Vision, Chapter 65).

EFFERENT PROJECTIONS TO THE OLFACTORY BULB

In addition to primary afferent inputs and intrinsic bulbar circuitry, axonal endings in the bulb are received from each of the central areas to which mitral and tufted cells project. These contrifugal olfactory controls influence mitral and tufted cells indirectly by modifying activity in periglomerular and granule cells. This enables higher nervous centers to control olfactory sensory input by sharpening discriminations and shaping incoming sensory data according to memories, internal needs, etc. Rapid adaptation to repetition of the same stimulus is virtually abolished when descending (efferent) axons to the bulb are sectioned (Moulton, 1963).

Efferent control of olfaction is analagous to other centrifugal sensory control systems which modulate primary receptors, e.g., cochlea, muscle spindle, etc., and central sensory relays. Thus, efferent control is also exerted on each successive olfactory relay as signals travel centralward. Shaping of olfactory perception is introduced into olfactory channels well before the incoming sensory data can be presumed to reach consciousness.

Two additional centrifugal olfactory control projections arise in the brainstem and release aminergic neurotransmitters in the bulb. *Serotonin* (5-HT) fibers originate in the raphe nuclei, and *norepinephrine* (NE) fibers in the locus coeruleus. Among other contributions, these aminergic projections modify bulbar activity during behavioral arousal and attention.

MECHANISMS OF PATTERN DISCRIMINATION IN OLFACTION

Intrinsic inhibition affecting bulbar reception and relay subserve functions analagous to lateral inhibition in the retina. In the visual system, lateral inhibition has been shown to contribute to contrast enhancement, directional selectivity, detection of motion, adaptation, etc. Differential receptor selectivites and bulbar organization suggest that olfactory patterns in the bulb may be of even greater complexity than those in the retina. Like retinal receptors, olfactory neuroepithelium consists of a sheet of receptors. Differential regional sensitivity

Bulb

Retina

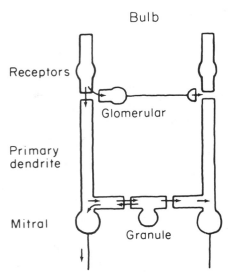
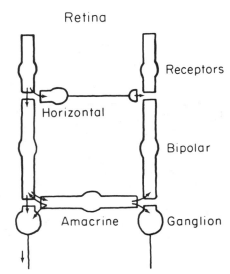

Figure 9.57. Basic circuit of bulb and retina compared. On the *left* is a schematic diagram of synaptic connections within the olfactory bulb for comparison with a similar diagram on the *right,* of the retina. Receptor-neurons project directly to the olfactory bulb where they excite primary dendrites of mitral (and tufted) cells. They also affect short-axon periglomerular (glomerular) cells that provide lateral inhibition which modulates mitral and tufted cell inputs. Output of mitral and tufted cells is modulated by reciprocal synaptic microcircuitry involving axonless granule cells. In the retina, receptors activate both bipolar and horizontal cells. Horizontal cells provide lateral inhibition that modulates retinal output. Bipolar output and that of ganglion cells are further modulated by the axonless amacrine cells. The organization of longitudinal and horizontal connections in a two-tiered system of sensory control is closely similar. Note that in each case throughput of sensory signals is being affected at the first synaptic relay and that refinement and analysis of sensory information is already taking place within the bulb and retina. (From Shepherd, 1970.)

of receptor cells is enhanced by controlled movement of odor-laden air across the sheet (as in gas chromatography). Sniffing relates to olfactory perception in the same fashion that actions of the iris, lens, and extraocular muscles regulate exposure of the retinal receptor sheet to odorant stimulus patterns.

Bulbar excitatory and inhibitory circuits are implicated in rhythmic activities in the bulb which are largely under centrifugal control. At times, the bulb undergoes seizure-like activity, perhaps analagous to the special susceptibility of olfactory cortex to seizure activity (see below).

Central Connections of the Olfactory Bulb

Central systems for the analysis of olfaction are both orderly and complex. Heimer (1976) provided a useful critical reveiw of olfactory projection systems and their wider relations. In brief, central olfactory representation forms a rough triangle that extends posteriorly from the bulb at the apex, widening along the posterior medial orbital surface of the frontal lobe and widening still farther laterally as the base of the triangle moves posteriorly onto the anterior and medial surfaces of the pole of the temporal lobe. Allen (1941) showed that homologous canine cortical projections are necessary and sufficient, as compared with all other cortical areas, to establishment and retention of olfactory conditioned responses.

Organization of the olfactory system has been schematized by Shepherd (1983) as illustrated in Figure 9.58. The olfactory bulb is properly represented as a part of the brain. Incoming primary afferents enter the bulbar glomerular layer where they contact mitral and tufted neurons (designated *m*). The glomerular relay is modulated by periglomerular and granule cell activity which, in term is under the influence of centrifugal neurons *(dashed lines)* projecting from areas of higher olfactory representation and from the brainstem reticular formation. Major projections centralward are depicted in order. The distribution of tufted cell axons is limited to the more rostral olfactory projection areas: anterior olfactory nucleus, medial pyriform cortex, and lateral olfactory tubercle. Mitral cell axons project farther and distribute terminals to all regions of olfactory representation (Price, 1977).

Mitral and tufted cells send axons along the olfactory tract. This tract flattens posteriorly and separates into medial and lateral olfactory striae. Divergence of the two striae exposes a triangular area, the olfactory trigone. The striae distribute fibers to the anterior olfactory nucleus, olfactory tubercle, septal nuclei, subdivisions of the amygdaloid complex, parts of the hypothalamus, pyriform cortex, and transitional entorhinal cortex.

The *anterior olfactory nucleus* represents an indirect relay station; it projects to the same structures that receive direct input from the bulb. The

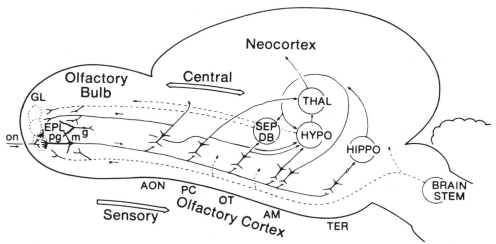

Figure 9.58. Central olfactory projections. Primary receptor-neurons enter the olfactory bulb glomerular layer (*GL*) where they are affected by periglomerular (*pg*) cells in the external plexiform layer (*EPL*) and by granule (*g*) cells deeper within the bulb. Abbreviations for olfactory projections: *AON*, anterior olfactory nucleus; *PC*, pyriform cortex; *OT*, olfactory tubercle; *AM*, amygdala; *TER*, transitional entorhinal cortex (mesocortex); *Hippo*, hippocampus (archicortex). Abbreviations for subcortical regions: *SEP-DB*, septum and diagonal band; *HYPO*, hypo-thalamus; *THAL*, thalamus, which projects to precentral cortex in the frontal lobes (neocortex). Brainstem aminergic projections (serotonin and norepinephrine to the bulb and dopamine to the olfactory tubercle) are indicated by *dashed lines*. All destinations of mitral and tufted cell axons send back reciprocal projections which are schematically dia-grammed by *upper dashed lines*. These modulate afferent throughput indirectly by influence on granule and periglomerular cells. (From Shep-herd, 1983.)

olfactory tubercle corresponds to the anterior per-forated space, grey matter which lies immediately posterior to the olfactory trigone. The tubercle receives direct bulbar projections and projections relayed through the anterior olfactory nucleus. It also receives dense dopaminergic (DA) innervation from the substantia nigra and associated reticular formation in the brainstem (Ungerstedt, 1974; Bjöerklund and Lindevall, 1976).

The olfactory tubercle is important because of its extensive projections to additional limbic structures which lie beyond those which receive direct olfac-tory contributions from the bulb. Because of the close association of the olfactory tubercle with both the basal ganglia and the limbic system, the olfac-tory tubercle may be involved in important dopa-mine-related psychological and schizoaffective types of behavior disorders (Ungerstedt, 1974). Olfactory tubercle projections to limbic mesocortex, and olfactory cortical association fibers projecting thence to hippocampus—where olfactory signals can be integrated with information from other sen-sory modalities—probably contribute to olfactory perceptions and memories during emotional experi-ence and behavior (MacLean, 1975).

The *pyriform lobe* is the primary cortical region engaged in olfactory discrimination. It has three principal parts: (1) the *prepyriform area*, which extends along the medial posterior orbital surface of the frontal lobe from pyriform cortex itself almost to the olfactory bulb—hence its designation; (2) *pyriform cortex* which covers the anterior and anteromedial surface of the pole of the temporal lobe; and (3) *entorhinal cortex*, bordering the hip-pocampus (Fig. 9.58).

Parts of the medial and posterior cortical nuclei of the amygdala receive powerful olfactory projec-tions from the bulb. These nuclei project to the ven-tromedial nucleus of the hypothalamus, which has importantly to do with feeding, appetite, and sati-ety. The amygdala is involved in fundamental aspects of social behavior; olfactory projectons to the amygdaloid complex contribute to this behavior (see below).

Olfactory signals also reach an intermediate zone of cortex that bounds the margins of the hemi-sphere, *mesocortex* (paleocortex). Mesocortex is three-layered and thus intermediate in complexity between *archicortex* which is one-layered, and *neo-cortex*, which is six-layered. Neocortex, as its name implies, is the newest as well as the most elabo-rately layered cortex. In evolution, neocortex makes its first appearance in mammals.

Other sensory modalities—auditory-vestibular, visual, and somesthetic—all project via the thala-mus to broad expanses of neocortex. Not only does olfaction have a more primitive neuroepithelium but its direct central projections are confined to primitive mesocortex.

Functions of the Accessory Olfactory Bulb

The accessory olfactory bulb receives olfactory afferents carrying olfactory pheromone signals from the vomeronasal organ. It projects to exclusive areas of prepyriform and pyriform cortex. It also projects to parts of the medial and posterior cortical nuclei of the amygdala, exclusive areas of the amygdala that do not receive terminals from the principal olfactory bulb. Thus, the accessory olfactory bulb projects to central regions separate from those receiving signals from the main olfactory bulb (Scalia and Winans, 1975; Broadwell, 1975). This independence persists also in next-order projections from the amygdala to the hypothalamus (Raisman, 1972). Like the olfactory bulb, the accessory olfactory bulb receives centrifugal (efferent) projections from all regions to which it projects.

Organization of Olfactory Mesocortex

Olfactory mesocortex receives axons from olfactory bulb mitral and tufted cells. These enter via the most superficial layer of cortex and terminate on apical dendrites of an upper layer of pyramidal cells. (This contrasts with sensory projections to neocortex where afferent input enters from below and terminates in midcortical layers.)

Mesocortical pyramidal cells are first strongly excited by olfactory signals and then inhibited by local intracortical circuits. Most of the pyramidal cells have extraordinarily long axon collaterals which arborize extensively throughout olfactory cortex and distribute the initial excitatory activity widely. Perhaps because of this excitatory feedforward circuit, olfactory cortex is especially susceptible to seizure activity. Mesocortex, including extensive nonolfactory mesocortex which completes a ring around the hilus of the hemisphere on each side, is nearly uniformly susceptible to seizure activity.

Archicortex forms a parallel ring adjacent to the mesocortical ring, but it is narrow except in the temporal lobe where it is greatly expanded and is identified as *hippocampus*. Seizures induced anywhere in mesocortex and hippocampus tend to spread throughout this entire double ring of primitive cortex and into directly connected subcortical structures.

Both archi- and meso-cortex are especially vulnerable to the effects of psychotropic drugs, narcotic substances, alcohol, and anoxia. Chronic alcoholism can induce signs of neuronal degeneration throughout the limbic system. The combination of temporal cortical and subcortical seizure susceptibility (including amygdala) accounts for a major category of epilepsy, *temporal lobe seizures*. Such seizure activity readily spreads throughout the entire ring of primitive cortex but does not usually spread to neocortex. When neocortex is activated, however, seizure activity manifests as a classical grand mal seizure, with gross tonic-clonic, rhythmic skeletal muscle contractions, breath-holding, tongue-biting, frothing at the mouth, sphincter relaxation, etc. Temporal lobe seizures, confined to limbic circuits, on the other hand, manifest simply as *altered states of consciousness* (see below).

Olfactory Associations

Finally, the cortical and subcortical regions that receive input from the olfactory bulb and those that receive input from the accessory olfactory bulb (the pheromone analysis system), which have been kept separate to this stage, exchange fibers with one another.

Olfactory association fibers from the pyriform lobe project to the large medial dorsal nucleus of the thalamus. From this important station they are systematically reprojected onto neocortex of the frontal lobes (prefrontal cortex) which is involved in planning behavior. Olfactory corticocortical fibers also extend to neocortical areas adjacent to olfactory cortex on the medial side of the hemisphere, and to ventral posterior cortex of the insula (island of Reil).

A massive fiber bundle, known as the *olfactohypothalamic system* connects the olfactory bulb and adjacent retrobulbar areas with the *septum* and *anterior, lateral, and posterior hypothalamus*. By way of the anterior and lateral hypothalamus, olfaction contributes to neuroendocrine function and to behavior relating to reproduction (Scott and Leonard, 1971; Powell et al, 1965). Via projections to posterior diencephalon, olfaction has a strong influence on approach-avoidance behavior.

Additional olfactory projections from pyriform and prepyriform cortex go to the ventral putamen (part of the basal ganglia, which have to do with postural expression), and to the lateral hypothalamus. These olfactory cortical efferent projections arise from an evolutionarily recent, additional deep layer of mesocortex, exclusive to this region of olfactory cortex. It is as though this area of mesocortex were in the process of an evolutionary transition, adding more layers, perhaps functionally akin to layers V and VI of neocortex.

Olfaction and Social Behavior

Ethological studies show that olfactory signals have importance in determining affective attitudes and behavior (Schultze-Westrum, 1969; Ralls, 1971). Olfactory memories in humans are powerful in commanding attention and they are highly resistant to decay (Engen et al., 1973a; 1973b). Complex perceptions and perceptual memories project to the amygdala. In decision-making contexts, the amygdala generates commands, executed through hypothalamic and limbic projections to posterior diencephalon and midbrain, that control approach and avoidance behaviors (see Chapter 71, Cortical and Subcortical Integrative Mechanisms).

The amygdala receives highly complex derivations from activities in sensory projection and association fields—auditory, visual, and somesthetic, and directly, olfactory projections and olfactory associations. This is the first place that olfaction and other sense modalities meet. Gloor (1976) suggested that the amygdala contributes an affective bias which we call "mood," in relation to sensory stimuli. Mood greatly influences hypothalamic appetitive, neuroendocrine, and behavioral channels. It contributes centrally to limbic integration of emotional experience and emotional expression (see Limbic Controls, Chapter 9).

Animals in which both amygdalae have been destroyed are seriously handicapped in their social behavior. They evidently are unable adequately to integrate social signals emitted by other animals, signals that should guide them through appropriate social transactions. Kling (1972; 1975) showed that wild monkeys in whom bilateral amygdalectomy was performed shun conspecifics, even a troop which they may have dominated, and fail to respond to socially significant signals emitted by other monkeys. After amygdalectomy, they become social isolates, but manage solitary existence effectively.

The amygdala, according to Gloor's interpretation, contributes to deciding what is suitable approach/avoidance behavior (i.e., for eating, socializing, and reproduction). Moods which represent superordinate integrations of perceptions and memories, contribute to the control of highly consequential constructive or destructive social behaviors (Gloor, 1976). It may be that sociopathy, which is by far the most expensive and tragic of all behavioral problems, is related to disorders in the amygdala and its connections.

Temporal Lobe Seizures

Temporal lobe seizures present useful insights into the physiological organization of the brain and form a bridge from neurology to psychiatry. The first clinical case description is one of the best, published a century ago by Jackson (1888). This was followed by autopsy finding of a cyst in the anteromedial part of the temporal lobe (Jackson and Colman, 1898). The patient, a physician, described "intellectual auras" which occurred just prior to fugue states which lasted from a few minutes to an hour, and which he could never remember. During

the fugue states, he maintained a tolerable general and even technical medical capability. He was experiencing long-lasting temporal lobe seizures which allowed unconscious, overlearned behavior, but which completely obstructed memory storage.

Another form of temporal lobe seizure (called "petit mal," to contrast with neocortical generalized "grand mall" seizures), may be accompanied by brief loss of consciousness, lip smacking, incoherent speech, mental "absence," fugue states, and periods of amnesia lasting from seconds to many minutes, with clouding of consciousness following the attack. Temporal lobe epilepsies commonly result from degeneration, cortical scar formation (e.g., from birth injury to the temporal lobes), drug abuse, and alcohol abuse. They present special medicolegal problems inasmuch as patients have no recollection or sense of responsibility for whatever they may do during the amnestic state.

Seizure Induction

Local direct electrical stimulation of cortex over the exposed hemispheres of monkeys reveals marked regional differences in susceptibility to seizure induction (French et al., 1956). Some cortical areas have low thresholds—being very seizure-prone, others are quite refractory. The differences are evidently due to differences in subcortical connections. A "reticular activating system" in the brainstem (Moruzzi and Magoun, 1949) relays impulses through the basal forebrain nucleus, which sends cholinergic impulses that arouse and activate the entire forebrain. Arousal is associated with behavioral alertness and readiness for action, and the cortical neurons exhibit rapid, asynchronous firing. Another powerful forebrain system has a complementary influence. This originates in the reticularis nucleus of the thalamus. These neurons deliver rhythmic inhibitory (GABA) impulses to thalamocortical neurons which, in turn, induce generalized rhythmic synchronous firing of cortical neurons. The electroencephalogram (EEG) shows repeated (2–15) rapid rhythmic, synchronous, high-voltage discharges which wax and wane—presenting a spindle-shaped profile on the EEG record. During arousal, the basal forebrain nucleus inhibits the reticularis-thalamocortical projections. The two systems thus provide reciprocal alertness and what appears to be behavioral disengagement or clinical "absence" (Buzsaki et al., 1988).

KINDLING

Olfactory and limbic cortex—and the underlying amygdala—are among cortical regions that are especially susceptible to seizure induction.

Repeated, weak stimulation of these regions also provides valuable clues to brain physiology. The process is called "kindling." Kindling is a term borrowed from the expression for using small pieces of wood to ignite a fire. By analogy, repeated weak excitation of certain brain structures can, over time, kindle full-blown seizures.

Cortical or subcortical kindling is induced when short bursts of low-intensity electical stimulation are applied daily. With daily mild stimulation, the threshold for after-discharge is progressively reduced (Racine, 1972a). Kindling manifests itself by a spreading of local afterdischarge and subtle behavioral changes (e.g., pausing in activity, engaging in sterotypic behavior).

Once kindling has been established, the threshold for convulsions typically continues to decline even without further stimulation. Convulsions occur spontaneously and recur with increasing frequency (Wada et al., 1976). Seizure activity continues to increase and spread to adjacent structures, then to the contralateral mirror-image site, and ultimately, to neocortical structures, giving rise to grand mal seizures (Goddard, 1967; Goddard et al., 1969; review by Goddard and Douglas, 1976).

After temporal lobe seizures or grand mal seizures have been established in this way, destruction of all neurons directly activated during the kindling process does not prevent spontaneous seizure activation by neurons on the periphery of the kindled site, nor by spontaneous seizure activity among neurons in the mirror-image focus in the contralateral hemisphere.

The olfactory bulb and all olfactory projection systems are readily kindled. Temporal olfactory and limbic cortex and the amygdala are particularly susceptible to kindling. Although notably less readily kindled, other primary sensory pathways are also subject to kindling, provided that daily kindling stimulation has been repeated often enough (Cain, 1979). Kindling has been observed in a wide range of mammals including primates, and has been obtained using a wide variety of stimulants in addition to mild, local electrical excitation: focal chemical (cholinergic) stimulation as well as stimulation by a variety of pharmacological agents—pentylene tetrazol, fluorothyl, cocaine, repeated alcohol withdrawal, repeated transcranial electroshock, and even repeated auditory stimulation (Goddard et al., 1976). Temporal lobe seizures and mechanisms of kindling have been linked to a variety of memory, mental and neurological disorders (Livingston and Hornykiewicz, 1978; Doane and Livingston, 1986).

Sutula et al. (1988) observed axonal and synaptic growth and reorganization in the hippocampus induced early in the course of kindling before the development of generalized seizures. They observed similar phenomena following simple synchronous stimulation of related pathways. Since similar strengthening and elaboration of circuits is predicted by a number of theories of learning and memory, it has been conjectured that some of the underlying synaptic enhancement that takes place in kindling and in learning may be similar (Goddard et al., 1976).

OLFACTORY-NEUROENDOCRINE INFLUENCES

A typical sequence of reproductive behavior in mammals involves several sensory modalities which play separate but interrelated roles. They usually come into action sequentially, as when olfaction and vision are followed by acoustic exchange, touching, and tasting. *Primer pheromones* bring about slow, long-term responses of a developmental nature or a change of physiological state. *Releaser pheromones* bring about more immediate behavioral responses. *Alarm pheromones* cause prompt general excitement and defensive or escape behavior.

A wide variety of neuroendocrine responses can be elicited by olfactory activation by pheromones. These include prolonged effects involving release of growth hormone (GH), luteinizing hormone (LH), and follicle stimulating hormone (FSH), and more immediate effects involving release of adrenal corticotrophic hormone (ACTH), and other factors, each with particular tissue targets, cellular and tissue effects, and functional schedules. Pheromones not only activate, they also habituate; heavy exposure may be followed by habituation lasting weeks (see Mueller-Schwarze and Mozell, 1976).

TASTE

Sensations of Taste

FAMILIAR TASTE CATEGORIES

The simplest categories of taste are sweet, salt, sour or acid, and bitter. Taste buds exhibit preferences along these lines in different regions of the tongue (Fig. 9.54*A*). The four fundamental qualities of taste, combined with tactile and olfactory cues in various combinations, account for taste perception. Taste ordinarily involves a wide mixture of chemical stimuli. Many different kinds of receptors are activated simultaneously. Each receptor presumably responds best to certain chemical phenomena and is also sensitive to others.

Taste is influenced by the recent past history of taste receptors (adaptation) and by longer-standing past history (memory and conditioning) which may associate special savoring or revulsion for particular tastes. There are conspicuous cultural differences in gustatory appreciation. Taste in this context is presumably acquired. Some constituents in the blood not only influence taste but alter an individuals' sensitivity to certain tastes. A broad overlap of representations of taste sensitivities obtains. Integration doubtless involves both peripheral and central efferent controls and lateral inhibitory influences.

Taste provides a vital gateway for perceptual judgment as to what may be appropriate or inappropriate to swallow. In this way, taste plays a decisive role in maintaining appropriate nutritional balances (Pfaffman, 1970). For example, animals with a salt deficiency show a definite taste preference for salt, and carbohydrate deficiency may be associated with an increased appetite for sugars. Some deficiencies may not create specific taste preferences but may increase food searching activities and preferences for novel foods (Nachman and Cole, 1971).

Taste requires chemical discrimination by touch. Taste thresholds are remarkably low as shown in Table 9.2. The process of moving food about in the mouth by motion of lips, cheeks, jaws, tongue, and gullet, aided by the secretion of salivary juices, serves to distribute food over the taste neuroepithelia in much the way the eye moves over visual objects to improve visual perception. Chewing alters the physical bulk and consistency of food and mechanically releases various tastes. The enzymatic actions of salivary secretions break down some compounds into tastier and sometimes less tasty components. Liquification of food from salivation and from drinking allows material to find readier access to taste buds. A dry mouth is disadvantageous for tasting purposes; it is partly for this reason that we reflexly salivate while eating.

Combinations of taste tend to enhance one another, e.g., sweets without salt tend to be relatively less attractive. Mixtures of several different sugars taste better than only a few. It is for this reason that fruit tastes better in spring and autumn when there are greater diurnal temperature fluctu-

Table 9.2.
Absolute Thresholds

Substance	% Concentration (approx.)	Molar Concentration
Sucrose	7×10^{-1}	2×10^{-2}
Sodium Chloride	2×10^{-1}	3.5×10^{-2}
Hydrochloric acid	7×10^{-3}	2×10^{-3}
Crystallose (sodium salt of saccharin)	5×10^{-4}	2×10^{-5}
Quinine sulphate	3×10^{-5}	4×10^{-7}

(From Stevens, 1951).

ations which influence deposition of different fruit sugars during ripening.

Taste sensations are generated by complex transactions among chemicals and receptors in taste buds together with subsequent activities occurring along the taste pathways and areas of central representation. There is much sensory processing, centrifugal (efferent) control, convergence, and global integration among related systems contributing to gustatory experiences.

Olfaction can attract us toward or repel us from chemical sources that are airborne. Chemicals important for taste are rarely airborne. Yet while we are tasting something, olfaction continues to present strong cues as to what may or may not be acceptable for ingestion. Foul or repulsive odors present a barrier to ingestion, even against great force of will. What has been taken into the mouth may be forcibly ejected. What has been swallowed may be propulsively vomited. We may have to "hold our noses" in order to taste and swallow something that has an offensive odor even though we know it to be tasty. When we know something to be important for nutrition and health, it may be necessary to sugar-coat or provide an agreeable vehicle to enable a patient to swallow something that is malodorous or distasteful.

ORGANIZATION OF TASTE BUDS AND PAPILLAE

Taste is a contact chemical sense, innervated by sensory components of the VIIth, IXth, and Xth cranial nerves (Fig. 9.59). Taste neuroepithelium consists of a host of taste buds widely distributed over tongue, pharynx, and larynx. They are aggregated in relation to three kinds of papillae: fungiform, foliate, and circumvallate. Taste buds appear to be generally similar among these supporting tissues.

Fungiform papillae are widely distributed on the top and sides of the tongue. They look like blunt pegs which stick up from the surface of the tongue with 1–5 taste buds mounted on top. It is readily feasible to obtain electrical recordings from single fibers, and integrated records from the whole branch of the VIIth cranial nerve devoted to taste, during application of solutions to fungiform papillae on the anterior two-thirds of the tongue of the monkey (Zotterman, 1967). For example: There are active neuronal responses to Ringer's solution, to sucrose, and to water alone; but the response to sucrose in Ringer's is smaller than that to sucrose in water because Ringer's inhibits the response to water.

Foliate papillae resemble such fungiform papillary pegs which have been submerged in a surrounding moat filled with serous fluid, so that the

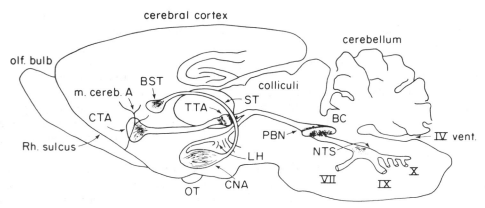

Figure 9.59. Basic pathways of the central gustatory system superimposed on a parasagittal section of the rat brain. *BC,* brachium conjunctivum; *BST,* bed nucleus of stria terminalis; *CNA,* central nucleus of the amygdala; *CTA,* cortical taste area; *IV vent.,* fourth ventricle; *XI,* glossopharyngeal nerve; *LH,* lateral hypothalamus; *m. cereb A,* middle cerebral artery; *NTS,* nucleus of the solitary tract; *olf. bulb,* olfactory bulb; *OT,* optic tract; *PPB,* parabrachial nuclei; *Rh. sulcus,* rhinal sulcus; *ST,* stria terminalis; *TTA,* thalamic taste area; *VII,* intermediate (facial) nerve; *X,* vagus nerve. (From Pfaffman; from Norgren, 1977.)

top of each peg is approximately even with the surrounding tongue surface. Thousands of taste buds lie on the sides of the foliate pegs and face the fluid. Foliate papillae are found on the posterior and lateral surfaces of the tongue, in a region characterized by bitter taste perception. They are innervated by both the chorda tympani (VIIth) and the glossopharyngeal (IXth) cranial nerves (Fig. 9.54*A* and *B*).

Circumvallate papillae, innervated by the glossopharyngeal (IXth) nerve, are much larger organs. They contain thousands of taste buds which face out from the sides of stout central stalks into large-scale circumferential serous-filled moats.

Each taste bud in any of these papillae consists of 40–50 modified epithelial cells, grouped in a barrel-shaped aggregate beneath a small pore which opens onto the epithelial surface. *Receptor cells* crowd their narrow, specialized apical surfaces together beneath the side walls of the pore, forming the bottom of a cylindrical pit. Their exposed surfaces, dense with microvilli, lie in contact with the acidophillic contents of the pit. Within this pit chemicals exert their effects on receptor molecules borne on membranes of the receptor cells.

Each taste nerve arborizes before it innervates several taste buds as well as many indiviual cells within each taste bud. Each afferent axon thus responds to activity among many receptor cells, providing convergent integration in the first-order neuron.

The effects of chemoactivation of a receptor cell are to generate simultaneously a receptor potential and a generator potential which, in turn, activates nerve endings that are linked synaptically to the base of the receptor cell. Receptor cells are surrounded by supporting cells. Individual cells in each taste bud differentiate, function, then degenerate and are replaced on about a weekly basis, from a stock of basal cells lying in the floor of each taste bud. This means that taste nerves continually remodel synapses on newly generated receptor cells.

Whereas the first-order olfactory neuron, the chemoselective unit, dies and must be replaced, in the case of taste, only the taste receptor cells and taste buds are recycled. The first-order taste neuron does not die but instead provides trophic influences essential for regeneration of taste receptor cells and taste buds.

TROPHIC INFLUENCES OF GUSTATORY NERVES

Outward transport of substances manufactured in the nerve cell body is required on a continuing basis for functioning of gustatory nerve and taste receptors. Recordings can be made from the gustatory nerves, but if their ends are severed centrally, chemically activated nerve impulses die out within minutes to hours (Bergland et al., 1977). Afferent innervation is also essential for regeneration of receptor cells and taste buds and for maintaining nerve-receptor synapses. Efferent fibers, which travel in the IXth nerve, are not necessary for peripheral trophic effects. All taste afferents convey impulses from taste receptors to the *nucleus of the solitary tract* in the brainstem (see *NTS,* Fig. 9.59).

TASTE BUD INNERVATION

Sensory innervation of the tongue and other surfaces of the oral cavity is provided by the trigeminal (Vth) cranial nerve. Taste buds on the anterior two-thirds of the tongue are innervated by terminals of the intermediate nerve, a branch of the facial (VIIth) nerve, with cell bodies in the *geniculate ganglion* (Fig. 9.54*B*). The glossopharyngeal (IXth) nerve, in addition to its efferent elements, includes afferents which serve taste buds on the posterior third of the tongue. Its afferent cell bodies lie in the *superior petrosal ganglion*. Branches of the vagus (Xth) nerve, with cell bodies in the *nodosal ganglion*, innervate taste buds found on the soft palate and palatal arches, the extreme posterior part of the dorsum of the tongue, and on the superior surfaces of the epiglottis and larynx (Fig. 9.54*A*).

Taste buds on the larynx and epiglottis are abundant at birth but begin to disappear during infancy. Finally, the tongue is left as the major organ of taste, with separate regional areas of innervation. The number of taste buds on the anterior two-thirds of the tongue is not great, and they tend to disappear progressively with age.

Central Gustatory Pathways

The principal components of central representation of taste are illustrated in the rat brain in Figure 9.59. The first relay of taste impulses occurs in the *cephalic half of the nucleus of the solitary tract* (*NTS*). There, second-order neurons receive centrally branching axons from bipolar cells of the VIIth, IXth, and Xth cranial nerves which are activated by taste buds distributed in the oral cavity (Fig. 9.54). NTS neurons project directly to the most medial part of the *ventral posterior thalamus* (*VPM*), thalamic taste area (*TTA* in Fig. 9.59), adjacent to oral fibers of the trigeminal (Vth cranial nerve) in juxtaposition to facial, oral, and tongue representation. Third-order axons project from the thalamus to the *taste area of somesthetic (primary sensory) cortex*, cortical taste area (*CTA* in Fig. 9.59), in the area of face representation in the postcentral gyrus, near the Sylvian fissure. Another thalamocortical projection goes to temporal lobe anterior *limbic* (central nucleus of the amygdala,

CNA in Fig. 9.59), and *entorhinal* cortex and cortex on the anterior portion of the *insula* (island of Reil). From the amygdala are phylogenetically very old projections via the stria terminalis to the bed nucleus of the stria terminalis (*BST* in Fig. 9.59) in the medial forebrain, closely connected with the hypothalamus.

The taste portion of the nucleus of the solitary tract also sends axons or collaterals to many regions of brainstem and diencephalon (lateral hypothalamus, *LH* in Fig. 9.59), including the reticular formation, median raphe nuclei, and diffusely projecting systems (Ricardo and Koh, 1978). The caudal (nontaste) part of the nucleus of the solitary tract receives impulses from the nodosal ganglion of the vagus (Xth cranial nerve) and projects to the brainstem and spinal cord. The NTS thereby contributes to visceral as well as to gustatory reflexes.

A "pontine taste area" has been located which projects bilaterally to thalamus, lateral hypothalamus, central nucleus of the amygdala, and basal forebrain. Hypothalamic projections include paraventricular, dorsomedial, and arcuate nuclei and the medial preoptic area. This provides taste information to areas in the amygdala and other temporal lobe structures juxtaposed to olfactory representations. Both limbic and basal forebrain structures probably contribute to affective qualities associated with taste, and the hypothalamic projections contribute to regulation of appetite and satiety.

Brodal (1981, p. 503) cities clinical observations which confirm much of the basic physiological and anatomical evidence: contralateral impairment of taste following tumors involving VPM of the thalamus; subjective sensations of taste in unanesthetized patients on stimulation of the lowermost portion of the postcentral gyrus, where it approaches the insula; gustatory auras associated with epileptic convulsions originating in the region of taste representation in mesocortex; and impairment of taste following cortical lesions in that vicinity.

Touch, Pain, and Temperature

NEURAL MECHANISMS OF SOMESTHESIS

Somesthesis and somatic sensibility are synonymous terms which include *touch-pressure, pain, temperature, and proprioception.* Proprioception includes sensibility of motion and tension in skin, muscles, tendons, and joints. Somesthesis includes all perceived events referable to body surface—except taste, and all sensed events referable to body wall—except vestibular. Somesthesis is usually considered as if it were a sensory system independent of autonomic mechanisms. This excludes central chemoreceptors, sensors associated with blood vessels and viscera—concerned with visceral functions of which we are generally unaware—and sensed functions, partly somesthetic, partly visceral, relating to swallowing, coughing, defecating, urinating, and coitus.

Somesthesis excludes mechanisms relating to outwardly projected perceptions—hearing, vision, and olfaction. With hearing and vision, physical transductions take place deep in tissues: photoreception in the layer of retina farthest from outside, and mechanoreception in the cochlea—embedded in densest bone. These are distance sensory systems because the events of perceptual interest are outside. The corresponding perceptions refer to an "outside world." Indeed, we create and project a unique "outside world" based on individual sensory experiences. Functional appropriateness of these projections is tested and affirmed through other sensory and motor experiences. With body movements we measure our way to sources of nearby sights, sounds, and smells, and extrapolate to more remote sense data sources.

BODY IMAGERY

Our "inmost self" consists of feeling-states: energy levels, moods, emotions, appetites, motivations, and the like. These guide our behavior for obtaining internal satisfactions: oxygen in/carbon dioxide out, water and salt balance, food intake and waste elimination, social relations, reproductive activities, sleep, etc. Self-awareness—perception of our existence and state of being—combined with visceral, vestibular, and somesthetic sensations establishes a *body image.* Somesthesis contributes sense data from body wall, cutaneous surface, and immediate surround—self-reference of our physical existence. This includes a sense of body fitness and availability for action: limb mobilities, positions, tensions, inertial forces, etc. Building on our immediate sense of mood, feeling, and energy state, somesthesis contributes the bulk of sense data required for conscious modeling of behavior.

Body imagery entails an internal map or model of our self to which all sensations are referred. Vestibular and distance sense data orient our body image with respect to three-dimensional space. Sensations of hearing, vision, and olfaction, which inform us about the outside world, we project outward with reference to our body image. We experience ourselves as bundles of feeling states and appetites, occupying generally competent yet vulnerable bodies which we move about in a dynamic physical and social environment. We perceive the outside world as real, and our feelings as referred, but it is quite the other way around. What we know by first intention and directly are only our own feelings. What we know about the "real" world is derived and projected according to elaborate inferences and assumptions. By social custom, that which we create and which we accept as "reality," as an outside world we can only sketchily compare to our own internal experiences. By custom, we behave as though our feeling states were less accessible, less real.

TOUCH

Perhaps the most biologically valuable and socially cherished sensation is touch. Hearing, which is equally socially cherished, is an evolutionary derivative of touch. A newborn deprived of varied, gentle touchings fails to thrive and may die without any other deprivation. Proprioception provides immediate experience of our body, but self-touching and active and receptive social touching generates even more vivid bodily sensations. Gentle touch conveys an intimate, gratifying signal; the handshake is a ritual social assurance by touch. Except when we are asleep or otherwise unconscious, we experience touch continuously, from some fetal origin of consciousness to our last glimmer.

A Sense of Touch

The customary surfaces on which touch stimuli are applied are cutaneous, but touch is also elicitable from lips, mouth, tongue, gums, throat, esophagus, trachea, and other ports, as well as periosteum, tendons, muscles, and deep tissues that can respond to movement, pressure, tension, and shearing forces. All hollow viscera and many blood vessels are sensitive to ballooning pressure from within and shearing forces, but can be touched, squeezed, cut, or burned, without eliciting any sensation. All else, including the brain, is quite insensitive.

ADEQUATE TOUCH STIMULI

Sensory endings for touch are known to respond to pressures that result in deformation of skin or movement of hairs. Sensitivity to touch varies in different regions of the body but may be elicited by very slight skin deformation or deflection of a single hair. A wisp of cotton moving over a hairy skin surface or over bare skin is a highly arousing sequence of touches.

Tactile sensibility combines discrimination of the physical characteristics of the stimulus, plus location, time, and intensity. Touch sensation elicited by mechanical stimulation of the skin varies from a faint, barely discernible tickle to painful pressure. Information about shape and temperature of the stimulus, depth of indentation, degree of flexion, stretching or shearing deformation of the skin, two-point sensibility, stimulus movement across skin, the rate at which stimuli are delivered, and spatial and temporal characteristics of moving and non-moving stimuli are accurately discriminated by fingertip and lip, and somewhat less accurately by skin elsewhere. Skin with a comparatively high threshold for touch has a comparably reduced capacity for detailed discrimination. Hairy skin reception includes discrimination of hair movements, including unidirectional motion of single guard hairs, and patterned movement of downy hair, not dependent on skin touch.

A few touch receptor endings are characterized schematically in Figure 9.60. Specialized touch receptors include Meissner's corpuscles, found in abundance in hand, foot, nipple, lips, and tip of

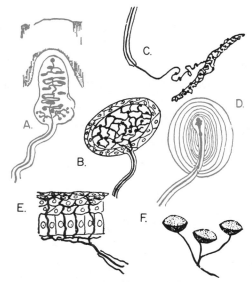

Figure 9.60.   Various elaborations of cutaneous receptor endings. *A*, Meissner's corpuscle. *B*, Krause's end bulb. *C*, Ruffini end-organ. *D*, Pacinian corpuscle. *E*, Bare nerve endings in the cornea (pain). *F*, Merkel's disks (touch). (From Bainbridge and Menzie, 1925.)

tongue; Merkel's disks found in fingertips (Fig. 9.61), lips and mouth, and basket-like endings surrounding hair follicles. Pacinian corpuscles have elongated, multilayered capsular end-organs. These layers focus the stimulus energy onto a nerve terminal that lies along the capsule's mechanical centerline. They send only a transient response to sustained mechanical displacement, and the axon follows stimuli optimally between 250 and 300 Hz (Lowenstein and Mendelson, 1965).

Threshold for touch is quantitatively measured in terms of pressure required to bend standard hair bristles against the skin. Surfaces of the body sparsely covered with hair tend to be especially sensitive to direct touch. Shaving the hair results in generally reduced sensitivity to touch. Minimum pressures required to elicit sensations of touch vary by a factor of more than 1:20, depending on body region. Tip of nose, lips, and fingertips require only 2–3 g/mm^2. On the backs of fingers and upper arm, higher pressures are required, and outer surface of thigh may require about 50 g/mm^2.

Touch Receptors

All cutaneous receptors are embedded in skin or subcutaneous tissue. Responses depend on the physical characteristics of the receptor and on the elasticity, deformability, hysteresis, thermal and electrical conductile properties, etc., of the surrounding tissue. Multiple cutaneous receptors that stem from a single axon may serve a less than 1 mm^2 region of skin, as in a fingertip. Others may be distributed over several cm^2. Scattered distributions

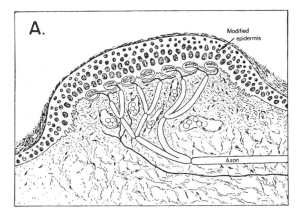

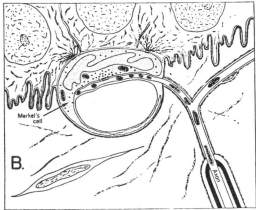

Figure 9.61. *A,* Cross-section of an epidermal dome with layer of nerve endings just inside the basement membrane. Endings are derived from a single myelinated axon. *B,* Detail of Merkel's cell at one of the nerve endings. (From Iggo, 1966.)

may leave wide, unresponsive gaps between receptor sites.

The simplest kinds of receptors consist of bare nerve terminals which arborize among cells in the superficial epidermis and across the corneal surface, and are also found in deeper layers of cutaneous and subcutaneous tissues. Pain sensations are thought to originate from such bare nerve endings. However, both pain and discriminative touch can be elicited from stimulation of free nerve endings, e.g., in the cornea where these are the only kinds of nerve terminals present. It is established that both modalities, touch and pain, are served by similar-looking undifferentiated bare nerve endings. In contrast, temperature sensibility, which might be expected to represent a continuum, is coded by separate warm and cold receptors.

There is usually considerable overlap in distribution of cutaneous receptors among adjacent neurons, and overlap between fibers from neighboring nerve bundles. Thus the boundaries of anesthetized or denervated skin are broad and fuzzy, a few millimeters in width rather than being sharply

bounded. From studies of regeneration following peripheral nerve injury, it is clear that there is active competition for fields of representation between neighboring nerve fibers and between fibers belonging to neighboring nerve bundles. If a nerve bundle or fascicle is cut, one or more neighboring bundles or fasicles may send out axon collateral sprouts which may successfully reinnervate the denervated zone. This reinforces evidence that (a) there is normally a trophic barrier which inhibits such cross-territorial invasion, and that therefore, (b) without meticulous anesthetic nerve block, clinical evidence for reinnervation does not prove that the reinnervation was restored through regeneration of the severed nerves (Livingston, 1947).

As with taste and olfactory neuroepithelium, there is a dynamic turnover of sensory end-organs even in areas not subjected to wear-and-tear. When skin is abraded, superficial nerve endings are also regenerated. The cutaneous scene is one of dynamic morphological as well as functional activity.

VELOCITY GROUPING OF MECHANORECEPTORS

A *sensory unit* includes the end-organ, corresponding axon and dorsal root cell body, and the distribution of axon terminals in the spinal cord or brainstem. Units are grouped according to axonal conduction velocities which range from about 2 m/s (equivalent to leisurely walking velocity) to 75 m/s (equivalent to propeller-driven aircraft velocity). Generally, the higher velocity sensory units have relatively more elaborate end-organs which respond at lower thresholds and adapt more rapidly. They reliably detect motion and timing of stimulus onset and duration, contributing to sensations of flutter and vibrations. They also send terminal axon collaterals to higher relay stations in the central nervous system as well as to the level of entry.

Slowly Conducting Fibers and Pain Sensation

C-fibers are slowest conducting, at 2 m/s; they cannot follow repetitive stimuli faster than about 2–5/s. A-δ units are thinly myelinated and conduct about 10 fold more rapidly than C-fibers. Both fiber types have unmyelinated nerve endings which arborize freely in skin and subcutaneous tissues. A-δ units have endings in the vicinity of hair follicles and are abundant in tooth pulp. Both C-fibers and A-δ fibers are capable of responding to touch and also of subserving pain perception. They have relatively high thresholds and do not respond to brief stimuli. With adequate stimulation they discharge after a slight delay and continue to respond to continuing stimulation. They generally deliver a burst of impulses, an "after-discharge," when stimulation is discontinued.

The central distribution of C-fibers axons is to grey matter in the dorsal horn at the spinal level of dorsal root entry and via bifurcating branches that terminate in the same dorsal horn region in adjacent segments above and below. A-δ fiber terminals serve these same segments and also send collateral axons which ascend the dorsal columns a few segments higher, where they also end in grey matter of the dorsal horn.

Hair Innervation—Mechanoreceptors

Three additional types of afferents innervate hairy skin. They have elaborate end-organs which are spear-like, fence-like, or basket-like, all distributed in the vicinity of hair follicles. They vary in conduction velocity from 50–75 m/s and, beside distributing terminals to local spinal segments, ascend the dorsal columns most or all the way to the bulb (medulla oblongata) before terminating in high spinal dorsal horn or bulbar dorsal column nuclei. These hairy skin units respond promptly and have stronger "on" than "off" responses. They generate impulses during slow or rapid hair movement. They may respond only to unidirectional movement of guard hairs.

Nonhairy Skin Mechanoreceptors

These sensory units have specialized receptor end-organs which manifest a variety of forms, including lamination, like Meissner's corpuscles, and encapsulation, like Krause's end-bulbs (Fig. 9.60). Conduction velocities range from 50–65 m/s. These units are rapid in onset and follow repetitive stimuli at moderately high rates. They respond best to deformation of skin and continue to discharge for several seconds following stimulus discontinuation.

Rapidly Responding Afferent Units

There are sensory units with even more elaborate receptors, consisting of Ruffini end-organs, Pacinian corpuscles, and Merkel's disks (Fig. 9.60). Sensory units with Ruffini end-organs may or may not show a resting discharge. They characteristically respond promptly with a regular firing rate to sustained light touch. Pacinian corpuscles, however, adapt rapidly and follow high-frequency rates of stimulation, optimally at 250–300 Hz.

Pacinian corpuscles are rare in hairy skin and common in glabrous skin. They are prevalent throughout subcutaneous and intramuscular connective tissues and periosteum. They are also found in mesentery where they can be readily isolated for physiological study. They respond to extremely slight perturbations: the pulse, respiratory movements, finger tremor, even a breath of air across an exposed corpuscle.

A myelinated fiber enters the Pacinian corpuscle and a half-node of Ranvier gives rise to a bare nerve terminal which lies along the center mechanical focus of the multilayered capsule (Fig. 9.60). Slight pressure on the capsule leads to a generator potential in the nerve terminal which, in turn, leads to depolarization of the node of Ranvier and an action potential conducted along the parent axon.

Physiological Responses of Cutaneous Afferents

UNITS IN GLABROUS SKIN

Units in glabrous skin on the palmar surfaces of the hand and foot match special features such as dermal ridges which contribute to mechanoreception (Fig. 9.61). C-fibers are sparse. Ruffini end-organs and Merkel's disks exhibit irregular tonic discharges which contribute to reception of velocity and acceleration as well as displacement. Four types of sensory units innervate glabrous skin: two are rapidly adapting (RA) and two are slowly adapting (SA). RA units respond only during skin movement, not during static displacement; SA units continue to respond during sustained skin displacement (Fig. 9.62).

Rapidly Adapting (RA) Sensory Units

Fibers associated with *Meissner's corpuscles* discharge a brief burst of impulses when skin is indented, adapt rapidly to steady pressure, and deliver a brief burst on removal of the stimulus. Their rate of adaptation makes them selectively sensitive to stimulus frequencies in the range of 30–40 Hz. They contribute to perception of what is called *flutter*. Another type of RA unit is associated with *Pacinian corpuscles*. These subserve sensations of *vibration* and can follow 10 fold higher frequencies.

Slowly Adapting (SA) Afferents

Units associated with *Merkel's disks* (Fig. 9.61) are slowly adapting and very persistent. They are quiescent in the absence of mechanical stimulation. A mechanical stimulus of 1- or 2-ms duration elicits a burst of impulses lasting 10 ms during physical recovery of surrounding tissue. The frequency of discharge varies with the amount of displacement. When a smooth probe is drawn across skin, successive Merkel's disk units can respond at rates in excess of 1000/s (Iggo, 1966).

Ruffini end-organ units show a regular resting discharge in absence of stimulation. Response to stimulation yields a rapid discharge which slows to a rate above the resting level, corresponding to the degree of static skin displacement.

In Figure 9.62, we see responses of two cutaneous afferents, one RA and the other SA, in the same area

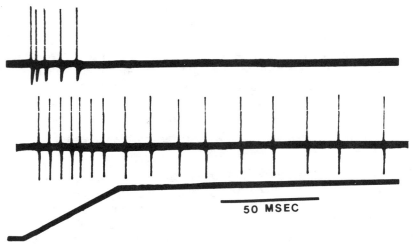

50 MSEC

Figure 9.62. Touch reception from a single rapidly adapting fiber *(upper trace)* and a slowly adapting type I fiber *(middle trace)* to the same mechanical displacement of glabrous skin in the hand of a rac- coon. Displacement ramp velocity 10 μm/s, final displacement, 480 μm/s *(lower trace)*. (From Pubols and Pubols, 1963.)

of glabrous skin, responding to the same mechanical stimulation delivered to the paw of a raccoon. The RA unit responds very promptly and at a high rate. It ceases to discharge even though the mechanical displacement continues. The SA sensory unit fires at a relatively rapid, steady rate during ramp onset of mechanical displacement and then settles down to a slow rate of discharge which continues as long as skin indentation is maintained.

Figure 9.63 depicts schematic maps of 45 human glabrous skin receptive fields. On the *left* (*A*) are outlined fields associated with rapidly adapting (RA) units and on the *right* (*B*) are fields associated with slowly adapting (SA) units in the same hand of the same subject. To the *right* are action potentials recorded from a single ulnar nerve afferent fiber in the same individual. This fiber innervated a recep- tive field on the volar surface of the middle phalanx of the fifth finger. The unit adapted slowly to each of three stimulating forces depicted below each tracing (Knibestol and Vallbo, 1970).

AFFERENT UNITS IN HAIRY SKIN

In addition to endings similar to those found in glabrous skin, hairy skin has a variety of endings associated with hairs and structures allied with hairs. In hairy skin where innervation is less dense (i.e., on the back of the forearm), one can identify localized "spots" which may have a dominant sin- gle-modality characteristic of sensory experience evoked by stimulation, no matter how the tempo and intensity of the stimulus is modified. Individual spots can be identified as "cold," "warm," "pain," or "touch" spots. This implies that in sparsely

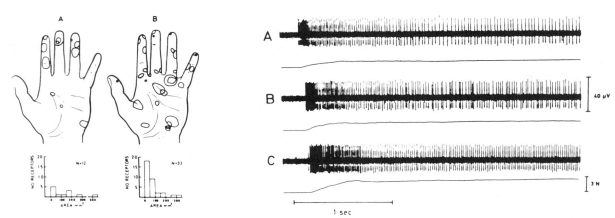

Figure 9.63. Schematic drawing of 45 cutaneous receptor fields of human glabrous skin *(above)* with histograms showing the distribtuion of receptive field sizes *(below)*. *A,* Rapidly adapting units. *B,* Slowly adapting units. *Right, A, B, C,* Responses of a slowly adapting unit which showed distinctive receptive field borders to stimuli of different intensities. The unit was located on the volar aspect of the 5th finger on the middle phalanx. The approximate stimulating force, indicated by the lower traces, and the mean impulse frequency in the steady state were in *A,* 1.2 N and 28 impulses/s; *B,* 1.6 N and 37 impulses/s; and in *C,* 2.4 N and 48 impulses/s. (From Knibestol and Vallbo, 1970.)

innervated areas, firing of one or another specialized sensory unit can dominate subjective experience.

Hair receptors may be usefully divided into those associated with short, fine downy hairs and those associated with long, thick guard hairs. The former are innervated by thinly myelinated axons having conduction velocities in the A-δ range. They tend not to be very sensitive to the direction of hair bending, but a number of them can signal patterns of downy hair bending.

Guard hair receptors have a wider range of conduction velocities and tend to be specifically sensitive to direction and velocity of hair deflection. Figure 9.64 illustrates a unit with Merkel's disk end-organs which responds with directional sensitivity to bending a single hair. It responds briskly during displacement of the hair in one direction, and emits a steady train of impulses while the hair is stati-

cally displaced. It ceases firing the instant the hair is released from displacement. When the same hair is displaced in the opposite direction, there is silence until the hair is released from that direction of displacement, whereupon the unit manifests a rapid train of impulses while the hair recovers its normal position. This unit shows remarkable fidelity to dynamic vibration, following reliably to 1200 Hz (Gottschaldt and Vahle-Hinz, 1981).

Structure-Function Relations in Finely Myelinated Pain Fibers

A-δ nociceptive fibers have "bare" nerve endings. Kruger et al. (1981) confirmed that the receptive fields of such nociceptive units may leave numerous large gaps between spots of local innervation (Fig. 9.65). A single unit can serve up to a score of separate nociceptive spots. Stimulation of gaps between spots evokes no response, but vigorous responses

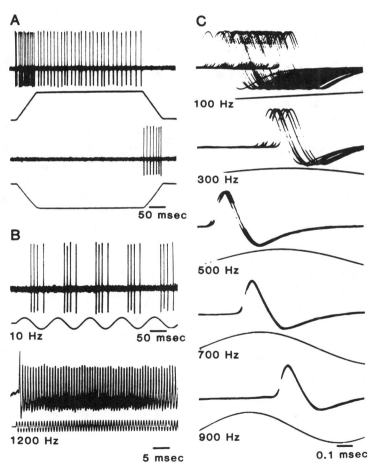

Figure 9.64. Merkel's cell transducer function. *A,* Directionally sensitive, slowly adapting nerve fiber during bending of a sinus hair in opposite directions. The stimulus signal is displayed below the spike records. *B,* Responses to hair movement at 10 and 1200 Hz. *C,* Phase jitter of at least 10 superimposed impulses in response to the same cycle of repeated vibratory stimuli at different frequencies. The elec-tronic sine wave signal driving the stimulator is pictured below the spike records. The mechanoelectric transduction process is faster than that for any other mechanoreceptor, responding to vibration as well as displacement. It is the nerve endings themselves and not the Merkel's cells that are the critical mechanoelectric transducer elements in these receptor complexes. (From Gottschaldt and Vahle-Hinz, 1981.)

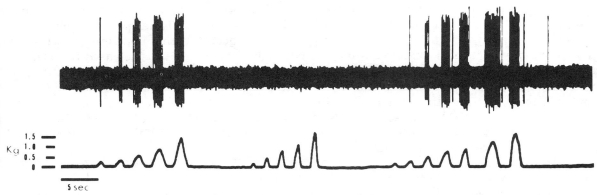

Figure 9.65. Discontinuity of receptive field of a single nociception fiber. This unit had a 17-spot receptive field. *Upper trace,* Recording from two adjacent responsive points. *Lower trace,* Force applied with a 1-mm probe to first spot *(left),* an unresponsive intermediate point, and the second spot 5 mm from the first. (From Kruger et al., 1981.)

are elicited when stimuli are applied to each innervated spot.

Mapping nociceptive spots by physiological means, Kruger et al. examined the same loci for innervation patterns, using light and electron microscopy. Beneath each nociceptive spot they found a thinly myelinated axon which, when it entered the epidermal layer lost its myelin sheath but continued to be associated with a thin Schwann cell shroud throughout its terminal ramifications. These "unmyelinated" terminal axons contained abundant, clear round vesicles, and sparse, large dense-core vesicles. The coapted Schwann cell processes exhibited many pinocytotic vesicles. The vesicles and pinocytosis indicate active communication between axonal terminal and glia. The combination of unmyelinated axon terminal and companion Schwann cell process are presumed to constitute the receptive apparatus for A-δ endings.

Elaborate corpuscle-laden end-organs which are commonly associated with sensory units are thought to be of Schwann cell origin. The boundary between presumably glial-fashioned end-organs and the axon terminals they surround resembles the shroud found on A-δ terminals. The relationship of terminal axons to Schwann cell processes constitute a fundamental synergistic junction for simple as well as elaborate end-organs. The rest of the elaborate end-organ apparatus contributes selective physical advantages.

Temperature Receptor Mechanisms

Specific end-organs have not been established for thermoreceptors, but warm and cold sensations are mediated by two physiologically distinctive systems. It is likely that both types of endings belong to the category of "bare" nerve endings, unencapsulated except for the shroud of Schwann cell membrane. Both types of thermoreceptor are responsive to the rate of change of temperature, linear at least for changes of intensity introduced by cold pulses. Perception of warmth is not dependent on inhibition of cold receptors but involves an independent receptor population.

Thermosensitive receptor fields have been found to consist of dispersed, identifiable warm and cold spots which are persistent over time. Warming a cold spot is ineffective or yields paradoxical cold sensations.

Physiological Responses of Cutaneous Afferents in Human Subjects

Hagbarth and Vallbo (1967) pioneered research by inserting microelectrodes through skin into nerve trunks in humans and recording from single sensory and motor axons. The peripheral fields of sensory units could be explored for distribution, modality specificity, response characteristics, and conduction velocity while at the same time the benefits of subjective testimony could be obtained while the unit was being tested. By stimulating as well as recording through such percutaneous microelectrodes in human subjects, Torebjoerk and Ochoa (1980) succeeded in identifying correspondences between single median nerve axons and their peripheral receptive fields established by referred sensory localization and the characteristics of their responses to mechanical stimulation.

Figure 9.66 illustrates microelectrode stimulation and recording from single myelinated cutaneous sensory units in man. Microstimulation of the unit at the elbow created a sensation referred to the thumb. Trains of stimuli were interpreted as flutter or vibration, depending on stimulus frequency. (1) Activation of RA units yielded painless "tapping" sensations which become "flutter" or "buzzing" at successively higher stimulus frequencies. (2) Painless continuous pressure was evoked by repetitive

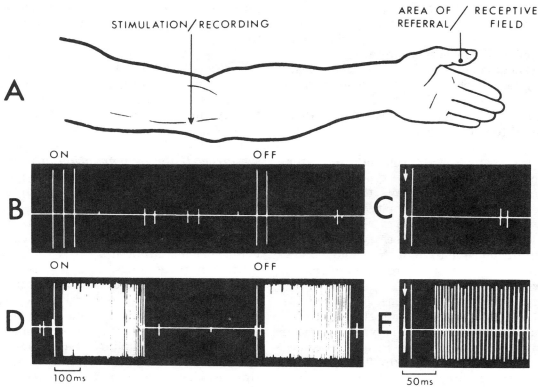

Figure 9.66. Microelectrode stimulation and recording in man from a single sensory nerve fiber, with perceptual referral to distal receptive field. *A,* Intraneural microstimulation in the median nerve at the elbow is referred as flutter-vibration to a focal area of referral in the thumb. *B,* Intraneural recording detects "on" and "off" responses to a single rapidly adapting unit responding to skin indentation in the area of referral. *C,* Electrical stimulation in the receptive field of unit excites a fiber with a conduction velocity of 45 m/s (*white arrow* indicates stimulus artifact). *D–E,* Prolonged intraneural microstimulation at high frequency induces posttetanic hyperexcitability, whereby mechanical (*D*) and electrical (*E*) stimulation elicits stereotyped bursts of impulses. Time bases given for *B–D* and *C–E,* respectively. (From Torebjork and Ochoa, 1980.)

impulses delivered to SA units that have fast conduction velocities. (3) Sharp, stinging pain sensation was elicited from high rate threshold units with conduction velocities in the A-δ range. (4) Poorly localized dull pain sensation was experienced when slower-conducting C-fibers were excited.

Trophic Influences of Afferent Neurons on Innervated Tissues

Most afferent units contribute trophic influences that affect the tissues they innervate, including other nerve cells, receptors, end-organs, etc. The effects apparently depend on protein synthesis in the cell body and axonal transport to the site of innervation. The effects may be reciprocal, i.e., tissues and end-organs can modify structural and functional characteristics of the nerves that innervate them.

Merkel's disks depend on such trophic influences; if the afferent nerve is sectioned, the disk degenerates. The disk regenerates on reinnervation. Pacinian corpuscles, however, do not degenerate and may be reinnervated by another afferent neuron, whether or not that neuron had previously served Pacinian end-organs (Schiff and Loewenstein, 1972). The Pacinian corpuscle contributes trophic effects which dictate the physiological characteristics of the reinnervating nerve fiber.

Small diameter afferents associated with blood vessels may release substance P which contributes neurogenically to inflammatory responses (Kruger, 1983). Activation of one branch of an afferent nerve can generate an *axon reflex* in adjoining branches of the same neuron. Signals ascend the local afferent fiber to the nearest bifurcation and then conduct in a direction opposite to ordinary afferent conduction. Impulses travel out to the periphery along adjoining branches of the same parent fiber. These are called *axon reflexes* and they contribute to vasodilation and inflammatory reactions.

PERIPHERAL PAIN RECEPTION

Pain is ordinarily aroused by stimuli that damage or risk tissue destruction. Pain can be elicited by stimulation of most tissues except the brain. Subjective pain experiences consist of: (a) *sensations* of pain, experienced as extreme hurting, and (b) *reactions* to pain. The latter may include somatic and autonomic reflex responses, voluntary efforts to remove the stimulus and to escape, emotional reactions, and "suffering" that may be burdened by previous experiences associated with pain.

There are two principal types of cutaneous afferents considered to constitute *pain units:* a population of A-δ, small diameter, finely myelinated nociceptive fibers, and still smaller diameter, unmyelinated C-fibers. Both these types of cutaneous afferents include sensory units that are fully capable of responding discriminatively to light touch. But with each of these two types of neurons, pain sensation may dominate. Both types of units are especially responsive to high threshold, tissue-damaging stimuli which usually result in definitely painful perceptions.

Activation of A-δ nociceptive sensory units yields sharp, pricking pain such as that elicited by an abrupt prick with a sharp pin. This type of unit can be activated by introducing a not-adequately-sharp hypodermic needle through the skin. It results in an immediate sharp pricking pain, accurately localized, which shortly abates.

Activation of C-fiber nociceptive units gives rise to a slower onset, a "second" pain that is more intense and longer lasting. Both types of nociceptive sensory units can be readily and effectively activated by high intensity tissue-damaging stimuli, using any mechanical, chemical, electrical, or thermal means. Threshold for thermal excitation of a pricking heat sensation is at a skin temperature of 45°. If such heat is prolonged, it can cause tissue damage.

Collins et al. (1966) stimulated C-fibers in conscious human subjects after compression block of myelinated fibers. They found that a single C-fiber volley did not reach consciousness, whereas if stimuli were repeated at 3/s, subjects invariably felt pain. Maximum response frequency stimulation (ca. 5/s) quickly became unbearable pain. Temporal summation along pain pathways apparently contributes importantly to the intensity of pain perception.

Characteristically, C-fiber pain has a burning quality. Strong activation of C-fiber units gives rise to a subjective feeling of *intense hurting*. An aching, paralyzing hurt can extend rapidly and inexorably to affect a whole limb or entire side of the body, even including temporary paralysis, after which the pain may slowly subside. While the pain sensation is expanding, it can engulf all other sensations and absolutely transfix attention with suf-

fering. C-fiber pain can be long persisting and, if so, it is likely to generate even more of a burning quality.

Sensitization of Denervation

Denervation (particularly partial denervation) upsets normal tissue functions by inducing tissue hypertrophy, hyperactivity, and hypersensitivity. *Partial destruction of any nerve pathway, peripheral or central, can result in "supersensitivity" of the target tissue or population of neurons.* Supersensitive neurons show exaggerated responses to both excitatory and inhibitory synaptic influences from other sources of innervation. A stimulating neurotransmitter or drug is more stimulating, and an inhibitory neurotransmitter or depressing drug is likewise more inhibitory.

Other nerve endings, not customarily associated with pain, if sensitized, can also contribute to pain perception. If both peripheral and central circuits become hypersensitive, a condition of intractable, searing, burning pain may ensue—called *causalgia* (Greek for "burning pain"). Patients with causalgia describe the feeling "as if a red hot poker were continuously pressing into" the affected parts. In this context, any kind of stimulation, even a gentle breath across the skin, may be perceived as excruciatingly painful (Livingston, 1943).

In chronic pain syndromes like causalgia, there may be obvious trophic changes in the distribution of the affected peripheral nerves. The skin may show marked loss of subcutaneous fat, elimination of normal creases in glabrous skin, reddened, mottled parchment-like appearance of skin, and overgrowth of nails—giving the appearance of claws, increased pigmentation, and coarseness of hair. There may be rapid dripping of sweat from the affected region. There may be local osteoporosis, greater in extent than expected from disuse atrophy. Segmental reflexes are likely to be affected, manifesting excessive sweating, heightened withdrawal (flexion) reflexes, spontaneous painful spasms, and spontaneous or reactive tonic-clonic contractions. It is important to realize that although causalgia is rare, lesser symptoms of burning pain, and less obvious trophic changes constitute an important class of "minor causalgias."

Severe, refractory pain is more likely with partial than with complete severance of a peripheral nerve. Importantly, the degree of pathophysiological changes, including disabling pain and trophic changes, does not depend on the severity of initial injury. A relatively minor trauma may give rise to full-blown causalgia (Livingston, 1943). Why most instances of acute painful trauma fade away without residual effects, whereas, occasionally, a seemingly similar acutely painful trauma may give rise to enduring, overwhelming signs and symptoms of intractable burning pain, we shall consider further.

SPINAL MECHANISMS OF SOMESTHESIS

Entrance of Somesthetic Information

AFFERENT NEURON INPUTS

Afferent fibers from skin, body wall, and viscera enter the spinal cord by way of the dorsal roots. Their cell bodies lie in the dorsal root ganglia. These units originate as bipolar cells. The two extending processes fuse into a single axon which bifurcates, sending one branch to a sensory receptive field, visceral or somatic, and the other to a central destination. Afferents innervating the face, the cranial analogues of spinal afferents, have cell bodies in the Gasserian ganglion (Vth cranial nerve) and enter the brainstem by way of the trigeminal nerve. Visceral afferents arriving by way of the vagus (Xth cranial nerve) have cell bodies which lie in the nodosal ganglia likewise homologous to dorsal root ganglia.

CATEGORIZING AFFERENT UNITS ACCORDING TO FIBER DIAMETER

Afferent nerves are grouped into four categories, I–IV, taking into account fiber diameter, physiological role, and peripheral end-organ. The largest diameter afferents, I-A, and many in category II, arise from muscles. Group I A-α, 12–22 μm in diameter, arise exclusively from tendon organs. Group II includes A-β units from muscle spindles. These are described in Chapter 70 in relation to the segmental control of skeletal muscles.

Group II also includes the largest cutaneous afferents, which range from 5–12 μm in diameter, and relate to touch, pressure, flutter and vibration. These cutaneous afferents include units associated with slowly adapting Merkel's disks, rapidly adapting Meissner's corpuscles, high-frequency Pacinian corpuscles, rapidly adapting hair receptors, and units relating to slowly adapting Merkel's and Ruffini's end-organs.

Group III consists exclusively of A-δ sensory units, 1–5 μm in diameter. These include quickly adapting mechanoreceptors which sense touch contact, and slowly adapting thermoceptor-nociceptive units. Group IV consists solely of C-fiber units, 0.2–1.5 μm in diameter, which constitute *slowly adapting* receptors concerned with thermosensitivity and/or nociception, and which also possess discriminative mechanoreception.

TEMPORAL DISTRIBUTION OF AFFERENT SIGNALS

Cutaneous sensory units which belong to all three groups, II, III, and IV, have representation in both glabrous and hairy skin areas. Considering differences in conduction distances, in conduction velocities (ranging between 75 m/s and 0.5 m/s), and latencies of response, plus the differences in time of adaptation, there is a very broad distribution of arrival times and a prolonged temporal spread of responses following the termination of stimuli. There is a spread of a few to many milliseconds between onset of stimuli and arrival of signals at spinal (dorsal root) and brainstem (trigeminal spinal root) terminals. There are even greater time differences following discontinuation of stimuli, and during the restoration of activity to steady-state conditions.

Information entering the spinal cord and brainstem continues through successions of relay neurons. As nerve impulses approach their termination, they markedly slow in velocity by virtue of branching (accompanied by reductions in fiber diameters) and by narrowing of terminal branches. A considerable portion of the approximately 0.5-ms "synaptic delay" is taken up by the slowing of impulses in presynaptic terminals. The range of diameters in the afferent fibers is usually repeated in fiber diameters of the corresponding central relay neurons. What has been conveyed relatively slowly in the periphery is relayed and transmitted relatively slowly along central pathways as well, contributing to a further temporal spread of central signals.

BIOLOGICAL ADVANTAGES OF RAPIDLY CONDUCTING SYSTEMS

Information contributing to perception is less urgent than information that can save us from falling. Rapidly conducting muscle spindle afferents and Golgi tendon organs serve to maintain, and when necessary, to inhibit postural muscle "tone." Rapidly conducting systems also control equilibrium, an urgent problem for all land-dwelling creatures. Both afferent and efferent arcs of the vestibular system as well as their relays to brainstem and cerebellar mechanisms involved in the control of balance are rapidly conducting.

CONTRIBUTIONS OF SLOWER CONDUCTING AFFERENT SYSTEMS

Other body wall afferents, cutaneous and visceral, do not possess monosynaptic access to motoneurons. They serve polysynaptic, multineuronal spinal reflexes. They are responsible for flexor reflexes. These involve excitation of flexor muscles and inhibition of extensor muscles on the ipsilateral side, and excitation of extensor muscles (the crossed-extensor reflex) and inhibition of flexor muscles, on the contralateral side of the body. This is made possible by segmental crossing of relayed, second-order sensory axons. The effect is to help maintain balance while retreating from the stimulus source.

Strong or prolonged excitation of any small diameter afferents contributes to ipsilateral withdrawal and protective, guarding reflexes. Flexor reflexes may involve all flexor muscle groups in the correspondings limb and ipsilateral side of the body. If there is, for example, a prolonged visceral pain stimulus originating in the right lower quadrant of the abdomen, as may accompany an inflamed appendix, or an ectopic pregnancy, skeletal muscles in the abdominal wall relating to that quadrant provide a flexion—abdomen-guarding, splinting—reflex. The resulting reflex muscular contractions may be strong enough to introduce an additional source of pain.

SPINAL AND BULBAR CONNECTIONS OF CUTANEOUS SENSORY UNITS

Incoming afferent axons characteristically send collateral branches ascending and descending for one or two segments above and below the level of dorsal root entry. These longitudinal, first-order neuronal branchings contribute to *intersegmental sensorimotor integration.*

Many incoming fibers send collaterals into the spinal grey matter while sending their main axons via dorsal column white matter to first central relays in the caudal brainstem, the dorsal column nuclei. Primary afferent axons can thus ascend the whole length of the spinal cord. This makes these the longest nerve fibers in the body, the maxima extending from toes to above the foramen magnum. Dorsal column fibers convey sensations of light discriminative touch, two-point sensibility, flutter and vibratory sense, and position sense and displacement of limbs and joints.

When syphilis was more prevalent (before the introduction of penicillin), there were many cases of tertiary syphilis affecting the spinal cord, *tabes dorsalis.* Persons afflicted with this disorder had a dorsal root radiculopathy which caused intermittent lancinating, excruciating but brief, "lightning" pains referred to the limb and body wall dermatomes of the affected dorsal roots. Their pathology revealed a drastic loss of fibers in the dorsal columns. As a consequence of the dorsal column loss, patients could not sense the position of their legs. They were obliged to use two canes and to rely on

vision to place their feet. They were therefore unable to get about in the dark.

After synaptic relay in the dorsal column nuclei, second-order neurons cross the midline and ascend the brainstem contralateral to the peripheral sensory fields of afferent representation. There they assemble into a flat ribbon-like bundle of fibers (hence *lemniscus,* meaning "ribbon"). The medial lemniscus ascends through the pons and midbrain to the *ventral posterior thalamus.*

The lemniscal system is a phylogenetically new, rapidly conducting, three-neuron (two synapses) path which projects to sites of primary somesthetic representation in neocortex. Before lemniscal fibers enter the thalamus, they co-mingle with fibers which ascended the ventrolateral and lateral spinothalamic pathways. These latter crossed the midline at their spinal levels of entry or higher. The ventral posterior thalamus receives an all-crossed, multimodality sensory input from cutaneous, body wall, and visceral sources.

At the spinal level, the uncrossed dorsal columns and the predominantly crossed spinothalamic tracts give rise, on lateral hemisection of the spinal cord, to loss of discriminative touch, two-point sensibility, position and vibratory sense, ipsilateral to the hemisection, with contralateral loss of pain and temperature sensation. (This bears the eponym Brown-Sequard syndrome.)

EVENT-RELATED POTENTIALS RECORDED OVER SPINAL PATHWAYS

Far-field potentials elicited by sensory stimulation can be detected from skin electrodes (Jewett et al., 1970). Because of feebleness of the potentials, many signals must be repeated and averaged to obtain a reasonable signal-to-noise ratio. Event-related potentials have been most thoroughly investigated in relation to auditory pathways, as discussed in Chapter 63 (Hearing).

Far-field potentials can be recorded from electrodes placed at intervals along the skin overlying the dorsal spinal processes. Repeated strong stimuli to the peroneal nerve will elicit low-amplitude responses which can be averaged over many stimulus events. Despite attenuation and dispersal of signals as impulses ascend the spinal cord, the ascending time displacement of the elicited potentials will provide a reasonable approximation of velocities of conduction along ascending spinal pathways. Such analyses can be used to detect interference with spinal conduction systems, such as may occur with multiple sclerosis affecting the spinal cord.

Organization of Neurons in Spinal Grey Matter

From a functional point of view, neurons in spinal grey matter can be conveniently divided into four categories:

1. *Somatic Motoneurons.* These constitute groups of motoneurons neatly arranged according to anatomical groupings of muscles, which occupy the large ventral horns. Their axons exit via ventral roots to innervate skeletal muscles.

2. *Visceral Motoneurons.* Small preganglionic sympathetic neurons cluster in the intermediolateral part of spinal grey matter, the "lateral horns," throughout thoracic and upper lumbar levels. They send axons via the corresponding ventral roots to sympathetic ganglia. There they activate sympathetic neurons which innervate visceral effectors, smooth muscles, and glands. Another kind of visceral motoneuron projects to *parasympathetic craniosacral autonomic ganglia.* They originate in efferent nuclei of the IIIrd, VIIth, IXth, and Xth cranial nerves in the brainstem, and in the intermediolateral nuclear clusters in the S_3 and S_4 sacral nerves (occasionally also S_2 and S_5).

3. *Propriospinal Interneurons.* An important type of intrinsic spinal neuron occupies the broad base of the dorsal horn and sends long ascending and descending axons which communicate with grey matter throughout most other spinal segments and caudal brainstem. Propriospinal interneurons functionally stitch the spinal cord together horizontally (segmentally) as well as longitudinally. They have elaborate dendrites which penetrate spinal grey matter for several segments up and down the spinal cord. Their axons enter neighboring white matter to ascend and descend for varying distances before sending collateral branches to reenter grey matter at higher and lower levels.

   Propriospinal interneurons derive information from several neighboring spinal segments and contribute integrated signals to all spinal segments and caudal brainstem. They integrate spinal afferent signals with intrinsic spinal activities and relay the resulting information toward the forebrain. They contribute to propriospinal reflex mechanisms which coordinate interlimb spinal and brainstem reflexes (Shimamura and Livingston, 1963).

   Propriospinal interneurons are interneurons with a far wider ranging influence than classic Golgi type II interneurons. Golgi type II interneurons have only local dendrites and short axons; hence their influence is limited to the vicinity of their cell bodies. There are relatively few Golgi type II neurons anywhere in the spinal cord; their place is taken over by these farther reaching, longitudinally integrating, propriospinal interneurons.

4. *First-Order Sensory Relay and Sensory Control Neurons.* A fourth type of intrinsic spinal neuron occupies the tapering dorsal part of the dorsal horn. They lie closest to afferent input. They constitute sensory relay and sensory control nuclei that are organized in a series of layers (Rexed, 1952). The layering of these neurons (resembling sensory cortex) implies that the primary integrative role here is *sensory analysis.* The ventral horn, on the ventral (motor) side, in contrast, is organized more like reticular formation, indicating signal convergence and *motor synthesis.*

   Organization of Second-Order Sensory Neurons. The most superficial dorsal horn lamina contains large marginal neurons and horizontal fibers. The most notable part of the dorsal horn lies directly beneath these. It has a gelatinous appearance and is called the *substantia gelatinosa (SG).* It continues without interruption throughout the length of the neuraxis from caudal medulla to coccygeal spinal cord. Apparently no cells in the SG contribute directly to ascending projections. Instead, they provide a spinal relay and control station relating to sensory input, especially to pain signals (see below). Deeper layers of the dorsal horn constitute the *nucleus proprius,* containing large relay neurons which contribute axons to the crossed ventral and lateral spinothalamic projections.

AFFERENT NERVE TERMINATIONS IN SPINAL GREY MATTER

Second-order sensory neurons in the most superficial (dorsal) layer of the dorsal horn are activated mainly by A-δ primary afferent fibers. These convey signals from cold thermoreception, nociceptive mechanoreceptors, and low-threshold, long-latency mechanoreceptors that appear to be particularly responsive to slowly moving stimuli contributing to sensations of tickle and itch. *Substance P* is the principal neurotransmitter among primary nociceptive afferents, A-δ and C-fibers alike. Cells of the SG are mainly activated by afferent impulses delivered by C-fibers.

SG neurons send axon collaterals that terminate on dendrites of neurons in deeper layers of the dorsal horn. They influence interneurons along the pain pathway. Many of them are enkaphalinergic and have strongly inhibitory effects. They contribute to central control over the first sensory relay along the pain pathway. Other cells in the SG contribute ascending fibers which relay upward toward thalamic stations, lateral spinothalamic projections which convey mainly pain and temperature signals but which do not deliver their signals directly to the thalamus. Pain modality projections are relayed one or more times during their ascent to the thalamus.

Some central projections that descend the spinal cord terminate in the SG and bear *enkephalin* terminals which end on the cell body and proximal dendrites of neurons that receive direct signals from cutaneous nociceptive units (Ruda, 1982). Other descending projections directed to this part of the dorsal horn come from the nucleus raphe magnus in the bulb (medulla oblongata) and contain *serotonin*. All three, the local SG interneuron and the descending enkephalin and serotonin projections, are inhibitory. They modulate and inhibit pain transmission at the first synaptic relay of impulses generated by noxious stimuli. They exercise control over A-δ and C-fiber pain signals before these are relayed to the spinothalamic tract.

PAIN MECHANISMS AT SPINAL AND BRAINSTEM LEVELS

Search for Endogenous Mechanisms for Control of Pain

Morphine and morphine-like substances have been used for the relief of pain since antiquity. Recently it was found that cells in certain regions of the brain bind opiates stereospecifically and that the analgesic potency of a drug correlated directly with its binding affinity for these receptors (see Snyder and Matthysse, 1975). This led to a search for naturally occurring endogenous morphine-like compounds. Two naturally occurring groups of compounds were discovered, *leucine enkephalin* and *methionine enkephalin* (Hughes et al., 1975), and the *endorphins* (Loh et al., 1976), including β-endorphin, a part of pituitary β-lipoprotein.

A few years earlier, it was found that focal electrical stimulation of certain parts of the central nervous system of the rat produces analgesia without general behavioral depression (Reynolds, 1969). The brain regions most responsive to electrical induction of analgesia coincided with sites found to exhibit powerful opiate binding. Moreover, opiates injected locally contributed as effectively as did electrical stimulation. Further, an opiate antagonist, *nalaxone*, reversed the analgesia produced by electrical stimulation (Adams, 1976; Hosobuchi et al., 1977). Attention then focused on midbrain *peri-*

aqueductal grey matter because cells in this region showed great affinity for opiates and opiate agonist binding. This region produces endogenous morphine-like *enkephalins*. This region is also prepotent among sites for electrically induced analgesia.

DESCENDING PAIN CONTROL MECHANISMS

Opiate administration activates pain suppression which is organized at three levels of the neuraxis: midbrain, bulb, and spinal cord. Activation of neurons in the midbrain periaqueductal grey matter is accomplished by direct electrical stimulation, by local introduction of small quantities of opiates, by larger doses of opiates administered systemically, and by limbic activation associated with strong emotions. The limbic system has a widespread distribution of opiate receptors.

A result of activating neurons in the periaqueductal grey is direct and indirect activation of *serotonergic neurons* in the rostral bulb, in the *nucleus raphe magnus* (see Basbaum and Fields, 1978). These serotonergic cells project to the trigeminal spinal root, equivalent of the spinal substantia gelatinosa and relay for the trigeminal (Vth cranial) nerve. They also project all along the spinal cord, via the dorsolateral funiculus of the spinal cord, going to nuclear (SG) relays in the dorsal horn. Interruption of this pathway produces a loss of analgesia that depends on centrally introduced opiates or electrical stimulation. This interruption affects only ipsilateral segments below the level of cutting of the dorsolateral fibers (Basbaum et al., 1976).

The influence of this descending pain control mechanism is to interfere with transmission between nociceptive afferents and second-order pain pathway neurons. These are relays which are presumably ordinarily activated by the release of substance P by the nociceptive afferents. Second-order neurons, if activated by pain stimuli and insufficiently inhibited by descending pain control mechanisms, send signals up the contralateral lateral spinothalamic tract. Collaterals from this tract activate periaqueductal grey matter and the bulbar raphe nucleus. This, in turn, activates the feedback loop which augments activity in the descending pain control system.

BRAINSTEM PATHWAYS ACTIVATED BY TOOTH PULP STIMULATION

Popular consensus attributes nociceptive receptive capacities to tooth pulp stimulation. The tooth pulp is served by A-δ and C-fiber touch, pain, and temperature receptors. Stimulation of tooth pulp activates crossed trigeminal lemniscal and trigeminothalamic pathways. It activates at least five different ascending brainstem projections, which can be distinguished by differences in latency, velocity, and other dynamic response characteristics, and by differential susceptibilities to anesthetic agents (Melzack et al., 1958).

Cats with bilateral lesions of a single descending tract, the central tegmental tract (CTT), respond as though they had become hypersensitive over all of their bodies. It is, therefore, logical to assume that defective central grey and (CTT) systems may contribute to chronic intractable pain syndromes, like causalgia.

The consequences of activating periaqueductal neurons rich in enkephalins is to activate circuits involved in the endogenous control of pain. It is this systems control of pain processes which presumably ordinarily permits a painful injury to subside. And, the corollary proposition is that disorders affecting the periaqueductal grey and associated central controls of pain transmission may account for the minor and major causalgias.

SUMMING UP SPINAL AND BRAINSTEM SENSORY MECHANISMS

Afferent Contributions to Reflexes

The fastest afferent inputs, from muscle and tendon end-organs, contribute mainly to skeletal muscle stretch reflexes and skeletal muscle motor "tone." The next fastest cutaneous and body wall afferents contribute mainly to flexion reflexes. These latter are accompanied by ipsilateral inhibition of stretch reflexes. Contralaterally, there are mirror image reflex responses.

Incoming sensory information contributes to four main ascending sensory systems:

1. *Dorsal column fibers to dorsal column nuclei in the caudal brainstem (the "lemniscal" system).* This represents an ipsilateral upward pathway of axons of first-order afferent neurons—the longest nerve cells in the body. This phylogenetically recent tract represents the shortest—minimum of two synaptic links—to cortex. The dorsal columns convey light discriminative touch, two-point sensibility, position, and vibratory sense.
2. *Lateral spinothalamic tract serving crude touch, pain, and temperature sensations (the "spinothalamic" system).* This is relayed by cells in the dorsal horn whose axons largely cross the midline and ascend in lateral and ventrolateral white matter. There are one to a few additional synapses along this route to thalamus.

The spinothalamic system is neither so dynamically accurate, modality specific, nor synaptically reliable as is the lemniscal system.

The spinothalamic pathway communicates in the bulb with the large raphe nucleus and in the midbrain with the periaqueductal grey matter. These serve as central stations responsible for descending pain control. The periaqueductal neurons are opiate receptive and produce endogenous opioids. The raphe nucleus, to which the periaqueductal cells project, is serotonergic. This latter nucleus projects to the trigeminal spinal root in the brainstem and the equivalent spinal substantia gelatinosa where it inhibits relay of nociceptive signals. Lateral spinothalamic tract collaterals contribute to this central pain control feedback loop thus augmenting descending traffic in the control of pain signals.

3. *Ascending visceral projections ("archispinothalamic tract")* are phylogenetically oldest, slowest, and most peurisynaptic, most diffused in space and time, least specific in modality characteristics, and are somewhat more than half crossed. This pathway undergoes repeated synaptic relays at successively higher spinal and brainstem levels on its way to diencephalon, primarily to hypothalamus.

Each of these three systems gives off axon collaterals to the spinal cord and brainstem grey matter thereby contributing messages which relate to general states of visceral feeling, readiness for action, and arousal.

4. Propriospinal projections, far-reaching spinal interneurons which vertically stitch spinal and brainstem segments together and contribute to intersegmental and interlimb reflexes. Afferent and intrinsic spinal signals contribute to these vertically oriented spinal interneurons which have dendrites that reach upward and downward several segments, and axons that go up and down to reach most spinal segments and caudal brainstem.

RELAY OF SOMESTHETIC INFORMATION THROUGH THALAMUS

Ventral Posterior Nuclei

As noted earlier, when the lemniscal system and spinothalamic tracts approach the thalamus, their fibers converge. Nevertheless, *the two systems maintain separate channels* during their thalamic relay. There appears to be a great deal of independence and segregation of information transmitted through the thalamus. The *ventral posterolateral nucleus* (VPL) of the thalamus receives ascending fibers from the *medial lemniscus*. Information concerning peripheral stimulus location, form, modality, and dynamic characteristics are transmitted in an orderly fashion. Precision in sensory processing along the lemniscal pathway derives from afferent neurons that have limited receptive fields, prefer single-modality reception, show dynamic responses, and transmit through conservative synaptic junctions to reliable third-order (thalamocortical) neurons. Lateral inhibition increases the sharpness of spatial and temporal patterns at each synaptic relay.

Lemniscal information concerns light discriminative touch, two-point sensibility, position sense, flutter, and vibratory sense. Lemniscal relay in VPL is not without corollary influences, however. VPL receives substantial inputs from the same somatosensory cortical areas to which it projects. It also receives projections from the median forebrain bundle (relating to frontal lobe and hypothalamic functions), and from midbrain reticular formation and midline, intralaminar, and reticular nuclei in the thalamus (concerned with arousal).

The neighboring *ventral posteromedial nucleus (VPM)* receives axons from the main sensory nucleus (trigeminal lemniscus relay) and spinal root (trigeminothalamic relay). *The combination of VPL and VPM therefore comprises a comprehensive topological map of the entire contralateral side of the body.* This is organized in the ventral posterior part of the thalamus at its boundary with midbrain. Face representation is lowermost and medial, with the rest of the body positioned progressively more laterally and tilted foot-upward. Somatotopic representation in VPL and VPM contains partitioned distributions of neurons reflecting different sensory modalities. This three-dimensional somatotopy is conserved two-dimensionally in primary somatosensory cortex.

COMPARISON BETWEEN LEMNISCAL AND SPINOTHALAMIC SENSORY PROCESSING

Spinothalamic cortical representation is less abundant and less specific than the lemniscal system. Spinothalamic representation in the thalamus is relatively widely distributed and emphasizes activation of diffusely projecting thalamic nuclei (midline and intralaminar) which participate in forebrain arousal. Spinothalamic projections contribute general features of sensations and qualitative experience relating to crude touch, pain, and temperature. Their information is less precise concerning locus, spatial organization, and dynamics of stimuli. Their projections to spinal and bulbar reticular formation contribute important information as to readiness and competence of visceral mechanisms.

Evolution, in establishing a lemniscal system in mammals, not only shortened the pathways to higher centers but delivered vastly refined information concerning the body and environmental events impinging on the body. Information transmitted by the lemniscal systems travels in first-order neurons by way of ipsilateral dorsal columns most or all the way to the dorsal column nuclei. From those nuclei at the bulbar level, second-order neurons cross the midline to form the *medial lemniscus* on the contralateral side. *Trigeminal* second-order fibers, from the *nucleus proprius V*, similarly cross the midline to form the *trigeminal lemniscus*. Second-order trigeminal crude touch, pain, and temperature signals are relayed through the *spinal root of the trigeminal* (comparable to the spinal SG) as mainly crossed, trigeminothalamic projections.

Powers of discrimination relating to the somatosensory system are remarkable. As in other sensory systems, stimulation of neighboring cutaneous fields results in powerful *lateral inhibition*. This pattern is reinforced from peripheral nerves through dorsal column and thalamic relays. Thus, edges and points tend to be sharpened in their localization. Complex textures are similarly made more precise.

As with eye movements, skin surfaces can be moved about to enhance the combination of spatial and temporal discrimination. Thus, by applying fingertip movements, an individual who is adept at reading braille can discriminate braille points separated by distances that are equivalent to the distances between adjacent cutaneous afferent stems. When an ordinary sighted person runs a fingertip across a line of braille, he feels an irregular pattern of protuberances, like goose-skin bumps. The blind braille reader can interpret their meaning at a rate of 60–120 words per minute, extracting language from something like 2500 protuberances in 60 s (Critchley, 1986). In this respect, cutaneous searching is able to define discriminable touch-contact differences with skill similar to visual search in detecting vernier displacement differences.

SOMATIC REPRESENTATION IN CEREBRAL CORTEX

Thalamic relay of somesthetic projections is predominantly to the postcentral gyrus. Somatotopy is preserved. Body representation projected onto the hemisphere is contralateral, with the foot represented at the top of the postcentral gyrus, the leg, pelvis, trunk, arm, hand, and face all arranged in sequence laterally, ventrally, and inclining anteriorly, following the gyral pattern. Following face projections onto the superior bank of the Sylvian fissure are represented successively, mouth, tongue, throat, with visceral representation still more deeply enfolded into the Sylvian fissure and refolded onto the surface of the insula.

This extended representation of visceral sensation approaches closure of a loop because, superiorly, on the medial surface of the hemisphere, over the crest of the postcentral gyrus, there is continuous representation of lower limb, foot, buttock and anus, in that order. A ring of somatic and visceral structures are looped in a nearly completed circle, interrupted medially by the hylus of the hemisphere. The somatovisceral map lies like a waistband not quite clasped around the middle of each

hemisphere. The representational gap is approximately at the pelvic floor.

Just as there are differences in density of distribution of cutaneous afferents in the periphery, there are even more spectacular differences in scale of representation of body parts in cortex. Hand and fingertip representation, lip, and other specially discriminative parts have greatly enlarged representations. The somatotopic representation of the contralateral cutaneous surface of the monkey in Figure 9.67, from the work of Nelson et al. (1980), shows the remarkably expanded cortical representations for foot and hand. These two parts occupy about half of the total primary cortical representation of the contralateral side of the body. The smallest area representations relate to dorsal hairy skin

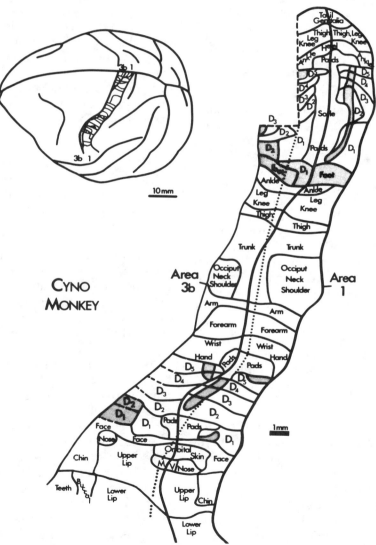

Figure 9.67. Unfolding of bodily representation in *areas 3b* and *1* on the medial and superior surface of the right hemisphere. The *dotted line* indicates the position of the top of the central sulcus. *Shaded areas* indicate representations of hairy dorsum of foot and digits. (From Nelson et al., 1980.)

areas. Glabrous skin areas are relatively more expanded.

Cortical representation is banded horizontally—exhibiting an anteroposterior representation of the same parts of the body, but organized in relation to different modalities. In this depth of the central sulcus, rising up along the posterior bank of the central sulcus *(area 3a)* is representation of muscle stretch afferents, immediately across the bank of the central (Rolandic) sulcus in motor cortex, and representing the same muscles. Posteriorly are found *(3b* in Fig. 9.67) slowly adapting cutaneous representation, followed by *(area 1)* rapidly adapting cutaneous representation and *(area 2)* periosteal, fascial, and joint sensibility (Kaas et al., 1981). These cortical bands can be traced back to distinctive cellular arrangements in the thalamus where these modalities are similarly segregated. It is clear that distinctive but related *somesthetic signals are projected along a series of parallel channels.*

Dynamics of Cortical Organization

EFFECTS OF DEAFFERENTATION

These precise cortical maps are also dynamic. Apparent stability is the result of balancing among dynamic influences. When the median nerve is sectioned, for example, there is only temporary inactivation of parts of the cortical map for innervation of the now-denervated glabrous skin of the palm. Neurons in this "deafferented" cortical field after a short interval are activated by stimulating parts of the hand outside the peripheral distribution of the median nerve. The new receptive fields are not random; they represent new encroachments which enlarge maps of the neighboring dorsum of the hand (Kaas et al., 1981). The result of interruptions of input from part of the periphery is expansion of cortical representation of other parts. There is constant competition for representation of adjacent skin areas, for space in the peripheral sensory fields, along central sensory pathways, and at cortical sites.

EFFECTS OF REMOVAL OF VIBRISSAE

Many mammals have vibrissae (whiskers) which constitute marvelous sweeping probes (like a blind person's wand) which are used to explore the environment of face and snout. The base of these long whiskers is sensitive to motion, and, like guard hairs, they are innervated by a number of directionally sensitive receptors. Vibrissae can be moved about, even somewhat independently as well as en masse. Woolsey and van der Loos (1970) found that each vibrissa has a large central cortical representation, a nest of cells that constitutes a cortical "barrel" for each vibrissa.

If vibrissae are destroyed in a young animal, the corresponding cortical barrels die back and disappear. The consequences of destroying a row of such vibrissae are shown in Figure 9.68. Retrogressive changes are seen in primary trigeminal neurons, second-order trigeminal lemniscal neurons, third-order thalamocortical neurons, and the cortical barrel itself. Retrogression induced by peripheral loss obviously affects the entire nth-order sensory path. This is a remarkable feature of the ability of the nervous system to match its body parts' internal mapping in accordance with the receptive mechanisms available for exploring the "outside world."

Additional Major Somatosensory Representation

There are many cortical areas from which potentials can be elicited following stimulation of somatic receptors. A very large area of cortex can be assigned somesthetic representation, but the postcentral gyrus deserves priority as the *primary somesthetic cortex,* thus, it is designated *S I.* Additional areas warrant inclusion: *sensory II* and *supplementary motor cortex.* Because these include motor as well as sensory representation, they are appropriately referred to as *sensorimotor cortices (Fig. 9.69).*

Columnar Organization of Somesthetic Cortex

The basic functional organization of cortex is in columns of cells, vertically oriented through the thickness of cortex, perpendicular to the cortical surface. Cortical columns were first discovered by Mountcastle (1957) in S I and have subsequently been found in all areas of cortex. Columns represent a vertically disposed chain of neurons which organizes locally specialized tasks between cortical input and output. A single cortical column constitutes a modular unit for analytic operations. Input arrives from thalamocortical projections with terminals which have ready access to columnar output via cells in layers V and VI. Incoming signals contribute to and are influenced by transactions among cells in more superficial layers of the same column. These receive inputs from several other cortical fields and from mechanisms for arousal and attention.

Somesthetic cortical columns receive sensory signals from thalamocortical fibers VPL (body) and VPM (face) representations. Adjacent columns have closely overlapping receptive fields which convey the same sensory modalities. Each cortical column sends reciprocal impulses to the same nuclear populations that provided their thalamocortical input. Cortical outputs also contribute to descending projections that can influence sensory relay nuclei in a downward cascade, improving central control over sensory transmission. Other cortical outputs go to

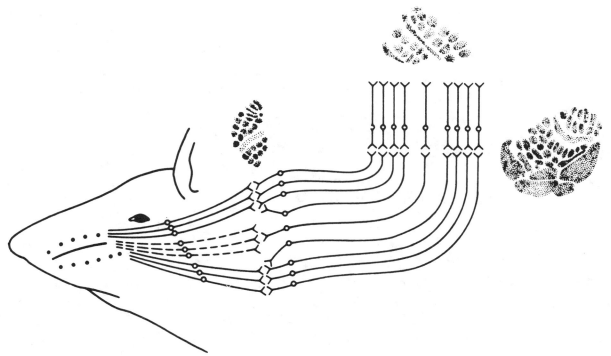

Figure 9.68. Summary of effects of vibrissae damage on pattern formation in the trigeminal distribution to somatosensory cortex in the rat. Following removal of the middle row (only three of the five rows are illustrated in this schematic), changes are detected in the termination of the primary neurons in the brainstem, secondary neurons in the thalamus, and tertiary neruons in the cortex. At all three levels, the effect is a band rather than discrete clusters. (From Killackey, 1982.)

motor control systems, somatic and visceral, in basal ganglia, diencephalon, cerebellum, and brainstem, with relays to the spinal cord. Still other cortical outputs go to particular motor and sensory fields in cortex, thereby contributing to higher order motor command and to multimodality, sensorimotor global mappings.

We have seen that lateral transactions contribute to sensory analysis in each of the other sensory systems. This principle holds equally well for somesthesis. Directionally sensitive neurons in S I were analyzed in response to moving edges applied to the skin of waking monkeys (Gardner and Costanzo, 1980). Directional sensitivity is provided by an asymmetric distribution of lateral inhibition and lateral facilitation. Clearly, lateral transactions play a significant role in analysis of spatial and temporal patterns in somesthesis.

SENSORY ASSOCIATION CORTEX

Evidence from the study of parietal cortex in waking monkeys indicates that this region is neither dominated by afferent input, as are primary sensory areas, nor dominating to peripheral effectors, as are primary motor areas. It is relatively remote from both input and output. The specific activities to which parietal cortex (areas 5 and 7 of Brodmann) respond share a common characteristic.

They link pertinent sensory information to particular motivations and particular opportunities for action. They are concerned with spatial relations of body parts with one another in respect to the gravitational field and the near environment (Mountcastle, 1975; 1976).

Within area 7 there are sets of columns which are active during movement of the arm toward an object of interest. Other columns are active during manipulation of objects and others during the direction of visual attention toward objects. Some columns are activated only when eye movements are attracted by the sight of objects in the environment, during slow pursuit movements of the eyes, and during visual searching for an object. There are columns which are activated preferentially with the appearance of objects of special interest in peripheral vision.

Association areas appear to be organized so as to pool sensory information from a wide variety of sources, olfactory, visual, auditory, tactile, body wall, etc., and to use pooled, evaluated information to specify actions in accordance with particular bodily needs and motivations. For example, units in a given column may be activated only on the appearance of an interesting (food) object, suitable (banana slice) in contralateral space, only if the animal is hungry, and only if the object is within reach.

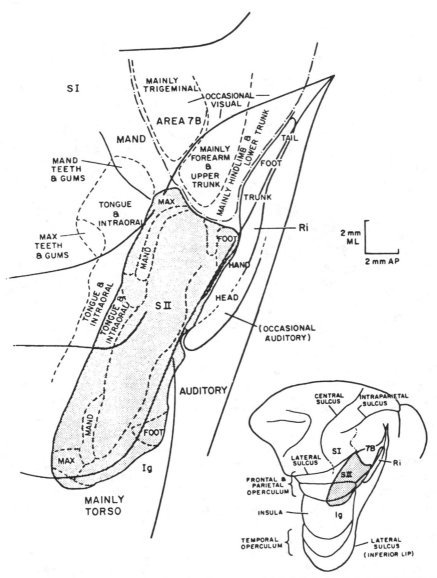

Figure 9.69. Somatosensory areas in monkey showing functional organization of secondary sensory area *(SII)*, including exposed view of cortical surfaces normally buried within the lateral (Sylvian) fissure. *Lower right* of figure shows SII in relation to the left hemisphere and in more detail, *left center.* Note somatotopic arrangements, including parallel-processing representations and representations of additional sensory modalities, auditory and visual at the margins. (From Robinson and Burton, 1980.)

This reflects the significance of global mappings relating to the individual and his/her extrapersonal space (see Edelman, 1987; 1989).

Entities of the nervous system at various levels of the neuraxis, known as dorsal horn, dorsal column nuclei, reticular formation, thalamic relays, neocortex, basal ganglia, cerebellum, brainstem, and spinal motor nuclei, etc., are composed of local circuits which make them functional subsystems. These, in turn, are grouped vertically into larger functional entities according to their extrinsic connections, fields of topographic representation, and kinds of transactional connections with other functional entities (Mountcastle, 1976). Connections run in parallel bundles which preserve topological relations with other bundles, thus ensuring competitive, precisely connected topographic representations at different levels of the neuraxis. These are differently affected by requirements for central magnification, and for modality partitioning among multiple, distributed parallel-processing systems. Reassortments of representations are utilized to assemble higher-order functional organizations (Crick, 1979; "global maps" of Edelman, 1987).

PARIETAL LOBE DEFICITS IN HUMANS

Humans with parietal lobe lesions may manifest peculiar disturbances of perception and behavior.

They may show alterations in perception of their own body and its relation to surrounding space. Such a patient may deny owning half of his/her body: if you show the patient his/her own hand and arm and its attachment to the shoulder, he/she will nevertheless insist that it belongs to someone else.

Directed attention to extrapersonal space constitutes a compound function that is based on transactions among at least three major cortical areas, and depends also on mechanisms essential for maintaining attention and arousal. Quite precise clinical neglect syndromes include: *anosognosia* (denial of one's own body parts), *dressing apraxia* (neglect to wash, dry, groom, and dress parts of the body), and *construction apraxia* (inability to assemble simple components or to draw, for example, a complete clock face).

Neglect follows disruption of mechanisms relating to spatial organization of attention. Unilateral neglect syndromes are more common and more severe after lesions in the right hemisphere. The right hemisphere appears to be involved in the execution of most attentional tasks. It appears capable of directing attention to both sides of the body, whereas the left hemisphere seems able to direct attention only contralaterally. It is assumed that the right hemisphere, at least in dextrals, is specialized for the direction of attention globally throughout extrapersonal space.

Mesulam (1981) has analyzed unilateral neglect syndromes into four categories relating to four brain locations. Each site has a unique functional role that reflects its profile of connections in the absence of which perceptual and behavioral neglect syndromes become manifest.

Mesulam's four brain regions represent what amount to four systems of global mappings: parietal, limbic, reticular, and frontal. With *parietal neglect*, the patient directs bodily action only into the ipsilateral extrapersonal space, i.e., neglects to reach for or approach objects in the contralateral extrapersonal space. With *frontal neglect*, the patient does not look into or visually examine the contralateral hemispace. It is not paralysis of gaze but neglect of visual interest and search, for the eyes move there readily enough in visual pursuit reflexes and with passive head turning. With *limbic neglect*, following lesions in the cingulate cortex, the patient lacks directed motivation and hence neglects purposes and objects that would otherwise be motivating if they were not located in contralateral space. *Reticular neglect* interferes with central state conditions relating to attention and arousal directed to extrapersonal space. This occurs with lesions in the posterior margins of the aqueduct, presumably affecting the *locus coeruleus*.

Problems of Interpreting Somesthetic Experiences

What is perceived as a consequence of stimuli applied to skin and deeper tissues in various parts of the body depends on (a) general density of innervation, (b) differences in distribution of specific types of sensory units, (c) physical characteristics of surrounding tissues and sensory end-organs, and (d) spatial and temporal patterns of activation and adaptation among different sensory units. If, for example, the skin is under tension, being stretched, deformed from outside pressure or pressure from within, tumescence, hemorrhage, inflammation, or irritated from scarification, blistering, histamine release, etc., sensations will be correspondingly modified, sensitized, numbed, etc. An example of an intervening variable is that single touch receptor activity in isolated skin of the frog is facilitated by stimulation of the sympathetic nerve supply to that region (Lowenstein, 1956). Moreover, circulating epinephrine increases the excitability of many end-organs, reflexes, and sensory relays. This indicates how sensory experiences may be disturbed by emotional excitement.

Physiological consequences of learning, early during development, undoubtedly improve the analytic capabilities and reliabilities of sensory circuits. The experiences of musicians, artists, and athletes while perfecting their performance skills through disciplined practice suggest that this is the case. It is not practice that makes perfect—it is practice, the consequences of which are taken into account, that contributes to skill acquisition. One role of teachers is to identify and exemplify pertinent feedback. It is likely that skill acquisition results in improvement of accuracy, reliability, and shortening of access time to appropriate circuits, by neuronal group selection (Edelman, 1987; 1988; 1989).

CHAPTER 68

Nervous System Communication and Control

MOTOR CONTROL

The following chapters portray how various effector systems, glands and muscles—smooth, cardiac, and skeletal—come under integrated control according to the overall purposes of the individual. Adult behavior requires competent central nervous system control over three aspects of behavior:

1. Visceral activities must accurately regulate, e.g., pH, body temperature, oxygen and energy demands, nutritional fitness, waste disposal, etc. By and large, these programs are subservient to, but are necessary for support of other aspects of behavior.
2. Outward expressions of the internal state, e.g., bodily attitudes and movements, gestures and emotions, must be appropriately adapted to individual and societal requirements. These constitute behavioral tools to establish and maintain social rapport. Unreasonable expressions erode social relations. But, adequate visceral and expressive behavior alone does not provide behavior essential for independent survival.
3. Therefore, consciously goal-directed behaviors must be organized and pursued in accordance with both immediate and long-range individual plans. These require cognitive skill and discipline to enable individual survival. In an adult, anything less than competence in all three of these categories of behavior implies nervous system disorder or deficit in some degree.

Behavioral control can originate in a variety of widely distributed "motor maps" represented in different ways in different parts of forebrain, brainstem, and spinal cord. It involves focusing central nervous system signals on about 500,000 executive neurons. Executive neurons send messages out from brainstem and spinal cord: indirectly—via peripheral autonomic ganglia—and directly, to a variety of effector tissues. With 50 billion central neurons, there are a hundred thousand central neurons for each executive neuron, an immense convergence, implying extraordinary discretionary potential. Discretionary control is available to behavior, but conscious discretionary control is largely limited to the control of skeletal muscles.

Higher levels of control have only indirect access to executive neurons serving cardiac and smooth muscles and glands. Visceral functions are therefore relatively less under voluntary control. Visceral mechanisms, and to a lesser extent, mechanisms involved in expressing bodily attitudes and emotions, are governed nearly automatically, semiautonomously. This is all to the good; if we had to regulate visceral activities consciously, we would probably do a poor job of it, and have correspondingly less to work with consciously. Expressive behavior is considerably under voluntary control, but the more this behavior is consciously directed, the more wooden and artificial expressive behavior becomes, and the less it appears to represent (express) the internal state validly.

Visceral and expressive behaviors are managed by circuits that are the oldest and next-oldest motor control systems. They are both well-tested in evolution. Visceral control is organized by nervous system circuits belonging to the phylogenetically oldest (archiontologic, inmost lining of the neuraxis). Behavioral expression and emotion are organized by intermediately old parts (paleo-ontologic, mid-thickness of the wall of the neuraxis). Thus, there is a step-wise progression from oldest (innermost)—most autonomous to intermediately old (midwall of the neuraxis—semiautonomous, yet largely spontaneous) to newest cortical surface (outermost fiber bundles which enclose the nuclear masses and innermost grey matter of the neuraxis). These newest controls have relatively the fullest representation in consciousness and facultative access to voluntary control (MacLean, 1990; Edelman, 1989).

No matter how refined our upbringing and sublime our thoughts, all we can do, ultimately, is to contract and relax muscles and to secrete. Extensive fields of neocortex can influence almost any skeletal motoneurons thereby providing considerable discretionary control over voluntary activity. We are capable of linking almost any skeletal muscle action to almost any sensory or mnemonic cue. With some muscle groups, controlling eye movements, lips, tongue, vocal cords, fingers, etc., we can develop fine, selective control of individual muscle fascicles and organize and manipulate at will complex patterns of muscular contraction; this is part of what is required for great musical performance.

SENSORIMOTOR INTEGRATION

We have seen in previous chapters how sensory systems contribute not only to ascending signals but to reflex mechanisms and to successively higher levels of sensorimotor integration, processes that are greatly elaborated at cortical levels. At lower levels, sensory input contributes to relatively automatic processes, i.e., without evident conscious participation. At successively higher levels, increasing awareness may be involved, enabling increasingly purposeful control of behavior. In ways that are not yet understood, higher levels of perception contribute to the organization of sensorimotor experiences into memory. These experiences can be retrieved from memory stores and compared for imaginative behavioral rehearsal, evaluation, and the development of consciously organized newly adaptive behavior (Squire, 1987).

Motor control functions are repeatedly and variously represented and rerepresented; "motor maps" and multimodal "global maps" are distributed throughout extensive regions of each hemisphere. Most such maps possess some large pyramidal cells whose axons descend through the internal capsule to innervate brainstem and spinal levels of motor control. Among these maps, the precentral gyrus is considered the *primary motor cortex*. It sends many long descending axons, some of which make direct monosynaptic contact with executive neurons, the anterior motor horn cells.

About in the middle of each hemisphere, the primary motor cortex occupies a roughly 1 cm-wide strip that extends like a belt across the medial, superior, and lateral surfaces and descends into the Sylvian fissure. The motor strip lies immediately anterior to the central sulcus, i.e., directly across the sulcus from primary somesthetic representation in the postcentral gyrus, the organization of which is discussed in Chapter 67.

Motor and sensory representations on either side of the central sulcus relate to corresponding body parts. Representation begins with perineum and lower limb on the medial surface of the hemisphere and extends lateralward with successive representations of pelvis, trunk, arm, hand, and face in that order. Complementary motor and sensory representations match one another precisely, anterior and posterior to the central sulcus. Corticocortical U-fibers pass in both directions to foster detailed integration.

Although most descending motor controls act through interneurons, many of the long pyramidal axons from the precentral gyrus obtain the most direct possible, i.e., monosynaptic linkages, between motor cortex and motorneurons which directly control skeletal muscles. Whereas the shortest ascending sensory path to cortex involves two synapses, the shortest motor path from cortical command to executive neurons is monosynaptic. Direct corticospinal projections, like the most direct sensory projections (the lemniscal systems), is a recent evolutionary development.

MECHANISMS OF BEHAVIOR

At all levels of the neuraxis, motor systems benefit from sensory and perceptually organized information. But behavior does not depend exclusively on sensory initiatives. Motor command operates largely on the basis of central initiatives, visceral and somatic, motivational mechanisms which are spontaneously active and which may only secondarily relate to memory mechanisms, body images, etc., and on immediate perceptual experience. Most smooth muscles and glands are rhythmically and automatically active, while innervation provides their functional integration for the body as a whole. Cardiac muscle contracts rhythmically and vascular muscles contract tonically without requiring innervation, although innervation plays a role in overall integration of cardiovascular performance, e.g., startling perceptual experiences, memory mechanisms, emotional perturbations, anticipation of exercise, etc.

Motivations for action can originate in peripheral sensory mechanisms (e.g., cutaneous thermal receptors) as well as central sensory machanisms (e.g., gas exchange, salt, water, and nutrient balance, etc.). At lower levels, these modulate ongoing spontaneous activity, rhythmicity, etc., and in conjunction with appetitive systems and other sensorimotor mechanisms react to fluctuating bodily needs for air, water, food, sleep, sex, waste disposal, etc. Neuroendocrine rhythmicities also occur at longer intervals such as diurnal and menstrual cycles, and even longer developmental epochs. Neuroendocrine mechanisms, for example, trigger birth of the individual, and neuroendocrine changes are associated with puberty, menopause, and the like.

Skeletal muscles, on the contrary, are quiescent unless contraction is commanded. Nonetheless, skeletal muscle tone, and both tonic and phasic and dynamic skeletal postural reflexes, including interlimb reflexes and head-on-body reflexes basic to walking, running, and swimming activities are largely generated and maintained by segmental and longitudinal spinal and brainstem sensorimotor control systems. Higher nervous centers are thereby spared excessive preoccupation with the lower-level control mechanisms which underlie coordinated skeletal muscle performance (see Chapters 69, Visceral Control Mechanisms, and 70, Segmental Control of Skeletal Muscle).

Lower-level controls provide reflexive support of the body against gravity and muscular readiness for action. Many important integrative spinal and brainstem functions are organized relatively autonomously. Relative autonomy obtains segmentally (horizontally) and also along the neuraxis (longitudinally), with the brainstem and spinal cord being organized functionally for fully integrated interlimb and head-on-body motor control.

In previous chapters we described analogous mechanisms with respect to sensory integration. Immediate sensory analysis is achieved by remarkably adaptive features of receptors and first-order afferent axon integration, and by central control of receptors and cental sensory transmission systems. Using centrifugal control, the central nervous system can select adaptive sensory signals and model them according to past experiences, expectations, and purposes. This control of sensory input is exerted as far "downstream" as the receptors.

By similar token, the central nervous system exercises "downstream" control over effector systems through diencephalospinal and corticospinal control of executive neurons. These analogies underline two principles: (1) that the forebrain gains control over sensory processing from the outside all the way in, beginning with the receptor; (2) that the forebrain exerts control over skeletal motor functions all the way down the neuraxis to the final neuronal output.

The relatively involuntary and autonomous aspects of visceral functioning is indicated by the fact that neocortical projections have only indirect access to visceral innervation. Visceral autonomy is further buffered by peripheral ganglia of the autonomic nervous system. Nevertheless, some neurons exert visceral control directly from the forebrain. A few long-axon visceral control fibers pass directly from hypothalamus and allied limbic structures to brainstem and lumbosacral parasympathetic and spinal sympathetic output neurons. By direct diencephalofugal projections to visceral systems, and by monosynaptic neocorticifugal projections to skeletal motor activity, the forebrain obtains relatively direct command and control. Higher centers thus enjoy as intimate two-way communication with the "outside world" as is possible.

General principles emerge: (1) that the nervous system can obtain information about the "outside world" quite directly; (2) that it has quite direct access to motor control; and (3) that the forebrain can employ vastly more elaborate circuits to make longer range behavioral adaptations. There is more to higher nervous system physiology than simply sensory and motor systems controls (see Chapter 72, Higher Neural Functions.)

COMMUNICATION

Sensory systems operate in a generally *ascending* direction, with abundant horizontal and descending feedback controls; and motor systems operate in a generally *descending* direction, with abundant horizontal and ascending feedback controls. As noted previously, sensory systems involve three distinctive, interrelated ascending systems: (a) *visceral* ("archispinothalamic"), (b) *spinothalamic* (including additional visceral modalities plus somesthetic crude touch, pain, and temperature modalities), and (c) *lemniscal* (medial lemniscus for bodily sensations; trigeminal lemniscus for facial sensations; lateral lemniscus for auditory sensations). In analogous fashion, motor systems involve three distinctive, interrelated descending systems which severally govern (a) *visceral behavior*, (b) *expressive behavior*, and (c) *manipulative behavior*. Like the three sensory systems, the three motor systems have distinctive central representations and projection pathways.

Bidirectionality of Neuronal Signaling

Fibers entering via dorsal roots appear to be strictly *afferent* (sensory). Fibers leaving via ventral roots appear to be strictly *efferent* (motor). Central neurons are ambiguous as to whether they are strictly sensory or strictly motor. A spinal interneuron lying directly between an afferent and an efferent neuron is clearly functionally intermediate. In a strict sense, all central neurons are functionally intermediate. It is noteworthy that the winnowing effects of evolutionary selection "pay-off" exclusively on the basis of *behavior*, regardless of what occurs along any of the sensory, perceptual, ideational, motivational, or motor control circuits. There is "no pay-off" except through behavior, e.g., action (or inaction).

The nervous system is "built for action": afferent,

reflex, ascending, central integrating, and descending systems develop and operate to provide behavior suitable to the purposes of the individual. Longer looping projections, and elaborate forebrain representations support their biological costs by contributing to favorable behavioral adaptations. They do this on a broad time-scale that ranges from instantaneous turnaround to lifelong preparations for appropriate action. Central circuits enable representation and rerepresentation thereby obtaining a succession of adaptive analytical procedures for interpretation of sensory signals. These procedures are concerned with dynamic aspects of individual experiences within a dynamic environment.

Forebrain mechanisms enable vast neuronal communication, with options for access to memory, imaginative improvement of behavior, and further behavioral selection and elaboration through learning. More important, the forebrain allows mechanisms concerned with imperative biological values of appetite, emotion, and short-term motivation to compete with higher values which have been built up from more primitive ones through individual and social experiences, to involve long-circuiting pathways so that deferred gratification and long-range plans can be organized around superordinate motivations (e.g., for studying medicine).

We are taught to assume that there are strictly afferent and efferent neurons, and ascending "sensory" and descending "motor" mechanisms and that this is the way the nervous system works. But, we have to reckon with several functionally complicating factors such as bidirectional axoplasmic flow of chemical messages, including trophic factors; axon reflexes whereby afferent nerves generate efferent signals that contribute to local inflammatory reactions; central efferent control of sensory receptors; descending projections that control ascending synaptic relays; ascending projections that control descending synaptic relays, etc. All central representations seem to be facultatively linked by reentry paths so that not only is there parallel processing but what might be recognized as crisscrossing processing as well. It is obvious all the way from molecular transport to global mapping systems, that bidirectionally closely coupled control of communications is the rule. This contributes to correspondences in the mapping business and how that contributes in concatenating ways to the development of categories relating to temporal and spatial things in the world (Edelman, 1987).

Neuronal Decision-Making

Individual neurons are decision-makers. Nervous systems are assemblages for multilevel, multiplex decision-making. By virtue of complex, dynamic, centrally distributed representations of the individual and a dynamic environment, and likewise centrally distributed representations of motor control, the human brain possesses extraordinary degrees of freedom of action. This is self-reflected as consciously organized and guided volition, "will." To the extent that we experience consciousness of a given behavioral context, we bear responsibility for our behavior.

Evolutionary advantages of nervous systems in higher organisms derive from greatly expanded and elaborated (better analyzed) sensory representations, memory stores, capacities for learning adaptive behavioral strategies and refined skeletal motor skills. Nervous systems of mammals are crowned by neocortex and directly associated subcortical structures. These have expanded roughly threefold between the great apes and humankind. This is a saltatory evolutionary step, perhaps the most remarkable and consequential since the invention of sex. (This is more than a bon mot; sex accelerated the rate of evolutionary change by a factor of about 3000. Furthermore, sex introduced natural death. Previously, replication by fission or budding maintained continuity of that protoplasm. When specialized sex cells were evolved, the rest of the organism was left to die.)

In humans, neocortex provides decision-making capabilities in more than 600 million systematically but variously connected cortical columns (Edelman and Mountcastle, 1978). This enormous endowment of corticocortical and corticosubcortical communication is kept reasonably coherent by mutual reentry signals and adaptively variable by neuronal population selection. The consequence is that integrated representations of self, objects, space, and time result in categorization in an otherwise unlabeled existence (Edelman, 1987).

Some decisions require prompt fulfillment; others permit self-consciously prolonged reflection. Swiftly acting spinal and brainstem reflexes are in physiological continuity with longer routing, ascending, central, and descending sensorimotor control systems. Forebrain decision-making requires longer pathways and hence longer turnaround time, but this time is reduced by the recent evolutionary provision of larger diameter fibers, fewer synaptic relays, and special access to cortical fields (via lemniscal channels) and prompt corticocortical routing, and cortical access to executive neurons (via pyramidal and other very direct motor controls). Somewhat more reflective decision-making can take advantage of the parallel processing, multiple reentry cortical representations, including memory stores. Within promptly accessible memory stores are motivations and value attachments, repository of decades of experiences and conscientious thinking.

FUNCTIONAL INTEGRATION OF SENSORY AND MOTOR SYSTEMS

Sensory systems are organized into multiple, distributed, distinctive representations and rerepresentations of self and "outside world." There exists no single comprehensive map of all sensory sources relating to body image or environment; nor is there any single neuronal field for representation of transactions between self and world. Perception thus can be likened to a multi-tabled feast (for sensory intake).

Motor systems operate with widely distributed controls, involving motor patterns that are many times represented and rerepresented in various ways. There is no single arena for behavioral representation. Behavior, thus, can be likened to a multi-ringed circus (for motor output).

As populations among sensory representations are selected for functional adequacy and proficiency, they acquire more adaptive perceptual coherence. Thus, with practice, we recognize things more rapidly and reliably. Acquired recognition (categorization) of the "outside world" is naturally constructed and behaviorally focused according to individual past experiences. We increasingly stabilize perceptions with continuing experience. Perceptions become increasingly attached to specific motivations and memories. Perceptual coherence, of course, can become stabilized whether it involves realistic representations or not. Only behavior can put perceptual representations to test. At the same time, by virtue of the requirement to perform in the "outside world," sensory representations become increasingly stable in respect to categorization and in relation to sensorimotor options.

Perceptions frequently become so stable that rectification does not occur even in the context of obvious mistakes. Thus, we may continue "blindly" to make perceptual errors in the face of evidence to the contrary and although others try to persuade us that we are mistaken. Groucho Marx asked, "Am I to believe you, or my own eyes?"

Thus are sensory and motor, emotional and cognitive systems brought by experience to relate to behavior in a real world—a sometimes-bruising, sometimes-rewarding world. In various ways and at different levels of the neuraxis, these systems become increasingly integrated and adapted to rendering satisfying motor performances. This involves an individualistic composition of variables among sensory input, memory stores of previous sensory experiences, and the consequences of performances in a given "outside world" that could not have been programmed genetically or epigenetically. Sensory and motor circuits have to be selected in accordance with reality-driven experiences.

There certainly is brain design according to individual experiences. Evidence indicates that much of this "plasticity" resides in the forebrain. Individuality of brain design may be achieved by neuronal population selection among abundant circuit possibilities in the forebrain. And, as motor performance becomes increasingly overlearned, the selection process may move downstream to reside more and more autonomously in cerebellum, brainstem, and spinal cord. "Practice" *(the consequences of which are taken into account)* "makes perfect."

Like sensory mechanisms, motor mechanisms are functionally increasingly elaborated from lower levels upward. For performance of urgent behaviors, motor responses do not wait for sensory signals to proceed to conscious perceptions. On arrival of sensory signals that have been well modeled by experience, motor mechanims may immediately release selected lower-level action programs. Motor mechanisms, nevertheless, continue to make use of incoming information while sensory signals ascend the neuraxis. As described in Chapter 67 (Touch, Pain, and Temperature), higher-order cortical organization involves multiple arrangements for perceptual processing, deficits of which demonstrate, for example, several distinctive neglect syndromes which combine sensory and motor features.

MECHANISMS OF BEHAVIOR

Toward Understanding Behavior

We need to consider the physiological basis for planning and pursuit of behavior. How does the nervous system control glands, smooth, cardiac, and skeletal muscles, and visceral mechanisms in orderly and coherent ways as required for complicated behaviors. How does the nervous system select among intentions, and organize, carry out, and learn from experience? How do we choose among options? How can we simultaneously control viscerosomatic and expressive behaviors (as in singing, gesture-vocalization, and language) which are critical for social communication? How do we exercise conscious dominion over our effectors?

We tend to think of behavior as a straightforward process, "a thing in itself." We think of the environment as a concrete reality, a three-dimensional space containing various physical and organic "objects." We think of time as flowing through this self-environment combination. The outcome is a combined *internal experience* of outwardly projected dynamic space, time, things, events, and personal actions.

THREE FUNDAMENTAL BEHAVIORS

Visceral Behavior: Visceration

This involves nervous system activities relating to respiration, circulation, ingestion, secretion, digestion, excretion, reproduction, etc. We control the throughput of energy in relation to the environment and selectively distribute that energy internally. Yakovlev (1948) called this "motility of *visceration.*"

Expressive Behavior: Expression

This expresses outwardly the internal state: emotion, e-motion, ex-motion, ex-pression, expression. Expression manifests as spontaneous responses to air-hunger, temperature imbalance (sweating, shivering, goose-flesh), exhaustion, thirst, hunger, fear, rage, grief, pain, horror, etc.; and responses to satisfactions—happiness, joy, ecstasy, laughter, satiety, and the like. Expression may involve struggling, screaming, grimacing, laughing, crying, sighing, or satisfactions expressed by thirst-slaking exclamations, feast-repletion groaning, joy of biological or social accomplishment, orgasm, etc. Yakovlev called this "motility of *expression.*"

Internal experiences tend to be expressed outwardly and spontaneously. Expressions are generally without enduring effects on the environment except for their reception in social circumstances. Charlie Chaplin, the great pantomimist, revolutionized cinema with his expertness at subtle as well as slapstick expression of his internal state. Actors say they must genuinely experience the internal state they seek to manifest; otherwise their actions are manifestly feigned.

Manipulative Behavior: Effectuation

This motility changes the world of matter by moving, shaping, manipulating, or otherwise altering structures. Human artifacts accrue mainly from cooperative, consciously directed "motility of *effectuation*" (Yakovlev, 1948).

Other forms of motility do not fundamentally change the world, but the physical products of human activity—buildings, roads, bridges, mines, factories, power lines, ships, aircraft, missiles, technology, and industry in general, the arts, statuary, musical instruments, implements of science and education, computers, paper, books, these printed words, are all evidences of human effectuation.

INTEGRATION OF THREE MODES OF BEHAVIOR

The three modes of behavior function in interdependent harmony: normal behavior for any individual is all of a piece, a coherent composition of the three. Otherwise, there is play-acting or manifest neurological or psychiatric disorder. Every heart beat, peristaltic wave, contraction of skeletal muscle, is part of a continuous comprehensive integrative scheme of living, in which nervous and endocrine systems play superordinate, integrative roles, melding three distinctive categories of behavior.

The nervous system exercises command and control over the three behavioral domains. It integrates them according to internally organized biological and social values that selectively motivate different behaviors. Many biological values originated far back in evolution. Other values are individuated according to particular experiences, and many are incorporated through social conditioning. Individual development and personal experiences shape several kinds of values, give them priorities and attach them selectively to different environmental contexts and different aspects of individually styled behavior.

With visceration and expression, only relatively superficial aspects of biological and socially attached values reach consciousness. With effectuation, the entire enterprise is consciously controlled. Most values, including socially acquired values, become deeply embedded and subconsciously integrated into behavior.

NERVOUS MECHANISMS GOVERNING BEHAVIOR

All organisms continuously integrate visceral motility. For higher forms, this involves integration of activities throughout grey matter that surrounds the spinal canal and aqueduct, lines the floor of the fourth ventricle, and the ventral walls and floor of the third ventricle. Most visceration is expressed by means of the autonomic nervous system. Visceration is also to some extent expressed via the basal ganglia, as changes in bodily attitude during temperature regulation, and in linked viscerosomatic functions, e.g., respiration, speech singing, ingestion, excretion, coitus, parturition, etc.

Expression invokes manifestations of emotions, postural attitudes, facial gesturing, gesture vocalization, etc. This constitutes a complementary, higher-order mode of behavior which for its normal operations is strictly dependent on continuing successful visceration. Expression involves the intermediate layer of grey and white reticular formation which surrounds the central visceral core. Controlling expression involves the limbic system, basal ganglia, and thalamic and brainstem mechanisms that govern sleep-wakefulness. Expression contributes evolutionary advantages by providing means for communicating social signals, interspecific and particularly conspecific.

Expressive behavior is socially and culturally limited by arbitrary rules that require discretion in public for some expressions, and strict privacy for others. Yet appropriate public manifestations are almost obligatory for expressions of empathy, compassion, grief, shared excitement and joy, and commonplace gestures and entreaties that attend decorum and cultural communication. Rules for expressive behavior are considerably adjusted according to environmental and social circumstances, and differ notably among different cultures. People with Parkinsonism, who may have markedly reduced neurological means for expression, may experience social reserve on the part of others because of personal inadequacy in reciprocal social expression.

Controls for Visceration

The phylogenetically oldest, most central neuronal components—common to all vertebrate nervous systems—are engaged in the control of visceral motility. Later phylogenetic developments are assembled around the central core of visceral controls. Visceral control mechanisms are first to be established. They derive from cells that line the walls of the primitive neural tube. These cells do not migrate far. They establish a vertical core of nucleated grey matter, distributed the length of the neuraxis. The two later-developing motor control systems, expression and effectuation, migrate in that order from the walls of the central canal, aqueduct, and the ventricular walls through the visceral command system and establish phylogenetically newer, successively more external mechanisms for motor control.

The primordial nature of visceration is reflected in its being composed mostly of relatively small neurons which have a great range of lengths of dendritic, and particularly, of axonal processes. A few axons reach all the way from hypothalamus to lower spinal centers (Swanson et al., 1983), but the vast majority are quite short. In general, they provide a locally signaling, diffusely projecting neuronal feltwork. Typically these neurons lack myelin. This made it difficult to trace central visceral pathways until the recent development of immunohistochemical, monoclonal antibody methods for specific labeling.

Conduction in visceral control pathways tends to be slow, pleurisynaptic, and relatively diffusely projecting. This means that visceral motility—cardiovascular, gastrointestinal, respiratory, genitourinary, etc.—is controlled in a relatively gross manner as compared with the precision of activating individual skeletal muscle fibers during discrete manipulative behaviors.

Peripheral visceral control components include distributed chains of ganglia and plexuses of ganglia, and nerve networks disposed along the walls of viscera (see Chapter 69, Visceral Control Mechanisms). These peripherally distributed ganglionic constellations resemble invertebrate nervous systems. We may suppose they constitute an evolutionary retention persisting from prevertebrate ancestry. Even the neurotransmitters employed in their central and peripheral synaptic junctions are characteristic of evolutionarily primitive organisms. Of course, every neuronal network, primitive or not, has continued to evolve through all the intervening millennia. But evolution is conservative, especially with respect to systems that work well, that satisfy biological requirements. Accordingly, visceral controls probably constitute the best-tested elements in the human nervous systems. Visceral peripheral ganglionic systems and nerve nets exert control largely locally, and with considerable autonomy. They are therefore referred to as the *autonomic nervous system.*

Control Mechanisms for Expression

Neural mechanisms involved in the motility of expression are organized paracentrally along the neuraxis, surrounding visceral controls. This intermediate system is responsible for outward expression of the internal state. It is not surprising, therefore, that the *limbic system,* which is involved in emotional experience and expression, and the *basal ganglia,* which are involved in postural expression and bodily attitude, are major components of this system. They operate closely together, structurally and functionally, and are richly interconnected and integrated with visceral control mechanisms. Limbic mechanisms and the basal ganglia communicate reciprocally with hypothalamus, brainstem, and spinal grey and white reticular formation.

Control Mechanisms for Effectuation

The greatest architectural structures designed and built by cooperative enterprise are enormous coral reefs. The greatest biological achievement of all time, the conversion of earth's atmosphere from reducing to oxidizing, mostly accomplished by phytoplankton, is a less tangible but nonetheless monumental example of concerted if not cooperative behavior. There are only a few large-scale evidences of biological effectuation, including blanketing the continents with grasses, flowering plants, and trees.

Effectuation by individuals awaited mammals; still the bulk of human effectuation is achieved through cooperation. Effectuation is generated and controlled consciously and fulfilled almost entirely by corticifugal motor controls operating on skeletal muscles. Effectual behavior relies utterly on underlying stratified physiological stabilities provided by appropriate visceration and expression. Human effectuation has festooned the earth with cities, monuments and tracts, and recently traces on the moon and other planets. Lately, some human artifacts have left the solar system to sail beyond the solar winds.

COMPONENTS GOVERNING BEHAVIOR

Physiology of Motoneurons

Peripheral neural controls are diverse and complex, but they are conveyed by only a few types of neurons which govern only a few types of tissues: glands as well as smooth, cardiac, and skeletal muscles. Control involves changing rates of secretion in glands and changing lengths and/or tensions in muscles. The sum of these activities is behavior.

Neurons exerting motor control, whose axons leave the ventral surface of the brainstem and spinal cord, are known as *efferent neurons* inasmuch as they convey signals that travel away from the central nervous system. They are also called *motor neurons* or *motoneurons,* although these terms seem inappropriate for control of glands. Efferent neurons present three characteristic patterns of organization as they project axons from the central nervous system to glands and muscles (see Fig. 9.70).

1. For cardiac and smooth muscles, exocrine glands, and certain endocrine glands, efferent axons terminate in autonomic ganglia where their endings activate ganglion-based neurons. Ganglionic neurons relay signals to the effector tissues which they control by release of neurotransmitters. Cardiac muscle is spontaneously rhythmically active and has a specialized muscular conduction system which coordinates cardiac muscular contractions. Cardiac innervation influences ongoing activity of cardiac muscle fibers and triggers or blocks spontaneous initiatives generated at nodal points along the cardiac conduction system. Smooth muscle also has a considerable degree of autonomy of contractile function and is also controlled by neurotransmitters released by peripheral autonomic ganglionic innervation.

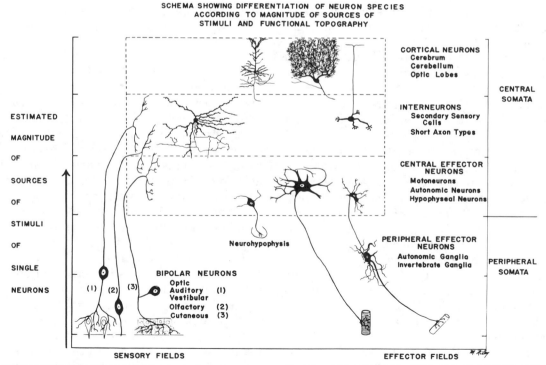

SCHEMA SHOWING DIFFERENTIATION OF NEURON SPECIES
ACCORDING TO MAGNITUDE OF SOURCES OF
STIMULI AND FUNCTIONAL TOPOGRAPHY

Figure 9.70. Major neuron types in mammalian central nervous system, classified according to general function hierarchical level, probable diversity, and magnitude of sources of synaptic connections. (From Bodian, 1967.)

2. For some endocrine glands there is no direct innervation. In this case, the control is neuro-humoral. A neurotransmitter, neurohormone, or neurohumor is conveyed to the gland by the bloodstream. This blood-borne neuroendocrine signal is delivered to the bloodstream by nerve endings or by glands controlled by neuronal signals originating elsewhere. Cells of the neuro-hypophysis are controlled in this way by cells lying in the anterior hypothalamus. Control of transmission of neurohormonal signals in this instance is made more secure by the fact that the circulation between hypothalamus and anterior hypophysis is a confined, portal circulation. Thus, the neurohumoral signal does not get too distributed or diluted before it reaches appropriate receptors.

3. Skeletal muscles receive direct innervation from neurons whose cell bodies lie in motor nuclei of brainstem and spinal cord. Their axons, in contrast to those for glands as well as smooth and cardiac muscles, course directly to the muscle which they activate by release of a locally acting neurotransmitter—acetylcholine. This is a system of precise control. The neuromuscular junction is highly reliable. The efferent axon and the muscle it innervates, taken together, constitute a single *motor unit*.

Convergence onto Efferent Neurons

Important aspects of motor integration result from the confluence of multiple control messages which originate from a wide variety of sources, local and distributed. Motor systems in general and central executive neurons in particular are characterized by having extensive, highly branched dendritic arborizations which allow a great deal of signal convergence within the spatial range of that cell.

Every neuron that receives multiple convergent signals from distributed sources engages in important decision-making in the process of generating its output. The locus of axon origin, the axon hillock, represents the final cellular site of decision-making. Impulses initiated or held in check by motoneurons represent biologically important go/no go contractile decisions. Neuronal decision-making is affected by several factors: (1) the tendency for that neuron to exhibit spontaneous activity, (2) the recent past history of that neuron, (3) the level of excitation and inhibition taking place in its local environment, including specifically within local (ultrastructural) neuronal circuits, and (4) the precise geometry (spatial summation) and timing (temporal summation) of signals (excitatory, inhibitory, and modulatory) delivered to various specific parts of that cell's dendritic arborizations and cell body.

Signals terminating on remote dendritic branches are less likely to exert effective control over that neuron. Messages generated in remote dendritic branches are more likely to decrement as they pass toward the cell body, or to be overwhelmed by signals which may be synaptically more powerful or which lie closer to the cell body. Signals which impinge on the major dendritic trunks and cell body are likely to exert more commanding control over neuronal decision-making. It is instructive to contrast this convergent decision-making strategy with the nearly one-to-one reliability, the minimal modification of throughput seen at relays along the lemniscal sensory pathways. There, speed and reliability of signaling is important for higher centers. In final motor control, however, decisions are postponed to the last possible instant, while information from local, specific remote, and nonspecific remote sources of control are all taken into account.

Sherrington (1906) called the motoneuron the *"final common path"* because all neural paths lead, ultimately, to the efferent neuron. What happens among prior circuits is relatively less important for immediate behavior. The critical event is action, or nonaction. Therefore, cells in cortical and subcortical systems that are engaged in the convergent integration of motor control, and especially brainstem and spinal interneurons, are providing biologically important information for decision-making by the motor neuron.

Physiology of Interneurons

In brainstem and spinal motor nuclei, there are many *internuncial cells* or *interneurons* (see Fig. 9.70). Most centrifugal controls communicate indirectly by terminating on interneurons which then communicate with efferent neurons. Descending control systems provide a vast confluence of excitatory and inhibitory neurotransmitter releases at synaptic junctions with interneurons. Some terminals can directly and indirectly influence the junction between interneurons and motoneurons. Interneurons themselves may be excitatory or inhibitory. The organization of descending controls provides a variety of different regulatory combinations by which spontaneous activity among descending projections and spinal interneurons convey important influences.

Neurons are not only decision-makers and commanders of action among other neurons, effectors, etc., and functionally very adroit at these responsibilities, but they also have remarkably various ways of reaching out for incoming information. Their dendritic arborizations are selective, to the point, for example, of having relations with only a single inputting neuron (e.g., the calyx junction in the trapezoid body of the auditory pathway). At the other extreme, they may distribute almost endlessly branching dendrites which are covered with synaptic terminals from tens of thousands inputting axons from at least a few thousand distinctive sources of information. Some of the variations

among dendritic patterns illustrated in Figure 9.71 are worthy of consideration with respect to the receptive-discriminative control that is available through dendritic outreach. Recall that these protoplasmic processes are capable of growth, branching out or trimming back, in response to neuronal messages, and in response to neuromodulators (hormones, neurohumors, and some neurotransmitters such as norepinephrine as it may invoke the "now print" order described below). Spatial and temporal variations that can be introduced by dendritic arbo-

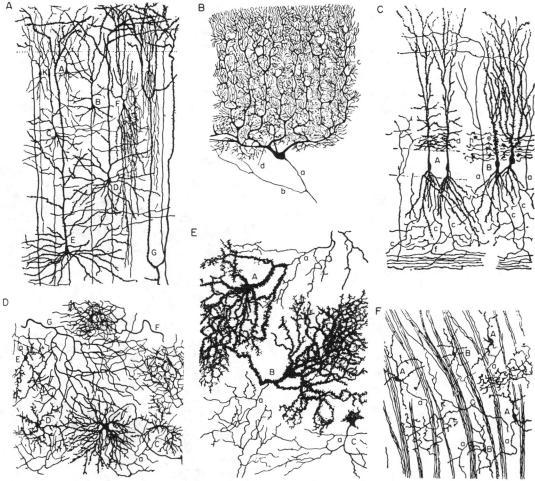

Figure 9.71. Examples of dendritic patterns of neurons which constitute the major cell types of elements in different structures of the mammalian brain. A, Outermost three layers of frontal cortex of a 1-month-old human infant. *Small* (A, B, C,) and *large* (D and E) medium pyramidal neurons as well as the dendritic process of a large pyramidal cell (G) of the fourth layer are shown. Cells with double dendritic bouquets (F) and fusiform appearance (K) are also identified. B, Purkinje cell of the cerebellum from adult human. Axon (a) and axon-collateral (b), capillary spaces (c), and spaces occupied by basket cells (d) are indicated. C, Pyramidal cells of hippocampus from a 1-month-old rabbit. A, small pyramidal cells of the superior region. B, large pyramidal

cells of the inferior region. Large ascending collaterals (a), axons (c), and sites of contact of mossy fibers (h) are to be noted. D, Frontal section of the thalamic somatic sensory nucleus of the cat a few days old. A, cell with a long axon; C, D, and E, cells with short axons; F, sensory fibers; G, axons of cortical origin terminating in the sensory nucleus. E, Cells of the pons from a human infant, a few days old. A and B, Cells with axons arising from dendrites; C, cell with a bifurcated axon; F, Saggital section of the caudate nucleus from a newborn rat. A, Cells with a long axon; B, Cells with short axon; C, ascending afferent fibers. (From Ramón y Cajal, cited by Purpura, 1967.)

rizations make all the difference in the decision-making and command and control options available to a given neuron. They are very likely involved in mechanisms of memory acquisition and forgetting.

Local Circuit Neurons (LCNs)

Mammalian brains exhibit a greater variety of cells than all other organs and tissues of the body combined. Bullock (1981) estimated that there are probably more than 50 million distinctly different *types* of neurons in the human brain. Yet it is conceptually useful to divide all neurons into two major classes: (1) cells with long axons which project from one structure to one or more other structures; these are considered to be *extrinsically projecting neurons;* (2) cells with short dendrites and short axons (or no axons) with connections confined to a small group of neurons; these are considered to be *intrinsic, local circuit neurons (LCNs) (Rakic, 1975).* LCNs may be very numerous and densely packed. In the caudate nucleus, for example, one type of LCN,

the common spiny neuron, dominates the cell population, constituting 95% of all caudate cells. Of special interest is the fact that LCNs are involved in intricate local circuits.

The synaptic organization of circuits appears to be highly plastic at the ultrastructural level (Akert and Livingston, 1973). As illustrated in Figure 9.72, there are distinctive morphological differences, borne out statistically, between synaptic junctions in the spinal cord of anesthetized and unanesthetized rats. The synaptic cleft in the anesthetized condition is flatter and in freeze-fracture electron micrographs the presynaptic membrane manifests few if any pitted indentations. In contrast, in the waking state, the synaptic cleft shows greater curvature and increased frequency of pitted indentations which indicate fusion of synaptic vesicles to the presynaptic membrane and therefore imply active release of neurotransmitter. "Omega" figures, seen in thin-section electron micrographs in spinal cords of waking rats, confirm this assumption

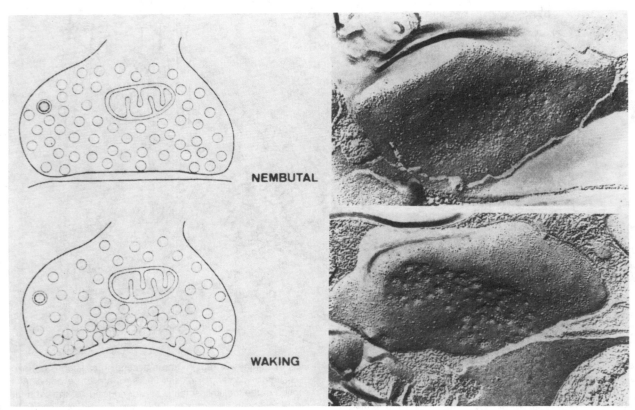

Figure 9.72. Distinctions revealed by freeze-etching techniques between anesthetized and unanesthetized junctions in the rat. Diagram illustrates *left (above)* a synaptic junction: presynaptic terminal containing numerous synaptic vesicles, synaptic cleft, and postsynaptic membrane. Effects of general Nembutal anesthesia include flattening of synaptic cleft, reduction of crowding of synaptic vesicles in the vicinity of presynaptic membrane, and reduction or elimination of numbers of vesicles fused to presynaptic membrane having open access of vesicle contents to synaptic cleft. In waking state, *left (below),* vesicles crowd toward presynaptic membrane and "omega" figures are seen resulting from vesicle fusion and opening of vesicle contents to synaptic cleft. On *right* are representative electron micrographs of an anesthetized *(above)* and unanesthetized presynaptic terminal *(below).* The former shows only rare pitted indentations associated with vesicle fusion to presynaptic membrane, the latter shows numerous such indentations. (From Akert and Livingston, 1973.)

and are consistently more prevalent in the waking state. Two are shown diagrammatically to the left in Figure 9.72.

Bodian (1972) categorized various synaptic junctions according to a schema presented in Figure 9.73. This depicts only axonal endings and does not begin to exhaust the varieties of structurally distinguished synapses. The figure illustrates a few varieties which have been found to possess distinctive operational characteristics. Such "microcircuits" are organized into systematically different "tools for management of information" in local circuits. Local circuits contain detailed microcircuitry which provides local microcontrol of information processing.

"Serial synapses" (Fig. 9.73) occur where two to five or more terminals are stacking in serial succession, linked together in potentially avalanching, or quenching, or ambiguous interdependence. Other configurations enable LCN terminals to interfere with the terminal release of neurotransmitter by neurons coming from disparate parts of the brain. Such presynaptic terminals can interfere with axonal signals at the last instant before those signals would otherwise have been synaptically effective.

Anatomical evidence that such local circuits are widespread is instructive in light of results from bioelectric studies that demonstrate special response properties of larger-scale, functionally comparable circuits: electrotonic coupling, effects of weak electric fields on local ionic flux in nervous tissue, etc. (Adey, 1981). Through such external influences, intraneuronal and interneuronal transport of substances that contributes to trophic effects of

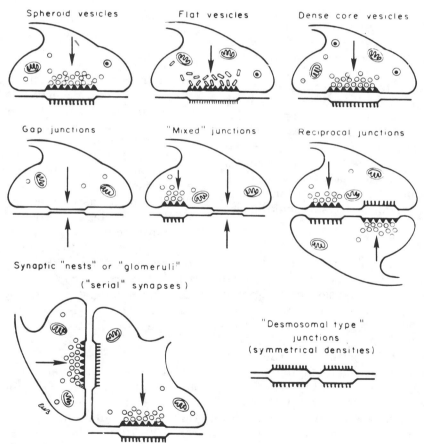

Figure 9.73. Schematic representation of some major variations of synaptic structures and topography, presumably associated with functional diversity. *Upper row,* "Asymmetric" contacts characteristic of most central synapses, with uniform cleft of about 16 nm width, dense pre- and postsynaptic materials and in the cleft (not shown). The presynaptic bulbs contain mitochondria and various kinds of microvesicles of about 20–40 nm. Spherical vesicles are ordinarily associated with known excitatory synapses and flattened vesicles with inhibitory synapses. Microvesicles with electron-dense cores are characteristic of adrenergic nerve terminals; those in the CNS have larger, 60–80 nm vesicles. *Middle row,* Neuron clefts of about 2 nm (gap junctions) characteristic of electrotonic transmission, devoid of junctional densities; they may occur on the same synaptic interface with chemical synapses. Reciprocal junctions are facing in opposite directions in the same interface. *Lower row,* Specialized interneuronal junctions may be arranged in configurations such as nest, glomeruli, and serial synapses which may be multiple. Desmosomal junctions are symmetric, with a cleft of 20 nm. (From Bodian, 1972.)

neurons on other neurons and on non-neural tissues are interwoven with intrinsic biochemical and bioelectric phenomena. The obvious implication, at present unverifiable, is that local circuits may be susceptible to a variety of weak outside electrical field forces.

Dimensions of the Neuropil

There is probably no structure in the mammalian central nervous system that lacks LCNs. The relative proportion of LCNs to extrinsically projecting neurons is probably greater than 3:1. Both the proportion of LCNs and the complexity of local circuits increases with phylogeny. Another phyletic sign of importance is the ratio of volume of grey matter to volume of nerve cell bodies, the *G/C (grey to cell volume) coefficient*. This gives a rough measure of the relative amount of neuropil available for information processing. This may be thought of as *the local transactional space. Neuropil* is a term that implies a feltwork of interacting neuronal terminals, as in the "pile" of a rug. *Transaction* implies multiple, mutually interdependent components in simultaneous action.

For primary visual cortex, Haug (1958) estimated G/C coefficients in a variety of species: man, 47; chimpanzee, 31; macaque, 19; rabbit, 15.4; guinea pig, 11.5; and mouse, 8.5. Cell dimensions in the different species were not substantially different, but the space available for local neuropil increased dramatically going up the phylogenetic scale. Expanding neuropil undoubtedly provides increasing degrees of freedom for local decision-making processes.

LOCAL CIRCUITS

Integrative processes in local circuits are enriched by the presence of numerous somatosomatic and dendrodendritic synapses and "gap junctions" among neurons (Peters et al., 1970; Sotelo and Llinas, 1972). This suggests that a great exchange of information between neurons may be taking place on the receiving surfaces (dendrites and somas) of neurons, decision-making regions, between neighboring neurons, prior to the arrival of signals at the critical zone—the axon hillock, where action potentials are generated. Gap junctions link cells electronically (by allowing relatively free exchange of ions), and biochemically (by allowing exchange of chemicals below about 1000 mW). Thus, both synaptic and nonsynaptic processes contribute to integration in local circuits. Elsewhere, gap junctions are rare between neurons.

Electron microscopy has revealed many other aspects of local circuit complexity which imply additional functional precision and flexibility (see Peters et al., 1970, for exemplary thin-section electron microscopy, and Sandri et al., 1977, for exemplary freeze-fracture electron microscopy). Local circuits exhibit an astonishing flow and counterflow of information among neighboring junctions facing upstream and downstream along the presumed directions of general information delivery (Schmitt et al., 1976). It is implied by fine-structure morphology that there are many local feedback and feedforward "microcontrol" circuits which may contribute extraordinary functional plasticity. It is therefore reasonable to assume that distinctive properties of local circuits contribute integratively at all levels of nervous system organization including higher cognitive functions (e.g., perception, thought processing, language, and the like).

PHYSIOLOGY OF NEUROGLIA

Nerve cells are almost completely surrounded by satellite cells, *neuroglia (glia)*. They probably outnumber neurons 10:1 and they contribute about half the volume of the nervous system. Glia are divided into astrocytes, microglia, and oligodendrocytes in the central nervous system, and Schwann cells in peripheral nerves. Astrocytes are divided into *fibrous astrocytes*, which contain many filaments and are abundant among nerve fibers in white matter, and *protoplasmic astrocytes*, which have fewer filaments and are commonplace in grey matter, among nerve cell bodies and neuropil.

Both types of astrocytes have extended processes which are applied to the outside of brain capillaries. These astrocytic "foot processes" provide a route for bailing excess potassium out of the brain, and contribute to the "blood brain barrier." They respond expressly to injury of the brain by closing gaps and repairing wounds—providing scar formation. Astrocytes also contribute to phagocytosis. Microglia, in contrast, are primarily phagocytotic.

Electrophysiology of Neuroglia

Glia respond passively to electric current and do not conduct impulses. The glial membrane potential (about 90 mV) depends primarily on the distribution of potassium and is greater than that of neurons (60–70 mV). In contrast to neurons, when current is passed through glial cell membranes, the membrane behaves passively (Fig. 9.74). Potassium is the principal cation inside glia. Glia apparently contribute as a source and sink for this ion. Impulse conduction by neurons releases potassium into extracellular spaces. This provides a *nonsynaptic signal to glia*, causing them to depolarize. The glial

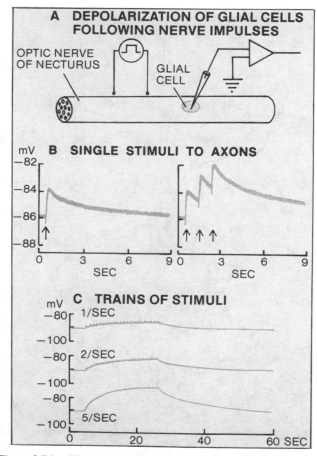

Figure 9.74. Effect of neural activity on glial cells in the optic nerve of the mudpuppy. Synchronous impulses in nerve fibers cause glial cells to become depolarized. Each volley of impulses leads to a depolarization that takes seconds to decline. The amplitude of the potential depends on the number of axons activated and the frequency of stimulation as shown in *B* and *C*. (From Kuffler and Nicholls, 1967.)

response is a straightforward index of the amount of neuronal activity: glia equilibrate potassium ions over large volumes and deliver excess potassium to capillaries.

Gap Junctions and Neuroglia

Glial cells are linked by numerous low-resistance gap junctions which permit the direct passage of ions and both neutral and negatively charged molecules up to a molecular weight of about 1000 (Bennett and Goodenough, 1978). Neurons share gap junctions with other neurons much less frequently and such gap junctions are confined primarily to local circuits. There appear to be no gap junctions between glia and neurons. By virtue of the widespread prevalence of low resistance gap junctions linking glia, depolarization among glia following neuronal release of potassium is widespread. If neuron activity is further increased, glial depolariza-

tion is correspondingly augmented. Such glial potential shifts affect the electroencephalogram and electroretinogram and may have weak "tidal" effects on neuronal activity.

Myelin Sheathing of Axons by Glia

Myelin sheaths increase reliability of axonal conduction by a factor of about 2 and increase velocity of conduction by a factor of about 6. In the central nervous system, myelin is wrapped around and around axons repeatedly (like window-shade winding) by *oligodendrocytes* or *oligos (oligo = few, dendro = branches)*. In peripheral nerves, Schwann cells perform provide analogous wrappings. In the central nervous system single oligos may be responsible for myelinating several segments of myelin in as many as 30 different axons (Bunge, 1968). In the periphery, single Schwann cells wrap only a single segment of myelin around a single axon.

These central/peripheral differences in myelination may reflect central integrative processes among axons, as contrasted with a strict preservation of signal independence in peripheral axons. Both types of glial wrappings form glial-axonal junctions along the paranodal fringes of the spirally wound glial wrappings at the margins of the nodes of Ranvier (Livingston et al., 1973).

DEVELOPMENT OF THE NERVOUS SYSTEM

Most neurons arise from stem cells lining the neural tube. Peripheral sensory and autonomic neurons and cells of the adrenal medulla arise from the neural crest which itself originates from the lining of the neural tube. Neurons are said to have their birthday at the time of their final cell division. They then migrate from the vicinity of the inner wall of the neural tube to some intermediate or final station. From the time of their birth, neurons seem to be committed to proceed to certain locations (in relation to other neurons). After they reach their destiny, they send out processes to make certain kinds of connections on certain dendrite and cell body regions of specific target neurons, near and remote (Edelman, 1984; 1987).

The sequence of cell migration and formation of connections is an orderly process with specific spatial and temporal patterns (Schmitt et al., 1970; 1981). Levi-Montalcini (1966) discovered a nerve growth factor that selectively influences the growth of sympathetic and sensory neurons. Presumably there are several other such specific growth factors. Glial guides assure appropriate radial migration and contribute to the establishment of orderly stations for neurons in subcortical and cortical regions. Cells arising from a single location on the ventricular surface end up in the same radially oriented cortical column. Those generated later take up positions external to the station occupied by their predecessors (Rakic, 1981).

DYNAMIC FEATURES OF STRUCTURAL ORGANIZATION

Prenatal Development

During embryogenesis, neuroblasts divide a number of times in rapid succession. This provides an enormous numerical increase. The human brain must produce, on average, on the order of one-third of a million nerve cells per minute during the whole 9 months of gestation. This proliferation, of course, accelerates as more and more cell divisions proceed, and it continues at a prodigiously high rate during the first months of postnatal life, tapering off gradually to a roughly zero base at about the end of the first year of life. *The brain doubles in size during the first 6 months of life and doubles again by about the fourth birthday.* Most of the latter growth is accounted for by increase in neuronal size and arborization, and by increase in number of glia and the fabrication of myelin.

Brain growth and development is affected by a number of genetic, nutritional, and other environmental factors (Brazier, 1975). The brain is the first organ to differentiate, the fastest growing, and therefore the largest organ throughout gestation and early childhood. Because of rapid growth and extraordinary plasticity, involving vast neuronal population selection and ''pruning'' of circuits, the brain of a child, until adolescence, has a metabolic rate per unit volume that is twice that of an adult. The plasticity of development and the high metabolic rate may account for the remarkable learning capacity of the preadolescent brain.

Postnatal Development

Much of the ultimate selection of circuitry and the dying back (nonselection?) of synapses, processes, and cell bodies takes place after birth. In this way, the environment and behavioral transactions with the environment play an important role in determining the abundance and specific architectural organization of neuronal circuits (Edelman, 1987). Figure 9.75 illustrates the orderliness of such early developmental processes in normal corticocortical projections compared with corticocortical projections following removal of the region of normal cortical destiny. Cortical projections deprived of their normal target seek out comparable targets and

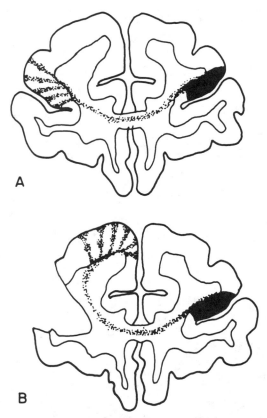

Figure 9.75. Plasticity of association cortex in primates. *A*, Drawing of prefrontal cortex in the left and right hemispheres of normal rhesus monkey whose superior bank of the principal sulcus in the right hemisphere was injected with a mixture of radioactive leucine and proline 3 weeks prior to sacrifice. The drawing illustrates the pattern of alternating callosal fiber bundles in the homotopic cortex of the left hemisphere. *B*, A monkey in which the dorsolateral prefrontal cortex was resected in the left hemisphere at 8 weeks of age. Two months later, intracortical injections of the radioactive amino acids were placed in the superior bank of the principal sulcus of the right hemisphere 3 weeks prior to sacrifice. Callosal fibers, deprived of their normal target in the cortex, project to areas dorsal and medial to the resected area, but the anomalous pathway retains its pattern of spatial periodicity. (From Goldman-Rakic, 1981; Rakic and Goldman-Rakic, 1982.)

retain their orderly pattern of interhemispheric innervation, even though it is to an abnormal region of cortex (Goldman-Rakic, 1982).

Architectural dynamics of brain growth can be appreciated at the light microscopic level by inspection of Figure 9.76 from the work of Conel (1959). The three sections illustrated are from the same region of primary visual cortex. Other cortical regions were shown by Conel to undergo similar progressive elaborations, although according to somewhat different age-correlated schedules.

It is evident that throughout early postnatal life, neurons are rapidly elaborating processes and that the entire neuropil becomes greatly expanded. Other anatomical methods indicate that additional

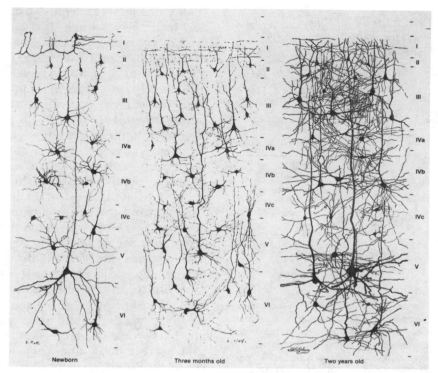

Figure 9.76. Development of dendritic arborizations in human visual cortex, as revealed by Golgi stain techniques, showing continuing development during early life, from newborn, 3-month-old, and 2-year- old infants. Depth of cortex has been arbitrarily expanded for newborn and 3-month-old (almost double for the newborn) in order to match layers as seen in the older infant. (From Conel, 1959.)

neurons are migrating into the cortex postnatally, especially into the superficial layers. In monkeys, it is implied that these late-arriving neurons establish connections and contribute to circuitry in accordance with ongoing experiences (Altman, 1966).

ARE NEURONS ADDED DURING ADULT LIFE?

By use of thymidine autoradiography, it was found that dentate granule cells in the hippocampus of the rat continue to be produced during adulthood (Altman 1963; Altman and Das, 1965; Bayer and Altman, 1975). There is an absolute numerical increase in dentate neuron population during adult life (Bayer et al., 1982). Thus, in addition to powerful genetic and early environmental effects on brain architecture and continuing modeling of circuit architecture by selection, and, at least in the hippocampus, new neurons may also be contributing to the phenomena of neuronal plasticity.

NEURONAL SPROUTING

It has long been known that following damage to the central nervous system, traumatic or vascular, severe functional deficits may be reduced and considerable restoration of functions may take place. It is possible that undamaged parts of the brain take over the functional role of damaged parts, i.e., that remaining parts may take on additional or altered functional responsibilities. One evidence of plastic reorganization is *collateral sprouting* of central neurons. Such collateral sprouting has been observed in the spinal cord (Liu and Chambers, 1958; Murray and Goldberger, 1974), hippocampus (Lynch et al., 1973; Zimmer, 1973), and superior colliculus (Lund and Lund, 1971).

Sprouting may begin within hours of injury and may be completed within a few days or may continue longer (Zimmer, 1974). Central sprouting can establish new synaptic connections at vacated synaptic sites (Raisman, 1969; Raisman and Field, 1973). Sprouting collaterals appear to be selective with respect to areas that they do or do not invade and the kinds of synaptic connections they establish.

Various responses of axons to damage to nervous and other tissue they innervate are illustrated diagrammatically in Figure 9.77. Typically, slight damage to skeletal muscle or to the neuromuscular junction is followed by sprouting from the nerve terminal itself and even from nodes of Ranvier farther back along the axon (*A*). If a myelinated fiber is cut, there is likely to be sprouting from the cut end and from proximal nodes (*B*). If cells innervated by a neuron are damaged, the innervating

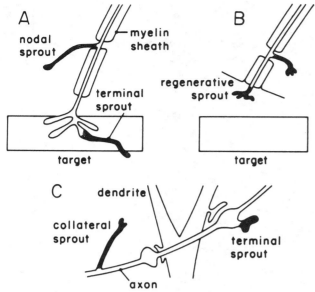

Figure 9.77. Axonal sprouting with consequences for regeneration, reinnervation, and synaptic restoration. Types of sprouting illustrated: *A*, Intact myelinated axon showing a collateral or nodal sprout arising from a node of Ranvier between myelin segments, and a terminal sprout. Most typical cases occur in the peripheral nervous system, e.g., neuromuscular junctions. *B*, Damaged myelinated axon showing regenerative sprouting. *C*, Axons with boutons en passant showing collateral and terminal sprouting, a situation arising frequently in the central nervous system. Growing sprouts are shaded. (From Cotman et al., 1981.)

neuron may sprout collaterals from more proximal regions of the axon, including where the neuron has synaptic junctions en passant (*C*).

GRAFTING FETAL BRAIN TISSUE INTO ADULT BRAINS

As indicated, the hippocampus is a region that possesses a potential for plastic morphological and functional reconstitution. Moore (1975) found that monoamine neurons innervating the hippocampus and septum undergo extensive reorganization in response to injury. Bjöerklund et al. (1976) successfully transplanted monoaminergic neurons from rat embryos into the hippocampus of adult rats. Adrenergic, serotonergic, and dopaminergic neurons from the embryonic tissue each have distinctive patterns of innervation in the adult hippocamus. The transplants develop different specific connections according to the areas damaged in the host brain. This means that considerable power for reestablishment of fiber pathways and connections are retained by the adult brain (Björklund and Stenevi, 1984).

Two principles emerged: (1) early fetal tissue survives better than older fetal transplants; and (2) transplants into young brains are more successful than transplants into older brains. Thus both donor and host tissues have time-diminishing potentialities for reorganization. Tissue from different parts of the fetal brain retain some of the developmental patterns that they would have exhibited if they had been left to develop in situ in the donor brain. Embryonic retinas retain retina-like organization when transplanted into adult brains. Septal and striatal cholinergic tissues transplanted into the hippocampus form normal patterns of cholinergic innervation there despite the fact that the septal region normally does and the striatal region normally does not innervate the hippocampus.

The brain is an immunologically protected region and may therefore be less likely to reject grafts than are other tissues. There has been some success with cross-species grafting of fetal tissue into adult brains. Experimental grafting in animals has involved grafts explicitly introduced to correct damage resulting in movement disorders, memory loss, hyperactivity, and overresponsiveness to stimuli. This work has succeeded in maintaining grafts that secrete dopamine, acetylcholine, norepinephrine, serotonin, vasopressin, and gonadotropin-releasing hormone. Even genetically altered neurons have been successfully implanted. Implantations can be facilitated by utilization of nerve growth factors (see Gage and Björklund, 1986a; 1986b; Gage et al., 1987; Azmitia and Björklund, 1987; Gash and Sladek, 1988).

Rather than transplanting fetal tissue, Stenevi and Björklund found it effective to inject 1–2 μl of brain cells suspensions into adult rat brain. Hippocampal damage in the rat results in short-term memory deficits, hyperactivity, and overreaction to stimuli. By infusing cells from the septum, deficits in short-term memory can be ameliorated, while the animals remain hyperactive and startle-prone. When cells from the locus coeruleus are infused in other animals, the rats are no longer hyperactive, but retain their short-term memory deficits and exaggerated startle responses. Suspensions containing dopamine-secreting cells introduced into the striatum of aging rats resulted in significant improvement in their motor coordination (Gage et al., 1983). Using infection with a retroviral vector, it is possible to genetically modify fibroblasts of the host to secrete nerve growth factor (NGF) which protects injured cells from dying back; instead they sprout axons that project in the direction of the cellular source of NGF. Grafts are similarly supported by this means. The combined approach of retroviral gene transfer and neural grafting may have considerable reparative effect in selected nervous system disorders (Rosenberg et al., 1988).

COMMUNICATION

The "business" of the nervous system is communication. Communication has the biological goal of improving behavior in regard to securing internal satisfactions. In general, communication in the nervous system involves the organization of appropriate input and output relations with respect to the environment and with respect to internal control mechanisms of viscera. Specialized elements in the nervous system are particularly sensitive to various mechanical and physical perturbations, electrical, chemical, gaseous, and ionic shifts; but most neurons only excite or inhibit one another. Communication is accomplished mainly by activation of systems of receptors and effectors—external and internal—between which are organized extensive decision-making networks.

The nervous system processes information that is conveyed by networks of metabolically and architecturally self-organizing neurons. Transactions are based on detailed electrophysical chemistry and physiology of receptors and junctions, membranes, and intracellular events relating to signaling between cell membranes and second-messenger systems which can influence the cell's genome, thus altering the activity bias or having trophic influences affecting axoplasmic transport, release of neurotransmitters, neurohormones, etc.

NEUROTRANSMITTERS AND NEUROPHARMACOLOGY

Synaptic Transmission

Synaptic transmission differs from axonal transmission because signals travel only in one direction across chemical synaptic junctions. In axons, impulses can travel in either direction, although, because of the directionality of junctions, axons ordinarily conduct orthodromically. There is junctional delay in synaptic conduction, but no similar delay in axonal conduction. Axonal conduction can follow up to high stimulation rates, from a few to more than 1000/s, but synaptic communication can follow only slower rates of transmission, below about 50/s. Action potentials are all-or-nothing in axons, synaptic potentials summate algebraically, adding excitatory and subtracting inhibitory potentials, i.e., providing interactive, graded responses.

Most nervous system communication involves synaptic transmission. In general, excitatory or inhibitory neurotransmitters are released by the presynaptic terminal. These have predictable consequences with respect to the postsynaptic membrane, depending on the characteristics of the neurotransmitter and also on the specifics of the postsynaptic receptor. A given neurotransmitter may have opposite effects in different synapses, depending on the receptor involved.

Synaptic communications are complicated by the fact that a given neuron can synthesize and release more than one neurotransmitter. Furthermore, local circuit configurations imply logical consequences for neuronal decision-making, e.g., "and; but; either-or; only-if; when-if; if-then, etc.," blocking excitation; inhibiting inhibition, etc. The more complicated of these may have concatenating circuits in extremely small space, and these can have significant dynamic consequences for communication and control.

Postsynaptic Potentials

Synaptic sites are not directly excitable; neurotransmitter is released across the junction and this induces more or less prolonged postsynaptic potentials. The results depend on rates of release and diffusion, rates of neurotransmitter binding and release by receptors on the postsynaptic membrane, rates of enzyme action to inactivate the neurotransmitter, and rates of neurotransmitter reuptake by the presynaptic terminal. Reuptake mechanisms are both energy- and sodium-dependent. Almost all neurons that release neurotransmitters actively reaccumulate the transmitter(s) they release. An exception is acetylcholine which is rapidly dissociated, leaving choline for reuptake.

Excitatory neurotransmitters released by presynaptic terminals activate postsynaptic membranes by increasing ionic conductance. This drives the membrane potential toward the sodium equilibrium potential which results in *depolarization*. If depolarization is sufficient, an all-or-nothing action potential is generated. Inhibitory neurotransmitters, similarly released, selectively activate chloride ion conductance which gives rise to an inward flow of chloride and an increased *polarization*. This increases stability of the membrane and makes it less likely that a spike impulse will be discharged.

Most postsynaptic potentials are of relatively short duration (less than about 20 ms), but some postsynaptic potentials may last for 1 to 2 s. Slow postsynaptic potentials may occur as a result of neurotransmitter inactivation of a tonically active postsynaptic membrane, one that has a slow sodium or potassium resting conductance. Slow postsynaptic potentials occur when tonically active excitatory or inhibitory terminals are themselves inactivated by *presynaptic inhibition*. They may also occur as a result of neurotransmitter activation of cyclic nucleotides which, in turn, bring about phosphorylation of membrane proteins which alter ion permeability. In this way, the set point or bias toward activity or inactivity on the part of the postsynaptic neuron may be controlled.

Neuropharmacology

Thousands of drugs and agents affect the nervous system. Aside from local anesthetics that can block transmission in axons, most drug effects involve synaptic events. These include analgesics, narcotics, hypnotics, sedatives, anesthetics, anticonvulsants, psychotropic drugs, and a number of drugs that specifically affect the autonomic nervous system. Most drugs appear to act either pre- or postsynaptically with respect to the action of neurotransmitters. Drugs may interfere with metabolism of the neurotransmitter or with its synthesis, storage, release, or reuptake. Or they may act postsynaptically by activating or blocking neurotransmitter receptors.

An important aspect of neuropharmacology relates to blood-brain permeability barriers. Some of the barriers are physical: (1) lack of fenestra in brain capillary endothelium, (2) tight junctions that provide a double seal joining cerebral endothelial cells to each other in capillary walls, and (3) the foot-processes of protoplasmic astrocytes. There are also chemical barriers, including charges borne by ion membrane transport channels and across solvent partitions. Variations in kinds and rates of metabolic activity of the local neuropil, differences in distance between capillaries and differences in diffusion barriers because of altered dimensions of extracellular space and changes in the ionic make-up of the glycoprotein "fuzz" between brain cells.

Release of Neurotransmitters

Action potentials usually lead to neurotransmitter release. Transmitters are also released from short-axon neurons, and axonless neurons, and from dendrites and cell bodies where graded potentials rather than action potentials occur. Spikeless release of neurotransmitters may occur with interneurons and between cells involved in local circuits.

Axonal excitation is coupled to neurotransmitter release at synaptic junctions. There is a shift from sodium ion conductance control of the axonal membrane to calcium ion control of neurotransmitter release. The linkage between these two ionic controls can be separated by using *tetrodotoxin* which selectively blocks sodium ion channels.

Calcium plays an instrumental role in all forms of neurotransmitter release. It is now possible to observe and measure the dynamics of release or admission of ionized calcium in cell terminals, using luminescence of aequorin which reacts with Ca^{2+} by

emitting light, by the use of X-ray or proton probe spectroscopy, and by the use of metallochromic dyes such as arsenazo III which changes its absorption spectrum when it complexes with Ca^{2+}. Most of the calcium in neurons is membrane-bound. Because the level of ionized calcium is normally very low (about 10^{-7}) in cytoplasm, the cell is strongly affected by very small shifts in local Ca^{2+} levels, thus yielding large physiological effects. The invasion of a nerve terminal by an action potential is associated with an influx of Ca^{2+} from extracellular space. Neurons then utilize calcium shifts from free to bound states to control secretion of neurotransmitters.

Similar Ca^{2+} control also accounts for axoplasmic flow, many enzymatic functions, and membrane permeability. Mechanisms involved in the control of calcium availability are also coupled to neurotransmitter synthesis and to the setting of regional metabolic activity by mitochondria. By participating in these long-lasting processes, calcium contributes to synaptic plasticity.

Receptors and Regulation

The process of synaptic transmission involves not only secretion of neurotransmitter but also depends on the special affinity of the neurotransmitter to postsynaptic receptors. A specific transmitter substance is not limited to having only one specific postsynaptic effect. The same transmitter can be used for different effects, both excitation and inhibition, and both excitation and inhibition can be elicited by different transmitters. Acetylcholine, for example, released by the motoneuron, is excitatory at the neuromuscular junction, whereas the same substance released by the vagus is inhibitory at the cardiac nerve-muscle contact. Dopamine and epinephrine can both be excitatory or inhibitory, depending on local receptors.

Kandel summarized the evidence as follows:

1. The sign of synaptic action is not determined by the transmitter but by the properties of the receptors on the postsynaptic cell.
2. The receptors in the follower (postsynaptic) cells of a single presynaptic neuron can be pharmacologically distinctive and can control different ionic channels.
3. A single follower cell may have more than one kind of receptor for a given transmitter, with each receptor controlling a different ionic conductance mechanism.

As a consequence of these three features, cells can mediate opposite synaptic actions to different follower cells or to a single follower cell.

Presynaptic terminal uptake of norepinephrine (NE) from the synaptic cleft and neighboring extracellular space is affected by the amount of NE present. The concentration of reuptake sites on the presynaptic terminal varies according to the local concentration of NE. Depletion of NE by administration of reserpine reduces the number of reuptake sites. Increase of concentration of NE by administration of monoamine oxidase inhibitors increases the number of binding sites. When NE is scarce, reuptake is retarded; when NE is abundant, reuptake is accelerated. These phenomena provide homeostatic regulation of NE availability at the synapse.

Neurotransmitters

The main categories of neurotransmitters include acetylcholine, biogenic amines, a few amino acids, and a large number of peptides. All neurotransmitters have low molecular weight and are water-soluble compounds which are synthesized from precursor molecules. The amino acids and peptides are formed ultimately from circulating glucose. It is evident that these transmitters have great utility because the same categories are found in invertebrates as well as vertebrates despite great differences in organization of the respective nervous systems.

PROBLEM OF IDENTIFYING NEUROTRANSMITTERS

To establish whether a putative neurotransmitter can be categorized as a bona fide neurotransmitter, convincing evidence requires: (1) presence and activity of pertinent enzymes necessary for the synthesis of the substance in the presynaptic terminals; (2) presence and activity of enzymes for inactivation or removal or reuptake of the substance at the synapse; (3) release of the substance by stimulation of the presynaptic terminals; (4) reproduction of physiological consequences by iontophoresis of the substance at the synaptic site; (5) appropriate influence of drugs that affect the enzymes synthesizing or inactivating the substance; (6) appropriate high-affinity receptors, release and uptake actions, inactivation by agonists or antagonists of the substance, its analogues, and related pharmacological evidence.

With respect to the identification of neuropeptides that serve as neurotransmitters, use is made of immunological methods. The putative oligopeptide is injected into animals as an antigen which produces an antibody specific to the peptide. It may be necessary to resort to monoclonal antibodies to obtain a sufficient specificity. The antibody can then be applied to the histological sections and a suitable antibody to the antibody then applied. The second-echelon antibody carries a fluorescent molecule, metallic label, or horseradish peroxidase which can be seen in electron microscopy and light microscopy.

Neuroactive peptides constitute an expanding list that includes: carnosine, thyrotropin-releasing hormone (TRH), met-enkephalin, leu-enkephalin, angiotensin II, cholecystokinine-like peptide (CCK), oxytocin, vasopressin, luteinizing-hormone releasing hormone (LHRH), substance P, neurotensin, bombesin, somatostatin, vasoactive intestinal polypeptide (VIP), β-endorphin, adrenocorticotropin hormone (ACTH), and others. The list ranges from single amino acids to lengthy hormone molecules. The list is expanding rapidly and may include hundreds of neuroactive peptides that might contribute to synaptic communication and control.

INDIVIDUAL NEURONS MAY CONTAIN MORE THAN ONE NEUROTRANSMITTER

Evidence for this phenomenon, to the extent of at least four neurotransmitters, is available in invertebrates: 5-hydroxytryptamine (5-HT), thyrotropin-releasing hormone (TRH), and substance P, in the pons and in the ventral horn of spinal cord of rats (Hökfelt et al., 1980). In some autonomic junctions in the cat, exocrine glands are innervated by two populations of axons. One contains acetylcholine plus the vasoactive intestinal polypeptide (VIP); the other contains norepinephrine plus a polypeptide that blocks vasodilation. Each of the peptides potentiates the transmitter with which it is associated. The two neuronal populations provide physiological responses of opposite nature.

VESICLE HYPOTHESIS FOR NEUROTRANSMITTER RELEASE

Freeze-fracture electron microscopy, combined with thin section analysis, has provided essential proof that synaptic vesicles fuse with the axolemma when secretion of transmitter is elicited (see Fig. 9.72). Rapid freezing shows that fusion of the vesicle and quantal release of acetylcholine occur within milliseconds. Extracellular markers indicate that stimulated nerve terminals contain many vesicles that have formed from axolemma, suggesting that the vesicle membrane is recycled after fusion with the terminal membrane (Heuser and Reese, 1973). Additional vesicles are presumably formed from smooth endoplasmic reticulum and from other membranous components, including preformed vesicles transported into the terminal by axoplasic flow.

In summarizing the evidence for quantal release

of acetylcholine, Ceccarelli and Hurlbut (1980) state that the preponderance of evidence indicates that quantal release comes from the vesicles. Sites for vesicular fusion in the grid of presynaptic dense projections appear to align with counterpart sites on the postsynaptic membrane where the highest density of acetylcholine receptors are aggregated.

DOES NONVESICULAR RELEASE OF NEUROTRANSMITTERS EXIST?

Because of the small dimensions and high speed of neurotransmission, alternative hypotheses for the release of neurotransmitters remain alive, including exclusive alternatives and supplementary release mechanisms (Tauc, 1982). It may be that there are several mechanisms for neurotransmitter release. Considerable acetylcholine leaves terminals in a nonquantal manner, and part of this release is Ca^{2+} dependent. Moreover, both types of release appear to be related to excitation-secretion ion coupling.

ANALYZING NEUROLOGICAL DEFECTS

Within the large and important category of *genetic defects*, methods of analysis include securing detailed history and lineage of carriers as well as of individuals manifesting the disorder. Genetic expression can vary considerably and long-range interactions among chromosomes and genes can add complications even in persons having identical point gene defects (Rosenberg, 1980). *Congenital defects* can be traced to injuries, infections, nutritional deficiencies, immune disorders, induced sensitivities, and toxic influences, e.g., tobacco, alcohol, cocaine—leading to striking syndromes in the newborn.

The nervous system exists in dynamic equilibrium in health and disease with restitutional changes that occur promptly after insult to any part of the nervous system and lasting indefinitely, presumably being stabilized eventually on some utilitarian selection basis. A great variety of diagnostic opportunities is available involving physical (e.g., nuclear magnetic resonance imaging; positron emission tomography; computed axial tomography; radioisotope tracing of tumors; electroretinography, electroencephalography, evoked potentials, echoencephalography, etc.) and, of course, biochemical analyses of blood, urine, and cerebrospinal fluid for particular drugs, metabolic products, toxic agents, etc.

It is important to remember that destruction of part of the nervous system does not reveal the function of that part—even in a substractive sense. It reveals only what the remainder of the nervous system can do in the absence of that part.

Neuropsychologists have developed powerful tests for neurological and psychological evaluation of patients with nervous system disorders. These are becoming increasingly specific and accurate in their implications for neuroanatomical and physiological localization. It is evident that subtle pathophysiological processes may have as much effect on a patient's mental status as anatomical defects. Note especially when processes that are slow in onset, considerable compensation may take place without any signs or symptoms appearing. For example, a child with a glial tumor in the cerebellum may lose almost all of the cerebellum, and yet the first signs may appear only when mass effects cause elevated intracranial pressure. Children with infiltrating blood discrasias and carcinomatosis may similarly have presenting signs of headache and vomiting and eye signs after the brain has been rather completely infiltrated. Moreover, a focal lesion may give rise to remote symptoms, e.g., a frontal tumor may present initially with cerebellar signs.

Structure and function are two sides of the same coin. If examination of one does not reveal something about the other, it has not been examined adequately. Undertake the examination of any patient presenting a neurological problem with a sacred respect for the complexities of the nervous system. Communication and control is what the nervous system is all about. The patient is, of course, the most valuable witness. Careful history and examination of the patient is likely to reveal information that can fill in gaps in the frontiers of neurosciences. If death eventuates, by all means obtain permission for a thorough postmortem examination of the nervous system to disabuse yourself of illusions, and to add greater perspective to your clinical experiences.

CHAPTER 69

Visceral Control Mechanisms

Multicellular existence depends on continuing appropriate distribution of nutrients and simultaneous adequate disposal of wastes for all cells of the body. This must be achieved in a context of widely differing local and regional physiological demands on cells, in a context of a wide range of environmental stresses on the body as a whole. It requires differential control of smooth muscles, cardiac muscle, and glands among several different interdependently controlled visceral systems. It also mandates integration of these functional systems with appropriate skeletal muscle activities. Moreover, visceral systems have so directly to do with overall appetite and satiety that their control is bound up with biological purposes and goal-directed activities of the whole organism.

A comparison is useful between *somatic nervous system mechanisms,* generally governing skeletal musculature and voluntary activity, and *visceral nervous system mechanisms,* generally governing biological housekeeping. Visceral nervous controls maintain a dynamic internal environment essential for proper functioning of all cells, tissues, and organ systems, and the individual as a whole. As with somatic nervous system organization, visceral organization includes sensory, motor, and sensorimotor mechanisms such as reflexes and, at higher levels, drives and integrative processes that include mood and feeling states, motivation, and emotion. Coordination is maintained between somatic and visceral controls, longitudinal as well as segmental, involving spinal cord, brainstem, and forebrain.

ORGANIZATION OF VISCERAL NERVOUS MECHANISMS

Autonomic Nervous System

Smooth muscles, cardiac muscle, and glands have a propensity to be active on their own initiative, in contrast to skeletal muscles which, in mammals, are activated strictly by nerve impulse. Nervous system components governing smooth and cardiac muscles and glands operate largely unconsciously and generally function independently of volition. Nonetheless, some visceral mechanisms which have been considered inaccessible to conscious control can be brought under partial voluntary influence (Miller, 1969; Miller and Dworkin, 1980).

CONTRAST BETWEEN SOMATIC AND VISCERAL SYSTEMS CONTROLS

Because of their general autonomy, visceral neural mechanisms are referred to as being *autonomic.* The *efferent* components are collectively called the *autonomic nervous system.* Note that afferent neurons which send impulses to the central nervous system from visceral organs are *not* considered part of the autonomic nervous system, even though they accompany efferent autonomic axons peripherally and pass through autonomic ganglia. They do not synapse in the autonomic ganglia but pass directly from peripheral receptors to dorsal horn grey matter of the spinal cord.

The autonomic nervous system is divided into two major divisions: the *sympathetic* or thoracolumbar division, and the *parasympathetic,* or craniosacral division (Fig. 9.78). The autonomic nervous system is distinguished by having relays between the central nervous system and the viscera they serve. This takes the form of numerous, variously distributed *peripheral ganglia,* complete with interneurons. This provides control loops close to the visceral organs they control; ganglia are organized and disposed for functional autonomy. There is no such peripheral relay in the somatic nervous system.

The last neurons in the central nervous system which direct efferent impulses to control smooth muscles, cardiac muscle, and glands do not represent a "final common path" such as skeletal motor neurons do. Somatic motor neurons generate final neural commands for execution by skeletal muscles, and they can alter signals up to the last possible moment. They provide final convergent integration for a wide variety of peripheral and central influ-

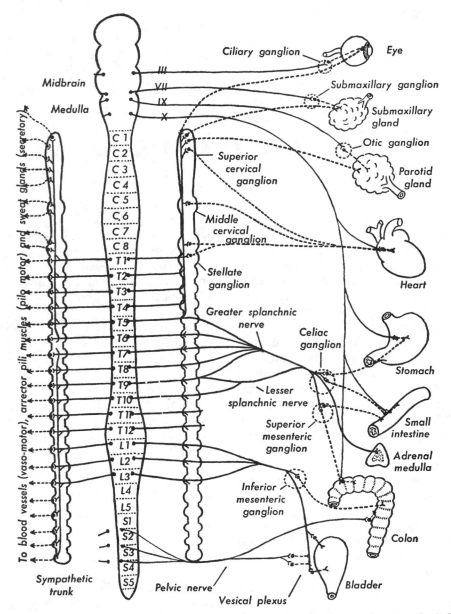

Figure 9.78. Diagram of connections of autonomic or visceromotor system. Craniosacral components are in *black;* thoracolumbar connections are in *color.* Course of postganglionic fibers is shown by *broken lines.* (From Carpenter, 1976.)

ences from sources remote as well as neighboring. Further, somatic motor neurons have a precise and strictly confined sphere of influence, affecting only one muscle or even a single fascicle of muscle fibers within a single muscle. The central nervous system is organized so that we can consciously control skeletal muscle contractions down to single motor units. This contrasts with the autonomic nervous system where voluntary control is rarely achieved and requires training and discipline. Moreover, voluntary visceral control has widespread, unrefinable effects.

The final central neurons to generate visceral commands influence visceral effectors only indirectly. Such commands become widely distributed as they pass through peripheral ganglionic relays. The relays are dominated by locally autonomous visceral activities.

Visceral mechanisms control *respiratory exchange* in lungs and tissues. In strategic locations, visceral systems monitor blood and other tissue concentrations of oxygen, carbon dioxide, and acid-base balance. Visceral systems control *cardiovascular performance* throughout the body, with local need-regulated blood flow; *alimentation,* ingestion, digestion involving gastrointestinal, liver, gallbladder, and pancreatic functions; differential *local metabolism; elimination* of wastes; *water balance,* drinking, renal excretion and bladder function; and *reproduction,* from start to finish. Visceral governance is continuous—awake and asleep—and quite refined at tissue levels.

Central neurons that contribute executive messages to the autonomic system have cell bodies located along an interrupted string of grey matter in brainstem and spinal cord. Interruptions occur in the brainstem at the level of the pons, between midbrain and bulb, throughout all spinal cervical levels, and between lumbar 4 and sacral 1. Cranial midbrain and bulbar visceral efferents contribute to the IIIrd, VIIth, IXth, and Xth nerves. These, and three sacral segmental efferents (S2–4) constitute *parasympathetic outflow.* Thoracolumbar visceral outflow, from T1 to L3, is *sympathetic.* In the left side of Figure 9.78, the "sympathetic trunk" includes numerous ganglionic relays, depicted as bulges. More peripheral ganglia are depicted as *dashed circles,* intermediate between the sympathetic trunk and the viscera. Parasympathetic outflow, as depicted, goes directly to the peripheral ganglia or directly to nerve nets which are disposed in the walls of viscera.

Sympathetic and parasympathetic divisions of the autonomic nervous system differ in morphological, pharmacological, and functional respects. Although they often operate reciprocally or antagonistically, their activities are thoroughly integrated. Hess (1932; 1949; 1956), winner of a Nobel Prize for his research on visceral physiology, emphasized that sympathetic commands enable the organism to mobilize and expend energy. Cardiac acceleration, for example, is accompanied by relaxation of coronary artery walls and contraction of peripheral arterioles, elevating blood perfusion pressures; increased ventilation is fostered by relaxation of bronchial musculature; increased blood volume for use by skeletal muscles is provided by contraction of splanchnic veins; widening of the pupils admits more light to the eye, etc. Such sympathetic contributions enable peak performance by effectors needed during emergencies. Hess called these concerted sympathetic functions *ergotropic* (ergo = energy; tropic = releasing).

Parasympathetic functions, reduction of the pupils, gastrointestinal secretions and peristalsis with relaxation of intestinal sphincters, pyloric and jejunal, slowing of the heart, reduction of arterial blood pressure, etc., contribute to conservation of energy and operation of vegetative functions to restore resources for the organism's general well-being. Hess characterized parasympathetic predominantly

as a *trophotropic* function (trophos = nourishment).

Distribution of neurons and fibers in the *sympathetic* division provides for widely distributed controls. Preganglionic fibers usually project to several sympathetic ganglia. Postganglionic fibers often project to more than one visceral organ. The predominant neurotransmitter, *norepinephrine,* has prolonged, generalized effects. Neuronal reuptake of norepinephrine is slow and incomplete. Large-scale sympathetic discharge is accompanied by a spill-over of norepinephrine into the general circulation.

Sympathetic preganglionic fibers also activate the medulla of the adrenal gland and release a mixture of *epinephrine* and *norepinephrine* into the adrenal veins. These two neurotransmitter-neurohumors add to generalized energy-releasing activities by increasing strength of contraction of skeletal muscles, increasing vasomotor tone, raising blood sugar, increasing metabolic rate, etc.

In contrast, *parasympathetic preganglionic* axons go directly to the vicinity of visceral organs where local peripheral autonomic ganglia, e.g., ciliary, submaxillary, otic ganglia, etc. reside. Or, they go directly to ganglia in the walls of the viscera. *Parasympathetic postganglionic neurons* control visceral effectors, locally, discretely. The predominant neurotransmitter is *acetylcholine* which is rapidly dissociated locally by cholinacetylase. Contrasting autonomic functions are schematized in Figure 9.79. Both sympathetic and parasympathetic postganglionic axonal endings do not have elaborate junctions with smooth muscles and glands, but end with vesicle-filled sacculations.

Visceral Afferent Innervation

Visceral afferents are similar to somatic afferents. Their cell bodies lie in the dorsal root ganglia. The axons are generally larger in diameter than autonomic efferents and they are usually myelinated. Although visceral afferent fibers travel with autonomic efferents, they pass through the autonomic ganglia without relay. They reach the central nervous system via the vagus as well as the pelvic, splanchnic, and other spinal nerves.

Afferent fibers, conveyed along the *thoracolumbar* (sympathetic) division and *sacral* (parasympathetic) division of the autonomic nervous system, after passing through the autonomic ganglia, join somatic afferents by way of *white rami.* They enter the dorsal roots from T1 to L3 or L4, and S2–4. Central terminals of visceral afferents penetrate the dorsal horn and end on interneurons in the vicinity of the lateral horn where visceral efferents take origin (Petras and Cummings, 1972).

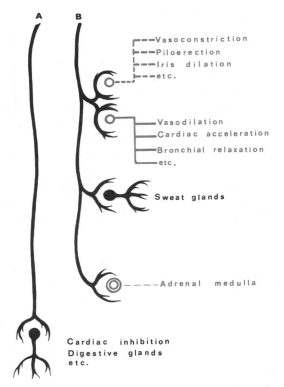

Figure 9.79. Distribution of cholinergic and adrenergic fibers, in parasympathetic *(left)* and sympathetic divisions of the autonomic system. Cholinergic mechanisms are in *black;* adrenergic neurons are in *color.*

Afferent fibers of the *cranial division* are axons whose cell bodies reside in the equivalent of dorsal root ganglia, for the VIIth, IXth, and Xth cranial nerves. Their central processes end in the *nucleus of the solitary tract (NTS) (see Chapter 65, Fig. 9.59).* Ninth cranial nerve afferents project to an intermediate zone of the NTS, the parahypoglossal area, and nucleus ambiguus, in the neighboring medullary reticular formation. These centers are involved, along with the suprameduallary areas in the pons and hypothalamus, in modulating the important *carotid sinus reflex* (Kostreva, 1983). *Carotid body chemoreceptors* project to the medial part of NTS. *Aortic depressor afferents* project to the caudal part of NTS, initiating reflexes that are mediated by the nucleus ambiguus.

NTS receives *gustatory afferents* from the Vth, VIIth, IXth, and Xth cranial nerves. It receives *lung stretch receptor respiratory afferents, aortic baroreceptors, carotid body chemoreceptors, carotid sinus baroreceptors,* and *cardiopulmonary baroreceptors.* NTS also receives collaterals from the vestibular nuclei, fastigial nucleus (a cerebellar roof nucleus), medullary reticular formation, amygdala, and cerebral cortex (Fig. 9.59).

NTS projects to (a) spinal sympathetic outflow,

including that to the adrenal medulla, (b) inhibitory sympathetic loci in the medullary reticular formation, (c) hypothalamic preoptic area (endocrine functions), (d) olfactory grey matter in the basal forebrain, (e) amygdala, (f) diffusely projecting nuclei in thalamus, (g) inferior olive, vermis, and flocculus of cerebellum, and (h) medullary efferents, including dorsal motor nucleus of the vagus, (i) visceral efferents to the heart and lungs, (j) somatic efferents to the larynx and pharynx in nucleus ambiguus, and other bulbar controls.

Afferent fibers coming from the visceral distribution of the pelvic nerves, serving bladder, urethra, and secondary sex organs, stem from cell bodies in the dorsal root ganglia of *sacral segments 2, 3, and 4.* Their reflexes are partly mediated locally and they have ascending projections to medulla, diencephalon, limbic and neocortex.

SYMPATHETIC VISCERAL CONTROLS

Cells of origin of the *sympathetic division* of the autonomic nervous system are located in *lateral horn* spinal grey matter from 8th cervical and 1st thoracic through 3rd lumbar segmental levels. Their axons, almost all myelinated, exit the spinal cord by way of local ventral roots and pass by *white rami* to *sympathetic ganglia.* Sympathetic ganglia are arranged in three groups: (1) *paravertebral;* (2) *collateral;* and (3) *peripheral* (Fig. 9.78).

Preganglionic sympathetic axons divide and pass up and down the sympathetic ganglionic chain, gaining access to ganglia close to their destined visceral influence. In these ganglia, sympathetic preganglionic fibers synapse with ganglionic neurons and interneurons.

Postganglionic axons are almost all unmyelinated. Those innervating smooth muscles and glands in skin and body wall join peripheral nerves by way of grey rami. Many investigators have demonstrated that sympathetic ganglia can mediate visceral reflexes without dependence on central nervous circuits (see Bosnjak and Kampine, 1982). In addition to local reflexes organized by neurons residing in the wall of the viscera, some autonomic neurons and autonomic ganglia effect reflex actions without signaling the central nervous system.

Sympathetic Ganglia and Ganglionic Chain

Sympathetic ganglia closest to spinal sympathetic outflow lie alongside the vertebral bodies, underlying pleura in thoracic regions and peritoneum in abdominal regions. These ganglia are interconnected by fibers in transit and by interneurons linking the ganglionic chain. Ganglia are spread out from their segmental origins from upper cervical to coccygeal levels. The lower four sacral ganglia receive sacral parasympathetic outflow.

CERVICAL SYMPATHETIC GANGLIA

Superior, middle, and *inferior cervical sympathetic ganglia* are relatively large and represent fusion of two or more smaller ganglia each. The superior cervical ganglion, located close to the base of the skull, is largest. It receives preganglionic fibers from upper thoracic spinal segments and sends postganglionic axons to upper cervical peripheral nerves serving blood vessels, glands, and smooth muscles in head and neck. It sends other postganglionic fibers to the *superior cardiac nerves.* The middle cervical ganglion innervates smooth muscles and glands in the distribution of the 5th and 6th cervical nerves, thyroid and parathyroid glands, and the *middle cardiac nerve.* The inferior cervical ganglion is usually fused with the first thoracic ganglion and occasionally with the second as well, forming a large stellate ganglion. It serves visceral control in the distribution of the 7th and 8th cervical nerves, 1st and 2nd thoracic nerves, and the *inferior cardiac nerve.* Postganglionic fibers from the stellate ganglion form plexuses on the subclavian artery and its branches whereby sympathetic innervation is supplied to vertebral, axillary, and brachial arteries.

THORACIC, LUMBAR, AND SACRAL GANGLIA

There are usually 10–12 sympathetic ganglia in chains on each side corresponding to 12 thoracic spinal segments. There are four lumbar and usually four or five sacral ganglia the latter of which may be fused with counterpart ganglia across the midline (Fig. 9.78).

COLLATERAL SYMPATHETIC GANGLIA

These lie in thorax, abdomen, and pelvis and are closely related to the aorta and its branches. The largest is the *celiac* (solar or semilunar) ganglion (see Fig. 9.78), related to the celiac artery. *Superior* and *inferior mesenteric* ganglia lie just caudal to the superior and inferior mesenteric arteries, respectively. *Terminal sympathetic ganglia* lie closer to visceral organs and especially close to rectum and bladder.

Distribution of Preganglionic Sympathetic Fibers

A preganglionic fiber arriving at a sympathetic ganglion may pursue one of three courses: (1) form synapses with cells in that ganglion; (2) traverse the ganglionated chain without interruption to relay in ganglia higher or lower than the level of entry; or (3) relay synaptically in a collateral sympathetic ganglion. A single preganglionic fiber typically contributes synaptic junctions by way of collateral axon branches to from 5–9 ganglia; at every ganglion, preganglionic axons usually serve several postganglionic cells, each of which has a large number of widespread axonal branches, thus broadcasting the effects of sympathetic excitation.

Distribution of Postganglionic Sympathetic Fibers

Sympathetic ganglia contribute postganglionic fibers via grey rami to each peripheral nerve, to serve skin and body wall blood vessels, smooth muscles, and glands. Preganglionic fibers which serve the upper limb arise from thoracic 2–7, reach corresponding ganglia by way of white rami, and pass up and down the chain. Postganglionic fibers exit via grey rami and enter the limb nerves. Thoracic and lumbar preganglionics similarly travel along the sympathetic chain serving several neurons in each of several ganglia. Postganglionic fibers exit via grey rami to join peripheral nerves emerging for thoracic, lumbar, sacral, and coccygeal segments.

Excision of the stellate and upper four thoracic ganglia deprives upper limb, head, and neck of sympathetic innervation and disables most postganglionic sympathetic innervation to the heart. Middle and superior cervical sympathetic ganglia are thereby orphaned from the central nervous system.

PARASYMPATHETIC CRANIOSACRAL VISCERAL CONTROLS

Cells providing central parasympathetic outflow originate from midbrain and medulla, and midsacral spinal segments. Parasympathetic ganglia, as previously noted, lie within or closely neighboring the visceral organs. Preganglionic parasympathetic fibers are myelinated, postganglionics are unmyelinated.

Midbrain Outflow

Cells included in the Edinger-Westphal nucleus of the oculomotor system reside in the midbrain equivalent of the lateral (visceral) motor column of the spinal cord. Preganglionic axons pass through short ciliary nerves to innervate the ciliary muscle for lens accommodation and the iris sphincter for pupillar contraction.

Medullary Outflow

Efferent preganglionic parasympathetic fibers leave the brainstem via the VIIth (facial), IXth (glossopharyngeal), and Xth (vagal) nerves.

1. Parasympathetic vasodilatory and secretory fibers arise from the *superior salivatory nucleus.* They emerge from brainstem along with other

fibers of the VIIth cranial nerve, as the *intermediate nerve*, and pass through the facial canal of the temporal bone. Thereafter, they follow either of two paths: (1) to the submaxillary (submandibular) ganglion to innervate the *submaxillary* and *submandibular salivary glands* and *mucous membranes of mouth*, and (2) to the sphenopalatine ganglion to innervate the *lachrymal gland, mucous membranes of soft palate, nasopharynx, and pharynx*, and send *vasodilator fibers* into the cranium via the middle meningeal artery and its branches.

2. Secretory and vasodilatory fibers of the IXth cranial nerve arise from *inferior salivatory nucleus*. They leave the brainstem with the glossopharyngeal nerve but separate from it to join fibers from the facial nerve and internal carotid plexus which pass as the *superficial petrosal nerve* to the otic ganglion. Postganglionic fibers pass to the *parotid gland*. Some preganglionic fibers bypass the otic ganglion and innervate the *mucous membranes of the tympanic cavity, mastoid air cells, eustachian tube*, and *internal ear*.

3. *The dorsal motor nucleus of the vagus* sends large numbers of preganglionic fibers through several branches of thoracic and abdominal viscera. The vagus innervates all viscera from throat to transverse colon. Vagal ganglia are located within visceral organs. Vagal impulses to the heart have synaptic relay through ganglionic cells in walls of the heart; bronchi through intrinsic bronchial ganglia; esophagus, stomach, and intestine through ganglion cells of *myenteric plexus* of Auerbach and *submucous plexus* of Meissner.

Sacral Outflow

Cells of origin lie along lateral columns of spinal segments *sacral 2, 3, and 4* (Fig. 9.78). They exit the spinal cord by way of corresponding ventral roots. Some preganglionic fibers accompany peripheral nerves to skin and body wall. Others enter the pelvis and form the *pelvic nerve* and *pelvic plexus*. They terminate with synaptic contacts on ganglion cells distributed in walls of the descending colon, rectum, and bladder. They carry inhibitory impulses to *internal anal, vesical*, and *uterine sphincters*, and dilator impulses to *blood vessels of bladder, rectum*, and *genitalia*. By relaxing vasoconstrictor tone, these postganglionics control tumescence of labia and clitoris and penile erection.

AUTONOMIC NERVOUS SYSTEM DEVELOPMENT

The autonomic nervous system (ANS), including adrenal gland medulla, derives from the neural crest which also gives rise to dorsal root ganglia including analogous cranial sensory nerve nuclei, olfactory and taste neuroepithelia, parts of pharynx and jaw, and squamous bones covering the forebrain. Neural crest derivatives form unusually pleomorphic and pleochemic neuronal tissues (Potter et al., 1983). Neurons destined for the ANS make a number of "decisions" during emergence from neuroblasts to mature cells—neural rather than osseous tissue—neuronal rather than glial cells—autonomic rather than sensory functions, and choices of neurotransmitters (Bunge et al., 1978).

In contrast to cells stemming from the neural tube, neural crest derivatives retain a capacity to divide even after phenotypic expression is established, and to switch neurotransmitters depending on environmental conditions (see Fed. Proc. Symposium, 1983). ANS neurotransmitter diversity includes: acetylcholine (ACh), norepinephrine (NE), 5-hydroxytryptamine (5-HT), dopamine (DA), and neuropeptides such as vasoactive intestinal polypeptide (VIP), substance P, somatostatin, enkaphalin, cholecystokinin (CCK), neurotensin, and bombesin.

Enteric Nervous System

Myenteric neurons stemming from the neural crest colonize the intestinal walls before sympathetic and parasympathetic neurons send out processes. Peripheral myenteric plexuses have a disproportion of postganglionic to preganglionic neurons, roughly 10^5:1. Cell division may continue even after phenotypic expression has been established (Gershon et al., 1983). The intestinal microenvironment influences organization of the enteric plexus, e.g., establishing proximodistal direction of peristaltic reflexes from esophagus to rectum.

Enteric reflexes involve adrenergic and cholinergic axo-axonal synapses which are widely distributed in the ANS. Internal gut pressure releases 5-HT (Gershon, 1981; Gershon et al., 1983). Descending inhibition necessary for peristalsis is controlled by somatostatin. Neurons in the wall of intestine, lung, bladder, and some blood vessels use 5-HT, DA, GABA, even ATP as neurotransmitters; some store and release more than one transmitter.

Functions of the Autonomic Nervous System

Autonomic activity controls the functional milieu for cells and tissues throughout the body, making adjustments differentially depending on global as well as local demands, and in support of somatic behavior. Autonomic service is unremitting and continuous. Through integration relating to emotional behavior, the ANS prepares the body for anticipated somatic and visceral requirements. For example, it increases cardiac output, funnels blood flow to safeguard emergency metabolic demands of heart, brain, and skeletal muscles, and in other ways prepares the individual for urgent vigorous activity, by both supporting and anticipating needs. But autonomic actions are not inevitably secondary to somatic functions; they can absolutely dominate—as everyone knows who has experienced an irresistable peristaltic wave.

Most preganglionic and postganglionic nerves are tonically active. Peripheral ganglia, e.g., those in the wall of the gut, are likewise tonically active. Tonic activity enables greater precision of control. The rate of tonic activity depends on dynamics originating in the visceral organs, in peripheral and central reflex activities, and in central nervous system controls engaged in patterns of visceral drives, appetite, and satiety, emotions, and forward planning for visceral housekeeping.

Cholinergic Transmission

Acetylcholine is a principal transmitter for all autonomic ganglia; for the sympathetic division, its actions are nicotinic. At many postganglionic endings of the parasympathetic division ACh actions are muscarinic. ACh is rapidly hydrolyzed by acetylcholinesterase. This enzyme can be inhibited by physostigmine, neostigmine, diisopropylfluorophosphate (DFP), and tetraethyl pyrophosphate (TEPP). The latter two combine irreversibly with the enzyme, so ACh must be newly synthesized for renewed activity.

Catecholamine Transmission

Epinephrine and norepinephrine are examples of catecholamines synthesized in the body from tyrosine (Fig. 9.80). Other examples include L-tyrosine, L-dopa, dopamine, and isoproterenol. These compounds are synthesized by a series of enzymes which catelyze reactions according to numbered arrows in Figure 9.80:

1. Tyrosine hydroxylase
2. L-dopa decarboxylase
3. Dopamine β-hydroxylase
4. Phenylethanolamine *N*-methyltransferase.

The synthetic pathways are known (Iverson, 1967; Blaschko, 1973; Sharman, 1973). Different actions by the same transmitter are accounted for by different postganglionic or effector cell receptors.

Vesicles in adrenergic nerve endings which store epinephrine also contain the enzyme β-hydroxylase which converts dopamine to norepinephrine. Much of the norepinephrine released is taken up into the same nerve endings and vesicles again. Reuptake is

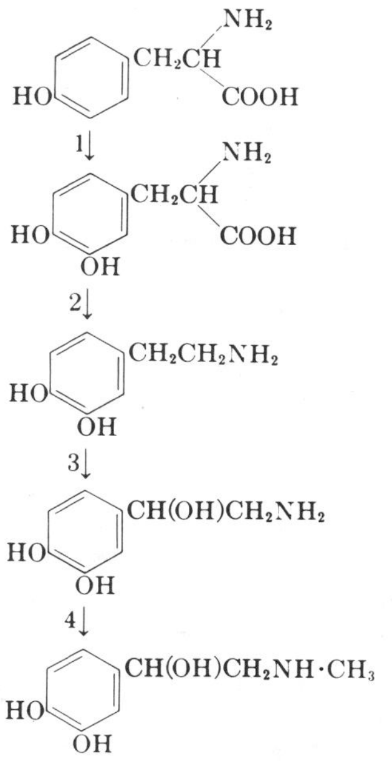

Figure 9.80. Biosynthesis of catecholamines from L-tyrosine (i.e., hydroxyphenylalanine). The enzyme for the fourth step, phenylethanolamine *N*-methyltransferase, is present in adrenal medulla and chromaffin tissues, but not in typical postganglionic adrenergic endings (From Iversen, 1967.)

the principal mechanism for norepinephrine inactivation. Primary amines can be oxidized by the enzyme *monoamine oxidase*. Circulating epinephrine and norepinephrine are metabolized by the methylating enzyme, *O*-methyltransferase (Fig. 9.81).

Examples of Sympathetic Actions

When the whole sympathetic system is activated, combined sympathoadrenal responses resemble severe stress or anger. The individual looks and feels angry or frightened and exhibits cardiac acceleration, elevated blood pressure, elevated blood glucose, piloerection, sweating, pupillary dilation and other responses which prepare the individual for "fight" or "flight." Generalized venous and arteriolar constriction reduces the pool of blood in veins and raises blood pressure. Blood vessels to skeletal muscles are dilated. Mechanical and secretory activities of the stomach and intestines are inhibited, and intestinal sphincters are constricted. Nonetheless, visceral representations have enough discretionary control to obtain local, even subtle, reflex actions.

Stimulation of the superior cervical sympathetic ganglion results in pupillary dilatation, exophthalmos (eye protrusion), and sweating on the ipsilateral side of the head. This combination of signs is known as *Horner's syndrome*. Stimulation of the stellate (inferior cervical) sympathetic ganglion

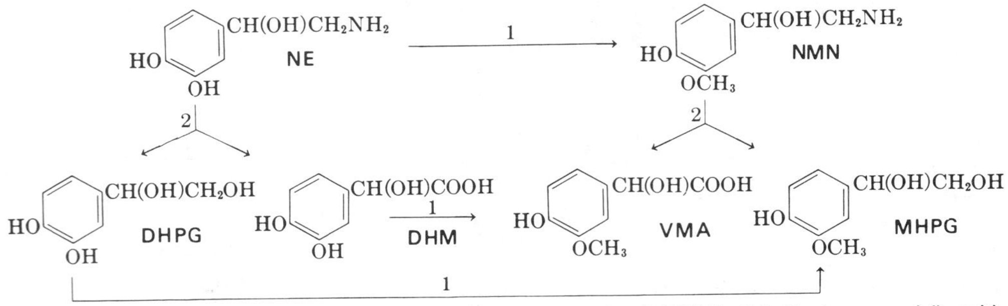

Figure 9.81. Pathways by which catecholamines are inactivated by catechol-*O*-methyltransferase (*1*) or by monoamine oxidase (*2*). *NE* is norepinephrine; *NMN* is normetanephrine; *DHPG* is 3,4-dihydroxyphen-ylethylglycol; *DHM* is 3,4-dihydroxymandelic acid; *VMA* is vanillymandelic acid; *MHPG* is 3-methoxy-4-hydroxymandelic acid. (From Iversen, 1967.)

results in increased rate of the cardiac pacemaker, increased rate of intracardiac conduction, increased cardiac force of contraction and increased stroke volume, bronchial dilatation, and moderate pulmonary vasocontriction.

Activation of sympathetic fibers innervating liver elicits vasoconstriction and inhibits gallbladder contractions. Activation of sympathetic fibers to the spleen causes splenic contraction, discharging red and white blood cells into the general circulation. Thoracic sympathetic release of epinephrine and norepinephrine from the medulla of the adrenal gland produces glycogenolysis and liberation of glucose, augments glucagon output, inhibits insulin secretion, increases metabolic rate, and reduces blood clotting time.

Sympathetic activation of sweat glands induces secretion by local release of ACh (Fig. 9.79). Sweating takes place in small bursts of activity at about 6–7 minute. This reflects a widespread rhythmic discharge of central origin; consequently widespread skin areas show synchronous sweating bursts. Apocrine sweat glands in the axilla are especially activated by mental stress.

Piloerection, associated with cold, fear, and anger, is adrenergic. Sympathetic activation of the bladder elicits relaxation of the bladder wall and internal sphincter contraction. Most pelvic structures undergo vasoconstriction, but vasodilation occurs in the external sex organs.

Denervation and localized central lesions can eliminate tonic sympathetic activity to visceral organs. Although innervation of abdominal viscera is orderly, elimination of a few sympathetic ganglia does not ordinarily induce clinical signs, presumably because several ganglia are served by each spinal level.

Examples of Parasympathetic Actions

Third cranial nerve parasympathetic innervation provides tonic and phasic pupillary constriction. Denervation leaves a dilated pupil which dilates more with sympathetic activation. Stimulation of the lachrymal branches of the VIIth nerve causes tear production. Activation of the VIIth and IXth cranial nerves yields salivation. Parasympathetic innervation of the heart inhibits cardiac pacemakers, decreases the rate of depolarization, slows cardiac conduction, and interferes with transmission at the atrioventricular node, resulting in reduction of blood pressure and decrease of cardiac output. Accompanying bronchiolar constriction and increased secretion, with increased viscosity of mucous, affects ventilation.

The gastrointestinal tract responds with increased peristaltic activity, relaxed intestinal sphincters, reduced gastric emptying time, and increased secretion of gastrin and other digestive juices. Parasympathetic activation also elicits contraction of gallbladder, secretion of pancreatic digestive juices, and release of insulin. Parasympathetic discharge from sacral levels induces relaxation of sphincter muscles during urination and defecation and contributes to clitoral and labial tumescence and penile erection.

Effects of Spinal Cord Transection on Autonomic Functions

The absence of two-way central connections for temperature regulation results in *poikilothermia below the level of cord transection*. Visceral reflexes below the transection tend to be hyperactive. Orthostatic hypotension is exaggerated, and cold immersion of the foot leads to reflex hypertension. Various stimuli, principally bladder and rectal in origin, and most marked in cases of high spinal transection, can initiate paroxysmal bouts of arterial hypertension. Bladder distention usually elicits excessive sweating, including sweating above the level of sensory loss. Such mass action responses from below the level of transection can initiate headache, facial flushing, and pilomotor hyperactivity. Characteristic signs of this kind can be recognized by the patient and utilized as an indication that the bladder or rectum or some other visceral organ needs attention.

Spinally transected individuals, male and female, may be capable of reproduction even though satisfactions are denied for want of sensory and motor communication to higher centers. Spinalized man can remain sexually potent, capable of ejaculation, and fertile, if sympathetic outflow from T6 to L3 is preserved below the level of transection. Even if potency or ejaculation is precluded by nerve or cord damage below the level of transection, fertility may still be realized by means of artificial insemination.

HIGHER LEVELS OF VISCERAL CONTROL

Autonomic controls are a major function of the brainstem. Medullary loci are referred to as cardiovascular vasopressor or vasodepressor "centers"; centers for respiratory inspiration and expiration; a "pneumotactic center"; etc. The terminology, however, is misleading. These are only lower-level neuronal circuits within more extensive systems.

Systematic stimulation of the hypothalamus by Ranson (1936; 1937) and by Hess (1932; 1949) revealed functional representations of autonomic functions relating to *temperature control, eating, drinking, reproduction, and sleep* (not necessarily in that order). Later it became clear that the limbic cortex has elaborate representations and rerepresentations of visceral functions. As with sensory and motor systems, autonomic representations become more varied in pattern and organization of options for functional control, more operationally comprehensive and holistic in compass—going from peripheral to spinal to brainstem to diencephalic to limbic representations. Not only are visceral functions multiply distributed through parallel processing and specialized recombinatorial mappings, but they are ultimately related to conditionable neocortical global mappings for motivations and anticipated behaviors.

Whereas brainstem representations emphasize reflex controls, especially controls that achieve integration essential for survival ("vital centers") the hypothalamus reflects representations that relate to more interdependent, global, and longer time-scale control combinations. Both levels are goal-oriented, but hypothalamic goals are more inclusively integrated. For example, the hypothalamus combines control of thermoregulation and nutritional intake with neuroendocrine and bodily activities for the individual as a whole, and these controls are integrated by the hypothalamus over extended time periods.

Visceral foundations for human fulfillment provide even more encompassing options. Frontal lobe and limbic mechanisms combine long-range planning with biological and cognitive development. This incorporates personal motivations with inherited neuroendocrine schedules for visceral and psychological organization of mating, bearing, and rearing of offspring.

MacLean (1958) introduced an appealing generalization that the anteromedial region of the limbic system, especially the septum, represents a cluster of functions relating to *survival of the species.* The anterolateral part of the temporal lobe, represents a cluster of functions relating to *survival of the individual.* Stimulation of either of these regions can elicit grooming, searching for food and defensive actions. But in the septal region, these functions relate to finding and grooming a mate, and breeding, feeding and protecting a mate and offspring. In the temporal lobe, these same responses are directed toward serving individualistic purposes. No one yet knows enough about the organization of higher levels of cerebral representation to be able to elicit or simulate elaborate emotional experiences in their fullest complexities, but careful analysis of visceral systems capabilities and various clinical disorders associated with limbic and frontal lobe dysfunctions suggests that biologically appropriate representations of this kind relate to higher nervous functions. This evidence provides useful insight into the field of psychosomatic medicine (MacLean, 1949; 1990).

REGULATION OF VISCERAL MECHANISMS

We have considered the control of visceral systems, but *regulation* implies more. Regulation involves maintenance of relative equilibria, what Cannon called *homeostasis*, in respect to physiological variables that must be separately maintained within certain strict ranges of tolerance (1932). Homeostasis is complex because of the interdependence of these primary physiological requirements. It is understandably difficult to satisfy requirements for individual survival in a harsh, dynamic environment.

Our inheritance has rehearsed these mechanisms well. Throughout evolution, none of our direct ancestors died before reaching reproductive age. All of them reproduced something that possessed survival and reproductive capabilities. Our survivorship goes all the way back to some fateful catalyst in a primeval sea.

All successful multicellular organisms require satisfactory performance by a number of homeostatic mechanisms, all of which must be successfully integrated in service of the organism as a whole. Evolutionary continuity is evidence that homeostasis was tolerably successful at all levels of physiological organization: molecular, cellular, tissue, organ system, individual, even interindividual. Each level regulates relatively stable equilibria on which successively higher levels of evolutionary organization predicated additional levels of complexity. Gradual ascent of the evolutionary ladder achieved increasing degrees of freedom for individual action. Many higher evolutionary stages are accomplished by finesse of neuroendocrine integration in favor of consciously integrated control, e.g., diet and exercise. Each level of integrative achievement depends upon what Bronowski called *stratified stabilities* (1973).

Physiological Regulation

For any visceral regulation, three questions are pertinent: (1) What is the process? (2) What is its rate? (3) What initiates the process and governs the kinetics? These questions can be asked about functions of molecular components, cells, and complex functions like digestion, circulation, respiration, endocrine and central nervous system functions. The physiologist addresses a given level of stratified stability in terms of *its* regulation, assuming that other levels regulate themselves.

All living systems, at any level of complexity, are open systems, i.e., they depend on energy intake, temperature regulation, waste disposal, etc. Regulation depends on a *detector* (interoceptor or exteroceptor) by which to gauge a variable or some derivative, and negative feedback from the detector by which to regulate the system. Feedback, which closes the loop, is the sine qua non of regulation. Feedback may lead to *adaptation* which is more complicated than regulation. An example of adaptation is provided by the vestibulo-ocular reflex (VOR; see Chapters 63, Vestibular Function, and 64, Vision). The VOR does not depend, strictly speaking, on feedback, but on outside signals that indicate whether or not the VOR has functioned appropriately. In the VOR, "error messages" are detected in the visual system. These feedback messages call for the generation of variations in the regulatory pathway. From among newly generated variations in the VOR three-neuron pathway, favorable variations may be "selected" in the manner suggested by Edelman (1987).

In maintaining physiological systems within tolerable ranges the endocrine system is an important contributor. Hormones not only have specific feedback actions on certain control systems, they also have metabolic and other effects, including direct effects on neuronal levels of activity. In the absence of thyroxine, for example, the nervous system manifests pathognomonic abnormalities; sex steroids exert feedback control on specific hypothalamic cell receptors, and manifest effects on behavior.

Mechanisms of visceral control can be roughly schematized by three statements: (1) The nervous system exerts control over relatively autonomous visceral effectors via direct and indirect neuronal actions, and it exerts slower, more prolonged adjustments via neuroendocrine actions. (2) The nervous system gathers information quickly via neuronal visceral receptors and analytic systems, and more slowly via neuroendocrine feedback. (3) The nervous system adjusts to bodily and environmental demands on visceral integrity by altering neuroendocrine tides and by adapting neuronal circuitry and synaptic organization throughout different levels of the neuraxis, presumably by means of neuronal group selection (Edelman, 1987).

CIRCULATION

Quantities regulated are (1) arterial blood pressure, monitored by stretch receptors in walls of major arteries, and (2) adequacy of blood supply to tissues, monitored as CO_2 partial pressure, and pH. Non-neural mechanisms alter smooth muscle tone in small blood vessels in relation to local carbon dioxide. General arterial blood pressure is maintained by mechanisms that control: (1) stroke volume and frequency of heart beat (cardiac output); (2) degree of relaxation of arterioles controlling outflow of blood (peripheral resistance); (3) tension in walls of large arteries and veins; and (4) mechanical forces such as bodily acceleration, direction and force of gravity, tension in skeletal muscles, tissue turgor, and intrathoracic pressure, all of which affect return of blood to the heart. Blood gas partial pressure reflexes, with sensors in the medulla and carotid and aortic bodies, affect circulation as well as pulmonary ventilation.

Electrical stimulation of limbic cortex, parts of hypothalamus and medullary reticular formation elicits circulatory changes. Hypothalamic regulation is active in control of circulatory changes during exposure to heat and cold and during exercise (Folkow and Rubinstein, 1965). Limbic cortex and hypothalamus are active in circulatory control during emotional expression. In effect, circulation is controlled by a complex of integrative visceral mechanisms distributed throughout the neuraxis, with final global integration occurring in the medullary reticular formation (Korner, 1971).

RESPIRATION

Central and peripheral detectors of partial pressures of carbon dioxide and oxygen introduce changes in respiratory frequency and depth which control respiratory minute volume (von Euler et al., 1970). Pulmonary stretch receptors send messages by way of the vagus and reflexly control excursion of the chest wall and contraction of the diaphragm. A combination of gas and mechanical monitoring is integrated in the medullary reticular formation, mainly activating inspiratory neurons. Respiratory regulation is not only reflexive and autonomic but controlled in relation to a variety of behavioral patterns, gesture vocalization, singing, language, and securing truncal stability as in pushing and lifting, and abdominal pressure as in defecation and childbirth, when respiratory muscles serve nonrespiratory functions.

WATER BALANCE

Hypothalamic mechanisms control rates of water ingestion and water loss through the kidneys. Both hypothalamus and pituitary are involved in a disorder known as *diabetes insipidus*. The patient drinks copiously to keep up with copious excretion of very dilute (i.e., "insipid") urine (Fischer et al., 1938). Antidiuretic hormone secretion by the posterior lobe of the pituitary is controlled by neurons in the supraoptic hypothalamic nucleus. Brooks and associates (Koizumi et al., 1964) recorded single neuron activity in the supraoptic nucleus in response to changes in extracellular osmolarity. Animals in which the hypothalamus is removed except for the supraopticohypophyseal system are spared diabetes insipidus (Woods and Bard, 1960). Control involves supraoptic neurons responding to water concentration and secreting antidiuretic hormone (Dreifuss et al., 1971). Central venous pressure acting through receptors in major blood vessels, particularly the left atrium, contributes to water regulation via hypothalamic control of antidiuretic hormone (Share, 1968).

WATER INTAKE

Control of water intake involves receptors in mouth and throat responding to a range of moistness-dryness; detectors of water concentration in various body fluids; pressure-sensitive neurons in the great vessels involved in secretion of antidiuretic hormone (as above); and kidney release of renin, an enzyme that produces angiotensin II, which acts through the hypothalamus to induce water drinking.

Lesions in lateral hypothalamus abolish drinking except that generated by mouth dryness (Epstein and Teitelbaum, 1964). Among priorities for visceral needs, water is more vital than temperature regulation, energy balance, or reproduction, but neither circulation nor respiration can be greatly modified to preserve water balance.

ENERGY BALANCE

Energy is gained by food intake. Energy is lost as heat or physical work, or retained in the body as protein during growth and recovery from undernutrition, as carbohydrate during recovery from a brief fast, and as fat in potentially almost limitless quantities. Stimulation of the lateral hypothalamus induces feeding behavior and can be used as a surrogate reward in conditioning experiments. It makes sense then that medial lesions in the hypothalamus induce hyperphagia and obesity (Tepperman et al., 1941) and lateral lesions induce failure of feeding (Anand and Brobeck, 1951). Damage to the hypothalamus following basal brain injury may result in misregulated hyperthermia. Removal of a small patch of limbic cortex in the posterior orbital surface, the only cortical representation of the vagus nerve, induces general hyperactivity balanced by increased food intake.

The normal human body temperature recorded orally varies from 35°C to 37.8°C during the day with an early morning minimum and an early evening maximum. Heat storage is a function of thermal capacity, which varies with body mass; heat loss varies with surface area. Stability of body temperature depends on body size. In a baby, with a high metabolic rate and relatively great surface/mass ratio, rectal temperature may vary through a range twice that of an adult. In an infant's smaller body, heat "stored" is a relatively smaller fraction of metabolic regulation. In large persons, stored heat has "inertia" which tends to dampen thermal oscillations.

At basal conditions, with no external work being done, all metabolic energy appears as heat and amounts to about 1 kcal/kg body weight/hour. Since body composition is mostly water with a specific heat of slightly less than 1, body temperature will rise about 1°C/hour if no heat is lost externally. With strenuous exercise more than three-quarters of increased metabolism appears as body heat, the remainder as a combination of work and dissipated heat. Strenuous exercise can elevate body temperature to 40.0°C. Fever increases metabolic rate about 13% for each degree (C) rise in mean body temperature and needs to be compensated by increased food intake. Reducing body temperature below about 23°C may result in loss of heat-regulation by depression of the neural controls. This is near the low temperature lethal limit. The upper temperature lethal limit is probably about 43°C. Death in either event is due to cardiac failure.

HEAT LOSS

Body heat is lost through (1) radiation, conduction, and surface convection; (2) evaporating sweat and insensible perspiration; (3) warming and humidifying inspired air; and (4) voiding urine and feces. The first two are subject to already considered physiological controls; animals that pant exhibit physiological control of the third. Roughly, for somebody doing light work, with an overall energy exchange of about 3000 kcal/day, radiation is responsible for half of the total heat loss; convection for 15%; evaporation for 30%, and loss with excreta for 2%.

Increased cardiac output allows rapid blood flow through dilated cutaneous vessels, promoting transport of heat to the body surface and increased heat loss by radiation. At ambient temperatures of 34°C, blood circulating through the skin may approach 12% of cardiac output, roughly a tenfold increase. Finger pads have arteriovenous shunts that permit rapid blood flow and augmented heat loss. During heat exposure, circulating blood volume may increase 10%, diluted by fluid from skin, muscles, liver, and other tissues, and blood cells from splenic contraction. Cold exposure brings about a reversal of these changes. Cutaneous vasoconstriction begins at skin temperature of about 15°C. Cold exposure is followed by marked diuresis and rapid lowering of circulating blood volume.

Convection is promoted by movement of air across body surfaces. For any given air temperature heat losses from a naked body increase with the square of wind velocity up to 60 mph; beyond this there is little further increase because of limitations on heat movement through tissues. Chill factor in cold weather winds can be appreciated by the rapidity of convection heat loss, with the wind velocity value being squared.

EVAPORATION OF WATER

The closer environmental temperatures approach that of blood the less heat can be lost by radiation. At about 38°C, of course, the body begins gaining heat by radiation from the environment. Heat can still be dissipated, nevertheless, by evaporation of water. One ml of water absorbs 0.58 kcal during evaporation. At room temperature, with insensible perspiration, heat loss by evaporation from lungs and skin amounts to about one-fourth of basal heat production, about 17 kcal/hour. Two-thirds of this loss is through insensible perspiration; the remainder by respiratory heat exchange. At higher temperatures, the proportion of heat loss by evaporation increases markedly, mainly from the skin, so that at ambient temperatures above 35°C evaporation accounts for nearly all heat loss.

The rate of evaporation is limited, of course, by the relative humidity. Sweat drips instead of evaporating when the air is humid. In dry air a normal body temperature can be maintained in an ambient temperature of over 100°C, but in a humid atmosphere, a temperature of 50°C causes the body temperature to rise.

Sweat is a weak solution of sodium chloride in water, together with urea and small quantities of potassium and other electrolytes and lactic acid. It has a specific gravity of 1.002 and a pH that varies from 4.2–7.5. The concentration of NaCl varies from 50–100 mEq/liter. Profuse sweating loses electrolytes at higher values, hence excessive sweating can lead to serious water and salt deprivation. If water is replaced without salt, cramps of muscles of the limbs and abdominal wall ensue. This can be prevented or relieved by administration of NaCl tablets and by recovery of electrolytes as found naturally in most foods.

Sweat glands, numbering over 2.5 million, are cholinergic although the nerves are postganglionic fibers of the (sympathetic) division of the autonomic nervous system. Sweating is controlled at many levels throughout the neuraxis. At the beginning of exercise, sweat is initiated by anticipatory activation of limbic cortex before there is any appreciable heat load. Spinal centers exert segmental control of sweating; a quadriplegic or paraplegic person sweats reflexly in parts of the body that are innervated below the level of transection. Sweat is secreted rather than filtered and can produce pressures of 250 mm Hg or more. The rate of sweating can exceed 1.5 liter/hour, representing an evaporative heat loss of 900 kcal/hour.

HEAT PRODUCTION

At environmental temperatures between 28°C and 31°C, basal heat production in a naked person can be dissipated to the environment by radiation, convection, and insensible vaporization. If the temperature goes below 28°C, loss by radiation and convection increases, and skin and mean body temperatures fall. Increased heat production, mainly from increased tension in skeletal muscles, begins even before shivering starts. At ambient temperatures of about 23°C, shivering begins. With intense shivering, heat production can increase to three times the basal rate. Shivering serves an unconscious, automatic function of temperature regulation, executed by the somatic division of the nervous system.

CENTRAL MECHANISMS OF TEMPERATURE REGULATION

Integration of visceral mechanisms that control heat loss and somatic mechanisms that govern heat production, posture, and shivering behavior is controlled by the hypothalamus, with finer control added by limbic cortex. As noted, hypothalamic injury can lead to temperature misregulation. Rostral lesions may lead to hyperthermia, with fever leading to death. This is apparently a "release" phenomenon, i.e., release of heat production from a suppressor mechanism in the anterior hypothalamus. Barbiturates can be used to decrease activity of fever-producing mechanisms. More caudal hypothalamic lesions, extending laterally, lead to hypothermia if the environment is cool.

Experimental analysis of hypothalamic temperature regulation by implanting stimulating electrodes and thermodes, and by analyzing neural responses to changes in ambient and central temperatures has led to the following general conclusions. The body does not respond to a change in skin temperature if there is no central change. Likewise, it does not respond to a central change in the absence of pheripheral changes. Central integration therefore involves integration of some functions or derivations of temperatures in both locations.

Hardy (1965) and Hammell (1965) established that exercise leads to a change in hypothalamic set point that originates from neural adjustments that accompany exercise. The set point is also adjusted in relation to different levels of sleep (Glotzbach and Heller, 1976). Set point changes may account for fevers due to infections (Eisenman, 1969). Toxins known as *pyrogens* are liberated by bacteria and by white blood cells (Moore et al., 1970; Dinarello and Wolff, 1978). A chemical mediary of pyrogens may be prostaglandins (Vane, 1971). Prostaglandins injected directly into the hypothalamus induce fever. Aspirin, which has pronounced antipyretic action, inhibits the synthesis of prostaglandins.

ENDOCRINE GLANDS IN THERMOREGULATION

Thyroid and adrenal glands have significant roles in thermoregulation. Rats exposed to low temperature (7–12°C) show hyperplasia of the thyroid with metabolic rates increasing as much as 16%. Thyroidectomized rats show little rise in metabolic rate with similar cold exposure. Thyrotropic hormone secreted by the anterior pituitary on hormonal command from the hypothalamic thyrotropin-releasing factor is responsible for thyroid responses to hypothermia.

HYPOTHALAMIC-ENDOCRINE INTEGRATION

We have seen how the hypothalamus integrates visceral mechanisms through neuronal connections. We now turn to evidence of hypothalamic influences on neuroendocrine regulation. Supraoptic and para-

ventricular nuclei (Fig. 9.82) each contain cells that synthesize either *oxytocin* or *vasopressin* (antidiuretic hormone), nonapeptides produced exclusively in these nuclei. Each of the hormones is found in separate neurons (Swanson and Swachenko, 1983). Descending axons deliver oxytocin and vasopressin directly to the bloodstream from neurohypophyseal axonal terminals. Within both hypothalamic nuclei, there are many somatosomatic gap junctions and there is conspicuous electrotonic coupling among the neurons. Impulse propagation in the predominantly unmyelinated, small-diameter hypothalamoneurohypophyseal fibers causes hormone release which is calcium-dependent (Lincoln and Wakerly, 1975).

In contrast to the supraoptic nucleus which apparently projects solely to the neurohypophysis, certain axons from the paraventricular nucleus also project to cells in the median eminence which, in turn, secrete into the portal circulation, delivering secondary signals to the anterior pituitary. The paraventricular nucleus also sends certain axons directly to autonomic nervous system (ANS) cells in the brainstem and spinal cord. Neurons in the paraventricular nucleus can thereby influence hormonal release from anterior as well as posterior pituitary and activate preganglionic sympathetic and parasympathetic neurons (Swanson and Swachenko, 1983). Individual paraventricular neurons contrib-

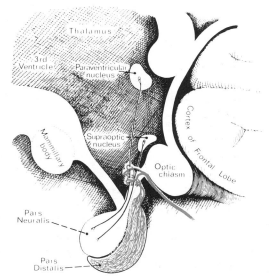

Figure 9.82. Schematic representation of key relations between hypothalamus and hypophysis (pituitary). The hypthalamohypophyseal tract carries neurosecretory granules between the paraventricular and supraoptic nuclei and the neurohypophysis (posterior lobe). The figure also depicts the hypothalamic nuclei, mostly in the region of the median eminence, which send axon terminals to discharge releasing factors into the blood vessels of the portal circulatory path to the anterior hypophysis. (From Noback, 1967.)

ute separately to each of these destinations, and are segregated within the nucleus, indicating that there may be synchronized autonomic and endocrine control, as implied by electrotonic gap junction linkages, and including more specific neuronal control.

Oxytocin is liberated reflexly following distension of the uterus and vagina in childbirth and the vagina during sexual intercourse, and during stimulation of the nipples as in suckling. Oxytocin causes the uterus to secrete and contract, and breast secretory epithelium to secrete, and the contractile ducts to "let down" milk. Late in pregnancy, the uterus becomes increasingly sensitive to oxytocin. The influence of oxytocin on the sensitized uterus ensures strong rhythmical contractions and easier parturition. Oxytocin effects on the breasts ensure postpartum lactation. Repeated suckling induces further oxytocin secretion and repeated uterine contractions. Both nuclei are involved in the release of oxytocin from the neurohypophysis in response to dilatation of the birth canal and suckling.

Vasopressin, an antidiuretic hormone, is released as a result of an increase in plasma osmotic pressure from 290 mosmol/liter by as little as 2%. This can be induced by injecting hypertonic saline solutions into the carotid artery which is followed by an increase in firing rate of supraoptic neurons and an increase in plasma vasopressin (Koizumi et al., 1964; Koizumi and Yamashita, 1978). The paraventricular nucleus responds to baroreceptor activity from aortic, carotid sinus, and renal nerves and contributes to cardiovascular regulation.

MECHANISMS OF REINFORCEMENT

Visceral control mechanisms are organized to provide a continuing, relatively satisfactory internal milieu. But organisms, especially higher animals, are engaged in additional pursuits. Particularly for humans, there are many activities which depend on internal satisfactions being met which cannot be directly or simply attributed to appetites and other simple visceral motivations. Some behaviors relate to long-range goals that may be attenuated and postponed for years, as in studying medicine. How can brain circuits establish delayed, higher purpose satisfactions as compared with immediate visceral satisfactions? This requires consideration of mechanisms that underlie learning, memory, and motivation.

Reinforcement subsumes processes involved in what are commonly called *reward* and *punishment*, experiences liked and disliked. Reinforcement yields behaviors directed toward *(approach)* or away from objects and experiences *(avoidance)*. The term reinforcement is taken from more elementary neurophysiological phenomena concerned with strengthening a unitary response to one stimulus by concurrent application of another stimulus. Here, reinforcement refers to whatever underlies the shaping of perception, memory, judgment, or behavior, in accordance with experience, as in learning.

Pavlovian (type I) conditioning involves pairing somatic or visceral stimuli with somatic or visceral responses so that a response is more likely to recur to a stimulus that had not previously been effective. Reinforcement, e.g., food for a hungry animal, is one procedure by which conditioning (new or strengthened "conditional" connections) can be established.

Type II, operant conditioning, usually associated with Konorski (1948; 1967) and, more recently, with Skinner (1966), involves reinforcement that is applied after a desired or desirable behavior has been emitted by the animal. In the Pavlovian case, the experimenter controls when a conditional stimulus (CS) will be presented; in the Konorskian case, the animal controls the desired behavior which is then reinforced.

Reinforcement, both positive (rewarding) and negative (punishing) has been a principal means for shaping behavior since time immemorial, but until 1954, it was understood only in terms of external manipulations. In that year, independent discoveries by Olds and Milner, and by Delgado, Roberts and Miller, located central nervous regions that represent positive and negative reinforcement, respectively. In general, negative reinforcement can be elicited from lateral midbrain and posterolateral diencephalon extending into ventrolateral thalamus and ventral thalamus. Positive reinforcement is more widespread, including the whole of the limbic system and most of the median forebrain bundle, sweeping from septum through medial hypothalamus into medial midbrain reticular formation and periaqueductal grey. There is no question that stimulation to certain areas yields negative reinforcement and that similar stimulation in other areas yields positive reinforcement. The brain volume of the positive reinforcement system, fortunately, is relatively much larger, roughly by tenfold (Livingston, 1967; 1978).

Mechanisms for central reinforcement relate closely to systems involved in the experience of *feeling states and moods;* indeed, they may consist of the same or overlapping circuits. They are similarly closely related to mechanisms concerned with motivation. Moreover, it is evident that central stimulation in areas of positive or negative reinforcement can be used to modify appetitive and motivational states. The close relationship—anatomical, physiological, and subjective—between drive and reinforcement has been firmly established.

To survive, an organism must be able to modify its behavior so that its basic needs can be met despite changes taking place in a dynamic environment. The nervous system is organized so that behavior is not only *goal-directed,* but it is guided by the *consequences* of both spontaneous and responsive behavior. Actions are initiated by internal feeling states and drives. They are maintained, altered, or discontinued in accordance with success or failure in fulfilling satisfactions sought. Central reinforcement and drive mechanisms, affected by the consequences of behavior, modify subsequent behavior. The nervous system is built for action. Systems for satisfaction and dissatisfaction are built into the circuits of brainstem, diencephalon, and limbic systems. Perceptual processes, memory, learning, and cognitive processes are all directed toward biologically improved internal satisfactions.

In the centralmost portions of the forebrain, neighboring the ventricular passages, crowned on each side by a ring of cortical and subcortical structures surrounding the hylus of the hemispheres, are organized systems involved in feeling, mood, appetite, satiety, approach, and avoidance. Lesions in alimentary appetitive loci reduce drive for food; lesions in food aversion loci yield inadequacy of the restraint that is normally exerted by satiety. The surrounding limbic system is less imperious, less compelling, and more modifiable through experience than are the underlying core mechanisms. In the limbic system, there seems to be more flexibility for the exercise of a range of judgments as to whether an ongoing event is biologically significant, and whether it should be approached or avoided. It is important to note that the limbic system can modulate or control the core system.

Out beyond the limbic system is phylogenetically newer *neocortex.* Neocortex appears to be even less imperious and perhaps entirely neutral with respect to reward or punishment, approach or avoidance. Neocortex is apparently also inessential for the exercise of judgments regarding biological significance. Stimulation of neocortex does not give rise to bar-pressing for central reward, for satisfaction of local neuronal activation.

Positive and negative reinforcement mechanisms exist without obvious biological advantage beyond the consequences of activation of particular primitive brain regions. Reinforcement is involved directly when there is actual biological gain, such as food for a hungry animal. Yet it is not essential that there be an obvious biological gain for reinforcement to be manifested through approach or avoidance behavior, including reinforcement induced by direct central brain stimulation. Behavior associated with reinforcement can also exist without any obvious drive reduction.

There is an interface between innate, reflexive behavior and learned behavior, and between unconscious and conscious processes, which cannot be defined in static terms. That interface appears to involve an adaptive, dynamic boundary. For behavior to be learned, for improvement in biological directedness and goal satisfaction, the intervening steps do not need to be consciously experienced. And, as William James emphasized, even a consciously learned response can become so habitual that it submerges to a nonconscious level and thereafter is carried out quite automatically by the nervous system, more or less on the same plane as genetically endowed behavior.

CHAPTER 70

Segmental Control of Skeletal Muscle

ORGANIZATION OF SPINAL SEGMENTS

Spinal control of skeletal muscles is ultimately expressed by an uninterrupted column of ventral horn motoneurons. Their dendrites arborize widely in ventral grey matter. Their axons leave the ventral surface of the cord on each side in a continuous longitudinal stream of *ventral roots*. These and the corresponding axons of dorsal root ganglion fibers, which lie on the dorsolateral aspect of the spinal cord, gather laterally to form a succession of *spinal nerves*. Intervals between successive intervertebral foramina divide the spinal cord into a series of *spinal segments* which vary in length. Spinal segments innervate a succession of particular derivatives of embryonic dermatomes and myotomes.

Injury to the Spinal Cord

Spinal cord injury occurs frequently and the effects are usually devastating. There are more than 200,000 people in the United States who are paralyzed as a consequence of spinal cord injury. They are mostly young people, most of them victims of automobile or motorcycle accidents. When spinal injury occurs below the second lumbar vertebra, it is likely to damage spinal nerves in the cauda equina rather than the spinal cord. In this event, losses of sensory and motor functions are apt to be uneven and incomplete. But peripheral nerve recovery is more promising than is recovery from spinal cord damage. Spinal injury at higher levels often results in a complete transection with an abrupt level of interruption of sensory and motor functions. If good care is provided, there is likely to be healthy survival of the spinal segments below the level of the lesion.

PARAPLEGIA AND QUADRIPLEGIA

Loss of power in the lower limbs as a consequence of spinal transection identifies *paraplegia*. Lesions at cervical levels result in *quadriplegia*. People with transections as high as the second cervical segment can survive if they are immediately provided adequate artificial respiration. Inasmuch as the respiratory center is cut off from both intercostal and phrenic nerves, permanent artificial respiration must be maintained. Insight into what spinal cord injury entails from a physiological, psychological, and social point of view should be understood by all physicians. Read "Sitting on a Basketball: How it Feels to be a Paraplegic," by the distinguished neurosurgeon, Paul C. Bucy (1974).

ORGANIZATION OF SPINAL MOTOR CONTROL

Development of Spinal Motor Organization

Motoneurons connect with specific muscles according to regular myomeric patterns. Early specification of motoneurons is based on their position in the wall of the neural tube. Medially located motoneurons project to ventral muscles, ventrally located motoneurons project to lateral muscles, and laterally located motoneurons project to dorsal muscles. A specific trophic recognition process takes place between the motoneuron and muscle fibers once the axons contact the muscle.

Specialization of Spinal Interneurons

The spinal cord is, in effect, a continuous, stretched-out column of sensory, sensorimotor, visceromotor, and somatic motor nuclei. Spinal interneurons gather information from many spinal segments and from different sensory and central motor control sources. They accomplish this by means of elongated, abundantly branched dendrites, and axons which descend and ascend to serve practically all spinal cord levels, and reach into the brainstem to coordinate spinal with cranial nerve reflex systems. These elongated interneurons solve some integrative problems of the spinal cord and brainstem by linking intersegmental and interlimb reflexes and reflexes serving head and body. Inhibition in the spinal cord is dominated by one neurotransmitter, *glycine*, which is characteristic of the spinal cord and is not prevalent elsewhere.

Spinal interneurons are divided into functional groups each of which participates in the integration of somewhat different spinal functions. Their role is determined in part by impulses arriving from supraspinal systems involved in motoneuron control and from particular peripheral afferent neurons involved in reflexes (Kostyuk and Vasilenko, 1979). Interneuron functional specificities also derive from the array of muscles to which they become indirectly associated functionally by second-order retrograde trophic influences.

Shorter and Longer Propriospinal Interneurons

Relatively shorter propriospinal interneurons reside in the deeper laminae of spinal grey matter. Their axons pass into the lateral and ventral cord funiculi to influence motoneurons over distances of five or so segments. They form short propriospinal paths, limited to the spinal cord, with conduction velocities that range from 35–40 m/s. They coordinate intralimb and interlimb reflexes. Surrounding these neurons are interneurons with longer processes which cover greater longitudinal distances and have higher conduction velocities (110–120 m/s). These provide the swiftest interconnections between lumbar and cervical levels for interlimb reflexes, and between spinal and brainstem motor control systems for coordination of limbs, trunk, and head.

The shorter propriospinal interneurons receive direct, monosynaptic descending corticospinal, rubrospinal, tectospinal, reticulospinal, and vestibulospinal collateral terminals. Their direct effects are reinforced by even more abundant pleurisynaptic connections. These interneurons provide discriminative integrative among the various descending motor control signals. They are only weakly affected by segmental afferents by which input signals they provide mild excitation of flexors and inhibition of extensors, thereby contributing to readiness of flexor reflexes. Powerful afferent stimulation of propriospinal interneurons can override descending motor commands and elicit flexion (protective, avoidance) responses at any time. This local spinal loop integrates higher command signals and yet can override those signals for protective purposes.

Specializations of Motoneurons

HOW MOTOR CONTROLS CONVERGE ON MOTONEURONS

Spinal and cranial motoneurons have unusually extended shaggily branched dendrites which reach out three-dimensionally into neighboring spinal and brainstem levels. They reach interneurons, spinal reticular formation, and motor and sensory nuclei as well as receiving axon terminals of interneurons and axons belonging to the major descending motor control systems.

The only comparably widespread dendritic outreach is typical of large neurons in the brainstem and spinal reticular formation. In functional terms, these latter neurons serve as interneurons for interneurons. They receive signals from just about everywhere in the nervous system including all descending forebrain and brainstem motor control systems. Widespread motoneuron outreach and high velocity interneurons—which have access to brainstem reticular formation activity—combine to give motoneurons the benefit of a triple-tiered synthesis which integrates spinal, brainstem, and forebrain activities. By this arrangement almost everything that can generate or shape somatic behavior can obtain nearly immediate access to brainstem and spinal motoneurons.

Organization of Brainstem Motor Control

SKELETAL MOTOR CONTROL

As noted, spinal motor neurons governing skeletal muscle are somatotopically organized in a continuous longitudinal column in the ventral horn. The most ventral cranial motor column in the brainstem is discontinuous, but it likewise innervates somatic skeletal muscles which derive embryologically from myotomes. They are thus functionally similar to motoneurons in the ventral horn of the spinal cord. The brainstem skeletal muscle control column includes motor nuclei of the oculomotor (IIIrd), trochlear (IVth), abducens (VIth), and hypoglossal (XIIth) nerves.

SPECIAL VISCERAL MOTOR CONTROL

In the brainstem, a unique motor column governing special kinds of striated muscle is added: this more lateral cranial motor column innervates special visceral striated muscles which derive embryologically from branchial arches, and ultimately from the neural crest. These special visceral striated muscles are innervated by axons from motor nuclei of the trigeminal (Vth) and facial (VIIth) cranial nerves, and the nucleus ambiguus. The nucleus ambiguus contributes motor fibers to the glossopharyngeal (IXth), vagus (Xth), and accessory (XIth) cranial nerves, which together innervate striated muscles of the pharynx and larynx. Since this effector system is unique to the head, there is no spinal homology (see Chapter 69, Visceral Control Mechanisms).

VISCERAL MOTOR CONTROL

Still more laterally in the brainstem is a column of cells which controls smooth muscles and glands, the dorsal motor nucleus of the vagus (Xth) cranial nerve. This is homologous to visceral preganglionic neurons in the intermediolateral column of the spinal cord. These autonomic mechanisms control smooth muscle and glands: pupillary dilatation (Edinger-Westphal nucleus of the IIIrd) and salivation (superior salivatory nucleus, VIIth, and inferior salivatory nucleus, IXth cranial nerves) (see Chapter 69).

These three motoric systems (skeletal motor, special visceral motor, and visceral autonomic) function in close harmony with one another. Eating, for example, requires integration of all three brainstem motor control systems governing skeletal muscles, special visceral striated muscles, and visceral smooth muscles and glands.

INTEGRATIVE ORGANIZATION OF SPINAL AND BRAINSTEM REFLEXES

Swallowing, coughing, and vomiting are extremely important reflexes. Stimulation of the pharynx and back of the tongue initiates *swallowing*. This activates a linked chain of closely coordinated rapid reflexes that delivers food and fluids across the glottal lid, over the trachea, to the esophagus. Commonplace experience demonstrates that breathing and swallowing are incompatible. They are reciprocally controlled during eating and drinking.

Swallowing begins with signals carried in the glossopharyngeal and vagues nerves to the nucleus of the solitary tract (NTS) and neighboring reticular formation, a region that coordinates swallowing. Boluses of food and drink are virtually thrust into the esophagus by coordinated and exquisitely timed powerful movements of the tongue (hypoglossal XIIth), palate, and pharynx (glossopharyngeal IXth and vagus Xth). It is instructive to watch this swiftly coordinated, vigorous thrusting reflex in slow motion film of a fluoroscopic examination of swallowing. The action needs to be slowed down to be able to appreciate the deftness and accuracy of this vital performance.

When swallowing falters, and when food, fluids, or other foreign objects get into the windpipe, irritation of the tracheal and bronchial mucosa elicits *coughing*. Afferent impulses travel in the glossopharyngeal (IXth) and vagus (Xth) nerves to the nucleus of the solitary tract (NTS) (see Chapter 65, Fig. 9.59). Motor impulses are mediated through the IXth, Xth, and XIIth nerves to the pharynx, larynx, and tongue. Other essential impulses control coordinated expulsive movements for tracheal and bronchial clearance which are executed by diaphragm, intercostals, and abdominal muscles.

Vomiting is another propulsive performance. In this event, there is rapid ejection of gastric contents, effected by coordinated contraction of abdominal muscles and diaphragm at the same time that the cardiac sphincter, throat, and esophagus are opened widely. Concomitant closure of the glottis guards against aspiration of vomitus. Vomiting is preceded and accompanied by copious salivation which lubricates bulk flow and somewhat dilutes the acid in vomitus.

Vomiting can be initiated by the gag reflex, by physical stimulation of the back of the throat. More generally, however, vomiting is caused by irritation of mucosa of the upper digestive tract. This initiates impulses that reach the NTS by way of (IXth) and (Xth) cranial nerves. The coordinated response is controlled by a "vomiting center" in the medulla. Vomiting can also be triggered by centrally acting agents (e.g., apomorphine). These initiate responses from a chemosensitive zone in or near the *area postrema*, a brainstem area outside the blood-brain permeability barrier and therefore accessible to activation by blood-borne substances. The area postrema then triggers the vomiting center.

These reflexes—swallowing, coughing, and vomiting—are counted among vital reflexes. Their integration is achieved mainly by spinal and brainstem interneurons which interlink brainstem and spinal levels. If any of the three fails to function in nearly perfect order, the individual is likely to succumb, sooner than later, from aspiration pneumonia. Temporary bypass of the respiratory channels protects individuals from the consequences of temporary failure of these reflexes, but permanent tracheostomy is difficult to manage.

MOTONEURONS AND CONTROL OF SKELETAL MUSCLES

Functional Plasticity of Motor Units

Chapter 4 depicts the excitation and contraction of muscle along with the function of the neuromuscular junction. Recall that a combination of the motoneuron and the muscle fibers under its command constitutes a *motor unit*. Functional unity is bolstered by contributions of trophic factors which are exchanged both ways between muscle and motoneuron. *Note* (a) there is no spontaneous activity in normal mammalian skeletal muscles; (b) the effects of motoneuron activity are always excitatory; and (c) the neurotransmitter is uniformly acetylcholine.

Skeletal muscle fibers are divided functionally into three types: "red," "white," and "intermediate." Most muscles consist of a mixture of the three, but proportions vary in different muscles. *Red fibers* are smaller, more granular, and darker. They are adapted for sustained posture-maintenance and antigravity actions. Capillary density and mitochondrial and myoglobin contents in red fibers are increased, indicating that these fibers are especially suited for high levels of sustained metabolic activity. They are activated by relatively low-frequency, steady trains of motoneuron impulses.

White fibers are adapted to swift, phasic contractions. Their motoneurons generate intermittent bursts of relatively high-frequency impulses. *Intermediate fibers* are structurally and functionally intermediate in respect to these characteristics.

Relatively speaking, larger motoneurons have larger diameter fibers which subdivide into more numerous branches and terminals and command larger aggregations of muscle fibers. They serve white, or white and mixed white and intermediate type muscles. Thus, they tend to discharge phasically, with infrequent bursts at higher discharge rates, than do other motoneurons.

Fast and Slow Muscles

Fast and slow muscles are functionally adaptive. Adaptation is dictated by the pattern of nerve impulses by which they are driven. Reversing innervation between fast and slow muscles converts the characteristics of the muscles to the type of reinnervation: slow becomes fast and vice versa (Buller et al., 1960). Moreover, if implanted electrodes are applied to a fast motor unit axon and a persistent low-frequency stimulus is applied, the fast muscle assumes the morphologial and functional characteristics of a slow muscle. A slow muscle given intermittent high-frequency stimulation assumes the characteristics of a fast muscle.

Twitch and Tetanic Contractions

A refractory period is necessary for repolarization of the muscle membrane. Repetition of nerve impulses reestablishes or sustains muscle tension. During the briefest contractions of individual muscles, twitch contractions, extraocular muscles show the fastest onset and briefest duration of contraction. This contrasts with the gastrocnemius, soleus, and interosseus plantar muscles, which are relatively slow. The rate at which twitches overlap and summate (to develop tetanus) depends on the combined speeds of contraction and relaxation of muscle fibers. Inasmuch as these rates vary greatly muscle to muscle, it follows that the frequency of nerve impulse discharges necessary to achieve smooth maximum contractions varies greatly.

Much of what is achieved by athletic conditioning and training in dexterity, e.g., playing a musical instrument, relates to mutual functional improvements that occur between motoneurons and muscle fibers, i.e., a conditioning of motor units. Other conditioning effects relate to peripheral sensory contributions, especially the γ motoneuron-muscle spindle system, and functionally related circuits in the spinal cord and brainstem. In sum, a great deal of motor conditioning and functional adaptation is accomplished "downstream" among components which are immediately involved in effector activities.

SKELETAL NEUROMUSCULAR JUNCTION

Each motoneuron terminal ends in a specialized region called the *neuromuscular junction*, a localized adaptation compounded of nerve terminal and sarcolemma. Transmission of nerve impulse to muscle takes place here, resulting in the generation of a muscle action current which initiates contraction.

A nerve impulse induces a prolonged negative potential at the neuromuscular junction. This is not propagated but it generates a muscle spike potential by depolarizing the muscle membrane surrounding the junction. The muscle spike potential sweeps over the muscle surface and commands muscle contraction. The neuromuscular junction is important clinically because it is the site of several important neuromuscular disorders.

Myasthenia Gravis

Myasthenia gravis is a neuromuscular disorder characterized by weakness and easy fatigability of skeletal muscles. Symptoms include weakness of the levator palpebrae which results in eyelid lowering, weakness of extraocular muscles, impairment and weakness of voice, speech, and ability to chew and swallow. Generalized weakness occurs, especially of proximal limb muscles and neck support of the head. Respiratory weakness may be life-threatening.

This disorder involves failure of neuromuscular transmission and resembles neuromuscular block by curare. Myasthenic symptoms respond to treatment with anticholinesterases whereby the effects of acetylcholine are enhanced and prolonged.

Isolation and characterization of snake venom neurotoxins by Lee (1972) led to the isolation of acetylcholine receptors (AChR). Almost immediately, it was shown that myasthenia gravis is associated with a shortage of AChR (Fambrough et al.,

1973). It was then demonstrated that myasthenia gravis is the result of an antibody-mediated autoimmune disorder (Lindstrom et al., 1976). Immune globulin (IgG) and AChR crosslink. This is followed by endocytosis and degradation of AChR within the muscle cell. Neural functions remain normal.

Treatment of myasthenia gravis involves suppression of IgG with steroids, removal of the thymus gland, depletion of the antibody by plasmopheresis, and specific immunotherapy. Solving the puzzle of myasthenia gravis provides a good example of combined basic and clinical neurosciences research (Drachman, 1981).

Neuromuscular Changes Following Denervation

If the axon of a motoneuron is severed, three dynamic effects follow: (a) the neuromuscular junction becomes simplified, (b) the AChR receptors spread out over the sarcolemma, and (c) the proximal portions of the motoneuron terminals begin to sprout new axonal processes. Nerve sprouting takes place at the point of severance of the axon and at more proximal nodes of Ranvier (see Fig. 9.77). Terminal sprouting is in response to degenerative changes in the muscle. Nodal sprouting is in response to processes associated with retrograde degeneration of the axon (Brown et al., 1981). When sprouts reinnervate muscle fibers, the neuromuscular junction is reconstituted, the excess sprouts are pruned back, and normal function may be restored.

CENTRAL CONTROL OF MOTOR UNITS

The Central Excitatory State

Motor units are most directly affected by what Sherrington called the *"central excitatory state"* (CES). In terms of reflexes, the CES increases the amplitude and duration of single muscle twitches, the rate at which individual muscle contractions reach sustained (tetanic) levels, and the thresholds at which afferent stimuli are effective in initiating reflex responses. With increased CES, any given motoneuron will discharge more impulses at higher frequencies, and motoneurons otherwise unresponsive will be recruited (Sherrington, 1906).

The CES represents convergent activity generated by a variety of spontaneously active neurons situated in different parts of the neuraxis, especially along its central core. Several motor control loci in the telencephalon show spontaneous activity as do appetitive regions in the hypothalamus. Mainly, the CES is attributed to spontaneous activity in brainstem and spinal reticular formation, in particular, that of the raphe nuclei. These generate widespread excitatory and inhibitory patterns which influence the CES along ascending as well as descending trajectories (Magoun, 1950). Widespread inhibition of motoneurons, for example, occurs during rapid eye movement (REM) sleep as a consequence of widespread descending pontine inhibitory influences (Jouvet, 1973). This, in turn, prevents us from acting out our dreams.

Afferent activity also makes contributions to the CES. Many afferent neurons are spontaneously active. But the CES does not depend on afferent input as it persists after virtually all sensory nerves have been severed (Taub et al., 1980). Certain drugs have the effect of heightening or reducing the CES. In the waking state, consciousness plays a role, from barely sustaining wakefulness (with a diminished CES) to powerful agitation, palpitation, exophthalmia, insomnia, and tremor characteristic of anxiety (with greatly heightened CES).

Characteristics of Spinal Reflex Activities

The anatomical basis for reflex activity is the *reflex arc*. Essential elements include: afferents, interneurons (internuncials), and efferent neurons (Fig. 9.83). At each segmental level, some reflex connections between Ia muscle afferents and motoneurons are direct, i.e., *monosynaptic*, with no interneurons intervening. The vast majority of reflex connections, however, are *pleurisynaptic*, involving a few to several synaptic relays among interneurons. Characteristically, afferent neurons contribute to reflexes through numerous collaterals and via numerous interneurons whereby, in an organized way, they influence a very large number of motoneurons.

Superficial reflexes are elicited by stimulation of mucous membranes and skin. They are all pleurisynaptic, usually involving a moving stimulus, and a coordinated ensemble of multiple interneurons and effectors. Superficial reflexes include corneal, snout, rooting, sucking, abdominal, plantar, cremasteric, sphincter, and other reflexes. When the cor-

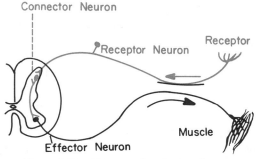

Figure 9.83. Diagram of a simple reflex arc.

nea is lightly stroked with a wisp of cotton, the eyelids close (blink) reflexly, unilaterally. Stronger stimulation results in bilateral blinking accompanied by strong eye closure and secretion of tears. Any painful (nociceptive) stimulus can engage both superficial and deep reflexes and may result in ipsilateral withdrawal accompanied by crossed extensor muscle contractions.

Snout, rooting, and sucking reflexes are functionally important in infants to ensure successful ingestion. They are pleasurable in their own right and reinforce thumb suckling, for example. They disappear in childhood, but may reappear in senile patients. It is presumed that the reflexes remained intact but were suppressed until they reappeared, being "released," especially with diminution of frontal lobe control.

If one quadrant of the relaxed abdomen is slowly, gently stroked, the abdominal muscles reflexly, transiently, pull in that direction. If the abdominal wall near the groin or the skin inside the thigh is stroked, the cremaster muscle (which is a transposed abdominal muscle) draws the testicle up on that side. These superficial reflexes may be weak or absent after a stroke, a sign of "upper (suprasegmental) motor neuron lesions."

When the sole of the foot is slowly, firmly stroked, the foot and all the toes flex, curling downward. This is a normal *plantar reflex*. This reflex changes to a pathological *"sign of Babinski"* when there is an "upper motor neuron lesion." With this "sign," the small toes fan outward and the big toe points upward. The "sign of Babinski" is seen in patients with functional damage to the contralateral sensorimotor cortex, internal capsule (common in cases of stroke), or interference with any of several descending motor control pathways.

The question whether the loss of abdominal reflexes and pathological conversion of the plantar response constitutes positive evidence of damage to the pyramidal tract is moot (see Brodal, 1981). Interestingly, the "sign of Babinski" is normally present in infants until about the age when they begin to walk. The infantile reflex is thought to represent a normal developmental absence of tonic descending suprasegmental inhibition.

Deep reflexes are stretch reflexes in which abrupt short extension of a muscle is obtained by striking the tendon of that muscle. Stretch reflexes consist of combined monosynaptic and pleurisynaptic reflex actions which lead to a brisk concerted contraction of the muscle that is attached to the tendon that was struck. These include the jaw jerk, biceps, triceps, knee, and ankle jerks.

CLASSIFICATION OF REFLEXES

Reflexes may be classified as segmental, intersegmental, interlimb, and suprasegmental on the basis of their horizontal and longitudinal connections within the spinal cord and brainstem. A typical *segmental reflex* is the deep tendon or stretch reflex such as the patellar tendon evoked knee jerk. *Interlimb reflexes* engage multisegmental reflex patterns and involve ascending and descending pathways. For example, pinching the hind limb of a cat yields reflex flexion of the contralateral forelimb.

Suprasegmental reflexes involve reflex interconnections extending upward and downward between spinal and brainstem levels. These involve important reflex head and body coordination controls, and are important in semiautomatic somatovisceral skeletal and autonomic reflexes related to swallowing, coughing, vomiting, etc., as described above. An example of suprasegmental reflexes is seen in decerebrate rigidity (seen in patients with functional transections at midbrain or higher levels). Turning the head to one side activates vestibulospinal responses which alter muscle tone in all four extremities. Limbs ipsilateral to the side to which the face is turned extend and contralateral limbs flex. This yields a "posture of an archer." When the head position is reversed, limb reflex posture is reversed.

Reflexes of Cutaneous Origin

FLEXOR REFLEXES

The most thoroughly studied cutaneous reflexes are flexor reflexes which involve contraction of functional flexors and relaxation of functional extensors. The pattern of reflex flexion varies with the nerve stimulated. Light stimulation of the skin elicits weak contraction of one or more flexors, but without much effect on the skeleton. Painful stimulation of skin elicits widespread contraction of flexors throughout the limb and generates brisk limb withdrawal. With nociceptive shocks of increasing intensity, withdrawal responses may expand regionally, generalize, and become long-lasting.

CROSSED EXTENSOR REFLEXES

When a "spinal" animal or spinal-transected human withdraws a limb in response to local noxious stimulation, the contralateral limb is simultaneously extended. Collaterals of interneurons that excite ipsilateral flexors cross to the contralateral side of the spinal cord at the same level and excite extensor and inhibit flexor motoneurons. The function of this response is to support the weight of the body when flexion is withdrawing a limb from its contribution to weight bearing. This crossed extensor reflex underlies the initiation of walking and running.

Reflexes of Muscular Origin

MUSCLE SPINDLES AND STRETCH REFLEXES

Stretching a muscle causes it to contract reflexly. This contributes to recovery of any given posture or movement from any direction of displacement. All skeletal muscles exhibit stretch reflexes. Appropriate stretch of flexor muscles yields a brief but unsustained flexor reflex contraction. Stretch of extensors elicits a more powerful and longer lasting reflex, especially in antigravity muscles.

Muscle Spindle Afferents

The stretch reflex is the most rapid of all reflexes. Ia afferents are the largest diameter (fastest conducting) of any afferent nerves. They readily respond to stretch (less than 1-mm displacement may be sufficient). They have monosynaptic connections with motoneurons. Motoneurons respond promptly to Ia afferent signals, yielding abrupt contraction. The reflex response is very specific: it results in contraction of the same muscle from which the Ia afferent originated.

Ia fibers constitute the primary afferent component of muscle spindles, the peripheral stretch reflex control system for skeletal muscles (Fig. 9.84). Each of these fibers serves as a specialized detector that has its own intrinsic motor innervation, functionally distinct from that of surrounding host muscle. γ-*Motoneurons* induce contraction of striated muscle fibers in the muscle spindle. Their contraction induces activity in Ia afferents. The receptors involved are annulospiral endings which coil around the equatorial region of intrafusal (intraspindle) muscle fibers. Ia primary afferents have to do mainly with phasic aspects of muscle contraction. Group II afferents, with secondary endings, have to do with measuring static forces relating to slow or sustained muscle contractions.

γ Efferents to Muscle Spindles

As noted, muscle spindles contain motor as well as sensory elements. The motor parts consist of two

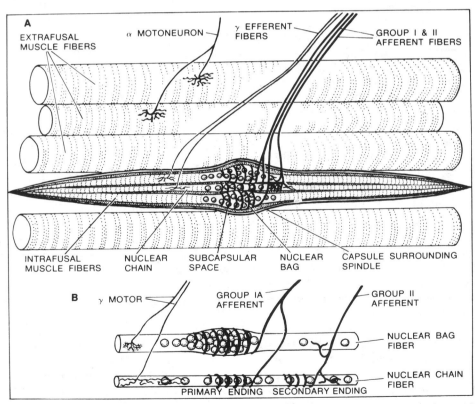

Figure 9.84. Mammalian muscle spindle. *A*, Schema of mammalian spindle innervation. The spindle is embedded in the bulk of the muscle made up of large extrafusal muscle fibers and supplied by α-motoneu- rons. *B*, Diagram of the two intrafusal muscle fiber types and their afferent innervation. (*B* from Matthews, 1964; from Kuffler and Nicholls, 1976.)

types of skeletal muscle fibers, both of which are contained within the spindle, that is, they are located *intrafusally*. Multiple nuclei in the muscle fibers are collected centrally in the case of "nuclear bag" fibers, and are longitudinally distributed in the case of the "nuclear chain" fibers. Both types of spindle muscle fiber are innervated by γ motoneurons.

Spindle Afferent Activation and Control

Mechanically sensitive primary afferents arise as annulospiral endings within the equatorial zone of nuclear bag cells. They are large diameter Ia fibers, greater than 12 μm in diameter, with conduction velocities greater than 90 m/s. Secondary afferents, "flower-spray" type, originate outside the nuclear bag region, in the myotubular region. Their axons belong to group II afferents, and exhibit somewhat slower conduction velocities, less than 80 m/s.

For each spindle, there is usually one type Ia afferent and from 0–5 type II afferents. Both type Ia and type II spindle afferents are excitatory to the muscle containing the spindle. Type Ia spindle afferents are especially sensitive to abrupt, short stretch, hence they are responsive to vibratory stimuli applied to the muscle tendon. Type II afferents are more sensitive to static stretch phenomena (Matthews, 1982; Harris and Henneman, 1980).

Spindle Servocontrol of Stretch Reflexes

Muscle spindles are attached at each end to extrafusal muscle fibers, to other spindles, or to tendons. Muscle spindles are anatomically rigged "in parallel" to extrafusal (host) muscle fibers. Contraction (hence shortening) of the host muscle relieves tension of the muscle spindles. If it were not for γ-motoneuron control, spindle afferents would automatically decrease their rate of firing when the spindle is slackened; in some instances there is a corresponding "pause" in Ia afferent activity at the start of host muscle contracton. But γ-motoneurons, which are fired by the same signals that fire α-motoneurons to the host muscle, accelerate firing of Ia afferents so that they keep up with or exceed the slackening effects of contraction of the host muscle.

Of course, either passive or active stretching of a muscle will lengthen extrafusal fibers and exert tension on spindles. γ-Motoneurons adjust their firing rate to correspond to the "intention" for the movement under control.

Servocontrol provided by γ-motoneurons and spindle afferents has the net effect of allowing the muscle spindle system to utilize the full range of its effectiveness regardless of dynamic changes in length of the host muscle. The effect of spindle feedback control at the segmental level is to increase excitation—hence to increase the range under control by the α-motoneurons, in accordance with whatever load may be imposed on that muscle. Golgi tendon organ afferents, on the other hand, operate to prevent muscles and tendons from being overloaded.

GOLGI TENDON ORGAN INHIBITORY CONTROL

Golgi tendon organs reside in tendons attached to skeletal muscles. These sense organs are anatomically "in series" rather than "in parallel" with the muscle. When a muscle is passively stretched, Golgi tendon organs are stretched and thereby their afferent fibers are activated. When the host muscle itself contracts, Golgi tendon organs are likewise stretched-activated.

Golgi apparatus afferents are classified as Ib, having a modal conduction velocity of about 80 m/s. The length of stretch necessary to elicit a response from Golgi tendon organs is greater than that to activate the spindle annulospiral endings. The reflex effects of Golgi tendon organ afferents are inhibitory. Therefore, their servocontrol operates to prevent too powerful passive or active muscle loading.

γ *EFFERENT EXCITATORY CONTROL*

Muscle spindles provide information about the state of muscles: the length and the rate at which the length is changing. Skeletal muscles are directly innervated by large α-motoneurons in the ventral horn. The ventral horn also contains small γ-motoneurons, concerned with regulation of muscle spindle contraction and hence only indirectly with overall muscle contraction. γ-Motoneurons constitute one-third of all motoneurons. Central motor commands that converge on α-motoneurons also converge on γ-motoneurons and both types are ordinarily co-activated.

γ Efferent Control of Spindle Afferents

Figure 9.85 illustrates the feedforward sensory control and feedback excitatory muscular control generated by γ-motoneurons. A single γ-motoneuron is electrically activated to exert a standard response in a single muscle spindle. Responses of the primary spindle afferent (Ia) from that spindle are recorded. Extrafusal muscle efferents are eliminated. A brief train of stimuli to the α-motoneuron axon yields a different response depending on load (0–20 g) imposed on the muscle.

The finer the movements subject to control, the

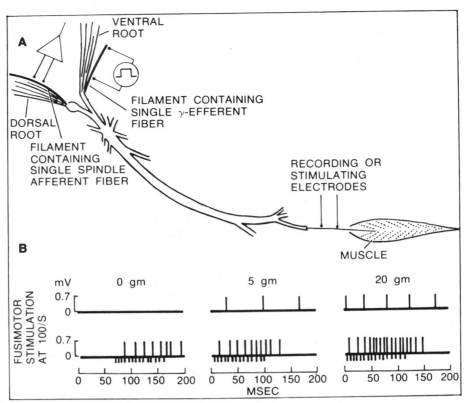

Figure 9.85. Centrifugal control of muscle spindle. *A,* Recordings from a single dorsal root axon arising in a muscle spindle; single fusimotor (γ) axon in a ventral root innervating the same spindle is stimulated. *B,* Upper records show sensory discharges when the muscle is slack (0-g tension) or lightly stretched (5- and 20-g tension). A brief train of 14 or 15 stimuli at 100/s either initiates sensory discharges *(lower left)* or accelerates them. Extrafusal muscle fibers remain inactive. (Adapted from Kuffler et al.; redrawn from Kuffler and Nicholls, 1976.)

greater the concentration of muscle spindles per host muscle unit mass, and hence of γ-motoneurons and of group Ia and group II afferents. Intrinsic muscles of the hand, for example, are far more densely supplied with this regulatory apparatus than are antigravity muscles.

Muscle Spindle Control in Human Subjects

Responses of muscle spindle afferents during voluntary contraction of extensor muscles indicate that γ-motoneurons are recruited in an orderly manner along with α-motoneurons. Thus, individual spindle endings become activated at specific contraction strengths. This is evidence that in voluntary contraction, as in reflex contraction, muscle spindles are activated in parallel with α-motoneurons.

There is a linear relationship between spindle activity and strength of contraction during voluntary isometric contractions. Different spindle afferents have widely different thresholds. Recruitment threshold for a given spindle afferent appears to depend on its fusimotor innervation and on the recruitment threshold for that γ-motoneuron. The recruitment threshold depends on desired muscle force rather than on muscle length (Burke et al., 1978a). Even more significant is the fact that muscle spindle afferents respond promptly and accurately to changes in load during voluntary accurate position maintenance (Burke et al., 1978b).

Muscle spindle afferents contribute to body imagery. When vibrations of 100 Hz are applied to the tendons of biceps or triceps muscles, subjects make systematic misjudgments as to the angle at the elbow. The subject perceives that the elbow is in a position that it would have assumed if the vibrated muscle had been stretched.

REGULATION OF STIFFNESS BY SKELETOMOTOR REFLEXES

Skeletomotor reflexes regulate length and force variables simultaneously: the ratio of force change to length change regulates what is called "stiffness." Spindle feedback tends to maintain stable muscle length; tendon organ feedback tends to keep muscle tension steady. These two feedbacks somewhat oppose one another. The physiological requirement is to control the balance of these two variables when the body or limb experiences variable external loads. These may be imposed by changes of loading, by alterations of the position of the body and limbs in the gravitational field, by weightlessness, or by centrifugation which increases the gravitational force.

Ordinarily, we successfully adjust to shifting loads, and we readily lift, load, and swing things in different postural contexts, even in a moving environment. Skeletomotor reflexes combine to provide *stiffness regulation* during various static and dynamic behaviors (see review by Houk, 1979). The net result is that higher motor command does not need to cope individually with either length or force but only with the intention of the movement. This simplifies higher command and provides a spring-like linkage between our body and its encumberances. Operation of this segmental control loop enables higher centers to guide movement in dynamic circumstances, and serves to absorb the impact of unexpected shifting of loads.

INTERSEGMENTAL AND SUPRASEGMENTAL REFLEXES

Spinal and Cranial Intersegmental Reflexes

PROPRIOSPINAL AND SPINO-BULBO-SPINAL REFLEXES

Two physiologically distinctive reflex systems interconnect the long sequence of spinal segments. One system, *propriospinal* (PS), is carried by spinal interneurons, and is entirely sufficient within the spinal cord. That is, PS reflexes remain intact when the spinal cord is isolated from the brainstem. The other system, *spino-bulbo-spinal* (SBS), involves a more circuitous path that goes through the brainstem reticular formation and reenters the spinal cord from above downward (Shimamura and Livingston, 1963). The brainstem relay takes place in the caudal part of the medullary (bulbar) reticular formation, near the midline. Because the predominant and shortest-latency responses to the SBS reflex system arise from cutaneous afferents, it is clear that by this means, cutaneous sensory signals are promptly integrated into motor control patterns (Shimamura and Akert, 1965).

PROPRIOCRANIAL AND BULBAR RELAYED CRANIAL REFLEXES

In ways that parallel spinal PS and SBS reflexes, when electrical stimuli are applied to cranial nerves, direct (propriocranial, PC) and indirect (cranio-bulbo-cranial, CBC) reflex responses occur among all cranial motor neurons (Shimamura, 1963). Cutaneous and muscular afferent and motor response patterns similar to the spinal and brainstem relays obtain with these cranial reflexes. The same medial medullary reticular formation that relays the spino-bulbo-spinal reflexes also relays cranio-bulbo-cranial reflexes in both ascending and descending directions.

Figure 9.86 illustrates (*A*) how spinal input generates a medullary relay that elicits cranial motor responses, and (*B*) how cranial input generates a medullary relay that elicits spinal motor responses, both being relayed through the same medullary reticular neurons that relay spino-bulbo-spinal reflexes. This represents an important reflex linkage between spinal and cranial sensorimotor controls. As in the case of spinal projections to the bulbar relay, cranial stimuli do not elicit motor responses until after they have relayed in the reticular formation. There they generate an ascending (bulbo-cranial) and a descending (bulbo-spinal) response that manifests throughout the entire motor column.

Vestibular and Local Limb Supporting Reflexes

It is clear that proprioceptive and vestibular systems both actively contribute to posture and locomotion. Impulses arising from receptors in each of these systems converge to influence motoneurons. The effects of their mutual interaction have been analyzed by recording ventral root responses following stimulation of the vestibular nerve and appropriate proprioceptive sources.

Vestibular stimulation alone elicits a rapidly conducted two-peaked ventral root response seen bilaterally in both spinal and cranial motor roots all along the neuraxis (Fig. 9.87). The first of the two peaks is facilitatory to skeletal motoneurons. This is conveyed by direct vestibulospinal and vestibulocranial impulses. The second peak results from relays through reticular neurons in the medial medulla. It conveys both facilitatory and inhibitory motor influences.

One kind of peripheral stimulation found to facilitate these vestibular ventral root responses powerfully is to move one or more limbs into weight-bearing positions. Upward displacement of the foot and extension of the tarsal-metatarsal joints markedly enhances the vestibular-evoked response. This activates both flexors and extensors throughout the limb. The consequence is establishment of pillar-like stability for that extremity.

Vestibular ventral root enhancement of the weight-bearing posture takes place only in the ventral roots belonging to the same limb and only when

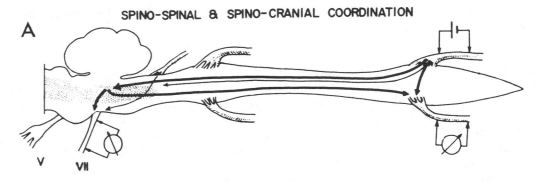

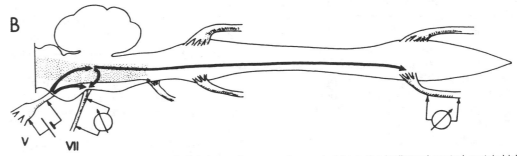

Figure 9.86. Longitudinal coordination systems of the neuraxis. *Above,* spinospinal and spinocranial coordination. *Below,* craniocranial and craniospinal coordination. Each dorsal root volley elicits motor discharges along ascending (and descending) propriospinal pathway continuous with propriocranial coordination pathway to all spinal and cranial motor nuclei. Additional influences, relayed in the bulbar reticular formation course both upward to cranial motor nuclei (spinobulbocranial relays) and upward then downward to spinal motor nuclei (spinobulbospinal relays). In *A,* stimuli are introduced into a lumbar dorsal root and recorded from the ipsilateral ventral root (which reveals the segmental and appropriately delayed spinobulbospinal discharge) and from the facial nerve (VIIth) (which reveals a propriospinal-cranial discharge as well as a relayed spinobulbocranial discharge). In *B,* stimuli are introduced into the trigeminal nerve and recorded as propriocranial and relayed craniobulbocranial discharge recorded from the facial nerve, and a craniobulbospinal discharge recorded from the lumbar ventral root. (From Shimamura, 1963.)

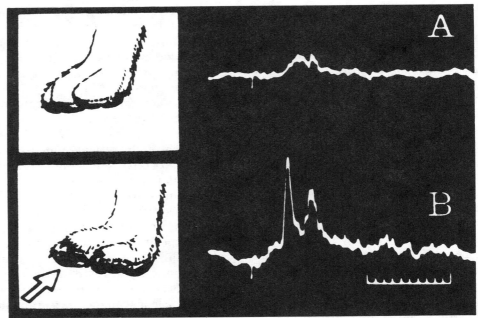

Figure 9.87. Effect of weight-bearing displacement of foot of cat on local ventral root responses to vestibular stimulation. In *A,* vestibular ventral root (L$_7$) responses are seen beginning about 4 ms following stimulus artefact and lasting for 4–5 ms with the foot in a nonweight-supporting position. In *B,* the cat's paw is pushed gently upwards as in light weight-bearing. The local vestibular responses are greatly augmented. The effect is to increase contraction of both flexors and extensors and hence to increase weight-bearing capability of the limb. (From Gernandt et al., 1963.)

that limb is in a position to bear weight. Therefore, standing, stepping, springing, or landing maneuvers which will displace the foot appropriately provide an adequate stimulus for powerful augmentation of this vestibulospinal, limb-stabilizing response. Given background vestibular excitation, there will be a corresponding effective increase in the weight-bearing stability of that limb. This is advantageous for an individual proceeding in the dark on uneven ground. There is obvious benefit from a vestibulospinal signal that can provide automatic limb support whenever a weight-bearing surface becomes available. This phenomenon also accounts for the positive supporting reactions—the so-called "magnet reaction" of Magnus (1924).

CHAPTER 71

Cortical and Subcortical Integrative Mechanisms

Sensory, motor, and central control systems depicted earlier summarize nervous system input and output, together with their central representations and controls. Two questions emerge to be answered briefly, then partially explicated in this and the final chapter.

1. *What lies beyond sensory and motor mechanisms and sensorimotor controls?* Nervous and endocrine systems have almost unlimited architectural and functional capacities to create higher order associations among various perceptual, judgmental, and behavioral representations, dealing with freedom and contingencies of experience and performance. It is believed that this is how the nervous system can create branching cognitive processes to support speech, language, and inclusive environmental social and cultural relationships, all within a self-organizing system by which we categorize objects and create "reality."

2. *How can diverse neuronal mechanisms provide coherent subjective experiences and volition?* The answer is *integration*. But how does subjective experience emerge? How do neuronal systems operate with conscious intent? Of course, every living creature—beginning with the most primitive—is fully integrated; without integration, life is not possible. How does consciousness enter the picture, and how can consciousness contribute to adaptive mental and behavioral skills? Partial answers are available.

Consciousness is an emergent, expanding evolutionary function of some utility. In humans, it enables subjective, reflective, and anticipatory memory, imagination, and disciplined learning to create innovative biological adaptations, e.g., engineering, medicine. As Bertrand Russell emphasized, consciousness is what we experience by first intention, that is, directly. All else that we experience, concerning our bodies and the "outside world," is derivative, second intentional. What we develop in

the course of our individual acquaintance with life experiences is, in fact, an individually created hypothetical "story"—an account of ourself in relation to an "outside world"—what we call "reality." Because of its behavioral contributions through perception, judgment, and volition, consciousness "pays its own way" in the same sense that binocular vision, upright gait, and the prehensile thumb contribute advantageous evolutionary adaptations.

Earliest emergence of consciousness in evolution probably started with tenuous and feeble self-reflective bodily feelings, e.g., of pressure, tension, duration, etc. (Herrick, 1949). Primitive consciousness may have gained some evolutionary advantage when it first generated tenuous, self-projective influences, e.g., biasing of kinetics, relaxing of tensions, altering of rhythms, etc. In any case, a fuller blown consciousness is available to us now and obviously involves both reflective and directive functions, including consciously directed behaviors—culminating, for example, in social altruism which can confer exemplary evolutionary advantages (Herrick, 1956).

Throughout its evolutionary development, consciousness, along with other nervous improvements, contributes to improving internal satisfactions. Now, we need desperately to sustain human survival. For that, it is required that we cultivate self-conscious thinking about our collective long-range interests, and cooperative strategies that will ensure an increasingly satisfying future for all humankind. Collective evolutionary advantages that are achievable through consciously directed constructive adaptations are essentially unlimited.

PHYSIOLOGICAL ADAPTATIONS SERVE DYNAMIC BIOLOGICAL PURPOSES

The nervous system functions to achieve harmony between inside visceral requirements and outside rewards and risks. The nervous system orchestrates a flexible, adaptive program which matches dynamic biological needs with dynamic environmental options. In developing this program, the nervous system evaluates and integrates biological needs in relation to behavioral options with respect to the individual as a whole. *Assigning values to both needs and options generates a matrix, a nervous system program, a plan*, probably integrated in the frontal lobes, *by which to channel behavior toward satisfying biological needs in an orderly sequence and in a timely fashion.* When this cannot be accomplished with resources available to the individual, outside help, including medical help, may be required.

Short-term and long-term behavioral goals are "shingled" in succession beneath one another. Survival and reproduction require sustained flexibility and ingenuity to "prioritize" the pursuit of multiple goals. It necessitates temporal integrations spread over seconds to decades. It requires the incorporation of lasting nervous system adaptations that suit altered needs and capabilities throughout a lifetime, and to meet the challenges of a dynamic environment. In the course of life's trajectory, each individual acquires an almost unlimited repertoire of individuated behaviors.

MOST OF THE HUMAN BRAIN MAY BE ENGAGED IN HIGHER NERVOUS PROCESSING

Sensory and motor mechanisms considered in previous chapters involve only a modest fraction of the total bulk of the nervous system. Measured in computer graphic representations of a normal adult human brain, the combined cortical and subcortical representations of the principal sensory, motor, and central control functions apparently embody less than one-third of the total brain volume. What do the remaining two-thirds contribute? The neurophysiology of this great volume of brain undoubtedly involves much additional decision-making activity. How might this "extra" activity be integrated into the whole?

Toward a Definition of Integration

ORGANISMS ARE INTERDEPENDENT WITH THEIR ENVIRONMENT AND WITH OTHER ORGANISMS

Behavior of any organism reflects its organizational and integrative capacities. Both visceral and somatic behavior are integrated according to overall individual purposes. The nervous system integrates signals informative about the internal state with signals informative about the environment and assigns value functions in relation to past experiences, expectations, and purposes.

Well-being depends on satisfaction of biological needs at each level of integrative organization, among cells, tissues, organ systems, and the whole

organism. An anoxic cardiac muscle, a hyperactive reflex, an epileptic focus, or a memory lapse represent deficient operations of integrative mechanisms at different levels of biological organization.

All organisms are "open" systems, dependent on throughput from the environment. All are perpetually engaged in a resourceful interplay with a dynamic environment, and interdependently with many other organisms. Survival depends on behavior being sufficiently individually integrated, and in a larger sense, behavior must be integrated with respect to the surrounding environment—including other organisms on which individual lives depend. Mankind has only recently begun to understand the vital interdependence of all living systems, and to appreciate some of the mutual benefits that depend on the preservation of physical, biological, psychological, social, and cultural diversity. This diversity is essential for a bioregion, indeed, for the whole planet.

We have lately been obliged to reconsider what "survival of the fittest" really means: *It means that those are fittest who contribute most constructively to the whole interdependent life-support system, globally as well as locally.* This requires unprecedented self-conscious human restraint, cooperation, and wisdom.

Considering the number of components and the complexity of interdependent physiological regulatory mechanisms—nervous and endocrine—we must inquire: How can so many dynamic control processes stay coordinated in guiding behavior toward successive successful satisfactions in a never-completely-fulfilled nexus of biological requirements? Since biological needs keep shifting, sometimes swiftly and radically, *no static rules for integration can suffice.*

We previously described mechanisms that control patterns of integration for a wide variety of goal-seeking subordinate activities. An overall, concatenating systems requirement demands overall, concatenating integration. Overall integration can perhaps be appreciated best by contrast: cancer is not integrative, it violates rules for restraint of its own growth; it can be said to be disintegrative. Aggregate behavior of many human populations likewise violates rules for restraint of their growth and exploitation of essential resources and thereby threatens disintegration of global life-support systems.

EVOLUTIONARY CONTINUITY OF NERVOUS AND ENDOCRINE INTEGRATION

Nervous and endocrine systems contribute most of the integrative mechanisms required for multicellular existence. Nervous and endocrine systems evolved early, together, and survived cooperatively, evolving interdependent capacities to provide integration for increasingly complex individual and collective organisms. Evolutionary persistence of successful integrative signal-response systems is demonstrated by the fact that neurotransmitters, peptides, neurohumors and hormones employed in nervous, endocrine, and neuroendocrine signaling are among the widest distributed, hence, most primitive as well as successful families of compounds encountered in comparative biochemistry (Dayhoff, 1976).

Clinical Significance of Integative Mechanisms

Knowledge about integration provides an important basis for physiological thinking. It presents an evocative question for consideration in relation to any clinical problem: *Which principal physiological mechanisms are insufficiently integrated in this individual's history and present condition?*

Fundamental integrative processing is intrinsic to neurons. Integrative capacities of single-celled organisms are not abandoned by multicelled organisms; rather, cellular integrative mechanisms are capitalized by neighboring cells (tissue integration) and built into specialized integrative systems, e.g., respiratory, cardiovascular, and endocrine systems, etc. Neurons make use of intrinsic integrative capacities of neuroglia and other functionally associated cells, such as other neurons, sensory receptors, and motor effectors. Further, we have seen that trophic influences are frequently reciprocally expressed among neurons and between neurons and many other cell types, receptors, effectors, glia, etc.

Individual Neurons are Decision-Makers

Neurons function as receivers, decision-makers, and distributors of modified information. Figure 9.88 provides a schematic representation of successive transformations of information from one neu-

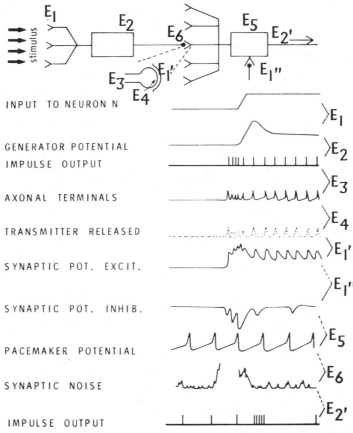

Figure 9.88. Schematic diagram and summary of events plotted against time (duration of tracings about 0.1 s). This illustrates several successive stages in the transfer of information from one nerve cell to the next. *Es* represent independent excitabilities. *Boxes* in the schematic represent two nerve cells of order N and N + 1; the synaptic contact between them is enlarged to show the location of the *Es* involved. Release of neurotransmitter is shown by a *broken line* to indicate that the recording trace is hypothetical. *Dashed lines* to *Es* in the bottom four records indicate that the events immediately above those *Es* are not the input for that particular *E*. (From Bullock, 1968, 1977.)

ron to another. Different parts of two neurons involved in this schema display different excitabilities, as represented in the circuit diagram and as depicted physiologically in the accompanying tracings (Bullock, 1968; Bullock et al., 1977).

Each neuron recieves and relays signals from one to many, including a variety of sources, and relays to one, many, and a variety of recipient cells. Sources and recievers of neuronal information may be neighboring or remote. *No neuron behaves like an all-or-nothing unit. Its membrane consists of a mosaic of discriminatively reactive parts.* It is capable of adjusting its internal bias to favor spontaneous activity and altered responsivity. Such biases can be introduced from the outside of the individual neuron by hormones, neurohumors, neurotransmitters, and other agents, some of which act by way of adenyl cyclase across the membrane. Pacemaker neurons characteristically generate information on their own as well as being influenced in ways similar to other neurons.

Localized functional individuation is expressed by different membrane potentials (Fig. 9.88, E_1s and E_6s) for both receiving and sending signals. Local membrane effects may be continuously graded while other parts of the same cell may convey all-or-nothing signals. Input and output signals may be inhibitory or excitatory, or inhibitory at one type of junction and excitatory at another, using the same neurotransmitter or more than one type of neurotransmitter. To those features are added important integrative contributions by Local Neuron Circuits (see Chapter 68, Nervous System Communication and Control).

All neurons are affected by blood-borne agents as well as by their extracellular milieu, including all the milieux to which their protoplasmic processes may extend. They are also affected by a variety of functional contacts and trophic influences with nonneurons, and other cells.

HOW DOES OVERALL INTEGRATION OPERATE?

Stratified Stabilities

Functional ensembles in the nervous system constitute *"stratified stabilities,"* a term used by Bronowski (1970; 1973) to explain increasing functional complexities observed throughout inorganic as well as organic evolution. Stable organization that constitutes one level of complexity provides the structural-functional foundation for the stable organization of higher order complexities.

Higher level nervous system integration is based on the organization of such stratified stabilities. Nervous system integration is distributed rather than centralized. Organization at every level of complexity of the nervous system, e.g., motor unit, spinal segment, interlimb spinal organization, midbrain sensorimotor control, etc., is functionally relatively stable. Each such complex can be controlled as a modifiable ensemble by other reciprocating neuronal circuits. These may be composed of larger scale relatively stable ensembles that exert more elaborate functional control over more widely distributed populations of neurons. Large-scale functional ensembles are themselves relatively stable, as they gain and reciprocate stability through their relations with stratified stabilities elsewhere. *By functionally linking combinations of stratified stabilities, the nervous system provides superordinate integration for the individual as a whole.* By virtue of this organization of interdependent stabilities, when there is a defect or loss of parts of the system, the remainder retains considerable stability. Assuming neuronal population selection in accordance with the consequences of behavior, the remainder of the nervous system not only rebalances integrative processes as a whole, but it optimizes usefulness of the remainder.

Stratified stabilities do not organize themselves simply horizontally, segmentally, although that may be a simple way to think about them. Nor do they organize themselves simply vertically, longitudinally. Such inordinately abstract assumptions derive from the way we examine nervous tissue, using microscopically thin sections. These obscure the most important potentialities of neuronal organization. Indeed, all neurons and neuronal circuits comprise three-dimensional dynamic protoplasmic configurations, delicately related in a complicated neurogeometry to numerous other three-dimensional dynamic protoplasmic configurations. Multiple and various connections may be both local and remote.

Dynamic fluctuations of protoplasmic activities and electrical and chemical processes, considered along with the crucial past history of the circuits, require that conceptions be "four-dimensional."

Stratified Stability Exemplified by a Brainstem Reflex

As described in Chapter 64, Vestibular Functions, the vestibulo-ocular reflex (VOR) adapts, after a few days of error accumulation, by reestablishing stability of vision in the presence of any of a variety of systematic, arbitrary displacements of retinal images by means of spectacles. VOR adaptation is challenging because the reflex pathway is direct (two synaptic relays) and rapid (12 ms). It is particularly challenging because the error messages by which precise adjustments are made along the pathway arise from outside the reflex circuit itself.

Error messages from the visual system have to be analyzed in relation to sensory inputs from five vestibular neuroepithelia, bilaterally, and applied to six extraocular muscles, bilaterally. Some of the brainstem interneurons involved in the normal and reorganized VOR pathway have been analyzed recently by Lisberger and Pavelko (1988) (see also Chapter 65, Vision, and Chapter 68, Nervous System Communication and Control).

Precise VOR adaptation takes place along the direct reflex path, in accordance with specifications from outside that path. The reconstructed reflex pathway works well, as effectively as before, for skills such as bicycle riding and fencing. The integrative utility of this and similar examples of motor learning is obvious. The fact that they take place far "downstream" means that the controls are swift and expeditious and that higher centers are relieved of having to cope with horrendous perceptual and motor control problems. This provides a good example of integration by reorganizing a stratified stability.

MOTIVATION PROPELS INTEGRATION

The power for integration derives from spontaneously active grey matter in the medial diencephalon and brainstem, floor of the fourth ventricle, and spinal central grey matter. Neurons in the floor and walls of ventricular channels, constituting the central core of the neuraxis, provide respiratory, cardiovascular, appetitive, neuroendocrine, sleep and arousal, and visceral and somatic regulations.

Visceral needs are motivating and remain restless until satisfied. Thus, core neurons generate rhythmical activity, tidally fluctuating central state conditions which represent appetites and satieties, feeling tones, anxiety levels, moods, sensations of fitness, readiness, and directedness for action—including substrates of consciousness. They radiate influences outward from the central core of the neuraxis to orchestrate activities among neuronal circuits that represent sensory, motor, and central control mechanisms, and higher nervous functions.

It is noteworthy that many basic physiological mechanisms that generate motivation are bound up with central systems for reward and punishment (see below). Feedback loops indicate whether internal needs are being met and celebrate visceral satisfactions. More complex, higher level circuits, involved in elaborate neuronal transactions, including conscious cognitive phenomena, are stage-by-stage erected from elementary stratified stabilities. Upward-bound activations serve to evaluate and "prioritize" incoming signals, judgments, and

behavioral options. Limbic and basal ganglia systems provide outward expressions of the internal state, and the frontal lobes organize plans for subsequent behavior (Livingston, 1978; Edelman, 1990).

Thus nervous and endocrine systems are poised to release such behaviors as have in evolutionary and individual past history succeeded in securing visceral satisfactions, survival, and reproduction. For example, hypothalamic guidance of pituitary hormone release (Axelrod and Reisine, 1984) affects the immune system (Berczi, 1988). Further, the immune system influences neoplasia (Riley, 1981). There is emerging a new field of psychoneuroimmunology (Cotman et al., 1987).

Hippocampal Pyramidal Cells Correlate Interoceptive and Exteroceptive States

There is good evidence that the limbic system, specifically the hippocampus, plays an importnat role in biological and psychological integration. Figure 9.89 presents a diagrammatic neuronal linkage between information coming from the internal milieu and the external milieu. Exteroceptive input comes from olfactory sources and is distributed to apical dendrites of hippocampal pyramid cells. Activation of this input elicits excitatory postsynaptic potentials (EPSPs) which ordinarily do not yield neuronal discharge.

Interoceptive information arises from the hypothalamus and is relayed to the hippocampal pyramids via the septum. Septal input is distributed to basal dendrites of the same pyramidal cells that receive apical olfactory information. Hypothalamic projections inform the hippocampal pyramids about conditions relating to the internal state. The activating effect of this input excites pyramidal EPSPs which are usually followed by pyramidal axon discharge. Olfactory input, also excitatory, contributes conditionally (see Fig. 9.89), making the pyramidal cell more likely to fire more decisively (MacLean, 1970).

EXAMPLE OF MAMMALIAN INTERGENERATIONAL INTEGRATION: BREAST FEEDING

Nervous and endocrine systems provide several familiar examples of integrative activities. Breast feeding, for example, involves the infant's olfactory guidance, suckling, and rooting activities which stimulate reflex suffusion and erection of the mother's areolae and nipples. Simultaneously, this activity elicits impulses that ascend the spinal cord and brainstem and penetrate the diencephalon to activate, inter alia, the *hypothalamic supraoptic and*

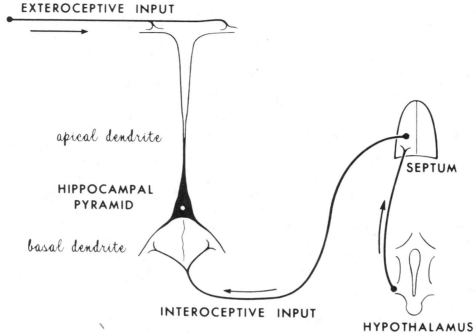

EXTEROCEPTIVE INPUT

apical dendrite

HIPPOCAMPAL
PYRAMID

basal dendrite

INTEROCEPTIVE INPUT

SEPTUM

HYPOTHALAMUS

Figure 9.89. Sketch of principal anatomical details of hippocampal pyramidal cells and their exteroceptive and interoceptive inputs, Note that the exteroceptive input (olfactory) goes to the outermost apical dendrites. The interoceptive input, from hypothalamus via septum, goes to the basal dendrites. Pyramidal cell discharge will be conditional by comparing information between internal state and external opportunity. (From MacLean, 1969, 1970.)

paraventricular nuclei. Discharge from both nuclei releases *oxytocin* from the neurohypophysis, which elicits contraction of collecting ducts in the breasts and active release of milk.

Each occasion of suckling results in periodic increases of *prolactin* through indirect influences by the paraventricular nucleus of the anterior pituitary (see Chapter 69, Visceral Control Mechanisms, and Fig. 9.82). This takes place through inhibition of a normal hypothalamic dopamine restraint of lactotroph growth and prolactin synthesis. The consequences of inhibition of this inhibitory influence maintains milk production for subsequent feedings.

STRESS RESPONSES AND SUSCEPTIBILITY TO NEOPLASM

Not all integrative mechanisms are entirely successful. Stress can elicit neuroendocrine initiatives that result in release of adrenal corticosteroids. Increased plasma concentrations of adrenal corticosteroids may have adverse effects on thymus and thymus-dependent T cells, elements of the immune system which defend against a variety of neoplastic processes and other pathologies. This may leave an individual vulnerable to actions of latent oncogenic viruses, newly transformed cancer cells, metastatic spread of cancers, and other pathological processes including infections, such as AIDS, which may be normally partially held in check by an intact immune system.

EXAMPLE OF BRAINSTEM INTEGRATIVE MECHANISMS

A decerebrated cat, with the brainstem completely transected transversely at midbrain level, between the superior and inferior colliculi, with all brain tissue forward of the transection removed, retains some behavioral capabilities commonly observed in the intact cat.

A decerebrated cat displays normal-appearing skin ruffling with its extensive platysma muscle, in response to being groomed. Brushing hair the "wrong way" increases decerebrate rigidity. Hair smoothing results in partial relaxation of decerebrate rigidity and typical feline slow tail switching. Electrical simulation of "flea motion" across the skin results in well-directed scratch reflexes.

Milk touched to its lips, nose, or tongue is adroitly licked clean and swallowed, but weak acid is forcibly rejected. Light stimulation of guard hairs on the margins of its external ear elicits a brisk ear flick. Gentle probing of hairs inside of its pinna results in a vigorously repeated shaking of the head.

If the cat is inverted a few feet above a mattress, and let fall freely, it rotates in the air and lands on its extended feet. If its four feet are placed beneath its center of gravity, a decerebrated cat will reflexly support its own weight against gravity. If it is pushed gently off balance, it falls without attempting to right itself. If a treadmill moves backward under its feet while it is supported upright, it will walk, run, and gallop with appropriate gaits, fitting to the speed of the treadmill. If it is supported in a millrace stream, it will swim at different speeds according to the rate of water flow.

There are many examples of brainstem and spinal mechanisms which obviously underlie behavior in an intact individual. Even in the absence of higher neuronal activity, lower level circuits show remarkably stable and elaborate integrative performance. This does not mean that the lower levels are unmodified in the intact state; an animal is always integrated insofar as integrative circuitry is available. Decerebrated animals behave like complex, "mindless" automatons. They exhibit no social initiatives and do not learn in the ordinary sense of that term.

NEURONAL SEX DIFFERENCES INTEGRATED BY STEROIDS

An integrative strategem of evolution has been to modify the functions of nerve cells by exposing them briefly to hormones. Particular brain regions contain abundant cells with cytoplasmic receptors specific for androgens, estrogens, or progestins. A complex of receptor-steroid moves across the cell membrane, through the cytoplasm, into the nucleus of the cell where it interacts with DNA to regulate specific RNA and protein manufacture and to alter cell functions (McEwen, 1976).

The "natural" mammalian brain is female. Absence of testosterone at a critical perinatal period allows development of female sexual behavior patterns and cyclic release of luteinizing hormone necessary for ovulation.

Gonadal steroids exert powerful effects on neuromorphogenesis and survival of specialized neurons in the hypothalamus and spinal cord. The timing of hormone exposure is critical. The "critical period" for sex steroid sensitivity begins a few days before birth and lasts a few days postnatally. Gorski et al. (1978; 1980) found that natural release of testosterone by the testis at birth enduringly commits a sexually dimorphic nucleus in the medial preoptic area of the rat hypothalamus (SDN-POA). This nucleus is important for control of masculine sexual behavior and for cyclic release of gonadotropin. SDN-POA grows three to seven times larger in the male than in the female. Castration of the newborn male rat produces a reduction to about half the normal adult male volume of SDN-POA; this effect

can be prevented by a single administration of exogenous androgen one day postnatal (Jacobson et al., 1981). Testosterone administered to a female newborn rat significantly increases the adult volume of SDN-POA.

Although steroids given at birth can determine the adult size of the SDN-POA nucleus, permanent effects from a single injection disappear if the injection is given later than about the 10th postnatal day. Hormones administered to adult gonadectomized male and female rats restore copulatory behavior but do not influence SDN-POA volumes.

Testosterone that induces sexual dimorphism is metabolized in the brain to form estradiol. This both masculinizes and defeminizes (Goy and McEwen, 1980). In fact, both estrogens and androgens contribute independently to masculinization of the nervous system. Steroids have a neurotrophic effect: both testosterone and estrogen enhance outgrowth of neural processes in explants of newborn median preoptic tissue. Needle biopsy transplants of male SDN-POA tissue in the newborn do not much change adult secretory patterns of hormones in recipient neonate females, but male SDN-POA transplants do have conspicuous masculinizing effects on female behavior (Arendash and Gorski, 1982; 1983). *Thus, steroid actions are involved in alterations of CNS structure, hormonal secretory patterns, and sexually distinctive behaviors.*

CORTICAL AND SUBCORTICAL CONTRIBUTIONS TO INTEGRATION

Harmon (1957) published an important study showing that expansion of cortical volume during primate evolution—from lemurs to man—is in direct proportion to expansion of the volume of the brain as a whole. Szentagothai (1984) pointed out that expansion of cortex is accompanied by relatively less expansion of subcortical grey matter, the remainder being made up by increases in white matter necessary to support additional connections. Relatively, then, cortical volume is gaining, and with it, potentialities for cortical and subcortical integration.

CORTICAL INTEGRATION

Cortical columns (see Chapter 65, Vision) constitute functional modules, organized for critical analytic functions and decision making. There are about 600 million columns in the human cortex, and the total cortical cell count approaches 50 billion cells (Edelman and Mountcastle, 1978).

Cells in a given cortical area originate from the same locality in the embryonic neural tube. They share similar input and output connections which are distinctive in different cortical areas. Intracolumnar circuits are basically similar within a given functional area but differ in different areas (Szentagothai and Arbib, 1975).

Evidence suggests that all neurons within a column respond to the same locus and modality-specific signal source. This is true whether the source is outside the body, on the body surface, in the body wall, or in the viscera. Since cortical architecture is only partially specified genetically, epigenetic variations in detail abound. Because of epigenetic differences, there are consequently many original circuit variations. When the individual begins to act out in the world, some circuits prove to be more favorably adaptive in reference to the consequences, or outcomes, of behavior. Outcomes of behavior that are successful in obtaining internal (visceral) satisfactions are likely to be associated with a generalized "now print" order (see below) that will thereby "select" the most suitable circuits and consolidate their connections in ways that may be useful for future behavior (Edelman, 1987).

Distinctive functional cortical areas are widely reciprocally interconnected, transcortically and subcortically. Subcortical structures are organized for systematic correlation and coordination with cortical areas, having a shared pattern of connections. As previously noted, there is functional partitioning within subcortical structures which links them to particular cortical areas and to particular cortical (modality specific) representations within those areas. The processes of representation, rerepresentation, and rererepresentation, cortical and subcortical, provide precisely connected, distributed maps, global maps, for correlative coordination contributing to consciousness (Edelman, 1990).

BASAL GANGLIA INTEGRATION

Neostriatum (Caudate and Putamen Nuclei)

Previous chapters showed that the basal ganglia contribute to posture and locomotion. Interactions among cortex, neostriatum, brainstem, and cerebellum are involved in going from static posture to motor performance. How are these circuits integrated for the initiation and control of movement?

Groves (1983) divided neostriatal neurons into two principal functional groups, a dominant population of common spiny cells which exert lateral inhibition, and a group of far less numerous aspiny cells which are excitatory. Each spiny neuron is excited by coincidence of firing from a wide variety of different cortical and brainstem sources, all specific and topologically orderly in their origins and destinations. Both cell types have long axons which

provide the principal striatal outflow. Dynamic excitatory signals, provided by aspiny neurons, accompanied by lateral inhibition, provided by spiny neurons, contributes to smooth motor performance. These striatal neurons are activated in anticipation of movement.

Substantia Nigra

Circuits within the basal ganglia involve different neurotransmitters. This accounts for some disabilites experienced by patients with Parkinson's disease, Huntington's chorea, and other basal ganglia disorders where degenerative processes predominantly affect specific neuron types. The substantia nigra distributes dopaminergic axons to the neostriatum. When these neurons degenerate, Parkinson's disease results. Parkinson patients have to exert special volition to prevent tremor at rest and especially to initiate movements. Patients may benefit from replacement therapy with L-dopa, a precursor of dopamine. Collateral axons of neurons in the substantia nigra also innervate the ventromedial (VM) thalamus and the superior colliculus (SC). These collateral axons contribute through both cortical and collicular circuits to the control of facial, head, and eye movements (Anderson and Yoshida, 1980), which are prominently affected in basal ganglia disorders.

Many neurons in the globus pallidus of the monkey discharge phasically in relation to "voluntary" and "operant" (instrumentally conditioned) contralateral limb movements (DeLong, 1971). Interference with the pallidothalamocortical pathway results in "release" of contralateral "hemiballismic" movements. These flail-like, ballistic gestures of limbs and face characterize vascular damage to the *subthalamic nucleus* which exclusively projects to and receives axons from the globus pallidus (see Chapter 70, Segmental Control of Skeletal Muscle).

LIMBIC INTEGRATION

Limbic cortex is not only phylogenetically old, it is organizationally distinctive from neocortex, both in its internal architecture and in its extrinsic connections (Swanson et al., 1981). Studying complicated circuitry has been advanced by Kuypers and colleagues who introduced retrograde and orthograde axonal transport markers, mostly fluorescent dyes, which can be utilized for identification of cells doubly and multiply labeled according to their various destinations (Kuypers et al., 1977; Van der Kooy et al., 1978). This allows explicit identification of neuronal branchings which serve multiple anatomical destinations.

Individual pyramidal neurons in the hippocampus project by way of collaterals to both the septum and the entorhinal cortex. Commissural, ipsilateral associational, septal, and subicular projections from the hippocampus all come from collaterals of individual hippocampal pyramidal cells. At least 80% of the cells in deep layers of the dentate gyrus of the hippocampus give rise to both ipsilateral associational and commissural crossed projections to dentate granule cells in the contralateral hippocampus. For a general survey of the neurobiology of the hippocampus, see Seifert, 1983.

FRONTAL INTEGRATION

By double-labeling neurons, Goldman-Rakic and Schwartz (1982) found that in frontal lobe association cortex of monkeys, associational projections from the parietal lobe of one hemisphere interdigitate with callosal projections from the opposite frontal lobe, forming adjacent slabs or bands 300–750 μm wide. Similarly orderly interdigitation has been identified as well for frontal projections from primary visual and auditory as well as somatosensory areas of cortex. In association cortical areas as well as in primary sensory and motor systems, vertical 'columns' or 'bands' related to input serving one class of sensory receptors alternate with input from another group of receptors within the same modality. Such side-by-side registration of inputs from one hemisphere and between the two hemispheres contributes to intrahemispheric as well as interhemispheric integration.

Analysis of autoradiographic data reveals that radioactively labeled callosal fibers from the contralateral prefrontal cortex are distributed in 300- to 750-μm bands that extend through all layers of the cortex and alternate with spaces which receive ipsilateral parietal association fibers. Callosal and association axons characteristically occupy mutually exclusive interdigitating columnar territories. Apparently each fiber system is represented across the prefrontal cortical surface as maps of regularly interpenetrating stripes, reminiscent of alternating ocular dominance stripes in primary visual cortex (see Chapter 65, Vision).

A patient whose frontal eye fields have been removed on one side can move his eyes perfectly well in ordinary circumstances, but he cannot direct his gaze in a voluntary searching mode toward the contralateral side. Bilateral frontal eye field removal handicaps both sides for voluntary contralateral conjugate gaze. It is believed that the frontal eye fields do not deliver motor command—they deliver plans for the desired eye movement. If a monkey is trained to move his eyes in a certain direction, specific cells in the frontal eye fields fire just prior to that movement. Yet these "eye movement planning cells" do not fire if the same eye movements are incidental to what the monkey is otherwise doing.

Higher Frontal Integration

Following removal of part of their frontal lobes in order to reduce intractable epileptic seizures, patients have been found unable to shift criteria for card sorting tasks. Milner and Petrides (1984) developed tests to distinguish between left and right frontal lobe lesions. Patients with left frontal lobe surgery have difficulty keeping track of word lists. Patients with right frontal lobe surgery have difficulty keeping track of distinctive drawings. To the researchers' surprise, patients with surgery in the left frontal lobe also have difficulty with the drawing tests. They interpret this to mean that "the left hemisphere as a whole may be dominant for the initiation, programming, and monitoring of sequential tasks," and that perhaps all strategies involve to some degree, verbal formulations.

PARIETAL LOBE INTEGRATION

Inferior parietal cortex (area 7) consititutes a major region for sensorimotor integration. There, visual association signals are linked to motivational messages, contributing to control of the direction of visual attention. Impulse discharge of neurons in the inferior parietal region in alert, behaving monkeys, trained to fixate and follow visual targets, indicate three classes of parietal visual task neurons: (1) cells sensitive to visual stimuli, with large contralateral receptive fields, maximally sensitive in remote temporal quadrants; (2) cells involved in establishing visual fixation; (3) cells relating to the initiation of saccades. These three classes of cells are neither sensory nor motor in the usual sense. They are not activated during spontaneous saccades nor with fixations that the monkey makes when casually exploring the visual environment.

MOTIVATIONAL INTEGRATION

Central Positive Reinforcement

Olds and Milner (1954) discovered localized regions in brainstem, diencephalon, and limbic system which, on stiumulation, provide *central positive reinforcement*. These contain pathways and grey matter and have relatively sharply defined boundaries. The most powerful reinforcing effects are obtained in the median forebrain bundle, a massive throughway of neurons that have mixed diameters, mixed origins, and mixed destinations, and bear the heaviest activity traffic in the brain.

The median forebrain bundle interconnects frontal lobes, limbic system, hypothalamus, and midbrain. The limbic system is positively reinforcing throughout its entire extent. Neuronal outflow from frontal lobes and limbic system pass via the median forebrain bundle to a region in the midbrain known as the *frontal-limbic midbrain area*. This area itself is powerfully positively reinforcing.

This important discovery of centrally rewarding mechanisms was made when rats with implanted electrodes were allowed to roam freely, tethered only by long, flexible leads suspended from overhead. Whenever weak electrical stimulation was applied to animals with electrodes in limbic structures such as the septum or in the median forebrain bundle, the animals would stop whatever they were doing.

It dawned on Olds and Milner that the animals actually "liked" receiving electrical stimulation in certain locations in their brains. That insight provided a conceptual revolution. They went on to establish that rats will cross an open field in any direction the experimenters wish if a brief reinforcement is given each time the animal moves in a desired direction. Rats will learn a complex maze for no reward other than to receive a few electrical stimuli in this positively reinforcing region. Animals will reverse their course through a maze to receive another limited number of electrical stimuli and they quickly learn to run the maze back and forth for reinforcement at each end of the maze. Rats will cross an electrified grid to obtain central stimulation, a grid that no hungry animal will cross for food. Hungry animals and sex-deprived animals will neglect feeding or access to receptive sexual partners in favor of self-stimulation (Olds, 1958).

Central Negative Reinforcement

This was discovered the same year by Delgado, Roberts, and Miller (1954). These authors found that rats will perform work and learn complicated tasks to avoid receiving mild central electrical stimulation. Such negative reinforcement is consistently obtained in a midbrain region overlapping with pain pathway projections. It continues into the ventral posterior hypothalamus. Another negatively reinforcing one passes dorsally, between hypothalamus and thalamus. The negative reinforcement region forks upward from brainstem to diencephalon like a serpent's tongue. The negative reinforcement system narrows and disappears abruptly in the diencephalon in close juxtaposition to positive reinforcement areas.

Electrical stimulation of other regions of the brain, such as the whole of neocortex and the basal ganglia, results in neither positive nor negative reinforcement. These vast regions seem to be with-out intrinsic motivation or motivational relations. Electrical stimulation in such motivationally neutral areas can be linked to behavior as signaling (cueing) events, but they are ineffectual for reinforcement purposes, either positive or negative.

Previously, it was supposed that behavior is controlled by *sensations* originating outside the nervous system. It is now clear that what makes certain sensations pleasant and others unpleasant depends especially on the activation of certain central neuronal systems which convey intrinsically pleasant and unpleasant feelings and attitudes.

Motivations stemming from feelings and appetites are related to these same central reinforcement systems. Feelings and appetites depend on central neuronal systems which are responsible for reward and punishment. Reinforcement systems contribute to approach and avoidance behavior, motivate learning, and modify higher neural systems.

OVERALL NERVOUS SYSTEM INTEGRATION

Figure 9.90 is a diagrammatic schema to account for global integration in the nervous system. A receptor, resembling a pacinian corpuscle, is intended to represent all sensory neuroepithelia connected to all sensory pathways. A skeletal muscle fasicle is intended to represent all effectors, secretory as well as contractile. On the left are depicted phylogenetically new (paucisynaptic) and phylogenetically old (pleurisynaptic) sensory pathways which contribute to spinal, brainstem (and cerebellar), diencephalic, and forebrain levels of integration. On the right are depicted phylogenetically old (pleurisynaptic) and new (paucisynaptic, in this case, monosynaptic) projections from cortex to brainstem, spinal cord, and motor effectors.

Broad ascending and descending arrows represent projections of spinal, brainstem, and diencephalic central core neurons which project cephalically and caudally to control central and peripheral state conditions that establish forebrain alertness and readiness for action along the entire neuraxis.

Note that central state control systems project downward onto *sensory* as well as motor relays, and upward onto association as well as sensory and motor cortex. It should be remembered that reticular activating systems are more discriminating than simply inducing gross arousal and readiness (Brazier and Hobson, 1980). They also contribute to focusing excitabilities and readiness for particular actions in accordance with peripheral and central circuits, visceral, sensory and motor, that are, in turn, impelled by motivational imperatives. This selectivity of activation includes the influences of corresponding memory stores.

Figure 9.91 depicts theoretical functional conver-

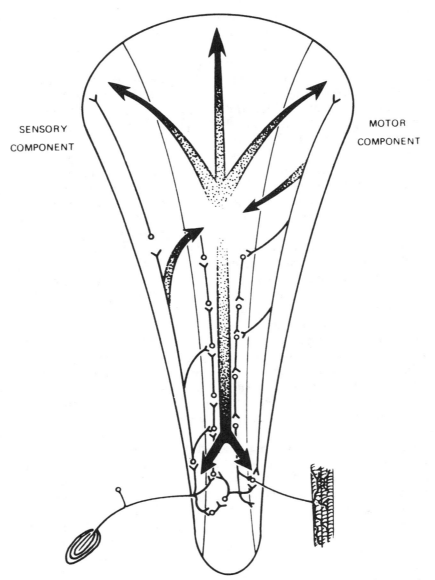

SENSORY
COMPONENT

MOTOR
COMPONENT

Figure 9.90. Diagrammatic representation of central neural signaling for forebrain arousal and arousal of brainstem and spinal sensory and motor background activity. The *left side* represents an ascending sensory column characterizing all sensory systems as having sequences of first-order, second-order, etc. neurons that may be directly and indirectly involved in sensorimotor reflexes and ascending relays to cortex. All sensory trajectories except olfactory send relatively direct signals into the brainstem reticular formation; olfactory arousal is powerful but less direct. The consequences of activation of brainstem reticular formation is ascending arousal affecting the whole forebrain and descending impulses which increase the central excitatory state all along the brainstem and spinal cord. The centrifugal effects influence all sensory as well as motor relays and also efferents such as muscle spindles that directly affect sensory input. The *right side* depicts a descending motor column representing all motor pathways from cortex to final motor neurons and effector organs. The motor as well as sensory columns contribute directly and by way of collaterals to brainstem and spinal reticular formation interneurons and associated diffusely projecting systems. (From Livingston, 1978.)

gence of central motivational and decision-making processes on systems that release or prevent the release of behavior. We have seen that during sleep there is a generalized inhibition of skeletal muscle activation under control of a localized pontine brainstem nucleus (Jouvet, 1973). Evidence from electrical recordings in awake animals during choice discrimination indicates that "go" and "no go" decisions evidently occur in the brainstem approximately between posterior diencephalon and midbrain (Grastyan et al., 1956; Adey, 1974).

At that level, there is a convergence of motivational information represented by the "frontal-limbic midbrain area" where approach and avoidance initiatives can be linked with motor control systems to release behavior ("go") or inhibit its release ("no go"). The consequences of decision-making at this level affect sensory processing as well as behavior

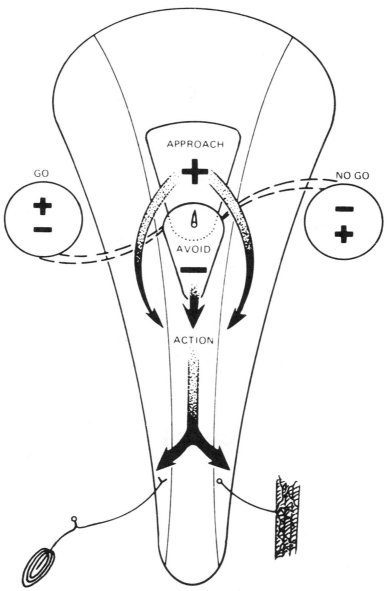

Figure 9.91. Diagrammatic representation of theoretical motivation-driven decision-making systems in cephalic diencephalon and brainstem. Neural activities relating to the status of appetite and satiety mechanisms and any perceived opportunities or threats in the environment relating to biological needs result in increased arousal followed by approach or avoidance behavior, "go," or freeze "no go." Posterior diencephalon and cephalic brainstem exhibit changes in electrical activity associated with switching on and off of both approach and avoidance behavior. Ambiguous stimuli lead to oscillatory electrical activities in this region which closely correlates with the animal's oscillatory behavior. (From Livingston, 1978.)

(depicted by *downward projecting arrows*, Fig. 9.91).

Figure 9.92 represents a theoretical model for memory, based on suggestions by Livingston (1966; 1967a and b), reinforced and made explicit in terms of the likely role of biogenic amines by Kety (1967; 1974). Previous theories of memory assumed synaptic enhancements would be limited to synapses along specific circuits engaged in associating unconditional and conditional stimulus-response events. The "now print" order implies that *all recently active circuits are enhanced ("printed")* whenever a biologically important (punishing or rewarding) event occurs. Accordingly, all recently active circuits are reinforced. This accounts for phenomena of stimulus and response generalization. Ultimately, only those central signals which are uniquely and regularly followed by biologically meaningful events will be consistently reinforced, "selected."

Figure 9.93 illustrates the upward paths from brainstem levels that may induce generalized "now

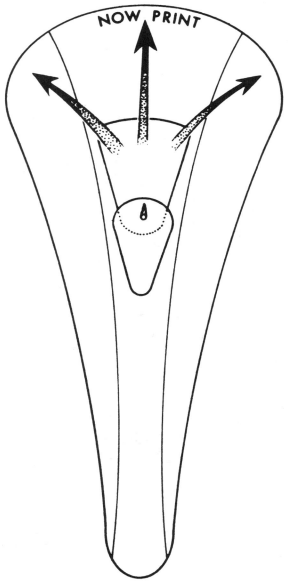

Figure 9.92. Diagrammatic representation of a theoretical mechanism for imprinting of memory. A schematic CNS shows the brainstem reticular mechanism for arousal and theoretical "now print" order for consolidating memory. Theory assumes that when a "biologically meaningful event" occurs, there is activation of cerebral arousal (by known reticular activating mechanisms) and also activation of an additional widely distributed message (presumably norepinephrine from the pontine raphé nucleus, locus coeruleus). It is presumed that widely distributed norepinephrine might effect synaptic consolidation at all recently active synaptic junctions (Kety, 1969). Such a sequence would provide that any significant positive or negative reinforcement, relating to any peripheral or central signaling system or combination of systems, would be followed by strengthening of those (and other) recently active and neural circuits. Repeated pavlovian or operant conditioning would continue consolidation of only those circuits which are instrumental to the reinforcement process. Other adventitious brain events would not be equivalently synaptically reinforced. (From Livingston, 1969.)

print" memory storage throughout the forebrain. Nuclei such as the locus coeruleus, located dorsally near the end of the aqueduct, at the junction of midbrain and pons, project directly to all regions of cerebellum and forebrain and release norepinephrine (NE) with extremely widespread distribution. NE and possibly other aminergic neurotransmitters are postulated to strengthen synaptic connections throughout all recently active circuits. Release of norepinephrine would stimulate protein synthesis through the medium of adenylcyclase. This effect is potentiated by magnesium and potassium ions and inhibited by calcium, indicating that the adrenergic effect would be prepotent for recently active rather than for recently inactive synapses (Kety, 1974).

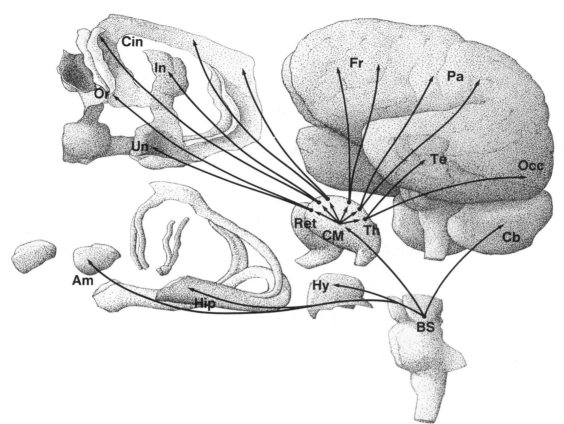

Figure 9.93.   Sketch by Dana R. Livingston from three-dimensional graphics display of normal human brain with selected brainstem, diencephalic, and limbic structures expanded *below* and to the *left*. Activation of the reticular activating system in the tegmentum of the midbrain is followed by arousal of interneurons and other state-related neurons in the hypothalamus, hippocampus, and amygdala. This is by way of the midbrain, centrum medianum, intralaminar and reticularis nuclei in the thalamus. Impulses radiate to all areas of the forebrain, including the limbic system indicated by the double ring of archicortex and mesocortex in *upper and lower left*.

INTEGRATION OF CONSCIOUSNESS

Consciously experienced learning may become so habituated that it consolidates below the conscious level into neuronal circuits by the process of "selection" so that the behavior no longer needs reinforcement and performs like genetically endowed behavior. Conditioned visceral behavior may be of this nature and capable of contributing to shaping and emotionally coloring conscious experiences.

The boundary between unconscious and conscious experience is difficult to imagine; the interface is apparently moving dynamically in time and three-dimensionally in brain-space. It occupies finely scaled neurogeometric regions, and spreads, dilates, thins, retracts, outreaches, etc. (Think of a large, collapsing parachute cloth that is made up of many interwoven pieces and surfaces that are irregularly connected, and are characterized by dynamic ribbons and fenestra. This is occasionally being puffed up and collapsed, undulating, ruffling, rising and sinking throughout delicate brain architecture.)

This "dynamic subjective frontier" of consciousness may be as fine-grained as cortical columns and cortical columnar bands (see Figs. 9.46, 9.49, and 9.50). Consider, for example, the alternating strabismus involved in "seeing alternately with one or the other eye" (see Chapter 65, Vision). Study the "zebra pattern" which interlaces left-eye and right-eye ocular dominance patterns as represented in occipital cortex in each hemisphere. Consciousness must alternate instantaneously and reciprocally, between the immediately adjacent left- and right-eyed cortical representations. In general, conscious experience and the direction of attention seems to be organized throughout cortical and subcortical regions in this fleetingly dynamic way.

CHAPTER 72

Higher Neural Functions

SLEEP

Sleep introduces remarkable changes in neurophysiological activity. The neurological examination of a normal adult, asleep, shows many "abnormalities." Thresholds for event-related potentials (ERPs) are greatly elevated. Muscle tone and reflex activity are extraordinarily reduced; superficial reflexes are depressed and deep tendon responses are practically absent. Metabolic rate is low, as is core temperature; skin temperature is slightly elevated. Heart rate and blood pressure are greatly reduced, as is respiratory function. Similar findings in a waking subject would be disturbing to the physician. Taken together they signify profound neurological depression, except that they "normalize" when the subject is awake.

Although perceptual mechanisms are markedly obtunded during sleep, they are nonetheless selective; they may be capable of initiating arousal with specific, quite subtle stimuli. For example, a child's whimper may be highly arousing even though that sound is less loud than many ambient sounds.

Sleep cycles are regulated by a biological clock (see Suprachiasmatic Circadian Control in Chapter 64). Circadian rhythm has a powerful effect on sleep patterns. This is obvious to travelers flying east and west. Flying east is somewhat more troublesome, and it takes about as many days to recover as time zones traversed.

Sleep is divided into stages I–IV (Fig. 9.94). The electroencephalogram (EEG) of a normal relaxed adult in the waking state shows a random pattern of low-voltage rapid electrical activity (20–30 Hz) over all scalp regions. With eyes closed, the α *rhythm* (8–13 Hz), always present, becomes more clear-cut. If the person is sleepy, sleep spindles (14–16 Hz) may appear, correlated with brief lapses of consciousness. Early sleep usually discloses a rapid transit to stage IV, a deeply relaxed slow wave sleep (SWS), manifested by highly synchronized brain activity.

After a substantial period of SWS, the EEG record may change to low voltage fast (20–30 Hz) asynchronous activity resembling the waking record. However, muscle relaxation is profound, limbs are flaccid, stretch reflexes are weak or absent. This stage is referred to as paradoxical sleep (sleep with waking EEG) or REM (rapid eye movement) sleep. Most dreaming occurs during REM sleep.

During REM sleep, a region in the pontine reticular formation establishes generalized spinal inhibition which prevents acting out during dreams. As sleep continues, there may be transitions to other sleep stages, with REM sleep occurring at approximately 90-min intervals (Fig. 9.94). REM sleep is important because it apparently contributes to memory consolidation. Figure 9.95 shows EEG recordings from an intact cat with implanted electrodes. Deep structures as well as eye movement (yeux) and an electromyogram (EMG) illustrate waking, SWS, and REM sleep. Note the complete muscular relaxation and the rapid eye movements during REM sleep.

Brainstem Reticular Activating System

In the 1950s, Magoun, Lindsley, and their colleagues (see Magoun, 1950; 1952a; 1952b; 1958), discovered that mechanisms other than the classical sensory pathways are necessary for arousal and for maintaining wakefulness. They made massive electrolytic lesions in the upper brainstem in cats and monkeys, interrupting the major sensory pathways (except for olfaction and vision which do not pass through the brainstem), but sparing the central core of the neuraxis. Other experiments involved interrupting the brainstem core while sparing the major sensory pathways. If the central brainstem reticular formation in the midbrain and its upward projection pathways are spared, even though the major sensory pathways are destroyed, the animals continue regular sleep-wakefulness cycles, seek and consume food, and crudely care for themselves, although they exhibit obviously inappropriate reactions to their environment.

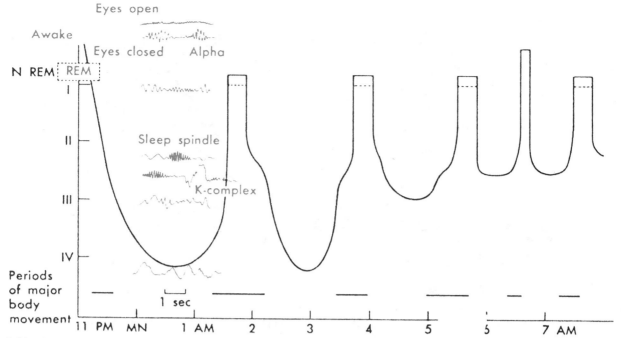

Figure 9.94. Progression of states of sleep during one night's sleep as recorded in a young adult; awake, eyes open; eyes closed with prominent alpha waves, sleep spindles, K-complex and deep slow-wave sleep, characterizing sleep stages I–IV of nonrapid-eye-move- ment (non-REM) sleep. REM sleep occurred four times during the night. One awakening took place shortly before 7 a.m. (From Willis and Grossman, 1981.)

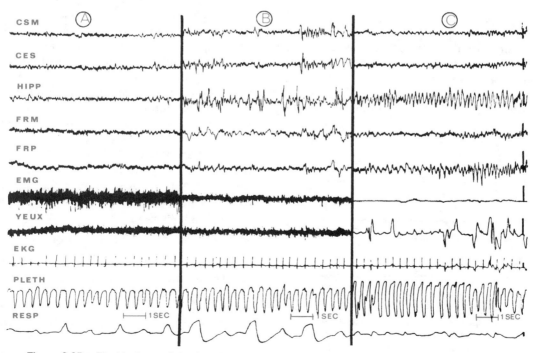

Figure 9.95. Electrical recordings that show differences in waking state (*A*), slowwave sleep (*B*), and paradoxical (REM) sleep (*C*). (From Jouvet, 1967.)

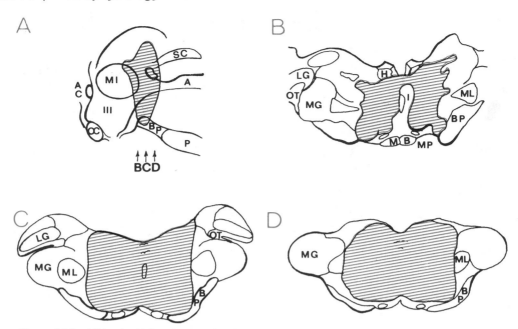

Figure 9.96.   Midsagittal (*A*) and cross-section *(B–D)* through brainstem of cat with destruction of central portions of reticular formation. (From Lindsley et al., 1950.)

Animals with lesions in the midbrain reticular formation, sparing the main sensory paths, are lastingly comatose (Fig. 9.96). EEG tracings (Fig. 9.97) from the cat with the massive lesion above show continuous high amplitude slow waves which are only briefly interrupted by forceful rousing stimuli. There was no spontaneous behavioral or electrographic arousal. Such animals resembled patients in coma and, like patients, had to be conscientiously nursed and tube-fed. Patients in lasting coma are likely to show either a similar massive midbrain reticular lesion or widespread cortical and subcortical lesions throughout the forebrain.

Moruzzi and Magoun (1949) demonstrated that

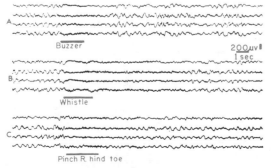

Figure 9.97.   Electroencephalogram of cat with lesion shown in Figure 9.96. Record shows persistent sleep pattern, with brief transition to waking pattern during and just after stimulation by sound or pain. Four different cortical regions show similar wave forms. (From Lindsley et al., 1950.)

electrical stimulation of the midbrain reticular formation transforms SWS sleep into low-voltage fast wave activity typical of the waking EEG pattern. In animals with implanted electrodes, EEG activation and behavioral arousal occurs with such stimulation. The neurophysiological basis for arousal is the *reticular activating system,* RAS (see Fig. 9.98). Access to this system for sensory arousal is by axon collaterals from the primary ascending sensory pathways (lemniscal, spinothalamic, trigeminolemniscal, trigeminothalamic, lateral lemniscal, etc.).

Forebrain arousal induced by the RAS is transmitted via the nucleus basalis and thence by cholinergic projections to cortical and subcortical structures in each hemisphere (Buzsaki et al., 1988). Certain particular areas of cerebral cortex project to the RAS and contribute to arousal and maintenance of arousal (French et al., 1955; Adey et al., 1957). Some of these same corticifugal projection systems were shown to be especially susceptible to seizure activation (French et al., 1956).

Induction of Sleep

Hess (1932) originated techniques for implanting electrodes to enable localized brain stimulation in the waking state after the animal has fully recovered from the anesthetic. Hess discovered that slow frequency stimulation of midline thalamus induces waking cats to sleep. It is neither catelepsy nor adynamia but naturally appearing sleep from which the animal can be readily wakened. This work was rein-

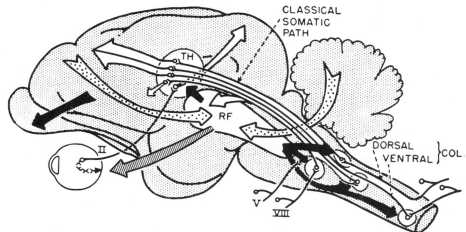

Figure 9.98. Systems for central control of sensory receptors and central sensory transmission. About 10% of efferent neurons are directed to sense receptors and to sensory organs like olfactory bulb, retina, cochlea, taste buds, etc. Sensory relay nuclei characteristically send feedback signals to lower stations, e.g., dorsal column nuclei (second-order neurons) to spinal sensory relay throughout the spinal cord, and cortical sensory areas project downstream onto the thalamic (TH) sources of their input, e.g., primary visual cortex onto lateral geniculate neurons. Moreover, projections from cerebrum and cerebellum to brainstem reticular formation *(RF)* generate diffusely projecting influences which alter sensory input, presumably in more general, nonspecific ways. These sensory control systems are conditionable; hence the brain controls its own input in accordance with past experiences. The full biological significance with respect to sensory signal processing is only beginning to be understood. (From Hernández-Péon and Livingston, 1978.)

forced by Akert et al. (1952) and R. Hess, Jr. (1953), who demonstrated a close correspondence between an animal's sleep-waking behavior and its EEG.

Lesions in the suprachiasmatic and supraoptic area of the hypothalamus in rats produced locomotor hyperactivity and failure to sleep (Nauta, 1946). Generalized hyperactivity, 8–10 fold above normal in monkeys, was induced by removal of a small patch of cortex (area 13) on posterior orbital surface of the frontal lobes bilaterally (Ruch and Shenken, 1943; Livingston et al., 1947a; 1947b). The vagus nerve has its only cortical representation in this area. By destroying a serotonin-rich brainstem raphe nuclei in the pons, a region that induces sleep, Jouvet (1973) produced prolonged wakefulness. This wakefulness controlling site is itself under control of cholinergic neurons in the neighboring brainstem.

HUMORAL MECHANISMS IN SLEEP AND WAKEFULNESS

Jouvet (1967) reviewed evidence for humoral influences affecting sleep mechanisms and concluded that the neurotransmitter serotonin is involved in normal sleep induction. The existence of an endogenous substance that activates behavioral arousal was demonstrated by Dell (1952) and extended by Purpura (1956). Purpura cross-connected circulatory systems in cats and found that stimulation of the reticular activating system in one animal induced prompt EEG arousal of that animal and, following delay of about 1 min, EEG arousal in the cross-circulated cat. It is likely, according to

subsequent work, that both norepinephrine and dopamine are released by such stimulation: NE activates the cross-circulated cortex and DA organizes behavioral readiness for action.

DISORDERS OF SLEEP

Somnolence, disordered sleep patterns, and coma are encountered frequently in medical practice. Higher neural functions deteriorate following REM sleep deprivation. Selective deprivation of REM sleep in animals shows that they need REM sleep and that, given opportunity, they will recover a roughly equivalent amount of REM sleep to the REM sleep deprivation. If deprivation is prolonged, more automatic, subcortical functions begin to deteriorate: tremors, nystagmus, ptosis, dysarthria, and affective disorders appear. As Hobson (1969; 1988) found, certain illnesses are likely to occur when patients are asleep: nocturnal angina, asthma, emphysematous anoxia, and periodic breathing. Some forms of epilepsy occur only during sleep and can be prevented by pharmocological regulation of the depth of sleep.

EMOTION

Physiological Basis of Emotion

Self-awareness relating to our moods and feeling states is directly linked to underlying neural substrates. We are thus continually, directly aware of these states when we are awake, and at least partially aware of them when we are asleep. Moods and feeling states clarify in our consciousness as we awaken, they attend all our waking hours, and they will likely accompany our final flicker of consciousness.

Moods and feeling states are the most direct, intimate, and valid of all our experiences. What we know otherwise—about the rest of the world, even about the rest of our own bodies—is less direct. Moods and feeling states are incompletely shareable socially through emotional expressions and bodily attitudes; even less adequately, by verbal expressions. As subjective states, they are completely private. They are perhaps easier to simulate than to mask. The physician needs to recognize that they are virtually impossible to transform by willpower.

A depressed patient is not relieved by being told to cheer up or to desist from outward expression of his internal state. It is nearly as difficult to hide unbounded joy. We describe the partially publicly sharable expressions of mood and feeling states in another person as emotion. We are the only ones who know what it really feels like, inside ourselves. That is first-intentional experience. All else that we perceive and know is a cognitive construct of second-intentional experiences and speculation.

Emotional Expression

Emotional expression is to a considerable extent involuntary, although from childhood we learn a systematic code for emotional constraint relating to exactly what is, and what is not, socially permissible, for intrafamily and more so for public expression, and on what occasions. Learning to maneuver with appropriate social-emotional finesse is one of life's most challenging and elaborate feats of acculturation. These acquired codes show considerable cross-cultural variability. Socially defined measures of maturity, and mental health, as well, are codified for almost uniform compliance (within a given culture), with respect to what is to be permitted in emotional expression. The fact that expressions differ widely, according to culture, emphasizes the adventitious nature of such viscerosomatic skills.

CATEGORIES OF AFFECT

MacLean (1990) divides affect into three categories: basic, specific, and general. *Basic affects* are subjectively recognized as hunger, thirst, sexual desire, urge to defecate, urinate, regulate body temperature, and in the shortest time-frame, the urge to breathe. *Specific affects* are those introduced by activation of specific sensory systems (e.g., enthusiasm aroused on smelling food cooking, pain on noxious stimulation). *General affects* include traditionally regarded emotions such as love, anger, fear,

disappointment, happiness, and so forth. General affects are distinguished from basic or specific affects inasmuch as they are projected outward, connecting oneself with individuals, institutions, and situations. Moreover, general affects can persist and return after the inciting occasions have passed (see Chapters 67–69 for related neurophysiological insights).

Temporal Lobe Contributions to Vividness of Experience

The temporal lobe can generate, even totally out of context, what is subjectively experienced as real and important. The temporal lobe establishes feelings of deja vu (already seen), an illusion in which a new situation is mistakenly experienced as repetition of previous experience, and jamais vu (never seen), mistakenly experienced as never seen although, in fact, previously experienced. These states occasionally occur in the experience of normal individuals, but in persons with temporal lobe disorder they may compellingly dominate subjective experiences.

Temporal lobe seizures can create feelings of depersonalization—or objects may be seen as abnormally large or small, close or remote; one's body parts may be distorted, time as well as space may be warped, and so on. Very commonly, there may be interference with remembering experiences. These subjective phenomena are associated with toxic psychoses, use of psychedelic drugs, chronic alcoholism, and so forth. Patients with lesions in the temporal lobe may undergo a form of epilepsy where the symptoms include autonomic discharges and emotional experiences. Bursts of anger have been elicited by mild electrical stimulation of the amygdala. Other electrically evoked subjective states include pleasant sensations, elation, deep thoughtful concentration, "odd feelings," utter relaxation, and colored visions (Delgado, 1970).

SPEECH AND ITS DISORDERS

Speech, reading, and writing are elaborate forms of communication. The cortex in the left cerebral hemisphere is usually responsible for speech, even in left-handed persons where the right hemisphere is said to be "dominant" for actions such as writing and other skilled motor performance (Penfield and Roberts, 1966). Speech localization in the right hemisphere, among right-handed individuals, is rare (Milner, 1974).

Aphasia

Aphasia is a term applied to disorders of expression in speech, writing, and sign language, and use of symbols, as well as to disabilities in comprehension of spoken, written, and signed language. Lesions causing aphasia are usually in the cerebral cortex and typically are found in cortical "association" areas.

Using electrical stimulation, it has been possible to interfere with speech production and linguistic thinking. Four critical areas have been found in the left hemisphere: (1) lower prefrontal cortex, Brodmann's area 44 (known as Broca's area); (2) upper frontal cortex on the mesial surface of the hemisphere, anterior to motor representation of the foot; (3) parietal cortex, posterior to the lower part of the postcentral gyrus; and (4) temporal lobe cortex, posteriorly (Wernicke's area). Damage to these areas can result in persisting aphasia.

Four types of aphasias (Head, 1926) include: (*1*) *verbal aphasia,* practical loss of the power to express ideas in words; reading is difficult; may understand written or oral commands; (*2*) *jargon dysphasia,* showing syntactical deficiencies, speaking jargon volubly, agrammatically; (*3*) *naming defects,* unable to name objects or express meaning, confused evaluation of coins; (*4*) *semantic defects,* little difficulty in speech and naming but does not comprehend the meaning of speech, misses jokes in speech and writing.

Other disabilities of speech and perception include: (*1*) *dysarthria,* loss or difficulty of expression with internal speech and psychical aspect of speech unimpaired; (*2*) *agnosia* (not knowing), failure to interpret sensory impressions which enable recognition and categorization of objects or symbols; "word blindness," "word deafness," are forms of visual and auditory agnosia; (*3*) *agraphia,* inability or difficulty in writing language; (*4*) *astereognosis,* loss of ability with eyes closed to recognize objects placed in the hand, or letters or numbers drawn on the skin; cannot recognize commonplace objects by feel; (*5*) *apraxia* (unable to act) inability to perform purposeful movements at will—by command or in imitation, although there is no paralysis.

FRONTAL LOBE FUNCTIONS

Basic stratified stabilities underlying aggression, sexual behavior, drinking, feeding, and elimination activities can be identified at different levels of the brainstem and spinal cord. At the level of the hypothalamus and limbic system, internal and external signals converge to enable integration of endocrine, autonomic, and somatic functions into organized behavioral expressions which correspond to inner emotional states and motives. Neocortex lacks representation for affect. But rich, extensive relations of neocortex, particularly in the frontal lobes, to limbic and hypothalamic and other related mechanisms contribute importantly to limbic and hypothalamic motivations, e.g., rationalizing them with respect to goal-directed behaviors, etc.

Attention and Selective Attention

We have previously concentrated on what enters higher level transactions that needs to be discriminated in order to match information coming in from the outside world. This requires decision-making as exemplified by the example of a hippocampal pyramidal cell depicted in Figure 9.89. In that context, individual cells receive interoceptive information on basal dendrites and exteroceptive olfactory information on apical dendrites. Inputs relating to internal needs are potentially compelling, contingent on arrival of signals indicating a suitable external biological opportunity. Signal coincidence increases the probability of axonal firing, a "decision" being made by that cell in that context.

By extrapolation, a population of such cells is able to make a concerted decision that will be reflected in messages, presumably to the "frontal-limbic midbrain area" where "go" and "no go" decisions are made in a behavioral context (see Fig. 9.91). By virtue of experiential memory, this transaction will be reinforced, more readily available, more prone to "decide" similarly in a similar context that combines internal needs with external opportunities.

How Is Incoming Information Selectively Discriminated?

We observe in examining Figure 9.99 that stratified stabilities can organize behavior on a very large scale to account for eating, drinking, eliminating, reproducing, and defending reactions. How can "higher neural functions" influence such basic behavioral repertoires? How does the nervous system go from being "mainly an automaton" to being highly discriminatory—in accordance with past experiences, expectations and purposes? How do "thinking processes" influence this inventory of stratified stabilities?

In previous chapters we considered how information enters the nervous system, how it is affected in various ways that integrate perceptions and synthesize motor programs, but we largely overlooked: how does a conscious person selectively attend to particular incoming information? Earlier, we established that attention and discrimination occurs, but, now, how?

Figure 9.100, from Desmedt (1981), shows early and middle range event-related potentials (ERPs) recorded during selective attention. The task is

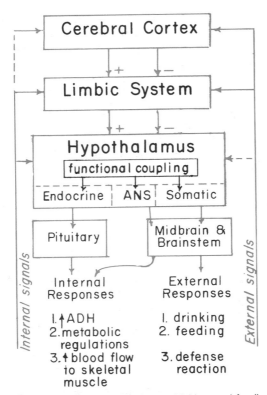

Figure 9.99. Basic reflexes for defense, drinking, and feeding operate through several levels of brainstem and spinal cord. Superimposed are facilitatory and inhibitory controls by limbic system and cerebral cortex. The hypothalamus integrates internal and external signals, as well as input from other neural structures, and thus provides a "functional coupling" of the endocrine, autonomic *(ANS),* and somatic efferent systems. (From Mogenson and Huang, 1973.)

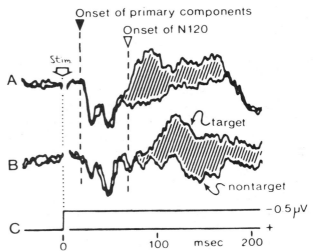

Figure 9.100. Early and middle range event-related potentials (ERPs) in selective attention. Experiment involves a bisensory paradigm with random sequences of acoustic clicks and brief electrical square pulses to one finger. All stimuli are near threshold and difficult to detect. Intervals between stimuli varied randomly between 1 and 15 s. *A* and *B,* ERPs to contralateral finger stimuli recorded from parietal scalp focus with earlobe reference. When the stimuli are targets to be mentally counted, they result in the superimposed traces showing a large N140. *A* and *B* correspond to two experiments on different subjects. The onset of primary components occurs at about 20 ms *(first vertical dashed line).* The primary components are not changed by the task. The N140 is elicited when the finger stimuli are selectively attended. The tracings diverge from control at about 70 ms *(second vertical dashed line).* (From Desmedt, 1981.)

demanding because it involves bisensory stimulation (acoustic and somesthetic) with intensities near threshold—difficult to detect. Moreover, intervals between stimuli are varied randomly over several seconds. When the finger is receiving signals that are to be counted, as compared with such signals that are not to be counted, there is a large amplitude negative far field potential that is maximal about 140 ms after stimulus onset. It is noteworthy that selective attention to the channel in question makes a detectable difference in averaged ERPs. The subject is able, by selective attention, to pick out and selectively literally "signify" cognitively identified events. It is noteworthy that selective attention does not seem to influence far field potentials earlier than about 70 ms after stimulus onset. The cognitive processing must occur within this interval.

Behavior: Higher Command Authority of Parietal Cortex

Figure 9.101 (from Mountcastle et al., 1975) illustrates what remarkable control can be obtained by using well-trained, intelligent animals in circum-

stances in which they draw positive reinforcement as much through individual excitement and challenge as from visceral satisfactions designed into the experience. The posterior parietal cortex is involved with how an animal explores and coordinates eye-hand movements relating to immediate extrapersonal space. The experiment involves establishing the temporal pattern of responses of a single parietal cortical neuron found to be engaged in the course of deliberate reaching of the contralateral hand to touch a moving target. The animal is trained visually to fixate a standing target and then to track that target as it moves, to detect a change in a light on the target, and successfully project the arm and hand to the target in order to receive a reward presented through the drinking tube.

Neurons in area 5 of parietal cortex show consistent distinctive functional responses relating to cutaneous, muscular, other deep tissues, joint rotation, visual target properties and events, including arm projection and hand manipulation. One of the active projection units is documented in Figure 9.101. The alignment of histograms shows that neuronal activity begins to accelerate, on average, prior to the animal's release of the detect key, reaching a peak during motion of the arm toward the target,

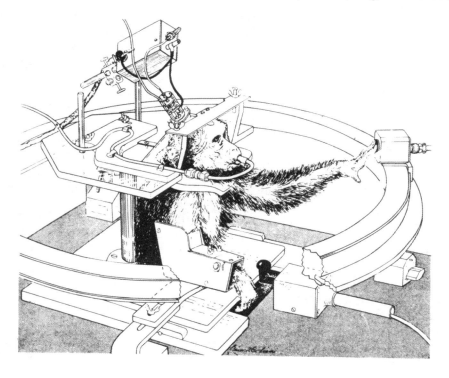

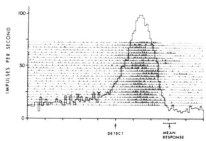

Figure 9.101. *Upper,* Drawing of monkey working in test apparatus used for study of posterior parietal cortex. The head fixation, implanted microelectrode drive, cathode follower, and reward tube are shown *upper left.* The signal key is shown through a cutaway of the circular race. The animal has just released the key with his left hand and projected that arm and hand forward and to his left to contact the lighted switch mounted on the moving carriage, shown *upper right.* The carriage can be moved from any present position in either direction at speed of 12 or 21°/s for preset distances. *Lower,* Replicas of original records, pre-, intra-, and postresponse histograms, made during study of an active projection neuron. The cell never responded to any passively delivered mechanical stimulus to the arm, to visual or auditory stimuli, or during aversive-aggressive movements of the contralateral arm. Each *horizontal line* is the time course of a single trial; each upstroke is the instant at which a nerve impulse occurred. *Left,* records and histograms aligned to the instant of detection *(arrow);* the bar indicates mean response time ± 1 sd. *Right,* the same records and histograms, now aligned by the instant closure of target switch by projected contralateral hand; the bar indicates mean instant detection ± 1 sd. Neuronal activity began to accelerate before release of detect key, reached a peak as arm moved toward target switch, and declined virtually to zero before the hand contacted the switch. (From Mountcastle et al., 1975.)

and declining to below resting rates before the hand actually reaches the target switch.

Parietal cortical neurons, in contrast to those in the precentral motor cortex, do not show much difference in discharge patterns when the projected movements differ greatly (by as much as 60°), or when the arm is forced to go under or over an obstacle to reach the target, or whether the arm is permitted to project to a visually fixed target only after vision has been occluded. The discharge patterns of parietal neurons are also independent of the sensory channels used for cueing signals, somesthetic or visual.

The authors concluded that

"projection and hand manipulation cells . . . compose a command apparatus for manual exploration of extrapersonal space. These command signals do not contain the detailed specification for the movements commanded, matters left . . . to the precentral motor field. The parietal command for projection is holistic in nature; *it is a gestalt.*"

MEMORY AND LEARNING

One of the early speculations concerning the nature of memory is attributed to Müller and Pilzecker (1900), that the establishment of a memory trace takes place in two stages. There is an initial stage in which the memory trace is fragile and easily lost, and a later, more stable state which constitutes a "permanent" memory trace.

In 1977, E. Roy John summarized information relating to memory consolidation. His interpretation of the timing of different phases of memory storage is presented in Figure 9.102. A temporary holding phase that is capable of mediating retrieval decays in a few hours. It appears that learning does not depend on protein synthesis but that retention involves protein synthesis.

Habituation as Negative Learning

If a specific stimulus is repeated several times and then test stimuli are presented, the test responses are depressed, although other unrepeated, i.e., novel, signals elicit full amplitude responses. Apparently little significant habituation occurs in first-order afferent neurons. In hippocampal pyramidal neurons, which show convergence of signals from all sensory channels, instead of cells habituating to characteristics of the stimulus, there is habituation of cells to the orienting response to the stimulus.

Neocortical neurons, in contrast, are relatively stable. After 6 hours of repeated stimuli, little or no habituation was found among responsive neocortical units. The filters appear to be established in interneurons along the incoming sensory pathways

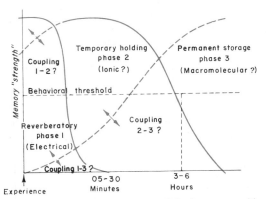

Figure 9.102. Relation of time to "strength" of memory, with *curves* showing hypothetical duration of three phases in memory process. (From John, 1970.)

and in hippocampal neurons (Sokolov, 1976). Clearly, learning depends on neuronal plasticity, and the hippocampus is engaged in perhaps the most plastic of processes involved in learning.

Apparently, decisions about what constitute biologically important events are generated in the hippocampus. The outcome is a discharge to the "frontal-limbic area in the midbrain." From this region, some kind of "now print" order signals *reinforcement or consolidation among all recently active neurons* (Fig. 9.92). This may provide a conceptual link to Edelman's neuronal group selection theory (Edelman, 1987). According to Vingradova (1975; 1980) the hippocampus plays a double role: (a) to establish biologically important signals and (b) to block or unblock brainstem reticular mechanisms involved in comparing novel vs. familiar signals. The outcome of these two functions might well be, "to print, or not print!"

The consolidation of a memory trace for a particular experience involves a complex process that has critical time parameters. Immediately following the learning experience, there is a phase of consolidation. This phase continues during the first subsequent episode of rapid eye movement (REM) sleep. REM sleep may play a role in *enhancing conditions needed for memory consolidation.* These conditions seem to depend upon large-scale forebrain activation, "now print," by locus coeruleus activation during REM sleep (Bloch, 1976).

Squire's View of Memory and Learning

Squire has written an authoritative and judicious review of present perspectives on learning and memory (1987). Whereas we have repeatedly dealt with various aspects of learning and memory, Squire's insight is particularly pertinent here:

"Memory is presumed to depend on the cooperative participation of assemblies of neurons . . . which are specialized to process different kinds of information . . . Each specialized system . . . stores the product of its own processing. Long-term memory of even a single event depends on synaptic change in a distributed ensemble of neurons, which themselves belong to many different processing systems . . . [R]epresentations of events in memory are subject to competition and dynamic change. The strengthening of some connections . . . are the synaptic reflections of rehearsal, relearning, normal forgetting . . . [resulting] in a resculpting of the neural circuitry that originally represented the stored information . . .

"*Declarative memory* . . . is stored as events, facts, images, and propositions . . . explicit knowledge. *Procedural memory* . . . is stored as procedures, or as changes in the facility to perform specific cognitive operations. The ability to find quickly a face that is hidden in a picture is a perceptual skill, but the ability to recognize the face as familiar after it is found depends on declarative representation in memory . . . Information concerning where memory is localized is available in a few well-studied cases . . . the vestibulo-ocular reflex . . . visual discrimination learning, and visual object recognition (see Chapters 64, Vestibular Functions; and 65, Vision). Changes in these pathways appear to store memory . . .

"Complex learning, such as that required for mastery of a visual discrimination problem, is almost certainly stored—at least partially—in neocortex. . . . Certain hormones and transmitters influence the strength of learning and can modulate memory if given close to the time of learning. Some of these effects may involve the action of mechanisms that subserve attention or reinforcement . . . Memory also depends on an interaction established at the time of learning between the cortical [and other] regions that represent experience and a medial temporal/midline diencephalic brain system [involving hippocampus and its descending discharge to the limbic-frontal midbrain area]. This system is the one damaged in amnesia. It provides the capacity for . . . [declarative] memory (Squire, 1987, pp. 241–245).

HIGHER CONSCIOUSNESS

The most primitive stages of consciousness in phylogeny may have arisen, as proposed by Herrick and Coghill, when the customary automatisms of animals prove inadequate to gain internal satisfaction. The basic ingredients of subjective experiences are some degree of self-awareness, feeling, and mood, i.e., the feelings of "self-in-action." An emergent affect or feeling, perhaps a feeling of appetite, tension or fatigue associated with movement, may be the most elementary subjective experiences in early evolution and in individual development (Herrick, 1949).

Many features characteristic of human mentation are observable in lower animals (Griffin, 1981). What we are concerned with is not the search for a liaison between brain as a physical "thing" and some other entity, which we call "mind." We need to understand brain as a functioning tissue that generates feelings, moods, and self-awareness, that constitutes the basic ingredients of mind (Herrick, 1956, Edelman, 1990).

Predictable behavior which might be explained on a rather mechanistic basis, dependent on stratified stabilities, prevails only briefly during vertebrate embryonic life. The individual as a whole soon exhibits unpredictable, individualistic behaviors. This reflects the vast internal degrees of freedom in the human brain, latitudes of freedom accumulating and concatenating during early development, and

expanding degrees of freedom acquired during the course of developing perceptual, judgmental, and behavioral repertoires in a given, dynamic environment.

In the case of humanity, where self-consciousness is obvious, and where we have access to our own subjective experiences, freedom becomes moral freedom when the social implications of our actions are taken into account. As Herrick wrote:

"We have at our command all the physical and mental resources that are needed for further advance [toward more efficient techniques for mentation, for widening the range of experiences and rational interpretations of life] . . . if only we do not squander them in senseless and suicidal rivalry and conflict" (Herrick, 1956, pp. 220–221).

Higher Perceptual Processes

Figure 9.103, from Bach-y-Rita (1972) illustrates a method for sensory substitution, in this case, skin stimulation as a substitute for retinal stimulation. A portable television camera scans the environment and the circuit applies stimuli to the skin of the abdomen in a corresponding but greatly simplified pattern. Thus, patterns of skin stimulation corresponding to images established by the video system are employed as a somesthetic substitute for visual mechanisms lost to blind persons.

After a score of hours of actively directing the camera attached to the moving head, the subject loses the feeling of abdominal skin stimulation and instead perceives forms that are *projected* into the space under exploration by the "gaze" of the camera. Mechanisms of lateral inhibition which operate along somesthetic channels from skin to forebrain establish somesthetic representations, images of objects bounded in three-dimensional space within a dynamic environment. The perception, as projected, is well enough defined so that it can guide behavior. After about 50 hours of actively "seeing" with skin, the individual makes correct automatic averting movements when the projected "visual field" conveys a threat occasioned by motion of the subject or something in the environment.

Bach-y-Rita learned from a young woman, blind from birth, who undertook such sensory substitution that, "You sighted people live in a strange world!" When he inquired how so, she replied, "All your right angles are not right angles when perceived by vision." Her experience with tactile explorations of the world had always affirmed rectilinearity in given architectural contexts, but right angles projected to her by camera perspectives were only rarely rectilinear. The truth of this insight tells

Figure 9.103. A blind subject with a 16-line portable electrical system which includes a TV camera attached to a pair of spectacle frames. Wires lead to an electrical stimulus drive circuitry (held in the right hand) which drives the matrix of 256 concentric silver electrodes which provide skin stimulation in patterns related to video scan density differences. When the blind subject moves the camera across a field or an object, he obtains an image that moves across receptors in his skin. Mechanisms similar to those in the retina, such as lateral inhibition, integrate skin receptor activity to produce edge enhancement. After some score or so hours of active experience while *moving* head with camera attached, blind subject begins to "project" objects perceived through the skin as being "out there" and related spatially to the subject in accordance with the position and direction of the head. (From Bach-y-Rita, 1972.)

us how much of what we experience is reorganized to conform to acquired mental constructions. Stable forms of "reality"—which we have created out of our experiences and assumptions establish "what we know to be true."

SPECIALIZATION OF THE CEREBRAL HEMISPHERES

Until relatively recently, asymmetric brain damage served as the primary source of knowledge concerning functional difference between the two hemispheres. Recently, evidence has been harvested through examination of patients who have had complete or partial section of the corpus callosum.

The operations are designed to help control epileptic seizures that are disabling and refractory to other forms of treatment.

The *corpus callosum* contains more than 200 million axons which interconnect neocortical systems, transferring information from one hemisphere to be integrated with information in the other. However, the corpus callosum does not transfer learning activity or consolidated memory traces. The corpus callosum apparently deals with information from ongoing unilateral experiences which crosses the midline to the contralateral hemisphere for comparison purposes (Doty and Negrão, 1973). This is an advantageous means to store twice the information in two hemispheres.

In both animal and human studies it has been established that different aspects of experience can be represented in the two hemispheres at the same time, even involving stable contradictory conditioning to the same stimuli (Gazzaniga, 1970; Sperry, 1974). Sperry and colleagues, studying humans with section of the corpus callosum, found two distinctive modes of perception which involve complementary specializations in the two sides of the brain (Fig. 9.104).

The *anterior commissure* interconnects temporal lobe structures. In contrast to the corpus callosum, the anterior commissure sends information that instructs and stabilizes memory experiences between the two hemispheres (Doty and Overman, 1977).

The right and left halves of the visual field are divided down the midline and present information to the contralateral hemispheres. The same is true for cerebral representation of somesthetic information. Auditory functions are projected more bilaterally, yet, as we shall see, the auditory contributions to the two hemispheres are distinctive.

Surgical sectioning of the corpus callosum effectively separates neocortex of the two hemispheres and makes much of what goes on in them more independent. Individuals with total section of the corpus callosum, 2 years after surgery, show so few effects of the separation of the hemispheres that the effects would not likely be recognized in a routine medical examination. All cerebral functions, speech and language, verbal intelligence, calculation, motor coordination, reasoning and recall, including personality and temperament remain essentially intact.

There is also a surprising lack of symptoms in persons born with agenesis of the corpus callosum. Some of these individuals are essentially without functional stigmata. They may be able to perform as well as do normal subjects when tested in apparatus that brings out defects in the performance of com-

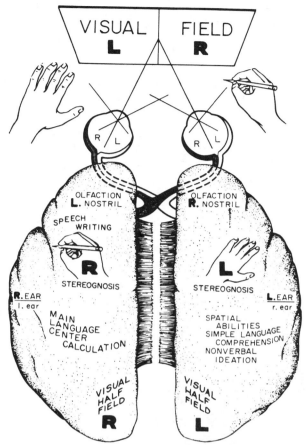

Figure 9.104. Section of the corpus callosum in humans, as a means to prevent spread of epileptic seizure activity, enabled Sperry and collaborators to demonstrate differences between the two hemispheres. This summarizing diagram shows functions known from neuroanatomy, cortical lesion data, and postoperative testing that are separated by the surgery. (From Sperry, 1974.)

missurotomized patients. Intracarotid anesthesia of one hemisphere at a time in one of these subjects indicated that speech had developed on both sides.

Bisected Brain

Sperry and others have demonstrated that despite the apparent normalcy of people with complete callosal section, appropriate tests reveal many distinctive impairments. Even though brainstem and anterior commissure interhemispheric connections remain, left and right sides of the brain function independently in most conscious activities. Each hemisphere is autonomous with respect to sensations, perceptions, ideas, in isolation from the experiences in the contralateral hemisphere. Stored memories in neocortex are unavailable for recall by the other hemisphere. Odors identified only through one nostril are not recognized when exposed to the other nostril (Gordon and Sperry, 1969). There is a

general inability to name objects in the left field of vision, or objects felt by the left hand or foot, or smelled via the left nostril.

These deficits are not very obvious in everyday life because people are usually exploring with eye movements, feeling things with both hands, and using language cues to bilateralize sensory information. Since the retinae are intact and the eyes move conjugately, there is less separation of visual experiences in the two hemispheres than would otherwise be the case. Sperry reports that emotional reactions to stimuli presented to one hemisphere tend to spread easily into the contralateral hemisphere, presumably through intact brainstem projections and perhaps through the anterior and hippocampal commissures which interconnect the two halves of the limbic system.

Figure 9.105 indicates that the two hemispheres, in the absence of the corpus callosum, cannot transfer information necessary for recognition of hand and finger positions either when gestured by the subject or imposed by the examiner. Figure 9.106 shows that the two hemispheres deal independently with composite figures which have different and conflicting images joined in the middle. As a consequence, the hemispheres are each unaware of the discordance between two sides of composite pictures.

In trials of verbal and manual matching options, if linguistic processing is required, the response is dominated by the left hemisphere. If manual matching only is required, the right hemisphere is dominant. An interesting observation by Preilowski and Sperry (1972) indicates that in patients with section of the corpus callosum, either hemisphere may occasionally capture motor control for either side of the body, depending on which hemisphere is dominant for the task. Nonetheless, commissurotomized patients are impaired in learning new motor acts that require coordination between the right and left hands, although habitual tasks such as tying shoelaces and neckties are not noticeably affected.

The right ear, which projects predominantly to the left hemisphere, is employed more adaptively for speech. The left ear, which projects to the contralateral hemisphere, is better at discriminating music. Milner (1974) showed that the right hemisphere is also superior to the left in discriminating and remembering spatial patterns. Right frontal lobe deficits lead to faulty performance in the temporal ordering of nonverbal events.

Children appear to utilize both hemispheres in language function during early development, prior to commitment of language to its characteristic adult location in the left hemisphere. Early damage

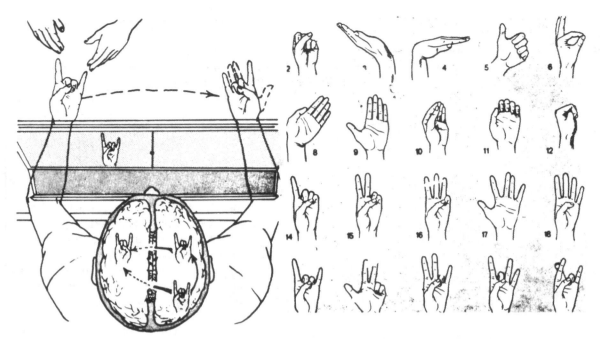

Figure 9.105. Tests used to detect lack of interhemispheric integration following section of the corpus callosum. Subject attempts to replicate complex hand and finger postures flashed to left and right visual half-fields or imposed directly on one hand by examiner. Interhemispheric combinations are performed successfully, but crossed combinations fail. (From Sperry, 1974.)

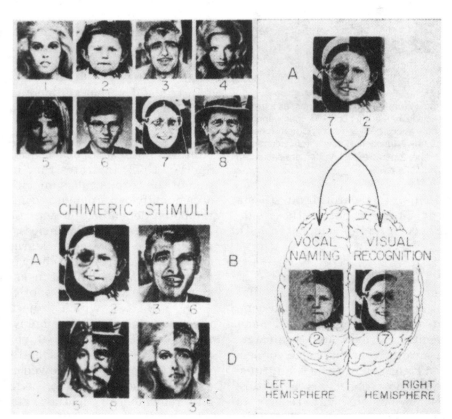

Figure 9.106. Perception with hemispheric disconnection to tachistoscopic presentations of chimeric faces viewed on center apparently involves two separate percepts for which the two test subjects could make correct verbal and manual responses. Manual responses were directed to a match for the left half of the stimulus and the verbal responses described the right half. With such double responses, a conflict between the verbal and the manual responses became evident to the subjects and resulted in considerable perplexity and confusion, lessened by brevity of the exposure. The perplexity evidently arises "as a result of responses by one hemisphere that were recognized to be incorrect by the other hemisphere." (From Levy et al., 1972.)

to protospeech areas in the child's left hemisphere may induce representation of speech in the *right* hemisphere. Berlucchi (1974) found asymmetries by examining normal subjects with respect to visual discriminations, using very brief exposures to the two visual half-fields of noncorresponding, conflicting images. Broadbent (1974) conducted similar studies with normal subjects in relation to conflicting or competing information introduced separately in the two ears.

BIBLIOGRAPHY

ADAMS, J. E. Naloxone reversal of analgesia produced by brain stimulation in the human. *Pain* 2: 161–166, 1976.

ADEY, W. R. Spontaneous electrical brain rhythms accompanying learned responses. In: *The Neurosciences: Second Study Program*, edited by F. O. Schmitt. New York: The Rockefeller University Press, 1974, p. 224–243.

ADEY, W. R. Tissue interaction with nonionizing electromagnetic fields. *Physiol. Rev.* 61: 435–514, 1981.

ADEY, W. R., J. P. SEGUNDO, AND R. B. LIVINGSTON. Corticifugal influences on intrinsic brain-stem conduction in cat and monkey. *J. Neurophysiol.* 20: 1–16, 1957.

ADLER, F. H. *Physiology of the Eye*, 4th ed. St. Louis: Mosby, 1965.

AKERT, K., AND B. E. GERNANDT. Neurophysiological study of vestibular and limbic influences upon vagal outflow. *Electroencephal. Clin. Neurophysiol.* 14: 904–914, 1962.

AKERT, K., W. P. KOELLA, AND R. HESS, JR. Sleep produced by electrical stimulation of the thalamus. *Am. J. Physiol.* 168: 260–267, 1952.

AKERT, K., AND R. B. LIVINGSTON. Morphological plasticity of the synapse. In: *Surgical Approaches in Psychiatry*, edited by L. V. Laitinen and K. E. Livingston. Lancaster, UK: Medical and Technical Publishing Co. Ltd, 1973, p. 315–330.

ALLEN, W. F. Effect of ablating the pyriform-amygdaloid areas and hippocampi on positive and negative olfactory conditioned reflexes and on conditioned olfactory differentiation. *Am. J. Physiol.* 132: 81–91, 1941.

ALLMAN, J. M. Evolution of the visual system in the early primates. In: *Progress in Psychobiology and Physiological Psychology*, edited by J. Sprague and A. Epstein. New York, Academic Press, 1977, vol. 7, p. 1–53.

ALLMAN, J. M., J. H. KAAS, R. H. LANE, AND F. M. MIEZEN. A representation of the visual field in the inferior nucleus of the pulvinar in the owl monkey (Aotus trivirgatus). *Brain Res.* 40: 291–302, 1972.

ALTMAN, J. Autoradiographic and histological studies of postnatal neurogenesis. II. A longitudinal investigation of the kinetics, migration and transformation of cells incorporating tritiated thymidine in infant rats, with special references to postnatal neurogenesis in some brain regions. *J. Comp. Neurol.* 128: 431–473, 1963.

ALTMAN, J. *Organic Foundations of Animal Behavior*. New York: Holt, Rinehart and Winston, 1966.

ALTMAN, J., AND G. D. DAS. Autoradiographic and histological studies of postnatal neurogenesis. *J. Comp. Neurol.* 126: 337–390, 1966.

ALTMAN, J., AND C. D. DAS. Post-natal origin of microneurons in the rat brain. *Nature (London)* 207: 953–965, 1965.

AMOORE, J. E., J. W. JOHNSTON, AND M. RUBIN. The stereochemical theory of odor. *Sci. Am.* 210 (Suppl 2): 42–49, 1964.

ANAND, B. K., AND J. R. BROBECK. Hypothalamic control of food intake in rats and cats. *Yale J. Bio. Med.* 24: 123–140, 1951.

ANDERSON, M. E., AND M. YOSHIDA. Axonal branching patterns and location of nigrothalamic and nigrocollicular neurons in the cat. *J. Neurophysiol.* 43: 883–895, 1980.

ANDERSEN, R. A., AND V. B. MOUNTCASTLE. The influence of the angle of gaze upon the excitability of the light-sensitive neurons of the posterior parietal cortex. *J. Neurosci.* 3: 532–548, 1982.

ARENDASH, G., AND R. GORSKI. Enhancement of sexual behavior in female rats by neonatal transplantation of brain tissue from males. *Science* 217: 1276–1278, 1982.

ARENDASH, G., AND R. GORSKI. Effects of discrete lesions of sexually dimorphic nucleus of the preoptic area or other medial preoptic regions on the sexual behavior of male rats. *Brain Res. Bull.* 10: 147–154, 1983.

AXELROD, J., AND REISINE, T. D. Stress hormones: Their interaction and regulation. *Science* 224: 452–459, 1984.

AZMITIA, E. C., AND A. BJOERKLUND (eds.) *Cell and Tissue Transplantation into the Adult Brain.* New York: New York Academy of Sciences, 1987.

BACH-Y-RITA, P. *Brain Mechanisms in Sensory Substitution.* New York: Academic Press, 1972.

BAINBRIDGE, F. A., AND J. A. MENZIE. *Essentials of Physiology*, 5th ed., edited and revised by C. L. Evans. New York: Longmans, 1925.

BARD, P., AND V. B. MOUNTCASTLE. Some forebrain mechanisms involved in expression of rage with special reference to suppression of angry behavior. *Res. Publ. Ass. Res. Nerv. Ment. Dis.* 27: 362–404, 1948.

BARD, P. Delimitation of central nervous mechanisms involved in motion sickness. *Fed. Proc.* 6: 72, 1947.

BARKER, J. L., AND T. G. SMITH, JR. (eds.) *The Role of Peptides in Neuronal Function.* New York: Dekker, 1980.

BARLOW, H. B., C. BLAKEMORE, AND J. D. PETTIGREW. The neural mechanisms of binocular depth discrimination. *J. Physiol. (London)* 193: 327–342, 1967.

BASBAUM, A. I., C. H. CLANTON, AND H. L. FIELDS. Ascending projections of nucleus raphe magnus in the cat. An autoradiographic study. *Anat. Rec.* 184: 354, 1976.

BASBAUM, A. I., AND H. L. FIELDS. Endogenous pain control mechanisms: review and hypothesis. *Ann. Neurol.* 4: 451–462, 1978.

BAYER, S. A., AND J. ALTMAN. The effects of x-irradiation in the postna-tally-forming granule cell population in the olfactory bulb, hippocampus, and cerebellum of the rat. *Exp. Neurol.* 48: 167–174, 1975.

BAYER, S. A., J. W. YACKEL, AND P. S. PURI. Neurons in the rat dentate gyrus granular layer substantially increase during juvenile and adult life. *Science* 216: 890–892, 1982.

BELLUGI, U., AND E. S. KLIMA. Language: perspectives from another modality. Ciba Foundation Symposium, 69: 99–117, 1979 (See also KLIMA AND BELLUGI, 1979).

BENNETT, M. V. L., AND D. A. GOODENOUGH. Gap junctions, electrotonic coupling, and intercellular communication. *Neurosci. Prog. Bull.* 16: 373–486, 1978.

BERCZI, I. The influence of pituitary hormones and neurotransmitters on the immune system. *J. Immunol. Immunopharmacol.* 8: 186–194, 1988.

BERGLAND, D. W., J. S. CHU, M. A. HOLSEY, L. B. JONES, J. M. KALISZEWSKI, AND B. OAKLEY. New approaches to the problem of the trophic function of neurons. In: *Proceedings of the Sixth International Symposium on Olfaction and Taste*, edited by J. Le Magnen and P. MacLeod. London: Information Retrieval Ltd., 1977, p. 217–224.

BERLUCCHI, G. Cerebral dominance and interhemispheric communication in normal man. In: *The Neurosciences Third Study Program*, edited by F. O. Schmitt and W. G. Worden. Cambridge, MA: The MIT Press, 1974, p. 65–69.

BIZZI, E. The coordination of eye-head movements. *Sci. Am.* 231: 100–106, 1974.

BIZZI, E., AND P. H. SCHILLER. Single unit activity in the frontal eye fields of unanesthetized monkeys during eye and head movement. *Exp. Brain Res.* 10: 150–158, 1970.

BJÖERKLUND, A., AND O. LINDEVALL. The meso-telencephalic dopamine neuron system: a review of its anatomy. In: *Limbic Mechanisms: The Continuing Evolution of the Limbic System Concept*, edited by K. E. Livingston and O. Hornykiewicz. New York: Plenum Press, 1976, p. 307–331.

BJÖERKLUND, A., U. STENEVI, AND N. A. SVENDGAARD. Growth of transplanted monoaminergic neurones into the adult hippocampus along the perforant path. *Nature* 262: 787–790, 1976.

BJÖERKLUND, A., AND U. STENEVI. Intracerebral neural implants: neuronal replacement and reconstruction of damaged circuits. *Ann. Rev. Neurosci.* 7: 279–308, 1984.

BLAKEMORE, C. Developmental factors in the formation of feature extracting neurons. In: *The Neurosciences Third Study Program*, edited by F. O. Schmitt and F. G. Worden. Cambridge, MA: The MIT Press, 1974, p. 105–113.

BLAKEMORE, C., AND J. D. PETTIGREW. Eye dominance in the visual cortex. *Nature (London)* 225: 426–429, 1970.

BLASCHKO, H. Catecholamine biosynthesis. *Brain Med. Bull.* 29: 105–109, 1973.

BLOCH, V. Brain activation and memory consolidation. In: *Neural Mechanisms of Learning and Memory*, edited by M. R. Rosenzweig and E. L. Bennett. Cambridge, MA: The MIT Press, 1976, p. 583–590.

BLOOM, F. E., B. J. HOFFER, AND G. R. SIGGINS. Studies on norepinephrine-containing afferents to Purkinje cells of rat cerebellum. I. Localization of the fibers and their synapses. *Brain Res.* 25: 501–521, 1971.

BLUM, P. S., L. D. ABRAHAM, AND S. GILMAN. Vestibular, auditory and somatic input to the posterior thalamus of the cat. *Exp. Brain Res.* 34: 1–9, 1979.

BODIAN, D. Neuron junctions: a revolutionary decade. *Anat. Rec.* 174: 73–82, 1972.

BODIAN, D. Neurons, circuits, and neuroglia. In: *The Neurosciences. A Study Program*, edited by G. C. Quarton, R. Melnechuk, and F. O. Schmitt. New York: Rockefeller Univ. Press, 1967, p. 6–24.

BORG, E., AND J. E. JACKRISSON. Stapedius reflex and speech features. *J. Acoust. Soc. Am.* 54: 525–527, 1973.

BOSNJAK, Z. B., AND J. P. KAMPINE. Intracellular recordings from the stellate ganglion of the cat. *J. Physiol. (London)* 324: 273–283, 1982.

BRAZIER, M. A. B. (ed.). *Growth and Development of the Brain: Nutritional, Genetic, and Environmental Factors.* International Brain Research Organization Series, Vol. I. New York: Raven Press, 1975.

BRAZIER, M. A. B., AND J. A. HOBSON. *The Reticular Formation Revisited: Specifying Functions for a Nonspecific System.* New York: Raven Press, 1980.

BROADBENT, D. E. Division of function and integration of behavior. In: *The Neurosciences Third Study Program*, edited by F. O. Schmitt and F. G. Worden, Cambridge, MA: The MIT Press, 1974, p. 31–41.

BROADWELL, R. D. Olfactory relationships of the telencephalon and diencephalon in the rabbit. I: An autoradiographic study of the efferent connections of the main and accessory olfactory bulbs. *J. Comp. Neurol.* 163: 329–345, 1975.

BRODAL, A. *The Cranial Nerves: Anatomy and Anatomicoclinical Correlations.* Springfield, IL: Charles C Thomas, 1959.

BRODAL, A. *Neurological Anatomy in Relation to Clinical Medicine*, 3rd ed. New York: Oxford Univ. Press, 1981.

BRONOWSKI, J. New concepts in the evolution of complexity. *Synthese* 21: 228–246, 1970.

BRONOWSKI, J. *The Ascent of Man.* Boston: Little, Brown, 1973.

BROWN, M. C., R. L. HOLLAND, AND W. G. HOPKINS. Motor nerve sprouting. *Ann. Rev. Neurosci.* 4:17–42, 1981.

BUCY, P. C. Sitting on a basketball: how it feels to be a paraplegic. *Perspect. Biol. Med.* 17: 151–163, 1974.

BULLER, A., J. C. ECCLES, AND R. M. ECCLES. Differentiation of fast and slow muscles in the cat hind limb. *J. Physiol.* 150: 399–416, 1960.

BULLOCK, T. H. Representation of information in neurons and sites for molecular participation. *Proc. Natl. Acad. Sci.* 60: 1058–1068, 1968.

BULLOCK, T. H. AND G. A. HORRIDGE. *Structure and Function of the Nervous Systems of Invertebrates.* San Francisco: Freeman and Company, 1965, 2 vols.

BULLOCK, T. H., R. ORKLAND, AND A. GRINNELL. *Introduction to Nervous Systems.* San Francisco: W. H. Freeman and Company, 1977.

BULLOCK, T. H. Reassessment of neuronal connectivity and its specificity. In: Pinsker, H. M., and Willis, W. D. (eds.) *Information Processing in the Nervous System.* New York: Raven Press, 1980, pp. 199–220.

BUNGE, R. P. Glial cells and central myelin sheath. *Physiol. Rev.* 48: 197–251, 1968.

BUNGE, R., M. JOHNSON, AND C. D. ROSS. Nature and nurture in development of the autonomic neuron. *Science* 199: 1409–1416, 1978.

BURKE, D., K-E. HAGBARTH, AND L. LÖFSTEDT. Muscle spindle responses in man to change in load during accurate position maintenance. *J. Physiol.* 276: 159–164, 1978a.

BURKE, D., K-E. HAGBARTH, AND N. F. SKUSE. Recruitment order of human spindle endings in isometric voluntary contractions. *J. Physiol.* 285: 101–112, 1978b.

BUZSAKI, G., R. BICKFORD, G. PONOMAREFF, L. J. THAL, R. MANDEL AND F. H. GAGE. Nucleus basalis and thalamic control of neocortical activity in the freely moving rat. *J. Neurosci.* 8: 4007–4026, 1988.

CAIN, D. P. Sensory kindling: implications for development of sensory prosthesis. *Neurol.* 29: 1595–1599, 1979.

CAJAL, S. R. *Textura del Sistema Nerviosa del Hombre y de los Vertebrados.* Madrid: Moya, 1904.

CALLAWAY, E., P. TEUTING, AND S. H. KOSLOW (eds.). *Event-related Brain Potentials in Man.* New York: Academic Press, 1978.

CANNON, W. B. *The Wisdom of the Body.* New York: Norton, 1932.

CARMEL, P. W., AND A. STARR. Acoustic and nonacoustic factors modifying middle-ear muscle activity in waking cats. *J. Neurophysiol.* 26: 598–616, 1963.

CARPENTER, M. B. *Human Neuroanatomy*, 7th ed. Baltimore: Williams & Wilkins, 1976.

CAVENESS, W. F. Ontogeny of focal seizures. In: *Basic Mechanisms of the Epilepsies*, edited by H. H. Jasper, A. A. Ward, and A. Pope. Boston: Little, Brown, 1969, p. 517–534.

CECCARELLI, B. AND W. P. HURLBUT. Vesicle hypothesis of the release of quanta of acetylcholine. *Physiol. Rev.* 60: 396–434, 1980.

CHANGEUX, J. P., AND S. G. DENNIS. Signal transduction across cellular membranes. *Neurosci. Res. Prog. Bull.* 20: 267–426, 1982.

CLARK, B. Thresholds for the perception of angular acceleration in man. *Aerosp. Med.* 38: 443–450, 1967.

CLARK, B., AND J. D. STEWART. Comparison of sensitivity for the perception of bodily rotation and the oculogyral illusion. *Percept. Psychophys.* 3: 253–256, 1968.

COLLINS, W. E. Vestibular responses from figure skaters. *Aerosp. Med.* 37: 1098–1104, 1966.

COLLINS, W. F., F. E. NULSEN, AND C. N. SHEALY. Electrophysiological studies of peripheral and central pathways conducting pain. In: *Pain*, edited by R. S. Knighton and P. R. Dumke. Boston: Little, Brown, and Company, 1966.

CONEL, J. R. *The Postnatal Development of the Human Cerebral Cortex.* 6 vols. Cambridge, MA: Harvard University Press, 1939–1959.

CONSTANTINE-PATON, M., AND M. LAW. The development of maps and stripes in the brain. *Sci. Am.* 247: 62–70, 1982.

COTMAN, C. W., R. A. BRINTON, A. GALABURDA, B. McEWEN, AND D. M. SCHNEIDER (eds). *The Neuroimmune Endocrine Connection.* New York: Raven Press, 1987.

COTMAN, C. W., M. NIETO-SAMPEDRO, AND E. W. HARRIS. Synapse replacement in the nervous system of adult vertebrates. *Physiol. Rev.* 61: 684–784, 1981.

COWEY, A. Sensory and non-sensory visual disorders in man and monkey. *Philos. Trans. Roy. Soc. Lond. (Biol.)* 25: 3–13, 1982.

CRICK, F. H. A. Thinking about the brain. *Sci. Am.* 241: 219–232, 1979.

CRICK, F. H. C., D. C. MARR, AND T. POGGIO. An information-processing approach to understanding the visual cortex. In: *The Organization of the Cerebral Cortex*, edited by F. O. Schmitt, F. G. Worden, G. Adelman, and S. G. Dennis. Cambridge, MA: The MIT Press, 1981, p. 505–533.

CRITCHLEY, M. *The Citadel of the Senses and Other Essays.* New York: Raven Press, 1986.

CROSBY, E. C., T. HUMPHREY, AND E. W. LAVER. *Correlative Neuroanatomy of the Nervous System.* New York: Macmillan, 1962.

DAVIS, H. Psychophysiology of hearing and deafness. In: *Handbook of Experimental Psychology*, edited by S. S. Stevens. New York: Wiley, 1951, p. 1116–1142.

DAVIS, H. A mechano-electrical theory of cochlear action. *Ann. Otol. Rhin. Laryngol.* 67: 789–801, 1958.

DAVIS, H. Guide for the classification

and evaluation of hearing handicaps in relation to the international audiometric zero. *Trans. Am. Acad. Ophthal. Otolaryngol.* 69: 740–751, 1965b.

DAYHOFF, M. O. The origin of protein superfamilies. *Fed. Proc.* 35: 2132–2138, 1976.

DeCASPER, A. J., AND W. P. FIFER. Of human bonding: newborns prefer their mother's voices. *Science* 208: 1174–1176, 1980.

DELGADO, J. M. R. Modulation of emotions by cerebral radio stimulation. In: *Physiological Correlates of Emotion*, edited by P. Black. New York: Academic Press, 1970, p. 189–202.

DELGADO, J. M. R., W. M. ROBERTS, AND N. E. MILLER. Learning motivated by electrical stimulation of the brain. *Am. J. Physiol.* 179: 587–593, 1954.

DeLONG, M. R. Activity of pallidal neurons during movements. *J. Neurophysiol.* 34: 414–427, 1971.

DELL, P. Correlations entrè le système nerveux vègètatif et le système de la vie de relation: mèsencèphale, diencèphale et cortex cèrèbrale. *J. Physiol. (Paris)* 44: 471, 1952.

DENSERT, O. Adrenergic innervation in the rabbit cochlea. *Acta. Ototlaryngol. (Stockholm)* 78: 345–356, 1974.

DESMEDT, J. E. Auditory-evoked potentials from cochlear to cortex as influenced by activation of the efferent olivocochlear bundle. *J. Acoust. Soc. Am.* 34: 1478–1496, 1962.

DESMEDT, J. E. Scalp-recorded cerebral event-related potentials in man as point of entry into the analysis of cognitive processing. In: *The Organization of the Cerebral Cortex*, edited by F. O. Schmitt. Cambridge, MA: The MIT Press, 1981, p. 441–473.

DeVALOIS, R. L. Central mechanisms of color vision. In: *Handbook of Sensory Physiology*, vol. VII/3A: Central Visual Information, edited by R. Jung. Heidelberg, FRG: Springer, 1973, p. 209–253.

DEWSON, J. H., III. Efferent olivocochlear bundle: some relationships to noise masking and to stimulus attenuation. *J. Neurophysiol.* 30: 817–832, 1967.

DEWSON, J. H., III. Efferent olivocochlear bundle: some relationships to stimulus discrimination in noise. *J. Neurophysiol.* 31: 122–130, 1968.

DINARELLO, C. A., AND S. M. WOLFF. Pathogenesis of fever in man. *N. Engl. J. Med.* 298: 607–612, 1978.

DOANE, B. K. AND K. E. LIVINGSTON. The Limbic System: Functional Organization and Clinical Disorders. New York: Raven Press, 1986, xiv–349.

DONALDSON, I. M. L. The properties of some human thalamic units. Some new observations and a critical review of the localization of thalamic nuclei. *Brain* 96: 419, 1973.

DOTY, R. W., AND N. NEGRÃO. Forebrain commissures and vision. *Handbook of Sensory Physiology*, edited by R. Jung.

Berlin: Springer Verlag, 1973, vol. 7, p. 543–582.

DOTY, R. W., AND W. H. OVERMAN, JR. Mnemonic role of forebrain commissures in macaques. In: *Lateralization in the Nervous System*, edited by S. Harnand, R. W. Doty, J. Jaynes, L. Goldstein, and G. Krauthamer. New York: Academic Press, 1977, p. 75–88.

DOVING, K. B., AND A. J. PINCHING. Selective degeneration of neurons in the olfactory bulb following prolonged odour exposure. *Brain Res.* 52: 115–129, 1973.

DOWNER, J. L. Changes in visualgnostic functions and emotional behavior following unilateral temporal pole damage in the "split-brain" monkey. *Nature (London)* 191: 50–51, 1961.

DRACHMAN, D. B. The biology of myasthenia gravis. *Ann. Rev. Neurosci.* 4: 195–225, 1981.

DREIFUSS, J. J., I. KALNINS, J. S. KELLY, AND K. B. RUF. Action potentials and release of neurohypophysial hormones *in vitro. J. Physiol (London)* 215: 805–817, 1971.

EASON, R. G. Visual evoked potential correlates of early neural filtering during selective attention. *Bull. Psychonomic. Soc.* 18: 203–206, 1981.

EASON, R. G., M. OAKLEY, AND L. FLOWERS. Central neural influences on the human retina during selective attention. *Physiol. Psychol.* 11: 18–28, 1983.

ECONOMO, C., AND L. HORN. Ueber Windungsrelief, Masse und Rindenarchitektonik der Supratemporalflaeche, ihre individuellen und ihre Seil Unterschiede. *Zeitschr. Ges. Neurol. Psychiat.* 130: 678–757, 1930.

EDELMAN, G. M. Cell adhesion and morphogenesis: the regulatory hypothesis. *Proc. Natl. Acad. Sci.* 81: 1460, 1984.

EDELMAN, G. M. *Neural Darwinism: The Theory of Neuronal Group Selection.* New York: Basic Books, 1987.

EDELMAN, G. M. *The Remembered Present: A Biological Theory of Consciousness.* New York: Basic Books, 1989.

EDELMAN, G. M. *Topobiology: An Introduction to Molecular Embryology.* New York: Basic Books, 1988.

EDELMAN, G. M. AND V. B. MOUNTCASTLE. *The Mindful Brain: Cortical Organization and Group-Selective Theory of Higher Brain Function.* Cambridge, MA: The MIT Press, 1978.

EISENMAN, J. S. Pyrogen-induced changes in the thermosensitivity of septal and preoptic neurons. *Am. J. Physiol.* 216: 330–334, 1969.

ENGEN, T., J. E. KUISMA, AND P. E. EIMAS. Short term memory for odors. *J. Exp. Psychol.* 99: 222–225, 1973a.

ENGEN, T., AND B. M. ROSS. Long term memory for odors with and without verbal descriptions. *J. Exp. Psychol.* 100: 221–227, 1973b.

ENGSTRÖM, H., H. W. ADES, AND J. E. HAWKINS. The vestibular sensory cells and their innervation. (In: *Modern Trends in Neuromorphology,*

edited by J. Szentagothai.) *Symp. Biol. Hungary* 5: 21–41, 1965.

EPSTEIN, A. N., AND P. TEITELBAUM. Severe and persistent deficits in thirst produced by lateral hypothalamic damage. In: *Thirst*, edited by M. J. Wayner. Oxford, UK: symposium Publications Division, Permagon Press, 1964, p. 395–410.

ERULKAR, S. D. Comparative aspects of spatial localization sound. *Physiol. Rev.* 52: 237–360, 1972.

EVANS, E. F. Effects of hypoxia on the tuning of single cochlear nerve fibers. *J. Physiol.* 238: 65–67, 1974.

EVANS, E. F. The dynamic range problem: place and time-coding at the level of cochlear nerve and nucleus. In: *Neuronal Mechanisms of Hearing*, edited by J. Syka and L. Aitkin. New York: Plenum Press, 1981, p. 69–85.

FAMBROUGH, D. M., D. B. DRACHMAN, AND S. SATYAMURTI. Neuromuscular junction in myasthenia gravis: decreased acetlycholine receptors. *Science* 182: 293–295, 1973.

FEDERATION PROCEEDINGS SYMPOSIUM. Development of the autonomic nervous system, chaired by P. M. Gootman. *Fed. Proc.* 42: 1619–1655, 1983.

FISCHER, B., AND G. F. POGGIO. Depth sensitivity of binocular cortical neurons of behaving monkeys. *Proc. Roy. Soc. (Biol.)* 204: 409–414, 1979.

FISCHER, C., W. R. INGRAM, AND S. W. RANSON. *Diabetes Insipidus and the Neuro-hormonal Control of Water Balance: A Contribution to the Structure and Function of the Hypothalamico-Hypophyseal System.* Ann Arbor, MI: Edwards Brothers, 1938.

FOLKOW, F., AND E. RUBINSTEIN. Behavioural and autonomic patterns evoked by stimulation of the lateral hypothalamic area in the cat. *Acta Physiol. Scand.* 65: 292–299, 1965.

FRENCH, J. D., R. HERNANDEZ-PEON, AND R. B. LIVINGSTON. Projections from cortex to cephalic brain stem (re formation) in monkey. *J. Neurophysiol.* 18: 74–95, 1955.

FRENCH, J. D., B. E. GERNANDT, AND R. B. LIVINSGTON. Regional differences in seizure susceptibility in monkey cortex. *Arch. Neurol. Psychiat. (Chicago)* 75: 260–274, 1956.

FRENCH, N. R., AND J. C. STEINBERG. Factors governing the intelligibility of speech sounds. *J. Acoust. Soc. Am.* 19: 90–119, 1947.

FURUKAWA, T., AND Y. ISHII. Neurophysiological studies on hearing in gold fish. *J. Neurophysiol.* 30: 1377–1403, 1967a.

FURUKAWA, T., AND Y. ISHII. Effects of static bending of sensory hair cells on sound reception in the goldfish. *Jpn. Physiol.* 17: 572–588, 1967b.

FUSTER, J. M., AND J. P. JERVEY. Neuronal firing in the inferotemporal cortex of monkey in a visual memory task. *Neurosci.* 2: 361–375, 1982.

GACEK, R. R., AND M. LYON. The localization of vestibular efferent neurons in the kitten with horseradish peroxidase *Acta Otolaryngol. (Stockholm)* 77: 92–101, 1974.

GAGE, F. H. AND A. BJOERKLUND. Denervation-induced enhacement of graft survival and growth. Cell and Tissue Transplantation into the Adult Brain. *Ann. N.Y. Acad. Sci.* 495: 378–395, 1987.

GAGE, F. H. AND A. BJOERKLUND. Compensatory collateral sprouting of aminergic systems in the hippocampal formation following partial deafferentation. *The Hippocampus* 3, 1986b.

GAGE, F. H., AND A. BJOERKLUND. Neural grafting in the aged rat brain. *Ann. Rev. Physiol.* 8: 447–459, 1986a.

GAGE, F. H., G. BUZSAKI, O. NILSSON, AND A. BJOERKLUND. Grafts of fetal cholinergic neurons to the deafferented hippocampus. In: F. J. Seil, E. Herbert and B. M. Carlson, (eds.), *Progress in Brain Research* 71: 335–347, 1987.

GAGE, F. H., S. B. DUNNETT, U. STENEVI, AND A. BJOERKLUND. Aged rats: recovery of motor impairments by intrastriatal nigral drafts. *Science* 221: 966–968, 1983.

GAGE, F. H., U. STENEVI, T. CARLSTEDT, G. FOSTER, A. BJOERKLUND, AND A. J. AGUAYO. Anatomical and functional consequences of grafting mesencephalic neurons in a peripheral nerve "bridge" connected to the denervated striatum. *Exper. Brain Res.* 60: 584–589, 1985.

GAGE, F. H., J. A. WOLFF, M. B. ROSENBERG, L. XU, J.-K. YEE, C. SHULTS, AND T. FRIEDMANN. Grafting genetically modified cells to the brain: Possibilities for the future. *Neuroscience* 3: 795–807, 1987.

GALABURDA, A. M., F. SANIDES, AND N. GESCHWIND. Human brain, cytoarchitectonis, left-right assymetries in the temporal speech regions. *Arch. Neurol.* 35: 812–817, 1978.

GALAMBOS, R. Processing of auditory information. In: *Brain and Behavior*, edited by M. A. B. Brazier. Washington, D.C.: Amer. Instit. Biol. Sci., 1961, vol. 1, p. 171–203.

GALAMBOS, R. Suppression of auditory nerve activity by stimulation of efferent fibers to cochlea. *J. Neurophysiol.* 19: 424–437, 1956.

GALAMBOS, R. The human auditory evoked response. In: *Sensation and Movement*, edited by H. R. Moskowitz. Dordrecht, Netherlands: Reidel Publishing Co, 1974, p. 215–221.

GALAMBOS, R., AND S. A. HILLYARD. Electrophysiological approaches to human cognitive processing. *Neurosci. Res. Prog. Bull.* 20: 141–265, 1981.

GALAMBOS, R., J. SCHWARTZKOPF, AND A. RUPERT. A microelectrode study of superior olivary nuclei. *Am. J. Physiol.* 197: 527–536, 1959.

GALAMBOS, R., G. SHEATZ, AND V. G. VERNIER. Electrophyiological correlates of a conditioned response in cats. *Science*, 123: 376–377, 1956.

GARDNER, E. P. AND R. M. COSTANZO. Neuronal mechanisms underlying direction sensitivity of somatosensory cortical neurons in awake monkeys. *J. Neurophysiol.* 43: 1342–1354, 1980.

GAZZANIGA, M. S. *The Bisected Brain*, New York: Appleton-Century-Crofts, 1970.

GERNANDT, B. E., M. IRANYI, AND R. B. LIVINGSTON. Vestibular influences on spinal mechanisms. *Exp. Neurol.* 1: 248–273, 1959.

GERNANDT, B. E., Y. KATSUKI, AND R. B. LIVINGSTON. Functional organization of descending vestibular influences. *J. Neurophysiol.* 20: 453–469, 1957.

GERNANDT, B. E., AND S. GILMAN. Vestibular and propriospinal interactions and protracted spinal inhibition by brain stem activation. *J. Neurophysiol.* 23: 269–287, 1960.

GERSHON, M. D. The enteric nervous system. *Ann. Rev. Neurosci.* 4: 227–272, 1981.

GERSHON, M. D., R. F. PAYETTE, AND R. P. ROTHMAN. Development of the enteric nervous system. *Fed. Proc.* 42: 1620–1625, 1983.

GESTELAND, R. C., J. Y. LETTVIN, AND W. H. PITTS. Chemical transmission in the nose of the frog. *J. Physiol.* 181: 525–559, 1965.

GIBSON, J. J. *The Senses Considered as Perceptual Systems*. Boston: Houghton Mifflin, 1966.

GLOOR, P. Inputs and outputs of the amygdala: what the amygdala is trying to tell the rest of the brain. In: *Limbic Mechanisms: The Continuing Evolution of the Limbic System Concept*, edited by K. E. Livingston and O. Hornykiewicz. New York: Plenum Press, 1976, p. 189–210.

GLOTZBACH, S. F., AND H. C. HELLER. Central nervous regulation of body temperature during sleep. *Science* 194: 537–539, 1976.

GODDARD, G. V. Development of epileptic seizures through brain stimulation at low intensity. *Nature (London)* 214: 1020–1021, 1967.

GODDARD, G. V., AND R. M. DOUGLAS. Does the engram of kindling model the engram of normal long term memory? In: *Kindling*, edited by J. A. Wada. New York: Raven Press, 1976, p. 1–18.

GODDARD, G. V., D. C. McINTYRE, AND C. K. LEECH. A permanent change in brain function from daily electrical stimulation. *Exp. Neurol.* 25: 295–330, 1969.

GODDARD, G. V., B. L. McNAUGHTON, R. M. DOUGLAS, AND C. A. BARNES. Synaptic change in the limbic system: evidence from studies using electrical stimulation with and without seizure activity. In: *Limbic Mechanisms: The Continuing Evolution of the Limbic System Concept*, edited by K. E. Livingston and O. Hornykiewicz. New York: Plenum Press, 1976, p. 355–368.

GOLDBERG, J. M., AND P. B. BROWN. Responses of binaural neurons of dog superior olivary complex to dichotic tonal stimuli: some physiological mechanisms of sound localization. *J. Neurophysiol.* 32: 613–636, 1969.

GOLDMAN-RAKIC, P. S. Neuronal development and plasticity of association cortex in primates. *Neurosci. Res. Prog. Bull.* 20: 471–479, 1982.

GOLDMAN-RAKIC, P. S., AND M. L. SCHWARTZ. Interdigitation of contralateral and ipsilateral columnar projections to frontal association cortex in primates. *Science* 216: 755–757, 1982.

GOODHILL, V. *Beethoven: Triumph over Silence*. Composite 37 min, film. Los Angeles: UCLA Film Library, 1983.

GORDON, H. W., AND R. W. SPERRY. Lateralization of olfactory perception in the surgically separated hemispheres in man. *Neuropsychologia* 7: 111–120, 1969.

GORSKI, R. A., J. H. GORDON, J. E. SHRYNE, AND A. M. SOUTHAM. Evidence for a morphological sex difference within the medial preoptic area of the rat brain. *Brain Res.* 148: 333–346, 1978.

GORSKI, R., R. HARLAN, C. JACOBSON, J. SHRYNE, AND A. SOUTHAM. Evidence for the existence of a sexually dimorphic nucleus in the preoptic area of the rat. *J. Comp. Neurol.* 193: 529–539, 1980.

GOTTSCHALDT, K-M., AND C. VAHLEHINZ. Merkel cell receptors: structure and transducer function. *Science* 214: 183–186, 1981.

GOY, R. W., AND B. S. McEWEN. *Sexual Differentiation of the Brain*. Cambridge, MA: M.I.T. Press, 1980.

GRASTYAN, E., G. KARMOS, L. VERECZKEY, AND L. KELLENYI. The hippocampal electrical correlates of the homeostatic regulation of motivation. *Electroencephal. Clin. Neurophysiol.* 21: 34–53, 1966.

GRAZIADEI, P. P. C. Functional anatomy of the mammalian chemoreceptor system. In: *Chemical Signals in Vertebrates*, edited by D. Mueller-Schwarze and M. M. Mozell. New York: Plenum Press, 1976.

GREER, C. A., W. B. STEWART, M. T. TEICHER, AND G. M. SHEPHERD. Functional development of the olfactory bulb and a unique glomerular complex in the neonatal rat. *J. Neurosci.* 2: 1744–1759, 1982.

GRIFFIN, D. R. *The Question of Animal Awareness: Evolutionary Continuity of Mental Experience*. New York: Rockefeller University Press, 1981.

GROVES, P. M. A theory of functional organization of the neostriatum and the neostriatal control of voluntary movement. *Brain Res. Rev.* 5: 109–132, 1983.

GUEDRY, F. E. Psychophysiological

studies of vestibular function. In: *Contributions to Sensory Physiology*, edited by W. D. Neff, New York: Academic Press, 1965.

HABER, R. N., AND L. S. NATHANSON. Processing of sequentially presented letters. *Percept. Psychophys.* 5: 359–361, 1969.

HAGBARTH, K.-E., AND A. B. VALLBO. Mechanoreceptor activity recorded percutaneously with semimicroelectrodes in human peripheral nerves. *Acta Physiol. Scand.* 69: 121–122, 1967.

HAMMEL, H. T. Neurons and temperature regulation. In: *Physiological Controls and Regulations*, edited by W. S. Yamamoto and J. R. Brobeck. Philadelphia: Saunders, 1965, p. 71–97.

HAMMES, G. G. Receptor biophysics and biochemistry: enzymes. *Neurosci. Res. Prog. Bull.* 11: 164–175, 1973.

HARDY, J. D. The "set-point" concept in physiological temperature regulation. In: *Physiological Controls and Regulations*, edited by W. S. Yamamoto and J. R. Brobeck. Philadelphia: Saunders, 1965, p. 98–116.

HARMAN, P. J. *Paleoneurologic, Neoneurologic, and Ontogenetic Aspects of Brain Phylogeny. James Arthur Lecture on the Evolution of the Human Brain*. New York: American Museum of Natural History, 1957, 24 pp.

HARRIS, D. A., AND E. HENNEMAN. Feedback signals from muscle and their efferent control. In: *Medical Physiology*, 14th ed., edited by V. B. Mountcastle. St. Louis: C. V. Mosby, 1980, vol. 1, p. 703–717.

HAUG, H. Remarks on the determination and significance of the gray cell coefficient. *J. Comp. Neurol.* 104: 473–492, 1956.

HEAD, H. *Aphasia and Kindred Disorders of Speech*. Cambridge, UK: Cambridge Univ. Press, 1926, 2 vols.

HECHT, S., S. SHLAER, AND M. H. PIRENNE. Energy, quanta and vision. *J. Gen. Physiol.* 25: 819–840, 1942.

HEIMER, L. The olfactory cortex and the ventral striatum. In: *Limbic Mechanisms: The Continuing Evolution of the Limbic System Concept*, edited by K. E. Livingston and O. Hornykiewicz. New York: Plenum Press, 1976, p. 95–187.

HEISTAD, D. D. The blood-brain barrier. *Fed. Proc. (Symp.)* 43: 185–219, 1984.

HELD, R., AND J. A. BAUER, JR. Visually-guided reaching in infant monkeys after restricted rearing. *Science* 155: 718–720, 1967.

HERNANDEZ-PEON, R. Neurophysiological correlates of habituation and other manifestations of plastic inhibition (internal inhibition). *Electroencephal. Clin. Neurophysiol.* 13 (Suppl): 101–114, 1960.

HERNANDEZ-PEON, R., H. SCHERRER, AND M. JOUVET. Modification of electric activity in cochlear nucleus during "attention" in unanesthetized cats. *Science* 123: 331–332, 1956.

HERRICK, C. J. *George Ellett Coghill, Naturalist and Philosopher*. Chicago: Univ. of Chicago Press, 1949.

HERRICK, C. J. *The Evolution of Human Nature*. Austin, TX: Univ. of Texas Press, 1956.

HESS, R., JR., W. P. KOELLA, AND K. AKERT. Cortical and subcortical recordings in natural and artificially induced sleep in cats. *Electroenceph. Clin. Neurophysiol.* 5: 75–90, 1953.

HESS, W. R. *Die Methodik der lokalisierten Reizung und Auschaltung subkortikaler Hirnabschnitte*. Leipzig, E. Germany: Georg Thieme, 1932.

HESS, W. R. *Das Zwischenhirn, Syndrome, Lokalisationen, Funktionen*. Basel, Switzerland: Schwabe, 1949, p. 187.

HESS, W. R. *Hypothalamus and Thalamus, Experimental Documentation*. Stuttgart, FRG: Georg Thieme, 1956.

HEUSER, J. E., AND T. S. REESE. Evidence for recycling synaptic vesicle membrane during transmitter release at the frog neuromuscular junction. *J. Cell Biol.* 57: 315–344, 1973.

HOBSON, J. A. Sleep: Physiologic aspects. *N. Engl. J. Med* 281: 1343–1345, 1969.

HOBSON, J. A. *The Dreaming Brain*. New York: Basic Books, 1988.

HOEKFELT, T., O. JOHANSSON, A. LJUNGDAHL, J. M. LUNDBERG, AND M. SCHULTZBERG. Peptidergic neurons. *Nature (London)* 284: 515–521, 1980.

HOLLEY, A., AND DOVING. Receptor sensitivity, acceptor distribution, convergence and neural coding in the olfactory system. In: *Olfaction and Taste*, edited by J. LeMagnen and P. MacLeod. London: Information Retrieval Ltd., 1977, vol. 6, p. 113–123.

HONRUBIA, V. Temporal and spatial distribution of the CM and SP of the cochlea. In: *Frequency Analysis and Periodicity Detection in Hearing*, edited by R. Plomp and G. F. Smoorenburg, Leiden, Netherlands: Sijthoff, 1970, p. 94–106.

HORTON, E. W. Hypotheses on physiological roles of prostaglandins. *Physiol. Rev.* 39: 122–161, 1969.

HOSOBUCHI, Y., J. E. ADAMS, AND R. LIPSHITZ. Pain relief by electrical stimulation of the central gray matter in humans and its reversal by naloxone. *Science* 197: 183–186, 1977.

HOUK, J. D. Regulation of stiffness by skeletomotor reflexes. *Ann. Rev. Physiol.* 41: 99–114, 1979.

HUBBELL, W. L., AND M. D. BOWNDS. Visual transduction in vertebrate photoreceptors. *Ann. Rev. Neurosci.* 2: 17–34, 1979.

HUBEL, D. H., AND T. N. WIESEL. Cells sensitive to binocular depth in area 18 of the macaque monkey cortex. *Nature (London)* 225: 41–42, 1970.

HUBEL, D. H., AND T. N. WIESEL. Brain mechanisms of vision. *Sci. Am.* 241: 150–162, 1979.

HUDSPETH, A. J. The hair cells of the inner ear. *Sci. Am.* 248: 54–56, 1983.

HUDSPETH, A. J., AND R. JACOBS. Stereocilia mediate transduction in vertebrate hair cells. *Proc. Natl. Acad. Sci. U.S.A.* 76: 1506–1509, 1979.

HUGHES, J., T. W. SMITH, H. W. KOSTERLITZ, L. A. FOTHERGILL, B. A. MORGAN, AND H. R. MORRIS. Leuenkephalin and Met-enkephalin-endogenous ligands for opiate receptors. *Nature (London)* 258: 577–579, 1975.

IGGO, A. Cutaneous receptors with a high sensitivity to mechanical displacement. In: *Touch, Heat and Pain*. Ciba Foundation Symposium. Boston: Little Brown, 1966.

IMIG, T. J., AND H. D. ADRIAN. Binaural columns in the primary field (AI) of cat auditory cortex. *Brain Res.* 138: 241–257, 1977.

IMIG, T. J., AND R. A. REALE. Ipsilateral corticocortical projections related to binaural columns in cat primary auditory cortex. *J. Comp. Neurol.* 203: 1–14, 1981.

INGVAR, D. H., AND L. PHILIPSON. Distribution of cerebral blood flow in the dominant hemisphere during motor ideation and motor performance. *Ann. Neurol.* 2: 230–237, 1977.

ITO, M. The control mechanisms of cerebellar motor systems. In: *The Neurosciences Third Study Program*, edited by F. O. Schmitt and F. G. Worden. Cambridge, MA: The MIT Press, 1974, p. 293–303.

IVERSON, L. L. The catecholamines. *Nature (London)* 214: 8–14, 1967.

IVERSEN, L. L., R. A. NICOLL, AND W. W. VALE. Neurobiology of peptides. *Neurosci. Res. Prog. Bull.* 16: 209–330, 1978.

JACKSON, J. H. On a particular variety of epilepsy ("intellectual aura"), one case with symptoms of organic brain disease. *Brain* 11: 179–207, 1888.

JACKSON, J. H., AND W. S. COLMAN. Case of epilepsy with tasting movements and "dreamy state"—very small patch of softening in the left uncinate gyrus. *Brain* 21: 580–590, 1898.

JACOBSON, C. D., V. J. CSERNUS, J. E. SHRYNE, AND R. A GORSKI. The influence of gonadectomy, androgen exposure, or a gonadal graft in the neonatal rat on the volume of the sexually dimorphic nucleus of the preoptic area. *J. Neurosci.* 1: 1142–1147, 1981.

JANTSCH, E. *The Self-Organizing Universe: Scientific and Human Implications of the Emerging Paradigm of Evolution*, edited by E. Laszlo. New York: Permagon Press, 1980.

JEWETT, D. L. Volume-conducted potentials in response to auditory stimuli as detected by averaging in the cat. *Electroencephalogr. Clin. Neurophysiol.* 28: 609–618, 1970.

JEWETT, D. L., AND J. S. WILLISTON. Auditory-evoked far fields averaged from the scalp of humans. *Brain* 94: 681–696, 1971.

JOHN, E. R. *Mechanisms of Memory*. New York: Academic Press, 1967; See also THATCHER, R. W., AND E. R. JOHN: *Foundations of Cognitive Processes*. Hillsdale, NJ: Erlbaum; New York: distributed by Halsted Press, 1977.

JOHN, E. R. Summary symposium on memory transfer, American Association for the Advancement of Science, New York, December 1976. In: *Molecular Approaches to Learning and Memory*, edited by W. L. Byrne. New York: Academic Press, 1970, pp 335–342.

JOHNSON, W. H., R. A. STUBBS, G. F. KELK, AND W. R. FRANKS. Stimulus required to produce motion sickness. *J. Aviat. Med.* 22: 363–374, 1951.

JOUVET, M. Neurophysiology of the states of sleep. *Physiol. Rev.* 47: 117–177, 1967.

JOUVET, M. A possible role of catecholamine-containing neurons of the brainstem of the cat in the sleep-waking cycle. *Acta Physiol. Pol.* 24: 5–19, 1973.

KAAS, J. H., R. J. NELSON, M. SUR, AND M. M. MEZENICH. Multiple representations of the body in the postcentral somatosensory cortex of the primates. In: *Cortical Sensory Organization*, vol. 1: *Multiple Somatic Areas*, edited by N. Woolsey. Clifton, NJ: Humana, 1981, p. 29–45.

KANEKO, A. Physiology of the retina. *Ann. Rev. Neurosci.* 2: 169–191, 1979.

KANDEL, E., AND J. H. SCHWARTZ. *Principles of Neural Science*. New York: Elsevier, 1985.

KATSUKI, Y. Neural mechanisms of auditory sensation in cats. In: *Sensory Communication*, edited by W. A. Rosenblith. Cambridge, MA: The MIT Press, 1961, p. 561–583.

KATZ, B. *Nerve, Muscle, and Synapse*. New York: McGraw-Hill Book Co., 1966.

KEEGAN, J. J., AND F. D. GARRETT. The segmental distribution of the cutaneous nerves in the limbs of man. *Anat. Rec.* 102: 409–437, 1948.

KENNEDY, C., M. DES ROSIERS, J. W. JEHLE. M. REIVICH, F. SHARP, AND L. SOKOLOFF. Mapping of functional neural pathways by autoradiographic survey of local metabolic rates with [14]C. *Science* 187: 850–853, 1975.

KETY, S. S. The central physiological and pharmacological effects of the biogenic amines and their correlations with behavior. In: *The Neurosciences: A Study Program*, edited by G. C. Quarton, T. Melnechuk, and F. O. Schmitt. New York: The Rockefeller University Press, 1967, p. 444–451.

KETY, S. S. The biogenic amines in the central nervous system: Their possible roles in arousal, emotion, and learning. In: *The Neurosciences: Second Study Program*, edited by F. O. Schmitt. New York: The Rockefeller University Press. 1974, p. 324–336.

KHANNA, S. M., AND D. G. B. LEONARD. Basilar membrane tuning in the cat cochlea. *Science* 215: 305–306, 1982.

KILLACKEY, H. P. Development and plasticity of somatosensory cortex. *Neurosci. Res. Prog. Bull.* 20: 507–513, 1982.

KIMURA, D. Cerebral dominance and the perception of verbal stimuli. *Can. J. Physiol.* 15: 166–171, 1961b.

KIMURA, D. Left-right differences in the perception of melodies. *Q. J. Exp. Psychol.* 16: 355–368, 1964.

KIMURA, D. Some effects of temporal lobe damage on auditory perception. *Can. J. Psychol.* 15: 156–171, 1961a.

KITAHATA, L. M., Y. AMAKATA, AND R. GALAMBOS. Effects of halothane upon auditory recovery function in cats. *J. Pharmacol. Exp. Therapeut.* 167: 14–25, 1969.

KLIMA, E. S., AND U. BELLUGI. *The Signs of Language*. Cambridge, MA: Harvard University Press, 1979.

KLING, A. Brain lesions and aggressive behavior of monkeys in free living groups. In: *Neural Bases of Violence and Aggression*, edited by W. S. Fields and W. H. Sweet. St. Louis: Warren H. Green, Inc., 1975.

KLING, A. Effects of amygdalectomy on social-affective behaviour in non-human primates. In: *Neurobiology of the Amygdala*, edited by B. E. Eleftheriou. New York: Plenum Press, 1972, p. 511–536.

KLUVER, H., AND P. C. BUCY. An analysis of certain effects of bilateral temporal lobectomy in the rhesus monkey, with special reference to "psychic blindness." *J. Psychol.* 5: 33–54, 1938.

KNIBESTOL, M., AND A. B. VALLBO. Single unit analysis of mechanoreceptors activity from the human glabrous skin. *Acta Physiol. Scand.* 80: 178–195, 1970.

KOIZUMI, K., T. ISIKAWA, AND C. McC. BROOKS. Control of activity of neurons in the supraoptic nucleus. *J. Neurophysiol.* 27: 878–892, 1964.

KOIZUMI, K., AND H. YAMASHITA. Influence of atrial stretch receptors on hypothalamic neurosecretory neurons. *J. Physiol. (Lond.)* 285: 341–358, 1978.

KONORSKI, J. *Conditioned Reflexes and Neuron Organization*. Cambridge, UK: Cambridge University Press, 1948.

KONORSKI, J. *Integrative Activity of the Brain. An Interdisciplinary Approach*. Chicago: University of Chicago Press, 1967.

KORNER, P. I. Integrative neural cardiovascular control. *Physiol. Rev.* 51: 312–367, 1971.

KOSHLAND, D. E., JR. Bacterial chemotaxis in relation to neurobiology. *Ann. Rev. Neurosci.* 3: 43–75, 1980.

KOSHLAND, D. E., JR., A. GOLDBETER, AND J. B. STOCK. Amplification and adaptation in regulatory and sensory system. *Science* 217: 220–225, 1982.

KOSTREVA. D. R. Functional mappings

of cardiovascular reflexes and the heart using 2-D deoxyglucose. *Physiologist* 26: 333–350, 1983.

KOSTYUK, P. G., AND D. A. VASILENKO. Spinal interneurons. *Ann. Rev. Physiol.* 41: 115–126, 1979.

KRUGER, L. Information processing in cutaneous mechanoreceptors: feature extraction at the periphery. Introduction: an overview. *Fed. Proc.* 42: 2519–2520, 1983; whole Symposium, *Fed. Proc., 42: 2519–2552, 1983.*

KRUGER, L., E. R. PERL, AND M. J. SEDIVEC. Fine structure of myelinated mechanical nociceptor endings in cat hairy skin. *J. Comp. Neurol.* 198: 137–154, 1981.

KUENZLE, H., AND A. AKERT. Efferent connections of cortical area 8 (frontal eye field) in *Macaca fascicularis*. A reinvestigation using the autoradiographic technique. *J. Comp. Neurol.* 17: 147–164, 1972.

KUFFLER, S. W., AND J. G. NICHOLLS. *From Neuron to Brain: A Cellular Approach to the Function of the Nervous System*. Sunderland: MA: Sinauer Assoc., 1976.

KUHL, D. E., J. ENGEL, JR., M. E. PHELPS, AND C. SELIN. Epileptic patterns of local cerebral metabolism and perfusion in humans determined by emission computed tomography of 18 FDG and 13 NH3. *Ann. Neurol.* 8: 348–360, 1980.

KUPPERMAN, R. The dynamic DC potentials in the cochlea of the guinea pig. *Acta Otolaryngol.* 62: 465–480, 1966.

KUPPERMAN, R. The SP connection with movements of the basilar membrane. In: *Frequency Analysis and Periodicity Detection in Hearing*, edited R. Plomp and G. F. Smoorenburg. Leiden, Netherlands: Sijthoff, 1970, p. 126–133.

KUTAS, M., AND S. A. HILLYARD. Reading senseless sentences: brain potentials reflect semantic incongruity. *Science* 207: 203–205, 1980.

KUYPERS, H. G., C. E. CATSMAN-BERREVOETS, AND R. PADT. Retrograde axonal transport of fluorescent substances in the rat's forebrain. *Neurosci. Lett.* 6: 127–135, 1977.

LEE, C. Y. *Chemistry and Pharmacology of Polypeptide Toxins in Snake Venoms*. Pharmacological Institute, College of Medicine, National Taiwan University, Taipei, Taiwan, China, 1972, p. 265–286.

LeVAY, S., AND H. SHERK. The visual claustrum of the cat. I. Structure and connections. *J. Neurosci.* 1: 956–980, 1981.

LEVI-MONTALCINI, R. The nerve growth factor: its mode of action on sensory and sympathetic cells. *Harvey Lectures Ser.* 60: 217–259, 1966.

LEVY, J., C. TREVARTHEN, AND R. W. SPERRY. Perception of bilateral chimeric figures following deconnection. *Brain* 95: 61–78, 1972.

LIBERMAN, M. C. Auditory nerve

responses from cats raised in a low-noise chamber. *J. Acoust. Soc. Amer.* 63: 442–455, 1978.

LICKLIDER, J. C. R. Basic correlates of the auditory stimulus. In: *Handbook of Experimental Psychology,* edited by S. S. Stevens. New York: Wiley, 1951, p. 985–1039.

LINCOLN, D. W., AND J. B. WAKERLY. Factors governing the periodic activation of supraoptic and paraventricular neurosecretory cells during suckling in the rat. *J. Physiol. (Lond.)* 250: 443–461, 1975.

LINDSLEY, D. B., J. W. BOWDEN, AND H. W. MAGOUN. Effect upon the EEG of acute injury to the brain stem activating system. *Electroencephal. Clin. Neurophysiol.* 1: 475–486, 1949.

LINDSLEY, D. B., L. H. SCHREINER, W. B. KNOWLES, AND H. W. MAGOUN. Behavioral and EEG changes following brain stem lesions in the cat. *Electroencephal. Clin. Neurophysiol.* 2: 483–498, 1950.

LINDSTROM, J. M., M. E. SEYBOLD, V. A. LENNON, S. WHITTINGHAM, AND D. D. DUANE. Antibody to acetylcholine receptor in myasthenia gravis: prevalence, clinical correlates, and diagnostic value. *Neurology* 26: 1054–1059, 1976.

LISBERGER, S. G., AND T. A. PAVELKO. Brain stem neurons in modified pathways for motor learning in the primate vestibulo-ocular reflex. *Science* 242: 771–773, 1988.

LISBERGER, S. G., F. A. MILES, AND L. M. OPTICAN. Frequency-selective adaptations: evidence for channels in the vestibulo-ocular reflex? *J. Neurosci.* 3: 1234–1244, 1983.

LIU, C.-N., AND W. W. CHAMBERS. Intraspinal sprouting of dorsal root axons. *Arch. Neurol. Psychiat.* 79: 46–61, 1958.

LIVINGSTON, K. E., AND O. HORNYKIEW-ICZ (eds.). *Limbic Mechanisms: The Continuing Evolution of the Limbic Concept.* New York: Plenum Press, 1978, 542 pp.

LIVINGSTON, R. B. Brain circuitry relating to complex behavior. In: *The Neurosciences Third Study Program,* edited by G. C. Quarton, T. Melnechuk, and F. O. Schmitt. New York: Rockefeller Univ. Press, 1967b, p. 449–515.

LIVINGSTON, R. B. Brain mechanisms in conditioning and learning. *Neurosci. Res. Prog. Bull.* 4: 235–354, 1966.

LIVINGSTON, R. B. Reinforcement. In: *The Neurosciences Third Study Program,* edited by G. C. Quarton, T. Melnichuk, and F. O. Schmitt. New York: Rockefeller Univ. Press, 1967a, p. 449–515.

LIVINGSTON, R. B. *Sensory Processing, Perception, and Behavior.* New York: Raven Press, 1978.

LIVINGSTON, R. B., J. F. FULTON, J. M. R. DELGADO, E. SACHS, JR., S. J. BRENDLER, AND G. D. DAVIS. Stimulation and regional ablation of orbital surface of frontal lobe. *Res. Publ. Assoc. Res. Nerv. Ment. Dis.* 27: 405–420, 1947a.

LIVINGSTON, R. B., W. P. CHAPMAN, K. E. LIVINGSTON, AND L. KRAINTZ. Stimulation of orbital surface of man prior to frontal lobotomy. *Res. Publ. Assoc. Res. Nerv. Ment. Dis.* 27: 421–432, 1947b.

LIVINGSTON, R. B., K. PFENNINGER, H. MOOR, AND K. AKERT. Specialized paranodal and interperinodal glial-axonal junctions in the peripheral and central nervous system: a freeze-etching study. *Brain Res.* 58: 1–24, 1973.

LIVINGSTON, W. K. *Pain Mechanisms.* New York: MacMillan, 1943; Republished, with a Forward by R. Melzack. New York: Plenum Press, 1976.

LIVINGSTON, W. K. Evidence of active invasion of denervated areas by sensory fibers from neighboring nerves in man. *J. Neurosurg.* 4: 140–145, 1947.

LIVINGSTONE, M. S. AND D. H. HUBEL. Anatomy and physiology of a color system in the primate visual system. *J. Neurosci.* 4: 309–356, 1984.

LOH, H. H., L. F. TSENG, E. WEI, AND C. H. LI. Beta-endorphin is a potent analgesic agent. *Proc. Natl. Acad. Sci. USA* 83: 2895–2898, 1976.

LOWENSTEIN, W. R. Modulation of cutaneous mechanoreceptors by sympathetic stimulation. *J. Physiol.* 132: 40–60, 1956.

LOWENSTEIN, W. R., AND M. MENDELSON. Components of receptor adaptation in a Pacinian corpuscle. *J. Physiol.* 177: 377–397, 1965.

LUND, R. D., AND J. S. LUND. Synaptic adjustment after deafferentation of the superior colliculus of the rat. *Science* 171: 804–807, 1971.

LYNCH, G., S. DEADWYLER, AND C. COTMAN. Post lesion axonal growth produces permanent functional connections. *Science* 180: 1364–1366, 1973.

MacLEAN, P. D. The limbic system with respect to self-preservation and the preservation of the species. *J. Nerv. Ment. Dis.* 127: 1–11, 1958.

MacLEAN, P. D. The co-evolution of the brain and family. In: *Anthroquest,* The L. S. B. Leakey Foundation News, No. 24, p. 1, 14–15. Pasadena, CA: The L. S. B. Leakey Foundation, Winter 1982.

MacLEAN, P. D. The limbic brain in relation to the psychoses. In: *Physiological Correlates of Emotion,* edited by P. Black. New York: Academic Press, 1970, p. 130–146.

MacLEAN, P. D. An ongoing analysis of hippocampal inputs and outputs: microelectrode and anatomic findings in the squirrel monkey. In: *The Hippocampus,* vol. I: *Structure and Development,* edited by R. L. Isaacson and K. H. Pribram. New York: Plenum Press, 1975, p. 177–211.

MacLEAN, P. D. *The Triune Brain in Evolution: Role in Paleocerebral Functions.* New York: Plenum Press, 1990.

MACRIDES, F. Dynamic aspects of central olfactory processing. In: *Chemical Signals in Vertebrates,* edited by D. Mueller-Schwarze and M. M. Mozell. New York: Plenum Press, 1976, p. 499–514.

MAGNUS, R. *Koerperstellung,* Berlin: Springer, 1924.

MAGOUN, H. W. Caudal and cephalic influences of the brain stem reticular formation. *Physiol. Rev.* 30: 459–474, 1950.

MAGOUN, H. W. The ascending reticular activating system. *Organization in the Central Nervous System* 30: 480–492, 1952a.

MAGOUN, H. W. An ascending reticular activating system in the brain stem. *Arch. Neurol. Psychiatr.* 67: 145–154, 1952b.

MAGOUN, H. W. *The Waking Brain.* Springfield, IL: Charles C Thomas, 1958.

MARR, D. *Vision.* San Francisco: W. H. Freeman, 1982.

MASERA, R. Sul'esistenza di un particolare organo olfatico nel setto nasale della cavia e di altri roditori. *Arch. Ital. Anat. Embriol.* 48: 157–212, 1943.

MATHEWS, D. F. Response patterns of single neurons in the tortoise olfactory epithelium and olfactory bulb. *J. Gen. Physiol.* 60: 166–180, 1972.

MATTHEWS, P. B. C. Where does Sherrington's "muscular sense" originate? Muscles, joints, corollary discharges? *Ann. Rev. Neurosci.* 5: 189–218, 1982.

MATURNA, H. R., J. Y. LETTVIN, W. S. McCULLOCH, AND W. H. PITTS. Anatomy and physiology of vision in the frog (Rana pipens). *J. Gen. Physiol.* 43: 129–175, 1960.

MAYNE, R. A systems concept of the vestibular organs. In: *Handbook of Sensory Physiology VI/2,* edited by H. H. Kornhuber. Berlin: Springer Verlag, 1974, p. 494–580.

McEWEN, B. S. Steroid receptors in neuroendocrine tissues: topography, subcellular distribution and functional implications. In: *International Symposium on Subcellular Mechanisms in Reproductive Neuroendocrinology,* edited by F. Naftolin, K. J. Ryan, and J. Davies. Amsterdam: Elsevier, 1976, p. 277–304.

McKINLEY, R. W. (ed.). *IES Lighting Handbook,* New York: Illuminating Engineering Society, 1947.

MELLONI, B. *The Internal Ear, An Atlas of Some Pathological Conditions of the Eye, Ear and Throat.* Chicago: Abbott Laboratories, 1957, p. 26–31.

MELVILL JONES, G. Plasticity in the adult vestibulo-ocular reflex arc. *Phil. Trans. Roy. Soc. (Biol)* 278: 319–334, 1977.

MELZACK, R., W. A. STOTLER, AND W. K. LIVINGSTON. Effects of discrete brainstem lesions in cats on perception of noxious stimulation. *J. Neurophysiol.* 21: 353–367, 1958.

MEREDITH, M., D. M. MARQUES, R. D. O'CONNELL, AND F. L. STERN. Vomeronasal pump: significance for male hamster sexual behavior. *Science* 207: 1224–1226, 1980.

MERZENICH, M. M., AND J. H. KAAS. Principles of organization of sensory-perceptual systems in mammals. In: *Progress in Psychobiology and Physiological Psychology*, edited by J. M. Sprague and A. N. Epstein. New York: Academic Press, 1975, p. 2–43.

MESULAM, M-M. A cortical network for directed attention and unilateral neglect. *Ann. Neurol.* 10: 309–325, 1981.

MEYER, J. S., L. A. HAYMAN, T. AMANO, S. NAKAJIMA, T. SHAW, P. LAUZON, S. DERMAN, E. KARACAN, AND Y. HARATI. Mapping local blood flow of human brain by CT scanning during stable xenon inhalation. *Stroke* 12: 426–435, 1981.

MILES, F. A. Centrifugal control of the avian retina. IV. Effects of reversible cold block of the isthmooptic tract on the receptive field properties of cells in the retina and isthmooptic nucleus. *Brain Res.* 48: 131–145, 1972.

MILES, F. A., AND S. G. LISBERGER. Plasticity in the vestibulo-ocular reflex: a new hypothesis. *Ann. Rev. Neurosci.* 4: 273–299, 1981.

MILLER, E. F., AND A. GRAYBIEL. Otolith function as measured by ocular counterrolling. In: *The Role of the Vestibular Organs in Space Exploration*. National Aeronautics Space Administration, NASA SP-77: 121–130, 1965.

MILLER, E. F., JR., A. GRAYBIEL, AND R. S. KELLOG. Otolith organ activity within earth standard, one-half standard and zero gravity environments. *Aerosp. Med.* 37: 399–403, 1966.

MILLER, N. E. Learning of visceral and glandular responses. *Science* 163: 434–445, 1969.

MILLER, N. E., AND B. R. DWORKIN. Different ways in which learning is involved in homeostasis. In: *Neural Mechanisms of Goal-Directed Behavior and Learning*. New York: Academic Press, Inc., 1980.

MILLS, E., AND P. G. SMITH. Functional development of the cervical sympathetic pathway in the neonatal rat. *Fed. Proc.* 42: 1639–1642, 1983.

MILNER, B. Hemispheric specialization: scope and limits. In: *The Neuroscience Third Study Program*, edited by F. O. Schmitt and F. G. Worden. Cambridge, MA: The MIT Press, 1974, p. 75–89.

MILNER, B., AND M. PETRIDES. Behavioral effects of frontal lobe lesions in man. *Trends Neuro.* 7: 403–407, 1984.

MOGENSON, G. J., AND Y. H. HUANG. The neurobiology of motivated behavior. In: *Progress in Neurobiology*, 1973, vol. I, p. 53–83.

MOORE, D. M., S. F. CHEUK, J. D. MORTON, R. D. BERLIN, AND W. B. WOOD, JR. Studies on the pathogenesis of fever. XVII. Activation of leucocytes for pyrogen production. *J. Exp. Med.* 131: 179–188, 1970.

MOORE, E. J. (Ed.). *Bases of Auditory Brain Stem Evoked Responses*. New York: Grune & Stratton, 1983.

MOORE, M. J., AND D. M. CASPAERY. Strychnine blocks binaural inhibition in lateral superior olivary neurons. *J. Neurosci.* 3: 237–242, 1983.

MOORE, R. Y. Visual pathways and the central neural control of diurnal rhythms. In: *The Neurosciences Third Study Program*, edited by F. O. Schmitt and F. G. Worden. Cambridge, MA: The MIT Press, 1974, p. 537–542.

MOORE, R. Y. Monoamine neurons innervating the hippocampal formation and septum: organization and response to injury. In: *Hippocampus*, edited by R. L. Issacson and K. H. Pribram. New York: Plenum Press, 5: 215–237, 1975.

MOREST, D. K. Auditory neurons of the brain stem. *Adv. Otol. Rhinol. Laryngol.* 20: 337–356, 1973.

MORUZZI, G., AND H. W. MAGOUN. Brain stem reticular formation and activation of the E. E. G. *Electroencephal. Clin. Neurophysiol.* 1: 455–473, 1949.

MOTTER, B. C., AND MOUNTCASTLE, V. B. The functional properties of the light-sensitive neurons of the posterior parietal cortex studied in waking monkeys: foveal sparing and opponent vector organization. *J. Neurosci.* 1: 3–26, 1981.

MOULTON, D. G. Electrical activity in the olfactory system of rabbits with indwelling electrodes. In: *Olfaction and Taste*. New York: Macmillan, 1963.

MOULTON, D. G. Minimum odorant concentrations detectable by the dog and their implications for olfactory receptor sensitivity. In: *Chemical Signals in Vertebrates*, edited by D. Mueller-Schwarze and M. M. Mozell. New York: Plenum Press, 1976, p. 455–464.

MOUNTCASTLE, V. B. Modality and topographic properties of single neurons of cat's somatic sensory cortex. *J. Neurophysiol.* 20: 408–434, 1957.

MOUNTCASTLE, V. B. The view from within: pathways to the study of perception. *Johns Hopkins Med. J.* 136: 109–131, 1975.

MOUNTCASTLE, V. B. The world around us: neural command functions for selective attention. The F. O. Schmitt Lecture in Neuroscience for 1975. *Neurosci. Res. Prog. Bull.* 14: 1–47, 1976.

MOUNTCASTLE, V. B., J. C. LYNCH, A. GEORGOPOULOS, H. SAKATA, AND C. ACUNA. Posterior parietal association cortex of the monkey: command functions for operations within extrapersonal space. *J. Neurophysiol.* 38: 871–908, 1975.

MOVSHON, J. A., AND R. C. VAN SLYTERS. Visual neural development. *Ann. Rev. Psychol.* 32: 477–522, 1981.

MUELLER-SCHWARZE, D., AND M. M. MOZELL (eds.). *Chemical Signals in Vertebrates*. New York: Plenum Press, 1976.

MÜLLER, G. E., AND A. PILZECKER. Experimentelle Beitraege zur Lehre vom Gedaechtnis. *Z. Psychol. Physiol. Sinnesorg. (Suppl. 1):* 1–300, 1900.

MURRAY, M., AND M. E. GOLDBERGER. Restitution of function and collateral sprouting in the cat spinal cord: the partially hemisected animal. *J. Comp. Neurol.* 155: 19–36, 1974.

NACHMAN, M., AND L. P. COLE. Role of taste in specific hungers. *Handbook of Sensory Physiology*, vol. IV/2: *Chemical Senses*, edited by L. M. Beidler. Heidelberg, FRG: Springer-Verlag, 1971, p. 337–362.

NAUTA, W. J. H. Hypothalamic regulation of sleep in rats. An experimental study. *J. Neurophysiol.* 9: 285–316, 1946.

NELSON, R. J., M. SUR, D. J. FELDMAN, AND J. H. KAAS. Representations of the body surface in postcentral parietal cortex of Macaca fascicularis. *J. Comp. Neurol.* 192: 611–643, 1980.

NIEMYER, W., AND I. STARLINGER. Do the blind hear better? *Audiology* 20: 510–515, 1981.

NOBACK, C. R., AND R. J. DEMAREST. *The Human Nervous System: Basic Principles of Neurobiology*. New York: McGraw Hill, 1980.

NORGEN, R. A synopsis of gustatory neuroanatomy. In: *Olfaction and Taste VI*, edited by J. LeMagnen and P. MacLeod. London: Information Retrieval, Ltd., 1977, p. 225–232.

OLDS, J. Self-stimulation of the brain. *Science* 127: 315–324, 1958.

OLDS, J. AND P. MILNER. Positive reinforcement produced by electrical stimulation of septal area and other regions of the brain. *J. Comp. Physiol. Psychol.* 47: 419–427, 1954.

OSTERBERG, G. A. Topography of the layer of rods and cones in the human retina. *Acta Ophthalmol.* 13 (Suppl. VI): 103 pp., 1935.

PENFIELD, W., AND H. H. JASPER. *Epilepsy and the Functional Anatomy of the Human Brain*. Boston: Little, Brown, 1954, 896 pp.

PENFIELD, W., AND L. ROBERTS. *Speech and Brain Mechanisms*. Princeton, NJ: Princeton University Press, 1959; reprinted 1966.

PERKEL, D.H., AND T. H. BULLOCK. Neural coding. *Neurosci. Res. Prog. Bull.* 6: 221–348, 1968.

PETERS, A., S. L. PALAY, AND H. DE F. WEBSTER. *The Fine Structure of the Nervous System*. New York: Harper & Row, 1970.

PETRAS, J. M., AND J. F. CUMMINGS. Autonomic neurons in the spinal cord of the rhesus monkey: a correlation of the findings of cytoarchitectonics and sympathectomy with fiber degeneration following dorsal rhizotomy. *J. Comp. Neurol.* 146: 189–218, 1972.

PFAFFMAN, C. Physiological and behavioural process of the sense of taste. In: *Taste and Smell in Vertebrates*. Ciba Foundation. London: J. & A. Churchill, 1970, p. 31–50.

PHELPS, M. E., J. C. MAZZIOTA, AND D.

E. KUHL. Metabolic mapping of the brain's response to visual stimulation: studies in man. *Science* 211: 1445–1448, 1981.

PICTON, T. W., S. A. HILLYARD, H. I. KRAUS, AND R. GALAMBOS. Human auditory evoked potentials. I: Evaluation of components. *Electroencephal. Clin. Neurophysiol.* 36: 170–190, 1974.

POGGIO, G. F., F. H. BAKER, Y. LAMARRE, AND E. RIVA SANSEVER-INO. Afferent inhibition at input to visual cortex of the cat. *J. Neurophysiol.* 32: 892–915, 1969.

POGGIO, G. F., AND B. FISCHER. Binocular interaction and depth sensitivity of striate and prestriate cortical neurons of the behaving rhesus monkey. *J. Neurophysiol.* 40: 1392–1405, 1977.

POTTER, D. D., E. J. FURSHPAN, AND S. C. LANDIS. Transmitter status in cultured rat sympathetic neurons; plasticity and multiple function. *Fed. Proc.* 42: 1626–1632, 1983.

POWELL, T. P. S., W. M. COWAN, AND G. RAISMAN. The central olfactory connexions. *J. Anat. (London)* 99: 791–813, 1965.

POWERS, J. B., R. B. FIELDS, AND S. S. WILLIAMS. Olfactory and vomeronasal system participates in male hamsters' attraction to female vaginal secretions. *Physiol. Behav.* 22: 77–84, 1979.

PRECHT, W. Vestibular mechanisms. *Ann. Rev. Neurosci.* 2: 265–289, 1979.

PREILOWSKI, U., AND R. W. SPERRY. Minor hemisphere dominance in a bilateral competitive tactual word recognition task. *Biol. Ann. Rep. (California Institute of Technology)* p. 83–84, 1972.

PRICE, J. L. Structural organization of the olfactory pathways. In: *Olfaction and Taste VI*, edited by J. LeMagnen and P. MacLeod. London: Information Retrieval, Ltd., 1977, p. 87–96.

PRIGOGINE, I. *From Being to Becoming: Time and Complexity in the Physical Sciences.* San Francisco: W. H. Freeman, 1980.

PUBOLS, B. H., JR., AND L. M. PUBOLS. Tactile receptor discharge and mechanical properties of glabrous skin. *Fed. Proc.* 42: 2528–2535, 1983.

PURPURA, D. P. A neurohumoral mechanism of reticulocortical activation. *Am. J. Physiol.* 186: 250–254, 1956.

PURPURA, D. P. Comparative physiology of dendrites. In: *The Neurosciences. A Study Program*, edited by G. C. Quarton, T. Melnechuk, and F. O. Schmitt. New York: Rockefeller University Press, 1967, p. 372–393.

RACINE, R. J. Modification of seizure activity by electrical stimulation: I. After-discharge threshold. *Electroencephal. Clin. Neurol.* 32: 269–279, 1972a.

RACINE, R. J. Modification of seizure activity by electrical stimulation: II. Motor seizure. *Electroencephal. Clin. Neurol.* 32: 281–294, 1972b.

RAISMAN, G. An experimental study of the projections of the amygdala to the accessory olfactory bulb and its relationship to the concept of a dual olfactory system. *Exp. Brain.* 14: 395–408, 1972.

RAISMAN, G. Neural plasticity in the septal nuclei of the adult rat. *Brain Res.* 14: 25–48, 1969.

RAISMAN, G., AND P. M. FIELD. A quantitative investigation of the development of collateral reinnervation after partial denervation of the septal nuclei. *Brain Res.* 50: 241–264, 1973.

RAKIC, P. Developmental events leading to laminar and areal organization of the neocortex. In: *The Organization of the Cerebral Cortex*, edited by F. O. Schmitt, F. G. Worden, G. Adelman, and S. G. Dennis. Cambridge, MA: The MIT Press, 1981.

RAKIC, P. *Local Circuit Neurons.* Cambridge, MA: The MIT Press, 1975.

RAKIC, P., AND P. S. GOLDMAN-RAKIC. The development and modifiability of the cerebral cortex. *Neurosci. Res. Prog. Bull.* 20: 433–438, 1982.

RALLS, K. Mammalian scent marking. *Science* 171: 443–449, 1971.

RALSTON, H. J. III. The synaptic organization of lemniscal projections to the ventrobasal thalamus of the cat. *Brain Res.* 14: 99–115, 1969.

RANSON, S. W. Some functions of the hypothalamus. *Harvey Lectures Ser.* 32: 92–121, 1936.

RANSON, S. W. Some functions of the hypothalamus. *Bull. N.Y. Acad. Med.* 13: 241–271, 1937.

RASMUSSEN, G. L. Efferent fibers in the cochlear nerve and cochlear nucleus. In: *Neural Mechanisms of the Auditory and Vestibular System*, edited by G. L. Rasmussen and W. Windle. Springfield, IL: Charles C Thomas, 1960, p. 105–115.

REIVICH, M., D. KUHL, A. WOLF, J. GREENBERG, M. PHELPS, T. IDO, V. CASSELLA, J. FOWLER, E. HOFFMAN, A. ALAVI, P. SOM, AND L. SOKOLOFF. The [18F] fluoro-deoxyglucose method for the measurement of local cerebral glucose utilization in man. *Circ. Res.* 44: 127–137, 1979.

REYNOLDS, D. V. Surgery in the rat during electrical analgesia induced by focal brain stimulation. *Science* 164: 444–445, 1969.

REXED, B. The cytoarchitectonic organization of the spinal cord in the cat. *J. Comp. Neurol.* 96: 415–495, 1952.

RICARDO, J. A., AND E. T. KOH. Anatomical evidence of direct projections from the nucleus of the solitary tract to the hypothalamus, amygdala, and other forebrain structures in rat. *Brain Res.* 153: 1–26, 1978.

RIGGS, L. A. Light as a stimulus for vision. In: *Vision and Visual Perception*, edited by C. H. Graham. New York: Wiley, 1965.

RILEY, V. Psychoneuroendocrine influences on immunocompetence and regulation. *Science* 224: 452–459, 1984.

ROBINSON, C. J., AND H. BURTON. Somatotopographic organization in the second somatosensory receptive fields in cortical areas, 76, retroinsula, postauditory and granular insula of M. fascicularis. *J. Comp. Neurol.* 192: 69–92, 1980.

ROCKEL, A. J., R. W. HIORNS, AND T. P. S. POWELL. Numbers of neurons through full depth of neocortex. *J. Anat.* 118: 371, 1974.

ROSENBERG, M. B., T. FRIEDMANN, R. C. ROBERSON, M. TUSZYNSKI, J. A. WOLFF, X. O. BREAKEFIELD, AND F. H. GAGE. Grafting genetically modified cells to the damaged brain: Restorative effects of NGF expression. *Science* 242: 1575–1578, 1988.

ROSENBERG, R. N. Genetic variation in neurological disease. *Trends Neurosci.* 3: 144–148, 1980.

RUCH, T. C. AND H. A. SHENKIN. The relation of area 13 on the orbital surface of the frontal lobes to hyperactivity and hyperphagia in monkeys. *J. Neurophysiol.* 6: 349–360, 1943.

RUDA, M. A. Opiates and pain pathways: demonstration of enkephalin synapses and dorsal horn projection neurons. *Science* 215: 1523–1524, 1982.

RUSHTON, W. A. H. Visual pigments and color blindness. *Sci. Am.* 232: 64–74, 1975.

RUSSO, A. F., AND D. E. KOSHLAND, JR. Separation of signal transduction and adaptation functions of the aspartate receptor in bacterial sensing. *Science* 220: 1016–1020, 1983.

SANDRI, C., J. VAN BUREN, AND K. AKERT. *Membrane Morphology of the Vertebrate Nervous System: A Study with Freeze-Etch Techniques.* Amsterdam: Elsevier, 1977.

SALOMON, G., AND A. STARR. Electromyography of middle ear muscles in man during motor activities. *Acta Neurol. Scand.* 39: 161–168, 1963.

SCALIA, F., AND S. S. WINANS. The differential projections of the olfactory bulb and accessory olfactory bulb in mammals. *J. Comp. Neurol.* 161: 31–56, 1975.

SCHELL, J. *The Fate of the Earth.* New York: Knopf, 1982.

SCHIFF, J., AND W. R. LOEWENSTEIN. Development of a receptor of a foreign nerve fiber in a pacinian corpuscle. *Science* 177: 712–715, 1972.

SCHMIDT, R. F. (ed.). *Fundamentals of Neurophysiology.* New York: Springer, 1978.

SCHMITT, F. O., P. DEV, AND B. H. SMITH. Electronic processing of information by brain cells. *Science* 193: 114–120, 1976.

SCHMITT, F. O., G. C. QUARTON, T. MELNECHUK, AND G. ADELMAN. *The Neurosciences: Second Study Program.* New York: The Rockefeller University Press, 1970.

SCHMITT, F. O., F. G. WORDEN, G.

ADELMAN, AND S. G. DENNIS. *The Organization of the Cerebral Cortex.* Cambridge, MA: The MIT Press, 1981.

SCHULMAN-GALAMBOS, C., AND R. GALAMBOS. Brainstem auditory-evoked responses in premature infants. *J. Speech Hearing Res.* 18: 456–465, 1975.

SCHULTZE-WESTRUM, T. G. Social communication by chemical signals in flying phalanger. In: *Olfaction and Taste III: Proceedings of the Third International Symposium*, edited by C. Pfaffman. New York: Rockefeller University Press, 1969, p. 269–277.

SCOTT, J. W., AND C. M. LEONARD. The olfactory connections of the lateral hypothalamus in the rat, mouse and hamster. *J. Comp. Neurol.* 141: 331–344, 1971.

SEIFERT, W. (ed.). *Neurobiology of the Hippocampus.* New York: Academic Press, 1983.

SEMPLE, M. N., AND L. M. AITKIN. Representation of sound frequency and laterality by units in central nucleus of cat inferior colliculus. *J. Neurophysiol.* 42: 1626–1638, 1979.

SHARE, L. Control of plasma ADH titer in hemorrhage: role of atrial and arterial receptors. *Am. J. Physiol.* 215: 1384–1389, 1968.

SHARMAN, D. F. The catabolism of catecholamines: recent studies. *Brain Med. Bull.* 29: 110–115, 1973.

SHARP, F., J. KAUER, AND G. M. SHEPHERD. Local sites of activity related glucose metabolism in the rat olfactory bulb during olfactory stimulation. *Brain Res.* 98: 596–600, 1975.

SHEPHERD, G. M. The olfactory bulb as a simple cortical system: experimental analysis and functional implications. In: *The Neurosciences Second Study Program*, edited by F. O. Schmitt, G. C. Quarton, T. Melnechuk, and G. Adelman. New York: Rockefeller University Press, 1970, p. 539–552.

SHEPHERD, G. M. *Neurobiology.* Oxford, UK: Oxford Univ. Press, 1983.

SHERMAN, S. M., AND P. D. SPEAR. Organization of visual pathways in normal and visually deprived cats. *Physiol. Rev.* 62: 738–855, 1982.

SHERRINGTON, C. S. *The Integrative Action of the Nervous System.* London: Constable, 1906; Reprinted with a new Forward, New Haven: Yale University Press, 1947.

SHIMAMURA, M. Longitudinal coordination between spinal and cranial reflex systems. *Exp. Neurol.* 8: 505–521, 1963.

SHIMAMURA, M., AND K. AKERT. Peripheral nervous relations of propriospinal and spino-bulbo-spinal reflex systems. *Jpn. J. Physiol.* 15: 638–647, 1965.

SHIMAMURA, M., AND R. B. LIVINGSTON. Longitudinal conduction systems serving spinal and brain-stem coordination. *J. Neurophysiol.* 26: 258–272, 1963.

SIMMONS, F. B., R. GALAMBOS, AND A. RUPERT. Conditioned response of middle ear muscles. *Am. J. Physiol.* 197: 537–538, 1959.

SKINNER, B. F. The phylogeny and ontogeny of behavior. *Science* 153: 1205–1213, 1966.

SMITH, C. A., AND G. L. RASMUSSEN. Recent observations on the olivocochlear bundle. *Ann. Otol. Rhinol. Laryngol.* 72: 489–506, 1963.

SMITH, C. A., AND G. L. RASMUSSEN. Nerve endings in the maculae and cristae of the chinchilla vestibule, with a special reference to the efferents. In: *Third Symposium on the Role of the Vestibular Organs in Space Exploration.* National Aeronautic Space Administration, NASA SP-152: 183–200, 1967.

SMITH, H. D. Audiometric effects of voluntary contraction of the tensor typmani muscle. *Arch. Otolaryngol.* 38: 369–372, 1943.

SNYDER, S. H., AND S. MATTHYSSE. Opiate receptor mechanisms. *Neurosci. Res. Prog. Bull.* 13: 3–166, 1975.

SOKOLOFF, L. Local cerebral circulation at rest and during altered cerebral activity induced by anesthesia or visual stimulation. In: *Regional Neurochemistry*, edited by S. S. Kety. Oxford, UK: Pergamon Press, 1961, p. 107–117.

SOKOLOFF, L., M. REIVICH, C. KENNEDY, M. H. DES ROSIERS, C. S. PATLAK, K. D. PETTIGREW, O. SAKURADA, AND M. SHINOHARA. The [^{14}C]deoxyglucose method for the measurement of local cerebral glucose utilization: theory, procedure, and normal values in the conscious and anesthetized albino rat. *J. Neurochem.* 28: 897–916, 1977.

SOKOLOV, E. N. Learning and memory: habituation as negative learning. In: *Neural Mechanisms of Learning*, edited by M. R. Rosenzweig and E. L. Bennett. Cambridge, MA: The MIT Press, 1976, p. 475–479.

SOTELO, C., AND R. LLINAS. Specialized membrane junctions between neurons in the vertebrate cerebellar cortex. *J. Cell. Biol.* 53: 271–289, 1972.

SPERRY, R. W. Lateral specialization in the surgically separated hemispheres. In: *The Neurosciences Third Study Program*, edited by F. O. Schmitt and F. G. Worden. Cambridge, MA: The MIT Press, 1974, p. 5–19.

SPOENDLIN, H. H. Ultrastructural studies of the labyrinth in squirrel monkeys. In: *The Role of Vestibular Organs in the Exploration of Space.* National Aeronautics Space Administration, NASA SP-77, 7–22, 1965.

SPRAGUE, J. M., H. C. HUGHES, AND G. BERLUCCHI. Cortical mechanisms in pattern and form perception. In: *Brain Mechanisms and Perceptual Awareness*, edited by O. Pompeiano and C. Ajmone Marsan. New York: Raven Press, 1981, p. 107–132.

SPRINGER, A. D. Retinopetal cells in the goldfish olfactory bulb. *Investig. Ophthal. Vis. Sci.* 22: 246, 1982.

SQUIRE, L. R. *Memory and Brain.* New York: Oxford University Press, 1987.

STARR, A. Auditory brainstorm responses in brain death. *Brain* 99: 543–554, 1976.

STARR, A. Regulatory mechanisms of the auditory pathway. In: *Modern Neurology.* Boston: Little Brown, 1969, p. 101–114.

STENEVI, U., A. BJOERKLUND, AND N.-AA. SVENDGAARD. Transplantation of central and peripheral monoamine neurons to the adult rat brain: Techniques and conditions for survival. *Brain Res.* 114: 1–20, 1976.

STEVENS S. S. *Handbook of Experimental Psychology.* New York: Wiley, 1951, ch. 29, p. 1143–1171.

STEVENS, S. S. Some similarities between hearing and seeing. *Laryngoscope* 68: 508–527, 1958.

SUTULA, T., X-X. HE, J. CAVAZOS, AND G. SCOTT. Synaptic reorganization in the hippocampus induced by abnormal functional activity. *Science* 239: 1147–1150, 1988.

SWANSON, L. W., P. E. SAWCHENKO, AND W. M. COWAN. Evidence for collateral projections by neurons in ammon's horn, the dentate gyrus, and the subiculum: a multiple retrograde labeling study in the rat. *J. Neurosci.* 1: 548–559, 1981.

SWANSON, L. W., AND P. E. SAWCHENKO. Hypothalamic integration: organization of the paraventricular and supraoptic nuclei. *Ann. Rev. Neurosci.* 6: 269–324, 1983.

SZENTAGOTHAI, J. Downward causation? *Ann. Rev. Neurosci.* 7: 1–11, 1984.

SZENTAGOTHAI, J., AND M. A. ARBIB. *Conceptual Models of Neural Organization.* Cambridge, MA: The MIT Press, 1975.

TASAKI, I. Nerve impulses in individual auditory nerve fibers of guinea pig. *J. Neurophysiol.* 17: 97–122, 1954.

TAUB, E., M. HARGER, H. C. GRIER, AND W. HODOS. Some anatomical observations following chronic dorsal rhizotomy in monkeys. *Neuroscience* 5: 389–401, 1980.

TAUC, L. Nonvesicular release of neurotransmitter. *Physiol. Rev.* 62: 857–887, 1982.

TEPPERMAN, J., J. R. BROBECK, AND C. N. H. LONG. A study of experimental hypothalamic obesity in the rat. *Am. J. Physiol.* 133: P468–P469, 1941.

TOMITA, T. Electrical response of single photoreceptors. *Proc. IEEE* 56: 1015–1023, 1968.

TOREBJOERK, H. E., AND J. L. OCHOA. Specific sensations evoked by activity in single identified sensory units in man. *Acta Physiol. Scand.* 110: 445–447, 1980.

TRUEX, R. C., AND M. B. CARPENTER. *Human Neuroanatomy.* Baltimore: Williams & Wilkins, 1969.

TYLER, D. B., AND P. BARD. Motion sickness. *Physiol. Rev.* 29: 311–369, 1949.

UNGERSTEDT, U. Stereotaxic mapping of the monoamine pathways in the rat brain. *Acta Physiol. Scand. (Suppl. 367)* 82: 1–48, 1971.

UNDERSTEDT, U. Brain dopamine neurons and behavior. In: *The Neurosciences Third Study Program*, edited by F. O. Schmitt and F. G. Worden. Cambridge, MA: The MIT Press, 1974, p. 695–703.

UOTILA, U. U. On the role of the pituitary stalk in the regulation of the anterior pituitary, with special reference to the thyrotropic hormone. *Endocrinology* 25: 605–612, 1939.

VAN DER KOOY, D., H. G. J. M. KUYPERS, AND C. E. CATSMAN-BERREVOETS. Single mammillary cells with axon collaterals. Demonstration by a simple fluorescent retrograde double-labeling technique in the rat. *Brain Res.* 158: 189–196, 1978.

VAN ESSEN, D. C. Visual areas of the mammalian cerebral cortex. *Ann. Rev. Neurosci.* 2: 227–263, 1979.

VANE, J. R. Inhibition of prostaglandin synthesis as a mechanism of action for aspirin-like drugs. *Nature New Biol.* 231: 232–235, 1971.

VINGRADOVA, O. S., E. S. BRAZHNIK, A. M. KARANOV, AND S. D. ZHADINA. Neuronal activity of the septum following various types of deafferentation. *Brain Res.* 187: 353–368, 1980; See also, VINGRADOVA, O. S. *Hippocampus and Memory.* Moscow: Nauka, 1975 (in Russian).

VON BEKESY, G. The variation of phase along the basilar membrane with sinusoidal vibrations. *J. Acoust. Soc. Am.* 19: 452–460, 1947.

VON EULER, C., F. HERRERO, AND I. WEXLER. Control mechanisms determining rate and depth of respiratory movements. *Respir. Physiol.* 10: 93–108, 1970.

WADA, J. A., T. OSAWA, AND T. MIZOGUCHI. Recurrent spontaneous seizure state induced by prefrontal kindling in Senegalese baboons, Papio papio. In: *Kindling*, edited by J. A. Wada. New York: Raven Press, 1976, p. 173–202.

WALD, G. Visual pigments and photoreceptors—review and outlook. *Exp. Eye Res.* 18: 333–343, 1974.

WALLS, G. L. *The Vertebrate Eye.* Bloomfield Hills, MI: Cranbrook Institute of Science, 1942.

WARR, W. B. Olivocochlear and vestibular efferent neurons of the feline brain stem; their location, morphology and number determined by retrograde axonal transport and acetylcholine-esterase histochemistry. *J. Comp. Neurol.* 161: 159–182, 1975.

WEBSTER, W. R. The effects of repetitive stimulation on auditory evoked potentials. *Electroencephal. Clin. Neurophysiol.* 30: 318–330, 1971.

WEISKRANTZ, L. The interaction between occipital and temporal cortex in vision: an overview. In: *The Neurosciences Third Study Program*, edited by F. O. Schmitt and F. G. Worden. Cambridge, MA: The MIT Press, 1974, p. 189–204.

WERBLIN, F. S. The control of sensitivity in the retina. *Sci. Am.* 228: 70–79, 1973.

WERNICK, J. S., AND A. STARR. Binaural interaction in the superior olivary complex of the cat: an analysis of field potentials evoked by binaural-beat stimuli. *J. Neurophysiol.* 31: 428–441, 1968.

WERSÄLL, J. S. Epithelium of the cristae ampullares of the guinea pig. *Acta Otolaryngol. (Stockh.) (Suppl. 126):* 1–85, 1956.

WEYMOUTH, F. W. The eye as an optical instrument. In: *A Textbook of Physiology*, 17th ed., edited by J. F. Fulton, Philadelphia: Saunders, 1955, ch. 23.

WIESEL, T. N., D. H. HUBEL, AND D. M. K. LAM. Autoradiographic demonstration of ocular-dominance columns in the monkey striate cortex by means of transneuronal transport. *Brain Res.* 79: 273–279, 1974.

WILLIS, W. D., AND R. G. GROSSMAN. *Medical Neurobiology.* St. Louis: C. V. Mosby Co., 1981.

WILLOT, J. F., AND S-M. LU. Noise-induced hearing loss can alter neural coding and increase excitability in the central nervous system. *Science* 216: 1331–1332, 1982.

WITELSON, S. F., AND W. PALLIE. Left hemisphere specialization for language in the newborn. Neuroanatomical evidence of asymmetry. *Brain* 96: 641–646, 1973.

WOLFE, J. M. Hidden visual processes. *Sci. Am.* 248: 94–103, 1983.

WOLTER, J. R. The centrifugal nerves in human optic tract, chiasm, optic nerve, and retina. *Trans. Am. Ophthalmol. Soc.* 63: 678–707, 1965.

WOLTER, J. R., AND O. E. LUND. Reaction of centrifugal nerves in the human retina. *Am. J. Ophthalmol.* 66: 221–232, 1968.

WOODS, J. W., AND P. BARD. Antidiuretic hormone secretion in the cat with a chronically denervated hypothalamus. Copenhagen: Proc. 1st Int. Congr. Endocr., 1960, p. 113.

WOODSON, W. E. *Human Engineering Guide for Equipment Designers.* Los Angeles: University of California Press, 1954.

WOOLSEY, T. A., AND H. VAN DER LOOS. The structural organization of layer IV in the somatosensory region (SI) of mouse cerebral cortex. The description of a cortical field composed of discrete cytoarchitectonic units. *Brain Res.* 17: 205–242, 1970.

WRIGHT, J. W., AND J. W. HARDING. Recovery of olfactory function after bilateral bulbectomy. *Science* 216: 322–324, 1982.

YAKOVLEV, P. I. Motility, behavior and the brain. *J. Nerv. Ment. Dis.* 107: 313–335, 1948.

YIN, T. C., AND V. B. MOUNTCASTLE. Mechanisms of neural integration in the parietal lobe for visual attention. *Fed. Proc.* 37: 2251–2257, 1978.

YIN, T. C., AND V. B. MOUNTCASTLE. Visual input to the visuomotor mechanisms of the monkey's parietal lobe. *Science* 197: 1381–1383, 1977.

ZEKI, S. M. Colour coding in rhesus monkey prestriate cortex. *Brain Res.* 53: 422–427, 1973.

ZEKI, S. M. Cells responding to changing image size and disparity in the cortex of the rhesus monkey. *J. Physiol.* 242: 827–841, 1974.

ZEKI, S. M. Colour coding in the superior temporal sulcus of the rhesus monkey visual cortex. *Proc. Roy. Soc. (Biol)* 197: 195–223, 1977.

ZEKI, S. M. The representation of colours in the cerebral cortex. *Nature* 284: 412–418, 1980.

ZIMMER, J. Extended commissural and ipsilateral projections in postnatally deentorhinated hippocampus and fascia dentata demonstrated in rats by silver impregnation. *Brain Res.* 64: 293–311, 1973.

ZIMMER, J. Proximity as a factor in the regulation of aberrant axonal growth in postnatally deafferented fascia dentata. *Brain Res.* 72: 137–142, 1974.

ZOTTERMAN, Y. The neural mechanisms of taste. In: *Sensory Mechanisms*, edited by Y. Zotterman. Amsterdam: Elsevier, 1967, p. 139–154.

Index

Note: Page numbers in *italics* indicate figures; those followed by *t* or *n* indicate tables or footnotes, respectively.

A

A band, 62–64, *63–65*, 197
Abdominal muscles, 580
Abdominal reflex, 1072–1073
Abdominal viscera, innervation of, 1060
Abdominal wall, expiratory muscles, 561
Abducens nerve (cranial nerve VI), 965, 1069
ABO blood group substances, 359
Absence seizures, 1007
Absolute (Kelvin) temperature, 521
Absolute refractory period, 45–46, 50–51, 165
Absorption, 693
  of calcium, 508, 705, 716, 839, 841
  of carbohydrate, 695–697
  of cholesterol, 700–701
  of drugs, 705–706
  of electrolytes, 707, 715
  intestinal, 693, 718t
  of iron, 371, 705
  methods of study, 695
  role of water in, 707
Absorption atelectasis, 593–594
Accelerated idioventricular rhythm, 190
Accessory muscle(s), 580
  of inspiration, 561
Accessory nerve (cranial nerve XI), 1069
Accessory olfactory bulb, functions of, 1005–1006
Accessory optic system, 982
  integration of visual-vestibular reflexes, 983
Accommodation, 39, 50–51
Accommodation reflex, 972, 975
ACE. *See* Angiotensin-converting enzyme
A cell(s), 754, 981
Acetazolamide, 497
  and CO_2 transport, 541
  mechanism of action, 506
Acetoacetate. *See* Ketone bodies
Acetoacetic acid, 487
  as urinary buffer, 495
Acetoacetyl coenzyme A, 767
Acetone, 729. *See also* Ketone bodies
Acetylcholine, 54, 82, 799, 831, 1040, 1055, 1058, 1070
  and action potential, in cardiac muscle, 164
  actions of, 1059
  cardiac effects of, 168–169
  drugs that mimic, 84
  effect on intestinal secretion, 717t, 718
  effect on vascular resistance, 532
  mechanism of action, 170
  quanta, 84
  quantal release of, 1051–1052
  regulation of, 1050

  in regulation of cellular transport pathways, 719
  storage and release, in nerve, 59–60
  as transmitter in central nervous system, 59–60
Acetylcholine analogues, receptors, on acinar cell, 670
Acetylcholine esterase, of red cell membrane, 375
Acetylcholine receptors, 60, 82–83, 1050
  on acinar cell, 670
  in myasthenia gravis, 1071–1072
Acetylcholinesterase, 60, 82, *83*, 84, 1059
  inactivation, 84
Acetyl coenzyme A, 92, 739, 762, 767
  metabolism, *94*
Acetyl glycerol ether phosphorylcholine, *365*, 366–367
  structure of, *367*
N-Acetylprocainamide, indications for, 195
Acetylstrophanthidin, and chemoreceptor response, 286
Acid(s)
  fixed, 486, 490
  metabolic, 486
  volatile, 486, 490
  weak
    as buffers, 487–488
    as urinary buffer, 495
Acid–base balance, 486
  and potassium secretion, 507
  regulation of, 469
Acid–base buffer systems, 487–490. *See also* Buffer systems
Acid–base disturbance(s), 498–502. *See also* Acidosis; Alkalosis; Diabetic ketoacidosis; Lactic acidosis; Metabolic acidosis; Metabolic alkalosis; Respiratory acidosis; Respiratory alkalosis
  acute, 499
  chronic, 499
    and potassium secretion, 507
  compensated phase, 499
  primary, 498
  uncompensated phase, 499
Acid–base status, 542
Acid excretion, quantitation of, 496
Acid hydrolases, in lysosomes, 10
Acidic isoferritins, 343
Acidosis, 383, 487, 504, 830, 836. *See also* Lactic acidosis; Metabolic acidosis; Respiratory acidosis
  acute, and potassium secretion, 507
  and calcium balance, 833
  cause of, 496
  chronic, 495
  definition of, 486, 498

  and H_2CO_3-derived hydrogen secretion, 496
  in shock, 319
Acinar cell(s), pancreatic, 645, 665
  digestive enzymes in, 668t
  enzyme secretory response of, agonists, 671
  gap junction, 665
  junctional complexes, 665
  receptors on, 670
  secretion, stimulation of, 666–667
  stimulus-secretion coupling in, 669–671
  ultrastructure of, *666*
Acoustic neuroepithelium, 936–937, 956–957
Acoustic neurons, 956, 960–961
Acoustic power, 937
Acoustic responses, plasticity of, 946–947
Acquired immune deficiency syndrome, 361
Acromegaly, 801, 803
Acrosome, 852
Acrosome reaction, 874
ACTH. *See* Adrenocorticotropin
Actin, 67–68, 85. *See also* F—actin; G—actin
  cardiac, 197, *198*
  erythrocyte, 374–375
  microfilaments, 12–13
  in muscle contraction and relaxation, 71–74
  regulated, 70
α-Actinin, 68
Actin–myosin interaction, 70
Action currents, 44
Action potential(s), 25, 28, 33, 38, 41, 46, 82, 1049. *See also* Nerve impulse
  all-or-nothing response, 45, 77, *77*, 163, 1049, 1083
  atrial, 172
  in auditory nerve, 941–943, *942*
  cardiac, 161–168, 172, 174–175, 206–207
    in animal species, 168
    conduction of, 166–168
  classical mechanism of, 46
  compound, 931
  fast-response, 162–164, 171
  in His-Purkinje system, 171
  ionic events in, 48–51
  mechanism of, 47–48
  of mixed nerve trunks, compound nature of, 42–43
  modern concept of, 47–52
  movement of sodium and potassium during, *44*
  of muscle, 41
  origin, 46
  overshoot, 46–48, 163
  plateau of, 163
  propagated, 162, 165
  recording, 41–43
  self-propagating activity, 44–45